TABLE OF ATOMIC MASSES (WEIGHTS) BASED ON CARBON-12

Name	Symbol	Atomic No.	Atomic Mass
Actinium	Ac	89	(227)[a]
Aluminum	Al	13	26.98154
Americium	Am	95	(243)[a]
Antimony	Sb	51	121.75
Argon	Ar	18	39.948
Arsenic	As	33	74.9216
Astatine	At	85	(210)[a]
Barium	Ba	56	137.34
Berkelium	Bk	97	(247)[a]
Beryllium	Be	4	9.01218
Bismuth	Bi	83	208.9804
Boron	B	5	10.81
Bromine	Br	35	79.904
Cadmium	Cd	48	112.40
Calcium	Ca	20	40.08
Californium	Cf	98	(251)[a]
Carbon	C	6	12.011
Cerium	Ce	58	140.12
Cesium	Cs	55	132.9054
Chlorine	Cl	17	35.453
Chromium	Cr	24	51.996
Cobalt	Co	27	58.9332
Copper	Cu	29	63.546
Curium	Cm	96	(247)[a]
Dysprosium	Dy	66	162.50
Einsteinium	Es	99	(252)[a]
Erbium	Er	68	167.26
Europium	Eu	63	151.96
Fermium	Fm	100	(257)[a]
Fluorine	F	9	18.99840
Francium	Fr	87	(223)[a]
Gadolinium	Gd	64	157.25
Gallium	Ga	31	69.72
Germanium	Ge	32	72.59
Gold	Au	79	196.9665
Hafnium	Hf	72	178.49
Helium	He	2	4.00260
Holmium	Ho	67	164.9304
Hydrogen	H	1	1.0079
Indium	In	49	114.82
Iodine	I	53	126.9045
Iridium	Ir	77	192.22
Iron	Fe	26	55.847
Krypton	Kr	36	83.80
Lanthanum	La	57	138.9055
Lawrencium	Lr	103	(260)[a]
Lead	Pb	82	207.2
Lithium	Li	3	6.941
Lutetium	Lu	71	174.97
Magnesium	Mg	12	24.305
Manganese	Mn	25	54.9380
Mendelevium	Md	101	(258)[a]
Mercury	Hg	80	200.59
Molybdenum	Mo	42	95.94
Neodymium	Nd	60	144.24
Neon	Ne	10	20.179
Neptunium	Np	93	237.0482[b]
Nickel	Ni	28	58.71
Niobium	Nb	41	92.9064
Nitrogen	N	7	14.0067
Nobelium	No	102	(259)[a]
Osmium	Os	76	190.2
Oxygen	O	8	15.9994
Palladium	Pd	46	106.4
Phosphorus	P	15	30.97376
Platinum	Pt	78	195.09
Plutonium	Pu	94	(244)[a]
Polonium	Po	84	(210)[a]
Potassium	K	19	39.098
Praseodymium	Pr	59	140.9077
Promethium	Pm	61	(145)[a]
Protactinium	Pa	91	231.0359[b]
Radium	Ra	88	226.0254[b]
Radon	Rn	86	(222)[a]
Rhenium	Re	75	186.2
Rhodium	Rh	45	102.9055
Rubidium	Rb	37	85.4678
Ruthenium	Ru	44	101.07
Samarium	Sm	62	150.4
Scandium	Sc	21	44.9559
Selenium	Se	34	78.96
Silicon	Si	14	28.086
Silver	Ag	47	107.868
Sodium	Na	11	22.98977
Strontium	Sr	38	87.62
Sulfur	S	16	32.06
Tantalum	Ta	73	180.9479
Technetium	Tc	43	98.9062[b]
Tellurium	Te	52	127.60
Terbium	Tb	65	158.9254
Thallium	Tl	81	204.37
Thorium	Th	90	232.0381[b]
Thulium	Tm	69	168.9342
Tin	Sn	50	118.69
Titanium	Ti	22	47.90
Tungsten	W	74	183.85
Unnilhexium	Unh	106	(263)[a]
Unnilpentium	Unp	105	(262)[a]
Unnilquadium	Unq	104	(261)[a]
Uranium	U	92	238.029
Vanadium	V	23	50.9414
Xenon	Xe	54	131.30
Ytterbium	Yb	70	173.04
Yttrium	Y	39	88.9059
Zinc	Zn	30	65.38
Zirconium	Zr	40	91.22

[a]Mass number of most stable or best-known isotope

[b]Mass number of the isotope of longest half-life

ntroduction to General, Organic, and Biochemistry

Sixth Edition

HV
286209
VALVE INSPECTION

Introduction to General, Organic, and Biochemistry

Sixth Edition

Morris Hein
Mount San Antonio College

Leo R. Best
Mount San Antonio College

Scott Pattison
Ball State University

Susan Arena
University of Illinois, Urbana–Champaign

Brooks/Cole Publishing Company
I(T)P ®An International Thomson Publishing Company

Pacific Grove • Albany • Belmont • Bonn • Boston • Cincinnati • Detroit • Johannesburg • London • Madrid • Melbourne
Mexico City • New York • Paris • Singapore • Tokyo • Toronto • Washington

Publisher: *Harvey C. Pantzis*
Marketing Team: *Kathleen Sharp, Christine Davis, Carrie Beckwith*
Assistant Editor: *Beth Wilbur*
Editorial Assistant: *Leigh Hamilton*
Production Service: *Ex Libris/Julie Kranhold*
Production Services Coordinator: *Jamie Sue Brooks*
Interior Design: *Nancy Benedict*
Illustrations: *Lotus Art; Pat Rogondino*
Photo Researcher: *Julie Kranhold*
Cover Design: *Vernon T. Boes*
Cover Illustration: *Atsuchi Tsunoda, Photonica*
Typesetting: *Monotype Composition*
Cover Printing: *Phoenix Color Corp.*
Printing and Binding: *Quebecor Printing/Hawkins*
Credits continue on page A-50.

FRONTIS PHOTOGRAPHS

- *Green leaves turn red and gold in autumn. As the days grow shorter and the nights grow chilly, the chlorophyll decomposes. The green fades away while other pigments remain, especially the reds and yellows.*
- *A worker in a pharmaceutical plant monitors the complex and precise machinery used in pill production.*
- *Chemistry is at work in every aspect of our lives. Scientists can actually see matter at the molecular level by using electron microscopes, and atomic-force microscopes—the first page of this book shows a neuron photographed by an electron microscope.*
- *Under inspection, a computer chip creates a colorful and intricate maze. These chips are made from the metalloid silicon.*

For more information, contact:

BROOKS/COLE PUBLISHING COMPANY
511 Forest Lodge Road
Pacific Grove, CA 93950
USA

International Thomson Publishing Europe
Berkshire House 168–173
High Holborn
London WC1V 7AA
England

Thomas Nelson Australia
102 Dodds Street
South Melbourne, 3205
Victoria, Australia

Nelson Canada
1120 Birchmount Road
Scarborough, Ontario
Canada M1K 5G4

International Thomson Editores
Seneca 53
Col. Polanco
11560 México, D. F., México

International Thomson Publishing GmbH
Königswinterer Strasse 418
53227 Bonn
Germany

International Thomson Publishing Asia
221 Henderson Road
#05-10 Henderson Building
Singapore 0315

International Thomson Publishing Japan
Hirakawacho Kyowa Building, 3F
2-2-1 Hirakawacho
Chiyoda-ku, Tokyo 102
Japan

Printed in the United States of America.

THIS BOOK IS PRINTED ON ACID-FREE RECYCLED PAPER

10 9 8 7 6 5 4 3 2

Library of Congress Cataloging-in-Publication Data

Introduction to general, organic, and biochemistry / Morris Hein ... [et al.]
p. cm.
Rev. ed. of: College chemistry. 5th ed. ©1993
Includes index.
ISBN 0-534-25878-6
1. Chemistry. I. Hein, Morris. II. Hein, Morris. College chemistry.
QD31.2.H43 1997 96-27546
540—dc20 CIP

To Morris Hein
Our esteemed colleague and coauthor
whose guidance and dedication
continue to be an inspiration to us.

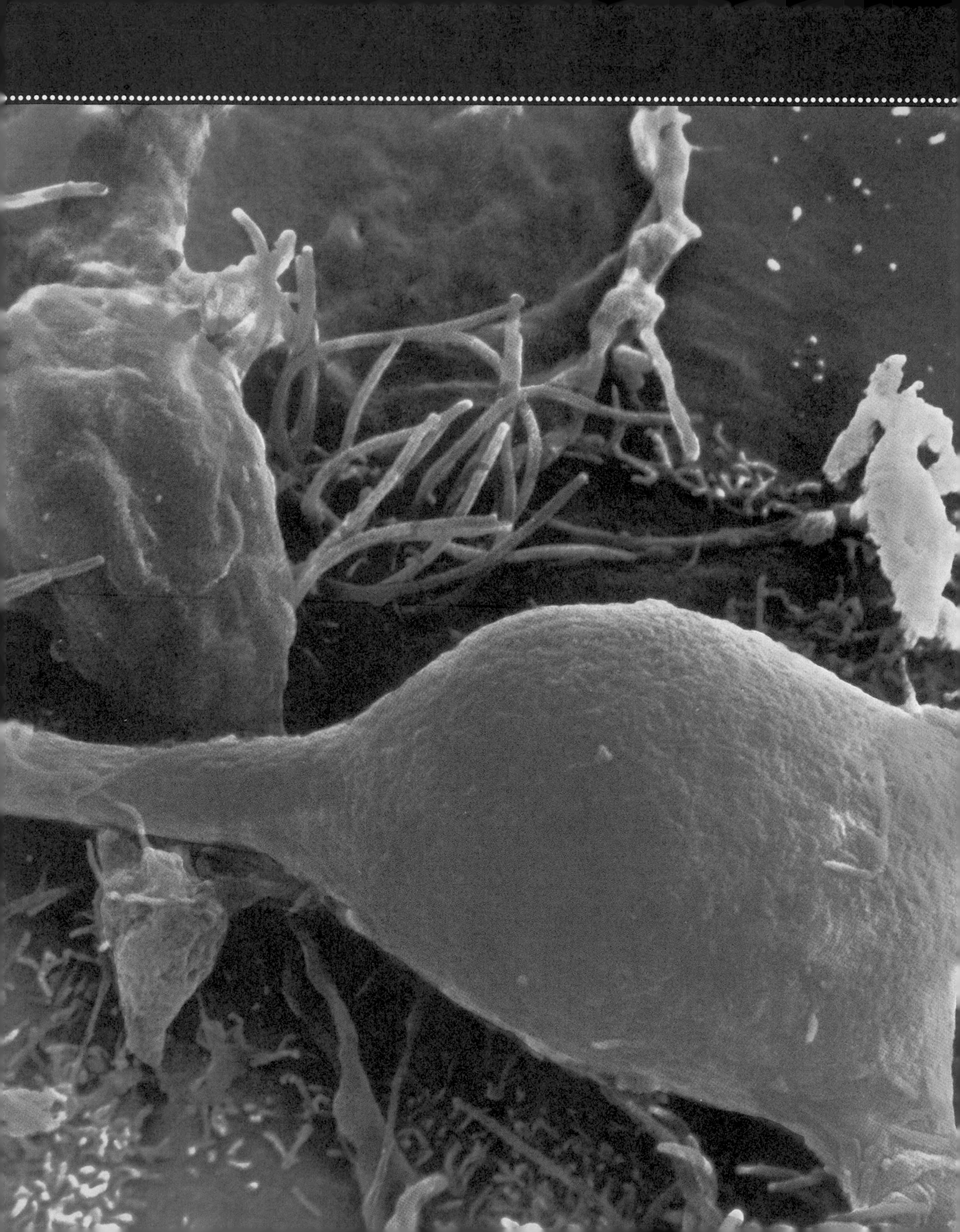

Our primary aim in writing *Introduction to General, Organic, and Biochemistry* has been steadfast throughout these six editions—to present chemistry in a clear, engaging manner that will stimulate students to further their scientific knowledge as they prepare for health sciences, nursing, and other careers.

This book is written for students who have not taken chemistry before and may have limited mathematical background and various career objectives. Even though we are constantly and intimately involved in a wide variety of chemical processes, occurring both within our bodies and in our surroundings, chemistry is often considered to be an esoteric subject—not easily accessible to students. Thus, we have portrayed the "everyday" nature of chemistry in photos, illustrations, examples, and in our "Chemistry in Action" series.

In preparing the Sixth Edition we considered the comments and suggestions of students, instructors, and researchers in the chemical education community to design a revision that builds on the strengths of previous editions and presents chemistry as a vital, coherent, and interesting subject. We have especially tried to relate chemistry to the lives of our students as we introduce and develop the principles that form the foundation for the study of general, organic, and biochemistry.

Development of Problem-Solving Skills

We all want our students to develop real skills in solving problems. We believe that the problem-solving approach we take works for students. This problem-solving approach (sometimes called a *dimensional analysis* approach) allows students to use units and show the change from one unit to the next. Students can learn most easily from defining and demonstrating concepts and problems step by step. In this edition we continue to show many examples, beginning with simple substitutions, progressing to the use of algorithms, and moving toward more complex problems. The examples show how to incorporate fundamental mathematical skills, scientific notation, and significant figures by following the rules consistently. Painstaking care has been taken to show each step in the problem-solving process (see pp. 107, 108, 138) and to give *alternative methods for solution* (ratio/proportion, algebraic, for example) where appropriate. These alternate methods give students flexibility in choosing the method that works the best for them. In this edition we have used four significant figures for atomic and molar masses for consistency and for rounding off answers appropriately. We have been careful to follow the rules set down in providing answers, correctly rounded so that students who have difficulty with mathematics do not become confused.

Dimensional analysis, or factor-label analysis, is explained in section 2.8, p. 23. Beginning students are encouraged to use this approach until they become comfortable with the terms used in calculations.

Alternative methods for solution: See, for example, pp. 38–39 and 178–179.

Fostering Student Skills

Attitude plays a critical role in problem-solving. We encourage students to learn that a systematic approach to solving problems is better than simple memorization. We begin to establish this attitude in Chapter 2. Throughout the book we encourage students to begin by writing down what is given (see p. 160) and to think their way through the problem to an answer, which is then checked to see if it makes sense. Once we have laid the foundations of concepts we highlight the steps in blue so students can locate them easily. Important rules and equations are highlighted in colored boxes for emphasis and ready reference.

Problem-solving steps are printed in blue: see pages 149–150, 183–184, and 235 for examples.

Boxed rules and equations: see pages 58, 202, 363, and 745 for examples.

Student Practice

Practice Problems: see p. 113 for an example and p. 123 for answers.

Questions review key terms, concepts, figures, and tables—see pp. 142–143.

Paired Exercises: pp. 143–145.

Additional Exercises: pp. 145–146.

Practice problems follow most of the examples in the text. Answers are provided at the end of each chapter for all of the practice problems. We have expanded and updated the number of end-of-chapter exercises. Each exercise set begins with a *Questions* section that helps students review key terms and concepts, as well as material presented in tables and figures. These are followed by a *Paired Exercises* section, where two similar exercises are presented side by side. These paired exercises cover concepts as well as numerical exercises. The section called *Additional Exercises* includes further practice on chapter concepts, presented in a more random order. Challenging questions and exercises are denoted with an asterisk. Answers for *all* even-numbered questions and exercises appear in Appendix V at the end of the book.

Organization

We emphasize the less theoretical aspects of chemistry early in the book, leaving the more abstract theory for later. Atoms, molecules, and reactions are all an integral part of the chemical nature of matter. A sound understanding of these topics will allow the student to develop a basic understanding of chemical properties and vocabulary.

We build toward a basic knowledge of organic and biochemistry for the health science student. Thus, we stress the nomenclature, structure, and reactivity of each major organic functional group. In turn, the basic biochemical concepts rest on this foundation. We encourage the students to apply their understanding to examples drawn from medicine, nutrition, agriculture, etc.

nitinol: pp. 3–5; risks and benefits: pp. 9–10

Chapter 1 is completely rewritten in this edition, to give students a better understanding of the scientific process by introducing the course with a narrative account of the discovery of nitinol, often called memory metal. Also included in this chapter is material on the benefits and risks of science in our high tech world.

Chapter 2 presents the basic mathematics and language of chemical calculations, including an explanation of the metric system and significant figures. Chapter 3 introduces the vocabulary of chemical substances, defining matter and the systems of naming and classifying elements. In Chapter 4 we present chemical properties—the ability of a substance to form new substances. Then, in Chapter 5, students encounter the history and terms of basic atomic theory.

We continue to present new material at a level appropriate for the beginning student by emphasizing nomenclature, composition of compounds, and reactions in Chapters 6 through 9 before moving into the details of modern atomic theory (Chapters 10 and 11). The entire text has been reexamined and the prose updated and rewritten to improve its clarity. Chapter 10, Modern Atomic Theory, has been extensively revised. The fifth edition chapter on the Periodic Table has been integrated into the revised Modern Atomic Theory chapter and into Chapter 11, Chemical Bonds. Those instructors who feel it is essential to cover atomic theory and bonding early in the course can assign Chapters 10 and 11 immediately following Chapter 5.

In Chapter 19 we study the chemistry of selected elements from the viewpoint of the periodic table. Then in Chapters 20–27 we introduce organic chemistry, and finally look into the principles of biochemistry in Chapters 28–36.

We have added current, relevant examples to most organic and biochemistry chapters. Some extended additions include:

- a section on polymers and recycling (Chapter 26);
- a section on micelles, liposomes and lipoproteins (Chapter 29);
- a section on new sources for and uses of industrial enzymes (Chapter 31);
- a section on the Human Genome Project and gene therapy (Chapter 32);
- a section on the Nutrition Facts Labels found on packaged foods (Chapter 33).

We have reviewed and carefully selected organic reactions to illustrate the reactivities of each important functional group. Chapter 20 now introduces three general categories of organic reactions (substitution, elimination, addition). Where possible, subsequent chapters present reactions within this conceptual framework.

IUPAC nomenclature is emphasized in this edition, but we have also specifically considered how organics are named in everyday usage. Thus, we present a common name if it continues to be widely used.

Biochemistry has become an increasingly visual science. Molecular pictures are often essential to the understanding of biochemical functions. Chapter 30 (Amino Acids, Polypeptides and Proteins) has been rewritten to emphasize a three-dimensional structure-to-function relationship. Chapter 31 (Enzymes) stresses a qualitative and visual approach to enzymes.

Learning Aids

In revising *Introduction to General, Organic, and Biochemistry* we have included new features to enhance the presentation and clarity as well as reinforce the practical, everyday nature of chemistry. The new design uses color to identify study aids, and the illustrations have been chosen to emphasize chemistry in familiar surroundings. We include numerous learning aids to help students develop a growing confidence with technical and abstract scientific content.

- Important **terms** are set off in boldface type where they are defined, and are printed in blue in the margin. These terms are listed alphabetically under the heading **Key Terms** at the end of each chapter with section references to assist in review of new vocabulary, and are also printed in boldface type in the index.

terms: pp. 47, 310–311, 756

Key Terms: pp. 80, 209, 850

- **Marginal Notations** have been added to help students in understanding basic concepts and problem-solving techniques. These are printed in magenta ink to clearly distinguish them from text and vocabulary terms.

marginal notations: pp. 262, 575, 963

- **Important statements,** equations, and laws are boxed and highlighted for emphasis.

boxed statements and equations: pp. 58–59, 745

- **Steps for solving problems** are printed in blue for easy reference.

steps for problem solving: pp. 149–150, 235, 539

- Worked **examples** with all steps included show students the how of problem solving before they are asked to tackle problems on their own.

worked examples and practice problems: pp. 17, 113, 696

- **Practice problems** permit immediate reinforcement of a skill shown in the example problem. Answers are provided at the end of the chapter to encourage students to check their problem solving immediately.

answers to practice problems: pp. 396, 855

end of chapter exercises: pp. 80–82, 651–655

- **End of chapter exercises** have been significantly revised, with approximately 200 new exercises, many emphasizing concepts and applications. Many of the existing problems have been shortened to fewer parts.

Concepts in Review: pp. 40, 209, 625

- A list of **Concepts in Review** given at the end of each chapter guides students in determining the most important concepts in the chapter.

paired exercises: pp. 121–122, 813–814

- This edition features **paired exercises** at the end of most chapters. Two parallel exercises are given, side by side, so the student can use the same problem-solving skills with two sets of similar information. Answers to the even-numbered paired exercises are given in Appendix V.

additional exercises: pp. 190–192, 872

- **Additional Exercises** are provided at the end of most chapters. They are arranged in a more random order, to encourage students to review the chapter material.

Chapter 1 begins with a garden photo as the metaphor for the diversity of the material world, which chemistry seeks to understand, explain, and utilize (pp. 1–2). Chapter 14 opens with an illustration of a surfer in the ocean, which is an aqueous solution.

- A **Review of Mathematics** is provided in Appendix I. (see p. A-1)
- **Units of measurement** are shown in table format in Appendix III and in the endpapers. (see p. A-12)
- **Answers** to the even-numbered exercises are given in Appendix V. (see p. A-15)
- **Each chapter opens** with a color photograph relating the chapter to our daily life. A chapter preview list assists students in viewing the topics covered in the chapter, and the introductory paragraph further connects the chapter topic to everyday life.

Chemistry in Action: See p. 135, The Taste of Chemistry, and p. 700, Coffee Talk

- Each chapter contains at least one special **Chemistry in Action** section that shows the impact of chemistry in a variety of practical applications. These essays cover such topical information as controlling graffiti and the fat content of fast food. Other Chemistry in Action essays introduce experimental information on new chemical discoveries and applications. Over twenty new essays have been added and the others carefully revised.

A Complete Ancillary Package

The following teaching materials have been developed to accompany this text.

For the Student

Study Guide by Peter Scott of Linn-Benton Community College and Rachel Porter of University of Illinois, Urbana-Champaign is a carefully revised self-study guide. A self-evaluation section presents a variety of exercises to the student, followed by answers and solutions. A recap section then concisely summarizes chapter concepts.

Solutions Manual by Morris Hein, Leo R. Best, Scott Pattison, and Susan Arena includes answers and solutions to all end-of-chapter questions and exercises.

Introduction to General, Organic, and Biochemistry in the Laboratory, 6th Edition, by Morris Hein, Leo R. Best, and Robert L. Miner, and James M. Ritchey includes 42 experiments for a laboratory program that may accompany the lecture course. Featuring updated information on waste disposal and emphasizing safe laboratory procedures, the lab manual also includes study aids and exercises.

A Basic Math Approach to Concepts of Chemistry, 6th Edition, by Leo Michels is a self-paced paperbound workbook that has proven itself an excellent resource for students needing help with mathematical aspects of chemistry. Evaluation tests are provided for each unit and the test answers are given in the back of the book. A glossary is also included.

Brooks/Cole Exerciser (BCX) 2.0, by Laurel Technical Services is a text-specific software tutorial for general chemistry, with exercises from the main text and the Study Guide. The program monitors student progress and generates reports. This software is available for DOS, DOS/Windows, and Macintosh platforms.

Alchemist: A Chemical Equation Balancer for Macintosh, by Steve Townsend and Joyce Brockwell is a software tool for balancing complex chemical equations.

Beaker is a sophisticated, yet easy to use program for exploring organic chemistry principles, for studying and solving, sketching and analyzing molecular structures, for constructing NMR spectra, for performing reactions, and more. This software is available for Macintosh *(Beaker 2.1)* and DOS/Windows *(Beaker 2.2).*

Organic Chemistry Toolbox, by Norbert J. Pienta is a text-specific software tool for constructing molecular models, drawing Lewis dot structures, creating animations of reactions, solving chemistry problems, and studying structures. Available for DOS/Windows and Macintosh platforms.

For the Instructor

Printed Test Items with Chapter Tests for *Introduction to General, Organic, and Biochemistry, 6th Edition,* includes a copy of the test questions provided electronically in *EXP-Test,* Review Exercise Worksheets, answers to the test item questions, and answers to the Review Exercise Worksheets.

Instructor's Manual for Introduction to General, Organic, and Biochemistry in the Laboratory, 6th Edition, includes information on the management of the lab, evaluation of experiments, notes for individual experiments, and answer keys to each experiment's report form and to all exercises.

EXP-Test, a computerized test generation system, is available for IBM PCs or compatibles. A Macintosh version, *ESATEST III,* is also available.

Transparencies in full color include illustrations from the text, enlarged for use in the classroom and lecture halls.

Acknowledgments

It is with great pleasure that we begin these acknowledgments by thanking our colleagues and students for many helpful comments and suggestions. These are the people who have made this book possible and it is for them we write.

We are especially thankful for the support, friendship, and constant encouragement of our spouses, Edna, Louise, Joan, and Steve, who have been patient and understanding through the long hours of this process. Their optiimism and good humor have given us a sense of balance and emotional stability.

Several colleagues have been instrumental in preparing this new edition. We appreciate the careful checking of answers and solutions to all exercises by Iraj Behbahani, Mt. San Antonio College, and Rachel Porter, University of Illinois.

Special thanks to the talented staff at Brooks/Cole. The careful attention and gentle encouragement from Assistant Editor Beth Wilbur was vital through much of the planning and writing. Julie Kranhold of Ex Libris has again earned our gratitude for her attention to detail as this book moved through various production stages. Finally, we appreciate the guidance we have received from Jamie Sue Brooks,

Editorial Production Supervisor, Harvey Pantzis, Publisher, and Leigh Hamilton, Editorial Assistant.

We gratefully acknowledge the following reviewers who were kind enough to read and give their professional comments: Kathleen Ashworth, Yakima Valley Community College; David Ball, Cleveland State University; Anne Barber, Manatee Community College; Harry Baxter, Fairmont State College; Mark Bishop, Monterey Peninsula College; Eugene Boney, Ocean County College; John Chapin, St. Petersburg Junior College; Mitchell Fedak, Community College of Allegheny County, Boyce Campus; Tom Frazee, Shawnee State University; William Hausler, Madison Area Technical College; Margaret Holzer, California State University, Northridge; James Jacobs, University of Rhode Island; Margaret Kimble, Indiana University and Purdue University; William Nickels, Schoolcraft College; Frank Ohene, Grambling State University; Jeffrey Schneider, State University of New York, Oswego; Donald Wink, University of Illinois, Chicago; and Donald Young, Ashville-Buncombe Community College.

About the Authors

Morris Hein is professor emeritus of chemistry at Mt. San Antonio College, where he regularly taught general and organic chemistry. His name is synonymous with clarity, meticulous accuracy, and a step-by-step approach that students can follow. Over the years, more than two million students have learned chemistry using a text by Morris Hein. In addition to *Introduction to General, Organic, and Biochemistry, 6th Edition,* he is co-author of *Foundations of College Chemistry, 9th Edition.* He has also co-authored *Foundations of Chemistry in the Laboratory* and *College Chemistry in the Laboratory.*

Leo R. Best taught chemistry for twenty-four years at Mt. San Antonio College. He has been a collaborator and co-author with Morris Hein since the first edition of *Introduction to General, Organic, and Biochemistry.* He is also co-author of *Foundations of Chemistry in the Laboratory* and *College Chemistry in the Laboratory.*

Scott Pattison lives in Muncie, Indiana, and is a dedicated teacher of general, organic, and biochemistry at Ball State University. He has been a co-author with Morris Hein since the third edition of *Introduction to General, Organic, and Biochemistry.*

Susan Arena currently teaches general chemistry and is director of the Merit Program for Emerging Scholars at the University of Illinois, Urbana–Champaign. She has worked with Morris Hein since the fifth edition of *Introduction to General, Organic, and Biochemistry,* and is co-author of *Foundations of College Chemistry, 9th Edition.*

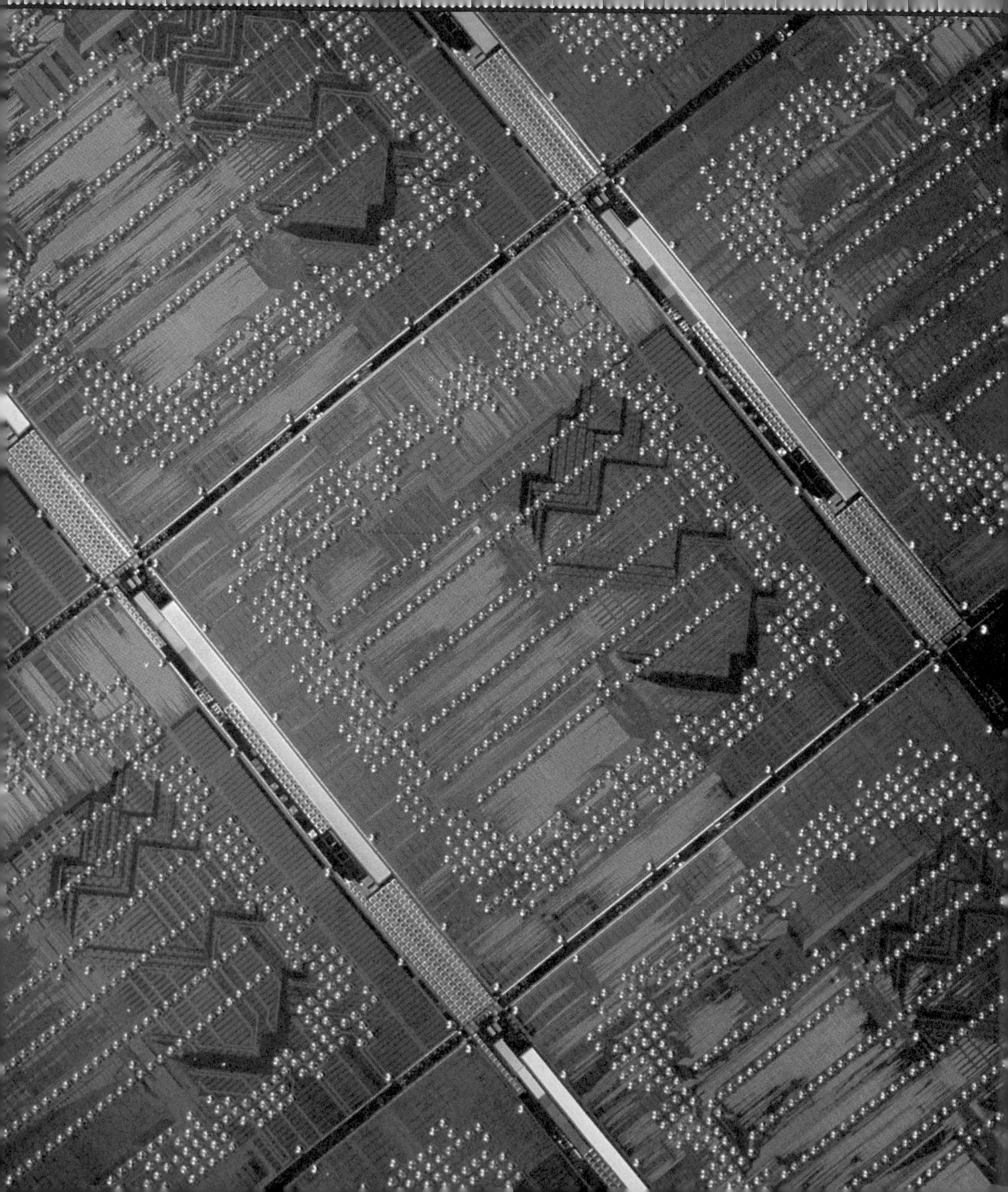

Chemistry in Action

Chemistry in Action

Brief Contents

Brief Contents

Contents

Contents

Contents

Contents

An Introduction to Chemistry

Chapter 1

Have you ever strolled through a spring garden and been amazed at the diversity of colors in the flowers? Or perhaps you have curled up in front of a winter fire and become fascinated watching the flames. And think of those times when you have dropped a beverage container on a hard floor, and were relieved to find that it was plastic instead of glass. All of these phenomena are the result of chemistry—not in the laboratory but rather in our everyday lives. Chemical changes can bring us beautiful colors, warmth and light, or new and exciting products. Chemists seek to understand, explain, and utilize the diversity of materials we find around us.

1.1 Why Study Chemistry?

Chemistry is a subject that is fascinating to many people. Learning about the composition of the world around us can lead to interesting and useful inventions and new technology. More than likely you are taking this chemistry course because someone has decided that it is an important part of your career goals. The field of chemistry is central to an understanding of many fields including agriculture, astronomy, animal science, geology, medicine, applied health technology, fire science, biology, molecular biology, and materials science. Even if you are not planning to work in any of these fields, chemistry is used by each of us every day in our struggle to cope with our technological world. Learning about the benefits and risks associated with chemicals will help you to be an informed citizen, able to make intelligent choices concerning the world around you. Studying chemistry teaches you to solve problems and communicate with others in an organized and logical manner. These skills will be helpful in college and throughout your career.

1.2 The Nature of Chemistry

chemistry

Key words are highlighted in bold and color in the margin to alert you to new terms defined in the text.

What is chemistry? A popular dictionary gives this definition: "**Chemistry** is the science of the composition, structure, properties, and reactions of matter, especially of atomic and molecular systems." Another, somewhat simpler definition is: "Chemistry is the science dealing with the composition of *matter* and the changes in composition that matter undergoes." Neither of these definitions is entirely adequate. Chemistry, along with the closely related science of physics, is a fundamental branch of knowledge. Chemistry is also closely related to biology, not only because living organisms are made of material substances but also because life itself is essentially a complicated system of interrelated chemical processes.

The scope of chemistry is extremely broad. It includes the whole universe and everything, animate and inanimate, in it. Chemistry is concerned not only with the composition and changes in the composition of matter, but also with the energy and energy changes associated with matter. Through chemistry we seek to learn and to understand the general principles that govern the behavior of all matter.

◀ Chapter Opening Photo: Chemistry can help us to understand nature, its beauty, and its complexity. The colors of this spring garden result from a series of chemical reactions.

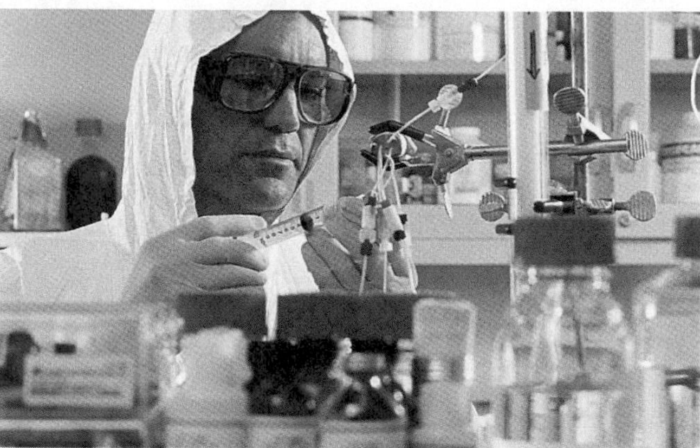

The field of chemistry is multi-layered and complex as you can see from the work of these two men. *Right:* This chemical operator is monitoring the process for production of a new medicine for postmenopausal osteoporosis at a plant in Northern Ireland. *Left:* This research chemist is developing a manufacturing process for hepatitis A vaccine.

The chemist, like other scientists, observes nature and attempts to understand its secrets: What makes a rose red? Why is sugar sweet? What is occurring when iron rusts? Why is carbon monoxide poisonous? Why do people wither with age? Problems such as these—some of which have been solved, some of which are still to be solved—are part of what we call chemistry.

A chemist may interpret natural phenomena, devise experiments that will reveal the composition and structure of complex substances, study methods for improving natural processes, or, sometimes, synthesize substances unknown in nature. Ultimately, the efforts of successful chemists advance the frontiers of knowledge and at the same time contribute to the well-being of humanity.

1.3 The Process of Chemistry

How does a chemist discover a new compound and study its behavior? How are these new substances developed into useful items for our everyday life? A combination of hard work, accident, and luck produced "memory metal," an advanced substance now used to make toys, braces for teeth, and to treat problems such as blood clots and reattachment of tendons. None of these applications was the target of the search for a new material. They all developed from the investigation of the properties of memory metal and the creativity of people associated with the work.

The discovery of "memory metal" began with William J. Buehler, a physical metallurgist at the Naval Ordnance Laboratory (NOL) in Maryland. He had the task of finding a substance composed of two or more metals that would be suitable for the nose cone of the Navy's underwater missile launchers. Buehler was experiencing difficulties at this particular time in his life and so he became particularly immersed in his work. He was tireless in his experimenting and eventually eliminated all but twelve compounds. He then began measuring impact resistance on them. His test was simple but effective: He made a button of the compound to be tested and hit it with a hammer. One of the substances, a 50%-50% mixture of nickel and titanium, produced a substance with greater impact resistance, elasticity, malleability, and resistance to fatigue than the others. Buehler named the substance *nitinol,* from *N*ickel *T*itanium *N*aval *O*rdnance *L*aboratory.

To begin testing, Buehler and his staff varied the percentages of nickel and titanium to determine the effect of composition on the properties of the compound. They made a series of bars in a furnace and, after cooling them, they smoothed them on a shop grinder. Buehler accidentally dropped one of the bars and noticed that it made a dull thud—much like that of a bar of lead. This aroused his curiosity and he began dropping other bars. To his surprise, he found that the cooled bars made a dull sound whereas the warm ones made a bell-like tone. Fascinated, he then began reheating and cooling the bars, discovering that the sound changed consistently back and forth from dull (cool) to bell-like (warm)—variances that indicated a change in the atomic structure of the metal.

During a project review at NOL, Buehler demonstrated the fatigue resistance of nitinol by bending a long strip of wire into accordion folds. He passed it around, letting the directors flex it and straighten it. One of them wanted to see what would happen when it was heated and held the pleated nitinol over a flame. To everyone's amazement it stretched out into a straight wire. Buehler recognized that this behavior related to the different sounds made when nitinol was heated or cooled.

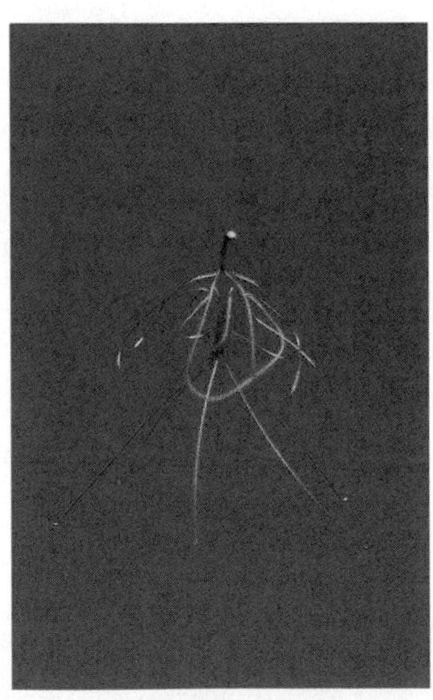

This nitinol filter can trap potentially fatal blood clots. When cooled below body temperature, it is collapsed into a straight bundle of wire. Then, with minor surgery, it is inserted into a large vein where it springs back into shape as it reaches body temperature.

Buehler recruited Frederick Wang, a crystallographer, to define the "memory" properties of the metal. Wang determined the atomic structural changes that give nitinol its unique characteristic of memory—changes that involve the rearrangement of the position of particles within the solid. Changes between solids and liquids or liquids and gases are well known (e.g., boiling water or melting ice), but the same sort of changes can occur between two solids. In the case of nitinol, a *parent shape* (the shape to which you want it to return) has to be defined. To fix the parent shape of nitinol it must be heated. No changes are apparent, but when it cools, the atoms in the solid rearrange into a slightly different structure. Thus, whenever the nitinol is heated the atoms rearrange themselves back to the structure necessary to produce the parent shape.

Buehler and Wang then began the process of moving nitinol from the experimental world of the laboratory to the commercial world of applications. By the late sixties nitinol was being used in pipe couplings in the aircraft industry.

In 1968, George Andreason, a dentist, began experimenting with nitinol in his metal-working shop where he made jewelry as a hobby. He developed a fine wire that could be molded to fit a patient's mouth (parent position), which when cooled could be bent to fit the misaligned teeth. When warmed to body temperature, it would exert a gentle constant pressure on the teeth. This was a major breakthrough in orthodontics, cutting the treatment time in half from that of steel braces.

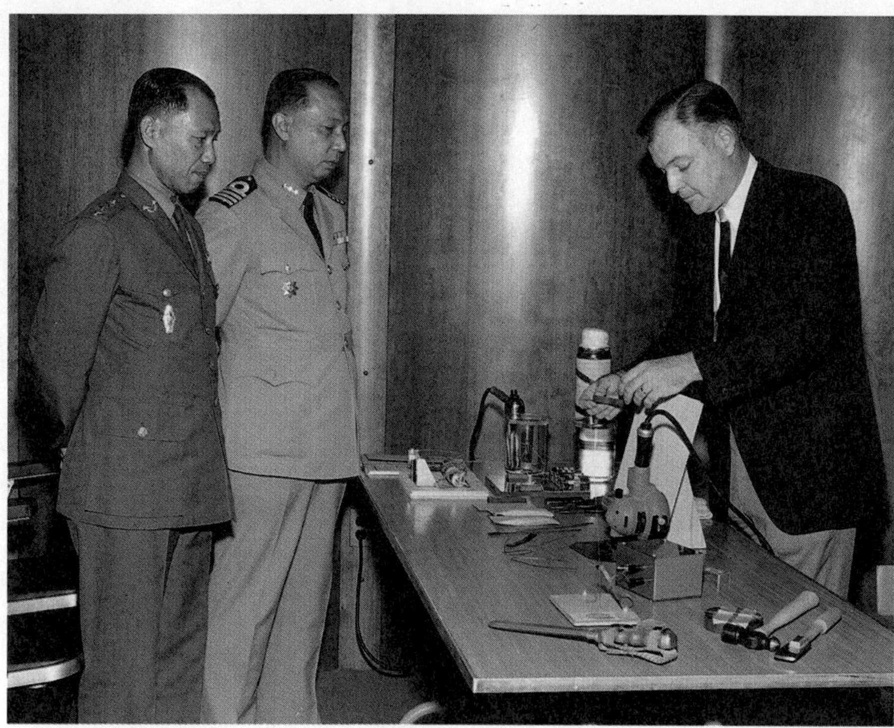

◀ **William Buehler demonstrating the unique properties of nitinol at the Naval Ordnance Laboratory in 1969.**

Dr. Wang left NOL in 1980 to become a supplier of nitinol to the many and varied manufacturers who wanted this exciting new product for such things as eyeglass frames, which can take extreme abuse (e.g., bending them, sitting on them, or twisting them) and return to their original shape; antiscalding devices for showerheads and faucets, which automatically shut off if the water approaches 120°F; and toys such as blinking movie posters and tail-swishing dinosaurs.

The uses of nitinol are varied and growing. Today it has many applications in the fields of medicine, engineering, and safety; in housewares; and even in the lingerie business in underwire brassieres. Nitinol was the first of the "intelligent" materials—ones that respond to changes in the environment.

1.4 The Scientific Method

The nitinol story illustrates how chemists and other scientists work together to solve a specific problem. Chemists, metallurgists, physicists, engineers, and a wide variety of technicians were involved in the development of nitinol. As scientists conduct studies they ask many questions, and their questions often lead in directions that are not part of the original problem. The amazing developments from chemistry and technology usually result from the use of the scientific method. Although complete agreement is lacking on exactly what is meant by "using the scientific method," the general approach is as follows.

1. **Collect the facts or data** that are relevant to the problem or question at hand. This is usually done by planned experimentation. In the nitinol story, Buehler compared the properties of various substances until he selected nitinol. He also investigated the properties of nitinol, including its ability to produce different sounds when warm and cool, and its unique ability to return to its original shape. Analyze the data to find trends or regularities that are pertinent to the problem.
2. **Formulate a hypothesis** that will account for the data that have been accumulated and that can be tested by further experimentation. Wang and Buehler investigated the properties and proposed a hypothesis regarding the two phases of nitinol. Further tests supported their model.
3. **Plan and do additional experiments to test the hypothesis.** The nitinol group continued its testing process until the mechanism was well understood.
4. **Modify the hypothesis** as necessary so that it is compatible with all the pertinent data.

hypothesis

theory

scientific laws

Confusion sometimes arises regarding the exact meanings of the words *hypothesis, theory,* and *law.* A **hypothesis** is a tentative explanation of certain facts that provides a basis for further experimentation. A well-established hypothesis is often called a **theory**. Thus a theory is an explanation of the general principles of certain phenomena with considerable evidence or facts to support it. Hypotheses and theories explain natural phenomena whereas **scientific laws** are simple statements of natural phenomena to which no exceptions are known under the given conditions.

Although these four steps are a broad outline of the general procedure that is followed in most scientific work, they are not a recipe for doing chemistry or any other science (Figure 1.1). But chemistry is an experimental science, and much of its progress has been due to application of the scientific method through systematic research.

Many theories and laws are studied in chemistry, which make the study of any science easier because they summarize particular aspects of that science. Some of the theories advanced by great scientists in the past have since been substantially altered and modified. Such changes do not mean that the discoveries of the past are less significant than those of today. Modification of existing theories in the light of new experimental evidence is essential to the growth and evolution of scientific knowledge.

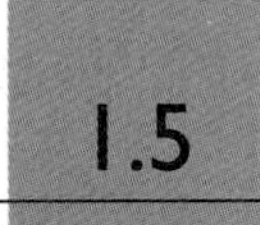

1.5 Relationship of Chemistry to Other Sciences and Industry

Besides being a science in its own right, chemistry is the servant of other sciences and industry. Chemical principles contribute to the study of physics, biology, agriculture, engineering, medicine, space research, oceanography, and many other sciences. Chemistry and physics are overlapping sciences, since both are based on the properties and behavior of matter. Biological processes are chemical in nature. The metabolism of food to provide energy to living organisms is a chemical process. Knowledge of molecular structure of proteins, hormones, enzymes, and the nucleic acids is assisting biologists in their investigations of the composition, development, and reproduction of living cells.

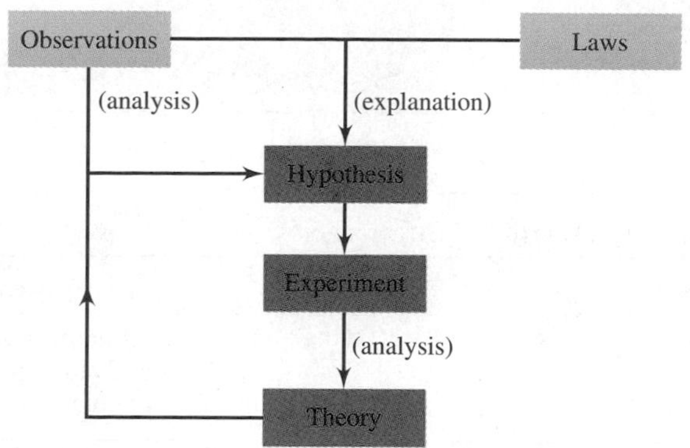

FIGURE 1.1
The scientific method.

Chemistry is playing an important role in alleviating the growing shortage of food in the world. Agricultural production has been increased with the use of chemical fertilizers, pesticides, and improved varieties of seeds. Chemical refrigerants make possible the frozen food industry, which preserves large amounts of food that might otherwise spoil. Chemistry is also producing synthetic nutrients, but much remains to be done as the world population increases relative to the land available for cultivation. Expanding energy needs have brought about difficult environmental problems in the form of air and water pollution. Chemists and other scientists are working diligently to alleviate these problems.

Scientists employ the scientific method every day in their laboratory work.

Advances in medicine and chemotherapy, through the development of new drugs, have contributed to prolonged life and the relief of human suffering. More than 90% of the drugs and pharmaceuticals being used in the United States today have been developed commercially within the past 50 years. The plastics and polymer industry, unknown 60 years ago, has revolutionized the packaging and textile industries and is producing durable and useful construction materials. Energy derived from chemical processes is used for heating, lighting, and transportation. Virtually every industry is dependent on chemicals—for example, the petroleum, steel, rubber, pharmaceutical, electronic, transportation, cosmetic, space, polymer, garment, aircraft, and television industries.

People outside of science usually have the perception that science is an intensely logical field. They picture the white-coated chemist going from hypothesis to experiment and then to laws and theories without error or emotion. Quite often scientific discoveries are the result of trial and error. In the nitinol story the memory nature of the material was discovered because someone wanted to see what would happen if a flame was applied to compressed nitinol. Buehler's creativity and insight and his ability to relate this chance discovery to his experiments enabled him to make the connection. This is an excellent example of what is meant by serendipity in science. Buehler was searching for a material to use as a nose cone in missiles, but the results of his experiments were quite unexpected. He and his coworkers discovered a material that has applications in medicine, engineering, and everyday life.

The nitinol story also illustrates the fact that scientists do not work alone in their laboratories. Buehler, an engineer, was joined by Wang, a crystallographer, and many other chemists, engineers, dentists, and physicians in developing the applications of nitinol. Each of them had a contribution to make to the body of

Serendipity in Science

Discoveries in the world of chemistry are for the most part made by people who are applying the scientific method in their work. Occasionally, important discoveries are made by chance, or through serendipity. But even when serendipity is involved, a discovery is more likely to be made by someone with a good knowledge of the field. Louis Pasteur summed this up in a statement made long ago: "Chance favors the prepared mind." In chemistry serendipity often can lead to whole new fields and technology.

The synthetic dye industry began in 1856 when William Perkin, an 18-year-old student at the Royal College of Chemistry in London, was attempting to synthesize quinine, a drug used to treat malaria. He reacted two chemicals, aniline sulfate and potassium dichromate, and obtained a black paste. Perkin then extracted the paste with alcohol. Upon evaporating the alcohol, violet crystals appeared that, when dissolved in water, made a beautiful purple solution. He so enjoyed the color he began investigating the solution; he then determined the purple color had a strong affinity for silk. Perkin had discovered the first synthetic aniline dye. Recognizing the commercial possibilities, he immediately left school and, with his father and an older brother, went into the dye manufacturing business. His dye, known as mauve, quickly became a success and inspired other research throughout Europe. By 1870, cloth could be purchased in more and brighter synthetic colors than were ever available with natural dyes.

A second, more recent, account of chance events in chemistry also led to a multimillion dollar industry (see table). In 1965 James Schlatter was researching anti-ulcer drugs for the pharmaceutical firm G. D. Searle. In the course of his work he accidentally ingested a small amount of his preparation and found, to his surprise, it had an extremely sweet taste. (*Note:* Tasting chemicals of any kind in the laboratory is not a safe procedure.) When purified, the sweet-tasting substance turned out to be aspartame, a molecule consisting of two amino acids joined together. Since only very small quantities are necessary to produce sweetness, it proved to be an excellent low calorie artificial sweetener. Today, under the trade names of "Equal" and "Nutrasweet," aspartame is one of the cornerstones of the artificial sweetener industry.

The Discovery of Artificial Sweeteners

Sweetener	Date	Discoverer
Saccharin	1878	I. Remsen and C. Fahlberg
Cyclamate	1937	M. Sveda
Aspartame	1965	J. Schlatter

The wide array of beautiful colors seen in this fabric shop come from synthetic dyes, which have been available for more than 100 years.

knowledge surrounding the substance nitinol. Chemistry is a field in which teamwork and cooperation play a vital role in understanding complex systems.

1.6 Risks and Benefits

DNA analysis is playing an ever-important role in such fields as genetics, disease control, and crime.

The nitinol story is only one example of the many problems confronting us today that rely on science for an answer. Virtually every day we read or hear about stories such as:

- developing an AIDS vaccine
- banning the use of herbicides and pesticides
- analyzing DNA to determine genetic disease, biological parents, or to place a criminal at the scene of a crime
- removing asbestos from public buildings
- removing lead from drinking water
- the danger of radon in our homes
- global warming
- the hole in the ozone layer
- health risks associated with coffee, margarine, saturated fats, and other foods
- burning of tropical rain forests and the effect on global ecology

Which of these risks present true danger to us and which pose no great problem? All of these problems will be around for many years and new ones will be continually added to the list. Wherever we live and whatever our occupation, each of us is exposed to chemicals and chemical hazards every day. The question we must answer is: Do the risks outweigh the benefits?

Risk assessment is a process that brings together professionals from the fields of chemistry, biology, toxicology, and statistics in order to determine the risk associated with exposure to a certain chemical. Assessment of risk involves determining both the probability of exposure and the severity of that exposure. Once this is done an estimate can be made of the overall risk. Studies have shown how people perceive various risks. The perception of risk depends on some rather interesting factors. Voluntary risks, such as smoking or flying, are much more easily accepted than involuntary ones, such as herbicides on apples or asbestos in buildings. People also often conclude that anything "synthetic" is bad while anything "organic" is good. Risk assessment may provide information on the degree of risk but not on whether the chemical is "safe." Safety is a qualitative judgment based on many personal factors including beliefs, preferences, benefits, and costs.

Once a risk has been assessed the next step is to manage it. This involves ethics, economics, and equity as well as government and politics. For example, some things are perceived as low risk by scientists (such as asbestos in buildings) but are classified as high risk by the general public. This inconsistency may result in the expenditure of millions of dollars to rid the public of a perceived threat that is much lower than they believe.

Risk management involves value judgments that integrate social, economic, and political issues. These risks must be weighed against the benefits of new technol-

Asbestos, once a widely used building material, was banned by the EPA in 1986 because of its health hazards. Today, the removal of asbestos from our homes and our public buildings has become a big concern. ▶

ogy and products in order to make the decisions required at home, in our local communities, and around the world. We use both risk assessment and risk management to decide whether to buy a certain product (such as a pesticide), take a certain drug (such as a pain reliever), or eat certain foods (such as hot dogs). We must realize that all risks can never be eliminated. Our goal is to minimize unnecessary risks and to make responsible decisions regarding the risks in our lives and our environment.

The theories and models used in risk assessment are based on concepts learned in chemistry—they are based on assumptions and therefore contain uncertainties. By improving your understanding of the concepts of chemistry you will be better able to understand the capabilities and limitations of science. You can then intelligently question the process of risk assessment and make decisions that will lead to a better understanding of our world and our responsibilities to each other.

Key Terms

The terms listed here have all been defined within the chapter. The section number is given for each term.

These terms can also be found in the Glossary.

chemistry (1.2)
hypothesis (1.4)
scientific laws (1.4)
theory (1.4)

Standards for Measurement

2

Chapter 2

Doing an experiment in chemistry is very much like cooking a meal in the kitchen. It is important to know the ingredients *and* the amounts of each in order to have a tasty product. Working on your car requires specific tools, in exact sizes. Buying new carpeting or draperies is an exercise in precise and accurate measurement for a good fit. A small difference in the concentration or amount of medication a pharmacist gives you may have significant effects on your well-being. In all of these cases, the ability to measure accurately and a strong foundation in the language and use of numbers provide the basis for success. In chemistry, we begin by learning the metric system and the proper units for measuring mass, length, volume, and temperature.

2.1 Mass and Weight

Chemistry is an experimental science. The results of experiments are usually determined by making measurements. In elementary experiments the quantities that are commonly measured are mass, length, volume, pressure, temperature, and time. Measurements of electrical and optical quantities may also be needed in more sophisticated experimental work.

mass

Although mass and weight are often used interchangeably, the two words have quite different meanings. The **mass** of a body is defined as the amount of matter in that body. The mass of an object is a fixed and unvarying quantity that is independent of the object's location. The mass of an object can be measured on a balance by comparison with other known masses.

An everyday example of a balance is a child's seesaw, shown in Figure 2.1. If children of equal mass sit on opposite ends the seesaw balances (2.1a). If one child is heavier than the other the seesaw sinks on the side holding the heavier child (2.1b). To bring the seesaw back into balance additional mass must be added to the side holding the lighter child (2.1c). Another common example of a balance is shown in Figure 2.2. These balances were used by assayers during the gold rush to determine the mass of gold brought in by the prospectors. The gold was placed on one pan of the balance and known masses on the other side until the pans were level with each other. (Several modern balances are shown in Figure 2.5, page 28.)

weight

The **weight** of a body is the measure of the earth's gravitational attraction for that body. Weight is measured on a device called a scale, which measures force against a spring. Unlike mass, weight varies in relation to the position of an object on earth or its distance from the earth.

Consider an astronaut of mass 70.0 kilograms (154 pounds) who is being shot into orbit. At the instant before blast-off the weight of the astronaut is also 70.0 kilograms. As the distance from the earth increases and the rocket turns into an orbiting course, the gravitational pull on the astronaut's body decreases until a state of weightlessness (zero weight) is attained. However, the mass of the astronaut's body has remained constant at 70.0 kilograms during the entire event.

◀ **Chapter Opening Photo: Measuring instruments come in various shapes and sizes, all necessary to allow components to fit together, and permit us to quantify our world.**

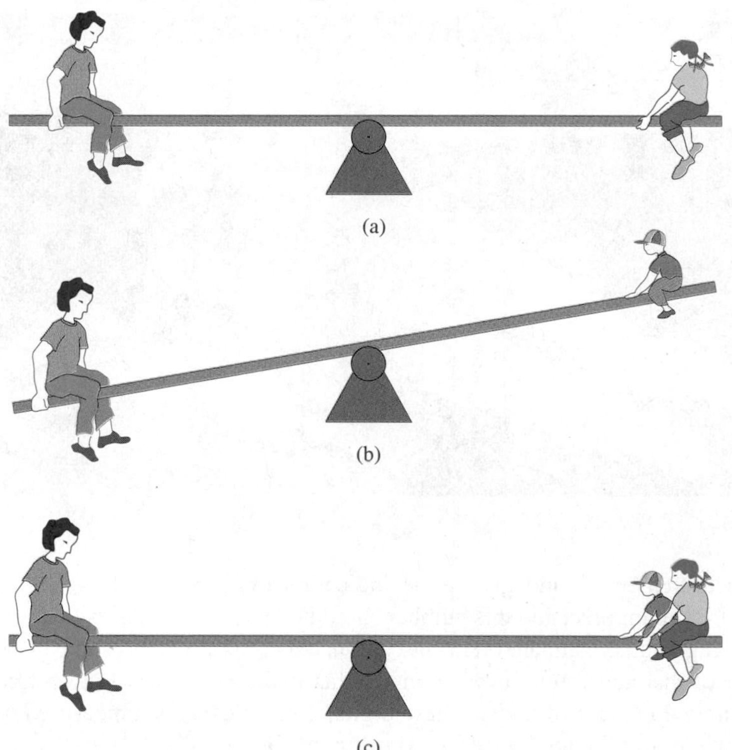

◀ **FIGURE 2.1**
We experience our first lessons in balancing from playing on a seesaw.

2.2 Measurement and Significant Figures

To understand certain aspects of chemistry it is necessary to set up and solve problems. Problem solving requires an understanding of the elementary mathematical operations used to manipulate numbers. Numerical values or data are obtained from measurements made in an experiment. A chemist may use these data to calculate the extent of the physical and chemical changes occurring in the substances that are being studied. By appropriate calculations the results of an experiment can be compared with those of other experiments and summarized in ways that are meaningful.

The result of a measurement is expressed by a numerical value together with a unit of that measurement. For example,

numerical value
70.0 kilograms = 154 pounds
unit

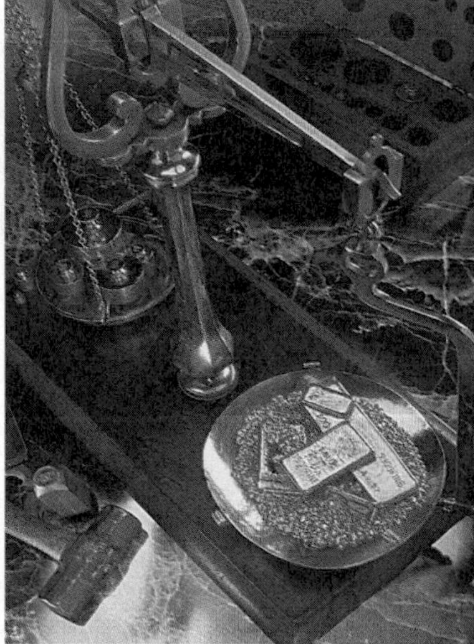

FIGURE 2.2
An assayer's balance is used for weighing gold.

Numbers obtained from a measurement are never exact values. They always have some degree of uncertainty due to the limitations of the measuring instrument and the skill of the individual making the measurement. The numerical value recorded

Floating some 130 nautical miles above the earth, this astronaut has attained weightlessness even though the mass of his body remains constant. ▶

for a measurement should give some indication of its reliability (precision). To express maximum precision this number should contain all the digits that are known plus one digit that is estimated. This last estimated digit introduces some uncertainty. Because of this uncertainty every number that expresses a measurement can have only a limited number of digits. These digits, used to express a measured quantity, are known as **significant figures**, or **significant digits**.

significant figures

Suppose we measure temperature on a thermometer calibrated in degrees and observe that the mercury stops between 21 and 22 (see Figure 2.3a). We then know that the temperature is at least 21 degrees and is less than 22 degrees. To express the temperature with greater precision, we estimate that the mercury is about two-tenths the distance between 21 and 22. The temperature is, therefore, 21.2 degrees. The last digit (2) has some uncertainty because it is an estimated value. The recorded temperature, 21.2 degrees, is said to have three significant figures. If the mercury stopped exactly on the 22 (Figure 2.3b), the temperature would be recorded as 22.0 degrees. The zero is used to indicate that the temperature was estimated to a precision of one-tenth degree. Finally, look at Figure 2.3c. On this thermometer, the temperature is recorded as 22.11°C (four significant figures). Since the thermometer is calibrated to tenths of a degree, the first estimated digit is the hundredths.

Some numbers are exact and have an infinite number of significant figures. Exact numbers occur in simple counting operations; when you count 25 dollars, you have exactly 25 dollars. Defined numbers, such as 12 inches in 1 foot, 60 minutes in 1 hour, and 100 centimeters in 1 meter, are also considered to be exact numbers. Exact numbers have no uncertainty.

Evaluating Zero

In any measurement all nonzero numbers are significant. However, zeros may or may not be significant depending on their position in the number. Here are some rules for determining when zero is significant.

1. Zeros between nonzero digits are significant:
   205 has three significant figures

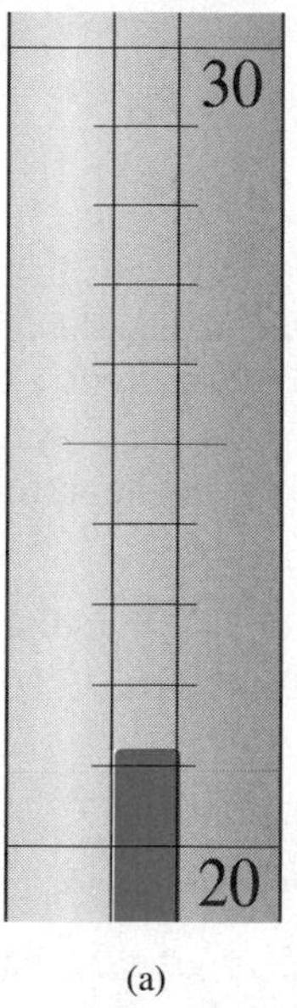

(a)

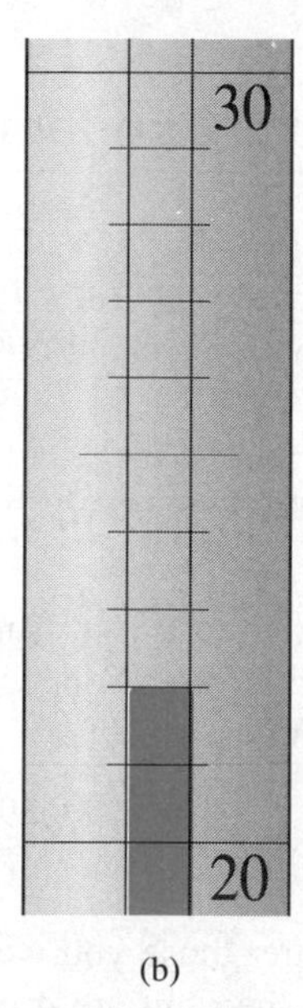

(b)

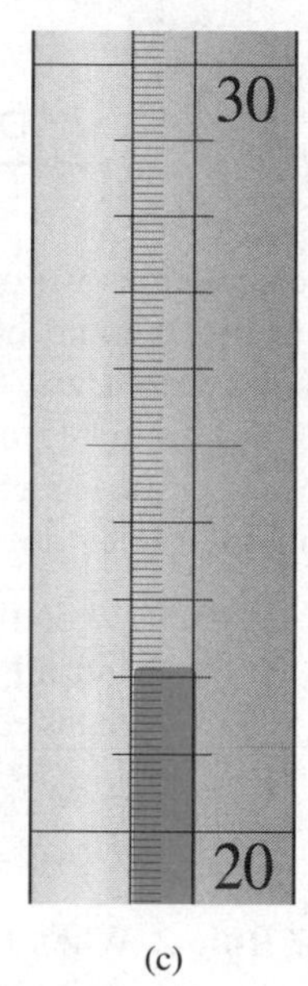

(c)

◀ **FIGURE 2.3**
Measuring temperature with various degrees of precision.

2.05 has three significant figures
61.09 has four significant figures

2. Zeros that precede the first nonzero digit are not significant. These zeros are used to locate a decimal point:
   0.0025 has two significant figures (2, 5)
   0.0108 has three significant figures (1, 0, 8)
3. Zeros at the end of a number that include a decimal point are significant:
   0.500 has three significant figures (5, 0, 0)
   25.160 has five significant figures
   3.00 has three significant figures
   20. has two significant figures
4. Zeros at the end of a number without a decimal point are not considered significant:
   1000 has one significant figure
   590 has two significant figures

   One way of indicating that these zeros are significant is to write the number using a decimal point and a power of 10. Thus, if the value 1000 has been determined to four significant figures, it is written as 1.000×10^3. If 590 has only two significant figures, it is written as 5.9×10^2.

Rules for significant figures should be memorized for use throughout the text.

Practice 2.1

How many significant figures are in each of these numbers?

(a) 4.5 inches	(e) 25.0 grams
(b) 3.025 feet	(f) 12.20 liters
(c) 125.0 meters	(g) 100,000 people
(d) 0.001 mile	(h) 205 birds

Answers to Practice Exercises are found at the end of each chapter.

2.3 Rounding Off Numbers

In calculations we often obtain answers that have more digits than we are justified in using. It is necessary, therefore, to drop the nonsignificant digits in order to express the answer with the proper number of significant figures. When digits are dropped from a number, the value of the last digit retained is determined by a process known as **rounding off numbers**. Two rules will be used in this book for rounding off numbers:

rounding off numbers

Not all schools use the same rules for rounding. Check with your instructor for variations in these rules.

Rule 1 When the first digit after those you want to retain is 4 or less, that digit and all others to its right are dropped. The last digit retained is not changed. The following examples are rounded off to four digits:

74.693 = 74.69 (This digit is dropped.)

1.00629 = 1.006 (These two digits are dropped.)

Rule 2 When the first digit after those you want to retain is 5 or greater, that digit and all others to the right are dropped and the last digit retained is increased by one. These examples are rounded off to four digits:

1.026868 = 1.027 (These three digits are dropped. This digit is changed to 7.)

18.02500 = 18.03 (These three digits are dropped. This digit is changed to 3.)

12.899 = 12.90 (This digit is dropped. These two digits are changed to 90.)

Practice 2.2

Round off these numbers to the number of significant digits indicated:

(a) 42.246 (four digits) (d) 0.08965 (two digits)
(b) 88.015 (three digits) (e) 225.3 (three digits)
(c) 0.08965 (three digits) (f) 14.150 (three digits)

2.4 Scientific Notation of Numbers

The age of the earth has been estimated to be about 4,500,000,000 (4.5 billion) years. Because this is an estimated value, say to the nearest 0.1 billion years, we are justified in using only two significant figures to express it. Thus, we write it, using a power of 10, as 4.5×10^9 years.

Very large and very small numbers are often used in chemistry and can be simplified and conveniently written using a power of 10. Writing a number as a power of 10 is called **scientific notation**.

scientific notation

To write a number in scientific notation, move the decimal point in the original number so that it is located after the first nonzero digit. This new number is multiplied by 10 raised to the proper power (exponent). The power of 10 is equal to the number of places that the decimal point has been moved. If the decimal is moved to the left, the power of 10 will be a positive number. If the decimal is moved to the right, the power of 10 will be a negative number.

The scientific notation of a number is the number written as a factor between 1 and 10 multiplied by 10 raised to a power. For example,

$2468 = 2.468 \times 10^3$

number — scientific notation of the number

Examples show you problem-solving techniques in a step-by-step form. Study each one and then try the Practice Exercises.

Example 2.1

Write 5283 in scientific notation.

Solution

5283. (3) Place the decimal between the 5 and the 2. Since the decimal was moved three places to the left the power of 10 will be 3, and the number 5.283 is multiplied by 10^3.

5.283×10^3

Example 2.2

Write 4,500,000,000 in scientific notation (two significant figures).

Solution

4 500 000 000. (9) Place the decimal between the 4 and the 5. Since the decimal was moved nine places to the left the power of 10 will be 9, and the number 4.5 is multiplied by 10^9.

4.5×10^9

Example 2.3

Write 0.000123 in scientific notation.

Solution

0.000123 (4) Place the decimal between the 1 and the 2. Since the decimal was moved four places to the right the power of 10 will be -4, and the number 1.23 is multiplied by 10^{-4}.

1.23×10^{-4}

Practice 2.3

Write the following numbers in scientific notation:

(a) 1200 (four digits)
(c) 0.0468
(b) 6,600,000 (two digits)
(d) 0.00003

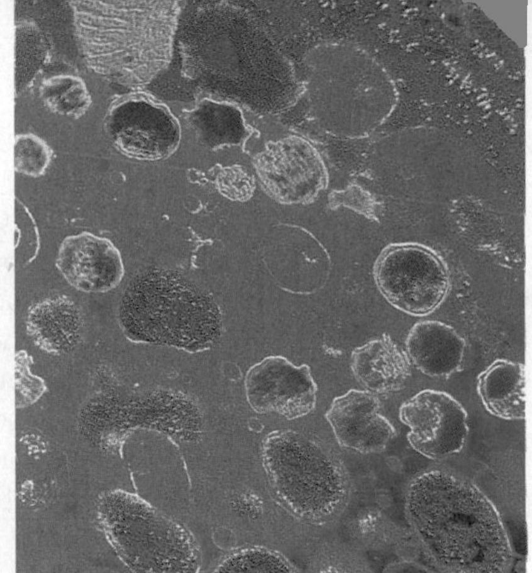

Left: Distances between the moon and stars are often so great they must be expressed in scientific notation. Right: These *Chlamydia* bacteria are so small, on the other hand, they need great magnification in order to be seen (18,000X).

2.5 Significant Figures in Calculations

The results of a calculation based on measurements cannot be more precise than the least precise measurement.

Use your calculator to check your work in the examples. Compare your results to be sure you understand the mathematics.

Multiplication or Division

In calculations involving multiplication or division, the answer must contain the same number of significant figures as in the measurement that has the least number of significant figures. Consider the following examples:

Example 2.4 $190.6 \times 2.3 = 438.38$

Solution The value 438.38 was obtained with a calculator. The answer should have two significant figures, because 2.3, the number with the fewest significant figures, has only two significant figures.

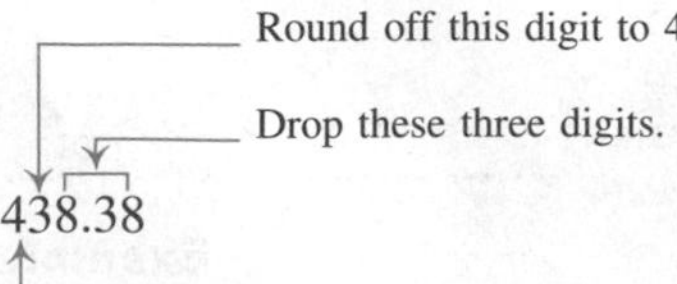

Round off this digit to 4.
Drop these three digits.
438.38
Move the decimal 2 places to the left to express in scientific notation.

The correct answer is 440 or 4.4×10^2.

Example 2.5 $\dfrac{13.59 \times 6.3}{12} = 7.13475$

Solution The value 7.13475 was obtained with a calculator. The answer should contain two significant figures because 6.3 and 12 have only two significant figures.

Drop these four digits.

7.13475

This digit remains the same.

The correct answer is 7.1.

Practice 2.4

134 in. × 25 in. = ?

Practice 2.5

$$\frac{213 \text{ miles}}{4.20 \text{ hours}} = ?$$

Practice 2.6

$$\frac{2.2 \times 273}{760} = ?$$

Addition or Subtraction

The results of an addition or a subtraction must be expressed to the same precision as the least precise measurement. This means the result must be rounded to the same number of decimal places as the value with the fewest decimal places.

Example 2.6

Add 125.17, 129, and 52.2.

Solution

```
 125.17
 129.
  52.2
 ------
 306.37   (306)
```

The number with the least precision is 129. Therefore the answer is rounded off to the nearest unit: 306.

Example 2.7

Subtract 14.1 from 132.56.

Solution

```
  132.56
 − 14.1
 -------
  118.46   (118.5)
```

14.1 is the number with the least precision. Therefore, the answer is rounded off to the nearest tenth: 118.5.

Example 2.8

Subtract 120 from 1587.

Solution

```
  1587
 − 120
 ------
  1467   (1.47 × 10^3)
```

120 is the number with least precision. The zero is not considered significant; therefore, the answer must be rounded to the nearest ten: 1470 or 1.47×10^3.

Example 2.9 Add 5672 and 0.00063.

Solution

$$\begin{array}{r} 5672 \\ +\ 0.00063 \\ \hline 5672.00063 \end{array} \quad (5672)$$

The number with least precision is 5672. Therefore, the answer is rounded off to the nearest unit: 5672.

Example 2.10 $\dfrac{1.039 - 1.020}{1.039} = 0.018286814$

Solution The value 0.018286814 was obtained with a calculator. When the subtraction in the numerator is done,

$$1.039 - 1.020 = 0.019$$

the number of significant figures changes from four to two. Therefore, the answer should contain two significant figures after the division is carried out:

0.018286814 — Drop these six digits (286814); the digit 8 (in 0.018) remains the same.

The correct answer is 0.018, or 1.8×10^{-2}.

Practice 2.7

How many significant figures should the answer contain in each of these calculations?

(a) 14.0×5.2

(b) 0.1682×8.2

(c) $\dfrac{160 \times 33}{4}$

(d) $8.2 + 0.125$

(e) $119.1 - 3.44$

(f) $\dfrac{94.5}{1.2}$

(g) $1200 + 6.34$

(h) $1.6 + 23 - 0.005$

If you need to brush up on your math skills refer to the "Mathematical Review" in Appendix I.

Additional material on mathematical operations is given in Appendix I, "Mathematical Review." Study any portions that are not familiar to you. You may need to do this at various times during the course when additional knowledge of mathematical operations arises.

2.6 The Metric System

metric system or SI

The **metric system**, or **International System** (**SI**, from *Système International*), is a decimal system of units for measurements of mass, length, time, and other physical quantities. It is built around a set of standard units and uses factors of 10 to express

TABLE 2.1 Prefixes and Numerical Values for SI Units*

Prefix	Symbol	Numerical value	Power of 10 equivalent
exa	E	1,000,000,000,000,000,000	10^{18}
peta	P	1,000,000,000,000,000	10^{15}
tera	T	1,000,000,000,000	10^{12}
giga	G	1,000,000,000	10^{9}
mega	M	1,000,000	10^{6}
kilo	k	1,000	10^{3}
hecto	h	100	10^{2}
deka	da	10	10^{1}
—	—	1	10^{0}
deci	d	0.1	10^{-1}
centi	c	0.01	10^{-2}
milli	m	0.001	10^{-3}
micro	μ	0.000001	10^{-6}
nano	n	0.000000001	10^{-9}
pico	p	0.000000000001	10^{-12}
femto	f	0.000000000000001	10^{-15}
atto	a	0.000000000000000001	10^{-18}

* The more commonly used prefixes are in color.

Most products today list both systems of measurement on their labels.

larger or smaller numbers of these units. To express quantities that are larger or smaller than the standard units, prefixes are added to the names of the units. These prefixes represent multiples of 10, making the metric system a decimal system of measurements. Table 2.1 shows the names, symbols, and numerical values of the prefixes. Some examples of the more commonly used prefixes are:

1 *kilo*meter = 1000 meters
1 *kilo*gram = 1000 grams
1 *milli*meter = 0.001 meter
1 *micro*second = 0.000001 second

The seven standard units in the International System, their abbreviations, and the quantities they measure are given in Table 2.2. Other units are derived from these units.

The metric system, or International System, is currently used by most of the countries in the world, not only in scientific and technical work but also in commerce and industry.

TABLE 2.2 International System's Standard Units of Measurement

Quantity	Name of unit	Abbreviation
Length	meter	m
Mass	kilogram	kg
Temperature	kelvin	K
Time	second	s
Amount of substance	mole	mol
Electric current	ampere	A
Luminous intensity	candela	cd

2.7 Measurement of Length

Standards for the measurement of length have an interesting history. The Old Testament mentions such units as the *cubit* (the distance from a man's elbow to the tip of his outstretched hand). In ancient Scotland the inch was once defined as a distance equal to the width of a man's thumb.

Reference standards of measurements have undergone continuous improvements in precision. The standard unit of length in the metric system is the **meter**. When the metric system was first introduced in the 1790s, the meter was defined as one ten-millionth of the distance from the equator to the North Pole, measured along the meridian passing through Dunkirk, France. In 1889 the meter was redefined as the distance between two engraved lines on a platinum–iridium alloy bar maintained at 0° Celsius. This international meter bar is stored in a vault at Sèvres near Paris. Duplicate meter bars have been made and are used as standards by many nations.

meter (m)

By the 1950s length could be measured with such precision that a new standard was needed. Accordingly, the length of the meter was redefined in 1960 and again in 1983. The latest definition is: A meter is the distance that light travels in a vacuum during 1/299,792,458 of a second.

A meter is 39.37 inches, a little longer than 1 yard. One meter contains 10 decimeters, 100 centimeters, or 1000 millimeters (see Figure 2.4). A kilometer contains 1000 meters. Table 2.3 shows the relationships of these units.

TABLE 2.3 Metric Units of Length

Unit	Abbreviation	Meter equivalent	Exponential equivalent
kilometer	km	1000 m	10^3 m
meter	m	1 m	10^0 m
decimeter	dm	0.1 m	10^{-1} m
centimeter	cm	0.01 m	10^{-2} m
millimeter	mm	0.001 m	10^{-3} m
micrometer	μm	0.000001 m	10^{-6} m
nanometer	nm	0.000000001 m	10^{-9} m
angstrom	Å	0.0000000001 m	10^{-10} m

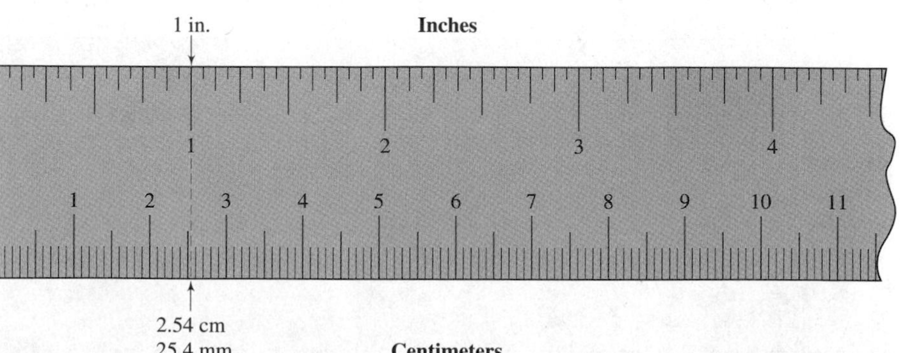

FIGURE 2.4 ▶ Comparison of the metric and American systems of length measurement: 2.54 cm = 1 in.

The nanometer (10^{-9} m) is used extensively in expressing the wavelength of light as well as in atomic dimensions. See inside back cover for a complete table of common conversions. Other important relationships are:

See inside back cover for a table of conversion.

$$1\ \text{m} = 100\ \text{cm} = 1000\ \text{mm} = 10^6\ \mu\text{m} = 10^{10}\ \text{Å}$$
$$1\ \text{cm} = 10\ \text{mm} = 0.01\ \text{m}$$
$$1\ \text{in.} = 2.54\ \text{cm}$$
$$1\ \text{mile} = 1.609\ \text{km}$$

2.8 Problem Solving

Many chemical principles are illustrated by mathematical concepts. Learning how to set up and solve numerical problems in a systematic fashion is *essential* in the study of chemistry. This skill, once acquired, is also very useful in other study areas. A calculator will save you much time in computation.

Usually a problem can be solved by several methods. But in all methods it is best, especially for beginners, to use a systematic, orderly approach. The *dimensional analysis,* or *factor-label, method* is stressed in this book because

1. It provides a systematic, straightforward way to set up problems.
2. It gives a clear understanding of the principles involved.
3. It trains you to organize and evaluate data.
4. It helps to identify errors because unwanted units are not eliminated if the setup of the problem is incorrect.

The basic steps for solving problems are

Steps for solving problems are highlighted in color for easy reference.

1. Read the problem very carefully to determine what is to be solved for, and write it down.
2. Tabulate the data given in the problem. Even in tabulating data it is important to label all factors and measurements with the proper units.
3. Determine which principles are involved and which unit relationships are needed to solve the problem. Sometimes it is necessary to refer to tables for needed data.
4. Set up the problem in a neat, organized, and logical fashion, making sure that unwanted units cancel. Use sample problems in the text as guides for making setups.
5. Proceed with the necessary mathematical operations. Make certain that the answer contains the proper number of significant figures.
6. Check the answer to see if it is reasonable.

Just a few more words about problem solving. Don't allow any formal method of problem solving to limit your use of common sense and intuition. If a problem is clear to you and its solution seems simpler by another method, by all means use it. But in the long run you should be able to solve many otherwise difficult problems by using the dimensional analysis method.

The dimensional analysis method of problem solving converts one unit to another unit by the use of conversion factors.

Important equations are boxed or highlighted in color.

$$\textbf{unit}_1 \times \textbf{conversion factor} = \textbf{unit}_2$$

If you want to know how many millimeters are in 2.5 meters, you need to convert meters (m) to millimeters (mm). Therefore, you start by writing

$$\text{m} \times \text{conversion factor} = \text{mm}$$

This conversion factor must accomplish two things. It must cancel (or eliminate) meters, and it must introduce millimeters—the unit wanted in the answer. Such a conversion factor will be in fractional form and have meters in the denominator and millimeters in the numerator:

$$\cancel{\text{m}} \times \frac{\text{mm}}{\cancel{\text{m}}} = \text{mm}$$

We know that 1 m = 1000 mm. From this relationship we can write two conversion factors—1 m per 1000 mm and 1000 mm per 1 m:

$$\frac{1\ \text{m}}{1000\ \text{mm}} \quad \text{and} \quad \frac{1000\ \text{mm}}{1\ \text{m}}$$

Using the conversion factor 1000 mm/1 m, we can set up the calculation for the conversion of 2.5 m to millimeters,

$$2.5\ \cancel{\text{m}} \times \frac{1000\ \text{mm}}{1\ \cancel{\text{m}}} = 2500\ \text{mm} \quad \text{or} \quad 2.5 \times 10^3\ \text{mm}$$

(two significant figures)

Note that, in making this calculation, units are treated as numbers; meters in the numerator are canceled by meters in the denominator.

Now suppose you need to change 215 centimeters to meters. We start with

$$\text{cm} \times \text{conversion factor} = \text{m}$$

The conversion factor must have centimeters in the denominator and meters in the numerator:

$$\cancel{\text{cm}} \times \frac{\text{m}}{\cancel{\text{cm}}} = \text{m}$$

From the relationship 100 cm = 1 m, we can write a factor that will accomplish this conversion:

$$\frac{1\ \text{m}}{100\ \text{cm}}$$

Now set up the calculation using all the data given.

$$215\ \cancel{\text{cm}} \times \frac{1\ \text{m}}{100\ \cancel{\text{cm}}} = \frac{215\ \text{m}}{100} = 2.15\ \text{m}$$

Some problems may require a series of conversions to reach the correct units in the answer. For example, suppose we want to know the number of seconds in 1 day. We need to go from the unit of days to seconds in this manner:

day → hours → minutes → seconds

Units are emphasized in problems by using color and flow diagrams to help visualize the steps in the process.

This series requires three conversion factors, one for each step. We convert days to hours (hr), hours to minutes (min), and minutes to seconds (s). The conversions can be done individually or in a continuous sequence:

$$\text{day} \times \frac{\text{hr}}{\text{day}} \rightarrow \text{hr} \times \frac{\text{min}}{\text{hr}} \rightarrow \text{min} \times \frac{\text{s}}{\text{min}} = \text{s}$$

$$\text{day} \times \frac{\text{hr}}{\text{day}} \times \frac{\text{min}}{\text{hr}} \times \frac{\text{s}}{\text{min}} = \text{s}$$

Inserting the proper factors we calculate the number of seconds in 1 day to be

$$1\ \text{day} \times \frac{24\ \text{hr}}{1\ \text{day}} \times \frac{60\ \text{min}}{1\ \text{hr}} \times \frac{60\ \text{s}}{1\ \text{min}} = 86{,}400.\ \text{s}$$

All five digits in 86,400 are significant, since all the factors in the calculation are exact numbers.

The dimensional analysis, or factor-label, method used in the preceding work shows how unit conversion factors are derived and used in calculations. After you become more proficient with the terms, you can save steps by writing the factors directly in the calculation. The problems that follow give examples of the conversion from American to metric units.

Label all factors with the proper units.

Example 2.11

How many centimeters are in 2.00 ft?

Solution

The stepwise conversion of units from feet to centimeters may be done in this manner. Convert feet to inches; then convert inches to centimeters:

ft → in. → cm

The conversion factors needed are

$$\frac{12\ \text{in.}}{1\ \text{ft}} \quad \text{and} \quad \frac{2.54\ \text{cm}}{1\ \text{in.}}$$

$$2.00\ \text{ft} \times \frac{12\ \text{in.}}{1\ \text{ft}} = 24.0\ \text{in.}$$

$$24.0\ \text{in.} \times \frac{2.54\ \text{cm}}{1\ \text{in.}} = 61.0\ \text{cm}$$

Since 1 ft and 12 in. are exact numbers, the number of significant figures allowed in the answer is three, based on the number 2.00.

Example 2.12

How many meters are in a 100.-yd football field?

Solution

The stepwise conversion of units from yards to meters may be done in this manner, using the proper conversion factors:

$$\text{yd} \rightarrow \text{ft} \rightarrow \text{in.} \rightarrow \text{cm} \rightarrow \text{m}$$

$$100.\ \cancel{\text{yd}} \times \frac{3\ \text{ft}}{1\ \cancel{\text{yd}}} = 300.\ \text{ft} \qquad (3\ \text{ft/yd})$$

$$300.\ \cancel{\text{ft}} \times \frac{12\ \text{in.}}{1\ \cancel{\text{ft}}} = 3600\ \text{in.} \qquad (12\ \text{in./ft})$$

$$3600\ \cancel{\text{in.}} \times \frac{2.54\ \text{cm}}{1\ \cancel{\text{in.}}} = 9144\ \text{cm} \qquad (2.54\ \text{cm/in.})$$

$$9144\ \cancel{\text{cm}} \times \frac{1\ \text{m}}{100\ \cancel{\text{cm}}} = 91.4\ \text{m} \qquad (1\ \text{m}/100\ \text{cm}) \quad \text{(three significant figures)}$$

Examples 2.11 and 2.12 may be solved using a linear expression and writing down conversion factors in succession. This method often saves one or two calculation steps and allows numerical values to be reduced to simpler terms, leading to simpler calculations. The single linear expressions for Examples 2.11 and 2.12 are

$$2.00\ \cancel{\text{ft}} \times \frac{12\ \cancel{\text{in.}}}{1\ \cancel{\text{ft}}} \times \frac{2.54\ \text{cm}}{1\ \cancel{\text{in.}}} = 61.0\ \text{cm}$$

$$100.\ \cancel{\text{yd}} \times \frac{3\ \cancel{\text{ft}}}{1\ \cancel{\text{yd}}} \times \frac{12\ \cancel{\text{in.}}}{1\ \cancel{\text{ft}}} \times \frac{2.54\ \cancel{\text{cm}}}{1\ \cancel{\text{in.}}} \times \frac{1\ \text{m}}{100\ \cancel{\text{cm}}} = 91.4\ \text{m}$$

Using the units alone (Example 2.12), we see that the stepwise cancellation proceeds in succession until the desired unit is reached.

$$\cancel{\text{yd}} \times \frac{\cancel{\text{ft}}}{\cancel{\text{yd}}} \times \frac{\cancel{\text{in.}}}{\cancel{\text{ft}}} \times \frac{\cancel{\text{cm}}}{\cancel{\text{in.}}} \times \frac{\text{m}}{\cancel{\text{cm}}} = \text{m}$$

Practice 2.8

How many meters are in 10.5 miles?

Example 2.13 How many cubic centimeters (cm^3) are in a box that measures 2.20 in. by 4.00 in. by 6.00 in.?

Solution First we need to determine the volume of the box in cubic inches (in.^3) by multiplying the length times the width times the height:

$$2.20\ \text{in.} \times 4.00\ \text{in.} \times 6.00\ \text{in.} = 52.8\ \text{in.}^3$$

Now we need to convert in.^3 to cm^3, which can be done by using the inches and centimeters relationship three times:

$$\cancel{\text{in.}^3} \times \frac{\text{cm}}{\cancel{\text{in.}}} \times \frac{\text{cm}}{\cancel{\text{in.}}} \times \frac{\text{cm}}{\cancel{\text{in.}}} = \text{cm}^3$$

$$52.8\ \cancel{\text{in.}^3} \times \frac{2.54\ \text{cm}}{1\ \cancel{\text{in.}}} \times \frac{2.54\ \text{cm}}{1\ \cancel{\text{in.}}} \times \frac{2.54\ \text{cm}}{1\ \cancel{\text{in.}}} = 865\ \text{cm}^3$$

Example 2.14

A driver of a car is obeying the speed limit of 55 miles per hour. How fast is it traveling in kilometers per second?

Solution

Two conversions are needed to solve this problem:

mi → km
hr → min → s

To convert mi → km,

$$\frac{55\ \cancel{\text{mi}}}{\text{hr}} \times \frac{1.609\ \text{km}}{1\ \cancel{\text{mi}}} = 88\ \frac{\text{km}}{\text{hr}}$$

Next we must convert hr → min → s. Notice that hours is in the denominator of our quantity, so the conversion factor must have hours in the numerator:

$$\frac{88\ \text{km}}{\cancel{\text{hr}}} \times \frac{1\ \cancel{\text{hr}}}{60\ \cancel{\text{min}}} \times \frac{1\ \cancel{\text{min}}}{60\ \text{s}} = 0.024\ \frac{\text{km}}{\text{s}}$$

Practice 2.9

How many cubic meters are in a room measuring 8 ft × 10 ft × 12 ft?

2.9 Measurement of Mass

The gram is used as a unit of mass measurement, but it is a tiny amount of mass; for instance, a nickel has a mass of about 5 grams. Therefore the *standard unit* of mass in the SI system is the **kilogram** (equal to 1000 g). The amount of mass in a kilogram is defined by international agreement as exactly equal to the mass of a platinum–iridium cylinder (international prototype kilogram) kept in a vault at Sèvres, France. Comparing this unit of mass to 1 lb (16 oz), we find that 1 kg is equal to 2.205 lb. A pound is equal to 453.6 g (0.4536 kg). The same prefixes used in length measurement are used to indicate larger and smaller gram units (see Table 2.4).

kilogram

TABLE 2.4 Metric Units of Mass

Unit	Abbreviation	Gram equivalent	Exponential equivalent
kilogram	kg	1000 g	10^3 g
gram	g	1 g	10^0 g
decigram	dg	0.1 g	10^{-1} g
centigram	cg	0.01 g	10^{-2} g
milligram	mg	0.001 g	10^{-3} g
microgram	μg	0.000001 g	10^{-6} g

FIGURE 2.5 ▶
(a) A quadruple beam balance with a precision of 0.01 g; (b) a single pan, top-loading balance with a precision of 0.001 g (1 mg); (c) a digital electronic analytical balance with a precision of 0.0001 g; and (d) a digital electronic balance with a precision of 0.001 g.

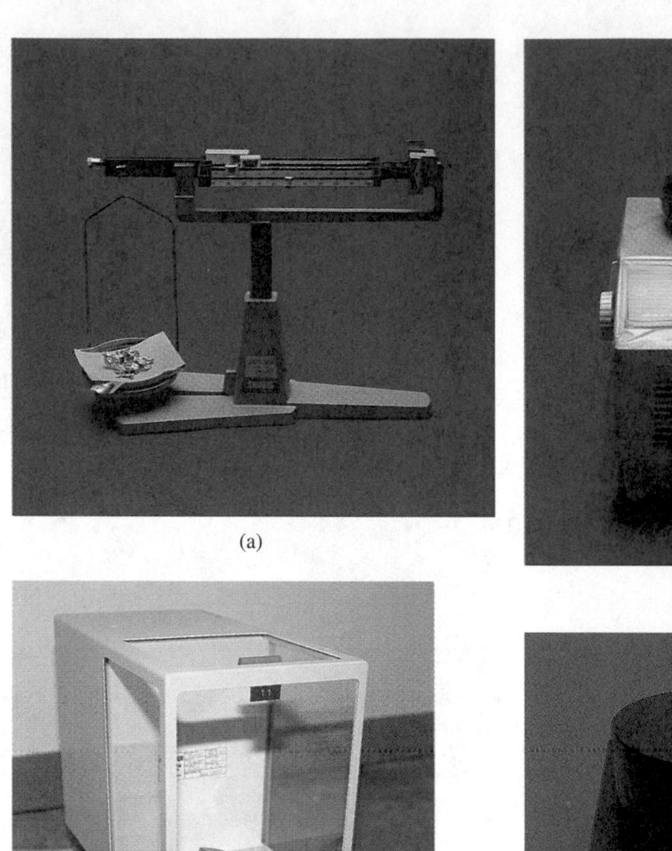

(a)

(b)

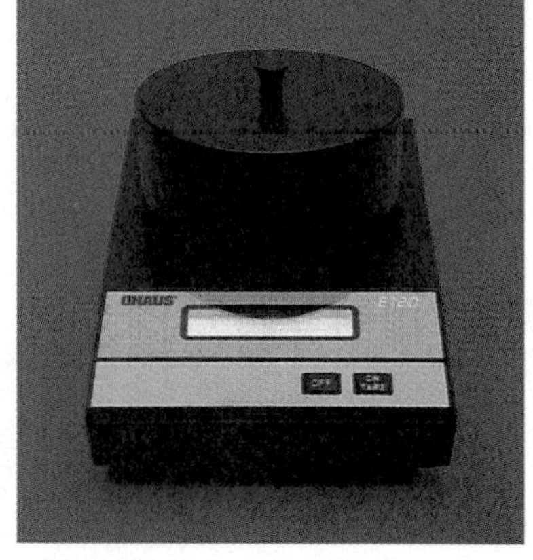

(c)

(d)

A balance is used to measure mass. Some balances will determine the mass of objects to the nearest microgram. The choice of balance depends on the precision required and the amount of material. Several balances are shown in Figure 2.5.

It is convenient to remember that

$$1 \text{ g} = 1000 \text{ mg}$$
$$1 \text{ kg} = 1000 \text{ g}$$
$$1 \text{ kg} = 2.205 \text{ lb}$$
$$1 \text{ lb} = 453.6 \text{ g}$$

To change grams to milligrams, multiply grams by the conversion factor 1000 mg/g. The setup for converting 25 g to milligrams is

$$25 \cancel{\text{g}} \times \frac{1000 \text{ mg}}{1 \cancel{\text{g}}} = 25{,}000 \text{ mg} \qquad (2.5 \times 10^4 \text{ mg})$$

Note that multiplying a number by 1000 is the same as multiplying the number by 10^3 and can be done simply by moving the decimal point three places to the right:

$$6.428 \times 1000 = 6428 \qquad (6.\underbrace{428}_{3})$$

To change milligrams to grams, multiply milligrams by the conversion factor 1 g/1000 mg. For example, to convert 155 mg to grams:

$$155\ \cancel{\text{mg}} \times \frac{1\ \text{g}}{1000\ \cancel{\text{mg}}} = 0.155\ \text{g}$$

Mass conversions from American to metric units are shown in Examples 2.15 and 2.16.

Example 2.15

A 1.50-lb package of baking soda costs 80 cents. How many grams of this substance are in this package?

Solution

We are solving for the number of grams equivalent to 1.50 lb. Since 1 lb = 453.6 g, the factor to convert pounds to grams is 453.6 g/lb:

$$1.50\ \cancel{\text{lb}} \times \frac{453.6\ \text{g}}{1\ \cancel{\text{lb}}} = 680.\ \text{g}$$

Note: The cost of the baking soda has no bearing on the question asked in this problem.

Example 2.16

Suppose four ostrich feathers weigh 1.00 lb. Assuming that each feather is equal in mass, how many milligrams does a single feather weigh?

Solution

The unit conversion in this problem is from 1 lb/4 feathers to milligrams per feather. Since the unit *feathers* occurs in the denominator of both the starting unit and the desired unit, the unit conversions needed are

$$\text{lb} \to \text{g} \to \text{mg}$$

$$\frac{1.00\ \cancel{\text{lb}}}{4\ \text{feathers}} \times \frac{453.6\ \cancel{\text{g}}}{1\ \cancel{\text{lb}}} \times \frac{1000\ \text{mg}}{1\ \cancel{\text{g}}} = \frac{113{,}400}{1\ \text{feather}} \qquad (1.13 \times 10^5\ \text{mg/feather})$$

Practice 2.10

You are traveling in Europe and wake up one morning to find your mass is 75.0 kg. Determine the American equivalent to see whether you need to go on a diet before you return home.

Practice 2.11

A tennis ball has a mass of 65 g. Determine the American equivalent in pounds.

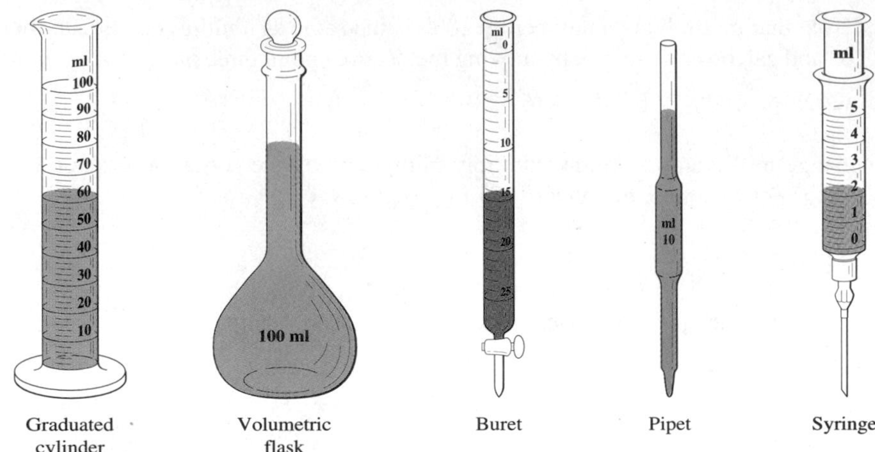

FIGURE 2.6 ▶ **Calibrated glassware for measuring the volume of liquids.**

2.10 Measurement of Volume

volume **Volume**, as used here, is the amount of space occupied by matter. The SI unit of volume is the *cubic meter* (m^3). However, the liter (pronounced *leeter* and abbreviated L) and the milliliter (abbreviated mL) are the standard units of volume used in most chemical laboratories. A **liter** is usually defined as 1 cubic decimeter (1 kg) of water at 4°C.

liter

The most common instruments or equipment for measuring liquids are the graduated cylinder, volumetric flask, buret, pipet, and syringe, which are illustrated in Figure 2.6. These pieces are usually made of glass and are available in various sizes.

It is convenient to remember that

$$1\ \text{L} = 1000\ \text{mL} = 1000\ \text{cm}^3$$
$$1\ \text{mL} = 1\ \text{cm}^3$$
$$1\ \text{L} = 1.057\ \text{qt}$$
$$946.1\ \text{mL} = 1\ \text{qt}$$

The volume of a cubic or rectangular container can be determined by multiplying its length × width × height. Thus a 10-cm-square box has a volume of $10\ \text{cm} \times 10\ \text{cm} \times 10\ \text{cm} = 1000\ \text{cm}^3$. Let's try some examples.

Example 2.17 How many milliliters are contained in 3.5 liters?

Solution The conversion factor to change liters to milliliters is 1000 mL/L:

$$3.5\ \cancel{\text{L}} \times \frac{1000\ \text{mL}}{\cancel{\text{L}}} = 3500\ \text{mL} \qquad (3.5 \times 10^3\ \text{mL})$$

Liters may be changed to milliliters by moving the decimal point three places to the right and changing the units to milliliters:

$$1.500\ \text{L} = 1500.\ \text{mL}$$

Example 2.18

How many cubic centimeters are in a cube that is 11.1 inches on a side?

Solution

First change inches to centimeters. The conversion factor is 2.54 cm/in.:

$$11.1\ \text{in.} \times \frac{2.54\ \text{cm}}{1\ \text{in.}} = 28.2\ \text{cm on a side}$$

Then change to cubic volume (length × width × height):

$$28.2\ \text{cm} \times 28.2\ \text{cm} \times 28.2\ \text{cm} = 22{,}426\ \text{cm}^3 \qquad (2.24 \times 10^4\ \text{cm}^3)$$

Practice 2.12

A bottle of excellent chianti holds 750. mL. Determine the volume in quarts.

Practice 2.13

Milk is often purchased by the half gallon. Determine the number of liters necessary to equal this amount.

When doing problems with multiple steps you should round only at the end of the problem. We are rounding at the end of each step in example problems to illustrate the proper significant figures.

2.11 Measurement of Temperature

heat

Heat is a form of energy associated with the motion of small particles of matter. The term *heat* refers to the quantity of energy within a system or to a quantity of energy added to or taken away from a system. *System* as used here simply refers to the entity that is being heated or cooled. Depending on the amount of heat energy present, a given system is said to be hot or cold. **Temperature** is a measure of the intensity of heat, or how hot a system is, regardless of its size. Heat always flows from a region of higher temperature to one of lower temperature. The SI unit of temperature is the kelvin. The common laboratory instrument for measuring temperature is a thermometer (see Figure 2.7).

temperature

The temperature of a system can be expressed by several different scales. Three commonly used temperature scales are the Celsius scale (pronounced *sell-see-us*), the Kelvin (absolute) scale, and the Fahrenheit scale. The unit of temperature on the Celsius and Fahrenheit scales is called a *degree,* but the size of the Celsius and the Fahrenheit degree is not the same. The symbol for the Celsius and Fahrenheit degrees is °, and it is placed as a superscript after the number and before the symbol for the scale. Thus, 100°C means 100 *degrees Celsius.* The degree sign is not used with Kelvin temperatures.

degrees Celsius = °C
Kelvin (absolute) = K
degrees Fahrenheit = °F

On the Celsius scale the interval between the freezing and boiling temperatures of water is divided into 100 equal parts, or degrees. The freezing point of water is

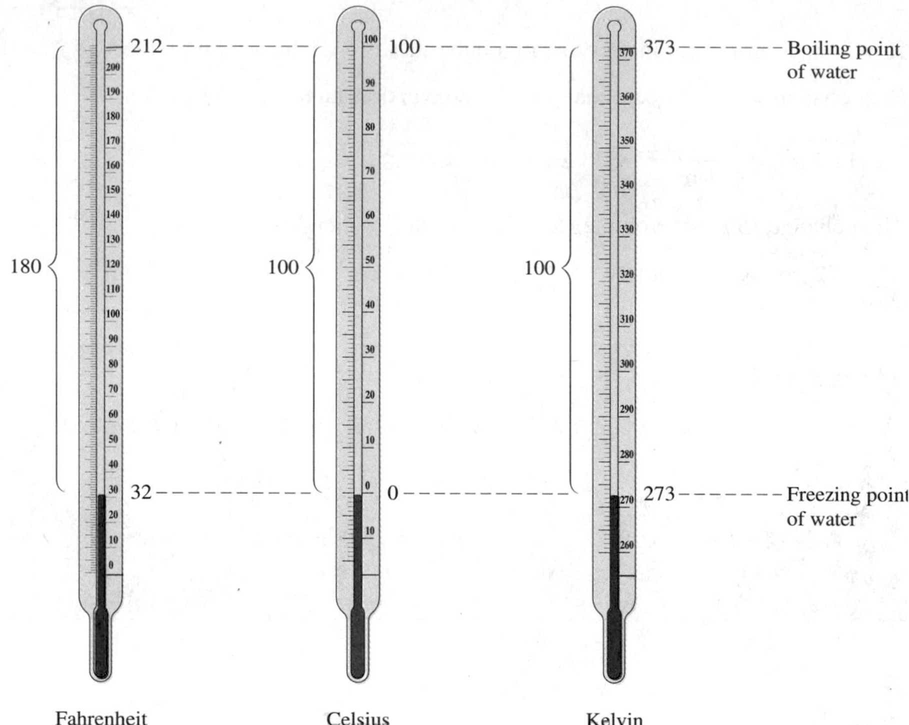

FIGURE 2.7 ▶
Comparison of Celsius, Kelvin, and Fahrenheit temperature scales.

assigned a temperature of 0°C and the boiling point of water a temperature of 100°C. The Kelvin temperature scale is known as the absolute temperature scale, because 0 K is the lowest temperature theoretically attainable. The Kelvin zero is 273.15 degrees below the Celsius zero. (A kelvin is equal in size to a Celsius degree.) The freezing point of water on the Kelvin scale is 273.15 K. The Fahrenheit scale has 180 degrees between the freezing and boiling temperatures of water. On this scale the freezing point of water is 32°F and the boiling point is 212°F.

$$0°\text{C} \cong 273 \text{ K} \cong 32°\text{F}$$

The three scales are compared in Figure 2.7. Although absolute zero (0 K) is the lower limit of temperature on these scales, temperature has no upper limit. (Temperatures of several million degrees are known to exist in the sun and in other stars.)

By examining Figure 2.7 we can see that there are 100 Celsius degrees and 100 kelvins between the freezing and boiling points of water, but there are 180 Fahrenheit degrees between these two temperatures. Hence, the size of a degree on the Celsius scale is the same as the size of 1 kelvin, but 1 Celsius degree corresponds to 1.8 degrees on the Fahrenheit scale.

$$\frac{180}{100} = 1.8$$

From these data, mathematical formulas have been derived to convert a temperature on one scale to the corresponding temperature on another scale:

$$\text{K} = °\text{C} + 273.15 \qquad (1)$$

$$°\text{F} = (1.8 \times °\text{C}) + 32 \qquad (2)$$

$$°\text{C} = \frac{°\text{F} - 32}{1.8} \qquad (3)$$

Example 2.19

The temperature at which table salt (sodium chloride) melts is 800.°C. What is this temperature on the Kelvin and Fahrenheit scales?

Solution

We need to calculate K from °C, so we use formula (1). We also need to calculate °F from °C; for this calculation we use formula (2).

$$K = °C + 273.15$$

$$K = 800.°C + 273.15 = 1073 K$$

$$°F = (1.8 \times °C) + 32$$

$$°F = (1.8 \times 800.°C) + 32$$

$$°F = 1440 + 32 = 1472°F$$

$$800.°C = 1073 K = 1472°F$$

Remember, since the original measurement of 800.°C was to the units' place, the converted temperature is also to the units' place.

Example 2.20

The temperature for December 1 in Honolulu, Hawaii, was 110.°F, a new record. Convert this temperature to °C.

Solution

Formula (3) applies here.

$$°C = \frac{°F - 32}{1.8}$$

$$°C = \frac{110. - 32}{1.8} = \frac{78}{1.8} = 43°C$$

Example 2.21

What temperature on the Fahrenheit scale corresponds to −8.0°C? (Notice the minus sign in this problem.)

Solution

$$°F = (1.8 \times °C) + 32$$

$$°F = [1.8 \times (-8.0)] + 32 = -14 + 32$$

$$°F = 18°F$$

Since the original measurement is to the tenth the converted temperature is also to the tenth.

Temperatures used throughout this book are in degrees Celsius (°C) unless specified otherwise. The temperature after conversion should be expressed to the same precision as the original measurement.

Practice 2.14

Helium boils at 4 K. Convert this temperature to °C and then to °F.

Practice 2.15

"Normal" human body temperature is 98.6°F. Convert this to °C and K.

How Hot Is Hot?

How hot is it today? Does my son have a fever? What is the best temperature for storing mayonnaise? What is the optimum temperature for oil in my car? People have many reasons to determine the temperature of objects around them. In all cases an instrument for measuring the temperature is required. These instruments, called *thermometers,* come in a variety of shapes and sizes.

The Galileo thermometer is one of the more interesting and decorative room thermometers (see photo). These bulbs floating in the colored liquid are filled with liquids at different densities. As the temperature of the surrounding liquid (and the room) changes, the density of the globes change relative to the surrounding liquid, and they sink or float. Each globe is calibrated to sink at a particular temperature. The tag on the globe that is floating highest in the solution tells the temperature.

Traditional thermometers are calibrated tubes filled with a liquid that expands as it becomes warm. Many different liquids are used in thermometers, two of the most common being mercury and alcohol (usually dyed red or green). Mercury is the liquid of choice because it has a wide temperature range in the liquid phase and it is compact so the thermometer doesn't have to be very long. One problem with the mercury thermometer is that it could have toxic effects on the patient if were to break in the mouth. Also a mercury spill is tricky to clean up. Another drawback with this type of oral thermometer is that it takes approximately 3 minutes to stabilize and register accurately—a long time to keep your mouth closed if you have a severe cold or are congested.

Traditional home thermometers need to be shaken down in order to take a new temperature. Why do they require this process while room thermometers do not? Because mercury begins to fall immediately upon being removed from a heat source, allowing no time for an accurate reading of the thermometer. To solve this problem a small kink is placed above the bulb in the tube (see diagram). The mercury is forced through this kink as it is being warmed, but it cannot fall back into the bulb as it cools. It remains in place until the thermometer is shaken, which forces the mercury through the kink and prepares the thermometer for its next use. Alcohol thermometers are more generally used because they are inexpensive, relatively safe, and work well at or near room temperature.

Today many hospitals, physicians' offices, and homes are equipped with electronic thermometers. These thermometers contain thermistors that are sensitive to temperature. A voltage reading is taken and associated with a number (calibrated into the thermometer). The temperature reading appears as a digital readout on the thermometer. These thermometers have several advantages, including the lack of toxic liquids, speed (they require much less time to register an accurate reading), convenience (they can be used with sterile cover to prevent the spread of infection), and size.

Engineers at Johns Hopkins Applied Physics Laboratory have built a battery-powered transmitting thermometer the size of an aspirin capsule that works after it is swallowed. It is capable of transmitting temperature measurements to within 0.01 degree until it passes out of the body (usually 1–2 days). This is useful in monitoring temperature patterns in the treatment of hypothermia, during which the body must be warmed at a slow, constant rate with continual temperature monitoring. The capsule can also be used to prevent hyperthermia in athletes, race drivers, or even in those taking routine treadmill tests.

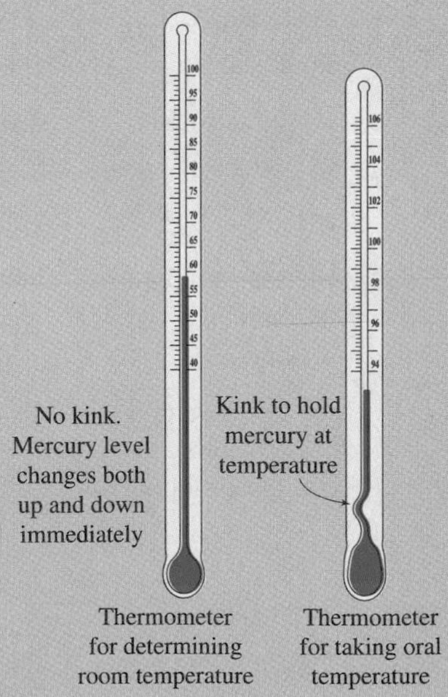

2.12 Density

density

Density *(d)* is the ratio of the mass of a substance to the volume occupied by that mass; it is the mass per unit of volume and is given by the equation

$$d = \frac{\text{mass}}{\text{volume}}$$

Density is a physical characteristic of a substance and may be used as an aid to its identification. When the density of a solid or a liquid is given, the mass is usually expressed in grams and the volume in milliliters or cubic centimeters.

$$d = \frac{\text{mass}}{\text{volume}} = \frac{\text{g}}{\text{mL}} \quad \text{or} \quad d = \frac{\text{g}}{\text{cm}^3}$$

Since the volume of a substance (especially liquids and gases) varies with temperature, it is important to state the temperature along with the density. For example, the volume of 1.0000 g of water at 4°C is 1.0000 mL, at 20°C it is 1.0018 mL, and at 80°C it is 1.0290 mL. Density therefore also varies with temperature.

The density of water at 4°C is 1.0000 g/mL, but at 80°C the density of water is 0.9718 g/mL.

$$d^{4°\text{C}} = \frac{1.0000\text{ g}}{1.0000\text{ mL}} = 1.0000\text{ g/mL}$$

$$d^{80°\text{C}} = \frac{1.0000\text{ g}}{1.0290\text{ mL}} = 0.97182\text{ g/mL}$$

The density of iron at 20°C is 7.86 g/mL.

$$d^{20°\text{C}} = \frac{7.86\text{ g}}{1.00\text{ mL}} = 7.86\text{ g/mL}$$

The densities of a variety of materials are compared in Figure 2.8.

Densities for liquids and solids are usually represented in terms of grams per milliliter (g/mL) or grams per cubic centimeter (g/cm^3). The density of gases, however, is expressed in terms of grams per liter (g/L). Unless otherwise stated, gas densities are given for 0°C and 1 atmosphere pressure (discussed further in Chapter 13). Table 2.5 lists the densities of some common materials.

Suppose that water, Karo syrup, and vegetable oil are successively poured into a graduated cylinder. The result is a layered three-liquid system (Figure 2.9). Can we predict the order of the liquid layers? Yes, by looking up the densities in Table 2.5. Karo syrup has the greatest density (1.37 g/mL), and vegetable oil has the lowest density (0.91 g/mL). Karo syrup will be the bottom layer and vegetable oil will be the top layer. Water, with a density between the other two liquids, will form the middle layer. This information can also be determined by experiment. Vegetable oil, being less dense than water, will float when added to the graduated cylinder.

The density of air at 0°C is approximately 1.293 g/L. Gases with densities less than this value are said to be "lighter than air." A helium-filled balloon will rise rapidly in air because the density of helium is only 0.178 g/L.

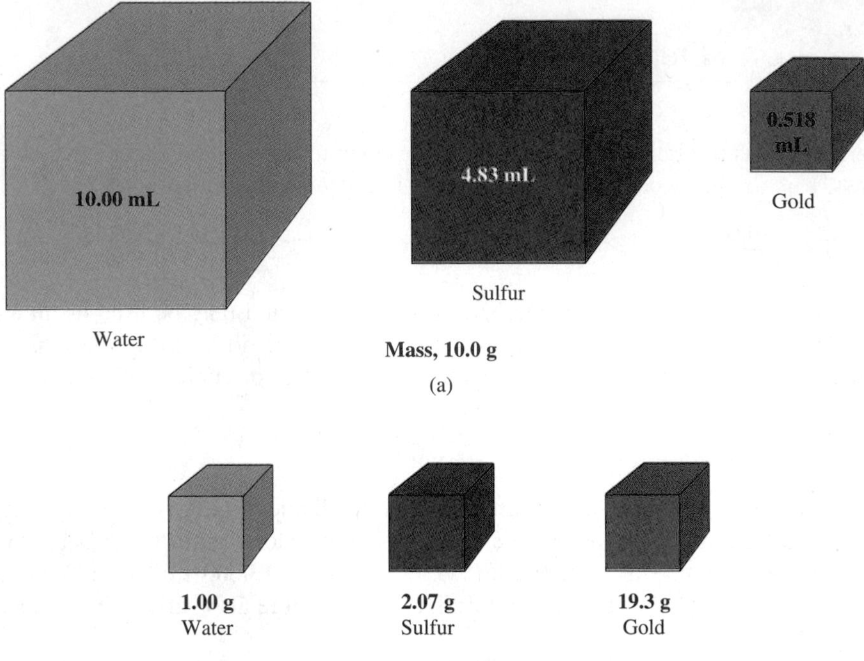

FIGURE 2.8 ▶
(a) Comparison of the volumes of equal masses (10.0 g) of water, sulfur, and gold.
(b) Comparison of the masses of equal volumes (1.00 cm³) of water, sulfur, and gold. (Water is at 4°C; the two solids, at 20°C.)

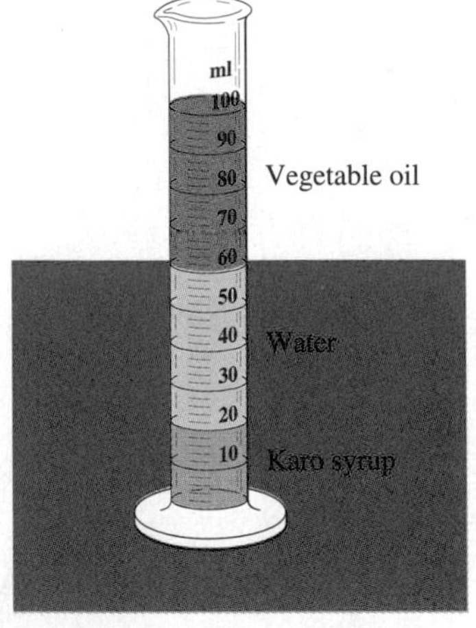

▲
FIGURE 2.9
Relative density of liquids. When three liquids are poured together, the liquid with the highest density will be the bottom layer. In the case of vegetable oil, water, and Karo syrup, vegetable oil is the top layer.

TABLE 2.5 Densities of Some Selected Materials*

Liquids and solids		Gases	
Substance	**Density (g/mL at 20°C)**	**Substance**	**Density (g/L at 0°C)**
Wood (Douglas fir)	0.512	Hydrogen	0.090
Ethyl alcohol	0.789	Helium	0.178
Vegetable oil	0.91	Methane	0.714
Water (4°C)	**1.000**	Ammonia	0.771
Sugar	1.59	Neon	0.90
Glycerin	1.26	Carbon monoxide	1.25
Karo syrup	1.37	Nitrogen	1.251
Magnesium	1.74	**Air**	**1.293**
Sulfuric acid	1.84	Oxygen	1.429
Sulfur	2.07	Hydrogen chloride	1.63
Salt	2.16	Argon	1.78
Aluminum	2.70	Carbon dioxide	1.963
Silver	10.5	Chlorine	3.17
Lead	11.34		
Mercury	13.55		
Gold	19.3		

*For comparing densities the density of water is the reference for solids and liquids; air is the reference for gases.

Healthy Measurements

Obesity, a known health hazard, is among the top ten factors in cardiovascular disease. Simply measuring the weight of an individual is not a good indicator of leanness or obesity. A person may appear to be thin, yet have a high percentage of body fat. Someone else may appear "overweight" in comparison to published height-weight charts, but actually be especially lean as a result of a large percentage of muscle mass. To assess the body composition of an individual requires a measurement of the percent body fat.

Body fat is defined by health science professionals as the percentage of weight attributable to fat. It is the sum of the essential fat, surrounding and cushioning the internal organs, and the storage fat, which acts as a reservoir for energy in the body. A variety of techniques are currently in use to measure the body composition, including skin-fold tests, bioelectrical impedance, and hydrostatic weighing.

Hydrostatic weighing is considered to be one of the most accurate methods for determining body density. The individual is weighed in air, then seated on a chair suspended from a scale and lowered into a tank of warm water. After exhaling as much as possible the individual is submerged in the water and remains under the surface for 5–7 seconds. The underwater weight is recorded during this time. A series of calculations can then be made to determine percent body fat. The basis for these calculations lies in the variation in density of different tissues shown in the above table. A person with more bone and muscle mass will weigh more in water and have a higher body density.

As in all measurements errors in measured values result in variations in calculated results. A 100-g error in underwater weight results in a 1% body fat error. Various values for percent body fat calculations are shown in the table above.

Percent Body Fat

Classification	Male	Female
Lean	<8%	<15%
Healthy	8–15%	15–22%
Plump	16–19%	23–27%
Fat	20–24%	28–33%
Obese	>24%	>33%
Average college-age	15%	25%
Average middle-age	23%	32%
Distance runner	4–9%	6–15%
Tennis player	14–17%	19–22%

Density of Various Tissues in the Human Body

Tissue	Density
Bone	3.0 g/cm^3
Muscle	1.06 g/cm^3
Water	1.00 g/cm^3
Fat	0.9 g/cm^3

This person is being weighed hydrostatically to determine body composition.

When an insoluble solid object is dropped into water, it will sink or float, depending on its density. If the object is less dense than water, it will float, displacing a *mass* of water equal to the mass of the object. If the object is more dense than water, it will sink, displacing a *volume* of water equal to the volume of the object. This information can be utilized to determine the volume (and density) of irregularly shaped objects.

specific gravity

The **specific gravity** (sp gr) of a substance is the ratio of the density of that substance to the density of another substance, usually water at 4°C. Specific gravity has no units because the density units cancel. The specific gravity tells us how many times as heavy a liquid, a solid, or a gas is as compared to the reference material. Since the density of water at 4°C is 1.00 g/mL, the specific gravity of a solid or liquid is the same as its density in g/mL without the units.

$$\text{sp gr} = \frac{\text{density of a liquid or solid}}{\text{density of water}}$$

Sample calculations of density problems follow.

Example 2.22 What is the density of a mineral if 427 g of the mineral occupy a volume of 35.0 mL?

Solution We need to solve for density, so we start by writing the formula for calculating density:

$$d = \frac{\text{mass}}{\text{volume}}$$

Then we substitute the data given in the problem into the equation and solve:

$$\text{mass} = 427 \text{ g} \qquad \text{volume} = 35.0 \text{ mL}$$

$$d = \frac{\text{mass}}{\text{volume}} = \frac{427 \text{ g}}{35.0 \text{ mL}} = 12.2 \text{ g/mL}$$

Example 2.23 The density of gold is 19.3 g/mL. What is the mass of 25.0 mL of gold?

Solution Two ways to solve this problem are: (a) Solve the density equation for mass, then substitute the density and volume data into the new equation and calculate. (b) Solve by dimensional analysis.

When alternative methods of solution are available, more than one is shown in the example. Choose the method you are most comfortable using to solve the problem.

Method 1 (a) Solve the density equation for mass:

$$d = \frac{\text{mass}}{\text{volume}} \qquad d \times \text{volume} = \text{mass}$$

(b) Substitute the data and calculate.

$$\text{mass} = \frac{19.3 \text{ g}}{\cancel{\text{mL}}} \times 25.0 \cancel{\text{mL}} = 483 \text{ g}$$

Method 2 Dimensional analysis: Use density as a conversion factor, converting

mL → g

The conversion of units is

$$\text{mL} \times \frac{\text{g}}{\text{mL}} = \text{g}$$

$$25.0\ \cancel{\text{mL}} \times \frac{19.3\ \text{g}}{\cancel{\text{mL}}} = 483\ \text{g}$$

Example 2.24

Calculate the volume (in mL) of 100. g of ethyl alcohol.

Solution

From Table 2.5 we see that the density of ethyl alcohol is 0.789 g/mL. This density also means that 1 mL of the alcohol has a mass of 0.789 g (1 mL/0.789 g).

Method 1: This problem may be done by solving the density equation for volume and then substituting the data in the new equation.

$$d = \frac{\text{mass}}{\text{volume}}$$

$$\text{volume} = \frac{\text{mass}}{d}$$

$$\text{volume} = \frac{100.\ \cancel{\text{g}}}{0.789\ \cancel{\text{g}}/\text{mL}} = 127\ \text{mL}$$

Method 2: Dimensional analysis. For a conversion factor, we can use either

$$\frac{\text{g}}{\text{mL}} \quad \text{or} \quad \frac{\text{mL}}{\text{g}}$$

In this case the conversion is from g → mL, so we use mL/g. Substituting the data,

$$100.\ \cancel{\text{g}} \times \frac{1\ \text{mL}}{0.789\ \cancel{\text{g}}} = 127\ \text{mL of ethyl alcohol}$$

Example 2.25

The water level in a graduated cylinder stands at 20.0 mL before and at 26.2 mL after a 16.74-g metal bolt is submerged in the water. (a) What is the volume of the bolt? (b) What is the density of the bolt?

Solution

(a) The bolt will displace a volume of water equal to the volume of the bolt. Thus the increase in volume is the volume of the bolt.

$$\begin{aligned} 26.2\ \text{mL} &= \text{volume of water plus bolt} \\ -20.0\ \text{mL} &= \text{volume of water} \\ \hline 6.2\ \text{mL} &= \text{volume of bolt} \end{aligned}$$

(b) $d = \dfrac{\text{mass of bolt}}{\text{volume of bolt}} = \dfrac{16.74\ \text{g}}{6.2\ \text{mL}} = 2.7\ \text{g/mL}$

Practice 2.16

Pure silver has a density of 10.5 g/mL. A ring sold as pure silver has a mass of 25.0 g. When placed in a graduated cylinder the water level rises 2.0 mL. Determine whether the ring is actually pure silver or if the customer should see the Better Business Bureau.

Practice 2.17

The water level in a metric measuring cup is 0.75 L before the addition of 150. g of shortening. The water level after submerging the shortening is 0.92 L. Determine the density of the shortening.

Concepts in Review

The major concepts of the chapter are listed in this section to help you review the chapter.

1. Differentiate between mass and weight. Indicate the instruments used to measure each.
2. Know the metric units of mass, length, and volume.
3. Know the numerical equivalent for the metric prefixes *deci, centi, milli, micro, nano, kilo,* and *mega.*
4. Express any number in scientific notation.
5. Express answers to calculations to the proper number of significant figures.
6. Set up and solve problems utilizing the method of dimensional analysis (factor-label method).
7. Convert measurements of mass, length, and volume from American units to metric units, and vice versa.
8. Make temperature conversions among Fahrenheit, Celsius, and Kelvin scales.
9. Differentiate between heat and temperature.
10. Calculate the density, mass, or volume of an object from the appropriate data.

Key Terms

The terms listed here have been defined within this chapter. Section numbers are referenced in parentheses for each term.

density (2.12)
heat (2.11)
kilogram (2.9)
liter (2.10)
mass (2.1)
meter (2.7)
metric system (SI) (2.6)
rounding off numbers (2.3)
scientific notation (2.4)
significant figures (2.2)
specific gravity (2.12)
temperature (2.11)
volume (2.10)
weight (2.1)

Questions

Questions refer to tables, figures, and key words and concepts defined within the chapter. A particularly challenging question or exercise is indicated with an asterisk.

1. Determine how many centimeters make up 1 km. (Table 2.3)
2. Determine the metric equivalent of 3 in. (Figure 2.4)
3. Why is the neck of a 100-mL volumetric flask narrower than the top of a 100-mL graduated cylinder? (Figure 2.6)
4. Describe the order of the following substances (top to bottom) if these three substances were placed in a 100-mL graduated cylinder: 25 mL glycerin, 25 mL mercury, and a cube of magnesium 2.0 cm on an edge. (Table 2.5)
5. Arrange these materials in order of increasing density: salt, vegetable oil, lead, and ethyl alcohol. (Table 2.5)
6. Ice floats in vegetable oil and sinks in ethyl alcohol. The density of ice must lie between what numerical values? (Table 2.5)
7. Distinguish between heat and temperature.
8. Distinguish between density and specific gravity.
9. Why is measuring the weight of a person a poor indication of leanness or obesity?
10. How is body density determined? What is the basis for this determination?
11. What are some of the important advantages of the metric system over the American system of weights and measurements?
12. State the rules used in this text for rounding off numbers.
13. Compare the number of degrees between the freezing point of water and its boiling point on the Fahrenheit, Kelvin, and Celsius temperature scales. (Figure 2.7)
14. Which of the following statements are correct? Rewrite the incorrect statements to make them correct:
    - **(a)** The prefix *micro* indicates one-millionth of the unit expressed.
    - **(b)** The length 10 cm is equal to 1000 mm.
    - **(c)** The number 383.263 rounded to four significant figures becomes 383.3.
    - **(d)** The number of significant figures in the number 29,004 is five.
    - **(e)** The number 0.00723 contains three significant figures.
    - **(f)** The sum of 32.276 + 2.134 should contain four significant figures.
    - **(g)** The product of 18.42 cm $\times$ 3.40 cm should contain three significant figures.
    - **(h)** One microsecond is 10^{-6} second.
    - **(i)** One thousand meters is a longer distance than 1000 yards.
    - **(j)** One liter is a larger volume than 1 quart.
    - **(k)** One centimeter is longer than 1 inch.
    - **(l)** One cubic centimeter (cm^3) is equal to 1 milliliter.
    - **(m)** The number 0.0002983 in exponential notation is 2.983×10^{-3}.
    - **(n)** $(3.0 \times 10^4)(6.0 \times 10^6) = 1.8 \times 10^{11}$
    - **(o)** Temperature is a form of energy.
    - **(p)** The density of water at 4°C is 1.00 g/mL.
    - **(q)** A pipet is a more accurate instrument for measuring 10.0 mL of water than is a graduated cylinder.

Paired Exercises

These exercises are paired. Each odd-numbered exercise is followed by a similar even-numbered exercise. Answers to the even-numbered exercises are given in Appendix V.

Metric Abbreviations

15. State the abbreviation for each of the following units:
    - **(a)** gram
    - **(b)** microgram
    - **(c)** centimeter
    - **(d)** micrometer
    - **(e)** milliliter
    - **(f)** deciliter
16. State the abbreviation for each of the following units:
    - **(a)** milligram
    - **(b)** kilogram
    - **(c)** meter
    - **(d)** nanometer
    - **(e)** angstrom
    - **(f)** microliter

Significant Figures, Rounding, Exponential Notation

17. For the following numbers, tell whether the zeros are significant:
(a) 503
(b) 0.007
(c) 4200
(d) 3.0030
(e) 100.00
(f) 8.00×10^2

18. Are the zeros significant in these numbers?
(a) 63,000
(b) 6.004
(c) 0.00543
(d) 8.3090
(e) 60.
(f) 5.0×10^{-4}

19. How many significant figures are in each of the following numbers?
(a) 0.025
(b) 22.4
(c) 0.0404
(d) 5.50×10^3

20. State the number of significant figures in each of the following numbers:
(a) 40.0
(b) 0.081
(c) 129,042
(d) 4.090×10^{-3}

21. Round each of the following numbers to three significant figures:
(a) 93.246
(b) 0.02857
(c) 4.644
(d) 34.250

22. Round each of the following numbers to three significant figures:
(a) 8.8726
(b) 21.25
(c) 129.509
(d) 1.995×10^6

23. Express each of the following numbers in exponential notation:
(a) 2,900,000
(b) 0.587
(c) 0.00840
(d) 0.0000055

24. Write each of the following numbers in exponential notation:
(a) 0.0456
(b) 4082.2
(c) 40.30
(d) 12,000,000

25. Solve the following problems, stating answers to the proper number of significant figures:
(a) $12.62 + 1.5 + 0.25 =$
(b) $(2.25 \times 10^3)(4.80 \times 10^4) =$
(c) $\dfrac{452 \times 6.2}{14.3} =$
(d) $0.0394 \times 12.8 =$
(e) $\dfrac{0.4278}{59.6} =$
(f) $10.4 + 3.75 \times (1.5 \times 10^4) =$

26. Evaluate each of the following expressions. State the answer to the proper number of significant figures:
(a) $15.2 - 2.75 + 15.67$
(b) 4.68×12.5
(c) $\dfrac{182.6}{4.6}$
(d) $1986 + 23.84 + 0.012$
(e) $\dfrac{29.3}{284 \times 415}$
(f) $(2.92 \times 10^{-3})(6.14 \times 10^5)$

27. Change these fractions into decimals. Express each answer to three significant figures:
(a) $\dfrac{5}{6}$ **(b)** $\dfrac{3}{7}$ **(c)** $\dfrac{12}{16}$ **(d)** $\dfrac{9}{18}$

28. Change each of the following decimals to fractions in lowest terms:
(a) 0.25 **(b)** 0.625 **(c)** 1.67 **(d)** 0.8888

29. Solve each of these equations for x:
(a) $3.42x = 6.5$
(b) $\dfrac{x}{12.3} = 7.05$
(c) $\dfrac{0.525}{x} = 0.25$

30. Solve each equation for the variable:
(a) $x = \dfrac{212 - 32}{1.8}$
(b) $8.9 \dfrac{\text{g}}{\text{mL}} = \dfrac{40.90 \text{ g}}{x}$
(c) $72°\text{F} = 1.8x + 32$

Unit Conversions

31. Complete the following metric conversions using the correct number of significant figures:
(a) 28.0 cm to m
(b) 1000. m to km
(c) 9.28 cm to mm
(d) 10.68 g to mg
(e) 6.8×10^4 mg to kg
(f) 8.54 g to kg
(g) 25.0 mL to L
(h) 22.4 L to μL

32. Complete the following metric conversions using the correct number of significant figures:
(a) 4.5 cm to Å
(b) 12 nm to cm
(c) 8.0 km to mm
(d) 164 mg to g
(e) 0.65 kg to mg
(f) 5.5 kg to g
(g) 0.468 L to mL
(h) 9.0 μL to mL

33. Complete the following American/metric conversions using the correct number of significant figures:
(a) 42.2 in. to cm
(b) 0.64 mi to in.
(c) 2.00 $in.^2$ to cm^2
(d) 42.8 kg to lb
(e) 3.5 qt to mL
(f) 20.0 gal to L

34. Make the following conversions using the correct number of significant figures:
(a) 35.6 m to ft
(b) 16.5 km to mi
(c) 4.5 $in.^3$ to mm^3
(d) 95 lb to g
(e) 20.0 gal to L
(f) 4.5×10^4 ft^3 to m^3

35. An automobile traveling at 55 miles per hour is moving at what speed in kilometers per hour?

36. A cyclist is traveling downhill at 55 km/hr. How fast is she moving in feet per second?

37. Carl Lewis, a sprinter in the 1988 Olympic Games, ran the 100.-m dash in 9.92 s. What was his speed in feet per second?

38. Al Unser, Jr., qualified for the pole position at the 1994 Indianapolis 500 at a speed of 229 mph. What was his speed in kilometers per second?

39. When the space probe *Galileo* reached Jupiter in 1995, it was traveling at an average speed of 27,000 miles per hour. What was its speed in kilometers per second?

40. The sun is approximately 93 million miles from the earth. How many seconds will it take for light from the sun to travel to the earth if the velocity of light is 3.00×10^8 m/s?

41. How many kilograms does a 176-lb man weigh?

42. The average mass of the heart of a human baby is about 1 oz. What is its mass in milligrams?

43. A regular aspirin tablet contains 5.0 grains of aspirin. How many grams of aspirin are in one tablet (1 grain = 1/7000. lb)?

44. An adult ruby-throated hummingbird has an average mass of 3.2 g, while an adult California condor may attain a weight of 21 lb. How many hummingbirds would it take to equal the mass of one condor?

45. A bag of pretzels has a mass of 283.5 g and costs $1.49. If a bag contains 18 pretzels determine the cost of a pound of pretzels?

46. The price of gold varies greatly and has been as high as $875 per ounce. What is the value of 250 g of gold at $350 per ounce? Gold is priced by troy ounces (14.58 troy ounces = 1 lb).

47. At 35¢/L how much will it cost to fill a 15.8-gal tank with gasoline?

48. How many liters of gasoline will be used to drive 525 miles in a car that averages 35 miles per gallon?

***49.** Assuming that there are 20. drops in 1.0 mL, how many drops are in 1.0 gallon?

50. How many liters of oil are in a 42-gal barrel of oil?

***51.** Calculate the number of milliliters of water in a cubic foot of water.

***52.** Oil spreads in a thin layer on water called an "oil slick." How much area in m^3 will 200 cm^3 of oil cover if it forms a layer 0.5 nm thick?

53. A textbook is 27 cm long, 21 cm wide, and 4.4 cm thick. What is the volume in:
(a) cubic centimeters?
(b) liters?
(c) cubic inches?

54. An aquarium measures 16 in. × 8 in. × 10 in. How many liters of water does it hold? How many gallons?

Temperature Conversions

55. Normal body temperature for humans is 98.6°F. What is this temperature on the Celsius scale?

56. Driving to the grocery store you notice the temperature is 45°C. Determine what this temperature is on the Fahrenheit scale and what season of the year it might be.

57. Make the following conversions and include an equation for each one:
(a) 162°F to °C
(b) 0.0°F to K
(c) −18°C to °F
(d) 212 K to °C

58. Make the following conversions and include an equation for each one:
(a) 32°C to °F
(b) −8.6°F to °C
(c) 273°C to K
(d) 100 K to °F

***59.** At what temperature are the Fahrenheit and Celsius temperatures exactly equal?

***60.** At what temperature are Fahrenheit and Celsius temperatures the same in value but opposite in sign?

Density

61. Calculate the density of a liquid if 50.00 mL of the liquid has a mass of 78.26 g.

62. A 12.8-mL sample of bromine has a mass of 39.9 g. What is the density of bromine?

63. When a 32.7-g piece of chromium metal was placed into a graduated cylinder containing 25.0 mL of water, the water level rose to 29.6 mL. Calculate the density of the chromium.

64. An empty graduated cylinder has a mass of 42.817 g. When filled with 50.0 mL of an unknown liquid it has a mass of 106.773 g. What is the density of the liquid?

65. Concentrated hydrochloric acid has a density of 1.19 g/mL. Calculate the mass of 250.0 mL of this acid.

66. What mass of mercury (density 13.6 g/mL) will occupy a volume of 25.0 mL?

Additional Exercises

These exercises are not paired or labeled by topic and provide additional practice on the concepts covered in this chapter.

67. One liter of homogenized whole milk has a mass of 1032 g. What is the density of the milk in grams per milliliter? In kilograms per liter?

68. The volume of blood plasma in adults is 3.1 L. Its density is 1.03 g/cm^3. Approximately how many pounds of blood plasma are there in your body?

***69.** The dashed lane markers on an interstate highway are 2.5 ft long and 4.0 in. wide. One (1.0) qt of paint covers 43 ft^2. How many dashed lane markers can be painted with 15 gal of paint?

70. Will a hollow cube with sides of length 0.50 m hold 8.5 L of solution? Depending on your answer, how much additional solution would be required to fill the container or how many times would the container need to be filled to measure the 8.5 L?

***71.** The accepted toxic dose of mercury is 300 μg/day. Dental offices sometimes contain as much as 180 μg of mercury per cubic meter of air. If a nurse working in the office ingests 2×10^4 L of air per day, is he or she at risk for mercury poisoning?

72. Which is the higher temperature, 4.5°F or −15°C?

***73.** A flask containing 100. mL of alcohol (d = 0.789 g/mL) is placed on one pan of a two-pan balance. A larger container, with a mass of 11.0 g more than the empty flask, is placed on the other pan of the balance. What volume of turpentine (d = 0.87 g/mL) must be added to this container to bring the two pans into balance?

74. Suppose you have samples of two metals, A and B. Use the data below to determine which sample occupies the larger volume

	A	B
Mass	25 g	65 g
Density	10 g/mL	4 g/mL

***75.** As a solid substance is heated its volume increases but its mass remains the same. Sketch a graph of density vs. temperature showing the trend you expect. Briefly explain.

76. A 35.0-mL sample of ethyl alcohol (density 0.789 g/mL) is added to a graduated cylinder that has a mass of 49.28 g. What will be the mass of the cylinder plus the alcohol?

77. You are given three cubes, A, B, and C; one is magnesium, one is aluminum, and the third is silver. All three cubes have the same mass, but cube A has a volume of 25.9 mL, cube B has a volume of 16.7 mL, and cube C has a volume of 4.29 mL. Identify cubes A, B, and C.

***78.** A cube of aluminum has a mass of 500. g. What will be the mass of a cube of gold of the same dimensions?

79. A 25.0-mL sample of water at 90°C has a mass of 24.12 g. Calculate the density of water at this temperature.

80. The mass of an empty container is 88.25 g. The mass of the container when filled with a liquid (d = 1.25 g/mL) is 150.50 g. What is the volume of the container?

81. Which liquid will occupy the greater volume, 50 g of water or 50 g of ethyl alcohol? Explain.

82. A gold bullion dealer advertised a bar of pure gold for sale. The gold bar had a mass of 3300 g and measured 2.00 cm × 15.0 cm × 6.00 cm. Was the bar pure gold? Show evidence for your answer.

83. The largest nugget of gold on record was found in 1872 in New South Wales, Australia, and had a mass of 93.3 kg. Assuming the nugget is pure gold, what is its volume in cubic centimeters? What is it worth by today's standards if gold is $345/oz? (14.58 troy oz = 1 lb.)

***84.** Forgetful Freddie placed 25.0 mL of a liquid in a graduated cylinder with a mass of 89.450 g when empty. When Freddie placed a metal slug with a mass of 15.454 g into the cylinder, the volume rose to 30.7 mL. Freddie was asked to calculate the density of the liquid and of the metal slug from his data, but he forgot to obtain the mass of the liquid. He was told that if he found the mass of the cylinder containing the liquid and the slug, he would have enough data for the calculations. He did so and found its mass to be 125.934 g. Calculate the density of the liquid and of the metal slug.

Answers to Practice Exercises

2.1 (a) 2, (b) 4, (c) 4, (d) 1, (e) 3, (f) 4, (g) 1, (h) 3

2.2 (a) 42.25 (Rule 2), (b) 88.0 (Rule 1), (c) 0.0897 (Rule 2), (d) 0.090 (Rule 2), (e) 225 (Rule 1), (f) 14.2 (Rule 2)

2.3 (a) 1200 = 1.200×10^3 (left means positive exponent), (b) 6,600,000 = 6.6×10^6 (left means positive exponent), (c) 0.0468 = 4.68×10^{-2} (right means negative exponent), (d) 0.00003 = 3×10^{-5} (right means negative exponent)

2.4 3350 in.^{2} = 3.4×10^3 in.^{2}

2.5 50.7 mph

2.6 0.79

2.7 (a) 2, (b) 2, (c) 1, (d) 2, (e) 4, (f) 2, (g) 2, (h) 2

2.8 1.69×10^4 m

2.9 30 m^3 or 3×10^1 m^3

2.10 165 lb

2.11 0.14 lb

2.12 0.793 qt

2.13 1.89 L (the number of significant figures is arbitrary)

2.14 −269°C, −452°F

2.15 37.0°C, 310.2 K

2.16 The density is 13 g/mL; therefore the ring is *not* pure silver.

2.17 0.88 g/mL

Classification of Matter

3

Chapter 3

Throughout our lives we seek to bring order into the chaos that surrounds us. To do this, we classify things according to their similarities. In the library, we find books grouped according to the subject, and then by author. Our local department store organizes its merchandise by the size and style of clothing, as well as by the type of customer. The ball park or theater classifies its seats by price and location. The biologist divides the living world into plants and animals; this broad classification is further simplified into various phyla and on to specific genera. In chemistry, this classification process begins with mixtures (such as air or vinegar) and then on to pure substances (such as water or mercury). This process continues and ultimately leads us to the fundamental building blocks of matter—the elements.

3.1 Matter Defined

The entire universe consists of matter and energy. Every day we come into contact with countless kinds of matter. Air, food, water, rocks, soil, glass, and this book are all different types of matter. Broadly defined, **matter** is *anything* that has mass and occupies space.

matter

Matter may be quite invisible. For example, if an apparently empty test tube is submerged mouth downward in a beaker of water, the water rises only slightly into the tube. The water cannot rise further because the tube is filled with invisible matter: air (see Figure 3.1).

To the eye, matter appears to be continuous and unbroken. However, it is actually discontinuous and is composed of discrete, tiny particles called *atoms*. The particulate nature of matter will become evident when we study atomic structure and the properties of gases.

3.2 Physical States of Matter

Matter exists in three physical states: solid, liquid, and gas. A **solid** has a definite shape and volume, with particles that cohere rigidly to one another. The shape of a solid can be independent of its container. For example, a crystal of sulfur has the same shape and volume whether it is placed in a beaker or simply laid on a glass plate.

solid

Most commonly occurring solids, such as salt, sugar, quartz, and metals, are *crystalline*. The particles that form crystalline materials exist in regular, repeating, three-dimensional, geometric patterns. Because their particles do not have any regular, internal geometric pattern, such solids as plastics, glass, and gels are called **amorphous** solids. (*Amorphous* means without shape or form.)

amorphous

A **liquid** has a definite volume but not a definite shape, with particles that cohere firmly but not rigidly. Although the particles are held together by strong attractive forces and are in close contact with one another, they are able to move freely. Particle mobility gives a liquid fluidity and causes it to take the shape of the container in which it is stored.

liquid

◀Chapter Opening Photo: Apples are classified by color, taste, and variety before they are shipped to consumers.

FIGURE 3.1
An apparently empty test tube is submerged, mouth downward, in water. Only a small volume of water rises into the tube, which is actually filled with air.

TABLE 3.1 Common Materials in the Solid, Liquid and Gaseous States of Matter

Solids	Liquids	Gases
Aluminum	Alcohol	Acetylene
Copper	Blood	Air
Gold	Gasoline	Butane
Polyethylene	Honey	Carbon dioxide
Salt	Mercury	Chlorine
Sand	Oil	Helium
Steel	Vinegar	Methane
Sulfur	Water	Oxygen

TABLE 3.2 Physical Properties of Solids, Liquids, and Gases

State	Shape	Volume	Particles	Compressibility
Solid	Definite	Definite	Rigidly cohering; tightly packed	Very slight
Liquid	Indefinite	Definite	Mobile; cohering	Slight
Gas	Indefinite	Indefinite	Independent of each other and relatively far apart	High

gas

A **gas** has indefinite volume and no fixed shape, with particles that move independently of one another. Particles in the gaseous state have gained enough energy to overcome the attractive forces that held them together as liquids or solids. A gas presses continuously in all directions on the walls of any container. Because of this quality a gas completely fills a container. The particles of a gas are relatively far apart compared with those of solids and liquids. The actual volume of the gas particles is usually very small in comparison with the volume of the space occupied by the gas. A gas therefore may be compressed into a very small volume or expanded almost indefinitely. Liquids cannot be compressed to any great extent, and solids are even less compressible than liquids.

If a bottle of ammonia solution is opened in one corner of the laboratory, we can soon smell its familiar odor in all parts of the room. The ammonia gas escaping from the solution demonstrates that gaseous particles move freely and rapidly and tend to permeate the entire area into which they are released.

Although matter is discontinuous, attractive forces exist that hold the particles together and give matter its appearance of continuity. These attractive forces are strongest in solids, giving them rigidity; they are weaker in liquids but still strong enough to hold liquids to definite volumes. In gases the attractive forces are so weak that the particles of a gas are practically independent of one another. Table 3.1 lists a number of common materials that exist as solids, liquids, and gases. Table 3.2 summarizes comparative properties of solids, liquids, and gases.

◀ **Water can exist as a solid (snow), a liquid (water), and a gas (steam) as shown here at Yellowstone National Park.**

3.3 Substances and Mixtures

The term *matter* refers to all materials or material things that make up the universe. Many thousands of different and distinct kinds of matter or substances exist. A **substance** is a particular kind of matter with a definite, fixed composition. A substance, sometimes known as a *pure substance,* is either an element or a compound. Familiar examples of elements are copper, gold, and oxygen. Familiar compounds are salt, sugar, and water.

substance

We can classify a sample of matter as either *homogeneous* or *heterogeneous* by examining it. **Homogeneous** matter is uniform in appearance and has the same properties throughout. Matter consisting of two or more physically distinct phases is **heterogeneous.** A **phase** is a homogeneous part of a system separated from other parts by physical boundaries. A **system** is simply the body of matter under consideration. Whenever we have a system in which visible boundaries exist between the parts or components, that system has more than one phase and is heterogeneous. It does not matter whether these components are in the solid, liquid, or gaseous states.

homogeneous

heterogeneous
phase
system

A pure substance may exist as different phases in a heterogeneous system. Ice floating in water, for example, is a two-phase system made up of solid water and liquid water. The water in each phase is homogeneous in composition, but because two phases are present, the system is heterogeneous.

A **mixture** is a material containing two or more substances and can be either heterogeneous or homogeneous. Mixtures are variable in composition. If we add a

mixture

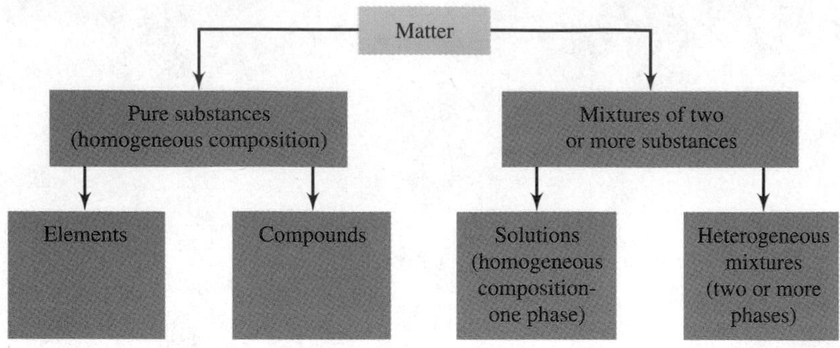

FIGURE 3.2 ▶ Classification of matter. A pure substance is always homogeneous in composition, whereas a mixture always contains two or more substances and may be either homogeneous or heterogeneous.

spoonful of sugar to a glass of water, a heterogeneous mixture is formed immediately. The two phases are a solid (sugar) and a liquid (water). But upon stirring the sugar dissolves to form a homogeneous mixture or solution. Both substances are still present: All parts of the solution are sweet and wet. The proportions of sugar and water can be varied simply by adding more sugar and stirring to dissolve.

Flow charts can help you to visualize the connections between concepts.

Many substances do not form homogeneous mixtures. If we mix sugar and fine white sand, a heterogeneous mixture is formed. Careful examination may be needed to decide that the mixture is heterogeneous because the two phases (sugar and sand) are both white solids. Ordinary matter exists mostly as mixtures. If we examine soil, granite, iron ore, or other naturally occurring mineral deposits, we find them to be heterogeneous mixtures. Air is a homogeneous mixture (solution) of several gases. Figure 3.2 illustrates the relationships of substances and mixtures.

3.4 Elements

element

All the words in the English dictionary are formed from an alphabet consisting of only 26 letters. All known substances on earth—and most probably in the universe, too—are formed from a sort of "chemical alphabet" consisting of 111 presently known elements. An **element** is a fundamental or elementary substance that cannot be broken down by chemical means to simpler substances. Elements are the building blocks of all substances. The elements are numbered in order of increasing complexity beginning with hydrogen, number 1. Of the first 92 elements, 88 are known to occur in nature. The other four—technetium (43), promethium (61), astatine (85), and francium (87)—either do not occur in nature or have only transitory existences during radioactive decay. With the exception of number 94, plutonium, elements above number 92 are not known to occur naturally but have been synthesized, usually in very small quantities, in laboratories. The discovery of trace amounts of element 94 (plutonium) in nature has been reported recently. The syntheses of elements 110 and 111 were reported in 1994. No elements other than those on the earth have been detected on other bodies in the universe.

Most substances can be decomposed into two or more simpler substances. Water can be decomposed into hydrogen and oxygen. Sugar can be decomposed into carbon, hydrogen, and oxygen. Table salt is easily decomposed into sodium and chlorine. An element, however, cannot be decomposed into simpler substances by ordinary chemical changes.

If we could take a small piece of an element, say copper, and divide it and subdivide it into smaller and smaller particles, we would finally come to a single unit of copper that we could no longer divide and still have copper. This smallest particle of an element that can exist, is called an **atom**, which is also the smallest unit of an element that can enter into a chemical reaction. Atoms are made up of still smaller subatomic particles. But these subatomic particles (described in Chapter 5) do not have the properties of elements.

atom

3.5 Distribution of Elements

Elements are distributed very unequally in nature, as shown in Figure 3.3. At normal room temperature two of the elements, bromine and mercury, are liquids. Eleven elements, hydrogen, nitrogen, oxygen, fluorine, chlorine, helium, neon, argon, krypton, xenon, and radon, are gases. All the other elements are solids.

Ten elements make up about 99% of the mass of the earth's crust, seawater, and atmosphere. Oxygen, the most abundant of these, constitutes about 50% of this mass. The distribution of the elements shown in Table 3.3 includes the earth's crust to a depth of about 10 miles, the oceans, fresh water, and the atmosphere but does not include the mantle and core of the earth, which are believed to consist of metallic iron and nickel. Because the atmosphere contains relatively little matter, its inclusion has almost no effect on the distribution shown in Table 3.3. But the inclusion of fresh and salt water does have an appreciable effect since water contains about 11.2% hydrogen. Nearly all of the 0.87% hydrogen shown in the table is from water.

The average distribution of the elements in the human body is shown in Figure 3.4. Note again the high percentage of oxygen.

FIGURE 3.3
Distribution of elements in nature.

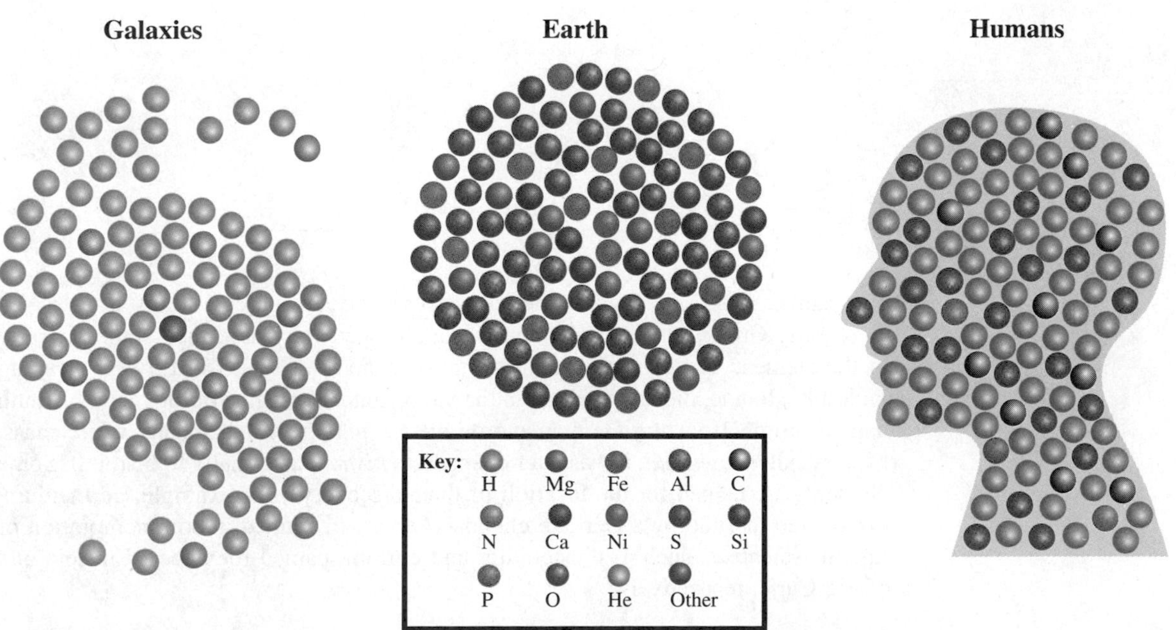

TABLE 3.3 Distribution of the Elements in the Earth's Crust, Seawater, and Atmosphere

Element	Mass percent	Element	Mass percent
Oxygen	49.20	Chlorine	0.19
Silicon	25.67	Phosphorus	0.11
Aluminum	7.50	Manganese	0.09
Iron	4.71	Carbon	0.08
Calcium	3.39	Sulfur	0.06
Sodium	2.63	Barium	0.04
Potassium	2.40	Nitrogen	0.03
Magnesium	1.93	Fluorine	0.03
Hydrogen	0.87		
Titanium	0.58	All others	0.47

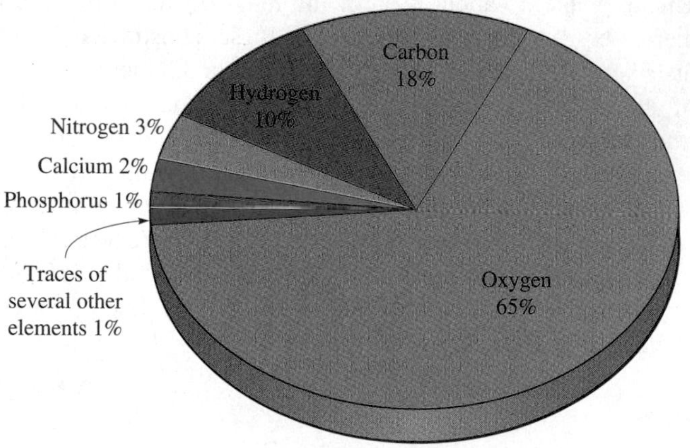

FIGURE 3.4 ▶ Average elemental composition of the human body.

3.6 Names of the Elements

The names of the elements come to us from various sources. Many are derived from early Greek, Latin, or German words that generally described some property of the element. For example, iodine is taken from the Greek word *iodes,* meaning violetlike. Iodine, indeed, is violet in the vapor state. The name of the metal bismuth had its origin from the German words *weisse masse,* which means white mass. Miners called it *wismat*; it was later changed to *bismat,* and finally to bismuth. Some elements are named for the location of their discovery—for example, germanium, discovered in 1886 by a German chemist. Others are named in commemoration of famous scientists, such as einsteinium and curium, named for Albert Einstein and Marie Curie, respectively.

3.7 Symbols of the Elements

We all recognize Mr., N.Y., and Ave. as abbreviations for mister, New York, and avenue. In a like manner chemists have assigned an abbreviation to each element; these are called **symbols** of the elements. Fourteen of the elements have a single letter as their symbol, six have three-letter symbols, and the rest have two letters. A symbol stands for the element itself, for one atom of the element, and (as we shall see later) for a particular quantity of the element.

symbol

Controversy exists over the names for the elements with three-letter symbols. (See Chemistry in Action, p. 103.)

Rules governing symbols of elements are as follows:

1. Symbols are composed of one, two, or three letters.
2. If one letter is used, it is capitalized.
3. If two or three letters are used, only the first is capitalized.

Examples: Sulfur S Barium Ba

The symbols and names of all the elements are given in the table on the inside front cover of this book. Table 3.4 lists the more commonly used symbols. If we examine this table carefully, we note that most of the symbols start with the same letter as the name of the element that is represented. A number of symbols, however, appear to have no connection with the names of the elements they represent (see Table 3.5). These symbols have been carried over from earlier names (usually in Latin) of the elements and are so firmly implanted in the literature that their use is continued today.

Special care must be taken in writing symbols. Begin each with a capital letter and use a lowercase second letter if needed. For example, consider Co, the symbol for the element cobalt. If through error CO (capital C and capital O) is written, the two elements carbon and oxygen (the *formula* for carbon monoxide) are represented

TABLE 3.4 Symbols of the Most Common Elements

Element	Symbol	Element	Symbol	Element	Symbol
Aluminum	Al	Fluorine	F	Phosphorus	P
Antimony	Sb	Gold	Au	Platinum	Pt
Argon	Ar	Helium	He	Potassium	K
Arsenic	As	Hydrogen	H	Radium	Ra
Barium	Ba	Iodine	I	Silicon	Si
Bismuth	Bi	Iron	Fe	Silver	Ag
Boron	B	Lead	Pb	Sodium	Na
Bromine	Br	Lithium	Li	Strontium	Sr
Cadmium	Cd	Magnesium	Mg	Sulfur	S
Calcium	Ca	Manganese	Mn	Tin	Sn
Carbon	C	Mercury	Hg	Titanium	Ti
Chlorine	Cl	Neon	Ne	Tungsten	W
Chromium	Cr	Nickel	Ni	Uranium	U
Cobalt	Co	Nitrogen	N	Zinc	Zn
Copper	Cu	Oxygen	O		

This tiny and powerful computer chip is made of silicon, a metalloid.

J. J. Berzelius (1779–1848) devised the chemical symbol system in use today.

TABLE 3.5 Symbols of the Elements Derived from Early Names*

Present name	Symbol	Former name
Antimony	Sb	Stibium
Copper	Cu	Cuprum
Gold	Au	Aurum
Iron	Fe	Ferrum
Lead	Pb	Plumbum
Mercury	Hg	Hydrargyrum
Potassium	K	Kalium
Silver	Ag	Argentum
Sodium	Na	Natrium
Tin	Sn	Stannum
Tungsten	W	Wolfram

* These symbols are in use today even though they do not correspond to the current name of the element.

instead of the single element cobalt. Another example of the need for care in writing symbols is with the symbol Ca for calcium versus Co for cobalt. The letters must be distinct, or else the symbol for the element may be misinterpreted.

Knowledge of symbols is essential for writing chemical formulas and equations, and will be used extensively in the remainder of this book and in any future chemistry courses you may take. One way to learn the symbols is to practice a few minutes a day by making side-by-side lists of names and symbols and then covering each list alternately and writing the corresponding name or symbol. Initially it is a good plan to learn the symbols of the most common elements shown in Table 3.4.

3.8 Metals, Nonmetals, and Metalloids

Three primary classifications of the elements are metals, nonmetals, and metalloids. Most of the elements are metals. We are familiar with them because of their widespread use in tools, materials of construction, automobiles, and so on. But nonmetals are equally useful in our everyday life as major components of clothing, food, fuel, glass, plastics, and wood. Metalloids are often used in the electronics industry.

metal

The **metals** are solids at room temperature (mercury is an exception). They have high luster, are good conductors of heat and electricity, are *malleable* (can be rolled or hammered into sheets), and are *ductile* (can be drawn into wires). Most metals have a high melting point and high density. Familiar metals are aluminum, chromium, copper, gold, iron, lead, magnesium, mercury, nickel, platinum, silver, tin, and zinc. Less familiar but still important metals are calcium, cobalt, potassium, sodium, uranium, and titanium.

Metals have little tendency to combine with each other to form compounds. But many metals readily combine with nonmetals such as chlorine, oxygen, and sulfur to form ionic compounds such as metallic chlorides, oxides, and sulfides. In nature the more reactive metals are found combined with other elements as minerals.

Chemistry in Action

Hydrogen: Fuel of the Future?

Hydrogen, the lightest element on the periodic table, could provide the basis for a society powered by a fuel that is nearly inexhaustible, is environmentally benign, and is available domestically. The space program has used hydrogen for more than 20 years to power rockets as well as to provide electrical power and even drinking water. How can we harness the energy found in hydrogen and use it to form a hydrogen-based fuel economy?

A major hurdle to using hydrogen as a primary fuel source is the difficulty in storing it. As a gas, it has a low density and requires a bulky container; liquid hydrogen (b.p. −253°C) requires intensive refrigeration and well-insulated containers. The best solution to the storage problem may be to generate hydrogen as it is needed for fuel.

The most promising, simple, and inexpensive scheme for generating hydrogen is based on sponge iron. The most remarkable quality of sponge iron is its ability to rust easily. In the rusting process water is split into hydrogen and oxygen while iron is converted to iron oxide (rust). Usually rusting takes place very slowly so that the hydrogen is formed in amounts too small to be useful. However, the H Power Corp of Belleville, N. J. has developed a process that increases the speed of rusting and cycles the iron oxide back to the sponge-iron state so it can produce more hydrogen.

This sponge-iron cycle could then be linked to fuel cells to eliminate the need for hydrogen storage units in transportation vehicles. The sponge-iron cycle could also be used in residential heating or in industrial processes. Since sponge iron is inexpensive and recyclable, it could be transported by ship and rail to many locations. The overall cost should be competitive to fossil fuels.

Changing over to hydrogen as a primary fuel source could be done gradually by using this type of technology. To use hydrogen effectively requires an infrastructure, which could take from 20 to 25 years to develop. Studies indicate that the transportation sector could be the first place to begin the use of hydrogen. Hydrogen-powered vehicles could have significant impact in places such as California where new regulations call for lower emission vehicles after 1998. Conventional cars could be converted to run on hydrogen, but the major difficulty is in getting enough hydrogen on board to provide a reasonable driving range. Another area of development for hydrogen fuels is for residential use. Studies at the University of Miami show that hydrogen can be used for cooking and heating.

The use of hydrogen as a fuel could have a fundamental impact on the way our society is organized. Instead of massive power plants, our energy needs would be met by on-site fuel cells that provide all of our heating, cooling, lighting, and other power requirements. There would be no losses over transmission lines and little capital investment. Best of all, the system would reduce the environmental impact of energy generation and take the cost of energy from a government-regulated commercial venture and put it into the hands of the individual consumer.

The LaserCel I, unveiled in 1991, is more economical to operate on hydrogen than on gasoline. This is accomplished by replacing the internal combustion engine with a hydrogen fuel cell.

A few of the less reactive ones such as copper, gold, and silver are sometimes found in a native, or free, state.

nonmetal

Nonmetals, unlike metals, are not lustrous, have relatively low melting points and densities, and are generally poor conductors of heat and electricity. Carbon, phosphorus, sulfur, selenium, and iodine are solids; bromine is a liquid; the rest of the nonmetals are gases. Common nonmetals found uncombined in nature are carbon (graphite and diamond), nitrogen, oxygen, sulfur, and the noble gases (helium, neon, argon, krypton, xenon, and radon).

Nonmetals combine with one another to form molecular compounds such as carbon dioxide (CO_2), methane (CH_4), butane (C_4H_{10}), and sulfur dioxide (SO_2). Fluorine, the most reactive nonmetal, combines readily with almost all other elements.

Samples of various metals, including aluminum, copper, mercury, titanium, beryllium, cadmium, calcium, and nickel.

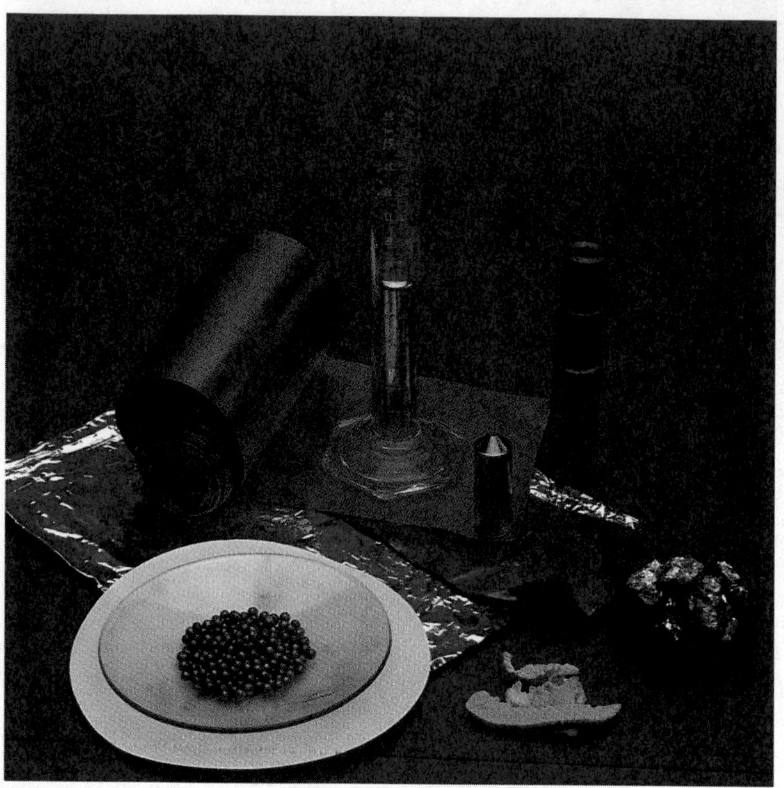

metalloid

Several elements (boron, silicon, germanium, arsenic, antimony, tellurium, and polonium) are classified as **metalloids** and have properties that are intermediate between those of metals and those of nonmetals. The intermediate position of these elements is shown in Table 3.6, which lists and classifies all the elements as metals, nonmetals, or metalloids. Certain metalloids, such as boron, silicon, and germanium, are the raw materials for the semiconductor devices that make our modern electronics industry possible.

3.9 Compounds

compound

A **compound** is a distinct substance containing two or more elements chemically combined in definite proportions by mass. Compounds, unlike elements, can be decomposed chemically into simpler substances—that is, into simpler compounds and/or elements. Atoms of the elements in a compound are combined in whole-number ratios, never as fractional parts. Compounds fall into two general types, *molecular* and *ionic.*

molecule

A **molecule** is the smallest uncharged individual unit of a compound formed by the union of two or more atoms. Water is a typical molecular compound. If we divide a drop of water into smaller and smaller particles, we finally obtain a single molecule of water consisting of two hydrogen atoms bonded to one oxygen atom. This molecule is the ultimate particle of water; it cannot be further subdivided without destroying the water molecule and forming hydrogen and oxygen.

TABLE 3.6 Classification of the Elements into Metals, Metalloids, and Nonmetals

Legend: Metals; Metalloids; Nonmetals

1 H																	2 He
3 Li	4 Be											5 B	6 C	7 N	8 O	9 F	10 Ne
11 Na	12 Mg											13 Al	14 Si	15 P	16 S	17 Cl	18 Ar
19 K	20 Ca	21 Sc	22 Ti	23 V	24 Cr	25 Mn	26 Fe	27 Co	28 Ni	29 Cu	30 Zn	31 Ga	32 Ge	33 As	34 Se	35 Br	36 Kr
37 Rb	38 Sr	39 Y	40 Zr	41 Nb	42 Mo	43 Tc	44 Ru	45 Rh	46 Pd	47 Ag	48 Cd	49 In	50 Sn	51 Sb	52 Te	53 I	54 Xe
55 Cs	56 Ba	57 La*	72 Hf	73 Ta	74 W	75 Re	76 Os	77 Ir	78 Pt	79 Au	80 Hg	81 Tl	82 Pb	83 Bi	84 Po	85 At	86 Rn
87 Fr	88 Ra	89 Ac†	104 Unq	105 Unp	106 Unh	107 Uns	108 Uno	109 Une									

*	58 Ce	59 Pr	60 Nd	61 Pm	62 Sm	63 Eu	64 Gd	65 Tb	66 Dy	67 Ho	68 Er	69 Tm	70 Yb	71 Lu
†	90 Th	91 Pa	92 U	93 Np	94 Pu	95 Am	96 Cm	97 Bk	98 Cf	99 Es	100 Fm	101 Md	102 No	103 Lr

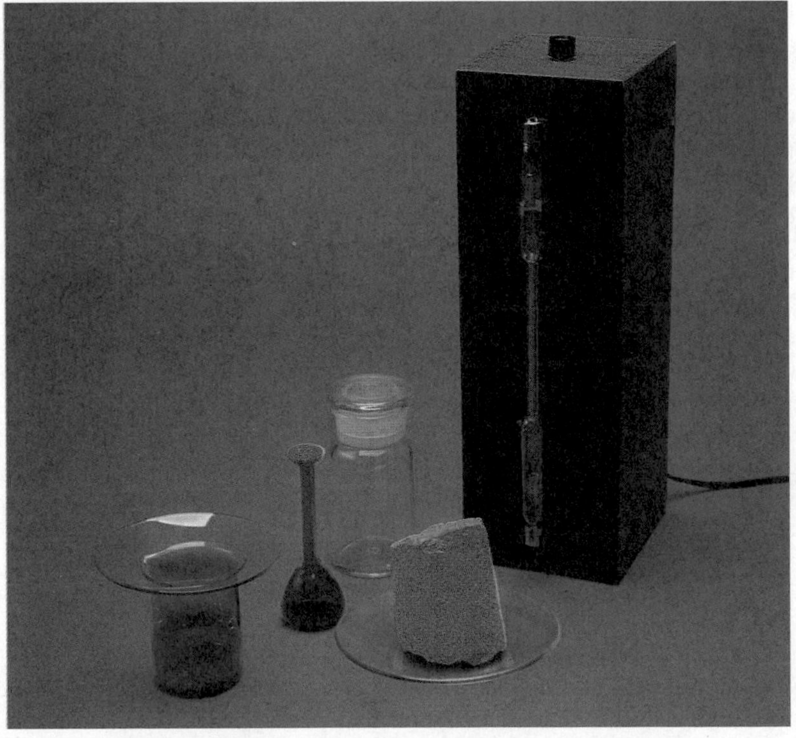

◀ Samples of various nonmetals, including iodine, bromine, oxygen, neon, and sulfur.

ion
cation
anion

An **ion** is a positively or negatively charged atom or group of atoms. An ionic compound is held together by attractive forces that exist between positively and negatively charged ions. A positively charged ion is called a **cation** (pronounced *cat-eye-on*); a negatively charged ion is called an **anion** (pronounced *an-eye-on*).

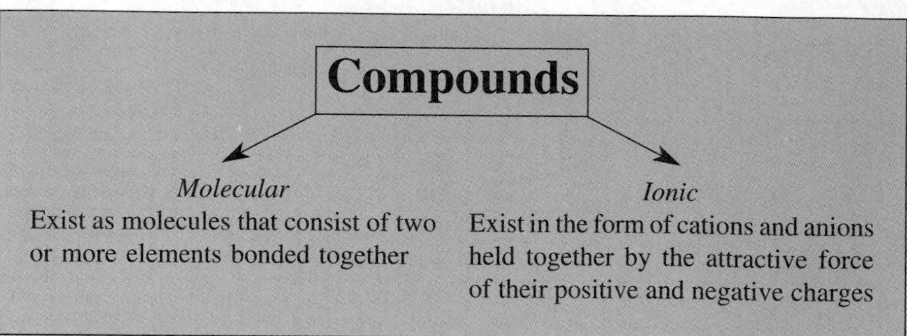

Compounds
Molecular
Exist as molecules that consist of two or more elements bonded together
Ionic
Exist in the form of cations and anions held together by the attractive force of their positive and negative charges

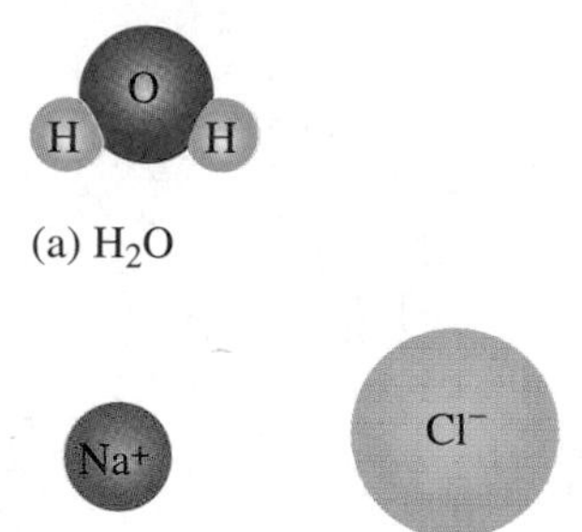

▲
FIGURE 3.5
Representation of molecular and ionic (nonmolecular) compounds. (a) Two hydrogen atoms combined with an oxygen atom to form a molecule of water. (b) A positively charged sodium ion and a negatively charged chloride ion form the compound sodium chloride.

Sodium chloride is a typical ionic compound. The ultimate particles of sodium chloride are positively charged sodium ions and negatively charged chloride ions. Sodium chloride is held together in a crystalline structure by the attractive forces existing between these oppositely charged ions. Although ionic compounds consist of large aggregates of cations and anions, their formulas are normally represented by the simplest possible ratio of the atoms in the compound. For example, in sodium chloride the ratio is one sodium ion to one chlorine ion, and the formula is NaCl. The two types of compounds, molecular and ionic, are illustrated in Figure 3.5.

There are more than 9 million known registered compounds, with no end in sight as to the number that will be prepared in the future. Each compound is unique and has characteristic properties. Let us consider two compounds, water and sodium chloride, in some detail. Water is a colorless, odorless, tasteless liquid that can be changed to a solid (ice) at 0°C and to a gas (steam) at 100°C. Composed of two atoms of hydrogen and one atom of oxygen per molecule, water is 11.2% hydrogen and 88.8% oxygen by mass. Water reacts chemically with sodium to produce hydrogen gas and sodium hydroxide, with lime to produce calcium hydroxide, and with sulfur trioxide to produce sulfuric acid. No other compound has all these exact physical and chemical properties; they are characteristic of water alone.

Sodium chloride is a colorless crystalline substance with a ratio of one atom of sodium to one atom of chlorine. Its composition by mass is 39.3% sodium and 60.7% chlorine. It does not conduct electricity in its solid state; it dissolves in water to produce a solution that conducts electricity. When a current is passed through molten sodium chloride, solid sodium and gaseous chlorine are produced. These specific properties belong to sodium chloride and to no other substance. Thus, a compound may be identified and distinguished from all other compounds by its characteristic properties.

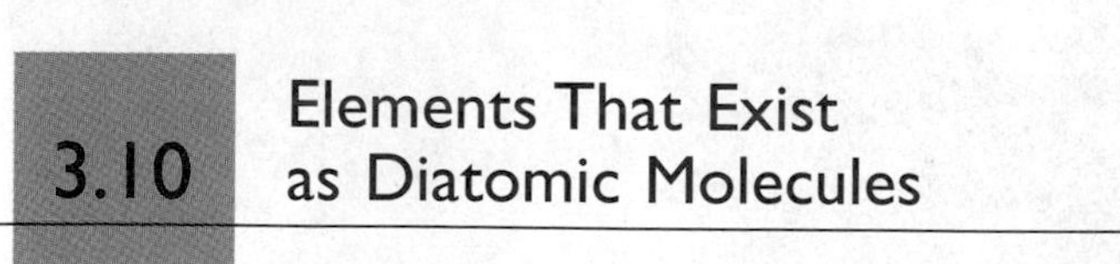

3.10 Elements That Exist as Diatomic Molecules

diatomic molecules

Seven of the elements (all nonmetals) occur as **diatomic molecules**. These elements and their symbols, formulas, and brief descriptions are listed in Table 3.7. Whether found free in nature or prepared in the laboratory, the molecules of these elements

TABLE 3.7 Elements That Exist as Diatomic Molecules

Element	Symbol	Molecular formula	Normal state
Hydrogen	H	H_2	Colorless gas
Nitrogen	N	N_2	Colorless gas
Oxygen	O	O_2	Colorless gas
Fluorine	F	F_2	Pale yellow gas
Chlorine	Cl	Cl_2	Yellow-green gas
Bromine	Br	Br_2	Reddish-brown liquid
Iodine	I	I_2	Bluish-black solid

always contain two atoms. The formulas of the free elements are therefore always written to show this molecular composition: H_2, N_2, O_2, F_2, Cl_2, Br_2, and I_2.

It is important to see how symbols are used to designate either an atom or a molecule of an element. Consider hydrogen and oxygen. Hydrogen gas is present in volcanic gases and can be prepared by many chemical reactions. Regardless of their source, all samples of free hydrogen gas consist of diatomic molecules. Free hydrogen is designated by the formula H_2, which also expresses its composition. Oxygen makes up about 21% by volume of the air that we breathe. This free oxygen is constantly being replenished by photosynthesis; it can also be prepared in the laboratory by several reactions. The majority of free oxygen is diatomic and is designated by the formula O_2. Now consider water, a compound designated by the formula H_2O (sometimes HOH). Water contains neither free hydrogen (H_2) nor free oxygen (O_2). The H_2 part of the formula H_2O simply indicates that two atoms of hydrogen are combined with one atom of oxygen to form water.

Symbols are used to designate elements, show the composition of molecules of elements, and give the elemental composition of compounds.

3.11 Chemical Formulas

Chemical formulas are used as abbreviations for compounds. A **chemical formula** shows the symbols and the ratio of the atoms of the elements in a compound. Sodium chloride contains one atom of sodium per atom of chlorine; its formula is NaCl. The formula for water is H_2O; it shows that a molecule of water contains two atoms of hydrogen and one atom of oxygen.

chemical formula

The formula of a compound tells us which elements it is composed of and how many atoms of each element are present in a formula unit. For example, a molecule of sulfuric acid is composed of two atoms of hydrogen, one atom of sulfur, and four atoms of oxygen. We could express this compound as HHSOOOO, but the usual formula for writing sulfuric acid is H_2SO_4. The formula may be expressed verbally as "H-two-S-O-four." Numbers that appear partially below the line and to

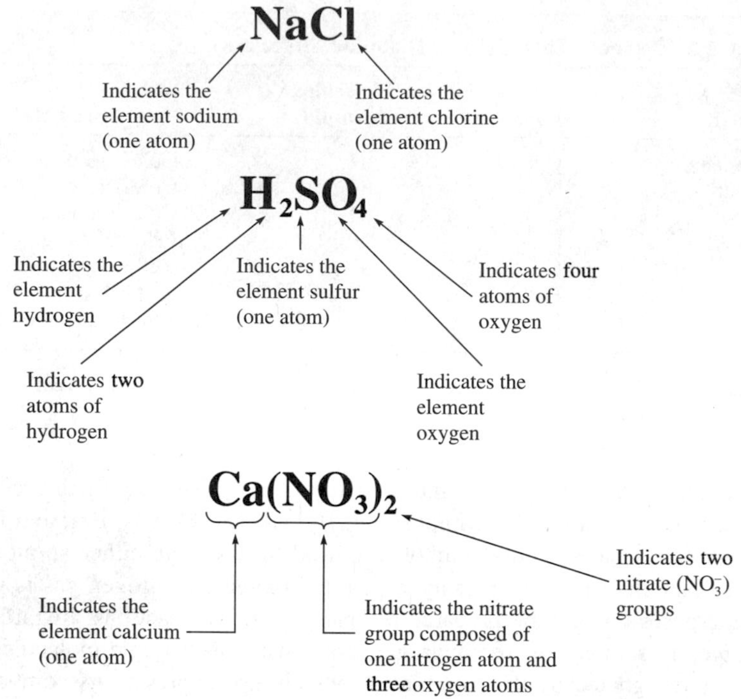

FIGURE 3.6 ▶ Explanation of the formulas NaCl, H_2SO_4, and $Ca(NO_3)_2$.

subscript

the right of a symbol of an element are called **subscripts.** Thus the 2 and the 4 in H_2SO_4 are subscripts (see Figure 3.6). Characteristics of chemical formulas are

1. The formula of a compound contains the symbols of all the elements in the compound.
2. When the formula contains one atom of an element, the symbol of that element represents that one atom. The number one (1) is not used as a subscript to indicate one atom of an element.
3. When the formula contains more than one atom of an element, the number of atoms is indicated by a subscript written to the right of the symbol of that atom. For example, the two (2) in H_2O indicates two atoms of H in the formula.
4. When the formula contains more than one of a group of atoms that occurs as a unit, parentheses are placed around the group, and the number of units of the group are indicated by a subscript placed to the right of the parentheses. Consider the nitrate group, NO_3. The formula for sodium nitrate, $NaNO_3$, has only one nitrate group; therefore no parentheses are needed. Calcium nitrate, $Ca(NO_3)_2$, has two nitrate groups, as indicated by the use of parentheses and the subscript 2. $Ca(NO_3)_2$ has a total of nine atoms: one Ca, two N, and six O atoms. The formula $Ca(NO_3)_2$ is read as "C-A [pause] N-O-three taken twice."
5. Formulas written as H_2O, H_2SO_4, $Ca(NO_3)_2$, and $C_{12}H_{22}O_{11}$ show only the number and kind of each atom contained in the compound; they do not show the arrangement of the atoms in the compound or how they are chemically bonded to one another.

Carbon—The Chameleon

One of the most diverse elements in the periodic table is carbon. Although it is much less abundant than many other elements, it is readily available. Carbon is found free in three forms called **allotropes**—as the mineral graphite, as diamonds, and as buckminsterfullerene, a form only recently discovered. The physical properties of the allotropes are quite distinct. Diamond consists of transparent, octahedral crystals that are colorless when pure—but may range from pale blue to jet black due to impurities. It is the hardest known substance and an excellent heat conductor. When a diamond has certain impurities added to the crystal intentionally it becomes an electrical semiconductor. Diamonds for cutting tools are both mined and produced synthetically. The majority of gem diamonds are mined in South Africa but some also come from South America. Graphite, on the other hand, consists of layers or sheets of carbon atoms, which are very soft and which easily slip over one another. It is an excellent conductor of electricity. Graphite is mined as massive crystals, or is obtained from heating coal and pitch in very high temperature furnaces. Buckminsterfullerene is composed of clusters of carbon atoms arranged in the shape of a soccer ball. The cage-like structure permits capture of other atoms leading to some interesting applications. More information on buckminsterfullerene can be found in the Chemistry in Action on page 232.

Carbon is an essential constituent of plant and animal life. For example, coal is formed by the gradual decay of plant life enriched in carbon through the loss of carbon dioxide and methane gas during the decaying process. If coal is heated without air present, *destructive distillation* occurs. This process decomposes the carbon compounds and produces coke, which is 90–95% graphite. Charcoal, consisting of tiny crystals of graphite, is prepared by the destructive distillation of wood. Bone black, also consisting of tiny graphite crystals, is made from the destructive distillation of bones and wastes from packing houses. Carbon black is formed when natural gas is burned with an insufficient quantity of air—a residue of graphite carbon is formed on a cold surface and then scraped off.

Free carbon finds a variety of uses in our society. Diamonds are collected and displayed for their gem quality, used in jewelry, and held as investments. They are also used in cutting and drilling tools. Graphite has many uses as a lubricant. Mixtures of clay and graphite are molded into the "lead" used in pencils. The higher the clay content, the harder the "lead." Graphite is also used as electrodes in dry cells and is found in some paints and stove polish.

Charcoal can absorb large quantities of substances onto its surface. For this reason it is very useful in water purification systems, in the manufacture of gas masks, and in removing color from solutions (as in the refining of sugar). Coke is an important fuel. It is also used to reduce iron from its ore. Alloys of iron and carbon form the major industrial product we call *steel.* Carbon black is used in making carbon paper, printer's ink, and some shoe polish. It is the additive in rubber that makes tires black.

In addition, carbon combines chemically with other elements to form a myriad of useful *compounds.* **Hydrocarbons** containing carbon and hydrogen are commonly found in petroleum and natural gas. There are so many hydrocarbons and their derivatives that an entire branch of chemistry, *organic chemistry,* is dedicated to studying and understanding them. Carbon is also found as carbon dioxide, CO_2, in our atmosphere. As such, it is one of the greenhouse gases that contributes to global warming and is of great concern to scientists (see Section 8.7). Carbon monoxide, CO, is an important fuel gas. It is colorless, odorless, tasteless, and extremely poisonous. Breathing even small quantities of carbon monoxide can be fatal (see Chemistry in Action, p. 422). Carbon can combine with chlorine and fluorine to make a group of compounds known as **chlorofluorocarbons**. These compounds are widely used as refrigerants. They are definite contributors to the destruction of the ozone layer in our atmosphere and are the topic of debate and regulation worldwide (see Section 12.17).

Diamonds and graphite.

Example 3.1 Write formulas for the following compounds, the atom composition of which is given. (a) Hydrogen chloride: 1 atom hydrogen + 1 atom chlorine; (b) Methane: 1 atom carbon + 4 atoms hydrogen; (c) Glucose: 6 atoms carbon + 12 atoms hydrogen + 6 atoms oxygen.

Solution

(a) First write the symbols of the atoms in the formula: H Cl. Since the ratio of atoms is one to one, we merely bring the symbols together to give the formula for hydrogen chloride as HCl.
(b) Write the symbols of the atoms: C H. Now bring the symbols together and place a subscript 4 after the hydrogen atom. The formula is CH_4.
(c) Write the symbols of the atoms: C H O. Now write the formula, bringing together the symbols followed by the correct subscripts according to the data given (six C, twelve H, six O). The formula is $C_6H_{12}O_6$.

3.12 Mixtures

Single substances—elements or compounds—seldom occur naturally in a pure state. Air is a mixture of gases; seawater is a mixture of a variety of dissolved minerals; ordinary soil is a complex mixture of minerals and various organic materials.

How is a mixture distinguished from a pure substance? A mixture always contains two or more substances that can be present in varying concentrations. Let us consider an example of a homogeneous mixture and an example of a heterogeneous mixture. Homogeneous mixtures (solutions) containing either 5% or 10% salt in water can be prepared simply by mixing the correct amounts of salt and water. These mixtures can be separated by boiling away the water, leaving the salt as a residue. The composition of a heterogeneous mixture of sulfur crystals and iron filings can be varied by merely blending in either more sulfur or more iron filings. This mixture can be separated physically by using a magnet to attract the iron or by adding carbon disulfide to dissolve the sulfur.

(a) When iron and sulfur exist as pure substances, only the iron is attracted to a magnet. (b) A mixture of iron and sulfur can be separated by using the difference in magnetic attraction. (c) The compound iron(II) sulfide cannot be separated into its elements with a magnet.

(a)

(b)

(c)

TABLE 3.8 Comparison of Mixtures and Compounds

	Mixture	Compound
Composition	May be composed of elements, compounds, or both in variable composition.	Composed of two or more elements in a definite, fixed proportion by mass.
Separation of components	Separation may be made by physical or mechanical means.	Elements can be separated by chemical changes only.
Identification of components	Components do not lose their identity.	A compound does not resemble the elements from which it is formed.

Iron(II) sulfide (FeS) contains 63.5% Fe and 36.5% S by mass. If we mix iron and sulfur in this proportion, do we have iron(II) sulfide? No, it is still a mixture; the iron is still attracted by a magnet. But if this mixture is heated strongly, a chemical change (reaction) occurs in which the reacting substances, iron and sulfur, form a new substance, iron(II) sulfide. Iron(II) sulfide, FeS, is a compound of iron and sulfur and has properties that are different from those of either iron or sulfur: It is neither attracted by a magnet nor dissolved by carbon disulfide. The differences between the iron and sulfur *mixture* and the iron(II) sulfide *compound* are as follows:

Iron(II) sulfide is the correct name for the compound formed from iron and sulfur. We will discuss the reason for the (II) in Chapter 6 when we learn to name compounds.

	Mixture of iron and sulfur	Compound of iron and sulfur
Formula	Has no definite formula; consists of Fe and S.	FeS
Composition	Contains Fe and S in any proportion by mass.	63.5% Fe and 36.5% S by mass.
Separation	Fe and S can be separated by physical means.	Fe and S can be separated only by chemical change.

The general characteristics of mixtures and compounds are compared in Table 3.8.

Concepts in Review

1. Identify the three physical states of matter.
2. Distinguish between substances and mixtures.
3. Classify common materials as elements, compounds, or mixtures.
4. Write the symbols when given the names, or write the names when given the symbols, of the common elements listed in Table 3.4.

5. Understand how symbols, including subscripts and parentheses, are used to write chemical formulas.
6. Differentiate among atoms, molecules, and ions.
7. List the characteristics of metals, nonmetals, and metalloids.
8. List the elements that occur as diatomic molecules.

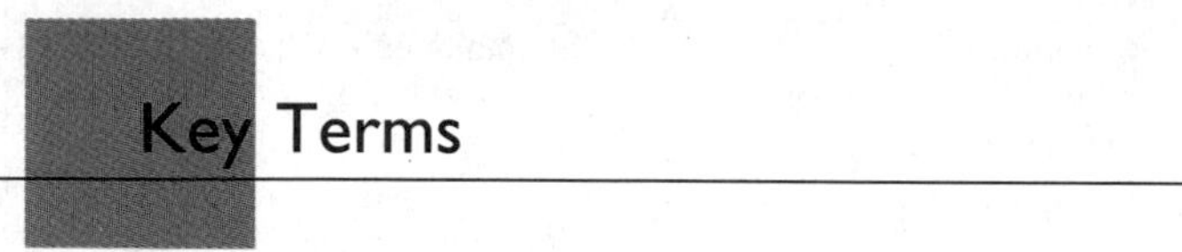

Key Terms

The terms listed here have been defined within this chapter. Section numbers are referenced in parentheses for each term.

CIA is used to indicate a term found in a "Chemistry in Action" essay.

allotropes (CIA)
amorphous (3.2)
anion (3.9)
atom (3.4)
cation (3.9)
chemical formula (3.11)
chlorofluorocarbons (CIA)
compound (3.9)
diatomic molecules (3.10)
element (3.4)
gas (3.2)
heterogeneous (3.3)
homogeneous (3.3)
hydrocarbons (CIA)
ion (3.9)
liquid (3.2)
matter (3.1)
metal (3.8)
metalloid (3.8)
mixture (3.3)
molecule (3.9)
nonmetal (3.8)
phase (3.3)
solid (3.2)
subscripts (3.11)
substance (3.3)
symbol (3.7)
system (3.3)

Questions

Questions refer to tables, figures, and key words and concepts defined within the chapter. A particularly challenging question or exercise is indicated with an asterisk.

1. List four different substances in each of the three states of matter.
2. In terms of the properties of the ultimate particles of a substance, explain
   (a) why a solid has a definite shape but a liquid does not.
   (b) why a liquid has a definite volume but a gas does not.
   (c) why a gas can be compressed rather easily but a solid cannot be compressed appreciably.
3. What evidence can you find in Figure 3.1 that gases occupy space?
4. Which liquids listed in Table 3.1 are not mixtures?
5. Which of the gases listed in Table 3.1 are not pure substances?
6. When the stopper is removed from a partly filled bottle containing solid and liquid acetic acid at 16.7°C, a strong vinegarlike odor is noticeable immediately. How many acetic acid phases must be present in the bottle? Explain.
7. Is the system enclosed in the bottle of Question 6 homogeneous or heterogeneous? Explain.
8. Is a system that contains only one substance necessarily homogeneous? Explain.
9. Is a system that contains two or more substances necessarily heterogeneous? Explain.
10. Are there more atoms of silicon or hydrogen in the earth's crust, seawater, and atmosphere? Use Table 3.3 and the fact that the mass of a silicon atom is about 28 times that of a hydrogen atom.
11. What does the symbol of an element stand for?
12. Write down what you believe to be the symbols for the elements phosphorus, aluminum, hydrogen, potassium,

magnesium, sodium, nitrogen, nickel, silver, and plutonium. Now look up the correct symbols and rewrite them, comparing the two sets.

13. Interpret the difference in meanings for each of these pairs: (a) Si and SI (b) Pb and PB (c) 4 P and P_4

14. List six elements and their symbols in which the first letter of the symbol is different from that of the name. (Table 3.5)

15. Write the names and symbols for the 14 elements that have only one letter as their symbol. (See table on inside front cover.)

16. Distinguish between an element and a compound.

17. How many metals are there? Nonmetals? Metalloids? (Table 3.6)

18. Of the ten most abundant elements in the earth's crust, seawater, and atmosphere, how many are metals? Nonmetals? Metalloids? (Table 3.3)

19. Of the six most abundant elements in the human body, how many are metals? Nonmetals? Metalloids? (Figure 3.4)

20. Why is the symbol for gold Au rather than G or Go?

21. Give the names of (a) the solid diatomic nonmetal and (b) the liquid diatomic nonmetal. (Table 3.7)

22. Distinguish between a compound and a mixture.

23. What are the two general types of compounds? How do they differ from each other?

24. What is the basis for distinguishing one compound from another?

25. How many atoms are contained in (a) one molecule of hydrogen, (b) one molecule of water, and (c) one molecule of sulfuric acid?

26. What is the major difference between a cation and an anion?

27. Write the names and formulas of the elements that exist as diatomic molecules. (Table 3.7)

28. Distinguish between homogeneous and heterogeneous mixtures.

29. Tabulate the properties that characterize metals and nonmetals.

30. Which of the following are diatomic molecules?
(a) H_2 (c) HCl (e) NO (g) $MgCl_2$
(b) SO_2 (d) H_2O (f) NO_2

31. What is the major difficulty in using hydrogen as a primary source of fuel?

32. How can sponge iron be used to create a recyclable source of hydrogen for fuel?

33. Name the three allotropes of carbon.

34. List a minimum of three forms of graphite crystals and indicate a source for each.

35. List four uses of graphite in daily life.

36. Which of the following statements are correct? Rewrite the incorrect statements to make them correct. (Try to answer this question without referring to the text.)

(a) Liquids are the least compact state of matter.
(b) Liquids have a definite volume and a definite shape.
(c) Matter in the solid state is discontinuous; that is, it is made up of discrete particles.
(d) Wood is homogeneous.
(e) Wood is a substance.
(f) Dirt is a mixture.
(g) Seawater, although homogeneous, is a mixture.
(h) Any system made up of only one substance is homogeneous.
(i) Any system containing two or more substances is heterogeneous.
(j) A solution, although it contains dissolved material, is homogeneous.
(k) The smallest unit of an element that can exist and enter into a chemical reaction is called a molecule.
(l) The basic building blocks of all substances that cannot be decomposed into simpler substances by ordinary chemical change are compounds.
(m) The most abundant element in the earth's crust, seawater, and atmosphere by mass is oxygen.
(n) The most abundant element in the human body, by mass, is carbon.
(o) Most of the elements are represented by symbols consisting of one or two letters.
(p) The symbol for copper is Co.
(q) The symbol for sodium is Na.
(r) The symbol for potassium is P.
(s) The symbol for lead is Le.
(t) Early names for some elements led to unlikely symbols, such as Fe for iron.
(u) A compound is a distinct substance that contains two or more elements combined in a definite proportion by mass.
(v) The smallest uncharged individual unit of a compound formed by the union of two or more atoms is called a substance.
(w) An ion is a positive or negative electrically charged atom or group of atoms.
(x) Bromine is an element that occurs as a diatomic molecule, Br_2.
(y) The formula Na_2CO_3 indicates a total of six atoms, including three oxygen atoms.
(z) A general property of nonmetals is that they are good conductors of heat and electricity.

Paired Exercises

These exercises are paired. Each odd-numbered exercise is followed by a similar even-numbered exercise. Answers to the even-numbered exercises are given in Appendix V.

37. Given the following list of compounds and their formulas, what elements are present in each compound?
(a) Potassium iodide KI
(b) Sodium carbonate Na_2CO_3
(c) Aluminum oxide Al_2O_3
(d) Calcium bromide $CaBr_2$
(e) Acetic acid $HC_2H_3O_2$

38. Given the following list of compounds and their formulas, what elements are present in each compound?
(a) Magnesium bromide $MgBr_2$
(b) Carbon tetrachloride CCl_4
(c) Nitric acid HNO_3
(d) Barium sulfate $BaSO_4$
(e) Aluminum phosphate $AlPO_4$

39. Write the formula for each of the following compounds (the composition of the compound is given after each name):
(a) Zinc oxide 1 atom Zn, 1 atom O
(b) Potassium chlorate 1 atom K, 1 atom Cl, 3 atoms O
(c) Sodium hydroxide 1 atom Na, 1 atom O, 1 atom H
(d) Ethyl alcohol 2 atoms C, 6 atoms H, 1 atom O

40. Write the formula for each of the following compounds (the composition of the compound is given after each name):
(a) Aluminum bromide 1 atom Al, 3 atoms Br
(b) Calcium fluoride 1 atom Ca, 2 atoms F
(c) Lead(II) chromate 1 atom Pb, 1 atom Cr, 4 atoms O
(d) Benzene 6 atoms C, 6 atoms H

41. Explain the meaning of each symbol and number in the following formulas:
(a) H_2O
(b) Na_2SO_4
(c) $HC_2H_3O_2$

42. Explain the meaning of each symbol and number in the following formulas:
(a) $AlBr_3$
(b) $Ni(NO_3)_2$
(c) $C_{12}H_{22}O_{11}$ (sucrose)

43. How many atoms are represented in each of these formulas?
(a) KF
(b) $CaCO_3$
(c) $K_2Cr_2O_7$
(d) $NaC_2H_3O_2$
(e) $(NH_4)_2C_2O_4$

44. How many atoms are represented in each of these formulas?
(a) $NaCl$
(b) N_2
(c) $Ba(ClO_3)_2$
(d) CCl_2F_2 (Freon)
(e) $Al_2(SO_4)_3$

45. How many atoms of oxygen are represented in each formula?
(a) H_2O
(b) $CuSO_4$
(c) H_2O_2
(d) $Fe(OH)_3$
(e) $Al(ClO_3)_3$

46. How many atoms of hydrogen are represented in each formula?
(a) H_2
(b) $Ba(C_2H_3O_2)_2$
(c) $C_6H_{12}O_6$
(d) $HC_2H_3O_2$
(e) $(NH_4)_2Cr_2O_7$

47. Classify each of the following materials as an element, compound, or mixture:
(a) Air
(b) Oxygen
(c) Sodium chloride
(d) Wine

48. Classify each of the following materials as an element, compound, or mixture:
(a) Platinum
(b) Sulfuric acid
(c) Iodine
(d) Crude oil

49. Classify each of the following materials as an element, compound, or mixture:
(a) Paint
(b) Salt
(c) Copper
(d) Beer

50. Classify each of the following materials as an element, compound, or mixture:
(a) Hydrochloric acid
(b) Silver
(c) Milk
(d) Sodium hydroxide

51. Reduce each of the following chemical formulas to the smallest whole-number relationship among atoms. (In chemistry this is called the empirical formula. You will learn more about it later.)
(a) $C_6H_{12}O_6$ glucose
(b) C_8H_{18} octane
(c) $C_{25}H_{52}$ paraffin wax

52. Reduce each of the following chemical formulas to the smallest whole-number relationship among atoms. (In chemistry this is called the empirical formula. You will learn more about it later.)
(a) H_2O_2 hydrogen peroxide
(b) C_2H_6O ethyl alcohol
(c) $Na_2Cr_2O_7$ sodium dichromate

53. Is there a pattern to the location of the gaseous elements on the periodic table? If so describe it.

54. Is there a pattern to the location of the liquid elements on the periodic table? If so describe it.

55. Of the first 36 elements on the periodic table what percent are metals?

56. Of the first 36 elements on the periodic table what percent are solids?

Additional Exercises

These exercises are not paired or labeled by topic and provide additional practice on the concepts covered in this chapter.

57. Vitamin B_{12} has a formula that is $C_{63}H_{88}CoN_{14}O_{14}P$.
(a) How many atoms make up one molecule of vitamin B_{12}?
(b) What percentage of the total atoms are carbon?
(c) What fraction of the total atoms are metallic?

58. How many total atoms are there in seven dozen molecules of nitric acid, HNO_3?

59. The following formulas look similar but represent different things. Compare and contrast them. How are they alike? How are they different?

8 S S_8

60. Calcium dihydrogen phosphate is an important fertilizer. How many atoms of hydrogen are there in ten formula units of $Ca(H_2PO_4)_2$?

61. How many total atoms are there in one molecule of $C_{145}H_{293}O_{168}$?

62. Name:
(a) three elements, all metals, beginning with the letter M.
(b) four elements, all solid nonmetals.
(c) five elements, all solids in the first five rows of the periodic table, whose symbols start with letters different than the element name.

63. It has been estimated that there is 4×10^{-4} mg of gold/L of sea water. At a price of \$19.40/g, what would be the value of the gold in 1 km^3 (1×10^{15} cm^3) of the ocean?

64. Make a graph using the data below. Plot the density of air in grams per liter along the *x*-axis and temperature along the *y*-axis.

Temperature (°C)	Density (g/L)
0	1.29
10	1.25
20	1.20
40	1.14
80	1.07

(a) What is the relationship between density and temperature according to your graph?
(b) From your plot find the density of air at these temperatures:

5°C 25°C 70°C

Properties of Matter

4

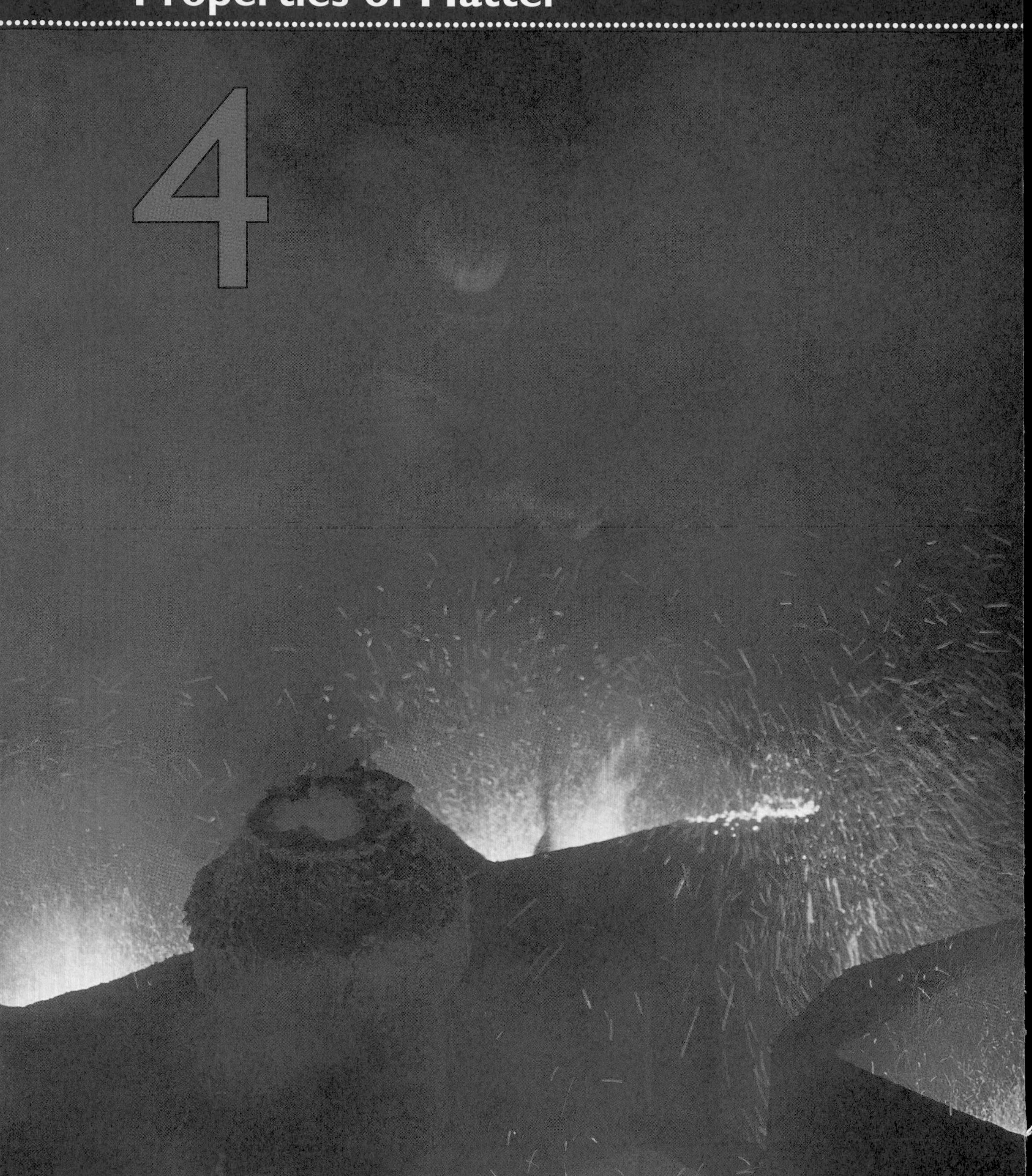

The world we live in is a myriad of sights, sounds, smells, and tastes. Our senses help us to describe these objects in our lives. For example, the smell of freshly baked cinnamon rolls creates a mouth-watering desire to gobble down a fresh-baked sample. And so it is with each substance—its own unique properties allow us to identify it and predict its interactions.

These interactions produce both physical and chemical changes. When you eat an apple, the ultimate metabolic result is carbon dioxide and water. These same products are achieved by burning logs. Not only does a chemical change occur in these cases, but an energy change as well. Some reactions release energy (as does the apple or the log) whereas others require energy, such as the production of steel or the melting of ice. Over 90% of our current energy comes from chemical reactions.

4.1 Properties of Substances

properties

How do we recognize substances? Each substance has a set of **properties** that is characteristic of that substance and gives it a unique identity. Properties are the personality traits of substances and are classified as either physical or chemical. **Physical properties** are the inherent characteristics of a substance that can be determined without altering its composition; they are associated with its physical existence. Common physical properties are color, taste, odor, state of matter (solid, liquid, or gas), density, melting point, and boiling point. **Chemical properties** describe the ability of a substance to form new substances, either by reaction with other substances or by decomposition.

physical properties

chemical properties

We can select a few of the physical and chemical properties of chlorine as an example. Physically, chlorine is a gas about 2.4 times heavier than air. It is yellowish-green in color and has a disagreeable odor. Chemically, chlorine will not burn but will support the combustion of certain other substances. It can be used as a bleaching agent, as a disinfectant for water, and in many chlorinated substances such as refrigerants and insecticides. When chlorine combines with the metal sodium, it forms a salt called sodium chloride. These properties, among others, help to characterize and identify chlorine.

Substances, then, are recognized and differentiated by their properties. Table 4.1 lists four substances and several of their common physical properties. Information about common physical properties, such as that given in Table 4.1, is available in handbooks of chemistry and physics. Scientists do not pretend to know all the answers or to remember voluminous amounts of data, but it is important for them to know where to look for data in the literature. Handbooks are one of the most widely used resources for scientific data.*

No two substances have identical physical and chemical properties.

*Two such handbooks are David R. Lide, ed., *Handbook of Chemistry and Physics,* 75th Ed. (Cleveland: Chemical Rubber Company, 1995) and Norbert A. Lange, comp., *Handbook of Chemistry,* 13th ed. (New York: McGraw-Hill, 1985).

◀ **Chapter Opening Photo: Iron can be melted at very high temperatures and then cast into a variety of shapes.**

TABLE 4.1 Physical Properties of Chlorine, Water, Sugar, and Acetic Acid

Substance	Color	Odor	Taste	Physical state	Melting point (°C)	Boiling point (°C)
Chlorine	Yellowish-green	Sharp, suffocating	Sharp, sour	Gas	−101.6	−34.6
Water	Colorless	Odorless	Tasteless	Liquid	0.0	100.0
Sugar	White	Odorless	Sweet	Solid	—	Decomposes 170–186
Acetic acid	Colorless	Like vinegar	Sour	Liquid	16.7	118.0

4.2 Physical Changes

physical change

Matter can undergo two types of changes, physical and chemical. **Physical changes** are changes in physical properties (such as size, shape, and density) or changes in the state of matter without an accompanying change in composition. The changing of ice into water and water into steam are physical changes from one state of matter into another (Figure 4.1). No new substances are formed in these physical changes.

When a clean platinum wire is heated in a burner flame, the appearance of the platinum changes from silvery metallic to glowing red. This change is physical because the platinum can be restored to its original metallic appearance by cooling and, more importantly, because the composition of the platinum is not changed by heating and cooling.

4.3 Chemical Changes

chemical change

In a **chemical change**, new substances are formed that have different properties and composition from the original material. The new substances need not in any way resemble the initial material.

When a clean copper wire is heated in a burner flame, the appearance of the copper changes from coppery metallic to glowing red. Unlike the platinum previously mentioned, the copper is not restored to its original appearance by cooling but becomes instead a black material. This black material is a new substance called copper(II) oxide. It was formed by chemical change when copper combined with oxygen in the air during the heating process. The unheated wire was essentially 100% copper, but the copper(II) oxide is 79.9% copper and 20.1% oxygen. One gram of copper will yield 1.251 g of copper(II) oxide (see Figure 4.2). The platinum is changed only physically when heated, but the copper is changed both physically and chemically when heated.

When 1.00 g of copper reacts with oxygen to yield 1.251 g of copper(II) oxide, the copper must have combined with 0.251 g of oxygen. The percentage of copper

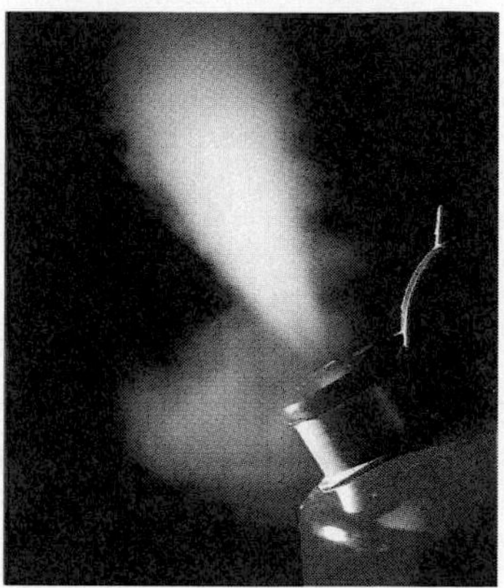

FIGURE 4.1
Ice melting into water or water turning into steam are physical changes from one state of matter to another.

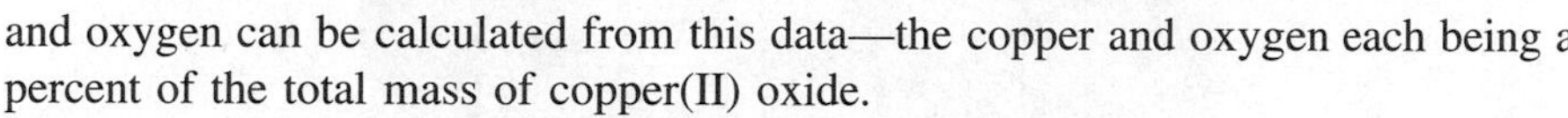

and oxygen can be calculated from this data—the copper and oxygen each being a percent of the total mass of copper(II) oxide.

$$1.00 \text{ g copper} + 0.251 \text{ g oxygen} \longrightarrow 1.251 \text{ g copper(II) oxide}$$

$$\frac{1.00 \text{ g copper}}{1.251 \text{ g copper(II) oxide}} \times 100 = 79.9\% \text{ copper}$$

$$\frac{0.251 \text{ g oxygen}}{1.251 \text{ g copper(II) oxide}} \times 100 = 20.1\% \text{ oxygen}$$

Water can be decomposed chemically into hydrogen and oxygen. This is usually accomplished by passing electricity through the water in a process called *electrolysis.* Hydrogen collects at one electrode while oxygen collects at the other (see Figure 4.3). The composition and the physical appearance of the hydrogen and the oxygen are quite different from that of water. They are both colorless gases, but each behaves differently when a burning splint is placed into the sample: The hydrogen explodes with a pop while the flame brightens considerably in the oxygen (oxygen supports and intensifies the combustion of the wood). From these observations we conclude that a chemical change has taken place.

Before heating,
wire is copper-colored
Copper wire: 1.00 g
(100% copper)

Copper and oxygen
from the air combine
chemically on heating
After heating,
wire is black
Copper (II) oxide: 1.251 g
79.9% copper: 1.00 g
20.1% oxygen: 0.251 g

FIGURE 4.2
Chemical change: formation of copper(II) oxide from copper and oxygen.

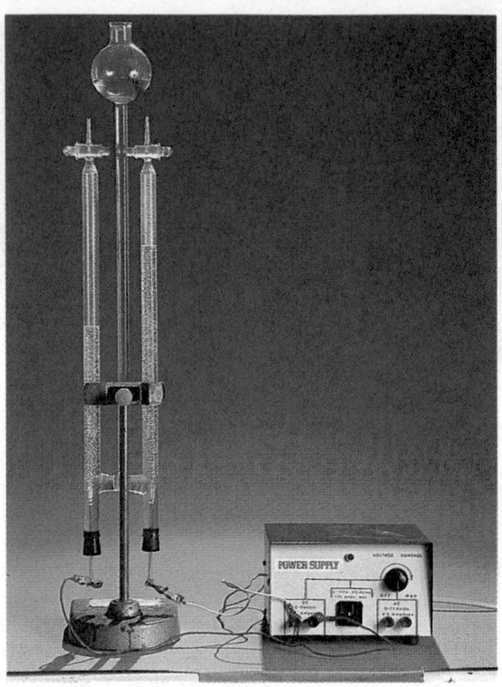

FIGURE 4.3 ▶
Electrolysis of water is forming hydrogen in the left tube and oxygen in the right tube.

chemical equations

Chemists have devised a shorthand method for expressing chemical changes in the form of **chemical equations**. The two previous examples of chemical changes can be represented by the following word equations:

$$\text{water} \xrightarrow[\text{energy}]{\text{electrical}} \text{hydrogen} + \text{oxygen}$$

$$\text{copper} + \text{oxygen} \xrightarrow{\Delta} \text{copper(II) oxide}$$

Equation (1) states that water decomposes into hydrogen and oxygen when electrolyzed. Equation (2) states that copper plus oxygen when heated produce copper(II) oxide. The arrow means "produces," and it points to the products. The greek letter delta (Δ) represents heat. The starting substances (water, copper, and oxygen) are called the **reactants** and the substances produced (hydrogen, oxygen, and copper(II) oxide) are called the **products**. These chemical equations can be presented in still more abbreviated form by using symbols to represent the substances:

reactants
products

$$2\,H_2O \xrightarrow[\text{energy}]{\text{electrical}} 2\,H_2 + O_2$$

$$2\,Cu + O_2 \xrightarrow{\Delta} 2\,CuO$$

We will learn more about writing chemical equations in later chapters.

Physical change usually accompanies a chemical change. Table 4.2 lists some common physical and chemical changes. In the examples given in the table, you will note that wherever a chemical change occurs, a physical change occurs also. However, wherever a physical change is listed, only a physical change occurs.

TABLE 4.2 Examples of Processes Involving Physical or Chemical Changes

Process taking place	Type of change	Accompanying observations
Rusting of iron	Chemical	Shiny, bright metal changes to reddish-brown rust.
Boiling of water	Physical	Liquid changes to vapor.
Burning of sulfur in air	Chemical	Yellow, solid sulfur changes to gaseous, choking sulfur dioxide.
Boiling an egg	Chemical	Liquid white and yolk change to solids.
Combustion of gasoline	Chemical	Liquid gasoline burns to gaseous carbon monoxide, carbon dioxide, and water.
Digesting food	Chemical	Food changes to liquid nutrients and partially solid wastes.
Sawing of wood	Physical	Smaller pieces of wood and sawdust are made from a larger piece of wood.
Burning of wood	Chemical	Wood burns to ashes, gaseous carbon dioxide, and water.
Heating of glass	Physical	Solid becomes pliable during heating, and the glass may change its shape.

4.4 Conservation of Mass

Law of Conservation of Mass

The **Law of Conservation of Mass** states that no change is observed in the total mass of the substances involved in a chemical change. This law, tested by extensive laboratory experimentation, is the basis for the quantitative mass relationships among reactants and products.

The decomposition of water into hydrogen and oxygen illustrates this law. One hundred grams of water decomposes into 11.2 g of hydrogen and 88.8 g of oxygen. For example,

$$\underset{100.0\text{ g}}{\text{water}} \longrightarrow \underset{11.2\text{ g}}{\text{hydrogen}} + \underset{88.8\text{ g}}{\text{oxygen}}$$

100.0 g Reactant

100.0 g Products

mass of reactants = mass of products

4.5 Energy

From the prehistoric discovery that fire could be used to warm shelters and cook food to the modern-day discovery that nuclear reactors can be used to produce vast amounts of controlled energy, our technical progress has been directed by our ability to produce, harness, and utilize energy. **Energy** is the capacity of matter to do work. Energy exists in many forms; some of the more familiar forms are mechanical, chemical, electrical, heat, nuclear, and radiant or light energy. Matter can have both potential and kinetic energy.

energy

potential energy

Potential energy is stored energy, or energy an object possesses due to its relative position. For example, a ball located 20 ft above the ground has more potential energy than when located 10 ft above the ground and will bounce higher when allowed to fall. Water backed up behind a dam represents potential energy that can be converted into useful work in the form of electrical or mechanical energy. Gasoline is a source of chemical potential energy. When gasoline burns (combines with oxygen), the heat released is associated with a decrease in potential energy. The new substances formed by burning have less chemical potential energy than the gasoline and oxygen did.

kinetic energy

Kinetic energy is energy that matter possesses due to its motion. When the water behind the dam is released and allowed to flow, its potential energy is changed into kinetic energy, which can be used to drive generators and produce electricity. All moving bodies possess kinetic energy. The pressure exerted by a confined gas is due to the kinetic energy of rapidly moving gas particles. We all know the results when two moving vehicles collide: Their kinetic energy is expended in the crash that occurs.

Energy can be converted from one form to another form. Some kinds of energy can be converted to other forms easily and efficiently. For example, mechanical energy can be converted to electrical energy with an electric generator at better than 90% efficiency. On the other hand, solar energy has thus far been directly converted to electrical energy at an efficiency of only about 15%. In chemistry, energy is most frequently expressed as heat.

4.6 Heat: Quantitative Measurement

The SI-derived unit for energy is the joule (pronounced *jool* and abbreviated J). Another unit for heat energy, which has been used for many years, is the calorie (abbreviated cal). The relationship between joules and calories is

$$4.184 \text{ J} = 1 \text{ cal (exactly)}$$

joule

calorie

To give you some idea of the magnitude of these heat units, 4.184 **joule** or one **calorie** is the quantity of heat energy required to change the temperature of one g of water by 1°C, usually measured from 14.5°C to 15.5°C.

Since joule and calorie are rather small units, kilojoules (kJ) and kilocalories (kcal) are used to express heat energy in many chemical processes:

TABLE 4.3 Specific Heat of Selected Substances

Substance	Specific heat J/g°C	Specific heat cal/g°C
Water	4.184	1.00
Ethyl alcohol	2.138	0.511
Ice	2.059	0.492
Aluminum	0.900	0.215
Iron	0.473	0.113
Copper	0.385	0.0921
Gold	0.131	0.0312
Lead	0.128	0.0305

The mechanical energy of falling water is converted to electrical energy at this hydroelectric plant in the North Alps of Japan.

$$1 \text{ kJ} = 1000 \text{ J}$$
$$1 \text{ kcal} = 1000 \text{ cal}$$

The kilocalorie is also known as the nutritional or large Calorie (spelled with a capital C and abbreviated Cal). In this book heat energy will be expressed in joules with parenthetical values in calories.

The difference in the meanings of the terms *heat* and *temperature* can be seen by this example: Visualize two beakers, A and B. Beaker A contains 100 g of water at 20°C, and beaker B contains 200 g of water also at 20°C. The beakers are heated until the temperature of the water in each reaches 30°C. The temperature of the water in the beakers was raised by exactly the same amount, 10°C. But twice as much heat (8368 J or 2000 cal) was required to raise the temperature of the water in beaker B as was required in beaker A (4184 J or 1000 cal).

In the middle of the 18th century Joseph Black, a Scottish chemist, was experimenting with the heating of elements. He heated and cooled equal masses of iron and lead through the same temperature range. Black noted that much more heat was needed for the iron than for the lead. He had discovered a fundamental property of matter, namely, that every substance has a characteristic heat capacity. Heat capacities may be compared in terms of specific heats. The **specific heat** of a substance is the quantity of heat (lost or gained) required to change the temperature of one g of that substance by 1°C. It follows then that the specific heat of liquid water is 4.184 J/g°C (1.000 cal/g°C). The specific heat of water is high compared with that of most substances. Aluminum and copper, for example, have specific heats of 0.900 and 0.385 J/g°C, respectively (see Table 4.3). The relation of mass, specific heat, temperature change (Δt), and quantity of heat lost or gained by a system is expressed by this general equation:

specific heat

$$\left(\begin{array}{c}\textbf{mass of}\\ \textbf{substance}\end{array}\right) \times \left(\begin{array}{c}\textbf{specific heat}\\ \textbf{of substance}\end{array}\right) \times \boldsymbol{\Delta t} = \textbf{energy (heat)} \qquad (1)$$

Thus, the amount of heat needed to raise the temperature of 200. g of water by 10.0°C can be calculated as follows:

$$200.\ \cancel{\text{g}} \times \frac{4.184 \text{ J}}{\cancel{\text{g}}\cancel{°\text{C}}} \times 10.0\cancel{°\text{C}} = 8.37 \times 10^3 \text{ J}$$

Examples of specific heat problems follow.

Chemistry in Action

Fast Energy or Fast Fat?

Fast food restaurants have a special aroma. That aroma is the smell of fat. One whiff can stir the appetite for a juicy burger and fries.

Enjoying fat in our diets today has become a major problem since we now live longer than our ancestors. Diets that are high in fats have been connected with heart disease and to a variety of cancers. In the average American diet, fat supplies about 40% of the Calories (kcal). Nutritionists suggest that fat should be not more than 30% of our total daily Calories. For the average person eating about 2000 Calories per day, this is about 67 grams, 600 Calories, or 15 teaspoons of fat (e.g., a double hamburger with cheese). A small amount of fat is necessary in the diet to provide a natural source of vitamins A, D, E, and K, as well as polyunsaturated fats necessary for maintaining growth and good health.

Fats also supply us with energy. One gram of fat supplies 9 Calories of energy; the same amount of protein or carbohydrate supplies only 4 Calories. Human beings utilize energy in a variety of ways but one of the most important is to maintain body temperature, which for healthy individuals is around 37°C. The body tends to lose heat to the surroundings since heat flows from an area of higher temperature to an area of lower temperature. Additional heat energy is used to evaporate moisture and cool our bodies as we perspire. Still more energy is demanded by our daily physical activities. The source for all this energy is the chemical oxidation of the food we eat.

The energy content of food is determined by burning it in a calorimeter and measuring the heat released. Since the initial substances and the final products of the combustion in the calorimeter are the same as those accomplished in the human body, a calorie content can be assigned to each food. These calorie values are now found on food packages along with nutritional information regarding the contents of the food we eat.

Fats remain in the stomach longer giving us the "full" feeling we like after a meal. But too much of a good thing can lead to weight and health problems.

How does fast food stack up in the nutrition department? Some sample meals are shown here from fast food restaurants that get less than 30% of their calories from fat. In general, chicken and turkey sandwiches that are not fried have less fat than hamburgers or roast beef. Salads with the lowest fat have little or no cheese. The best way to reduce fat in salads is to eliminate the dressing or to opt for a lowfat dressing.

Believe it or not, most of the milkshakes and frozen desserts served in fast food restaurants have less than 30% Calories from fat per serving. This is primarily because they are made with skim milk. Fast food does not necessarily mean fast fat if the consumer is careful in selecting the particular food he or she consumes. Balancing a full day's diet is more important than worrying about each and every food eaten, although choosing foods that have less than 30% Calories from fat makes maintaining the balance easier.

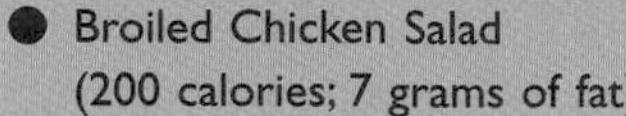

- Broiled Chicken Salad (200 calories; 7 grams of fat)
- Newman's Own Light Italian dressing (30 calories; 1 gram of fat)
- Vanilla shake (310 calories; 7 grams of fat)

Total—540 calories; 15 grams of fat 25% calories from fat

- McLean Deluxe (340 calories; 12 grams of fat)
- Diet Coke, 12 ounces (1 calorie; 0 grams of fat)
- Hot Fudge Lowfat Frozen Yogurt Sundae (290 calories; 5 grams of fat)

Total—651 calories; 17 grams of fat 24% calories from fat

- Grilled Chicken Sandwich (290 calories; 7 grams of fat)
- Black coffee (0 calories; 0 grams fat)
- Chocolate Frosty Dairy Dessert (medium) (460 calories; 13 grams of fat)

Total—750 calories; 20 grams of fat 24% calories from fat

Example 4.1

Calculate the specific heat of a solid in J/g°C and cal/g°C if 1638 J raise the temperature of 125 g of the solid from 25.0°C to 52.6°C.

Solution

First solve equation (1) to obtain an equation for specific heat:

$$\text{specific heat} = \frac{\text{energy}}{\text{g} \times \Delta t}$$

Now substitute in the data:

$$\text{energy} = 1638\ \text{J} \qquad \text{mass} = 125\ \text{g} \qquad \Delta t = 52.6°\text{C} - 25.0°\text{C} = 27.6°\text{C}$$

$$\text{specific heat} = \frac{1638\ \text{J}}{125\ \text{g} \times 27.6°\text{C}} = 0.475\ \text{J/g°C}$$

Now convert joules to calories using 1.000 cal/4.184 J:

$$\text{specific heat} = \frac{0.475\ \cancel{\text{J}}}{\text{g°C}} \times \frac{1.000\ \text{cal}}{4.184\ \cancel{\text{J}}} = 0.114\ \text{cal/g°C}$$

Example 4.2

A sample of a metal with a mass of 212 g is heated to 125.0°C and then dropped into 375 g water at 24.0°C. If the final temperature of the water is 34.2°C, what is the specific heat of the metal? (Assume no heat losses to the surroundings.)

Solution

When the metal enters the water it begins to cool, losing heat to the water. At the same time the temperature of the water rises. This process continues until the temperature of the metal and the temperature of the water are equal, at which point (34.2°C) no net flow of heat occurs.

The heat lost or gained by a system is given by equation (1). We use this equation first to calculate the heat gained by the water and then to calculate the specific heat of the metal:

$$\text{temperature rise of the water } (\Delta t) = 34.2°\text{C} - 24.0°\text{C} = 10.2°\text{C}$$

$$\text{heat gained by the water} = 375\ \cancel{\text{g}} \times \frac{4.184\ \text{J}}{\cancel{\text{g°C}}} \times 10.2\cancel{°\text{C}} = 1.60 \times 10^4\ \text{J}$$

The metal dropped into the water must have a final temperature the same as the water (34.2°C):

$$\text{temperature drop by the metal } (\Delta t) = 125.0°\text{C} - 34.2°\text{C} = 90.8°\text{C}$$

$$\text{heat lost by the metal} = \text{heat gained by the water} = 1.60 \times 10^4\ \text{J}$$

Rearranging equation (1) we get

$$\text{specific heat} = \frac{\text{energy}}{\text{g} \times \Delta t}$$

$$\text{specific heat of the metal} = \frac{1.60 \times 10^4\ \text{J}}{212\ \text{g} \times 90.8°\text{C}} = 0.831\ \text{J/g°C}$$

Practice 4.1

Calculate the quantity of energy needed to heat 8.0 g of water from 42.0°C to 45.0°C.

Energy from the sun is used to produce the chemical changes that occur during photosynthesis in the rain forests.

Practice 4.2
A 110.0-g sample of iron at 55.5°C raises the temperature of 150.0 mL of water from 23.0°C to 25.5°C. Determine the specific heat of the iron in cal/g°C.

4.7 Energy in Chemical Changes

In all chemical changes matter either absorbs or releases energy. Chemical changes can produce different forms of energy. For example, electrical energy to start automobiles is produced by chemical changes in the lead storage battery. Light energy is produced by the chemical change that occurs in a light stick. Heat and light energies are released from the combustion of fuels. All the energy needed for our life processes—breathing, muscle contraction, blood circulation, and so on—is produced by chemical changes occurring within the cells of our bodies.

Conversely, energy is used to cause chemical changes. For example, a chemical change occurs in the electroplating of metals when electrical energy is passed through a salt solution in which the metal is submerged. A chemical change also occurs when radiant energy from the sun is used by green plants in the process of photosynthesis. And, as we saw, a chemical change occurs when electricity is used to decompose water into hydrogen and oxygen. Chemical changes are often used primarily to produce energy rather than to produce new substances. The heat or thrust generated by the combustion of fuels is more important than the new substances formed.

4.8 Conservation of Energy

An energy transformation occurs whenever a chemical change occurs (see Figure 4.4). If energy is absorbed during the change, the products will have more chemical potential energy than the reactants. Conversely, if energy is given off in a chemical change, the products will have less chemical potential energy than the reactants. Water can be decomposed in an electrolytic cell by absorbing electrical energy. The products, hydrogen and oxygen, have a greater chemical potential energy level than that of water (see Figure 4.4a). This potential energy is released in the form of heat and light when the hydrogen and oxygen are burned to form water again (see Figure 4.4b). Thus, energy can be changed from one form to another or from one substance to another, and therefore is not lost.

The energy changes occurring in many systems have been thoroughly studied by many investigators. No system has been found to acquire energy except at the expense of energy possessed by another system. This principle is stated in other words as the **Law of Conservation of Energy**: Energy can be neither created nor destroyed, though it can be transformed from one form to another.

Law of Conservation of Energy

Hot to Go!

Instant hot and cold packs utilize the properties of matter to release or absorb energy. When a chemical dissolves in water, energy can be released or absorbed. In a cold pack a small sealed package of ammonium nitrate is placed in a separate sealed pouch containing water. As long as the substances remain separated, nothing happens. When the small package is broken (when the pack is activated), the substances mix and, as the temperature of the solution falls, the solution absorbs heat from the surroundings.

Instant hot packs work in one of two ways. The first type depends on a spontaneous chemical reaction that releases heat energy. In one product an inner paper bag perforated with tiny holes is contained in a plastic envelope. The inner bag contains a mixture of powdered iron, salt, activated charcoal, and dampened sawdust. The heat pack is activated by removing the inner bag, shaking to mix the chemicals, and replacing it in the outer envelope. The heat is the result of a chemical change—that of iron rusting (oxidizing) very rapidly. In the second type of heat pack a physical property of matter is responsible for the production of the heat. The hot pack consists of a tough sealed plastic envelope containing a solution of sodium acetate or sodium thiosulfate. A small crystal is added by squeezing a corner of the pack, or by bending a small metal activator. Crystals then form throughout the solution and release heat to the surroundings. This type of hot pack has the advantages of not being able to overheat and of being reusable. To reuse, it is simply heated in boiling water until the crystals dissolve, and then cooled and stored away until needed again.

Immediate application of hot and cold packs have helped reduce injuries to athletes.

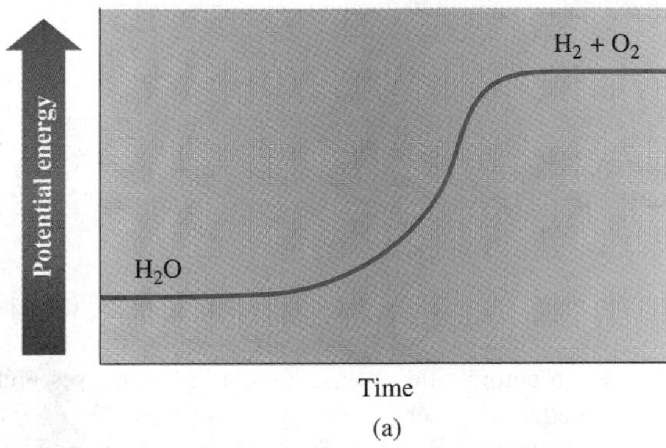

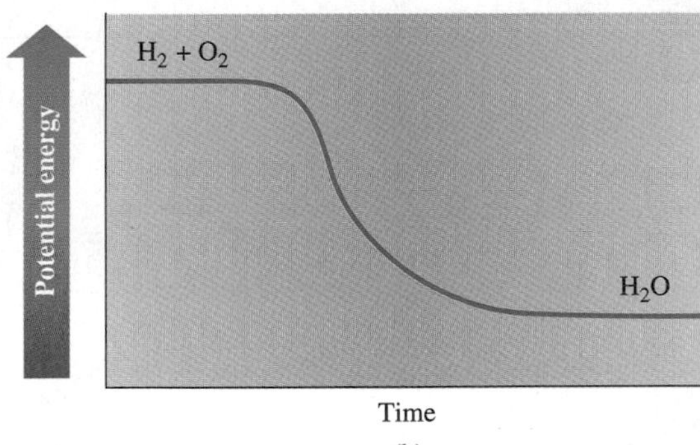

FIGURE 4.4
(a) In electrolysis of water, energy is absorbed by the system so the products H_2 and O_2 have a higher potential energy. (b) When hydrogen is burned (in O_2) energy is released and the product (H_2O) has lower potential energy.

Concepts in Review

1. List the physical properties used to characterize a substance.
2. Distinguish between the physical and chemical properties of matter.

3. Classify changes undergone by matter as either physical or chemical.
4. Distinguish between kinetic and potential energy.
5. State the Law of Conservation of Mass.
6. State the Law of Conservation of Energy.
7. Differentiate clearly between heat and temperature.
8. Make calculations using the equation:

$$\text{energy} = (\text{mass}) \times (\text{specific heat}) \times (\Delta t)$$

Key Terms

The terms listed here have been defined within this chapter. Section numbers are referenced in parentheses for each term.

calorie (4.6)
chemical change (4.3)
chemical equations (4.3)
chemical properties (4.1)
energy (4.5)
joule (4.6)
kinetic energy (4.5)
Law of Conservation of Energy (4.8)
Law of Conservation of Mass (4.4)
physical change (4.2)
physical properties (4.1)
potential energy (4.5)
products (4.3)
properties (4.1)
reactants (4.3)
specific heat (4.6)

Questions

Questions refer to tables, figures, and key words and concepts defined within the chapter. A particularly challenging question or exercise is indicated with an asterisk.

1. In what physical state does acetic acid exist at 10°C? (Table 4.1)
2. In what physical state does chlorine exist at 102 K? (Table 4.1)
3. What evidence of chemical change is visible when electricity is run through water? (Figure 4.2)
4. What physical changes occur during the electrolysis of water?
5. Distinguish between chemical and physical properties.
6. What is the fundamental difference between a chemical change and a physical change?
7. Explain how foods are assigned a specific energy (caloric) value.
8. Describe how a "reusable hot pack" works.
9. In a chemical change why can we consider that mass is neither gained nor lost (for practical purposes)?
10. Distinguish between potential and kinetic energy.
11. Calculate the boiling point of acetic acid in
    (a) Kelvins
    (b) degrees Fahrenheit. (Table 4.1)
12. Which of the following statements are correct? Rewrite the incorrect ones to make them correct.
    (a) An automobile rolling down a hill possesses both kinetic and potential energy.
    (b) When heated in the air, a platinum wire gains mass.
    (c) When heated in the air, a copper wire loses mass.
    (d) 4.184 cal is the equivalent of 1.000 J of energy.
    (e) Boiling water represents a chemical change, because a change of state occurs.
    (f) All the following represent chemical changes: baking a cake, frying an egg, leaves changing color, rusting iron.
    (g) All of the following represent physical changes: breaking a stick, melting wax, folding a napkin, burning hydrogen to form water.
    (h) Chemical changes can produce electrical energy.
    (i) Electrical energy can produce chemical changes.
    (j) A stretched rubber band possesses kinetic energy.

Paired Exercises

These exercises are paired. Each odd-numbered exercise is followed by a similar even-numbered exercise. Answers to the even-numbered exercises are given in Appendix V.

13. Classify each of the following as being primarily a physical or primarily a chemical change:
(a) Formation of a snowflake
(b) Freezing ice cream
(c) Boiling water
(d) Churning cream to make butter
(e) Boiling an egg
(f) Souring milk

15. Cite the evidence that indicates that only physical changes occur when a platinum wire is heated in a Bunsen burner flame.

17. Identify the reactants and products for heating a copper wire in a Bunsen burner flame.

19. What happens to the kinetic energy of a speeding car when the car is braked to a stop?

21. Indicate with a plus sign (+) any of these processes that require energy, and a negative (−) sign for any that release energy:
(a) melting ice
(b) relaxing a taut rubber band
(c) a rocket launching
(d) striking a match
(e) a Slinky toy (spring) "walking" down stairs

23. How many joules of energy are required to raise the temperature of 75 g of water from 20.0°C to 70.0°C?

25. A 250.0-g metal bar requires 5.866 kJ to change its temperature from 22°C to 100.0°C. What is the specific heat of the metal?

***27.** A 325-g piece of gold at 427°C is dropped into 200.0 mL of water at 22.0°C. The specific heat of gold is 0.131 J/g°C. Calculate the final temperature of the mixture. (Assume no heat loss to the surroundings.)

14. Classify each of the following as being primarily a physical or primarily a chemical change:
(a) Lighting a candle
(b) Stirring cake batter
(c) Dissolving sugar in water
(d) Decomposition of limestone by heat
(e) A leaf turning yellow
(f) Formation of bubbles in a pot of water long before the water boils.

16. Cite the evidence that indicates that both physical and chemical changes occur when a copper wire is heated in a Bunsen burner flame.

18. Identify the reactants and products for the electrolysis of water.

20. What energy transformation is responsible for the fiery reentry of the space shuttle?

22. Indicate with a plus sign (+) any of these processes that require energy and a negative (−) sign for any that release energy:
(a) boiling water
(b) releasing a balloon full of air with the neck open
(c) a race car crashing into the wall
(d) cooking a potato in a microwave oven
(e) ice cream freezing in an ice cream maker

24. How many joules of energy are required to raise the temperature of 65 g of iron from 25°C to 95°C?

26. A 1.00-kg sample of antimony absorbed 30.7 kJ, thus raising the temperature of the antimony from 20.0°C to its melting point (630.0°C). Calculate the specific heat of antimony.

***28.** A 500.0-g iron bar at 212°C is placed in 2.0 L of water at 24.0°C. What will be the change in temperature of the water? (Assume no heat is lost to the surroundings.)

Additional Exercises

These exercises are not paired or labeled by topic and provide additional practice on the concepts covered in this chapter.

***29.** The specific heat of zinc is 0.096 cal/g°C. Determine the energy required to raise the temperature of 250.0 g of zinc from room temperature (24°C) to 150.0°C.

***30.** If 40.0 kJ of energy is absorbed by 500.0 g of water at 10.0°C what would be the final temperature of the water?

***31.** The heat of combustion of a sample of coal is 5500 cal/g. What quantity of this coal must be burned to heat 500.0 g of water from 20.0°C to 90.0°C?

32. One gram of anthracite coal gives off 7000. cal when burned. How many joules is this? If 4.0 L of water is heated from 20.0°C to 100.0°C, how many grams of anthracite are needed?

33. A 100.0-g sample of copper is heated from 10.0°C to 100.0°C.

(a) Determine the number of calories needed. (Specific heat of copper is 0.0921 cal/g°C.)

(b) The same amount of heat is added to 100.0 g of Al at 10.0°C. (Specific heat of Al is 0.215 cal/g°C.) Which metal gets hotter, the copper or the aluminum?

34. A 500.0-g piece of iron is heated in a flame and dropped into 400.0 g of water at 10.0°C. The temperature of the water rises to 90.0°C. How hot was the iron when it was first removed from the flame? (Specific heat of iron is 0.113 cal/g°C.)

***35.** A 20.0-g piece of metal at 203°C is dropped into 100.0 g of water at 25.0°C. The water temperature rises to 29.0°C. Calculate the specific heat of the metal (J/g°C). Assume that all of the heat lost by the metal is transferred to the water and no heat is lost to the surroundings.

***36.** Assuming no heat loss by the system, what will be the final temperature when 50.0 g of water at 10.0°C are mixed with 10.0 g of water at 50.0°C?

37. Three 500.0-g pans of iron, aluminum, and copper are each used to fry an egg. Which pan fries the egg (105°C) the quickest? Explain.

38. At 6:00 P.M. you put a 300.0-g copper pan containing 800.0 mL of water (all at room temperature, which is 25°C) on the stove. The stove supplies 150 cal/s. When will the water reach the boiling point? (Assume no heat loss.)

39. Why does blowing gently across the surface of a hot cup of coffee help to cool it? Why does inserting a spoon into the coffee do the same thing?

40. If you are boiling some potatoes in a pot of water, will they cook faster if the water is boiling vigorously than if the water is only gently boiling? Explain your reasoning.

41. Homogenized whole milk contains 4% butterfat by volume. How many milliliters of fat are there in a glass (250 mL) of milk? How many grams of butterfat (d = 0.8 g/mL) are in this glass of milk?

42. A 100.0-mL volume of mercury (density = 13.6 g/mL) is put into a container along with 100.0 g of sulfur. The two substances react when heated and result in 1460 g of a dark, solid matter. Is that material an element or a compound? Explain. How many grams of mercury were in the container? How does this support the Law of Conservation of Matter?

Answers to Practice Exercises

4.1 1.0×10^2 J = 24 cal

4.2 0.114 cal/g°C

5 Early Atomic Theory and Structure

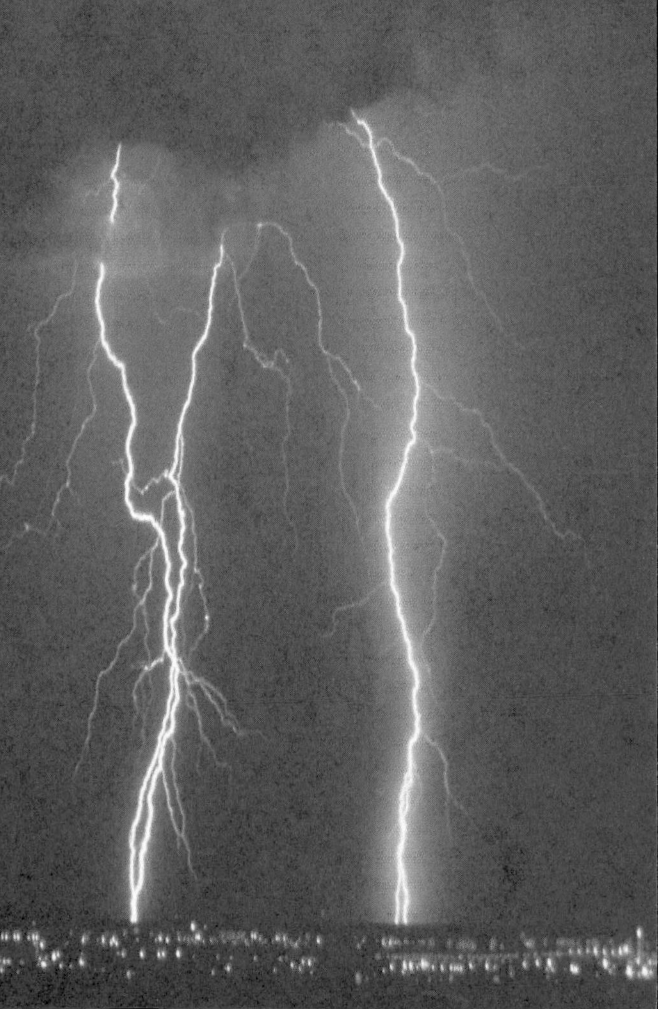

Chapter 5

◀ **Lightning occurs when electrons move to neutralize a charge difference between the clouds and the earth.**

Pure substances are classified into elements and compounds. But just what makes a substance possess its unique properties? Salt tastes salty, but how small a piece of salt will retain this property? Carbon dioxide puts out fires, is used by plants to produce oxygen, and forms "dry" ice when solidified. But how small a mass of this material still behaves like carbon dioxide? When substances finally reach the atomic, ionic, or molecular level they are in their simplest identifiable form. Further division produces a loss of characteristic properties.

What particles lie within an atom or ion? How are these tiny particles alike? How do they differ? How far can we continue to divide them? Alchemists began the quest, early chemists laid the foundation, and the modern chemist continues to build and expand on models of the atom.

5.1 Early Thoughts

The structure of matter has long intrigued and engaged the minds of people. The seed of modern atomic theory was sown during the time of the ancient Greek philosophers. About 440 B.C. Empedocles stated that all matter was composed of four "elements"—earth, air, water, and fire. Democritus (about 470–370 B.C.), one of the early atomic philosophers, thought that all forms of matter were divisible into invisible particles, which he called atoms, derived from the Greek word *atomos,* meaning indivisible. He held that atoms were in constant motion and that they combined with one another in various ways. This purely speculative hypothesis was not based on scientific observations. Shortly thereafter, Aristotle (384–322 B.C.) opposed the theory of Democritus and instead endorsed and advanced the Empedoclean theory. So strong was the influence of Aristotle that his theory dominated the thinking of scientists and philosophers until the beginning of the 17th century.

5.2 Dalton's Atomic Theory

Dalton's atomic theory

More than 2000 years after Democritus, the English schoolmaster John Dalton (1766–1844) revived the concept of atoms and proposed an atomic theory based on facts and experimental evidence. This theory, described in a series of papers published from 1803 to 1810, rested on the idea of a different kind of atom for each element. The essence of **Dalton's atomic theory** may be summed up as follows:

1. Elements are composed of minute, indivisible particles called atoms.
2. Atoms of the same element are alike in mass and size.
3. Atoms of different elements have different masses and sizes.
4. Chemical compounds are formed by the union of two or more atoms of different elements.
5. Atoms combine to form compounds in simple numerical ratios, such as one to one, two to two, two to three, and so on.

John Dalton (1766–1844)

6. Atoms of two elements may combine in different ratios to form more than one compound.

Dalton's atomic theory stands as a landmark in the development of chemistry. The major premises of his theory are still valid, but some of the statements must be modified or qualified because later investigations have shown that (1) atoms are composed of subatomic particles; (2) not all the atoms of a specific element have the same mass; and (3) atoms, under special circumstances, can be decomposed.

5.3 Composition of Compounds

A large number of experiments extending over a long period of time have established the fact that a particular compound always contains the same elements in the same proportions by mass. For example, water will always contain 11.2% hydrogen and 88.8% oxygen by mass. The fact that water contains hydrogen and oxygen in this particular ratio does not mean that hydrogen and oxygen cannot combine in some other ratio, but that a compound with a different ratio would not be water. In fact, hydrogen peroxide is made up of two atoms of hydrogen and two atoms of oxygen per molecule and contains 5.9% hydrogen and 94.1% oxygen by mass; its properties are markedly different from those of water.

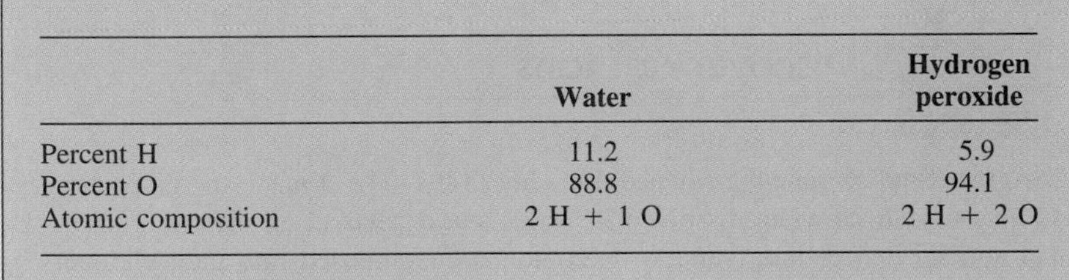

	Water	Hydrogen peroxide
Percent H	11.2	5.9
Percent O	88.8	94.1
Atomic composition	2 H + 1 O	2 H + 2 O

Law of Definite Composition

The **Law of Definite Composition** states:

A compound always contains two or more elements combined in a definite proportion by mass.

Let us consider two elements, oxygen and hydrogen, that form more than one compound. In water there are 8.0 g of oxygen for each gram of hydrogen. In hydrogen peroxide there are 16.0 g of oxygen for each gram of hydrogen. The masses of oxygen are in the ratio of small whole numbers, 16:8 or 2:1. Hydrogen peroxide has twice as much oxygen (by mass) as does water. Using Dalton's atomic theory, we deduce that hydrogen peroxide has twice as many oxygens per hydrogen as water. In fact, we now write the formulas for water as H_2O and for hydrogen peroxide as H_2O_2.

Law of Multiple Proportions

The **Law of Multiple Proportions** states:

Atoms of two or more elements may combine in different ratios to produce more than one compound.

Michael Faraday (1791–1867)

The reliability of this law and the Law of Definite Composition is the cornerstone of the science of chemistry. In essence these laws state that (1) the composition of a particular substance will always be the same no matter what its origin or how it is formed, and (2) the composition of different compounds formed from the same elements will always be unique.

5.4 The Nature of Electric Charge

Many of us have received a shock after walking across a carpeted area on a dry day. We have also experienced the static associated with combing our hair, and have had our clothing cling to us. All of these phenomena result from an accumulation of *electric charge.* This charge may be transferred from one object to another. The properties of electric charge follow:

1. Charge may be of two types, positive and negative.
2. Unlike charges attract (positive attracts negative) and like charges repel (negative repels negative and positive repels positive).
3. Charge may be transferred from one object to another, by contact or induction.
4. The less the distance between two charges, the greater the force of attraction between unlike charges (or repulsion between identical charges).

5.5 Discovery of Ions

The great English scientist Michael Faraday (1791–1867) made the discovery that certain substances when dissolved in water could conduct an electric current. He also noticed that certain compounds could be decomposed into their elements by passing an electric current through the compound. Atoms of some elements were attracted to the positive electrode, while atoms of other elements were attracted to the negative electrode. Faraday concluded that these atoms were electrically charged. He called them *ions* after the Greek word meaning "wanderer."

Any moving charge is an electric current. The electrical charge must travel through a substance known as a conducting medium. The most familiar conducting media are metals that are used in the form of wires.

The Swedish scientist Svante Arrhenius (1859–1927) extended Faraday's work. Arrhenius reasoned that an ion was an atom carrying a positive or negative charge. When a compound such as sodium chloride (NaCl) was melted, it conducted electricity. Water was unnecessary. Arrhenius's explanation of this conductivity was that upon melting, the sodium chloride dissociated, or broke up, into charged ions, Na^+ and Cl^-. The Na^+ ions moved toward the negative electrode (cathode) whereas the Cl^- migrated toward the positive electrode (anode). Thus, positive ions are *cations,* and negative ions are *anions.*

From Faraday's and Arrhenius's work with ions, Irish physicist G. J. Stoney (1826–1911) realized there must be some fundamental unit of electricity associated with atoms. He named this unit the *electron* in 1891. Unfortunately he had no means of supporting his idea with experimental proof. Evidence remained elusive until 1897 when English physicist Joseph Thomson (1856–1940) was able to show experimentally the existence of the electron.

Chemistry in Action

Triboluminescence

Some substances give off light when they are rubbed, crushed, or broken in a phenomenon called triboluminescence. Examples of substances exhibiting this property include crystals of quartz, sugar cubes, adhesive tape torn off certain surfaces, and Wintergreen Lifesavers.

Linda M. Sweeting, a chemist at Towson State University in Maryland, investigated the sparks created by a Wintergreen Lifesaver when crushed inside one's mouth in a dark room. When a sugar crystal is fractured, separate areas of positive and negative charge form on opposite sides of the crack. Electrons tend to leap the gap and neutralize the charge. When the electrons collide with nitrogen molecules in the air, small amounts of light are emitted. (Lightning is a somewhat similar phenomenon but on a much grander scale.) The addition of wintergreen molecules changes the outcome though because they absorb some of the light energy from the electron leap and reemit it as bright blue-green flashes. These are the sparks we see as we crush a Lifesaver in the dark.

Triboluminescence occurs when Wintergreen Lifesavers are crushed in the mouth.

5.6 Subatomic Parts of the Atom

The concept of the atom—a particle so small that until recently it could not be seen even with the most powerful microscope—and the subsequent determination of its structure stand among the very greatest creative intellectual human achievements.

Any visible quantity of an element contains a vast number of identical atoms. But when we refer to an atom of an element, we isolate a single atom from the multitude in order to present the element in its simplest form. Figure 5.1 shows individual atoms as we can see them today.

Let us examine this tiny particle we call the atom. The diameter of a single atom ranges from 0.1 to 0.5 nanometers (1 nm $= 1 \times 10^{-9}$ m). Hydrogen, the smallest atom, has a diameter of about 0.1 nm. To arrive at some idea of how small an atom is, consider this dot (•), which has a diameter of about 1 mm, or 1×10^6 nm. It would take 10 million hydrogen atoms to form a line of atoms across this dot. As inconceivably small as atoms are, they contain smaller particles, the **subatomic particles**, such as electrons, protons, neutrons.

subatomic particles

The development of atomic theory was helped in large part by the invention of new instruments. For example, the Crookes tube, developed by Sir William Crookes in 1875, opened the door to the subatomic structure of the atom (Figure 5.2). The emissions generated in a Crookes tube are called *cathode rays.* Joseph Thomson demonstrated in 1897 that cathode rays (1) travel in straight lines, (2) are negative in charge, (3) are deflected by electric and magnetic fields, (4) produce sharp shadows, and (5) are capable of moving a small paddle wheel. This was the experimental discovery of the fundamental unit of charge—the electron.

electron

The **electron** (e^-) is a particle with a negative electrical charge and a mass of 9.110×10^{-28} g. This mass is 1/1837 the mass of a hydrogen atom and corresponds to 0.0005486 atomic mass unit (amu) (defined in Section 5.11). One atomic mass unit has a mass of 1.6606×10^{-24} g. Although the actual charge of an electron is known, its value is too cumbersome for practical use and therefore has been assigned a relative electrical charge of -1. The size of an electron has not been determined exactly, but its diameter is believed to be less than 10^{-12} cm.

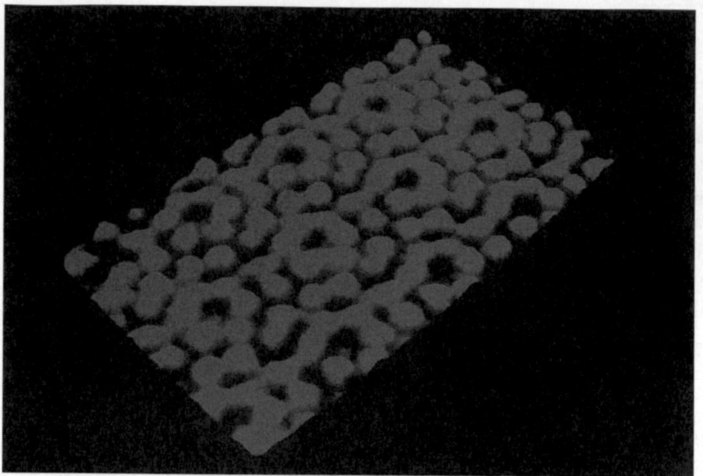

FIGURE 5.1
A scanning tunneling microscope shows an array of copper atoms.

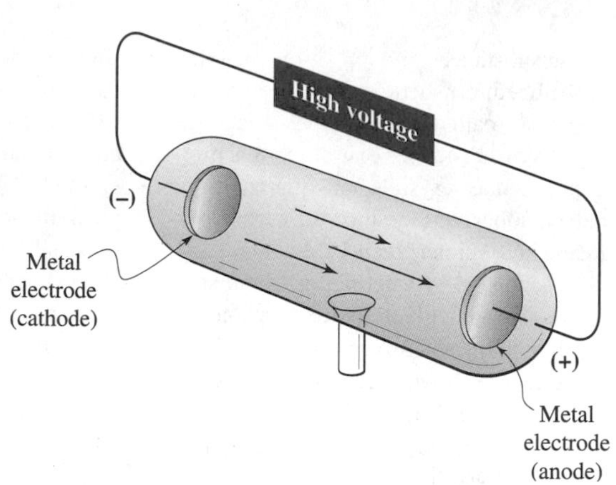

FIGURE 5.2
Crookes tube. In a Crookes tube a gas is contained at very low pressure. When a current is applied across electrodes within the tube, a stream of electrons travels from the cathode toward the anode.

Protons were first observed by German physicist E. Goldstein (1850–1930) in 1886. However, it was Thomson who discovered the nature of the proton. He showed that the proton was a particle, and he calculated its mass to be about 1837 times that of an electron. The **proton** (p) is a particle with a relative mass of 1 amu and an actual mass of 1.673×10^{-24} g. Its relative charge ($+1$) is equal in magnitude, but of opposite sign, to the charge on the electron. The mass of a proton is only very slightly less than that of a hydrogen atom.

proton

Thomson had shown that atoms contained both negatively and positively charged particles. Clearly, the Dalton model of the atom was no longer acceptable. Atoms were not indivisible. They were composed of smaller parts. Thomson proposed a new model of the atom.

Thomson model of the atom

In the **Thomson model of the atom** the electrons were negatively charged particles embedded in the atomic sphere. Since atoms were electrically neutral the sphere also contained an equal number of protons, or positive charges. A neutral atom could become an ion by gaining or losing electrons.

Positive ions were explained by assuming that the neutral atom had lost electrons. An atom with a net charge of $+1$ (for example, Na^+ or Li^+) would have lost one electron. An atom with a net charge of $+3$ (for example, Al^{3+}) would have lost three electrons.

Negative ions were explained by assuming that extra electrons had been added to the atoms. A net charge of -1 (for example, Cl^- or F^-) would be produced by the addition of one electron. A net charge of -2 (for example, O^{2-} or S^{2-}) would be explained as the addition of two electrons.

neutron

The third major subatomic particle was discovered in 1932 by James Chadwick (1891–1974). This particle, the **neutron** (n), bears neither a positive nor a negative charge and has a relative mass of about 1 amu. Its actual mass (1.675×10^{-24} g) is only very slightly greater than that of a proton. The properties of these three subatomic particles are summarized in Table 5.1.

Nearly all the ordinary chemical properties of matter can be explained in terms of atoms consisting of electrons, protons, and neutrons. The discussion of atomic

◀ Joseph Thomson (1856–1940)

structure that follows is based on the assumption that atoms contain only these principal subatomic particles. Many other subatomic particles, such as mesons, positrons, neutrinos, and antiprotons, have been discovered, but it is not yet clear whether all these particles are actually present in the atom or whether they are produced by reactions occurring within the nucleus. The fields of atomic and particle or high-energy physics have produced a long list of subatomic particles. Descriptions of the properties of many of these particles are to be found in recent physics textbooks, in various articles appearing in *Scientific American,* and in several of Isaac Asimov's books.

TABLE 5.1 Electrical Charge and Relative Mass of Electrons, Protons, and Neutrons

Particle	Symbol	Relative electrical charge	Relative mass (amu)	Actual mass (g)
Electron	e^-	-1	$\frac{1}{1837}$	9.110×10^{-28}
Proton	p	$+1$	1	1.673×10^{-24}
Neutron	n	0	1	1.675×10^{-24}

Example 5.1

The mass of a helium atom is 6.65×10^{-24} g. How many atoms are in a 4.0-g sample of helium?

Solution

$$4.0 \text{ g} \times \frac{1 \text{ atom He}}{6.65 \times 10^{-24} \text{ g}} = 6.0 \times 10^{23} \text{ atoms He}$$

Practice 5.1

The mass of an atom of hydrogen is 1.673×10^{-24} g. How many atoms are in a 10.0-g sample of hydrogen?

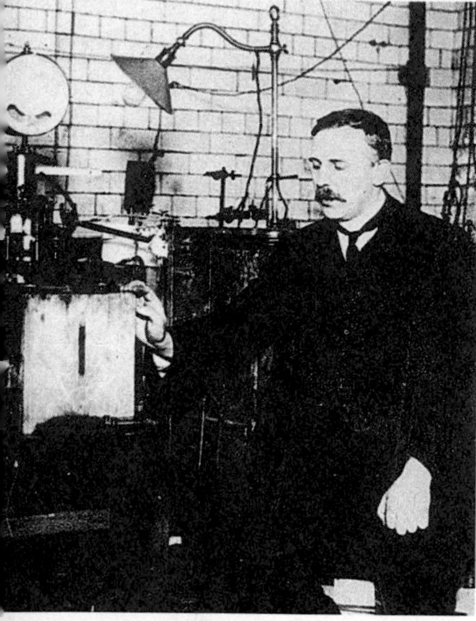

Ernest Rutherford (1871–1937)

5.7 The Nuclear Atom

The discovery that positively charged particles were present in atoms came soon after the discovery of radioactivity by Henri Becquerel in 1896. Radioactive elements spontaneously emit alpha particles, beta particles, and gamma rays from their nuclei (see Chapter 18).

Ernest Rutherford had, by 1907, established that the positively charged alpha particles emitted by certain radioactive elements were ions of the element helium. Rutherford used these alpha particles to establish the nuclear nature of atoms. In experiments performed in 1911, he directed a stream of positively charged helium ions (alpha particles) at a very thin sheet of gold foil (about 1000 atoms thick). See Figure 5.3(a). He observed that most of the alpha particles passed through the foil with little or no deflection; but a few of the particles were deflected at large angles, and occasionally one even bounced back from the foil (Figure 5.3b). It was known that like charges repel each other and that an electron with a mass of 1/1837 amu could not possibly have an appreciable effect on the path of a 4-amu alpha particle, which is about 7350 times more massive than an electron. Rutherford therefore reasoned that each gold atom must contain a positively charged mass occupying a relatively tiny volume and that, when an alpha particle approached close enough to this positive mass, it was deflected. Rutherford spoke of this positively charged mass as the *nucleus* of the atom. Because alpha particles have relatively high masses, the extent of the deflections (some actually bounced back) indicated to Rutherford that the nucleus is very heavy and dense. (The density of the nucleus of a hydrogen atom is about 10^{12} g/cm^3—about one trillion times the density of water.) Because most of the alpha particles passed through the thousand or so gold atoms without any apparent deflection, he further concluded that most of an atom consists of empty space.

When we speak of the mass of an atom, we are, for practical purposes, referring primarily to the mass of the nucleus. The nucleus contains all the protons and neutrons, which represent more than 99.9% of the total mass of any atom (see Table 5.1). By way of illustration, the largest number of electrons known to exist in an

FIGURE 5.3
(a) Diagram representing Rutherford's experiment on alpha-particle scattering. (b) Positive alpha particles (α), emanating from a radioactive source, were directed at a thin gold foil. Diagram illustrates the deflection and repulsion of the positive alpha particles by the positive nuclei of the gold atoms.
▼

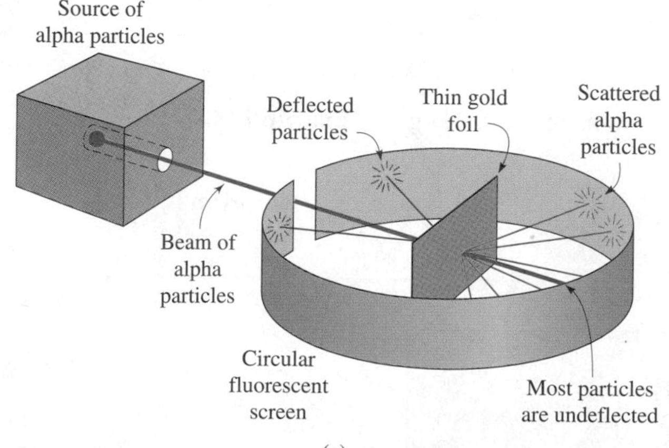

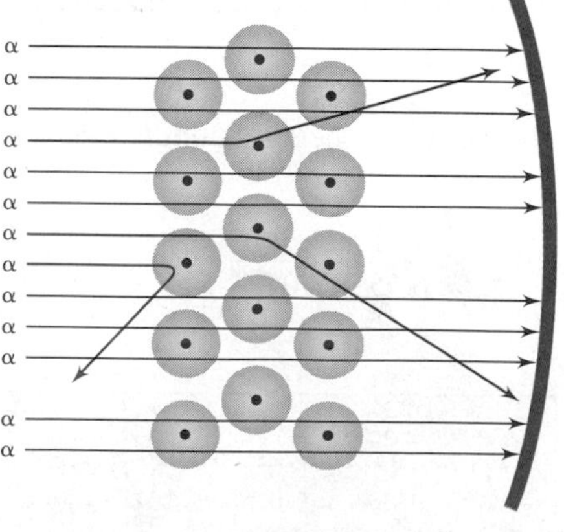

atom is 111. The mass of even 111 electrons is only about 1/17 of the mass of a single proton or neutron. The mass of an atom, therefore, is primarily determined by the combined masses of its protons and neutrons.

5.8 General Arrangement of Subatomic Particles

The alpha-particle scattering experiments of Rutherford established that the atom contains a dense, positively charged nucleus. The later work of Chadwick demonstrated that the atom contains neutrons, which are particles with mass but no charge. He also noted that light, negatively charged electrons are present and offset the positive charges in the nucleus. Based on this experimental evidence, a general description of the atom and the location of its subatomic particles was devised. Each atom consists of a **nucleus** surrounded by electrons. The nucleus contains protons and neutrons but not electrons. In a neutral atom the positive charge of the nucleus (due to protons) is exactly offset by the negative electrons. Because the charge of an electron is equal but of opposite sign to the charge of a proton, a neutral atom must contain exactly the same number of electrons as protons. However, this generalized picture of atomic structure provides no information on the arrangement of electrons within the atom.

nucleus

A neutral atom contains the same number of protons and electrons.

5.9 Atomic Numbers of the Elements

The **atomic number** of an element is the number of protons in the nucleus of an atom of that element. The atomic number determines the identity of an atom. For example, every atom with an atomic number of 1 is a hydrogen atom; it contains one proton in its nucleus. Every atom with an atomic number of 8 is an oxygen atom; it contains 8 protons in its nucleus. Every atom with an atomic number of 92 is a uranium atom; it contains 92 protons in its nucleus. The atomic number tells us not only the number of positive charges in the nucleus, but also the number of electrons in the neutral atom.

atomic number

The number of protons *defines* the element.

atomic number = number of protons in the nucleus

There is no need to memorize the atomic numbers of the elements because a periodic table is commonly provided in texts, laboratories, and on examinations. The atomic numbers of all elements are shown in the periodic table on the inside front cover of this book and are also listed in the table of atomic masses on the inside front cover.

5.10 Isotopes of the Elements

Shortly after Rutherford's conception of the nuclear atom, experiments were performed to determine the masses of individual atoms. These experiments showed that the masses of nearly all atoms were greater than could be accounted for by simply adding up the masses of all the protons and electrons that were known to be present in an atom. This fact led to the concept of the neutron, a particle with no charge but with a mass about the same as that of a proton. Because this particle has no charge, it was very difficult to detect, and the existence of the neutron was not proven experimentally until 1932. All atomic nuclei, except that of the simplest hydrogen atom, are now believed to contain neutrons.

All atoms of a given element have the same number of protons. Experimental evidence has shown that, in most cases, all atoms of a given element do not have identical masses. This is because atoms of the same element may have different numbers of neutrons in their nuclei.

isotopes

Atoms of an element having the same atomic number but different atomic masses are called **isotopes** of that element. Atoms of the various isotopes of an element, therefore, have the same number of protons and electrons but different numbers of neutrons.

Three isotopes of hydrogen (atomic number 1) are known. Each has one proton in the nucleus and one electron. The first isotope (protium), without a neutron, has a mass number of 1; the second isotope (deuterium), with one neutron in the nucleus, has a mass number of 2; the third isotope (tritium), with two neutrons, has a mass number of 3 (see Figure 5.4).

mass number

The three isotopes of hydrogen may be represented by the symbols $^{1}_{1}H$, $^{2}_{1}H$, $^{3}_{1}H$, indicating an atomic number of 1 and mass numbers of 1, 2, and 3, respectively. This method of representing atoms is called *isotopic notation.* The subscript (Z) is the atomic number; the superscript (A) is the **mass number**, which is the sum of the number of protons and the number of neutrons in the nucleus. The hydrogen isotopes may also be referred to as hydrogen-1, hydrogen-2, and hydrogen-3.

The *mass number* of an element is the sum of the protons and neutrons in the nucleus.

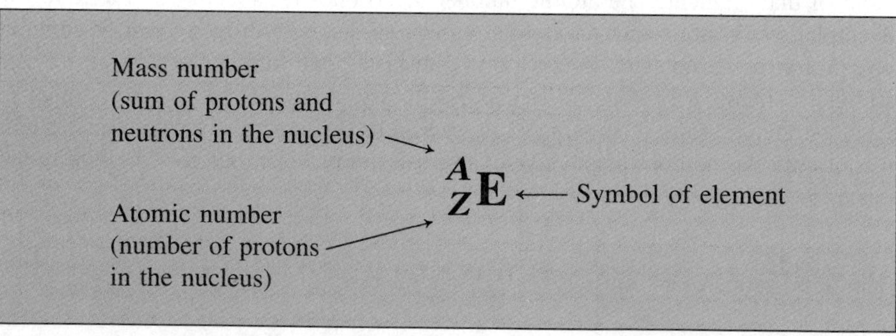

Most of the elements occur in nature as mixtures of isotopes. However, not all isotopes are stable; some are radioactive and are continuously decomposing to form other elements. For example, of the seven known isotopes of carbon, only two, carbon-12 and carbon-13, are stable. Of the seven known isotopes of oxygen, only three, $^{16}_{8}O$, $^{17}_{8}O$, and $^{18}_{8}O$, are stable. Of the fifteen known isotopes of arsenic, $^{75}_{33}As$ is the only one that is stable.

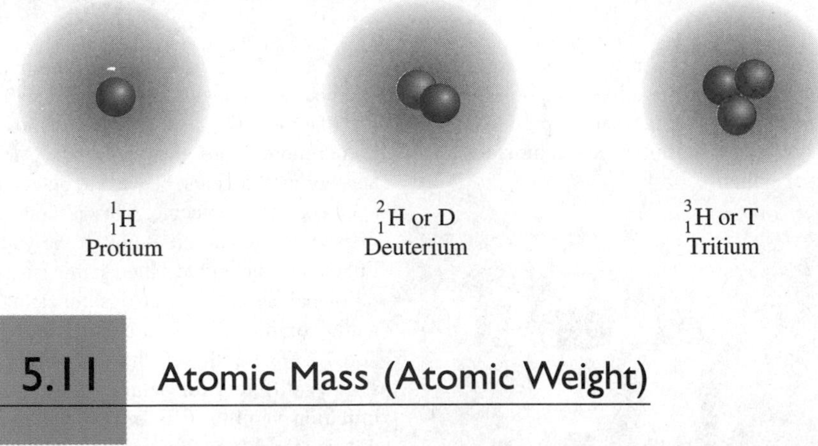

FIGURE 5.4
Schematic of the isotopes of hydrogen. The number of protons (purple) and neutrons (blue) are shown within the nucleus. The electron (e^-) exists outside the nucleus (gray).

5.11 Atomic Mass (Atomic Weight)

The mass of a single atom is far too small to measure individually on a balance. But fairly precise determinations of the masses of individual atoms can be made with an instrument called a *mass spectrometer* (see Figure 5.5). The mass of a single hydrogen atom is 1.673×10^{-24} g. However, it is neither convenient nor practical to compare the actual masses of atoms expressed in grams; therefore, a table of relative atomic masses using *atomic mass units* was devised. (The term *atomic weight* is often used instead of atomic mass.) The carbon isotope having six protons and six neutrons and designated carbon-12, or $^{12}_{6}C$, was chosen as the standard for atomic masses. This reference isotope was assigned a value of exactly 12 atomic mass units (amu). Thus one **atomic mass unit** is defined as equal to exactly 1/12 of the mass of a carbon-12 atom. The actual mass of a carbon-12 atom is 1.9927×10^{-23} g, and that of one atomic mass unit is 1.6606×10^{-24} g. In the table of atomic masses all elements then have values that are relative to the mass assigned to the reference isotope carbon-12.

atomic mass unit

A table of atomic masses is given on the inside front cover of this book. Hydrogen atoms, with a mass of about 1/12 that of a carbon atom, have an average atomic mass of 1.00797 amu on this relative scale. Magnesium atoms, which are about twice as heavy as carbon, have an average mass of 24.305 amu. The average atomic mass of oxygen is 15.9994 amu.

Since most elements occur as mixtures of isotopes with different masses, the atomic mass determined for an element represents the average relative mass of all the naturally occurring isotopes of that element. The atomic masses of the individual isotopes are approximately whole numbers, because the relative masses of the protons and neutrons are approximately 1.0 amu each. Yet we find that the atomic masses given for many of the elements deviate considerably from whole numbers.

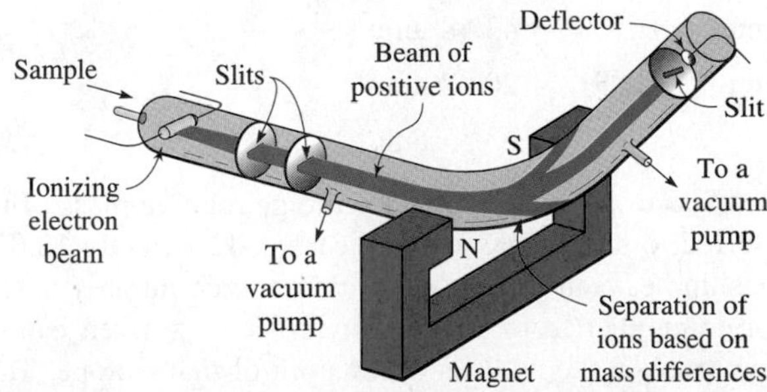

FIGURE 5.5
Schematic of a modern mass spectrometer. A beam of positive ions is produced from the sample as it enters the chamber. The positive ions are then accelerated as they pass through slits in an electric field. When the ions enter the magnetic field, they are deflected differently, depending on mass and charge. The ions are then detected at the end of the tube. From intensity and position of the lines on the mass spectrogram, the different isotopes of the elements and their relative amounts can be determined.

Chemistry in Action

Chemical Fraud

The bottle of vanilla flavoring found in most kitchens may be labeled either as "vanilla extract" or as "imitation vanilla extract." Both substances taste the same because they contain the same molecules, and taste is related to molecular structure. What is the difference then between real vanilla extract and imitation vanilla extract?

Vanilla extract is an alcohol and water solution of materials extracted from vanilla beans. Imitation vanilla extract is made from lignin, a waste product in the wood pulp industry, which is converted to vanillin—the same molecule found in natural vanilla extract. The advantage of this process is monetary. The cost of producing natural vanilla is approximately \$125 per gallon, whereas production of synthetic vanillin is approximately \$60 per gallon.

One of the roles of a government chemist is to prevent fraudulent mislabeling of products. Since the compounds in both products are chemically the same, another method must be used to detect the source of the molecule. This is done by inspecting the carbon atoms in the vanillin, using a technique called stable isotope ratio analysis (SIRA). Two isotopes in organic compounds are carbon-12 and carbon-13. The ratio of these isotopes is slightly different for natural vanillin and imitation vanillin. It is from these slight differences that government scientists distinguish whether the vanillin is from the vanilla bean or from the lignin.

For example, the atomic mass of rubidium is 85.4678 amu, that of copper is 63.546 amu, and that of magnesium is 24.305 amu. The deviation of an atomic mass from a whole number is due mainly to the unequal occurrence of the various isotopes of an element. For example, the two principal isotopes of copper are $^{63}_{29}Cu$ and $^{65}_{29}Cu$. It is apparent that copper-63 atoms are the more abundant isotope, since the atomic mass of copper, 63.546 amu, is closer to 63 than to 65 amu (see Figure 5.6). The actual values of the copper isotopes observed by mass spectra determination are shown here:

Isotope	Isotopic mass (amu)	Abundance (%)	Average atomic mass (amu)
$^{63}_{29}Cu$	62.9298	69.09	63.55
$^{65}_{29}Cu$	64.9278	30.91	

The average atomic mass can be calculated by multiplying the atomic mass of each isotope by the fraction of each isotope present and adding the results. The calculation for copper is

$$\begin{aligned} 62.9298 \text{ amu} \times 0.6909 &= 43.48 \text{ amu} \\ 64.9278 \text{ amu} \times 0.3091 &= \underline{20.07} \text{ amu} \\ & 63.55 \text{ amu} \end{aligned}$$

atomic mass

The **atomic mass** of an element is the average relative mass of the isotopes of that element referred to the atomic mass of carbon-12 (exactly 12.0000 . . . amu).

The relationship between mass number and atomic number is such that, if we subtract the atomic number from the mass number of a given isotope, we obtain the number of neutrons in the nucleus of an atom of that isotope. Table 5.2 shows

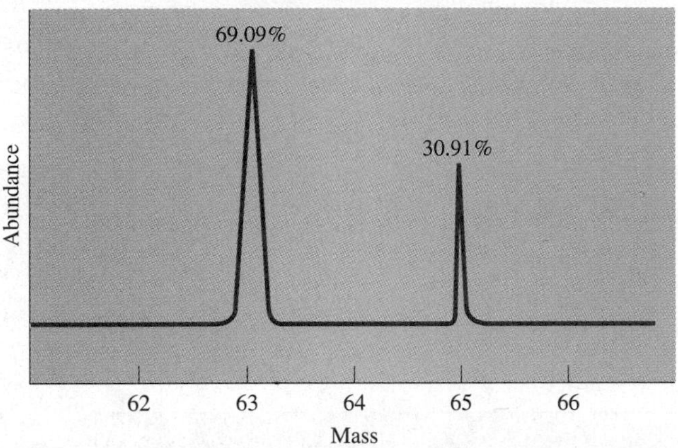

FIGURE 5.6 A typical reading from a mass spectrometer. The two principal isotopes of copper are shown with the abundance (%) given.

TABLE 5.2 Determination of the Number of Neutrons in an Atom by Subtracting Atomic Number from Mass Number

	Hydrogen (^1_1H)	Oxygen ($^{16}_8O$)	Sulfur ($^{32}_{16}S$)	Fluorine ($^{19}_9F$)	Iron ($^{56}_{26}Fe$)
Mass number	1	16	32	19	56
Atomic number	(−)1	(−)8	(−)16	(−)9	(−)26
Number of neutrons	0	8	16	10	30

the application of this method of determining the number of neutrons. For example, the fluorine atom ($^{19}_9F$), atomic number 9, having a mass of 19 amu, contains 10 neutrons:

Mass number	−	Atomic number	=	Number of neutrons
19	−	9	=	10

The atomic masses given in the table on the inside front cover of this book are values accepted by international agreement. You need not memorize atomic masses. In most of the calculations needed in this book, the use of atomic masses to four significant figures will give results of sufficient accuracy.

Example 5.2

How many protons, neutrons, and electrons are found in an atom of $^{14}_6C$?

Solution

The element is carbon, atomic number 6. The number of protons or electrons equals the atomic number and is 6. The number of neutrons is determined by subtracting the atomic number from the mass number: 14 − 6 = 8.

Practice 5.2

How many protons, neutrons, and electrons are in each of the following?
(a) $^{16}_8O$, (b) $^{80}_{35}Br$, (c) $^{235}_{92}U$, (d) $^{64}_{29}Cu$

Example 5.3 Chlorine is found in nature as two isotopes, $^{37}_{17}Cl$ (24.47%) and $^{35}_{17}Cl$ (75.53%). The atomic masses are 36.96590 and 34.96885 amu, respectively. Determine the average atomic mass of chlorine.

Solution Multiply each mass by its percentage and add the results to find the average.

$$(0.2447) \times (36.96590 \text{ amu}) + (0.7553) \times (34.96885 \text{ amu})$$
$$= 35.4575 \text{ amu}$$
$$= 35.46 \text{ amu (4 significant figures)}$$

Practice 5.3

Silver is found in two isotopes with atomic masses of 106.9041 and 108.9047 amu, respectively. The first isotope represents 51.82% and the second 48.18%. Determine the average atomic mass of silver.

Concepts in Review

1. State the major provisions of Dalton's atomic theory.
2. State the Law of Definite Composition and indicate its significance.
3. State the Law of Multiple Proportions and indicate its significance.
4. Give the names, symbols, and relative masses of the three principal subatomic particles.
5. Describe the Thomson model of the atom.
6. Describe the atom as conceived by Ernest Rutherford after his alpha-scattering experiment.
7. Determine the atomic number, mass number, or number of neutrons of an isotope when given the values of any two of these three items.
8. Name and distinguish among the three isotopes of hydrogen.
9. Calculate the average atomic mass of an element, given the isotopic masses and the abundance of its isotopes.
10. Determine the number of protons, neutrons, and electrons from the atomic number and atomic mass of an atom.

Key Terms

atomic mass (5.11)
atomic mass unit (5.11)
atomic number (5.9)
Dalton's atomic theory (5.2)
electron (5.6)
isotopes (5.10)
Law of Definite Composition (5.3)
Law of Multiple Proportions (5.3)
mass number (5.10)
neutron (5.6)
nucleus (5.8)
proton (5.6)
subatomic particles (5.6)
Thomson model of the atom (5.6)

Questions

Questions refer to tables, figures, and key words and concepts defined within the chapter. A particularly challenging question is indicated with an asterisk.

1. What are the atomic numbers of (a) copper, (b) nitrogen, (c) phosphorus, (d) radium, and (e) zinc?

2. A neutron is approximately how many times heavier than an electron?

3. From the point of view of a chemist, what are the essential differences among a proton, a neutron, and an electron?

4. Distinguish between an atom and an ion.

5. What letters are used to designate atomic number and mass number in isotopic notation of atoms?

6. Indicate the relationship between triboluminescence and Wintergreen Lifesavers.

7. Differentiate between synthetic and natural vanilla extract. Explain why this is important to our society.

8. Describe and indicate the value of the SIRA technique.

9. In what ways are isotopes alike? In what ways different?

10. Which of the following statements are correct? Rewrite the incorrect statements to make them correct.
- **(a)** John Dalton developed an important atomic theory in the early 1800s.
- **(b)** Dalton said that elements are composed of minute indivisible particles called atoms.
- **(c)** Dalton said that when atoms combine to form compounds, they do so in simple numerical ratios.
- **(d)** Dalton said that atoms are composed of protons, neutrons, and electrons.
- **(e)** All of Dalton's theory is still considered valid today.
- **(f)** Hydrogen is the smallest atom.
- **(g)** A proton is about 1837 times as heavy as an electron.
- **(h)** The nucleus of an atom contains protons, neutrons, and electrons.

11. Which of the following statements are correct? Rewrite the incorrect statements to make them correct.
- **(a)** An element with an atomic number of 29 has 29 protons, 29 neutrons, and 29 electrons.
- **(b)** An atom of the isotope $^{60}_{26}Fe$ has 34 neutrons in its nucleus.
- **(c)** $^{2}_{1}H$ is a symbol for the isotope deuterium.
- **(d)** An atom of $^{31}_{15}P$ contains 15 protons, 16 neutrons, and 31 electrons.
- **(e)** In the isotope $^{6}_{3}Li$, $Z = 3$ and $A = 3$.
- **(f)** Isotopes of a given element have the same number of protons but differ in the number of neutrons.
- **(g)** The three isotopes of hydrogen are called protium, deuterium, and tritium.
- **(h)** $^{23}_{11}Na$ and $^{24}_{11}Na$ are isotopes.
- **(i)** $^{24}_{11}Na$ has one more electron than $^{23}_{11}Na$.
- **(j)** $^{24}_{11}Na$ has one more proton than $^{23}_{11}Na$.
- **(k)** $^{24}_{11}Na$ has one more neutron than $^{23}_{11}Na$.
- **(l)** Only a few of the elements exist in nature as mixtures of isotopes.

12. Which of the following statements are correct? Rewrite the incorrect statements to make them correct.
- **(a)** One atomic mass unit has a mass 12 times as much as one carbon-12 atom.
- **(b)** The atomic masses of protium and deuterium differ by about 100%.
- **(c)** $^{23}_{11}Na$ and $^{24}_{11}Na$ have the same atomic masses.
- **(d)** The atomic mass of an element represents the average relative atomic mass of all the naturally occurring isotopes of that element.

Paired Exercises

These exercises are paired. Each odd-numbered exercise is followed by a similar even-numbered exercise. Answers to even-numbered exercises are given in Appendix V.

13. Explain why, in Rutherford's experiments, some alpha particles were scattered at large angles by the gold foil or even bounced back?

14. What experimental evidence led Rutherford to conclude each of the following?
- **(a)** The nucleus of the atom contains most of the atomic mass.
- **(b)** The nucleus of the atom is positively charged.
- **(c)** The atom consists of mostly empty space.

15. Describe the general arrangement of subatomic particles in the atom.

16. What part of the atom contains practically all its mass?

17. What contribution did each of the following scientists make to atomic theory?
(a) Dalton
(b) Thomson
(c) Rutherford

18. Consider each of the following models of the atom: (a) Dalton, (b) Thomson, (c) Rutherford. How does the location of the electrons in an atom vary? How does the location of the atom's positive matter compare?

19. Explain why the atomic masses of elements are not whole numbers.

20. Is the isotopic mass of a given isotope ever an exact whole number? Is it always? In answering, consider the masses of $^{12}_{6}C$ and $^{63}_{29}Cu$.

21. What special names are given to the isotopes of hydrogen?

22. List the similarities and differences in the three isotopes of hydrogen.

23. What is the symbol and name of the element that has an atomic number of 24 and a mass number of 52?

24. An atom of an element has a mass number of 201 and has 121 neutrons in its nucleus.
(a) What is the electrical charge of the nucleus?
(b) What is the symbol and name of the element?

25. What is the nuclear composition of the six naturally occurring isotopes of calcium having mass numbers of 40, 42, 43, 44, 46, and 48?

26. Which of the isotopes of calcium in the previous exercise is the most abundant isotope? Can you be sure? Explain your choice.

27. Write isotopic notation symbols for the following:
(a) $Z = 26, \quad A = 55$
(b) $Z = 12, \quad A = 26$
(c) $Z = 3, \quad A = 6$
(d) $Z = 79, \quad A = 188$

28. Give the nuclear composition and isotopic notation for:
(a) An atom containing 27 protons, 32 neutrons, and 27 electrons
(b) An atom containing 15 protons, 16 neutrons, and 15 electrons
(c) An atom containing 110 neutrons, 74 electrons, and 74 protons
(d) An atom containing 92 electrons, 143 neutrons, and 92 protons

29. Naturally occurring lead exists as four stable isotopes: ^{204}Pb with a mass of 203.973 amu (1.480%); ^{206}Pb, 205.974 amu (23.60%); ^{207}Pb, 206.9759 amu (22.60%); and ^{208}Pb, 207.9766 amu (52.30%). Calculate the average atomic mass of lead.

30. Naturally occurring magnesium consists of three stable isotopes: ^{24}Mg, 23.985 amu (78.99%); ^{25}Mg, 24.986 amu (10.00%); and ^{26}Mg, 25.983 amu (11.01%). Calculate the average atomic mass of magnesium.

31. 68.9257 amu is the mass of 60.4% of the atoms of an element with only two naturally occurring isotopes. The atomic mass of the other isotope is 70.9249 amu. Determine the average atomic mass of the element. Identify the element.

32. A sample of enriched lithium contains 30.00% ^{6}Li (6.015 amu) and 70.00% ^{7}Li (7.016 amu). What is the average atomic mass of the sample?

33. An aluminum atom has an average diameter of about 3.0×10^{-8} cm. The nucleus has a diameter of about 2.0×10^{-13} cm. Calculate the ratio of these diameters.

***34.** An average dimension for the radius of an atom is 1.0×10^{-8} cm, and the average radius of the nucleus is 1.0×10^{-13} cm. Determine the ratio of atomic volume to nuclear volume. Assume that the atom is spherical ($V = (4/3)\pi r^3$ for a sphere).

Additional Exercises

These exercises are not paired or labeled by topic and provide additional practice on the concepts covered in this chapter.

35. What experimental evidence supports each of the following statements?
(a) The nucleus of an atom is small.
(b) The atom consists of both positive and negative charges.
(c) The nucleus of the atom is positive.

36. What is the relationship between the following two atoms:
(a) one atom with 10 protons, 10 neutrons, and 10 electrons; and another atom with 10 protons, 11 neutrons, and 10 electrons
(b) one atom with 10 protons, 11 neutrons, and 10 electrons; and another atom with 11 protons, 10 neutrons, and 11 electrons.

37. How will an atom's nucleus differ if it loses an alpha particle?

38. Suppose somewhere in the universe another kind of matter is discovered and your job is to determine the characteristics of this new matter. You first place a gaseous sample of this matter in a cathode-ray tube and find that, as with ordinary matter, cathode rays are seen. When they are deflected with a magnet, the rays emitted from the anode (+) are deflected thousands of times more than those emitted from the cathode (−).
(a) What conclusions can you draw from this experiment? When a metal foil of the new matter is bombarded by alpha particles, most of them pass right on through, but one every few billion or so goes in and never comes out!
(b) What further conclusions can you draw from this new information?

39. The radius of a carbon atom in many compounds is 0.77×10^{-8} cm. If the radius of a Styrofoam ball used to represent the carbon atom in a molecular model is 1.5 cm, how much of an enlargement is this?

40. How is it possible for there to be more than one kind of atom of the same element?

41. Which of the following contains the largest number of neutrons per atom: ^{210}Bi, ^{210}Po, ^{210}At, ^{211}At

***42.** An unknown element Q has two known isotopes: ^{60}Q and ^{63}Q. If the average atomic mass is 61.5 amu, what are the relative percentages of the isotopes?

***43.** The actual mass of one atom of an unknown isotope is 2.18×10^{-22} g. Calculate the atomic mass of this isotope.

44. The mass of an atom of argon is 6.63×10^{-24} g. How many atoms are in a 40.0-g sample of argon?

45. Using the periodic table inside the front cover of the book, determine which of the first twenty elements have isotopes that you would expect to have the same number of protons, neutrons, and electrons.

46. Complete the following table with the appropriate data for each isotope given:

Atomic number	Mass number	Symbol of element	Number of protons	Number of neutrons
(a) 8	16			
(b)		Ni		30
(c)	199		80	

47. Complete the following table (all are neutral atoms):

Element	Symbol	Atomic number	Number of protons	Number of neutrons	Number of electrons
(a) Platinum					
(b)	^{30}P				
(c)		53			
(d)			36		
(e)				45	34
(f)	^{40}Ca				

Answers to Practice Exercises

5.1 5.98×10^{24} atoms

5.2

	protons	neutrons	electrons
(a)	8	8	8
(b)	35	45	35
(c)	92	143	92
(d)	29	35	29

5.3 107.9 amu

Nomenclature of Inorganic Compounds

As children, we begin to communicate with other people in our lives by learning the names of objects around us. As we continue to develop, we learn to speak and use language to complete a wide variety of tasks. As we enter school, we begin to learn of other languages—the languages of mathematics, of other cultures, of computers. In each case, we begin by learning the names of the building blocks, and then proceed to more abstract concepts. In chemistry, a new language beckons also—a whole new way of describing the objects so familiar to us in our daily lives—the nomenclature of compounds. Only after learning the language are we able to understand the complexities of the modern model of the atom and its applications to the various fields we have chosen as our careers.

6.1 Common and Systematic Names

Chemical nomenclature is the system of names that chemists use to identify compounds. When a new substance is formulated, it must be named in order to distinguish it from all other substances (see Figure 6.1). In this chapter, we will restrict our discussion to the nomenclature of inorganic compounds—compounds that do not generally contain carbon. The naming of organic compounds (carbon-containing compounds) will be covered separately in Chapter 19.

Common names are arbitrary names that are not based on the chemical composition of compounds. Before chemistry was systematized, a substance was given a name that generally associated it with one of its outstanding physical or chemical properties. For example, *quicksilver* was a common name for mercury, and nitrous oxide (N_2O), used as an anesthetic in dentistry, has been called *laughing gas* because it induces laughter when inhaled. Water and ammonia also are common names because neither provides any information about the chemical composition of the compounds. If every substance were assigned a common name, the amount of memorization required to learn over nine million names would be astronomical.

Water (H_2O) and ammonia (NH_3) are almost always referred to by their common names.

Common names have distinct limitations, but they remain in frequent use. Often common names continue to be used in industry because the systematic name is too long or too technical for everyday use. For example, calcium oxide (CaO) is called *lime* by plasterers; photographers refer to *hypo* rather than sodium thiosulfate

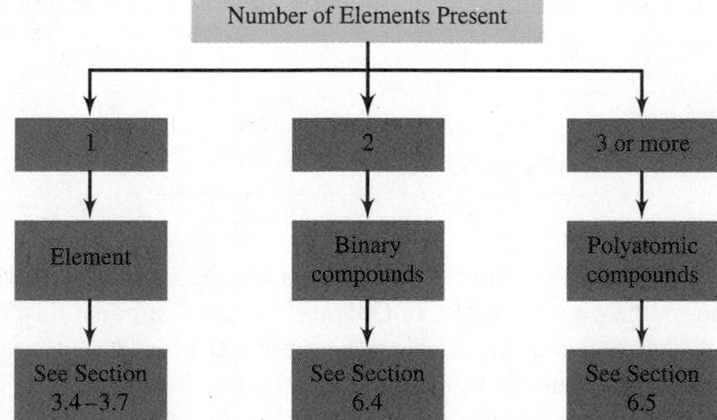

FIGURE 6.1 Flow chart summarizing the text locations of rules for naming inorganic substances.

Chapter Opening Photo: This exquisite chambered nautilus shell is formed from calcium carbonate, commonly called limestone.

TABLE 6.1 Common Names, Formulas, and Chemical Names of Familiar Substances

Common name	Formula	Chemical name
Acetylene	C_2H_2	Ethyne
Lime	CaO	Calcium oxide
Slaked lime	$Ca(OH)_2$	Calcium hydroxide
Water	H_2O	Water
Galena	PbS	Lead(II) sulfide
Alumina	Al_2O_3	Aluminum oxide
Baking soda	$NaHCO_3$	Sodium hydrogen carbonate
Cane or beet sugar	$C_{12}H_{22}O_{11}$	Sucrose
Borax	$Na_2B_4O_7 \cdot 10\ H_2O$	Sodium tetraborate decahydrate
Brimstone	S	Sulfur
Calcite, marble, limestone	$CaCO_3$	Calcium carbonate
Cream of tartar	$KHC_4H_4O_6$	Potassium hydrogen tartrate
Epsom salts	$MgSO_4 \cdot 7\ H_2O$	Magnesium sulfate heptahydrate
Gypsum	$CaSO_4 \cdot 2\ H_2O$	Calcium sulfate dihydrate
Grain alcohol	C_2H_5OH	Ethanol, ethyl alcohol
Hypo	$Na_2S_2O_3$	Sodium thiosulfate
Laughing gas	N_2O	Dinitrogen monoxide
Lye, caustic soda	$NaOH$	Sodium hydroxide
Milk of magnesia	$Mg(OH)_2$	Magnesium hydroxide
Muriatic acid	HCl	Hydrochloric acid
Plaster of Paris	$CaSO_4 \cdot \frac{1}{2}\ H_2O$	Calcium sulfate hemihydrate
Potash	K_2CO_3	Potassium carbonate
Pyrite (fool's gold)	FeS_2	Iron disulfide
Quicksilver	Hg	Mercury
Saltpeter (chile)	$NaNO_3$	Sodium nitrate
Table salt	$NaCl$	Sodium chloride
Vinegar	$HC_2H_3O_2$	Acetic acid
Washing soda	$Na_2CO_3 \cdot 10\ H_2O$	Sodium carbonate decahydrate
Wood alcohol	CH_3OH	Methanol, methyl alcohol

($Na_2S_2O_3$); and nutritionists use the name *vitamin* D_3, instead of 9,10-secocholesta-5,7,10(19)-trien-3-β-ol ($C_{27}H_{44}O$). Table 6.1 lists the common names, formulas, and systematic names of some familiar substances.

Chemists prefer systematic names that precisely identify the chemical composition of chemical compounds. The system for inorganic nomenclature was devised by the International Union of Pure and Applied Chemistry (IUPAC), which was founded in 1921. They continue to meet regularly and constantly review and update the system.

6.2 Elements and Ions

In Chapter 3 you learned the names and symbols for the elements as well as their general location on the periodic table. In Chapter 5 we investigated the composition of the atom and learned that all atoms are composed of protons, electrons, and neutrons; that a particular element is defined by the number of protons it contains; and that atoms are uncharged because they contain equal numbers of protons and electrons.

What's in a Name?

Sometimes it takes an extremely long time to name a new substance. Such was the case with element 106. It was discovered in 1974 by a team of scientists in California at Lawrence Berkeley Laboratories at the same time it was discovered by a team of Russian scientists at the Joint Institute for Nuclear Research in Dubna. Names for new elements are generally provided by its discoverers, but in this case neither team could reproduce their discovery since element 106 is a substance that disappears within a few seconds. Only in 1993 after almost 20 years did the group at Berkeley succeed in once more producing element 106, therefore giving Berkeley the claim for its discovery and the freedom to name it.

The eight-member team of scientists discussed naming the element after Sir Isaac Newton or Luis Alvarez (a Berkeley scientist who played a major role in developing the theory that the dinosaurs were killed by the impact of a comet). Seven members of the team finally met and suggested the name seaborgium after the eighth member, Dr. Glenn T. Seaborg. This is the first time that an element has ever been named for a living person.

Seaborg, now 81, is a former chancellor of UC Berkeley and a former chairman of the Atomic Energy Commission (now called the Department of Energy). He was first recognized as the prime mover in the discovery of plutonium, the radioactive element that is best known as the key component of atomic bombs. Since then, he has worked with a team of scientists to discover nine of the transuranium elements (those coming after uranium on the periodic table). They are created by firing streams of particles and nuclei into heavy elements in a nuclear accelerator. Elements credited to his team from the 1940s to the 1960s include neptunium (named after neptune), americium (America), fermium (Enrico Fermi, a famous physicist), mendelevium (Dimitri Mendeleev, creator of the modern periodic table), nobelium (Alfred Nobel, inventor of dynamite and founder of the Nobel prize), berkelium (Berkeley, California, home of the group's lab), californium (California), curium (Marie Curie, discoverer of several elements), and einsteinium (Albert Einstein). In 1951, Seaborg received the Nobel prize for his role in the discoveries of these elements.

Glenn T. Seaborg, discoverer of element 106.

Since Berkeley's last discovery (element 106) in 1974, the creation of new elements has been led by a team in Germany that has discovered elements 107, 108, and 109. They suggested neilsbohrium (Niels Bohr, a Danish physicist), hassium (Hesse, a German state), and meitnerium (Lise Meitner, a German atomic fission pioneer).

All of these names must still be approved by the IUPAC before becoming official and placed into common use on the periodic table. The IUPAC group met in August 1994 and assigned different names to several of the elements above 100. The names for elements 101, 102, and 103 remained the same as they had been, but the name for 106 was changed because Glenn Seaborg is still alive. The names for elements 99 and 100 were also proposed (but not approved) when the scientists were still living. The proposed IUPAC names for 104 and 105 would change names that have been in wide use for over 20 years. The latest decisions by IUPAC have rekindled controversy over the naming of elements. If you look at the periodic table today you will find each of the elements (104–109) named only by the Latin version of its number. This was a provisional recommendation until names could be agreed upon. It appears that the newest elements will have to wait a little longer to be officially named.

Some of the Names Proposed for Elements Discovered in Recent Years

Number	American	German	Russian	IUPAC
101				mendelevium
102			joliotium	nobelium
103				lawrencium
104	rutherfordium			dubnium
105	hahnium		neilsbohrium	joliotium
106	seaborgium			rutherfordium
107		neilsbohrium		bohrium
108		hassium		hahnium
109		meitnerium		meitnerium

The formula for most elements is simply the symbol of the element. In chemical reactions or mixtures the element behaves as though it were a collection of individual particles. A small number of elements have formulas that are not single atoms at normal temperatures. Seven of the elements are *diatomic* molecules—that is, two atoms bonded together to form a molecule. These diatomic elements are hydrogen, H_2; oxygen, O_2; nitrogen, N_2; fluorine, F_2; chlorine, Cl_2; bromine, Br_2; and iodine, I_2. Two other elements that are commonly polyatomic are S_8, sulfur; and P_4, phosphorus.

Elements Occurring as Molecules

Hydrogen	H_2	Chlorine	Cl_2	Sulfur	S_8
Oxygen	O_2	Fluorine	F_2	Phosphorus	P_4
Nitrogen	N_2	Bromine	Br_2		
		Iodine	I_2		

A charged atom, known as an *ion,* can be produced by adding or removing one or more electrons from a neutral atom. For example, potassium atoms contain 19 protons and 19 electrons. To make a potassium ion we remove one electron leaving 19 protons and only 18 electrons. This gives an ion with a positive one (+1) charge (see diagram).

Written in the form of an equation, $K \rightarrow K^+ + e^-$. A positive ion is called

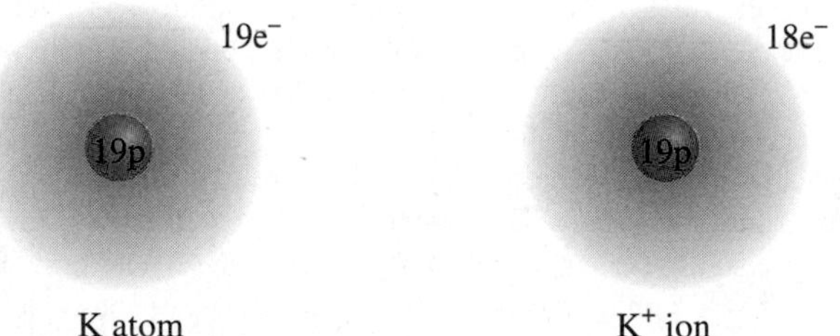

a *cation*. Any neutral atom that *loses* an electron will form a cation. Sometimes an atom may lose one electron as we have seen in our potassium example. Other atoms may lose more than one electron as shown below:

$$Mg \rightarrow Mg^{2+} + 2e^-$$

or

$$Al \rightarrow Al^{3+} + 3e^-$$

Cations are named the same as their parent atoms, as shown below:

Atom		Ion	
K	potassium	K^+	potassium ion
Mg	magnesium	Mg^{2+}	magnesium ion
Al	aluminum	Al^{3+}	aluminum ion

Ions can also be formed by adding electrons to a neutral atom. For example, the chlorine atom contains 17 protons and 17 electrons. The equal number of positive charges and negative charges results in a net charge of zero for the atom. If one electron is added to the chlorine atom it now contains 17 protons and 18 electrons resulting in a net charge of negative one (−1) on the ion.

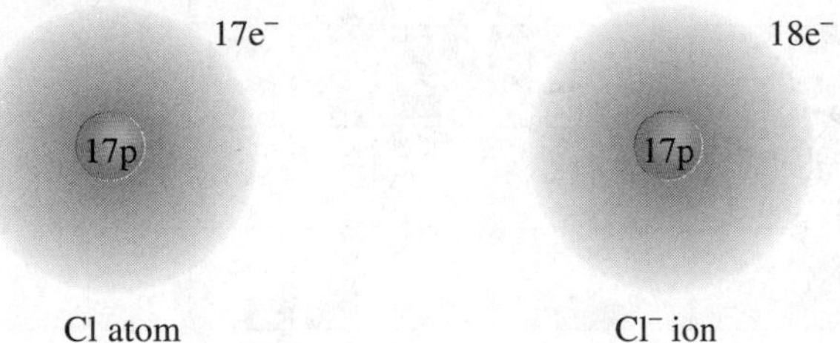

Summarizing the process in a chemical equation produces $Cl + e^- \rightarrow Cl^-$. A negative ion is called an *anion.* Any neutral atom that *gains* an electron will form an anion. Atoms may gain more than one electron to form anions with different charges as shown in the following examples:

$$O + 2e^- \rightarrow O^{2-}$$

$$N + 3e^- \rightarrow N^{3-}$$

Anions are named differently than cations. To name an anion consisting of only one element, use the stem part of the parent element name and change the ending to *-ide.* For example, the Cl^- ion is named in the following way: use the *chlor-* from chlorine and add *-ide* to form chloride ion. Other examples are shown below:

Symbol	Name of atom	Ion	Name of ion
F	fluorine	F^-	fluoride ion
Br	bromine	Br^-	bromide ion
Cl	chlorine	Cl^-	chloride ion
I	iodine	I^-	iodide ion
O	oxygen	O^{2-}	oxide ion
N	nitrogen	N^{3-}	nitride ion

Ions are always formed by adding or removing electrons from an atom. Atoms do not form ions on their own. Most often ions are formed when metals combine with nonmetals.

The charge on an ion can often be predicted from the position of the element on the periodic table. Figure 6.2 shows the charges of selected ions from several groups on the periodic table. Notice that all the metals in the far left column (Group IA) are (1+), all those in the next column (Group IIA) are (2+), and the metals in the next tall column (Group IIIA) form (3+) ions. The elements in the lower center part of the table are called *transition metals.* These elements tend to form

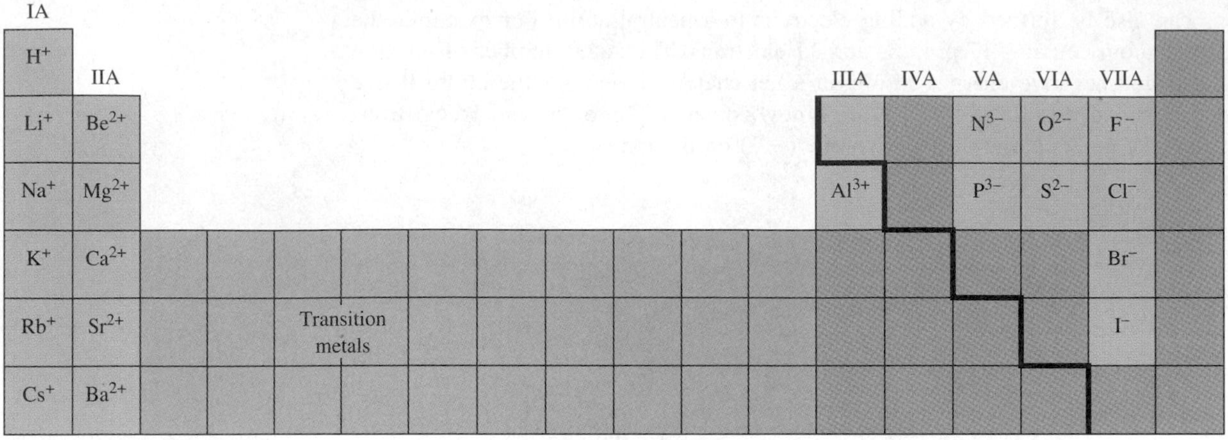

IA	IIA											IIIA	IVA	VA	VIA	VIIA	
H^+																	
Li^+	Be^{2+}													N^{3-}	O^{2-}	F^-	
Na^+	Mg^{2+}											Al^{3+}		P^{3-}	S^{2-}	Cl^-	
K^+	Ca^{2+}															Br^-	
Rb^+	Sr^{2+}			Transition metals												I^-	
Cs^+	Ba^{2+}																

▲ **FIGURE 6.2**
Charges of selected ions in the periodic table.

FIGURE 6.3 ▶
Salt (NaCl) is a strong electrolyte as can be seen from the brightly glowing light bulb.

cations with various positive charges. There is no easy way to predict the charges on these cations. All metals lose electrons to form positive ions.

In contrast, the nonmetals form anions by gaining electrons. On the right side of the periodic table in Figure 6.2 you can see that the atoms in Group VIIA form (1−) ions. The nonmetals in Group VIA form (2−) ions. It is important to learn the charges on the ions shown in Figure 6.2 and their relationship to the group number at the top of the column. We will learn more about why these ions carry their particular charges later in the course.

6.3 Writing Formulas from Names of Compounds

In Chapters 3 and 5 we learned that compounds can be composed of ions. These substances are called *ionic compounds* and will conduct electricity when dissolved in water. An excellent example of an ionic compound is ordinary table salt. It is composed of crystals of sodium chloride. When dissolved in water, sodium chloride conducts electricity very well as shown in Figure 6.3.

A chemical compound must have a net charge of zero. If it contains ions, the charges on the ions must add up to zero in the formula for the compound. This is relatively easy in the case of sodium chloride. The sodium ion (1+) and the chloride ion (1−) add to zero, resulting in the formula NaCl. Now consider an ionic compound containing calcium (Ca^{2+}) and fluoride (F^-) ions. How can we write a formula with a net charge of zero? To do this we need one Ca^{2+} and two F^- ions. The correct formula is CaF_2. The subscript 2 indicates two fluoride ions are needed for each calcium ion. Aluminum oxide is a bit more complicated since it consists of Al^{3+} and O^{2-} ions. Since 6 is the least common multiple of 3 and 2, we have $2(3+) = 3(2-) = 0$ or a formula containing 2 Al^{3+} ions and 3 O^{2-} ions for Al_2O_3. The following table gives more examples of formula writing for ionic compounds.

Name of compound	Ions	Lowest common multiple	Sum of charges on ions	Formula
Sodium bromide	Na^+, Br^-	1	$(+1) + (-1) = 0$	NaBr
Potassium sulfide	K^+, S^{2-}	2	$2(+1) + (-2) = 0$	K_2S
Zinc sulfate	Zn^{2+}, SO_4^{2-}	2	$(+2) + (-2) = 0$	$ZnSO_4$
Ammonium phosphate	NH_4^+, PO_4^{3-}	3	$3(+1) + (-3) = 0$	$(NH_4)_3PO_4$
Aluminum chromate	Al^{3+}, CrO_4^{2-}	6	$2(+3) + 3(-2) = 0$	$Al_2(CrO_4)_3$

Example 6.1

Write formulas for (a) calcium chloride, (b) magnesium oxide, (c) barium phosphide.

Solution

(a) Use the following steps for calcium chloride.

Step 1 From the name we know that calcium chloride is composed of calcium and chloride ions. First write down the formulas of these ions:

Ca^{2+} Cl^-

Step 2 To write the formula of the compound, combine the smallest numbers of Ca^{2+} and Cl^- ions to give the charge sum equal to zero. In this case the lowest common multiple of the charges is 2:

$$(Ca^{2+}) + 2(Cl^-) = 0$$

$$(2+) + 2(1-) = 0$$

Therefore, the formula is $CaCl_2$.

(b) Use the same procedure for magnesium oxide.

Step 1 From the name we know that magnesium oxide is composed of magnesium and oxide ions. First write down the formulas of these ions:

Mg^{2+} O^{2-}

Step 2 To write the formula of the compound, combine the smallest numbers of Mg^{2+} and O^{2-} ions to give the charge sum equal to zero:

1. Cobalt(II) chloride, $CoCl_2$
2. Sodium chloride, NaCl
3. Lead(II) sulfide, PbS
4. Sulfur, S
5. Zinc, Zn
6. Marble chips, $CaCO_3$
7. Logwood chips
8. Charcoal, C
9. Mercury(II) iodide, HgI_2
10. Pyrite, FeS
11. Chromium(III) oxide, Cr_2O_3
12. Iron(II) sulfate, $FeSO_4$
13. Sodium sulfite, Na_2SO_3
14. Rosin
15. Sodium thiosulfate, $Na_2S_2O_3$
16. Iron, Fe
17. Aluminum, Al
18. Potassium hexacyanoferrate, $K_3Fe(CN)_6$
19. Potassium chromium(III) sulfate, $KCr(SO_4)_2$
20. Menthol, $C_{10}H_{19}OH$
21. Potassium permanganate, $KMnO_4$
22. Ammonium nickel(II) sulfate, $(NH_4)_2Ni(SO_4)_2$
23. Copper(II) sulfate pentahydrate, $CuSO_4{\cdot}5H_2O$
24. Sodium chromate, Na_2CrO_4
25. Trilead tetraoxide, Pb_3O_4
26. Hydroquinone, $C_6H_4(OH)_2$
27. Copper, Cu

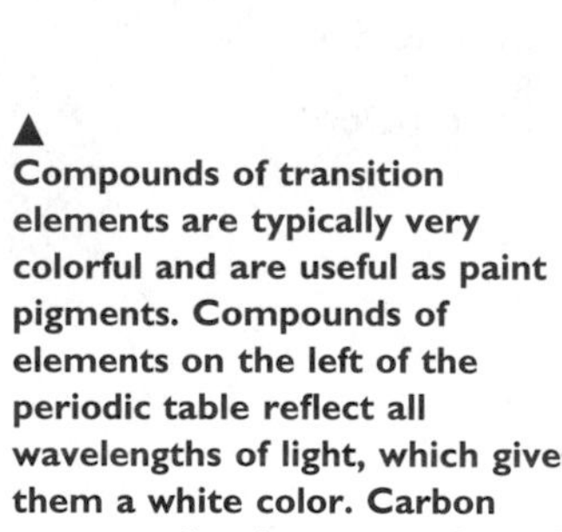

▲ **Compounds of transition elements are typically very colorful and are useful as paint pigments. Compounds of elements on the left of the periodic table reflect all wavelengths of light, which gives them a white color. Carbon compounds take on a variety of colors.**

$$(Mg^{2+}) + (O^{2-}) = 0$$

$$(2+) \quad + (2-) \quad = 0$$

Therefore, the formula is MgO.

(c) Use the same procedure for barium phosphide.

Step 1 From the name we know that barium phosphide is composed of barium and phosphide ions. First write down the formulas of these ions:

$$Ba^{2+} \qquad P^{3-}$$

Step 2 To write the formula of the compound, combine the smallest numbers of Ba^{2+} and P^{3-} ions to give the charge sum equal to zero. In this case the lowest common multiple of the charges is 6:

$$3(Ba^{2+}) + 2(P^{3-}) = 0$$

$$3(2+) \quad + 2(3-) \quad = 0$$

Therefore, the formula is Ba_3P_2.

Practice 6.1

Write formulas for compounds containing the following ions:

(a) K^+ and F^-
(b) Ca^{2+} and Br^-
(c) Mg^{2+} and N^{3-}
(d) Na^+ and S^{2-}
(e) Ba^{2+} and O^{2-}

6.4 Binary Compounds

Binary compounds contain only two different elements. Many binary compounds are formed when a metal combines with a nonmetal to form a *binary ionic compound.* The metal loses one or more electrons to become a cation while the nonmetal gains one or more electrons to become an anion. The cation is written first in the formula and is followed by the anion.

A. Binary Ionic Compounds Containing a Metal Forming Only One Type of Cation

The chemical name is composed of the name of the metal followed by the name of the nonmetal, which has been modified to an identifying stem plus the suffix *-ide.*

For example, sodium chloride, NaCl, is composed of one atom each of sodium and chlorine. The name of the metal, sodium, is written first and is not modified. The second part of the name is derived from the nonmetal, chlorine, by using the stem *chlor-* and adding the ending *-ide;* it is named *chloride.* The compound name is *sodium chloride.*

NaCl

Elements:	Sodium (metal)
	Chlorine (nonmetal)
	name modified to the stem *chlor-* + *-ide*
Name of compound:	Sodium chloride

To name these compounds:
1. **Write the name of the cation.**
2. **Write the stem for the anion with the suffix *-ide*.**

Stems of the more common negative-ion-forming elements are shown in Table 6.2. Table 6.3 shows some compounds with names ending in *-ide.*

Compounds may contain more than one atom of the same element, but as long as they contain only two different elements and only one compound of these two elements exists, the name follows the rules for binary compounds:

Examples:

$CaBr_2$	Mg_3N_2	Li_2O
Calcium bromide	Magnesium nitride	Lithium oxide

TABLE 6.2 Examples of Elements Forming Anions

Symbol	Element	Stem	Anion name
Br	Bromine	Brom	Bromide
Cl	Chlorine	Chlor	Chloride
F	Fluorine	Fluor	Fluoride
H	Hydrogen	Hydr	Hydride
I	Iodine	Iod	Iodide
N	Nitrogen	Nitr	Nitride
O	Oxygen	Ox	Oxide
P	Phosphorus	Phosph	Phosphide
S	Sulfur	Sulf	Sulfide

Example 6.6 Name the compound CaS.

Solution

Step 1 From the formula it is a two-element compound and follows the rules for binary compounds.

Step 2 The compound is composed of Ca, a metal, and S, a nonmetal. Elements in the IIA column form only one type of cation. Thus, we name the positive part of the compound *calcium.*

Step 3 Modify the name of the second element to the identifying stem *sulf-* and add the binary ending *-ide* to form the name of the negative part, *sulfide.*

Step 4 The name of the compound, therefore, is *calcium sulfide.*

Practice 6.2

Write formulas for the following compounds:
(a) strontium chloride, (b) potassium iodide, (c) aluminum nitride, (d) calcium sulfide, (e) sodium oxide

B. Binary Ionic Compounds Containing a Metal That Can Form Two or More Types of Cations

The metals in the center of the periodic table (including the transition metals) often form more than one type of cation. For example, iron can form Fe^{2+} and Fe^{3+} ions, and copper can form Cu^{+} and Cu^{2+} ions. This can lead to confusion when naming compounds. For example, copper chloride could be $CuCl_2$ or CuCl. To resolve this difficulty the IUPAC devised a system to name these compounds, which is known as the **Stock System.** This system is currently recognized as the official system to name these compounds although another older system is sometimes used. In the Stock System, when a compound contains a metal that can form more than one type of cation, the charge on the cation of the metal is designated by a Roman numeral placed in parentheses immediately following the name of the metal. The negative element is treated in the usual manner for binary compounds.

Stock System

TABLE 6.3 Examples of Compounds with Names Ending in *-ide*

Formula	Name	Formula	Name
$AlCl_3$	Aluminum chloride	BaS	Barium sulfide
Al_2O_3	Aluminum oxide	LiI	Lithium iodide
CaC_2	Calcium carbide	$MgBr_2$	Magnesium bromide
HCl	Hydrogen chloride	NaH	Sodium hydride
HI	Hydrogen iodide	Na_2O	Sodium oxide

To name these compounds:
1. Write the name of the cation.
2. Write the charge on the cation as a Roman numeral in parentheses.
3. Write the stem of the anion with the suffix *-ide*.

Cation charge	+1	+2	+3
Roman numeral	(I)	(II)	(III)

Examples: $FeCl_2$ Iron(II) chloride Fe^{2+}
$FeCl_3$ Iron(III) chloride Fe^{3+}
CuCl Copper(I) chloride Cu^{+}
$CuCl_2$ Copper(II) chloride Cu^{2+}

The fact that $FeCl_2$ has two chloride ions, each with a -1 charge, establishes that the charge of Fe is $+2$. To distinguish between the two iron chlorides, $FeCl_2$ is named iron(II) chloride and $FeCl_3$ is named iron(III) chloride.

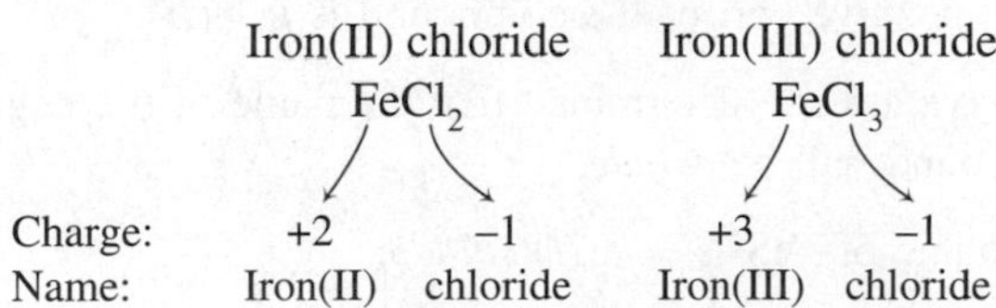

When a metal forms only one possible cation, we need not distinguish one cation from another, so Roman numerals are not needed. Thus we do not say calcium(II) chloride for $CaCl_2$, but rather calcium chloride, since the charge of calcium is understood to be $+2$.

In classical nomenclature, when the metallic ion has only two cation types, the name of the metal (usually the Latin name) is modified with the suffixes *-ous* and *-ic* to distinguish between the two. The lower charge cation is given the *-ous* ending, and the higher one, the *-ic* ending.

Examples: $FeCl_2$ Ferrous chloride Fe^{2+} (lower charge cation)
$FeCl_3$ Ferric chloride Fe^{3+} (higher charge cation)
CuCl Cuprous chloride Cu^{+} (lower charge cation)
$CuCl_2$ Cupric chloride Cu^{2+} (higher charge cation)

Table 6.4 lists some common metals that have more than one type of cation.

Notice that the *ous–ic* naming system does not give the charge of the cation of an element but merely indicates that at least two types of cations exist. The Stock System avoids any possible uncertainty by clearly stating the charge on the cation.

In this book we will use only the Stock System.

Example 6.7

Name the compound FeS.

Solution

Step 1 This compound follows the rules for a binary compound and, like CaS, must be a sulfide.

TABLE 6.4 Names and Charges of Some Common Metal Ions That Have More Than One Type of Cation

Formula	Stock System name	Classical name
Cu^{1+}	Copper(I)	Cuprous
Cu^{2+}	Copper(II)	Cupric
Hg^{1+} $(Hg_2)^{2+}$	Mercury(I)	Mercurous
Hg^{2+}	Mercury(II)	Mercuric
Fe^{2+}	Iron(II)	Ferrous
Fe^{3+}	Iron(III)	Ferric
Sn^{2+}	Tin(II)	Stannous
Sn^{4+}	Tin(IV)	Stannic
Pb^{2+}	Lead(II)	Plumbous
Pb^{4+}	Lead(IV)	Plumbic
As^{3+}	Arsenic(III)	Arsenous
As^{5+}	Arsenic(V)	Arsenic
Ti^{3+}	Titanium(III)	Titanous
Ti^{4+}	Titanium(IV)	Titanic

Step 2 It is a compound of Fe, a metal, and S, a nonmetal, and Fe is a transition metal that has more than one type of cation. In sulfides, the charge on the S is -2. Therefore, the charge on Fe must be $+2$, and the name of the positive part of the compound is *iron(II)*.

Step 3 We have already determined that the name of the negative part of the compound will be *sulfide*.

Step 4 The name of FeS is *iron(II) sulfide*.

Practice 6.3

Write the name for each of the following compounds using the official Stock System:
(a) PbI_2, (b) SnF_4, (c) Fe_2O_3, (d) CuO

C. Binary Compounds Containing Two Nonmetals

In a compound formed between two nonmetals, the element that occurs first in the series is written and named first.

Si, B, P, H, C, S, I, Br, N, Cl, O, F

The name of the second element retains the *-ide* ending as though it were an anion. A Latin or Greek prefix (*mono-*, *di-*, *tri-*, and so on) is attached to the name of each element to indicate the number of atoms of that element in the molecule. The prefix *mono-* is never used for naming the first element. Some common prefixes and their numerical equivalences follow.

mono = 1	*tetra* = 4	*hepta* = 7	*nona* = 9
di = 2	*penta* = 5	*octa* = 8	*deca* = 10
tri = 3	*hexa* = 6		

Here are some examples of compounds that illustrate this system:

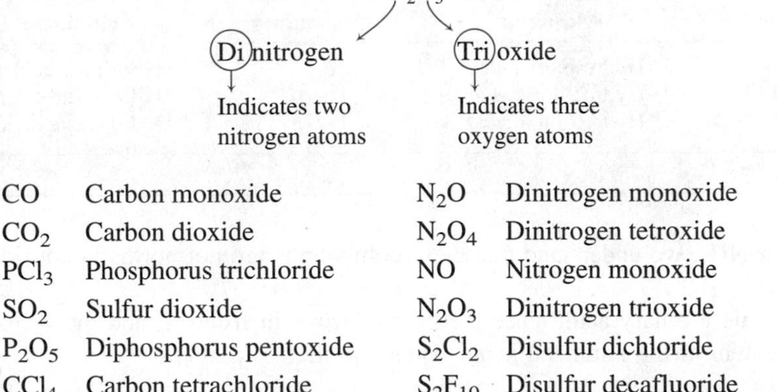

CO	Carbon monoxide	N_2O	Dinitrogen monoxide
CO_2	Carbon dioxide	N_2O_4	Dinitrogen tetroxide
PCl_3	Phosphorus trichloride	NO	Nitrogen monoxide
SO_2	Sulfur dioxide	N_2O_3	Dinitrogen trioxide
P_2O_5	Diphosphorus pentoxide	S_2Cl_2	Disulfur dichloride
CCl_4	Carbon tetrachloride	S_2F_{10}	Disulfur decafluoride

To name these compounds:
1. Write the name for the first element using a prefix if there is more than one atom of this element.
2. Write the stem for the second element with the suffix *-ide*. Use a prefix to indicate the number of atoms for the second element.

The example above illustrates that we sometimes drop the final *o*(mono) or *a*(penta) of the prefix when the second element is oxygen. This avoids creating a name that is awkward to pronounce. For example, CO is carbon monoxide instead of carbon mon*oo*xide.

Example 6.8

Name the compound PCl_5.

Solution

Step 1 Phosphorus and chlorine are nonmetals, so the rules for naming binary compounds containing two nonmetals apply. Phosphorus is named first. Therefore, the compound is a chloride.

Step 2 No prefix is needed for phosphorus because each molecule has only one atom of phosphorus. The prefix *penta-* is used with chloride to indicate the five chlorine atoms. (PCl_3 is also a known compound.)

Step 3 The name for PCl_5 is *phosphorus pentachloride.*

Practice 6.4

Name the following compounds:
(a) Cl_2O, (b) SO_2, (c) CBr_4, (d) N_2O_5, (e) NH_3

D. Acids Derived from Binary Compounds

Certain binary hydrogen compounds, when dissolved in water, form solutions that have *acid* properties. Because of this property, these compounds are given acid names in addition to their regular *-ide* names. For example, HCl is a gas and is called *hydrogen chloride,* but its water solution is known as *hydrochloric acid.* Binary acids are composed of hydrogen and one other nonmetallic element. However, not all binary hydrogen compounds are acids. To express the formula of a binary acid it is customary to write the symbol of hydrogen first, followed by the symbol of the second element (for example, HCl, HBr, H_2S). When we see formulas such

TABLE 6.5 Names and Formulas of Selected Binary Acids

Formula	Acid name	Formula	Acid name
HF	Hydrofluoric acid	HI	Hydroiodic acid
HCl	Hydrochloric acid	H_2S	Hydrosulfuric acid
HBr	Hydrobromic acid	H_2Se	Hydroselenic acid

as CH_4 or NH_3, we understand that these compounds are not normally considered to be acids.

To name a binary acid, place the prefix *hydro-* in front of, and the suffix *-ic* after, the stem of the nonmetal name. Then add the word *acid.*

To name these compounds:
1. **Write the prefix *hydro-* with the stem of the second element and add the suffix *-ic*.**
2. **Write the word *acid*.**

	HCl	H_2S
Examples:	*Hydro*-chlor-*ic acid* (hydrochloric acid)	*Hydro*-sulfur-*ic acid* (hydrosulfuric acid)

Acids are hydrogen-containing substances that liberate hydrogen ions when dissolved in water. The same formula is often used to express binary hydrogen compounds, such as HCl, regardless of whether or not they are dissolved in water. Table 6.5 shows several examples of binary acids.

A summary of the approach to naming binary compounds is shown in Figure 6.4.

Practice 6.5

Name each of the following binary compounds:
(a) KBr, (b) Ca_3N_2, (c) SO_3, (d) SnF_2, (e) $CuCl_2$, (f) N_2O_4

6.5 Naming Compounds Containing Polyatomic Ions

polyatomic ion

A **polyatomic ion** is an ion that contains two or more elements. Compounds containing polyatomic ions are composed of three or more elements and usually consist of one or more cations combined with a negative polyatomic ion. In general, naming compounds containing polyatomic ions is similar to naming binary compounds. The cation is named first, followed by the name for the negative polyatomic ion.

To name these compounds:
1. **Write the name of the cation.**
2. **Write the name of the anion.**

In order to name these compounds you must learn to recognize the common polyatomic ions (Table 6.6) and know their charges. Consider the formula $KMnO_4$. You must be able to recognize that it consists of two parts K | MnO_4. These parts are composed of a K^+ ion and a MnO_4^- ions. The correct name for this compound is potassium permanganate. Many polyatomic ions that contain oxygen are called *oxy-anions,* and generally have the suffix *-ate* or *-ite*. Unfortunately, the suffix does not indicate the number of oxygen atoms present. The *-ate* form contains more oxygen atoms than the *-ite* form. Examples include sulfate (SO_4^{2-}), sulfite (SO_3^{2-}), nitrate (NO_3^-), and nitrite (NO_2^-).

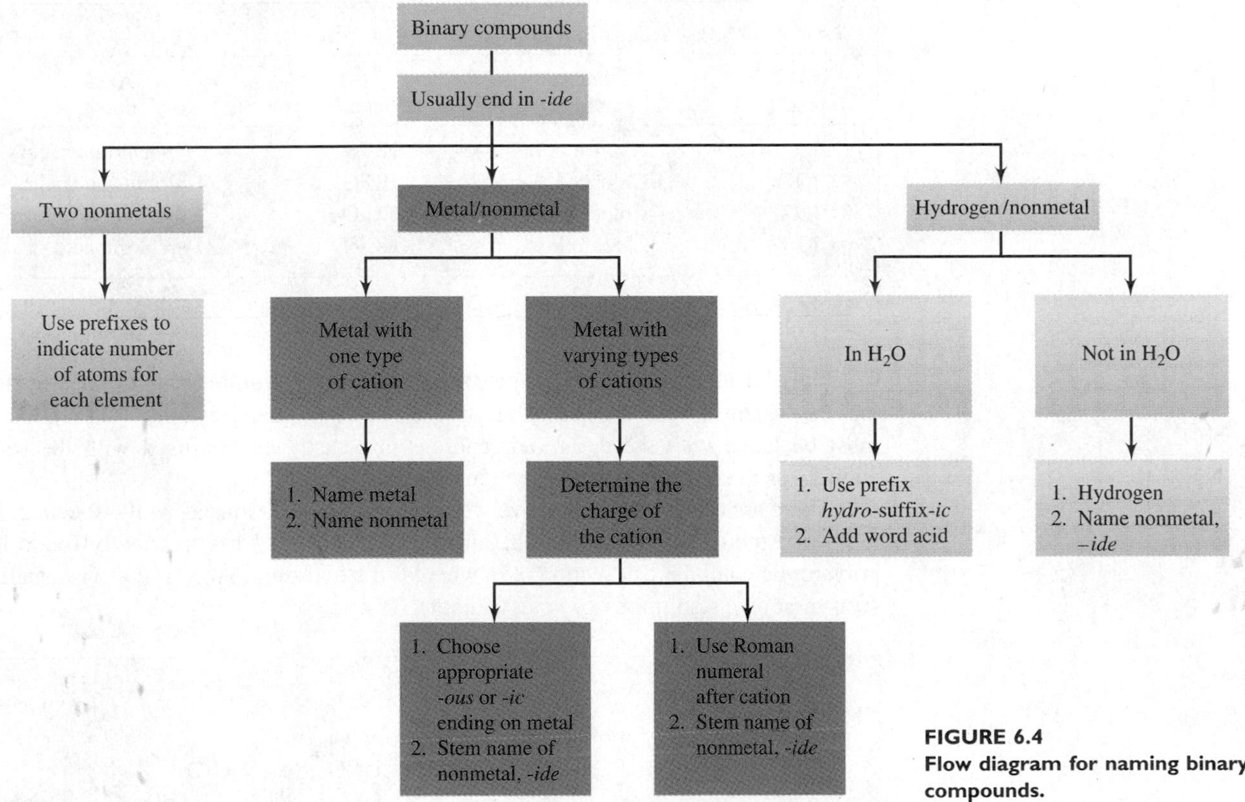

FIGURE 6.4
Flow diagram for naming binary compounds.

TABLE 6.6 Names, Formulas, and Charges of Some Common Polyatomic Ions

Name	Formula	Charge	Name	Formula	Charge
Acetate	$C_2H_3O_2^-$	−1	Cyanide	CN^-	−1
Ammonium	NH_4^+	+1	Dichromate	$Cr_2O_7^{2-}$	−2
Arsenate	AsO_4^{3-}	−3	Hydroxide	OH^-	−1
Hydrogen carbonate	HCO_3^-	−1	Nitrate	NO_3^-	−1
Hydrogen sulfate	HSO_4^-	−1	Nitrite	NO_2^-	−1
Bromate	BrO_3^-	−1	Permanganate	MnO_4^-	−1
Carbonate	CO_3^{2-}	−2	Phosphate	PO_4^{3-}	−3
Chlorate	ClO_3^-	−1	Sulfate	SO_4^{2-}	−2
Chromate	CrO_4^{2-}	−2	Sulfite	SO_3^{2-}	−2

Some elements form more than two different polyatomic ions containing oxygen. To name these ions, prefixes are used in addition to the suffix. To indicate more oxygen than in the *-ate* form we add the prefix *per-*, which is a short form of *hyper-*, meaning more. The prefix *hypo-*, meaning less (oxygen in this case), is used for the ion containing less oxygen than the *-ite* form. An example of this system is shown for the polyatomic ions containing chlorine and oxygen in Table 6.7. The prefixes are also used with other similar ions, such as iodate (IO_3^-), bromate (BrO_3^-), and phosphate (PO_4^{3-}).

TABLE 6.7 Oxy-Anions and Oxy-Acids of Chlorine

Anion formula	Anion name	Acid formula	Acid name
ClO^-	*Hypo*chlor*ite*	$HClO$	*Hypo*chlor*ous* acid
ClO_2^-	Chlor*ite*	$HClO_2$	Chlor*ous* acid
ClO_3^-	Chlor*ate*	$HClO_3$	Chlor*ic* acid
ClO_4^-	*Per*chlor*ate*	$HClO_4$	*Per*chlor*ic* acid

Only two of the common negatively charged polyatomic ions do not use the *ate–ite* system. These exceptions are hydroxide (OH^-) and cyanide (CN^-). Care must be taken with these, as their endings can easily be confused with the *-ide* ending for binary compounds (Section 6.4).

There are two common positively charged polyatomic ions as well—the ammonium and the hydronium ions. The ammonium ion (NH_4^+) is frequently found in polyatomic compounds (Section 6.5), whereas the hydronium ion (H_3O^+) is usually seen in aqueous solutions of acids (Chapter 15).

Practice 6.6

Name each of the following compounds:
(a) $NaNO_3$, (b) $Ca_3(PO_4)_2$, (c) KOH, (d) Li_2CO_3, (e) $NaClO_3$

Inorganic compounds are also formed from more than three elements (see Table 6.8). In these cases one or more of the ions is often a polyatomic ion. Once you have learned to recognize the polyatomic ions, naming these compounds follows the patterns we have already learned. First identify the ions. Name the cations in the order given, and follow them with the names of the anions. Study the following examples:

Compound	Ions	Name
$NaHCO_3$	Na^+; HCO_3^-	Sodium hydrogen carbonate
NaHS	Na^+; HS^-	Sodium hydrogen sulfide
$MgNH_4PO_4$	Mg^{2+}; NH_4^+; PO_4^{3-}	Magnesium ammonium phosphate
$NaKSO_4$	Na^+; K^+; SO_4^{2-}	Sodium potassium sulfate

6.6 Acids

While we will learn much more about acids later (see Chapter 15), it is helpful to be able to recognize and name common acids both in the laboratory and in class. The simplest way to recognize many acids is to know that acid formulas often begin with hydrogen. Naming binary acids was covered in Section 6.4D. Inorganic

TABLE 6.8 Names of Selected Compounds That Contain More Than One Kind of Positive Ion

Formula	Name of Compound
$KHSO_4$	Potassium hydrogen sulfate
$Ca(HSO_3)_2$	Calcium hydrogen sulfite
NH_4HS	Ammonium hydrogen sulfide
$MgNH_4PO_4$	Magnesium ammonium phosphate
NaH_2PO_4	Sodium dihydrogen phosphate
Na_2HPO_4	Disodium hydrogen phosphate
KHC_2O_4	Potassium hydrogen oxalate
$KAl(SO_4)_2$	Potassium aluminum sulfate
$Al(HCO_3)_3$	Aluminum hydrogen carbonate

TABLE 6.9 Comparison of Acid and Anion Names for Selected Oxy-Acids

Acid	Anion	Acid	Anion
H_2SO_4 Sulfuric acid	SO_4^{2-} Sulfate ion	H_3PO_4 Phosphoric acid	PO_4^{3-} Phosphate ion
H_2SO_3 Sulfurous acid	SO_3^{2-} Sulfite ion	H_3PO_3 Phosphorous acid	PO_3^{3-} Phosphite ion
HNO_3 Nitric acid	NO_3^- Nitrate ion	HIO_3 Iodic acid	IO_3^- Iodate ion
HNO_2 Nitrous acid	NO_2^- Nitrite ion	$HC_2H_3O_2$ Acetic acid	$C_2H_3O_2^-$ Acetate ion
H_2CO_3 Carbonic acid	CO_3^{2-} Carbonate ion	$H_2C_2O_4$ Oxalic acid	$C_2O_4^{2-}$ Oxalate ion

compounds containing hydrogen, oxygen, and one other element are called *oxy-acids.* The element other than hydrogen or oxygen in these acids is often a nonmetal, but it can also be a metal.

The first step in naming these acids is to determine that the compound in question is really an oxy-acid. The keys to identification are that (1) hydrogen is the first element in the compound's formula; and (2) the second portion of the formula consists of a polyatomic ion containing oxygen.

Hydrogen in an oxy-acid is not specifically designated in the acid name. The presence of hydrogen in the compound is indicated by the use of the word *acid* in the name of the substance. To determine the particular type of acid, the polyatomic ion following hydrogen must be examined. The name of the polyatomic ion is modified in the following manner: (1) *-ate* changes to an *-ic* ending; (2) *-ite* changes to an *-ous* ending. (See Table 6.9.) The compound with the *-ic* ending contains more oxygen than the one with the *-ous* ending. Consider the following examples:

H_2SO_4 sulf*ate* ⟶ sulfur*ic* acid (contains 4 oxygens)

H_2SO_3 sulf*ite* ⟶ sulfur*ous* acid (3 oxygens)

HNO_3 nitr*ate* ⟶ nitr*ic* acid (3 oxygens)

HNO_2 nitr*ite* ⟶ nitr*ous* acid (2 oxygens)

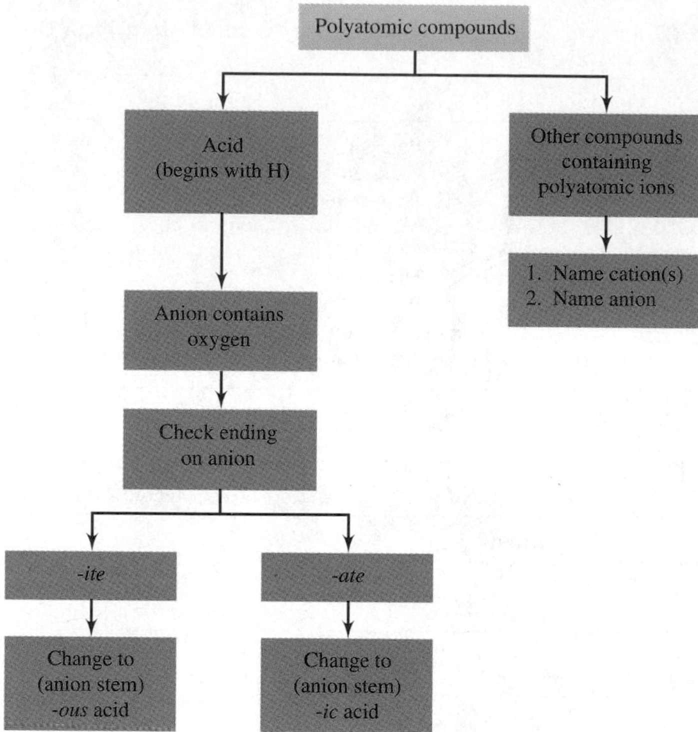

FIGURE 6.5 ▶
Flow diagram for naming polyatomic compounds.

The complete system for naming oxy-acids is shown in Table 6.7 for the various acids containing chlorine. Examples of other oxy-acids and their names are shown in Table 6.10.

A summary of the approach to naming polyatomic compounds is shown in Figure 6.5. We have now looked at ways of naming a variety of inorganic compounds—binary compounds consisting of a metal and a nonmetal and of two nonmetals, and binary acids. These compounds are just a small part of the classified chemical compounds. Most of the remaining classes are in the broad field of organic chemistry under such categories as hydrocarbons, alcohols, ethers, aldehydes, ketones, phenols, and carboxylic acids.

TABLE 6.10 Formulas and Names of Selected Oxy-Acids

Formula	Acid name	Formula	Acid name
H_2SO_3	Sulfurous acid	$HC_2H_3O_2$	Acetic acid
H_2SO_4	Sulfuric acid	$H_2C_2O_4$	Oxalic acid
HNO_2	Nitrous acid	H_2CO_3	Carbonic acid
HNO_3	Nitric acid	$HBrO_3$	Bromic acid
H_3PO_3	Phosphorous acid	HIO_3	Iodic acid
H_3PO_4	Phosphoric acid	H_3BO_3	Boric acid

Charges in Your Life

Ions are used in living organisms to perform many important functions. For example, electrical neutrality must be maintained both inside and outside the body cells. Within the cell, neutrality is maintained by potassium ions (K^+) and hydrogen phosphate ions (HPO_4^{2-}). Outside the cell in the intercellular fluid the ions responsible for neutrality are sodium (Na^+) and chloride (Cl^-).

Another important ion within organisms is magnesium (Mg^{2+}), found in chlorophyll and used during nerve and muscle activity, as well as in conjunction with certain enzymes. Iron ions (Fe^{2+}) are incorporated within the hemoglobin molecule and are an integral part of the oxygen transport system within the body. Calcium ions (Ca^{2+}) are part of the matrix of both bones and teeth and play a significant role in the clotting of blood.

Ions also have a significant role in the detergent industry. Water is said to be "hard" when it contains relatively high concentrations of Ca^{2+} and Mg^{2+} ions. In solution, soaps combine with these ions to form an insoluble scum. This material forms the common bathtub ring and, in laundry, settles on the clothes leaving them gray and dingy. Although soaps and detergents are similar in their cleansing actions, detergents are less likely to form this scum. For this reason, many people who live in areas with "hard" water use detergents instead of soaps.

Ions form an insoluble scum when combined with some products.

Concepts in Review

1. Write the formulas of compounds formed by combining the ions from Figure 6.2 (or from the inside back cover of this book) in the correct ratios.
2. Write the names or formulas for inorganic binary compounds in which the metal has only one type of cation.
3. Write the names or formulas for inorganic binary compounds that contain metals with multiple types of cations, using the Stock System.
4. Write the names or formulas for inorganic binary compounds that contain two nonmetals.
5. Write the names or formulas for binary acids.
6. Write the names or formulas for oxy-acids.
7. Write the names or formulas for compounds that contain polyatomic ions.

Key Terms

The terms listed here have been defined within this chapter. Section numbers are referenced in parenthesis for each term.

polyatomic ion (6.5) Stock System (6.4)

Questions

Questions refer to tables, figures, and key words and concepts defined within the chapter. A particularly challenging question or exercise is indicated with an asterisk.

1. Use the common ion table on the inside back cover of your text to determine the formulas for compounds composed of the following ions:
(a) Sodium and chlorate
(b) Hydrogen and sulfate
(c) Tin(II) and acetate
(d) Copper(I) and oxide
(e) Zinc and hydrogen carbonate
(f) Iron(III) and carbonate

2. Does the fact that two elements combine in a one-to-one atomic ratio mean that the charges on their ions are both 1? Explain.

3. Explain how "hard water" may cause white clothing to become dingy and gray.

4. Why did the IUPAC vote not to use the name seaborgium for element 106?

5. What elements did Glenn Seaborg help to discover?

6. Which of these statements are correct? Rewrite each incorrect statement to make it correct.
(a) The formula for calcium hydride is CaH_2.
(b) The ions of all the following metals have a charge of +2: Ca, Ba, Sr, Cd, Zn.
(c) The formulas for nitrous and sulfurous acids are HNO_2 and H_2SO_3.
(d) The formula for the compound between Fe^{3+} and O^{2-} is Fe_3O_2.
(e) The name for $NaNO_2$ is sodium nitrite.
(f) The name for $Ca(ClO_3)_2$ is calcium chlorate.
(g) The name for CuO is copper(I) oxide.
(h) The name for SO_4^{2-} is sulfate ion.
(i) The name for N_2O_4 is dinitrogen tetroxide.
(j) The name for Na_2O is disodium oxide.
(k) If the name of an anion ends with *-ide,* the name of the corresponding acid will start with *hydro-*.
(l) If the name of an anion ends with *-ite,* the corresponding acid name will end with *-ic.*
(m) If the name of an acid ends with *-ous,* the corresponding polyatomic ion will end with *-ate.*
(n) In FeI_2, the iron is iron(II) because it is combined with two I^- ions.
(o) In Cu_2SO_4, the copper is copper(II) because there are two copper ions.
(p) When two nonmetals combine, prefixes of *di-, tri-, tetra-,* and so on, are used to specify how many atoms of each element are in a molecule.
(q) N_2O_3 is called dinitrogen trioxide.
(r) $Sn(CrO_4)_2$ is called tin dichromate.

Paired Exercises

These exercises are paired. Each odd-numbered exercise is followed by a similar even-numbered exercise. Answers to the even-numbered exercises are given in Appendix V.

7. Write the formula of the compound that would be formed between the given elements:
(a) Na and I
(b) Ba and F
(c) Al and O
(d) K and S
(e) Cs and Cl
(f) Sr and Br

8. Write the formula of the compound that would be formed between the given elements:
(a) Ba and O
(b) H and S
(c) Al and Cl
(d) Be and Br
(e) Li and Se
(f) Mg and P

9. Write formulas for the following cations (do not forget to include the charges): sodium, magnesium, aluminum, copper(II), iron(II), iron (III), lead(II), silver, cobalt(II), barium, hydrogen, mercury(II), tin(II), chromium(III), tin(IV), manganese(II), bismuth(III).

10. Write formulas for the following anions (do not forget to include the charges): chloride, bromide, fluoride, iodide, cyanide, oxide, hydroxide, sulfide, sulfate, hydrogen sulfate, hydrogen sulfite, chromate, carbonate, hydrogen carbonate, acetate, chlorate, permanganate, oxalate.

11. Complete the table, filling in each box with the proper formula.

Cations \ Anions	Br^-	O^{2-}	NO_3^-	PO_4^{3-}	CO_3^{2-}
K^+	KBr				
Mg^{2+}					
Al^{3+}					
Zn^{2+}				$Zn_3(PO_4)_2$	
H^+					

12. Complete the table, filling in each box with the proper formula.

Cations \ Anions	SO_4^{2-}	Cl^-	AsO_4^{3-}	$C_2H_3O_2^-$	CrO_4^{2-}
NH_4^+			$(NH_4)_3AsO_4$		
Ca^{2+}					
Fe^{3+}	$Fe_2(SO_4)_3$				
Ag^+					
Cu^{2+}					

13. Write formulas for the following binary compounds, all of which are composed of nonmetals:
(a) Carbon monoxide
(b) Sulfur trioxide
(c) Carbon tetrabromide
(d) Phosphorus trichloride
(e) Nitrogen dioxide
(f) Dinitrogen pentoxide
(g) Iodine monobromide
(h) Silicon tetrachloride
(i) Phosphorus pentaiodide
(j) Diboron trioxide

14. Name the following binary compounds, all of which are composed of nonmetals:
(a) CO_2
(b) N_2O
(c) PCl_5
(d) CCl_4
(e) SO_2
(f) N_2O_4
(g) P_2O_5
(h) OF_2
(i) NF_3
(j) CS_2

15. Write formulas for the following compounds:
(a) Sodium nitrate
(b) Magnesium fluoride
(c) Barium hydroxide
(d) Ammonium sulfate
(e) Silver carbonate
(f) Calcium phosphate
(g) Potassium nitrite
(h) Strontium oxide

16. Name the following compounds:
(a) K_2O
(b) NH_4Br
(c) CaI_2
(d) $BaCO_3$
(e) Na_3PO_4
(f) Al_2O_3
(g) $Zn(NO_3)_2$
(h) Ag_2SO_4

17. Name each of the following compounds by the Stock (IUPAC) System:
(a) $CuCl_2$
(b) $CuBr$
(c) $Fe(NO_3)_2$
(d) $FeCl_3$
(e) SnF_2
(f) $HgCO_3$

18. Write formulas for the following compounds:
(a) Tin(IV) bromide
(b) Copper(I) sulfate
(c) Iron(III) carbonate
(d) Mercury(II) nitrite
(e) Titanium(IV) sulfide
(f) Iron(II) acetate

19. Write formulas for the following acids:
(a) Hydrochloric acid
(b) Chloric acid
(c) Nitric acid
(d) Carbonic acid
(e) Sulfurous acid
(f) Phosphoric acid

20. Write formulas for the following acids:
(a) Acetic acid
(b) Hydrofluoric acid
(c) Hypochlorous acid
(d) Boric acid
(e) Nitrous acid
(f) Hydrosulfuric acid

21. Name the following acids:
(a) HNO_2
(b) H_2SO_4
(c) $H_2C_2O_4$
(d) HBr
(e) H_3PO_3
(f) $HC_2H_3O_2$
(g) HF
(h) $HBrO_3$

22. Name the following acids:
(a) H_3PO_4
(b) H_2CO_3
(c) HIO_3
(d) HCl
(e) HClO
(f) HNO_3
(g) HI
(h) $HClO_4$

23. Write formulas for the following compounds:
(a) Silver sulfite
(b) Cobalt(II) bromide
(c) Tin(II) hydroxide
(d) Aluminum sulfate
(e) Manganese(II) fluoride
(f) Ammonium carbonate
(g) Chromium(III) oxide
(h) Copper(II) chloride
(i) Potassium permanganate
(j) Barium nitrite
(k) Sodium peroxide
(l) Iron(II) sulfate
(m) Potassium dichromate
(n) Bismuth(III) chromate

24. Write formulas for the following compounds:
(a) Sodium chromate
(b) Magnesium hydride
(c) Nickel(II) acetate
(d) Calcium chlorate
(e) Lead(II) nitrate
(f) Potassium dihydrogen phosphate
(g) Manganese(II) hydroxide
(h) Cobalt(II) hydrogen carbonate
(i) Sodium hypochlorite
(j) Arsenic(V) carbonate
(k) Chromium(III) sulfite
(l) Antimony(III) sulfate
(m) Sodium oxalate
(n) Potassium thiocyanate

25. Write the name of each compound.
(a) $ZnSO_4$
(b) $HgCl_2$
(c) $CuCO_3$
(d) $Cd(NO_3)_2$
(e) $Al(C_2H_3O_2)_3$
(f) CoF_2
(g) $Cr(ClO_3)_3$
(h) Ag_3PO_4
(i) NiS
(j) $BaCrO_4$

26. Write the name of each compound.
(a) $Ca(HSO_4)_2$
(b) $As_2(SO_3)_3$
(c) $Sn(NO_2)_2$
(d) $FeBr_3$
(e) $KHCO_3$
(f) $BiAsO_4$
(g) $Fe(BrO_3)_2$
(h) $(NH_4)_2HPO_4$
(i) NaClO
(j) $KMnO_4$

27. Write the chemical formula for each of the following substances:
(a) Baking soda
(b) Lime
(c) Epsom salts
(d) Muriatic acid
(e) Vinegar
(f) Potash
(g) Lye

28. Write the chemical formula for each of the following substances:
(a) Fool's gold
(b) Saltpeter
(c) Limestone
(d) Cane sugar
(e) Milk of magnesia
(f) Washing soda
(g) Grain alcohol

Additional Exercises

These exercises are not paired or labeled by topic and provide additional practice on the concepts covered in this chapter.

29. Name the following compounds:

(a) $Ba(NO_3)_2$	**(d)** $MgSO_4$	**(g)** NiS
(b) $NaC_2H_3O_2$	**(e)** $CdCrO_4$	**(h)** $Sn(NO_3)_4$
(c) PbI_2	**(f)** $BiCl_3$	**(i)** $Ca(OH)_2$

30. State how each of the following is used in naming inorganic compounds: *ide, ous, ic, hypo, per, ite, ate,* Roman numerals.

31. Translate each of the following formula sentences into unbalanced chemical equations:
(a) Silver nitrate and sodium chloride react to form silver chloride and sodium nitrate.
(b) Iron(III) sulfate and calcium hydroxide react to form iron(III) hydroxide and calcium sulfate.
(c) Potassium hydroxide and sulfuric acid react to form potassium sulfate and water.

32. How many of each type of subatomic particle (protons and electrons) is in:
(a) an atom of tin
(b) an Sn^{2+} ion
(c) an Sn^{4+} ion

33. The compound X_2Y_3 is a stable solid. What ionic charge do you expect for *X* and *Y*? Explain.

***34.** The ferricyanide ion has the formula $Fe(CN)_6^{3-}$. Write the formula for the compounds that ferricyanide would form with the cations of elements 3, 13, and 30.

35. Compare and contrast the formulas of:
(a) nitride with nitrite
(b) nitrite with nitrate
(c) nitrous acid with nitric acid

36. In the beaker below there is a pool of ions. Write all possible formulas for ionic compounds that could form using these ions. Write the name of each compound next to its formula.

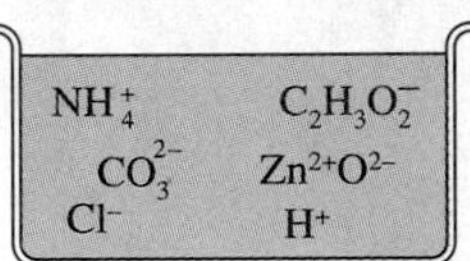

Answers to Practice Exercises

6.1 (a) KF, (b) $CaBr_2$, (c) Mg_3N_2, (d) Na_2S, (e) BaO
6.2 (a) $SrCl_2$, (b) KI, (c) AlN, (d) CaS, (e) Na_2O
6.3 (a) lead(II) iodide, (b) tin(IV) fluoride, (c) iron(III) oxide, (d) copper(II) oxide
6.4 (a) dichlorine monoxide, (b) sulfur dioxide, (c) carbon tetrabromide, (d) dinitrogen pentoxide, (e) ammonia
6.5 (a) potassium bromide, (b) calcium nitride, (c) sulfur trioxide, (d) tin(II) fluoride, (e) copper(II) chloride, (f) dinitrogen tetroxide
6.6 (a) sodium nitrate, (b) calcium phosphate, (c) potassium hydroxide, (d) lithium carbonate, (e) sodium chlorate

7 Quantitative Composition of Compounds

Chapter 7

Knowing the substances contained in a product is not sufficient information to produce the product. An artist can create an incredible array of colors from a limited number of pigments. A pharmacist can combine the same drugs in various amounts to produce different effects in patients. Cosmetics, cereals, cleaning products, and pain remedies all show a list of ingredients on the labels of the packages.

In each of these products the key to successful production lies in the quantity of each ingredient. The pharmaceutical industry maintains strict regulations on the amounts of the ingredients in the medicines we purchase. The formulas for soft drinks and most cosmetics are trade secrets. Small deviations in the composition of these products can result in large losses or lawsuits for these organizations.

The composition of compounds is an important concept in chemistry. The numerical relationships among elements within compounds and the measurement of exact quantities of particles are fundamental tasks for the chemist.

7.1 The Mole

In the laboratory we normally determine the mass of substances on a balance. When we run a chemical reaction, the reaction occurs between atoms and molecules. For example, in the reaction between magnesium and sulfur, one atom of sulfur reacts with one atom of magnesium:

$$\text{Mg} + \text{S} \longrightarrow \text{MgS}$$

However, when we measure the masses of these two elements, we find that 24.31 g of Mg are required to react with 32.06 g of S. Because magnesium and sulfur react in a 1:1 atom ratio, we can conclude from this experiment that 24.31 g of Mg contain the same number of atoms as 32.06 g of S. How many atoms are in 24.31 g of Mg or 32.06 g of S? These two amounts each contain 1 *mole* of atoms.

The mole (abbreviated mol) is one of the seven base units in the International System and is the unit for an amount of substance. The mole is a counting unit as in other things that we count, such as a dozen (12) eggs or a gross (144) of pencils. The mole is sometimes called the chemist's dozen. But a mole is a much larger number of things, namely 6.022×10^{23}. Thus 1 mole contains 6.022×10^{23} entities of anything. In reference to our reaction between magnesium and sulfur, 1 mol Mg (24.31 g) contains 6.022×10^{23} atoms of magnesium and 1 mol S (32.06 g) contains 6.022×10^{23} atoms of sulfur.

Avogadro's number

The number 6.022×10^{23} is known as **Avogadro's number** in honor of Amedeo Avogadro (1776–1856), an Italian physicist. Avogadro's number is an important constant in chemistry and physics and has been experimentally determined by several independent methods.

Avogadro's number = 6.022×10^{23}

It is difficult to imagine how large Avogadro's number really is, but the following analogy will help to express it: If 10,000 people started to count Avogadro's number and each counted at the rate of 100 numbers per minute each minute of the day, it

◀ **Chapter Opening Photo: For these chefs, preparing gourmet meals is an exercise in combining precise quantities of various ingredients to produce a delightful experience for their patrons.**

would take them over 1 trillion (10^{12}) years to count the total number. So, even the most minute amount of matter contains extremely large numbers of atoms. For example, 1 mg (0.001 g) of sulfur contains 2×10^{19} atoms of sulfur.

mole

Avogadro's number is the basis for the amount of substance used to express a particular number of chemical species, such as atoms, molecules, formula units, ions, or electrons. This amount of substance is the mole. We define a **mole** as an amount of a substance containing the same number of formula units as there are atoms in exactly 12 g of carbon-12. (Recall that carbon-12 is the reference isotope for atomic masses.) Other definitions of the mole are used, but they all relate to Avogadro's number of formula units of a substance. A **formula unit** is the atom or molecule indicated by the formula of the substance under consideration—for example, Mg, MgS, H_2O, O_2.

formula unit

From the definition, we can say that the atomic mass in grams of any element contains 1 mol of atoms. The term *mole* is so commonplace in chemical jargon that chemists use it as freely as the words *atom* and *molecule*. The mole is used in conjunction with many different particles, such as atoms, molecules, ions, and electrons, to represent Avogadro's number of these particles. If we can speak of a mole of atoms, we can also speak of a mole of molecules, a mole of electrons, a mole of ions, understanding that in each case we mean 6.022×10^{23} formula units of these particles.

1 mol of atoms = 6.022×10^{23} atoms
1 mol of molecules = 6.022×10^{23} molecules
1 mol of ions = 6.022×10^{23} ions

molar mass

The atomic mass, in grams of an element, contains Avogadro's number of atoms and is defined as the **molar mass** of that element. To determine the molar mass of an element, change the units of the atomic mass found in the periodic table from atomic mass units to grams. For example, sulfur has an atomic mass of 32.06 amu. One molar mass of sulfur has a mass of 32.06 grams and contains 6.022×10^{23} atoms of sulfur. See the following table.

Element	Atomic mass	Molar mass	Number of atoms
H	1.008 amu	1.008 g	6.022×10^{23}
Mg	24.31 amu	24.31 g	6.022×10^{23}
Na	22.99 amu	22.99 g	6.022×10^{23}

1 molar mass (g) = 1 mol of atoms
= Avogadro's number (6.022×10^{23}) of atoms

Amedeo Avogadro (1776–1856).

In chemistry, we frequently encounter problems that require conversions involving quantities of mass (g), moles, and the number of atoms of an element. From the boxed equivalence statement above, many different conversion factors can be written. The key to writing a conversion factor is to remember that both sides of an equivalence statement represent equal amounts. This means that, when one of these amounts is placed in the numerator and another is placed in the denominator of a conversion factor, the resulting ratio has a value equal to one. To determine which amount should be placed in the numerator, consider the units that are desired in the result. The denominator should contain the units you are trying to eliminate. Study the following examples.

Example 7.1

How many moles of iron does 25.0 g of iron, Fe, represent?

Solution

The problem requires that we change grams of Fe to moles of Fe. We look up the atomic mass of Fe in the table of atomic masses on the periodic table or table of atomic masses and find it to be 55.85. Then we use the proper conversion factor to obtain moles:

$$\text{grams Fe} \longrightarrow \text{moles Fe} \qquad \text{grams Fe} \times \frac{1 \text{ mol Fe}}{1 \text{ molar mass Fe}}$$

$$25.0 \cancel{\text{g Fe}} \times \frac{1 \text{ mol Fe}}{55.85 \cancel{\text{g Fe}}} = 0.448 \text{ mol Fe}$$

Example 7.2

How many magnesium, Mg, atoms are contained in 5.00 g of Mg?

Solution

The problem requires that we change grams of Mg to atoms of Mg.

$$\text{grams Mg} \longrightarrow \text{atoms Mg}$$

We find the atomic mass of magnesium to be 24.31 and set up the calculation using a conversion factor between atoms and molar mass:

$$\text{grams Mg} \times \frac{6.022 \times 10^{23} \text{ atoms Mg}}{1 \text{ molar mass Mg}}$$

$$5.00 \cancel{\text{g Mg}} \times \frac{6.022 \times 10^{23} \text{ atoms Mg}}{24.31 \cancel{\text{g Mg}}} = 1.24 \times 10^{23} \text{ atoms Mg}$$

An alternative solution is first to convert grams of Mg to moles of Mg, which are then changed to atoms of Mg.

$$\text{grams Mg} \longrightarrow \text{moles Mg} \longrightarrow \text{atoms Mg}$$

Use conversion factors for each step. The calculation setup is

$$5.00 \cancel{\text{g Mg}} \times \frac{\cancel{1 \text{ mol Mg}}}{24.31 \cancel{\text{g Mg}}} \times \frac{6.022 \times 10^{23} \text{ atoms Mg}}{\cancel{1 \text{ mol Mg}}} = 1.24 \times 10^{23} \text{ atoms Mg}$$

Thus 1.24×10^{23} atoms of Mg are contained in 5.00 g of Mg.

Example 7.3 What is the mass, in grams, of one atom of carbon, C?

Solution The molar mass of C is 12.01 g. Create a conversion factor between molar mass and atoms.

$$\text{atoms C} \longrightarrow \text{grams C}$$

$$\text{atoms C} \times \frac{1 \text{ molar mass}}{6.022 \times 10^{23} \text{ atoms C}}$$

$$1 \cancel{\text{atom C}} \times \frac{12.01 \text{ g C}}{6.022 \times 10^{23} \cancel{\text{atoms C}}} = 1.994 \times 10^{-23} \text{ g C}$$

Example 7.4 What is the mass of 3.01×10^{23} atoms of sodium (Na)?

Solution The information needed to solve this problem is the molar mass of Na (22.99 g) and a conversion factor between molar mass and atoms.

$$\text{atoms Na} \longrightarrow \text{grams Na}$$

$$\text{atoms Na} \times \frac{1 \text{ molar mass Na}}{6.022 \times 10^{23} \text{ atoms Na}}$$

$$3.01 \times 10^{23} \cancel{\text{atoms Na}} \times \frac{22.99 \text{ g Na}}{6.022 \times 10^{23} \cancel{\text{atoms Na}}} = 11.5 \text{ g Na}$$

Example 7.5 What is the mass of 0.252 mol of copper (Cu)?

Solution The information needed to solve this problem is the molar mass of Cu (63.55 g) and a conversion factor between molar mass and moles.

$$\text{moles Cu} \longrightarrow \text{grams Cu}$$

$$\text{moles Cu} \times \frac{1 \text{ molar mass Cu}}{1 \text{ mol Cu}}$$

$$0.252 \cancel{\text{mol Cu}} \times \frac{63.55 \text{ g Cu}}{1 \cancel{\text{mol Cu}}} = 16.0 \text{ g Cu}$$

Example 7.6 How many oxygen atoms are present in 1.00 mol of oxygen molecules?

Solution Oxygen is a diatomic molecule with the formula O_2. Therefore a molecule of oxygen contains two atoms of oxygen.

$$\frac{2 \text{ atoms O}}{1 \text{ molecule } O_2}$$

The sequence of conversions is

$$\text{moles } O_2 \longrightarrow \text{molecules } O_2 \longrightarrow \text{atoms O}$$

Two conversion factors are needed; they are

$$\frac{6.022 \times 10^{23} \text{ molecules } O_2}{1 \text{ mol } O_2} \quad \text{and} \quad \frac{2 \text{ atoms O}}{1 \text{ molecule } O_2}$$

The calculation is

$$1.00 \cancel{\text{mol } O_2} \times \frac{6.022 \times 10^{23} \cancel{\text{molecules } O_2}}{1 \cancel{\text{mol } O_2}} \times \frac{2 \text{ atoms O}}{1 \cancel{\text{molecule } O_2}} = 1.20 \times 10^{24} \text{ atoms O}$$

Practice 7.1

What is the mass of 2.50 mol of helium (He)?

Practice 7.2

How many atoms are present in 0.025 mol of iron?

7.2 Molar Mass of Compounds

One mole of a compound contains Avogadro's number of formula units of that compound. The terms *molecular weight, molecular mass, formula weight,* and *formula mass* have been used in the past to refer to the mass of one mole of a compound. However, the term *molar mass* is more inclusive, because it can be used for all types of compounds.

If the formula of a compound is known, its molar mass may be determined by adding the molar masses of all the atoms in the formula. If more than one atom of any element is present, its mass must be added as many times as it is used.

Example 7.7

The formula for water is H_2O. What is its molar mass?

Solution

Proceed by looking up the molar masses of H (1.008 g) and O (16.00 g) and adding the masses of all the atoms in the formula unit. Water contains two atoms of H and one atom of O. Thus,

$$\begin{aligned} 2\text{ H} &= 2 \times 1.008 \text{ g} = 2.016 \text{ g} \\ 1\text{ O} &= 1 \times 16.00 \text{ g} = \underline{16.00 \text{ g}} \\ & \qquad\qquad\quad 18.02 \text{ g} = \text{molar mass} \end{aligned}$$

Example 7.8

Calculate the molar mass of calcium hydroxide, $Ca(OH)_2$.

Solution

The formula of this substance contains one atom of Ca and two atoms each of O and H. Proceed as in Example 7.7. Thus,

$$\begin{aligned} 1\text{ Ca} &= 1 \times 40.08 \text{ g} = 40.08 \text{ g} \\ 2\text{ O} &= 2 \times 16.00 \text{ g} = 32.00 \text{ g} \\ 2\text{ H} &= 2 \times 1.008 \text{ g} = \underline{2.016 \text{ g}} \\ & \qquad\qquad\quad 74.10 \text{ g} = \text{molar mass} \end{aligned}$$

Each sample contains one mole. *Clockwise from lower left:* magnesium, carbon, copper(II) sulfate, copper, mercury, potassium permanganate, cadmium, and sodium chloride *(center).*

Practice 7.3

Calculate the molar mass of KNO_3.

Remember in this text we round all molar masses to four significant figures, although you may need to use a different number of significant figures for other work (perhaps in the lab).

The mass of 1 mol of a compound contains Avogadro's number of formula units or molecules. Now consider the compound hydrogen chloride (HCl). One atom of H combines with one atom of Cl to form one molecule of HCl. When 1 mol of H (1.008 g of H representing 1 mol or 6.022×10^{23} H atoms) combines with 1 mol of Cl (35.45 g of Cl representing 1 mol or 6.022×10^{23} Cl atoms), 1 mol of HCl (36.46 g of HCl representing 1 mol or 6.022×10^{23} HCl molecules) is produced. These relationships are summarized in the following table:

H	Cl	HCl
6.022×10^{23} H *atoms*	6.022×10^{23} Cl *atoms*	6.022×10^{23} HCl *molecules*
1 mol H *atoms*	1 mol Cl *atoms*	1 mol HCl *molecules*
1.008 g H	35.45 g Cl	36.46 g HCl
1 molar mass H	1 molar mass Cl	1 molar mass HCl

In dealing with diatomic elements (H_2, O_2, N_2, F_2, Cl_2, Br_2, and I_2), we must take special care to distinguish between a mole of atoms and a mole of molecules.

For example consider 1 mol of oxygen molecules, which has a mass of 32.00 g. This quantity is equal to 2 mol of oxygen atoms. The key concept is that 1 mol represents Avogadro's number of the particular chemical entity—atoms, molecules, formula units, and so forth—that is under consideration.

$$1 \text{ mol } H_2O = 18.02 \text{ g } H_2O = 6.022 \times 10^{23} \text{ molecules}$$
$$1 \text{ mol NaCl} = 58.44 \text{ g NaCl} = 6.022 \times 10^{23} \text{ formula units}$$
$$1 \text{ mol } H_2 = 2.016 \text{ g } H_2 = 6.022 \times 10^{23} \text{ molecules}$$
$$1 \text{ mol } HNO_3 = 63.02 \text{ g } HNO_3 = 6.022 \times 10^{23} \text{ molecules}$$
$$1 \text{ mol } K_2SO_4 = 174.3 \text{ g } K_2SO_4 = 6.022 \times 10^{23} \text{ formula units}$$

1 mol = 6.022×10^{23} formula units or molecules
= 1 molar mass of a compound

Remember to create the appropriate conversion factor by using the equivalence statement. Place the unit desired in the numerator and the unit to be eliminated in the denominator.

Example 7.9

What is the mass of 1 mol of sulfuric acid (H_2SO_4)?

Solution

One mole of H_2SO_4 is 1 molar mass of H_2SO_4. The problem, therefore, is solved in a similar manner to Examples 7.7 and 7.8. Look up the molar masses of hydrogen, sulfur, and oxygen, and solve.

$$2\text{ H} = 2 \times 1.008 \text{ g} = 2.016 \text{ g}$$
$$1\text{ S} = 1 \times 32.06 \text{ g} = 32.06 \text{ g}$$
$$4\text{ O} = 4 \times 16.00 \text{ g} = 64.00 \text{ g}$$
$$98.08 \text{ g} = \text{mass of 1 mol of } H_2SO_4$$

Example 7.10

How many moles of sodium hydroxide, NaOH, are there in 1.00 kg of sodium hydroxide?

Solution

First we know that

$$1 \text{ molar mass} = (22.99 \text{ g} + 16.00 \text{ g} + 1.008 \text{ g}) \text{ or } 40.00 \text{ g NaOH}$$

To convert grams to moles we use the conversion factor

$$\frac{1 \text{ mol}}{1 \text{ molar mass}} \quad \text{or} \quad \frac{1 \text{ mol NaOH}}{40.00 \text{ g NaOH}}$$

Use this conversion sequence:

$$\text{kg NaOH} \longrightarrow \text{g NaOH} \longrightarrow \text{mol NaOH}$$

The calculation is

$$1.00 \text{ } \cancel{\text{kg NaOH}} \times \frac{1000 \text{ } \cancel{\text{g NaOH}}}{\cancel{\text{kg NaOH}}} \times \frac{1 \text{ mol NaOH}}{40.00 \text{ g NaOH}} = 25.0 \text{ mol NaOH}$$

$$1.00 \text{ kg NaOH} = 25.0 \text{ mol NaOH}$$

Example 7.11 What is the mass of 5.00 mol of water?

Solution First we know that

$$1 \text{ mol } H_2O = 18.02 \text{ g (Example 7.7)}$$

The conversion is

$$\text{mol } H_2O \longrightarrow \text{g } H_2O$$

To convert moles to grams use the conversion factor

$$\frac{1 \text{ molar mass } H_2O}{1 \text{ mol } H_2O} \quad \text{or} \quad \frac{18.02 \text{ g } H_2O}{1 \text{ mol } H_2O}$$

The calculation is

$$5.00 \cancel{\text{mol } H_2O} \times \frac{18.02 \text{ g } H_2O}{1 \cancel{\text{mol } H_2O}} = 90.1 \text{ g } H_2O$$

Example 7.12 How many molecules of hydrogen chloride, HCl, are there in 25.0 g of hydrogen chloride?

Solution From the formula, we find that the molar mass of HCl is 36.46 g (1.008 g + 35.45 g). The sequence of conversions is

$$\text{g HCl} \longrightarrow \text{mol HCl} \longrightarrow \text{molecules HCl}$$

Using the conversion factors

$$\frac{1 \text{ mol HCl}}{36.46 \text{ g HCl}} \quad \text{and} \quad \frac{6.022 \times 10^{23} \text{ molecules HCl}}{1 \text{ mol HCl}}$$

$$25.0 \cancel{\text{g HCl}} \times \frac{1 \cancel{\text{mol HCl}}}{36.46 \cancel{\text{g HCl}}} \times \frac{6.022 \times 10^{23} \text{ molecules HCl}}{1 \cancel{\text{mol HCl}}} =$$

$$4.13 \times 10^{23} \text{ molecules HCl}$$

Practice 7.4

What is the mass of 0.150 mol of Na_2SO_4?

Practice 7.5

How many moles are there in 500.0 g of $AlCl_3$?

7.3 Percent Composition of Compounds

percent composition of a compound

Percent means parts per 100 parts. Just as each piece of pie is a percent of the whole pie, each element in a compound is a percent of the whole compound. The **percent composition of a compound** is the *mass percent* of each element in the compound. The molar mass represents the total mass, or 100%, of the compound. Thus, the percent composition of water, H_2O, is 11.19% H and 88.79% O by mass. According to the Law of Definite Composition, the percent composition must be the same no matter what size sample is taken.

The percent composition of a compound can be determined (1) from knowing its formula or (2) from experimental data.

Percent Composition from Formula

If the formula is known, it is essentially a two-step process to determine the percent composition.

Step 1 Calculate the molar mass (Section 7.2).

Step 2 Divide the total mass of each element in the formula by the molar mass and multiply by 100. This gives the percent composition.

$$\frac{\text{total mass of the element}}{\text{molar mass}} \times 100 = \text{percent of the element}$$

Example 7.13

Calculate the percent composition of sodium chloride, NaCl.

Solution

Step 1 Calculate the molar mass of NaCl:

$$1\ \text{Na} = 1 \times 22.99\ \text{g} = 22.99\ \text{g}$$
$$1\ \text{Cl} = 1 \times 35.45\ \text{g} = \underline{35.45\ \text{g}}$$
$$58.44\ \text{g (molar mass)}$$

Step 2 Now calculate the percent composition. We know there are 22.99 g Na and 35.45 g Cl in 58.44 g NaCl.

$$\text{Na:}\quad \frac{22.99\ \text{g Na}}{58.44\ \text{g NaCl}} \times 100 = 39.34\%\ \text{Na}$$

$$\text{Cl:}\quad \frac{35.45\ \text{g Cl}}{58.44\ \text{g NaCl}} \times 100 = \underline{60.66\%\ \text{Cl}}$$
$$100.00\%\ \text{total}$$

In any two-component system, if the percent of one component is known, the other is automatically defined by the difference; that is, if Na is 39.34%, then Cl is 100% − 39.34% = 60.66%. However, the calculation of the percent of each component should be carried out, since this provides a check against possible error. The percent composition data should add up to 100 ± 0.2%.

Example 7.14

Calculate the percent composition of potassium chloride, KCl.

Solution

Step 1 Calculate the molar mass of KCl:

$$1\ \text{K} = 1 \times 39.10\ \text{g} = 39.10\ \text{g}$$
$$1\ \text{Cl} = 1 \times 35.45\ \text{g} = \underline{35.45\ \text{g}}$$
$$74.55\ \text{g (molar mass)}$$

Step 2 Now calculate the percent composition. We know there are 39.10 g K and 35.45 g Cl in 74.55 g KCl.

$$\text{K:} \quad \frac{39.10 \text{ g K}}{74.55 \text{ g KCl}} \times 100 = 52.45\% \text{ K}$$

$$\text{Cl:} \quad \frac{35.45 \text{ g Cl}}{74.55 \text{ g KCl}} \times 100 = 47.55\% \text{ Cl}$$

$$100.00\% \text{ total}$$

Comparing the data calculated for NaCl and for KCl, we see that NaCl contains a higher percentage of Cl by mass, although each compound has a one-to-one atom ratio of Cl to Na and Cl to K. The reason for this mass percent difference is that Na and K do not have the same atomic masses.

It is important to realize that, when we compare 1 mol of NaCl with 1 mol of KCl, each quantity contains the same number of Cl atoms—namely, 1 mol of Cl atoms. However, if we compare equal masses of NaCl and KCl, there will be more Cl atoms in the mass of NaCl since NaCl has a higher mass percent of Cl.

1 mol NaCl contains	100.00 g NaCl contains
1 mol Na	39.34 g Na
1 mol Cl	60.66 g Cl
60.66%Cl	

1 mol KCl contains	100.00 g KCl contains
1 mol K	52.45 g K
1 mol Cl	47.55 g Cl
47.55%Cl	

Example 7.15 Calculate the percent composition of potassium sulfate, K_2SO_4.

Solution

Step 1 Calculate the molar mass of K_2SO_4:

$$2\text{ K} = 2 \times 39.10 \text{ g} = 78.20 \text{ g}$$

$$1\text{ S} = 1 \times 32.06 \text{ g} = 32.06 \text{ g}$$

$$4\text{ O} = 4 \times 16.00 \text{ g} = 64.00 \text{ g}$$

$$174.3 \text{ g (molar mass)}$$

Step 2 Now calculate the percent composition. We know there are 78.20 g of K, 32.06 g of S, and 64.00 g of O in 174.3 g of K_2SO_4.

$$\text{K:} \quad \frac{78.20 \text{ g K}}{174.3 \text{ g } K_2SO_4} \times 100 = 44.87\% \text{ K}$$

$$\text{S:} \quad \frac{32.06 \text{ g S}}{174.3 \text{ g } K_2SO_4} \times 100 = 18.39\% \text{ S}$$

$$\text{O:} \quad \frac{64.00 \text{ g O}}{174.3 \text{ g } K_2SO_4} \times 100 = 36.72\% \text{ O}$$

$$99.98\% \text{ total}$$

The Taste of Chemistry

Flavorings, seasonings, and preservatives have been added to foods since ancient civilizations. Spices were originally added as preservatives when refrigeration was unavailable. These spices contained mild antiseptics and antioxidants and were effective in prolonging the time during which food could be eaten. Over the course of time a great variety of substances (food additives) came to be used in foods—preservatives, colorings, flavorings, antioxidants, sweeteners, and so on. Many of these substances are now regarded as virtual necessities for processing foods. However, there is genuine concern that some additives, particularly those that are not naturally present in foods, may be detrimental to one's health when consumed. This concern has led to the passage of federal and state laws that regulate the food industry.

These are a few foods that contain additives.

The United States Food and Drug Administration (FDA) is the principal agency charged with enforcing federal laws concerning food and must approve the use of all food additives. The FDA divides food additives into several categories. These include a classification known as GRAS (Generally Regarded As Safe). The GRAS designation includes substances that were in use in 1958 and that met certain specifications for safety. All substances introduced after 1958 have been approved on an individual basis.

Before commercial use of a new food additive, a company must provide the FDA with satisfactory evidence that the chemical is safe for the proposed usage. Providing this evidence of safety is complex and expensive. Although mainly concerned with chemistry, this task requires the services of many people trained in a variety of disciplines—biochemistry, microbiology, medicine, physiology, and so on. Research and testing may have to be done in several laboratories before final FDA approval (or rejection) is obtained.

The amount of an additive that is considered to be safe in foods is determined by the maximum tolerable daily intake (MTDI). This is the amount of the food additive that can be eaten daily for a lifetime without adverse effects. It is calculated on the basis of body mass (mg/kg/day). If there is any doubt regarding the safety of an additive, a time limit is determined and a conditional MTDI is issued and reviewed with further testing at the end of the initial time period. To establish the MTDI, experiments are run on animals, increasing the quantities of the additive in successive experiments until acute and chronic toxicity occurs. A minimum of two species of animals must be tested, the most sensitive species forming the basis for the appropriate level for the additive. This quantity is then divided by 100 (for additional safety) to set the MTDI.

Once the MTDI is established, all foods in which the use of the additive is proposed must be considered. An estimate is made of the maximum amount of the additive that could be ingested. On the basis of this information the use may then be restricted to only certain foods (as in the exclusion of sulfites from most meats) or broadened to include additional foods.

Unfortunately, even with all the quantitative testing it is extremely difficult to determine safe levels of additives since the majority of toxicological data is from animal testing. The results are often not the same as in humans—chemicals toxic to one species may not be so for humans.

Children add a further complication. They cannot be considered simply as small adults. A child has a much greater energy demand per kilogram than an adult. Often children's chemical defense mechanisms are much different than those of an adult.

Flavorings are the largest class of food additives. There are 1100–1400 natural or synthetic flavorings used in foods. The task of checking for effects of these additives is herculean. Flavorings are complex mixtures of chemicals used in very small amounts making them difficult to analyze. Distinctions between artificial and natural flavorings must also be monitored (see Chemistry in Action, page 94). Chemists rely on quantitative analysis techniques (such as gas–liquid chromatography) to separate and identify components in these mixtures. It is possible to detect compounds in amounts as low as 10 μg/kg (parts per billion).

Practice 7.6

Calculate the percent composition of $Ca(NO_3)_2$.

Practice 7.7

Calculate the percent composition of K_2CrO_4.

Percent Composition from Experimental Data

The percent composition can be determined from experimental data without knowing the formula of a compound.

Step 1 Calculate the mass of the compound formed.

Step 2 Divide the mass of each element by the total mass of the compound and multiply by 100.

Example 7.16 When heated in the air, 1.63 g of zinc, Zn, combine with 0.40 g of oxygen, O_2, to form zinc oxide. Calculate the percent composition of the compound formed.

Solution **Step 1** First, calculate the total mass of the compound formed.

$$\begin{array}{l} 1.63 \text{ g Zn} \\ \underline{0.40 \text{ g } O_2} \\ 2.03 \text{ g} \quad = \text{total mass of product} \end{array}$$

Step 2 Divide the mass of each element by the total mass (2.03 g) and multiply by 100.

$$\frac{1.63 \text{ g}}{2.03 \text{ g}} \times 100 = 80.3\% \text{ Zn} \qquad \frac{0.40 \text{ g}}{2.03 \text{ g}} \times 100 = \underline{19.7\% \text{ O}}$$

$$100.0\% \text{ total}$$

The compound formed contains 80.3% Zn and 19.7% O.

Practice 7.8

Aluminum chloride is formed by reacting 13.43 g aluminum with 53.18 g chlorine. What is the percent composition of the compound?

7.4 Empirical Formula Versus Molecular Formula

empirical formula

The **empirical formula**, or *simplest formula,* gives the smallest whole-number ratio of the atoms that are present in a compound. This formula gives the relative number of atoms of each element in the compound.

molecular formula

The **molecular formula** is the true formula, representing the total number of atoms of each element present in one molecule of a compound. It is entirely possible that two or more substances will have the same percent composition yet be distinctly different compounds. For example, acetylene, C_2H_2, is a common gas used in welding; benzene, C_6H_6, is an important solvent obtained from coal tar and is used in the synthesis of styrene and nylon. Both acetylene and benzene contain 92.3% C and 7.7% H. The smallest ratio of C and H corresponding to these percentages is CH (1:1). Therefore the empirical formula for both acetylene and benzene is CH, even though it is known that the molecular formulas are C_2H_2 and C_6H_6, respectively. Often the molecular formula is the same as the empirical formula. If the molecular formula is not the same, it will be an integral (whole number) multiple of the empirical formula. For example,

CH = empirical formula

$(CH)_2 = C_2H_2$ = acetylene (molecular formula)

$(CH)_6 = C_6H_6$ = benzene (molecular formula)

Table 7.1 summarizes the data concerning these CH formulas. Table 7.2 shows empirical and molecular formula relationships of other compounds.

TABLE 7.1 Molecular Formulas of Two Compounds Having an Empirical Formula with a 1:1 Ratio of Carbon and Hydrogen Atoms

	Composition		
Formula	% C	% H	Molar mass
CH (empirical)	92.3	7.7	13.02 (empirical)
C_2H_2 (acetylene)	92.3	7.7	26.04 (2 × 13.02)
C_6H_6 (benzene)	92.3	7.7	78.12 (6 × 13.02)

TABLE 7.2 Some Empirical and Molecular Formulas

Compound	Empirical formula	Molecular formula	Compound	Empirical formula	Molecular formula
Acetylene	CH	C_2H_2	Diborane	BH_3	B_2H_6
Benzene	CH	C_6H_6	Hydrazine	NH_2	N_2H_4
Ethylene	CH_2	C_2H_4	Hydrogen	H	H_2
Formaldehyde	CH_2O	CH_2O	Chlorine	Cl	Cl_2
Acetic acid	CH_2O	$C_2H_4O_2$	Bromine	Br	Br_2
Glucose	CH_2O	$C_6H_{12}O_6$	Oxygen	O	O_2
Hydrogen chloride	HCl	HCl	Nitrogen	N	N_2
Carbon dioxide	CO_2	CO_2			

7.5 Calculation of Empirical Formulas

It is possible to establish an empirical formula because (1) the individual atoms in a compound are combined in whole-number ratios, and (2) each element has a specific atomic mass.

In order to calculate the empirical formula we need to know (1) the elements that are combined, (2) their atomic masses, and (3) the ratio by mass or percentage in which they are combined. If elements *A* and *B* form a compound, we may represent the empirical formula as A_xB_y, where *x* and *y* are small whole numbers that represent the atoms of *A* and *B*. To write the empirical formula we must determine *x* and *y*. Three or four steps are required to do this.

Step 1 Assume a definite starting quantity (usually 100.0 g) of the compound, if not given, and express the mass of each element in grams.

Step 2 Convert the grams of each element into moles using the molar mass of each element. This conversion gives the number of moles of atoms of each element in the quantity assumed in Step 1. At this point these numbers will usually not be whole numbers.

Step 3 Divide each of the values obtained in Step 2 by the smallest of these values. If the numbers obtained by this procedure are whole numbers, use them as subscripts in writing the empirical formula. If the numbers obtained are not whole numbers, go on to Step 4.

Step 4 Multiply the values obtained in Step 3 by the smallest number that will convert them to whole numbers. Use these whole numbers as the subscripts in the empirical formula. For example, if the ratio of *A* to *B* is 1.0:1.5, multiply both numbers by 2 to obtain a ratio of 2:3. The empirical formula then is A_2B_3.

Some common fractions and their decimal equivalents are
$1/4 = 0.25$
$1/3 = 0.333\ldots$
$2/3 = 0.666\ldots$
$1/2 = 0.5$
$3/4 = 0.75$
Multiply the decimal equivalent by the number in the denominator of the fraction to give a whole number: $4(0.75) = 3$.

In many of these calculations results will vary somewhat from an exact whole number; this can be due to experimental errors in obtaining the data or from rounding off numbers. Calculations that vary by no more than ± 0.1 from a whole number usually can be rounded off to the nearest whole number. Deviations greater than about 0.1 unit usually mean that the calculated ratios need to be multiplied by a factor to make them all whole numbers. For example, an atom ratio of 1:1.33 should be multiplied by 3 to make the ratio 3:4.

Example 7.17 Calculate the empirical formula of a compound containing 11.19% hydrogen, H, and 88.79% oxygen, O.

Solution

Step 1 Express each element in grams. If we assume that there are 100.00 g of material, then the percent of each element is equal to the grams of each element in 100.00 g, and the percent sign can be omitted:

H = 11.19 g

O = 88.79 g

Step 2 Convert the grams of each element to moles:

$$\text{H:}\quad 11.19\ \text{g H} \times \frac{1\ \text{mol H atoms}}{1.008\ \text{g H}} = 11.10\ \text{mol H atoms}$$

$$\text{O:}\quad 88.79\ \text{g O} \times \frac{1\ \text{mol O atoms}}{16.00\ \text{g O}} = 5.549\ \text{mol O atoms}$$

The formula could be expressed as $H_{11.10}O_{5.549}$. However, it is customary to use the smallest whole-number ratio of atoms. This ratio is calculated in Step 3.

Step 3 Change these numbers to whole numbers by dividing each of them by the smaller number.

$$\text{H} = \frac{11.10\ \text{mol}}{5.549\ \text{mol}} = 2.000 \qquad \text{O} = \frac{5.549\ \text{mol}}{5.549\ \text{mol}} = 1.000$$

In this step, the ratio of atoms has not changed, because we divided the number of moles of each element by the same number.

The simplest ratio of H to O is 2:1.
Empirical formula = H_2O

Example 7.18

The analysis of a salt shows that it contains 56.58% potassium, K, 8.68% carbon, C, and 34.73% oxygen, O. Calculate the empirical formula for this substance.

Solution

Steps 1 and 2 After changing the percentage of each element to grams, find the relative number of moles of each element by multiplying by the proper mole/molar mass factor:

$$\text{K:}\quad 56.58\ \text{g K} \times \frac{1\ \text{mol K atoms}}{39.10\ \text{g K}} = 1.447\ \text{mol K atoms}$$

$$\text{C:}\quad 8.68\ \text{g C} \times \frac{1\ \text{mol C atoms}}{12.01\ \text{g C}} = 0.723\ \text{mol C atoms}$$

$$\text{O:}\quad 34.73\ \text{g O} \times \frac{1\ \text{mol O atoms}}{16.00\ \text{g O}} = 2.171\ \text{mol O atoms}$$

Step 3 Divide each number of moles by the smallest value:

$$\text{K} = \frac{1.447\ \text{mol}}{0.723\ \text{mol}} = 2.00$$

$$\text{C} = \frac{0.723\ \text{mol}}{0.723\ \text{mol}} = 1.00$$

$$\text{O} = \frac{2.171\ \text{mol}}{0.723\ \text{mol}} = 3.00$$

The simplest ratio of K:C:O is 2:1:3.
Empirical formula = K_2CO_3

Example 7.19 A sulfide of iron was formed by combining 2.233 g of iron, Fe, with 1.926 g of sulfur, S. What is the empirical formula of the compound?

Solution

Steps 1 and 2 The grams of each element are given, so we use them directly in our calculations. Calculate the relative number of moles of each element by multiplying the grams of each element by the proper mole/molar mass factor:

$$\text{Fe:}\quad 2.233\ \cancel{\text{g Fe}} \times \frac{1\ \text{mol Fe atoms}}{55.85\ \cancel{\text{g Fe}}} = 0.03998\ \text{mol Fe atoms}$$

$$\text{S:}\quad 1.926\ \cancel{\text{g S}} \times \frac{1\ \text{mol S atoms}}{32.06\ \cancel{\text{g S}}} = 0.06007\ \text{mol S atoms}$$

Step 3 Divide each number of moles by the smaller of the two numbers:

$$\text{Fe} = \frac{0.03998\ \text{mol}}{0.03998\ \text{mol}} = 1.000 \qquad \text{S} = \frac{0.06007\ \text{mol}}{0.03998\ \text{mol}} = 1.503$$

Step 4 We still have not reached a ratio that will give a formula containing whole numbers of atoms, so we must double each value to obtain a ratio of 2.000 atoms of Fe to 3.000 atoms of S. Doubling both values does not change the ratio of Fe and S atoms.

Fe: $1.000 \times 2 = 2.000$

S: $1.503 \times 2 = 3.006$

Empirical formula $= Fe_2S_3$

Practice 7.9

Calculate the empirical formula of a compound containing 52.14% C, 13.12% H, and 34.73% O.

Practice 7.10

Calculate the empirical formula of a compound that contains 43.7% phosphorus and 56.3% O by mass.

7.6 Calculation of the Molecular Formula from the Empirical Formula

In addition to data for calculating the empirical formula, the molecular formula can be calculated from the empirical formula if the molar mass is known. The molecular formula, as stated in Section 7.4, will be either equal to or some multiple of the empirical formula. For example, if the empirical formula of a compound of hydrogen and fluorine is HF, the molecular formula can be expressed as $(HF)_n$, where $n =$ 1, 2, 3, 4, ... This n means that the molecular formula could be HF, H_2F_2, H_3F_3, H_4F_4, and so on. To determine the molecular formula, we must evaluate n.

$$n = \frac{\text{molar mass}}{\text{mass of empirical formula}} = \text{number of empirical formula units}$$

What we actually calculate is the number of units of the empirical formula contained in the molecular formula.

Example 7.20

A compound of nitrogen and oxygen with a molar mass of 92.00 g was found to have an empirical formula of NO_2. What is its molecular formula?

Solution

Let n be the number of NO_2 units in a molecule; then the molecular formula is $(NO_2)_n$. Each NO_2 unit has a mass of [14.01 g + (2 × 16.00 g)] or 46.01 g. The molar mass of $(NO_2)_n$ is 92.00 g and the number of 46.01 units in 92.00 is 2:

$$n = \frac{92.00\ \text{g}}{46.01\ \text{g}} = 2 \quad \text{(empirical formula units)}$$

The molecular formula is $(NO_2)_2$ or N_2O_4.

Example 7.21

The hydrocarbon propylene has a molar mass of 42.00 g and contains 14.3% H and 85.7% C. What is its molecular formula?

Solution

First find the empirical formula:

$$\text{C:} \quad 85.7\ \cancel{\text{g C}} \times \frac{1\ \text{mol C atoms}}{12.01\ \cancel{\text{g C}}} = 7.14\ \text{mol C atoms}$$

$$\text{H:} \quad 14.3\ \cancel{\text{g H}} \times \frac{1\ \text{mol H atoms}}{1.008\ \cancel{\text{g H}}} = 14.2\ \text{mol H atoms}$$

Divide each value by the smaller number of moles:

$$\text{C} = \frac{7.14\ \text{mol}}{7.14\ \text{mol}} = 1.00$$

$$\text{H} = \frac{14.2\ \text{mol}}{7.14\ \text{mol}} = 1.99$$

Empirical formula = CH_2

Then determine the molecular formula from the empirical formula and the molar mass:

Molecular formula = $(CH_2)_n$

Molar mass = 42.00 g

Each CH_2 unit has a mass of (12.01 g + 2.016 g) or 14.03 g. The number of CH_2 units in 42.00 g is 3:

$$n = \frac{42.00\ \text{g}}{14.03\ \text{g}} = 3 \quad \text{(empirical formula units)}$$

The molecular formula is $(CH_2)_3$ or C_3H_6.

Practice 7.11

Calculate the empirical and molecular formulas of a compound that contains 80.0% C, 20.0% H, and has a molar mass of 30.00 g.

Concepts in Review

1. Explain the meaning of the mole.
2. Discuss the relationship between a mole and Avogadro's number.
3. Convert grams, atoms, molecules, and molar masses to moles, and vice versa.
4. Determine the molar mass of a compound from the formula.
5. Calculate the percent composition of a compound from its formula.
6. Calculate the percent composition of a compound from experimental data.
7. Explain the relationship between an empirical formula and a molecular formula.
8. Determine the empirical formula for a compound from its percent composition.
9. Calculate the molecular formula of a compound from its percent composition and molar mass.

Key Terms

The terms listed here have been defined within this chapter. Section numbers are referenced in parenthesis for each term.

Avogadro's number (7.1)
empirical formula (7.4)
formula unit (7.1)
molar mass (7.1)
mole (7.1)
molecular formula (7.4)
percent composition of a compound (7.3)

Questions

Questions refer to tables, figures, and key words and concepts defined within the chapter. A particularly challenging question or exercise is indicated with an asterisk.

1. What is a mole?
2. Which would have a higher mass: a mole of K atoms or a mole of Au atoms?
3. Which would contain more atoms: a mole of K atoms, or a mole of Au atoms?
4. Which would contain more electrons: a mole of K atoms, or a mole of Au atoms?

* 5. If the atomic mass scale had been defined differently, with an atom of $^{12}_{6}C$ being defined as a mass of 50 amu, would this have any effect on the value of Avogadro's number? Explain.

6. What is the numerical value of Avogadro's number?
7. What is the relationship between Avogadro's number and the mole?

8. Complete the following statements, supplying the proper quantity.
   (a) A mole of O atoms contains ________ atoms.
   (b) A mole of O_2 molecules contains ________ molecules.
   (c) A mole of O_2 molecules contains ________ atoms.
   (d) A mole of O atoms has a mass of ________ grams.
   (e) A mole of O_2 molecules has a mass of ________ grams.

9. How many molecules are present in 1 molar mass of sulfuric acid (H_2SO_4)? How many atoms are present?

10. In calculating the empirical formula of a compound from its percent composition, why do we choose to start with 100.0 g of the compound?

11. List four characteristics of food additives.

12. Explain how the maximum tolerable daily intake of an additive is determined.

13. Why is it difficult to determine safe levels of food additives?

14. What is the difference between an empirical formula and a molecular formula?

15. Which of the following statements are correct? Rewrite the incorrect statements to make them correct.
   (a) One atomic mass of any element contains 6.022×10^{23} atoms.
   (b) The mass of 1 atom of Cl is $\frac{35.45\text{ g}}{6.022 \times 10^{23}\text{ atoms}}$.
   (c) A mole of Mg atoms (24.31 g) contains the same number of atoms as a mole of Na atoms (22.99 g).
   (d) A mole of bromine atoms contains 6.022×10^{23} atoms of bromine.
   (e) A mole of Cl_2 molecules contains 6.022×10^{23} atoms of Cl.
   (f) A mole of Al atoms has the same mass as a mole of tin (Sn) atoms.
   (g) A mole of H_2O contains 6.022×10^{23} atoms.
   (h) A mole of H_2 molecules contains 1.204×10^{24} electrons.

16. Which of the following statements are correct? Rewrite the incorrect statements to make them correct.
   (a) A mole of Na and a mole of NaCl contain the same number of Na atoms.
   (b) One mole of nitrogen gas (N_2) has a mass of 14.01 g.
   (c) The percent of oxygen is higher in K_2CrO_4 than it is in Na_2CrO_4.
   (d) The number of Cr atoms is the same in a mole of K_2CrO_4 as it is in a mole of Na_2CrO_4.
   (e) Both K_2CrO_4 and Na_2CrO_4 contain the same percent by mass of Cr.
   (f) A molar mass of sucrose ($C_{12}H_{22}O_{11}$) contains 1 mol of sucrose molecules.
   (g) Two moles of nitric acid (HNO_3) contain 6 mol of O atoms.
   (h) The empirical formula of sucrose ($C_{12}H_{22}O_{11}$) is CH_2O.
   (i) A hydrocarbon that has a molar mass of 280 and an empirical formula of CH_2 has a molecular formula of $C_{22}H_{44}$.
   (j) The empirical formula is often called the simplest formula.
   (k) The empirical formula of a compound gives the smallest whole-number ratio of the atoms that are present in the compound.
   (l) If the molecular formula and the empirical formula of a compound are not the same, the empirical formula will be an integral multiple of the molecular formula.
   (m) The empirical formula of benzene (C_6H_6) is CH.
   (n) A compound having an empirical formula of CH_2O and a molar mass of 60 has a molecular formula of $C_3H_6O_3$.

Paired Exercises

These exercises are paired. Each odd-numbered exercise is followed by a similar even-numbered exercise. Answers to the even-numbered exercises are given in Appendix V.

Molar Masses

17. Determine the molar masses of the following compounds:
   (a) KBr
   (b) Na_2SO_4
   (c) $Pb(NO_3)_2$
   (d) C_2H_5OH
   (e) $HC_2H_3O_2$
   (f) Fe_3O_4
   (g) $C_{12}H_{22}O_{11}$
   (h) $Al_2(SO_4)_3$
   (i) $(NH_4)_2HPO_4$

18. Determine the molar masses of the following compounds:
   (a) NaOH
   (b) Ag_2CO_3
   (c) Cr_2O_3
   (d) $(NH_4)_2CO_3$
   (e) $Mg(HCO_3)_2$
   (f) C_6H_5COOH
   (g) $C_6H_{12}O_6$
   (h) $K_4Fe(CN)_6$
   (i) $BaCl_2 \cdot 2\ H_2O$

Moles and Avogadro's Number

19. How many moles of atoms are contained in the following?
(a) 22.5 g Zn
(b) 0.688 g Mg
(c) 4.5×10^{22} atoms Cu
(d) 382 g Co
(e) 0.055 g Sn
(f) 8.5×10^{24} molecules N_2

20. How many moles are contained in the following?
(a) 25.0 g NaOH
(b) 44.0 g Br_2
(c) 0.684 g $MgCl_2$
(d) 14.8 g CH_3OH
(e) 2.88 g Na_2SO_4
(f) 4.20 lb ZnI_2

21. Calculate the number of grams in each of the following:
(a) 0.550 mol Au
(b) 15.8 mol H_2O
(c) 12.5 mol Cl_2
(d) 3.15 mol NH_4NO_3

22. Calculate the number of grams in each of the following:
(a) 4.25×10^{-4} mol H_2SO_4
(b) 4.5×10^{22} molecules CCl_4
(c) 0.00255 mol Ti
(d) 1.5×10^{16} atoms S

23. How many molecules are contained in each of the following:
(a) 1.26 mol O_2
(b) 0.56 mol C_6H_6
(c) 16.0 g CH_4
(d) 1000. g HCl

24. How many molecules are contained in each of the following:
(a) 1.75 mol Cl_2
(b) 0.27 mol C_2H_6O
(c) 12.0 g CO_2
(d) 100. g CH_4

25. Calculate the mass in grams of each of the following:
(a) 1 atom Pb
(b) 1 atom Ag
(c) 1 molecule H_2O
(d) 1 molecule $C_3H_5(NO_3)_3$

26. Calculate the mass in grams of each of the following:
(a) 1 atom Au
(b) 1 atom U
(c) 1 molecule NH_3
(d) 1 molecule $C_6H_4(NH_2)_2$

27. Make the following conversions:
(a) 8.66 mol Cu to grams Cu
(b) 125 mol Au to kilograms Au
(c) 10 atoms C to moles C
(d) 5000 molecules CO_2 to moles CO_2

28. Make the following conversions:
(a) 28.4 g S to moles S
(b) 2.50 kg NaCl to moles NaCl
(c) 42.4 g Mg to atoms Mg
(d) 485 mL Br_2 (d = 3.12 g/mL) to moles Br_2

29. Exactly 1 mol of carbon disulfide contains
(a) how many carbon disulfide molecules?
(b) how many carbon atoms?
(c) how many sulfur atoms?
(d) how many total atoms of all kinds?

30. One mole of ammonia contains
(a) how many ammonia molecules?
(b) how many nitrogen atoms?
(c) how many hydrogen atoms?
(d) how many total atoms of all kinds?

31. How many atoms of oxygen are contained in each of the following?
(a) 16.0 g O_2
(b) 0.622 mol MgO
(c) 6.00×10^{22} molecules $C_6H_{12}O_6$

32. How many atoms of oxygen are contained in each of the following?
(a) 5.0 mol MnO_2
(b) 255 g $MgCO_3$
(c) 5.0×10^{18} molecules H_2O

33. Calculate the number of
(a) grams of silver in 25.0 g AgBr
(b) grams of nitrogen in 6.34 mol $(NH_4)_3PO_4$
(c) grams of oxygen in 8.45×10^{22} molecules SO_3

34. Calculate the number of
(a) grams of chlorine in 5.0 g $PbCl_2$
(b) grams of hydrogen in 4.50 mol H_2SO_4
(c) grams of iodine in 5.45×10^{22} molecules CaI_2

Percent Composition

35. Calculate the percent composition by mass of the following compounds:
(a) NaBr
(b) $KHCO_3$
(c) $FeCl_3$
(d) $SiCl_4$
(e) $Al_2(SO_4)_3$
(f) $AgNO_3$

36. Calculate the percent composition of the following compounds:
(a) $ZnCl_2$
(b) $NH_4C_2H_3O_2$
(c) MgP_2O_7
(d) $(NH_4)_2SO_4$
(e) $Fe(NO_3)_3$
(f) ICl_3

37. Calculate the percent of iron in the following compounds:
(a) FeO
(b) Fe_2O_3
(c) Fe_3O_4
(d) $K_4Fe(CN)_6$

38. Which of the following chlorides has the highest and which has the lowest percentage of chlorine, by mass, in its formula?
(a) KCl
(b) $BaCl_2$
(c) $SiCl_4$
(d) LiCl

39. A 6.20-g sample of phosphorus was reacted with oxygen to form an oxide with a mass of 14.20 g. Calculate the percent composition of the compound.

40. A sample of ethylene chloride was analyzed to contain 6.00 g of C, 1.00 g of H, and 17.75 g of Cl. Calculate the percent composition of ethylene chloride.

41. Answer the following by examining the formulas. Check your answers by calculation if you wish. Which compound has the
(a) higher percent by mass of hydrogen, H_2O or H_2O_2?
(b) lower percent by mass of nitrogen, NO or N_2O_3?
(c) higher percent by mass of oxygen, NO_2 or N_2O_4?

42. Answer the following by examining the formulas. Check your answers by calculation if you wish. Which compound has the
(a) lower percent by mass of chlorine, $NaClO_3$ or $KClO_3$?
(b) higher percent by mass of sulfur, $KHSO_4$ or K_2SO_4?
(c) lower percent by mass of chromium, Na_2CrO_4 or $Na_2Cr_2O_7$?

Empirical and Molecular Formulas

43. Calculate the empirical formula of each compound from the percent compositions given:
(a) 63.6% N, 36.4% O
(b) 46.7% N, 53.3% O
(c) 25.9% N, 74.1% O
(d) 43.4% Na, 11.3% C, 45.3% O
(e) 18.8% Na, 29.0% Cl, 52.3% O
(f) 72.02% Mn, 27.98% O

44. Calculate the empirical formula of each compound from the percent compositions given:
(a) 64.1% Cu, 35.9% Cl
(b) 47.2% Cu, 52.8% Cl
(c) 51.9% Cr, 48.1% S
(d) 55.3% K, 14.6% P, 30.1% O
(e) 38.9% Ba, 29.4% Cr, 31.7% O
(f) 3.99% P, 82.3% Br, 13.7% Cl

45. A sample of tin having a mass of 3.996 g was oxidized and found to have combined with 1.077 g of oxygen. Calculate the empirical formula of this oxide of tin.

46. A 3.054-g sample of vanadium (V) combined with oxygen to form 5.454 g of product. Calculate the empirical formula for this compound.

47. Hydroquinone is an organic compound commonly used as a photographic developer. It has a molar mass of 110.1 g/mol and a composition of 65.45% C, 5.45% H, and 29.09% O. Calculate the molecular formula of hydroquinone.

48. Fructose is a very sweet natural sugar that is present in honey, fruits, and fruit juices. It has a molar mass of 180.1 g/mol and a composition of 40.0% C, 6.7% H, and 53.3% O. Calculate the molecular formula of fructose.

Additional Exercises

These problems are not paired or labeled by topic and provide additional practice on the concepts covered in this chapter.

49. White phosphorus is one of several forms of phosphorus and exists as a waxy solid consisting of P_4 molecules. How many atoms are present in 0.350 mol of P_4?

50. How many grams of sodium contain the same number of atoms as 10.0 g of potassium?

51. One atom of an unknown element is found to have a mass of 1.79×10^{-23} g. What is the molar mass of this element?

*52. If a stack of 500 sheets of paper is 4.60 cm high, what will be the height, in meters, of a stack of Avogadro's number of sheets of paper?

53. There are about 5.0 billion (5.0×10^9) people on earth. If exactly 1 mol of dollars were distributed equally among these people, how many dollars would each person receive?

*54. If 20. drops of water equal 1.0 mL (1.0 cm^3),
(a) how many drops of water are there in a cubic mile of water?
(b) what would be the volume in cubic miles of a mole of drops of water?

***55.** Silver has a density of 10.5 g/cm^3. If 1.00 mol of silver were shaped into a cube,
(a) what would be the volume of the cube?
(b) what would be the length of one side of the cube?

***56.** A sulfuric acid solution contains 65.0% H_2SO_4 by mass and has a density of 1.55 g/mL. How many moles of the acid are present in 1.00 L of the solution?

***57.** A nitric acid solution containing 72.0% HNO_3 by mass has a density of 1.42 g/mL. How many moles of HNO_3 are present in 100. mL of the solution?

58. Given 1.00-g samples of each of the following compounds, CO_2, O_2, H_2O, and CH_3OH,
(a) which sample will contain the largest number of molecules?
(b) which sample will contain the largest number of atoms? Show proof for your answers.

59. How many grams of Fe_2S_3 will contain a total number of atoms equal to Avogadro's number?

60. How many grams of lithium will combine with 20.0 g of sulfur to form the compound Li_2S?

61. Calculate the percentage of
(a) mercury in $HgCO_3$
(b) oxygen in $Ca(ClO_3)_2$
(c) nitrogen in $C_{10}H_{14}N_2$ (nicotine)
(d) Mg in $C_{55}H_{72}MgN_4O_5$ (chlorophyll)

***62.** Zinc and sulfur react to form zinc sulfide, ZnS. If we mix 19.5 g of zinc and 9.40 g of sulfur, have we added sufficient sulfur to fully react all the zinc? Show evidence for your answer.

63. Aspirin is well known as a pain reliever (analgesic) and as a fever reducer (antipyretic). It has a molar mass of 180.2 g/mol and a composition of 60.0% C, 4.48% H, and 35.5% O. Calculate the molecular formula of aspirin.

64. How many grams of oxygen are contained in 8.50 g $Al_2(SO_4)_3$?

65. Gallium arsenide is one of the newer materials used to make semiconductor chips for use in supercomputers. Its composition is 48.2% Ga and 51.8% As. What is the empirical formula?

66. Listed below are the compositions of four different compounds of carbon and chlorine. Determine both the empirical formula and the molecular formula for each.

Percent C	Percent Cl	Molar mass (g)
(a) 7.79	92.21	153.8
(b) 10.13	89.87	236.7
(c) 25.26	74.74	284.8
(d) 11.25	88.75	319.6

67. How many years is a mole of seconds?

68. A normal penny has a mass of about 2.5 g. If we assume the penny to be pure copper (which means the penny is very old since newer pennies are a mixture of copper and zinc), how many atoms of copper does it contain?

69. What would be the mass (in grams) of one thousand trillion molecules of glycerin, $C_3H_8O_3$?

70. If we assume there are 5.0 billion people on the earth, how many moles of people is this?

71. An experimental catalyst used to make polymers has the following composition: Co, 23.3%; Mo, 25.3%; and Cl, 51.4%. What is the empirical formula for this compound?

72. If a student weighs 18 g of aluminum and needs twice as many atoms of magnesium as she has of aluminum, how many grams of Mg does she need?

***73.** If 10.0 g of an unknown compound composed of carbon, hydrogen, and nitrogen contains 17.7% N and 3.8×10^{23} atoms of hydrogen, what is its empirical formula?

74. A substance whose formula is A_2O (A is a mystery element) is 60.0% A and 40.0% O. Identify the element A.

75. For the following compounds whose molecular formulas are given, indicate the empirical formula:
(a) $C_6H_{12}O_6$ glucose
(b) C_8H_{18} octane
(c) $C_3H_6O_3$ lactic acid
(d) $C_{25}H_{52}$ paraffin
(e) $C_{12}H_4Cl_4O_2$ dioxin (a powerful poison)

Answers to Practice Exercises

7.1 10.0 g helium
7.2 1.5×10^{22} atoms
7.3 101.1 g KNO_3
7.4 21.3 g Na_2SO_4
7.5 3.751 mol $AlCl_3$
7.6 24.42% Ca; 17.07% N; 58.50% O
7.7 40.27% K; 26.78% Cr; 32.96% O
7.8 20.16% Al; 79.84% Cl
7.9 C_2H_6O
7.10 P_2O_5
7.11 The empirical formula is CH_3; the molecular formula is C_2H_6.

Chemical Equations

8

Chapter 8

In the world today, much of our energy is directed toward expressing information in a concise, useful manner. From our earliest days in childhood, we are taught to translate ideas and desires into sentences. In mathematics, we learn to translate numerical relationships and situations into mathematical expressions and equations. A historian translates a thousand years of history into a 500-page textbook. A secretary translates an entire letter or document into a few lines of shorthand. A film maker translates an entire event, such as the Olympics, into several hours of entertainment.

And so it is with chemistry. A chemist uses a chemical equation to translate reactions that are observed in the laboratory or in nature. Chemical equations provide the necessary means (1) to summarize the reaction, (2) to display the substances that are reacting, (3) to show the products, and (4) to indicate the amounts of all component substances in the reaction.

8.1 The Chemical Equation

In a chemical reaction, the substances entering the reaction are called *reactants* and the substances formed are called the *products*. During a chemical reaction, atoms, molecules, or ions interact and rearrange themselves to form the products. In this process, chemical bonds are broken and new bonds are formed. The reactants and products may be in the solid, liquid, or gaseous state, or in solution.

chemical equation

A **chemical equation** is a shorthand expression for a chemical change or reaction. A chemical equation uses the chemical symbols and formulas of the reactants and products and other symbolic terms to represent a chemical reaction. The equations are written according to this general format:

1. The reactants are separated from the products by an arrow ($\rightarrow$) that indicates the direction of the reaction. The reactants are placed to the left and the products to the right of the arrow. A plus sign (+) is placed between reactants and between products when needed.
2. Coefficients (whole numbers) are placed in front of substances (e.g., 2 H_2O) to balance the equation and to indicate the number of units (atoms, molecules, moles, ions) of each substance reacting or being produced. When no number is shown, it is understood that one unit of the substance is indicated.
3. Conditions required to carry out the reaction may, if desired, be placed above or below the arrow or equality sign. For example, a delta sign placed over the arrow ($\xrightarrow{\Delta}$) indicates that heat is supplied to the reaction.
4. The physical state of a substance is indicated by the following symbols: (*s*) for solid state: (*l*) for liquid state; (*g*) for gaseous state; and (*aq*) for substances in aqueous solution. States are not always given in chemical equations.

The symbols commonly used in equations are given in Table 8.1.

Chapter Opening Photo: The thermite reaction is a reaction between elemental aluminum and iron oxide. This reaction produces so much energy that the iron becomes molten. The thermite reaction is used to weld railroad rails.

TABLE 8.1 Symbols Commonly Used in Chemical Equations

Symbol	Meaning
$\rightarrow$	Yields; produces (points to products)
(s)	Solid state (written after a substance)
(l)	Liquid state (written after a substance)
(g)	Gaseous state (written after a substance)
(aq)	Aqueous solution (substance dissolved in water)
Δ	Heat (written above arrow)
$+$	Plus or added to (placed between substances)

8.2 Writing and Balancing Equations

To represent the quantitative relationships of a reaction, the chemical equation must be balanced. A **balanced equation** contains the same number of each kind of atom on each side of the equation. The balanced equation, therefore, obeys the Law of Conservation of Mass.

balanced equation

Every chemistry student must learn to *balance* equations. Simple equations are easy to balance, but care and attention to detail are required. The way to balance an equation is to adjust the number of atoms of each element so that it is the same on each side of the equation, but a correct formula must not be changed in order to balance an equation. The following is a general procedure for balancing equations. Study this outline and refer to it as needed when working examples.

Step 1. Identify the Reaction for Which the Equation Is to Be Written. Formulate a description or word equation for the reaction if needed (e.g., mercury(II) oxide decomposes yielding mercury and oxygen).

Step 2. Write the Unbalanced, or Skeleton, Equation. Make sure that the formula for each substance is correct and that the reactants are written to the left and the products to the right of the arrow (e.g., $HgO \rightarrow Hg + O_2$). The correct formulas must be known or ascertained from the periodic table, lists of ions, or experimental data.

Step 3. Balance the Equation. Use the following steps as necessary:

(a) Count and compare the number of atoms of each element on each side of the equation and determine those that must be balanced.

(b) Balance each element, one at a time, by placing whole numbers (coefficients) in front of the formulas containing the unbalanced element. It is usually best to balance metals first, then nonmetals, then hydrogen and oxygen. Select the smallest coefficients that will give the same number of atoms of the element on each side. A coefficient placed before a formula multiplies every atom in the formula by that number (e.g., 2 H_2SO_4 means two molecules of sulfuric acid and also means four H atoms, two S atoms, and eight O atoms.)

It is often helpful to leave elements that are in two or more formulas on the same side of the equation until just before balancing hydrogen and oxygen.

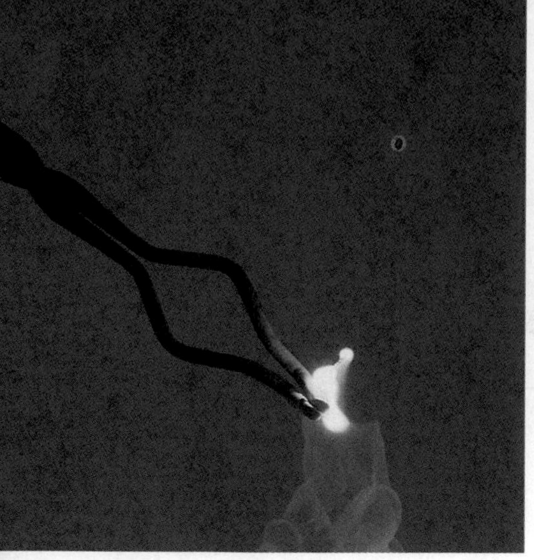

A burst of UV light is emitted as metallic magnesium combusts in air, producing magnesium oxide.

(c) Check all other elements after each individual element is balanced to see whether, in balancing one element, other elements have become unbalanced. Make adjustments as needed.
(d) Balance polyatomic ions such as SO_4^{2-}, which remain unchanged from one side of the equation to the other, in the same way as individual atoms.
(e) Do a final check, making sure that each element and/or polyatomic ion is balanced and that the smallest possible set of whole-number coefficients has been used.

$$4\ HgO \longrightarrow 4\ Hg + 2\ O_2 \quad \text{(incorrect form)}$$

$$2\ HgO \longrightarrow 2\ Hg + O_2 \quad \text{(correct form)}$$

Not all chemical equations can be balanced by the simple method of inspection just described. The following examples show *stepwise* sequences leading to balanced equations. Study each one carefully.

Example 8.1 Write the balanced equation for the reaction that takes place when magnesium metal is burned in air to produce magnesium oxide.

Solution

Step 1 *Word equation:*

magnesium + oxygen ⟶ magnesium oxide

Step 2 *Skeleton equation:*

$$Mg + O_2 \longrightarrow MgO \quad \text{(unbalanced)}$$

Step 3 *Balance:*

(a) Mg is balanced.
(b) Oxygen is not balanced. Two O atoms appear on the left side and one on the right side.
Place the coefficient 2 before MgO:

$$Mg + O_2 \longrightarrow 2\ MgO \quad \text{(unbalanced)}$$

(c) Now Mg is not balanced. One Mg atom appears on the left side and two on the right side. Place a 2 before Mg:

$$2\ Mg + O_2 \longrightarrow 2\ MgO \quad \text{(balanced)}$$

(d) *Check:* Each side has two Mg and two O atoms.

Example 8.2 When methane, CH_4, undergoes complete combustion, it reacts with oxygen to produce carbon dioxide and water. Write the balanced equation for this reaction.

Solution

Step 1 *Word equation:*

methane + oxygen ⟶ carbon dioxide + water

Step 2 *Skeleton equation:*

$$CH_4 + O_2 \longrightarrow CO_2 + H_2O \quad \text{(unbalanced)}$$

Step 3 *Balance:*

(a) Carbon is balanced.

(b) Hydrogen and oxygen are not balanced. Balance H atoms by placing a 2 before H_2O:

$$CH_4 + O_2 \longrightarrow CO_2 + 2\ H_2O \quad \text{(unbalanced)}$$

Each side of the equation has four H atoms; oxygen is still not balanced. Place a 2 before O_2 to balance the oxygen atoms:

$$CH_4 + 2\ O_2 \longrightarrow CO_2 + 2\ H_2O \quad \text{(balanced)}$$

(c) *Check:* The equation is correctly balanced; it has one C, four O, and four H atoms on each side.

Example 8.3

Oxygen and potassium chloride are formed by heating potassium chlorate. Write a balanced equation for this reaction.

Solution

Step 1 *Word equation:*

$$\text{potassium chlorate} \xrightarrow{\Delta} \text{potassium chloride} + \text{oxygen}$$

Step 2 *Skeleton equation:*

$$KClO_3 \xrightarrow{\Delta} KCl + O_2 \quad \text{(unbalanced)}$$

Step 3 *Balance:*

(a) Potassium and chlorine are balanced.

(b) Oxygen is unbalanced (three O atoms on the left and two on the right side).

(c) How many oxygen atoms are needed? The subscripts of oxygen (3 and 2) in $KClO_3$ and O_2 have a least common multiple of 6. Therefore, coefficients for $KClO_3$ and O_2 are needed to give six O atoms on each side. Place a 2 before $KClO_3$ and a 3 before O_2 to give six O atoms on each side:

$$2\ KClO_3 \xrightarrow{\Delta} KCl + 3\ O_2 \quad \text{(unbalanced)}$$

Now K and Cl are not balanced. Place a 2 before KCl, which balances both K and Cl at the same time:

$$2\ KClO_3 \xrightarrow{\Delta} 2\ KCl + 3\ O_2 \quad \text{(balanced)}$$

(d) *Check:* Each side now contains two K, two Cl, and six O atoms.

Example 8.4

Silver nitrate reacts with hydrogen sulfide to produce silver sulfide and nitric acid. Write a balanced equation for this reaction.

Solution

Step 1 *Word equation:*

$$\text{silver nitrate} + \text{hydrogen sulfide} \longrightarrow \text{silver sulfide} + \text{nitric acid}$$

Step 2 *Skeleton equation:*

$$AgNO_3 + H_2S \longrightarrow Ag_2S + HNO_3 \quad \text{(unbalanced)}$$

Step 3 *Balance:*

(a) Ag and H are unbalanced.

(b) Place a 2 in front of $AgNO_3$ to balance Ag:

$$2\ AgNO_3 + H_2S \longrightarrow Ag_2S + HNO_3 \quad \text{(unbalanced)}$$

(c) H and NO_3^- are still unbalanced. Balance by placing a 2 in front of HNO_3:

$$2\ AgNO_3 + H_2S \longrightarrow Ag_2S + 2\ HNO_3 \quad \text{(balanced)}$$

(d) In this example N and O atoms are balanced by balancing the NO_3^- ion as a unit.

(e) *Check:* Each side has two Ag, two H, and one S atom. Also, each side has two NO_3^- ions.

Example 8.5 When aluminum hydroxide is mixed with sulfuric acid the products are aluminum sulfate and water. Write a balanced equation for this reaction.

Solution

Step 1 *Word equation:*

aluminum hydroxide + sulfuric acid ⟶ aluminum sulfate + water

Step 2 *Skeleton equation:*

$$Al(OH)_3 + H_2SO_4 \longrightarrow Al_2(SO_4)_3 + H_2O \quad \text{(unbalanced)}$$

Step 3 *Balance:*

(a) All elements are unbalanced.

(b) Balance Al by placing a 2 in front of $Al(OH)_3$. Treat the unbalanced SO_4^{2-} ion as a unit and balance by placing a 3 before H_2SO_4:

$$2\ Al(OH)_3 + 3\ H_2SO_4 \longrightarrow Al_2(SO_4)_3 + H_2O \quad \text{(unbalanced)}$$

(c) Balance the unbalanced H and O by placing a 6 in front of H_2O:

$$2\ Al(OH)_3 + 3\ H_2SO_4 \longrightarrow Al_2(SO_4)_3 + 6\ H_2O \quad \text{(balanced)}$$

(d) *Check:* Each side has two Al, twelve H, three S, and eighteen O atoms.

Example 8.6 When the fuel in a butane gas stove undergoes complete combustion, it reacts with oxygen to form carbon dioxide and water. Write the balanced equation for this reaction.

Solution

Step 1 *Word equation:*

butane + oxygen ⟶ carbon dioxide + water

Step 2 *Skeleton equation:*

$$C_4H_{10} + O_2 \longrightarrow CO_2 + H_2O \quad \text{(unbalanced)}$$

Step 3 *Balance:*

(a) All elements are unbalanced.

(b) Balance C by placing a 4 in front of CO_2:

$$C_4H_{10} + O_2 \longrightarrow 4\ CO_2 + H_2O \quad \text{(unbalanced)}$$

Balance H by placing a 5 in front of H_2O:

$$C_4H_{10} + O_2 \longrightarrow 4\ CO_2 + 5\ H_2O \quad \text{(unbalanced)}$$

Oxygen remains unbalanced. The oxygen atoms on the right side are fixed, because 4 CO_2 and 5 H_2O are derived from the single C_4H_{10} molecule on the left. When we try to balance the O atoms, we find that there is no whole number that can be placed in front of O_2 to bring about a balance. The equation can be balanced if we use $6\frac{1}{2}\ O_2$ and then double the coefficients of each substance, including the $6\frac{1}{2}\ O_2$, to obtain the balanced equation:

$$C_4H_{10} + 6\tfrac{1}{2}\,O_2 \longrightarrow 4\ CO_2 + 5\ H_2O \quad \text{(balanced—incorrect form)}$$

$$2\ C_4H_{10} + 13\ O_2 \longrightarrow 8\ CO_2 + 10\ H_2O \quad \text{(balanced)}$$

(c) *Check:* Each side now has eight C, twenty H, and twenty-six O atoms.

Cooking with gas. The fuel in this butane stove undergoes complete combustion as it reacts with oxygen to form CO_2 and water.

Practice 8.1

Balance the following word equation.

aluminum + oxygen ⟶ aluminum oxide

Practice 8.2

Balance the following word equation.

magnesium hydroxide + phosphoric acid ⟶ magnesium phosphate + water

8.3 What Information Does an Equation Tell Us?

Depending on the particular context in which it is used, a formula can have different meanings. The meanings refer either to an individual chemical entity (atom, ion, molecule, or formula unit) or to a mole of that chemical entity. For example, the formula H_2O can be used to indicate any of the following:

1. 2 H atoms and 1 O atom
2. 1 molecule of water
3. 1 mol of water
4. 6.022×10^{23} molecules of water
5. 18.02 g of water

Chemistry in Action

Autumn Leaf Color

Chemical reactions can be the source of much natural beauty, even as they contribute to the metabolic cycle in plants. The wondrous array of color seen in the leaf display of autumn is the result of chemical reactions.

Chlorophylls are responsible for the common green color in plants and are necessary for the plant to produce food in the process called *photosynthesis.* Photosynthesis involves many chemical reactions but can be summarized by the following equation:

$$\underset{\text{carbon dioxide}}{6\,CO_2} + \underset{\text{water}}{6\,H_2O} \xrightarrow[\text{sunlight}]{\text{chlorophyll}} \underset{\text{glucose}}{C_6H_{12}O_6} + \underset{\text{oxygen}}{6\,O_2}$$

There are several types of chlorophyll, including chlorophyll *a*, chlorophyll *b*, and chlorophyll *c*. All photosynthetic plants contain chlorophyll *a*, and some contain chlorophyll *b* and chlorophyll *c*, which are called *accessory pigments.* The additional pigments extend the range of colored light that plants can use for photosynthesis. In addition, plants also contain carotenoids (orange, yellow, and red pigments), which protect the plant from the destructive potential of chlorophyll. The carotenoids absorb the high-energy oxygen that is released when the chlorophyll absorbs light energy, and they release the oxygen when it can be used constructively. The least prevalent pigments are *anthocyanins* (red or blue). These pigments are responsible for the great diversity of combinations and concentrations of colors in the leaves. The carotenoids and anthocyanins are masked by the chlorophylls and can only be seen as the chlorophyll disintegrates in the autumn.

Chilly nights cause chlorophyll to decompose in leaves. As the green fades away, other pigments emerge, giving us our beautiful fall colors.

To produce brilliant autumn color, the proper set of conditions must exist. The trees need a long, vigorous growing season with plenty of water for photosynthesis. The autumn days should be bright to provide a longer photosynthetic period, with plenty of sugar production. When less intense sunlight and cold nights signal the leaves to stop photosynthesis, the protein bound to the chlorophyll begins to release from it. The protein breaks down into amino acids that go to the roots for storage. The chlorophyll decomposes and the green color fades away. The other pigments, especially the yellows and reds of the carotenoids, can now be seen. Sugar, trapped in the leaves during the very chilly nights, is changed through a set of complex reactions into anthocyanins to produce a variety of colors, mostly reds. The more productive a leaf has been, the greater the concentration of the other pigments and the more brilliant the color.

Formulas used in equations can be expressed in units of individual chemical entities or as moles, the latter being more commonly used. For example, in the reaction of hydrogen and oxygen to form water,

$$2\,H_2 + O_2 \longrightarrow 2\,H_2O$$

We usually use moles in equations because molecules are so small that we generally work with large collections at once.

the 2 H_2 can represent 2 molecules or 2 mol of hydrogen; the O_2, 1 molecule or 1 mol of oxygen; and the 2 H_2O, 2 molecules or 2 mol of water. In terms of moles, this equation is stated: 2 mol of H_2 react with 1 mol of O_2 to give 2 mol of H_2O.

As indicated earlier, a chemical equation is a shorthand description of a chemical reaction. Interpretation of a balanced equation gives us the following information:

1. What the reactants are and what the products are
2. The formulas of the reactants and products

3. The number of molecules or formula units of reactants and products in the reaction
4. The number of atoms of each element involved in the reaction
5. The number of moles of each substance
6. The number of grams of each substance used or produced

Consider the equation

$$H_2(g) + Cl_2(g) \longrightarrow 2\ HCl(g)$$

Here, hydrogen gas reacts with chlorine gas to produce hydrogen chloride, also a gas. Let's summarize all the information that can be stated about the relative amount of each substance, with respect to all other substances in the balanced equation:

Hydrogen		Chlorine		Hydrogen chloride
$H_2(g)$	+	$Cl_2(g)$	→	$2\ HCl(g)$
1 molecule 2 atoms H 1 molar mass 1 mol 2.016 g		1 molecule 2 atoms Cl 1 molar mass 1 mol 70.90 g		2 molecules 2 atoms H + 2 atoms Cl 2 molar masses 2 mol 2×36.46 g or 72.92 g

These data are very useful in calculating quantitative relationships that exist among substances in a chemical reaction. For example, if we react 2 mol of hydrogen (twice as much as is indicated by the equation) with 2 mol of chlorine, we can expect to obtain 4 mol, or 145.8 g, of hydrogen chloride as a product. We will study this phase of using equations in more detail in the next chapter.

Let us try another equation. When propane gas (C_3H_8) is burned in air, the products are carbon dioxide, CO_2, and water, H_2O. The balanced equation and its interpretation are as follows:

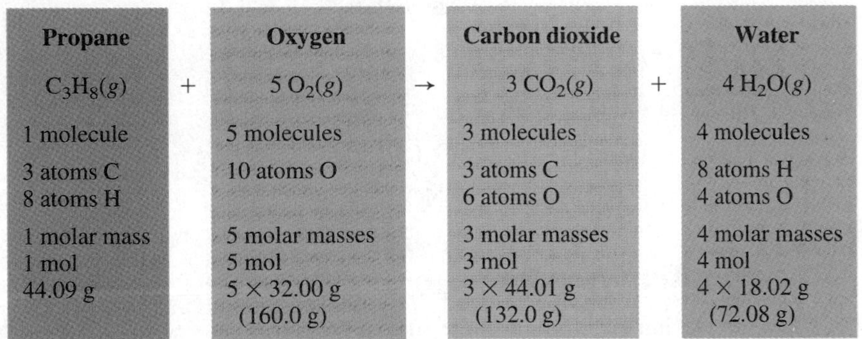

Propane		Oxygen		Carbon dioxide		Water
$C_3H_8(g)$	+	$5\ O_2(g)$	→	$3\ CO_2(g)$	+	$4\ H_2O(g)$
1 molecule		5 molecules		3 molecules		4 molecules
3 atoms C 8 atoms H		10 atoms O		3 atoms C 6 atoms O		8 atoms H 4 atoms O
1 molar mass 1 mol 44.09 g		5 molar masses 5 mol 5×32.00 g (160.0 g)		3 molar masses 3 mol 3×44.01 g (132.0 g)		4 molar masses 4 mol 4×18.02 g (72.08 g)

8.4 Types of Chemical Equations

Chemical equations represent chemical changes or reactions. Reactions are classified into types to assist in writing equations and in predicting other reactions. Many chemical reactions fit one or another of the four principal reaction types that we discuss in the following paragraphs. Reactions are also classified as oxidation–reduction. Special methods are used to balance complex oxidation–reduction equations.

Combination, or Synthesis, Reaction

combination reaction

In a **combination reaction**, two reactants combine to give one product. The general form of the equation is

$$A + B \longrightarrow AB$$

in which A and B are either elements or compounds and AB is a compound. The formula of the compound in many cases can be determined from a knowledge of the ionic charges of the reactants in their combined states. Some reactions that fall into this category are given here.

(a) metal + oxygen $\longrightarrow$ metal oxide

$$2\ Mg(s) + O_2(g) \xrightarrow{\Delta} 2\ MgO(s)$$
$$4\ Al(s) + 3\ O_2(g) \xrightarrow{\Delta} 2\ Al_2O_3(s)$$

(b) nonmetal + oxygen $\longrightarrow$ nonmetal oxide

$$S(s) + O_2(g) \xrightarrow{\Delta} SO_2(g)$$
$$N_2(g) + O_2(g) \xrightarrow{\Delta} 2\ NO(g)$$

(c) metal + nonmetal $\longrightarrow$ salt:

$$2\ Na(s) + Cl_2(g) \longrightarrow 2\ NaCl(s)$$
$$2\ Al(s) + 3\ Br_2(l) \longrightarrow 2\ AlBr_3(s)$$

(d) metal oxide + water $\longrightarrow$ metal hydroxide

$$Na_2O(s) + H_2O(l) \longrightarrow 2\ NaOH(aq)$$
$$CaO(s) + H_2O(l) \longrightarrow Ca(OH)_2(aq)$$

(e) nonmetal oxide + water $\longrightarrow$ oxy-acid

$$SO_3(g) + H_2O(l) \longrightarrow H_2SO_4(aq)$$
$$N_2O_5(s) + H_2O(l) \longrightarrow 2\ HNO_3(aq)$$

Decomposition Reaction

decomposition reaction

In a **decomposition reaction**, a single substance is decomposed or broken down to give two or more different substances. The reaction may be considered the reverse of combination. The starting material must be a compound, and the products may be elements or compounds. The general form of the equation is

$$AB \longrightarrow A + B$$

Predicting the products of a decomposition reaction can be difficult and requires an understanding of each individual reaction. Heating oxygen-containing compounds often results in decomposition. Some reactions that fall into this category are:

(a) Metal oxides. Some metal oxides decompose to yield the free metal plus oxygen; others give another oxide, and some are very stable, resisting decomposition by heating:

$$2\ HgO(s) \xrightarrow{\Delta} 2\ Hg(l) + O_2(g)$$

$$2\ PbO_2(s) \xrightarrow{\Delta} 2\ PbO(s) + O_2(g)$$

(b) Carbonates and hydrogen carbonates decompose to yield CO_2 when heated:

$$CaCO_3(s) \xrightarrow{\Delta} CaO(s) + CO_2(g)$$

$$2\ NaHCO_3(s) \xrightarrow{\Delta} Na_2CO_3(s) + H_2O(g) + CO_2(g)$$

(c) Miscellaneous:

$$2\ KClO_3(s) \xrightarrow{\Delta} 2\ KCl(s) + 3\ O_2(g)$$

$$2\ NaNO_3(s) \xrightarrow{\Delta} 2\ NaNO_2(g) + O_2(g)$$

$$2\ H_2O_2(l) \xrightarrow{\Delta} 2\ H_2O(l) + O_2(g)$$

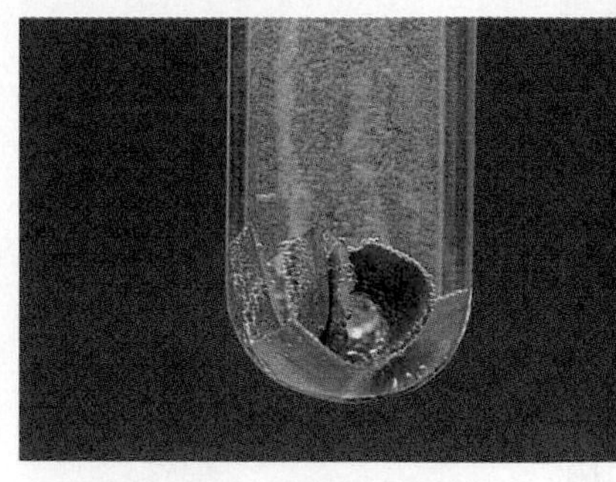

A single-displacement reaction occurs as elemental zinc reacts with hydrochloric acid to produce bubbles of hydrogen.

Single-Displacement Reaction

single-displacement reaction

In a **single-displacement reaction** one element reacts with a compound to take the place of one of the elements of that compound, yielding a different element and a different compound. The general form of the equation is

$$A + BC \longrightarrow B + AC \quad \text{or} \quad A + BC \longrightarrow C + BA$$

If A is a metal, A will replace B to form AC, provided A is a more reactive metal than B. If A is a halogen, it will replace C to form BA, provided A is a more reactive halogen than C.

A brief activity series of selected metals (and hydrogen) and halogens are shown in Table 8.2. This series is listed in descending order of chemical reactivity, with the most active metals and halogens at the top. From such series it is possible to predict many chemical reactions. In an activity series, the atoms of any element in the series will replace the atoms of those elements below it. For example, zinc metal will replace hydrogen from a hydrochloric acid solution. But copper metal, which is underneath hydrogen on the list and thus less reactive than hydrogen, will not replace hydrogen from a hydrochloric acid solution. Some reactions that fall into this category follow:

TABLE 8.2 Activity Series (increasing activity ↑)

Metals	Halogens
K	F_2
Ca	Cl_2
Na	Br_2
Mg	I_2
Al	
Zn	
Fe	
Ni	
Sn	
Pb	
H	
Cu	
Ag	
Hg	
Au	

(a) metal + acid ⟶ hydrogen + salt

$$Zn(s) + 2\ HCl(aq) \longrightarrow H_2(g) + ZnCl_2(aq)$$

$$2\ Al(s) + 3\ H_2SO_4(aq) \longrightarrow 3\ H_2(g) + Al_2(SO_4)_3(aq)$$

(b) metal + water ⟶ hydrogen + metal hydroxide or metal oxide

$$2\ Na(s) + 2\ H_2O \longrightarrow H_2(g) + 2\ NaOH(aq)$$

$$Ca(s) + 2\ H_2O \longrightarrow H_2(g) + Ca(OH)_2(aq)$$

$$3\ Fe(s) + \underset{\text{steam}}{4\ H_2O(g)} \longrightarrow 4\ H_2(g) + Fe_3O_4(s)$$

(c) metal + salt ⟶ metal + salt

$Fe(s) + CuSO_4(aq) \longrightarrow Cu(s) + FeSO_4(aq)$

$Cu(s) + 2\ AgNO_3(aq) \longrightarrow 2\ Ag(s) + Cu(NO_3)_2(aq)$

(d) halogen + halide salt ⟶ halogen + halide salt

$Cl_2(g) + 2\ NaBr(aq) \longrightarrow Br_2(l) + 2\ NaCl(aq)$

$Cl_2(g) + 2\ KI(aq) \longrightarrow I_2(s) + 2\ KCl(aq)$

A common chemical reaction is the displacement of hydrogen from water or acids. This reaction is a good illustration of the relative reactivity of metals and the use of the activity series. For example,

- K, Ca, and Na displace hydrogen from cold water, steam, and acids.
- Mg, Al, Zn, and Fe displace hydrogen from steam and acids.
- Ni, Sn, and Pb displace hydrogen only from acids.
- Cu, Ag, Hg, and Au do not displace hydrogen.

Example 8.7 Will a reaction occur between (a) nickel metal and hydrochloric acid and (b) tin metal and a solution of aluminum chloride? Write balanced equations for the reactions.

Solution (a) Nickel is more reactive than hydrogen, so it will displace hydrogen from hydrochloric acid. The products are hydrogen gas and a salt of Ni^{2+} and Cl^- ions:

$Ni(s) + 2\ HCl(aq) \longrightarrow H_2(g) + NiCl_2(aq)$

(b) According to the activity series, tin is less reactive than aluminum, so no reaction will occur:

$Sn(s) + AlCl_3(aq) \longrightarrow$ no reaction

Practice 8.3

Write balanced equations for the reactions:
(a) iron metal and a solution of magnesium chloride
(b) zinc metal and a solution of lead(II) nitrate

Double-Displacement, or Metathesis, Reaction

double-displacement reaction

In a **double-displacement reaction**, two compounds exchange partners with each other to produce two different compounds. The general form of the equation is

$$AB + CD \longrightarrow AD + CB$$

This reaction can be thought of as an exchange of positive and negative groups, in which A combines with D and C combines with B. In writing formulas for the products, we must account for the charges of the combining groups.

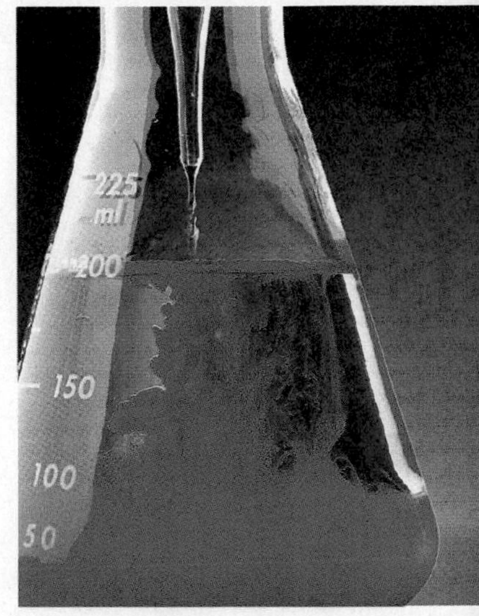

A double-displacement reaction results from pouring a clear, colorless solution of $Pb(NO_3)_2$ into a clear, colorless solution of KI forming a yellow precipitate of PbI_2.

It is also possible to write an equation in the form of a double-displacement reaction when a reaction has not occurred. For example, when solutions of sodium chloride and potassium nitrate are mixed, the following equation can be written:

$$NaCl(aq) + KNO_3(aq) \longrightarrow NaNO_3(aq) + KCl(aq)$$

When the procedure is carried out, no physical changes are observed, indicating that no chemical reaction has taken place.

A double-displacement reaction is accompanied by evidence of such reactions as the evolution of heat, the formation of an insoluble precipitate, or the production of gas bubbles. We will now look at some of these reactions.

Neutralization of an acid and a base. The production of a molecule of water from an H^+ and an OH^- ion is accompanied by a release of heat, which can be detected by touching the reaction container. For neutralization reactions, $H^+ + OH^- \longrightarrow H_2O$.

$$\text{acid} + \text{base} \longrightarrow \text{salt} + \text{water}$$

$$HCl(aq) + NaOH(aq) \longrightarrow NaCl(aq) + H_2O(l)$$

$$H_2SO_4(aq) + Ba(OH)_2(aq) \longrightarrow BaSO_4(s) + 2\ H_2O(l)$$

Formation of an insoluble precipitate. The solubilities of the products can be determined by consulting the Solubility Table in Appendix IV. One or both of the products may be insoluble.

$$BaCl_2(aq) + 2\ AgNO_3(aq) \longrightarrow 2\ AgCl(s) + Ba(NO_3)_2(aq)$$

$$FeCl_3(aq) + 3\ NaOH(aq) \longrightarrow Fe(OH)_3(s) + 3\ NaCl(aq)$$

Metal oxide + acid. Heat is released by the production of a molecule of water.

$$\text{metal oxide} + \text{acid} \longrightarrow \text{salt} + \text{water}$$

$$CuO(s) + 2\ HNO_3(aq) \longrightarrow Cu(NO_3)_2(aq) + H_2O(l)$$

$$CaO(s) + 2\ HCl(aq) \longrightarrow CaCl_2(aq) + H_2O(l)$$

Formation of a gas. A gas such as HCl or H_2S may be produced directly, as in these two examples:

$$H_2SO_4(l) + NaCl(s) \longrightarrow NaHSO_4(s) + HCl(g)$$

$$2\ HCl(aq) + ZnS(s) \longrightarrow ZnCl_2(aq) + H_2S(g)$$

A gas can also be produced indirectly. Some unstable compounds formed in a double-displacement reaction, such as H_2CO_3, H_2SO_3, and NH_4OH, will decompose to form water and a gas:

$$2\ HCl(aq) + Na_2CO_3(aq) \longrightarrow 2\ NaCl(aq) + H_2CO_3(aq) \longrightarrow 2\ NaCl(aq) + H_2O(l) + CO_2(g)$$

$$2\ HNO_3(aq) + K_2SO_3(aq) \longrightarrow 2\ KNO_3(aq) + H_2SO_3(aq) \longrightarrow 2\ KNO_3(aq) + H_2O(l) + SO_2(g)$$

$$NH_4Cl(aq) + NaOH(aq) \longrightarrow NaCl(aq) + NH_4OH(aq) \longrightarrow NaCl(aq) + H_2O(l) + NH_3(g)$$

Example 8.8 Write the equation for the reaction between aqueous solutions of hydrobromic acid and potassium hydroxide.

Solution First write the formulas for the reactants. (They are HBr and KOH.) Then classify the type of reaction that would occur between them. Because the reactants are compounds, one an acid and the other a base, the reaction will be of the neutralization type:

$$\text{acid} + \text{base} \longrightarrow \text{salt} + \text{water}$$

Now rewrite the equation using the formulas for the known substances:

$$HBr(aq) + KOH(aq) \longrightarrow \text{salt} + H_2O$$

In this reaction, which is a double-displacement type, the H^+ from the acid combines with the OH^- from the base to form water. The ionic compound must be composed of the other two ions, K^+ and Br^-. We determine the formula of the ionic compound to be KBr from the fact that K is a $+1$ cation and Br is a -1 anion. The final balanced equation is

$$HBr(aq) + KOH(aq) \longrightarrow KBr(aq) + H_2O(l)$$

Example 8.9 Complete and balance the equation for the reaction between aqueous solutions of barium chloride and sodium sulfate.

Solution First determine the formula for the reactants. (They are $BaCl_2$ and Na_2SO_4.) Then classify these substances as acids, bases, or ionic compounds. (Both substances are salts.) Since both substances are compounds, the reaction will be of the double-displacement type. Start writing the equation with the reactants:

$$BaCl_2(aq) + Na_2SO_4(aq) \longrightarrow$$

If the reaction is double-displacement, Ba^{2+} will be written combined with SO_4^{2-}, and Na^+ with Cl^- as the products. The balanced equation is

$$BaCl_2(aq) + Na_2SO_4(aq) \longrightarrow BaSO_4 + 2\ NaCl$$

The final step is to determine the nature of the products, which controls whether or not the reaction will take place. If both products are soluble, we may merely have a mixture of all the ions in solution. But if an insoluble precipitate is formed, the reaction will definitely occur. We know from experience that NaCl is fairly soluble in water, but what about $BaSO_4$? The Solubility Table in Appendix IV can give us this information. From this table we see that $BaSO_4$ is insoluble in water, so it will be a precipitate in the reaction. Thus the reaction will occur, forming a precipitate. The equation is

$$BaCl_2(aq) + Na_2SO_4(aq) \longrightarrow BaSO_4(s) + 2\ NaCl(aq)$$

A double-displacement reaction occurs between solutions of barium chloride and sodium sulfate to form a sodium chloride solution (colorless) and a precipitate of barium sulfate (white).

Practice 8.4

Complete and balance the equations for the reactions:
(a) potassium phosphate + barium chloride
(b) hydrochloric acid + nickel carbonate
(c) ammonium chloride + sodium nitrate

Some of the reactions we attempt may fail because the substances are not reactive or because the proper conditions for reaction are not present. For example, mercury(II) oxide does not decompose until it is heated; magnesium does not burn in air or oxygen until the temperature reaches a certain point. When silver is placed in a solution of copper(II) sulfate, no reaction occurs. When a strip of copper is placed in a solution of silver nitrate, a single-displacement reaction takes place, because copper is a more reactive metal than silver.

In this single-displacement reaction between a strip of copper and a solution of silver nitrate, crystals of silver metal are formed and the solution turns blue, indicating the presence of copper ions.

The successful prediction of the products of a reaction is not always easy. The ability to predict products correctly comes with knowledge and experience. Although you may not be able to predict many reactions at this point, as you continue to experiment you will find that reactions can be categorized, and that prediction of the products thereby becomes easier, if not always certain.

We have a great deal yet to learn about which substances react with each other, how they react, and what conditions are necessary to bring about their reaction. It is possible to make accurate predictions concerning the occurrence of proposed reactions, but they require, in addition to appropriate data, a good knowledge of thermodynamics—a subject usually reserved for advanced courses in chemistry and physics. Even without the formal use of thermodynamics, your knowledge of the four general reaction types, the periodic table, atomic structure, and charges of ions can be put to good use in predicting reactions and in writing equations.

8.5 Heat in Chemical Reactions

Energy changes always accompany chemical reactions. One reason why reactions occur is that the products attain a lower, more stable energy state than the reactants. When the reaction leads to a more stable state, energy is released to the surroundings as heat (or as heat and work). When a solution of a base is neutralized by the addition of an acid, the liberation of heat energy is signaled by an immediate rise in the temperature of the solution. For example, when an automobile engine burns gasoline, heat is certainly liberated; at the same time, part of the liberated energy does the work of moving the automobile.

exothermic reaction
endothermic reaction

Reactions are either exothermic or endothermic. **Exothermic reactions** liberate heat; **endothermic reactions** absorb heat. In an exothermic reaction, heat is a product and may be written on the right side of the equation for the reaction. In an endothermic reaction, heat can be regarded as a reactant and is written on the left side of the equation. Here are two:

$$H_2(g) + Cl_2(g) \longrightarrow 2\,HCl(g) + 185\text{ kJ} \quad (\textit{exothermic})$$

$$N_2(g) + O_2(g) + 181\text{ kJ} \longrightarrow 2\,NO(g) \quad (\textit{endothermic})$$

heat of reaction

The quantity of heat produced by a reaction is known as the **heat of reaction**. The units used can be kilojoules or kilocalories. Consider the reaction represented by this equation:

$$C(s) + O_2(g) \longrightarrow CO_2(g) + 393\text{ kJ}$$

When the heat liberated is expressed as part of the equation, the substances are expressed in units of moles. Thus, when 1 mol (12.01 g) of C combines with 1 mol (32.00 g) of O_2, 1 mol (44.01 g) of CO_2 is formed and 393 kJ of heat are liberated.

This cornfield is a good example of the endothermic reactions happening through photosynthesis in plants.

In this reaction, as in many others, the heat energy is more useful than the chemical products.

Aside from relatively small amounts of energy from nuclear processes, the sun is the major provider of energy for life on earth. The sun maintains the temperature necessary for life and also supplies light energy for the endothermic photosynthetic reactions of green plants. In photosynthesis, carbon dioxide and water are converted to free oxygen and glucose:

$$6\ CO_2 + 6\ H_2O + 2519\ kJ \longrightarrow \underset{\text{glucose}}{C_6H_{12}O_6} + 6\ O_2$$

Nearly all of the chemical energy used by living organisms is obtained from glucose or compounds derived from glucose.

The major source of energy for modern technology is fossil fuel, such as coal, petroleum, and natural gas. The energy is obtained from the combustion (burning) of these fuels, which are converted to carbon dioxide and water. Fossil fuels are mixtures of **hydrocarbons**, compounds containing only hydrogen and carbon.

hydrocarbon

Natural gas is primarily methane, CH_4. Petroleum is a mixture of hydrocarbons (compounds of carbon and hydrogen). Liquefied petroleum gas (LPG) is a mixture of propane (C_3H_8) and butane (C_4H_{10}).

The combustion of these fuels releases a tremendous amount of energy, but reactions won't occur to a significant extent at ordinary temperatures. A spark or a flame must be present before methane will ignite. The amount of energy that must be supplied to start a chemical reaction is called the **activation energy**. Once this activation energy is provided, enough energy is then generated to keep the reaction going. Here are some examples:

activation energy

$$CH_4(g) + 2\ O_2(g) \longrightarrow CO_2(g) + 2\ H_2O(g) + 890\ kJ$$

$$C_3H_8(g) + 5\ O_2(g) \longrightarrow 3\ CO_2(g) + 4\ H_2O(g) + 2200\ kJ$$

$$2\ C_8H_{18}(l) + 25\ O_2(g) \longrightarrow 16\ CO_2(g) + 18\ H_2O(g) + 10{,}900\ kJ$$

Be careful not to confuse an exothermic reaction that merely requires heat (activation energy) to get it started with a truly endothermic process. The combustion of magnesium, for example, is highly exothermic, yet magnesium must be heated to a fairly high temperature in air before combustion begins. Once started, however, the combustion reaction goes very vigorously until either the magnesium or the available supply of oxygen is exhausted. The electrolytic decomposition of water to hydrogen and oxygen is highly endothermic. If the electric current is shut off when this process is going on, the reaction stops instantly. The relative energy levels of reactants and products in exothermic and in endothermic processes are presented graphically in Figure 8.1.

In reaction (a) of Figure 8.1, the products are at a lower potential energy than the reactants. Energy (heat) is given off, producing an exothermic reaction. In reaction (b) the products are at a higher potential energy than the reactants. Energy has therefore been absorbed, and the reaction is endothermic.

Examples of endothermic and exothermic processes can be easily demonstrated in the laboratory. In Figure 8.2, solid $Ba(OH)_2$ and solid NH_4SCN are mixed in a beaker, which is standing in a puddle of water. The solids liquefy and absorb heat from the surroundings causing the beaker to freeze to the board. In another demonstration, potassium chlorate ($KClO_3$) and sugar are mixed and placed into a pile. A drop of concentrated sulfuric acid is added, creating a spectacular exothermic reaction (Figure 8.2).

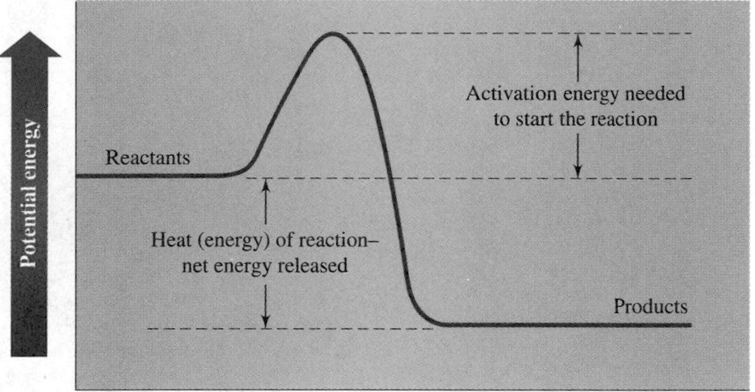

(a) Exothermic reaction

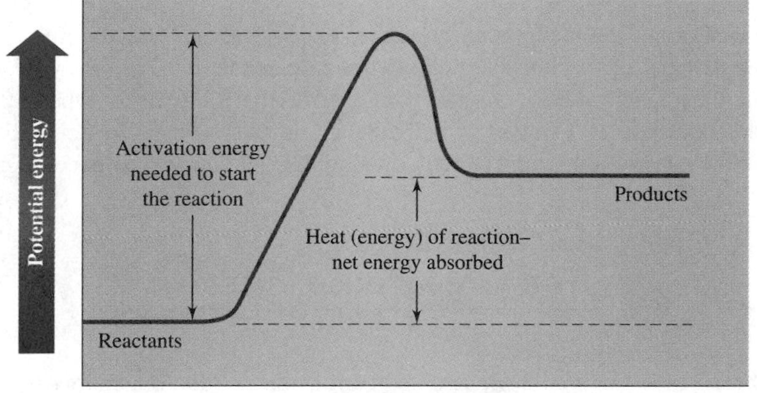

(b) Endothermic reaction

FIGURE 8.1
Energy changes in exothermic and endothermic reactions.

FIGURE 8.2
Left: **$Ba(OH)_2$ and NH_4SCN are mixed (endothermic reaction).** ***Right:*** **Sugar and $KClO_3$ ignite in a spectacular exothermic reaction.**

Carbon dioxide dissolves in the ocean, forming carbonates and hydrogen carbonates.

Outside the laboratory you can experience an endothermic process when applying a cold pack to an injury. In this case ammonium chloride (NH_4Cl) dissolves in water. Temperature changes from 24.5°C to 18.1°C result when 10 g of NH_4Cl are added to 100 mL of water. Energy, in the form of heat, is taken from the immediate surroundings (water) causing the salt solution to become cooler.

8.6 Global Warming: The Greenhouse Effect

Fossil fuels, derived from coal and petroleum, provide the energy we use to power our industries, heat and light our homes and workplaces, and run our cars. As these fuels are burned they produce carbon dioxide and water, releasing over 50 billion tons of carbon dioxide into our atomosphere each year.

The concentration of carbon dioxide has been monitored by scientists since 1958. Analysis of the air trapped in a core sample of snow from Antartica provides data on carbon dioxide levels for the past 160,000 years. The results of this study show that as the carbon dioxide increased, the global temperature increased as well. The levels of carbon dioxide remained reasonably constant from the last ice age, 100,000 years ago, until the industrial revolution. Since then the concentration of carbon dioxide in our atmosphere has risen 15% to an all-time high.

Carbon dioxide is a minor component in our atmosphere and is not usually considered to be a pollutant. The concern expressed by scientists arises from the dramatic increase occurring in the earth's atmosphere. Without the influence of man in the environment, the exchange of carbon dioxide between plants and animals would be relatively balanced. Our continued use of fossil fuels has led to an increase of 7.4% in carbon dioxide between 1900 and 1970 and an additional 3.5% increase during the 1980s. See Figure 8.3.

In addition to the larger consumption of fossil fuels, there are still other factors that increase carbon dioxide levels in the atmosphere: Rain forests are being destroyed by cutting and burning to make room for increased population and agricul-

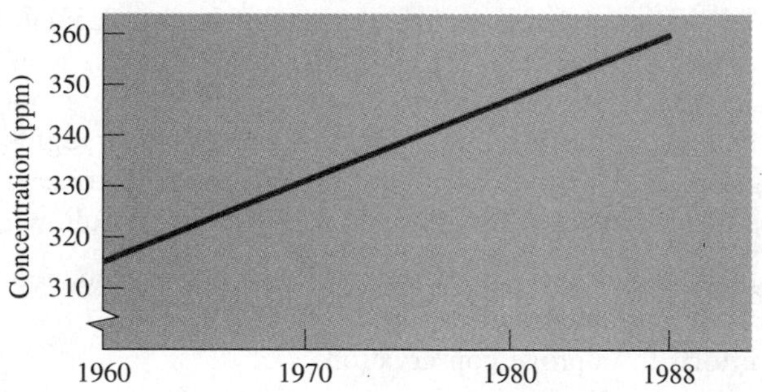

FIGURE 8.3 Concentration of carbon dioxide in the atmosphere.

tural needs. Carbon dioxide is added to the atmosphere during the burning, and the loss of trees diminishes the uptake of carbon dioxide by plants.

About half of all the carbon dioxide released into our atmosphere each year remains there, thus increasing its concentration. The other half is absorbed by plants during photosynthesis or is dissolved in the ocean to form hydrogen carbonates and carbonates.

Carbon dioxide and other greenhouse gases, such as methane and water, act to warm our atmosphere by trapping heat near the surface of the earth. Solar radiation strikes the earth and warms the surface. The warmed surface then reradiates this energy as heat. The greenhouse gases absorb some of this heat energy from the surface, which warms our atmosphere. A similar principle is illustrated in a greenhouse where sunlight comes through the glass yet heat cannot escape. The air in the greenhouse warms, producing a climate considerably different than in nature. In the atmosphere these greenhouse gases are acting to warm our air and produce changes in our climate. See Figure 8.4.

FIGURE 8.4 Elements of a global temperature warming are caused by the greenhouse effect.

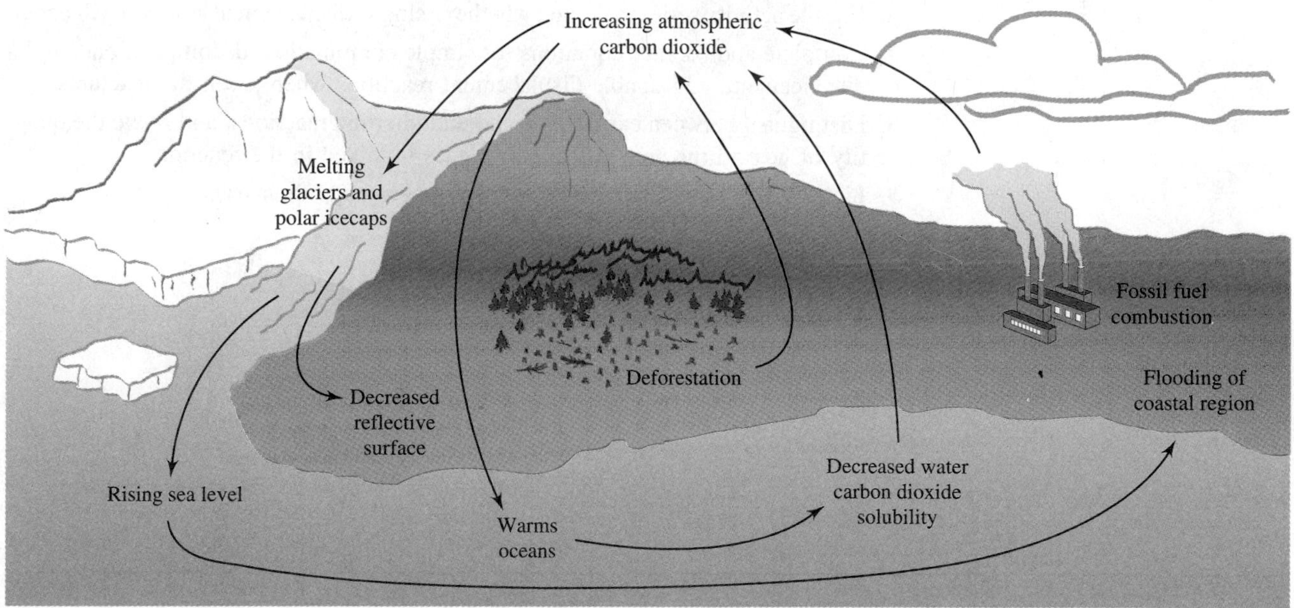

Long-term effects of global warming are still a matter of speculation and debate. One consideration is whether the polar ice caps would melt; this would cause a rise in sea level and lead to major flooding on the coasts of our continents. Further effects could include shifts in rainfall patterns, producing droughts and extreme seasonal changes in such major agricultural regions as California.

To reverse these trends will require major efforts in the following areas:

- The development of new energy sources to cut our dependence on fossil fuels
- An end to deforestation worldwide
- Intense efforts to improve conservation

On an individual basis each of us can play a significant role. For example, the simple conversion of a 100-watt incandescent light bulb to a compact fluorescent bulb can reduce the electrical consumption for that light 20%, and the bulb can last 10 times longer. Recycling, switching to more fuel-efficient cars, and energy-efficient applicances, heaters, and air conditioners all would result in decreased energy consumption and less carbon dioxide being released into our atmosphere.

Concepts in Review

1. Know the format used in setting up chemical equations.
2. Recognize the various symbols commonly used in writing chemical equations.
3. Be able to balance simple chemical equations.
4. Interpret a balanced equation in terms of the relative numbers or amounts of molecules, atoms, grams, or moles of each substance represented.
5. Classify equations as combination, decomposition, single-displacement, or double-displacement reactions.
6. Use the activity series to predict whether a single-displacement reaction will occur.
7. Complete and balance equations for simple combination, decomposition, single-displacement, and double-displacement reactions when given the reactants.
8. Distinguish between exothermic and endothermic reactions, and relate the quantity of heat to the amounts of substances involved in the reaction.
9. Identify the major sources of chemical energy and their uses.

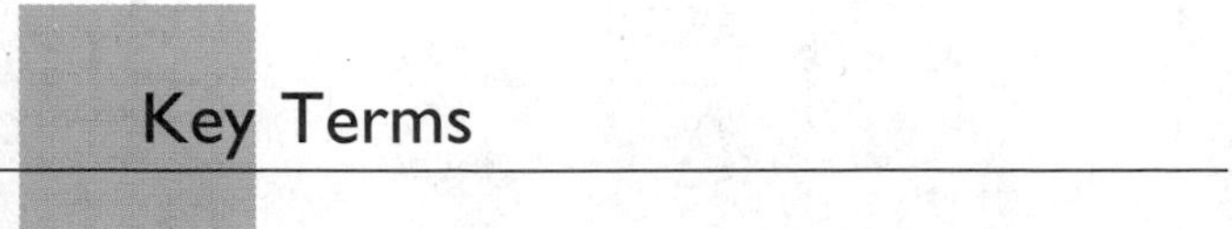

Key Terms

The terms listed here have all been defined within the chapter. Section numbers are referenced in parenthesis for each term.

activation energy (8.5)
balanced equation (8.2)
chemical equation (8.1)
combination reaction (8.4)
combustion (8.5)
decomposition reaction (8.4)
double-displacement reaction (8.4)
endothermic reaction (8.5)
exothermic reaction (8.5)
heat of reaction (8.5)
hydrocarbons (8.5)
single-displacement reaction (8.4)

Questions

Questions refer to tables, figures, and key words and concepts defined within the chapter. A particularly challenging question or exercise is indicated with an asterisk.

1. What is the purpose of balancing equations?
2. What is represented by the numbers (coefficients) that are placed in front of the formulas in a balanced equation?
3. In a balanced chemical equation:
   (a) are atoms conserved?
   (b) are molecules conserved?
   (c) are moles conserved?
   Explain your answers briefly.
4. Which of the following statements are correct? Rewrite the incorrect ones to make them correct.
   (a) The coefficients in front of the formulas in a balanced chemical equation give the relative number of moles of the reactants and products in the reaction.
   (b) A balanced chemical equation is one that has the same number of moles on each side of the equation.
   (c) In a chemical equation, the symbol $\xrightarrow{\Delta}$ indicates that the reaction is exothermic.
   (d) A chemical change that absorbs heat energy is said to be endothermic.
   (e) In the reaction $H_2 + Cl_2 \longrightarrow 2\,HCl$, 100 molecules of HCl are produced for every 50 molecules of H_2 reacted.
   (f) The symbol (aq) after a substance in an equation means that the substance is in a water solution.
   (g) The equation $H_2O \longrightarrow H_2 + O_2$ can be balanced by placing a 2 in front of H_2O.
   (h) In the equation $3\,H_2 + N_2 \longrightarrow 2\,NH_3$, there are fewer moles of product than there are moles of reactants.
   (i) The total number of moles of reactants and products represented by this equation is 5 mol:
   $Mg + 2\,HCl \longrightarrow MgCl_2 + H_2$
   (j) One mole of glucose, $C_6H_{12}O_6$, contains 6 mol of carbon atoms.
   (k) The reactants are the substances produced by the chemical reaction.
   (l) In a balanced equation, each side of the equation contains the same number of atoms of each element.
   (m) When a precipitate is formed in a chemical reaction, it can be indicated in the equation with an (s) immediately before the formula of the substance precipitated.
   (n) When a gas is involved in a chemical reaction, it can be indicated in the equation with a (g) immediately following the formula of the gas.
   (o) According to the equation $3\,H_2 + N_2 \longrightarrow 2\,NH_3$, 4 mol of NH_3 will be formed when 6 mol of H_2 and 2 mol of N_2 react.
   (p) The products of an exothermic reaction are at a lower potential energy than the reactants.
   (q) The combustion of hydrocarbons produces carbon dioxide and water as products.

Paired Exercises

These exercises are paired. Each odd-numbered exercise is followed by a similar even-numbered exercise. Answers to the even-numbered exercises are given in Appendix V.

5. Balance the following equations:
   (a) $H_2 + O_2 \longrightarrow H_2O$
   (b) $C + Fe_2O_3 \longrightarrow Fe + CO$
   (c) $H_2SO_4 + NaOH \longrightarrow H_2O + Na_2SO_4$
   (d) $Al_2(CO_3)_3 \xrightarrow{\Delta} Al_2O_3 + CO_2$
   (e) $NH_4I + Cl_2 \longrightarrow NH_4Cl + I_2$
6. Balance the following equations:
   (a) $H_2 + Br_2 \longrightarrow HBr$
   (b) $Al + C \xrightarrow{\Delta} Al_4C_3$
   (c) $Ba(ClO_3)_2 \xrightarrow{\Delta} BaCl_2 + O_2$
   (d) $CrCl_3 + AgNO_3 \longrightarrow Cr(NO_3)_3 + AgCl$
   (e) $H_2O_2 \longrightarrow H_2O + O_2$
7. Classify the reactions in Exercise 5 as combination, decomposition, single displacement, or double displacement.
8. Classify the reactions in Exercise 6 as combination, decomposition, single displacement, or double displacement.

9. Balance the following equations:
(a) $SO_2 + O_2 \longrightarrow SO_3$
(b) $Al + MnO_2 \xrightarrow{\Delta} Mn + Al_2O_3$
(c) $Na + H_2O \longrightarrow NaOH + H_2$
(d) $AgNO_3 + Ni \longrightarrow Ni(NO_3)_2 + Ag$
(e) $Bi_2S_3 + HCl \longrightarrow BiCl_3 + H_2S$
(f) $PbO_2 \xrightarrow{\Delta} PbO + O_2$
(g) $LiAlH_4 \xrightarrow{\Delta} LiH + Al + H_2$
(h) $KI + Br_2 \longrightarrow KBr + I_2$
(i) $K_3PO_4 + BaCl_2 \longrightarrow KCl + Ba_3(PO_4)_2$

11. Change the following word equations into formula equations and balance them:
(a) copper + sulfur $\xrightarrow{\Delta}$ copper(I) sulfide
(b) phosphoric acid + calcium hydroxide $\longrightarrow$ calcium phosphate + water
(c) silver oxide $\xrightarrow{\Delta}$ silver + oxygen
(d) iron(III) chloride + sodium hydroxide $\longrightarrow$ iron(III) hydroxide + sodium chloride
(e) nickel(II) phosphate + sulfuric acid $\longrightarrow$ nickel(II) sulfate + phosphoric acid
(f) zinc carbonate + hydrochloric acid $\longrightarrow$ zinc chloride + water + carbon dioxide
(g) silver nitrate + aluminum chloride $\longrightarrow$ silver chloride + aluminum nitrate

13. Use the activity series to predict which of the following reactions will occur. Complete and balance the equations. Where no reaction will occur, write "no reaction" as the product.
(a) $Ag(s) + H_2SO_4(aq) \longrightarrow$
(b) $Cl_2(g) + NaBr(aq) \longrightarrow$
(c) $Mg(s) + ZnCl_2(aq) \longrightarrow$
(d) $Pb(s) + AgNO_3(aq) \longrightarrow$

15. Complete and balance the equations for these reactions. All reactions yield products.
(a) $H_2 + I_2 \longrightarrow$
(b) $CaCO_3 \xrightarrow{\Delta}$
(c) $Mg + H_2SO_4 \longrightarrow$
(d) $FeCl_2 + NaOH \longrightarrow$

17. Complete and balance the equations for the following reactions. All reactions yield products.
(a) $Ba + O_2 \longrightarrow$
(b) $NaHCO_3 \xrightarrow{\Delta} Na_2CO_3 +$
(c) $Ni + CuSO_4 \longrightarrow$
(d) $MgO + HCl \longrightarrow$
(e) $H_3PO_4 + KOH \longrightarrow$

19. Interpret the following chemical reactions in terms of the number of moles of each reactant and product:
(a) $MgBr_2 + 2\ AgNO_3 \longrightarrow Mg(NO_3)_2 + 2\ AgBr$
(b) $N_2 + 3\ H_2 \longrightarrow 2\ NH_3$
(c) $2\ C_3H_7OH + 9\ O_2 \longrightarrow 6\ CO_2 + 8\ H_2O$

10. Balance the following equations:
(a) $MnO_2 + CO \longrightarrow Mn_2O_3 + CO_2$
(b) $Mg_3N_2 + H_2O \longrightarrow Mg(OH)_2 + NH_3$
(c) $C_3H_5(NO_3)_3 \longrightarrow CO_2 + H_2O + N_2 + O_2$
(d) $FeS + O_2 \longrightarrow Fe_2O_3 + SO_2$
(e) $Cu(NO_3)_2 \longrightarrow CuO + NO_2 + O_2$
(f) $NO_2 + H_2O \longrightarrow HNO_3 + NO$
(g) $Al + H_2SO_4 \longrightarrow Al_2(SO_4)_3 + H_2$
(h) $HCN + O_2 \longrightarrow N_2 + CO_2 + H_2O$
(i) $B_5H_9 + O_2 \longrightarrow B_2O_3 + H_2O$

12. Change the following word equations into formula equations and balance them:
(a) water $\longrightarrow$ hydrogen + oxygen
(b) acetic acid + potassium hydroxide $\longrightarrow$ potassium acetate + water
(c) phosphorus + iodine $\longrightarrow$ phosphorus triiodide
(d) aluminum + copper(II) sulfate $\longrightarrow$ copper + aluminum sulfate
(e) ammonium sulfate + barium chloride $\longrightarrow$ ammonium chloride + barium sulfate
(f) sulfur tetrafluoride + water $\longrightarrow$ sulfur dioxide + hydrogen fluoride
(g) chromium(III) carbonate $\xrightarrow{\Delta}$ chromium(III) oxide + carbon dioxide

14. Use the activity series to predict which of the following reactions will occur. Complete and balance the equations. Where no reaction will occur, write "no reaction" as the product.
(a) $Cu(s) + FeCl_3(aq) \longrightarrow$
(b) $H_2(g) + Al_2O_3(aq) \longrightarrow$
(c) $Al(s) + HBr(aq) \longrightarrow$
(d) $I_2(s) + HCl(aq) \longrightarrow$

16. Complete and balance the equations for these reactions. All reactions yield products.
(a) $SO_2 + H_2O \longrightarrow$
(b) $SO_3 + H_2O \longrightarrow$
(c) $Ca + H_2O \longrightarrow$
(d) $Bi(NO_3)_3 + H_2S \longrightarrow$

18. Complete and balance the equations for the following reactions. All reactions yield products.
(a) $C + O_2 \longrightarrow$
(b) $Al(ClO_3)_3 \xrightarrow{\Delta} O_2 +$
(c) $CuBr_2 + Cl_2 \longrightarrow$
(d) $SbCl_3 + (NH_4)_2S \longrightarrow$
(e) $NaNO_3 \xrightarrow{\Delta} NaNO_2 +$

20. Interpret each of the following equations in terms of the relative number of moles of each substance involved and indicate whether the reaction is exothermic or endothermic:
(a) $2\ Na + Cl_2 \longrightarrow 2\ NaCl + 822\ kJ$
(b) $PCl_5 + 92.9\ kJ \longrightarrow PCl_3 + Cl_2$

21. Write balanced equations for each of these reactions, including the heat term:
   (a) Lime (CaO) is converted to slaked lime $Ca(OH)_2$ by reaction with water. The reaction liberates 65.3 kJ of heat for each mole of lime reacted.
   (b) The industrial production of aluminum metal from aluminum oxide is an endothermic electrolytic process requiring 1630 kJ per mole of Al_2O_3. Oxygen is also a product.

22. Write a balanced equation for each of the following descriptions. Include a heat term on the appropriate side of the equation.
   (a) Powdered aluminum will react with crystals of iodine when moistened with dishwashing detergent. The reaction produces violet sparks and flaming aluminum. The major product is aluminum iodide (AlI_3). The detergent is not a reactant.
   (b) Copper(II) oxide (CuO), a black powder, can be decomposed to produce pure copper by heating the powder in the presence of methane gas (CH_4). The products are copper, carbon dioxide and water vapor.
   (c) A form of rust, iron(III) oxide (Fe_2O_3), reacts with powdered aluminum to produce molten iron and aluminum oxide in the spectacular reaction shown in the opening photo for this chapter.

Additional Exercises

These problems are not paired or labeled by topic and provide additional practice on the concepts covered in this chapter.

23. Name one piece of evidence that a chemical reaction is actually taking place in each of these situations:
   (a) making a piece of toast
   (b) frying an egg
   (c) striking a match

24. Balance this equation, using the smallest possible whole numbers. Then determine how many atoms of oxygen appear on each side of the equation:

$$P_4O_{10} + HClO_4 \longrightarrow Cl_2O_7 + H_3PO_4$$

25. Suppose that in a balanced equation the term $7\ Al_2(SO_4)_3$ appears.
   (a) How many atoms of aluminum are represented?
   (b) How many atoms of sulfur are represented?
   (c) How many atoms of oxygen are represented?
   (d) How many atoms of all kinds are represented?

26. Name two pieces of information that can be obtained from a balanced chemical equation. Name two pieces of information that the reaction does not provide.

27. Make a drawing to show six molecules of ammonia gas decomposing to form hydrogen and nitrogen gases.

28. Explain briefly why the following single-displacement reaction will not take place:

$$Zn + Mg(NO_3)_2 \longrightarrow \text{no reaction}$$

29. A student does an experiment to determine where titanium metal should be placed on the activity series chart. He places newly cleaned pieces of titanium into solutions of nickel(II) nitrate, lead(II) nitrate, and magnesium nitrate. He finds that the titanium reacts with the nickel(II) nitrate and lead(II) nitrate solutions, but not with the magnesium nitrate solution. From this information place titanium in the activity series in a position relative to these ions.

30. Complete and balance the equations for these combination reactions:
   (a) $K + O_2 \longrightarrow$
   (b) $Al + Cl_2 \longrightarrow$
   (c) $CO_2 + H_2O \longrightarrow$
   (d) $CaO + H_2O \longrightarrow$

31. Complete and balance the equations for these decomposition reactions:
   (a) $HgO \xrightarrow{\Delta}$
   (b) $NaClO_3 \xrightarrow{\Delta}$
   (c) $MgCO_3 \xrightarrow{\Delta}$
   (d) $PbO_2 \xrightarrow{\Delta} PbO +$

32. Complete and balance the equations for these single-displacement reactions:
   (a) $Zn + H_2SO_4 \longrightarrow$
   (b) $AlI_3 + Cl_2 \longrightarrow$
   (c) $Mg + AgNO_3 \longrightarrow$
   (d) $Al + CoSO_4 \longrightarrow$

33. Complete and balance the equations for these double-displacement reactions:
   (a) $ZnCl_2 + KOH \rightarrow$
   (b) $CuSO_4 + H_2S \rightarrow$
   (c) $Ca(OH)_2 + H_3PO_4 \rightarrow$
   (d) $(NH_4)_3PO_4 + Ni(NO_3)_2 \rightarrow$
   (e) $Ba(OH)_2 + HNO_3 \rightarrow$
   (f) $(NH_4)_2S + HCl \rightarrow$

34. Predict which of the following double-displacement reactions will occur. Complete and balance the equations. Where no reaction will occur, write "no reaction" as the product.
   (a) $AgNO_3(aq) + KCl(aq) \longrightarrow$
   (b) $Ba(NO_3)_2(aq) + MgSO_4(aq) \longrightarrow$
   (c) $H_2SO_4(aq) + Mg(OH)_2(aq) \longrightarrow$
   (d) $MgO(s) + H_2SO_4(aq) \longrightarrow$
   (e) $Na_2CO_3(aq) + NH_4Cl(aq) \longrightarrow$

35. Write balanced equations for the combustion of the following hydrocarbons:
(a) ethane, C_2H_6
(b) benzene, C_6H_6
(c) heptane, C_7H_{16}

36. List the various pigments found in plants and state the function of each.

37. State the four necessary requirements for successful photosynthesis.

38. Draw a flowchart illustrating how leaves "change color" in the autumn of the year.

39. List the factors that contribute to an increase in carbon dioxide in our atmosphere.

40. List three gases considered to be greenhouse gases. Explain why they are given this name.

41. How can the effects of global warming be reduced?

42. What happens to carbon dioxide released into our atmosphere?

Answers to Practice Exercises

8.1 $4\ Al + 3\ O_2 \longrightarrow 2\ Al_2O_3$

8.2 $3\ Mg(OH)_2 + 2\ H_3PO_4 \longrightarrow Mg_3(PO_4)_2 + 6\ H_2O$

8.3 (a) $Fe + MgCl_2 \longrightarrow$ no reaction
(b) $Zn(s) + Pb(NO_3)_2(aq) \longrightarrow Pb(s) + Zn(NO_3)_2(aq)$

8.4 (a) $2\ K_3PO_4(aq) + 3\ BaCl_2(aq) \longrightarrow Ba_3(PO_4)_2(s) + 6\ KCl(aq)$
(b) $2\ HCl(aq) + NiCO_3(aq) \longrightarrow NiCl_2(aq) + H_2O(l) + CO_2(g)$
(c) $NH_4Cl(aq) + NaNO_3(aq) \longrightarrow$ no reaction

Calculations from Chemical Equations

Chapter 9

◀ Chapter Opening Photo: A chemist must measure exact quantities of each reactant to produce new chemical compounds for use in our complex society.

The old adage "waste not, want not" is appropriate in our daily life and in the laboratory. Determining correct amounts comes into play in most all professions. For example, a hostess determines the quantity of food and beverage necessary to serve her guests. These amounts are defined by specific recipes and a knowledge of the particular likes and dislikes of the guests. A seamstress determines the amount of material, lining, and trim necessary to produce a gown for her client by relying on a pattern or her own experience to guide the selection. A carpet layer determines the correct amount of carpet and padding necessary to recarpet a customer's house by calculating the floor area. The IRS determines the correct deduction for federal income taxes from your paycheck based on your expected annual income.

The chemist also finds it necessary to calculate amounts of products or reactants by using a balanced chemical equation. With these calculations, the chemist can control the amount of product by scaling the reaction up or down to fit the needs of the laboratory, and can thereby minimize waste or excess materials formed during the reaction.

9.1 A Short Review

Molar Mass. The molar mass is the sum of the atomic masses of all the atoms in a molecule. The molar mass also applies to the mass of a mole of any formula unit—atoms, molecules, or ions; it is the atomic mass of an atom, or the sum of the atomic masses in a molecule or an ion (in grams).

Relationship Between Molecule and Mole. A molecule is the smallest unit of a molecular substance (e.g., Cl_2), and a mole is Avogadro's number (6.022×10^{23}) of molecules of that substance. A mole of chlorine (Cl_2) has the same number of molecules as a mole of carbon dioxide, a mole of water, or a mole of any other molecular substance. When we relate molecules to molar mass, 1 molar mass is equivalent to 1 mol, or 6.022×10^{23} molecules.

In addition to referring to molecular substances, the term *mole* may refer to any chemical species. It represents a quantity (6.022×10^{23} particles) and may be applied to atoms, ions, electrons, and formula units of nonmolecular substances. In other words,

$$1 \text{ mole} = \begin{cases} 6.022 \times 10^{23} \text{ molecules} \\ 6.022 \times 10^{23} \text{ formula units} \\ 6.022 \times 10^{23} \text{ atoms} \\ 6.022 \times 10^{23} \text{ ions} \end{cases}$$

Other useful mole relationships are

$$\text{molar mass} = \frac{\text{grams of a substance}}{\text{number of moles of the substance}}$$

$$\text{molar mass} = \frac{\text{grams of a monatomic element}}{\text{number of moles of the element}}$$

$$\text{number of moles} = \frac{\text{number of molecules}}{6.022 \times 10^{23} \text{ molecules/mole}}$$

Balanced Equations. When using chemical equations for calculations of mole–mass–volume relationships between reactants and products, the equations must be balanced. Remember that the number in front of a formula in a balanced chemical equation can represent the number of moles of that substance in the chemical reaction.

9.2 Introduction to Stoichiometry: The Mole-Ratio Method

It is often necessary to calculate the amount of a substance that is produced from or needed to react with a given quantity of another substance. The area of chemistry that deals with the quantitative relationships among reactants and products is known as **stoichiometry** (*stoy-key-ah-meh-tree*). Although several methods are known, we firmly believe that the *mole* or *mole-ratio* method is generally best for solving problems in stoichiometry.

stoichiometry

A **mole ratio** is a ratio between the number of moles of any two species involved in a chemical reaction. For example, in the reaction

mole ratio

$$\underset{2\ \text{mol}}{2\,H_2} + \underset{1\ \text{mol}}{O_2} \longrightarrow \underset{2\ \text{mol}}{2\,H_2O}$$

only six mole ratios apply to this reaction. They are

$$\frac{2\ \text{mol}\ H_2}{1\ \text{mol}\ O_2} \qquad \frac{2\ \text{mol}\ H_2}{2\ \text{mol}\ H_2O} \qquad \frac{1\ \text{mol}\ O_2}{2\ \text{mol}\ H_2}$$

$$\frac{1\ \text{mol}\ O_2}{2\ \text{mol}\ H_2O} \qquad \frac{2\ \text{mol}\ H_2O}{2\ \text{mol}\ H_2} \qquad \frac{2\ \text{mol}\ H_2O}{1\ \text{mol}\ O_2}$$

The mole ratio is a conversion factor used to convert the number of moles of one substance to the corresponding number of moles of another substance in a chemical reaction. For example, if we want to calculate the number of moles of H_2O that can be obtained from 4.0 mol of O_2, we use the mole ratio 2 mol H_2O/1 mol O_2:

$$4.0\ \cancel{\text{mol}\ O_2} \times \frac{2\ \text{mol}\ H_2O}{1\ \cancel{\text{mol}\ O_2}} = 8.0\ \text{mol}\ H_2O$$

Since stoichiometric problems are encountered throughout the entire field of chemistry, it is prudent to master this general method for their solution. The mole-ratio method makes use of three basic operations:

1. Convert the quantity of starting substance to moles (if it is not given in moles).
2. Convert the moles of starting substance to moles of desired substance.
3. Convert the moles of desired substance to the units specified in the problem.

Like learning to balance chemical equations, learning to make stoichiometric calculations requires practice. A detailed step-by-step description of the general method, together with a variety of worked examples, is given in the following paragraphs. Study this material and apply the method to the problems at the end of this chapter.

Use a balanced equation.

You may need to write the equation before beginning the problem.

Step 1 Determine the number of moles of starting substance.
Identify the starting substance from the data given in the statement of the problem. If it is not in moles, convert the quantity of the starting substance to moles.

$$\text{moles of starting substance} = \text{grams of starting substance} \times \frac{1 \text{ mol of starting substance}}{\text{molar mass of starting substance}}$$

As in all problems with units, the desired quantity is in the numerator, and the quantity to be eliminated is in the denominator.

Step 2 Determine the mole ratio of the desired substance to the starting substance.
The number of moles of each substance in the balanced equation is indicated by the coefficient in front of each substance. Use these coefficients to set up the mole ratio:

$$\text{mole ratio} = \frac{\text{moles of desired substance in the equation}}{\text{moles of starting substance in the equation}}$$

Multiply the number of moles of starting substance (from Step 1) by the mole ratio to obtain the number of moles of desired substance:

Units of moles of starting substance cancel in the numerator and denominator.

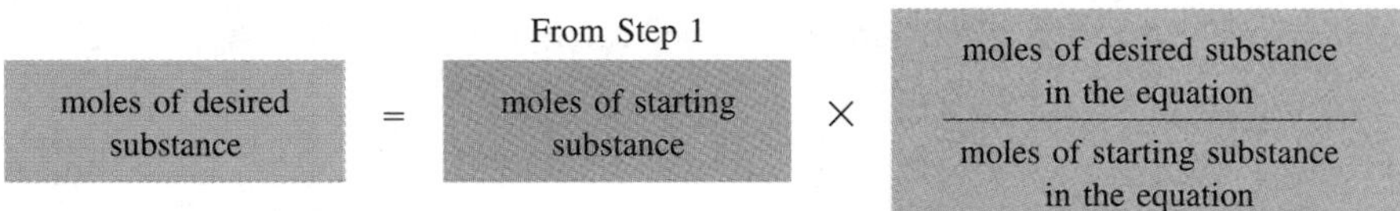

Step 3 Calculate the desired substance in the units specified in the problem.
If the answer is to be in moles, the problem is finished. If units other than moles are wanted, multiply the moles of the desired substance (from Step 2) by the appropriate factor to convert moles to the units required. For example, if grams of the desired substance are wanted,

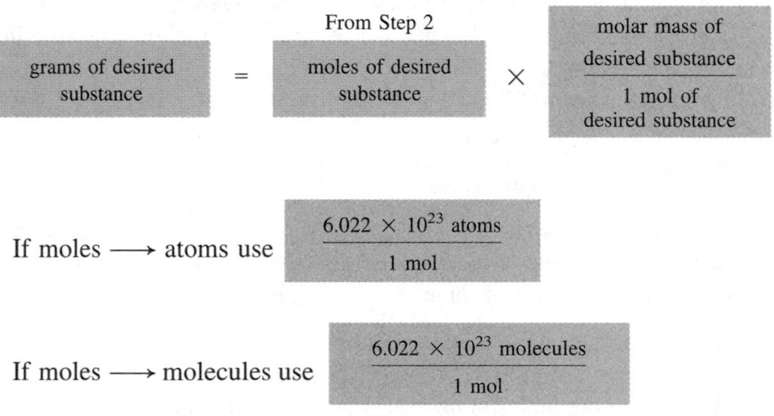

The steps for converting the mass of a starting substance *A* to either the mass, atoms, or molecules of desired substance *B* are summarized in Figure 9.1.

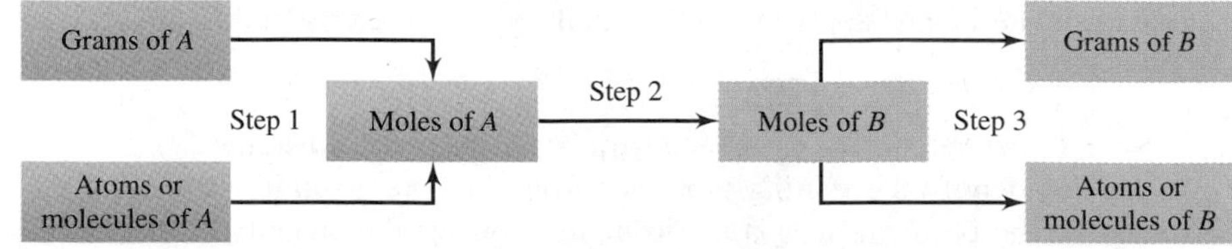

FIGURE 9.1 ▶ Steps for converting starting substance *A* to mass, atoms, or molecules of desired substance *B*.

9.3 Mole–Mole Calculations

The first application of the mole-ratio method of solving stoichiometric problems is that of mole–mole calculations.

The quantity of starting substance is given in moles and the quantity of desired substance is requested in moles. Understanding the use of mole ratios is very helpful in solving many problems. Some examples follow.

Example 9.1

How many moles of carbon dioxide will be produced by the complete reaction of 2.0 mol of glucose ($C_6H_{12}O_6$), according to the following reaction?

$$\underset{1\ \text{mol}}{C_6H_{12}O_6} + \underset{6\ \text{mol}}{6\ O_2} \longrightarrow \underset{6\ \text{mol}}{6\ CO_2} + \underset{6\ \text{mol}}{6\ H_2O}$$

Solution

The balanced equation states that 6 mol of CO_2 will be produced from 1 mol of $C_6H_{12}O_6$. Even though we can readily see that 12 mol of CO_2 will be formed from 2.0 mol of $C_6H_{12}O_6$, the mole-ratio method of solving the problem is shown below.

Step 1 The number of moles of starting substance is 2.0 mol $C_6H_{12}O_6$.
Step 2 The conversion needed is

$$\text{moles } C_6H_{12}O_6 \longrightarrow \text{moles } CO_2$$

Multiply 2.0 mol of glucose (given in the problem) by this mole ratio:

$$2.0\ \cancel{\text{mol}\ C_6H_{12}O_6} \times \frac{6\ \text{mol}\ CO_2}{1\ \cancel{\text{mol}\ C_6H_{12}O_6}} = 12\ \text{mol}\ CO_2$$

The numbers in the mole ratio are absolute and do not affect the number of significant figures in the answer.

Again note the use of units. The moles of $C_6H_{12}O_6$ cancel, leaving the answer in units of moles of CO_2.

Example 9.2

How many moles of ammonia can be produced from 8.00 mol of hydrogen reacting with nitrogen? The balanced equation is

$$3\ H_2 + N_2 \longrightarrow 2\ NH_3$$

Solution

Step 1 The starting substance is 8.00 mol of H_2.
Step 2 The conversion needed is

$$\text{moles } H_2 \longrightarrow \text{moles } NH_3$$

The balanced equation states that we get 2 mol of NH_3 for every 3 mol of H_2 that react. Set up the mole ratio of desired substance (NH_3) to starting substance (H_2):

$$\text{mole ratio} = \frac{2\ \text{mol}\ NH_3}{3\ \text{mol}\ H_2} \quad \text{(from equation)}$$

Multiply the 8.00 mol H_2 by the mole ratio:

$$8.00\ \cancel{\text{mol}\ H_2} \times \frac{2\ \text{mol}\ NH_3}{3\ \cancel{\text{mol}\ H_2}} = 5.33\ \text{mol}\ NH_3$$

Example 9.3 Given the balanced equation

$$\underset{1\ \text{mol}}{K_2Cr_2O_7} + \underset{6\ \text{mol}}{6\ KI} + 7\ H_2SO_4 \longrightarrow Cr_2(SO_4)_3 + 4\ K_2SO_4 + \underset{3\ \text{mol}}{3\ I_2} + 7\ H_2O$$

calculate (a) the number of moles of potassium dichromate ($K_2Cr_2O_7$) that will react with 2.0 mol of potassium iodide (KI); (b) the number of moles of iodine (I_2) that will be produced from 2.0 mol of potassium iodide.

Solution Since the equation is balanced, we are concerned only with $K_2Cr_2O_7$, KI, and I_2, and we can ignore all the other substances. The equation states that 1 mol of $K_2Cr_2O_7$ will react with 6 mol of KI to produce 3 mol of I_2.

(a) Calculate the number of moles of $K_2Cr_2O_7$.

Step 1 The starting substance is 2.0 mol of KI.

Step 2 The conversion needed is

$$\text{moles KI} \longrightarrow \text{moles } K_2Cr_2O_7$$

Set up the mole ratio of desired substance to starting substance:

$$\text{mole ratio} = \frac{1\ \text{mol}\ K_2Cr_2O_7}{6\ \text{mol KI}} \quad \text{(from equation)}$$

Multiply the moles of starting material by this ratio:

$$2.0\ \cancel{\text{mol KI}} \times \frac{1\ \text{mol}\ K_2Cr_2O_7}{6\ \cancel{\text{mol KI}}} = 0.33\ \text{mol}\ K_2Cr_2O_7$$

(b) Calculate the number of moles of I_2.

Step 1 The moles of starting substance are 2.0 mol KI as in part (a).

Step 2 The conversion needed is

$$\text{moles KI} \longrightarrow \text{moles } I_2$$

Set up the mole ratio of desired substance to starting substance:

$$\text{mole ratio} = \frac{3\ \text{mol}\ I_2}{6\ \text{mol KI}} \quad \text{(from equation)}$$

Multiply the moles of starting material by this ratio:

$$2.0\ \cancel{\text{mol KI}} \times \frac{3\ \text{mol}\ I_2}{6\ \cancel{\text{mol KI}}} = 1.0\ \text{mol}\ I_2$$

Example 9.4 How many molecules of water can be produced by reacting 0.010 mol of oxygen with hydrogen?

Solution The balanced equation is $2\ H_2 + O_2 \rightarrow 2\ H_2O$.
The sequence of conversions needed in the calculation is

$$\text{moles } O_2 \longrightarrow \text{moles } H_2O \longrightarrow \text{molecules } H_2O$$

◀ **The space shuttle is powered by H_2 and O_2, which react to produce H_2O.**

Step 1 The starting substance is 0.010 mol O_2.

Step 2 The conversion needed is moles $O_2 \longrightarrow$ moles H_2O. Set up the mole ratio of desired substance to starting substance:

$$\text{mole ratio} = \frac{2 \text{ mol } H_2O}{1 \text{ mol } O_2} \quad \text{(from equation)}$$

Multiply 0.010 mol O_2 by the mole ratio:

$$0.010 \cancel{\text{mol } O_2} \times \frac{2 \text{ mol } H_2O}{1 \cancel{\text{mol } O_2}} = 0.020 \text{ mol } H_2O$$

Step 3 Since the problem asks for molecules instead of moles of H_2O, we must convert moles to molecules. Use the conversion factor (6.022×10^{23} molecules)/mole:

$$0.020 \cancel{\text{mol}} \, H_2O \times \frac{6.022 \times 10^{23} \text{ molecules}}{1 \cancel{\text{mol}}} = 1.2 \times 10^{22} \text{ molecules } H_2O$$

Note that 0.020 mol is still quite a large number of water molecules.

Practice 9.1

How many moles of aluminum oxide will be produced from 0.50 mol of oxygen?

$$4\ Al + 3\ O_2 \longrightarrow 2\ Al_2O_3$$

Practice 9.2

How many moles of aluminum hydroxide are required to produce 22.0 mol of water?

$$2\ Al(OH)_3 + 3\ H_2SO_4 \longrightarrow Al_2(SO_4)_3 + 6\ H_2O$$

9.4 Mole–Mass Calculations

The object of this type of problem is to calculate the mass of one substance that reacts with or is produced from a given number of moles of another substance in a chemical reaction. If the mass of the starting substance is given, it is necessary to convert it to moles. The mole ratio is used to convert from moles of starting substance to moles of desired substance. Moles of desired substance can then be changed to mass if required. Each example is solved in two ways:

- **Method 1:** Step-by-Step.
- **Method 2:** Continuous Calculation where all the individual steps are combined in a single line.

Select the method that is easiest for you and focus your attention on solving problems that way.

Example 9.5 What mass of hydrogen can be produced by reacting 6.0 mol of aluminum with hydrochloric acid?

Solution The balanced equation is $2\ Al(s) + 6\ HCl(aq) \longrightarrow 2\ AlCl_3(aq) + 3\ H_2(g)$.

Method 1: Step-by-Step
First calculate the moles of hydrogen produced, using the mole-ratio method; then calculate the mass of hydrogen by multiplying the moles of hydrogen by its grams per mole. The sequence of conversions in the calculation is

$$\text{moles Al} \longrightarrow \text{moles } H_2 \longrightarrow \text{grams } H_2$$

Step 1 The starting substance is 6.0 mol of aluminum.
Step 2 Calculate moles of H_2 by the mole-ratio method:

$$6.0\ \cancel{\text{mol Al}} \times \frac{3\ \text{mol } H_2}{2\ \cancel{\text{mol Al}}} = 9.0\ \text{mol } H_2$$

Step 3 Convert moles of H_2 to grams [g = mol × (g/mol)]:

$$9.0 \cancel{\text{mol H}_2} \times \frac{2.016 \text{ g H}_2}{1 \cancel{\text{mol H}_2}} = 18 \text{ g H}_2 \quad \text{(Answer)}$$

We see that 18 g of H_2 can be produced by reacting 6.0 mol of Al with HCl.

Method 2: Continuous Calculation

$$6.0 \cancel{\text{mol Al}} \times \frac{3 \cancel{\text{mol H}_2}}{2 \cancel{\text{mol Al}}} \times \frac{2.016 \text{ g H}_2}{1 \cancel{\text{mol H}_2}} = 18 \text{ g H}_2$$

Example 9.6

How many moles of water can be produced by burning 325 g of octane (C_8H_{18})? The balanced equation is $2\ C_8H_{18}(g) + 25\ O_2(g) \rightarrow 16\ CO_2(g) + 18\ H_2O(g)$.

Solution

Method 1: Step-by-Step

The sequence of conversions in the calculation is

$$\text{grams } C_8H_{18} \longrightarrow \text{moles } C_8H_{18} \longrightarrow \text{moles } H_2O$$

Step 1 The starting substance is 325 g C_8H_{18}. Convert 325 g of C_8H_{18} to moles:

$$325 \cancel{\text{g C}_8\text{H}_{18}} \times \frac{1 \text{ mol C}_8\text{H}_{18}}{114.2 \cancel{\text{g C}_8\text{H}_{18}}} = 2.85 \text{ mol C}_8\text{H}_{18}$$

Step 2 Calculate the moles of water by the mole-ratio method:

$$2.85 \cancel{\text{mol C}_8\text{H}_{18}} \times \frac{18 \text{ mol H}_2\text{O}}{2 \cancel{\text{mol C}_8\text{H}_{18}}} = 25.7 \text{ mol H}_2\text{O} \quad \text{(Answer)}$$

Method 2: Continuous Calculation

$$325 \cancel{\text{g C}_8\text{H}_{18}} \times \frac{1 \cancel{\text{mol C}_8\text{H}_{18}}}{114.2 \cancel{\text{g C}_8\text{H}_{18}}} \times \frac{18 \text{ mol H}_2\text{O}}{2 \cancel{\text{mol C}_8\text{H}_{18}}} = 25.6 \text{ mol H}_2\text{O}$$

Notice the answers for the different methods vary in the last digit. This results from rounding off at different times in the calculation. Check with your instructor to find the appropriate rules for your course.

Practice 9.3

How many moles of potassium chloride can be produced from 100.0 g of potassium chlorate?

$$2\ KClO_3 \longrightarrow 2\ KCl + 3\ O_2$$

Practice 9.4

How many grams of silver nitrate are required to produce 0.25 mol of silver sulfide?

$$2\ AgNO_3 + H_2S \longrightarrow Ag_2S + 2\ HNO_3$$

9.5 Mass–Mass Calculations

Remember to select either step-by-step method or continuous calculation for solving these problems.

Solving mass–mass stoichiometry problems requires all the steps of the mole-ratio method. The mass of starting substance is converted to moles. The mole ratio is then used to determine moles of desired substance, which in turn is converted to mass.

Example 9.7 What mass of carbon dioxide is produced by the complete combustion of 100. g of the hydrocarbon pentane, C_5H_{12}? The balanced equation is $C_5H_{12} + 8\ O_2 \rightarrow 5\ CO_2 + 6\ H_2O$.

Solution **Method 1: Step-by-Step**

The sequence of conversions in the calculation is

$$\text{grams } C_5H_{12} \longrightarrow \text{moles } C_5H_{12} \longrightarrow \text{moles } CO_2 \longrightarrow \text{grams } CO_2$$

Step 1 The starting substance is 100. g of C_5H_{12}. Convert 100. g of C_5H_{12} to moles:

$$100.\ \text{g } C_5H_{12} \times \frac{1\ \text{mol } C_5H_{12}}{72.15\ \text{g } C_5H_{12}} = 1.39\ \text{mol } C_5H_{12}$$

Step 2 Calculate the moles of CO_2 by the mole-ratio method:

$$1.39\ \text{mol } C_5H_{12} \times \frac{5\ \text{mol } CO_2}{1\ \text{mol } C_5H_{12}} = 6.95\ \text{mol } CO_2$$

Step 3 Convert moles of CO_2 to grams:

$$\text{mol } CO_2 \times \frac{\text{molar mass } CO_2}{1\ \text{mol } CO_2} = \text{grams } CO_2$$

$$6.95\ \text{mol } CO_2 \times \frac{44.01\ \text{g } CO_2}{1\ \text{mol } CO_2} = 306\ \text{g } CO_2$$

Method 2: Continuous Calculation

Remember to round off as appropriate for your particular course.

$$100.\ \text{g } C_5H_{12} \times \frac{1\ \text{mol } C_5H_{12}}{72.15\ \text{g } C_5H_{12}} \times \frac{5\ \text{mol } CO_2}{1\ \text{mol } C_5H_{12}} \times \frac{44.01\ \text{g } CO_2}{1\ \text{mol } CO_2} = 305\ \text{g } CO_2$$

Example 9.8 How many grams of nitric acid, HNO_3, are required to produce 8.75 g of dinitrogen monoxide (N_2O) according to the following equation?

$$4\ Zn(s) + \underset{10\ \text{mol}}{10\ HNO_3(aq)} \longrightarrow 4\ Zn(NO_3)_2(aq) + \underset{1\ \text{mol}}{N_2O(g)} + 5\ H_2O(l)$$

Solution **Method 1: Step-by-Step**

The sequence of conversions in the calculation is

$$\text{grams } N_2O \longrightarrow \text{moles } N_2O \longrightarrow \text{moles } HNO_3 \longrightarrow \text{grams } HNO_3$$

A Shrinking Technology

In the high-tech world of computers, the microchip has miniaturized the field of electronics. In order to produce ever smaller computers, calculators, and even microbots (microsized robots) precise quantities of chemicals in exact proportions are required.

Engineers at Bell Laboratories, Massachusetts Institute of Technology, University of California, and Stanford University are busy trying to produce parts for tiny machines and robots. New techniques are allowing them to produce gears smaller than a grain of sand, and motors lighter than a speck of dust.

The secret behind these tiny circuits is to print the entire circuit or blueprint at one time. Computers are used to draw a design of how the chip will look. This image is then transferred into a pattern, or mask, with details finer than a human hair. Light is then shone through the mask onto a silicon-coated surface. The process is similar to photography. The areas created on the silicon exhibit high or low resistance to chemical etching. Chemicals are then applied to etch away the silicon.

Micromachinery is produced in the same way. First a thin layer of silicon dioxide is applied (sacrificial material). Then a layer of polysilicon is carefully applied (structural material). A mask is then applied and the whole structure is covered with plasma (excited gas). The plasma acts as a tiny sandblaster removing everything the mask doesn't protect. This process is repeated as the entire machine is constructed. When the entire assembly is complete, the whole machine is placed in hydrofluoric acid, which dissolves all the sacrificial material and permits the various parts of the machine to move.

More development is necessary to turn these micromachines into true microbots, but research is currently underway to design methods of locomotion and sensing imaging systems. Possible uses for these microbots include "smart" pills, which could contain sensors (see Chemistry in Action, page 34), or drug reservoirs (currently done for birth control). Tiny pumps, once inside the body, will be able to dispense the proper amount of medication at precisely the correct site. These microbots are currently in production for the treatment of diabetes (to release insulin).

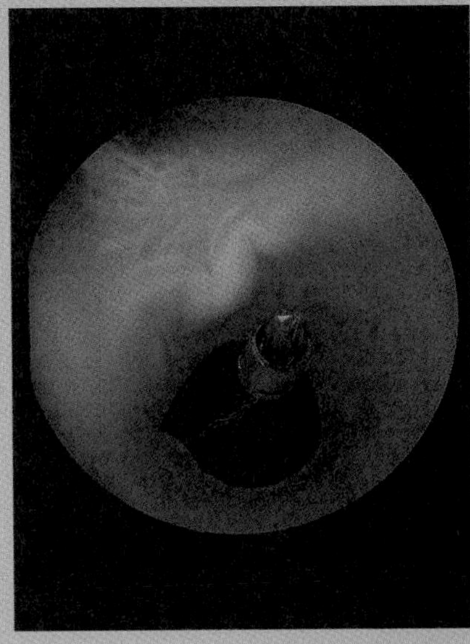

An endoscopic image of a micromotor in an artery.

Step 1 The starting substance is 8.75 g of N_2O. Convert 8.75 g of N_2O to moles:

$$8.75\ \cancel{\text{g N}_2\text{O}} \times \frac{1\ \text{mol N}_2\text{O}}{44.02\ \cancel{\text{g N}_2\text{O}}} = 0.199\ \text{mol N}_2\text{O}$$

Step 2 Calculate the moles of HNO_3 by the mole-ratio method:

$$0.199\ \cancel{\text{mol N}_2\text{O}} \times \frac{10\ \text{mol HNO}_3}{1\ \cancel{\text{mol N}_2\text{O}}} = 1.99\ \text{mol HNO}_3$$

Step 3 Convert moles of HNO_3 to grams:

$$1.99\ \cancel{\text{mol HNO}_3} \times \frac{63.02\ \text{g HNO}_3}{1\ \cancel{\text{mol HNO}_3}} = 125\ \text{g HNO}_3$$

Method 2: Continuous Calculation

$$8.75\ \cancel{\text{g N}_2\text{O}} \times \frac{1\ \cancel{\text{mol N}_2\text{O}}}{44.02\ \cancel{\text{g N}_2\text{O}}} \times \frac{10\ \cancel{\text{mol HNO}_3}}{1\ \cancel{\text{mol N}_2\text{O}}} \times \frac{63.02\ \text{g HNO}_3}{1\ \cancel{\text{mol HNO}_3}} = 125\ \text{g HNO}_3$$

Practice 9.5

How many grams of chromium(III) chloride are required to produce 75.0 g of silver chloride?

$$CrCl_3 + 3\ AgNO_3 \longrightarrow Cr(NO_3)_3 + 3\ AgCl$$

Practice 9.6

What mass of water is produced by the complete combustion of 225.0 g of butane (C_4H_{10})?

$$2\ C_4H_{10} + 13\ O_2 \longrightarrow 8\ CO_2 + 10\ H_2O$$

9.6 Limiting-Reactant and Yield Calculations

In many chemical processes the quantities of the reactants used are such that the amount of one reactant is in excess of the amount of a second reactant in the reaction. The amount of the product(s) formed in such a case will depend on the reactant that is not in excess. Thus the reactant that is not in excess is known as the **limiting reactant** (sometimes called the limiting reagent), because it limits the amount of product that can be formed.

limiting reactant

Consider the case illustrated in Figure 9.2. How many bicycles can be assembled from the parts shown?

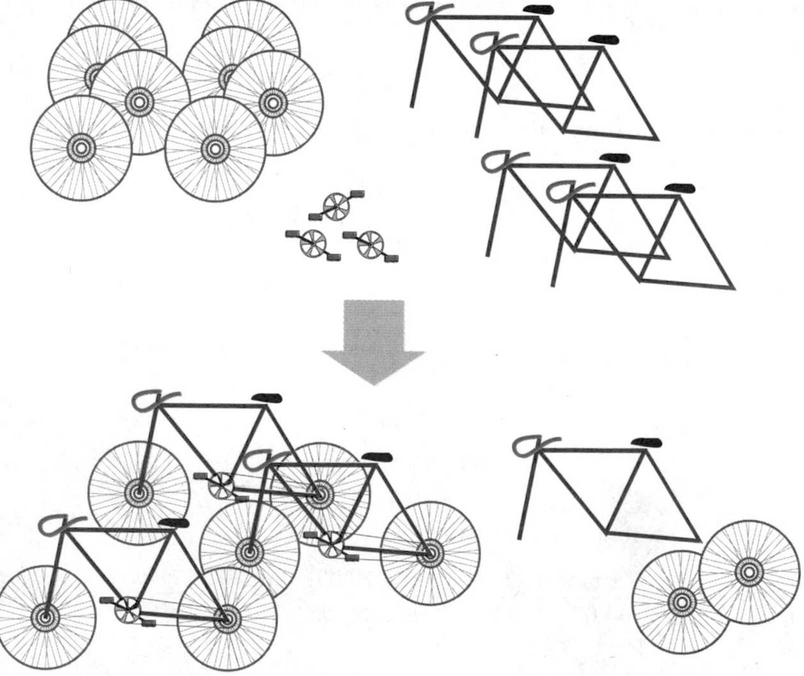

The number of bicycles that can be built from this pile of parts is determined by the limiting reactant (the pedal assemblies).

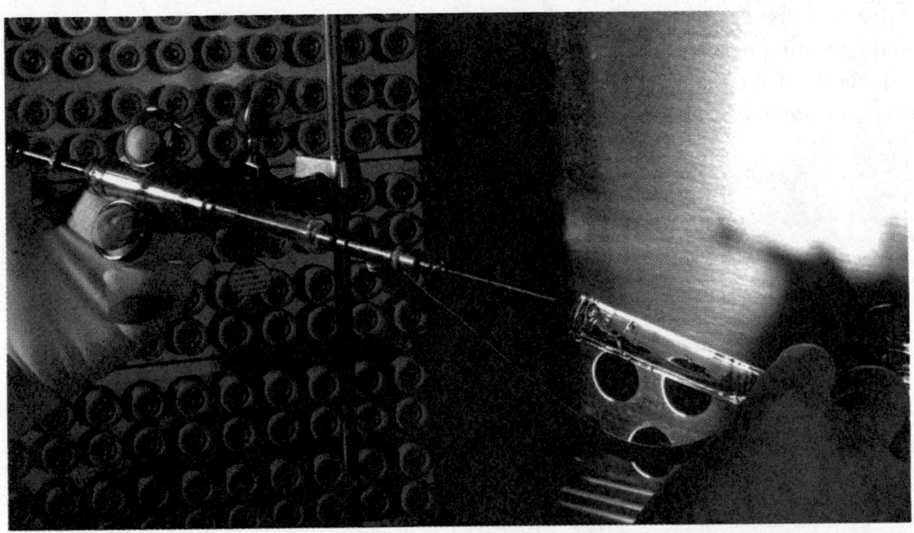

The amount of product formed into each medicine is determined by the limiting reactant.

The limiting reactant in this case is the number of pedal assemblies. This part limits the number of bicycles that can be built. The wheels and frames are reactants in excess.

Now let's consider a chemical situation in which solutions containing 1.0 mol of sodium hydroxide and 1.5 mol of hydrochloric acid are mixed:

$$\underset{1\ \text{mol}}{NaOH} + \underset{1\ \text{mol}}{HCl} \longrightarrow \underset{1\ \text{mol}}{NaCl} + \underset{1\ \text{mol}}{H_2O}$$

According to the equation it is possible to obtain 1.0 mol of NaCl from 1.0 mol of NaOH, and 1.5 mol of NaCl from 1.5 mol of HCl. However, we cannot have two different yields of NaCl from the reaction. When 1.0 mol of NaOH and 1.5 mol of HCl are mixed, there is insufficient NaOH to react with all of the HCl. Therefore, HCl is the reactant in excess and NaOH is the limiting reactant. Since the amount of NaCl formed is dependent on the limiting reactant, only 1.0 mol of NaCl will be formed. Because 1.0 mol of NaOH reacts with 1.0 mol of HCl, 0.5 mol of HCl remains unreacted:

$$\left.\begin{array}{l}1.0\ \text{mol NaOH}\\ 1.5\ \text{mol HCl}\end{array}\right\} \longrightarrow \begin{array}{l}1.0\ \text{mol NaCl}\\ 1.0\ \text{mol}\ H_2O\end{array} + 0.5\ \text{mol HCl unreacted}$$

Problems giving the amounts of two reactants are generally of the limiting-reactant type, and there are several methods used to identify them in a chemical reaction. In the most direct method two steps are needed to determine the limiting reactant and the amount of product formed.

1. Calculate the amount of product (moles or grams, as needed) that can be formed from each reactant.
2. Determine which reactant is limiting. (The reactant that gives the least amount of product is the limiting reactant; the other reactant is in excess. The limiting reactant will determine the amount of product formed in the reaction.)

Sometimes, however, it is necessary to calculate the amount of excess reactant.

3. This can be done by first calculating the amount of excess reactant required to react with the limiting reactant. Then subtract the amount that reacts from the starting quantity of the reactant in excess. This result is the amount of that substance that remains unreacted.

Example 9.9 How many moles of Fe_3O_4 can be obtained by reacting 16.8 g Fe with 10.0 g H_2O? Which substance is the limiting reactant? Which substance is in excess?

Solution

$$3\ Fe(s) + 4\ H_2O(g) \xrightarrow{\Delta} Fe_3O_4(s) + 4\ H_2(g)$$

Step 1 Calculate the moles of Fe_3O_4 that can be formed from each reactant:

$$\text{g reactant} \longrightarrow \text{mol reactant} \longrightarrow \text{mol } Fe_3O_4$$

The continuous calculation method is shown here. You can also use the step-by-step method to determine the requested substance.

$$16.8\ \cancel{\text{g Fe}} \times \frac{1\ \cancel{\text{mol Fe}}}{55.85\ \cancel{\text{g Fe}}} \times \frac{1\ \text{mol } Fe_3O_4}{3\ \cancel{\text{mol Fe}}} = 0.100\ \text{mol } Fe_3O_4$$

$$10.0\ \cancel{\text{g } H_2O} \times \frac{1\ \cancel{\text{mol } H_2O}}{18.02\ \cancel{\text{g } H_2O}} \times \frac{1\ \text{mol } Fe_3O_4}{4\ \cancel{\text{mol } H_2O}} = 0.139\ \text{mol } Fe_3O_4$$

Step 2 Determine the limiting reactant. The limiting reactant is Fe because it produces less Fe_3O_4; the H_2O is in excess. The yield of product is 0.100 mol of Fe_3O_4.

Example 9.10 How many grams of silver bromide (AgBr) can be formed when solutions containing 50.0 g of $MgBr_2$ and 100.0 g of $AgNO_3$ are mixed together? How many grams of the excess reactant remain unreacted?

Solution

$$MgBr_2(aq) + 2\ AgNO_3(aq) \longrightarrow 2\ AgBr(s) + Mg(NO_3)_2(aq)$$

Step 1 Calculate the grams of AgBr that can be formed from each reactant.

$$\text{g reactant} \longrightarrow \text{mol reactant} \longrightarrow \text{mol AgBr} \longrightarrow \text{g AgBr}$$

$$50.0\ \cancel{\text{g } MgBr_2} \times \frac{1\ \cancel{\text{mol } MgBr_2}}{184.1\ \cancel{\text{g } MgBr_2}} \times \frac{2\ \cancel{\text{mol AgBr}}}{1\ \cancel{\text{mol } MgBr_2}} \times \frac{187.8\ \text{g AgBr}}{1\ \cancel{\text{mol AgBr}}} = 102\ \text{g AgBr}$$

$$100.0\ \cancel{\text{g } AgNO_3} \times \frac{1\ \cancel{\text{mol } AgNO_3}}{169.9\ \cancel{\text{g } AgNO_3}} \times \frac{2\ \cancel{\text{mol AgBr}}}{2\ \cancel{\text{mol } AgNO_3}} \times \frac{187.8\ \text{g AgBr}}{1\ \cancel{\text{mol AgBr}}} = 110.5\ \text{g AgBr}$$

Step 2 Determine the limiting reactant. The limiting reactant is $MgBr_2$ because it gives less AgBr; $AgNO_3$ is in excess. The yield is 102 g AgBr.

Step 3 Calculate the grams of unreacted $AgNO_3$. Calculate the grams of $AgNO_3$ that will react with 50.0 g of $MgBr_2$:

$$\text{g } MgBr_2 \longrightarrow \text{mol } MgBr_2 \longrightarrow \text{mol } AgNO_3 \longrightarrow \text{g } AgNO_3$$

$$50.0\ \cancel{\text{g } MgBr_2} \times \frac{1\ \cancel{\text{mol } MgBr_2}}{184.1\ \cancel{\text{g } MgBr_2}} \times \frac{2\ \cancel{\text{mol } AgNO_3}}{1\ \cancel{\text{mol } MgBr_2}} \times \frac{169.9\ \text{g } AgNO_3}{1\ \cancel{\text{mol } AgNO_3}} = 92.3\ \text{g } AgNO_3$$

Thus, 92.3 g of $AgNO_3$ react with 50.0 g of $MgBr_2$. The amount of $AgNO_3$ that remains unreacted is

$$100.0 \text{ g } AgNO_3 - 92.3 \text{ g } AgNO_3 = 7.7 \text{ g } AgNO_3 \text{ unreacted}$$

The final mixture will contain 102 g $AgBr(s)$, 7.7 g $AgNO_3$ and an undetermined amount of $Mg(NO_3)_2$ in solution.

Practice 9.7

How many grams of hydrogen chloride can be produced from 0.490 g of hydrogen and 50.0 g of chlorine?

$$H_2(g) + Cl_2(g) \longrightarrow 2\ HCl(g)$$

Practice 9.8

How many grams of barium sulfate will be formed from 200.0 g of barium nitrate and 100.0 grams of sodium sulfate?

$$Ba(NO_3)_2(aq) + Na_2SO_4(aq) \longrightarrow BaSO_4(s) + 2\ NaNO_3(aq)$$

The quantities of the products we have been calculating from equations represent the maximum yield (100%) of product according to the reaction represented by the equation. Many reactions, especially those involving organic substances, fail to give a 100% yield of product. The main reasons for this failure are the side reactions that give products other than the main product and the fact that many reactions are reversible. In addition, some product may be lost in handling and transferring from one vessel to another. The **theoretical yield** of a reaction is the calculated amount of product that can be obtained from a given amount or reactant, according to the chemical equation. The **actual yield** is the amount of product that we finally obtain.

theoretical yield

actual yield

percent yield

The **percent yield** is the ratio of the actual yield to the theoretical yield multiplied by 100. Both the theoretical and the actual yields must have the same units to obtain a percent:

$$\frac{\text{actual yield}}{\text{theoretical yield}} \times 100 = \text{percent yield}$$

For example, if the theoretical yield calculated for a reaction is 14.8 g and the amount of product obtained is 9.25 g, the percent yield is

$$\text{percent yield} = \frac{9.25 \text{ g}}{14.8 \text{ g}} \times 100 = 62.5\%$$

Remember to round off as appropriate for your particular course.

Example 9.11

Carbon tetrachloride (CCl_4) was prepared by reacting 100. g of carbon disulfide and 100. g of chlorine. Calculate the percent yield if 65.0 g of CCl_4 was obtained from the reaction:

$$CS_2 + 3\ Cl_2 \longrightarrow CCl_4 + S_2Cl_2$$

Solution In this problem we need to determine the limiting reactant in order to calculate the quantity of CCl_4 (theoretical yield) that can be formed. Then we can compare that amount with the 65.0 g CCl_4 (actual yield) to calculate the percent yield.

Step 1 Determine the theoretical yield. Calculate the grams of CCl_4 that can be formed from each reactant:

$$\text{g reactant} \longrightarrow \text{mol reactant} \longrightarrow \text{mol } CCl_4 \longrightarrow \text{g } CCl_4$$

$$100.\ \cancel{\text{g } CS_2} \times \frac{1\ \cancel{\text{mol } CS_2}}{76.13\ \cancel{\text{g } CS_2}} \times \frac{1\ \cancel{\text{mol } CCl_4}}{1\ \cancel{\text{mol } CS_2}} \times \frac{153.8\ \text{g } CCl_4}{1\ \cancel{\text{mol } CCl_4}} = 202\ \text{g } CCl_4$$

$$100.\ \cancel{\text{g } Cl_2} \times \frac{1\ \cancel{\text{mol } Cl_2}}{70.90\ \cancel{\text{g } Cl_2}} \times \frac{1\ \cancel{\text{mol } CCl_4}}{3\ \cancel{\text{mol } Cl_2}} \times \frac{153.8\ \text{g } CCl_4}{1\ \cancel{\text{mol } CCl_4}} = 72.3\ \text{g } CCl_4$$

Step 2 Determine the limiting reactant. The limiting reactant is Cl_2 because it gives less CCl_4. The CS_2 is in excess. The theoretical yield is 72.3 g CCl_4.

Step 3 Calculate the percent yield. According to the equation, 72.3 g of CCl_4 is the maximum amount or theoretical yield of CCl_4 possible from 100. g of Cl_2. Actual yield is 65.0 g of CCl_4:

$$\text{percent yield} = \frac{65.0\ \cancel{\text{g}}}{72.3\ \cancel{\text{g}}} \times 100 = 89.9\%$$

Example 9.12 Silver bromide was prepared by reacting 200.0 g of magnesium bromide and an adequate amount of silver nitrate. Calculate the percent yield if 375.0 g of silver bromide was obtained from the reaction:

$$MgBr_2 + 2\ AgNO_3 \longrightarrow Mg(NO_3)_3 + 2\ AgBr$$

Solution **Step 1** Determine the theoretical yield. Calculate the grams of AgBr that can be formed:

$$200.0\ \text{g } \cancel{MgBr_2} \times \frac{1\ \text{mol } \cancel{MgBr_2}}{184.1\ \text{g } \cancel{MgBr_2}} \times \frac{2\ \text{mol } \cancel{AgBr}}{1\ \text{mol } \cancel{MgBr_2}} \times \frac{187.8\ \text{g AgBr}}{1\ \text{mol } \cancel{AgBr}} = 408.0\ \text{g AgBr}$$

The theoretical yield is 408.0 g AgBr.

Step 2 Calculate the percent yield. According to the equation, 408.0 g AgBr is the maximum amount of AgBr possible from 200.0 g $MgBr_2$. Actual yield is 375.0 g AgBr:

$$\text{Percent yield} = \frac{375.0\ \cancel{\text{g AgBr}}}{408.0\ \cancel{\text{g AgBr}}} \times 100 = 91.91\%$$

Practice 9.9

Aluminum oxide was prepared by heating 225 g of chromium(II) oxide with 125 g of aluminum. Calculate the percent yield if 100.0 g of aluminum oxide was obtained:

$$2\ Al + 3\ CrO \longrightarrow Al_2O_3 + 3\ Cr$$

When solving problems, you will achieve better results if you work in an organized manner.

1. Write the data and numbers in a logical, orderly manner.
2. Make certain that the equations are balanced and that the computations are accurate and expressed to the correct number of significant figures.
3. Remember that units are very important; a number without units has little meaning.

Concepts in Review

1. Write mole ratios for any two substances involved in a chemical reaction.
2. Outline the mole or mole-ratio method for making stoichiometric calculations.
3. Calculate the number of moles of a desired substance obtainable from a given number of moles of a starting substance in a chemical reaction (mole–mole calculations).
4. Calculate the mass of a desired substance obtainable from a given number of moles of a starting substance in a chemical reaction, and vice versa (mole–mass and mass–mole calculations).
5. Calculate the mass of a desired substance involved in a chemical reaction from a given mass of a starting substance (mass–mass calculation).
6. Deduce the limiting reactant or reagent when given the amounts of starting substances, and then calculate the moles or mass of desired substance obtainable from a given chemical reaction (limiting reactant calculation).
7. Apply theoretical yield or actual yield to any of the foregoing types of problems, or calculate theoretical and actual yields of a chemical reaction.

Key Terms

The terms listed here have all been defined within the chapter. Section numbers are referenced in parenthesis for each term.

actual yield (9.6)
limiting reactant (9.6)
mole ratio (9.2)
percent yield (9.6)
stoichiometry (9.2)
theoretical yield (9.6)

Questions

Questions refer to tables, figures, and key words and concepts defined within the chapter. A particularly challenging question or exercise is indicated with an asterisk.

1. Phosphine (PH_3) can be prepared by the hydrolysis of calcium phosphide, Ca_3P_2:

$$Ca_3P_2 + 6\ H_2O \longrightarrow 3\ Ca(OH)_2 + 2\ PH_3$$

Based on this equation, which of the following statements are correct? Show evidence to support your answer.

(a) One mole of Ca_3P_2 produces 2 mol of PH_3.

(b) One gram of Ca_3P_2 produces 2 g of PH_3.

(c) Three moles of $Ca(OH)_2$ are produced for each 2 mol of PH_3 produced.

(d) The mole ratio between phosphine and calcium phosphide is $\frac{2\text{ mol }PH_3}{1\text{ mol }Ca_3P_2}$

(e) When 2.0 mol of Ca_3P_2 and 3.0 mol of H_2O react, 4.0 mol of PH_3 can be formed.

(f) When 2.0 mol of Ca_3P_2 and 15.0 mol of H_2O react, 6.0 mol of $Ca(OH)_2$ can be formed.

(g) When 200. g of Ca_3P_2 and 100. g of H_2O react, Ca_3P_2 is the limiting reactant.

(h) When 200. g of Ca_3P_2 and 100. g of H_2O react, the theoretical yield of PH_3 is 57.4 g.

2. The equation representing the reaction used for the commercial preparation of hydrogen cyanide is

$$2\ CH_4 + 3\ O_2 + 2\ NH_3 \longrightarrow 2\ HCN + 6\ H_2O$$

Based on this equation, which of the following statements are correct? Rewrite incorrect statements to make them correct.

(a) Three moles of O_2 are required for 2 mol of NH_3.

(b) Twelve moles of HCN are produced for every 16 mol of O_2 that react.

(c) The mole ratio between H_2O and CH_4 is $\frac{6\text{ mol }H_2O}{2\text{ mol }CH_4}$

(d) When 12 mol of HCN are produced, 4 mol of H_2O will be formed.

(e) When 10 mol of CH_4, 10 mol of O_2, and 10 mol of NH_3 are mixed and reacted, O_2 is the limiting reactant.

(f) When 3 mol each of CH_4, O_2, and NH_3 are mixed and reacted, 3 mol of HCN will be produced.

Paired Exercises

These exercises are paired. Each odd-numbered exercise is followed by a similar even-numbered exercise. Answers to the even-numbered exercises are given in Appendix V.

Mole Review Exercises

3. Calculate the number of moles in each of the following quantities:

(a) 25.0 g KNO_3

(b) 56 millimol NaOH

(c) 5.4×10^2 g $(NH_4)_2C_2O_4$

(d) 16.8 mL H_2SO_4 solution (d = 1.727 g/mL, 80.0% H_2SO_4 by mass)

4. Calculate the number of moles in each of the following quantities:

(a) 2.10 kg $NaHCO_3$

(b) 525 mg $ZnCl_2$

(c) 9.8×10^{24} molecules CO_2

(d) 250 mL ethyl alcohol, C_2H_5OH (d = 0.789 g/mL)

5. Calculate the number of grams in each of the following quantities:

(a) 2.55 mol $Fe(OH)_3$

(b) 125 kg $CaCO_3$

(c) 10.5 mol NH_3

(d) 72 millimol HCl

(e) 500.0 mL of liquid Br_2 (d = 3.119 g/mL)

6. Calculate the number of grams in each of the following quantities:

(a) 0.00844 mol $NiSO_4$

(b) 0.0600 mol $HC_2H_3O_2$

(c) 0.725 mol Bi_2S_3

(d) 4.50×10^{21} molecules glucose, $C_6H_{12}O_6$

(e) 75 mL K_2CrO_4 solution (d = 1.175 g/mL, 20.0% K_2CrO_4 by mass)

7. Which contains the larger number of molecules, 10.0 g H_2O or 10.0 g H_2O_2?

8. Which contains the larger numbers of molecules, 25.0 g HCl or 85.0 g $C_6H_{12}O_6$?

Mole-Ratio Exercises

9. Given the equation for the combustion of isopropyl alcohol

$$2\ C_3H_7OH + 9\ O_2 \longrightarrow 6\ CO_2 + 8\ H_2O$$

what is the mole ratio of
(a) CO_2 to C_3H_7OH
(b) C_3H_7OH to O_2
(c) O_2 to CO_2
(d) H_2O to C_3H_7OH
(e) CO_2 to H_2O
(f) H_2O to O_2

10. For the reaction

$$3\ CaCl_2 + 2\ H_3PO_4 \longrightarrow Ca_3(PO_4)_2 + 6\ HCl$$

set up the mole ratio of
(a) $CaCl_2$ to $Ca_3(PO_4)_2$
(b) HCl to H_3PO_4
(c) $CaCl_2$ to H_3PO_4
(d) $Ca_3(PO_4)_2$ to H_3PO_4
(e) HCl to $Ca_3(PO_4)_2$
(f) H_3PO_4 to HCl

11. How many moles of Cl_2 can be produced from 5.60 mol HCl?

$$4\ HCl + O_2 \longrightarrow 2\ Cl_2 + 2\ H_2O$$

12. How many moles of CO_2 can be produced from 7.75 mol C_2H_5OH? (Balance the equation first.)

$$C_2H_5OH + O_2 \longrightarrow CO_2 + H_2O$$

13. Given the equation

$$Al_4C_3 + 12\ H_2O \longrightarrow 4\ Al(OH)_3 + 3\ CH_4$$

(a) How many moles of water are needed to react with 100. g of Al_4C_3?
(b) How many moles of $Al(OH)_3$ will be produced when 0.600 mol of CH_4 is formed?

14. An early method of producing chlorine was by the reaction of pyrolusite (MnO_2) and hydrochloric acid. How many moles of HCl will react with 1.05 mol of MnO_2? (Balance the equation first.)

$$MnO_2(s) + HCl(aq) \longrightarrow Cl_2(g) + MnCl_2(aq) + H_2O(l)$$

15. How many grams of sodium hydroxide can be produced from 500 g of calcium hydroxide according to this equation?

$$Ca(OH)_2 + Na_2CO_3 \longrightarrow 2\ NaOH + CaCO_3$$

16. How many grams of zinc phosphate, $Zn_3(PO_4)_2$, are formed when 10.0 g of Zn are reacted with phosphoric acid?

$$3\ Zn + 2\ H_3PO_4 \longrightarrow Zn_3(PO_4)_2 + 3\ H_2$$

17. In a blast furnace, iron(III) oxide reacts with coke (carbon) to produce molten iron and carbon monoxide:

$$Fe_2O_3 + 3\ C \longrightarrow 2\ Fe + 3\ CO$$

How many kilograms of iron would be formed from 125 kg of Fe_2O_3?

18. How many grams of steam and iron must react to produce 375 g of magnetic iron oxide, Fe_3O_4?

$$3\ Fe(s) + 4\ H_2O(g) \longrightarrow Fe_3O_4(s) + 4\ H_2(g)$$

19. Ethane gas, C_2H_6, burns in air (i.e., reacts with the oxygen in air) to form carbon dioxide and water:

$$2\ C_2H_6 + 7\ O_2 \longrightarrow 4\ CO_2 + 6\ H_2O$$

(a) How many moles of O_2 are needed for the complete combustion of 15.0 mol of ethane?
(b) How many grams of CO_2 are produced for each 8.00 g of H_2O produced?
(c) How many grams of CO_2 will be produced by the combustion of 75.0 g of C_2H_6?

20. Given the equation

$$4\ FeS_2 + 11\ O_2 \longrightarrow 2\ Fe_2O_3 + 8\ SO_2$$

(a) How many moles of Fe_2O_3 can be made from 1.00 mol of FeS_2?
(b) How many moles of O_2 are required to react with 4.50 mol of FeS_2?
(c) If the reaction produces 1.55 mol of Fe_2O_3, how many moles of SO_2 are produced?
(d) How many grams of SO_2 can be formed from 0.512 mol of FeS_2?
(e) If the reaction produces 40.6 g of SO_2, how many moles of O_2 were reacted?
(f) How many grams of FeS_2 are needed to produce 221 g of Fe_2O_3?

Limiting-Reactant and Percent-Yield Exercises

21. In the following equations, determine which reactant is the limiting reactant and which reactant is in excess. The amounts mixed together are shown below each reactant. Show evidence for your answers.

(a) $KOH + HNO_3 \longrightarrow KNO_3 + H_2O$
16.0 g 12.0 g

(b) $2\ NaOH + H_2SO_4 \longrightarrow Na_2SO_4 + 2\ H_2O$
10.0 g 10.0 g

22. In the following equations, determine which reactant is the limiting reactant and which reactant is in excess. The amounts mixed together are shown below each reactant. Show evidence for your answers.

(a) $2\ Bi(NO_3)_3 + 3\ H_2S \longrightarrow Bi_2S_3 + 6\ HNO_3$
50.0 g 6.00 g

(b) $3\ Fe + 4\ H_2O \longrightarrow Fe_3O_4 + 4\ H_2$
40.0 g 16.0 g

23. The reaction for the combustion of propane is

$$C_3H_8 + 5\ O_2 \longrightarrow 3\ CO_2 + 4\ H_2O$$

(a) If 5.0 mol of C_3H_8 and 5.0 mol of O_2 are reacted, how many moles of CO_2 can be produced?
(b) If 3.0 mol of C_3H_8 and 20.0 mol of O_2 are reacted, how many moles of CO_2 are produced?
(c) If 20.0 mol of C_3H_8 and 3.0 mol of O_2 are reacted, how many moles of CO_2 can be produced?

24. The reaction for the combustion of propane is

$$C_3H_8 + 5\ O_2 \longrightarrow 3\ CO_2 + 4\ H_2O$$

(a) If 20.0 g of C_3H_8 and 20.0 g of O_2 are reacted, how many moles of CO_2 can be produced?
(b) If 20.0 g of C_3H_8 and 80.0 g of O_2 are reacted, how many moles of CO_2 are produced?
(c) If 2.0 mol of C_3H_8 and 14.0 mol of O_2 are placed in a closed container and they react to completion (until one reactant is completely used up), what compounds will be present in the container after the reaction, and how many moles of each compound are present?

***25.** When a particular metal *X* reacts with HCl, the resulting products are XCl_2 and H_2. Write and balance the equation. When 78.5 g of the metal completely react, 2.42 g of hydrogen gas results. Identify the element *X*.

***26.** When a certain nonmetal whose formula is X_8 combusts in air, XO_3 forms. Write a balanced equation for this reaction. If 120.0 g of oxygen gas are consumed completely, along with 80.0 g of X_8, identify element *X*.

27. Aluminum reacts with bromine to form aluminum bromide:

$$2\ Al + 3\ Br_2 \longrightarrow 2\ AlBr_3$$

If 25.0 g of Al and 100. g of Br_2 are reacted, and 64.2 g of $AlBr_3$ product are recovered, what is the percent yield for the reaction?

28. Iron was reacted with a solution containing 400. g of copper(II) sulfate. The reaction was stopped after 1 hour, and 151 g of copper were obtained. Calculate the percent yield of copper obtained:

$$Fe(s) + CuSO_4(aq) \longrightarrow Cu(s) + FeSO_4(aq)$$

***29.** Carbon disulfide, CS_2, can be made from coke, C, and sulfur dioxide, SO_2:

$$3\ C + 2\ SO_2 \longrightarrow CS_2 + 2\ CO_2$$

If the actual yield of CS_2 is 86.0% of the theoretical yield, what mass of coke is needed to produce 950 g of CS_2?

***30.** Acetylene (C_2H_2) can be manufactured by the reaction of water and calcium carbide, CaC_2:

$$CaC_2(s) + 2\ H_2O(l) \longrightarrow C_2H_2(g) + Ca(OH)_2(s)$$

When 44.5 g of commercial grade (impure) calcium carbide is reacted, 0.540 mol of C_2H_2 is produced. Assuming that all of the CaC_2 was reacted to C_2H_2, what is the percent of CaC_2 in the commercial grade material?

Additional Exercises

These problems are not paired or labeled by topic and provide additional practice on the concepts covered in this chapter.

31. A tool set contains 6 wrenches, 4 screwdrivers, and 2 pliers. The manufacturer has 1000 pliers, 2000 screwdrivers, and 3000 wrenches in stock. Can an order for 600 tool sets be filled? Explain briefly.

32. What is the difference between using a number as a subscript and using a number as a coefficient in a chemical equation?

33. Oxygen masks for producing O_2 in emergency situations contain potassium superoxide (KO_2). It reacts according to the following equation:

$$4\ KO_2 + 2\ H_2O + 4\ CO_2 \longrightarrow 4\ KHCO_3 + 3\ O_2$$

(a) If a person wearing such a mask exhales 0.85 g of CO_2 every minute, how many moles of KO_2 are consumed in 10.0 minutes?
(b) How many grams of oxygen are produced in 1.0 hour?

34. Ethyl alcohol is the result of fermentation of sugar, $C_6H_{12}O_6$:

$$C_6H_{12}O_6 \longrightarrow 2\ C_2H_5OH + 2\ CO_2$$

(a) How many grams of ethyl alcohol and how many grams of carbon dioxide can be produced from 750 g of sugar?
(b) How many milliliters of alcohol (d = 0.79 g/mL) can be produced from 750 g of sugar?

35. Phosphoric acid, H_3PO_4, can be synthesized from phosphorus, oxygen, and water according to these two reactions

$$4\ P + 5\ O_2 \longrightarrow P_4O_{10}$$

$$P_4O_{10} + 6\ H_2O \longrightarrow 4\ H_3PO_4$$

Starting with 20.0 g P, 30.0 g O_2, and 15.0 g H_2O, what is the mass of phosphoric acid that can be formed?

36. The methyl alcohol (CH_3OH) used in alcohol burners combines with oxygen gas to form carbon dioxide and water. How many grams of oxygen are required to burn 60.0 mL of methyl alcohol (d = 0.72 g/mL)?

37. Hydrazine (N_2H_4) and hydrogen peroxide (H_2O_2) have been used as rocket propellants. They react according to the following equation:

$$7\ H_2O_2 + N_2H_4 \longrightarrow 2\ HNO_3 + 8\ H_2O$$

(a) How many moles of HNO_3 are formed from 0.33 mol of hydrazine?
(b) How many moles of hydrogen peroxide are required if 2.75 mol of water are to be produced?
(c) How many moles of water are produced if 8.72 mol of HNO_3 are also produced?
(d) How many grams of hydrogen peroxide are needed to completely react with 120 g of hydrazine?

38. Silver tarnishes in the presence of hydrogen sulfide (which smells like rotten eggs) and oxygen because of the reaction:

$$4\ Ag + 2\ H_2S + O_2 \longrightarrow 2\ Ag_2S + 2\ H_2O$$

How many grams of silver sulfide can be formed from a mixture of 1.1 g Ag, 0.14 g H_2S, and 0.080 g O_2?

39. After 180.0 g of zinc were dropped into a beaker of hydrochloric acid and the reaction ceased, 35 g of unreacted zinc remained in the beaker:

$$Zn + HCl \longrightarrow ZnCl_2 + H_2$$

Balance the equation first.
(a) How many moles of hydrogen gas were produced?
(b) How many grams of HCl were reacted?

40. Given the following equation, answer (a) and (b) below:

$$Fe(s) + CuSO_4(aq) \longrightarrow Cu(s) + FeSO_4(aq)$$

(a) When 2.0 mol of Fe and 3.0 mol of $CuSO_4$ are reacted, what substances will be present when the reaction is over? How many moles of each substance are present?
(b) When 20.0 g of Fe and 40.0 g of $CuSO_4$ are reacted, what substances will be present when the reaction is over? How many grams of each substance are present?

41. Methyl alcohol (CH_3OH) is made by reacting carbon monoxide and hydrogen in the presence of certain metal oxide catalysts. How much alcohol can be obtained by reacting 40.0 g of CO and 10.0 g of H_2? How many grams of excess reactant remain unreacted?

$$CO(g) + 2\ H_2(g) \longrightarrow CH_3OH(l)$$

*42. Ethyl alcohol (C_2H_5OH), also called grain alcohol, can be made by the fermentation of sugar, which often comes from starch in grain:

$$\underset{\text{glucose}}{C_6H_{12}O_6} \longrightarrow 2\ \underset{\text{ethyl alcohol}}{C_2H_5OH} + 2\ CO_2$$

If an 84.6% yield of ethyl alcohol is obtained,
(a) what mass of ethyl alcohol will be produced from 750 g glucose?
(b) what mass of glucose should be used to produce 475 g of C_2H_5OH?

*43. Both $CaCl_2$ and $MgCl_2$ react with $AgNO_3$ to precipitate AgCl. When solutions containing equal masses of $CaCl_2$ and $MgCl_2$ are reacted with $AgNO_3$ solutions that will produce the larger amount of AgCl? Show proof.

*44. An astronaut excretes about 2500 g of water a day. If lithium oxide (Li_2O) is used in the spaceship to absorb this water, how many kilograms of Li_2O must be carried for a 30-day space trip for three astronauts?

$$Li_2O + H_2O \longrightarrow 2\ LiOH$$

*45. Much commercial hydrochloric acid is prepared by the reaction of concentrated sulfuric acid with sodium chloride:

$$H_2SO_4 + 2\ NaCl \longrightarrow Na_2SO_4 + 2\ HCl$$

How many kilograms of concentrated H_2SO_4, 96% H_2SO_4 by mass, are required to produce 20.0 L of concentrated hydrochloric acid (d = 1.20 g/mL, 42.0% HCl by mass)?

*46. Gastric juice contains about 3.0 g HCl per liter. If a person produces about 2.5 L of gastric juice per day, how many antacid tablets, each containing 400. mg of $Al(OH)_3$, are needed to neutralize all the HCl produced in 1 day?

$$Al(OH)_3(s) + 3\ HCl(aq) \longrightarrow AlCl_3(aq) + 3\ H_2O(l)$$

*47. When 12.82 g of a mixture of $KClO_3$ and NaCl is heated strongly, the $KClO_3$ reacts according to the following equation:

$$2\ KClO_3(s) \longrightarrow 2\ KCl(s) + 3\ O_2(g)$$

The NaCl does not undergo any reaction. After the heating, the mass of the residue (KCl and NaCl) is 9.45 g. Assuming that all the loss of mass represents loss of oxygen gas, calculate the percent of $KClO_3$ in the original mixture.

48. What is the purpose of a mask in the manufacture of tiny circuits and machines?

49. Why is silicon dioxide called a sacrificial material in the production of micromachinery?

50. What are some possible uses for micromachines in our society? Indicate some of the difficulties that must be overcome to implement these uses.

Answers to Practice Exercises

9.1 0.33 mol Al_2O_3
9.2 7.33 mol $Al(OH)_3$
9.3 0.8157 mol KCl
9.4 85 g $AgNO_3$
9.5 27.6 g $CrCl_3$
9.6 348.8 g H_2O
9.7 17.7 g HCl
9.8 164.4 g $BaSO_4$
9.9 88.5% yield

Modern Atomic Theory and the Periodic Table

10

Chapter 10

How do we go about studying an object that is too small to see? Think back to that birthday when you had a present you could look at but not yet open. Simply judging from the wrapping and size of the box was not very useful. Shaking, turning, and lifting the package all gave clues, indirectly, to the contents. After all the experiments were done on the package, a fairly good hypothesis of the contents could be made. But was it correct? The only means of absolute verification would be to open the package.

The same is true for chemists in their study of the atom. Atoms are so very small that it is not possible to use the normal senses and rules to describe them. Chemists are essentially working in the dark with this package we call the atom. Yet, as instruments (x-ray machines and scanning-tunneling microscopes) and measuring devices (spectrophotometers and magnetic resonance imaging, MRI) have progressed, along with our skill in mathematics and probability, the secrets of the atom are beginning to unravel.

10.1 A Brief History

In the last 200 years, vast amounts of data have been accumulated to support atomic theory. When atoms were originally suggested by the early Greeks, no evidence existed to physically support their ideas. Faraday, Arrhenius, and others did a variety of experiments, which culminated in Dalton's theory of the atom. Because of the limitations of Dalton's model, modifications were proposed by Thomson and then by Rutherford, which eventually led to our modern concept of the nuclear atom. These early models of the atom work reasonably well—in fact, we continue to use them today to visualize a variety of chemical concepts. There remain, however, questions that these models cannot answer, including an explanation of how atomic structure relates to the periodic table. In this chapter, we will look at our modern model of the atom to see how it varies from and improves upon the earlier atomic theories.

10.2 Electromagnetic Radiation

Scientists have studied energy and light for centuries, and several models have been proposed to explain how energy is transferred from place to place. One of the ways that energy travels through space is by *electromagnetic radiation.* Examples of electromagnetic radiation include light from the sun, X rays in your dentist's office, microwaves from your microwave oven, radio and television waves, and radiant heat from your fireplace. While these examples seem quite different, they are all similar in some important ways. Each shows wave-like behavior, and all travel at the same speed in a vacuum (3.00×10^8 m/s).

wavelength

The study of wave behavior is a topic for another course, but we shall need some basic terminology in order to understand atoms. Waves have three basic characteristics: wavelength, frequency, and speed. **Wavelength** (lambda, λ) is the distance between consecutive peaks (or troughs) in a wave, as shown in Figure 10.1.

◀ **Chapter Opening Photo: The brilliance of these neon signs is the result of electrons being transferred between energy levels.**

Chemistry in Action

Atomic Clocks

Imagine a clock that keeps time to within one second over a million years. The National Institute of Standards and Technology in Boulder, Colorado, has an atomic clock that does just that—a little better than your average alarm, grandfather, or cuckoo clock! This atomic clock serves as the international standard for time and frequency. How does it work?

Within the glistening case are several layers of magnetic shielding. In the heart of the clock is a small oven that heats cesium metal to release cesium atoms, which are collected into a narrow beam (1 mm wide). The beam of atoms passes down a long evacuated tube while being excited by a laser until all the cesium atoms are in the same electron state.

The atomic clock at the National Institute of Standards and Technology in Boulder, CO, loses only 1 second in a million years.

The atoms then enter another chamber filled with reflecting microwaves. The frequency of the microwaves (9,192,631,770 cycles per second) is exactly the same frequency required to excite a cesium atom from its ground state to the next higher energy level. These excited cesium atoms then release electromagnetic radiation in a process known as fluorescence. Electronic circuits maintain the microwave frequency at precisely the right level to keep the cesium atoms moving from one level to the next. One second is equal to 9,192,631,770 of these vibrations. The clock is set to this frequency and can keep accurate time for over a million years.

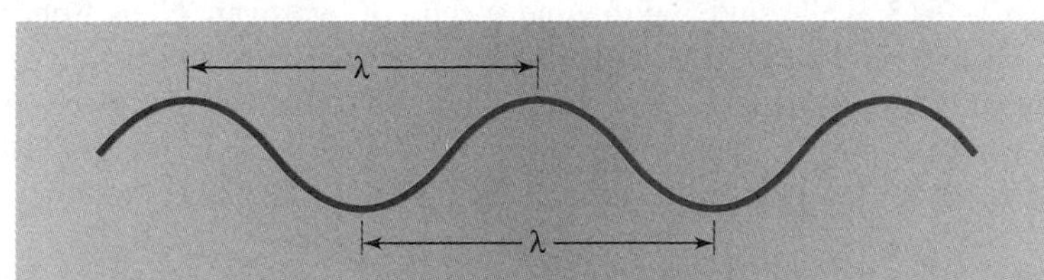

FIGURE 10.1
The wavelength of this wave is shown by λ. It can be measured from peak to peak or trough to trough.

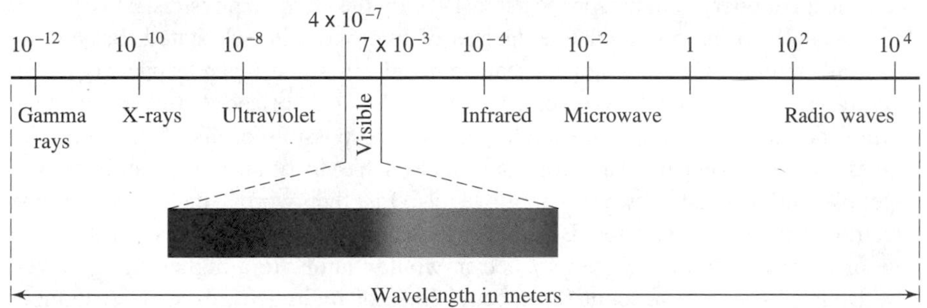

FIGURE 10.2
The electromagnetic spectrum.

Frequency (nu, ν) tells how many waves pass a particular point per second. **Speed** (v) tells how fast a wave moves through space.

frequency
speed

Light is one form of electromagnetic radiation and is usually classified by its wavelength, as shown in Figure 10.2. Notice that visible light is only a tiny part of the electromagnetic spectrum. Some examples of electromagnetic radiation involved in energy transfer outside the visible region are hot coals in your backyard grill, which transfer infrared radiation to cook your food; and microwaves, which transfer energy to water molecules in the food, causing them to move more quickly and thus raise the temperature of your food.

photons

Scientists have evidence for the wave-like nature of light. They also know that a beam of light behaves like a stream of tiny packets of energy called **photons**. So what is light exactly? Is it a particle? Is it a wave? Scientists have agreed to explain the properties of electromagnetic radiation by using both wave and particle properties. Neither explanation is ideal, but currently these are our best models.

10.3 The Bohr Atom

Niels Bohr (1885–1962).

As scientists struggled to understand the properties of electromagnetic radiation, evidence began to accumulate that atoms could radiate light. At high temperatures, or when subjected to high voltages, elements in the gaseous state give off colored light. Brightly colored neon signs illustrate this property of matter very well. When the light emitted by a gas is passed through a prism or diffraction grating, a set of brightly colored lines called a **line spectrum** results. These colored lines indicate that the light is being emitted only at certain wavelengths, or frequencies, that correspond to specific colors. Each element possesses a unique set of these spectral lines that is different from the sets of all the other elements. This is illustrated in Figure 10.3.

In 1912–1913, while studying the line spectra of hydrogen, Niels Bohr (1885–1962), a Danish physicist, made a significant contribution to the rapidly growing knowledge of atomic structure. His research led him to believe that electrons in an atom exist in specific regions at various distances from the nucleus. He also visualized the electrons as revolving in orbits around the nucleus, like planets rotating around the sun.

line spectrum

quanta

Bohr's first paper in this field dealt with the hydrogen atom, which he described as a single electron revolving in an orbit about a relatively heavy nucleus. He applied the concept of energy quanta, proposed in 1900 by the German physicist Max Planck (1858–1947), to the observed line spectra of hydrogen. Planck stated that energy is never emitted in a continuous stream but only in small discrete packets called **quanta** (Latin, *quantus,* how much). From this, Bohr theorized that electrons have several possible energies corresponding to several possible orbits at different distances from the nucleus. Therefore, an electron has to be in one specific energy level; it cannot exist between energy levels. In other words, the energy of the electron is said to be quantized. Bohr also stated that when a hydrogen atom absorbed one or more quanta of energy, its electron would "jump" to a higher energy level.

ground state

Bohr was able to account for spectral lines of hydrogen this way. A number of energy levels are available, the lowest of which is called the **ground state**. When an electron falls from a high-energy level to a lower one (say, from the fourth to the second), a quantum of energy is emitted as light at a specific frequency, or wavelength. This light corresponds to one of the lines visible in the hydrogen spectrum (see Figure 10.3). Several lines are visible in this spectrum, each one corresponding to a specific electron energy-level shift within the hydrogen atom.

The chemical properties of an element and its position in the periodic table depend on electron behavior within the atoms. In turn, much of our knowledge of the behavior of electrons within atoms is based on spectroscopy. Niels Bohr contributed a great deal to our knowledge of atomic structure by (1) suggesting quantized energy levels for electrons and (2) showing that spectral lines result from the radiation of small increments of energy (Planck's quanta) when electrons shift from one energy level to another. Bohr's calculations succeeded very well in correlating the experimentally observed spectral lines with electron energy levels for the hydrogen atom. However, Bohr's methods of calculation did not succeed for heavier atoms. More theoretical work on atomic structure was needed.

This stamp commemorates the work of Max Planck (1858–1947).

In 1924, the French physicist Louis de Broglie suggested a surprising hypothesis: All objects have wave properties. De Broglie used sophisticated mathematics to show that the wave properties for an object of ordinary size, such as a baseball, are too small to be observed. But for small objects, such as an electron, the wave properties become significant. Other scientists confirmed de Broglie's hypothesis, showing that electrons do exhibit wave properties. In 1926, Erwin Schrödinger, an Austrian physicist, created a mathematical model that described electrons as waves. Using Schrödinger's wave mechanics, it is possible to determine the probability of finding an electron in a certain region around the atom.

This treatment of the atom led to a new branch of physics called *wave mechanics* or *quantum mechanics,* which forms the basis for the modern understanding of atomic structure. Although the wave-mechanical description of the atom is mathematical it can be translated, at least in part, into a visual model. It is important to recognize that we cannot locate an electron precisely within an atom; however, it is clear that electrons are not revolving around the nucleus in *orbits* as Bohr postulated. The electrons are instead found in *orbitals.* An **orbital** can be pictured as a region in space around the nucleus where there is a high probability of finding a given electron. We will have more to say about the meaning of orbitals in the next section.

orbital

10.4 Energy Levels of Electrons

One of the ideas Bohr contributed to the modern concept of the atom was that the energy of the electron is quantized—that is, the electron is restricted to only certain allowed energies. The wave-mechanical model of the atom also predicts discrete **principal energy levels** within the atom. These energy levels are designated by the letter *n,* where *n* is a positive integer. The lowest principal energy level corresponds to $n = 1$, the next to $n = 2$, and so on. As *n* increases, the energy of the electron increases, and the electron is found on average farther from the nucleus.

principal energy levels

◀ FIGURE 10.3
Line spectrum of hydrogen. Each line corresponds to the wavelength of the energy emitted when the electron of a hydrogen atom, which has absorbed energy, falls back to a lower principal energy level.

Chemistry in Action

Ripples on the Surface

Modern physicists have astounded us with the idea that particles have wave properties. Protons, electrons, and all other elementary particles sometimes behave like waves and sometimes like particles. Donald Eigler at the IBM Almaden Research Center in San Jose, California, used a scanning-tunneling microscope at 4 kelvins to produce the picture shown here. The surface of this copper crystal surprised even the physicists. The waves in the photo are produced by electrons moving around on the surface of the crystal and bouncing off impurities (the two pits in the photo) in the copper. Since each electron behaves like a wave, it interferes with itself after reflecting off an impurity in the copper. The interference pattern is called a standing wave, which is a wave that vibrates up and down without visible transverse movement, similar to a plucked violin string. The crests of the waves represent regions where the electron is most probably in its particle form.

Metal atoms easily lose one or more electrons, which move about freely within the metal crystal, forming what is often called an "electron sea." At the surface of the metal these loose electrons are confined to a single layer, which is free to move in only two dimensions. Within these constraints the particles also behave as waves. The electron layer responsible for this beautiful pattern is only 0.02 Å thick. Similar images of standing electron waves have also been produced with gold at room temperature.

This STM of the surface of a copper crystal beautifully illustrates the wave nature of matter. Magnification is 215,000,000×.

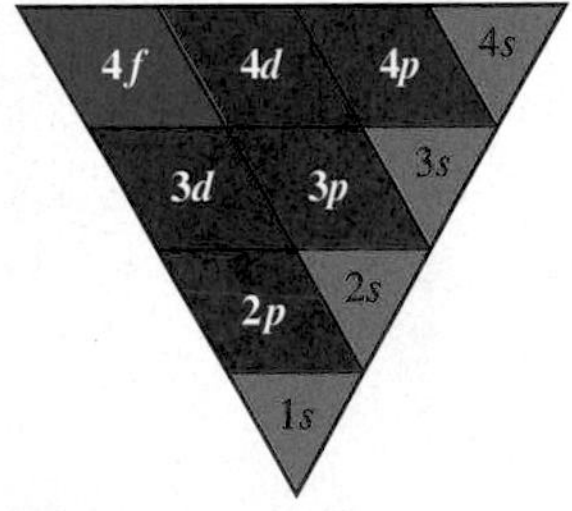

FIGURE 10.4
The types of orbitals on each of the first four principal energy levels.

Corresponding to each energy level are orbitals (the regions of space where the electrons are found). The first principal energy level has one type of **orbital or sublevel**. The second principal energy level has two sublevels, and so on. A good way to visualize this is to use an inverted triangle as shown in Figure 10.4.

orbital or sublevel

In each sublevel the electrons are found within orbitals. Let's consider each principal energy level in turn. The first principal energy level ($n = 1$) has one sublevel. This orbital is spherical in shape and is designated as $1s$. It is important to understand what the spherical shape of the $1s$ orbital means. The electron does *not* move around on the surface of the sphere, but rather the surface encloses a space where there is a 90% probability that the electron may be found. It might help to consider these orbital shapes in the same way we consider the atmosphere. There is no distinct dividing line between the atmosphere and "space." The boundary is quite fuzzy. The same is true for atomic orbitals. Each has a region of highest density roughly corresponding to its shape. The probability of finding the electron outside this region drops rapidly but never quite reaches zero. Scientists often speak of orbitals as electron "clouds" to emphasize the fuzzy nature of their boundaries.

spin

How many electrons can fit into a $1s$ orbital? To answer this question we need to consider one more property of electrons. This property is called **spin**. Each electron appears to be spinning on an axis, like a globe. It can only spin in two directions. We represent this spin with an arrow, ↑ or ↓. In order to occupy the same orbital, electrons must have *opposite* spins. That is, two electrons with the same spin cannot occupy the same orbital. This gives us the answer to our question: An atomic orbital can hold a maximum of two electrons, which must have opposite spins. This rule is called the **Pauli exclusion principle**. To summarize, the first principal energy level contains one type of orbital ($1s$) that holds a maximum of two electrons.

Pauli exclusion principle

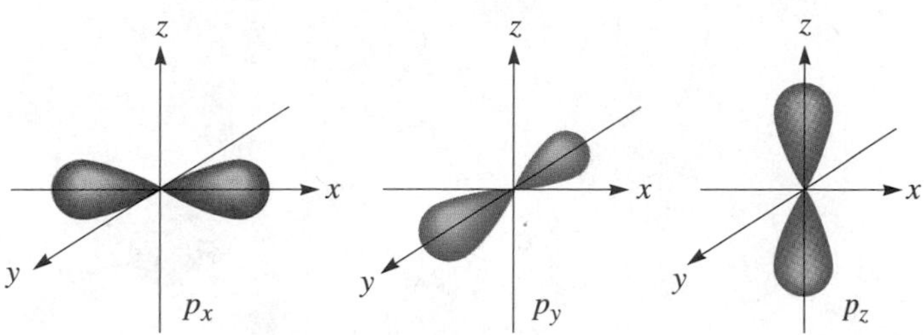

FIGURE 10.5
Perspective representation of the p_x, p_y, and p_z atomic orbitals.

What happens with the second principal energy level ($n = 2$)? Here we find two types of orbitals (sublevels), 2*s* and 2*p*. The 2*s* orbital is spherical in shape as it is for the first principal energy level but is larger in size and higher in energy than the 1*s*. It can also hold a maximum of two electrons. The second type of orbital is designated by 2*p*. The shape of these orbitals is quite different from the *s* orbitals. The shapes of the *p* orbitals are shown in Figure 10.5.

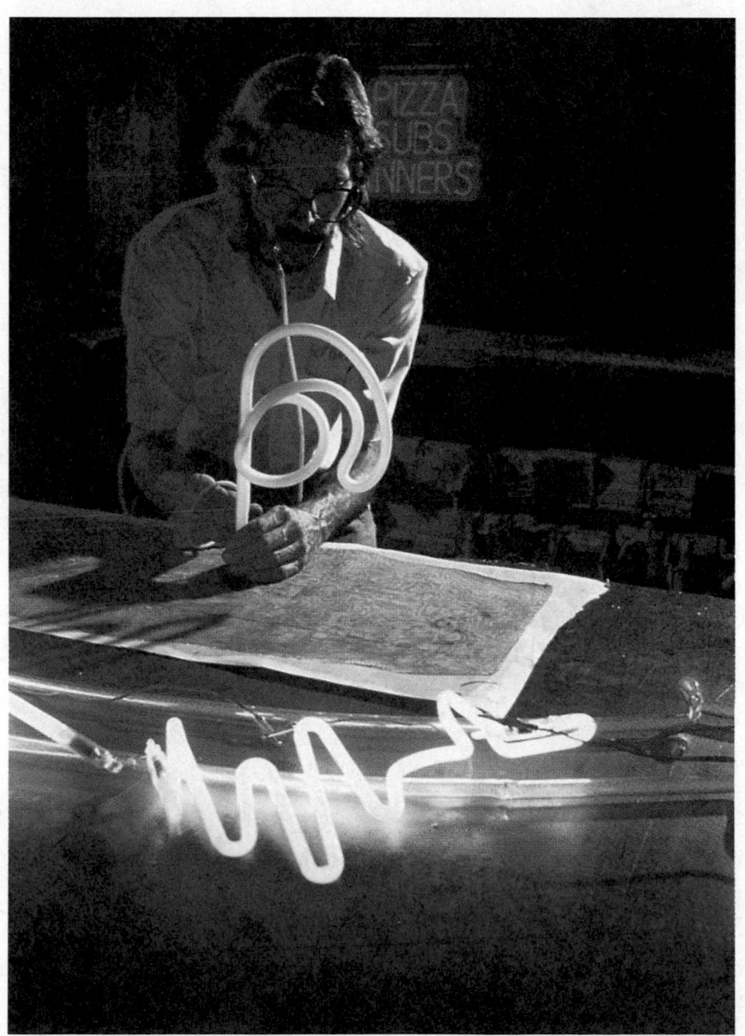

The light in this neon sign is the result of electrons falling from one principal energy level to another.

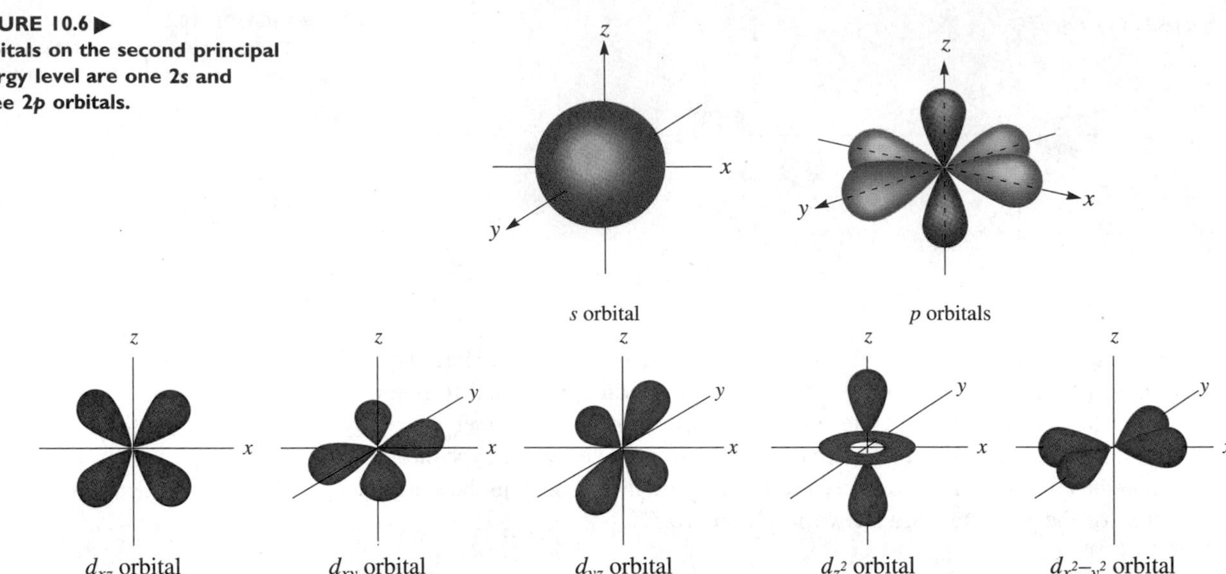

FIGURE 10.6 ▶
Orbitals on the second principal energy level are one 2*s* and three 2*p* orbitals.

FIGURE 10.7
The five *d* orbitals are found in the third principal energy level along with one 3*s* orbital and three 3*p* orbitals.

Each *p* orbital has two "lobes." Remember, the space enclosed by these surfaces represents the regions of probability for finding the electrons 90% of the time. Notice there are three separate *p* orbitals, each oriented in a different direction. Each *p* orbital can hold a maximum of two electrons. Thus the total number of electrons that can reside in all three *p* orbitals is six. To summarize once again, the first principal energy level of an atom has a 1*s* orbital. The second principal energy level has a 2*s* and three 2*p* orbitals labeled $2p_x$, $2p_y$, and $2p_z$ as shown in Figure 10.6.

The third principal energy level has three types of orbitals (sublevels) labeled 3*s*, 3*p*, and 3*d*. The 3*s* orbital is spherical and larger than the 1*s* and 2*s* orbitals. The $3p_x$, $3p_y$, $3p_z$ orbitals are shaped like those of the second level, only larger. There are five 3*d* orbitals with the shapes shown in Figure 10.7. You do not need to memorize these shapes, but notice that they look different from the *s* or *p* orbitals.

Each time a new principal energy level is added, we also add a new type of orbital. This makes sense when we realize that each energy level corresponds to a larger average distance from the nucleus, which provides more room on each level for new orbitals.

The pattern continues with the fourth principal energy level. It has 4*s*, 4*p*, 4*d*, and 4*f* orbitals. There are one 4*s*, three 4*p*, five 4*d*, and seven 4*f* orbitals. The shapes of the *s*, *p*, and *d* orbitals are the same as those for lower levels, only larger. We will not consider the shapes of the *f* orbitals. To summarize atomic structure, we can show the types of orbitals associated with each principal energy level:

$n = 1$	1*s*			
$n = 2$	2*s*	2*p* 2*p* 2*p*		
$n = 3$	3*s*	3*p* 3*p* 3*p*	3*d* 3*d* 3*d* 3*d* 3*d*	
$n = 4$	4*s*	4*p* 4*p* 4*p*	4*d* 4*d* 4*d* 4*d* 4*d*	4*f* 4*f* 4*f* 4*f* 4*f* 4*f* 4*f*

FIGURE 10.8
The modern concept of a hydrogen atom is shown here. It consists of an electron in an *s* orbital as a cloud of negative charge surrounding the proton in the nucleus.

The hydrogen atom consists of a nucleus (containing one proton) and an electron moving around the outside of the nucleus. In its ground state the electron occupies a 1*s* orbital, but by absorbing energy the electron can become *excited* and move to a higher energy level.

Yes, We Can See Atoms!

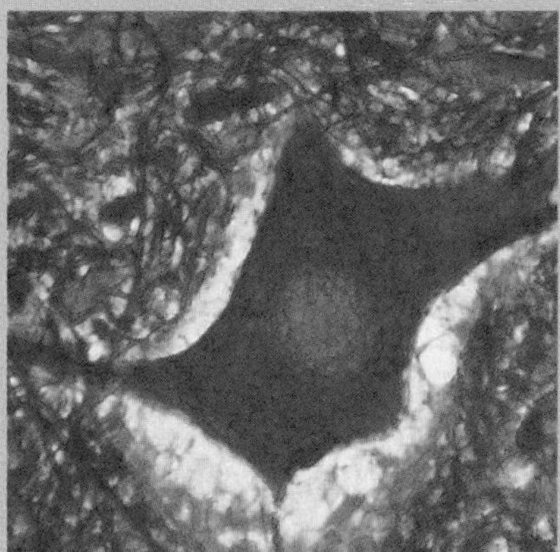

Neurons, optical microscope.

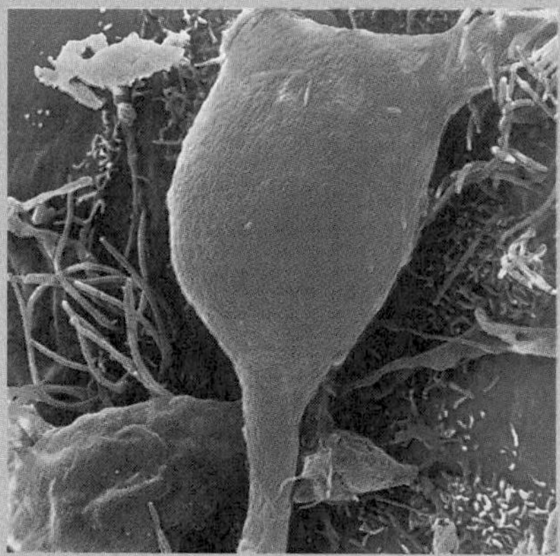

Neurons, electron microscope.

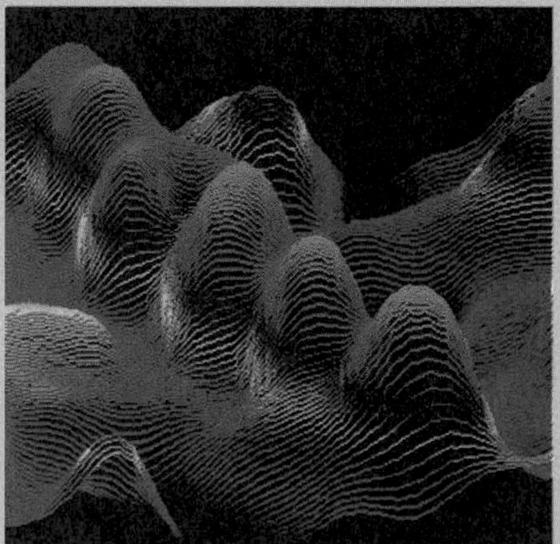

DNA molecule, scanning-tunneling microscope.

For centuries, scientists have argued and theorized over the nature and existence of atoms. Today, physicists and chemists can produce pictures of atoms and even move them individually from place to place. This new-found ability to see atoms, molecules, and even watch chemical reactions occur is the direct result of the evolution of the microscope.

An optical microscope is capable of viewing objects as small as the size of a cell. To see smaller objects an electron microscope is necessary. Since the eye cannot respond to a beam of electrons, the image is produced on a fluorescent screen or on film. These microscopes have been used for some time to photograph large molecules. To see tiny objects, however, the objects must be placed in a vacuum and the electrons must be in a high energy state. If the sample is fragile, as are most molecules, it can be destroyed before the image is formed.

In 1981, Gerd Binnig and Heinrich Rohrer, two scientists from IBM, invented the first scanning-probe microscope. These instruments are fundamentally different from previous microscopes. In a scanning-probe microscope a probe is placed near the surface of a sample and a parameter of some sort (voltage, magnetic field, etc.) is measured. As the probe is moved across the surface, an image is produced—in the same manner a child would determine the identity of an object sealed in an opaque bag. The first of the scanning-probe instruments was called a scanning-tunneling microscope. It produced the first clear pictures of silicon atoms in January 1983. The greatest limitation for the scanning-tunneling microscope is that, in order to be viewed, organic molecules must be given a thin coating of metal so that electrons are free to jump from the surface to the probe.

In 1985, a team of physicists from Stanford University and IBM found a solution to this problem. In a new instrument known as an atomic-force microscope, the probe measures tiny electric forces between electrons instead of the actual movement of electrons from surface to probe. The greatest advantage of this approach is that the probe is so gentle that even very fragile molecules remain intact. The probe is a tiny shard of a diamond attached to a tiny piece of silicon, and it works like a phonograph needle. At University of California, Santa Barbara, a group of scientists has even succeeded in making a movie of the formation of a blood clot on the molecular level.

In industry, another type of scanning-probe instrument has been developed to check the quality of microelectronic equipment. Researchers at IBM have developed a laser-force microscope in which a tiny wire probe measures small attractive forces (surface tension of water on the sample) to show imperfections as small as 25 atoms across.

The hydrogen atom can be represented as shown in Figure 10.8. The diameter of the nucleus is about 10^{-13} cm, and the diameter of the electron orbital is about 10^{-8} cm. The diameter of the electron cloud of a hydrogen atom is about 100,000 times greater than the diameter of the nucleus.

10.5 Atomic Structures of the First 18 Elements

We have seen that hydrogen has one electron that can occupy a variety of orbitals in different principal energy levels. Now we must consider the structure of atoms with more than one electron. We have learned that all atoms contain orbitals similar to those found in hydrogen. Therefore, we can describe the structures of atoms beyond hydrogen by systematically placing electrons in these hydrogen-like orbitals. The following guidelines apply to this process:

1. No more than two electrons can occupy one orbital.
2. Electrons occupy the lowest energy orbitals available. They enter a higher energy orbital only when the lower orbitals are filled. For the atoms beyond hydrogen, orbital energies vary as $s < p < d < f$ for a given value of n.
3. Each one of a given type of orbital is occupied by a single electron before a second electron enters. For example, all three p orbitals must contain one electron before a second electron enters a p orbital.

We can use several methods to represent the atomic structures of atoms, depending on what we are trying to illustrate. When we want to show both the nuclear make-up and the electron structure of each principal energy level (without orbital detail), we can use a diagram such as Figure 10.9.

n = 1 2
9p 10n 2e^- 7e^-

Fluorine atom

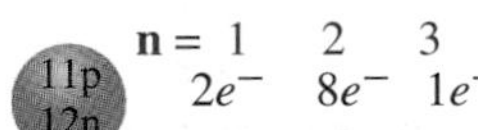

Sodium atom

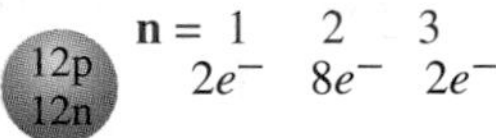

Magnesium atom

FIGURE 10.9
Atomic structure diagrams of fluorine, sodium, and magnesium atoms. The number of protons and neutrons is shown in the nucleus. The number of electrons is shown in each principal energy level outside the nucleus. ▶

electron configuration

Often we are interested in showing the arrangement of the electrons in an atom in their orbitals. There are two ways to show this. The first method is called the **electron configuration**. In this method we list each type of orbital, showing the number of electrons in it as an exponent. The following diagram shows how to read these:

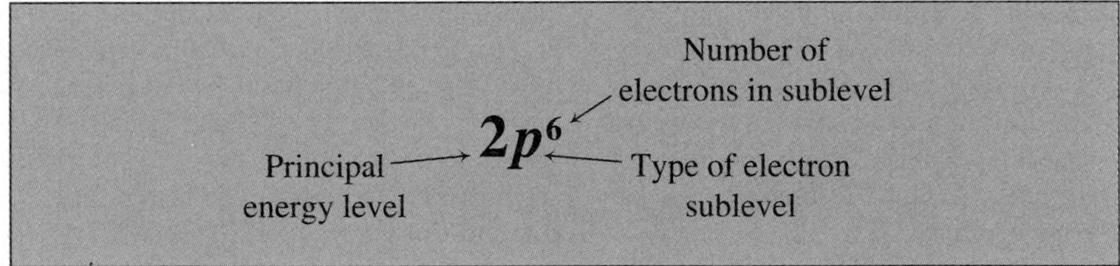

orbital diagram

We can also represent this configuration by using an **orbital diagram** in which the orbitals are represented by boxes grouped by sublevel with small arrows indicating the electrons. When the orbital contains one electron, an arrow, pointing upward (↑), is placed in the box. A second arrow, pointing downward (↓), indicates the second electron in that orbital.

Let's consider each of the first 20 elements on the periodic table in turn. The order of filling for the orbitals in these elements is 1s, 2s, 2p, 3s, 3p, and 4s.

Gerd Bennig and Heinrich Rohrer won the Nobel Prize in 1986 for inventing the scanning-tunneling microscope, which measures electron patterns.

Hydrogen, the first element, has only one electron. The electron will be in the 1*s* orbital since this is the most favorable position (where it will have the greatest attraction for the nucleus). Both representations for hydrogen are shown below:

H	[↑]	$1s^1$
	Orbital diagram	Electron configuration

Helium with two electrons can be shown as

He	[↑↓]	$1s^2$
	Orbital diagram	Electron configuration

The first energy level is now full. An atom with three electrons will have its third electron in the second energy level. Thus in lithium (atomic number 3), the third electron is in the 2*s* orbital of the second energy level. Lithium has the following structure:

Li [↑↓] [↑] $1s^2 2s^1$

All four electrons of beryllium are *s* electrons:

Be [↑↓] [↑↓] $1s^2 2s^2$

The next six elements illustrate the filling of the *p* orbitals. Boron has the first *p* electron. Because all of the *p* orbitals have the same energy, it does not matter which of these orbitals fills first:

B [↑↓] [↑↓] [↑][ ][ ] $1s^2 2s^2 2p^1$

Carbon is the sixth element. It has two electrons in the 1*s* orbital, 2 electrons in the 2*s* orbital and 2 electrons to place in 2*p* orbitals. Because it is energetically more difficult for the *p* electrons to pair up than to occupy a second *p* orbital, the second *p* electron is located in a different *p* orbital. We could show this by writing $2p_x^1 2p_y^1$, but usually it is written as $2p^2$ and it is *understood* that the electrons are in different *p* orbitals. The spins on these electrons are alike for reasons we will not explain here.

C [↑↓] [↑↓] [↑][↑][ ] $1s^2 2s^2 2p^2$

Nitrogen has 7 electrons. They are in 1*s*, 2*s*, and 2*p* orbitals. The third *p* electron in nitrogen is still unpaired and is found in the $2p_z$ orbital:

N [↑↓] [↑↓] [↑][↑][↑] $1s^2 2s^2 2p^3$

Oxygen is the eighth element. It has 2 electrons in both the 1*s* and 2*s* orbitals, and 4 electrons in the 2*p* orbitals. One of the 2*p* orbitals is now occupied by a second electron, which has a spin opposite the electron already in that orbital:

O [↑↓] [↑↓] [↑↓][↑][↑] $1s^2 2s^2 2p^4$

The next two elements are fluorine with 9 electrons and neon with 10 electrons:

F [↑↓] [↑↓] [↑↓][↑↓][↑] $1s^2 2s^2 2p^5$

Ne [↑↓] [↑↓] [↑↓][↑↓][↑↓] $1s^2 2s^2 2p^6$

With neon, the first and second energy levels are filled.

TABLE 10.1 Orbital Diagrams and Electron Configurations for Elements 11–18

Number	Element	1s	2s	2p	3s	3p	Electron configuration
11	Na	↑↓	↑↓	↑↓ ↑↓ ↑↓	↑		$1s^22s^22p^63s^1$
12	Mg	↑↓	↑↓	↑↓ ↑↓ ↑↓	↑↓		$1s^22s^22p^63s^2$
13	Al	↑↓	↑↓	↑↓ ↑↓ ↑↓	↑↓	↑	$1s^22s^22p^63s^23p^1$
14	Si	↑↓	↑↓	↑↓ ↑↓ ↑↓	↑↓	↑ ↑	$1s^22s^22p^63s^23p^2$
15	P	↑↓	↑↓	↑↓ ↑↓ ↑↓	↑↓	↑ ↑ ↑	$1s^22s^22p^63s^23p^3$
16	S	↑↓	↑↓	↑↓ ↑↓ ↑↓	↑↓	↑↓ ↑ ↑	$1s^22s^22p^63s^23p^4$
17	Cl	↑↓	↑↓	↑↓ ↑↓ ↑↓	↑↓	↑↓ ↑↓ ↑	$1s^22s^22p^63s^23p^5$
18	Ar	↑↓	↑↓	↑↓ ↑↓ ↑↓	↑↓	↑↓ ↑↓ ↑↓	$1s^22s^22p^63s^23p^6$

Sodium, element 11, has 2 electrons in the first energy level and 8 electrons in the second energy level, with the remaining electron occupying the 3*s* orbital in the third energy level:

	1s	2s	2p	3s	
Na	↑↓	↑↓	↑↓ ↑↓ ↑↓	↑	$1s^22s^22p^63s^1$

Magnesium (12), aluminum (13), silicon (14), phosphorus (15), sulfur (16), chlorine (17), and argon (18) follow in order. Table 10.1 summarizes the filling of the orbitals for elements 11–18.

valence electrons

The electrons in the outermost (highest) energy level of an atom are called the **valence electrons**. For example, oxygen, which has the electron configuration of $1s^22s^22p^4$, has electrons in the first and second energy levels. Therefore, the second level (2) is the valence level for oxygen. The 2*s* and 2*p* electrons are the valence electrons. In the case of magnesium ($1s^22s^22p^63s^2$), the valence electrons are in the 3*s* orbital since the outermost level containing electrons is the third energy level. The valence electrons are involved in bonding atoms together to form compounds and are of particular interest to chemists, as we will see in Chapter 11.

10.6 Electron Structures and the Periodic Table

We have seen how the electrons are assigned for the atoms of elements 1–18. How do the electron structures of these atoms relate to their position on the periodic table? To answer this question we need to look at the periodic table more closely.

The periodic table represents the efforts of chemists to organize the elements logically. Chemists of the early 19th century had sufficient knowledge of the properties of elements to recognize similarities among groups of elements. In 1869 Dimitri Mendeleev (1834–1907) of Russia and Lothar Meyer (1830–1895) of Germany independently published periodic arrangements of the elements that were based on increasing atomic masses. Mendeleev's arrangement is the precursor to the modern periodic table and his name is associated with it. The periodic table is shown in Figure 10.10.

Atomic number / Symbol: 9 F

Period	IA	IIA	IIIB	IVB	VB	VIB	VIIB	VIII	VIII	VIII	IB	IIB	IIIA	IVA	VA	VIA	VIIA	Noble gases
1	1 **H**																	2 **He**
2	3 **Li**	4 **Be**											5 **B**	6 **C**	7 **N**	8 **O**	9 **F**	10 **Ne**
3	11 **Na**	12 **Mg**											13 **Al**	14 **Si**	15 **P**	16 **S**	17 **Cl**	18 **Ar**
4	19 **K**	20 **Ca**	21 **Sc**	22 **Ti**	23 **V**	24 **Cr**	25 **Mn**	26 **Fe**	27 **Co**	28 **Ni**	29 **Cu**	30 **Zn**	31 **Ga**	32 **Ge**	33 **As**	34 **Se**	35 **Br**	36 **Kr**
5	37 **Rb**	38 **Sr**	39 **Y**	40 **Zr**	41 **Nb**	42 **Mo**	43 **Tc**	44 **Ru**	45 **Rh**	46 **Pd**	47 **Ag**	48 **Cd**	49 **In**	50 **Sn**	51 **Sb**	52 **Te**	53 **I**	54 **Xe**
6	55 **Cs**	56 **Ba**	57–71 **La–Lu**	72 **Hf**	73 **Ta**	74 **W**	75 **Re**	76 **Os**	77 **Ir**	78 **Pt**	79 **Au**	80 **Hg**	81 **Tl**	82 **Pb**	83 **Bi**	84 **Po**	85 **At**	86 **Rn**
7	87 **Fr**	88 **Ra**	89–103 **Ac–Lr**	104 **Unq**	105 **Unp**	106 **Unh**	107 **Uns**	108 **Uno**	109 **Une**									

FIGURE 10.10
The periodic table of the elements.

Each horizontal row in the periodic table is called a **period** as shown in Figure 10.10. The number of each period corresponds with the number of the outermost energy level that contains electrons for elements in that period. Those in the first row contain electrons only in energy level 1, while those in the second period contain electrons in levels 1 and 2. In the third row, electrons are found in levels 1, 2, and 3, and so on.

period

Elements that behave in a similar manner are found in **groups** or **families**. These form the vertical columns on the periodic table. There are several systems used for numbering the groups. In one system the columns are numbered from left to right using the numbers 1–18. We use a system that numbers the columns with Roman numerals and the letters A and B, as shown in Figure 10.10. The A groups are known as the **representative elements**. The B groups and Group VIII are called the **transition elements**. In this book we will focus on the representative elements. In addition, the groups (columns) of the periodic table often have family names. For example, the group on the far right side of the periodic table (He, Ne, Ar, Kr, Xe, and Rn) is called the *noble gases.* Group IA is called the *alkali metals,* Group IIA the *alkaline earth metals,* and Group VIIA the *halogens.*

groups, families

representative elements

transition elements

How does the structure of the periodic table connect to the atomic structures of the elements? We've just seen that the periods of the periodic table are associated with the energy level of the outermost electrons of the atoms in that row. Let's summarize the valence electron configurations for the elements we have examined so far. Figure 10.11 gives the valence electron configurations for the first eighteen elements. Notice that a pattern is emerging. The valence configuration for the elements in columns is the same. Only the number for the energy level is different. This is expected since each new period is associated with a different energy level for the valence electrons. The chemical behavior and properties of elements in a particular family must therefore be associated with the electronic configuration of the elements.

The electron configurations for elements beyond these eighteen become long and tedious to write. We often abbreviate the electron configuration by using the following notation:

Na $[Ne]3s^1$

FIGURE 10.11 ▶
Valence shell electron configurations for the first 18 elements.

IA	IIA	IIIA	IVA	VA	VIA	VIIA	Noble gases
1 **H** $1s^1$							2 **He** $1s^2$
3 **Li** $2s^1$	4 **Be** $2s^2$	5 **B** $2s^22p^1$	6 **C** $2s^22p^2$	7 **N** $2s^22p^3$	8 **O** $2s^22p^4$	9 **F** $2s^22p^5$	10 **Ne** $2s^22p^6$
11 **Na** $3s^1$	12 **Mg** $3s^2$	13 **Al** $3s^23p^1$	14 **Si** $3s^23p^2$	15 **P** $3s^23p^3$	16 **S** $3s^23p^4$	17 **Cl** $3s^23p^5$	18 **Ar** $3s^23p^6$

Look carefully at Figure 10.11. Notice that the *p* orbitals are full at the noble gases on the periodic table. By placing the symbol for the noble gas in square brackets we can abbreviate the complete electron configuration and focus our attention on the valence electrons (the electrons we will be interested in when we discuss bonding in Chapter 11). To write the abbreviated electron configuration for any element, go back to the previous noble gas and place it in square brackets. Then list the valence electrons. Several examples are given below:

B	$1s^2\ 2s^2\ 2p^1$	$[\text{He}]2s^22p^1$
Cl	$1s^22s^22p^63s^23p^5$	$[\text{Ne}]3s^23p^5$
Na	$1s^22s^22p^63s^1$	$[\text{Ne}]3s^1$

The sequence for filling the orbitals is exactly as expected up through the 3*p* orbitals. The third energy level might logically be expected to fill with 3*d* electrons before electrons enter the 4*s* orbital, but this is not the case. The behavior and properties of the next two elements, potassium and calcium, are very similar to the elements in Groups IA and IIA. They clearly belong in these groups. The other elements in Group IA and Group IIA have electron configurations that indicate valence electrons in the *s* orbitals. For example, since the electron configuration is connected to the element's properties, we should place the last electrons for potassium and calcium in the 4*s* orbital. Their electron configurations are

K	$1s^22s^22p^63s^23p^64s^1$	or	$[\text{Ar}]4s^1$
Ca	$1s^22s^22p^63s^23p^64s^2$	or	$[\text{Ar}]4s^2$

Elements 21–30 belong to the elements known as *transition elements* on the periodic table. Electrons are placed in the 3*d* orbitals for each of these elements. When the 3*d* orbitals are full, the electrons fill the 4*p* orbitals to complete the fourth row of the periodic table. Let's consider the overall relationship between filling of the orbitals and the periodic table. Figure 10.12 illustrates the type of orbital filling and its location on the periodic table. The tall columns on the table (labeled IA–VIIA, and noble gases) are often called the *representative elements.* Valence electrons in these elements occupy *s* and *p* orbitals. The row number of the periodic table corresponds to the energy level of the valence electrons. The elements in the center of the periodic table (shown in ■) are the transition elements where the *d* orbitals are being filled. Notice that the number for the *d* orbitals is one behind the row number on the periodic table. The two rows shown at the bottom of the table in Figure 10.12 are called the *inner transition elements.* The last electrons in these elements are placed in the *f* orbitals. The number for the *f* orbitals is always two behind that of the *s* and *p* orbitals. A periodic table is almost always available to

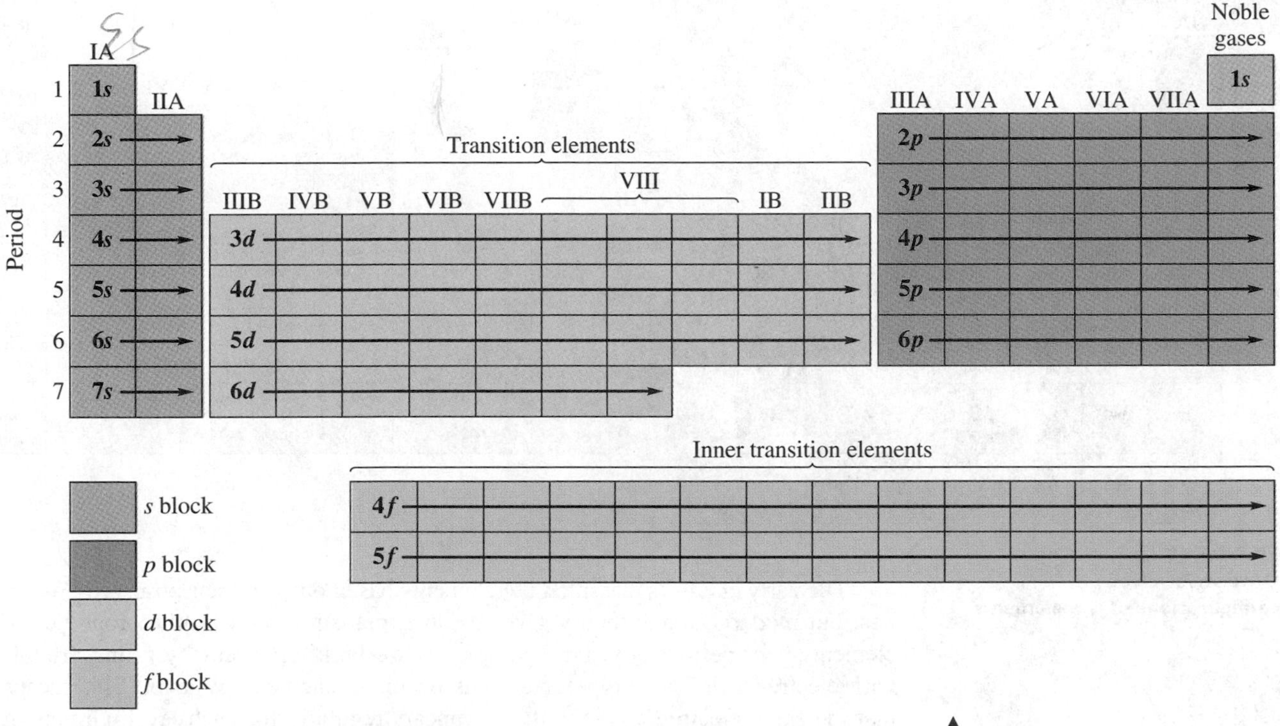

▲
FIGURE 10.12
Arrangement of elements according to the sublevel of electrons being filled in their atomic structure.

you, so if you understand the relationship between the orbitals and the periodic table you can write the electron configuration for any element. There are several minor variations to these rules. We shall not concern ourselves with these variations in this course.

Example 10.1

Use the periodic table to write the electron configuration for phosphorus and tin.

Solution

Phosphorus is element 15 and is located in Period 3, Group VA. The electron configuration must have a full first and second energy level:

P $1s^22s^22p^63s^23p^3$

You can determine the electron configuration by looking across the period and counting the element blocks.

Tin is element 50 in Period 5, Group IVA, two places after the transition metals. It must have two electrons in the $5p$ series. Its electron configuration is

Sn $1s^22s^22p^63s^23p^64s^23d^{10}4p^65s^24d^{10}5p^2$

Notice that the d series of electrons is always one energy level behind the period number.

Practice 10.1

Use the periodic table to write the electron configuration for (a) O, (b) Ca, and (c) Ti.

Period	Group number IA	IIA	IIIB	IVB	VB	VIB	VIIB	VIII	VIII	VIII	IB	IIB	IIIA	IVA	VA	VIA	VIIA	Noble gases
1	1 **H** $1s^1$																	2 **He** $1s^2$
2	3 **Li** $2s^1$	4 **Be** $2s^2$											5 **B** $2s^22p^1$	6 **C** $2s^22p^2$	7 **N** $2s^22p^3$	8 **O** $2s^22p^4$	9 **F** $2s^22p^5$	10 **Ne** $2s^22p^6$
3	11 **Na** $3s^1$	12 **Mg** $3s^2$											13 **Al** $3s^23p^1$	14 **Si** $3s^23p^2$	15 **P** $3s^23p^3$	16 **S** $3s^23p^4$	17 **Cl** $3s^23p^5$	18 **Ar** $3s^23p^6$
4	19 **K** $4s^1$	20 **Ca** $4s^2$	21 **Sc** $4s^23d^1$	22 **Ti** $4s^23d^2$	23 **V** $4s^23d^3$	24 **Cr** $4s^13d^5$	25 **Mn** $4s^23d^5$	26 **Fe** $4s^23d^6$	27 **Co** $4s^23d^7$	28 **Ni** $4s^23d^8$	29 **Cu** $4s^13d^{10}$	30 **Zn** $4s^23d^{10}$	31 **Ga** $4s^24p^1$	32 **Ge** $4s^24p^2$	33 **As** $4s^24p^3$	34 **Se** $4s^24p^4$	35 **Br** $4s^24p^5$	36 **Kr** $4s^24p^6$
5	37 **Rb** $5s^1$	38 **Sr** $5s^2$	39 **Y** $5s^24d^1$	40 **Zr** $5s^24d^2$	41 **Nb** $5s^14d^4$	42 **Mo** $5s^14d^5$	43 **Te** $5s^14d^6$	44 **Ru** $5s^14d^7$	45 **Rh** $5s^1 4d^8$	46 **Pd** $5s^04d^{10}$	47 **Ag** $5s^14d^{10}$	48 **Cd** $5s^24d^{10}$	49 **In** $5s^25p^1$	50 **Sn** $5s^25p^2$	51 **Sb** $5s^25p^3$	52 **Te** $5s^25p^4$	53 **I** $5s^25p^5$	54 **Xe** $5s^25p^6$
6	55 **Cs** $6s^1$	56 **Ba** $6s^2$	57 **La** $6s^25d^1$	72 **Hf** $6s^25d^2$	73 **Ta** $6s^25d^3$	74 **W** $6s^25d^4$	75 **Re** $6s^25d^5$	76 **Os** $6s^25d^6$	77 **Ir** $6s^25d^7$	78 **Pt** $6s^15d^9$	79 **Au** $6s^15d^{10}$	80 **Hg** $6s^25d^{10}$	81 **Tl** $6s^26p^1$	82 **Pb** $6s^26p^2$	83 **Bi** $6s^26p^3$	84 **Po** $6s^26p^4$	85 **At** $6s^26p^5$	86 **Rn** $6s^26p^6$
7	87 **Fr** $7s^1$	88 **Ra** $7s^2$	89 **Ac** $7s^26d^1$	104 **Unq** $7s^26d^2$	105 **Unp** $7s^26d^3$	106 **Unh** $7s^26d^4$	107 **Uns** $7s^26d^5$	108 **Uno** $7s^26d^6$	109 **Une** $7s^26d^7$									

FIGURE 10.13
Outermost electron configurations of the elements.

The early chemists classified the elements based only on their observed properties, but modern atomic theory gives us an explanation for why the properties of elements vary periodically. For example, as we build up atoms by filling orbitals with electrons, the same type of orbitals occur on each energy level. This means that the same electron configuration reappears regularly for each level. Groups of elements show similar chemical properties because of these outermost electron similarities.

Consider the periodic table shown in Figure 10.13. For the representative elements in this table only the electron configuration of the outer shell electrons is given. This table illustrates the following important points:

1. In each row the number of the period corresponds with the highest energy level occupied by electrons.
2. The group numbers for the representative elements are equal to the total number of outer shell electrons in the atoms of the group. For example, the elements in Group VIIA always have the electron configuration ns^2np^5. The d and f electrons are always in a lower energy level than the highest energy level and so are not considered as outermost electrons.
3. The elements of a family have the same outer shell electron configuration except that the electrons are in different energy levels.
4. The elements within each of the s, p, d, f blocks are filling the s, p, d, f orbitals, as shown in Figure 10.12.
5. Within the transition elements some discrepancies in the order of filling occur. (Explanation of these discrepancies and the similar ones that occur in the inner transition elements are beyond the scope of this book.)

Example 10.2 Write the electron configuration of a zinc atom and a rubidium atom.

Solution The atomic number of zinc is 30; therefore it has 30 protons and 30 electrons in a neutral atom. Using Figure 10.10, the electron configuration of a zinc atom is $1s^22s^22p^63s^23p^64s^23d^{10}$. Check by adding the superscripts, which should equal 30.

The atomic number of rubidium is 37; therefore it has 37 protons and 37 electrons in a neutral atom. With a little practice using a periodic table, the electron configuration may be written directly. The electron configuration of a rubidium atom is $1s^22s^22p^63s^23p^64s^23d^{10}4p^65s^1$. Check by adding the superscripts, which should equal 37.

Concepts in Review

1. Describe the atom as conceived by Niels Bohr.
2. Discuss the contributions to atomic theory made by Dalton, Thomson, Rutherford, Bohr, and Schrödinger. (See Chapter 5 as well.)
3. Explain what is meant by an electron orbital.
4. Explain how an electron configuration can be determined from the periodic table.
5. Write the electron configuration for any of the first 56 elements.
6. Indicate the locations of the metals, nonmetals, metalloids, and noble gases in the periodic table. (See Chapter 3 for help.)
7. Indicate the areas in the periodic table where the *s, p, d,* and *f* orbitals are being filled.
8. Determine the number of valence electrons in any atom in the Group A elements.
9. Distinguish between representative elements and transition elements.
10. Identify groups of elements by their special family names.
11. Describe the changes in outer shell electron structure (a) when moving from left to right in a period, and (b) when going from top to bottom in a group.
12. Explain the relationship between group number and the number of outer shell electrons for the representative elements.

Key Terms

The terms listed here have been defined within this chapter. Section numbers are referenced in parenthesis for each term.

electron configuration (10.5)
frequency (10.2)
ground state (10.3)
groups, families (10.6)
line spectrum (10.3)
orbital (10.3), (10.4)
orbital diagram (10.5)
Pauli exclusion principle (10.4)
period (10.6)
photons (10.2)
principal energy levels (10.4)
quanta (10.3)
representative elements (10.6)
speed (10.2)
spin (10.4)
transition elements (10.6)
valence electrons (10.5)
wavelength (10.2)

Questions

Questions refer to tables, figures, and key words and concepts defined within the chapter. A particularly challenging question or exercise is indicated with an asterisk.

1. What is an electron orbital?

2. Under what conditions can a second electron enter an orbital already containing one electron?

3. What is meant when we say that the electron structure of an atom is in its ground state?

4. How do 1*s* and 2*s* orbitals differ? How are they alike?

5. What letters are used to designate the types of orbitals?

6. List the following sublevels in order of increasing energy: 2*s*, 2*p*, 4*s*, 1*s*, 3*d*, 3*p*, 4*p*, 3*s*.

7. How many *s* electrons, *p* electrons, and *d* electrons are possible in any energy level?

8. What is the major difference between an orbital and a Bohr orbit?

9. Explain how and why Bohr's model of the atom was modified to include the cloud model of the atom.

10. Sketch the s, p_x, p_y, and p_z orbitals.

11. In the designation $3d^7$, give the significance of 3, *d*, and 7.

12. Describe the difference between transition and representative elements.

13. From the standpoint of electron structure, what do the elements in the *s* block have in common?

14. Write symbols for elements with atomic numbers 8, 16, 34, 52, and 84. What do these elements have in common?

15. Write the symbols of the family of elements that have seven electrons in their outermost energy level.

16. What is the greatest number of elements to be found in any period? Which periods have this number?

17. From the standpoint of energy level, how does the placement of the last electron in the Group A elements differ from that of the Group B elements?

18. Find the places in the modern periodic table where elements are not in proper sequence according to atomic mass. (See inside of front cover for periodic table.)

19. Which of the following statements are correct? Rewrite each incorrect statement to make it correct.

(a) In the ground state, electrons tend to occupy orbitals having the lowest possible energy.

(b) The maximum number of *p* electrons in the first energy level is six.

(c) A 2*s* electron is in a lower energy state than a 2*p* electron.

(d) The electron structure for a carbon atom is $1s^22s^22p^2$.

(e) The $2p_x$, $2p_y$, and $2p_z$ electron orbitals are all in the same energy state.

(f) The energy level of a 3*d* electron is higher than that of a 4*s* electron.

(g) The electron structure for a calcium atom is $1s^22s^22p^63s^23p^63d^2$.

(h) The third energy level can have a maximum of 18 electrons.

(i) The number of possible *d* electrons in the third energy level is ten.

(j) The first *f* electron occurs in the fourth principal energy level.

(k) Atoms of all the noble gases (except helium) have eight electrons in their outermost energy level.

(l) A *p* orbital is spherically symmetrical around the nucleus.

(m) An atom of nitrogen has two electrons in a 1*s* orbital, two electrons in a 2*s* orbital, and one electron in each of three different 2*p* orbitals.

(n) When an orbital contains two electrons, the electrons have parallel spins.

(o) The Bohr theory proposed that electrons move around the nucleus in circular orbits.

(p) Bohr concluded from his experiment that the positive charge and almost all the mass were concentrated in a very small nucleus.

20. Which of the following statements are correct? Rewrite the incorrect statements to make them correct.

(a) Properties of the elements are periodic functions of their atomic numbers.

(b) There are more nonmetallic elements than metallic elements.

(c) Calcium is a member of the alkaline earth metal family.

(d) Iron belongs to the alkali metal family.

(e) Bromine belongs to the halogen family.

(f) Neon is a noble gas.

(g) Group A elements do not contain partially filled *d* or *f* sublevels.

(h) An atom of aluminum (Group IIIA) has five electrons in its outer shell.

(i) The element $[Ar]4s^23d^{10}4p^5$ is a halogen.

(j) The element $[Kr]5s^2$ is a nonmetal.

(k) The element with $Z = 12$ forms compounds similar to the element with $Z = 37$.

(l) Nitrogen, fluorine, neon, gallium, and bromine are all nonmetals.

(m) The atom having an outer shell electron structure of $5s^25p^2$ would be in period 6, Group IVA.

(n) The yet-to-be-discovered element with an atomic number of 118 would be a noble gas.

Paired Exercises

These exercises are paired. Each odd-numbered exercise is followed by a similar even-numbered exercise. Answers to the even-numbered exercises are given in Appendix V.

21. How many protons are in the nucleus of an atom of each of these elements?
(a) H
(b) B
(c) Sc
(d) U

22. How many protons are in the nucleus of an atom of each of these elements?
(a) F
(b) Ag
(c) Br
(d) Sb

23. Give the electron configuration for the following:
(a) B
(b) Ti
(c) Zn
(d) Br
(e) Sr

24. Give the electron configuration for the following:
(a) chlorine
(b) silver
(c) lithium
(d) iron
(e) iodine

25. Explain how the spectral lines of hydrogen occur.

26. Explain how Bohr used the data from the hydrogen spectrum to support his model of the atom.

27. How many orbitals exist in the third energy level? What are they?

28. How many electrons can be present in the fourth energy level?

29. Write orbital diagrams for the following:
(a) N
(b) Cl
(c) Zn
(d) Zr
(e) I

30. Write orbital diagrams for the following:
(a) Si
(b) S
(c) Ar
(d) V
(e) P

31. Which atoms have the following electron configurations?
(a) $1s^22s^22p^63s^2$
(b) $1s^22s^22p^63s^23p^1$
(c) $1s^22s^22p^63s^23p^64s^23d^8$
(d) $1s^22s^22p^63s^23p^64s^23d^5$

32. Which atoms have the following electron configurations?
(a) $[Ar]4s^23d^1$
(b) $[Ar]4s^23d^{10}$
(c) $[Kr]5s^24d^{10}5p^2$
(d) $[Xe]6s^1$

33. Show the electron configurations for elements with atomic numbers of:
(a) 8
(b) 11
(c) 17
(d) 23
(e) 28
(f) 34

34. Show the electron configurations for elements with atomic numbers of:
(a) 9
(b) 26
(c) 31
(d) 39
(e) 52
(f) 10

35. Identify these atoms from their atomic structure diagrams:
(a) (16p 16n) 2e⁻ 8e⁻ 6e⁻
(b) (28p 32n) 2e⁻ 8e⁻ 16e⁻ 2e⁻

36. Diagram the atomic structures (as you see in Exercise 35) for these atoms:
(a) $^{27}_{13}Al$
(b) $^{48}_{22}Ti$

37. Why is the eleventh electron of the sodium atom located in the third energy level rather than in the second energy level?

38. Why is the last electron in potassium located in the fourth energy level rather than in the third energy level?

39. What electron structure do the noble gases have in common?

40. What is unique about the noble gases, from an electron point of view?

41. How are elements in a period related to one another?

42. How are elements in a group related to one another?

43. What is common about the electron structures of the alkali metals?

44. Why would you expect the elements zinc, cadmium, and mercury to be in the same chemical family?

45. Pick the electron structures below that represent elements in the same chemical family:
(a) $1s^22s^1$
(b) $1s^22s^22p^4$
(c) $1s^22s^22p^2$
(d) $1s^22s^22p^63s^23p^4$
(e) $1s^22s^22p^63s^23p^6$
(f) $1s^22s^22p^63s^23p^64s^2$
(g) $1s^22s^22p^63s^23p^64s^1$
(h) $1s^22s^22p^63s^23p^64s^23d^1$

46. Pick the electron structures below that represent elements in the same chemical family:
(a) $[He]2s^22p^6$
(b) $[Ne]3s^1$
(c) $[Ne]3s^2$
(d) $[Ne]3s^23p^3$
(e) $[Ar]4s^23d^{10}$
(f) $[Ar]4s^23d^{10}4p^6$
(g) $[Ar]4s^23d^5$
(h) $[Kr]5s^24d^{10}$

47. In the periodic table, calcium, element 20, is surrounded by elements 12, 19, 21, and 38. Which of these have physical and chemical properties most resembling calcium?

48. In the periodic table, phosphorus, element 15, is surrounded by elements 14, 7, 16, and 33. Which of these have physical and chemical properties most resembling phosphorus?

49. Classify each of the following elements as metals, nonmetals, or metalloids (review Chapter 3 if you need help):
(a) potassium
(b) plutonium
(c) sulfur
(d) antimony

50. Classify each of the following elements as metals, nonmetals, or metalloids (review Chapter 3 if you need help):
(a) iodine
(b) tungsten
(c) molybdenum
(d) germanium

51. In which period and group does an electron first appear in a *d* orbital?

52. In which period and group does an electron first appear in an *f* orbital?

53. How many electrons occur in the valence level of Group IIIA and IIIB elements? Why are they different?

54. How many electrons occur in the valence level of Group VIIA and VIIB elements? Why are they different?

Additional Exercises

These exercises are not paired or labeled by topic and provide additional practice on the concepts covered in this chapter.

55. If all the orbitals within an atom could hold three electrons rather than two, what would be the atomic numbers and identities of the first three noble gases?

56. Why does the emission spectrum for nitrogen reveal many more spectral lines than that for hydrogen?

57. Among the first 100 elements on the periodic table, how many have
(a) at least one *s* electron?
(b) at least one *p* electron?
(c) at least one *d* electron?
(d) at least one *f* electron?

58. For each of the following elements, what percent of their electrons are *s* electrons?
(a) He
(b) Be
(c) Xe
(d) Se
(e) Cs

59. How many pairs of valence electrons do these elements have?
(a) O
(b) P
(c) I
(d) Xe
(e) Rb

60. Suppose we use a foam ball to represent a typical atom. If the radius of the ball is 1.5 cm and the radius of a typical atom is 1.0×10^{-8} cm, how much of an enlargement is this? Use a ratio to express your answer.

61. List the first element on the periodic table that satisfies each of these conditions:
(a) a completed set of *p* orbitals
(b) two 4*p* electrons
(c) seven valence electrons
(d) three unpaired electrons

62. Explain how scanning-probe microscopes are different from earlier microscopes.

63. Describe the functioning of a scanning-tunneling microscope.

64. What is the difference between a scanning-tunneling microscope and an atomic-force microscope?

65. Oxygen is a gas. Sulfur is a solid. What is it about their electron structures that causes them to be grouped in the same chemical family?

66. In which groups are the transition elements located?

67. How do the electron structures of the transition elements differ from the representative elements?

Try to answer Exercises 68–71 without referring to the periodic table.

68. The atomic numbers of the noble gases are 2, 10, 18, 36, 54, and 86. What are the atomic numbers for the elements with six electrons in their outer electron shells?

69. Element number 87 is in Group IA, Period 7. Describe its outermost energy level. How many energy levels of electrons does it have?

70. If element 36 is a noble gas, in which groups would you expect elements 35 and 37 to occur?

71. Write a paragraph describing the general features of the periodic table.

72. Some scientists have proposed the existence of element 117. If it were to exist:
(a) what would its electron configuration be?
(b) how many valence electrons would it have?
(c) what element would it likely resemble?
(d) to what family and period would it belong?

73. What is the relationship between two elements if
(a) one of them has 10 electrons, 10 protons, and 10 neutrons and the other has 10 electrons, 10 protons, and 12 neutrons?
(b) one of them has 23 electrons, 23 protons, and 27 neutrons and the other has 24 electrons, 24 protons, and 26 neutrons?

74. Is there any pattern to the location of gases on the periodic table? Is there a pattern for the location of liquids? Is there a pattern for the location of solids?

Answers to Practice Exercises

10.1 (a) O $1s^22s^22p^4$
(b) Ca $1s^22s^22p^63s^23p^64s^2$
(c) Ti $1s^22s^22p^63s^23p^64s^23d^2$

Chemical Bonds: The Formation of Compounds from Atoms

Chapter 11

For centuries, we have been aware that certain metals cling to a magnet. Balloons may stick to a wall. Why? The activity of a superconductor floating in air has been the subject of television commercials. High-speed levitation trains are heralded to be the wave of the future. How do they function? In each case, forces of attraction and repulsion are at work.

Interestingly, human interactions also suggest that "opposites attract" and "likes repel." Attractions draw us into friendships and significant relationships, whereas repulsive forces may produce debate and antagonism. We form and break apart interpersonal bonds throughout our lives.

In chemistry, we also see this phenomenon. Substances form chemical bonds as a result of electrical attractions. These bonds provide the tremendous diversity of compounds found in nature.

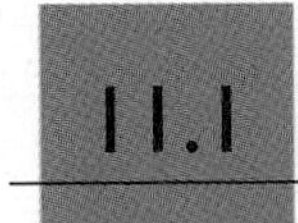

11.1 Periodic Trends in Atomic Properties

Although atomic theory and electron configuration can help us to understand the arrangement and behavior of the elements, it is important to remember that the design of the periodic table is based on the observation of properties of the elements. Before we proceed to use the concept of atomic structure to explain how and why atoms combine to form compounds, we need to understand the characteristic properties of the elements and the trends that occur in these properties on the periodic table. These trends in observed properties allow us to use the periodic table to accurately predict properties and reactions of a diversity of substances without needing to possess the substance or complete the reaction.

Metals and Nonmetals

In Section 3.8 we began our study of the periodic table by classifying elements as metals, nonmetals, or metalloids. The heavy stair-step line beginning at boron and running diagonally down the periodic table separates the elements into metals and nonmetals. Metals are usually lustrous, malleable, and good conductors of heat and electricity. Nonmetals are just the opposite—nonlustrous, brittle, and poor conductors. Metalloids are found bordering the heavy diagonal line and may have properties of both metals and nonmetals.

Most elements are classified as metals (see Figure 11.1). Metals are found on the left side of the stair-step line, while the nonmetals are located toward the upper right of the table. Note that hydrogen does not fit into the division of metals and nonmetals. It displays nonmetallic properties under normal conditions, even though it has only one outermost electron like the alkali metals. Hydrogen is considered to be a unique element.

It is the chemical properties of metals and nonmetals that are most interesting. Metals tend to lose electrons and form positive ions, while nonmetals tend to gain electrons and form negative ions. When a metal reacts with a nonmetal a transfer of electrons from the metal to the nonmetal frequently occurs.

◀ **Chapter Opening Photo: This colorfully lighted limestone cave reveals dazzling stalactites and stalagmites, formed from calcium carbonate.**

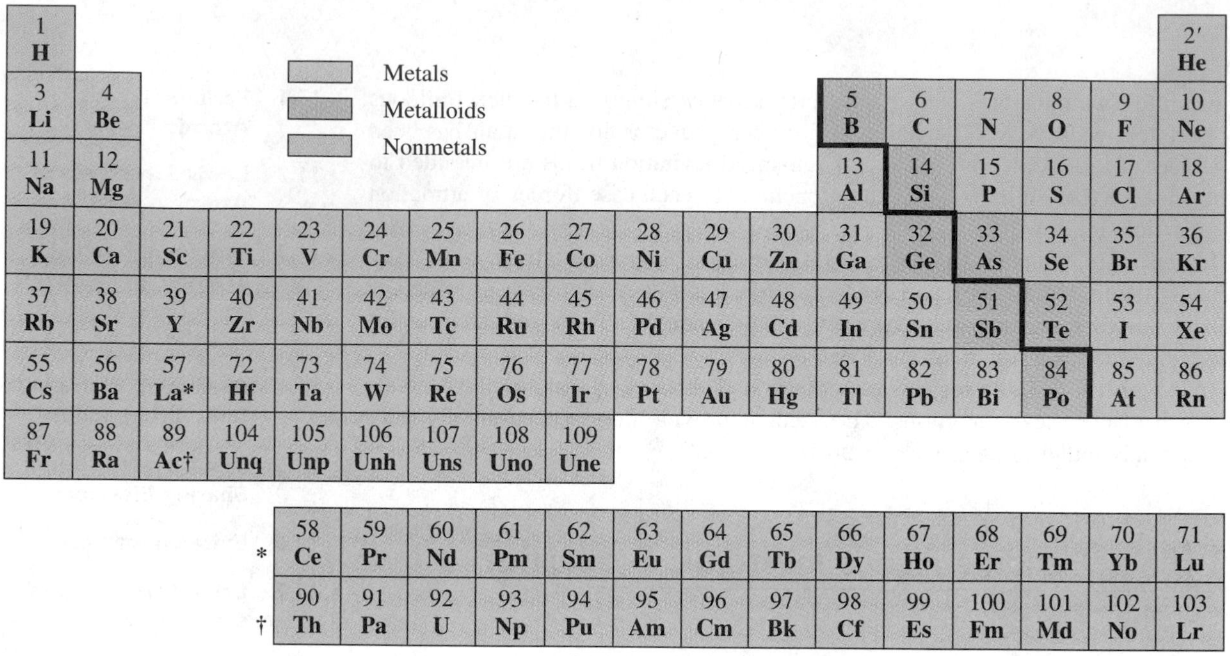

Metals
Metalloids
Nonmetals

1 H																	2′ He
3 Li	4 Be											5 B	6 C	7 N	8 O	9 F	10 Ne
11 Na	12 Mg											13 Al	14 Si	15 P	16 S	17 Cl	18 Ar
19 K	20 Ca	21 Sc	22 Ti	23 V	24 Cr	25 Mn	26 Fe	27 Co	28 Ni	29 Cu	30 Zn	31 Ga	32 Ge	33 As	34 Se	35 Br	36 Kr
37 Rb	38 Sr	39 Y	40 Zr	41 Nb	42 Mo	43 Tc	44 Ru	45 Rh	46 Pd	47 Ag	48 Cd	49 In	50 Sn	51 Sb	52 Te	53 I	54 Xe
55 Cs	56 Ba	57 La*	72 Hf	73 Ta	74 W	75 Re	76 Os	77 Ir	78 Pt	79 Au	80 Hg	81 Tl	82 Pb	83 Bi	84 Po	85 At	86 Rn
87 Fr	88 Ra	89 Ac†	104 Unq	105 Unp	106 Unh	107 Uns	108 Uno	109 Une									

*	58 Ce	59 Pr	60 Nd	61 Pm	62 Sm	63 Eu	64 Gd	65 Tb	66 Dy	67 Ho	68 Er	69 Tm	70 Yb	71 Lu
†	90 Th	91 Pa	92 U	93 Np	94 Pu	95 Am	96 Cm	97 Bk	98 Cf	99 Es	100 Fm	101 Md	102 No	103 Lr

▲
FIGURE 11.1
Classification of the elements into metals, nonmetals, and metalloids.

Atomic Radius

The relative radii of the representative elements are shown in Figure 11.2. Notice that the radii of the atoms tend to increase down each group and that they tend to decrease from left to right across a period.

The increase in radius down a column can be understood if we consider the electron structure of the atoms. For each step down a column an additional energy level is added to the atom. The average distance from the nucleus to the outside edge of the atom must increase as each new energy level is added. The atoms get bigger as electrons are placed in these new higher energy levels.

Understanding the decrease in atomic radius across a row requires more thought, however. As we move from left to right across a period electrons within the same block are being added to the same energy level. Within a given energy level we expect the orbitals to have about the same size. We would then expect the atoms to be about the same size across the period. But each time an electron is added a proton is added to the nucleus as well. The increase in positive charge (in the nucleus) pulls the electrons closer to the nucleus resulting in a gradual decrease in atomic radius across a row.

Ionization Energy

ionization energy

The **ionization energy** of an atom is the energy required to remove an electron from the atom. For example,

$$\text{Na} + \text{ionization energy} \longrightarrow \text{Na}^+ + e^-$$

The first ionization energy is the amount of energy required to remove the first electron from an atom, the second is the amount required to remove the second electron from that atom, and so on.

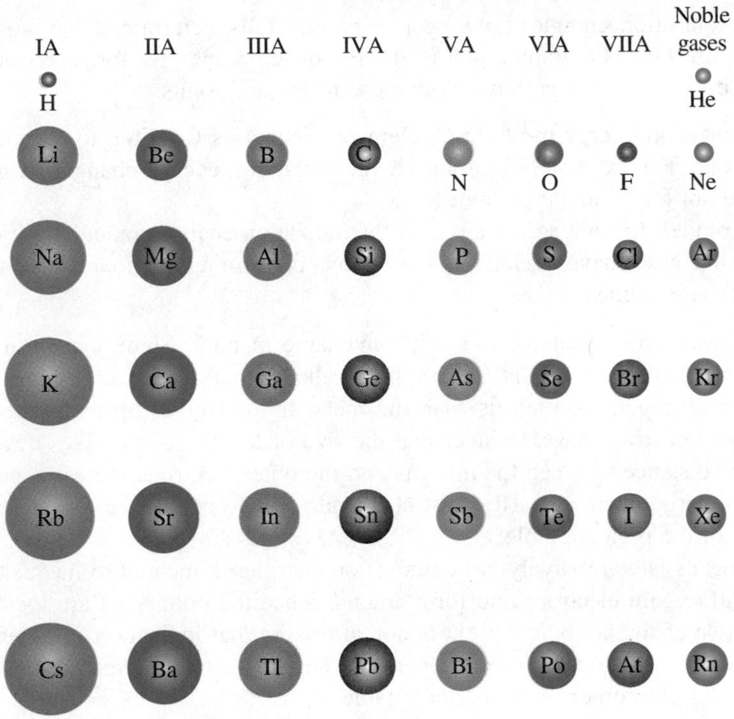

◀ **FIGURE 11.2**
Relative atomic radii for representative elements. Atomic radius decreases across a period and increases down a group.

TABLE 11.1 Ionization Energies for Selected Elements*

	Required amounts of energy (kJ/mol)				
Element	1st e^-	2nd e^-	3rd e^-	4th e^-	5th e^-
H	1,314				
He	2,372	5,247			
Li	520	7,297	11,810		
Be	900	1,757	14,845	21,000	
B	800	2,430	3,659	25,020	32,810
C	1,088	2,352	4,619	6,222	37,800
Ne	2,080	3,962	6,276	9,376	12,190
Na	496	4,565	6,912	9,540	13,355

* Values are expressed in kilojoules per mole, showing energies required to remove 1 to 5 electrons per atom. Color indicates the energy needed to remove an electron from a noble-gas electron structure.

Table 11.1 gives the ionization energies for the removal of one to five electrons from several elements. The table shows that even higher amounts of energy are needed to remove the second, third, fourth, and fifth electrons. This makes sense because removing electrons does not decrease the positive charge in the nucleus; the remaining electrons are held even more tightly. The data in Table 11.1 also show that an extra large ionization energy (red) is needed when an electron is removed from a noble gas structure, clearly showing the stability of this configuration.

First ionization energies have been experimentally determined for most elements. Figure 11.3 is a graphic plot of these ionization energies for representative elements in the first four periods. Note these important points:

1. Ionization energy in Group A elements decreases from top to bottom in a group. For example, in Group IA the ionization energy changes from 520 kJ/mol for Li to 419 kJ/mol for K.
2. From left to right across a period the ionization energy gradually increases. Noble gases have a relatively high value, confirming the nonreactive nature of these elements.

All metals do not behave in exactly the same manner. Some metals give up electrons much more easily than others. In the alkali metal family, cesium gives up its 6*s* electron much more easily than the metal lithium gives up its 2*s* electron. This makes sense when we consider that the size of the atoms increases down the group. The distance between the nucleus and the outer electrons increases and the ionization energy decreases. The most chemically active metals are located on the lower left of the periodic table.

Nonmetals have relatively large ionization energies compared to metals. Nonmetals tend to gain electrons and form anions. Since the nonmetals are located at the right side of the periodic table it is not surprising that ionization energies tend to increase from left to right across a period. The most active nonmetals are found in the *upper* right corner of the periodic table.

11.2 Lewis Structures of Atoms

Metals tend to form cations (positively charged ions) and nonmetals form anions (negatively charged ions) in order to attain a stable valence electron structure. For

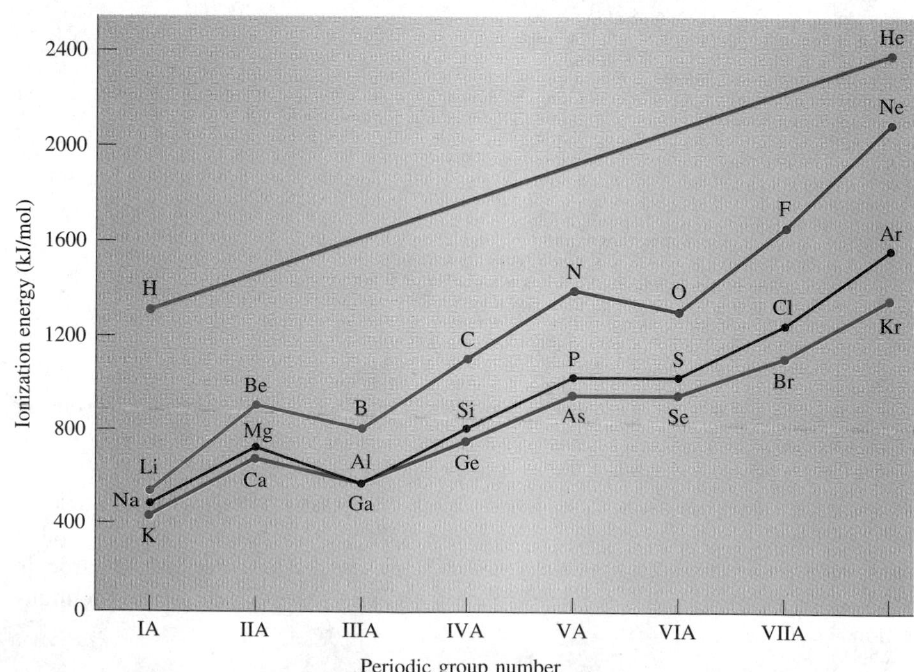

FIGURE 11.3 ▶ Periodic relationship of the first ionization energy for representative elements in the first four periods.

◀ **Gilbert N. Lewis (1875–1946) in his laboratory.**

many elements this stable valence level contains eight electrons (two *s* and six *p*), identical to the valence electron configuration of the noble gases. Atoms undergo rearrangements of electron structure to lower their chemical potential energy (or become more stable). These rearrangements are accomplished by losing, gaining, or sharing electrons with other atoms. For example, a hydrogen atom could accept a second electron and attain an electron structure the same as the noble gas helium. A fluorine atom could gain an electron and attain an electron structure like neon. A sodium atom could lose one electron to attain an electron structure like neon.

The valence electrons in the outermost energy level of an atom are responsible for the electron activity that occurs to form chemical bonds. The **Lewis structure** of an atom is a representation that shows the valence electrons for that atom. American chemist, Gilbert N. Lewis (1875–1946) proposed using the symbol for the element and dots for electrons. The number of dots placed around the symbol equals the number of *s* and *p* electrons in the outermost energy level of the atom. Paired dots represent paired electrons; unpaired dots represent unpaired electrons. For example, **H·** is the Lewis symbol for a hydrogen atom, $1s^1$; $:\dot{\mathbf{B}}$ is the Lewis symbol for a boron atom, with valence electrons $2s^2 2p^1$. In the case of boron, the symbol B represents the boron nucleus and the $1s^2$ electrons; the dots represent only the $2s^2 2p^1$ electrons.

Lewis structure

Paired electrons → $:\dot{\mathbf{B}}$ ← Unpaired electron (dot); ← Symbol of the atom

The Lewis method is often used not only because of its simplicity of expression but also because much of the chemistry of the atom is directly associated with the electrons in the outermost energy level. Figure 11.4 shows Lewis structures for the elements hydrogen through calcium.

FIGURE 11.4 ▶ **Lewis structures of the first 20 elements. Dots represent electrons in the outermost energy level only.**

IA	IIA	IIIA	IVA	VA	VIA	VIIA	Noble Gases
H·							He:
Li·	Be:	:Ḃ	:Ċ·	:N·	·Ö:	:F:	:Ne:
Na·	Mg:	:Al	:Si·	:P·	·S:	:Cl:	:Ar:
K·	Ca:						

Example 11.1 Write the Lewis structure for a phosphorus atom.

Solution First establish the electron structure for a phosphorus atom. It is $1s^22s^22p^63s^23p^3$. Note that there are five electrons in the outermost energy level; they are $3s^23p^3$. Write the symbol for phosphorus and place the five electrons as dots around it.

A quick way to determine the correct number of dots (electrons) for a Lewis structure is to use the group number. For the A groups on the periodic table, the roman numeral is the same as the number of electrons in the Lewis structure.

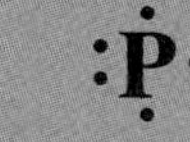

The $3s^2$ electrons are paired and are represented by the paired dots. The $3p^3$ electrons, which are unpaired, are represented by the single dots.

Practice 11.1

Write the Lewis structure for the following elements:
(a) N (b) Al (c) Sr (d) Br

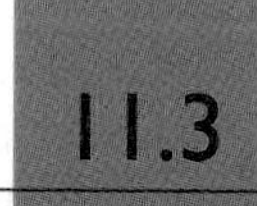

11.3 The Ionic Bond: Transfer of Electrons from One Atom to Another

The chemistry of many elements, especially the representative ones, is to attain an outer shell electron structure like that of the chemically stable noble gases. With the exception of helium, this stable structure consists of eight electrons in the outer shell (see Table 11.2).

Let us look at the electron structures of sodium and chlorine to see how each element can attain a structure of eight electrons in its outer shell. A sodium atom has eleven electrons: two in the first energy level, eight in the second energy level, and one in the third energy level. A chlorine atom has seventeen electrons: two in the first energy level, eight in the second energy level, and seven in the third energy level. If a sodium atom transfers or loses its 3*s* electron, its third energy level becomes vacant, and it becomes a sodium ion with an electron configuration identical to that of the noble gas neon. This process requires energy:

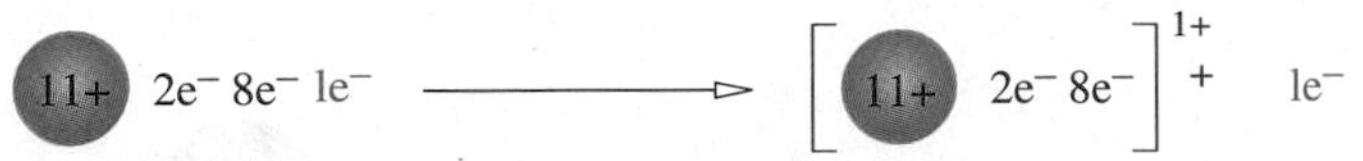

$$\text{(11+) } 2e^- \ 8e^- \ 1e^- \longrightarrow \left[\text{(11+) } 2e^- \ 8e^-\right]^{1+} + 1e^-$$

Na atom ($1s^22s^22p^63s^1$) $\qquad$ Na^+ ion ($1s^22s^22p^6$)

An atom that has lost or gained electrons will have a positive or negative charge, depending on which particles, protons or electrons are in excess. Remember that a charged atom or group of atoms is called an *ion.*

By losing a negatively charged electron, the sodium atom becomes a positively charged particle known as a sodium ion. The charge, +1, results because the nucleus still contains eleven positively charged protons, and the electron orbitals contain only ten negatively charged electrons. The charge is indicated by a plus sign (+) and is written with a superscript after the symbol of the element (Na^+).

A chlorine atom with seven electrons in the third energy level needs one electron to pair up with its one unpaired $3p$ electron to attain the stable outer shell electron structure of argon. By gaining one electron the chlorine atom becomes a chloride ion (Cl^-), a negatively charged particle containing seventeen protons and eighteen electrons. This process releases energy:

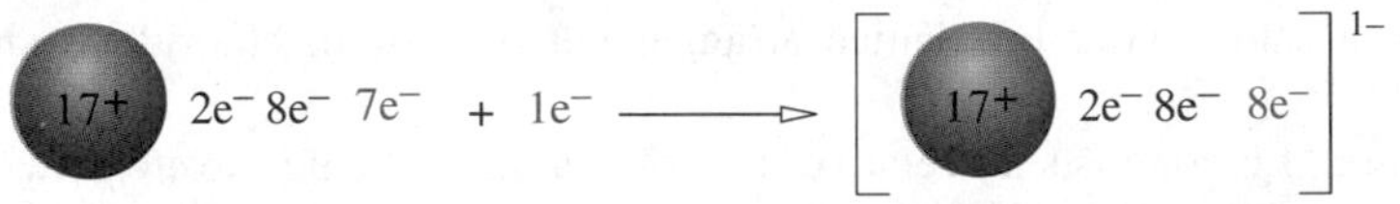

$$\text{(17+) } 2e^- \ 8e^- \ 7e^- + 1e^- \longrightarrow \left[\text{(17+) } 2e^- \ 8e^- \ 8e^-\right]^{1-}$$

Cl atom ($1s^22s^22p^63s^23p^5$) $\qquad$ Cl^- ion ($1s^22s^22p^63s^23p^6$)

Consider sodium and chlorine atoms reacting with each other. The $3s$ electron from the sodium atom transfers to the half-filled $3p$ orbital in the chlorine atom to form a positive sodium ion and a negative chloride ion. The compound sodium chloride results because the Na^+ and Cl^- ions are strongly attracted to each other

TABLE 11.2 Arrangement of Electrons in the Noble Gases*

		Electron structure					
Noble gas	**Symbol**	$n = 1$	2	3	4	5	6
Helium	He	$1s^2$					
Neon	Ne	$1s^2$	$2s^22p^6$				
Argon	Ar	$1s^2$	$2s^22p^6$	$3s^23p^6$			
Krypton	Kr	$1s^2$	$2s^22p^6$	$3s^23p^63d^{10}$	$4s^24p^6$		
Xenon	Xe	$1s^2$	$2s^22p^6$	$3s^23p^63d^{10}$	$4s^24p^64d^{10}$	$5s^25p^6$	
Radon	Rn	$1s^2$	$2s^22p^6$	$3s^23p^63d^{10}$	$4s^24p^64d^{10}4f^{14}$	$5s^25p^65d^{10}$	$6s^26p^6$

* Each gas except helium has eight electrons in its outermost energy level.

by their opposite electrostatic charges. The force holding the oppositely charged ions together is an ionic bond:

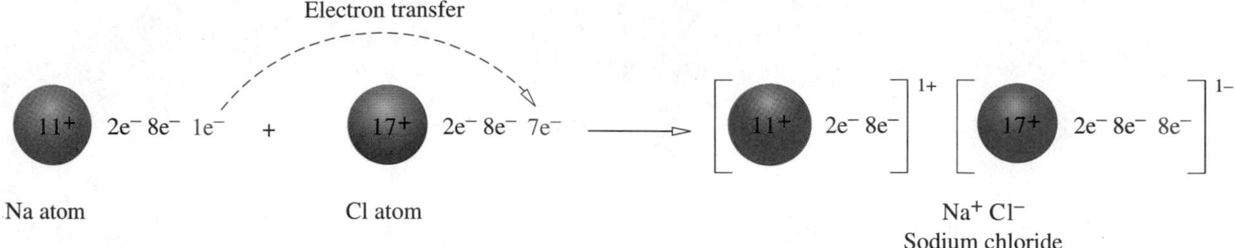

The Lewis representation of sodium chloride formation is shown here:

$$\text{Na}\cdot + \cdot\ddot{\underset{..}{\text{Cl}}}\text{:} \longrightarrow [\text{Na}]^+\left[\text{:}\ddot{\underset{..}{\text{Cl}}}\text{:}\right]^-$$

The chemical reaction between sodium and chlorine is a very vigorous one, producing considerable heat in addition to the salt formed. When energy is released in a chemical reaction, the products are more stable than the reactants. Note that in NaCl both atoms attain a noble-gas electron structure.

Sodium chloride is made up of cubic crystals in which each sodium ion is surrounded by six chloride ions and each chloride ion by six sodium ions, except at the crystal surface. A visible crystal is a regularly arranged aggregate of millions of these ions, but the ratio of sodium to chloride ions is 1:1, hence the formula NaCl. The cubic crystalline lattice arrangement of sodium chloride is shown in Figure 11.5.

It is helpful to remember that a cation is always smaller than its parent atom whereas an anion is always larger than its parent atom.

Figure 11.6 contrasts the relative sizes of sodium and chlorine atoms with those of their ions. The sodium ion is smaller than the atom due primarily to two factors: (1) The sodium atom has lost its outer shell of one electron, thereby reducing its size; and (2) the ten remaining electrons are now attracted by eleven protons and are thus drawn closer to the nucleus. Conversely, the chloride ion is larger than the atom because (1) it has eighteen electrons but only seventeen protons; and (2) the nuclear attraction on each electron is thereby decreased, allowing the chlorine atom to expand as it forms an ion.

We have seen that when sodium reacts with chlorine, each atom becomes an ion. Sodium chloride, like all ionic substances, is held together by the attraction existing between positive and negative charges. An **ionic bond** is the attraction between oppositely charged ions.

ionic bond

Ionic bonds are formed whenever one or more electrons are transferred from one atom to another. Metals, which have relatively little attraction for their valence electrons, tend to form ionic bonds when they combine with nonmetals.

It is important to recognize that substances with ionic bonds do not exist as molecules. In sodium chloride, for example, the bond does not exist solely between a single sodium ion and a single chloride ion. Each sodium ion in the crystal attracts six near-neighbor negative chloride ions; in turn, each negative chloride ion attracts six near-neighbor positive sodium ions (see Figure 11.5).

A metal will usually have one, two, or three electrons in its outer energy level. In reacting, metal atoms characteristically lose these electrons, attain the electron

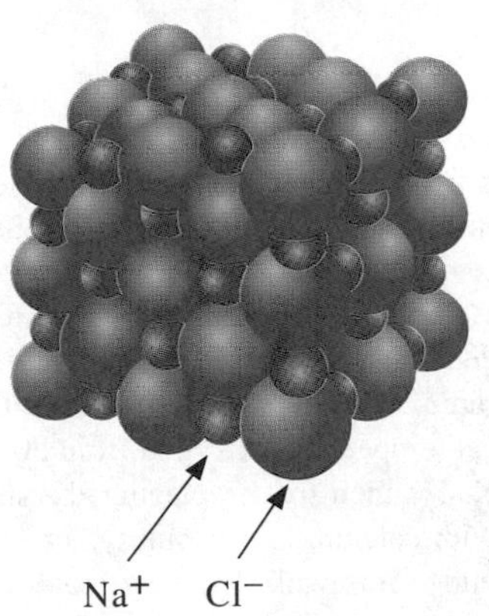

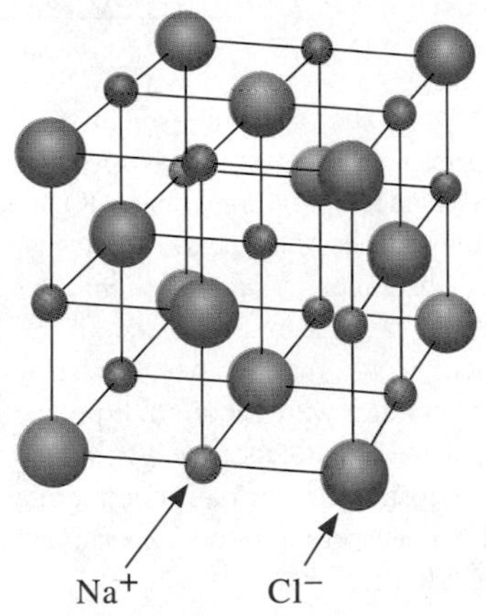

FIGURE 11.5 Sodium chloride crystal. Diagram represents a small fragment of sodium chloride, which forms cubic crystals. Each sodium ion is surrounded by six chloride ions, and each chloride ion is surrounded by six sodium ions.

structure of a noble gas, and become positive ions. A nonmetal, on the other hand, is only a few electrons short of having a noble gas electron structure in its outer energy level and thus has a tendency to gain electrons. In reacting with metals, nonmetal atoms characteristically gain one, two or three electrons, attain the electron structure of a noble gas, and become negative ions. The ions formed by loss of electrons are much smaller than the corresponding metal atoms; the ions formed by gaining electrons are larger than the corresponding nonmetal atoms. The actual dimensions of the atomic and ionic radii of several metals and nonmetals are given in Table 11.3.

Study the following examples. Note the loss and gain of electrons between atoms; also note that the ions in each compound have a noble-gas electron structure.

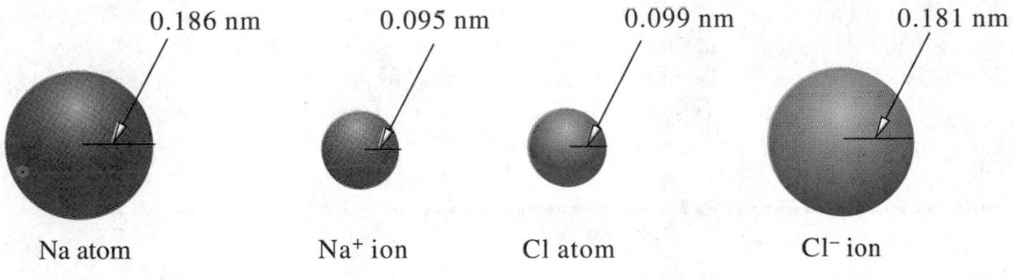

FIGURE 11.6 Relative radii of sodium and chlorine atoms and their ions.

TABLE 11.3 Change in Atomic Radii of Selected Metals and Nonmetals*

Atomic radius (nm)	Ionic radius (nm)	Atomic radius (nm)	Ionic radius (nm)
Li 0.152	Li^+ 0.060	F 0.071	F^- 0.136
Na 0.186	Na^+ 0.095	Cl 0.099	Cl^- 0.181
K 0.227	K^+ 0.133	Br 0.114	Br^- 0.195
Mg 0.160	Mg^{2+} 0.065	O 0.074	O^{2-} 0.140
Al 0.143	Al^{3+} 0.050	S 0.103	S^{2-} 0.184

* The metals shown lose electrons to become positive ions. The nonmetals gain electrons to become negative ions.

Chemistry in Action

Superconductors—A New Frontier

When electric current flows through a wire, resistance slows the current and heats the wire. In order to keep the current flowing this electrical "friction" must be overcome by adding energy to the system. In fact the existence of electrical resistance limits the efficiency of all electrical devices.

When chilled in liquid nitrogen, the superconductor acts as a perfect mirror to the magnet, causing it to levitate as it "sees" its reflection in the superconductor.

In 1911, a Dutch scientist, Heike Kamerlingh Onnes, discovered that at very cold temperatures (near 0 K), electrical resistance disappears. Onnes named this phenomenon *superconductivity.* Scientists have been fascinated by it ever since. Unfortunately, because such low temperatures were required, liquid helium was necessary to cool the wires. With helium costing $7 per liter, commercial applications were far too expensive to be considered.

For many years scientists were convinced that superconductivity was not possible at higher temperatures (not even as high as 77 K, the boiling point of liquid nitrogen, a bargain at $0.17 per liter). The first high-temperature superconductor, developed in 1986, was superconducting at 30 K. This material is a complex metal oxide that has a sandwich-like crystal structure with copper and oxygen atoms on the inside and barium and lanthanum atoms on the outside.

Scientists immediately tried to develop materials that would be superconducting at even higher temperatures. To do this they relied on their knowledge of the periodic table and the properties of chemical families. Paul Chu at the University of Houston, Texas, found the critical temperature could be raised by compressing the superconducting oxide. The pressure was too intense to be useful commercially, so Chu looked for another way to bring the layers closer together. He recognized that this could be accomplished by replacing the barium with strontium, an element in the same family with similar chemical properties and a smaller ionic radius. The idea was successful—the critical temperature changed from 30 K to 40 K. He then tried replacing the strontium with calcium (same family, smaller still) but to no avail. The new material had a lower critical temperature! But Chu persisted, and on January 12, 1987, by substituting yttrium for lanthanum (same family, smaller radius) he produced a new superconductor having a critical temperature of 95 K, well above the 77 K boiling point of liquid nitrogen. This material has the formula $YBa_2Cu_3O_7$ and is a good candidate for commercial applications.

There are several barriers to cross before superconductors are in wide use. The material is brittle and easily broken, nonmalleable, and does not carry as high a current per unit cross section as conventional conductors. Many researchers are currently working to surmount these problems and develop potential uses for superconductors, including high-speed levitation trains, tiny efficient electric motors, and smaller, faster computers.

Example 11.2 Explain how magnesium and chlorine combine to form magnesium chloride, $MgCl_2$.

Solution A magnesium atom of electron structure $1s^22s^22p^63s^2$ must lose two electrons or gain six to reach a stable electron structure. If magnesium reacts with chlorine and each chlorine atom can accept only one electron, two chlorine atoms will be needed for the two electrons from each magnesium atom. The compound formed will contain one magnesium ion and two chloride ions. The magnesium atom, having lost two electrons, becomes a magnesium ion with a $+2$ charge. Each chloride ion will have a -1 charge. The transfer of electrons from a magnesium atom to two chlorine atoms is shown in the following illustration:

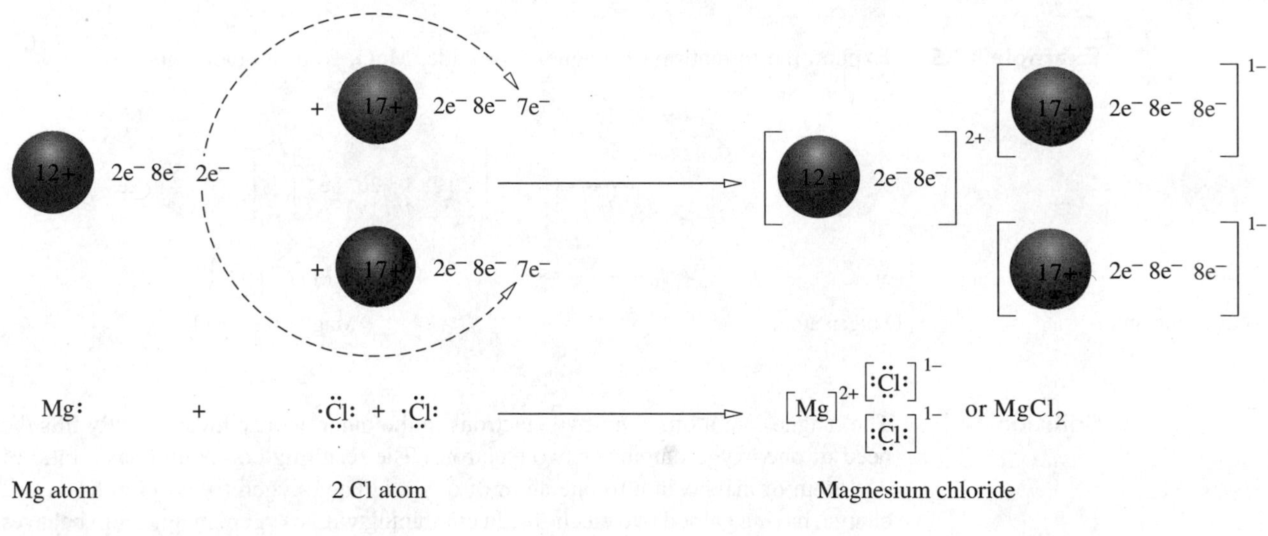

Example 11.3

Explain the formation of sodium fluoride, NaF, from its elements.

$$(11+)\ 2e^-\ 8e^-\ 1e^- \quad + \quad (9+)\ 2e^-\ 7e^- \longrightarrow [(11+)\ 2e^-\ 8e^-]^{1+}\ [(9+)\ 2e^-\ 8e^-]^{1-}$$

$$\text{Na}\cdot \quad + \quad \cdot\ddot{\underset{..}{\text{F}}}: \longrightarrow [\text{Na}]^+[:\ddot{\underset{..}{\text{F}}}:]^-$$

Sodium atom — Fluorine atom — Sodium fluoride

Solution

The fluorine atom, with seven electrons in its outer shell, behaves similarly to the chlorine atom.

Example 11.4

Explain the formation of aluminum fluoride, AlF_3, from its elements.

Solution

$$\dot{\text{Al}}\cdot \quad + \quad 3\ \cdot\ddot{\underset{..}{\text{F}}}: \longrightarrow [\text{Al}]^{3+}\,3[:\ddot{\underset{..}{\text{F}}}:]^- \quad \text{or} \quad \text{AlF}_3$$

aluminum atom — fluorine atoms — aluminum flouride

Each fluorine atom can accept only one electron. Therefore three fluorine atoms are needed to combine with the three outer shell electrons of one aluminum atom. The aluminum atom has lost three electrons to become an aluminum ion, Al^{3+}, with a +3 charge.

Example 11.5 Explain the formation of magnesium oxide, MgO, from its elements.

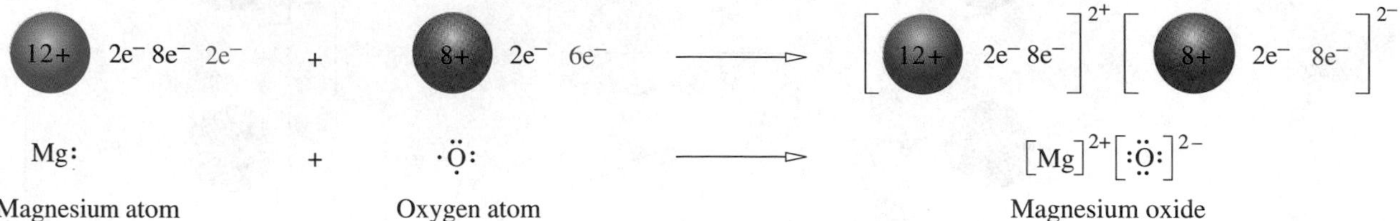

Solution The magnesium atom, with two electrons in the outer energy level, exactly fills the need of one oxygen atom for two electrons. The resulting compound has a ratio of one atom of magnesium to one atom of oxygen. The oxygen (oxide) ion has a -2 charge, having gained two electrons. In combining with oxygen, magnesium behaves the same way as when combining with chlorine; it loses two electrons.

Example 11.6 Explain the formation of sodium sulfide, Na_2S, from its elements.

Solution

Na· + ·S̈: ⟶ $[Na]^+$ $[:\ddot{S}:]^{2-}$ or Na_2S
Na· $[Na]^+$

sodium atoms, sulfur atom, sodium sulfide

Two sodium atoms supply the electrons that one sulfur atom needs to make eight in its outer shell.

Example 11.7 Explain the formation of aluminum oxide, Al_2O_3, from its elements.

Solution

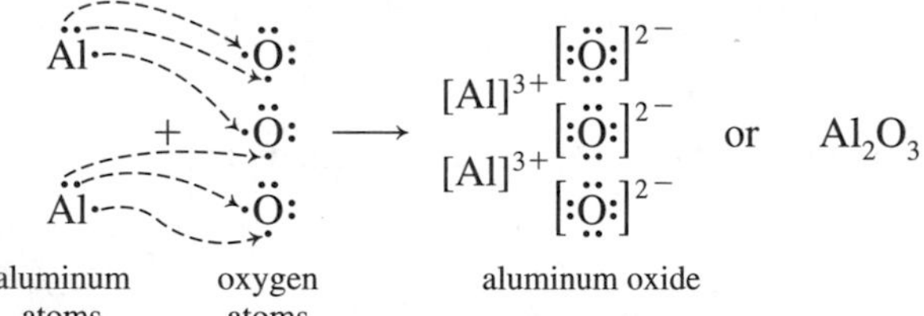

One oxygen atom, needing two electrons, cannot accommodate the three electrons from one aluminum atom. One aluminum atom falls one electron short of the four electrons needed by two oxygen atoms. A ratio of two atoms of aluminum to three atoms of oxygen, involving the transfer of six electrons (two to each oxygen atom), gives each atom a stable electron configuration.

Note that in each of the examples above, outer shells containing eight electrons were formed in all the negative ions. This formation resulted from the pairing of all the *s* and *p* electrons in these outer shells.

11.4 Predicting Formulas of Ionic Compounds

In the previous examples we have seen that when a metal and a nonmetal react to form an ionic compound, the metal loses one or more electrons to the nonmetal. In Chapter 6, where we learned to name compounds and write formulas, we saw that Group IA metals always formed $+1$ cations, whereas Group IIA formed $+2$ cations. Group VIIA elements formed -1 anions and Group VIA formed -2 anions.

We now realize that this pattern is directly related to the stability of the noble gas configuration. Metals lose electrons to attain the electron configuration of a noble gas (the previous one on the periodic table). A nonmetal forms an ion by gaining enough electrons to achieve the electron configuration of the noble gas following it on the periodic table. These observations lead us to an important chemical principle: In almost all stable chemical compounds of representative elements, each atom attains a noble gas electron configuration. This concept forms the basis for our understanding of chemical bonding.

We can apply this principle in predicting the formulas of ionic compounds. To predict the formula of an ionic compound we must recognize that chemical compounds are always electrically neutral. In addition, the metal will lose electrons to achieve noble gas configuration and the nonmetal will gain electrons to achieve noble gas configuration. Consider the compound formed between barium and sulfur. Barium has two valence electrons whereas sulfur has six valence electrons:

Ba $[Xe]6s^2$ S $[Ne]3s^23p^4$

If barium loses two electrons it will achieve the configuration of Xenon. By gaining two electrons sulfur achieves the configuration of argon. Consequently a pair of electrons is transferred between atoms. Now we have Ba^{2+} and S^{2-}. Since compounds are electrically neutral there must be a ratio of one Ba to one S, giving the empirical formula BaS.

The same principle works for many other cases. Since the key to the principle lies in the electron configuration, the periodic table can be used to extend the prediction even further. Because of similar electron structures, the elements in a family generally form compounds with the same atomic ratios. In general, if we know the atomic ratio of a particular compound, say NaCl, we can predict the atomic ratios and formulas of the other alkali metal chlorides. These formulas are LiCl, KCl, RbCl, CsCl, and FrCl (see Table 11.4).

In a similar way, if we know that the formula of the oxide of hydrogen is H_2O, we can predict that the formula of the sulfide will be H_2S, because sulfur has the same valence electron structure as oxygen. It must be recognized, however, that these are only predictions; it does not necessarily follow that every element in a group will behave like the others or even that a predicted compound will actually exist. Knowing the formulas for potassium chlorate, bromate, and iodate to be $KClO_3$, $KBrO_3$, and KIO_3, we can correctly predict the corresponding sodium compounds to have the formulas $NaClO_3$, $NaBrO_3$, and $NaIO_3$. Fluorine belongs to the same family of elements (Group VIIA) as chlorine, bromine, and iodine. So we could predict that potassium and sodium fluorates would have the formulas

TABLE 11.4 Formulas of Compounds Formed by Alkali Metals

Lewis structure	Monoxides	Chlorides	Bromides	Sulfates
Li·	Li_2O	LiCl	LiBr	Li_2SO_4
Na·	Na_2O	NaCl	NaBr	Na_2SO_4
K·	K_2O	KCl	KBr	K_2SO_4
Rb·	Rb_2O	RbCl	RbBr	Rb_2SO_4
Cs·	Cs_2O	CsCl	CsBr	Cs_2SO_4

KFO_3 and $NaFO_3$. But this prediction would not be correct, because potassium and sodium fluorates are not known to exist. However, if they did exist, the formulas could very well be correct, for these predictions are based on comparisons with known formulas and similar electron structures.

In the discussion in this section we refer only to representative metals (Groups IA, IIA, IIIA). The transition metals (Group B) show more complicated behavior (they form multiple ions) and their formulas are not as easily predicted.

Example 11.8 The formula for calcium sulfide is CaS and that for lithium phosphide is Li_3P. Predict formulas for (a) magnesium sulfide, (b) potassium phosphide, and (c) magnesium selenide.

Solution

(a) Look in the periodic table for calcium and magnesium. They are both in Group IIA. Since the formula for calcium sulfide is CaS, it is reasonable to predict that the formula for magnesium sulfide is MgS.
(b) Find lithium and potassium in the periodic table. They are in Group IA. Since the formula for lithium phosphide is Li_3P, it is reasonable to predict that K_3P is the formula for potassium phosphide.
(c) Find selenium in the periodic table. It is in Group VIA just below sulfur. Therefore, it is reasonable to assume that selenium forms selenide in the same way that sulfur forms sulfide. Since MgS was the predicted formula for magnesium sulfide in part (a), it is reasonable to assume that the formula for magnesium selenide is MgSe.

Practice 11.2

The formula for sodium oxide is Na_2O. Predict the formula for
(a) sodium sulfide
(b) rubidium oxide

Practice 11.3

The formula for barium phosphide is Ba_3P_2. Predict the formula for
(a) magnesium nitride
(b) barium arsenide

11.5 The Covalent Bond: Sharing Electrons

Some atoms do not transfer electrons from one atom to another to form ions. Instead they form a chemical bond by sharing pairs of electrons between them.

covalent bond

A **covalent bond** consists of a pair of electrons shared between two atoms. This bonding concept was introduced in 1916 by G. N. Lewis. In the millions of compounds that are known, the covalent bond is the predominant chemical bond.

True molecules exist in substances in which the atoms are covalently bonded. It is proper to refer to molecules of such substances as hydrogen, chlorine, hydrogen chloride, carbon dioxide, water, or sugar. These substances contain only covalent bonds and exist as aggregates of molecules. We do not use the term *molecule* when talking about ionically bonded compounds such as sodium chloride, because such substances exist as large aggregates of positive and negative ions, not as molecules.

A study of the hydrogen molecule gives us an insight into the nature of the covalent bond and its formation. The formation of a hydrogen molecule, H_2, involves the overlapping and pairing of $1s$ electron orbitals from two hydrogen atoms. This overlapping and pairing is shown in Figure 11.7. Each atom contributes one electron of the pair that is shared jointly by two hydrogen nuclei. The orbital of the electrons now includes both hydrogen nuclei, but probability factors show that the most likely place to find the electrons (the point of highest electron density) is between the two nuclei. The two nuclei are shielded from each other by the pair of electrons, allowing the two nuclei to be drawn very close to each other.

The formula for chlorine gas is Cl_2. When the two atoms of chlorine combine to form this molecule, the electrons must interact in a manner that is similar to that shown in the preceding example. Each chlorine atom would be more stable with eight electrons in its outer shell. But chlorine atoms are identical, and neither is able to pull an electron away from the other. What happens is this: The unpaired $3p$ electron orbital of one chlorine atom overlaps the unpaired $3p$ electron orbital of the other atom, resulting in a pair of electrons that are mutually shared between the two atoms. Each atom furnishes one of the pair of shared electrons. Thus, each atom attains a stable structure of eight electrons by sharing an electron pair with the other atom. The pairing of the p electrons and the formation of a chlorine molecule are illustrated in Figure 11.8. Neither chlorine atom has a positive or negative charge, since both contain the same number of protons and have equal attraction for the pair of electrons being shared. Other examples of molecules in which electrons are equally shared between two atoms are hydrogen, H_2; oxygen,

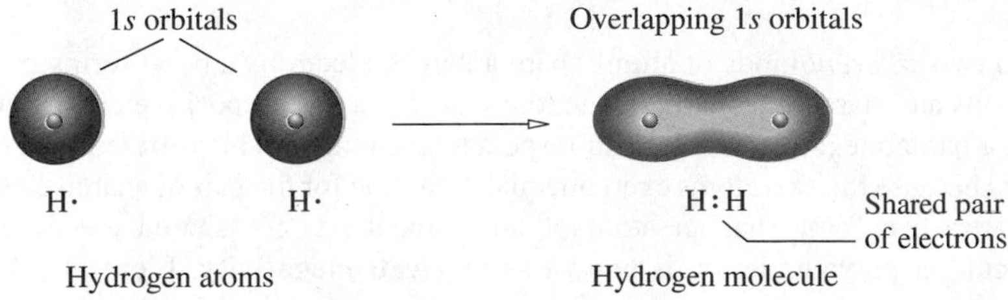

FIGURE 11.7
The formation of a hydrogen molecule from two hydrogen atoms. The two $1s$ orbitals overlap, forming the H_2 molecule. In this molecule the two electrons are shared between the atoms, forming a covalent bond.

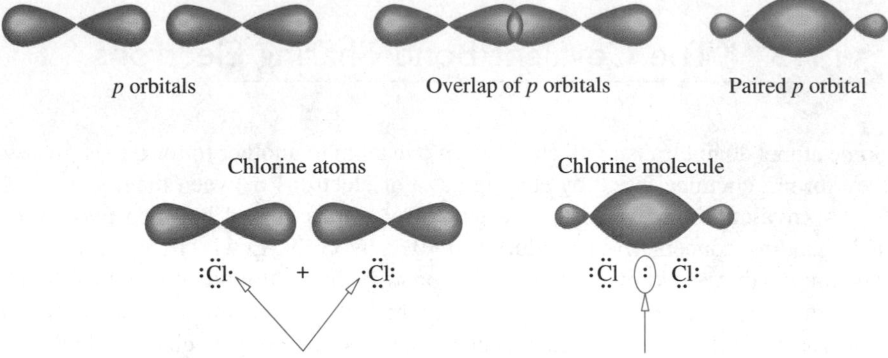

FIGURE 11.8 ▶ **Pairing of *p* electrons in the formation of a chlorine molecule.**

O_2; nitrogen, N_2; fluorine, F_2; bromine, Br_2; and iodine, I_2. Note that more than one pair of electrons may be shared between atoms:

H:H	:F:F:	:Br:Br:	:I:I:	:O::O:	:N:::N:
hydrogen	fluorine	bromine	iodine	oxygen	nitrogen

The Lewis structure given for oxygen does not adequately account for all the properties of the oxygen molecule. Other theories explaining the bonding in oxygen molecules have been advanced, but they are complex and beyond the scope of this book.

Remember that a dash represents a shared pair of electrons.

A common practice in writing structures is to replace the pair of dots used to represent a shared pair of electrons by a dash (—). One dash represents a single bond; two dashes, a double bond; and three dashes, a triple bond. The six structures just shown may be written thus:

H—H   :F—F: :Br—Br: :I—I: :O=O: :N≡N:

polar covalent bond

The ionic bond and the covalent bond represent two extremes. In ionic bonding the atoms are so different that electrons are transferred between them, forming a charged pair of ions. In covalent bonding two identical atoms share electrons equally. The bond is the mutual attraction of the two nuclei for the shared electrons. Between these extremes lie many cases in which the atoms are not different enough for a transfer of electrons, but are different enough that the electrons cannot be shared equally. This unequal sharing of electrons results in the formation of a **polar covalent bond**.

11.6 Electronegativity

electronegativity

When two *different* kinds of atoms share a pair of electrons a bond forms in which electrons are shared unequally. One atom assumes a partial positive charge and the other a partial negative charge with respect to each other. This difference in charge occurs because the two atoms exert unequal attraction for the pair of shared electrons. The attractive force that an atom of an element has for shared electrons in a molecule or polyatomic ion is known as its **electronegativity**. Elements differ in

◀ **Nobel laureate Linus Pauling (1901–1994).**

their electronegativities. For example, both hydrogen and chlorine need one electron to form stable electron configurations. They share a pair of electrons in hydrogen chloride, HCl. Chlorine is more electronegative and therefore has a greater attraction for the shared electrons than does hydrogen. As a result, the pair of electrons is displaced toward the chlorine atom, giving it a partial negative charge and leaving the hydrogen atom with a partial positive charge. It should be understood that the electron is not transferred entirely to the chlorine atom, as in the case of sodium chloride, and no ions are formed. The entire molecule, HCl, is electrically neutral. A partial charge is usually indicated by the Greek letter delta, δ. Thus, a partial positive charge is represented by δ+ and a partial negative charge by δ−.

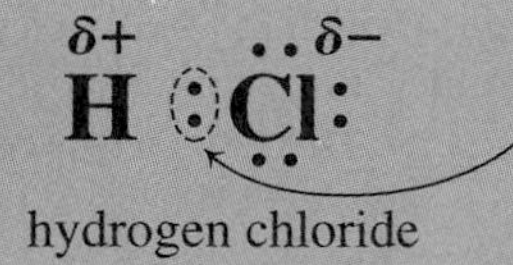

The pair of shared electrons in HCl is closer to the more electronegative chlorine atom than to the hydrogen atom, giving chlorine a partial negative charge with respect to the hydrogen atom.

A scale of relative electronegativities, in which the most electronegative element, fluorine, is assigned a value of 4.0, was developed by the Nobel laureate (1954 and 1962) Linus Pauling (1901–1994). Table 11.5 shows that the relative electronegativity of the nonmetals is high and that of the metals is low. These electronegativities indicate that atoms of metals have a greater tendency to lose electrons than do atoms of nonmetals and that nonmetals have a greater tendency to gain electrons than do metals. The higher the electronegativity value, the greater the attraction for electrons. Notice that electronegativity generally increases from left to right across a period and decreases down a group for the representative elements. The highest electronegativity is 4.0 for fluorine, and the lowest is 0.7 for francium and cesium. It is important to remember that the higher the electronegativity the stronger an atom holds electrons.

The polarity of a bond is determined by the difference in electronegativity

Goal! A Spherical Molecule

One of the most diverse elements in the periodic table is carbon. Although it is much less abundant than many other elements, it is readily available. Two well-known forms of elemental carbon are graphite and diamond. Both contain extended arrays of carbon atoms. In graphite, the carbon atoms are arranged in sheets, the bonds forming hexagons that look like chicken wire (see figure). The bonding between the sheets is very weak and the atoms can slide past each other. This slippery property makes graphite useful as a lubricant. Diamond consists of transparent octahedral crystals in which each carbon atom is bonded to four other carbon atoms. This three-dimensional network of bonds gives diamond the property of hardness for which it is noted. In the 1980s, a new form of carbon was discovered in which the atoms are arranged in relatively small clusters.

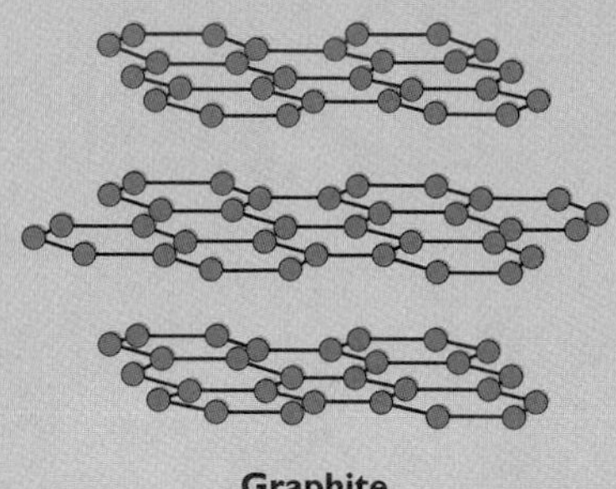

Graphite

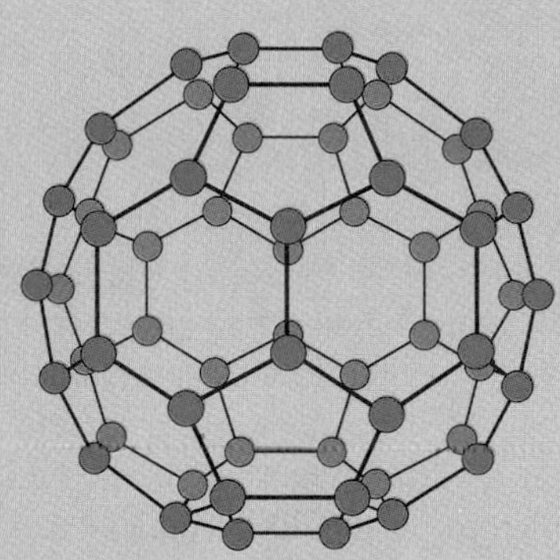

Buckminsterfullerene

Just how do scientists discover a new form of an element? Harold Kroto of the University of Sussex, England, and Richard Smalley of Rice University, Texas, were investigating the effect of light on the surface of graphite. As they analyzed the surface clusters with a mass spectrometer, they discovered a strange molecule whose formula is C_{60}. What could be the structure of such a molecule? They deduced that the most stable arrangement for the atoms would be in the shape of a soccer ball. In thinking about possible arrangements, the scientists considered the geodesic domes designed by R. Buckminster Fuller in the 1960s. The C_{60} form of carbon was named *buckminsterfullerene.* Its structure is shown in the diagram.

The amounts of buckminsterfullerene (now known as buckyballs or fullerenes) prepared by laser were very small; thus much effort went into finding a way to prepare larger amounts of the new allotrope. In 1990, a group led by Donald Huffman of the University of Arizona discovered that by vaporizing graphite electrodes, fullerenes could be manufactured in large quantities. Now that relatively large amounts of C_{60} are available, buckyballs have captured the imagination of a variety of chemists.

Research on fullerenes has led to a host of possible applications for the molecules. If metals are bound to the carbon atoms the fullerenes become superconducting; that is, they conduct electrical current without resistance, at very low temperature. Scientists are now able to make buckyball compounds that superconduct at temperatures of 45 K. Other fullerenes are being used in lubricants and optical materials.

Chemists at Yale University have managed to trap helium and neon inside buckyballs. This is the first time chemists have ever observed helium or neon in a compound of any kind. They found that at temperatures from 1000 °F to 1500 °F one of the covalent bonds linking neighboring carbon atoms in the buckyball breaks. This opens a window in the fullerene molecule through which a helium or neon atom can enter the buckyball. When the fullerene is allowed to cool, the broken bond between carbon atoms re-forms, shutting the window and trapping the helium or neon atom inside the buckyball. Since the visiting helium or neon cannot react or share electrons with its host, the resulting compound has forced scientists to invent a new kind of chemical formula to describe the compound. The relationship between the "prisoner" helium or neon and the host buckyball is shown with an @ sign. Therefore, a helium fullerene containing 60 carbon atoms would be $He@C_{60}$.

The noble-gas fullerenes could be used to "label" crude oil and other pollutants in order to identify and track down the polluter. Two different isotopes of helium could be trapped with buckyballs in a specific ratio to create a coding system for each manufacturer. This ratio of fullerenes could then be detected even in a small quantity of oil recovered from an oil slick at sea and could be used to identify the source of the oil.

Buckyballs can also be tailored to fit a particular size requirement. Raymond Schinazi of the Emory University School of Medicine made a buckyball to fit the active site of a key HIV enzyme that paralyzes the virus, making it noninfectious

continued →

Goal! A Spherical Molecule continued

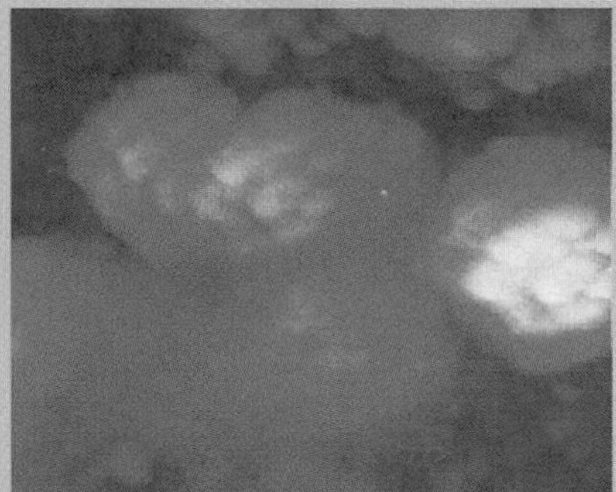

"Raspberries" of worn fullerene lubricant.

in human cells. The key to making this compound was preparing a water-soluble buckyball that would fit in the active site of the enzyme. Eventually, scientists created a fullerene molecule that is water soluble and has two charged arms to grasp the binding site of the enzyme. It is also toxic to the virus but does not appear to harm the host cells.

Fullerenes also are being tested as lubricants to protect surfaces under conditions present in space. Bharat Bhushan of Ohio State University is depositing thin fullerene films on silicon surfaces, and then testing the film by sliding steel balls across the surface while measuring the friction. Engineers have evaluated the films under a nitrogen atmosphere and in vacuum. The film breaks down slightly as the fullerene molecules cluster together to form larger balls resembling raspberries (see photo). These larger clusters roll like ball bearings between the silicon surface and the steel ball. Fullerene lubricants seem to work best at 110°C, in low humidity, and inert environments.

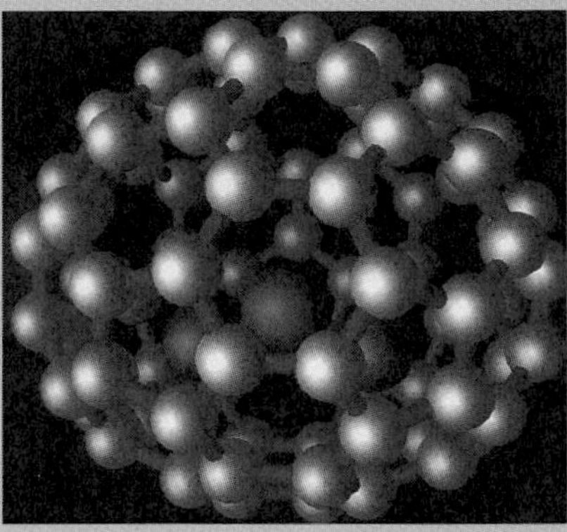

Scandium atoms trapped in a buckyball.

TABLE 11.5 Relative Electronegativity of the Elements*

Key: 9 — Atomic number; F — Symbol; 4.0 — Electronegativity

1 **H** 2.1																	2 **He**
3 **Li** 1.0	4 **Be** 1.5											5 **B** 2.0	6 **C** 2.5	7 **N** 3.0	8 **O** 3.5	9 **F** 4.0	10 **Ne**
11 **Na** 0.9	12 **Mg** 1.2											13 **Al** 1.5	14 **Si** 1.8	15 **P** 2.1	16 **S** 2.5	17 **Cl** 3.0	18 **Ar**
19 **K** 0.8	20 **Ca** 1.0	21 **Sc** 1.3	22 **Ti** 1.4	23 **V** 1.6	24 **Cr** 1.6	25 **Mn** 1.5	26 **Fe** 1.8	27 **Co** 1.8	28 **Ni** 1.8	29 **Cu** 1.9	30 **Zn** 1.6	31 **Ga** 1.6	32 **Ge** 1.8	33 **As** 2.0	34 **Se** 2.4	35 **Br** 2.8	36 **Kr**
37 **Rb** 0.8	38 **Sr** 1.0	39 **Y** 1.2	40 **Zr** 1.4	41 **Nb** 1.6	42 **Mo** 1.8	43 **Tc** 1.9	44 **Ru** 2.2	45 **Rh** 2.2	46 **Pd** 2.2	47 **Ag** 1.9	48 **Cd** 1.7	49 **In** 1.7	50 **Sn** 1.8	51 **Sb** 1.9	52 **Te** 2.1	53 **I** 2.5	54 **Xe**
55 **Cs** 0.7	56 **Ba** 0.9	57–71 **La–Lu** 1.1–1.2	72 **Hf** 1.3	73 **Ta** 1.5	74 **W** 1.7	75 **Re** 1.9	76 **Os** 2.2	77 **Ir** 2.2	78 **Pt** 2.2	79 **Au** 2.4	80 **Hg** 1.9	81 **Tl** 1.8	82 **Pb** 1.8	83 **Bi** 1.9	84 **Po** 2.0	85 **At** 2.2	86 **Rn**
87 **Fr** 0.7	88 **Ra** 0.9	89–103 **Ac–Lr** 1.1–1.7	104 **Unq**	105 **Unp**	106 **Unh**	107 **Uns**	108 **Uno**	109 **Une**									

*The electronegativity value is given below the symbol of each element.

FIGURE 11.9 ▶
Nonpolar, polar covalent, and ionic compounds.

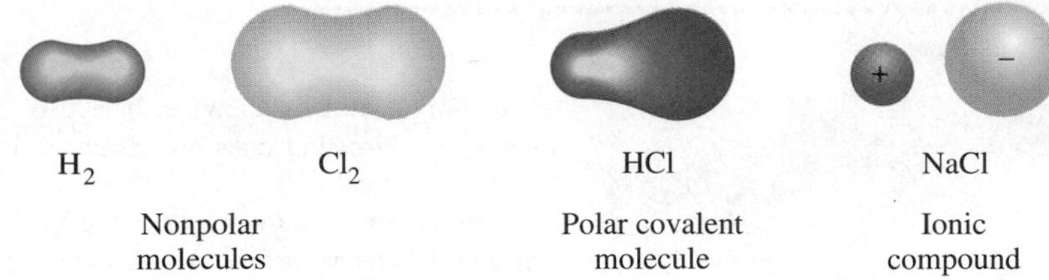

values of the atoms forming the bond (see Figure 11.9). If the electronegativities are the same the bond is **nonpolar** and the electrons are shared equally. If the atoms have greatly different electronegativities the bond is very *polar*. At the extreme, one or more electrons are actually transferred and an ionic bond results.

nonpolar bond

dipole

A **dipole** is a molecule that is electrically asymmetrical, causing it to be oppositely charged at two points. A dipole is often written as (+−). A hydrogen chloride molecule is polar and behaves as a small dipole. The HCl dipole may be written as H ⟼ Cl. The arrow points toward the negative end of the dipole. Molecules of H_2O, HBr, and ICl are polar:

H ⟼ Cl   H ⟼ Br   I ⟼ Cl   H ⟼ O ⟻ H (bent)

How do we know whether a bond between two atoms is ionic or covalent? The difference in electronegativity between two atoms determines the character of the bond formed between them. As the difference in electronegativity increases, the polarity of the bond (or percent ionic character) increases. As a rule, if the electronegativity difference between two bonded atoms is greater than 1.7–1.9, the bond will be more ionic than covalent. If the electronegativity difference is greater than 2.0, the bond is strongly ionic. If the electronegativity difference is less than 1.5, the bond is strongly covalent.

Care must be taken to distinguish between polar bonds and polar molecules. A covalent bond between different kinds of atoms is always polar. But a molecule containing different kinds of atoms may or may not be polar, depending on its shape or geometry. Molecules of HF, HCl, HBr, HI, and ICl are all polar because each contains a single polar bond. However, CO_2, CH_4, and CCl_4 are nonpolar molecules despite the fact that all three contain polar bonds. The carbon dioxide molecule O=C=O is nonpolar because the carbon–oxygen dipoles cancel each other by acting in opposite directions.

⟻ ⟼
O=C=O
dipoles in opposite directions

Methane (CH_4) and carbon tetrachloride (CCl_4) are nonpolar because the four C—H and C—Cl polar bonds are identical, and, because these bonds emanate from the center to the corners of a tetrahedron in the molecule, the effect of their polarities cancel one another. We shall discuss the shapes of molecules later in this chapter.

We have said that water is a polar molecule. If the atoms in water were linear like those in carbon dioxide, the two O—H dipoles would cancel each other, and the molecule would be nonpolar. However, water is definitely polar and has a nonlinear (bent) structure with an angle of 105° between the two O—H bonds.

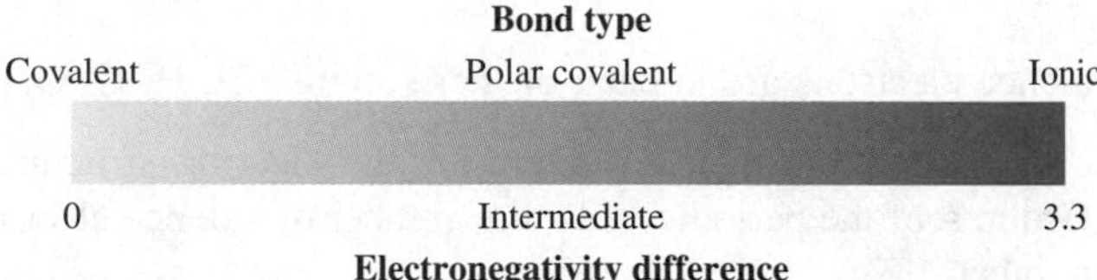

FIGURE 11.10 Relating bond type to electronegativity difference between atoms.

The relationships among types of bonds are summarized in Figure 11.10. It is important to realize that bonding is a continuum, the difference between ionic and covalent is a gradual change.

11.7 Lewis Structures of Compounds

As we have seen, Lewis structures are a convenient way of showing the covalent bonds in many molecules or ions of the representative elements. In writing Lewis structures the most important consideration for forming a stable compound is that the atoms attain a noble gas configuration.

The most difficult part of writing Lewis structures is determining the arrangement of the atoms in a molecule or an ion. In simple molecules with more than two atoms, one atom will be the central atom surrounded by the other atoms. Thus Cl_2O has two possible arrangements, Cl—Cl—O or Cl—O—Cl. Usually, but not always, the single atom in the formula (except H) will be the central atom.

Athough Lewis structures for many molecules and ions can be written by inspection, the following procedure is helpful for learning to write these structures:

Remember, the number of valence electrons of Group A elements is the same as their group number in the periodic table.

Step 1. Obtain the total number of valence electrons to be used in the structure by adding the number of valence electrons in all of the atoms in the molecule or ion. If you are writing the structure of an ion, add one electron for each negative charge or subtract one electron for each positive charge on the ion.

Step 2. Write down the skeletal arrangement of the atoms and connect them with a single covalent bond (two dots or one dash). Hydrogen, which contains only one bonding electron, can form only one covalent bond. Oxygen atoms are not normally bonded to each other, except in compounds known to be peroxides. Oxygen atoms normally have a maximum of two covalent bonds, two single bonds or one double bond.

Step 3. Subtract two electrons for each single bond you used in Step 2 from the total number of electrons calculated in Step 1. This calculation gives you the net number of electrons available for completing the structure.

Step 4. Distribute pairs of electrons (pairs of dots) around each atom (except hydrogen) to give each atom a noble gas structure.

Step 5. If there are not enough electrons to give these atoms eight electrons, change single bonds between atoms to double or triple bonds by shifting unbonded pairs of electrons as needed. Check to see that each atom has a noble gas electron structure (two electrons for hydrogen and eight for the others). A double bond counts as four electrons for each atom to which it is bonded.

Example 11.9 How many valence electrons are in each of these atoms: Cl, H, C, O, N, S, P, I?

Solution You can look at the periodic table to determine the electron structure, or, if the element is in Group A of the periodic table, the number of valence electrons is equal to the group number:

Atom	Periodic group	Valence electrons
Cl	VIIA	7
H	IA	1
C	IVA	4
O	VIA	6
N	VA	5
S	VIA	6
P	VA	5
I	VIIA	7

Example 11.10 Write the Lewis structure for water, H_2O.

Solution

Step 1 The total number of valence electrons is eight, two from the two hydrogen atoms and six from the oxygen atom.

Step 2 The two hydrogen atoms are connected to the oxygen atom. Write the skeletal structure:

```
H O      or      H O H
  H
```

Place two dots between the hydrogen and oxygen atoms to form the covalent bonds:

```
H:O      or      H:O:H
  ..
  H
```

Step 3 Subtract the four electrons used in Step 2 from eight to obtain four electrons yet to be used.

Step 4 Distribute the four electrons around the oxygen atom. Hydrogen atoms cannot accommodate any more electrons:

```
   ..                 ..
H—O:       or     H—O—H
   |                  ..
   H
```

This arrangement is the Lewis structure. Each atom has a noble-gas electron structure. Notice the shape of the molecule is not shown by the Lewis structure.

Example 11.11 Write Lewis structures for a molecule of (a) methane, CH_4, and (b) carbon tetrachloride, CCl_4.

Solution

Part A

Step 1 The total number of valence electrons is eight, one from each hydrogen atom and four from the carbon atom.

Step 2 The skeletal structure contains four H atoms around a central C atom. Place two electrons between the C and each H.

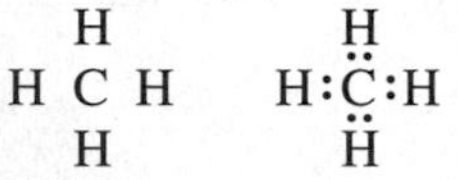

```
  H         H
                ..
H C H     H:C:H
                ..
  H         H
```

Step 3 Subtract the eight electrons used in Step 2 from eight to obtain zero electrons yet to be placed. Therefore, the Lewis structure must be as written in Step 2:

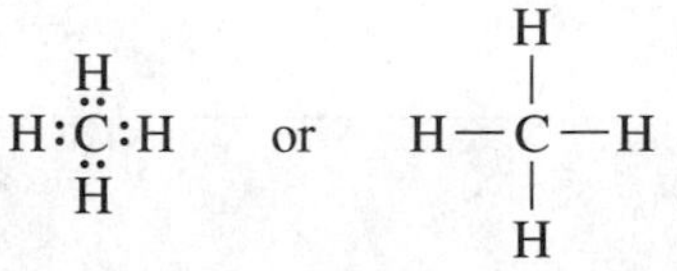

```
                    H
  H                 |
  ..
H:C:H    or    H—C—H
  ..                |
  H                 H
```

Part B

Step 1 The total number of valence electrons to be used is 32, four from the carbon atom and seven from each of the four chlorine atoms.

Step 2 The skeletal structure contains the four Cl atoms around a central C atom. Place two electrons between the C and each Cl:

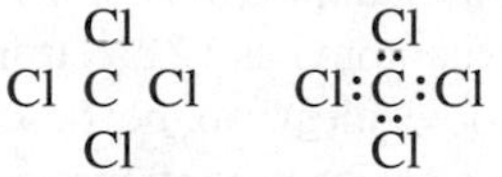

```
   Cl          Cl
                ..
Cl C Cl    Cl:C:Cl
                ..
   Cl          Cl
```

Step 3 Subtract the eight electrons used in Step 2 from 32, to obtain 24 electrons yet to be placed.

Step 4 Distribute the 24 electrons (12 pairs) around the Cl atoms so that each Cl atom has eight electrons around it:

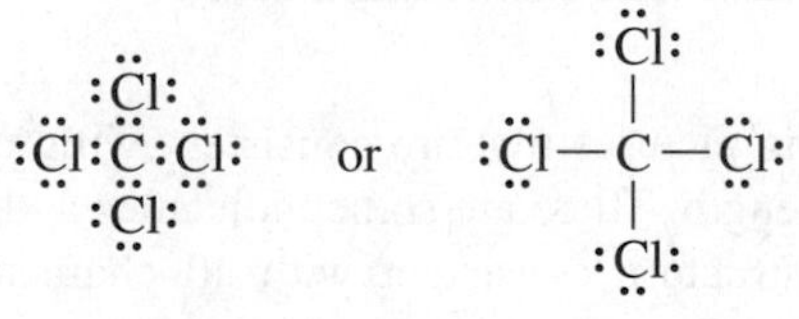

```
                          ..
      ..                 :Cl:
     :Cl:                 |
  ..  ..  ..          ..      ..
 :Cl:C:Cl:    or    :Cl—C—Cl:
  ..  ..  ..          ..      ..
     :Cl:                 |
      ..                 :Cl:
                          ..
```

This arrangement is the Lewis structure; CCl_4 contains four covalent bonds.

Example 11.12

Write a Lewis structure for CO_2.

Solution

Step 1 The total number of valence electrons is 16, four from the C atom and six from each O atom.

Step 2 The two O atoms are bonded to a central C atom. Write the skeletal structure and place two electrons between the C and each O atom.

O:C:O

Step 3 Subtract the four electrons used in Step 2 from 16 to obtain 12 electrons yet to be placed.

Step 4 Distribute the 12 electrons around the C and O atoms. Several possibilities exist:

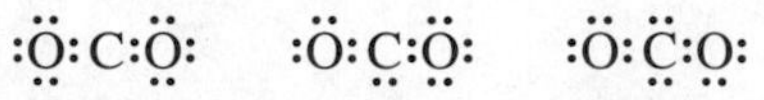

```
 ..   ..       ..   ..        ..  ..
:O:C:O:       :O:C:O:        :O:C:O:
 ..   ..          ..  ..      ..  ..
   I              II            III
```

Step 5 All the atoms do not have eight electrons around them (noble gas structure). Remove one pair of unbonded electrons from each O atom in structure I and place one pair between each O and the C atom forming two double bonds:

$$:\ddot{O}::C::\ddot{O}: \quad \text{or} \quad :\ddot{O}{=}C{=}\ddot{O}:$$

Each atom now has eight electrons around it. The carbon is sharing four pairs of electrons, and each oxygen is sharing two pairs. These bonds are known as double bonds since each involves sharing two pairs of electrons.

Practice 11.4

Write the Lewis structure for each of the following:
(a) PBr_3, (b) $CHCl_3$, (c) HF, (d) H_2CO.

Although many compounds attain a noble gas structure in covalent bonding, there are numerous exceptions. Sometimes it is impossible to write a structure in which each atom has eight electrons around it. For example, in BF_3 the boron atom has only six electrons around it, and in SF_6 the sulfur atom has 12 electrons around it.

Even though there are exceptions, many molecules can be described using Lewis structures where each atom has a noble gas electron configuration. This is a useful model for understanding chemistry.

11.8 Complex Lewis Structures

Most Lewis structures give bonding pictures that are consistent with experimental information on bond strength and length. There are some molecules and polyatomic ions for which no single Lewis structure consistent with all characteristics and bonding information can be written. For example, consider the nitrate ion, NO_3^-. To write a Lewis structure for this polyatomic ion we would use the following steps.

Step 1 The total number of valence electrons is 24, five from the nitrogen atom, and six from each oxygen atom, plus one electron from the −1 charge.

Step 2 The three O atoms are bonded to a central N atom. Write the skeletal structure and place two electrons between each pair of atoms:

$$\left[\begin{array}{c} O \\ O:\ddot{N}:O \end{array}\right]^-$$

Step 3 Subtract the six electrons used in Step 2 from 24 to obtain eighteen electrons yet to be placed.

Step 4 Distribute the eighteen electrons around the N and O atoms:

$$\begin{array}{c} :\ddot{O} \leftarrow \text{electron deficient} \\ :\ddot{\underset{..}{O}}:\underset{..}{\ddot{N}}:\ddot{\underset{..}{O}}: \end{array}$$

Step 5 One pair of electrons is still needed to give all the N and O atoms a noble gas structure. Move the unbonded pair of electrons from the N atom and place it between the N and the electron-deficient O atom, making a double bond.

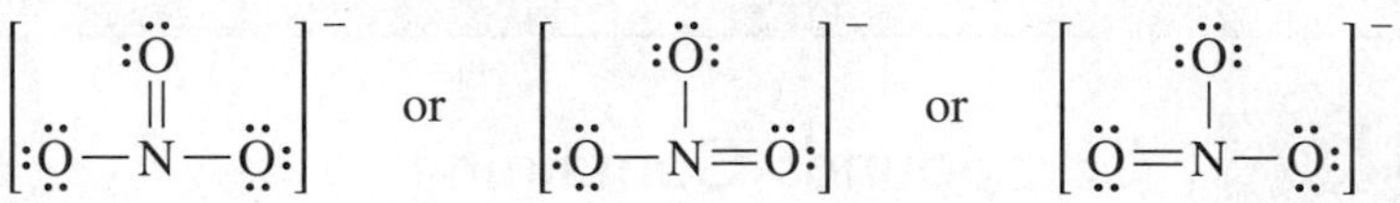

```
[    :Ö      ]-        [     :Ö:     ]-        [    :Ö:     ]-
[     ||     ]    or   [      |      ]    or   [     |      ]
[ :Ö—N—Ö:    ]         [ :Ö—N=Ö:     ]         [ Ö=N—Ö:     ]
```

Are these all valid Lewis structures? Yes, so there are really three possible Lewis structures for NO_3^-.

A molecule or ion that has multiple correct Lewis structures shows *resonance*. Each of these Lewis structures is called a **resonance structure**. In this book, however, we will not be concerned with how to choose the correct resonance structure for a molecule or ion. Therefore, any of the possible resonance structures may be used to represent the ion or molecule.

resonance structure

Example 11.13

Write the Lewis structure for a carbonate ion, CO_3^{2-}.

Solution

Step 1 These four atoms have 22 valence electrons plus two electrons from the -2 charge, which makes 24 electrons to be placed.

Step 2 In the carbonate ion the carbon is the central atom surrounded by the three oxygen atoms. Write the skeletal structure and place two electrons between each pair of atoms:

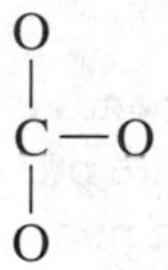

```
O
|
C—O
|
O
```

Step 3 Subtract the six electrons used in Step 2 from 24 to give 18 electrons yet to be placed.

Step 4 Distribute the 18 electrons around the three oxygen atoms and indicate that the carbonate ion has a -2 charge:

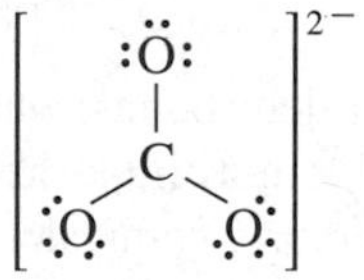

```
[     :Ö:     ]2-
[      |      ]
[      C      ]
[    /   \    ]
[  :Ö.   .Ö:  ]
```

The difficulty with this structure is that the carbon atom has only six electrons around it instead of a noble gas octet.

Step 5 Move one of the nonbonding pairs of electrons from one of the oxygens and place them between the carbon and the oxygen. These Lewis structures are possible:

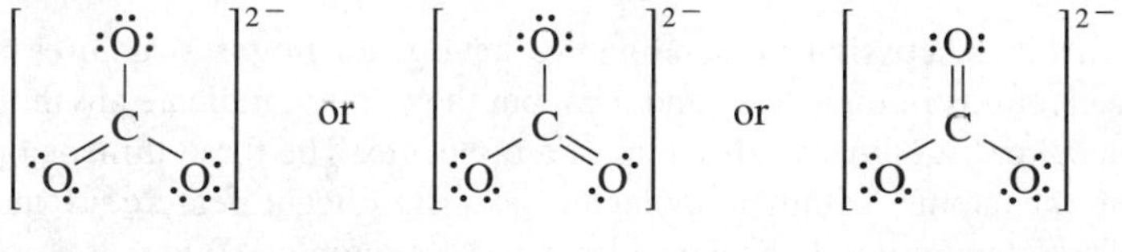

```
[    :Ö:    ]2-        [    :Ö:    ]2-        [    :O:    ]2-
[     |     ]          [     |     ]          [     ||    ]
[     C     ]    or    [     C     ]    or    [     C     ]
[   //  \   ]          [   /  \\   ]          [   /   \   ]
[  O.    .Ö: ]         [ :Ö.    O  ]          [ :Ö.   .Ö: ]
```

Practice 11.5

Write the Lewis structure for each of the following: (a) NH_3, (b) H_3O^+, (c) NH_4^+, (d) HCO_3^-.

11.9 Compounds Containing Polyatomic Ions

A polyatomic ion is a stable group of atoms that has either a positive or a negative charge and behaves as a single unit in many chemical reactions. Sodium carbonate, Na_2CO_3, contains two sodium ions and a carbonate ion. The carbonate ion, CO_3^{2-}, is a polyatomic ion composed of one carbon atom and three oxygen atoms and has a charge of -2. One carbon and three oxygen atoms have a total of 22 electrons in their outer shells. The carbonate ion contains 24 outer shell electrons and therefore has a charge of -2. In this case, the two additional electrons come from the two sodium atoms, which are now sodium ions:

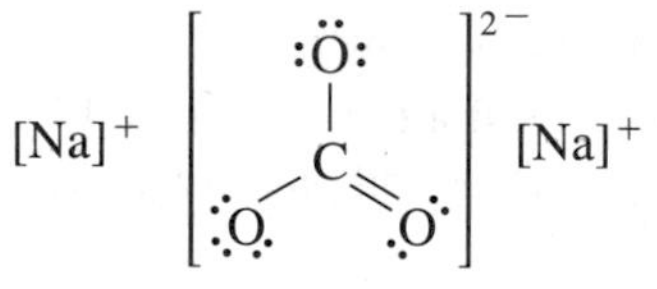

[Na]$^+$ [:Ö: | C (—:O: , =O:)]$^{2-}$ [Na]$^+$

sodium carbonate

[:Ö: | C (—:O: , =O:)]$^{2-}$

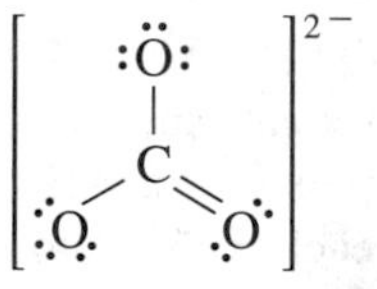

carbonate ion

Sodium carbonate has both ionic and covalent bonds. Ionic bonds exist between each of the sodium ions and the carbonate ion. Covalent bonds are present between the carbon and oxygen atoms within the carbonate ion. One important difference between the ionic and covalent bonds in this compound can be demonstrated by dissolving sodium carbonate in water. It dissolves in water forming three charged particles, two sodium ions and one carbonate ion, per formula unit of Na_2CO_3:

$$\underset{\text{sodium carbonate}}{Na_2CO_3} \xrightarrow{\text{water}} \underset{\text{sodium ions}}{2\,Na^+} + \underset{\text{carbonate ion}}{CO_3^{2-}}$$

The CO_3^{2-} ion remains as a unit, held together by covalent bonds; but where the bonds are ionic, dissociation of the ions takes place. Do not think, however, that polyatomic ions are so stable that they cannot be altered. Chemical reactions by which polyatomic ions can be changed to other substances do exist.

11.10 Molecular Structure

So far in our discussion of bonding we have used Lewis structures to represent valence electrons in molecules and ions, but they do not indicate anything regarding the molecular or geometric structure of a molecule. The three-dimensional arrangement of the atoms within a molecule is a significant feature in understanding molecular interactions. Let's consider several examples. Water is known to have the geometric structure:

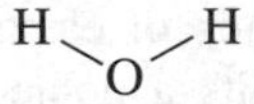

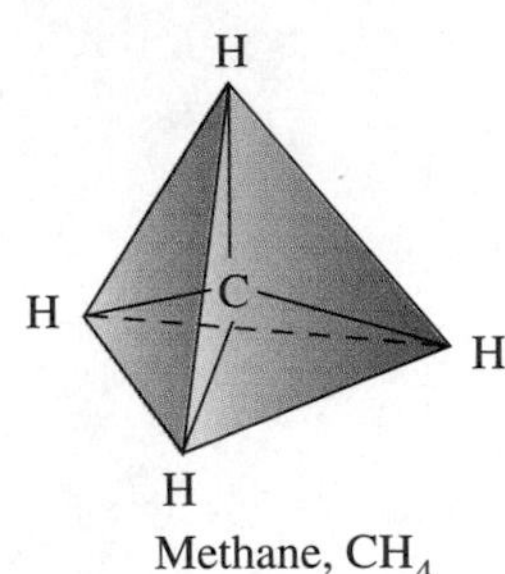

Methane, CH_4

H H
 O

known as "bent" or "V-shaped." Carbon dioxide exhibits a linear shape:

O=C=O

whereas BF_3 forms a third molecular structure:

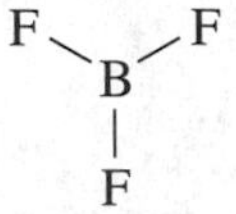

F   F
  B
  |
  F

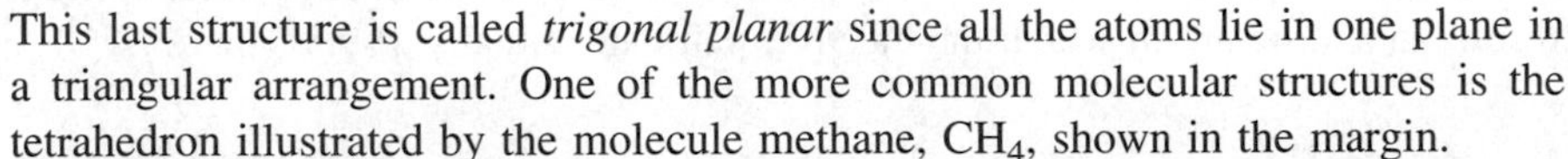

This last structure is called *trigonal planar* since all the atoms lie in one plane in a triangular arrangement. One of the more common molecular structures is the tetrahedron illustrated by the molecule methane, CH_4, shown in the margin.

How can chemists predict the geometric structure of a molecule? We will now study a model developed to assist in making predictions from the Lewis structure.

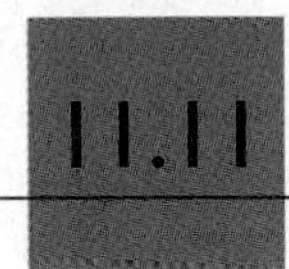

11.11 The Valence Shell Electron Pair Repulsion (VSEPR) Model

The chemical properties of a substance are closely related to the structure of its molecules. A change in a single site on a large biomolecule can make a difference in whether or not a particular reaction occurs.

Instrumental analysis can be used to determine exact spatial arrangements of atoms. Quite often, though, we only need to be able to predict the approximate structure of a molecule. A relatively simple model has been developed by chemists to allow us to make predictions of shape from Lewis structures.

The VSEPR model is based on the idea that electron pairs will repel each other electrically and will seek to minimize this repulsion. In order to accomplish this minimization, the electron pairs will be arranged around a central atom as far apart as possible. Consider $BeCl_2$, a molecule with only two pairs of electrons surrounding the central atom. These electrons are arranged 180° apart for maximum separation:

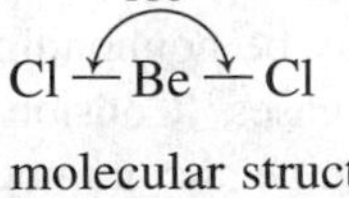

180°
Cl—Be—Cl

The molecular structure can now be labeled as a **linear structure**. When only two pairs of electrons surround an atom they should be placed 180° apart to give a linear structure.

linear structure

What occurs when there are three pairs of electrons on the central atom? Consider the BF_3 molecule. The greatest separation of electron pairs occurs when the angles between atoms are 120°:

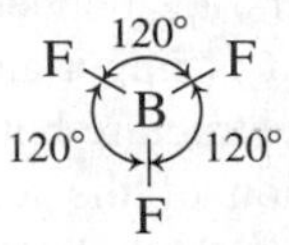

F 120° F
    B
120°   120°
    F

This arrangement of atoms is flat (planar) and is usually called **trigonal planar**. When three pairs of electrons surround an atom they should be placed 120° apart to show the trigonal planar structure.

trigonal planar

Now consider the most common situation, CH_4, with four pairs of electrons on the central carbon atom. In this case, the central atom exhibits a noble gas electron structure. What arrangement will best minimize the electron pair repulsions? At first it seems that an obvious choice is a 90° angle with all the atoms in a single plane:

```
        H
  90°  -|-  90°
  H ---  C  --- H
  90°  -|-  90°
        H
```

However, we must consider that molecules are three-dimensional. This concept results in a structure in which the electron pairs are actually 109.5° apart:

```
     109.5°
  H  <-->  H
      C
  H        H
```

In this diagram the wedged line seems to protrude from the page whereas the dashed line recedes. Two examples showing representations of this arrangement, known as **tetrahedral structure**, are illustrated in Figure 11.11. When four pairs of electrons surround an atom they should be placed 109.5° apart to give them a tetrahedral structure.

tetrahedral structure

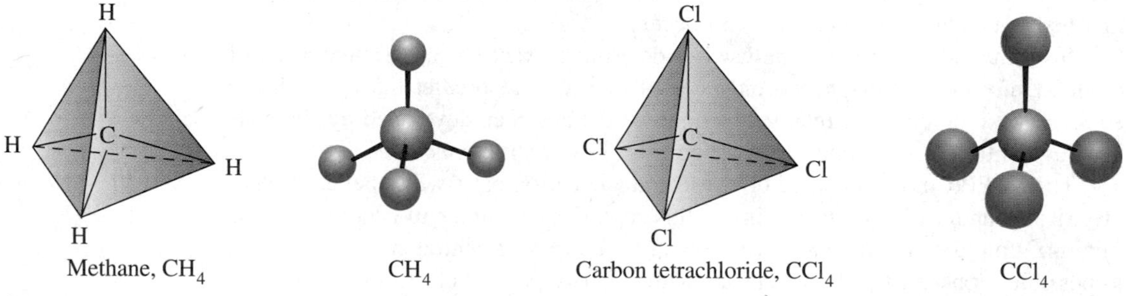

FIGURE 11.11 ▲ Ball-and-stick models of methane and carbon tetrachloride. Methane and carbon tetrachloride are nonpolar molecules because their polar bonds cancel each other in the tetrahedral arrangement of their atoms. The carbon atoms are located in the centers of the tetrahedrons.

The VSEPR model is based on the premise that we are counting electron pairs. It is quite possible that one or more of these electron pairs may be nonbonding or lone pairs. What happens to the molecular structure in these cases? Consider the ammonia molecule. First draw the Lewis structure to determine the number of electron pairs around the central atom:

```
   ..
 H:N:H
   ..
   H
```

Since there are four pairs of electrons the arrangement of electrons around the central atom will be tetrahedral. However, only three of the pairs are bonded to another atom, so the shape of the molecule itself is pyramidal. It is important to understand that the placement of the electron pairs determines the structure but the name of the shape of the molecule is determined by the position of the atoms themselves. Therefore, ammonia has a pyramidal shape, not a tetrahedral shape. See Figure 11.12.

Now consider the effect of two unbonded pairs of electrons in the water molecule. The Lewis structure for water is

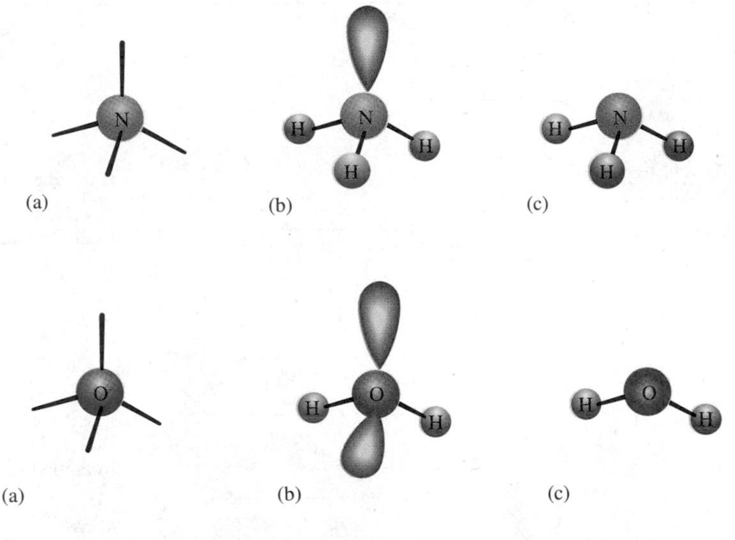

FIGURE 11.12
(a) The tetrahedral arrangement of electron pairs around the N atom in the NH_3 molecule. (b) Three pairs are shared and one is unshared. (c) The NH_3 molecule has a trigonal pyramidal structure.

FIGURE 11.13
(a) The tetrahedral arrangement of the four electron pairs around oxygen in the H_2O molecule. (b) Two of the pairs are shared and two are unshared. (c) The H_2O molecule has a bent molecular structure.

```
H—Ö:
  |
  H
```

The four electron pairs indicate a tetrahedral arrangement is necessary (see Figure 11.11). The molecule is not tetrahedral because two of the electron pairs are unbonded pairs. The atoms in the water molecule form a "bent" shape as shown in Figure 11.13. Using the VSEPR model helps to explain some of the unique properties of the water molecule. Because it is bent and not linear we can see that the molecule is polar.

The properties of water that cause it to be involved in so many interesting and important roles are largely a function of its shape and polarity. We will consider water in greater detail in Chapter 13.

Example 11.14

Predict the molecular structure for each molecule: H_2S, CCl_4, AlF_3.

Solution

1. Draw the Lewis structure.
2. Count the electron pairs and determine the arrangement that will minimize repulsions.
3. Determine the positions of the atoms and name the structure.

Molecule	Lewis structure	Number of electron pairs	Electron pair arrangement	Molecular structure
H_2S	H:S̤̈:H	4	Tetrahedral	Bent
CCl_4	:C̤̈l: above; :C̤̈l:C:C̤̈l: ; :C̤̈l: below	4	Tetrahedral	Tetrahedral
AlF_3	:F̈: above; :F̤̈:Al:F̤̈:	3	Trigonal planar	Trigonal planar

Liquid Crystals

What do color-changing pens or toy cars, bullet-resistant vests, wristwatches, calculators, and color-changing thermometers have in common? The chemicals that make each of them work are liquid crystals. In a normal crystal the molecules have an orderly arrangement. In a liquid crystal the molecules can flow and maintain an orderly arrangement at the same time.

The molecules in all types of liquid crystals are linear and polar. The atoms of linear molecules tend to lie in a relatively straight line and the molecules are generally much longer than they are wide. Polar molecules are attracted to each other, and certain linear ones are able to line up in an orderly fashion, without solidifying, to form liquid crystals. Such a substance can be used as a liquid crystal display (LCD), to change color with changing temperature, or to make a super-strong synthetic fiber.

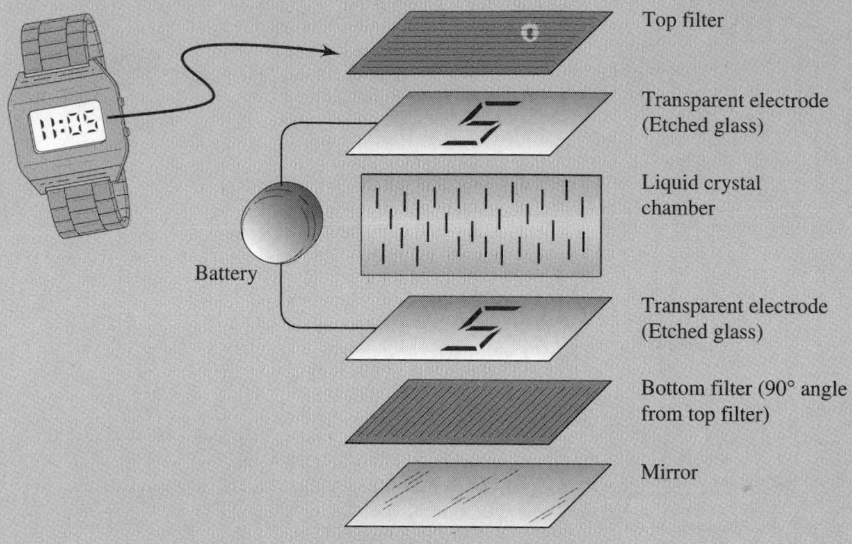

Adapted from *Chem Matters Magazine*.

The key to these products that change color with temperature is in the twisted arrangement of the molecules. The linear molecules form a generally flat surface. In the liquid crystal the molecules lie side by side in a nearly flat layer. The next layer is similar but at an angle to the one below. These closely packed flat layers have a special effect on light. As the light strikes the surface, some of it is reflected from the top layer and more from lower layers. When the same wavelength is reflected from many layers a color is observed. This is similar to the rainbow of colors formed by oil in a puddle on the street or the film of a soap bubble. As the temperature increases the molecules move faster, causing a change in the angle and the space between the layers. These changes result in a color change in the reflected light. Different compounds change color within different temperature ranges, allowing a variety of practical and amusing applications.

In liquid crystal displays (LCDs) in watches and calculators the process is similar. Normally the LCD acts as a mirror reflecting the light that enters it. The display is created by a series of layers, consisting of a filter, a glass etched with tiny lines, a liquid crystal chamber, a second etched glass, a bottom filter, and a mirror (see diagram). The molecules at the top of the liquid crystal chamber align with the lines on the top layer of glass whereas those at the bottom align with the grooves on that glass (90-degree turn from the top). In between the molecules line up as closely as possible with neighboring molecules to form a twisted spiral. To display a number a tiny current is sent to the proper SiO_2 segments on the etched piece of glass. The plates become charged, and the polar molecules of the liquid crystal are attracted to the charged segments, thus destroying the carefully arranged spirals. The pattern of reflected light is changed and a numeral appears.

Another type of liquid crystal (nematic) contains molecules that all point in the same direction. These liquid crystals are used to manufacture very strong synthetic fibers. The molecules in nematic crystals all line up in the same direction but are free to slide past one another.

Perhaps the best example of these liquid crystals is the manufacture of Kevlar, a synthetic fiber used in bullet-resistant vests, canoes, and parts of the space shuttle. Kevlar is a synthetic polymer, like nylon or polyester, that gains strength by passing through a liquid crystal state during its manufacture. In a typical polymer the long molecular chains are jumbled together, somewhat like spaghetti. The strength of the material is limited by the disorderly arrangement.

The trick is to get the molecules to line up parallel to each other. Once the giant molecules have been synthesized they are dissolved in sulfuric acid. At the proper concentration the molecules align, and the solution is forced through tiny holes in a nozzle and further aligned. The sulfuric acid is removed in a water bath forming solid fibers in near perfect alignment. One strand of Kevlar is stronger than an equal-sized strand of steel. It has a much lower density as well, making it a material of choice in bullet-resistant vests.

Practice 11.6

Predict the shape for CF_4, NF_3, and BeF_2.

Concepts in Review

1. Describe how atomic radii vary (a) from left to right in a period, and (b) from top to bottom in a group.
2. Describe how the ionization energies of the elements vary with respect to (a) the position in the periodic table and (b) the removal of successive electrons.
3. Write Lewis structures for the representative elements from their position in the periodic table.
4. Describe (a) the formation of ions by electron transfer and (b) the nature of the chemical bond formed by electron transfer.
5. Show by means of Lewis structures the formation of an ionic compound from atoms.
6. Describe a crystal of sodium chloride.
7. Predict the relative sizes of an atom and a monatomic ion for a given element.
8. Describe the covalent bond and predict whether a given covalent bond will be polar or nonpolar.
9. Draw Lewis structures for the diatomic elements.
10. Identify single, double, and triple covalent bonds.
11. Describe the changes in electronegativity in (1) moving across a period and (2) moving down a group in the periodic table.
12. Predict formulas of simple compounds formed between the representative (Group A) elements using the periodic table.
13. Describe the effect of electronegativity on the type of chemical bonds in a compound.
14. Draw Lewis structures for (a) the molecules of covalent compounds and (b) polyatomic ions.
15. Describe the difference between polar and nonpolar bonds.
16. Distinguish clearly between ionic and molecular substances.
17. Predict whether the bonding in a compound will be primarily ionic or covalent.
18. Describe the VSEPR model for molecular shape.
19. Use the VSEPR model to determine molecular structure from Lewis structure for given compounds.

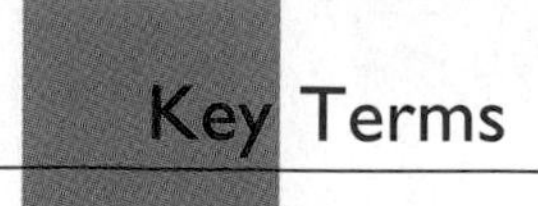

Key Terms

The terms listed here have been defined within this chapter. Section numbers are referenced in parenthesis for each term.

covalent bond (11.5)
dipole (11.6)
electronegativity (11.6)
ionic bond (11.3)
ionization energy (11.1)
Lewis structure (11.2)
linear structure (11.11)
nonpolar bond (11.6)
polar covalent bond (11.5)
resonance structure (11.8)
tetrahedral structure (11.11)
trigonal planar (11.11)

Questions

Questions refer to tables, figures, and key words and concepts defined within the chapter. A particularly challenging question or exercise is indicated with an asterisk.

1. Rank the following five elements according to the radii of their atoms, from smallest to largest: Na, Mg, Cl, K, and Rb. (Figure 11.2)

2. Explain why much more ionization energy is required to remove the first electron from neon than from sodium. (Table 11.1)

3. Explain the large increase in ionization energy needed to remove the third electron from beryllium compared with that needed for the second electron. (Table 11.1)

4. Does the first ionization energy increase or decrease from top to bottom in the periodic table for the alkali metal family? Explain. (Figure 11.3)

5. Does the first ionization energy increase or decrease from top to bottom in the periodic table for the noble gas family? Explain. (Figure 11.3)

6. Why does barium (Ba) have a lower ionization energy than beryllium (Be)? (Figure 11.3)

7. Why is there such a large increase in the ionization energy required to remove the second electron from a sodium atom as opposed to the first? (Table 11.1)

8. Which element in each of the following pairs has the larger atomic radius? (Figure 11.2)
(a) Na or K **(c)** O or F **(e)** Ti or Zr
(b) Na or Mg **(d)** Br or I

9. Which element in each of Groups IA–VIIA has the smallest atomic radius? (Figure 11.2)

10. Why does the atomic size increase in going down any family of the periodic table?

11. All the atoms within each Group A family of elements can be represented by the same Lewis structure. Complete the table below, expressing the Lewis structure for each group. (Use E to represent the elements.) (Figure 11.4)

Group	IA	IIA	IIIA	IVA	VA	VIA	VIIA
	$E\cdot$						

12. Draw the Lewis structure for Cs, Ba, Tl, Pb, Po, At, and Rn. How do these structures correlate with the group in which each element occurs?

13. In which general areas of the periodic table are the elements with (a) the highest and (b) the lowest electronegativities found?

14. What are valence electrons?

15. Explain why potassium usually forms a K^+ ion but not a K^{2+} ion.

16. Why does an aluminum ion have a +3 charge?

17. Which of these statements are correct? (Try to answer this exercise using only the periodic table.) Rewrite each incorrect statement to make it correct.
(a) If the formula for calcium iodide is CaI_2, then the formula for cesium iodide is CsI_2.
(b) Metallic elements tend to have relatively low electronegativities.
(c) If the formula for aluminum oxide is Al_2O_3, then the formula for gallium oxide is Ga_2O_3.
(d) Sodium and chlorine react to form molecules of NaCl.
(e) A chlorine atom has fewer electrons than a chloride ion.
(f) The noble gases have a tendency to lose one electron to become positively charged ions.
(g) The chemical bonds in a water molecule are ionic.
(h) The chemical bonds in a water molecule are polar.

(i) Valence electrons are those electrons in the outermost shell of an atom.
(j) An atom with eight electrons in its outer shell has all its *s* and *p* orbitals filled.
(k) Fluorine has the lowest electronegativity of all the elements.
(l) Oxygen has a greater electronegativity than carbon.
(m) A cation is larger than its corresponding neutral atom.
(n) Cl_2 is more ionic in character than HCl.
(o) A neutral atom with eight electrons in its valence shell will likely be an atom of a noble gas.
(p) A nitrogen atom has four valence electrons.
(q) An aluminum atom must lose three electrons to become an aluminum ion, Al^{3+}.
(r) A stable group of atoms that has either a positive or a negative charge and behaves as a single unit in many chemical reactions is called a *polyatomic ion.*
(s) Sodium sulfate, Na_2SO_4, has covalent bonds between sulfur and the oxygen atoms, and ionic bonds between the sodium ions and the sulfate ion.
(t) The water molecule is a dipole.
(u) In an ethylene molecule, C_2H_4,

$$\begin{matrix} H & & & & H \\ & \diagdown & & \diagup & \\ & & C{=}C & & \\ & \diagup & & \diagdown & \\ H & & & & H \end{matrix}$$

two pairs of electrons are shared between the carbon atoms.
(v) When electrons are transferred from one atom to another, the resulting compound contains ionic bonds.
(w) A phosphorus atom, $\cdot\ddot{\text{P}}\cdot$, needs three additional electrons to attain a noble-gas electron structure.
(x) The simplest compound between oxygen, $\cdot\ddot{\underset{\cdot\cdot}{\text{O}}}\cdot$, and fluorine, $:\ddot{\underset{\cdot}{\text{F}}}\cdot$, atoms is FO_2.
(y) The $\text{H}:\ddot{\underset{\cdot\cdot}{\text{Cl}}}:$ molecule has three unshared pairs of electrons.

18. Which of these statements are correct? (Try to answer this exercise using only the periodic table.) Rewrite each incorrect statement to make it correct.

(a) The smaller the difference in electronegativity between two atoms, the more ionic the bond between them will be.
(b) Lewis structures are mainly useful for the representative elements.
(c) The correct Lewis structure for NH_3 is

$$\text{H}:\ddot{\underset{\cdot\cdot}{\text{N}}}:\text{H} \\ \text{H}$$

(d) The correct Lewis structure for CO_2 is

$$:\ddot{\underset{\cdot\cdot}{\text{O}}}:\text{C}:\ddot{\underset{\cdot\cdot}{\text{O}}}:$$

(e) The Lewis structure for the noble gas helium is

$$:\ddot{\underset{\cdot\cdot}{\text{He}}}:$$

(f) In Period 5 of the periodic table, the element having the lowest ionization energy is Xe.
(g) An atom having an electron structure of $1s^22s^22p^63s^23p^2$ has four valence electrons.
(h) When an atom of bromine becomes a bromide ion, its size increases.
(i) The structures that show that H_2O is a dipole and that CO_2 is not a dipole are

$$\text{H}-\ddot{\underset{\cdot\cdot}{\text{O}}}-\text{H} \quad \text{and} \quad :\ddot{\text{O}}{=}\text{C}{=}\ddot{\text{O}}:$$

(j) The Cl^- and S^{2-} ions have the same electron structure.
(k) A molecule with the central atom surrounded by two pairs of electrons has a linear shape.
(l) A molecule with a central atom having three bonding pairs and one nonbonding electron pair will have a tetrahedral electron structure and a tetrahedral shape.
(m) A molecule in which the central atom is surrounded by three bonding pairs of electrons will have a trigonal planar shape.
(n) A molecule with two bonding and two nonbonding pairs of electrons will have a bent shape.
(o) The Lewis structure for nitrogen is $:\dot{\underset{\cdot}{\text{N}}}\cdot$.
(p) The Lewis structure for potassium is P·.

Paired Exercises

These exercises are paired. Each odd-numbered exercise is followed by a similar even-numbered exercise. Answers to the even-numbered exercises are given in Appendix V.

19. Which is larger, a magnesium atom or a magnesium ion? Explain.

20. Which is smaller, a bromine atom or a bromide ion? Explain.

21. Using the table of electronegativity values (Table 11.5), indicate which element is more positive and which is more negative in the following compounds:
(a) H_2O (c) NH_3 (e) NO
(b) NaF (d) PbS (f) CH_4

22. Using the table of electronegativity values (Table 11.5), indicate which element is more positive and which is more negative in the following compounds:
(a) HCl (c) CCl_4 (e) MgH_2
(b) LiH (d) IBr (f) OF_2

23. Classify the bond between the following pairs of elements as principally ionic or principally covalent (use Table 11.5):
(a) sodium and chlorine
(b) carbon and hydrogen
(c) chlorine and carbon
(d) calcium and oxygen

24. Classify the bond between the following pairs of elements as principally ionic or principally covalent (use Table 11.5):
(a) hydrogen and sulfur
(b) barium and oxygen
(c) fluorine and fluorine
(d) potassium and fluorine

25. Explain what happens to the electron structures of Mg and Cl atoms when they react to form $MgCl_2$.

26. Write an equation representing
(a) the change of a fluorine atom to a fluoride ion and
(b) the change of a calcium atom to a calcium ion

27. Use Lewis structures to show the electron transfer for the formation of the following ionic compounds from the atoms:
(a) MgF_2 (b) K_2O

28. Use Lewis structures to show the electron transfer for the formation of the following ionic compounds from the atoms:
(a) CaO (b) NaBr

29. How many valence electrons are in each of these atoms? H, K, Mg, He, Al

30. How many valence electrons are in each of these atoms? Si, N, P, O, Cl

31. How many electrons must be gained or lost for each of the following to achieve a noble-gas electron structure?
(a) a calcium atom
(b) a sulfur atom
(c) a helium atom

32. How many electrons must be gained or lost for each of the following to achieve a noble-gas electron structure?
(a) a chloride ion
(b) a nitrogen atom
(c) a potassium atom

33. Which is larger? Explain.
(a) a potassium atom or a potassium ion
(b) a bromine atom or a bromide ion

34. Which is larger? Explain.
(a) a magnesium ion or an aluminum ion
(b) Fe^{2+} or Fe^{3+}

35. Let *E* be any representative element. Following the pattern in the table, write the formulas for the hydrogen and oxygen compounds of
(a) Na (c) Al
(b) Ca (d) Sn

Group IA	IIA	IIIA	IVA	VA	VIA	VIIA
EH	EH_2	EH_3	EH_4	EH_3	H_2E	HE
E_2O	EO	E_2O_3	EO_2	E_2O_5	EO_3	E_2O_7

36. Let *E* be any representative element. Following the pattern in the table, write the formulas for the hydrogen and oxygen compounds of
(a) Sb (c) Cl
(b) Se (d) C

Group IA	IIA	IIIA	IVA	VA	VIA	VIIA
EH	EH_2	EH_3	EH_4	EH_3	H_2E	HE
E_2O	EO	E_2O_3	EO_2	E_2O_5	EO_3	E_2O_7

37. The formula for sodium sulfate is Na_2SO_4. Write the names and formulas for the other alkaline earth metal sulfates.

38. The formula for calcium bromide is $CaBr_2$. Write the names and formulas for the other alkaline earth metal bromides.

39. Write Lewis structures for
(a) Na (b) Br^- (c) O^{2-}

40. Write Lewis structures for
(a) Ga (b) Ga^{3+} (c) Ca^{2+}

41. Classify the bonding in each compound as ionic or covalent:
(a) H_2O (c) MgO
(b) NaCl (d) Br_2

42. Classify the bonding in each compound as ionic or covalent:
(a) HCl (c) NH_3
(b) $BaCl_2$ (d) SO_2

43. Predict the type of bond that would be formed between each of the following pairs of atoms:
(a) Na and N
(b) N and S
(c) Br and I

44. Predict the type of bond that would be formed between each of the following pairs of atoms:
(a) H and Si
(b) O and F
(c) Ca and I

45. Draw Lewis structures for
(a) H_2 (b) N_2 (c) Cl_2

46. Draw Lewis structures for
(a) O_2 (b) Br_2 (c) I_2

47. Draw Lewis structures for
(a) NCl_3 (c) C_2H_6
(b) H_2CO_3 (d) $NaNO_3$

48. Draw Lewis structures for
(a) H_2S (c) NH_3
(b) CS_2 (d) NH_4Cl

49. Draw Lewis structures for
(a) Ba^{2+} (d) CN^-
(b) Al^{3+} (e) HCO_3^-
(c) SO_3^{2-}

50. Draw Lewis structures for
(a) I^- (d) ClO_3^-
(b) S^{2-} (e) NO_3^-
(c) CO_3^{2-}

51. Classify the following molecules as polar or nonpolar:
(a) H_2O
(b) HBr
(c) CF_4

52. Classify the following molecules as polar or nonpolar:
(a) F_2
(b) CO_2
(c) NH_3

53. Give the number and arrangement of the electron pairs around the central atom:
(a) C in CCl_4
(b) S in H_2S
(c) Al in AlH_3

54. Give the number and arrangement of the electron pairs around the central atom:
(a) Be in BeF_2
(b) N in NF_3
(c) Cl in HCl

55. Use VSEPR theory to predict the structure of the following polyatomic ions:
(a) sulfate ion
(b) chlorate ion
(c) periodate ion

56. Use VSEPR theory to predict the structure of the following polyatomic ions:
(a) ammonium ion
(b) sulfite ion
(c) phosphate ion

57. Use VSEPR theory to predict the shape of the following molecules:
(a) SiH_4
(b) PH_3
(c) SeF_2

58. Use VSEPR theory to predict the shape of the following molecules:
(a) SiF_4
(b) OF_2
(c) Cl_2O

59. Identify this element, from the following description: Element X reacts with sodium to form the compound Na_2X and is in the second period on the periodic table.

60. Identify this element from the following description: Element Y reacts with oxygen to form the compound Y_2O and has the lowest ionization energy of any fourth-period element on the periodic table.

Additional Exercises

These exercises are not paired or labeled by topic and provide additional practice on concepts covered in this chapter.

61. Identify the element on the periodic table that satisfies each of the following descriptions:
(a) the transition metal with the largest atomic radius
(b) the alkaline earth metal with the greatest ionization energy
(c) the least dense member of the nitrogen family
(d) the alkali metal with the greatest ratio of neutrons to protons
(e) the most electronegative transition metal

62. Choose the element that fits each of the following descriptions:
(a) the lower electronegativity — As or Zn
(b) the lower chemical reactivity — Ba or Be
(c) the fewer valence electrons — N or Ne

63. Identify two reasons for the fact that fluorine has a much higher electronegativity than neon.

64. When one electron is removed from an atom of Li, it has two left. Helium atoms also have two electrons. Why is more energy required to remove the second electron from Li than to remove the first from He?

65. Group IB elements (see the periodic table on the inside cover of your book) have one electron in their outer shell, as do Group IA elements. Would you expect them to form compounds such as CuCl, AgCl, and AuCl? Explain.

66. The formula for lead(II) bromide is $PbBr_2$: predict formulas for tin(II) and germanium(II) bromides.

67. Why is it not proper to speak of sodium chloride molecules?

68. What is a covalent bond? How does it differ from an ionic bond?

69. Briefly comment on the structure Na:Ö:Na for the compound Na_2O.

70. What are the four most electronegative elements?

71. Rank these elements from highest electronegativity to lowest: Mg, S, F, H, O, Cs.

72. Is it possible for a molecule to be nonpolar even though it contains polar covalent bonds? Explain.

73. Why is CO_2 a nonpolar molecule, whereas CO is a polar molecule?

74. Estimate the bond angle between atoms in each of these molecules.

(a) H_2S
(b) NH_3
(c) NH_4^+
(d) $SiCl_4$

75. What is meant by the term *superconductor*?

76. What is the important relationship between 77 K and superconductivity?

77. What was the relationship on the periodic table that enabled Chu to find a high temperature superconductor?

78. Indicate the limitations of current superconducting material.

79. Indicate three uses for liquid crystals.

80. Explain how a liquid crystal display works.

81. What is Kevlar? How does it attain the property of superstrength?

82. Consider the two molecules BF_3 and NF_3. Compare and contrast these two in terms of

(a) the valence level orbitals on the central atom that are used for bonding
(b) the shape of the molecule
(c) the number of lone electron pairs on the central atom
(d) the type and number of bonds found in the molecule

83. With respect to electronegativity, why is fluorine such an important atom? What combination of atoms on the periodic table results in the most ionic bond?

84. Why does the Lewis structure of each element in a given group of representative elements on the periodic table have the same number of dots?

85. A sample of an air pollutant composed of sulfur and oxygen was found to contain 1.40 g sulfur and 2.10 g oxygen. What is the empirical formula for this compound? Draw a Lewis structure to represent it.

86. A dry-cleaning fluid composed of carbon and chlorine was found to have the composition 14.5% carbon and 85.5% chlorine. Its known molar mass is 166 g/mol. Draw a Lewis structure to represent it.

Answers to Practice Exercises

11.1 (a) :Ṅ· (b) :Ȧl (c) Sr: (d) :B̈r·

11.2 (a) Na_2S (b) Rb_2O

11.3 (a) Mg_3N_2 (b) Ba_3As_2

11.4 (a)
```
 :B̈r:
:P̈:B̈r:
 :B̈r:
```
(b)
```
    H
:C̈l:C:C̈l:
   :C̈l:
```
(c) H:F̈:
(d)
```
   Ö:
   ::
 H:C:H
```

11.5 (a)
```
H:N̈:H
   H
```
(b)
```
[H:Ö:H]+
    H
```
(c)
```
[   H   ]+
[ H:N̈:H ]
[   H   ]
```
(d)
```
[      :Ö:   ]−
[ H:Ö:C::Ö:  ]
```

11.6 CF_4 tetrahedral, NF_3 pyramidal, BeF_2 linear

The Gaseous State of Matter

12

Chapter 12

Our atmosphere is composed of a mixture of gases, including nitrogen, oxygen, carbon dioxide, ozone, and trace amounts of other gases. But there are cautions to be observed and respected when dealing with our atmosphere. For example, carbon dioxide is valuable as it is taken in by plants and converted to carbohydrates, but it also is associated with the potentially hazardous greenhouse effect. Ozone surrounds the earth at high altitudes and protects us from harmful ultraviolet rays, but it also destroys rubber and plastics. We require air to live, yet scuba divers must be concerned about oxygen poisoning, and the "bends."

In chemistry, the study of the behavior of gases allows us to lay a foundation for understanding our atmosphere and the effects that gases have on our lives.

12.1 General Properties

In Chapter 3, solids, liquids, and gases were described briefly. In this chapter we shall consider the behavior of gases in greater detail.

Gases are the least dense and most mobile of the three states of matter. A solid has a rigid structure, and its particles remain in essentially fixed positions. When a solid absorbs sufficient heat, it melts and changes into a liquid. Melting occurs because the molecules (or ions) have absorbed enough energy to break out of the rigid crystal lattice structure of the solid. The molecules or ions in the liquid are more energetic than they were in the solid, as indicated by their increased mobility. Molecules in the liquid state cling to one another. When the liquid absorbs additional heat, the more energetic molecules break away from the liquid surface and go into the gaseous state—the most mobile state of matter. Gas molecules move at very high velocities and have high kinetic energy (KE). The average velocity of hydrogen molecules at 0°C is over 1600 meters (1 mile) per second. Mixtures of gases are uniformly distributed within the container in which they are confined.

A quantity of a substance occupies a much greater volume as a gas than it does as a liquid or a solid. For example, 1 mol of water (18.02 g) has a volume of 18 mL at 4°C. This same amount of water would occupy about 22,400 mL in the gaseous state—more than a 1200-fold increase in volume. We may assume from this difference in volume that (1) gas molecules are relatively far apart, (2) gases can be greatly compressed, and (3) the volume occupied by a gas is mostly empty space.

12.2 The Kinetic-Molecular Theory

Kinetic-Molecular Theory (KMT)

Careful scientific studies of the behavior and properties of gases were begun in the 17th century by Robert Boyle (1627–1691). This work was carried forward by many investigators, and the accumulated data were used in the second half of the 19th century to formulate a general theory to explain the behavior and properties of gases. This theory is called the **Kinetic-Molecular Theory (KMT)**. The KMT has since been extended to cover, in part, the behavior of liquids and solids. It ranks today with the atomic theory as one of the greatest generalizations of modern science.

◀ **Chapter Opening Photo: The colorful art of hot air ballooning illustrates the multiple properties of gases.**

The KMT is based on the motion of particles, particularly gas molecules. A gas that behaves exactly as outlined by the theory is known as an **ideal gas**. Actually no ideal gases exist, but under certain conditions of temperature and pressure, gases approach ideal behavior, or at least show only small deviations from it. Under extreme conditions, such as very high pressure and low temperature, real gases may deviate greatly from ideal behavior. For example, at low temperature and high pressure many gases become liquids.

ideal gas

The principal assumptions of the Kinetic-Molecular Theory are:

1. Gases consist of tiny (submicroscopic) particles.
2. The distance between particles is large compared with the size of the particles themselves. The volume occupied by a gas consists mostly of empty space.
3. Gas particles have no attraction for one another.
4. Gas particles move in straight lines in all directions, colliding frequently with one another and with the walls of the container.
5. No energy is lost by the collision of a gas particle with another gas particle or with the walls of the container. All collisions are perfectly elastic.
6. The average kinetic energy for particles is the same for all gases at the same temperature, and its value is directly proportional to the Kelvin temperature.

The kinetic energy (KE) of a particle is one-half its mass times its velocity squared. It is expressed by the equation

$$KE = \frac{1}{2}mv^2$$

where m is the mass and v is the velocity of the particle.

All gases have the same kinetic energy at the same temperature. Therefore, from the kinetic energy equation we can see that, if we compare the velocities of the molecules of two gases, the lighter molecules will have a greater velocity than the heavier ones. For example, calculations show that the velocity of a hydrogen molecule is four times the velocity of an oxygen molecule.

Due to their molecular motion, gases have the property of **diffusion**, the ability of two or more gases to mix spontaneously until they form a uniform mixture. The diffusion of gases may be illustrated by use of the apparatus shown in Figure 12.1. Two large flasks, one containing reddish brown bromine vapors and the other dry air, are connected by a side tube. When the stopcock between the flasks is opened, the bromine and air will diffuse into each other. After standing awhile, both flasks will contain bromine and air.

diffusion

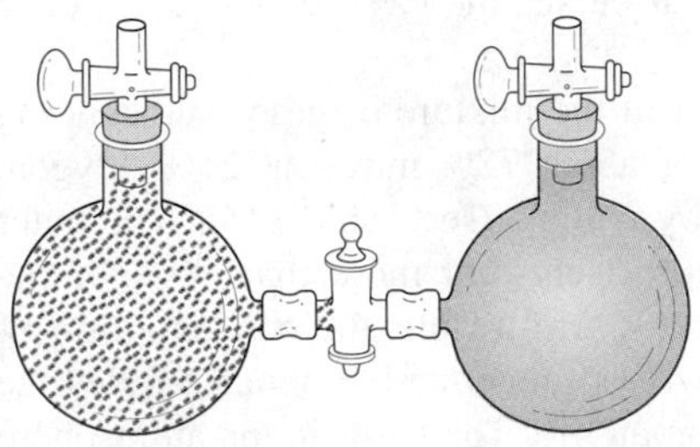

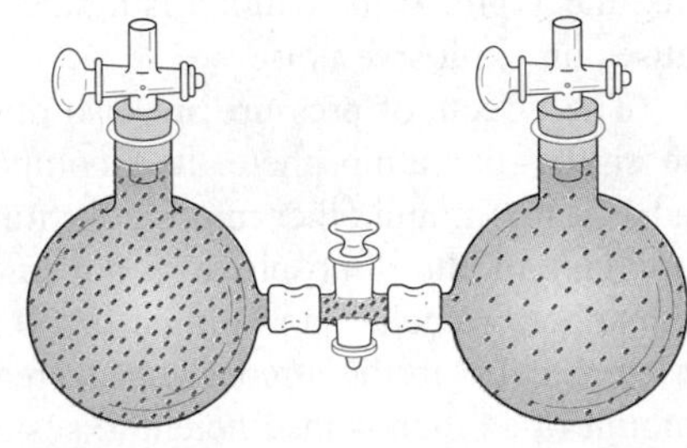

FIGURE 12.1
Diffusion of gases. When the stopcock between the two flasks is opened, colored bromine molecules can be seen diffusing into the flask containing air.

effusion

If we put a pinhole in a balloon, the gas inside will effuse or flow out of the balloon. **Effusion** is a process by which gas molecules pass through a very small orifice (opening) from a container at higher pressure to one at lower pressure.

Graham's law of effusion

Thomas Graham (1805–1869), a Scottish chemist, observed that the rate of effusion was dependent on the density of a gas. This observation led to **Graham's law of effusion**.

The rates of effusion of two gases at the same temperature and pressure are inversely proportional to the square roots of their densities or molar masses:

$$\frac{\text{rate of effusion of gas } A}{\text{rate of effusion of gas } B} = \sqrt{\frac{dB}{dA}} = \sqrt{\frac{\text{molar mass } B}{\text{molar mass } A}}$$

A major application of Graham's law occurred during World War II with the separation of the isotopes of uranium-235 (U-235) and uranium-238 (U-238). Naturally occurring uranium consists of 0.7% U-235, 99.3% U-238, and a trace of U-234. However, only U-235 is useful as fuel for nuclear reactors and atomic bombs, so the concentration of U-235 in the mixture of isotopes had to be increased.

Uranium was first changed to uranium hexafluoride, UF_6, a white solid that readily goes into the gaseous state. The gaseous mixture of $^{235}UF_6$ and $^{238}UF_6$ was then allowed to effuse through porous walls. Although the effusion rate of the lighter gas is only slightly faster than that of the heavier one,

$$\frac{\text{effusion rate } ^{235}UF_6}{\text{effusion rate } ^{238}UF_6} = \sqrt{\frac{\text{molar mass } ^{238}UF_6}{\text{molar mass } ^{235}UF_6}} = \sqrt{\frac{352}{349}} = 1.0043$$

the separation and enrichment of U-235 was accomplished by subjecting the gaseous mixture to several thousand stages of effusion.

12.3 Measurement of Pressure of Gases

pressure

Pressure is defined as force per unit area. When a rubber balloon is inflated with air, it stretches and maintains its larger size because the pressure on the inside is greater than that on the outside. Pressure results from the collisions of gas molecules with the walls of the balloon (see Figure 12.2). When the gas is released, the force or pressure of the air escaping from the small neck propels the balloon in a rapid, irregular flight. If the balloon is inflated until it bursts, the gas escaping all at once causes an explosive noise.

The effects of pressure are also observed in the mixture of gases surrounding the earth—our atmosphere. It is composed of about 78% nitrogen, 21% oxygen, and 1% argon, and other minor constituents by volume (see Table 12.1). The outer boundary of the atmosphere is not known precisely, but more than 99% of the atmosphere is below an altitude of 20 miles (32 km). Thus, the concentration of gas molecules in the atmosphere decreases with altitude, and at about 4 miles the amount of oxygen is insufficient to sustain human life. The gases in the atmosphere

atmospheric pressure

exert a pressure known as **atmospheric pressure**. The pressure exerted by a gas depends on the number of molecules of gas present, the temperature, and the volume in which the gas is confined. Gravitational forces hold the atmosphere relatively close to the earth and prevent air molecules from flying off into outer space. Thus, the atmospheric pressure at any point is due to the mass of the atmosphere pressing downward at that point.

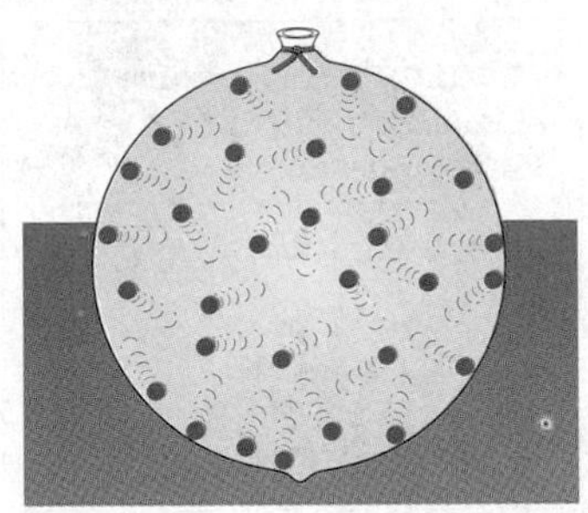

FIGURE 12.2
Here we see the pressure resulting from the collisions of gas molecules with the walls of the balloon. This pressure keeps the balloon inflated.

The pressure of the gases in the atmosphere can be measured with a **barometer**. A mercury barometer may be prepared by completely filling a long tube with pure, dry mercury and inverting the open end into an open dish of mercury. If the tube is longer than 760 mm, the mercury level will drop to a point at which the column of mercury in the tube is just supported by the pressure of the atmosphere. If the tube is properly prepared, a vacuum will exist above the mercury column. The weight of mercury, per unit area, is equal to the pressure of the atmosphere. The column of mercury is supported by the pressure of the atmosphere, and the height of the column is a measure of this pressure (see Figure 12.3). The mercury barometer was invented in 1643 by the Italian physicist E. Torricelli (1608–1647), for whom the unit of pressure *torr* was named.

barometer

1 atmosphere

Air pressure is measured and expressed in many units. The standard atmospheric pressure, or simply **1 atmosphere** (atm), is the pressure exerted by a column of mercury 760 mm high at a temperature of 0°C. The normal pressure of the atmosphere at sea level is 1 atm or 760 torr or 760 mm Hg. The SI unit for pressure is the pascal (Pa), where 1 atm = 101,325 Pa or 101.3 kPa. Other units for expressing

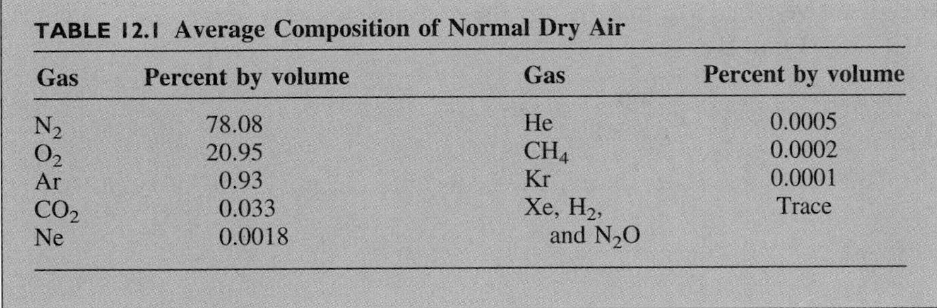

TABLE 12.1 Average Composition of Normal Dry Air

Gas	Percent by volume	Gas	Percent by volume
N_2	78.08	He	0.0005
O_2	20.95	CH_4	0.0002
Ar	0.93	Kr	0.0001
CO_2	0.033	Xe, H_2, and N_2O	Trace
Ne	0.0018		

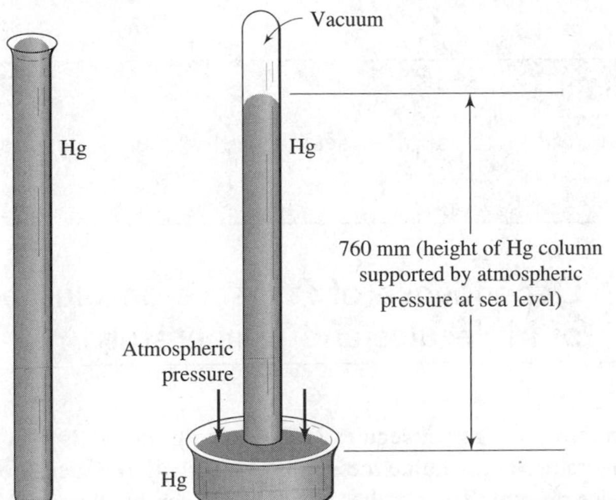

FIGURE 12.3
Preparation of a mercury barometer. The full tube of mercury at the left is inverted and placed in a dish of mercury.

TABLE 12.2 Pressure Units Equivalent to 1 Atmosphere

1 atm
760 torr
760 mm Hg
76 cm Hg
101,325 kPa
1013 mbar
29.9 in. Hg
14.7 lb/in.^{2}

pressure are inches of mercury, centimeters of mercury, the millibar (mbar), and pounds per square inch (lb/in.^{2} or psi). The meteorologist uses inches of mercury in reporting atmospheric pressure. The values of these units equivalent to 1 atm are summarized in Table 12.2 (1 atm ≡ 760 torr ≡ 760 mm Hg ≡ 76 cm Hg ≡ 101,325 Pa ≡ 1013 mbar ≡ 29.9 in. Hg ≡ 14.7 lb/in.^{2}). (The symbol ≡ means *identical with.*)

Atmospheric pressure varies with altitude. The average pressure at Denver, Colorado, 1.61 km (1 mile) above sea level, is 630 torr (0.83 atm). Atmospheric pressure is 0.5 atm at about 5.5 km (3.4 miles) altitude.

Pressure is often measured by reading the height of a mercury column in millimeters on a barometer. Thus pressure may be recorded as mm Hg. But in many applications the torr is superseding mm Hg as a unit of pressure. In problems dealing with gases it is necessary to make interconversions among the various pressure units. Since atm, torr, and mm Hg are common pressure units, we use illustrative problems involving all three of these units.

$$1 \text{ atm} = 760 \text{ torr} = 760 \text{ mm Hg}$$

Example 12.1 The average atmospheric pressure at Walnut, California, is 740. mm Hg. Calculate this pressure in (a) torr and (b) atmospheres.

Solution This problem can be solved using conversion factors, relating one unit of pressure to another.

(a) To convert mm Hg to torr, use the conversion factor 760 torr/760 mm Hg (1 torr/1 mm Hg):

$$740. \cancel{\text{mm Hg}} \times \frac{1 \text{ torr}}{1 \cancel{\text{mm Hg}}} = 740. \text{ torr}$$

(b) To convert mm Hg to atm, use the conversion factor 1 atm/760. mm Hg:

$$740. \cancel{\text{mm Hg}} \times \frac{1 \text{ atm}}{760. \cancel{\text{mm Hg}}} = 0.974 \text{ atm}$$

Practice 12.1

A barometer reads 1.12 atm. Calculate the corresponding pressure in (a) torr and (b) mm Hg.

12.4 Dependence of Pressure on Number of Molecules and Temperature

Pressure is produced by gas molecules colliding with the walls of a container. At a specific temperature and volume the number of collisions depends on the number of gas molecules present. The number of collisions can be increased by increasing the number of gas molecules present. If we double the number of molecules, the frequency of collisions and the pressure should double. We find, for an ideal gas,

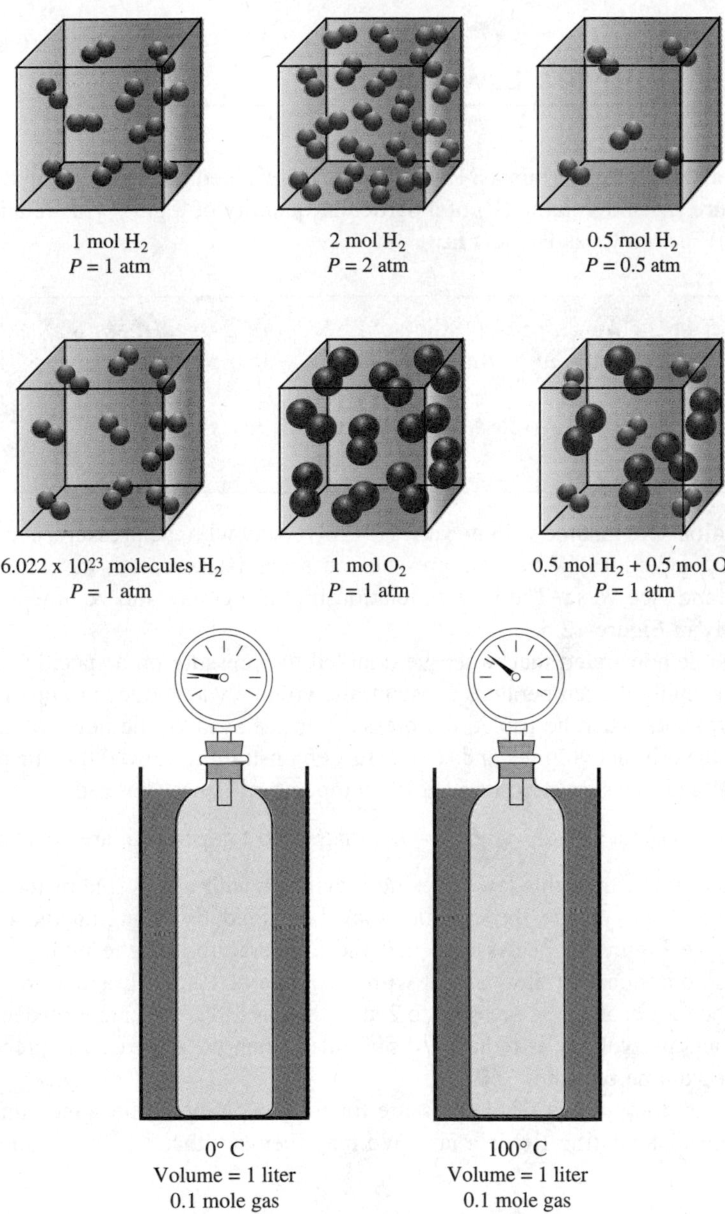

FIGURE 12.4
The pressure exerted by a gas is directly proportional to the number of molecules present. In each case shown, the volume is 22.4 L and the temperature is 0°C.

FIGURE 12.5
The pressure of a gas in a fixed volume increases with increasing temperature. The increased pressure is due to more frequent and more energetic collisions of the gas molecules with the walls of the container at the higher temperature.

kept constant, the pressure is directly proportional to the number of moles or molecules of gas present. Figure 12.4 illustrates this concept.

A good example of this molecule–pressure relationship may be observed in an ordinary cylinder of compressed gas that is equipped with a pressure gauge. When the valve is opened, gas escapes from the cylinder. The volume of the cylinder is constant, and the decrease in quantity (moles) of gas is registered by a drop in pressure indicated on the gauge.

The pressure of a gas in a fixed volume also varies with temperature. When the temperature is increased, the kinetic energy of the molecules increases, causing more frequent and more energetic collisions of the molecules with the walls of the container. This increase in collision frequency and energy results in a pressure increase (see Figure 12.5).

Robert Boyle (1627–1691).

12.5 Boyle's Law

Through a series of experiments, Robert Boyle determined the relationship between the pressure (P) and volume (V) of a particular quantity of a gas. This relationship of P and V is known as **Boyle's law**.

At constant temperature (T), the volume (V) of a fixed mass of a gas is inversely proportional to the pressure (P), which may be expressed as:

$$V \propto \frac{1}{P} \quad \text{(mass and temperature are constant)}$$

This equation says that the volume varies ($\propto$) inversely with the pressure, at constant mass and temperature. When the pressure on a gas is increased, its volume will decrease, and vice versa. The inverse relationship of pressure and volume is shown graphically in Figure 12.6.

Boyle's law

Boyle demonstrated that, when he doubled the pressure on a specific quantity of a gas, keeping the temperature constant, the volume was reduced to one-half the original volume; when he tripled the pressure on the system, the new volume was one-third the original volume; and so on. His demonstration showed that the product of volume and pressure is constant if the temperature is not changed:

$$PV = \text{constant} \qquad \text{or} \qquad PV = k \quad \text{(mass and temperature are constant)}$$

Let us demonstrate this law by using a cylinder with a movable piston so that the volume of gas inside the cylinder may be varied by changing the external pressure (see Figure 12.7). Assume that the temperature and the number of gas molecules do not change. Let us start with a volume of 1000 mL and a pressure of 1 atm. When we change the pressure to 2 atm, the gas molecules are crowded closer together, and the volume is reduced to 500 mL. When we increase the pressure to 4 atm, the volume becomes 250 mL.

Note that the product of the pressure times the volume is the same number in each case, substantiating Boyle's law. We may then say that

$$P_1V_1 = P_2V_2$$

where P_1V_1 is the pressure–volume product at one set of conditions, and P_2V_2 is the product at another set of conditions. In each case the new volume may be calculated by multiplying the starting volume by a ratio of the two pressures involved. Of course, the ratio of pressures used must reflect the direction in which the volume should change. When the pressure is changed from 1 atm to 2 atm, the ratio to be used is 1 atm/2 atm. Now we can verify the results given in Figure 12.7:

(a) Starting volume, 1000 mL; pressure change, 1 atm $\longrightarrow$ 2 atm

$$1000 \text{ mL} \times \frac{1 \cancel{\text{atm}}}{2 \cancel{\text{atm}}} = 500 \text{ mL}$$

(b) Starting volume, 1000 mL; pressure change, 1 atm $\longrightarrow$ 4 atm

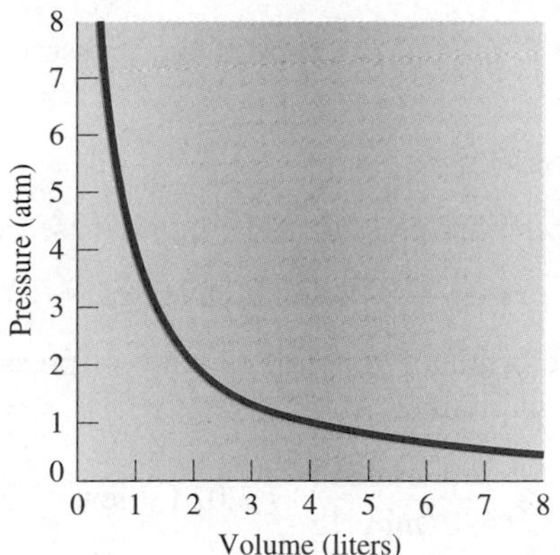

FIGURE 12.6
Graph of pressure versus volume showing the inverse *PV* relationship of an ideal gas.

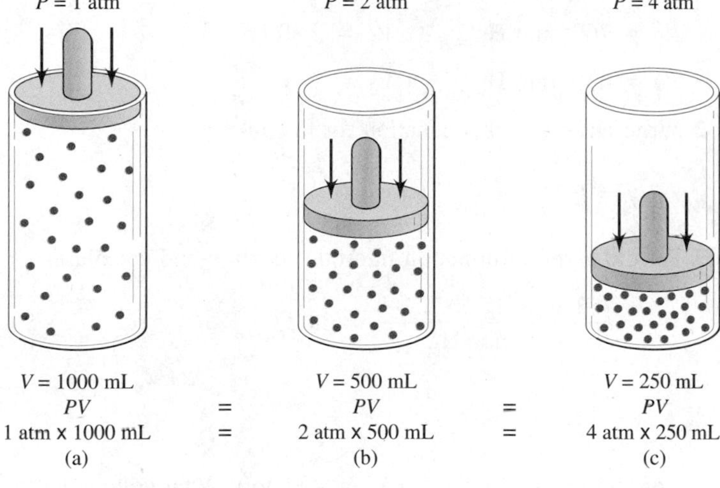

FIGURE 12.7
The effect of pressure on the volume of a gas.

$$1000 \text{ mL} \times \frac{1 \text{ atm}}{4 \text{ atm}} = 250 \text{ mL}$$

(c) Starting volume, 500 mL; pressure change, 2 atm $\longrightarrow$ 4 atm

$$500 \text{ mL} \times \frac{2 \text{ atm}}{4 \text{ atm}} = 250 \text{ mL}$$

In summary, a change in the volume of a gas due to a change in pressure can be calculated by multiplying the original volume by a ratio of the two pressures. If the pressure is increased, the ratio should have the smaller pressure in the numerator and the larger pressure in the denominator. If the pressure is decreased, the larger pressure should be in the numerator and the smaller pressure in the denominator.

new volume = original volume $\times$ ratio of pressures

The following examples are problems based on Boyle's law. If no mention is made of temperature, assume that it remains constant.

Example 12.2 What volume will 2.50 L of a gas occupy if the pressure is changed from 760. mm Hg to 630. mm Hg?

Solution

Method A: Conversion Factors

Step 1 Determine whether pressure is being increased or decreased:

pressure decreases ⟶ volume increases

Step 2 Multiply the original volume by a ratio of pressures that will result in an increase in volume:

$$V = 2.50\text{ L} \times \frac{760.\ \cancel{\text{mm Hg}}}{630.\ \cancel{\text{mm Hg}}} = 3.02\text{ L (new volume)}$$

Decide which method is the best for you and stick with it.

Method B: Algebraic Equation

Step 1 Organize the given information:

$P_1 = 760.$ mm Hg $\qquad V_1 = 2.50$ L

$P_2 = 630.$ mm Hg $\qquad V_2 = ?$

Step 2 Write and solve this equation for the unknown:

$$P_1V_1 = P_2V_2 \qquad V_2 = \frac{P_1V_1}{P_2}$$

Step 3 Put the given information into this equation and calculate:

$$V_2 = \frac{760.\ \cancel{\text{mm Hg}} \times 2.50\text{ L}}{630.\ \cancel{\text{mm Hg}}} = 3.02\text{ L}$$

Example 12.3 A given mass of hydrogen occupies 40.0 L at 700. torr. What volume will it occupy at 5.00 atm pressure?

Solution

Method A: Conversion Factors

Step 1 Determine whether the pressure is being increased or decreased. Notice that in order to compare the values the units must be the same.

$$700.\ \cancel{\text{torr}} \times \frac{1\text{ atm}}{760\ \cancel{\text{torr}}} = 0.921\text{ atm}$$

pressure increases ⟶ volume decreases

Step 2 Multiply the original volume by a ratio of pressures that will result in a decrease in volume:

$$V = 40.0\text{ L} \times \frac{0.921\ \cancel{\text{atm}}}{5.00\ \cancel{\text{atm}}} = 7.37\text{ L}$$

Method B: Algebraic Equation

Step 1 Organize the given information. Remember to make the pressure units the same.

$P_1 = 700.\text{ torr} = 0.921\text{ atm} \qquad V_1 = 40.0\text{ L}$

$P_2 = 5.00\text{ atm} \qquad V_2 = ?$

Step 2 Write and solve this equation for the unknown:

$$P_1V_1 = P_2V_2 \qquad V_2 = \frac{P_1V_1}{P_2}$$

Step 3 Put the given information into this equation and calculate:

$$V_2 = \frac{0.921\text{ atm} \times 40.0\text{ L}}{5.00\text{ atm}} = 7.37\text{ L}$$

Example 12.4

A gas occupies a volume of 200. mL at 400. torr pressure. To what pressure must the gas be subjected in order to change the volume to 75.0 mL?

Method A: Conversion Factors

Step 1 Determine whether volume is being increased or decreased:

volume decreases $\longrightarrow$ pressure increases

Step 2 Multiply the original pressure by a ratio of volumes that will result in an increase in pressure:

new pressure = original pressure × ratio of volumes

$$P = 400.\text{ torr} \times \frac{200.\text{ mL}}{75.0\text{ mL}} = 1067\text{ torr} \quad \text{or} \quad 1.07 \times 10^3\text{ torr (new pressure)}$$

Method B: Algebraic Equation

Step 1 Organize the given information. Remember to make units the same.

$P_1 = 400.\text{ torr} \qquad V_1 = 200.\text{ mL}$

$P_2 = ? \qquad V_2 = 75.0\text{ mL}$

Step 2 Write and solve this equation for the unknown:

$$P_1V_1 = P_2V_2 \qquad P_2 = \frac{P_1V_1}{V_2}$$

Step 3 Put the given information into the equation and calculate:

$$P_2 = \frac{P_1V_1}{V_2} = \frac{400.\text{ torr} \times 200.\text{ mL}}{75.0\text{ mL}} = 1.07 \times 10^3\text{ torr}$$

Practice 12.2

A gas occupies a volume of 3.86 L at 0.750 atm. At what pressure will the volume be 4.86 L?

12.6 Charles' Law

The effect of temperature on the volume of a gas was observed in about 1787 by the French physicist J. A. C. Charles (1746–1823). Charles found that various gases expanded by the same fractional amount when they underwent the same change in temperature. Later it was found that if a given volume of any gas initially at 0°C was cooled by 1°C, the volume decreased by $\frac{1}{273}$; if cooled by 2°C, it decreased by $\frac{2}{273}$; if cooled by 20°C, by $\frac{20}{273}$; and so on. Since each degree of cooling reduced the volume by $\frac{1}{273}$, it was apparent that any quantity of any gas would have zero volume if it could be cooled to −273°C. Of course, no real gas can be cooled to −273°C for the simple reason that it would liquefy before that temperature is reached. However, −273°C (more precisely −273.15°C) is referred to as **absolute zero**; this temperature is the zero point on the Kelvin (absolute) temperature scale—the temperature at which the volume of an ideal, or perfect, gas would become zero.

absolute zero

The volume–temperature relationship for methane is shown graphically in Figure 12.8. Experimental data show the graph to be a straight line that, when extrapolated, crosses the temperature axis at −273.15°C, or absolute zero. This is characteristic for all gases.

Charles' law

In modern form, **Charles' law** is as follows:

At *constant pressure* the volume of a fixed mass of any gas is directly proportional to the absolute temperature, which may be expressed as:

$$V \propto T \quad (P \text{ is constant})$$

A capital T is usually used for absolute temperature, K, and a small t for °C.

Mathematically this states that the volume of a gas varies directly with the absolute temperature when the pressure remains constant. In equation form Charles' law may be written as

$$V = kT \quad \text{or} \quad \frac{V}{T} = k \quad \text{(at constant pressure)}$$

where k is a constant for a fixed mass of the gas. If the absolute temperature of a gas is doubled, the volume will double.

To illustrate, let us return to the gas cylinder with the movable or free-floating piston (see Figure 12.9). Assume that the cylinder labeled (a) contains a quantity of gas and the pressure on it is 1 atm. When the gas is heated, the molecules move faster, and their kinetic energy increases. This action should increase the number of collisions per unit of time and therefore increase the pressure. However, the increased internal pressure will cause the piston to rise to a level at which the internal and external pressures again equal 1 atm, as we see in cylinder (b). The net result is an increase in volume due to an increase in temperature.

Don't forget to change temperature to Kelvin when doing gas law calculations.

Another equation relating the volume of a gas at two different temperatures is

$$\frac{V_1}{T_1} = \frac{V_2}{T_2} \quad (\text{constant } P)$$

where V_1 and T_1 are one set of conditions and V_2 and T_2 are another set of conditions.

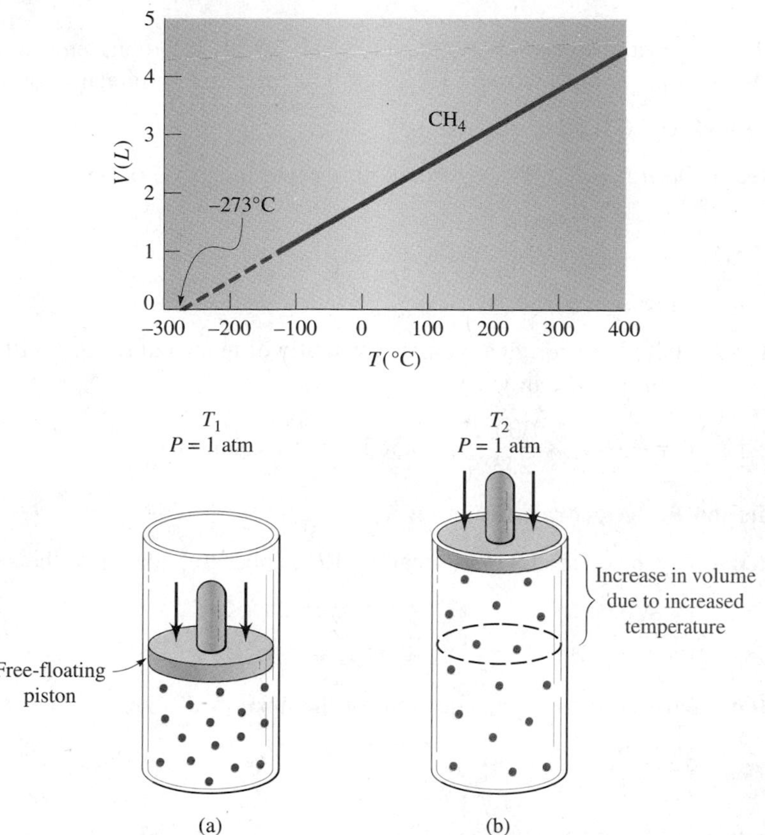

FIGURE 12.8
Volume–temperature relationship of methane (CH_4). Extrapolated portion of the graph is shown by the broken line.

FIGURE 12.9
The effect of temperature on the volume of a gas. The gas in cylinder (a) is heated from T_1 to T_2. With the external pressure constant at 1 atm, the free-floating piston rises, resulting in an increased volume, as shown in cylinder (b).

A simple experiment showing the variation of the volume of a gas with temperature is illustrated in Figure 12.10. A balloon is placed in a beaker and liquid N_2 is poured over it. The volume is reduced, as shown by the collapse of the balloon; when the balloon is removed from the liquid N_2, the gas expands and the balloon increases in size.

(a)

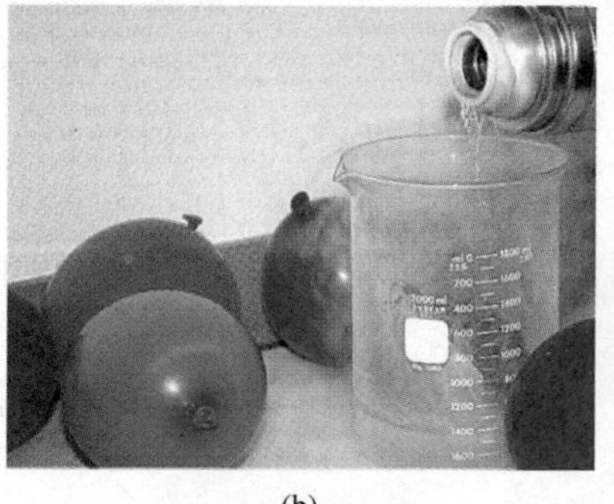

(b)

(c)

FIGURE 12.10
The air-filled balloons in (a) are placed in liquid nitrogen (b). The volume of the air decreases tremendously at this temperature. In (c) the balloons are removed from the beaker and are beginning to return to their original volume as they warm back to room temperature.

Example 12.5 Three liters of hydrogen at $-20.°C$ are allowed to warm to a room temperature of 27°C. What is the volume at room temperature if the pressure remains constant?

Solution

Method A: Conversion Factors

Step 1 Determine whether temperature is being increased or decreased.

Remember temperature must be changed to Kelvin in gas problems. Note that we use 273 to convert instead of 273.15 since our original measurements are to the nearest degree.

$$-20.°C + 273 = 253\ K$$
$$27°C + 273 = 300.\ K$$

temperature increases ⟶ volume increases

Step 2 Multiply the original volume by a ratio of temperatures that will result in an increase in volume.

$$V = 3.00\ L \times \frac{300.\ \cancel{K}}{253\ \cancel{K}} = 3.56\ L \quad \text{(new volume)}$$

Method B: Algebraic Equation

Step 1 Organize the given information. Remember to make units the same.

$V_1 = 3.00\ L$ $\qquad T_1 = 20.°C = 253\ K$

$V_2 = ?$ $\qquad T_2 = 27°C = 300.\ K$

Step 2 Write and solve the equation for the unknown:

$$\frac{V_1}{T_1} = \frac{V_2}{T_2} \qquad V_2 = \frac{V_1 T_2}{T_1}$$

Step 3 Put the given information into the equation and calculate:

$$V_2 = \frac{V_1 T_2}{T_1} = \frac{3.00\ L \times 300.\ \cancel{K}}{253\ \cancel{K}} = 3.56\ L$$

Example 12.6 If 20.0 L of oxygen are cooled from 100°C to 0°C, what is the new volume?

Solution Since no mention is made of pressure, assume that pressure does not change.

Method A: Conversion Factors

Step 1 Change °C to K:

$$100°C + 273 = 373\ K$$
$$0°C + 273 = 273\ K$$

Step 2 The ratio of temperature to be used is 273 K/373 K, because the final volume should be smaller than the original volume. The calculation is

$$V = 20.0\ L \times \frac{273\ \cancel{K}}{373\ \cancel{K}} = 14.6\ L \quad \text{(new volume)}$$

Method B: Algebraic Equation

Step 1 Organize the given information. Remember to make units coincide.

$V_1 = 20.0\ L$ $\qquad T_1 = 100°C = 373\ K$

$V_2 = ?$ $\qquad T_2 = 0°C = 273\ K$

Step 2 Write and solve equation for the unknown:

$$\frac{V_1}{T_1} = \frac{V_2}{T_2} \qquad V_2 = \frac{V_1 T_2}{T_1}$$

Step 3 Put the given information into the equation and calculate:

$$V_2 = \frac{V_1 T_2}{T_1} = \frac{20.0\ \text{L} \times 273\ \cancel{\text{K}}}{373\ \cancel{\text{K}}} = 14.6\ \text{L}$$

Practice 12.3

A 4.50-L container of nitrogen at 28.0°C is heated to 56.0°C. Assuming the volume of the container can vary, what is the new volume of the gas?

Joseph Louis Gay-Lussac (1778–1850).

12.7 Gay-Lussac's Law

J. L. Gay-Lussac was a French chemist involved in the study of volume relationships of gases. The three variables (pressure, P; volume, V; and temperature, T) are needed to describe a fixed amount of a gas. Boyle's law, $PV = k$, relates pressure and volume at constant temperature; Charles' law, $V = kT$, relates volume and temperature at constant pressure. A third relationship involving pressure and temperature at constant volume is a modification of Charles' law and is sometimes called **Gay-Lussac's law**:

Gay-Lussac's law

The pressure of a fixed mass of a gas, at constant volume, is directly proportional to the Kelvin temperature:

$$P = kT \quad \text{(at constant volume)} \qquad \frac{P_1}{T_1} = \frac{P_2}{T_2}$$

Example 12.7

The pressure of a container of helium is 650. torr at 25°C. If the sealed container is cooled to 0°C, what will the pressure be?

Solution

Method A: Conversion Factors

Step 1 Determine whether temperature is being increased or decreased.

temperature decreases ⟶ pressure decreases

Step 2 Multiply the original pressure by a ratio of Kelvin temperatures that will result in a decrease in pressure:

$$650.\ \text{torr} \times \frac{273\ \cancel{\text{K}}}{298\ \cancel{\text{K}}} = 595\ \text{torr}$$

Method B: Algebraic Equation

Step 1 Organize the given information. Remember to make units the same.

$P_1 = 650.\text{ torr}$ $\quad T_1 = 25°\text{C} = 298\text{ K}$

$P_2 = ?$ $\quad T_2 = 0°\text{C} = 273\text{ K}$

Step 2 Write and solve equation for the unknown:

$$\frac{P_1}{T_1} = \frac{P_2}{T_2} \qquad P_2 = \frac{P_1 T_2}{T_1}$$

Step 3 Put given information into equation and calculate:

$$P_2 = \frac{650.\text{ torr} \times 273\text{ K}}{298\text{ K}} = 595\text{ torr}$$

Practice 12.4

A gas cylinder contains 40.0 L of gas at 45.0°C and has a pressure of 650. torr. What will the pressure be if the temperature is changed to 100.°C?

We may summarize the effects of changes in pressure, temperature, and quantity of a gas as follows:

1. In the case of a constant volume,
   (a) when the temperature is increased, the pressure increases.
   (b) when the quantity of a gas is increased, the pressure increases (T remaining constant).
2. In the case of a variable volume,
   (a) when the external pressure is increased, the volume decreases (T remaining constant).
   (b) when the temperature of a gas is increased, the volume increases (P remaining constant).
   (c) when the quantity of a gas is increased, the volume increases (P and T remaining constant).

12.8 Standard Temperature and Pressure

standard conditions
standard temperature and pressure (STP)

In order to compare volumes of gases, common reference points of temperature and pressure were selected and called **standard conditions** or **standard temperature and pressure** (abbreviated **STP**). Standard temperature is 273.15 K (0°C), and standard pressure is 1 atm or 760 torr or 760 mm Hg or 101.325 kPa. For purposes of comparison, volumes of gases are usually changed to STP conditions.

standard temperature = 273.15 K or 0°C
standard pressure = 1 atm or 760 torr or 760 mm Hg or 101.325 kPa

In this text we'll use 273 K for temperature conversions and calculations. Check with your instructor for rules in your class.

12.9 Combined Gas Laws: Simultaneous Changes in Pressure, Volume, and Temperature

When temperature and pressure change at the same time, the new volume may be calculated by multiplying the initial volume by the correct ratios of both pressure and temperature, as follows:

$$\text{final volume} = \text{initial volume} \times \begin{pmatrix}\text{ratio of}\\ \text{pressures}\end{pmatrix} \times \begin{pmatrix}\text{ratio of}\\ \text{temperatures}\end{pmatrix}$$

This equation combines Boyle's and Charles' laws, and the same considerations for the pressure and temperature ratios should be used in the calculation. The four possible variations are as follows:

1. Both T and P cause an increase in volume.
2. Both T and P cause a decrease in volume.
3. T causes an increase and P causes a decrease in volume.
4. T causes a decrease and P causes an increase in volume.

The P, V, and T relationships for a given mass of any gas, in fact, may be expressed as a single equation, $PV/T = k$. For problem solving this equation is usually written

$$\frac{P_1V_1}{T_1} = \frac{P_2V_2}{T_2}$$

where P_1, V_1, and T_1 are the initial conditions and P_2, V_2, and T_2 are the final conditions.

This equation can be solved for any one of the six variables and is useful in dealing with the pressure–volume–temperature relationships of gases. Note that when T is constant ($T_1 = T_2$), Boyle's law is represented; when P is constant ($P_1 = P_2$), Charles' law is represented; and when V is constant ($V_1 = V_2$), Gay-Lussac's law is represented.

Note, in the examples below, the use of 273 K does not change the number of significant figures in the temperature.
The converted temperature is expressed to the same precision as the original measurement.

Example 12.8

Given 20.0 L of ammonia gas at 5°C and 730. torr, calculate the volume at 50.°C and 800. torr.

Solution

Step 1 Organize the given information, putting temperatures in Kelvin:

$P_1 = 730.$ torr $\qquad P_2 = 800.$ torr

$V_1 = 20.0$ L $\qquad V_2 = ?$

$T_1 = 5°C = 278$ K $\qquad T_2 = 50.°C = 323$ K

Method A: Conversion Factors

Step 2 Set up ratios of T and P:

$$T \text{ ratio} = \frac{323 \text{ K}}{278 \text{ K}} \quad \text{(increase in } T \text{ should increase } V\text{)}$$

$$P \text{ ratio} = \frac{730. \text{ torr}}{800. \text{ torr}} \quad \text{(increase in } P \text{ should decrease } V\text{)}$$

Step 3 Multiply the original pressure by the ratios:

$$V_2 = 20.0 \text{ L} \times \frac{730. \cancel{\text{torr}}}{800. \cancel{\text{torr}}} \times \frac{323 \cancel{\text{K}}}{278 \cancel{\text{K}}} = 21.2 \text{ L}$$

Method B: Algebraic Equation

Step 2 Write and solve the equation for the unknown. Solve

$$\frac{P_1V_1}{T_1} = \frac{P_2V_2}{T_2}$$

for V_2 by multiplying both sides of the equation by T_2/P_2 and rearranging to obtain

$$V_2 = \frac{V_1P_1T_2}{P_2T_1}$$

Step 3 Put the given information into the equation and calculate:

$$V_2 = \frac{20.0 \text{ L} \times 730. \cancel{\text{torr}} \times 323 \cancel{\text{K}}}{800. \cancel{\text{torr}} \times 278 \cancel{\text{K}}} = 21.2 \text{ L}$$

Example 12.9 To what temperature (°C) must 10.0 L of nitrogen at 25°C and 700. torr be heated in order to have a volume of 15.0 L and a pressure of 760. torr?

Solution

Step 1 Organize the given information, putting temperatures in Kelvin:

$P_1 = 700.$ torr $\qquad P_2 = 760.$ torr

$V_1 = 10.0$ L $\qquad V_2 = 15.0$ L

$T_1 = 25°\text{C} = 298$ K $\qquad T_2 = ?$

Method A: Conversion Factors

Step 2 Set up ratios of V and P.

$$P \text{ ratio} = \frac{760. \text{ torr}}{700. \text{ torr}} \quad \text{(increase in } P \text{ should increase } T\text{)}$$

$$V \text{ ratio} = \frac{15.0 \text{ L}}{10.0 \text{ L}} \quad \text{(increase in } V \text{ should increase } T\text{)}$$

Step 3 Multiply the original temperature by the ratios:

$$T_2 = 298 \text{ K} \times \frac{760. \cancel{\text{torr}} \times 15.0 \cancel{\text{L}}}{700. \cancel{\text{torr}} \times 10.0 \cancel{\text{L}}} = 485 \text{ K}$$

Method B: Algebraic Equation

Step 2 Write and solve the equation for the unknown:

$$\frac{P_1V_1}{T_1} = \frac{P_2V_2}{T_2} \qquad T_2 = \frac{T_1P_2V_2}{P_1V_1}$$

Step 3 Put the given information into the equation and calculate:

$$T_2 = \frac{298\text{ K} \times 760.\text{ torr} \times 15.0\text{ L}}{700.\text{ torr} \times 10.0\text{ L}} = 485\text{ K}$$

In either method, since the problem asks for °C, we must subtract 273 from the Kelvin answer:

485 K − 273 = 212°C

Example 12.10

The volume of a gas-filled balloon is 50.0 L at 20.°C and 742 torr. What volume will it occupy at standard temperature and pressure (STP)?

Solution

Step 1 Organize the given information, putting temperatures in Kelvin.

P_1 = 742 torr $\qquad$ P_2 = 760. torr (standard pressure)

V_1 = 50.0 L $\qquad$ V_2 = ?

T_1 = 20.°C = 293 K $\qquad$ T_2 = 273 K (standard temperature)

Method A: Conversion Factors

Step 2 Set up ratios of T and P:

$$T \text{ ratio} = \frac{273\text{ K}}{293\text{ K}} \quad \text{(decrease in } T \text{ should decrease } V\text{)}$$

$$P \text{ ratio} = \frac{742\text{ torr}}{760.\text{ torr}} \quad \text{(increase in } P \text{ should decrease } V\text{)}$$

Step 3 Multiply the original volume by the ratios:

$$V_2 = 50.0\text{ L} \times \frac{273\text{ K}}{293\text{ K}} \times \frac{742\text{ torr}}{760.\text{ torr}} = 45.5\text{ L}$$

Method B: Algebraic Equation

Step 2 Write and solve the equation for the unknown:

$$\frac{P_1V_1}{T_1} = \frac{P_2V_2}{T_2} \qquad V_2 = \frac{P_1V_1T_2}{P_2T_1}$$

Step 3 Put the given information into the equation and calculate:

$$V_2 = \frac{742\text{ torr} \times 50.0\text{ L} \times 273\text{ K}}{760.\text{ torr} \times 293\text{ K}} = 45.5\text{ L}$$

Messenger Molecules

Nitrogen monoxide, NO, and carbon monoxide, CO, are small, gaseous molecules, often considered toxic to the body and the environment because of their contribution to photochemical smog and acid rain. But now, biochemists have found that these molecules are important players in a newly discovered group of biological messengers called *neurotransmitter molecules.*

Traditional neurotransmitter molecules, such as acetylcholine, norepinephrine, and epinephrine, are biogenic amines. In the 1950s, scientists discovered that amino acids could signal nerve cells, and in the 1970s researchers produced another class of transmitter molecules known as peptides—short strings of amino acids commonly called *endorphins* and *enkephalins* (associated with "runner's high"). In the late 1980s, nitrogen monoxide was discovered to be the first gas that could act as a neurotransmitter.

Until recently, neurotransmitter molecules were thought to be specific. That is, each neurotransmitter would fit into the next cell like a key in a lock. These transmitters were thought to be stored in tiny pouches in the cells where they were manufactured and released through transporters or channels only when needed. Nitrogen monoxide and carbon monoxide gases break the rules for explaining neurotransmission because gases are volatile and nonspecific. A gas freely diffuses into any nearby cell and must be made only when needed. The actions of gas molecules depend on their chemical properties rather than their molecular shape, as the lock-and-key model for neurotransmitters had suggested.

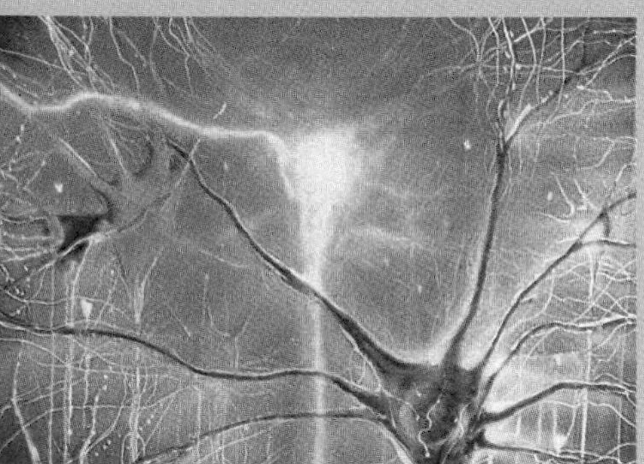

Incoming synaptic transmission ignites a neuron on impulse neural pathway.

Nitrogen monoxide has been used for nearly a century to dilate blood vessels and increase blood flow, thereby lowering blood pressure.

In the late 1980s, scientists discovered that biologically derived NO is an important signaling molecule in both the central and peripheral nervous system. Nitrogen monoxide was found to be a mediator of certain neurons that do not respond to the normal chemical neurotransmitters (acetylcholine or norepinephrine). These NO-sensitive neurons are in the cardiovascular, respiratory, digestive, and urogenital systems. One study suggests that nitrogen monoxide may be a way to treat impotence by mediating the relaxation of the smooth muscle of the major erectile tissue in the penis, causing erection. Current research indicates that nitrogen monoxide plays a role in regulating blood pressure, blood clotting, and neurotransmission, and it may play a role in the ability of the immune system to kill tumor cells.

Researchers at Johns Hopkins Medical School reasoned that if one gas could act as a transmitter molecule, others might also be found. They began searching for other gases that might exhibit this biological activity. Carbon monoxide was proposed as a possible transmitter because the enzyme used to make it is localized within specific parts of the brain, such as the olfactory neurons and those cells thought to play a role in long-term memory. The enzyme used to make an intracellular messenger was found in exactly the same locations. Researchers then showed that nerve cells made the intracellular messenger when stimulated with carbon monoxide. An inhibitor for carbon monoxide blocked the production of the same messenger molecule.

Dr. Charles Stevens, from Salk Institute in La Jolla, California, thinks carbon monoxide might be an important factor in understanding long-term memory. Researchers think that in order to lay down a memory, a nerve cell is signaled repeatedly in a process called *long-term potentiation*. An important feature of the potentiation process is that the cell receiving the signal sends a message back to the cell that first signaled it. This message stimulates the originating cell to release its neurotransmitter molecule more easily. Investigators have never been able to find the message that the receiving cell sends to the originating cell.

When nitrogen monoxide was discovered to be a neurotransmitter, scientists thought that it might be the return signal. But the enzymes necessary to produce nitrogen monoxide were not found in the cells where long-term potentiation occurs. The enzyme that makes carbon monoxide, however, is abundant in these cells, and so Stevens began testing his theory. He found that inhibitors for carbon monoxide prevented long-term potentiation, and when CO was blocked from these cells, memories that were already there were erased. Further research may uncover still other roles for these gaseous molecules.

Practice 12.5

15.00 L of gas at 45.0°C and 800. torr is heated to 400.°C, and the pressure changed to 300. torr. What is the new volume?

Practice 12.6

To what temperature must 5.00 L of oxygen at 50.°C and 600. torr be heated in order to have a volume of 10.0 L and a pressure of 800. torr?

12.10 Dalton's Law of Partial Pressures

If gases behave according to the Kinetic-Molecular Theory, there should be no difference in the pressure–volume–temperature relationships whether the gas molecules are all the same or different. This similarity in the behavior of gases is the basis for an understanding of **Dalton's law of partial pressures**:

Dalton's law of partial pressures

The total pressure of a mixture of gases is the sum of the partial pressures exerted by each of the gases in the mixture.

Each gas in the mixture exerts a pressure that is independent of the other gases present. These pressures are called **partial pressures**. Thus, if we have a mixture of three gases, *A*, *B*, and *C*, exerting partial pressures of 50 torr, 150 torr, and 400 torr, respectively, the total pressure will be 600 torr:

partial pressure

$$P_{\text{Total}} = p_A + p_B + p_C$$

$$P_{\text{Total}} = 50 \text{ torr} + 150 \text{ torr} + 400 \text{ torr} = 600 \text{ torr}$$

We can see an application of Dalton's law in the collection of insoluble gases over water. When prepared in the laboratory, oxygen is commonly collected by the downward displacement of water. Thus, the oxygen is not pure but is mixed with water vapor (see Figure 12.11). When the water levels are adjusted to the same height inside and outside the bottle, the pressure of the oxygen plus water vapor inside the bottle is equal to the atmospheric pressure:

$$P_{\text{atm}} = p_{O_2} + p_{H_2O}$$

To determine the amount of O_2 or any other gas collected over water, we must subtract the pressure of the water vapor from the total pressure of the gas. The vapor pressure of water at various temperatures is tabulated in Appendix II.

$$p_{O_2} = P_{\text{atm}} - p_{H_2O}$$

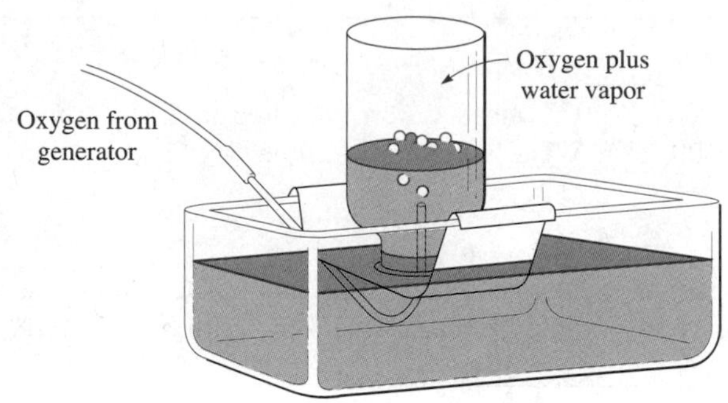

FIGURE 12.11
Oxygen collected over water. ▶

Example 12.11 A 500.-mL sample of oxygen was collected over water at 23°C and 760. torr. What volume will the dry O_2 occupy at 23°C and 760. torr? The vapor pressure of water at 23°C is 21.2 torr.

Solution To solve this problem, we must first determine the pressure of the oxygen alone, by subtracting the pressure of the water vapor present.

Step 1 Determine the pressure of dry O_2:

$$P_{\text{Total}} = 760.\ \text{torr} = p_{O_2} + p_{H_2O}$$

$$p_{O_2} = 760.\ \text{torr} - 21.2\ \text{torr} = 739\ \text{torr} \quad (\text{dry } O_2)$$

Step 2 Organize the given information:

$P_1 = 739$ torr $P_2 = 760.$ torr

$V_1 = 500.$ mL $V_2 = ?$

T is constant

Step 3 Solve as a Boyle's law problem:

$$V = \frac{500.\ \text{mL} \times 739\ \cancel{\text{torr}}}{760.\ \cancel{\text{torr}}} = 486\ \text{mL dry } O_2$$

Practice 12.7

Hydrogen gas was collected by downward displacement of water. A volume of 600.0 mL of gas was collected at 25.0°C and 740.0 torr. What volume will the dry hydrogen occupy at STP?

12.11 Avogadro's Law

Gay-Lussac's law of combining volumes

Early in the 19th century Gay-Lussac studied the volume relationships of reacting gases. His results, published in 1809, were summarized in a statement known as **Gay-Lussac's law of combining volumes**:

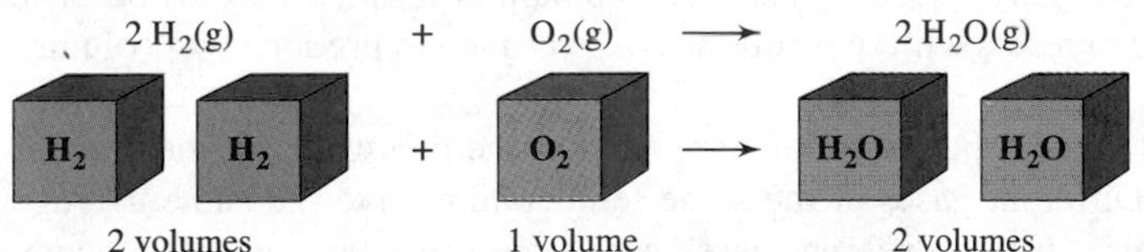

FIGURE 12.12
Gay-Lussac's law of combining volumes of gases applied to the reaction of hydrogen and oxygen. When measured at the same temperature and pressure, hydrogen and oxygen react in a volume ratio of 2:1.

When measured at the same temperature and pressure, the ratios of the volumes of reacting gases are small whole numbers.

Thus, H_2 and O_2 combine to form water vapor in a volume ratio of 2:1 (Figure 12.12); H_2 and Cl_2 react to form HCl in a volume ratio of 1:1; and H_2 and N_2 react to form NH_3 in a volume ratio of 3:1.

Two years later, in 1811, Amedeo Avogadro used the law of combining volumes of gases to make a simple but significant and far-reaching generalization concerning gases. **Avogadro's law** states:

Avogadro's law

Equal volumes of different gases at the same temperature and pressure contain the same number of molecules.

This law was a real breakthrough in understanding the nature of gases.

1. It offered a rational explanation of Gay-Lussac's law of combining volumes of gases and indicated the diatomic nature of such elemental gases as hydrogen, chlorine, and oxygen.
2. It provided a method for determining the molar masses of gases and for comparing the densities of gases of known molar mass (see Sections 12.12 and 12.13).
3. It afforded a firm foundation for the development of the Kinetic-Molecular Theory.

By Avogadro's law, equal volumes of hydrogen and chlorine contain the same number of molecules. On a volume basis, hydrogen and chlorine react thus:

hydrogen +	chlorine ⟶	hydrogen chloride
1 volume	1 volume	2 volumes

Therefore, hydrogen molecules react with chlorine molecules in a 1:1 ratio. Since two volumes of hydrogen chloride are produced, one molecule of hydrogen and one molecule of chlorine must produce two molecules of hydrogen chloride. Therefore, each hydrogen molecule and each chlorine molecule must be made up of two atoms. The coefficients of the balanced equation for the reaction give the correct ratios for volumes, molecules, and moles of reactants and products:

H_2 +	Cl_2 ⟶	2 HCl
1 volume	1 volume	2 volumes
1 molecule	1 molecule	2 molecules
1 mol	1 mol	2 mol

By like reasoning, oxygen molecules also must contain at least two atoms because one volume of oxygen reacts with two volumes of hydrogen to produce two volumes of water vapor.

The volume of a gas depends on the temperature, the pressure, and the number of gas molecules. Different gases at the same temperature have the same average kinetic energy. Hence, if two different gases are at the same temperature, occupy equal volumes, and exhibit equal pressures, each gas must contain the same number of molecules. This statement is true because systems with identical *PVT* properties can be produced only by equal numbers of molecules having the same average kinetic energy.

12.12 Mole–Mass–Volume Relationships of Gases

Because a mole contains 6.022×10^{23} molecules (Avogadro's number), a mole of any gas will have the same volume as a mole of any other gas at the same temperature and pressure. It has been experimentally determined that the volume occupied by a mole of any gas is 22.4 L at STP. This volume, 22.4 L, is known as the **molar volume** of a gas. The molar volume is a cube about 28.2 cm (11.1 in.) on a side. The molar masses of several gases, each occupying 22.4 L at STP, are shown in Figure 12.13.

molar volume

As with many constants, the molar volume is known more exactly to be 22.414 L. We use 22.4 L in our calculations since the extra figures do not often affect the result, given the other measurements in the calculation.

One mole of a gas occupies 22.4 L at STP.

The molar volume is useful for determining the molar mass of a gas or of substances that can be easily vaporized. If the mass and the volume of a gas at STP are known, we can calculate its molar mass. For example, 1 L of pure oxygen at STP has a mass of 1.429 g. The molar mass of oxygen may be calculated by

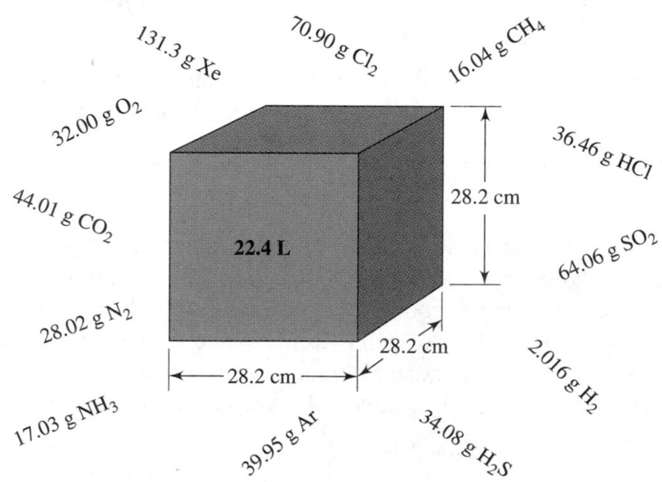

FIGURE 12.13 ▶ One mole of a gas occupies 22.4 L at STP. The mass given for each gas is the mass of 1 mol.

multiplying the mass of 1 L by 22.4 L/mol:

$$\frac{1.429\ \text{g}}{1\ \cancel{\text{L}}} \times \frac{22.4\ \cancel{\text{L}}}{1\ \text{mol}} = 32.00\ \text{g/mol} \quad \text{(molar mass)}$$

If the mass and volume are at other than standard conditions, we change the volume to STP and then calculate the molar mass.

Standard conditions apply only to pressure, temperature, and volume. Mass is not affected.

The molar volume, 22.4 L/mol, is used as a conversion factor to convert grams per liter to grams per mole (molar mass) and also to convert liters to moles. The two conversion factors are

$$\frac{22.4\ \text{L}}{1\ \text{mol}} \quad \text{and} \quad \frac{1\ \text{mol}}{22.4\ \text{L}}$$

These conversions must be done at STP except under certain special circumstances. Examples follow.

Example 12.12

If 2.00 L of a gas measured at STP have a mass of 3.23 g, what is the molar mass of the gas?

Solution

The unit of molar mass is g/mol; the conversion is from

$$\frac{\text{g}}{\text{L}} \longrightarrow \frac{\text{g}}{\text{mol}}$$

The starting amount is $\frac{3.23\ \text{g}}{2.00\ \text{L}}$. The conversion factor is $\frac{22.4\ \text{L}}{1\ \text{mol}}$.

The calculation is $\frac{3.23\ \text{g}}{2.00\ \cancel{\text{L}}} \times \frac{22.4\ \cancel{\text{L}}}{1\ \text{mol}} = 36.2\ \text{g/mol} \quad \text{(molar mass)}$

Example 12.13

Measured at 40°C and 630. torr, the mass of 691 mL of ethyl ether is 1.65 g. Calculate the molar mass of ethyl ether.

Solution

Step 1 Organize the given information, converting temperatures to Kelvin. Note that we must change to STP in order to determine molar mass.

$P_1 = 630.$ torr $P_2 = 760.$ torr

$V_1 = 691$ mL $V_2 = ?$

$T_1 = 313$ K (40.°C) $T_2 = 273$ K

Step 2 Use either the conversion factor method or the algebraic method and the combined gas law to correct the volume (V_2) to STP:

$$V_2 = \frac{691\ \text{mL} \times 273\ \cancel{\text{K}} \times 630.\ \cancel{\text{torr}}}{313\ \cancel{\text{K}} \times 760.\ \cancel{\text{torr}}} = 500.\ \text{mL} = 0.500\ \text{L} \quad \text{(at STP)}$$

Step 3 In the example, V_2 is the volume for 1.65 g of the gas, so we can now find the molar mass by converting g/L to g/mol:

$$\frac{1.65\ \text{g}}{0.500\ \cancel{\text{L}}} \times \frac{22.4\ \cancel{\text{L}}}{\text{mol}} = 73.9\ \text{g/mol}$$

Practice 12.8

A gas with a mass of 86 g occupies 5.00 L at 25°C and 3.00 atm pressure. What is the molar mass of the gas?

12.13 Density of Gases

The density, d, of a gas is its mass per unit volume, which is generally expressed in grams per liter as follows:

$$d = \frac{\text{mass}}{\text{volume}} = \frac{\text{g}}{\text{L}}$$

Because the volume of a gas depends on temperature and pressure, both should be given when stating the density of a gas. The volume of a solid or liquid is hardly affected by changes in pressure and is changed only slightly when the temperature is varied. Increasing the temperature from 0°C to 50°C will reduce the density of a gas by about 18% if the gas is allowed to expand, whereas a 50°C rise in the temperature of water (0°C $\longrightarrow$ 50°C) will change its density by less than 0.2%.

The density of a gas at any temperature and pressure can be determined by calculating the mass of gas present in 1 L. At STP, in particular, the density can be calculated by multiplying the molar mass of the gas by 1 mol/22.4 L:

$$d\text{ (at STP)} = \text{molar mass} \times \frac{1\text{ mol}}{22.4\text{ L}}$$

$$\text{molar mass} = d\text{ (at STP)} \times \frac{22.4\text{ L}}{1\text{ mol}}$$

Table 12.3 lists the densities of some common gases.

TABLE 12.3 Density of Common Gases at STP

Gas	Molar mass (g/mol)	Density (g/L at STP)	Gas	Molar mass (g/mol)	Density (g/L at STP)
H_2	2.016	0.0900	H_2S	34.08	1.52
CH_4	16.04	0.716	HCl	36.45	1.63
NH_3	17.03	0.760	F_2	38.00	1.70
C_2H_2	26.04	1.16	CO_2	44.01	1.96
HCN	27.03	1.21	C_3H_8	44.09	1.97
CO	28.01	1.25	O_3	48.00	2.14
N_2	28.02	1.25	SO_2	64.06	2.86
air	**(28.9)**	**(1.29)**	Cl_2	70.90	3.17
O_2	32.00	1.43			

Chemistry in Action

Physiological Effects of Pressure Changes

The human body has a variety of methods for coping with the changes in atmospheric pressure. As we travel to the mountains, fly in an airplane, or take a high-speed elevator to the top of a skyscraper, the pressure around us decreases. Our ears are sensitive to this because the eardrum (tympanic membrane) has air on both sides of it. The difference in pressure is relieved by yawning or moving the jaw to open the (Eustachian) tubes that connect the middle ear and throat and allow the pressure inside the eardrum to equalize with the outside.

Divers must also contend with the effects of pressure, most notably in body cavities containing air, such as the lungs, ears, and sinuses. Scuba divers do not experience a crushing effect of pressure at increased depths because the tank regulators deliver air at the same pressure as that of the surroundings. The diver must always breathe out regularly while ascending to the surface. Failure to do so may cause the lungs to expand, thus rupturing some of the alveoli, and resulting in loss of consciousness, brain damage, or heart attack. This is a clear application of Boyle's law.

Divers are also affected by consequences of **Henry's law,** which states that the amount of gas that will dissolve in a liquid varies directly with the pressure above the liquid. This means that during a dive the gases entering the lungs are absorbed into the blood to a greater extent than at the water's surface. If the diver returns too rapidly to the surface, the swift pressure reduction may cause dissolved gases to produce bubbles in the blood, resulting in a condition known as *decompression sickness* or "the bends." The only successful method of treatment for this involves the use of a decompression chamber to increase the pressure once again and slowly decompress the diver back to normal pressure.

In the medical field, *hyperbaric units* are used to treat patients who have cells starved for oxygen. In these units the whole room may be placed at high pressure (2 or 3 atm), and the entire staff as well as the patient undergo gradual compression and, following treatment, decompression. These units are widely used to treat carbon monoxide poisoning. Oxygen is dissolved directly into the plasma giving the tissues temporary relief from oxygen deprivation. Hyperbaric units are also effective in treating other problems such as skin grafts, severe thermal burns, and radiation tissue damage.

If a diver returns too quickly to the surface, the pressure reduction may produce bubbles in the blood—a condition known as "the bends."

Example 12.14

Calculate the density of Cl_2 at STP.

Solution

First calculate the molar mass of Cl_2. It is 70.90 g/mol. Since $d = \text{g/L}$, the conversion is

$$\frac{\text{g}}{\text{mol}} \longrightarrow \frac{\text{g}}{\text{L}}$$

The conversion factor is $\dfrac{1 \text{ mol}}{22.4 \text{ L}}$:

$$d = \frac{70.90 \text{ g}}{1 \text{ mol}} \times \frac{1 \text{ mol}}{22.4 \text{ L}} = 3.165 \text{ g/L}$$

Practice 12.9

The molar mass of a gas is 20. g/mol. Calculate the density of the gas at STP.

12.14 Ideal Gas Equation

We have used four variables in calculations involving gases: the volume, V; the pressure, P; the absolute temperature, T; and the number of molecules or moles, (abbreviated n). Combining these variables into a single expression, we obtain

$$V \propto \frac{nT}{P} \quad \text{or} \quad V = \frac{nRT}{P}$$

where R is a proportionality constant known as the *ideal gas constant*. The equation is commonly written as

$$PV = nRT$$

ideal gas equation

and is known as the **ideal gas equation**. This equation states in a single expression what we have considered in our earlier discussions: The volume of a gas varies directly with the number of gas molecules and the absolute temperature, and varies inversely with the pressure. The value and units of R depend on the units of P, V, and T. We can calculate one value of R by taking 1 mol of a gas at STP conditions. Solve the equation for R:

$$R = \frac{PV}{nT} = \frac{1 \text{ atm} \times 22.4 \text{ L}}{1 \text{ mol} \times 273 \text{ K}} = 0.0821 \frac{\text{L-atm}}{\text{mol-K}}$$

The units of R in this case are liter-atmospheres (L-atm) per mole Kelvin (mol-K). When the value of $R = 0.0821$ L-atm/mol-K, P is in atmospheres, n is in moles, V is in liters, and T is in Kelvin.

The ideal gas equation can be used to calculate any one of the four variables when the other three are known.

Example 12.15 What pressure will be exerted by 0.400 mol of a gas in a 5.00-L container at 17.0°C?

Solution

Step 1 Organize the given information, putting temperatures in Kelvin:

$P = ?$

$V = 5.00$ L

$T = 290.$ K

$n = 0.400$ mol

Step 2 Write and solve the ideal gas equation for the unknown:

$$PV = nRT \quad \text{or} \quad P = \frac{nRT}{V}$$

Step 3 Put given information into the equation and calculate:

$$P = \frac{0.400 \text{ mol} \times 0.0821 \text{ L-atm/mol-K} \times 290. \text{ K}}{5.00 \text{ L}} = 1.90 \text{ atm}$$

Example 12.16

How many moles of oxygen gas are in a 50.0-L tank at 22.0°C if the pressure gauge reads 2000. lb/in.^{2}?

Solution

Step 1 Organize the given information, putting the temperature in Kelvin and changing pressure to atmospheres:

$$P = \frac{2000. \text{ lb}}{\text{in.}^2} \times \frac{1 \text{ atm}}{14.7 \text{ lb/in.}^2} = 136.1 \text{ atm}$$

$V = 50.0$ L

$T = 295$ K

$n = ?$

Step 2 Write and solve the equation for the unknown:

$$PV = nRT \qquad \text{or} \qquad n = \frac{PV}{RT}$$

Step 3 Put given information into the equation and calculate:

$$n = \frac{136.1 \text{ atm} \times 50.0 \text{ L}}{(0.0821 \text{ L-atm/mol-K}) \times 295 \text{ K}} = 281 \text{ mol } O_2$$

Practice 12.10

A 23.8-L cylinder contains oxygen gas at 20.0°C and 732 torr. How many moles of oxygen are in the cylinder?

The molar mass of a gaseous substance can be determined using the ideal gas equation. Since molar mass = g/mol, then mol = g/molar mass. Using M for molar mass, we can substitute g/M for n (moles) in the ideal gas equation to get

This form of the gas equation is most useful in problems containing mass instead of moles.

$$PV = \frac{g}{M}RT \qquad \text{or} \qquad M = \frac{gRT}{PV} \quad \text{(modified ideal gas equation)}$$

which will allow us to calculate the molar mass, M, for any substance in the gaseous state.

Example 12.17

Calculate the molar mass of butane gas, if 3.69 g occupy 1.53 L at 20.0°C and 1.00 atm.

Solution

Change 20°C to 293 K and substitute the data into the modified ideal gas equation:

$$M = \frac{gRT}{PV} = \frac{3.69 \text{ g} \times 0.0821 \text{ L-atm/mol-K} \times 293 \text{ K}}{1.00 \text{ atm} \times 1.53 \text{ L}} = 58.0 \text{ g/mol}$$

Practice 12.11

A sample of 0.286 g of a certain gas occupies 50.0 mL at standard temperature and 76.0 cm Hg. Determine the molar mass of the gas.

12.15 Stoichiometry Involving Gases

Mole–Volume and Mass–Volume Calculations

Stoichiometric problems involving gas volumes can be solved by the general mole-ratio method outlined in Chapter 9. The factors 1 mol/22.4 L and 22.4 L/1 mol are used for converting volume to moles and moles to volume, respectively. (See Figure 12.14.) These conversion factors are used under the assumption that the gases are at STP and that they behave as ideal gases. In actual practice, gases are measured at other than STP conditions, and the volumes are converted to STP for stoichiometric calculations.

In a balanced equation, the number preceding the formula of a gaseous substance represents the number of moles or molar volumes (22.4 L at STP) of that substance.

The following are examples of typical problems involving gases and chemical equations.

Example 12.18 What volume of oxygen (at STP) can be formed from 0.500 mol of potassium chlorate?

Solution

Step 1 Write the balanced equation:

$$\underset{2\ \text{mol}}{2\ KClO_3} \longrightarrow 2\ KCl + \underset{3\ \text{mol}}{3\ O_2(g)}$$

Step 2 The starting amount is 0.500 mol $KClO_3$. The conversion is from

$$\text{moles } KClO_3 \longrightarrow \text{moles } O_2 \longrightarrow \text{liters } O_2$$

Step 3 Calculate the moles of O_2, using the mole-ratio method:

$$0.500\ \cancel{\text{mol } KClO_3} \times \frac{3\ \text{mol } O_2}{2\ \cancel{\text{mol } KClO_3}} = 0.750\ \text{mol } O_2$$

Step 4 Convert moles of O_2 to liters of O_2. The moles of a gas at STP are converted to liters by multiplying by the molar volume, 22.4 L/mol:

$$0.750\ \cancel{\text{mol}}\ O_2 \times \frac{22.4\ \text{L}}{1\ \cancel{\text{mol}}} = 16.8\ \text{L } O_2$$

Setting up a continuous calculation, we obtain

$$0.500\ \cancel{\text{mol } KClO_3} \times \frac{3\ \cancel{\text{mol}}\ O_2}{2\ \cancel{\text{mol } KClO_3}} \times \frac{22.4\ \text{L}}{1\ \cancel{\text{mol}}} = 16.8\ \text{L } O_2$$

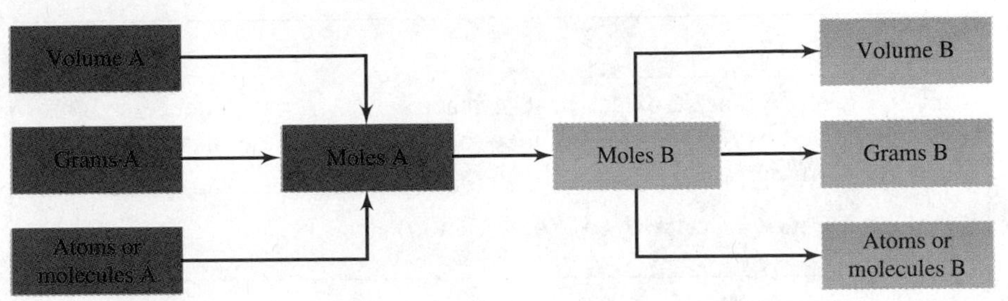

FIGURE 12.14 Summary of the primary conversions involved in stoichiometry. The conversion for volumes of gases is included.

Example 12.19

How many grams of aluminum must react with sulfuric acid to produce 1.25 L of hydrogen gas at STP?

Solution

Step 1 The balanced equation is

$$\underset{2\ \text{mol}}{2\ \text{Al}(s)} + 3\ H_2SO_4(aq) \longrightarrow Al_2(SO_4)_3(aq) + \underset{3\ \text{mol}}{3\ H_2(g)}$$

Step 2 We first convert liters of H_2 to moles of H_2. Then the familiar stoichiometric calculation from the equation is used. The conversion is

$$\text{L } H_2 \longrightarrow \text{mol } H_2 \longrightarrow \text{mol Al} \longrightarrow \text{g Al}$$

$$1.25\ \cancel{\text{L}}\ \cancel{H_2} \times \frac{1\ \cancel{\text{mol}}}{22.4\ \cancel{\text{L}}} \times \frac{2\ \cancel{\text{mol Al}}}{3\ \cancel{\text{mol}\ H_2}} \times \frac{26.98\ \text{g Al}}{1\ \cancel{\text{mol Al}}} = 1.00\ \text{g Al}$$

Example 12.20

What volume of hydrogen, collected at 30°C and 700. torr, will be formed by reacting 50.0 g of aluminum with hydrochloric acid?

$$\underset{2\ \text{mol}}{2\ \text{Al}(s)} + 6\ \text{HCl}(aq) \longrightarrow 2\ AlCl_3(aq) + \underset{3\ \text{mol}}{3\ H_2(g)}$$

Solution

In this problem the conditions are not at STP, so we cannot use the method shown in Example 12.18. Either we need to calculate the volume at STP from the equation and then convert this volume to the conditions given in the problem, or we can use the ideal gas equation. Let's use the ideal gas equation.

First calculate the moles of H_2 obtained from 50.0 g of Al. Then, using the ideal gas equation, calculate the volume of H_2 at the conditions given in the problem.

Step 1 Moles of H_2: The conversion is

$$\text{grams Al} \longrightarrow \text{moles Al} \longrightarrow \text{moles } H_2$$

$$50.0\ \cancel{\text{g Al}} \times \frac{1\ \cancel{\text{mol Al}}}{26.98\ \cancel{\text{g Al}}} \times \frac{3\ \text{mol}\ H_2}{2\ \cancel{\text{mol Al}}} = 2.78\ \text{mol}\ H_2$$

Step 2 Liters of H_2: Solve $PV = nRT$ for V and substitute the data into the equation.

Convert °C to K: 30°C + 273 = 303 K.

Convert torr to atm: 700. $\cancel{\text{torr}}$ × 1 atm/760. $\cancel{\text{torr}}$ = 0.921 atm.

$$V = \frac{nRT}{P} = \frac{2.78\ \cancel{\text{mol}}\ H_2 \times 0.0821\ \text{L-}\cancel{\text{atm}} \times 303\ \cancel{\text{K}}}{0.921\ \cancel{\text{atm}} \times \cancel{\text{mol-K}}} = 75.1\ \text{L}\ H_2$$

Note: The volume at STP is 62.3 L H_2.

Practice 12.12

If 10.0 g of sodium peroxide, Na_2O_2, react with water to produce sodium hydroxide and oxygen, how many liters of oxygen will be produced at 20°C and 750. torr?

$$2\ Na_2O_2(s) + 2\ H_2O(l) \longrightarrow 4\ NaOH(aq) + O_2(g)$$

Volume–Volume Calculations

When all substances in a reaction are in the gaseous state, simplifications in the calculation can be made. These are based on Avogadro's law, which states that gases under identical conditions of temperature and pressure contain the same number of molecules and occupy the same volume. Using this same law, we can also state that, under the same conditions of temperature and pressure, the volumes of gases reacting are proportional to the numbers of moles of the gases in the balanced equation. Consider the reaction:

$$H_2(g) + Cl_2(g) \longrightarrow 2\ HCl(g)$$

$H_2(g)$	$Cl_2(g)$	$2\ HCl(g)$
1 mol	1 mol	2 mol
22.4 L	22.4 L	2 × 22.4 L
1 volume	1 volume	2 volumes
Y volume	*Y* volume	2 *Y* volumes

In this reaction 22.4 L of hydrogen will react with 22.4 L of chlorine to give 2 × 22.4 = 44.8 L of hydrogen chloride gas. This statement is true because these volumes are equivalent to the number of reacting moles in the equation. Therefore, *Y* volume of H_2 will combine with *Y* volume of Cl_2 to give 2 *Y* volumes of HCl. For example, 100 L of H_2 react with 100 L of Cl_2 to give 200 L of HCl; if the 100 L of H_2 and of Cl_2 are at 50°C, they will give 200 L of HCl at 50°C. When the temperature and pressure before and after a reaction are the same, volumes can be calculated without changing the volumes to STP.

For reacting gases at constant temperature and pressure: Volume–volume relationships are the same as mole–mole relationships.

Example 12.21 What volume of oxygen will react with 150 L of hydrogen to form water vapor? What volume of water vapor will be formed?

Solution Assume that both reactants and products are measured at the same conditions. Calculate by using reacting volumes:

$$2\ H_2(g) + O_2(g) \longrightarrow 2\ H_2O(g)$$

$2\ H_2(g)$	$O_2(g)$	$2\ H_2O(g)$
2 mol	1 mol	2 mol
2 × 22.4 L	22.4 L	2 × 22.4 L
2 volumes	1 volume	2 volumes
150 L	75 L	150 L

For every two volumes of H_2 that react, one volume of O_2 reacts and two volumes of $H_2O(g)$ are produced:

$$150 \text{ L } H_2 \times \frac{1 \text{ volume } O_2}{2 \text{ volumes } H_2} = 75 \text{ L } O_2$$

$$150 \text{ L } H_2 \times \frac{2 \text{ volume } H_2O}{2 \text{ volume } H_2} = 150 \text{ L } H_2O$$

Example 12.22

The equation for the preparation of ammonia is

$$3H_2(g) + N_2(g) \xrightarrow{400°C} 2\,NH_3(g)$$

Assuming that the reaction goes to completion.

(a) What volume of H_2 will react with 50.0 L of N_2?
(b) What volume of NH_3 will be formed from 50.0 L of N_2?
(c) What volume of N_2 will react with 100. mL of H_2?
(d) What volume of NH_3 will be produced from 100. mL of H_2?
(e) If 600. mL of H_2 and 400. mL of N_2 are sealed in a flask and allowed to react, what amounts of H_2, N_2, and NH_3 are in the flask at the end of the reaction?

Solution

The answers to parts (a)–(d) are shown in the boxes and can be determined from the equation by inspection, using the principle of reacting volumes:

	$3\,H_2(g)$	+ $N_2(g)$	$\longrightarrow$ $2\,NH_3(g)$
	3 volumes	1 volume	2 volumes
(a)	**[150. L]**	50.0 L	
(b)		50.0 L	**[100. L]**
(c)	100. mL	**[33.3 mL]**	
(d)	100. mL		**[66.7 mL]**

(e) Volume ratio from the equation $= \dfrac{3 \text{ volumes } H_2}{1 \text{ volume } N_2}$

$$\text{Volume ratio used} = \frac{600.\text{ mL } H_2}{400.\text{ mL } N_2} = \frac{3 \text{ volumes } H_2}{2 \text{ volumes } N_2}$$

Comparing these two ratios, we see that an excess of N_2 is present in the gas mixture. Therefore, the reactant limiting the amount of NH_3 that can be formed is H_2:

$3\,H_2(g)$	+ $N_2(g)$	$\longrightarrow$ $2\,NH_3(g)$
600 mL	200 mL	400 mL

To have a 3:1 ratio of volumes reacting, 600 mL of H_2 will react with 200 mL of N_2 to produce 400 mL of NH_3, leaving 200 mL of N_2 unreacted. At the end of the reaction the flask will contain 400 mL of NH_3 and 200 mL of N_2.

Practice 12.13

What volume of oxygen will react with 15.0 L of propane (C_3H_8) to form carbon dioxide and water? What volume of carbon dioxide will be formed? What volume of water vapor?

$$C_3H_8(g) + 5\,O_2(g) \longrightarrow 3\,CO_2(g) + 4\,H_2O(g)$$

12.16 Real Gases

All the gas laws are based on the behavior of an ideal gas—that is, a gas with a behavior that is described exactly by the gas laws for all possible values of P, V, and T. Most real gases actually do behave very nearly as predicted by the gas laws over a fairly wide range of temperatures and pressures. However, when conditions are such that the gas molecules are crowded closely together (high pressure and/or low temperature), they show marked deviations from ideal behavior. Deviations occur because molecules have finite volumes and also have intermolecular attractions, which result in less compressibility at high pressures and greater compressibility at low temperatures than predicted by the gas laws. Many gases become liquids at high pressure and low temperature.

12.17 Air Pollution

Chemical reactions occur among the gases that are emitted into our atmosphere. In recent years, there has been growing concern over the effects these reactions have on our environment and our lives.

The outer portion (stratosphere) of the atmosphere plays a significant role in determining the conditions for life at the surface of the earth. This stratosphere protects the surface from the intense radiation and particles bombarding our planet. Some of the high-energy radiation from the sun acts upon oxygen molecules in the stratosphere, converting them into ozone, O_3. Different molecular forms of an element are called **allotropes** of that element. Thus oxygen and ozone are allotropic forms of oxygen:

allotrope

$$O_2 \xrightarrow{\text{sunlight}} \underset{\text{oxygen atoms}}{O + O}$$

$$O_2 + O \longrightarrow O_3$$

Ultraviolet radiation from the sun is highly damaging to living tissues of plants and animals. The ozone layer, however, shields the earth by absorbing ultraviolet radiation and thus prevents most of this lethal radiation from reaching the earth's surface. The reaction that occurs is the reverse of the preceding one:

$$O_3 \xrightarrow[\text{radiation}]{\text{ultraviolet}} O_2 + O + \text{heat}$$

Scientists have become concerned about a growing hazard to the ozone layer. Chlorofluorocarbon propellants, such as the Freons, CCl_3F and CCl_2F_2, which were used in aerosol spray cans and are used in refrigeration and air-conditioning units, are stable compounds and remain unchanged in the lower atmosphere. But when these chlorofluorocarbons are carried by convection currents to the stratosphere, they absorb ultraviolet radiation and produce chlorine atoms (chlorine free radicals), which in turn react with ozone. The following reaction sequence involving free

A free radical is a species containing an odd number of electrons. Free radicals are highly reactive.

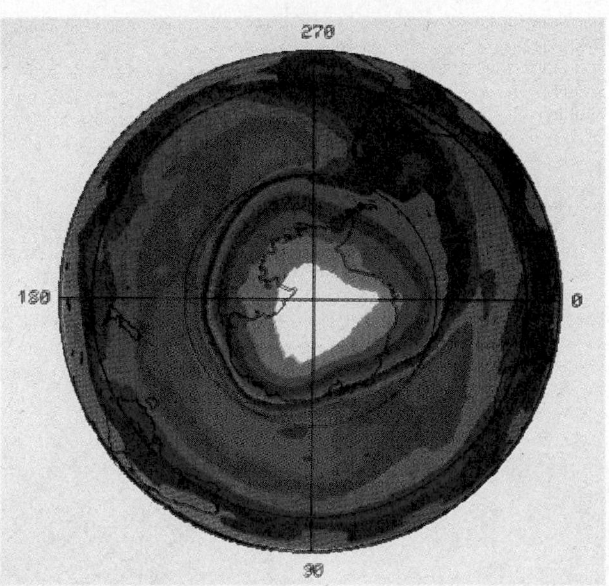

◀ **FIGURE 12.15**
Satellite map showing a severe depletion or "hole" in the ozone layer over Antarctica in October, 1990. The hole is believed to be due to pollution of the atmosphere by chlorofluorocarbons used in aerosols and refrigerants.

radicals has been proposed to explain the partial destruction of the ozone layer by chlorofluorocarbons.

$$\underset{\text{fluorocarbon molecule}}{CCl_3F} \xrightarrow[\text{radiation}]{\text{ultraviolet}} \underset{\text{fluorocarbon free radical}}{\cdot CCl_2F} + \underset{\text{chlorine free radical (atom)}}{Cl\cdot} \quad (1)$$

$$Cl\cdot + O_3 \longrightarrow ClO\cdot + O_2 \quad (2)$$

$$ClO\cdot + O \longrightarrow O_2 + Cl\cdot \quad (3)$$

Because a chlorine atom is generated for each ozone molecule that is destroyed (reactions 2 and 3 can proceed repeatedly), a single chlorofluorocarbon molecule can be responsible for the destruction of many ozone molecules. During the past decade, scientists have discovered an annual thinning in the ozone layer over Antarctica. This is what we call the "hole" in the ozone layer. If this hole were to occur over populated regions of the world, severe effects would result, including a rise in the cancer rate, increased climatic temperatures, and vision problems. See Figure 12.15.

Ozone can be prepared by passing air or oxygen through an electrical discharge:

$$3\,O_2(g) + 286\text{ kJ} \xrightarrow[\text{discharge}]{\text{electrical}} 2\,O_3(g)$$

The characteristic pungent odor of ozone is noticeable in the vicinity of electrical machines and power transmission lines. Ozone is formed in the atmosphere during electrical storms and by the photochemical action of ultraviolet radiation on a mixture of nitrogen dioxide and oxygen. Areas with high air pollution are subject to high atmospheric ozone concentrations.

Ozone is not a desirable low-altitude constituent of the atmosphere because it is known to cause extensive plant damage, cracking of rubber, and the formation of eye-irritating substances. Concentrations of ozone greater than 0.1 part per million (ppm) of air cause coughing, choking, headache, fatigue, and reduced resistance to respiratory infection. Concentrations between 10 and 20 ppm are fatal to humans.

A smoggy day in Los Angeles, California. High concentrations of ozone near the surface of the earth can cause plant damage, headaches, and deterioration of rubber, among other things.

In addition to ozone, the air in urban areas contains nitrogen oxides, which are components of smog. The term *smog* refers to air pollution in urban environments. Often the chemical reactions occur as part of a *photochemical process*. Nitrogen monoxide (NO) is oxidized in the air or in automobile engines to produce nitrogen dioxide (NO_2). In the presence of light,

$$NO_2 \xrightarrow{\text{light}} NO + O$$

In addition to nitrogen oxides, combustion of fossil fuels releases CO_2, CO, and sulfur oxides. Incomplete combustion releases unburned and partially burned hydrocarbons.

Society is continually attempting to discover, understand, and control emissions that contribute to this sort of atmospheric chemistry. It is a problem that each one of us faces as we look to the future if we want to continue to support life as we know it on our planet.

Concepts in Review

1. State the principal assumptions of the Kinetic-Molecular Theory.
2. Estimate the relative rates of effusion of two gases of known molar mass.
3. Sketch and explain the operation of a mercury barometer.
4. List two factors that determine gas pressure in a vessel of fixed volume.
5. State Boyle's, Charles' and Gay-Lussac's laws. Use all of them in problems.
6. State the combined gas law. Indicate when it is used.
7. Use Dalton's law of partial pressures and the combined gas law to determine the dry STP volume of a gas collected over water.

8. State Avogadro's law.
9. Understand the mole–mass–volume relationship of gases.
10. Determine the density of any gas at STP.
11. Determine the molar mass of a gas from its density at a known temperature and pressure.
12. Solve problems involving the ideal gas equation.
13. Make mole–volume, mass–volume, and volume–volume stoichiometric calculations from balanced chemical equations.
14. State two reasons why real gases may deviate from the behavior predicted for an ideal gas.

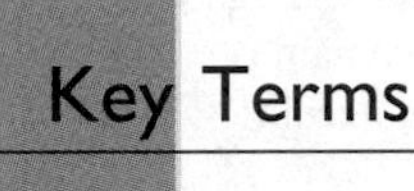

Key Terms

The terms listed here have been defined within this chapter. Section numbers are referenced in parenthesis for each term.

absolute zero (12.6)
allotrope (12.17)
1 (one) atmosphere (12.3)
atmospheric pressure (12.3)
Avogadro's law (12.11)
barometer (12.3)
Boyle's law (12.5)
Charles' law (12.6)
Dalton's law of partial pressures (12.10)
diffusion (12.2)
effusion (12.2)
Gay Lussac's law (12.7)
Gay Lussac's law of combining volumes (12.11)
Graham's law of effusion (12.2)
Henry's law (CIA)
ideal gas (12.2)
ideal gas equation (12.14)
Kinetic-Molecular Theory (KMT) (12.2)
molar volume (12.12)
partial pressure (12.10)
pressure (12.3)
standard conditions (12.8)
standard temperature and pressure (STP) (12.8)

Questions

Questions refer to tables, figures, and key words and concepts defined within the chapter. A particularly challenging question or exercise is indicated with an asterisk.

1. What evidence is used to show diffusion in Figure 12.1? If H_2 and O_2 were in the two flasks, how could you prove that diffusion had taken place?
2. How does the air pressure inside the balloon shown in Figure 12.2 compare with the air pressure outside the balloon? Explain.
3. According to Table 12.1, what two gases are the major constituents of dry air?
4. How does the pressure represented by 1 torr compare in magnitude to the pressure represented by 1 mm Hg? See Table 12.2.
5. In which container illustrated in Figure 12.5 are the molecules of gas moving faster? Assume both gases to be hydrogen.
6. In Figure 12.6, what gas pressure corresponds to a volume of 4 L?
7. How do the data illustrated in Figure 12.6 substantiate Boyle's law?
8. What effect would you observe in Figure 12.9 if T_2 were lower than T_1?

9. In the diagram shown in Figure 12.11, is the pressure of the oxygen plus water vapor inside the bottle equal to, greater than, or less than the atmospheric pressure outside the bottle? Explain.
10. List five gases in Table 12.3 that are more dense than air. Explain the basis for your selection.
11. What are the basic assumptions of the Kinetic-Molecular Theory?
12. Arrange the following gases, all at standard temperature, in order of increasing relative molecular velocities: H_2, CH_4, Rn, N_2, F_2, He. What is your basis for determining the order?
13. List, in descending order, the average kinetic energies of the molecules in Question 12.
14. What are the four parameters used to describe the behavior of a gas?
15. What are the characteristics of an ideal gas?
16. Under what condition of temperature, high or low, is a gas least likely to exhibit ideal behavior? Explain.
17. Under what condition of pressure, high or low, is a gas least likely to exhibit ideal behavior? Explain.
18. Compare, at the same temperature and pressure, equal volumes of H_2 and O_2 as to
    - **(a)** number of molecules
    - **(b)** mass
    - **(c)** number of moles
    - **(d)** average kinetic energy of the molecules
    - **(e)** rate of effusion
    - **(f)** density
19. How does the Kinetic-Molecular Theory account for the behavior of gases as described by
    - **(a)** Boyle's law?
    - **(b)** Charles' law?
    - **(c)** Dalton's law of partial pressures?
20. Explain how the reaction

    $N_2(g) + O_2(g) \xrightarrow{\Delta} 2\ NO(g)$ proves that nitrogen and oxygen are diatomic molecules.
21. What is the reason for referring gases to STP?
22. Is the conversion of oxygen to ozone an exothermic or endothermic reaction? How do you know?
23. Write formulas for an oxygen atom, an oxygen molecule, and an ozone molecule. How many electrons are in an oxygen molecule?
24. When constant pressure is maintained, what effect does heating a mole of N_2 gas have on
    - **(a)** its density?
    - **(b)** its mass?
    - **(c)** the average kinetic energy of its molecules?
    - **(d)** the average velocity of its molecules?
    - **(e)** the number of N_2 molecules in the sample?
25. State Henry's law and indicate its significance to a scuba diver.
26. Assuming ideal gas behavior, which of the following statements are correct? (Try to answer without referring to your text.) Rewrite each incorrect statement to make it correct.
    - **(a)** The pressure exerted by a gas at constant volume is independent of the temperature of the gas.
    - **(b)** At constant temperature, increasing the pressure exerted on a gas sample will cause a decrease in the volume of the gas sample.
    - **(c)** At constant pressure, the volume of a gas is inversely proportional to the absolute temperature.
    - **(d)** At constant temperature, doubling the pressure on a gas sample will cause the volume of the gas sample to decrease to one-half its original volume.
    - **(e)** Compressing a gas at constant temperature will cause its density and mass to increase.
    - **(f)** Equal volumes of CO_2 and CH_4 gases at the same temperature and pressure contain
        - (1) the same number of molecules
        - (2) the same mass
        - (3) the same densities
        - (4) the same number of moles
        - (5) the same number of atoms
    - **(g)** At constant temperature, the average kinetic energy of O_2 molecules at 200 atm pressure is greater than the average kinetic energy of O_2 molecules at 100 atm pressure.
    - **(h)** According to Charles' law, the volume of a gas becomes zero at $-273°C$.
    - **(i)** One liter of O_2 gas at STP has the same mass as 1 L of O_2 gas at 273°C and 2 atm pressure.
    - **(j)** The volume occupied by a gas depends only on its temperature and pressure.
    - **(k)** In a mixture containing O_2 molecules and N_2 molecules, the O_2 molecules, on the average, are moving faster than the N_2 molecules.
    - **(l)** $PV = k$ is a statement of Charles' law.
    - **(m)** If the temperature of a sample of gas is increased from 25°C to 50°C, the volume of the gas will increase by 100%.
    - **(n)** One mole of chlorine, Cl_2, at 20°C and 600 torr pressure contains 6.022×10^{23} molecules.
    - **(o)** One mole of H_2 plus 1 mol of O_2 in an 11.2-L container exert a pressure of 4 atm at 0°C.
    - **(p)** When the pressure on a sample of gas is halved, with the temperature remaining constant, the density of the gas is also halved.
    - **(q)** When the temperature of a sample of gas is increased at constant pressure, the density of the gas will decrease.
    - **(r)** According to the equation

      $$2\ KClO_3(s) \xrightarrow{\Delta} 2\ KCl(s) + 3\ O_2(g)$$

      1 mol of $KClO_3$ will produce 67.2 L of O_2 at STP.
    - **(s)** $PV = nRT$ is a statement of Avogadro's law.
    - **(t)** STP conditions are 1 atm and 0°C.

Paired Exercises

These exercises are paired. Each odd-numbered exercise is followed by a similar even-numbered exercise. Answers to the even-numbered exercises are given in Appendix V.

Pressure Units

27. The barometer reads 715 mm Hg. Calculate the corresponding pressure in
(a) atmospheres
(b) inches of Hg
(c) lb/in.^{2}

28. The barometer reads 715 mm Hg. Calculate the corresponding pressure in
(a) torrs
(b) millibars
(c) kilopascals

29. Express the following pressures in atmospheres:
(a) 28 mm Hg
(b) 6000. cm Hg
(c) 795 torr
(d) 5.00 kPa

30. Express the following pressures in atmospheres:
(a) 62 mm Hg
(b) 4250. cm Hg
(c) 225 torr
(d) 0.67 kPa

Boyle's and Charles' laws

31. A gas occupies a volume of 400. mL at 500. mm Hg pressure. What will be its volume, at constant temperature, if the pressure is changed to (a) 760 mm Hg? (b) 250 torr?

32. A gas occupies a volume of 400. mL at 500. mm Hg pressure. What will be its volume, at constant temperature, if the pressure is changed to (a) 2.00 atm? (b) 325 torr?

33. A 500.-mL sample of a gas is at a pressure of 640. mm Hg. What must be the pressure, at constant temperature, if the volume is changed to 855 mL?

34. A 500.-mL sample of a gas is at a pressure of 640. mm Hg. What must be the pressure, at constant temperature, if the volume is changed to 450. mL?

35. Given 6.00 L of N_2 gas at -25°C, what volume will the nitrogen occupy at (a) 0.0°C? (b) 100. K? (Assume constant pressure.)

36. Given 6.00 L of N_2 gas at -25°C, what volume will the nitrogen occupy at (a) 0.0°F? (b) 345. K? (Assume constant pressure.)

Combined Gas Laws

37. A gas occupies a volume of 410 mL at 27°C and 740 mm Hg pressure. Calculate the volume the gas would occupy at STP.

38. A gas occupies a volume of 410 mL at 27°C and 740 mm Hg pressure. Calculate the volume the gas would occupy at 250.°C and 680 mm Hg pressure.

39. An expandable balloon contains 1400. L of He at 0.950 atm pressure and 18°C. At an altitude of 22 miles (temperature 2.0°C and pressure 4.0 torr), what will be the volume of the balloon?

40. A gas occupies 22.4 L at 2.50 atm and 27°C. What will be its volume at 1.50 atm and -5.00°C?

Dalton's Law of Partial Pressures

41. What would be the partial pressure of N_2 gas collected over water at 20°C and 720. torr pressure? (Check Appendix II for the vapor pressure of water.)

42. What would be the partial pressure of N_2 gas collected over water at 25°C and 705. torr pressure? (Check Appendix II for the vapor pressure of water.)

43. A mixture contains H_2 at 600. torr pressure, N_2 at 200. torr pressure, and O_2 at 300. torr pressure. What is the total pressure of the gases in the system?

44. A mixture contains H_2 at 325. torr pressure, N_2 at 475. torr pressure, and O_2 at 650. torr pressure. What is the total pressure of the gases in the system?

45. A sample of methane gas, CH_4, was collected over water at 25.0°C and 720. torr. The volume of the wet gas is 2.50 L. What will be the volume of the dry methane at standard pressure?

46. A sample of propane gas, C_3H_8, was collected over water at 22.5°C and 745 torr. The volume of the wet gas is 1.25 L. What will be the volume of the dry propane at standard pressure?

Mole–Mass–Volume Relationships

47. What volume will 2.5 mol of Cl_2 occupy at STP?

48. What volume will 1.25 mol of N_2 occupy at STP?

49. How many grams of CO_2 are present in 2500 mL of CO_2 at STP?

50. How many grams of NH_3 are present in 1.75 L of NH_3 at STP?

51. What volume will each of the following occupy at STP?
   (a) 1.0 mol of NO_2
   (b) 17.05 g of NO_2
   (c) 1.20×10^{24} molecules of NO_2

52. What volume will each of the following occupy at STP?
   (a) 0.50 mol of H_2S
   (b) 22.41 g of H_2S
   (c) 8.55×10^{23} molecules of H_2S

53. How many molecules of NH_3 gas are present in a 1.00-L flask of NH_3 gas at STP?

54. How many molecules of CH_4 gas are present in a 1.00-L flask of CH_4 gas at STP?

Density of Gases

55. Calculate the density of the following gases at STP:
   (a) Kr
   (b) SO_3

56. Calculate the density of the following gases at STP:
   (a) He
   (b) C_4H_8

57. Calculate the density of:
   (a) F_2 gas at STP
   (b) F_2 gas at 27°C and 1.00 atm pressure

58. Calculate the density of:
   (a) Cl_2 gas at STP
   (b) Cl_2 gas at 22°C and 0.500 atm pressure

Ideal Gas Equation and Stoichiometry

59. At 27°C and 750 torr pressure, what will be the volume of 2.3 mol of Ne?

60. At 25°C and 725 torr pressure, what will be the volume of 0.75 mol of Kr?

61. What volume will a mixture of 5.00 mol of H_2 and 0.500 mol of CO_2 occupy at STP?

62. What volume will a mixture of 2.50 mol of N_2 and 0.750 mol of HCl occupy at STP?

63. Given the equation:

$$4\,NH_3(g) + 5\,O_2(g) \longrightarrow 4\,NO(g) + 6\,H_2O(g)$$

   (a) How many moles of NH_3 are required to produce 5.5 mol of NO?
   (b) How many liters of NO can be made from 12 L of O_2 and 10. L of NH_3 at STP?
   (c) At constant temperature and pressure, what is the maximum volume, in liters, of NO that can be made from 3.0 L of NH_3 and 3.0 L of O_2?

64. Given the equation:

$$4\,NH_3(g) + 5\,O_2(g) \longrightarrow 4\,NO(g) + 6\,H_2O(g)$$

   (a) How many moles of NH_3 will react with 7.0 mol of O_2?
   (b) At constant temperature and pressure, how many liters of NO can be made by the reaction of 800. mL of O_2?
   (c) How many grams of O_2 must react to produce 60. L of NO measured at STP?

65. Given the equation:

$$4\,FeS(s) + 7\,O_2(g) \xrightarrow{\Delta} 2\,Fe_2O_3(s) + 4\,SO_2(g)$$

how many liters of O_2, measured at STP, will react with 0.600 kg of FeS?

66. Given the equation:

$$4\,FeS(s) + 7\,O_2(g) \xrightarrow{\Delta} 2\,Fe_2O_3(s) + 4\,SO_2(g)$$

how many liters of SO_2, measured at STP, will be produced from 0.600 kg of FeS?

Additional Exercises

These exercises are not paired or labeled by topic and provide additional practice on concepts covered in this chapter.

67. Sketch a graph to show each of the following relationships:
(a) P vs V at constant temperature and number of moles
(b) T vs V at constant pressure and number of moles
(c) T vs P at constant volume and number of moles
(d) n vs V at constant temperature and pressure

68. Why is it dangerous to incinerate an aerosol can?

69. What volume does 1 mol of an ideal gas occupy at standard conditions?

70. Which of these occupies the greatest volume?
(a) 0.2 mol of chlorine gas at 48°C and 80 cm Hg
(b) 4.2 g of ammonia at 0.65 atm and −112°C
(c) 21 g of sulfur trioxide at room temperature and 110 kPa

71. Which of these has the greatest density?
(a) SF_6 at STP
(b) C_2H_6 at room conditions
(c) He at −80°C and 2.15 atm

72. A chemist carried out a chemical reaction that produced a gas. It was found that the gas contained 80.0% carbon and 20.0% hydrogen. It was also noticed that 1500 mL of the gas at STP had a mass of 2.01 g.
(a) What is the empirical formula of the compound?
(b) What is the molecular formula of the compound?
(c) What Lewis structure fits this compound?

* **73.** Three gases were added to the same 2.0-L container. The total pressure of the gases was 790 torr at room temperature (25.0°C). If the mixture contained 0.65 g of oxygen gas, 0.58 g of carbon dioxide, and an unknown amount of nitrogen gas, determine:
(a) the total number of moles of gas in the container
(b) the number of grams of nitrogen in the container
(c) the partial pressure of each gas in the mixture

* **74.** When carbon monoxide and oxygen gas react, carbon dioxide results. If 500. mL of O_2 at 1.8 atm and 15°C are mixed with 500. mL of CO at 800 mm Hg and 60°C, how many milliliters of CO_2 at STP could possibly result?

75. One of the methods for estimating the temperature at the center of the sun is based on the ideal gas equation. If the center is assumed to be a mixture of gases whose average molar mass is 2.0 g/mol, and if the density and pressure are 1.4 g/cm^3 and 1.3×10^9 atm, respectively, calculate the temperature.

76. A soccer ball of constant volume 2.24 L is pumped up with air to a gauge pressure of 13 lb/in.^{2} at 20.0°C. The molar mass of air is about 29 g/mol.
(a) How many moles of air are in the ball?
(b) What mass of air is in the ball?
(c) During the game, the temperature rises to 30.0°C. What mass of air must be allowed to escape to bring the gauge pressure back to its original value?

77. A balloon will burst at a volume of 2.00 L. If it is partially filled at a temperature of 20.0°C and a pressure of 65 cm Hg to occupy 1.75 L, at what temperature will it burst if the pressure is exactly 1 atm at the time that it breaks?

78. At constant temperature, what pressure would be required to compress 2500 L of hydrogen gas at 1.0 atm pressure into a 25-L tank?

79. Given a sample of a gas at 27°C, at what temperature would the volume of the gas sample be doubled, the pressure remaining constant?

80. A gas sample at 22°C and 740 torr pressure is heated until its volume is doubled. What pressure would restore the sample to its original volume?

81. A gas occupies 250 mL at 700. torr and 22°C. When the pressure is changed to 500. torr, what temperature (°C) is needed to maintain the same volume?

82. Hydrogen stored in a metal cylinder has a pressure of 252 atm at 25°C. What will be the pressure in the cylinder when the cylinder is lowered into liquid nitrogen at −196°C?

83. The tires on an automobile were filled with air to 30. psi at 71.0°F. When driving at high speeds, the tires become hot. If the tires have a bursting pressure of 44 psi, at what temperature (°F) will the tires "blow out"?

84. What volume would 5.30 L of H_2 gas at STP occupy at 70°C and 830 torr pressure?

85. What pressure will 800. mL of a gas at STP exert when its volume is 250. mL at 30°C?

86. How many gas molecules are present in 600. mL of N_2O at 40°C and 400. torr pressure? How many atoms are present? What would be the volume of the sample at STP?

87. 5.00 L of CO_2 at 500. torr and 3.00 L of CH_4 at 400. torr are put into a 10.0-L container. What is the pressure exerted by the gases in the container?

88. A steel cylinder contains 60.0 mol of H_2 at a pressure of 1500 lb/in.^{2}.
(a) How many moles of H_2 are in the cylinder when the pressure reads 850 lb/in.^{2}?
(b) How many grams of H_2 were initially in the cylinder?

89. At STP, 560. mL of a gas have a mass of 1.08 g. What is the molar mass of the gas?

***90.** How many moles of Cl_2 are in one cubic meter ($1.00\ m^3$) of Cl_2 gas at STP?

91. A gas has a density at STP of 1.78 g/L. What is its molar mass?

***92.** At what temperature (°C) will the density of methane, CH_4, be 1.0 g/L at 1.0 atm pressure?

93. Using the ideal gas equation, $PV = nRT$, calculate:
(a) the volume of 0.510 mol of H_2 at 47°C and 1.6 atm pressure
(b) the number of grams in 16.0 L of CH_4 at 27°C and 600. torr pressure
***(c)** the density of CO_2 at 4.00 atm pressure and −20.0°C
***(d)** the molar mass of a gas having a density of 2.58 g/L at 27°C and 1.00 atm pressure.

94. What is the molar mass of a gas if 1.15 g occupy 0.215 L at 0.813 atm and 30.0°C?

95. What is the Kelvin temperature of a system in which 4.50 mol of a gas occupies 0.250 L at 4.15 atm?

96. How many moles of N_2 gas occupy 5.20 L at 250 K and 0.500 atm?

97. What volume of hydrogen at STP can be produced by reacting 8.30 mol of Al with sulfuric acid? The equation is

$$2\ Al(s) + 3\ H_2SO_4(aq) \longrightarrow Al_2(SO_4)_3(aq) + 3\ H_2(g)$$

***98.** Acetylene, C_2H_2, and hydrogen fluoride, HF, react to give difluoroethane.

$$C_2H_2(g) + 2\ HF(g) \longrightarrow C_2H_4F_2(g)$$

When 1.0 mol of C_2H_2 and 5.0 mol of HF are reacted in a 10.0-L flask, what will be the pressure in the flask at 0°C when the reaction is complete?

99. What are the relative rates of effusion of N_2 and He?

***100.** **(a)** What are the relative rates of effusion of CH_4 and He?
(b) If these two gases are simultaneously introduced into opposite ends of a 100.-cm tube and allowed to diffuse toward each other, at what distance from the helium end will molecules of the two gases meet?

***101.** A gas has a percent composition by mass of 85.7% carbon and 14.3% hydrogen. At STP the density of the gas is 2.50 g/L. What is the molecular formula of the gas?

***102.** Assume that the reaction

$$2\ CO(g) + O_2(g) \longrightarrow 2\ CO_2(g)$$

goes to completion. When 10. mol of CO and 8.0 mol of O_2 react in a closed 10.-L vessel,
(a) how many moles of CO, O_2, and CO_2 are present at the end of the reaction?
(b) what will be the total pressure in the flask at 0°C?

***103.** 250 mL of O_2, measured at STP, were obtained by the decomposition of the $KClO_3$ in a 1.20-g mixture of KCl and $KClO_3$:

$$2\ KClO_3(s) \longrightarrow 2\ KCl(s) + 3\ O_2(g)$$

What is the percent by mass of $KClO_3$ in the mixture?

***104.** Look at the apparatus below. When a small amount of water is squirted into the flask containing ammonia gas (by squeezing the bulb of the medicine dropper), water from the beaker fills the flask through the long glass tubing. Explain this phenomenon. (Remember that ammonia dissolves in water.)

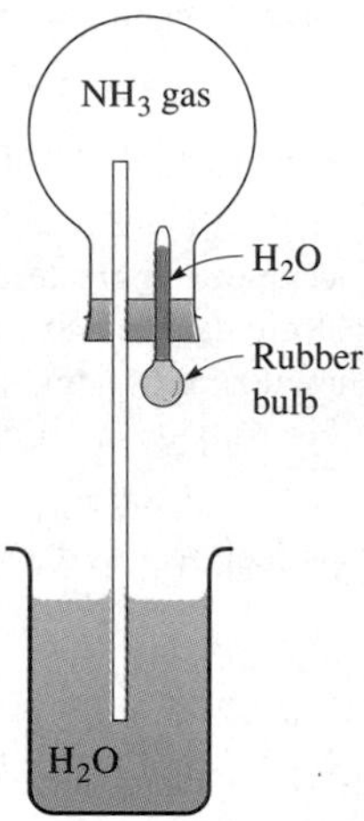

***105.** Determine the pressure of the gas in each of the figures below:

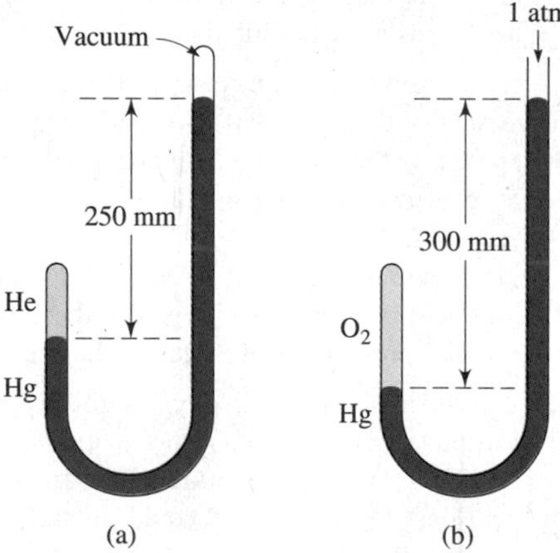

***106.** Consider the arrangement of gases shown below. If the valve between the gases is opened and the temperature is held constant:

(a) determine the pressure of each gas

(b) determine the total pressure in the system

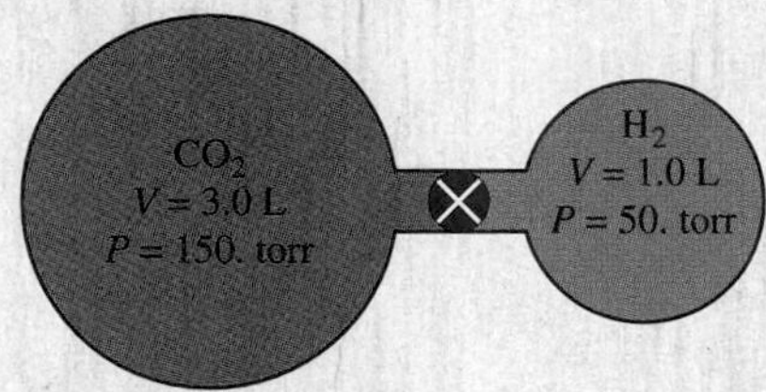

***107.** Air has a density of 1.29 g/L at STP. Calculate the density of air on Pikes Peak, where the pressure is 450 torr and the temperature is 17°C.

***108.** A room is 16 ft × 12 ft × 12 ft. Would air enter or leave the room and how much, if the temperature in the room changed from 27°C to −3°C with the pressure remaining constant? Show evidence.

***109.** A steel cylinder contained 50.0 L of oxygen gas under a pressure of 40.0 atm and a temperature of 25°C. What was the pressure in the cylinder during a storeroom fire that caused the temperature to rise 152°C? (Be careful!)

Answers to Practice Exercises

12.1 (a) 851 torr, (b) 851 mm Hg
12.2 0.596 atm
12.3 4.92 L
12.4 762 torr
12.5 84.7 L
12.6 861 K (588°C)
12.7 518 mL
12.8 1.4×10^2 g/mol
12.9 0.89 g/L
12.10 0.953 mol
12.11 128 g/mol
12.12 1.56 L O_2
12.13 75.0 L O_2, 45.0 L CO_2, 60.0 L H_2O

Water and the Properties of Liquids

13

Planet earth, that magnificent blue sphere we enjoy viewing from the space shuttle, is spectacular. Over 75% of the earth is covered with water. We are born from it, drink it, bathe in it, cook with it, enjoy its beauty in waterfalls and rainbows, and stand in awe of the majesty of icebergs. Water supports and enhances life.

In chemistry, water provides the medium for numerous reactions. The shape of the water molecule is the basis for hydrogen bonds. These bonds determine the unique properties and reactions of water. The tiny water molecule holds the answers to many of the mysteries of chemical reactions.

13.1 Liquids and Solids

In the last chapter, we found that a gas can be a substance containing particles that are far apart, in rapid random motion, and independent of each other. The Kinetic-Molecular Theory, along with the ideal gas equation, summarizes the behavior of most gases at relatively high temperatures and low pressures.

Solids are obviously very different from gases. A solid contains particles that are very close together, has a high density, compresses negligibly, and maintains its shape regardless of container. These characteristics indicate large attractive forces between particles. The model for solids is very different from the one for gases.

Liquids, on the other hand, lie somewhere in between the extremes of gases and solids. A liquid contains particles that are close together, is essentially incompressible, and has a definite volume. These properties are very similar to solids. But, a liquid also takes the shape of its container, which is closer to the model of a gas.

Although liquids and solids show similar properties, they differ tremendously from gases. No simple mathematical relationship, like the ideal gas equation, works well for liquids or solids. Instead, these models are directly related to the forces of attraction between molecules. With these general statements in mind, let us consider some of the specific properties of liquids.

13.2 Evaporation

When beakers of water, ethyl ether, and ethyl alcohol are allowed to stand uncovered in an open room, their volumes gradually decrease. The process by which this change takes place is called *evaporation.*

Attractive forces exist between molecules in the liquid state. Not all of these molecules, however, have the same kinetic energy. Molecules that have greater than average kinetic energy can overcome the attractive forces and break away from the surface of the liquid to become a gas. **Evaporation** or **vaporization** is the escape of molecules from the liquid state to the gas or vapor state.

evaporation
vaporization

In evaporation, molecules of higher-than-average kinetic energy escape from a liquid, leaving it cooler than it was before they escaped. For this reason, evaporation of perspiration is one way the human body cools itself and keeps its temperature

◀ **Chapter Opening Photo: Water on planet earth as viewed from an Apollo spacecraft.**

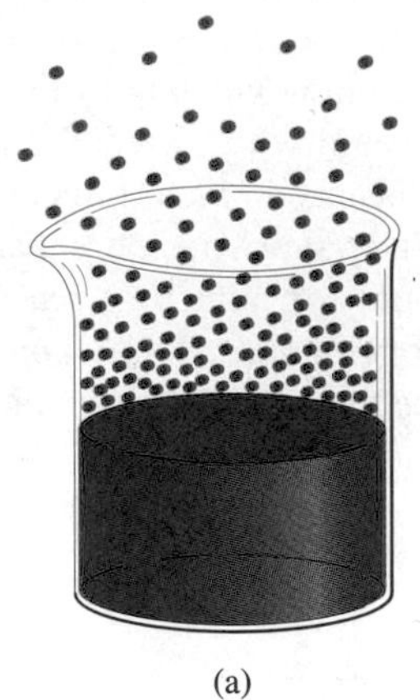
(a)

(b)

FIGURE 13.1 ▶
(a) Molecules in an open beaker evaporate from the liquid and disperse into the atmosphere. Evaporation will continue until all the liquid is gone.
(b) Molecules leaving the liquid are confined to a limited space. With time, the concentration in the vapor phase will increase until an equilibrium between liquid and vapor is established.

constant. When volatile liquids such as ethyl chloride, C_2H_5Cl, are sprayed on the skin, they evaporate rapidly, cooling the area by removing heat. The numbing effect of the low temperature produced by evaporation of ethyl chloride allows it to be used as a local anesthetic for minor surgery.

sublimation

Solids such as iodine, camphor, naphthalene (moth balls), and, to a small extent, even ice will go directly from the solid to the gaseous state, bypassing the liquid state. This change is a form of evaporation and is called **sublimation**:

$$\text{liquid} \xrightarrow{\text{evaporation}} \text{vapor}$$

$$\text{solid} \xrightarrow{\text{sublimation}} \text{vapor}$$

13.3 Vapor Pressure

condensation

When a liquid vaporizes in a closed system as shown in Figure 13.1, part (b), some of the molecules in the vapor or gaseous state strike the surface and return to the liquid state by the process of **condensation**. The rate of condensation increases until it is equal to the rate of vaporization. At this point, the space above the liquid is said to be saturated with vapor, and an equilibrium, or steady state, exists between the liquid and the vapor. The equilibrium equation is

$$\text{liquid} \underset{\text{condensation}}{\overset{\text{vaporization}}{\rightleftarrows}} \text{vapor}$$

This equilibrium is dynamic; both processes—vaporization and condensation—are taking place, even though one cannot see or measure a change. The number of molecules leaving the liquid in a given time interval is equal to the number of molecules returning to the liquid.

vapor pressure

At equilibrium the molecules in the vapor exert a pressure like any other gas. The pressure exerted by a vapor in equilibrium with its liquid is known as the **vapor pressure** of the liquid. The vapor pressure may be thought of as a measure of the "escaping" tendency of molecules to go from the liquid to the vapor state. The

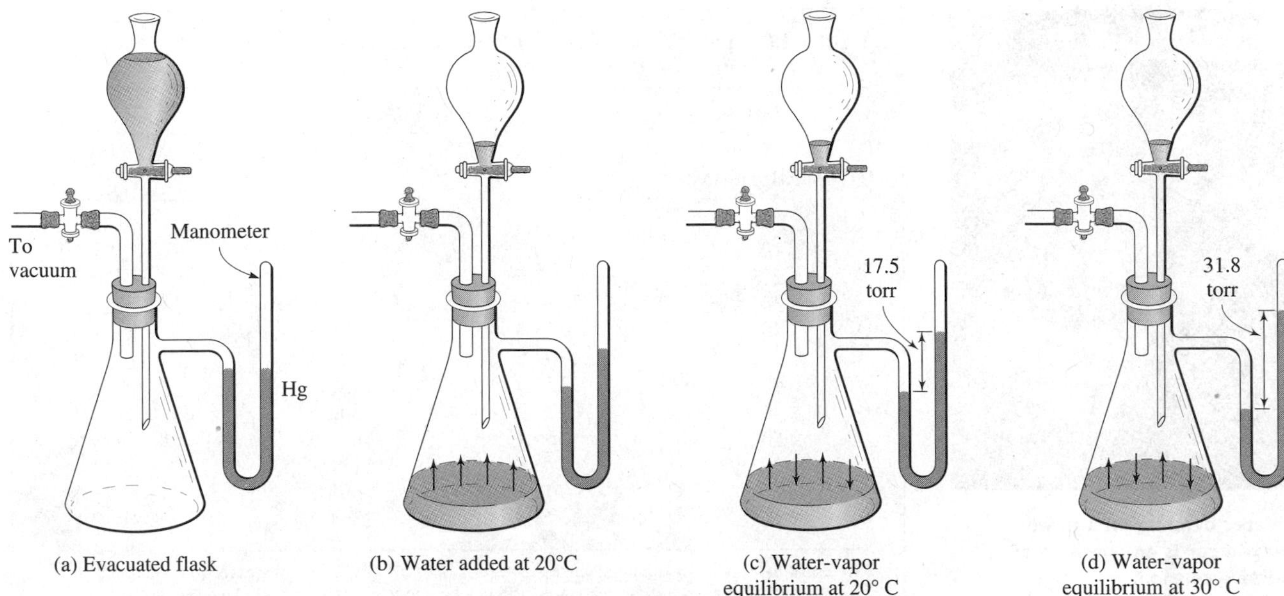

▲
FIGURE 13.2
Measurement of the vapor pressure of water at 20°C and 30°C. In flask (a) the system is evacuated. The mercury manometer attached to the flask shows equal pressure in both legs. In (b) water has been added to the flask and begins to evaporate, exerting pressure as indicated by the manometer. In (c), when equilibrium is established, the pressure inside the flask remains constant at 17.5 torr. In (d) the temperature is changed to 30°C, and equilibrium is reestablished with the vapor pressure at 31.8 torr.

vapor pressure of a liquid is independent of the amount of liquid and vapor present, but it increases as the temperature rises. Figure 13.2 illustrates a liquid–vapor equilibrium and the measurement of vapor pressure.

When equal volumes of water, ethyl ether, and ethyl alcohol are placed in separate beakers and allowed to evaporate at the same temperature, we observe that the ether evaporates faster than the alcohol, which evaporates faster than the water. This order of evaporation is consistent with the fact that ether has a higher vapor pressure at any particular temperature than ethyl alcohol or water. One reason for this higher vapor pressure is that the attraction is less between ether molecules than between alcohol molecules or between water molecules. The vapor pressures of these three compounds at various temperatures are compared in Table 13.1.

volatile

Substances that evaporate readily are said to be **volatile**. A volatile liquid has a relatively high vapor pressure at room temperature. Ethyl ether is a very volatile liquid, water is not too volatile, and mercury, which has a vapor pressure of 0.0012 torr at 20°C, is essentially a nonvolatile liquid. Most substances that are normally in a solid state are nonvolatile (solids that sublime are exceptions).

13.4 Surface Tension

Have you ever observed water and mercury in the form of small drops? These liquids occur as drops due to *surface tension* of liquids. A droplet of liquid that is not falling or under the influence of gravity (as on the space shuttle) will form a sphere. Minimum surface area is found in the geometrical form of the sphere. The molecules within the liquid are attracted to the surrounding liquid molecules. However, at the surface of the liquid, the attraction is nearly all inward, pulling the surface into a spherical shape. The resistance of a liquid to an increase in its surface

Water beading on a newly waxed car is an example of surface tension.

TABLE 13.1 The Vapor Pressure of Water, Ethyl Alcohol, and Ethyl Ether at Various Temperatures

	Vapor pressure (torr)		
Temperature (°C)	Water	Ethyl alcohol	Ethyl ether*
0	4.6	12.2	185.3
10	9.2	23.6	291.7
20	17.5	43.9	442.2
30	31.8	78.8	647.3
40	55.3	135.3	921.3
50	92.5	222.2	1276.8
60	152.9	352.7	1729.0
70	233.7	542.5	2296.0
80	355.1	812.6	2993.6
90	525.8	1187.1	3841.0
100	760.0	1693.3	4859.4
110	1074.6	2361.3	6070.1

* Note that the vapor pressure of ethyl ether at temperatures of 40°C and higher exceeds standard pressure, 760 torr, which indicates that the substance has a low boiling point and therefore should be stored in a cool place in a tightly sealed container.

surface tension

area is called the **surface tension** of the liquid. Substances with large attractive forces between molecules have high surface tensions. The effect of surface tension in water is illustrated by the phenomenon of floating a needle on the surface of still water. Other examples include the movement of the water strider insect across a calm pond, or the beading of water on a freshly waxed car.

capillary action

Liquids also exhibit a phenomenon known as **capillary action**, which is the spontaneous rising of a liquid in a narrow tube. This action results from the *cohesive forces* within the liquid, and the *adhesive forces* between the liquid and the walls of the container. If the forces between the liquid and the container are greater than those within the liquid itself, the liquid will climb the walls of the container. For example, consider the California sequoia tree, which can be over 200 feet in height. Under atmospheric pressure, water will only rise 33 feet in a glass tube, but capillary action will cause water to rise from the roots to all parts of the tree.

meniscus

The meniscus in liquids is further evidence of these cohesive and adhesive forces. When a liquid is placed in a glass cylinder the surface of the liquid shows a curve called the **meniscus** (see Figure 13.3). The concave shape of the meniscus of water shows that the adhesive forces between the glass and the liquid are stronger than the cohesive forces within the liquid. In a nonpolar substance such as mercury, the meniscus is convex, indicating that the cohesive forces within the mercury are greater than the adhesive forces between the glass wall and the mercury.

13.5 Boiling Point

The boiling temperature of a liquid is associated with its vapor pressure. We have seen that the vapor pressure increases as the temperature increases. When the internal or vapor pressure of a liquid becomes equal to the external pressure, the liquid boils. (By external pressure we mean the pressure of the atmosphere above the

TABLE 13.2 Physical Properties of Ethyl Chloride, Ethyl Ether, Ethyl Alcohol, and Water

Substance	Boiling point (°C)	Melting point (°C)	Heat of vaporization J/g (cal/g)	Heat of fusion J/g (cal/g)
Ethyl chloride	13	−139	387 (92.5)	—
Ethyl ether	34.6	−116	351 (83.9)	—
Ethyl alcohol	78.4	−112	855 (204.3)	104 (24.9)
Water	100.0	0	2259 (540)	335 (80)

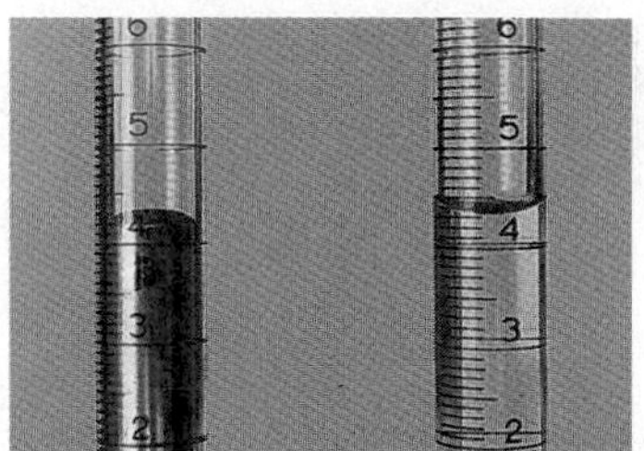

FIGURE 13.3
The meniscus is the characteristic curve of the surface of a liquid in a narrow capillary tube. Here we see the meniscus of mercury (left) and water (right)

liquid.) The boiling temperature of a pure liquid remains constant as long as the external pressure does not vary.

The boiling point (bp) of water is 100°C at 1 atm pressure. Table 13.1 shows that the vapor pressure of water at 100°C is 760 torr, a figure we have seen many times before. The significant fact here is that the boiling point is the temperature at which the vapor pressure of the water or other liquid is equal to standard, or atmospheric, pressure at sea level. These relationships lead to the following definition: The **boiling point** is the temperature at which the vapor pressure of a liquid is equal to the external pressure above the liquid.

boiling point

We can readily see that a liquid has an infinite number of boiling points. When we give the boiling point of a liquid, we should also state the pressure. When we express the boiling point without stating the pressure, we mean it to be the **normal boiling point** at standard pressure (760 torr). Using Table 13.1 again, we see that the normal boiling point of ethyl ether is between 30°C and 40°C, and for ethyl alcohol it is between 70°C and 80°C, because, for each compound, 760 torr pressure lies within these stated temperature ranges. At the normal boiling point, 1 g of a liquid changing to a vapor (gas) absorbs an amount of energy equal to its heat of vaporization (see Table 13.2).

normal boiling point

The boiling point at various pressures may be evaluated by plotting the data of Table 13.2 on the graph in Figure 13.4, where temperature is plotted horizontally along the *x* axis and vapor pressure is plotted vertically along the *y* axis. The resulting curves are known as **vapor-pressure curves**. Any point on these curves represents a vapor–liquid equilibrium at a particular temperature and pressure. We may find the boiling point at any pressure by tracing a horizontal line from the designated pressure to a point on the vapor-pressure curve. From this point we draw a vertical line to obtain the boiling point on the temperature axis. Four such points are shown in Figure 13.4; they represent the normal boiling points of the four compounds at 760 torr pressure. By reversing this process, you can ascertain at what pressure a substance will boil at a specific temperature. The boiling point is one of the most commonly used physical properties for characterizing and identifying substances.

vapor-pressure curves

Practice 13.1

Use the graph in Figure 13.4 to determine the boiling points of ethyl chloride, ethyl ether, ethyl alcohol, and water at 600 torr.

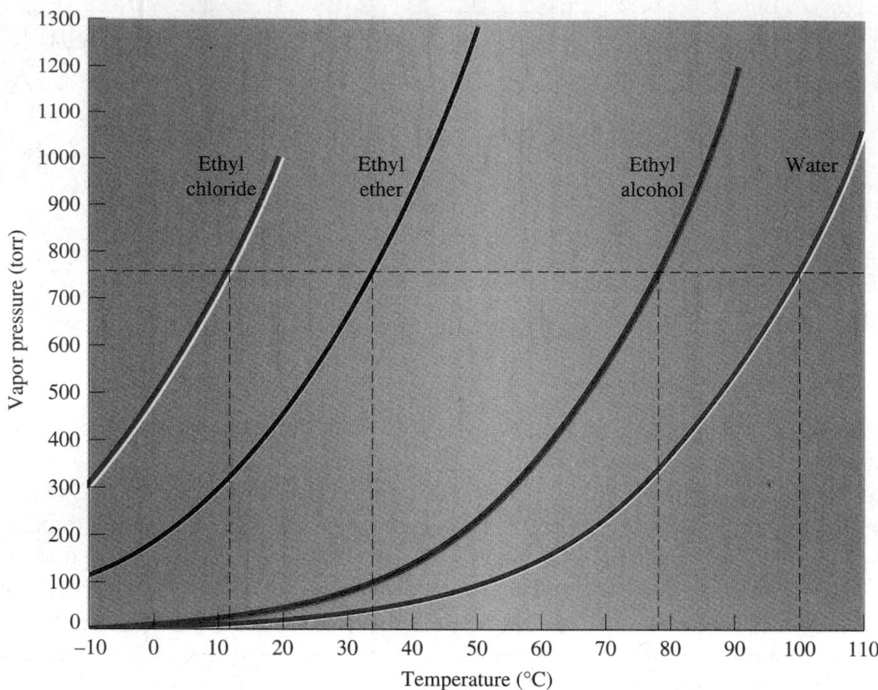

FIGURE 13.4 ▶ **Vapor-pressure–temperature curves for ethyl chloride, ethyl ether, ethyl alcohol, and water.**

Practice 13.2

The average atmospheric pressure in Denver is 0.83 atm. What is the boiling point of water in Denver?

13.6 Freezing Point or Melting Point

As heat is removed from a liquid, the liquid becomes colder and colder, until a temperature is reached at which it begins to solidify. A liquid changing into a solid is said to be *freezing,* or *solidifying.* When a solid is heated continuously, a temperature is reached at which the solid begins to liquefy. A solid that is changing into a liquid is said to be *melting.* The temperature at which the solid phase of a substance is in equilibrium with its liquid phase is known as the **freezing point** or **melting point** of that substance. The equilibrium equation is

freezing or melting point

$$\text{solid} \underset{\text{freezing}}{\overset{\text{melting}}{\rightleftarrows}} \text{liquid}$$

When a solid is slowly and carefully heated so that a solid–liquid equilibrium is achieved and then maintained, the temperature will remain constant as long as both phases are present. The energy is used solely to change the solid to the liquid. The melting point is another physical property that is commonly used for characterizing substances.

The most common example of a solid–liquid equilibrium is ice and water. In a well-stirred system of ice and water, the temperature remains at 0°C as long as both phases are present. The melting point changes with pressure but only slightly unless the pressure change is very large.

13.7 Changes of State

The majority of solids undergo two changes of state upon heating. A solid changes to a liquid at its melting point, and a liquid changes to a gas at its boiling point. This warming process can be represented by a graph called a *heating curve* (Figure 13.5). This figure shows ice being heated at a constant rate. As energy flows into the ice, the vibrations within the crystal increase and the temperature rises ($A \longrightarrow B$). Eventually, the molecules begin to break free from the crystal and melting occurs ($B \longrightarrow C$). During the melting process all energy goes into breaking down the crystal structure; the temperature remains constant.

The energy required to change 1 g of a solid at its melting point into a liquid is called the **heat of fusion**. When the solid has completely melted, the temperature once again rises ($C \longrightarrow D$); the energy input is increasing the molecular motion within the water. At 100°C, the water reaches its boiling point; the temperature remains constant while the added energy is used to vaporize the water to steam ($D \longrightarrow E$). The **heat of vaporization** is the energy required to change 1 g of liquid to vapor at its normal boiling point. The attractive forces between the liquid molecules are overcome during vaporization. Beyond this temperature, all the water exists as steam and is being heated further ($E \longrightarrow F$).

heat of fusion

heat of vaporization

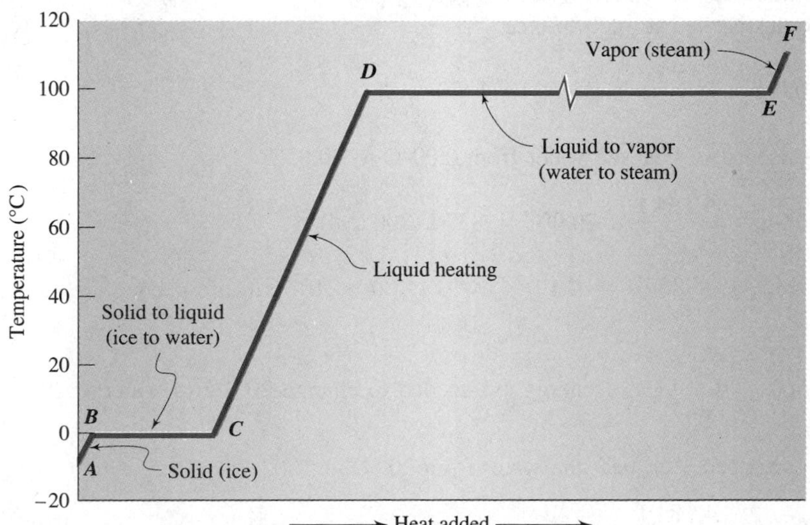

FIGURE 13.5
Heating curve for a pure substance—the absorption of heat by a substance from the solid state to the vapor state. Using water as an example, the *AB* interval represents the ice phase; *BC* interval, the melting of ice to water; *CD* interval, the elevation of the temperature of water from 0°C to 100°C; *DE* interval, the boiling of water to steam; and *EF* interval, the heating of steam.

Chemistry in Action

Comet Power

When a comet is near the sun, some of the ice on its surface sublimates into the vacuum of space. This rapidly moving water vapor drags comet dust along with it, creating two of the characteristic features of a comet, its tail and its dusty shroud (usually called a coma). Scientists understand that the production of a tail on a comet traveling near the sun is the result of the rapid conversion of ice to water vapor from the heat of the sun. But why would a comet traveling far from the sun have a tail when there is not enough heat to transform ice to water vapor? This is a mystery. How can these comets produce such celestial wonders?

New observations from the University of Hawaii, Honolulu, suggest that the release of carbon monoxide provides the spectacle. Astronomers Matthew C. Senay and David Jewitt used two short-wavelength radio telescopes to detect the emissions of gases from Comet Schwassmann-Wachmann 1. This particular comet never gets closer to the sun than Jupiter. Their telescopes revealed that the comet releases a startling amount of carbon monoxide. Carbon monoxide sublimes at a temperature of 25 K. The carbon monoxide coming from the comet has about the same speed as the comet but moves in general toward the sun. Senay and Jewitt suggest that the gas comes from the surface of the comet most illuminated by the sun. The sun sublimes the carbon monoxide at a rate of nearly 2000 kg/sec, producing the tail of gas headed toward the sun.

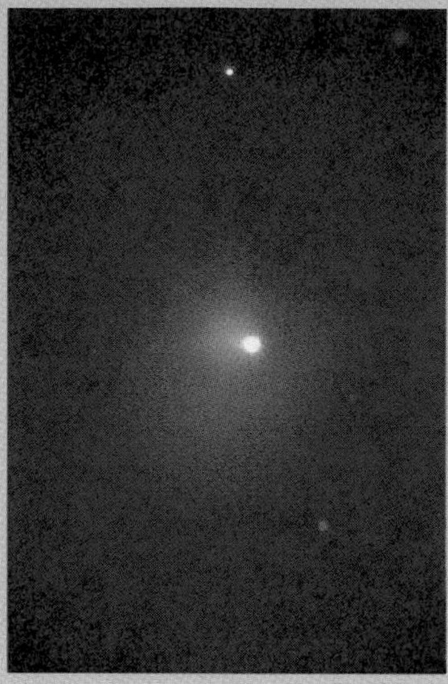

Example 13.1 How many joules of energy are needed to change 10.0 g of ice at 0.00°C to water at 20.0°C?

Solution Ice will absorb 335 J/g (heat of fusion) in going from a solid at 0°C to a liquid at 0°C. An additional 4.184 J/g°C (specific heat of water) are needed to raise the temperature of the water for each 1°C.

Joules needed to melt the ice:

$$10.0\ \cancel{g} \times \frac{335\ \text{J}}{1\ \cancel{g}} = 3.35 \times 10^3\ \text{J (801 cal)}$$

Joules needed to heat the water from 0.00°C to 20.0°C:

$$10.0\ \cancel{g} \times \frac{4.184\ \text{J}}{1\ \cancel{g}\cancel{^\circ C}} \times 20.0\cancel{^\circ C} = 837\ \text{J (200. cal)}$$

Thus, 3350 J + 837 J = 4.19×10^3 J (1.00×10^3 cal) are needed.

Example 13.2 How many kilojoules of energy are needed to change 20.0 g of water at 20.°C to steam at 100.°C?

Solution Kilojoules needed to heat the water from 20.°C to 100.°C:

$$20.0\ \cancel{g} \times \frac{4.184\ \cancel{J}}{1\ \cancel{g}\cancel{^\circ C}} \times \frac{1\ \text{kJ}}{1000\ \cancel{J}} \times 80.\cancel{^\circ C} = 6.7\ \text{kJ}$$

Kilojoules needed to change water at 100.°C to steam at 100.°C:

$$20.0\ \text{g} \times \frac{2.26\ \text{kJ}}{1\ \text{g}} = 45.2\ \text{kJ}$$

Thus, 6.7 kJ + 45.2 kJ = 51.9 kJ are needed.

Practice 13.3

How many kilojoules of energy are required to change 50.0 g of ethyl alcohol from 60.0°C to vapor at 78.4°C? The specific heat of ethyl alcohol is 2.138 J/g°C.

13.8 Occurrence of Water

Water is our most common natural resource. It covers about 75% of the earth's surface. Not only is it found in the oceans and seas, in lakes, rivers, streams, and in glacial ice deposits, it is also always present in the atmosphere and in cloud formations.

About 97% of the earth's water is in the oceans. This *saline* water contains vast amounts of dissolved minerals. More than 70 elements have been detected in the mineral content of seawater. Only four of these—chlorine, sodium, magnesium, and bromine—are now commercially obtained from the sea. The world's *fresh* water comprises the other 3%, of which about two-thirds is locked up in polar ice caps and glaciers. The remaining fresh water is found in groundwater, lakes, and the atmosphere.

Water is an essential constituent of all living matter. It is the most abundant compound in the human body, making up about 70% of total body mass. About 92% of blood plasma is water; about 80% of muscle tissue is water; and about 60% of a red blood cell is water. Water is more important than food in the sense that a person can survive much longer without food than without water.

13.9 Physical Properties of Water

Water is a colorless, odorless, tasteless liquid with a melting point of 0°C and a boiling point of 100°C at 1 atm. The heat of fusion of water is 335 J/g (80 cal/g). The heat of vaporization of water is 2.26 kJ/g (540 cal/g). The values for water for both the heat of fusion and the heat of vaporization are high compared with those for other substances; these high values indicate that strong attractive forces are acting between the molecules.

Ice and water exist together in equilibrium at 0°C, as shown in Figure 13.6. When ice at 0°C melts, it absorbs 335 J/g (80 cal/g) in changing into a liquid; the temperature remains at 0°C. In order to refreeze the water we have to remove 335 J/g (80 cal/g) from the liquid at 0°C.

FIGURE 13.6 ▶
Water equilibrium systems. In the beaker on the left, ice and water are in equilibrium at 0°C; in the flask on the right, boiling water and steam are in equilibrium at 100°C.

In Figure 13.6 both boiling water and steam are shown to have a temperature of 100°C. It takes 418 J (100 cal) to heat 1 g of water from 0°C to 100°C, but water at its boiling point absorbs 2.26 kJ/g (540 cal/g) in changing to steam. Although boiling water and steam are both at the same temperature, steam contains considerably more heat per gram and can cause more severe burns than hot water. In Table 13.3 the physical properties of water are tabulated and compared with those of other hydrogen compounds of Group VIA elements.

The maximum density of water is 1.000 g/mL at 4°C. Water has the unusual property of contracting in volume as it is cooled to 4°C and then expanding when cooled from 4°C to 0°C. Therefore 1 g of water occupies a volume greater than 1 mL at all temperatures except 4°C. Although most liquids contract in volume all the way down to the point at which they solidify, a large increase (about 9%) in volume occurs when water changes from a liquid at 0°C to a solid (ice) at 0°C. The density of ice at 0°C is 0.917 g/mL, which means that ice, being less dense than water, will float in water.

TABLE 13.3 Physical Properties of Water and Other Hydrogen Compounds of Group VIA Elements

Formula	Color	Molar mass (g/mol)	Melting point (°C)	Boiling point, 1 atm (°C)	Heat of fusion J/g (cal/g)	Heat of vaporization J/g (cal/g)
H_2O	Colorless	18.02	0.00	100.0	335 (80.0)	2.26×10^3 (540)
H_2S	Colorless	34.08	−85.5	−60.3	69.9 (16.7)	548 (131)
H_2Se	Colorless	80.98	−65.7	−41.3	31 (7.4)	238 (57.0)
H_2Te	Colorless	129.6	−49	−2	—	179 (42.8)

13.10 Structure of the Water Molecule

A single water molecule consists of two hydrogen atoms and one oxygen atom. Each hydrogen atom is attached to the oxygen atom by a single covalent bond. This bond is formed by the overlap of the 1*s* orbital of hydrogen with an unpaired 2*p* orbital of oxygen. The average distance between the two nuclei is known as the *bond length.* The O—H bond length in water is 0.096 nm. The water molecule is nonlinear and has a bent structure with an angle of about 105 degrees between the two bonds (see Figure 13.7).

Oxygen is the second most electronegative element. As a result, the two covalent OH bonds in water are polar. If the three atoms in a water molecule were aligned in a linear structure, such as H⟶O⟵H, the two polar bonds would be acting in equal and opposite directions and the molecule would be nonpolar. However, water is a highly polar molecule. Therefore, it does not have a linear structure. When atoms are bonded together in a nonlinear fashion, the angle formed by the bonds is called the *bond angle.* In water, the HOH bond angle is 105°. The two polar covalent bonds and the bent structure result in a partial negative charge on the oxygen atom and a partial positive charge on each hydrogen atom. The polar nature of water is responsible for many of its properties, including its behavior as a solvent.

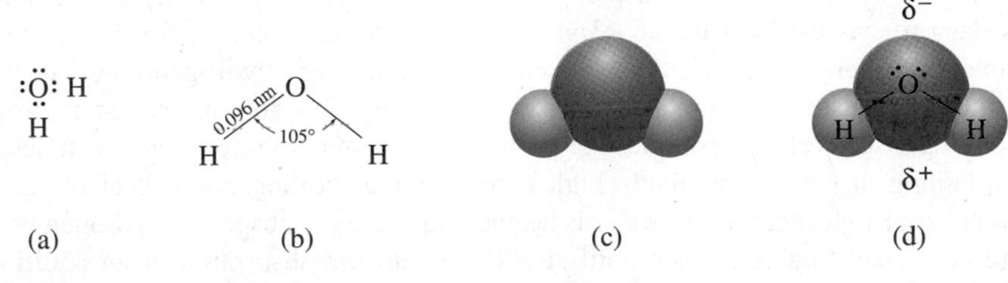

FIGURE 13.7
Diagrams of a water molecule: (a) electron distribution, (b) bond angle and O—H bond length, (c) molecular orbital structure, and (d) dipole representation.

13.11 The Hydrogen Bond

Table 13.3 compares the physical properties of H_2O, H_2S, H_2Se, and H_2Te. From this comparison it is apparent that four physical properties of water—melting point, boiling point, heat of fusion, and heat of vaporization—are extremely high and do not fit the trend relative to the molar masses of the four compounds. If the properties of water followed the progression shown by the other three compounds, we would expect the melting point of water to be below −85°C and the boiling point to be below −60°C.

Why does water have these anomalous physical properties? It is because liquid water molecules are held together more strongly than other molecules in the same family. The intermolecular force acting between water molecules is called a **hydrogen bond**, which acts like a very weak bond between two polar molecules. A

hydrogen bond

hydrogen bond is formed between polar molecules that contain hydrogen covalently bonded to a small, highly electronegative atom such as fluorine, oxygen, or nitrogen (F—H, O—H, N—H). A hydrogen bond is actually the dipole–dipole attraction between polar molecules containing these three types of polar bonds.

Elements that have significant hydrogen-bonding ability are F, O, and N.

Because a hydrogen atom has only one electron, it can form only one covalent bond. When it is attached to a strong electronegative atom such as oxygen, a hydrogen atom will also be attracted to an oxygen atom of another molecule, forming a dipole–dipole attraction (H-bond) between the two molecules. Water has two types of bonds: covalent bonds that exist between hydrogen and oxygen atoms within a molecule and hydrogen bonds that exist between hydrogen and oxygen atoms in *different* water molecules.

Hydrogen bonds are *intermolecular* bonds; that is, they are formed between atoms in different molecules. They are somewhat ionic in character because they are formed by electrostatic attraction. Hydrogen bonds are much weaker than the ionic or covalent bonds that unite atoms to form compounds. Despite their weakness, they are of great chemical importance.

The oxygen atom in water can form two hydrogen bonds—one through each of the unbonded pairs of electrons. Figure 13.8 shows (a) two water molecules linked by a hydrogen bond and (b) six water molecules linked by hydrogen bonds. A dash (—) is used for the covalent bond and a dotted line (••••) for the hydrogen bond. In water each molecule is linked to others through hydrogen bonds to form a three-dimensional aggregate of water molecules. This intermolecular hydrogen bonding effectively gives water the properties of a much larger, heavier molecule, explaining in part its relatively high melting point, boiling point, heat of fusion, and heat of vaporization. As water is heated and energy is absorbed, hydrogen bonds are continually being broken until at 100°C, with the absorption of an additional 2.26 kJ/g (540 cal/g), water separates into individual molecules, going into the gaseous state. Sulfur, selenium, and tellurium are not sufficiently electronegative for their hydrogen compounds to behave like water. The lack of hydrogen bonding is one reason why H_2S is a gas and not a liquid at room temperature.

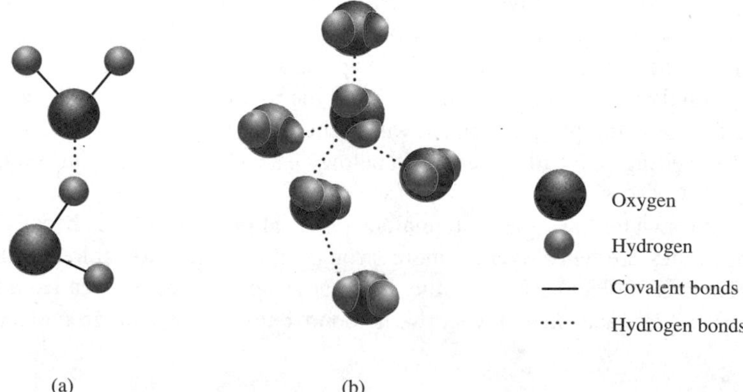

FIGURE 13.8 ▶ Hydrogen bonding. Water in the liquid and solid states exists as aggregates in which the water molecules are linked together by hydrogen bonds.

Fluorine, the most electronegative element, forms the strongest hydrogen bonds. This bonding is strong enough to link hydrogen fluoride molecules together as *dimers,* H_2F_2, or as larger $(HF)_n$ molecular units. The dimer structure may be represented in this way:

$$\mathrm{H{-}\ddot{\underset{\cdot\cdot}{F}}\!:\cdots H{-}\ddot{\underset{\cdot\cdot}{F}}\!:}$$

H-bond

Hydrogen bonding can occur between two different atoms that are capable of forming H-bonds. Thus we may have an O····H—N or O—H····N linkage in which the hydrogen atom forming the H-bond is between an oxygen and a nitrogen atom. This form of the H-bond exists in certain types of protein molecules and many biologically active substances.

Example 14.3

Would you expect hydrogen bonding to occur between molecules of the following substances?

$$\text{(a)}\ \begin{array}{ccccccc} & & \mathrm{H} & & \mathrm{H} & & \\ & & | & & | & & \\ \mathrm{H} & - & \mathrm{C} & - & \mathrm{C} & - & \mathrm{\ddot{\underset{\cdot\cdot}{O}}{-}H} \\ & & | & & | & & \\ & & \mathrm{H} & & \mathrm{H} & & \end{array} \qquad \text{(b)}\ \begin{array}{ccccccccc} & & \mathrm{H} & & & & \mathrm{H} & & \\ & & | & & & & | & & \\ \mathrm{H} & - & \mathrm{C} & - & \mathrm{\ddot{\underset{\cdot\cdot}{O}}} & - & \mathrm{C} & - & \mathrm{H} \\ & & | & & & & | & & \\ & & \mathrm{H} & & & & \mathrm{H} & & \end{array}$$

ethyl alcohol — dimethyl ether

Solution

(a) Hydrogen bonding should occur in ethyl alcohol because one hydrogen atom is bonded to an oxygen atom:

$$\begin{array}{ccccccccccccc} & & \mathrm{H} & & \mathrm{H} & & & & & \mathrm{H} & \mathrm{H} & \mathrm{H} & \\ & & | & & | & & & & & | & | & | & \\ \mathrm{H} & - & \mathrm{C} & - & \mathrm{C} & - & \mathrm{\ddot{\underset{\cdot\cdot}{O}}} & - & \mathrm{H}\cdots & \mathrm{:\!\underset{\cdot\cdot}{O}} & -\mathrm{C} & -\mathrm{C} & -\mathrm{H} \\ & & | & & | & & & & \uparrow & & | & | & \\ & & \mathrm{H} & & \mathrm{H} & & & & \text{H-bond} & & \mathrm{H} & \mathrm{H} & \end{array}$$

(b) There is no hydrogen bonding in dimethyl ether because all the hydrogen atoms are bonded only to carbon atoms.

Both ethyl alcohol and dimethyl ether have the same molar mass (46.07). Although both compounds have the same molecular formula, C_2H_6O, ethyl alcohol has a much higher boiling point (78.4°C) than dimethyl ether (−23.7°C) because of hydrogen bonding between the alcohol molecules.

Practice 13.4

Would you expect hydrogen bonding to occur between molecules of the following substances?

$$\text{(a)}\ \begin{array}{ccccccccc} & & \mathrm{H} & & \mathrm{H} & & \mathrm{H} & & \\ & & | & & | & & | & & \\ \mathrm{H} & - & \mathrm{C} & - & \mathrm{C} & - & \mathrm{\underset{\cdot\cdot}{N}} & - & \mathrm{H} \\ & & | & & | & & & & \\ & & \mathrm{H} & & \mathrm{H} & & & & \end{array} \quad \text{(b)}\ \begin{array}{ccccccccc} & & \mathrm{H} & & \mathrm{H} & & \mathrm{H} & & \\ & & | & & | & & | & & \\ \mathrm{H} & - & \mathrm{C} & - & \mathrm{\underset{\cdot\cdot}{N}} & - & \mathrm{C} & - & \mathrm{H} \\ & & | & & & & | & & \\ & & \mathrm{H} & & & & \mathrm{H} & & \end{array} \quad \text{(c)}\ \begin{array}{ccccccc} & & & & & & \mathrm{H} \\ & & & & & & | \\ & & \mathrm{H} & & \mathrm{H} & & \mathrm{H{-}C{-}H} \\ & & | & & | & & | \\ \mathrm{H} & - & \mathrm{C} & - & \mathrm{C} & \text{———} & \mathrm{N\!:} \\ & & | & & | & & | \\ & & \mathrm{H} & & \mathrm{H} & & \mathrm{H{-}C{-}H} \\ & & & & & & | \\ & & & & & & \mathrm{H} \end{array}$$

How Sweet It Is!

Intermolecular forces provide the chemical basis for the sweet taste of many consumer products. The artificial sweetener industry was built on the sweet taste of chemicals discovered quite by accident in the chemistry laboratory. In 1878, Ira Remsen was working late in his laboratory and glanced at the clock to discover he was about to miss a dinner with friends. In his haste to leave the lab he forgot to wash his hands. Later that evening at dinner he broke a piece of bread and tasted it only to discover that it was very sweet. He realized the sweet taste was the result of the chemical he had been working with in the lab and rushed back to the lab where he isolated saccharin—the first of the artificial sweeteners.

In 1937, while working in his laboratory, Michael Sveda was smoking a cigarette (a very dangerous practice to say the least!). He touched the cigarette to his lips and was surprised by the exceedingly sweet taste. When he purified the chemical he had on his hands it was found to be cyclamate, a staple of the artificial sweetener industry for many years.

In 1965, James Schlatter was researching anti-ulcer drugs for the pharmaceutical firm G. D. Searle. In the course of his work, he accidentally ingested a small amount of his preparation and found to his surprise that it had an extremely sweet taste. When purified, the sweet-tasting substance turned out to be aspartame, a molecule consisting of two amino acids joined together. Since only very small quantities of aspartame are necessary to produce sweetness, it proved to be an excellent low-calorie artificial sweetener. Today under the trade names of "Equal" and "Nutrasweet" aspartame is one of the cornerstones of the artificial sweetener industry.

There are more than 50 different molecules that have a sweet taste and all of them have similar molecular shapes. The triangle of sweetness theory, developed by Lamont Kier (Massachusetts College of Pharmacy) indicates that these molecules contain three sites that produce the proper structure to attach to the taste bud and trigger the response that registers "sweet" in our brains.

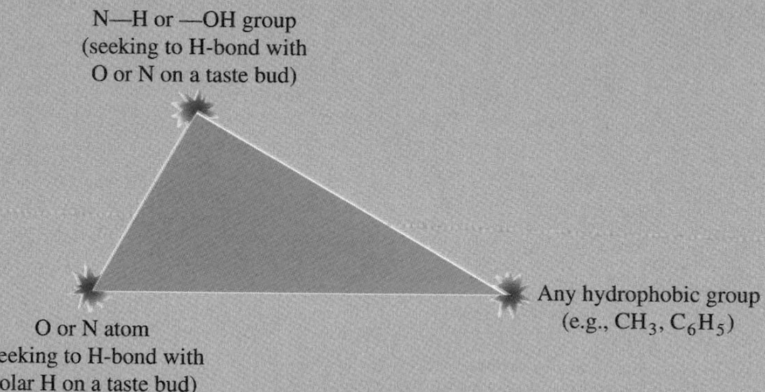

Triangle of sweetness

Taste buds are composed of proteins that can form hydrogen bonds with other molecules. The proteins contain —N—H and —OH groups (with hydrogen available to bond) as well as C=O groups (providing oxygen for hydrogen bonding). Molecules that are sweet also contain H-bonding groups including —OH, $—NH_2$, and O or N. The molecules must not only have the proper atoms to form hydrogen bonds, but must also contain a region that is hydrophobic (repels H_2O). The triangle in the diagram shows the three necessary sites. The molecule must contain the three sites located at just the proper distances.

The search for new and better sweeteners is continuing. The perfect sweetener would have the following qualities: (1) be as sweet or sweeter than sucrose (table sugar), (2) be nontoxic, (3) be quick to register sweet on the taste buds, (4) be easy to release so the taste doesn't linger, (5) have no calories, (6) be stable when cooked or dissolved, and (7) of course, be inexpensive. Scientists are continuing to search for different molecules that possess these qualities.

13.12 Formation and Chemical Properties of Water

Water is very stable to heat; it decomposes to the extent of only about 1% at temperatures up to 2000°C. Pure water is a nonconductor of electricity. But when a small amount of sulfuric acid or sodium hydroxide is added, the solution is readily decomposed into hydrogen and oxygen by an electric current. Two volumes of hydrogen are produced for each volume of oxygen:

$$2\,H_2O(l) \xrightarrow[H_2SO_4 \text{ or } NaOH]{\text{electrical energy}} 2\,H_2(g) + O_2(g)$$

Formation

Water is formed when hydrogen burns in air. Pure hydrogen burns very smoothly in air, but mixtures of hydrogen and air or oxygen explode when ignited. The reaction is strongly exothermic:

$$2\,H_2(g) + O_2(g) \longrightarrow 2\,H_2O(g) + 484\text{ kJ}$$

Water is produced by a variety of other reactions, especially by (1) acid–base neutralizations, (2) combustion of hydrogen-containing materials, and (3) metabolic oxidation in living cells:

1. $HCl(aq) + NaOH(aq) \longrightarrow NaCl(aq) + H_2O(l)$

2. $\underset{\text{acetylene}}{2\,C_2H_2(g)} + 5\,O_2(g) \longrightarrow 4\,CO_2(g) + 2\,H_2O(g) + 1212\text{ kJ}$

$\underset{\text{methane}}{CH_4(g)} + 2\,O_2(g) \longrightarrow CO_2(g) + 2\,H_2O(g) + 803\text{ kJ}$

3. $\underset{\text{glucose}}{C_6H_{12}O_6(aq)} + 6\,O_2(g) \xrightarrow{\text{enzymes}} 6\,CO_2(g) + 6\,H_2O(l) + 2519\text{ kJ}$

The combustion of acetylene shown in (2) is strongly exothermic and is capable of producing very high temperatures. It is used in oxygen–acetylene torches to cut and weld steel and other metals. Methane is known as natural gas and is commonly used as fuel for heating and cooking. The reaction of glucose with oxygen shown in (3) is the reverse of photosynthesis. It is the overall reaction by which living cells obtain needed energy by metabolizing glucose to carbon dioxide and water.

Reactions of Water with Metals and Nonmetals

The reactions of metals with water at different temperatures show that these elements vary greatly in their reactivity. Metals such as sodium, potassium, and calcium react with cold water to produce hydrogen and a metal hydroxide. A small piece of sodium added to water melts from the heat produced by the reaction, forming a silvery metal ball, which rapidly flits back and forth on the surface of the water. Caution must be used when experimenting with this reaction, because the hydrogen produced is frequently ignited by the sparking of the sodium, and it will explode, spattering sodium. Potassium reacts even more vigorously than sodium. Calcium sinks in water and liberates a gentle stream of hydrogen. The equations for these reactions are

$$2\ Na(s) + 2\ H_2O(l) \longrightarrow H_2(g) + 2\ NaOH(aq)$$

$$2\ K(s) + 2\ H_2O(l) \longrightarrow H_2(g) + 2\ KOH(aq)$$

$$Ca(s) + 2\ H_2O(l) \longrightarrow H_2(g) + Ca(OH)_2(aq)$$

Zinc, aluminum, and iron do not react with cold water but will react with steam at high temperatures, forming hydrogen and a metallic oxide. The equations are

$$Zn(s) + H_2O(g) \longrightarrow H_2(g) + ZnO(s)$$

$$2\ Al(s) + 3\ H_2O(g) \longrightarrow 3\ H_2(g) + Al_2O_3(s)$$

$$3\ Fe(s) + 4\ H_2O(g) \longrightarrow 4\ H_2(g) + Fe_3O_4(s)$$

Copper, silver, and mercury are examples of metals that do not react with cold water or steam to produce hydrogen. We conclude that sodium, potassium, and calcium are chemically more reactive than zinc, aluminum, and iron, which are more reactive than copper, silver, and mercury.

Certain nonmetals react with water under various conditions. For example, fluorine reacts violently with cold water, producing hydrogen fluoride and free oxygen. The reactions of chlorine and bromine are much milder, producing what is commonly known as "chlorine water" and "bromine water," respectively. Chlorine water contains HCl, HOCl, and dissolved Cl_2; the free chlorine gives it a yellow-green color. Bromine water contains HBr, HOBr, and dissolved Br_2; the free bromine gives it a reddish-brown color. Steam passed over hot coke (carbon) produces a mixture of carbon monoxide and hydrogen that is known as "water gas." Since water gas is combustible, it is useful as a fuel. It is also the starting material for the commercial production of several alcohols. The equations for these reactions are

$$2\ F_2(g) + 2\ H_2O(l) \longrightarrow 4\ HF(aq) + O_2(g)$$

$$Cl_2(g) + H_2O(l) \longrightarrow HCl(aq) + HOCl(aq)$$

$$Br_2(l) + H_2O(l) \longrightarrow HBr(aq) + HOBr(aq)$$

$$C(s) + H_2O(g) \xrightarrow{1000^\circ C} CO(g) + H_2(g)$$

Reactions of Water with Metal and Nonmetal Oxides

basic anhydride

Metal oxides that react with water to form hydroxides are known as **basic anhydrides**. Examples are

$$CaO(s) + H_2O(l) \longrightarrow \underset{\text{calcium hydroxide}}{Ca(OH)_2(aq)}$$

$$Na_2O(s) + H_2O(l) \longrightarrow \underset{\text{sodium hydroxide}}{2\ NaOH(aq)}$$

Certain metal oxides, such as CuO and Al_2O_3, do not form solutions containing OH^- ions because the oxides are insoluble in water.

acid anhydride

Nonmetal oxides that react with water to form acids are known as **acid anhydrides**. Examples are

$$CO_2(g) + H_2O(l) \rightleftarrows H_2CO_3(aq)$$
carbonic acid

$$SO_2(g) + H_2O(l) \rightleftarrows H_2SO_3(aq)$$
sulfurous acid

$$N_2O_5(s) + H_2O(l) \longrightarrow 2\ HNO_3(aq)$$
nitric acid

The word *anhydrous* means "without water." An anhydride is a metal oxide or a nonmetal oxide derived from a base or an oxy-acid by the removal of water. To determine the formula of an anhydride, the elements of water are removed from an acid or base formula until all the hydrogen is removed. Sometimes more than one formula unit is needed to remove all the hydrogen as water. The formula of the anhydride then consists of the remaining metal or nonmetal and the remaining oxygen atoms. In calcium hydroxide, removal of water as indicated leaves CaO as the anhydride:

$$Ca\begin{matrix}OH \\ OH\end{matrix} \xrightarrow{\Delta} CaO + H_2O$$

In sodium hydroxide, H_2O cannot be removed from one formula unit, so two formula units of NaOH must be used, leaving Na_2O as the formula of the anhydride:

$$\begin{matrix}NaOH \\ NaOH\end{matrix} \xrightarrow{\Delta} Na_2O + H_2O$$

The removal of H_2O from H_2SO_4 gives the acid anhydride SO_3:

$$H_2SO_4 \xrightarrow{\Delta} SO_3 + H_2O$$

The foregoing are examples of typical reactions of water but are by no means a complete list of the known reactions of water.

13.13 Hydrates

When certain solutions containing ionic compounds are allowed to evaporate, some water molecules remain as part of the crystalline compound that is left after evaporation is complete. Solids that contain water molecules as part of their crystalline structure are known as **hydrates**. Water in a hydrate is known as **water of hydration**, or **water of crystallization**.

hydrate
water of hydration
water of crystallization

Formulas for hydrates are expressed by first writing the usual anhydrous (without water) formula for the compound and then adding a dot followed by the number of water molecules present. An example is $BaCl_2 \cdot 2\ H_2O$. This formula tells us that each formula unit of this compound contains one barium ion, two chloride ions, and two water molecules. A crystal of the compound contains many of these units in its crystalline lattice.

As a solution of $CuSO_4$ evaporates beautiful blue crystals of $CuSO_4 \cdot 5\ H_2O$ are formed.

In naming hydrates, we first name the compound exclusive of the water and then add the term *hydrate,* with the proper prefix representing the number of water molecules in the formula. For example, $BaCl_2 \cdot 2\ H_2O$ is called *barium chloride dihydrate.* Hydrates are true compounds and follow the Law of Definite Composition. The molar mass of $BaCl_2 \cdot 2\ H_2O$ is 244.2 g/mol; it contains 56.22% barium, 29.03% chlorine, and 14.76% water.

Water molecules in hydrates are bonded by electrostatic forces between polar water molecules and the positive or negative ions of the compound. These forces are not as strong as covalent or ionic chemical bonds. As a result, water of crystallization can be removed by moderate heating of the compound. A partially dehydrated or completely anhydrous compound may result. When $BaCl_2 \cdot 2\ H_2O$ is heated, it loses its water at about 100°C:

$$BaCl_2 \cdot 2\ H_2O(s) \xrightarrow{100°C} BaCl_2(s) + 2\ H_2O(g)$$

When a solution of copper(II) sulfate ($CuSO_4$) is allowed to evaporate, beautiful blue crystals containing 5 moles of water per mole of $CuSO_4$ are formed. The formula for this hydrate is $CuSO_4 \cdot 5\ H_2O$; it is called copper(II) sulfate pentahydrate. When $CuSO_4 \cdot 5\ H_2O$ is heated, water is lost, and a pale green-white powder, anhydrous $CuSO_4$, is formed:

$$CuSO_4 \cdot 5\ H_2O(s) \xrightarrow{250°C} CuSO_4(s) + 5\ H_2O(g)$$

When water is added to anhydrous copper(II) sulfate, the foregoing reaction is reversed, and the compound turns blue again. Because of this outstanding color change, anhydrous copper(II) sulfate has been used as an indicator to detect small amounts of water. The formation of the hydrate is noticeably exothermic.

The formula for plaster of paris is $(CaSO_4)_2 \cdot H_2O$. When mixed with the proper quantity of water, plaster of paris forms a dihydrate and sets to a hard mass. It is, therefore, useful for making patterns for the reproduction of art objects, molds, and surgical casts. The chemical reaction is

$$(CaSO_4)_2 \cdot H_2O(s) + 3\ H_2O(l) \longrightarrow 2\ CaSO_4 \cdot 2\ H_2O(s)$$

Table 13.4 lists a number of common hydrates.

13.14 Hygroscopic Substances

hygroscopic substance

Many anhydrous compounds and other substances readily absorb water from the atmosphere. Such substances are said to be **hygroscopic**. This property can be observed in the following simple experiment: Spread a 10–20 g sample of anhydrous copper(II) sulfate on a watch glass and set it aside so that the compound is exposed to the air. Then determine the mass of the sample periodically for 24 hours, noting the increase in mass and the change in color. Over time, water is absorbed from the atmosphere, forming the blue pentahydrate $CuSO_4 \cdot 5\ H_2O$.

deliquescence

Some compounds continue to absorb water beyond the hydrate stage to form solutions. A substance that absorbs water from the air until it forms a solution is said to be **deliquescent**. A few granules of anhydrous calcium chloride or pellets of sodium hydroxide exposed to the air will appear moist in a few minutes, and within an hour will absorb enough water to form a puddle of solution. Diphosphorus

TABLE 13.4 Selected Hydrates

Hydrate	Name	Hydrate	Name
$CaCl_2 \cdot 2\ H_2O$	Calcium chloride dihydrate	$Na_2CO_3 \cdot 10\ H_2O$	Sodium carbonate decahydrate
$Ba(OH)_2 \cdot 8\ H_2O$	Barium hydroxide octahydrate	$(NH_4)_2C_2O_4 \cdot H_2O$	Ammonium oxalate monohydrate
$MgSO_4 \cdot 7\ H_2O$	Magnesium sulfate heptahydrate	$NaC_2H_3O_2 \cdot 3\ H_2O$	Sodium acetate trihydrate
$SnCl_2 \cdot 2\ H_2O$	Tin(II) chloride dihydrate	$Na_2B_4O_7 \cdot 10\ H_2O$	Sodium tetraborate decahydrate
$CoCl_2 \cdot 6\ H_2O$	Cobalt(II) chloride hexahydrate	$Na_2S_2O_3 \cdot 5\ H_2O$	Sodium thiosulfate pentahydrate

pentoxide (P_2O_5) picks up water so rapidly that its mass cannot be determined accurately except in an anhydrous atmosphere.

Compounds that absorb water are useful as drying agents (desiccants). Refrigeration systems must be kept dry with such agents or the moisture will freeze and clog the tiny orifices in the mechanism. Bags of drying agents are often enclosed in packages containing iron or steel parts to absorb moisture and prevent rusting. Anhydrous calcium chloride, magnesium sulfate, sodium sulfate, calcium sulfate, silica gel, and diphosphorus pentoxide are some of the compounds commonly used for drying liquids and gases that contain small amounts of moisture.

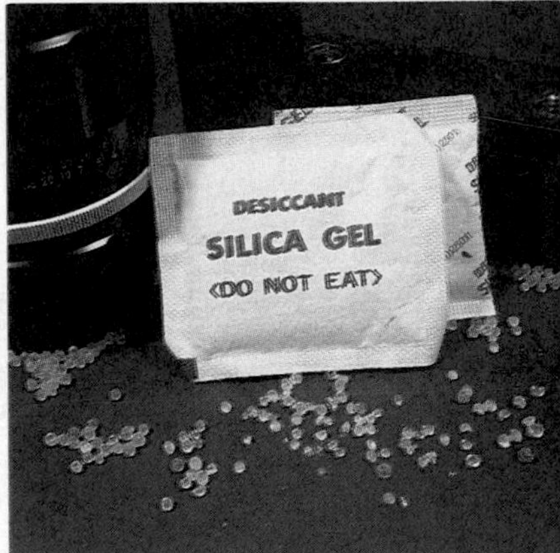

This common hygroscopic substance is packaged with many consumer products containing metals.

13.15 Natural Waters

Natural fresh waters are not pure, but contain dissolved minerals, suspended matter, and sometimes harmful bacteria. The water supplies of large cities are usually drawn from rivers or lakes. Such water is generally unsafe to drink without treatment. To make such water safe to drink it is treated by some or all of the following processes (See Figure 13.9).

1. *Screening.* Removal of relatively large objects, such as trash, fish, and so on.
2. *Flocculation and sedimentation.* Chemicals, usually lime, CaO, and alum (aluminum sulfate), $Al_2(SO_4)_3$, are added to form a flocculent jelly-like precipitate of aluminum hydroxide. This precipitate traps most of the fine suspended matter in the water and carries it to the bottom of the sedimentation basin.
3. *Sand filtration.* Water is drawn from the top of the sedimentation basin and passed downward through fine sand filters. Nearly all the remaining suspended matter and bacteria are removed by the sand filters.
4. *Aeration.* Water is drawn from the bottom of the sand filters and is aerated by spraying. The purpose of this process is to remove objectionable odors and tastes.
5. *Disinfection.* In the final stage chlorine gas is injected into the water to kill harmful bacteria before the water is distributed to the public. Ozone is also used in some countries to disinfect water. In emergencies water may be disinfected by simply boiling it for a few minutes.

Chemistry in Action

Moisturizers

Moisturizers have long been used to protect and rehydrate the skin. These products contain humectants and emollients that increase the water content of the skin in different ways. The emollients cover the skin with a layer of material that is immiscible with water thus preventing water from within the skin from evaporating. In contrast, humectants add water to the skin by attracting water vapor from the air.

The most common humectants are sorbitol, glycerin, and polypropylene glycol. Each molecule is polar, containing multiple —OH groups (see formulas).

The oxygen atom in each —OH group is considerably more electronegative than the hydrogen atom. This electronegativity difference results in a partial negative charge on the oxygen, whereas the hydrogen carries a partial positive charge. This polarity is the basis for the attraction between the humectant and the water molecules (also polar).

Emollients are composed of hydrophobic (water-insoluble) molecules. These products are made of nonpolar molecules. There is a great diversity of compounds in this category, including animal oils, vegetable oils, exotic oils (such as jojoba and aloe vera), and synthetic oils. In each case the molecules form a water-insoluble layer on the skin, which traps the skin's own moisture and feels smooth to the touch.

$$\begin{array}{c}
\mathrm{H} \\
| \\
\mathrm{H-C-OH} \\
| \\
\mathrm{H-C-OH} \\
| \\
\mathrm{H-C-OH} \\
| \\
\mathrm{H}
\end{array}
\qquad
\begin{array}{c}
\mathrm{H} \\
| \\
\mathrm{H-C-OH} \\
| \\
\mathrm{H-C-OH} \\
| \\
\mathrm{H-C-H} \\
| \\
\mathrm{H}
\end{array}$$

Glycerin — Propylene glycol

$$\begin{array}{c}
\mathrm{H} \\
| \\
\mathrm{H-C-OH} \\
| \\
\mathrm{H-C-OH} \\
| \\
\mathrm{HO-C-H} \\
| \\
\mathrm{H-C-OH} \\
| \\
\mathrm{H-C-OH} \\
| \\
\mathrm{H-C-OH} \\
| \\
\mathrm{H}
\end{array}$$

Sorbitol

Many consumers believe that alcohols can dry the skin and should not be used in skin care products. Is this concern justified? What is the purpose of alcohols in skin care products? The problem is not so simple as we might expect. There are two different categories of alcohol with distinctly different properties. Fatty acid alcohols, such as caprylic, isocetyl, oleyl, and stearyl, are made from long-chain fatty acids. These molecules are essentially nonpolar and behave like emollients. They form a water-insoluble layer on the skin that traps the moisture. The second type of alcohol is the simple alcohol, which includes ethyl, isopropyl, and methyl alcohols. These substances act as solvents in the skin care product and can be drying. They absorb excess oil, dissolve one ingredient into another and make products evaporate. Some also act to keep products from spoiling and separating. The problem with alcohol in skin care products results from overuse or use by people who don't need them.

One of the best tests for whether a skin care product will tend to dry the skin is to examine the texture. Liquids are typically the most drying formulations and are best used on oily skin. Gels containing lightweight emollients are best for skin with varying degrees of oiliness. Creams tend to contain heavier moisturizers and are best for normal to dry skin. Ointments are very heavy, creamy products that form a barrier to the skin, acting in the same manner as a humectant. Ointments are for use on severely dry or damaged skin. The key to selecting the proper moisturizer lies in understanding the way in which the product functions.

If the drinking water of children contains an optimum amount of fluoride ion, their teeth will be more resistant to decay. Therefore, in many communities NaF or Na_2SiF_6 is added to the water supply to bring the fluoride ion concentration up to the optimum level of about 1.0 ppm. Excessively high concentrations of fluoride ion can cause brown mottling of the teeth.

Water that contains dissolved calcium and magnesium salts is called *hard water*. One drawback of hard water is that ordinary soap does not lather well in it; the soap reacts with the calcium and magnesium ions to form an insoluble greasy scum. However, synthetic soaps, known as detergents, are available that have excellent

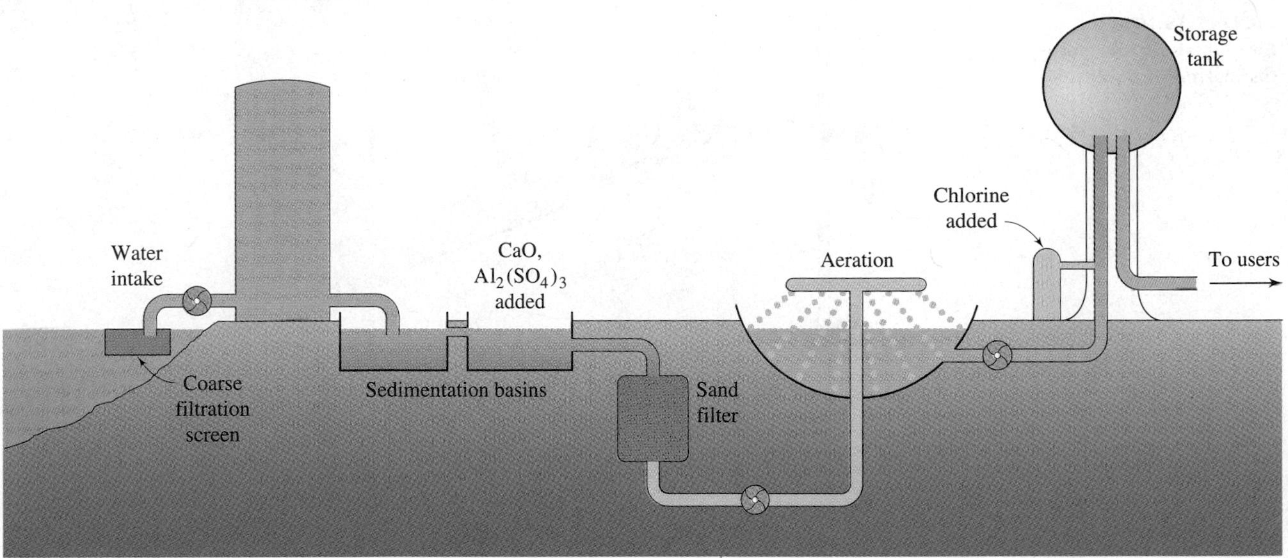

FIGURE 13.9
Typical municipal water treatment plant.

cleaning qualities and do not form precipitates with hard water. Hard water is also undesirable because it causes "scale" to form on the walls of water heaters, teakettles, coffee pots, and steam irons, which greatly reduces their efficiency.

Four techniques used to "soften" hard water are:

1. **Distillation** The water is boiled, and the steam formed is condensed into a liquid again, leaving the minerals behind in the distilling vessel. Figure 13.10 illustrates a simple laboratory distillation apparatus. Commercial stills are capable of producing hundreds of liters of distilled water per hour.
2. **Calcium precipitation** Calcium and magnesium ions are precipitated from hard water by adding sodium carbonate and lime. Insoluble calcium carbonate and magnesium hydroxide are precipitated and are removed by filtration or sedimentation.
3. **Ion-exchange** Hard water is effectively softened as it is passed through a bed or tank of zeolite—a complex sodium aluminum silicate. In this process, sodium ions replace objectionable calcium and magnesium ions, and the water is thereby softened:

   $$Na_2(\text{zeolite})(s) + Ca^{2+}(aq) \longrightarrow Ca(\text{zeolite})(s) + 2\,Na^+(aq)$$

   The zeolite is regenerated by back-flushing with concentrated sodium chloride solution, reversing the foregoing reaction.
4. **Demineralization** Both cations and anions are removed by a two-stage ion-exchange system. Special synthetic organic resins are used in the ion-exchange beds. In the first stage metal cations are replaced by hydrogen ions. In the second stage anions are replaced by hydroxide ions. The hydrogen and hydroxide ions react, and essentially pure, mineral-free water leaves the second stage.

The sodium ions present in water softened either by chemical precipitation or by the zeolite process are not objectionable to most users of soft water.

The oceans are an enormous source of water, but seawater contains about 3.5 lb of salts per 100 lb of water. This 35,000 ppm of dissolved salts makes seawater unfit for agricultural and domestic uses. Water that contains less than 1000 ppm of

FIGURE 13.10 ▶
Simple laboratory setup for distillation of liquids.

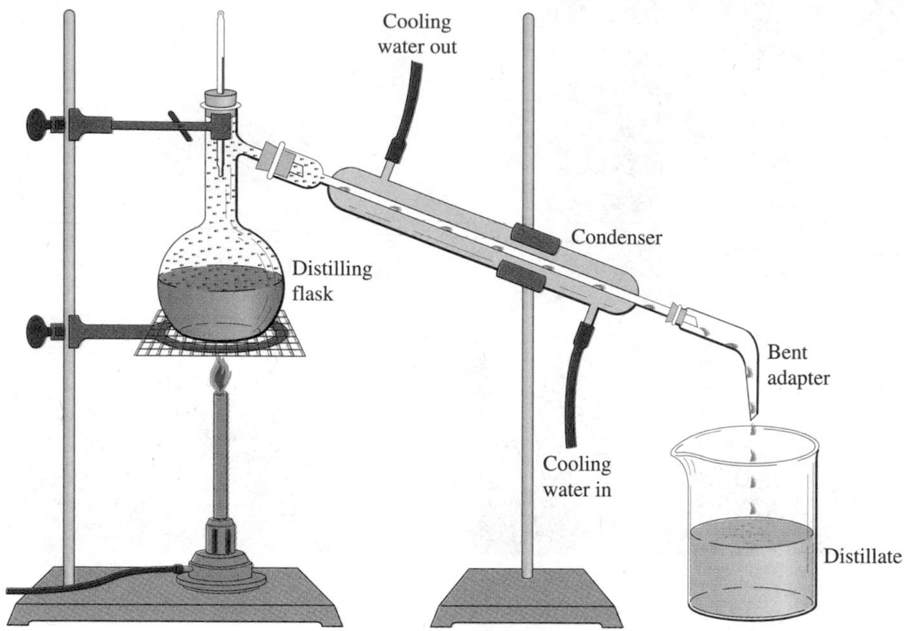

salts is considered reasonably good for drinking, and safe drinking water is already being obtained from the sea in many parts of the world. Continuous research is being done in an effort to make usable water from the oceans more abundant and economical. See Figure 13.11.

13.16 Water Pollution

Polluted water was formerly thought of as water that was unclear, had a bad odor or taste, and contained disease-causing bacteria. However, such factors as increased population, industrial requirements for water, atmospheric pollution, toxic waste dumps, and use of pesticides have greatly expanded the problem of water pollution.

Many of the newer pollutants are not removed or destroyed by the usual water-treatment processes. For example, among the 66 organic compounds found in the drinking water of a major city on the Mississippi River, 3 are labeled slightly toxic, 17 moderately toxic, 15 very toxic, 1 extremely toxic, and 1 supertoxic. Two are known carcinogens (cancer-producing agents), 11 are suspect, and 3 are metabolized to carcinogens. The U.S. Public Health Service classifies water pollutants under eight broad categories. These categories are shown in Table 13.5.

Many outbreaks of disease or poisoning, such as typhoid, dysentery, and cholera have been attributed directly to drinking water. Rivers and streams are an easy means for municipalities to dispose of their domestic and industrial waste products. Much of this water is used again by people downstream, and then discharged back into the water source. Then another community still farther downstream draws the same water and discharges its own wastes. Thus, along waterways such as the

FIGURE 13.11
Catalina Island, California (left), gets most of its water from a desalinization plant. Below is a series of filtration units used to treat the sea water, which is then pumped into the main reservoir.

TABLE 13.5 Classification of Water Pollutants

Type of pollutant	Examples
Oxygen-demanding wastes	Decomposable organic wastes from domestic sewage and industrial wastes of plant and animal origin
Infectious agents	Bacteria, viruses, and other organisms from domestic sewage, animal wastes, and animal process wastes
Plant nutrients	Principally compounds of nitrogen and phosphorus
Organic chemicals	Large numbers of chemicals synthesized by industry, pesticides, chlorinated organic compounds
Other minerals and chemicals	Inorganic chemicals from industrial operations, mining, oil field operations, and agriculture
Radioactive substances	Waste products from mining and processing radioactive materials, airborne radioactive fallout, increased use of radioactive materials in hospitals and research
Heat from industry	Large quantities of heated water returned to water bodies from power plants and manufacturing facilities after use for cooling
Sediment from land erosion	Solid matter washed into streams and oceans by erosion, rain, and water runoff

Mississippi and Delaware rivers, water is withdrawn and discharged many times. If this water is not properly treated, harmful pollutants build up, causing epidemics of various diseases.

The disposal of hazardous waste products adds to the water pollution problem. These products are unavoidable in the manufacture of many products that we use in everyday life. One common way to dispose of these wastes is to place them in toxic waste dumps. What has been found after many years of disposing of wastes in this manner is that toxic substances have seeped into the groundwater deposits. As a result many people have become ill, and water wells have been closed until satisfactory methods of detoxifying this water are found. This problem is serious, because one-half the United States population gets its drinking water from groundwater. To clean up the thousands of industrial dumps and to find and implement new and safe methods of disposing of wastes is ongoing and costly.

Many major water pollutants have been recognized and steps have been taken to eliminate them. Three that pose serious problems are lead, detergents, and chlorine-containing organic compounds. Lead poisoning, for example, has been responsible for many deaths in past years. One major toxic action of lead in the body is the inhibition of the enzyme necessary for the production of hemoglobin in the blood. The usual intake of lead into the body is through food. However, extraordinary amounts of lead can be ingested from water running through lead pipes and by using lead-containing ceramic containers for storage of food and beverages.

It has been clearly demonstrated that waterways rendered so polluted that the water is neither fit for human use nor able to sustain marine life can be successfully restored. However, keeping our lakes and rivers free from pollution is a very costly and complicated process.

Concepts in Review

1. List the common properties of liquids and solids. Explain how they are different from gases.
2. Explain the process of evaporation from the standpoint of kinetic energy.
3. Relate vapor-pressure data or vapor-pressure curves of different substances to their relative rates of evaporation and to their relative boiling points.
4. Explain the forces involved in surface tension of a liquid. Give two examples.
5. Explain why a meniscus forms on the surface of liquids in a container.
6. Explain what is occurring throughout the heating curve for water.
7. Describe a water molecule with respect to the Lewis structure, bond angle, and polarity.
8. Make sketches showing hydrogen bonding (a) between water molecules, (b) between hydrogen fluoride molecules, and (c) between ammonia molecules.
9. Explain the effect of hydrogen bonding on the physical properties of water.
10. Determine whether a compound will or will not form hydrogen bonds.
11. Identify metal oxides as basic anhydrides and write balanced equations for their reactions with water.

12. Identify nonmetal oxides as acid anhydrides and write balanced equations for their reactions with water.
13. Deduce the formula of the acid anhydride or basic anhydride when given the formula of the corresponding acid or base.
14. Identify the product, name each reactant and product, and write equations for the complete dehydration of hydrates.
15. Outline the processes necessary to prepare safe drinking water from a contaminated river source.
16. Describe how water may be softened by distillation, chemical precipitation, ion exchange, and demineralization.
17. Complete and balance equations for (a) the reactions of water with Na, K, and Ca; (b) the reaction of steam with Zn, Al, Fe, and C; and (c) the reaction of water with halogens.

Key Terms

The terms listed here have been defined within this chapter. Section numbers are referenced in parenthesis for each term.

acid anhydride (13.12)
basic anhydride (13.12)
boiling point (13.15)
capillary action (13.4)
condensation (13.3)
deliquescence (13.14)
evaporation (13.2)
freezing or melting point (13.6)
heat of fusion (13.7)
heat of vaporization (13.7)
hydrate (13.13)
hydrogen bond (13.11)
hygroscopic substance (13.14)
meniscus (13.4)
nonvolatile (13.3)
normal boiling point (13.5)
sublimation (13.2)
surface tension (13.4)
vapor pressure (13.3)
vapor pressure curves (13.5)
vaporization (13.2)
volatile (13.3)
water of crystallization (13.13)
water of hydration (13.13)

Questions

Questions refer to tables, figures, and key words and concepts defined within the chapter. A particularly challenging question or exercise is indicated with an asterisk.

1. Compare the potential energy of the two states of water shown in Figure 13.6.
2. In what state (solid, liquid, or gas) would H_2S, H_2Se, and H_2Te be at 0°C? (See Table 13.3.)
3. The two thermometers in the flask on the hotplate (Figure 13.6) read 100°C. What is the pressure of the atmosphere?
4. Draw a diagram of a water molecule and point out the areas that are the negative and positive ends of the dipole.
5. If the water molecule were linear, with all three atoms in a straight line rather than in the shape of a V, as shown in Figure 13.7, what effect would this have on the physical properties of water?
6. Based on Table 13.4, how do we specify 1, 2, 3, 4, 5, 6, 7, and 8 molecules of water in the formulas of hydrates?
7. Would the distillation setup in Figure 13.10 be satisfactory for separating salt and water? Ethyl alcohol and water? Explain.
8. If the liquid in the flask in Figure 13.10 is ethyl alcohol and the atmospheric pressure is 543 torr, what temperature will show on the thermometer? (Use Figure 13.4.)

9. If water were placed in both containers in Figure 13.1, would both have the same vapor pressure at the same temperature? Explain.
10. In Figure 13.1, in which case, (a) or (b), will the atmosphere above the liquid reach a point of saturation?
11. Suppose that a solution of ethyl ether and ethyl alcohol were placed in the closed bottle in Figure 13.1. Use Figure 13.4 for information on the substances.
    (a) Would both substances be present in the vapor?
    (b) If the answer to part (a) is yes, which would have more molecules in the vapor?
12. In Figure 13.2, if 50% more water had been added in part (b), what equilibrium vapor pressure would have been observed in (c)?
13. At approximately what temperature would each of the substances listed in Table 13.2 boil when the pressure is 30 torr? (See Figure 13.4.)
14. Use the graph in Figure 13.4 to find the following:
    (a) The boiling point of water at 500 torr
    (b) The normal boiling point of ethyl alcohol
    (c) The boiling point of ethyl ether at 0.50 atm
15. Consider Figure 13.5.
    (a) Why is line *BC* horizontal? What is happening in this interval?
    (b) What phases are present in the interval *BC*?
    (c) When heating is continued after point *C*, another horizontal line, *DE*, is reached at a higher temperature. What does this line represent?
16. List six physical properties of water.
17. What condition is necessary for water to have its maximum density? What is its maximum density?
18. Account for the fact that an ice–water mixture remains at 0°C until all the ice is melted, even though heat is applied to it.
19. Which contains less heat, ice at 0°C or water at 0°C? Explain.
20. Why does ice float in water? Would ice float in ethyl alcohol (d = 0.789 g/mL)? Explain.
21. If water molecules were linear instead of bent, would the heat of vaporization be higher or lower? Explain.
22. The heat of vaporization for ethyl ether is 351 J/g (83.9 cal/g) and that for ethyl alcohol is 855 J/g (204.3 cal/g). Which of these compounds has hydrogen bonding? Explain.
23. Would there be more or less H-bonding if water molecules were linear instead of bent? Explain.
24. Which would show hydrogen bonding, ammonia, NH_3, or methane, CH_4? Explain.
25. In which condition are there fewer hydrogen bonds between molecules: water at 40°C or water at 80°C?
26. Which compound,

    $H_2NCH_2CH_2NH_2$ or $CH_3CH_2CH_2NH_2$,

    would you expect to have the higher boiling point? Explain your answer. (Both compounds have similar molar masses.)
27. Explain why rubbing alcohol which has been warmed to body temperature still feels cold when applied to your skin.
28. The vapor pressure at 20°C is given for the following compounds:

methyl alcohol	96 torr
acetic acid	11.7 torr
benzene	74.7 torr
bromine	173 torr
water	17.5 torr
carbon tetrachloride	91 torr
mercury	0.0012 torr
toluene	23 torr

    (a) Arrange these compounds in their order of increasing rate of evaporation.
    (b) Which substance listed would have the highest boiling point? The lowest?
29. Suggest a method whereby water could be made to boil at 50°C.
30. Explain why a higher temperature is obtained in a pressure cooker than in an ordinary cooking pot.
31. What is the relationship between vapor pressure and boiling point?
32. On the basis of the Kinetic-Molecular Theory, explain why vapor pressure increases with temperature.
33. Why does water have such a relatively high boiling point?
34. The boiling point of ammonia, NH_3, is −33.4°C and that of sulfur dioxide, SO_2, is −10.0°C. Which has the higher vapor pressure at −40°C?
35. Explain what is occurring physically when a substance is boiling.
36. Explain why HF (bp = 19.4°C) has a higher boiling point than HCl (bp = −85°C), whereas F_2 (bp = −188°C) has a lower boiling point than Cl_2 (bp = −34°C).
37. Why does a boiling liquid maintain a constant temperature when heat is continuously being added?
38. At what specific temperature will ethyl ether have a vapor pressure of 760 torr?
39. Why does a lake freeze from the top down?
40. What water temperature would you theoretically expect to find at the bottom of a very deep lake? Explain.
41. Write equations to show how the following metals react with water: aluminum, calcium, iron, sodium, zinc. State the conditions for each reaction.

42. Is the formation of hydrogen and oxygen from water an exothermic or an endothermic reaction? How do you know?
43. (a) What is an anhydride?
(b) What type of compound will be an acid anhydride?
(c) What type of compound will be a basic anhydride?
44. Which of the following statements are correct? Rewrite each incorrect statement to make it correct.
(a) The process of a substance changing directly from a solid to a gas is called sublimation.
(b) When water is decomposed, the volume ratio of H_2 to O_2 is 2:1, but the mass ratio of H_2 to O_2 is 1:8.
(c) Hydrogen sulfide is a larger molecule than water.
(d) The changing of ice into water is an exothermic process.
(e) Water and hydrogen fluoride are both nonpolar molecules.
(f) Hydrogen bonding is stronger in H_2O than in H_2S because oxygen is more electronegative than sulfur.
(g) $H_2O_2 \longrightarrow 2\ H_2O + O_2$ represents a balanced equation for the decomposition of hydrogen peroxide.
(h) Steam at 100°C can cause more severe burns than liquid water at 100°C.
(i) The density of water is independent of temperature.
(j) Liquid A boils at a lower temperature than liquid B. This fact indicates that liquid A has a lower vapor pressure than liquid B at any particular temperature.
(k) Water boils at a higher temperature in the mountains than at sea level.
(l) No matter how much heat you put under an open pot of pure water on a stove, you cannot heat the water above its boiling point.
(m) The vapor pressure of a liquid at its boiling point is equal to the prevailing atmospheric pressure.
(n) The normal boiling temperature of water is 273°C.
(o) The pressure exerted by a vapor in equilibrium with its liquid is known as the vapor pressure of the liquid.
(p) Sodium, potassium, and calcium each react with water to form hydrogen gas and a metal hydroxide.
(q) Calcium oxide reacts with water to form calcium hydroxide and hydrogen gas.
(r) Carbon dioxide is the hydride of carbonic acid.
(s) Water in a hydrate is known as water of hydration or water of crystallization.
(t) A substance that absorbs water from the air until it forms a solution is deliquescent.
(u) Distillation is effective for softening water because the minerals boil away, leaving soft water behind.
(v) Disposal of toxic industrial wastes in toxic waste dumps has been found to be a very satisfactory long-term solution to the problem of what to do with these wastes.
(w) The amount of heat needed to change 1 mol of ice at 0°C to a liquid at 0°C is 6.02 kJ (1.44 kcal).
(x) $BaCl_2 \cdot 2\ H_2O$ has a higher percentage of water than does $CaCl_2 \cdot 2\ H_2O$.

Paired Exercises

These exercises are paired. Each odd-numbered exercise is followed by a similar even-numbered exercise. Answers to the even-numbered exercises are given in Appendix V.

45. Write the formulas for the anhydrides of the following acids:
H_2SO_3, H_2SO_4, HNO_3
46. Write the formulas for the anhydrides of the following acids:
$HClO_4$, H_2CO_3, H_3PO_4
47. Write the formulas for the anhydrides of the following bases:
LiOH, NaOH, $Mg(OH)_2$
48. Write the formulas for the anhydrides of the following bases:
KOH, $Ba(OH)_2$, $Ca(OH)_2$
49. Complete and balance the following equations:
(a) $Ba(OH)_2 \xrightarrow{\Delta}$
(b) $CH_3OH + O_2 \longrightarrow$
methyl alcohol
(c) $Rb + H_2O \longrightarrow$
(d) $SnCl_2 \cdot 2\ H_2O \xrightarrow{\Delta}$
(e) $HNO_3 + NaOH \longrightarrow$
(f) $CO_2 + H_2O \longrightarrow$
50. Complete and balance the following equations:
(a) $Li_2O + H_2O \longrightarrow$
(b) $KOH \xrightarrow{\Delta}$
(c) $Ba + H_2O \longrightarrow$
(d) $Cl_2 + H_2O \longrightarrow$
(e) $SO_3 + H_2O \longrightarrow$
(f) $H_2SO_3 + KOH \longrightarrow$

51. Name each of the following hydrates:
(a) $BaBr_2 \cdot 2\ H_2O$
(b) $AlCl_3 \cdot 6\ H_2O$
(c) $FePO_4 \cdot 4\ H_2O$

52. Name each of the following hydrates:
(a) $MgNH_4PO_4 \cdot 6\ H_2O$
(b) $FeSO_4 \cdot 7\ H_2O$
(c) $SnCl_4 \cdot 5\ H_2O$

53. Distinguish between deionized water and
(a) hard water
(b) soft water

54. Distinguish between deionized water and
(a) distilled water
(b) natural water

55. How many moles of compound are in 100. g of $CoCl_2 \cdot 6\ H_2O$?

56. How many moles of compound are in 100. g of $FeI_2 \cdot 4\ H_2O$?

57. How many moles of water can be obtained from 100. g of $CoCl_2 \cdot 6\ H_2O$?

58. How many moles of water can be obtained from 100. g of $FeI_2 \cdot 4\ H_2O$?

59. When a person purchases epsom salts, $MgSO_4 \cdot 7\ H_2O$, what percent of the compound is water?

60. Calculate the mass percent of water in the hydrate $Al_2(SO_4)_3 \cdot 18\ H_2O$.

61. Sugar of lead, a hydrate of lead acetate, $Pb(C_2H_3O_2)_2$, contains 14.2% H_2O. What is the empirical formula for the hydrate?

62. A 25.0-g sample of a hydrate of $FePO_4$ was heated until no more water was driven off. The mass of anhydrous sample is 16.9 g. What is the empirical formula of the hydrate?

63. How many joules are needed to change 120. g of water at 20.°C to steam at 100.°C?

64. How many joules of energy must be removed from 126 g of water at 24°C to form ice at 0°C?

***65.** Suppose 100. g of ice at 0°C are added to 300. g of water at 25°C. Is this sufficient ice to lower the temperature of the system to 0°C and still have ice remaining? Show evidence for your answer.

***66.** Suppose 35.0 g of steam at 100.°C are added to 300. g of water at 25°C. Is this sufficient steam to heat all the water to 100.°C and still have steam remaining? Show evidence for your answer.

***67.** If 75 g of ice at 0.0°C were added to 1.5 L of water at 75°C, what would be the final temperature of the mixture?

***68.** If 9560 J of energy were absorbed by 500. g of ice at 0.0°C, what would be the final temperature?

69. How many grams of water will react with each of the following?
(a) 1.00 mol K
(b) 1.00 mol Ca
(c) 1.00 mol SO_3

70. How many grams of water will react with each of the following?
(a) 1.00 g Na
(b) 1.00 g MgO
(c) 1.00 g N_2O_5

Additional Exercises

These exercises are not paired or labeled by topic and provide additional practice on concepts covered in this chapter.

71. Which would cause a more severe burn, liquid water at 100°C or steam at 100°C? Why?

72. Imagine a shallow dish of alcohol set into a tray of water. If one were to blow across the tray, the alcohol would evaporate, while the water would cool significantly and eventually freeze. Explain why.

73. Regardless of how warm the outside temperature may be, one always feels cool when stepping out of a swimming pool, the ocean, or a shower. Why is this so?

74. Sketch a heating curve for a substance *X* whose melting point is 40°C and whose boiling point is 65°C.

(a) Describe what you will observe as a 60. g sample of *X* is warmed from 0°C to 100°C.

(b) If the heat of fusion of *X* is 80. J/g, the heat of vaporization is 190. J/g, and if 3.5 J are required to warm 1 g of *X* each degree, how much energy will be needed to accomplish the change in (a)?

75. Why does the vapor pressure of a liquid increase as the temperature of it is increased?

76. At the top of Mount Everest, which is just about 29,000 feet above sea level, the atmospheric pressure is about

270 torr. Using Figure 13.4 determine the approximate boiling temperature of water on Mt. Everest.

77. Explain how anhydrous copper(II) sulfate, $CuSO_4$, can act as an indicator for moisture.

78. Write formulas of magnesium sulfate heptahydrate and disodium hydrogen phosphate dodecahydrate.

79. How can soap function to make soft water from hard water? What objections are there to using soap for this purpose?

80. What substance is commonly used to destroy bacteria in water?

81. What chemical, other than chlorine or chlorine compounds, can be used to disinfect water for domestic use?

82. Some organic pollutants in water can be oxidized by dissolved molecular oxygen. What harmful effect can result from this depletion of oxygen in the water?

83. Why should you not drink liquids that are stored in ceramic containers, especially unglazed ones?

84. Write the chemical equation showing how magnesium ions are removed by a zeolite water softener.

85. Write an equation to show how hard water containing calcium chloride, $CaCl_2$, is softened by using sodium carbonate, Na_2CO_3.

86. How is the chemical structure of a humectant different from that of an emollient?

87. Explain how humectants and emollients help to moisturize the skin.

88. Explain the triangle theory of sweetness.

89. What are the characteristics of a good sweetener?

90. How many calories are required to change 225 g of ice at 0°C to steam at 100.°C?

91. The molar heat of vaporization is the number of joules required to change one mole of a substance from liquid to vapor at its boiling point. What is the molar heat of vaporization of water?

***92.** The specific heat of zinc is 0.096 cal/g°C. Determine the energy required to raise the temperature of 250. g of zinc from room temperature (20.0°C) to 150.°C.

***93.** Suppose 150. g of ice at 0.0°C is added to 0.120 L of water at 45°C. If the mixture is stirred and allowed to cool to 0.0°C, how many grams of ice remain?

94. How many joules of energy would be liberated by condensing 50.0 mol of steam at 100.0°C and allowing the liquid to cool to 30.0°C?

95. How many kilojoules of energy are needed to convert 100. g of ice at −10.0°C to water at 20.0°C. (The specific heat of ice at −10.0°C is 2.01 J/g°C.)

96. What mass of water must be decomposed to produce 25.0 L of oxygen at STP?

97. Compare the volume occupied by 1.00 mol of liquid water at 0°C and 1.00 mol of water vapor at STP.

***98.** Suppose 1.00 mol of water evaporates in 1.00 day. How many water molecules, on the average, leave the liquid each second?

***99.** A quantity of sulfuric acid is added to 100. mL of water. The final volume of the solution is 122 mL and has a density of 1.26 g/mL. What mass of acid was added? Assume the density of the water is 1.00 g/mL.

100. A mixture of 80.0 mL of hydrogen and 60.0 mL of oxygen is ignited by a spark to form water.
(a) Does any gas remain unreacted? Which one, H_2 or O_2?
(b) What volume of which gas (if any) remains unreacted? (Assume the same conditions before and after the reaction.)

101. A student (with slow reflexes) puts his hand in a stream of steam at 100.°C until 1.5 g of water have condensed. If the water then cools to room temperature (20.0°C), how many joules have been absorbed by the student's hand?

Answers to Practice Exercises

13.1 8.5°C, 28°C, 73°C, 93°C
13.2 approximately 95°C
13.3 44.8 kJ
13.4 (a) yes, (b) yes, (c) no

14

Most of the substances we encounter in our daily lives are mixtures. Often they are homogeneous mixtures, which are called *solutions*. When you think of a solution, juices, blood plasma, shampoo, soft drinks, or wine may come to mind. These solutions all have water as a main component, but many common items, such as air, gasoline, and steel, are also solutions that do not contain water. What are the necessary components of a solution? Why do some substances mix while others do not? What effect does a dissolved substance have on the properties of the solution? Answering these questions is the first step in understanding the solutions we encounter in our daily lives.

14.1 General Properties of Solutions

The term **solution** is used in chemistry to describe a system in which one or more substances are homogeneously mixed or dissolved in another substance. A simple solution has two components, a solute and a solvent. The **solute** is the component that is dissolved or is the least abundant component in the solution. The **solvent** is the dissolving agent or the most abundant component in the solution. For example, when salt is dissolved in water to form a solution, salt is the solute and water is the solvent. Complex solutions containing more than one solute and/or more than one solvent are common.

solution

solute

solvent

From the three states of matter—solid, liquid, and gas—it is possible to have nine different types of solutions: solid dissolved in solid, solid dissolved in liquid, solid dissolved in gas, liquid dissolved in liquid, and so on. Of these, the most common solutions are solid dissolved in liquid, liquid dissolved in liquid, gas dissolved in liquid, and gas dissolved in gas. Some common types of solutions are listed in Table 14.1.

A true solution is one in which the particles of dissolved solute are molecular or ionic in size, generally in the range of 0.1 to 1 nm (10^{-8} to 10^{-7} cm). The properties of a true solution are as follows:

1. It is a homogeneous mixture of two or more components—solute and solvent—and has a variable composition; that is, the ratio of solute to solvent may be varied.
2. The dissolved solute is molecular or ionic in size.
3. It may be either colored or colorless but it is usually transparent.
4. The solute remains uniformly distributed throughout the solution and will not settle out with time.
5. The solute generally can be separated from the solvent by purely physical means (e.g., by evaporation).

These properties are illustrated by water solutions of sugar and of potassium permanganate. Suppose that we prepare two sugar solutions, the first containing 10 g of sugar added to 100 mL of water and the second containing 20 g of sugar added to 100 mL of water. Each solution is stirred until all the solute dissolves, demonstrating that we can vary the composition of a solution. Every portion of the

◀ **Chapter Opening Photo: The ocean is a salt solution covering the majority of the earth's surface.**

Note the beautiful purple trails of $KMnO_4$ as the crystals dissolve.

TABLE 14.1 Common Types of Solutions

Phase of solution	Solute	Solvent	Example
Gas	Gas	Gas	Air
Liquid	Gas	Liquid	Soft drinks
Liquid	Liquid	Liquid	Antifreeze
Liquid	Solid	Liquid	Salt water
Solid	Gas	Solid	H_2 in Pt
Solid	Solid	Solid	Brass

solution has the same sweet taste because the sugar molecules are uniformly distributed throughout. If confined so that no solvent is lost, the solution will taste and appear the same a week or a month later. The properties of the solution are unaltered after the solution is passed through filter paper. But by carefully evaporating the water, we can recover the sugar from the solution.

To observe the dissolving of potassium permanganate, $KMnO_4$, we affix a few crystals of it to paraffin wax or rubber cement at the end of a glass rod and submerge the entire rod, with the wax-permanganate end up, in a cylinder of water. Almost at once the beautiful purple color of dissolved permanganate ions, MnO_4^-, appears at the top of the rod and streams to the bottom of the cylinder as the crystals dissolve. The purple color at first is mostly at the bottom of the cylinder because $KMnO_4$ is denser than water. But after a while the purple color disperses until it is evenly distributed throughout the solution. This dispersal demonstrates that molecules and ions move about freely and spontaneously (diffuse) in a liquid or solution.

Solution permanency is explained in terms of the Kinetic-Molecular Theory (see Section 12.2). According to the KMT both the solute and solvent particles (molecules and/or ions) are in constant random motion. This motion is energetic enough to prevent the solute particles from settling out under the influence of gravity.

14.2 Solubility

solubility

The term **solubility** describes the amount of one substance (solute) that will dissolve in a specified amount of another substance (solvent) under stated conditions. For example, 36.0 g of sodium chloride will dissolve in 100 g of water at 20°C. We say, then, that the solubility of NaCl in water is 36.0 g/100 g H_2O at 20°C.

Solubility is often used in a relative way. For instance, we say that a substance is very soluble, moderately soluble, slightly soluble, or insoluble. Although these terms do not accurately indicate how much solute will dissolve, they are frequently used to describe the solubility of a substance qualitatively.

miscible
immiscible

Two other terms often used to describe solubility are *miscible* and *immiscible*. Liquids that are capable of mixing and forming a solution are **miscible**; those that do not form solutions or are generally insoluble in each other are **immiscible**. Methyl alcohol and water are miscible in each other in all proportions. Oil and water are immiscible, forming two separate layers when they are mixed, as illustrated in Figure 14.1.

The general guidelines for the solubility of common ionic compounds (salts) are given in Figure 14.2. These guidelines have some exceptions but they provide a solid foundation for the compounds considered in this course. The solubilities of over 200 compounds are given in the Solubility Table in Appendix IV. Solubility data for thousands of compounds can be found by consulting standard reference sources.*

The term *salt* is used interchangeably with *ionic compound* by many chemists.

The quantitative expression of the amount of dissolved solute in a particular quantity of solvent is known as the **concentration of a solution**. Several methods of expressing concentration will be described in Section 14.6.

concentration of a solution

14.3 Factors Related to Solubility

Predicting solubilities is complex and difficult. Many variables, such as size of ions, charge on ions, interaction between ions, interaction between solute and solvent, and temperature, complicate the problem. Because of the factors involved, the general rules of solubility given in Figure 14.2 have many exceptions. However, the rules are very useful, because they do apply to many of the more common compounds that we encounter in the study of chemistry. Keep in mind that these are rules, not laws, and are therefore subject to exceptions. Fortunately the solubility of a solute is relatively easy to determine experimentally. We will now discuss factors related to solubility.

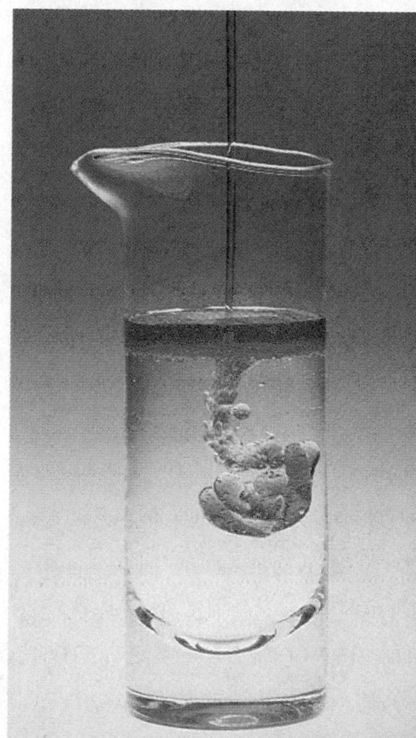

FIGURE 14.1
An immiscible mixture of oil and water.

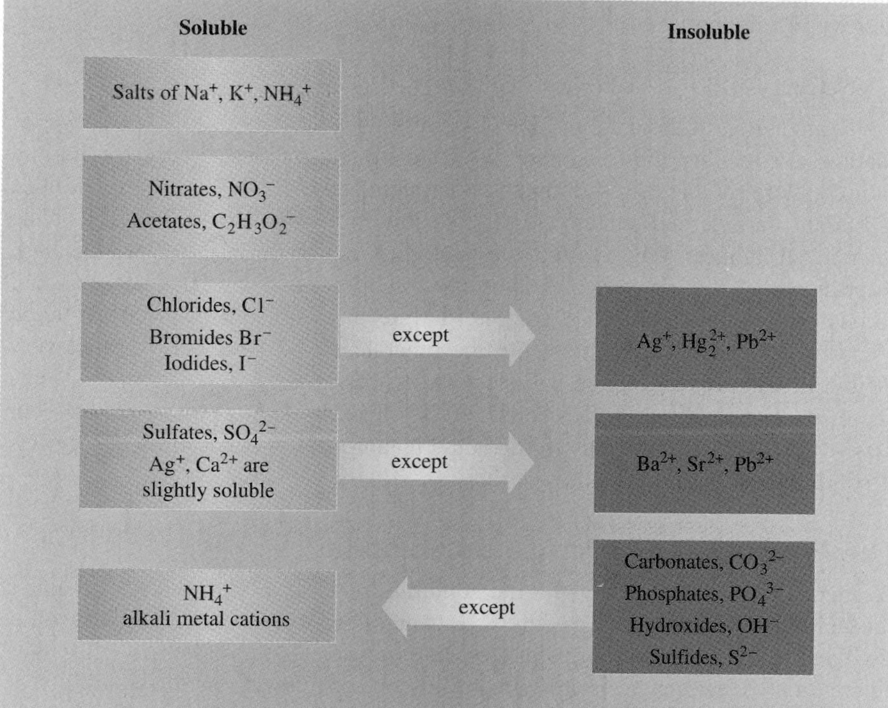

FIGURE 14.2
The solubility of various common ions. Substances containing the ions on the left are generally soluble in cold water, while those substances containing the ions on the right are insoluble in cold water. The arrows point to the exceptions.

*Two commonly used handbooks are *Lange's Handbook of Chemistry*, 13th ed. (New York: McGraw-Hill, 1985), and *Handbook of Chemistry and Physics*, 76th ed. (Cleveland: Chemical Rubber Co., 1996).

FIGURE 14.3
Dissolution of sodium chloride in water. Polar water molecules are attracted to Na^+ and Cl^- ions in the salt crystal, weakening the attraction between the ions. As the attraction between the ions weakens, the ions move apart and become surrounded by water dipoles. The hydrated ions slowly diffuse away from the crystal to become dissolved in solution.

The Nature of the Solute and Solvent

The old adage "like dissolves like" has merit, in a general way. Polar or ionic substances tend to be more miscible, or soluble, with other polar substances. Nonpolar substances tend to be miscible with other nonpolar substances and less miscible with polar substances. Thus, ionic compounds, which are polar, tend to be much more soluble in water, which is polar, than in solvents such as ether, hexane, or benzene, which are essentially nonpolar. Sodium chloride, an ionic substance, is soluble in water, slightly soluble in ethyl alcohol (less polar than water), and insoluble in ether and benzene. Pentane, C_5H_{12}, a nonpolar substance, is only slightly soluble in water but is very soluble in benzene and ether.

At the molecular level the formation of a solution from two nonpolar substances, such as hexane and benzene, can be visualized as a process of simple mixing. The nonpolar molecules, having little tendency to either attract or repel one another, easily intermingle to form a homogeneous mixture.

Solution formation between polar substances is much more complex. See for example, the process by which sodium chloride dissolves in water (Figure 14.3). Water molecules are very polar and are attracted to other polar molecules or ions. When salt crystals are put into water, polar water molecules become attracted to the sodium and chloride ions on the crystal surfaces and weaken the attraction between Na^+ and Cl^- ions. The positive end of the water dipole is attracted to the Cl^- ions, and the negative end of the water dipole to the Na^+ ions. The weakened attraction permits the ions to move apart, making room for more water dipoles. Thus, the surface ions are surrounded by water molecules, becoming hydrated ions, $Na^+(aq)$ and $Cl^-(aq)$, and slowly diffuse away from the crystals and dissolve in solution:

$$\text{NaCl(crystal)} \xrightarrow{H_2O} Na^+(aq) + Cl^-(aq)$$

Examination of the data in Table 14.2 reveals some of the complex questions relating to solubility. For example: Why are lithium halides, except for lithium fluoride, more soluble than sodium and potassium halides? Why are the solubilities of lithium fluoride and sodium fluoride so low in comparison with those of the other metal halides? Why doesn't the solubility of LiF, NaF, and NaCl increase proportionately with temperature, as do the solubilities of the other metal halides? Sodium chloride is appreciably soluble in water but is insoluble in concentrated hydrochloric acid, HCl, solution. On the other hand, LiF and NaF are not very soluble in water but are quite soluble in hydrofluoric acid, HF, solution—Why? These questions will not be answered directly here, but are meant to arouse your curiosity to the point that you will do some reading and research on the properties of solutions.

The Effect of Temperature on Solubility

Temperature has major effects on the solubility of most substances. Most solutes have a limited solubility in a specific solvent at a fixed temperature. For most solids dissolved in a liquid, an increase in temperature results in increased solubility (see Figure 14.4). However, no completely valid general rule governs the solubility of solids in liquids with change in temperature. Some solids increase in solubility only slightly with increasing temperature (see NaCl in Figure 14.4); other solids decrease in solubility with increasing temperature (see Li_2SO_4 in Figure 14.4).

On the other hand, the solubility of a gas in water usually decreases with

TABLE 14.2 Solubility of Alkali Metal Halides in Water

Salt	Solubility (g salt/100 g H_2O) 0°C	100°C
LiF	0.12	0.14 (at 35°C)
LiCl	67	127.5
LiBr	143	266
LiI	151	481
NaF	4	5
NaCl	35.7	39.8
NaBr	79.5	121
NaI	158.7	302
KF	92.3 (at 18°C)	Very soluble
KCl	27.6	57.6
KBr	53.5	104
KI	127.5	208

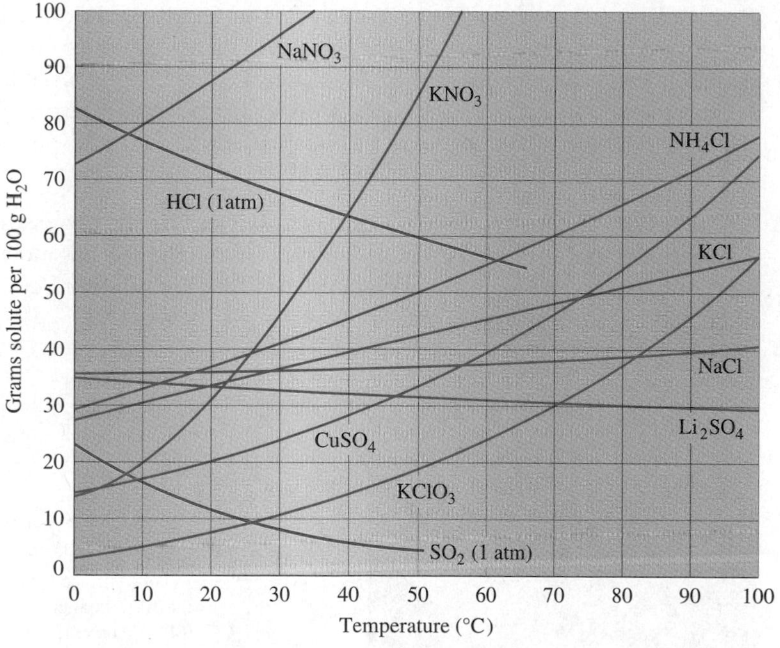

FIGURE 14.4 Solubility of various compounds in water. Solids are shown in red and gases are shown in blue.

increasing temperature (see HCl and SO_2 in Figure 14.4). The tiny bubbles that form when water is heated are due to the decreased solubility of air at higher temperatures. The decreased solubility of gases at higher temperatures is explained in terms of the KMT by assuming that, in order to dissolve, the gas molecules must form bonds of some sort with the molecules of the liquid. An increase in temperature decreases the solubility of the gas because it increases the kinetic energy (speed) of the gas molecules and thereby decreases their ability to form "bonds" with the liquid molecules.

Pouring root beer into a glass illustrates the effect of pressure on solubility. The escaping CO_2 produces the foam.

The Effect of Pressure on Solubility

Small changes in pressure have little effect on the solubility of solids in liquids or liquids in liquids but have a marked effect on the solubility of gases in liquids. The solubility of a gas in a liquid is directly proportional to the pressure of that gas above the solution. Thus, the amount of a gas that is dissolved in solution will double if the pressure of that gas over the solution is doubled. For example, carbonated beverages contain dissolved carbon dioxide under pressures greater than atmospheric pressure. When a bottle of carbonated soda is opened, the pressure is immediately reduced to the atmospheric pressure, and the excess dissolved carbon dioxide bubbles out of the solution.

14.4 Rate of Dissolving Solids

The rate at which a solid dissolves is governed by (1) the size of the solute particles, (2) the temperature, (3) the concentration of the solution, and (4) agitation or stirring. Let's look at each of these conditions.

1. *Particle Size.* A solid can dissolve only at the surface that is in contact with the solvent. Because the surface-to-volume ratio increases as size decreases, smaller crystals dissolve faster than large ones. For example, if a salt crystal 1 cm on a side (6-cm^2 surface area) is divided into 1000 cubes, each 0.1 cm on a side, the total surface of the smaller cubes is 60 cm^2—a tenfold increase in surface area (see Figure 14.5).

2. *Temperature.* In most cases the rate of dissolving of a solid increases with temperature. This increase is due to kinetic effects. The solvent molecules move more rapidly at higher temperatures and strike the solid surfaces more often and harder, causing the rate of dissolving to increase.

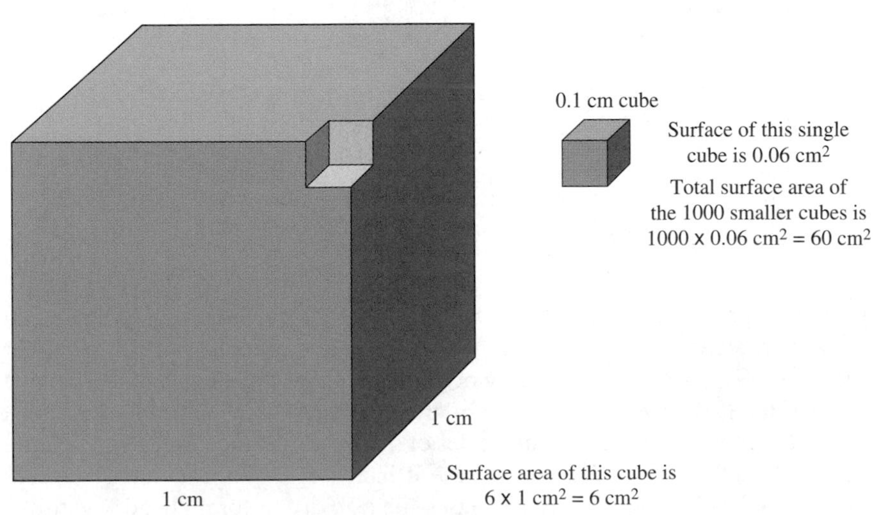
0.1 cm cube
Surface of this single cube is 0.06 cm²
Total surface area of the 1000 smaller cubes is 1000 x 0.06 cm² = 60 cm²
1 cm
1 cm
Surface area of this cube is 6 x 1 cm² = 6 cm²

FIGURE 14.5 ▶
Surface area of crystals. A crystal 1 cm on a side has a surface area of 6 cm^2. Subdivided into 1000 smaller crystals, each 0.1 cm on a side, the total surface area is increased to 60 cm^2.

Killer Lakes

In a tiny African nation called Cameroon, two towns border on lakes that are people killers. The townspeople do not die by drowning in the lakes, or by drinking contaminated water. Instead they die from carbon dioxide asphyxiation. The lakes give off clouds of carbon dioxide at irregular intervals. Thirty-seven people died near Lake Monoun in August, 1984. Just two years later 1700 people died at nearby Lake Nyos.

Scientists studying these volcanic crater lakes have found that CO_2 percolates upward from groundwater into the bottom of these lakes. The CO_2 accumulates to dangerous levels because the water is naturally stratified into layers that do not mix. A boundary called a *chemocline* separates the layers, keeping fresh water at the surface of the lake. The lower layers of the lake contain dissolved minerals and gases (including CO_2).

The disasters occur when something disturbs the layers. An earthquake, a landslide, or even winds can trigger the phenomenon. As waves form and move across the lake, the layers within the lake are mixed. When the deep water containing the CO_2 rises near the top of the lake, the dissolved CO_2 is released from the solution (similar to the bubbles released on opening a can of soda). At Lake Nyos, where 1700 people died, the cloud of CO_2 spilled over the edge of the crater and traveled down a river valley. Since CO_2 is denser than air, the cloud stayed near the ground. It traveled at an amazing speed of 45 mph and killed people as far away as 25 miles.

Although scientists do not know precisely what causes the water layers to turnover, they have succeeded in measuring the rate at which gas seeps into the lake bottoms. The rate is so fast that some scientists think the bottom waters of Lake Nyos could be saturated in less than 20 years, and Lake Monoun could be saturated in less than 10 years.

Scientists and engineers are working to lower gas concentrations in both lakes. In Lake Monoun, water is being pumped through pipes from the lake bottom to the surface to release the gas slowly. Lake Nyos is a larger lake and represents a more difficult problem. One end of the lake is supported by a weak natural dam. If the dam were to break, the water from the lake would spill into a valley with about 10,000 residents and could trigger a CO_2 release. Water could be pumped out of the lake to release the CO_2 and lower the level of the lake to accommodate the dam, but funding for these projects is uncertain. Until money is found to relieve the build-up of gases, these lakes will remain disasters waiting to happen.

3. *Concentration of the Solution.* When the solute and solvent are first mixed, the rate of dissolving is at its maximum. As the concentration of the solution increases and the solution becomes more nearly saturated with the solute, the rate of dissolving decreases greatly. The rate of dissolving is pictured graphically in Figure 14.6. Note that about 17 g dissolve in the first 5-minute interval, but only about 1 g dissolves in the fourth 5-minute interval. Although

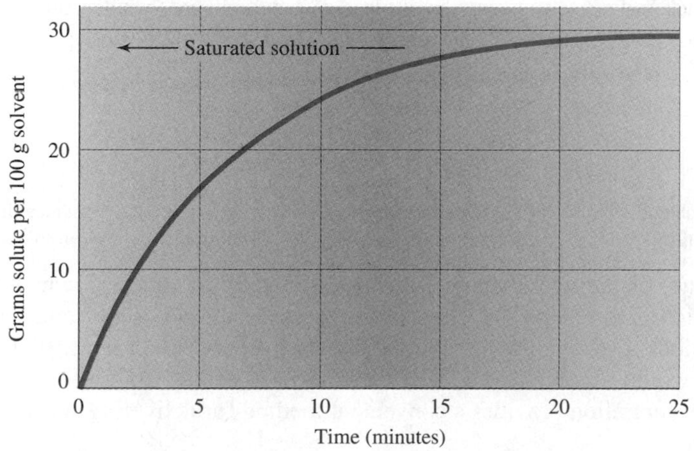

FIGURE 14.6
Rate of dissolution of a solid solute in a solvent. The rate is maximum at the beginning and decreases as the concentration approaches saturation.

different solutes show different rates, the rate of dissolving always becomes very slow as the concentration approaches the saturation point.

4. *Agitation or Stirring.* The effect of agitation or stirring is kinetic. When a solid is first put into water, the only solvent with which it comes in contact is in the immediate vicinity. As the solid dissolves, the amount of dissolved solute around the solid becomes more and more concentrated, and the rate of dissolving slows down. If the mixture is not stirred, the dissolved solute diffuses very slowly through the solution; weeks may pass before the solid is entirely dissolved. Stirring distributes the dissolved solute rapidly through the solution, and more solvent is brought into contact with the solid, causing it to dissolve more rapidly.

14.5 Solutions: A Reaction Medium

Many solids must be put into solution in order to undergo appreciable chemical reaction. We can write the equation for the double displacement reaction between sodium chloride and silver nitrate:

$$NaCl + AgNO_3 \longrightarrow AgCl + NaNO_3$$

But suppose we mix solid NaCl and solid $AgNO_3$ and look for a chemical change. If any reaction occurs, it is slow and virtually undetectable. In fact, the crystalline structures of NaCl and $AgNO_3$ are so different that we could separate them by tediously picking out each kind of crystal from the mixture. But if we dissolve the NaCl and $AgNO_3$ separately in water and mix the two solutions, we observe the immediate formation of a white, curd-like precipitate of silver chloride.

Molecules or ions must come into intimate contact or collide with one another in order to react. In the foregoing example, the two solids did not react because the ions were securely locked within their crystal structures. But when the NaCl and $AgNO_3$ are dissolved, their crystal lattices are broken down and the ions become mobile. When the two solutions are mixed, the mobile Ag^+ and Cl^- ions come into contact and react to form insoluble AgCl, which precipitates out of solution. The soluble Na^+ and NO_3^- ions remain mobile in solution but form the crystalline salt $NaNO_3$ when the water is evaporated:

$$NaCl(aq) + AgNO_3(aq) \longrightarrow AgCl(s) + NaNO_3(aq)$$

$$\underbrace{Na^+(aq) + Cl^-(aq)}_{\text{sodium chloride solution}} + \underbrace{Ag^+(aq) + NO_3^-(aq)}_{\text{silver nitrate solution}} \longrightarrow \underbrace{AgCl(s)}_{\text{silver chloride}} + \underbrace{Na^+(aq) + NO_3^-(aq)}_{\text{sodium nitrate in solution}}$$

The mixture of the two solutions provides a medium or space in which the Ag^+ and Cl^- ions can react. (See Chapter 15 for further discussion of ionic reactions.)

Solutions also function as diluting agents in reactions in which the undiluted reactants would combine with each other too violently. Moreover, a solution of known concentration provides a convenient method for delivering specific amounts of reactants.

14.6 Concentration of Solutions

The concentration of a solution expresses the amount of solute dissolved in a given quantity of solvent or solution. Because reactions are often conducted in solution, it is important to understand the methods of expressing concentration and to know how to prepare solutions of particular concentrations. The concentration of a solution may be expressed qualitatively or quantitatively. Let's begin with a look at the qualitative methods of expressing concentration.

Dilute and Concentrated Solutions

When we say that a solution is *dilute* or *concentrated*, we are expressing, in a relative way, the amount of solute present. One gram of a compound and 2 g of a compound in solution are both dilute solutions when compared with the same volume of a solution containing 20 g of a compound. Ordinary concentrated hydrochloric acid contains 12 mol of HCl per liter of solution. In some laboratories the dilute acid is made by mixing equal volumes of water and the concentrated acid. In other laboratories the concentrated acid is diluted with two or three volumes of water, depending on its use. The term **dilute solution**, then, describes a solution that contains a relatively small amount of dissolved solute. Conversely, a **concentrated solution** contains a relatively large amount of dissolved solute.

dilute solution

concentrated solution

Saturated, Unsaturated, and Supersaturated Solutions

At a specific temperature there is a limit to the amount of solute that will dissolve in a given amount of solvent. When this limit is reached, the resulting solution is said to be *saturated*. For example, when we put 40.0 g of KCl into 100 g of H_2O at 20°C, we find that 34.0 g of KCl dissolves and 6.0 g of KCl remains undissolved. The solution formed is a saturated solution of KCl.

Two processes are occurring simultaneously in a saturated solution. The solid is dissolving into solution and, at the same time, the dissolved solute is crystallizing out of solution. This may be expressed as

$$\text{solute (undissolved)} \rightleftharpoons \text{solute (dissolved)}$$

When these two opposing processes are occurring at the same rate, the amount of solute in solution is constant, and a condition of equilibrium is established between dissolved and undissolved solute. Therefore, a **saturated solution** contains dissolved solute in equilibrium with undissolved solute.

saturated solution

It is important to state the temperature of a saturated solution, because a solution that is saturated at one temperature may not be saturated at another. If the temperature of a saturated solution is changed, the equilibrium is disturbed, and the amount of dissolved solute will change to reestablish equilibrium.

A saturated solution may be either dilute or concentrated, depending on the solubility of the solute. A saturated solution can be conveniently prepared by dissolving a little more than the saturated amount of solute at a temperature somewhat higher than room temperature. Then the amount of solute in solution will be in

TABLE 14.3 Saturated Solutions at 20°C and 50°C

Solute	Solubility (g solute/100 g H_2O) 20°C	50°C
NaCl	36.0	37.0
KCl	34.0	42.6
$NaNO_3$	88.0	114.0
$KClO_3$	7.4	19.3
$AgNO_3$	222.0	455.0
$C_{12}H_{22}O_{11}$	203.9	260.4

excess of its solubility at room temperature, and, when the solution cools, the excess solute will crystallize, leaving the solution saturated. (In this case, the solute must be more soluble at higher temperatures and must not form a supersaturated solution.) Examples expressing the solubility of saturated solutions at two different temperatures are given in Table 14.3.

unsaturated solution

An **unsaturated solution** contains less solute per unit of volume than does its corresponding saturated solution. In other words, additional solute can be dissolved in an unsaturated solution without altering any other conditions. Consider a solution made by adding 40 g of KCl to 100 g of H_2O at 20°C (see Table 14.3). The solution formed will be saturated and will contain about 6 g of undissolved salt, because the maximum amount of KCl that can dissolve in 100 g of H_2O at 20°C is 34 g. If the solution is now heated and maintained at 50°C, all the salt will dissolve and, in fact, even more can be dissolved. Thus the solution at 50°C is unsaturated.

supersaturated solution

In some circumstances, solutions can be prepared that contain more solute than that needed for a saturated solution at a particular temperature. These solutions are said to be **supersaturated**. However, we must qualify this definition by noting that a supersaturated solution is unstable. Disturbances, such as jarring, stirring, scratching the walls of the container, or dropping in a "seed" crystal, cause the supersaturation to return to saturation. When a supersaturated solution is disturbed, the excess solute crystallizes out rapidly, returning the solution to a saturated state.

Supersaturated solutions are not easy to prepare but may be made from certain substances by dissolving, in warm solvent, an amount of solute greater than that needed for a saturated solution at room temperature. The warm solution is then allowed to cool very slowly. With the proper solute and careful work, a supersaturated solution will result.

Example 14.1 Will a solution made by adding 2.5 g of $CuSO_4$ to 10 g of H_2O be saturated or unsaturated at 20°C?

Solution To answer this question we first need to know the solubility of $CuSO_4$ at 20°C. From Figure 14.4 we see that the solubility of $CuSO_4$ at 20°C is about 21 g per 100 g of H_2O. This amount is equivalent to 2.1 g of $CuSO_4$ per 10 g of H_2O.

Since 2.5 g per 10 g of H_2O is greater than 2.1 g per 10 g of H_2O, the solution will be saturated and 0.4 g of $CuSO_4$ will be undissolved.

Practice 14.1

Will a solution made by adding 9.0 g NH_4Cl to 20 g of H_2O be saturated or unsaturated at 50°C?

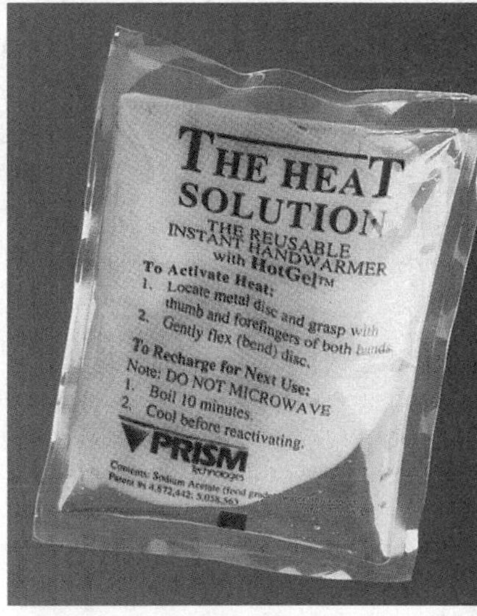

The heat released in this hot pack results from the crystallization of a supersaturated solution of sodium acetate.

Mass Percent Solution

The mass percent method expresses concentration of the solution as the percent of solute in a given mass of solution. It says that for a given mass of solution a certain percent of that mass is solute. Suppose that we take a bottle from the reagent shelf that reads "sodium hydroxide, NaOH, 10%." This statement means that for every 100 g of this solution, 10 g will be NaOH and 90 g will be water. (Note that this amount of solution is 100 g and not 100 mL.) We could also make this same concentration of solution by dissolving 2.0 g of NaOH in 18 g of water. Mass percent concentrations are most generally used for solids dissolved in liquids:

$$\text{mass percent} = \frac{\text{g solute}}{\text{g solute} + \text{g solvent}} \times 100 = \frac{\text{g solute}}{\text{g solution}} \times 100$$

Note that mass percent is independent of the formula for the solute.

As instrumentation advances are made in chemistry, our ability to measure the concentration of dilute solutions is increasing as well. Instead of mass percent, chemists now commonly use **parts per million (ppm)**:

parts per million (ppm)

$$\text{parts per million} = \frac{\text{g solute}}{\text{g solute} + \text{g solvent}} \times 1{,}000{,}000$$

Currently, air and water contaminants, drugs in the human body, and pesticide residues are some substances measured in parts per million.

Example 14.2

What is the mass percent of sodium hydroxide in a solution that is made by dissolving 8.00 g of NaOH in 50.0 g of H_2O?

Solution

$$\text{grams of solute (NaOH)} = 8.00 \text{ g}$$

$$\text{grams of solvent } (H_2O) = 50.0 \text{ g}$$

$$\frac{8.00 \text{ g NaOH}}{8.00 \text{ g NaOH} + 50.0 \text{ g } H_2O} \times 100 = 13.8\% \text{ NaOH solution}$$

Example 14.3

What masses of potassium chloride and water are needed to make 250. g of 5.00% solution?

Solution

The percent expresses the mass of the solute:

$$250.\text{ g} = \text{total mass of solution}$$

$$5.00\% \text{ of } 250.\text{ g} = 0.0500 \times 250.\text{ g} = 12.5 \text{ g KCl (solute)}$$

$$250.\text{ g} - 12.5 \text{ g} = 237.5 \; H_2O$$

Dissolving 12.5 g of KCl in 237.5 g of H_2O gives a 5.00% KCl solution.

Example 14.4 A 34.0% sulfuric acid solution has a density of 1.25 g/mL. How many grams of H_2SO_4 are contained in 1.00 L of this solution?

Solution Since H_2SO_4 is the solute, we first solve the mass percent equation for grams of solute:

$$\text{mass percent} = \frac{\text{g solute}}{\text{g solution}} \times 100$$

$$\text{g solute} = \frac{\text{mass percent} \times \text{g solution}}{100}$$

The mass percent is given in the problem. We need to determine the grams of solution. The mass of the solution can be calculated from the density data. Convert density (g/mL) to grams:

$$1.00 \text{ L} = 1.00 \times 10^3 \text{ mL}$$

$$\frac{1.25 \text{ g}}{\text{mL}} \times 1.00 \times 10^3 \text{ mL} = 1250 \text{ g} \quad \text{(mass of solution)}$$

Now we have all the figures to calculate the grams of solute:

$$\text{g solute} = \frac{34.0 \text{ g} \times 1250 \text{ g}}{100} = 425 \text{ g } H_2SO_4$$

Thus, 1.00 L of 34.0% H_2SO_4 solution contains 425 g of H_2SO_4.

Practice 14.2

What is the mass percent of Na_2SO_4 in a solution that is made by dissolving 25.0 g of Na_2SO_4 in 225.0 g of H_2O?

Mass/Volume Percent (m/v)

This method expresses concentration as grams of solute per 100 mL of solution. With this system, a 10.0% (m/v) glucose solution is made by dissolving 10.0 g of glucose in water, diluting to 100 mL, and mixing. The 10.0% (m/v) solution could also be made by diluting 20.0 g to 200 mL, 50.0 g to 500 mL, and so on. Of course, any other appropriate dilution ratio may be used:

$$\text{mass/volume percent} = \frac{\text{g solute}}{\text{mL solution}} \times 100$$

Volume Percent

Solutions that are formulated from two liquids are often expressed as *volume percent* with respect to the solute. The volume percent is the volume of a liquid in 100 mL of solution. The label on a bottle of ordinary rubbing alcohol reads "isopropyl alcohol, 70% by volume." Such a solution could be made by mixing 70 mL of alcohol with water to make a total volume of 100 mL, but we cannot use 30 mL of water because the two volumes are not necessarily additive:

$$\text{volume percent} = \frac{\text{volume of liquid in question}}{\text{total volume of solution}} \times 100$$

Volume percent is used to express the concentration of alcohol in beverages. Wines generally contain 12% alcohol by volume. This translates into 12 mL of alcohol in each 100 mL of wine. The beverage industry also uses the concentration unit of *proof* (twice the volume percent). Pure alcohol is 100%, therefore 200 proof. Scotch whiskey is 86 proof or 43% alcohol.

Molarity

Mass percent solutions do not equate or express the molar masses of the solute in solution. For example, 1000. g of 10.0% NaOH solution contains 100. g of NaOH; 1000. g 10.0% KOH solution contains 100. g of KOH. In terms of moles of NaOH and KOH, these solutions contain

$$\text{mol NaOH} = 100.\ \text{g NaOH} \times \frac{1\ \text{mol NaOH}}{40.00\ \text{g NaOH}} = 2.50\ \text{mol NaOH}$$

$$\text{mol KOH} = 100.\ \text{g KOH} \times \frac{1\ \text{mol KOH}}{56.11\ \text{g KOH}} = 1.78\ \text{mol KOH}$$

From these figures we see that the two 10.0% solutions do not contain the same number of moles of NaOH and KOH. Yet 1 mol of each of these bases will neutralize the same amount of acid. As a result we find that a 10.0% NaOH solution has more reactive alkali than a 10.0% KOH solution.

We need a method of expressing concentration that will easily indicate how many moles of solute are present per unit volume of solution. For this purpose the molar method of expressing concentration is used.

A 1 molar solution contains 1 mol of solute per liter of solution. For example, to make a 1 molar solution of sodium hydroxide, NaOH, we dissolve 40.00 g of NaOH (1 mol) in water and dilute the solution with more water to a volume of 1 L. The solution contains 1 mol of the solute in 1 L of solution and is said to be 1 molar in concentration. Figure 14.7 illustrates the preparation of a 1 molar solution. Note that the volume of the solute and the solvent together is 1 L.

The concentration of a solution can, of course, be varied by using more or less solute or solvent; but in any case the **molarity** of a solution is the number of moles of solute per liter of solution. The abbreviation for molarity is *M*. The units of molarity are moles per liter. The expression "2.0 *M* NaOH" means a 2.0 molar solution of NaOH (2.0 mol, or 80.00 g, of NaOH dissolved in 1 L of solution).

molarity (*M*)

$$\textbf{molarity} = M = \frac{\text{number of moles of solute}}{\text{liter of solution}} = \frac{\text{moles}}{\text{liter}}$$

Flasks that are calibrated to contain specific volumes at a particular temperature are used to prepare solutions of a desired concentration. These *volumetric flasks* have a calibration mark on the neck to indicate accurately the measured volume. Molarity is based on a specific volume of solution and therefore will vary slightly

FIGURE 14.7
Preparation of a 1 *M* solution. ▶

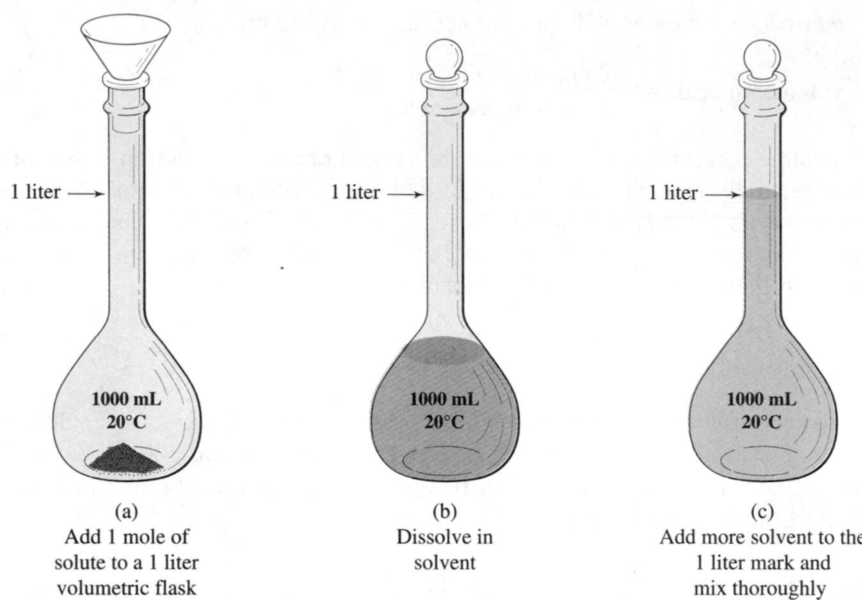

(a)
Add 1 mole of solute to a 1 liter volumetric flask

(b)
Dissolve in solvent

(c)
Add more solvent to the 1 liter mark and mix thoroughly

with temperature because volume varies with temperature (1000 mL of H_2O at 20°C = 1001 mL at 25°C).

Suppose we want to make 500 mL of 1 *M* solution. This solution can be prepared by determining the mass of 0.5 mol of the solute and diluting with water in a 500-mL (0.5-L) volumetric flask. The molarity will be

$$M = \frac{0.5 \text{ mol solute}}{0.5 \text{ L solution}} = 1 \text{ molar}$$

Thus you can see that it is not necessary to have a liter of solution to express molarity. All we need to know is the number of moles of dissolved solute and the volume of solution. Thus 0.001 mol of NaOH in 10 mL of solution is 0.1 *M*:

$$\frac{0.001 \text{ mol}}{10 \cancel{\text{mL}}} \times \frac{1000 \cancel{\text{mL}}}{1 \text{ L}} = 0.1\ M$$

When we stop to think that a balance is not calibrated in moles but in grams, we can incorporate grams into the molarity formula. We do so by using the relationship:

$$\text{moles} = \frac{\text{grams of solute}}{\text{molar mass}}$$

The molarities of the concentrated acids commonly used in the laboratory are

HCl	**12 *M***
$HC_2H_3O_2$	**17 *M***
HNO_3	**16 *M***
H_2SO_4	**18 *M***

Substituting this relationship into our expression for molarity, we get

$$M = \frac{\text{mol}}{\text{L}} = \frac{\text{g solute}}{\text{molar mass solute} \times \text{L solution}}$$

$$= \frac{\text{g}}{\text{molar mass} \times \text{L}}$$

We can now determine the mass of any amount of a solute that has a known formula, dilute it to any volume, and calculate the molarity of the solution using this formula.

Example 14.5

What is the molarity of a solution containing 1.4 mol of acetic acid, $HC_2H_3O_2$, in 250. mL of solution?

Solution

Substitute the data, 1.4 mol and 250. mL (0.250 L), directly into the equation for molarity:

$$M = \frac{\text{mol}}{\text{L}} = \frac{1.4\ \text{mol}}{0.250\ \text{L}} = \frac{5.6\ \text{mol}}{1\ \text{L}} = 5.6\ M$$

By the unit conversion method we note that the concentration given in the problem statement is 1.4 mol per 250. mL (mol/mL). Since molarity = mol/L, the needed conversion is

$$\frac{\text{mol}}{\text{mL}} \longrightarrow \frac{\text{mol}}{\text{L}} = M$$

$$\frac{1.4\ \text{mol}}{250.\ \cancel{\text{mL}}} \times \frac{1000\ \cancel{\text{mL}}}{1\ \text{L}} = \frac{5.6\ \text{mol}}{1\ \text{L}} = 5.6\ M$$

Example 14.6

What is the molarity of a solution made by dissolving 2.00 g of potassium chlorate in enough water to make 150. mL of solution?

Solution

This problem can be solved using the unit conversion method. The steps in the conversions must lead to units of moles/liter:

$$\frac{\text{g KClO}_3}{\text{mL}} \longrightarrow \frac{\text{g KClO}_3}{\text{L}} \longrightarrow \frac{\text{mol KClO}_3}{\text{L}} = M$$

The data are

$$\text{g} = 2.00\ \text{g} \qquad \text{molar mass KClO}_3 = 122.6\ \text{g/mol} \qquad \text{volume} = 150.\ \text{mL}$$

$$\frac{2.00\ \cancel{\text{g KClO}_3}}{150.\ \cancel{\text{mL}}} \times \frac{1000\ \cancel{\text{mL}}}{1\ \text{L}} \times \frac{1\ \text{mol KClO}_3}{122.6\ \cancel{\text{g KClO}_3}} = \frac{0.109\ \text{mol}}{1\ \text{L}} = 0.109\ M$$

Example 14.7

How many grams of potassium hydroxide are required to prepare 600. mL of 0.450 M KOH solution?

Solution

The conversion is

$$\text{milliliters} \longrightarrow \text{liters} \longrightarrow \text{moles} \longrightarrow \text{grams}$$

The data are

$$\text{volume} = 600.\ \text{mL} \qquad M = \frac{0.450\ \text{mol}}{\text{L}} \qquad \text{molar mass KOH} = \frac{56.11\ \text{g KOH}}{\text{mol}}$$

The calculation is

$$600.\ \cancel{\text{mL}} \times \frac{1\ \cancel{\text{L}}}{1000\ \cancel{\text{mL}}} \times \frac{0.450\ \cancel{\text{mol}}}{\cancel{\text{L}}} \times \frac{56.11\ \text{g KOH}}{\cancel{\text{mol}}} = 15.1\ \text{g KOH}$$

Practice 14.3

What is the molarity of a solution made by dissolving 7.50 g of magnesium nitrate, $Mg(NO_3)_2$, in enough water to make 25.0 mL of solution?

Practice 14.4

How many grams of sodium chloride are needed to prepare 125 mL of a 0.037 *M* NaCl solution?

Example 14.8 How many milliliters of 2.00 *M* HCl will react with 28.0 g of NaOH?

Solution

Step 1 Write and balance the equation for the reaction:

$$HCl(aq) + NaOH(aq) \longrightarrow NaCl(aq) + H_2O(aq)$$

The equation states that 1 mol of HCl reacts with 1 mol of NaOH.

Step 2 Find the number of moles of NaOH in 28.0 g of NaOH:

$$\text{g NaOH} \longrightarrow \text{mol NaOH}$$

$$28.0\ \text{g NaOH} \times \frac{1\ \text{mol}}{40.00\ \text{g}} = 0.700\ \text{mol NaOH}$$

$$28.0\ \text{g NaOH} = 0.700\ \text{mol NaOH}$$

Step 3 Solve for moles and volume of HCl needed. From Steps 1 and 2 we see that 0.700 mol of HCl will react with 0.700 mol of NaOH, because the ratio of moles reacting is 1:1. We know that 2.00 *M* HCl contains 2.00 mol of HCl per liter, and so the volume that contains 0.700 mol of HCl will be less than 1 L:

$$\text{mol NaOH} \longrightarrow \text{mol HCl} \longrightarrow \text{L HCl} \longrightarrow \text{mL HCl}$$

$$0.700\ \cancel{\text{mol NaOH}} \times \frac{1\ \cancel{\text{mol HCl}}}{1\ \cancel{\text{mol NaOH}}} \times \frac{1\ \text{L HCl}}{2.00\ \cancel{\text{mol HCl}}} = 0.350\ \text{L HCl}$$

$$0.350\ \cancel{\text{L}}\ \text{HCl} \times \frac{1000\ \text{mL}}{1\ \cancel{\text{L}}} = 350.\ \text{mL HCl}$$

Therefore, 350. mL of 2.00 *M* HCl contains 0.700 mol HCl and will react with 0.700 mol, or 28.0 g, of NaOH.

Example 14.9 What volume of 0.250 *M* solution can be prepared from 16.0 g of potassium carbonate?

Solution We start with 16.0 g of K_2CO_3 and need to find the volume of 0.250 *M* solution that can be prepared from this K_2CO_3. The conversion, therefore, is

$$\text{g } K_2CO_3 \longrightarrow \text{mol } K_2CO_3 \longrightarrow \text{L solution}$$

The data are

$$16.0 \text{ g } K_2CO_3 \qquad M = \frac{0.250 \text{ mol}}{1 \text{ L}} \qquad \text{molar mass } K_2CO_3 = \frac{138.2 \text{ g } K_2CO_3}{1 \text{ mol}}$$

$$16.0 \text{ g } \cancel{K_2CO_3} \times \frac{1 \cancel{\text{mol } K_2CO_3}}{138.2 \cancel{\text{g } K_2CO_3}} \times \frac{1 \text{ L}}{0.250 \cancel{\text{mol } K_2CO_3}} = 0.463 \text{ L } (463 \text{ mL})$$

Thus, 463 mL of 0.250 M solution can be made from 16.0 g K_2CO_3.

Example 14.10

Calculate the number of moles of nitric acid in 325 mL of 16 M HNO_3 solution.

Solution

Use the equation

$$\text{moles} = \text{liters} \times M$$

Substitute the data given in the problem and solve:

$$\text{moles} = 0.325 \cancel{\text{L}} \times \frac{16 \text{ mol } HNO_3}{1 \cancel{\text{L}}} = 5.2 \text{ mol } HNO_3$$

Practice 14.5

What volume of 0.035 M $AgNO_3$ can be made from 5.0 g of $AgNO_3$?

Practice 14.6

How many milliliters of 0.50 M NaOH are required to react completely with 25.00 mL of 1.5 M HCl?

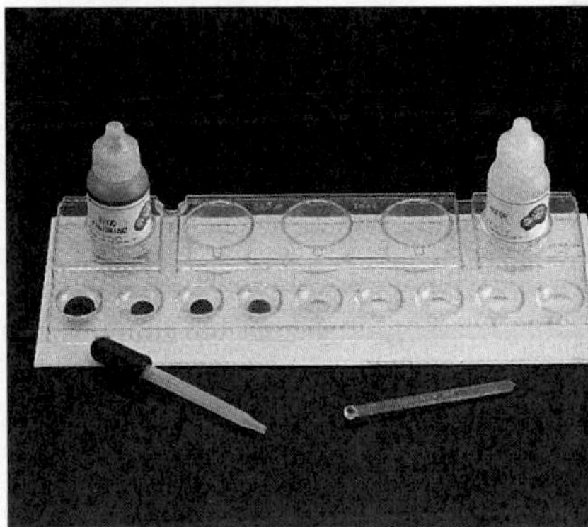

A serial dilution. The concentration of food coloring in cup 1 is 1 part per 10 (by weight), cup 2 is 1 part per 100; cup 3 is 1 part per 1000, and so on. The concentration in cup 6 is 1 part per million (ppm).

Dilution Problems

Chemists often find it necessary to dilute solutions from one concentration to another by adding more solvent to the solution. If a solution is diluted by adding pure solvent, the volume of the solution increases, but the number of moles of solute in the solution remains the same. Thus, the moles/liter (molarity) of the solution decreases. It is important to read a problem carefully to distinguish between (1) how much solvent must be added to dilute a solution to a particular concentration and (2) to what volume a solution must be diluted to prepare a solution of a particular concentration.

Example 14.11

Calculate the molarity of a sodium hydroxide solution that is prepared by mixing 100. mL of 0.20 M NaOH with 150. mL of water. Assume the volumes are additive.

Solution

This problem is a dilution problem. If we double the volume of a solution by adding water, we cut the concentration in half. Therefore, the concentration of the above solution should be less than 0.10 M. In the dilution, the moles of NaOH remain constant; the molarity and volume change. The final volume is (100. mL + 150. mL) or 250. mL.

To solve this problem, (1) calculate the moles of NaOH in the original solution, and (2) divide the moles of NaOH by the final volume of the solution to obtain the new molarity.

Step 1 Calculate the moles of NaOH in the original solution:

$$M = \frac{\text{mol}}{\text{L}} \qquad \text{mol} = \text{L} \times M$$

$$0.100\ \cancel{\text{L}} \times \frac{0.20\ \text{mol NaOH}}{1\ \cancel{\text{L}}} = 0.020\ \text{mol NaOH}$$

Step 2 Solve for the new molarity, taking into account that the total volume of the solution after dilution is 250. mL (0.250 L):

$$M = \frac{0.020\ \text{mol NaOH}}{0.250\ \text{L}} = 0.080\ M\ \text{NaOH}$$

Alternative Solution

When the moles of solute in a solution before and after dilution are the same, then the moles before and after dilution may be set equal to each other:

$$\text{mol}_1 = \text{mol}_2$$

where mol_1 = moles before dilution, and mol_2 = moles after dilution. Then

$$\text{mol}_1 = \text{L}_1 \times M_1 \qquad \text{mol}_2 = \text{L}_2 \times M_2$$

$$\text{L}_1 \times M_1 = \text{L}_2 \times M_2$$

When both volumes are in the same units, a more general statement can be made:

$$V_1 \times M_1 = V_2 \times M_2$$

For this problem

$$V_1 = 100.\ \text{mL} \qquad M_1 = 0.20\ M$$
$$V_2 = 250.\ \text{mL} \qquad M_2 = \text{(unknown)}$$

Then

$$100.\ \text{mL} \times 0.20\ M = 250.\ \text{mL} \times M_2$$

Solving for M_2, we get

$$M_2 = \frac{100.\ \cancel{\text{mL}} \times 0.20\ M}{250.\ \cancel{\text{mL}}} = 0.080\ M\ \text{NaOH}$$

Practice 14.7

Calculate the molarity of a solution prepared by diluting 125 mL of 0.400 M $K_2Cr_2O_7$ with 875 mL of water.

Example 14.12

How many grams of silver chloride will be precipitated by adding sufficient silver nitrate to react with 1500. mL of 0.400 *M* barium chloride solution?

$$2\ AgNO_3(aq) + BaCl_2(aq) \longrightarrow 2\ AgCl(s) + Ba(NO_3)_2(aq)$$

1 mol (BaCl₂) 2 mol (AgCl)

Solution

This problem is a stoichiometry problem. The fact that $BaCl_2$ is in solution means that we need to consider the volume and concentration of the solution in order to determine the number of moles of $BaCl_2$ reacting.

Step 1 Determine the number of moles of $BaCl_2$ in 1500. mL of 0.400 *M* solution:

$$M = \frac{\text{mol}}{\text{L}} \qquad \text{mol} = \text{L} \times M \qquad 1500.\ \text{mL} = 1.500\ \text{L}$$

$$1.500\ \cancel{\text{L}} \times \frac{0.400\ \text{mol BaCl}_2}{\cancel{\text{L}}} = 0.600\ \text{mol BaCl}_2$$

Step 2 Use the mole-ratio method to calculate the moles and grams of AgCl:

$$\text{mol BaCl}_2 \longrightarrow \text{mol AgCl} \longrightarrow \text{g AgCl}$$

$$0.600\ \cancel{\text{mol BaCl}_2} \times \frac{2\ \cancel{\text{mol AgCl}}}{1\ \cancel{\text{mol BaCl}_2}} \times \frac{143.4\ \text{g AgCl}}{\cancel{\text{mol AgCl}}} = 172\ \text{g AgCl}$$

Practice 14.8

How many grams of lead(II) iodide will be precipitated by adding sufficient $Pb(NO_3)_2$ to react with 750 mL of 0.250 *M* KI solution?

$$2\ KI(aq) + Pb(NO_3)_2(aq) \longrightarrow PbI_2(s) + 2\ KNO_3(aq)$$

Normality

Normality is another way of expressing the concentration of a solution. It is based on an alternative chemical unit of mass called the *equivalent mass*. The **normality** of a solution is the concentration expressed as the number of equivalent masses (equivalents, abbreviated equiv) of solute per liter of solution. A 1 normal (1 *N*) solution contains 1 equivalent mass of solute per liter of solution. Normality is widely used in analytical chemistry because it simplifies many of the calculations involving solution concentration:

normality

$$\text{normality} = N = \frac{\text{number of equivalents of solute}}{1\ \text{liter of solution}} = \frac{\text{equivalents}}{\text{liter}}$$

where

$$\text{number of equivalents of solute} = \frac{\text{grams of solute}}{\text{equivalent mass of solute}}$$

Every substance may be assigned an equivalent mass. The equivalent mass may be equal either to the molar mass of the substance or to an integral fraction of the molar mass (i.e., the molar mass divided by 2, 3, 4, and so on). To gain an understanding of the meaning of equivalent mass, let us start by considering these two reactions:

$$\underset{\substack{1\text{ mole}\\(36.46\text{ g})}}{HCl(aq)} + \underset{\substack{1\text{ mol}\\(40.00\text{ g})}}{NaOH(aq)} \longrightarrow NaCl(aq) + H_2O(l)$$

$$\underset{\substack{1\text{ mole}\\(98.08\text{ g})}}{H_2SO_4(aq)} + \underset{\substack{2\text{ mol}\\(80.00\text{ g})}}{2\ NaOH(aq)} \longrightarrow Na_2SO_4(aq) + 2\ H_2O(l)$$

We note first that 1 mol of hydrochloric acid reacts with 1 mol of sodium hydroxide and 1 mol of sulfuric acid reacts with 2 mol of NaOH. If we make 1 *M* solutions of these substances, 1 L of 1 *M* HCl will react with 1 L of 1 *M* NaOH, and 1 L of 1 *M* H_2SO_4 will react with 2 L of 1 *M* NaOH. From this reaction, we can see that H_2SO_4 has twice the chemical capacity of HCl when reacting with NaOH. We can, however, adjust these acid solutions to be equivalent in reactivity by dissolving only 0.5 mol of H_2SO_4 per liter of solution. By doing so, we find that we are required to use 49.04 g of H_2SO_4/L (instead of 98.08 g of H_2SO_4/L) to make a solution that is equivalent to one made from 36.46 g of HCl/L. These masses, 49.04 g of H_2SO_4 and 36.46 g of HCl, are chemically equivalent and are known as the equivalent masses of these substances, because each will react with the same amount of NaOH (40.00 g). The equivalent mass of HCl is equal to its molar mass, but that of H_2SO_4 is one-half its molar mass.

Thus, 1 L of solution containing 36.46 g of HCl would be 1 *N*, and 1 L of solution containing 49.04 g of H_2SO_4 would also be 1 *N*. A solution containing 98.08 g of H_2SO_4 (1 mol per liter) would be 2 *N* when reacting with NaOH in the given equation.

equivalent mass

The **equivalent mass** is the mass of a substance that will react with, combine with, contain, replace, or in any other way be equivalent to 1 mol of hydrogen atoms or hydrogen ions.

Normality and molarity can be interconverted in the following manner:

$$N = \frac{\text{equiv}}{\text{L}} \qquad M = \frac{\text{mol}}{\text{L}}$$

$$N = M \times \frac{\text{equiv}}{\text{mol}} = \frac{\cancel{\text{mol}}}{\text{L}} \times \frac{\text{equiv}}{\cancel{\text{mol}}} = \frac{\text{equiv}}{\text{L}}$$

$$M = N \times \frac{\text{mol}}{\text{equiv}} = \frac{\cancel{\text{equiv}}}{\text{L}} \times \frac{\text{mol}}{\cancel{\text{equiv}}} = \frac{\text{mol}}{\text{L}}$$

Thus a 2.0 *N* H_2SO_4 solution is 1.0 *M*.

One application of normality and equivalents is in acid–base neutralization reactions. An equivalent of an acid is that mass of the acid that will furnish 1 mol of H^+ ions. An equivalent of a base is that mass of base that will furnish 1 mol of OH^- ions. Using concentrations in normality, 1 equivalent of acid (A) will react with 1 equivalent of base (B):

$$N_A = \frac{\text{equiv}_A}{L_A} \quad \text{and} \quad N_B = \frac{\text{equiv}_B}{L_B}$$

$$\text{equiv}_A = L_A \times N_A \quad \text{and} \quad \text{equiv}_B = L_B \times N_B$$

Since $\text{equiv}_A = \text{equiv}_B$,

$$L_A \times N_A = L_B \times N_B$$

When both volumes are in the same units, we can write a more general equation:

$$V_A N_A = V_B N_B$$

which states that the volume of acid times the normality of the acid equals the volume of base times the normality of the base.

Example 14.13

(a) What is the normality of an H_2SO_4 solution if 25.00 mL of the solution requires 22.48 mL of 0.2018 N NaOH for complete neutralization? (b) What is the molarity of the H_2SO_4 solution?

Solution

(a) Solve for N_A by substituting the data into

$$V_A N_A = V_B N_B$$

$$25.00\text{ mL} \times N_A = 22.48\text{ mL} \times 0.2018\ N$$

$$N_A = \frac{22.48\ \cancel{\text{mL}} \times 0.2018\ N}{25.00\ \cancel{\text{mL}}} = 0.1815\ N\ H_2SO_4$$

(b) When H_2SO_4 is completely neutralized it furnishes 2 equivalents of H^+ ions per mole of H_2SO_4. The conversion from N to M is

$$\frac{\text{equiv}}{\text{L}} \longrightarrow \frac{\text{mol}}{\text{L}}$$

$$H_2SO_4 = 0.1815\ N$$

$$\frac{0.1815\text{ equiv}}{1\text{ L}} \times \frac{1\text{ mol}}{2\text{ equiv}} = 0.09075\text{ mol/L}$$

The H_2SO_4 solution is 0.09075 M.

Practice 14.9

What is the normality of a NaOH solution if 50.0 mL of the solution requires 23.72 mL of 0.0250 N H_2SO_4? What is the molarity of the NaOH?

The equivalent mass of a substance may be variable; its value is dependent on the reaction that the substance is undergoing. Consider the reactions represented by these equations:

$$NaOH + H_2SO_4 \longrightarrow NaHSO_4 + H_2O$$

$$2\ NaOH + H_2SO_4 \longrightarrow Na_2SO_4 + 2\ H_2O$$

In the first reaction 1 mol of H_2SO_4 furnishes 1 mol of hydrogen atoms. Therefore the equivalent mass of H_2SO_4 is the molar mass, namely 98.08 g. But in the

Microencapsulation

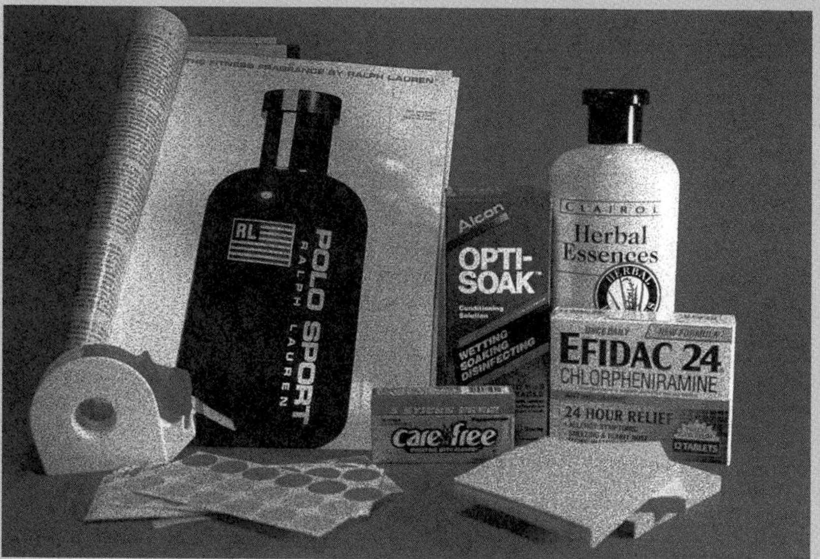

Many items we come across in our daily lives are microencapsulated. These are just a few examples.

Producing chemical reactions that will occur at precisely the correct moment is one of the tasks facing the chemist in industry. For this to happen, one or more of the reactants must be stored separately and released under controlled conditions precisely when the reaction is desired. A technique developed to accomplish this is microencapsulation, in which reactive chemicals—solids, liquids, or gases—are sealed inside tiny capsules. The material forming the wall of the capsule is carefully chosen so that the encapsulated chemicals can be released, at the appropriate time, by one of several methods. This release can be accomplished in a variety of ways, which include dissolving the capsules, diffusion through the capsule walls, and by mechanical, thermal, electrical, or chemical disruption of the capsules.

In one type of microencapsulation, water diffuses into the capsule and forms a solution which then diffuses out into the surroundings at a constant rate. Some types of capsules contain materials that dissolve at a certain level of acidity and form pores in the capsule through which the encapsulated materials escape. Still other types of capsules dissolve completely over a given period of time, releasing their contents into the system.

Applications of microencapsulation are found everywhere. Carbonless paper, often used in receipts makes use of pressure-sensitive microcapsules containing colorless dye precursors. Another reactive substance is present and converts the precursor to the colored form when pressure is applied by a pen or printer.

Adhesives are frequently encapsulated to prevent them from becoming tacky too soon. The active surfaces on pressure-sensitive labels and certain self-sealing envelopes are coated with encapsulated adhesives that are released by pressure. Heat-sensitive encapsulation is used for the adhesives in iron-on patches for clothing.

Many microencapsulated products are to be found in the kitchen. Flavorings are encapsulated to make them easier to store in a powdered state, cut evaporation, and reduce reactions with the air. These advantages increase the shelf life of products. Flavoring microcapsules may also be heat sensitive and release their contents during cooking or pressure sensitive (as in chewing gum) and release their contents upon chewing.

Still other encapsulated products are found in our bathrooms. Time-release microencapsulation is used in deodorants, moisturizers, colognes, and perfumes. The encapsulation process prevents evaporation, decomposition, and unwanted reactions with the air and other ingredients. Drugs and medications are frequently encapsulated to dissolve slowly over a long period of time in the body. These medications generally work in the intestinal tract, but some may also be given by injection to work within other tissues.

Fragrances have undergone microencapsulation in such products as cosmetics, health care products, detergents, and even foods. Gas companies use encapsulated propyl mercaptan, $CH_3CH_2CH_2SH$, to teach children how to detect a gas leak (natural gas without this substance is odorless). Encapsulated fragrances are responsible for the ever-present scratch-and-sniff labels found in children's books and fashion magazines. When the paper is scratched or pulled open, the fragrance is released into the air.

Other applications of microencapsulation include time-release pesticides and neutralizer for contact lenses, as well as special additives in detergents, cleaners, and paints.

TABLE 14.4 Concentration Units for Solutions

Units	Symbol	Definition
Mass percent	% m/m	$\frac{\text{Mass solute}}{\text{Mass solution}} \times 100$
Parts per million	ppm	$\frac{\text{Mass solute}}{\text{Mass solution}} \times 1{,}000{,}000$
Mass/volume percent	% m/v	$\frac{\text{Mass solute}}{\text{mL solution}} \times 100$
Volume percent	% v/v	$\frac{\text{mL solute}}{\text{mL solution}} \times 100$
Molarity	*M*	$\frac{\text{Moles solute}}{\text{L solution}}$
Normality	*N*	$\frac{\text{Equivalents solute}}{\text{L solution}}$
Molality	*m*	$\frac{\text{Moles solute}}{\text{kg solvent}}$

second reaction 1 mol of H_2SO_4 furnishes 2 mol of hydrogen atoms. Therefore, the equivalent mass of the H_2SO_4 is one-half the molar mass, or 49.04 g. A summary of the quantitative concentration units is found in Table 14.4.

14.7 Colligative Properties of Solutions

Two solutions—one containing 1 mol (60.06 g) of urea, NH_2CONH_2, and the other containing 1 mol (342.3 g) of sucrose, $C_{12}H_{22}O_{11}$, in 1 kg of water—both have a freezing point of $-1.86°C$, not 0°C as for pure water. Urea and sucrose are distinctly different substances, yet they lower the freezing point of the water by the same amount. The only thing apparently common to these two solutions is that each contains 1 mol (6.022×10^{23} molecules) of solute and 1 kg of solvent. In fact, when we dissolve 1 mol of any nonionizable solute in 1 kg of water, the freezing point of the resulting solution is $-1.86°C$.

These results lead us to conclude that the freezing-point depression for a solution containing 6.022×10^{23} solute molecules (particles) and 1 kg of water is a constant, namely, 1.86°C. Freezing-point depression is a general property of solutions. Furthermore, the amount by which the freezing point is depressed is the same for all solutions made with a given solvent; that is, each solvent shows a characteristic *freezing-point depression constant*. Freezing-point depression constants for several solvents are given in Table 14.5.

The solution formed by the addition of a nonvolatile solute to a solvent has a lower freezing point, a higher boiling point, and a lower vapor pressure than that of the pure solvent. All these effects are related and are known as colligative

TABLE 14.5 Freezing-Point Depression and Boiling-Point Elevation Constants of Selected Solvents

Solvent	Freezing point of pure solvent (°C)	Freezing-point depression constant, K_f $\left(\frac{°C\ kg\ solvent}{mol\ solute}\right)$	Boiling point of pure solvent (°C)	Boiling-point elevation constant, K_b $\left(\frac{°C\ kg\ solvent}{mol\ solute}\right)$
Water	0.00	1.86	100.0	0.512
Acetic acid	16.6	3.90	118.5	3.07
Benzene	5.5	5.1	80.1	2.53
Camphor	178	40	208.2	5.95

colligative properties

properties. The **colligative properties** are properties that depend only on the number of solute particles in a solution and not on the nature of those particles. Freezing-point depression, boiling-point elevation, and vapor-pressure lowering are colligative properties of solutions.

The colligative properties of a solution can be considered in terms of vapor pressure. The vapor pressure of a pure liquid depends on the tendency of molecules to escape from its surface. Thus, if 10% of the molecules in a solution are nonvolatile solute molecules, the vapor pressure of the solution is 10% lower than that of the pure solvent. The vapor pressure is lower because the surface of the solution contains 10% nonvolatile molecules and 90% of the volatile solvent molecules. A liquid boils when its vapor pressure equals the pressure of the atmosphere. Thus, we can see that the solution just described as having a lower vapor pressure will have a higher boiling point than the pure solvent. The solution with a lowered vapor pressure does not boil until it has been heated above the boiling point of the solvent (see Figure 14.8(a)). Each solvent has its own characteristic boiling-point elevation constant (Table 14.5). The boiling-point elevation constant is based on a solution that contains 1 mol of solute particles per kilogram of solvent. For example, the boiling-point elevation constant for a solution containing 1 mol of solute particles per kilogram of water is 0.512°C, which means that this water solution will boil at 100.512°C.

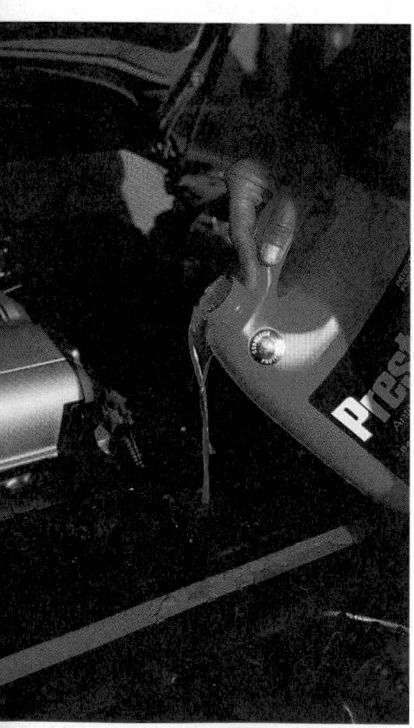

Engine coolant is an example of the use of colligative properties. The additon of coolant to the water in a radiator raises its boiling point and lowers its freezing point.

The freezing behavior of a solution can also be considered in terms of lowered vapor pressure. Figure 14.8(b) shows the vapor-pressure relationships of ice, water, and a solution containing 1 mol of solute per kilogram of water. The freezing point of water is at the intersection of the water and ice vapor-pressure curves (i.e., at the point where water and ice have the same vapor pressure). Because the vapor pressure of water is lowered by the solute, the vapor-pressure curve of the solution does not intersect the vapor-pressure curve of ice until the solution has been cooled below the freezing point of pure water. Thus, it is necessary to cool the solution below 0°C in order for it to freeze.

The foregoing discussion dealing with freezing-point depressions is restricted to *un-ionized* substances. The discussion of boiling-point elevations is restricted to *nonvolatile* and un-ionized substances. The colligative properties of ionized substances are not under consideration at this point, but are discussed in Chapter 15.

Some practical applications involving colligative properties are (1) use of salt–ice mixtures to provide low freezing temperatures for homemade ice cream,

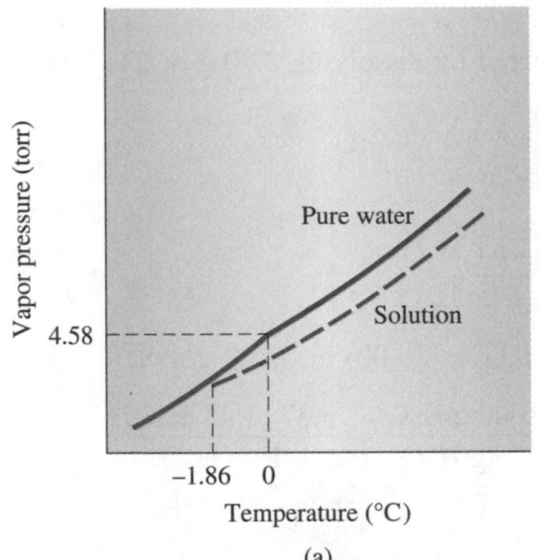

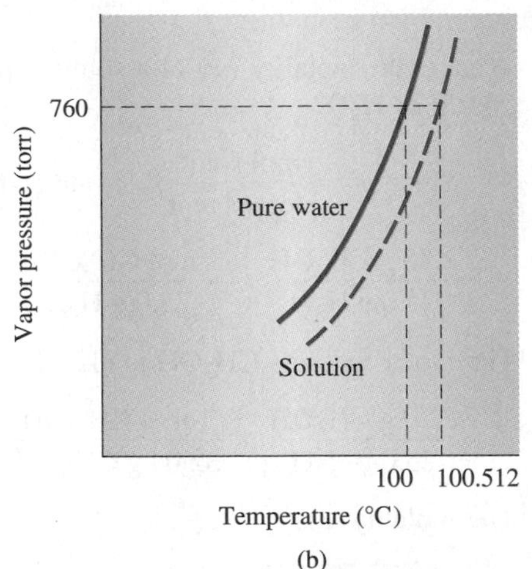

FIGURE 14.8
Vapor-pressure curves of pure water and water solutions, showing (a) freezing-point depression and (b) boiling-point elevation effects (concentration: 1 mol solute/1 kg water).

(2) use of sodium chloride or calcium chloride to melt ice from streets, and (3) use of ethylene glycol and water mixtures as antifreeze in automobile radiators (ethylene glycol also raises the boiling point of radiator fluid, thus allowing the engine to operate at a higher temperature).

Both the freezing-point depression and the boiling-point elevation are directly proportional to the number of moles of solute per kilogram of solvent. When we deal with the colligative properties of solutions, another concentration expression, *molality*, is used. The **molality (m)** of a solute is the number of moles of solute per kilogram of solvent:

molality (m)

$$m = \frac{\text{mol solute}}{\text{kg solvent}}$$

Note that a lowercase m is used for molality concentrations and a capital M for molarity. The difference between molality and molarity is that molality refers to moles of solute *per kilogram of solvent*, whereas molarity refers to moles of solute *per liter of solution*. For un-ionized substances, the colligative properties of a solution are directly proportional to its molality.

Molality is independent of volume. It is a mass-to-mass relationship of solute to solvent and allows for experiments, such as freezing-point depression and boiling-point elevation, to be conducted at variable temperatures.

The following equations are used in calculations involving colligative properties and molality:

$$\Delta t_f = mK_f \qquad \Delta t_b = mK_b \qquad m = \frac{\text{mol solute}}{\text{kg solvent}}$$

m = molality; mol solute/kg solvent

Δt_f = freezing-point depression; °C

Δt_b = boiling-point elevation; °C

K_f = freezing-point depression constant; °C kg solvent/mol solute

K_b = boiling-point elevation constant; °C kg solvent/mol solute

Sodium chloride or calcium chloride is used to melt ice on snowy streets and highways.

Example 14.14 What is the molality (m) of a solution prepared by dissolving 2.70 g of CH_3OH in 25.0 g of H_2O?

Solution

Since $m = \dfrac{\text{mol solute}}{\text{kg solvent}}$, the conversion is

$$\frac{2.70 \text{ g } CH_3OH}{25.0 \text{ g } H_2O} \longrightarrow \frac{\text{mol } CH_3OH}{25.0 \text{ g } H_2O} \longrightarrow \frac{\text{mol } CH_3OH}{1 \text{ kg } H_2O}$$

The molar mass of CH_3OH is (12.01 + 4.032 + 16.00) or 32.04 g/mol:

$$\frac{2.70 \text{ g } CH_3OH}{25.0 \text{ g } H_2O} \times \frac{1 \text{ mol } CH_3OH}{32.04 \text{ g } CH_3OH} \times \frac{1000 \text{ g } H_2O}{1 \text{ kg } H_2O} = \frac{3.37 \text{ mol } CH_3OH}{1 \text{ kg } H_2O}$$

The molality is 3.37 m.

Practice 14.10

What is the molality of a solution prepared by dissolving 150.0 g of $C_6H_{12}O_6$ in 600.0 g of H_2O?

Example 14.15 A solution is made by dissolving 100. g of ethylene glycol ($C_2H_6O_2$) in 200. g of water. What is the freezing point of this solution?

Solution

To calculate the freezing point of the solution, we first need to calculate Δt_f, the change in freezing point. Use the equation

$$\Delta t_f = mK_f = \frac{\text{mol solute}}{\text{kg solvent}} \times K_f$$

K_f (for water): $\dfrac{1.86°\text{C kg solvent}}{\text{mol solute}}$ (from Table 14.5)

mol solute: $100. \text{ g } C_2H_6O_2 \times \dfrac{1 \text{ mol } C_2H_6O_2}{62.07 \text{ g } C_2H_6O_2} = 1.61 \text{ mol } C_2H_6O_2$

kg solvent: $200. \text{ g } H_2O \times \dfrac{1 \text{ kg}}{1000 \text{ g}} = 0.200 \text{ kg } H_2O$

$$\Delta t_f = \frac{1.61 \text{ mol } C_2H_6O_2}{0.200 \text{ kg } H_2O} \times \frac{1.86°\text{C kg } H_2O}{1 \text{ mol } C_2H_6O_2} = 15.0°\text{C}$$

The freezing-point depression, 15.0°C, must be subtracted from 0°C, the freezing point of the pure solvent (water):

$$\text{freezing point of solution} = \text{freezing point of solvent} - \Delta t_f$$
$$= 0.0°\text{C} - 15.0°\text{C} = -15.0°\text{C}$$

Therefore, the freezing point of the solution is −15.0°C. The calculation can also be done using the equation

$$\Delta t_f = K_f \times \frac{\text{g solute}}{\text{molar mass solute}} \times \frac{1}{\text{kg solvent}}$$

Example 14.16

A solution made by dissolving 4.71 g of a compound of unknown molar mass in 100.0 g of water has a freezing point of $-1.46°C$. What is the molar mass of the compound?

Solution

First substitute the data in $\Delta t_f = mK_f$ and solve for m:

$$\Delta t_f = +1.46 \text{ (since the solvent, water, freezes at 0°C)}$$

$$K_f = \frac{1.86°\text{C kg H}_2\text{O}}{\text{mol solute}}$$

$$1.46°\text{C} = mK_f = m \times \frac{1.86°\text{C kg H}_2\text{O}}{\text{mol solute}}$$

$$m = \frac{1.46\cancel{°\text{C}} \times \text{mol solute}}{1.86\cancel{°\text{C}} \times \text{kg H}_2\text{O}} = \frac{0.785 \text{ mol solute}}{\text{kg H}_2\text{O}}$$

Now convert the data, 4.71 g solute/100.0 g H_2O, to g/mol:

$$\frac{4.71 \text{ g } \cancel{\text{solute}}}{100.0 \text{ g } \cancel{\text{H}_2\text{O}}} \times \frac{1000 \text{ g } \cancel{\text{H}_2\text{O}}}{1 \text{ kg } \cancel{\text{H}_2\text{O}}} \times \frac{1 \cancel{\text{kg H}_2\text{O}}}{0.785 \text{ mol } \cancel{\text{solute}}} = 60.0 \text{ g/mol}$$

The molar mass of the compound is 60.0 g/mol.

Practice 14.11

What is the freezing point of the solution in Practice exercise 14.10? What is the boiling point?

14.8 Osmosis and Osmotic Pressure

When red blood cells are put into distilled water, they gradually swell and, in time, may burst. If red blood cells are put in a 5% urea (or a 5% salt) solution, they gradually shrink and take on a wrinkled appearance. The cells behave in this fashion because they are enclosed in semipermeable membranes. A **semipermeable membrane** allows the passage of water (solvent) molecules through it in either direction but prevents the passage of larger solute molecules or ions. When two solutions of different concentrations (or water and a water solution) are separated by a semipermeable membrane, water diffuses through the membrane from the solution of lower concentration into the solution of higher concentration. The diffusion of water, either from a dilute solution or from pure water, through a semipermeable membrane into a solution of higher concentration is called **osmosis**.

semipermeable membrane

osmosis

A 0.90% (0.15 M) sodium chloride solution is known as a *physiological saline solution* because it is *isotonic* with blood plasma; that is, it has the same osmotic pressure as blood plasma. Because each mole of NaCl yields about 2 mol of ions when in solution, the solute particle concentration in physiological saline solution is nearly 0.30 M. Five percent glucose solution (0.28 M) is also approximately isotonic with blood plasma. Blood cells neither swell nor shrink in an isotonic solution. The cells described in the first paragraph of this section swell in water

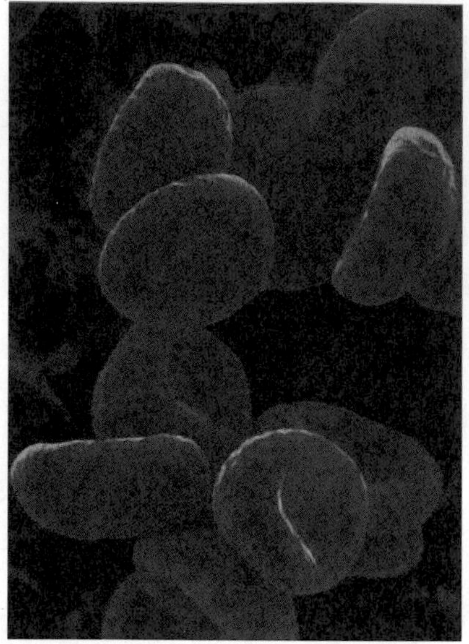

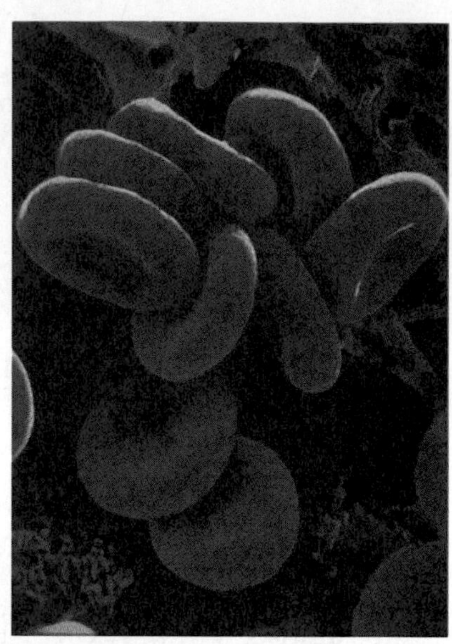

▲
Human red blood cells. *Left:* In a hypotonic solution (0.2% saline) the cells swell as water moves into the cell center. *Center:* In a hypertonic solution (1.6% saline) water leaves the cells causing them to crenate (shrink). *Right:* In an isotonic solution the concentration is the same inside and outside the cell (0.9% saline). Cells do not change in size. Magnification is 260,000 X.

because water is *hypotonic* to cell plasma. The cells shrink in 5% urea solution because the urea solution is *hypertonic* to the cell plasma. In order to prevent possible injury to blood cells by osmosis, fluids for intravenous use are usually made up at approximately isotonic concentration.

All solutions exhibit *osmotic pressure*, which is another colligative property. Osmotic pressure is dependent only on the concentration of the solute particles and is independent of their nature. The osmotic pressure of a solution can be measured by determining the amount of counterpressure needed to prevent osmosis; this pressure can be very large. The osmotic pressure of a solution containing 1 mol of solute particles in 1 kg of water is about 22.4 atm, which is about the same as the pressure exerted by 1 mol of a gas confined in a volume of 1 L at 0°C.

Osmosis has a role in many biological processes, and semipermeable membranes occur commonly in living organisms. An example is the roots of plants, which are covered with tiny structures called root hairs; soil water enters the plant by osmosis, passing through the semipermeable membranes covering the root hairs. Artificial or synthetic membranes can also be made.

Osmosis can be demonstrated with the simple laboratory setup shown in Figure 14.9. As a result of osmotic pressure, water passes through the cellophane membrane into the thistle tube, causing the solution level to rise. In osmosis the net transfer of water is always from a less concentrated to a more concentrated solution; that is, the effect is toward equalization of the concentration on both sides of the membrane. It should also be noted that the effective movement of water in osmosis is always from the region of *higher water concentration* to the region of *lower water concentration*.

Osmosis can be explained by assuming that a semipermeable membrane has passages that permit water molecules and other small molecules—to pass in either direction. Both sides of the membrane are constantly being struck by water molecules in random motion. The number of water molecules crossing the membrane is proportional to the number of water molecule-to-membrane impacts per unit of time.

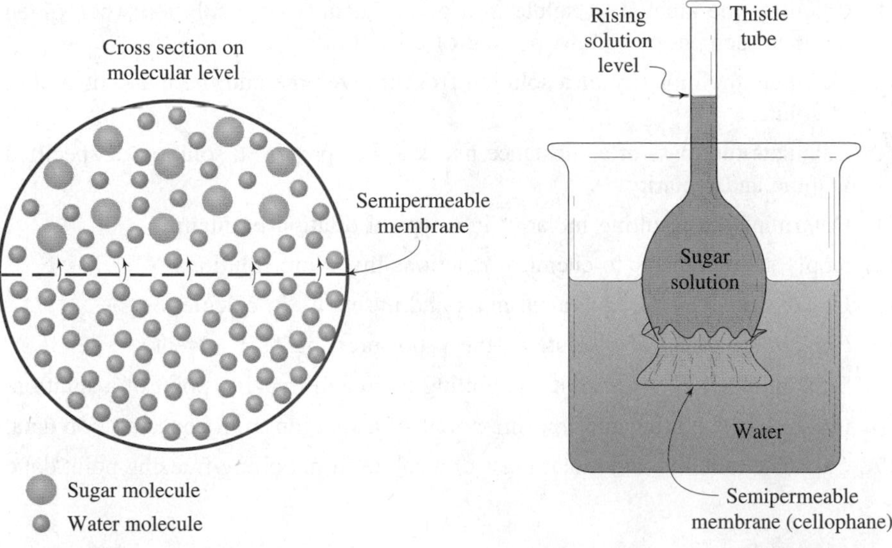

FIGURE 14.9
Laboratory demonstration of osmosis: As a result of osmosis, water passes through the membrane causing the solution to rise in the thistle tube.

Because the solute molecules or ions reduce the concentration of water, there are more water molecules and more water-molecule impacts on the side with the lower solute concentration (more dilute solution). The greater number of water molecule-to-membrane impacts on the dilute side thus causes a net transfer of water to the more concentrated solution. Again, note that the overall process involves the net transfer, by diffusion through the membrane, of water molecules from a region of higher water concentration (dilute solution) to one of lower water concentration (more concentrated solution).

This explanation is a simplified picture of osmosis. No one has ever seen the hypothetical passages that allow water molecules and other small molecules or ions to pass through them. Alternative explanations have been proposed, but our discussion has been confined to water solutions. Osmotic pressure is a general colligative property, however, and is known to occur in nonaqueous systems.

Concepts in Review

1. Describe the types of solutions.
2. List the general properties of solutions.
3. Describe and illustrate the process by which an ionic substance dissolves in water.
4. Indicate the effects of temperature and pressure on the solubility of solids and gases in liquids.
5. Identify and explain the factors affecting the rate at which a solid dissolves in a liquid.
6. Use a solubility table or graph to determine whether a solution is saturated, unsaturated, or supersaturated at a given temperature.
7. Calculate the mass percent or volume percent for a solution.

8. Calculate the amount of solute in a given quantity of a solution when given the mass percent or volume percent of a solution.
9. Calculate the molarity of a solution from the volume and the mass, or moles, of solute.
10. Calculate the mass of a substance necessary to prepare a solution of specified volume and molarity.
11. Determine the resulting molarity in a typical dilution problem.
12. Apply stoichiometry to chemical reactions involving solutions.
13. Use the concepts of equivalent mass and normality in calculations.
14. Explain the effect of a solute on the vapor pressure of a solvent.
15. Explain the effect of a solute on boiling point and freezing point of a solution.
16. Calculate the boiling and freezing points of a solution from concentration data.
17. Calculate molality and molar mass of a solute from boiling/freezing point data.

Key Terms

The terms listed here have been defined within this chapter. Section numbers are referenced in parenthesis for each term. More detailed definitions are given in the Glossary.

colligative properties (14.7)
concentrated solution (14.6)
concentration of a solution (14.2)
dilute solution (14.6)
equivalent mass (14.6)
immiscible (14.2)
miscible (14.2)
molality (*m*) (14.7)
molarity (*M*) (14.6)
normality (14.6)
osmosis (14.8)
parts per million (ppm) (14.6)
saturated solution (14.6)
semipermeable membrane (14.8)
solubility (14.2)
solute (14.1)
solution (14.1)
solvent (14.1)
supersaturated solution (14.6)
unsaturated solution (14.6)

Questions

Questions refer to tables, figures, and key words and concepts defined within the chapter. A particularly challenging question or exercise is indicated with an asterisk.

1. Make a sketch indicating the orientation of water molecules (a) about a single sodium ion and (b) about a single chloride ion in solution.
2. Estimate the number of grams of sodium fluoride that would dissolve in 100 g of water at 50°C. (See Table 14.2.)
3. What is the solubility at 25°C of each of the substances listed below? (See Figure 14.4.)
   (a) potassium chloride (c) potassium nitrate
   (b) potassium chlorate
4. What is different in the solubility trend of the potassium halides compared with that of the lithium halides and the sodium halides? (See Table 14.2.)
5. What is the solubility, in grams of solute per 100 g of H_2O of (a) $KClO_3$ at 60°C, (b) HCl at 20°C, (c) Li_2SO_4 at 80°C, and (d) KNO_3 at 0°C? (See Figure 14.4.)
6. Which substance, KNO_3 or NH_4Cl, shows the greater increase in solubility with increased temperature? (See Figure 14.4.)
7. Does a 2 molal solution in benzene or a 1 molal solution in camphor show the greater freezing-point depression? (See Table 14.5.)

8. What would be the total surface area if the 1-cm cube in Figure 14.5 were cut into cubes 0.01 cm on a side?
9. At which temperatures—10°C, 20°C, 30°C, 40°C, or 50°C—would you expect a solution made from 63 g of ammonium chloride and 150 g of water to be unsaturated? (See Figure 14.4.)
10. Explain why the rate of dissolving decreases as shown in Figure 14.6.
11. Would the volumetric flasks in Figure 14.7 be satisfactory for preparing normal solutions? Explain.
12. Assume that the thistle tube in Figure 14.9 contains 1.0 *M* sugar solution and that the water in the beaker has just been replaced by a 2.0 *M* solution of urea. Would the solution level in the thistle tube continue to rise, remain constant, or fall? Explain.
13. Name and distinguish between the two components of a solution.
14. Is it always apparent in a solution which component is the solute, for example, in a solution of a liquid in a liquid?
15. Explain why the solute does not settle out of a solution.
16. Is it possible to have one solid dissolved in another? Explain.
17. An aqueous solution of KCl is colorless, $KMnO_4$ is purple, and $K_2Cr_2O_7$ is orange. What color would you expect of an aqueous solution of $Na_2Cr_2O_7$? Explain.
18. Explain why hexane will dissolve benzene but will not dissolve sodium chloride.
19. Some drinks like tea are consumed either hot or cold, whereas others like Coca Cola are drunk only cold. Why?
20. Why is air considered to be a solution?
21. In which will a teaspoonful of sugar dissolve more rapidly, 200 mL of iced tea or 200 mL of hot coffee? Explain in terms of the KMT.
22. What is the effect of pressure on the solubility of gases in liquids? Solids in liquids?
23. Why do smaller particles dissolve faster than large ones?
24. In a saturated solution containing undissolved solute, solute is continuously dissolving, but the concentration of the solution remains unchanged. Explain.
25. Explain why there is no apparent reaction when crystals of $AgNO_3$ and NaCl are mixed, but a reaction is apparent immediately when solutions of $AgNO_3$ and NaCl are mixed.
26. What do we mean when we say that concentrated nitric acid, HNO_3, is 16 molar?
27. Will 1 L of 1 *M* NaCl contain more chloride ions than 0.5 L of 1 *M* $MgCl_2$? Explain.
28. Champagne is usually cooled in a refrigerator prior to opening. It is also opened very carefully. What would happen if a warm bottle of champagne is shaken and opened quickly and forcefully?
29. Explain how a supersaturated solution of $NaC_2H_3O_2$ can be prepared and proven to be supersaturated.
30. Explain in terms of the KMT how a semipermeable membrane functions when placed between pure water and a 10% sugar solution.
31. Which has the higher osmotic pressure, a solution containing 100 g of urea, NH_2CONH_2, in 1 kg of H_2O or a solution containing 150 g of glucose, $C_6H_{12}O_6$, in 1 kg of H_2O?
32. Explain why a lettuce leaf in contact with salad dressing containing salt and vinegar soon becomes wilted and limp whereas another lettuce leaf in contact with plain water remains crisp.
33. A group of shipwreck survivors floated for several days on a life raft before being rescued. Those who had drunk some seawater were found to be suffering the most from dehydration. Explain.
34. Which of the following statements are correct? Rewrite the incorrect statements to make them correct.
   (a) A solution is a homogeneous mixture.
   (b) It is possible for the same substance to be the solvent in one solution and the solute in another.
   (c) A solute can be removed from a solution by filtration.
   (d) Saturated solutions are always concentrated solutions.
   (e) If a solution of sugar in water is allowed to stand undisturbed for a long time, the sugar will gradually settle to the bottom of the container.
   (f) It is not possible to prepare an aqueous 1.0 *M* AgCl solution.
   (g) Gases are generally more soluble in hot water than in cold water.
   (h) It is impossible to prepare a two-phase liquid mixture from two liquids that are miscible with each other in all proportions.
   (i) A solution that is 10% NaCl by mass always contains 10 g of NaCl.
   (j) Small changes in pressure have little effect on the solubility of solids in liquids but a marked effect on the solubility of gases in liquids.
   (k) How fast a solute dissolves depends mainly on the size of the solute particles, the temperature of the solvent, and the degree of agitation or stirring taking place.
   (l) In order to have a 1 *M* solution you must have 1 mol of solute dissolved in sufficient solvent to give 1 L of solution.
   (m) Dissolving 1 mol of NaCl in 1 L of water will give a 1 *M* solution.
   (n) One mole of solute in 1 L of solution has the same concentration as 0.1 mol of solute in 100 mL of solution.
   (o) When 100 mL of 0.200 *M* HCl is diluted to 200-mL volume by the addition of water, the resulting solution is 0.100 *M* and contains one-half the number of moles of HCl as were in the original solution.

(p) Fifty milliliters of 0.1 M H_2SO_4 will neutralize the same volume of 0.1 M NaOH as 100 mL of 0.1 M HCl.
(q) Fifty milliliters of 0.1 N H_2SO_4 will neutralize the same volume of 0.1 M NaOH as 100 mL of 0.1 M HCl.
(r) The molarity of a solution will vary slightly with temperature.
(s) The equivalent mass of $Ca(OH)_2$ is one-half its molar mass.
(t) Gram for gram, methyl alcohol, CH_3OH, is more effective than ethyl alcohol, C_2H_5OH, in lowering the freezing point of water.
(u) An aqueous solution that freezes below 0°C will have a normal boiling point below 100°C.
(v) The colligative properties of a solution depend on the number of solute particles dissolved in solution.
(w) A solution of 1.00 mol of a nonionizable solute and 1000 g of water will freeze at −1.86°C and will boil at 99.5°C at atmospheric pressure.
(x) Water will diffuse from a 0.1 M sugar solution to a 0.2 M sugar solution when these two solutions are separated by a semipermeable membrane.
(y) An isotonic salt solution has the same osmotic pressure as blood plasma.
(z) Red blood cells will neither swell nor shrink when placed into an isotonic salt solution.

35. What disadvantages are there in expressing the concentration of solutions as dilute or concentrated?

36. Explain how concentrated H_2SO_4 can be both 18 M and 36 N in concentration.

37. Describe how you would prepare 750 mL of 5 M NaCl solution.

38. Arrange the following bases (in descending order) according to the volume of each that will react with 1 L of 1 M HCl; (a) 1 M NaOH, (b) 1.5 M $Ca(OH)_2$, (c) 2 M KOH, and (d) 0.6 M $Ba(OH)_2$.

***39.** Explain in terms of vapor pressure why the boiling point of a solution containing a nonvolatile solute is higher than that of the pure solvent.

40. Explain why the freezing point of a solution is lower than the freezing point of the pure solvent.

41. Which would be colder, a glass of water and crushed ice or a glass of Seven-Up and crushed ice? Explain.

42. When water and ice are mixed, the temperature of the mixture is 0°C. But, if methyl alcohol and ice are mixed, a temperature of −10°C is readily attained. Explain why the two mixtures show such different temperature behavior.

43. Which would be more effective in lowering the freezing point of 500. g of water?
(a) 100. g of sucrose ($C_{12}H_{22}O_{11}$) or 100. g of ethyl alcohol (C_2H_5OH)
(b) 100. g of sucrose or 20.0 g of ethyl alcohol
(c) 20.0 g of ethyl alcohol or 20.0 g of methyl alcohol (CH_3OH)

44. Is the molarity of a 5 molal aqueous solution of NaCl greater or less than 5 molar? Explain.

45. What is microencapsulation?

46. Explain how a scratch-and-sniff label works.

47. What are the purposes for timed-release microencapsulation?

48. State three types of microencapsulation systems and give a practical application of each.

Paired Exercises

These exercises are paired. Each odd-numbered exercise is followed by a similar even-numbered exercise. Answers to the even-numbered exercises are given in Appendix V.

49. Which of the substances listed below are reasonably soluble and which are insoluble in water? (See Figure 14.2 or Appendix IV.)
(a) KOH
(b) $NiCl_2$
(c) ZnS
(d) $AgC_2H_3O_2$
(e) Na_2CrO_4

50. Which of the substances listed below are reasonably soluble and which are insoluble in water? (See Figure 14.2 or Appendix IV.)
(a) PbI_2
(b) $MgCO_3$
(c) $CaCl_2$
(d) $Fe(NO_3)_3$
(e) $BaSO_4$

Percent Solutions

51. Calculate the mass percent of the following solutions:
(a) 25.0 g NaBr + 100.0 g H_2O
(b) 1.20 g K_2SO_4 + 10.0 g H_2O

52. Calculate the mass percent of the following solutions.
(a) 40.0 g $Mg(NO_3)_2$ + 500.0 g H_2O
(b) 17.5 g $NaNO_3$ + 250.0 g H_2O

53. How many grams of a solution that is 12.5% by mass $AgNO_3$ would contain 30.0 g of $AgNO_3$?

54. How many grams of a solution that is 12.5% by mass $AgNO_3$ would contain 0.400 mol of $AgNO_3$?

55. Calculate the mass percent of the following solutions:
(a) 60.0 g NaCl + 200.0 g H_2O
(b) 0.25 mol $HC_2H_3O_2$ + 3.0 mol H_2O

56. Calculate the mass percent of the following solutions:
(a) 145.0 g NaOH in 1.5 kg H_2O
(b) 1.0 *m* solution of $C_6H_{12}O_6$ in water

57. How much solute is present in 65 g of 5.0% KCl solution?

58. How much solute is present in 250. g of 15.0% K_2CrO_4 solution?

59. Calculate the mass/volume percent of a solution made by dissolving 22.0 g of CH_3OH (methanol) in C_2H_5OH (ethanol) to make 100. mL of solution.

60. Calculate the mass/volume percent of a solution made by dissolving 4.20 g of NaCl in H_2O to make 12.5 mL of solution.

61. What is the volume percent of 10.0 mL of CH_3OH (methanol) dissolved in water to a volume of 40.0 mL?

62. What is the volume percent of 2.0 mL of hexane, C_6H_{14}, dissolved in benzene, C_6H_6, to a volume of 9.0 mL?

Molarity Problems

63. Calculate the molarity of the following solutions:
(a) 0.10 mol of solute in 250 mL of solution
(b) 2.5 mol of NaCl in 0.650 L of solution
(c) 53.0 g of Na_2CrO_4 in 1.00 L of solution
(d) 260 g of $C_6H_{12}O_6$ in 800. mL of solution

64. Calculate the molarity of the following solutions:
(a) 0.025 mol of HCl in 10. mL of solution
(b) 0.35 mol $BaCl_2 \cdot H_2O$ in 593 mL of solution
(c) 1.50 g of $Al_2(SO_4)_3$ in 2.00 L of solution
(d) 0.0282 g of $Ca(NO_3)_2$ in 1.00 mL of solution

65. Calculate the number of moles of solute in each of the following solutions:
(a) 40.0 L of 1.0 *M* LiCl
(b) 25.0 mL of 3.00 *M* H_2SO_4

66. Calculate the number of moles of solute in each of the following solutions:
(a) 349 mL of 0.0010 *M* NaOH
(b) 5000. mL of 3.1 *M* $CoCl_2$

67. Calculate the grams of solute in each of the following solutions:
(a) 150 L of 1.0 *M* NaCl
(b) 260 mL of 18 *M* H_2SO_4

68. Calculate the grams of solute in each of the following solutions:
(a) 0.035 L of 10.0 *M* HCl
(b) 8.00 mL of 8.00 *M* $Na_2C_2O_4$

69. How many milliliters of 0.256 *M* KCl solution will contain the following?
(a) 0.430 mol of KCl
(b) 20.0 g of KCl

70. How many milliliters of 0.256 *M* KCl solution will contain the following?
(a) 10.0 mol of KCl
***(b)** 71.0 g of chloride ion, Cl^-

Dilution Problems

71. What will be the molarity of the resulting solutions made by mixing the following? Assume volumes are additive.
(a) 200. mL of 12 *M* HCl + 200.0 mL H_2O
(b) 60.0 mL of 0.60 *M* $ZnSO_4$ + 500. mL H_2O

72. What will be the molarity of the resulting solutions made by mixing the following? Assume volumes are additive.
(a) 100. mL 1.0 *M* HCl + 150 mL 2.0 *M* HCl
(b) 25.0 mL 12.5 *M* NaCl + 75.0 mL 2.00 *M* NaCl

73. Calculate the volume of concentrated reagent required to prepare the diluted solutions indicated:
(a) 12 *M* HCl to prepare 400. mL of 6.0 *M* HCl
(b) 16 *M* HNO_3 to prepare 100. mL of 2.5 *N* HNO_3

74. Calculate the volume of concentrated reagent required to prepare the diluted solutions indicated:
(a) 15 *M* NH_3 to prepare 50. mL of 6.0 *M* NH_3
(b) 18 *M* H_2SO_4 to prepare 250 mL of 10.0 *N* H_2SO_4

75. What will be the molarity of each of the solutions made by mixing 250 mL of 0.75 *M* H_2SO_4 with (a) 150 mL of H_2O? (b) 250 mL of 0.70 *M* H_2SO_4?

76. What will be the molarity of each of the solutions made by mixing 250 mL of 0.75 *M* H_2SO_4 with (a) 400. mL of 2.50 *M* H_2SO_4? (b) 375 mL of H_2O?

Stoichiometry Problems

77. $BaCl_2(aq) + K_2CrO_4(aq) \rightarrow BaCrO_4(s) + 2\ KCl(aq)$
Using the above equation, calculate
(a) The grams of $BaCrO_4$ that can be obtained from 100.0 mL of 0.300 *M* $BaCl_2$
(b) The volume of 1.0 *M* $BaCl_2$ solution needed to react with 50.0 mL of 0.300 *M* K_2CrO_4 solution

78. $3\ MgCl_2(aq) + 2\ Na_3PO_4(aq) \rightarrow Mg_3(PO_4)_2(s) + 6\ NaCl(aq)$
Using the above equation, calculate:
(a) The milliliters of 0.250 *M* Na_3PO_4 that will react with 50.0 mL of 0.250 *M* $MgCl_2$.
(b) The grams of $Mg_3(PO_4)_2$ that will be formed from 50.0 mL of 0.250 *M* $MgCl_2$.

79. Given the balanced equation
$6\ FeCl_2(aq) + K_2Cr_2O_7(aq) + 14\ HCl(aq) \rightarrow 6\ FeCl_3(aq) + 2\ CrCl_3(aq) + 2\ KCl(aq) + 7\ H_2O(l)$
(a) How many moles of KCl will be produced from 2.0 mol of $FeCl_2$?
(b) How many moles of $CrCl_3$ will be produced from 1.0 mol of $FeCl_2$?
(c) How many moles of $FeCl_2$ will react with 0.050 mol of $K_2Cr_2O_7$?
(d) How many milliliters of 0.060 *M* $K_2Cr_2O_7$ will react with 0.025 mol of $FeCl_2$?
(e) How many milliliters of 6.0 *M* HCl will react with 15.0 mL of 6.0 *M* $FeCl_2$?

80. $2\ KMnO_4(aq) + 16\ HCl(aq) \rightarrow 2\ MnCl_2(aq) + 5\ Cl_2(g) + 8\ H_2O(l) + 2\ KCl(aq)$
Calculate the following using the above equation:
(a) The moles of Cl_2 produced from 0.050 mol of $KMnO_4$
(b) The moles of HCl required to react with 1.0 L of 2.0 *M* $KMnO_4$
(c) The milliliters of 6.0 *M* HCl required to react with 200. mL of 0.50 *M* $KMnO_4$
(d) The liters of Cl_2 gas at STP produced by the reaction of 75.0 mL of 6.0 *M* HCl

Equivalent Mass and Normality Problems

81. Calculate the equivalent mass of the acid and base in each of the following reactions:
(a) $HCl + NaOH \rightarrow NaCl + H_2O$
(b) $2\ HCl + Ba(OH)_2 \rightarrow 2\ H_2O + BaCl_2$
(c) $H_2SO_4 + Ca(OH)_2 \rightarrow CaSO_4 + 2\ H_2O$

82. Calculate the equivalent mass of the acid and base in each of the following reactions:
(a) $H_2SO_4 + KOH \rightarrow KHSO_4 + H_2O$
(b) $H_3PO_4 + 2\ LiOH \rightarrow Li_2HPO_4 + 2\ H_2O$
(c) $HNO_3 + NaOH \rightarrow NaNO_3 + H_2O$

83. What is the normality of the following solutions? Assume complete neutralization.
(a) 4.0 *M* HCl
(b) 0.243 *M* HNO_3
(c) 3.0 *M* H_2SO_4

84. What is the normality of the following solutions? Assume complete neutralization.
(a) 1.85 *M* H_3PO_4
(b) 0.250 *M* $HC_2H_3O_2$
(c) 1.25 *M* NaOH

85. What volume of 0.2550 *N* NaOH is required to neutralize
(a) 20.22 mL of 0.1254 *N* HCl?
(b) 14.86 mL of 0.1246 *N* H_2SO_4?

86. What volume of 0.2550 *N* NaOH is required to neutralize
(a) 21.30 mL of 0.1430 *M* HCl?
(b) 18.00 mL of 0.1430 *M* H_2SO_4?

Molality and Colligative Properties Problems

87. Calculate the molality of these solutions:
(a) 14.0 g of CH_3OH in 100.0 g of H_2O
(b) 2.50 mol of benzene (C_6H_6) in 250 g of hexane (C_6H_{14})

88. Calculate the molality of these solutions:
(a) 1.0 g of $C_6H_{12}O_6$ in 1.0 g of H_2O
(b) 0.250 mol of iodine in 1.0 kg of H_2O

89. **(a)** What is the molality of a solution containing 100.0 g of ethylene glycol, $C_2H_6O_2$, in 150.0 g of water?
(b) What is the boiling point of this solution?
(c) What is the freezing point of this solution?

90. What is (a) the molality, (b) the freezing point, and (c) the boiling point of a solution containing 2.68 g of naphthalene, $C_{10}H_8$, in 38.4 g of benzene, C_6H_6?

*91. The freezing point of a solution of 8.00 g of an unknown compound dissolved in 60.0 g of acetic acid is 13.2°C. Calculate the molar mass of the compound.

*92. What is the molar mass of a compound if 4.80 g of the compound dissolved in 22.0 g of H_2O gives a solution that freezes at −2.50°C?

Additional Exercises

These exercises are not paired or labeled by topic and provide additional practice on concepts covered in this chapter.

93. How many grams of solution, 10.0% NaOH by mass, are required to neutralize 150 mL of a 1.0 *M* HCl solution?

***94.** How many grams of solution, 10.0% NaOH by mass, are required to neutralize 250.0 g of a 1.0 *m* solution of HCl?

***95.** A sugar syrup solution contains 15.0% sugar, $C_{12}H_{22}O_{11}$, by mass and has a density of 1.06 g/mL.
(a) How many grams of sugar are in 1.0 L of this syrup?
(b) What is the molarity of this solution?
(c) What is the molality of this solution?

***96.** A solution of 3.84 g of C_4H_2N (empirical formula) in 250.0 g of benzene depresses the freezing point of benzene 0.614°C. What is the molecular formula for the compound?

***97.** Hydrochloric acid, HCl, is sold as a concentrated aqueous solution (12.0 mol/L). If the density of the solution is 1.18 g/mL, determine the molality of the solution.

***98.** How many grams of KNO_3 must be used to make 450 mL of a solution that is to contain 5.5 mg/mL of potassium ion? Calculate the molarity of the solution.

99. What mass of 5.50% solution can be prepared from 25.0 g KCl?

100. Physiological saline, NaCl, solutions used in intravenous injections have a concentration of 0.90% NaCl (mass/volume).
(a) How many grams of NaCl are needed to prepare 500.0 mL of this solution?
***(b)** How much water must evaporate from this solution to give a solution that is 9.0% NaCl (mass/volume)?

***101.** A solution is made from 50.0 g of KNO_3 and 175 g of H_2O. How many grams of water must evaporate to give a saturated solution of KNO_3 in water at 20°C? (See Figure 14.4.)

102. What volume of 70.0% rubbing alcohol can you prepare if you have only 150 mL of pure isopropyl alcohol on hand?

103. At 20°C an aqueous solution of HNO_3 that is 35.0% HNO_3 by mass has a density of 1.21 g/mL.
(a) How many grams of HNO_3 are present in 1.00 L of this solution?
(b) What volume of this solution will contain 500. g of HNO_3?

***104.** What is the molarity of a nitric acid solution, if the solution is 35.0% HNO_3 by mass and has a density of 1.21 g/mL?

105. To what volume must a solution of 80.0 g of H_2SO_4 in 500.0 mL of solution be diluted to give a 0.10 *M* solution?

106. How many milliliters of water must be added to 300.0 mL of 1.40 *M* HCl to make a solution that is 0.500 *M* HCl?

107. A 10.0-mL sample of 16 *M* HNO_3 is diluted to 500.0 mL. What is the molarity of the final solution?

108. Given a 5.00 *M* KOH solution, how would you prepare 250.0 mL of 0.625 *M* KOH?

109. (a) How many moles of hydrogen will be liberated from 200.0 mL of 3.00 *M* HCl reacting with an excess of magnesium? The equation is

$$Mg(s) + 2\ HCl(aq) \rightarrow MgCl_2(aq) + H_2(g)$$

(b) How many liters of hydrogen gas, H_2, measured at 27°C and 720 torr, will be obtained? (*Hint*: Use the ideal gas equation.)

***110.** What is the molarity of an HCl solution, 150.0 mL of which, when treated with excess magnesium, liberates 3.50 L of H_2 gas measured at STP?

111. What is the normality of an H_2SO_4 solution if 36.26 mL are required to neutralize 2.50 g of $Ca(OH)_2$?

112. Which will be more effective in neutralizing stomach acid, HCl: a tablet containing 12.0 g of $Mg(OH)_2$ or a tablet containing 10.0 g of $Al(OH)_3$? Show evidence for your answer.

113. Which would be more effective as an antifreeze in an automobile radiator? A solution containing
(a) 10 kg of methyl alcohol, CH_3OH, or 10 kg of ethyl alcohol, C_2H_5OH?
(b) 10 *m* solution of methyl alcohol or 10 *m* solution of ethyl alcohol?

114. Automobile battery acid is 38% H_2SO_4 and has a density of 1.29 g/mL. Calculate the molality and the molarity of this solution.

***115.** A sugar solution made to feed hummingbirds contains 1.00 lb of sugar, $C_{12}H_{22}O_{11}$, to 4.00 lb of water. Can this solution be put outside without freezing where the temperature falls to 20.0°F at night? Show evidence for your answer.

***116.** What would be (a) the molality and (b) the boiling point of an aqueous sugar, $C_{12}H_{22}O_{11}$, solution that freezes at −5.4°C?

117. A solution of 6.20 g of $C_2H_6O_2$ in water has a freezing point of $-0.372°C$. How many grams of H_2O are in the solution?

118. What (a) mass and (b) volume of ethylene glycol ($C_2H_6O_2$, density = 1.11 g/mL) should be added to 12.0 L of water in an automobile radiator to protect it fom freezing at $-20°C$? (c) To what temperature Fahrenheit will the radiator be protected?

119. Can a saturated solution ever be a dilute solution? Explain.

120. What volume of 0.65 *M* HCl is needed to completely neutralize 12 g of NaOH?

***121.** If 150 mL of 0.055 *M* HNO_3 are needed to completely neutralize 1.48 g of an *impure* sample of sodium hydrogen carbonate (baking soda), what percent of the sample is baking soda?

122. **(a)** How much water must be added to concentrated sulfuric acid, H_2SO_4 (17.8 *M*), to prepare 8.4 L of 1.5 *M* sulfuric acid solution?
(b) How many moles of H_2SO_4 would be in each milliliter of the original concentrate?
(c) How many moles would be in each milliliter of the diluted solution?

123. An aqueous solution freezes at $-3.6°C$. What is its boiling temperature?

***124.** How would you prepare a 6.00 *M* HNO_3 solution if only 3.00 *M* and 12.0 *M* solutions of the acid were available for mixing?

***125.** A 20.0-mL portion of an HBr solution of unknown strength was diluted to exactly 240 mL. Now if 100.0 mL of this diluted solution required 88.4 mL of 0.37 *M* NaOH to achieve complete neutralization, what was the strength of the original HBr solution?

126. When 80.5 mL of 0.642 *M* $Ba(NO_3)_2$ is mixed with 44.5 mL of 0.743 *M* KOH, a precipitate of $Ba(OH)_2$ forms. How many grams of $Ba(OH)_2$ do you expect?

127. Exactly three hundred grams of a 5.0% sucrose solution is to be prepared. How many grams of a 2.0% solution of sucrose would contain the same number of grams of sugar?

128. Lithium carbonate, Li_2CO_3, is a drug used to treat manic depression. A 0.25 *M* solution of it is prepared.
(a) How many moles of Li_2CO_3 are present in 45.8 mL of the solution?
(b) How many grams of Li_2CO_3 are in 750 mL of the same solution?
(c) How many milliliters of the solution would be needed to supply 6.0 g of the solute?
(d) If the solution has a density of 1.22 g/mL, what is its mass percent?

129. If a student accidentally mixed 400.0 mL of 0.35 *M* HCl with 1100 mL of 0.65 *M* HCl, what would be the final molarity of the hydrochloric acid solution?

130. Suppose you start with 100. mL of distilled water. Then suppose you add one drop (20 drops/mL) of concentrated acetic acid, $HC_2H_3O_2$ (17.8 *M*). What is the molarity of the resulting solution?

Answers to Practice Exercises

14.1 unsaturated
14.2 10.0% Na_2SO_4 solution
14.3 2.02 *M*
14.4 0.27 g NaCl
14.5 0.84 L (840 mL)
14.6 75 mL NaOH
14.7 5.00×10^{-2} *M*
14.8 43 g
14.9 1.19×10^{-2} *N* NaOH, 1.19×10^{-2} *M* NaOH
14.10 1.387 *m*
14.11 freezing point = $-2.58°C$, boiling point = $100.71°C$

Acids, Bases, and Salts

15

Chapter 15

Chapter Opening Photo: As lava flows into the sea, a slightly acidic solution is formed.

Water is the medium for the reactions that provide the foundation for life on this planet. Many significant chemical reactions occur in aqueous solutions. Volcanos and hot springs yield acidic solutions formed from hydrochloric acid and sulfur dioxide. These acids mix with bases (such as calcium carbonate) dissolved from rocks and microscopic animals to form the salts of our oceans and produce the beauty of stalactites and stalagmites in limestone caves. Photosynthesis and respiration reactions cannot occur without an adequate balance between acids and bases. Even small deviations in this balance can be detrimental to living organisms. Human activity has resulted in changes in this delicate acid–base balance. For example, in recent years we have become very concerned about the effects of acid rain upon our environment. We are beginning to examine ways in which we can maintain the balance as we continue to change our planet.

15.1 Acids and Bases

The word *acid* is derived from the Latin *acidus,* meaning "sour" or "tart," and is also related to the Latin word *acetum,* meaning "vinegar." Vinegar has been known since antiquity as a product of the fermentation of wine and apple cider. The sour constituent of vinegar is acetic acid, $HC_2H_3O_2$. Some of the characteristic properties commonly associated with acids are the following:

1. sour taste
2. change the color of litmus, a vegetable dye, from blue to red
3. react with
   - metals such as zinc and magnesium to produce hydrogen gas
   - hydroxide bases to produce water and an ionic compound (salt)
   - carbonates to produce carbon dioxide

These properties are due to the hydrogen ions, H^+, released by acids in a water solution.

Classically, a *base* is a substance capable of liberating hydroxide ions, OH^-, in water solution. Hydroxides of the alkali metals (Group IA) and alkaline earth metal (Group IIA), such as LiOH, NaOH, KOH, $Ca(OH)_2$, and $Ba(OH)_2$, are the most common inorganic bases. Water solutions of bases are called *alkaline solutions* or *base solutions.* Some of the characteristic properties commonly associated with bases are the following:

1. bitter or caustic taste
2. a slippery, soapy feeling
3. the ability to change litmus from red to blue
4. the ability to interact with acids

Several theories have been proposed to answer the question "What is an acid and a base?" One of the earliest, most significant of these theories was advanced in a doctoral thesis in 1884 by Svante Arrhenius (1859–1927), a Swedish scientist, who stated that "an acid is a hydrogen-containing substance that dissociates to produce hydrogen ions, and that a base is a hydroxide-containing substance that dissociates to produce hydroxide ions in aqueous solutions." Arrhenius postulated that the hydrogen ions were produced by the dissociation of acids in water, and that the hydroxide ions were produced by the dissociation of bases in water:

$$\underset{\text{acid}}{HA} \longrightarrow H^+ + A^-$$

$$\underset{\text{base}}{MOH} \longrightarrow M^+ + OH^-$$

Thus, an acid solution contains an excess of hydrogen ions and a base an excess of hydroxide ions.

In 1923, the Brønsted–Lowry proton transfer theory was introduced by J. N. Brønsted (1897–1947), a Danish chemist, and T. M. Lowry (1847–1936), an English chemist. This theory states that an acid is a proton donor and a base is a proton acceptor.

A Brønsted–Lowry acid is a proton (H^+) donor.
A Brønsted–Lowry base is a proton (H^+) acceptor.

Consider the reaction of hydrogen chloride gas with water to form hydrochloric acid:

$$HCl(g) + H_2O(l) \longrightarrow H_3O^+(aq) + Cl^-(aq) \qquad (1)$$

In the course of the reaction, HCl donates, or gives up, a proton to form a Cl^- ion, and H_2O accepts a proton to form the H_3O^+ ion. Thus, HCl is an acid and H_2O is a base, according to the Brønsted–Lowry theory.

A hydrogen ion, H^+, is nothing more than a bare proton and does not exist by itself in an aqueous solution. In water a proton combines with a polar water molecule to form a hydrated hydrogen ion, H_3O^+, commonly called a **hydronium ion**. The proton is attracted to a polar water molecule, forming a bond with one of the two pairs of unshared electrons:

hydronium ion

$$H^+ + H{:}\ddot{\underset{\cdot\cdot}{O}}{:} \longrightarrow \left[H{:}\ddot{\underset{\cdot\cdot}{O}}{:}H\right]^+$$
$$\qquad\quad H \qquad\qquad\quad H$$

hydronium ion

Note the electron structure of the hydronium ion. For simplicity of expression in equations, we often use H^+ instead of H_3O^+, with the explicit understanding that H^+ is always hydrated in solution.

As illustrated in Equation (1) when a Brønsted–Lowry acid donates a proton, it forms the conjugate base of that acid. When a base accepts a proton, it forms the conjugate acid of that base. A conjugate acid and base are produced as products. The formulas of a conjugate acid–base pair differ by one proton (H^+). Consider what happens when HCl(g) is bubbled through water, as shown by the equation below:

conjugate acid–base pair

$$HCl(g) + H_2O(l) \longrightarrow Cl^-(aq) + H_3O^+(aq)$$

conjugate acid–base pair

acid base base acid

The conjugate acid–base pairs are HCl—Cl^- and H_3O^+—H_2O. Cl^- is the conjugate base of HCl, and HCl is the conjugate acid of Cl^-. H_2O is the conjugate base of H_3O^+, and H_3O^+ is the conjugate acid of H_2O.

Another example of conjugate acid–base pairs is observed in the following equation:

$$NH_4^+ + H_2O \longrightarrow H_3O^+ + NH_3$$

acid base acid base

In this equation the conjugate acid–base pairs are NH_4^+—NH_3 and H_3O^+—H_2O.

Example 15.1 Write the formula for (a) the conjugate base of H_2O and of HNO_3, and (b) the conjugate acid of SO_4^{2-} and of $C_2H_3O_2^-$.

Solution

(a) To write the conjugate base of an acid, remove one proton from the acid formula. Thus,

Remember, the difference between an acid or a base and its conjugate is one proton, H^+.

$$H_2O \xrightarrow{-H^+} OH^- \quad \text{(conjugate base)}$$

$$HNO_3 \xrightarrow{-H^+} NO_3^- \quad \text{(conjugate base)}$$

Note that, by removing an H^+, the conjugate base becomes more negative than the acid by one minus charge.

(b) To write the conjugate acid of a base, add one proton to the formula of the base. Thus,

$$SO_4^{2-} \xrightarrow{+H^+} HSO_4^- \quad \text{(conjugate acid)}$$

$$C_2H_3O_2^- \xrightarrow{+H^+} HC_2H_3O_2 \quad \text{(conjugate acid)}$$

In each case the conjugate acid becomes more positive than the base by one positive charge due to the addition of H^+.

Practice 15.1

Indicate the conjugate base for each of the following acids:
(a) H_2CO_3 (b) HNO_2 (c) $HC_2H_3O_2$

Practice 15.2

Indicate the conjugate acid for each of the following bases:
(a) HSO_4^- (b) NH_3 (c) OH^-

A more general concept of acids and bases was introduced by Gilbert N. Lewis. The Lewis theory deals with the way in which a substance with an unshared pair of electrons reacts in an acid–base type of reaction. According to this theory a base is any substance that has an unshared pair of electrons (electron-pair donor), and an acid is any substance that will attach itself to or accept a pair of electrons.

A Lewis acid is an electron pair acceptor.
A Lewis base is an electron pair donor.

In the reaction

$$\underset{\text{acid}}{H^+} + \underset{\text{base}}{\begin{array}{c} H \\ \ddot{}\\ :\!N\!:\!H \\ \ddot{} \\ H \end{array}} \longrightarrow \left[\begin{array}{c} H \\ \ddot{} \\ H\!:\!N\!:\!H \\ \ddot{} \\ H \end{array}\right]^+$$

The H^+ is a Lewis acid and $:NH_3$ is a Lewis base. According to the Lewis theory, substances other than proton donors (e.g., BF_3) behave as acids:

$$\underset{\text{acid}}{\begin{array}{c} F \\ \ddot{} \\ F\!:\!B \\ \ddot{} \\ F \end{array}} + \underset{\text{base}}{\begin{array}{c} H \\ \ddot{} \\ :\!N\!:\!H \\ \ddot{} \\ H \end{array}} \longrightarrow \begin{array}{c} F\ H \\ \ddot{}\ \ \ddot{} \\ F\!:\!B\!:\!N\!:\!H \\ \ddot{}\ \ \ddot{} \\ F\ H \end{array}$$

The Lewis and Brønsted–Lowry bases are identical because, to accept a proton, a base must have an unshared pair of electrons.

The three theories are summarized in Table 15.1. These theories explain how acid–base reactions occur. We will generally use the theory that best explains the reaction that is under consideration. Most of our examples will refer to aqueous solutions. It is important to realize that in an aqueous acidic solution the H^+ ion concentration is always greater than OH^- ion concentration. And, vice versa, in an aqueous basic solution the OH^- ion concentration is always greater than the H^+ ion concentration. When the H^+ and OH^- ion concentrations in a solution are equal, the solution is neutral; that is, it is neither acidic nor basic.

TABLE 15.1 Summary of Acid–Base Definitions

Theory	Acid	Base
Arrhenius	A hydrogen-containing substance that produces hydrogen ions in aqueous solution	A hydroxide-containing substance that produces hydroxide ions in aqueous solution
Brønsted-Lowry	A proton (H^+) donor	A proton (H^+) acceptor
Lewis	Any species that will bond to an unshared pair of electrons (electron-pair acceptor)	Any species that has an unshared pair of electrons (electron-pair donor)

15.2 Reactions of Acids

In aqueous solutions the H^+ or H_3O^+ ions are responsible for the characteristic reactions of acids. The following reactions are in an aqueous medium.

Reaction with Metals Acids react with metals that lie above hydrogen in the activity series of elements to produce hydrogen and an ionic compound (salt) (see Section 17.5):

$$\text{acid} + \text{metal} \longrightarrow \text{hydrogen} + \text{ionic compound}$$

$$2\ HCl(aq) + Ca(s) \longrightarrow H_2(g) + CaCl_2(aq)$$

$$H_2SO_4(aq) + Mg(s) \longrightarrow H_2(g) + MgSO_4(aq)$$

$$6\ HC_2H_3O_2(aq) + 2\ Al(s) \longrightarrow 3\ H_2(g) + 2\ Al(C_2H_3O_2)_3(aq)$$

Acids such as nitric acid (HNO_3) are oxidizing substances (see Chapter 17) and react with metals to produce water instead of hydrogen. For example:

$$3\ Zn(s) + 8\ HNO_3(\text{dilute}) \longrightarrow 3\ Zn(NO_3)_2(aq) + 2\ NO(g) + 4\ H_2O(l)$$

Reaction with Bases The interaction of an acid and a base is called a *neutralization reaction.* In aqueous solutions, the products of this reaction are a salt and water:

$$\text{acid} + \text{base} \longrightarrow \text{salt} + \text{water}$$

$$HBr(aq) + KOH(aq) \longrightarrow KBr(aq) + H_2O(l)$$

$$2\ HNO_3(aq) + Ca(OH)_2(aq) \longrightarrow Ca(NO_3)_2(aq) + 2\ H_2O(l)$$

$$2\ H_3PO_4(aq) + 3\ Ba(OH)_2(aq) \longrightarrow Ba_3(PO_4)_2(s) + 6\ H_2O(l)$$

Reaction with Metal Oxides This reaction is closely related to that of an acid with a base. With an aqueous acid, the products are a salt and water:

$$\text{acid} + \text{metal oxide} \longrightarrow \text{salt} + \text{water}$$

$$2\ HCl(aq) + Na_2O(s) \longrightarrow 2\ NaCl(aq) + H_2O(l)$$

$$H_2SO_4(aq) + MgO(s) \longrightarrow MgSO_4(aq) + H_2O(l)$$

$$6\ HCl(aq) + Fe_2O_3(s) \longrightarrow 2\ FeCl_3(aq) + 3\ H_2O(l)$$

Reaction with Carbonates Many acids react with carbonates to produce carbon dioxide, water, and an ionic compound:

Carbonic acid (H_2CO_3) is not the product, because it is unstable and decomposes into water and carbon dioxide.

$$H_2CO_3(aq) \longrightarrow CO_2(g) + H_2O(l)$$

$$\text{acid} + \text{carbonate} \longrightarrow \text{salt} + \text{water} + \text{carbon dioxide}$$

$$2\ HCl(aq) + Na_2CO_3(aq) \longrightarrow 2\ NaCl(aq) + H_2O(l) + CO_2(g)$$

$$H_2SO_4(aq) + MgCO_3(s) \longrightarrow MgSO_4(aq) + H_2O(l) + CO_2(g)$$

15.3 Reactions of Bases

The OH^- ions are responsible for the characteristic reactions of bases. All the following reactions are in an aqueous medium.

Reaction with Acids Bases react with acids to produce a salt and water. See reaction of acids with bases in Section 15.2.

Amphoteric Hydroxides Hydroxides of certain metals, such as zinc, aluminum, and chromium, are **amphoteric**—that is, they are capable of reacting as either an acid or a base. When treated with a strong acid, they behave like bases; when reacted with a strong base, they behave like acids:

amphoteric

$$Zn(OH)_2(s) + 2\ HCl(aq) \longrightarrow ZnCl_2(aq) + 2\ H_2O(l)$$

$$Zn(OH)_2(s) + 2\ NaOH(aq) \longrightarrow Na_2Zn(OH)_4(aq)$$

Reaction of NaOH and KOH with Certain Metals Some amphoteric metals react directly with the strong bases sodium hydroxide and potassium hydroxide to produce hydrogen:

$$\text{base} + \text{metal} + \text{water} \longrightarrow \text{salt} + \text{hydrogen}$$

$$2\ NaOH(aq) + Zn(s) + 2\ H_2O(l) \longrightarrow Na_2Zn(OH)_4(aq) + H_2(g)$$

$$2\ KOH(aq) + 2\ Al(s) + 6\ H_2O(l) \longrightarrow 2\ KAl(OH)_4(aq) + 3\ H_2(g)$$

15.4 Salts

Salts are very abundant in nature. Most of the rocks and minerals of the earth's mantle are salts of one kind or another. Huge quantities of dissolved salts also exist in the oceans. Salts can be considered to be compounds that have been derived from acids and bases. They consist of positive metal or ammonium ions (H^+ excluded) combined with negative nonmetal ions (OH^- and O^{2-} excluded). The positive ion is the base counterpart and the nonmetal ion is the acid counterpart:

Chemists use the terms *ionic compound* and *salt* interchangeably.

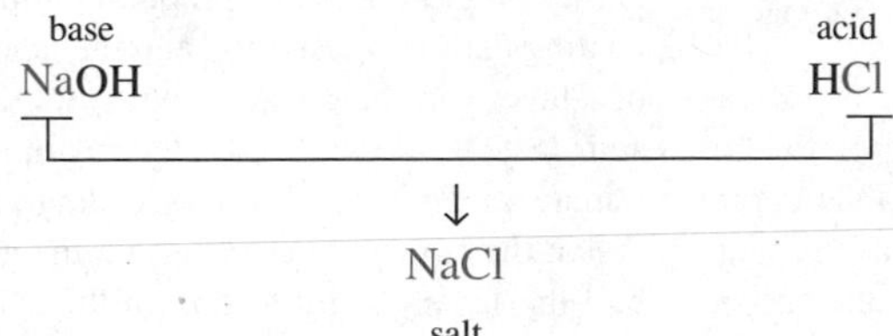

Salts are usually crystalline and have high melting and boiling points.

Pucker Power

The candy counter is filled with a variety of treats designed to cause your lips to pucker and to stimulate your tongue to send signals to your brain saying "SOUR." The human tongue has four types of taste receptors (known as "taste buds"). Sweet, bitter, salty, and sour taste buds are each concentrated in a different area of the tongue. These receptors are molecules that fit together with molecules in the candy (or other food), sending a signal to the brain, which is interpreted as a taste.

Substances that taste sour are acids. The substances that produce this sour taste in candy and other confections are often malic acid and/or citric acid.

$$\begin{array}{l} \mathrm{H-\overset{H}{\underset{|}{C}}-\overset{O}{\overset{\|}{C}}-O-H} \\ \mathrm{HO-\underset{|}{C}-\overset{O}{\overset{\|}{C}}-O-H} \\ \mathrm{H-\underset{H}{\underset{|}{C}}-\overset{O}{\overset{\|}{C}}-O-H} \end{array}$$

Citric acid
mp 153°C

$$\begin{array}{l} \mathrm{H-\overset{H}{\underset{|}{C}}-\overset{O}{\overset{\|}{C}}-O-H} \\ \mathrm{HO-\underset{H}{\underset{|}{C}}-\overset{O}{\overset{\|}{C}}-O-H} \end{array}$$

Malic acid
mp 100°C

Citric acid tastes more sour than malic acid. When these substances are mixed with other ingredients such as sugar, corn syrup, flavorings, and preservatives, the result is a candy both sweet and tart! The same acids can be mixed with synthetic rubber or chicle (dried latex from the sapodilla tree) to produce bubble gum with real pucker power. The most sour of this type of gum is called Face Slammers.™ Try some on your favorite 12-year-old.

$$\begin{array}{l} \mathrm{CH_2-\overset{O}{\overset{\|}{C}}-O-O-H} \\ \mathrm{HO-C-\overset{O}{\overset{\|}{C}}-O-H} \\ \mathrm{CH_2-\overset{O}{\overset{\|}{C}}-O-O-H} \end{array} \longrightarrow \begin{array}{l} \mathrm{CH_2-\overset{O}{\overset{\|}{C}}-O^-} \\ \mathrm{HO-C-\overset{O}{\overset{\|}{C}}-OH} \\ \mathrm{CH_2-\overset{O}{\overset{\|}{C}}-O-H} \end{array} + H^+(aq)$$

Citric acid → Citrate ion

Gum manufacturers have gone a step further in incorporating acid–base chemistry into their products for an even greater surprise. In Mad Dawg™ gum the initial taste is sour. But after a couple of minutes of chewing, brightly colored foam begins to accumulate in your mouth and ooze out over your lips. What is going on here?

The foam is a mixture of sugar and saliva mixed with carbon dioxide bubbles released when several of the gum's ingredients are mixed in the watery environment of your mouth. The citric and malic acids dissociate to form hydrogen ions while sodium hydrogen carbonate (baking soda) dissolves into sodium and hydrogen carbonate ions:

$$NaHCO_3(s) \longrightarrow Na^+(aq) + HCO_3^-(aq)$$

The hydrogen ions from the acids mix with the hydrogen carbonate ions from the baking soda to produce water and carbon dioxide gas:

$$H^+(aq) + HCO_3^-(aq) \longrightarrow H_2O(l) + CO_2(g)$$

The acids stimulate production of saliva and the food coloring adds color to the foamy mess. The major problem for chemists in creating this treat was keeping the reactants apart until the consumer pops the gum into his or her mouth. As solids, sodium hydrogen carbonate and citric acid do not react. But with the slightest amount of water present the process begins. When a tablet such as Alka Seltzer™ is manufactured, the ingredients (solid citric acid, sodium hydrogen carbonate, aspirin, and flavoring) are compressed into a tablet that is sealed into a dry foil packet. When opened and dropped into water the reaction (and the relief) begins and the bubbles are released.

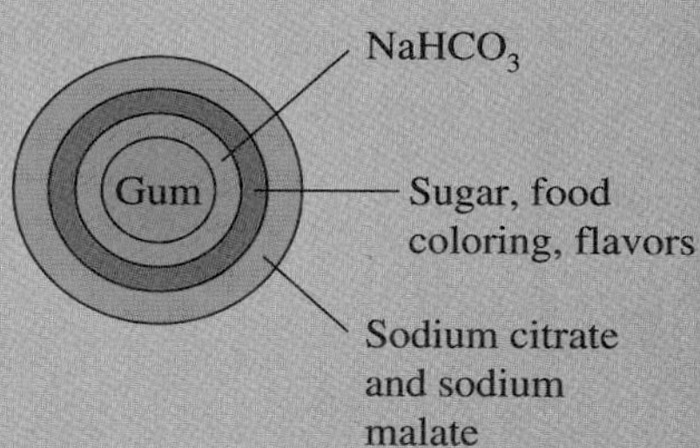

Cross-section of a Mad Dawg™ gum ball.

In Mad Dawg™ gum balls, the center core is moist rubber and the coating is applied in solution. Some of the early versions of these gum balls actually exploded as they were removed from the candy machine. To eliminate this problem multiple coatings are now used to keep the acid in one layer (on the outside to give the first sour taste) and the sodium hydrogen carbonate in an inner layer (see diagram). When the consumer crunches on the gum ball the layers begin to mix in the saliva and the fun begins!

◀ **These strange mineral formations called "tufa" exist at Mono Lake in California. Tufa is formed by water bubbling through sand saturated with NaCl, Na_2CO_3, and Na_2SO_4.**

From a single acid, such as hydrochloric acid (HCl), we can produce many chloride compounds by replacing the hydrogen with metal ions (e.g., NaCl, KCl, RbCl, $CaCl_2$, $NiCl_2$). Hence, the number of known salts greatly exceeds the number of known acids and bases. If the hydrogen atoms of a binary acid are replaced by a nonmetal, the resulting compound has covalent bonding and is therefore not considered to be ionic (e.g., PCl_3, S_2Cl_2, Cl_2O, NCl_3, ICl).

You may want to review Chapter 6 for nomenclature of acids, bases, and salts.

15.5 Electrolytes and Nonelectrolytes

We can show that solutions of certain substances are conductors of electricity by using a simple conductivity apparatus consisting of a pair of electrodes connected to a voltage source through a light bulb and switch (see Figure 15.1). If the medium between the electrodes is a conductor of electricity, the light bulb will glow when the switch is closed. When chemically pure water is placed in the beaker and the switch is closed, the light does not glow, indicating that water is a virtual nonconductor. When we dissolve a small amount of sugar in the water and test the solution, the light still does not glow, showing that a sugar solution is also a nonconductor. But, when a small amount of salt, NaCl, is dissolved in water and this solution is tested, the light glows brightly. Thus, the salt solution conducts electricity. A fundamental difference exists between the chemical bonding of sugar and that of salt. Sugar is a covalently bonded (molecular) substance; common salt is a substance with ionic bonds.

Substances whose aqueous solutions are conductors of electricity are called **electrolytes**. Substances whose solutions are nonconductors are known as **nonelectrolytes**. The classes of compounds that are electrolytes are acids, bases, and other ionic compounds (salts). Solutions of certain oxides also are conductors because the oxides form an acid or a base when dissolved in water. One major difference between

electrolyte
nonelectrolyte

FIGURE 15.1
A simple conductivity apparatus for testing electrolytes and nonelectrolytes in solution. The light bulb glows because the solution contains an electrolyte (acetic acid solution).

electrolytes and nonelectrolytes is that electrolytes are capable of producing ions in solution, whereas nonelectrolytes do not have this property. Solutions that contain a sufficient number of ions will conduct an electric current. Although pure water is essentially a nonconductor, many city water supplies contain enough dissolved ionic matter to cause the light to glow dimly when the water is tested in a conductivity apparatus. Table 15.2 lists some common electrolytes and nonelectrolytes.

Acids, bases, and salts are electrolytes.

15.6 Dissociation and Ionization of Electrolytes

Arrhenius received the 1903 Nobel Prize in chemistry for his work on electrolytes. He stated that a solution conducts electricity because the solute dissociates immediately upon dissolving into electrically charged particles (ions). The movement of these ions toward oppositely charged electrodes causes the solution to be a conductor. According to his theory, solutions that are relatively poor conductors contain electrolytes that are only partly dissociated. Arrhenius also believed that ions exist in solution whether or not an electric current is present. In other words, the electric current does not cause the formation of ions. Remember that positive ions are cations; negative ions are anions.

dissociation

We have seen that sodium chloride crystals consist of sodium and chloride ions held together by ionic bonds. **Dissociation** is the process by which the ions of a salt separate as the salt dissolves. When placed in water, the sodium and chloride ions are attracted by the polar water molecules, which surround each ion as it dissolves. In water, the salt dissociates, forming hydrated sodium and chloride ions (see Figure 15.2). The sodium and chloride ions in solution are surrounded by a specific number of water dipoles and have less attraction for each other than they had in the crystalline state. The equation representing this dissociation is

$$NaCl(s) + (x + y)\ H_2O \longrightarrow Na^+(H_2O)_x + Cl^-(H_2O)_y$$

A simplified dissociation equation in which the water is omitted but understood to be present is

$$NaCl(s) \longrightarrow Na^+(aq) + Cl^-(aq)$$

TABLE 15.2 Representative Electrolytes and Nonelectrolytes

Electrolytes		Nonelectrolytes	
H_2SO_4	$HC_2H_3O_2$	$C_{12}H_{22}O_{11}$ (sugar)	CH_3OH (methyl alcohol)
HCl	NH_3	C_2H_5OH (ethyl alcohol)	$CO(NH_2)_2$ (urea)
HNO_3	K_2SO_4	$C_2H_4(OH)_2$ (ethylene glycol)	O_2
$NaOH$	$NaNO_3$	$C_3H_5(OH)_3$ (glycerol)	H_2O

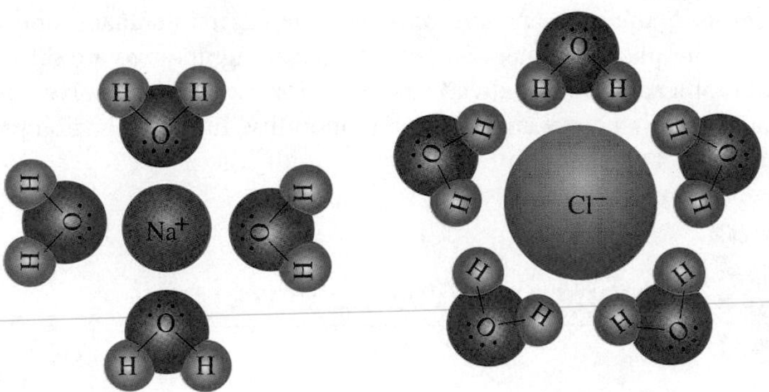

FIGURE 15.2
Hydrated sodium and chloride ions. When sodium chloride dissolves in water, each Na^+ and Cl^- ion becomes surrounded by water molecules. The negative end of the water dipole is attracted to the Na^+ ion, and the positive end is attracted to the Cl^- ion.

It is important to remember that sodium chloride exists in an aqueous solution as hydrated ions and not as NaCl units, even though the formula NaCl or $Na^+ + Cl^-$ is often used in equations.

The chemical reactions of salts in solution are the reactions of their ions. For example, when sodium chloride and silver nitrate react and form a precipitate of silver chloride, only the Ag^+ and Cl^- ions participate in the reaction. The Na^+ and NO_3^- remain as ions in solution:

$$Ag^+(aq) + Cl^-(aq) \longrightarrow AgCl(s)$$

ionization

Ionization is the formation of ions; it occurs as a result of a chemical reaction of certain substances with water. Glacial acetic acid (100% $HC_2H_3O_2$) is a liquid that behaves as a nonelectrolyte when tested by the method described in Section 15.5. But a water solution of acetic acid conducts an electric current (as indicated by the dull-glowing light of the conductivity apparatus). The equation for the reaction with water, which forms hydronium and acetate ions, is

$$\underset{\text{acid}}{HC_2H_3O_2} + \underset{\text{base}}{H_2O} \rightleftharpoons \underset{\text{acid}}{H_3O^+} + \underset{\text{base}}{C_2H_3O_2^-}$$

or, in the simplified equation,

$$HC_2H_3O_2 \rightleftharpoons H^+ + C_2H_3O_2^-$$

In this ionization reaction, water serves not only as a solvent but also as a base according to the Brønsted–Lowry theory.

Hydrogen chloride is predominantly covalently bonded, but when dissolved in water it reacts to form hydronium and chloride ions:

$$HCl(g) + H_2O(l) \longrightarrow H_3O^+(aq) + Cl^-(aq)$$

When a hydrogen chloride solution is tested for conductivity, the light glows brilliantly, indicating many ions in the solution.

Ionization occurs in each of the above two reactions with water, producing ions in solution. The necessity for water in the ionization process can be demonstrated by dissolving hydrogen chloride in a nonpolar solvent such as hexane, and testing the solution for conductivity. The solution fails to conduct electricity, indicating that no ions are produced.

The terms *dissociation* and *ionization* are often used interchangeably to describe processes taking place in water. But, strictly speaking, the two are different. In the dissociation of a salt, the salt already exists as ions; when it dissolves in water, the ions separate, or dissociate, and increase in mobility. In the ionization process, ions are produced by the reaction of a compound with water.

15.7 Strong and Weak Electrolytes

strong electrolyte
weak electrolyte

Electrolytes are classified as strong or weak depending on the degree, or extent, of dissociation or ionization. **Strong electrolytes** are essentially 100% ionized in solution; **weak electrolytes** are much less ionized (based on comparing 0.1 *M* solutions). Most electrolytes are either strong or weak, with a few classified as moderately strong or weak. Most salts are strong electrolytes. Acids and bases that are strong electrolytes (highly ionized) are called *strong acids* and *strong bases*. Acids and bases that are weak electrolytes (slightly ionized) are called *weak acids* and *weak bases*.

For equivalent concentrations, solutions of strong electrolytes contain many more ions than do solutions of weak electrolytes. As a result, solutions of strong electrolytes are better conductors of electricity. Consider the two solutions, 1 *M* HCl and 1 *M* $HC_2H_3O_2$. Hydrochloric acid is almost 100% ionized; acetic acid is about 1% ionized. Thus HCl is a strong acid, and $HC_2H_3O_2$ is a weak acid. Hydrochloric acid has about 100 times as many hydronium ions in solution as acetic acid, making the HCl solution much more acidic.

One can distinguish between strong and weak electrolytes experimentally using the apparatus described in Section 15.5. A 1 *M* HCl solution causes the light to glow brilliantly, but a 1 *M* $HC_2H_3O_2$ solution causes only a dim glow. In a similar fashion the strong base sodium hydroxide, NaOH, may be distinguished from the weak base ammonia, NH_3. The ionization of a weak electrolyte in water is represented by an equilibrium equation showing that both the un-ionized and ionized forms are present in solution. In the equilibrium equation of $HC_2H_3O_2$ and its ions, we say that the equilibrium lies "far to the left" because relatively few hydrogen and acetate ions are present in solution:

$$HC_2H_3O_2(aq) \rightleftharpoons H^+(aq) + C_2H_3O_2^-(aq)$$

We have previously used a double arrow in an equation to represent reversible processes in the equilibrium between dissolved and undissolved solute in a saturated solution. A double arrow ($\rightleftharpoons$) is also used in the ionization equation of soluble weak electrolytes to indicate that the solution contains a considerable amount of the un-ionized compound in equilibrium with its ions in solution. (See Section 16.1 for a discussion of reversible reactions.) A single arrow is used to indicate that the electrolyte is essentially all in the ionic form in the solution. For example, nitric acid is a strong acid; nitrous acid is a weak acid. Their ionization equations in water may be indicated as

$$HNO_3(aq) \xrightarrow{H_2O} H^+(aq) + NO_3^-(aq)$$

$$HNO_2(aq) \overset{H_2O}{\rightleftharpoons} H^+(aq) + NO_2^-(aq)$$

TABLE 15.3 Strong and Weak Electrolytes

Strong electrolytes		Weak electrolytes	
Most soluble salts	$HClO_4$	$HC_2H_3O_2$	$H_2C_2O_4$
H_2SO_4	$NaOH$	H_2CO_3	H_3BO_3
HNO_3	KOH	HNO_2	$HClO$
HCl	$Ca(OH)_2$	H_2SO_3	NH_3
HBr	$Ba(OH)_2$	H_2S	HF

Practically all soluble salts, acids (such as sulfuric, nitric, and hydrochloric acids), and bases (such as sodium, potassium, calcium, and barium hydroxides) are strong electrolytes. Weak electrolytes include numerous other acids and bases such as acetic acid, nitrous acid, carbonic acid, and ammonia. The terms *strong acid, strong base, weak acid,* and *weak base* refer to whether an acid or base is a strong or weak electrolyte. A brief list of strong and weak electrolytes is given in Table 15.3.

Electrolytes yield two or more ions per formula unit upon dissociation—the actual number being dependent on the compound. Dissociation is complete or nearly complete for nearly all soluble ionic compounds and for certain other strong electrolytes, such as those given in Table 15.3. The following are dissociation equations for several strong electrolytes. In all cases the ions are actually hydrated:

$$NaOH \xrightarrow{H_2O} Na^+(aq) + OH^-(aq) \qquad \text{2 ions in solution per formula unit}$$

$$Na_2SO_4 \xrightarrow{H_2O} 2\ Na^+(aq) + SO_4^{2-}(aq) \qquad \text{3 ions in solution per formula unit}$$

$$Fe_2(SO_4)_3 \xrightarrow{H_2O} 2\ Fe^{3+}(aq) + 3\ SO_4^{2-}(aq)$$
5 ions in solution per formula unit

One mole of NaCl will give 1 mol of Na^+ ions and 1 mol of Cl^- ions in solution, assuming complete dissociation of the salt. One mole of $CaCl_2$ will give 1 mol of Ca^{2+} ions and 2 mol of Cl^- ions in solution:

$$\underset{1\ \text{mol}}{NaCl} \xrightarrow{H_2O} \underset{1\ \text{mol}}{Na^+(aq)} + \underset{1\ \text{mol}}{Cl^-(aq)}$$

$$\underset{1\ \text{mol}}{CaCl_2} \xrightarrow{H_2O} \underset{1\ \text{mol}}{Ca^{2+}(aq)} + \underset{2\ \text{mol}}{2\ Cl^-(aq)}$$

Example 15.2

What is the molarity of each ion in a solution of (a) 2.0 *M* NaCl, and (b) 0.40 *M* K_2SO_4? Assume complete dissociation.

Solution

(a) According to the dissociation equation,

$$\underset{1\ \text{mol}}{NaCl} \xrightarrow{H_2O} \underset{1\ \text{mol}}{Na^+(aq)} + \underset{1\ \text{mol}}{Cl^-(aq)}$$

the concentration of Na^+ is equal to that of NaCl (1 mol NaCl $\longrightarrow$ 1 mol Na^+), and the concentration of Cl^- is also equal to that of NaCl. Therefore, the concentrations of the ions in 2.0 *M* NaCl are 2.0 *M* Na^+ and 2.0 *M* Cl^-.

(b) According to the dissociation equation,

$$\underset{1\ \text{mol}}{K_2SO_4} \xrightarrow{H_2O} \underset{2\ \text{mol}}{2\ K^+(aq)} + \underset{1\ \text{mol}}{SO_4^{2-}(aq)}$$

the concentration of K^+ is twice that of K_2SO_4 and the concentration of SO_4^{2-} is equal to that of K_2SO_4. Therefore, the concentrations of the ions in 0.40 *M* K_2SO_4 are 0.80 *M* K^+ and 0.40 *M* SO_4^{2-}.

Practice 15.3

What is the molarity of each ion in a solution of (a) 0.050 *M* $MgCl_2$, and (b) 0.070 *M* $AlCl_3$?

Colligative Properties of Electrolyte Solutions

We have learned that when 1 mol of sucrose, a nonelectrolyte, is dissolved in 1000 g of water, the solution freezes at $-1.86°C$. When 1 mol of NaCl is dissolved in 1000 g of water, the freezing point of the solution is not $-1.86°C$, as might be expected, but is closer to $-3.72°C$ (-1.86×2). The reason for the lower freezing point is that 1 mol of NaCl in solution produces 2 mol of particles ($2 \times 6.022 \times 10^{23}$ ions) in solution. Thus, the freezing-point depression produced by 1 mol of NaCl is essentially equivalent to that produced by 2 mol of a nonelectrolyte. An electrolyte such as $CaCl_2$, which yields three ions in water, gives a freezing-point depression of about three times that of a nonelectrolyte. These freezing-point data provide additional evidence that electrolytes dissociate when dissolved in water. The other colligative properties are similarly affected by substances that yield ions in aqueous solutions.

15.8 Ionization of Water

The more we study chemistry, the more intriguing the water molecule becomes. Two equations commonly used to show how water ionizes are

$$\underset{\text{acid}}{H_2O} + \underset{\text{base}}{H_2O} \rightleftharpoons \underset{\text{acid}}{H_3O^+} + \underset{\text{base}}{OH^-}$$

and

$$H_2O \rightleftharpoons H^+ + OH^-$$

The first equation represents the Brønsted–Lowry concept, with water reacting as both an acid and a base, forming a hydronium ion and a hydroxide ion. The second equation is a simplified version, indicating that water ionizes to give a hydrogen and a hydroxide ion. Actually, the proton, H^+, is hydrated and exists as a hydronium ion. In either case equal molar amounts of acid and base are produced so that water is neutral, having neither H^+ nor OH^- ions in excess. The ionization of water at 25°C produces an H^+ ion concentration of 1.0×10^{-7} mol/L and an OH^- ion concentration of 1.0×10^{-7} mol/L. Square brackets, [ ], indicate that the concentration is in moles per liter. Thus $[H^+]$ means the concentration of H^+ is in moles per liter. These concentrations are usually expressed as

$$[H^+] \text{ or } [H_3O^+] = 1.0 \times 10^{-7} \text{ mol/L}$$

$$[OH^-] = 1.0 \times 10^{-7} \text{ mol/L}$$

These figures mean that about two out of every billion water molecules are ionized. This amount of ionization, small as it is, is a significant factor in the behavior of water in many chemical reactions.

15.9 Introduction to pH

The acidity of an aqueous solution depends on the concentration of hydrogen or hydronium ions. The pH scale of acidity was devised to fill the need for a simple, convenient numerical way to state the acidity of a solution. Values on the pH scale are obtained by mathematical conversion of H^+ ion concentrations to pH by the expression:

$$\text{pH} = -\log[H^+]$$

where $[H^+] = H^+$ or H_3O^+ ion concentration in moles per liter. **pH** is defined as the *negative* logarithm of the H^+ or H_3O^+ concentration in moles per liter: **pH**

$$\text{pH} = -\log[H^+] = -\log(1 \times 10^{-7}) = -(-7) = 7$$

For example, the pH of pure water at 25°C is 7 and is said to be neutral; that is, it is neither acidic nor basic, because the concentrations of H^+ and OH^- are equal. Solutions that contain more H^+ ions than OH^- ions have pH values less than 7, and solutions that contain less H^+ ions than OH^- ions have values greater than 7.

pH $<$ 7.00 is an acidic solution
pH = 7.00 is a neutral solution
pH $>$ 7.00 is a basic solution

When $[H^+] = 1 \times 10^{-5}$ mol/L, pH = 5 (acidic)

When $[H^+] = 1 \times 10^{-9}$ mol/L, pH = 9 (basic)

Instead of saying that the hydrogen ion concentration in the solution is 1×10^{-5} mol/L, it is customary to say that the pH of the solution is 5. The smaller the pH value, the more acidic the solution (see Figure 15.3).

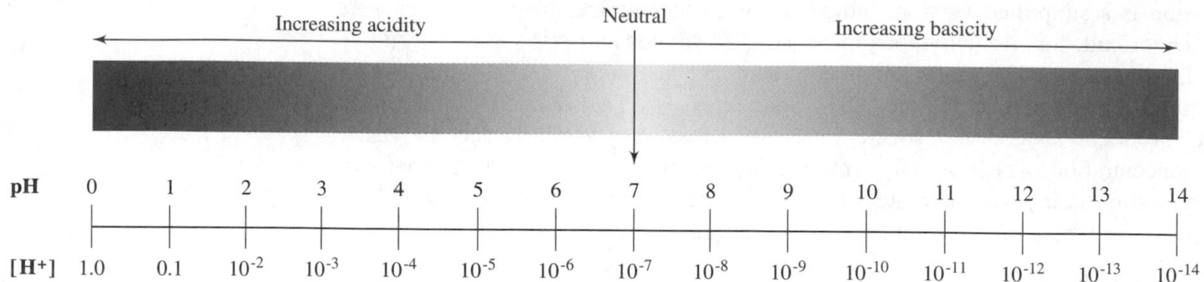

▲
FIGURE 15.3
The pH scale of acidity and basicity.

The pH scale, along with its interpretation, is given in Table 15.4. Table 15.5 lists the pH of some common solutions. Note that a change of only 1 pH unit means a tenfold increase or decrease in H^+ ion concentration. For example, a solution with a pH of 3.0 is ten times more acidic than a solution with a pH of 4.0. A simplified method of determining pH from $[H^+]$ follows:

$$[H^+] = 1 \times 10^{-5} \longleftarrow$$ pH = this number (5)
pH = 5

when this number is exactly 1

$$[H^+] = 2 \times 10^{-5} \longleftarrow$$ pH is between this number and next lower number (4 and 5)
pH = 4.7

when this number is between 1 and 10

Calculation of the pH value corresponding to any H^+ ion concentration requires the use of logarithms, which are exponents. The **logarithm** (log) of a number is

logarithm

TABLE 15.4 The pH Scale for Expressing Acidity

$[H^+]$ (mol/L)	pH	
1×10^{-14}	14	↑
1×10^{-13}	13	
1×10^{-12}	12	Increasing basicity
1×10^{-11}	11	
1×10^{-10}	10	
1×10^{-9}	9	
1×10^{-8}	8	
1×10^{-7}	7	Neutral
1×10^{-6}	6	
1×10^{-5}	5	
1×10^{-4}	4	
1×10^{-3}	3	Increasing acidity
1×10^{-2}	2	
1×10^{-1}	1	↓
1×10^{0}	0	

TABLE 15.5 The pH of Some Common Solutions

Solution	pH
Gastric juice	1.0
0.1 *M* HCl	1.0
Lemon juice	2.3
Vinegar	2.8
0.1 *M* $HC_2H_3O_2$	2.9
Orange juice	3.7
Tomato juice	4.1
Coffee, black	5.0
Urine	6.0
Milk	6.6
Pure water (25°C)	7.0
Blood	7.4
Household ammonia	11.0
1 *M* NaOH	14.0

simply the power to which 10 must be raised to give that number. Thus the log of 100 is 2 ($100 = 10^2$), and the log of 1000 is 3 ($1000 = 10^3$). The log of 500 is 2.70, but you cannot easily determine this value without a scientific calculator.

Let us determine the pH of a solution with $[H^+] = 2 \times 10^{-5}$. The exponent (-5) indicates that the pH is between 4 and 5. Enter 2×10^{-5} into your calculator and press the log key. The number $-4.69\ldots$ will be displayed. The pH is then

$$pH = -\log[H^+] = -(-4.69\ldots) = 4.7$$

Remember to change the sign on your calculator since $pH = -\log[H^+]$.

Next we must determine the correct number of significant figures in the logarithm. The rules for logs are different from those we use in other math operations. The number of decimal places for a log must equal the number of significant figures in the original number. Since 2×10^{-5} has one significant figure, we should round the log to one decimal place $(4.69\ldots) = 4.7$.

Example 15.3

What is the pH of a solution with an $[H^+]$ of (a) 1.0×10^{-11}, and (b) 6.0×10^{-4}, and (c) 5.47×10^{-8}?

Solution

(a) $[H^+] = 1.0 \times 10^{-11}$
(2 significant figures)
$pH = -\log(1.0 \times 10^{-11})$
$pH = 11.00$
(2 decimal places)

(b) $[H^+] = 6.0 \times 10^{-4}$
(2 significant figures)
$\log 6.0 \times 10^{-4} = -3.22$
$pH = -\log[H^+]$
$pH = -(-3.22) = 3.22$
(2 decimal places)

(c) $[H^+] = 5.47 \times 10^{-8}$
(3 significant figures)
$\log 5.47 \times 10^{-8} = -7.262$
$pH = -\log[H^+]$
$pH = -(-7.262) = 7.262$
(3 decimal places)

Practice 15.4

What is the pH of a solution with $[H^+]$ of (a) 3.9×10^{-12} *M*, (b) 1.3×10^{-3} *M*, and (c) 3.72×10^{-6} *M*?

The measurement and control of pH is extremely important in many fields of science and technology. The proper soil pH is necessary to grow certain types of plants successfully. The pH of certain foods is too acidic for some diets. Many biological processes are delicately controlled pH systems. The pH of human blood is regulated to very close tolerances through the uptake or release of H^+ by mineral ions, such as HCO_3^-, HPO_4^{2-}, and $H_2PO_4^-$. Changes in the pH of the blood by as little as 0.4 pH unit result in death.

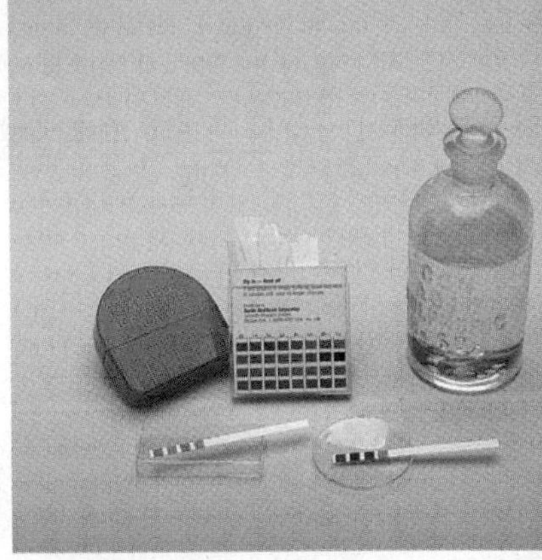

FIGURE 15.4
pH test paper for determining the approximate acidity of solutions.

Compounds with colors that change at particular pH values are used as indicators in acid–base reactions. For example, phenolphthalein, an organic compound, is colorless in acid solution and changes to pink at a pH of 8.3. When a solution of sodium hydroxide is added to a hydrochloric acid solution containing phenolphthalein, the change in color (from colorless to pink) indicates that all the acid is neutralized. Commercially available pH test paper, such as shown in Figure 15.4, contains chemical indicators. The indicator in the paper takes on different colors when wetted with solutions of different pH. Thus the pH of a solution can be estimated by placing a drop on the test paper and comparing the color of the test

Chemistry in Action

Hair Care and pH—A Delicate Balance

Hair shampoo advertisements often proclaim the proper pH for their products, but does controlling the pH of hair care products really make hair cleaner, shiny, or stronger?

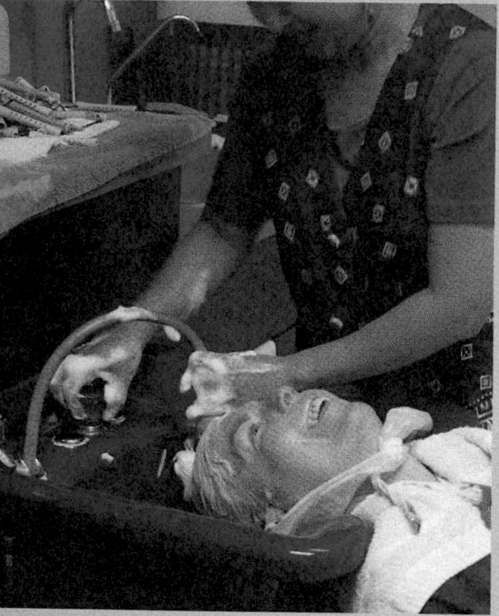

The chemistry of shampoo is ever improving to suit different hair conditions.

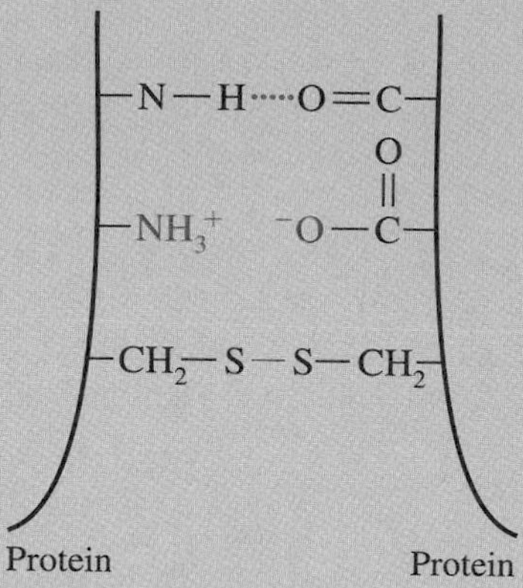

Interaction between strands of hair protein.

Each strand of hair is composed of many long chains of amino acid linked together as polymers called *proteins.* The individual chains can connect with other chains in one of three ways: (1) hydrogen bonds, red; (2) salt bridges (the result of acid–base interactions), green; and (3) disulfide bonds, blue. These interactions are shown in the diagram.

When hair is wet with water, the hydrogen bonds are broken. As the wet hair is shaped, set, or dried, the hydrogen bonds form at new positions and hold the hair in the style desired. If an acidic solution (pH 1.0–2.0) is used on the hair, the hydrogen bonds and the salt bridges are both broken, leaving only the disulfide bonds to hold the chains together. In a mildly alkaline solution (pH 8.5) some of the disulfide bonds are also broken. The outer surface of the hair becomes rough, and light does not reflect evenly from the surface, making the hair look dull. Using an alkaline shampoo will cause damage by continued breakage of the disulfide bonds, which results in "split ends." If the pH is increased further to approximately 12.0 the hair dissolves as all types of bonds break. This is the working basis for depilatories (hair removers), such as Neet™ and Nair™.

Hair has its maximum strength at pH 4.0–5.0. Shampooing tends to leave the hair slightly alkaline, so an acid rinse is sometimes used to bring the pH back into the normal range. Lemon juice or vinegar are common household products that are used for this purpose. The shampoo may also be "acid-balanced," containing a weak acid (such as citric acid) to counteract the alkalinity of the solution formed when the detergent interacts with water.

paper with a color chart calibrated at different pH values. Common applications of pH test indicators are the kits used to measure and adjust the pH of swimming pools and hot tubs. Electronic pH meters are used for making rapid and precise pH determinations.

15.10 Neutralization

neutralization The reaction of an acid and a base to form a salt and water is known as **neutralization**. We have seen this reaction before, but now with our knowledge about ions and ionization, let us reexamine the process of neutralization.

Consider the reaction that occurs when solutions of sodium hydroxide and hydrochloric acid are mixed. The ions present initially are Na^+ and OH^- from the base and H^+ and Cl^- from the acid. The products, sodium chloride and water, exist as Na^+ and Cl^- ions and H_2O molecules. A chemical equation representing this reaction is

$$HCl(aq) + NaOH(aq) \longrightarrow NaCl(aq) + H_2O(l)$$

This equation, however, does not show that HCl, NaOH, and NaCl exist as ions in solution. The following total ionic equation gives a better representation of the reaction:

$$(H^+ + Cl^-) + (Na^+ + OH^-) \longrightarrow Na^+ + Cl^- + H_2O(l)$$

This equation shows that the Na^+ and Cl^- ions did not react. These ions are called **spectator ions** because they were present but did not take part in the reaction. The only reaction that occurred was that between the H^+ and OH^- ions. Therefore, the equation for the neutralization can be written as this net ionic equation:

spectator ion

$$\underset{\text{acid}}{H^+(aq)} + \underset{\text{base}}{OH^-(aq)} \longrightarrow \underset{\text{water}}{H_2O(l)}$$

This simple net ionic equation represents not only the reaction of sodium hydroxide and hydrochloric acid, but also the reaction of any strong acid with any water-soluble hydroxide base in an aqueous solution. The driving force of a neutralization reaction is the ability of an H^+ ion and an OH^- ion to react and form a molecule of un-ionized water.

The amount of acid, base, or other species in a sample may be determined by **titration**, which is the process of measuring the volume of one reagent that is required to react with a measured mass or volume of another reagent.

titration

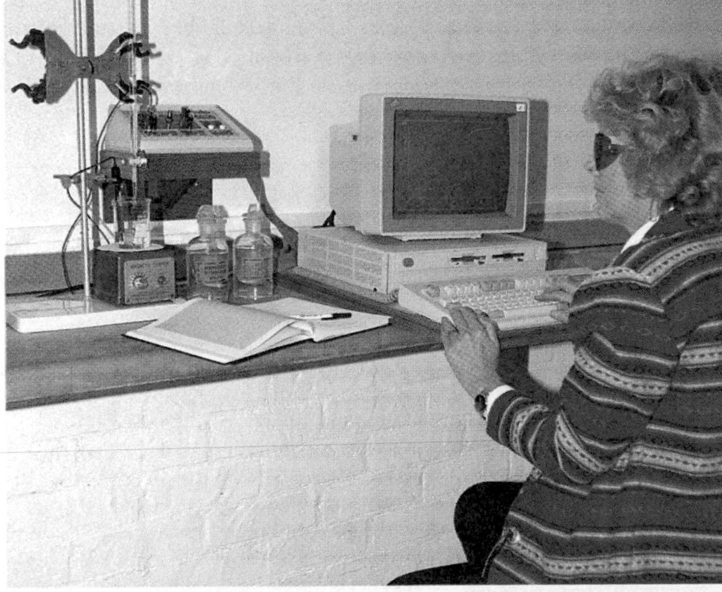

◀ **The progress of a titration can be monitored by computer graphing.**

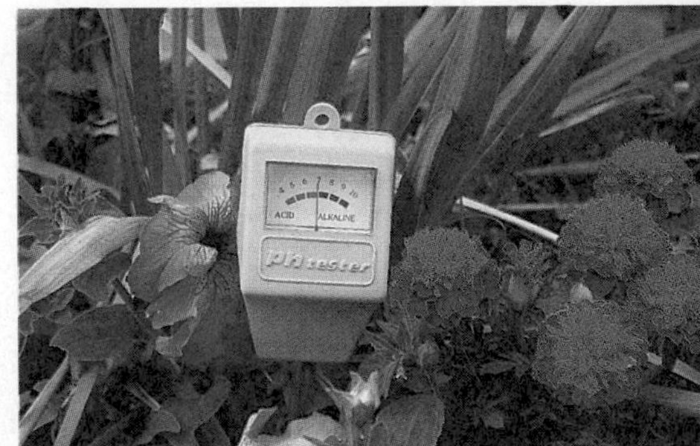

This pH meter measures the acidity and alkalinity of garden soil.

Let us consider the titration of an acid with a base. A measured volume of acid of unknown concentration is placed in a flask, and a few drops of an indicator solution are added. Base solution of known concentration is slowly added from a buret to the acid until the indicator changes color The indicator selected is one that changes color when the stoichiometric quantity (according to the equation) of base has been added to the acid. At this point, known as the *end point of the titration,* the titration is complete, and the volume of base used to neutralize the acid is read from the buret. The concentration or amount of acid in solution can be calculated from the titration data and the chemical equation for the reaction. Let's look at some illustrative examples.

Example 15.4 Suppose that 42.00 mL of 0.150 *M* NaOH solution is required to neutralize 50.00 mL of hydrochloric acid solution. What is the molarity of the acid solution?

Solution The equation for the reaction is

$$NaOH(aq) + HCl(aq) \longrightarrow NaCl(aq) + H_2O(l)$$

In this neutralization NaOH and HCl react in a 1:1 mole ratio. Therefore, the moles of HCl in solution are equal to the moles of NaOH required to react with it. First we calculate the moles of NaOH used, and from this value we determine the moles of HCl:

Data: 42.00 mL of 0.150 *M* NaOH   50.00 mL of HCl
Molarity of acid = *M* (unknown)

Determine the moles of NaOH:

$$M = \text{mol/L} \qquad 42.00\ \text{mL} = 0.04200\ \text{L}$$

$$0.04200\ \cancel{\text{L}} \times \frac{0.150\ \text{mol NaOH}}{1\ \cancel{\text{L}}} = 0.00630\ \text{mol NaOH}$$

Since NaOH and HCl react in a 1:1 ratio, 0.00630 mol of HCl was present in the 50.00 mL of HCl solution. Therefore, the molarity of the HCl is

$$M = \frac{\text{mol}}{\text{L}} = \frac{0.00630\ \text{mol HCl}}{0.05000\ \text{L}} = 0.126\ M\ \text{HCl}$$

Example 15.5 Suppose that 42.00 mL of 0.150 *M* NaOH solution is required to neutralize 50.00 mL of H_2SO_4 solution. What is the molarity of the acid solution?

Solution The equation for the reaction is

$$2\ NaOH(aq) + H_2SO_4(aq) \longrightarrow Na_2SO_4(aq) + 2\ H_2O(l)$$

The same amount of base (0.00630 mol of NaOH) is used in this titration as in Example 15.4, but the mole ratio of acid to base in the reaction is 1:2. The moles of H_2SO_4 reacted can be calculated by using the mole-ratio method:

Data: 42.00 mL of 0.150 *M* NaOH = 0.00630 mol NaOH

$$0.00630\ \cancel{\text{mol NaOH}} \times \frac{1\ \text{mol}\ H_2SO_4}{2\ \cancel{\text{mol NaOH}}} = 0.00315\ \text{mol}\ H_2SO_4$$

Therefore, 0.00315 mol of H_2SO_4 was present in 50.00 mL of H_2SO_4 solution. The molarity of the H_2SO_4 is

$$M = \frac{\text{mol}}{\text{L}} = \frac{0.00315 \text{ mol } H_2SO_4}{0.05000 \text{ L}} = 0.0630\ M\ H_2SO_4$$

Example 15.6

A 25.00-mL sample of H_2SO_4 solution required 14.26 mL of 0.2240 N NaOH for complete neutralization. What is the normality and the molarity of the sulfuric acid?

Solution

The equation for the reaction is

$$2\ NaOH(aq) + H_2SO_4(aq) \longrightarrow Na_2SO_4(aq) + 2\ H_2O(l)$$

The normality of the acid can be calculated from

$$V_A N_A = V_B N_B$$

Substitute the data into the equation and solve for N_A:

$$25.00 \text{ mL} \times N_A = 14.26 \text{ mL} \times 0.2240\ N$$

$$N_A = \frac{14.26 \cancel{\text{mL}} \times 0.2240\ N}{25.00 \cancel{\text{mL}}} = 0.1278\ N\ H_2SO_4$$

The normality of the acid is 0.1278 N.

Because H_2SO_4 furnishes 2 equivalents of H^+ per mole, the conversion to molarity is

$$\frac{\text{equiv}}{\text{L}} \times \frac{\text{mol}}{\text{equiv}}$$

$$\frac{0.1278 \cancel{\text{equiv } H_2SO_4}}{1 \text{ L}} \times \frac{1 \text{ mol } H_2SO_4}{2 \cancel{\text{equiv } H_2SO_4}} = 0.06390 \text{ mol/L } H_2SO_4$$

The H_2SO_4 solution is 0.06390 M.

Practice 15.5

A 50.0-mL sample of HCl required 24.81 mL of 0.1250 M NaOH for neutralization. What is the molarity of the acid?

15.11 Acid Rain

Acid rain is defined as any atmospheric precipitation that is more acidic than normal. The increase in acidity might be from natural or man-made sources. Rain acidity varies throughout the world and across the United States. The pH of rain is generally lower in the eastern United States and higher in the west. Unpolluted rain has a pH of 5.6, and so is slightly acidic. This acidity results from the dissolution of carbon dioxide in the water producing carbonic acid:

$$CO_2(g) + H_2O(l) \longrightarrow H_2CO_3(aq) \rightleftharpoons H^+(aq) + HCO_3^-(aq)$$

Marble masterpieces, sculpted to last forever, are slowly disappearing as acid rain dissolves the calcium carbonate ($CaCO_3$). ▶

Although the details of acid-rain formation are not yet fully understood, chemists know the general process involves the following steps:

1. emission of nitrogen and sulfur oxides into the air
2. transportation of these oxides throughout the atmosphere
3. chemical reactions between the oxides and water forming sulfuric acid (H_2SO_4) and nitric acid (HNO_3)
4. rain or snow, which carries the acids to the surface

The oxides may also be deposited directly on a dry surface and become acidic when normal rain falls on them.

Acid rain is not a new phenomenon. Rain was probably acidic in the early days of our planet as volcanic eruptions, fires, and decomposition of organic matter released large volumes of nitrogen and sulfur oxides into the atmosphere. Use of fossil fuels, especially since the industrial revolution about 250 years ago, has made significant changes in the amounts of pollutants being released into the atmosphere. As increasing amounts of fossil fuels have been burned, more and more sulfur and nitrogen oxides have poured into the atmosphere, thus increasing the acidity of rain.

Acid rain affects a variety of factors in our environment. For example, fresh water plants and animals decline significantly when rain is acidic; large numbers of fish and plants die when acidic water from spring thaws enters the lakes. Aluminum is leached from the soil into lakes by acidic rain water where the aluminum compounds adversely affect the gills of fish. In addition to leaching aluminum from the soil, acid rain also causes other valuable minerals, such as magnesium and calcium, to dissolve and run into lakes and streams. It can also dissolve the waxy protective coat on plant leaves making them vulnerable to attack by bacteria and fungi.

In our cities, acid rain is responsible for extensive and continuing damage to buildings, monuments, and statues. It may also reduce the durability of paint and promote the deterioration of paper, leather, and cloth. In short, we are just beginning to explore the effects of acid rain on human beings and on our food chain.

15.12 Writing Net Ionic Equations

In Section 15.10 we wrote the reaction of hydrochloric acid and sodium hydroxide in three different equations:

(1) $HCl(aq) + NaOH(aq) \longrightarrow NaCl(aq) + H_2O(l)$

(2) $(H^+ + Cl^-) + (Na^+ + OH^-) \longrightarrow Na^+ + Cl^- + H_2O(l)$

(3) $H^+ + OH^- \longrightarrow H_2O$

un-ionized equation
total ionic equation

In the **un-ionized equation** (1), compounds are written in their molecular, or normal, formula expressions. In the **total ionic equation** (2), compounds are written to show the form in which they are predominantly present: strong electrolytes as ions in solution; and nonelectrolytes, weak electrolytes, precipitates, and gases in their molecular (or un-ionized) forms. In the **net ionic equation** (3), only those molecules or ions that have changed are included in the equation; ions or molecules that do not change (spectators) are omitted.

net ionic equation

Up to this point, when balancing an equation we have been concerned only with the atoms of the individual elements. Because ions are electrically charged, ionic equations often end up with a net electrical charge. A balanced equation must have the same net charge on each side, whether that charge is positive, negative, or zero. Therefore, when balancing an ionic equation, we must make sure that both the same number of each kind of atom and the same net electrical charge are present on each side.

Following is a list of rules for writing ionic equations:

1. Strong electrolytes in solution are written in their ionic form.
2. Weak electrolytes are written in their molecular (un-ionized) form.
3. Nonelectrolytes are written in their molecular form.
4. Insoluble substances, precipitates, and gases are written in their molecular forms.
5. The net ionic equation should include only substances that have undergone a chemical change. Spectator ions are omitted from the net ionic equation.
6. Equations must be balanced, both in atoms and in electrical charge.

Study the following examples. In each one the un-ionized equation is given. Write the total ionic equation and the net ionic equation for each.

Example 15.7

$HNO_3(aq) + KOH(aq) \longrightarrow KNO_3(aq) + H_2O(l)$

Solution

$$(H^+ + NO_3^-) + (K^+ + OH^-) \longrightarrow (K^+ + NO_3^-) + H_2O$$

total ionic equation

$$H^+ + OH^- \longrightarrow H_2O$$

net ionic equation

The HNO_3, KOH, and KNO_3 are soluble, strong electrolytes. The K^+ and NO_3^- ions are spectator ions, have not changed, and are not included in the net ionic equation. Water is a nonelectrolyte and is written in the molecular form.

Example 15.8 $2\ AgNO_3(aq) + BaCl_2(aq) \longrightarrow 2\ AgCl(s) + Ba(NO_3)_2(aq)$

un-ionized equation

$(2\ Ag^+ + 2\ NO_3^-) + (Ba^{2+} + 2\ Cl^-) \longrightarrow 2\ AgCl(s) + (Ba^{2+} + 2\ NO_3^-)$

total ionic equation

$Ag^+ + Cl^- \longrightarrow AgCl(s)$

net ionic equation

Solution Although AgCl is an ionic compound, it is written in the un-ionized form on the right side of the ionic equations because most of the Ag^+ and Cl^- ions are no longer in solution but have formed a precipitate of AgCl. The Ba^{2+} and NO_3^- ions are spectator ions.

Example 15.9 $Na_2CO_3(aq) + H_2SO_4(aq) \longrightarrow Na_2SO_4(aq) + H_2O(l) + CO_2(g)$

un-ionized equation

$(2\ Na^+ + CO_3^{2-}) + (2\ H^+ + SO_4^{2-}) \longrightarrow (2\ Na^+ + SO_4^{2-}) + H_2O(l) + CO_2(g)$

total ionic equation

$CO_3^{2-} + 2\ H^+ \longrightarrow H_2O(l) + CO_2(g)$

net ionic equation

Solution Carbon dioxide, CO_2, is a gas and evolves from the solution; Na^+ and SO_4^{2-} are spectator ions.

Example 15.10 $HC_2H_3O_2(aq) + NaOH(aq) \longrightarrow NaC_2H_3O_2(aq) + H_2O(l)$

un-ionized equation

$HC_2H_3O_2 + (Na^+ + OH^-) \longrightarrow (Na^+ + C_2H_3O_2^-) + H_2O$

total ionic equation

$HC_2H_3O_2 + OH^- \longrightarrow C_2H_3O_2^- + H_2O$

net ionic equation

Solution Acetic acid, $HC_2H_3O_2$, a weak acid, is written in the molecular form, but sodium acetate, $NaC_2H_3O_2$, a soluble salt, is written in the ionic form. The Na^+ ion is the only spectator ion in this reaction. Both sides of the net ionic equation have a -1 electrical charge.

Example 15.11 $Mg(s) + 2\ HCl(aq) \longrightarrow MgCl_2(aq) + H_2(g)$

un-ionized equation

$Mg + (2\ H^+ + 2\ Cl^-) \longrightarrow (Mg^{2+} + 2\ Cl^-) + H_2(g)$

total ionic equation

$Mg + 2\ H^+ \longrightarrow Mg^{2+} + H_2(g)$

net ionic equation

Solution The net electrical charge on both sides of the equation is $+2$.

Example 15.12

$$H_2SO_4(aq) + Ba(OH)_2(aq) \longrightarrow BaSO_4(s) + 2\ H_2O(l)$$

un-ionized equation

$$(2\ H^+ + SO_4^{2-}) + (Ba^{2+} + 2\ OH^-) \longrightarrow BaSO_4(s) + 2\ H_2O(l)$$

total ionic equation

$$2\ H^+ + SO_4^{2-} + Ba^{2+} + 2\ OH^- \longrightarrow BaSO_4(s) + 2\ H_2O(l)$$

net ionic equation

Solution

Barium sulfate, $BaSO_4$, is a highly insoluble salt. If we conduct this reaction using the conductivity apparatus described in Section 15.5, the light glows brightly at first but goes out when the reaction is complete because almost no ions are left in solution. The $BaSO_4$ precipitates out of solution, and water is a nonconductor of electricity.

Practice 15.6

Write the net ionic equation for
$3\ H_2S(aq) + 2\ Bi(NO_3)_3(aq) \longrightarrow Bi_2S_3(s) + 6\ HNO_3(aq)$

15.13 Colloids: An Introduction

When we add sugar to a flask of water and shake it, the sugar dissolves and forms a clear homogeneous *solution.* When we do the same experiment with very fine sand and water, the sand particles form a *suspension,* which settles when the shaking stops. When we repeat the experiment again using ordinary cornstarch, we find that the starch does not dissolve in cold water. But if the mixture is heated and stirred, the starch forms a cloudy, opalescent *dispersion.* This dispersion does not appear to be clear and homogeneous like the sugar solution, yet it is not obviously heterogeneous and does not settle like the sand suspension. In short, its properties are intermediate between those of the sugar solution and those of the sand suspension. The starch dispersion is actually a *colloid,* a name derived from the Greek *kolla,* meaning "glue," and was coined by the English scientist Thomas Graham in 1861. Graham classified solutes as crystalloids if they diffused through a parchment membrane and as colloids if they did not diffuse through the membrane.

colloid

As it is now used, the word **colloid** means a dispersion in which the dispersed particles are larger than the solute ions or molecules of a true solution and smaller than the particles of a mechanical suspension. The term does not imply a glue-like quality, although most glues are colloidal materials. The size of colloidal particles ranges from a lower limit of about 1 nm (10^{-7} cm) to an upper limit of about 1000 nm (10^{-4} cm).

The fundamental difference between a colloidal dispersion and a true solution is the size not the nature of the particles. The solute particles in a solution are usually single ions or molecules that may be hydrated to varying degrees. Colloidal particles are usually aggregations of ions or molecules. However, the molecules of

TABLE 15.6 Types of Colloidal Dispersions

Type	Name	Examples
Gas in liquid	Foam	Whipped cream, soap suds
Gas in solid	Solid foam	Styrofoam, foam rubber, pumice
Liquid in gas	Liquid aerosol	Fog, clouds
Liquid in liquid	Emulsion	Milk, vinegar in oil salad dressing, mayonnaise
Liquid in solid	Solid emulsion	Cheese, opals, jellies
Solid in gas	Solid aerosol	Smoke, dust in air
Solid in liquid	Sol	India ink, gold sol
Solid in solid	Solid Sol	Tire rubber, certain gems (e.g., rubies)

some polymers, such as proteins, are large enough to be classified as colloidal particles when in solution. To appreciate fully the differences in relative sizes, the volumes (not just the linear dimensions) of colloidal particles and solute particles must be compared. The difference in volumes can be approximated by assuming that the particles are spheres. A large colloidal particle has a diameter of about 500 nm, whereas a fair-sized ion or molecule has a diameter of about 0.5 nm. Thus, the diameter of the colloidal particle is about 1000 times that of the solute particle. Because the volumes of spheres are proportional to the cubes of their diameters, we can calculate that the volume of a colloidal particle can be up to a billion ($10^3 \times 10^3 \times 10^3 = 10^9$) times greater than that of a solution particle.

Colloids are mixtures in which one component, the *dispersal phase,* exists as discrete particles in the other component, the *dispersing phase* or *dispersing medium.* The components of a colloidal dispersion are also sometimes called the *discontinuous phase* and the *continuous phase.* Each component, or phase, can exist as a solid, a liquid, or a gas. The components cannot be mutually soluble, nor can both phases of the dispersion be gases, because such conditions would result in an ordinary solution. Hence, only eight types of colloidal dispersions, based on the possible physical states of the phases, are known. The eight types, with specific examples, are listed in Table 15.6.

15.14 Preparation of Colloids

Colloidal dispersions can be prepared by two methods: (1) *dispersion,* the breaking down of larger particles to colloidal size, and (2) *condensation,* the formation of colloidal particles from solutions.

Homogenized milk is a good example of a colloid prepared by dispersion. Milk, as drawn from the cow, is an unstable emulsion of fat in water. The fat globules are so large that they rise and form a cream layer in a few hours. To avoid separation of the cream, the milk is homogenized by pumping it through very small holes, or orifices, at high pressure. The violent shearing action of this treatment breaks the fat globules into particles well within the colloidal size range. The butterfat in homogenized milk remains dispersed indefinitely. Colloid mills, which reduce particles to colloidal size by grinding or shearing, are used in preparing many commercial products such as paints, cosmetics, and salad dressings.

The preparation of colloids by condensation frequently involves a precipitation reaction in a dilute solution. A colloidal dispersion in a liquid is called a **sol**. For example, a good colloidal sulfur sol can be made by bubbling hydrogen sulfide into a solution of sulfur dioxide. Solid sulfur is formed and dispersed as a colloid:

sol

$$SO_2 + 2\ H_2S \longrightarrow \underset{\substack{\text{colloidal}\\ \text{sulfur}}}{3\ S} + 2\ H_2O$$

A colloidal dispersion is also easily made by adding iron(III) chloride solution to boiling water. The reddish-brown colloidal dispersion that is formed probably consists of iron(III) hydroxide and hydrated iron(III) oxide:

$$FeCl_3 + 3\ HOH \xrightarrow{H_2O} Fe(OH)_3 + 3\ HCl$$

$$Fe(OH)_3 \xrightarrow{H_2O} Fe_2O_3 \cdot xH_2O$$

A great many products for home use (insecticides, insect repellents, and deodorants, to name a few) are packaged as aerosols. The active ingredient, either a liquid or a solid, is dissolved in a liquefied gas and sealed under pressure in a container fitted with a release valve. When this valve is opened, the pressurized solution is ejected. The liquefied gas vaporizes, and the active ingredient is converted to a colloidal aerosol almost instantaneously.

15.15 Properties of Colloids

In 1827 Robert Brown (1773–1858), while observing a strongly illuminated aqueous suspension of pollen under a high-powered microscope, noted that the pollen grains appeared to have a trembling, erratic motion. He determined later that this erratic motion was not confined to pollen but was characteristic of colloidal particles in general. This random motion of colloidal particles is called **Brownian movement**. It can be readily observed by confining cigarette smoke in a small transparent chamber and illuminating it with a strong beam of light at right angles to the optical axis of the microscope. The smoke particles appear as tiny randomly moving lights, because the light is reflected from their surfaces. This motion is due to the continual bombardment of the smoke particles by air molecules. Since Brownian movement can be seen when colloidal particles are dispersed in either a gaseous or a liquid medium, it affords nearly direct visual proof that matter at the molecular level actually is moving randomly, as postulated by the Kinetic-Molecular Theory.

Brownian movement

When an intense beam of light is passed through an ordinary solution and is viewed at an angle, the beam passing through the solution is hardly visible. A beam of light, however, is clearly visible and sharply outlined when it is passed through a colloidal dispersion (see Figure 15.5). This phenomenon is known as the **Tyndall effect**. It was first described by Michael Faraday in 1857 and later amplified by John Tyndall. The Tyndall effect, like the Brownian movement, can be observed

Tyndall effect

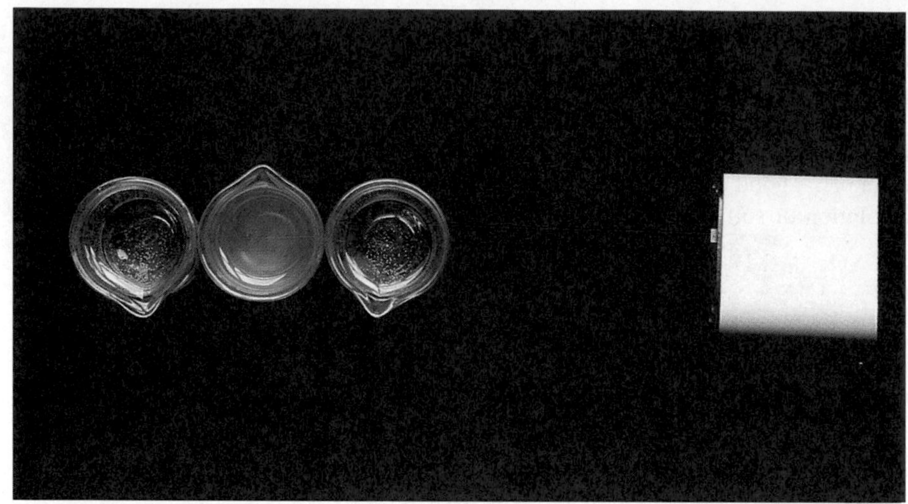

FIGURE 15.5
The Tyndall effect. The beakers on the left and right each contain a true solution, whereas the beaker in the center contains a collodial starch solution. A laser beam, emitted by the laser on the far left, scatters in the colloid but is invisible in the true solution. ▶

in nearly all colloidal dispersions. It occurs because the colloidal particles are large enough to scatter the rays of visible light. The ions or molecules of true solutions are too small to scatter light and, therefore, do not exhibit a noticeable Tyndall effect.

Another important characteristic of colloids is that the particles have relatively huge surface areas. We saw in Section 14.6 that the surface area is increased tenfold when a 1-cm cube is divided into 1000 cubes with sides of 0.1 cm. When a 1-cm cube is divided into colloidal-size cubes measuring 10^{-6} cm, the combined surface area of all the particles becomes a million times greater than that of the original cube.

Colloidal particles become electrically charged when they adsorb ions on their surfaces. *Adsorption* should not be confused with *absorption.* Adsorption refers to the adhesion of molecules or ions to a surface, whereas absorption refers to the taking in of one material by another material. Adsorption occurs because the atoms or ions at the surface of a particle are not completely surrounded by other atoms or ions as are those in the interior. Consequently these surface atoms or ions attract and adsorb ions or polar molecules from the dispersion medium onto the surfaces of the colloidal particles. This property is directly related to the large surface area presented by the many tiny particles.

The particles in a given dispersion tend to adsorb only ions having one kind of charge. For example, cations are primarily adsorbed on iron(III) hydroxide sol, resulting in positively charged colloidal particles. On the other hand, the particles of an arsenic(III) sulfide sol primarily adsorb anions, resulting in negatively charged colloidal particles. The properties of true solutions, colloidal dispersions, and mechanical suspensions are summarized and compared in Table 15.7.

15.16 Stability of Colloids

The stability of different dispersions varies with the properties of the dispersed and dispersing phases. We have noted that nonhomogenized milk is a colloid, yet it will separate after standing for a few hours. However, the particles of a good colloidal dispersion will remain in suspension indefinitely. As a case in point, a ruby-red

TABLE 15.7 Comparison of the Properties of True Solutions, Colloidal Dispersions, and Suspensions

	Particle size (nm)	Ability to pass through filter paper	Ability to pass through parchment	Exhibits Tyndall effect	Exhibits Brownian movement	Settles out on standing	Appearance
True solution	<1	Yes	Yes	No	No	No	Transparent, homogeneous
Colloidal dispersion	1–1000	Yes	No	Yes	Yes	Generally does not	Usually not transparent, but may appear to be homogeneous
Suspension	>1000	No	No	—	No	Yes	Not transparent, heterogeneous

gold sol has been kept in the British Museum for more than a century without noticeable settling. (This specimen is kept for historical interest; it was prepared by Michael Faraday.) The particles of a specific colloid remain dispersed for two reasons:

1. They are bombarded by the molecules of the dispersing phase, which keeps the particles in motion (Brownian movement) so that gravity does not cause them to settle out.
2. Since the colloidal particles have the same kind of electrical charge, they repel each other. This mutual repulsion prevents the dispersed particles from coalescing to larger particles, which would settle out of suspension.

In certain types of colloids, the presence of a material known as a protective colloid is necessary for stability. Egg yolk, for example, acts as a stabilizer, or protective colloid, in mayonnaise. The yolk adsorbs on the surfaces of the oil particles and prevents them from coalescing.

15.17 Applications of Colloidal Properties

Activated charcoal has an enormous surface area, approximately 1 million square centimeters per gram in some samples. Hence, charcoal is very effective in selectively adsorbing the polar molecules of some poisonous gases and is therefore used in gas masks. Charcoal can be used to adsorb impurities from liquids as well as from gases, and large amounts are used to remove substances that have objectionable tastes and odors from water supplies. In sugar refineries activated charcoal is used to adsorb colored impurities from the raw sugar solutions.

A process widely used for dust and smoke control in many urban and industrial areas was devised by an American, Frederick Cottrell (1877–1948). The Cottrell process takes advantage of the fact that the particulate matter in dust and smoke is electrically charged. Air to be cleaned of dust or smoke is passed between electrode plates charged with a high voltage. Positively charged particles are attracted to,

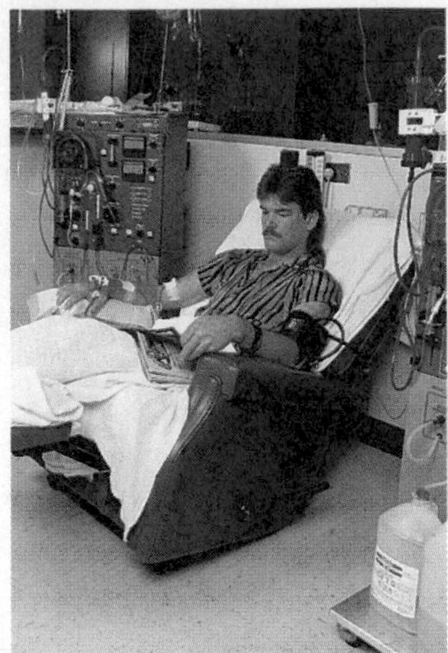

This dialysis machine acts as artificial kidneys by removing soluble waste products from the blood.

dialysis

neutralized, and thereby precipitated at the negative electrodes. Negatively charged particles are removed in the same fashion at the positive electrodes. Large Cottrell units are fitted with devices for automatic removal of precipitated material. In some installations, particularly at cement mills and smelters, the value of the dust collected may be sufficient to pay for the precipitation equipment. Small units, designed for removing dust and pollen from air in the home, are now on the market. Unfortunately, Cottrell units remove only particulate matter; they cannot remove gaseous pollutants such as carbon monoxide, sulfur dioxide, and nitrogen oxides.

Thomas Graham found that a parchment membrane would allow the passage of true solutions but would prevent the passage of colloidal dispersions. Dissolved solutes can be removed from colloidal dispersions through the use of such a membrane by a process called **dialysis**. The membrane itself is called a *dialyzing membrane.* Artificial membranes are made from such materials as parchment paper, collodion, or certain kinds of cellophane. Dialysis can be demonstrated by putting a colloidal starch dispersion and some copper(II) sulfate solution in a parchment paper bag and suspending it in running water. In a few hours the blue color of the copper(II) sulfate has disappeared, and only the starch dispersion remains in the bag.

A life-saving application of dialysis has been the development of artificial kidneys. The blood of a patient suffering from partial kidney failure is passed through the artificial kidney machine for several hours, during which time the soluble waste products are removed by dialysis.

Concepts in Review

1. State the general characteristics of acids and bases.
2. Define an acid and base in terms of the Arrhenius, Brønsted–Lowry, and Lewis theories.
3. Identify acid–base conjugate pairs in a reaction.
4. When given the reactants, complete and balance equations for the reactions of acids with bases, metals, metal oxides, and carbonates.
5. When given the reactants, complete and balance equations of the reaction of an amphoteric hydroxide with either a strong acid or a strong base.
6. Write balanced equations for the reaction of sodium hydroxide or potassium hydroxide with zinc and with aluminum.
7. Classify common compounds as electrolytes or nonelectrolytes.
8. Distinguish between strong and weak electrolytes.
9. Explain the process of dissociation and ionization. Indicate how they differ.
10. Write equations for the dissociation and/or ionization of acids, bases, and salts in water.
11. Describe and write equations for the ionization of water.
12. Explain how pH expresses hydrogen-ion concentration or hydronium-ion concentration.
13. Given pH as an integer, calculate the H^+ molarity, and vice versa.
14. Use a calculator to calculate pH values from corresponding H^+ molarities.

15. Explain the process of acid–base neutralization.
16. Calculate the molarity, normality, or volume of an acid or base solution from appropriate titration data.
17. Write un-ionized, total ionic, and net ionic equations for neutralization equations.
18. Discuss colloids and describe methods for their preparation.
19. Describe the characteristics that distinguish true solutions, colloidal dispersions, and mechanical suspensions.
20. Explain (a) how colloidal dispersions can be stabilized and (b) how they can be precipitated.

Key Terms

The terms listed here have been defined within this chapter. Section numbers are referenced in parenthesis for each term.

amphoteric (15.2)
Brownian movement (15.15)
colloid (15.13)
dialysis (15.17)
dissociation (15.6)
electrolyte (15.5)
hydronium ion (15.1)
ionization (15.6)
logarithm (15.9)
net ionic equation (15.12)
neutralization (15.10)
nonelectrolyte (15.5)
pH (15.9)
sol (15.14)
spectator ion (15.10)
strong electrolyte (15.7)
titration (15.10)
total ionic equation (15.12)
Tyndall effect (15.15)
un-ionized equation (15.12)
weak electrolyte (15.7)

Questions

Questions refer to tables, figures, and key words and concepts defined within the chapter. A particularly challenging question or exercise is indicated with an asterisk.

1. Since a hydrogen ion and a proton are identical, what differences exist between the Arrhenius and Brønsted–Lowry definitions of an acid? (See Table 15.1.)
2. According to Figure 15.1, what type of substance must be in solution in order for the bulb to light?
3. Which of the following classes of compounds are electrolytes: acids, alcohols, bases, salts? (See Table 15.2.)
4. What two differences are apparent in the arrangement of water molecules about the hydrated ions as depicted in Figure 15.2?
5. The pH of a solution with a hydrogen ion concentration of 0.003 M is between what two whole numbers? (See Table 15.4.)
6. Which is the more acidic, tomato juice or blood? (See Table 15.5.)
7. Using each of the three acid–base theories (Arrhenius, Brønsted–Lowry, and Lewis), define an acid and a base.
8. For each of the acid–base theories referred to in Exercise 7, write an equation illustrating the neutralization of an acid with a base.
9. Write the Lewis structure for the (a) bromide ion, (b) hydroxide ion, and (c) cyanide ion. Why are these ions considered to be bases according to the Brønsted–Lowry and Lewis acid–base theories?

10. Into what three classes of compounds do electrolytes generally fall?

11. Name each compound listed in Table 15.3.

12. A solution of HCl in water conducts an electric current, but a solution of HCl in hexane does not. Explain this behavior in terms of ionization and chemical bonding.

13. How do ionic compounds exist in their crystalline structure? What occurs when they are dissolved in water?

14. An aqueous methyl alcohol, CH_3OH, solution does not conduct an electric current, but a solution of sodium hydroxide, NaOH, does. What does this information tell us about the OH group in the alcohol?

15. Why does molten NaCl conduct electricity?

16. Explain the difference between dissociation of ionic compounds and ionization of molecular compounds.

17. Distinguish between strong and weak electrolytes.

18. Explain why ions are hydrated in aqueous solutions.

19. What is the main distinction between water solutions of strong and weak electrolytes?

20. What are the relative concentrations of $H^+(aq)$ and $OH^-(aq)$ in (a) a neutral solution, (b) an acid solution, and (c) a basic solution?

21. Write the net ionic equation for the reaction of a strong acid with a water-soluble hydroxide base in an aqueous solution.

22. The solubility of HCl gas in water, a polar solvent, is much greater than its solubility in hexane, a nonpolar solvent. How can you account for this difference?

23. Pure water, containing equal concentrations of both acid and base ions, is neutral. Why?

24. Which of the following statements are correct? Rewrite each incorrect statement to make it correct.

(a) The Arrhenius theory of acids and bases is restricted to aqueous solutions.

(b) The Brønsted–Lowry theory of acids and bases is restricted to solutions other than aqueous solutions.

(c) All substances that are acids according to the Brønsted–Lowry theory will also be acids by the Lewis theory.

(d) All substances that are acids according to the Lewis theory will also be acids by the Brønsted–Lowry theory.

(e) An electron-pair donor is a Lewis acid.

(f) All Arrhenius acid–base neutralization reactions can be represented by a single net ionic equation.

(g) When an ionic compound dissolves in water, the ions separate; this process is called *ionization.*

(h) In the auto-ionization of water,

$$2\ H_2O \rightleftharpoons H_3O^+ + OH^-$$

the H_3O^+ and the OH^- constitute a conjugate acid–base pair.

(i) In the reaction in part (h), H_2O is both the acid and the base.

(j) Most common Na^+, K^+, and NH_4^+ salts are soluble in water.

(k) A solution of pH 3 is 100 times more acidic than a solution of pH 5.

(l) In general, ionic substances when placed in water will give a solution capable of conducting an electric current.

(m) The terms *dissociation* and *ionization* are synonymous.

(n) A solution of $Mg(NO_3)_2$ contains three ions per formula unit in solution.

(o) The terms *strong acid, strong base, weak acid,* and *weak base* refer to whether an acid or base solution is concentrated or dilute.

(p) pH is defined as the negative logarithm of the molar concentration of H^+ ions (or H_3O^+ ions).

(q) All reactions may be represented by net ionic equations.

(r) One mole of $CaCl_2$ contains more anions than cations.

(s) It is possible to boil seawater at a lower temperature than that required to boil pure water (both at the same pressure).

(t) It is possible to have a neutral aqueous solution whose pH is not 7.

(u) The size of colloidal particles ranges from 1 mm to 1000 mm.

(v) The Tyndall effect is observable in both colloidal and true solutions.

25. Indicate the fundamental difference between a colloidal dispersion and a true solution.

26. List the two methods used to prepare colloidal dispersions and briefly explain how each is accomplished.

27. Explain the Tyndall effect and how it may be used to distinguish between a colloidal dispersion and a true solution.

28. Distinguish between adsorption and absorption.

29. Why do particles in a specific colloid remain dispersed?

30. Explain the process of dialysis, giving a practical application in society.

31. Diagram and label the types of interactions occurring between strands of protein in the hair.

32. What is the difference between a shampoo and a depilatory? Include a discussion of the types of bonds broken in your answer.

Paired Exercises

These exercises are paired. Each odd-numbered exercise is followed by a similar even-numbered exercise. Answers to the even-numbered exercises are given in Appendix V.

33. Identify the conjugate acid–base pairs in the following equations:

(a) $HCl + NH_3 \longrightarrow NH_4^+ + Cl^-$

(b) $HCO_3^- + OH^- \rightleftharpoons CO_3^{2-} + H_2O$

(c) $HCO_3^- + H_3O^+ \rightleftharpoons H_2CO_3 + H_2O$

(d) $HC_2H_3O_2 + H_2O \rightleftharpoons H_3O^+ + C_2H_3O_2^-$

34. Identify the conjugate acid–base pairs in the following equations:

(a) $HC_2H_3O_2 + H_2SO_4 \rightleftharpoons H_2C_2H_3O_2^+ + HSO_4^-$

(b) The two-step ionization of sulfuric acid,

$H_2SO_4 + H_2O \longrightarrow H_3O^+ + HSO_4^-$

$HSO_4^- + H_2O \rightleftharpoons H_3O^+ + SO_4^{2-}$

(c) $HClO_4 + H_2O \longrightarrow H_3O^+ + ClO_4^-$

(d) $CH_3O^- + H_3O^+ \longrightarrow CH_3OH + H_2O$

35. Complete and balance the following equations:

(a) $Mg(s) + HCl(aq) \longrightarrow$

(b) $BaO(s) + HBr(aq) \longrightarrow$

(c) $Al(s) + H_2SO_4(aq) \longrightarrow$

(d) $Na_2CO_3(aq) + HCl(aq) \longrightarrow$

(e) $Fe_2O_3(s) + HBr(aq) \longrightarrow$

(f) $Ca(OH)_2(aq) + H_2CO_3(aq) \longrightarrow$

36. Complete and balance the following equations:

(a) $NaOH(aq) + HBr(aq) \longrightarrow$

(b) $KOH(aq) + HCl(aq) \longrightarrow$

(c) $Ca(OH)_2(aq) + HI(aq) \longrightarrow$

(d) $Al(OH)_3(s) + HBr(aq) \longrightarrow$

(e) $Na_2O(s) + HClO_4(aq) \longrightarrow$

(f) $LiOH(aq) + FeCl_3(aq) \longrightarrow$

37. Which of the following compounds are electrolytes? Consider each substance to be mixed with water.

(a) HCl

(b) CO_2

(c) $CaCl_2$

(d) $C_{12}H_{22}O_{11}$ (sugar)

(e) C_3H_7OH (rubbing alcohol)

(f) CCl_4 (insoluble)

38. Which of the following compounds are electrolytes? Consider each substance to be mixed with water.

(a) $NaHCO_3$ (baking soda)

(b) N_2 (insoluble gas)

(c) $AgNO_3$

(d) HCOOH (formic acid)

(e) RbOH

(f) K_2CrO_4

39. Calculate the molarity of the ions present in each of the following salt solutions. Assume each salt to be 100% dissociated:

(a) 0.015 *M* NaCl

(b) 4.25 *M* $NaKSO_4$

(c) 0.20 *M* $CaCl_2$

(d) 22.0 g KI in 500. mL of solution

40. Calculate the molarity of the ions present in each of the following salt solutions. Assume each salt to be 100% dissociated:

(a) 0.75 *M* $ZnBr_2$

(b) 1.65 *M* $Al_2(SO_4)_3$

(c) 900. g $(NH_4)_2SO_4$ in 20.0 L of solution

(d) 0.0120 g $Mg(ClO_3)_2$ in 1.00 mL of solution

41. In Exercise 39, how many grams of each ion would be present in 100. mL of each solution?

42. In Exercise 40, how many grams of each ion would be present in 100. mL of each solution?

43. What is the molar concentration of all ions present in a solution prepared by mixing the following? Neglect the concentration of H^+ and OH^- from water. Also, assume volumes of solutions are additive:

(a) 30.0 mL of 1.0 *M* NaCl and 40.0 mL of 1.0 *M* NaCl

(b) 30.0 mL of 1.0 *M* HCl and 30.0 mL of 1.0 *M* NaOH

*(c) 100.0 mL of 0.40 *M* KOH and 100.0 mL of 0.80 *M* HCl

44. What is the molar concentration of all ions present in a solution prepared by mixing the following? Neglect the concentration of H^+ and OH^- from water. Also, assume volumes of solutions are additive.

(a) 100.0 mL of 2.0 *M* KCl and 100.0 mL of 1.0 *M* $CaCl_2$

(b) 35.0 mL of 0.20 *M* $Ba(OH)_2$ and 35.0 mL of 0.20 *M* H_2SO_4

(c) 1.00 L of 1.0 *M* $AgNO_3$ and 500. mL of 2.0 *M* NaCl

45. Given the data for the following separate titrations, calculate the molarity of the HCl:

	mL HCl	Molarity HCl	mL NaOH	Molarity NaOH
(a)	40.13	*M*	37.70	0.728
(b)	19.00	*M*	33.66	0.306
(c)	27.25	*M*	18.00	0.555

46. Given the data for the following separate titrations, calculate the molarity of the NaOH:

	mL HCl	Molarity HCl	mL NaOH	Molarity NaOH
(a)	37.19	0.126	31.91	*M*
(b)	48.04	0.482	24.02	*M*
(c)	13.x13	1.425	39.39	*M*

47. Rewrite the following unbalanced equations, changing them into balanced net ionic equations. All reactions are in water solution:
(a) $K_2SO_4(aq) + Ba(NO_3)_2(aq) \longrightarrow KNO_3(aq) + BaSO_4(s)$
(b) $CaCO_3(s) + HCl(aq) \longrightarrow CaCl_2(aq) + CO_2(g) + H_2O(l)$
(c) $Mg(s) + HC_2H_3O_2(aq) \longrightarrow Mg(C_2H_3O_2)_2(aq) + H_2(g)$

48. Rewrite the following unbalanced equations, changing them into balanced net ionic equations. All reactions are in water solution:
(a) $H_2S(g) + CdCl_2(aq) \longrightarrow CdS(s) + HCl(aq)$
(b) $Zn(s) + H_2SO_4(aq) \longrightarrow ZnSO_4(aq) + H_2(g)$
(c) $AlCl_3(aq) + Na_3PO_4(aq) \longrightarrow AlPO_4(s) + NaCl(aq)$

49. In each of the following pairs which solution is more acidic? All are water solutions. Explain your answer.
(a) 1 molar HCl or 1 molar H_2SO_4?
(b) 1 molar HCl or 1 molar $HC_2H_3O_2$?

50. In each of the following pairs which solution is more acidic? All are water solutions. Explain your answer.
(a) 1 molar HCl or 2 molar HCl?
(b) 1 normal H_2SO_4 or 1 molar H_2SO_4?

51. What volume (in milliliters) of 0.245 *M* HCl will neutralize 50.0 mL of 0.100 *M* $Ca(OH)_2$? The equation is
$2\ HCl(aq) + Ca(OH)_2(aq) \longrightarrow CaCl_2(aq) + 2\ H_2O(l)$.

52. What volume (in milliliters) of 0.245 *M* HCl will neutralize 10.0 g of $Al(OH)_3$? The equation is
$3\ HCl(aq) + Al(OH)_3(s) \longrightarrow AlCl_3(aq) + 3\ H_2O(l)$.

*53. A 0.200-g sample of impure NaOH requires 18.25 mL of 0.2406 *M* HCl for neutralization. What is the percent of NaOH in the sample?

*54. A batch of sodium hydroxide was found to contain sodium chloride as an impurity. To determine the amount of impurity, a 1.00-g sample was analyzed and found to require 49.90 mL of 0.466 *M* HCl for neutralization. What is the percent of NaCl in the sample?

*55. What volume of H_2 gas, measured at 27°C and 700. torr pressure, can be obtained by reacting 5.00 g of zinc metal with 100. mL of 0.350 *M* HCl? The equation is
$Zn(s) + 2\ HCl(aq) \longrightarrow ZnCl_2(aq) + H_2(g)$

*56. What volume of H_2 gas, measured at 27°C and 700. torr, can be obtained by reacting 5.00 g of zinc metal with 200. mL of 0.350 *M* HCl? The equation is
$Zn(s) + 2\ HCl(aq) \longrightarrow ZnCl_2(aq) + H_2(g)$

57. Calculate the pH of solutions having the following H^+ ion concentrations:
(a) 0.01 *M*
(b) 1.0 *M*
(c) 6.5×10^{-9} *M*

58. Calculate the pH of solutions having the following H^+ ion concentrations:
(a) 1×10^{-7} *M*
(b) 0.50 *M*
(c) 0.00010 *M*

59. Calculate the pH of the following:
(a) orange juice, 3.7×10^{-4} *M* H^+
(b) vinegar, 2.8×10^{-3} *M* H^+

60. Calculate the pH of the following:
(a) black coffee, 5.0×10^{-5} *M* H^+
(b) limewater, 3.4×10^{-11} *M* H^+

61. How many milliliters of 0.325 *N* HNO_3 are required to neutralize 32.8 mL of 0.225 *N* NaOH?

62. How many milliliters of 0.325 *N* H_2SO_4 are required to neutralize 32.8 mL of 0.225 *N* NaOH?

63. The equivalent masses of organic acids are often determined by titration with a standard solution of a base. Determine the equivalent mass of benzoic acid, if 0.305 g of it requires 25 mL of 0.10 *N* NaOH for neutralization.

*64. Determine the equivalent mass of succinic acid, if 0.738 g is required to neutralize 125 mL of 0.10 *N* base.

Additional Exercises

These exercises are not paired or labeled by topic and provide additional practice on concepts covered in this chapter.

65. What is the concentration of Ca^{2+} ions in a solution of CaI_2 having an I^- ion concentration of 0.520 *M*?

66. How many milliliters of 0.40 *M* HCl can be made by diluting 100. mL of 12 *M* HCl with water?

67. If 29.26 mL of 0.430 *M* HCl neutralizes 20.40 mL of $Ba(OH)_2$ solution, what is the molarity of the $Ba(OH)_2$ solution? The reaction is

$$Ba(OH)_2(aq) + 2\ HCl(aq) \longrightarrow BaCl_2(aq) + 2\ H_2O(l)$$

68. A 1 molal solution of acetic acid, $HC_2H_3O_2$, in water freezes at a lower temperature than a 1 molal solution of ethyl alcohol, C_2H_5OH, in water. Explain.

69. At the same cost per pound, which alcohol, CH_3OH or C_2H_5OH, would be more economical to purchase as an antifreeze for your car? Why?

70. How does a hydronium ion differ from a hydrogen ion?

71. Arrange, in decreasing order of freezing points, 1 molal aqueous solutions of HCl, $HC_2H_3O_2$, $C_{12}H_{22}O_{11}$ (sucrose), and $CaCl_2$. (List the one with the highest freezing point first.)

72. At 100°C the H^+ concentration in water is about 1×10^{-6} mol/L, about ten times that of water at 25°C. At which of these temperatures is
(a) the pH of water the greater?
(b) the hydrogen ion (hydronium ion) concentration the higher?
(c) the water neutral?

73. What is the relative difference in H^+ concentration in solutions that differ by 1 pH unit?

74. A sample of pure sodium carbonate with a mass of 0.452 g was dissolved in water and neutralized with 42.4 mL of hydrochloric acid. Calculate the molarity of the acid:

$$Na_2CO_3(aq) + 2\ HCl(aq) \longrightarrow 2\ NaCl(aq) + CO_2(g) + H_2O(l)$$

75. What volume (mL) of 0.1234 *M* HCl is needed to neutralize 2.00 g $Ca(OH)_2$?

76. How many grams of KOH are required to neutralize 50.00 mL of 0.240 *M* HNO_3?

77. Two drops (0.1 mL) of 1.0 *M* HCl are added to water to make 1.0 L of solution. What is the pH of this solution if the HCl is 100% ionized?

78. What volume of concentrated (18.0 *M*) sulfuric acid must be used to prepare 50.0 L of 5.00 *M* solution?

79. Three (3.0) grams of NaOH are added to 500. mL of 0.10 *M* HCl. Will the resulting solution be acidic or basic? Show evidence for your answer.

80. A 10.00-mL sample of base solution requires 28.92 mL of 0.1240 *N* H_2SO_4 for neutralization. What is the normality of the base?

81. What is the normality and the molarity of a 25.00-mL sample of H_3PO_4 solution that requires 22.68 mL of 0.5000 *N* NaOH for complete neutralization?

82. How many milliliters of 0.10 *N* NaOH is required to neutralize 60.0 mL of 0.20 *N* H_2SO_4?

***83.** A 25-mL sample of H_2SO_4 requires 40. mL of 0.20 *N* NaOH for neutralization.
(a) What is the normality of the sulfuric acid solution?
(b) How many grams of sulfuric acid are contained in the 25 mL sample?

***84.** A solution of 40.0 mL of HCl is neutralized by 20.0 mL of NaOH solution. The resulting neutral solution is evaporated to dryness, and the residue is found to have a mass of 0.117 g. Calculate the normality of the HCl and NaOH solutions.

85. Compare and contrast a concentrated solution of a weak electrolyte with a dilute solution of a strong electrolyte.

86. Under what circumstances will a solution's normality be equal to its molarity?

87. How many milliliters of water must be added to 85.00 mL of 1.000 *N* H_3PO_4 to give a solution that is 0.6500 *N* H_3PO_4?

***88.** If 380 mL of 0.35 *M* $Ba(OH)_2$ is added to 500.0 mL of 0.65 *M* HCl, will the mixture be acidic or basic? Find the pH of the resulting solution.

89. Fifty milliliters (50.00 mL) of 0.2000 *M* HCl is titrated with 0.2000 *M* NaOH. Find the pH of the solution after:
(a) 0.000 mL of base have been added
(b) 10.00 mL of base have been added
(c) 25.00 mL of base have been added
(d) 49.00 mL of base have been added
(e) 49.90 mL of base have been added
(f) 49.99 mL of base have been added
(g) 50.00 mL of base have been added

Plot your answers on a graph with pH on the *y*-axis and mL NaOH on the *x*-axis.

90. NaOH reacts with sulfuric acid:
(a) Write a balanced equation for the reaction producing Na_2SO_4.
(b) How many milliliters of 0.10 *M* NaOH are needed to react with 0.0050 mol H_2SO_4?
(c) How many grams of Na_2SO_4 will also form?

***91.** Lactic acid (found in sour milk) has an empirical formula of $HC_3H_5O_3$. A 1.0-g sample of lactic acid required 17.0 mL of 0.65 *M* NaOH to reach the end point of a titration. What is the molecular formula for lactic acid?

92. A 10.0-mL sample of HNO_3 was diluted to a volume of 100.00 mL. Then, 25 mL of that diluted solution was needed to neutralize 50.0 mL of 0.60 *M* KOH. What was the concentration of the original nitric acid?

93. The pH of a solution of a strong acid was determined to be 3. If water is then added to dilute this solution, would the pH change? Why or why not? Could enough water ever be added to raise the pH of an acid solution above 7?

94. Solution *X* has a pH of 2. Solution *Y* has a pH of 4. On the basis of this information, which of the following is true?
(a) the $[H^+]$ of *X* is one half that of *Y*
(b) the $[H^+]$ of *X* is twice that of *Y*
(c) the $[H^+]$ of *X* is 100 times that of *Y*
(d) the $[H^+]$ of *X* is 1/100 that of *Y*

95. A student is given three solutions: an acid, a base, and one that is neither acidic nor basic. The student performs tests on these solutions and records their properties. For each result, tell whether it is the property of an acid, a base, or whether you cannot decide.
(a) the solution has $[H^+] = 1 \times 10^{-7}$ *M*
(b) the solution has $[OH^-] = 1 \times 10^{-2}$ *M*
(c) the solution turns litmus red
(d) the solution is a good conductor of electricity

Answers to Practice Exercises

15.1 (a) HCO_3^-, (b) NO_2^-, (c) $C_2H_3O_2^-$
15.2 (a) H_2SO_4, (b) NH_4^+, (c) H_2O
15.3 (a) 0.050 *M* Mg^{2+}, 0.10 *M* Cl^-, (b) 0.070 *M* Al^{3+}, 0.21 *M* Cl^-
15.4 (a) 11.41, (b) 2.89, (c) 5.429
15.5 (c) 0.0620 *M* HCl
15.6 $3\ H_2S(aq) + 2Bi^{3+}(aq) \longrightarrow Bi_2S_3(s) + 6\ H^+(aq)$

Chemical Equilibrium

16

Chapter 16

Thus far, we have considered chemical change to proceed from reactants to products. Does that mean that reactions stop? No. A solute dissolves until the solution becomes saturated. Once a solid remains undissolved in a container, the system appears to be at rest. The human body is a marvelous chemical factory, yet from day to day it appears to be quite the same. For example, the blood remains at a constant pH, even though all sorts of chemical reactions are taking place. Another example is a terrarium, which can be watered and sealed for long periods of time with no ill effects. Or an antacid, which absorbs excess stomach acid and *does not* change the pH of the stomach. In all of these cases, reactions are proceeding, even though visible signs of chemical change are absent. When the system is at equilibrium, chemical reactions are dynamic. As the Frenchman Alphonse Karr so succinctly stated in 1849, "The more things change the more they stay the same."

16.1 Reversible Reactions

In the preceding chapters we have treated chemical reactions mainly as reactants changing to products. However, many reactions do not go to completion. Some reactions do not go to completion because they are reversible; that is, when the products are formed, they react to produce the starting reactants.

We have encountered reversible systems before. One is the vaporization of a liquid by heating and its subsequent condensation by cooling:

$$\text{liquid} + \text{heat} \longrightarrow \text{vapor}$$
$$\text{vapor} + \text{cooling} \longrightarrow \text{liquid}$$

The interconversion of nitrogen dioxide, NO_2, and dinitrogen tetroxide, N_2O_4, offers visible evidence of the reversibility of a reaction. The NO_2 is a reddish-brown gas that changes with cooling to N_2O_4, a yellow liquid that boils at 21.2°C, and then to a colorless solid, N_2O_4, that melts at −11.2°C. The reaction is reversible by heating N_2O_4:

$$2\ NO_2(g) \xrightarrow{\text{cooling}} N_2O_4(l)$$

$$N_2O_4(l) \xrightarrow{\text{heating}} 2\ NO_2(g)$$

These two reactions may be represented by a single equation with a double arrow, $\rightleftharpoons$ to indicate that the reactions are taking place in both directions at the same time:

$$2\ NO_2(g) \rightleftharpoons N_2O_4(l)$$

This reversible reaction can be demonstrated by sealing samples of NO_2 in two tubes and placing one tube in warm water and the other in ice water (see Figure 16.1). Heating promotes disorder or randomness in a system, so we would expect more NO_2 (a gas) to be present at higher temperatures.

◀ **Chapter Opening Photo: A coral reef is a system in ecological equilibrium with the ocean surrounding it.**

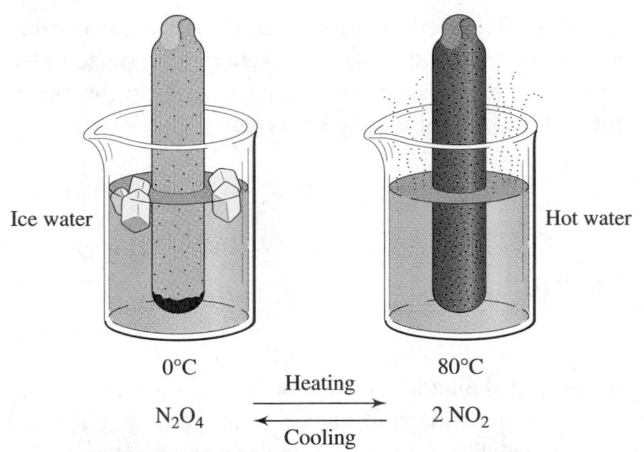

FIGURE 16.1
Reversible reaction of NO_2 and N_2O_4. More of the reddish-brown NO_2 molecules are visible in the tube that is heated than in the tube that is cooled.

Fog forming on a bathroom mirror is a reversible reaction familiar to all of us.

reversible chemical reaction

A **reversible chemical reaction** is one in which the products formed react to produce the original reactants. Both the forward and reverse reactions occur simultaneously. The forward reaction is called *the reaction to the right,* and the reverse reaction is called *the reaction to the left.* A double arrow is used in the equation to indicate that the reaction is reversible.

16.2 Rates of Reaction

chemical kinetics

Every reaction has a rate, or speed, at which it proceeds. Some are fast and some are extremely slow. The study of reaction rates and reaction mechanisms is known as **chemical kinetics.**

The rate of a reaction is variable and depends on the concentration of the reacting species, the temperature, the presence or absence of catalytic agents, and the nature of the reactants. Consider the hypothetical reaction

$$A + B \longrightarrow C + D \quad \text{(forward reaction)}$$

$$C + D \longrightarrow A + B \quad \text{(reverse reaction)}$$

in which a collision between A and B is necessary for a reaction to occur. The rate at which A and B react depends on the concentration or the number of A and B molecules present; it will be fastest, for a fixed set of conditions, when they are first mixed. As the reaction proceeds, the number of A and B molecules available for reaction decreases, and the rate of reaction slows down. If the reaction is reversible, the speed of the reverse reaction is zero at first and gradually increases as the concentrations of C and D increase. As the number of A and B molecules decreases, the forward rate slows down because A and B cannot find one another as often in order to accomplish a reaction. To counteract this diminishing rate of reaction, an excess of one reagent is often used to keep the reaction from becoming impractically slow. Collisions between molecules may be likened to be space video games. When many objects are on the screen, collisions occur frequently; but if only a few objects are present, collisions can usually be avoided.

16.3 Chemical Equilibrium

equilibrium

Any system at **equilibrium** represents a dynamic state in which two or more opposing processes are taking place at the same time and at the same rate. A chemical equilibrium is a dynamic system in which two or more opposing chemical reactions are going on at the same time and at the same rate. When the rate of the forward reaction is exactly equal to the rate of the reverse reaction, a condition of **chemical equilibrium** exists (see Figure 16.2). The concentrations of the products and the reactants are not changing, and the system appears to be at a standstill because the products are reacting at the same rate at which they are being formed.

chemical equilibrium

Chemical equilibrium:
rate of forward reaction = rate of reverse reaction

A saturated salt solution is in a condition of equilibrium:

$$NaCl(s) \rightleftharpoons Na^+(aq) + Cl^-(aq)$$

At equilibrium, salt crystals are continuously dissolving, and Na^+ and Cl^- ions are continuously crystallizing. Both processes are occurring at the same rate.

The ionization of weak electrolytes is another chemical equilibrium system:

$$HC_2H_3O_2(aq) + H_2O(l) \rightleftharpoons H_3O^+(aq) + C_2H_3O_2^-(aq)$$

In this reaction, the equilibrium is established in a 1 *M* solution when the forward reaction has gone about 1%—that is, when only 1% of the acetic acid molecules in solution have ionized. Therefore, only a relatively few ions are present, and the acid behaves as a weak electrolyte. In any acid–base equilibrium system the position of equilibrium is toward the weaker conjugate acid and base. In the ionization of acetic acid, $HC_2H_3O_2$ is a weaker acid than H_3O^+, and H_2O is a weaker base than $C_2H_3O_2^-$.

The reactions represented by

$$H_2(g) + I_2(g) \rightleftharpoons 2\ HI(g)$$

provide another example of chemical equilibrium. Theoretically, 1.00 mol of hydrogen should react with 1.00 mol of iodine to yield 2.00 mol of hydrogen iodide. Actually, when 1.00 mol of H_2 and 1.00 mol of I_2 are reacted at 700 K, only 1.58 mol of HI are present when equilibrium is attained. Since 1.58 is 79% of the theoretical yield of 2.00 mol of HI, the forward reaction is only 79% complete at

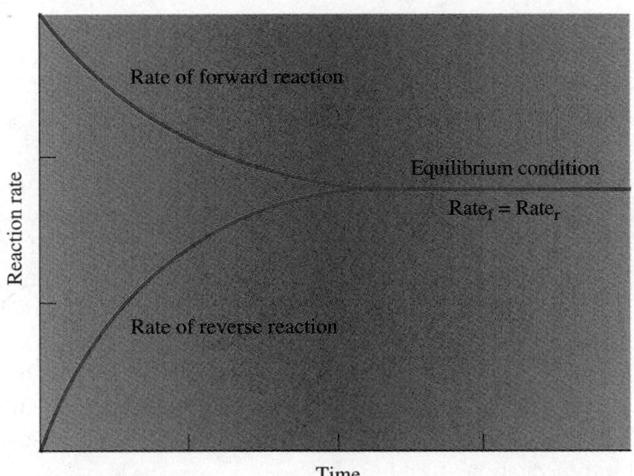

◀ **FIGURE 16.2**
The graph illustrates that the rates of the forward and reverse reactions become equal at some point in time. The forward reaction rate decreases as a result of decreasing amounts of reactants. The reverse reaction rate starts at zero and increases as the amount of product increases. When the two rates become equal, a state of chemical equilibrium has been reached.

equilibrium. The equilibrium mixture will also contain 0.21 mol each of unreacted H_2 and I_2 (1.00 mol − 0.79 mol = 0.21 mol):

$$H_2(g) + I_2(g) \xrightarrow{700K} 2HI(g)$$

1.00 mol	1.00 mol	2.00 mol

(This equation represents the condition if the reaction were 100% complete; 2.00 mol of HI would be formed and no H_2 and I_2 would be left unreacted.)

$$H_2(g) + I_2(g) \overset{700K}{\rightleftharpoons} 2HI(g)$$

0.21 mol	0.21 mol	1.58 mol

(This equation represents the actual equilibrium attained starting with 1.00 mol each of H_2 and I_2. It shows that the forward reaction is only 79% complete.)

16.4 Principle of Le Chatelier

In 1888, the French chemist Henri Le Chatelier (1850–1936) set forth a simple, far-reaching generalization on the behavior of equilibrium systems. This generalization, known as the **principle of Le Chatelier**, states:

principle of Le Chatelier

If a stress is applied to a system in equilibrium, the system will respond in such a way as to relieve that stress and restore equilibrium under a new set of conditions.

The application of Le Chatelier's principle helps us to predict the effect of changing conditions in chemical reactions. We will examine the effect of changes in concentration, temperature, and pressure.

16.5 Effect of Concentration on Reaction Rate and Equilibrium

The way in which the rate of a chemical reaction depends on the concentration of the reactants must be determined experimentally. Many simple, one-step reactions result from a collision between two molecules or ions. The rate of such one-step reactions can be altered by changing the concentration of the reactants or products. An increase in concentration of the reactants provides more individual reacting species for collisions and results in an increase in the rate of reaction.

An equilibrium is disturbed when the concentration of one or more of its components is changed. As a result the concentration of all species will change, and a new equilibrium mixture will be established. Consider the hypothetical equilibrium represented by the equation

$$A + B \rightleftharpoons C + D$$

where A and B react in one step to form C and D. When the concentration of B is increased, the following occurs:

1. The rate of the reaction to the right (forward) increases. This rate is proportional to the concentration of A times the concentration of B.
2. The rate to the right becomes greater than the rate to the left.
3. Reactants A and B are used faster than they are produced; C and D are produced faster than they are used.
4. After a period of time, rates to the right and left become equal, and the system is again in equilibrium.
5. In the new equilibrium, the concentration of A is less and the concentrations of B, C, and D are greater than in the original equilibrium.

Conclusion: The equilibrium has shifted to the right.

Applying this change in concentration to the equilibrium mixture of 1.00 mol of hydrogen and 1.00 mol of iodine from Section 16.3, we find that, when an additional 0.20 mol of I_2 is added, the yield of HI (based on H_2) is 85% (1.70 mol) instead of 79%. A comparison of the two systems, after the new equilibrium mixture is reached, follows.

Original equilibrium	**New equilibrium**
1.00 mol H_2 + 1.00 mol I_2 Yield: 79% HI Equilibrium mixture contains: 1.58 mol HI 0.21 mol H_2 0.21 mol I_2	1.00 mol H_2 + 1.20 mol I_2 Yield: 85% HI (based on H_2) Equilibrium mixture contains: 1.70 mol HI 0.15 mol H_2 0.35 mol I_2

Analyzing this new system, we see that, when the 0.20 mol of I_2 was added, the equilibrium shifted to the right in order to counteract the increase in I_2 concentration. Some of the H_2 reacted with added I_2 and produced more HI, until an equilibrium mixture was established again. When I_2 was added, the concentration of I_2 increased, the concentration of H_2 decreased, and the concentration of HI increased. What do you think would be the effects of adding (a) more H_2 or (b) more HI?

The equation

$$\underset{\text{pale yellow}}{Fe^{3+}(aq)} + \underset{\text{colorless}}{SCN^-(aq)} \rightleftharpoons \underset{\text{red}}{Fe(SCN)^{2+}(aq)}$$

represents an equilibrium that is used in certain analytical procedures as an indicator because of the readily visible, intense red color of the complex $Fe(SCN)^{2+}$ ion. A very dilute solution of iron(III), Fe^{3+}, and thiocyanate, SCN^-, is light red. When the concentration of either Fe^{3+} or SCN^- is increased, the equilibrium shift to the right is observed by an increase in the intensity of the color, resulting from the formation of additional $Fe(SCN)^{2+}$.

If either Fe^{3+} or SCN^- is removed from solution, the equilibrium will shift to the left, and the solution will become lighter in color. When Ag^+ is added to the solution, a white precipitate of silver thiocyanate (AgSCN) is formed, thus removing SCN^- ion from the equilibrium:

$$Ag^+(aq) + SCN^-(aq) \rightleftharpoons AgSCN(s)$$

The system accordingly responds to counteract the change in SCN^- concentration by shifting the equilibrium to the left. This shift is evident by a decrease in the intensity of the red color due to a decreased concentration of $Fe(SCN)^{2+}$.

Let us now consider the effect of changing the concentrations in the equilibrium mixture of chlorine water. The equilibrium equation is

$$Cl_2(aq) + 2\ H_2O(l) \rightleftharpoons HOCl(aq) + H_3O^+(aq) + Cl^-(aq)$$

The variation in concentrations and the equilibrium shifts are tabulated in the following table. An X in the second or third column indicates that the reagent is increased or decreased. The fourth column indicates the direction of the equilibrium shift.

	Concentration		
Reagent	Increase	Decrease	Equilibrium shift
Cl_2	—	X	Left
H_2O	X	—	Right
HOCl	X	—	Left
H_3O^+	—	X	Right
Cl^-	X	—	Left

Consider the equilibrium in a 0.100 *M* acetic acid solution:

$$HC_2H_3O_2(aq) + H_2O(l) \rightleftharpoons H_3O^+(aq) + C_2H_3O_2^-(aq)$$

In this solution the concentration of the hydronium ion, H_3O^+, which is a measure of the acidity, is 1.34×10^{-3} mol/L, corresponding to a pH of 2.87. What will happen to the acidity when 0.100 mol of sodium acetate, $NaC_2H_3O_2$, is added to 1 L of 0.100 *M* acetic acid, $HC_2H_3O_2$? When $NaC_2H_3O_2$ dissolves, it dissociates into sodium ions, Na^+, and acetate ions, $C_2H_3O_2^-$. The acetate ion from the salt is a common ion to the acetic acid equilibrium system and increases the total acetate ion concentration in the solution. As a result the equilibrium shifts to the left, decreasing the hydronium ion concentration and lowering the acidity of the solution. Evidence of this decrease in acidity is shown by the fact that the pH of a solution that is 0.100 *M* in $HC_2H_3O_2$ and 0.100 *M* in $NaC_2H_3O_2$ is 4.74. The pH of several different solutions of $HC_2H_3O_2$ and $NaC_2H_3O_2$ is shown in the table that follows. Each time the acetate ion is increased, the pH increases, indicating a further shift in the equilibrium toward un-ionized acetic acid.

Solution	pH
1 L 0.100 *M* $HC_2H_3O_2$	2.87
1 L 0.100 *M* $HC_2H_3O_2$ + 0.100 mol $NaC_2H_3O_2$	4.74
1 L 0.100 *M* $HC_2H_3O_2$ + 0.200 mol $NaC_2H_3O_2$	5.05
1 L 0.100 *M* $HC_2H_3O_2$ + 0.300 mol $NaC_2H_3O_2$	5.23

In summary, we can say that, when the concentration of a reagent on the left side of an equation is increased, the equilibrium shifts to the right. When the concentration of a reagent on the right side of an equation is increased, the equilibrium shifts to the left. In accordance with Le Chatelier's principle the equilibrium always shifts in the direction that tends to reduce the concentration of the added reactant.

Practice 16.1

Aqueous chromate ion, CrO_4^{2-}, exists in equilibrium with aqueous dichromate ion, $Cr_2O_7^{2-}$ in an acidic solution. What effect will (a) increasing the dichromate ion and (b) adding HCl have on the equilibrium?

$$2\ CrO_4^{2-}(aq) + 2\ H^+(aq) \rightleftharpoons Cr_2O_7^{2-}(aq) + H_2O(l)$$

16.6 Effect of Pressure on Reaction Rate and Equilibrium

Changes in pressure significantly affect the reaction rate only when one or more of the reactants or products is a gas. In these cases the effect of increasing the pressure of the reacting gases is equivalent to increasing their concentrations. In the reaction

$$CaCO_3(s) \overset{\Delta}{\rightleftharpoons} CaO(s) + CO_2(g)$$

calcium carbonate decomposes into calcium oxide and carbon dioxide when heated above 825°C. Increasing the pressure of the equilibrium system by adding CO_2 or by decreasing the volume of the container, speeds up the reverse reaction and causes the equilibrium to shift to the left. The increased pressure gives the same effect as that caused by increasing the concentration of CO_2, the only gaseous substance in the reaction.

If the volume of a gas is decreased, the pressure of the gas will increase. In a system composed entirely of gases, an increase in pressure will cause the reaction

▲ **Gaseous ammonia is often used to add nitrogen to the fields before planting and during early growth.**

and the equilibrium to shift to the side that contains the smaller number of moles. This shift occurs because fewer moles of gas at the same temperature and volume will exert less pressure.

Prior to World War I, Fritz Haber (1868–1934) invented the first major process for the fixation of nitrogen in Germany. For this process, Haber received the Nobel Prize in chemistry in 1918. In the Haber process nitrogen and hydrogen are reacted together in the presence of a catalyst at moderately high temperature and pressure to produce ammonia. The catalyst consists of iron and iron oxide with small amounts of potassium and aluminum oxides:

$$\underset{\substack{1\text{ mol}\\1\text{ volume}}}{N_2(g)} + \underset{\substack{3\text{ mol}\\3\text{ volumes}}}{3\,H_2(g)} \rightleftharpoons \underset{\substack{2\text{ mol}\\2\text{ volumes}}}{2\,NH_3(g)} + 92.5\text{ kJ} \qquad \text{(at 25°C)}$$

The left side of the equation in the Haber process represents 4 mol of gas combining to give 2 mol of gas on the right side of the equation. An increase in the total pressure on the system shifts the equilibrium to the right. This increase in pressure results in a higher concentration of both reactants and products. The equilibrium shifts to the right when the pressure is increased, because fewer moles of NH_3 than moles of N_2 and H_2 are in the equilibrium reaction.

Ideal conditions for the Haber process are 200°C and 1000 atm pressure. However, at 200°C the rate of reaction is very slow, and at 1000 atm extraordinarily heavy equipment is required. As a compromise the reaction is run at 400–600°C and 200–350 atm pressure, which gives a reasonable yield at a reasonable rate. The effect of pressure on the yield of ammonia at one particular temperature is shown in Table 16.1.

TABLE 16.1 The Effect of Pressure on the Conversion of H_2 and N_2 to NH_3 at 450°C*

Pressure (atm)	Yield of NH_3(%)	Pressure (atm)	Yield of NH_3(%)
10	2.04	300	35.5
30	5.8	600	53.4
50	9.17	1000	69.4
100	16.4		

*The starting ratio of H_2 to N_2 is 3 mol to 1 mol.

When the total number of gaseous molecules on both sides of an equation is the same, a change in pressure does not cause an equilibrium shift. The following reaction is an example:

$N_2(g)$	+	$O_2(g)$	$\rightleftarrows$	$2\ NO(g)$
1 mol		1 mol		2 mol
1 volume		1 volume		2 volumes
6.022×10^{23} molecules		6.022×10^{23} molecules		$2(6.022 \times 10^{23})$ molecules

When the pressure on this system is increased, the rate of both the forward and the reverse reactions will increase because of the higher concentrations of N_2, O_2, and NO. But the equilibrium will not shift, because the increase in concentration of molecules is the same on both sides of the equation and the decrease in volume is the same on both sides of the equation.

Example 16.1 What effect would an increase in pressure have on the position of equilibrium in the following reactions?

(a) $2\ SO_2(g) + O_2(g) \rightleftarrows 2\ SO_3(g)$

(b) $H_2(g) + Cl_2(g) \rightleftarrows 2\ HCl(g)$

(c) $N_2O_4(l) \rightleftarrows 2\ NO_2(g)$

Solution

(a) The equilibrium will shift to the right because the substance on the right has a smaller volume than those on the left.
(b) The equilibrium position will be unaffected because the volumes (or moles) of gases on both sides of the equation are the same.
(c) The equilibrium will shift to the left because $N_2O_4(l)$ occupies a much smaller volume than does $2\ NO_2(g)$.

Practice 16.2

What effect would an increase in pressure have on the position of the equilibrium in the following reactions?

(a) $2\ NO(g) + Cl_2(g) \rightleftarrows 2\ NOCl(g)$

(b) $COBr_2(g) \rightleftarrows CO(g) + Br_2(g)$

16.7 Effect of Temperature on Reaction Rate and Equilibrium

When the temperature of a system is raised, the rate of reaction increases because of increased kinetic energy and more frequent collisions of the reacting species. In a reversible reaction the rate of both the forward and the reverse reactions is increased by an increase in temperature; however, the reaction that absorbs heat increases to a greater extent, and the equilibrium shifts to favor that reaction.

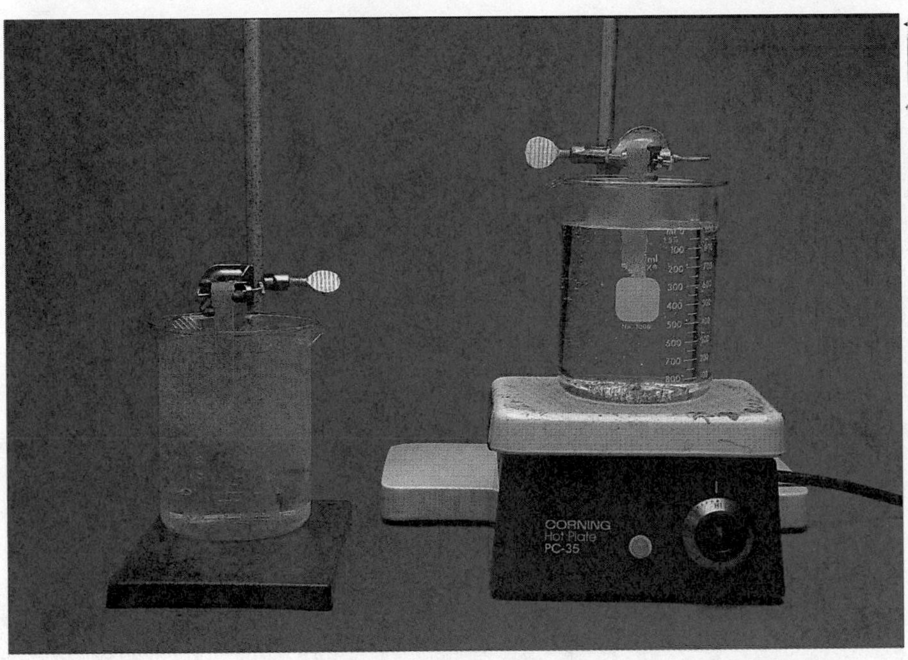

◀ **Light sticks. The chemical reaction that produces light in these light sticks is endothermic. Placing the light stick in hot water (right) favors this reaction, producing a brighter light than when the light stick is in ice water (left).**

High temperatures may cause the destruction or decomposition of the reactants or products.

An increase in temperature generally increases the rate of reaction. Molecules at elevated temperatures are more energetic and have more kinetic energy; thus, their collisions are more likely to result in a reaction. However, we cannot assume that the rate of a desired reaction will increase indefinitely as the temperature is raised.

When heat is applied to a system in equilibrium, the reaction that absorbs heat is favored. When the process, as written, is endothermic, the forward reaction is increased. When the reaction is exothermic, the reverse reaction is favored. In this sense heat may be treated as a reactant in endothermic reactions or as a product in exothermic reactions. Therefore, temperature is analogous to concentration when applying Le Chatelier's principle to heat effects on a chemical reaction.

Hot coke (C) is a very reactive element. In the reaction

$$C(s) + CO_2(g) + \text{heat} \rightleftharpoons 2\,CO(g)$$

very little if any CO is formed at room temperature. At 1000°C the equilibrium mixture contains about an equal number of moles of CO and CO_2. At higher temperatures the equilibrium shifts to the right, increasing the yield of CO. The reaction is endothermic, and, as can be seen, the equilibrium is shifted to the right at higher temperatures.

When phosphorus trichloride reacts with dry chlorine gas to form phosphorus pentachloride, the reaction is exothermic:

$$PCl_3(l) + Cl_2(g) \rightleftharpoons PCl_5(s) + 88\text{ kJ}$$

Heat must continuously be removed during the reaction to obtain a good yield of the product. According to the principle of Le Chatelier, heat will cause the product, PCl_5, to decompose, re-forming PCl_3 and Cl_2. The equilibrium mixture at 200°C contains 52% PCl_5, and at 300°C it contains 3% PCl_5, verifying that heat causes the equilibrium to shift to the left. The following example illustrates these effects.

Example 16.2 What effect would an increase in temperature have on the position of equilibrium in the following reactions?

$$4\ HCl(g) + O_2(g) \rightleftharpoons 2\ H_2O(g) + 2\ Cl_2(g) + 95.4\ kJ \quad (1)$$

$$H_2(g) + Cl_2(g) \rightleftharpoons 2\ HCl(g) + 185\ kJ \quad (2)$$

$$CH_4(g) + 2\ O_2(g) \rightleftharpoons CO_2(g) + 2\ H_2O(g) + 890\ kJ \quad (3)$$

$$N_2O_4(l) + 58.6\ kJ \rightleftharpoons 2\ NO_2(g) \quad (4)$$

$$2\ CO_2(g) + 566\ kJ \rightleftharpoons 2\ CO(g) + O_2(g) \quad (5)$$

$$H_2(g) + I_2(g) + 51.9\ kJ \rightleftharpoons 2\ HI(g) \quad (6)$$

Solution Reactions (1), (2), and (3) are exothermic; an increase in temperature will cause the equilibrium to shift to the left. Reactions (4), (5), and (6) are endothermic; an increase in temperature will cause the equilibrium to shift to the right.

Practice 16.3

What effect would an increase in temperature have on the position of the equilibrium in the following reactions?

(a) $2\ SO_2(g) + O_2(g) \rightleftharpoons 2\ SO_3(g) + 198\ kJ$

(b) $H_2(g) + CO_2(g) + 41\ kJ \rightleftharpoons H_2O(g) + CO(g)$

16.8 Effect of Catalysts on Reaction Rate and Equilibrium

catalyst

A **catalyst** is a substance that influences the rate of a reaction and can be recovered essentially unchanged at the end of the reaction. A catalyst does not shift the equilibrium of a reaction; it affects only the speed at which the equilibrium is reached. If a catalyst does not affect the equilibrium, then it follows that it must affect the rate of both the forward and the reverse reactions equally.

The reaction between phosphorus trichloride and sulfur is highly exothermic, but it is so slow that very little product, thiophosphoryl chloride is obtained, even after prolonged heating. When a catalyst, such as aluminum chloride is added, the reaction is complete in a few seconds:

$$PCl_3(l) + S(s) \xrightarrow{AlCl_3} PSCl_3(l)$$

The lab preparation of oxygen uses manganese dioxide as a catalyst to increase the rates of decomposition of both potassium chlorate and hydrogen peroxide:

$$2\ KClO_3(s) \xrightarrow[\Delta]{MnO_2} 2\ KCl(s) + 3\ O_2(g)$$

$$2\ H_2O_2(aq) \xrightarrow{MnO_2} 2\ H_2O(l) + O_2(g)$$

Catalysts are extremely important to industrial chemistry. Hundreds of chemical reactions that are otherwise too slow to be of practical value have been put to commercial use once a suitable catalyst was found. But in the area of biochemistry, catalysts are of supreme importance because nearly all chemical reactions in all forms of life are completely dependent on biochemical catalysts known as *enzymes*.

16.9 Equilibrium Constants

In a reversible chemical reaction at equilibrium, the concentrations of the reactants and products are constant; that is, they are not changing. The rates of the forward and reverse reactions are constant, and an equilibrium constant expression can be written relating the products to the reactants. For the general reaction

$$aA + bB \rightleftharpoons cC + dD$$

at constant temperature, the following equilibrium constant expression can be written:

$$K_{eq} = \frac{[C]^c[D]^d}{[A]^a[B]^b}$$

where K_{eq} is constant at a particular temperature and is known as the **equilibrium constant**. The quantities in brackets are the concentrations of each substance in moles per liter. The superscript letters *a*, *b*, *c*, and *d* are the coefficients of the substances in the balanced equation. The convention is to place the concentrations of the products (the substances on the right side of the equation as written) in the numerator and the concentrations of the reactants in the denominator.

equilibrium constant, K_{eq}

Note that the exponents are the same as the coefficients in the balanced equation.

Example 16.3

Write equilibrium constant expressions for

(a) $3\ H_2(g) + N_2(g) \rightleftharpoons 2\ NH_3(g)$

(b) $CO(g) + 2\ H_2(g) \rightleftharpoons CH_3OH(g)$

Solution

(a) The only product, NH_3, has a coefficient of 2. Therefore, the numerator of the equilibrium constant will be $[NH_3]^2$. Two reactants are present, H_2 with a coefficient of 3 and N_2 with a coefficient of 1. Therefore, the denominator will be $[H_2]^3[N_2]$. The equilibrium constant expression is

$$K_{eq} = \frac{[NH_3]^2}{[H_2]^3[N_2]}$$

(b) For this equation, the numerator is $[CH_3OH]$ and the denominator is $[CO][H_2]^2$. The equilibrium constant expression is

$$K_{eq} = \frac{[CH_3OH]}{[CO][H_2]^2}$$

Practice 16.4

Write the equilibrium constant expressions for

(a) $2\ N_2O_5(g) \rightleftarrows 4\ NO_2(g) + O_2(g)$

(b) $4\ NH_3(g) + 3\ O_2(g) \rightleftarrows 2\ N_2(g) + 6\ H_2O(g)$

The magnitude of an equilibrium constant indicates the extent to which the forward and reverse reactions take place. When K_{eq} is greater than 1, the amount of the products at equilibrium is greater than the amount of the reactants. When K_{eq} is less than 1, the amount of reactants at equilibrium is greater than the amount of the products. A very large value for K_{eq} indicates that the forward reaction goes essentially to completion. A very small K_{eq} means that the reverse reaction goes nearly to completion and that the equilibrium is far to the left (toward the reactants). Two examples follow:

$$H_2(g) + I_2(g) \rightleftarrows 2\ HI(g) \qquad K_{eq} = 54.8 \text{ at } 425°C$$

This K_{eq} indicates that more product than reactant exists at equilibrium.

$$COCl_2(g) \rightleftarrows CO(g) + Cl_2(g) \qquad K_{eq} = 7.6 \times 10^{-4} \text{ at } 400°C$$

This K_{eq} indicates that $COCl_2$ is stable and that very little decomposition to CO and Cl_2 occurs at 400°C. The equilibrium is far to the left.

When the molar concentrations of all the species in an equilibrium reaction are known, the K_{eq} can be calculated by substituting the concentrations into the equilibrium constant expression.

Example 16.4 Calculate the K_{eq} for the following reaction based on concentrations of: PCl_5 = 0.030 mol/L; PCl_3 = 0.97 mol/L; Cl_2 = 0.97 mol/L at 300°C.

$$PCl_5(g) \rightleftarrows PCl_3(g) + Cl_2(g)$$

Solution First write the K_{eq} expression; then substitute the respective concentrations into this equation and solve:

$$K_{eq} = \frac{[PCl_3][Cl_2]}{[PCl_5]} = \frac{(0.97)(0.97)}{(0.030)} = 31$$

This K_{eq} is considered to be a fairly large value, indicating that at 300°C the decomposition of PCl_5 proceeds far to the right.

Practice 16.5

Calculate the K_{eq} for the following reaction. Is the forward or the reverse reaction favored?

$$2\ NO(g) + O_2(g) \rightleftarrows 2\ NO_2(g)$$

when [NO] = 0.050 *M*, $[O_2]$ = 0.75 *M*, $[NO_2]$ = 0.25 *M*

16.10 Ion Product Constant for Water

We have seen that water ionizes to a slight degree. This ionization is represented by these equilibrium equations:

$$H_2O + H_2O \rightleftharpoons H_3O^+ + OH^- \qquad (1)$$

$$H_2O \rightleftharpoons H^+ + OH^- \qquad (2)$$

Equation (1) is the more accurate representation of the equilibrium, since free protons (H^+) do not exist in water. Equation (2) is a simplified and often-used representation of the water equilibrium. The actual concentration of H^+ produced in pure water is minute and amounts to only 1.00×10^{-7} mol/L at 25°C. In pure water,

$$[H^+] = [OH^-] = 1.00 \times 10^{-7} \text{ mol/L}$$

since both ions are produced in equal molar amounts, as shown in Equation (2).

The $H_2O \rightleftharpoons H^+ + OH^-$ equilibrium exists in water and in all water solutions. A special equilibrium constant called the **ion product constant for water, K_w,** applies to this equilibrium. The constant K_w is defined as the product of the H^+ ion concentration and the OH^- ion concentration, each in moles per liter:

ion product constant for water, K_w

$$K_w = [H^+][OH^-]$$

The numerical value of K_w is 1.00×10^{-14}, since for pure water at 25°C,

$$K_w = [H^+][OH^-] = (1.00 \times 10^{-7})(1.00 \times 10^{-7}) = 1.00 \times 10^{-14}$$

The value of K_w for all water solutions at 25°C is the constant 1.00×10^{-14}. It is important to realize that, as the concentration of one of these ions, H^+ or OH^-, increases, the other decreases. However, the product of $[H^+]$ and $[OH^-]$ always equals the constant 1.00×10^{-14}. This relationship can be seen in the examples shown in Table 16.2. If the concentration of one ion is known, the concentration of the other can be calculated from the K_w expression.

$$K_w = [H^+][OH^-] \qquad [H^+] = \frac{K_w}{[OH^-]} \qquad [OH^-] = \frac{K_w}{[H^+]}$$

Example 16.5

What is the concentration of (a) H^+ and (b) OH^- in a 0.001 *M* HCl solution? Remember that HCl is 100% ionized.

Solution

(a) Since all the HCl is ionized, $H^+ = 0.001$ mol/L or 1×10^{-3} mol/L:

$$\underset{0.001\,M}{HCl} \longrightarrow \underset{0.001\,M}{H^+} + \underset{0.001\,M}{Cl^-}$$

$$[H^+] = 1 \times 10^{-3} \text{ mol/L}$$

(b) To calculate the $[OH^-]$ in this solution, use the following equation and substitute the values for K_w and $[H^+]$:

$$[OH^-] = \frac{K_w}{[H^+]}$$

$$[OH^-] = \frac{1.00 \times 10^{-14}}{1 \times 10^{-3}} = 1 \times 10^{-11} \text{ mol/L}$$

Practice 16.6

Determine the $[H^+]$ and $[OH^-]$ in
(a) 5.0×10^{-5} M HNO_3 (b) 2.0×10^{-6} M KOH

Example 16.6 What is the pH of a 0.010 M NaOH solution? Assume that NaOH is 100% ionized.

Solution Since all the NaOH is ionized, $[OH^-] = 0.010$ mol/L or 1.0×10^{-2} mol/L.

$$\underset{0.00\,M}{\text{NaOH}} \longrightarrow \underset{0.010\,M}{\text{Na}^+} + \underset{0.010\,M}{\text{OH}^-}$$

To find the pH of the solution, we first calculate the H^+ ion concentration. Use the following equation and substitute the values for K_w and $[OH^-]$:

$$[H^+] = \frac{K_w}{[OH^-]} = \frac{1.00 \times 10^{-14}}{1.0 \times 10^{-2}} = 1.0 \times 10^{-12} \text{ mol/L}$$

$$\text{pH} = -\log[H^+] = -\log(1.0 \times 10^{-12}) = 12.00$$

Practice 16.7

Determine the pH for the following solutions:
(a) 5.0×10^{-5} M HNO_3 (b) 2.0×10^{-6} M KOH

Just as pH is used to express the acidity of a solution, pOH is used to express the basicity of an aqueous solution. The pOH is related to the OH^- ion concentration in the same way that the pH is related to the H^+ ion concentration.

$$\text{pOH} = -\log[OH^-]$$

Thus, a solution in which $[OH^-] = 1.0 \times 10^{-2}$, as in Example 16.6, will have pOH = 2.00.

In pure water, where $[H^+] = 1 \times 10^{-7}$ and $[OH^-] = 1 \times 10^{-7}$, the pH is 7.0, and the pOH is 7.0. The sum of the pH and pOH is always 14.0:

$$\text{pH} + \text{pOH} = 14.00$$

In Example 16.6 the pH can also be found by first calculating the pOH from the OH^- ion concentration and then subtracting from 14.00.

TABLE 16.2 Relationship of H^+ and OH^- Concentrations in Water Solutions

$[H^+]$	$[OH^-]$	K_w	pH	pOH
1.00×10^{-2}	1.00×10^{-12}	1.00×10^{-14}	2.00	12.00
1.00×10^{-4}	1.00×10^{-10}	1.00×10^{-14}	4.00	10.00
2.00×10^{-6}	5.00×10^{-9}	1.00×10^{-14}	5.70	8.30
1.00×10^{-7}	1.00×10^{-7}	1.00×10^{-14}	7.00	7.00
1.00×10^{-9}	1.00×10^{-5}	1.00×10^{-14}	9.00	5.00

$$\text{pH} = 14.00 - \text{pOH} = 14.00 - 2.00 = 12.00$$

Table 16.2 summarizes the relationship between $[H^+]$ and $[OH^-]$ in water solutions.

16.11 Ionization Constants

As a first application of an equilibrium constant, let us consider the constant for acetic acid in solution. Because it is a weak acid, an equilibrium is established between molecular $HC_2H_3O_2$ and its ions in solution. The constant is called the **acid ionization constant,** K_a, a special type of equilibrium constant. The concentration of water in the solution is large compared to other concentrations and does not change appreciably, so we use this simplified equation to set up the constant:

acid ionization constant, K_a

$$HC_2H_3O_2(aq) \rightleftharpoons H^+(aq) + C_2H_3O_2^-(aq)$$

The ionization constant expression is the concentration of the products divided by the concentration of the reactants:

$$K_a = \frac{[H^+][C_2H_3O_2^-]}{[HC_2H_3O_2]}$$

It states that the ionization constant, K_a, is equal to the product of the hydrogen ion $[H^+]$ concentration and the acetate ion $[C_2H_3O_2^-]$ concentration divided by the concentration of the un-ionized acetic acid $[HC_2H_3O_2]$.

At 25°C, a 0.100 *M* $HC_2H_3O_2$ solution is 1.34% ionized and has an $[H^+]$ of 1.34×10^{-3} mol/L. From this information we can calculate the ionization constant for acetic acid.

A 0.100 *M* solution initially contains 0.100 mol of acetic acid per liter. Of this, 0.100 mol, only 1.34%, or 1.34×10^{-3} mol, is ionized, which gives an $[H^+] = 1.34 \times 10^{-3}$ mol/L. Because each molecule of acid that ionizes yields one H^+ and one $C_2H_3O_2^-$, the concentration of $C_2H_3O_2^-$ ions is also 1.34×10^{-3} mol/L. This ionization leaves $0.100 - 0.00134 = 0.0987$ mol/L of un-ionized acetic acid.

Percent ionization of an acid, HA, is determined by dividing the concentration of H^+ or A^- ions at equilibrium by the initial concentration of HA.

Acid	Initial concentration (mol/L)	Equilibrium concentration (mol/L)
$[HC_2H_3O_2]$	0.100	0.0987
$[H^+]$	0	0.00134
$[C_2H_3O_2^-]$	0	0.00134

TABLE 16.3 Ionization Constants (K_a) of Weak Acids at 25°C

Acid	Formula	K_a	Acid	Formula	K_a
Acetic	$HC_2H_3O_2$	1.8×10^{-5}	Hydrocyanic	HCN	4.0×10^{-10}
Benzoic	$HC_7H_5O_2$	6.3×10^{-5}	Hypochlorous	HClO	3.5×10^{-8}
Carbolic (phenol)	HC_6H_5O	1.3×10^{-10}	Nitrous	HNO_2	4.5×10^{-4}
Cyanic	HCNO	2.0×10^{-4}	Hydrofluoric	HF	6.5×10^{-4}
Formic	$HCHO_2$	1.8×10^{-4}			

Substituting these concentrations in the equilibrium expression, we obtain the value for K_a:

$$K_a = \frac{[H^+][C_2H_3O_2^-]}{[HC_2H_3O_2]} = \frac{(1.34 \times 10^{-3})(1.34 \times 10^{-3})}{(0.0987)} = 1.82 \times 10^{-5}$$

The K_a for acetic acid, 1.82×10^{-5}, is small and indicates that the position of the equilibrium is far toward the un-ionized acetic acid. In fact, a 0.100 *M* acetic acid solution is 98.7% un-ionized.

Once the K_a for acetic acid is established, it can be used to describe other systems containing H^+, $C_2H_3O_2^-$, and $HC_2H_3O_2$ in equilibrium at 25°C. The ionization constants for several other weak acids are listed in Table 16.3.

Example 16.7 What is the $[H^+]$ in a 0.50 *M* $HC_2H_3O_2$ solution? The ionization constant, K_a, for $HC_2H_3O_2$ is 1.8×10^{-5}.

Solution To solve this problem, first write the equilibrium equation and the K_a expression:

$$HC_2H_3O_2 \rightleftharpoons H^+ + C_2H_3O_2^- \qquad K_a = \frac{[H^+][C_2H_3O_2^-]}{[HC_2H_3O_2]} = 1.8 \times 10^{-5}$$

We know that the initial concentration of $HC_2H_3O_2$ is 0.50 *M*. We also know from the ionization equation that one $C_2H_3O_2^-$ is produced for every H^+ produced; that is, the $[H^+]$ and the $[C_2H_3O_2^-]$ are equal. To solve, let $Y = [H^+]$, which also equals the $[C_2H_3O_2^-]$. The un-ionized $[HC_2H_3O_2]$ remaining will then be $0.50 - Y$, the starting concentration minus the amount that ionized:

$$[H^+] = [C_2H_3O_2^-] = Y \qquad [HC_2H_3O_2] = 0.50 - Y$$

Substituting these values into the K_a expression, we obtain

$$K_a = \frac{(Y)(Y)}{0.50 - Y} = \frac{Y^2}{0.50 - Y} = 1.8 \times 10^{-5}$$

The quadratic equation is

$$Y = \frac{-b \pm \sqrt{b^2 - 4ac}}{2a}$$

for the equation $ay^2 + by + c = 0$

An exact solution of this equation for Y requires the use of a mathematical equation known as the *quadratic equation*. However, an approximate solution is obtained if we assume that Y is small and can be neglected compared with 0.50. Then $0.50 - Y$ will be equal to approximately 0.50. The equation now becomes

$$\frac{Y^2}{0.50} = 1.8 \times 10^{-5}$$

$$Y^2 = 0.50 \times 1.8 \times 10^{-5} = 0.90 \times 10^{-5} = 9.0 \times 10^{-6}$$

Taking the square root of both sides of the equation, we obtain

$$Y = \sqrt{9.0 \times 10^{-6}} = 3.0 \times 10^{-3} \text{ mol/L}$$

Thus, the $[H^+]$ is approximately 3.0×10^{-3} mol/L in a 0.50 *M* $HC_2H_3O_2$ solution. The exact solution to this problem, using the quadratic equation, gives a value of 2.99×10^{-3} mol/L for $[H^+]$, showing that we were justified in neglecting Y compared with 0.50.

Practice 16.8

Calculate the hydrogen ion concentration in (a) 0.100 *M* hydrocyanic acid (HCN) solution and (b) 0.0250 *M* carbolic acid (HC_6H_5O) solution.

Example 16.8

Calculate the percent ionization in a 0.50 *M* $HC_2H_3O_2$ solution.

Solution

The percent ionization of a weak acid, $HA(aq) \rightleftharpoons H^+(aq) + A^-(aq)$, is found by dividing the concentration of the H^+ or A^- ions at equilibrium by the initial concentration of HA. For acetic acid,

$$\frac{\text{concentration of } [H^+] \text{ or } [C_2H_3O_2^-]}{\text{initial concentration of } [HC_2H_3O_2]} \times 100 = \text{percent ionized}$$

To solve this problem we first need to calculate the $[H^+]$. This calculation has already been done in Example 16.7 for a 0.50 *M* solution:

$$[H^+] = 3.0 \times 10^{-3} \text{ mol/L in a 0.50 } M \text{ solution} \qquad \text{(from Example 16.7)}$$

This $[H^+]$ represents a fractional amount of the initial 0.50 *M* $HC_2H_3O_2$. Therefore,

$$\frac{3.0 \times 10^{-3} \text{ mol/L}}{0.50 \text{ mol/L}} \times 100 = 0.60\% \text{ ionized}$$

A 0.50 *M* $HC_2H_3O_2$ solution is 0.60% ionized.

Practice 16.9

Calculate the percent ionization for
(a) 0.100 *M* hydrocyanic acid (HCN)
(b) 0.0250 *M* carbolic acid (HC_6H_5O)

16.12 Solubility Product Constant

solubility product constant, K_{sp}

The **solubility product constant**, abbreviated K_{sp}, is another application of the equilibrium constant. It is the equilibrium constant of a slightly soluble salt. The following example illustrates how K_{sp} is evaluated.

The solubility of AgCl in water is 1.3×10^{-5} mol/L at 25°C. The equation for the equilibrium between AgCl and its ions in solution is

$$AgCl(s) \rightleftharpoons Ag^+(aq) + Cl^-(aq)$$

The equilibrium constant expression is

$$K_{eq} = \frac{[Ag^+][Cl^-]}{[AgCl(s)]}$$

The amount of solid AgCl does not affect the equilibrium system provided that some is present. In other words, the concentration of solid AgCl is constant whether 1 mg or 10 g of the salt is present. Therefore, the product obtained by multiplying the two constants K_{eq} and $[AgCl(s)]$ is also a constant. This constant is the solubility product constant, K_{sp}:

$$K_{eq} \times [AgCl(s)] = [Ag^+][Cl^-] = K_{sp}$$

$$K_{sp} = [Ag^+][Cl^-]$$

The K_{sp} is equal to the product of the $[Ag^+]$ and the $[Cl^-]$, each in moles per liter. When 1.3×10^{-5} mol/L of AgCl dissolves, it produces 1.3×10^{-5} mol/L each of Ag^+ and Cl^-. From these concentrations the K_{sp} can be evaluated:

$$[Ag^+] = 1.3 \times 10^{-5} \text{ mol/L} \qquad [Cl^-] = 1.3 \times 10^{-5} \text{ mol/L}$$

$$K_{sp} = [Ag^+][Cl^-] = (1.3 \times 10^{-5})(1.3 \times 10^{-5}) = 1.7 \times 10^{-10}$$

Once the K_{sp} value for AgCl is established, it can be used to describe other systems containing Ag^+ and Cl^-.

The K_{sp} expression does not have a denominator. It consists only of the concentrations (mol/L) of the ions in solution. As in other equilibrium expressions, each of these concentrations is raised to a power that is the same number as its coefficient in the balanced equation. The equilibrium equations and the K_{sp} expressions for several other substances follow:

$$AgBr(s) \rightleftharpoons Ag^+(aq) + Br^-(aq) \qquad K_{sp} = [Ag^+][Br^-]$$

$$BaSO_4(s) \rightleftharpoons Ba^{2+}(aq) + SO_4^{2-}(aq) \qquad K_{sp} = [Ba^{2+}][SO_4^{2-}]$$

$$Ag_2CrO_4(s) \rightleftharpoons 2Ag^+(aq) + CrO_4^{2-}(aq) \qquad K_{sp} = [Ag^+]^2[CrO_4^{2-}]$$

$$CuS(s) \rightleftharpoons Cu^{2+}(aq) + S^{2-}(aq) \qquad K_{sp} = [Cu^{2+}][S^{2-}]$$

$$Mn(OH)_2(s) \rightleftharpoons Mn^{2+}(aq) + 2OH^-(aq) \qquad K_{sp} = [Mn^{2+}][OH^-]^2$$

$$Fe(OH)_3(s) \rightleftharpoons Fe^{3+}(aq) + 3OH^-(aq) \qquad K_{sp} = [Fe^{3+}][OH^-]^3$$

Table 16.4 lists K_{sp} values for these and several other substances.

When the product of the molar concentration of the ions in solution (each raised to its proper power) is greater than the K_{sp} for that substance, precipitation should occur. If the ion product is less than the K_{sp} value, no precipitation will occur.

Example 16.9 Write K_{sp} expressions for AgI and PbI_2, both of which are slightly soluble salts.

Solution First write the equilibrium equations:

$$AgI(s) \rightleftharpoons Ag^+(aq) + I^-(aq)$$

$$PbI_2(s) \rightleftharpoons Pb^{2+}(aq) + 2\,I^-(aq)$$

TABLE 16.4 Solubility Product Constants (K_{sp}) at 25°C

Compound	K_{sp}	Compound	K_{sp}
$AgCl$	1.7×10^{-10}	CaF_2	3.9×10^{-11}
$AgBr$	5×10^{-13}	CuS	9×10^{-45}
AgI	8.5×10^{-17}	$Fe(OH)_3$	6×10^{-38}
$AgC_2H_3O_2$	2×10^{-3}	PbS	7×10^{-29}
Ag_2CrO_4	1.9×10^{-12}	$PbSO_4$	1.3×10^{-8}
$BaCrO_4$	8.5×10^{-11}	$Mn(OH)_2$	2.0×10^{-13}
$BaSO_4$	1.5×10^{-9}		

Since the concentration of the solid crystals is constant, the K_{sp} equals the product of the molar concentrations of the ions in solution. In the case of PbI_2, the $[I^-]$ must be squared:

$$K_{sp} = [Ag^+][I^-] \qquad K_{sp} = [Pb^{2+}][I^-]^2$$

Example 16.10

The K_{sp} value for lead sulfate is 1.3×10^{-8}. Calculate the solubility of $PbSO_4$ in grams per liter.

Solution

First write the equilibrium equation and the K_{sp} expression:

$$PbSO_4 \rightleftharpoons Pb^{2+}(aq) + SO_4^{2-}(aq)$$

$$K_{sp} = [Pb^{2+}][SO_4^{2-}] = 1.3 \times 10^{-8}$$

Since the lead sulfate that is in solution is completely dissociated, the $[Pb^{2+}]$ or $[SO_4^{2-}]$ is equal to the solubility of $PbSO_4$ in moles per liter. Let

$$Y = [Pb^{2+}] = [SO_4^{2-}]$$

Substitute Y into the K_{sp} equation and solve:

$$[Pb^{2+}][SO_4^{2-}] = (Y)(Y) = 1.3 \times 10^{-8}$$

$$Y^2 = 1.3 \times 10^{-8}$$

$$Y = 1.1 \times 10^{-4} \text{ mol/L}$$

The solubility of $PbSO_4$, therefore, is 1.1×10^{-4} mol/L. Now convert mol/L to g/L:

1 mol of $PbSO_4$ has a mass of (207.2 g + 32.06 g + 64.00 g) or 303.3 g

$$\frac{1.1 \times 10^{-4} \cancel{\text{mol}}}{\text{L}} \times \frac{303.3 \text{ g}}{\cancel{\text{mol}}} = 3.3 \times 10^{-2} \text{ g/L}$$

The solubility of $PbSO_4$ is 3.3×10^{-2} g/L.

Practice 16.10

Write the K_{sp} expression for
(a) $Cr(OH)_3$ (b) $Cu_3(PO_4)_2$

Practice 16.11

The K_{sp} value for CuS is 9.0×10^{-45}. Calculate the solubility of CuS in grams per liter.

An ion added to a solution that already contains that ion is called a *common ion.* When a common ion is added to an equilibrium solution of a weak electrolyte or a slightly soluble salt, the equilibrium shifts according to Le Chatelier's principle. For example, when silver nitrate, $AgNO_3$, is added to a saturated solution of AgCl, ($AgCl(s) \rightleftharpoons Ag^+ + Cl^-$), the equilibrium shifts to the left due to the increase in the $[Ag^+]$. As a result, the $[Cl^-]$ and the solubility of AgCl decreases. The AgCl and $AgNO_3$ have the common ion Ag^+. A shift in the equilibrium position upon addition of an ion already contained in the solution is known as the **common ion effect**.

common ion effect

Example 16.11 Silver nitrate is added to a saturated AgCl solution until the $[Ag^+]$ is 0.10 *M*. What will be the $[Cl^-]$ remaining in solution?

Solution This problem is an example of the common ion effect. The addition of $AgNO_3$ puts more Ag^+ in solution; the Ag^+ combines with Cl^- and causes the equilibrium to shift to the left, reducing the $[Cl^-]$ in solution. After the addition of Ag^+ to the mixture, the $[Ag^+]$ and $[Cl^-]$ in solution are no longer equal.

We use the K_{sp} to calculate the $[Cl^-]$ remaining in solution. The K_{sp} is constant at a particular temperature and remains the same no matter how we change the concentration of the species involved:

$$K_{sp} = [Ag^+][Cl^-] = 1.7 \times 10^{-10} \qquad [Ag^+] = 0.10 \text{ mol/L}$$

We then substitute the $[Ag^+]$ into the K_{sp} expression and calculate the $[Cl^-]$:

$$[0.10][Cl^-] = 1.7 \times 10^{-10}$$

$$[Cl^-] = \frac{1.7 \times 10^{-10}}{0.10} = 1.7 \times 10^{-9} \text{ mol/L}$$

This calculation shows a 10,000-fold reduction of Cl^- ions in solution. It illustrates that Cl^- ions may be quantitatively removed from solution with an excess of Ag^+ ions.

Practice 16.12

Sodium sulfate, Na_2SO_4, is added to a saturated solution of $BaSO_4$ until the concentration of sulfate ion is 2.0×10^{-2} *M*. What will be the concentration of the Ba^{2+} ions remaining in solution?

TABLE 16.5 Ionic Composition of Salts and the Nature of the Aqueous Solutions They Form

Type of salt	Nature of aqueous solution	Examples
Weak base–strong acid	Acidic	NH_4Cl, NH_4NO_3
Strong base–weak acid	Basic	$NaC_2H_3O_2$, K_2CO_3
Weak base–weak acid	Depends on the salt	$NH_4C_2H_3O_2$, NH_4NO_2
Strong base–strong acid	Neutral	NaCl, KBr

16.13 Hydrolysis

Hydrolysis is the term used for the general reaction in which a water molecule is split. For example, the net ionic hydrolysis reaction for a sodium acetate solution is

hydrolysis

$$C_2H_3O_2^-(aq) + H_2O(l) \rightleftharpoons HC_2H_3O_2(aq) + OH^-(aq)$$

In this reaction the water molecule is split, with the H^+ combining with $C_2H_3O_2^-$ to give the weak acid $HC_2H_3O_2$ and the OH^- going into solution, making the solution more basic.

Salts that contain an ion of a weak acid undergo hydrolysis. For example, a 0.10 *M* NaCN solution has a pH of 11.1. The hydrolysis reaction that causes this solution to be basic is

$$CN^-(aq) + H_2O(l) \rightleftharpoons HCN(aq) + OH^-(aq)$$

If a salt contains the ion of a weak base, the ion hydrolyzes to produce an acidic solution. An example is ammonium chloride, which produces the NH_4^+ and Cl^- in solution. The NH_4^+ hydrolyzes to produce an acidic solution:

$$NH_4^+(aq) + H_2O(l) \longrightarrow NH_3(aq) + H_3O^+(aq)$$

The ions of a salt derived from a strong acid and a strong base, such as NaCl, do not undergo hydrolysis and thus form neutral solutions. Table 16.5 lists the ionic composition of various salts and the nature of the aqueous solutions that they form.

Practice 16.13

Indicate whether each of the following salts would produce an acidic, basic, or neutral aqueous solution.
(a) KCN (b) $NaNO_3$ (c) NH_4Br

16.14 Buffer Solutions: The Control of pH

The control of pH within narrow limits is critically important in many chemical applications and vitally important in many biological systems. For example, human

A saltwater aquarium is a buffer system. ▶

blood must be maintained between pH 7.35 and 7.45 for the efficient transport of oxygen from the lungs to the cells. This narrow pH range is maintained by buffer systems in the blood.

buffer solution

A **buffer solution** resists changes in pH when diluted or when small amounts of acid or base are added. Two common types of buffer solutions are (1) a weak acid mixed with its conjugate base and (2) a weak base mixed with its conjugate acid.

The action of a buffer system can be understood by considering a solution of acetic acid and sodium acetate. The weak acid, $HC_2H_3O_2$, is mostly un-ionized and is in equilibrium with its ions in solution. The sodium acetate is completely ionized:

$$HC_2H_3O_2(aq) \rightleftarrows H^+(aq) + C_2H_3O_2^-(aq)$$

$$NaC_2H_3O_2(aq) \longrightarrow Na^+(aq) + C_2H_3O_2^-(aq)$$

Because the sodium acetate is completely ionized, the solution contains a much higher concentration of acetate ions than would be present if only acetic acid were in solution. The acetate ion represses the ionization of acetic acid and also reacts with water, causing the solution to have a higher pH (be more basic) than an acetic acid solution (see Section 16.5). Thus, a 0.1 *M* acetic acid solution has a pH of 2.87, but a solution that is 0.1 *M* in acetic acid and 0.1 *M* in sodium acetate has a pH of 4.74. This difference in pH is the result of the common ion effect.

A buffer solution has a built-in mechanism that counteracts the effect of adding acid or base. Consider the effect of adding HCl or NaOH to an acetic acid–sodium acetate buffer. When a small amount of HCl is added, the acetate ions of the buffer combine with the H^+ ions from HCl to form un-ionized acetic acid, thus neutralizing the added acid and maintaining the approximate pH of the solution. When NaOH is added, the OH^- ions react with acetic acid to neutralize the added base and thus maintain the approximate pH. The equations for these reactions are

$$H^+(aq) + C_2H_3O_2^-(aq) \rightleftarrows HC_2H_3O_2(aq)$$

$$OH^-(aq) + HC_2H_3O_2(aq) \rightleftarrows H_2O(l) + C_2H_3O_2^-(aq)$$

Data comparing the changes in pH caused by adding HCl and NaOH to pure water and to an acetic acid–sodium acetate buffer solution are shown in Table 16.6.

The human body has a number of buffer systems. One of these, the hydrogen carbonate–carbonic acid buffer, $HCO_3^- - H_2CO_3$, maintains the blood plasma at a pH of 7.4. The phosphate system, $HPO_4^{2-} - H_2PO_4^-$, is an important buffer in the red blood cells as well as in other places in the body.

Exchange of Oxygen and Carbon Dioxide in the Blood

The transport of oxygen and carbon dioxide between the lungs and tissues is a complex process that involves several reversible reactions, each of which behaves in accordance with Le Chatelier's principle.

The binding of oxygen to hemoglobin is a reversible reaction. The oxygen molecule must attach to the hemoglobin (Hb) and then later detach. The equilibrium equation for this reaction can be written:

$$Hb + O_2 \rightleftharpoons HbO_2$$

In the lungs, the concentration of oxygen is high and favors the forward reaction. Oxygen quickly binds to the hemoglobin until it is saturated with oxygen.

In the tissues the concentration of oxygen is lower and in accordance with Le Chatelier's principle: The equilibrium position shifts to the left and the hemoglobin releases oxygen to the tissues. Approximately 45% of the oxygen diffuses out of the capillaries into the tissues where it may be picked up by *myoglobin,* another carrier molecule.

Myoglobin functions as an oxygen storage molecule, holding the oxygen until it is required in the energy-producing portions of the cell. The reaction between myoglobin (Mb) and oxygen can be written as an equilibrium reaction:

$$Mb + O_2 \rightleftharpoons MbO_2$$

Since both the hemoglobin and myoglobin equations are so similar, what accounts for the transfer of the oxygen from the hemoglobin to the myoglobin? Although both equilibria involve similar interactions the affinity between oxygen and hemoglobin is different from the affinity

Oxygen and carbon dioxide are exchanged in the red blood cells when they are in capillaries.

between myoglobin and oxygen. In the tissues, the position of the hemoglobin equilibrium is such that it is 55% saturated with oxygen, whereas the myoglobin is at 90% oxygen saturation. Under these conditions hemoglobin will release oxygen while myoglobin will bind oxygen. Thus, oxygen is loaded onto hemoglobin in the lungs and unloaded in the tissue's cells.

Carbon dioxide produced in the cells must be removed from the tissues. Oxygen-depleted hemoglobin molecules accomplish this by becoming carriers of carbon dioxide. The carbon dioxide does not bind at the heme site as the oxygen does, but rather at one end of the protein chain. When carbon dioxide dissolves in water some of the CO_2 reacts to release hydrogen ions:

$$CO_2 + H_2O \rightleftharpoons HCO_3^- + H^+$$

In order to facilitate the removal of CO_2 from the tissues this equilibrium needs to be moved toward the right. This shift is accomplished by the removal of H^+ from the tissues by the hemoglobin molecule. The deoxygenated hemoglobin molecule can bind H^+ ions as well as CO_2. In the lungs this whole process is reversed so the CO_2 is removed from the hemoglobin and exhaled.

Molecules that are similar in structure to the oxygen molecule can become involved in competing equilibria. Hemoglobin is capable of binding with carbon monoxide, CO, nitrogen monoxide, NO, and cyanide, CN^-. The extent of the competition depends on the affinity. Since these molecules have a greater affinity for hemoglobin than oxygen they will effectively displace oxygen from hemoglobin. For example,

$$HbO_2 + CO \rightleftharpoons HbCO + O_2$$

Since the affinity of hemoglobin for CO is 150 times stronger than its affinity for oxygen, the equilibrium position lies far to the right. This explains why CO is a poisonous substance and why oxygen is administered to victims of CO poisoning. The hemoglobin molecules can only transport oxygen if the CO is released and the oxygen shifts the equilibrium toward the left.

TABLE 16.6 Changes in pH Caused by the Addition of HCl and NaOH

Solution	pH	Change in pH
H_2O (1000 mL)	7	—
H_2O + 0.010 mol HCl	2	5
H_2O + 0.010 mol NaOH	12	5
Buffer solution (1000 mL)		
0.10 *M* $HC_2H_3O_2$ + 0.10 *M* $NaC_2H_3O_2$	4.74	—
Buffer + 0.010 mol HCl	4.66	0.08
Buffer + 0.010 mol NaOH	4.83	0.09

16.15 Mechanism of Reactions

mechanism of a reaction

How a reaction occurs—that is, the manner in which it proceeds—is known as the **mechanism of the reaction**. The mechanism is the path, or route, the atoms and molecules take to arrive at the products. Our aim here is not to study the mechanisms themselves but to show that chemical reactions occur by specific routes.

activated complex

activation energy

When hydrogen and iodine are mixed at room temperature, we observe no appreciable reaction. In this case, the reaction takes place as a result of collisions between H_2 and I_2 molecules, but at room temperature the collisions do not result in reaction because the molecules lack sufficient energy to react. We say that an energy barrier to reaction exists. If heat is added, the kinetic energy of the molecules increases. When molecules of H_2 and I_2 with sufficient energy collide, an intermediate product, known as the **activated complex**, is formed. The amount of energy needed to form the activated complex is known as the **activation energy**. The activated complex, H_2I_2, is in a metastable form and has an energy level higher than that of the reactants or the product. It can decompose to form either the reactants or the product. Three steps constitute the mechanism of the reaction: (1) collision of an H_2 and an I_2 molecule; (2) formation of the activated complex, H_2I_2; and (3) decomposition to the product, HI. The various steps in the formation of HI are shown in Figure 16.3. Figure 16.4 illustrates the energy relationships in this reaction.

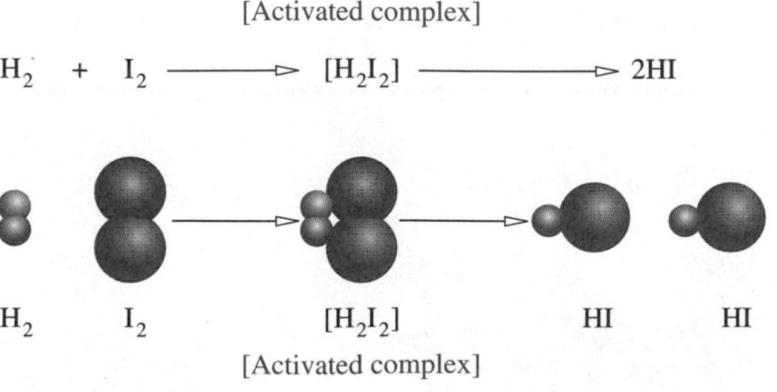

FIGURE 16.3 ▶ Mechanism of the reaction between hydrogen and iodine. H_2 and I_2 molecules of sufficient energy unite, forming the intermediate activated complex that decomposes to the product, HI.

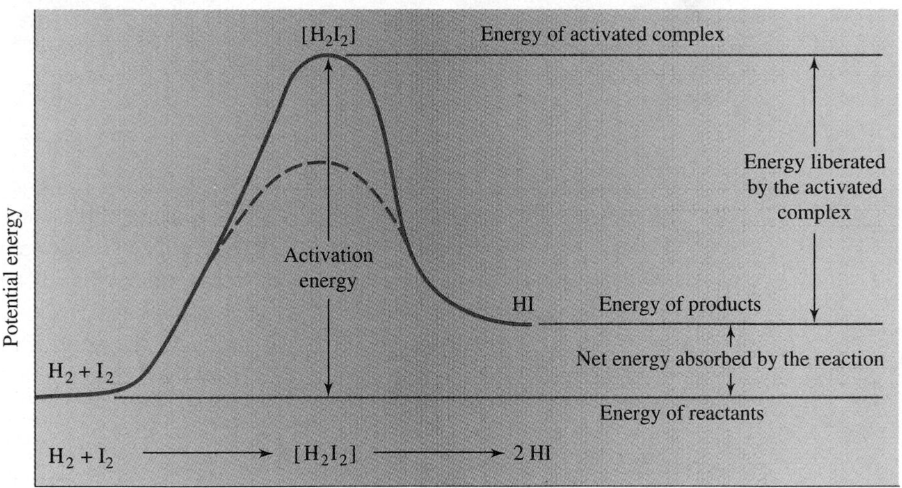

FIGURE 16.4
Relative energy diagram for the reaction between hydrogen and iodine. Energy equal to the activation energy is put into the system to form the activated complex, H_2I_2. When this complex decomposes, it liberates energy, forming the product. In this case, the product is at a higher energy level than the reactants, indicating that the reaction is endothermic and that energy is absorbed during the reaction. The dashed line represents the effect that a catalyst would have on the reaction. The catalyst lowers the activation energy, thereby increasing the rate of the reaction.

The reaction of hydrogen and chlorine proceeds by a different mechanism. When H_2 and Cl_2 are mixed and kept in the dark, essentially no product is formed. But, if the mixture is exposed to sunlight or ultraviolet radiation, it reacts very rapidly. The overall reaction is

$$H_2(g) + Cl_2(g) \longrightarrow 2\ HCl(g)$$

free radical

This reaction proceeds by what is known as a *free radical mechanism.* A **free radical** is a neutral atom or group of atoms containing one or more unpaired electrons. Both atomic chlorine ($:\ddot{\underset{..}{Cl}}\cdot$) and atomic hydrogen ($H\cdot$) have an unpaired electron and are free radicals. The reaction occurs in three steps.

Step 1 *Initiation:*

$$:\ddot{\underset{..}{Cl}}:\ddot{\underset{..}{Cl}}: + h\nu \longrightarrow :\ddot{\underset{..}{Cl}}\cdot + :\ddot{\underset{..}{Cl}}\cdot$$

chlorine free radicals

In this step a chlorine molecule absorbs energy in the form of a photon, $h\nu$, of light or ultraviolet radiation. The energized chlorine molecule then splits into two chlorine free radicals.

Step 2 *Propagation:*

$$:\ddot{\underset{..}{Cl}}\cdot + H:H \longrightarrow HCl + H\cdot$$

hydrogen free radical

$$H\cdot + :\ddot{\underset{..}{Cl}}:\ddot{\underset{..}{Cl}}: \longrightarrow HCl + :\ddot{\underset{..}{Cl}}\cdot$$

This step begins when a chlorine free radical reacts with a hydrogen molecule to produce a molecule of hydrogen chloride and a hydrogen free radical. The hydrogen radical then reacts with another chlorine molecule to form hydrogen chloride and another chlorine free radical. This chlorine free radical can repeat the process by reacting with another hydrogen molecule, and the reaction continues to propagate itself in this manner until one or both of the reactants are used up. Almost all of the product is formed in this step.

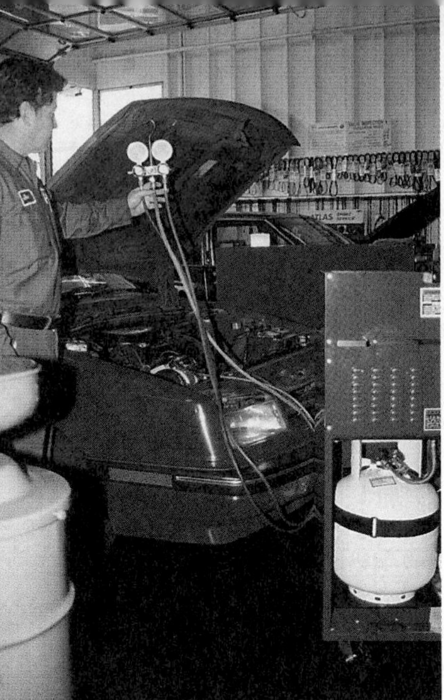

The freon used in automobile air conditioners can be converted into chlorine free radicals in the upper atmosphere. These free radicals play a key role in creating the hole in the ozone layer.

Step 3 *Termination:*

$$:\ddot{\underset{..}{Cl}}\cdot + :\ddot{\underset{..}{Cl}}\cdot \longrightarrow Cl_2$$

$$H\cdot + H\cdot \longrightarrow H_2$$

$$H\cdot + :\ddot{\underset{..}{Cl}}\cdot \longrightarrow HCl$$

Hydrogen and chlorine free radicals can react in any of the three ways shown. Unless further activation occurs, the formation of hydrogen chloride will terminate when the free radicals form molecules. In an exothermic reaction such as that between hydrogen and chlorine, usually enough heat and light energy is available to maintain the supply of free radicals, and the reaction will continue until at least one reactant is exhausted.

Concepts in Review

1. Describe a reversible reaction.
2. Explain why the rate of the forward reaction decreases and the rate of the reverse reaction increases as a chemical reaction approaches equilibrium.
3. Describe the qualitative effect of Le Chatelier's principle.
4. Predict how the rate of a chemical reaction is affected by (a) changes in the concentration of reactants, (b) changes in pressure of gaseous reactants, (c) changes in temperature, and (d) the presence of a catalyst.
5. Write the equilibrium constant expression for a chemical reaction from a balanced chemical equation.
6. Explain the meaning of the numerical constant, K_{eq}, when given the concentration of the reactants and products in equilibrium.
7. Calculate the concentration of one substance in equilibrium when given the equilibrium constant and the concentrations of all the other substances.
8. Calculate the equilibrium constant, K_{eq}, when given the concentration of reactants and products in equilibrium.
9. Calculate the ionization constant for a weak acid from appropriate data.
10. Calculate the concentrations of all the chemical species in a solution of a weak acid when given the percent ionization or the ionization constant.
11. Compare the relative strengths of acids by using their ionization constants.
12. Use the ion product constant for water, K_w, to calculate $[H^+]$, $[OH^-]$, pH, and pOH when given any one of these quantities.
13. Calculate the solubility product constant, K_{sp}, of a slightly soluble salt when given its solubility, or vice versa.
14. Compare relative solubilities of salts if solubility products are known.
15. Discuss the common ion effect on a system at equilibrium.
16. Explain hydrolysis and why some salts form acidic or basic aqueous solutions.

17. Explain how a buffer solution is able to counteract the addition of small amounts of either H^+ or OH^- ions.
18. Draw the relative energy diagram of an exothermic or endothermic reaction. Label the activation energy, and show the effect of a catalyst.

Key Terms

The terms listed here have been defined within this chapter. Section numbers are referenced in parenthesis for each term.

acid ionization constant, K_a (16.11)
activated complex (16.15)
activation energy (16.15)
buffer solution (16.14)
catalyst (16.8)
chemical equilibrium (16.3)
chemical kinetics (16.2)
common ion effect (16.12)
equilibrium (16.3)
equilibrium constant, K_{eq} (16.9)
free radical (16.15)
hydrolysis (16.13)
ion product constant for water, K_w (16.10)
mechanism of a reaction (16.15)
principle of Le Chatelier (16.4)
reversible chemical reaction (16.1)
solubility product constant, K_{sp} (16.12)

Questions

Questions refer to tables, figures, and key words and concepts defined within the chapter. A particularly challenging question or exercise is indicated with an asterisk.

1. How would you expect the two tubes in Figure 16.1 to appear if both are at 25°C?
2. Is the reaction $N_2O_4 \rightleftarrows 2\,NO_2$ exothermic or endothermic? (See Figure 16.1)
3. At equilibrium how do the forward and reverse reaction rates compare? (See Figure 16.2)
4. Would the reaction of 30 mol of H_2 and 10 mol of N_2 produce a greater yield of NH_3 if carried out in a 1-L or a 2-L vessel? (Table 16.1)
5. For each of the solutions in Table 16.2, what is the sum of the pH plus the pOH? What would be the pOH of a solution whose pH was -1?
6. Of the acids listed in Table 16.3, which ones are stronger than acetic acid and which are weaker?
7. Tabulate the relative order of molar solubilities of AgCl, AgBr, AgI, $AgC_2H_3O_2$, $PbSO_4$, $BaSO_4$, $BaCrO_4$, and PbS. List the most soluble first. (Use Table 16.4.)
8. Which compound in each of the following pairs has the greater molar solubility? (See Table 16.4)
   (a) $Mn(OH)_2$ or Ag_2CrO_4
   (b) $BaCrO_4$ or Ag_2CrO_4
9. Using Table 16.6, explain how the acetic acid–sodium acetate buffer system maintains its pH when 0.010 mol of HCl is added to 1 L of the buffer solution.
10. How would Figure 16.4 be altered if the reaction were exothermic?
11. Explain why a precipitate of NaCl forms when HCl gas is passed into a saturated aqueous solution of NaCl.
12. Why does the rate of a reaction usually increase when the concentration of one of the reactants is increased?
13. If pure hydrogen iodide, HI, is placed in a vessel at 700 K, will it decompose? Explain.
14. Why does an increase in temperature cause the rate of reaction to increase?
15. Give a word description of how equilibrium is reached when the substances A and B are first mixed and react as

$$A + B \rightleftarrows C + D$$

16. With dilution, aqueous solutions of acetic acid, $HC_2H_3O_2$, show increased ionization. For example, a 1.0 *M* solution of acetic acid is 0.42% ionized, whereas a 0.10 *M* solution is 1.34% ionized. Explain the behavior using the ionization equation and equilibrium principles.
17. A 1.0 *M* solution of acetic acid ionizes less and has a higher concentration of H^+ ions than a 0.10 *M* acetic acid solution. Explain this behavior. (See Question 16 for data.)
18. What would cause two separate samples of pure water to have slightly different pH values?

19. Why are the pH and pOH equal in pure water?

20. Explain why silver acetate is more soluble in nitric acid than in water. [*Hint:* Write the equilibrium equation first and then consider the effect of the acid on the acetate ion.] What would happen if hydrochloric acid were used in place of nitric acid?

21. Dissolution of sodium acetate, $NaC_2H_3O_2$, in pure water gives a basic solution. Why? [*Hint:* A small amount of $HC_2H_3O_2$ is formed.]

22. Describe why the pH of a buffer solution remains almost constant when a small amount of acid or base is added to it.

23. How does Le Chatelier's principle explain the transport of oxygen from the lungs to the tissues?

24. State the similarities and differences between hemoglobin and myoglobin.

25. Explain how the removal of carbon dioxide from the tissues is facilitated and accomplished by the hemoglobin molecule.

26. Which of the following statements are correct? Rewrite the incorrect statements to make them correct.

(a) In a reaction at equilibrium the concentrations of reactants and products are equal.

(b) A catalyst increases the concentration of products present at equilibrium.

(c) Enzymes are the catalysts in living systems.

(d) A catalyst lowers the activation energy of a reaction by equal amounts for both the forward and the reverse reactions.

(e) If an increase in temperature causes an increase in the concentration of products present at equilibrium, the reaction is exothermic.

(f) The magnitude of an equilibrium constant is independent of the reaction temperature.

(g) A large equilibrium constant for a reaction indicates that the reaction, at equilibrium, favors products over reactants.

(h) The amount of product obtained at equilibrium is proportional to how fast equilibrium is attained.

(i) The study of reaction rates and reaction mechanisms is known as chemical kinetics.

(j) For the reaction

$$CaCO_3(s) \overset{\Delta}{\rightleftharpoons} CaO(s) + CO_2(g)$$

increasing the pressure of CO_2 present at equilibrium will cause the reaction to shift left.

(k) The larger the value of the equilibrium constant, the greater the proportion of products present at equilibrium.

(l) At chemical equilibrium, the rate of the reverse reaction is equal to the rate of the forward reaction.

Statements (m)–(s) pertain to the following equilibrium system:

$$2\,NO(g) + O_2(g) \rightleftharpoons 2\,NO_2(g) + \text{heat}$$

(m) The reaction as shown is endothermic.

(n) Increasing the temperature will cause the equilibrium to shift left.

(o) Increasing the temperature will increase the magnitude of the equilibrium constant, K_{eq}.

(p) Decreasing the volume of the reaction vessel will shift the equilibrium to the right and decrease the concentrations of the reactants.

(q) Removal of some of the O_2 will cause an increase in the concentration of NO.

(r) High temperatures and pressures favor increased yields of NO_2.

(s) The equilibrium constant expression for the reaction is

$$K_{eq} = \frac{[NO_2]^2}{[NO]^2[O_2]}$$

(t) A solution with an $[H^+]$ of 1×10^{-5} mol/L has a pOH of 9.

(u) An aqueous solution that has an $[OH^-]$ of 1×10^{-4} mol/L has an $[H^+]$ of 1×10^{-10} mol/L.

(v) $K_w = [H^+][OH^-] = 1.00 \times 10^{-14}$, and pH + pOH = 14.

(w) As solid $BaSO_4$ is added to a saturated solution of $BaSO_4$, the magnitude of its K_{sp} increases.

(x) A solution of pOH 10 is basic.

(y) The pH increases as the $[H^+]$ increases.

(z) The pH of 0.0500 *M* $Ca(OH)_2$ is 13.

Paired Exercises

These exercises are paired. Each odd-numbered exercise is followed by a similar even-numbered exercise. Answers to the even-numbered exercises are given in Appendix V.

27. Express the following reversible systems in equation form:
(a) a mixture of ice and liquid water at 0°C
(b) crystals of Na_2SO_4 in a saturated aqueous solution of Na_2SO_4

28. Express the following reversible systems in equation form:
(a) liquid water and vapor at 100°C in a pressure cooker
(b) a closed system containing boiling sulfur dioxide, SO_2

29. Consider the following system at equilibrium:

$$4\ NH_3(g) + 3\ O_2(g) \rightleftharpoons 2\ N_2(g) + 6\ H_2O(g) + 1531\ \text{kJ}$$

(a) Is the reaction exothermic or endothermic?
(b) If the system's state of equilibrium is disturbed by the addition of O_2, in which direction, left or right, must the reaction occur to reestablish equilibrium? After the new equilibrium has been established, how will the final molar concentrations of NH_3, O_2, N_2, and H_2O compare (increase or decrease) with their concentrations before the addition of the O_2?

30. Consider the following system at equilibrium:

$$4\ NH_3(g) + 3\ O_2(g) \rightleftharpoons 2\ N_2(g) + 6\ H_2O(g) + 1531\ \text{kJ}$$

(a) If the system's state of equilibrium is disturbed by the addition of N_2, in which direction, left or right, must the reaction occur to reestablish equilibrium? After the new equilibrium has been established, how will the final molar concentrations of NH_3, O_2, N_2, and H_2O compare (increase or decrease) with their concentrations before the addition of the N_2?
(b) If the system's state of equilibrium is disturbed by the addition of heat, in which direction will the reaction occur, left or right, to reestablish equilibrium?

31. Consider the following system at equilibrium:

$$N_2(g) + 3\ H_2(g) \rightleftharpoons 2\ NH_3(g) + 92.5\ \text{kJ}$$

Complete the following table. Indicate changes in moles by entering I, D, N, or ? in the table. (I = increase, D = decrease, N = no change, ? = insufficient information to determine.)

Change of stress imposed on the system at equilibrium	Direction of reaction, left or right, to reestablish equilibrium	Change in number of moles		
		N_2	H_2	NH_3
(a) Add N_2				
(b) Remove H_2				
(c) Decrease volume of reaction vessel				
(d) Increase temperature				

32. Consider the following system at equilibrium:

$$N_2(g) + 3\ H_2(g) \rightleftharpoons 2\ NH_3(g) + 92.5\ \text{kJ}$$

Complete the following table. Indicate changes in moles by entering I, D, N, or ? in the table. (I = increase, D = decrease, N = no change, ? = insufficient information to determine.)

Change of stress imposed on the system at equilibrium	Direction of reaction, left or right, to reestablish equilibrium	Change in number of moles		
		N_2	H_2	NH_3
(a) Add NH_3				
(b) Increase volume of reaction vessel				
(c) Add catalyst				
(d) Add both H_2 and NH_3				

33. For each of the equations that follow, tell in which direction, left or right, the equilibrium will shift when the following changes are made: The temperature is increased; the pressure is increased by decreasing the volume of the reaction vessel; a catalyst is added.
(a) $3\ O_2(g) + 271\ \text{kJ} \rightleftharpoons 2\ O_3(g)$
(b) $CH_4(g) + Cl_2(g) \rightleftharpoons CH_3Cl(g) + HCl(g) + 110\ \text{kJ}$
(c) $2\ NO(g) + 2\ H_2(g) \rightleftharpoons N_2(g) + 2\ H_2O(g) + 665\ \text{kJ}$

34. For each of the equations that follow, tell in which direction, left or right, the equilibrium will shift when the following changes are made: The temperature is increased; the pressure is increased by decreasing the volume of the reaction vessel; a catalyst is added.
(a) $2\ SO_3(g) + 197\ \text{kJ} \rightleftharpoons 2\ SO_2(g) + O_2(g)$
(b) $4\ NH_3(g) + 3\ O_2(g) \rightleftharpoons 2\ N_2(g) + 6\ H_2O(g) + 1531\ \text{kJ}$
(c) $OF_2(g) + H_2O(g) \rightleftharpoons O_2(g) + 2\ HF(g) + 318\ \text{kJ}$

35. Utilizing Le Chatelier's principle, indicate the shift (if any) that would occur to

$C_2H_6(g) + \text{heat} \rightleftharpoons C_2H_4(g) + H_2(g)$

(a) if the concentration of hydrogen gas is decreased
(b) if the temperature is lowered
(c) if a catalyst is added

36. Utilizing Le Chatelier's principle, indicate the shift (if any) that would occur to

$C_2H_6(g) + \text{heat} \rightleftharpoons C_2H_4(g) + H_2(g)$

(a) if C_2H_6 is removed from the system
(b) if the volume of the container is increased
(c) if the temperature is raised

37. Write the equilibrium constant expression for each of the following reactions:
(a) $4\ HCl(g) + O_2(g) \rightleftharpoons 2\ Cl_2(g) + 2\ H_2O(g)$
(b) $N_2(g) + 3\ H_2(g) \rightleftharpoons 2\ NH_3(g)$
(c) $PCl_5(g) \rightleftharpoons PCl_3(g) + Cl_2(g)$

38. Write the equilibrium constant expression for each of the following reactions:
(a) $HClO_2(aq) \rightleftharpoons H^+(aq) + ClO_2^-(aq)$
(b) $HC_2H_3O_2(aq) \rightleftharpoons H^+(aq) + C_2H_3O_2^-(aq)$
(c) $4\ NH_3(g) + 5\ O_2(g) \rightleftharpoons 4\ NO(g) + 6\ H_2O(g)$

39. Write the solubility product expression, K_{sp}, for each of the following substances:
(a) CuS **(c)** $PbBr_2$
(b) $BaSO_4$ **(d)** Ag_3AsO_4

40. Write the solubility product expression, K_{sp}, for each of the following substances:
(a) $Fe(OH)_3$ **(c)** CaF_2
(b) Sb_2S_5 **(d)** $Ba_3(PO_4)_2$

41. What effect will decreasing the $[H^+]$ of a solution have on (a) pH, (b) pOH, (c) $[OH^-]$, and (d) K_w?

42. What effect will increasing the $[H^+]$ of a solution have on (a) pH, (b) pOH, (c) $[OH^-]$, and (d) K_w?

43. Decide whether each of the following salts forms an acidic, basic, or neutral aqueous solution when dissolved in water:
(a) KCl **(c)** K_2SO_4
(b) Na_2CO_3 **(d)** $(NH_4)_2SO_4$

44. Decide whether each of the following salts forms an acidic, basic, or neutral aqueous solution when dissolved in water:
(a) $Ca(CN)_2$ **(c)** $NaNO_2$
(b) $BaBr_2$ **(d)** NaF

45. Write hydrolysis equations for aqueous solutions of these salts:
(a) KNO_2
(b) $Mg(C_2H_3O_2)_2$

46. Write hydrolysis equations for aqueous solutions of these salts:
(a) NH_4NO_3
(b) Na_2SO_3

47. Write hydrolysis equations for the following ions:
(a) HCO_3^-
(b) NH_4^+

48. Write hydrolysis equations for the following ions:
(a) OCl^-
(b) ClO_2^-

49. One of the important pH-regulating systems in the blood consists of a carbonic acid–sodium hydrogen carbonate buffer:

$H_2CO_3(aq) \rightleftharpoons H^+(aq) + HCO_3^-(aq)$
$NaHCO_3\ (aq) \longrightarrow Na^+(aq) + HCO_3^-(aq)$

Explain how this buffer resists changes in pH when excess acid, H^+, gets into the bloodstream.

50. One of the important pH-regulating systems in the blood consists of a carbonic acid–sodium hydrogen carbonate buffer:

$H_2CO_3(aq) \rightleftharpoons H^+(aq) + HCO_3^-(aq)$
$NaHCO_3(aq) \longrightarrow Na^+(aq) + HCO_3^-(aq)$

Explain how this buffer resists changes in pH when excess base, OH^-, gets into the bloodstream.

51. Calculate (a) the $[H^+]$, (b) the pH, and (c) the percent ionization of a 0.25 *M* solution of $HC_2H_3O_2$. ($K_a = 1.8 \times 10^{-5}$)

52. Calculate (a) the $[H^+]$, (b) the pH, and (c) the percent ionization of a 0.25 *M* solution of phenol, HC_6H_5O. ($K_a = 1.3 \times 10^{-10}$)

53. A 1.000 *M* solution of a weak acid, HA, is 0.52% ionized. Calculate the ionization constant, K_a, for the acid.

54. A 0.15 *M* solution of a weak acid, HA, has a pH of 5. Calculate the ionization constant, K_a, for the acid.

55. Calculate the percent ionization and pH of solutions of $HC_2H_3O_2$ ($K_a = 1.8 \times 10^{-5}$) having the following molarities: (a) 1.0 *M*, (b) 0.10 *M*, and (c) 0.010 *M*.

56. Calculate the percent ionization and pH of solutions of HClO ($K_a = 3.5 \times 10^{-8}$) having the following molarities: (a) 1.0 *M*, (b) 0.10 *M*, and (c) 0.010 *M*.

***57.** A 0.37 *M* solution of a weak acid, HA, has a pH of 3.7. What is the K_a for this acid?

***58.** A 0.23 *M* solution of a weak acid, HA, has a pH of 2.89. What is the K_a for this acid?

59. A common laboratory reagent is 6.0 *M* HCl. Calculate the $[H^+]$, $[OH^-]$, pH, and pOH of this solution.

60. A common laboratory reagent is 1.0 *M* NaOH. Calculate the $[H^+]$, $[OH^-]$, pH, and pOH of this solution.

61. Calculate the pH and the pOH of the following solutions:
(a) 0.00010 M HCl
(b) 0.010 M NaOH

62. Calculate the pH and the pOH of the following solutions:
(a) 0.0025 M NaOH
(b) 0.10 M HClO ($K_a = 3.5 \times 10^{-8}$)
*(c) Saturated $Fe(OH)_2$ solution ($K_{sp} = 8.0 \times 10^{-16}$)

63. Calculate the $[OH^-]$ in each of these solutions:
(a) $[H^+] = 1.0 \times 10^{-4}$
(b) $[H^+] = 2.8 \times 10^{-6}$

64. Calculate the $[OH^-]$ in each of these solutions:
(a) $[H^+] = 4.0 \times 10^{-9}$
(b) $[H^+] = 8.9 \times 10^{-2}$

65. Calculate the $[H^+]$ in each of these solutions:
(a) $[OH^-] = 6.0 \times 10^{-7}$
(b) $[OH^-] = 1 \times 10^{-8}$

66. Calculate the $[H^+]$ in each of these solutions:
(a) $[OH^-] = 4.5 \times 10^{-6}$
(b) $[OH^-] = 7.3 \times 10^{-4}$

67. Given the following solubility data, calculate the solubility product constant for each substance:
(a) $BaSO_4$, 3.9×10^{-5} mol/L
(b) Ag_2CrO_4, 7.8×10^{-5} mol/L
(c) $CaSO_4$, 0.67 g/L
(d) AgCl, 0.0019 g/L

68. Given the following solubility data, calculate the solubility product constant for each substance:
(a) ZnS, 3.5×10^{-12} mol/L
(b) $Pb(IO_3)_2$, 4.0×10^{-5} mol/L
(c) Ag_3PO_4, 6.73×10^{-3} g/L
(d) $Zn(OH)_2$, 2.33×10^{-4} g/L

69. Calculate the molar solubility for each of the following substances:
(a) $BaCO_3$, $K_{sp} = 2.0 \times 10^{-9}$
(b) $AlPO_4$, $K_{sp} = 5.8 \times 10^{-19}$

70. Calculate the molar solubility for each of the following substances:
(a) Ag_2SO_4, $K_{sp} = 1.5 \times 10^{-5}$
(b) $Mg(OH)_2$, $K_{sp} = 7.1 \times 10^{-12}$

71. Calculate for each of the substances in Question 69, the solubility in grams per 100. mL of solution.

72. Calculate for each of the substances in Question 70, the solubility in grams per 100. mL of solution.

*73. Solutions containing 100 mL of 0.010 M Na_2SO_4 and 100 mL of 0.001 M $Pb(NO_3)_2$ are mixed. Show by calculation whether or not a precipitate will form. Assume the volumes are additive. (K_{sp} $PbSO_4 = 1.3 \times 10^{-8}$)

*74. Solutions containing 50.0 mL of 1.0×10^{-4} M $AgNO_3$ and 100. mL of 1.0×10^{-4} M NaCl are mixed. Show by calculation whether or not a precipitate will form. Assume the volumes are additive. (K_{sp} AgCl $= 1.7 \times 10^{-10}$)

75. How many moles of AgBr will dissolve in 1.0 L of 0.10 M NaBr? ($K_{sp} = 5.0 \times 10^{-13}$ for AgBr)

76. How many moles of AgBr will dissolve in 1.0 L of 0.10 M $MgBr_2$? ($K_{sp} = 5.0 \times 10^{-13}$ for AgBr)

77. Calculate the $[H^+]$ and the pH of a buffer solution that is 0.20 M in $HC_2H_3O_2$ and contains sufficient sodium acetate to make the $[C_2H_3O_2^-]$ equal to 0.10 M. (K_a for $HC_2H_3O_2 = 1.8 \times 10^{-5}$)

78. Calculate the $[H^+]$ and the pH of a buffer solution that is 0.20 M in $HC_2H_3O_2$ and contains sufficient sodium acetate to make the $[C_2H_3O_2^-]$ equal to 0.20 M. (K_a for $HC_2H_3O_2 = 1.8 \times 10^{-5}$)

79. When 1.0 mL of 1.0 M HCl is added to 50. mL of 1.0 M NaCl, the $[H^+]$ changes from 1×10^{-7} M to 2.0×10^{-2} M. Calculate the initial pH and the pH change in the solution.

80. When 1.0 mL of 1.0 M HCl is added to 50. mL of a buffer solution that is 1.0 M in $HC_2H_3O_2$ and 1.0 M in $NaC_2H_3O_2$, the $[H^+]$ changes from 1.8×10^{-5} M to 1.9×10^{-5} M. Calculate the initial pH and the pH change in the solution.

Additional Exercises

These exercises are not paired or labeled by topic and provide additional practice on concepts covered in this chapter.

81. What is the maximum number of moles of HI that can be obtained from a reaction mixture containing 2.30 mol of I_2 and 2.10 mol of H_2?

82. (a) How many moles of HI will be produced when 2.00 mol of H_2 and 2.00 mol of I_2 are reacted at 700 K? (Reaction is 79% complete.)
(b) Addition of 0.27 mol of I_2 to the system increases the yield of HI to 85%. How many moles of H_2, I_2, and HI are now present?
(c) From the data in part (a), calculate K_{eq} for the reaction at 700 K.

***83.** 6.00 g of H_2, and 200. g of I_2 are reacted at 500. K. After equilibrium is reached, analysis shows that the flask contains 64.0 g of HI. How many moles of H_2, I_2, and HI are present in this equilibrium mixture?

84. What is the equilibrium constant of the reaction below

$$PCl_3(g) + Cl_2(g) \rightleftharpoons PCl_5(g)$$

if a 20. L flask contains 0.10 mol of PCl_3, 1.50 mol of Cl_2, and 0.22 mol of PCl_5?

85. If the rate of a reaction doubles for every 10°C rise in temperature, how much faster will the reaction go at 100°C than at 30°C?

86. Calculate the ionization constant for each of the following acids. Each acid ionizes as follows: $HA \rightleftharpoons H^+ + A^-$.

Acid	Acid concentration	$[H^+]$
Hypochlorous, HOCl	0.10 M	5.9×10^{-5} mol/L
Propanoic, $HC_3H_5O_2$	0.15 M	1.4×10^{-3} mol/L
Hydrocyanic, HCN	0.20 M	8.9×10^{-6} mol/L

87. The K_{sp} of CaF_2 is 3.9×10^{-11}. Calculate (a) the molar concentrations of Ca^{2+} and F^- in a saturated solution, and (b) the grams of CaF_2 that will dissolve in 500. mL of water.

88. The following pairs of solutions are mixed. Show by calculation whether or not a precipitate will form:
(a) 100 mL of 0.010 M Na_2SO_4 and 100 mL of 0.001 M $Pb(NO_3)_2$
(b) 50.0 mL of 1.0×10^{-4} M $AgNO_3$ and 100. mL of 1.0×10^{-4} M NaCl
(c) 1.0 g $Ca(NO_3)_2$ in 150 mL H_2O and 250 mL of 0.01 M NaOH
K_{sp} $PbSO_4 = 1.3 \times 10^{-8}$
K_{sp} $AgCl = 1.7 \times 10^{-10}$
K_{sp} $Ca(OH)_2 = 1.3 \times 10^{-6}$

89. $BaCl_2$ is added to a saturated $BaSO_4$ solution until the $[Ba^{2+}]$ is 0.050 M.
(a) What concentration of SO_4^{2-} remains in solution?
(b) How many grams of $BaSO_4$ remain dissolved in 100. mL of the solution? ($K_{sp} = 1.5 \times 10^{-9}$ for $BaSO_4$)

90. The concentration of a solution is 0.10 M Ba^{2+} and 0.10 M Sr^{2+}. Which sulfate, $BaSO_4$ or $SrSO_4$, will precipitate first when a dilute solution of H_2SO_4 is added dropwise to the solution? Show evidence for your answer. ($K_{sp} = 1.5 \times 10^{-9}$ for $BaSO_4$ and $K_{sp} = 3.5 \times 10^{-7}$ for $SrSO_4$)

91. The K_{sp} for $PbCl_2$ is 2.0×10^{-5}. Will a precipitate form when 0.050 mol of $Pb(NO_3)_2$ and 0.010 mol of NaCl are dissolved in 1.0 L H_2O? Show evidence for your answer.

92. Calculate the K_{eq} for the reaction

$$SO_2(g) + O_2(g) \rightleftharpoons SO_3(g)$$

when the equilibrium concentrations of the gases are at 530°C: $[SO_3] = 11.0$ M, $[SO_2] = 4.20$ M, and $[O_2] = 0.60 \times 10^{-3}$ M

93. If it takes 0.048 g of BaF_2 to saturate 15.0 mL of water, what is the K_{sp} of BaF_2?

94. The K_{eq} for the formation of ammonia gas from its elements is 4.0. If the equilibrium concentrations of nitrogen gas and hydrogen gas are both 2.0 M, what is the equilibrium concentration of the ammonia gas?

95. The K_{sp} of $SrSO_4$ is 7.6×10^{-7}. Should precipitation occur when 25.0 mL of 1.0×10^{-3} M $SrCl_2$ solution is mixed with 15.0 mL of 2.0×10^{-3} M Na_2SO_4? Show proof.

***96.** The solubility of Hg_2I_2 in H_2O is 3.04×10^{-7} g/L. The reaction $Hg_2I_2 \rightleftharpoons Hg_2^{2+} + 2\,I^-$ represents the equilibrium. Calculate the K_{sp}.

97. Under certain circumstances, when oxygen gas is heated, it can be converted into ozone according to the following reaction equation:

$$3\,O_2(g) + \text{heat} \rightleftharpoons 2\,O_3(g)$$

Name three different ways that you could increase the production of the ozone.

98. One day in a laboratory, some water spilled on a table. In just a few minutes the water had evaporated. Some days later, a similar amount of water spilled again. This time, the water remained on the table after 7 or 8 hours. Name three conditions that could have changed in the lab to cause this difference.

***99.** All a snowbound skier had to eat were walnuts! He was carrying a bag holding 12 dozen nuts. With his mittened hands, he cracked open the shells. Each nut that was opened resulted in one kernel and two shell halves. When he tired of the cracking and got ready to do some eating, he discovered he had 194 total pieces (whole nuts, shell halves, and kernels). What is the K_{eq} for this reaction?

100. For the reaction $CO(g) + H_2O(g) \rightleftharpoons CO_2(g) + H_2(g)$ at a certain temperature, K_{eq} is 1. At equilibrium would you expect to find:
(a) only CO and H_2
(b) mostly CO_2 and H_2
(c) about equal concentrations of CO and H_2O, compared to CO_2 and H_2
(d) mostly CO and H_2O
(e) only CO and H_2O
Explain your answer briefly.

101. Write the equilibrium constant expressions for each of the following reactions:
(a) $3\,O_2(g) \rightleftharpoons 2\,O_3(g)$
(b) $H_2O(g) \rightleftharpoons H_2O(l)$
(c) $MgCO_3(s) \rightleftharpoons MgO(s) + CO_2(g)$
(d) $2\,Bi^{3+}(aq) + 3\,H_2S(aq) \rightleftharpoons Bi_2S_3(s) + 6\,H^+(aq)$

102. Reactants A and B are mixed, each initially at a concentration of 1.0 *M*. They react to produce C according to this equation:

$$2\,A + B \rightleftharpoons C$$

When equilibrium is established, the concentration of C is found to be 0.30 *M*. Find the value of K_{eq}.

103. At a certain temperature, K_{eq} is 2.2×10^{-3} for the reaction

$$2\,ICl(g) \rightleftharpoons I_2(g) + Cl_2(g)$$

Now calculate the K_{eq} value for the reaction

$$I_2(g) + Cl_2(g) \rightleftharpoons 2\,ICl(g)$$

104. One drop of 1 *M* OH^- ion is added to a 1 *M* solution of HNO_2. What will be the effect of this addition on the equilibrium concentration of each of the following:
(a) $[OH^-]$
(b) $[H^+]$
(c) $[NO_2^-]$
(d) $[HNO_2]$

***105.** At 500°C the reaction

$$SO_2(g) + NO_2(g) \rightleftharpoons NO(g) + SO_3(g)$$

has $K_{eq} = 90$. What will be the equilibrium concentrations of the four gases if the two reactants begin with equal concentrations of 0.50 *M*?

106. How many grams of $CaSO_4$ will dissolve in 600. mL of water? ($K_{sp} = 2.0 \times 10^{-4}$ for $CaSO_4$)

107. A student found that 0.098 g of PbF_2 was dissolved in 400. mL of saturated PbF_2. What is the K_{sp} for the lead(II) fluoride?

Answers to Practice Exercises

16.1 (a) Equilibrium shifts left; (b) equilibrium shifts right
16.2 (a) Equilibrium shifts right; (b) equilibrium shifts left
16.3 (a) Equilibrium shifts left; (b) equilibrium shifts right
16.4 (a) $K_{eq} = \dfrac{[NO_2]^4[O_2]}{[N_2O_5]^2}$; (b) $K_{eq} = \dfrac{[N_2]^2[H_2O]^6}{[NH_3]^4[O_2]^3}$
16.5 $K_{eq} = 33$; the forward reaction is favored
16.6 (a) $[H^+] = 5.0 \times 10^{-5}$ $[OH^-] = 2.0 \times 10^{-10}$
(b) $[H^+] = 5.0 \times 10^{-9}$ $[OH^-] = 2.0 \times 10^{-6}$
16.7 (a) 4.30; (b) 8.30
16.8 (a) 6.3×10^{-6}; (b) 1.8×10^{-6}
16.9 (a) 6.3×10^{-3}% ionized; (b) 7.2×10^{-3}% ionized
16.10 (a) $K_{sp} = [Cr^{3+}][OH^-]^3$;
(b) $K_{sp} = [Cu^{2+}]^3[PO_4^{3-}]^2$
16.11 9.1×10^{-21} g/L
16.12 7.5×10^{-8} mol/L
16.13 (a) basic; (b) neutral; (c) acidic

17 Oxidation–Reduction

Chapter 17

The variety of chemical reactions that occur in our daily lives is amazing. Our society seems to run on batteries—in calculators, cars, toys, lights, thermostats, radios, televisions, and more. We polish sterling silver, paint iron railings, and galvanize nails to combat corrosion. Jewelry and computer chips are electroplated with very thin coatings of gold or silver. Clothes are bleached, and photographs are developed in solutions using chemical reactions that involve electron transfer. Tests for glucose in urine, or alcohol in the breath show vivid color changes. Plants turn energy into chemical compounds through a series of reactions called the *electron transport chain.* All of these reactions involve the transfer of electrons between substances in a chemical process called *oxidation–reaction.*

17.1 Oxidation Number

The oxidation number of an atom (sometimes called its *oxidation state*) can be considered to represent the number of electrons lost, gained, or unequally shared by the atom. Oxidation numbers can be zero, positive, or negative. When the oxidation number of an atom is zero, the atom has the same number of electrons assigned to it as there are in the free neutral atom. When the oxidation number is positive, the atom has fewer electrons assigned to it than there are in the neutral atom. When the oxidation number is negative, the atom has more electrons assigned to it than there are in the neutral atom.

The oxidation number of an atom that has lost or gained electrons to form an ion is the same as the plus or minus charge of the ion. (See Table 17.1.) In the ionic compound NaCl, the oxidation numbers are clearly established to be +1 for the Na^+ ion and −1 for the Cl^- ion. The Na^+ ion has one less electron than the neutral Na atom; and the Cl^- ion has one more electron than the neutral Cl atom. In $MgCl_2$ two electrons have transferred from the Mg atom to the Cl atoms; thus, the oxidation number of Mg is +2.

TABLE 17.1 Oxidation Numbers for Common Ions

Ion	Oxidation number
Na^+	+1
K^+	+1
Li^+	+1
Ag^+	+1
Ca^{2+}	+2
Ba^{2+}	+2
Mg^{2+}	+2
Al^{3+}	+3
Cl^-	−1
Br^-	−1
F^-	−1
I^-	−1
S^{2-}	−2
O^{2-}	−2

In covalently bonded substances, where electrons are shared between two atoms, oxidation numbers are assigned by a somewhat arbitrary system based on relative electronegativities. For symmetrical covalent molecules, such as H_2 and Cl_2, each atom is assigned an oxidation number of zero because the bonding pair of electrons is shared equally between two like atoms, neither of which is more electronegative than the other:

H:H :C̈l:C̈l:

When the covalent bond is between two unlike atoms, the bonding electrons are shared unequally because the more electronegative element has a greater attraction for them. In this case the oxidation numbers are determined by assigning both electrons to the more electronegative element.

Thus, in compounds with covalent bonds, such as NH_3 and H_2O,

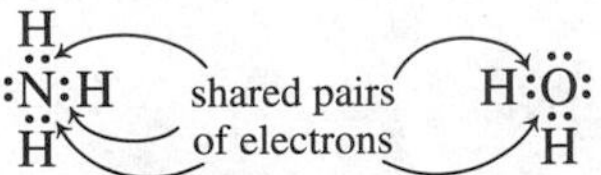

◀ **Chapter Opening Photo: Continuous painting of the Golden Gate Bridge is necessary to control the corrosion of the salt air.**

TABLE 17.2 Rules for Assigning Oxidation Number

1. All elements in their free state (uncombined with other elements) have an oxidation number of zero (e.g., Na, Cu, Mg, H_2, O_2, Cl_2, N_2).
2. H is $+1$, except in metal hydrides, where it is -1 (e.g., NaH, CaH_2).
3. O is -2, except in peroxides, where it is -1, and in OF_2, where it is $+2$.
4. The metallic element in an ionic compound has a positive oxidation number.
5. In covalent compounds the negative oxidation number is assigned to the most electronegative atom.
6. The algebraic sum of the oxidation numbers of the elements in a compound is zero.
7. The algebraic sum of the oxidation numbers of the elements in a polyatomic ion is equal to the charge of the ion.

the pairs of electrons are unequally shared between the atoms and are attracted toward the more electronegative elements, N and O. This unequal sharing causes the N and O atoms to be relatively negative with respect to the H atoms. At the same time, it causes the H atoms to be relatively positive with respect to the N and O atoms. In H_2O, both pairs of shared electrons are assigned to the O atom, giving it two electrons more than the neutral O atom. At the same time, each H atom is assigned one electron less than the neutral H atom. Therefore, the O atom is assigned an oxidation number of -2, and each H atom is assigned an oxidation number of $+1$. In NH_3, the three pairs of shared electrons are assigned to the N atom, giving it three electrons more than the neutral N atom. At the same time, each H atom has one electron less than the neutral atom. Therefore, the N atom is assigned an oxidation number of -3, and each H atom is assigned an oxidation number of $+1$.

The assignment of correct oxidation numbers to elements is essential for balancing oxidation–reduction equations.

oxidation number
oxidation state

The **oxidation number** or **oxidation state** of an element is an integer value assigned to each element in a compound or ion. This value allows us to keep track of electrons associated with each atom. Oxidation numbers have a variety of uses in chemistry—from writing formulas, to predicting properties of compounds, and assisting in the balancing of oxidation–reduction reactions in which electrons are transferred.

As a starting point, the oxidation number of an uncombined element, regardless of whether it is monatomic or diatomic, is zero. Rules for assigning oxidation numbers are summarized in Table 17.2.

Use the following steps to find the oxidation number for an element within a compound.

Step 1 Write the oxidation number of each known atom below the atom in the formula.

Step 2 Multiply each oxidation number by the number of atoms of that element in the compound.

Step 3 Write an equation indicating the sum of all the oxidation numbers in the compound. Remember that the sum of all the oxidation numbers in a compound must equal zero.

Example 17.1

Determine the oxidation number of carbon in carbon dioxide:

$$CO_2$$

Step 1 -2
Step 2 $(-2)2$
Step 3 $C + (-4) = 0$
$C = +4$ (oxidation number for carbon)

Example 17.2

Determine the oxidation number for sulfur in sulfuric acid:

$$H_2SO_4$$

Step 1 $+1 \quad -2$
Step 2 $2(+1) = +2 \qquad 4(-2) = -8$
Step 3 $+2 + S + (-8) = 0$
$S = +6$

Practice 17.1

Determine the oxidation number of (a) S in Na_2SO_4, (b) As in K_3AsO_4, (c) C in $CaCO_3$.

Oxidation numbers in a polyatomic ion (ions containing more than one atom) are determined in a similar fashion, remembering that in a polyatomic ion the sum of the oxidation numbers must equal the charge on the ion instead of zero.

Example 17.3

Determine the oxidation number of manganese in the permanganate ion MnO_4^-:

$$MnO_4^-$$

Step 1 -2
Step 2 $(-2)4$
Step 3 $Mn + (-8) = -1$ (the charge on the ion)
$Mn = +7$ (oxidation number for manganese)

Example 17.4 Determine the oxidation number of carbon in the oxalate ion $C_2O_4^{2-}$:

$$C_2O_4^{2-}$$

-2

Step 1 $(-2)4$

Step 2 $2C + (-8) = -2$ (the charge on the ion)

Step 3 $2C = +6$

$C = +3$ (oxidation number for C)

Practice 17.2

Determine the oxidation numbers of (a) N in NH_4^+, (b) Cr in $Cr_2O_7^{2-}$, (c) P in PO_4^{3-}.

Example 17.5 Determine the oxidation number of each element in (a) KNO_3 and (b) SO_4^{2-}.

Solution

(a) Potassium is a Group IA metal; therefore it has an oxidation number of $+1$. The oxidation number of each O atom is -2 (Table 17.2, Rule 3). Using these values and the fact that the sum of the oxidation numbers of all the atoms in a compound is zero, we can determine the oxidation number of N:

$$KNO_3$$

$$+1 + N + 3(-2) = 0$$

$$N = +6 - 1 = +5$$

The oxidation numbers are K, $+1$; N, $+5$; O, -2.

(b) SO_4^{2-} is an ion; therefore, the sum of oxidation numbers of the S and the O atoms must be -2, the charge of the ion. The oxidation number of each O atom is -2 (Table 17.2, Rule 3). Then

$$SO_4^{2-}$$

$$S + 4(-2) = -2, \quad S - 8 = -2$$

$$S = -2 + 8 = +6$$

The oxidation numbers are S, $+6$; O, -2.

Practice 17.3

Determine the oxidation number of each element in the following species:
(a) $BeCl_2$ (b) HClO (c) H_2O_2 (d) NH_4^+ (e) BrO_3^-

17.2 Oxidation–Reduction

Oxidation–reduction, also known as **redox**, is a chemical process in which the oxidation number of an element is changed. The process may involve the complete transfer of electrons to form ionic bonds or only a partial transfer or shift of electrons to form covalent bonds.

oxidation–reduction
redox

Oxidation occurs whenever the oxidation number of an element increases as a result of losing electrons. Conversely, **reduction** occurs whenever the oxidation number of an element decreases as a result of gaining electrons. For example, a change in oxidation number from +2 to +3 or from −1 to 0 is oxidation; a change from +5 to +2 or from −2 to −4 is reduction (see Figure 17.1). Oxidation and reduction occur simultaneously in a chemical reaction; one cannot take place without the other.

oxidation
reduction

Many combination, decomposition, and single-displacement reactions involve oxidation–reduction. Let us examine the combustion of hydrogen and oxygen from this point of view:

$$2\ H_2 + O_2 \longrightarrow 2\ H_2O$$

Both reactants, hydrogen and oxygen, are elements in the free state and have an oxidation number of zero. In the product (water), hydrogen has been oxidized to +1 and oxygen reduced to −2. The substance that causes an increase in the oxidation state of another substance is called an **oxidizing agent**. The substance that causes a decrease in the oxidation state of another substance is called a **reducing agent**. In this reaction the oxidizing agent is free oxygen, and the reducing agent is free hydrogen. In the reaction,

oxidizing agent
reducing agent

$$Zn(s) + H_2SO_4(aq) \longrightarrow ZnSO_4(aq) + H_2(g)$$

metallic zinc is oxidized, and hydrogen ions are reduced. Zinc is the reducing agent, and hydrogen ions, the oxidizing agent. Electrons are transferred from the zinc metal to the hydrogen ions. The reaction is better expressed as

$$Zn^0 + 2\ H^+ + SO_4^{2-} \longrightarrow Zn^{2+} + SO_4^{2-} + H_2^0$$

Oxidation: **Increase in oxidation number**
Loss of electrons

Reduction: **Decrease in oxidation number**
Gain of electrons

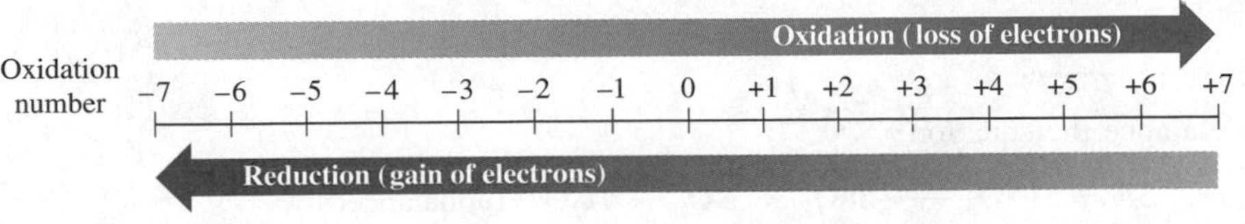

FIGURE 17.1 **Oxidation and reduction. Oxidation results in an increase in the oxidation number, and reduction results in a decrease in the oxidation number.**

The oxidizing agent is reduced and gains electrons. The reducing agent is oxidized and loses electrons. The transfer of electrons is characteristic of all redox reactions.

17.3 Balancing Oxidation–Reduction Equations

Many simple redox equations can be balanced readily by inspection, or by trial and error:

$$Na + Cl_2 \longrightarrow NaCl \quad \text{(unbalanced)}$$

$$2\,Na + Cl_2 \longrightarrow 2\,NaCl \quad \text{(balanced)}$$

Balancing this equation is certainly not complicated. But as we study more complex reactions and equations, such as

$$P + HNO_3 + H_2O \longrightarrow NO + H_3PO_4 \quad \text{(unbalanced)}$$

$$3\,P + 5\,HNO_3 + 2\,H_2O \longrightarrow 5\,NO + 3\,H_3PO_4 \quad \text{(balanced)}$$

the trial-and-error method of finding the proper numbers to balance the equation would take an unnecessarily long time.

One systematic method for balancing oxidation–reduction equations is based on the transfer of electrons between the oxidizing and reducing agents. Consider the first equation again:

$$Na^0 + Cl_2^0 \longrightarrow Na^+Cl^- \quad \text{(unbalanced)}$$

In this reaction sodium metal loses one electron per atom when it changes to a sodium ion. At the same time chlorine gains one electron per atom. Because chlorine is diatomic, two electrons per molecule are needed to form a chloride ion from each atom. These electrons are furnished by two sodium atoms. Stepwise, the reaction may be written as two half-reactions, the oxidation half-reaction and the reduction half-reaction:

$$2\,Na^0 \longrightarrow 2\,Na^+ + 2e^- \qquad \text{oxidation half-reaction}$$

$$Cl_2^0 + 2e^- \longrightarrow 2\,Cl^- \qquad \text{reduction half-reaction}$$

$$2\,Na^0 + Cl_2^0 \longrightarrow 2\,Na^+Cl^-$$

When the two half-reactions, each containing the same number of electrons, are added together algebraically, the electrons cancel out. In this reaction there are no excess electrons; the two electrons lost by the two sodium atoms are utilized by chlorine. In all redox reactions the loss of electrons by the reducing agent must equal the gain of electrons by the oxidizing agent. Here, sodium is oxidized and chlorine is reduced. Chlorine is the oxidizing agent; sodium is the reducing agent.

The following examples illustrate a systematic method of balancing more complicated redox equations by the change-in-oxidation-number method.

Example 17.6 Balance the equation

$$Sn + HNO_3 \longrightarrow SnO_2 + NO_2 + H_2O \quad \text{(unbalanced)}$$

Solution

Step 1 Assign oxidation numbers to each element to identify the elements that are being oxidized and those that are being reduced. Write the oxidation numbers below each element in order to avoid confusing them with the charge on an ion:

$$\underset{0}{Sn} + \underset{+1}{H}\ \underset{+5}{N}\ \underset{-2}{O_3} \longrightarrow \underset{+4}{Sn}\ \underset{-2}{O_2} + \underset{+4}{N}\ \underset{-2}{O_2} + \underset{+1}{H_2}\ \underset{-2}{O}$$

Note that the oxidation numbers of Sn and N have changed.

Step 2 Now write two new equations, using only the elements that change in oxidation number. Then add electrons to bring the equations into electrical balance. One equation represents the oxidation step; the other represents the reduction step. The oxidation step produces electrons; the reduction step uses electrons.

$$Sn^0 \longrightarrow Sn^{4+} + 4e^- \qquad \text{oxidation}$$

Sn^0 loses 4 electrons

$$N^{5+} + 1e^- \longrightarrow N^{4+} \qquad \text{reduction}$$

N^{5+} gains 1 electron

Step 3 Now multiply the two equations by the smallest whole numbers that will make the loss of electrons by the oxidation step equal to the number of electrons gained in the reduction step. In this reaction, the oxidation step is multiplied by 1 and the reduction step by 4. The equations become

$$Sn^0 \longrightarrow Sn^{4+} + 4e^- \qquad \text{oxidation}$$

Sn^0 loses 4 electrons

$$4\,N^{5+} + 4e^- \longrightarrow 4\,N^{4+} \qquad \text{reduction}$$

$4\,N^{5+}$ gain 4 electrons

We have now established the ratio of the oxidizing to the reducing agent as being four atoms of N to one atom of Sn.

Step 4 Now transfer the coefficient that appears in front of each substance in the balanced oxidation–reduction equations to the corresponding substance in the original equation. We need to use $1\,Sn$, $1\,SnO_2$, $4\,HNO_3$, and $4\,NO_2$:

$$Sn + 4\,HNO_3 \longrightarrow SnO_2 + 4\,NO_2 + H_2O \quad \text{(unbalanced)}$$

Step 5 In the usual manner, balance the remaining elements that are not oxidized or reduced to give the final balanced equation:

$$Sn + 4\,HNO_3 \longrightarrow SnO_2 + 4\,NO_2 + 2\,H_2O \quad \text{(balanced)}$$

In balancing the final elements, we must not change the ratio of the elements that were oxidized and reduced. Make a final check to ensure that both sides of the equation have the same number of atoms of each element. The final balanced equation contain 1 atom of Sn, 4 atoms of N, 4 atoms of H, and 12 atoms of O on each side.

Because each new equation may present a slightly different problem and because proficiency in balancing equations requires practice, we will work through two more examples.

Example 17.7 Balance the equation

$$I_2 + Cl_2 + H_2O \longrightarrow HIO_3 + HCl \quad \text{(unbalanced)}$$

Solution

Step 1 Assign oxidation numbers:

$$\underset{0}{I_2} + \underset{0}{Cl_2} + \underset{+1\ -2}{H_2O} \longrightarrow \underset{+1\ +5\ -2}{HIO_3} + \underset{+1\ -1}{HCl}$$

The oxidation numbers of I_2 and Cl_2 have changed, I_2 from 0 to +5, and Cl_2 from 0 to −1.

Step 2 Write oxidation and reduction steps. Balance the number of atoms and then balance the electrical charge using electrons:

$$I_2 \longrightarrow 2\,I^{5+} + 10e^- \qquad \text{oxidation}$$

I_2 loses 10 electrons

$$Cl_2 + 2e^- \longrightarrow 2\,Cl^- \qquad \text{reduction}$$

Cl_2 gains 2 electrons

Step 3 Adjust loss and gain of electrons so that they are equal. Multiply the oxidation step by 1 and the reduction step by 5:

$$I_2 \longrightarrow 2\,I^{5+} + 10e^- \qquad \text{oxidation}$$

I_2 loses 10 electrons

$$5\,Cl_2 + 10e^- \longrightarrow 10\,Cl^- \qquad \text{reduction}$$

5 Cl_2 gain 10 electrons

Step 4 Transfer the coefficients from the balanced redox equations into the original equation. We need to use 1 I_2, 2 HIO_3, 5 Cl_2, and 10 HCl:

$$I_2 + 5\,Cl_2 + H_2O \longrightarrow 2\,HIO_3 + 10\,HCl \quad \text{(unbalanced)}$$

Step 5 Balance the remaining elements, H and O:

$$I_2 + 5\,Cl_2 + 6\,H_2O \longrightarrow 2\,HIO_3 + 10\,HCl \quad \text{(balanced)}$$

Check: The final balanced equation contains 2 atoms of I, 10 atoms of Cl, 12 atoms of H, and 6 atoms of O on each side.

Example 17.8 Balance the equation

$$K_2Cr_2O_7 + FeCl_2 + HCl \longrightarrow CrCl_3 + KCl + FeCl_3 + H_2O \quad \text{(unbalanced)}$$

Solution

Step 1 Assign oxidation numbers (Cr and Fe have changed):

$$\underset{+1+6\ -2}{K_2Cr_2O_7} + \underset{+2-1}{FeCl_2} + \underset{+1-1}{HCl} \longrightarrow \underset{+3-1}{CrCl_3} + \underset{+1-1}{KCl} + \underset{+3-1}{FeCl_3} + \underset{+1-2}{H_2O}$$

Step 2 Write the oxidation and reduction steps. Balance the number of atoms and then balance the electrical charge using electrons:

$$Fe^{2+} \longrightarrow Fe^{3+} + 1e^{-} \quad \text{oxidation}$$

Fe^{2+} loses 1 electron

$$2\ Cr^{6+} + 6e^{-} \longrightarrow 2\ Cr^{3+} \quad \text{reduction}$$

$2\ Cr^{6+}$ gain 6 electrons

Step 3 Balance the loss and gain of electrons. Multiply the oxidation step by 6 and the reduction step by 1 to equalize the transfer of electrons.

$$6\ Fe^{2+} \longrightarrow 6\ Fe^{3+} + 6e^{-} \quad \text{oxidation}$$

$6\ Fe^{2+}$ lose 6 electrons

$$2\ Cr^{6+} + 6e^{-} \longrightarrow 2\ Cr^{3+} \quad \text{reduction}$$

$2\ Cr^{6+}$ gain 6 electrons

Step 4 Transfer the coefficients from the balanced redox equations into the original equation. (Note that one formula unit of $K_2Cr_2O_7$ contains two Cr atoms.) We need to use 1 $K_2Cr_2O_7$, 2 $CrCl_3$, 6 $FeCl_2$, and 6 $FeCl_3$:

$$K_2Cr_2O_7 + 6\ FeCl_2 + HCl \longrightarrow 2\ CrCl_3 + KCl + 6\ FeCl_3 + H_2O \quad \text{(unbalanced)}$$

Step 5 Balance the remaining elements in this order: K, Cl, H, O.

$$K_2Cr_2O_7 + 6\ FeCl_2 + 14\ HCl \longrightarrow 2\ CrCl_3 + 2\ KCl + 6\ FeCl_3 + 7\ H_2O \quad \text{(balanced)}$$

Check: The final balanced equation contains 2 K atoms, 2 Cr atoms, 7 O atoms, 6 Fe atoms, 26 Cl atoms, and 14 H atoms on each side.

Practice 17.4

Balance the following equations using the change in oxidation number method:

(a) $HNO_3 + S \longrightarrow NO_2 + H_2SO_4 + H_2O$

(b) $CrCl_3 + MnO_2 + H_2O \longrightarrow MnCl_2 + H_2CrO_4$

(c) $KMnO_4 + HCl + H_2S \longrightarrow KCl + MnCl_2 + S + H_2O$

17.4 Balancing Ionic Redox Equations

The main difference between balancing ionic redox equations and molecular redox equations is in the handling of ions. In addition to having the same number of each kind of element on both sides of the final equation, the net charges must also be equal. In assigning oxidation numbers we must be careful to consider the charge

Chemistry in Action

Sensitive Sunglasses

Oxidation–reduction reactions are the basis for many interesting and useful applications in technology. One such application is photochromic glass, which is used for the lenses in light-sensitive glasses. Lenses manufactured by the Corning Glass Company can change from transmitting 85% of light to only transmitting 22% of light when exposed to bright sunlight.

Photochromic glass is composed of linked tetrahedrons of silicon and oxygen atoms jumbled together in a disorderly array, with crystals of silver chloride caught in between the silica tetrahedrons. When the glass is clear, the visible light passes right through the molecules. The glass absorbs ultraviolet light, however, and this energy triggers an oxidation–reduction reaction between Ag^+ and Cl^-:

$$Ag^+ + Cl^- \xrightarrow{\text{UV light}} Ag^0 + Cl^0$$

An oxidation–reduction reaction causes these photochromic glasses to change from light to dark in bright sunlight.

To prevent the reaction from reversing itself immediately, a few ions of Cu^+ are incorporated into the silver chloride crystal. These Cu^+ ions react with the newly formed chlorine atoms:

$$Cu^+ + Cl^0 \longrightarrow Cu^{2+} + Cl^-$$

The silver atoms move to the surface of the crystal and form small colloidal clusters of silver metal. This metallic silver absorbs visible light, making the lens appear dark (colored).

As the glass is removed from the light, the Cu^{2+} ions slowly move to the surface of the crystal where they interact with the silver metal:

$$Cu^{2+} + Ag^0 \longrightarrow Cu^+ + Ag^+$$

The glass clears as the silver ions rejoin chloride ions in the crystals.

on the ions. In many respects, balancing ionic equations is much simpler than balancing molecular equations.

Several methods can be used to balance ionic redox equations. These methods include, with slight modification, the oxidation-number method just shown for molecular equations. But the most popular method is probably the ion–electron method, which we will now discuss.

The ion–electron method uses ionic charges and electrons to balance ionic redox equations. Oxidation numbers are not formally used, but it is necessary to determine what is being oxidized and what is being reduced. The method is as follows:

1. Write the two half-reactions that contain the elements being oxidized and reduced using the entire formula of the ion or molecule.
2. Balance the elements other than oxygen and hydrogen.
3. Balance oxygen and hydrogen.
   Acidic Solution: For reactions that occur in acidic solution, use H^+ and H_2O to balance oxygen and hydrogen. For each oxygen needed, use one H_2O. Then add H^+ as needed to balance the hydrogen atoms.
   Basic Solution: Balancing equations that occur in alkaline solutions is a bit more complicated. For reactions that occur in alkaline solutions, first balance as though the reaction were in an acidic solution, using Steps 1, 2, and 3. Then add as many OH^- ions to each side of the equation as there are H^+ ions in the equation. Now combine the H^+ and OH^- ions into water (for

example, 4 H^+ and 4 OH^- give 4 H_2O). Rewrite the equation, canceling equal numbers of water molecules that appear on opposite sides of the equation.

4. Add electrons (e^-) to each half-reaction to bring them into electrical balance.
5. Since the loss and gain of electrons must be equal, multiply each half-reaction by the appropriate number to make the number of electrons the same in each half-reaction.
6. Add the two half-reactions together, canceling electrons and any other identical substances that appear on opposite sides of the equation.

Example 17.9

Balance this equation using the ion–electron method:

$$MnO_4^- + S^{2-} \longrightarrow Mn^{2+} + S^0 \quad \text{(acidic solution)}$$

Solution

Step 1 Write two half-reactions, one containing the element being oxidized and the other the element being reduced (use the entire molecule or ion):

$$S^{2-} \longrightarrow S^0 \qquad \text{oxidation}$$
$$MnO_4^- \longrightarrow Mn^{2+} \qquad \text{reduction}$$

Step 2 Balance elements other than oxygen and hydrogen (accomplished in Step 1 in this example—1S and 1Mn on each side).

Step 3 Balance O and H. Remember the solution is acidic. The oxidation requires neither O nor H, but the reduction equation needs 4 H_2O on the right and 8 H^+ on the left.

$$S^{2-} \longrightarrow S^0$$
$$8\,H^+ + MnO_4^- \longrightarrow Mn^{2+} + 4\,H_2O$$

Step 4 Balance each half-reaction electrically with electrons:

$$S^{2-} \longrightarrow S^0 + 2e^-$$

net charge = −2 on each side

$$5e^- + 8\,H^+ + MnO_4^- \longrightarrow Mn^{2+} + 4\,H_2O$$

net charge = +2 on each side

Step 5 Equalize loss and gain of electrons. In this case multiply the oxidation equation by 5 and the reduction equation by 2:

$$5\,S^{2-} \longrightarrow 5\,S^0 + 10e^-$$
$$10e^- + 16\,H^+ + 2\,MnO_4^- \longrightarrow 2\,Mn^{2+} + 8\,H_2O$$

Step 6 Add the two half-reactions together, canceling the $10e^-$ from each side, to obtain the balanced equation:

$$5\,S^{2-} \longrightarrow 5\,S^0 + \cancel{10e^-}$$
$$\cancel{10e^-} + 16\,H^+ + 2\,MnO_4^- \longrightarrow 2\,Mn^{2+} + 8\,H_2O$$
$$16\,H^+ + 2\,MnO_4^- + 5\,S^{2-} \longrightarrow 2\,Mn^{2+} + 5\,S^0 + 8\,H_2O \quad \text{(balanced)}$$

Check: Both sides of the equation have a charge of +4 and contain the same number of atoms of each element.

Example 17.10 Balance the following equation:

$$CrO_4^{2-} + Fe(OH)_2 \longrightarrow Cr(OH)_3 + Fe(OH)_3 \quad \text{(basic solution)}$$

Solution

Step 1 Write the two half-reactions:

$$Fe(OH)_2 \longrightarrow Fe(OH)_3 \quad \text{oxidation}$$

$$CrO_4^{2-} \longrightarrow Cr(OH)_3 \quad \text{reduction}$$

Step 2 Balance elements other than H and O (accomplished in Step 1).

Step 3 Remember the solution is basic. Balance O and H as though the solution were acidic. Use H_2O and H^+. To balance O and H in the oxidation equation, add 1 H_2O on the left and 1 H^+ on the right side of the equation:

$$Fe(OH)_2 + H_2O \longrightarrow Fe(OH)_3 + H^+$$

Add 1 OH^- to each side:

$$Fe(OH)_2 + H_2O + OH^- \longrightarrow Fe(OH)_3 + H^+ + OH^-$$

Combine H^+ and OH^- as H_2O and rewrite, canceling H_2O on each side:

$$Fe(OH)_2 + \cancel{H_2O} + OH^- \longrightarrow Fe(OH)_3 + \cancel{H_2O}$$

$$\boxed{Fe(OH)_2 + OH^- \longrightarrow Fe(OH)_3}$$

To balance O and H in the reduction equation, add 1 H_2O on the right and 5 H^+ on the left:

$$CrO_4^{2-} + 5\,H^+ \longrightarrow Cr(OH)_3 + H_2O$$

Add 5 OH^- to each side:

$$CrO_4^{2-} + 5\,H^+ + 5\,OH^- \longrightarrow Cr(OH)_3 + H_2O + 5\,OH^-$$

Combine $5\,H^+ + 5\,OH^- \longrightarrow 5\,H_2O$:

$$CrO_4^{2-} + 5\,H_2O \longrightarrow Cr(OH)_3 + H_2O + 5\,OH^-$$

Rewrite, canceling 1 H_2O from each side:

$$\boxed{CrO_4^{2-} + 4\,H_2O \longrightarrow Cr(OH)_3 + 5\,OH^-}$$

Step 4 Balance each half-reaction electrically with electrons:

$$Fe(OH)_2 + OH^- \longrightarrow Fe(OH)_3 + e^- \quad \text{(balanced oxidation equation)}$$

$$CrO_4^{2-} + 4\,H_2O + 3e^- \longrightarrow Cr(OH)_3 + 5\,OH^- \quad \text{(balanced reduction equation)}$$

Step 5 Equalize the loss and gain of electrons. Multiply the oxidation reaction by 3:

$$3\,Fe(OH)_2 + 3\,OH^- \longrightarrow 3\,Fe(OH)_3 + 3e^-$$

$$CrO_4^{2-} + 4\,H_2O + 3e^- \longrightarrow Cr(OH)_3 + 5\,OH^-$$

Step 6 Add the two half-reactions together, canceling the $3e^-$ and $3\ OH^-$ from each side of the equation:

$$3\ Fe(OH)_2 + 3\ OH^- \longrightarrow 3\ Fe(OH)_3 + 3e^-$$
$$CrO_4^{2-} + 4\ H_2O + 3e^- \longrightarrow Cr(OH)_3 + 5\ OH^-$$
$$CrO_4^{2-} + 3\ Fe(OH)_2 + 4\ H_2O \longrightarrow Cr(OH)_3 + 3\ Fe(OH)_3 + 2\ OH^- \quad \text{(balanced)}$$

Check: Each side of the equation has a charge of -2 and contains the same number of atoms of each element.

Practice 17.5

Balance the following equations using the ion–electron method:

(a) $I^- + NO_2^- \longrightarrow I_2 + NO$ (acidic solution)
(b) $Cl_2 + IO_3^- \longrightarrow IO_4^- + Cl^-$ (basic solution)
(c) $AuCl_4^- + Sn^{2+} \longrightarrow Sn^{4+} + AuCl + Cl^-$

Ionic equations can also be balanced by using the change-in-oxidation-number method shown in Example 17.6. Let us examine the same equation as in Example 17.10 to illustrate this method.

Example 17.11

Balance the following equation using the change-in-oxidation-number method:

$$CrO_4^{2-} + Fe(OH)_2 \longrightarrow Cr(OH)_3 + Fe(OH)_3 \quad \text{(basic solution)}$$

Solution

Steps 1 and 2 Assign oxidation numbers and balance the charges with electrons:

$$Cr^{6+} + 3e^- \longrightarrow Cr^{3+} \quad \text{reduction}$$

Cr^{6+} gains $3e^-$

$$Fe^{2+} \longrightarrow Fe^{3+} + e^- \quad \text{oxidation}$$

Fe^{2+} loses $1e^-$

Step 3 Equalize the loss and gain of electrons, and then multiply the oxidation step by 3:

$$Cr^{6+} + 3e^- \longrightarrow Cr^{3+}$$

Cr^{6+} gains $3e^-$

$$3\ Fe^{2+} \longrightarrow 3\ Fe^{3+} + 3e^-$$

$3\ Fe^{2+}$ lose $3e^-$

Step 4 Transfer coefficients back to the original equation:

$$CrO_4^{2-} + 3\ Fe(OH)_2 \longrightarrow Cr(OH)_3 + 3\ Fe(OH)_3$$

Step 5 Balance electrically. Because the solution is basic, use OH^- to balance charges. The charge on the left side is -2, and on the right side is 0. Add $2\ OH^-$ ions to the right side of the equation:

$$CrO_4^{2-} + 3\ Fe(OH)_2 \longrightarrow Cr(OH)_3 + 3\ Fe(OH)_3 + 2\ OH^-$$

Adding 4 H_2O to the left side balances the equation:

$$CrO_4^{2-} + 3\ Fe(OH)_2 + 4\ H_2O \longrightarrow Cr(OH)_3 + 3\ Fe(OH)_3 + 2\ OH^- \quad \text{(balanced)}$$

Check: Each side of the equation has a charge of -2 and contains the same number of atoms of each element.

Practice 17.6

Balance each of the following equations using the change-in-oxidation-number method:

(a) $Zn \longrightarrow Zn(OH)_4^{2-} + H_2$ (basic solution)
(b) $H_2O_2 + Sn^{2+} \longrightarrow Sn^{4+}$ (acidic solution)
(c) $Cu + Cu^{2+} \longrightarrow Cu_2O$ (basic solution)

17.5 Activity Series of Metals

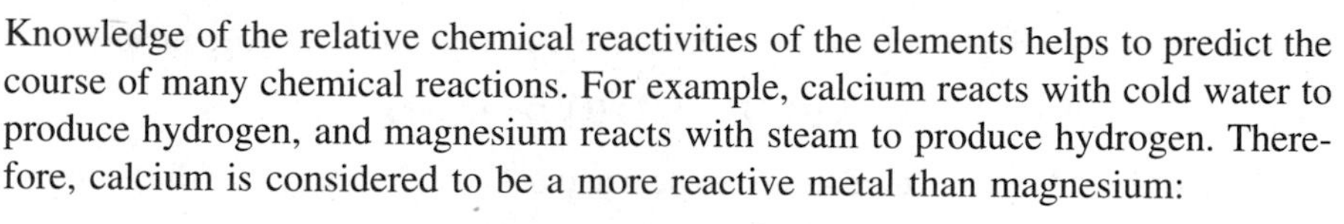

Knowledge of the relative chemical reactivities of the elements helps to predict the course of many chemical reactions. For example, calcium reacts with cold water to produce hydrogen, and magnesium reacts with steam to produce hydrogen. Therefore, calcium is considered to be a more reactive metal than magnesium:

$$Ca(s) + 2\ H_2O(l) \longrightarrow Ca(OH)_2(aq) + H_2(g)$$
$$Mg(s) + \underset{\text{steam}}{H_2O(g)} \longrightarrow MgO(s) + H_2(g)$$

FIGURE 17.2
A coil of copper placed in a silver nitrate solution forms silver crystals on the wire. The pale blue of the solution indicates the presence of copper ions.

The difference in their activity is attributed to the fact that calcium loses its two valence electrons more easily than does magnesium and is therefore more reactive and/or more readily oxidized than magnesium.

When a coil of copper is placed in a solution of silver nitrate ($AgNO_3$), free silver begins to plate out on the copper. See Figure 17.2. After the reaction has continued for some time, we can observe a blue color in the solution, indicating the presence of copper(II) ions. The equations are

$$Cu^0(s) + 2\ AgNO_3(aq) \longrightarrow 2\ Ag^0(s) + Cu(NO_3)_2(aq)$$
$$Cu^0(s) + 2\ Ag^+(aq) \longrightarrow 2\ Ag^0(s) + Cu^{2+}(aq) \quad \text{net ionic equation}$$
$$Cu^0(s) \longrightarrow Cu^{2+}(aq) + 2e^- \quad \text{oxidation of } Cu^0$$
$$Ag^+(aq) + e^- \longrightarrow Ag^0(s) \quad \text{reduction of } Ag^+$$

If a coil of silver is placed in a solution of copper(II) nitrate, $Cu(NO_3)_2$, no reaction is visible.

$$Ag^0(s) + Cu(NO_3)_2(aq) \longrightarrow \text{no reaction}$$

In the reaction between Cu and $AgNO_3$, electrons are transferred from Cu^0 atoms to Ag^+ ions in solution. Copper has a greater tendency than silver to lose electrons, so an electrochemical force is exerted upon silver ions to accept electrons from copper atoms. When an Ag^+ ion accepts an electron, it is reduced to an Ag^0 atom and is no longer soluble in solution. At the same time, Cu^0 is oxidized and goes into solution as Cu^{2+} ions. From this reaction we can conclude that copper is more reactive than silver.

Metals such as sodium, magnesium, zinc, and iron, that react with solutions of acids to liberate hydrogen are more reactive than hydrogen. Metals such as copper, silver, and mercury, that do not react with solutions of acids to liberate hydrogen are less reactive than hydrogen. By studying a series of reactions such as those given above, we can list metals according to their chemical activity, placing the most active at the top and the least active at the bottom. This list is called the **activity series of metals**. Table 17.3 shows some of the common metals in the series. The arrangement corresponds to the ease with which the elements are oxidized or lose electrons, with the most easily oxidizable element listed first. More extensive tables are available in chemistry reference books.

TABLE 17.3 Activity Series of Metals

Ease of oxidation ↑			
K	$\longrightarrow$	K^+	$+ e^-$
Ba	$\longrightarrow$	Ba^{2+}	$+ 2e^-$
Ca	$\longrightarrow$	Ca^{2+}	$+ 2e^-$
Na	$\longrightarrow$	Na^+	$+ e^-$
Mg	$\longrightarrow$	Mg^{2+}	$+ 2e^-$
Al	$\longrightarrow$	Al^{3+}	$+ 3e^-$
Zn	$\longrightarrow$	Zn^{2+}	$+ 2e^-$
Cr	$\longrightarrow$	Cr^{3+}	$+ 3e^-$
Fe	$\longrightarrow$	Fe^{2+}	$+ 2e^-$
Ni	$\longrightarrow$	Ni^{2+}	$+ 2e^-$
Sn	$\longrightarrow$	Sn^{2+}	$+ 2e^-$
Pb	$\longrightarrow$	Pb^{2+}	$+ 2e^-$
$\mathbf{H_2}$	$\longrightarrow$	$\mathbf{2\,H^+}$	$\mathbf{+ 2e^-}$
Cu	$\longrightarrow$	Cu^{2+}	$+ 2e^-$
As	$\longrightarrow$	As^{3+}	$+ 3e^-$
Ag	$\longrightarrow$	Ag^+	$+ e^-$
Hg	$\longrightarrow$	Hg^{2+}	$+ 2e^-$
Au	$\longrightarrow$	Au^{3+}	$+ 3e^-$

The general principles governing the arrangement and use of the activity series are as follows:

1. The reactivity of the metals listed decreases from top to bottom.
2. A free metal can displace the ion of a second metal from solution, provided that the free metal is above the second metal in the activity series.
3. Free metals above hydrogen react with nonoxidizing acids in solution to liberate hydrogen gas.
4. Free metals below hydrogen do not liberate hydrogen from acids.
5. Conditions such as temperature and concentration may affect the relative position of some of these elements.

Two examples of the application of the activity series of metals are given in the following examples.

activity series of metals

Example 17.12

Will zinc metal react with dilute sulfuric acid?

Solution

From Table 17.3 we see that zinc is above hydrogen; therefore zinc atoms will lose electrons more readily than hydrogen atoms. Hence, zinc atoms will reduce hydrogen ions from the acid to form hydrogen gas and zinc ions. In fact, these reagents are commonly used for the laboratory preparation of hydrogen. The equation is

$$Zn(s) + H_2SO_4(aq) \longrightarrow ZnSO_4(aq) + H_2(g)$$

$$Zn(s) + 2\,H^+(aq) \longrightarrow Zn^{2+}(aq) + H_2(g) \qquad \text{net ionic equation}$$

Example 17.13

Will a reaction occur when copper metal is placed in an iron(II) sulfate solution?

Solution

No, copper lies below iron in the series, loses electrons less easily than iron, and therefore will not displace iron(II) ions from solution. In fact, the reverse is true. When an iron nail is dipped into a copper(II) sulfate solution, it becomes coated with free copper. The equations are

$$Cu(s) + FeSO_4(aq) \longrightarrow \text{no reaction}$$

$$Fe(s) + CuSO_4(aq) \longrightarrow FeSO_4(aq) + Cu(s)$$

From Table 17.3 we may abstract the following pair in their relative position to each other:

$$Fe \longrightarrow Fe^{2+} + 2e^-$$

$$Cu \longrightarrow Cu^{2+} + 2e^-$$

According to the second principle listed above on the use of the activity series, we can predict that free iron will react with copper(II) ions in solution to form free copper metal and iron(II) ions in solution:

$$Fe(s) + Cu^{2+}(aq) \longrightarrow Fe^{2+}(aq) + Cu(s) \quad \text{net ionic equation}$$

Practice 17.7

Indicate whether the following reactions will occur:
(a) Sodium metal is placed in dilute hydrochloric acid.
(b) A piece of lead is placed in magnesium nitrate solution.
(c) Mercury is placed in a solution of silver nitrate.

17.6 Electrolytic and Voltaic Cells

electrolysis
electrolytic cell

The process in which electrical energy is used to bring about chemical change is known as **electrolysis**. An **electrolytic cell** uses electrical energy to produce a nonspontaneous chemical reaction. The use of electrical energy has many applications in the chemical industry—for example, in the production of sodium, sodium hydroxide, chlorine, fluorine, magnesium, aluminum, and pure hydrogen and oxygen, and in the purification and electroplating of metals.

cathode
anode

What happens when an electric current is passed through a solution? Let us consider a hydrochloric acid solution in a simple electrolytic cell, as shown in Figure 17.3. The cell consists of a source of direct current (a battery) connected to two electrodes that are immersed in a solution of hydrochloric acid. The negative electrode is called the **cathode** because cations are attracted to it. The positive electrode is called the **anode** because anions are attracted to it. The cathode is attached to the negative pole and the anode to the positive pole of the battery. The battery supplies electrons to the cathode.

When the electric circuit is completed, positive hydronium ions (H_3O^+) migrate to the cathode where they pick up electrons and evolve as hydrogen gas. At the same time the negative chloride ions (Cl^-) migrate to the anode, where they lose electrons and evolve as chlorine gas.

Reaction at the cathode:

$$H_3O^+ + 1e^- \longrightarrow H^0 + H_2O \qquad \text{(reduction)}$$

$$H^0 + H^0 \longrightarrow H_2$$

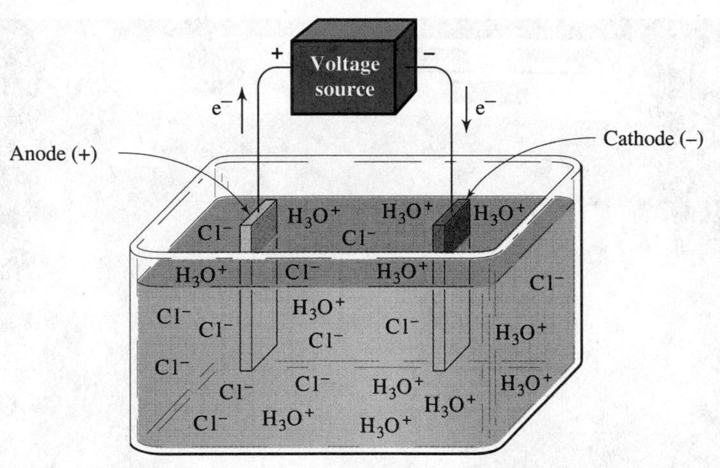

FIGURE 17.3 During the electrolysis of a hydrochloric acid solution, positive hydronium ions are attracted to the cathode, where they gain electrons and form hydrogen gas. Chloride ions migrate to the anode, where they lose electrons and form chlorine gas. The equation for this process is $2\ HCl(aq) \longrightarrow H_2(g) + Cl_2(g)$

Reaction at the anode:

$$Cl^- \longrightarrow Cl^0 + 1e^- \qquad \text{(oxidation)}$$

$$Cl^0 + Cl^0 \longrightarrow Cl_2$$

$$2\ HCl(aq) \xrightarrow{\text{electrolysis}} H_2(g) + Cl_2(g) \qquad \text{net reaction}$$

Note that oxidation–reduction has taken place. Chloride ions lost electrons (were oxidized) at the anode, and hydronium ions gained electrons (were reduced) at the cathode.

Oxidation always occurs at the anode and reduction at the cathode.

When concentrated sodium chloride solutions (brines) are electrolyzed, the products are sodium hydroxide, hydrogen, and chlorine. The overall reaction is

$$2\ Na^+(aq) + 2\ Cl^-(aq) + 2\ H_2O(l) \xrightarrow{\text{electrolysis}}$$
$$2\ Na^+(aq) + 2\ OH^-(aq) + H_2(g) + Cl_2(g)$$

The net ionic equation is

$$2\ Cl^-(aq) + 2\ H_2O(l) \longrightarrow 2\ OH^-(aq) + H_2(g) + Cl_2(g)$$

During the electrolysis, Na^+ ions move toward the cathode and Cl^- ions move toward the anode. The anode reaction is similar to that of hydrochloric acid; the chlorine is liberated:

$$2\ Cl^-(aq) \longrightarrow Cl_2(g) + 2e^-$$

Even though Na^+ ions are attracted by the cathode, the facts show that hydrogen is liberated there. No evidence of metallic sodium is found, but the area around the cathode tests alkaline from the accumulated OH^- ions. The reaction at the cathode is

$$2\ H_2O(l) + 2e^- \longrightarrow H_2(g) + 2\ OH^-(aq)$$

Electroplating is a process ▶ involving oxidation–reduction reactions. It can be used to protect an object or for decorative effect.

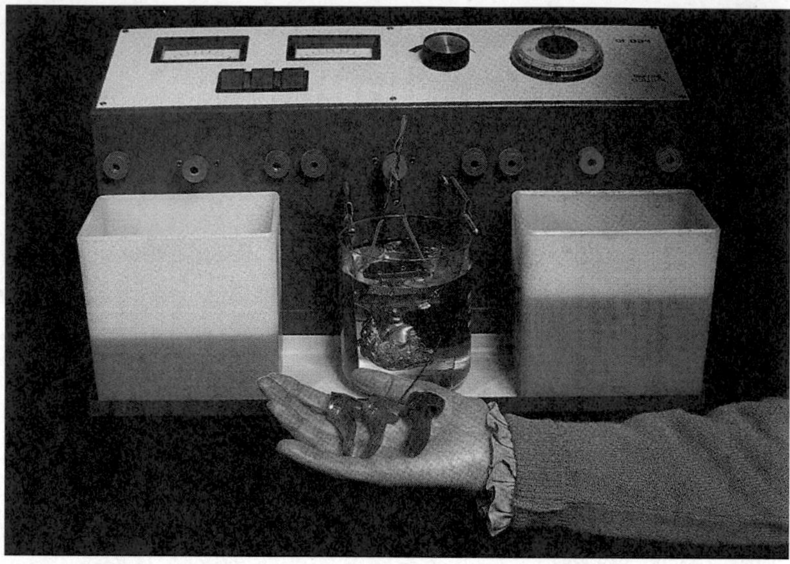

If the electrolysis is allowed to continue until all the chloride is reacted the solution remaining will contain only sodium hydroxide, which on evaporation yields solid NaOH. Large tonnages of sodium hydroxide and chlorine are made by this process.

When molten sodium chloride (without water) is subjected to electrolysis, metallic sodium and chlorine gas are formed:

$$2\ Na^+(l) + 2\ Cl^-(l) \xrightarrow{\text{electrolysis}} 2\ Na(l) + Cl_2(g)$$

An important electrochemical application is the electroplating of metals. Electroplating is the art of covering a surface or an object with a thin adherent electrodeposited metal coating. Electroplating is done for protection of the surface of the base metal or for a purely decorative effect. The layer deposited is surprisingly thin, varying from as little as 5×10^{-5} cm to 2×10^{-3} cm, depending on the metal and the intended use. The object to be plated is set up as the cathode and is immersed in a solution containing ions of the metal to be plated. When an electric current passes through the solution, metal ions that migrate to the cathode are reduced, depositing on the object as the free metal. In most cases the metal deposited on the object is replaced in the solution by using an anode of the same metal. The following equations show the chemical changes in the electroplating of nickel:

Reaction at the cathode:	$Ni^{2+}(aq) + 2e^- \longrightarrow Ni(s)$	Ni plated out on an object
Reaction at the anode:	$Ni(s) \longrightarrow Ni^{2+}(aq) + 2e^-$	Ni replenished in solution

Metals commonly used in commercial electroplating are copper, nickel, zinc, lead, cadmium, chromium, tin, gold, and silver.

In the electrolytic cell shown in Figure 17.3, electrical energy from the voltage source is used to bring about nonspontaneous redox reactions. The hydrogen and chlorine produced have more potential energy than was present in the hydrochloric acid before electrolysis.

Conversely, some spontaneous redox reactions can be made to supply useful

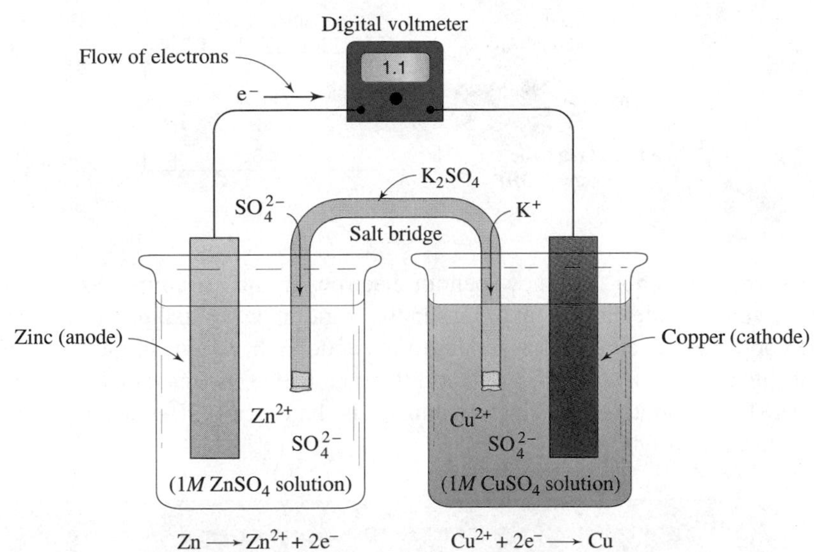

FIGURE 17.4
Zinc–copper voltaic cell. The cell has a potential of 1.1 volts when $ZnSO_4$ and $CuSO_4$ solutions are 1.0 *M*. The salt bridge provides electrical contact between the two half-cells.

amounts of electrical energy. When a piece of zinc is put in a copper(II) sulfate solution, the zinc quickly becomes coated with metallic copper. We expect this coating to happen because zinc is above copper in the activity series; copper(II) ions are therefore reduced by zinc atoms:

$$Zn^0(s) + Cu^{2+}(aq) \longrightarrow Zn^{2+}(aq) + Cu^0(s)$$

This reaction is clearly a spontaneous redox reaction, but simply dipping a zinc rod into a copper(II) sulfate solution will not produce useful electric current. However, when we carry out this reaction in the cell shown in Figure 17.4, an electric current is produced. The cell consists of a piece of zinc immersed in a zinc sulfate solution and connected by a wire through a voltmeter to a piece of copper immersed in copper(II) sulfate solution. The two solutions are connected by a salt bridge. Such a cell produces an electric current and a potential of about 1.1 volts when both solutions are 1.0 *M* in concentration. A cell that produces electric current from a spontaneous chemical reaction is called a **voltaic cell**. A voltaic cell is also known as a *galvanic cell.*

voltaic cell

The driving force responsible for the electric current in the zinc–copper cell originates in the great tendency of zinc atoms to lose electrons relative to the tendency of copper(II) ions to gain electrons. In the cell shown in Figure 17.4, zinc atoms lose electrons and are converted to zinc ions at the zinc electrode surface; the electrons flow through the wire (external circuit) to the copper electrode. Here copper(II) ions pick up electrons and are reduced to copper atoms, which plate out on the copper electrode. Sulfate ions flow from the $CuSO_4$ solution via the salt bridge into the $ZnSO_4$ solution (internal circuit) to complete the circuit. The equations for the reactions of this cell are

anode	$Zn^0(s) \longrightarrow Zn^{2+}(aq) + 2e^-$	(oxidation)
cathode	$Cu^{2+}(aq) + 2e^- \longrightarrow Cu^0(s)$	(reduction)
net ionic	$Zn^0(s) + Cu^{2+}(aq) \longrightarrow Zn^{2+}(aq) + Cu^0(s)$	
overall	$Zn(s) + CuSO_4(aq) \longrightarrow Cu(s) + ZnSO_4(aq)$	

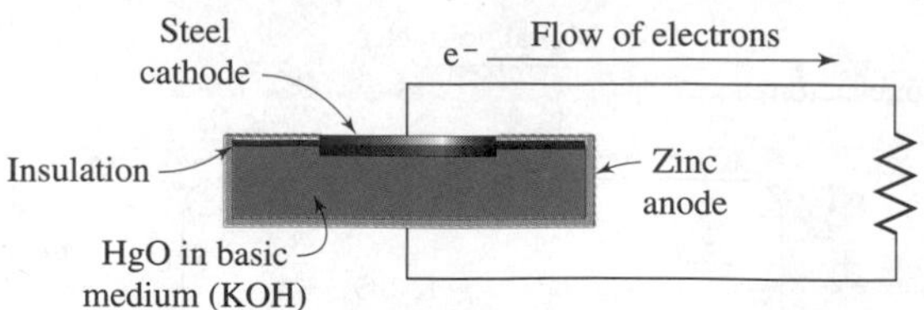

FIGURE 17.5 ▶ Diagram of an alkaline zinc–mercury cell.

The redox reaction, the movement of electrons in the metallic or external part of the circuit, and the movement of ions in the solution or internal part of the circuit of the copper–zinc cell are very similar to the actions that occur in the electrolytic cell of Figure 17.3. The only important difference is that the reactions of the zinc–copper cell are spontaneous. This spontaneity is the crucial difference between all voltaic and electrolytic cells.

Voltaic cells use chemical reactions to produce electrical energy, and electrolytic cells use electrical energy to produce chemical reactions.

Although the zinc–copper voltaic cell is no longer used commercially, it was used to energize the first transcontinental telegraph lines. Such cells were the direct ancestors of the many different kinds of "dry" cells that operate portable radio and television sets, automatic cameras, tape recorders, and so on.

One such "dry" cell, the alkaline zinc–mercury cell, is diagrammed in Figure 17.5. The reactions occurring in this cell are

anode	$Zn^0 + 2\ OH^- \longrightarrow ZnO + H_2O + 2e^-$	(oxidation)
cathode	$HgO + H_2O + 2e^- \longrightarrow Hg^0 + 2\ OH^-$	(reduction)
net ionic	$Zn^0 + Hg^{2+} \longrightarrow Zn^{2+} + Hg^0$	
overall	$Zn^0 + HgO \longrightarrow ZnO + Hg^0$	

To offset the relatively high initial cost, this cell (a) provides current at a very steady potential of about 1.5 volts, (b) has an exceptionally long service life—that is, high energy output to weight ratio, (c) is completely self-contained, and (d) can be stored for relatively long periods of time when not in use.

An automobile storage battery is an energy reservoir. The charged battery acts as a voltaic cell and through chemical reactions furnishes electrical energy to operate the starter, lights, radio, and so on. When the engine is running, a generator, or alternator, produces and forces an electric current through the battery and, by electrolytic chemical action, restores it to the charged condition.

The cell unit consists of a lead plate filled with spongy lead and a lead dioxide plate, both immersed in dilute sulfuric acid solution, which serves as the electrolyte (see Figure 17.6). When the cell is discharging, or acting as a voltaic cell, these reactions occur:

Pb plate (anode):	$Pb^0 \longrightarrow Pb^{2+} + 2e^-$	(oxidation)
PbO_2 plate (cathode):	$PbO_2 + 4\ H^+ + 2e^- \longrightarrow Pb^{2+} + 2\ H_2O$	(reduction)

A New Look for a Great Lady

Corrosion has been a problem for the world since the earliest use of metals. One familiar example of the complexities of corrosion has been the renovation of our famous Statue of Liberty. The beautiful lady was given to the United States by France in 1886. She is constructed of a skeleton of iron bars (ribs) that were bent to precisely conform to her shape and covered with a thin skin of copper. Her skin is connected to the skeleton with special copper "saddles" or bands that overlap the iron bars and are riveted in place.

Iron will corrode (oxidize) in moist air and, if in contact with copper and an electrolyte (seawater), the corrosion rate can increase one-hundred-fold. The harbor environment of New York fosters accelerated corrosion, and within less than a century the iron ribs had corroded to a point where the lady was in serious danger of collapse.

In order to repair and reinforce the statue each of the iron bars (over 1300 in all) were replaced one at a time with stainless steel, which corrodes much less easily. At the same time, the copper saddles were coated with Teflon to insulate them from contact with the iron and further reduce future corrosion.

The Statue of Liberty was in serious danger of collapse before renovation.

The external appearance of the statue was left essentially unchanged. The bluish green patina is a natural product formed as copper undergoes atmospheric corrosion. Patina may take several forms and exists as a protective film that adheres tightly to the surface of the copper and reduces the rate of corrosion underneath.

The patina of the Statue of Liberty is primarily $CuSO_4 \cdot 3Cu(OH)_2$. It has protected the lady well—during her hundred years, the copper skin has thinned only about 4%. Unfortunately, acid rain appears to be changing the patina to $CuSO_4 \cdot 2Cu(OH)_2$, which doesn't bond as closely to the copper surface. Chemists are concerned that this change will result in the loss of patina and increased corrosion of the copper skin.

Net ionic redox reaction: $Pb^0 + PbO_2 + 4\,H^+ \longrightarrow 2\,Pb^{2+} + 2\,H_2O$

Precipitation reaction on plates: $Pb^{2+}(aq) + SO_4^{2-}(aq) \longrightarrow PbSO_4(s)$

Because lead(II) sulfate is insoluble, the Pb^{2+} ions combine with SO_4^{2-} ions to form a coating of $PbSO_4$ on each plate. The overall chemical reaction of the cell is

$$Pb(s) + PbO_2(s) + 2\,H_2SO_4(aq) \xrightarrow[\text{cycle}]{\text{discharge}} 2\,PbSO_4(s) + 2\,H_2O(l)$$

The cell can be recharged by reversing the chemical reaction. This reversal is accomplished by forcing an electric current through the cell in the opposite direction. Lead sulfate and water are reconverted to lead, lead (IV) oxide, and sulfuric acid:

$$2\,PbSO_4(s) + 2\,H_2O(l) \xrightarrow[\text{cycle}]{\text{charge}} Pb(s) + PbO_2(s) + 2\,H_2SO_4(aq)$$

The electrolyte in a lead storage battery is a 38% by mass sulfuric acid solution having a density of 1.29 g/mL. As the battery is discharged, sulfuric acid is removed, thereby decreasing the density of the electrolyte solution. The state of charge or

FIGURE 17.6 ▶
Cross-sectional diagram of a lead storage battery cell.

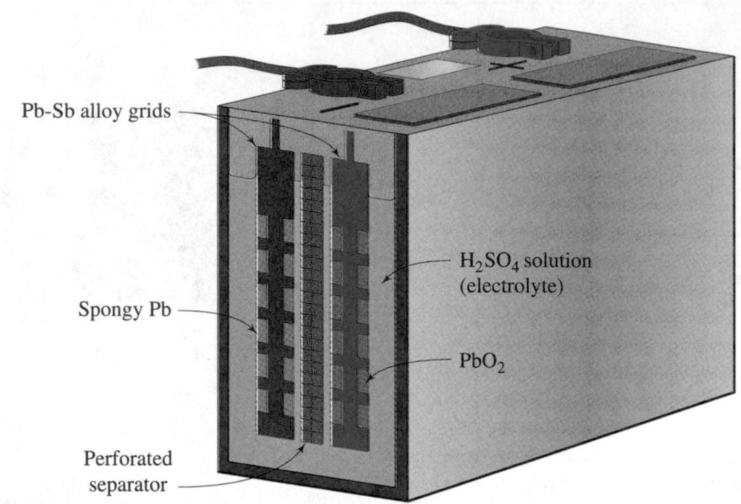

discharge of the battery can be estimated by measuring the density (or specific gravity) of the electrolyte solution with a hydrometer. When the density has dropped to about 1.05 g/mL, the battery needs recharging.

In a commercial battery, each cell consists of a series of cell units of alternating lead–lead (IV) oxide plates separated and supported by wood, glass wool, or fiberglass. The energy storage capacity of a single cell is limited, and its electrical potential is only about 2 volts. Therefore, a bank of six cells is connected in series to provide the 12-volt output of the usual automobile battery.

Concepts in Review

1. Assign oxidation numbers to all the elements in a given compound or ion.
2. Determine which element is being oxidized and which element is being reduced in an oxidation–reduction reaction.
3. Identify the oxidizing agent and the reducing agent in an oxidation–reduction reaction.
4. Balance oxidation–reduction equations in molecular and ionic forms.
5. Outline the general principles concerning the activity series of the metals.
6. Use the activity series to determine whether a proposed single-displacement reaction will occur.
7. Distinguish between an electrolytic and a voltaic cell.
8. Draw a voltaic cell that will produce electric current from an oxidation–reduction reaction involving two metals and their salts.
9. Identify the anode reaction and the cathode reaction in a given electrolytic or voltaic cell.

Key Terms

The terms listed here have been defined within this chapter. Section numbers are referenced in parenthesis for each term.

activity series of metals (17.5)
anode (17.6)
cathode (17.6)
electrolysis (17.6)
electrolytic cell (17.6)
oxidation (17.2)
oxidation number (17.1)
oxidation–reduction (17.2)
oxidation state (17.1)
oxidizing agent (17.2)
redox (17.2)
reducing agent (17.2)
reduction (17.2)
voltaic cell (17.6)

Questions

Questions refer to tables, figures, and key words and concepts defined within the chapter. A particularly challenging question or exercise is indicated with an asterisk.

1. In the equation

$$I_2 + 5\ Cl_2 + 6\ H_2O \longrightarrow 2\ HIO_3 + 10\ HCl$$

(a) has iodine been oxidized or has it been reduced?
(b) has chlorine been oxidized or has it been reduced? (Figure 17.1)

2. Based on Table 17.3, which element of each pair is more active?
(a) Ag or Al
(b) Na or Ba
(c) Ni or Cu

3. Based on Table 17.3, will the following combinations react in aqueous solution?
(a) $Zn + Cu^{2+}$
(b) $Ag + H^+$
(c) $Sn + Ag^+$
(d) $As + Mg^{2+}$
(e) $Ba + FeCl_2$
(f) $Pb + NaCl$
(g) $Ni + Hg(NO_3)_2$
(h) $Al + CuSO_4$

4. The reaction between powdered aluminum and iron(III) oxide (in the thermite process) producing molten iron is very exothermic.
(a) Write the equation for the chemical reaction that occurs.
(b) Explain in terms of Table 17.3 why a reaction occurs.
(c) Would you expect a reaction between powdered iron and aluminum oxide?
(d) Would you expect a reaction between powdered aluminum and chromium(III) oxide?

5. Write equations for the chemical reaction of each of the following metals with dilute solutions of (a) hydrochloric acid and (b) sulfuric acid: aluminum, chromium, gold, iron, copper, magnesium, mercury, and zinc. If a reaction will not occur, write "no reaction" as the product. (See Table 17.3)

6. An $NiCl_2$ solution is placed in the apparatus shown in Figure 17.3, instead of the HCl solution shown. Write equations for
(a) the anode reaction
(b) the cathode reaction
(c) the net electrochemical reaction

7. What is the major distinction between the reactions occurring in Figure 17.3 and those in Figure 17.4?

8. In the cell shown in Figure 17.4,
(a) what would be the effect of removing the voltmeter and connecting the wires shown coming to the voltmeter?
(b) what would be the effect of removing the salt bridge?

9. Why are oxidation and reduction said to be complementary processes?

10. When molten $CaBr_2$ is electrolyzed, calcium metal and bromine are produced. Write equations for the two half-reactions that occur at the electrodes. Label the anode half-reaction and the cathode half-reaction.

11. Why is direct current used instead of alternating current in the electroplating of metals?

12. What property of lead IV oxide and lead(II) sulfate makes it unnecessary to have salt bridges in the cells of a lead storage battery?

13. Explain why the density of the electrolyte in a lead storage battery decreases during the discharge cycle.

14. In one type of alkaline cell used to power devices such as portable radios, Hg^{2+} ions are reduced to metallic mercury when the cell is being discharged. Does this reduction occur at the anode or the cathode? Explain.

15. Differentiate between an electrolytic cell and a voltaic cell.

16. Why is a porous barrier or a salt bridge necessary in some voltaic cells?

17. Explain the chemical basis for the corrosion and deterioration of the Statue of Liberty.

18. What is patina? How does it act to protect an object from further corrosion?

19. What is photochromic glass? How does it work?

20. Which of the following statements are correct? Rewrite the incorrect statements to make them correct.
    (a) An atom of an element in the uncombined state has an oxidation number of zero.
    (b) The oxidation number of molybdenum in Na_2MoO_4 is +4.
    (c) The oxidation number of an ion is the same as the electrical charge on the ion.
    (d) The process in which an atom or an ion loses electrons is called reduction.
    (e) The reaction $Fe^{3+} + e^- \longrightarrow Fe^{2+}$ is a reduction reaction.
    (f) In the reaction

$$2\ Al + 3\ CuCl_2 \longrightarrow 2\ AlCl_3 + 3\ Cu$$

    aluminum is the oxidizing agent.
    (g) In a redox reaction the oxidizing agent is reduced and the reducing agent is oxidized.
    (h) $Cu^0 \longrightarrow Cu^{2+}$ is a balanced oxidation half-reaction.
    (i) In the electrolysis of sodium chloride brine (solution), Cl_2 gas is formed at the cathode, and hydroxide ions are formed at the anode.
    (j) In any cell, electrolytic or voltaic, reduction takes place at the cathode, and oxidation occurs at the anode.
    (k) In the Zn–Cu voltaic cell, the reaction at the anode is $Zn \longrightarrow Zn^{2+} + 2e^-$.

    The statements in (l) through (o) pertain to this activity series:

    Ba Mg Zn Fe H Cu Ag

    (l) The reaction $Zn + MgCl_2 \longrightarrow Mg + ZnCl_2$ is a spontaneous reaction.
    (m) Barium is more active than copper.
    (n) Silver metal will react with acids to liberate hydrogen gas.
    (o) Iron is a better reducing agent than zinc.
    (p) Oxidation and reduction occur simultaneously in a chemical reaction; one cannot take place without the other.
    (q) A free metal can displace from solution the ions of a metal that lie below the free metal in the activity series.
    (r) In electroplating, the piece to be electroplated with a metal is attached to the cathode.
    (s) In an automobile lead storage battery, the density of the sulfuric acid solution decreases as the battery discharges.
    (t) In an electrolytic cell, chemical energy is used to produce electrical energy.

Paired Exercises

These exercises are paired. Each odd-numbered exercise is followed by a similar even-numbered exercise. Answers to the even-numbered exercises are given in Appendix V.

21. What is the oxidation number of the underlined element in each compound?
    (a) $\underline{Na}Cl$
    (b) $Fe\underline{Cl}_3$
    (c) $\underline{Pb}O_2$
    (d) $Na\underline{N}O_3$
    (e) $H_2\underline{S}O_3$
    (f) $\underline{N}H_4Cl$

22. What is the oxidation number of the underlined element in each compound?
    (a) $K\underline{Mn}O_4$
    (b) $\underline{I}_2$
    (c) $\underline{N}H_3$
    (d) $K\underline{Cl}O_3$
    (e) $K_2\underline{Cr}O_4$
    (f) $K_2\underline{Cr}_2O_7$

23. What is the oxidation number of the underlined elements?
    (a) $\underline{S}^{2-}$
    (b) $\underline{N}O_2^-$
    (c) $Na_2\underline{O}_2$
    (d) $\underline{Bi}^{3+}$

24. What is the oxidation number of the underlined elements?
    (a) $\underline{O}_2$
    (b) $\underline{As}O_4^{3-}$
    (c) $Fe(\underline{O}H)_3$
    (d) $\underline{I}O_3^-$

25. In the following half-reactions, which element is changing oxidation state? Is the half-reaction an oxidation or a reduction? Supply the proper number of electrons to the proper side to balance each equation.
(a) $Zn^{2+} \longrightarrow Zn$
(b) $2\ Br^- \longrightarrow Br_2$
(c) $MnO_4^- + 8\ H^+ \longrightarrow Mn^{2+} + 4\ H_2O$
(d) $Ni \longrightarrow Ni^{2+}$

26. In the following half-reactions, which element is changing oxidation state? Is the half-reaction an oxidation or a reduction? Supply the proper number of electrons to the proper side to balance each equation:
(a) $SO_3^{2-} + H_2O \longrightarrow SO_4^{2-} + 2\ H^+$
(b) $NO_3^- + 4\ H^+ \longrightarrow NO + 2\ H_2O$
(c) $S_2O_4^{2-} + 2\ H_2O \longrightarrow 2\ SO_3^{2-} + 4\ H^+$
(d) $Fe^{2+} \longrightarrow Fe^{3+}$

27. In the following unbalanced equations,
(a) identify the element that is oxidized and the element that is reduced.
(b) identify the oxidizing agent and the reducing agent:
(1) $Cr + HCl \longrightarrow CrCl_3 + H_2$
(2) $SO_4^{2-} + I^- + H^+ \longrightarrow H_2S + I_2 + H_2O$

28. In the following unbalanced equations,
(a) identify the element that is oxidized and the element that is reduced.
(b) identify the oxidizing agent and the reducing agent:
(1) $AsH_3 + Ag^+ + H_2O \longrightarrow H_3AsO_4 + Ag + H^+$
(2) $Cl_2 + NaBr \longrightarrow NaCl + Br_2$

29. Balance these equations by the change-in-oxidation-number method:
(a) $Zn + S \longrightarrow ZnS$
(b) $AgNO_3 + Pb \longrightarrow Pb(NO_3)_2 + Ag$
(c) $Fe_2O_3 + CO \longrightarrow Fe + CO_2$
(d) $H_2S + HNO_3 \longrightarrow S + NO + H_2O$
(e) $MnO_2 + HBr \longrightarrow MnBr_2 + Br_2 + H_2O$

30. Balance these equations by the change-in-oxidation-number method:
(a) $Cl_2 + KOH \longrightarrow KCl + KClO_3 + H_2O$
(b) $Ag + HNO_3 \longrightarrow AgNO_3 + NO + H_2O$
(c) $CuO + NH_3 \longrightarrow N_2 + Cu + H_2O$
(d) $PbO_2 + Sb + NaOH \longrightarrow PbO + NaSbO_2 + H_2O$
(e) $H_2O_2 + KMnO_4 + H_2SO_4 \longrightarrow$
$O_2 + MnSO_4 + K_2SO_4 + H_2O$

31. Balance the following ionic redox equations using the ion–electron method. All of these reactions occur in acidic solution.
(a) $Zn + NO_3^- \longrightarrow Zn^{2+} + NH_4^+$
(b) $NO_3^- + S \longrightarrow NO_2 + SO_4^{2-}$
(c) $PH_3 + I_2 \longrightarrow H_3PO_2 + I^-$
(d) $Cu + NO_3^- \longrightarrow Cu^{2+} + NO$
*(e) $ClO_3^- + Cl^- \longrightarrow Cl_2$

32. Balance the following ionic redox equations using the ion–electron method. All of these reactions occur in acidic solution.
(a) $ClO_3^- + I^- \longrightarrow I_2 + Cl^-$
(b) $Cr_2O_7^{2-} + Fe^{2+} \longrightarrow Cr^{3+} + Fe^{3+}$
(c) $MnO_4^- + SO_2 \longrightarrow Mn^{2+} + SO_4^{2-}$
(d) $H_3AsO_3 + MnO_4^- \longrightarrow H_3AsO_4 + Mn^{2+}$
*(e) $Cr_2O_7^{2-} + H_3AsO_3 \longrightarrow Cr^{3+} + H_3AsO_4$

33. Balance the following ionic redox equations using the ion–electron method. All of these reactions occur in basic solutions.
(a) $Cl_2 + IO_3^- \longrightarrow Cl^- + IO_4^-$
(b) $MnO_4^- + ClO_2^- \longrightarrow MnO_2 + ClO_4^-$
(c) $Se \longrightarrow Se^{2-} + SeO_3^{2-}$
*(d) $Fe_3O_4 + MnO_4^- \longrightarrow Fe_2O_3 + MnO_2$
*(e) $BrO^- + Cr(OH)_4^- \longrightarrow Br^- + CrO_4^{2-}$

34. Balance the following ionic redox equations using the ion–electron method. All of these reactions occur in basic solutions.
(a) $MnO_4^- + SO_3^{2-} \longrightarrow MnO_2 + SO_4^{2-}$
(b) $ClO_2 + SbO_2^- \longrightarrow ClO_2^- + Sb(OH)_6^-$
(c) $Al + NO_3^- \longrightarrow NH_3 + Al(OH)_4^-$
*(d) $P_4 \longrightarrow HPO_3^{2-} + PH_3$
*(e) $Al + OH^- \longrightarrow Al(OH)_4^- + H_2$

Additional Exercises

These exercises are not paired or labeled by topic and provide additional practice on concepts covered in this chapter.

35. The chemical reactions taking place during discharge in a lead storage battery are

$$Pb + SO_4^{2-} \longrightarrow PbSO_4$$
$$PbO_2 + SO_4^{2-} + 4\,H^+ \longrightarrow PbSO_4 + 2\,H_2O$$

(a) Complete each half-reaction by supplying electrons.
(b) Which reaction is oxidation and which is reduction?
(c) Which reaction occurs at the anode of the battery?

36. Use the following unbalanced redox equation to indicate

$$KMnO_4 + HCl \longrightarrow KCl + MnCl_2 + H_2O + Cl_2$$

(a) the oxidizing agent
(b) the reducing agent
(c) the number of electrons that are transferred per mole of oxidizing agent

37. How many moles of NO gas will be formed by the reaction of 25.0 g of silver with nitric acid?

$$Ag + HNO_3 \longrightarrow AgNO_3 + NO + H_2O \quad \text{(unbalanced)}$$

38. What volume of chlorine gas, measured at STP, is required to react with excess KOH to form 0.300 mol of $KClO_3$?

$$Cl_2 + KOH \longrightarrow KCl + KClO_3 + H_2O \quad \text{(unbalanced)}$$

39. What mass of $KMnO_4$ would be needed to react with 100. mL of H_2O_2 solution ($d = 1.031$ g/mL, 9.0% H_2O_2 by mass)?

$$H_2O_2 + KMnO_4 + H_2SO_4 \longrightarrow O_2 + MnSO_4 + K_2SO_4 + H_2O \quad \text{(unbalanced)}$$

***40.** What volume of 0.200 M $K_2Cr_2O_7$ will be required to oxidize 5.00 g of H_3AsO_3?

$$Cr_2O_7^{2-} + H_3AsO_3 \longrightarrow Cr^{3+} + H_3AsO_4 \quad \text{(unbalanced)}$$

***41.** What volume of 0.200 M $K_2Cr_2O_7$ will be required to oxidize the Fe^{2+} ion in 60.0 mL of 0.200 M $FeSO_4$ solution?

$$Cr_2O_7^{2-} + Fe^{2+} \longrightarrow Cr^{3+} + Fe^{3+} \quad \text{(unbalanced)}$$

***42.** A sample of crude potassium iodide was analyzed using the following reaction (not balanced):

$$I^- + SO_4^{2-} \longrightarrow I_2 + H_2S \quad \text{(acid solution)}$$

If a 4.00-g sample of crude KI produced 2.79 g of iodine, what is the percent purity of the KI?

***43.** What volume of NO gas, measured at 28°C and 744 torr, will be formed by the reaction of 0.500 mol of Ag reacting with excess nitric acid?

$$Ag + HNO_3 \longrightarrow AgNO_3 + NO + H_2O \quad \text{(unbalanced)}$$

44. How many moles of H_2 can be produced from 100.0 g of Al according to the following reaction?

$$Al + OH^- \longrightarrow Al(OH)_4^- + H_2 \quad \text{(basic solution)}$$

45. There is something incorrect about each of the following half-reactions:

(a) $Cu^+ + e^- \longrightarrow Cu^{2+}$ **(b)** $Pb^{2+} + e^{2-} \longrightarrow Pb$

Identify what is wrong, and correct each half-reaction.

46. Why can oxidation *never* occur without reduction?

47. The following observations were made concerning four different metals: A, B, C, and D.

(a) When a strip of metal A is placed in a solution of B^{2+} ions, no reaction is observed.
(b) Similarly, A in a solution containing C^+ ions produces no reaction.
(c) When a strip of metal D is placed in a solution of C^+ ions, black metallic C deposits on the surface of D, and the solution tests positively for D^{2+} ions.
(d) When a piece of metallic B is placed in a solution of D^{2+} ions, metallic D appears on the surface of B and B^{2+} ions are found in the solution.

Arrange the ions, A^+, B^{2+}, C^+, and D^{2+} in order of their ability to attract electrons. List them in order of increasing ability.

48. Tin normally has oxidation numbers of 0, +2, and +4. Which of these species can be an oxidizing agent, which can be a reducing agent, and which can be both? In each case, what product would you expect as the tin reacts?

49. Manganese is an element that can take on numerous oxidation states. In each of these compounds, identify the oxidation number of the manganese. Which compound would you expect to be the best oxidizing agent and why?

(a) $Mn(OH)_2$
(b) MnF_3
(c) MnO_2
(d) K_2MnO_4
(e) $KMnO_4$

50. Which of these equations represent oxidations?

(a) $Mg \longrightarrow Mg^{2+}$
(b) $SO_2 \longrightarrow SO_3$
(c) $KMnO_4 \longrightarrow MnO_2$
(d) $Cl_2O_3 \longrightarrow Cl^-$

***51.** In the equation that follows, one can see the reaction between manganese (IV) oxide and bromide ions:

$$MnO_2 + Br^- \longrightarrow Br_2 + Mn^{2+}$$

(a) Balance this redox reaction in acidic solution.
(b) How many grams of MnO_2 would be needed to produce 100.0 mL of 0.05 M Mn^{2+}?
(c) How many liters of bromine vapor at 50°C and 1.4 atm would also result?

52. Use the table shown below to complete the following reactions. If no reaction occurs, write NR:
(a) $F_2 + Cl^- \longrightarrow$
(b) $Br_2 + Cl^- \longrightarrow$
(c) $I_2 + Cl^- \longrightarrow$
(d) $Br_2 + I^- \longrightarrow$

Activity
F_2
Cl_2
Br_2
I_2

(↑ ease of reduction)

53. Manganese metal reacts with HCl to give hydrogen gas and the Mn^{2+} ion in solution. Write a balanced equation for the reaction.

54. If zinc is allowed to react with dilute nitric acid, zinc is oxidized to the +2 ion, while the nitrate ion can be reduced to ammonium, NH_4^+. Write a balanced equation for the reaction in acidic solution.

55. In each of the following equations, identify
(a) the atom or ion oxidized
(b) the atom or ion reduced
(c) the oxidizing agent
(d) the reducing agent
(e) the change in oxidation number associated with each oxidizing process
(f) the change in oxidation number associated with each reducing process
(1) $C_3H_8 + O_2 \longrightarrow CO_2 + H_2O$
(2) $HNO_3 + H_2S \longrightarrow NO + S + H_2O$
(3) $CuO + NH_3 \longrightarrow N_2 + H_2O + Cu$
(4) $H_2O_2 + Na_2SO_3 \longrightarrow Na_2SO_4 + H_2O$
(5) $H_2O_2 \longrightarrow H_2O + O_2$

56. In the galvanic cell shown in the diagram, a strip of silver is placed in a solution of silver nitrate, and a strip of lead placed in a solution of lead(II) nitrate. The two beakers are connected with a salt bridge. Determine
(a) the anode
(b) the cathode
(c) where oxidation occurs
(d) where reduction occurs
(e) which direction electrons flow through the wire
(f) which direction ions flow through the solution

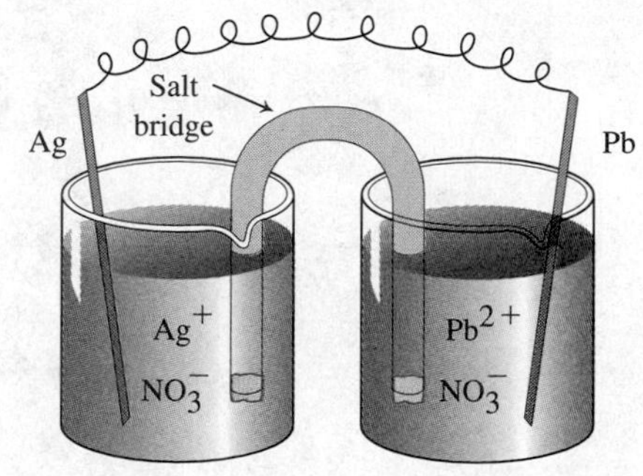

Answers to Practice Exercises

17.1 (a) S = +6, (b) As = +5, (c) C = +4
17.2 (*Note*: H = +1 even though it comes second in the formula; N is a nonmetal.)
(a) N = −3, (b) Cr = +6, (c) P = +5
17.3 (a) Be = +2; Cl = −1, (b) H = +1; Cl = +1; O = −2, (c) H = +1; O = −1, (d) N = −3; H = +1, (e) Br = +5; O = −2
17.4 (a) $6\,HNO_3 + S \longrightarrow 6\,NO_2 + H_2SO_4 + 2H_2O$,
(b) $2\,CrCl_3 + 3\,MnO_2 + 2\,H_2O \longrightarrow 3\,MnCl_2 + 2\,H_2CrO_4$,
(c) $2\,KMnO_4 + 6\,HCl + 5\,H_2S \longrightarrow 2\,KCl + 2\,MnCl_2 + 5\,S + 8\,H_2O$
17.5 (a) $4\,H^+ + 2\,I^- + 2\,NO_2^- \longrightarrow I_2 + 2\,NO + 2\,H_2O$,
(b) $2\,OH^- + Cl_2 + IO_3^- \longrightarrow IO_4^- + H_2O + 2\,Cl^-$,
(c) $AuCl_4^- + Sn^{2+} \longrightarrow Sn^{4+} + AuCl + 3Cl^-$
17.6 (a) $Zn + 2\,H_2O + 2\,OH^- \longrightarrow Zn(OH)_4^{2-} + H_2$,
(b) $H_2O_2 + Sn^{2+} + 2\,H^+ \longrightarrow Sn^{4+} + 2\,H_2O$,
(c) $Cu + Cu^{2+} + 2\,OH^- \longrightarrow Cu_2O + H_2O$
17.7 (a) yes, (b) no, (c) no

Nuclear Chemistry

18

Chapter 18

An unusual telephone conversation originated from the University of Chicago in December of 1942. The caller was American physicist Arthur Compton. "Jim," he said, "the Italian navigator has just landed in the new world." The reference was to the great physicist Enrico Fermi; the landing was the first controlled atomic chain reaction; and the new world was the dawning of the nuclear age.

A new world indeed! Locked within the nucleus of an atom is a source of tremendous energy. Its power can be devastating, as seen in the effects of nuclear weapons, or accidents such as Three Mile Island or Chernobyl. But nuclear energy can also be harnessed to perform such useful tasks as generating power, treating cancer, and preserving food. Isotopes are used in medicine to diagnose illness and detect minute quantities of drugs or hormones. Researchers are using radioactive tracers to sequence the human genome. The applications of nuclear chemistry are important in medicine, industry, history, and research. Its impact both threatens and enhances our lives and our future.

18.1 Discovery of Radioactivity

One of the most important steps leading to the discovery of radioactivity was made in 1895 by Wilhelm Konrad Roentgen (1845–1923). Roentgen discovered X rays when he observed that a vacuum discharge tube, enclosed in a thin, black cardboard box, caused a nearby piece of paper coated with barium platinocyanide to glow with a brilliant phosphorescence. From this and other experiments he concluded that certain rays, which he called X rays, were emitted from the discharge tube, penetrated the box, and caused the salt to glow. Roentgen also showed that X rays could penetrate other bodies and affect photographic plates. This observation led to the development of X-ray photography.

Shortly after this discovery, Antoine Henri Becquerel (1852–1908) attempted to show a relationship between X rays and the phosphorescence of uranium salts. In one of his experiments he wrapped a photographic plate in black paper, placed a sample of uranium salt on it, and exposed it to sunlight. The developed photographic plate showed that rays emitted from the salt had penetrated the paper. Later Becquerel prepared to repeat the experiment, but, because the sunlight was intermittent, he placed the entire setup in a drawer. Several days later he developed the photographic plate, expecting to find it only slightly affected. He was amazed to observe an intense image on the plate. He repeated the experiment in total darkness and obtained the same results, proving that the uranium salt emitted rays that affected the photographic plate without its being exposed to sunlight. In this way Becquerel discovered radioactivity, but the actual name radioactivity was given to this phenomenon two years later (in 1898) by Marie Curie. **Radioactivity** is the spontaneous emission of particles and/or rays from the nucleus of an atom. Elements having this property are said to be radioactive. Becquerel later showed that the rays coming from uranium were able to ionize air and were also capable of penetrating thin sheets of metal.

radioactivity

In 1898, Marie Sklodowska Curie (1867–1934) and her husband Pierre Curie (1859–1906) turned their research interests to radioactivity. In a short time the Curies discovered two new elements, polonium and radium, both of which are radio-

◀ Chapter Opening Photo: The rings on a bristlecone pine provide a means of calibrating the carbon-14 system used for dating artifacts.

Marie Curie (1867–1934).

active. To confirm their work on radium they processed 1 ton of pitch-blende residue ore to obtain 0.1 g of pure radium chloride, which they used to make further studies on the properties of radium and to determine its atomic mass.

In 1899, Ernest Rutherford began to investigate the nature of the rays emitted from uranium. He found two particles, which he called *alpha* and *beta particles*. Soon he realized that uranium, while emitting these particles, was changing into another element. By 1912, over 30 radioactive isotopes were known, and many more are known today. The *gamma ray*, a third type of emission from radioactive materials and similar to an X ray, was discovered by Paul Villard in 1900. After the description of the nuclear atom by Rutherford, the phenomenon of radioactivity was attributed to reactions taking place in the nuclei of atoms.

The symbolism and notation described for isotopes in Chapter 5 is very useful in nuclear chemistry and is briefly reviewed here:

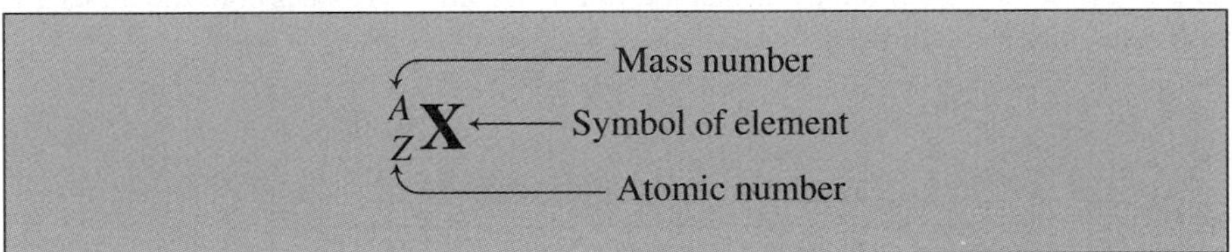

For example, $^{238}_{92}U$ represents a uranium isotope with an atomic number of 92 and a mass number of 238. This isotope is also designated as U-238 or uranium-238 and contains 92 protons and 146 neutrons. The protons and neutrons collectively

nucleon

are known as **nucleons**. The mass number is the total number of nucleons in the nucleus. Table 18.1 shows the isotopic notations for several particles associated with nuclear chemistry.

When we speak of isotopes, we generally mean atoms of the same element with different masses, such as $^{16}_{8}O$, $^{17}_{8}O$, $^{18}_{8}O$. In nuclear chemistry, we use the term

nuclide

nuclide to mean any isotope of any atom. Thus $^{16}_{8}O$ and $^{235}_{92}U$ are referred to as nuclides. Nuclides that spontaneously emit radiation are referred to as *radionuclides*.

TABLE 18.1 Symbols in Isotopic Notation for Several Particles (and Small Isotopes) Associated with Nuclear Chemistry

Particle	Symbol	Atomic number Z	Mass number A
Neutron	$^{1}_{0}n$	0	1
Proton	$^{1}_{1}H$	1	1
Beta particle (electron)	$^{0}_{-1}e$	−1	0
Positron (positive electron)	$^{0}_{+1}e$	1	0
Alpha particle (helium nucleus)	$^{4}_{2}He$	2	4
Deuteron (heavy hydrogen nucleus)	$^{2}_{1}H$	1	2

18.2 Natural Radioactivity

radioactive decay

Radioactive elements continuously undergo **radioactive decay**, or disintegration, to form different elements. The chemical properties of an element are associated with its electronic structure, but radioactivity is a property of the nucleus. Therefore, neither ordinary changes of temperature and pressure nor the chemical or physical state of an element has any effect on its radioactivity.

The principal emissions from the nuclei of radionuclides are known as alpha rays (or particles), beta rays (or particles), and gamma rays. Upon losing an alpha or beta particle, the radioactive element changes into a different element. We will explain this process in detail later.

half-life

Each radioactive nuclide disintegrates at a specific and constant rate, which is expressed in units of half-life. The **half-life** ($t_{1/2}$) is the time required for one-half of a specific amount of a radioactive nuclide to disintegrate. The half-lives of the elements range from a fraction of a second to billions of years. For example, $^{238}_{92}U$ has a half-life of 4.5×10^9 years, $^{226}_{88}Ra$ has a half-life of 1620 years, and $^{15}_{6}C$ has a half-life of 2.4 seconds. To illustrate, if we start today with 1.0 g of $^{226}_{88}Ra$, we will have 0.50 g of $^{226}_{88}Ra$ remaining at the end of 1620 years; at the end of another 1620 years, 0.25 g will remain; and so on.

$$1.0\text{ g }^{226}_{88}\text{Ra} \xrightarrow[\text{1620 years}]{t_{1/2}} 0.50\text{ g }^{226}_{88}\text{Ra} \xrightarrow[\text{1620 years}]{t_{1/2}} 0.25\text{ g }^{226}_{88}\text{Ra}$$

The half-lives of the various radioisotopes of the same element are different from one another. Half-lives for some isotopes of radium, carbon, and uranium are listed in Table 18.2. A radioactive decay curve is illustrated in Figure 18.1.

Example 18.1

The half-life of $^{131}_{53}I$ is 8 days. How much $^{131}_{53}I$ will be left of a 32-g sample after five half-lives?

Solution

This problem can be solved by using the graph in Figure 18.1. To find the number of grams of ^{131}I left after one half-life, trace a perpendicular line from 8 days on the x axis to the line on the graph. Now trace a horizontal line from this point on the plotted line to the y axis and read the corresponding grams of ^{131}I. This process continues for each half-life, adding 8 days to the previous value on the x axis.

Half-lives	0	1	2	3	4	5
Number of days		8	16	24	32	40
Amount remaining	32 g	16 g	8 g	4 g	2 g	1 g

Starting with 32 g, 1 g of ^{131}I will be left after five half-lives (40 days).

TABLE 18.2 Half-Lives for Radium, Carbon and Uranium Isotopes

Isotope	Half-life	Isotope	Half-life
Ra-223	11.7 days	C-14	5668 years
Ra-224	3.64 days	C-15	2.4 seconds
Ra-225	14.8 days	U-235	7.1×10^8 years
Ra-226	1620 years	U-238	4.5×10^9 years
Ra-228	6.7 years		

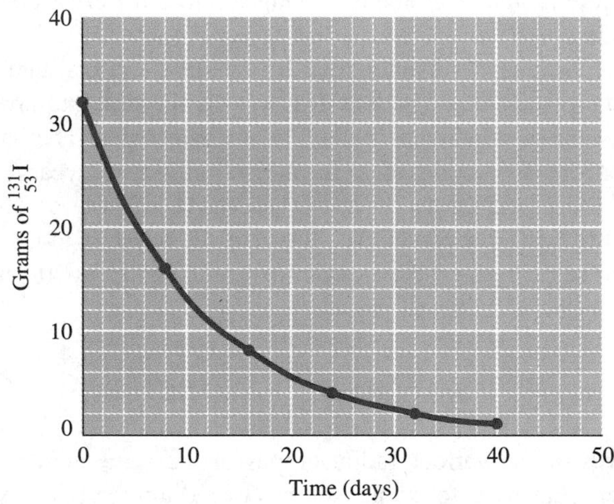

FIGURE 18.1 ▶ **Radioactive decay curve for $^{131}_{53}I$, which has a half-life of 8 days:**

$$^{131}_{53}I \longrightarrow {}^{131}_{54}Xe + {}^{0}_{-1}e$$

Example 18.2 In how many half-lives will 10.0 g of a radioactive nuclide decay to less than 10% of its original value?

Solution Ten percent of the original amount is 1.0 g. After the first half-life, half of the original material remains and half has decayed. After the second half-life, one-fourth of the original material remains, which is one-half of the starting amount at the end of the first half-life. This progression continues, reducing the quantity remaining by half for each half-life that passes.

Half-lives	0	1	2	3	4
Percent remaining	100%	50%	25%	12.5%	6.25%
Amount remaining	10.0 g	5.00 g	2.50 g	1.25 g	0.625 g

Therefore, the amount remaining will be less than 10% sometime between the third and the fourth half-lives.

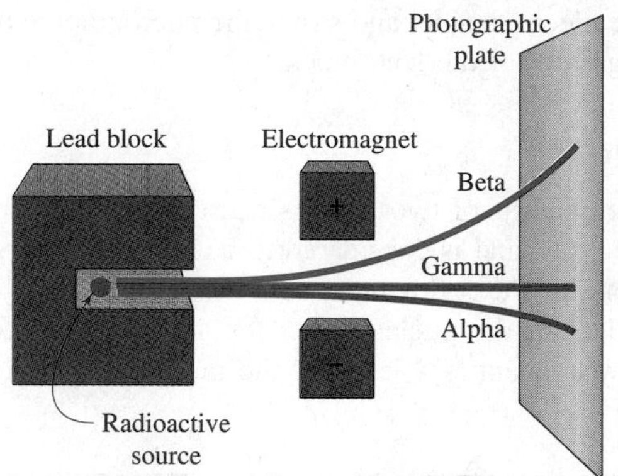

FIGURE 18.2
The effect of an electromagnetic field on alpha, beta, and gamma rays. Lighter beta particles are deflected considerably more than alpha particles. Alpha and beta particles are deflected in opposite directions. Gamma radiation is not affected by the electromagnetic field.

Practice 18.1

The half-life of $^{14}_{6}C$ is 5668 years. How much $^{14}_{6}C$ will remain after six half-lives in a sample that initially contains 25.0 g?

Nuclides are said to be either *stable* (nonradioactive) or *unstable* (radioactive). All elements that have atomic numbers greater than 83 (bismuth) are naturally radioactive, although some of the nuclides have extremely long half-lives. Some of the naturally occurring nuclides of elements 81, 82, and 83 are radioactive, and some are stable. Only a few naturally occurring elements that have atomic numbers less than 81 are radioactive. However, no stable isotopes of element 43 (technetium) or of element 61 (promethium) are known.

Radioactivity is believed to be a result of an unstable ratio of neutrons to protons in the nucleus. Stable nuclides of elements up to about atomic number 20 generally have about a 1:1 neutron-to-proton ratio. In elements above number 20 the neutron-to-proton ratio in the stable nuclides gradually increases to about 1.5:1 in element number 83 (bismuth). When the neutron-to-proton ratio is too high or too low, alpha, beta, or other particles are emitted to achieve a more stable nucleus.

18.3 Properties of Alpha Particles, Beta Particles, and Gamma Rays

The classical experiment proving that alpha and beta particles are oppositely charged was performed by Marie Curie (see Figure 18.2). She placed a radioactive source in a hole in a lead block and positioned two poles of a strong electromagnet so that the radiations that were given off passed between them. The paths of three different kinds of radiation were detected by means of a photographic plate placed some distance beyond the electromagnet. The lighter beta particles were strongly deflected toward the positive pole of the electromagnet; the heavier alpha particles were less strongly deflected and in the opposite direction. The uncharged gamma rays were not affected by the electromagnet and struck the photographic plates after traveling along a path straight out of the lead block.

Alpha Particle

alpha particle

An **alpha particle** consists of two protons and two neutrons, has a mass of about 4 amu, a charge of +2, and is considered to be a doubly charged helium atom. It is usually given one of the following symbols: α, He^{2+}, or ${}^{4}_{2}He$. When an alpha particle is emitted from the nucleus, a different element is formed. The atomic number of the new element is 2 less and the mass is 4 amu less than that of the starting element.

Loss of an alpha particle from the nucleus results in
loss of 4 in the mass number (*A*)
loss of 2 in the atomic number (*Z*)

For example, when ${}^{238}_{92}U$ loses an alpha particle, ${}^{234}_{90}Th$ is formed, because two neutrons and two protons are lost from the uranium nucleus. This disintegration may be written as a nuclear equation:

$${}^{238}_{92}U \longrightarrow {}^{234}_{90}Th + \alpha \qquad \text{or} \qquad {}^{238}_{92}U \longrightarrow {}^{234}_{90}Th + {}^{4}_{2}He$$

For the loss of an alpha particle from ${}^{226}_{88}Ra$, the equation is

$${}^{226}_{88}Ra \longrightarrow {}^{222}_{86}Rn + {}^{4}_{2}He \qquad \text{or} \qquad {}^{226}_{88}Ra \longrightarrow {}^{222}_{86}Rn + \alpha$$

A nuclear equation, like a chemical equation, consists of reactants and products and must be balanced. To have a balanced nuclear equation, the sum of the mass numbers (superscripts) on both sides of the equation must be equal, and the sum of the atomic numbers (subscripts) on both sides of the equation must be equal:

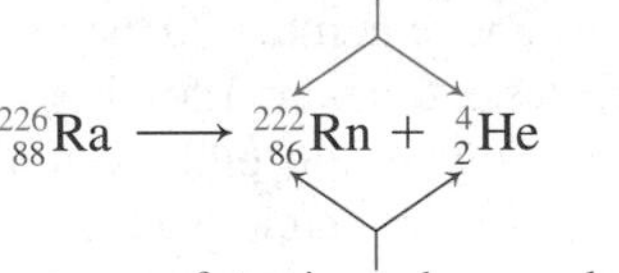

What new nuclide will be formed when ${}^{230}_{90}Th$ loses an alpha particle? This loss is equivalent to two protons and two neutrons. The new nuclide will have a mass of (230 − 4) or 226 amu and will contain (90 − 2) or 88 protons, so its atomic number is 88. Locate the corresponding element on the periodic chart. It is ${}^{226}_{88}Ra$ or radium-226.

Beta Particle

beta particle

The **beta particle** is identical in mass and charge to an electron; its charge is −1. Both a beta particle and a proton are produced by the decomposition of a neutron:

$${}^{1}_{0}n \longrightarrow {}^{1}_{1}p + {}^{\ 0}_{-1}e$$

The beta particle leaves, and the proton remains in the nucleus. When an atom loses a beta particle from its nucleus, a different element is formed that has essentially the same mass but an atomic number that is 1 greater than that of the starting element. The beta particle is written as β or ${}_{-1}^{0}e$.

> **Loss of a beta particle from the nucleus results in**
> **no change in the mass number (*A*)**
> **increase of 1 in the atomic number (*Z*)**

Examples of equations in which a beta particle is lost are

$${}_{90}^{234}Th \longrightarrow {}_{91}^{234}Pa + \beta$$

$${}_{91}^{234}Pa \longrightarrow {}_{92}^{234}U + {}_{-1}^{0}e$$

$${}_{82}^{210}Pb \longrightarrow {}_{83}^{210}Bi + \beta$$

Gamma Ray

Gamma rays are photons of energy. A gamma ray is similar to an X ray, but is more energetic. They have no electrical charge and no measurable mass. Gamma rays emanate from the nucleus in many radioactive changes along with either alpha or beta particles. The designation for a gamma ray is γ. Gamma radiation does not result in a change of atomic number or the mass of an element.

gamma ray

> **Loss of a gamma ray from the nucleus results in**
> **no change in mass number (*A*) or atomic number (*Z*)**

Example 18.3

(a) Write an equation for the loss of an alpha particle from the nuclide ${}_{78}^{194}Pt$.
(b) What nuclide is formed when ${}_{88}^{228}Ra$ loses a beta particle from its nucleus?

Solution

(a) Loss of an alpha particle, ${}_{2}^{4}He$, means the loss of two neutrons and two protons. This change results in a decrease of 4 in the mass number and a decrease of 2 in the atomic number:

Mass of new nuclide: $A - 4$ or $194 - 4 = 190$

Atomic number of new nuclide: $Z - 2$ or $78 - 2 = 76$

Looking up element number 76 on the periodic table, we find it to be osmium, Os. The equation, then, is

$${}_{78}^{194}Pt \longrightarrow {}_{76}^{190}Os + {}_{2}^{4}He$$

(b) The loss of a beta particle from a ${}_{88}^{228}Ra$ nucleus means a gain of 1 in the atomic number with no essential change in mass.
The new nuclide will have an atomic number of $(Z + 1)$ or 89, which is actinium, Ac. The nuclide formed is ${}_{89}^{228}Ac$:

$${}_{88}^{228}Ra \longrightarrow {}_{89}^{228}Ac + {}_{-1}^{0}e$$

Example 18.4 What nuclide will be formed when $^{214}_{82}Pb$ successively emits β, β, and α particles from its nucleus? Write successive equations showing these changes.

Solution The changes brought about in the three steps outlined are as follows:

β loss: Increase of 1 in the atomic number; no change in mass

β loss: Increase of 1 in the atomic number; no change in mass

α loss: Decrease of 2 in the atomic number; decrease of 4 in the mass

The equations are

$$^{214}_{82}Pb \longrightarrow {}^{214}_{83}X + \beta \longrightarrow {}^{214}_{84}X + \beta \longrightarrow {}^{210}_{82}X + \alpha$$

where X stands for the new nuclide formed. Looking up each of these elements by their atomic numbers, we rewrite the equations

$$^{214}_{82}Pb \xrightarrow[\beta]{} {}^{214}_{83}Bi \xrightarrow[\beta]{} {}^{214}_{84}Po \xrightarrow[\alpha]{} {}^{210}_{82}Pb$$

Practice 18.2

What nuclide will be formed when $^{230}_{90}Th$ emits an alpha particle?

The ability of radioactive rays to pass through various objects is in proportion to the speed at which they leave the nucleus. Gamma rays travel at the velocity of light (186,000 miles per second) and are capable of penetrating several inches of lead. The velocities of beta particles are variable, the fastest being about nine-tenths the velocity of light. Alpha particles have velocities less than one-tenth the velocity of light. Figure 18.3 illustrates the relative penetrating power of these rays. A few sheets of paper will stop alpha particles; a thin sheet of aluminum will stop both alpha and beta particles; and a 5-cm block of lead will reduce, but not completely stop, gamma radiation. In fact, it is difficult to stop all gamma radiation. Table 18.3 summarizes the properties of alpha, beta, and gamma radiation.

TABLE 18.3 Characteristics of Nuclear Radiation

Radiation	Symbol	Mass (amu)	Electrical charge	Velocity	Composition	Ionizing power
Alpha	α $^{4}_{2}He$	4	+2	Variable, less than 10% the speed of light	Identical to He^{2+}	High
Beta	β $^{0}_{-1}e$	$\frac{1}{1837}$	−1	Variable, up to 90% the speed of light	Identical to an electron	Moderate
Gamma	γ	0	0	Speed of light	Photons or electromagnetic waves of energy	Almost none

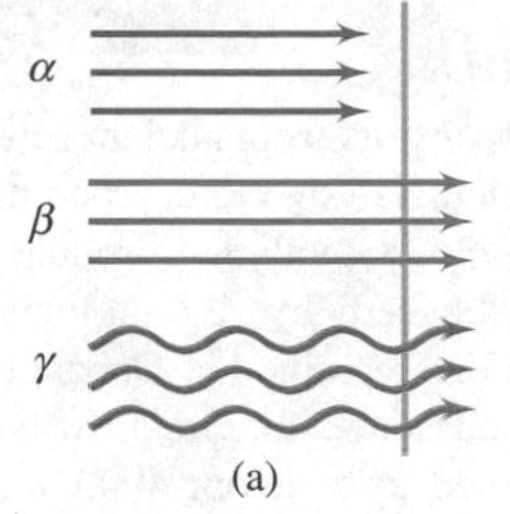

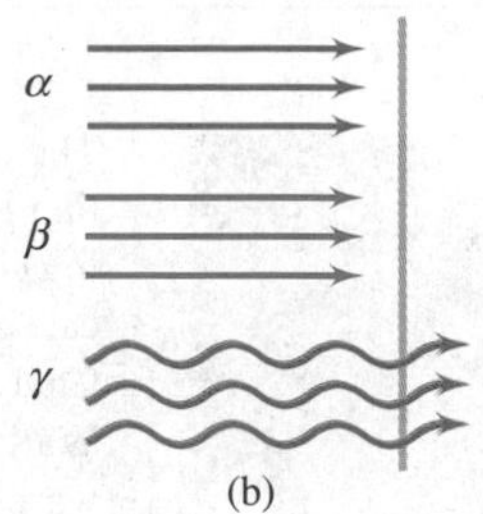

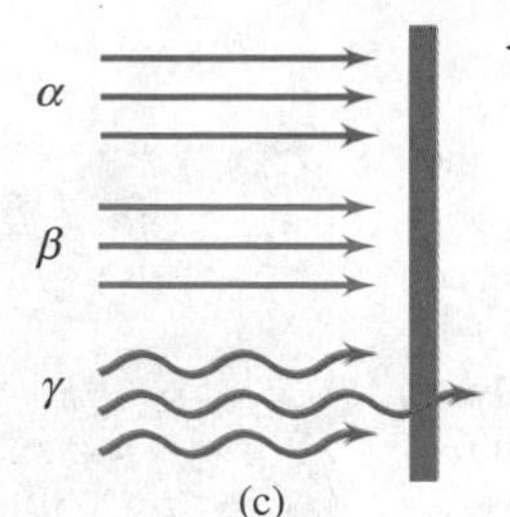

FIGURE 18.3
Relative penetrating ability of alpha, beta, and gamma radiation. (a) Thin sheet of paper; (b) thin sheet of aluminum; (c) 5-cm lead block.

18.4 Radioactive Disintegration Series

The naturally occurring radioactive elements with a higher atomic number than lead (Pb) fall into three orderly disintegration series. Each series proceeds from one element to the next by the loss of either an alpha or a beta particle, finally ending in a nonradioactive nuclide. The uranium series starts with $^{238}_{92}U$ and ends with $^{206}_{82}Pb$. The thorium series starts with $^{232}_{90}Th$ and ends with $^{208}_{82}Pb$. The actinium series starts with $^{235}_{92}U$ and ends with $^{207}_{82}Pb$. A fourth series, the neptunium series, starts with the synthetic element $^{241}_{94}Pu$ and ends with the stable bismuth nuclide $^{209}_{83}Bi$. The uranium series is shown in Figure 18.4; gamma radiation, which accompanies alpha and beta radiation, is not shown in the figure.

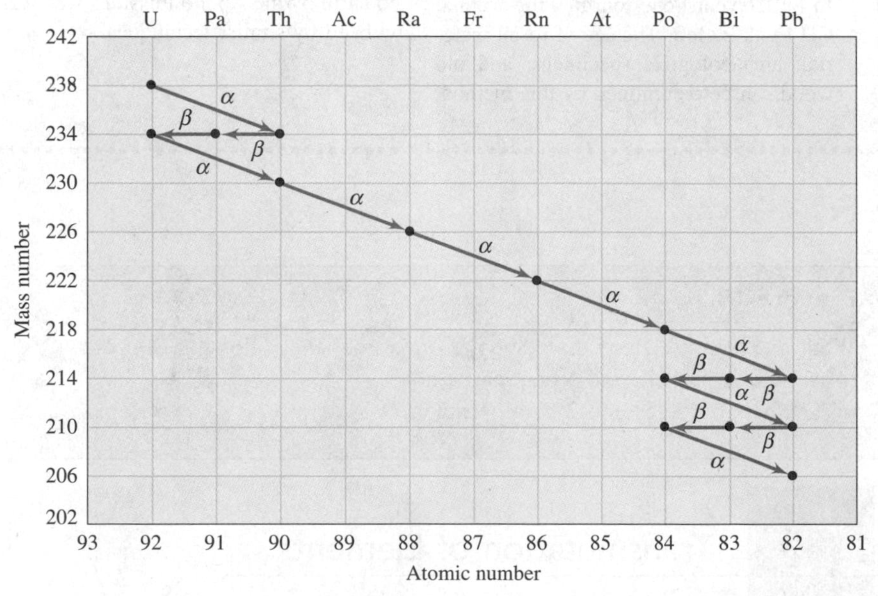

FIGURE 18.4
The uranium disintegration series. $^{238}_{92}U$ decays by a series of alpha (α) and beta (β) emissions to the stable nuclide $^{206}_{82}Pb$.

By using such a series and the half-lives of its members, scientists have been able to approximate the age of certain geologic deposits. This approximation is done by comparing the amount of $^{238}_{92}U$ with the amount of $^{206}_{82}Pb$ and other nuclides in the series that are present in a particular geologic formation. Rocks found in Canada and Finland have been calculated to be about 3.0×10^9 (3 billion) years old. Some meteorites have been determined to be 4.5×10^9 years old.

Chemistry in Action

Artifacts and Geologic Formations

An interesting outgrowth of the use of the radionuclide techniques is *radiocarbon dating*. The method is based on the decay rate of C-14 and was devised by the American chemist W. F. Libby, who received the Nobel Prize in chemistry in 1960 for this work. The principle of radiocarbon dating is as follows: Carbon dioxide in the atmosphere contains a fixed ratio of radioactive C-14 to ordinary C-12, because C-14 is produced at a steady rate in the atmosphere by bombardment of N-14 by neutrons from cosmic ray sources:

$$^{14}_{7}N + ^{1}_{0}n \longrightarrow ^{14}_{6}C + ^{1}_{1}H$$

Plants that consume carbon dioxide during photosynthesis and animals that eat the plants contain the same proportion of C-14 to C-12 as long as they are alive. When an organism dies, the amount of C-12 remains fixed, but the C-14 content diminishes according to its half-life (5668 years). By comparing the ratio of C-14 to C-12 in an object to the same ratio in living plants, one can estimate the age of the object being evaluated. In 5668 years, one-half of the radiocarbon initially present will have undergone decomposition.

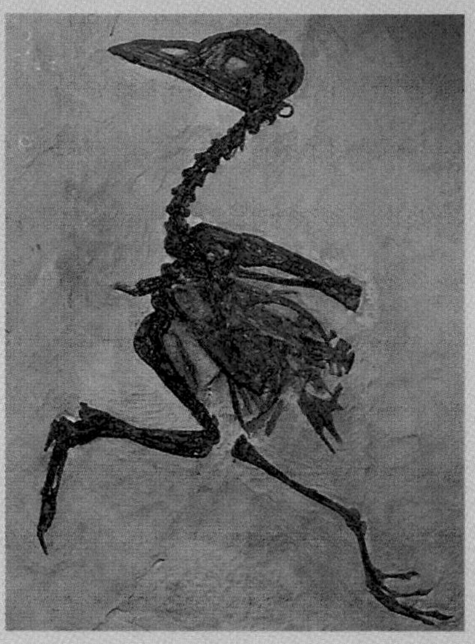

Radiocarbon dating is used to verify the age of artifacts and fossils. This "bird" is about 40 million years old (Eocene Period).

In 11,336 years, one-fourth of the original C-14 will be left. The age of fossil material, archaeological specimens, and old wood can be determined by this method.

The age of specimens from ancient Egyptian tombs calculated by radiocarbon dating correlates closely with the chronological age established by Egyptologists. Charcoal samples obtained at Darrington Walls, a wood-henge in Great Britain, were determined to be about 4000 years old. Radiocarbon dating instruments currently in use enable researchers to date specimens back as far as 70,000 years. This technique was used recently to estimate the age of the shroud of Turin.

Radioactive decay has been used to date samples other than those containing carbon. For example, the age of rock formations containing uranium has been approximated by determining the ratio of U-238 to Pb-206. Lead-206 is the last isotope formed in the U-238 disintegration series. Thus, a geologic deposit containing a 1:1 ratio of U-238 to Pb-206 would correspond to a time lapse of one half-life of U-238, which is 4.5×10^9 years, assuming that all the lead came from the decay of U-238. The age of moon rocks returned to earth by the Apollo missions were calculated by similar techniques.

Practice 18.3

What nuclides are formed when $^{238}_{92}U$ undergoes the following decays?
(a) an alpha particle and a beta particle
(b) 3 alpha particles and 2 beta particles

18.5 Transmutation of Elements

transmutation

Transmutation is the conversion of one element into another by either natural or artificial means. Transmutation occurs spontaneously in natural radioactive disintegrations. Alchemists tried for centuries to convert lead and mercury into gold by artificial means. But transmutation by artificial means was not achieved until 1919, when Ernest Rutherford succeeded in bombarding the nuclei of nitrogen atoms with alpha particles and produced oxygen nuclides and protons. The nuclear equation for this transmutation can be written as

$$^{14}_{7}N + \alpha \longrightarrow {}^{17}_{8}O + {}^{1}_{1}H \qquad \text{or} \qquad {}^{14}_{7}N + {}^{4}_{2}He \longrightarrow {}^{17}_{8}O + {}^{1}_{1}H$$

It is believed that the alpha particle enters the nitrogen nucleus, forming $^{18}_{9}F$ as an intermediate, which then decomposes into the products.

Rutherford's experiments opened the door to nuclear transmutations of all kinds. Atoms were bombarded by alpha particles, neutrons, protons, deuterons ($^{2}_{1}H$), electrons, and so forth. Massive instruments were developed for accelerating these particles to very high speeds and energies to aid their penetration of the nucleus. Some of these instruments are the famous cyclotron, developed by E. O. Lawrence at the University of California; the Van de Graaf electrostatic generator; the betatron; and the electron and proton synchrotrons. With these instruments many nuclear transmutations became possible. Equations for a few of these follow:

$$^{7}_{3}Li + {}^{1}_{1}H \longrightarrow 2\,{}^{4}_{2}He$$

$$^{40}_{18}Ar + {}^{1}_{1}H \longrightarrow {}^{40}_{19}K + {}^{1}_{0}n$$

$$^{23}_{11}Na + {}^{1}_{1}H \longrightarrow {}^{23}_{12}Mg + {}^{1}_{0}n$$

$$^{114}_{48}Cd + {}^{2}_{1}H \longrightarrow {}^{115}_{48}Cd + {}^{1}_{1}H$$

$$^{2}_{1}H + {}^{2}_{1}H \longrightarrow {}^{3}_{1}H + {}^{1}_{1}H$$

$$^{209}_{83}Bi + {}^{2}_{1}H \longrightarrow {}^{210}_{84}Po + {}^{1}_{0}n$$

$$^{16}_{8}O + {}^{1}_{0}n \longrightarrow {}^{13}_{6}C + {}^{4}_{2}He$$

$$^{238}_{92}U + {}^{12}_{6}C \longrightarrow {}^{244}_{98}Cf + 6\,{}^{1}_{0}n$$

Aerial view of Stanford University's Linear Accelerator.

18.6 Artificial Radioactivity

Irene Joliot-Curie (daughter of Pierre and Marie Curie) and her husband Frederic Joliot-Curie observed that when aluminum-27 was bombarded with alpha particles, neutrons and positrons (positive electrons) were emitted as part of the products. When the source of alpha particles was removed, neutrons ceased to be produced, but positrons continued to be emitted. This observation suggested that the neutrons and positrons were coming from two separate reactions. It also indicated that one of the products of the first reaction was radioactive. After further investigation they discovered that, when aluminum-27 is bombarded with alpha particles, phosphorus-30 and neutrons are produced. Phosphorus-30 is radioactive, has a half-life of 2.5 minutes, and decays to silicon-30 with the emission of a positron. The equations for these reactions follow:

$$^{27}_{13}Al + {}^{4}_{2}He \longrightarrow {}^{30}_{15}P + {}^{1}_{0}n$$

$$^{30}_{15}P \longrightarrow {}^{30}_{14}Si + {}^{0}_{+1}e$$

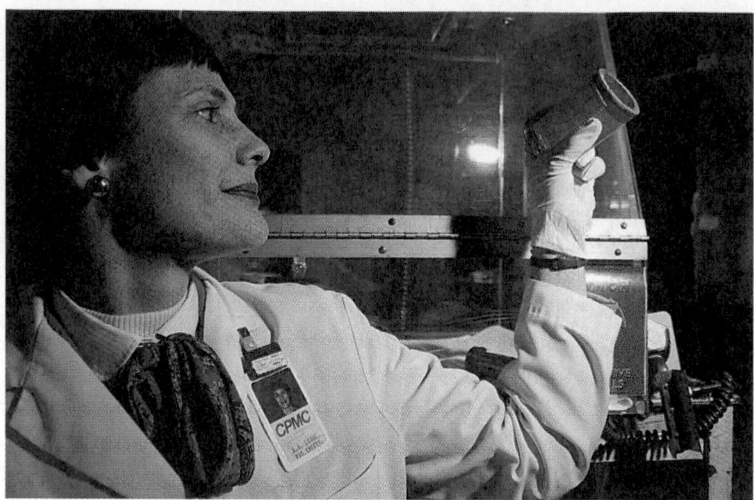

FIGURE 18.5
Left: Geiger–Müller survey meter.
Right: Worker wearing film badge (above I.D. tag) to measure radioactivity.

The radioactivity of nuclides produced in this manner is known as **artificial radioactivity** or **induced radioactivity**. Artificial radionuclides behave like natural radioactive elements in two ways: They disintegrate in a definite fashion and they have a specific half-life. The Joliot-Curies received the Nobel Prize in chemistry in 1935 for the discovery of artificial, or induced, radioactivity.

artificial radioactivity
induced radioactivity

18.7 Measurement of Radioactivity

Radiation from radioactive sources is so energetic that it is called *ionizing radiation*. When it strikes an atom or a molecule, one or more electrons are knocked off, and an ion is created. One of the common instruments used to detect and measure radioactivity, the Geiger counter, depends on this fact. The instrument consists of a Geiger–Müller detecting tube and a counting device. The detector tube is a pair of oppositely charged electrodes in an argon gas-filled chamber fitted with a thin window. When radiation, such as a beta particle, passes through the window into the tube, some argon is ionized, and a momentary pulse of current (discharge) flows between the electrodes. These current pulses are electronically amplified in the counter and appear as signals in the form of audible clicks, flashing lights, meter defections, or numerical readouts (Figure 18.5 left).

The amount of radiation that an individual encounters can be measured by a film badge (Figure 18.5 right). This badge contains a piece of photographic film in a light-proof holder, and is worn in areas where radiation might be encountered. The silver grains in the film will darken when exposed to radiation. The badges are processed after a predetermined time interval to determine the amount of radiation the wearer has been exposed to.

A scintillation counter is used to measure radioactivity for biomedical applications. A scintillator is composed of molecules that emit light when exposed to

Chemistry in Action

Isotopes in Agriculture

Agricultural research scientists use gamma radiation from Co-60 or other sources to develop disease-resistant and highly productive grains and other crops. The seeds are exposed to gamma radiation to induce mutations. The most healthy and vigorous plants grown from the irradiated seed are then selected and propagated to obtain new and improved varieties for commercial use. Preservation of foodstuffs by radiation is another beneficial application. Food is exposed to either gamma radiation or a beam of beta particles supplied by Co-60 or Cs-137. Microorganisms that can cause food spoilage are destroyed, but the temperature of the food is raised only slightly. The food does not become radioactive as a result of this process, but the shelf-life is extended significantly.

Radioactivity has been used to control and, in some areas, to eliminate the screw-worm fly. The larvae of this obnoxious insect pest burrow into wounds in livestock. The female fly, like a queen bee, mates only once. When large numbers of gamma ray–sterilized male flies are released at the proper time in an area infested with screw-worm flies, the majority of the females mate with sterile males. As a consequence, the flies fail to reproduce sufficiently to maintain their numbers. This technique has also been used to eradicate the Mediterranean fruit fly in some areas.

The mushrooms on the right were treated with radiation.

ionizing radiation. A light-sensitive detector counts the flashes and converts them to a numerical readout.

The *curie* is the unit used to express the amount of radioactivity produced by an element. One **curie** (ci) is defined as the quantity of radioactive material giving 3.7×10^{10} disintegrations per second. The basis for this figure is pure radium, which has an activity of 1 curie per gram. Because the curie is such a large quantity, the millicurie and microcurie, representing one-thousandth and one-millionth of a curie, respectively, are more practical and more commonly used.

curie

The curie only measures radioactivity emitted by a radionuclide. Different units are required to measure exposure to radiation. The **Roentgen (R)** is the unit that quantifies exposure to gamma or X rays. One Roentgen is defined as the amount of radiation required to produce 2.1×10^9 ions per cm^3 of dry air. The **rad** (***r**adiation **a**bsorbed **d**ose*) is defined as the amount of radiation that provides 0.01 J of energy per kilogram of matter. The amount of radiation absorbed will change depending on the type of matter. The Roentgen and the rad are numerically similar. One Roentgen of gamma radiation provides 0.92 rad in bone tissue.

Roentgen (R)

rad (radiation absorbed dose)

Neither the rad nor the Roentgen indicates the biological damage caused by radiation. One rad of alpha particles has the ability to cause ten times more damage than one rad of gamma rays or beta particles. Another unit, **rem (*r*oentgen *e*quivalent to *m*an)** takes into account the degree of biological effect caused by the type of radiation exposure. One rem is equal to the dose in rads multiplied by a factor specific to the form of radiation. The factor is 10 for alpha particles and 1 for both beta particles and gamma rays. Units of radiation are summarized in Table 18.4.

rem (roentgen equivalent to man)

TABLE 18.4 Radiation Units

Curie (ci)	A unit of radioactivity indicating the rate of decay of a radioactive substance. $1\ Ci = 3.7 \times 10^{10}$ disintegrations/sec
Roentgen (R)	A unit of exposure of gamma or X radiation based on the quantity of ionization produced in air. $1\ R = 2.1 \times 10^{9}$ ions/cm^3
Rad	A unit of absorbed dose of radiation indicating the energy absorbed from any ionizing radiation. 1 rad = 0.01 J/kg matter
Rem	A unit of radiation dose equivalent. 1 rem is equal to the dose in rads multiplied by a factor dependent on the particular type of radiation. 1 rem = 1 rad $\times$ factor
Gray (Gy) (SI unit)	Energy absorbed by tissue. 1 Gy = 1 J/kg tissue (1 Gy = 100 rad)

18.8 Nuclear Fission

nuclear fission

In **nuclear fission** a heavy nuclide splits into two or more intermediate-sized fragments when struck in a particular way by a neutron. The fragments are called *fission products*. As the atom splits, it releases energy and two or three neutrons, each of which can cause another nuclear fission. The first instance of nuclear fission was reported in January 1939 by the German scientists Otto Hahn and F. Strassmann. Detecting isotopes of barium, krypton, cerium, and lanthanum after bombarding uranium with neutrons, the scientists were led to believe that the uranium nucleus had been split.

Characteristics of nuclear fission are as follows:

1. Upon absorption of a neutron, a heavy nuclide splits into two or more smaller nuclides (fission products).
2. The mass of the nuclides formed range from about 70 to 160 amu.
3. Two or more neutrons are produced from the fission of each atom.
4. Large quantities of energy are produced as a result of the conversion of a small amount of mass into energy.
5. Most nuclides produced are radioactive and continue to decay until they reach a stable nucleus.

One process by which this fission takes place is illustrated in Figure 18.6. When a heavy nucleus captures a neutron, the energy increase may be sufficient to cause deformation of the nucleus until the mass finally splits into two fragments, releasing energy and usually two or more neutrons.

In a typical fission reaction, a $^{235}_{92}U$ nucleus captures a neutron and forms unstable $^{236}_{92}U$. This $^{236}_{92}U$ nucleus undergoes fission, quickly disintegrating into two

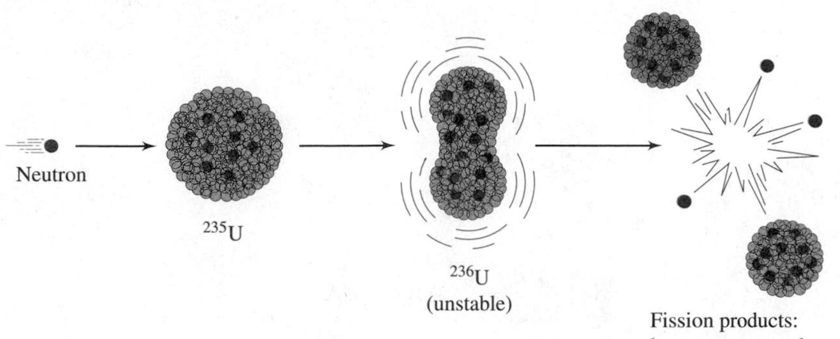

FIGURE 18.6
The fission process. When a neutron is captured by a heavy nucleus, the nucleus becomes more unstable. The more energetic nucleus begins to deform, resulting in fission. Two nuclear fragments and three neutrons are produced by this fission process.

fragments, such as $^{139}_{56}Ba$ and $^{94}_{36}Kr$, and three neutrons. The three neutrons in turn may be captured by three other $^{235}_{92}U$ atoms, each of which undergoes fission, producing nine neutrons, and so on. A reaction of this kind, in which the products cause the reaction to continue or magnify, is known as a **chain reaction**. For a chain reaction to continue, enough fissionable material must be present so that each atomic fission causes, on the average, at least one additional fission. The minimum quantity of an element needed to support a self-sustaining chain reaction is called the **critical mass**. Since energy is released in each atomic fission, chain reactions constitute a possible source of a steady supply of energy. (A chain reaction is illustrated in Figure 18.7.) Two of the many possible ways in which $^{235}_{92}U$ may fission are shown by the following equations.

chain reaction

critical mass

$$^{235}_{92}U + ^{1}_{0}n \longrightarrow ^{139}_{56}Ba + ^{94}_{36}Kr + 3\,^{1}_{0}n$$

$$^{235}_{92}U + ^{1}_{0}n \longrightarrow ^{144}_{54}Xe + ^{90}_{38}Sr + 2\,^{1}_{0}n$$

18.9 Nuclear Power

Nearly all electricity for commercial use is produced by machines consisting of a turbine linked by a drive shaft to an electrical generator. The energy required to run the turbine can be supplied by falling water, as in hydroelectric power plants, or by steam as in thermal power plants.

The world's demand for energy, largely from fossil fuels, has continued to grow at an ever-increasing rate for about 250 years. Even at the present rates of consumption, the estimated world supply of fossil fuels is sufficient for only a few centuries. Although the United States has large coal and oil shale deposits, it is currently importing about 40% of its oil supply. We clearly need to develop alternative energy sources. At present, uranium is the most productive alternative energy source, and about 17% of the electrical energy used in the United States is generated from power plants using uranium fuel.

A nuclear power plant is a thermal power plant in which heat is produced by a nuclear reactor instead of by combustion of fossil fuel. The major components of a nuclear reactor are as follows:

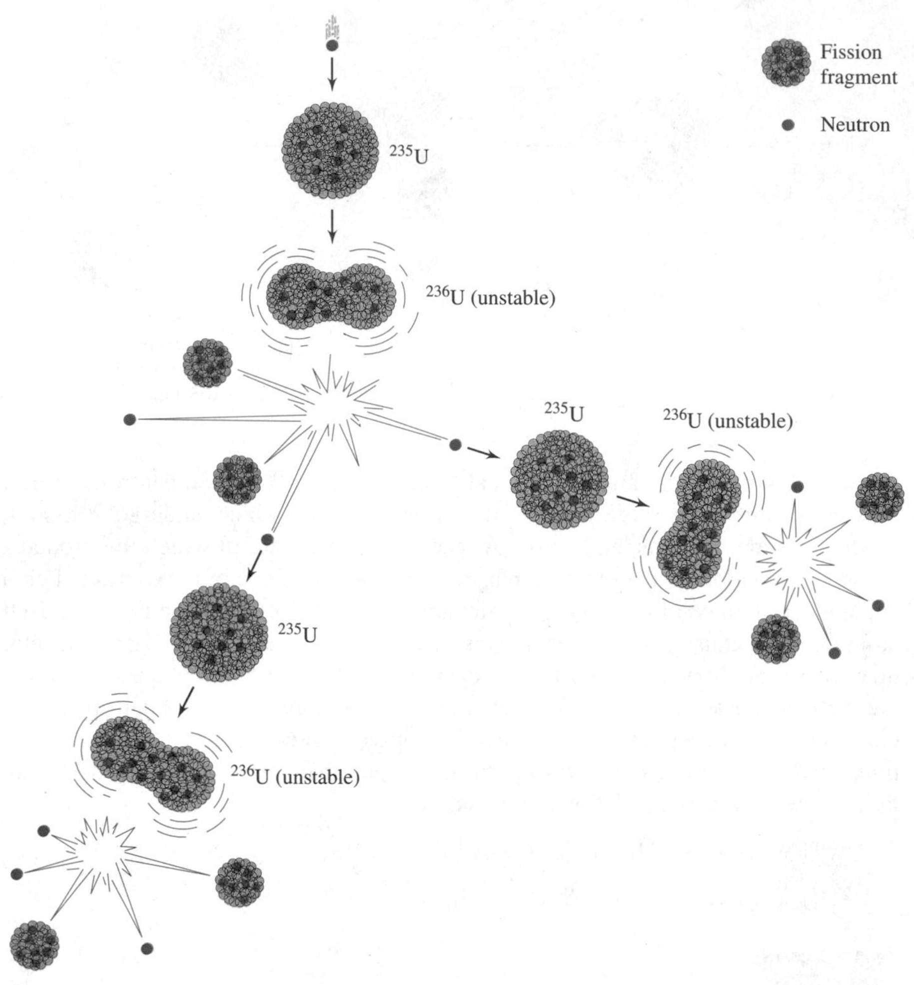

FIGURE 18.7 ▶ **Fission and chain reaction of $^{235}_{92}U$. Each fission produces two major fission fragments and three neutrons, which may be captured by other $^{235}_{92}U$ nuclei, continuing the chain reaction.**

1. an arrangement of nuclear fuel, called the reactor core
2. a control system, which regulates the rate of fission and thereby the rate of heat generation
3. a cooling system, which removes the heat from the reactor and also keeps the core at the proper temperature

One type of reactor uses metal slugs containing uranium enriched from the normal 0.7% U-235 to about 3% U-235. The self-sustaining fission reaction is moderated, or controlled, by adjustable control rods containing substances that slow down and capture some of the neutrons produced. Ordinary water, heavy water, and molten sodium are typical coolants used. Energy obtained from nuclear reactions in the form of heat is used in the production of steam to drive turbines for generating electricity. (See Figure 18.8.)

Two events that demonstrate the potential dangers of nuclear power are the accidents at Three Mile Island, Pennsylvania (1979) and Chernobyl, U.S.S.R. (1986). Both of these nuclear accidents resulted from the loss of coolant to the reactor core.

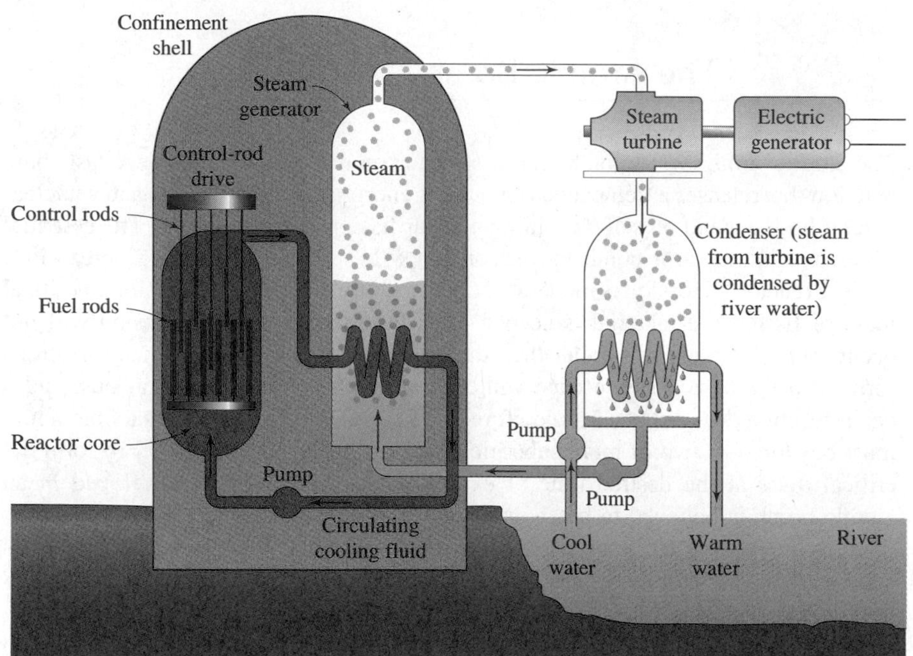

FIGURE 18.8
Diagram of a nuclear power plant. Heat produced by the reactor core is carried by a cooling fluid to a steam generator, which turns a turbine that produces electricity.

The reactors at Three Mile Island were covered by concrete containment buildings and therefore released a relatively small amount of radioactive material into the atmosphere. But because the Soviet Union did not require containment structures on nuclear power plants, the Chernobyl accident resulted in 31 deaths and the resettlement of 135,000 people. The release of large quantities of I-131, Cs-134, and Cs-137 may cause long-term health problems in that exposed population.

Another major disadvantage of nuclear-fueled power plants is that they produce highly radioactive waste products, some of which have half-lives of thousands of years. As yet, no agreement has been reached on how to safely dispose of these dangerous wastes.

In the United States, reactors designed for commercial power production use uranium oxide, U_3O_8, that is enriched with the relatively scarce fissionable U-235 isotope. Because the supply of U-235 is limited, a new type of reactor known as the *breeder reactor* has been developed. Breeder reactors are designed to produce additional fissionable material at the same time that the fission reaction is occurring. In a breeder reactor, excess neutrons convert nonfissionable isotopes, such as U-238 or Th-232, to fissionable isotopes, Pu-239 or U-233:

$$^{238}_{92}\text{U} + ^{1}_{0}\text{n} \longrightarrow ^{239}_{92}\text{U} \xrightarrow[\beta]{} ^{239}_{93}\text{Np} \xrightarrow[\beta]{} ^{239}_{94}\text{Pu}$$

$$^{232}_{90}\text{Th} + ^{1}_{0}\text{n} \longrightarrow ^{233}_{90}\text{Th} \xrightarrow[\beta]{} ^{233}_{91}\text{Pa} \xrightarrow[\beta]{} ^{233}_{92}\text{U}$$

These transmutations make it possible to greatly extend the supply of fuel for nuclear reactors. No breeder reactors are presently in commercial operation in the United States, but a number of them are being operated in Europe and Great Britain.

Interior view of a reactor hall at a nuclear power plant in France. The core of the reactor is situated in the water pool (blue area).

18.10 The Atomic Bomb

The atomic bomb is a fission bomb; it operates on the principle of a very fast chain reaction that releases a tremendous amount of energy. An atomic bomb and a nuclear reactor both depend on self-sustaining nuclear fission chain reactions. The essential difference is that in a bomb the fission is "wild," or uncontrolled, whereas in a nuclear reactor the fission is moderated and carefully controlled. A minimum critical mass of fissionable material is needed for a bomb, or a major explosion will not occur. When a quantity smaller than the critical mass is used, too many neutrons formed in the fission step escape without combining with another nucleus, and a chain reaction does not occur. Therefore, the fissionable material of an atomic bomb must be stored as two or more subcritical masses and brought together to form the critical mass at the desired time of explosion. The temperature developed in an atomic bomb is believed to be about 10 million degrees Celsius.

The nuclides used in atomic bombs are U-235 and Pu-239. Uranium deposits contain about 0.7% of the U-235 isotope, the remainder being U-238. Uranium-238 does not undergo fission except with very high energy neutrons. It was discovered, however, that U-238 captures a low-energy neutron without undergoing fission and that the product, U-239, changes to Pu-239 (plutonium) by a beta-decay process. Plutonium-239 readily undergoes fission upon capture of a neutron and is therefore useful for nuclear weapons. The equations for the nuclear transformations are

$$^{238}_{92}\text{U} + {}^{1}_{0}\text{n} \longrightarrow {}^{239}_{92}\text{U} \xrightarrow{\ \ \beta\ \ } {}^{239}_{93}\text{Np} \xrightarrow{\ \ \beta\ \ } {}^{239}_{94}\text{Pu}$$

The hazards of an atomic bomb explosion include not only shock waves from the explosive pressure and tremendous heat, but also intense radiation in the form of alpha particles, beta particles, gamma rays, and ultraviolet rays. Gamma rays and X rays can penetrate deeply into the body, causing burns, sterilization, and mutation of the genes, which may have an adverse effect on future generations. Both radioactive fission products and unfissioned material are present after the explosion. If the bomb explodes near the ground, many tons of dust are lifted into the air. Radioactive material adhering to this dust, known as *fallout*, is spread by air currents over wide areas of the land and constitutes a lingering source of radiation hazard.

Today nuclear war is probably the most awesome threat facing civilization. Only two rather primitive fission-type atom bombs were used to destroy the Japanese cities of Hiroshima and Nagasaki and bring World War II to an early end. The threat of nuclear war is increased by the fact that the number of nations possessing nuclear weapons is steadily increasing.

18.11 Nuclear Fusion

nuclear fusion

The process of uniting the nuclei of two light elements to form one heavier nucleus is known as **nuclear fusion**. Such reactions can be used for producing energy, because the mass of the two individual nuclei that are fused into a single nucleus is greater than the mass of the nucleus formed by their fusion. The mass differential is liberated in the form of energy. Fusion reactions are responsible for the tremendous energy output of the sun. Thus, aside from relatively small amounts from nuclear fission and radioactivity, fusion reactions are the ultimate source of our energy, even the energy from fossil fuels. They are also responsible for the devastating power of the thermonuclear, or hydrogen, bomb.

Fusion reactions require temperatures on the order of tens of millions of degrees for initiation. Such temperatures are present in the sun but have been produced only momentarily on earth. For example, the hydrogen, or fusion, bomb is triggered by the temperature of an exploding fission bomb. Two typical fusion reactions are

$$\underset{\text{tritium}}{{}^{3}_{1}\text{H}} + \underset{\text{deuterium}}{{}^{2}_{1}\text{H}} \longrightarrow {}^{4}_{2}\text{He} + {}^{1}_{0}\text{n} + \text{energy}$$

$$\underset{\substack{3.0150\\ \text{amu}}}{{}^{3}_{1}\text{H}} + \underset{\substack{1.0079\\ \text{amu}}}{{}^{1}_{1}\text{H}} \longrightarrow \underset{\substack{4.0026\\ \text{amu}}}{{}^{4}_{2}\text{He}} + \text{energy}$$

The total mass of the reactants in the second equation is 4.0229 amu, which is 0.0203 amu greater than the mass of the product. This difference in mass is manifested in the great amount of energy liberated.

During the past 40 to 45 years, a large amount of research on controlled nuclear fusion reactions has been done in the United States and in other countries, especially the former Soviet Union. So far, the goal of controlled nuclear fusion has not been attained, although the required ignition temperature has been reached in several devices. Evidence to date leads us to believe that it is possible to develop a practical fusion power reactor. Fusion power, if it can be developed, will be far superior to fission power for the following reasons:

1. Virtually infinite amounts of energy are to be had from fusion. Uranium supplies for fission power are limited, but heavy hydrogen, or deuterium (the most likely fusion fuel), is relatively abundant. It is estimated that the deuterium in a cubic mile of seawater is an energy resource (as fusion fuel) greater than the petroleum reserves of the entire world.
2. From an environmental viewpoint, fusion power is much "cleaner" than fission power because fusion reactions, in contrast to uranium and plutonium fission reactions, do not produce large amounts of long-lived and dangerously radioactive isotopes.

18.12 Mass–Energy Relationship in Nuclear Reactions

Because large amounts of energy are released in nuclear reactions, significant amounts of mass are converted to energy. We stated earlier that the amount of mass converted to energy in chemical changes is considered insignificant. In fission reactions, about 0.1% of the mass is converted into energy. In fusion reactions as much as 0.5% of the mass may be changed into energy. The Einstein equation, $E = mc^2$, can be used to calculate the energy liberated, or available, when the mass loss is known. For example, in the reaction

$$\underset{7.016\text{ g}}{{}^{7}_{3}\text{Li}} + \underset{1.008\text{ g}}{{}^{1}_{1}\text{H}} \longrightarrow \underset{4.003\text{ g}}{{}^{4}_{2}\text{He}} + \underset{4.003\text{ g}}{{}^{4}_{2}\text{He}} + \text{energy}$$

the mass difference between the reactants and products (8.024 g − 8.006 g) is 0.018 g. The energy equivalent to this amount of mass is 1.62×10^{12} J. By comparison, this is more than 4 million times greater than the 3.9×10^{5} J of energy obtained from the complete combustion of 12.01 g (1 mol) of carbon.

The mass of a nucleus is actually less than the sum of the masses of the protons and neutrons that make up that nucleus. The difference between the mass of the protons and the neutrons in a nucleus and the mass of the nucleus is known as the **mass defect**. The energy equivalent to this difference in mass is known as the **nuclear binding energy**. This energy is the amount that would be required to break up a nucleus into its individual protons and neutrons. The higher the binding energy, the more stable is the nucleus. Elements of intermediate atomic masses have high binding energies. For example, iron (element number 26) has a very high binding energy and therefore has a very stable nucleus. Just as electrons attain less energetic and more stable arrangements through ordinary chemical reactions, neutrons and protons attain less energetic and more stable arrangements through nuclear fission or fusion reactions. Thus, when uranium undergoes fission, the products have less mass (and greater binding energy) than the original uranium. In like manner, when hydrogen and lithium fuse to form helium, the helium has less mass (and greater binding energy) than the hydrogen and lithium. It is this conversion of mass to energy that accounts for the very large amounts of energy associated with both nuclear fission and fusion reactions.

mass defect

nuclear binding energy

18.13 Transuranium Elements

The elements following uranium on the periodic table and having atomic numbers greater than 92 are known as the **transuranium elements**. All of them are synthetic radioactive elements; none of them occur naturally.

transuranium elements

The first transuranium element, number 93, was discovered in 1939 by Edwin M. McMillan (1907–1991) at the University of California while he was investigating the fission of uranium. He named it neptunium for the planet Neptune. In 1941, element 94, plutonium, was identified as a beta-decay product of neptunium:

$${}^{238}_{93}\text{Np} \longrightarrow {}^{238}_{94}\text{Pu} + {}^{0}_{-1}\text{e}$$

$${}^{239}_{93}\text{Np} \longrightarrow {}^{239}_{94}\text{Pu} + {}^{0}_{-1}\text{e}$$

Plutonium is one of the most important fissionable elements known today.

Since 1964, the discoveries of eight new transuranium elements, numbers 104–111, have been announced. All of these elements were produced in minute quantities by high-energy particle accelerators.

A Window into Living Organisms

Imagine being able to view a living process as it is occurring. This was the dream of many a scientist in the past as they tried to extract this knowledge from dead tissue. Today, because of innovations in nuclear chemistry, this dream is a common, everyday occurrence.

Compounds containing a radionuclide are described as being *labeled* or *tagged.* These compounds undergo their normal chemical reactions, but their location can be detected because of their radioactivity. When such compounds are given to a plant or an animal, the movement of the nuclide can be traced through the organism by the use of a Geiger counter or other detecting device.

One important use of the tracer technique was determining the pathway by which CO_2 becomes fixed into carbohydrate ($C_6H_{12}O_6$) during photosynthesis. The net equation for the photosynthesis is

$$6\,CO_2 + 6\,H_2O \longrightarrow C_6H_{12}O_6 + 6\,O_2$$

In this technique, radioactive $^{14}CO_2$ was injected into a colony of green algae. The algae was then placed in the dark, killed at selected time intervals, and the radioactive compounds separated by paper chromatography and analyzed. The results elucidated a series of light-independent photosynthetic reactions.

Some other examples of biological research in which tracer techniques have been employed include determining

1. the rate of phosphate uptake by plants, using radiophosphorus
2. the flow of nutrients in the digestive tract using radioactive barium compounds
3. the accumulation of iodine in the thyroid gland, using radioactive iodine
4. the absorption of iron by the hemoglobin of the blood, using radioactive iron

In chemistry, the uses for tracers are unlimited. The study of reaction mechanisms, the measurement of the rates of chemical reactions, and the determination of physical constants are just a few of the areas of application.

Radioactive tracers are commonly used in medical diagnosis. Because radiation must be detected outside of the body, radionuclides that emit gamma rays are usually chosen. The radionuclide must also be effective at a low concentration and have a short half-life to reduce the possibility of damage to the patient.

Radioactive iodine (I-131) is used to determine thyroid function, where the body concentrates iodine. In this process, a small amount of radioactive potassium or sodium iodide is ingested. A detector is focused on the thyroid gland and measures the amount of iodine in the gland. This picture can be compared to that of a normal thyroid to detect any differences.

Doctors can examine the heart's pumping performance and check for evidence of obstruction in the coronary arteries by *nuclear scanning*. The radionuclide Tl-201, when injected into the bloodstream, lodges in healthy heart muscle. Thallium-201 emits gamma radiation, which is detected by a special imaging device called a *scintillation camera*. The data obtained are simultaneously translated into pictures by a computer. With this technique doctors can observe whether heart tissue has died after a heart attack and whether blood is flowing freely through the coronary passages.

One of the newest applications of nuclear chemistry is the use of positron emission tomography (PET) in the measurement of dynamic processes in the body, such as oxygen use or blood flow. In this application a compound is made that contains a positron-emitting nuclide such as C-11, O-15, or N-13. The compound is injected into the body, and the patient is placed in an instrument that detects the positron emission. A computer produces a three-dimensional image of the area.

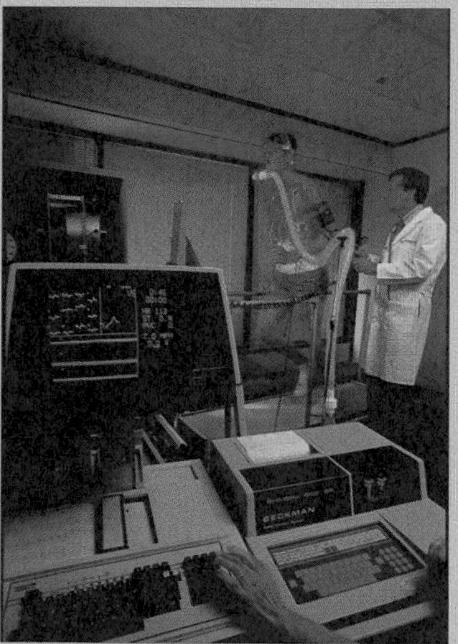

During a stress test, radionuclides help doctors to determine cardiovascular problems in their patients.

PET scans have been used to locate the areas of the brain involved with epileptic seizures. Glucose tagged with C-11 is injected, and an image of the brain is produced. Since the brain uses glucose almost exclusively for energy, diseased areas that use glucose at a rate different than normal tissue can then be identified.

18.14 Biological Effects of Radiation

ionizing radiation

Radiation that has enough energy to dislocate bonding electrons and create ions when passing through matter is classified as **ionizing radiation**. Alpha, beta, and gamma rays, along with X rays, fall into this classification. Ionizing radiation can damage or kill living cells. This damage is particularly devastating when it strikes the cell nuclei and affects molecules involved in cell reproduction. The overall effects of radiation on living organisms fall into these general categories: (1) acute or short-term, effects; (2) long-term effects, and (3) genetic effects.

Acute Radiation Damage

High levels of radiation, especially of gamma or X rays, produce nausea, vomiting, and diarrhea. The effect has been likened to a sunburn throughout the body. If the dosage is high enough, death will occur in a few days. The damaging effects of radiation appear to be centered in the nuclei of the cells, and cells that are undergoing rapid cell division are most susceptible to damage. It is for this reason that cancers are often treated with gamma radiation from a Co-60 source. Cancerous cells multiply rapidly and are destroyed by a level of radiation that does not seriously damage normal cells.

Long-Term Radiation Damage

Protracted exposure to low levels of any form of ionizing radiation can weaken an organism and lead to the onset of malignant tumors, even after fairly long time delays. The largest exposure to man-made sources of radiation is from X rays. Evidence suggests that a number of early workers in radioactivity and X-ray technology may have had their lives shortened by long-term radiation damage.

Strontium-90 isotopes are present in the fallout from atmospheric testing of nuclear weapons. Strontium is in the same periodic-table group as calcium, and its chemical behavior is similar to that of calcium. Hence, when foods contaminated with Sr-90 are eaten, Sr-90 ions are laid down in the bone tissue along with ordinary calcium ions. Strontium-90 is a beta emitter with a half-life of 28 years. Blood cells that are manufactured in bone marrow are affected by the radiation from Sr-90. Hence, there is concern that Sr-90 accumulation in the environment may cause an increase in the incidence of leukemia and bone cancers. Currently no countries are testing nuclear weapons in the atmosphere.

Genetic Effects

All the information needed to create an individual of a particular species, be it a bacterial cell or a human being, is contained within the nucleus of a cell. This genetic information is encoded in the structure of DNA (deoxyribonucleic acid) molecules, which make up genes. The DNA molecules form precise duplicates of themselves when cells divide, thus passing genetic information from one generation to the next. Radiation can damage DNA molecules. If the damage is not severe enough to prevent the individual from reproducing, a mutation may result. Most mutation-induced traits are undesirable. Unfortunately, if the bearer of the altered genes survives to reproduce, these traits are passed along to succeeding generations. In other words, the genetic effects of increased radiation exposure are found in future generations, not in the present generation.

Because radioactive rays are hazardous to health and living tissue, special precautions must be taken in designing laboratories and nuclear reactors, in disposing of waste materials, and in monitoring the radiation exposure of people working in this field. For example, personnel working in areas of hazardous radiation wear film badges or pocket dosimeters to provide them with an accurate indication of cumulative radiation exposure.

Containment shell around nuclear reactor.

Concepts in Review

1. Outline the historical development of nuclear chemistry, including the major contributions of Henri Becquerel, Marie Curie, Ernest Rutherford, Irene Joliet-Curie, Otto Hahn, Fritz Strassmann, and Edwin McMillen.
2. Write balanced nuclear chemical equations using isotopic notation.
3. Determine the amount of radionuclide remaining after a given period of time when the starting amount and half-life are given.
4. List the characteristics that distinguish alpha, beta, and gamma rays from the standpoint of mass, charge, relative velocities, and penetrating power.
5. Describe the effect of a magnetic field on alpha particles, beta particles, and gamma rays.
6. Describe a radioactive disintegration series, and predict which isotope would be formed by the loss of specified numbers of alpha and beta particles from a given radionuclide.
7. Discuss the transmutation of elements.
8. Indicate the methods used for the detection of radiation.
9. Distinguish between radioactive disintegration and nuclear fission reactions.
10. Explain how the fission of U-235 can lead to a chain reaction and why a critical mass is necessary.
11. Explain how the energy from nuclear fission is converted to electrical energy.
12. Explain the difference between fission reactions in a nuclear reactor and those of an atomic bomb.
13. Explain what is meant by the term *nuclear fusion* and why a massive effort to develop controlled nuclear fusion is in progress.
14. Indicate the significance of mass defect and nuclear binding energy.
15. Indicate the major effects of radiation on living organisms.
16. Explain how the age of objects can be determined using radioactivity.
17. Indicate several current uses for radioactive tracers.

Key Terms

The terms listed here have been defined within this chapter. Section numbers are referenced in parenthesis for each term.

alpha particle (18.3)
artificial radioactivity (18.6)
beta particle (18.3)
chain reaction (18.8)
critical mass (18.8)
curie (18.7)
gamma ray (18.3)
half-life (18.2)
induced radioactivity (18.6)
ionizing radiation (18.14)
mass defect (18.12)
nuclear binding energy (18.12)
nuclear fission (18.8)
nuclear fusion (18.11)
nucleon (18.1)
nuclide (18.1)
rad (radiation absorbed dose) (18.7)
radioactive decay (18.2)
radioactivity (18.1)
rem (roentgen equivalent to man) (18.7)
Roentgen (R) (18.7)
transmutation (18.5)
transuranium elements (18.13)

Questions

Questions refer to tables, figures, and key words and concepts defined within the chapter. A particularly challenging question or exercise is indicated with an asterisk.

1. To afford protection from radiation injury, which kind of radiation requires (a) the most shielding? (b) the least shielding?
2. Why is an alpha particle deflected less than a beta particle in passing through an electromagnetic field?
3. Name three pairs of nuclides that might be obtained by fissioning U-235 atoms.
4. Identify the following people and their associations with the early history of radioactivity.
   (a) Antoine Henri Becquerel
   (b) Marie and Pierre Curie
   (c) Wilhelm Roentgen
   (d) Ernest Rutherford
   (e) Otto Hahn and Fritz Strassmann
5. Why is the radioactivity of an element unaffected by the usual factors that affect the rate of chemical reactions, such as ordinary changes of temperature and concentration?
6. Distinguish between the terms *isotope* and *nuclide*.
7. The half-life of Pu-244 is 76 million years. If the age of the earth is about 5 billion years, discuss the feasibility of this nuclide being found as a naturally occurring nuclide.
8. Tell how alpha, beta, and gamma radiation are distinguished from the standpoint of
   (a) charge
   (b) relative mass
   (c) nature of particle or ray
   (d) relative penetrating power
9. Distinguish between natural and artificial radioactivity.
10. What is a radioactive disintegration series?
11. Briefly discuss the transmutation of elements.
12. Stable Pb-208 is formed from Th-232 in the thorium disintegration series by successive, α, β, β, α, α, α, α, β, β, α particle emissions. Write the symbol (including mass and atomic number) for each nuclide formed in this series.
13. The nuclide Np-237 loses a total of seven alpha particles and four beta particles. What nuclide remains after these losses?
14. Bismuth-211 decays by alpha emission to give a nuclide that in turn decays by beta emission to yield a stable nuclide. Show these two steps with nuclear equations.
15. Describe the use of a Geiger counter in measuring radioactivity.
16. What was the contribution to nuclear physics of Otto Hahn and Fritz Strassmann?
17. What is a breeder reactor? Explain how it accomplishes the "breeding"?
18. What is the essential difference between the nuclear reactions in a nuclear reactor and those in an atomic bomb?
19. Why must a certain minimum amount of fissionable material be present before a self-supporting chain reaction can occur?

20. What is mass defect and nuclear binding energy?
21. Explain why radioactive rays are classified as ionizing radiation.
22. Give a brief description of the biological hazards associated with radioactivity.
23. Strontium-90 has been found to occur in radioactive fallout. Why is there so much concern about this radionuclide being found in cow's milk? (Half-life of Sr-90 is 28 years.)
24. What is a radioactive tracer? How is it used?
25. Describe the radiocarbon method for dating archaeological artifacts.
26. How might radioactivity be used to locate a leak in an underground pipe?
27. Anthropologists have found bones whose age suggests that the human line may have emerged in Africa as much as 4 million years ago. If wood or charcoal were found with such bones, would C-14 dating be useful in dating the bones? Explain.
28. Which of the following statements are correct? Rewrite the incorrect statements to make them correct.
    (a) Radioactivity was discovered by Marie Curie.
    (b) An atom of $^{59}_{28}Ni$ has 59 neutrons.
    (c) The loss of a beta particle by an atom of $^{75}_{33}As$ forms an atom of increased atomic number.
    (d) The emission of an alpha particle from the nucleus of an atom lowers its atomic number by 4 and lowers its mass number by 2.
    (e) Emission of gamma radiation from the nucleus of an atom leaves both the atomic number and mass number unchanged.
    (f) Relatively few naturally occurring radioactive nuclides have atomic numbers below 81.
    (g) The longer the half-life of a radionuclide, the more slowly it decays.
    (h) The beta ray has the greatest penetrating power of all the rays emitted from the nucleus of an atom.
    (i) Radioactivity is due to an unstable ratio of neutrons to nucleons in an atom.
    (j) The half-life of different nuclides can vary from a fraction of a second to millions of years.
    (k) The symbol $^{0}_{+1}e$ is used to indicate a positron, which is a positively charged particle with the mass of an electron.
    (l) The disintegration of $^{226}_{88}Ra$ into $^{214}_{83}Po$ involves the loss of 3 alpha particles and 2 beta particles.
    (m) If 1.0 g of a radionuclide has a half-life of 7.2 days, the half-life of 0.50 g of that nuclide is 3.6 days.
    (n) If the mass of a radionuclide is reduced by radioactive decay from 12 g to 0.75 g in 22 hours, the half-life of the isotope is 5.5 hours.
    (o) A very high temperature is required to initiate nuclear fusion reactions.
    (p) Radiocarbon dating of archaeological artifacts is based on an increase in the ratio of C-14 to C-12 in the object.
    (q) Cancers are often treated by radiation from Co-60, which destroys the rapidly growing cancer cells.
    (r) High levels of radiation produce nausea, vomiting, and diarrhea.
    (s) Carbon-14, used in radiocarbon dating, is produced in living matter by the beta decay of N-14.
    (t) Radioactive tracers are small amounts of radioactive nuclides of selected elements whose progress in chemical or biological processes can be followed using a radiation detection instrument.

Paired Exercises

These exercises are paired. Each odd-numbered exercise is followed by a similar even-numbered exercise. Answers to the even-numbered exercises are given in Appendix V.

29. Indicate the number of protons, neutrons, and nucleons in each of the following nuclei:
    (a) $^{35}_{17}Cl$ (b) $^{226}_{88}Ra$
30. Indicate the number of protons, neutrons, and nucleons in each of the following nuclei:
    (a) $^{235}_{92}U$ (b) $^{82}_{35}Br$
31. How are the mass and the atomic number of a nucleus affected by the loss of an alpha particle?
32. How are the mass and the atomic number of a nucleus affected by the loss of a beta particle?
33. Write nuclear equations for the alpha decay of
    (a) $^{218}_{85}At$ (b) $^{221}_{87}Fr$
34. Write nuclear equations for the alpha decay of
    (a) $^{192}_{78}Pt$ (b) $^{210}_{84}Po$
35. Write nuclear equations for the beta decay of
    (a) $^{14}_{6}C$ (b) $^{137}_{55}Cs$
36. Write nuclear equations for the beta decay of
    (a) $^{239}_{93}Np$ (b) $^{90}_{38}Sr$

37. Write nuclear equations for the conversion of ${}^{13}_{6}C$ to ${}^{14}_{6}C$.

38. Write nuclear equations for the conversion of ${}^{30}_{15}P$ to ${}^{30}_{14}Si$.

39. Complete and balance the following nuclear equations by supplying the missing particles:
(a) ${}^{27}_{13}Al + {}^{4}_{2}He \longrightarrow {}^{30}_{15}P + ?$
(b) ${}^{27}_{14}Si \longrightarrow {}^{0}_{+1}e + ?$
(c) $? + {}^{2}_{1}H \longrightarrow {}^{13}_{7}N + {}^{1}_{0}n$
(d) $? \longrightarrow {}^{82}_{36}Kr + {}^{0}_{-1}e$

40. Complete and balance the following nuclear equations by supplying the missing particles:
(a) ${}^{66}_{29}Cu \longrightarrow {}^{66}_{30}Zn + ?$
(b) ${}^{0}_{-1}e + ? \longrightarrow {}^{7}_{3}Li$
(c) ${}^{27}_{13}Al + {}^{4}_{2}He \longrightarrow {}^{30}_{14}Si + ?$
(d) ${}^{85}_{37}Rb + ? \longrightarrow {}^{82}_{35}Br + {}^{4}_{2}He$

41. Strontium-90 has a half-life of 28 years. If a 1.00-mg sample were stored for 112 years, what mass of Sr-90 would remain?

42. Strontium-90 has a half-life of 28 years. If a sample was tested in 1980 and found to be emitting 240 counts/minute, in what year would the same sample be found to be emitting 30 counts/minute? How much of the original Sr-90 would be left?

***43.** Consider the fission reaction

$${}^{235}_{92}U + {}^{1}_{0}n \longrightarrow {}^{94}_{38}Sr + {}^{139}_{54}Xe + 3\,{}^{1}_{0}n + \text{energy}$$

Calculate the following using the mass data (1.0 g is equivalent to 9.0×10^{13} J):
U-235 = 235.0439 amu Sr-94 = 93.9154 amu
Xe-139 = 138.9179 amu n = 1.0087 amu
(a) the energy released in joules for a single event (one uranium atom splitting)
(b) the energy released in joules per mole of uranium splitting
(c) the percentage of the mass that is lost in the reaction

44. Consider the fusion reaction

$${}^{1}_{1}H + {}^{2}_{1}H \longrightarrow {}^{3}_{2}He + \text{energy}$$

Calculate the following using the mass data (1.0 g is equivalent to 9.0×10^{13} J):
${}^{1}_{1}H$ = 1.00794 amu
${}^{2}_{1}H$ = 2.01410 amu
${}^{3}_{2}He$ = 3.01603 amu
(a) the energy released in joules per mole of He-3 formed
(b) the percentage of the mass that is lost in the reaction

Additional Exercises

These exercises are not paired or labeled by topic and provide additional practice on concepts covered in this chapter.

45. If radium costs \$50,000 a gram, how much will 0.0100 g of ${}^{226}RaCl_2$ cost if the price is based only on the radium content?

46. An archaeological specimen was analyzed and found to be emitting only 25% as much C-14 radiation per gram of carbon as newly cut wood. How old is this specimen?

47. Barium-141 is a beta emitter. What is the half-life if a 16.0-g sample of the nuclide decays to 0.500 g in 90 minutes?

***48.** Calculate (a) the mass defect and (b) the binding energy of ${}^{7}_{3}Li$ using the mass data:
${}^{7}_{3}Li$ = 7.0160 g n = 1.0087 g
p = 1.0073 g e^- = 0.00055 g
1.0 g ≡ 9.0×10^{13} J (from $E = mc^2$).

***49.** In the disintegration series ${}^{235}_{92}U \longrightarrow {}^{207}_{82}Pb$, how many alpha and beta particles are emitted?

50. List three devices used for radiation detection and explain their operation.

51. The half-life of I-123 is 13 hours. If 10 mg of I-123 is administered to a patient, how much I-123 remains after 3 days and 6 hours?

52. Clearly distinguish between fission and fusion. Give an example of each.

53. Starting with 1 g of a radioactive isotope whose half-life is 10 days, sketch a graph showing the pattern of decay for that material. On the *x* axis, plot time (you may want to simply show multiples of the half-life), and on the *y* axis, plot mass of material remaining. Then, after completing the graph, explain why one never really gets to the point where *all* of a sample's radioactivity is considered to be gone.

54. Identify each of the missing products (name the element and give its atomic number and mass number) by balancing each of the following nuclear equations:

(a) $^{235}U + {}^{1}_{0}n \longrightarrow {}^{143}Xe + 3\,{}^{1}_{0}n +$ _____
(b) $^{235}U + {}^{1}_{0}n \longrightarrow {}^{102}Y + 3\,{}^{1}_{0}n +$ _____
(c) $^{14}N + {}^{1}_{0}n \longrightarrow {}^{1}H +$ _____

55. Consider the three reactions listed below.
(a) $H_2O(l) \longrightarrow H_2O(g)$
(b) $2\,H_2(g) + O_2(g) \longrightarrow 2\,H_2O(g)$
(c) ${}^{2}_{1}H + {}^{2}_{1}H \longrightarrow {}^{3}_{1}H + {}^{1}_{1}H$

Each of the following energy values belong to one of these equation:

$energy_1$	115.6 kcal released
$energy_2$	10.5 kcal absorbed
$energy_3$	7.5×10^7 kcal released

Match the equation to the energy value and explain briefly your choices.

56. When ${}^{235}_{92}U$ is struck by a neutron, the unstable isotope ${}^{236}_{92}U$ results. When that daughter isotope undergoes fission, there are numerous possible products. If strontium-90 and 3 neutrons are the results of one such fission, what is the other product?

57. Write balanced nuclear equations for
(a) the beta emission by ${}^{29}_{12}Mg$
(b) the alpha emission by ${}^{150}_{60}Nd$
(c) the positron emission by ${}^{72}_{33}As$

58. Rubidium-87, a beta emitter, is the product of positron emission. Identify
(a) the product of rubidium decay
(b) the precursor of rubidium-87

59. Potassium-42 is used to locate brain tumors. Its half-life is 12.5 hours. Starting with 15.4 mg, what fraction will remain after 100 hours? If it was necessary to have at least 1 microgram for a particular procedure, could you hold the original sample for 200 hours?

60. How much of a sample of cesium-137 ($t_{1/2}$ = 30 years) must have been present originally if, after 270 years, 15.0 g remain?

61. Suppose that the existence of element 114 were confirmed and reported. What element would this new substance fall beneath in the periodic table? Would it be a metal? What typical ion might you expect it to form in solution?

62. Cobalt-60 has a half-life of 5.26 years. If 1.00 g of ^{60}Co was allowed to decay, how many grams would be left after
(a) one half-life?
(b) two half-lives?
(c) four half-lives?
(d) ten half-lives?

63. Write balanced equations to show each of the following changes:
(a) alpha emission by boron-11
(b) beta emission by strontium-88
(c) neutron absorption by silver-107
(d) proton emission by potassium-41
(e) electron absorption by antimony-116

64. The ${}^{14}_{6}C$ content of an ancient piece of wood was found to be one-sixteenth of that in living trees. How many years old is this piece of wood, if the half-life of carbon-14 is 5668 years?

65. The curie is equal to 3.7×10^{10} disintegrations/second and the becqueral is equivalent to just one disintegration per second. Suppose that a hospital has a 150-g radioactive source that has an activity of 1.24 curies. What is its activity in becquerals?

***66.** When an electron and a positron (a positively charged electron) encounter each other, they destroy each other and release energy. How much energy, in joules, would you expect from such a collision if the mass of the electron is 9.1096×10^{-31} kg?

Answers to Practice Exercises

18.1 0.391 g

18.2 ${}^{226}_{88}Ra$

18.3 (a) ${}^{234}_{91}Pa$
(b) ${}^{226}_{88}Ra$

Chemistry of Selected Elements

19

Chapter 19

It is amazing to consider the vast number of substances and products that surround us. Even more astounding is the fact that all forms of matter are combinations of only slightly more than 100 elements. The human body is a complex of billions of cells, but all the cells are primarily combinations of carbon, hydrogen, oxygen, and nitrogen. This is typical of many substances, either manufactured or naturally occurring. A typical cell contains countless numbers of atoms and molecules but very few elements. In an earlier chapter the design of the periodic table was considered. Each element belongs to a particular family whose members are similar in electronic structure. The chemical behavior of elements in each family is similar as well. In this chapter we will take a brief tour of the periodic table as we discuss the chemistry of selected elements and discover some of their uses and characteristics that make them an integral part of our lives.

19.1 Periodic Trends: A Quick Review

A number of important properties of elements correspond directly to the position of the elements on the periodic table. Several of these properties are summarized in Figure 19.1.

One of the most useful of these characteristics is the division of the elements into metals and nonmetals. More than three-fourths of the elements are metals. The outstanding properties of metals and nonmetals are summarized in Table 19.1.

The first member of a family often exhibits chemistry that is significantly different from that of other members of the same family. The differences are partially the result of the smaller size and greater electronegativity of the first member. Hydrogen is so small and unique that it is often classified as a family of one.

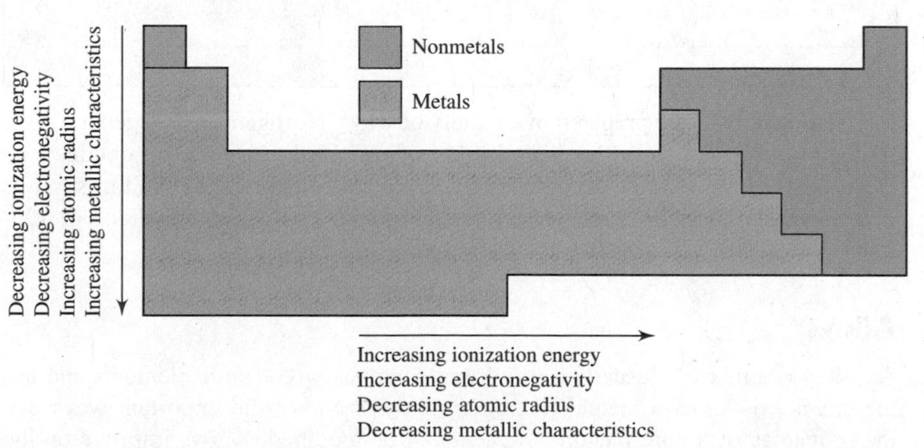

FIGURE 19.1
A summary of periodic trends.

Chapter Opening Photo: Alloys of steel, chrome, and bronze were used to create this sculpture, which stands in front of the YMCA in Los Angeles.

TABLE 19.1 Physical Properties and Chemical Properties of Metals and Nonmetals

Metals	Nonmetals
Lustrous	Nonlustrous
Good thermal conductors	Poor thermal conductors
Good electrical conductors	Poor electrical conductors
Malleable	Brittle
Ductile	Shatter easily
Form positive ions in compounds	Form negative ions in compounds
Form basic oxides	Form acidic oxides

19.2 Preparation of the Elements

metallurgy

ore

Only about a quarter of the elements found in nature exist in the free state. Most are found in compounds. **Metallurgy** is the science of obtaining and refining metals from their ores and the study of the properties and applications of metals. An **ore** is the commercial mineral source of a metal. Since the metals in ores are usually in the form of positive ions (cations), the processing of an ore involves the *reduction* of the ion to the free metal (oxidation number 0). Reduction may be accomplished by the following methods:

Method	Example
Carbon reduction	$2\ PbO + C \xrightarrow{\Delta} 2Pb + CO_2$
Active metal (Mg, Zn) reduction	$TiCl_4 + 2\ Mg \xrightarrow{\Delta} Ti + 2\ MgCl_2$
Hydrogen reduction	$WO_3 + 3\ H_2 \xrightarrow{\Delta} W + 3\ H_2O$
Electrolysis	$MgCl_2 \xrightarrow[\text{current}]{\text{electric}} Mg + Cl_2$
Heating	$2\ Ag_2O \xrightarrow{\Delta} 4\ Ag + O_2$

The nonmetals are prepared in a variety of ways. Nitrogen and oxygen are most frequently obtained from the *liquefication* of air. Hydrogen can be obtained by electrolysis of water or the decomposition of methane (natural gas). Sulfur is mined in its elemental form, and halogens are produced by the oxidation of halide salt solutions.

Alloys

alloy

An **alloy** is a mixture (usually a solid) that contains two or more elements and has the characteristics of a metal. Making an alloy is one of the important ways that the properties of a pure metallic element can be modified. Alloys improve on the

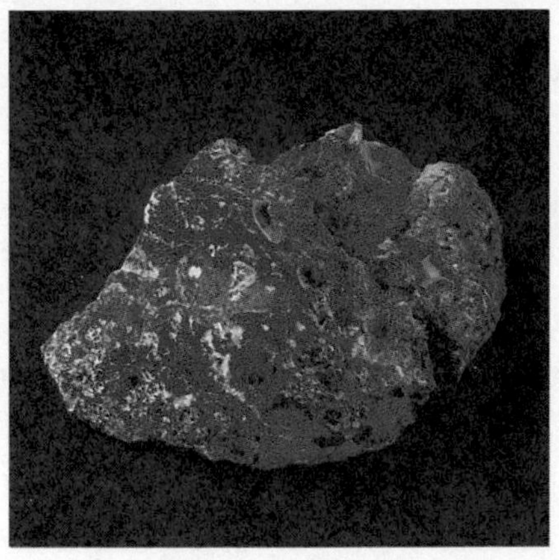

◀ **Copper ore has a much different appearance and color than refined copper metal.**

tensile strength, wearability, corrosion resistance, or conductivity of pure metals. **Steel**, for example, is an alloy of iron that contains small amounts of carbon, and it is much stronger and harder than pure iron. Pure gold is too soft to use in jewelry, but alloys of gold and copper or silver form beautiful decorative pieces.

steel

Alloys can be classified as *solution alloys* or *intermetallic compounds.* In a **solution alloy** the metals are combined in the molten state and the components mix randomly and uniformly. Examples of this type of alloy include brass, pewter, and nitinol (a nickel–titanium alloy that "remembers" its shape). Dental **amalgam** (alloys of mercury) dates back to early China. **Intermetallic compounds** are homogeneous alloys with definite composition. In modern society the uses of these intermetallic compounds are continually expanding. Razor blades are advertised with edges of Cr_3Pt. Stereo headphones contain permanent lightweight magnets of Co_5Sm (a cobalt–samarium alloy). Most recently, intermetallic compounds have been important in the development of superconductors. These materials are utilized as superconducting magnets in magnetic resonance imaging (MRI), a noninvasive technique for observing tissues in living organisms. The number of alloys and their uses are almost limitless. Table 19.2 gives several more examples of the thousands of alloys now available.

solution alloy

amalgam

intermetallic compound

TABLE 19.2 Composition of Selected Alloys

Alloy	Percent composition
Stainless steel	74 Fe, 18 Cr, 8 Ni, 0.18 C (Many others are known).
Storage battery plates	94 Pb, 6 Sb
Coinage silver (U.S.)	90 Ag, 10 Cu
Plumber's solder	67 Pb, 33 Sn
Babbitt metal	90 Sn, 7 Sb, 3 Cu
Yellow brass	67 Cu, 33 Zn
10 carat gold	42 Au, 38–46 Cu, 12–20 Ag
18 carat gold	75 Au, 10–20 Ag, 5–15 Cu
Nichrome	60 Ni, 40 Cr
Stellite	55 Co, 15 W, 25 Cr, 5 Mo
Spring steel	98.6 Fe, 1 Cr, 0.4 C

Liquefied hydrogen gas is stored in huge tanks at Kennedy Space Center before being loaded onto the space shuttle as fuel.

19.3 Hydrogen: A Unique Element

Hydrogen is the most abundant element in the universe. In the early days of chemistry, hydrogen was considered to be the precursor to all the other elements. Hydrogen combines chemically with most of the elements, and its history includes many of the scientists we've encountered before. Jacques Charles, in 1783, first used hydrogen to float his balloon above France. By 1928, a dirigible designed to carry passengers was built in Germany, soon to be followed by the infamous *Hindenburg*. Unfortunately, the flammable nature of hydrogen resulted in the explosion and destruction of the *Hindenburg* in 1939, killing many passengers. This accident caused hydrogen to be classified by many people as a dangerous substance. Today the use of hydrogen is once again expanding.

Hydrogen is used commercially in the production of ammonia, NH_3, by the Haber process:

$$N_2 + 3\ H_2 \rightleftharpoons 2\ NH_3$$

Most of the ammonia manufactured this way is converted to fertilizers. Hydrogen is also used to produce methanol, CH_3OH, by reaction with carbon monoxide.

$$2\ H_2 + CO \xrightarrow[\Delta]{\text{catalyst}} CH_3OH$$

The methanol produced this way can be used as a gasoline additive or converted to gasoline. Methanol is used as an alternative fuel and in the manufacture of adhesives and plastics.

A less obvious but extensive use of hydrogen is found in our food. Fats and oils, from peanut butter to solid shortening or margarine, are hardened by the addition of hydrogen to carbon–carbon double bonds in the molecules. Hydrogenation converts inexpensive oils to popular cooking products.

Hydrogen has also been proposed as the fuel of the future (see Chemistry in Action, p. 55). It is currently used as a fuel for the space shuttle; the large external tanks are filled with hydrogen and oxygen, which combine to form water vapor and thus provide energy to lift the shuttle into orbit. On board the shuttle are hydrogen–oxygen fuel cells that provide power for the shuttle during flight.

19.4 The Alkali Metals: Group IA

Physical Properties

alkali metal

The elements of Group IA in the periodic table are known as the **alkali metals**. They are named alkali metals because all of them form strong basic aqueous solutions. The Group IA metals are silvery-white in color when freshly cut, are soft and ductile, and are good conductors of electricity. Their physical properties are summarized in Table 19.3.

TABLE 19.3 Physical Properties of the Alkali Metals

Element	Atomic number	Atomic mass	Electron configuration	Metallic atomic radius (nm)	Radius of M^+ ion (nm)	Density (20°C, g/ml)	Melting point (°C)	Boiling point (°C)
Li	3	6.941	$[He]2s^1$	0.158	0.060	0.53	186.0	1336
Na	11	22.99	$[Ne]3s^1$	0.186	0.095	0.97	97.7	880
K	19	39.10	$[Ar]4s^1$	0.238	0.133	0.86	63.6	760
Rb	37	85.47	$[Kr]5s^1$	0.253	0.148	1.53	39.0	700
Cs	55	132.9	$[Xe]6s^1$	0.272	0.169	1.90	28.4	670

The alkali metals are among the most reactive elements and are never found in the free state in nature. Their chemical activity increases with size and atomic number. Because these metals react readily with oxygen and moisture, they are stored in a vacuum or under an inert liquid such as kerosene, benzene, or toluene. The similarity in the chemical properties of this family arises because of their similar electron configurations. An atom of an alkali metal has only one valence electron. Therefore, these elements form ions generally having a $+1$ oxidation number.

Chemical Properties

The chemical properties of alkali metals depend on the ease with which they each lose one outer shell electron. The ion formed when an electron is lost from an alkali metal has a noble-gas electron configuration and therefore has great stability. The energy required to remove an electron from an alkali metal (ionization energy, see Chapter 11) is lower than that for other metals. The alkali metals are excellent reducing agents. A number of these oxidation–reduction reactions are summarized in Table 19.4.

Sodium metal is so soft it can be cut with a knife.

Applications of Alkali Metal Compounds

Uses of the alkali metals come primarily from their ability to lose electrons. The formation of basic solutions in reactions with water is a sometimes spectacular

TABLE 19.4 Chemical Reactions of Alkali Metals

Reacts with	Products	Example
H_2O	Hydroxide, basic solution	$2\,Na(s) + 2\,H_2O(l) \longrightarrow 2\,NaOH(aq) + H_2(g)$
H_2	Hydride	$2\,Rb(s) + H_2(g) \longrightarrow 2\,RbH(s)$
Halogens	Metal halides (salts)	$2\,Na(s) + Cl_2(g) \longrightarrow 2\,NaCl(s)$
Oxygen	Oxide	$4\,Li\,(s) + O_2(g) \longrightarrow 2\,Li_2O(s)$
	Peroxide	$2\,Na(s) + O_2(g) \longrightarrow Na_2O_2(s)$
	Superoxide	$K(s) + O_2(g) \longrightarrow KO_2(s)$

Firefighters use tanks of potassium superoxide as a source of oxygen in breathing equipment. ▶

reaction. Sodium hydroxide, NaOH (lye), is widely used in the manufacturing of other chemicals and in processing pulp in the paper industry. Sodium hydroxide is also used to treat cotton to produce mercerized cotton, often used in thread. Some household cleaners contain lye. Soap is formed when sodium hydroxide is reacted with fats.

In reactions with oxygen, the alkali metals produce a diversity of useful compounds. Sodium peroxide can be reacted with water to produce hydrogen peroxide:

$$Na_2O_2 + 2\ H_2O \longrightarrow 2\ NaOH + H_2O_2$$

Hydrogen peroxide is a good oxidizing agent often used as a bleach for hair or as a disinfectant. Potassium, rubidium, and cesium react with oxygen to form superoxides. These compounds (especially KO_2) are used as a source of oxygen in emergency breathing equipment. The carbon dioxide and water vapor exhaled by the wearer react with the superoxide to form oxygen.

$$4\ KO_2 + 4\ CO_2 + 2\ H_2O \longrightarrow 4\ KHCO_3 + 3\ O_2$$

The salts formed by the alkali metals also find many uses. Sodium nitrate (Chile saltpeter) is found in large deposits formed by bacteria acting on saltwater organisms in shallow water. Sodium nitrate can be converted into potassium nitrate in a double-displacement reaction:

$$NaNO_3 + KCl \longrightarrow KNO_3 + NaCl$$

Potassium nitrate has long been used in gunpowder and in pyrotechnics. Sodium bicarbonate, $NaHCO_3$ (commonly called baking soda), is a component of baking powder and antacids. Lithium carbonate is currently prescribed as a treatment for manic depression.

TABLE 19.5 Physical Properties of the Alkaline Earth Metals

Element	Atomic number	Atomic mass	Electron configuration	Metallic atomic radius (nm)	Radius of M^{2+} ion (nm)	Density (20°C, g/mL)	Melting point (°C)	Boiling point (°C)
Be	4	9.012	$[He]2s^2$	0.112	0.031	1.85	1278	2970
Mg	12	24.31	$[Ne]3s^2$	0.116	0.065	1.74	651	1107
Ca	20	40.08	$[Ar]4s^2$	0.197	0.099	1.55	843	1487
Sr	38	87.62	$[Kr]5s^2$	0.215	0.113	2.6	769	1384
Ba	56	137.3	$[Xe]6s^2$	0.222	0.135	3.6	725	1140
Ra	88	226.0	$[Rn]7s^2$	0.246	—	5.5	700	<1737

Biological Applications

In living organisms sodium and potassium ions play a vital role in the functioning of biochemical systems. Inside living cells the concentration of potassium ions is very high compared to the concentration of sodium ions, while outside the cell the reverse is true. To maintain these levels, an elaborate mechanism is required to transport these ions through the cell membrane. Potassium ions help the cell to maintain intercellular osmotic pressure and to regulate pH. Potassium ions promote metabolic reactions and play an important role in membrane polarization. This polarization is necessary for proper nerve impulse conduction and muscle contraction. Sources of potassium in the diet include avocados, peanut butter, potatoes, and bananas.

Sodium ions are absorbed from foods by a transport system. The blood concentration of sodium ions is regulated by the kidneys. The major role of sodium ions in the extracellular fluid is to maintain osmotic pressure and regulate water balance. Sodium ions are also significant in the regulation of pH, the maintenance of membrane polarization, and the transportation of substances through cell membranes.

19.5 Alkaline Earth Metals: Group IIA

Physical Properties

alkaline earth metal

The **alkaline earth metals** were named for the alkaline properties of their oxides and because early chemists called all insoluble nonmetallic substances that were not affected by fire "earth." Group IIA metals are too reactive to be found free in nature but are generally harder and less reactive than the Group IA metals. At ordinary temperatures beryllium and magnesium are stable in air, but barium and strontium are so reactive that they are usually stored under an inert solvent such as kerosene. Radium is highly radioactive. Some of the physical properties of the alkaline earth metals are shown in Table 19.5.

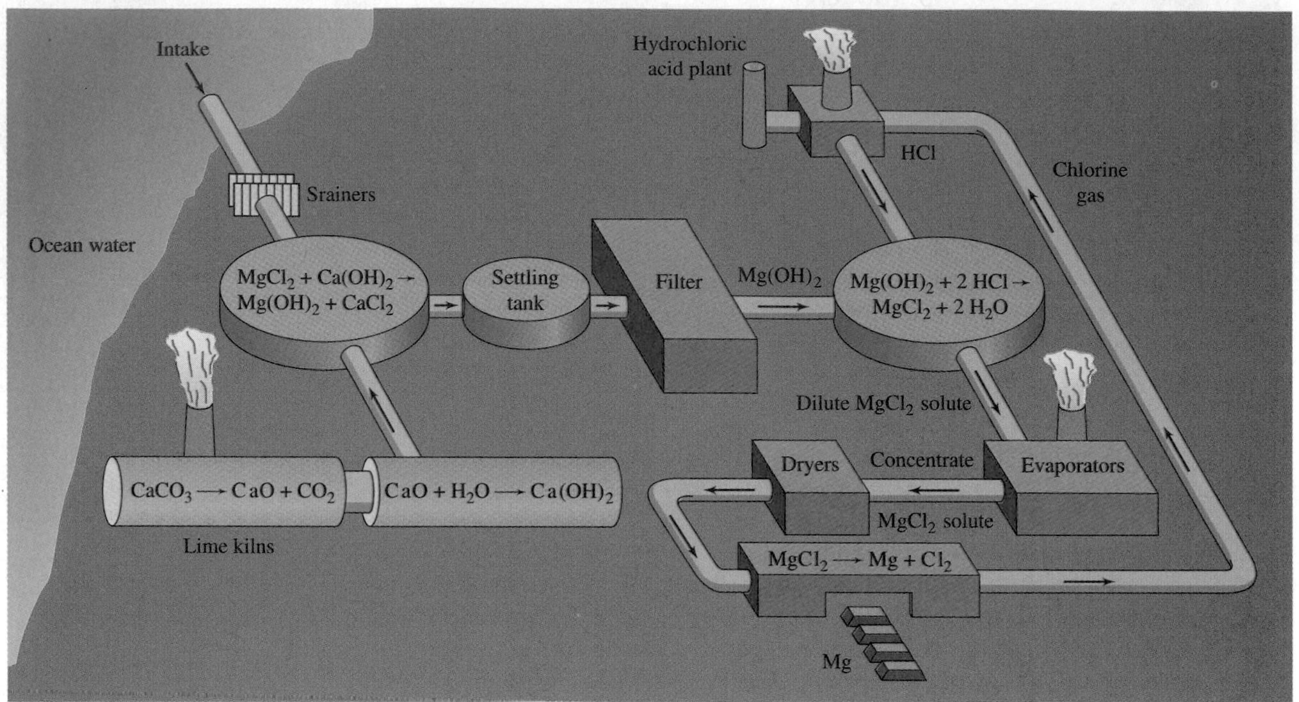

▲
FIGURE 19.2
Plant for processing Mg^{2+} ions from seawater.

Magnesium is the only metal that is obtained commercially from seawater (Figure 19.2). The process involves the following steps:

1. Precipitation of magnesium ions from seawater (which contains approximately 0.13% Mg^{2+}) as magnesium hydroxide by adding calcium oxide.

$$CaO(s) + H_2O(l) \longrightarrow Ca^{2+}(aq) + 2\ OH^-(aq)$$

$$Mg^{2+}(aq) + 2\ OH^-(aq) \longrightarrow Mg(OH)_2(s)$$

2. Filtering out the $Mg(OH)_2$ and converting it to magnesium chloride by reaction with hydrochloric acid.

$$Mg(OH)_2(s) + 2\ HCl(aq) \longrightarrow MgCl_2(aq) + 2\ H_2O(l)$$

3. Drying and electrolyzing the $MgCl_2$ to obtain magnesium and chlorine.

$$MgCl_2(l) \xrightarrow{\text{electrolysis}} Mg(s) + Cl_2(g)$$

Chemical Properties and Applications

All the Group IIA elements react by losing the outermost pair of *s* electrons to form +2 ions. Hence, the formulas of their compounds are similar ($BeSO_4$, $MgSO_4$, $CaSO_4$, and so on). However, these elements differ widely in reactivity.

Beryllium is the least reactive of the alkaline earth metals. It is resistant to oxidation and stable in air at ordinary temperatures but oxidizes rapidly at about 800°C.

$$2\ Be + O_2 \longrightarrow 2\ BeO$$

Because of its many special properties, beryllium is useful in technology. For example, it is highly permeable to X rays and is therefore used in X-ray tube windows. In nuclear technology, beryllium serves as a moderator to slow down neutrons; and when bombarded with alpha particles, it is a good source of neutrons.

Alloys of beryllium and copper form a metal that conducts electricity almost as well as copper. Because of its strength it does not generate sparks, and this alloy can be made into tools for use in areas where explosive gases are present. Its light weight and strength also have made it a choice (an expensive one) for golf clubs.

The use of magnesium is also based on its property of high strength-to-weight ratios. When exposed to air, magnesium quickly becomes coated with a thin, transparent layer of MgO. This layer protects the surface from further corrosion at ordinary temperatures. The majority of magnesium produced is made into lightweight alloys, which are used in aircraft, automotive parts, lightweight tools, and luggage trim.

The oxides and carbonates of calcium are compounds of economic value and special interest. Calcium carbonate decomposes to CaO (lime) in a reaction that ranks among the oldest chemistry known to us. The primary use of lime until 1900 was as a building material in the form of mortar (a sand, lime, and water mixture). Mortar was used by the Greeks in building temples. The chemistry involved in the mortar hardening process is unique: The mortar slowly absorbs CO_2 from the air, and the lime reverts back to calcium carbonate.

$$Ca(OH)_2 + CO_2 \longrightarrow CaCO_3 + H_2O$$

The sand simply remains in the matrix, forming a hard substance resistant to wear. Calcium carbonate forms the beautiful stalactites and stalagmites in caverns and also marble, which we see so often in buildings and sculpture. When lime is added to water, an alkaline solution of calcium hydroxide is produced. This solution has been used in the removal of hair from hides in the leather tanning process.

The mineral gypsum is another common calcium compound, hydrous calcium sulfate. Although gypsum is much less common than calcium carbonate, the several hundred square miles of sand in the White Sands Desert in New Mexico are composed of finely divided gypsum. Plaster of Paris is made by heating gypsum:

$$\underset{\text{gypsum}}{2\,CaSO_4 \cdot 2\,H_2O} \xrightarrow{\Delta} \underset{\text{plaster of Paris}}{(CaSO_4)_2 \cdot H_2O} + 3\,H_2O$$

To make surgical casts or plaster molds and impressions, the foregoing reaction is reversed by wetting powdered plaster of Paris to form a solid cast of calcium sulfate dihydrate.

Calcium and magnesium salts present problems when dissolved in our water. Water supplies containing Ca^{2+}, Mg^{2+}, and Fe^{3+} are called "hard water." The walls of pipes and the insides of tea kettles and coffee pots often become coated with insoluble carbonates of these ions. To prevent this scaly buildup, the water is "softened" by removing the Ca^{2+} and Mg^{2+} ions (see Section 13.15).

Light weight and strength make berillium golf clubs a favorite though expensive choice of golfers.

Biological Applications

Calcium and magnesium are required for many essential biological processes in both animal and plant life. Magnesium is present in the chlorophyll of green plants and is required for the activity of some enzymes involved in animal metabolism.

It is involved in a variety of metabolic functions within the mitochondria and is associated with the production and use of ATP, the central energy transfer molecule of the cell. Calcium is necessary for nitrogen utilization by plants.

Ninety-nine percent of the calcium in humans is found as inorganic salts in the bones and teeth. In dissolved form calcium ions play a role in nerve impulses, muscle contraction, hormone regulation, and blood coagulation. The absorption of calcium by the body is dependent on a variety of factors, including the need for calcium. The presence of vitamin D and a high protein intake increase absorption of calcium, while excessive fat seems to lower absorption. The role of calcium in the prevention of osteoporosis has been well documented. Good sources of calcium in the diet include milk and leafy green vegetables.

The remaining Group IIA elements are toxic to living systems. Interestingly, physicians often use barium sulfate (a barium "cocktail") to observe the digestive tract by X-rays. This does not mean the doctor is poisoning the patient; rather, the $BaSO_4$ is *very* insoluble and passes through the digestive system without absorption. Strontium is believed to be a biologically nonessential element; however, the isotope Sr-90, present in nuclear fallout, can be absorbed by the body where it concentrates in bone tissue creating a radiation hazard. Strontium chloride is used as a dental desensitizer in some toothpastes.

19.6 Flame Tests and Spectroscopy

All Group IA elements and calcium, strontium, and barium from Group IIA can be detected by simple flame tests. A Bunsen burner and a piece of clean platinum wire are the only essential pieces of apparatus needed to make flame tests. When salts of the elements are volatilized in the burner flame, the characteristic colors are visible.

Element	Color
Li	Red
Na	Bright yellow (persistent)
K	Violet (short duration)
Rb	Violet
Ca	Brick red (short duration)
Sr	Crimson
Ba	Light green

Flame color has been exploited by spectroscopy both to identify elements and to provide detailed information on the structure of atoms. Electrons can exist only at certain definite quantized energy levels within atoms (Section 5.6). At ordinary temperatures the electrons in atoms are at their lowest, or ground-state, energy levels. When heated in a Bunsen flame (or electric arc for some elements), some ground-state electrons are promoted to higher excited-state energy levels. In falling back to the ground state, each electron releases the same amount of energy (as radiation) that was absorbed when it was promoted to the excited state. If the wavelength of this radiation falls in the visible spectrum (400–700 nm), it is seen as colored light corresponding to a particular wavelength. When such light is passed through a prism (or ruled grating) in a spectroscope, it is found to consist of narrow colored bands called *spectral lines.* Each band (or line) corresponds to the energy emitted when an electron falls from a specific excited state to a specific lower energy level. Since the number of electrons in each element is different, each element has a characteristic set of spectral lines that is different from that of every other element. This set of spectral lines identifies the element and provides a means of determining the energy levels that are occupied by electrons within the atoms of that element.

A spectacular example of the colors produced by the nitrate salts of Groups IA and IIA can be seen in fireworks displays. The colors of the flames from treated pine cones or commercial pressed sawdust logs also result from these salts.

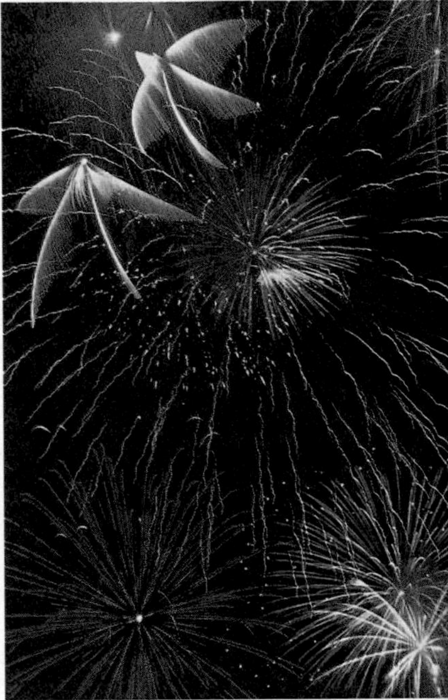

The brilliant colors in fireworks are produced by the burning of metal salts.

19.7 Group IIIA Elements

Physical Properties

The elements in Group IIIA include such various elements as the most abundant metal on earth, aluminum, and the first metalloid (boron) we've encountered in our survey. These elements do not occur free in nature but are generally found as oxides in many locations worldwide. The physical properties of these elements are summarized in Table 19.6.

Boron is relatively rare; it is usually found combined with oxygen in borates. Large deposits of borax, $Na_2B_4O_7 \cdot 10\ H_2O$, are mined in the Mojave Desert near Boron, California. In the 1800s, deposits of borate were mined in Death Valley using the famous 20-mule teams to haul the ore. Another source of boron is boric acid, H_3BO_3, found in the volcanic regions of Italy.

Applications

Although boron is not a common element, it finds many interesting uses. Diboron trioxide can be reduced with carbon to form tetraboron carbide.

$$2\ B_2O_3 + 4\ C \longrightarrow B_4C + 3\ CO_2$$

This material is extremely hard and has a low density. One of its major uses is in bulletproof armor. Borax is probably the most important boron compound. It is

TABLE 19.6 Physical Properties of the Group IIIA Elements

Element	Atomic number	Atomic mass	Electron configuration	Density (20°C, g/mL)	Melting point (°C)	Boiling point (°C)
B	5	10.81	$[He]2s^22p^1$	2.34	2300	2550
Al	13	26.98	$[Ne]3s^23p^1$	2.70	660	2327
Ga	31	69.72	$[Ar]4s^23d^{10}4p^1$	5.91	29.78	2403
In	49	114.8	$[Kr]5s^24d^{10}5p^1$	7.31	156.6	2000
Tl	81	204.4	$[Xe]6s^24f^{14}5d^{10}6p^1$	11.85	303.5	1457

mildly alkaline and works well as a cleansing agent or water softener. Borax can be reacted with sulfuric acid to form boric acid.

$$Na_2B_4O_7 \cdot 10\ H_2O + H_2SO_4 \longrightarrow 4\ H_3BO_3 + Na_2SO_4 + 5\ H_2O$$

Boric acid is a weak acid that is slightly toxic. It is used as a mild antiseptic in the home and, in dilute solution, as an eyewash. Boric acid is also used as a flame retardant in insulating materials made from cellulose. The dehydration of boric acid yields boric anhydride (diboron trioxide).

$$2\ H_3BO_3 \longrightarrow B_2O_3 + 3\ H_2O$$

This oxide is used in the manufacture of borosilicate glass, commonly known as Pyrex. This type of glass has a higher softening temperature and expands less on heating than other types of glass, so it can be used for freezer-to-microwave (or stove) cookware.

Aluminum (generally spelled *aluminium* in countries other than the United States) is the most abundant metal and the third most abundant element in the earth's crust, exceeded only by oxygen and silicon. Aluminum occurs mainly in clays, feldspars, and granite as complex aluminum silicates and in bauxite ores as hydrated aluminum oxide. Pure aluminum oxide (Al_2O_3) is known as *alumina.*

Alumina occurs naturally as the mineral corundum. Corundum is a very hard material, standing next to diamond on the hardness scale, and it is used as an abrasive in sandpaper and toothpaste. Emery, also widely used as an abrasive, is an impure corundum consisting of Al_2O_3 and Fe_3O_4. Rubies and sapphires are much sought after precious gemstones. They consist of corundum colored by small amounts of other metallic oxides. Artificial rubies and sapphires are produced that are practically indistinguishable from the natural stones. In addition to their use as gemstones and bearings in watches and other instruments, artificial rubies and sapphires are used in lasers for the production of intense beams of coherent light.

Although aluminum is the most abundant metal in the earth's crust, free aluminum metal was not known until 1825. In that year the Danish scientist Hans Christian Oersted succeeded in making some impure samples of aluminum metal by heating aluminum chloride with potassium–mercury amalgam. However, the reduction of aluminum by an alkali metal was an inherently costly process, and aluminum remained nearly as expensive as gold and platinum during most of the 19th century. Finally in 1886, a relatively cheap method for the electrolytic production of aluminum was devised. Since then, aluminum has been used in ever-increasing amounts. Aluminum has a density of only 2.7 g/mL, a value about one-third that of other common metals such as iron, copper, and zinc. Because it is a lightweight metal and has the ability to form a variety of useful alloys, aluminum has many important commercial applications.

Aluminum is an excellent conductor of heat. As an electrical conductor, it is surpassed only by silver and copper. However, when a comparison is made with wires having the same mass per unit of length, aluminum wire, being larger, surpasses both copper and silver as an electrical conductor. For this reason—and because of the increasing scarcity of copper—high-voltage transmission cables are now usually made of aluminum wires reinforced with a steel core for additional strength.

Aluminum metal has numerous uses for structural materials as well as in packaging. It is most commonly found as an alloy, since alone aluminum is soft and weak. When combined with copper and smaller amounts of silicon and magnesium, the alloy is used in large quantities to build airplanes. In combination with manganese, a softer, more corrosion-resistant alloy is formed that is used in kitchen utensils, lawn furniture, highway and directional signs, aluminum foil, and soda cans.

Bauxite or clay can be treated with sulfuric acid to produce aluminum sulfate. Double salts, such as $KAl(SO_4)_2 \cdot 12\ H_2O$ form a class of substances known as alums. These aluminum compounds are used in the paper industry to make the product stronger and less porous. In water treatment plants alums are used as clarifiers. The alum surrounds the suspended particles in the water and precipitates them.

The vibrant colors in fabrics are produced by adhering dye molecules to the cloth with a mordant such as aluminum hydroxide.

When aluminum salts react with bases, aluminum hydroxide forms as a gelatinous white precipitate. The ionic equation is

$$Al^{3+}(aq) + 3\ OH^-(aq) \longrightarrow Al(OH)_3(s)$$

This hydroxide is amphoteric and dissolves in either a strong base or a strong acid:

$$Al(OH)_3(s) + NaOH(aq) \longrightarrow NaAlO_2(aq) + 2\ H_2O(l)$$
$$2\ Al(OH)_3(s) + 6\ HCl(aq) \longrightarrow 2\ AlCl_3(aq) + 3\ H_2O(l)$$

Aluminum hydroxide is called a *mordant* in the textile industry. It binds to cloth and also to dye molecules, adhering the dye to the cloth. Some common antacid medications are also preparations of aluminum hydroxide.

Certain aluminum salts exert an astringent or constricting action on the skin and mucous membranes, a property exploited in many commercial antiperspirant preparations. The principal active agents in many of these formulations are aluminum chlorohydrates, actually aluminum hydroxy chlorides with formulas of the type $Al_2(OH)_5Cl \cdot 2\ H_2O$ or $Al(OH)_2Cl$. These substances prevent or retard perspiration by constricting the openings of the sweat glands.

19.8 Group IVA Elements

Physical Properties

The elements of Group IVA exhibit the full range of metallic character. Carbon is distinctly nonmetallic, silicon is classified as a metalloid, and the heavier elements, tin and lead, are most certainly metals. The physical properties of these elements are summarized in Table 19.7.

Occurrence and Applications

Carbon is widely distributed in many forms throughout the planet. As an element it commonly occurs in three allotropes, diamond, graphite, and buckminsterfullerene. The diamond form of carbon has long been of interest to us. The majority of the most common naturally occurring diamonds are mined in South Africa. The use of diamonds for abrasives and cutting tools in industry, in addition to their gem status, has led to the production of synthetic diamonds. These synthetic crystals are made by subjecting graphite to intense pressures and temperatures (70,000 atm and 2000°C). Buckminsterfullerene is the newest form of carbon to be discovered. Its structure and uses are described in the Chemistry in Action, Chapter 11, p. 232.

TABLE 19.7 Physical Properties of the Group IVA Elements

Element	Atomic number	Atomic mass	Electron configuration	Density (20°C, g/mL)	Melting point (°C)	Boiling point (°C)
C	6	12.01	$[He]2s^22p^2$	3.15	3550	Sublimes
Si	14	28.09	$[Ne]3s^23p^2$	2.33	1410	2355
Ge	32	77.59	$[Ar]4s^23d^{10}4p^2$	5.32	937	2830
Sn	50	118.7	$[Kr]5s^24d^{10}5p^2$	7.30	232	2270
Pb	82	207.2	$[Xe]6s^24f^{14}5d^{10}6p^2$	11.4	328	1744

Graphite fibers form high strength, lightweight tennis racquets, making a faster more arduous game. ▶

Graphite is by far the most common form of elemental carbon. It melts at extremely high temperatures (3550°C), so it is sometimes used to line electric furnaces. It is remarkably soft and sticks to paper, making graphite the common pencil "lead." Harder "lead" contains more clay mixed with the graphite. This allotrope is also used as a lubricant. There are a variety of other relatively pure forms of carbon, including charcoal and carbon black. Charcoal has remarkable properties as an adsorbant after it has been heated to remove volatile impurities in a process known as activation. The activated charcoal is used to remove colored impurities in the refining of sugar and can adsorb a variety of contaminants when used in gas masks or a water purification system. Carbon black is used as a reinforcing agent in rubber and is the reason that tires are black and not the natural color of rubber. Inks used on nonglossy productions (such as your newspaper) are made with carbon black. Graphite can also be mixed with plastics to produce highly flexible, strong products such as tennis racquets, golf clubs, and fishing rods.

Carbon is easily oxidized to either carbon monoxide or carbon dioxide, depending upon the amount of oxygen available. For this reason carbon is used industrially for the reduction of metal oxides to metals. Approximately half of the carbon dioxide produced in the United States is used as a refrigerant. When cooled and compressed into solid blocks, it is known as dry ice. About one-fourth of the CO_2 produced is used in the production of carbonated beverages. The carbon dioxide released into the atmosphere during combustion of fossil fuels has become a subject of controversy in recent years because of its potential effect on the global heat balance ("greenhouse effect").

Carbon monoxide is a toxic gas. It can interact with the iron in the hemoglobin of the blood, preventing proper transport of oxygen. Exposure to even a small amount of carbon monoxide can be serious and even fatal.

Carbon exhibits a tremendous ability to form compounds with hydrogen, nitrogen, and oxygen. These compounds are so numerous and diverse that entire branches of chemistry are devoted to their properties and reactions. Chapters 20–36 provide a closer look at carbon in the fields of organic chemistry and biochemistry.

Silicon is the second most abundant element in the crust of the earth. Its principal sources are in the form of silicon dioxide, found in quartz, sand, sandstone, agate, and amethyst. It also occurs in a wide variety of silicate minerals ranging from very hard garnet to very soft asbestos. Silicate minerals also include clay, mica, feldspar, talc, and zeolite.

Crystalline quartz is used commercially to control the frequency of virtually all radio and television transmissions. It is used in instruments and timepieces as well. Silica can be dissolved in hot sodium carbonate to produce sodium silicate.

$$Na_2CO_3 + SiO_2 \longrightarrow Na_2SiO_3 + CO_2$$

The sodium silicate produced is known as water glass. It dissolves in water, and the silicate ions can act as buffers to maintain the pH of the solution. They also allow animal and vegetable oils to be dispersed in water. For this reason water glass is frequently used in detergents. Sodium silicate can be treated with a dilute acid to form a gelatinous noncrystalline SiO_2. This substance when dried is known as silica gel. It has many uses as a drying agent or a dehumidifier. It is placed (in small packets) in manufacturers' boxes to protect moisture-susceptible products (like electronics).

Silicon has the ability to form long chains in combination with oxygen in molecules known as silicones. The silicones are a part of a group of compounds called polymers (see Chapter 26). Silicones are resistant to heat, light, and oxygen. They also have some interesting antisticking properties. Commercial applications of silicones are truly incredible, ranging from lipstick and lotions to Silly Putty, auto ignition systems, and car polish.

Silicon also behaves as a semiconductor. This property has made silicon the backbone of the computer and microchip industry. Microchips are used in the myriad of small electronic gadgets and personal computers so prevalent in our society. Silicon is also used in photoelectric cells to generate electricity.

Tin and lead are metals easily recovered from their ores. They have long been used in alloys such as bronze (tin and copper) for making tools and decorative objects. Tin is especially corrosion-resistant and is used to plate the surfaces of other metals. The best-known example of this is "tin cans," which are really steel cans coated with a layer of tin. Lead is primarily used in the manufacture of automotive batteries (see Section 17.6). Lead oxide is often used in rust-inhibiting paints and primers. Lead chromate is the brilliant yellow pigment found in high-

▲ **The brilliant yellow pigment found in highway markings contains lead chromate.**

way markings. The toxicity of lead has been known for centuries. Recent concern over the effects of lead poisoning has resulted in the removal of lead from paints and gasoline.

19.9 Group VA Elements

Physical Properties

Group VA of the periodic table includes the elements nitrogen, phosphorus, arsenic, antimony, and bismuth. Nitrogen is the most abundant element in the atmosphere, and phosphorus is the twelfth most abundant element in the earth's crust. Compounds of nitrogen and phosphorus are essential for all forms of life.

The electron structures of these elements show that they have five electrons in their outer shells, two *s* and three *p* electrons. The elements of this group become increasingly metallic from top to bottom. Nitrogen is a colorless gas with virtually no hint of metallic properties, but bismuth is distinctly metallic in character. Some of the physical properties of Group VA elements are listed in Table 19.8.

Details of the Haber process are given in Chapter 16

Occurrence and Applications of Nitrogen

Nitrogen is found in the atmosphere, living organic matter, decayed matter, ammonia, ammonium salts, and nitrate deposits. Elemental nitrogen, N_2, is useful in many ways. It is extremely inert and highly stable as a result of the triple bond between nitrogen atoms. Nitrogen gas is used to provide an inert atmosphere for sensitive chemical reactions. It also provides a nonoxidizing atmosphere for packaging foods and wine, or preserving articles of sentimental value (such as a wedding dress). Nitrogen is easily liquefied to produce a coolant used to freeze such things as termites, warts, and biological samples.

In the Haber process, ammonia is manufactured by passing nitrogen and hydrogen gases over a catalyst at elevated temperatures and pressures:

$$N_2(g) + 3\ H_2(g) \rightleftharpoons 2\ NH_3(g)$$

TABLE 19.8 Physical Properties of the Group VA Elements

Element	Atomic number	Atomic mass	Electron configuration	Density (20°C, g/mL)	Melting point (°C)	Boiling point (°C)
N	7	14.01	$[He]2s^22p^3$	0.81[a]	−210.1	−195.8
P	15	30.97	$[Ne]3s^23p^3$	1.82[b]	44.2	280
As	33	74.92	$[Ar]4s^23d^{10}4p^3$	5.7	817[c]	615[d]
Sb	51	121.8	$[Kr]5s^24d^{10}5p^3$	6.6	630.5	1380
Bi	83	209.0	$[Xe]6s^24f^{14}5d^{10}6p^3$	9.8	271.3	1450

[a]Liquid at boiling point
[b]White phosphorus
[c]At 36 atm
[d]Sublimes

The current production of ammonia in the United States is over 1.7×10^{10} kg annually. By far the greatest consumption of ammonia is in the manufacture of fertilizers or in its direct use as a fertilizer. Other industrial uses of ammonia or its derivatives include the manufacture of nitric acid, plastics, refrigerants, explosives, pulp and paper, and textiles.

Another important hydrogen compound of nitrogen is hydrazine, N_2H_4. It is a powerful reducing agent that has been used in liquid and solid form. The most spectacular use of hydrazine is in rocket fuel. A combination of monomethylhydrazine ($CH_3N_2H_3$) and dinitrogen tetroxide powers the booster rockets of the space shuttle. Hydrazine was also used during lunar landings. Here on earth hydrazine is most commonly used as a blowing agent in the production of plastics. It decomposes to nitrogen gas, producing the porous structure of the plastic.

Nitrogen combines with oxygen in various stoichiometric ratios to form a series of oxides. Several have interesting applications. Dinitrogen oxide, N_2O, is odorless and tasteless and is used as an anesthetic for minor surgery. It is commonly known as "laughing gas." Dinitrogen oxide also has the ability to easily dissolve in vegetable fats. This results in its use as a propellant in canned whipped cream. Nitrogen oxide, NO, is a colorless gas that can be produced in the laboratory by reacting copper with concentrated nitric acid. In the presence of air this oxide immediately undergoes oxidation to the brown nitrogen dioxide gas, which is a visible component in the air pollution plaguing our cities.

Nitric acid is an important industrial chemical used in the manufacture of many products. Most nitric acid is made by the Ostwald process. In this process, an ammonia–air mixture is passed over a platinum catalyst at about 800°C. Ammonia is first oxidized to nitrogen oxide, which in turn is further oxidized to nitrogen dioxide. The latter is then absorbed in water to form the acid. The equations are

$$4\ NH_3(g) + 5\ O_2(g) \longrightarrow 4\ NO(g) + 6\ H_2O(g)$$
$$2\ NO(g) + O_2(g) \longrightarrow 2\ NO_2(g)$$
$$3\ NO_2(g) + H_2O(l) \longrightarrow 2\ HNO_3(aq) + NO(g)$$

The acid produced is about 60% HNO_3. The nitrogen oxide formed in the last step is recycled back through the process. The annual production of nitric acid is more than 8.0×10^9 kg.

When nitric acid comes into contact with the skin, it reacts with the proteins to cause destruction of the tissues, producing a yellow discoloration of the skin.

The major uses for nitric acid are in the manufacture of ammonium nitrate fertilizers, explosives, plastics, pharmaceuticals, dye intermediates, and organic nitro compounds, and in pickling stainless steel in steel refining.

The Nitrogen Cycle

nitrogen cycle

The process by which nitrogen is circulated and recirculated from the atmosphere through living organisms and back to the atmosphere is known as the **nitrogen cycle**.

Nitrogen compounds are required by all living organisms. Despite the fact that the atmosphere is about 78% nitrogen, all animals as well as the higher plants are unable to utilize free nitrogen. Higher plants require inorganic nitrogen compounds, and animals must have nitrogen in the form of organic compounds.

Atmospheric nitrogen is *fixed*—that is, converted into chemical compounds that are useful to higher forms of life—by three general routes:

1. *Bacterial action* Certain soil bacteria are capable of converting N_2 into nitrates. Most of these bacteria live in the soil in association with legumes (e.g., peas, beans, clover). Some free-living soil bacteria and the blue-green algae, which live in water, are also capable of fixing nitrogen.
2. *High Temperature* The high temperature of lightning flashes causes the formation of substantial amounts of nitrogen oxide in the atmosphere. This NO is dissolved in rainwater and is eventually converted to nitrate ions in the soil. The combustion of fuels also provides temperatures high enough to form NO in the atmosphere. The total amount of nitrogen fixed by combustion is relatively insignificant on a worldwide basis. However, NO produced by combustion, especially in automobile engines, is a serious air pollution problem in some areas.
3. *Chemical fixation* Chemical processes have been devised for making nitrogen compounds directly from atmospheric nitrogen. By far the most important of these is the Haber process for making ammonia from nitrogen and hydrogen. This process is the major means of production for the millions of tons of nitrogen fertilizers that are produced synthetically each year.

A schematic diagram of the nitrogen cycle is shown in Figure 19.3. Starting with the atmosphere, the cycle begins with the fixation of nitrogen by any one of the three routes. In the soil, nitrogen or nitrogen compounds are converted to nitrates, taken up by higher plants, and converted to organic compounds. The plants eventually die or are eaten by animals. During the life of the animal, part of the nitrogen from plants in its diet is returned to the soil in the form of fecal and urinary excreta. Eventually after death, both plants and animals are decomposed by bacterial action. Part of the nitrogen from plant and animal tissues is returned to the atmosphere as free nitrogen and part is retained in the soil. The cycle thus continues.

Other Elements of the Nitrogen Family

Phosphorus is the twelfth most abundant element in the earth's crust. Free phosphorus is never found in nature. The element occurs mainly in phosphate minerals. Calcium phosphate is present in the bones and teeth of animals, and small amounts of

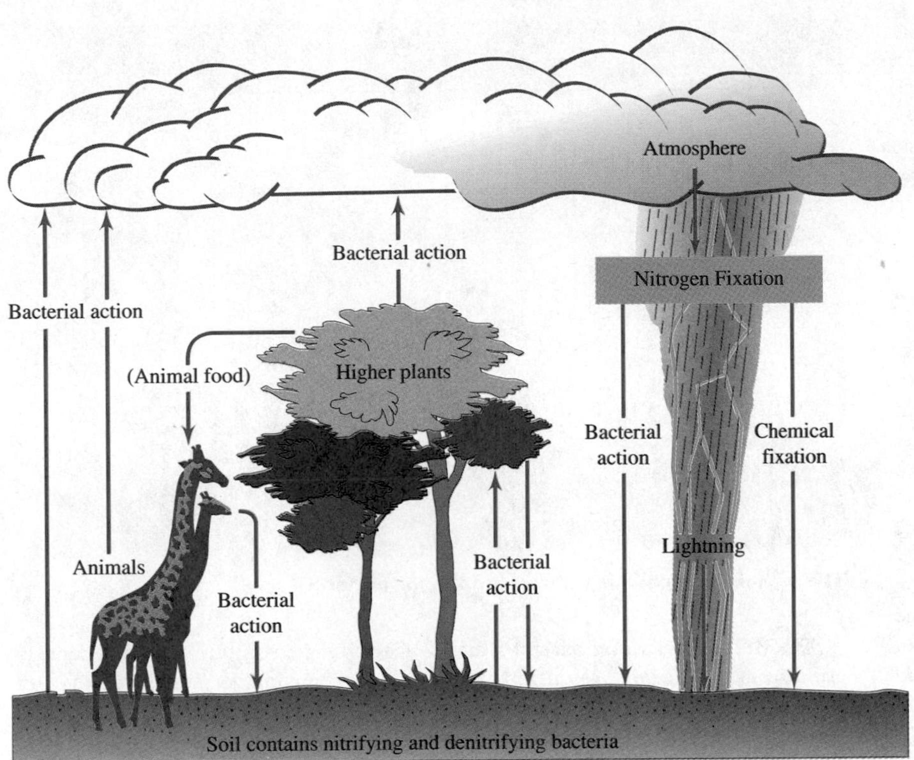

FIGURE 19.3
The nitrogen cycle. Arrows indicate the movement of nitrogen through the cycle.

phosphorus compounds are present in the cells of all plants and animals. The free element exists in several allotropic modifications—white, red, and black being the most common. The white form is tetratomic, P_4; the red and black forms are polymeric chains of phosphorus atoms.

Phosphorus is essential for plant growth. It is often present in soil as insoluble minerals. Phosphate rock can be treated with sulfuric or phosphoric acid to make fertilizer high in phosphates.

The sodium salts of phosphates are used in a variety of everyday products. Monosodium phosphate, NaH_2PO_4, forms an acidic solution and is a component in baking powder. Disodium phosphate, Na_2HPO_4, is used to soften boiler water (by precipitating the Ca^{2+} and Mg^{2+} ions). Trisodium phosphate, Na_3PO_4, is used as a common household cleaner (TSP).

In the human body, phosphorus makes up about 1% of the body's weight. It is an essential component of the bones, teeth, muscle, brain, and nervous tissue. Phosphorus plays a role in many metabolic reactions occurring in the body. Phosphorus is a component in nucleic acids and the phospholipids in the cell membrane. It is an essential component of ATP and is therefore important in much of the cell's metabolism. Good sources of phosphate in the diet include protein foods, whole-grain cereals, milk, and legumes.

Although nitrogen and phosphorus are nonmetals, arsenic, antimony, and bismuth become progressively more metallic; bismuth has properties approaching those of a true metal. Arsenic, antimony, and bismuth are mainly used in alloys to increase the hardness of metals. Lead shot is hardened by adding up to 0.5% arsenic; pewter contains 6% bismuth and 1.7% antimony. Low-melting alloys containing over 50%

Chemistry in Action

Arsenic for Dinner?

Most of us would shy away from a meal containing arsenic. It is a common poison; a large dose kills within hours while small doses cause a gradual wasting away. Arsenic interferes directly with the generation of energy in cells, stopping all life processes. Imagine how extraordinary it would be to discover an organism thriving on a diet containing arsenic.

Scientists at MIT have discovered such an organism, a bacterium that uses arsenic for respiration. These arsenic-using bacteria are called MIT-13 and were first identified in a watershed north of Boston, which was a dump for toxic wastes for nearly a century. When chemical plants dump arsenic, it is usually in the form of As(V), which binds to iron and manganese in the sediment and is trapped there. In this Boston watershed the majority of the arsenic was found dissolved in the water in the form of As(III). The newly discovered bacteria are able to convert As(V) into As(III) much faster than any previously known organisms.

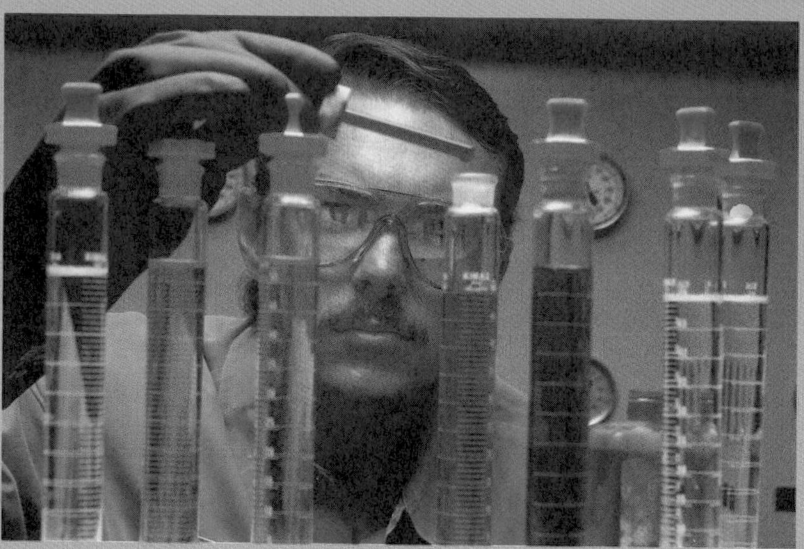

The technician is checking water samples for bacteria.

The MIT-13 bacteria are anaerobic and use arsenic in the same way aerobic organisms use oxygen. Arsenic is converted from As(V) to As(III) during the process of releasing energy from food. Scientists are now looking for similar bacteria in other contaminated places.

bismuth are used in safety sprinkler heads for fire protection. Arsenic and antimony compounds, although very toxic, have been successfully used as medicinals in controlled dosages.

19.10 Group VIA Elements

Physical Properties

The elements in Group VIA all have six electrons in their outermost shell. Each forms compounds with metals or hydrogen in which the oxidation number is -2. These elements are mainly nonmetals; yet all have the ability to form compounds with more electronegative elements as well. The physical properties of the Group VIA elements are summarized in Table 19.9.

Oxygen

Occurrence and Physical Properties Oxygen is the most abundant element on earth; it occurs both as free oxygen gas and combined with other elements. It forms compounds with all the elements except some noble gases. Water is 88.9% oxygen;

TABLE 19.9 Physical Properties of the Group VIA Elements

Element	Atomic number	Atomic mass	Electron configuration	Density (20°C, g/mL)	Melting point (°C)	Boiling point (°C)
O	8	16.00	$[He]2s^22p^4$	1.43×10^{-3}	−218	−183
S	16	32.06	$[Ne]3s^23p^4$	2.07	112	444
Se	34	78.96	$[Ar]4s^23d^{10}4p^4$	4.79	217	685
Te	52	127.6	$[Kr]5s^24d^{10}5p^4$	6.24	450	990
Po	84	210	$[Xe]6s^24f^{14}5d^{10}6p^4$	9.32	254	962

the human body is 65% oxygen by mass. The atmosphere contains about 21% oxygen by volume and about 23% oxygen by mass. The oxygen molecule is diatomic.

Oxygen is a colorless, odorless, tasteless gas. It is a pale blue solid or liquid, melting at −218.4°C and boiling at −183°C. The liquid is *paramagnetic*; that is, it is attracted by a magnet. The solubility of the gas is about 0.03 mL per gram of water at 20°C, a fact that is very important for most aquatic life.

Preparation of Oxygen We have seen that oxygen can be prepared by electrolysis of water (Section 4.3) and by decomposition of hydrogen peroxide. Many additional methods for the laboratory preparation are known.

Oxygen is obtained commercially from air. Air is first freed of carbon dioxide and moisture and then liquefied to give a mixture that is essentially liquid oxygen and nitrogen. Oxygen is separated from nitrogen by fractional distillation.

One of the outstanding properties of oxygen is its ability to support combustion. By combustion we generally mean the act of burning or, commonly, the union of oxygen with other substances accompanied by the evolution of light and heat. Fire, caused by the rapid reaction of oxygen with some other substance, is a familiar example. Oxygen is consumed during combustion but does not burn. The importance of combustion is underscored by our dependence on heat, electrical, and mechanical energy obtained from burning fossil fuels.

Animal life as we know it is dependent on oxygen. Absorbed in the blood, oxygen is carried by way of the arteries to all tissues of the body, where it is used in metabolism. Metabolism includes the oxidation of carbohydrates, fats, and proteins and provides the energy needed to carry on the normal life processes. The free oxygen in the atmosphere and that dissolved in fresh water and seawater is constantly replenished by the photosynthesis carried on by green plants (Section 34.9).

Sulfur

Occurrence and Physical Properties Sulfur has been known for thousands of years; evidence of its use dates back 16–20 centuries. It is mentioned in the Bible, where it is called *brimstone*, a name sometimes used today.

Sulfur occurs both in the free state and in compounds. It is found in various metal sulfides and sulfates, hydrogen sulfide, sulfur dioxide, and in a variety of compounds associated with petroleum and natural gas. Sulfur is an essential element in life processes and is present in all living organisms.

FIGURE 19.4 ▶
Diagrams of the S_8 molecule: (a) crown-shaped ring formation with bond angles of 105°; (b) Lewis structure; and (c) space-filling models. The latter most nearly represents the S_8 molecule.

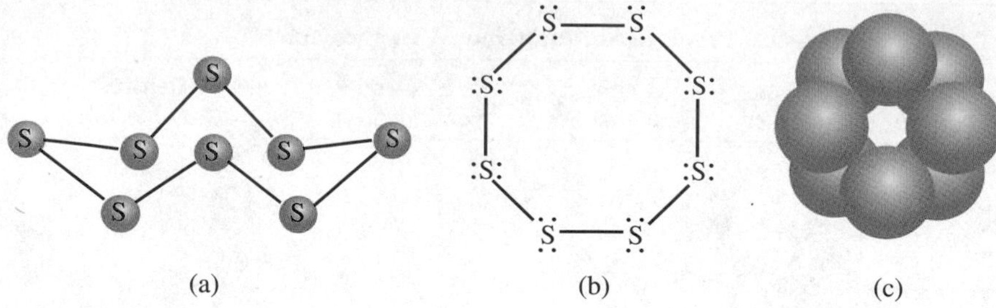

On a commercial basis, elemental sulfur is obtained from deposits in Sicily, Texas, Louisiana, and in volcanic regions of Japan and Mexico. Large quantities of sulfur and sulfur compounds are recovered as by-products from petroleum, natural gas, and metallurgical processing. Sulfur is widely used in many phases of agriculture and industry. More than 80% of all sulfur produced is made into sulfuric acid.

Sulfur is a bright yellow, odorless solid. In the solid form, sulfur exists in a crown-shaped ring structure composed of eight atoms connected by covalent bonds. The molecular aggregate is actually S_8, but in many of its reactions sulfur is represented by a single S. Figure 19.4 shows several ways of representing the S_8 molecule.

The element sulfur exists only as S_2 molecules in the gas phase at high temperatures. It is generally found as one of several allotropes containing the S_8 ring. As sulfur is heated beyond its melting point, a viscous (resistant to flowing) liquid of S_8 rings develops and the color darkens. If heating is continued, the rings rearrange into long chains, which are in turn broken down at high temperatures. If the liquid is cooled quickly, a substance called *plastic sulfur* is formed. Eventually this reverts back to the more stable rings.

Chemical Properties of Sulfur Since sulfur is less electronegative than oxygen, its compounds are less ionic and more covalent than those of oxygen. Ionic sulfur compounds are formed with only the most electropositive metals—the alkali metals (Group IA) and the alkaline earth metals (Group IIA). The oxidation states of sulfur range from −2 to +6. Sulfur combines directly with many elements, both metals and nonmetals. For example,

$$Zn(s) + S(s) \xrightarrow{\Delta} ZnS(s)$$

$$H_2(g) + S(l) \xrightarrow{200°C} H_2S(g)$$

$$C(s) + 2\ S(g) \xrightarrow{3000°C} CS_2(g)$$

There are at least ten known oxides of sulfur, two of which are well known—sulfur dioxide, SO_2, and sulfur trioxide, SO_3.

The prefix *thio* is used to indicate the presence of sulfur in a compound. More specifically, it means that sulfur is used in place of oxygen in an analogous compound. Two examples of thio compounds are sodium thiosulfate and thiourea. Their formulas and the formulas of their oxygen analogs are shown here.

◀ **Sulfur mine and production plant in Italy.**

```
              ..       2−
            [ :O: ]
               |
     Na+ [ :O—S—O: ] Na+
               |
            [ :O: ]
              ..
```

sodium sulfate, Na_2SO_4

```
              ..
            [ :O: ]
               |
     Na+ [ :O—S—O: ] Na+
               |
            [ :S: ]
              ..
```

sodium thiosulfate, $Na_2S_2O_3$

```
  H         H
   \       /
    N—C—N
   /  ||   \
  H  :O:    H
```

urea

```
  H         H
   \       /
    N—C—N
   /  ||   \
  H  :S:    H
```

thiourea

Application of Sulfur Compounds An important use of the thiosulfate ion is in photography. The $S_2O_3^{2-}$ ion reacts with the suspension of silver bromide, AgBr, which coats the photographic film. The AgBr is "activated" in areas exposed to light. This silver bromide is reduced to free Ag by the developing solution, and the film is treated with thiosulfate solution to remove the excess AgBr, thus "fixing" the image.

Sulfur burns in air or oxygen with a bluish flame to form sulfur dioxide:

$$S(l) + O_2(g) \longrightarrow SO_2(g)$$

Sulfur dioxide is also formed by reacting iron pyrites, FeS_2, or other metal sulfides with oxygen (air) at elevated temperatures:

$$4\ FeS_2(s) + 11\ O_2(g) \xrightarrow{\Delta} 2\ Fe_2O_3(s) + 8\ SO_2(g)$$

Sulfur dioxide is a colorless gas that is a throat and eye irritant, causing choking, coughing, and tearing. It also is an air pollutant known to damage plant life.

Sulfur dioxide is an acid anhydride and forms sulfurous acid, H_2SO_3, when dissolved in water. It is a weak acid, which also decomposes easily into sulfur dioxide and water.

$$SO_2(g) + H_2O(l) \rightleftharpoons H_2SO_3(aq)$$

Much industrial sulfur dioxide goes into the production of sulfuric acid. It is also used as a bleaching agent in the pulp and paper industry, in refining sugar, and in processing dried fruit.

Hydrogen sulfide, H_2S, is the sulfur analog of water, but that is as far as the resemblance goes. It is normally a gas; the liquid boils at $-60°C$. It has a disagreeable odor, like rotten eggs, and is very toxic. The gas is slightly soluble in water, forming an extremely weak acid solution called hydrosulfuric acid.

The uses of sulfuric acid, H_2SO_4, are numerous and varied. The current U.S. production is greater than 4.3×10^{10} kg annually. Large amounts are used to convert insoluble phosphates into soluble phosphates for fertilizers. Sulfuric acid is used in the production of hydrochloric acid, ammonium sulfate, and aluminum sulfate. It is also used in petroleum refining; the manufacture of rayon; the making of explosives (nitroglycerin and TNT), alcohols, detergents, and metal sulfates, textile printing, dyeing; and in scores of other applications. Dilute sulfuric acid is the electrolyte used in millions of ordinary automobile storage batteries.

Sulfuric acid is a colorless, oily liquid with a density of 1.84 g/mL. Its freezing point is 10.4°C; its boiling point is 338°C. The acid decomposes as it boils, liberating its anhydride, sulfur trioxide (SO_3):

$$H_2SO_4(l) \xrightarrow{\Delta} H_2O(g) + SO_3(g)$$

Ordinary concentrated sulfuric acid contains about 96–98% H_2SO_4; the remainder is water.

Other Group VI Elements

Selenium and tellurium are relatively rare elements. In recent years there has been a growing interest in the chemistry of selenium. It is used in the manufacture of red glass, such as that found in traffic lights. It has unusual properties of electrical conduction that make it useful in the switches of automatic light timers. The most familiar use of selenium is in xerography. Inside a photocopying machine is an aluminum plate coated with a thin film of selenium. Light coming in discharges the electric charge on the selenium. The black toner sticks only to the remaining charged areas, creating the photocopy as soon as the toner is transferred to paper.

19.11 The Halogens: Group VIIA Elements

Physical Properties

halogen

The **halogens,** Group VIIA of the periodic table, consist of five elements: fluorine, F; chlorine, Cl; bromine, Br; iodine, I; and astatine, At. None of these are found

TABLE 19.10 Physical Characteristics of the Halogens

Property	Fluorine	Chlorine	Bromine	Iodine
Molecular formula	F_2	Cl_2	Br_2	I_2
Atomic number	9	17	35	53
Atomic mass	19.00	35.45	79.90	126.9
Electron configuration	$[He]2s^22p^5$	$[Ne]3s^23p^5$	$[Ar]4s^23d^{10}4p^5$	$[Kr]5s^24d^{10}5p^5$
Atomic radius (nm)	0.072	0.099	0.114	0.133
Radius of X^- ion (nm)	0.136	0.181	0.195	0.216
Appearance	Pale yellow gas	Yellow-green gas	Reddish-brown liquid	Bluish-black solid
Density (g/mL)	1.11 (liquid at bp)	1.57 (liquid at bp)	3.14	4.94
Melting point (°C)	−219.6	−102.4	−7.2	113.6
Boiling point (°C)	−187.9	−34.0	58.8	184.5
Relative electronegativity	4.0	3.0	2.8	2.5

as free elements in nature. The term *halogen* is derived from the Greek words *halos* and *gen*, which mean "salt producer." Chlorine, bromine, and iodine are easily prepared in the laboratory; fluorine is much more difficult to prepare. Astatine is found only in very small amounts as a radioactive decay product of other elements.

All the halogen atoms have seven electrons in their outermost shell and are followed, in each case, in the periodic table by a noble gas. Consequently, their most stable oxidation state is −1. All the halogens except fluorine are known to exist in several oxidation states. A halogen atom has a great tendency to fill its outer-orbital electron vacancy either by gaining an electron, as in the formation of a sodium halide, Na^+X^- (where X = F, Cl, Br, I), or by sharing electrons and forming a covalent bond. In the free state, the halogens exist as diatomic molecules, F_2, Cl_2, Br_2, and I_2:

$$:\ddot{\underset{..}{F}}:\ddot{\underset{..}{F}}: \qquad :\ddot{\underset{..}{Cl}}:\ddot{\underset{..}{Cl}}: \qquad :\ddot{\underset{..}{Br}}:\ddot{\underset{..}{Br}}: \qquad :\ddot{\underset{..}{I}}:\ddot{\underset{..}{I}}:$$

The differences in reactivity of the halogens are due to differences in their electronegativities. Fluorine, as the most electronegative, has the greatest attraction for electrons and is the most reactive, followed in descending electronegativity by chlorine, bromine, and iodine (see Table 19.10).

Since fluorine has the highest electronegativity, it is the most characteristically nonmetallic of all the elements. The general trend of the elements to become less electronegative (from top to bottom within a periodic group), and hence more metallic in characteristics, is clearly evident among the halogens. Fluorine and chlorine are gases, bromine is a liquid, and iodine is a solid with a characteristic metallic luster.

Chemical Properties and Applications

The halogen family exhibits a very smooth transition in chemical properties as we progress down the group. Some of the chemical properties of the halogens are summarized in Table 19.11.

Over three-fourths of the fluorine produced in the United States is used for

TABLE 19.11 Chemical Properties of the Halogens

Characteristic	F_2	Cl_2	Br_2	I_2
General activity	Extreme	Very high	Less active	Least active
Activity with H_2	Violent	Slow (dark) Fast (light)	Requires heat	Slow
Hydrogen halide formula	HF	HCl	HBr	HI
Oxidizing strength	Highest	Second	Third	Lowest
Reducing strength of anion	Lowest	Third	Second	Highest

processing uranium for nuclear power plants. The UF_6 formed vaporizes at a fairly low temperature, allowing easy separation from other substances.

Fluorine has become an important industrial chemical. It is used in familiar compounds such as Teflon (nonstick coating on cookware) and Freon (used in refrigerants and formerly in aerosol cans as a propellant). Some of these fluorocarbons have been implicated in the destruction of the ozone layer.

Fluorine is also present in the bones and teeth of mammals. Some fluoride ions are necessary for the development of sound teeth in children, although too much results in mottled teeth. Controlled amounts of sodium fluoride are sometimes added to water supplies in areas deficient in fluoride.

Over 90% of all commercial chlorine is produced by the electrolysis of molten or aqueous sodium chloride. About half of this inorganic chlorine eventually ends up as vinyl chloride, C_2H_3Cl, used in the manufacture of the plastic polyvinyl chloride (PVC) as well as other chlorine-containing organic solvents. About 20% of the chlorine produced is used to bleach wood pulp for the paper industry and cotton and linen fabrics for the textile industry. Chlorine is used to oxidize and destroy bacteria in virtually all water treatment plants in urban areas in the United States. Swimming pools are chlorinated in a similar manner.

Neither bromine nor iodine is as widely found as chlorine and fluorine. Bromine is the only nonmetallic element that is a liquid at ordinary temperatures. It is used in the petroleum, pharmaceutical, and photographic industries. The compound 1,2-dibromoethane, $C_2H_4Br_2$, was used as an additive in leaded gasoline to reduce engine "knock." This has been replaced in unleaded fuels by methyl tertiary butyl ether (MTBE). A variety of drugs and dyes also contain bromine. Bromine compounds are known to have a soothing effect on the nerves. In photography light-sensitive silver bromide is coated on the surface of the film. Precautions must be taken when working with bromine because its fumes are toxic and it produces painful sores when spilled on the skin.

Iodine is a bluish-black solid that forms violet vapors when heated. Compounds of iodine are found in seawater, where they are selectively absorbed and concentrated by certain types of seaweed (kelp). The first commercial source was from ashes of seaweed. Today iodine is obtained commercially from certain oil well brines and from Chilean sodium nitrate deposits, where it occurs as sodium iodate ($NaIO_3$).

The brines are treated with chlorine to liberate iodine:

$$Cl_2(g) + 2\,I^-(aq) \longrightarrow I_2(s) + 2\,Cl^-(aq)$$

Sodium iodate is separated from other salts found in the nitrate deposits, then reacted with sodium hydrogen sulfite to liberate free iodine:

$$2\,NaIO_3 + 5\,NaHSO_3 \longrightarrow I_2 + 3\,NaHSO_4 + 2\,Na_2SO_4 + H_2O$$

Iodine is only slightly soluble in water, but it is very soluble in organic solvents such as alcohol, carbon tetrachloride, and carbon disulfide. The brown alcohol solution is known as *tincture* of iodine and is used as an antiseptic and disinfectant for cuts and scratches. The carbon tetrachloride and carbon disulfide solutions are violet. Iodine is soluble in aqueous potassium iodide because of the formation of the soluble complex triiodide ion, I_3^-. The solution is labeled "I_2 in KI" and is brown in color.

$$I_2(s) + KI(aq) \longrightarrow KI_3(aq)$$
$$I_2(s) + I^-(aq) \longrightarrow I_3^-(aq)$$

About half of all commercial chlorine produced today ends up as polyvinyl chloride (PVC).

An intense blue complex is formed when iodine is added to a starch suspension. This iodine–starch color is a very sensitive test for iodine; it is a useful indicator that allows iodine to be used in the analysis of other substances.

The thyroid gland is richer in iodine than any other part of the human body. A deficiency of iodine causes this gland to enlarge, creating the condition known as goiter. Iodine and iodides have been used for many years as an effective treatment for goiter. Iodine in the regular diet may be supplemented by iodized salt, which contains a small amount of a soluble iodide.

Silver iodide (AgI) is very sensitive to light; it is used in high-speed photographic films.

All of the halogens form stable compounds when reacted with hydrogen. The aqueous solutions of the compounds are all strongly acidic with the exception of hydrogen fluoride.

The melting and boiling points of HF are abnormally high as a direct result of hydrogen bonding. Hydrogen fluoride is dangerous to inhale since it causes edema of the lungs. The eyes are also quite sensitive to this compound.

A unique property of HF is its ability to etch glass. The art of etching glass with HF was practiced as early as the 16th century. The chemical reaction that takes place is the dissolving of silicon dioxide in the glass, forming silicon tetrafluoride:

$$SiO_2(s) + 4\,HF(g) \longrightarrow SiF_4(g) + 2\,H_2O(l)$$

The area to be etched is first coated with paraffin. The design is then scratched through the wax. The glass is dipped in the HF solution and then washed, removing the paraffin, which is not attacked by hydrogen fluoride. The glass in light bulbs is also "frosted" in the same manner.

Hydrochloric acid—the technical grade is called *muriatic acid*—is one of the most widely used acids. It is essential in the manufacture of textiles, soaps, glue, and in the cleaning of metals. Hydrochloric acid is also used for cleaning mortar from stone or brick structures. The acidity of gastric juice is the result of the presence of HCl. It is necessary for digestion and is also the source of heartburn and acid indigestion.

19.12 Noble Gases

The last column on the right-hand side of the periodic table consists of the group of elements known as the noble or inert gases. These elements are characterized by filled *s* and *p* orbitals in the outermost electron shell, which results in an almost complete lack of reactivity, a common characteristic for these gases. No noble-gas compounds were known until 1962.

19.13 Transition Metals

All the elements we have surveyed to this point have been representative elements, which show similarities in vertical columns and incredible diversity across a period. In contrast, the transition metals exhibit similarities across each period as well as in vertical columns. The last electron in these elements is added to a *d* or an *f* orbital that is not in the outermost shell. This results in more gradual changes than those seen in the representative elements.

Overall, the transition metals exhibit the typical metallic characteristics of luster and high electrical and thermal conductivity. The best heat conductor is silver. You may have practical knowledge of this if you have handled a hot sterling silver gravy ladle. Copper is also a very good conductor of heat and electricity and is much less expensive than silver. For this reason many electrical systems contain copper.

The transition metals also exhibit considerable diversity in other properties. Mercury is a liquid at room temperature, yet tungsten (used in light bulbs for the filament) does not melt until 3400°C. Some transition metals, such as silver, gold, and copper, are relatively soft, while others, like iron, are strong and used for structural materials. Some transition metals react with oxygen to form oxides. These oxides are of two types. In the first type, the oxide coating adheres to the metal surface, protecting it from further corrosion. Examples of this type include chromium and nickel. The second type of oxide is scaly, constantly exposing the metal underneath to further oxidation. The most common example of this type is the rusting of iron. A few of the transition metals, such as gold, silver, and platinum, form oxides only with difficulty.

Ionic transition metal compounds also exhibit specific characteristics. The metals are often found in multiple oxidation states. Many of these compounds are highly colored and quite often are paramagnetic. Let's consider several transition metals that are important in our daily lives.

Iron

Iron is the fourth most abundant element, following oxygen, silicon, and aluminum. Iron was known in prehistoric times, and its known use dates back almost 8000 years. The first iron to be used was probably obtained from meteorites, but iron was produced from ores as early as 1300 B.C.

Pure iron is a silvery-white, relatively soft, very ductile metal. It has a melting point of 1532°C and it boils at 3000°C. One of its most distinguishing characteristics

◀ **This Calder sculpture in downtown Los Angeles illustrates the strength and beauty of steel.**

is its magnetism, which it loses when it is heated to 770°C. In the presence of a magnetic field or electric current, iron shows greater magnetic properties than any other element. (Cobalt and nickel are also strongly magnetic.)

Very little iron is used in its pure form; most of it is used in the form of steel. Steel is an iron-carbon alloy (0.05–1.7% C). It may contain small amounts of other elements such as Mn, Si, Cr, Ni, V, W, Mo, Co, and Ti. These alloying elements enhance certain qualities of the steel such as toughness, hardness, corrosion resistance, and so on (see Table 19.12).

Steel making is a basic industry; the quantity of steel produced each year in the world far exceeds the combined quantities of all other metals. The first step in steel making is the reduction of iron ore to impure iron, commonly called *pig iron.* Reduction occurs in a huge reactor called a *blast furnace.* Furnace operation is continuous; ore, coke, and limestone are charged at the top while steady streams of preheated air are blown in through bottom nozzles (*tuyeres*). Furnace temperatures range from about 200°C near the top to about 1900°C near the bottom (the combustion zone). Coke is both the fuel and the source of the principal reducing agent, CO. The limestone reacts with impurities in the ore and coke to form a slag that coats and protects the reduced iron from reoxidation in the combustion zone. Molten pig iron and slag are removed at intervals from the bottom of the furnace. The following are equations for typical blast furnace reactions.

$$\underset{\text{coke}}{2\,C(s)} + O_2(g) \longrightarrow 2\,CO(g)$$

$$\underset{\text{ore}}{Fe_3O_4(s)} + CO(g) \longrightarrow 3\,FeO(s) + CO_2(g)$$

$$FeO(s) + CO(g) \longrightarrow Fe(l) + CO_2(g)$$

$$\underset{\text{limestone}}{CaCO_3(s)} \longrightarrow CaO(s) + CO_2(g)$$

$$CaO(s) + SiO_2(s) \longrightarrow \underset{\text{slag}}{CaSiO_3(l)}$$

Pig iron contains on the average about 1% Si, 0.03% S, 0.27% P, 2.4% Mn, and 4.6% C, the balance being Fe. These impurities must be removed or lowered to carefully controlled levels to convert pig iron into steel.

Iron is a fairly reactive metal and forms two principal series of compounds: iron(II) or ferrous compounds and iron(III) or ferric compounds. Pure iron reacts

TABLE 19.12 Effects of Selected Alloying Elements on the Properties of Steel

Alloying element	Effect on steel
Carbon	Makes tempering possible; increases tensile strength (up to 0.83% C); increases hardness
Cobalt	Imparts high-temperature strength
Chromium	Improves hardness, abrasion resistance, corrosion resistance, and high-temperature properties
Manganese	Imparts hardness, toughness, and resistance to wear and abrasion
Molybdenum	Imparts hardness, shock resistance, and high-temperature strength
Nickel	Imparts corrosion resistance; increases toughness and high-temperature strength; reduces brittleness at very low temperatures
Silicon	Increases strength without affecting ductility; modifies magnetic and electrical properties; high Si concentrations impart resistance to acid corrosion
Tungsten	Imparts hardness, which is retained at high temperatures
Vanadium	Increases strength and toughness; improves heat-treating characteristics

with dilute acids or steam (at high temperatures) to form iron(II) compounds. These can then be oxidized to iron(III) compounds.

In addition to its use as a structural material in a diverse array of objects, iron is also used because of its magnetic properties. Iron is found in magnets, telephones, electric motors, generators, and many other common devices.

Dipping iron into concentrated nitric acid (a powerful oxidizing agent) results in the formation of *passive iron*, which is coated with a very thin layer of oxide. Passive iron will not rust unless the film of oxide is broken.

Corrosion of iron, commonly called *rusting*, is a serious economic problem that causes losses of billions of dollars annually in the United States. Rusting occurs on the surface of the iron, where it is exposed to oxygen and moisture in the atmosphere. The iron is transformed into reddish-brown rust, a hydrated iron(III) oxide ($Fe_2O_3 \cdot x\ H_2O$). As the rust is formed, it flakes off the surface, allowing the corrosion to penetrate deeper into the iron. Both oxygen and water are needed for rusting. The process is summarized in the following equations:

$$2\ Fe + O_2 + 2\ H_2O \longrightarrow 2\ Fe(OH)_2$$
$$4\ Fe(OH)_2 + O_2 + 2\ H_2O \longrightarrow 4\ Fe(OH)_3$$
$$2\ Fe(OH)_3 \longrightarrow Fe_2O_3 \cdot x\ H_2O$$

The most common general methods for protecting iron from corrosion are (1) protective coatings, (2) alloying, and (3) cathodic protection. Protective coatings of paint, enamel, tar, grease, or metal are commonly used to prevent contact with the atmosphere. The term *galvanized* is applied to iron or steel that is protected by a coating of zinc metal. Highly corrosion-resistant iron alloys are produced industrially.

To provide *cathodic protection* to a steel pipeline or tank, magnesium (or zinc) stakes or rods are driven into the earth and connected by wires to the pipeline or tank. This in effect sets up an electrochemical cell, with the more easily oxidized magnesium metal acting as the anode. Electrons flow from the magnesium to the iron, causing the iron to become negatively charged or cathodic with respect to the magnesium. Oxidation does not occur at the iron cathode, and thus no corrosion occurs on the iron pipe or tank. The magnesium or zinc rods or stakes must be replaced periodically since they are consumed in affording protection to the iron or steel. The iron hulls of ships can be protected by fastening blocks of magnesium metal to them. The zinc coating on galvanized iron provides cathodic protection for the iron.

In the human body, iron is found primarily in the blood with reserves in the liver, spleen, and bone marrow. Iron is central in the hemoglobin molecule, where it is responsible for the transport of oxygen. Another complex molecule, known as myoglobin, also contains iron. Myoglobin is synthesized in the muscle cells, where it serves as a temporary storage depot for oxygen. Iron is also important in many enzymes that catalyze oxidation reactions, such as the conversion of β-carotene to vitamin A. Sources of iron in the diet include liver, lean meat, whole-grain cereals, apricots, raisins, and legumes.

Copper

Copper is a soft, extremely ductile and malleable metal with a reddish-brown color. Its excellent electrical and thermal conductivity make it the primary source for wires in electrical systems. Copper is not very active chemically, but when exposed to moist air over a period of time, it becomes coated with the green *patina* seen on statues. This patina consists of thin layers of two compounds, $CuSO_4 \cdot 3\ Cu(OH)_2$ and $CuSO_4 \cdot 2\ Cu(OH)_2$. A good deal of chemistry was applied in the restoration of the Statue of Liberty to transplant weathered copper onto the statue and affix the patina to the surface to match the old. See Chemistry in Action, Chapter 17, p. 453.

Copper forms alloys with a variety of metals. Table 19.13 indicates the uses of some of these alloys. Copper-containing alloys also constitute material used in coins today. The nickel coin is an alloy of 75% copper and 25% nickel. Most other coins are clad-metal. These coins are now constructed with a core metal, then coated with a thin layer of an alloy. The dime, quarter, half-dollar, and dollar coins contain 8.33% nickel and 91.7% copper. Other copper-containing alloys include white gold, used in jewelry, and dental gold.

Trace amounts of copper are essential for life. Copper is concentrated in the liver, heart, and brain in human beings. It is instrumental in the synthesis of hemoglobin, bone and connective tissue development, the production of melanin, and for the formation of myelin in the nervous system. Sources of copper in the diet include liver, oysters, crab, nuts, and whole-grain cereals.

TABLE 19.13 Uses of Copper Alloys

Alloy metal	Name	Use
Zinc	Brass	Cartridges, musical instruments, hardware
Tin, zinc	Bronze	Statues, medals, primitive tools
Tin	Bell metal	Bells

Zinc oxide (white) has been used for many years as a sun screen. In today's fashion conscious market, colors (e.g., blue and yellow) have been added to the protectant.

Copper in large amounts can have a toxic effect. Salts containing copper are used to kill bacteria, fungi, and algae. Copper sulfate pentahydrate, $CuSO_4 \cdot 5\ H_2O$ (commonly known as blue vitriol), is used to treat water in reservoirs and swimming pools. Paints containing copper are used on the hulls of ships to keep them free from marine organisms.

Zinc

Zinc is a silvery-white metal that tarnishes easily. It melts at 420°C and boils at 907°C. The surface oxide film protects zinc from further corrosion. Zinc is a very active metal and is an excellent reducing agent. Approximately 90% of the zinc produced is used in galvanizing steel. Galvanized objects include water pails, gutters, fencing, and nails. Zinc forms the case and the negative electrode in most dry-cell batteries. It is also an important component in alloys. In combination with aluminum, the alloy is used in radio and auto parts. The "copper" penny is minted from an alloy of 97.5% zinc and 2.5% copper.

Many chemists do not classify zinc as a transition metal because this element has a filled *d* subshell. In fact, the chemistry of zinc is more like that of the alkaline earth metals than the transition metals.

Compounds of zinc are also important in our lives. Zinc oxide, ZnO, is a white pigment used in paints. It is used in making tire rubber and as an antiseptic or skin protectant. A paste of zinc oxide and phosphoric acid solidifies into a hard cement used in dentistry.

Trace amounts of zinc are found in the human body. It is a constituent of many enzymes in the digestive and respiratory systems. Zinc also plays a role in bone and liver metabolism. It is necessary for healing wounds and plays a key role in skin integrity. Sources of zinc include seafood and meats.

Concepts in Review

1. Indicate the similarities and differences in the physical and chemical properties of metals and nonmetals.
2. Summarize the chemical process that occurs when an ore is converted to a free element.
3. Summarize the important trends on the periodic table.
4. State the definition of an alloy, indicate the types of alloy, and the reasons for making them.
5. List the alkali metals (Group IA) and compare their atomic radii, densities, melting points, and first ionization energies.
6. Write balanced chemical equations for the reaction of any alkali metal with (a) water, (b) any halogen, (c) hydrogen.
7. Describe the preparation of magnesium metal from seawater and write equations for the reactions involved.

An Ironic Solution to CO_2 Pollution?

Oceanographers have been struggling to explain why algae don't thrive in the nutrient rich areas of the sub-Arctic and the Pacific Ocean near the equator. One theory is that iron controls the rate at which plants grow and absorb carbon dioxide. If this is true, then an iron deficiency would prevent algae from using the abundant nitrates in these ocean areas.

To test this hypothesis, Kenneth Coale, from Moss Landing Marine Laboratories in California, seeded the Pacific Ocean near the Galapagos Islands with iron. The first iron seeding experiment (1993) resulted in an increase by four fold in chlorophyll but only a tiny decrease in carbon dioxide concentration in the water.

Coale persisted in his efforts and ran another experiment in 1995. This time the iron was infused into the ocean three times in much smaller amounts over the course of a week. This technique was used to mimic nature's way of distributing iron to the ocean (through windblown dust particles). It also prevented the suspended iron from coalescing into heavy particles, which dropped below the surface layer where the algae live. This time scientists saw a 30–40 fold increase in chlorophyll and a corresponding increase in CO_2 absorption from the ocean.

This result raises an intriguing question. Can we seed our oceans with iron to reduce the levels of CO_2 pollution in our atmosphere? Unfortunately, the answer is probably no. Adding iron to the ocean would certainly change the food web in unpredictable ways. Also, it has been shown that increasing iron levels cause an increase in the release of methane, an even more potent greenhouse gas than carbon dioxide. Studies of wetlands plants over the last 10 years have shown that plants that absorb higher amounts of CO_2 produce more methane than those with less CO_2 absorption. Because the methane is released by bacteria living on the organic matter, increased plant growth means more methane. Alas, iron isn't a quick fix for global warming.

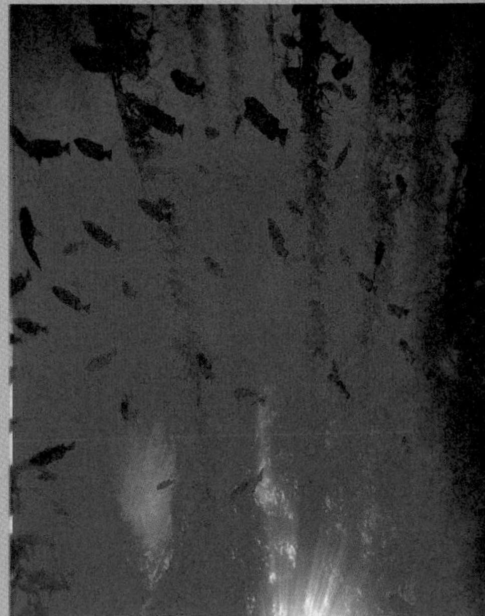

Fish and algae flourish in this kelp forest near Moss Landing, California.

8. Explain why colors appear when certain elements are heated strongly in a nonluminous flame, and list the specific colors obtained from Li, Na, K, Rb, Ca, Sr, and Ba.
9. Explain why aluminum is more expensive than iron or steel even though it is more abundant than iron in the earth's crust.
10. List the common allotropes of carbon and state two uses for each.
11. State which element of Group VA is the most metallic and which is the least metallic.
12. Diagram and discuss the nitrogen cycle.
13. Identify the halogen that has the highest electronegativity (and first ionization energy) and explain why its electronegativity is especially high.
14. Arrange the halogens in order of increasing strength as oxidizing agents and the halide ions in order of increasing strength as reducing agents.
15. State the essential difference between pig iron and steel.
16. Explain cathodic protection of metals.
17. Explain how a clad-metal coin is made and list several examples of this type of coin.

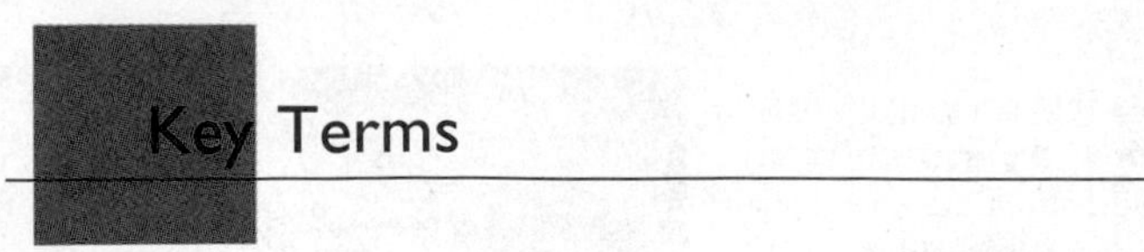

Key Terms

The terms listed here have been defined within the chapter. Section numbers are given in parenthesis for each term.

alkali metal (19.4)
alkaline earth metal (19.5)
alloy (19.2)
amalgam (19.2)
halogen (19.11)
intermetallic compound (19.2)
metallurgy (19.2)
nitrogen cycle (19.9)
ore (19.2)
solution alloy (19.2)
steel (19.2)

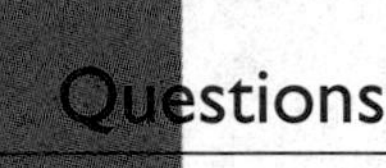

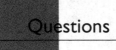

Questions

Questions refer to figures, tables, and key words and concepts defined within the chapter. A particularly challenging question or exercise is indicated with an asterisk.

1. Why is the chemistry of the first element in a family often significantly different from the other elements in the family?
2. Which element in Group IVA would have the least metallic character? Give two properties that support this conclusion (Table 19.7)
3. Why are the alkali metals of higher molar mass more active than those of lower molar mass, while in the halogens the reverse is true?
4. Identify the chemical process used in the conversion of an ore to a free metal.
5. Indicate two methods for the preparation of nonmetals.
6. Differentiate between a solution alloy and an intermetallic compound. Give an example of each.
7. Why are metals combined to form alloys?
8. Why is hydrogen often considered to be in a family or group of one?
9. Explain in terms of electronic structure why potassium reacts more violently with water than does lithium.
10. How would the ionization energy of francium be expected to compare with that of cesium?
11. Explain how potassium superoxide (KO_2) can be used as a source of oxygen in emergency breathing apparatus.
12. State two biologically significant uses of alkali metals.
13. On the basis of electron structure, propose an explanation for why a Group IIA metal is less reactive than a Group IA metal that precedes it in the periodic table.
14. Outline the process for isolating magnesium from seawater. Include equations in your answer.
15. Explain the chemical process that occurs as mortar hardens.
16. List two major functions of calcium and magnesium in living organisms.
17. Explain why it is possible to use barium sulfate, but not barium chloride, as an aid to X-ray photography of the gastrointestinal tract.
18. Briefly discuss the nature and origin of the colored flames that are used for the qualitative detection of certain Group IA and Group IIA elements.
19. State three uses for the element boron.
20. What is a mordant and how is it used commercially?
21. What is an allotrope? State an example from Group IVA.
22. State three commercial uses for the element silicon.
23. Does the tendency to become more metallic increase or decrease from top to bottom in Group VA elements? Explain.
24. Outline and discuss the nitrogen cycle in nature.
25. Which Group VA elements are absolutely essential to our health and well-being?
26. Explain why nitrogen is so inert in its free state. What commercial uses result from this property?
27. Explain in terms of electronic structure why the reactivity of Group VIIA elements decreases from top to bottom in the periodic table.
28. How are the compounds of sulfur different from the compounds of oxygen?
29. Why is the boiling point of HF abnormally high when compared to that of the other hydrogen halides?

30. Arrange the halogens in order of increasing strength as oxidizing agents. Justify the order in terms of electronic structure.

31. How do the transition metals differ from the representative elements?

32. What is the essential chemical difference between pig iron and steel?

33. Explain the process commonly known as the rusting of iron.

34. Bare aluminum and magnesium metal structures are not damaged by prolonged exposure to the atmosphere. Yet bare iron, generally less reactive than either aluminum or magnesium, is severely corroded by long exposure to the atmosphere. Explain.

35. Explain how cathodic protection can prevent oxidation of an iron storage tank or a ship's hull.

36. State four important uses for copper and its compounds in our daily lives.

37. Explain why a metal is galvanized. What element is used in this process?

38. Which of the following statements are correct? Rewrite each incorrect statement to make it correct.
   - **(a)** Most of the common metals are found in the free or uncombined state in nature.
   - **(b)** Metals may be alloyed to increase tensile strength, hardness, and electrical conductivity.
   - **(c)** The chemical reactivity of the alkali metals (Group IA) increases from top to bottom in the periodic table.
   - **(d)** In binary compounds between metals and nonmetals, the metals generally form positive ions.
   - **(e)** Chile saltpeter is another name for sodium nitrate.
   - **(f)** Both calcium and magnesium are essential elements for animals, but only calcium is essential for plants.
   - **(g)** Very pure iron is used in some industrial applications because of its great hardness and strength, especially at high temperatures.
   - **(h)** Copper and aluminum alloys are used for electrical wiring because of their excellent electrical conductivity.
   - **(i)** Galvanized steel is protected from rusting or corrosion by a zinc coating.
   - **(j)** Aluminum is the most abundant metal in the earth's crust.
   - **(k)** The electronegativity of the halogens (Group VIIA) increases from top to bottom in the periodic table.
   - **(l)** Iron plays a central role in bone development in the human body.
   - **(m)** The oxidation number of oxygen in oxygen difluoride, OF_2, is $+2$.
   - **(n)** The major use of fluorine in the United States is in the manufacture of toothpaste.
   - **(o)** Pure sulfur is a bright yellow, odorless solid.
   - **(p)** Sulfur occurs as molecules of S_2 in the gaseous state.
   - **(q)** The Ostwald process is used to prepare ammonia on an industrial scale.
   - **(r)** Phosphorus is indispensable to plants and animals.
   - **(s)** The copper patina on statues protects them from further oxidation.

Paired Exercises

These exercises are paired. Each odd-numbered exercise is followed by a similar even-numbered exercise. Answers to the even-numbered exercises are given in Appendix V.

39. Identify each of the following as a metal, nonmetal, or metalloid:
   - **(a)** strontium
   - **(b)** krypton
   - **(c)** silicon
   - **(d)** tungsten

40. Identify each of the following as a metal, nonmetal, or metalloid:
   - **(a)** bromine
   - **(b)** arsenic
   - **(c)** palladium
   - **(d)** hydrogen

41. Consider the following elements: Na, K, S, O, Ar, Kr. Choose the element with the
   - **(a)** highest electronegativity
   - **(b)** smallest atomic radius
   - **(c)** smallest ionization energy
   - **(d)** greatest metallic character

42. Consider the following elements: Mg, Ca, P, N, Cl, Br. Choose the element with the
   - **(a)** highest electronegativity
   - **(b)** smallest atomic radius
   - **(c)** smallest ionization energy
   - **(d)** greatest metallic character

43. Graph the melting points of the alkali metals versus atomic number. Predict the melting point for francium. Will it be a liquid or solid at room temperature?

44. Graph the densities of the alkali metals versus atomic number. Predict the density for francium. Will it float or sink in water at room temperature?

45. Write the formulas of an alkali metal oxide, hydroxide, acetate, hydrogen carbonate, and carbonate using a different metal for each compound.

46. Write the formulas of an alkaline earth metal oxide, hydroxide, acetate, hydrogen carbonate, and carbonate using a different metal for each compound.

47. Draw Lewis structures for
(a) a nitrogen molecule
(b) an ammonium ion
(c) an ammonia molecule

48. Draw Lewis structures for
(a) a nitrogen molecule
(b) a nitride ion
(c) a nitrate ion

49. What ill effect, if any, would occur from a lack of the following in one's diet?
(a) fluoride ions
(b) chloride ions
(c) bromide ions
(d) iodide ions

50. What ill effect, if any, would occur from a lack of the following in one's diet?
(a) copper
(b) zinc
(c) iron
(d) calcium

Additional Exercises

These exercises are not paired by topic and provide additional practice on the concepts covered in this chapter.

51. Write an equation for the reduction of a metal oxide by the following methods:
(a) reaction with carbon
(b) reaction with hydrogen
(c) heating

52. Use Lewis structures to illustrate how the halogen elements, designated X, can form both covalent bonds (X_2, HX) and ionic bonds (Na^+X^-).

53. Write the equations for the production of ammonia by the Haber process. What happens to the majority of ammonia manufactured in this way?

54. What is the oxidation number of hydrogen in a hydride? Write an equation for the formation of potassium hydride from its elements.

55. Write equations for the formation of the oxide, peroxide, and superoxide of potassium.

56. What compound is formed if carbon is burned
(a) in a limited supply of oxygen?
(b) in an excess of oxygen?

57. How many liters of SO_2 (at STP) can be obtained by combustion of 150 g of sulfur?

***58.** The water coming from a large reservoir is being treated with 0.20 ppm (parts per million) of chlorine. Calculate the number of grams of chlorine required each hour when water is being pumped from the reservoir at a rate of 300,000 L/hr.

59. Calculate the molarity of each of the following concentrated acids.
(a) sulfuric acid: d = 1.84 g/mL, containing 96% H_2SO_4 by mass
(b) nitric acid: d = 1.42 g/mL, containing 70.% HNO_3 by mass
(c) hydrochloric acid: d = 1.19 g/mL, containing 37% HCl by mass

60. Name two elements that you would expect to show the same kind of chemical reactivity as calcium. Explain your answer.

61. Carbon is the element that makes up the charcoal briquettes used in grills.
(a) Assuming access to an unlimited supply of air, write an equation to show the complete combustion of the carbon.
(b) Looking at the products in the equation in part (a), why is it inadvisable to cook with charcoal indoors? (In addition, incomplete charcoal combustion produces poisons such as carbon monoxide.)

62. Metal scouring pads used in the kitchen are composed largely of iron that has been drawn out into fine strings and woven into a pad. Often these pads are manufactured with a detergent inside. As the wet pads are exposed to the open air, the iron reacts with the oxygen in the air. Write the formula for the final product produced.

63. Chlorine, Cl_2, is an extremely hazardous and harmful chemical, yet in the compound sodium chloride, it is relatively harmless. What accounts for this difference between the elemental chlorine and the chloride in the compound? Name another element that is harmful in its elemental form.

Organic Chemistry: Saturated Hydrocarbons

Chapter 20

Many substances throughout nature contain silicon or carbon within their molecular structures. Silicon is the staple of the geologist—it combines with oxygen in a variety of ways to produce silica and a family of compounds known as the silicates. These compounds form the chemical foundation of most types of sand, rocks, and soil, essential materials of the construction industry.

In the living world, carbon, in combination with hydrogen, oxygen, nitrogen, and sulfur, provides the basis for millions of organic compounds. Carbon compounds provide us with energy sources in the form of hydrocarbons and their derivatives that allow us to heat and light our homes, drive our automobiles to work, and fly off to Hawaii for vacation. Small substitutions in these carbon molecules can produce chlorofluorocarbons, compounds used in plastics and refrigerants. An understanding of these molecules and their effect upon our global environment is vital in the continuing search to find ways to maintain our lifestyles while preserving the planet.

20.1 Organic Chemistry: History and Scope

During the late 18th and the early 19th centuries, chemists were baffled by the fact that compounds obtained from animal and vegetable sources defied the established rules for inorganic compounds—namely, that compound formation is due to a simple attraction between positively and negatively charged elements. In their experience with inorganic chemistry, groups of two or three elements formed only one, or at most a few, different compounds. However, they observed that one group—carbon, hydrogen, oxygen, and nitrogen—gave rise to a large number of different compounds that often were remarkably stable.

vital force theory

Because no organic compounds had been synthesized from inorganic substances and because there was no other explanation for the complexities of organic compounds, chemists believed that organic compounds were formed by some "vital force." The **vital force theory** held that organic substances could originate only from living material. In 1828, a German chemist Friedrich Wöhler (1800–1882) did a simple experiment that eventually proved to be the death blow to this theory. In attempting to prepare ammonium cyanate (NH_4CNO) by heating cyanic acid ($HCNO$) and ammonia, Wöhler obtained a white crystalline substance that he identified as urea. Wöhler knew that urea must be an authentic organic substance because it is a product of metabolism that can be isolated from urine. Although Wöhler's discovery was not immediately and generally recognized, the vital force theory was overthrown by this simple observation that an organic compound had been made from nonliving materials.

The formula for urea is:

$$H_2N-\overset{\overset{\displaystyle O}{\|}}{C}-NH_2$$

organic chemistry

After the work of Wöhler, it was apparent that no vital force other than skill and knowledge was needed to make organic chemicals in the laboratory and that inorganic as well as organic substances could be used as raw materials. Today, **organic chemistry** designates the branch of chemistry that deals with carbon compounds but does not imply that these compounds must originate from some form of life. A few special kinds of carbon compounds (e.g., carbon oxides, metal carbides, and metal carbonates) are often excluded from the organic classification because their chemistry is more conveniently related to that of inorganic substances.

◀ **Chapter Opening Photo: We are surrounded by organic compounds—the food we prepare, the clothes we wear, even our bodies are composed of organic molecules.**

The field of organic chemistry is vast, for it includes not only the composition of all living organisms, but also that of a great many other materials that we use daily. Examples of organic materials are nutrients (fats, proteins, carbohydrates); fuels; fabrics (cotton, wool, rayon, nylon); wood and paper products; paints and varnishes; plastics; dyes; soaps and detergents; cosmetics; medicinals; rubber products; and explosives.

What makes carbon special and different from the other elements in the periodic table? Carbon has the unique ability to bond to itself in long chains and rings of varying size. The greater the number of carbon atoms in a molecule, the more ways there are to link these atoms in different arrangements. This flexibility in the arrangement of atoms produces compounds with the same chemical composition and different structures. There is no theoretical limit on the number of organic compounds that can exist. In addition to this unique bonding property, carbon forms strong covalent bonds with a variety of elements, especially hydrogen, nitrogen, oxygen, sulfur, phosphorus, and the halogens. These are the elements most commonly found in organic compounds.

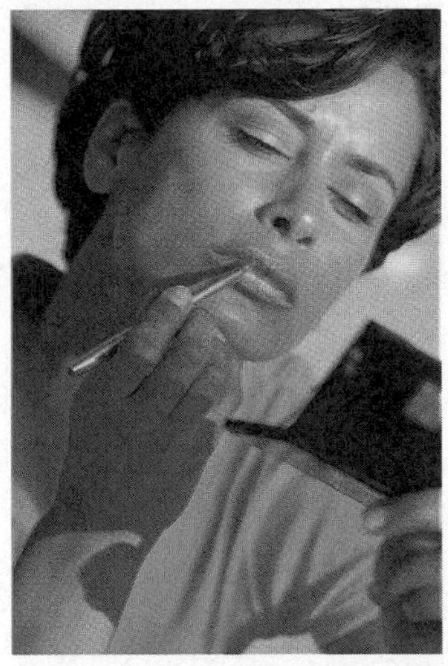

Organic molecules form the products used in cosmetics and perfumes.

20.2 The Carbon Atom: Bonding and Shape

The carbon atom is central to all organic compounds. The atomic number of carbon is 6, and its electron structure is $1s^22s^22p^2$. Two stable isotopes of carbon exist, ^{12}C and ^{13}C. In addition, there are several radioactive isotopes; ^{14}C is the most widely known of these because of its use in radiocarbon dating. With four electrons in its outer shell, carbon has oxidation numbers ranging from $+4$ to -4, and it forms predominantly covalent bonds. Carbon occurs as the free element in diamond, graphite, coal, coke, carbon black, charcoal, lampblack, and buckminsterfullerene.

A carbon atom usually forms four covalent bonds; each bond results from two atoms sharing a pair of electrons (see Section 11.5). The number of electron pairs that two atoms share determines whether the bond is single or multiple. In a single bond, only one pair of electrons is shared by the atoms. If both atoms have the same electronegativity (see Section 11.6) the bond is classified as nonpolar. In this type of bond there is no separation of positive and negative charge between the atoms.

Carbon can also form multiple bonds by sharing two or three pairs of electrons between two atoms. The double bond formed by sharing two electron pairs is stronger than a single bond but not twice as strong. It is also shorter than a single bond. Similarly, the triple bond formed by sharing three electron pairs is stronger and shorter than a double bond. An organic compound is classified as **saturated** if it contains only single bonds and as **unsaturated** if the molecules possess one or more multiple carbon–carbon bonds.

saturated compound
unsaturated compound

Lewis structures are useful in representing the bonding between atoms in a molecule, but these representations tell us little about the geometry of the molecules. There are a number of bonding theories that can be used to predict the shape of molecules. One of the common theories is called the valence shell electron pair repulsion (VSEPR) theory (see Section 11.11). This is a fairly simple, yet accurate method for determining the shape of a molecule.

VSEPR theory states that electron pairs repel each other since they have like charges. The electron pairs will therefore try to spread out as far as possible around

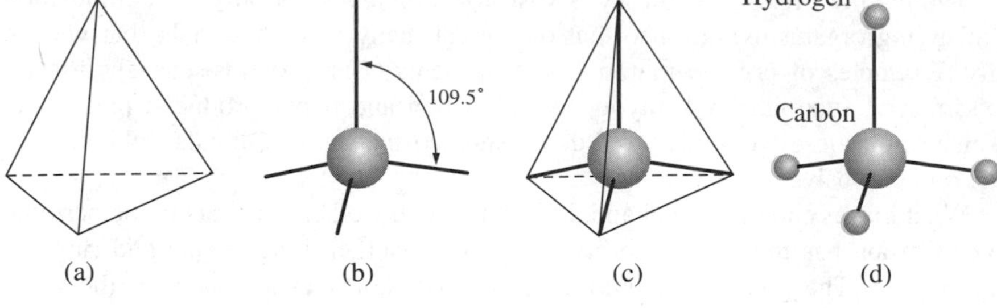

FIGURE 20.1 ▶
Tetrahedral structure of carbon: (a) a regular tetrahedron; (b) a carbon atom with tetrahedral bonds; (c) a carbon atom within a regular tetrahedron; (d) a methane molecule, CH_4.

```
(a) H:C::C:H
      ..  ..
      H   H

(b) H   120°  H
     \       /
      C ═ C
     /       \
    H         H
```

▲
FIGURE 20.2
(a) Lewis structure for C_2H_4.
(b) Shape of a molecule with a carbon–carbon double bond. The hydrogens and carbon form the vertices of a triangle. The bond angles around the carbon are 120°.

an atom. In addition, unshared pairs of electrons occupy more space than shared electron pairs.

What does VSEPR theory tell us about the shapes of carbon-containing compounds? Consider the simplest organic molecule, methane, CH_4. It contains one carbon atom with four single bonds to hydrogen atoms. The four shared electron pairs must be placed as far apart as possible in three dimensions. This results in the hydrogen atoms forming the corners of a tetrahedron with the carbon atom in the center (see Figure 20.1). The angle between these tetrahedral bonds is 109.5°, which is the least strained of the possible angles.

Double or triple bonds have a significant effect on the shape of the molecule. Because of increased repulsion between the pairs, additional electron pairs that are in close proximity take up more space than those in a single bond. Consider the Lewis structure for C_2H_4 (see Figure 20.2a). Each carbon atom has three separate regions for shared electrons. To place them as far apart as possible requires placing each atom at the corner of a triangle (see Figure 20.2b). The bond angles around the carbon atoms are 120°.

In a triple bond the carbon has only two regions for shared electrons. To be placed as far apart as possible, a linear arrangement is required. The bond angle is 180°.

```
                        180°
H:C:::C:H       H—C≡C—H
```

20.3 Molecular Models

Models are often used in organic chemistry to illustrate molecules. There are many ways to represent molecules, and each has its own advantages and disadvantages. Two common methods are ball-and-stick models and space-filling models. Both types of models are illustrated in Figure 20.3 for the methane molecule.

In the ball-and-stick model, different atoms are represented by different colored balls, and bonds are represented by sticks. The space-filling model gives a more accurate representation of the actual molecule but is not as clear in representing the chemical bonds between the atoms. Since most organic molecules contain many atoms, it is often difficult or inconvenient to draw either of these types of model on paper every time we wish to represent a molecule. Chemists often translate a three-dimensional model into a two-dimensional representation (a Lewis structure

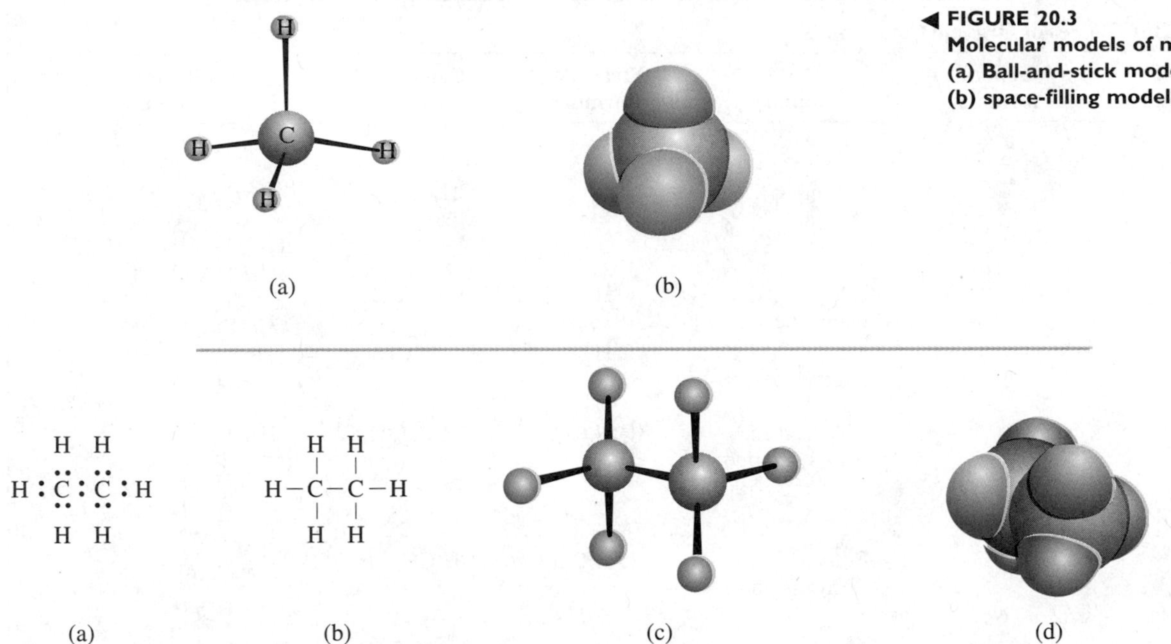

◀ **FIGURE 20.3**
Molecular models of methane. (a) Ball-and-stick model; (b) space-filling model.

▲ **FIGURE 20.4**
Models of ethane: (a) Lewis structure; (b) bond-line drawing; (c) ball-and-stick model; (d) space-filling model.

or bond-line drawing). An example of all four models is shown in Figure 20.4. In both Lewis and bond-line drawings, it is important to remember that the bond angle is not 90° the way it appears in the drawing. Rather it is 109.5° because the molecule is actually three-dimensional.

Formulas of organic molecules are also represented differently than those of inorganic compounds. A formula gives information about the composition of a compound. Inorganic compounds contain relatively few groups of atoms and are frequently represented by *empirical formulas*, which give only the simplest ratio of the atoms in a molecule. Larger molecules are often represented by a *molecular formula*, which gives more information. The molecular formula gives the actual number of atoms in a molecule. For inorganic compounds the empirical and molecular formulas are often the same. The models we've just been considering represent yet another even more informative type of formula. In a *structural formula*, the arrangement of the atoms within the molecule is clearly shown. Organic chemists often shorten these structural formulas into a final type called *condensed structural formulas*. In these formulas each carbon is grouped with the hydrogens bonded to it and then written as a formula—for example, CH_3CH_3 or $CH_3CH_2CH_2CH_3$. The diversity of methods for writing organic formulas is illustrated in Table 20.1.

20.4 Classifying Organic Compounds

It is impossible for anyone to study the properties of each of the millions of organic compounds. Organic compounds with similar structures are grouped into classes as shown in Table 20.1. The members of each class of compounds contain a characteristic atom or group of atoms called a **functional group** shown as the red portion

functional group

TABLE 20.1 Classes of Organic Compounds

Class of compound	General formula[a]	IUPAC name[b, c]	Molecular formula	Condensed formula	Structural formula
Alkane	RH	Eth*ane* (Ethane)	C_2H_6	CH_3CH_3	$\begin{array}{cccc} & H & H & \\ & \vert & \vert & \\ H- & C- & C- & H \\ & \vert & \vert & \\ & H & H & \end{array}$
Alkene	$R-CH=CH_2$	Eth*ene* (Ethylene)	C_2H_4	$H_2C=CH_2$	$\begin{array}{ccc} H & & H \\ & C=C & \\ H & & H \end{array}$
Alkyne	$R-C\equiv C-H$	Eth*yne* (Acetylene)	C_2H_2	$HC\equiv CH$	$H-C\equiv C-H$
Alkyl halide	RX	Chloroethane (Ethyl chloride)	C_2H_5Cl	CH_3CH_2Cl	$\begin{array}{cccc} & H & H & \\ & \vert & \vert & \\ H- & C- & C- & Cl \\ & \vert & \vert & \\ & H & H & \end{array}$
Alcohol	ROH	Ethan*ol* (Ethyl alcohol)	C_2H_6O	CH_3CH_2OH	$\begin{array}{cccc} & H & H & \\ & \vert & \vert & \\ H- & C- & C- & OH \\ & \vert & \vert & \\ & H & H & \end{array}$
Ether	$R-O-R$	Methoxymethane (Dimethyl ether)	C_2H_6O	CH_3OCH_3	$\begin{array}{ccccc} & H & & H & \\ & \vert & & \vert & \\ H- & C- & O- & C- & H \\ & \vert & & \vert & \\ & H & & H & \end{array}$
Aldehyde	$\begin{array}{ccc} R- & C & =O \\ & \vert & \\ & H & \end{array}$	Ethan*al* (Acetaldehyde)	C_2H_4O	CH_3CHO	$\begin{array}{cccc} & H & & \\ & \vert & & \\ H- & C- & C- & H \\ & \vert & \Vert & \\ & H & O & \end{array}$
Ketone	$\begin{array}{ccc} R- & C & -R \\ & \Vert & \\ & O & \end{array}$	Propan*one* (Dimethyl ketone)	C_3H_6O	CH_3COCH_3	$\begin{array}{ccccc} & H & & H & \\ & \vert & & \vert & \\ H- & C- & C- & C- & H \\ & \vert & \Vert & \vert & \\ & H & O & H & \end{array}$
Carboxylic acid	$\begin{array}{ccc} R- & C & -OH \\ & \Vert & \\ & O & \end{array}$	Ethan*oic acid* (Acetic acid)	$C_2H_4O_2$	CH_3COOH	$\begin{array}{cccc} & H & & \\ & \vert & & \\ H- & C- & C- & OH \\ & \vert & \Vert & \\ & H & O & \end{array}$
Ester	$\begin{array}{ccc} R- & C & -OR \\ & \Vert & \\ & O & \end{array}$	Methyl ethan*oate* (Methyl acetate)	$C_3H_6O_2$	CH_3COOCH_3	$\begin{array}{cccccc} & H & & & H & \\ & \vert & & & \vert & \\ H- & C- & C- & O- & C- & H \\ & \vert & \Vert & & \vert & \\ & H & O & & H & \end{array}$
Amide	$\begin{array}{ccc} R- & C & -NH_2 \\ & \Vert & \\ & O & \end{array}$	Ethan*amide* (Acetamide)	C_2H_5ON	CH_3CONH_2	$\begin{array}{ccccc} & H & & & H \\ & \vert & & & / \\ H- & C- & C- & N & \\ & \vert & \Vert & & \backslash \\ & H & O & & H \end{array}$
Amine	$R-CH_2-NH_2$	Aminoethane (Ethylamine)	C_2H_7N	$CH_3CH_2NH_2$	$\begin{array}{ccccc} & H & H & & \\ & \vert & \vert & & \\ H- & C- & C- & N- & H \\ & \vert & \vert & \vert & \\ & H & H & H & \end{array}$

[a] The letter R is used to indicate any of the many possible alkyl groups.
[b] Class name ending in italic.
[c] Common name in parentheses.

Fossil fuels are mined from many sources: Here we see an oil exploration rig in Texas, a coal miner in Pennsylvania, and a natural gas operation in Washington state.

of the structural formula in Table 20.1. Organic compounds from different classes may have the same molecular formulas but completely different chemical and physical properties.

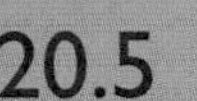

20.5 Hydrocarbons

hydrocarbon

aliphatic

Hydrocarbons are compounds that are composed entirely of carbon and hydrogen atoms bonded to each other by covalent bonds. Hydrocarbons are classified into two major categories, aliphatic and aromatic. The term *aromatic* refers to compounds that contain benzene rings. All hydrocarbons that are not aromatic are often described as **aliphatic** (from the Greek word *aleiphar*, meaning "fat"). The aliphatic hydrocarbons include the alkanes, alkenes, and alkynes (see Figure 20.5).

Fossil fuels—natural gas, petroleum, and coal—are the principal sources of hydrocarbons. Natural gas is primarily methane with small amounts of ethane, propane, and butane. Petroleum is a mixture of hydrocarbons from which gasoline, kerosene, fuel oil, lubricating oil, paraffin wax, and petrolatum (themselves mixtures of hydrocarbons) are separated. Coal tar, a volatile by-product of the process of

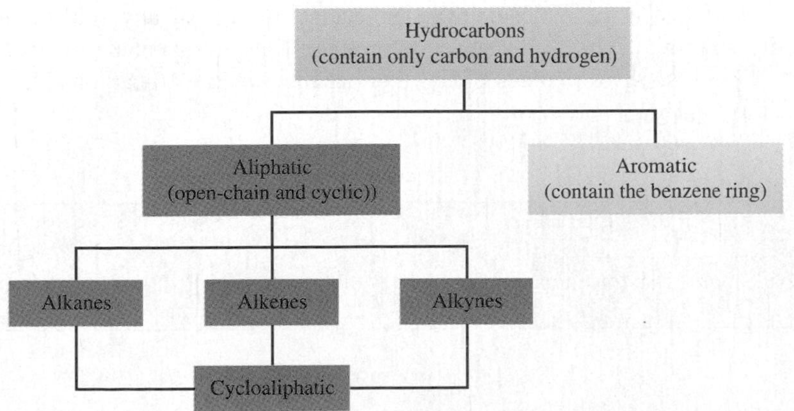

◀ **FIGURE 20.5**
Classes of hydrocarbons.

making coke from coal for use in the steel industry, is the source of many valuable chemicals including the aromatic hydrocarbons benzene, toluene, and naphthalene.

The fossil fuels provide a rich resource of hydrocarbons for human society. In the past these resources have been used primarily as a source of heat (via combustion). We now see that extensive combustion can have severe environmental consequences (e.g., air pollution and global warming). Fossil fuels also serve as the raw materials for much of today's chemical industry. One of the themes that runs through the study of organic chemistry is the synthetic relationship between compounds; for example, acids are often formed from alcohols. Fossil fuels are the starting materials for many of these synthetic sequences. And, in the long run, the fossil fuels may well prove to be more valuable to us as a source of organic chemicals than as a source of heat.

20.6 Saturated Hydrocarbons: Alkanes

alkane

The **alkanes**, also known as *paraffins* or *saturated hydrocarbons*, are straight- or branched-chain hydrocarbons with only single covalent bonds between the carbon atoms. We shall study the alkanes in some detail because many other classes of organic compounds can be considered as derivatives of these substances. For example, it is necessary to learn the names of the first ten members of the alkane series because these names are used as a basis for naming other classes of compounds.

Methane, CH_4, is the first member of the alkane series. Alkanes with two-, three-, and four-carbon atoms are ethane, propane, and butane, respectively. The names of the first four alkanes are of common or trivial origin and must be memorized; but the names beginning with the fifth member, pentane, are derived from Greek numbers and are relatively easy to recall. The names and formulas of the first ten members of the series are given in Table 20.2.

homologous series

Successive compounds in the alkane series differ from each other in composition by one carbon and two hydrogen atoms. When each member of a series differs from the next member by a CH_2 group, the series is called a **homologous series**. The members of a homologous series are similar in structure but have a regular difference in formula. All common classes of organic compounds exist in homologous series. Each homologous series can be represented by a general formula. For all open-chain alkanes, the general formula is C_nH_{2n+2}, where n corresponds to the number of carbon atoms in the molecule. The molecular formula of any specific alkane is easily determined from this general formula. Thus, for pentane, $n = 5$ and $2n + 2 = 12$, so the formula is C_5H_{12}. For hexadecane, the 16-carbon alkane, the formula is $C_{16}H_{34}$.

Practice 20.1

Write molecular formulas for the alkanes that contain 12, 14, and 20 carbons.

TABLE 20.2 Names, Formulas, and Physical Properties of Straight-Chain Alkanes

Name	Molecular formula C_nH_{2n+2}	Condensed structural formula	Boiling point (°C)	Melting point (°C)
Methane	CH_4	CH_4	−161	−183
Ethane	C_2H_6	CH_3CH_3	−88	−172
Propane	C_3H_8	$CH_3CH_2CH_3$	−45	−187
Butane	C_4H_{10}	$CH_3CH_2CH_2CH_3$	−0.5	−138
Pentane	C_5H_{12}	$CH_3CH_2CH_2CH_2CH_3$	36	−130
Hexane	C_6H_{14}	$CH_3CH_2CH_2CH_2CH_2CH_3$	69	−95
Heptane	C_7H_{16}	$CH_3CH_2CH_2CH_2CH_2CH_2CH_3$	98	−90
Octane	C_8H_{18}	$CH_3CH_2CH_2CH_2CH_2CH_2CH_2CH_3$	125	−57
Nonane	C_9H_{20}	$CH_3CH_2CH_2CH_2CH_2CH_2CH_2CH_2CH_3$	151	−54
Decane	$C_{10}H_{22}$	$CH_3CH_2CH_2CH_2CH_2CH_2CH_2CH_2CH_2CH_3$	174	−30

20.7 Carbon Bonding in Alkanes

A carbon atom is capable of forming single covalent bonds with one, two, three, or four other atoms. To understand this remarkable bonding ability, we must look at the electron structure of carbon. The valence electrons of carbon in their ground state are $2s^22p_x^12p_y^1$. All of carbon's valence electrons can be shared to make a total of four bonds. When a carbon atom is bonded to other atoms by single bonds (e.g., to four hydrogen atoms in CH_4), it would appear at first that there should be two different types of bonds—bonds involving the $2s$ electrons and bonds involving the $2p$ electrons of the carbon atom. However, this is not the case. All four carbon–hydrogen bonds are identical.

If the carbon atom is to form four equivalent bonds, its electrons in the $2s$ and $2p$ orbitals must rearrange to four equivalent orbitals. To form the four equivalent orbitals, imagine that a $2s$ electron is promoted to a $2p$ orbital, giving carbon an outer shell electron structure of $2s^12p_x^12p_y^12p_z^1$. The $2s$ orbital and the three $2p$ orbitals then hybridize to form four equivalent hybrid orbitals, which are designated sp^3 orbitals. The orbitals formed (sp^3) are neither s orbitals nor p orbitals, but are instead a hybrid of those orbitals, having one-fourth s orbital character and three-fourths p orbital character. This process is illustrated in Figure 20.6. It is these sp^3 orbitals that are directed toward the corners of a regular tetrahedron (see Figure 20.7).

A single bond is formed when one of the sp^3 orbitals overlaps an orbital of another atom. Thus each C—H bond in methane is the result of the overlapping of a carbon sp^3 orbital and a hydrogen s orbital, Figure 20.7(c). Once the bond is formed, the pair of bonding electrons constituting it are said to be in a molecular orbital. In a similar way, a C—C single bond results from the overlap of sp^3 orbitals between two carbon atoms. This type of bond is called a sigma (σ) bond. A **sigma bond** exists if the electron cloud formed by the pair of bonding electrons lies on a straight line drawn between the nuclei of the bonded atoms.

sigma bond

FIGURE 20.6 ▶ **Schematic hybridization of $2s^22p_x^12p_y^1$ orbitals of carbon to form four sp^3 electron orbitals.**

$$2p\ \underline{\uparrow}\ \underline{\uparrow}\ \underline{\ \ }\qquad 2s\ \underline{\uparrow\downarrow}\quad \xrightarrow[\text{is promoted to a }2p\text{ orbital}]{\text{A }2s\text{ electron}}\quad 2p\ \underline{\uparrow}\ \underline{\uparrow}\ \underline{\uparrow}\qquad 2s\ \underline{\uparrow}\quad \xrightarrow{\text{Hybridization}}\quad sp^3\ \underline{\uparrow}\ \underline{\uparrow}\ \underline{\uparrow}\ \underline{\uparrow}$$

Four carbon electrons in their ground-state orbitals

Four equivalent sp^3 orbitals—each contains one electron

FIGURE 20.7 ▶ **Tetrahedral nature of sp^3 orbitals: (a) a single sp^3-hybridized orbital; (b) four sp^3-hybridized orbitals in tetrahedral arrangement; (c) sp^3 and s orbitals overlapping to form C—H bonds in methane.**

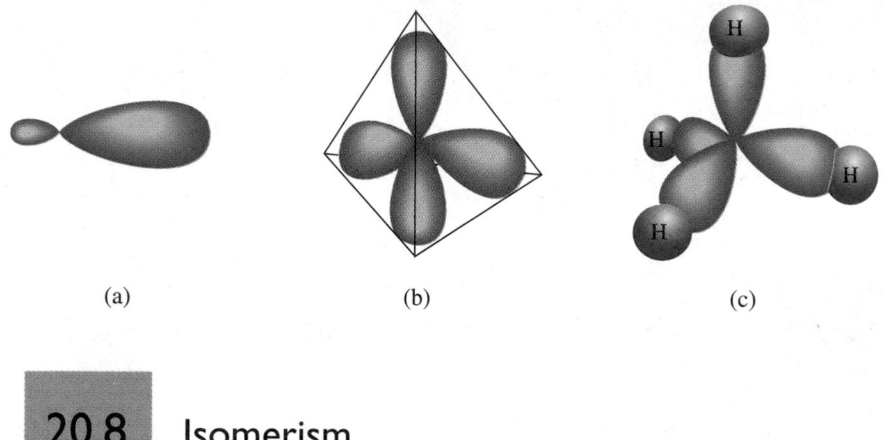

20.8 Isomerism

The properties of an organic substance are dependent on its molecular structure. The majority of organic compounds are made from relatively few elements—carbon, hydrogen, oxygen, nitrogen, and the halogens. The valence bonds or points of attachment may be represented in structural formulas by a corresponding number of dashes attached to each atom:

—C— (with bonds above and below) H— —O— —N— (with bond below) Cl— Br— I— F—

Thus carbon has four bonds to each atom, nitrogen three bonds, oxygen two bonds, and hydrogen and the halogens one bond to each atom.

In an alkane, each carbon atom is joined to four other atoms by four single covalent bonds. These bonds are separated by angles of 109.5° (the angles correspond to those formed by lines drawn from the center of a regular tetrahedron to its corners). Alkane molecules contain only carbon–carbon and carbon–hydrogen bonds and are essentially nonpolar. Alkane molecules are nonpolar because (1) carbon–carbon bonds are nonpolar since they are between like atoms; (2) carbon–hydrogen bonds are only slightly polar since there is only a small difference in electronegativity between carbon and hydrogen atoms; and (3) the bonds in an alkane are symmetrically directed toward the corners of a tetrahedron. Because of their low polarity, alkane molecules have very little intermolecular attraction and therefore have relatively low boiling points compared with other organic compounds of similar molar mass.

Without the use of models or perspective drawings, the three-dimensional character of atoms and molecules is difficult to portray accurately. However, concepts of structure can be conveyed to some extent by structural formulas.

To write the correct structural formula for propane, C_3H_8, we must determine how to place each atom in the molecule. An alkane contains only single bonds. Therefore, each carbon atom must be bonded to four other atoms by either C—C or C—H bonds. Hydrogen must be bonded to only one carbon atom by a C—H bond, since C—H—C bonds do not occur and an H—H bond would simply represent a hydrogen molecule. Applying this information, we find that the only possible structure for propane is

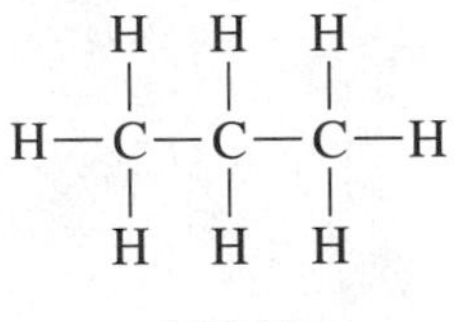

```
    H   H   H
    |   |   |
H — C — C — C — H
    |   |   |
    H   H   H
```

propane

However, it is possible to write two structural formulas corresponding to the molecular formula C_4H_{10}:

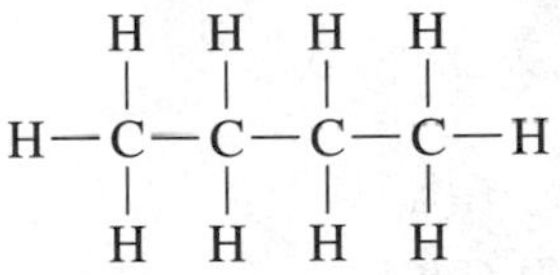

```
                                       H
                                    H  |  H
                                     \ | /
                                       C
    H   H   H   H                  H   |   H
    |   |   |   |                  |   |   |
H — C — C — C — C — H    and   H — C — C — C — H
    |   |   |   |                  |   |   |
    H   H   H   H                  H   H   H
```

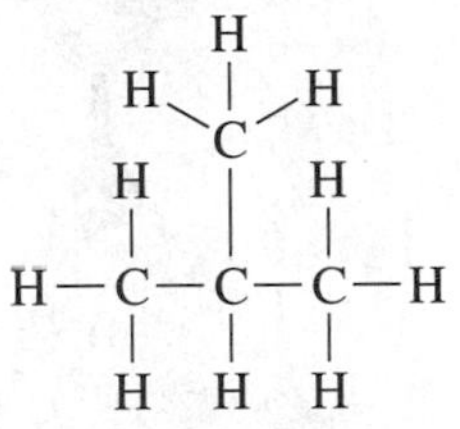

butane isobutane

"Isobutane" is a common name; a more systematic nomenclature will be introduced in Section 20.9.

Two C_4H_{10} compounds with the structural formulas shown here actually exist. The butane with the unbranched carbon chain is called *butane* or *normal butane* (abbreviated *n*-butane); it boils at $-0.5°C$ and melts at $-138.3°C$. The common name for the branched-chain butane is *isobutane*; it boils at $-11.7°C$ and melts at $-159.5°C$. These differences in physical properties are sufficient to establish that the two compounds are different substances even though they have the same molecular formula. Models illustrating the structural arrangement of the atoms in methane, ethane, propane, butane, and isobutane are shown in Figure 20.8.

This phenomenon of two or more compounds having the same molecular formula but different structural arrangements of their atoms is called **isomerism**. The various individual compounds are called **isomers**. For example, there are two isomers of butane, C_4H_{10}. Isomerism is very common among organic compounds and is another reason for the large number of known compounds. There are 3 isomers of pentane, 5 isomers of hexane, 9 isomers of heptane, 18 isomers of octane, 35 isomers of nonane, and 75 isomers of decane. The phenomenon of isomerism is a compelling reason for the use of structural formulas.

isomerism

isomer

Isomers are compounds that have the same molecular formula but different structural formulas.

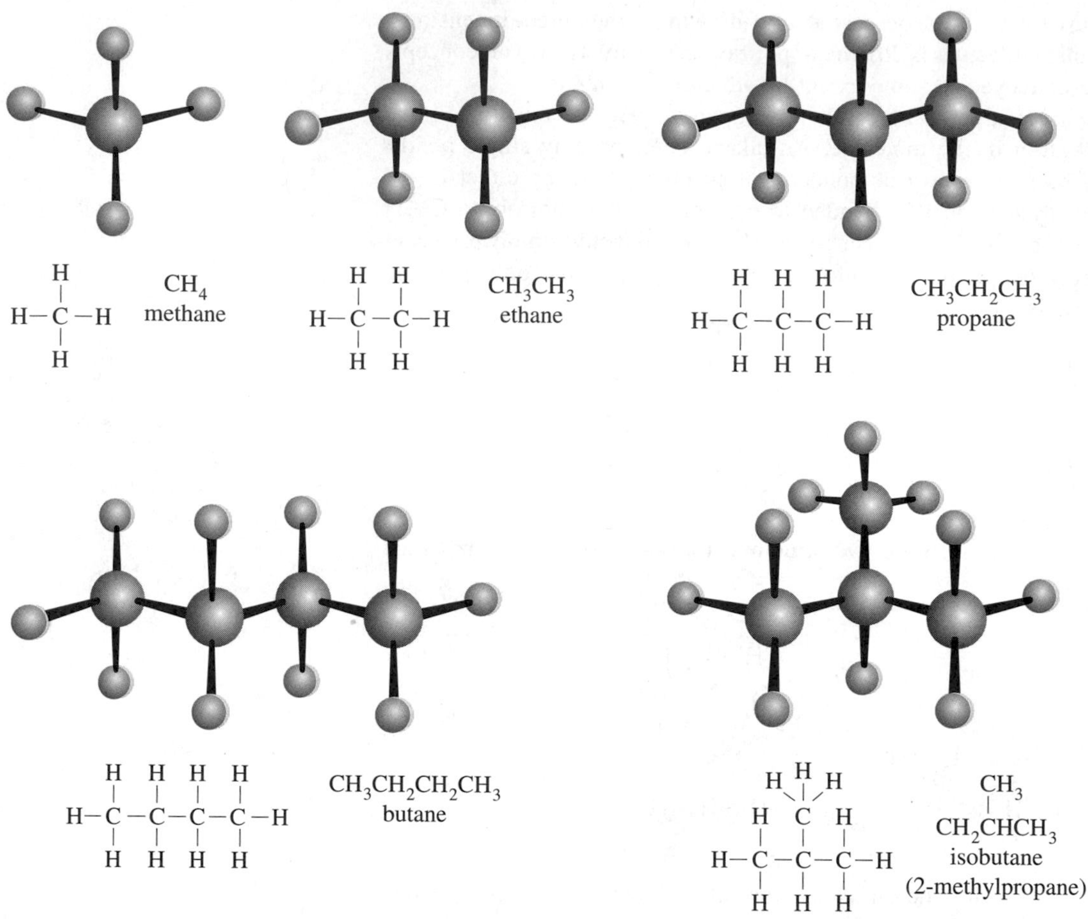

▲
FIGURE 20.8
Ball-and-stick models illustrating structural formulas of methane, ethane, propane, butane, and isobutane.

To save time and space in writing, condensed structural formulas are often used. In the condensed structural formulas, atoms and groups that are attached to a carbon atom are generally written to the right of that carbon atom.

Let's interpret the condensed structural formula for propane:

$$\overset{1}{C}H_3-\overset{2}{C}H_2-\overset{3}{C}H_3 \quad \text{or} \quad CH_3CH_2CH_3$$

Carbon 1 has three hydrogen atoms attached to it and is bonded to carbon 2, which has two hydrogen atoms on it and is bonded to carbon 3, which has three hydrogen atoms bonded to it.

When structures are written as in butane, $CH_3CH_2CH_2CH_3$, the bonds (long dashes) are often not shown. But when a group is shown above or below the main carbon chain, as in 2-methylpropane (isobutane), a vertical dash must be used to indicate the point of attachment of the group to the main carbon chain.

Example 20.1 There are three isomers of pentane, C_5H_{12}. Write structural formulas and condensed structural formulas for these isomers.

Solution

In a problem of this kind it is best to start by writing the carbon skeleton of the compound containing the longest continuous carbon chain. In this case, it is five carbon atoms:

C—C—C—C—C

Now complete the structure by attaching hydrogen atoms around each carbon atom so that each carbon atom has four bonds attached to it. The carbon atoms at each end of the chain need three hydrogen atoms. The three inner carbon atoms each need two hydrogen atoms to give them four bonds.

```
   H   H   H   H   H
   |   |   |   |   |
H—C—C—C—C—C—H
   |   |   |   |   |
   H   H   H   H   H
```

$CH_3CH_2CH_2CH_2CH_3$

For the next isomer, start by writing a four-carbon chain and attach the fifth carbon atom to either of the middle carbon atoms—do not use the end ones.

```
        C             C
        |             |
C—C—C—C      C—C—C—C
```

Both of these structures represent the same compound.

Now add the 12 hydrogen atoms to complete the structure:

```
           H
        H  |  H
          \|/
           C
   H   H   |   H
   |   |   |   |
H—C—C—C—C—H
   |   |   |   |
   H   H   H   H
```

$CH_3CH_2CH(CH_3)CH_3$ (written with CH_3 above the CH of $CH_3CH_2CHCH_3$) or $CH_3CH_2CH(CH_3)_2$

For the third isomer, write a three-carbon chain, attach the other two carbon atoms to the central carbon atom, and complete the structure by adding the 12 hydrogen atoms:

```
                          H
                       H  |  H
                         \|/
    C                     C
    |               H     |     H
    |               |     |     |
C—C—C           H—C—C—C—H
    |               |     |     |
    C               H     |     H
                          C
                         /|\
                       H  |  H
                          H
```

CH_3CCH_3 (with CH_3 above and below the central C) or $C(CH_3)_4$

Practice 20.2

Write structural and condensed formulas for the isomers of hexane, C_6H_{14}.

20.9 Naming Organic Compounds

In the early development of organic chemistry, each new compound was given a name, usually by the person who had isolated or synthesized it. Names were not systematic but often conveyed some information—usually about the origin of the substance. Wood alcohol (methanol), for example, was so named because it was obtained by destructive distillation or pyrolysis of wood. Methane, formed during underwater decomposition of vegetable matter in marshes, was originally called marsh gas. A single compound was often known by several names. For example, the active ingredient in alcoholic beverages has been called alcohol, ethyl alcohol, methyl carbinol, grain alcohol, spirit, and ethanol.

A meeting in Geneva in 1892 initiated the development of an international system for naming compounds. In its present form the method recommended by the International Union of Pure and Applied Chemistry is systematic, generally unambiguous, and internationally accepted. It is called the **IUPAC System**. Despite the existence of the official IUPAC System, a great many well-established common, or trivial, names and abbreviations (e.g., TNT and DDT) are used because of their brevity and/or convenience. It is thus necessary to have a knowledge of both the IUPAC System and many common names.

IUPAC System

In order to name organic compounds systematically, you must be able to recognize certain common alkyl groups. **Alkyl groups** have the general formula C_nH_{2n+1} (one less hydrogen atom than the corresponding alkane). The missing H atom may have been detached from any carbon in the alkane. The name of the group is formed from the name of the corresponding alkane by simply dropping *-ane* and substituting a *-yl* ending. The names and formulas of selected alkyl groups up to and including four carbon atoms are given in Table 20.3. The letter R is often used in formulas to mean any of the many possible alkyl groups.

alkyl group

$$R = C_nH_{2n+1} \qquad \text{Any alkyl group}$$

Three prefixes are commonly used to indicate structural information. They are *iso-*, *sec-* (for secondary) and *tert-* or *t-* (for tertiary). *Iso-* is used to indicate that a compound or alkyl group has the following structure at one end of the carbon chain:

```
        CH3
        |
  CH3—CH—
```

(See Table 20.3.) A carbon atom bonded to only one other carbon is a primary carbon atom. A carbon atom bonded to two other carbons is a secondary (*sec-*) carbon atom. Thus, the name *sec*-butyl indicates that this group contains four carbons (butyl) and the carbon that connects this group to other atoms is bonded to two other carbon atoms. A carbon atom bonded to three other carbons is a tertiary carbon atom. So the name *tert*-butyl indicates that this group contains four carbons (butyl) and the carbon that connects this group to other atoms is bonded to three other carbons.

TABLE 20.3 Names and Formulas of Selected Alkyl Groups

Formula	Name	Formula	Name
$CH_3—$	Methyl	$CH_3CH(CH_3)—$	Isopropyl
$CH_3CH_2—$	Ethyl	$CH_3CH(CH_3)CH_2—$	Isobutyl
$CH_3CH_2CH_2—$	Propyl	$CH_3CH_2CH(CH_3)—$	*sec*-Butyl (secondary butyl)
$CH_3CH_2CH_2CH_2—$	Butyl	$CH_3C(CH_3)_2—$	*tert*-Butyl or t-Butyl (tertiary butyl)
$CH_3(CH_2)_3CH_2—$	Pentyl		
$CH_3(CH_2)_4CH_2—$	Hexyl		
$CH_3(CH_2)_5CH_2—$	Heptyl		
$CH_3(CH_2)_6CH_2—$	Octyl		
$CH_3(CH_2)_7CH_2—$	Nonyl		
$CH_3(CH_2)_8CH_2—$	Decyl		

Practice 20.3

Locate the primary, secondary, and tertiary carbon atoms in this compound:

$$CH_3-CH(CH_3)-CH_2-CH_2-CH(CH_3)-CH_2-C(CH_3)(CH_3)-CH_2-CH(CH_2CH_3)-CH_3$$

A few relatively simple rules are all that are needed to name a great many alkanes according to the IUPAC System. In later sections these rules will be extended to cover other classes of compounds, but advanced texts or references must be consulted for the complete system.

IUPAC Nomenclature for Alkanes

1. Select the longest continuous chain of carbon atoms as the parent compound, and consider all alkyl groups attached to it as side chains that have replaced hydrogen atoms of the parent hydrocarbon. The name of the alkane consists of the name of the parent compound prefixed by the names of the side-chain alkyl groups attached to it.
2. Number the carbon atoms in the parent carbon chain starting from the end closest to the first carbon atom that has an alkyl or other group substituted for a hydrogen atom.
3. Name each side-chain alkyl group and designate its position on the parent carbon chain by a number (e.g., 2-methyl means a methyl group attached to carbon 2).

Adding numerical prefixes to the alkyl name does *not* change the alphabetical order.

4. When the same alkyl-group side chain occurs more than once, indicate this by a prefix (*di-*, *tri-*, *tetra-*, etc.) written in front of the alkyl-group name (e.g., *dimethyl* indicates two methyl groups). The numbers indicating the positions of these alkyl groups are separated by a comma and followed by a hyphen and are placed in front of the name (e.g., 2,3-dimethyl).
5. When several different alkyl groups are attached to the parent compound, list them in alphabetical order (e.g., *ethyl* before *methyl* in 3-ethyl-4-methyloctane).

The compound shown here is commonly called isopentane. Let's name it using the IUPAC System:

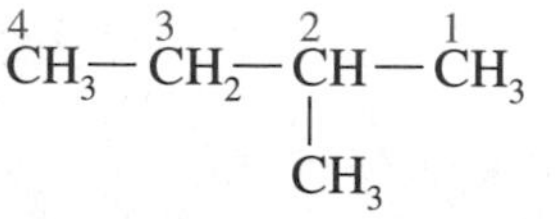

$$\overset{4}{C}H_3-\overset{3}{C}H_2-\overset{2}{C}H(CH_3)-\overset{1}{C}H_3$$

2-methylbutane
(isopentane)

The longest continuous chain contains four carbon atoms. Therefore, the parent compound is butane and the compound is named as a butane. The methyl group (CH_3—) attached to carbon 2 is named as a prefix to butane, the "2-" indicating the point of attachment of the methyl group on the butane chain.

How would we write the structural formula for 2-methylpentane? An analysis of its name gives us this information.

1. The parent compound, pentane, contains five carbons. Write and number the five-carbon skeleton of pentane:

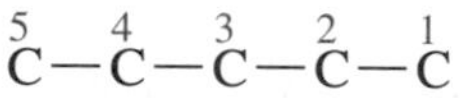

$$\overset{5}{C}-\overset{4}{C}-\overset{3}{C}-\overset{2}{C}-\overset{1}{C}$$

2. Put a methyl group on carbon 2 ("2-methyl" in the name gives this information):

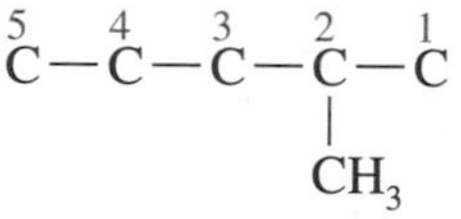

$$\overset{5}{C}-\overset{4}{C}-\overset{3}{C}-\overset{2}{C}(CH_3)-\overset{1}{C}$$

3. Add hydrogens to give each carbon four bonds. The structural formula is

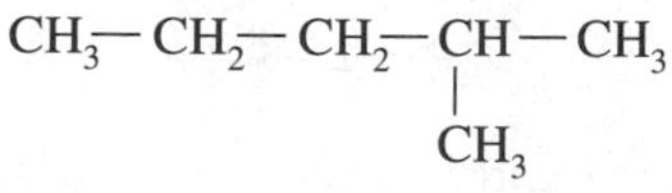

$$CH_3-CH_2-CH_2-CH(CH_3)-CH_3$$

2-methylpentane

Should this compound be called 4-methylpentane? No, the IUPAC System specifically states that the parent carbon chain is numbered starting from the end nearest to the side or branch chain.

It is important to understand that it is the sequence of atoms and groups that determines the name of a compound, and not the way the sequence is written. Each of the following formulas represents 2-methylpentane:

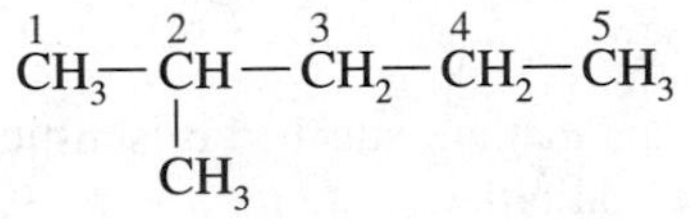

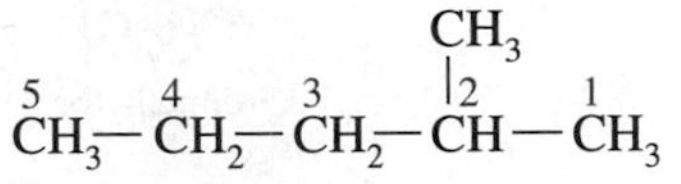

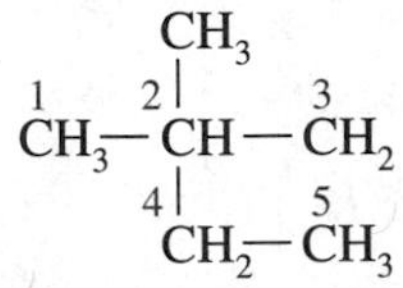

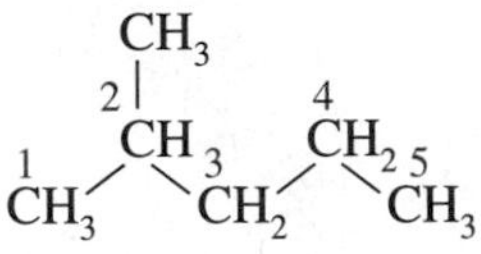

$$\overset{1}{CH_3}-\underset{\underset{\displaystyle CH_3}{|}}{\overset{2}{CH}}-\overset{3}{CH_2}-\overset{4}{CH_2}-\overset{5}{CH_3} \qquad \overset{5}{CH_3}-\overset{4}{CH_2}-\overset{3}{CH_2}-\overset{\displaystyle CH_3}{\overset{|2}{CH}}-\overset{1}{CH_3}$$

$$\begin{array}{l} \quad\; CH_3 \\ \overset{1}{} \quad\;\; |{\scriptstyle 2} \quad\;\; {\scriptstyle 3} \\ CH_3-CH-CH_2 \\ \quad\;\; |{\scriptstyle 4} \quad\; {\scriptstyle 5} \\ \quad CH_2-CH_3 \end{array} \qquad \overset{1}{CH_3}\diagup\overset{\displaystyle CH_3}{\overset{|2}{CH}}\diagdown\overset{3}{CH_2}\diagup\overset{4}{CH_2}\diagdown\overset{5}{CH_3}$$

The following formulas and names demonstrate other aspects of the official nomenclature system:

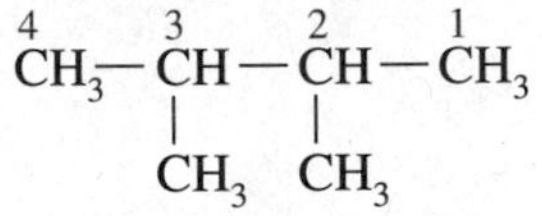

$$\overset{4}{CH_3}-\underset{\underset{\displaystyle CH_3}{|}}{\overset{3}{CH}}-\underset{\underset{\displaystyle CH_3}{|}}{\overset{2}{CH}}-\overset{1}{CH_3}$$

2,3-dimethylbutane

The name of this compound is 2,3-dimethylbutane. The longest carbon atom chain is four, indicating butane; "dimethyl" indicates two methyl groups; "2,3-" means that one CH_3 is on carbon 2 and one is on carbon 3.

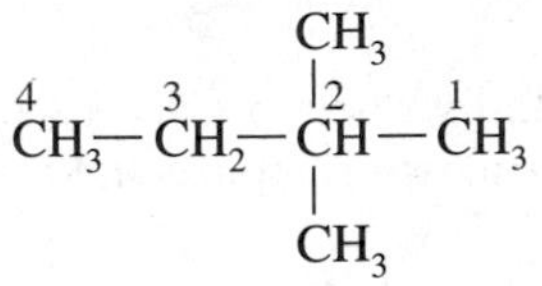

$$\overset{4}{CH_3}-\overset{3}{CH_2}-\underset{\underset{\displaystyle CH_3}{|}}{\overset{\displaystyle CH_3}{\overset{|2}{CH}}}-\overset{1}{CH_3}$$

2,2-dimethylbutane

In this compound, both methyl groups are on the same carbon atom; both numbers are required.

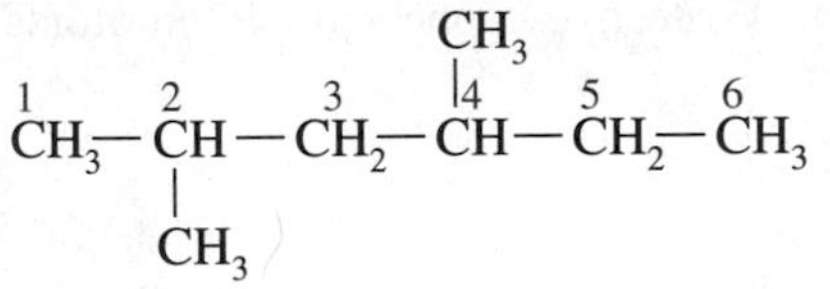

$$\overset{1}{CH_3}-\underset{\underset{\displaystyle CH_3}{|}}{\overset{2}{CH}}-\overset{3}{CH_2}-\overset{\displaystyle CH_3}{\overset{|4}{CH}}-\overset{5}{CH_2}-\overset{6}{CH_3}$$

2,4-dimethylbutane

Here the molecule is numbered from left to right.

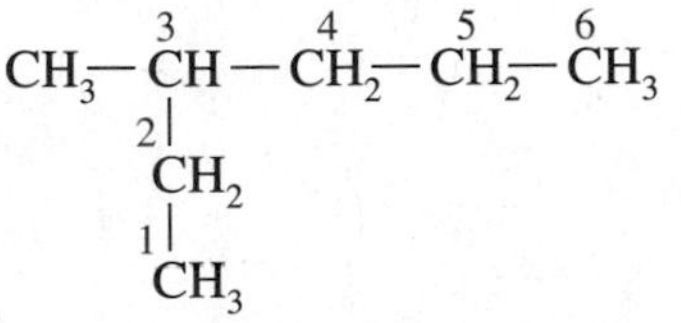

$$CH_3-\underset{\underset{\underset{\underset{\displaystyle CH_3}{|{\scriptstyle 1}}}{\displaystyle CH_2}}{|{\scriptstyle 2}}}{\overset{3}{CH}}-\overset{4}{CH_2}-\overset{5}{CH_2}-\overset{6}{CH_3}$$

3-methylhexane

In this compound, there are six carbons in the longest continuous chain.

$$\begin{array}{ccccccccccccccc}
 & & & & & & & & CH_2{-}CH_3 & & & & & & \\
 & & & & & & & & | & & & & & & \\
\overset{8}{CH_3} & - & \overset{7}{CH_2} & - & \overset{6}{CH_2} & - & \overset{5}{CH_2} & - & \overset{4}{C} & - & \overset{3}{CH} & - & \overset{2}{CH} & - & \overset{1}{CH_3} \\
 & & & & & & & & | & & | & & | & & \\
 & & & & & & & & CH_3 & & Cl & & CH_3 & &
\end{array}$$

3-chloro-4-ethyl,2,4-dimethyloctane

Here the longest carbon chain is eight. The groups that are attached or substituted for H on the octane chain are named in alphabetical order.

Example 20.2 Write the formulas for

(a) 3-ethylpentane (b) 2,2,4-trimethylpentane

Solution

(a) The name *pentane* indicates a five-carbon chain. Write five connected carbon atoms and number them:

$$\overset{1}{C}-\overset{2}{C}-\overset{3}{C}-\overset{4}{C}-\overset{5}{C}$$

An ethyl group is written as CH_3CH_2—. Attach this group to carbon 3:

$$\begin{array}{ccccccccc}
\overset{1}{C} & - & \overset{2}{C} & - & \overset{3}{C} & - & \overset{4}{C} & - & \overset{5}{C} \\
 & & & & | & & & & \\
 & & & & CH_2CH_3 & & & &
\end{array}$$

Now add hydrogen atoms to give each carbon atom four bonds. Carbons 1 and 5 each need three H atoms; carbons 2 and 4 each need two H atoms; and carbon 3 needs one H atom. The formula is complete:

$$\begin{array}{c}
CH_3CH_2CHCH_2CH_3 \\
| \\
CH_2CH_3
\end{array}$$

(b) Pentane indicates a five-carbon chain. Write five connected carbon atoms and number them:

$$\overset{1}{C}-\overset{2}{C}-\overset{3}{C}-\overset{4}{C}-\overset{5}{C}$$

There are three methyl groups (CH_3—) in the compound (*trimethyl*), two attached to carbon 2 and one attached to carbon 4. Attach these three methyl groups to their respective carbon atoms:

$$\begin{array}{ccccccccc}
 & & CH_3 & & & & CH_3 & & \\
 & & | & & & & | & & \\
\overset{1}{C} & - & \overset{2}{C} & - & \overset{3}{C} & - & \overset{4}{C} & - & \overset{5}{C} \\
 & & | & & & & & & \\
 & & CH_3 & & & & & &
\end{array}$$

Now add H atoms to give each carbon atom four bonds. Carbons 1 and 5 each need three H atoms; carbon 2 does not need any H atoms; carbon 3 needs two H atoms; and carbon 4 needs one H atom. The formula is complete:

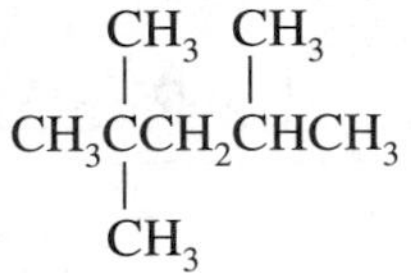

```
     CH3  CH3
     |    |
  CH3CCH2CHCH3
     |
     CH3
```

Example 20.3

Name the following compounds:

(a)

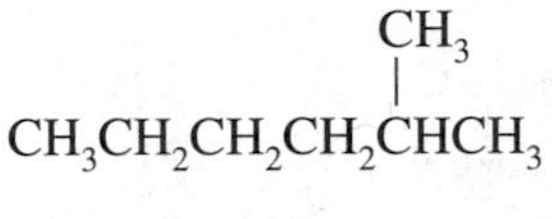

```
              CH3
              |
CH3CH2CH2CH2CHCH3
```

(b)

```
             CH2CH3
             |
CH3CH2CH2CHCH2CHCH3
                |
                CH2CH3
```

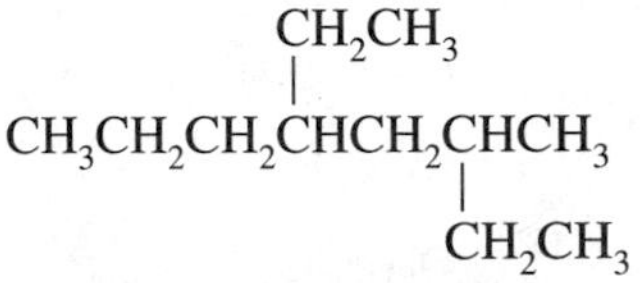

Solution

(a) The longest continuous carbon chain contains six carbon atoms (Rule 1). Thus the parent name of the compound is hexane. Number the carbon chain from right to left so that the methyl group attached to carbon 2 is given the lowest possible number (Rule 2). With a methyl group on carbon 2, the name of the compound is 2-methylhexane (Rule 3).

(b) The longest continuous carbon chain contains eight carbon atoms:

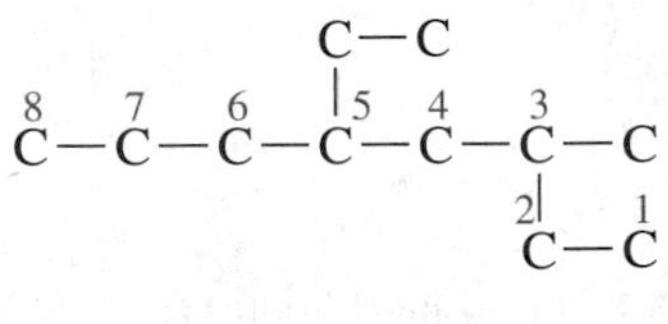

```
         C—C
         |
8  7  6  |5  4  3
C—C—C—C—C—C—C
               2|  1
               C—C
```

Thus the parent name is octane. As the chain is numbered, there is a methyl group on carbon 3 and an ethyl group on carbon 5, so the name of the compound is 5-ethyl-3-methyloctane. Note that ethyl is named before methyl (alphabetical order, Rule 5).

Practice 20.4

Write the formula for (a) 2-methylhexane, (b) 3,4-dimethylheptane, (c) 2-chloro-3-ethylpentane

Practice 20.5

Name the following compounds:

(a)

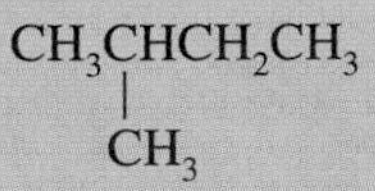

```
CH3CHCH2CH3
   |
   CH3
```

(b)

```
            CH2CH3
            |
CH3CH2CHCH2CHCH3
      |
      CH3
```

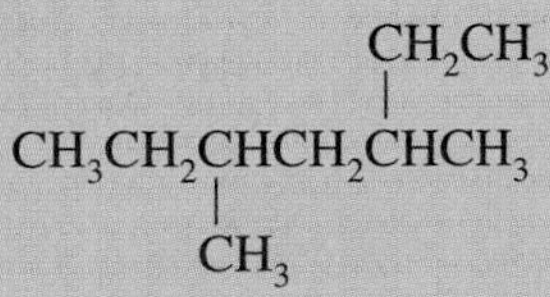

20.10 Introduction to the Reactions of Carbon

Carbon, like most elements, can undergo oxidation–reduction reactions. When carbon atoms are oxidized, they often form additional bonds to oxygen. In the following example, a methane carbon is oxidized to form carbon dioxide:

$$\underset{\text{methane}}{\begin{array}{c} \text{H} \\ | \\ \text{H}-\text{C}-\text{H} \\ | \\ \text{H} \end{array}} + 2\,O_2 \longrightarrow \underset{\text{carbon dioxide}}{O=C=O} + 2\,H_2O$$

When carbon atoms are reduced, they often form additional bonds to hydrogen. During this reaction, the carbon may decrease the number of bonds to oxygen and increase the number of bonds to hydrogen. The following reaction shows the reduction of a carbon compound:

$$\underset{\text{carbon dioxide}}{O=C=O} + 3\,H_2 \longrightarrow \underset{\text{methanol}}{\begin{array}{c} \text{H} \\ | \\ \text{H}-\text{C}-\text{OH} \\ | \\ \text{H} \end{array}} + H_2O$$

Each class of organic compounds can undergo important oxidation–reduction reactions.

Although countless different carbon-containing molecules exist, there are relatively few different types of organic reactions, and only a few carbon atoms, at most, are involved in any common reaction. These changes can be usefully categorized by counting the number of atoms bonded to a reactive carbon. Carbon can be bonded to a maximum of four atoms.

substitution reaction

1. If in a reaction one atom in a molecule is exchanged by another atom or group of atoms, a **substitution reaction** has taken place. For example,

$$\underset{\text{methane, colorless gas}}{\begin{array}{c} \text{H} \\ | \\ \text{H}-\text{C}-\text{H} \\ | \\ \text{H} \end{array}} + I_2 \longrightarrow \underset{\text{iodomethane, colorless liquid}}{\begin{array}{c} \text{H} \\ | \\ \text{H}-\text{C}-\text{I} \\ | \\ \text{H} \end{array}} + \text{HI}$$

elimination reaction

2. An **elimination reaction** is a reaction in which a single reactant is split into two products, and one of the products is eliminated. For example,

$$\underset{\text{bromoethane, colorless liquid}}{\begin{array}{c} \text{H}\;\;\text{H} \\ |\;\;\;| \\ \text{H}-\text{C}-\text{C}-\text{Br} \\ |\;\;\;| \\ \text{H}\;\;\text{H} \end{array}} \longrightarrow \underset{\text{ethene, colorless gas}}{\begin{array}{c} \text{H}\;\;\text{H} \\ |\;\;\;| \\ \text{C}=\text{C} \\ |\;\;\;| \\ \text{H}\;\;\text{H} \end{array}} + \text{HBr}$$

Elimination reactions form multiple bonds.

The combustion of hydrocarbons is spectacular but can be a huge problem, as shown in this oil well fire in Wyoming.

3. Two reactants adding together to form a single product is called an **addition reaction**. An addition reaction can be thought of as the reverse of an elimination reaction.

addition reaction

$$\underset{\substack{\text{ethene,}\\\text{colorless gas}}}{\mathrm{H_2C{=}CH_2}} + \mathrm{HBr} \longrightarrow \underset{\substack{\text{bromoethane,}\\\text{colorless liquid}}}{\mathrm{H{-}CH_2{-}CH_2{-}Br}}$$

20.11 Reactions of Alkanes

A single type of alkane reaction has inspired people to explore equatorial jungles, endure the heat and sandstorms of the deserts of Africa and the Middle East, mush across the frozen Arctic, and drill holes—some more than 30,000 feet deep—on land and on the ocean floor! These strenuous and expensive activities have been undertaken because alkanes, as well as other hydrocarbons, undergo combustion with oxygen with the evolution of large amounts of heat energy. Methane, for example, reacts with oxygen:

$$CH_4(g) + 2\ O_2(g) \longrightarrow CO_2(g) + 2\ H_2O(g) + 802.5\ \text{kJ}\ (191.8\ \text{kcal})$$

When carbon dioxide is formed, the alkane has undergone *complete* oxidation. The resulting thermal energy can be converted to mechanical and electrical energy. Combustion reactions overshadow all other reactions of alkanes in economic importance. But combustion reactions are not usually of great interest to organic chemists because carbon dioxide and water are the only chemical products of complete combustion.

Aside from their combustibility, alkanes are relatively sluggish and limited in reactivity. But with proper activation, such as high temperature and/or catalysts, alkanes can be made to react in a variety of ways. Some important noncombustion reactions of alkanes include the following:

1. **Halogenation** (a *substitution* reaction). A halogen is substituted for a hydrogen in halogenation. When a specific halogen such as chlorine is used, the reaction is called chlorination; RH is an alkane (alkyl group + H atom) that reacts with halogens in this manner.

   Example:
   $$RH + X_2 \longrightarrow RX + HX \qquad (X = Cl \text{ or } Br)$$
   $$CH_3CH_3 + Cl_2 \longrightarrow \underset{\text{chloroethane}}{CH_3CH_2Cl} + HCl$$

   This reaction yields alkyl halides, RX, which are useful as intermediates for the manufacture of other substances.
2. **Dehydrogenation** (an *elimination* reaction). Hydrogen is lost from an organic compound during dehydrogenation.

   Example:
   $$C_nH_{2n+2} \xrightarrow{700\text{–}900°C} C_nH_{2n} + H_2$$
   $$CH_3CH_2CH_3 \xrightarrow{\Delta} \underset{\text{propene}}{CH_3CH{=}CH_2} + H_2$$

   This reaction yields alkenes, which, like alkyl halides, are useful chemical intermediates. Hydrogen is a valuable by-product.
3. **Cracking** (breaking up large molecules to form smaller ones).

   Example: $\underset{\text{alkane}}{C_{16}H_{34}} \xrightarrow{\Delta} \underset{\text{alkane}}{C_8H_{18}} + \underset{\text{alkene}}{C_8H_{16}}$ (one of several possibilities)
4. **Isomerization** (rearrangement of molecular structures).

   Example:
   $$CH_3{-}CH_2{-}CH_2{-}CH_2{-}CH_3 \longrightarrow CH_3{-}CH_2{-}\underset{\displaystyle CH_3}{\underset{|}{CH}}{-}CH_3$$

Halogenation is used extensively in the manufacture of petrochemicals (chemicals derived from petroleum and used for purposes other than fuels). The other three reactions—dehydrogenation, cracking, and isomerization—singly or in combination, are of great importance in the production of motor fuels and petrochemicals.

A well-known reaction of methane and chlorine is shown by the equation

$$CH_4 + Cl_2 \longrightarrow \underset{\substack{\text{chloromethane}\\ \text{(methyl chloride)}}}{CH_3Cl} + HCl$$

The reaction of methane and chlorine gives a mixture of mono-, di-, tri-, and tetra-substituted chloromethanes.

$$CH_4 \xrightarrow{Cl_2} CH_3Cl \xrightarrow{Cl_2} CH_2Cl_2 \xrightarrow{Cl_2} CHCl_3 \xrightarrow{Cl_2} CCl_4 + 4\ HCl$$

TABLE 20.4 Chlorination Products of Methane

Formula	IUPAC name	Common name
CH_3Cl	Chloromethane	Methyl chloride
CH_2Cl_2	Dichloromethane	Methylene chloride
$CHCl_3$	Trichloromethane	Chloroform
CCl_4	Tetrachloromethane	Carbon tetrachloride

However, if an excess of chlorine is used, the reaction can be controlled to give all tetrachloromethane (carbon tetrachloride). On the other hand, if a large ratio of methane to chlorine is used, the product will be predominantly chloromethane (methyl chloride). Table 20.4 lists the formulas and names for all the chloromethanes. The names for the other halogen-substituted methanes follow the same pattern as for the chloromethanes; for example, CH_3Br is bromomethane, or methyl bromide, and CHI_3 is triiodomethane, or iodoform.

monosubstitution

Chloromethane is a monosubstitution product of methane. The term **monosubstitution** refers to the fact that one hydrogen atom in an organic molecule is substituted by another atom or by a group of atoms. In hydrocarbons, for example, when we substitute one chlorine atom for a hydrogen atom, the new compound is a monosubstitution (monochlorosubstitution) product. In a like manner we can have di-, tri-, tetra-, and so on, substitution products.

This kind of chlorination (or bromination) is general with alkanes. There are nine different chlorination products of ethane. See if you can write the structural formulas for all of them.

When propane is chlorinated, two isomeric monosubstitution products are obtained because a hydrogen atom can be replaced on either the first or second carbon as shown here:

$$CH_3CH_2CH_3 + Cl_2 \xrightarrow[25°C]{\text{light}} \underset{\text{1-chloropropane}}{CH_3CH_2CH_2Cl} + \underset{\text{2-chloropropane}}{CH_3CHClCH_3} + HCl$$

alkyl halide

The letter X is commonly used to indicate a halogen atom. The formula RX indicates a halogen atom attached to an alkyl group and represents the class of compounds known as the **alkyl halides**. When R is CH_3, then CH_3X can be CH_3F, CH_3Cl, CH_3Br, or CH_3I.

Alkyl halides are named systematically in the same general way as alkanes. Halogen atoms are identified as *fluoro-*, *chloro-*, *bromo-*, or *iodo-* and are named as substituents like side-chain alkyl groups. Study these examples:

$$\underset{\text{2-chlorobutane}}{CH_3{-}CHCl{-}CH_2{-}CH_3} \qquad \underset{\text{2-bromo-1-chloropropane}}{CH_2Cl{-}CHBr{-}CH_3}$$

$$\underset{\text{2-chloro-3-methylpentane}}{CH_3{-}CH_2{-}\underset{\substack{|\\ CH_3}}{CH}{-}CHCl{-}CH_3}$$

Example 20.4 How many monochlorosubstitution products can be obtained from pentane?

Solution First write the formula for pentane:

$$\overset{5}{C}H_3\overset{4}{C}H_2\overset{3}{C}H_2\overset{2}{C}H_2\overset{1}{C}H_3$$

Now rewrite the formula five times substituting a Cl atom for an H atom on each C atom:

I	$CH_3CH_2CH_2CH_2\overset{1}{C}H_2Cl$	Cl on carbon 1
II	$CH_3CH_2CH_2\overset{2}{C}HClCH_3$	Cl on carbon 2
III	$CH_3CH_2\overset{3}{C}HClCH_2CH_3$	Cl on carbon 3
IV	$CH_3\overset{4}{C}HClCH_2CH_2CH_3$	Cl on carbon 4
V	$\overset{5}{C}H_2ClCH_2CH_2CH_2CH_3$	Cl on carbon 5

Compounds I and V are identical. By numbering compound V from left to right, we find that both compounds (I and V) are 1-chloropentane. Compounds II and IV are identical. By numbering compound IV from left to right, we find that both compounds (II and IV) are 2-chloropentane. Thus there are three monochlorosubstitution products of pentane: 1-chloropentane, 2-chloropentane, and 3-chloropentane.

Practice 20.6

How many dichlorosubstitution products can be obtained from hexane?

20.12 Sources of Alkanes

The two main sources of alkanes are natural gas and petroleum. Natural gas is formed by the anaerobic decay of plants and animals. The composition of natural gas varies in different locations. Its main component is methane (80–95%), the balance being varying amounts of other hydrocarbons, hydrogen, nitrogen, carbon monoxide, carbon dioxide, and in some locations, hydrogen sulfide. Economically significant amounts of methane are now obtained by the decomposition of sewage, garbage, and other organic waste products.

Petroleum, also called *crude oil*, is a viscous black liquid consisting of a mixture of hydrocarbons with smaller amounts of nitrogen and sulfur-containing organic compounds. Petroleum is formed by the decomposition of plants and animals over millions of years. The composition of petroleum varies widely from one locality to another. Crude oil is refined into such useful products as gasoline, kerosene, diesel fuel, jet fuel, lubricating oil, heating oil, paraffin wax, petroleum jelly (petrolatum), tars, and asphalt.

◀ **Petroleum is converted to the starting materials for many chemicals at a refinery.**

At the rate that natural gas and petroleum are being used, these sources of hydrocarbons are destined to be in short supply and virtually exhausted in the not-too-distant future. Alternative sources of fuels must be developed.

20.13 Gasoline: A Major Petroleum Product

A large fraction of all petroleum is burned as fuel (see Figure 20.9). Historically, fuels were commonly used for heating and cooking; today most fuels provide power for transportation. Because different engines require different fuels, there are many different fuels. However, befitting the importance of the automobile in transportation, gasoline is the fuel most produced by the oil companies.

Gasoline, aside from the additives put into it, consists primarily of hydrocarbons. Gasoline, as it is distilled from crude oil, causes "knocking" when burned in high-compression automobile engines. Knocking, which is caused by a too-rapid combustion or detonation of the air–gasoline mixture, is a severe problem in high-compression engines. The knock-resistance of gasolines, a quality that varies widely, is usually expressed in terms of *octane number*, or *octane rating*.

Because of its highly branched chain structure, isooctane (2,2,4-trimethylpentane), is a motor fuel that is resistant to knocking. Mixtures of isooctane and heptane, a straight-chain alkane that knocks badly, have been used as standards to establish octane ratings of gasolines. Isooctane is arbitrarily assigned an octane number of 100 and heptane an octane number of 0. To determine the octane rating, a gasoline is compared with mixtures of isooctane and heptane in a test engine. The octane number of the gasoline corresponds to the percentage of isooctane present in the isooctane and heptane mixture that matches the knocking characteristics of the

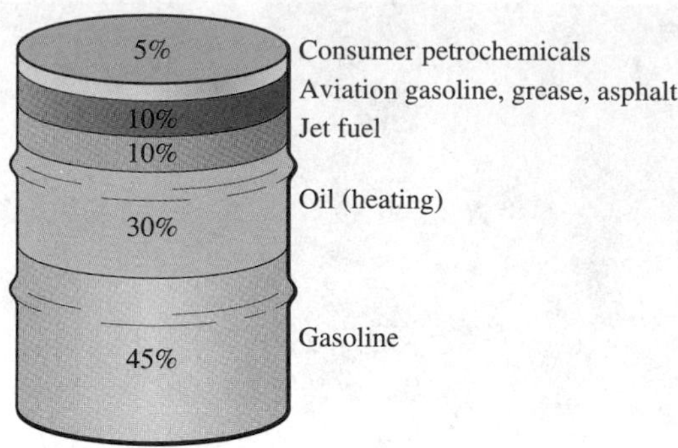

FIGURE 20.9 ▶ **Uses of petroleum.**

gasoline being tested. Thus a 90-octane gasoline has knocking characteristics matching those of a mixture of 90% isooctane and 10% heptane.

When first used to establish octane numbers, isooctane was the most knock-resistant substance available. However, because technological advances have resulted in engines with greater power and compression ratios, higher-quality fuels were subsequently necessary. Fuels containing more highly branched hydrocarbons, unsaturated hydrocarbons, or aromatic hydrocarbons burn more smoothly than isooctane and have a higher octane rating than 100.

An alternative method to boost octane rating and minimize engine knocking involves adding small amounts of an additive to the fuel. One such additive commonly used in gasoline was tetraethyllead, $(C_2H_5)_4Pb$. Adding only 3 mL per gallon of gasoline can increase the octane rating by 15 units. The presence of tetraethyllead prevents the premature explosions that constitute knocking. However, the use of tetraethyllead additives poses a serious environmental hazard. Lead becomes yet another air pollutant, in addition to the others (carbon monoxide, hydrocarbons, and nitrogen oxides) already produced by the automobile. Since lead is a toxic substance that accumulates in living organisms, legal restrictions on the use of tetraethyllead are constantly being tightened. In addition, most cars are now equipped with a catalytic converter, which reduces the emission of pollutants into the atmosphere. These catalytic converters are deactivated by lead.

Environmental problems are causing changes in the formulation of gasoline. Major oil companies have changed the formulation of gasoline in recent years to eliminate leaded gasoline and reduce emissions into the atmosphere. Current additives in unleaded gasoline include aromatic compounds such as toluene and xylene (Chapter 21), methyl *tert*-butyl ether (MTBE) or ethanol:

$$CH_3{-}O{-}\underset{\displaystyle CH_3}{\overset{\displaystyle CH_3}{\underset{|}{\overset{|}{C}}}}{-}CH_3 \qquad CH_3CH_2{-}OH$$

MTBE   ethanol

Chemical conversion of straight-chain alkanes into branched or cyclic compounds is also a method for "reformulating" gasoline. Fuels are also changed from season to season as well as from one region of the country to another. These seasonal and regional changes reflect variations in air pollution and environmental standards.

◀ **Unleaded gasolines are formulated to different octane ratings so they can be used in different engines.**

Although scientists have worked long and hard to improve gasoline to make a better fuel and to decrease air pollution, it is an unavoidable fact that even the cleanest burning gasoline increases atmospheric carbon dioxide, which can contribute to the "greenhouse effect." Furthermore, combustion uses hydrocarbons in a nonrecyclable way. Burning gasoline pollutes the air and consumes a valuable resource. As petroleum chemists look to the future, they recommend (1) development of nonpolluting energy sources, and (2) shifting the major use of petroleum from fuel to the manufacture of other organic compounds.

20.14 Cycloalkanes

Cyclic, or closed-chain, alkanes also exist. These substances, called **cycloalkanes**, *cycloparaffins*, or *naphthenes*, have the general formula C_nH_{2n}. Note that this series of compounds has two fewer hydrogen atoms than the open-chain alkanes. The bonds for the two missing hydrogen atoms are accounted for by an additional carbon–carbon bond in the cyclic ring of carbon atoms. Structures for the four smallest cycloalkanes are shown in Figure 20.10.

cycloalkane

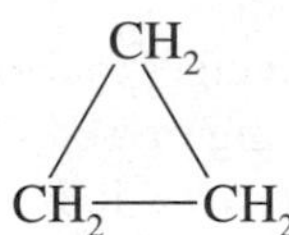

CH_2—CH_2 / CH_2—CH_2 (four-membered ring)

CH_2, CH_2, CH_2, CH_2—CH_2 (five-membered ring)

CH_2, CH_2, CH_2, CH_2, CH_2, CH_2 (six-membered ring)

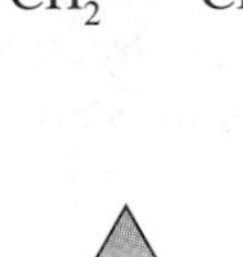

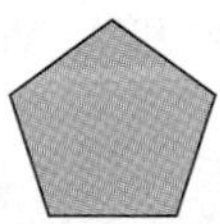
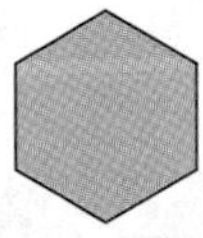

cyclopropane C_3H_6 | cyclobutane C_4H_8 | cyclopentane C_5H_{10} | cyclohexane C_6H_{12}

◀ **FIGURE 20.10** **Cycloalkanes. In the line representations, each corner of the diagram represents a CH_2 group.**

With the following exception noted for cyclopropane and cyclobutane, cycloalkanes are generally similar to open-chain alkanes in both physical properties and chemical reactivity. Cycloalkanes are saturated hydrocarbons; they contain only single bonds between carbon atoms.

The reactivity of cyclopropane, and to a lesser degree that of cyclobutane, is greater than that of other alkanes. This greater reactivity exists because the carbon–carbon bond angles in these substances deviate substantially from the normal tetrahedral angle. The carbon atoms form a triangle in cyclopropane, and in cyclobutane they approximate a square. Cyclopropane molecules therefore have carbon–carbon bond angles of 60°, and in cyclobutane the bond angles are about 90°. In the open-chain alkanes and in larger cycloalkanes, the carbon atoms are in a three-dimensional zigzag pattern in space and have normal (tetrahedral) bond angles of about 109.5°.

Bromine adds to cyclopropane readily and to cyclobutane to some extent. In this reaction the ring breaks and an open-chain dibromopropane is formed:

$$\underset{\text{cyclopropane}}{\overset{\quad CH_2}{CH_2{-}CH_2}} + Br_2 \longrightarrow \underset{\text{1,3-dibromopropane}}{BrCH_2CH_2CH_2Br}$$

Cyclopropane and cyclobutane react in this way because their carbon–carbon bonds are strained and therefore weakened. Cycloalkanes whose rings have more than four carbon atoms do not react in this way because their molecules take the shape of nonplanar puckered rings. These rings can be considered to be formed by simply joining the end carbon atoms of the corresponding normal alkanes. The resulting cyclic molecules are nearly strain-free, with carbon atoms arranged in space so that the bond angles are close to 109.5° (Figure 20.11).

Molecular models show that cyclohexane can assume two distinct nonplanar conformations. One form is shaped like a chair, while the other is shaped like a boat (see Figure 20.12). In the chair form the hydrogen atoms are separated as effectively as possible, so this is the more stable conformation. Six of the hydrogens in "chair" cyclohexane lie approximately in the same plane as the carbon ring. These are called **equatorial hydrogens**. The other six hydrogen atoms are approximately at right angles above or below the plane of the ring. These are called **axial hydrogens**. Substituent groups are usually found in equatorial positions where they are farthest from hydrogens and other groups. Five- and six-membered rings are a common occurrence in organic chemistry and biochemistry. These conformations and isomers are evident in carbohydrates as well as in nucleic acids.

equatorial hydrogen

axial hydrogen

Cyclopropane is a useful general anesthetic and, along with certain other cycloalkanes, is used as an intermediate in some chemical syntheses. The high reactivity of cyclopropane requires great care in its use as an anesthetic because it is an extreme fire and explosion hazard. The cyclopentane and cyclohexane ring structures are present in many naturally occurring molecules such as prostaglandins, steroids (e.g., cholesterol and sex hormones), and some vitamins.

Practice 20.7

Draw the structures for the two cycloalkanes that have the molecular formula, C_4H_8.

cyclopropane

hexane $CH_3CH_2CH_2CH_2CH_2CH_3$

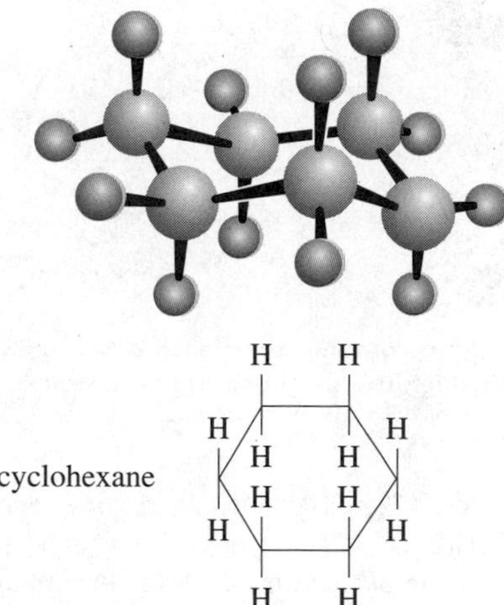

FIGURE 20.11
Ball-and-stick models illustrating cyclopropane, hexane, and cyclohexane. In cyclopropane all the carbon atoms are in one plane. The angle between the carbon atoms is 60°, not the usual 109.5°; therefore, the cyclopropane ring is strained. In cyclohexane the carbon–carbon bonds are not strained. This is because the molecule is puckered (as shown in the chair conformation) with carbon–carbon bond angles of 109.5°, as found in hexane.

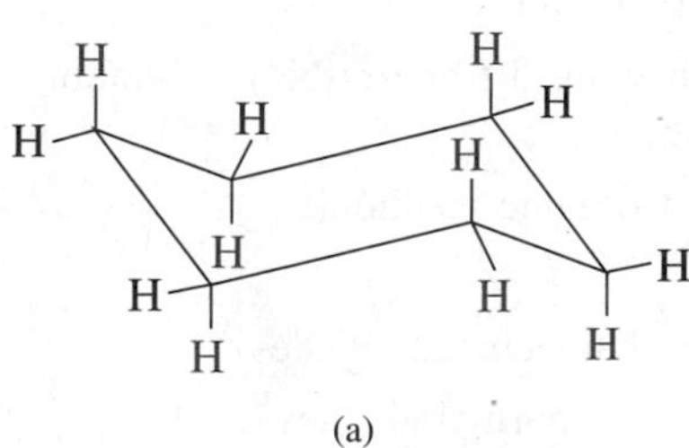

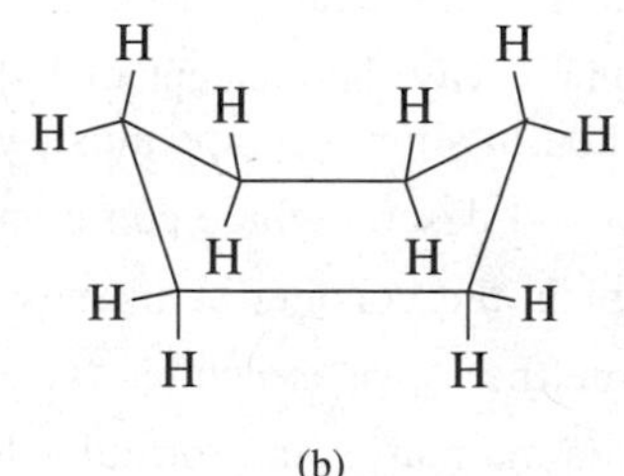

FIGURE 20.12
Conformations of cyclohexane: (a) chair conformation; (b) boat conformation. Axial hydrogens are shown in blue in the chair conformation.

Chemistry in Action

Molecules to Communicate, Refrigerate, and Save Lives

By far the most important use of alkanes in our world is as a source of energy for industrial and consumer use. Hydrocarbons do not generally function as an energy source in the physiological world of living organisms. The compounds that supply fuel necessary for most life contain oxygen as well as carbon and hydrogen.

Alkanes do find other significant uses in our lives. High-molar-mass alkanes are used to soften or curtail evaporation from the skin (see Section 13.15). Petroleum jelly and mineral oil are both mixtures of hydrocarbons used as skin protectants and as lubricants.

Certain insects use hydrocarbons (as well as other organic compounds) as chemical communication devices. These compounds, called pheromones, are secreted by an insect and recognized by other members of the species as a message. The meaning of the pheromone varies with its composition. It can be a sex attractant, an alarm, or an indication of the path to a source of food. When ants are disturbed, they release alarm pheromones that have been identified as undecane, $CH_3(CH_2)_9CH_3$, and tridecane, $CH_3(CH_2)_{11}CH_3$. Our growing understanding of these molecules is beginning to lead to their use in insect abatement. Commercial traps are baited with sex pheromone, and the insects are captured without the use of pesticides.

Ants communicate to each other through the secretion of pheromones.

Freons and other chlorofluorocarbons (CFCs) are useful because they are nontoxic, nonflammable, and noncorrosive. Many of these compounds have low boiling points and make excellent refrigerants. They have also been used as propellants in aerosol sprays and in the production of some fast-food containers. Unfortunately, the use of CFCs is a major factor in the destruction of the ozone layer (see Section 12.17). Use of these compounds is currently being discontinued by major industries in an effort to protect our atmosphere.

Some fluorinated hydrocarbons are used as blood substitutes. All the hydrogens in the hydrocarbon are replaced with fluorine, producing a compound known as a fluorocarbon. Dispersions of these compounds in water can absorb nearly three times the oxygen per unit as whole blood. Organisms receiving blood substitutes continue to produce whole blood. The substitute permits the tissues to absorb oxygen while it remains chemically inert and is excreted over a period of time.

Concepts in Review

1. Describe the tetrahedral nature of the carbon atom.
2. Explain why the concept of hybridization is used to describe the bonding of carbon in simple compounds such as methane.
3. List and describe three common classes of organic reactions.
4. Explain the bonding in alkanes.
5. Write the Lewis structures for alkanes and halogenated alkanes.
6. Write the names and formulas for the first ten normal alkanes.
7. Understand the concept of isomerization.
8. Write structural formulas and IUPAC names for the isomers of an alkane or a halogenated alkane.

9. Give the IUPAC name of a hydrocarbon or a halogenated hydrocarbon when given the structural formula, and vice versa.
10. Write equations for the halogenation of an alkane, giving all possible monohalo-substitution products.
11. Write structural formulas and names for simple cycloalkanes.
12. Draw the two major conformations for cyclohexane.
13. Discuss the advantages and disadvantages of gasoline as a fuel, including octane rating.

Key Terms

The terms listed here have been defined within this chapter. Section numbers are referenced in parenthesis for each term.

addition reaction (20.10)
aliphatic (20.5)
alkane (20.6)
alkyl group (20.9)
alkyl halide (20.11)
axial hydrogen (20.14)
cycloalkane (20.14)
elimination reaction (20.10)
equatorial hydrogen (20.14)
functional group (20.4)
homologous series (20.6)
hydrocarbon (20.5)
isomerism (20.8)
isomer (20.8)
IUPAC System (20.9)
monosubstitution (20.11)
organic chemistry (20.1)
sigma bond (20.7)
saturated compound (20.2)
substitution reaction (20.10)
unsaturated compound (20.2)
vital force theory (20.1)

Questions

Questions refer to tables, figures, and key words and concepts defined within the chapter. A particularly challenging question is indicated with an asterisk.

1. What are the main reasons for the large number of organic compounds?
2. Why is it believed that a carbon atom must form hybrid electron orbitals when it bonds to hydrogen atoms to form methane?
3. Write the names and formulas for the first ten normal alkanes.
4. How many sigma bonds are in a molecule of
   (a) ethane **(b)** butane **(c)** 2-methylpropane
5. What is a major advantage and a major disadvantage to the industrial use of the Freons?
6. Which of these statements are correct? Rewrite the incorrect statements to make them correct.
   (a) Alkane hydrocarbons are essentially nonpolar.
   (b) The C—H sigma bond in methane is made from an overlap of an *s* electron orbital and an sp^3 electron orbital.
   (c) The valence electrons of every carbon atom in an alkane are in sp^3-hybridized orbitals.
   (d) The four carbon–hydrogen bonds in methane are equivalent.
   (e) Hydrocarbons are composed of carbon and water.
   (f) In the alkane homologous series, the formula of each member differs from its preceding member by CH_3.
   (g) Carbon atoms can form single, double, and triple bonds with other carbon atoms.
   (h) The name for the alkane C_5H_{12} is propane.
   (i) There are eight carbon atoms in a molecule of 2,3,3-trimethylpentane.
   (j) The general formula for an alkyl halide is RX.
   (k) The IUPAC name for $CH_3CH_2CH_2CHClCH_3$ is 4-chloropentane.
   (l) Isopropyl chloride is also called 2-chloropropane.
   (m) The molecular formula for chlorocyclohexane is $C_6H_{11}Cl$.
   (n) Chlorocyclohexane and 1-chlorohexane are isomers.
   (o) The products of complete combustion of a hydrocarbon are carbon monoxide and water.

(p) When pentane is chlorinated, three monochlorosubstitution products can be obtained.

(q) Isobutane, 2-methylpropane, and 1,1-dimethylethane are all correct names for the same compound.

(r) Only one monochlorosubstituted product results from the chlorination of butane.

(s) Cycloalkanes have the general formula C_nH_{2n}.

Paired Exercises

These exercises are paired. Each odd-numbered exercise is followed by a similar even-numbered exercise. Answers to the even-numbered exercises are given in Appendix V.

7. Write the Lewis structures for
(a) CCl_4 **(b)** C_2Cl_6 **(c)** $CH_3CH_2CH_3$

9. Which of these formulas represent isomers?

(a) $CH_3CH_2CH_2CH_3$

(b) $CH_3CH_2CH_2CH_2CH_3$

(c) CH_3CHCH_3
$\quad\;\; CH_3CH_2$ (attached to the central CH)

(d) $CH_3CH_2CH_2CH_2CH_2CH_3$

(e) $CH_3 \quad\quad CH_3$
$\quad CH—CH$
$CH_3 \quad\quad CH_3$

(f) $CH_2—CH_2$
$CH_2 \quad CH_2$
$\quad CH_2$ (five-membered ring)

(g) $CH_3CHCH_2CH_2CH_3$
$\quad\;\; CH_3$ (attached to the second carbon)

(h) CH_2
$\quad CHCH_2CH_3$
CH_2 (three-membered ring)

(i) CH_3CH_2
$\quad CH_2CH_3$

8. Write the Lewis structures for
(a) methane **(b)** propane **(c)** pentane

10. Which of these formulas represent the same compound?

(a) $CH_3CHCH_2CHCH_3$
$\quad\quad CH_3 \quad CH_3$

(b) $\quad CH_3$
$CH_2CHCH_2CHCH_3$
$CH_3 \quad\quad CH_3$

(c) $\quad\quad\quad\quad CH_3$
$CH_3CHCH_2CH_2CHCH_3$
$\quad CH_3$

(d) $\quad CH_3$
$CH_3CHCHCH_2CH_2$
$\quad CH_3 \quad\quad CH_3$

(e) CH_3
CH_3CHCH_2
$CH_3CH_2CHCH_3$

(f) CH_3
CH_3CH
$\quad CH_2CHCH_3$
$\quad\quad CH_2CH_3$

11. How many methyl groups are in each formula in Exercise 9?

12. How many methyl groups are in each formula of Exercise 10?

13. Write the condensed structural formulas for the nine isomers of heptane.

14. Write the condensed structural formulas for the five isomers of hexane.

15. Draw structural formulas for all the isomers of
(a) CH_3Br **(b)** C_2H_5Cl
(c) C_4H_9I **(d)** C_3H_6BrCl

16. Draw structural formulas for all the isomers of
(a) CH_2Cl_2 **(b)** C_3H_7Br
(c) $C_3H_6Cl_2$ **(d)** $C_4H_8Cl_2$

17. Give IUPAC names for the following:
(a) $CH_3CH_2CH_2Cl$
(b) $CH_3CHClCH_3$
(c) $(CH_3)CCl$
(d) $CH_3CH_2CHCH_3$ with CH_3 on the third carbon
(e) $CH_3CH_2CH_2CHCH_3$ with CH_3CHCH_3 on the fourth carbon

18. Give the IUPAC name for each of the following compounds:
(a) CH_3CH_2Cl
(b) $(CH_3)_2CHCH_2Cl$
(c) $CH_3CHClCH_2CH_3$
(d) a three-membered ring of CH_2, CH_2 and $CHCH_3$
(e) $(CH_3)_2CHCH_2CH(CH_3)_2$

19. Draw structural formulas for the following compounds:
(a) 2,4-dimethylpentane
(b) 2,2-dimethylpentane
(c) 3-isopropyloctane
(d) 5,6-diethyl-2,7-dimethyl-5-propylnonane

20. Draw structural formulas for the following compounds:
(a) 4-ethyl-2-methylhexane
(b) 4-*tert*-butylheptane
(c) 4-ethyl-7-isopropyl-2,4,8-trimethyldecane
(d) 3-ethyl-2,2-dimethyloctane

21. The following names are incorrect. Explain why the name is wrong and give the correct name.
(a) 3-methylbutane
(b) 2-ethylbutane
(c) 2-dimethylpentane
(d) 1,4-dimethylcyclopentane

22. The following names are incorrect. Explain why the name is wrong and give the correct name.
(a) 3-methyl-5-ethyloctane
(b) 3,5,5-triethylhexane
(c) 4,4-dimethyl-3-ethylheptane
(d) 1,6-dimethylcyclohexane

23. Draw the structures for the ten dichlorosubstituted isomers, $C_5H_{10}Cl_2$, of 2-methylbutane.

24. The structure for hexane is

$$CH_3CH_2CH_2CH_2CH_2CH_3$$

Draw the structural formulas for all the monochlorohexanes, $C_6H_{13}Cl$, that have the same linear carbon structure as hexane.

25. Complete the equations for (a) the monochlorination, and (b) the complete combustion of butane.

(a) $CH_3CH_2CH_2CH_3 + Cl_2 \xrightarrow{hv}$

(b) $CH_3CH_2CH_2CH_3 + O_2 \xrightarrow{\Delta}$

26. Complete the equations for (a) the monobromination, and (b) the complete combustion of propane.

(a) $CH_3CH_2CH_3 + Br_2 \xrightarrow{hv}$

(b) $CH_3CH_2CH_3 + O_2 \xrightarrow{\Delta}$

Additional Exercises

These exercises are not paired or labeled by topic and provide additional practice on concepts covered in this chapter.

27. A hydrocarbon sample of formula C_4H_{10} is brominated, and four different monobromo compounds of formula C_4H_9Br are isolated. Is the sample a pure compound or a mixture of compounds? Explain your answer.

28. The name of the compound of formula $C_{11}H_{24}$ is undecane. What is the formula for dodecane, the next higher homologue in the alkane series?

29. The newer, more environmentally safe substitutes for chlorofluorocarbons (CFCs) are termed hydrochlorofluorocarbons (HCFCs). Show the equation for monochlorination of 1,2-difluoroethane to produce an HCFC.

*30. High-efficiency cars use gasoline at about the rate of 60 miles per gallon. One gallon of gasoline contains about 19 mol of octane. How many moles of carbon dioxide will be exhausted to travel 60 miles (assuming complete combustion of the octane)? How many liters of carbon dioxide will be exhausted (at 20.°C and 1.0 atm pressure)?

31. Write the structures and names for all the cycloalkanes that have the molecular formula C_5H_{10}.

32. The following reactions are important in biochemistry. Classify them as either substitution, elimination, or addition reactions.

(a) $HO-\overset{O}{\overset{\|}{C}}-\underset{PO_4}{\underset{|}{C}}HCH_2OH \longrightarrow HO-\overset{O}{\overset{\|}{C}}-\underset{PO_4}{\underset{|}{C}}=CH_2 + H_2O$

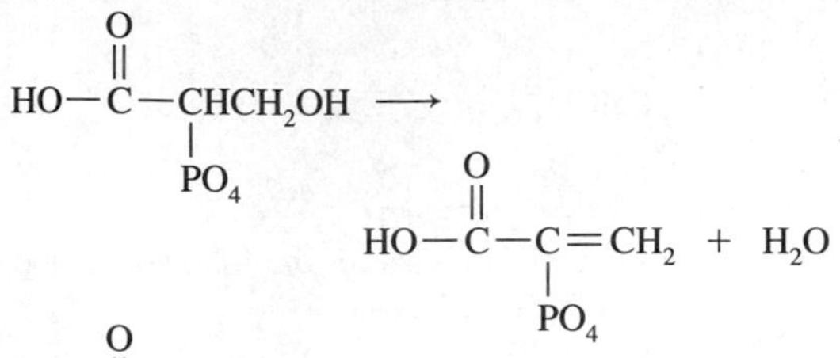

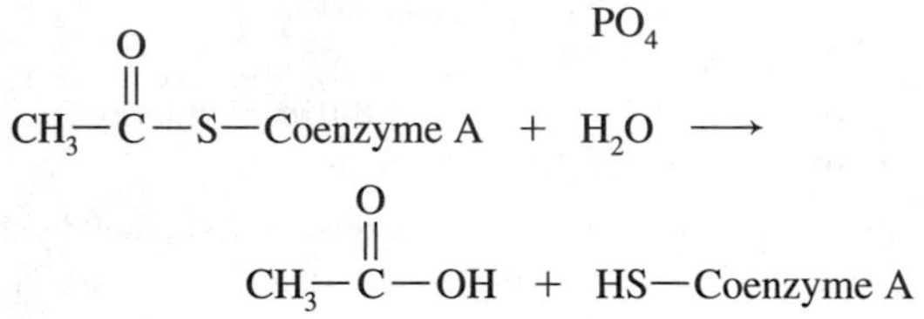

(b) $CH_3-\overset{O}{\overset{\|}{C}}-S-\text{Coenzyme A} + H_2O \longrightarrow CH_3-\overset{O}{\overset{\|}{C}}-OH + HS-\text{Coenzyme A}$

(c) $HO-\overset{O}{\overset{\|}{C}}-CH-\underset{\underset{O}{\underset{\|}{C-OH}}}{\underset{|}{C}}CH_2-\overset{O}{\overset{\|}{C}}-OH + H_2O \longrightarrow HO-\overset{O}{\overset{\|}{C}}-\underset{OH}{\underset{|}{C}H}-\underset{\underset{O}{\underset{\|}{C-OH}}}{\underset{|}{C}}HCH_2-\overset{O}{\overset{\|}{C}}-OH$

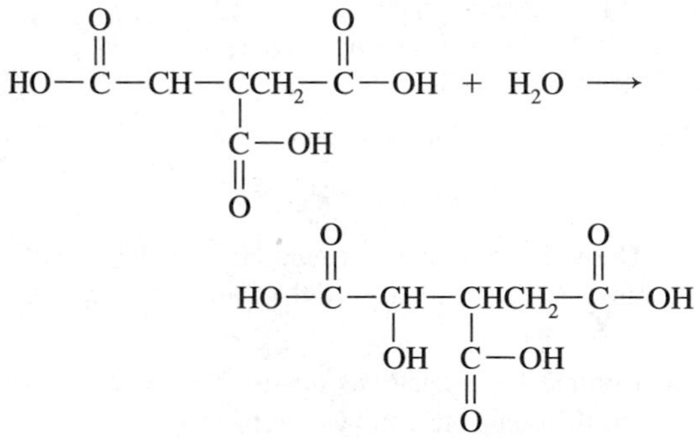

33. Assume that you have two test tubes, one containing hexane and the other containing 3-methylheptane. Can you tell which one is in which tube by testing their solubility in water? Why or why not?

34. Gasoline is a mixture of hydrocarbons. The components must be blended so that the gasoline has the correct volatility, and gasoline must have enough highly volatile compounds so that sufficient vapors are available to burn when the spark plug fires. In general, the lower the boiling point of a substance, the higher the volatility. Which of these compounds would you expect to be the most volatile? Butane, propane, or hexane?

35. In each of the following, state whether the compounds given are (i) isomers of one another, (ii) the same compound, or (iii) neither isomers nor the same compound.

(a) $CH_3CH_2CHBrCH_2Br$ and $CH_2BrCH_2CH_2CH_2Br$

(b) 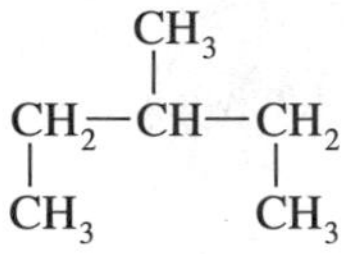 $\underset{CH_3}{\underset{|}{CH_2}}-\overset{CH_3}{\overset{|}{C}H}-\underset{CH_3}{\underset{|}{CH_2}}$ and 2-methylhexane

(c) 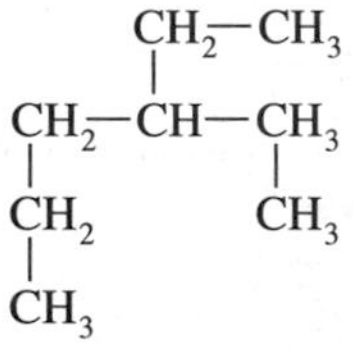 $\underset{\underset{CH_3}{\underset{|}{CH_2}}}{\underset{|}{CH_2}}-\overset{CH_2-CH_3}{\overset{|}{C}H}-\underset{CH_3}{\underset{|}{CH_3}}$ and $CH_3(CH_2)_5CH_3$

(d) Pentane and cylcopentane

36. Freon-12 is nontoxic, nonflammable, and noncorrosive. It has a boiling point of $-30°C$ and is commonly used as a refrigerant. Its formula is CCl_2F_2.

(a) What types of hybrid orbitals would you expect to be present in this compound?

(b) Would you expect Freon-12 to exhibit structural isomerism? Why or why not?

(c) Give an appropriate IUPAC name for Freon-12.

(d) Based on Table 20.1, to which class of compounds does Freon belong?

37. Pheromones are chemical communicators. When an ant is disturbed, it can release undecane, $C_{11}H_{24}$, and tridecane, $C_{13}H_{28}$, to send an alarm signal to other ants.

(a) Write the structural formulas for undecane and tridecane.

(b) Are either of these compounds alkanes? How can you tell?

Answers to Practice Exercises

20.1 $C_{12}H_{26}$, $C_{14}H_{30}$, $C_{20}H_{42}$

20.2

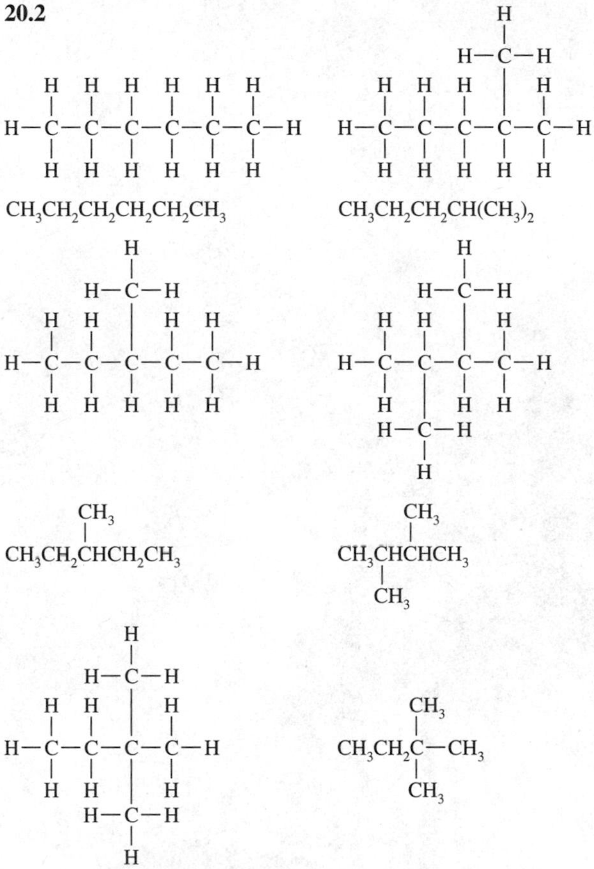

```
    H  H  H  H  H  H
    |  |  |  |  |  |
 H—C—C—C—C—C—C—H
    |  |  |  |  |  |
    H  H  H  H  H  H
```

$CH_3CH_2CH_2CH_2CH_2CH_3$

```
                   H
                   |
                H—C—H
    H  H  H        |  H
    |  |  |        |  |
 H—C—C—C—C—C—H
    |  |  |  |  |
    H  H  H  H  H
```

$CH_3CH_2CH_2CH(CH_3)_2$

```
          H
          |
       H—C—H
    H  H  |  H  H
    |  |  |  |  |
 H—C—C—C—C—C—H
    |  |  |  |  |
    H  H  H  H  H
```

```
          CH3
          |
 CH3CH2CHCH2CH3
```

```
          H
          |
       H—C—H
    H  H  |  H
    |  |  |  |
 H—C—C—C—C—H
    |  |  |  |
    H  |  H  H
    H—C—H
       |
       H
```

```
     CH3
     |
 CH3CHCHCH3
     |
     CH3
```

```
          H
          |
       H—C—H
    H  H  |  H
    |  |  |  |
 H—C—C—C—C—H
    |  |  |  |
    H  H  |  H
       H—C—H
          |
          H
```

```
           CH3
           |
 CH3CH2C—CH3
           |
           CH3
```

20.3

```
  16          15         13      12  11
  CH3         CH3        CH3     CH2CH3
  |           |          |       |
CH3CHCH2CH2CHCH2CCH2CHCH3
 1   2   3   4   5   6  |7  8  9   10
                        CH3
                        14
```

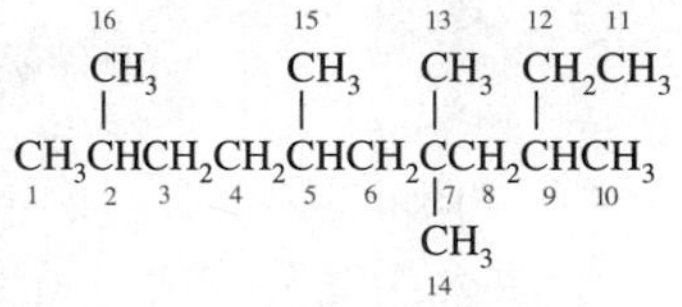

Primary carbons: 1 10 11 13 14 15 16

Secondary carbons: 3 4 6 8 12

Tertiary carbons: 2 5 9

20.4

(a)
```
 CH3CHCH2CH2CH2CH3
     |
     CH3
```

(c)
```
         Cl
         |
 CH3CHCHCH2CH3
       |
       CH2CH3
```

(b)
```
          CH3
          |
 CH3CH2CHCHCH2CH2CH3
            |
            CH3
```

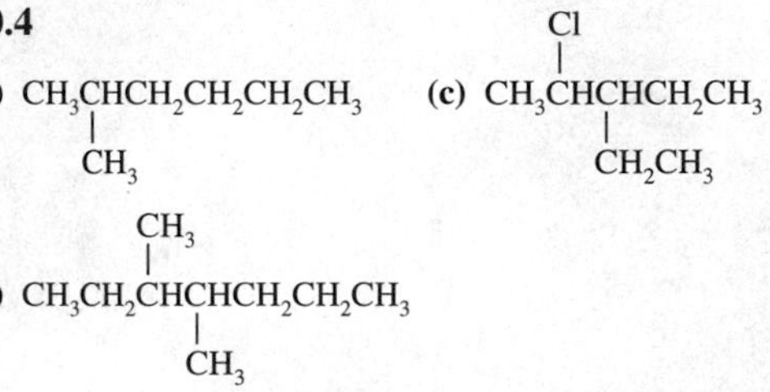

20.5 (a) 2-methylbutane (b) 3,5-dimethylheptane

20.6 There are twelve dichloro products of hexane

20.7 [cyclobutane] C_4H_8 [methylcyclopropane, —CH_3] C_4H_8

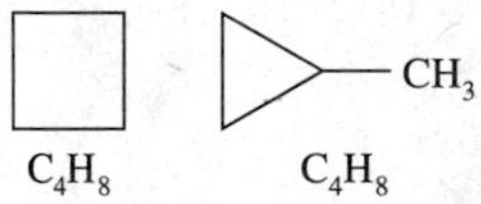

Unsaturated Hydrocarbons

21

Chapter 21

Think of the many images a particular fragrance or smell can evoke. A favorite perfume may provide memories of a romantic evening, while the aroma of a light-bodied red wine may remind us of a favorite restaurant. The compounds responsible for such evocative odors are the essential oils in plants. Many of these molecules are classified as unsaturated hydrocarbons.

Unsaturated hydrocarbons enhance our lives in many ways. There are the polymers, manufactured from unsaturated molecules, that would be difficult to do without. There are large numbers of these: you are probably familiar with polyethylene plastic bags and bottles and polystyrene styrofoam cups, and plastic wrap. The molecules associated with the essential oils in plants often contain multiple bonds beween carbon atoms. These oils are widely used in cosmetics, medicines, flavorings, and perfumes.

Hydrocarbons also often form into rings. These ring molecules are the basis for many consumer products such as detergents, insecticides, and dyes. Such molecules are known as aromatic carbon compounds and are also found in living organisms.

21.1 Bonding in Unsaturated Hydrocarbons

The unsaturated hydrocarbons consist of three families of homologous compounds that contain multiple bonds between carbon atoms. In each family every compound contains fewer hydrogens than the alkane with the corresponding number of carbons. Compounds in the first family, known as the **alkenes**, contain carbon–carbon double bonds. Those in the second family, known as the **alkynes**, contain carbon–carbon triple bonds, and those in the third family, known as **aromatic compounds**, contain benzene rings.

alkene
alkyne
aromatic compound

The double bonds in alkenes are different from the sp^3 hybrid bonds found in the alkanes. The hybridization of the carbons that form the double bond in alkenes may be visualized in the following way. One of the $2s$ electrons of carbon is promoted to a $2p$ orbital to form the four half-filled orbitals, $2s^1 2p_x^1 2p_y^1 2p_z^1$. Three of these orbitals $(2s^1 2p_x^1 2p_y^1)$ hybridize, thereby forming three equivalent orbitals designated as sp^2. Thus the four orbitals available for bonding are three sp^2orbitals and one p orbital. This process is illustrated in Figure 21.1.

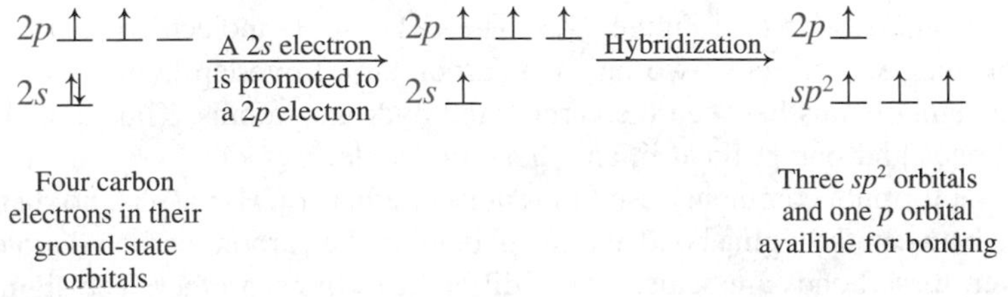

$2p$ ↿ ↿ —	A $2s$ electron is promoted to a $2p$ electron →	$2p$ ↿ ↿ ↿	Hybridization →	$2p$ ↿
$2s$ ⇅		$2s$ ↿		sp^2 ↿ ↿ ↿
Four carbon electrons in their ground-state orbitals				Three sp^2 orbitals and one p orbital availible for bonding

FIGURE 21.1 Schematic hybridization of $2s^2 2p_x^1 2p_y^1$ *orbitals of carbon to form three* sp^2 electron orbitals and one p electron orbital.

The three sp^2 hybrid orbitals form angles of 120° with each other and lie in a single plane. The remaining $2p$ orbital is oriented perpendicular to this plane, with one lobe above and one lobe below the plane (see Figure 21.2). In the formation

Chapter Opening Photo: Perfumes and colognes interact with chemicals in our skin to produce unique fragrances.

FIGURE 21.2 ▶
(a) A single sp^2 electron orbital and (b) a side view of three sp^2 orbitals all lying in the same plane with a p orbital perpendicular to the three sp^2 orbitals.

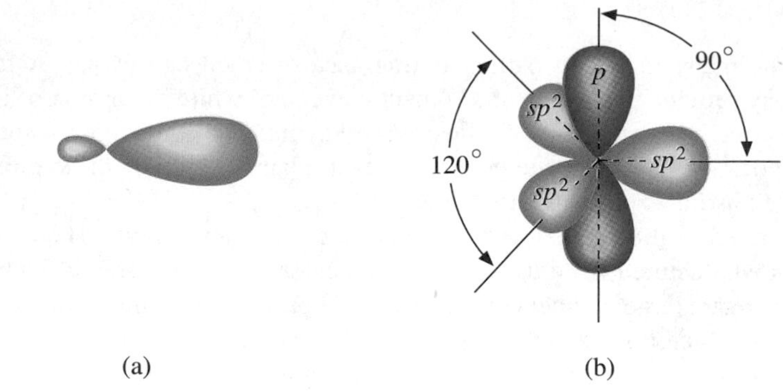

FIGURE 21.3 ▶
Pi (π) and sigma (σ) bonding in ethene.

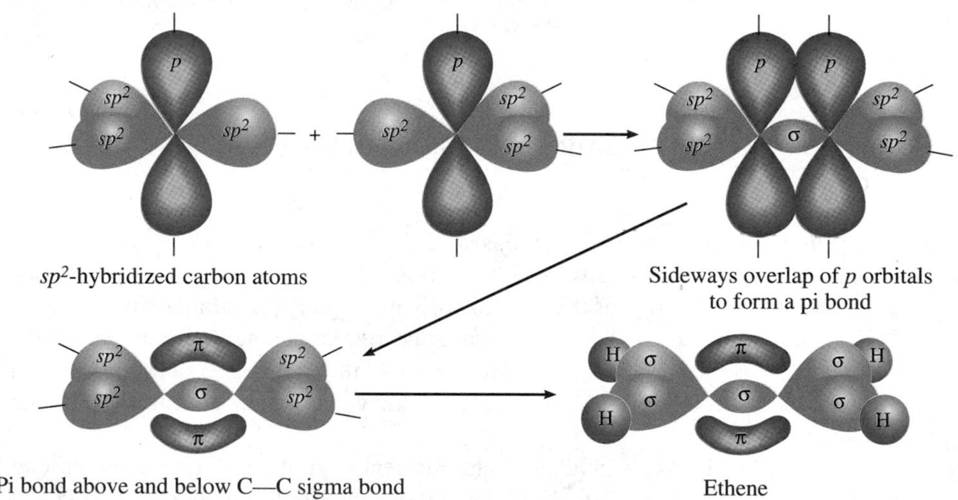

of a double bond, an sp^2 orbital of one carbon atom overlaps an identical sp^2 orbital of another carbon to form a sigma bond. At the same time the two perpendicular p orbitals (one on each carbon atom) overlap to form a **pi (π) bond** between the two carbon atoms. This pi bond consists of two electron clouds, one above and one below the sigma bond (see Figure 21.3) The $CH_2{=}CH_2$ molecule is completed as the remaining sp^2 orbitals (two on each carbon atom) overlap hydrogen s orbitals to form sigma bonds between the carbon and hydrogen atoms. Thus there are five sigma bonds and one pi bond in an ethene molecule.

pi bond

In the formula commonly used to represent ethene, $CH_2{=}CH_2$, no distinction is made between the sigma bond and the pi bond in the carbon–carbon double bond. However, these bonds are actually very different from each other. The sigma bond is formed by the overlap of sp^2 orbitals; the pi bond is formed by the overlap of p orbitals. The sigma bond electron cloud is distributed about a line joining the carbon nuclei, but the pi bond electron cloud is distributed above and below the sigma bond region (see Figure 21.3). The carbon–carbon pi bond is much weaker and, as a consequence, much more reactive than the carbon–carbon sigma bond.

The formation of a triple bond between carbon atoms, as in acetylene, H≡CH, may be visualized as follows:

1. A carbon atom 2*s* electron is promoted to a 2*p* orbital ($2s^12p_x^12p_y^12p_z^1$).
2. The 2*s* orbital hybridizes with one of the 2*p* orbitals to form two equivalent orbitals known as *sp* orbitals. These two hybrid orbitals lie on a straight line that passes through the center of the carbon atom. The remaining two unhybridized 2*p* orbitals are oriented at right angles to these *sp* orbitals and to each other.
3. In forming carbon–carbon bonds, one carbon *sp* orbital overlaps an identical *sp* orbital on another carbon atom to establish a sigma bond between the two carbon atoms.
4. The remaining *sp* orbitals (one on each carbon atom) overlap *s* orbitals on hydrogens to form sigma bonds and establish the H—C—C—H bond sequence. Because all the atoms forming this sequence lie in a straight line, the acetylene molecule is linear.
5. The two 2*p* orbitals on each carbon overlap simultaneously to form two pi bonds. These two pi bond orbitals occupy sufficient space to overlap each other and form a continuous tubelike electron cloud surrounding the sigma bond between the carbon atoms (Figure 21.4). These pi bond electrons (as in ethene) are not as tightly held by the carbon nuclei as the sigma bond electrons. Acetylene, consequently, is a very reactive substance.

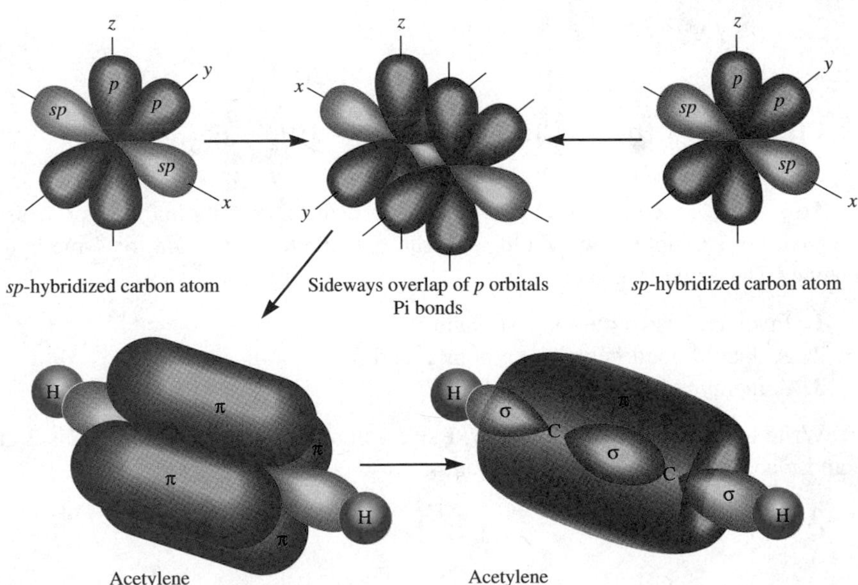

FIGURE 21.4
Pi (π) and sigma (σ) bonding in acetylene.

21.2 Nomenclature of Alkenes

The names of alkenes are derived from the names of corresponding alkanes.

IUPAC Nomenclature for Alkenes

1. Select the longest carbon–carbon chain that contains the double bond.
2. Name this parent compound as you would an alkane but change the *-ane* ending to *-ene*; for example, propane is changed to propene.

$$\underset{\text{propane}}{CH_3CH_2CH_3} \qquad \underset{\text{propene}}{CH_3CH{=}CH_2}$$

3. Number the carbon chain of the parent compound starting with the end nearer to the double bond. Use the smaller of the two numbers on the double-bonded carbon atoms to indicate the position of the double bond. Place this number in front of the alkene name; for example, 2-butene means that the carbon–carbon double bond is between carbons 2 and 3.
4. Side chains and other groups are treated as in naming alkanes, by numbering and assigning them to the carbon atom to which they are bonded.

Study the following examples of named alkenes:

$$\underset{\text{1-butene}}{\overset{4}{C}H_3\overset{3}{C}H_2\overset{2}{C}H{=}\overset{1}{C}H_2} \qquad \underset{\text{2-butene}}{\overset{1}{C}H_3\overset{2}{C}H{=}\overset{3}{C}H\overset{4}{C}H_3}$$

$$\underset{\text{3-methyl-1-butene}}{\overset{4}{C}H_3\overset{\overset{CH_3}{|}\,3}{C}H\overset{2}{C}H{=}\overset{1}{C}H_2} \qquad \underset{\text{3-propyl-1-hexene}}{\overset{6}{C}H_3\overset{5}{C}H_2\overset{4}{C}H_2\overset{\overset{CH_2CH_2CH_3}{|}\,3}{C}H\overset{2}{C}H{=}\overset{1}{C}H_2}$$

To write a structural formula from a systematic name, the naming process is reversed. For example, how would we write the structural formula for 4-methyl-2-pentene? The name indicates

1. Five carbons in the longest chain
2. A double bond between carbons 2 and 3
3. A methyl group on carbon 4

Write five carbon atoms in a row. Place a double bond between carbons 2 and 3, and place a methyl group on carbon 4:

$$\overset{1}{C}-\overset{2}{C}{=}\overset{3}{C}-\underset{\underset{CH_3}{|}}{\overset{4}{C}}-\overset{5}{C}$$

carbon skeleton

Now add hydrogen atoms to give each carbon atoms four bonds. Carbons 1 and 5 each need three H atoms; carbons 2, 3, and 4 each need one H atom. The complete formula is

$$\begin{array}{l} CH_3CH{=}CH\underset{\displaystyle CH_3}{\underset{|}{C}}HCH_3 \end{array}$$

4-methyl-2-pentene

Example 21.1

Write structural formulas for (a) 2-pentene and (b) 7-methyl-2-octene.

Solution

(a) The stem *pent-* indicates a five-carbon chain; the suffix *-ene* indicates a carbon–carbon double bond; the number 2 locates the double bond between carbons 2 and 3. Write five carbon atoms in a row and place a double bond between carbons 2 and 3:

$$\overset{1}{C}-\overset{2}{C}=\overset{3}{C}-\overset{4}{C}-\overset{5}{C}$$

Add hydrogen atoms to give each carbon atom four bonds. Carbons 1 and 5 each need three H atoms; carbons 2 and 3 each need one H atom; carbon 4 needs two H atoms. The complete formula is

$$CH_3CH{=}CHCH_2CH_3$$

2-pentene

(b) Octene, like octane, indicates an eight-carbon chain. The chain contains a double bond between carbons 2 and 3 and a methyl group on carbon 7. Write eight carbon atoms in a row, place a double bond between carbons 2 and 3, and place a methyl group on carbon 7:

$$\overset{1}{C}-\overset{2}{C}=\overset{3}{C}-\overset{4}{C}-\overset{5}{C}-\overset{6}{C}-\underset{\displaystyle CH_3}{\underset{|}{\overset{7}{C}}}-\overset{8}{C}$$

Now add hydrogen atoms to give each carbon atom four bonds. The complete formula is

$$CH_3CH{=}CHCH_2CH_2CH_2\underset{\displaystyle CH_3}{\underset{|}{C}}HCH_3$$

7-methyl-2-octene

Example 21.2

Name this compound.

$$CH_3CH_2\underset{\displaystyle CH_2}{\underset{||}{C}}CH_2CH_2CH_3$$

Solution

The longest carbon chain contains six carbons. However, since the compound is an alkene, we must include the double bond in the chain. The longest carbon chain containing the double bond has five carbons. Therefore, the compound is named as a pentene.

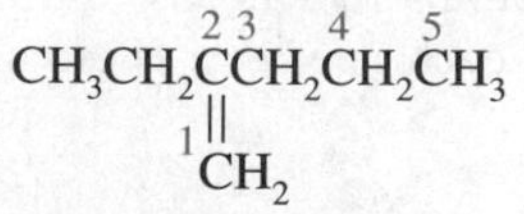

$$\overset{}{CH_3CH_2}\overset{2}{C}\overset{3}{C}H_2\overset{4}{C}H_2\overset{5}{C}H_3$$
$$\overset{1}{C}H_2 \text{ (double-bonded to C2)}$$

Attached to carbon 2 is an ethyl group. the name is 2-ethyl-1-pentene.

Practice 21.1

Write structural formulas for (a) 3-hexene, (b) 4-ethyl-2-heptene, and (c) 3,4-dimethyl-2-pentene.

Practice 21.2

Name these compounds.

(a) 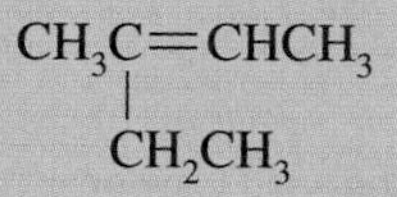 $CH_3C{=}CHCH_3$ with CH_2CH_3 on the second carbon   (b) 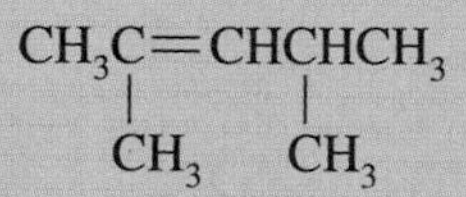 $CH_3C{=}CHCHCH_3$ with CH_3 on the second and fourth carbons

The common name for a compound is often more widely used than the IUPAC name. For example, "ethylene" is the common name for ethene. Ethylene is used to make many important products such as the plastic polyethylene and the major component of automobile antifreeze, ethylene glycol.

21.3 Geometric Isomerism in Alkenes

There are only two dichloroethanes: 1,1-dichloroethane, $CHCl_2CH_3$, and 1,2-dichloroethane, CH_2ClCH_2Cl. But surprisingly there are three dichloroethenes—namely, 1,1-dichloroethene, $CCl_2{=}CH_2$, and *two* isomers of 1,2-dichloroethene, $CHCl{=}CHCl$. There is only one 1,2-dichloroethane because carbon atoms can rotate freely about a single bond. Thus the structural formulas I and II that follow represent the same compound. The chlorine atoms are simply shown in different relative positions in the two formulas due to rotation of the CH_2Cl group about the carbon–carbon single bond.

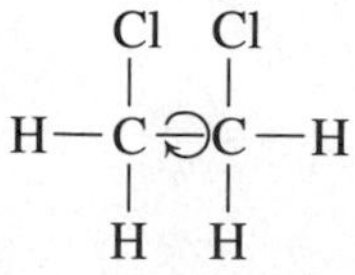

```
 Cl  Cl            Cl  H
 |   |             |   |
H—C—C—H         H—C—C—H
 |   |             |   |
 H   H             H   Cl

   I                 II
```

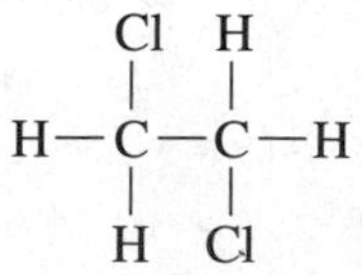

1,2-dichloroethane

Compounds containing a carbon–carbon double bond have restricted rotation about that double bond. Rotation cannot occur unless the pi bond breaks—a difficult process. Restricted rotation in a molecule gives rise to a type of isomerism known as *geometric isomerism.* Isomers that differ from each other only in the geometry of the molecules and not in the order of their atoms are known as **geometric isomers**. They are also called cis–trans isomers. Two isomers of 1,2-dichloroethene exist because of geometric isomerism.

geometric isomer

For a further explanation, let's look at the geometry of an ethene molecule. This molecule is planar, or flat, with all six atoms lying in a single plane as in a rectangle:

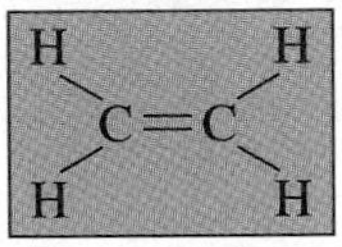

```
H       H
 \     /
  C=C
 /     \
H       H
```

Because the hydrogen atoms are identical, only one structural arrangement is possible for ethene. But if one hydrogen atom on each carbon atom is replaced by chlorine, for example, two different geometric isomers are possible:

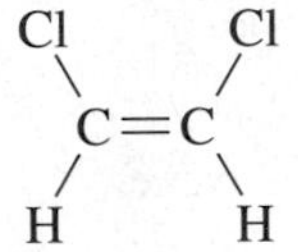

```
Cl      Cl              H       Cl
  \    /                  \    /
   C=C          and        C=C
  /    \                  /    \
H       H              Cl       H
```

cis-1,2-dichloroethene (bp = 60.1°C) and *trans*-1,2-dichloroethene (bp = 48.4°C)

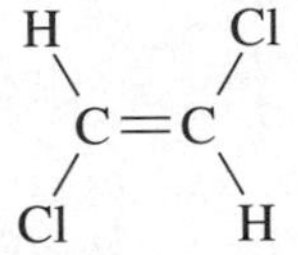

Both of these isomers are known and have been isolated. The fact that they have different boiling points as well as other different physical properties is proof that they are not the same compound. Note that, in naming these geometric isomers, the prefix *cis-* is used to designate the isomer having the substituent groups (chlorine atoms) on the same side of the double bond, and the prefix *trans-* is used to designate the isomer having the substituent groups on opposite sides of the double bond.

Molecules of *cis-* and *trans-*1,2-dichloroethene are not superimposable. That is, we cannot pick up one molecule and place it over the other in such a way that all the atoms in each molecule occupy the same relative positions in space. Nonsuperimposability is a general test for isomerism that all kinds of isomers must meet.

An alkene shows cis–trans isomerism when each of the carbon atoms of the double bond has two different kinds of groups attached to it.

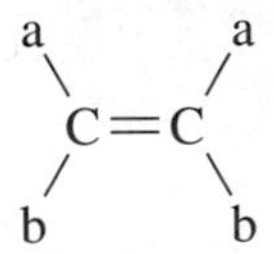

```
a       a              a       b
 \     /                \     /
  C=C                    C=C
 /     \                /     \
b       b              b       a
```

cis isomer and trans isomer

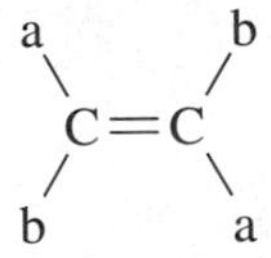

An alkene does not show cis–trans isomerism if one of the carbon atoms of the double bond has two identical groups attached to it. Thus there are no geometric isomers of ethene or propene.

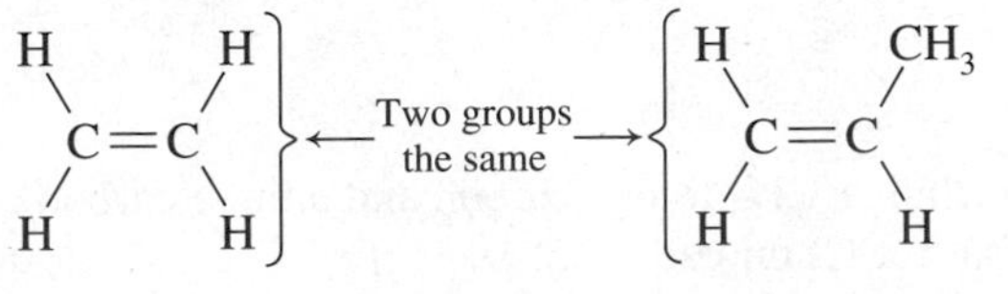

```
H      H ]                      [ H      CH3
 \    /  |                      |   \    /
  C=C    } <-- Two groups   --> {    C=C
 /    \  |     the same         |   /    \
H      H ]                      [ H       H
```

Four structural isomers of butene, C_4H_8, are known. Two of these, 1-butene and 2-methylpropene, do not show geometric isomerism.

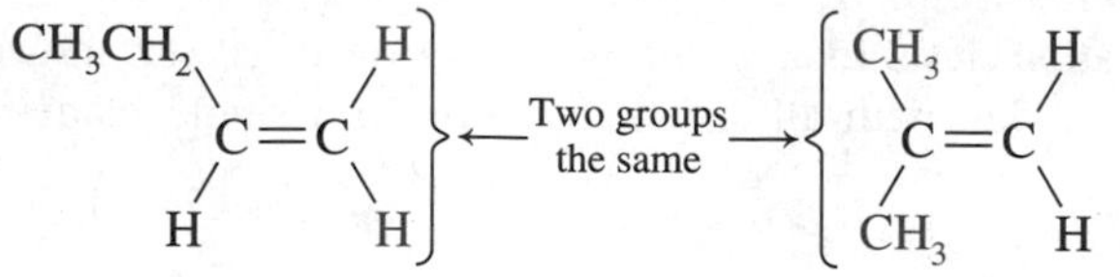

```
CH3CH2       H  }                        { CH3     H
      \     /   }                        {    \   /
       C=C      } <-- Two groups   -->   {     C=C
      /     \   }     the same           {    /   \
     H       H  }                        { CH3     H

     1-butene                              2-methylpropene
```

The other two butenes are the cis–trans isomers shown here.

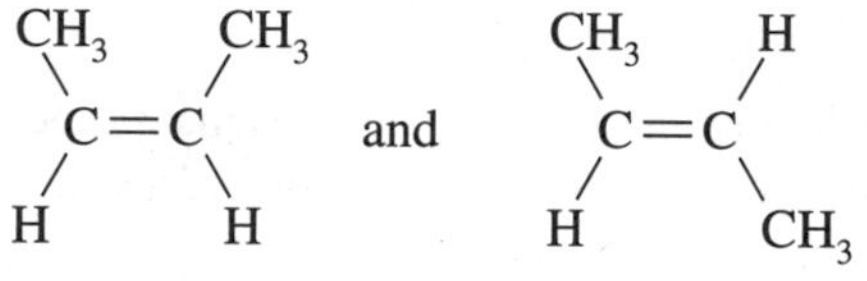

```
CH3     CH3                 CH3      H
   \   /                       \    /
    C=C          and            C=C
   /   \                       /    \
  H     H                     H      CH3

 cis-2-butene                trans-2-butene
```

Example 21.3 Draw structures for (a) *trans*-3-heptene and (b) *cis*-5-chloro-2-hexene.

Solution

(a) The compound contains seven carbon atoms with a double bond between carbons 3 and 4. First draw a C=C double bond in a planar arrangement:

```
 \   /
  C=C
 /   \
```

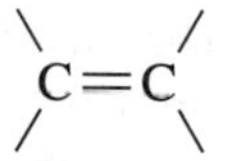

In the trans positions attach a two-carbon chain to one carbon atom and a three-carbon chain to the other carbon atom:

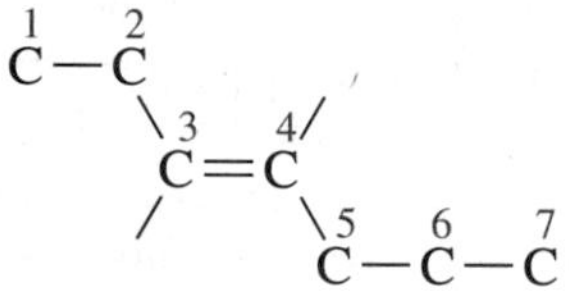

```
1   2
C — C
     \3  4/
      C=C
     /   \5  6   7
          C — C — C
```

Now attach hydrogen atoms to give each carbon atom four bonds. Carbons 3 and 4 need only one hydrogen atom apiece: these two hydrogen atoms are also trans to each other. The structure is

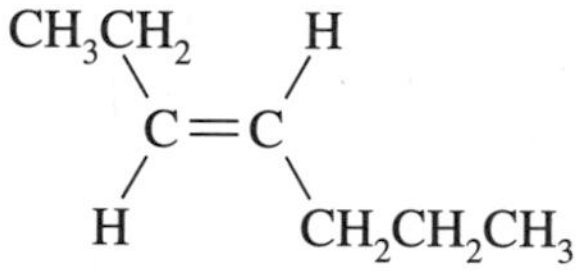

```
CH3CH2      H
      \    /
       C=C
      /    \
     H      CH2CH2CH3
```

(b) The compound contains six carbons, a double bond between carbons 2 and 3, and a Cl atom on carbon 5. Draw a C=C double bond in a planar arrangement:

```
 \   /
  C=C
 /   \
```

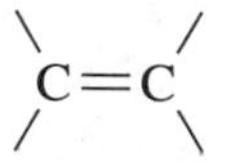

In the cis positions attach a $—CH_3$ to one carbon and a three-carbon chain to the other carbon. Place a Cl on carbon 5.

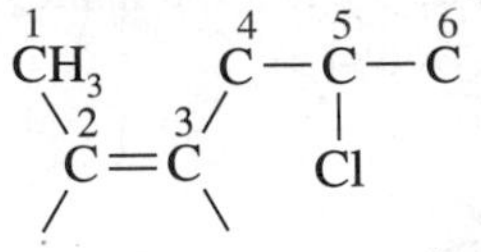

```
 1        4   5   6
CH3      C — C — C
  \2   3/      |
   C=C        Cl
  /    \
```

Now add H atoms to give each carbon four bonds. Carbons 2, 3, and 5 each need only one H atom. The structure is

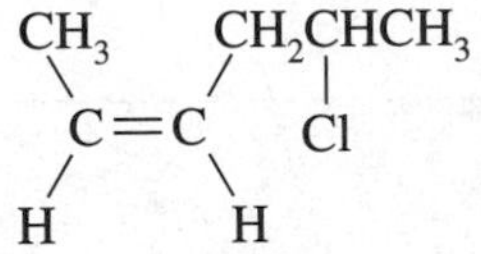

```
CH3      CH2CHCH3
  \     /     |
   C=C       Cl
  /    \
 H      H
```

Example 21.4

Draw structural formulas and names for all the isomers of pentene, C_5H_{10}. Identify all geometric isomers.

Solution

Start by drawing the isomers of the five-carbon chain, placing the C=C in all possible positions (there are two possible isomers). Then proceed to the four-carbon chains with a methyl group side chain. Locate the C=C and the CH_3 group in all possible positions. Check for duplications from the names of the compounds.

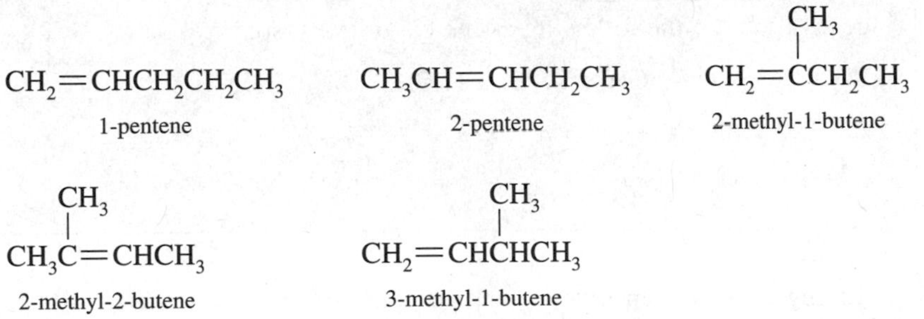

```
                                                   CH3
                                                   |
CH2=CHCH2CH2CH3        CH3CH=CHCH2CH3         CH2=CCH2CH3
   1-pentene              2-pentene          2-methyl-1-butene

   CH3                        CH3
   |                          |
CH3C=CHCH3              CH2=CHCHCH3
2-methyl-2-butene       3-methyl-1-butene
```

Of these five compounds, only 2-pentene can have cis–trans isomers. All of the others have two identical groups on one of the carbon atoms of the double bond. Draw the cis–trans isomers.

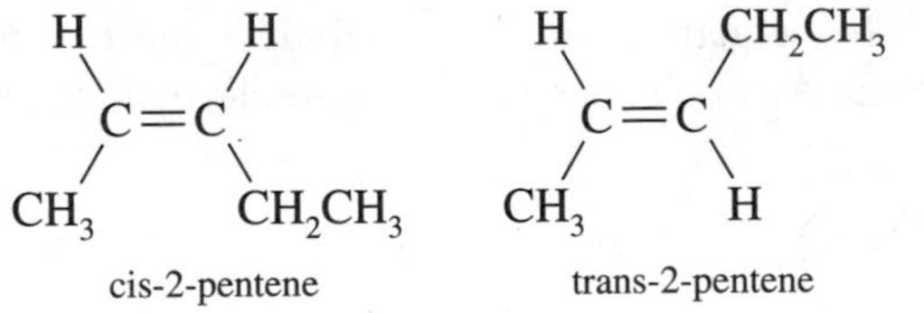

```
 H       H              H       CH2CH3
  \     /                \     /
   C=C                    C=C
  /    \                 /    \
CH3     CH2CH3         CH3     H
   cis-2-pentene          trans-2-pentene
```

Example 21.5

Is the compound below the cis or the trans isomer?

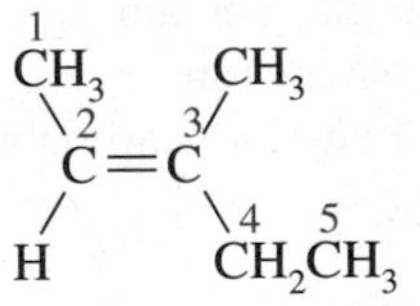

```
 1
CH3     CH3
  \2   3/
   C=C
  /    \4  5
 H      CH2CH3
```

Solution

In branched-chain alkenes, the cis–trans designation is ordinarily given to the structure containing the longest carbon chain that includes the carbon–carbon double bond. In this case the longest chain is the 2-pentene, in which the methyl group on carbon 2 and the ethyl group on carbon 3 are trans to each other. Thus the name of the compound shown is *trans*-3-methyl-2-pentene. The cis isomer is

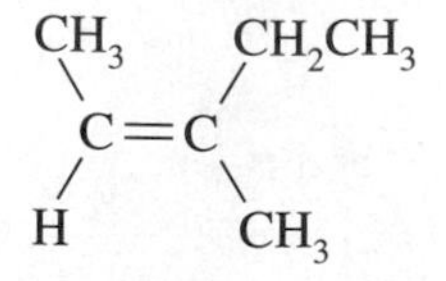

$$\begin{array}{ccc} CH_3 & & CH_2CH_3 \\ & \diagdown\quad\diagup & \\ & C{=}C & \\ & \diagup\quad\diagdown & \\ H & & CH_3 \end{array}$$

Practice 21.3

Draw structures (a) cis-3-methyl-2-pentene and (b) trans-2-bromo-2-pentene.

Practice 21.4

Determine whether geometric isomers exist for the following compounds. Draw structures for the cis and trans isomers.
(a) 1-chloro-3-methyl-2-pentene (b) 3-hexene (c) 2,3-dimethyl-2-pentene

Practice 21.5

Identify each of the following as the cis or trans isomer.

(a) 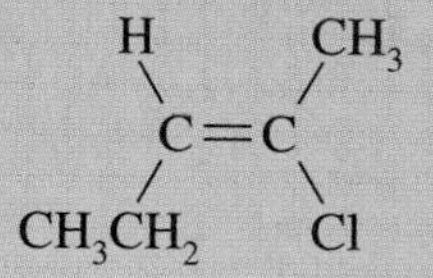
$$\begin{array}{ccc} H & & CH_3 \\ & \diagdown\quad\diagup & \\ & C{=}C & \\ & \diagup\quad\diagdown & \\ CH_3CH_2 & & Cl \end{array}$$

(b) 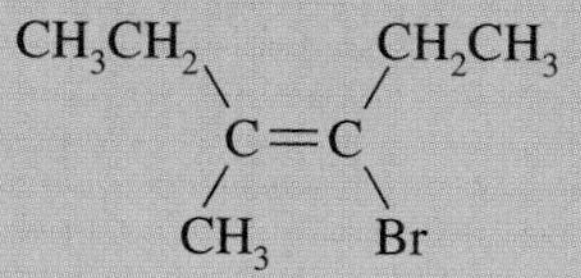
$$\begin{array}{ccc} CH_3CH_2 & & CH_2CH_3 \\ & \diagdown\quad\diagup & \\ & C{=}C & \\ & \diagup\quad\diagdown & \\ CH_3 & & Br \end{array}$$

Living cells are sensitive to, and respond to, specific isomers. The shape of these molecules determines their biological activity. Components of our fat tissue, unsaturated fatty acids, are commonly metabolized as the cis isomer. Vitamin A is only biologically active as an all-trans isomer. The effectiveness of drugs is also strongly dependent on their shape. Cisplatin (a cis isomer) is an effective anticancer agent; the trans isomer is not. The penicillin family of antibiotics must be the cis isomers to effectively kill bacteria. As these examples show, molecular shape is of vital importance in biology.

21.4 Cycloalkenes

cycloalkene

As the name implies, **cycloalkenes** are cyclic compounds that contain a carbon–carbon double bond in the ring. The two most common cycloalkenes are cyclopentene and cyclohexene. The double bond may be placed between any two carbon atoms:

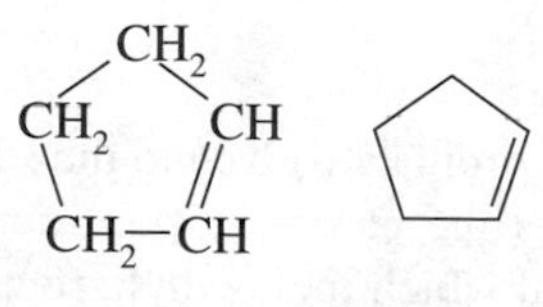

$$\begin{array}{ccc} & CH_2 & \\ CH_2 & & CH \\ & CH_2{-}CH & \end{array}$$

cyclopentene (C_5H_8)

$$\begin{array}{ccc} & CH_2 & \\ CH_2 & & CH \\ | & & \| \\ CH_2 & & CH \\ & CH_2 & \end{array}$$

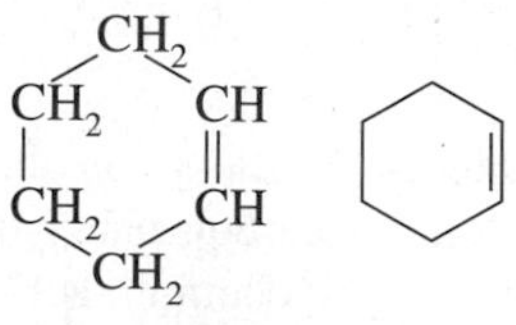

cyclohexene (C_6H_{10})

In cycloalkenes the carbons of the double bond are assigned numbers 1 and 2. Thus the positions of the double bond need not be indicated in the name of the compound. Other substituents on the ring are named in the usual manner, and their positions on the ring are indicated with the smallest possible numbers.

In the following examples, note that a Cl or a CH_3 has replaced one H atom in the molecule. The ring is numbered either clockwise or counterclockwise starting with the carbon–carbon double bond so that the substituted group(s) have the smallest possible numbers.

3-chlorocyclohexene (C_6H_9Cl)

1-methylcyclopentene ($C_5H_7CH_3$)

1, 3-dimethylcyclohexene [$C_6H_8(CH_3)_2$]

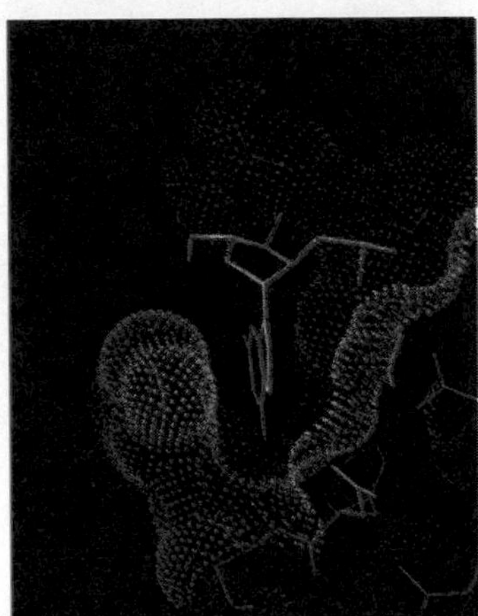

The shape of the molecule in green is just right to fit into the active site of the ribonuclease molecule (shown in red).

No cis–trans designation is necessary for cycloalkenes that contain up to seven carbon atoms in the ring. Cyclooctene has been shown to exist in both cis and trans-forms.

21.5 Preparation and Physical Properties of Alkenes

Common preparation methods for alkenes start with saturated organic molecules; each carbon in these molecules is bonded to four other atoms. Then, to form double bonds, atoms must be removed. The product contains double-bonded carbons that are connected to only three atoms each. Alkene synthesis commonly means "getting rid" of some atoms—an elimination reaction. Watch for evidence of elimination in the following two examples of alkene preparation.

Cracking

cracking

Ethene can be produced by the cracking of petroleum. **Cracking**, or pyrolysis, is the process in which saturated hydrocarbons are heated to very high temperatures in the presence of a catalyst (usually silica–alumina). This results in large molecules breaking into smaller ones, with the elimination of hydrogen, to form alkenes and small hydrocarbons like methane and ethane. Unfortunately, cracking always results in mixtures of products and is therefore not used often in the laboratory.

$$\text{Alkane } (C_nH_{2n+2}) \xrightarrow[\text{catalyst}]{\Delta} \text{Mixture of alkenes} + \text{Alkanes} + \text{Hydrogen gas}$$

$$2CH_3CH_2CH_3 \xrightarrow{\sim 500°C} CH_3CH{=}CH_2 + CH_2{=}CH_2 + CH_4 + H_2$$

Dehydration of Alcohols

dehydration

Dehydration involves the elimination of a molecule of water from a reactant molecule. Dehydration reactions are very common in organic chemistry as well as in biochemistry.

To produce an alkene by dehydration, an alcohol is heated in the presence of concentrated sulfuric acid.

$$CH_3\underset{\underset{H}{|}}{\overset{\overset{H}{|}}{C}}-\underset{\underset{OH}{|}}{\overset{\overset{H}{|}}{C}}CH_3 \xrightarrow[\Delta]{\text{conc. } H_2SO_4} CH_3\overset{\overset{H}{|}}{C}=\overset{\overset{H}{|}}{C}CH_3 + H_2O$$

The alkene is formed as a result of elimination of H and OH (shown in color) from adjacent carbon atoms.

Physical Properties

Alkenes have physical properties very similar to the corresponding alkanes. This is not surprising since the difference between an alkane and an alkene is simply two hydrogen atoms.

General formula for alkanes	C_nH_{2n+2}
General formula for alkenes	C_nH_{2n}

Since alkenes have slightly smaller molar masses, their boiling points are slightly lower than the corresponding alkanes. The smaller alkenes (to 5 carbons) are gases at room temperature. As the chain lengthens (5–17 carbons), the alkenes are liquid, and above 17 carbons they are solid. The alkenes are nonpolar, like the other hydrocarbons, and so are insoluble in water but soluble in organic solvents. The densities of most alkenes are much less than water. Table 21.1 shows the properties of some alkenes. Notice that isomers (C_4H_8) have similar boiling points although they differ significantly in melting points. Melting-point differences occur because the isomers have different shapes and therefore fit into their crystalline structures in significantly different ways.

21.6 Chemical Properties of Alkenes

What type of reaction might be expected for an alkene (or an alkyne)? Note that ethene ($CH_2{=}CH_2$) has only three atoms bonded to each carbon; acetylene ($CH{\equiv}CH$) has only two atoms bonded to each carbon. Both alkenes and alkynes have fewer than the maximum of four atoms bonded per carbon. These molecules are more reactive than the corresponding alkanes and readily undergo addition reactions.

Addition

Addition at the carbon–carbon double bond is the most common reaction of alkenes. Hydrogen, halogens (Cl_2 or Br_2), hydrogen halides, sulfuric acid, and water are

TABLE 21.1 Physical Properties of Alkenes

Molecular formula	Structural formula	IUPAC name	Density (g/mL)	Melting point (°C)	Boiling point (°C)
C_2H_4	$CH_2{=}CH_2$	Ethene	—	−169	−104
C_3H_6	$CH_3CH{=}CH_2$	Propene	—	−185	−48
C_4H_8	$CH_3CH_2CH{=}CH_2$	1-Butene	0.595	−185	−6
C_4H_8	$(CH_3)_2C{=}CH_2$	2-Methylpropene	0.594	−14	−7
C_5H_{10}	$CH_3(CH_2)_2CH{=}CH_2$	1-Pentene	0.641	−138	30

some of the reagents that can be added to unsaturated hydrocarbons. Ethene, for example, reacts in the presence of a platinum catalyst in this fashion:

$$\underset{\text{ethene}}{CH_2{=}CH_2} + H_2 \xrightarrow[\text{1 atm}]{\text{Pt, 25°C}} \underset{\text{ethane}}{CH_3{-}CH_3}$$

The double bond is broken, and unsaturated alkene molecules become saturated by an addition reaction.

$$\underset{\text{colorless}}{CH_2{=}CH_2} + \underset{\text{reddish brown}}{Br{-}Br} \longrightarrow \underset{\substack{\text{1,2-dibromoethane}\\ \text{colorless}}}{CH_2Br{-}CH_2Br}$$

The disappearance of the reddish-brown bromine color provides visible evidence of reaction. Other reactions of ethene include the following.

$$CH_2{=}CH_2 + HCl \longrightarrow \underset{\substack{\text{chloroethane}\\ \text{(ethyl chloride)}}}{CH_3CH_2Cl}$$

$$CH_2{=}CH_2 + \underset{\text{sulfuric acid}}{HOSO_3H\,(conc.)} \longrightarrow \underset{\text{ethyl hydrogen sulfate}}{CH_3CH_2OSO_3H}$$

$$CH_2{=}CH_2 + HOH \xrightarrow{H^+} \underset{\substack{\text{ethanol}\\ \text{(ethyl alcohol)}}}{CH_3CH_2OH}$$

The H^+ indicates that the reaction is carried out under acidic conditions.

Unsaturated alkenes are readily converted to saturated molecules by addition. Also, note that addition is the reverse of elimination. For example, by the addition of water, an alkene can be converted to an alcohol, and by the elimination of water, an alcohol can be converted to an alkene.

The preceding examples dealt with ethene, but reactions of this kind can be made to occur on almost any molecule that contains a carbon–carbon double bond. If a symmetrical molecule such as Cl_2 is added to propene, only one product, 1,2-dichloropropane, is formed:

$$CH_2{=}CH{-}CH_3 + Cl_2 \longrightarrow \underset{\text{1,2-dichloropropane}}{CH_2Cl{-}CHCl{-}CH_3}$$

Bromine has the characteristic orange color shown in the flask on the left, but when added to an alkene the color disappears, as shown in the flask on the right.

But if an unsymmetrical molecule such as HCl is added to propene, two products are theoretically possible, depending upon which carbon atom adds the hydrogen. The two possible products are 1-chloropropane and 2-chloropropane. Experimentally we find that 2-chloropropane is formed almost exclusively:

$$CH_3{-}CH{=}CH_2 + HCl \longrightarrow CH_3CHClCH_3 \text{ (about 100\%)}$$

$$CH_3{-}CH{=}CH_2 + HCl \not\longrightarrow CH_3CH_2CH_2Cl \text{ (trace)}$$

A single product is obtained because the reaction follows specific steps. The sum of these steps is termed a *reaction mechanism.* The reaction between propene and HCl proceeds by the following mechanism:

1. A proton (H^+) from HCl bonds to carbon 1 of propene utilizing the pi bond electrons. The intermediate formed is a positively charged alkyl group, or carbocation. The positive charge is localized on carbon 2 of this carbocation.

$$CH_2{=}CH{-}CH_3 + HCl \longrightarrow CH_3{-}\overset{+}{C}H{-}CH_3 + Cl^-$$

isopropyl carbocation

2. The chloride ion then adds to the positively charged carbon atom to form a molecule of 2-chloropropane:

$$CH_3{-}\overset{+}{C}H{-}CH_3 + Cl^- \longrightarrow CH_3{-}CHCl{-}CH_3$$

2-chloropropane

carbocation

An ion in which a carbon atom has a positive charge is known as a **carbocation.** There are four types of carbocations: methyl, primary (1°), secondary (2°), and tertiary (3°). Examples of these four types are as follows:

H_3C^+	$CH_3{-}CH_2^+$	$CH_3CH_2{-}CH_2^+$	$CH_3{-}\overset{+}{C}H{-}CH_3$	$(CH_3)_3C^+$
methyl carbocation	ethyl carbocation (primary)	*n*-propyl carbocation (primary)	isopropyl carbocation (secondary)	*t*-butyl carbocation (tertiary)

A carbon atom is designated as primary if it is bonded to one carbon atom, secondary if it is bonded to two carbon atoms, and tertiary if it is bonded to three carbon atoms. Thus in a primary carbocation the positive carbon atom is bonded to only one carbon atom. In a secondary carbocation the positive carbon atom is bonded to two carbon atoms. In a tertiary carbocation the positive carbon atom is bonded to three carbon atoms.

The order of stability of carbocations and hence the ease with which they are formed is tertiary > secondary > primary. Thus, in the reaction of propene and HCl, isopropyl carbocation (secondary) is formed as an intermediate in preference to *n*-propyl carbocation (primary).

Stability of carbocations: $3° > 2° > 1° > \overset{+}{C}H_3$

In the middle of the 19th century, a Russian chemist, V. Markovnikov, observed reactions of this kind, and in 1869 he formulated a useful generalization now known as **Markovnikov's rule**. This rule in essence states:

Markovnikov's rule

> **When an unsymmetrical molecule such as HX(HCl) adds to a carbon–carbon double bond, the hydrogen from HX goes to the carbon atom that has the greater number of hydrogen atoms.**

As you can see, the addition of HCl to propene, discussed previously, follows Markovnikov's rule. The addition of HI to 2-methylpropene is another example illustrating this rule:

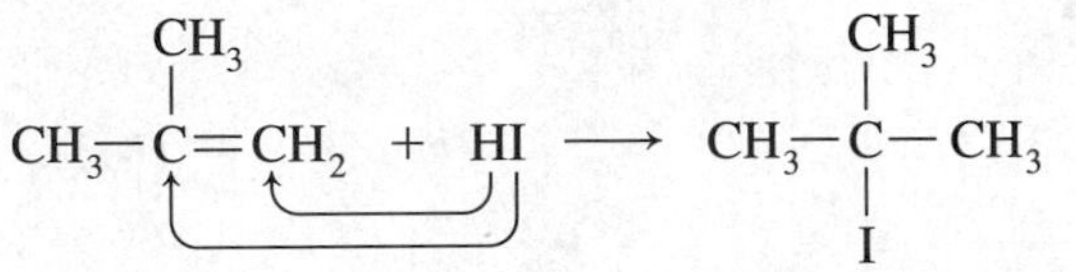

$$CH_3-\overset{\overset{\large CH_3}{|}}{C}=CH_2 + HI \longrightarrow CH_3-\underset{\underset{\large I}{|}}{\overset{\overset{\large CH_3}{|}}{C}}-CH_3$$

2-iodo-2-methylpropane
(*tert*-butyl iodide)

In the reaction to the left, the $=CH_2$ is the alkene carbon with the most hydrogen atoms (two); it thus adds the hydrogen atom followed by the addition of I^- to the other alkene carbon atom.

General rules of this kind are useful in predicting the products of reactions; however, exceptions are known for most such rules.

Example 21.6

Write formulas for the organic products formed when 2-methyl-1-butene reacts with (a) H_2, Pt/25°C; (b) Cl_2; (c) HCl; and (d) H_2O, H^+.

Solution

First write the formula for 2-methyl-1-butene:

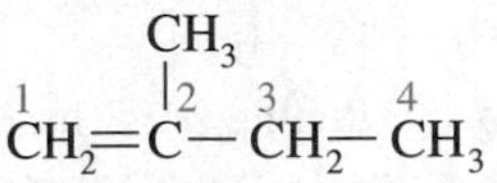

$$\overset{1}{C}H_2=\overset{\overset{\large CH_3}{|}}{\overset{2}{C}}-\overset{3}{C}H_2-\overset{4}{C}H_3$$

(a) The double bond is broken when a hydrogen molecule adds. One H atom adds to each carbon atom of the double bond. Platinum, Pt, is a necessary catalyst in this reaction. The product is

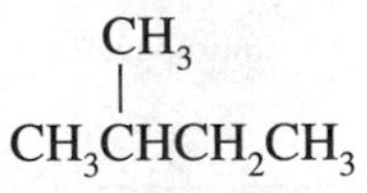

$$CH_3\overset{\overset{\large CH_3}{|}}{C}HCH_2CH_3$$

2-methylbutane

(b) The Cl_2 molecule adds to the carbons of the double bond. One Cl atom adds to each carbon atom of the double bond. The product is

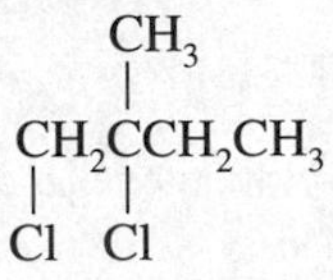

$$\underset{\underset{\large Cl}{|}}{CH_2}\underset{\underset{\large Cl}{|}}{\overset{\overset{\large CH_3}{|}}{C}}CH_2CH_3$$

1,2-dichloro-2-methylbutane

(c) HCl adds to the double bond according to Markovnikov's rule. The H^+ goes to carbon 1 (the more stable 3° carbocation is formed as an intermediate product), and the Cl^- goes to carbon 2. The product is

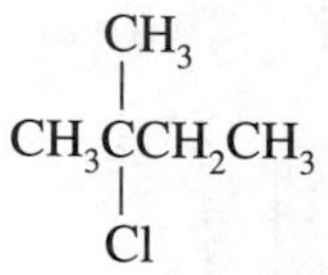

$$\begin{array}{c} CH_3 \\ | \\ CH_3CCH_2CH_3 \\ | \\ Cl \end{array}$$

2-chloro-2-methylbutane

(d) The net result of this reaction is a molecule of water added across the double bond. The H adds to carbon 1 (the carbon with the greater number of hydrogen atoms), and the OH adds to carbon 2 (the carbon of the double bond with the lesser number of hydrogen atoms). The product is

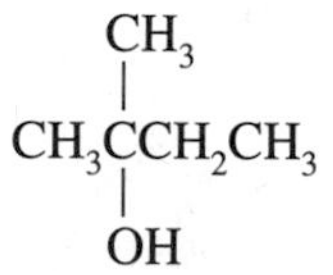

$$\begin{array}{c} CH_3 \\ | \\ CH_3CCH_2CH_3 \\ | \\ OH \end{array}$$

2-methyl-2-butanol
(an alcohol)

Example 21.7 Write equations for the addition of HCl to (a) 1-pentene and (b) 2-pentene.

Solution

(a) In the case of 1-pentene, $CH_3CH_2CH_2CH{=}CH_2$, the proton from HCl adds to carbon 1 to give the more stable secondary carbocation, followed by the addition of Cl^- to give the product 2-chloropentane. The addition is directly in accordance with Markovnikov's rule.

$$CH_3CH_2CH_2CH{=}CH_2 + HCl \longrightarrow \underset{\text{2-chloropentane}}{CH_3CH_2CH_2CHClCH_3}$$

(b) In 2-pentene, $CH_3CH_2CH{=}CHCH_3$, each carbon of the double bond has one hydrogen atom, and the addition of a proton to either one forms a secondary carbocation. After the addition of Cl^-, the result is two isomeric products that are formed in almost equal quantities:

$$CH_3CH_2CH{=}CHCH_3 + HCl \longrightarrow \underset{\text{2-chloropentane}}{CH_3CH_2CH_2CHClCH_3} + \underset{\text{3-chloropentane}}{CH_3CH_2CHClCH_2CH_3}$$

Practice 21.6

Write formulas for the organic products formed when 3-methyl-2-pentene reacts with (a) H_2/Pt, (b) Br_2, (c) HCl, and (d) H_2O, H^+.

Practice 21.7

Write equations for (a) addition of water to 1-methylcyclopentene and (b) addition of HI to 2-methyl-2-butene.

Oxidation

Another typical reaction of alkenes is oxidation at the double bond. For example, when shaken with a cold, dilute solution of potassium permanganate, $KMnO_4$, an alkene is converted to a glycol (glycols are dihydroxy alcohols). Ethene reacts in this manner:

$$\underset{\substack{\text{ethene}\\\text{(ethylene)}}}{CH_2{=}CH_2} + \underset{\text{(purple)}}{KMnO_4(aq)} + H_2O \longrightarrow \underset{\substack{\text{1,2 ethanediol}\\\text{(ethylene glycol)}}}{\underset{\text{OH}\quad\text{OH}}{CH_2{-}CH_2}} + \underset{\text{(brown)}}{MnO_2} + KOH$$

The *Baeyer test* makes use of this reaction to detect or confirm the presence of double (of triple) bonds in hydrocarbons. Evidence of reaction (positive Baeyer test) is the disappearance of the purple color of permanganate ions. The Baeyer test is not specific for detecting unsaturation in hydrocarbons because other classes of compounds may also give a positive Baeyer test.

Carbon–carbon double bonds are found in many different kinds of molecules. Most of these substances react with potassium permanganate and undergo somewhat similar reactions with other oxidizing agents including oxygen in the air and, especially, with ozone. Such reactions are frequently troublesome. For example, premature aging and cracking of automobile tires in smoggy atmospheres occur because ozone attacks the double bonds in rubber molecules. Cooking oils and fats sometimes develop disagreeable odors and flavors because the oxygen of the air reacts with the double bonds present in these materials. Potato chips, because of their large surface area, are especially subject to flavor damage caused by oxidation of the unsaturated cooking oils that they contain.

21.7 Alkynes: Nomenclature and Preparation

Nomenclature

IUPAC Nomenclature for Alkynes

The procedure for naming alkynes is the same as that for alkenes, but the ending used is *−yne* to indicate the presence of a triple bond. Table 21.2 lists names and formulas for some common alkynes.

Practice 21.8

Write formulas for
(a) 3-methyl-1-butyne
(b) 4-ethyl-4-methyl-2-hexyne

Preparation

Although triple bonds are very reactive, it is relatively easy to synthesize alkynes. Acetylene can be prepared inexpensively from calcium carbide and water.

$$CaC_2 + 2\ H_2O \longrightarrow HC{\equiv}CH + Ca(OH)_2$$

Acetylene is also prepared by the cracking of methane in an electric arc.

$$2\ CH_4 \xrightarrow{1500°C} HC{\equiv}CH + 3\ H_2$$

TABLE 21.2 Nomenclature for Some Common Alkynes

Molecular formula	Structural formula	IUPAC name
C_2H_2	$H{-}C{\equiv}C{-}H$	Ethyne*
C_3H_4	$CH_3{-}C{\equiv}C{-}H$	Propyne
C_4H_6	$CH_3CH_2{-}C{\equiv}C{-}H$	1-Butyne
C_4H_6	$CH_3{-}C{\equiv}C{-}CH_3$	2-Butyne

*Ethyne is commonly known as acetylene.

21.8 Physical and Chemical Properties of Alkynes

Physical Properties

Acetylene is a colorless gas, with little odor when pure. The disagreeable odor we associate with it is the result of impurities (usually PH_3). Acetylene is insoluble in water and is a gas at normal temperature and pressure (bp $= -84°C$). As a liquid, acetylene is very sensitive and may decompose violently (explode), spontaneously, or from a slight shock.

$$HC{\equiv}CH \longrightarrow H_2 + 2\ C + 227\ kJ\ (54.3\ kcal)$$

To eliminate the danger of explosions, acetylene is dissolved under pressure in acetone and is packed in cylinders that contain a porous inert material.

Chemical Properties

Acetylene is used mainly (1) as fuel for oxyacetylene cutting and welding torches and (2) as an intermediate in the manufacture of other substances. Both uses are dependent upon the great reactivity of acetylene. Acetylene and oxygen mixtures produce flame temperatures of about 2800°C. Acetylene readily undergoes addition reactions rather similar to those of ethene. It reacts with chlorine and bromine and decolorizes a permanganate solution (Baeyer's test). Either one or two molecules of bromine or chlorine can be added:

$$HC{\equiv}CH + Br_2 \longrightarrow \underset{\text{1,2-dibromoethene}}{CHBr{=}CHBr}$$

Superabsorbants are used in disposable diapers to increase the amount of liquid the diaper will absorb without leaking.

or

$$HC\equiv CH + 2\ Br_2 \longrightarrow CHBr_2\text{—}CHBr_2$$
1,1,2,2-tetrabromoethane

It is apparent that either unsaturated or saturated compounds can be obtained as addition products of acetylene. Often, unsaturated compounds capable of undergoing further reactions are made from acetylene. For example, vinyl chloride, which is used to make the plastic polyvinyl chloride (PVC), can be made by simple addition to HCl to acetylene:

$$CH\equiv CH + HCl \longrightarrow CH_2\text{=}CHCl$$
chloroethene
(vinyl chloride)

(*Note:* The common name for the $CH_2\text{=}CH\text{—}$ group is *vinyl.*) If the reaction is not properly controlled, another HCl adds to the chloroethene. This reaction follows Markovinkov's rule.

$$CH_2\text{=}CHCl + HCl \longrightarrow CH_3CHCl_2$$
1,1-dichloroethane

Addition reactions are common for all alkynes.

Hydrogen chloride reacts with other alkynes in a similar fashion to form substituted alkenes. The addition follows Markovnikov's rule. Alkynes can react with 1 or 2 mol of HCl. Consider the reaction of propyne with HCl:

$$CH_3C\equiv CH + HCl \longrightarrow CH_3CCl\text{=}CH_2$$
2-chloropropene

$$CH_3CCl\text{=}CH_2 + HCl \longrightarrow CH_3CCl_2CH_3$$
2,2-dichloropropane

There are certain unique reactions for the alkynes. They are capable of reacting at times when alkenes will not. Acetylene, with certain catalysts, reacts with HCN to form $CH_2\text{=}CHCN$ (acrylonitrile). This chemical is used industrially to manufacture Orlon, a polymer commonly found in clothing. It is also used to form the superabsorbants, which are capable of retaining up to 2000 times their mass of water. These superabsorbants are used in disposable diapers as well as in soil additives to retain water.

The reactions of other alkynes are similar to those of acetylene. Although many other alkynes are known, acetylene is by far the most important industrially.

Practice 21.9

Write formulas for the products formed by the reaction of HCl with (a) 1-butyne and (b) 2-pentyne to form saturated compounds.

21.9 Aromatic Hydrocarbons: Structure

Benzene and all substances that have structures and chemical properties resembling benzene are classified as *aromatic compounds.* The word *aromatic* originally referred to the rather pleasant odor possessed by many of these substances, but this meaning has been dropped. Benzene, the parent substance of the aromatic hydrocarbons, was first isolated by Michael Faraday in 1825; its correct molecular formula, C_6H_6, was established a few years later. The establishment of a reasonable structural formula that would account for the properties of benzene was a difficult problem for chemists in the mid-19th century.

Finally, in 1865, August Kekulé proposed that the carbon atoms in a benzene molecule are arranged in a six-membered ring with one hydrogen atom bonded to each carbon atom and with three carbon–carbon double bonds, as shown in the following formula:

```
          H
          |
  H\    / C \\    /H
     C          C
     ||         |
     C          C
  H/    \ C //    \H
          |
          H
```

Kekulé soon realized that there should be two dibromobenzenes, based on double- and single-bond positions relative to the two bromine atoms:

```
          Br                            Br
          |                             |
  H\    / C \\    /Br           H\   // C \    /Br
     C          C                  C          C
     ||         |       and        |          ||
     C          C                  C          C
  H/    \ C //    \H            H/   \\ C /    \H
          |                             |
          H                             H
```

Since only one dibromobenzene (with bromine atoms on adjacent carbons) could be produced, Kekulé suggested that the double bonds are in rapid oscillation within the molecule. He therefore proposed that the structure of benzene could be represented in this fashion:

```
        H                         H
        |                         |
  H\  / C \\ /H             H\  //C \  /H
    C        C                C        C
    ||       |      <==>      |        ||
    C        C                C        C
  H/  \ C // \H             H/  \\C /  \H
        |                         |
        H                         H
```

Kekulé's concepts are a landmark in the history of chemistry. They are the basis of the best representation of the benzene molecule devised in the 19th century, and they mark the beginning of our understanding of structure in aromatic compounds.

Kekulé's formulas have one serious shortcoming: They represent benzene and related substances as highly unsaturated compounds. Yet benzene does not react like a typical alkene; it does not decolorize bromine solutions rapidly, nor does it destroy the purple color of permanganate ions (Baeyer's test). Instead, the chemical behavior of benzene resembles that of an alkane. It's typical reactions are the substitution type, wherein a hydrogen atom is replaced by some other group; for example,

$$C_6H_6 + Cl_2 \xrightarrow{Fe} C_6H_5Cl + HCl$$

This problem was not fully resolved until the technique of X-ray diffraction, developed in the years following 1912, permitted us to determine the actual distances between the nuclei of carbon atoms in molecules. The center-to-center distances between carbon atoms in different kinds of hydrocarbon molecules are

Ethane (single bond)	0.154 nm
Ethene (double bond)	0.134 nm
Benzene	0.139 nm

Because only one carbon–carbon distance (bond length) is found in benzene, it is apparent that alternating single and double bonds do not exist in the benzene molecule.

Modern theory accounts for the structure of the benzene molecule in this way: the orbital hybridization of the carbon atoms is sp^2 (see the structure of ethene in Section 21.1). A planar hexagonal ring is formed by the overlapping of two sp^2 orbitals on each of six carbon atoms. The other sp^2 orbital on each carbon atom overlaps an s orbital of a hydrogen atom, bonding the carbon to the hydrogen by a sigma bond. The remaining six p orbitals, one on each carbon atom, overlap each other and form doughnut-shaped pi electron clouds above and below the plane of the ring (see Figure 21.5). The electrons composing these clouds are not attached to particular carbon atoms but are delocalized and associated with the entire molecule. This electronic structure imparts unusual stability to benzene and is responsible for many of the characteristic properties of aromatic compounds. Because of this stable electronic structure, benzene does not readily undergo addition or elimination reactions, but it does undergo substitution reactions.

For convenience, present-day chemists usually write the structure of benzene as one or the other of these abbreviated forms:

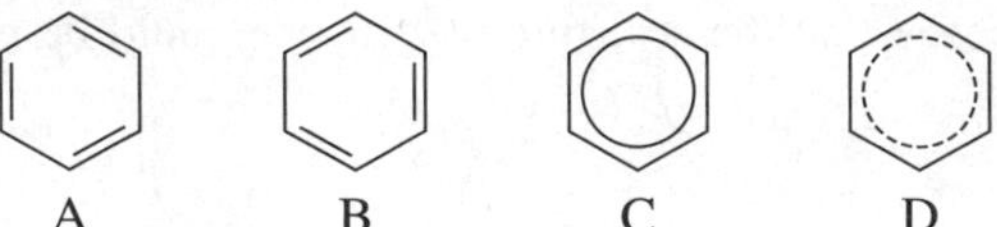

In all of these representations, it is understood that there is a carbon atom and a hydrogen atom at each corner of the hexagon. The classical Kekulé structures are represented by formulas A and B. However, neither of these Kekulé structures actually exists. The real benzene molecule is a hybrid of these structures and is commonly represented by either formula C or D. The circle or the dashed circle indicates the special nature of the benzene pi bonds, as shown in Figure 21.5. We will use the hexagon with the solid circle to represent a benzene ring.

Hexagons are used in representing the structural formulas of benzene derivatives—that is, substances in which one or more hydrogen atoms in the ring have been replaced by other atoms or groups. Chlorobenzene, for example, is written in this fashion:

Cl

chlorobenzene, C_6H_5Cl

This notation indicates that the chlorine atom has replaced a hydrogen atom and is bonded directly to a carbon atom in the ring. Thus the correct formula for chlorobenzene is C_6H_5Cl, not C_6H_6Cl.

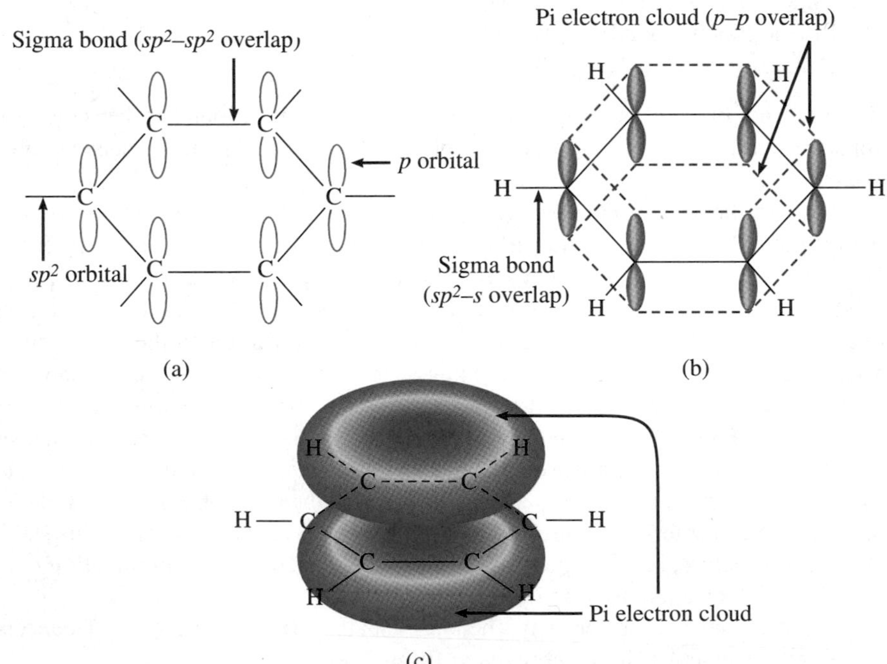

FIGURE 21.5 ▶
Bonding in a benzene molecule: (a) sp^2–sp^2 orbital overlap to form the carbon ring structure; (b) carbon–hydrogen bonds formed by sp^2–s orbital overlap and overlapping of p orbitals; (c) pi electron clouds above and below the plane of the carbon ring.

21.10 Naming Aromatic Compounds

Substituted benzenes are the most common benzene derivatives because substitution is the most common reaction type for benzene. A substituted benzene is derived by replacing one or more hydrogen atoms of benzene by another atom or group of atoms. Thus a monosubstituted benzene has the formula C_6H_5G, where G is the group replacing a hydrogen atom.

Monosubstituted Benzenes

Some monosubstituted benzenes are named by adding the name of the substitutent group as a prefix to the word *benzene*. The name is written as one word. Note that the position of the substituent is not important here as all the positions in the hexagon are equivalent. Examples include the following:

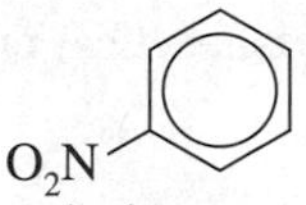

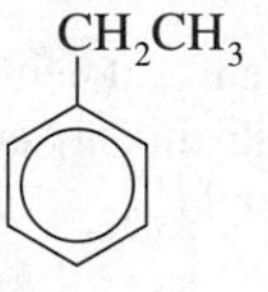

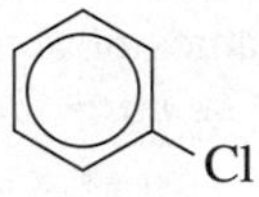

O_2N nitrobenzene; CH_2CH_3 ethylbenzene; Cl chlorobenzene; Br bromobenzene

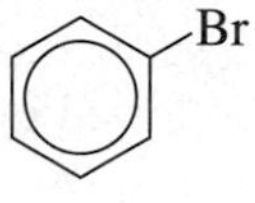

Certain monosubstituted benzenes have special names. These are used as parent names for further substituted compounds, so they should be memorized.

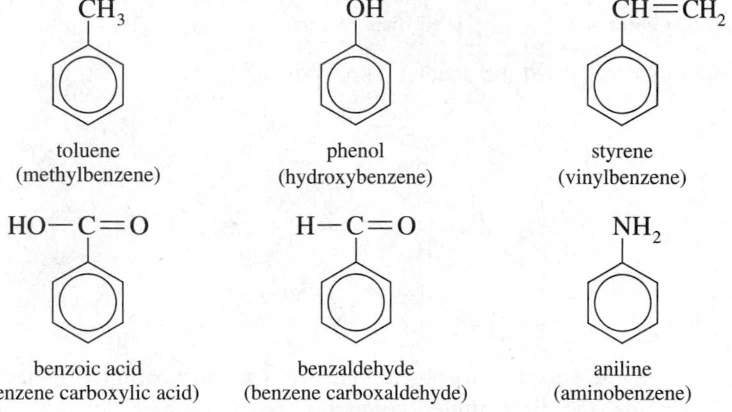

The C_6H_5— group is known as the phenyl group, and the name *phenyl* is used to name compounds that cannot easily be named as benzene derivatives. For example, the following compounds are named as derivatives of alkanes:

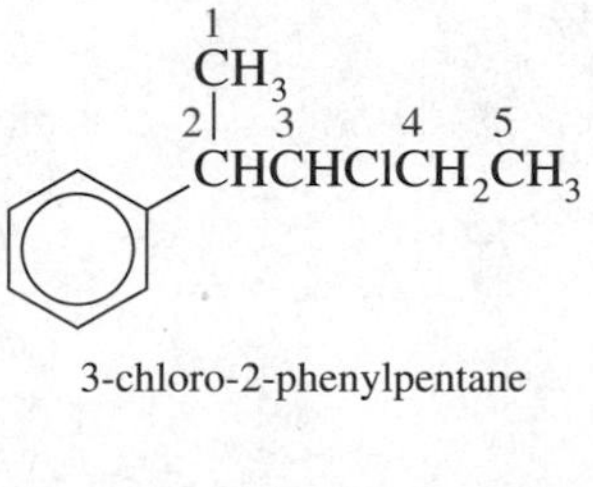

$\overset{1}{CH_3}$ — $\overset{2}{C}H\overset{3}{C}HCl\overset{4}{C}H_2\overset{5}{C}H_3$ (phenyl on C-2)

3-chloro-2-phenylpentane

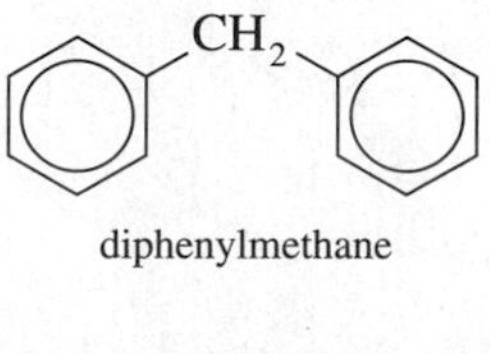

$C_6H_5CH_2C_6H_5$

diphenylmethane

Disubstituted Benzenes

There are two ways of naming disubstituted benzenes.

When two substituent groups replace two hydrogen atoms in a benzene molecule, three different isomeric compounds are possible. The prefixes *ortho-*, *meta-* and *para-* (abbreviated *o-*, *m-*, and *p-*) are used to name these disubstituted benzenes in one nomenclature system. Ortho designates 1,2 disubstitution, meta represents 1,3 disubstitution, and para designates 1,4 disubstitution.

Consider the dichlorobenzenes, $C_6H_4Cl_2$. Note that the three isomers have different physical properties, indicating that they are truly different substances:

Moth crystals or mothballs used to be made of naphthalene, but today most of them are made from para-dichlorobenzene.

ortho-dichlorobenzene	*meta*-dichlorobenzene	*para*-dichlorobenzene
(1,2-dichlorobenzene)	(1,3-dichlorobenzene)	(1,4-dichlorobenzene)
(mp –17.2°C, bp 180.4°C)	(mp –24.8°C, bp 172°C)	(mp 53.1°C, bp 174.4°C)

When the two substituents are different and neither is part of a compound with a special name, the names of the two substituents are given in alphabetical order, followed by the word *benzene.* For example,

o-bromochlorobenzene   *m*-ethylnitrobenzene

The dimethylbenzenes have the special name *xylene.*

ortho-xylene   *meta*-xylene   *para*-xylene

When one of the substituents corresponds to a monosubstituted benzene that has a special name, the disubstituted compound is named as a derivative of that parent compound. In the following examples the parent compounds are phenol, aniline, and toluene:

o-nitrophenol   *p*-bromoaniline   *m*-nitrotoluene

The numbering system as described under polysubstituted benzenes is also used to name disubstituted benzenes.

Polysubstituted Benzenes

When there are more than two substituents on a benzene ring, the carbon atoms in the ring are numbered starting at one of the substituted groups. Numbering may be either clockwise or counterclockwise but must be done in the direction that gives the lowest possible numbers to the substituent groups. When the compound is named as a derivative of one of the special parent compounds, the substituent of the parent compound is considered to be on carbon 1 of the ring (the CH_3 group is on carbon 1 in 2,4,6-trinitrotoluene). The following examples illustrate this system:

1,3,5-trinitrobenzene

1,2,4-tribromobenzene (not 1,4,6-)

2,4,6-trinitrotoluene (TNT)

5-bromo-2-chlorophenol

Example 21.8

Write formulas and names for all the possible isomers of (a) chloronitrobenzene, $C_6H_4Cl(NO_2)$, and (b) tribromobenzene, $C_6H_3Br_3$.

Solution

(a) The name and formula indicate a chloro group (Cl) and a nitro group (NO_2) attached to a benzene ring. There are six positions in which to place these two groups. They can be ortho, meta, or para to each other.

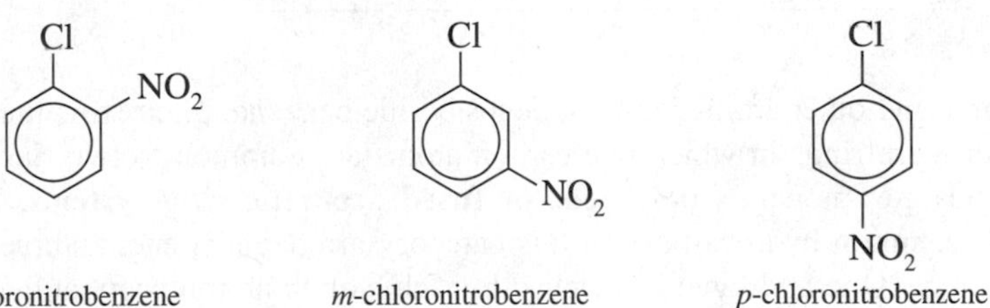

o-chloronitrobenzene *m*-chloronitrobenzene *p*-chloronitrobenzene

(b) For tribromobenzene, start by placing the three bromo groups in the 1-, 2-, and 3-positions; then the 1-, 2-, and 4-positions; and so on until all the possible isomers are formed. The name of each isomer will allow you to check that no duplication of formulas has been written.

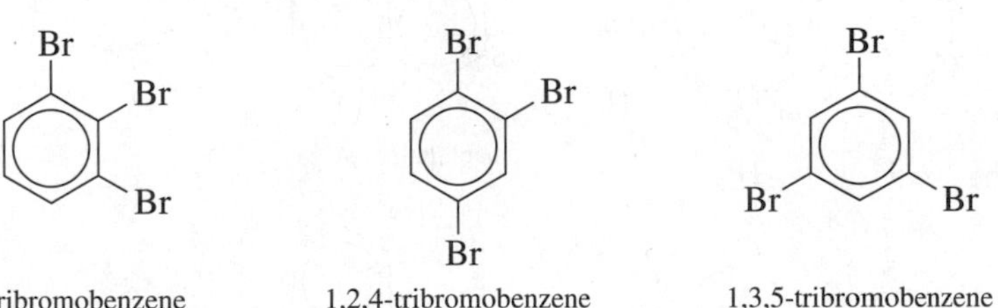

1,2,3-tribromobenzene 1,2,4-tribromobenzene 1,3,5-tribromobenzene

The dramatic collapse of this building is from the explosive TNT (common name for trinitroluene).

There are only three isomers of tribromobenzene. If one erroneously writes the 1,2,5- compound, a further check will show that, by numbering the rings as indicated, it is in reality the 1,2,4- isomer:

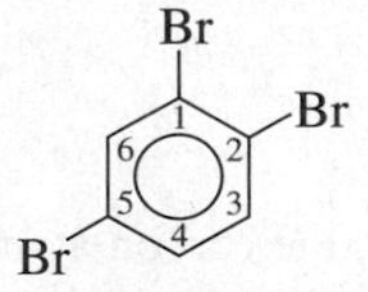

1,2,5-tribromobenzene (erroneous name)   1,2,4-tribromobenzene (correct name)

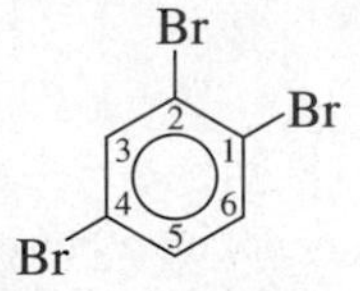

Practice 21.10

Write formulas and names for all possible isomers of chlorophenol.

Practice 21.11

Name the following:

(a) COOH, Br (b) NH_2, F (c) C(=O)—H, Cl, NO_2

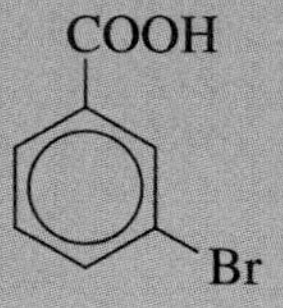

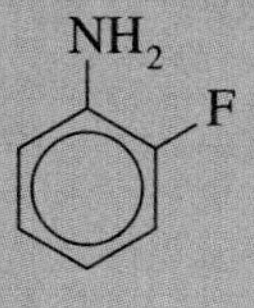

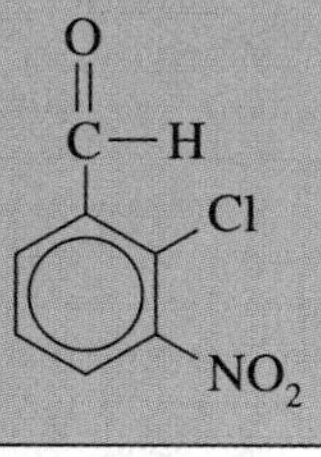

21.11 Polycyclic Aromatic Compounds

polycyclic or fused aromatic ring system

There are many other aromatic ring systems beside benzene. Their structures consist of two or more rings in which two carbon atoms are common to two rings. These compounds are known as **polycyclic or fused aromatic ring systems**. Three of the most common hydrocarbons in this category are naphthalene, anthracene, and phenanthrene. One hydrogen is attached to each carbon atom except at the carbons that are common to two rings.

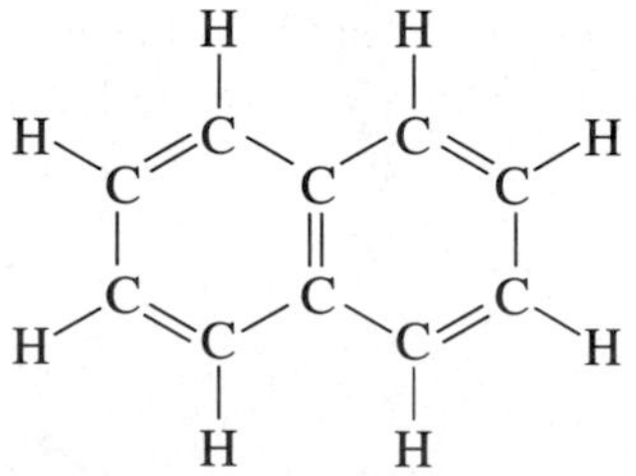

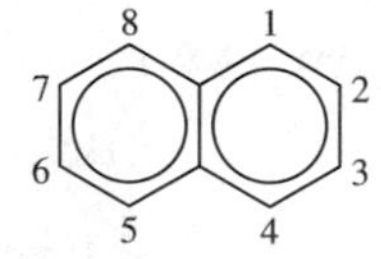

naphthalene, $C_{10}H_8$

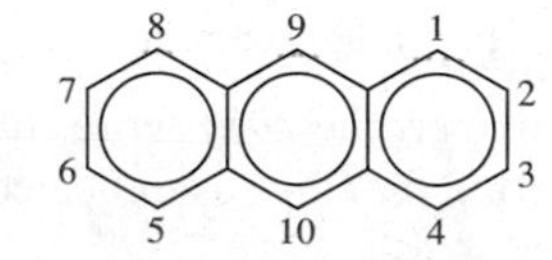

anthracene, $C_{14}H_{10}$ 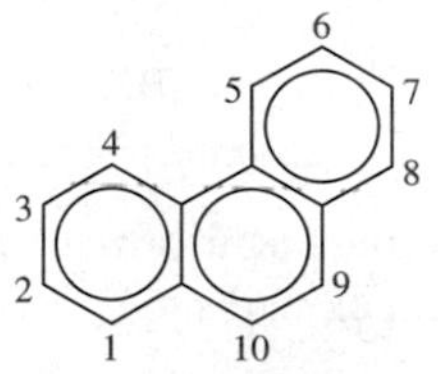   phenanthrene, $C_{14}H_{10}$

Chemistry in Action

Plants Reduce "PAH-lution"

Polycyclic aromatic hydrocarbons (PAHs) are by-products of combustion released into the air by traffic, residential heating, industrial processes, and other human activities. The bad news is that a number of these PAHs are known or suspected carcinogens and both our homes and offices appear to have significant accumulations of these chemicals. The good news is that houseplants remove significant amounts of these pollutants from the air.

A recent study (Staci Simonich and Ronald Hites, Indiana University) of the concentrations of ten representative PAHs in the vegetation and soil has concluded that houseplants absorb PAHs by accumulating the chemicals in their waxy outer tissues. Grasses, shrubs, and trees provide the same sort of cleanup outdoors. Apparently, the waxiness of a plant controls its PAH uptake; the plant's ability to absorb PAHs varies with season and temperature.

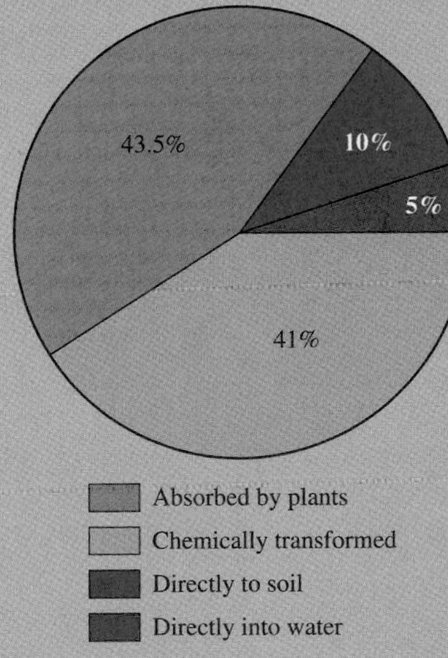

Simonich and Hites modeled the removal of PAHs by plants in the region of the United States encompassing Kansas, Missouri, Kentucky, and Virginia north to the Canadian border where about 52% of the U.S. population lives. They concluded that plants absorb a large fraction of this "PAH-lution," about 43.5% in this region of the country. Of the remainder, 10% falls directly on the soil, 5% into the water, and 41% is chemically transformed or drifts into Canada or the ocean.

Although plants are useful in controlling PAHs indoors and out, they don't completely solve the PAH-lution problem. John Roberts, a Seattle-based consulting engineer, found that carpets are a significant reservoir for PAHs. His team found that the levels of several PAH compounds in carpets equaled or exceeded the PAH levels in yard soils by as much as 20 times. Such levels are well above the PAH levels tolerated by hazardous waste sites in the state of Washington. Carpets shield the PAHs from sunlight, moisture, and microbes that would normally break down or transform them.

Thus far, there is no obvious connection between cigarette smoking or fireplace use and PAH levels in indoor air. Since PAH concentrations in the dust in outside doormats correlate with those found in carpet dust, one theory is that we track the pollutants in on our shoes. While adding plants to our houses can remove PAHs from the air, we may also need to consider removing carpets and/or removing our shoes at the door to lower the levels of these chemicals in our homes.

Indoor plants absorb polycyclic aromatic hydrocarbons in their waxy outer tissue.

All three of these substances can be obtained from coal tar. Naphthalene is known as moth balls and has been used as a moth repellant for many years. A number of the polycyclic aromatic hydrocarbons (and benzene) have been shown to be carcinogenic (cancer-producing). Formulas for some of the more notable ones, found in coal tar, tar from cigarette smoke, and soot in urban environments are as follows:

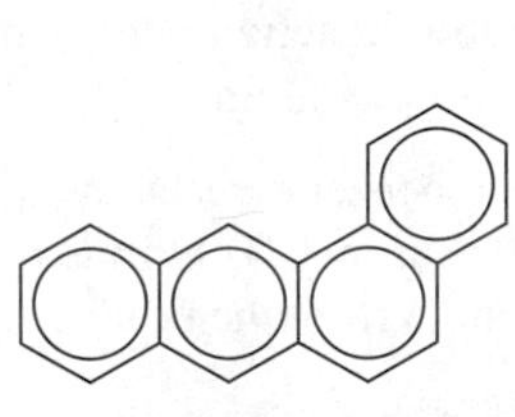

1,2-benzanthracene

1,2,5,6-dibenzanthracene

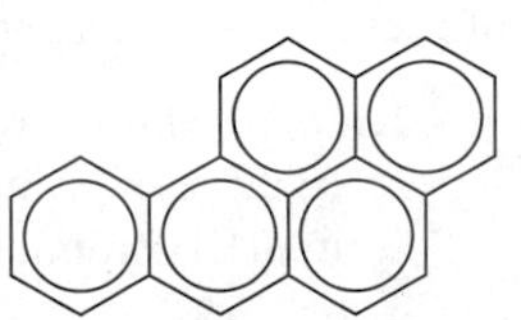

3,4-benzpyrene

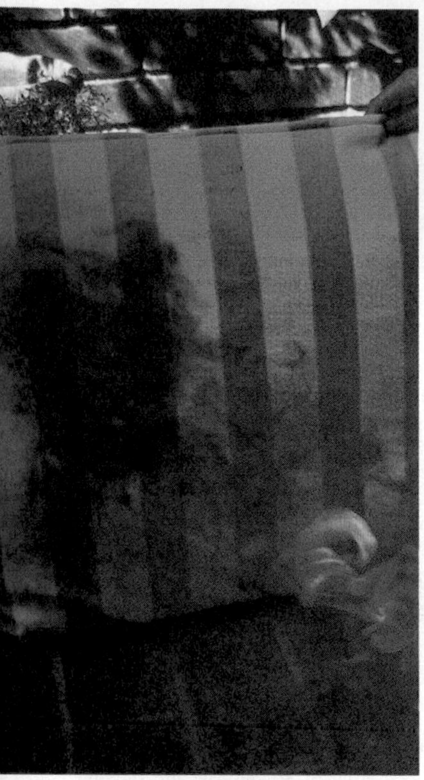
Yellow flames and smoke result from incomplete carbon combustion of burning gasoline.

These compounds are processed in living cells and become potent carcinogens. For example, the lung cells absorb polycyclic aromatic hydrocarbons from inhaled cigarette smoke. The aromatic rings are chemically modified, and the compounds become active carcinogens.

21.12 Sources and Physical Properties of Aromatic Hydrocarbons

When coal is heated to high temperatures (450–1200°C) in the absence of air to produce coke (C), coal gas and a complex mixture of condensable substances called *coal tar* are driven off:

$$\text{Coal} \xrightarrow{\Delta} \text{Coke} + \text{Coal gas} + \text{Coal tar}$$

The aromatic hydrocarbons, such as benzene, toluene, xylene, naphthalene, and anthracene, were first obtained in quantity from coal tar. Since coal tar itself is a by-product of the manufacture of coke, the total amount of aromatics that can be obtained from this source is limited. The demand for aromatic hydrocarbons, which are used in the production of a vast number of materials such as drugs, dyes, detergents, explosives, insecticides, plastics, and synthetic rubber, became too great to be obtained from coal tar alone. Processes were devised to make aromatic hydrocarbons from the relatively inexpensive alkanes found in petroleum. Currently, about one-third of our benzene supply and the greater portion of our toluene and xylene supplies are obtained from petroleum.

Aromatic hydrocarbons are essentially nonpolar substances, insoluble in water but soluble in many organic solvents. They are liquids or solids and usually have densities less than that of water. Aromatic hydrocarbons burn readily, usually with smoky yellow flames as a result of incomplete carbon combustion. Some are blended with other hydrocarbons to make good motor fuels with excellent antiknock properties.

21.13 Chemical Properties of Aromatic Hydrocarbons

Substitution

The most characteristic reactions of aromatic hydrocarbons involve the substitution of some group for a hydrogen on one of the ring carbons. Recall that a substitution reaction does not change the number of atoms bonded to each carbon. For example, each carbon in benzene will start and end a substitution reaction bonded to three other atoms.

The following are examples of typical aromatic substitution reactions. In each of these reactions, a functional group is substituted for a hydrogen atom.

1. *Halogenation* (chlorination or bromination) When benzene reacts with chlorine or bromine in the presence of a catalyst such as iron(III) chloride or iron(III) bromide, a Cl or a Br atom replaces an H atom to form the products.

$$\text{C}_6\text{H}_6 + X_2 \xrightarrow{FeX_3} \text{C}_6\text{H}_5X + HX$$

benzene | Chlorine or bromine | Chlorobenzene or bromobenzene

2. *Nitration* When benzene reacts with a mixture of concentrated nitric acid and concentrated sulfuric acid at about 50°C, nitrobenzene is formed. In this reaction a nitro group, —NO_2, is substituted for an H atom of benzene.

$$\text{C}_6\text{H}_6 + HO{-}NO_2 \xrightarrow{H_2SO_4} \text{C}_6\text{H}_5NO_2 + H_2O$$

benzene | nitric acid | nitrobenzene

3. *Alkylation* (Friedel–Crafts reaction) There are many variations of the Friedel–Crafts reaction. In this type the alkyl group from an alkyl halide (RX), in the presence of $AlCl_3$ catalyst, substitutes for an H atom on the benzene ring.

$$\text{C}_6\text{H}_6 + CH_3CH_2Cl \xrightarrow{AlCl_3} \text{C}_6\text{H}_5CH_2CH_3 + HCl$$

benzene | chloroethane | ethylbenzene

From about 1860 onward, especially in Germany, a great variety of useful substances such as dyes, explosives, and drugs were synthesized from aromatic hydrocarbons by reactions of the types just described. These early syntheses were developed by trial-and-error methods. A good picture of the reaction mechanism, or step-by-step sequence of intermediate stages in the overall reaction, was not obtained until about 1940.

It is now recognized that aromatic substitution reactions usually proceed by a mechanism called *electrophilic substitution.* Three steps are involved:

Step 1. An electrophile (electron-seeking group) is formed.

Step 2. The electrophile is attached to the benzene ring forming a positively charged carbocation intermediate.

Step 3. A hydrogen ion is lost from the carbocation to form the product. This reaction mechanism is illustrated in the chlorination of benzene catalyzed by iron (III) chloride.

Step 1 $FeCl_3 + Cl_2 \longrightarrow FeCl_4^- + Cl^+$

iron(III) chloride | chloronium ion (electrophile)

Step 2 $\text{C}_6\text{H}_6 + Cl^+ \longrightarrow [\text{C}_6\text{H}_6Cl]^+$ (H and Cl on the same ring carbon)

a carbocation

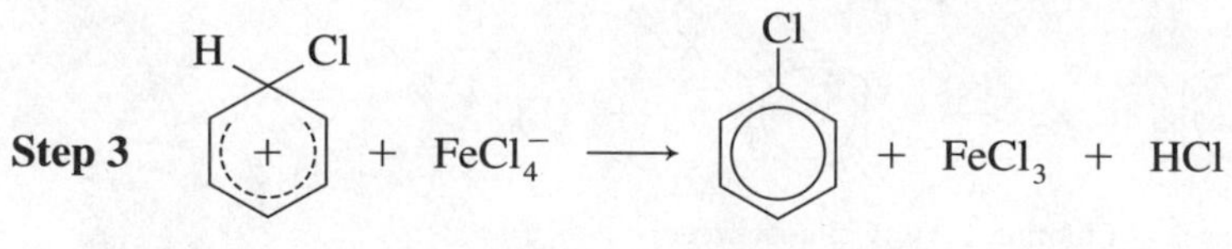

$$\text{Step 3} \quad \text{[H, Cl-substituted carbocation intermediate]} + FeCl_4^- \longrightarrow C_6H_5Cl + FeCl_3 + HCl$$

chlorobenzene

In Step 1 the electrophile (chloronium ion) is formed. In Step 2 the chloronium ion adds to benzene to form an intermediate carbocation, which loses a proton (H^+) in Step 3 to form the products, C_6H_5Cl and HCl. The catalyst, $FeCl_3$, is regenerated in Step 3.

This same mechanism is used by living organisms when aromatic rings gain or lose substituents. For example, the thyroxines (thyroid gland hormones) contain aromatic rings that are iodinated by following an electrophilic substitution mechanism. Iodination is a key step in producing these potent hormones. In general, scientists find that most of life's reactions follow mechanisms that have been elucidated in the organic chemist's laboratory.

Side-Chain Oxidation

Carbon chains attached to an aromatic ring are fairly easy to oxidize. Reagents most commonly used to accomplish this in the laboratory are $KMnO_4$ or $K_2Cr_2O_7$ + H_2SO_4. No matter how long the side chain is, the carbon atom attached to the aromatic ring is oxidized to a carboxylic acid group, —COOH. For example, toluene, ethylbenzene, and propylbenzene are all oxidized to benzoic acid:

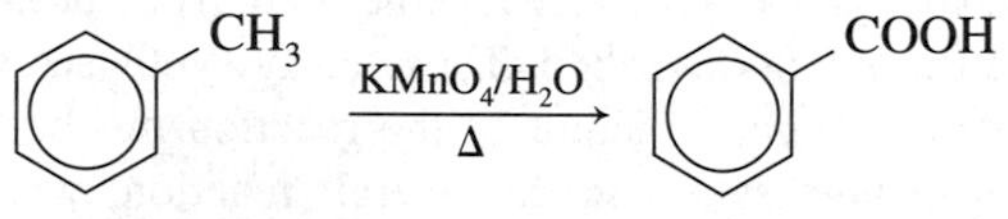

$$C_6H_5CH_3 \xrightarrow[\Delta]{KMnO_4/H_2O} C_6H_5COOH$$

toluene → benzoic acid

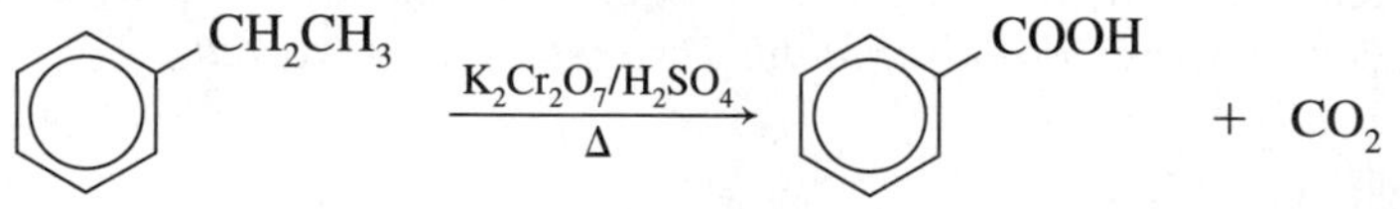

$$C_6H_5CH_2CH_3 \xrightarrow[\Delta]{K_2Cr_2O_7/H_2SO_4} C_6H_5COOH + CO_2$$

ethylbenzene

$$C_6H_5CH_2CH_2CH_3 \xrightarrow[\Delta]{K_2Cr_2O_7/H_2SO_4} C_6H_5COOH + CH_3COOH$$

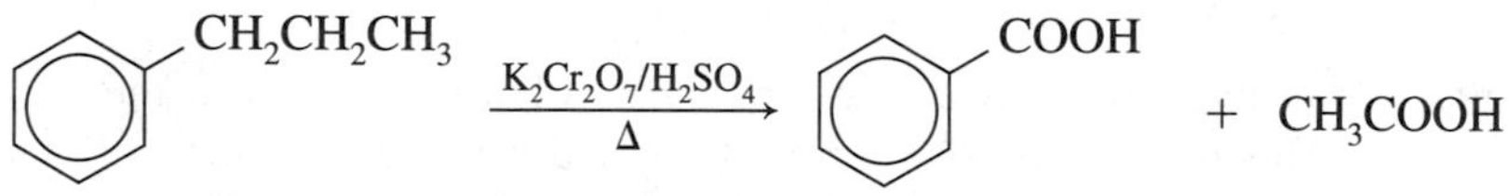

propylbenzene

Concepts in Review

1. Explain the sp^2 and *sp* hybridization of carbon atoms.
2. Explain the formation of a pi bond.
3. Explain the formation of double and triple bonds.
4. Distinguish, by formulas, the difference between saturated and unsaturated hydrocarbons.
5. Name and write structural formulas of alkenes, alkynes, cycloalkenes, and aromatic compounds.
6. Determine from structural formulas whether a compound can exist as geometric isomers.
7. Name geometric isomers by the cis–trans method.
8. Write equations for addition reactions of the alkenes and alkynes.
9. Explain the formation of carbocations and the role they play in chemical reactions.
10. Apply Markovnikov's rule to the addition of HCl, HBr, HI, and H_2O/H^+ to alkenes and alkynes.
11. Explain the Baeyer test for unsaturation.
12. Distinguish, using simple chemical tests, among alkanes, alkenes, and alkynes.
13. Describe the nature of benzene and how its properties differ from open-chain unsaturated compounds.
14. Name monosubstituted, disubstituted, and polysubstituted benzene compounds.
15. Draw structural formulas of substituted benzene compounds.
16. Recognize the more common fused aromatic ring compounds.
17. Write equations for the following reactions of benzene and substituted benzenes: halogenation (chlorination or bromination), nitration, alkylation (Friedel–Crafts reaction), and side-chain oxidation.
18. Describe and write equations for the mechanism by which benzene compounds are brominated in the presence of $FeBr_3$ or chlorinated in the presence of $FeCl_3$.

Key Terms

The terms listed here have been defined within this chapter. Section numbers are referenced in parenthesis for each term.

alkene (21.1)
alkyne (21.1)
aromatic compound (21.1)
carbocation (21.6)
cracking (21.5)
cycloalkene (21.4)
dehydration (21.5)
geometric isomer (21.3)
Markovnikov's rule (21.6)
pi bond (21.1)
polycyclic or fused aromatic ring system (21.11)

Questions

Questions refer to tables, figures, and key words and concepts defined within the chapter. A particularly challenging question or exercise is indicated with an asterisk.

1. The double bond in ethene, C_2H_4, is made up of a sigma bond and a pi bond. Explain how the pi bond differs from the sigma bond.

2. Why is it possible to obtain cis and trans isomers of 1,2-dichloroethene but not of 1,2-dichloroethane?

3. Why do many rubber products deteriorate rapidly in smog-ridden areas?

4. Explain the two different types of explosion hazards present when acetylene is being handled.

5. Why is an elimination reaction often considered to be the reverse of an addition reaction?

6. In terms of historical events, why did the major source of aromatic hydrocarbons shift from coal tar to petroleum during the 10-year period 1935–1945?

7. Explain how the reactions of benzene provide evidence that its structure does not include double bonds like those found in alkenes.

8. Which of the following statements are correct? Rewrite the incorrect statements to make them correct.

(a) The compound with the formula C_5H_{10} can be either an alkene or a cycloalkane.

(b) If C_8H_{10} is an open-chain compound with multiple double bonds, it needs an additional ten hydrogen atoms to become a saturated hydrocarbon.

(c) Propene and propane are isomers.

(d) The pi bond is formed from two sp^2 electron orbitals.

(e) A double bond consists of two equivalent bonds called pi bonds.

(f) A triple bond consists of one sigma bond and two pi bonds.

(g) The hybridized electron structure of a carbon atom in alkynes is *sp*, *sp*, *p*, *p*.

(h) The acetylene molecule is linear.

(i) A molecule of 2,3-dimethyl-1-pentene contains seven carbon atoms.

(j) When an alkene is reacted with cold $KMnO_4$, a glycol is formed.

(k) The compound C_6H_{10} can have in its structure: two carbon–carbon double bonds, or one carbon–carbon triple bond, or one cyclic ring and one carbon–carbon double bond.

(l) The disappearance of the purple color when $KMnO_4$ reacts with an alkene is known as the Markovnikov test for unsaturation.

(m) $CH_3CH_2CH_2^+$ is a primary carbocation.

(n) A secondary carbocation is more stable than a primary carbocation and less stable than a tertiary carbocation.

(o) After bromine has added to an alkene, the product is no longer unsaturated.

(p) Alkynes have the general formula C_nH_{2n-4}.

(q) Cis–trans isomerism occurs in alkenes and alkynes.

(r) Geometric isomers are superimposable on each other.

(s) All six hydrogen atoms in benzene are equivalent.

(t) The chemical behavior of benzene is similar to that of alkenes.

(u) Toluene and benzene are isomers.

(v) Two substituents on a benzene ring that are in the 1,2-position are ortho to each other.

(w) 1,4-Dichlorobenzene and *p*-dichlorobenzene are different names for the same compound.

(x) The oxidation of toluene with hot $KMnO_4$ yields benzoic acid.

(y) The oxidation of ethylbenzene with hot $KMnO_4$ yields benzoic acid.

(z) Toluene and benzene are homologues.

Paired Exercises

These exercises are paired. Each odd-numbered exercise is followed by a similar even-numbered exercise. Answers to the even-numbered exercises are given in Appendix V.

9. Draw Lewis structures to represent the following molecules:
(a) ethane **(b)** ethene **(c)** ethyne

10. Draw Lewis structures to represent the following molecules:
(a) propane **(b)** propene **(c)** propyne

11. There are eleven possible isomeric iodobutenes, C_4H_7I, including geometric isomers.
(a) Write the structural formula for each isomer.
(b) Name each isomer and include the prefix *cis-* or *trans-* where apppropriate.

12. There are four possible isomeric chloropropenes, C_3H_5Cl.
(a) Draw and name the structural formula for each isomer.
(b) There is another compound with this same molecular formula. What is its structure?

13. Draw structural formulas for the following:
(a) 2,5-dimethyl-3-hexene
(b) *cis*-4-methyl-2-pentene
(c) 3-pentene-1-yne
(d) *trans*-3-hexene
(e) 3-methyl-1-pentyne
(f) 3-methyl-2-phenylhexane

14. Draw structural formulas for the following:
(a) 3-ethyl-3-methyl-1-pentene
(b) *cis*-1,2-diphenylethene
(c) 3-phenyl-1-butyne
(d) cyclopentene
(e) 1-methylcyclohexene
(f) 3-isopropylcyclopentene

15. The following names are incorrect. State why each name is wrong and give the correct name.
(a) 3-methyl-3-butene
(b) 3-pentene
(c) *cis*-2-methyl-2-pentene

16. The following names are incorrect. State why each name is wrong and give the correct name.
(a) 3-ethyl-1-butene
(b) 2-chlorocyclohexene
(c) 4-hexene

17. Name the following compounds:

(a)
```
CH3CH2       CH3
       \    /
        C=C
       /    \
      H      CH2CH3
```

(b) (benzene ring)–$CH_2C{\equiv}CH$

(c) $CH_3CHBrCHBrC{\equiv}CCH_3$

18. Name the following compounds:

(a)
```
 H       H
  \     /
   C=C
  /     \
CH3      CHCH2CH3
         |
         CH3
```

(b)
```
CH3     CH3
   \   /
    C=C
   /   \
CH3     CH3
```

(c)
```
CH3CH2CHCH=CH2
      |
      CH
     /  \
  CH3    CH3
```

19. Write the structural formulas and IUPAC names for all the hexynes.

20. Write the structural formulas and IUPAC names for all the pentynes.

21. Which of the following molecules have structural formulas that permit cis–trans isomers to exist?
(a) $(CH_3)_2C{=}CHCH_3$
(b) $CH_3CH{=}CHCl$
(c) $CCl_2{=}CBr_2$

22. Which of the following molecules have structural formulas that permit cis–trans isomers to exist?
(a) $CH_2{=}CHCl$
(b) $CH_3CH_2C{\equiv}CCH_3$
(c) $CH_2ClCH{=}CHCH_2Cl$

23. Complete the following equations:

(a) $CH_3CH_2CH_2CH{=}CH_2 + Br_2 \longrightarrow$

(b) $CH_3CH_2\underset{\displaystyle CH_3}{\underset{|}{C}}{=}CHCH_3 + HI \longrightarrow$

(c) $CH_3CH_2CH{=}CH_2 + H_2O \xrightarrow{H^+}$

(d) $C_6H_5CH{=}CH_2 + H_2 \xrightarrow[1\ atm]{Pt,\ 25°C}$ (benzene ring bearing CH=CH₂)

(e) $CH_3CH{=}CHCH_3 + KMnO_4 \xrightarrow[cold]{H_2O}$

24. Complete the following equations:

(a) $CH_3CH_2CH_2CH{=}CH_2 + H_2O \xrightarrow{H^+}$

(b) $CH_3CH_2CH{=}CHCH_3 + HBr \longrightarrow$

(c) $CH_2{=}CHCl + Br_2 \longrightarrow$

(d) $C_6H_5CH{=}CH_2 + HCl$ (benzene ring bearing CH=CH₂)

(e) $CH_2{=}CHCH_2CH_3 + KMnO_4 \xrightarrow[cold]{H_2O}$

25. Complete the following equations:

(a) $CH_3C{\equiv}CCH_3 + Br_2\ (1\ mol) \longrightarrow$

(b) The two-step reaction

$$CH{\equiv}CH + HCl \longrightarrow \text{first product} \xrightarrow{HCl}$$

(c) $CH_3CH_2CH{\equiv}CH + H_2\ (1\ mole) \xrightarrow[25°C,\ 1\ atm]{Pt}$

26. Complete the following equations:

(a) $CH_3C{\equiv}CH + H_2\ (1\ mole) \xrightarrow[25°C,\ 1\ atm]{Pt}$

(b) $CH_3C{\equiv}CCH_3 + Br_2\ (2\ mol) \longrightarrow$

(c) The two-step reaction

$$CH_3C{\equiv}CH + HCl \longrightarrow \text{first product} \xrightarrow{HCl}$$

27. Write the formula and name for the product when cyclohexene reacts with

(a) Br_2 (c) H_2O, H^+

(b) HI (d) $KMnO_4(aq)$(cold)

28. Write the formula and name for the product when cyclopentene reacts with

(a) Cl_2 (c) H_2, Pt

(b) HBr (d) H_2O, H^+

29. Write equations to show how 2-butyne can be converted into

(a) 2,3-dibromobutane

(b) 2,2-dibromobutane

(c) 2,2,3,3-tetrabromobutane

30. Write equations to show how 2-pentyne can be converted into

(a) 2,3-dichloropentane

(b) 2,2-dichloropentane

(c) 2,2,3,3-tetrachloropentane

31. Write structural formulas for

(a) benzene (c) benzoic acid

(b) aniline (d) naphthalene

32. Write structural formulas for

(a) toluene (c) phenol

(b) ethylbenzene (d) anthracene

33. Write structural formulas for

(a) 1,3,5-tribromobenzene

(b) *o*-bromochlorobenzene

(c) *tert*-butylbenzene

(d) *p*-xylene

34. Write structural formulas for

(a) 1,3-dichloro-5-nitrobenzene

(b) *m*-dinitrobenzene

(c) 1,1-diphenylethane

(d) styrene

35. Write structural formulas and names for all the isomers of

(a) dichlorobromobenzene, $C_6H_3Cl_2Br$

(b) the toluene derivatives of formula C_9H_{12}

36. Write structural formulas and names for all the isomers of

(a) trichlorobenzene, $C_6H_3Cl_3$

(b) the benzene derivatives of formula C_8H_{10}

37. Write the structures and names for all the isomers that can be written by substituting an additional chlorine atom in *o*-chlorobromobenzene.

38. Write the structures and names for all the isomers that can be written by substituting a third chlorine atom in *o*-dichlorobenzene.

39. Name the following compounds:

(a) CH_2CH_3 on a benzene ring, with Cl para

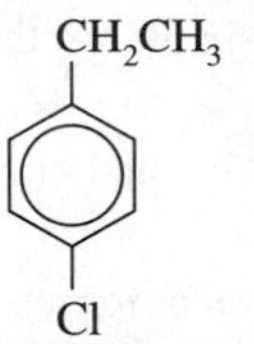

(b) benzene ring with $CH_2CH_2CH_3$

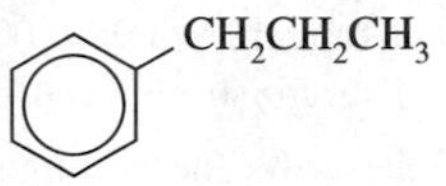

(c) benzene ring with NH_2 and NO_2

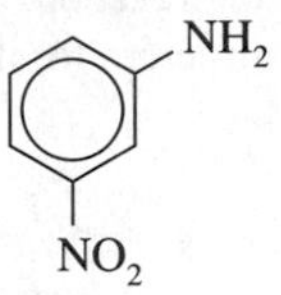

(d) benzene ring with OH and Br para

(e) CH with three benzene rings

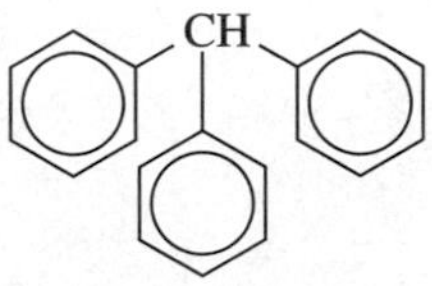

40. Name the following compounds:

(a) $CH{=}CH_2$ on a benzene ring

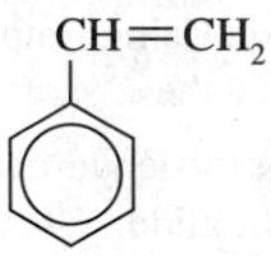

(b) benzene ring with CH_3 and NO_2

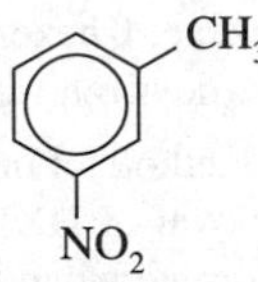

(c) benzene ring with COOH, Br, Br

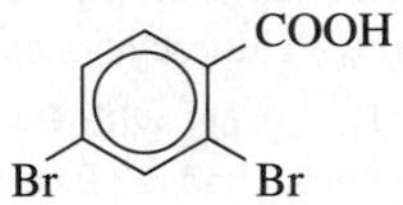

(d) benzene ring with $CH-CH_3$, CH_3 below the CH

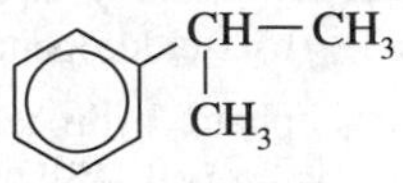

(e) benzene ring with Br, Br, Br and OH

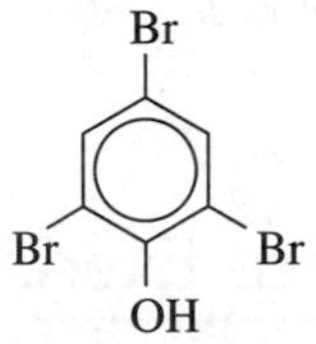

41. Complete the following equations and name the organic products:

(a) benzene $+ Br_2 \xrightarrow{FeBr_3}$

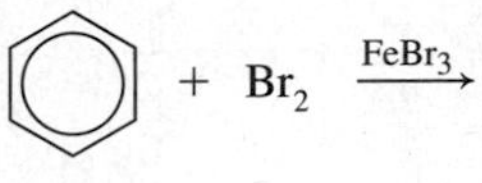

(b) benzene ring with two CH_3 (para) $+ HNO_3 \xrightarrow{H_2SO_4}$

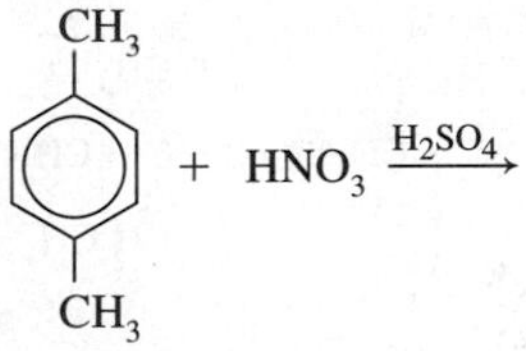

42. Complete the following equations and name the organic products:

(a) benzene $+ CH_3CHCH_3$ (Cl on the middle C) $\xrightarrow{AlCl_3}$

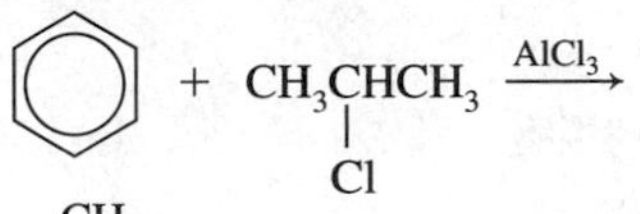

(b) benzene ring with CH_3 $+ KMnO_4 \xrightarrow{H_2O}$

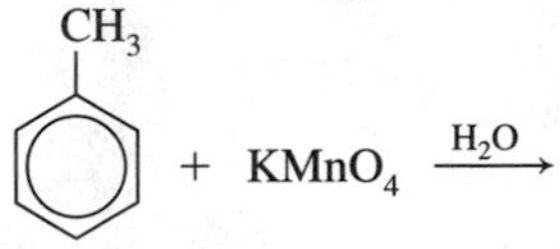

Additional Exercises

These exercises are not paired or labeled by topic and provide additional practice on concepts covered in this chapter.

***43.** Two alkyl bromides are possible when 2-methyl-1-pentene is reacted with HBr. Which one will predominate? Why?

***44.** Cyclohexane and 2-hexene both have the formula C_6H_{12}. How could you distinguish one from the other by chemical tests?

45. Write the structures and names of the methyl pentenes that show geometric isomerism.

46. Write structural formulas for (a) methyl carbocation, (b) propyl carbocation, (c) *tert*-butyl carbocation, and (d) pentyl carbocation.

47. Complete the following equations and name the products:

(a) cyclohexene $+ Br_2 \longrightarrow$

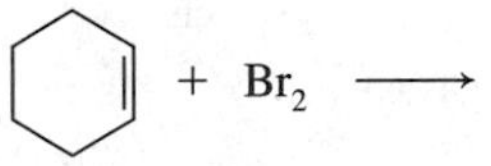

(b) cyclohexene $+ HCl \longrightarrow$

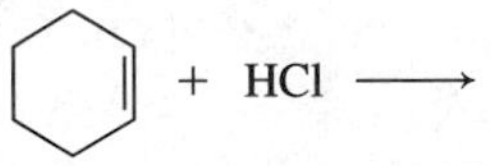

(c) cyclohexene with CH_3 $+ HCl \longrightarrow$

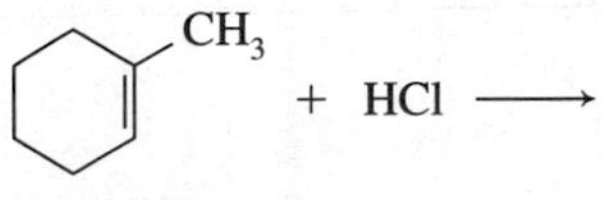

(d) cyclohexene with CH_3 $+ HCl \longrightarrow$

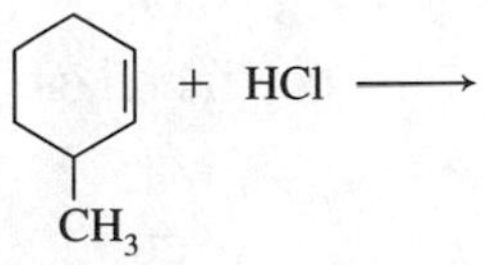

***48.** Describe the reaction mechanism by which benzene is brominated in the presence of $FeBr_3$. Show the equations.

*49. You are given three colorless liquid samples that are known to be benzene, 1-hexene, and 1-hexyne. Using simple chemical tests, devise a scheme to identify these samples.

50. Another method for making alkenes is by dehydrohalogenation (removal of HX) from an alkyl halide. What alkenes will be formed when HCl is removed from (a) 2-chlorobutane, (b) 1-chloropentane, and (c) chlorocyclohexane?

51. An aliphatic hydrocarbon, C_4H_8, reacts with Br_2, but no HBr is formed. Will there be a color change? Explain. Give possible structures for the original molecule.

52. Heptane and 1-heptene are both colorless liquids that boil in the range of 90–100°C. Describe a simple test that would distinguish between the two liquids.

53. Describe the bonding in ethane, ethene, and ethyne. Are they different or the same? Explain.

54. An unknown compound has the molecular formula C_5H_{10}. This compound does not show addition reactions with hydrogen or water. Write one possible structure for this compound.

Answers to Practice Exercises

21.1 (a) $CH_3CH_2CH{=}CHCH_2CH_3$

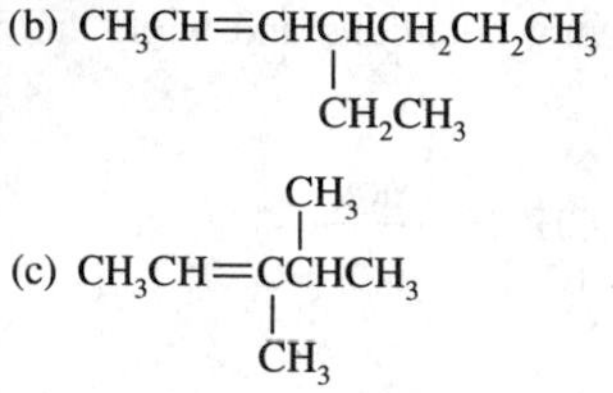

(b) $CH_3CH{=}CHCHCH_2CH_2CH_3$ (with CH_2CH_3 on the fourth carbon)

(c) $CH_3CH{=}CCHCH_3$ (with CH_3 above and CH_3 below)

21.2 (a) 3-methyl-2-pentene
(b) 2,4-dimethyl-2-pentene

21.3 (a) CH_3CH_2, CH_3 / C=C / CH_3, H (b) Br, CH_2CH_3 / C=C / CH_3, H

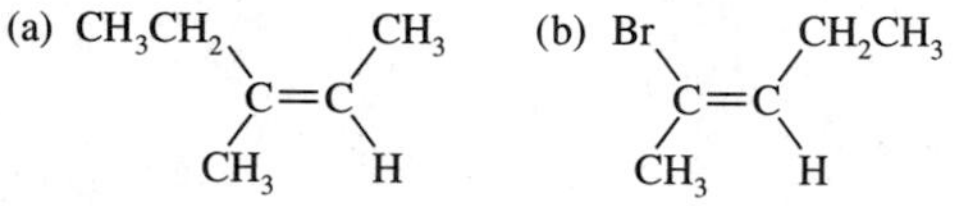

21.4 (a) Yes
CH_2Cl, CH_3 / C=C / H, CH_2CH_3 — trans
CH_2Cl, CH_2CH_3 / C=C / H, CH_3 — cis

(b) Yes
CH_3CH_2, CH_2CH_3 / C=C / H, H — cis
H, CH_2CH_3 / C=C / CH_3CH_2, H — trans

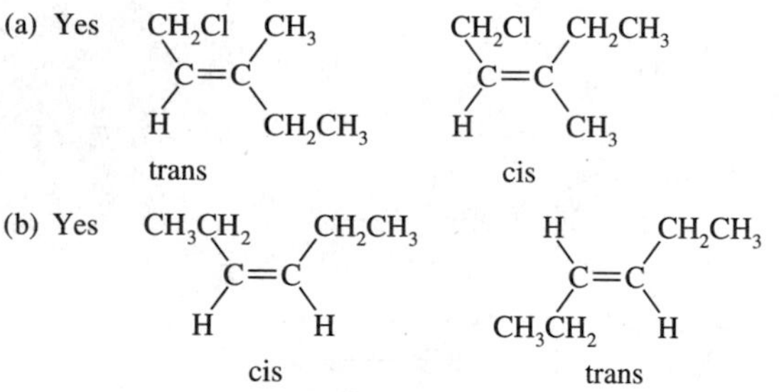

(c) No geometric isomers

21.5 (a) trans (b) cis

21.6 (a) $CH_3CH_2CHCH_2CH_3$ (with CH_3 below)

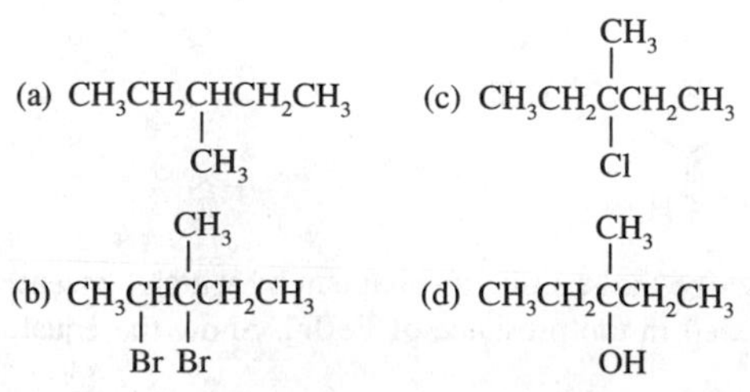

(b) $CH_3CHCCH_2CH_3$ (with CH_3 above; Br Br below)

(c) $CH_3CH_2CCH_2CH_3$ (with CH_3 above; Cl below)

(d) $CH_3CH_2CCH_2CH_3$ (with CH_3 above; OH below)

21.7 (a) 1-methylcyclopentene $+ H_2O \xrightarrow{H^+}$ cyclopentane ring bearing CH_3 and OH

(b) H, CH_3 / C=C / CH_3, CH_3 + HI ⟶ $CH_3CH_2CCH_3$ (with CH_3 above; I below)

21.8 (a) $CH_3CHC{\equiv}CH$ (with CH_3 below) (b) $CH_3C{\equiv}C{-}C{-}CH_2CH_3$ (with CH_3 above; CH_2CH_3 below)

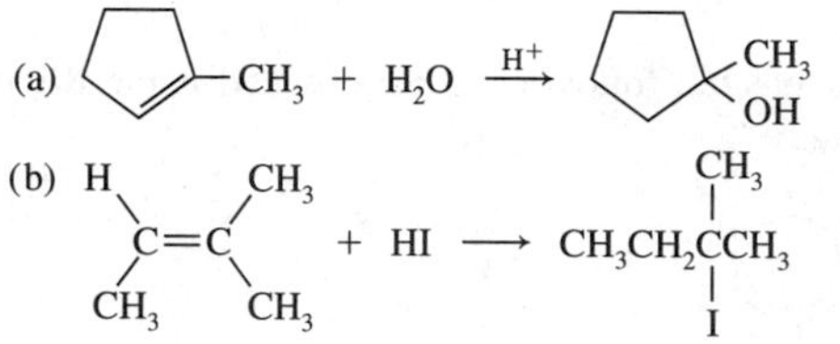

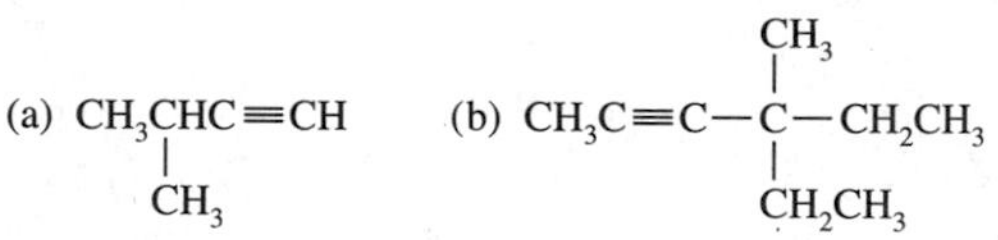

21.9 (a) $CH_3CH_2CCl_2CH_3$
(b) $CH_3CH_2CCl_2CH_2CH_3 + CH_3CCl_2CH_2CH_2CH_3$

21.10 OH, Cl (ortho) — *o*-chlorophenol; OH, Cl (meta) — *m*-chlorophenol; OH, Cl (para) — *p*-chlorophenol

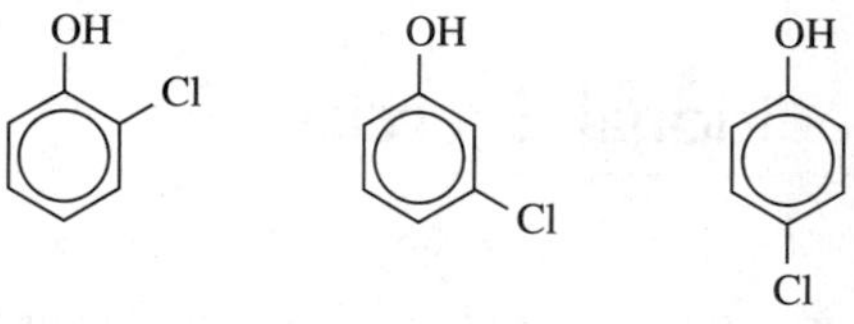

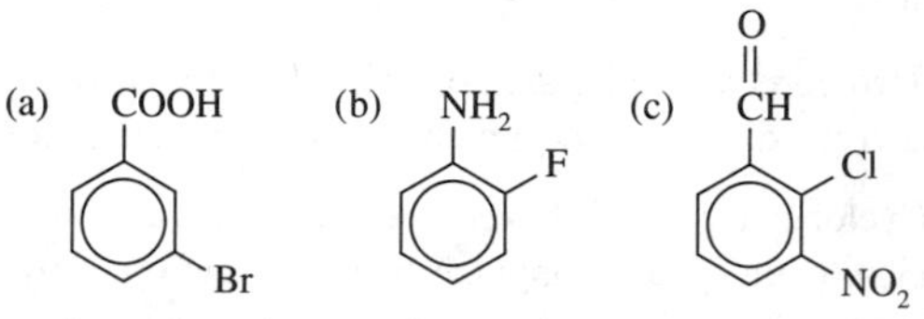

21.11 (a) COOH, Br — *m*-bromobenzoic acid (b) NH_2, F — *o*-fluoroaniline (c) CH=O, Cl, NO_2 — 2-chloro-3-nitrobenzaldehyde

Alcohols, Ethers, Phenols, and Thiols

22

Chapter 22

We tend to play as strenuously as we work. Our "vacations" often involve traveling by car to an area where we can indulge in vigorous physical activity like hiking, cycling, skiing, or swimming. After a hard day "torturing" our bodies, we relax and commiserate over our sore muscles. During all of these activities, organic molecules that contain an —OH group play a significant role. Ethylene glycol acts as a coolant in the radiator of our car, sugar and other carbohydrates provide the biochemical energy for our physical activities, ethyl alcohol is a component in any alcoholic beverage we might consume, and phenolic compounds are an active ingredient in the muscle rubs and analgesics that relieve our sore muscles.

Just what changes does the addition of an —OH group produce in the physical and chemical properties of an organic molecule? In this chapter we begin to examine the effect of various functional groups on organic molecules.

22.1 Functional Groups

Organic compounds were originally obtained from plants and animals, which, even today, are still the direct sources of many important chemicals. As a case in point, millions of tons of sucrose (table sugar) are obtained from sugar cane and sugar beet juices each year. However, as our knowledge of chemistry has increased, we have been able to synthesize many naturally occurring compounds, often at far less cost than the natural products. Of even greater significance than the cheaper manufacture of natural substances has been the synthesis of new substances totally unlike any natural product. These syntheses have been greatly aided by the realization that organic chemicals can be divided into a relatively small number of classes and studied on the basis of similar chemical properties (Table 20.1). The various classes of compounds are identified by the presence of certain characteristic groups called functional groups. For example, if a hydroxyl group (—OH) is substituted for a hydrogen atom in an alkane molecule, the resulting compound is an **alcohol**. Thus alcohols are a class of compounds in which the functional group is the hydroxyl group.

alcohol

Through the chemical reactions of functional groups, it is possible to create or synthesize new substances. The synthesis of new and possibly useful compounds or the more economical synthesis of known compounds is a main concern of modern organic chemistry. Most chemicals used today do not occur in nature but are synthesized from naturally occurring materials. The chemical and physical properties of an organic compound depend on (1) the kinds and number of functional groups present and (2) the shape and size of the molecule.

The structures of alcohols, ethers, and phenols may be derived from the structure of water by replacing the hydrogen atoms of water with alkyl groups (R) or aromatic rings:

H–O–H (water) R–O–H (alcohol) R–O–R (ether) C_6H_5–O–H (phenol)

The R— groups in ethers can be the same or different and can be alkyl groups or aromatic rings.

◀ **Chapter Opening Photo: These Indy cars have just refueled with methanol during the annual race at Elkart Lake, Wisconsin.**

TABLE 22.1 Names and Classification of Alcohols

Class	Formula	IUPAC name	Common name*	Boiling point (°C)
Primary	CH_3OH	Methanol	Methyl alcohol	65.0
Primary	CH_3CH_2OH	Ethanol	Ethyl alcohol	78.5
Primary	$CH_3CH_2CH_2OH$	1-Propanol	*n*-Propyl alcohol	97.4
Primary	$CH_3CH_2CH_2CH_2OH$	1-Butanol	*n*-Butyl alcohol	118
Primary	$CH_3(CH_2)_3CH_2OH$	1-Pentanol	*n*-Pentyl alcohol (*n*-amyl)	138
Primary	$CH_3CH(CH_3)CH_2OH$	2-Methyl-1-propanol	Isobutyl alcohol	108
Primary	$C_6H_5CH_2OH$	Phenylmethanol	Benzyl alcohol	205
Secondary	$CH_3CH(OH)CH_3$	2-Propanol	Isopropyl alcohol	82.5
Secondary	$CH_3CH_2CH(OH)CH_3$	2-Butanol	*sec*-Butyl alcohol	91.5
Tertiary	$CH_3C(CH_3)_2OH$	2-Methyl-2-propanol	*t*-Butyl alcohol	82.9
Dihydroxy	$HOCH_2CH_2OH$	1,2-Ethanediol	Ethylene glycol	197
Trihydroxy	$HOCH_2CH(OH)CH_2OH$	1,2,3-Propanetriol	Glycerol or glycerine	290

*The abbreviations *n*, *sec*, and *t* stand for normal, secondary, and tertiary, respectively.

22.2 Classification of Alcohols

Structurally, an alcohol is derived from an aliphatic hydrocarbon by the replacement of at least one hydrogen atom with a hydroxyl group (—OH). Alcohols are represented by the general formula ROH, with methanol (CH_3OH) being the first member of the homologous series (R represents an alkyl or substituted alkyl group). Models illustrating the structural arrangements of the atoms in methanol and ethanol are shown in Figure 22.1.

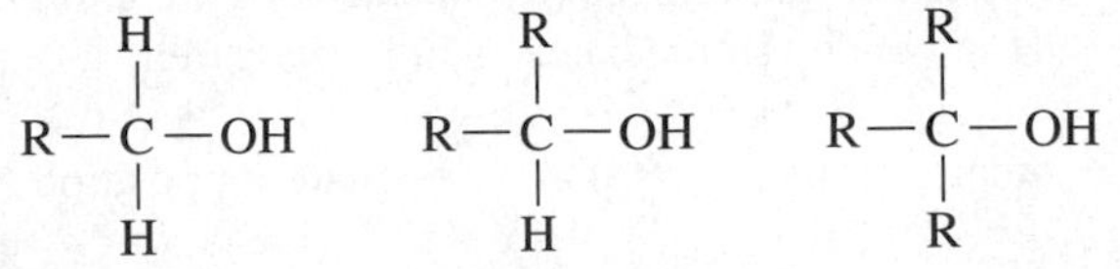

primary alcohol
secondary alcohol
tertiary alcohol

Alcohols are classified as **primary** (1°), **secondary** (2°), or **tertiary** (3°), depending on whether the carbon atom to which the —OH group is attached is directly bonded to one, two, or three other carbon atoms, respectively. (The terms secondary and tertiary are used as they were for alkyl groups in Chapter 20.) Generalized formulas for 1°, 2°, and 3° alcohols are as follows:

$$\underset{\text{primary alcohol}}{R{-}CH_2{-}OH} \qquad \underset{\text{secondary alcohol}}{R{-}CHR{-}OH} \qquad \underset{\text{tertiary alcohol}}{R{-}CR_2{-}OH}$$

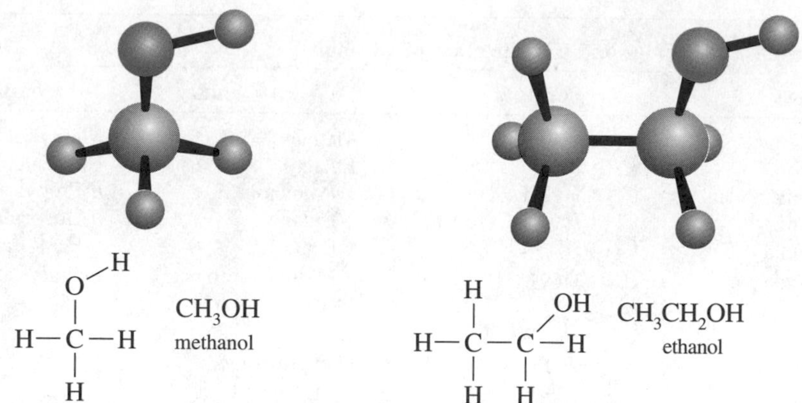

FIGURE 22.1 ▶ Ball-and-stick models illustrating structural formulas of methanol and ethanol.

Formulas of specific examples of these classes of alcohols are shown in Table 22.1. Methanol (CH_3OH) is grouped with the primary alcohols.

Molecular structures with more than one —OH group attached to a single carbon atom are generally not stable. But an alcohol molecule can contain two or more —OH groups if each —OH is attached to a different carbon atom. Accordingly, alcohols are also classified as monohydroxy, dihydroxy, trihydroxy, and so on, on the basis of the number of hydroxyl groups per molecule. **Polyhydroxy alcohols** and *polyols* are general terms for alcohols that have more than one —OH group per molecule. Polyhydroxy compounds are very important molecules in living cells as they include the carbohydrate class of biochemicals.

polyhydroxy alcohol

An alcohol such as 2-butanol can be written in a single-line formula by putting the —OH group in parentheses and placing it after the carbon to which it is bonded. For example, the following two formulas represent the same compound:

$$\underset{\textstyle OH}{CH_3CH_2\overset{}{C}HCH_3} \qquad CH_3CH_2CH(OH)CH_3$$

22.3 Naming Alcohols

If you know how to name alkanes, it is easy to name alcohols by the IUPAC System:

IUPAC Nomenclature for Alcohols

1. Select the longest continuous chain of carbon atoms containing the hydroxyl group.
2. Number the carbon atoms in this chain so that the one bearing the —OH group has the lowest possible number.
3. Form the parent alcohol name by replacing the final *e* of the corresponding alkane name by *ol*. When isomers are possible, locate the position of the —OH by placing the number (hyphenated) of the carbon atom to which the —OH is bonded immediately before the parent alcohol name.
4. Name each alkyl side chain (or other group) and designate its position by number.

For example, let's go through the steps above to name the alcohol $CH_3CH_2CH_2CH_2OH$.

Step 1 The longest carbon chain containing the —OH group has four carbons.

Step 2 Number the carbon atoms, giving the carbon bonded to the —OH the number 1.

$$\overset{4}{C}—\overset{3}{C}—\overset{2}{C}—\overset{1}{C}—OH$$

Step 3 The name of the four-carbon alkane is butane. Replace the final *e* in butane with *ol*, forming the name *butanol*. Since the —OH is on carbon 1, place a *1* before butanol to give the complete alcohol name 1-butanol.

Step 4 No groups of atoms other than hydrogen are attached to the butanol chain, so the name of this alcohol is 1-butanol.

Study the application of this naming system to these examples and to those shown in Table 22.1.

$$\overset{3}{CH_3}—\overset{2}{CH_2}—\overset{1}{CH_2OH}$$

1-propanol

$$\overset{1}{CH_3}—\underset{\displaystyle OH}{\overset{2}{CH}}—\overset{3}{CH_3}$$

2-propanol

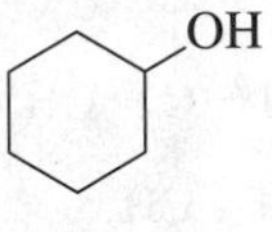

(cyclohexane ring with OH)

cyclohexanol

$$\overset{4}{CH_3}—\underset{\displaystyle CH_3}{\overset{3}{CH}}—\overset{2}{CH_2}—\overset{1}{CH_2OH}$$

3-methyl-1-butanol

$$HO\overset{2}{CH_2}—\overset{1}{CH_2OH}$$

1,2-ethanediol

Example 22.1

Name this alcohol by the IUPAC method.

$$CH_3CH_2\underset{\displaystyle CH_3}{CH}CH_2\underset{\displaystyle OH}{CH}CH_3$$

Solution

Step 1 The longest continuous carbon chain containing the —OH group has six carbon atoms.

Step 2 This carbon chain is numbered from right to left so that the —OH group has the lowest possible number:

$$\overset{6}{C}—\overset{5}{C}—\underset{\displaystyle CH_3}{\overset{4}{C}}—\overset{3}{C}—\underset{\displaystyle OH}{\overset{2}{C}}—\overset{1}{C}$$

In this case, the —OH is on carbon 2.

Step 3 The name of the six-carbon alkane is hexane. Replace the final *e* in hexane by *ol*, to form the name *hexanol*. Since the —OH is on carbon 2, place a *2* before hexanol to give the parent alcohol name 2-hexanol.

Step 4 A methyl group (—CH_3) is located on carbon 4. Therefore, the full name of the compound is 4-methyl-2-hexanol.

Example 22.2 Write the structural formula of 3,3-dimethyl-2-hexanol.

Solution

Step 1 The "2-hexanol" refers to a six-carbon chain with an —OH group on carbon 2. Write the skeleton structure as follows:

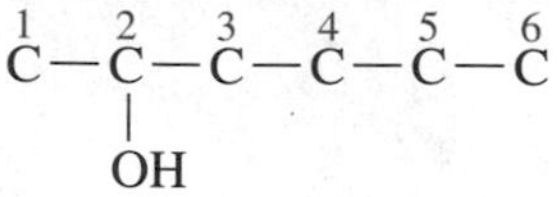

```
1   2   3   4   5   6
C — C — C — C — C — C
    |
    OH
```

Step 2 Place the two methyl groups ("3,3-dimethyl") on carbon 3:

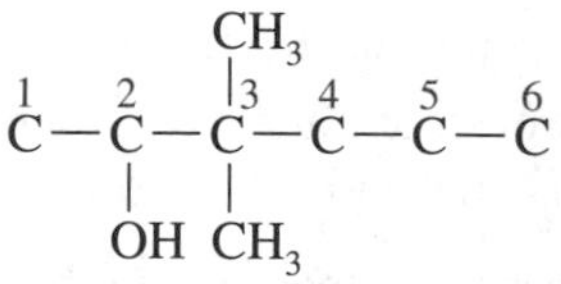

```
        CH3
        |3
1   2   |   4   5   6
C — C — C — C — C — C
    |   |
    OH  CH3
```

Step 3 Finally, add H atoms to give each carbon atom four bonds:

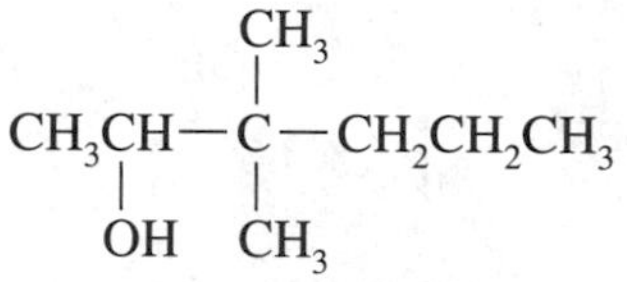

```
           CH3
           |
CH3CH — C — CH2CH2CH3
    |      |
    OH     CH3
```

3,3-dimethyl-2-hexanol

Practice 22.1

Write the IUPAC name for each of the following:

```
                                                      Br
                                                      |
(a) CH3CH2CHCH2CH2CH3      (b) CH3CH2CHCH2CCH3
             |                          |     |
             CH2OH                      OH    CH3
```

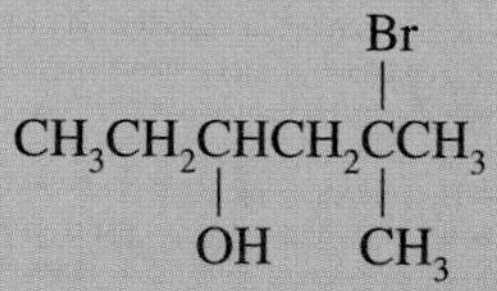

Practice 22.2

Write the structural formula for each of the following:
(a) 3-methylcyclohexanol (b) 4-ethyl-2-methyl-3-heptanol

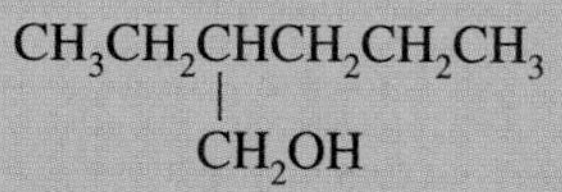

Several of the alcohols are generally known by their common names, so it may be necessary to know more than one name for a given alcohol. The common name is usually formed from the name of the alkyl group that is attached to the —OH group, followed by the word *alcohol*. (See the examples in Table 22.1.)

22.4 Physical Properties of Alcohols

The physical properties of alcohols are related to those of both water and alkane hydrocarbons. This is easily understandable if we recall certain facts about water and the alkanes. Water molecules are quite polar. The properties of water, such as its high boiling point and its ability to dissolve many polar substances, are largely due to the polarity of its molecules. Alkane molecules possess almost no polarity. The properties of the alkanes reflect this lack of polarity—for example, their relatively low boiling points and inability to dissolve water and other polar substances. An alcohol molecule is made up of a water-like hydroxyl group joined to a hydrocarbon-like alkyl group.

H–O–H (water) R–O–H (alcohol)

One striking property of alcohols is their relatively high boiling points. The simplest alcohol, methanol, boils at 65°C. But methane, the simplest hydrocarbon, boils at −162°C. The boiling points of the normal alcohols increase in a regular fashion with increasing number of carbon atoms. Branched-chain alcohols have lower boiling points than the corresponding straight-chain alcohols (see Table 22.1).

Alcohols containing up to three carbon atoms are infinitely soluble in water. With one exception (2-methyl-2-propanol), alcohols with four or more carbon atoms have limited solubility in water. In contrast, all hydrocarbons are essentially insoluble in water.

The hydroxyl group on the alcohol molecule is responsible for both the water solubility and the relatively high boiling points of the low molar-mass alcohols. Hydrogen bonding (see Section 13.11) between water and alcohol molecules accounts for the solubility, and hydrogen bonding between alcohol molecules accounts for their high boiling points.

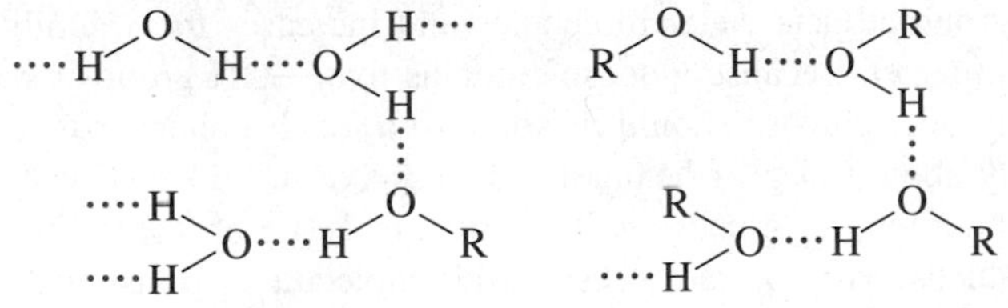

water–alcohol alcohol–alcohol

As the length of the hydrocarbon chain increases, the polar effect of the hydroxyl group becomes relatively less important. Alcohols with 5–11 carbons are oily liquids of slight water solubility, and in physical behavior they resemble the corresponding alkane hydrocarbons. However, the effect of the —OH group is still noticeable in that their boiling points are higher than those of alkanes with similar molar masses (see Table 22.2). Alcohols containing 12 or more carbons are wax-like solids that resemble solid alkanes in physical appearance.

TABLE 22.2 Comparison of Boiling Points of Alkanes and Monohydroxy Alcohols

Name	Boiling point (°C)	Name	Boiling point (°C)
Hexane	69	1-Hexanol	156
Octane	126	1-Octanol	195
Decane	174	1-Decanol	228

TABLE 22.3 Comparison of the Boiling Points of Ethanol, 1,2-Ethanediol, and 1-Propanol

Name	Formula	Molar mass	Boiling point (°C)
Ethanol	CH_3CH_2OH	46	78
1,2-Ethanediol	$CH_2(OH)CH_2OH$	62	197
1-Propanol	$CH_3CH_2CH_2OH$	60	97

In alcohols with two or more hydroxyl groups, the effect of the hydroxyl groups on intermolecular attractive forces is, as we might suspect, even more striking than in monohydroxy alcohols. Ethanol, CH_3CH_2OH, boils at 78°C, but the boiling point of 1,2-ethanediol, $CH_2(OH)CH_2OH$, is 197°C. Comparison of the boiling points of ethanol, 1-propanol, and 1,2-ethanediol shows that increased molar mass does not account for the high boiling point of 1,2-ethanediol (see Table 22.3). The higher boiling point is primarily a result of additional hydrogen bonding due to the two —OH groups in the 1,2-ethanediol molecule.

Example 22.3 Glucose is one of the most important carbohydrates in biochemistry. It has six carbons and five alcohol groups (molar mass = 180.2 g). How would you predict the water solubility of glucose will differ from that of hexanol?

Solution Each polar alcohol group attracts water molecules and increases the solubility of organic compounds in water. Because glucose contains five —OH groups whereas hexanol contains only one, glucose should dissolve to a much greater extent than hexanol. In fact, only about 0.6 g of hexanol will dissolve in 100 g of water (at 20°C). In contrast, about 95 g of glucose will dissolve in 100 g of water. The high water solubility of glucose is important because this molecule is transported in a water solution to the body's cells.

Practice 22.3

List the following alcohols in order of increasing solubility in water:
(a) 1-hexanol (b) 1-pentanol (c) 1,3-propanediol (d) 1-butanol

22.5 Chemical Properties of Alcohols

An alcohol contains the hydroxyl functional group (—OH). Chemists have chosen the term *functional group* to indicate that this group brings a "function" to an organic molecule. Just as adding new electronic chips to a calculator can add new functions, attaching a hydroxyl group to an alkyl chain can allow a molecule to function in new ways. As we have seen, the —OH group can interact with water so that when a molecule adds a hydroxyl group, its solubility in water will increase. In general, the "functions" of a hydroxyl group will help determine the properties of an alcohol.

Acidic and Basic Properties

Aliphatic alcohols are similar to water in their acidic/basic properties. If an alcohol is mixed with a strong acid, it will accept a proton (act as a Brønsted–Lowry base) to form a protonated alcohol or **oxonium ion**.

oxonium ion

$$CH_3{-}\ddot{O}H + H_2SO_4 \longrightarrow CH_3{-}\overset{+}{\ddot{O}}(H)_2 + HSO_4^-$$

Methanol and ethanol are approximately the same strength as water as an acid, while the larger alcohols are weaker acids than water. Both water and alcohols react with alkali metals to release hydrogen gas and an anion.

$$2\ H_2O + 2\ Na \longrightarrow \underset{\text{sodium hydroxide}}{2\ Na^+\ {}^-OH} + H_2(g)$$

$$2\ CH_3CH_2OH + 2\ Na \longrightarrow \underset{\text{sodium ethoxide}}{2\ Na^+\ {}^-OCH_2CH_3} + H_2(g)$$

The resulting anion in the alcohol reaction is known as an **alkoxide ion** (RO^-). Alkoxides are strong bases (stronger than hydroxide) and so are used in organic chemistry when a strong base is required in a nonaqueous solution.

alkoxide ion

The order of reactivity of alcohols with sodium or potassium is primary > secondary > tertiary. Alcohols do not react with sodium as vigorously as water. Reactivity decreases with increasing molar mass since the —OH group becomes a relatively smaller, less significant part of the molecule.

As you will see in later chapters, the —OH group is also part of the carboxylic acid functional group ($-\overset{O}{\overset{\|}{C}}-OH$). The —OH behaves much differently in the carboxylic acid group than it does in an alcohol.

Oxidation

We will consider only a few of the many reactions that alcohols are known to undergo. One important reaction is oxidation. We saw in Chapter 17 that the oxidation number of an element increases as a result of oxidation. Carbon can exist in several oxidation states, ranging from -4 to $+4$. In the -4 oxidation state, such as in methane, the carbon atom is considered to be completely reduced. In carbon dioxide the carbon atom is completely oxidized; that is, it is in its highest oxidation state ($+4$). In many cases, oxidation reactions in organic chemistry and biochemistry can be considered in a simple manner without the use of oxidation numbers. Oxidation is the loss of hydrogen or the gain of bonds to oxygen by the organic reactant. Table 22.4 illustrates the progression of oxidation states for various compounds containing one carbon atom.

Carbon atoms exist in progressively higher stages of oxidation in different functional-group compounds:

TABLE 22.4 Oxidation States of Carbon in One-Carbon Compounds

	Compound	Number of C—O bonds	Oxidation state
CH_4	Methane	0	−4
CH_3OH	Methanol	1	−2
$H_2C{=}O$	Methanal (formaldehyde)	2	0
$HC(=O){-}OH$	Methanoic acid (formic acid)	3	+2
$O{=}C{=}O$	Carbon dioxide	4	+4

$$\text{Alkanes} \longrightarrow \text{Alcohols} \longrightarrow \left\{\begin{matrix}\text{Aldehydes}\\ \text{Ketones}\end{matrix}\right\} \longrightarrow \begin{matrix}\text{Carboxylic}\\ \text{acids}\end{matrix} \longrightarrow \begin{matrix}\text{Carbon}\\ \text{dioxide}\end{matrix}$$

Increasing oxidation state ⟶

The synthesis of alcohols enables us to make many compounds that are more oxidized. In general, it is not practical to convert alkanes directly to alcohols, but once the —OH functional group has been attached, then further oxidations are possible. The hydroxyl group gives an organic compound the capability ("function") of forming an aldehyde, ketone, or carboxylic acid. Industrial chemists use this hydroxyl group function often; compounds as diverse as plastics, antibiotics, and fertilizers are synthesized starting with alcohols. The following equations represent generalized oxidation reactions in which the oxidizing agent is represented by [O].

$$\underset{\text{primary alcohol}}{R{-}CH(H){-}OH} \xrightarrow{[O]} \underset{\text{aldehyde}}{R{-}C(=O){-}H} + H_2O \xrightarrow{[O]} \underset{\text{carboxylic acid}}{R{-}C(=O){-}OH}$$

$$\underset{\text{secondary alcohol}}{R{-}C(H)(OH){-}R} \xrightarrow{[O]} \underset{\text{ketone}}{R{-}C(=O){-}R} + H_2O$$

$$R{-}C(R)(OH){-}R \xrightarrow{[O]} \text{No reaction}$$

Tertiary alcohols do not have a hydrogen on the —OH carbon and so cannot react with oxidizing agents except by such drastic procedures as combustion. Both primary and secondary alcohols are oxidized as shown in the equations by the loss of the colored hydrogen atoms.

Some common oxidizing agents used for specific reactions are potassium permanganate, $KMnO_4$, in an alkaline solution, potassium dichromate, $K_2Cr_2O_7$, in an acid solution, or oxygen of the air. A complete equation for an alcohol oxidation may be fairly complex. Since our main interest is in the changes that occur in the functional groups, we can convey this information in abbreviated form:

$$\underset{\text{ethanol}}{CH_3CH_2OH} \xrightarrow[\Delta]{K_2Cr_2O_7/H_2SO_4} \underset{\text{ethanal}}{CH_3\overset{\overset{\displaystyle O}{\|}}{C}{-}H} + H_2O$$

Although the abbreviated equation lacks some of the details, it does show the overall reaction involving the organic compounds. Additional information is provided by notations above and below the arrow, which indicate that this reaction is carried out in heated potassium dichromate–sulfuric acid solution. Abbreviated equations of this kind will be used frequently in the remainder of this book.

Example 22.4

What are the products when (a) 1-propanol, (b) 2-propanol, and (c) cyclohexanol are oxidized with $K_2Cr_2O_7/H_2SO_4$?

Solution

(a) The formula for 1-propanol is $CH_3CH_2CH_2OH$. Since it is a primary alcohol, it can be oxidized to an aldehyde or a carboxylic acid. The oxidation occurs at the carbon bonded to the —OH group, and this carbon atom becomes an aldehyde or a carboxylic acid. The rest of the molecule remains the same.

$$CH_3CH_2CH_2OH \nearrow \underset{\text{propanal}}{CH_3CH_2\overset{\overset{\displaystyle H}{|}}{C}{=}O}$$

$$CH_3CH_2CH_2OH \searrow \underset{\text{propanoic acid}}{CH_3CH_2COOH}$$

(b) 2-Propanol is a secondary alcohol and is oxidized to a ketone. Ketones resist further oxidation. The oxidation occurs at the carbon bonded to the —OH group.

$$CH_3\underset{\underset{\displaystyle OH}{|}}{C}HCH_3 \longrightarrow CH_3\underset{\underset{\displaystyle O}{\|}}{C}CH_3$$

propanone (acetone)

(c) Cyclohexanol is also a secondary alcohol and is oxidized to cyclohexanone, a ketone.

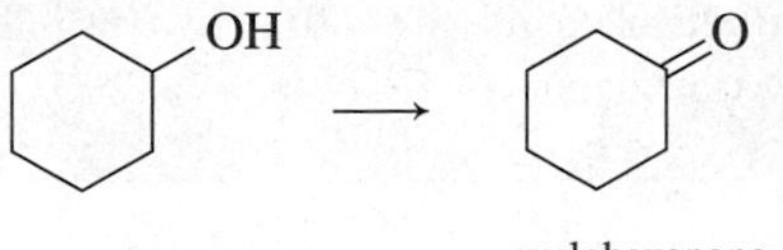

cyclohexanol (OH) ⟶ cyclohexanone (O)

cyclohexanone

Practice 22.4

Write the structure for the products of the oxidation of these alcohols with $K_2Cr_2O_7$ and H_2SO_4:
(a) 1-butanol (b) 2-methyl-2-butanol (c) cyclopentanol

Dehydration

The term *dehydration* implies the elimination of water. Alcohols can be dehydrated to form alkenes (Section 21.5) or ethers. One of the more effective dehydrating agents is sulfuric acid. Whether an ether or an alkene is formed depends on the ratio of alcohol to sulfuric acid, the reaction temperature, and the type of alcohol (1°, 2°, or 3°).

Alkenes: Intramolecular Dehydration The formation of alkenes from alcohols requires a relatively high temperature. Water is removed from within a *single* alcohol molecule and a carbon–carbon double bond forms as shown here:

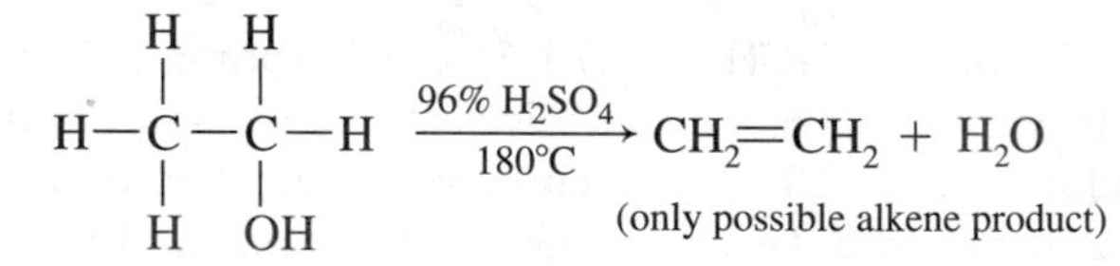

$$\begin{array}{c} \text{H}\quad\text{H} \\ |\quad\;| \\ \text{H}-\text{C}-\text{C}-\text{H} \\ |\quad\;| \\ \text{H}\quad\text{OH} \end{array} \xrightarrow[180°\text{C}]{96\%\ H_2SO_4} CH_2{=}CH_2 + H_2O$$

(only possible alkene product)

For many alcohols there are a number of ways to remove water. Therefore, the double bond can be located in different positions. The major product in these cases is the compound in which the double bond has the greatest number of alkyl substituents (or lesser number of hydrogens).

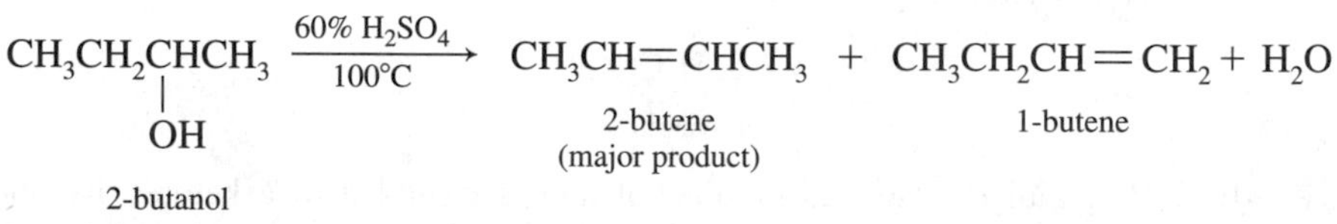

$$\underset{\text{2-butanol}}{\underset{\text{OH}}{CH_3CH_2\overset{}{C}HCH_3}} \xrightarrow[100°\text{C}]{60\%\ H_2SO_4} \underset{\substack{\text{2-butene}\\ \text{(major product)}}}{CH_3CH{=}CHCH_3} + \underset{\text{1-butene}}{CH_3CH_2CH{=}CH_2} + H_2O$$

To predict the major product in an intramolecular dehydration, follow these steps: (1) Remove H and OH from adjacent carbons forming a carbon–carbon double bond; (2) if there are choices (multiple hydrogen-containing neighboring carbon atoms adjacent to the —OH carbon), remove the hydrogen from the carbon with fewer hydrogens. This is known as **Saytzeff's rule**. This rule in essence states:

Saytzeff's rule

During intramolecular dehydration, if there is a choice of positions for the carbon–carbon double bond, the preferred location is the one that generally gives the more highly substituted alkene—that is, the alkene with the most alkyl groups on the double-bond carbons.

Ethers: Intermolecular Dehydration A dehydration reaction can take place between two alcohol molecules to produce an ether. However, the dehydration to make an ether is only useful for primary alcohols since secondary and tertiary alcohols predominantly yield alkenes.

$$\begin{array}{l} CH_3CH_2O\boxed{H} \\ CH_3CH_2\boxed{OH} \end{array} \xrightarrow[140°C]{96\%\ H_2SO_4} \underset{\text{diethyl ether}}{CH_3CH_2OCH_2CH_3} + H_2O$$

A reaction in which *two* molecules are combined by removing a small molecule is known as a **condensation reaction**. There are many examples of other condensation reactions in the formation of biochemical molecules.

condensation reaction

The type of dehydration that occurs depends on the temperature and the number of reactant molecules. Lower temperatures and *two* alcohol molecules produce ethers while higher temperatures and a *single* alcohol produce alkenes.

Dehydration reactions are often used in industry to make relatively expensive ethers from lower-priced alcohols. Alkenes are less expensive than alcohols and so are not produced industrially by this method.

Esterification (Conversion of Alcohols to Esters)

An alcohol can react with a carboxylic acid to form an ester and water. The reaction is represented as follows:

$$\underset{\text{carboxylic acid}}{R-\overset{\overset{\displaystyle O}{\|}}{C}-\boxed{OH}} + \underset{\text{alcohol}}{\boxed{H}-\boxed{OR'}} \overset{H^+}{\rightleftharpoons} \underset{\text{ester}}{R\overset{\overset{\displaystyle O}{\|}}{C}-OR'} + HOH$$

$$\underset{\text{acetic acid}}{CH_3\overset{\overset{\displaystyle O}{\|}}{C}-OH} + \underset{\text{ethanol}}{HOCH_2CH_3} \overset{H^+}{\rightleftharpoons} \underset{\text{ethyl acetate}}{CH_3\overset{\overset{\displaystyle O}{\|}}{C}-OCH_2CH_3} + HOH$$

Esterification is an important reaction of alcohols and is discussed in greater detail in Chapter 24.

Utility of the Hydroxyl Functional Group

The hydroxyl group is a particularly important functional group; it introduces a myriad of possible reactions leading to a variety of other valuable organic compounds. An organic chemist might term the hydroxyl group as a valuable intermediate, a "gateway" functional group:

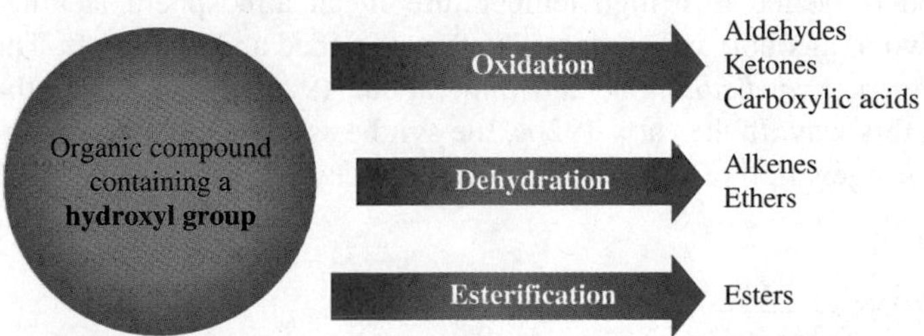

The common alcohols are often made by special reactions (see subsequent sections) from relatively inexpensive hydrocarbon sources. Once the hydroxyl functional group is introduced, the reactivity of the hydrocarbon has been changed, and the chemist has many options available. Note in the upcoming sections how many different uses are possible for the simple alcohols.

22.6 Common Alcohols

Three general methods for making alcohols are

1. *Hydrolysis of an ester*

$$\underset{\text{ester}}{\mathrm{R\overset{\displaystyle O}{\overset{\|}{C}}{-}OR'}} + \mathrm{HOH} \xrightarrow[\Delta]{\mathrm{H^+}} \underset{\text{carboxylic acid}}{\mathrm{R\overset{\displaystyle O}{\overset{\|}{C}}{-}OH}} + \underset{\text{alcohol}}{\mathrm{R'OH}}$$

2. *Alkaline hydrolysis of an alkyl halide* (1° and 2° alcohols only)

$$\underset{\text{alkyl halide}}{\mathrm{RX}} + \mathrm{NaOH}(aq) \longrightarrow \underset{\text{alcohol}}{\mathrm{ROH}} + \mathrm{NaX}$$

$$\mathrm{CH_3CH_2Cl} + \mathrm{NaOH}(aq) \longrightarrow \mathrm{CH_3CH_2OH} + \mathrm{NaCl}$$

Hydrolysis is a reaction of water with another species in which the water molecule is split (see Section 16.13). The hydrolysis of an ester is the reverse reaction of esterification. A carboxylic acid and an alcohol are formed as products. The reaction can be conducted in an acid or an alkaline medium.

3. *Catalytic reduction of aldehydes and ketones* to produce primary and secondary alcohols. These reactions are discussed in Chapter 23.

In theory, these general methods provide a way to make almost any desired alcohol, but they may not be practical for a specific alcohol because the necessary starting material—ester, alkyl halide, aldehyde, or ketone—cannot be obtained at a reasonable cost. Hence, for economic reasons, most of the widely used alcohols are made on an industrial scale by special methods that have been developed for specific alcohols. The preparation and properties of several of these alcohols are described in the following paragraphs.

Methanol

When wood is heated to a high temperature in an atmosphere lacking oxygen, methanol (wood alcohol) and other products are formed and driven off. The process is called *destructive distillation*, and until about 1925, nearly all methanol was obtained in this way. In the early 1920s, the synthesis of methanol by high-pressure catalytic hydrogenation of carbon monoxide was developed in Germany. The reaction is

$$\mathrm{CO} + 2\,\mathrm{H_2} \xrightarrow[\text{300–400°C, 200 atm}]{\mathrm{ZnO{-}Cr_2O_3}} \mathrm{CH_3OH}$$

◀ **Chemical tests that determine alcohol level can be administered to drivers at the roadside.**

The most economical nonpetroleum source of carbon monoxide for making methanol is coal. In addition to coal, burnable materials such as wood, agricultural wastes, and sewage sludge also are potential sources of methanol.

Methanol is a volatile (bp 65°C), highly flammable liquid. It is poisonous and capable of causing blindness or death if taken internally. Exposure to methanol vapors for even short periods of time is dangerous. Despite this danger, over 5.1×10^9 kg is manufactured annually. Methanol is used for

1. Conversion to formaldehyde (methanal), primarily for use in the manufacture of polymers
2. Manufacture of other chemicals, especially various kinds of esters
3. Denaturing ethyl alcohol (rendering it unfit as a beverage)
4. An industrial solvent

The use of about 10% methanol or ethanol in gasoline (*gasohol*) has shown promising results in reducing the amount of air pollutants emitted in automobile exhausts. Another benefit of using methanol in gasoline is that methanol can be made from nonpetroleum sources.

Ethanol

Ethanol is without doubt the earliest and most widely known alcohol. It is or has been known by a variety of other names such as ethyl alcohol, "alcohol," grain alcohol, and spirit. Huge quantities of this substance are prepared by fermentation. Starch and sugar are the raw materials. Starch is first converted to sugar by enzyme- or acid-catalyzed hydrolysis. (An enzyme is a biological catalyst, as discussed in Chapter 31.) Conversion of simple sugars to ethanol is accomplished by yeast:

$$\underset{\text{glucose}}{C_6H_{12}O_6} \xrightarrow{\text{yeast}} 2\ \underset{\text{ethanol}}{CH_3CH_2OH} + 2\ CO_2$$

Before bottling the wine, a vintner checks alcohol content, color, and sediment. ▶

For legal use in beverages, ethanol is made by fermentation, but a large part of the alcohol for industrial uses (5.9×10^8 kg annually) is made from petroleum-derived ethene:

$$CH_2{=}CH_2 + H_2O \xrightarrow{H^+} CH_3CH_2OH$$

Some of the economically significant uses of ethanol include the following:

1. Intermediate in the manufacture of other chemicals such as acetaldehyde, acetic acid, ethyl acetate, and diethyl ether
2. Solvent for many organic substances
3. Compounding ingredient for pharmaceuticals, perfumes, flavorings, and so on
4. Essential ingredient of alcoholic beverages

Ethanol acts physiologically as a food, as a drug, and as a poison. It is a food in the limited sense that the body is able to metabolize it to carbon dioxide and water with the production of energy. As a drug, ethanol is often mistakenly considered to be a stimulant, but it is in fact a depressant. In moderate quantities ethanol causes drowsiness and depresses brain functions so that activities requiring skill and judgment (such as automobile driving) are impaired. In larger quantities ethanol causes nausea, vomiting, impaired perception, and incoordination. If a very large amount is consumed, unconsciousness and ultimately death may occur.

Authorities maintain that the effects of ethanol on automobile drivers are a factor in about half of all fatal traffic accidents in the United States. The gravity of this problem can be grasped when you realize that traffic accidents are responsible for many thousands of deaths each year in the United States.

Heavy taxes are imposed on alcohol in beverages. A gallon of pure alcohol costs only a few dollars to produce, but in a distilled beverage it bears a U.S. federal tax of about 27 dollars.

Ethanol for industrial use is often denatured (rendered unfit for drinking) and thus is not taxed. Denaturing is done by adding small amounts of methanol and other denaturants that are extremely difficult to remove. Denaturing is required by

Hospital Supplies from an Alcohol

Polyvinyl alcohol (PVA) bags used for hospital supplies and chemicals dissolve in water. In this photo red dye is being released as the PVA bag dissolves.

Used surgical supplies are a tremendous waste disposal problem for hospitals. However, a new system may allow these used materials to be literally washed right down the drain. In a process developed by Travis Honeycutt, executive vice president of Medsurg Isolyzer, a wide range of surgical supplies, including drapes, sponges, hats, and even plastic basins are made from a material called polyvinyl alcohol (PVA). These supplies can be processed in a special washer at temperatures up to 203°F where they dissolve into degradable fibers. The fibers are then flushed down the drain. The PVA fiber is produced at a factory in Shanghai, made into fabric in the United States, and manufactured into surgical products in domestic and foreign factories.

Economic factors strongly favor the new system. Costs for the new PVA disposables are 3¢ per pound while incineration costs 10¢ per pound and off-site disposal averages 30¢ per pound. The PVA-based materials can save thousands of dollars in disposal costs per year for hospitals and eliminate a large waste disposal problem.

From its name you might expect polyvinyl alcohol (PVA) to be formed from molecules of vinyl alcohol. However, this is not possible because vinyl alcohol does not exist. Vinyl alcohol is unstable and immediately rearranges to form ethanal, a more stable molecule. To make PVA, acetic acid is reacted with acetylene to produce vinyl acetate:

$$\underset{\text{acetic acid}}{CH_3-\overset{\overset{\displaystyle O}{\|}}{C}-OH} + \underset{\text{acetylene}}{HC\equiv CH} \longrightarrow \underset{\text{vinyl acetate}}{CH_3-\overset{\overset{\displaystyle O}{\|}}{C}-O-CH=CH_2}$$

The vinyl acetate is polymerized to form polyvinyl acetate, which reacts with methanol to form PVA:

$$\underset{\text{polyvinyl acetate}}{\left[-CH_2-\underset{\underset{\displaystyle O-\underset{\underset{\displaystyle O}{\|}}{C}-CH_3}{|}}{\overset{\overset{\displaystyle H}{|}}{C}}-\right]_n} \xrightarrow{CH_3OH} \underset{\text{polyvinyl alcohol (PVA)}}{\left[-CH_2-\underset{\underset{\displaystyle OH}{|}}{\overset{\overset{\displaystyle H}{|}}{C}}-\right]_n} + CH_3\overset{\overset{\displaystyle O}{\|}}{C}-OCH_3$$

Hospitals are using PVA bags to hold laundry from patients with infectious diseases. The sealed bags are dropped in the washing machines and dissolve as the hot water kills the infectious organisms, and the workers never touch the microbe-covered materials. The use of PVA products is spreading; for example, farmers are using PVA bags to hold powdered insecticides and herbicides so they don't have to handle the chemicals directly. As in hospital laundry, the bags are dropped into water and dissolve along with the pesticide.

the federal government to protect the beverage-alcohol tax source. Special tax-free use permits are issued to scientific and industrial users who require pure ethanol for nonbeverage uses.

2-Propanol (Isopropyl Alcohol)

2-Propanol (isopropyl alcohol) is made from propene derived from petroleum. This synthesis is analogous to that used for making ethanol from ethene:

$$\underset{\text{propene}}{CH_3CH{=}CH_2} + H_2O \xrightarrow{H^+} \underset{\text{2-propanol}}{\underset{\displaystyle OH}{CH_3\overset{}{C}HCH_3}}$$

Note that 2-propanol, not 1-propanol, is produced. This is because, in the first step of the reaction, an H^+ adds to carbon 1 of propene according to Markovnikov's rule (Section 21.6). The —OH group then ends up on carbon 2 to give the final product.

2-Propanol is a relatively low-cost alcohol that is manufactured in large quantities, about 6.3×10^8 kg annually. It is not a potable alcohol, and merely breathing large quantities of the vapor may cause dizziness, headache, nausea, vomiting, mental depression, and coma. Isopropyl alcohol is used (1) to manufacture other chemicals (especially acetone), (2) as an industrial solvent, and (3) as the principal ingredient in rubbing alcohol formulations.

Ethylene Glycol (1,2-Ethanediol)

Ethylene glycol is the simplest alcohol containing two —OH groups. Like most other relatively cheap, low-molar-mass alcohols, it is commercially derived from petroleum. One industrial synthesis is from ethylene via ethylene oxide (oxirane).

$$\underset{\text{ethylene}}{2\,CH_2{=}CH_2} + O_2 \xrightarrow[200\text{–}300°C]{\text{Ag catalyst}} \underset{\text{(ethylene oxide)}}{\underset{\text{oxirane}}{2\,\overset{\displaystyle O}{CH_2{-}CH_2}}}$$

$$\overset{\displaystyle O}{CH_2{-}CH_2} + H_2O \xrightarrow{H^+} \underset{\text{(ethylene glycol)}}{\underset{\text{1,2-ethanediol}}{HOCH_2CH_2OH}}$$

This alcohol is commonly referred to as ethylene glycol in commercial products. Major uses of ethylene glycol are (1) in the preparation of the synthetic polyester fibers (Dacron) and film (Mylar), (2) as a major ingredient in "permanent-type" antifreeze for cooling systems, (3) as a solvent in the paint and plastics industries, and (4) in the formulations of printing ink and ink for ballpoint pens.

The low molar mass, complete water solubility, low freezing point, and high boiling point make ethylene glycol a nearly ideal antifreeze. A 58% by mass aqueous solution of this alcohol freezes at $-48°C$. Its high boiling point and high heat of vaporization prevent it from being boiled away and permit higher, and therefore more efficient, engine-operating temperatures than are possible with water alone. The U.S. production of ethylene glycol amounts to about 2.4×10^9 kg annually. Ethylene glycol is extremely toxic when ingested.

Glycerol (1,2,3-Propanetriol)

Glycerol, also known as *glycerine* or 1,2,3-propanetriol, is an important trihydroxy alcohol. Glycerol is a syrupy liquid with a sweet, warm taste. It is about 0.6 times as sweet as cane sugar. It is obtained as a by-product of processing animal and vegetable fats to make soap and other products, and it is also synthesized commercially from propene. The major uses of glycerol are (1) as a raw material in the manufacture of polymers and explosives, (2) as an emollient in cosmetics, (3) as a humectant in tobacco products, and (4) as a sweetener. Each use is directly related to the three —OH groups on glycerol.

The —OH groups provide sites through which the glycerol unit may be bonded to other molecules to form a polymer (Chapter 26). The explosive nitroglycerine, or glyceryltrinitrate, is made by reacting the —OH groups with nitric acid:

$$\begin{array}{c} CH_2OH \\ | \\ CHOH \\ | \\ CH_2OH \\ \text{glycerol} \end{array} + \underset{\text{nitric acid}}{3\,HONO_2} \longrightarrow \begin{array}{c} CH_2ONO_2 \\ | \\ CHONO_2 \\ | \\ CH_2ONO_2 \\ \text{glyceryltrinitrate} \\ \text{(nitroglycerine)} \end{array} + 3\,H_2O$$

The three polar —OH groups on the glycerol molecule are able to hold water molecules by hydrogen bonding. Consequently, glycerol is a hygroscopic substance; that is, it has the ability to take up water vapor from the air. It is therefore used as a skin moisturizer in cosmetic preparations. Glycerol is also used as an additive in tobacco products; by taking up moisture from the air, it prevents the tobacco from becoming excessively dry and crumbly.

22.7 Phenols

The term **phenol** is used for the class of compounds that have a hydroxy group attached to an aromatic ring. The parent compound is also called phenol, C_6H_5OH (pronounced *feenol*).

phenol

Naming Phenols

Many phenols are named as derivatives of the parent compound using the general methods for naming aromatic compounds. For example,

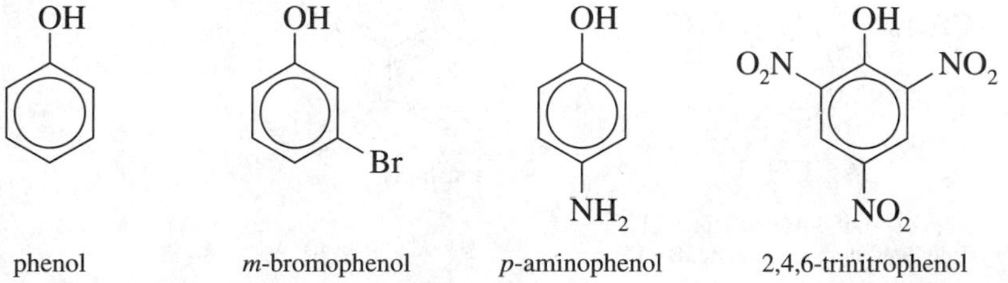

OH | Br | NH2 | O2N, NO2, NO2

phenol

m-bromophenol

p-aminophenol

2,4,6-trinitrophenol (picric acid)

Common Phenols

Many natural substances have phenolic groups in their structures. The formulas and brief descriptions of several examples include the following:

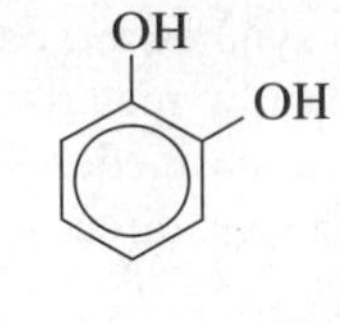

catechol
(*o*-dihydroxybenzene)

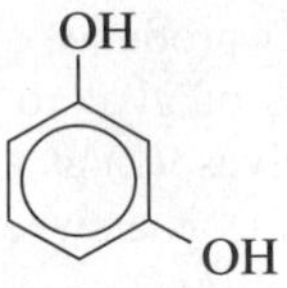

resorcinol
(*m*-dihydroxybenzene)

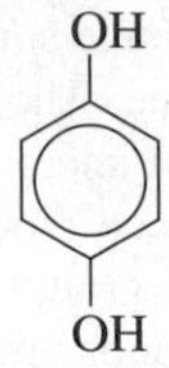

hydroquinone
(*p*-dihydroxybenzene)

The *ortho*-, *meta*-, and *para*-dihydroxybenzenes have the special names catechol, resorcinol, and hydroquinone, respectively. The catechol structure occurs in many natural substances, and hydroquinone, a manufactured product, is commonly used as a photographic reducer and developer.

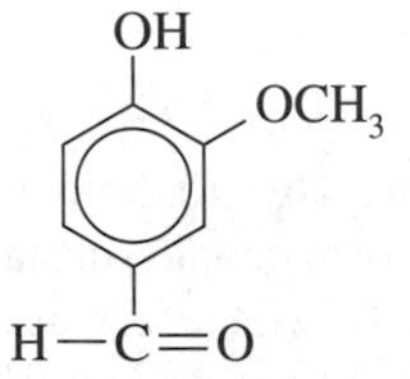

vanillin

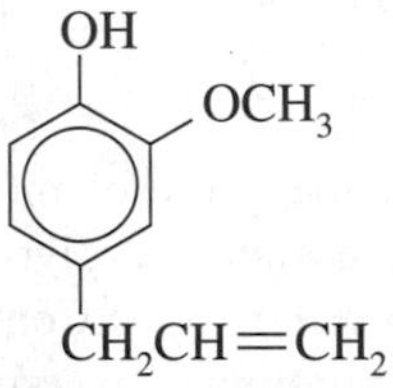

eugenol

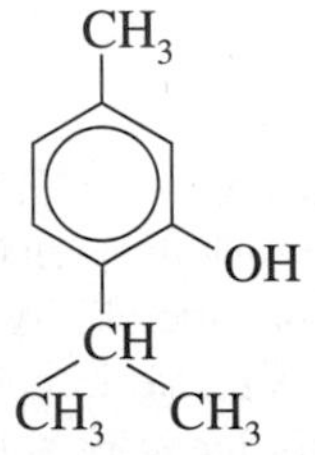

thymol

Vanillin is the principal odorous component of the vanilla bean. It is one of the most widely used flavorings and is also used for masking undesirable odors in many products such as paints.

Eugenol is the essence of oil of cloves. It is used in the manufacture of synthetic vanillin. Thymol occurs in the oil of thyme. It has a pleasant odor and flavor and is used as an antiseptic in preparations such as mouthwashes. Thymol is the starting material for the synthesis of menthol, the main constituent of oil of peppermint. Thymol is a widely used flavoring and pharmaceutical.

Butylated hydroxytoluene (BHT) is used in small amounts as an antioxidant preservative for food, synthetic rubber, vegetable oils, soap, and some plastics.

The active irritants in poison ivy and poison oak are called urushiols. They are catechol derivatives with an unbranched 15-carbon side chain in position 3 on the phenol ring.

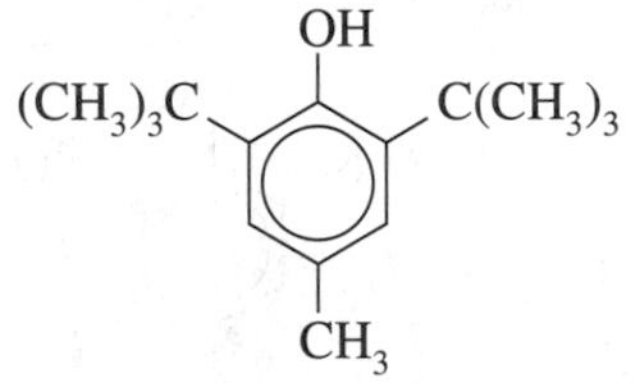

2,6-di-*t*-butyl-4-methylphenol
(butylated hydroxytoluene, BHT)

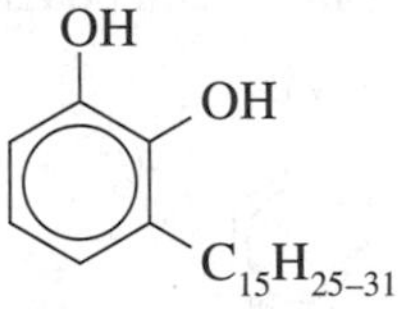

urushiols

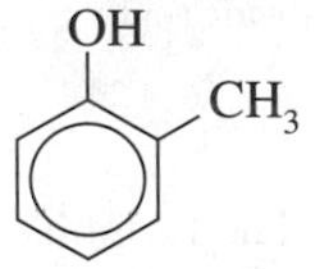

o-cresol

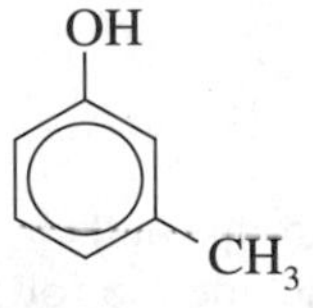

m-cresol

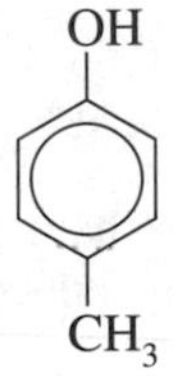

p-cresol

The *ortho-*, *meta-*, and *para-*methylphenols are present in coal tar and are known as cresols. They are all useful disinfectants.

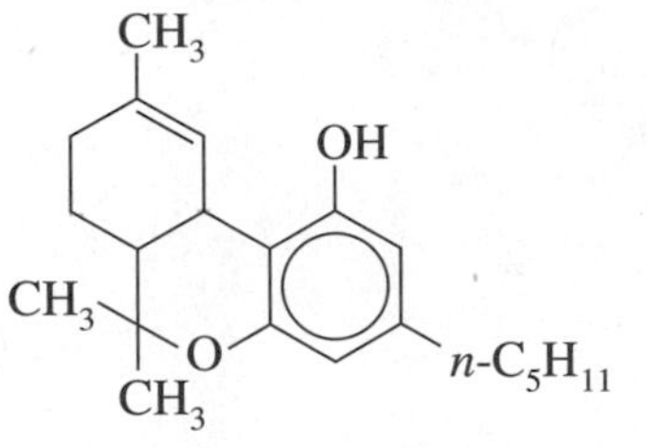

tetrahydrocannabinol (from marijuana)

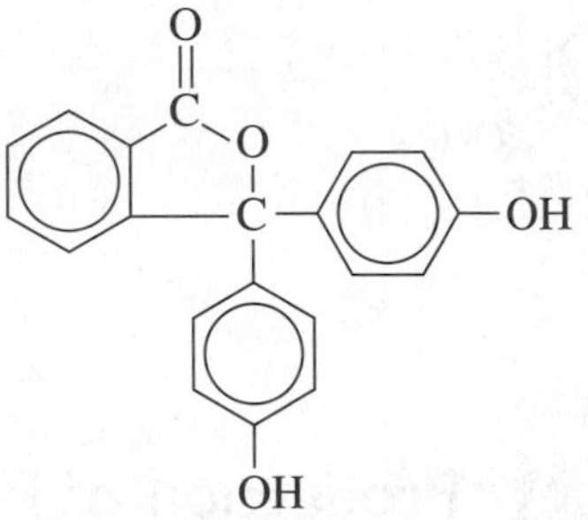

phenolphthalein

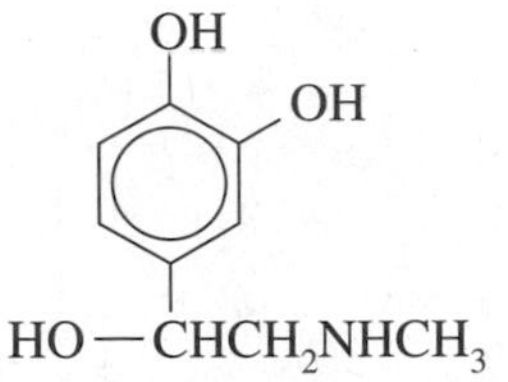

adrenalin (epinephrine)

The active principal component of marijuana is tetrahydrocannabinol. It is obtained from the dried leaves and flowering tops of the hemp plant and has been used since antiquity for its physiological effects. The common acid–base indicator phenolphthalein is a phenol derivative. Phenolphthalein is also used as a laxative. Epinephrine (adrenalin) is secreted by the adrenal gland in response to stress, fear, anger, or other heightened emotional states. It stimulates the conversion of glycogen to glucose in the body. Phenol is the starting material for the manufacture of aspirin, one of the most widely used drugs for self-medication.

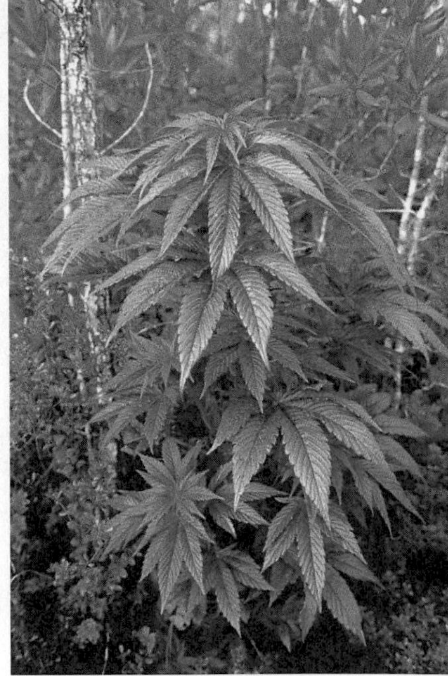

Marijuana is a member of the hemp family. It contains the physiologically active molecule tetrahydrocannabinal.

22.8 Properties of Phenols

In the pure state, phenol is a colorless crystalline solid with a melting point at about 41°C and a characteristic odor. Phenol is highly poisonous. Ingestion of even small amounts of it may cause nausea, vomiting, circulatory collapse, and death from respiratory failure.

Phenol is a weak acid; it is more acidic than alcohols and water but less acidic than acetic and carbonic acids. The pH values are as follows: 0.1 *M* acetic acid, 2.87; water, 7.0; 0.1 *M* phenol, 5.5. Thus phenol reacts with sodium hydroxide solution to form a salt but does not react with sodium hydrogen carbonate. The salt formed is called sodium phenoxide or sodium phenolate. Sodium hydroxide does not remove a hydrogen atom from an alcohol because alcohols are weaker acids than water.

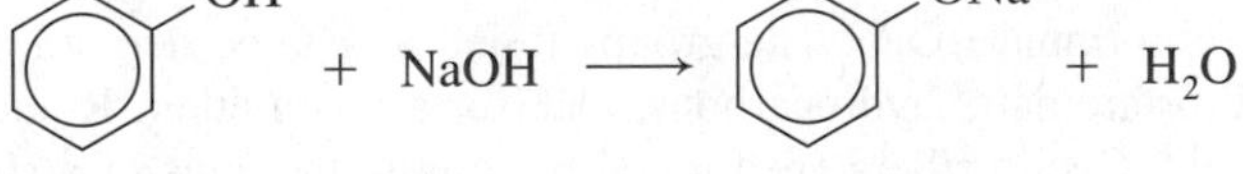

$$C_6H_5OH + NaOH \longrightarrow C_6H_5ONa + H_2O$$

sodium phenoxide

In general, the phenols are toxic to microorganisms. They are widely used as antiseptics and disinfectants. Phenol was the first compound to be used extensively as an operating room disinfectant. Joseph Lister (1827–1912) first used phenol for this purpose in 1867. The antiseptic power of phenols is increased by substituting alkyl groups (up to six carbons) in the benzene ring. For example, 4-hexylresorcinol is used as an antiseptic in numerous pharmaceuticals.

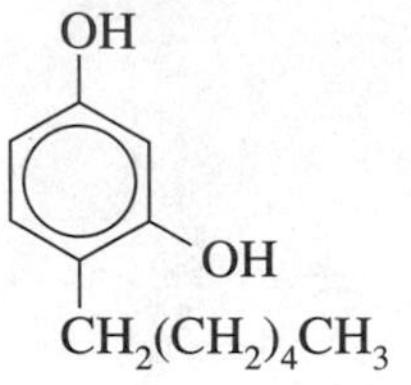

OH

OH

$CH_2(CH_2)_4CH_3$

4-hexylresorcinol

22.9 Production of Phenol

Phenol is obtained from coal tar. In addition, there are several commercial methods used to produce phenol synthetically. The most economical of these methods starts with benzene and propene, which react to form cumene. Cumene is then oxidized by air to cumene hydroperoxide, which is treated with dilute sulfuric acid to obtain phenol and acetone. The economic feasibility of the process is due to the fact that two important commercial products are produced. The equations for the reactions are

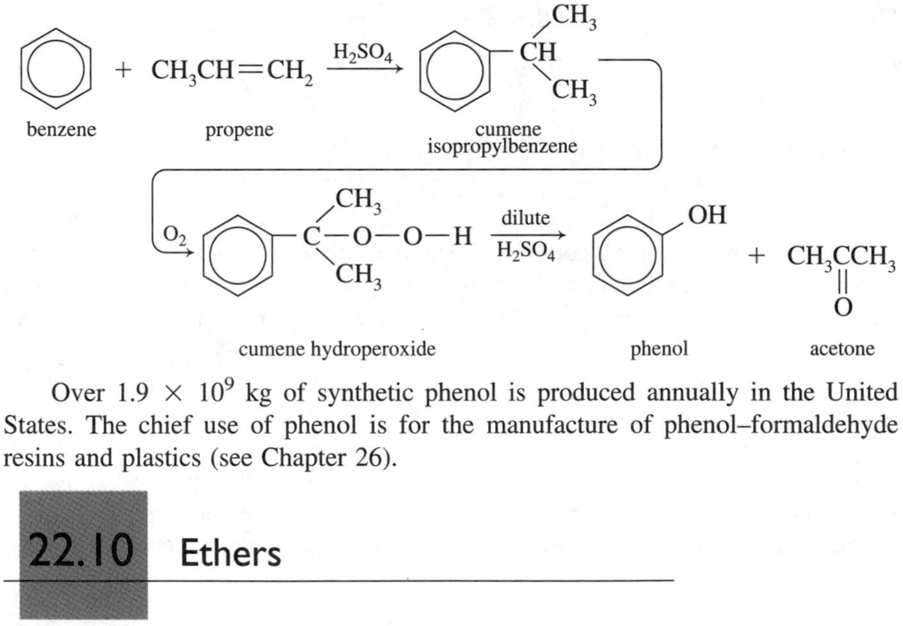

$$C_6H_6 + CH_3CH{=}CH_2 \xrightarrow{H_2SO_4} C_6H_5CH(CH_3)_2$$

benzene, propene, cumene (isopropylbenzene)

$$\xrightarrow{O_2} C_6H_5C(CH_3)_2{-}O{-}O{-}H \xrightarrow[H_2SO_4]{\text{dilute}} C_6H_5OH + CH_3\underset{\underset{O}{\|}}{C}CH_3$$

cumene hydroperoxide, phenol, acetone

Over 1.9×10^9 kg of synthetic phenol is produced annually in the United States. The chief use of phenol is for the manufacture of phenol–formaldehyde resins and plastics (see Chapter 26).

22.10 Ethers

ether **Ethers** have the general formula ROR′. The groups R and R′ can be derived from saturated, unsaturated, or aromatic hydrocarbons, and, for a given ether, R and R′ may be alike or different. Cyclic ethers are formed by joining the ends of a single hydrocarbon chain through an oxygen atom to form a ring structure. Table 22.5 shows structural formulas and names for some of the different kinds of ethers.

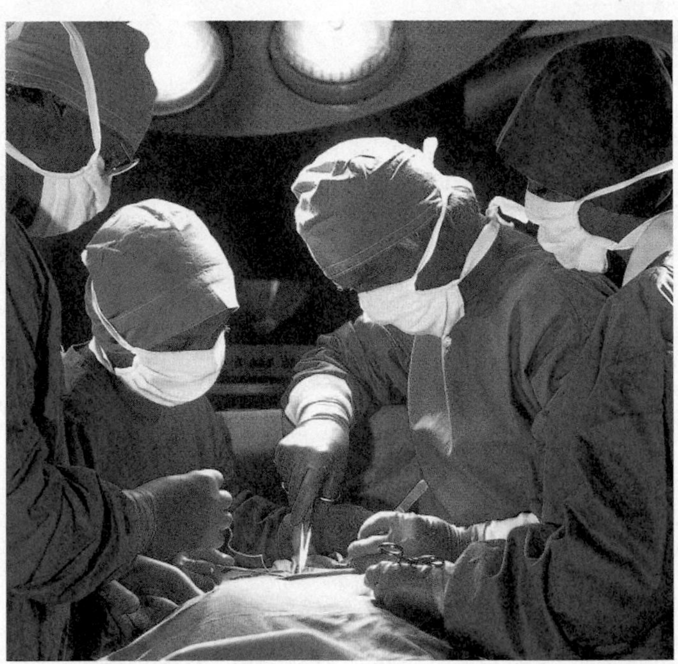

◀ **Phenols were among the first antiseptics to be used in operating rooms to prevent the spread of bacteria. Today, other antiseptics such as iodine solutions and germicidal soaps have replaced phenols.**

TABLE 22.5 Names and Structural Formulas of Ethers

Name*	Formula	Boiling point (°C)
Dimethyl ether (Methoxymethane)	$CH_3{-}O{-}CH_3$	−25
Methyl ethyl ether (Methoxyethane)	$CH_3CH_2{-}O{-}CH_3$	8
Diethyl ether (Ethoxyethane)	$CH_3CH_2{-}O{-}CH_2CH_3$	35
Ethyl isopropyl ether (2-Ethoxypropane)	$CH_3CH_2{-}O{-}CH(CH_3)CH_3$	54
Divinyl ether (Ethenyloxyethene)	$CH_2{=}CH{-}O{-}CH{=}CH_2$	39
Anisole (Methoxybenzene)	$C_6H_5{-}OCH_3$	154
Diphenyl ether (Phenoxy benzene)	$C_6H_5{-}O{-}C_6H_5$	259
Tetrahydrofuran (THF) (1-Epoxybutane)	$CH_2{-}CH_2{-}CH_2{-}CH_2$ ring closed by O	66

*The IUPAC name is in parentheses when given.

Naming Ethers

Individual ethers, like alcohols, may be known by several names. The ether having the formula $CH_3CH_2—O—CH_2CH_3$ and formerly widely used as an anesthetic is called diethyl ether, ethyl ether, ethoxyethane, or simply ether. Common names of ethers are formed from the names of the groups attached to the oxygen atom, followed by the word *ether*.

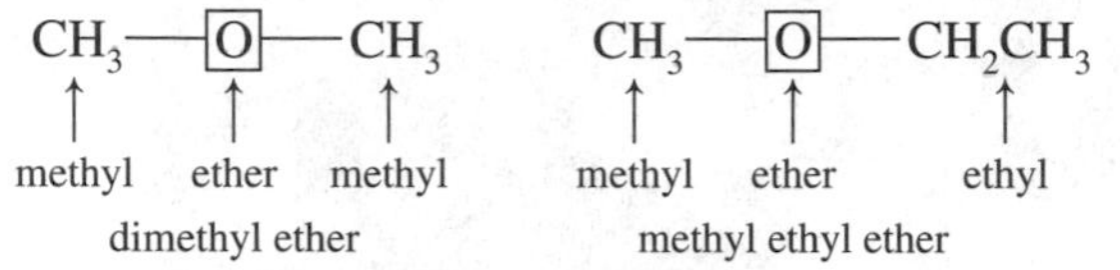

CH_3 —	O	— CH_3		CH_3 —	O	— CH_2CH_3
↑	↑	↑		↑	↑	↑
methyl	ether	methyl		methyl	ether	ethyl
	dimethyl ether				methyl ethyl ether	

In the IUPAC System, ethers are named as alkoxy (RO—) derivatives of the alkane corresponding to the longest carbon–carbon chain in the molecule. To name an ether by this system:

IUPAC Nomenclature for Ethers

1. Select the longest carbon–carbon chain and label it with the name of the corresponding alkane.
2. Change the *yl* ending of the other hydrocarbon group to *oxy* to obtain the alkoxy group name. For example, $CH_3O—$ is called *methoxy*.
3. Combine the two names from Steps 1 and 2, giving the alkoxy name first, to form the ether name.

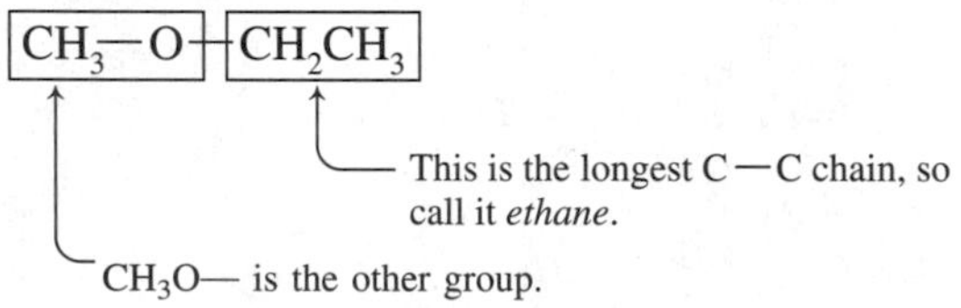

$CH_3—O—CH_2CH_3$

CH_2CH_3: This is the longest C—C chain, so call it *ethane*.

$CH_3O—$ is the other group.

Modify the name for the $CH_3O—$ group from *methyl* to *methoxy* and combine with *ethane* to obtain the name of the ether, methoxyethane. Thus

$CH_3CH_2—O—CH_2CH_3$ is ethoxyethane
$CH_3CH_2CH_2—O—CH_2CH_2CH_2CH_3$ is propoxybutane

Additional examples are found in Table 22.5.

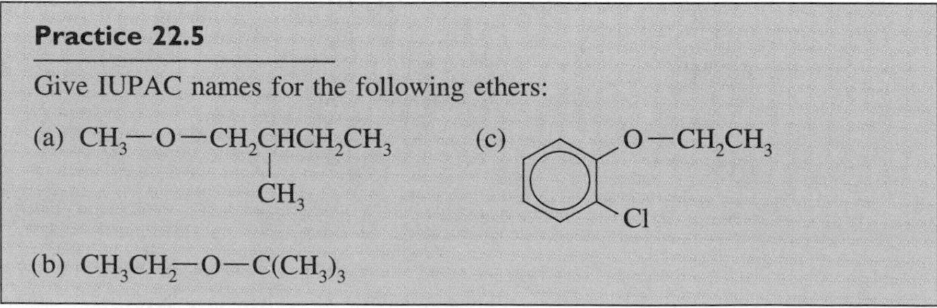
Practice 22.5
Give IUPAC names for the following ethers:
(a) $CH_3—O—CH_2CHCH_2CH_3$ with CH_3 on the CH
(c) O—CH_2CH_3
Cl
(b) $CH_3CH_2—O—C(CH_3)_3$

Practice 22.6

Give common names for the following ethers:

(a) $CH_3-O-CH_2CH_2CH_3$

(b) $(CH_3)_3C-O-C(CH_3)_3$

(c) 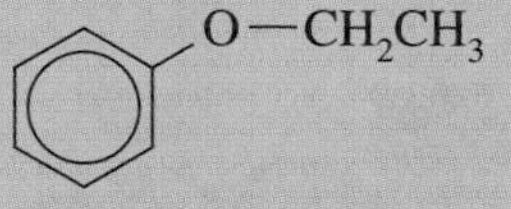 $C_6H_5-O-CH_2CH_3$

22.11 Structures and Properties of Ethers

An oxygen atom linking two carbon atoms is the key structural feature of an ether molecule. This oxygen atom causes ether molecules to have a bent shape somewhat like that of water and alcohol molecules:

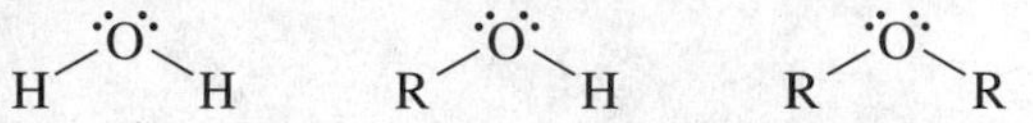

H–Ö–H (water) R–Ö–H (alcohol) R–Ö–R (ether)

Ethers are somewhat more polar than alkanes, because alkanes lack the oxygen atom with its exposed, nonbonded electrons. But ethers are much less polar than alcohols, since no hydrogen is attached to the oxygen atom in an ether. The solubility and boiling point (vapor pressure) characteristics of ethers are related to the C—O—C structure. Alkanes have virtually no solubility in water or acid. But about 7.5 g of diethyl ether will dissolve in 100 g of water at 20°C. Diethyl ether also dissolves in sulfuric acid. Hydrogen bonding between ether and water molecules and between ether and acid molecules is responsible for this solubility.

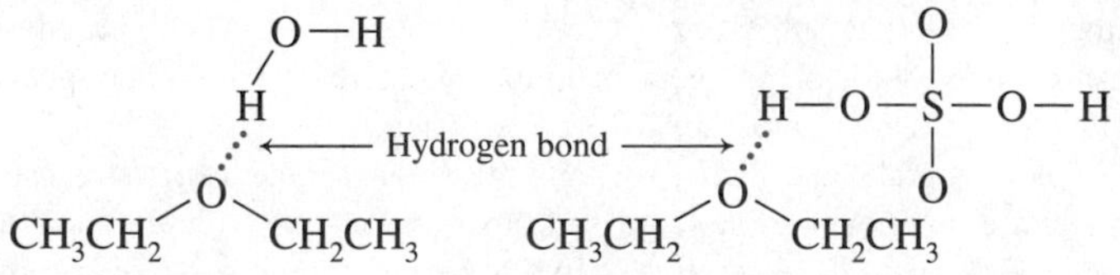

$CH_3CH_2-O(\cdots H-O-H)-CH_2CH_3$ (ether ••• water) ← Hydrogen bond → $CH_3CH_2-O(\cdots H-O-SO_2-O-H)-CH_2CH_3$ (ether ••• acid)

Because no —OH group is present, hydrogen bonding does not occur between ether molecules. This lack of hydrogen bonding can be seen by comparing the boiling points of a hydrocarbon, an ether, and an alcohol of similar molecular mass, as in Table 22.6. The boiling point of the ether is somewhat above that of the hydrocarbon but much lower than that of the more polar alcohol.

Ethers, especially diethyl ether, are exceptionally good solvents for organic compounds. Many polar compounds, including water, acids, alcohols, and other oxygenated organic compounds, dissolve, at least to some extent, in ethers. This solubility is a result of intermolecular attractions between the slightly polar ether molecules and the molecules of the other polar substance. Nonpolar compounds such as hydrocarbons and alkyl halides also dissolve in ethers. These substances dissolve because the ether molecules are not very polar and therefore are not strongly attracted either to one another or to the other kinds of molecules. Thus ether molecules

Chemistry in Action

Anesthetics

The most widely known use of ethyl ether has been as a general anesthetic for surgery. The introduction of ether for this purpose is one of the great landmarks of medicine. Two Americans, Crawford W. Long and William T. Morton, played important roles in this development. Long, a physician, used ether in a surgical operation as early as 1842 but did not publish his discovery until 1849. Morton, a dentist, used ether as an anesthetic for dental work in 1846. He publicly demonstrated its effectiveness by administering ether to a patient undergoing surgery at the Massachusetts General Hospital on October 16, 1846.

The word *anesthesia* is from the Greek, meaning insensibility, and was suggested to Morton by the poet and physician Oliver Wendell Holmes. A *general anesthetic* is a substance or combination of substances that produces both unconsciousness and insensitivity to pain.

Ether produces unconsciousness by depressing the activity of the central nervous system. Other ethers such as divinyl ether (Vinethene) and methoxyflurane (Penthrane) were also eventually used as general inhalation anesthetics.

$$CH_2{=}CH{-}O{-}CH{=}CH_2$$
divinyl ether

$$CHCl_2CF_2{-}O{-}CH_3$$
methoxyflurane

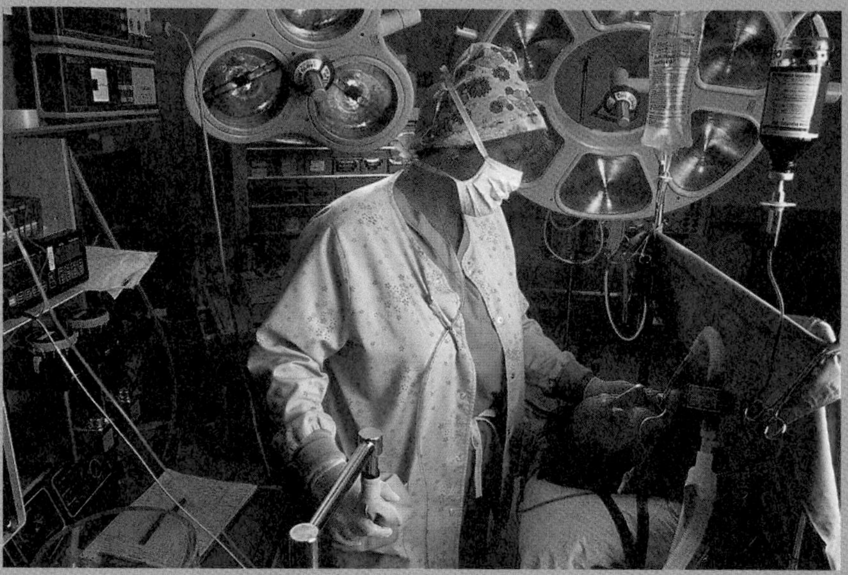

An anesthetist prepares to give a general anesthetic to her patient.

The major disadvantages of ether include flammability, irritation of respiratory passages, and occurrence of nausea and vomiting after its use.

These hazards posed by the ethers have resulted in the use of other substances (such as nitrous oxide, N_2O, or halogenated compounds like halothane, $CF_3CHClBr$) as general anesthetics. These substances are superior in some respects and, in recent years, have totally replaced the ethers; however, these substitute compounds pose other hazards and must also be used with caution.

A number of theories explaining the biochemical activity of anesthetics have been proposed though none has been proved conclusively. All are based on the nonpolar nature of the compounds. This property leads to high solubility in fats and membranes. One theory suggests that anesthetics interfere with electrical activity in nerve impulses by dissolving in brain cells.

are able to intermingle freely with the molecules of a nonpolar substance and form a solution by simple mixing.

In summary, ethers are polar enough to dissolve some polar substances, but their polarity is so slight that they act as nonpolar solvents for a great many nonpolar substances.

Ethers have little chemical reactivity, but because a great many organic substances dissolve readily in ethers, they are often used as solvents in laboratory and manufacturing operations. Their use may be dangerous, since low-molar-mass ethers

TABLE 22.6 Boiling Points of Ethers, Alkanes, and Alcohols

Name	Formula	Molar mass	Boiling point (°C)
Methoxymethane	CH_3OCH_3	46	−24
Propane	$CH_3CH_2CH_3$	44	−42
Ethanol	CH_3CH_2OH	46	78
Methoxyethane	$CH_3OCH_2CH_3$	60	8
Butane	$CH_3CH_2CH_2CH_3$	58	−0.6
1-Propanol	$CH_3CH_2CH_2OH$	60	97
2-Propanol	$CH_3CH(OH)CH_3$	60	83

are volatile, and their highly flammable vapors form explosive mixtures with air. Another hazard of ethers is that, despite their generally low chemical reactivity, oxygen of the air slowly reacts with them to form unstable peroxides that are subject to explosive decomposition:

$$CH_3CH_2{-}O{-}CH_2CH_3 + O_2 \longrightarrow CH_3\underset{\displaystyle O{-}O{-}H}{\underset{|}{C}H}{-}O{-}CH_2CH_3$$

diethyl ether hydroperoxide

22.12 Preparation of Ethers

We have seen that ethers can be made by intermolecular dehydration of alcohols by heating in the presence of an acid (see Section 22.5). Ethers are also made from alkyl halides and sodium alkoxides or sodium phenoxides via a substitution reaction (the Williamson synthesis).

Organic chemistry is replete with reactions named after scientists who discovered the reactions.

$$RX + R'ONa \longrightarrow ROR' + NaX$$

where R is a methyl, primary alkyl, or secondary alkyl group (but *not* a tertiary alkyl or an aromatic group).

Alkyl halides are especially capable of substitution reactions. Here, the halide is replaced by an alkoxide or phenoxide to form an ether. This synthesis is generally useful in the preparation of mixed ethers (where $R \neq R'$) and aromatic ethers:

$$\underset{\text{sodium ethoxide}}{CH_3CH_2ONa} + \underset{\text{bromomethane}}{CH_3Br} \longrightarrow \underset{\text{methoxyethane}}{CH_3CH_2{-}O{-}CH_3} + NaBr$$

$$\underset{\text{sodium phenoxide}}{C_6H_5ONa} + \underset{\text{bromomethane}}{CH_3Br} \longrightarrow \underset{\text{methoxybenzene (anisole)}}{C_6H_5OCH_3} + NaBr$$

Practice 22.7

Write equations for the preparation of 1-propoxybutane starting with an alcohol and an alkyl halide.

22.13 Thiols

Sulfur and oxygen are found next to each other in the same family on the periodic table. This indicates some similarity in the formulas of their compounds. Organic compounds that contain the —SH group are analogs of alcohols. The —SH-containing compounds are known as **thiols**, or mercaptans. Examples include:

thiol

CH_3SH
methanethiol
(methyl mercaptan)

$CH_3CH_2CHCH_3$ (with —SH on the third carbon)
2-butanethiol
(*sec*-butyl mercaptan)

Thiols have a higher molar mass than corresponding alcohols but boil at lower temperatures (ethanol, 78°C; ethanethiol, 36°C). The reason for this discrepancy lies in the fact that alcohols form hydrogen bonds while thiols do not.

The major important properties of thiols are summarized as follows:

1. Foul odors: some of these compounds smell so awful that companies make special labels to warn consumers. The odor given off by a frightened skunk has thiols as the active ingredient. Natural gas is odorized to be detectable by adding small amounts of methanethiol.
2. Oxidation to disulfides:

$$\underset{\text{thiol}}{2\ RSH} \xrightarrow{[O]} \underset{\text{disulfide}}{R—S—S—R}$$

This reaction can be accomplished using many oxidizing agents. The sulfur is being oxidized just as the carbon was being oxidized in alcohols.

Thiols serve important functions in living systems. The disulfide structure often binds proteins into biologically useful three-dimensional shapes (see Chapter 30). The thiol is also a key part of a molecule (coenzyme A) that plays a central role in the metabolism of carbohydrates, lipids, and proteins (see Chapter 35).

Concepts in Review

1. Name alcohols by common and IUPAC methods.
2. Write the structural formula when given the name of an alcohol.
3. Write the structural formulas for all the isomeric alcohols of a given molecular formula.
4. Recognize and identify primary, secondary, and tertiary alcohols.
5. Indicate a class of biochemicals that has many of the same properties as the polyhydroxy alcohols.
6. Summarize the acidic/basic properties of alcohols and alkoxides.
7. Write equations for the oxidation and dehydration of alcohols.
8. Understand the importance of the hydroxyl group as an intermediate for the synthesis of other organic molecules.
9. Differentiate between intramolecular and intermolecular dehydration, indicating proper conditions for each.
10. Write equations for the synthesis of alcohols from alkyl halides and from alkenes.
11. Explain the relative reactivities of primary, secondary, and tertiary alcohols.
12. Understand the common methods of preparing methyl alcohol, ethyl alcohol, isopropyl alcohol, ethylene glycol, and glycerol.
13. Name phenols and write their formulas.
14. Understand the differences in properties of the hydroxyl group when bonded to an aromatic ring (a phenol) and to an aliphatic group (an alcohol).
15. Be familiar with the general properties of phenols.
16. Name ethers and write their formulas.
17. Write equations for preparing ethers by the Williamson synthesis.
18. List major properties for thiols.
19. Discuss relative boiling points and water solubilities of comparable hydrocarbons, alcohols, dihydroxy alcohols, ethers, and thiols.
20. Discuss the hazards of using ethers in the laboratory.

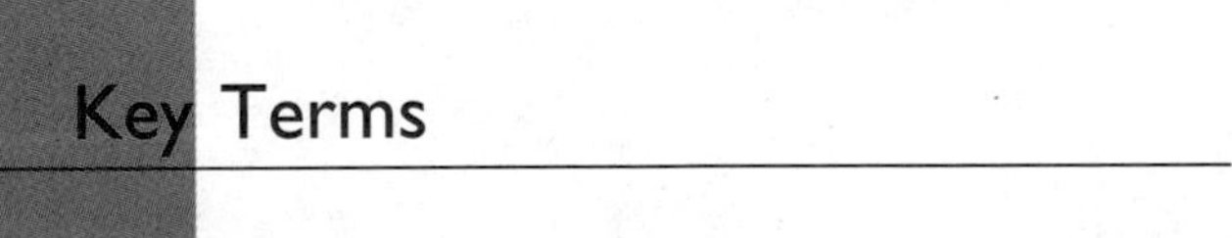

Key Terms

The terms listed here have been defined within this chapter. Section numbers are referenced in parenthesis for each term.

alcohol (22.1)
alkoxide ion (22.5)
condensation reaction (22.5)
ether (22.10)
oxonium ion (22.5)
polyhydroxy alcohol (22.2)
phenol (22.7)
primary alcohol (22.2)
Saytzeff's rule (22.5)
secondary alcohol (22.2)
tertiary alcohol (22.2)
thiol (22.13)

Questions

Questions refer to tables, figures, and key words and concepts defined within the chapter. A particularly challenging question or exercise is indicated with an asterisk.

1. Write the structural formulas for and give an example of
   (a) an alkyl halide
   (b) a phenol
   (c) an ether
   (d) an aldehyde
   (e) a ketone
   (f) a carboxylic acid
   (g) an ester
   (h) a thiol

2. Although it is possible to make alkenes from alcohols, alkenes are seldom, if ever, made in this way on an industrial scale. Why not?

3. Isopropyl alcohol is usually used in rubbing alcohol formulations. Why is this alcohol used in preference to normal propyl alcohol?

4. What classes of compounds can be formed by the oxidation of primary alcohols? Cite examples.

5. Why is 1,2-ethanediol (ethylene glycol) superior to methanol as an antifreeze for automobile radiators?

6. Briefly outline the physiological effects of (a) methanol and (b) ethanol.

7. Write equations for the cumene hydroperoxide synthesis of phenol and acetone.

8. What two hazards may be present when working with low-molar-mass ethers?

9. Explain, in terms of molecular structure, why ethanol, CH_3CH_2OH (molar mass = 46), is a liquid at room temperature and dimethyl ether, CH_3OCH_3 (molar mass = 46), is a gas.

10. Which of the following statements are correct? Rewrite the incorrect statements to make them correct.
   (a) Another name for isopropyl alcohol is 2-propanol.
   (b) Ethanol and dimethyl ether are isomers.
   (c) Alcohols and phenols are more acidic than water.
   (d) Sodium ethoxide can be prepared by reacting ethyl alcohol and sodium hydroxide solution.
   (e) Methyl alcohol is a very poisonous substance that can lead to blindness if ingested.
   (f) Tertiary alcohols are easier to oxidize than primary alcohols.
   (g) A correct name for $CH_3CH_2CH(OH)CH_3$ is *sec*-butyl alcohol.
   (h) $(CH_3)_3CCH_2OH$ is a primary alcohol.
   (i) When a secondary alcohol is oxidized, a ketone is formed.
   (j) Alcohols have higher boiling points than ethers with comparable molecular masses because of hydrogen bonding between the alcohol molecules.
   (k) The product formed when a molecule of water is split out between an alcohol and a carboxylic acid is called an ether.
   (l) When 1-butene is reacted with dilute H_2SO_4, the alcohol formed is 1-butanol.
   (m) Ethanol used for industrial purposes and rendered unfit for use in beverages is said to be denatured.
   (n) The common name for 1,2,3-propanetriol is ethylene glycol.
   (o) Although ethyl alcohol is used in beverages, it is still classified physiologically as a depressant and a poison.
   (p) Cyclohexanol is a primary alcohol.
   (q) Dihydroxy alcohols are more soluble in water than monohydroxy alcohols.
   (r) Aldehydes and ketones may be prepared by the oxidation of primary alcohols.
   (s) Thiols have a higher molar mass and lower boiling point than alcohols containing an equal number of carbons.
   (t) Coenzyme A is an alcohol that has a central role in metabolism.

Paired Exercises

These exercises are paired. Each odd-numbered exercise is followed by a similar even-numbered exercise. Answers to the even-numbered exercises are given in Appendix V.

11. Write structural formulas for the following:
(a) methanol
(b) 3-methyl-1-hexanol
(c) 1,2-propanediol
(d) 1-phenylethanol
(e) 2,3-butanediol
(f) 2-propanethiol

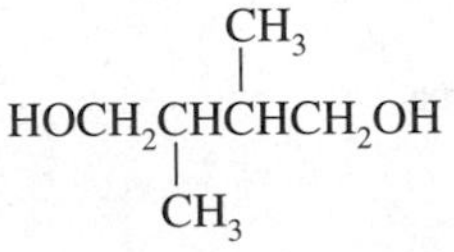

12. Write structural formulas for the following:
(a) 2-butanol
(b) 2-methyl-2-butanol
(c) 2-propanol
(d) cyclopentanol
(e) 1-pentanethiol
(f) 4-ethyl-3-hexanol

13. There are eight open-chain isomeric alcohols that have the formula $C_5H_{11}OH$. Write the structural formula and the IUPAC name for each of these alcohols.

14. Write structures for all the isomers (alcohols and ethers) with the formula
(a) C_3H_8O
(b) $C_4H_{10}O$

15. Which of the isomers in Exercise 13 are
(a) primary alcohols?
(b) secondary alcohols?
(c) tertiary alcohols?

16. Which of the isomers in Exercise 14 are
(a) primary alcohols?
(b) secondary alcohols?
(c) tertiary alcohols?

17. Name the following compounds:

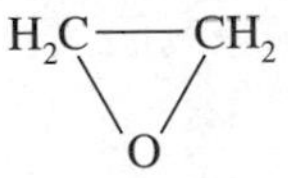

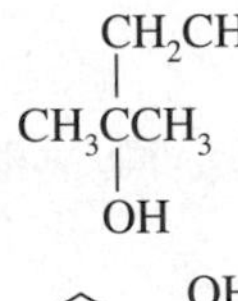

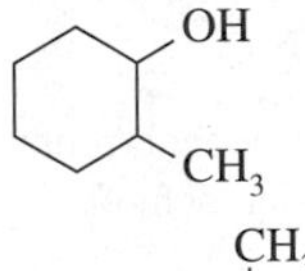

(a) $CH_3CH_2CH_2CH_2OH$

(b) $CH_3CH(OH)CH_3$

(c) (benzene ring)–CH_2CHCH_2OH with CH_3 on the CH

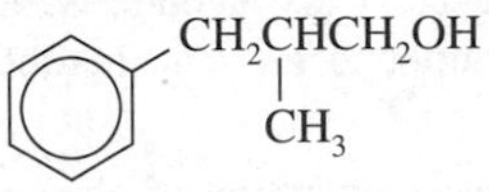

(d) $H_2C—CH_2$ bridged by O (epoxide ring)

(e) CH_3CCH_3 with CH_2CH_3 above and OH below the central C

(f) cyclohexane ring with OH and adjacent CH_3

(g) $HOCH_2CHCHCH_2OH$ with CH_3 above and CH_3 below

18. Name the following compounds:

(a) CH_3CH_2OH

(b) (benzene ring)–CH_2CH_2OH

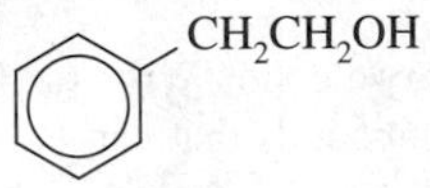

(c) $CH_3CCH_2CH_3$ with CH_2CH_3 above and OH below the central C

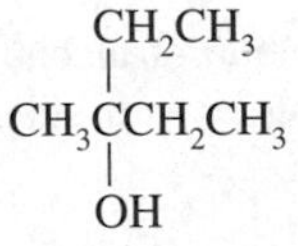

(d) cyclopentane ring with OH and CH_3 on the same carbon

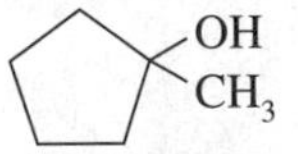

(e) $CH_3CH_2CH(OH)CH_2CH_3$

(f) CH_3CHCH_2 with HO and OH below

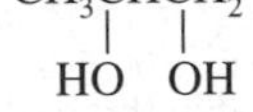

(g) $CH_3CH_2CHCH_2CHCH_3$ with OH above the second CH and CH_2CH_3 below the first CH

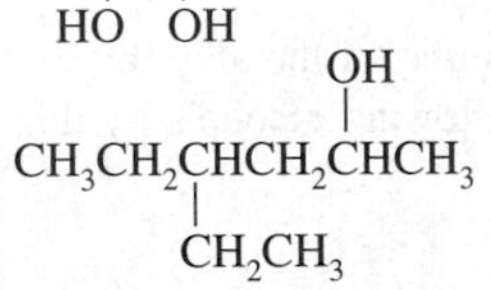

19. Write the formula and the name of the chief product when the following alcohols are dehydrated to alkenes:

$$\begin{array}{l} \quad\quad\; CH_3 \\ \quad\quad\;\; | \\ \textbf{(a)}\ CH_3CHCHCH_3 \\ \quad\quad\;\; | \\ \quad\quad\; OH \end{array}$$

$$\begin{array}{l} \textbf{(b)}\ CH_3CHCH_2CH_2CH_3 \\ \quad\quad\;\; | \\ \quad\quad\; OH \end{array}$$

(c) 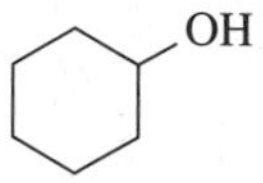 OH

20. Write the formula and the name of the chief product when the following alcohols are dehydrated to alkenes:

(a) 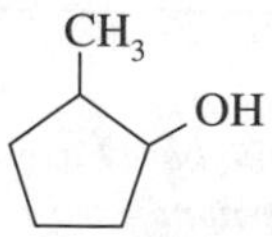 CH_3 OH

$$\begin{array}{l} \textbf{(b)}\ CH_3CH_2CHCH_3 \\ \quad\quad\quad\quad\;\; | \\ \quad\quad\quad\quad\; OH \end{array}$$

$$\begin{array}{l} \quad\quad\; CH_3 \\ \quad\quad\;\; | \\ \textbf{(c)}\ CH_3CHCHCH_2CH_3 \\ \quad\quad\quad\; | \\ \quad\quad\quad OH \end{array}$$

21. Write the formula and the name of the alcohol that, when dehydrated, gives rise to the following alkene <u>only</u>:

(a) $CH_3CH_2CH_2CH{=}CHCH_2CH_3$

(b)

(c) 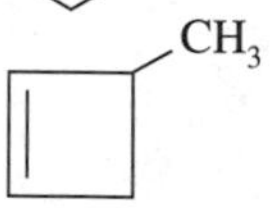 CH_3

22. Write the formula and the name of the alcohol that, when dehydrated, gives rise to the following alkene <u>only</u>:

(a)

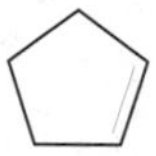

(b) 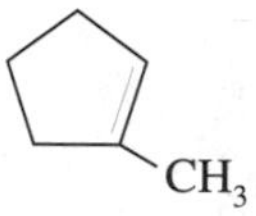 CH_3

$$\begin{array}{l} \textbf{(c)}\quad CH_3 \quad\;\; CH_3 \\ \quad\quad\quad | \quad\quad\quad\; | \\ \quad CH_3CHCH{=}CCH_3 \end{array}$$

23. 3-Ethyl-1-hexanol is a primary alcohol. Write the formulas of two different organic compounds that can be obtained by oxidizing this alcohol.

24. 3-Methyl-2-pentanol is a secondary alcohol. Write the formula of one organic compound that can be obtained by oxidizing this alcohol.

25. Write the equation for the preparation of alcohols by reacting each of the following alkenes with sulfuric acid and water:
(a) propene
(b) 1-butene
(c) 2-pentene

26. Write the equation for the preparation of alcohols by reacting each of the following alkenes with sulfuric acid and water:
(a) 2-butene
(b) 1-pentene
(c) 2-methyl-2-butene

27. Alcohols can be made by reacting alkyl halides with sodium hydroxide as follows:

$$RX + NaOH \longrightarrow ROH + NaX$$

Give the names and formulas of the alkyl bromides (RBr) needed to prepare the following alcohols by this method:
(a) 2-propanol
(b) cyclohexanol
(c) 3-methyl-1-butanol

28. Alcohols can be made by reacting alkyl halides with sodium hydroxide as follows:

$$RX + NaOH \longrightarrow ROH + NaX$$

Give the names and formulas of the alcohols produced from the following alkyl bromides by this method:
(a) 2-bromobutane
(b) 2-bromo-3-ethylpentane
(c) bromocyclopentane

29. Complete the following equations and name the principal organic product formed in each case:

(a) $2\ CH_3CH_2OH \xrightarrow[140°C]{96\%\ H_2SO_4}$

(b) $CH_3CH_2CH_2OH \xrightarrow[180°C]{96\%\ H_2SO_4}$

(c) $CH_3CH(OH)CH_2CH_3 \xrightarrow[\Delta]{K_2Cr_2O_7/H_2SO_4}$

(d) $CH_3CH_2\overset{O}{\overset{\|}{C}}-OCH_2CH_3 + H_2O \xrightarrow{H^+}$

30. Complete the following equations giving only the major organic products:

(a) $CH_3CH_2OH + Na \longrightarrow$

(b) $CH_3CH_2CH_2CH_2OH \xrightarrow[H_2SO_4]{K_2Cr_2O_7}$

(c) 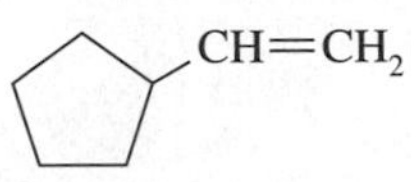 (cyclopentyl)$-CH{=}CH_2 \xrightarrow{H_2SO_4/H_2O}$

(d) $CH_3CH_2\overset{O}{\overset{\|}{C}}-OCH_3 + NaOH \longrightarrow$

31. Write structural formulas for each of the following:
(a) *o*-methylphenol
(b) 1,3-dihydroxybenzene
(c) 4-hydroxy-3-methoxybenzaldehyde (vanillin)

32. Write structural formulas for each of the following:
(a) *p*-nitrophenol
(b) 2,6-dimethylphenol
(c) 1,2-dihydroxybenzene

33. Name the following compounds:

(a) 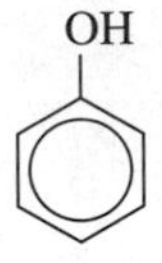 OH (on benzene ring)

(b) 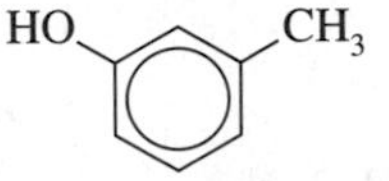 HO, CH_3 (on benzene ring)

(c) 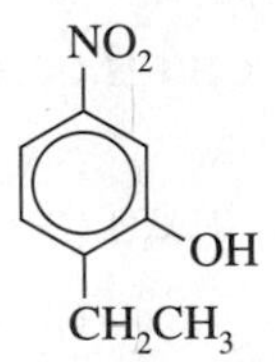 NO_2, OH, CH_2CH_3 (on benzene ring)

(d) 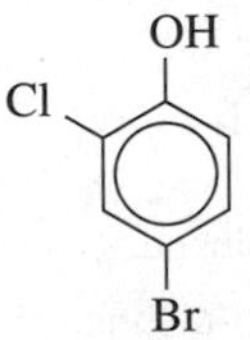 OH, Cl, Br (on benzene ring)

34. Name the following compounds:

(a) 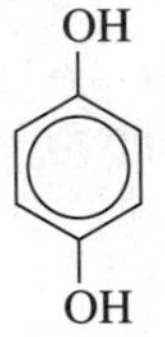 OH, OH (on benzene ring)

(b) 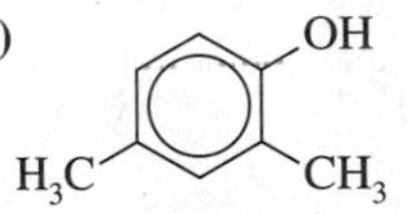 OH, H_3C, CH_3 (on benzene ring)

(c) 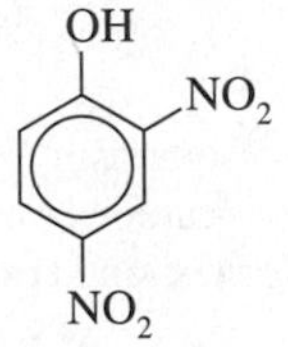 OH, NO_2, NO_2 (on benzene ring)

(d) 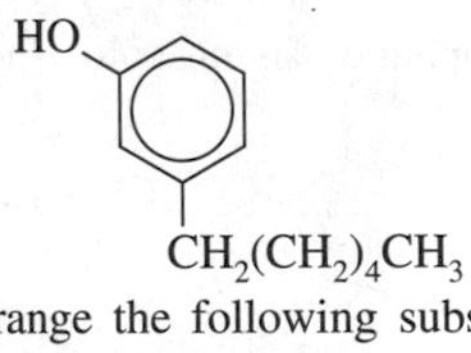 HO, $CH_2(CH_2)_4CH_3$ (on benzene ring)

35. Arrange the following substances in order of increasing solubility in water:
(a) $CH_3CH_2OCH_2CH_2CH_3$
(b) $CH_3CH(OH)CH_2CH_2CH_3$
(c) $CH_3CH_2CH_2CH_2CH_3$
(d) $CH_3CH(OH)CH(OH)CH_2CH_3$

36. Arrange the following substances in order of decreasing solubility in water:
(a) $CH_3CH(OH)CH(OH)CH_2OH$
(b) $CH_3CH_2OCH_2CH_3$
(c) $CH_3CH_2CH_2CH_2OH$
(d) $CH_3CH(OH)CH_2CH_2OH$

37. There are 14 isomeric saturated ethers that have the formula $C_6H_{14}O$. Write the structural formula for each of these ethers.

38. There are six isomeric saturated ethers that have the formula $C_5H_{12}O$. Write the structural formula and name for each of these ethers.

39. Write the balanced equation for the complete combustion of 2-butanol.

40. Write the balanced equation for the complete combustion of diethyl ether.

41. Write the formulas for all the possible combinations of RONa and RCl for making each of these ethers:

(a) $CH_3CH_2OCH_3$

(b) C_6H_5–$CH_2OCH_2CH_3$ (benzene ring attached to $CH_2OCH_2CH_3$)

42. Write the formulas for all the possible combinations of RONa and RCl for making each of these ethers:

(a) $CH_3CH_2CH_2OCH_2CH_2CH_3$

(b) CH_3–$\underset{CH_3}{HC}$–$OCH_2CH_2CH_3$ (HCOCH$_2$CH$_2$CH$_3$ with CH_3 above and CH_3 below the C)

Additional Exercises

These exercises are not paired or labeled by topic and provide additional practice on concepts covered in this chapter.

43. Phenol and its derivatives have been used as antiseptics since the middle 1800s. One common present-day antiseptic is related to phenol:

(benzene ring with OH at top, OH at adjacent position, and $CH_2CH_2CH_2CH_2CH_2CH_3$ at bottom)

Name this compound.

***44.** Cyclic glycols show cis–trans isomerism because of restricted rotation about the carbon–carbon bonds in the ring. Draw and label the structures for **cis**- and **trans**-1,2-cyclopentanediol.

45. Arrange these three compounds in order of increasing acidity:

HO–C_6H_4–CH_3 $\quad$ H_2O $\quad$ C_6H_5–CH_2OH

46. A deficiency in the production of dopamine by the brain is believed to be the cause of Parkinson's disease. Dopamine is part of a class of compounds called catecholamines. Draw the structure of the common phenol derivative that is a part of the dopamine molecule.

***47.** When 1-butanol is dehydrated to an alkene, it yields mainly 2-butene rather than 1-butene. This indicates that the dehydration to an alkene is at least a two-step reaction. Suggest a mechanism to explain the reaction. (*Hint*: a primary carbocation is formed initially.)

48. Methyl alcohol is toxic partly because it is metabolized to formaldehyde ($O{=}CH_2$) after ingestion. What type of reaction converts methanol to formaldehyde?

49. Write equations to show how each of the following transformations can be accomplished. Some conversions may require more than one step, and some reactions studied in previous chapters may be needed.

(a) $CH_3\underset{OH}{CH}CH_3 \longrightarrow CH_3\underset{\|\,O}{C}CH_3$

(b) $CH_3CH_2CH_2CH{=}CH_2 \longrightarrow CH_3CH_2CH_2\underset{OH}{CH}CH_3$

(c) $CH_3CH_2OH \longrightarrow CH_3CH_2O^-Na^+$

(d) $CH_3CH_2CH{=}CH_2 \longrightarrow CH_3CH_2\underset{\|\,O}{C}CH_3$

(e) $CH_3CH_2CH_2CH_2OH \longrightarrow CH_3CH_2\underset{Cl}{CH}CH_3$

(f) $CH_3CH_2CH_2Cl \longrightarrow CH_3CH_2\underset{H}{C}{=}O$

50. Starting with p-methylphenol and ethane, show the equations for the synthesis of p-ethoxytoluene.

***51.** Give a simple chemical test that will distinguish between the compounds in each of the following pairs:

(a) ethanol and methoxymethane
(b) 1-pentanol and 1-pentene
(c) *p*-methylphenol and methoxybenzene

52. In what important ways do phenols differ from alcohols?

53. Draw all the isomeric compounds that contain a benzene ring and have the molecular formula $C_8H_{10}O$.

54. Which of the following compounds would you expect to react with sodium hydroxide? Write equations for those that react.

(a) C_6H_5–OH $\quad$ **(b)** C_6H_5–CH_2OH $\quad$ **(c)** C_6H_5–O–C_6H_5

55. Arrange these compounds in order of *increasing* boiling points. Give reasons for your answer.
1-pentanol 1,2-pentanediol 1-octanol

Answers to Practice Exercises

22.1 (a) 2-ethyl-1-pentanol (b) 5-bromo-5-methyl-3-hexanol

22.2 (a) 3-methylcyclohexanol (ring with OH and CH_3) (b) $CH_3CH(CH_3)CH(OH)CH(CH_2CH_3)CH_2CH_2CH_3$

22.3 1-hexanol, 1-pentanol, 1-butanol, 1,3-propanediol

22.4 (a) $CH_3CH_2CH_2C(=O){-}H \xrightarrow{[O]} CH_3CH_2CH_2C(=O){-}OH$
(b) No reaction (2-methyl-2-butanol is a tertiary alcohol)
(c) cyclopentanone (cyclopentane ring with =O)

22.5 (a) 1-methoxy-2-methylbutane
(b) 2-ethoxy-2-methylpropane
(c) 2-chloro-1-ethoxybenzene

22.6 (a) methyl *n*-propyl ether
(b) di-*tert*-butyl ether
(c) ethyl phenyl ether

22.7 (1) $2\ CH_3CH_2CH_2OH + 2\ Na \longrightarrow 2\ CH_3CH_2CH_2ONa + H_2$
(2) $CH_3CH_2CH_2ONa + CH_3CH_2CH_2CH_2Br \longrightarrow CH_3CH_2CH_2{-}O{-}CH_2CH_2CH_2CH_3 + NaBr$

or

(1) $2\ CH_3CH_2CH_2CH_2OH + 2\ Na \longrightarrow 2\ CH_3CH_2CH_2CH_2ONa + H_2$
(2) $CH_3CH_2CH_2CH_2ONa + CH_3CH_2CH_2Br \longrightarrow CH_3CH_2CH_2CH_2{-}O{-}CH_2CH_2CH_3 + NaBr$

Aldehydes and Ketones

Many organic molecules contain a carbon atom that is connected to oxygen with a double bond. This grouping of atoms is particularly reactive and is present in both aldehydes and ketones. Formaldehyde is by far the most common aldehyde molecule. It is used commercially as a preservative for animal specimens and in the formation of formaldehyde polymers. Examples of these polymers are found in Melmac plastic dishes, Formica table tops, and in the adhesives used in the manufacture of plywood and fiberboard. Other aldehydes are found in nature as spices.

Acetone is the simplest and most common ketone in our lives. It is used in nail polish remover, paints, varnishes, and resins. Acetone is also produced in the body during lipid metabolism. It generally is metabolized, but in diabetic patients more is formed than can be oxidized. The presence of acetone in a urine sample or on the breath is a positive indicator of diabetes. An understanding of aldehydes and ketones forms the basis for a discussion of many of the important organic and biochemical reactions.

23.1 Structures of Aldehydes and Ketones

The aldehydes and ketones are closely related classes of compounds. Their structures contain the **carbonyl group**, $\gt C{=}O$, a carbon–oxygen double bond. **Aldehydes** have at least one hydrogen atom bonded to the carbonyl group, whereas **ketones** have only alkyl or aryl (aromatic, denoted Ar) groups bonded to the carbonyl group.

carbonyl group
aldehyde
ketone

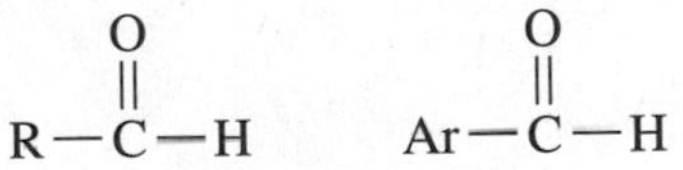

$$R-\overset{\overset{\displaystyle O}{\|}}{C}-H \qquad Ar-\overset{\overset{\displaystyle O}{\|}}{C}-H$$

aldehydes

$$R-\overset{\overset{\displaystyle O}{\|}}{C}-R \qquad R-\overset{\overset{\displaystyle O}{\|}}{C}-Ar \qquad Ar-\overset{\overset{\displaystyle O}{\|}}{C}-Ar$$

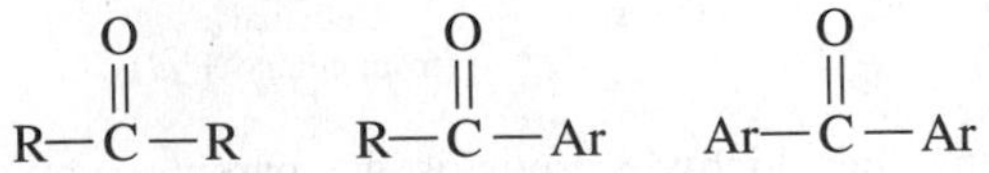

ketones

In a linear expression, the aldehyde group is often written as CHO. For example,

$$CH_3CHO \quad \text{is equivalent to} \quad CH_3\overset{\overset{\displaystyle O}{\|}}{C}-H$$

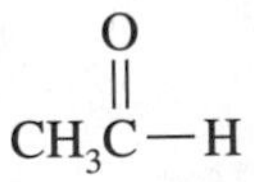

In the linear expression of a ketone, the carbonyl group is written as CO; for example,

$$CH_3COCH_3 \quad \text{is equivalent to} \quad CH_3\overset{\overset{\displaystyle O}{\|}}{C}CH_3$$

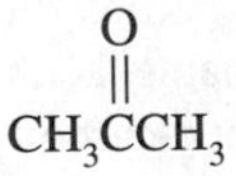

The general formula for the saturated homologous series of aldehydes and ketones is $C_nH_{2n}O$.

◀ Chapter Opening Photo: Many plastic products begin with molecules known as aldehydes.

23.2 Naming Aldehydes and Ketones

Aldehydes

IUPAC Nomenclature for Aldehydes

1. Select the longest continuous chain of carbon atoms that contains the aldehyde group.
2. The carbons of the parent chain are numbered starting with the aldehyde group. Since the aldehyde group is at the beginning (or end) of a chain, it is understood to be number 1.
3. Form the parent aldehyde name by dropping the *e* from the corresponding alkane name and adding the suffix *al*.
4. Other groups attached to the parent chain are named and numbered as we have done before.

The first member of the homologous series, $H_2C{=}O$, is methanal. The name *methanal* is derived from the hydrocarbon methane, which contains one carbon atom. The second member of the series is ethanal; the third member of the series is propanal; and so on.

$$\underset{\text{methane}}{CH_4} \qquad \underset{\substack{\text{methanal}\\ \text{(from methane̸ + } al\text{)}}}{H{-}\overset{\overset{\displaystyle O}{\|}}{C}{-}H} \qquad \underset{\text{ethane}}{CH_3CH_3} \qquad \underset{\substack{\text{ethanal}\\ \text{(from ethane̸ + } al\text{)}}}{CH_3\overset{\overset{\displaystyle O}{\|}}{C}{-}H}$$

The longest carbon chain containg the aldehyde group is the parent compound. Other groups attached to this chain are numbered and named as before. For example,

$$\underset{\text{4-methylhexanal}}{CH_3CH_2\overset{\overset{\displaystyle CH_3}{|}}{C}HCH_2CH_2\overset{\overset{\displaystyle O}{\|}}{C}{-}H}$$

The naming of aldehydes is illustrated in Table 23.1. The common names for some aldehydes are widely used. Common names for the aliphatic aldehydes are derived from the common names of the carboxylic acids (see Table 24.1). The *-ic acid* or *-oic acid* ending of the acid name is dropped and is replaced with the suffix *-aldehyde*. Thus the name of the one-carbon acid, formic acid, becomes formaldehyde for the one-carbon aldehyde; the name for the two-carbon acid, acetic acid, becomes acetaldehyde for the two-carbon aldehyde.

TABLE 23.1 IUPAC Names of Selected Aldehydes

Fomula	IUPAC name
$H{-}C({=}O){-}H$	Methanal
$CH_3C({=}O){-}H$	Ethanal
$CH_3CH_2C({=}O){-}H$	Propanal
$CH_3CH_2CH_2C({=}O){-}H$	Butanal
$CH_3CH(CH_3)C({=}O){-}H$	2-Methylpropanal

$$\underset{\text{formic acid}}{H{-}\overset{O}{\overset{\|}{C}}{-}OH} \qquad \underset{\text{formaldehyde}}{H{-}\overset{O}{\overset{\|}{C}}{-}H} \qquad \underset{\text{acetic acid}}{CH_3{-}\overset{O}{\overset{\|}{C}}{-}OH} \qquad \underset{\text{acetaldehyde}}{CH_3{-}\overset{O}{\overset{\|}{C}}{-}H}$$

Aromatic aldehydes contain an aldehyde group bonded to an aromatic ring. These aldehydes are named after the corresponding carboxylic acids. Thus the name benzaldehyde is derived from benzoic acid, and the name *p*-tolualdehyde is derived from *p*-toluic acid.

$$\underset{\text{benzaldehyde}}{C_6H_5{-}\overset{O}{\overset{\|}{C}}{-}H} \qquad \underset{\text{benzoic acid}}{C_6H_5{-}\overset{O}{\overset{\|}{C}}{-}OH} \qquad \underset{p\text{-tolualdehyde}}{CH_3{-}C_6H_4{-}\overset{O}{\overset{\|}{C}}{-}H} \qquad \underset{p\text{-toluic acid}}{CH_3{-}C_6H_4{-}\overset{O}{\overset{\|}{C}}{-}OH}$$

In dialdehydes the suffix *dial* is added to the corresponding hydrocarbon name; for example,

$$H{-}\overset{O}{\overset{\|}{C}}CH_2CH_2\overset{O}{\overset{\|}{C}}{-}H$$

is named butanedial.

Aldehydes are used in the ▶ manufacture of building materials.

Ketones

IUPAC Nomenclature for Ketones

1. Select the longest continuous chain of carbon atoms that contain the ketone group.
2. If the chain is longer than four carbons, it is numbered so that the carbonyl group has the smallest number possible; this number is prefixed to the parent name of the ketone.
3. Form the parent name by dropping the *e* from the corresponding alkane name and adding the suffix *one*.
4. Other groups attached to the parent chain are named and numbered as we have done before.

For example,

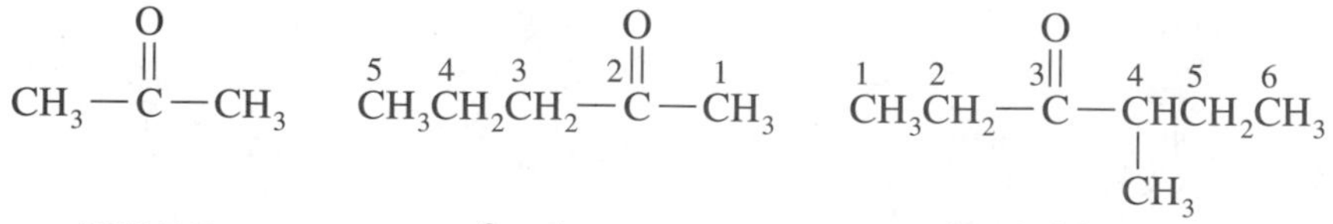

$$CH_3-\overset{\overset{O}{\|}}{C}-CH_3$$

propanone

$$\overset{5}{C}H_3\overset{4}{C}H_2\overset{3}{C}H_2-\overset{\overset{O}{\|}}{\overset{2}{C}}-\overset{1}{C}H_3$$

2-pentanone

$$\overset{1}{C}H_3\overset{2}{C}H_2-\overset{\overset{O}{\|}}{\overset{3}{C}}-\underset{\underset{CH_3}{|}}{\overset{4}{C}H}\overset{5}{C}H_2\overset{6}{C}H_3$$

4-methyl-3-hexanone

Note that in 4-methyl-3-hexanone the carbon chain is numbered from left to right to give the ketone group the lowest possible number.

An alternative non-IUPAC method commonly used to name simple ketones is to list the names of the alkyl or aromatic groups attached to the carbonyl carbon together with the word *ketone*. Thus butanone ($CH_3COCH_2CH_3$) is methyl ethyl ketone:

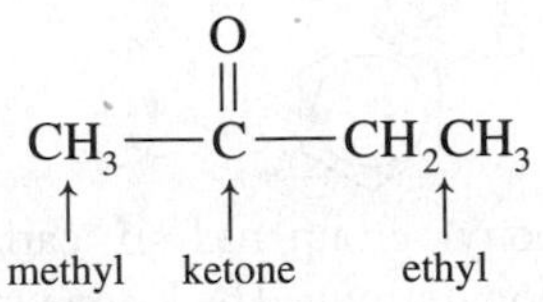

$$\underset{\text{methyl}}{CH_3}-\underset{\text{ketone}}{\overset{\overset{\displaystyle O}{\|}}{C}}-\underset{\text{ethyl}}{CH_2CH_3}$$

Two of the most widely used ketones have special common names: Propanone is called acetone, and butanone is known as methyl ethyl ketone, or MEK.

Aromatic ketones are named in a fashion similar to that for aliphatic ketones and often have special names as well.

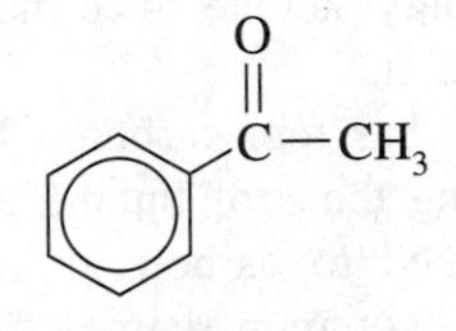

$$C_6H_5-\overset{\overset{\displaystyle O}{\|}}{C}-CH_3$$

1-phenylethanone (IUPAC)
(methyl phenyl ketone)
(acetophenone)

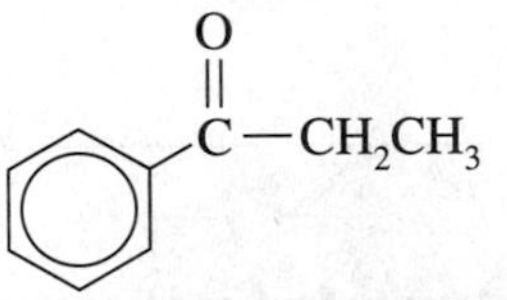

$$C_6H_5-\overset{\overset{\displaystyle O}{\|}}{C}-CH_2CH_3$$

1-phenyl-1-propanone (IUPAC)
(ethyl phenyl ketone)
(propiophenone)

Example 23.1

Write the formulas and the names for the straight-chain five- and six-carbon aldehydes.

Solution

The IUPAC names are based on the five- and six-carbon alkanes. Drop the *e* of the alkane name and add the suffix *al*. Pentane, C_5, becomes pentanal and hexane, C_6, becomes hexanal. (The aldehyde group does not need to be numbered; it is understood to be on carbon 1.) The common names are derived from valeric acid and caproic acid, respectively.

$$CH_3CH_2CH_2CH_2\overset{\overset{\displaystyle O}{\|}}{C}-H$$

pentanal (valeraldehyde)

$$CH_3CH_2CH_2CH_2CH_2\overset{\overset{\displaystyle O}{\|}}{C}-H$$

hexanal (caproaldehyde)

Example 23.2 Give two names for each of the following ketones:

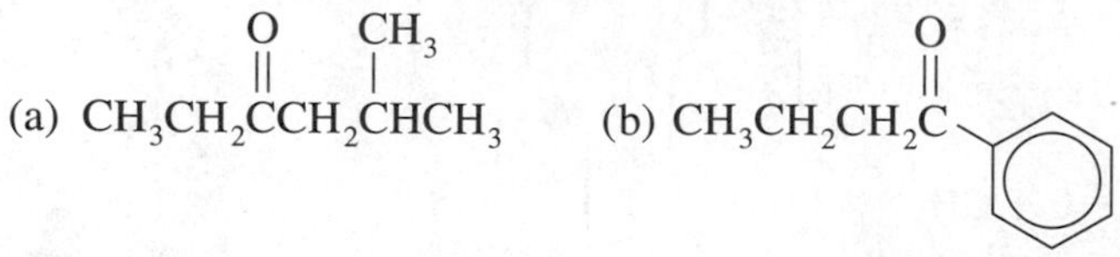

(a) $CH_3CH_2\overset{O}{\overset{\|}{C}}CH_2\overset{CH_3}{\overset{|}{C}}HCH_3$ (b) $CH_3CH_2CH_2\overset{O}{\overset{\|}{C}}-C_6H_5$

Solution

(a) The parent carbon chain that contains the carbonyl group has six carbons. Number this chain from the end nearer to the carbonyl group. The ketone group is on carbon 3, and a methyl group is on carbon 5. The six-carbon alkane is hexane. Drop the *e* from hexane and add *one* to give the parent name, hexanone. Prefix the name hexanone with a 3- to locate the ketone group and with 5-methyl- to locate the methyl group. The name is 5-methyl-3-hexanone. The common name is ethyl isobutyl ketone since the C=O has an ethyl group and an isobutyl group bonded to it.

(b) The longest aliphatic chain has four carbons. The parent ketone name is butanone, derived by dropping the *e* of butane and adding *one*. The butanone has a phenyl group attached to carbon 1. The IUPAC name is therefore 1-phenyl-1-butanone. The common name for this compound is phenyl *n*-propyl ketone, since the C=O group has a phenyl and an *n*-propyl group bonded to it.

Practice 23.1

Write structures for the following carbonyl compounds:

(a) 4-bromo-5-hydroxyhexanal (c) 3-buten-2-one
(b) phenylethanal (d) diphenylmethanone (diphenyl ketone)

Practice 23.2

Name each of the following compounds using the IUPAC system:

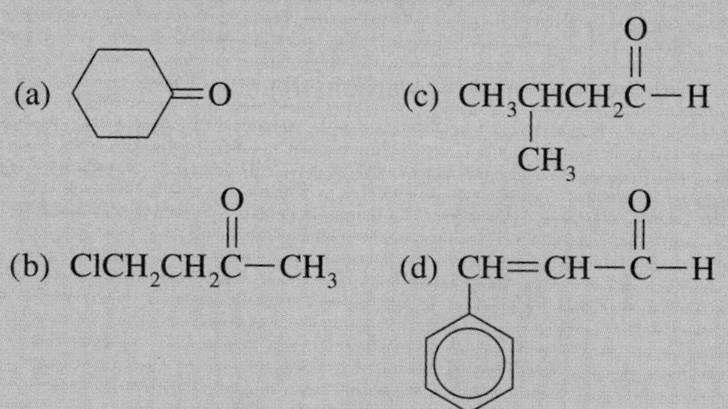

(a) cyclohexane ring =O

(c) $CH_3\underset{CH_3}{\underset{|}{C}}HCH_2\overset{O}{\overset{\|}{C}}-H$

(b) $ClCH_2CH_2\overset{O}{\overset{\|}{C}}-CH_3$

(d) $\underset{C_6H_5}{\underset{|}{C}}H=CH-\overset{O}{\overset{\|}{C}}-H$

23.3 Bonding and Physical Properties

The carbon atom of the carbonyl group is sp^2 hybridized and is joined to three other atoms by sigma bonds. The fourth bond is made by overlapping p electrons of carbon and oxygen to form a pi bond between the carbon and oxygen atoms. Since the carbon has a double bond, the possibility of adding another bonded atom (an addition reaction) exists for carbonyl carbon atoms.

Because the oxygen atom is considerably more electronegative than carbon, the carbonyl group is polar, with the electrons shifted toward the oxygen atom. This makes the oxygen atom partially negative (δ^-) and leaves the carbon atom partially positive (δ^+). Many of the chemical reactions of aldehydes and ketones are due to this polarity.

$$\overset{\delta^+}{C}=\overset{\delta^-}{\ddot{O}}:$$

polarity

Unlike alcohols, aldehydes and ketones cannot interact with themselves through hydrogen bonding, because there is no hydrogen atom attached to the oxygen atom of the carbonyl group. Aldehydes and ketones, therefore, have lower boiling points than alcohols of comparable molar mass (Table 23.2).

Low-molar-mass aldehydes and ketones are soluble in water, but for five or more carbons, the solubility decreases markedly. Ketones are very efficient organic solvents.

The lower-molar-mass aldehydes have a penetrating, disagreeable odor and are partially responsible for the taste of some rancid and stale foods. As the molar mass increases, the odor of both aldehydes and ketones, especially the aromatic ones, becomes more fragrant. Some are even used in flavorings and perfumes. A few of these and other selected aldehydes and ketones are shown in Figure 23.1.

TABLE 23.2 Boiling Points of Selected Aldehydes and Ketones and Corresponding Alcohols

Name	Molar mass	Boiling point (°C)
1-Propanol	60	97
Propanal	58	49
Propanone	58	56
1-Butanol	74	118
Butanal	72	76
Butanone	72	80
1-Pentanol	88	138
Pentanal	86	103
2-Pentanone	86	102

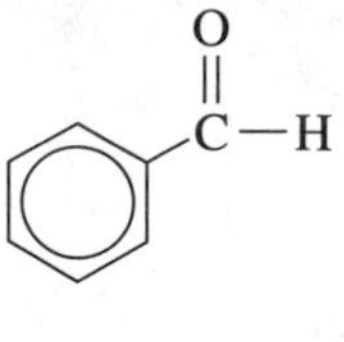

benzaldehyde
(oil of bitter almonds)

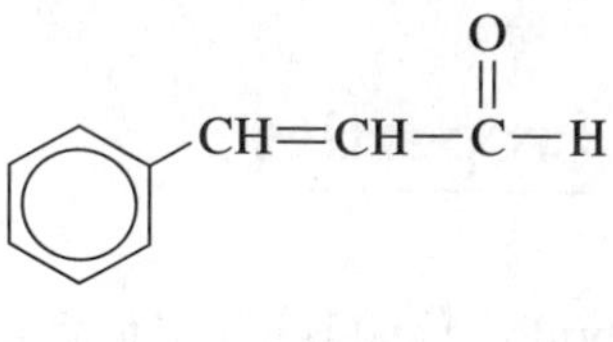

cinnamaldehyde
(oil of cinnamon)

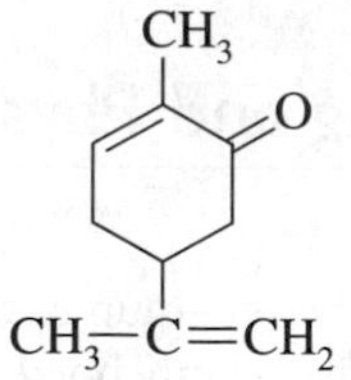

carvone
(chief component of spearmint oil)

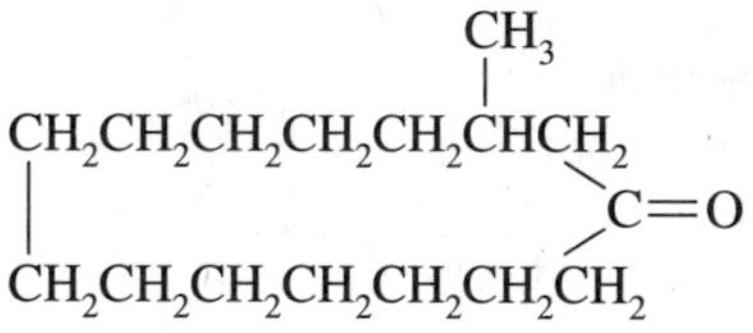

muscone
(gland of male musk deer, used in perfume)

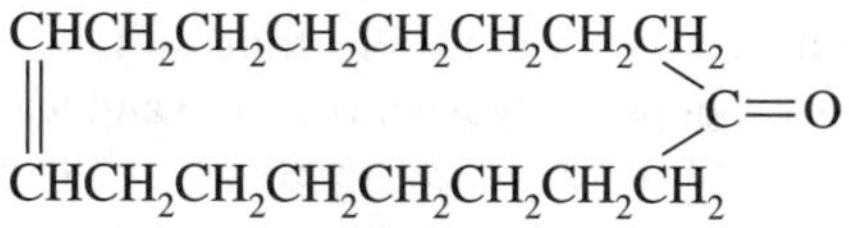

civetone
(secretion of the civet cat, used in perfume)

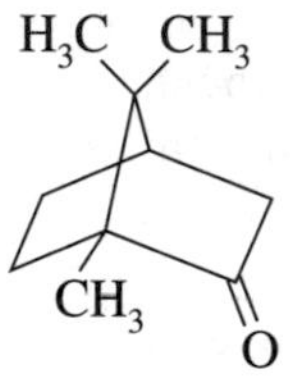

camphor
(from the camphor tree)

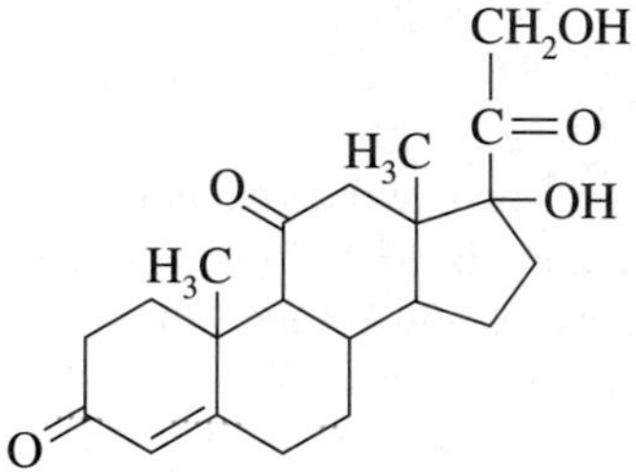

cortisone
(hormone; regulation of carbohydrate and protein metabolism; used to reduce inflammation)

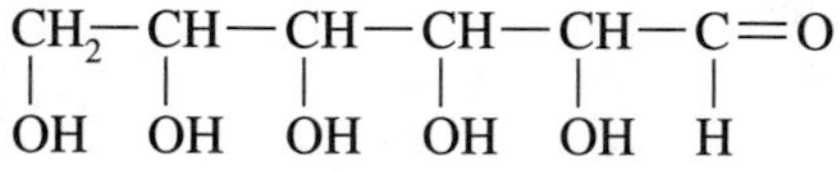

glucose
(sugar)

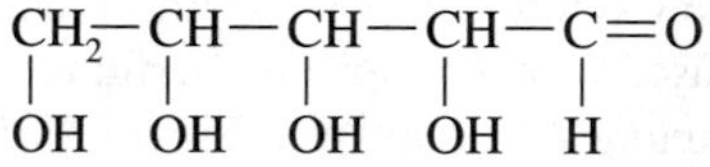

ribose
(sugar)

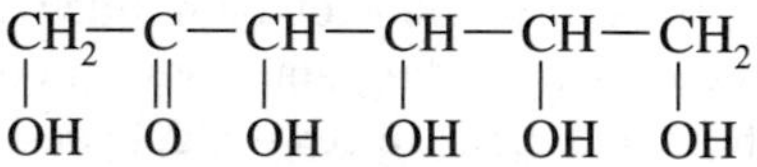

fructose
(sugar)

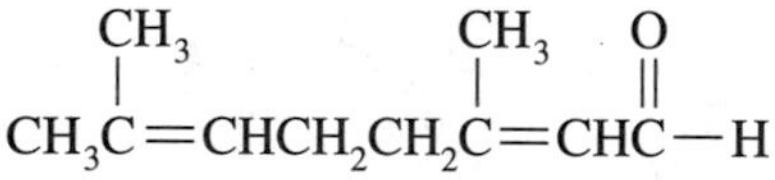

citral
(oil of lemon)

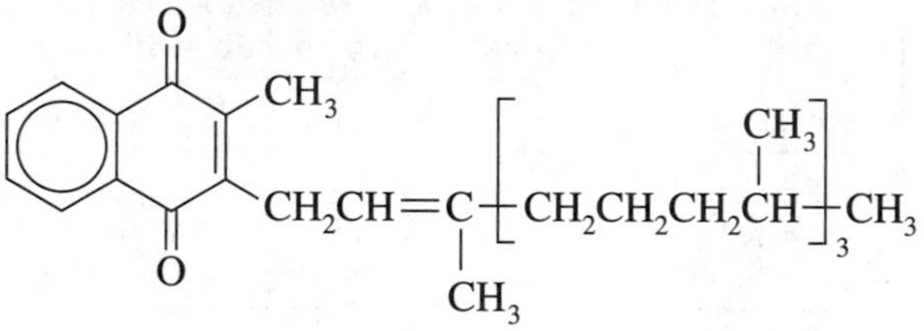

vitamin K_1
(antihemorrhagic vitamin)

FIGURE 23.1
Selected naturally occurring aldehydes and ketones

Chemistry in Action

Familiar Aldehydes and Ketones

Aldehydes and ketones are all around you. For example, aldehydes are used commercially in the production of plastics. The common polymers known as Formica, used on counters and in telephones; Melmac, used in plastic tableware; and Bakelite, found in handles on kitchen utensils, are all derived from formaldehyde.

Then there are the aromatic aldehydes. They are responsible for the characteristic flavor and aroma of many common flavorings and spices. Examples include vanillin (see Chemistry in Action, Chapter 5), cinnamaldehyde, citral, and benzaldehyde (see Figure 23.1).

Higher-molar-mass ketones are also useful because of their unique aromas. Camphor (Figure 23.1) is a moth repellant. Muscone, extracted from the musk glands of the male musk deer, is frequently used to give the musky aroma to perfumes. A tiny amount of these chemicals can be used to produce long-lasting aromas. Perfume manufacturers closely guard the secret of their ingredients to keep their formulations unique and selling well.

In living organisms aldehydes and ketones are found in many different compounds. Sugars are often classified as containing an aldehyde group, as in glucose or ribose, or a ketone group, as in fructose. The compounds known as steroids, belonging to the lipid family, also often contain aldehyde or ketone groups. These lipids may function as hormones, as in the case of cortisone.

Formica, one of the most familiar formaldehyde-derived polymers, comes in an array of colors and designs.

23.4 Chemical Properties of Aldehydes and Ketones

The carbonyl group ($\rangle C{=}O$) is the functional group of both aldehydes and ketones. Associated with every functional group are characteristic reactions. Aldehydes undergo both oxidation and reduction reactions; ketones undergo reduction reactions.

The carbon atom in the carbonyl group also has a double bond. What kind of reactivity does this suggest for aldehydes and ketones? This question can be answered by reviewing the chemistry of the alkenes in Chapter 21. Like alkenes, the aldehydes and ketones have a carbon atom that is bonded to three other atoms—one less than the usual maximum of four. Such a carbon atom can easily bond to one more atom. Alkenes readily undergo addition reactions. This suggests that the aldehydes and ketones also undergo addition reactions. In fact, addition is the characteristic reaction of aldehydes and ketones.

Oxidation

Aldehydes are easily oxidized to carboxylic acids by a variety of oxidizing agents, including (under some conditions) oxygen of the air. Oxidation is the reaction in which aldehydes differ most from ketones. In fact, aldehydes and ketones may be separated into classes by their relative susceptibilities to oxidation. Aldehydes are easily oxidized to carboxylic acids by $K_2Cr_2O_7$ + H_2SO_4 and by mild oxidizing agents such as Ag^+ and Cu^{2+} ions; ketones are unaffected by such reagents. Ketones can be oxidized under drastic conditions—for example, by treatment with hot potassium permanganate solution. However, under these conditions carbon–carbon bonds are broken, and a variety of products are formed. Equations for the oxidation of aldehydes by dichromate are

$$3\,R\overset{\overset{\large O}{\|}}{C}-H + Cr_2O_7^{2-} + 8\,H^+ \longrightarrow \underset{\text{carboxylic acid}}{3\,R\overset{\overset{\large O}{\|}}{C}-OH} + 2\,Cr^{3+} + 4\,H_2O$$

$$3\,CH_3\overset{\overset{\large O}{\|}}{C}-H + Cr_2O_7^{2-} + 8\,H^+ \longrightarrow \underset{\text{acetic acid}}{3\,CH_3\overset{\overset{\large O}{\|}}{C}-OH} + 2\,Cr^{3+} + 4\,H_2O$$

Tollens test

The **Tollens test** (silver-mirror test) for aldehydes is based on the ability of silver ions to oxidize aldehydes. The Ag^+ ions are thereby reduced to metallic silver. In practice, a little of the suspected aldehyde is added to a solution of silver nitrate and ammonia in a clean test tube. The appearance of a silver mirror on the inner wall of the tube is a positive test for the aldehyde group. The abbreviated equation is

$$R\overset{\overset{\large O}{\|}}{C}-H + 2\,Ag^+ \xrightarrow[H_2O]{NH_3} R\overset{\overset{\large O}{\|}}{C}-O^-NH_4^+ + 2\,Ag(s) \qquad \text{(general reaction)}$$

$$CH_3\overset{\overset{\large O}{\|}}{C}-H + 2\,Ag^+ \xrightarrow[H_2O]{NH_3} CH_3\overset{\overset{\large O}{\|}}{C}-O^-\,NH_4^+ + 2\,Ag(s)$$

Fehling and Benedict tests

Fehling and Benedict solutions contain Cu^{2+} ions in an alkaline medium. In the **Fehling and the Benedict tests**, the aldehyde group is oxidized to an acid by Cu^{2+} ions. The blue Cu^{2+} ions are reduced and form brick-red copper(I) oxide (Cu_2O), which precipitates during the reaction. These tests can be used for detecting carbohydrates that have an available aldehyde group. The abbreviated equation is

$$R\overset{\overset{\large O}{\|}}{C}-H + \underset{\text{(blue)}}{2\,Cu^{2+}} \xrightarrow[H_2O]{NaOH} R\overset{\overset{\large O}{\|}}{C}-O^-\,Na^+ + \underset{\text{(brick-red)}}{Cu_2O(s)}$$

Most ketones do not give a positive test with Tollens, Fehling, or Benedict solutions. These tests are used to distinguish between aldehydes and ketones.

$$R-\overset{\overset{\large O}{\|}}{C}-R + Ag^+ \xrightarrow[H_2O]{NH_3} \text{No reaction}$$

$$R-\overset{\overset{\large O}{\|}}{C}-R + Cu^{2+} \xrightarrow[H_2O]{OH^-} \text{No reaction}$$

Aldehydes and ketones are often involved in industrial syntheses because oxidation is a relatively simple reaction that can start with inexpensive reactants. Petroleum, natural gas, and the oxygen in the air are inexpensive and abundant materials; many aldehydes and ketones can be prepared from these starting materials. In turn, the aldehydes and ketones are useful for making even more valuable products. For example, simple hydrocarbons are oxidized to form acetaldehyde, which can be oxidized by air to acetic acid, the most widely used carboxylic acid. Air oxidation of 2-propanol yields acetone, an important commercial solvent. Methanol is air oxidized to formaldehyde, which is the precursor for a wide variety of plastics and glues.

The oxidation of aldehydes is an important reaction in biochemistry. When our cells "burn" carbohydrates, they take advantage of the aldehyde reactivity. The aldehyde group is oxidized to a carboxylic acid and is eventually converted to carbon dioxide, which is then exhaled. This stepwise oxidation provides some of the energy necessary to sustain life.

Practice 23.3

Write the structures of the products (or indicate "no reaction") for the following:
(a) pentanal in the Tollens test
(b) 4-methyl-2-hexanone in the Tollens test
(c) 3-pentanone in the Fehling test
(d) 4-methylhexanal in the Fehling test

Reduction

Aldehydes and ketones are easily reduced to alcohols, either by elemental hydrogen in the presence of a catalyst or by chemical reducing agents such as lithium aluminum hydride ($LiAlH_4$) or sodium borohydride ($NaBH_4$). Aldehydes yield primary alcohols; ketones yield secondary alcohols:

$$\text{R}-\overset{\overset{\displaystyle\text{O}}{\|}}{\text{C}}-\text{H} \xrightarrow[\Delta]{H_2/Ni} \underset{1^\circ\text{ alcohol}}{\text{RCH}_2\text{OH}} \qquad \text{(general reaction)}$$

$$\text{R}-\overset{\overset{\displaystyle\text{O}}{\|}}{\text{C}}-\text{R} \xrightarrow[\Delta]{H_2/Ni} \underset{2^\circ\text{ alcohol}}{\text{R}-\overset{\overset{\displaystyle\text{OH}}{|}}{\text{CH}}-\text{R}} \qquad \text{(general reaction)}$$

$$\text{CH}_3\overset{\overset{\displaystyle\text{O}}{\|}}{\text{C}}-\text{H} \xrightarrow[\Delta]{H_2/Ni} \text{CH}_3\text{CH}_2\text{OH}$$

$$\text{CH}_3\overset{\overset{\displaystyle\text{O}}{\|}}{\text{C}}\text{CH}_3 \xrightarrow[\Delta]{H_2/Ni} \text{CH}_3\overset{\overset{\displaystyle\text{OH}}{|}}{\text{C}}\text{HCH}_3$$

Addition Reactions

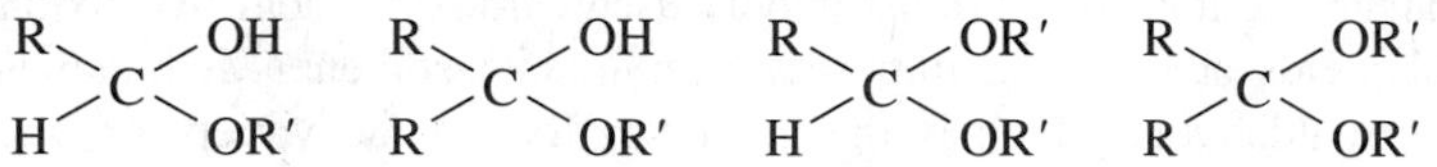

hemiacetal
hemiketal
acetal
ketal

Addition of Alcohols Compounds derived from aldehydes and ketones that contain an alkoxy and a hydroxy group on the same carbon atom are known as **hemiacetals** and **hemiketals**. In a like manner, compounds that have two alkoxy groups on the same carbon atom are known as **acetals** and **ketals**.

The alkoxy group in these compounds forms an ether linkage.

$$\underset{\text{hemiacetal}}{\mathrm{(R)(H)C(OH)(OR')}} \qquad \underset{\text{hemiketal}}{\mathrm{(R)(R)C(OH)(OR')}} \qquad \underset{\text{acetal}}{\mathrm{(R)(H)C(OR')(OR')}} \qquad \underset{\text{ketal}}{\mathrm{(R)(R)C(OR')(OR')}}$$

Most open-chain hemiacetals and hemiketals are so unstable that they cannot be isolated. On the other hand, acetals and ketals are stable in alkaline solutions but are unstable in acid solutions, in which they are hydrolyzed back to the original aldehyde or ketone.

In the following reactions we show only aldehydes in the equations, but keep in mind that ketones behave in a similar fashion, although they are not as reactive.

Aldehydes react with alcohols in the presence of a trace of acid to form hemiacetals:

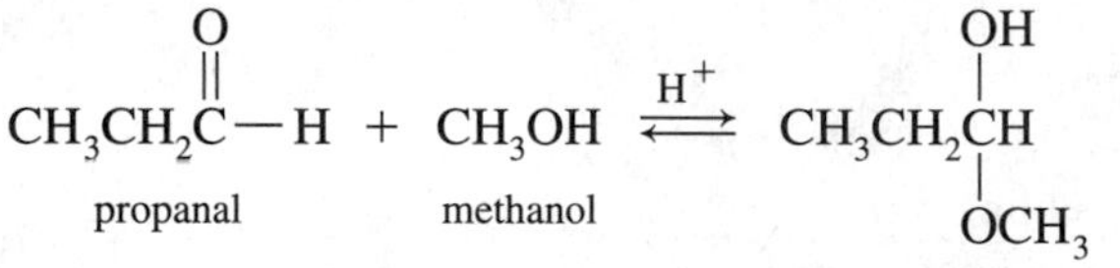

$$\underset{\text{propanal}}{\mathrm{CH_3CH_2\overset{O}{\overset{\|}{C}}{-}H}} + \underset{\text{methanol}}{\mathrm{CH_3OH}} \overset{H^+}{\rightleftharpoons} \mathrm{CH_3CH_2CH(OH)(OCH_3)}$$

1-methoxy-1-propanol
(propionaldehyde methyl hemiacetal)

In the presence of excess alcohol and a strong acid such as dry HCl, aldehydes or hemiacetals react with a second molecule of the alcohol to give an acetal:

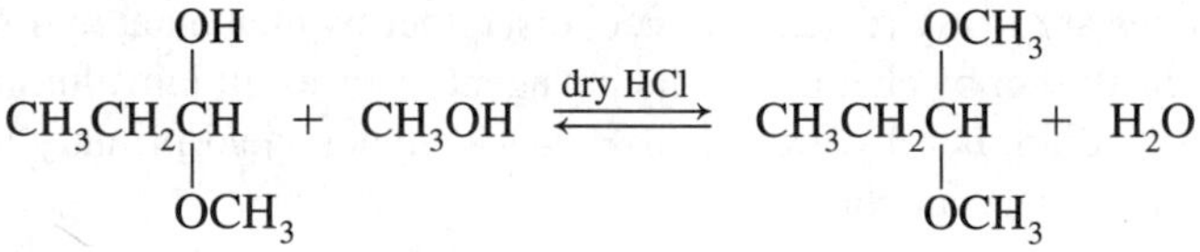

$$\mathrm{CH_3CH_2CH(OH)(OCH_3)} + \mathrm{CH_3OH} \overset{\text{dry HCl}}{\rightleftharpoons} \mathrm{CH_3CH_2CH(OCH_3)(OCH_3)} + \mathrm{H_2O}$$

1,1-dimethoxypropane
(propionaldehyde dimethyl acetal)

A hemiacetal has both an alcohol and an ether group attached to the aldehyde carbon. An acetal has two ether groups attached to the aldehyde carbon.

If the alcohol and carbonyl groups are within the same molecule, the result is the formation of a cyclic hemiacetal (or hemiketal). This *intramolecular* cyclization is particularly significant in carbohydrate chemistry during the study of monosaccharides (Chapter 28).

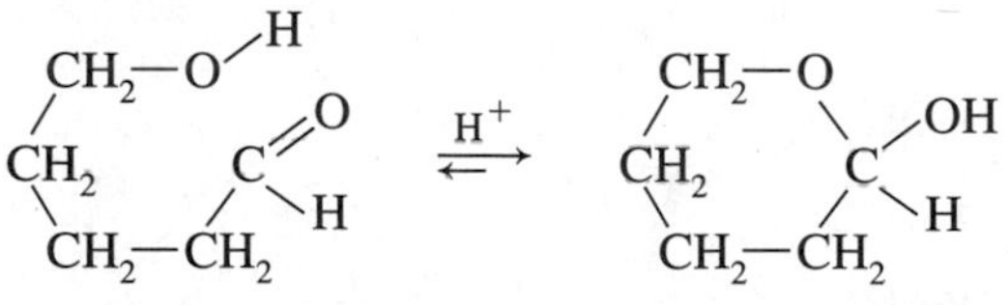

$$\underset{\text{5-hydroxypentanal}}{\mathrm{HO{-}CH_2CH_2CH_2CH_2{-}CHO}} \overset{H^+}{\rightleftharpoons} \underset{\text{stable hemiacetal}}{\text{cyclic } \mathrm{CH_2{-}O{-}CH(OH)}\text{ ring}}$$

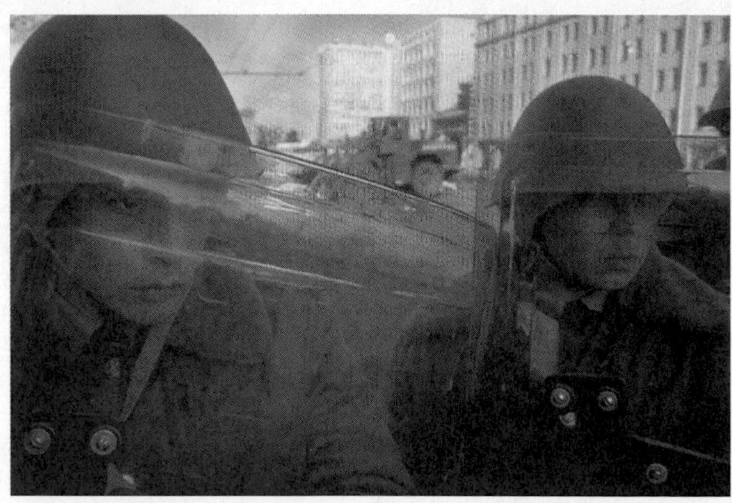

◀ **Riot control police use plastic Lucite shields for protection.**

> **Practice 23.4**
>
> Write structures and IUPAC names for the hemiacetal and acetal formed by reacting: (a) ethanal and ethanol, and (b) butanone and methanol.

Addition of Hydrogen Cyanide The addition of hydrogen cyanide, HCN, to aldehydes and ketones forms a class of compounds known as cyanohydrins. **Cyanohydrins** have a cyano (—CN) group and a hydroxyl group on the same carbon atom. The reaction is catalyzed by a small amount of base:

cyanohydrin

$$\underset{\text{acetaldehyde}}{CH_3\overset{\overset{\displaystyle O}{\|}}{C}-H} + HCN \xrightarrow{OH^-} \underset{\text{acetaldehyde cyanohydrin}}{CH_3\overset{\overset{\displaystyle OH}{|}}{C}HCN}$$

$$\underset{\text{acetone}}{CH_3\underset{\underset{\displaystyle O}{\|}}{C}CH_3} + HCN \xrightarrow{OH^-} \underset{\text{acetone cyanohydrin}}{CH_3\overset{\overset{\displaystyle CH_3}{|}}{\underset{\underset{\displaystyle OH}{|}}{C}}-CN}$$

In the cyanohydrin reaction, the more positive H atom of HCN adds to the oxygen of the carbonyl group, and the —CN group adds to the carbon atom of the carbonyl group. In the aldehyde addition, the length of the carbon chain is increased by one carbon. The ketone addition product also contains an additional carbon atom.

Cyanohydrins are useful intermediates for the synthesis of several important compounds. For example, the hydrolyses of cyanohydrins produce α-hydroxy acids:

$$CH_3\underset{\underset{\displaystyle OH}{|}}{C}H-CN + H_2O \xrightarrow{H^+} \underset{\text{lactic acid}}{CH_3\underset{\underset{\displaystyle OH}{|}}{C}HCOOH} + NH_4^+$$

Acetaldehyde can also be converted into other important biochemical compounds such as the amino acid alanine:

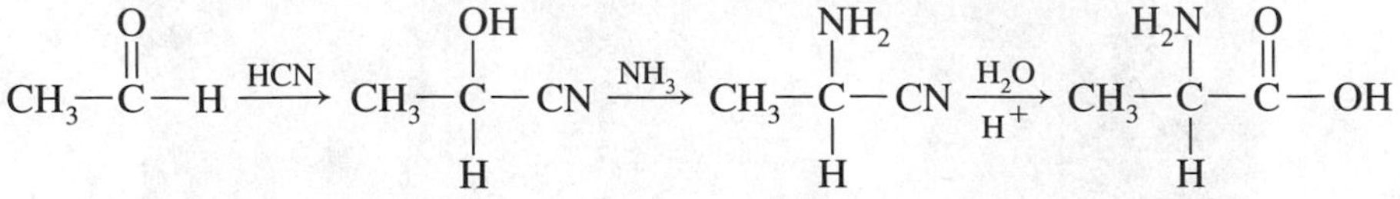

$$CH_3-\overset{\overset{O}{\|}}{C}-H \xrightarrow{HCN} CH_3-\underset{\underset{H}{|}}{\overset{\overset{OH}{|}}{C}}-CN \xrightarrow{NH_3} CH_3-\underset{\underset{H}{|}}{\overset{\overset{NH_2}{|}}{C}}-CN \xrightarrow[H^+]{H_2O} CH_3-\underset{\underset{H}{|}}{\overset{\overset{H_2N}{|}}{C}}-\overset{\overset{O}{\|}}{C}-OH$$

alanine

Some commercial reactions also involve the use of cyanohydrins. Acetone cyanohydrin can be converted to methyl methacrylate when refluxed with methanol and a strong acid. The methyl methacrylate can then be polymerized to Lucite or Plexiglas, both transparent plastics. See Chapter 26.

Aldol Condensation (Self-addition) In a carbonyl compound the carbon atoms are labeled alpha (α), beta (β), gamma (γ), delta (δ), and so on, according to their positions with respect to the carbonyl group. The α-carbon is adjacent to the carbonyl carbon, the β-carbon is next, the γ-carbon is third, and so forth. The hydrogens attached to the α-carbon atom are therefore called α-hydrogens, and so on, as shown here:

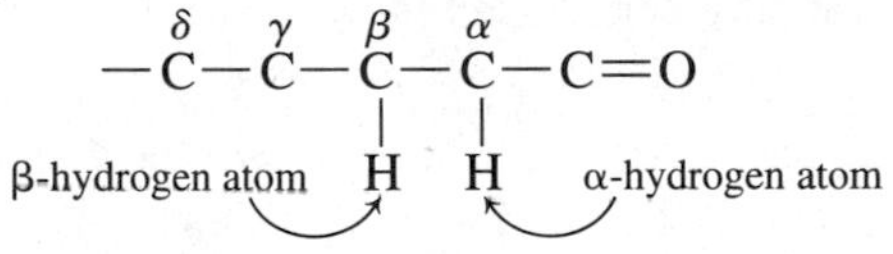

$$-\overset{\delta}{C}-\overset{\gamma}{C}-\underset{\underset{H}{|}}{\overset{\beta}{C}}-\underset{\underset{H}{|}}{\overset{\alpha}{C}}-C=O$$

β-hydrogen atom → H   H ← α-hydrogen atom

The hydrogen atoms attached to the α-carbon atom have the unique ability to be more easily released as protons than other hydrogens within the molecule.

Remember that a *condensation* reaction is one in which two smaller molecules combine to form a larger molecule.

aldol condensation

An aldehyde or ketone that contains α-hydrogens may add to itself or to another α-hydrogen containing aldehyde or ketone. The product of this reaction contains both a carbonyl group and an alcohol group within the same molecule. The reaction is known as an **aldol condensation** and is catalyzed by dilute base. The aldol condensation is similar to the other carbonyl addition reactions. An α-hydrogen adds to the carbonyl oxygen, and the remainder of the molecule adds to the carbonyl carbon.

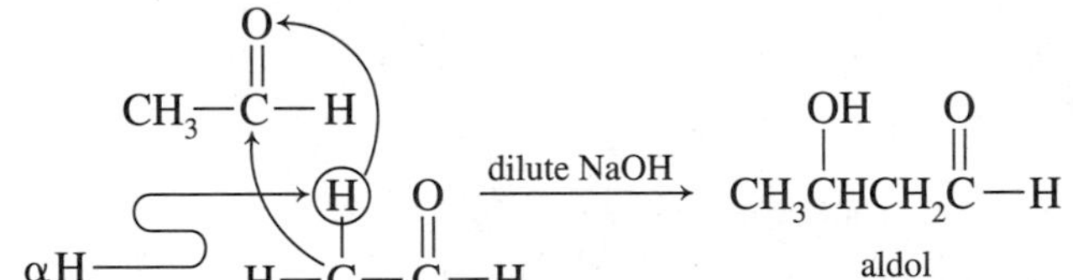

$$CH_3-\overset{\overset{O}{\|}}{C}-H \;+\; \alpha H\!:\; H-\underset{\underset{H}{|}}{\overset{\overset{H}{|}}{C}}-\overset{\overset{O}{\|}}{C}-H \xrightarrow{\text{dilute NaOH}} CH_3\overset{\overset{OH}{|}}{C}HCH_2\overset{\overset{O}{\|}}{C}-H$$

aldol
(3-hydroxybutanal)

In this reaction an alpha hydrogen first transfers from one molecule to the oxygen of the other molecule. This breaks the C=O pi bond, leaving intermediates in which a carbon atom of each molecule has three bonds.

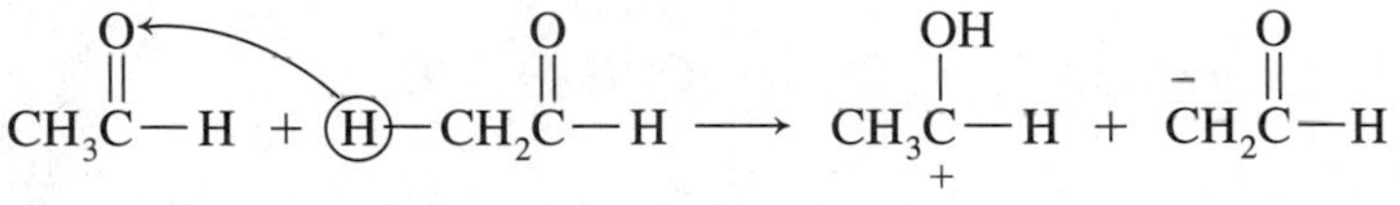

$$CH_3\overset{\overset{O}{\|}}{C}-H + H-CH_2\overset{\overset{O}{\|}}{C}-H \longrightarrow CH_3\underset{+}{\overset{\overset{OH}{|}}{C}}-H + \overset{-}{C}H_2\overset{\overset{O}{\|}}{C}-H$$

The two intermediates then bond to each other, forming the product.

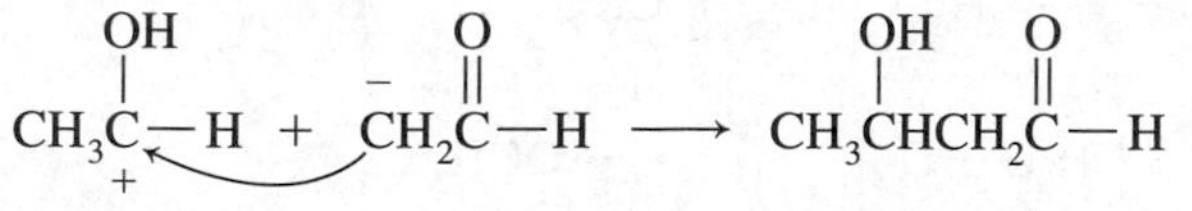

$$\underset{+}{\overset{\text{OH}}{\overset{|}{CH_3C}}}-H + \overset{-}{C}H_2\overset{\text{O}}{\overset{\|}{C}}-H \longrightarrow CH_3\overset{\text{OH}}{\overset{|}{C}}HCH_2\overset{\text{O}}{\overset{\|}{C}}-H$$

Acetone also undergoes the aldol condensation:

$$CH_3\overset{\text{O}}{\overset{\|}{C}}CH_3 + \text{(H)}-CH_2\overset{\text{O}}{\overset{\|}{C}}CH_3 \xrightarrow[\text{NaOH}]{\text{dilute}} CH_3\underset{CH_3}{\underset{|}{\overset{\text{OH}}{\overset{|}{C}}}}-CH_2\overset{\text{O}}{\overset{\|}{C}}CH_3$$

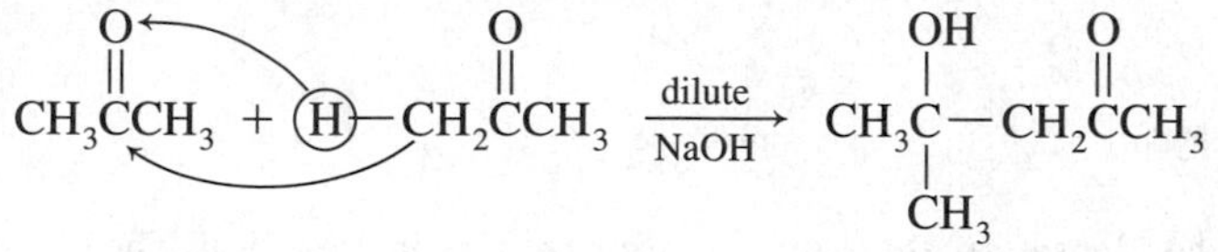

acetone acetone diacetone alcohol (4-hydroxy-4-methyl-2-pentanone)

Example 23.3

Write the equation for the aldol condensation of propanal.

Solution

First write the structure for propanal and locate the α-hydrogen atoms:

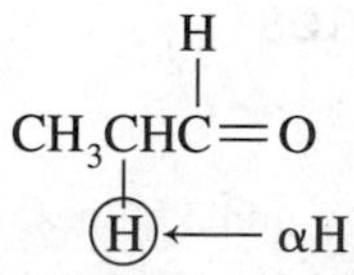

$$CH_3\underset{\text{(H)}\leftarrow\ \alpha H}{\underset{|}{C}}H\overset{\text{H}}{\overset{|}{C}}=O$$

Now write two propanal molecules and transfer an α-hydrogen from one molecule to the oxygen of the second molecule. After the π bond breaks, the two carbon atoms that are bonded to only three other atoms are attached to each other to form the product.

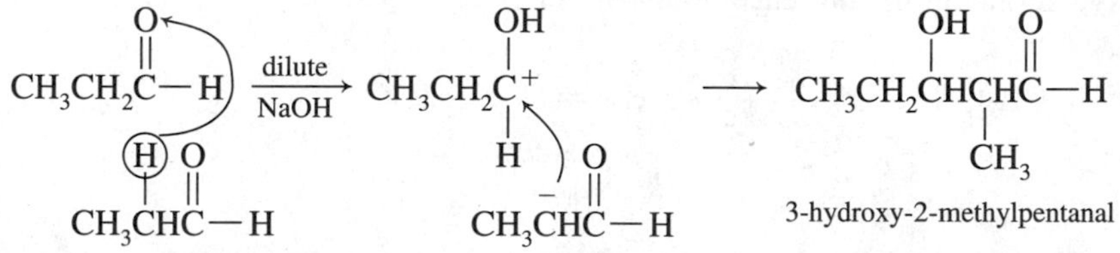

$$CH_3CH_2\overset{\text{O}}{\overset{\|}{C}}-H \quad CH_3\overset{\text{(H)}}{\overset{|}{C}}H\overset{\text{O}}{\overset{\|}{C}}-H \xrightarrow[\text{NaOH}]{\text{dilute}} CH_3CH_2\underset{H}{\underset{|}{\overset{\text{OH}}{\overset{|}{C^+}}}} \quad CH_3\overset{-}{C}H\overset{\text{O}}{\overset{\|}{C}}-H \longrightarrow CH_3CH_2\overset{\text{OH}}{\overset{|}{C}}H\underset{CH_3}{\underset{|}{C}}H\overset{\text{O}}{\overset{\|}{C}}-H$$

3-hydroxy-2-methylpentanal

Practice 23.5

(a) Write the equation for the aldol condensation of butanal.
(b) Write the equation for the aldol condensation of 3-pentanone.

Both the aldol condensation and the addition of hydrogen cyanide form new C—C bonds. These reactions are used in industry to build larger molecules from smaller precursors. For example, nifedipine (Procardia, Adalat) is an important heart medication (part of a class of drugs known as calcium-channel blockers) used to treat various forms of angina. Although nifedipine is a complex molecule, it can be synthesized from *o*-nitrobenzaldehyde (which contains a reactive aldehyde group) and two other reactants.

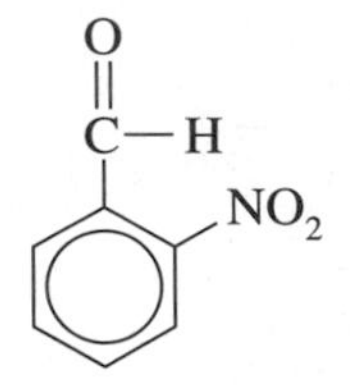

$$\text{o-nitrobenzaldehyde: } C_6H_4(NO_2)\text{—}CHO \qquad \text{nifedipine}$$

o-nitrobenzaldehyde 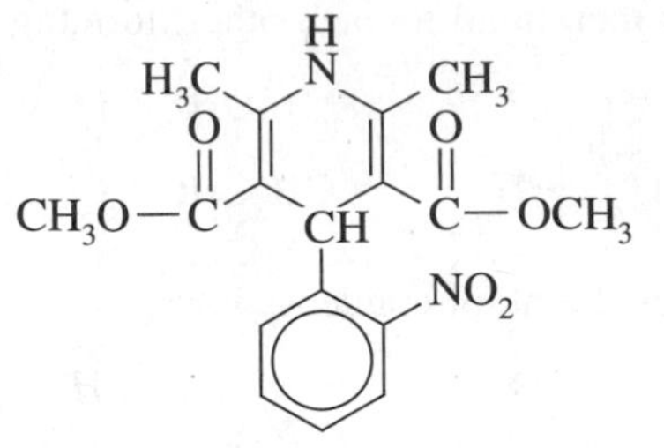 nifedipine

By using an aldol condensation, the drug is synthesized in one reaction. The carbonyl functional group provides the reactivity needed by the pharmaceutical chemist to create a complex molecule with important biological activity.

23.5 Common Aldehydes and Ketones

Numerous methods have been devised for making aldehydes and ketones. The oxidation of alcohols is a very general method. Special methods are often used for the commercial production of individual aldehydes and ketones.

Formaldehyde (Methanal) This aldehyde is made from methanol by reaction with oxygen (air) in the presence of a silver or copper catalyst:

$$2\ CH_3OH + O_2 \xrightarrow[400°C]{\text{Ag or Cu}} 2\ H_2C{=}O + 2\ H_2O$$

formaldehyde (methanal)

Formaldehyde is a poisonous, irritating gas that is very soluble in water. It is marketed as a 40% aqueous solution called *formalin*. By far the largest use of this chemical is in the manufacture of polymers (Chapter 26). About 1.33×10^9 kg of formaldehyde is manufactured annually in the United States.

Formaldehyde vapors are intensely irritating to the mucous membranes. Ingestion may cause severe abdominal pains, leading to coma and death.

It is of interest that formaldehyde may have had a significant role in chemical evolution. Formaldehyde is believed to have been a component of the primitive atmosphere of the earth. It is theorized that the reactivity of this single-carbon aldehyde enabled it to form more complex organic molecules—molecules that were precursors of the still more complicated substances that today are essential components of every living organism.

Acetaldehyde (Ethanal) Acetaldehyde is a volatile liquid (bp 21°C) with a pungent, irritating odor. It has a general narcotic action and in large doses may cause respiratory paralysis. Its principal use is as an intermediate in the manufacture of other chemicals such as acetic acid and 1-butanol. Acetic acid, for example, is made by air oxidation of acetaldehyde:

$$2\,CH_3\overset{\overset{\displaystyle O}{\|}}{C}{-}H + O_2 \xrightarrow[\Delta]{Mn^{2+}} 2\,CH_3\overset{\overset{\displaystyle O}{\|}}{C}{-}OH$$

Acetaldehyde undergoes reactions in which three or four molecules condense or polymerize to form the cyclic compounds paraldehyde and metaldehyde:

$$3\,CH_3{-}\overset{\overset{\displaystyle O}{\|}}{C}{-}H \xrightarrow[\Delta]{H^+} \text{paraldehyde}$$

paraldehyde
(bp 125°C)

Paraldehyde is a controlled substance and has been used as a sedative.

$$4\,CH_3\overset{\overset{\displaystyle O}{\|}}{C}{-}H \xrightarrow[-20°C]{Ca(NO_3)_2 + HBr} \text{metaldehyde}$$

metaldehyde
(mp 246°C)

Metaldehyde is very attractive to slugs and snails, and it is also very poisonous to them. For this reason, it is an active ingredient in some pesticides that are sold for lawn and garden use. Metaldehyde is also used as a solid fuel.

Acetone and Methyl Ethyl Ketone Ketones are widely used organic solvents. Acetone, in particular, is used in very large quantities for this purpose. The U.S. production of acetone is about 1.26×10^8 kg annually. It is used as a solvent in the manufacture of drugs, chemicals, and explosives; for removal of paints, varnishes, and fingernail polish; and as a solvent in the plastics industry. Methyl ethyl ketone (MEK) is also widely used as a solvent, especially for lacquers. Both acetone and MEK are made by oxidation (dehydrogenation) of secondary alcohols. Acetone is also a coproduct in the manufacture of phenol (see Section 22.9).

$$\underset{\text{2-propanol}}{CH_3\underset{\underset{\displaystyle OH}{|}}{C}HCH_3} \xrightarrow[250\text{–}300°C]{Cu} \underset{\text{acetone (propanone)}}{CH_3\underset{\underset{\displaystyle O}{\|}}{C}CH_3} + H_2$$

$$\underset{\text{2-butanol}}{CH_3CH_2\underset{\underset{\displaystyle OH}{|}}{C}HCH_3} \xrightarrow[250\text{–}300°C]{Cu} \underset{\text{methyl ethyl ketone (2-butanone)}}{CH_3CH_2\underset{\underset{\displaystyle O}{\|}}{C}CH_3} + H_2$$

Acetone is also formed in the human body as a by-product of lipid metabolism. Usually it is oxidized to carbon dioxide and water. Normal concentrations of acetone in the body are less than 1 mg/100 mL of blood volume. In patients with diabetes mellitus, the concentration of acetone may rise, and it is then excreted in the urine where it can be easily detected. Sometimes the odor of acetone can be detected on the breath of these patients.

Concepts in Review

1. Recognize aldehydes and ketones from their formulas.
2. Relate the general reactivity of aldehydes and ketones to the carbonyl functional group properties.
3. Give IUPAC and common names of aldehydes and ketones.
4. Write formulas of aldehydes and ketones when given their names.
5. Understand why aldehydes and ketones have lower boiling points than alcohols.
6. Write equations showing the oxidation of alcohols to aldehydes and ketones.
7. Write the structure of the alcohol formed when an aldehyde or ketone is reduced.
8. Discuss the Tollens, Benedict, and Fehling tests, including the reagents used, evidence of a positive test, and the equations of the reactions that occur in positive tests; distinguish between the reactions of aldehydes and ketones.
9. Recognize whether an aldehyde or ketone undergoes the aldol condensation.
10. Write equations showing the aldol condensation of aldehydes and ketones.
11. Write equations for the formation and hydrolysis of cyanohydrins.
12. Write equations for the formation and decomposition of hemiacetals, hemiketals, acetals, and ketals.

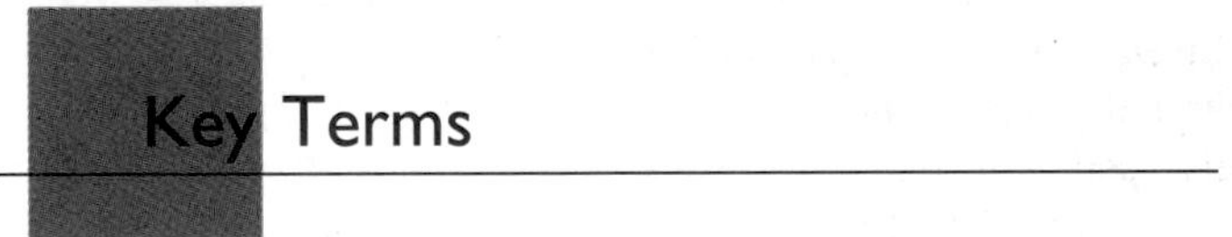

Key Terms

The terms listed here have been defined within this chapter. Section numbers are referenced in parenthesis for each term.

acetal (23.4)
aldehyde (23.1)
aldol condensation (23.4)
carbonyl group (23.1)
cyanohydrin (23.4)
Fehling and Benedict tests (23.4)
hemiacetal (23.4)
hemiketal (23.4)
ketal (23.4)
ketone (23.1)
Tollens test (23.4)

Questions

Questions refer to tables, figures, and key words and concepts defined within the chapter. A particularly challenging question or exercise is indicated with an asterisk.

1. Write generalized structures for
   (a) an aldehyde
   (b) a ketone
   (c) a dialdehyde
   (d) a hemiacetal
   (e) a hemiketal
   (f) an acetal
   (g) a ketal
   (h) a cyanohydrin

2. Explain, in terms of structure, why aldehydes and ketones have lower boiling points than alcohols of similar molar masses.

3. Write structural formulas for propanal and propanone. Judging from these formulas, do you think that aldehydes and ketones are isomeric with each other? Show evidence and substantiate your answer by testing with a four-carbon aldehyde and a four-carbon ketone.

4. Which of the following statements are correct? Rewrite the incorrect statements to make them correct.
   (a) The functional group that characterizes aldehydes and ketones is called a carboxyl group.
   (b) The carbonyl group contains a sigma and a pi bond.
   (c) The carbonyl group is polar, with the oxygen atom being more electronegative than the carbon atom.
   (d) The higher-molar-mass aldehydes and ketones are very soluble in water.
   (e) Ketones, like aldehydes, are easily oxidized to carboxylic acids.
   (f) A compound of formula $C_6H_{12}O$ can be either an aliphatic aldehyde or ketone.
   (g) Diethyl ketone has the same molecular formula as butyraldehyde.
   (h) In aldehydes and ketones, the hydrogen atoms that are bonded to carbon atoms adjacent to the carbonyl group (alpha position) are more reactive than other hydrogen atoms in the molecule.
   (i) In order for an aldehyde or a ketone to undergo the aldol condensation, it must have at least one alpha-hydrogen atom.
   (j) A hemiacetal has an alcohol and an ether group bonded to the same carbon atom.
   (k) Acetals are stable in acid solution but not in alkaline solution.
   (l) Formaldehyde is a gas, but it is usually handled in a solution.
   (m) The major use for formaldehyde is for making plastics.
   (n) Ethanal may be distinguished from propanal by use of Tollens reagent.
   (o) The general formula for the saturated homologous series of aldehydes and ketones is $C_nH_{2n}O$.
   (p) The compound C_2H_4O can be an aldehyde or a ketone.
   (q) The name for $O{=}\underset{|}{\underset{H}{C}}{-}\underset{|}{\underset{H}{C}}{=}O$ is ethanedial.
   (r) The oxidation product of 3-pentanol is 3-pentanone.
   (s) $C_6H_5CH_2CH(OC_2H_5)_2$ is a ketal.
   (t) When hydrolyzed, cyanohydrins form α-hydroxy acids.
   (u) Of the three compounds ethanal, propanal, and butanal, ethanal has the lowest vapor pressure.
   (v) Of the three compounds ethanal, propanal, and butanal, butanal has the highest boiling point.
   (w) Reduction of aldehydes yields secondary alcohols.

Paired Exercises

These exercises are paired. Each odd-numbered exercise is followed by a similar even-numbered exercise. Answers to the even-numbered exercises are given in Appendix V.

5. Give the IUPAC name for each of these aldehydes unless otherwise indicated:

(a) $H_2C{=}O$ (give both the IUPAC name and the common name)

(b) $CH_3CH(CH_3)CH_2C(=O){-}H$

(c) $H{-}C(=O)CH_2CH_2C(=O){-}H$

(d) $C_6H_5{-}CH{=}CHC(=O){-}H$

(e) $CH_3(H)C{=}C(H){-}C(=O){-}H$

6. Give the IUPAC name for each of these aldehydes unless otherwise indicated:

(a) $CH_3C(=O){-}H$ (give both the IUPAC name and the common name)

(b) $CH_3CH_2CH_2C(=O){-}H$

(c) $C_6H_5{-}C(=O){-}H$

(d) Benzene ring bearing $C(=O){-}H$, with Cl on the adjacent ring carbon and $CH(CH_3)_2$ ($H_3C{-}CH{-}CH_3$) para to the Cl

(e) $CH_3CH(OH)CH_2C(=O){-}H$

7. Give the IUPAC name for each of these ketones unless otherwise indicated:

(a) CH_3COCH_3 (three names)

(b) $C_6H_5{-}C(=O){-}CH_2CH_3$ (two names)

(c) cyclopentanone (five-membered ring with $=O$)

(d) $CH_3C(CH_3)(OH)CH_2C(=O)CH_3$

8. Give the IUPAC name for each of these ketones unless otherwise indicated:

(a) $CH_3CH_2COCH_3$ (two names)

(b) $CH_3C(=O){-}C(CH_3)_2CH_3$ (two names)

(c) $CH_3C(=O)CH_2CH_2C(=O)CH_3$

(d) $C_6H_5{-}CH_2C(=O)CH_3$ (two names)

9. Write the structural formulas for
(a) 1,3-dichloropropanone
(b) 3-butenal
(c) 4-phenyl-3-hexanone
(d) hexanal
(e) 3-ethyl-2-pentanone

11. Which compound in each of the following pairs has the higher boiling point? (Try to answer without consulting tables.)
(a) 2-hexanone or 2,5-hexanedione
(b) hexane or hexanal
(c) 2-pentanone or 2-pentanol
(d) propanone or butanone

13. Which compound in each of the following pairs has the higher aqueous solubility? (Try to answer without consulting tables.)
(a) 2-hexanone or 2,5-hexanedione
(b) propane or propanal
(c) heptanal or acetaldehyde

15. Write equations to show how each of the following is oxidized by (1) $K_2Cr_2O_7 + H_2SO_4$ and (2) air + Cu or Ag + heat:
(a) 3-pentanol
(b) 3-ethyl-1-hexanol

17. (a) What functional group is present in a compound that gives a positive Tollens test?
(b) What is the visible evidence for a positive Tollens test?
(c) Write an equation showing the reaction involved in a positive Tollens test.

19. Give the products of the reaction of the following with Tollens reagent:
(a) butanal
(b) benzaldehyde
(c) methyl ethyl ketone

21. Write equations showing the aldol condensation for the following compounds:
(a) butanol (b) phenylethanal

23. Complete the following equations:

(a) $CH_3C(=O)CH_3 + CH_2(OH)CH_2(OH) \xrightleftharpoons{\text{dry HCl}}$

(b) $CH_3CH_2C(=O){-}H + CH_3CH_2OH \xrightleftharpoons{H^+}$

(c) $CH_3CH(CH_3)CH_2CH(OCH_3)_2 \xrightarrow[H^+]{H_2O}$

25. Write equations for the following sequence of reactions:
(a) propanone + HCN ⟶
(b) product of part (a) + H_2O ⟶
(c) product of part (b) + acetaldehyde + dry HCl ⟶

10. Write the structural formulas for
(a) 3-hydroxypropanal
(b) 4-methyl-3-hexanone
(c) cyclohexanone
(d) 2,4,6-trichloroheptanal
(e) 3-pentenal

12. Which compound in each of the following pairs has the higher boiling point? (Try to answer without consulting tables.)
(a) pentane or pentanal
(b) benzaldehyde or benzyl alcohol
(c) 2-hexanone or butanone
(d) 1-butanol or butanal

14. Which compound in each of the following pairs has the higher aqueous solubility? (Try to answer without consulting tables.)
(a) acetaldehyde or ethane
(b) 2-pentanone or 2,4-pentanediol
(c) propanal or hexanal

16. Write equations to show how each of the following is oxidized by (1) $K_2Cr_2O_7 + H_2SO_4$ and (2) air + Cu or Ag + heat:
(a) 1-propanol
(b) 2,3-dimethyl-2-butanol

18. (a) What functional group is present in a compound that gives a positive Fehling test?
(b) What is the visible evidence for a positive Fehling test?
(c) Write an equation showing the reaction involved in a positive Fehling test.

20. Give the products of the reaction of the following with Fehling reagent:
(a) propanal
(b) acetone
(c) 3-methylpentanal

22. Write equations showing the aldol condensation for the following compounds:
(a) 3-pentanone (b) propanal

24. Complete the following equations:

(a) $CH_3CH_2C(=O){-}H + CH_3CH_2CH_2OH \xrightleftharpoons{\text{dry HCl}}$

(b) cyclohexanone $+ CH_3OH \xrightleftharpoons{H^+}$

(c) $CH_3CH_2CH_2CH(OCH_3)_2 \xrightarrow[H^+]{H_2O}$

26. Write equations for the following sequence of reactions:
(a) benzaldehyde + HCN ⟶
(b) product of part (a) + H_2O ⟶
(c) product of part (b) + $K_2Cr_2O_7 + H_2SO_4$ ⟶

Additional Exercises

These exercises are not paired or labeled by topic and provide additional practice on concepts covered in this chapter.

27. 3-Hydroxypropanal can form an intramolecular cyclic hemiacetal. What is the structure of the hemiacetal?

28. The following cyanohydrin is used in a multistep synthesis to make the artificial fiber Orlon (polyacrylonitrile):

$$CH_3CH(OH)C\equiv N$$

From what carbonyl-containing compound is this cyanohydrin synthesized (give both the structural formula and name)?

29. Write the structure for each aldol condensation product that is possible when a mixture of ethanal and propanal is reacted with dilute NaOH.

30. The following molecules ("ketone bodies") are used by the human body as an emergency energy supply primarily for the muscles:

$$\underset{\text{compound I}}{CH_3CH(OH)CH_2COOH} \qquad \underset{\text{compound II}}{CH_3C(=O)CH_2COOH}$$

These two molecules can be interconverted in a single biochemical reaction. What chemical change takes place as the compound I is converted to compound II?

31. Give a simple visible chemical test that will distinguish between the compounds in each of the following pairs:

(a) $CH_3CH_2C(=O)-H$ and $CH_3C(=O)CH_3$

(b) $CH_3CH_2C(=O)-H$ and $CH_2{=}CHC(=O)-H$

(c) $C_6H_5CH_2CH_2OH$ and $C_6H_5CH(OH)CH_3$

32. Write equations to show how you could prepare lactic acid, $CH_3CH(OH)COOH$, from acetaldehyde through a cyanohydrin intermediate.

33. The millipede carries the following cyanohydrin as a chemical defense weapon:

$$C_6H_5CH(OH)C\equiv N$$

When attacked, the millipede reverses the reaction that formed this cyanohydrin, and deadly hydrogen cyanide gas is released. One millipede is said to be able to release enough poisonous gas to kill a small mouse. Write the reaction that releases hydrogen cyanide and name the aldehyde product.

34. Pyruvic acid is formed during muscle exertion.

$$CH_3C(=O)COOH$$

If the muscles are working strenuously, this acid is converted to lactic acid.

$$CH_3CH(OH)COOH$$

What chemical change takes place as pyruvic acid is converted to lactic acid?

35. During the biological synthesis of glucose, two 3-carbon compounds are connected via a new carbon–carbon single bond to form a 6-carbon straight-chain molecule (a sugar). Given the reactants in a simplified version of this reaction, use your knowledge of the aldol condensation reaction to write the structure of the product.

$$CH_2(OH)CH(OH)C(=O)H + CH_2(OH)C(=O)CH_2(OH) \longrightarrow$$

36. Ketones are prepared by oxidation of secondary alcohols. Name the alcohol that should be used to prepare

(a) 3-pentanone

(b) methyl ethyl ketone

(c) 4-phenyl-2-butanone

37. Write structures for all isomeric aldehydes and ketones with the molecular formula of $C_6H_{12}O$.

38. Write structures for all the benzaldehyde isomers with the molecular formula of $C_9H_{10}O$.

Answers to Practice Exercises

23.1 (a) 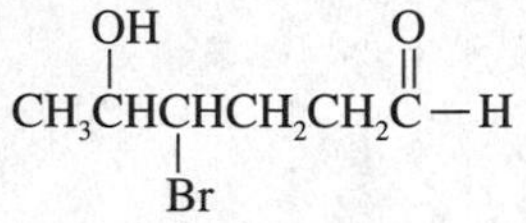 $CH_3CH(OH)CH(Br)CH_2CH_2C(=O)-H$ (c) 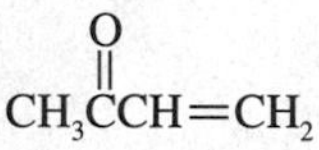 $CH_3C(=O)CH=CH_2$

(b) 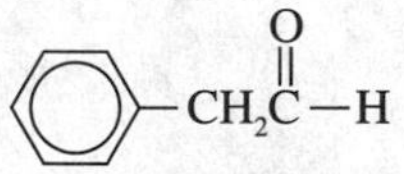 $C_6H_5-CH_2C(=O)-H$ (d) 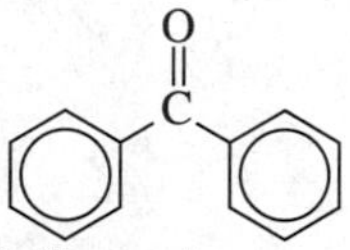 $C_6H_5C(=O)C_6H_5$

23.2 (a) cyclohexanone (c) 3-methylbutanal
(b) 4-chloro-2-butanone (d) 3-phenylpropenal

23.3 (a) 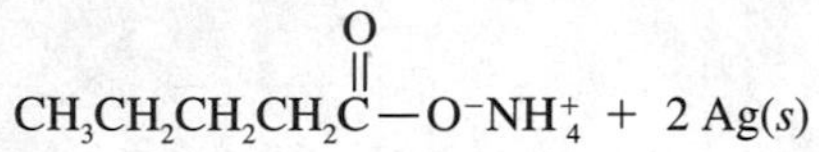 $CH_3CH_2CH_2CH_2C(=O)-O^-NH_4^+ + 2\,Ag(s)$

(b) No reaction

(c) No reaction

(d) 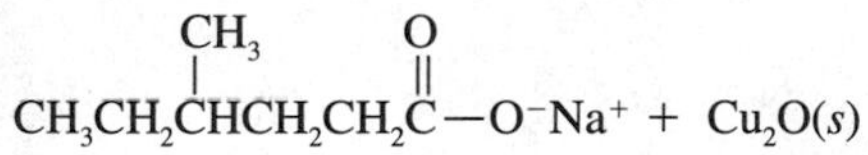 $CH_3CH_2CH(CH_3)CH_2CH_2C(=O)-O^-Na^+ + Cu_2O(s)$

23.4 (a) 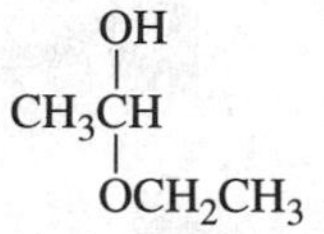 $CH_3CH(OH)(OCH_2CH_3)$ 1-ethoxyethanol; $CH_3CH(OCH_2CH_3)(OCH_2CH_3)$ 1,1-diethoxyethane

(b) 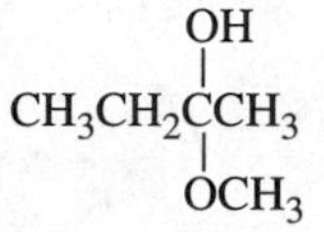 $CH_3CH_2C(OH)(OCH_3)CH_3$ 2-methoxy-2-butanol; $CH_3CH_2C(OCH_3)(OCH_3)CH_3$ 2,2-dimethoxybutane

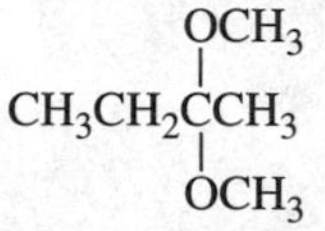

23.5 (a) 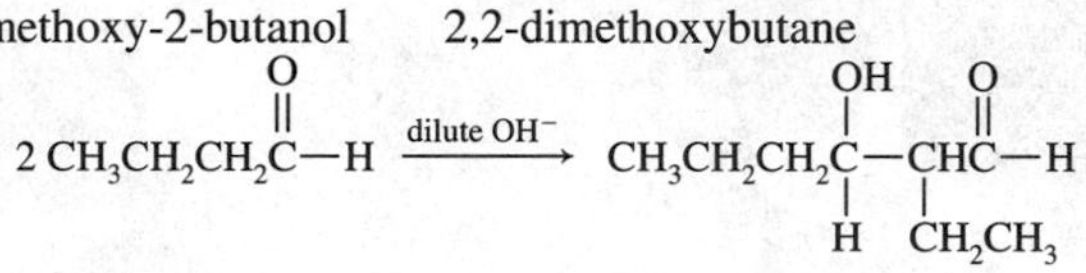 $2\,CH_3CH_2CH_2C(=O)-H \xrightarrow{\text{dilute OH}^-} CH_3CH_2CH_2C(OH)(H)-CH(CH_2CH_3)C(=O)-H$

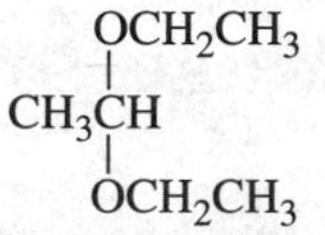

(b) 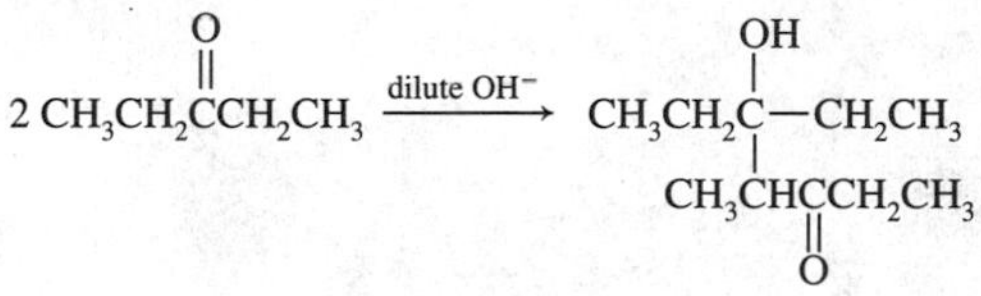 $2\,CH_3CH_2C(=O)CH_2CH_3 \xrightarrow{\text{dilute OH}^-} CH_3CH_2C(OH)(-CH_2CH_3)-CH(CH_3)C(=O)CH_2CH_3$

Carboxylic Acids and Esters

24

Whenever we eat food with a sour or tart taste, it is very likely that at least one carboxylic acid is present in that food. Lemons, for example, contain citric acid, vinegar contains acetic acid, and sour milk contains lactic acid. Carboxylic acids are also important compounds in biochemistry. Citric acid is also found in our blood, and lactic acid is produced in our muscles during the breakdown of glucose. In living systems, these acids are most often found in the form of salts or acid derivatives. In the world around us, carboxylic acid salts are commonly used as preservatives, especially in cheeses and breads. Some carboxylic acid salts are used to treat skin irritations like diaper rash and athlete's foot.

Carboxylic acid derivatives known as esters are also responsible for the sweet and pleasant odors and tastes of our food. These compounds are frequently used as artificial flavors in foods in place of more expensive natural extracts. In biochemistry ester-like molecules act as the energy carriers in many cells. As you will see, the properties of carboxylic acids are quite distinct from those of aldehydes and alcohols.

24.1 Carboxylic Acids

The functional group of the carboxylic acids is called a **carboxyl group** and is represented in the following ways:

carboxyl group

$$-\overset{\overset{\displaystyle O}{\|}}{C}-OH \quad \text{or} \quad -COOH \quad \text{or} \quad -CO_2H$$

Carboxylic acids can be either aliphatic or aromatic:

$$R\overset{\overset{\displaystyle O}{\|}}{C}-OH \qquad CH_3\overset{\overset{\displaystyle O}{\|}}{C}-OH \qquad Ar\overset{\overset{\displaystyle O}{\|}}{C}-OH \qquad C_6H_5-\overset{\overset{\displaystyle O}{\|}}{C}-OH$$

aliphatic (first two); aromatic (last two)

"R" is often used to symbolize an aliphatic group; "Ar" symbolizes an aromatic group.

24.2 Nomenclature and Sources of Aliphatic Carboxylic Acids

Aliphatic carboxylic acids form a homologous series. The carboxyl group is always at the beginning of a carbon chain, and the carbon atom in this group is understood to be carbon number 1 when the compound is named.

◀ Chapter Opening Photo: All citrus fruits contain a variety of carboxylic acids.

The sting of an ant bite is caused by formic acid.

IUPAC Nomenclature for Carboxylic Acids

1. Identify the longest carbon chain that includes the carboxyl group.
2. Drop the final *e* from the corresponding hydrocarbon name.
3. Add the suffix *-oic acid.*

Thus the names corresponding to the one-, two-, and three-carbon acids are methanoic acid, ethanoic acid, and propanoic acid. These names are derived from methane, ethane, and propane.

CH_4	methane	$HCOOH$	methanoic acid
CH_3CH_3	ethane	CH_3COOH	ethanoic acid
$CH_3CH_2CH_3$	propane	CH_3CH_2COOH	propanoic acid

Other groups bonded to the parent chain are numbered and named as we have done previously. For example,

$$\overset{5}{C}H_3\overset{4}{C}H_2\overset{3}{C}H(CH_3)\overset{2}{C}H_2\overset{1}{C}OOH$$

3-methylpentanoic acid

Unfortunately, the IUPAC method is not the only, nor the most used, method of naming acids. Organic acids are usually known by common names. Methanoic, ethanoic, and propanoic acids are called formic, acetic, and propionic acids, respectively. These names usually refer to a natural source of the acid and are not systematic. Formic acid was named from the Latin word *formica*, meaning ant. This acid contributes to the stinging sensation of ant bites. Acetic acid is found in vinegar and is so named from the Latin word for vinegar. The name of butyric acid is derived from the Latin term for butter, since it is a constituent of butterfat. The 6-, 8-, and 10-carbon acids are found in goat fat and have names derived from the Latin word for goat. These three acids—caproic, caprylic, and capric—along with butyric acid have characteristic and disagreeable odors. In a similar way the names of the 12-, 14-, and 16-carbon acids—lauric, myristic, and palmitic—come from plants from which the corresponding acid has been isolated. The name stearic acid is derived from a Greek word meaning beef fat or tallow, which is a good source of this acid. Many of the carboxylic acids, principally those with even numbers of carbon atoms ranging from 4 to about 20, exist in combined form in plant and animal fats. These are called *fatty acids* (see Chapter 29). Table 24.1 lists the common and IUPAC names, together with some of the physical properties, of the more important saturated aliphatic acids.

Common and IUPAC nomenclature systems should not be intermixed.

Another common nomenclature method uses letters of the Greek alphabet (α, β, γ, δ, . . .) to name certain acid derivatives, especially hydroxy, amino, and halogen acids. When Greek letters are used, the carbon atoms, beginning with the one adjacent to the carboxyl group, are labeled α, β, γ, δ, ... When numbers are used (the IUPAC System), the numbers begin with the carbon in the —COOH group.

$$\underset{5}{\overset{\delta}{C}}-\underset{4}{\overset{\gamma}{C}}-\underset{3}{\overset{\beta}{C}}-\underset{2}{\overset{\alpha}{C}}-\underset{1}{\overset{O}{\overset{\|}{C}}}-OH$$

TABLE 24.1 Names, Formulas, and Physical Properties of Saturated Aliphatic Carboxylic Acids

Common name (IUPAC name)	Formula	Melting point (°C)	Boiling point (°C)	Solubility in water[a]
Formic acid (Methanoic acid)	$HCOOH$	8.4	100.8	∞
Acetic acid (Ethanoic acid)	CH_3COOH	16.6	118	∞
Propionic acid (Propanoic acid)	CH_3CH_2COOH	−21.5	141.4	∞
Butyric acid (Butanoic acid)	$CH_3(CH_2)_2COOH$	−6	164	∞
Valeric acid (Pentanoic acid)	$CH_3(CH_2)_3COOH$	−34.5	186.4	3.3
Caproic acid (Hexanoic acid)	$CH_3(CH_2)_4COOH$	−3.4	205	1.1
Caprylic acid (Octanoic acid)	$CH_3(CH_2)_6COOH$	16.3	239	0.1
Capric acid (Decanoic acid)	$CH_3(CH_2)_8COOH$	31.4	269	Insoluble
Lauric acid (Dodecanoic acid)	$CH_3(CH_2)_{10}COOH$	44.1	225[b]	Insoluble
Myristic acid (Tetradecanoic acid)	$CH_3(CH_2)_{12}COOH$	54.2	251[b]	Insoluble
Palmitic acid (Hexadecanoic acid)	$CH_3(CH_2)_{14}COOH$	63	272[b]	Insoluble
Stearic acid (Octadecanoic acid)	$CH_3(CH_2)_{16}COOH$	69.6	287[b]	Insoluble
Arachidic acid (Eicosanoic acid)	$CH_3(CH_2)_{18}COOH$	77	298[b]	Insoluble

[a]Grams of acid per 100 g of water.
[b]Boiling point is given at 100 mm Hg pressure instead of atmospheric pressure because thermal decomposition occurs before this acid reaches its boiling point at atmospheric pressure.

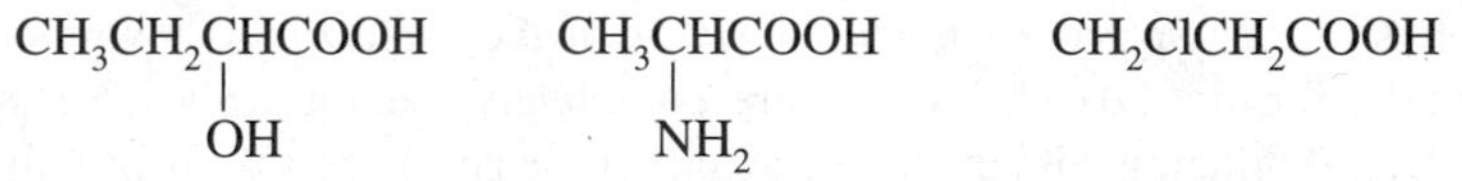

	$CH_3CH_2CH(OH)COOH$	$CH_3CH(NH_2)COOH$	CH_2ClCH_2COOH
Common name:	α-hydroxybutyric acid	α-aminopropionic acid	β-chloropropionic acid
IUPAC name:	2-hydroxybutanoic acid	2-aminopropanoic acid	3-chloropropanoic acid

Example 24.1

Write formulas for the following:

(a) 3-Chloropentanoic acid (b) γ-Hydroxybutyric acid (c) Phenylacetic acid

Solution

(a) This is an IUPAC name as indicated by the use of position numbers. *Pentanoic* indicates a five-carbon acid. Substituted on carbon 3 is a chlorine atom. Write five carbon atoms in a row. Make carbon 1 a carboxyl group, place a Cl on carbon 3, and add hydrogens to give each carbon four bonds. The formula is

$CH_3CH_2CHClCH_2COOH$

(b) This is a common name as indicated by the use of Greek letters. *Butyric* indicates a four-carbon acid. The γ-position is three carbons removed from the carboxyl group. Therefore, the formula is

$$\underset{\gamma \text{ carbon}}{HO-CH_2}CH_2CH_2COOH$$

(c) *Acetic acid* (a common name) is the familiar two-carbon acid. There is only one place to substitute the phenyl group and still call the compound an acid—that is, at the CH_3 group. Substitute a phenyl group for one of the three H atoms to give the formula:

$$C_6H_5-CH_2COOH$$

Practice 24.1

Write formulas for (a) 2-methylpropanoic acid, (b) β-chlorocaproic acid, and (c) cyclohexanecarboxylic acid.

24.3 Physical Properties of Carboxylic Acids

Each aliphatic carboxylic acid molecule is polar and consists of a carboxyl group and a hydrocarbon group (R). These two unlike parts have great bearing on the physical, as well as chemical, behavior of the molecule as a whole. The first four acids, formic through butyric, are completely soluble (miscible) in water (Table 24.1). Beginning with pentanoic acid (valeric acid), the water solubility falls sharply and is only about 0.1 g of acid per 100 g of water for octanoic acid (caprylic acid). Acids of this series with more than eight carbons are virtually insoluble in water. The water-solubility characteristics of the first four acids are determined by the highly soluble polar carboxyl group. Thereafter, the water insolubility of the nonpolar hydrocarbon chain is dominant.

The polarity due to the carboxyl group is evident in the boiling-point data. Formic acid (HCOOH) boils at about 101°C. Carbon dioxide, a nonpolar substance of similar molar mass, remains in the gaseous state until it is cooled to −78°C. In like manner, the boiling point of acetic acid (molar mass 60 g/mole) is 118°C, whereas nonpolar butane (molar mass 58 g/mole) boils at −0.6°C. The comparatively high boiling points for carboxylic acids are due to intermolecular attractions resulting from hydrogen bonding. In fact, molar mass determinations on gaseous acetic acid (near its boiling point) show a value of about 120 g/mol indicating that two molecules are joined together to form a *dimer,* $(CH_3COOH)_2$.

$$R{-}C\overset{\displaystyle O\cdots HO}{\underset{\displaystyle OH\cdots O}{\rightleftharpoons}}C{-}R \qquad CH_3{-}C\overset{\displaystyle O\cdots HO}{\underset{\displaystyle OH\cdots O}{\rightleftharpoons}}C{-}CH_3$$

hydrogen bonding in carboxylic acids — acetic acid dimer

Saturated monocarboxylic acids that have fewer than ten carbon atoms are liquids at room temperature, whereas those with more than ten carbon atoms are wax-like solids.

Carboxylic acids and phenols, like mineral acids such as HCl, ionize in water to produce hydronium ions and anions (Chapter 15). Carboxylic acids are generally weak acids; that is, they are only slightly ionized in water. Phenols are, in general, even weaker acids than carboxylic acids. For example, the ionization constant for acetic acid is 1.8×10^{-5} and that for phenol is 1.1×10^{-10}. Equations illustrating these ionizations are given here:

$$\underset{\text{hydrogen chloride}}{HCl} + H_2O \longrightarrow \underset{\text{hydronium ion}}{H_3O^+} + \underset{\text{chloride ion}}{Cl^-}$$

$$\underset{\text{acetic acid}}{CH_3\overset{\overset{\displaystyle O}{\|}}{C}{-}OH} + H_2O \rightleftharpoons \underset{\text{hydronium ion}}{H_3O^+} + \underset{\text{acetate ion}}{CH_3\overset{\overset{\displaystyle O}{\|}}{C}{-}O^-}$$

$$\underset{\text{phenol}}{C_6H_5OH} + H_2O \rightleftharpoons \underset{\text{hydronium ion}}{H_3O^+} + \underset{\text{phenoxide ion}}{C_6H_5O^-}$$

Carboxylic acids are very common in biological systems. Their tendency to ionize and form anions means that many biological molecules carry negative charges.

24.4 Classification of Carboxylic Acids

Thus far our discussion has dealt mainly with a single type of acid—that is, saturated monocarboxylic acids. But various other kinds of carboxylic acids are known. Some of the more important ones are discussed here.

Unsaturated Carboxylic Acids

An unsaturated acid contains one or more carbon–carbon double bonds. The first member of the homologous series of unsaturated carboxylic acids, containing one carbon–carbon double bond, is acrylic acid, $CH_2{=}CHCOOH$. The IUPAC name for $CH_2{=}CHCOOH$ is propenoic acid. Derivatives of acrylic acid are used to manufacture a class of synthetic polymers known as the acrylates (see Chapter 26). These polymers are widely used as textiles and in paints and lacquers. Unsaturated carboxylic acids undergo the reactions of both an unsaturated hydrocarbon and a carboxylic acid.

Even one carbon–carbon double bond in the molecule exerts a major influence on the physical and chemical properties of an acid. The effect of a double bond can be seen when comparing the two 18-carbon acids, stearic and oleic. Stearic acid, $CH_3(CH_2)_{16}COOH$, a solid that melts at 70°C, shows only the reactions of a carboxylic acid. On the other hand, oleic acid,

$$CH_3(CH_2)_7CH{=}CH(CH_2)_7COOH \text{ (mp 16°C)}$$

with one double bond, is a liquid at room temperature and shows the reactions of an unsaturated hydrocarbon as well as those of a carboxylic acid.

Aromatic Carboxylic Acids

In an aromatic carboxylic acid, the carbon of the carboxyl group (—COOH) is bonded directly to a carbon in an aromatic ring. The parent compound of this series is benzoic acid. Other common examples are the three isomeric toluic acids:

COOH CH$_3$ COOH CH$_3$ CH$_3$ COOH COOH

benzoic acid *o*-toluic acid *m*-toluic acid *p*-toluic acid

Dicarboxylic Acids

Acids of both the aliphatic and aromatic series that contain two or more carboxyl groups are known. Those with two —COOH groups are called dicarboxylic acids. The simplest member of the aliphatic series is oxalic acid. The next member in the homologous series is malonic acid. Several dicarboxylic acids and their names are listed in Table 24.2.

The IUPAC names for dicarboxylic acids are formed by modifying the corresponding hydrocarbon names to end in *dioic acid.* Thus the two-carbon acid is ethanedioic acid (derived from ethane). However, the common names for dicarboxylic acids are frequently used.

Oxalic acid is found in various plants including spinach, cabbage, and rhubarb. Among its many uses are bleaching straw and leather and removing rust and ink

TABLE 24.2 Names and Formulas of Selected Dicarboxylic Acids

Common name	IUPAC name	Formula
Oxalic acid	Ethanedioic acid	HOOCCOOH
Malonic acid	Propanedioic acid	$HOOCCH_2COOH$
Succinic acid	Butanedioic acid	$HOOC(CH_2)_2COOH$
Glutaric acid	Pentanedioic acid	$HOOC(CH_2)_3COOH$
Adipic acid	Hexanedioic acid	$HOOC(CH_2)_4COOH$
Fumaric acid	*trans*-2-Butenedioic acid	$HOOCCH{=}CHCOOH$
Maleic acid	*cis*-2-Butenedioic acid	$HOOCCH{=}CHCOOH$

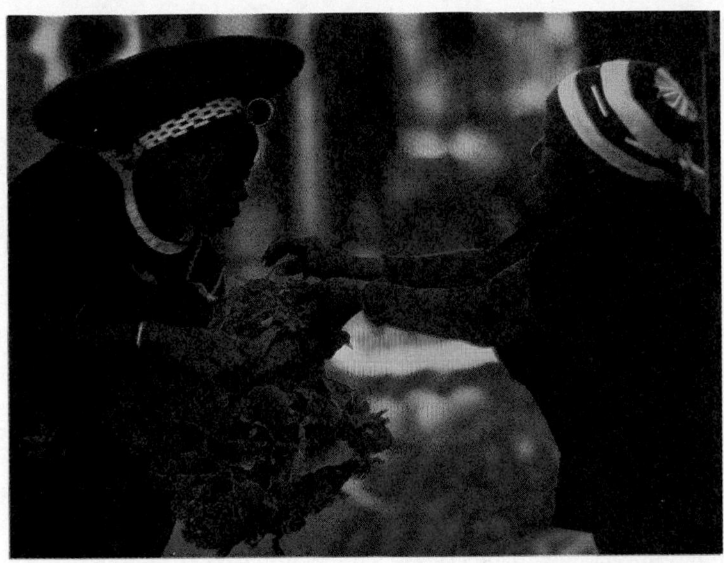

This South African vendor is selling fresh spinach, which contains oxalic acid in addition to important vitamins.

stains. Although oxalic acid is poisonous, the amounts present in the previously mentioned vegetables are usually not harmful.

Malonic acid is made synthetically but was originally prepared from malic acid, which is commonly found in apples and many fruit juices. Malonic acid is one of the major compounds used in the manufacture of the class of drugs known as barbiturates. When heated above their melting points, malonic acid and substituted malonic acids lose carbon dioxide to give monocarboxylic acids. Thus malonic acid yields acetic acid when strongly heated:

$$\begin{array}{c} \text{COOH} \\ | \\ \text{CH}_2 \\ | \\ \text{COOH} \\ \text{malonic acid} \end{array} \xrightarrow{150^\circ\text{C}} \underset{\text{acetic acid}}{\text{CH}_3\text{COOH}} + \text{CO}_2(g)$$

Malonic acid is used as the biological precursor for the synthesis of fatty acids. In living cells, when malonic acid loses carbon dioxide, the acetic acid units are linked together to begin formation of long chain fatty acids.

Succinic acid has been known since the 16th century, when it was obtained as a distillation product of amber. Succinic, fumaric, and citric acids are among the important acids in the energy-producing metabolic pathway known as the citric acid cycle (see Chapter 35). Citric acid is a *tricarboxylic acid* that is widely distributed in plant and animal tissue, especially in citrus fruits (lemon juice contains 5–8%). The formula for citric acid is

$$\begin{array}{c} \text{CH}_2\text{COOH} \\ | \\ \text{HO}-\text{C}-\text{COOH} \\ | \\ \text{CH}_2\text{COOH} \\ \text{citric acid} \end{array}$$

When succinic acid is heated, it loses water, forming succinic anhydride, an acid anhydride. Glutaric acid behaves similarly, forming glutaric anhydride.

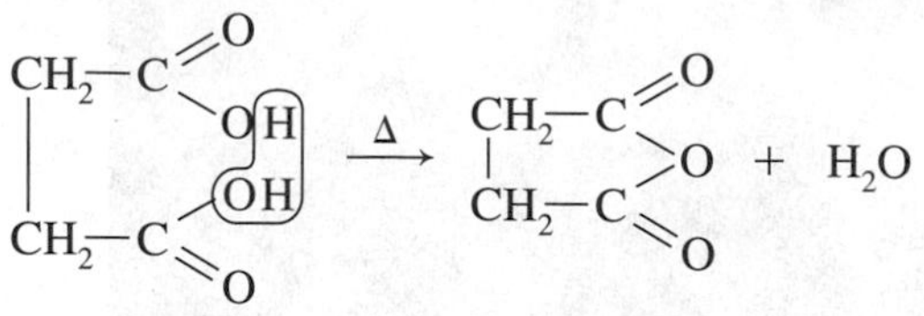

$$\underset{\text{succinic acid}}{\begin{array}{l} CH_2-C(=O)-OH \\ | \\ CH_2-C(=O)-OH \end{array}} \xrightarrow{\Delta} \underset{\text{succinic anhydride}}{\begin{array}{l} CH_2-C(=O) \\ | \qquad\qquad > O \\ CH_2-C(=O) \end{array}} + H_2O$$

Adipic acid is the most important commercial dicarboxylic acid. It is made from benzene by converting it first to cyclohexene and then by oxidation to adipic acid. About 8.2×10^8 kg of adipic acid is produced annually in the United States. Most of the adipic acid is used to produce nylon (Chapter 26). It is also used in polyurethane foams, plasticizers, and lubricating-oil additives.

Aromatic dicarboxylic acids contain two carboxyl groups attached directly to an aromatic ring.

Examples are the three isomeric phthalic acids, $C_6H_4(COOH)_2$:

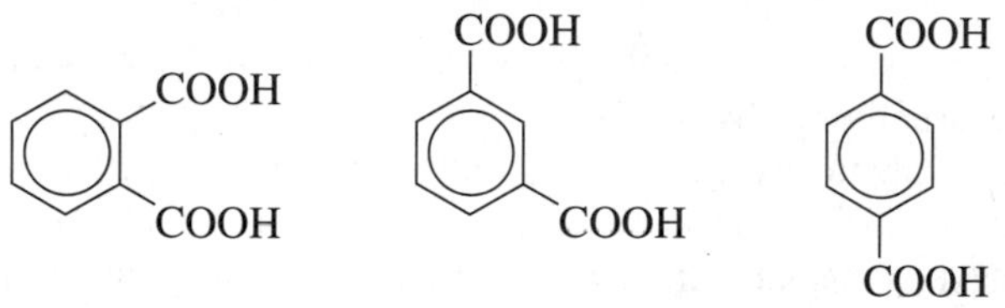

o-phthalic acid (phthalic acid) — benzene ring with COOH groups at positions 1,2

m-phthalic acid (isophthalic acid) — benzene ring with COOH groups at positions 1,3

p-phthalic acid (terephthalic acid) — benzene ring with COOH groups at positions 1,4

Dicarboxylic acids are *bifunctional*; that is, they have two sites where reactions can occur. Therefore, they are often used as monomers in the preparation of synthetic polymers such as Dacron polyester (Section 26.7).

Hydroxy Acids

Lactic acid, found in sour milk, sauerkraut, and dill pickles, has the functional groups of both a carboxylic acid and an alcohol. Lactic acid is the end product when our muscles use glucose for energy in the absence of oxygen, a process called *glycolysis* (see Chapter 35). Salicylic acid is both a carboxylic acid and a phenol. It is of special interest because a family of useful drugs—the salicylates—are derivatives of this acid. The salicylates include aspirin and function as *analgesics* (pain relievers) and as *antipyretics* (fever reducers). The structural formulas of several hydroxy acids are given here:

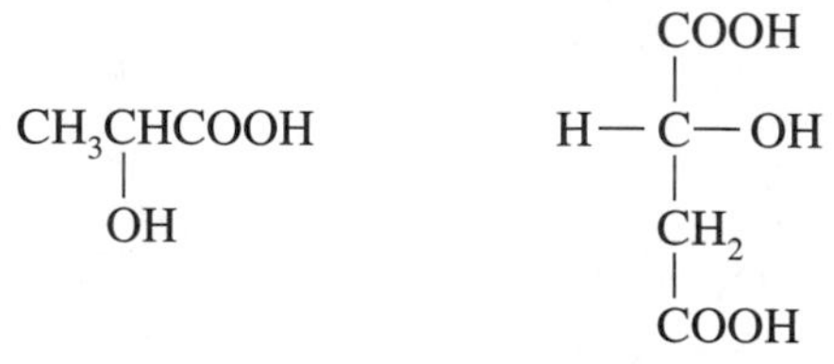

$$\underset{\substack{\text{lactic acid} \\ \text{(2-hydroxypropanoic acid)}}}{\begin{array}{l} CH_3CHCOOH \\ \quad\;\; | \\ \quad\; OH \end{array}} \qquad \underset{\substack{\text{malic acid} \\ \text{(2-hydroxybutanedioic acid)}}}{\begin{array}{c} COOH \\ | \\ H-C-OH \\ | \\ CH_2 \\ | \\ COOH \end{array}}$$

A carboxylic acid is one of the reactants used to form these acrylic nails.

salicylic acid (o-hydroxybenzoic acid): $C_6H_4(COOH)(OH)$ — benzene ring with COOH and OH on adjacent carbons

$$\begin{array}{l} HO-CHCOOH \\ \quad\quad\; | \\ HO-CHCOOH \end{array}$$

tartaric acid (2,3-dihydroxybutanedioic acid)

Amino Acids

Naturally occuring amino acids have this general formula; the amino group is in the alpha position:

$$\begin{array}{c} \quad\quad\quad\;\; O \\ \quad\quad\quad\;\; \| \\ R-\underset{\displaystyle NH_2}{\underset{|}{CH}}-C-OH \end{array}$$

Each amino acid molecule has a carboxyl group that acts as an acid and an amino group that acts as a base. About 20 biologically important amino acids, each with a different group represented by R, are found in nature. (In amino acids, R does not always represent an alkyl group.) The immensely complicated protein molecules, found in every form of life, are built from amino acids. Some protein molecules contain more than 10,000 amino acid units. Amino acids and proteins are discussed in more detail in Chapter 30.

Practice 24.2

Write a structural formula for the smallest molecule of each of the following acids:
(a) a carboxylic acid, (b) an α-amino acid, (c) an α-hydroxy acid,
(d) a β-bromo acid, (e) a dicarboxylic acid, (f) an aromatic carboxylic acid

24.5 Preparation of Carboxylic Acids

Many different methods of preparing carboxylic acids are known. We will consider only a few examples.

Oxidation of an Aldehyde or a Primary Alcohol This is a general method that can be used to convert an aldehyde or primary alcohol to the corresponding carboxylic acid:

$$R\overset{\displaystyle O}{\overset{\|}{C}}-H \xrightarrow{[O]} RCOOH$$

$$RCH_2OH \xrightarrow{[O]} RCOOH$$

Butyric acid (butanoic acid) can be obtained by oxidizing either 1-butanol or butanal with potassium dichromate in the presence of sulfuric acid. Aromatic acids may be prepared by the same general method. For example, benzoic acid is obtained by oxidizing benzyl alcohol.

$$\underset{\text{1-butanol}}{CH_3CH_2CH_2CH_2OH} \xrightarrow[H_2SO_4,\ \Delta]{Cr_2O_7^{2-}} \underset{\text{butanoic acid}}{CH_3CH_2CH_2COOH}$$

$$\underset{\text{butanal}}{CH_3CH_2CH_2\overset{O}{\overset{\|}{C}}-H} \xrightarrow[H_2SO_4,\ \Delta]{Cr_2O_7^{2-}} \underset{\text{butanoic acid}}{CH_3CH_2CH_2COOH}$$

$$\underset{\text{benzyl alcohol}}{C_6H_5CH_2OH} \xrightarrow[H_2SO_4,\ \Delta]{Cr_2O_7^{2-}} \underset{\text{benzoic acid}}{C_6H_5COOH}$$

Carboxylic acids can also be obtained by the hydrolysis or saponification of esters (see Section 24.9).

Practice 24.3

Name the compound formed when (a) 1-hexanol and (b) 2-hexanol are oxidized with $K_2Cr_2O_7/H_2SO_4$.

Oxidation of Alkyl Groups Attached to Aromatic Rings When reacted with a strong oxidizing agent (alkaline permanganate solution or potassium dichromate and sulfuric acid), alkyl groups bonded to aromatic rings are oxidized to carboxyl groups. Regardless of the size or length of the alkyl group, the carbon atom adjacent to the ring remains bonded to the ring and is oxidized to a carboxyl group. The remainder of the alkyl group goes either to carbon dioxide or to a salt of a carboxylic acid. Thus sodium benzoate is obtained when toluene, ethylbenzene, or propylbenzene is heated with alkaline permanganate solution:

$$\underset{\text{toluene}}{C_6H_5CH_3} \xrightarrow[NaOH,\ \Delta]{NaMnO_4} \underset{\text{sodium benzoate}}{C_6H_5\overset{O}{\overset{\|}{C}}-O^-\,Na^+}$$

$$\underset{\text{ethylbenzene}}{C_6H_5CH_2CH_3} \xrightarrow[NaOH,\ \Delta]{NaMnO_4} \underset{\text{sodium benzoate}}{C_6H_5\overset{O}{\overset{\|}{C}}-O^-\,Na^+} + CO_2(g)$$

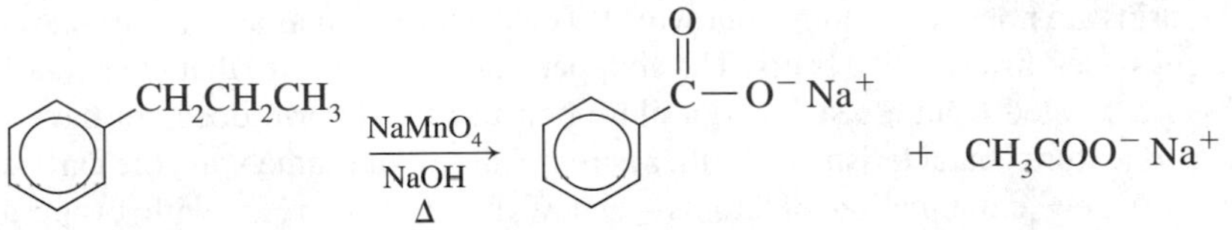

$$C_6H_5CH_2CH_2CH_3 \xrightarrow[\Delta]{NaMnO_4,\ NaOH} C_6H_5C(=O){-}O^- Na^+ + CH_3COO^- Na^+$$

propylbenzene — sodium benzoate — sodium acetate

Since the reaction is conducted in an alkaline medium, a salt of the carboxylic acid (sodium benzoate) is formed instead of the free acid. To obtain the free carboxylic acid, the reaction mixture is acidified with a strong mineral acid (HCl or H_2SO_4) in a second step.

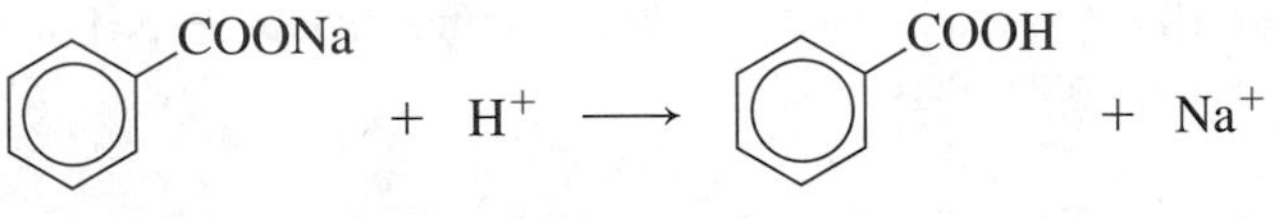

$$C_6H_5COONa + H^+ \longrightarrow C_6H_5COOH + Na^+$$

sodium benzoate — benzoic acid

Practice 24.4

What aromatic hydrocarbon of formula C_9H_{12}, when oxidized with $NaMnO_4$/NaOH, will give (a) benzoic acid and (b) *o*-phthalic acid?

Hydrolysis of Nitriles Nitriles, RCN, which can be prepared by adding HCN to aldehydes and ketones (Section 23.4) or by reacting alkyl halides with KCN, can be hydrolyzed to carboxylic acids:

$$\underset{\text{alkyl halide}}{RX} + KCN \longrightarrow \underset{\text{a nitrile}}{RCN} + KX$$

$$RCN + 2\ H_2O \xrightarrow{H^+} RCOOH + NH_4^+$$

$$CH_3CN + 2\ H_2O \xrightarrow{H^+} CH_3COOH + NH_4^+$$

Practice 24.5

Ethanal is reacted with HCN. Write equations for the nitrile addition product and the subsequent hydrolysis to the acid. Name the acid.

24.6 Chemical Properties of Carboxylic Acids

The structural formula of the carboxyl group consists of a carbonyl group and a hydroxyl group in combination. From such a combination, carboxylic acids might well be expected to have reactions characteristic of both aldehydes and alcohols. But such is not the case; carboxylic acids have their own unique properties.

Hybrids are not uncommon in our world. The nectarines available in our markets are large, sweet, firm-fleshed fruits. The shopper may not be aware that the nectarine is a hybrid created from a peach and a plum because this hybrid does not have the flavor and texture characteristics of either parent fruit. In the same way, the carboxyl group is a new combination of atoms—a new functional group with properties different from those of both the carbonyl and hydroxyl groups.

The carboxyl group is the most acidic of any organic functional group discussed so far. The —OH group acts as a proton donor, even more effectively than phenol.

The carboxyl group commonly is involved in substitution-type reactions. The C=O retains its double-bond character while the —OH is replaced by another group or atom. Thus carboxylic acids react with many substances to produce derivatives. In these reactions the hydroxyl group is replaced by a halogen (—Cl), an acyloxy group (—OOCR), an alkoxy group (—OR), or an amino group ($—NH_2$). The general reactions are summarized here:

$$\mathrm{R{-}\overset{\overset{\displaystyle O}{\|}}{C}{-}OH \xrightarrow{-Cl} R{-}\overset{\overset{\displaystyle O}{\|}}{C}{-}Cl} \quad \text{(acyl halide)}$$

$$\mathrm{R{-}\overset{\overset{\displaystyle O}{\|}}{C}{-}OH \xrightarrow{-OOCR'} R{-}\overset{\overset{\displaystyle O}{\|}}{C}{-}O{-}\overset{\overset{\displaystyle O}{\|}}{C}{-}R'} \quad \text{(acid anhydride)}$$

$$\mathrm{R{-}\overset{\overset{\displaystyle O}{\|}}{C}{-}OH \xrightarrow{-OR'} R{-}\overset{\overset{\displaystyle O}{\|}}{C}{-}OR'} \quad \text{(ester)}$$

$$\mathrm{R{-}\overset{\overset{\displaystyle O}{\|}}{C}{-}OH \xrightarrow{-NH_2} R{-}\overset{\overset{\displaystyle O}{\|}}{C}{-}NH_2} \quad \text{(amide)}$$

Acid–Base Reactions Because of their ability to form hydrogen ions in solution, acids in general have the following properties:

1. Sour taste
2. Change blue litmus to red and affect other suitable indicators
3. Form water solutions with pH values of less than 7
4. Undergo neutralization reactions with bases to form water and a salt

All of the foregoing general properties of an acid are readily seen in low-molar-mass carboxylic acids such as acetic acid. However, these general acid properties can be greatly influenced by the size of the hydrocarbon chain attached to the carboxyl group. In stearic acid, for example, taste, effect on indicators, and pH are not detectable because the large size of the hydrocarbon chain makes the acid insoluble in water. But stearic acid reacts with a base to form water and a salt. With sodium hydroxide, the equation for the reaction is

$$\underset{\text{stearic acid}}{C_{17}H_{35}COOH} + NaOH \longrightarrow \underset{\text{sodium stearate}}{C_{17}H_{35}COONa} + H_2O$$

The salts formed from this neutralization reaction have different properties than the acids. Salts are soluble in water and dissociate completely in solution. These properties assist in the separation of carboxylic acids from other nonpolar compounds. A base, like NaOH, is added to the mixture of compounds. The carboxylic acid reacts, forming its sodium salt and water. The salt dissolves in water while the remaining nonpolar molecules stay in the organic layer. Once the layers are separated, some mineral acid, HCl, can be added to the carboxylic acid salt to recover the acid.

Carboxylic acids generally react with sodium hydrogen carbonate to release carbon dioxide. This reaction can be used to distinguish a carboxylic acid from a phenol (also a weak acid). Phenols do not react with sodium hydrogen carbonate, although they will be neutralized by a strong base.

Acid Chloride Formation Thionyl chloride ($SOCl_2$) reacts with carboxylic acids to form acid chlorides. In this substitution reaction, a chlorine atom replaces an —OH group.

$$\underset{\text{acid}}{\mathrm{R\overset{\overset{\displaystyle O}{\|}}{C}{-}OH}} + \underset{\text{thionyl chloride}}{\mathrm{SOCl_2}} \longrightarrow \underset{\text{acid chloride}}{\mathrm{R\overset{\overset{\displaystyle O}{\|}}{C}{-}Cl}} + \mathrm{SO_2} + \mathrm{HCl}$$

$$\mathrm{CH_3\overset{\overset{\displaystyle O}{\|}}{C}{-}OH} + \mathrm{SOCl_2} \longrightarrow \underset{\text{acetyl chloride}}{\mathrm{CH_3\overset{\overset{\displaystyle O}{\|}}{C}{-}Cl}} + \mathrm{SO_2} + \mathrm{HCl}$$

Acid chlorides are extremely reactive substances. They must be kept away from moisture, or they will hydrolyze back to the acid:

Acid chlorides undergo substitution reactions.

$$\mathrm{R\overset{\overset{\displaystyle O}{\|}}{C}{-}Cl} + \mathrm{H_2O} \longrightarrow \mathrm{R\overset{\overset{\displaystyle O}{\|}}{C}{-}OH} + \mathrm{HCl}$$

Acid chlorides are more reactive than acids and can be used in place of acids to prepare esters and amides:

$$\mathrm{CH_3\overset{\overset{\displaystyle O}{\|}}{C}{-}Cl} + \mathrm{CH_3OH} \longrightarrow \underset{\text{methyl acetate}}{\mathrm{CH_3\overset{\overset{\displaystyle O}{\|}}{C}{-}OCH_3}} + \mathrm{HCl}$$

$$\mathrm{CH_3\overset{\overset{\displaystyle O}{\|}}{C}{-}Cl} + 2\,\mathrm{NH_3} \longrightarrow \underset{\text{acetamide}}{\mathrm{CH_3\overset{\overset{\displaystyle O}{\|}}{C}{-}NH_2}} + \mathrm{NH_4Cl}$$

Acid Anhydride Formation Inorganic anhydrides are formed by the elimination of a molecule of water from an acid or a base:

$$\mathrm{H_2SO_3} \longrightarrow \mathrm{SO_2} + \mathrm{H_2O}$$

$$\mathrm{Ba(OH)_2} \xrightarrow{\Delta} \mathrm{BaO} + \mathrm{H_2O}$$

An organic anhydride is formed by the elimination of a molecule of water from two molecules of acid.

$$\mathrm{R{-}\overset{\overset{\large O}{\|}}{C}{-}OH \;+\; HO{-}\overset{\overset{\large O}{\|}}{C}{-}R' \longrightarrow R{-}\overset{\overset{\large O}{\|}}{C}{-}O{-}\overset{\overset{\large O}{\|}}{C}{-}R' \;+\; H_2O}$$

The most commonly used organic anhydride is acetic anhydride. It can be prepared by the reaction of acetyl chloride with sodium acetate.

$$\mathrm{CH_3\overset{\overset{\large O}{\|}}{C}{-}Cl \;+\; Na^+\,{}^-O\overset{\overset{\large O}{\|}}{C}{-}CH_3 \longrightarrow \underset{\text{acetic anhydride}}{CH_3\overset{\overset{\large O}{\|}}{C}{-}O{-}\overset{\overset{\large O}{\|}}{C}{-}CH_3} \;+\; NaCl}$$

Acid anhydrides are very reactive and can be used to synthesize amides and esters. The anhydrides are not used as often as the acid chlorides in organic synthesis, however. In living cells, acid anhydrides are commonly used to activate carboxylic acids for further reaction.

ester

Ester Formation An **ester** is an organic compound formed by the reaction of an acid and an alcohol or a phenol. Water is also a product in this reaction.

$$\underset{\text{carboxylic acid}}{\mathrm{R\overset{\overset{\large O}{\|}}{C}{-}\boxed{OH}}} \;+\; \underset{\text{alcohol}}{\mathrm{R'O\boxed{H}}} \;\underset{}{\overset{H^+}{\rightleftharpoons}}\; \underset{\substack{\text{ester}\\ \text{(R can be H or Ar,}\\ \text{but R}'\text{ cannot be H)}}}{\mathrm{R\overset{\overset{\large O}{\|}}{C}{-}OR'}} \;+\; \mathrm{H_2O}$$

$$\underset{\substack{\text{formic acid}\\ \text{(methanoic acid)}}}{\mathrm{H{-}\overset{\overset{\large O}{\|}}{C}{-}OH}} \;+\; \mathrm{CH_3CH_2OH} \;\overset{H^+}{\rightleftharpoons}\; \underset{\substack{\text{ethyl formate}\\ \text{(ethyl methanoate)}}}{\mathrm{H{-}\overset{\overset{\large O}{\|}}{C}{-}OCH_2CH_3}} \;+\; \mathrm{H_2O}$$

At first glance this looks like the familiar acid–base neutralization reaction. But this is not the case, because the alcohol does not yield OH^- ions, and the ester, unlike a salt, is a molecular, not an ionic, substance. The forward reaction of an acid and an alcohol is called *esterification*; the reverse reaction of an ester with water is called *hydrolysis*. The work of a chemist may call for manipulating reaction conditions to favor the formation of either esters or their component parts, alcohols and acids.

Esterification is one of the most important reactions of carboxylic acids. Many biologically significant substances are esters.

24.7 Nomenclature of Esters

The general formula for an ester is RCOOR′, where R may be a hydrogen, alkyl group, or aryl group, and R′ may be an alkyl group, or aryl group, but *not* a hydrogen. Esters are found throughout nature. The ester linkage is particularly important in the study of fats and oils, both of which are esters. Esters of phosphoric acid are of vital importance to life as well.

◀ **The odor of pineapple is due to the ester ethyl butyrate, whereas the odor of ripe bananas is due to isoamyl acetate.**

Esters are alcohol derivatives of carboxylic acids. They are named in much the same way as salts. The alcohol part is named first, followed by the name of the acid modified to end in *ate*. The *ic* ending of the organic acid name is replaced by the ending *ate*. Thus in the IUPAC System, *ethanoic acid* becomes *ethanoate*. In the common names, *acetic acid* becomes *acetate*. To name an ester, it is necessary to recognize the portion of the ester molecule that comes from the acid and the portion that comes from the alcohol. In the general formula for an ester, the RC=O comes from the acid, and the R′O comes from the alcohol:

$$\underbrace{R-\overset{\displaystyle O}{\overset{\|}{C}}}_{\text{acid}}-\underbrace{O-R'}_{\text{alcohol}}$$

The R′ in R′O is named first, followed by the name of the acid modified by replacing *ic acid* with *ate*. The ester derived from ethyl alcohol and acetic acid is called ethyl acetate or ethyl ethanoate. Consider the ester formed from CH_3CH_2COOH and CH_3OH:

$$\underset{\substack{\text{propanoic acid}\\\text{(propionic acid)}}}{CH_3CH_2\overset{\displaystyle O}{\overset{\|}{C}}-\boxed{OH \;+\; H}}\underset{\substack{\text{methanol}\\\text{(methyl alcohol)}}}{-OCH_3} \overset{H^+}{\rightleftharpoons} \underset{\substack{\text{methyl propanoate}\\\text{(methyl propionate)}}}{CH_3CH_2\overset{\displaystyle O}{\overset{\|}{C}}-OCH_3} + H_2O$$

Esters of aromatic acids are named in the same general way as those of aliphatic acids. For example, the ester of benzoic acid and isopropyl alcohol is

$$C_6H_5-\overset{\displaystyle O}{\overset{\|}{C}}-O-CH(CH_3)_2$$

isopropyl benzoate

Formulas and names for additional esters are given in Table 24.3.

TABLE 24.3 Formulas, Names and Odors of Selected Esters

Formula	IUPAC name	Common name	Odor or flavor
$CH_3C(=O)-OCH_2CH_2CH(CH_3)CH_3$	Isopentyl ethanoate	Isoamyl acetate	Banana, pear
$CH_3CH_2CH_2C(=O)-OCH_2CH_3$	Ethyl butanoate	Ethyl butyrate	Pineapple
$HC(=O)-OCH_2CH(CH_3)CH_3$	Isobutyl methanoate	Isobutyl formate	Raspberry
$CH_3C(=O)-OCH_2(CH_2)_6CH_3$	Octyl ethanoate	*n*-Octyl acetate	Orange
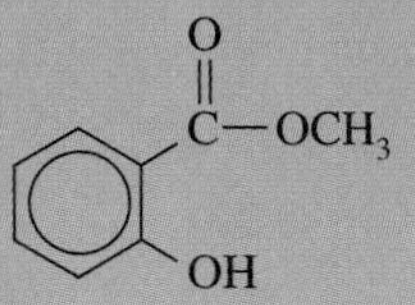 $C_6H_4(OH)C(=O)-OCH_3$	2-Hydroxymethylbenzoate	Methyl salicylate	Wintergreen

Example 24.2 Name the following esters:

(a) $H-C(=O)-OCH_2CH_2CH_3$

(b) 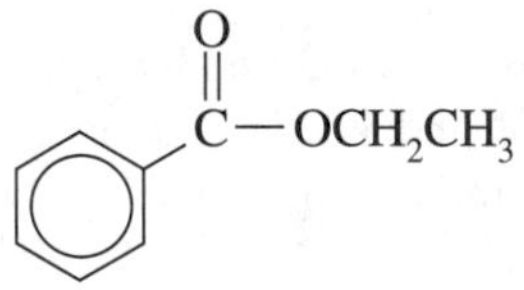 $C_6H_5-C(=O)-OCH_2CH_3$

(c) $O=C(-OCH_2CH_3)-CH_2-C(=O)-OCH_2CH_3$

Solution

(a) First identify the acid and alcohol components. The acid contains one carbon and is formic acid. The alcohol is propyl alcohol.

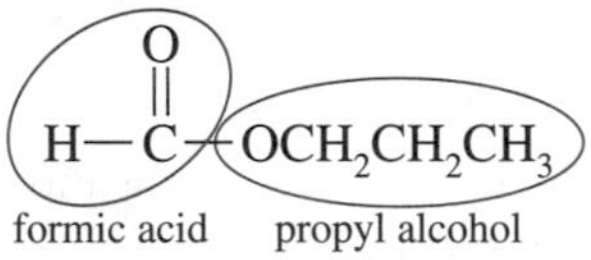

$H-C(=O) \,|\, OCH_2CH_2CH_3$

formic acid — propyl alcohol

Change the *ic* ending of the acid to *ate*, making the name formate or methanoate. The name of the ester then is propyl formate or propyl methanoate.

(b) The acid is benzoic acid; the alcohol is ethyl alcohol. Using the same procedure as in part (a), the name of the ester is ethyl benzoate.

(c) The acid is the three-carbon dicarboxylic acid, malonic acid. The alcohol is ethyl alcohol. Both acid groups are in the ester form. The name, therefore, is diethyl malonate.

Practice 24.6

Name the following esters:

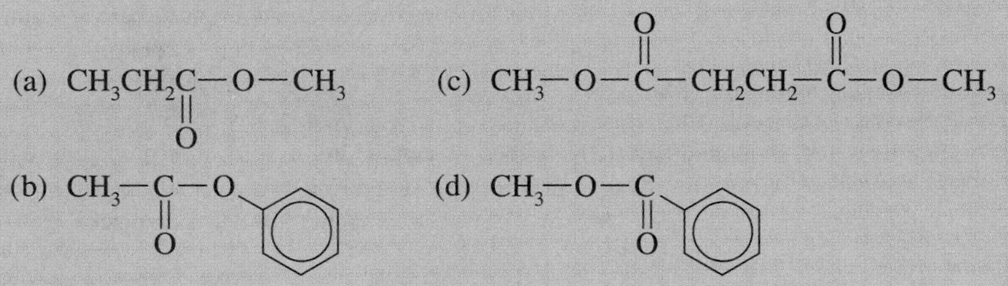

(a) $CH_3CH_2C(=O)-O-CH_3$

(b) $CH_3-C(=O)-O-C_6H_5$

(c) $CH_3-O-C(=O)-CH_2CH_2-C(=O)-O-CH_3$

(d) $CH_3-O-C(=O)-C_6H_5$

24.8 Occurrence and Physical Properties of Esters

Since many acids and many alcohols are known, the number of esters theoretically possible is very large. In fact, both natural and synthesized esters exist in almost endless variety. Simple esters derived from monocarboxylic acids and monohydroxy alcohols are colorless, generally nonpolar liquids or solids. The low polarity of ester molecules is substantiated by the fact that both their water solubility and boiling points are lower than those of either acids or alcohols of similar molar masses.

Low- and intermediate-molar-mass esters (from both acids and alcohols up to about ten carbons) are liquids with characteristic (usually fragrant or fruity) odors. The distinctive odor and flavor of many fruits are caused by one or more of these esters. The difference in properties between an acid and its esters is remarkable. For example, in contrast to the extremely unpleasant odor of butyric acid, ethyl butyrate has the pleasant odor of pineapple and methyl butyrate the odor of rum. Esters are used in flavoring and scenting agents (see Table 24.3). They are generally good solvents for organic substances, and those that have relatively low molar masses are volatile. Therefore, esters such as ethyl acetate, butyl acetate, and isoamyl acetate are extensively used in paints, varnishes, and lacquers.

High-molar-mass esters (formed from acids and alcohols of 16 or more carbons) are waxes and are obtained from various plants. They are used in furniture wax and automobile wax preparations; for example, carnauba wax contains esters of 24- and 28-carbon fatty acids and 32- and 34-carbon alcohols. Polyesters with very high molar masses, such as Dacron, are widely used in the textile industries (see Chapter 26).

24.9 Chemical Properties of Esters

The most important reaction of esters is *hydrolysis*. Hydrolysis is the splitting of molecules through the addition of water. The majority of organic and biochemical substances react only very slowly, if at all, with water. In order to increase the rate

Chemistry in Action

Aspirin

From the earliest days of medicine, people have obtained pain relief by chewing willow bark. In 1840, the active compound was isolated from the bark and identified as salicylic acid. Unfortunately, salicylic acid has several undesirable side effects, including a very sour taste and irritation of the stomach lining.

In 1883, an organic chemist reacted salicylic acid with acetic anhydride to form the ester acetylsalicylic acid.

$$\text{C}_6\text{H}_4(\text{COOH})\text{–O–}\overset{\overset{\text{O}}{\|}}{\text{C}}\text{–CH}_3$$

acetylsalicylic acid

The chemist happened to work for the Bayer Company, and the drug was aspirin.

Aspirin is the most widely used drug in the world and for good reason: it acts as a fever reducer (antipyretic), a pain reliever (analgesic), and an anti-inflammatory agent. In large doses (lethal is between 30 and 40 g), it is also a poison. Tablets are manufactured by mixing 0.32 g of aspirin with an inert binder (often starch) to hold the tablet together. Approximately half of the aspirin manufactured in the United States (more than 2.0×10^7 kg) is made into aspirin tablets. The remainder is used in combination pain relievers and cold remedies.

Until relatively recently it was not understood how aspirin acted within the body. Chemists have now determined that aspirin inhibits an enzyme necessary for the synthesis of prostaglandins (Chapter 29). Functions of prostaglandins include elevation of the blood pressure, tissue inflammation, and activation of the pain receptors in tissues. A reduction in the synthesis of prostaglandins produces both analgesic and anti-inflammatory effects.

Aspirin usage can have undesirable side effects as it can irritate the stomach lining, inhibit blood clotting, and prolong labor in childbirth. It is also associated with the development of Reyes syndrome, a brain disease that occurs in children recovering from chicken pox or flu. Some people are allergic to aspirin.

Publicity over Reyes syndrome and allergic reactions have resulted in the development of alternatives to aspirin. The most common alternatives are acetaminophen and ibuprofen. Acetaminophen acts an an analgesic and antipyretic but is not anti-inflammatory. Ibuprofen, a fairly recent addition to the non-prescriptive drug category, acts as a prostaglandin inhibitor in a manner similar to aspirin.

$$\text{HO–C}_6\text{H}_4\text{–NH–}\overset{\overset{\text{O}}{\|}}{\text{C}}\text{–CH}_3$$

acetaminophen

$$(\text{CH}_3)_2\text{CHCH}_2\text{–C}_6\text{H}_4\text{–CH(CH}_3)\text{–COOH}$$

ibuprofen

Other esters of salicylic acid also show medicinal effects. Methyl salicylate is an oil with the odor of wintergreen. It is commonly found as the active ingredient in pain-relieving liniments. It has the unusual property of penetrating the skin surface where it hydrolyzes, releasing salicylic acid and relieving the pain. Phenyl salicylate is an ester that is not hydrolyzed by acids. It is used to coat pills in order to permit them to pass through the stomach to the intestine before hydrolyzing and releasing their contents.

of these reactions, a catalyst is required. In the laboratory the chemist often employs an acid or base as a catalyst for hydrolysis. In living systems the role of catalyst is filled by enzymes.

Acid Hydrolysis

Hydrolysis of an ester involves reaction with water to form a carboxylic acid and an alcohol. The hydrolysis is catalyzed by strong acids (H_2SO_4 and HCl) or by certain enzymes.

$$\underset{\text{ester}}{\mathrm{R\overset{O}{\overset{\|}{C}}{-}OR'}} + H_2O \xrightarrow[\text{or enzyme}]{H^+} \underset{\text{acid}}{\mathrm{R\overset{O}{\overset{\|}{C}}{-}OH}} + \underset{\text{alcohol}}{\mathrm{R'OH}}$$

$$\underset{\text{methyl propanoate}}{\mathrm{CH_3CH_2\overset{O}{\overset{\|}{C}}{-}OCH_3}} + H_2O \xrightarrow{H^+} \underset{\text{propanoic acid}}{\mathrm{CH_3CH_2COOH}} + \underset{\text{methanol}}{\mathrm{CH_3OH}}$$

$$\underset{\text{methyl salicylate}}{\mathrm{C_6H_4(OH)\overset{O}{\overset{\|}{C}}{-}OCH_3}} + H_2O \xrightarrow{H^+} \underset{\text{salicylic acid}}{\mathrm{C_6H_4(OH)COOH}} + \underset{\text{methanol}}{\mathrm{CH_3OH}}$$

Alkaline Hydrolysis (Saponification)

Saponification is the hydrolysis of an ester by a strong base (NaOH or KOH) to produce an alcohol and a salt (or soap if the salt formed is from a high-molar-mass acid): **saponification**

$$\underset{\text{ester}}{\mathrm{R\overset{O}{\overset{\|}{C}}{-}OR'}} + NaOH \xrightarrow[\Delta]{H_2O} \underset{\text{salt}}{\mathrm{R\overset{O}{\overset{\|}{C}}{-}O^-Na^+}} + \underset{\text{alcohol}}{\mathrm{R'OH}}$$

$$\underset{\text{ethyl stearate}}{\mathrm{CH_3(CH_2)_{16}\overset{O}{\overset{\|}{C}}{-}OCH_2CH_3}} + NaOH \xrightarrow[\Delta]{H_2O} \underset{\text{sodium stearate}}{\mathrm{CH_3(CH_2)_{16}\overset{O}{\overset{\|}{C}}ONa}} + \mathrm{CH_3CH_2OH}$$

The carboxylic acid may be obtained by reacting the salt with a strong acid:

$$\mathrm{CH_3(CH_2)_{16}COONa} + \mathrm{HCl} \longrightarrow \underset{\text{stearic acid}}{\mathrm{CH_3(CH_2)_{16}COOH}} + \mathrm{NaCl}$$

Notice that in saponification the base is a reactant and not a catalyst.

Practice 24.7

Write equations showing (a) the acid hydrolysis and (b) the alkaline hydrolysis of isobutyl pentanoate.

24.10 Glycerol Esters

Fats and **oils** are esters of glycerol and predominantly long-chain fatty acids. Fats and oils are also called **triacylglycerols** or triglycerides, since each molecule is derived from one molecule of glycerol and three molecules of fatty acid.

fat, oil
triacylglycerol

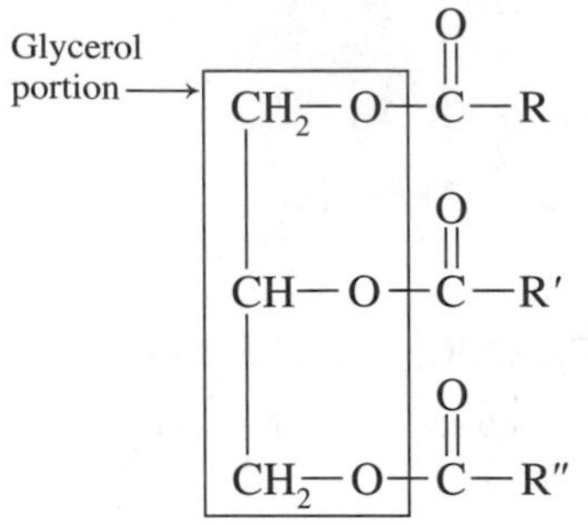

$$\begin{array}{l} \text{Glycerol portion} \rightarrow \boxed{\begin{array}{l} CH_2-O- \\ | \\ CH-O- \\ | \\ CH_2-O- \end{array}} \begin{array}{l} \overset{O}{\overset{\|}{C}}-R \\ \\ \overset{O}{\overset{\|}{C}}-R' \\ \\ \overset{O}{\overset{\|}{C}}-R'' \end{array} \end{array} \qquad \begin{array}{ll} CH_2-O-\overset{O}{\overset{\|}{C}}-C_{17}H_{35} & \text{(Stearic)} \\ | & \\ CH-O-\overset{O}{\overset{\|}{C}}-C_{15}H_{31} & \text{(Palmitic)} \\ | & \\ CH_2-O-\overset{O}{\overset{\|}{C}}-C_{11}H_{23} & \text{(Lauric)} \end{array}$$

General formula for a triacylglycerol

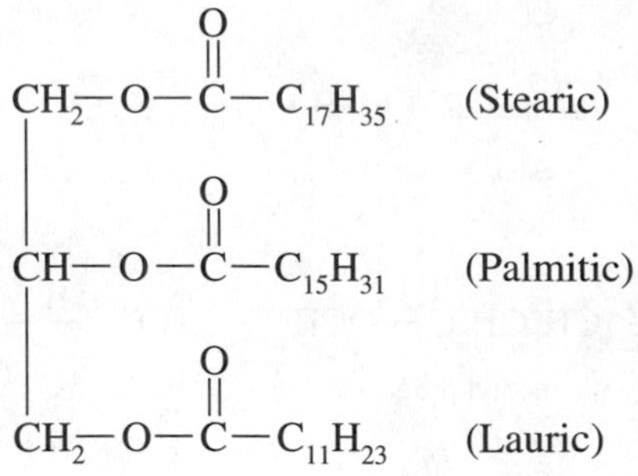

Typical triaclyglycerol containing three different fatty acids

The structural formulas of triacylglycerol molecules vary because

1. The length of the fatty acid chain may vary from 4 to 20 carbons, but the number of carbon atoms in the chain is nearly always even.
2. Each fatty acid may be saturated or may be unsaturated and contain one, two, or three carbon–carbon double bonds.
3. A triacylglycerol may, and frequently does, contain three different fatty acids.

The most abundant saturated fatty acids in fats and oils are lauric, myristic, palmitic, and stearic acids (Table 24.4). The most abundant unsaturated acids in fats and oils contain 18 carbon atoms and have one, two, or three carbon–carbon double bonds. In all of these naturally occurring unsaturated acids, the configuration about the double bond is cis. Their formulas are

$$CH_3(CH_2)_7CH{=}CH(CH_2)_7COOH$$
oleic acid

$$CH_3(CH_2)_4CH{=}CHCH_2CH{=}CH(CH_2)_7COOH$$
linoleic acid

$$CH_3CH_2CH{=}CHCH_2CH{=}CHCH_2CH{=}CH(CH_2)_7COOH$$
linolenic acid

The major physical difference between fats and oils is that fats are solid and oils are liquid at room temperature (see Section 29.2). Since the glycerol part of the structure is the same for a fat and an oil, the difference must be due to the fatty acid end of the molecule. Fats contain a larger proportion of saturated fatty acids, whereas oils contain greater amounts of unsaturated fatty acids. The term *polyunsaturated* has been popularized in recent years; it means that each molecule of fat in a particular product contains several double bonds.

Fats and oils are obtained from natural sources. In general, fats come from animal sources and oils from vegetable sources. Thus lard is obtained from hogs and tallow from cattle and sheep. Olive, cottonseed, corn, peanut, soybean, canola, linseed, and other oils are obtained from the fruit or seed of their respective vegetable sources. Table 24.4 shows the major constituents of several fats and oils.

The caloric value per unit mass for triacylglycerols is about 9 kcal/g, while proteins and carbohydrates yield about 4 kcal/g.

Triacylglycerols are the principal form in which energy is stored in the body. The caloric value per unit of mass is over twice as great as that for carbohydrates and proteins. As a source of energy, triacylglycerols can be completely replaced by

TABLE 24.4 Fatty Acid Composition of Selected Fats and Oils

	Fatty acid (%)				
Fat or oil	**Myristic acid**	**Palmitic acid**	**Stearic acid**	**Oleic acid**	**Linoleic acid**
Animal fat					
Butter[a]	7–10	23–26	10–13	30–40	4–5
Lard	1–2	28–30	12–18	41–48	6–7
Tallow	3–6	24–32	14–32	35–48	2–4
Vegetable oil					
Olive	0–1	5–15	1–4	49–84	4–12
Peanut	—	6–9	2–6	50–70	13–26
Corn	0–2	7–11	3–4	43–49	34–42
Cottonseed	0–2	19–24	1–2	23–33	40–48
Soybean	0–2	6–10	2–4	21–29	50–59
Linseed[b]	—	4–7	2–5	9–38	3–43

[a]Butyric acid, 3–4%
[b]Linolenic acid, 25–58%

either carbohydrates or proteins. However, some minimum amount of fat is needed in the diet, because fat supplies the nutritionally essential unsaturated fatty acids (linoleic, linolenic, and arachidonic) required by the body (see Section 29.2).

Hydrogenation of Glycerides Addition of hydrogen is a characteristic reaction of the carbon–carbon pi bonds. Industrially, low-cost vegetable oils are partially hydrogenated to obtain solid fats that are useful as shortening in baking or in making margarine. In this process, hydrogen gas is bubbled through hot oil that contains a finely dispersed nickel catalyst. The hydrogen adds to the carbon–carbon double bonds of the oil to saturate the double bonds and form fats:

$$H_2 + \underset{\text{in oil or fat}}{—CH{=}CH—} \xrightarrow{Ni} —CH_2—CH_2—$$

In actual practice, only some of the double bonds are allowed to become saturated. The degree of hydrogenation can be controlled to obtain a product of any desired degree of saturation. The products resulting from the partial hydrogenation of oils are marketed as solid shortening (Crisco, Spry, etc.) and are used for cooking and baking. Oils and fats are also partially hydrogenated to improve their keeping qualities. Rancidity in fats and oils results from air oxidation at points of unsaturation, producing low molar mass aldehydes and acids of disagreeable odor and flavor.

Hydrogenolysis Triacylglycerols can be split and reduced in a reaction called hydrogenolysis (splitting by hydrogen). Hydrogenolysis requires higher temperatures and pressures and a different catalyst (copper chromite) than does hydrogenation of double bonds. Each triacylglycerol molecule yields a molecule of glycerol and three primary alcohol molecules. The hydrogenolysis of glyceryl trilaurate is represented as follows:

$$
\begin{array}{l}
CH_2-O-\overset{O}{\overset{\|}{C}}-C_{11}H_{23} \\
| \\
CH-O-\overset{O}{\overset{\|}{C}}-C_{11}H_{23} \\
| \\
CH_2-O-\overset{O}{\overset{\|}{C}}-C_{11}H_{23}
\end{array}
+ 6\,H_2 \xrightarrow[\Delta,\ \text{pressure}]{\text{copper chromite}} 3\,CH_3(CH_2)_{10}CH_2OH + \begin{array}{c} CH_2OH \\ | \\ CHOH \\ | \\ CH_2OH \end{array}
$$

glyceryl trilaurate; lauryl alcohol (1-dodecanol); glycerol

Long-chain primary alcohols obtained by this reaction are important, since they are used to manufacture other products, especially synthetic detergents (see Section 24.11).

Hydrolysis Triacylglycerols can be hydrolyzed, yielding fatty acids and glycerol. The hydrolysis is catalyzed by digestive enzymes at room temperatures and by mineral acids at high temperatures:

$$
\begin{array}{l}
CH_2-O-\overset{O}{\overset{\|}{C}}-R \\
| \\
CH-O-\overset{O}{\overset{\|}{C}}-R' \\
| \\
CH_2-O-\overset{O}{\overset{\|}{C}}-R''
\end{array}
+ 3\,H_2O \xrightarrow[\text{enzymes}]{H^+ \text{ or}} \begin{array}{c} RCOOH \\ R'COOH \\ R''COOH \end{array} + \begin{array}{c} CH_2OH \\ | \\ CHOH \\ | \\ CH_2OH \end{array}
$$

triacylglycerol; fatty acids (3 molecules); glycerol

The enzyme-catalyzed reaction occurs in digestive reactions and in biological degradation (or metabolic) processes. The acid-catalyzed reaction is employed in the commercial preparation of fatty acids and glycerol.

Saponification Saponification of a fat or oil involves the alkaline hydrolysis of a triester. The products formed are glycerol and the alkali metal salts of fatty acids, which are called soaps. As a specific example, glyceryl tripalmitate reacts with sodium hydroxide to produce sodium palmitate and glycerol:

$$
\begin{array}{l}
CH_2-O-\overset{O}{\overset{\|}{C}}-C_{15}H_{31} \\
| \\
CH-O-\overset{O}{\overset{\|}{C}}-C_{15}H_{31} \\
| \\
CH_2-O-\overset{O}{\overset{\|}{C}}-C_{15}H_{31}
\end{array}
+ 3\,NaOH \xrightarrow{\Delta} 3\,C_{15}H_{31}COONa + \begin{array}{c} CH_2OH \\ | \\ CHOH \\ | \\ CH_2OH \end{array}
$$

glyceryl tripalmitate; sodium palmitate (a soap); glycerol

Glycerol, fatty acids, and soaps are valuable articles of commerce, and the processing of fats and oils to obtain these products is a major industry.

The cleansing action of soap is of little interest to this baby who is enjoying the water!

Practice 24.8

Write the structure for a triacylglycerol that has one unit each of myristic, palmitic, and oleic acids.

24.11 Soaps and Synthetic Detergents

In the broadest sense possible, a detergent is simply a cleansing agent. Soap has been used as a cleansing agent for at least 2000 years and thereby is classified as a detergent under this definition. However, beginning about 1930, a number of new cleansing agents that were superior in many respects to ordinary soap began to appear on the market. Because they were both synthetic organic products and detergents, they were called **synthetic detergents**, or syndets. A soap is distinguished from a synthetic detergent on the basis of chemical composition and not on the basis of function or usage.

synthetic deterent

Soaps

In former times soap-making was a crude operation. Surplus fats were boiled with wood ashes or with some other alkaline material. Today, soap is made in large manufacturing plants under controlled conditions. Salts of long-chain fatty acids are called **soaps**. However, only the sodium and potassium salts of carboxylic acids containing 12–18 carbon atoms are of great value as soaps, because of their abundance in fats.

soap

$$\text{Fat or oil} + \text{NaOH} \longrightarrow \text{Soap} + \text{Glycerol}$$

To understand how a soap works as a cleansing agent, let's consider sodium palmitate, $CH_3(CH_2)_{14}COONa$, as an example of a typical soap. In water this substance exists as sodium ions, Na^+, and palmitate ions, $CH_3(CH_2)_{14}COO^-$. The sodium ion is an ordinary hydrated metal ion. The cleansing property, then, must be centered in the palmitate ion. The palmitate ion contains both a **hydrophilic** (water-loving) and a **hydrophobic** (water-fearing) group. The hydrophilic end is the polar, negatively charged carboxylate group. The hydrophobic end is the long hydrocarbon group. The hydrocarbon group is soluble in oils and greases but not in water. The hydrophilic carboxylate group is soluble in water.

hydrophilic
hydrophobic

The cleansing action of a soap is explained in this fashion: When the soap comes in contact with grease on a soiled surface, the hydrocarbon end of the soap dissolves in the grease, leaving the negatively charged carboxylate end exposed on the grease surface. Because the negatively charged carboxylate groups are strongly attracted by water, small droplets are formed, and the grease is literally lifted or floated away from the soiled object (see Figure 24.1).

The cleansing property of a soap is due to its ability to act as an emulsifying

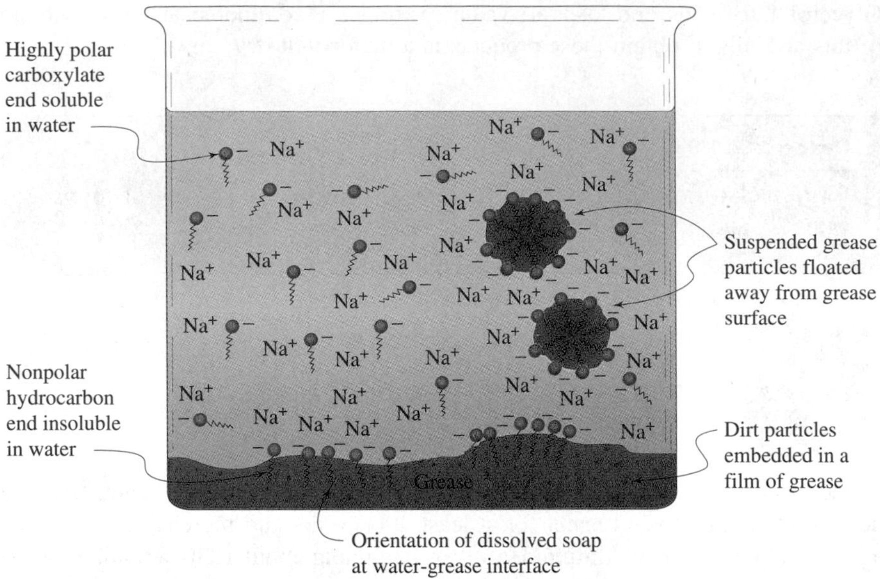

FIGURE 24.1 ▶ Cleansing action of soap: Dirt particles are embedded in a surface film of grease. The hydrocarbon ends of negative soap ions dissolve in the grease film, leaving exposed carboxylate groups. These carboxylate groups are attracted to water, and small droplets of grease-bearing dirt are formed and floated away from the surface.

agent between water and water-insoluble greases and oils. The grease–soap emulsion is stable because the oil droplets repel each other due to the negatively charged carboxyl groups on their surfaces. Some insoluble particulate matter is carried away with the grease; the remainder is wetted and mechanically washed away in the water. Synthetic detergents function in a similar way.

Ordinary soap is a good cleansing agent in soft water, but it is not satisfactory in hard water because insoluble calcium, magnesium, and iron(III) salts are formed. Palmitate ions, for example, are precipitated by calcium ions:

$$\underset{}{Ca^{2+}(aq)} + 2\ \underset{\text{palmitate ion}}{CH_3(CH_2)_{14}COO^-(aq)} \longrightarrow \underset{\text{calcium palmitate}}{[CH_3(CH_2)_{14}COO]_2Ca(s)}$$

These precipitates are sticky substances and are responsible for "bathtub ring" and the sticky feel of hair after being shampooed with soap in hard water.

Soaps are ineffective in acidic solutions because water-insoluble molecular fatty acids are formed:

$$CH_3(CH_2)_{14}COO^- + H^+ \longrightarrow \underset{\text{palmitic acid molecule}}{CH_3(CH_2)_{14}COOH}$$

Synthetic Detergents

Once it was recognized that the insoluble hydrocarbon radical joined to a highly polar group was the key to the detergent action of soaps, chemists set out to make new substances that would have similar properties. About 1930, synthetic detergents (syndets) began to replace soaps, and now about 4 lb of syndets are sold for each pound of soap.

Although hundreds of substances with detergent properties are known, an idea of their general nature can be obtained from consideration of sodium lauryl sulfate and sodium *p*-dodecylbenzene sulfonate.

$CH_3(CH_2)_{10}CH_2OSO_3^- Na^+$

sodium lauryl sulfate

$CH_3(CH_2)_{10}CH_2-C_6H_4-SO_3^- Na^+$

sodium *p*-dodecylbenzene sulfonate

Sodium lauryl sulfate and sodium *p*-dodecylbenzene sulfonate act in water in much the same way as sodium palmitate. Like the palmitate ion, the negative lauryl sulfate ion has a long hydrocarbon chain that is soluble in grease and a sulfate group that is attracted to water:

$$\underbrace{CH_3CH_2CH_2CH_2CH_2CH_2CH_2CH_2CH_2CH_2CH_2CH_2}_{\text{nonpolar hydrophobic end, grease soluble}}-\underbrace{OSO_3^-}_{\text{polar hydrophilic end, water soluble}}$$

The one great advantage these synthetic detergents have over soap is that their calcium, magnesium, and iron(III) salts, as well as their sodium salts, are soluble in water. Therefore, they are nearly as effective in hard water as in soft water.

The foregoing are anionic detergents, because they contain long chain negatively charged ions. Other detergents, both cationic and nonionic, have been developed for special purposes. A cationic detergent has a long hydrocarbon chain and a positive charge.

$$\underbrace{CH_3(CH_2)_{14}CH_2}_{\text{grease soluble, hydrophobic}}-\underbrace{\overset{+}{N}(CH_3)_3}_{\text{water soluble, hydrophilic}}$$

Nonionic detergents are molecular substances. The molecule of a nonionic detergent contains a grease-soluble component and a water-soluble component. Some of these substances are especially useful in automatic washing machines because they have good detergent but low sudsing properties. The structure of a representative nonionic detergent is

$$\underbrace{CH_3(CH_2)_{10}CH_2}_{\text{grease soluble, hydrophobic}}-O-\underbrace{(CH_2CH_2O)_7-CH_2CH_2OH}_{\text{water soluble, hydrophilic}}$$

Biodegradability

Organic substances that are readily decomposed by microorganisms in the environment are said to be **biodegradable**. All naturally occurring organic substances are eventually converted to simple inorganic molecules and ions such as CO_2, H_2O, N_2, Cl^-, and SO_4^{2-}. Most of these conversions are catalyzed by enzymes produced by microorganisms. These enzymes are capable of attacking only certain specific molecular configurations that are found in substances occurring in nature.

biodegradable

A number of years ago a serious environmental pollution problem arose in connection with synthetic detergents. Some of the early syndets, which contained highly branched chain hydrocarbons, had no counterparts in nature. Therefore, enzymes capable of degrading them did not exist, and the detergents were essentially nonbiodegradable and broke down very, very slowly. As a result, these syndets accumulated in water supplies, where they caused severe pollution problems due to excessive foaming and other undesirable effects.

Detergent manufacturers, acting on the recommendations of chemists and biologists, changed from a branched-chain alkyl benzene to a straight-chain alkyl benzene raw material. Detergents that contain the straight-chain alkyl groups are biodegradable.

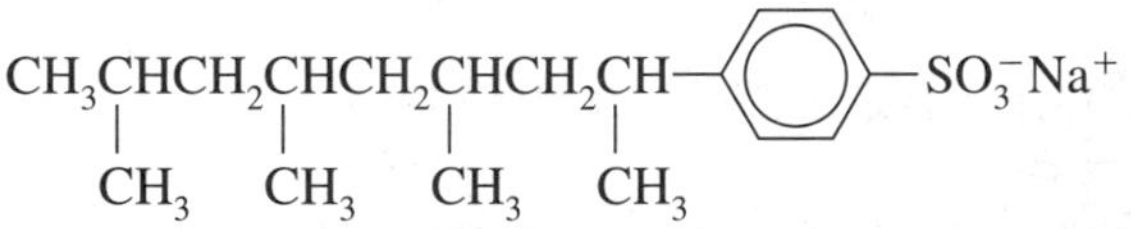

$$\underset{\displaystyle CH_3}{CH_3\underset{|}{C}H}\underset{\displaystyle CH_3}{CH_2\underset{|}{C}H}\underset{\displaystyle CH_3}{CH_2\underset{|}{C}H}\underset{\displaystyle CH_3}{CH_2\underset{|}{C}H}-C_6H_4-SO_3^-Na^+$$

a nonbiodegradable detergent

$$CH_3CH_2CH_2CH_2CH_2CH_2CH_2CH_2CH_2CH_2CH_2CH_2-C_6H_4-SO_3^-Na^+$$

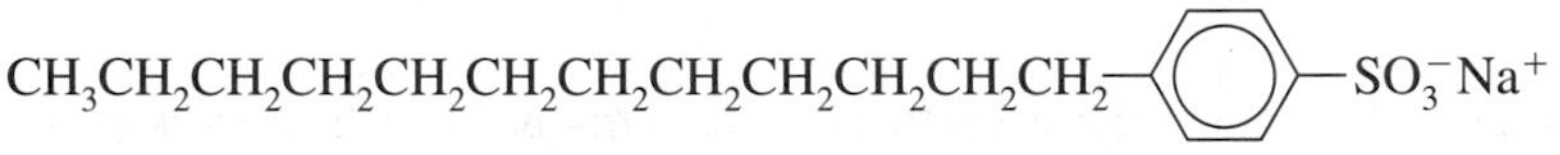

a biodegradable detergent

24.12 Esters and Anhydrides of Phosphoric Acid

Phosphoric acid has a Lewis structure similar to that of a carboxylic acid.

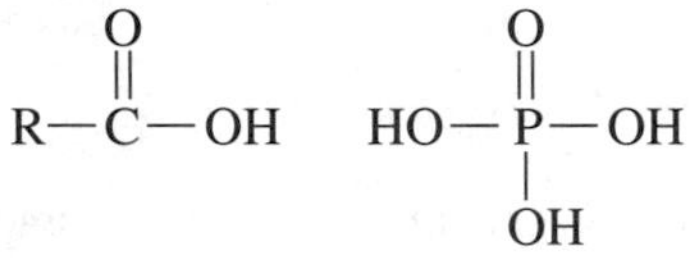

$$R-\overset{\displaystyle O}{\overset{\|}{C}}-OH \qquad HO-\overset{\displaystyle O}{\overset{\|}{\underset{\displaystyle \underset{OH}{|}}{P}}}-OH$$

carboxylic acid   phosphoric acid

In both molecules an —OH is attached to an element that is double-bonded to an oxygen. In fact, phosphoric acid has three such —OH groups. This similarity in structure permits phosphoric acid to behave as a carboxylic acid in reaction with an alcohol. The product of the esterification reaction is called a phosphate ester.

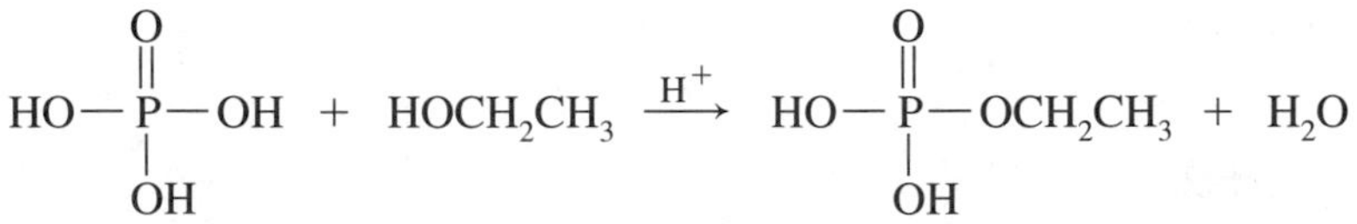

$$HO-\overset{\displaystyle O}{\overset{\|}{\underset{\displaystyle \underset{OH}{|}}{P}}}-OH + HOCH_2CH_3 \xrightarrow{H^+} HO-\overset{\displaystyle O}{\overset{\|}{\underset{\displaystyle \underset{OH}{|}}{P}}}-OCH_2CH_3 + H_2O$$

phosphoric acid   ethanol   monoethyl phosphate

This phosphate ester still has two —OH groups that can form additional esters. The result of one further esterification reaction is a molecule with two phosphate ester linkages. The diester then can contain two different alcohol groups in the same molecule. This structure is common in biochemistry specifically in nucleic acids (Chapter 32) and phospholipids (Chapter 29).

Anhydrides of phosphoric acid can be formed by bringing two molecules together and eliminating a water molecule.

$$\underset{\text{phosphoric acid}}{HO-\overset{\overset{\displaystyle O}{\|}}{\underset{\underset{\displaystyle OH}{|}}{P}}-OH} + \underset{\text{phosphoric acid}}{HO-\overset{\overset{\displaystyle O}{\|}}{\underset{\underset{\displaystyle OH}{|}}{P}}-OH} \longrightarrow \underset{\text{pyrophosphoric acid}}{HO-\overset{\overset{\displaystyle O}{\|}}{\underset{\underset{\displaystyle OH}{|}}{P}}-O-\overset{\overset{\displaystyle O}{\|}}{\underset{\underset{\displaystyle OH}{|}}{P}}-OH} + H_2O$$

In general, both phosphate esters and phosphoric acid anhydrides have significant biological importance. The phosphate esters "tag" many biochemicals, labeling them for specific biological purposes. The phosphoric acid anhydrides serve as a temporary store of metabolic energy (see Chapter 34).

Concepts in Review

1. Give the common and IUPAC names of carboxylic acids.
2. Write structural formulas for saturated carboxylic acids, unsaturated carboxylic acids, amino acids, hydroxy acids, aromatic carboxylic acids, and dicarboxylic acids.
3. Tell how water solubility of carboxylic acids varies with increasing molar mass.
4. Relate the boiling points of carboxylic acids to their structure.
5. Write equations for the preparation of carboxylic acids by (a) oxidation of alcohols and aldehydes, (b) hydrolysis or saponification of esters and fats, (c) oxidation of aromatic hydrocarbons, and (d) hydrolysis of nitriles.
6. Write equations showing the effect of heat on malonic, succinic, and glutaric acids.
7. Write equations for the reactions of carboxylic acids that form (a) salts, (b) esters, and (c) acid chlorides.
8. Write an equation to show the formation of an ester from an acid chloride.
9. Write common names, IUPAC names, and formulas of esters.
10. Identify the portion of an ester that is derived from a carboxylic acid and the portion derived from an alcohol.
11. Write the structure of a triacylglycerol (triglyceride) when given the fatty acid composition.
12. Explain the differences between a fat and an oil.
13. Write equations illustrating the (a) hydrogenation, (b) hydrogenolysis, (c) hydrolysis, and (d) saponification of a fat or oil.
14. Write the structural formulas for the three principal unsaturated carboxylic acids found in fats and oils.
15. Explain how a soap or synthetic detergent acts as a cleansing agent.
16. Explain why syndets are effective and soaps are not effective as cleansing agents in hard water.

17. Differentiate among cationic, anionic, and nonionic detergents.
18. Explain the similarities between a carboxylic acid and phosphoric acid.
19. Write equations showing the formation of a phosphate ester and the formation of a phosphoric acid anhydride.
20. Indicate the significant role of phosphate esters and anhydrides in living organisms.
21. Indicate the ways in which aspirin acts within the body.
22. Explain how aspirin and ibuprofen differ from acetaminophen in medicinal use within the body.

Key Terms

The terms listed here have been defined within this chapter. Section numbers are referenced in parenthesis for each term..

biodegradable (24.11)
carboxyl group (24.1)
ester (24.6)
fat (24.10)
hydrophilic (24.11)
hydrophobic (24.11)
oil (24.10)
saponification (24.9)
soap (24.11)
synthetic detergent (24.11)
triacylglycerol (24.10)

Questions

Questions refer to tables, figures, and key words and concepts defined within the chapter. A particularly challenging question or exercise is indicated with an asterisk.

1. Using a specific compound in each case, write structural formulas for the following:
   (a) an aliphatic carboxylic acid
   (b) an aromatic carboxylic acid
   (c) an α-hydroxy acid
   (d) an α-amino acid
   (e) a β-chloro acid
   (f) a dicarboxylic acid
   (g) an unsaturated carboxylic acid
   (h) an ester
   (i) a nitrile
   (j) a sodium salt of a carboxylic acid
   (k) an acid halide
   (l) a triacylglycerol
   (m) a soap

2. Which of the following would have the more objectionable odor? Briefly explain.
   (a) a 1% solution of butyric acid (C_3H_7COOH) or
   (b) a 1% solution of sodium butyrate (C_3H_7COONa)

3. Which has the greater solubility in water?
   (a) methyl propanoate or propanoic acid
   (b) sodium palmitate or palmitic acid
   (c) sodium stearate or barium stearate
   (d) phenol or sodium phenoxide

4. Explain the difference between
   (a) a fat and an oil
   (b) a soap and a syndet
   (c) hydrolysis and saponification

5. Explain the cleansing action of detergents.

6. Cite the principal advantages that synthetic detergents (syndets) have over soaps.

7. List the medicinal effects and the risks associated with aspirin.

8. Besides aspirin, give two examples of other esters of salicylic acid that have medicinal uses.

9. What substances are commonly substituted for aspirin? Indicate their medical effects.

10. Which of these statements are correct? Rewrite the incorrect statements to make them correct.
    (a) Carboxylic acids can be either aliphatic or aromatic.
    (b) The functional group —COOH is known as a carboxyl group.
    (c) The name for $CH_3CH_2CHBrCH_2COOH$ is γ-bromovaleric acid.
    (d) Acetic acid is a stronger acid than hydrochloric acid.
    (e) Benzoic acid is more soluble in water than sodium benzoate.
    (f) Carboxylic acids have relatively high boiling points because of hydrogen bonding between molecules.
    (g) The formula $C_{17}H_{33}COOH$ represents oleic acid.
    (h) Fumaric and maleic acids are cis–trans isomers.
    (i) If $CH_3CH_2CH_2COCl$ comes into contact with water, it is hydrolyzed to butyric acid.
    (j) Volatile esters generally have a pleasant odor.
    (k) Glycerol is a trihydroxy alcohol.
    (l) Fatty acids in fats usually have an even number of carbon atoms.
    (m) Fats and oils are esters of glycerol.
    (n) Oils are largely of vegetable origin.
    (o) The presence of unsaturation in the acid component of a fat tends to raise its melting point compared with the corresponding saturated compound.
    (p) Alkali metal salts of long-chain fatty acids are called soaps.
    (q) The chemical name for aspirin is acetylsalicylic acid.
    (r) In living cells, phosphate esters are most often used as a temporary store of metabolic energy.
    (s) The hydrophobic end of a detergent molecule is water soluble.
    (t) Saponification is the hydrolysis of a fat or an ester in an acid medium.
    (u) Methyl salicylate is a salt.

Paired Exercises

These exercises are paired. Each odd-numbered exercise is followed by a similar even-numbered exercise. Answers to the even-numbered exercises are given in Appendix V.

11. Name the following compounds:
    (a) $CH_3(CH_2)_4COOH$ (IUPAC name)
    (b) $CH_3CH{=}CHCOOH$ (common name for the trans isomer)
    (c) 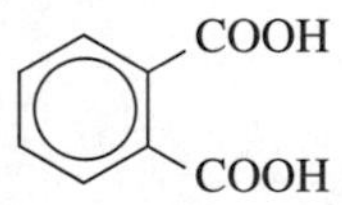

    (common name)
    (d) $CH_3(CH_2)_{16}COOH$ (common name)
    (e) $CH_3CH_2COO^-Na^+$ (IUPAC name)

12. Name the following compounds:
    (a) 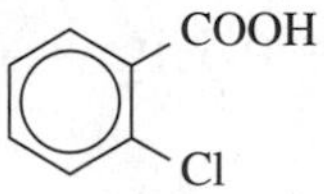

    (IUPAC name)
    (b) $CH_3(CH_2)_7CH{=}CH(CH_2)_7COOH$ (common name)
    (c) 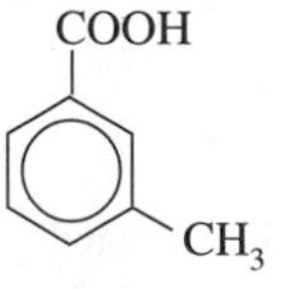

    (common name)
    (d) $CH_3CH_2CHCOOH$ (IUPAC name)
    with OH on the CH: $CH_3CH_2CH(OH)COOH$
    (e) $CH_3CH_2COO^-NH_4^+$ (IUPAC name)

13. Write structures for the following compounds:
    (a) hexanoic acid
    (b) malonic acid
    (c) sodium benzoate
    (d) *o*-toluic acid
    (e) stearic acid
    (f) 2-chloropropanoic acid
    (g) potassium butyrate

14. Write structures for the following compounds:
    (a) oxalic acid
    (b) pentanoic acid
    (c) *o*-phthalic acid
    (d) linolenic acid
    (e) sodium *p*-aminobenzoate
    (f) ammonium propanoate
    (g) β-hydroxybutyric acid

15. Assume that you have a 0.01 M solution of each of the following substances:

(a) NH_3 (c) NaCl (e) CH_3COOH
(b) HCl (d) NaOH (f) 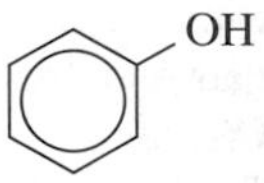 OH (phenol; benzene ring with OH)

Arrange them in order of increasing pH (list the most acidic solution first).

17. Give an IUPAC and a common name for each of the following:

(a) $CH_2{=}CH\overset{O}{\overset{\|}{C}}{-}OCH_3$

(b) $CH_3CH_2CH_2\overset{O}{\overset{\|}{C}}{-}OCH_2CH_3$

(c) 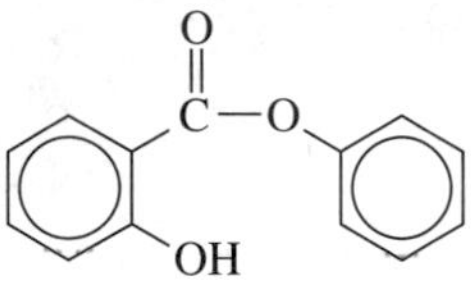 benzene ring bearing $\overset{O}{\overset{\|}{C}}{-}O{-}$ (bonded to a second benzene ring) and, on the adjacent carbon, OH

19. Write the structural formulas for each:
(a) methyl formate (c) ethyl benzoate
(b) butyl butanoate

21. Write the structural formula and name of the principal organic product for each of the following reactions:

(a) $CH_3(CH_2)_7CH{=}CH(CH_2)_7COOH + H_2 \xrightarrow{Ni}$

(b) 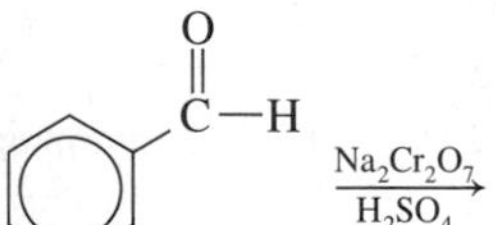 benzene ring bearing $\overset{O}{\overset{\|}{C}}{-}H$ $\xrightarrow[H_2SO_4]{Na_2Cr_2O_7}$

(c) $CH_3CH_2COOH + NaOH \longrightarrow$

(d) $CH_3(CH_2)_3CH_2\overset{O}{\overset{\|}{C}}{-}OCH_2CH_2CH_3 + NaOH \xrightarrow{\Delta}$

(e) 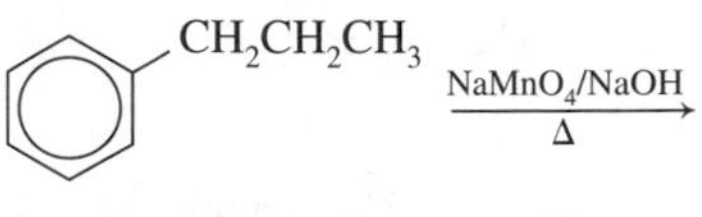 benzene ring bearing $CH_2CH_2CH_3$ $\xrightarrow[\Delta]{NaMnO_4/NaOH}$

23. Write the structural formula of the ester that when hydrolyzed would yield
(a) methanol and acetic acid
(b) ethanol and formic acid
(c) 2-propanol and benzoic acid

16. Assume that you have a 0.04 M solution of each of the following substances:

(a) NH_3 (c) HBr (e) KBr
(b) KOH (d) HCOOH (f) OH (phenol; benzene ring with OH)

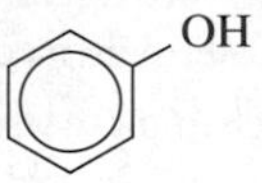

Arrange them in order of decreasing pH (list the most basic solution first).

18. Give an IUPAC and a common name for each of the following:

(a) $H\overset{O}{\overset{\|}{C}}{-}OCH_3$

(b) benzene ring bearing $\overset{O}{\overset{\|}{C}}{-}OCH_2CH_2CH_3$

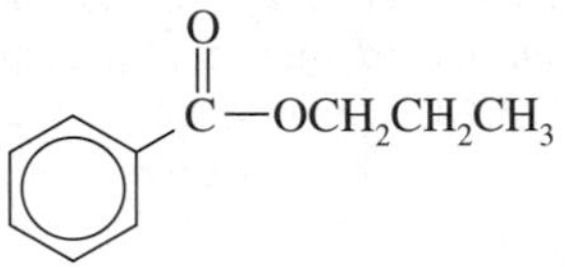

(c) $CH_3CH_2\overset{O}{\overset{\|}{C}}{-}OCH_2CH_3$

20. Write the structural formulas for each:
(a) propyl acetate (c) ethyl hexanoate
(b) methyl benzoate

22. Write the structural formula and name of the principal organic product for each of the following reactions:

(a) 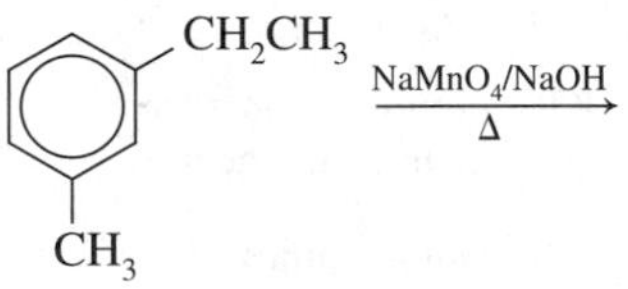 benzene ring bearing CH_2CH_3 and, in the meta position, CH_3 $\xrightarrow[\Delta]{NaMnO_4/NaOH}$

(b) $HOOCCH{=}CHCOOH + H_2 \xrightarrow{Ni}$

(c) $CH_3(CH_2)_7CH{=}CH(CH_2)_7COOH + NaOH \longrightarrow$

(d) $CH_3CH_2CH_2\overset{O}{\overset{\|}{C}}{-}H \xrightarrow[H_2SO_4]{Na_2Cr_2O_7}$

(e) 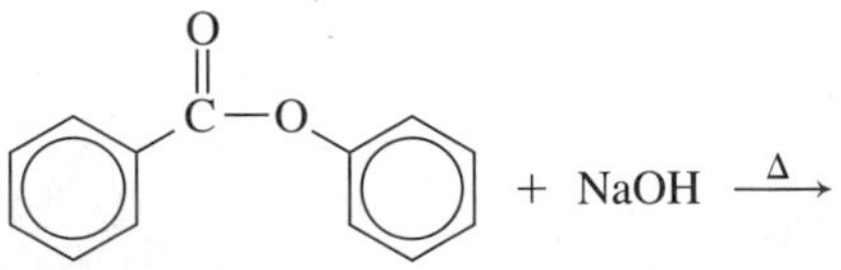 benzene ring bearing $\overset{O}{\overset{\|}{C}}{-}O{-}$ (bonded to a second benzene ring) $+ NaOH \xrightarrow{\Delta}$

24. Write the structural formula of the ester that when hydrolyzed would yield
(a) methanol and propanoic acid
(b) 1-octanol and acetic acid
(c) ethanol and butanoic acid

25. Write structural formulas for the reactants that will yield the following esters:
(a) methyl palmitate
(b) phenyl propionate
(c) dimethyl succinate

26. Write structural formulas for the reactants that will yield the following esters:
(a) isopropyl formate
(b) diethyl adipate
(c) benzyl benzoate

27. Write structural formulas for the organic products of the following reactions:

(a) 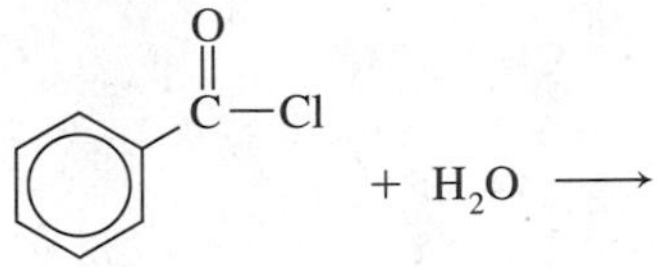 $C_6H_5C(=O)Cl + H_2O \longrightarrow$

(b) $CH_2{=}CHCOOH + Br_2 \longrightarrow$

(c) 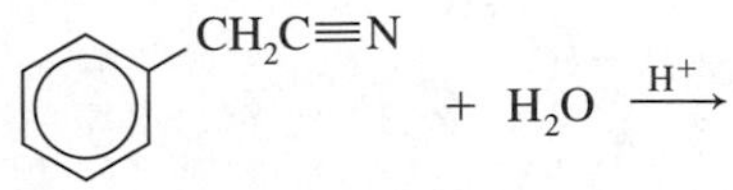 $C_6H_5CH_2C{\equiv}N + H_2O \xrightarrow{H^+}$

(d) $CH_3CH_2OH + CH_3C(=O){-}Cl \longrightarrow$

(e) $CH_3C(=O){-}Cl + NH_3 \longrightarrow$

28. Write structural formulas for the organic products of the following reactions:

(a) 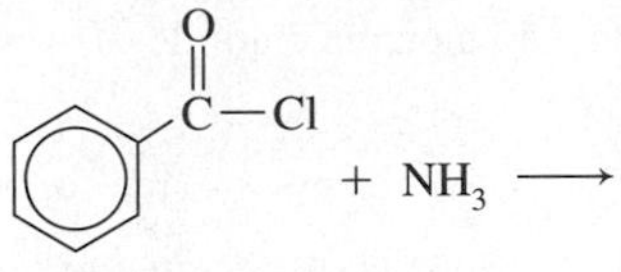 $C_6H_5C(=O){-}Cl + NH_3 \longrightarrow$

(b) $CH_3CH_2COOH + SOCl_2 \longrightarrow$

(c) $CH_3CH_2C(=O){-}Cl + CH_3OH \longrightarrow$

(d) $CH_3CH_2CH_2C{\equiv}N + H_2O \xrightarrow{H^+}$

(e) 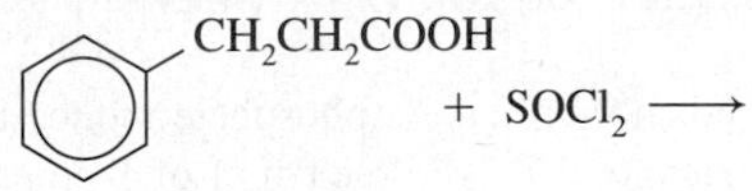 $C_6H_5CH_2CH_2COOH + SOCl_2 \longrightarrow$

29. What simple tests can be used to distinguish between the following pairs of compounds?
(a) benzoic acid and sodium benzoate
(b) maleic acid and malonic acid

30. What simple tests can be used to distinguish between the following pairs of compounds?
(a) benzoic acid and ethyl benzoate
(b) succinic acid and fumaric acid

31. Write structural formulas and names for the organic products of the following reactions:

(a) 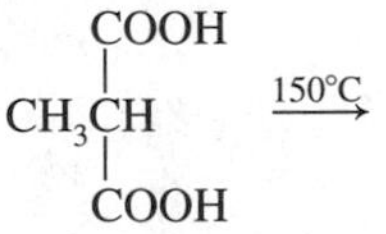 $CH_3CH(COOH)COOH \xrightarrow{150°C}$

(b) 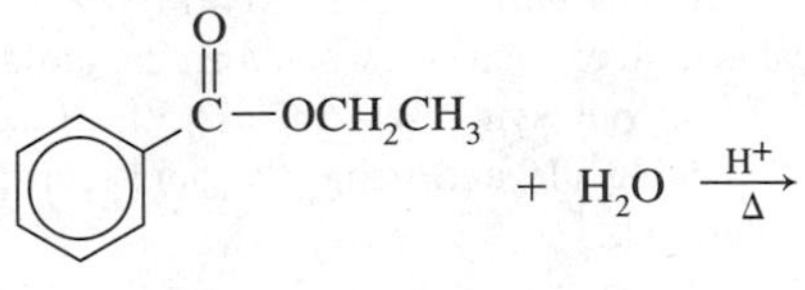 $C_6H_5C(=O){-}OCH_2CH_3 + H_2O \xrightarrow[\Delta]{H^+}$

(c) $CH_3CH_2COOH + C_6H_5OH \xrightarrow{H^+}$

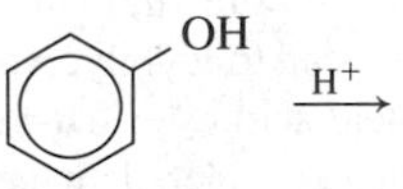

32. Write structural formulas and names for the organic products of the following reactions:

(a) 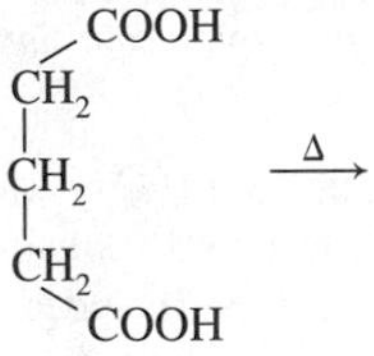 $HOOC{-}CH_2{-}CH_2{-}CH_2{-}COOH \xrightarrow{\Delta}$
glutaric acid

(b) 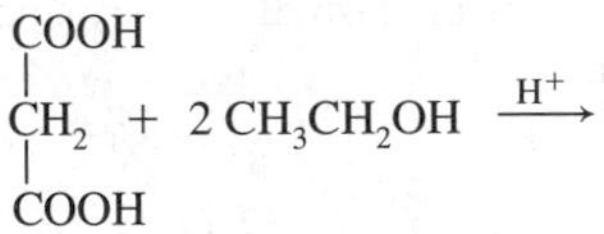 $HOOC{-}CH_2{-}COOH + 2\,CH_3CH_2OH \xrightarrow{H^+}$

(c) 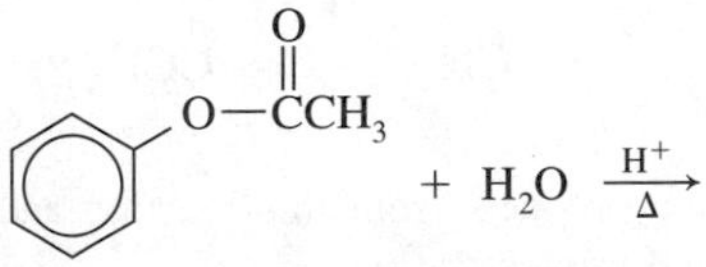 $C_6H_5O{-}C(=O)CH_3 + H_2O \xrightarrow[\Delta]{H^+}$

33. Trans isomers of naturally occurring fatty acids can act as metabolic inhibitors. Draw the structural formulas of *cis,cis,cis*-linolenic acid and its all-trans isomer.

34. The geometric configuration of naturally occurring unsaturated 18-carbon acids is all cis. Draw structural formulas for
(a) *cis*-oleic acid
(b) *cis,cis*-linoleic acid

35. Would $CH_3(CH_2)_{12}COOH$ or $CH_3(CH_2)_{12}COONa$ be the more useful cleansing agent in soft water? Explain.

36. Would $CH_3(CH_2)_{11}OSO_3Na$ (sodium lauryl sulfate) or $CH_3CH_2CH_2OSO_3Na$ (sodium propyl sulfate) be the more effective detergent in hard water? Explain.

37. Which one of the following substances is a good detergent in water? Is this substance a nonionic, anionic, or cationic detergent?

(a) $C_{16}H_{33}N(CH_3)_3^+Cl^-$
hexadecyltrimethyl ammonium chloride

(b) $C_{16}H_{34}$
hexadecane

(c) $C_{15}H_{31}COOH$
palmitic acid (hexadecanoic acid)

(d) $C_{15}H_{31}\overset{\overset{\displaystyle O}{\|}}{C}-OC_{16}H_{33}$
cetyl palmitate (hexadecyl hexadecanoate)

38. Which one of the following substances is a good detergent in water? Is this substance a nonionic, anionic, or cationic detergent?

(a) $CH_3(CH_2)_{11}O-\overset{\overset{\displaystyle O}{\|}}{C}(CH_2)_{14}CH_3$

(b) $HO-\overset{\overset{\displaystyle O}{\|}}{C}(CH_2)_{14}CH_3$

(c) $CH_3(CH_2)_{10}CH_2O(CH_2CH_2O)_7CH_2CH_2OH$

(d) 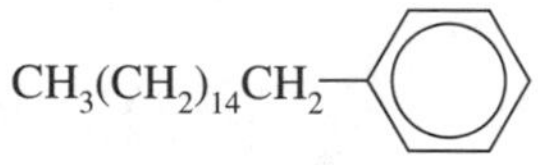 $CH_3(CH_2)_{14}CH_2-C_6H_5$

39. Show the products of the reaction of 1 mol of phosphoric acid with
(a) 1 mol of ethanol to make a phosphate monoester
(b) 1 mol of ethanol followed by 1 mol of 1-propanol to make a phosphate diester

40. Show the products of the reaction of 1 mol of phosphoric acid with
(a) 3 mol of methanol to make a phosphate ester
(b) 1 mol of methanol followed by 1 mol of 2-butanol to make a phosphate diester

Additional Exercises

These exercises are not paired or labeled by topic and provide additional practice on concepts covered in this chapter.

41. The FDA allows the addition of up to 0.1% by weight of sodium benzoate to baked goods to retard spoilage. How many grams of sodium benzoate can be included in 1 lb of dinner rolls (453.6 g = 1 lb)? How many moles?

42. Aspirin is synthesized on an industrial scale by combining the following reactants:

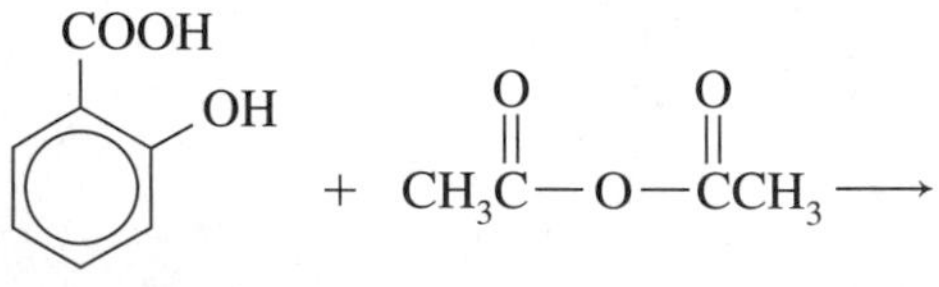

$$C_6H_4(COOH)(OH) + CH_3\overset{\overset{\displaystyle O}{\|}}{C}-O-\overset{\overset{\displaystyle O}{\|}}{C}CH_3 \longrightarrow$$

Show the organic products from this reaction and name both reactants and products.

43. Upon hydrolysis, an ester of formula $C_6H_{12}O_2$ yields an acid A and an alcohol B. When B is oxidized, it yields a product identical to A. What is the structure of the ester? Explain your answer.

44. The gas, phosgene,

$$Cl-\overset{\overset{\displaystyle O}{\|}}{C}-Cl$$

was used in chemical warfare during World War I. Today this chemical is used extensively when forming amide linkages during industrial syntheses. Identify the functional groups on this molecule and name the acid from which phosgene is derived.

45. Most plant oils are high in unsaturated fatty acids; the average soybean triacylglycerol has two linoleic acid esters and one oleic acid ester. However, there are several plant oils that are considered to be less healthy because they contain higher amounts of saturated fatty acids. For example, an average triacylglycerol from palm oil contains two palmitic acid esters and one oleic acid ester. Give structural formulas for both the soybean and the palm triacylglycerols.

46. Write the structural formula of a triacylglycerol that contains one unit each of lauric acid, palmitic acid, and oleic acid. How many other triacylglycerols, each containing all three of these acids, are possible?

47. Choose one of the triacylglycerols from Exercise 46. Write the names and formulas of all products expected when this triacylglycerol is
(a) reacted with water at high temperature and pressure in the presence of mineral acid
(b) reacted with hydrogen at relatively high pressure and temperature in the presence of a copper chromite catalyst
(c) boiled with potassium hydroxide
(d) reacted with hydrogen in the presence of Ni

48. Write a balanced chemical reaction for the reaction of KOH with glyceryl tristearate. Discuss any changes in solubility that might occur as a result of this reaction.

*49. Starting with ethyl alcohol as the only source of organic material and using any other reagents you desire, write equations to show the synthesis of
(a) acetic acid
(b) ethyl acetate
(c) β-hydroxybutyric acid

50. Waxes can be formed from the reaction between a fatty acid and a long-chain alcohol. A major component of beeswax is a combination of a 16-carbon acid and a 30-carbon alcohol. Write the condensed formula for this component of beeswax.

51. Compound A, C_7H_8, reacted with $KMnO_4$ to form compound B, $C_7H_6O_2$. Compound B then combined with methanol to yield a sweet smelling liquid, compound C. Compound A did not react with Br_2 in CCl_4. Write structural formulas of A, B, and C.

52. A good lipstick is uniform in color and sticks to the lips once it is applied. The stickiness of a lipstick can be controlled by the addition of a compound such as isopropyl myristate. Write the structural formula of isopropyl myristate.

53. An ester A of formula $C_{10}H_{12}O_2$ contains a benzene ring. When hydrolyzed, A gives an alcohol B, C_7H_8O, and an acid C, $C_3H_6O_2$. What are the structures of A, B, and C?

54. You have been given two unlabeled bottles, one of which contains butanoic acid and the other ethyl butanoate. Describe a simple test that you could do to determine which compound is in which bottle.

*55. If 1.00 kg of triolein (glycerol trioleate) is converted to tristearin (glycerol tristearate) by hydrogenation:
(a) How many liters of hydrogen (at STP) are required?
(b) What is the mass of the tristearin that is produced?

56. Consider this statement: "When methyl propanoate is hydrolyzed, formic acid and propanol are formed." If this statement is true, write out a balanced chemical reaction for it. If it is false, explain why.

57. Write structural formulas for all the esters that have a molecular formula of $C_5H_{10}O_2$.

58. What functional groups, if any, do aspirin, acetaminophen, and ibuprofen have in common?

Answers to Practice Exercises

24.1 (a) $CH_3CH(CH_3)COOH$ (b) $CH_3CH_2CH_2CHClCH_2COOH$ (c) cyclohexane–COOH

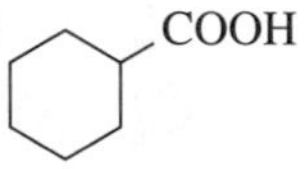

24.2 (a) $HC(=O)-OH$ (b) NH_2CH_2COOH (c) $HOCH_2COOH$ (d) CH_2BrCH_2COOH (e) $HO-C(=O)-C(=O)-OH$ (f) 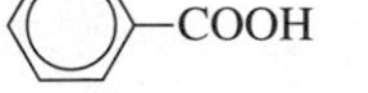 benzene–COOH

24.3 (a) caproic acid or hexanoic acid
(b) 2-hexanone

24.4 (a) 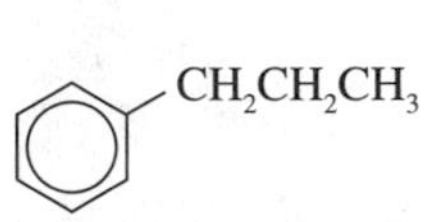 benzene–$CH_2CH_2CH_3$ (b) 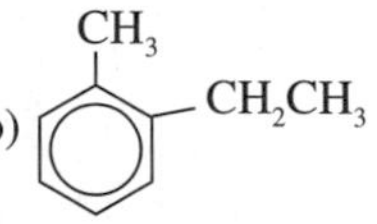 benzene with CH_3 and CH_2CH_3 (ortho)

24.5 $CH_3C(=O)-H + HCN \xrightarrow{OH^-} CH_3CH(OH)CN$

$CH_3CH(OH)CN + H_2O \xrightarrow{H^+} CH_3CH(OH)COOH + NH_4^+$
lactic acid

24.6 (a) methyl propanoate (b) phenyl ethanoate (phenyl acetate) (c) dimethyl succinate (d) methyl benzoate

24.7 (a) 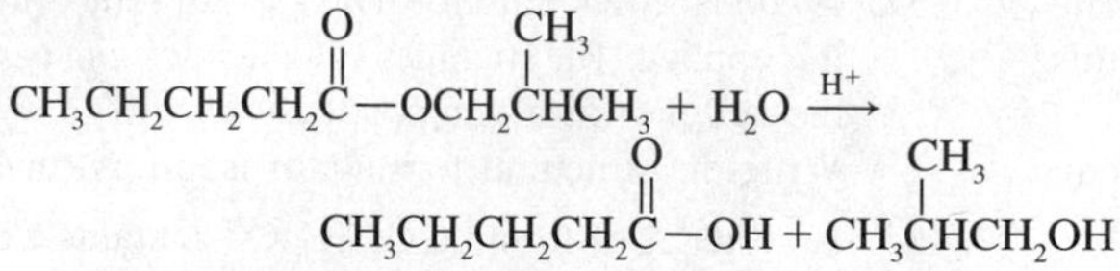 $CH_3CH_2CH_2CH_2\overset{\overset{\large O}{\|}}{C}-OCH_2\overset{\overset{\large CH_3}{|}}{C}HCH_3 + H_2O \xrightarrow{H^+}$

$$CH_3CH_2CH_2CH_2\overset{\overset{\large O}{\|}}{C}-OH + CH_3\overset{\overset{\large CH_3}{|}}{C}HCH_2OH$$

(b) 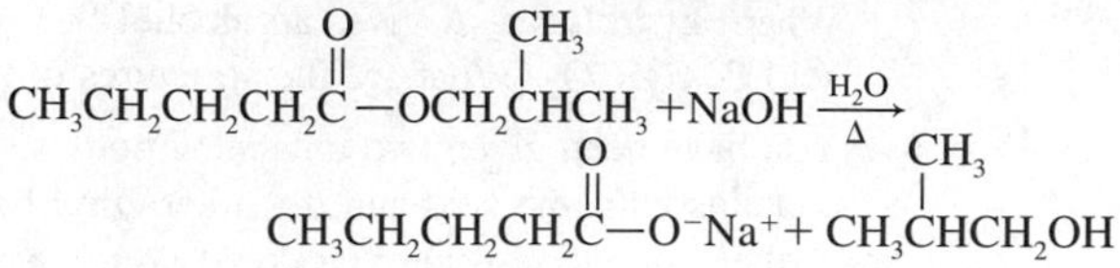 $CH_3CH_2CH_2CH_2\overset{\overset{\large O}{\|}}{C}-OCH_2\overset{\overset{\large CH_3}{|}}{C}HCH_3 + NaOH \xrightarrow[\Delta]{H_2O}$

$$CH_3CH_2CH_2CH_2\overset{\overset{\large O}{\|}}{C}-O^-Na^+ + CH_3\overset{\overset{\large CH_3}{|}}{C}HCH_2OH$$

24.8

$$\begin{array}{l} CH_2-O-\overset{\overset{\large O}{\|}}{C}(CH_2)_7CH{=}CH(CH_2)_7CH_3 \\ | \\ CH-O-\overset{\overset{\large O}{\|}}{C}(CH_2)_{12}CH_3 \\ | \\ CH_2-O-\overset{\overset{\large O}{\|}}{C}(CH_2)_{14}CH_3 \end{array}$$

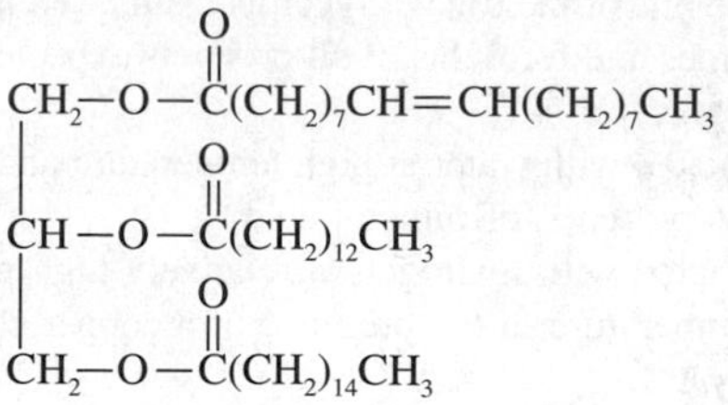

The sequence of the three esters in the triacylglycerol may vary.

Amides and Amines: Organic Nitrogen Compounds

25

Chapter 25

Many organic compounds contain nitrogen. Two major classes of nitrogen-containing compounds are amines and amides. Amines isolated from plants form a group of compounds called alkaloids. Thousands of alkaloids have been isolated. Many of these compounds exhibit physiological activity. Examples of common alkaloid compounds include quinine, used in the treatment of malaria; strychnine, a poison; morphine, a narcotic; and caffeine, a stimulant. Many other drugs are also nitrogen-containing compounds.

Amides are nitrogen derivatives of carboxylic acids. These compounds are found as polymers, both commercially as in nylon, and biologically as in proteins. An understanding of the chemistry of organic nitrogen compounds is the cornerstone of genetics and is essential to unlocking the chemical secrets of living organisms.

25.1 Amides: Nomenclature and Physical Properties

Carboxylic acids react with ammonia to form ammonium salts:

$$\underset{\text{carboxylic acid}}{\mathrm{R}\overset{\overset{\displaystyle O}{\|}}{C}-\mathrm{OH}} + \underset{\text{ammonia}}{\mathrm{NH_3}} \longrightarrow \underset{\text{ammonium salt}}{\mathrm{R}\overset{\overset{\displaystyle O}{\|}}{C}-\mathrm{O^-\,NH_4^+}}$$

$$\underset{\text{acetic acid}}{\mathrm{CH_3}\overset{\overset{\displaystyle O}{\|}}{C}-\mathrm{OH}} + \mathrm{NH_3} \longrightarrow \underset{\text{ammonium acetate}}{\mathrm{CH_3}\overset{\overset{\displaystyle O}{\|}}{C}-\mathrm{O^-\,NH_4^+}}$$

Ammonium salts of carboxylic acids are ionic substances. Ammonium acetate, for example, is ionized and exists as ammonium ions and acetate ions, both in the crystalline form and when dissolved in water.

When heated, ammonium salts of carboxylic acids lose a molecule of water and are converted to *amides*:

$$\underset{\text{ammonium salt}}{\mathrm{R}-\overset{\overset{\displaystyle O}{\|}}{C}-\mathrm{O^-\,NH_4^+}} \xrightarrow{\Delta} \underset{\text{amide}}{\mathrm{R}\overset{\overset{\displaystyle O}{\|}}{C}-\mathrm{NH_2}} + \mathrm{H_2O}$$

amide

$$\underset{\text{ammonium acetate}}{\mathrm{CH_3}\overset{\overset{\displaystyle O}{\|}}{C}-\mathrm{O^-\,NH_4^+}} \xrightarrow{\Delta} \underset{\substack{\text{ethanamide}\\ \text{(acetamide)}}}{\mathrm{CH_3}\overset{\overset{\displaystyle O}{\|}}{C}-\mathrm{NH_2}} + \mathrm{H_2O}$$

◀ **Chapter Opening Photo: Nylon is one of the materials used to give these colorful sails their strength and durability.**

Amides are neutral (nonbasic) molecular substances and exist as molecules (not ions) both in the crystalline form and when dissolved in water. An amide contains the following characteristic structure:

$$R-\overset{\overset{\displaystyle O}{\|}}{C}-NH_2$$

amide structure

In amides the carbon atom of a carbonyl group is bonded directly to a nitrogen atom of an $-NH_2$, $-NHR$, or $-NR_2$ group. The amide structure occurs in numerous substances, including proteins and some synthetic polymers such as nylon.

Two systems are used to name amides.

IUPAC Nomenclature for Amides

1. The IUPAC name is based on the longest carbon chain that includes the amide group.
2. Drop the *-oic acid* ending from the corresponding IUPAC acid name.
3. Add the suffix *-amide.*

methanoic acid becomes methanamide

$HCOOH$ $\qquad$ $H\overset{\overset{\displaystyle O}{\|}}{C}-NH_2$

ethanoic acid becomes ethanamide

CH_3COOH $\qquad$ $CH_3\overset{\overset{\displaystyle O}{\|}}{C}-NH_2$

In a like manner, the common names for amides are formed from the common names of the corresponding carboxylic acids by dropping the *ic* or *oic acid* ending and adding the suffix *amide*. Thus

formic acid becomes formamide,

$HCOOH$ $\qquad$ $H\overset{\overset{\displaystyle O}{\|}}{C}-NH_2$

butyric acid becomes butyramide,

$CH_3CH_2CH_2COOH$ $\qquad$ $CH_3CH_2CH_2\overset{\overset{\displaystyle O}{\|}}{C}-NH_2$

benzoic acid becomes benzamide,

C_6H_5-COOH $\qquad$ $C_6H_5-\overset{\overset{\displaystyle O}{\|}}{C}-NH_2$

When the nitrogen of an amide is connected to an alkyl or aryl group, the group is named as a prefix preceded by the letter *N*.

TABLE 25.1 Formulas and Names of Selected Amides

Formula	IUPAC name	Common name
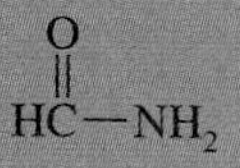 $HC(=O)-NH_2$	Methanamide	Formamide
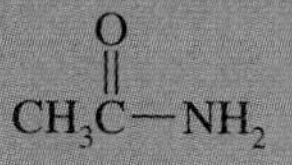 $CH_3C(=O)-NH_2$	Ethanamide	Acetamide
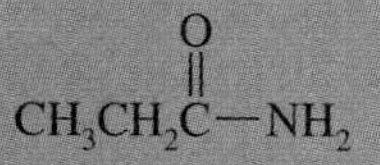 $CH_3CH_2C(=O)-NH_2$	Propanamide	Propionamide
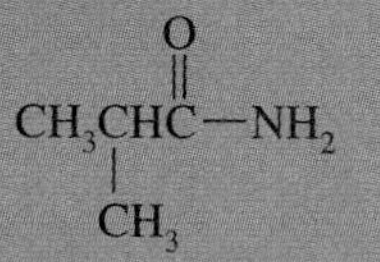 $CH_3CH(CH_3)C(=O)-NH_2$	2-Methylpropanamide	Isobutyramide
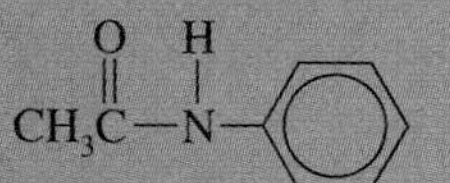 $CH_3C(=O)-N(H)-C_6H_5$	*N*-phenylethanamide	Acetanilide
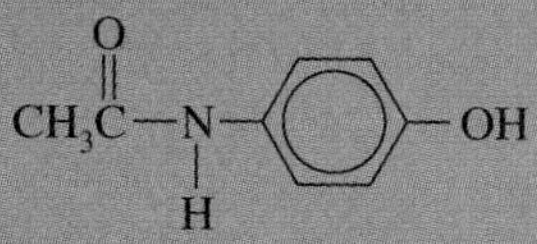 $CH_3C(=O)-N(H)-C_6H_4-OH$	*N*-*p*-hydroxyphenylethanamide	Acetaminophen
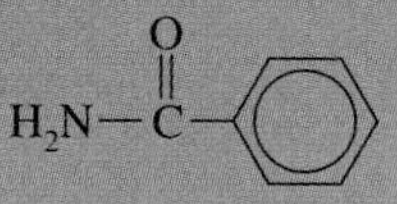 $H_2N-C(=O)-C_6H_5$	Benzamide	Benzamide
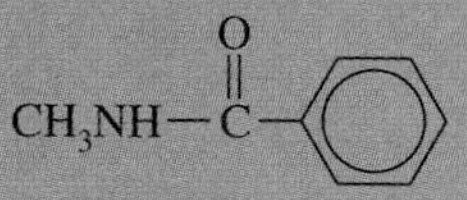 $CH_3NH-C(=O)-C_6H_5$	*N*-methylbenzamide	*N*-methylbenzamide

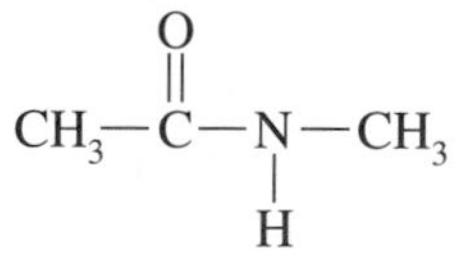

$CH_3-C(=O)-N(H)-CH_3$ is *N*-methylacetamide

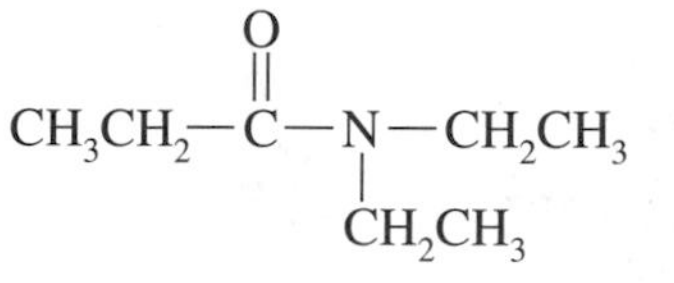

$CH_3CH_2-C(=O)-N(CH_2CH_3)-CH_2CH_3$ is *N,N*-diethylpropionamide

Formulas and names of selected amides are shown in Table 25.1.

Practice 25.1

Write formulas for

(a) *N*-isopropylmethanamide
(b) *N*-ethylbutanamide
(c) *N*-methyl-2-methylpropanamide
(d) *N,N*-diethylhexanamide
(e) *p*-aminobenzamide

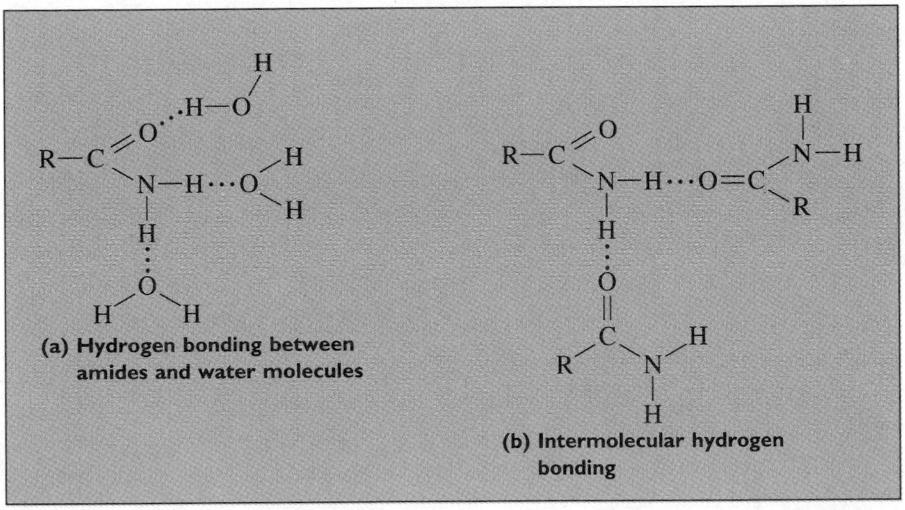

R—C(=O···H—O—H)—N(—H···OH₂)—H···OH₂

(a) Hydrogen bonding between amides and water molecules

R—C(=O)—N(—H···O=C(R)—NH₂)—H···O=C(R)—NH₂

(b) Intermolecular hydrogen bonding

FIGURE 25.1
Hydrogen bonding in amides.

Except for formamide, a liquid, all other unsubstituted amides are solids at room temperature. Many are odorless and colorless. Low-molar-mass amides are soluble in water, but solubility decreases quickly as molar mass increases. The amide functional group is polar, and nitrogen is capable of hydrogen bonding. The solubility of these molecules and their exceptionally high melting points and boiling points are the result of this polarity and hydrogen bonding between molecules, shown in Figure 25.1.

25.2 Chemical Properties of Amides

One of the more important reactions of amides is hydrolysis. This type of reaction is analogous to the hydrolysis of carboxylic acid esters. The amide is cleaved into two parts, the carboxylic acid portion and the nitrogen-containing portion. As in ester hydrolysis, this reaction requires the presence of a strong acid or a strong base for it to occur in the laboratory. Amide hydrolysis is accomplished in living systems during the degradation of proteins by enzymatic reactions under much milder conditions (Chapter 31). Hydrolysis of an unsubstituted amide in an acid solution produces a carboxylic acid and an ammonium salt.

$$CH_3CH_2CH_2\overset{\overset{\displaystyle O}{\|}}{C}{-}NH_2 + H_2O + HCl \xrightarrow{\Delta} CH_3CH_2CH_2\overset{\overset{\displaystyle O}{\|}}{C}{-}OH + NH_4Cl$$

Basic hydrolysis results in the production of ammonia and the salt of a carboxylic acid.

$$CH_3{-}\overset{\overset{\displaystyle O}{\|}}{C}{-}NH_2 + NaOH \longrightarrow CH_3\overset{\overset{\displaystyle O}{\|}}{C}{-}O^-\,Na^+ + NH_3$$

Example 25.1 Show the products of (a) the acid hydrolysis and (b) the basic hydrolysis of

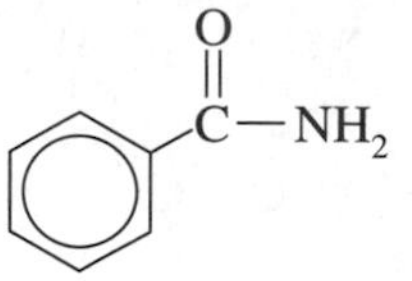

$$C_6H_5-\overset{O}{\overset{\|}{C}}-NH_2$$

Solution

(a) In acid hydrolysis the C—N bond is cleaved, and the carboxylic acid is formed. The $-NH_2$ group is converted into an ammonium ion.

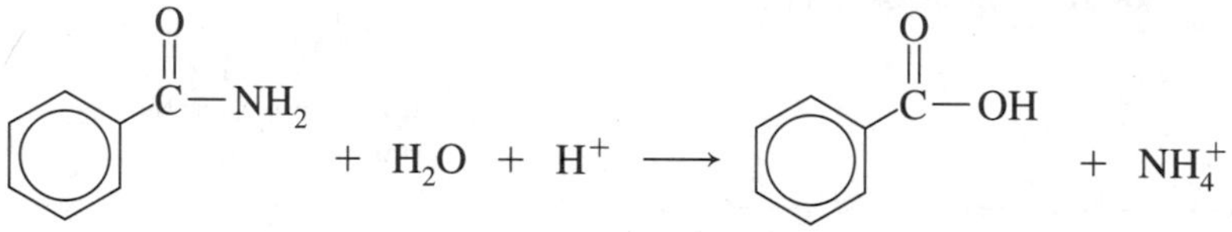

$$C_6H_5-\overset{O}{\overset{\|}{C}}-NH_2 + H_2O + H^+ \longrightarrow C_6H_5-\overset{O}{\overset{\|}{C}}-OH + NH_4^+$$

(b) In basic solution the C—N bond is also cleaved, but since the solution is basic, the salt of the carboxylic acid is formed along with ammonia.

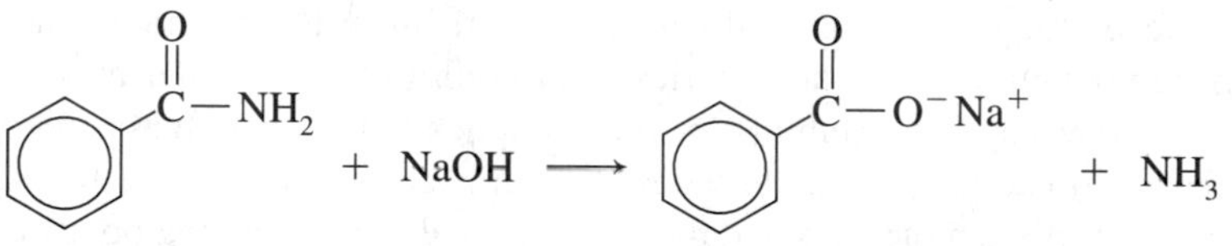

$$C_6H_5-\overset{O}{\overset{\|}{C}}-NH_2 + NaOH \longrightarrow C_6H_5-\overset{O}{\overset{\|}{C}}-O^-Na^+ + NH_3$$

Practice 25.2

Give the products of the acidic and basic hydrolysis of

(a) $CH_3CH_2\overset{O}{\overset{\|}{C}}-NH_2$ (b) $CH_3\overset{O}{\overset{\|}{C}}-N(CH_3)_2$

25.3 Urea

Hydrolysis of proteins in the digestion process involves cleavage of the amide linkage between adjacent amino acids (Chapter 30). The process begins in the acidic environment of the stomach, and hydrolysis continues in the small intestine.

The body disposes of nitrogen by the formation of a diamide known as urea.

$$H_2N-\overset{O}{\overset{\|}{C}}-NH_2$$

urea

Urea is a white solid that melts at 133°C. It is soluble in water and therefore is excreted from the body in the urine. The normal adult excretes about 30 g of urea daily.

Urea is a common commercial product as well. It is widely used in fertilizers to add nitrogen to the soil, or as a starting material in the production of plastics and barbiturates.

25.4 Amines: Nomenclature and Physical Properties

amine

An **amine** is a substituted ammonia molecule with basic properties and has the general formula RNH_2, R_2NH, or R_3N, where R is an alkyl or an aryl group. Amines are classified as primary (1°), secondary (2°), or tertiary (3°), depending on the number of hydrocarbon groups attached to the nitrogen atom. Some examples include the following:

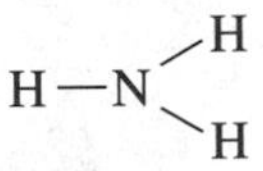

$H-NH_2$ (H—N bonded to two H)
ammonia

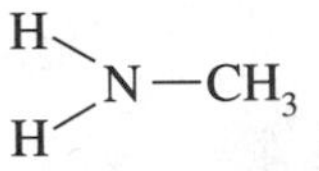

H_2N-CH_3
methylamine
(1° amine)

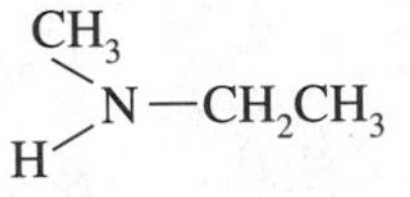

$(CH_3)(H)N-CH_2CH_3$
methylethylamine
(2° amine)

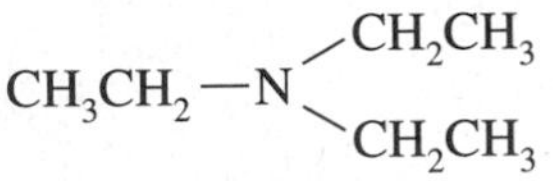

$CH_3CH_2-N(CH_2CH_3)_2$
triethylamine
(3° amine)

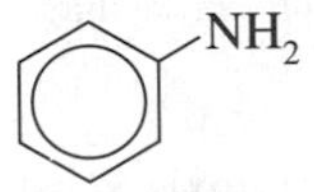

$C_6H_5-NH_2$
aniline
(1° amine)

In the IUPAC System, $—NH_2$ is called an amino group, and amines are named as amino-substituted hydrocarbons using the longest carbon chain as the parent compound (e.g., $CH_3CH_2CH_2NH_2$ is called 1-aminopropane).

Here are some examples:

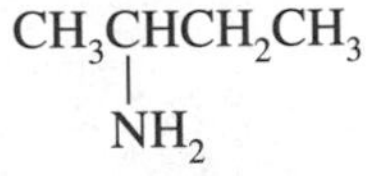

$CH_3CH(NH_2)CH_2CH_3$
2-aminobutane

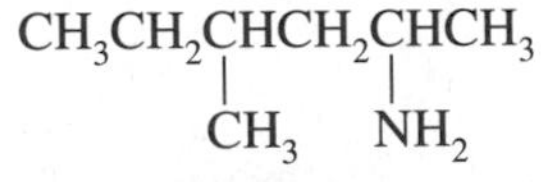

$CH_3CH_2CH(CH_3)CH_2CH(NH_2)CH_3$
2-amino-4-methylhexane

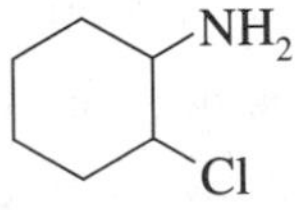

Cyclohexane ring with NH_2 and adjacent Cl
1-amino-2-chlorocyclohexane

Simple amines are most often referred to by their common names. The names for aliphatic amines are formed by naming the alkyl group or groups attached to the nitrogen atom followed by the ending *amine*. Thus CH_3NH_2 is methylamine, $(CH_3)_2NH$ is dimethylamine, and $(CH_3)_3N$ is trimethylamine.

The most important aromatic amine is aniline ($C_6H_5NH_2$). Derivatives are named as substituted anilines. To identify a substituted aniline in which the substituent group is attached to the nitrogen atom, an *N*- is placed before the group name to indicate that the substituent is bonded to the nitrogen atom and not to a carbon atom in the ring. For example, the following compounds are called aniline, *N*-methylaniline, and *N*,*N*-dimethylaniline:

Urea is one of the major sources of nitrogen in this rose food. It stimulates growth while supplying essential nitrogen.

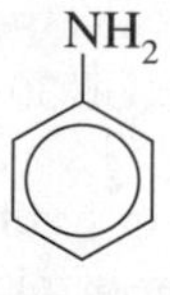

aniline

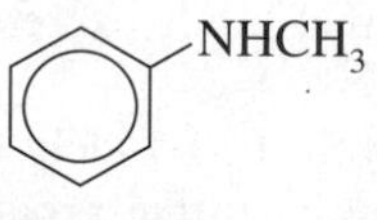

N-methylaniline

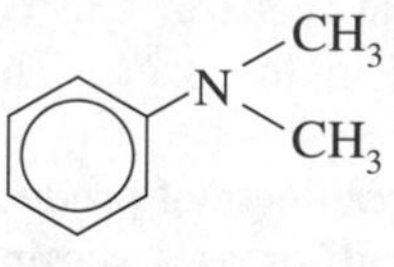

N,N-dimethylaniline

When a group is substituted for a hydrogen atom in the ring, the resulting ring-substituted aniline is named as we have previously done for naming aromatic compounds. The monomethyl ring-substituted anilines are known as toluidines. Study the names for the following substituted anilines:

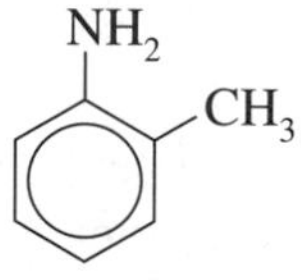

o-toluidine
(*o*-methylaniline)

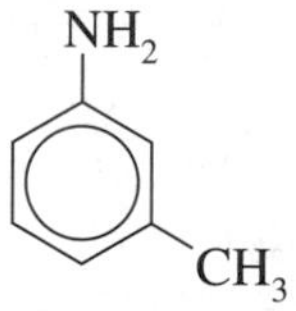

m-toluidine
(*m*-methylaniline)

p-toluidine
(*p*-methylaniline)

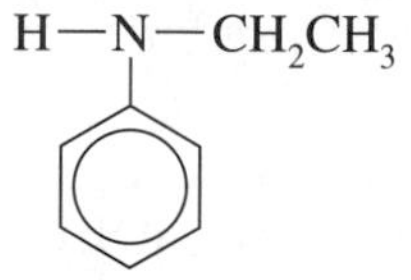

N-ethylaniline

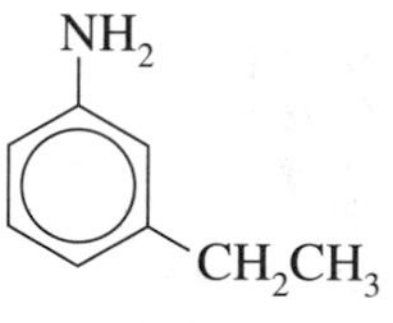

m-ethylaniline

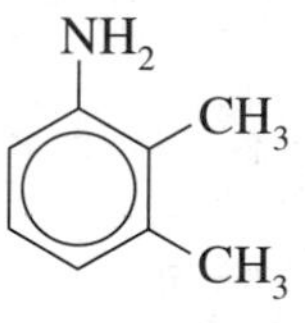

2,3-dimethylaniline

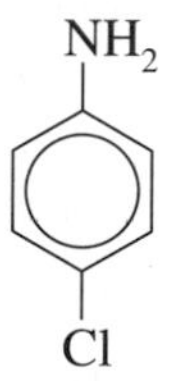

p-chloroaniline

Physiologically, aniline is a toxic substance. It is easily absorbed through the skin and affects both the blood and the nervous system. Aniline reduces the oxygen-carrying capacity of the blood by converting hemoglobin to methemoglobin. Methemoglobin is the oxidized form of hemoglobin in which the iron has gone from a +2 to a +3 oxidation state.

Example 25.2 Name the two compounds given:

(a) 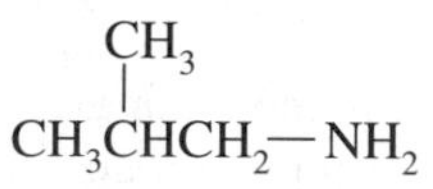

(b) 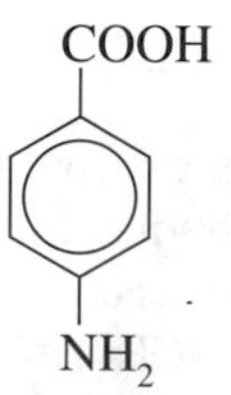

Solution

(a) The alkyl group attached to NH_2 is an isobutyl group. Thus the common name is isobutylamine. The longest chain containing the NH_2 has three carbons. Therefore, the parent carbon chain is propane, and the compound is called 1-amino-2-methylpropane.

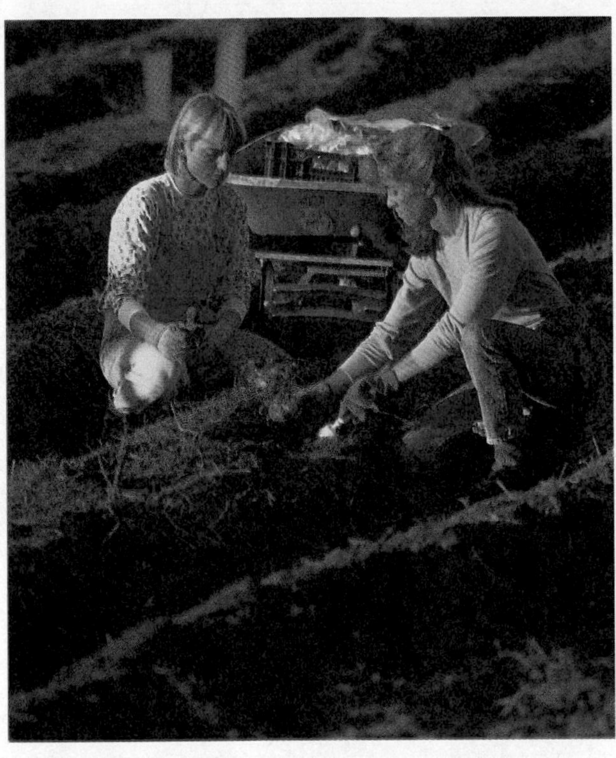

◀ The smell associated with fertilizers or decaying plant matter is characteristic of nitrogen-containing compounds. These women are adding nitrogen-containing compounds to the soil by mulching and fertilizing.

(b) The parent compound on which the name is based is benzoic acid. With an amino group in the para position, the name is *p*-aminobenzoic acid. The acronym for *p*-aminobenzoic acid is PABA. Esters of PABA are some of the most effective ultraviolet screening agents and are used in suntanning lotions. PABA is a component of folic acid, an essential vitamin for humans.

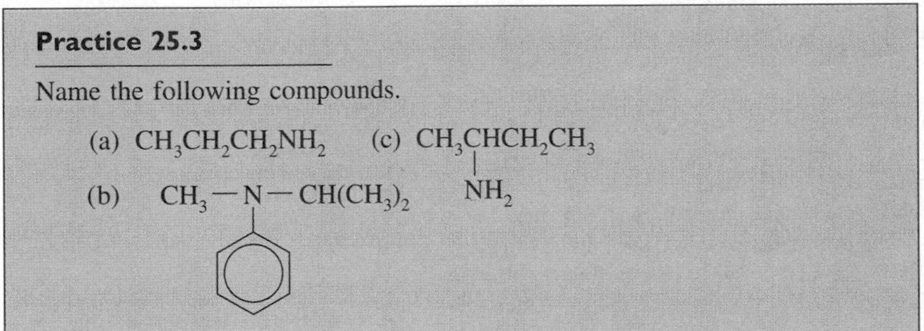

Practice 25.3

Name the following compounds.

(a) $CH_3CH_2CH_2NH_2$

(b) $CH_3-N-CH(CH_3)_2$ (with a benzene ring attached to N)

(c) $CH_3CHCH_2CH_3$ (with NH_2 attached to the second carbon)

Ring compounds in which all the atoms in the ring are not alike are known as **heterocyclic compounds**. The most common heteroatoms are oxygen, nitrogen, and sulfur. A number of the nitrogen-containing heterocyclic compounds are present in naturally occurring biological substances such as DNA, which controls heredity. The structural formulas of several nitrogen-containing heterocyclics are as follows:

heterocyclic compound

Coffee Talk

Coffee is one of the most widely consumed beverages in the world. Coffee drinkers rely on it to supply them a regular dose of caffeine, a stimulant sometimes called the world's most popular drug. Caffeine is an alkaloid.

H_3C O CH_3 N N N N O CH_3

caffeine

Caffeine is found in the seeds and leaves of certain plants especially trees or shrubs of the genus *Coffea*, widely cultivated in the tropics.

A number of foods and drugs including chocolate, soft drinks, tea, and pain relievers contain caffeine, but society mainly gets its caffeine from coffee.

The primary effect of caffeine occurs within 30–60 minutes after its consumption when caffeine levels peak in the bloodstream to produce alertness and heightened concentration. Caffeine also postpones exhaustion and augments the capacity for physical work. It is precisely this last effect that has caused the U.S. Olympic committee to consider caffeine a "performance enhancer" and to routinely screen athletes for the drug. Caffeine also constricts blood vessels, relieving some headaches, which is why it is found in some pain relievers. Too much caffeine results in "coffee nerves," which may include lightheadedness, irritability, unsteady hands, and diarrhea. Excess caffeine can also cause insomnia.

Caffeine used on a regular basis results in tolerance and dependence. Tolerance can develop very quickly, usually within days. Dependence is not a function of large amounts of caffeine. Researchers at Johns Hopkins University found signs of dependence in people who drank less than 3 cups of coffee per day. Caffeine-dependent people who try to quit the coffee habit are miserable; they suffer headaches, nausea, and runny noses, and, in the worst cases, vomiting.

Coffee houses provide an atmosphere for relaxation and conversation.

Studies over the last two decades have shown that coffee doesn't trigger heart attacks or raise blood cholesterol levels, at least for Americans. Americans tend to drink filtered coffee, which eliminates oils and other substances thought to trigger cholesterol problems. Caffeine does, however, raise blood pressure slightly in some people. In 1981, a study linked coffee to cancer of the pancreas, but since then several extensive studies have found no correlation between coffee and a variety of cancers.

Women who are pregnant or trying to become pregnant may want to carefully control their caffeine intake. Studies have found a link between the amount of caffeine consumed and infertility. A Yale University study (1993) showed 300 mg of caffeine per day reduced a woman's chance of conceiving by 25%.

Caffeine is not an innocuous drug. It acts powerfully on both the mind and the body, but it does not often pose a serious health risk.

HOW MUCH CAFFEINE?

	Caffeine Content (mg)	
Item	**Average**	**Range**
Hot drinks (per 8 fluid oz)		
Drip-brewed coffee	100	70–215
Instant coffee	70	35–169
Tea, U.S. brands	50	25–110
Cocoa	5	2–25
Decaf coffee (brewed and instant)	4	2–8
Soft drinks (per 12 fluid oz)		
Mountain Dew	54	
Coca-Cola (Diet, regular)	46	
Dr. Pepper	41	
Pepsi-Cola (Diet, regular)	35	
Chocolate (per 1 oz)		
Dark chocolate	20	5–35
Milk chocolate	6	1–15
Over-the-counter drugs (per tablet or capsule)		
Vivarin	200	
No Doz	100	
Excedrin	65	
Anacin	32	

Sources: U.S. Food and Drug Administration; soft drink and instant coffee manufacturers.

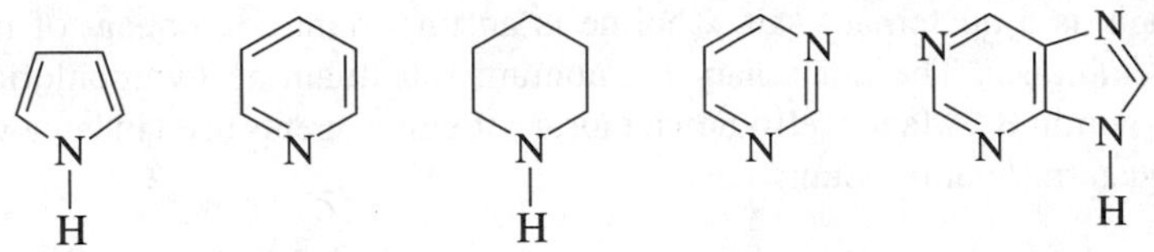

pyrrole (C_4H_5N)	pyridine (C_5H_5N)	piperidine ($C_5H_{11}N$)	pyrimidine ($C_4H_4N_2$)	purine ($C_5H_4N_4$)

Amines are capable of hydrogen bonding with water. As a result, the aliphatic amines with up to six carbons are quite soluble in water. Methylamine and ethylamine are flammable gases with a strong ammoniacal odor. Trimethylamine has a "fishy" odor. Higher-molar-mass amines have obnoxious odors. The foul odors arising from dead fish and decaying flesh are due to amines released by bacterial decomposition. Two of these compounds are diamines, 1,4-butanediamine and 1,5-pentanediamine. Each compound contains two amino groups:

$$H_2NCH_2CH_2CH_2CH_2NH_2 \qquad H_2NCH_2CH_2CH_2CH_2CH_2NH_2$$

1,4-butanediamine (putrescine) — 1,5-pentanediamine (cadaverine)

Simple aromatic amines are all liquids or solids. They are colorless or almost colorless when freshly prepared but become dark brown or red when exposed to air and light.

25.5 Preparation of Amines

Alkylation of Ammonia and Amines The substitution of alkyl groups for hydrogen atoms of ammonia can be done by reacting ammonia with alkyl halides. Thus, in successive reactions, a primary, a secondary, and a tertiary amine can be formed.

$$NH_3 \xrightarrow{CH_3Br} \underset{\substack{\text{methylamine} \\ (1°)}}{CH_3NH_2} \xrightarrow{CH_3Br} \underset{\substack{\text{dimethylamine} \\ (2°)}}{(CH_3)_2NH} \xrightarrow{CH_3Br} \underset{\substack{\text{trimethylamine} \\ (3°)}}{(CH_3)_3N}$$

Tertiary amines can be further alkylated so that there are four organic groups bonded to the nitrogen atom. Compounds of this type are called **quaternary ammonium salts**. For example,

quaternary ammonium salt

$$CH_3{-}\ddot{N}(CH_3){-}CH_3 + CH_3Br \longrightarrow CH_3{-}N^{+}(CH_3)_2{-}CH_3\ Br^{-}$$

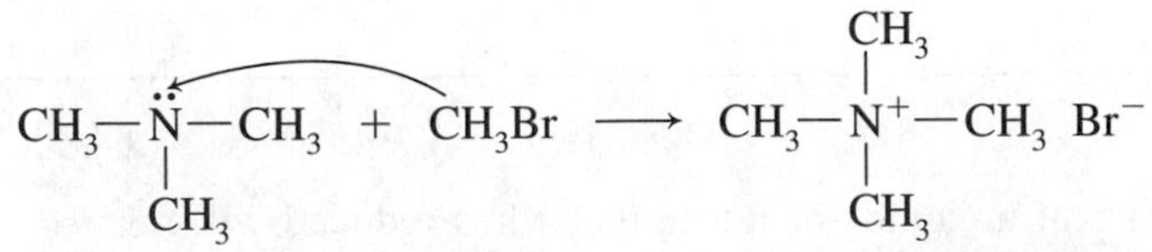

tetramethylammonium bromide

Quaternary ammonium salts are well known in biologically active compounds and in many popular medicinals. For example, acetylcholine, an active neurotransmitter in the brain, is a quaternary salt. Choline is an important component of many biological membranes. The quaternary ammonium salt thiamine hydrochloride is marketed as vitamin B_1. Many well-known fabric softening agents used in laundering clothes are quaternary ammonium salts.

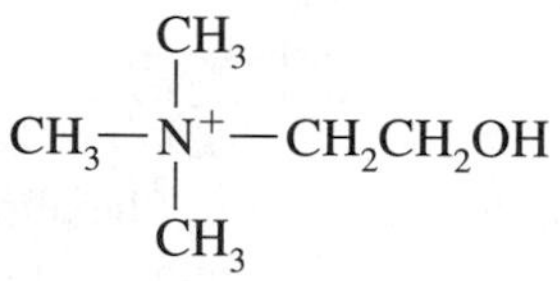

$$CH_3-\overset{\overset{\large CH_3}{|}}{\underset{\underset{\large CH_3}{|}}{N^+}}-CH_2CH_2OH \qquad CH_3-\overset{\overset{\large CH_3}{|}}{\underset{\underset{\large CH_3}{|}}{N^+}}-CH_2CH_2O\underset{\underset{\large O}{\|}}{C}CH_3$$

choline 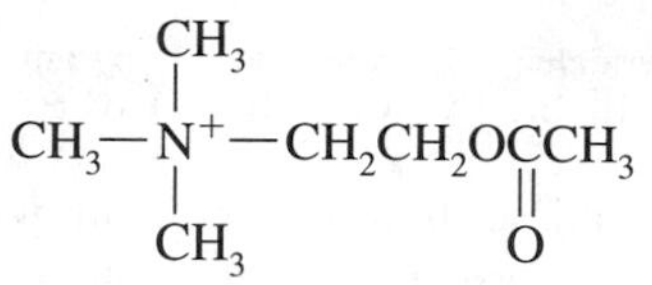 — acetylcholine

Reduction of Amides and Nitriles Amides can be reduced with lithium aluminum hydride to give amines. For example, acetamide can be reduced to ethylamine; and when *N,N*-diethylacetamide is reduced, triethylamine is formed:

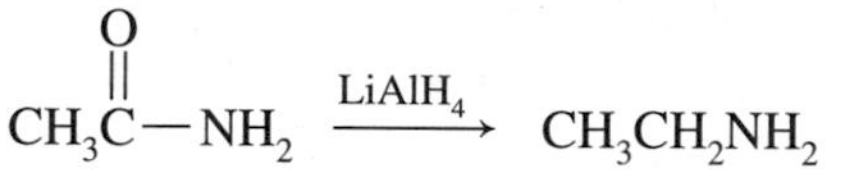

$$CH_3\overset{\overset{\large O}{\|}}{C}-NH_2 \xrightarrow{LiAlH_4} \underset{\text{ethylamine}}{CH_3CH_2NH_2}$$

$$CH_3\overset{\overset{\large O}{\|}}{C}-N\begin{matrix} \diagup CH_2CH_3 \\ \diagdown CH_2CH_3 \end{matrix} \xrightarrow{LiAlH_4} \underset{\text{triethylamine}}{(CH_3CH_2)_3N}$$

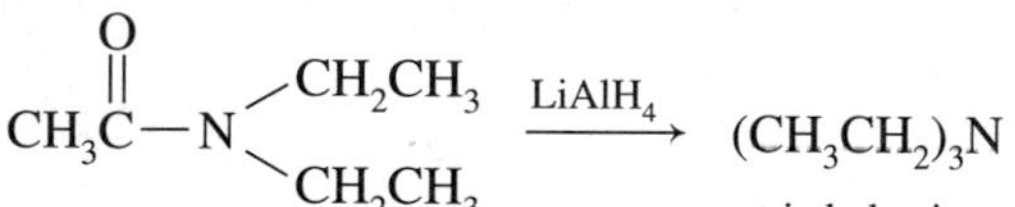

Nitriles, RCN, are also reducible to amines using hydrogen and a metal catalyst:

$$\underset{\text{propionitrile}}{CH_3CH_2C{\equiv}N} \xrightarrow{H_2/Ni} \underset{n\text{-propylamine}}{CH_3CH_2CH_2NH_2}$$

Reduction of Aromatic Nitro Compounds Aniline, the most widely used aromatic amine, is made by reducing nitrobenzene. The nitro group can be reduced by several reagents; Fe and HCl, or Sn and HCl, are commonly used.

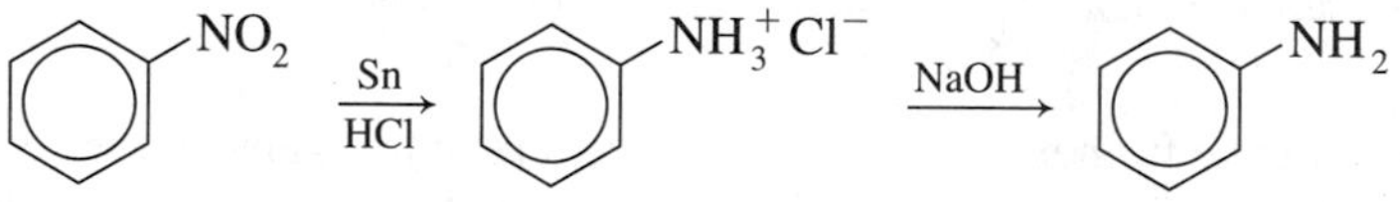

$$C_6H_5NO_2 \xrightarrow[HCl]{Sn} C_6H_5NH_3^+Cl^- \xrightarrow{NaOH} C_6H_5NH_2$$

Practice 25.4

Write the structure of an amide or nitrile that when reduced will form
(a) methylethylamine
(b) 1-amino-2-methylpropane
(c) 1-amino-3-ethyl-2-methylpentane

25.6 Chemical Properties of Amines

Alkaline Properties of Amines

In many respects, amines resemble ammonia in their reactions. Thus amines are bases and, like ammonia, produce OH^- ions in water:

$$\ddot{N}H_3 + (H)OH \rightleftharpoons NH_4^+ + OH^-$$

ammonia molecule ammonium ion hydroxide ion

Methylamine and aniline react in the same manner:

$$CH_3\ddot{N}H_2 + (H)OH \rightleftharpoons CH_3NH_3^+ + OH^-$$

methylamine molecule methylammonium ion hydroxide ion

$$C_6H_5\ddot{N}H_2 + (H)OH \rightleftharpoons C_6H_5\overset{+}{N}H_3 + OH^-$$

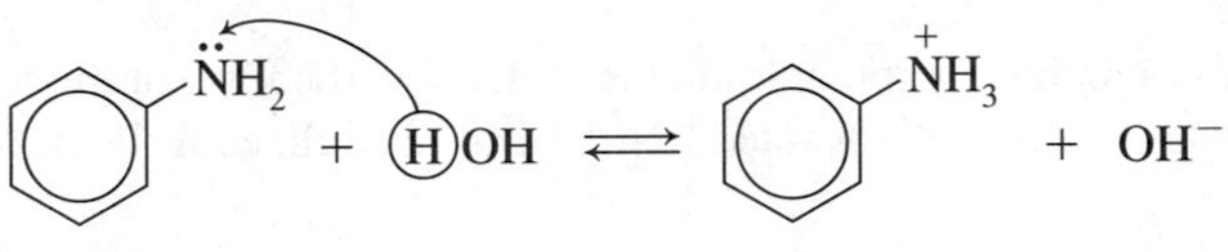

aniline molecule anilinium ion hydroxide ion

The ions formed are substituted ammonium ions. They are named by replacing the amine ending by ammonium and, for the aromatic amines, by replacing the aniline name by anilinium.

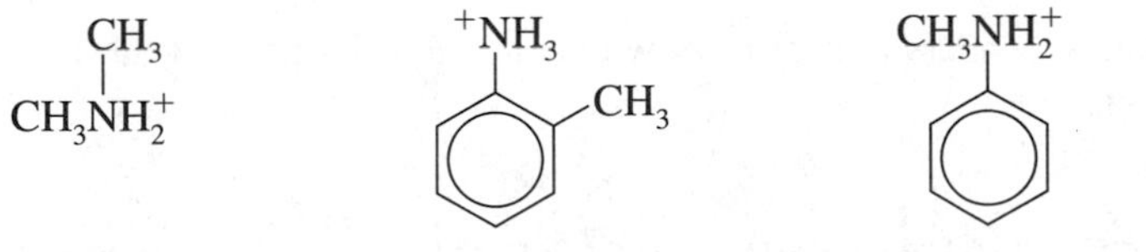

$(CH_3)CH_3NH_2^+$ — dimethylammonium ion

$^+NH_3$ and CH_3 on adjacent benzene ring positions — *o*-methylanilinium ion

$CH_3NH_2^+$ on benzene ring — *N*-methylanilinium ion

Like ammonia, amines are weak bases. Methylamine is a slightly stronger base than ammonia, and aniline is considerably weaker than ammonia. The pH values for 0.1 *M* solutions are: methylamine, 11.8; ammonia, 11.1; and aniline, 8.8.

Because amine groups form substituted ammonium ions under physiological conditions, they can provide the positive charge for biological molecules. For example, neurotransmitters are often positively charged. The structures of two such transmitter compounds, dopamine and serotonin, are shown here:

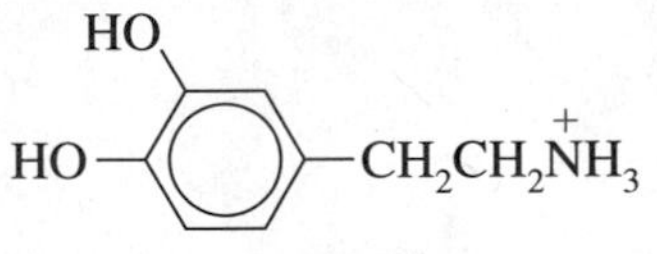

HO, HO on benzene ring — $CH_2CH_2\overset{+}{N}H_3$

dopamine

HO on indole ring (N—H) — $CH_2CH_2\overset{+}{N}H_3$

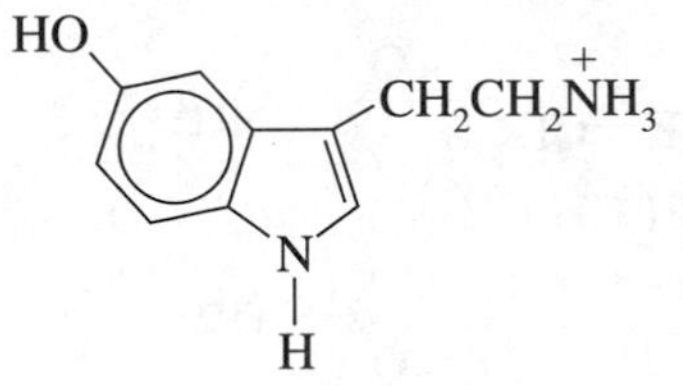

serotonin

Salt Formation

An amine reacts with a strong acid to form a salt; for example, methylamine and hydrogen chloride react in this fashion:

$$\underset{\text{methylamine molecule}}{CH_3NH_2(g)} + \underset{\text{hydrogen chloride molecule}}{HCl(g)} \longrightarrow \underset{\text{methylammonium chloride (salt)}}{CH_3\overset{+}{N}H_3\ Cl^-}$$

Methylammonium chloride is made up of methylammonium ions, $CH_3\overset{+}{N}H_3$, and chloride ions, Cl^-. It is a white crystalline salt that in physical appearance resembles ammonium chloride very closely.

Aniline reacts in a similar manner, forming anilinium chloride:

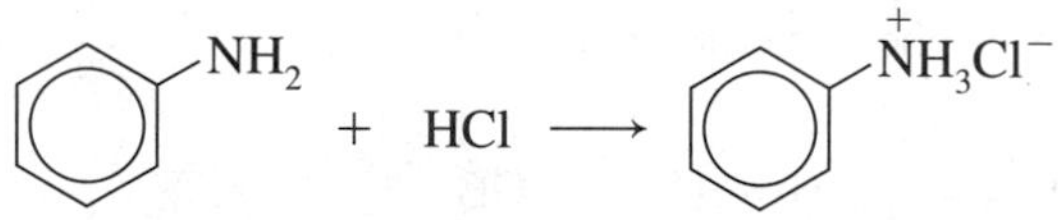

$$C_6H_5NH_2 + HCl \longrightarrow C_6H_5\overset{+}{N}H_3Cl^-$$

anilinium chloride
(aniline hydrochloride salt)

Many amines or amino compounds are more stable in the form of the hydrochloride salt. When the free amine is wanted, the HCl is neutralized to liberate the free amine. Thus,

$$\underset{\text{an amine hydrochloride salt}}{R\overset{+}{N}H_3\ Cl^-\ (\text{or } RNH_2 \cdot HCl)} + NaOH \longrightarrow \underset{\text{free amine}}{RNH_2} + NaCl + H_2O$$

Formation of Amides

Primary and secondary amines react with acid chlorides to form substituted amides. For example,

$$CH_3\overset{O}{\overset{\|}{C}}-Cl + 2\,(CH_3CH_2)_2NH \longrightarrow CH_3\overset{O}{\overset{\|}{C}}-N(CH_2CH_3)_2 + (CH_3CH_2)_2\overset{+}{N}H_2Cl^-$$

N, N-diethylacetamide

As seen from this example, amide formation creates larger molecules from smaller precursors. Building larger more complex molecules is often a goal in both industrial chemistry and biochemistry. Amide formation plays an important role in this process. The barbiturates provide a simple example from pharmaceutical chemistry. Amide bonds form the central six-membered ring that is characteristic of barbiturates (the amide bonds are shown in color).

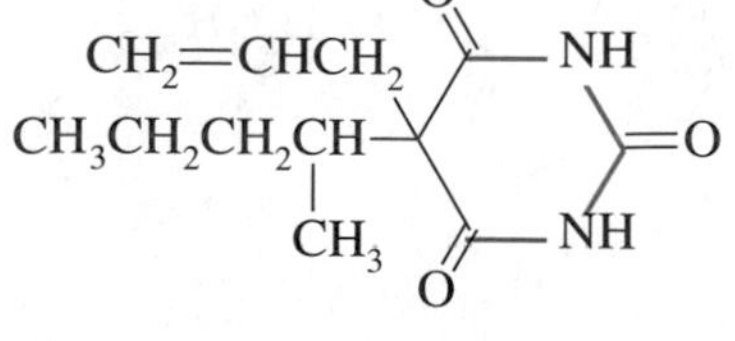

$CH_2{=}CHCH_2$, $CH_3CH_2CH_2CH(CH_3)$– substituted ring with C=O, NH, C=O, NH, C=O

Seconal

Living cells also use amide linkages to create proteins. In proteins small reactants, amino acids, are connected via amide bonds. It is not uncommon for several hundred amino acids to be linked together to form a single protein molecule.

25.7 Sources and Uses of Selected Amines

Nitrogen compounds are found throughout the plant and animal kingdoms. Amines, substituted amines, and amides occur in every living cell. Many of these compounds have important physiological effects. Several examples of well-known nitrogen compounds follow.

Many antibacterial agents contain nitrogen. Common examples include the synthetic sulfa drugs and penicillin-related antibiotics, which are synthesized by molds:

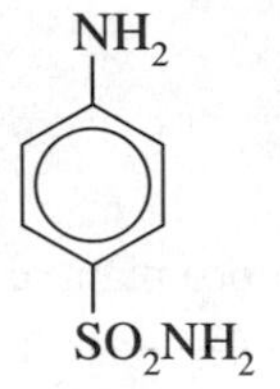

sulfanilamide

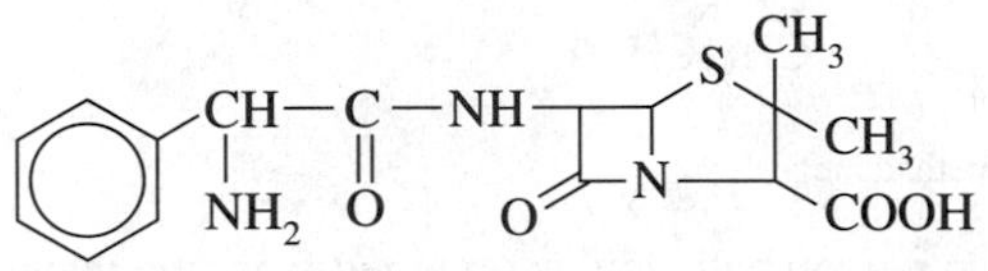

ampicillin

Amines and amides are part of the structures of the B-complex vitamins. Thiamine (vitamin B_1) and nicotinamide (niacin) are examples.

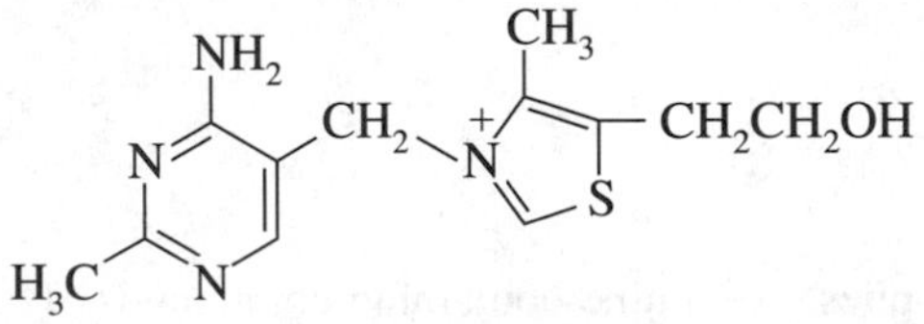

thiamine (vitamin B_1)

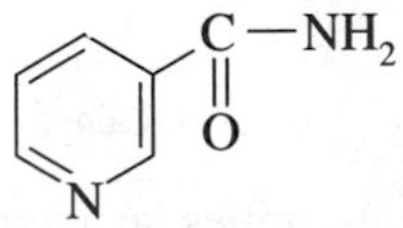

nicotinamide (niacin)

A wide variety of nitrogen-containing compounds can affect both the central and peripheral nervous system. Local anesthetics like the common drug, procaine, contain amines:

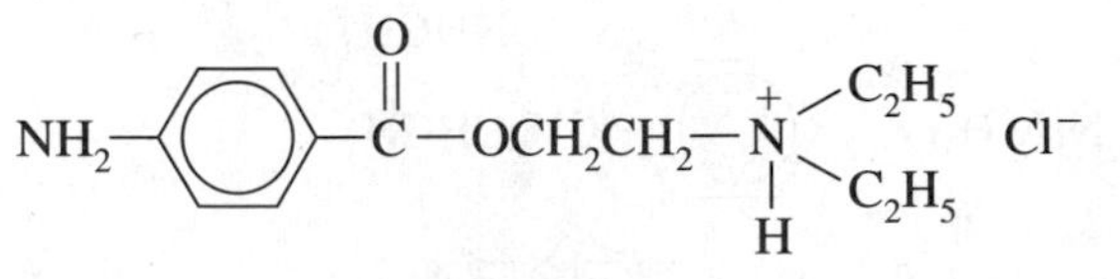

procaine hydrochloride (novocaine)

Basic compounds that are derived from plants and show physiological activity are known as **alkaloids**. These substances are usually amines. Nicotine, an alkaloid derived from tobacco leaves, acts to stimulate the nervous system: **alkaloid**

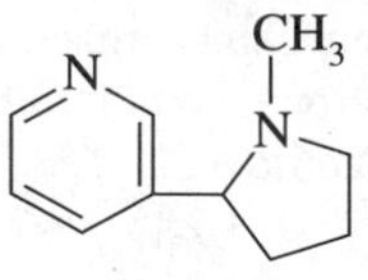

nicotine

The opium alkaloids are often called opiates and include both compounds derived from the opium poppy and synthetic compounds that have morphine-like activity (sleep-inducing and analgesic properties). These drugs are classified as narcotics because they produce physical addiction, and they are strictly regulated by federal law.

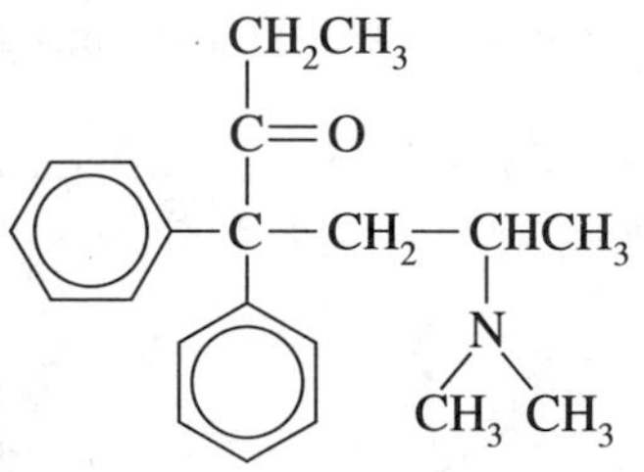

methadone

Another common narcotic alkaloid is obtained from the leaves of the coca plant, cocaine:

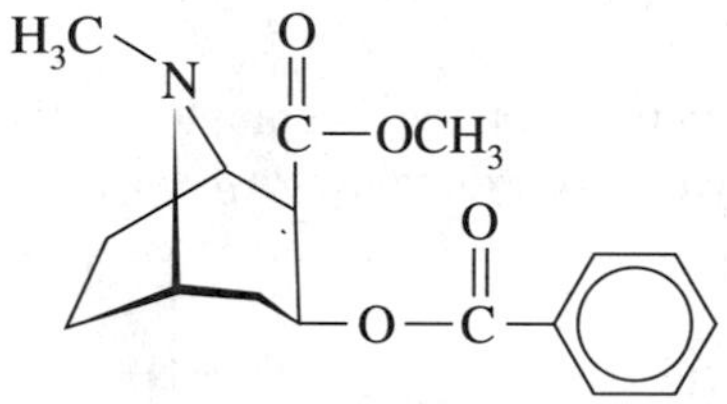

cocaine

Amphetamines, as the name implies, are amine-containing compounds. These drugs act to stimulate the central nervous system by mimicking the action of compounds such as epinephrine, which is produced naturally by the body. They are used to treat depression, narcolepsy, and obesity. Use of amphetamines produces a feeling of well-being, loss of fatigue, and increasing alertness. The most widely abused amphetamine is methamphetamine, commonly called "speed."

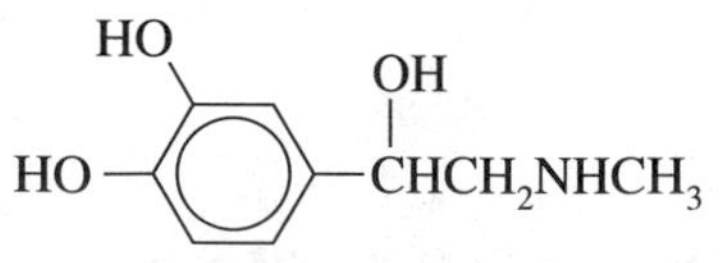

epinephrine (adrenalin)

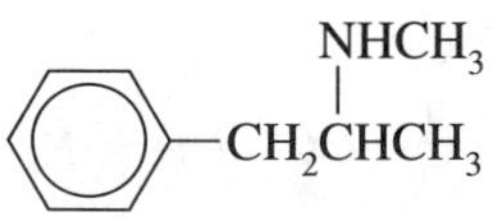

methamphetamine (Methedrine)

Barbiturates are synthetic drugs classified as sedatives. They are prepared from urea and substituted malonic acid, and they contain amide groups. Barbiturates act to depress the activity of brain cells. For this reason, they are often called "downers"; they are one of the more widely abused drugs.

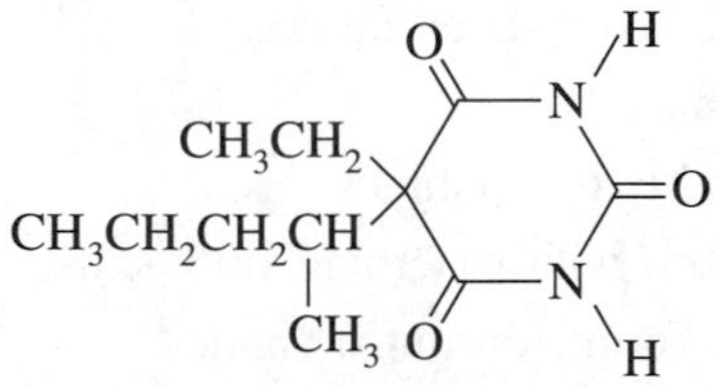

pentobarbital
(Nembutal)

Many common tranquilizers contain amines and amides. These drugs are used to modify psychotic behavior without inducing sleep, or to reduce anxiety or restlessness. Psychotic behavior is treated with strong tranquilizers such as Thorazine. Milder tranquilizers such as diazepam (Valium) are often used to relieve the pressure and anxiety of daily life.

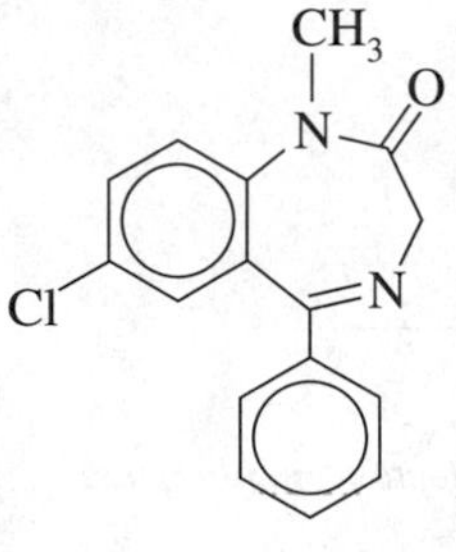

diazepam
(Valium)

Other amine-containing drugs are being developed as scientists learn more about the central nervous system. Fluoxetine (Prozac) is a widely used and very effective antidepressant.

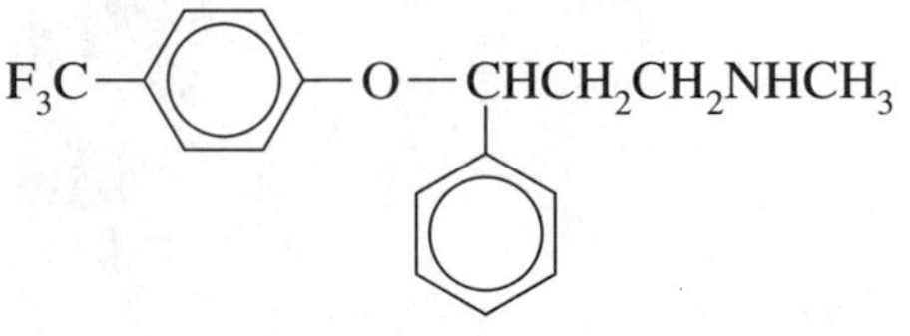

fluoxetine (Prozac)

Recently Prozac has been found useful in treating other central nervous system disorders.

Concepts in Review

1. Name and write structural formulas for amides.
2. Explain the high melting and boiling points of the amides, compared to alkanes of similar molar mass.

3. Write equations for both acidic and basic hydrolysis of amides.
4. Name and write structural formulas for amines.
5. Distinguish among primary, secondary, and tertiary amines.
6. Show that amines are bases in their reactions with water and with acids.
7. Write equations for reactions of amines to form substituted amides.
8. Write equations for the formation of amines from (a) alkyl halides plus ammonia, (b) reduction of amides with $LiAlH_4$, (c) reduction of nitriles with H_2 and Ni, and (d) reduction of aromatic nitro compounds.
9. Identify and write equations for the formation of quaternary ammonium salts.
10. Indicate the major physiological responses to barbiturates, tranquilizers, amphetamines, and alkaloids.
11. Identify amine and amide functional groups in selected naturally occurring nitrogen compounds.

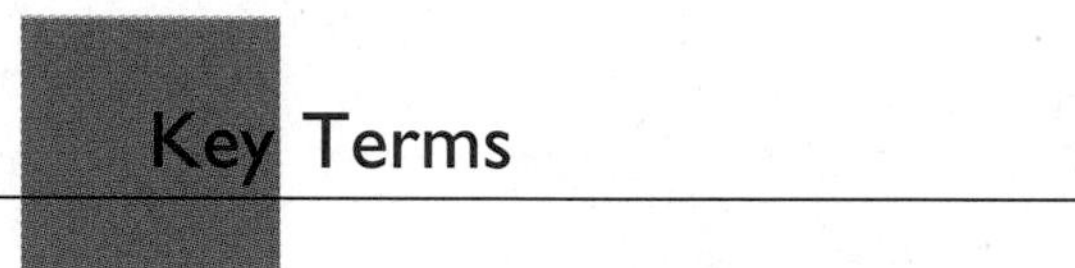

Key Terms

The terms listed here have been defined within this chapter. Section numbers are referenced in parenthesis for each term.

alkaloid (25.7)
amide (25.1)
amine (25.4)
heterocyclic compound (25.4)
quaternary ammonium salt (25.5)

Questions

Questions refer to tables, figures, and key words and concepts defined within the chapter. A particularly challenging question or exercise is indicated with an asterisk.

1. Write the structural formula and name of: (a) a typical amine and (b) a typical amide. Explain how the difference in functional groups affects the chemical properties of amines and amides.
2. Contrast the physical properties of amides with those of amines.
3. Explain why amines have approximately the same water solubility as alcohols of similar molar mass.
4. **(a)** What is a heterocyclic compound?
   (b) How many heterocyclic rings are present in (i) purine, (ii) ampicillin, (iii) methadone, and (iv) nicotine?
5. Indicate the functional groups present in
   (a) procaine hydrochloride
   (b) cocaine
   (c) nicotinamide
   (d) methamphetamine
6. Which of the following statements are correct? Rewrite the incorrect statements to make them correct.
   (a) The common name for CH_3CONH_2 is methanamide.
   (b) The general formula for a primary amine is

   (c) Uric acid is the primary way that the body loses nitrogen.
   (d) Heterocyclic compounds contain more than one ring.
   (e) Alkaloids are used to settle upset stomachs.
   (f) Barbiturates are stimulants.

(g) Amphetamines are also called tranquilizers.
(h) Water soluble amines react with water to produce basic solutions.
(i) The name for $[(CH_3)_2CH]_2NH$ is isopropyl amine.
(j) Aniline is soluble in dilute HCl because it forms a soluble salt.
(k) Most amines have pleasant odors.
(l) Lactic acid is an α-amino acid.
(m) Sulfanilamide is an aniline derivative.
(n) Isopropyl amine is a secondary amine.
(o) Butylamine and diethylamine are isomers.
(p) Aniline is made by the reduction of nitrobenzene.

Paired Exercises

These exercises are paired. Each odd-numbered exercise is followed by a similar even-numbered exercise. Answers to the even-numbered exercises are given in Appendix V.

7. Draw structural formulas for the following:
(a) *p*-methylaniline
(b) 2-aminobutane
(c) *N*-methyl-*p*-bromobenzamide

8. Draw structural formulas for the following:
(a) 2-amino-4-chloropentane
(b) *N,N*-diethylbenzamide
(c) *m*-chloroaniline

9. Name the following compounds:

(a) $CH_3C(=O)-NH_2$ (common name)

(b) $CH_3CH(OH)CH_2CH(NH_2)CH_3$ (IUPAC name)

(c) 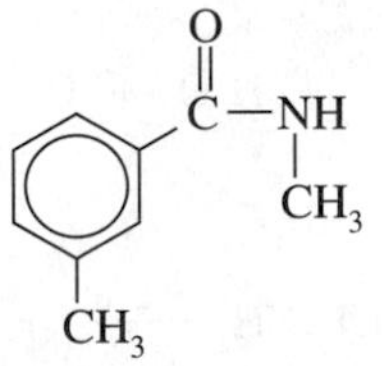 benzene ring bearing $-C(=O)-NH-CH_3$ with CH_3 meta to it (IUPAC name)

10. Name the following compounds:

(a) $CH_3CH(CH_3)CH_2CH_2NH_2$ (IUPAC name)

(b) $HC(=O)-NH_2$ (common name)

(c) 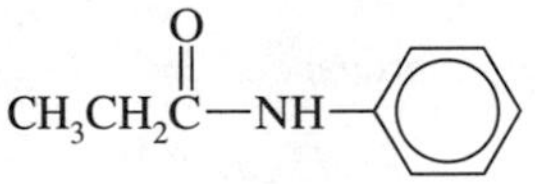 $CH_3CH_2C(=O)-NH-C_6H_5$ (IUPAC name)

11. Arrange this set of compounds in order of increasing solubility in water.

$CH_3CH_2C(=O)-NHCH_3$

$CH_3CH_2C(=O)-N(CH_3)_2$

$CH_3CH_2C(=O)-NH_2$

12. Arrange this set of compounds in order of increasing solubility in water.

$CH_3C(=O)-NH_2$

$CH_3(CH_2)_4C(=O)-NH_2$

$C_6H_5-C(=O)-NH_2$

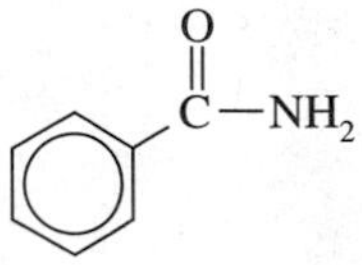

13. Acetamide is soluble in water. Show the possible hydrogen bonding that helps explain this property.

14. Urea is soluble in water. Show the possible hydrogen bonding that helps to explain this property.

15. Classify each of the following compounds as an acid, a base, or neither:

(a) CH_3CH_2OH

(b) $CH_3CH_2NH_2$

(c) 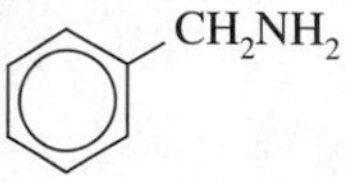 $C_6H_5CH_2NH_2$ (benzene ring with CH_2NH_2)

(d) 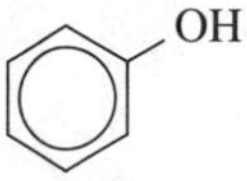 C_6H_5OH (benzene ring with OH)

(e) $CH_3CH_2\overset{O}{\overset{\|}{C}}{-}NH_2$

(f) $CH_3CH_2\overset{O}{\overset{\|}{C}}{-}OH$

16. Classify each of the following compounds as an acid, a base, or neither:

(a) $CH_3CH_2OCH_2CH_3$

(b) $CH_3CH_2NHCH_2CH_3$

(c) 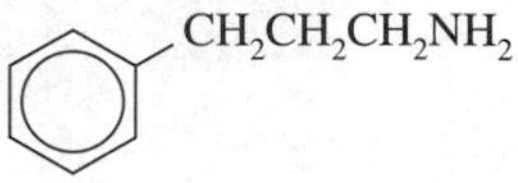 benzene ring with $CH_2CH_2CH_2NH_2$

(d) 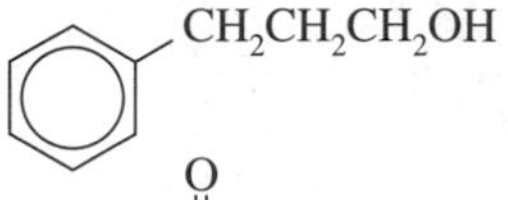 benzene ring with $CH_2CH_2CH_2OH$

(e) 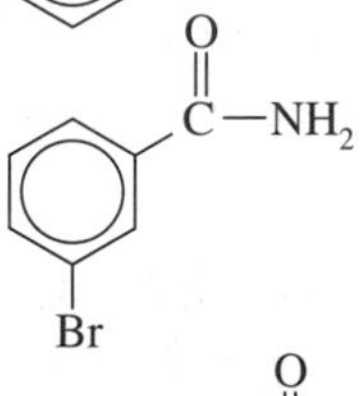 benzene ring with $\overset{O}{\overset{\|}{C}}{-}NH_2$ and Br

(f) 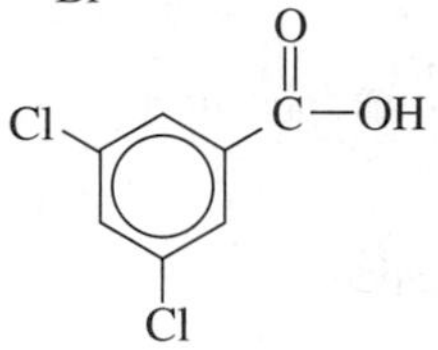 benzene ring with $\overset{O}{\overset{\|}{C}}{-}OH$, Cl, and Cl

17. Predict the organic products for the following reactions:

(a) $CH_3NH{-}\overset{O}{\overset{\|}{C}}CH_3 + H_2O + H^+ \xrightarrow{\Delta}$

(b) Acetic acid + 2-aminopropane $\xrightarrow{\Delta}$

(c) 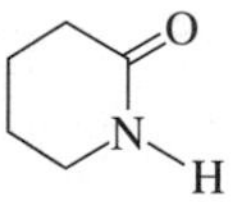 six-membered cyclic amide (ring C=O adjacent to N–H) + NaOH $\xrightarrow{\Delta}$

18. Predict the organic products for the following reactions:

(a) Acetic acid + diethylamine $\xrightarrow{\Delta}$

(b) 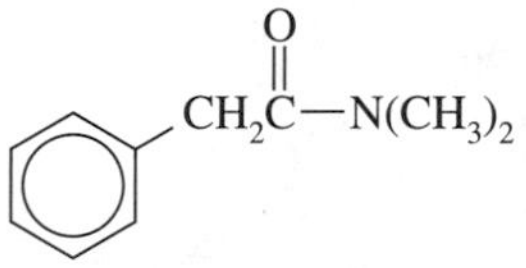 benzene ring with $CH_2\overset{O}{\overset{\|}{C}}{-}N(CH_3)_2$ $+ H_2O + H^+ \xrightarrow{\Delta}$

(c) $CH_3\underset{CH_3}{\underset{|}{C}}HCH_2\overset{O}{\overset{\|}{C}}{-}OH + CH_3NH_2 \xrightarrow{\Delta}$

19. Draw structural formulas for all the amines having the formula $C_4H_{11}N$. Classify each as either a primary amine, a secondary amine, or a tertiary amine.

20. Draw structural formulas for all the amines having the formula C_3H_9N. Classify each as either a primary amine, a secondary amine, or a tertiary amine.

21. Classify each of the following amines as primary, secondary, or tertiary:

(a) $CH_3CH_2CH_2NH_2$

(b) 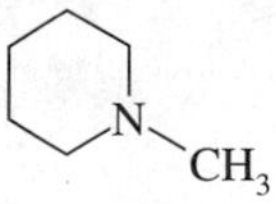 six-membered ring containing N bearing CH_3

(c) 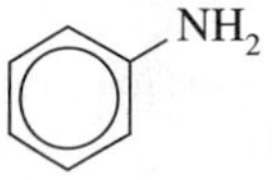 benzene ring with NH_2

22. Classify each of the following amines as primary, secondary, or tertiary:

(a) CH_3NHCH_3

(b) $CH_3CH_2N(CH_3)_2$

(c) 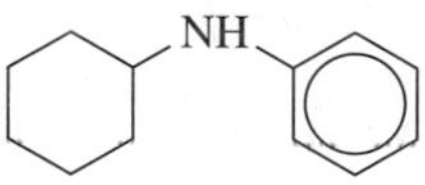 cyclohexane ring–NH–benzene ring

23. Low-molar-mass aliphatic amines generally have odors suggestive of ammonia and/or stale fish. Which of the following solutions would have the stronger odor? Explain.
(a) a 1% trimethylamine solution in 1.0 *M* sulfuric acid or
(b) a 1% trimethylamine solution in 1.0 *M* NaOH

24. Low-molar-mass aliphatic amines generally have odors suggestive of ammonia and/or stale fish. Which of the following solutions would have the stronger odor? Explain.
(a) a 1% 2-aminopropane solution in 1.0 *M* hydrochloric acid or
(b) a 1% 2-aminopropane solution in 1.0 *M* KOH

25. Name the following compounds:

(a) CH_3NHCH_3 (common name)

(b) 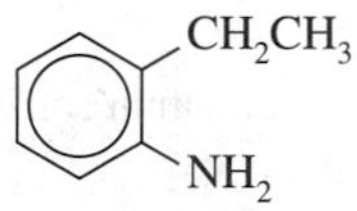 benzene ring with CH_2CH_3 and NH_2 on adjacent carbons (IUPAC name)

(c) 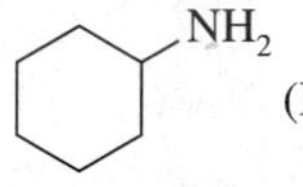 cyclohexane ring with NH_2 (IUPAC name)

(d) $(C_2H_5)_4N^+I^-$ (common name)

(e) 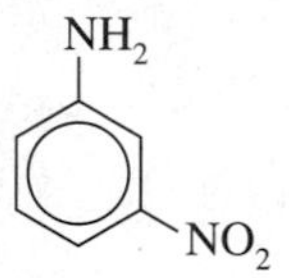 benzene ring with NH_2 and NO_2 (meta) (IUPAC name)

(f) 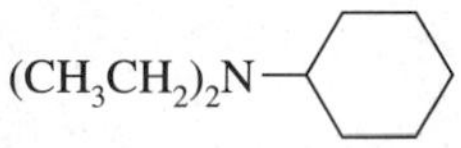 $(CH_3CH_2)_2N$—cyclohexyl (common name)

26. Name the following compounds:

(a) 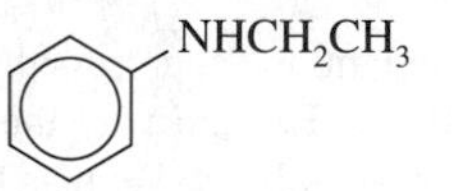 benzene ring with $NHCH_2CH_3$ (IUPAC name)

(b) $CH_3CH_2CHCH_3$ with NH_2 on C-2 (IUPAC name)

(c) 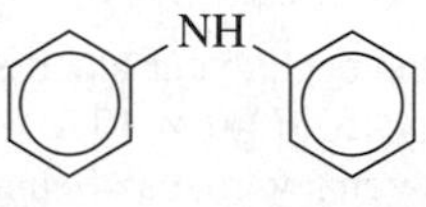 two benzene rings joined by NH (IUPAC name)

(d) $CH_3CH_2NH_3^+\ Br^-$ (common name)

(e) 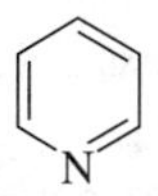 pyridine ring (N) (common name, which is also IUPAC name)

(f) $CH_3\overset{O}{\overset{\|}{C}}—NHCH_2CH_3$ (common name)

27. Draw structural formulas for the following compounds:
(a) *N*-methylaminoethane
(b) aniline
(c) 1,4-diaminobutane
(d) ethylisopropylmethylamine
(e) pyridine

28. Draw structural formulas for the following compounds:
(a) tributylamine
(b) *N*-methylanilinium chloride
(c) ethylammonium chloride
(d) 2-amino-1-pentanol
(e) 3,3-dimethyl-*N*-phenyl-2-aminohexane

29. Write equations to show how each conversion may be accomplished. (Some conversions may require more than one step.)

(a) $CH_3CH_2CH_2Br \longrightarrow CH_3CH_2CH_2NH_2$

(b) $CH_3CH_2CH_2Br \longrightarrow CH_3CH_2CH_2CH_2NH_2$

(c) 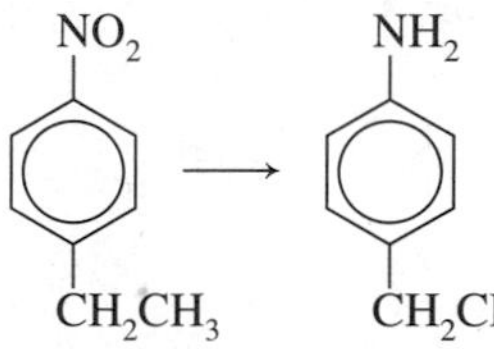 benzene ring with NO_2 and para CH_2CH_3 $\longrightarrow$ benzene ring with NH_2 and para CH_2CH_3

(d) $CH_3CH_2CH_2NH_2 \longrightarrow CH_3\overset{O}{\overset{\|}{C}}NHCH_2CH_2CH_3$

30. Show structural formulas for the organic products from each of the following reactions:

(a) $CH_3CH_2\overset{O}{\overset{\|}{C}}—O^-\ NH_4^+ \xrightarrow{\Delta}$

(b) $CH_3CH_2CH_2C{\equiv}N + H_2 \xrightarrow{Ni}$

(c) 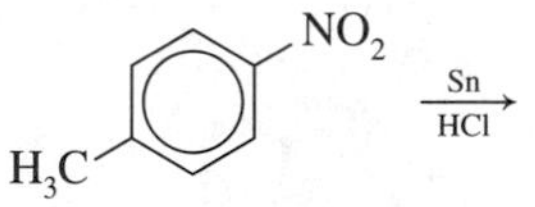 benzene ring with NO_2 and H_3C (meta) $\xrightarrow[HCl]{Sn}$

(d) 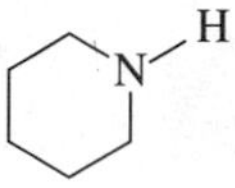 piperidine (ring N—H) $+ CH_3\overset{O}{\overset{\|}{C}}—Cl \longrightarrow$

Additional Exercises

These exercises are not paired or labeled by topic and provide additional practice on concepts covered in this chapter.

31. 1-Aminopropane, ethylmethylamine, and trimethylamine have the same formula, C_3H_9N. Explain why the boiling point of trimethylamine is considerably lower than the boiling points of the other two compounds.

32. List three classes of drugs that are also amines and indicate the major physiological effects of each.

33. When you experience a sudden fright, your blood epinephrine (adrenalin) level rises quickly to 1.0×10^{-7} *M*. Given a blood volume of 5.0 L, how many grams of epinephrine are in circulation when you are in this condition?

34. Show the structure known as an amide linkage. Indicate some important classes of biochemicals that contain this linkage.

35. After cleaning or packing fish, workers often use lemon juice to clean their hands. What is the purpose of the lemon juice? Explain.

36. Cephalosporins are a newer generation of antibiotics that are often effective against penicillin-resistant bacteria.

O
||
—CH_2—C—NH
S
S
O
||
N
O
CH_2O —CCH_3

(The colored ring is critical to the cephalosporin's antibiotic activity.) Identify the amide and ester groups.

37. Write a chemical reaction in which aniline acts as a base.

38. Putrescine, a product of decay of organic matter, is necessary for the growth of cells and occurs in animal tissues. Draw the structure and identify the type of nitrogen bonds that are present.

39. Why are many drugs given as an ammonium salt?

40. Aspartame is 200 times sweeter than sucrose. Circle any amine or amide in this structure.

O　　　CH_2
||　　　|
H_2NCHC—$NHCHC$ —OCH_3
|　　　　||
CH_2COOH　O

Answers to Practice Exercises

25.1

(a) $HC(=O)-NHCH(CH_3)CH_3$

(b) $CH_3CH_2CH_2C(=O)-NHCH_2CH_3$

(c) $CH_3CH(CH_3)C(=O)-NHCH_3$

(d) $CH_3CH_2CH_2CH_2CH_2C(=O)-N(CH_2CH_3)-CH_2CH_3$

(e) 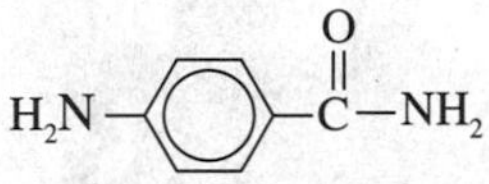 $H_2N-C_6H_4-C(=O)-NH_2$

25.2

(a) acid hydrolysis $CH_3CH_2C(=O)-OH + NH_4^+$

basic hydrolysis $CH_3CH_2C(=O)-O^- + NH_3$

(b) acid hydrolysis $CH_3C(=O)-OH + \overset{+}{N}H_2(CH_3)_2$

basic hydrolysis $CH_3C(=O)-O^- + NH(CH_3)_2$

25.3 (a) 1-aminopropane (propylamine)

(b) *N*-isopropyl-*N*-methylaniline

(c) 2-aminobutane (sec-butylamine)

25.4 (a) $CH_3C(=O)-NHCH_3$ or $HC(=O)-NHCH_2CH_3$

(b) $CH_3CH(CH_3)C\equiv N$ or $CH_3CH(CH_3)C(=O)-NH_2$

(c) $CH_3CH_2CH(CH_2CH_3)CH(CH_3)C\equiv N$ or $CH_3CH_2CH(CH_2CH_3)CH(CH_3)C(=O)-NH_2$

Polymers: Macromolecules

26

Chapter 26

What is a "mer"? The terms *polymer* and *monomer* are a part of our everyday speech, but we are usually familiar only with the prefixes, *poly* meaning "many" and *mono* meaning "one." Although "mer" sounds like something from a child's cartoon show, scientists have chosen this small three-letter root to convey an important concept. It is derived from the Greek *meros*, meaning "part." So, a monomer is a "one part" and a polymer is a "many part."

Look around you for a few minutes and note all the objects that could be described in these terms. A brick building could be described as a polymer, with each brick representing a monomer. A string of pearls is an elegant polymer, with each pearl a monomer. Starting with simple chemical monomers, chemists have learned to construct both utilitarian and elegant polymers. As you will see in this chapter, many of the polymers are of great commercial importance.

26.1 Macromolecules

Up to now we have dealt mainly with rather small organic molecules that contain up to 50 atoms and some (fats) that contain up to about 150 atoms. But there exist in nature some very large molecules (macromolecules) that contain tens of thousands of atoms. Some of these, such as starch, glycogen, cellulose, proteins, and DNA, have molar masses in the millions and are central to many of our life processes. Synthetic macromolecules touch every phase of our lives. Today it is hard to imagine a world without polymers. Textiles for clothing, carpeting, draperies, shoes, toys, automobile parts, construction materials, synthetic rubber, chemical equipment, medical supplies, cooking utensils, synthetic leather, recreational equipment—the list could go on and on. All these and a host of others that we consider to be essential in our daily lives are wholly or partly synthetic polymers. Until about 1930 most of these polymers were unknown. The vast majority of these polymeric materials are based on petroleum. Because petroleum is a nonreplaceable resource, our dependence on polymers is another good reason for not squandering the limited world supply of petroleum.

Polyethylene is an example of a synthetic polymer. Ethylene, derived from petroleum, is made to react with itself to form polyethylene (or polythene). Polyethylene is a long-chain hydrocarbon made from many ethylene units:

$$\underset{\text{ethylene}}{n\,CH_2{=}CH_2} \longrightarrow \underset{\text{polyethylene}}{—CH_2CH_2CH_2(CH_2CH_2)_nCH_2CH_2CH_2—}$$

A typical polyethylene molecule contains anywhere from 2500–25,000 ethylene units joined in a continuous chain.

The process of forming very large, high-molar-mass molecules from smaller units is called **polymerization**. The large molecule, or unit, is called the **polymer** and the small unit, the **monomer**. The term *polymer* is derived from the Greek word *polumerēs*, meaning "having many parts." Ethylene is a monomer, and polyethylene is a polymer. Because of their large size, polymers are often called *macromolecules*. Another commonly used term is *plastics*. The word *plastic* means "to be capable of being molded, or pliable." Although not all polymers are pliable and capable of being remolded, *plastics* has gained general acceptance and has come to mean any of a variety of polymeric substances.

polymerization
polymer
monomer

◀ **Chapter Opening Photo: River rafting through the Grand Canyon is often done on rafts made from polymers.**

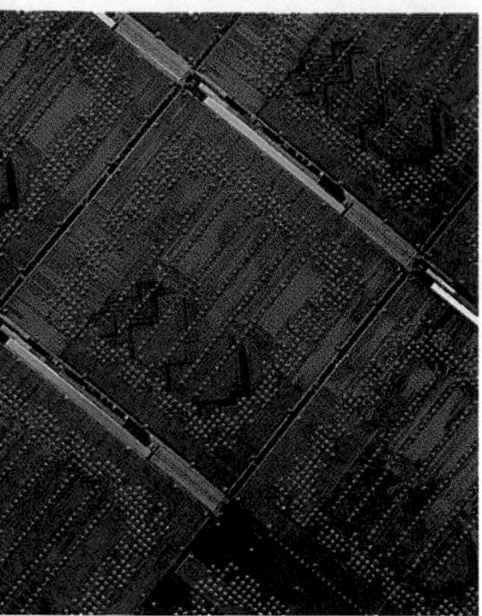

Synthetic polymers are used in the production of microchips.

26.2 Synthetic Polymers

Some of the early commercial polymers were merely modifications of naturally occurring substances. One chemically modified natural polymer, nitrated cellulose, was made and sold as Celluloid late in the 19th century. But the first commercially successful, fully synthetic polymer, Bakelite, was made from phenol and formaldehyde by Leo Baekeland in 1909. This was the beginning of the modern plastics industry. Chemists began to create many synthetic polymers in the late 1920s. Since then, ever-increasing numbers of synthetic macromolecular materials have become articles of commerce. Even greater numbers of polymers have been made and discarded for technical or economic reasons. Polymers are presently used extensively in nearly every industry. For example, the electronics industry uses "plastics" in applications ranging from microchip production to fabrication of heat and impact-resistant cases for the assembled products. In the auto industry, huge amounts of polymers are used as body, engine, transmission, and electrical system components, as well as for tires. Vast quantities of polymers with varied and sometimes highly specialized characteristics are used for packaging; for example, packaging for frozen foods must withstand subfreezing temperatures as well as those met in either conventional or microwave ovens.

Although there is a great variety of synthetic polymers on the market, they can be classified into the following general groups based on their properties and uses:

1. Rubber-like materials or elastomers
2. Flexible films
3. Synthetic textiles and fibers
4. Resins (or plastics) for casting, molding, and extruding
5. Coating resins for dip-, spray-, or solvent-dispersed applications
6. Miscellaneous (e.g., hydraulic fluids, foamed insulation, ion-exchange resins)

26.3 Polymer Types

Organic chemistry has had perhaps the greatest impact on our lives via the polymers (plastics) that we use. Although these polymers often have complex structures, the underlying chemistry of polymerization is relatively simple. Two classes of organic reactions are routinely used, the addition reaction and the substitution reaction. An **addition polymer** is one that is produced by successive *addition reactions.* Polyethylene is an example of an addition polymer. A **condensation polymer** is one that is formed when monomers combine and split out water or some other simple substance—a *substitution reaction.*

addition polymer
condensation polymer

Polymers are either thermoplastic or thermosetting. Those that soften on reheating are **thermoplastic polymers**; those that set to an infusible solid and do not soften on reheating are **thermosetting polymers**. Thermoplastic polymers are formed when monomer molecules join end to end in a linear chain with little or no cross-linking between the chains. Thermosetting polymers are macromolecules in which the polymeric chains are cross-linked to form a network structure. The structures of thermoplastic and thermosetting polymers are illustrated in Figure 26.1

thermoplastic polymer
thermosetting polymer

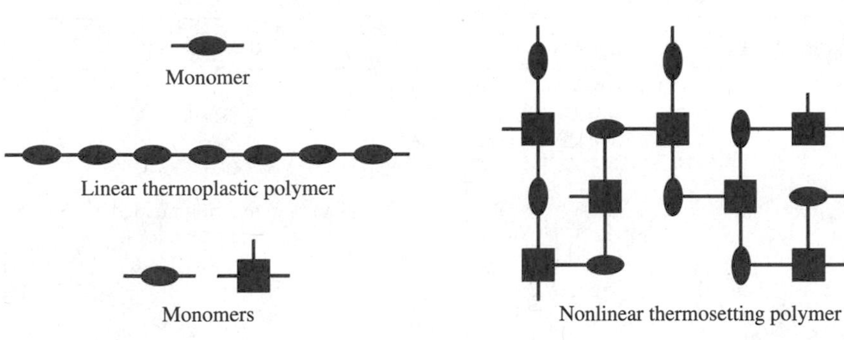

FIGURE 26.1
Schematic diagrams of thermoplastic and thermosetting polymer structures.

26.4 Addition Polymerization

Addition polymerization starts with monomers that contain C=C double bonds. When these double bonds react, each alkene carbon bonds to another monomer.

$$\mathrm{C{=}C} \quad \mathrm{C{=}C} \longrightarrow \mathrm{-C-C-C-C-}$$

This reaction is typical of many alkene addition reactions (see Chapter 21).

Ethylene polymerizes to form polyethylene according to the reaction shown in Section 26.1. Polyethylene is one of the most important and also the most widely used polymer on the market today. It is a tough, inert, but flexible thermoplastic material. Over 1.1×10^{10} kg of polyethylene is produced annually in the United States alone. Polyethylene is made into hundreds of different articles, many of which we use everyday. Low-density polyethylene is more stretchable and is used in objects such as food wrap and plastic squeeze bottles; high-density polyethylene is used in items that require more strength, such as plastic grocery bags and cases for electronic components.

The double bond is the key structural feature involved in the polymerization of ethylene. Ethylene derivatives, in which one or more hydrogen atoms have been replaced by other atoms or groups, can also be polymerized. This is often called *vinyl polymerization.* Many of our commercial synthetic polymers are made from such modified ethylene monomers. The names, structures, and uses of some of these polymers are given in Table 26.1.

Free radicals catalyze or initiate many addition polymerizations. Organic peroxides, ROOR, are frequently used for this purpose. The reaction proceeds as shown in the following three steps:

Step 1 *Free radical formation.* The peroxide splits into free radicals:

$$\mathrm{RO{:}OR} \longrightarrow \mathrm{2RO\cdot}$$

Step 2 *Propagation of polymeric chain.*

$$\mathrm{RO\cdot} + \mathrm{CH_2{=}CH_2} \longrightarrow \mathrm{ROCH_2CH_2\cdot}$$
$$\mathrm{ROCH_2CH_2\cdot} + \mathrm{CH_2{=}CH_2} \longrightarrow \mathrm{ROCH_2CH_2CH_2CH_2\cdot} \quad \text{(and so on)}$$

Polyethylene is used to wrap foods both in stores and at home.

TABLE 26.1 Polymers Derived from Modified Ethylene Monomers

Monomer	Polymer	Identification Code for Most Commonly Recycled Polymers	Uses
$CH_2{=}CH_2$ ethylene	$\text{-}(CH_2{-}CH_2)_n\text{-}$ polyethylene	4 LDPE (Low Density Polyethylene); 2 HDPE (High Density Polyethylene)	Packing material, molded articles, containers, toys
$CH_2{=}CH(CH_3)$ propylene	$\text{-}(CH_2{-}CH(CH_3))_n\text{-}$ polypropylene	5 PP	Textile fibers, molded articles, lightweight ropes, autoclavable biological equipment
$CH_2{=}C(CH_3)_2$ isobutylene	$\text{-}(CH_2{-}C(CH_3)_2)_n\text{-}$ polyisobutylene		Pressure-sensitive adhesives, butyl rubber (contains some isoprene as copolymer)
$CH_2{=}CH(Cl)$ vinyl chloride	$\text{-}(CH_2{-}CH(Cl))_n\text{-}$ polyvinyl chloride (PVC)	3 V	Garden hoses, pipes, molded articles, floor tile, electrical insulation, vinyl leather
$CH_2{=}CCl_2$ vinylidene chloride	$\text{-}(CH_2{-}CCl_2)_n\text{-}$ Saran		Food packaging, textile fibers, pipes, tubing (contains some vinyl chloride as copolymer)
$CH_2{=}CH(CN)$ acrylonitrile	$\text{-}(CH_2{-}CH(CN))_n\text{-}$ Orlon, Acrilan		Textile fibers
$CF_2{=}CF_2$ tetrafluoroethylene	$\text{-}(CF_2{-}CF_2)_n\text{-}$ Teflon		Gaskets, valves, insulation, heat-resistant and chemical-resistant coatings, linings for pots and pans
$CH_2{=}CH(C_6H_5)$ styrene	$\text{-}(CH_2{-}CH(C_6H_5))_n\text{-}$ polystyrene	6 PS	Molded articles, Styrofoam, insulation, toys, disposable food containers
$CH_2{=}CH{-}OC({=}O){-}CH_3$ vinyl acetate	$\text{-}(CH_2{-}CH(OC({=}O){-}CH_3))_n\text{-}$ polyvinyl acetate		Adhesives, paint, and varnish
$CH_2{=}C(CH_3){-}C({=}O){-}O{-}CH_3$ methylmethacrylate	$\text{-}(CH_2{-}C(CH_3)(C({=}O){-}O{-}CH_3))_n\text{-}$ Lucite, Plexiglas (acrylic resins)		Contact lenses, clear sheets for windows and optical uses, molded articles, automobile finishes

Slowing Soil Erosion

One of the least expensive ways to irrigate farm fields is by using furrows—ditches that run along each planted row. The major problem with furrow irrigation is that, as water runs along a ditch, it erodes valuable topsoil. Losses can run between 2.8 and 28 tons per acre. This erosion can also rob fields of fertilizers and pesticides, dumping them as pollutants into nearby lakes and streams. In addition, tiny clay particles eventually clog soil pores near the end of the furrow, thus reducing the amount of water that can filter through to the roots of the plants.

Soil scientists in Idaho have begun testing special crystals that can be added to the irrigation water to stop erosion. Just what are these magic crystals and how do they work? The crystals are polyacrylamide (PAM), a long-chain polymer commonly used to clean wastewater. There are thousands of different PAM molecules, varying by electrical charge, number of units, and cross-linking.

This field has been treated with polyacrylamide (PAM) crystals, which reduce erosion caused by irrigation.

$$-CH_2-\underset{\underset{\overset{\|}{O}}{C-NH_2}}{CH}-CH_2-\underset{C(=O)-NH_2}{CH}-CH_2-\underset{C(=O)-NH_2}{CH}-$$

short section of PAM polymer

The PAM polymer that interests Robert Sojka, of the Agricultural Research Service in Kimberly, Idaho, is one in which the molecules average 30,000 units in length and are negatively charged. It has very low toxicity and is primarily used in wastewater-treatment facilities to aggregate fine solids into particles large enough to settle out or be captured by filters. Scientists think the PAM seeks out and binds to the broken edges of crystalline clay particles, increasing their cohesiveness. This makes the dirt more resistant to the shearing forces exerted by water flowing down the furrow.

It takes very little PAM to dramatically reduce erosion. Just 10 ppm added to water during the first hour of irrigation can reduce erosion 70–99%. If the soil is not disturbed, the next irrigation can be done without additional PAM, giving about 50% erosion reduction. Or farmers can refresh their irrigation furrows by using 1 ppm PAM in the water. On steeply sloping fields, the use of PAM can improve water infiltration into soils so much that crop yields are dramatically boosted.

The new erosion treatment still has some drawbacks. The polymer works well in clean water, but when added to water clouded with sediment, it causes the particles to flocculate and settle out, filling the head of the furrow. It also requires extra steps to measure and mix proper concentrations of PAM. The cost of treating a field is about \$2 to \$3 per acre, which over a growing season can seriously affect profits.

Farmers view PAM treatment of soil as a long-term investment. By reducing topsoil loss, PAM preserves the value of their farm land. Scientists think PAM can help farmers all over the United States. For example, PAM could improve the absorption of water into soils in the Central Valley of California, increasing harvests by as much as 20%. This polymer gives farmers an opportunity to hold onto the earth and improve their crop yields at the same time.

Step 3 *Termination.* Polymerization stops when the free radicals are used up:

$$RO(CH_2CH_2)_n\cdot + \cdot OR \longrightarrow RO(CH_2CH_2)_nOR$$

$$RO(CH_2CH_2)_n\cdot + \cdot(CH_2CH_2)_nOR \longrightarrow RO(CH_2CH_2)_n(CH_2CH_2)_nOR$$

Addition (or vinyl) polymerization of ethylene and its substituted derivatives yields saturated polymers—that is, polymer chains without carbon–carbon double bonds. The pi bond is eliminated when a free radical adds to an ethylene molecule. One of the electrons of the pi bond pairs with the unpaired electron of the free radical, thus bonding the radical to the ethylene unit. The other pi bond electron remains unpaired, generating a new and larger free radical. This new free radical then adds another ethylene molecule, continuing the building of the polymeric chain. This process is illustrated by the following electron-dot diagram:

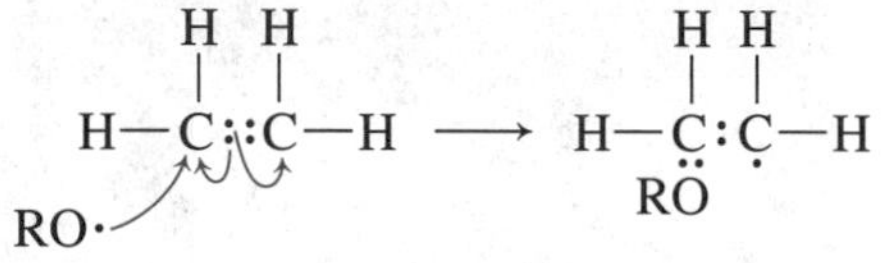

$$\mathrm{RO\cdot} + \mathrm{H{-}\overset{H}{\underset{}{C}}{:}{:}\overset{H}{C}{-}H} \longrightarrow \mathrm{H{-}\overset{H}{\underset{RO}{C}}{:}\overset{H}{\dot{C}}{-}H} \qquad (\mathrm{ROCH_2CH_2\cdot})$$

26.5 Addition Polymers: Use and Reuse (Recycling)

Over 50% of all polymers produced and used in the United States are made from ethylene and related compounds. These polymers are inexpensive to produce and can be shaped and molded into almost any form. Their very utility makes disposal a severe problem.

For millions of years, microorganisms have broken down large molecules to smaller ones, starting the biosynthetic process over again. Unfortunately, the C–C single bonds of common addition polymers, like polyethylene, are different from those of most natural polymers and cannot be metabolized by many microorganisms. Consequently, while paper and cardboard in our landfills are very slowly disintegrated by microorganisms, most of the plastic waste is not.

Recycling seems to be the best solution to the problem of disposing of these long-lived addition polymers. Reuse of polyethylene (or any other of these plastics in relatively pure form) is fairly easy—simply melt and re-form into the desired product. The big problems of recycling lie with separating the different types of plastic; some plastics are not compatible with others and must be separated before reuse. The symbols on plastic goods (see Table 26.1) are one step toward plastic identification and separation. Each plastic container used in commerce carries a number that identifies its constituent polymer. It is possible (although difficult) to manually separate containers based on this code. Current research has focused on machine-separation techniques that depend on properties of plastics such as density. We recycle about 1% of all plastics (about 20% of all soft-drink bottles) compared with about 30% of aluminum; we clearly could recycle much more.

26.6 Butadiene Polymers

A diene is a compound that contains two carbon–carbon double bonds. Another type of addition polymer is based on the compound 1,3-butadiene or its derivatives:

Plastic is ground up during the recycling process before it is used to make new articles such as these play structures.

$$\overset{1}{CH_2}{=}\overset{2}{CH}{-}\overset{3}{CH}{=}\overset{4}{CH_2}$$

1,3-butadiene

Natural rubber is a polymer of isoprene (2-methyl-1,3-butadiene). Many synthetic elastomers or rubber-like materials are polymers of isoprene or of butadiene. Unlike the saturated ethylene polymers, these polymers are unsaturated; that is, they have double bonds in their polymeric structures.

$$n\ CH_2{=}\underset{}{\overset{CH_3}{\overset{|}{C}}}{-}CH{=}CH_2 \longrightarrow {+}CH_2{-}\overset{CH_3}{\overset{|}{C}}{=}CH{-}CH_2{+}_n$$

isoprene — rubber polymer chain (polyisoprene)

$$n\ CH_2{=}CH{-}CH{=}CH_2 \longrightarrow {+}CH_2{-}CH{=}CH{-}CH_2{+}_n$$

1,3-butadiene — butadiene polymer chain

$$n\ CH_2{=}\overset{Cl}{\overset{|}{C}}{-}CH{=}CH_2 \longrightarrow {+}CH_2{-}\overset{Cl}{\overset{|}{C}}{=}CH{-}CH_2{+}_n$$

2-chloro-1,3-butadiene (chloroprene) — neoprene polymer chain (polychloroprene)

In this kind of polymerization, the free radical adds to the butadiene monomer at carbon 1 of the carbon–carbon double bond. At the same time, a double bond is formed between carbons 2 and 3, and a new free radical is formed at carbon 4. This process is illustrated in the following diagram:

$$H{-}\overset{H}{\overset{|1}{C}}{:}{:}\overset{H}{\overset{|2}{C}}{:}\overset{H}{\overset{|3}{C}}{:}{:}\overset{H}{\overset{|4}{C}}{-}H \quad RO\cdot \longrightarrow RO{:}\underset{H}{\underset{|}{\overset{H}{\overset{|}{C}}}}{:}\overset{H}{\overset{|}{C}}{:}{:}\overset{H}{\overset{|}{C}}{:}\underset{H}{\underset{|}{\overset{H}{\overset{|}{C}}}}\cdot$$

free radical — 1,3-butadiene — free radical chain lengthened by four carbon atoms

The tires of these cars are made of rubber that has been vulcanized. This process improves the temperature range for use and resistance to abrasion. ▶

One of the oustanding synthetic rubbers (styrene–butadiene rubber, SBR) is made from two monomers, styrene and 1,3-butadiene. These substances form a **copolymer**—that is, a polymer containing two different kinds of monomer units. Styrene and butadiene do not necessarily have to combine in a 1:1 ratio. In the actual manufacture of SBR polymers, about 3 moles of butadiene are used per mole of styrene. Thus the butadiene and styrene units are intermixed, but in a ratio of about 3:1.

copolymer

$$-CH_2CH{=}CHCH_2 \,|\, \underset{\displaystyle C_6H_5}{CH_2CH} \,|\, CH_2CH{=}CHCH_2 \,|\, CH_2CH{=}CHCH_2-$$

styrene unit — butadiene unit

segment of styrene–butadiene rubber (SBR)

The presence of double bonds at intervals along the chains of rubber and rubberlike synthetic polymers designed for use in tires is almost a necessity and, at the same time, a disadvantage. On the positive side, double bonds make vulcanization possible. On the negative side, double bonds afford sites where ozone, present especially in smoggy atmospheres, can attack the rubber, causing "age hardening" and cracking. Vulcanization extends the useful temperature range of rubber products and imparts greater abrasion resistance to them. The vulcanization process is usually accomplished by heating raw rubber with sulfur and other auxiliary agents. It consists of introducing sulfur atoms that connect or cross-link the long strands of polymeric chains. Vulcanization was devised through trial-and-error experimentation by the American inventor Charles Goodyear in 1839, long before any real understanding of the chemistry of the process was known. Goodyear's patent on "Improvement in India Rubber" was issued on June 15, 1844. In the segment of vulcanized rubber shown here, the chains of polymerized isoprene are cross-linked by sulfur–sulfur bonds, giving the polymer more strength and elasticity.

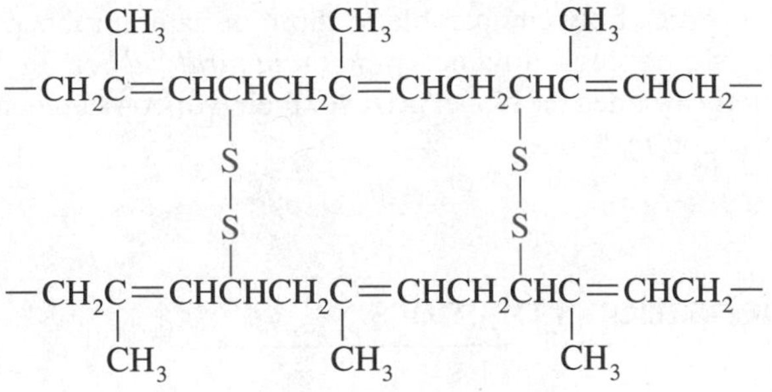

```
   CH3           CH3           CH3
   |             |             |
-CH2C=CHCHCH2C=CHCH2CHC=CHCH2-
          |                |
          S                S
          |                |
          S                S
          |                |
-CH2C=CHCHCH2C=CHCH2CHC=CHCH2-
   |             |             |
   CH3           CH3           CH3
```

segment of vulcanized rubber

26.7 Geometric Isomerism in Polymers

The recurring double bonds in isoprene and butadiene polymers make it possible to have polymers with specific spatial orientation as a result of cis–trans isomerism. Recall from Section 21.3 that two carbon atoms joined by a double bond are not free to rotate and thus give rise to cis–trans isomerism. An isoprene polymer can have all-cis, all-trans, or a random distribution of cis and trans configurations about the double bonds.

Natural rubber is *cis*-polyisoprene with an all-cis configuration about the carbon–carbon double bonds. Gutta-percha, also obtained from plants, is a *trans*-polyisoprene with an all-trans configuration. Although these two polymers have the same composition, their properties are radically different. The cis natural rubber is a soft, elastic material, whereas the trans gutta-percha is a tough, nonelastic, horn-like substance. Natural rubber has many varied uses. Some uses of gutta-percha include electrical insulation, dentistry, and golf balls.

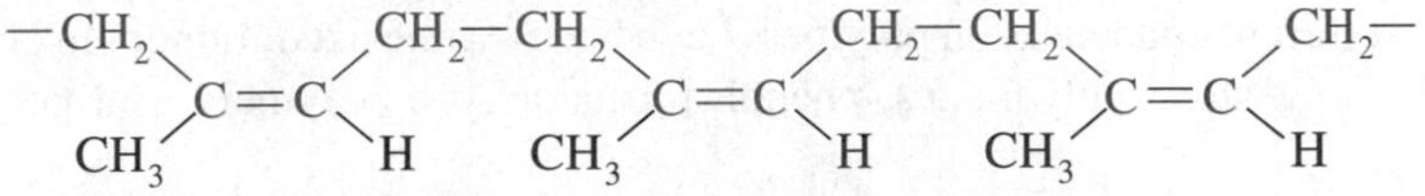

```
-CH2        CH2-CH2        CH2-CH2        CH2-
    \      /       \      /       \      /
     C=C            C=C            C=C
    /    \         /    \         /    \
 CH3      H     CH3      H     CH3      H
```

all-cis configuration of natural rubber

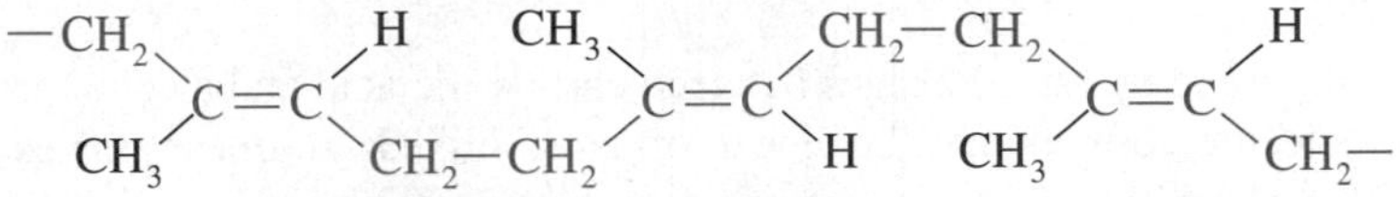

```
-CH2        H    CH3        CH2-CH2        H
    \      /        \      /       \      /
     C=C             C=C            C=C
    /    \          /    \         /    \
 CH3      CH2-CH2        H      CH3      CH2-
```

all-trans configuration of natural rubber

Chicle is another natural substance containing polyisoprenes. It is obtained by concentrating the latex from the sapodilla tree which contains about 5% *cis*-polyisoprene and 12% *trans*-polyisoprene. The chief use of chicle is in chewing gum.

Only random or nonstereospecific polymers are obtained by free radical polymerization. Synthetic polyisoprenes made by free radical polymerization are much inferior to natural rubber, since they contain both the cis and the trans isomers. But in the 1950s, Karl Ziegler (1898–1973) of Germany and Giulio Natta (1903–1979) of Italy developed catalysts [e.g., $(C_2H_5)_3Al/TiCl_4$] that allowed polymerization to proceed by an ionic mechanism, producing stereochemically controlled polymers. Ziegler–Natta catalysts made possible the synthesis of polyisoprene with an all-cis configuration and with properties fully comparable to those of natural rubber. This material is known by the odd but logical name *synthetic natural rubber.* In 1963, Natta and Ziegler were jointly awarded the Nobel prize for their work on stereochemically controlled polymerization reactions.

26.8 Condensation Polymers

Condensation polymers are formed by reactions between functional groups on adjacent monomers. As a rule, a smaller molecule, usually water, is eliminated in the reaction. Each monomer loses a small group of atoms and gains a larger group, a substitution reaction that allows the polymer to continue growing. Condensation polymerization uses functional groups that favor substitution reactions; thus carboxylic acids and their derivatives (see Chapter 24 for review) are found in many condensation polymerization reactions.

For the polymer to continue to grow, each monomer must undergo at least two substitution reactions and must thus be at least bifunctional. If cross-linking is to occur, there must be more than two functional groups on some monomer molecules.

Most biochemical polymers are the condensation type. Proteins are polyamides, whereas nucleic acids are polyesters. Just as with synthetic polymers, biological polymers can have varied physical properties and varied functions. The collagen protein is used to make durable structures such as feathers, hair, and hooves. The elastin protein can be used for structures that need to stretch, such as tendons. Other proteins provide the mucous coating that protects our nasal passages. These differences in physical properties are determined by (1) the functional groups that are incorporated in the polymer and (2) the structure into which the atoms have been bonded. This generalization applies to synthetic polymers as well.

Many different condensation polymers have been synthesized. Important classes include the polyesters, polyamides, phenol–formaldehyde polymers, and polyurethanes.

Polyesters

Polyesters are joined by ester linkages between carboxylic acid and alcohol groups; the macromolecule formed may be linear or cross-linked. A linear polyester is obtained from the bifunctional monomers terephthalic acid and ethylene glycol. Esterification occurs between the alcohol and acid groups on both ends of both monomers, forming long-chain macromolecules:

$$\mathrm{HOOC{-}C_6H_4{-}COOH} \qquad \mathrm{HOCH_2CH_2OH}$$

terephthalic acid ethylene glycol

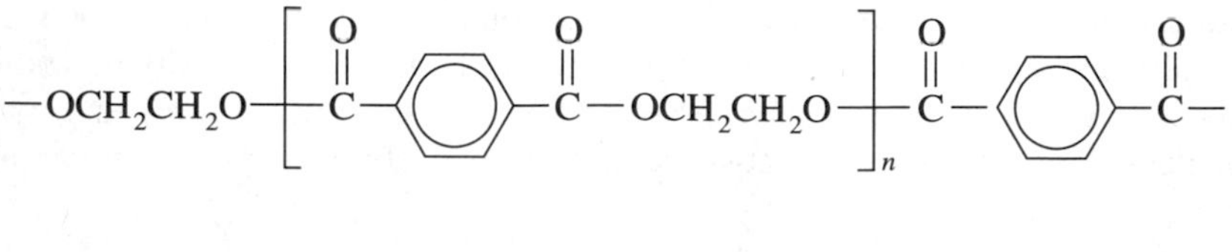

$$\mathrm{{-}OCH_2CH_2O}\left[\overset{\displaystyle O}{\overset{\|}{C}}{-}\mathrm{C_6H_4}{-}\overset{\displaystyle O}{\overset{\|}{C}}{-}\mathrm{OCH_2CH_2O}\right]_n\overset{\displaystyle O}{\overset{\|}{C}}{-}\mathrm{C_6H_4}{-}\overset{\displaystyle O}{\overset{\|}{C}}{-}$$

This polymer may be drawn into fibers or formed into transparent films of great strength. Dacron and Terylene synthetic textiles and Mylar films are made from this polyester, as is the familiar 2-L soft-drink bottle. The symbol on objects made of this material is shown in the margin. In actual practice the dimethyl ester of terephthalic acid is used, and the molecule split out is methyl alcohol instead of water.

When trifunctional acids or alcohols are used as monomers, cross-linked thermosetting polyesters are obtained (see Figure 26.1). One common example is the reaction of glycerol and *o*-phthalic acid. The polymer formed is one of a group of polymers known as alkyd resins. Glycerol has three functional —OH groups, and phthalic acid has two functional —COOH groups:

$$\underset{\text{glycerol}}{HOCH_2\overset{\displaystyle OH}{\overset{|}{C}}HCH_2OH} \qquad \underset{o\text{-phthalic acid}}{C_6H_4(COOH)_2}$$

A cross-linked macromolecular structure is formed that, with modifications, has proved to be one of the most outstanding materials used in the coatings industry. Alkyd resins have been used as "baked-on" finishes for automobiles and household appliances. Each year, more than 3.6×10^8 kg of alkyd resins are used in paints, varnishes, lacquers, electrical insulation, and so on.

Polyamides

Although there are several nylons, one of the best known and the first commercially successful polyamide is Nylon-66. This polymer was so named because it was made from two 6-carbon monomers, adipic acid, $HOOC(CH_2)_4COOH$, and 1,6-diaminohexane, $H_2N(CH_2)_6NH_2$. The polymer chains of polyamides contain recurring amide linkages. The amide linkage can be made by reacting a carboxylic acid group with an amine group:

$$\underset{\text{carboxylic acid group}}{R-\overset{\displaystyle O}{\overset{\|}{C}}-OH} + \underset{\text{amine group}}{H-\overset{\displaystyle H}{\overset{|}{N}}-CH_2-R'} \xrightarrow{\Delta} \underset{\text{amide linkage}}{R-\overset{\displaystyle O}{\overset{\|}{C}}-NH}-CH_2-R' + H_2O$$

The repeating structural unit of the Nylon-66 chain consists of one adipic acid unit and one 1,6-diaminohexane unit:

$$\underset{\text{adipic acid}}{HOOC-(CH_2)_4-COOH} \qquad \underset{\substack{\text{1,6-diaminohexane}\\ \text{(hexamethylenediamine)}}}{H_2N-(CH_2)_6-NH_2}$$

$$-NH(CH_2)_6-NH\left[\overset{\displaystyle O}{\overset{\|}{C}}(CH_2)_4-\overset{\displaystyle O}{\overset{\|}{C}}-NH(CH_2)_6-NH\right]_n\overset{\displaystyle O}{\overset{\|}{C}}(CH_2)_4-\overset{\displaystyle O}{\overset{\|}{C}}-$$

segment of Nylon-66 polyamide

Fibers made from polymers can be used to conduct light to a very specific location. This property is used extensively in the field of fiberoptics.

Nylon was developed as a synthetic fiber for the production of stockings and other wearing apparel. It was introduced to the public at the New York World's

Fair in 1939. Nylon is used to make fibers for clothing and carpeting, filaments for fishing lines and ropes, bristles for brushes, and molded objects such as gears and bearings. For the latter application, no lubrication is required because nylon surfaces are inherently slippery.

Phenol–Formaldehyde Polymers

As noted earlier, a phenol–formaldehyde condensation polymer (Bakelite) was first marketed over 85 years ago. Polymers of this type are still widely used, especially in electrical equipment, because of their insulating and fire-resistant properties.

Each phenol molecule can react with formaldehyde to lose an H atom from the para position and from each of the ortho positions (indicated by arrows):

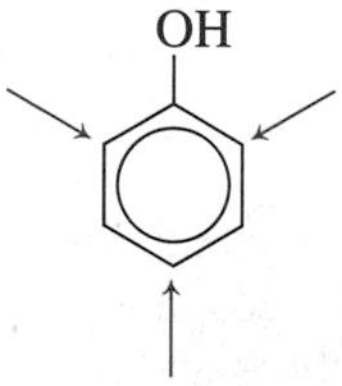

OH

Each formaldehyde molecule reacts with two phenol molecules to eliminate water:

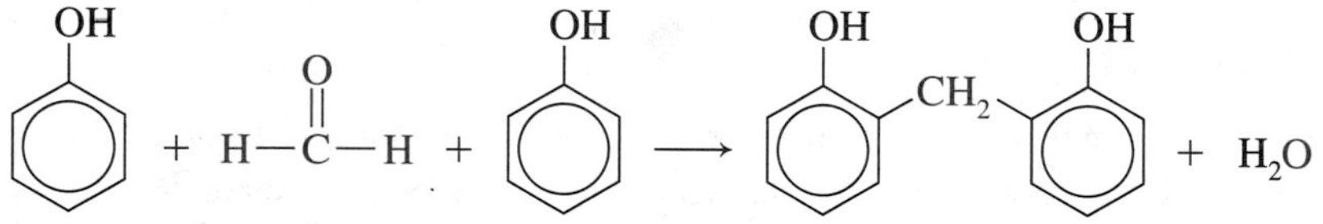

$$C_6H_5OH + H{-}\overset{O}{\overset{\|}{C}}{-}H + C_6H_5OH \longrightarrow HOC_6H_4{-}CH_2{-}C_6H_4OH + H_2O$$

Similar reactions can occur at the other two reactive sites on each phenol molecule, leading to the formation of the polymer. This polymer is thermosetting because it has an extensively cross-linked network structure. A typical section of this structure is illustrated as follows:

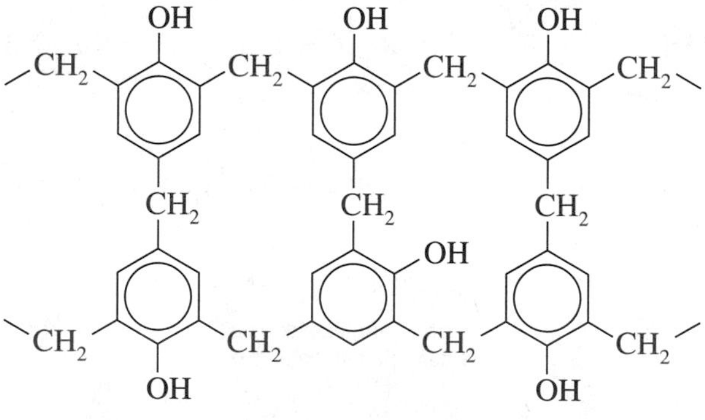

phenol–formaldehyde polymer

Polyurethanes

The compound urethane has structural features of both an ester and an amide. Its formula is

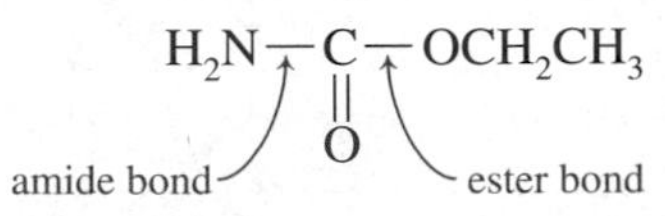

$$H_2N-\overset{\displaystyle O}{\overset{\|}{C}}-OCH_2CH_3$$

amide bond   ester bond

A substituted urethane can be made by reacting an isocyanate with an alcohol. For example,

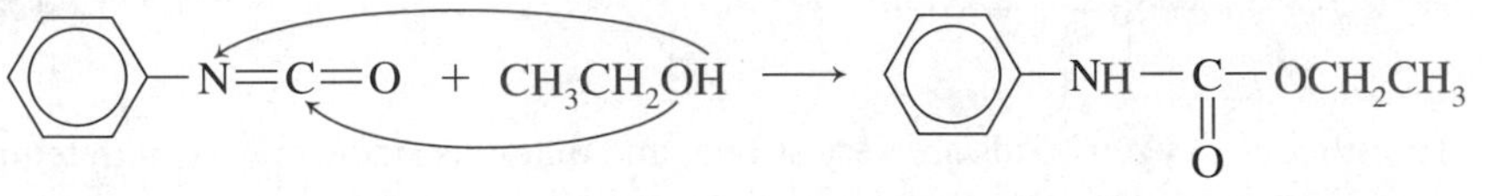

$$C_6H_5-N=C=O + CH_3CH_2OH \longrightarrow C_6H_5-NH-\underset{\displaystyle O}{\underset{\|}{C}}-OCH_2CH_3$$

phenyl isocyanate   *N*-phenyl urethane

Diisocyanates and diols are both bifunctional; therefore, they yield polymers called *polyurethanes*. The polyurethanes are classified as condensation polymers, although no water or other small molecule is split out when they are formed. From *p*-phenylene diisocyanate and ethylene glycol, we obtain a polyurethane with this structure:

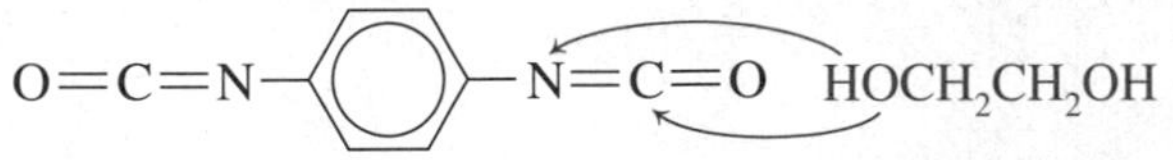

$$O=C=N-C_6H_4-N=C=O \quad HOCH_2CH_2OH$$

p-phenylene diisocyanate   ethylene glycol

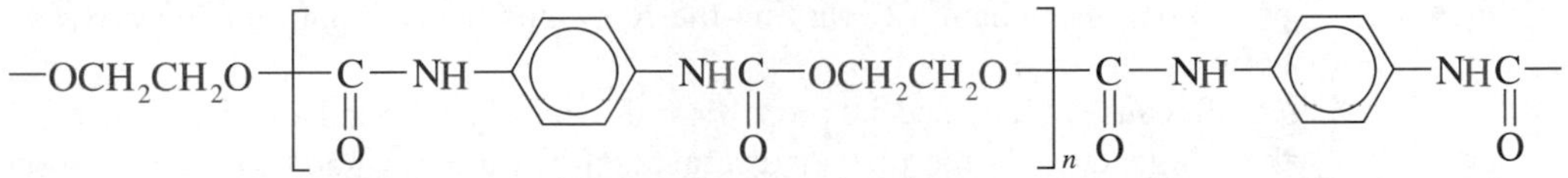

$$-OCH_2CH_2O-\left[\underset{\displaystyle O}{\underset{\|}{C}}-NH-C_6H_4-NH\underset{\displaystyle O}{\underset{\|}{C}}-OCH_2CH_2O\right]_n-\underset{\displaystyle O}{\underset{\|}{C}}-NH-C_6H_4-NH\underset{\displaystyle O}{\underset{\|}{C}}-$$

segment of a polyurethane

Some polyurethanes are soft elastic materials that are widely used as *foam rubber* in upholstery and similar applications. There are many other applications, including automobile safety padding, insulation, life preservers, elastic fibers, and semirigid or rigid foams. Polyurethane can be made into a foam or spongy polymer by adding water during the polymerization or during the molding process. Water reacts with the isocyanate to produce carbon dioxide, which causes the polymer to foam. The effect is similar to that of baking powder releasing carbon dioxide in dough, causing it to rise and become light. The result is a polymer containing innumerable tiny gas-filled cavities that give the product a spongelike quality.

26.9 Silicone Polymers

The silicones are an unusual group of polymers. They include oils and greases, molding resins, rubbers (elastomers), and Silly Putty, the latter being a remarkable material that bounces like a rubber ball when dropped but that can be shaped like putty! Silicones have properties common to both organic and inorganic compounds. The mineral quartz (found in igneous rocks and sand) has the empirical formula SiO_2 and is actually an inorganic polymer. Each silicon atom is bonded to four oxygen atoms, and each oxygen to two silicon atoms to form a three-dimensional structure:

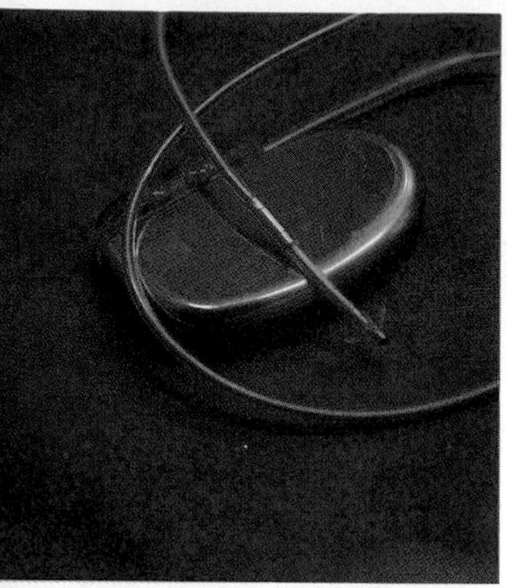

Steel pacemakers, used to stimulate the heart muscle, are encased in silicone.

```
               |         |
               O         O
      |        |         |         |
    —Si—O—Si—O—Si—O—Si—
      |        |         |         |
               O         O
               |         |
             —Si—      —Si—
               |         |
```

The silicon–oxygen bonds are very strong, and quartz is stable at very high temperatures. But it is also very hard and brittle and therefore difficult to form into useful shapes.

Linear silicone polymers, also called silicones or polysiloxanes, consist of silicon–oxygen chains with two alkyl groups attached to each silicon atom:

$$-O-\overset{R}{\underset{R}{\overset{|}{\underset{|}{Si}}}}-O-\left[\overset{R}{\underset{R}{\overset{|}{\underset{|}{Si}}}}-O\right]_n-\overset{R}{\underset{R}{\overset{|}{\underset{|}{Si}}}}-O-$$

The physical properties of silicones can be modified by (1) varying the length of the polymer chain, (2) varying the R groups, and (3) introducing cross-linking between the chains.

Because of their special properties, silicones have found a variety of uses despite their high cost. Some of their useful qualities are (1) good insulating properties (used in high-temperature applications), (2) little viscosity change over a wide temperature range (they can therefore be used as lubricating oils and greases at extreme temperatures), (3) excellent water repellency (used to waterproof many types of surfaces), and (4) good biological compatibility (used in medical and plastic surgery applications).

Because of their stability and tissue compatibility, silicones are frequently used for permanent surgical implants. For example, heart pacemakers are enclosed in protective casings made of silicone. These polymers are used in many reconstructive surgeries. Ears, noses, and chins may be rebuilt using silicones. The flexible nature of this polymer is important when it is used in finger joint reconstruction. Because of extensive litigation, breast implants are perhaps the most well-known of the silicone reconstructive surgeries.

Concepts in Review

1. Write formulas for addition polymers when given the monomer(s).
2. Write formulas for condensation polymers when given the monomers.
3. Describe the properties of a thermoplastic polymer and a thermosetting polymer.
4. Explain the free radical mechanism for polymer formation.

5. Identify polymers from their trade names (e.g., Dacron, Nylon, and Teflon).
6. Explain the effect of cross-linking in polymers.
7. Draw a segment of the structural formula of natural rubber to illustrate the all-cis configuration.
8. Explain how butadiene-type polymers are formed.
9. Explain vulcanization and its effect on rubber.
10. Identify polymers by type (e.g., vinyl, polyester, polyamide, or polyurethane).
11. Identify several useful qualities of silicones.

Key Terms

The terms listed here have been defined within this chapter. Section numbers are referenced in parenthesis for each term.

addition polymer (26.3)
condensation polymer (26.3)
copolymer (26.6)
monomer (26.1)
polymer (26.1)
polymerization (26.1)
thermoplastic polymers (26.3)
thermosetting polymers (26.3)

Questions

Questions refer to tables, figures, and key words and concepts defined within the chapter. A particularly challenging question or exercise is indicated with an asterisk.

1. How does condensation polymerization differ from addition polymerization?
2. What property distinguishes a thermoplastic polymer from a thermosetting polymer?
3. Why must at least some monomer molecules be trifunctional to form a thermosetting polymer?
4. How are rubber molecules modified by vulcanization?
5. Silicone polymers are more resistant to high temperatures than the usual organic polymers. Suggest an explanation for this property of the silicones.
6. Which of these statements are correct? Rewrite each incorrect statement to make it correct.
   (a) The process of forming macromolecules from small units is called polymerization.
   (b) The monomers in condensation polymerization must be at least bifunctional.
   (c) The monomers of Nylon-66 are a dicarboxylic acid and a diol.
   (d) Dacron and Mylar are both made from the same monomers.
   (e) 1,6-diaminohexane is a secondary amine.
   (f) Teflon is an addition polymer.
   (g) Vulcanization was invented by Charles Goodrich.
   (h) The monomer for polystyrene is

   $C_6H_5-CH_2CH{=}CH_2$

   (i) Polyurethanes have both ester and amide bonds in their structure.
   (j) Bakelite is a copolymer of phenol and ethylene glycol.
   (k) The monomer of natural rubber is 2-methyl-1,3-butadiene.

Paired Exercises

These exercises are paired. Each odd-numbered exercise is followed by a similar even-numbered exercise. Answers to the even-numbered exercises are given in Appendix V.

7. How many ethylene units are in a polyethylene molecule that has a molar mass of approximately 25,000?

8. Calculate the molar mass of a polystyrene molecule consisting of 3000 monomer units.

9. Write formulas showing the structure of
(a) Dacron
(b) Nylon-66
(c) Bakelite
(d) Polyurethane

10. Write structural formulas for the following polymers:
(a) Saran
(b) Orlon
(c) Teflon
(d) Polystyrene
(e) Lucite

11. Draw the structure of the free radical that is formed when the first two propylene units combine during propagation of the polypropylene chain.

12. Draw the structure of the free radical that is formed when the first two ethylene units combine during propagation of the polyethylene chain.

13. Write a structural formula showing the polymer that can be formed from
(a) propylene
(b) 2-methylpropene
(c) 2-butene

14. Write a structural formula showing the polymer that can be formed from
(a) ethylene
(b) chloroethene
(c) 1-butene

15. Write structures showing two possible ways in which vinyl chloride can polymerize to form polyvinyl chloride. Show four units in each structure.

16. Write structures showing two possible ways in which acrylonitrile can polymerize to form an "acrylic" fiber (Orlon, Acrilan). Show four units in each structure.

17. Natural rubber is the all-cis polymer of isoprene. Write the structure of the polymer showing at least three isoprene units.

18. Gutta-percha is the all-trans polymer of isoprene. Write the structure of the polymer showing at least three isoprene units.

19. Is the useful life of natural rubber shortened in smoggy atmospheres? Briefly explain your answer.

20. Is the useful life of styrene–butadiene rubber (SBR) shortened in smoggy atmospheres? Briefly explain your answer.

21. Write the chemical structures for the monomers of
(a) natural rubber
(b) synthetic (SBR) rubber

22. Write the chemical structures for the monomers of
(a) synthetic natural rubber
(b) neoprene rubber

23. Calculate the mass percent of
(a) nitrogen in Nylon-66
(b) chlorine in polyvinyl chloride

24. Calculate the mass percent of
(a) oxygen in Dacron
(b) fluorine in Teflon

25. Nitrile rubber (Buna N) is a copolymer of two parts of 1,3-butadiene and one part of acrylonitrile ($CH_2{=}CHCN$). Write a structure for this synthetic rubber, showing at least two units of the polymer.

26. Ziegler–Natta catalysts can orient the polymerization of propylene to form isotactic polypropylene—that is, polypropylene with all the methyl groups on the same side of the long carbon chain. Write the structure for (a) isotactic polypropylene and (b) another possible geometric form of polypropylene.

27. Lexan is a transparent plastic with high impact resistance. It is used in such sporting equipment as football helmets and hockey face masks. The two monomers that make up this polymer are

$$HO-C_6H_4-C(CH_3)_2-C_6H_4-OH \;+\; Cl-\overset{O}{\overset{\|}{C}}-Cl$$

Show the structure of the polymer (two units).

28. Kevlar is a polymer that is many times stronger than steel and is often used to make bulletproof vests. This polymer is formed by linking terephthalic acid with *p*-aminoaniline,

$$H_2N-C_6H_4-NH_2$$

via amide bonds. Show the structure of the polymer (two units).

Additional Exercises

These exercises are not paired or labeled by topic and provide additional practice on concepts covered in this chapter.

29. Using *p*-cresol (p-CH_3-C_6H_4-OH) in place of phenol to form a phenol–formaldehyde polymer results in a thermoplastic rather than a thermosetting polymer. Explain why this occurs.

30. How is "foam" introduced into foam or sponge rubber materials?

31. Of polyethylene, polypropylene, and polystyrene, which polymer contains the highest mass percent of carbon? Calculate the percentage.

32. Glyptal polyesters are made from glycerol and phthalic acid. Would this kind of polymer more likely be thermoplastic or thermosetting? Explain.

33. Nylon-6 (commonly used in tire cord and rope) is made by polymerizing the following molecule:

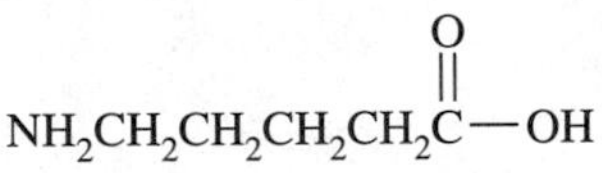

$$NH_2CH_2CH_2CH_2CH_2\overset{\overset{\displaystyle O}{\|}}{C}—OH$$

Show the polymer formed from this monomer.

34. Latex polymer has revolutionized the paint industry by allowing "easy to use" water-based paints that cover well and are not water soluble after drying. This polymer is made from the following monomer:

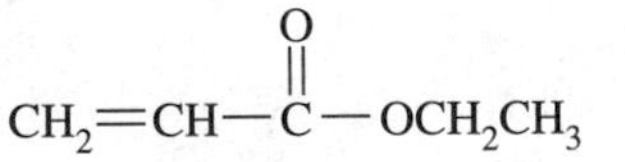

$$CH_2{=}CH—\overset{\overset{\displaystyle O}{\|}}{C}—OCH_2CH_3$$

cyaneacrylate ester

To what class of polymers does latex belong? Draw the polymer including at least three units of monomer.

35. Superglue contains the following monomer:

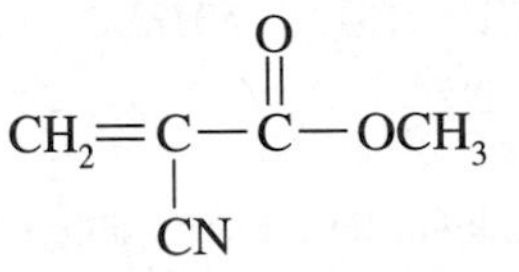

$$CH_2{=}\underset{\underset{\displaystyle CN}{|}}{C}—\overset{\overset{\displaystyle O}{\|}}{C}—OCH_3$$

cyanoacrylate ester

When this monomer is spread between two surfaces, it rapidly polymerizes to give a very strong bond. Write the structure of the polymer formed.

36. (a) Write the structure for a polymer that can be formed from 2,3-dimethyl-1,3-butadiene. (b) Can 2,3-dimethyl-1,3-butadiene form cis and trans isomers? Explain.

37. Would you expect it to be easier to recycle a thermosetting polymer or a thermoplastic polymer? Why?

38. A copolymer formed from ethylene and tetrafluoroethylene can be used to make wire insulation. Write out the structural formulas for the monomers. Write out a portion of the copolymer that you would expect from an addition polymerization reaction.

39. Write structures for the monomers of each of the polymers shown and classify each as polyvinyl, polyester, polyamide, or polyurethane.

(a) 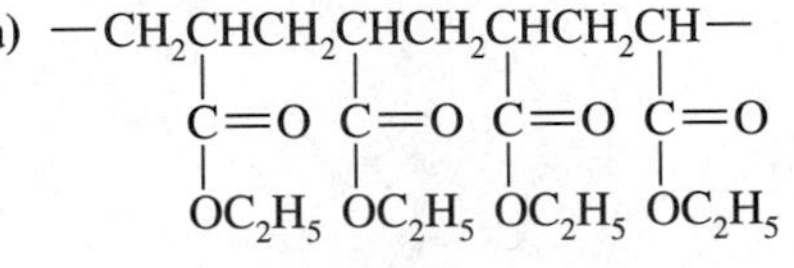
$$—CH_2\underset{\underset{\underset{\underset{\displaystyle OC_2H_5}{|}}{C=O}}{|}}{C}HCH_2\underset{\underset{\underset{\underset{\displaystyle OC_2H_5}{|}}{C=O}}{|}}{C}HCH_2\underset{\underset{\underset{\underset{\displaystyle OC_2H_5}{|}}{C=O}}{|}}{C}HCH_2\underset{\underset{\underset{\underset{\displaystyle OC_2H_5}{|}}{C=O}}{|}}{C}H—$$

(b) 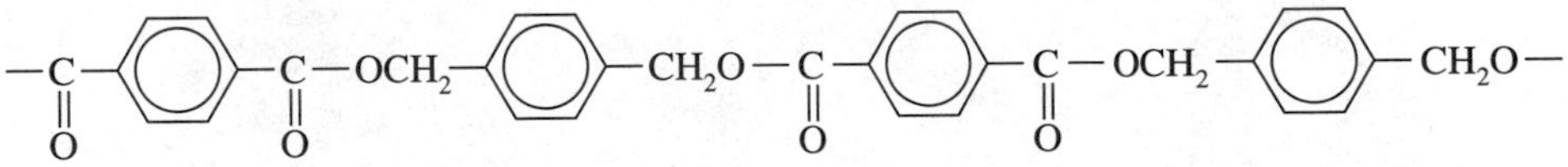
$$—\underset{\underset{\displaystyle O}{\|}}{C}—C_6H_4—\underset{\underset{\displaystyle O}{\|}}{C}—OCH_2—C_6H_4—CH_2O—\underset{\underset{\displaystyle O}{\|}}{C}—C_6H_4—\underset{\underset{\displaystyle O}{\|}}{C}—OCH_2—C_6H_4—CH_2O—$$

(c) 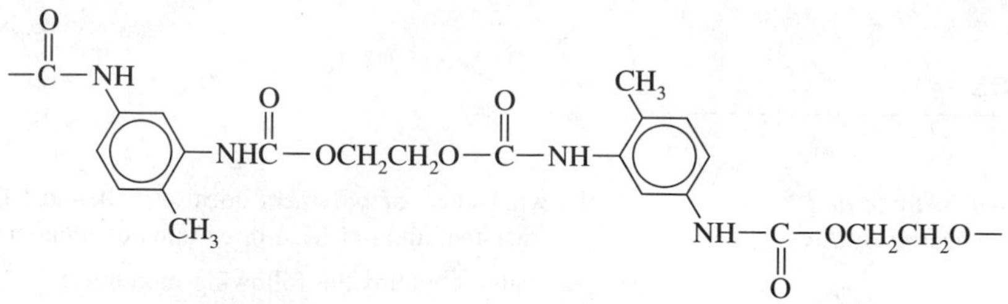 $-\overset{\displaystyle O}{\overset{\|}{C}}-NH$ CH_3

$\overset{\displaystyle O}{\overset{\|}{}}$ $\overset{\displaystyle O}{\overset{\|}{}}$

$-NH\overset{\|}{C}-OCH_2CH_2O-\overset{\|}{C}-NH-$

CH_3 $NH-\underset{\displaystyle O}{\underset{\|}{C}}-OCH_2CH_2O-$

(d) $-\underset{\displaystyle O}{\underset{\|}{C}}-(CH_2)_8-\underset{\displaystyle O}{\underset{\|}{C}}-NH-(CH_2)_6-NH-\underset{\displaystyle O}{\underset{\|}{C}}-(CH_2)_8-\underset{\displaystyle O}{\underset{\|}{C}}-NH-(CH_2)_6-NH-$

40. Quiana is a polymer that is spun into a soft, silky fabric. The structure of the polymer can be represented as

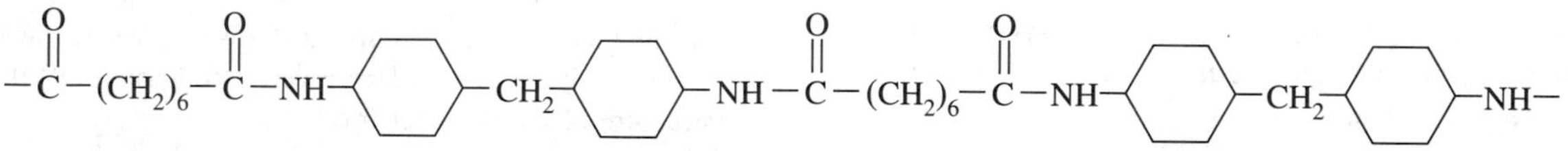

$-\overset{\displaystyle O}{\overset{\|}{C}}-(CH_2)_6-\overset{\displaystyle O}{\overset{\|}{C}}-NH-\langle\ \rangle-CH_2-\langle\ \rangle-NH-\overset{\displaystyle O}{\overset{\|}{C}}-(CH_2)_6-\overset{\displaystyle O}{\overset{\|}{C}}-NH-\langle\ \rangle-CH_2-\langle\ \rangle-NH-$

(a) To what class of polymers does Quiana belong? Be as specific as possible.

(b) Draw the structures of the two monomers used to make Quiana.

41. The polymer used in hard contact lenses is poly(methyl methacrylate). The structures of the monomer (methyl methacrylate) and a section of the polymer are shown. Would you expect this polymer to result from an addition reaction or a condensation reaction? Why?

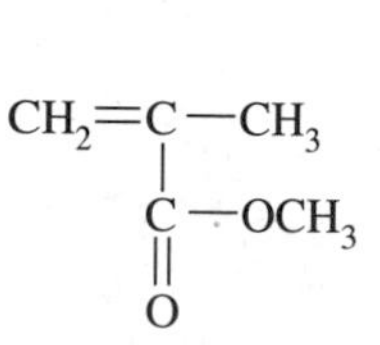

$CH_2{=}C(-CH_3)-\underset{\displaystyle O}{\underset{\|}{C}}-OCH_3$

methyl methacrylate

$-CH_2-\overset{CH_3}{\underset{C(=O)-OCH_3}{C}}-CH_2-\overset{CH_3}{\underset{C(=O)-OCH_3}{C}}-CH_2-\overset{CH_3}{\underset{C(=O)-OCH_3}{C}}-$

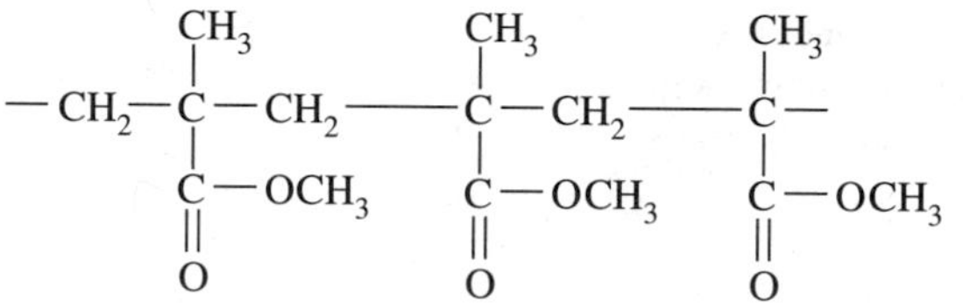

polymer

Stereoisomerism

27

Chapter 27

Many of us grew up hearing such comments as, "Can't you tell your left from your right?" Have you watched a small child try to differentiate between a right and left shoe? Not surprisingly, the distinction between right and left is difficult. After all, our bodies are reasonably symmetrical. For example, both hands are made up of the same components (four fingers, a thumb, and a palm) ordered in the same way (from thumb through little finger). Yet, there is a difference if we try to put a left-handed glove on our right hand or a right shoe on our left foot.

Molecules possess similar, subtle structural differences, which can have a major impact on chemical reactivity. For example, although there are two forms of blood sugar, related as closely as our left and right hands, only one of these structures can be used by our bodies for energy. Stereoisomerism is a subject that attempts to define these subtle differences in molecular structure.

27.1 Review of Isomerism

The phenomenon of two or more compounds having the same number and kinds of atoms is isomerism (see Section 20.9). There are two types of isomerism. In the first type, known as structural isomerism, the difference between isomers is due to different structural arrangements of the atoms that form the molecules. For example, butane and isobutane, ethanol and dimethyl ether, and 1-chloropropane and 2-chloropropane are structural isomers:

$CH_3CH_2CH_2CH_3$ butane	CH_3CH_2OH ethanol	$CH_3CH_2CH_2Cl$ 1-chloropropane
CH_3CHCH_3 with CH_3 on the central C 2-methylpropane (isobutane)	CH_3OCH_3 methoxymethane (dimethyl ether)	$CH_3CHClCH_3$ 2-chloropropane

stereoisomerism

stereoisomer

In the second type of isomerism, the isomers have the same structural formulas but differ in the spatial arrangement of the atoms. This type of isomerism is known as **stereoisomerism**. Thus compounds that have the same structural formulas but differ in their spatial arrangement are called **stereoisomers**. There are two types of stereoisomers: cis–trans or geometric isomers, which we have already considered, and optical isomers, the subject of this chapter. The outstanding feature of optical isomers is that they have the ability to rotate plane-polarized light.

plane-polarized light

27.2 Plane-Polarized Light

◀ **Chapter Opening Photo: These crystals of para toluene sulfonic acid are seen through a polarizing light microscope.**

Plane-polarized light is light that is vibrating in only one plane. Ordinary (unpolarized) light consists of electromagnetic waves vibrating in all directions (planes) perpendicular to the direction in which it is traveling. When ordinary light passes

(a) Ordinary (unpolarized) light

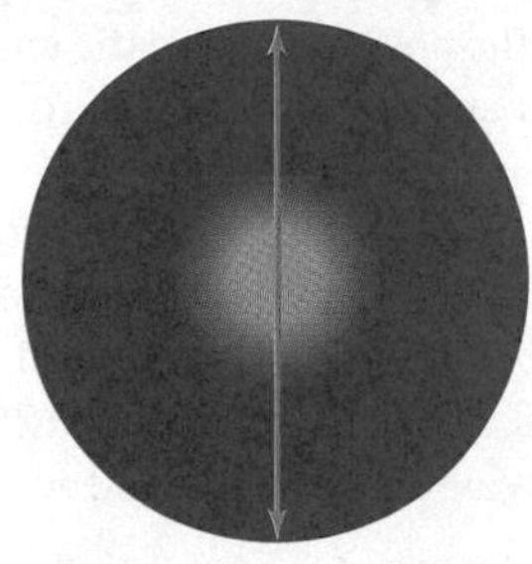
(b) Plane-polarized light

FIGURE 27.1
(a) Diagram of ordinary light vibrating in all possible directions (planes) and (b) diagram of plane-polarized light vibrating in a single plane. The beam of light is coming toward the reader.

through a polarizer, it emerges vibrating in only one plane and is called plane-polarized light (Figure 27.1).

Polarizers can be made from calcite or tourmaline crystals or from a Polaroid filter, which is a transparent plastic that contains properly oriented embedded crystals. Two Polaroid filters with parallel axes allow the passage of plane-polarized light. But when one filter is placed so that its axis is at right angles to that of the other filter, the passage of light is blocked and the filters appear black (Figure 27.2).

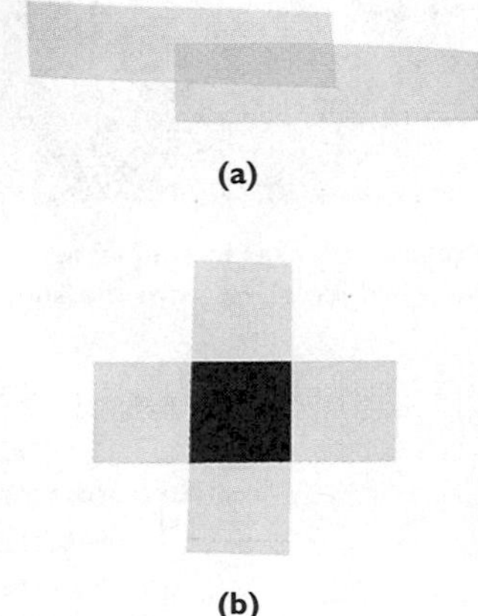

FIGURE 27.2
Two Polaroid filters (top) with axes parallel and (bottom) with axes at right angles. In (a), light passes through both filters and emerges polarized. In (b), the polarized light that emerges from one filter is blocked and does not pass through the second filter, which is at right angles to the first. With no light emerging, the filters appear black.

Specific Rotation

The rotation of plane-polarized light is quantitatively measured with a polarimeter. The essential features of this instrument are (1) a light source (usually a sodium lamp), (2) a polarizer, (3) a sample tube, (4) an analyzer (which is another matched polarizer), and (5) a calibrated scale (360°) for measuring the number of degrees the plane of polarized light is rotated. The calibrated scale is attached to the analyzer (see Figure 27.3). When the sample tube contains a solution of an optically inactive material, the axes of the polarizer and the analyzer are parallel and the scale is at zero degrees; the light passing through is at maximum intensity. When a solution of an optically active substance is placed in the sample tube, the plane in which the polarized light is vibrating is rotated through an angle (α). The analyzer is then

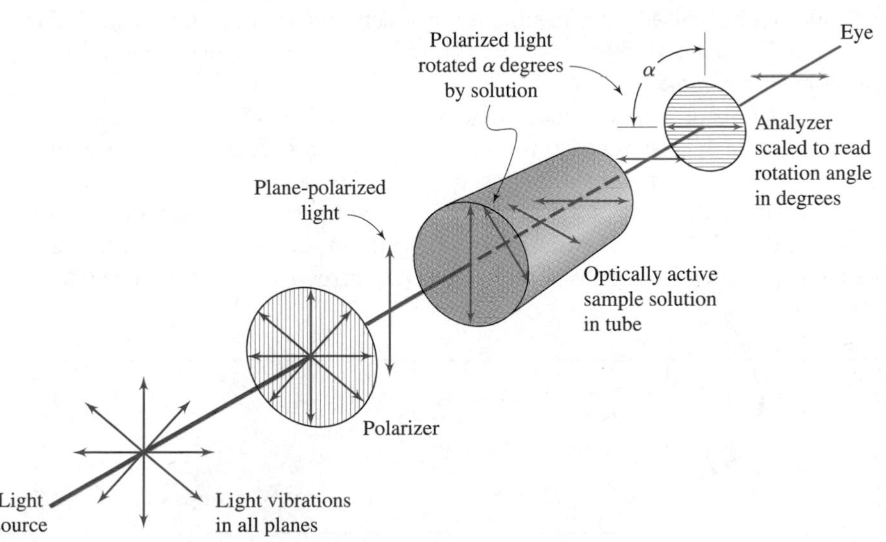

FIGURE 27.3
Schematic diagram of a polarimeter. This instrument is used to measure the angle α through which an optically active substance rotates the plane of polarized light.

Polarized lenses in sunglasses help reduce glare from the sun.

rotated to the position where the emerging light is at maximum intensity. The number of degrees and the direction of rotation by the solution are then read from the scale as the observed rotation.

specific rotation

The **specific rotation**, $[\alpha]$, of a compound is the number of degrees that polarized light would be rotated by passage through 1 decimeter (dm) of a solution of the substance at a concentration of 1 g/mL. The specific rotation of optically active substances is listed in chemical handbooks along with other physical properties. The following formula is used to calculate specific rotation from polarimeter data:

$$[\alpha] = \frac{\text{Observed rotation in degrees}}{\left(\begin{array}{c}\text{Length of}\\ \text{sample tube in decimeters}\end{array}\right)\left(\begin{array}{c}\text{Sample concentration in}\\ \text{grams per milliliter}\end{array}\right)}$$

27.3 Optical Activity

optical activity

Many naturally occurring substances are able to rotate the plane of polarized light. Because of this ability to rotate polarized light, such substances are said to be **optically active**. When plane-polarized light passes through an optically active substance, the plane of the polarized light is rotated. If the rotation is to the right (clockwise), the substance is said to be **dextrorotatory**; if the rotation is to the left (counterclockwise), the substance is said to be **levorotatory**.

dextrorotatory

levorotatory

Some minerals, notably quartz, rotate the plane of polarized light (in fact, optical activity was discovered in minerals). However, when such mineral crystals are melted, the optical activity disappears. This means that the optical activity of these crystals must be due to an ordered arrangement within the crystals.

In 1848, Louis Pasteur (1822–1895) observed that sodium ammonium tartrate, a salt of tartaric acid, exists as a mixture of two kinds of crystals. Pasteur carefully hand-separated the two kinds of crystals. Investigating their properties, he found that solutions made from either kind of crystal would rotate the plane of polarized light, but in opposite directions. Since this optical activity was present in a solution, it could not be caused by a specific arrangement within a crystal. Instead, Pasteur concluded that this difference in optical activity occurred because of a difference in molecular symmetry.

asymmetric carbon atom

The tetrahedral arrangement of single bonds around a carbon atom makes asymmetry (lack of symmetry) possible in organic molecules. When four different atoms or functional groups are bonded to a carbon atom, the molecule formed is asymmetric, and the carbon atom is called an **asymmetric carbon atom** (see Figure 27.4). In 1874, J. H. van't Hoff (1852–1911) and J. A. Le Bel (1847–1930) concluded that the presence of at least one asymmetric carbon atom in a molecule of an

FIGURE 27.4
Three-dimensional representation of an asymmetric carbon atom with four different groups bonded to it. The carbon atom is a sphere. Bonds to A and B project from the sphere toward the observer. Bonds to C and D project from the sphere away from the observer. ▶

D
A
B
C
○ = Carbon atom
A, B, C, D = Four different atoms or groups of atoms

optically active substance is a key factor for optical activity. The first Nobel prize in chemistry was awarded to van't Hoff in 1901.

Molecules of optically active substances must have at least one center of dissymmetry. Although there are optically active compounds that do not contain asymmetric carbon atoms (their center of dissymmetry is due to some other structural feature), most optically active organic substances contain one or more asymmetric carbon atoms.

Quartz is an optically active mineral.

27.4 Projection Formulas

Molecules of a compound that contain one asymmetric carbon atom occur in two optically active isomeric forms. This is because the four different groups bonded to the asymmetric carbon atom can be oriented in space in two different configurations. It is important to understand how we represent such isomers on paper.

Let us now consider the spatial arrangement of the lactic acid molecule, $CH_3CH(OH)COOH$, that contains one asymmetric carbon atom:

```
    ¹COOH
      |
   H—²C*—OH      C* = asymmetric carbon atom
      |
    ³CH₃
```

lactic acid

Three-dimensional models are the best means of representing such a molecule, but by adopting certain conventions and using imagination, we can formulate the images on paper. The geometrical arrangement of the four groups about the asymmetric carbon (carbon 2) is the key to the stereoisomerism of lactic acid. The four bonds attached to carbon 2 are separated by angles of about 109.5°. Diagram I in Figure 27.5 is a three-dimensional representation of lactic acid in which the asymmetric carbon atom is represented as a sphere, with its center in the plane of the paper. The —H and —OH groups are projected forward from the paper (toward the observer), and the —COOH and $—CH_3$ groups are projected back from the paper (away from the observer).

For convenience of expression, simpler diagrams such as II and III are used. These are much easier and faster to draw. In II, it is understood that the groups (—H and —OH) attached to the horizontal bonds are coming out of the plane of

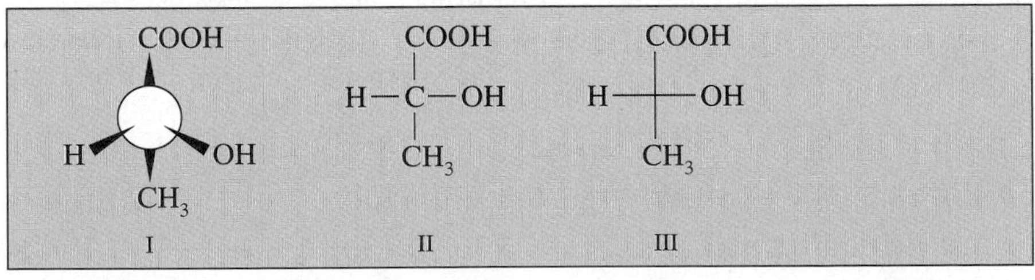

FIGURE 27.5
Methods of representing three-dimensional formulas of a compound that contains one asymmetric carbon atom. All three structures represent the same molecule.

the paper toward the observer, and the groups attached to the vertical bonds are projected back from the paper. The molecule represented by formula III is made by drawing a cross and attaching the four groups in their respective positions, as in formula II. The asymmetric carbon atom is understood to be located where the lines cross. Formulas II and III are called **projection formulas**.

projection formula

It is important to be careful when comparing projection formulas. Two rules apply: (1) Projection formulas must not be turned 90°; (2) projection formulas must not be lifted or flipped out of the plane of the paper. Projection formulas may, however, be turned 180° in the plane of the paper without changing the spatial arrangement of the molecule. Consider the following projection formulas:

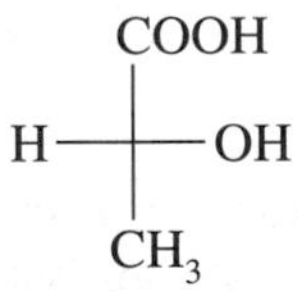

COOH
H—+—OH
CH_3
III

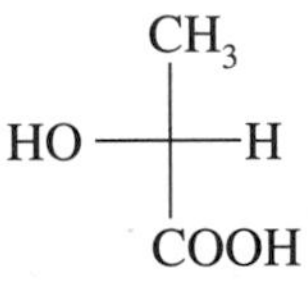

CH_3
HO—+—H
COOH
IV

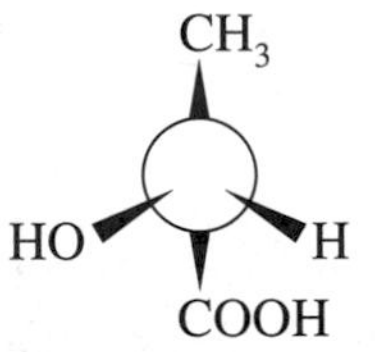

CH_3
HO ◯ H
COOH
V

Formulas I, II, III, IV, and V represent the same molecule. Formula IV was obtained by turning formula III 180°. Formula V is formula IV drawn in a three-dimensional representation. If formula III is turned 90°, the other stereoisomer of lactic acid is represented, as shown in formulas VI and VII.

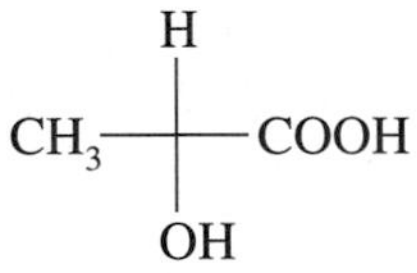

H
CH_3—+—COOH
OH
VI

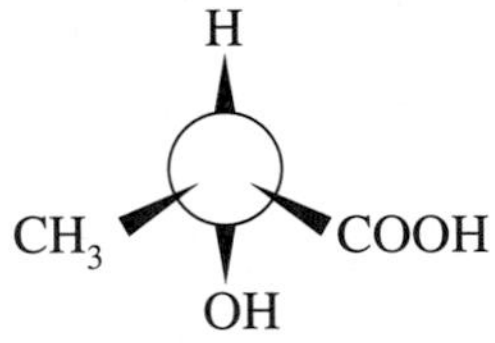

H
CH_3 ◯ COOH
OH
VII

Example 27.1 (a) Redraw the three-dimensional formula (A) as a projection formula. (b) Draw the three-dimensional formula represented by the projection formula (B).

(A) 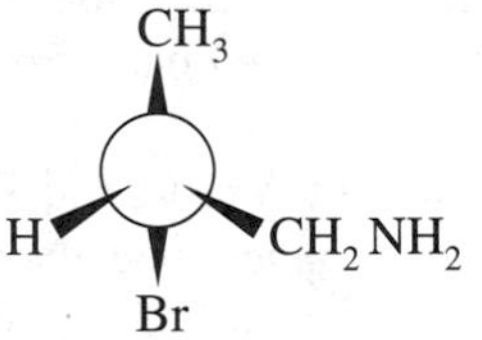
CH_3
H ◯ CH_2NH_2
Br
$CH_3CHBrCH_2NH_2$

(B) 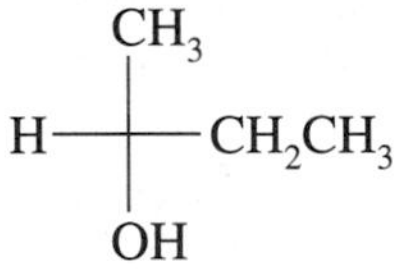
CH_3
H—+—CH_2CH_3
OH
$CH_3CH(OH)CH_2CH_3$

Solution (a) Draw a vertical and a horizontal line crossing each other. Place the CH_3 at the top and the Br at the bottom of the vertical line. Place the H at the left and the CH_2NH_2 at the right of the horizontal line to complete the projection formula:

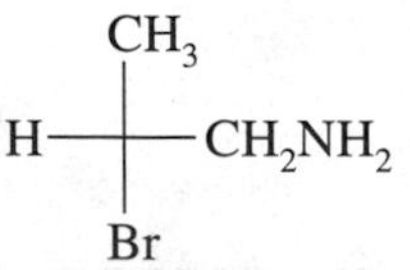

CH_3
H—+—CH_2NH_2 (projection formula of A)
Br

(b) Draw a small circle to represent the asymmetric carbon atom (carbon 2) that is located where the two lines cross in the projection formula. Draw a short line extending from the top and from the bottom of the circle. Place the CH_3 on the top and the OH on the bottom at the end of these lines. Now draw two short lines from within the circle coming toward your left and right arms. Place the H at the end of the left line and the CH_2CH_3 at the end of the right line. The finished formula should look like this:

CH_3

H ◯ CH_2CH_3 (three-dimensional formula of B)

OH

Practice 27.1

Redraw this projection formula as a three-dimensional formula.

Br

Cl—┼—CH_3

F

Practice 27.2

Redraw this three-dimensional formula as a projection formula.

COOH

H ◯ NH_2

CH_2CH_3

FIGURE 27.6
The left hand is the same as the mirror image of the right hand. Right and left hands are not superimposable; hence they are chiral.

27.5 Enantiomers

Your right and your left hands are mirror images of each other; that is, your left hand is a mirror reflection of your right hand, and vice versa. Furthermore, your hands are not superimposable on each other (see Figure 27.6). **Superimposable** means that, when we lay one object upon another, all parts of both objects coincide exactly.

superimposable

A molecule that is not superimposable on its mirror image is said to be **chiral**. The word *chiral* comes from the Greek word *cheir*, meaning hand. Chiral molecules have the property of "handedness"; that is, they are related to each other in the same manner as the right and left hands. An asymmetric carbon atom is often

chiral

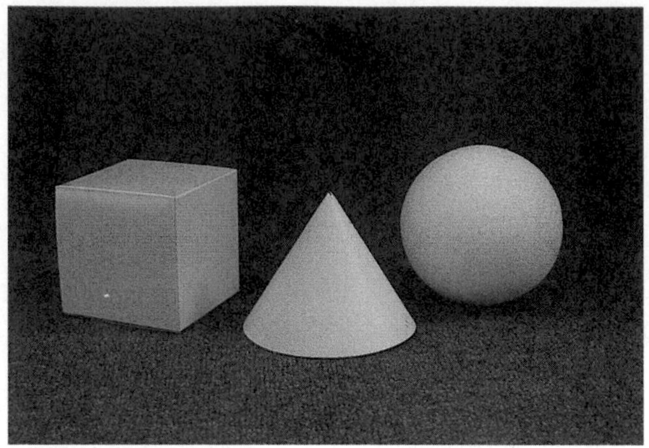

(a) achiral

(b) chiral

▲
FIGURE 27.7
Can you explain why these objects are achiral or chiral?

referred to as a chiral carbon or a chiral center. Molecules or objects that are superimposable are **achiral**. Some chiral and achiral objects are shown in Figure 27.7.

achiral

The formulas developed in Section 27.4 dealt primarily with a single kind of lactic acid molecule. But two stereoisomers of lactic acid are known, one that rotates the plane of polarized light to the right and one that rotates it to the left. These two forms of lactic acid are shown in Figure 27.8. If we examine these two structural formulas carefully, we can see that they are mirror images of each other. The reflection of either molecule in a mirror corresponds to the structure of the other molecule. Even though the two molecules have the same molecular formula and the same four groups attached to the central carbon atom, they are not superimposable. Therefore, the two molecules are not identical but are isomers. One molecule rotates the plane of polarized light to the left and is termed *levorotatory*; the other molecule rotates it to the right and is *dextrorotatory*. A plus (+) or a minus (−) sign written in parentheses and placed in front of a name or formula indicates the direction of rotation of polarized light and becomes part of the name of the compound. Plus (+) indicates rotation to the right and minus (−) to the left. Using projection formulas, we write the two lactic acids as shown on page 741.

FIGURE 27.8 ▶
Mirror-image isomers of lactic acid. Each isomer is the mirror reflection of the other. (−)-Lactic acid rotates plane-polarized light to the left, and (+)-lactic acid rotates plane-polarized light to the right.

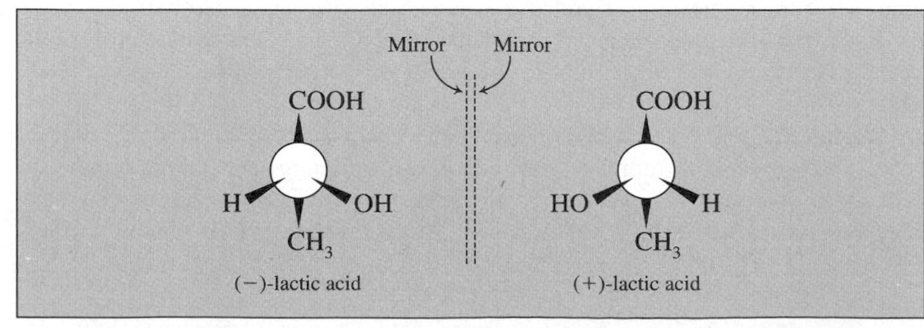

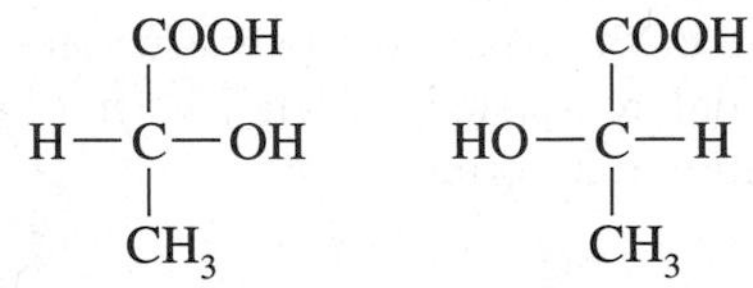

$$\begin{array}{c} COOH \\ | \\ H-C-OH \\ | \\ CH_3 \end{array} \qquad \begin{array}{c} COOH \\ | \\ HO-C-H \\ | \\ CH_3 \end{array}$$

(−)-lactic acid (+)-lactic acid

Originally, it was not known which lactic acid structure was the (+) or the (−) compound. However, it is now known that they are as shown. Isomers that are mirror images of each other are called **enantiomers**. The word *enantiomer* comes from the Greek word *enantios*, which means opposite.

enantiomer

Many, but not all, molecules that contain an asymmetric carbon are chiral. Most of the molecules we shall study that have an asymmetric carbon atom are chiral. To decide whether a molecule is chiral and has an enantiomer, make models of the molecule and of its mirror image and see if they are superimposable. This is the ultimate test, but instead of making models every time, first examine the formula to see if it has an asymmetric carbon atom. If an asymmetric carbon atom is found, draw a cross and attach the four groups on the asymmetric carbon to the four ends of the cross. The asymmetric carbon is understood to be located where the lines cross. Remember that an asymmetric carbon atom has four different groups attached to it. Let's test the compounds 2-butanol and 2-chloropropane:

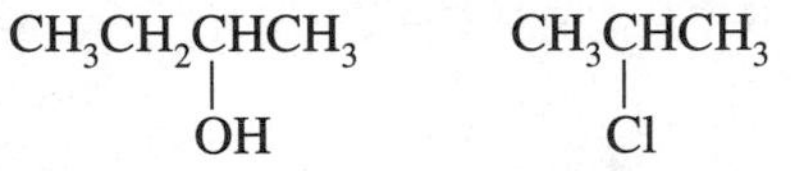

$$\begin{array}{c} CH_3CH_2CHCH_3 \\ \quad\;\; | \\ \quad\;\; OH \end{array} \qquad \begin{array}{c} CH_3CHCH_3 \\ | \\ Cl \end{array}$$

2-butanol 2-chloropropane

In 2-butanol, carbon 2 is asymmetric. The four groups attached to carbon 2 are H, OH, CH_3, and CH_2CH_3. Draw the structure and its mirror image:

Mirror

$$\begin{array}{c} CH_3 \\ H-\!\!+\!\!-OH \\ CH_2 \\ | \\ CH_3 \end{array} \quad \vdots \quad \begin{array}{c} CH_3 \\ HO-\!\!+\!\!-H \\ CH_2 \\ | \\ CH_3 \end{array} \qquad \begin{array}{c} CH_3 \\ | \\ CH_2 \\ H-\!\!+\!\!-OH \\ CH_3 \end{array}$$

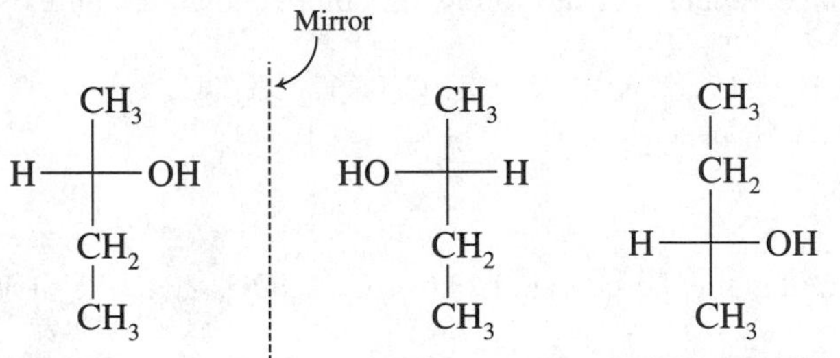

VIII IX X (IX turned 180°)

enantiomers

Turning structure IX 180° in the plane of the paper allows H and OH to coincide with their position in VIII, but CH_3 and CH_2CH_3 do not coincide. Therefore, we conclude that the mirror-image structures VIII and IX are enantiomers since they are not superimposable.

In 2-chloropropane, the four groups attached to carbon 2 are H, Cl, CH_3, and CH_3. Note that two groups are the same. Draw the structure and its mirror image:

Mirror

$$\begin{array}{c} CH_3 \\ H-\!\!+\!\!-Cl \\ CH_3 \end{array} \quad \vdots \quad \begin{array}{c} CH_3 \\ Cl-\!\!+\!\!-H \\ CH_3 \end{array} \qquad \begin{array}{c} CH_3 \\ H-\!\!+\!\!-Cl \\ CH_3 \end{array}$$

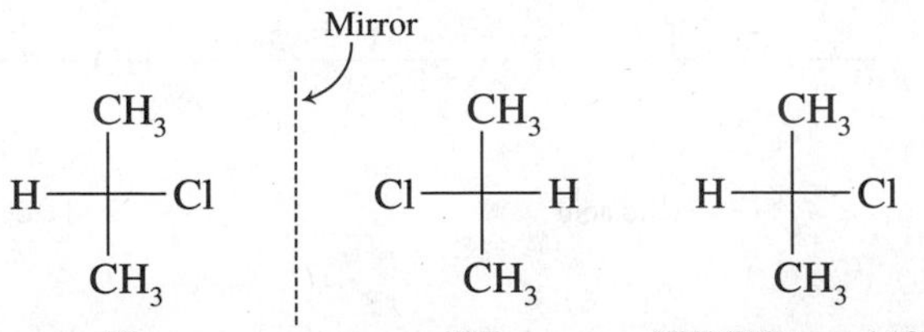

XI XII XIII (XII turned 180°)

When we turn structure XII 180° in the plane of the paper, the two structures XI and XIII are superimposable, proving that 2-chloropropane, which does not have an asymmetric carbon, does not exist in enantiomeric forms.

Example 27.2 Draw mirror-image isomers for any of the following compounds that can exist as enantiomers:

(a) $CH_3CH_2CH_2OH$
(b) $CH_3CH_2CHClCH_2CH_3$
(c) $CH_3CH_2CHClCH_2CH_2CH_3$

Solution First check each formula for asymmetric carbon atoms:

(a) No asymmetric carbon atoms; each carbon has at least two groups that are the same.
(b) No asymmetric carbon atoms; carbon 3 has H, Cl, and two CH_3CH_2 groups.
(c) Carbon 3 is asymmetric; the four groups on carbon 3 are H, Cl, CH_3CH_2, and $CH_3CH_2CH_2$. Draw mirror images:

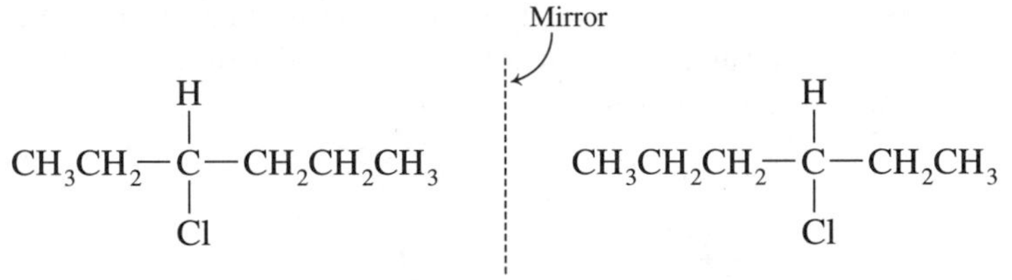

Mirror

$$CH_3CH_2-\overset{\displaystyle H}{\underset{\displaystyle Cl}{\overset{|}{\underset{|}{C}}}}-CH_2CH_2CH_3 \qquad \Big| \qquad CH_3CH_2CH_2-\overset{\displaystyle H}{\underset{\displaystyle Cl}{\overset{|}{\underset{|}{C}}}}-CH_2CH_3$$

Practice 27.3

Draw the mirror-image isomers for the following compounds that can exist as enantiomers:

(a) $NH_2CH_2CHBrCH_2OH$
(b) $NH_2CH_2CHBrCH_2NH_2$
(c) $NH_2CH_2CH_2CH_2OH$

Practice 27.4

Draw structural formulas for the pentyl alcohols, $C_5H_{11}OH$, that will show optical activity.

The relationship between enantiomers is such that if we change the positions of any two groups on a compound that contains only one asymmetric carbon atom, we obtain the structure of its enantiomer. If we make a second change, the structure of the original isomer is obtained again. In both cases shown here, (+)-lactic acid is formed by interchanging the positions of two groups on (−)-lactic acid:

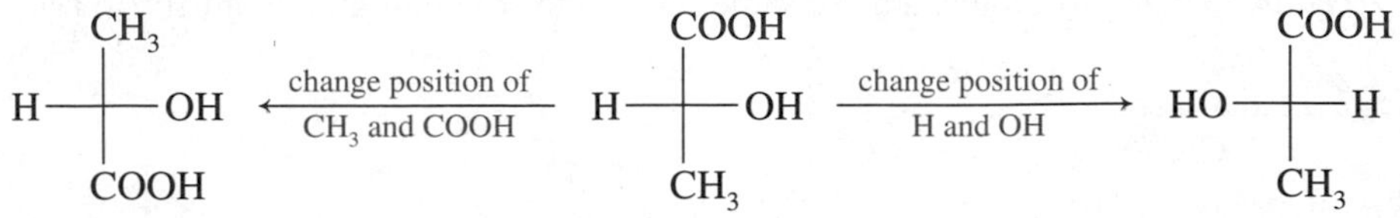

$$H-\overset{\displaystyle CH_3}{\underset{\displaystyle COOH}{\overset{|}{\underset{|}{+}}}}-OH \xleftarrow[\text{CH}_3\text{ and COOH}]{\text{change position of}} H-\overset{\displaystyle COOH}{\underset{\displaystyle CH_3}{\overset{|}{\underset{|}{+}}}}-OH \xrightarrow[\text{H and OH}]{\text{change position of}} HO-\overset{\displaystyle COOH}{\underset{\displaystyle CH_3}{\overset{|}{\underset{|}{+}}}}-H$$

(+)-lactic acid (−)-lactic acid (+)-lactic acid

To test whether two projection formulas are the same structure or are enantiomers, (1) turn one structure 180° in the plane of the paper and compare to see if they are superimposable, or (2) make successive group interchanges until the formulas are identical. If an odd number of interchanges are made, the two original formulas represent enantiomers; if an even number of interchanges are made, the two formulas represent the same compound. The following two examples illustrate this method. Are structures XIV and XV the same compound?

```
      CH3                H
       |                 |
 Br ---+--- H      Br ---+--- Cl
       |                 |
      Cl                CH3

      XIV                XV
```

```
       H                                    H                                  CH3
       |        interchange                 |         interchange               |
 Br ---+--- Cl  ------------>   Br ---+--- CH3  ------------>   Br ---+--- H
       |        CH3 and Cl                  |         CH3 and H                 |
      CH3                                  Cl                                  Cl

      XV                                                                       XIV
```

Two interchanges were needed to make structure XV identical to structure XIV. Therefore, structures XIV and XV represent the same compound.

Do structures XVI and XVII represent the same compound?

```
      CH3                       OH
       |                         |
 HO ---+--- H      CH3CH2 ---+--- CH3
       |                         |
    CH2CH3                       H

      XVI                      XVII
```

```
            OH                                       CH2CH3
             |          interchange                     |           interchange
 CH3CH2 ---+--- CH3   ---------------->   HO ---+--- CH3   ---------------->
             |          CH3CH2 and OH                   |           CH3 and H
             H                                          H

           XVII

        CH2CH3                                    CH3
           |            interchange                |
 --> HO ---+--- H   -------------------->   HO ---+--- H
           |            CH3 and CH3CH2             |
          CH3                                   CH2CH3

                                                  XVI
```

Three interchanges were needed to make structure XVII identical to structure XVI. Therefore, structures XVI and XVII do not represent the same compound; they are enantiomers.

Enantiomers ordinarily have the same chemical properties, and other than optical rotation, they also have the same physical properties (see Table 27.1). They rotate plane-polarized light the same number of degrees but in opposite directions.

Enantiomers usually differ in their biochemical properties. In fact, most living cells are able to use only one isomer of a mirror-image pair. For example, (+)-glucose ("blood sugar") can be used for metabolic energy, whereas (−)-glucose cannot. Enantiomers are truly different molecules and are treated as such by most organisms. The key factors of enantiomers and optical isomerism can be summarized as follows:

1. A carbon atom that has four different groups bonded to it is called an asymmetric or a chiral carbon atom.
2. A compound with one asymmetric carbon atom can exist in two isomeric forms called enantiomers.
3. Enantiomers are nonsuperimposable mirror-image isomers.
4. Enantiomers are optically active; that is, they are able to rotate plane-polarized light.
5. One isomer of an enantiomeric pair rotates polarized light to the left (counterclockwise). The other isomer rotates polarized light to the right (clockwise). The degree of rotation is the same but in opposite directions.
6. Rotation of polarized light to the right is indicated by (+), placed in front of the name of the compound. Rotation to the left is indicated by (−), for example, (+)-lactic acid and (−)-lactic acid.

27.6 Racemic Mixtures

racemic mixture

A mixture containing equal amounts of a pair of enantiomers is known as a **racemic mixture**. Such a mixture is optically inactive and shows no rotation of polarized light when tested in a polarimeter. Each of the enantiomers rotates the plane of polarized light by the same amount but in opposite directions. Thus the rotation by each isomer is canceled. The (±) symbol is often used to designate racemic mixtures. For example, a racemic mixture of lactic acid is written as (±)-lactic acid because this mixture contains equal molar amounts of (+)-lactic acid and (−)-lactic acid.

Racemic mixtures are usually obtained in laboratory syntheses of compounds in which an asymmetric carbon atom is formed. Thus catalytic reduction of pyruvic acid (an achiral compound) to lactic acid produces a racemic mixture containing equal amounts of (+)- and (−)-lactic acid:

$$\underset{\text{pyruvic acid}}{CH_3\underset{\underset{O}{\|}}{C}COOH} + H_2 \xrightarrow{Ni} \underset{(\pm)\text{-lactic acid}}{CH_3\underset{\underset{OH}{|}}{C}HCOOH}$$

As a general rule, in the biological synthesis of potentially optically active compounds, only one of the isomers is produced. For example, (+)-lactic acid is produced by reactions occurring in muscle tissue, and (−)-lactic acid is produced by lactic acid bacteria in the souring of milk. These stereospecific reactions occur because biochemical syntheses are enzyme catalyzed. The preferential production of one isomer over another is often due to the configuration (shape) of the specific enzyme involved. Returning to the hand analogy, if the "right-handed" enantiomer is produced, then the enzyme responsible for the product can be likened to a right-handed glove.

The mirror-image isomers (enantiomers) of a racemic mixture are alike in all ordinary physical properties except in their action on polarized light. It is possible to separate or resolve racemic mixtures into their optically active components. In fact, Pasteur's original work with sodium ammonium tartrate involved such a separation. But a general consideration of the methods involved in such separations is beyond the scope of our present discussion.

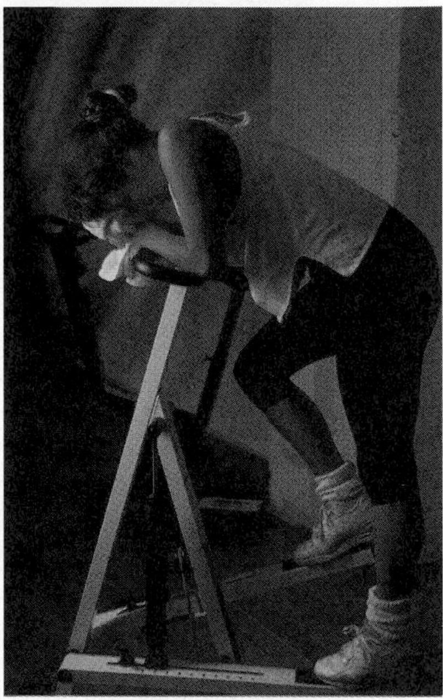

Overuse of muscle tissue results in the buildup of (+)-lactic acid.

27.7 Diastereomers and Meso Compounds

The enantiomers discussed in the preceding sections are stereoisomers. That is, they differ only in the spatial arrangement of the atoms and groups within the molecule. The number of stereoisomers increases as the number of asymmetric carbon atoms increases. The maximum number of stereoisomers for a given compound is obtained by the formula 2^n, where n is the number of asymmetric carbon atoms.

> **2^n = Maximum number of stereoisomers for a given chiral compound**
> **n = Number of asymmetric carbon atoms in a molecule**

As we have seen, there are two ($2^1 = 2$) stereoisomers of lactic acid. But for a substance with two nonidentical asymmetric carbon atoms, such as 2-bromo-3-chlorobutane ($CH_3CHBrCHClCH_3$), four stereoisomers are possible ($2^2 = 4$). These four possible stereoisomers are written as projection formulas in this way (carbons 2 and 3 are asymmetric):

$$\begin{array}{ccc} & CH_3 & \\ H & - & Br \\ H & - & Cl \\ & CH_3 & \end{array} \quad \begin{array}{ccc} & CH_3 & \\ Br & - & H \\ Cl & - & H \\ & CH_3 & \end{array} \qquad\qquad \begin{array}{ccc} & CH_3 & \\ H & - & Br \\ Cl & - & H \\ & CH_3 & \end{array} \quad \begin{array}{ccc} & CH_3 & \\ Br & - & H \\ H & - & Cl \\ & CH_3 & \end{array}$$

XVIII XIX XX XXI

enantiomers (XVIII, XIX) enantiomers (XX, XXI)

Remember that, for comparison, projection formulas may be turned 180° in the plane of the paper, but they cannot be lifted (flipped) out of the plane. Formulas XVIII and XIX, and formulas XX and XXI, represent two pairs of nonsuperimposable mirror-image isomers and are, therefore, two pairs of enantiomers. All four compounds are optically active. But the properties of XVIII and XIX differ from the properties of XX and XXI because they are not mirror-image isomers of each other. Stereoisomers that are not enantiomers (not mirror images of each other) are called **diastereomers**. There are four different pairs of diastereomers of 2-bromo-3-chlorobutane: They are XVIII and XX, XVIII and XXI, XIX and XX, and XIX and XXI.

diastereomer

The 2^n formula indicates that four stereoisomers of tartaric acid are possible. The projection formulas of these four possible stereoisomers are written in this way (carbons 2 and 3 are asymmetric):

$$\begin{array}{c} \mathrm{COOH} \\ \mathrm{HO}-\!\!\!+\!\!\!-\mathrm{H} \\ \mathrm{H}-\!\!\!+\!\!\!-\mathrm{OH} \\ \mathrm{COOH} \\ \text{XXII} \end{array} \qquad \begin{array}{c} \mathrm{COOH} \\ \mathrm{H}-\!\!\!+\!\!\!-\mathrm{OH} \\ \mathrm{HO}-\!\!\!+\!\!\!-\mathrm{H} \\ \mathrm{COOH} \\ \text{XXIII} \end{array}$$

$$\begin{array}{c} \mathrm{COOH} \\ \mathrm{H}-\!\!\!+\!\!\!-\mathrm{OH} \\ \mathrm{H}-\!\!\!+\!\!\!-\mathrm{OH} \\ \mathrm{COOH} \\ \text{XXIV} \end{array} \qquad \begin{array}{c} \mathrm{COOH} \\ \mathrm{HO}-\!\!\!+\!\!\!-\mathrm{H} \\ \mathrm{HO}-\!\!\!+\!\!\!-\mathrm{H} \\ \mathrm{COOH} \\ \text{XXV} \end{array}$$

Formulas XXII and XXIII represent nonsuperimposable mirror-image isomers and are therefore enantiomers. Formulas XXIV and XXV are also mirror images. But by turning XXV 180°, we see that it is exactly superimposable on XXIV. Therefore, XXIV and XXV represent the same compound, and only *three* stereoisomers of tartaric acid actually exist. Compound XXIV is achiral and does not rotate polarized light. A plane of symmetry can be passed between carbons 2 and 3 so that the top and bottom halves of the molecule are mirror images:

$$\begin{array}{c} \mathrm{COOH} \\ | \\ \mathrm{H-C-OH} \\ \text{-------------} \\ \mathrm{H-C-OH} \\ | \\ \mathrm{COOH} \end{array} \quad \leftarrow \text{Plane of symmetry}$$

Thus the molecule is internally compensated. The rotation of polarized light in one direction by half of the molecule is exactly compensated by an opposite rotation by the other half. Stereoisomers that contain asymmetric carbon atoms and are superimposable on their own mirror images are called **meso compounds**, or **meso structures**. All meso compounds are optically inactive.

meso compound
meso structure

The term *meso* comes from the Greek word *mesos*, meaning middle. It was first used by Pasteur to name a kind of tartaric acid that was optically inactive and could not be separated into different forms by any means. Pasteur called it *meso*-tartaric acid, because it seemed intermediate between the (+)- and (−)-tartaric acid. The three stereoisomers of tartaric acid are represented and designated in this fashion:

$$\begin{array}{c} \mathrm{COOH} \\ | \\ \mathrm{HO-C-H} \\ | \\ \mathrm{H-C-OH} \\ | \\ \mathrm{COOH} \\ (-)\text{-tartaric acid} \end{array} \qquad \begin{array}{c} \mathrm{COOH} \\ | \\ \mathrm{H-C-OH} \\ | \\ \mathrm{HO-C-H} \\ | \\ \mathrm{COOH} \\ (+)\text{-tartaric acid} \end{array} \qquad \begin{array}{c} \mathrm{COOH} \\ | \\ \mathrm{H-C-OH} \\ | \\ \mathrm{H-C-OH} \\ | \\ \mathrm{COOH} \\ \textit{meso}\text{-tartaric acid} \end{array}$$

TABLE 27.1 Properties of Tartaric Acid, HOOCCH(OH)CH(OH)COOH

Name	Specific gravity	Melting point (°C)	Solubility (g/100 g H_2O)	Specific rotation [α]
(+)-Tartaric acid	1.760	170	$147^{20°C}$	+12°
(−)-Tartaric acid	1.760	170	$147^{20°C}$	−12°
(±)-Tartaric acid (racemic mixture)	1.687	206	$20.6^{20°C}$	0°
meso-Tartaric acid	1.666	140	$125^{15°C}$	0°

The physical properties of tartaric acid stereoisomers are given in Table 27.1. Note that the properties of (+)-tartaric acid and (−)-tartaric acid are identical except for opposite rotation of polarized light. However, *meso*-tartaric acid has properties that are entirely different from those of the other isomers. But most surprising is the fact that the racemic mixture, though composed of equal parts of the (+) and (−) enantiomers, differs from them in specific gravity, melting point, and solubility. Why, for example, is the melting point of the racemic mixture higher than that of any of the other forms? The melting point of any substance is largely dependent on the attractive forces holding the ions or molecules together. The melting point of the racemic mixture is higher than that of either enantiomer. Therefore, we can conclude that the attraction between molecules of the (+) and (−) enantiomers in the racemic mixture is greater than the attraction between molecules of the (+) and (+) or the (−) and (−) enantiomers.

Example 27.3

How many stereoisomers can exist for the following compounds? Write their structures and label any pairs of enantiomers and meso compounds. Point out any diastereomers.

(a) $CH_3CHBrCHBrCH_2CH_3$ (b) $CH_2BrCHClCHClCH_2Br$

Solution

(a) Carbons 2 and 3 are asymmetric, so there can be a maximum of four stereoisomers ($2^2 = 4$). Write structures around the asymmetric carbons:

	I	II	III	IV
top	CH_3	CH_3	CH_3	CH_3
C-2	H—C—Br	Br—C—H	H—C—Br	Br—C—H
C-3	H—C—Br	Br—C—H	Br—C—H	H—C—Br
bottom	CH_2—CH_3	CH_2—CH_3	CH_2—CH_3	CH_2—CH_3

I and II: enantiomers; III and IV: enantiomers

There are four stereoisomers: two pairs of enantiomers (I and II, and III and IV) and no meso compounds. Structures I and III, I and IV, II and III, and II and IV are diastereomers.

(b) Carbons 2 and 3 are asymmetric, so there can be a maximum of four stereoisomers. Write structures around the asymmetric carbons:

$$
\begin{array}{cccc}
\begin{array}{c} CH_2Br \\ | \\ H-C-Cl \\ | \\ H-C-Cl \\ | \\ CH_2Br \\ \text{V} \end{array} &
\begin{array}{c} CH_2Br \\ | \\ Cl-C-H \\ | \\ Cl-C-H \\ | \\ CH_2Br \\ \text{VI} \end{array} &
\begin{array}{c} CH_2Br \\ | \\ Cl-C-H \\ | \\ H-C-Cl \\ | \\ CH_2Br \\ \text{VII} \end{array} &
\begin{array}{c} CH_2Br \\ | \\ H-C-Cl \\ | \\ Cl-C-H \\ | \\ CH_2Br \\ \text{VIII} \end{array}
\end{array}
$$

meso compound (V, VI) — enantiomers (VII, VIII)

There are three stereoisomers: one pair of enantiomers (VII and VIII) and one meso compound (V). Structures V and VI represent the meso compound because there is a plane of symmetry between carbons 2 and 3, and turning VI 180° makes it superimposable on V. Structures V and VII and V and VIII are diastereomers.

Practice 27.5

Write all stereoisomer structures and label any pairs of enantiomers and meso compounds for the following compound. Also, point out any diastereomers.

$HOOCCHClCH_2CH_2CHClCOOH$

Concepts in Review

1. Identify all asymmetric (chiral) carbon atoms in a given formula.
2. Explain the use of a polarimeter.
3. Explain how polarized light is obtained.
4. Explain the phenomenon of optical isomerism.
5. Determine whether a compound is chiral or achiral.
6. Draw projection formulas for all possible stereoisomers of a given compound. Label enantiomers, diastereomers, and meso compounds.
7. Given the formula of a compound, calculate the maximum possible number of optical isomers.
8. Understand the meaning of (+) and (−) relative to the optical activity of a compound.
9. Draw the mirror image of a given structure.
10. Compare projection formulas to ascertain whether they represent identical compounds or enantiomers.
11. Compare the physical properties of enantiomers, diastereomers, and racemic mixtures.
12. Explain why meso compounds are optically inactive.
13. Determine whether optical isomers are formed in simple chemical reactions.

Key Terms

The terms listed here have been defined within this chapter. Section numbers are referenced in parenthesis for each term.

achiral (27.5)
asymmetric carbon atom (27.3)
chiral (27.5)
dextrorotatory (27.3)
diastereomer (27.7)
enantiomer (27.5)
levorotatory (27.3)
meso compound (27.7)
meso structure (27.7)
optical activity (27.3)
plane-polarized light (27.2)
projection formula (27.4)
racemic mixture (27.6)
specific rotation (27.2)
stereoisomer (27.1)
stereoisomerism (27.1)
superimposable (27.5)

Questions

Questions refer to tables, figures, and key words and concepts defined within the chapter. A particularly challenging question or exercise is indicated with an asterisk.

1. What is an asymmetric carbon atom? Draw structural formulas of three different compounds that contain one asymmetric carbon atom and mark the asymmetric carbon in each with an asterisk.
2. How can you tell when the axes of two Polaroid filters are parallel? When one filter has been rotated by 90°?
3. What is a necessary and sufficient condition for a compound to show enantiomerism?
4. Differentiate between an enantiomer and a diastereomer.
5. The physical properties for (+)-2-methyl-1-butanol are specific rotation, +5.76°; bp, 129°C; density, 0.819 g/mL. What are these same properties for (−)-2-methyl-1-butanol?
6. Which of these statements are correct? Rewrite each incorrect statement to make it correct.
   (a) The polarizer and the analyzer of a polarimeter are made of the same material.
   (b) Very few natural products are optically active.
   (c) Cis–trans isomers are not considered to be stereoisomers.
   (d) J. A. Le Bel received the first Nobel prize in chemistry in 1901.
   (e) Molecules that contain only one asymmetric carbon atom are chiral, but not all chiral molecules contain an asymmetric carbon atom.
   (f) The compound $CH_3CHBrCHBrCH_2OH$ has eight optical isomers.
   (g) The compounds shown in these projection formulas are enantiomers:

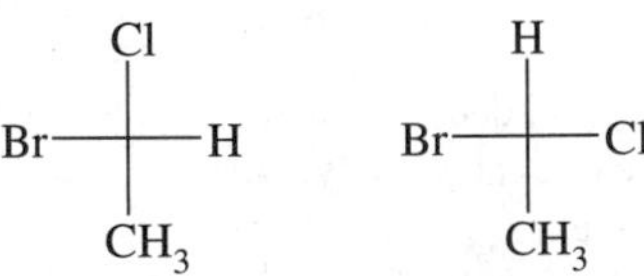

   (h) Diastereomers have identical melting points.
   (i) A molecule that contains two asymmetric carbon atoms may not be chiral.
   (j) A racemic mixture contains equal amounts of the dextrorotatory and levorotatory molecules of a compound.

Paired Exercises

These exercises are paired. Each odd-numbered exercise is followed by a similar even-numbered exercise. Answers to the even-numbered exercises are given in Appendix V.

7. Explain why it is not possible to separate enantiomers by ordinary chemical and physical means.

9. Which of these objects is chiral?
(a) your ear
(b) a pair of pliers
(c) a coiled spring
(d) the letter b

11. How many asymmetric carbon atoms are present in each of the following?

(a)
```
         H   Cl
         |   |
    Cl — C — C — Br
         |   |
         H   H
```

(b) $CH_3CH_2CH_2CHClCH_3$

(c)
```
H — C = O
    |
H — C — OH
    |
H — C — OH
    |
H — C — OH
    |
    H
```

(d)
```
         H   H   O
         |   |   ||
    HO — C — C — C — OH
         |   |
         H   NH2
```

13. Write the formulas and decide which of the following compounds will show optical activity:
(a) 1-chloropentane
(b) 3-chloropentane
(c) 2-chloro-2-methylpentane
(d) 4-chloro-2-methylpentane

15. Glucose, $C_6H_{12}O_6$, has four asymmetric carbon atoms. How many stereoisomers of glucose are theoretically possible?

17. Do structures (A) and (B) on next page represent enantiomers or the same compound? Justify your answer.

8. Explain why it is usually possible to separate diastereomers by ordinary chemical and physical means.

10. Which of these objects is chiral?
(a) a wood screw
(b) the letter o
(c) the letter g
(d) this textbook

12. How many asymmetric carbon atoms are present in each of the following?

(a)
```
        Br  Cl
        |   |
    H — C — C — H
        |   |
        H   H
```

(b)
```
              OH
              |
    CH3CH2CHCH2CH3
```

(c)
```
         H
         |
    H  — C — OH
         |
         C = O
         |
    HO — C — H
         |
    H  — C — OH
         |
    H  — C — OH
         |
    H  — C — OH
         |
         H
```

(d) $CH_3CH_2CHBrCHClCH_3$

14. Write the formulas and decide which of the following compounds will show optical activity:
(a) 2-chloropentane
(b) 1-chloro-2-methylpentane
(c) 3-chloro-2-methylpentane
(d) 3-chloro-3-methylpentane

16. Fructose, $C_6H_{12}O_6$, has three asymmetric carbon atoms. How many stereoisomers of fructose are theoretically possible?

18. Do structures (A) and (B) on next page represent enantiomers or the same compound? Justify your answer.

$$\text{(A)}\ \begin{array}{ccccc} & & Cl & & \\ & & | & & \\ Br & - & C & - & H \\ & & | & & \\ & & F & & \end{array} \qquad \text{(B)}\ \begin{array}{ccccc} & & Br & & \\ & & | & & \\ H & - & C & - & Cl \\ & & | & & \\ & & F & & \end{array}$$

19. Which of these projection formulas represent (−)-lactic acid and which represent (+)-lactic acid (see Section 27.5)?

$$\text{(a)}\ \begin{array}{ccc} & CH_3 & \\ HO & + & H \\ & COOH & \end{array} \qquad \text{(d)}\ \begin{array}{ccc} & CH_3 & \\ H & + & OH \\ & COOH & \end{array}$$

$$\text{(b)}\ \begin{array}{ccc} & OH & \\ H & + & COOH \\ & CH_3 & \end{array} \qquad \text{(e)}\ \begin{array}{ccc} & COOH & \\ CH_3 & + & H \\ & OH & \end{array}$$

$$\text{(c)}\ \begin{array}{ccc} & COOH & \\ H & + & CH_3 \\ & OH & \end{array} \qquad \text{(f)}\ \begin{array}{ccc} & H & \\ CH_3 & + & OH \\ & COOH & \end{array}$$

21. Draw projection formulas for all the possible stereoisomers of the following compounds. Label pairs of enantiomers and meso compounds.
(a) 1,2-dibromopropane
(b) 2-butanol
(c) 3-chlorohexane

23. Draw projection formulas for all the stereoisomers of 1,2,3-trihydroxybutane. Point out enantiomers, meso compounds, and diastereomers, where present.

***25.** Write structures for the four stereoisomers of 2-hydroxy-3-pentene.

27. **(a)** Draw the nine structural isomers of $C_4H_8Cl_2$.
(b) Identify which structures represent chiral molecules, and draw all possible pairs of enantiomers and meso compounds, if any.

29. (+)-2-Bromopentane is further brominated to give dibromopentanes, $C_5H_{10}Br_2$. Write structures for all the possible isomers formed and indicate which of these isomers will be optically active. [Remember that (+)-2-bromopentane is optically active.

$$\text{(A)}\ \begin{array}{ccccc} & & H & & \\ & & | & & \\ Br & - & C & - & F \\ & & | & & \\ & & CH_3 & & \end{array} \qquad \text{(B)}\ \begin{array}{ccccc} & & CH_3 & & \\ & & | & & \\ Br & - & C & - & H \\ & & | & & \\ & & F & & \end{array}$$

20. Which of these projection formulas represent the naturally occurring amino acid, (+)-alanine and which represent (−)-alanine?

$$\begin{array}{ccc} & COOH & \\ H_2N & + & H \\ & CH_3 & \end{array} \qquad \begin{array}{ccc} & COOH & \\ H & + & NH_2 \\ & CH_3 & \end{array}$$
$$\text{(+)-alanine} \qquad\qquad \text{(−)-alanine}$$

$$\text{(a)}\ \begin{array}{ccc} & CH_3 & \\ H_2N & + & H \\ & COOH & \end{array} \qquad \text{(d)}\ \begin{array}{ccc} & CH_3 & \\ H & + & NH_2 \\ & COOH & \end{array}$$

$$\text{(b)}\ \begin{array}{ccc} & NH_2 & \\ H & + & COOH \\ & CH_3 & \end{array} \qquad \text{(e)}\ \begin{array}{ccc} & COOH & \\ CH_3 & + & H \\ & NH_2 & \end{array}$$

$$\text{(c)}\ \begin{array}{ccc} & COOH & \\ H & + & CH_3 \\ & NH_2 & \end{array} \qquad \text{(f)}\ \begin{array}{ccc} & H & \\ CH_3 & + & NH_2 \\ & COOH & \end{array}$$

22. Draw projection formulas for all the possible stereoisomers of the following compounds. Label pairs of enantiomers and meso compounds.
(a) 2,3-dichlorobutane
(b) 2,4-dibromopentane
(c) 3-hexanol

24. Draw projection formulas for all the stereoisomers of 3,4-dichloro-2-methylpentane. Point out enantiomers, meso compounds, and diastereomers, where present.

***26.** Write structures for the four stereoisomers of 2-chloro-3-hexene.

28. **(a)** Draw the four structural isomers of $C_3H_6Br_2$.
(b) Identify which structures represent chiral molecules, and draw all possible pairs of enantiomers and meso compounds, if any.

30. (+)-2-Chlorobutane is further chlorinated to give dichlorobutanes, $C_4H_8Cl_2$. Write structures for all the possible isomers formed and indicate which of these isomers will be optically active. [Remember that (+)-2-chlorobutane is optically active.

31. In the chlorination of propane, 1-chloropropane and 2-chloropropane are obtained as products. After separation by distillation, neither product rotates the plane of polarized light. Explain these results.

33. Which, if any, of the following are meso compounds?

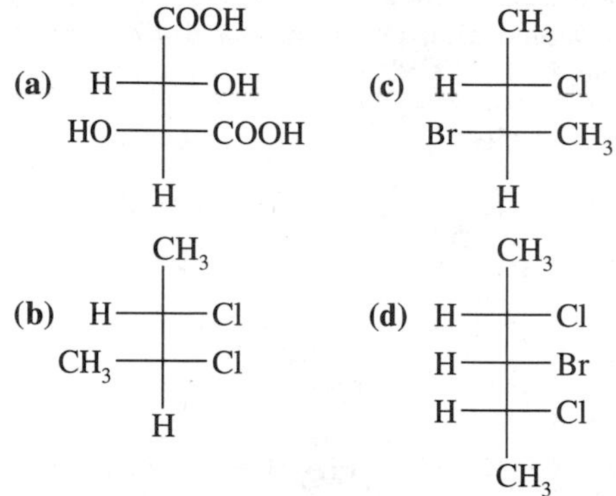

***35.** Some substituted cycloalkanes are chiral. Draw the structures and enantiomers for any of the following that are chiral:

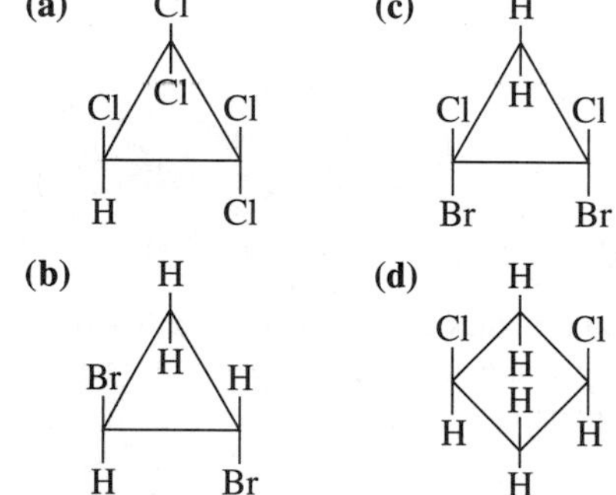

32. In the chlorination of butane, 1-chlorobutane and 2-chlorobutane are obtained as products. After separation by distillation, neither product rotates the plane of polarized light. Explain these results.

34. Which, if any, of the following are meso compounds?

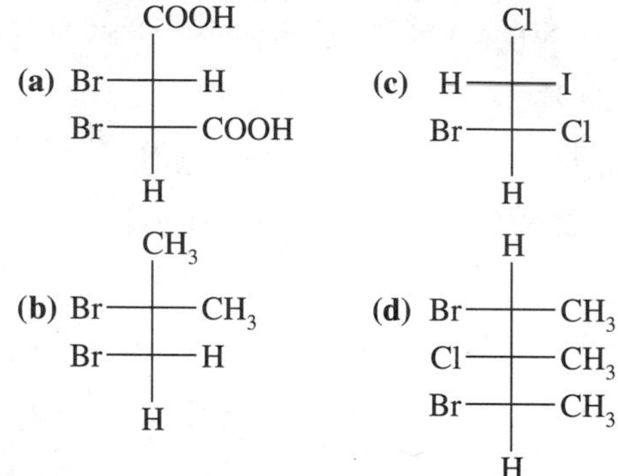

***36.** Some substituted cycloalkanes are chiral. Draw the structures and enantiomers for any of the following that are chiral:

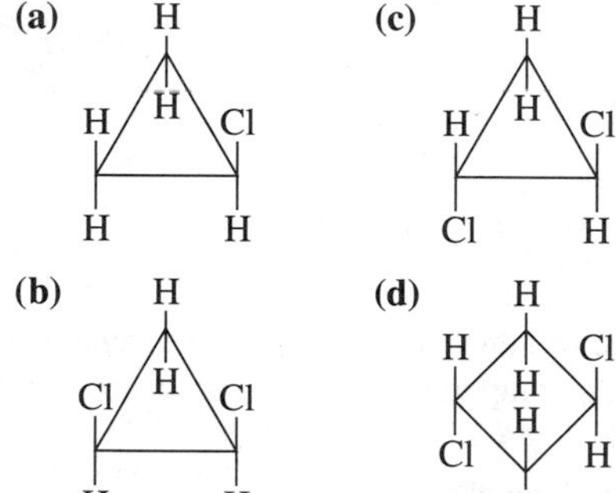

Additional Exercises

These exercises are not paired or labeled by topic and provide additional practice on concepts covered in this chapter.

***37.** Suppose a carbon atom is located at the center of a square with four different groups attached to the corners in a planar arrangement. Would the compound rotate polarized light? Explain.

38. Ibuprofen is a common pain reliever found in medications such as Motrin and Advil. It is sold as a racemic mixture. The structure of ibuprofen is as follows:

$$(CH_3)_2CHCH_2-C_6H_4-\underset{\displaystyle CH_3}{\overset{|}{C}}HCOOH$$

(a) Identify the one asymmetric carbon atom.
(b) Using projection formulas, write the two enantiomers of this drug.

*39. (a) A chiral substance was identified as a primary alcohol of the formula $C_5H_{12}O$. What is its structure?
(b) The compound $C_6H_{14}O_3$ has three primary alcohol groups and is chiral. What is its structure?

40. The following dextrorotatory isomer of carvone gives caraway seeds their distinctive smell.
(a) Identify the one asymmetric carbon in this molecule.
(b) The enantiomer of this molecule is responsible for the refreshing smell of spearmint. How does the "spearmint" molecule differ from the "caraway seed" molecule?

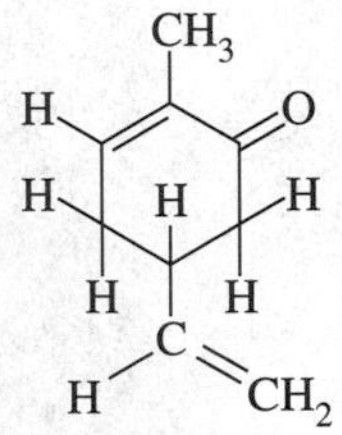

CH_3, H, O, H, H, H, H, H, C, H, CH_2

*41. What is the structure of a substance of formula $C_3H_8O_2$ that is (a) chiral and contains two —OH groups; (b) chiral and contains one —OH group; (c) achiral and contains two —OH groups? (Only one —OH group can be bonded to a carbon atom.)

42. Ephedrine is a potent drug originally found in a Chinese herbal medicine. It acts as a bronchial dilator and is used to treat asthma. The structure of ephedrine is

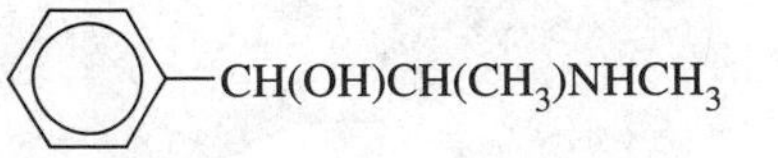

C_6H_5—$CH(OH)CH(CH_3)NHCH_3$

How many stereoisomers are possible for this drug? Briefly explain your reasoning.

43. One significant property of most amino acids is their chirality. Identify the chiral carbon in the amino acid alanine shown here:

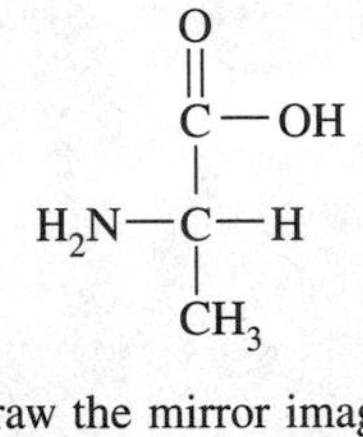

$H_2N-C(H)(CH_3)-C(=O)-OH$

Draw the mirror image of this alanine.

44. When biological interaction of a drug involves optical isomers, drug design becomes complicated because often only one isomer will have the desired property. For example, levomethorphan is an addictive opiate and dextromethorphan is a nonaddictive cough suppressant. What symbols would you use to differentiate between the two? Can you tell which one rotates light to the right? Explain why or why not.

45. Write structures for all stereoisomers and label enantiomers and meso compounds for (a) 2-bromo-3-chlorobutane and (b) 2,3,4-trichloro-1-pentanol.

Answers to Practice Exercises

27.1 Br, Cl, CH_3, F

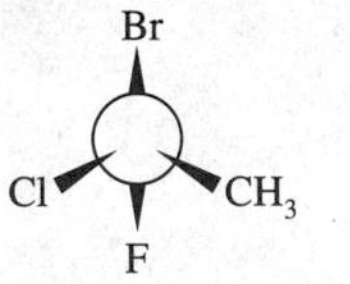

27.2 COOH; H—C—NH_2; CH_2CH_3

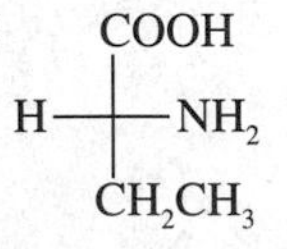

27.3 (a) $NH_2CH_2-C(H)(Br)-CH_2OH$ ← Mirror → $HOCH_2-C(H)(Br)-CH_2NH_2$

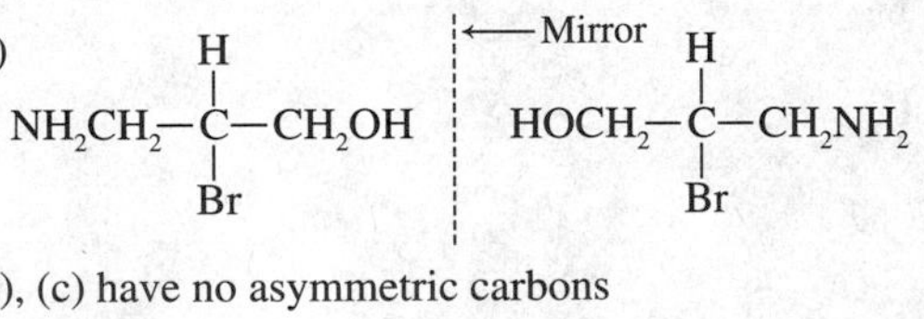

(b), (c) have no asymmetric carbons

27.4 $CH_3CH_2CH_2CH(OH)CH_3$

$CH_3CH_2CH(CH_3)CH_2OH$

$CH_3CH(OH)CH(CH_3)CH_3$

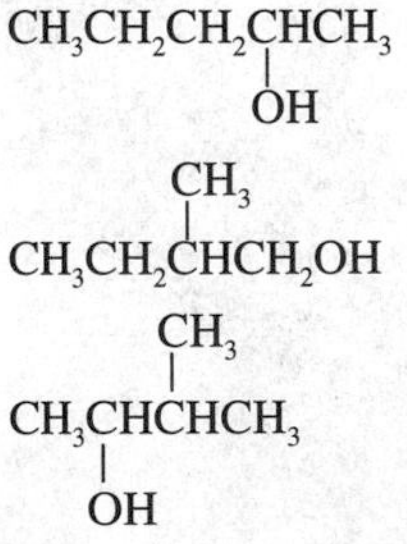

27.5

	I	II	III	IV
top	COOH	COOH	COOH	COOH
C2	H—Cl	Cl—H	H—Cl	Cl—H
C3	H—H	H—H	H—H	H—H
C4	H—H	H—H	H—H	H—H
C5	H—Cl	Cl—H	Cl—H	H—Cl
bottom	COOH	COOH	COOH	COOH

I, II: meso compound; III, IV: enantiomers

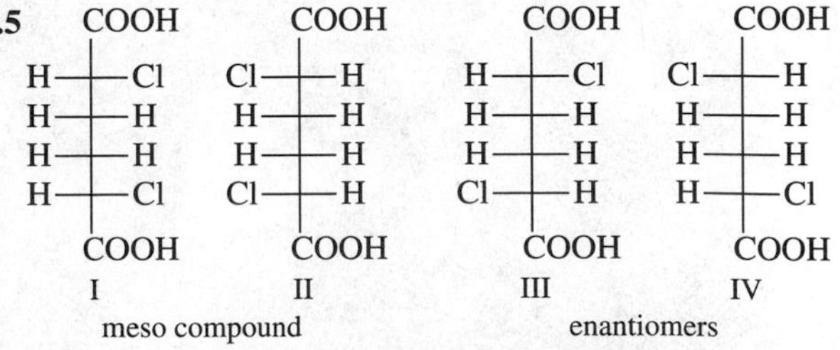

Structures I and III, I and IV, II and III, and II and IV are diastereomers.

28 Carbohydrates

Chapter 28

What is the most abundant organic chemical in the world? The answer is not petroleum products, plastics, or drugs. Rather, it is cellulose. An amazing 10 billion tons of cellulose are formed daily in the biosphere. Aggregates of cellulose allow the California redwoods to stretch hundreds of feet toward the sky and make a Brazil nut a "hard nut to crack." Products as diverse as the paper in this book and cotton in clothing are derived from cellulose. So, perhaps it is not surprising that this carbohydrate is the most widespread organic chemical in the world.

Carbohydrates are molecules of exceptional utility. These molecules provide a basic diet for many of us (starch and sugar), and a roof over our heads and clothes for our bodies (cellulose). They also thicken our ice cream, stick postage stamps to our letters, and provide biodegradable plastic trash sacks. Starting from relatively simple components (using carbon, hydrogen, and oxygen), nature has created one of the premier classes of biochemicals.

28.1 Carbohydrates: A First Class of Biochemicals

Carbohydrates are among the most widespread and important biochemicals. Most of the matter in plants, except water, consists of these substances. Carbohydrates are one of the three principal classes of energy-yielding nutrients; the other two are fats and proteins. Because of their widespread distribution and their role in many vital metabolic processes such as photosynthesis, carbohydrates have been subjected to a great deal of scientific study over the last 150 years.

The name *carbohydrates* was given to this class of compounds many years ago by French scientists who called them *hydrates de carbone* because their empirical formulas approximated $(C \cdot H_2O)_n$. It was found later that not all substances classified as carbohydrates conform to this formula (e.g., rhamnose, $C_6H_{12}O_5$, and deoxyribose, $C_5H_{10}O_4$). It seems clear that carbohydrates are not simply hydrated carbon; they are complex substances that contain from three to many thousands of carbon atoms. **Carbohydrates** are generally defined as polyhydroxy aldehydes or polyhydroxy ketones or substances that yield these compounds when hydrolyzed.

carbohydrate

The simplest carbohydrates are glyceraldehyde and dihydroxyacetone:

```
                            H
                            |
H—C=O                    H—C—OH
  |                         |
H—C—OH                     C=O
  |                         |
H—C—OH                   H—C—OH
  |                         |
  H                         H

glyceraldehyde         dihydroxyacetone
```

These substances are "polyhydroxy" because each molecule has more than one hydroxyl group. Glyceraldehyde contains a carbonyl carbon in a terminal position and is therefore an aldehyde. The internal carbonyl of dihydroxyacetone identifies

◀ Chapter Opening Photo: Green plants turn H_2O, CO_2 and sunlight into carbohydrates.

When sugar and sulfuric acid (left) are combined, water is removed from the sugar, leaving a black column of carbon (right). ▶

it as a ketone. Much of the chemistry and biochemistry of carbohydrates can be understood from a basic knowledge of the chemistry of the hydroxyl and carbonyl functional groups (see Chapters 22 and 23 for review).

28.2 Classification of Carbohydrates

A carbohydrate is classified as a monosaccharide, a disaccharide, an oligosaccharide, or a polysaccharide, depending on the number of monosaccharide units linked together to form a molecule. A **monosaccharide** is a carbohydrate that cannot be hydrolyzed to simpler carbohydrate units. The monosaccharide is the basic carbohydrate unit of cellular metabolism. A **disaccharide** yields two monosaccharides—either alike or different—when hydrolyzed:

monosaccharide

disaccharide

$$\text{Disaccharide} + \text{Water} \xrightarrow{\text{H}^+ \text{ or enzyme}} 2 \text{ Monosaccharides}$$

Disaccharides are often used by plants or animals to transport monosaccharides from one cell to another. The monosaccharides and disaccharides generally have names ending in *ose*—for example, glucose, sucrose, and lactose. These water-soluble carbohydrates, which have a characteristically sweet taste, are also called *sugars*.

oligosaccharide

An **oligosaccharide** has two to six monosaccharide units linked together. *Oligo* comes from the Greek word *oligos*, which means small or few. Free oligosaccharides that contain more than two monosaccharide units are rarely found in nature.

polysaccharide

A **polysaccharide** is a macromolecular substance that can be hydrolyzed to yield many monosaccharide units:

$$\text{Polysaccharide} + \text{Water} \xrightarrow{\text{H}^+ \text{ or enzyme}} \text{Many monosaccharide units}$$

Polysaccharides are important as structural supports (particularly in plants) and also serve as a storage depot for monosaccharides (which cells use for energy).

Carbohydrates can also be classified in other ways. A monosaccharide might be described with respect to several of these categories:

1. As a triose, tetrose, pentose, hexose, or heptose. Theoretically, a monosaccharide can have any number of carbons greater than three, but only monosaccharides of three to seven carbons are commonly found in the biosphere.

Trioses	$C_3H_6O_3$	Hexoses	$C_6H_{12}O_6$
Tetroses	$C_4H_8O_4$	Heptoses	$C_7H_{14}O_7$
Pentoses	$C_5H_{10}O_5$		

2. As an aldose or ketose, depending on whether an aldehyde group (—CHO) or keto group ($>$C=O) is present.
3. As a D or L isomer, depending on the spatial orientation of the —H and —OH groups attached to the carbon atom that is adjacent to the terminal primary alcohol group. When the —OH is written to the right of this carbon in the projection formula (see Section 27.4), the D isomer is represented. When the —OH is written to the left, the L isomer is represented. The reference compounds for this classification are the trioses D-glyceraldehyde and L-glyceraldehyde, whose formulas follow. Also shown are two aldohexoses (D- and L-glucose) and a ketohexose (D-fructose).

H—C=O
D-Configuration → H—C—OH
Terminal 1° ROH → CH_2OH
D-glyceraldehyde

H—C=O
HO—C—H ← L-Configuration
CH_2OH ← Terminal 1° ROH
L-glyceraldehyde

H—C=O
HO—C—H
H—C—OH
HO—C—H
L-Configuration → HO—C—H
Terminal 1° ROH → CH_2OH
L-glucose

H—C=O
H—C—OH
HO—C—H
H—C—OH
H—C—OH ← D-Configuration
CH_2OH ← Terminal 1° ROH
D-glucose

CH_2OH
C=O
HO—C—H
H—C—OH
D-Configuration → H—C—OH
Terminal 1° ROH → CH_2OH
D-fructose

The letters D and L do not in any way refer to the direction of optical rotation of a carbohydrate. The D and L forms of any specific compound are enantiomers (e.g., D- and L-glucose).

4. As a (+) or (−) isomer, depending on whether the monosaccharide rotates the plane of polarized light to the right (+) or to the left (−). (See Section 27.5.)
5. As a furanose or a pyranose, depending on whether the cyclic structure of the carbohydrate is related to that of the five- or six-membered heterocyclic ring compound furan or pyran (a heterocyclic ring contains more than one kind of atom in the ring):

Sugar cane is a major source for sucrose, or table sugar.

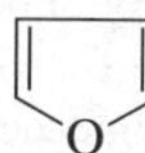

furan, C_4H_4O
(five-membered ring with oxygen in the ring)

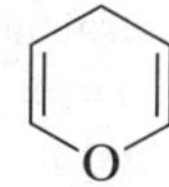

pyran, C_5H_6O
(six-membered ring with oxygen in the ring)

6. As having an alpha (α) or beta (β) configuration, based on the orientation of the —H and —OH groups about a specific asymmetric carbon in the cyclic form of the monosaccharide (Section 28.6).

Example 28.1 Write projection formulas for (a) an L-aldotriose, (b) a D-ketotetrose, and (c) a D-aldopentose.

Solution (a) Triose indicates a three-carbon carbohydrate; aldo indicates that the compound is an aldehyde; L- indicates that the —OH on carbon 2 (adjacent to the terminal CH_2OH) is on the left. The aldehyde group is carbon 1.

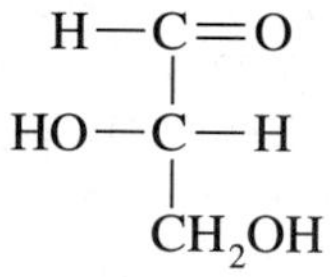

```
  H—C=O
    |
HO—C—H
    |
   CH2OH
```

an L-aldotriose

(b) Tetrose indicates a four-carbon carbohydrate; keto indicates a ketone group (on carbon 2); D- indicates that the —OH on carbon 3 (adjacent to the terminal CH_2OH) is on the right. Carbons 1 and 4 have the configuration of primary alcohols.

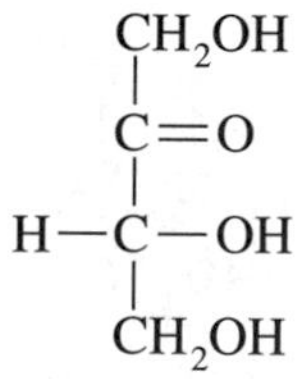

```
   CH2OH
   |
   C=O
   |
H—C—OH
   |
   CH2OH
```

a D-ketotetrose

(c) Pentose indicates a five-carbon carbohydrate; aldo indicates an aldehyde group (on carbon 1); D- indicates that the —OH on carbon 4 (adjacent to the terminal CH_2OH) is on the right. The orientation of the —OH groups on carbons 2 and 3 is not specified here and therefore can be written in either direction for this problem.

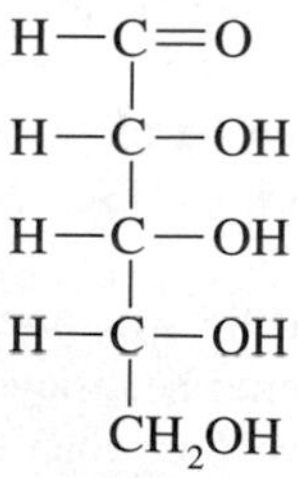

```
H—C=O
   |
H—C—OH
   |
H—C—OH
   |
H—C—OH
   |
   CH2OH
```

a D-aldopentose

Practice 28.1

Write the projection formula for an L-ketopentose.

Practice 28.2

In a projection formula for a D-aldotriose, is the —OH of the secondary alcohol carbon written on the right or the left side?

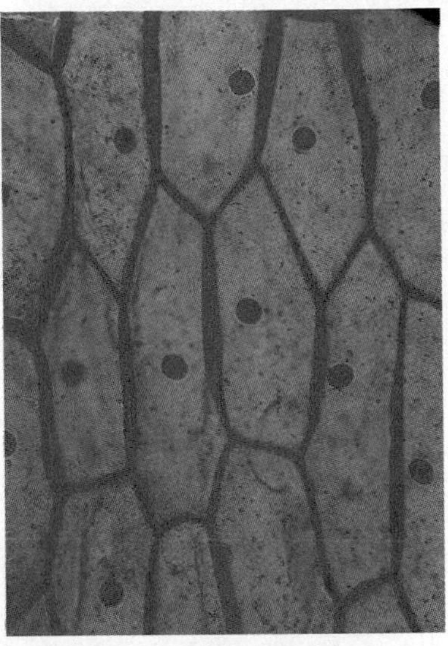

Carbohydrates are important building blocks in the formation of cell walls in higher plants. Shown here are the cell walls of an onion.

28.3 Importance of Carbohydrates

As we stated earlier, carbohydrates are the most abundant organic chemical—they must be molecules of exceptional utility. But what makes them so special? Why are they so important in our daily lives?

Carbohydrates are very effective energy-yielding nutrients. In general, a molecule can provide biological energy if it (1) contains carbon, (2) is relatively reactive (contains some polar bonds), and (3) contains carbons that are not completely oxidized. Carbohydrates fit this description very well. The average carbohydrate contains about 40% carbon by mass. Carbohydrates are polar molecules because they contain many polar groups (e.g., hydroxyl groups), and none of the carbons in a carbohydrate is completely oxidized. A useful estimate of a carbon's reduced state is the oxidation number (see Table 28.1). The lower the oxidation number, the more reduced the carbon is and the more energy it can supply. Carbohydrates are energy-yielding nutrients partly because they contain carbons in an intermediate state of reduction, with oxidation numbers ranging from -1 to $+2$.

Carbohydrates can serve as very effective building materials. These materials need to be strong, not brittle, and water insoluble. As you will see in Section 28.15, large, water-insoluble polymers (polysaccharides) are formed from carbohydrates. Cellulose polymers can form durable fibers by hydrogen bonding. These fibers provide building materials for strong, relatively rigid wood construction; for flexible, comfortable cotton clothing; and for flexible yet smooth paper writing material. Cellulose also forms at least part of the cell wall for all higher plants.

Carbohydrates are important water-soluble molecules. Many smaller carbohydrates (monosaccharides and disaccharides) are water soluble. These carbohydrates are often mixed into foods to give a sweet taste. Water solubility allows these molecules to pass easily from living cell to living cell.

28.4 Monosaccharides

Although a great many monosaccharides have been synthesized, only a very few appear to be of much biological significance. One pentose monosaccharide (ribose) and its deoxy derivative are essential components of ribonucleic acid (RNA) and of deoxyribonucleic acid (DNA) (see Chapter 32). However, the hexose monosaccharides are the most important carbohydrate sources of cellular energy. Three

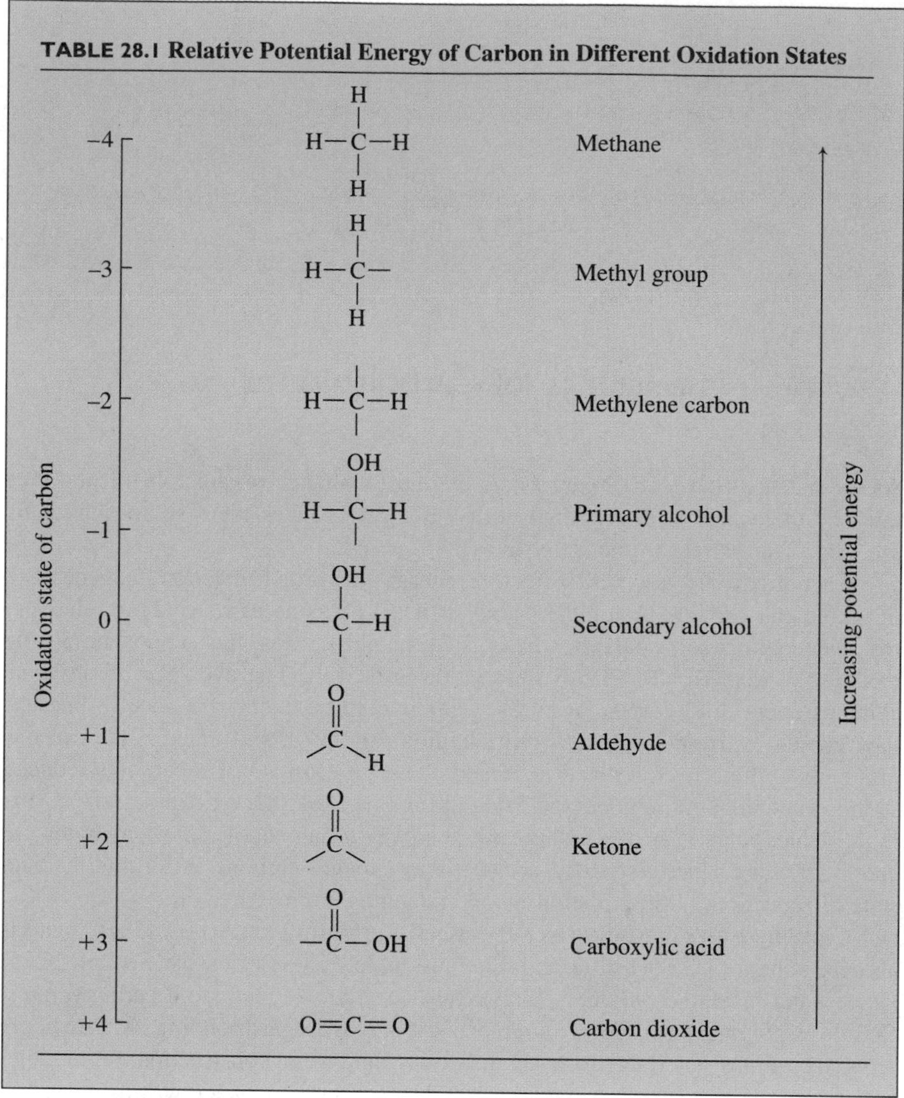

TABLE 28.1 Relative Potential Energy of Carbon in Different Oxidation States

Oxidation state of carbon	Structure	Name
−4	CH_4 (H—C—H with H above and below)	Methane
−3	H—C— (with H above and below)	Methyl group
−2	H—C—H	Methylene carbon
−1	H—C(OH)—H	Primary alcohol
0	—C(OH)—H	Secondary alcohol
+1	C(=O)H	Aldehyde
+2	C(=O)	Ketone
+3	—C(=O)—OH	Carboxylic acid
+4	O=C=O	Carbon dioxide

hexoses—glucose, galactose, and fructose—are of major significance in nutrition. All three have the same molecular formula, $C_6H_{12}O_6$, and thus contain an equal number of reduced carbons. They differ in structure but are biologically interconvertible. Glucose plays a central role in carbohydrate energy utilization. Other carbohydrates are usually converted to glucose before cellular utilization. The structure of glucose is considered in detail in Section 28.5.

Glucose

Glucose is the most important of the monosaccharides. It is an aldohexose and is found in the free state in plant and animal tissue. Glucose is also known as *dextrose* or *grape sugar.* It is a component of the disaccharides sucrose, maltose, and lactose, and is also the monomer of the polysaccharides starch, cellulose, and glycogen. Among the common sugars, glucose is of intermediate sweetness (see Section 28.12).

Glucose is the key sugar of the body and is carried by the bloodstream to all body parts. The concentration of glucose in the blood is normally 80–100 mg per 100 mL of blood. Because glucose is the most abundant carbohydrate in the blood, it is commonly called *blood sugar.* Glucose requires no digestion and may therefore be given intravenously to patients who cannot take food by mouth. Glucose is found in the urine of those who have diabetes mellitus (sugar diabetes). The condition in which glucose is excreted in the urine is called glycosuria.

Galactose

Galactose is also an aldohexose and occurs, along with glucose, in lactose and in many oligo- and polysaccharides such as pectin, gums, and mucilages. Galactose is an isomer of glucose, differing only in the spatial arrangement of the —H and —OH groups around carbon 4 (see Section 28.5). Galactose is synthesized in the mammary glands to make the lactose of milk. It is also a constituent of glycolipids and glycoproteins in many cell membranes such as those in nervous tissue. Galactose is less than half as sweet as glucose.

A severe inherited disease, called galactosemia, is the inability of infants to metabolize galactose. The galactose concentration increases markedly in the blood and also appears in the urine. Galactosemia causes vomiting, diarrhea, enlargement of the liver, and often mental retardation. If not recognized within a few days after birth, it can lead to death. If diagnosis is made early and lactose is excluded from the diet, the symptoms disappear and normal growth may be resumed.

Fructose

Fructose, also known as *levulose*, is a ketohexose that occurs in fruit juices, honey, and (along with glucose) as a constituent of sucrose. Fructose is the major constituent of the polysaccharide inulin, a starch-like substance present in many plants such as dahlia tubers, chicory roots, and Jerusalem artichokes. Fructose is the sweetest of all the common sugars, being about twice as sweet as glucose. This accounts for the sweetness of high-fructose corn syrup and honey. The enzyme invertase, present in bees, splits sucrose into glucose and fructose. Fructose is metabolized directly but is also readily converted to glucose in the liver.

28.5 Structure of Glucose and Other Aldoses

In one of the classic feats of research in organic chemistry, Emil Fischer (1852–1919), working in Germany, established the structural configuration of glucose along with that of many other sugars. He received the Nobel prize in chemistry in 1902. Fischer devised projection formulas that relate the structure of a sugar to one or the other of the two enantiomeric forms of glyceraldehyde. These projection formulas represent three-dimensional stereoisomers (see Chapter 27) in a two-dimensional plane. (Remember that stereoisomers cannot be interconverted without breaking and reforming covalent bonds. Each carbohydrate isomer has a different shape and thus reacts differently in biological systems.)

In Fischer projection formulas, the molecule is represented with the aldehyde (or ketone) group at the top. The —H and —OH groups attached to interior carbons are written to the right or to the left as they would appear when projected toward the observer. The two glyceraldehydes are represented thus:

A projection formula is commonly written in either of two ways, with or without the asymmetric carbon atoms.

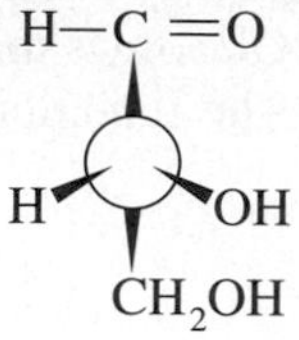

```
  H—C=O            H—¹C=O          H—¹C=O
                      |                |
H ◄—O—► OH        H—²C—OH        H——2——OH
                      |                |
  CH₂OH             ³CH₂OH          ³CH₂OH
```

D-glyceraldehyde (three-dimensional representation)

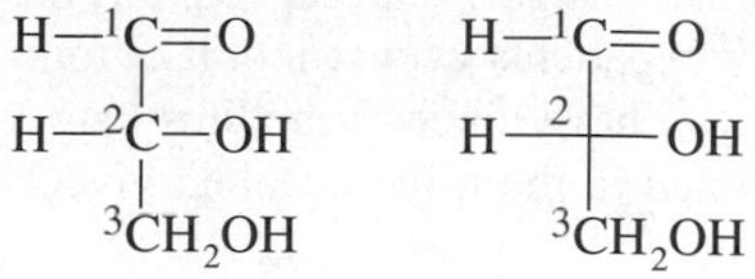

D-glyceraldehyde (Fischer projection formulas)

```
  H—C=O            H—¹C=O          H—¹C=O
                      |                |
HO ◄—O—► H        HO—²C—H        HO——2——H
                      |                |
  CH₂OH             ³CH₂OH          ³CH₂OH
```

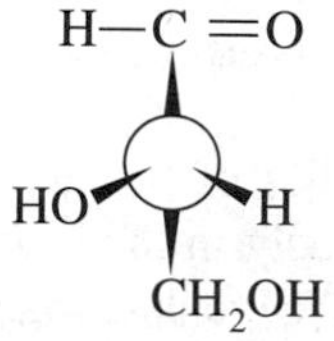

L-glyceraldehyde (three-dimensional representation)

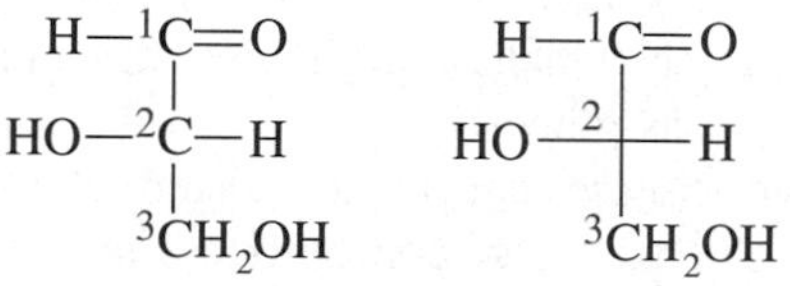

L-glyceraldehyde (Fischer projection formulas)

In the three-dimensional molecules represented by these formulas, the carbon 2 atoms are in the plane of the paper. The —H and —OH groups project forward (toward the observer); the —CHO and —CH_2OH groups project backward (away from the observer). Any two monosaccharides that differ only in the configuration around a single carbon atom are called **epimers**. Thus D- and L-glyceraldehyde are epimers.

epimer

Fischer recognized that there were two enantiomeric forms of glucose. To these forms he assigned the following structures and names:

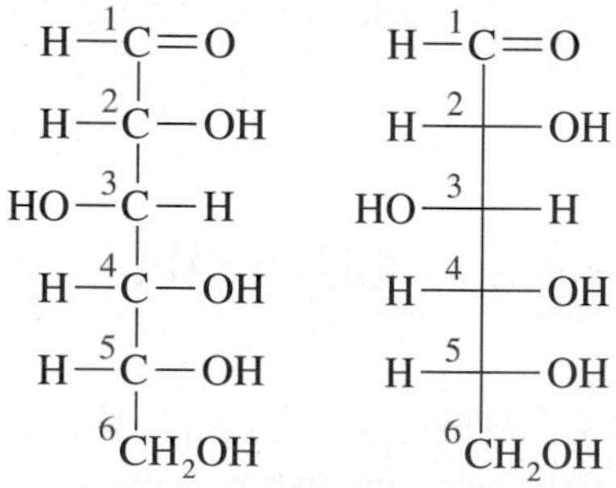

```
  H—¹C=O         H—¹C=O            H—¹C=O          H—¹C=O
     |               |                 |               |
  H—²C—OH      H——2——OH          HO—²C—H       HO——2——H
     |               |                 |               |
 HO—³C—H      HO——3——H            H—³C—OH        H——3——OH
     |               |                 |               |
  H—⁴C—OH      H——4——OH          HO—⁴C—H       HO——4——H
     |               |                 |               |
  H—⁵C—OH      H——5——OH          HO—⁵C—H       HO——5——H
     |               |                 |               |
   ⁶CH₂OH         ⁶CH₂OH            ⁶CH₂OH          ⁶CH₂OH
```

D-glucose 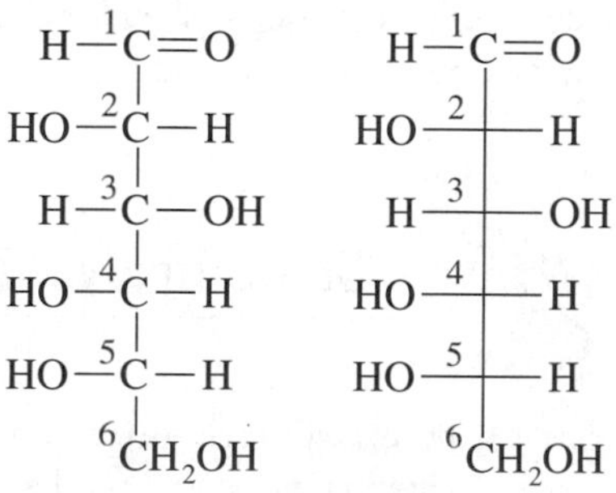 L-glucose

The structure called D-glucose is so named because the —H and —OH on carbon 5 are in the same configuration as the —H and —OH on carbon 2 in D-glyceraldehyde. The configuration of the —H and —OH on carbon 5 in L-glucose corresponds to the —H and —OH on carbon 2 in L-glyceraldehyde.

Fischer recognized that 16 different aldohexoses, 8 with the D configuration and 8 with the L configuration, were possible. This follows our formula 2^n for

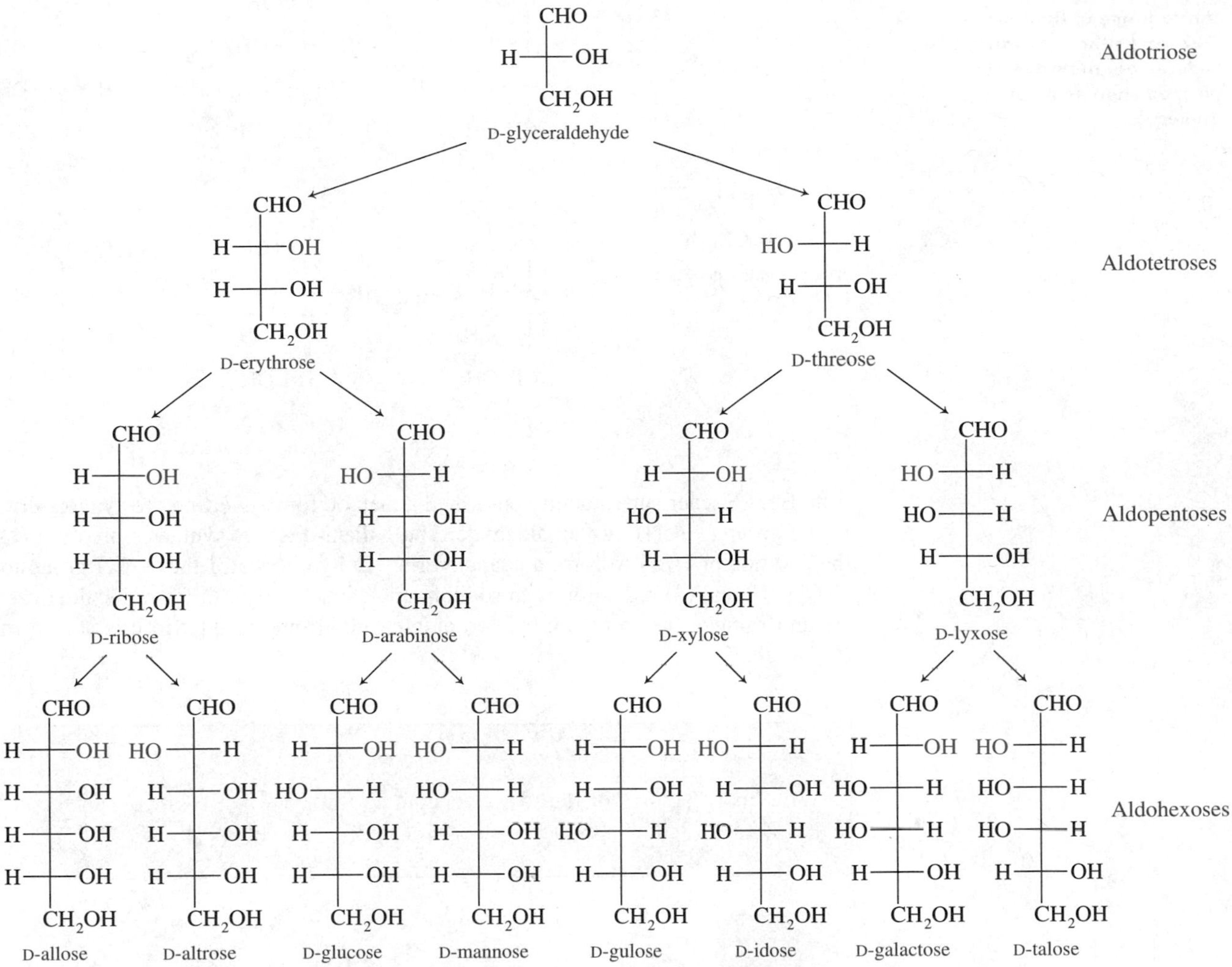

Aldotriose

CHO
H—OH
CH_2OH
D-glyceraldehyde

Aldotetroses

CHO
H—OH
H—OH
CH_2OH
D-erythrose

CHO
HO—H
H—OH
CH_2OH
D-threose

Aldopentoses

CHO
H—OH
H—OH
H—OH
CH_2OH
D-ribose

CHO
HO—H
H—OH
H—OH
CH_2OH
D-arabinose

CHO
H—OH
HO—H
H—OH
CH_2OH
D-xylose

CHO
HO—H
HO—H
H—OH
CH_2OH
D-lyxose

Aldohexoses

D-allose	D-altrose	D-glucose	D-mannose	D-gulose	D-idose	D-galactose	D-talose
CHO	CHO	CHO	CHO	CHO	CHO	CHO	CHO
H—OH	HO—H	H—OH	HO—H	H—OH	HO—H	H—OH	HO—H
H—OH	H—OH	HO—H	HO—H	H—OH	H—OH	HO—H	HO—H
H—OH	H—OH	H—OH	H—OH	HO—H	HO—H	HO—H	HO—H
H—OH	H—OH	H—OH	H—OH	H—OH	H—OH	H—OH	H—OH
CH_2OH	CH_2OH	CH_2OH	CH_2OH	CH_2OH	CH_2OH	CH_2OH	CH_2OH

▲
FIGURE 28.1
Configurations of the D-family of aldoses. The hydroxyl group on the new asymmetric carbon atom added in going from triose to tetrose to pentose to hexose is marked in color.

optical isomers (see Section 27.7). Glucose has four asymmetric carbon atoms and should have 16 stereoisomers (2^4). The configurations of the D-aldose family are shown in Figure 28.1. In this family, new asymmetric carbon atoms are formed as we go from triose to tetrose to pentose to hexose. Each time a new asymmetric carbon is added, a pair of epimers is formed that differ only in the structure around carbon 2. This sequence continues until eight D-aldohexoses are created. A similar series starting with L-glyceraldehyde is known, making a total of 16 aldohexoses.

All of the 16 aldohexoses have been synthesized, but only D-glucose and D-galactose appear to be of considerable biological importance. Since the metabolism of most living organisms revolves about D-glucose, our discussion will be centered on this substance.

The laboratory conversion of one aldose into another aldose containing one more carbon atom is known as the Kiliani–Fischer synthesis. This synthesis makes use of the aldehyde's ability to bond to an additional group (see Section 23.5). Heinrich Kiliani discovered that the addition of HCN to an aldose formed a cyanohy-

FIGURE 28.2 ▶ An example of the Kiliani–Fischer synthesis in which two aldotetrose molecules are formed from an aldotriose molecule.

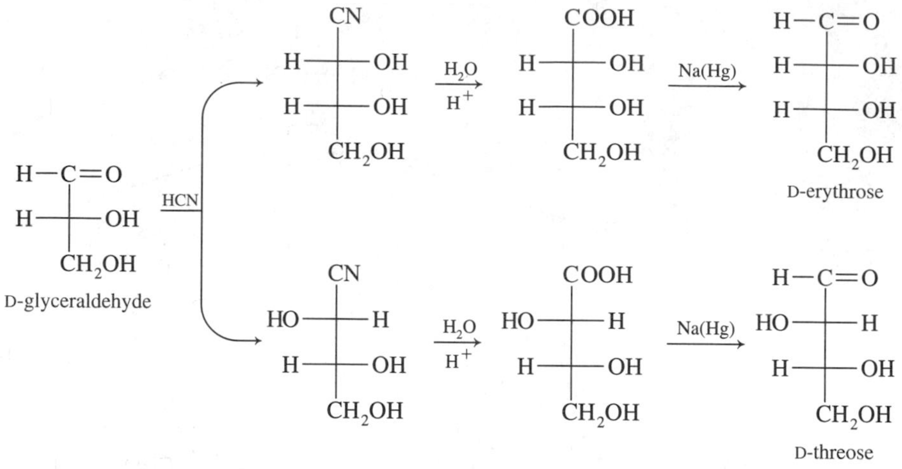

H—C=O
H—C—OH
CH_2OH
D-glyceraldehyde

HCN →

CN | H—C—OH | H—C—OH | CH_2OH —(H_2O, H^+)→ COOH | H—C—OH | H—C—OH | CH_2OH —Na(Hg)→ H—C=O | H—C—OH | H—C—OH | CH_2OH
D-erythrose

CN | HO—C—H | H—C—OH | CH_2OH —(H_2O, H^+)→ COOH | HO—C—H | H—C—OH | CH_2OH —Na(Hg)→ H—C=O | HO—C—H | H—C—OH | CH_2OH
D-threose

drin. Emil Fischer subsequently published a method for converting the cyanohydrin nitrile group (—CN) into an aldehyde. The Kiliani–Fischer synthesis involves (1) the addition of HCN to form a cyanohydrin; (2) hydrolysis of the —CN group to —COOH; and (3) reduction with sodium amalgam, Na(Hg), to form the aldehyde. As an example, the formation of two aldotetroses from an aldotriose is shown in Figure 28.2.

Practice 28.3

Write the structures of the two aldoheptoses formed when D-glucose is subjected to the Kiliani-Fischer synthesis.

28.6 Cyclic Structure of Glucose; Mutarotation

Straight open-chain D-glucose is so reactive that almost all molecules quickly rearrange their bonds to form two new structures. These two forms are diastereomers and differ with respect to their rotation of polarized light. One form, labeled α-D-glucopyranose, has a specific rotation, [α], of +112°; the other, labeled β-D-glucopyranose, has a specific rotation of +18.7°. An interesting phenomenon occurs when these two forms of glucose are put into separate solutions and allowed to stand for several hours. The specific rotation of each solution changes to +52.7°. This phenomenon is known as mutarotation. An explanation of mutarotation is that D-glucose exists in solution as an equilibrium mixture of two cyclic forms and the open-chain form (see Figure 28.3). The two cyclic molecules are optical isomers, differing only in the orientation of the —H and —OH groups about carbon 1. When dissolved, some α-D-glucopyranose molecules are transformed into β-D-glucopyranose, and vice versa, until an equilibrium is reached between the α and

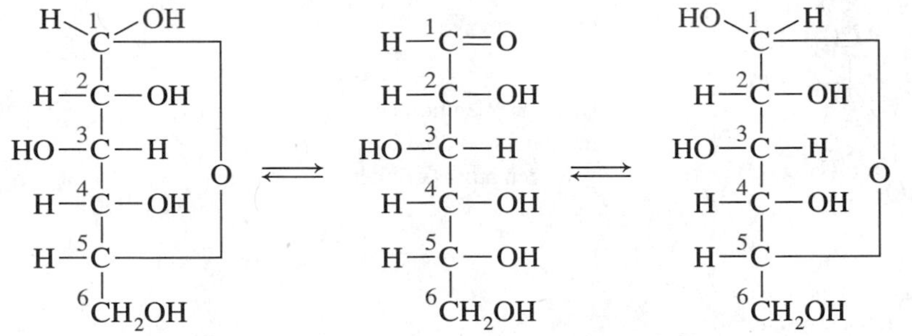

α-D-(+)-glucopyranose
$[\alpha] = +112°$

D-(+)-glucose
(open-chain form)

β-D-(+)-glucopyranose
$[\alpha] = +18.7°$

(a) Modified Fischer projection formulas

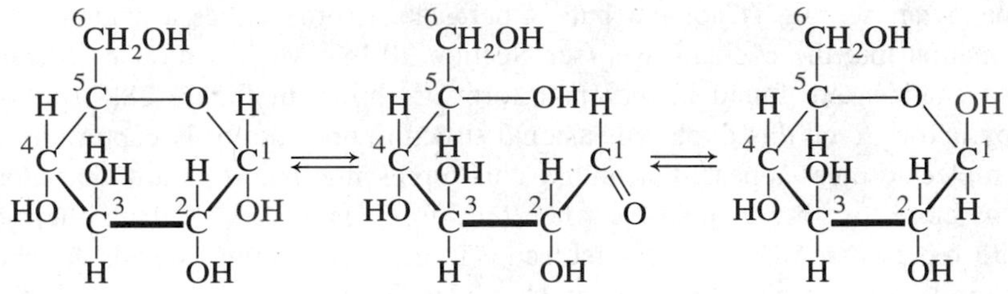

α-D-(+)-glucopyranose

D-(+)-glucose
(open-chain form)

β-D-(+)-glucopyranose

(b) Haworth perspective formulas

FIGURE 28.3
Mutarotation of D-glucose.

β forms. (Note that no other chiral centers in D-glucose are altered when this sugar is dissolved.) The equilibrium solution contains about 36% α molecules and 64% β molecules, with a trace of open-chain molecules. When two cyclic isomers differ only in their stereo arrangement about the carbon involved in mutarotation, they are called **anomers**. For example, α- and β-D-glucopyranose are anomers (see Figure 28.3). **Mutarotation** is the process by which anomers are interconverted.

anomer

mutarotation

The cyclic forms of D-glucose may be represented by either Fischer projection formulas or by Haworth perspective formulas. These structures are shown in Figure 28.3. In the cyclic Fischer projection formulas of the D-aldoses, the α form has the —OH on carbon 1 written to the right; in the β form the —OH on carbon 1 is on the left. The Haworth formula represents the molecule as a flat hexagon with the —H and —OH groups above and below the plane of the hexagon. In the α form the —OH on carbon 1 is written below the plane; in the β form the —OH on carbon 1 is above the plane. In converting the projection formula of a D-aldohexose to the Haworth formula, the —OH groups on carbons 2, 3, and 4 are written below the plane if they project to the right and above the plane if they project to the left. Carbon 6 is written above the plane.

Haworth formulas are sometimes shown in abbreviated schematic form. For example, α-D-(+)-glucopyranose is shown as

CH_2OH
O
OH
HO OH
OH

α-D-(+)-glucopyranose
(abbreviated form)

The four open bonds are understood to have hydrogen atoms on them.

Although both the Fischer projection formula and the Haworth formula provide useful representations of carbohydrate molecules, it is important to understand that these structures only approximate the true molecular shapes. We know, for example, that the pyranose ring is not flat but, rather, can assume either a chair or boat conformation like the cycloalkanes (see Section 20.14). Most naturally occurring monosaccharides are found in the chair form as shown in Figure 28.4 for α-D-glucopyranose. Even this three-dimensional structure does not truly capture how a sugar molecule must appear. For, unlike this representation, we know that atoms move as close together as possible when they form molecules. Perhaps the most accurate representation of a sugar molecule is the space-filling model. A space-filling model of α-D-glucopyranose is shown in Figure 28.4. At best, any two-dimensional representation is a compromise in portraying the three-dimensional configuration of such molecules. Structural models are much more effective, especially if constructed by the student.

The two cyclic forms of D-glucose differ only in the relative positions of the —H and —OH groups attached to carbon 1. Yet, this seemingly minor structural difference has important biochemical consequences because the physical shape of a molecule often determines its biological use. For example, the fundamental structural difference between starch and cellulose is that starch is a polymer of α-D-glucopyranose, whereas cellulose is a polymer of β-D-glucopyranose. As a consequence, starch is easily digested by humans, but we are totally unable to digest cellulose.

FIGURE 28.4 ▶ Three-dimensional representations of the chair form of α-D-glucopyranose.

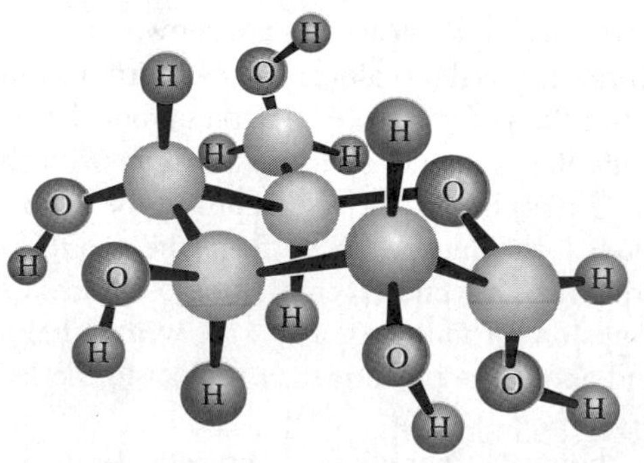

(a) Ball-and-stick model

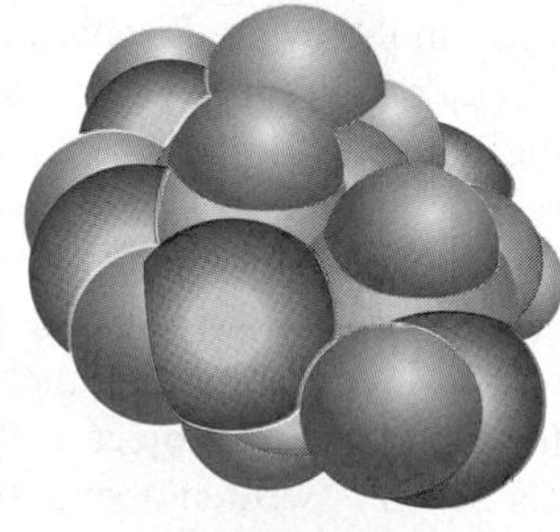

(b) Space-filling model

Example 28.2

Write the pyranose Haworth perspective formulas for the two anomers of D-mannose and name these isomers.

Solution

First, write the open-chain Fischer projection formula for D-mannose (you must memorize this structure or know where to find it in the text) and number the carbons from top (the aldehyde group) to bottom (the primary alcohol group).

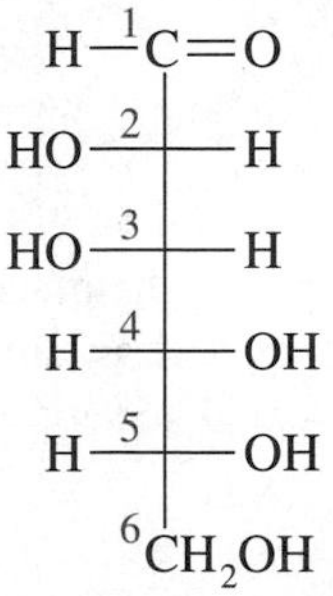

Next, draw the structure of the Haworth pyranose ring. Number the carbons from the right-hand point of the hexagon clockwise around the cyclic form, placing the CH_2OH (C6) group in the up position on the ring.

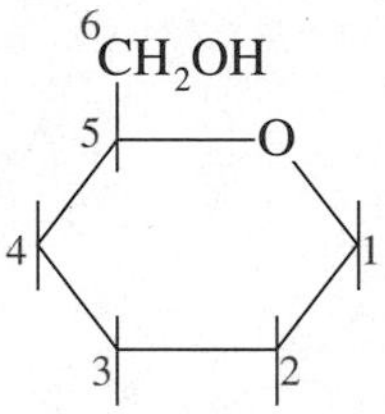

Then, refer to the open-chain Fischer projection formula. All the hydroxyl groups on the right of the open chain should be written *down* in the Haworth formula, and all the hydroxyl groups on the left should be written *up*. Since this rule only applies to chiral centers, we can ignore the hydroxyl on carbon 6.

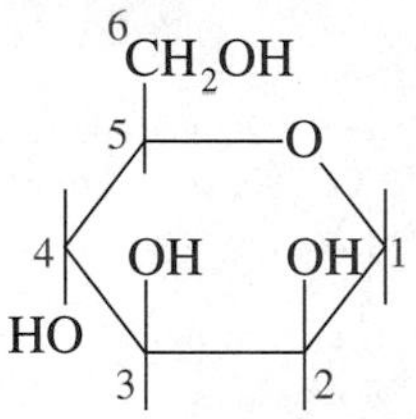

The carbon involved in mutarotation, carbon 1, can have either of two configurations, the α anomer when the hydroxyl is pointed down or the β anomer when pointed up. The last step is to add the hydroxyl group at carbon 1 and name the anomers.

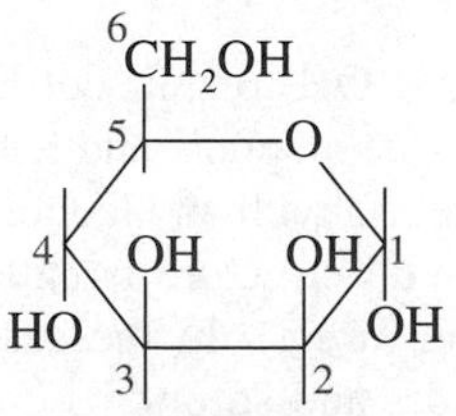

α-D-mannopyranose 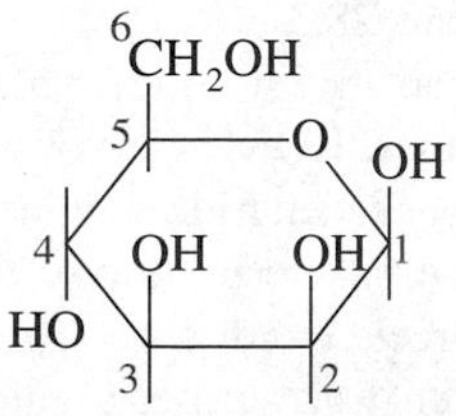   β-D-mannopyranose

Digitalis, a well-known heart stimulant, is produced in nature by the foxglove plant.

Practice 28.4

Draw the Haworth formulas for (a) α-D-galactopyranose and (b) β-D-glucopyranose.

28.7 Hemiacetals and Acetals

In Chapter 23 we studied the reactions of aldehydes (and ketones) to form hemiacetals and acetals. The hemiacetal structure consists of an ether linkage and an alcohol linkage on the same carbon atom (shown in red), whereas the acetal structure has two ether linkages to the same carbon atom:

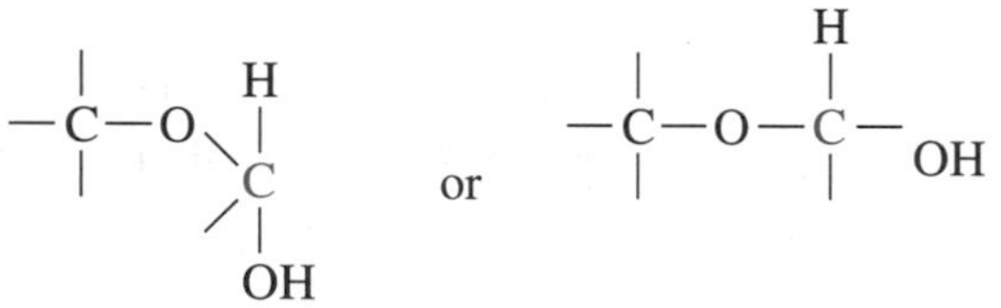

—C—O—CH(OH)— or —C—O—CH(OH)—

hemiacetal structures

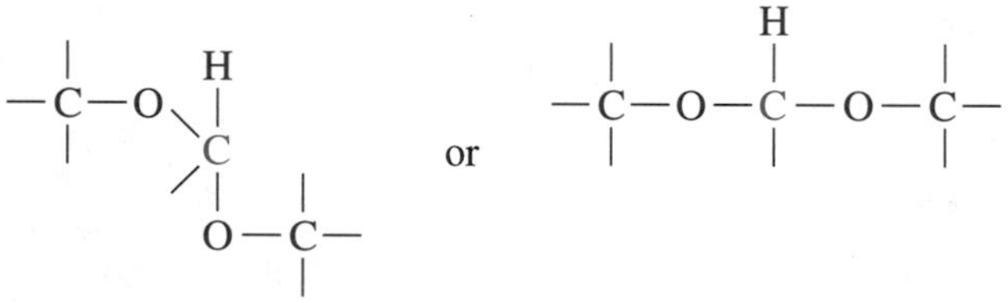

—C—O—CH(O—C—)— or —C—O—CH—O—C—

acetal structures

Cyclic structures of monosaccharides are intramolecular hemiacetals. Five- or six-membered rings are especially stable.

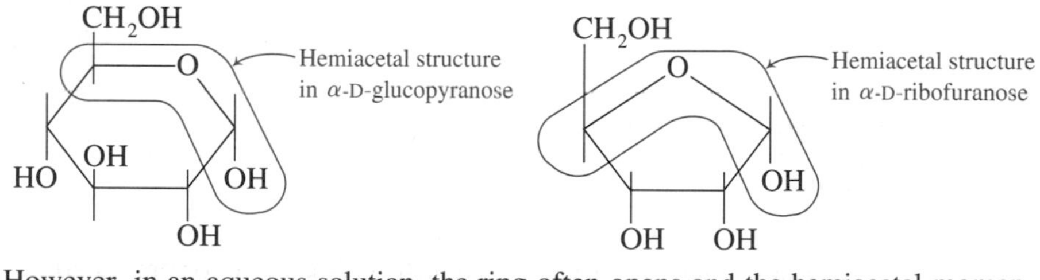

Hemiacetal structure in α-D-glucopyranose

Hemiacetal structure in α-D-ribofuranose

However, in an aqueous solution, the ring often opens and the hemiacetal momentarily reverts to the open-chain aldehyde. When the open chain closes, it forms either the α or the β anomer. Mutarotation results from this opening and closing of the hemiacetal ring (see Figure 28.3).

When an alcohol, ROH, reacts with another alcohol, R′OH to split out H_2O, the product formed can be an ether, ROR′. Carbohydrates are alcohols and behave in a similar manner. When a monosaccharide hemiacetal reacts with an alcohol, the product is an acetal. In carbohydrate terminology this acetal structure is called a **glycoside** (derived from the Greek word *glykys*, meaning sweet). In the case of glucose, it would be a glucoside; if galactose, a galactoside; and so on.

glycoside

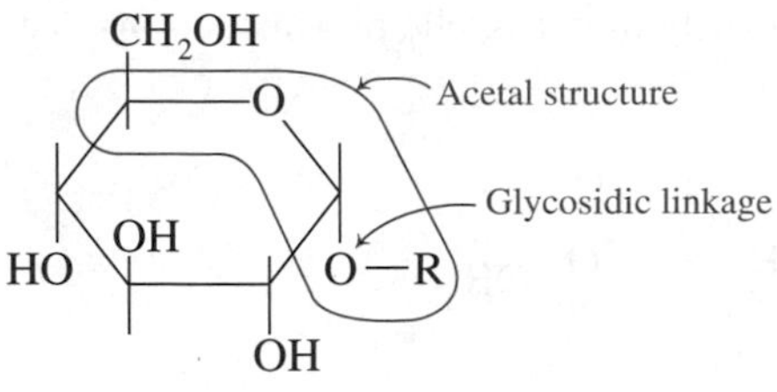

an α-glycoside
(R = a variety of groups)

A glycoside differs significantly from a monosaccharide with respect to chemical reactivity.

When α-D-glucopyranose is heated with methyl alcohol and a small quantity of hydrogen chloride is added, two optically active isomers are formed—methyl α-D-glucopyranoside and methyl β-D-glucopyranoside:

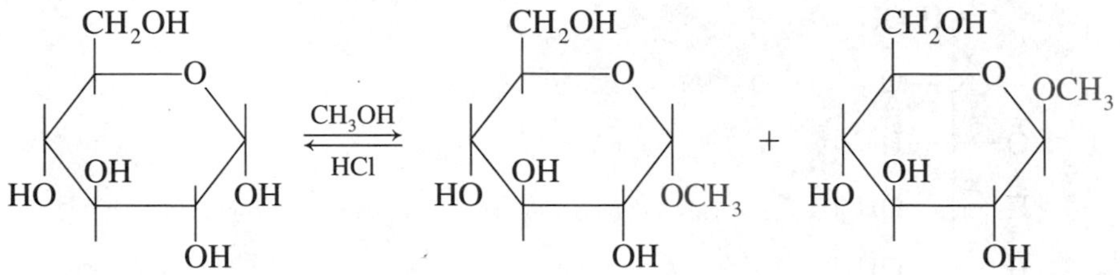

α-D-glucopyranose

methyl α-D-glucopyranoside
(mp 165°C, [α] = + 158°)

methyl β-D-glucopyranoside
(mp 107°C, [α] = − 33°)

Unlike D-glucose, the two glycoside products no longer undergo mutarotation. They do not form open-chain compounds in aqueous solution. Acetals tend to be more stable and less reactive than hemiacetals.

The glycosidic linkage occurs in a wide variety of natural substances. All carbohydrates other than monosaccharides are glycosides. Heart stimulants such as digitalis and ouabain are known as heart glycosides. Several antibiotics such as streptomycin and erythromycin are also glycosides.

28.8 Structures of Galactose and Fructose

Galactose, like glucose, is an aldohexose and differs structurally from glucose only in the configuration of the —H and —OH group on carbon 4:

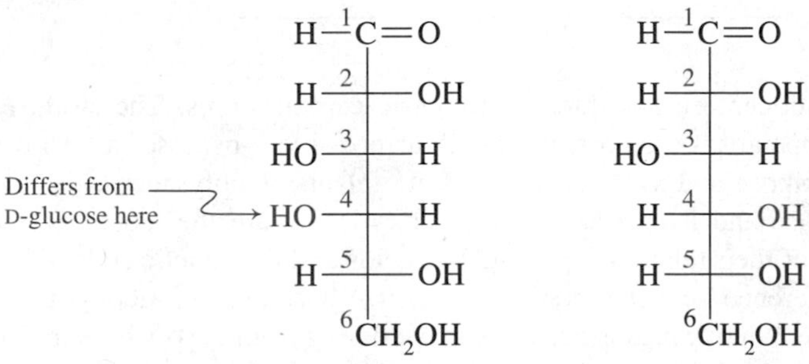

D-galactose

D-glucose

To identify anomers in the Haworth formula, focus on the carbon to the far right side of the ring. If the —OH on this carbon is written below the plane of the ring, the molecule is an α-anomer; if the —OH is written above the plane of the ring, the molecule is a β-anomer.

Galactose, like glucose, also exists primarily in two cyclic pyranose forms that have hemiacetal structures and undergo mutarotation:

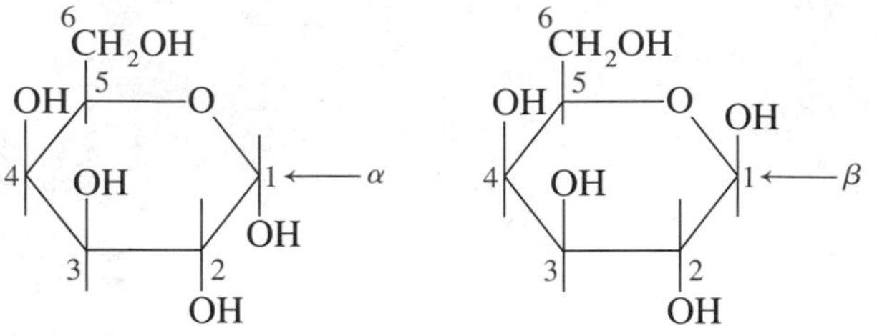

α-D-galactopyranose β-D-galactopyranose

Fructose is a ketohexose. The open-chain form may be represented in a Fischer projection formula:

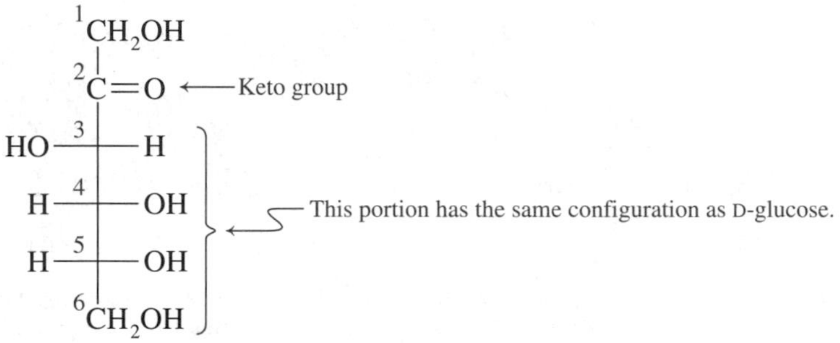

D-fructose

Like glucose and galactose, fructose exists in both cyclic and open-chain forms. One common cyclic structure is a five-membered furanose ring in the *β* configuration:

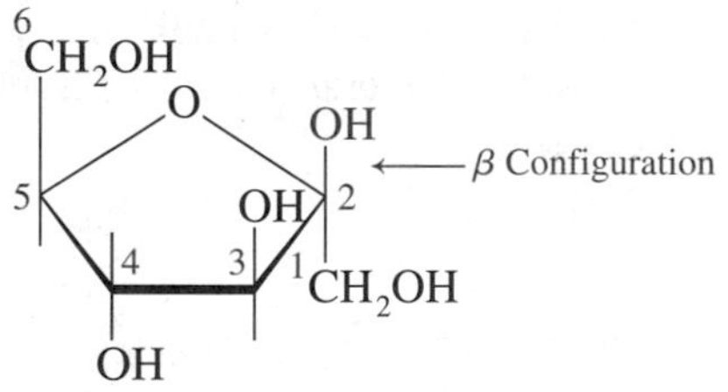

β-D-fructofuranose

28.9 Pentoses

An open-chain aldopentose has three asymmetric carbon atoms. Therefore, eight (2^3) isomeric aldopentoses are possible. The four possible D-pentoses are shown in Figure 28.1. Arabinose and xylose occur in some plants as polysaccharides called pentosans. D-Ribose and its derivative, D-2-deoxyribose, are the most interesting pentoses because of their relationship to nucleic acids and the genetic code (Chapter 32). Note the difference between the two names, D-ribose and D-2-deoxyribose. In the latter name, the 2-deoxy means that oxygen is missing from the D-ribose molecule at carbon 2. Check the formulas that follow to verify this difference.

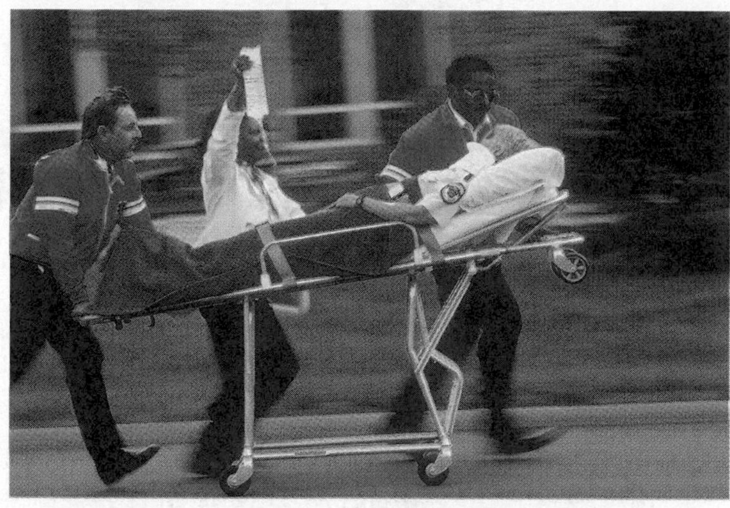

◀ **Glucose is often given intravenously to patients suffering from major trauma.**

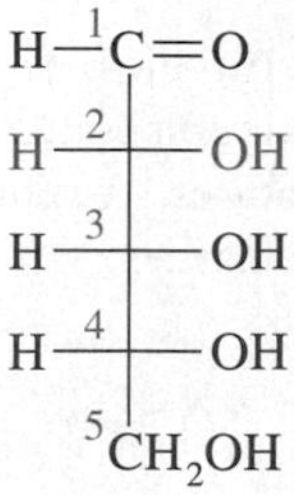

$$\begin{array}{c} H-\overset{1}{C}=O \\ H-\overset{2}{C}-OH \\ H-\overset{3}{C}-OH \\ H-\overset{4}{C}-OH \\ {}^{5}CH_2OH \end{array}$$

D-ribose

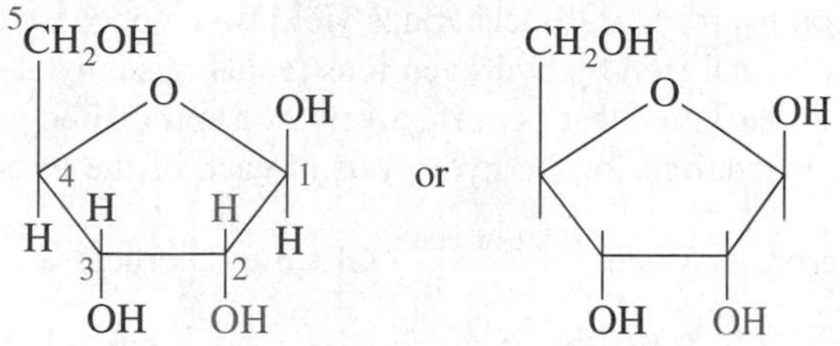

$^{5}CH_2OH$, O, OH, 4, 1, H, H, H, H, 3, 2, OH, OH — or — CH_2OH, O, OH, OH, OH

β-D-ribofuranose

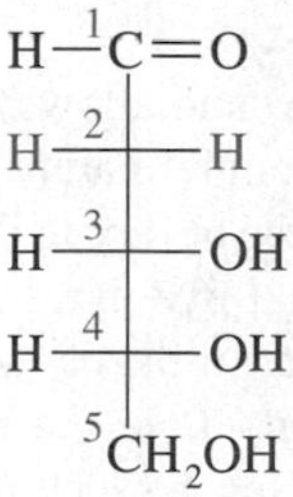

$$\begin{array}{c} H-\overset{1}{C}=O \\ H-\overset{2}{C}-H \\ H-\overset{3}{C}-OH \\ H-\overset{4}{C}-OH \\ {}^{5}CH_2OH \end{array}$$

D-2-deoxyribose

$^{5}CH_2OH$, O, OH, 4, 1, H, H, H, 3, H, 2, OH, H — or — CH_2OH, O, OH, OH

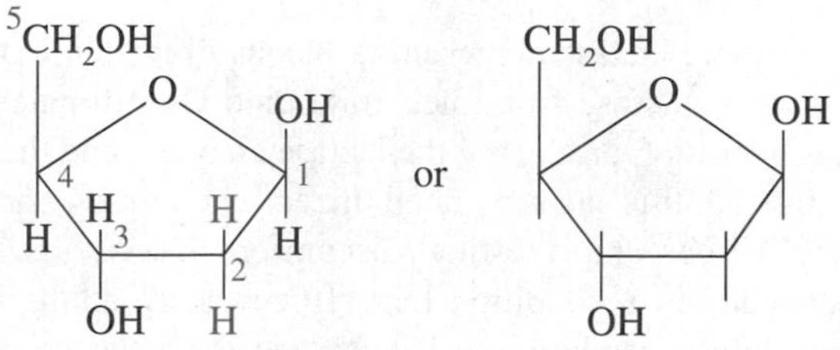

β-D-2-deoxyribofuranose

The ketose that is closely related to D-ribose is named D-ribulose. (Ketose names are often derived from the corresponding aldose name by modifying the suffix *-ose* to *-ulose*.) This ketose is an intermediate that allows cells to make many other monosaccharides. In photosynthetic organisms, D-ribulose is used to capture carbon dioxide and thus make new carbohydrates.

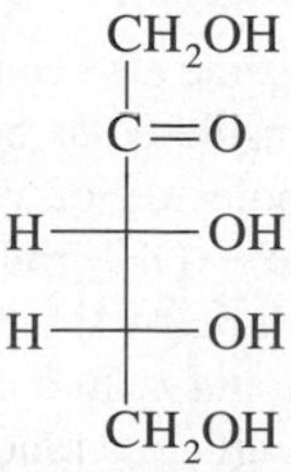

$$\begin{array}{c} CH_2OH \\ C=O \\ H-C-OH \\ H-C-OH \\ CH_2OH \end{array}$$

D-ribulose

Milk sugar, or lactose, is found mainly in the milk of mammals.

28.10 Disaccharides

Disaccharides are carbohydrates composed of two monosaccharide residues united by a glycosidic linkage. The two important disaccharides that are found in the free state in nature are sucrose and lactose ($C_{12}H_{22}O_{11}$). Sucrose, commonly known as *table sugar*, exists throughout the plant kingdom. Sugar cane contains 15–20% sucrose, and sugar beets 10–17%. Maple syrup and sorghum are also good sources of sucrose. Lactose, also known as *milk sugar*, is found free in nature mainly in the milk of mammals. Human milk contains about 6.7% lactose and cow milk about 4.5% lactose.

Unlike sucrose and lactose, several other important disaccharides are derived directly from polysaccharides by hydrolysis (see Section 28.15). For example, maltose, isomaltose, and cellobiose are formed when specific polysaccharides are hydrolyzed. Of this group, maltose is the most common and is found as a constituent of sprouting grain.

Upon hydrolysis, disaccharides yield two monosaccharide molecules. The hydrolysis is catalyzed by hydrogen ions (acids), usually at elevated temperatures, or by certain enzymes that act effectively at room or body temperatures. A different enzyme is required for the hydrolysis of each of the three disaccharides:

$$\text{Sucrose} + \text{Water} \xrightarrow{\text{H}^+ \text{ or sucrase}} \text{Glucose} + \text{Fructose}$$

$$\text{Lactose} + \text{Water} \xrightarrow{\text{H}^+ \text{ or lactase}} \text{Galactose} + \text{Glucose}$$

$$\text{Maltose} + \text{Water} \xrightarrow{\text{H}^+ \text{ or maltase}} \text{Glucose} + \text{Glucose}$$

The enzyme lactase is present in the small intestine of infants and allows them to easily digest lactose from their milk diet. Unfortunately, as people mature their intestines often stop producing the lactase enzyme, and they lose the ability to digest lactose. Instead, this sugar is metabolized by common bacteria that live in the large intestine. The gas and intestinal discomfort that results is termed milk or lactose intolerance and is a condition that afflicts many adults. Carefully note the small structural differences between lactose and the other common disaccharides in the next section. Small differences in molecular shape often determine what our bodies do with specific molecules.

28.11 Structures and Properties of Disaccharides

Disaccharides contain an acetal structure (glycosidic linkage), and some also contain a hemiacetal structure. The acetal structure in maltose may be considered as being derived from two glucose molecules by the elimination of a molecule of water between the —OH group on carbon 1 of one glucose unit and the —OH group of carbon 4 on the other glucose unit. This is an α-1,4-glycosidic linkage since the glucose units have the α configuration and are joined at carbons 1 and 4. In a more systematic nomenclature, this form of maltose is known as α-D-glucopyranosyl-

(1-4)-α-D-glucopyranose. If the structure of glucose is known, this name provides a complete description for drawing the maltose formula.

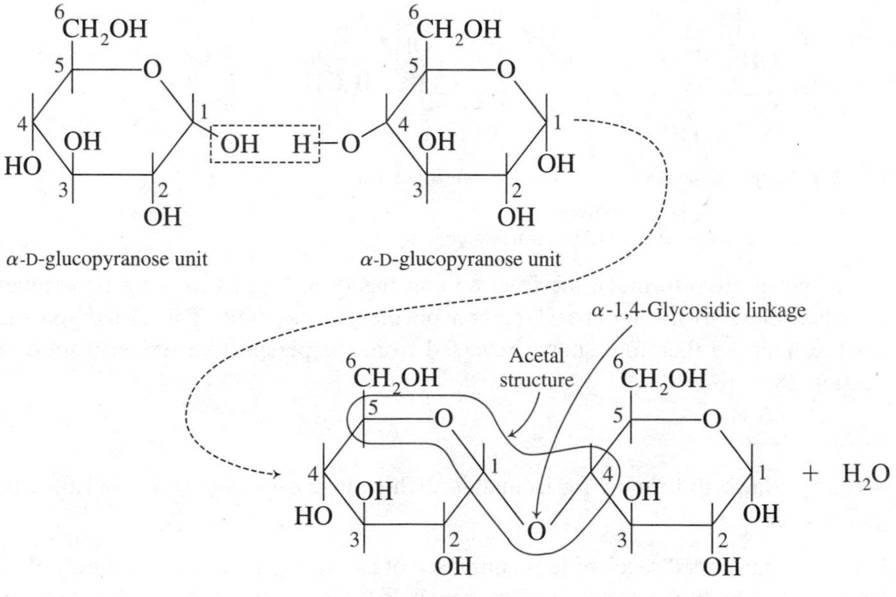

α-D-glucopyranose unit α-D-glucopyranose unit

α-1,4-Glycosidic linkage

Acetal structure

$+ \ H_2O$

maltose
[α-D-glucopyranosyl-(1,4)-α-D-glucopyranose]

Lactose consists of a β-D-galactopyranose unit linked to an α-D-glucopyranose unit. These are joined by a β-1,4-glycosidic linkage from carbon 1 on galactose to carbon 4 on glucose. The more systematic name for lactose is β-D-galactopyranosyl-(1,4)-α-D-glucopyranose. Note that, although glycosidic bonds are straight carbon–oxygen linkages, the structural formula represents them in a bent fashion to provide stereochemical information. Carbon 1 of the galactose is in a β configuration, so its bent bond initially points up. The oxygen on carbon 4 of the glucose is below the carbon in the Haworth formula, and thus the bent bond initially points down.

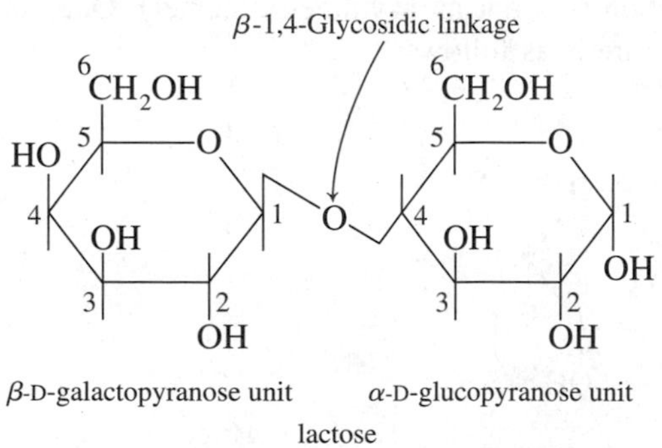

lactose
[β-D-galactopyranosyl-(1,4)-α-D-glucopyranose]

Sucrose consists of an α-D-glucopyranose unit and a β-D-fructofuranose unit. These monosaccharides are joined by an oxygen bridge from carbon 1 on glucose to carbon 2 on fructose—that is, by an α-1,2-glycosidic linkage.

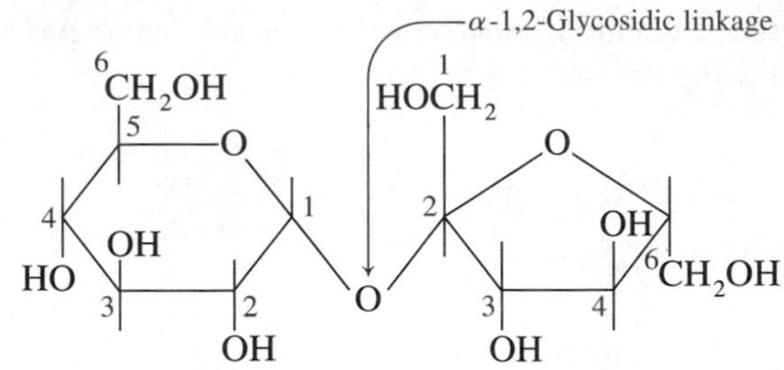

α-1,2-Glycosidic linkage

α-D-glucopyranose unit β-D-fructofuranose unit

sucrose
[α-D-glucopyranosyl-(1,2)-β-D-fructofuranose]

In this perspective formula, the fructose unit has been flipped to bring its number 2 carbon close to the number 1 carbon on the glucose unit. The groups on the fructose unit are therefore shown reversed from the perspective representation in Section 28.8.

Example 28.3 Write the Haworth formula for isomaltose, which is α-D-glucopyranosyl-(1,6)-α-D-glucopyranose.

Solution Recognize that this disaccharide is composed of two α-D-glucopyranose units linked between carbon 6 of one unit and carbon 1 of the other. First, write the Haworth formula for the monosaccharides and number their carbons.

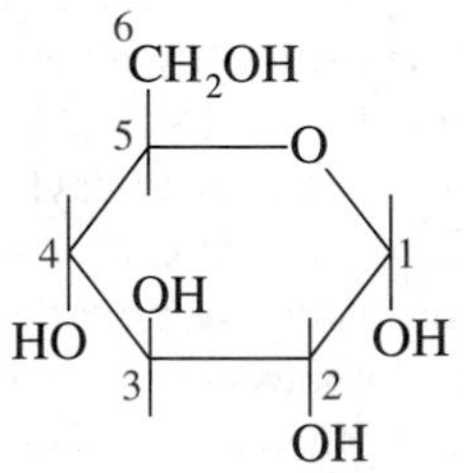

The two α-D-glucopyranose units must be linked in such a way that the stereochemistry at carbon 1 is preserved (carbon 6 is not an asymmetric center). One correct way to write the isomaltose structure is as follows:

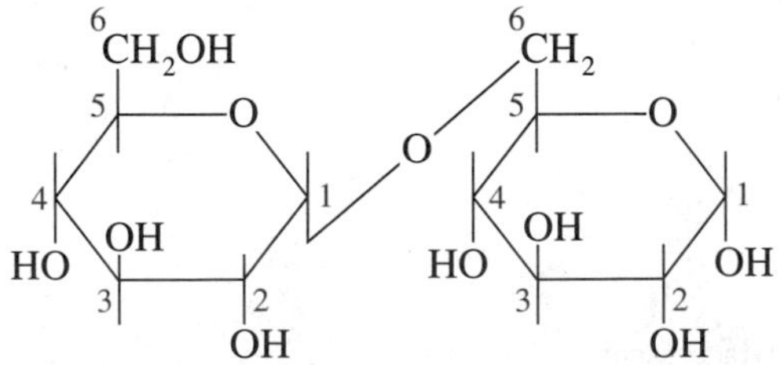

Practice 28.5

Write the structure for cellobiose (a disaccharide that can be derived from plants), β-D-glucopyranosyl-(1,4)-β-D-glucopyranose.

Lactose and maltose both show mutarotation, which indicates that one of the monosaccharide units has a hemiacetal ring that can open and close to interchange anomers. Sucrose has no hemiacetal structure and hence does not mutarotate.

The three disaccharides sucrose, lactose, and maltose have physical properties associated with large polar molecules. All three are crystalline solids and are quite soluble in water; the solubility of sucrose amounts to 200 g per 100 g of water at 0°C. Hydrogen bonding between the polar —OH groups on the sugar molecules and the water molecules is a major factor in this high solubility. These sugars are not easily melted. In fact, lactose is the only one with a clearly defined melting point (201.6°C). Sucrose and maltose begin to decompose when heated to 186°C and 103°C, respectively. When sucrose is heated to melting, it darkens and undergoes partial decomposition. The resulting mixture is known as caramel, or burnt sugar, and is used as coloring and as a flavoring agent in foods.

This sugar substitute is far sweeter than sucrose and is nonnutritive.

28.12 Sweeteners and Diet

Carbohydrates have long been valued for their ability to sweeten foods. Fructose is the sweetest of the common sugars (a scale of relative sweetness is given in Table 28.2), although sucrose (table sugar) is the most commonly used sweetener. Astonishingly large amounts of sucrose are produced from sugar beets and cane: World production is on the order of 90 million tons annually. There are no essential chemical differences between cane and beet sugar. In the United States, approximately 20–30% of the average caloric intake is sucrose (about 150 g/day per person). Low price and sweet taste are the major reasons for high sucrose consumption. Note that sucrose is only 58% as sweet as fructose (Table 28.2). However, because sucrose is inexpensive to produce and is amenable to a variety of food processing techniques, approximately 60–80% of all sweeteners is sucrose.

Sucrose has a tendency to crystallize from concentrated solutions or syrups. Therefore, in commercial food preparations (e.g., candies, jellies, and canned fruits) the sucrose is often hydrolyzed.

$$\text{Sucrose} + H_2O \xrightarrow{H^+} \text{Glucose} + \text{Fructose}$$

invert sugar

The resulting mixture of glucose and fructose, usually in solution, is called **invert sugar**. Invert sugar has less tendency to crystallize than sucrose, and it has greater sweetening power than an equivalent amount of sucrose. The nutritive value of the sucrose is not affected in any way by the conversion to invert sugar because the same hydrolysis reaction occurs in normal digestion.

High-fructose syrups, derived from cornstarch, are used as sweeteners in prod-

TABLE 28.2 Relative Sweetness of Sugars and Sugar Substitutes

Fructose	100	Galactose	19	Saccharin	1.7×10^4
Sucrose	58	Lactose	9.2	Aspartame	2.0×10^4
Glucose	43	Invert sugar	75		
Maltose	19				

ucts such as soft drinks. Starch is a polymer of glucose, contains no fructose, and is not sweet. But through biotechnology a method was developed to convert starch to a very sweet syrup of high fructose content. The starch polymers are hydrolyzed to glucose; then part of the glucose is enzymatically converted to fructose yielding the syrup. High-fructose corn syrups are used because they are more economical sweetening agents than either cane or beet sugar.

Unfortunately, high sugar consumption presents health problems. For many people, sucrose is a source of too many calories. Oral bacteria also find sucrose easy to metabolize, increasing the incidence of dental caries. Finally, because the monosaccharides, fructose and glucose, are quickly absorbed from the small intestine, sugar consumption leads to a rapid increase in blood sugar. Such a sharp rise can be dangerous for people with impaired carbohydrate metabolism—for example, with diabetes mellitus.

Scientists have searched for sugar substitutes. The ideal substitute might be sweeter than sucrose but lack the structural features and chemical reactivity that allow sugar to be absorbed and metabolized. Purely artificial, noncarbohydrate sweeteners have been developed starting with saccharin in 1879. This molecule is about 300 times sweeter than sucrose, cannot be metabolized, and so is nonnutritive. However, the search for other artificial sweeteners has continued because of saccharin's aftertaste and because of its potential health risks. Sodium cyclamate (20 times sweeter than fructose) was introduced in the early 1960s, but this molecule has been shown to cause cancer in laboratory animals. More recently, aspartame has become the artificial sweetener of choice. This molecule is about 200 times sweeter than fructose and poses no known health risks for most individuals, with the notable exception of people who suffer from phenylketonuria (PKU). Aspartame is composed of two amino acids (see Chapter 30) and can be metabolized to yield energy. However, foods sweetened with aspartame have many fewer calories than those containing sucrose because the same sweet taste is achieved with about 200 times less sweetener. The formulas for these sugar substitutes are:

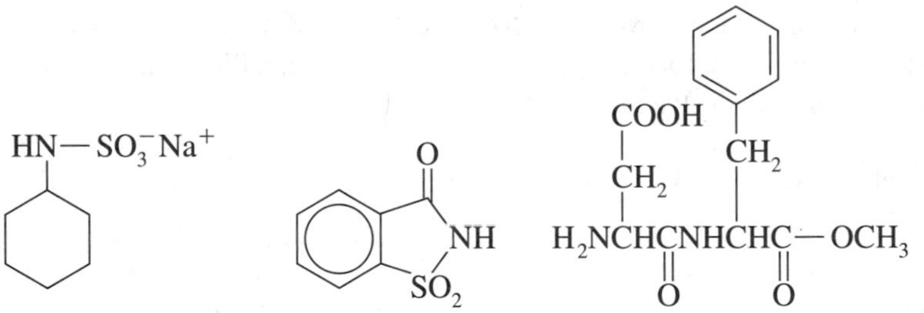

$HN{-}SO_3^-Na^+$ (on cyclohexane ring)

sodium cyclamate

saccharin (O, NH, SO_2)

$H_2NCH(CH_2COOH)C(=O)NHCH(CH_2C_6H_5)C(=O){-}OCH_3$

aspartame

28.13 Reducing Sugars

reducing sugar

Some sugars are capable of reducing silver ions to free silver, and copper(II) ions to copper(I) ions, under prescribed conditions. Such sugars are called **reducing sugars**. This reducing ability, which is useful in classifying sugars and in certain clinical tests, is dependent on the presence of (1) aldehydes, (2) α-hydroxyketone groups ($-CH_2COCH_2OH$) such as in fructose, or (3) hemiacetal structures in cyclic molecules such as maltose. These groups are easily oxidized to carboxylic acid (or carboxylate ion) groups; the metal ions are thereby reduced ($Ag^+ \longrightarrow Ag^0$; $Cu^{2+} \longrightarrow Cu^+$). Several different reagents, including Tollens, Fehling, Benedict, and Barfoed reagents, are used to detect reducing sugars (see Section 23.5). The Benedict, Fehling, and Barfoed tests depend on the formation of copper(I) oxide precipitate to indicate a positive reaction.

$$\underset{\text{aldehyde group}}{RC(=O)-H} + \underset{\text{(blue)}}{2\,Cu^{2+}} + 5\,OH^- \longrightarrow \underset{\text{carboxylate ion group}}{RC(=O)-O^-} + \underset{\text{copper(I) oxide (brick red)}}{Cu_2O(s)} + 3\,H_2O$$

Barfoed reagent contains Cu^{2+} ions in the presence of acetic acid. It is used to distinguish reducing monosaccharides from reducing disaccharides. Under the same reaction conditions, the reagent is reduced more rapidly by monosaccharides.

Glucose and galactose contain aldehyde groups; fructose contains an α-hydroxyketone group. Therefore, all three of these monosaccharides are reducing sugars.

A carbohydrate molecule need not have a free aldehyde or α-hydroxyketone group to be a reducing sugar. A hemiacetal structure (see below) is a potential aldehyde group. Maltose and the cyclic form of glucose are examples of molecules with the hemiacetal structure.

Hemiacetal structure that opens to form an aldehyde group

CH_2OH O OH HO OH OH — CH_2OH O OH HO OH O CH_2OH O OH OH OH

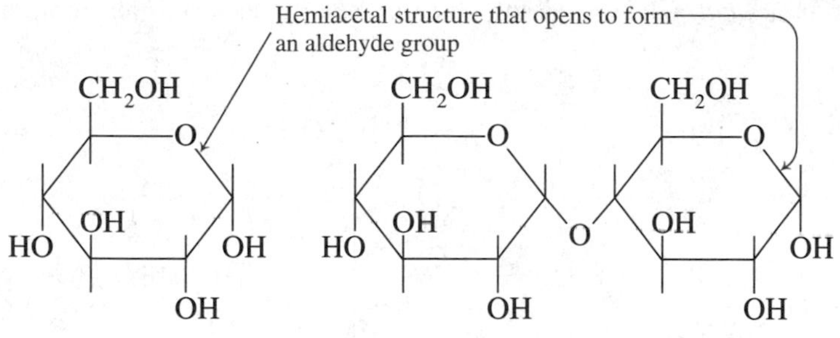

glucose maltose

Under mildly alkaline conditions, the rings open at the points indicated by the arrows to form aldehyde groups:

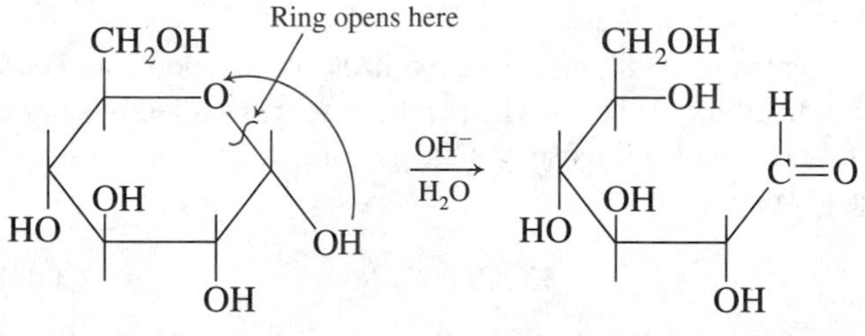

Ring opens here

CH_2OH O OH HO OH OH $\xrightarrow[H_2O]{OH^-}$ CH_2OH OH H C=O OH HO OH

glucose (ring structure) glucose (open-chain structure)

Any sugar that has the hemiacetal structure is classified as a reducing sugar. Among the disaccharides, lactose and maltose have hemiacetal structures and are therefore reducing sugars. Sucrose is not a reducing sugar because it does not have the hemiacetal structure.

Many clinical tests monitor glucose as a reducing sugar. For example, Benedict and Fehling reagents are used to detect the presence of glucose in urine. Initially, the reagents are deep blue in color. A positive test is indicated by a color change to greenish-yellow, yellowish-orange, or brick-red, corresponding to an increasing glucose (reducing sugar) concentration. These tests are used to estimate the amount of glucose in the urine of diabetics in order to adjust the amount of insulin needed for proper glucose utilization.

Alternatively, a clinical test (glucose oxidase test) makes use of an enzyme-catalyzed oxidation of glucose to test for urine sugar. The inclusion of an enzyme ensures a reaction that is specific for the glucose structure, allowing a more selective test for glucose in the urine.

28.14 Reactions of Monosaccharides

Oxidation

The oxidation of monosaccharides by copper ions is described in Section 28.13. The aldehyde groups in monosaccharides are also oxidized to monocarboxylic acids by other mild oxidizing agents such as bromine water. The carboxylic acid group is formed at carbon 1. The name of the resulting acid is formed by changing the *ose* ending to *onic acid*. Glucose yields gluconic acid; galactose, galactonic acid; and so on.

$$
\begin{array}{rcl}
 & \mathrm{H{-}C{=}O} & \\
\mathrm{H} & - & \mathrm{OH} \\
\mathrm{HO} & - & \mathrm{H} \\
\mathrm{H} & - & \mathrm{OH} \\
\mathrm{H} & - & \mathrm{OH} \\
 & \mathrm{CH_2OH} & \\
 & \text{D-glucose} &
\end{array}
\; + \mathrm{Br_2} + \mathrm{H_2O} \longrightarrow
\begin{array}{rcl}
 & \mathrm{COOH} & \\
\mathrm{H} & - & \mathrm{OH} \\
\mathrm{HO} & - & \mathrm{H} \\
\mathrm{H} & - & \mathrm{OH} \\
\mathrm{H} & - & \mathrm{OH} \\
 & \mathrm{CH_2OH} & \\
 & \text{D-gluconic acid} &
\end{array}
\; + 2\,\mathrm{HBr}
$$

Dilute nitric acid, a vigorous oxidizing agent, oxidizes both carbon 1 and carbon 6 of aldohexoses to form dicarboxylic acids. The resulting acid is named by changing the *ose* sugar suffix to *aric acid*. Glucose yields glucaric acid (saccharic acid); galactose, galactaric acid (mucic acid).

$$
\begin{array}{rcl}
 & \mathrm{H{-}C{=}O} & \\
\mathrm{H} & - & \mathrm{OH} \\
\mathrm{HO} & - & \mathrm{H} \\
\mathrm{H} & - & \mathrm{OH} \\
\mathrm{H} & - & \mathrm{OH} \\
 & \mathrm{CH_2OH} & \\
 & \text{D-glucose} &
\end{array}
\xrightarrow{\text{warm } \mathrm{HNO_3}}
\begin{array}{rcl}
 & \mathrm{COOH} & \\
\mathrm{H} & - & \mathrm{OH} \\
\mathrm{HO} & - & \mathrm{H} \\
\mathrm{H} & - & \mathrm{OH} \\
\mathrm{H} & - & \mathrm{OH} \\
 & \mathrm{COOH} & \\
 & \text{glucaric acid} &
\end{array}
\qquad
\begin{array}{rcl}
 & \mathrm{COOH} & \\
\mathrm{H} & - & \mathrm{OH} \\
\mathrm{HO} & - & \mathrm{H} \\
\mathrm{HO} & - & \mathrm{H} \\
\mathrm{H} & - & \mathrm{OH} \\
 & \mathrm{COOH} & \\
 & \text{galactaric acid} &
\end{array}
$$

```
H—C=O                         H—C=NNHC6H5                  H—C=NNHC6H5
H———OH                        H———OH                       C=O
HO———H     C6H5NHNH2 →        HO———H     C6H5NHNH2 →       HO———H
H———OH                        H———OH                       H———OH       C6H5NHNH2 ↘
H———OH                        H———OH                       H———OH
CH2OH                         CH2OH                        CH2OH
D-glucose                                                                    H—C=NNHC6H5
                                                                             C=NNHC6H5
                                                                             HO———H
                                                                C6H5NHNH2 ↗  H———OH
CH2OH                         CH2OH                        H—C=O             H———OH
C=O                           C=NNHC6H5                    C=NNHC6H5         CH2OH
HO———H     C6H5NHNH2 →        HO———H     C6H5NHNH2 →       HO———H            osazone
H———OH                        H———OH                       H———OH            (from either glucose
H———OH                        H———OH                       H———OH            or fructose)
CH2OH                         CH2OH                        CH2OH
D-fructose
```

◀ FIGURE 28.5
Reaction of glucose and fructose to form osazones. Common sugars form the same structure at carbons 1 and 2. Since D-glucose and D-fructose are identical at all other positions, these sugars yield the same osazone.

Osazone Formation

Phenylhydrazine ($C_6H_5NHNH_2$) reacts with carbons 1 and 2 of reducing sugars to form derivatives called osazones. The formation of these distinctive crystalline derivatives is useful for comparing the structures of sugars. Glucose and fructose react as shown in Figure 28.5.

Identical osazones are obtained from D-glucose and D-fructose. This demonstrates that carbons 3 through 6 of D-glucose and D-fructose molecules are identical. The same osazone is also obtained from D-mannose. This indicates that carbons 3 through 6 of the D-mannose molecule are the same as those of D-glucose and D-fructose molecules. In fact, D-mannose differs from D-glucose only in the configuration of the —H and —OH groups on carbon 2.

Reduction

Monosaccharides may be reduced to their corresponding polyhydroxy alcohols by reducing agents such as H_2/Pt or sodium amalgam, Na(Hg). For example, glucose yields sorbitol (glucitol), galactose yields galactitol (dulcitol), and mannose yields mannitol; all of these are hexahydric alcohols (containing six —OH groups).

```
H—C=O                       CH2OH
H———OH                      H———OH
HO———H      H2/Pt →         HO———H
H———OH                      H———OH
H———OH                      H———OH
CH2OH                       CH2OH
D-glucose                   D-glucitol
                            (sorbitol)
```

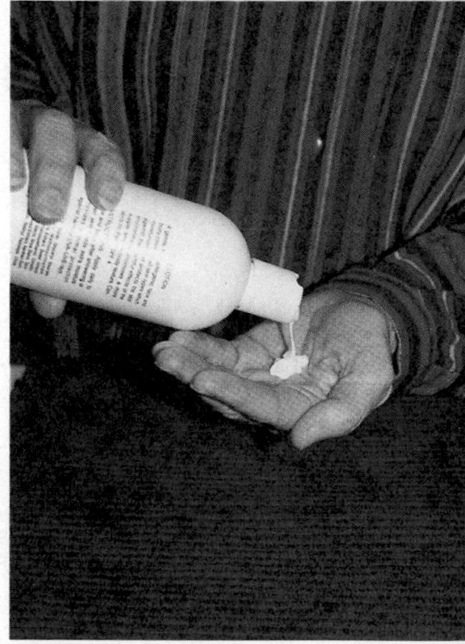

Sorbitol occurs naturally in plants and has affinity for water. It is an ingredient in many moisturizers and lotions.

Hexahydric alcohols have properties resembling those of glycerol (Section 22.7). Because of their affinity for water, they are used as moisturizing agents in food and cosmetics. Sorbitol, galactitol, and mannitol occur naturally in a variety of plants.

Example 28.4 Two samples labeled A and B are known to be D-threose and D-erythrose. Water solutions of each sample are optically active. However, when each solution was warmed with nitric acid, the solution from sample A became optically inactive while that from sample B was still optically active. Determine which sample (A or B) contains D-threose and which sample contains D-erythrose.

Solution In this problem we need to examine the structures of D-threose and D-erythrose, write equations for the reaction with nitric acid, and examine the products to see why one is optically active and the other optically inactive. Start by writing the formulas for D-threose and D-erythrose:

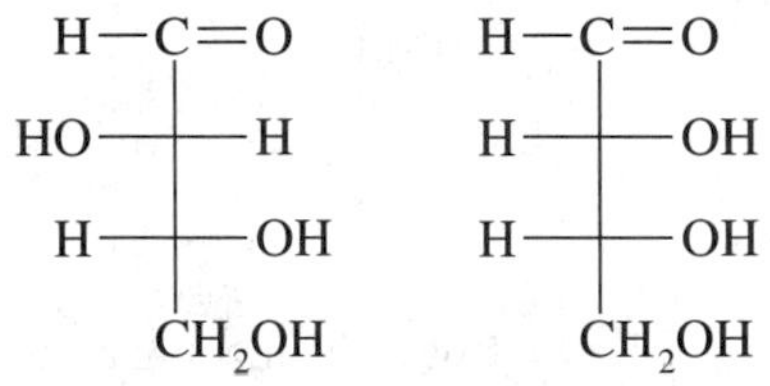

```
  H—C=O            H—C=O
HO——+——H         H——+——OH
 H——+——OH        H——+——OH
   CH2OH            CH2OH
 D-threose        D-erythrose
```

Oxidation of these tetroses with HNO_3 will yield dicarboxylic acids:

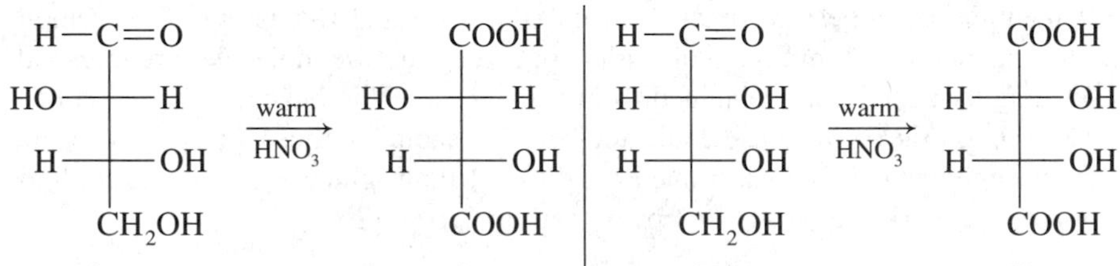

```
  H—C=O                  COOH   |   H—C=O                  COOH
HO——+——H     warm     HO——+——H  |  H——+——OH     warm     H——+——OH
 H——+——OH   ——————>    H——+——OH |  H——+——OH   ——————>    H——+——OH
             HNO3               |               HNO3
   CH2OH                 COOH   |    CH2OH                 COOH
 D-threose                 I    |  D-erythrose               II
```

Product I is a chiral molecule and is optically active. Product II is a meso compound and is optically inactive. Therefore, sample A is D-erythrose, since oxidation yields the meso acid. Sample B then must be D-threose.

Practice 28.6

A disaccharide yields no copper(I) oxide when treated in Benedict test. This carbohydrate is composed of two α-D-galactopyranose units. Identify the carbon from each monosaccharide involved in the acetal linkage.

Practice 28.7

Write the structure of the product formed when D-galactose is reduced with H_2/Pt. Is this compound optically active?

28.15 Polysaccharides Derived from Glucose

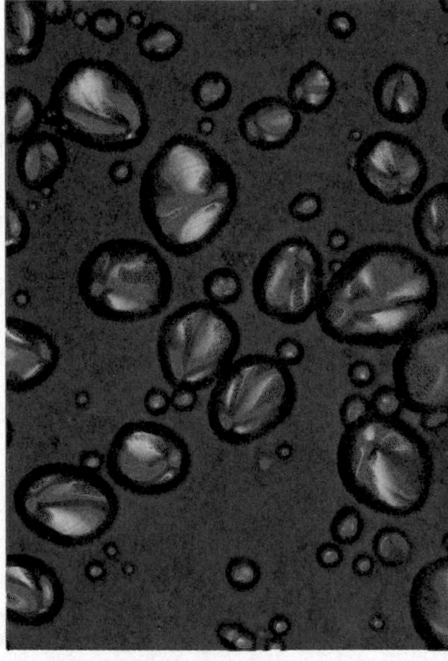

▲
FIGURE 28.6
Polarized light micrograph of starch granules in potato tuber cells.

Although many naturally occurring polysaccharides are known, three—starch, cellulose, and glycogen—are of outstanding importance. All three, when hydrolyzed, yield D-glucose as the only product, according to this approximate general equation:

$$\underset{\substack{\text{polysaccharide molecule}\\ \text{(approximate formula)}}}{(C_6H_{10}O_5)_n} + n\,H_2O \longrightarrow \underset{\text{D-glucose}}{n\,C_6H_{12}O_6}$$

This hydrolysis reaction establishes that all three polysaccharides are polymers made up of glucose monosaccharide units. It also means that the differences in properties among the three polysaccharides must be due to differences in the structures and/or sizes of these molecules.

Many years of research were required to determine the detailed structures for polysaccharide molecules. Consideration of all this work is beyond the scope of our discussion, but an abbreviated summary of the results is given in the following paragraphs.

Starch

Starch is found in plants, mainly in the seeds, roots, or tubers (see Figure 28.6). Corn, wheat, potatoes, rice, and cassava are the chief sources of dietary starch. The two main components of starch are amylose and amylopectin. Amylose molecules are unbranched chains composed of about 25–1300 α-D-glucose units joined by α-1,4-glycosidic linkages, as shown in Figure 28.7. The stereochemistry of the α anomer causes amylose to coil into a helical conformation. Partial hydrolysis of this linear polymer yields the disaccharide maltose.

Amylopectin is a branched-chain polysaccharide with much larger molecules than those of amylose. Amylopectin molecules consist on the average of several thousand α-D-glucose units with molar masses ranging up to 1 million or more. The main chain contains glucose units connected by α-1,4-glycosidic linkages. Branch chains are linked to the main chain through α-1,6-glycosidic linkages about every 25 glucose units, as shown in Figure 28.7. This molecule has a characteristic tree-like structure because of its many branch chains. Partial hydrolysis of amylopectin yields both maltose and the related disaccharide isomaltose, α-D-glucopyranosyl-(1-6)-α-D-glucopyranose.

Despite the presence of many polar —OH groups, starch molecules are insoluble in cold water, apparently because of their very large size. Starch readily forms colloidal dispersions in hot water. Such starch "solutions" form an intense blue-black color in the presence of free iodine. Hence, a starch solution can be used to detect free iodine, or a dilute iodine solution can be used to detect starch.

Starch is readily converted to glucose by heating with water and a little acid (e.g., hydrochloric or sulfuric acid). It is also readily hydrolyzed at room temperature by certain digestive enzymes. The hydrolysis of starch to maltose and glucose is shown in the following equation:

$$\text{Starch} \xrightarrow[\substack{\text{or salivary and}\\ \text{pancreatic amylase}}]{\text{acid } + \Delta} \text{Dextrins} + \text{Maltose} \xrightarrow[\substack{\text{or maltase and other}\\ \text{intestinal enzymes}}]{\text{acid } + \Delta} \text{D-Glucose}$$

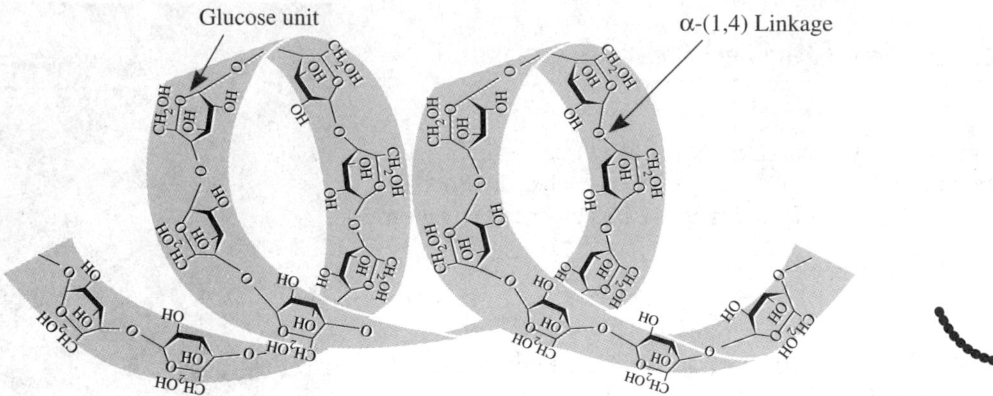

(a) Molecular structure of amylose chain

(b) Array of glucose units (dots) in amylose

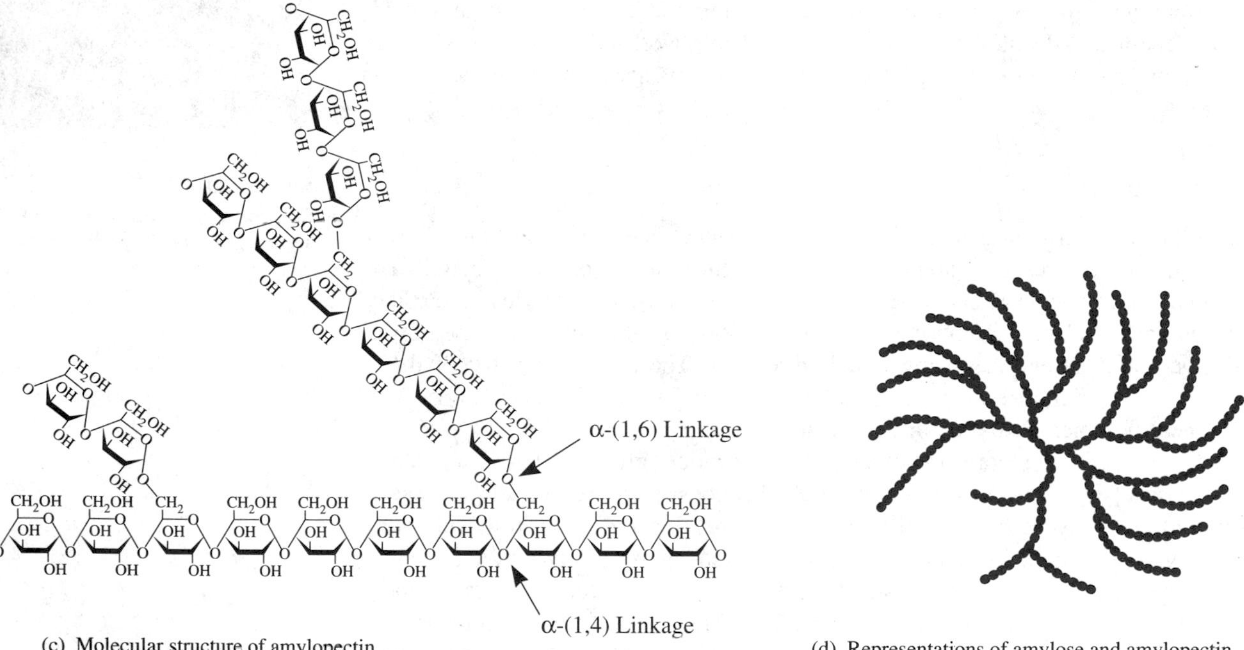

(c) Molecular structure of amylopectin

(d) Representations of amylose and amylopectin

▲
FIGURE 28.7
Representations of amylose and amylopectin.

The hydrolysis of starch can be followed qualitatively by periodically testing samples from a mixture of starch and saliva with a very dilute iodine solution. The change of color sequence is blue-black ⟶ blue ⟶ purple ⟶ pink ⟶ colorless as the starch molecules are broken down into smaller and smaller fragments.

Hydrolysis is a key chemical reaction in the digestion of starchy foods. If these foods are well chewed, salivary amylase normally decreases the starch polymer chain length from on the order of a thousand glucose units to about eight per chain. In the small intestine, pancreatic amylase continues digestion to form maltose. Enzymes in the small intestine membranes complete the conversion of starch to glucose, which is then absorbed into the bloodstream.

Starch is the most important energy storage carbohydrate of the plant kingdom.

FIGURE 28.8
Scanning electron micrograph of cellulose fibers in a plant cell wall.

In turn, humans and other animals consume huge quantities of starch. This polymer is an important food source because it has the appropriate structure to be readily broken down to D-glucose. The carbons from starch provide much of our daily energy needs as they are oxidized to carbon dioxide.

Glycogen

Glycogen is the energy storage carbohydrate of the animal kingdom. It is formed by polymerization of glucose and is stored in the liver and in muscle tissues. Structurally, it is very similar to the amylopectin fraction of starch except that it is more highly branched. The α-1,6-glycosidic linkages occur on one of every 12–18 glucose units.

Cellulose

Cellulose is the most abundant organic substance found in nature. It is the chief structural component of plants and wood (Figure 28.8). Cotton fibers are almost pure cellulose; wood, after removal of moisture, consists of about 50% cellulose. Cellulose is an important substance in the textile and paper industries.

Cellulose, like starch and glycogen, is a polymer of glucose. But cellulose differs from starch and glycogen because the glucose units are joined by β-1,4-glycosidic linkages instead of α-1,4-glycosidic linkages. The stereochemistry of the β anomer allows this polymer to form an extended chain that can hydrogen-bond to adjacent cellulose molecules. The large number of hydrogen bonds so formed partially accounts for the strength of the resulting plant cell walls. The cellulose structure is illustrated in Figure 28.9.

A partial hydrolysis of cellulose produces the disaccharide cellobiose, β-D-glucopyranosyl-(1-4)-β-D-glucopyranose. However, cellulose has greater resistance to hydrolysis than either starch or glycogen. It is not appreciably hydrolyzed when boiled in a 1% sulfuric acid solution. It does not show a color reaction with iodine. Humans cannot digest cellulose because they have no enzymes capable of catalyzing its hydrolysis. Fortunately, some microorganisms found in soil and in the digestive tracts of certain animals produce enzymes that do catalyze the breakdown of cellulose. The presence of these microorganisms explains why cows and other herbivorous animals thrive on grass—and also why termites thrive on wood.

The —OH groups of starch and cellulose can be reacted without destruction of the macromolecular structures. For example, nitric acid converts an —OH group to a nitrate group in this fashion:

$$\text{Cellulose}-\boxed{\text{OH} + \text{H}}\text{ONO}_2 \longrightarrow \text{Cellulose}-\text{ONO}_2 + \text{H}_2\text{O}$$

hydroxyl group on cellulose molecule; nitric acid; nitrate group on cellulose molecule; water

FIGURE 28.9 ▶
Two representations of cellulose. In the three-dimensional drawing, note the hydrogen bonding that links the extended cellulose polymers to form cellulose fibers.

(a) Haworth formulas

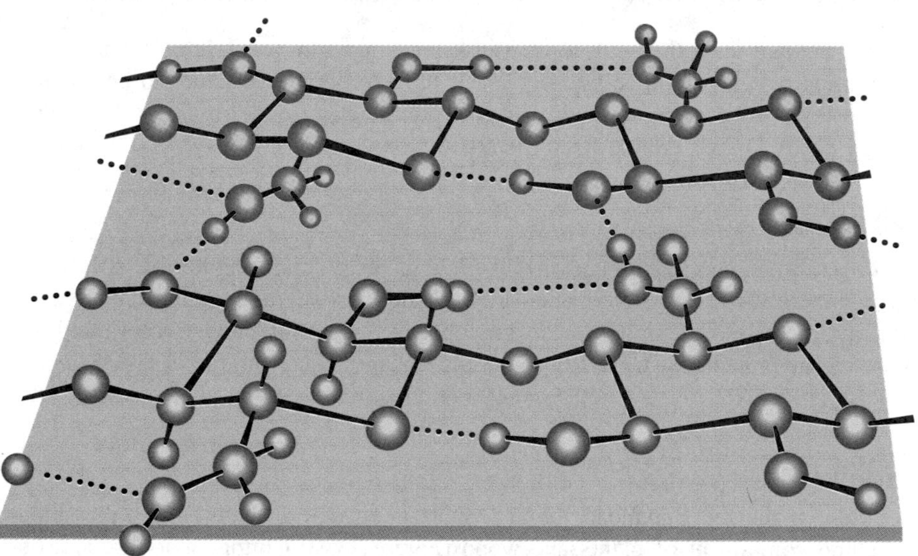

(b) Three-dimensional representation

If only a portion of the —OH groups on the cellulose molecule are nitrated, a plastic nitrocellulose material known as celluloid or pyroxylin is obtained. This material has been used to make such diverse articles as billiard balls, celluloid shirt collars, and photographic film. By nitration of nearly all —OH groups, a powerful high explosive is obtained. This highly nitrated cellulose, or "guncotton," is the basic ingredient in modern "smokeless" gunpowder.

Another modified cellulose, cellulose acetate, is made by esterification of —OH groups with acetic acid (acetic anhydride). About two-thirds of the —OH groups are esterified:

$$\underset{\text{hydroxyl group on cellulose molecule}}{\text{Cellulose}-\text{O}\boxed{\text{H}}} + \underset{\text{acetic anhydride}}{\text{CH}_3-\overset{\overset{\text{O}}{\|}}{\text{C}}-\text{O}-\overset{\overset{\text{O}}{\|}}{\text{C}}-\text{CH}_3} \longrightarrow \underset{\text{acetate group on cellulose molecule}}{\text{Cellulose}-\text{O}-\overset{\overset{\text{O}}{\|}}{\text{C}}-\text{CH}_3} + \text{CH}_3\text{COOH}$$

Chemistry in Action

Polysaccharides

Many carbohydrate polymers are composed of monosaccharides other than glucose and serve a variety of purposes for both plants and animals. Some polysaccharides are made up of only one type of monosaccharide. For example, D-mannose–containing polymers are found in many plants, including pine trees and orchid tubers. Polysaccharides composed of the pentose D-xylose are also found widely distributed in the vegetable kingdom.

More complex polysaccharides are often found in animals. These carbohydrates are linked together by glycosidic bonds at various positions on the monosaccharides. Thus a sugar may be bonded to its neighbor from carbon 1 to carbon 3, or carbon 1 to carbon 4, or carbon 1 to carbon 6, and so on. Because of these numerous linkages, the polysaccharides often have a complex, branching structure. Also, it is not uncommon for these molecules to contain many different kinds of monosaccharides. These complex carbohydrates serve a variety of functions in animals.

Mucopolysaccharides (or glycosaminoglycans) make up part of the connective tissue and are found in the joints and the skin. The slimy, mucus-like consistency of these molecules derives from their special chemical properties. About half of the sugar units are acidic in these carbohydrates. These acid groups become negatively charged under physiological conditions, causing the polymer chains to repel each other.

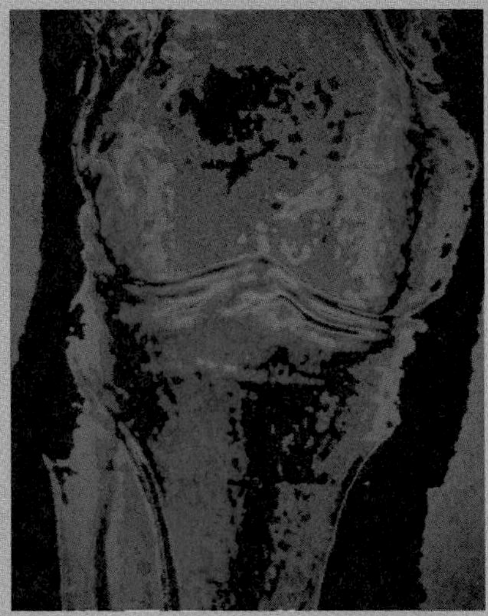

The connective tissue inside the knee contains mucopolysaccharides.

Water fills the space between the polymer chains and gives the mucopolysaccharide a spongelike consistency. One gram of some mucopolysaccharides can absorb up to 20 L of water. This natural shock absorber and lubricant is necessary for animal locomotion.

Even more complicated polysaccharides are found on the surfaces of almost all cells. These carbohydrates serve as "labels" or antigens allowing organisms to distinguish their own cells from invading bacteria, for example. Antigen recognition illustrates a very important biochemical principle: *molecular shape carries information that guides the reactions of life.*

In humans, the polysaccharides on the surface of the red blood cells give rise to many blood types, often classified by the ABO system. Cells of different blood types have surface polysaccharides with different structures. Cells carrying one carbohydrate structure are commonly not tolerated by an individual of another blood type. For example, if type AB blood is transfused into someone with type O red blood cells, the new cells will be attacked by the immunosystem. Red cell destruction can lead to serious injury or death.

Cellulose acetate, unlike the dangerously flammable cellulose nitrate, can be made to burn only with difficulty. For this reason, cellulose acetate has displaced cellulose nitrate in almost all kinds of photographic films. The textile known as acetate rayon is made from cellulose acetate. Cellulose acetate is also used as a clear, transparent packaging film. In another process, cellulose reacts with carbon disulfide in the presence of sodium hydroxide to form a soluble cellulose derivative called cellulose xanthate, from which cellulose can be regenerated. Viscose rayon textiles and cellophane packaging materials are made of regenerated cellulose prepared by this process.

Concepts in Review

1. List three important characteristics of an energy-providing food.
2. Compare the oxidation state of carbon in carbohydrates with that of carbon in carbon dioxide.
3. Classify carbohydrates as mono-, di-, oligo-, or polysaccharides.
4. Explain the use of D, L, (+), and (−) in naming carbohydrates.
5. Identify and write pyranose and furanose ring structures of carbohydrates.
6. Identify and write Fischer projection and Haworth formulas for carbohydrates.
7. Identify the structural feature of a carbohydrate that makes it a reducing sugar.
8. Explain the phenomenon of mutarotation.
9. Distinguish between hemiacetal and acetal structures in a carbohydrate.
10. Understand what a glycoside linkage is.
11. Understand disaccharide composition and the manner in which monosaccharides are linked together in sucrose, lactose, and maltose.
12. Identify monosaccharides that are epimers.
13. List the major sources of glucose, galactose, fructose, sucrose, lactose, and maltose.
14. Describe some disadvantages of a diet high in sucrose and some alternatives to this sweetener.
15. Identify three disaccharides that are derived from polysaccharides.
16. Describe the Benedict test and state what evidence must be seen to indicate a positive test.
17. Write chemical equations for the oxidation of monosaccharides by bromine and by nitric acid.
18. Write chemical equations for the reduction of monosaccharides.
19. Write chemical equations for the formation of osazones.
20. Identify monosaccharides that give identical osazones.
21. Identify monosaccharides that give optically inactive (meso) products when reduced to polyhydroxy alcohols or when oxidized to dicarboxylic acids.
22. Discuss the structural differences between amylose, amylopectin, and cellulose.

Key Terms

The terms listed here have been defined within this chapter. Section numbers are referenced in parenthesis for each term.

anomer (28.6)
carbohydrate (28.1)
disaccharide (28.2)
epimer (28.5)
glycoside (28.7)
invert sugar (28.12)
monosaccharide (28.2)
mutarotation (28.6)
oligosaccharide (28.2)
polysaccharide (28.2)
reducing sugar (28.13)

Questions

Questions refer to tables, figures, and key words and concepts defined within the chapter. A particularly challenging question or exercise is indicated with an asterisk.

1. Why is the oxidation state of a carbohydrate carbon important in metabolism?

2. What is the significance of the notations D and L in the name of a carbohydrate?

3. What is the significance of the notations (+) and (−) in the name of a carbohydrate?

4. What is galactosemia and what are its effects on humans?

5. Which of the D-aldohexoses in Figure 28.1 are epimers?

6. Explain how a carbohydrate with a pyranose structure differs from one with a furanose structure.

7. Explain how α-D-glucopyranose differs from β-D-glucopyranose.

8. Are the cyclic forms of monosaccharides hemiacetals or glycosides?

9. Explain the phenomenon of mutarotation.

10. What is (are) the major sources of each of these carbohydrates?
(a) sucrose (b) lactose (c) maltose

11. Consider the eight aldohexoses given in Figure 28.1 and answer the following:
(a) Which of these aldohexoses give the same osazone?
(b) The eight aldohexoses are oxidized by nitric acid to dicarboxylic acids. Which of these give meso (optically inactive) dicarboxylic acids?
(c) Write the structures and names for the enantiomers of D-altrose and D-idose.

12. Explain why invert sugar is sweeter than sucrose.

13. What are the structural differences between amylose and cellulose?

14. What visual difference would you expect between a dilute glucose solution and a concentrated glucose solution in the Benedict test?

15. What are the two main components of starch? How are they alike and how do they differ?

16. Which of these statements are correct? Rewrite each incorrect statement to make it correct.
(a) α-D-Glucopyranose and β-D-glucopyranose are enantiomers.
(b) The carbon of a secondary alcohol is more reduced than the carbon of a primary alcohol.
(c) D-Glyceraldehyde and L-glyceraldehyde are epimers.
(d) D-Threose and L-threose are epimers.
(e) There are eight stereoisomers of the aldopentoses.
(f) D-Glucose and L-glucose form identical osazones.
(g) Raffinose, which consists of one unit each of galactose, glucose, and fructose, is an oligosaccharide.
(h) Two aldohexoses that react with phenylhydrazine to yield identical osazones are epimers.
(i) Dextrose is another name for glucose.
(j) D-Mannitol is obtained from D-mannose by oxidation.
(k) Methyl glucosides are capable of reducing Fehling and Barfoed reagents.
(l) Humans are incapable of using cellulose directly as a food.
(m) Fructose can be classified as a hexose, a monosaccharide, and an aldose.
(n) The change in the specific rotation of a carbohydrate solution to an equilibrium value is called mutarotation.
(o) The disaccharide found in mammalian milk is galactose.
(p) The reserve carbohydrate of animals is glycogen.
(q) Starch consists of two polysaccharides known as amylose and amylopectin.
(r) Carbohydrates that are capable of reducing copper ions in Benedict reagent are called reducing sugars.
(s) The polysaccharides cellulose, starch, and glycogen are all composed of glucose units.
(t) Invert sugar is sweeter than fructose.
(u) Sucrose, glucose, galactose, and fructose are reducing sugars.
(v) Methyl glycosides do not undergo mutarotation.
(w) Oxidation of D-erythrose with dilute HNO_3 yields *meso*-tartaric acid.
(x) The amylose polysaccharide coils into a helical shape.
(y) Aspartame is a nonnutritive sweetener.
(z) Mucopolysaccharides have many sugar units that act as bases.

Paired Exercises

These exercises are paired. Each odd-numbered exercise is followed by a similar even-numbered exercise. Answers to the even-numbered exercises are given in Appendix V.

17. Dihydroxyacetone is the simplest ketose.
  (a) Write a Fischer projection formula for this ketose. Are there any asymmetric carbon atoms in this molecule?
  (b) Write the Fischer projection formula of the product obtained by reacting dihydroxyacetone with hydrogen in the presence of platinum.

18. Glyceraldehyde is the simplest aldose.
  (a) Write Fischer projection formulas for and identify the D and L forms of this aldose.
  (b) Write the Fischer projection formula of the product obtained by reacting D-glyceraldehyde with hydrogen in the presence of platinum.

19. Give the structure of an epimer of D-mannose.

20. Give the structure of an epimer of D-galactose.

21. Write structures and names for the enantiomers of D-galactose, D-mannose, and D-ribose.

22. Write structures and names for the enantiomers of D-glucose, D-fructose, and D-2-deoxyribose.

23. Write the cyclic structures for α-D-glucopyranose, β-D-galactopyranose, and α-D-mannopyranose.

24. Write the cyclic structures of β-D-glucopyranose, α-D-galactopyranose, and β-D-mannopyranose.

25. Write the structure of dihydroxyacetone and show, using oxidation numbers, which carbon is most oxidized.

26. Write the structure of L-glyceraldehyde and show, using oxidation numbers, which carbon is most reduced.

27. Starting with the proper D-tetrose, show the steps for the synthesis of D-glucose by the Kiliani–Fischer synthesis.

28. Starting with the proper D-triose, show the steps for the synthesis of D-ribose by the Kiliani–Fischer synthesis.

29. Is D-2-deoxymannose the same as D-2-deoxyglucose? Explain.

30. Is D-2-deoxygalactose the same as D-2-deoxyglucose? Explain.

31. What is the monosaccharide composition of each of the following?
  (a) sucrose
  (b) glycogen
  (c) amylose
  (d) maltose

32. What is the monosaccharide composition of each of the following?
  (a) lactose
  (b) amylopectin
  (c) cellulose
  (d) sucrose

33. How does the structure of cellobiose differ from that of isomaltose?

34. How does the structure of maltose differ from that of isomaltose?

35. Of the disaccharides sucrose and lactose, which show mutarotation? Explain why.

36. Of the disaccharides maltose and isomaltose, which show mutarotation? Explain why.

37. Draw structural formulas for cellobiose and isomaltose. Point out the portion of the structure that is responsible for making these sugars reducing.

38. Draw structural formulas for maltose and lactose. Point out the portion of the structure that is responsible for making these sugars reducing.

39. Give systematic names for isomaltose and cellobiose.

40. Give systematic names for maltose and lactose.

41. Describe the principal structural differences and similarities between the members of each of the following pairs:
  (a) D-glucose and D-fructose
  (b) maltose and sucrose
  (c) cellulose and glycogen

42. Describe the principal structural differences and similarities between the members of each of the following pairs:
  (a) D-ribose and D-2-dexoyribose
  (b) amylose and amylopectin
  (c) lactose and isomaltose

43. When D-glucose is oxidized with nitric acid, glucaric acid is formed. Write the structural formula and the name of the dicarboxylic acid that is formed when D-galactose is oxidized with nitric acid.

44. When D-glucose is oxidized with nitric acid, glucaric acid is formed. Write the structural formula and the name of the dicarboxylic acid that is formed when D-mannose is oxidized with nitric acid.

45. Write the formulas for the four L-ketohexoses. Indicate which pair (or pairs) of the L-ketohexoses give identical osazones with phenylhydrazine.

46. Write the formulas for the four L-aldopentoses. Indicate which pair (or pairs) of the L-aldopentoses give identical osazones with phenylhydrazine.

***47.** Draw the Haworth formulas for
  (a) β-D-mannopyranosyl-(1,4)-β-D-galactopyranose
  (b) β-D-galactopyranosyl-(1,6)-α-D-glucopyranose

***48.** Draw the Haworth formulas for
  (a) β-D-glucopyranosyl-(1,4)-α-D-galactopyranose
  (b) β-D-galactopyranosyl-(1,6)-β-D-mannopyranose

Additional Exercises

These exercises are not paired or labeled by topic and provide additional practice on concepts covered in this chapter.

49. Would D-mannose provide more metabolic energy than D-galactose? Explain.

50. Cite two advantages of aspartame, as a sweetener, over sucrose.

51. Trehalose is a disaccharide found in mushrooms and many insects. It is made up of two α-D-glucopyranose units and is a nonreducing sugar. Draw the two monosaccharides and star a carbon on each monosaccharide that must be involved in the trehalose glycosidic linkage.

52. How is the acidic nature of the mucopolysaccharides related to their biological function (refer to the Chemistry in Action for this Chapter)

53. Candy makers who want an especially sweet treat may add some lemon juice (an acid) to their boiling sucrose solution before letting the candy harden. What chemical reaction will the lemon juice aid? How will this chemical reaction make the candy sweeter? (See Section 28.12)

54. What changes must take place to convert cornstarch into the sweetener high-fructose corn syrup?

55. A thickener commonly used in ice cream, alginic acid, is a seaweed polymer composed of the following monomer:

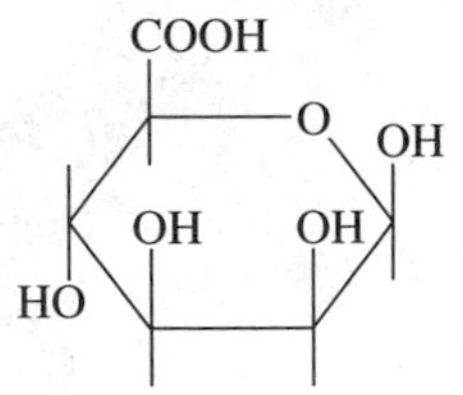

COOH, O, OH, OH, OH, HO

(a) From what monosaccharide might this monomer be derived?

(b) Alginic acid is made by linking these monomers together via β-(1,4)-glycosidic bonds. Draw a short section (three units) of this polymer.

56. A reddish color is obtained when compound A (a dissaccharide) is reacted with Benedict solution. Is this compound more likely to be maltose or sucrose?

57. Refer to this compound to answer the questions that follow:

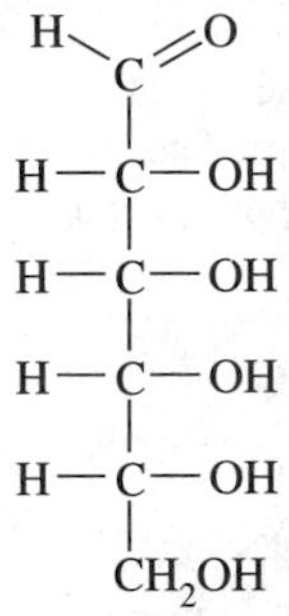

H—C=O
H—C—OH
H—C—OH
H—C—OH
H—C—OH
CH_2OH

(a) How many chiral carbons are present in this compound as it is written? Label them.

(b) Draw the α anomer pyranose ring form of this compound.

(c) How many chiral carbons are present in the structure you drew in part (b)? Label them.

58. If the compound shown below rotates light 25° to the right, draw the structure of a closely related compound that rotates light 25° to the left. Are the two compounds epimers of one another? Why or why not?

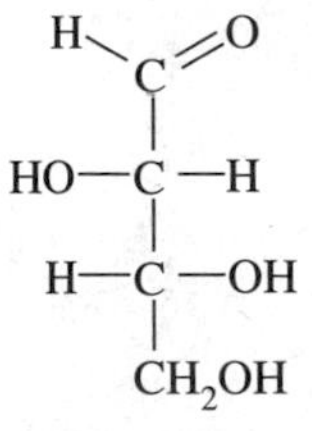

H—C=O
HO—C—H
H—C—OH
CH_2OH

59. Draw the structure of a nonreducing disaccharide composed of two molecules of α-D-galactopyranose.

60. Why is cellulose considered "fiber" in your diet but starch is not? Refer specifically to the structures of cellulose, amylose, and amylopectin in your answer.

61. When lactose is metabolized, the first step separates it into glucose and galactose. Galactose must be changed into glucose before it can be further metabolized in the body.

(a) What changes have to occur in the structure of galactose to make it into glucose?

(b) Which of these terms best describes the relationship of glucose to galactose: disastereomers, enantiomers, or epimers?

62. One of your classmates has told you that the optical rotation of D-glucose and D-mannose are equal and opposite because they are epimers of one another. Should you believe your classmate? Why or why not?

Answers to Practice Exercises

28.1

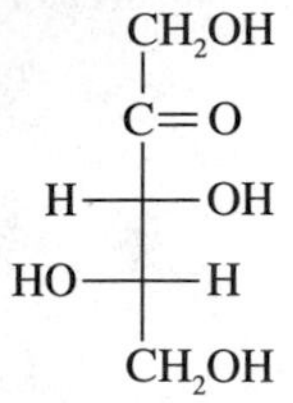

```
      CH2OH
       |
      C=O
       |
  H----+----OH
       |
 HO----+----H
       |
      CH2OH
```

The orientation of the —OH on carbon 3 is not specified and can be written in either direction.

28.2 The right side

28.3

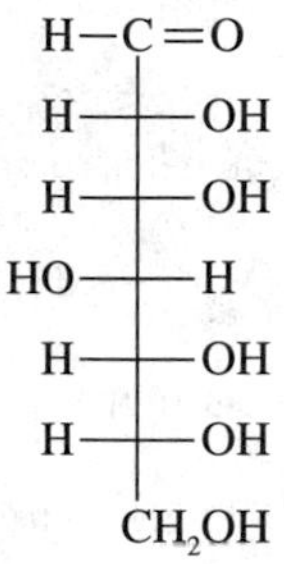

```
  H—C=O                    H—C=O
    |                        |
 H--+--OH               HO--+--H
    |                        |
 H--+--OH                H--+--OH
    |                        |
HO--+--H        and     HO--+--H
    |                        |
 H--+--OH                H--+--OH
    |                        |
 H--+--OH                H--+--OH
    |                        |
   CH2OH                    CH2OH
```

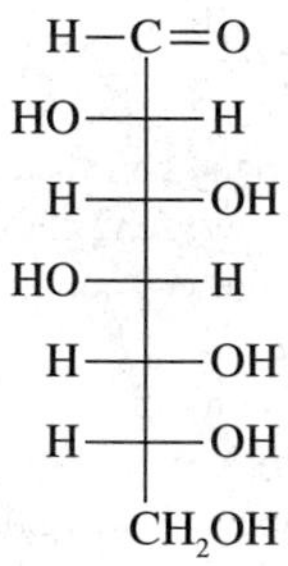

28.4 (a) 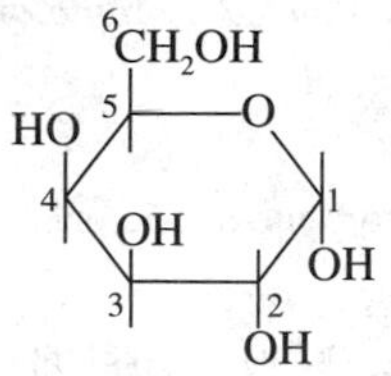 (b)

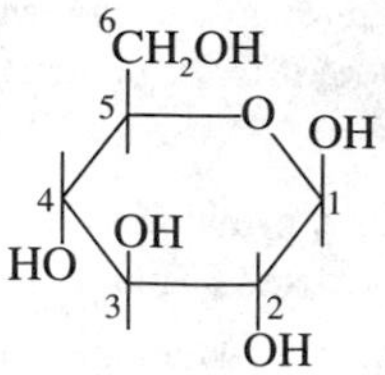

28.5

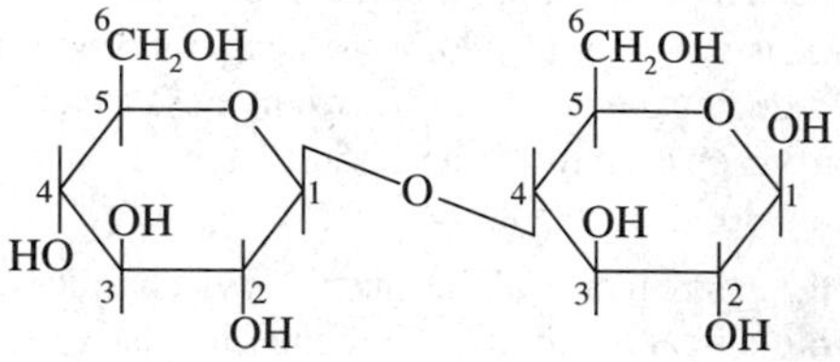

28.6 Carbon 1 from each α-D-galactopyranose

28.7

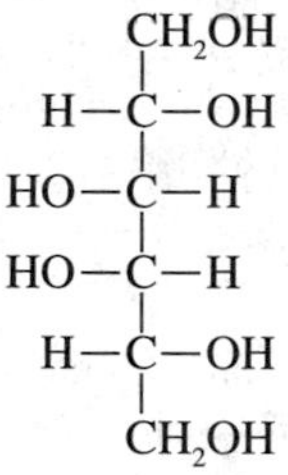

```
     CH2OH
      |
   H—C—OH
      |
  HO—C—H
      |
  HO—C—H
      |
   H—C—OH
      |
     CH2OH
```

This compound is not optically active, it is a meso compound.

Lipids

29

Chapter 29

Lipids provide the bad, the good, and the ugly of the cleaning industry. Splattering a drop of oil (a lipid) on a new shirt is bad—this spot just won't wash out with water. But thank goodness for soap, another form of lipid. And even the ugliest stack of greasy dishes is no match for a sudsy basin of hot water and soap.

The way lipids interact with water—the fact that they are insoluble—is a key to their importance in nature. An oil slick can spread for many square miles on the surface of the ocean partly because oil and water don't mix. Based on this same principle, cells surround themselves with a thin film of lipid, the cell membrane. We protect a fine wood floor with wax, another lipid, because we can depend on this material to adhere to the floor and not dissolve in water. A lipid's stickiness and water insolubility can also create diseases such as atherosclerosis, where arteries become partially clogged by cholesterol, another lipid. Lipids bring both benefits and problems—these molecules are truly a mixed blessing.

29.1 Lipids: Hydrophobic Molecules

lipid

Lipids are water insoluble, oily, or greasy biochemical compounds that can be extracted from cells by nonpolar solvents such as ether, chloroform, or benzene. Unlike carbohydrates, lipids share no common chemical structure. Still, these molecules must possess some structural similarities because of their shared water insolubility.

What makes a molecule such as a lipid insoluble in water? To answer this question, we must establish two important principles about water solutions: A compound may dissolve in water if (1) the water molecules bond well to the potential solute and if (2) the water molecules can still move relatively freely around the dissolved compound. For example, salt (sodium chloride) dissolves because it forms ions to which water molecules can bond *and* because these ions are small and so do not significantly impede the movement of the water molecules. Sugar (sucrose) dissolves because it can form hydrogen bonds with water and because it is still a relatively small molecule.

Lipid structures and solubilities differ from both salts and carbohydrates. Lipid molecules are big enough to substantially affect the free movement of water molecules. In addition, lipids cannot hydrogen-bond to the extent that carbohydrates can, nor do they form the large number of positive and negative charges found in a salt solution. Lipids are large and relatively nonpolar molecules and thus are water insoluble.

Compounds such as most carbohydrates and salts are said to be *hydrophilic* ("water loving"). In contrast, lipids are said to be *hydrophobic* ("water fearing").

Consider fatty acids, which are common components of lipids. As shown in Table 29.1, when the number of atoms in a fatty acid molecule increases, the water solubility of the fatty acid decreases dramatically. Water molecules can easily maneuver around smaller compounds like butyric acid, which is infinitely soluble in water. However, these same water molecules run into a huge barrier when they encounter the 18-carbon chain of stearic acid, and so only a little of this fatty acid dissolves in water (0.0003 g/100 g of water).

◀ **Chapter Opening Photo: Polar bears have a large reserve of lipids.**

The hydrophobic nature of lipids contributes significantly to the biological functions of these molecules. Their water insolubility allows lipids to serve as barriers to aqueous solutions. This property, as we shall see later, is of great importance when lipids form cellular membranes.

29.2 Classification of Lipids

Lipids are hydrophobic molecules; their structures are relatively large and nonpolar. Yet, within this broad description lipid structures vary markedly. The following classification scheme recognizes important structural differences.

1. **Simple lipids**
   (a) *Fats and oils:* esters of fatty acids and glycerol
   (b) *Waxes:* esters of high-molar-mass fatty acids and high-molar-mass alcohols
2. **Compound lipids**
   (a) *Phospholipids:* substances that yield glycerol, phosphoric acid, fatty acids, and a nitrogen-containing base upon hydrolysis
   (b) *Sphingolipids:* substances that yield an unsaturated amino alcohol (sphingosine), a long-chain fatty acid, and either a carbohydrate or phosphate and a nitrogen base upon hydrolysis
   (c) *Glycolipids:* substances that yield sphingosine, fatty acids, and a carbohydrate upon hydrolysis
3. **Steroids**
   Substances that possess the steroid nucleus, which is a 17-carbon structure consisting of four fused carbocyclic rings. Cholesterol and several hormones are in this class.
4. **Miscellaneous lipids**
   Substances that do not fit into the preceding classifications; these include the fat-soluble vitamins A, D, E, and K.

The most abundant lipids are the fats and oils. These substances constitute one of the three important classes of foods. The discussion that follows is centered on fats and oils. A more complete consideration of the properties and composition of various fats and oils is given in Section 24.10.

29.3 Simple Lipids

Fatty Acids

Fatty acids, which form a part of most lipids, are carboxylic acids with long, hydrophobic carbon chains. The formulas for some of the most common fatty acids are shown in Table 29.1. All these fatty acids are straight-chain compounds with an even number of carbon atoms. Five of the fatty acids in this table—palmitoleic, oleic, linoleic, linolenic, and arachidonic—are unsaturated, having carbon–carbon

TABLE 29.1 Some Naturally Occurring Fatty Acids

Fatty acid	Number of C atoms	Formula	Solubility (g/100 g water)	Melting point (°C)
Saturated acids				
Butyric acid	4	$CH_3CH_2CH_2COOH$	∞	−4.7
Caproic acid	6	$CH_3(CH_2)_4COOH$	1.08	−1.5
Caprylic acid	8	$CH_3(CH_2)_6COOH$	0.07	16
Capric acid	10	$CH_3(CH_2)_8COOH$	0.015	32
Lauric acid	12	$CH_3(CH_2)_{10}COOH$	0.006	48
Myristic acid	14	$CH_3(CH_2)_{12}COOH$	0.002	57
Palmitic acid	16	$CH_3(CH_2)_{14}COOH$	0.0007	63
Stearic acid	18	$CH_3(CH_2)_{16}COOH$	0.0003	70
Arachidic acid	20	$CH_3(CH_2)_{18}COOH$	—	77
Unsaturated acids				
Palmitoleic acid	16	$CH_3(CH_2)_5CH{=}CH(CH_2)_7COOH$	—	0.5
Oleic acid	18	$CH_3(CH_2)_7CH{=}CH(CH_2)_7COOH$	—	13
Linoleic acid	18	$CH_3(CH_2)_4CH{=}CHCH_2CH{=}CH(CH_2)_7COOH$	—	−5
Linolenic acid	18	$CH_3CH_2CH{=}CHCH_2CH{=}CHCH_2CH{=}CH(CH_2)_7COOH$	—	−11
Arachidonic acid	20	$CH_3(CH_2)_4(CH{=}CHCH_2)_4CH_2CH_2COOH$	—	−50

double bonds in their structures. Animal and higher plant cells produce lipids in which palmitic, oleic, linoleic, and stearic acids predominate. Over one-half of plant and animal fatty acids are unsaturated, plant lipids tending to be more unsaturated than their animal counterparts.

Double bonds impart some special characteristics to the unsaturated fatty acids. Remember that the presence of double bonds raises the possibility of geometric isomerism (Section 21.3). Unsaturated fatty acids may be either cis or trans isomers. To illustrate the effect of these double bonds on fatty acid structure, the following two fatty acids are portrayed in a simplified manner with each of the many $—CH_2—$ groups as an apex at the intersection between two single bonds.

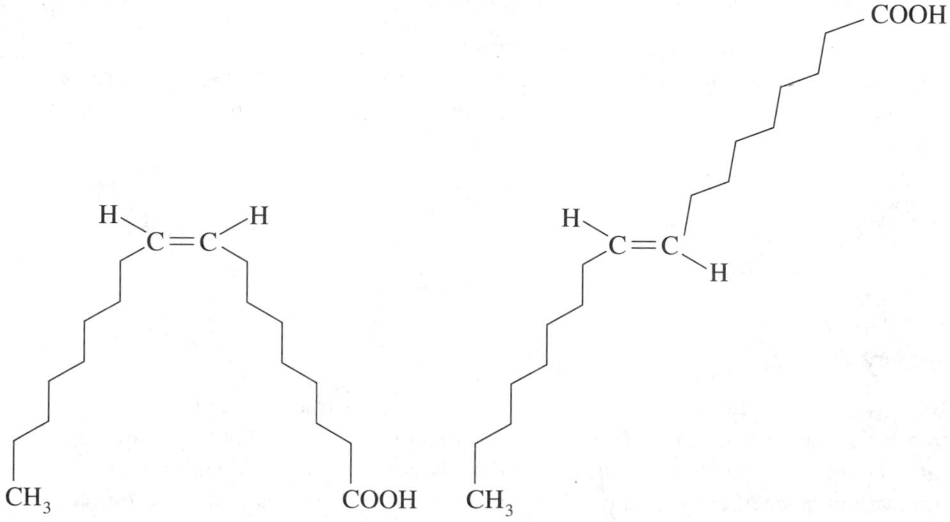

H C=C H CH$_3$ COOH

cis isomer

COOH H C=C H CH$_3$

trans isomer

Note that the trans isomer is almost a linear molecule while the double bond in the cis isomer introduces a kink in the fatty acid structure. Unsaturated fatty acids found in nature are almost always cis isomers. These kinked fatty acids cannot stack closely together and hence do not solidify easily. As shown in Table 29.1, unsaturated fatty acids have lower melting points than saturated fatty acids of a similar size. Cooking oils purchased from your market are liquids at room temperature because a high percentage of their fatty acids are unsaturated. In like manner, biological membranes are very fluid because of the presence of double bonds in their component fatty acids (see Section 29.6).

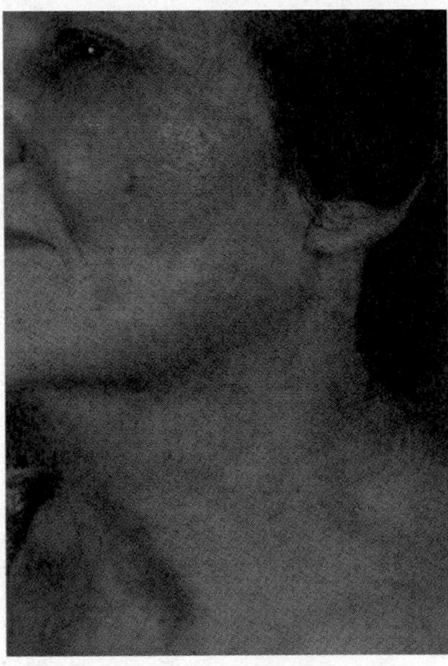

Prostaglandins are synthesized at the site in response to allergic reactions such as the dermatitis shown here.

Practice 29.1

Draw the cis and trans isomers for a fatty acid with the formula $CH_3(CH_2)_5CH{=}CH(CH_2)_7COOH$.

Three unsaturated fatty acids—linoleic, linolenic, and arachidonic—are essential for animal nutrition and must be present in the diet. Diets lacking these fatty acids lead to impaired growth and reproduction, and skin disorders such as eczema and dermatitis. A dermatitis disorder can be attributed to an unsaturated fatty acid deficiency if the symptoms clear up when that fatty acid is supplied in the diet.

Certain fatty acids, as well as other lipids, are biochemical precursors of several classes of hormones. The well-known steroid hormones are synthesized from cholesterol and will be discussed later in this chapter. Arachidonic acid and, to a lesser extent, linolenic acid are also used by the body to make hormone-like substances. The biochemicals derived from arachidonic acid are collectively termed eicosanoids using a derivative of the Greek word for twenty (*eikosi*) to indicate that these compounds have 20 carbon atoms. Prostaglandins are perhaps the best known of the eicosanoid class, which also includes the leukotrienes, prostacyclins, and thromboxanes. Cell membranes release arachidonic acid in response to a variety of circumstances, including infection and allergic reactions. In turn, enzymes in the surrounding fluid convert this fatty acid to specific eicosanoids by catalyzing the addition of oxygen to the arachidonic double bonds. Some examples of eicosanoids are shown in Figure 29.1.

Unlike true hormones, eicosanoids are not transported via the bloodstream to their site of action, but rather take effect where they are synthesized. Prostaglandins are a primary cause of the swelling, redness, and pain associated with tissue inflammation. Platelets in the bloodstream form the thromboxanes, which act as vasoconstrictors and stimulate platelet aggregation, as an initial step in blood clotting. Leukotrienes are formed by a variety of white blood cells as well as other tissues and cause many of the symptoms associated with an allergy attack. For example, asthma is thought to be mediated by the leukotrienes.

Nonsteroidal anti-inflammatory drugs (NSAIDs) block the oxidation of arachidonic acid to form prostaglandins and thromboxanes. These drugs include the common pain relievers aspirin; ibuprofen (e.g., Advil, Motrin, Nuprin); naproxon (e.g., Aleve); and ketoprofen (e.g., Orudis KT, Actron). Recent research indicates that low levels of aspirin may prevent heart attacks and strokes, possibly by blocking synthesis of the thromboxanes that participate in blood clotting. Cortisone acts as

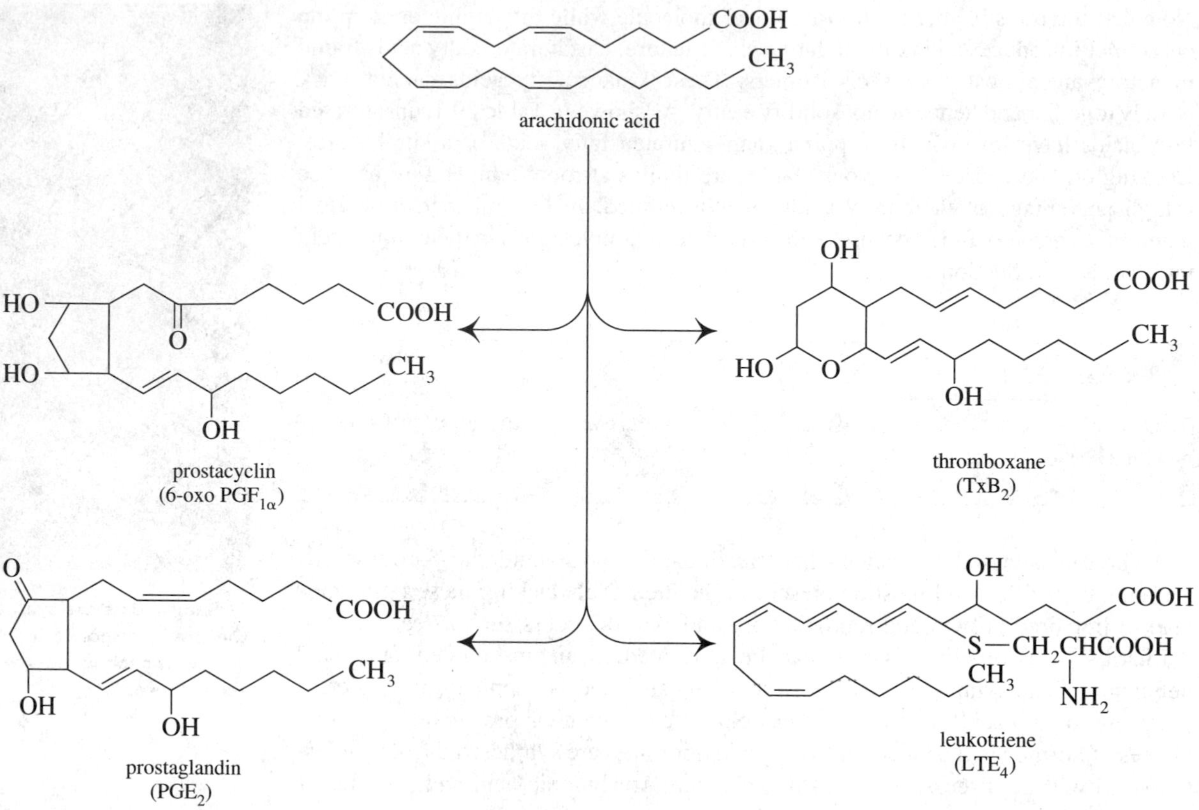

FIGURE 29.1
Several examples of eicosanoids. Each of these molecules is derived from arachidonic acid.

an anti-inflammatory drug by decreasing the release of arachidonic acid from the cell membranes.

Diet may have a significant effect on eicosanoid formation. Fish oils contain fatty acids that inhibit formation of the thromboxanes and lead to formation of the less potent leukotrienes. It has been suggested that cultures for which fish is a dietary staple (such as the Greenland Eskimos) have a low level of heart disease, possibly because of a decrease in thromboxane formation.

Fats and Oils

Chemically, fats and oils are esters of glycerol and the higher-molar-mass fatty acids. They have the general formula

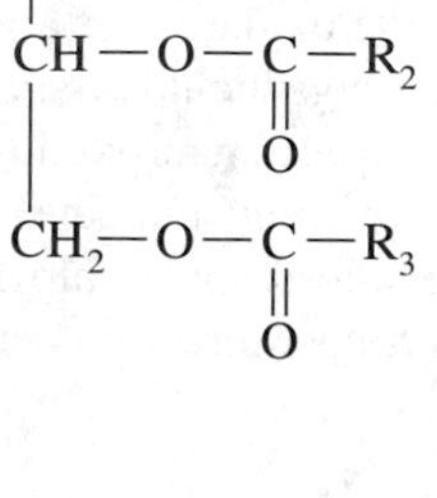

$$\begin{array}{l} CH_2-O-\overset{\displaystyle O}{\overset{\|}{C}}\!\!-R_1 \\ | \\ CH-O-\overset{\displaystyle O}{\overset{\|}{C}}\!\!-R_2 \\ | \\ CH_2-O-\overset{\displaystyle O}{\overset{\|}{C}}\!\!-R_3 \end{array}$$

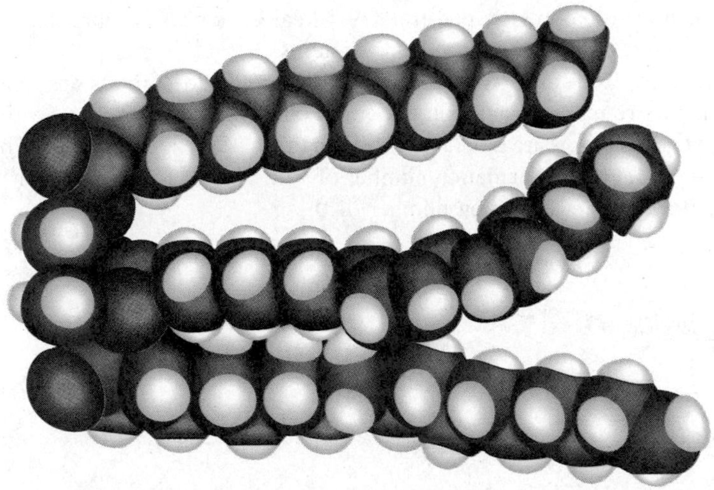

FIGURE 29.2
Space-filling model of a triaclyglycerol formed by reacting glycerol with one palmitic acid, one oleic acid, and one stearic acid. Note the kink introduced into oleic acid by the cis double bond.

where the R's can be either long-chain saturated or unsaturated hydrocarbon groups. Figure 29.2 shows a three-dimensional representation of a typical fat.

Fats may be considered to be triesters formed from the trihydroxy alcohol glycerol and three molecules of fatty acids. Most of the fatty acids in these esters have 14–18 carbons. Because there are three ester groups per glycerol, these molecules are called triacylglycerols, or triglycerides (an older name that is still commonly used). The three R groups are usually different.

$$
\begin{array}{lcll}
\mathrm{CH_2-O}\boxed{\mathrm{H + H-O}}\mathrm{-C(=O)-R_1} & & \mathrm{CH_2-O-C(=O)-R_1} & \\
\mathrm{CH-O}\boxed{\mathrm{H + H-O}}\mathrm{-C(=O)-R_2} & \longrightarrow & \mathrm{CH-O-C(=O)-R_2} & +\ 3\,\mathrm{H_2O} \\
\mathrm{CH_2-O}\boxed{\mathrm{H + H-O}}\mathrm{-C(=O)-R_3} & & \mathrm{CH_2-O-C(=O)-R_3} & \\
\text{glycerol} \qquad \text{fatty acids} & & \text{a triacylglycerol} &
\end{array}
$$

Practice 29.2

Write the balanced equation (showing structures of reactants and products) for the reaction between a molecule of glycerol, a molecule of stearic acid and two palmitic acid molecules to form a triacylglycerol.

Cultures that use fish as a dietary staple have a low level of heart disease.

Fats and oils fit the general description of a lipid. They are large molecules, averaging more than 50 carbon atoms per molecule, with many nonpolar, uncharged groups. Triacylglycerols are water insoluble.

Fats are an important food source for humans and normally account for about 25–50% of their caloric intake. When oxidized to carbon dioxide and water, fats

Nutritionists recommend that less than 30% of the daily dietary caloric intake be derived from fats.

supply about 40 kJ of energy per gram (9.5 kcal/g), which is more than twice the amount obtained from carbohydrates or proteins.

The energy from a fat is released when the reduced carbons are oxidized. In general, the more reduced a carbon, the more energy it contains (see Section 28.1). Triacylglycerol carbons are more reduced than those of most other foods. The typical carbon from a fat has an oxidation number of −2, whereas the typical carbon from a carbohydrate has an oxidation number of 0:

$$\underset{\text{in a fat}}{\mathrm{C{-}\overset{\overset{\Large H}{|}}{\underset{\underset{\Large H}{|}}{C}}{-}C}}\quad \text{(−2 Oxidation number)} \qquad\qquad \underset{\text{in a carbohydrate}}{\mathrm{C{-}\overset{\overset{\Large OH}{|}}{\underset{\underset{\Large H}{|}}{C}}{-}C}}\quad \text{(0 Oxidation number)}$$

This difference in oxidation numbers makes it clear that almost every carbon in a fat contains and can release more energy than a typical carbohydrate carbon. In addition, the average fat contains about 75% carbon by mass, whereas the average carbohydrate contains only about 40% carbon. Fats are indeed a rich source of biochemical energy.

Waxes

wax

Waxes are esters of high-molar-mass fatty acids and high-molar-mass alcohols. They have the general formula

$$\mathrm{R'{-}\overset{\overset{\Large O}{\|}}{C}{-}O{-}R}$$

The name, wax, derives from the old English word, "weax," which means "material from a honeycomb."

in which the alcohol (ROH) contributes up to about 30 carbons, and the fatty acid (R′COOH) also provides an equivalent number of carbons. Waxes are very large molecules with almost no polar groups. They represent one of the most hydrophobic lipid classes.

Their extreme water insolubility allows waxes to serve a protective function. Leaves, feathers, fruit, and fur are often naturally coated with a wax. Hardwood floors, cars, and leather goods are just a few of the many products that can be protected by a wax. Waxes tend to be the hardest of the lipids because their carbon chains are long and have very few double bonds. As with fats and oils, the size of the wax molecule and the number of double bonds contained in its carbon chains determine whether the wax will be a liquid or a solid.

Practice 29.3

Write the formula for a wax formed from stearic acid and $CH_3(CH_2)_{26}CH_2OH$.

Designing Low-Calorie Fats

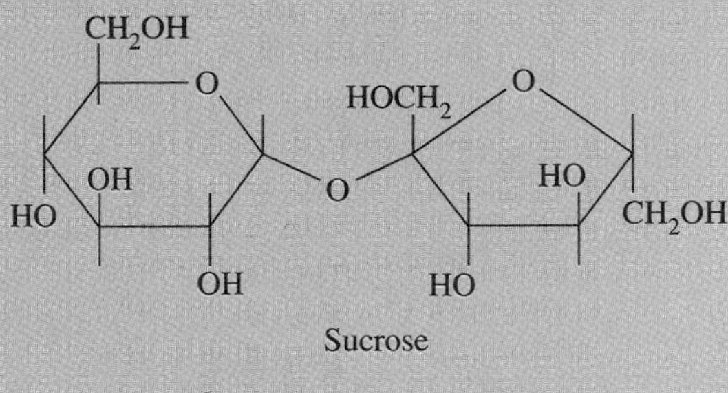

Sucrose

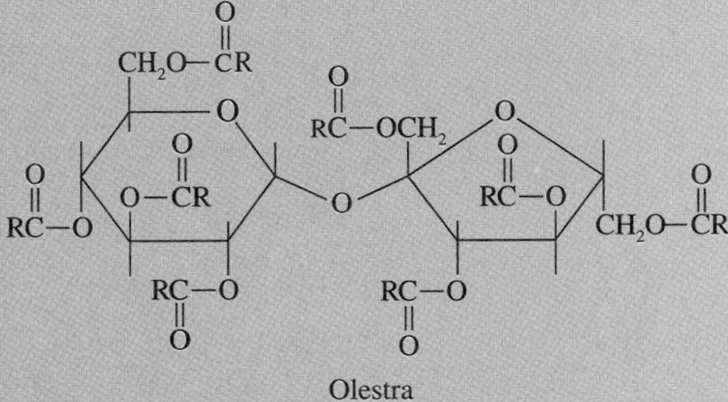

Olestra

Fatty foods provide eating enjoyment for many people. A baked potato just isn't the same without butter or sour cream. A doughnut or french fries shoud be deep-fried to be tasty. Yet, high-fat diets are a primary cause of obesity and heart disease. The choice of whether or not to eat fatty foods is a dilemma. A solution may be offered by a fat substitute called olestra. Approved by the U.S. Food and Drug Administration in 1996 for snack foods, this compound cooks and tastes like fat but is not digestable. Olestra is being marketed under the trade name, Olean.

Olestra is a synthetic compound that combines a sugar (sucrose) and six to eight fatty acids. The fatty acids bond to the sucrose through ester bonds. Like a triacylglycerol, olestra is hydrophobic. When olestra is mixed with water, this fat substitute aggregates just as a natural fat does. These aggregates make for the creamy flavor and texture that we have come to associate with fats.

However, the olestra molecule has a different structure than a common triacylglycerol. A triacylglycerol is formed from a small alcohol (glycerol) and three long-chain fatty acids.

$$\begin{array}{l} CH_2O-\overset{O}{\overset{\|}{C}}-R \\ | \\ CHO-\overset{O}{\overset{\|}{C}}-R \\ | \\ CH_2O-\overset{O}{\overset{\|}{C}}-R \end{array}$$

In olestra, sucrose provides the alcohol groups to which fatty acids are bonded.

The much greater number of fatty acid esters give olestra a significantly different shape than a triacylglycerol, and molecular shape is important in digestion. Digestive enzymes only break down molecules of specific shapes. The enzymes that digest fats look for a molecular shape similar to a triacylglycerol. Olestra, escapes digestion because it does not have that shape.

We can greatly reduce our fat intake by using an olestra product.

Olestra's nondigestability has some important consequences. This fat substitute isn't absorbed into the body. So, olestra doesn't provide energy to the cells, and it is rated as zero in caloric content. Also, olestra doesn't raise the level of lipids in the bloodstream and thus helps to avoid atherosclerosis.

Sound too good to be true? As always, there is a down side. Since olestra is not digested, aggregates of this fatty substance continue to pass through the digestive tract. These aggregates may trap fat-soluble nutrients (e.g., the fat-soluble vitamins A, D, E, and, K) leading to a nutritional deficiency. Also, for some people, these fatty aggregates cause lower intestinal problems such as diarrhea.

On balance, olestra may bring healthy and pleasurable eating for many people. Science has given us a new nutrient option. We can decide to avoid the excess calories and health risks of a high-fat diet by using olestra, but we must also be aware of potential side effects. As with all new food additives, we need to make an informed choice when we plan our diets.

29.4 Compound Lipids

Phosphoplipids

phospholipid

The **phospholipids** are a group of compounds that yield one or more fatty acid molecules, a phosphate group, and usually a nitrogenous base upon hydrolysis. The phosphate group and the nitrogenous base, which are found at one end of the phospholipid molecule, often have negative and positive charges. Consequently, in contrast to the triacylglycerols, phospholipids have hydrophobic ends that repel water and hydrophilic ends that interact with water.

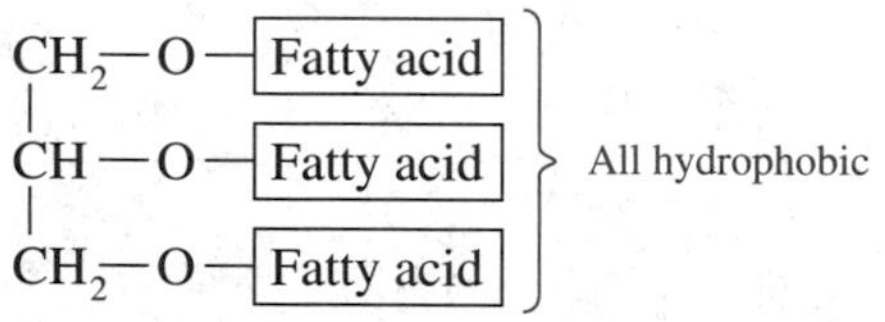

$$\left.\begin{array}{l} CH_2-O-\boxed{\text{Fatty acid}} \\ \;| \\ CH-O-\boxed{\text{Fatty acid}} \\ \;| \\ CH_2-O-\boxed{\text{Fatty acid}} \end{array}\right\} \text{All hydrophobic}$$

a triacylglycerol

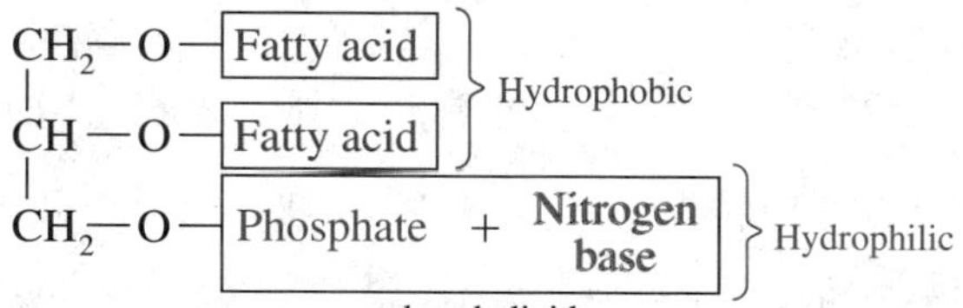

$$\begin{array}{l} \left.\begin{array}{l} CH_2-O-\boxed{\text{Fatty acid}} \\ \;| \\ CH-O-\boxed{\text{Fatty acid}} \end{array}\right\} \text{Hydrophobic} \\ \;| \\ CH_2-O-\boxed{\text{Phosphate} + \textbf{Nitrogen base}}\Big\} \text{Hydrophilic} \end{array}$$

a phospholipid

As will be seen later in this chapter, a lipid with both hydrophobic and hydrophilic character is needed to make membranes. It is not surprising that phospholipids are one of the most important membrane components.

Phospholipids are also involved in the metabolism of other lipids and nonlipids. Although they are produced to some extent by almost all cells, most of the phospholipids that enter the bloodstream are formed in the liver. Descriptions of representative phospholipids follow.

Phosphatidic Acids Phosphatidic acids are glyceryl esters of fatty acids and phosphoric acid. The phosphatidic acids are important intermediates in the synthesis of triacylglycerols and other phospholipids.

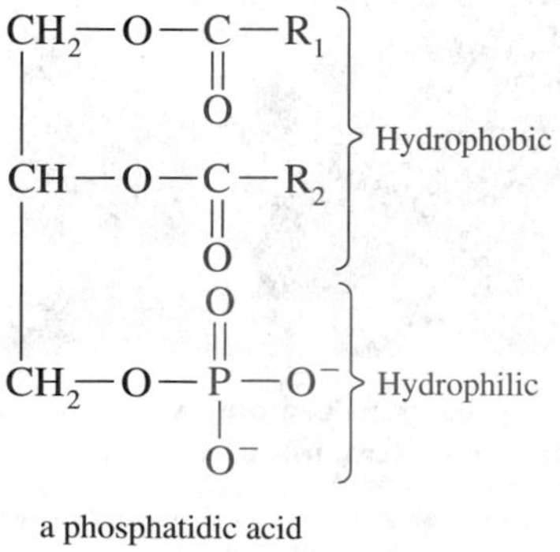

$$\begin{array}{l} \left.\begin{array}{l} CH_2-O-\underset{\underset{O}{\|}}{C}-R_1 \\ \;| \\ CH-O-\underset{\underset{O}{\|}}{C}-R_2 \end{array}\right\} \text{Hydrophobic} \\ \;| \\ \left.CH_2-O-\overset{\overset{O}{\|}}{\underset{\underset{O^-}{|}}{P}}-O^-\right\} \text{Hydrophilic} \end{array}$$

a phosphatidic acid

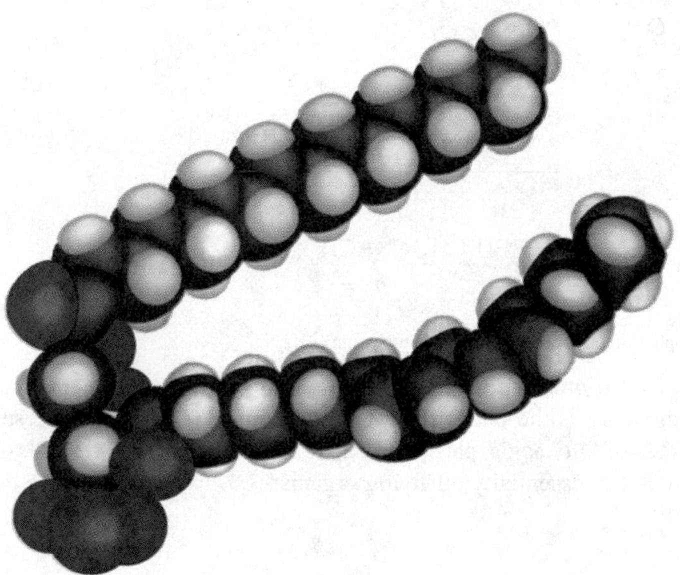

◀ **FIGURE 29.3**
Space-filling model of a phosphatidic acid that is esterified to palmitic acid at the top glycerol carbon, to oleic acid at the middle carbon, and to a phosphate group at the bottom glycerol carbon. Note the kink introduced into the oleic acid by the cis double bond.

A three-dimensional representation of a typical phosphatidic acid is given in Figure 29.3. As with all common phospholipids, the fatty acid chains are large relative to the rest of this molecule. Other phospholipids are formed from a given phosphatidic acid when specific nitrogen-containing compounds are linked to the phosphate group by an ester bond. Three commonly used nitrogen compounds are choline, ethanolamine, and L-serine:

$$HOCH_2CH_2\overset{\overset{\displaystyle CH_3}{|}}{\underset{\underset{\displaystyle CH_3}{|}}{N^+}}—CH_3 \qquad HOCH_2CH_2NH_3^+ \qquad HOCH_2\underset{\underset{\displaystyle NH_3^+}{|}}{C}HCOO^-$$

choline ethanolamine L-serine

Because other phospholipids are structurally related to phosphatidic acids, their names are also closely related.

Phosphatidyl Cholines (Lecithins) Phosphatidyl cholines (lecithins) are glyceryl esters of fatty acids, phosphoric acid, and choline. The synonym *lecithin* is an older term that is still used, particularly in commercial products that contain phosphatidyl choline. Phosphatidyl cholines are synthesized in the liver and are present in considerable amounts in nerve tissue and brain substance. Most commercial phosphatidyl choline is obtained from soybean oil and contains palmitic, stearic, palmitoleic, oleic, linoleic, linolenic, and arachidonic acids. Phosphatidyl choline is an edible and digestible emulsifying agent that is used extensively in the food industry. For example, chocolate and margarine are generally emulsified with phosphatidyl choline. Phosphatidyl choline is also used as an emulsifier in many pharmaceutical preparations.

The single most important biological function for phosphatidyl choline is as a membrane component. This phospholipid makes up between 10 and 20% of many membranes.

Soybean oil is the main commercial source of phosphatidyl choline.

$$
\begin{array}{l}
CH_2-O-\underset{\displaystyle \overset{\|}{O}}{C}-R_1 \\
| \\
CH-O-\underset{\displaystyle \overset{\|}{O}}{C}-R_2 \\
| \\
CH_2-O-\overset{\displaystyle O}{\overset{\|}{\underset{\displaystyle \underset{|}{O^-}}{P}}}-\overbrace{O-CH_2CH_2\overset{\displaystyle CH_3}{\overset{|}{\underset{\displaystyle \underset{|}{CH_3}}{N^+}}}-CH_3}^{\text{choline}}
\end{array}
\quad \begin{array}{l}\}\ \text{Hydrophobic} \\ \}\ \text{Hydrophilic}\end{array}
$$

a phosphatidyl choline (lecithin) molecule

Phosphatidyl Ethanolamines (Cephalins) Another important constituent of biological membranes is the phosphatidyl ethanolamines (cephalins). These lipids are glyceryl esters of fatty acids, phosphoric acid, and ethanolamine ($HOCH_2CH_2NH_2$). They are found in essentially all living organisms.

$$
\begin{array}{l}
CH_2-O-\underset{\displaystyle \overset{\|}{O}}{C}-R_1 \\
| \\
CH-O-\underset{\displaystyle \overset{\|}{O}}{C}-R_2 \\
| \\
CH_2-O-\overset{\displaystyle O}{\overset{\|}{\underset{\displaystyle \underset{|}{O^-}}{P}}}-\overbrace{OCH_2CH_2NH_3^+}^{\text{ethanolamine}}
\end{array}
\quad \begin{array}{l}\}\ \text{Hydrophobic} \\ \}\ \text{Hydrophilic}\end{array}
$$

a phosphatidyl ethanolamine (cephalin) molecule

Example 29.1 Write the formula for a phosphatidyl ethanolamine that contains two palmitic acid groups.

Solution Phosphatidyl ethanolamine is a phospholipid and so contains glycerol, phosphate, fatty acids, and a nitrogen base. First, write the structure for glycerol.

$$
\begin{array}{l}
CH_2OH \\
| \\
CHOH \\
| \\
CH_2OH
\end{array}
$$

The two palmitic acids are linked by ester bonds to the top two carbons of glycerol.

$$
\begin{array}{l}
CH_2-O-\underset{\displaystyle \overset{\|}{O}}{C}(CH_2)_{14}CH_3 \\
| \\
CH-O-\underset{\displaystyle \overset{\|}{O}}{C}(CH_2)_{14}CH_3 \\
| \\
CH_2OH
\end{array}
$$

The phosphate group is connected via an ester bond to the bottom glycerol carbon to form a phosphatidic acid.

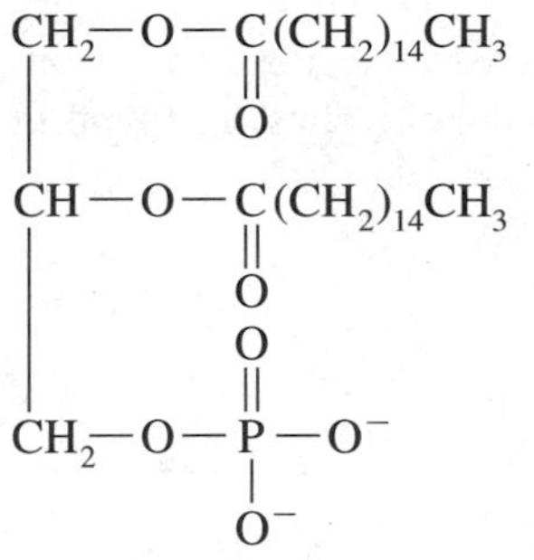

$$\begin{array}{l}
CH_2-O-\underset{\displaystyle \|\atop O}{C}(CH_2)_{14}CH_3 \\
\;| \\
CH-O-\underset{\displaystyle \|\atop O}{C}(CH_2)_{14}CH_3 \\
\;| \\
CH_2-O-\overset{\displaystyle O\atop \|}{\underset{\displaystyle |\atop O^-}{P}}-O^-
\end{array}$$

Finally, the ethanolamine is linked to the phosphate group to yield phosphatidyl ethanolamine.

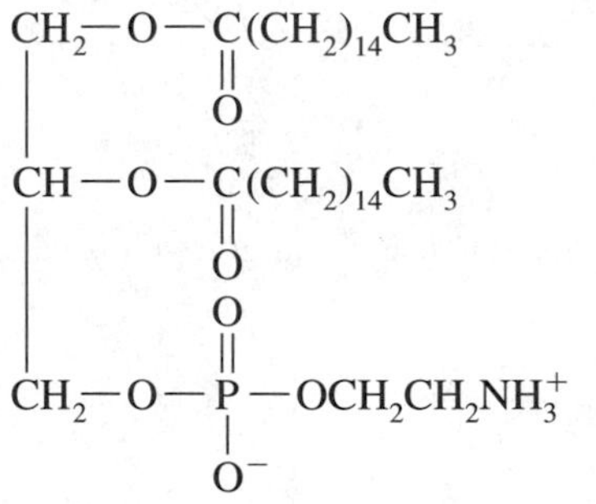

$$\begin{array}{l}
CH_2-O-\underset{\displaystyle \|\atop O}{C}(CH_2)_{14}CH_3 \\
\;| \\
CH-O-\underset{\displaystyle \|\atop O}{C}(CH_2)_{14}CH_3 \\
\;| \\
CH_2-O-\overset{\displaystyle O\atop \|}{\underset{\displaystyle |\atop O^-}{P}}-OCH_2CH_2NH_3^+
\end{array}$$

Practice 29.4

Write the structure of a phosphatidyl choline that contains palmitic acid and stearic acid.

Sphingolipids

sphingolipid

Sphingolipids are compounds that, when hydrolyzed, yield a hydrophilic group (either phosphate and choline or a carbohydrate), a long-chain fatty acid (18–26 carbons), and sphingosine (an unsaturated amino alcohol). When drawn as follows, sphingosine can be seen as similar to glycerol esterified to one fatty acid:

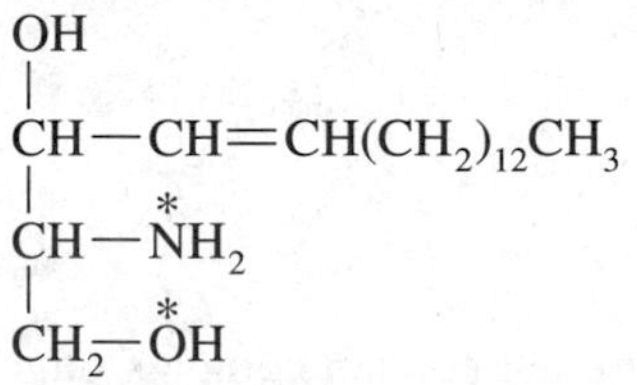

$$\begin{array}{l}
OH \\
\;| \\
CH-CH=CH(CH_2)_{12}CH_3 \\
\;| \\
CH-\overset{*}{N}H_2 \\
\;| \\
CH_2-\overset{*}{O}H
\end{array}
\qquad\qquad
\begin{array}{l}
CH_2-O-\overset{\displaystyle O\atop \|}{C}(CH_2)_nCH_3 \\
\;| \\
CH-\overset{*}{O}H \\
\;| \\
CH_2-\overset{*}{O}H
\end{array}$$

sphingosine 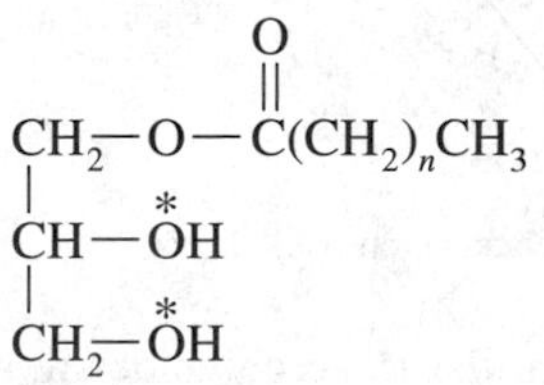 — glycerol esterified with one fatty acid

The starred atoms on sphingosine react further to make sphingolipids, just as the starred atoms on the glycerol compound react further to give triacylglycerols or phospholipids.

Sphingolipids are common membrane components because they have both hydrophobic and hydrophilic character. For example, sphingomyelins are found in the myelin sheath membranes that surround nerves:

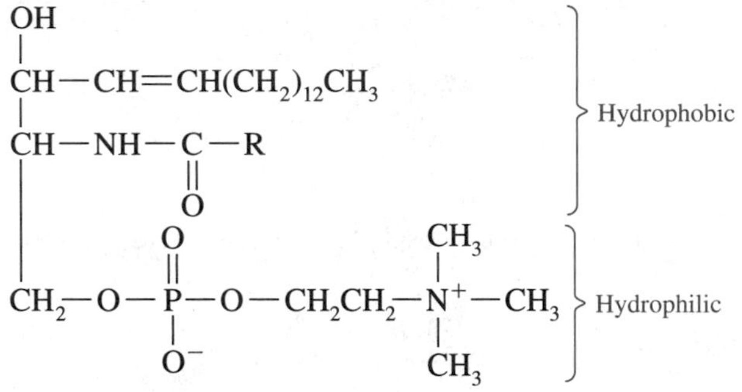

OH
|
$CH-CH{=}CH(CH_2)_{12}CH_3$
|
$CH-NH-\overset{\overset{\displaystyle O}{\|}}{C}-R$ } Hydrophobic
|
$CH_2-O-\overset{O}{\overset{\|}{P}}(O^-)-O-CH_2CH_2-N^+(CH_3)_2-CH_3$ } Hydrophilic

a sphingomyelin

Notice the hydrophobic and hydrophilic parts of this molecule. Sphingomyelins can also be classified as phospholipids.

Glycolipids

glycolipid Sphingolipids that contain carbohydrate groups are also known as **glycolipids**. The two most important classes of glycolipids are cerebrosides and gangliosides. These substances are found mainly in cell membranes of nerve and brain tissue. A cerebroside may contain either D-galactose or D-glucose. The following formula of a galactocerebroside shows the typical structure of cerebrosides:

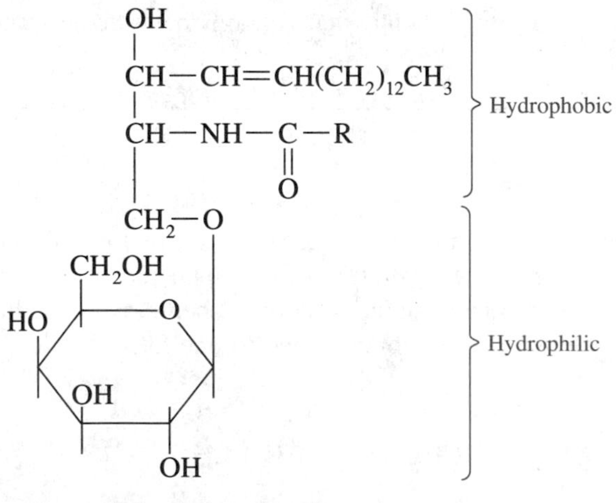

OH
|
$CH-CH{=}CH(CH_2)_{12}CH_3$
|
$CH-NH-\overset{\overset{\displaystyle O}{\|}}{C}-R$ } Hydrophobic
|
CH_2-O
(galactose ring: CH_2OH, HO, O, OH, OH) } Hydrophilic

A β-galactocerebroside

Gangliosides are similar to cerebrosides in structure but contain complex oligosaccharides instead of simple monosaccharides.

Practice 29.5

Write the structure of a sphingomyelin that contains stearic acid.

Muscle mass can be increased through the use of steroids.

29.5 Steroids

Steroids are compounds that have the steroid nucleus, which consists of four fused carbocyclic rings. This nucleus contains 17 carbon atoms in one five-membered and three six-membered rings. Modifications of this nucleus in the various steroid compounds include added side chains, hydroxyl groups, carbonyl groups, ring double bonds, and so on.

steroid ring nucleus

steroid

Steroids are closely related in structure but are highly diverse in function. Examples of steroids and steroid-containing materials are (1) cholesterol, the most abundant steroid in the body, which is widely distributed in all cells and serves as a major membrane component; (2) bile salts, which aid in the digestion of fats; (3) ergosterol, a yeast steroid, which is converted to vitamin D by ultraviolet radiation; (4) digitalis and related substances called cardiac glycosides, which are potent heart drugs; (5) the adrenal cortex hormones, which are involved in metabolism; and (6) the sex hormones, which control sexual characteristics and reproduction. The formulas for several steroids are given in Figure 29.4.

Cholesterol is the parent compound from which the steroid hormones are synthesized. As we will see, small changes in steroid structure can lead to large changes in hormonal action. Cholesterol is first converted to progesterone, a compound that helps control the menstrual cycle and pregnancy. This hormone is, in turn, the parent compound from which testosterone and the adrenal corticosteroids are produced. Notice that the long side chain on carbon 17 in cholesterol (Figure 29.4) is smaller in progesterone and is eliminated when testosterone is formed. Interestingly, testosterone is the precursor for the female sex hormones such as estradiol. These sex hormones are produced by the gonads, either the male testes or the female ovaries. The small structural differences between testosterone and estradiol trigger vastly different physiological responses. If the embryonic male gonads are surgically removed and testosterone is no longer available, the embryo develops as a female. In contrast, the female hormones seem to be important in sexual maturation and function but not in embryonic development. It appears that embryonic mammals are programmed to develop as females unless this program is overridden by the action of testosterone.

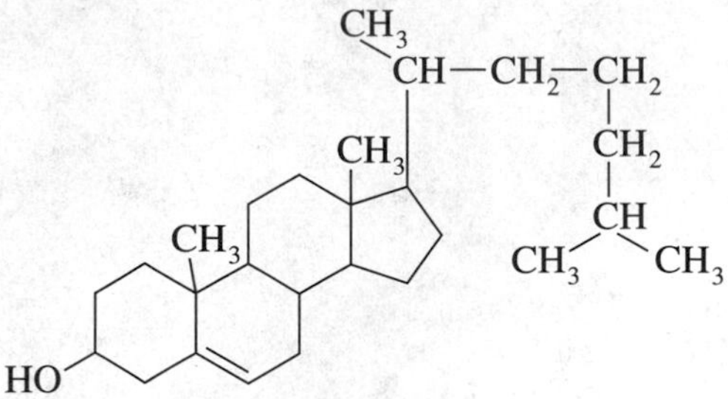

cholesterol
(parent steroid compound—precursor for steroid hormones and bile acids)

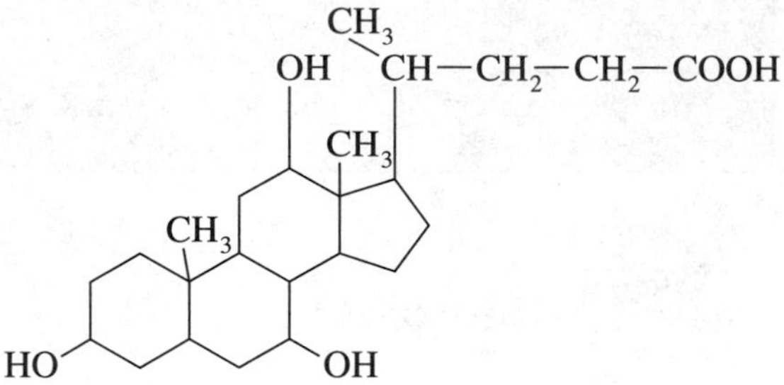

cholic acid
(a bile acid)

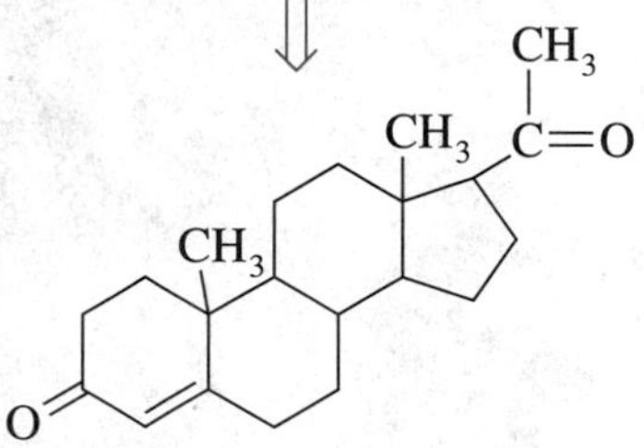

progesterone
(progestational hormone)

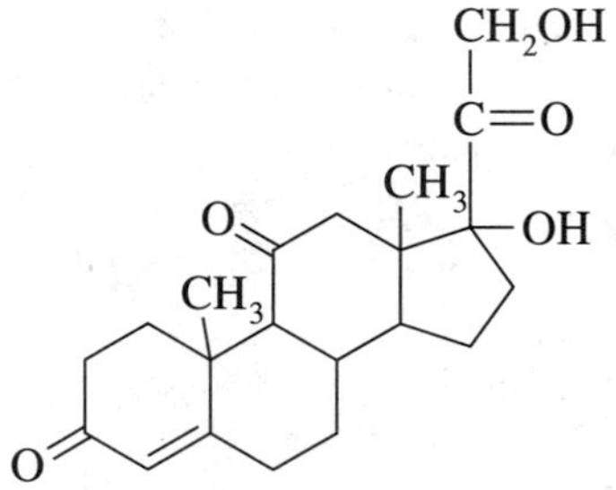

cortisone
(adrenal corticosteroid hormone)

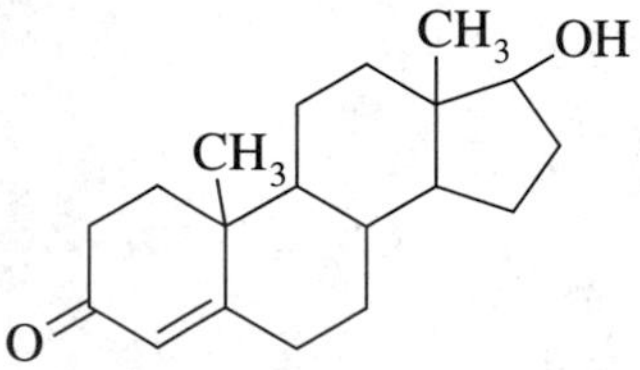

testosterone
(male sex hormone)

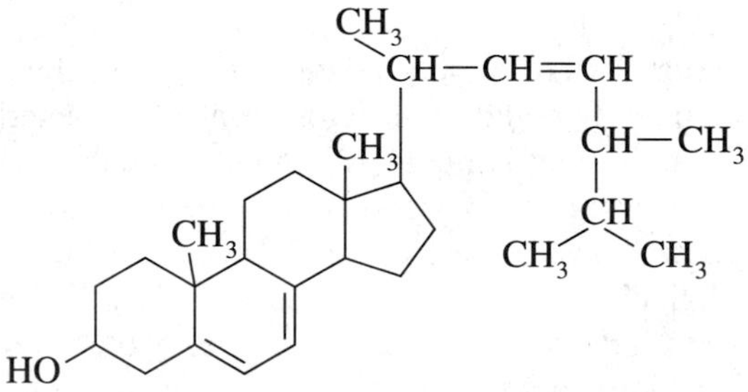

ergosterol
(precursor of vitamin D_2)

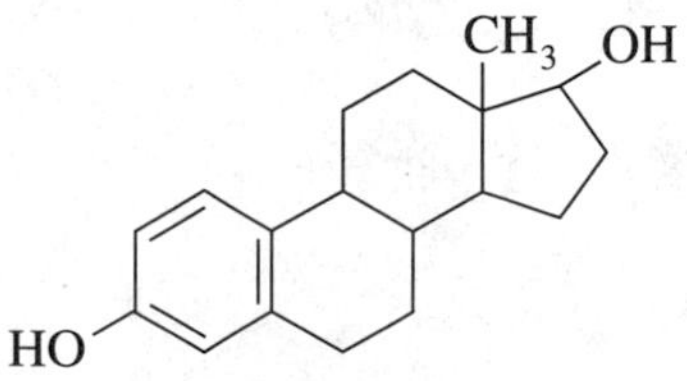

estradiol
(female sex hormone)

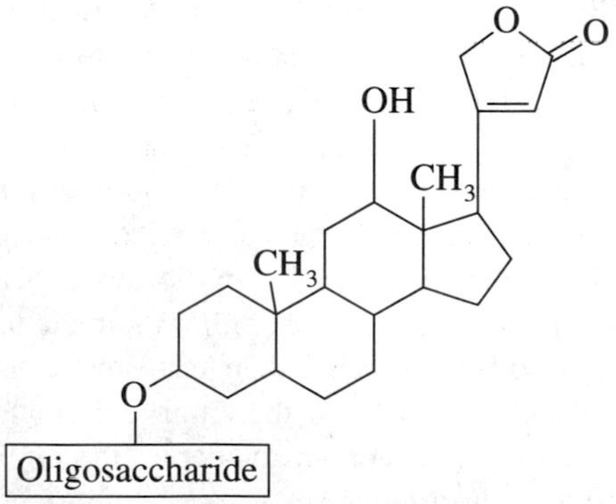

digoxin
(a cardiac glycoside)

▲
FIGURE 29.4
Structures of selected steroids. Arrows show the biosynthetic relationship between steroids derived from cholesterol. Ergosterol and digoxin are from plant sources.

Practice 29.6

Identify the functional groups in (a) progesterone, (b) cholic acid, and (c) estradiol.

Olive oil is a hydrophobic lipid. Notice how the oil molecules separate from the vinegar.

Cholesterol is also used to build cell membranes, many of which contain about 25% by mass of this steroid. In fact, often there is as much cholesterol as there is phospholipid, sphingolipid, or glycolipid. These latter three lipid classes cause the membrane to be more oily; cholesterol solidifies the membrane. This different behavior arises from an important structural difference. Cholesterol's four fused rings make it a rigid molecule while other membrane lipids are more flexible because of their fatty acid chains. When biological membranes are synthesized, their cholesterol level is adjusted to achieve an appropriate balance between a solid and liquid consistency.

29.6 Hydrophobic Lipids and Biology

The hydrophobic nature of lipids has many important biological consequences. The water insolubility of lipids results in (1) lipid aggregation that causes atherosclerosis, and (2) lipid aggregation that forms biological membranes. When a lipid is surrounded by water, it is in a hostile environment. The lipid molecules aggregate to minimize their contact with water. This explains why olive oil (a mixture of triacylglycerols) will separate from an Italian salad dressing if it is allowed to stand. This same process of separation occurs continuously in biological solutions.

Let's look at this process on a molecular level. In water, lipid molecules tend to aggregate and orient themselves in a definite manner. The hydrophilic part of lipid molecules is attracted to water and forms an interface with it, but the hydrophobic part distances itself from water molecules. Depending on the general shape of the lipid molecules, different-shaped aggregates will form. Smaller lipids like fatty acids will come together to make micelles. If the fatty acid is shown schematically as follows,

$CH_3(CH_2)_n$ | COOH

hydrophobic | hydrophilic

then a micelle can be visualized as in Figure 29.5. Notice that the hydrophilic carboxyl groups coat the aggregate and protect the hydrophobic alkyl groups from water.

More complex lipids such as phospholipids and sphingolipids are shaped differently than fatty acids and will aggregate differently in water solutions. These lipids have two hydrophobic alkyl groups and can be schematically represented as

hydrophobic hydrophilic

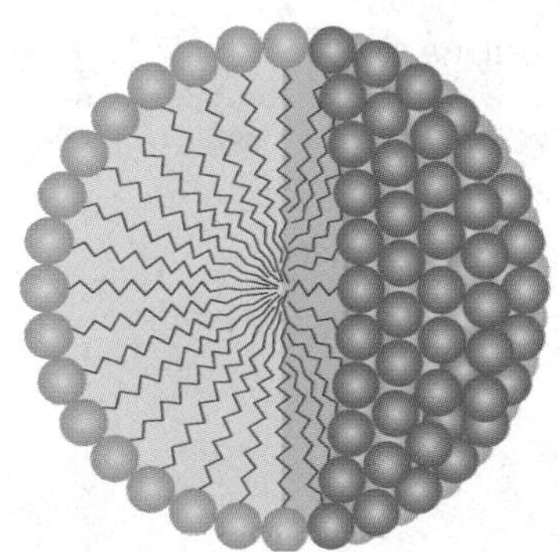

FIGURE 29.5
Schematic representation of a micelle. Note that the hydrophobic chains extend to the center of this aggregate.

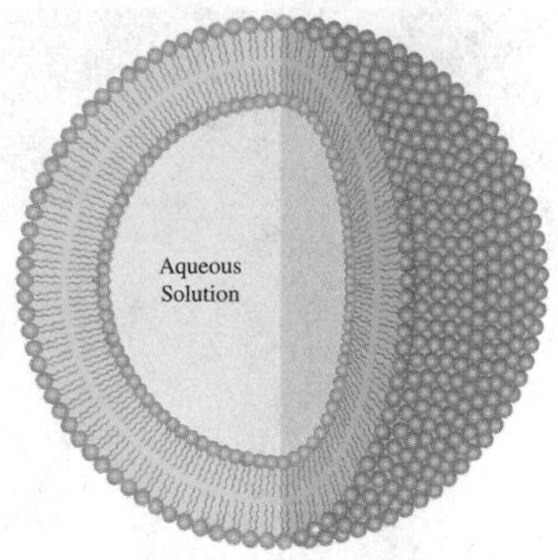

▲
FIGURE 29.6
Schematic representation of a single liposome. Note the bilayer shell and the water solution core.

Having two hydrophobic chains, these lipids are not as thin as fatty acids. They don't aggregate to form micelles in aqueous solution but rather form liposomes. As shown in Figure 29.6, a simple liposome is bounded by two layers of lipid. The hydrophobic alkyl chains are covered by hydrophilic groups on both the liposome's inside and outside. Thus, unlike micelles, liposomes have a water core.

Pharmaceutical companies are actively developing liposomes for use in drug delivery systems. These systems consist of specially prepared liposomes carrying drug molecules inside. If the liposome surface is modified appropriately, then the drug will be absorbed only by a specific organ or tissue. Current research aims at finding the correct lipids (and other hydrophobic molecules) to make liposomes that target a variety of tissues.

Atherosclerosis

Atherosclerosis is a metabolic disease that leads to deposits of cholesterol and other lipids on the inner walls of the arteries. The name *athere* is Greek for "mush" and nicely describes the appearance of these fatty deposits (called *plaque*). As plaque accumulates, the arterial passages become progressively narrower. The walls of the arteries also lose their elasticity and their ability to expand to accommodate the volume of blood pumped by the heart. Blood pressure increases as the heart works to pump sufficient blood through the narrowed passages; this may eventually lead to a heart attack. The accumulation of plaque also causes the inner walls to have a rough rather than a normal smooth surface. This condition may lead to coronary thrombosis (heart attack due to blood clots).

Plaque formation begins because of a lipid's natural tendency to aggregate. In fact, atherosclerosis might be described as a disease partly caused by unwanted lipid aggregation. Scientists have searched for the causes of this aggregation. High serum cholesterol levels often lead to plaque deposits. This makes sense since cholesterol is a large lipid molecule and higher lipid concentrations cause more aggregation. Physicians often recommend that serum cholesterol levels be lowered by restricting intake of major dietary sources such as red meat, liver, and eggs.

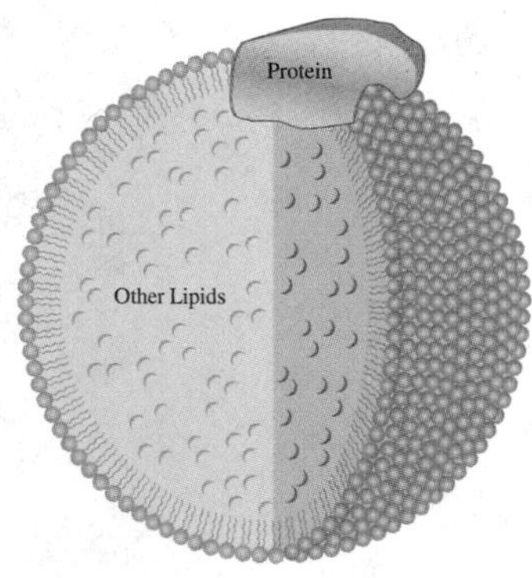

▲
FIGURE 29.7
Schematic representation of a VLDL. Note the single-layer shell and the lipid core.

Another cause of atherosclerosis is improper transportation of cholesterol through the blood. Cholesterol (and other lipids) must be packaged for transport since lipids aggregate in the aqueous bloodstream. The liver packages dietary lipid into aggregates known as *very-low-density lipoproteins* (VLDL). A schematic picture of a VLDL is shown in Figure 29.7. The VLDL surface is hydrophilic containing a single layer of lipids (and some protein) while the interior provides space for the more hydrophobic triacylglycerols and cholesterol. The VLDLs travel through the bloodstream delivering triacylglycerols to fat cells (the adipose tissue) (see Figure 29.8). The VLDL packages are converted to smaller *low-density lipoproteins* (LDL), which deliver cholesterol to peripheral tissues. Proper circulation of cholesterol also depends on a third lipoprotein, the *high-density lipoprotein* (HDL). The HDL acts as a cholesterol scavenger by collecting cholesterol and returning it to the liver, essentially opposing the action of LDL.

People with high plasma LDL concentrations are prone to atherosclerosis even though they may be on low-cholesterol diets. If LDLs are damaged by oxidation, they selectively accumulate in plaque causing the deposits to grow. In contrast, large amounts of plasma HDL seem to prevent plaque formation, and HDL levels can be increased by strenuous exercise, weight loss, and administration of estradiol.

However, the most common drug treatments to prevent atherosclerosis still involve the reduction of serum cholesterol levels, either by preventing absorption of cholesterol from the intestine or by inhibiting metabolic synthesis of this steroid. Treatments such as these can decrease serum cholesterol levels by as much as 50%.

This clogged artery shows plaque formation on the artery wall.

Biological Membranes

Biological membranes are thin, semipermeable cellular barriers. The general function of these barriers is to exclude dangerous chemicals from the cell while allowing nutrients to enter. Membranes also confine special molecules to specific sections of the cell. Because almost all the dangerous chemicals, nutrients, and special molecules are water soluble, the membranes can act as effective barriers only if they impede the movement of hydrophilic (water-soluble) molecules.

To act as such a barrier, a membrane must have some special properties. To exclude water and water solutes, the bulk of a membrane must be hydrophobic. But a membrane necessarily touches water both inside and outside the cell. Therefore, the surface of a membrane must be hydrophilic. Thus a membrane can be visualized as being layered much like a piece of laminated plywood.

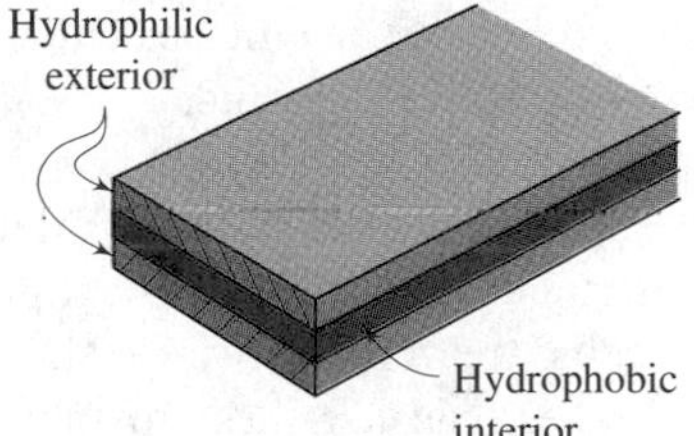

Plywood model of a simple membrane

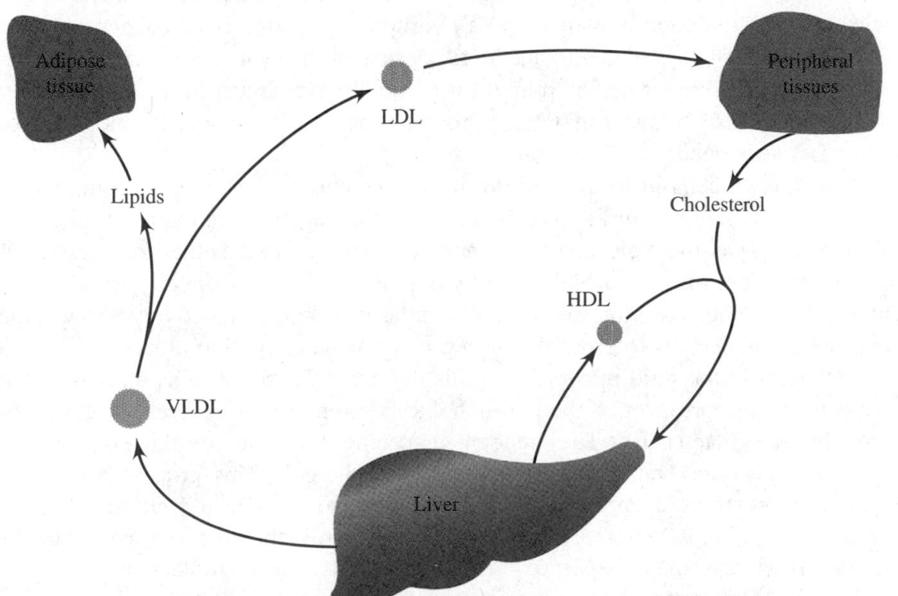

FIGURE 29.8
Schematic representation of the lipid distribution system through the bloodstream.

The interior provides the barrier while the exterior interacts with the aqueous environment.

The cell uses lipids to give the membrane its hydrophobic nature. In fact, by selecting the right lipids, both the hydrophobic and hydrophilic portions of a membrane can be assembled. There are several classes of membrane lipids. The most important of these are the phospholipids and sphingolipids. Remember that these lipids have both a hydrophobic and a hydrophilic section.

lipid bilayer

Membrane lipids naturally aggregate to form lipid bilayers. A **lipid bilayer** is composed of two adjoining layers of lipid molecules aligned so that their hydrophobic portions form the bilayer interior while their hydrophilic portions form the bilayer exterior. A lipid bilayer has the necessary properties of a membrane—a hydrophobic barrier and a hydrophilic surface.

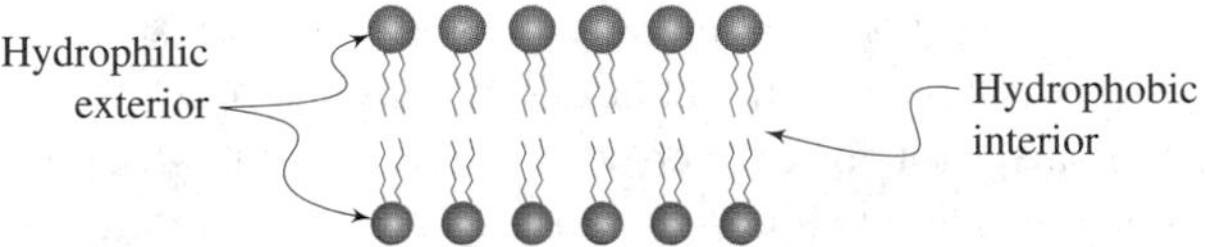

Lipids give this barrier an oily or fluid appearance. The more unsaturated fatty acids in the membrane, the more fluid it will be. Other lipids, most importantly cholesterol, cause the membrane to be less fluid.

Membrane fluidity can have significant effects on cell function. It is thought that general anesthetics (e.g., ether, halothane) are effective partly because they dissolve in membranes, altering the fluidity of the lipid bilayer. During severe cirrhosis of the liver, red cells are forced to take abnormally large amounts of cholesterol into their membranes, causing these membranes to be less fluid. The red cells become more rigid, have greater difficulty passing through narrow capillaries, and are destroyed more easily.

It is truly amazing that lipid bilayers form spontaneously, simply because of the ability of lipids to aggregate. With no complicated planning, a barrier forms with both a hydrophobic interior and a hydrophilic exterior. This barrier provides the basis of the biological membrane. In fact, scientists have proposed that once lipid-like molecules were formed on primal earth, lipid bilayers must have spontaneously aggregated. These bilayers may have provided the boundaries needed as primitive cells first developed.

All known cells in today's world need a membrane that is more complicated than a simple lipid bilayer. A membrane must function as more than just a barrier. Tasks such as passing molecules from one side of a bilayer to the other are an essential part of life. Yet many molecules are hydrophilic and have difficulty traversing a lipid bilayer. The cell is faced with a dilemma. How can it selectively allow some hydrophilic molecules to cross the lipid bilayer while excluding others?

Proteins in the fluid bilayer solve this dilemma. These proteins allow specific molecular transport through the hydrophobic interior. (For a general discussion of proteins, see Chapter 30.) They recognize specific molecules on the exterior of a cell membrane and shuttle these molecules into the cell. This process may be as simple as providing a tunnel through the membrane for selected nutrients. If the protein helps (facilitates) transport without using energy, the process is called **facilitated diffusion**. Some transport requires energy, such as when molecules are moved from areas of low concentration to areas of high concentration (the opposite direction

facilitated diffusion

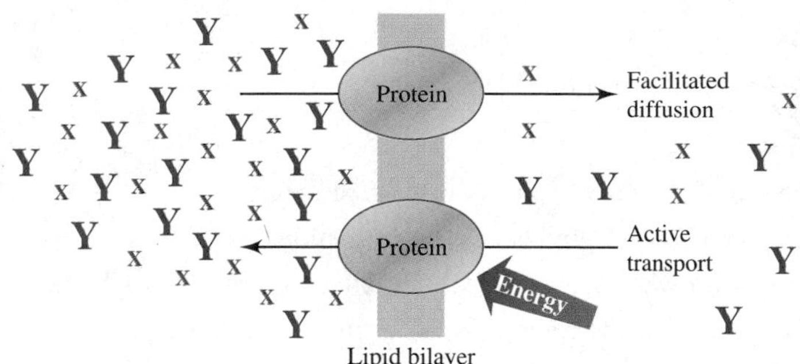

FIGURE 29.9 How membrane transport is aided by proteins: Molecules or ions (symbolized by Y and x) can move from high concentration to low concentration without energy (facilitated diffusion), but movement in the reverse direction requires energy (active transport).

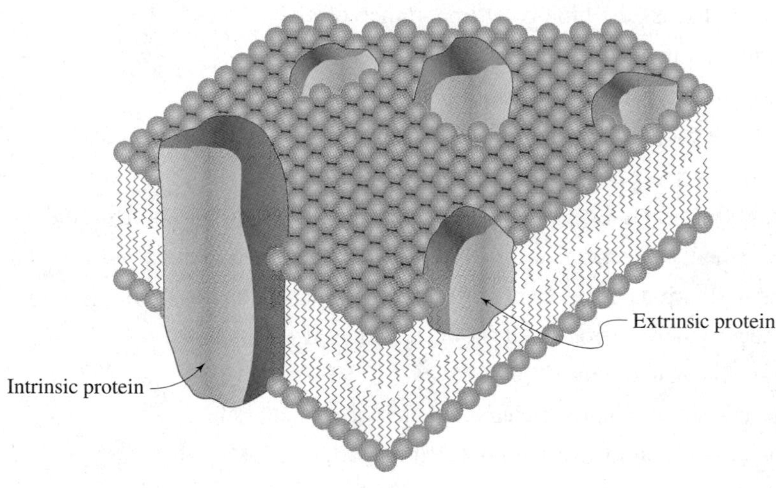

FIGURE 29.10 The fluid-mosaic model of a membrane.

from that of diffusion). This energy-requiring transport is termed **active transport** (see Figure 29.9).

active transport

Other proteins are located in the lipid bilayer to allow special reactions to occur. These special reactions, which often could be harmed by water, take place in the hydrophobic interior of the lipid bilayer. Much of the cellular energy production (oxidation–reduction) occurs in this environment.

Thus a complete cellular membrane must have both lipid and protein. A typical membrane includes about 60% protein, 25% phospholipid, 10% cholesterol, and 5% sphingolipid. The fluid lipid bilayer is studded with many solid proteins. The proteins form a random pattern on the outer surface of the oily lipid. This general membrane is called the **fluid-mosaic model** (see Figure 29.10). An *intrinsic membrane protein* is one in which the bulk of the protein is inside the lipid bilayer; a protein that is mainly on the surface is an *extrinsic membrane protein*.

fluid-mosaic model

Concepts in Review

1. Describe several general features of lipid structures.
2. Describe the classes of lipids and their functions.
3. Briefly explain why long carbon chain fatty acids are hydrophobic.
4. State which fatty acids commonly occur in fats and oils.
5. State which fatty acids are essential to human diets.
6. Explain why unsaturated fatty acids have lower melting points than the corresponding saturated fatty acids.
7. Briefly discuss the biological importance of eicosanoids.
8. Define the major biological function of waxes.
9. Describe the general structural makeup of phospholipids.
10. Describe the similarities between phospholipids and sphingolipids.
11. Draw the structural feature common to all steroids.
12. Draw the structures for cholesterol and several other common steroids.
13. Describe the biochemical relationships between common steroid hormones.
14. Discuss various components of the steroid distribution system.
15. Discuss atherosclerosis and the factors that affect it.
16. Draw a schematic of a lipid bilayer.
17. Discuss some important characteristics of a membrane lipid.
18. Describe two forms of membrane transport.

Key Terms

The terms listed here have been defined within this chapter. Section numbers are referenced in parenthesis for each term.

active transport (29.6)
facilitated diffusion (29.6)
fluid-mosaic model (29.6)
glycolipid (29.4)
lipid bilayer (29.6)
lipid (29.1)
phospholipid (29.4)
sphingolipid (29.4)
steroid (29.5)
wax (29.3)

Questions

Questions refer to tables, figures, and key words and concepts defined within the chapter. A particularly challenging question or exercise is indicated with an asterisk.

1. Why are the lipids, which are dissimilar substances, classified as a group?

2. Briefly, explain why caproic acid is more water soluble than stearic acid.

3. Briefly explain why arachidonic acid is of special biological importance.

4. What are the three essential fatty acids? What are the consequences of their being absent from the diet?

5. List two reasons why fats contain more biochemical energy than carbohydrates.

6. How is aspirin thought to relieve inflammation?

7. Wax serves a protective function on many types of leaves. How does it do this?

8. What two properties must a membrane lipid possess?

9. In what organ in the body are phospholipids mainly produced?

10. List the four classes of eicosanoids.

11. What two structural features allow sphingomyelin to serve as a membrane lipid?

12. What is atherosclerosis? In general, how is it produced and what are its symptoms?

13. Briefly describe the body's cholesterol distribution system.

14. How does a diet that contains a large amount of fish possibly decrease the risk of a heart attack?

15. Why is HDL a potential aid in controlling serum cholesterol levels?

16. Why can a lipid bilayer be described as a barrier?

17. What advantages does a liposome provide as a vehicle for drug delivery?

18. Distinguish active transport from facilitated diffusion.

19. What common structural feature is possessed by all steroids? Write the structural formulas for two steroids.

20. Which of these statements are correct? Rewrite each incorrect statement to make it correct.
- **(a)** A lipid will dissolve in water.
- **(b)** Lipids are often small molecules.
- **(c)** A triacylglycerol is high in biochemical energy because it contains many oxidized carbons.
- **(d)** Linoleic acid has a lower melting point than stearic acid.
- **(e)** Phosphatidyl choline is also known as lecithin.
- **(f)** A sphingolipid always contains carbohydrate.
- **(g)** Cholesterol is often found in membranes.
- **(h)** Facilitated diffusion moves molecules from areas of low concentration to areas of high concentration.
- **(i)** A lipid bilayer has a hydrophilic interior.
- **(j)** The "fluid" in the fluid-mosaic model refers to the lipid bilayer.
- **(k)** Estradiol is the precursor of testosterone.
- **(l)** Cortisone blocks inflammation by limiting the release of arachidonic acid from membranes.
- **(m)** Cells that need cholesterol are able to absorb the low density lipoprotein.
- **(n)** Thromboxane formation may be a factor in heart disease.

Paired Exercises

These exercises are paired. Each odd-numbered exercise is followed by a similar even-numbered exercise. Answers to the even-numbered exercises are given in Appendix V.

21. Write the structural formula of a triacylglycerol that contains one unit each of palmitic, stearic, and oleic acids. How many other triacylglycerols are possible using one unit of each of these acids?

22. Write the structural formula of a triacylglycerol that contains two units of palmitic acid and one unit of oleic acid. How many other triacylglycerols are possible using this same set of fatty acids?

23. Would a triacylglycerol that contains three units of lauric acid be more hydrophobic than a triacylglycerol that contains three units of stearic acid? Explain.

24. Would a triacylglycerol that contains three units of stearic acid be more hydrophobic than a triacylglycerol that contains three units of myristic acid? Explain.

25. Draw formulas for the products of a hydrolysis reaction involving a triacylglycerol that contains palmitic acid, oleic acid, and linoleic acid.

26. Draw formulas for the products of a hydrolysis reaction involving a triacylglycerol that contains oleic acid, stearic acid, and palmitic acid.

27. Give the structure of a phospholipid that would be formed from glycerol, phosphoric acid, palmitic acid, and ethanolamine.

28. Give the structure of a phospholipid that would be formed from glycerol, phosphoric acid, stearic acid, and choline.

29. Given the following components, draw the structure of a sphingolipid: sphingosine, oleic acid, phosphoric acid, choline.

30. Given the following components, draw the structure of a sphingolipid: sphingosine, stearic acid, phosphoric acid, ethanolamine.

31. Given the following components, draw the structure of a glycolipid: sphingosine, palmitic acid, D-glucose.

32. Given the following components, draw the structure of a glycolipid: sphingosine, oleic acid, D-galactose.

33. Sketch the portion of a micelle that forms from palmitic acid. Show the structural formula for one palmitic acid in your sketch.

34. Sketch the portion of a micelle that forms from myristic acid. Show the structural formula for one myristic acid in your sketch.

35. How does LDL differ in both structure and function from VLDL?

36. How does HDL differ in both structure and function from LDL?

37. In what ways is sphingosine similar to a glycerol molecule that has been esterified to one fatty acid?

38. In what ways is a diacylglycerol similar to a sphingosine that has been linked to a fatty acid by an amide bond?

39. If sodium ion is moved from a solution (0.1 *M*) on one side of a membrane to a solution (0.001 *M*) on the other side, is this process facilitated diffusion or active transport? Briefly explain.

40. If phosphate ion is moved from a solution (0.1 *M*) on one side of a membrane to a solution (0.5 *M*) on the other side, is this process facilitated diffusion or active transport? Briefly explain.

41. Differentiate between the function of a thromboxane and a prostaglandin.

42. Differentiate between the function of a thromboxane and a leukotriene.

Additional Exercises

These exercises are not paired or labeled by topic and provide additional practice on concepts covered in this chapter.

43. What is the biochemical effect of ibuprofen that decreases inflammation, redness, and swelling?

44. How does olestra differ structurally from natural fat?

45. **(a)** Draw structural formulas for the three essential fatty acids.
(b) What are the consequences of these fatty acids being absent from the diet?

46. Explain why olestra passes through the body without being digested.

47. Suppose a simple meal to be composed of 10 grams of fat, 17 grams of carbohydrate and 19 grams of proteins. The total energy from this meal is equal to approximately 234 Calories (kcal). Now suppose that the amount of fat intake is increased to 15 grams, what is the energy content now?

Answers to Practice Exercises

29.1 *cis*-isomer: $H(CH_3(CH_2)_5)C{=}C(H)(CH_2)_7COOH$ with both H on the same side

trans-isomer: $H(CH_3(CH_2)_5)C{=}C(H)(CH_2)_7COOH$ with H on opposite sides

$$\begin{array}{l}\text{H}\quad\quad\ \text{H}\\ \quad C{=}C\\ (CH_2)_5\quad(CH_2)_7\\ CH_3\qquad\quad COOH\end{array}\qquad\qquad\begin{array}{l}\qquad\qquad COOH\\ \text{H}\qquad (CH_2)_7\\ \quad C{=}C\\ (CH_2)_5\quad \text{H}\\ CH_3\end{array}$$

cis-isomer *trans*-isomer

29.2

$$\begin{array}{l}CH_2OH\\ |\\ CHOH\\ |\\ CH_2OH\end{array} + CH_3(CH_2)_{16}COOH + 2\ CH_3(CH_2)_{14}COOH \longrightarrow \begin{array}{l}CH_2O-\overset{\overset{\large O}{\|}}{C}(CH_2)_{16}CH_3\\ |\\ CHO-\overset{\overset{\large O}{\|}}{C}(CH_2)_{14}CH_3\\ |\\ CH_2O-\overset{\overset{\large O}{\|}}{C}(CH_2)_{14}CH_3\end{array} + 3\ H_2O$$

or

$$\begin{array}{l}CH_2O-\overset{\overset{\large O}{\|}}{C}(CH_2)_{14}CH_3\\ |\\ CHO-\overset{\overset{\large O}{\|}}{C}(CH_2)_{16}CH_3\\ |\\ CH_2O-\overset{\overset{\large O}{\|}}{C}(CH_2)_{14}CH_3\end{array} + 3\ H_2O$$

29.3 $CH_3(CH_2)_{16}\overset{\overset{\large O}{\|}}{C}-O(CH_2)_{27}CH_3$

29.4

$$\begin{array}{l}CH_2O-\overset{\overset{\large O}{\|}}{C}(CH_2)_{14}CH_3\\ |\\ CHO-\overset{\overset{\large O}{\|}}{C}(CH_2)_{16}CH_3\\ |\\ CH_2O-\overset{\overset{\large O}{\|}}{\underset{\underset{\large O^-}{|}}{P}}-OCH_2CH_2\overset{\overset{\large CH_3}{|}}{\underset{\underset{\large CH_3}{|}}{N^+}}-CH_3\end{array}\quad\text{or}\quad\begin{array}{l}CH_2O-\overset{\overset{\large O}{\|}}{C}(CH_2)_{16}CH_3\\ |\\ CHO-\overset{\overset{\large O}{\|}}{C}(CH_2)_{14}CH_3\\ |\\ CH_2O-\overset{\overset{\large O}{\|}}{\underset{\underset{\large O^-}{|}}{P}}-OCH_2CH_2\overset{\overset{\large CH_3}{|}}{\underset{\underset{\large CH_3}{|}}{N^+}}-CH_3\end{array}$$

29.5

$$\begin{array}{l}OH\\ |\\ CH-CH{=}CH(CH_2)_{12}CH_3\\ |\\ CH-NH-\underset{\underset{\large O}{\|}}{C}(CH_2)_{16}CH_3\\ |\\ CH_2-O-\overset{\overset{\large O}{\|}}{\underset{\underset{\large O^-}{|}}{P}}-OCH_2CH_2-\overset{\overset{\large CH_3}{|}}{\underset{\underset{\large CH_3}{|}}{N^+}}-CH_3\end{array}$$

29.6 (a) Progesterone: two ketone functional groups.

(b) Cholic acid: three alcohol functional groups, one carboxylic acid functional group

(c) Estradiol: one alcohol functional group, one phenol functional group

30 Amino Acids, Polypeptides, and Proteins

Chapter 30

Proteins are present in every living cell. Their very name, derived from the Greek word *proteios*, which means holding first place, signifies the importance of these substances. Think of the startling properties of these molecules. Spider-web protein is many times stronger than the toughest steel; hair, feathers, and hooves are all made from one related group of proteins; another protein provides the glass clear lens material needed for vision. If very small quantities (milligram amounts) of some proteins are missing from the blood, a person's metabolic processes may be out of control. Juvenile-onset diabetes mellitus results from a lack of the insulin protein. Dwarfism can arise when the growth hormone protein is lacking. A special "antifreeze" blood protein allows Antarctic fish to survive at body temperatures below freezing.

This list could go on and on, but what is perhaps most amazing is that this great variety of proteins is made from the same, relatively small, group of amino acids. By using various amounts of these amino acids in different sequences, nature has created biochemical compounds that are essential to the many functions needed to sustain life.

Another aspect of proteins is their importance in our nourishment. Proteins are one of the three major classes of foods. The other two, carbohydrates and fats, are needed for energy; proteins are needed for growth and maintenance of body tissue. Some common foods with high (over 10%) protein content are fish, beans, nuts, cheese, eggs, poultry, and meat. These foods tend to be scarce and relatively expensive. Therefore proteins are the class of foods that is least available to the undernourished people of the world. Hence the question of how to secure an adequate supply of high-quality protein for an ever-increasing population is one of the world's more critical problems (see Chapter 33).

30.1 The Structure–Function Connection

Proteins function as structural materials and as enzymes (catalysts) that regulate the countless chemical reactions taking place in every living organism, including the reactions involved in the decomposition and synthesis of proteins.

All proteins are polymeric substances that yield amino acids on hydrolysis. Those that yield only amino acids when hydrolyzed are classified as **simple proteins**; those that yield amino acids and one or more additional products are classified as **conjugated proteins**. There are approximately 200 different known amino acids in nature. Some are found in only one particular species of plant or animal, others in only a few life forms. But 20 of these amino acids are found in almost all proteins. Furthermore, these same 20 amino acids are used by all forms of life in the synthesis of proteins.

simple protein

conjugated protein

All proteins contain carbon, hydrogen, oxygen, and nitrogen. Some proteins contain additional elements, usually sulfur, phosphorus, iron, copper, or zinc. The significant presence of nitrogen in all proteins sets them apart from carbohydrates and lipids. The average nitrogen content is about 16%.

Proteins are highly specific in their functions. The amino acid units in a given protein molecule are arranged in a definite sequence. An amazing fact about proteins

◀ **Chapter Opening Photo: Bodybuilders sculpt their physiques by building muscles composed of long chains of proteins.**

is that in some cases if just one of the hundreds or thousands of amino acid units is missing or out of place, the biological function of that protein is seriously damaged or destroyed. The sequence of amino acids in a protein establishes the function of that protein.

This relationship between structure and function contrasts sharply with that for other classes of biochemicals. For example, carbohydrates can provide cellular energy because they contain one particular type of atom, reduced carbon, that is readily oxidizable. This important function does not depend directly on the sequence in which the atoms are arranged. On the other hand, an appropriate sequence of amino acids produces a protein strong enough to form a horse's hoof, a different sequence produces a protein capable of absorbing oxygen in the lungs and releasing it to needy cells; yet another sequence produces a hormone capable of directing carbohydrate metabolism for an entire organism. Full understanding of the function of a protein requires an understanding of its structure.

30.2 The Nature of Amino Acids

Each amino acid has two functional groups, an amino group (—NH_2) and a carboxyl group (—COOH). The amino acids found in proteins are called alpha (α) amino acids because the amino group is attached to the first or α-carbon atom adjacent to the carboxyl group. The beta (β) position is the next adjacent carbon; the gamma (γ) position the next carbon; and so on. The following formula represents an α-amino acid:

$$\overset{\gamma}{C}H_3\overset{\beta}{C}H_2\overset{\alpha}{C}HCOOH \quad (NH_2 \text{ on } \alpha\text{-carbon})$$

α-amino butyric acid

Amino acids are represented by this general formula:

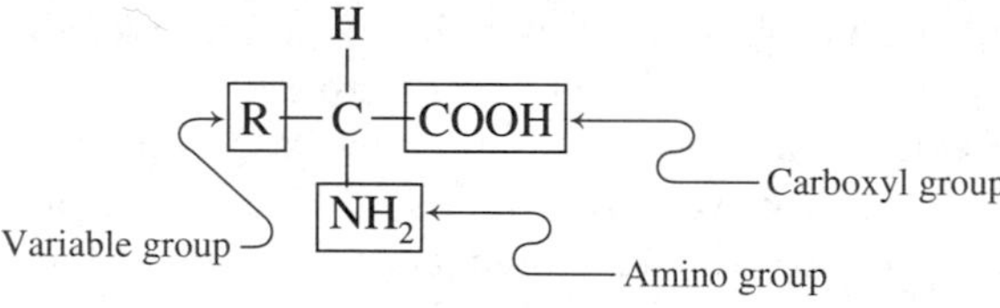

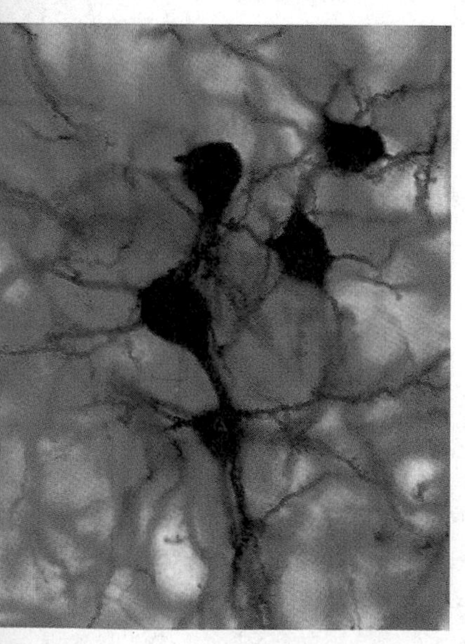

Cerebrum nerve synapse. Neurotransmitters, many of which are amino acids, cross these synapses to send messages between cells.

The portion of the molecule designated R is commonly referred to as the *amino acid side chain.* It is not restricted to alkyl groups and may contain (a) open-chain, cyclic, or aromatic hydrocarbon groups; (b) additional amino or carboxyl groups; (c) hydroxyl groups; or (d) sulfur-containing groups.

Amino acids are divided into three groups: neutral, acidic, and basic. They are classified as

1. *Neutral* when their molecules have the same number of amino and carboxyl groups
2. *Acidic* when their molecules have more carboxyl groups than amino groups
3. *Basic* when their molecules have more amino groups than carboxyl groups

The names, formulas, and abbreviations of the common amino acids are given in Table 30.1. Two of the these (aspartic acid and glutamic acid) are classified as acidic, three (lysine, arginine, and histidine) as basic, and the remainder are classified as neutral amino acids.

Perhaps the most important role played by amino acids is as the building blocks for proteins. However, selected amino acids also have physiological importance on their own. Many neurotransmitters are amino acids or their derivatives. Glycine and glutamic acid are known to act as chemical messengers between nerve cells in some organisms. Tyrosine is converted to the very important neurotransmitter dopamine. A deficiency of this amino acid derivative causes Parkinson's disease, which can be relieved by another compound formed from tyrosine, L-dopa. Tyrosine is also the parent compound for the "flight-or-fight" hormone, epinephrine (adrenalin), and the metabolic hormone thyroxine. Still another amino acid with an important physiological role is histidine, which is converted in the body to histamine. This derivative causes the stomach lining to secrete HCl but is probably best known for causing many of the symptoms associated with tissue inflammation and colds and is the reason antihistamines are such important over-the-counter medications.

Antihistamines lessen the symptoms of colds and allergies by reducing tissue imflammation.

Practice 30.1

Write the structures for the simplest α-amino acid and β-amino acid.

30.3 Essential Amino Acids

Protein is broken down during digestion into its constituent amino acids, which supply much of the body's need for amino acids (see Chapter 33). Eight of the amino acids are **essential amino acids** (see Table 30.2) because they are essential to the normal functioning of the human body. Since the body is not capable of synthesizing them, they must be supplied in our diets if we are to enjoy normal health. It is known that some animals require amino acids in addition to those listed for humans. Rats, for example, require two additional amino acids—arginine and histidine—in their diets.

essential amino acid

On a nutritional basis, proteins are classified as *complete* or *incomplete*. A complete protein supplies all the essential amino acids; an incomplete protein is deficient in one or more essential amino acids. Many proteins, especially those from vegetable sources, are incomplete. For example, protein from corn (maize) is deficient in lysine. The nutritional quality of such vegetable proteins can be greatly improved by supplementing them with the essential amino acids that are lacking, if these can be synthesized at reasonable costs. Lysine, methionine, and tryptophan are now used to enrich human food and livestock feed as a way to extend the world's limited supply of high-quality food protein. In still another approach to the problem of obtaining more high-quality protein, plant breeders have developed maize varieties with greatly improved lysine content. Genetic engineering may hold the key to further significant improvements in plant protein quality.

Vegetarians must carefully choose a combination of foods that include all essential amino acids.

Adding essential amino acids to livestock feed is a way to extend high quality protein.

TABLE 30.1 Common Amino Acids Derived from Proteins°

Name	Abbreviation	Formula
Alanine	Ala	$CH_3CH(NH_2)COOH$
Arginine	Arg	$NH_2-C(=NH)-NH-CH_2CH_2CH_2CH(NH_2)COOH$
Asparagine	Asn	$NH_2C(=O)-CH_2CH(NH_2)COOH$
Aspartic acid	Asp	$HOOCCH_2CH(NH_2)COOH$
Cysteine	Cys	$HSCH_2CH(NH_2)COOH$
Glutamic acid	Glu	$HOOCCH_2CH_2CH(NH_2)COOH$
Glutamine	Gln	$NH_2C(=O)CH_2CH_2CH(NH_2)COOH$
Glycine	Gly	$HCH(NH_2)COOH$
Histidine	His	imidazole ring (N=CH, HC, C, N–H) $-CH_2CH(NH_2)COOH$
Isoleucine*	Ile	$CH_3CH_2CH(CH_3)-CH(NH_2)COOH$
Leucine*	Leu	$(CH_3)_2CHCH_2-CH(NH_2)COOH$

Practice 30.2

Which of the following are essential amino acids? Gly, Ala, Leu, Lys, Tyr, Trp, His.

30.4 D-Amino Acids and L-Amino Acids

Refer to Chapter 27 to review stereoisomerism.

All amino acids, except glycine, have at least one asymmetric carbon atom. For example, two stereoisomers of alanine are possible:

TABLE 30.1 *(continued)*

Name	Abbreviation	Formula
Lysine*	Lys	$NH_2CH_2CH_2CH_2CH_2CH(NH_2)COOH$
Methionine*	Met	$CH_3SCH_2CH_2CH(NH_2)COOH$
Phenylalanine*	Phe	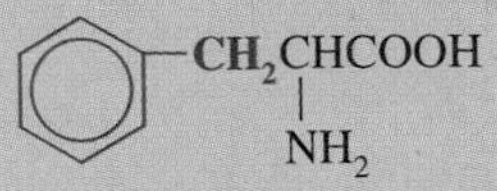 $C_6H_5{-}CH_2CH(NH_2)COOH$
Proline	Pro	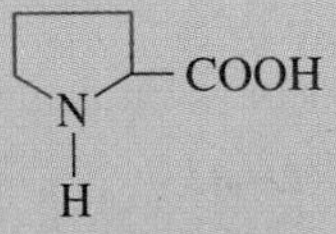 pyrrolidine ring (N–H) bearing COOH
Serine	Ser	$HOCH_2CH(NH_2)COOH$
Threonine*	Thr	$CH_3CH(OH){-}CH(NH_2)COOH$
Tryptophan*	Trp	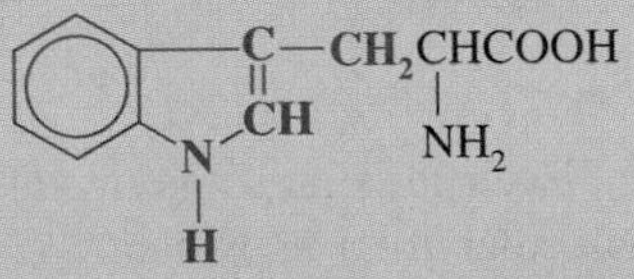 indole ring (N–H, C=CH) $C{-}CH_2CH(NH_2)COOH$
Tyrosine	Tyr	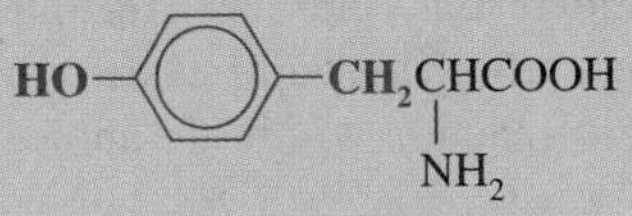 $HO{-}C_6H_4{-}CH_2CH(NH_2)COOH$
Valine*	Val	$(CH_3)_2CHCH(NH_2)COOH$

*Amino acids essential in human nutrition

TABLE 30.2 Essential Amino Acids for Humans

Isoleucine
Leucine
Lysine
Methionine
Phenylalanine
Threonine
Tryptophan
Valine

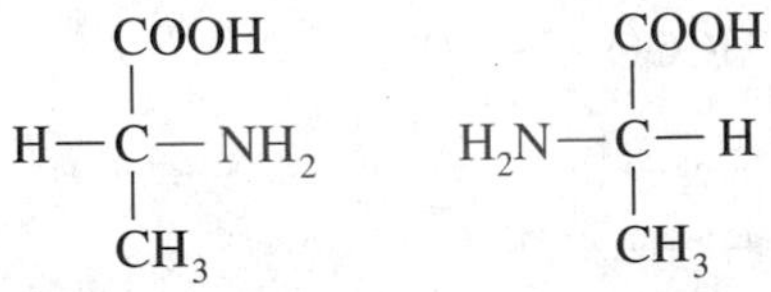

$$\begin{array}{c} COOH \\ | \\ H{-}C{-}NH_2 \\ | \\ CH_3 \end{array} \qquad \begin{array}{c} COOH \\ | \\ H_2N{-}C{-}H \\ | \\ CH_3 \end{array}$$

D-(−)-alanine L-(+)-alanine

Fischer projection formulas illustrate the D and L configurations of amino acids in the same way they illustrate the configurations of D- and L- glyceraldehyde (Section 28.2). The —COOH group is written at the top of the projection formula, and the D configuration is indicated by writing the alpha —NH_2 to the right of carbon 2. The L configuration is indicated by writing the alpha —NH_2 to the left of carbon 2. Although some D-amino acids occur in nature, only L-amino acids occur in proteins. The (+) and (−) signs in the name indicate the direction of rotation of plane-polarized light by the amino acid. Most amino acids have relatively complex structures that make use of the projection formula difficult. Thus unless stereochemical information is explicitly considered, amino acids will be shown using a condensed, structural formula.

30.5 Amphoterism

Amino acids are *amphoteric* (or *amphiprotic*); that is, they can react either as an acid or as a base. For example, with a strong base such as sodium hydroxide, alanine reacts as an acid:

$$\underset{\text{alanine}}{\underset{\displaystyle NH_2}{CH_3\underset{|}{C}HCOOH}} + NaOH \longrightarrow \underset{\text{sodium alanate}}{\underset{\displaystyle NH_2}{CH_3\underset{|}{C}HCOO^-Na^+}} + H_2O$$

With a strong acid such as HCl, alanine reacts as a base:

$$\underset{\text{alanine}}{\underset{\displaystyle NH_2}{CH_3\underset{|}{C}HCOOH}} + HCl \longrightarrow \underset{\text{alanyl ammonium chloride}}{\underset{\displaystyle NH_3^+Cl^-}{CH_3\underset{|}{C}HCOOH}}$$

zwitterion

Even in neutral biological solutions, amino acids do not actually exist in the molecular form shown in the preceding equations. Instead, they exist mainly as dipolar ions called **zwitterions**. Again, using alanine as an example, the proton on the carboxyl group transfers to the amino group, forming a zwitterion by an acid–base reaction within the molecule:

$$\underset{\text{alanine molecule}}{\underset{\displaystyle H_2N\!:}{CH_3\underset{|}{C}HCOO(H)}} \longrightarrow \underset{\text{alanine zwitterion}}{\underset{\displaystyle NH_3^+}{CH_3\underset{|}{C}HCOO^-}}$$

On an ionic basis, the reaction of alanine with OH^- and H^+ is

$$\underset{\text{alanine zwitterion}}{\underset{\displaystyle NH_3^+}{CH_3\underset{|}{C}HCOO^-}} + OH^- \longrightarrow \underset{\text{alanate anion}}{\underset{\displaystyle NH_2}{CH_3\underset{|}{C}HCOO^-}} + H_2O \qquad (1)$$

$$\underset{\text{alanine zwitterion}}{\underset{\displaystyle NH_3^+}{CH_3\underset{|}{C}HCOO^-}} + H^+ \longrightarrow \underset{\text{alanyl ammonium cation}}{\underset{\displaystyle NH_3^+}{CH_3\underset{|}{C}HCOOH}} \qquad (2)$$

Other amino acids behave like alanine. Together with protein molecules that contain —COOH and —NH_2 groups, they help to buffer or stabilize the pH of the blood at about 7.4. The pH is maintained close to 7.4 because any excess acid or base in the blood is neutralized by reactions such as shown in equations (1) and (2).

$$\underset{\substack{\text{cation form}\\ \text{II}}}{\underset{\displaystyle NH_3^+}{\overset{}{RCHCOOH}}} \underset{H^+}{\overset{OH^-}{\rightleftharpoons}} \underset{\substack{\text{zwitterion form}\\ \text{I}}}{\underset{\displaystyle NH_3^+}{RCHCOO^-}} \underset{H^+}{\overset{OH^-}{\rightleftharpoons}} \underset{\substack{\text{anion form}\\ \text{III}}}{\underset{\displaystyle NH_2}{RCHCOO^-}} \qquad (3)$$

When an amino acid in solution has equal positive and negative charges, as in formula I of equation (3), it is electrically neutral and does not migrate toward either the positive or negative electrode when placed in an electrolytic cell. The pH at which there is no migration toward either electrode is called the **isoelectric point** (see Table 30.3). If acid (H^+) is added to an amino acid at its isoelectric point, the equilibrium is shifted toward formula II, and the cation formed migrates toward the negative electrode. When base (OH^-) is added, the anion formed (formula III) migrates toward the positive electrode. Differences in isoelectric points are important in isolating and purifying amino acids and proteins, since their rates and directions of migration can be controlled in an electrolytic cell by adjusting the pH. This method of separation is called *electrophoresis.*

isoelectric point

TABLE 30.3 Isoelectric Points of Selected Amino Acids

Amino acid	pH at Isoelectric Point
Arginine	10.8
Lysine	9.7
Alanine	6.0
Glycine	6.0
Serine	5.7
Glutamic acid	3.2
Aspartic acid	2.9

Amino acids are classified as basic, neutral, or acidic depending on whether the ratio of $—NH_2$ to —COOH groups in the molecules is greater than 1:1, equal to 1:1, or less than 1:1, respectively. Furthermore, this ratio differs from 1:1 only if the amino acid side chain (R—) contains an additional amino or carboxyl group. For example, if the side chain contains a carboxyl group, the amino acid is considered acidic. Thus the R— group determines whether an amino acid is classified as basic, neutral, or acidic.

Isoelectric points are found at pH values ranging from 7.8 to 10.8 for basic amino acids, 4.8 to 6.3 for neutral amino acids, and 2.8 to 3.3 for acidic amino acids. It is logical that a molecule such as glutamic acid, with one amino group and two carboxyl groups, would be classified as acidic and that its isoelectric point would be at a pH lower than 7.0. It might also seem that the isoelectric point of an amino acid that is classified as neutral, such as alanine, with one amino and one carboxyl group, would have an isoelectric point of 7.0. However, the isoelectric point of alanine is 6.0, not 7.0. This is because the carboxyl group and the amino group are not equally ionized. The carboxyl group of alanine ionizes to a greater degree as an acid than the amino group ionizes as a base.

Example 30.1

Draw the structure of L-serine in a strongly acidic solution.

Solution

First, draw the structure for L-serine as the molecule would exist in a neutral solution. The α-amino and carboxylic acid groups form a zwitterion.

$$\underset{\displaystyle NH_3^+}{HOCH_2CHCOO^-}$$

When the solution is made acidic (the concentration of hydrogen ions is increased), the amine group is unaffected because it is already protonated. However, the carboxylate anion bonds to a hydrogen ion, resulting in an L-serine structure with a net positive charge.

$$\underset{\displaystyle NH_3^+}{HOCH_2CHCOOH}$$

Practice 30.3

Draw the structure of L-valine in a strongly basic solution.

30.6 Formation of Polypeptides

Proteins are polyamides consisting of amino acid units joined through amide structures. If we react two glycine molecules, with the elimination of a molecule of water, we form a compound containing the amide structure, also called the **peptide linkage**, or peptide bond. The elimination of water occurs between the carboxyl group of one amino acid and the α-amino group of a second amino acid (see Section 25.1). The product formed from two glycine molecules is called glycylglycine (abbreviated Gly-Gly). Because it contains two amino acid units, it is called a dipeptide.

peptide linkage

$$H_2N{-}CH_2{-}C(=O){-}OH + H{-}NH{-}CH_2{-}C(=O){-}OH \longrightarrow NH_2{-}CH_2{-}C(=O){-}NH{-}CH_2{-}C(=O){-}OH + H_2O$$

peptide linkage

glycylglycine (Gly-Gly)

If three amino acid units are included in a molecule, it is a tripeptide; if four, a tetrapeptide; if five, a pentapeptide; and so on. Peptides containing up to about 40–50 amino acid units in a chain are called **polypeptides**. The units making up the peptide are amino acids less the elements of water and are referred to as *amino acid residues* or, simply, residues. Still larger chains of amino acids are known as proteins.

polypeptide

When amino acids form a polypeptide chain, a carboxyl group and an α-amino group are involved in each peptide bond (amide bond). While these groups are joined in peptide bonds, they cannot ionize as acids or bases. Consequently, the properties of a polypeptide/protein are determined to a large extent by the side chains of the amino acid residues.

In linear peptides one end of the chain has a free amino group and the other end a free carboxyl group. The amino-group end is called the *N-terminal residue* and the other end the *C-terminal residue*:

1 2 3 4 5 6 7
Ala-Pro-Tyr-Met-Gly-Lys-Gly

N-Terminal residue — C-Terminal residue

The sequence of amino acids in a chain is numbered starting with the N-terminal residue, which is written to the left with the C-terminal residue at the right. Any segment of the sequence that is not specifically known is placed in parentheses. Thus in the heptapeptide above, if the order of tyrosine and methionine were not known, the structure would be written as

Ala-Pro-(Met, Tyr)-Gly-Lys-Gly

Peptides are named as acyl derivatives of the C-terminal amino acid with the C-terminal unit keeping its complete name. The *ine* ending of all but the C-terminal amino acid is changed to *yl*, and these listed in the order in which they appear, starting with the N-terminal amino acid.

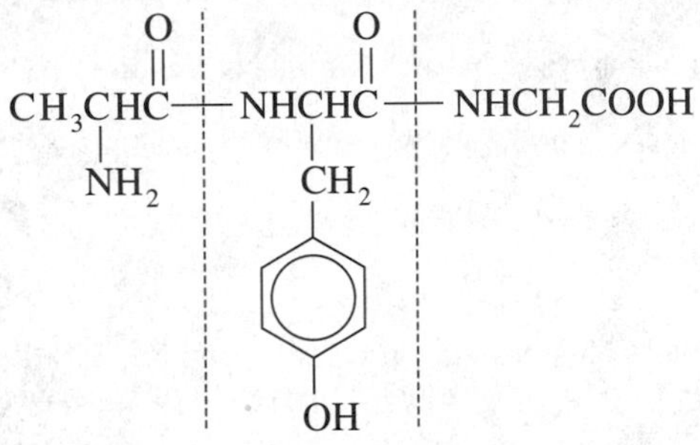

$$\underset{\text{alanyl}}{CH_3\underset{\underset{NH_2}{|}}{C}H\overset{\overset{O}{\|}}{C}}-\underset{\text{tyrosyl}}{NH\underset{\underset{CH_2C_6H_4OH}{|}}{C}H\overset{\overset{O}{\|}}{C}}-\underset{\text{glycine}}{NHCH_2COOH}$$

Ala-Tyr-Gly

Thus Ala-Tyr-Gly is called alanyltyrosylglycine. The name of Arg-Glu-His-Ala is arginylglutamylhistidylalanine.

Alanine and glycine can form two different dipeptides, Gly-Ala and Ala-Gly, using each amino acid only once:

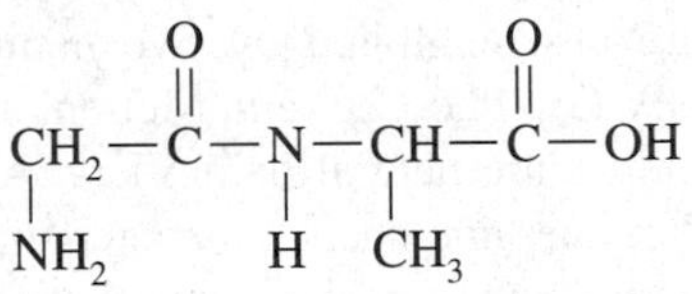

$$\underset{NH_2}{\underset{|}{CH_2}}-\overset{\overset{O}{\|}}{C}-\underset{H}{\underset{|}{N}}-\underset{CH_3}{\underset{|}{CH}}-\overset{\overset{O}{\|}}{C}-OH \quad \text{and} \quad CH_3\underset{NH_2}{\underset{|}{CH}}-\overset{\overset{O}{\|}}{C}-\underset{H}{\underset{|}{N}}-CH_2-\overset{\overset{O}{\|}}{C}-OH$$

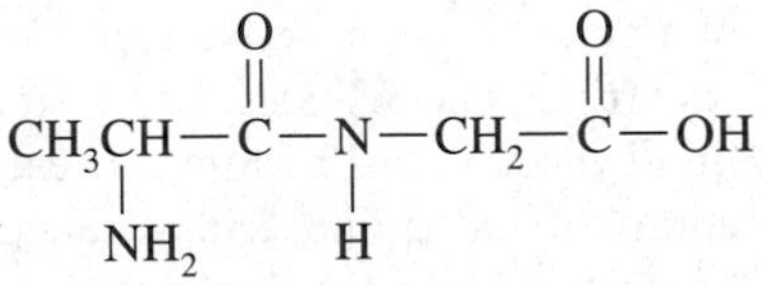

glycylalanine (Gly-Ala) alanylglycine (Ala-Gly)

If three different amino acids react—for example, glycine, alanine, and threonine—six tripeptides in which each amino acid appears only once are possible.

Gly-Ala-Thr	Ala-Thr-Gly	Thr-Ala-Gly
Gly-Thr-Ala	Ala-Gly-Thr	Thr-Gly-Ala

The number of possible peptides rises very rapidly as the number of amino acid units increases. For example, there are 120 ($1 \times 2 \times 3 \times 4 \times 5 = 120$) different ways to combine five different amino acids to form a pentapeptide, using each amino acid only once in each molecule. If the same constraints are applied to 15 different amino acids, the number of possible combinations is greater than 1 trillion (10^{12})! Since a protein molecule may contain several hundred amino acid units, with individual amino acids occurring several times, the number of possible combinations from 20 amino acids is simply beyond imagination.

There are a number of small, naturally occurring polypeptides with significant biochemical functions. In general, these substances serve as hormones or nerve transmitters. Their functions range from controlling pain and pleasure responses in the brain to controlling smooth muscle contraction or kidney fluid excretion rates (see Table 30.4).

The amino acid sequence and chain length give a polypeptide its biological effectiveness and specificity. For example, recent research has shown that the effects

TABLE 30.4 Primary Structures and Functions of Some Biological Polypeptides*

Name	Primary structure	General biological function
Substance P	Arg-Pro-Lys-Pro-Gln-Gln-Phe-Phe-Gly-Leu-Met-NH_2	Is a pain-producing agent
Bradykinin	Arg-Pro-Pro-Gly-Phe-Ser-Pro-Phe-Arg	Affects tissue inflammation and blood pressure
Angiotensin II	Asp-Arg-Val-Tyr-Val-His-Pro-Phe	Maintains water balance and blood pressure
Leu-enkephalin	Tyr-Gly-Gly-Phe-Leu	Relieves pain, produces sense of well-being
Met-enkephalin	Tyr-Gly-Gly-Phe-Met	
Vasopressin	Cys-Tyr-Phe-Gln-Asn-Cys-Pro-Arg-Gly-NH_2 (S—S bridge between the two Cys)	Increases blood pressure, decreases kidney water excretion
Oxytocin	Cys-Tyr-Ile-Gln-Asn-Cys-Pro-Leu-Gly-NH_2 (S—S bridge between the two Cys)	Initiates childbirth labor, causes mammary gland milk release, affects kidney excretion of water and sodium

*Where —NH_2 is indicated at the end of the sequence, the C-terminal amino acid has an amide structure rather than a free —COOH.

of opiates (opium derivatives) on the brain are also exhibited by two naturally occurring pentapeptides, Leu-enkephalin (Tyr-Gly-Gly-Phe-Leu) and Met-enkephalin (Tyr-Gly-Gly-Phe-Met). These two polypeptides are natural painkillers. Alterations of the amino acid sequence—which alter the side-chain characteristics—cause drastic changes in the analgesic effects of these pentapeptides. The substitution of L-alanine for either of the glycine residues in these compounds (simply changing one side-chain group from —H to —CH_3) causes an approximately 1000-fold decrease in effectiveness as a painkiller! The substitution of L-tyrosine for L-phenylalanine causes a comparable loss of activity. Even the substitution of D-tyrosine for the L-tyrosine residue causes a considerable loss in the analgesic effectiveness of the pentapeptides.

It is clearly evident that a particular sequence of amino acid residues is essential for proper polypeptide function. This sequence aligns the side-chain characteristics (large or small; polar or nonpolar; acidic, basic, or neutral) in the proper positions for a specific polypeptide function.

Oxytocin and vasopressin are similar nonapeptides, differing only at two positions in their primary structure (see Table 30.4). Yet their biological functions differ dramatically. Oxytocin controls uterine contractions during labor in childbirth and also causes contraction of the smooth muscles of the mammary glands, resulting in milk excretion. Vasopressin in high concentration raises the blood pressure and has been used in treatment of surgical shock for this purpose. Vasopressin is also an antidiuretic, regulating the excretion of fluid by the kidneys. The absence of vasopressin leads to diabetes insipidus. This condition is characterized by excretion of up to 30 L of urine per day but can be controlled by administration of vasopressin or its derivatives.

The isolation and synthesis of oxytocin and vasopressin was accomplished by Vincent du Vigneaud (1901–1978) and co-workers at Cornell University. Du Vig-

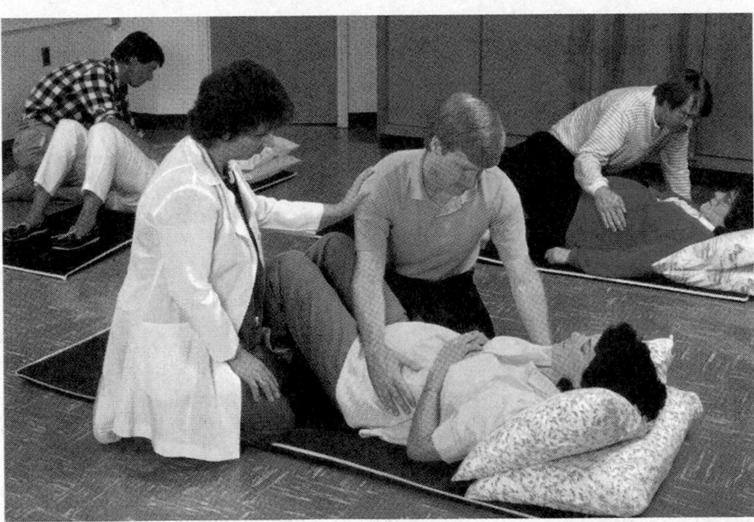

◀ **The nonapeptide oxytocin controls uterine contractions during labor.**

neaud was awarded the Nobel prize in chemistry in 1955 for this work. Synthetic oxytocin is indistinguishable from the natural material. It is available commercially and is used for the induction of labor in the late stages of pregnancy.

Example 30.2

Write the structure of the tripeptide Ser-Gly-Ala.

Solution

First, write the structures of the three amino acids in this tripeptide.

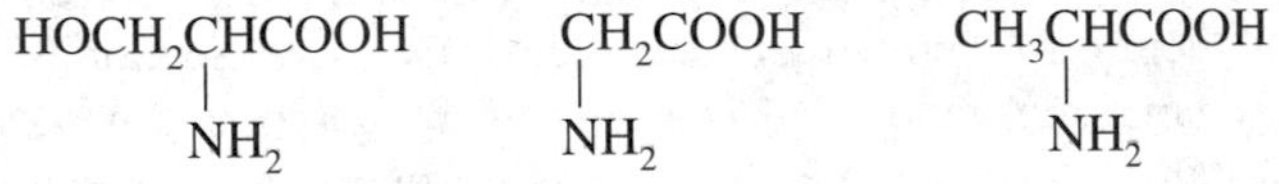

$$HOCH_2\underset{\underset{\displaystyle NH_2}{|}}{CH}COOH \qquad \underset{\underset{\displaystyle NH_2}{|}}{CH_2}COOH \qquad CH_3\underset{\underset{\displaystyle NH_2}{|}}{CH}COOH$$

serine (Ser) glycine (Gly) alanine (Ala)

By convention, the amino acid residue written at the left end of the tripeptide has a free amino group while the residue at the right end has a free carboxylic acid group. Now split out a water molecule in two places: between the carboxyl group of serine and the amino group of glycine, and between the carboxyl group of glycine and the amino group of alanine. When the amino acids are connected by peptide linkages, the following structure results:

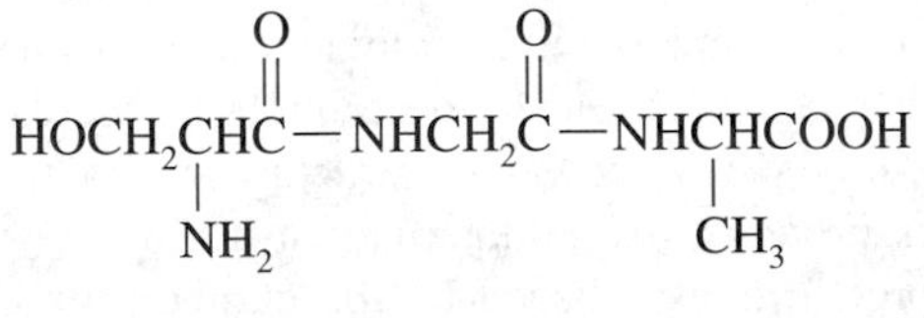

$$HOCH_2\underset{\underset{\displaystyle NH_2}{|}}{CH}\overset{\overset{\displaystyle O}{\|}}{C}-NHCH_2\overset{\overset{\displaystyle O}{\|}}{C}-NH\underset{\underset{\displaystyle CH_3}{|}}{CH}COOH$$

Practice 30.4

Write the structure of the pentapeptide Gly-Leu-Asp-Ser-Cys.

30.7 Protein Structure

By 1940, a great deal of information concerning proteins had been assembled. Their elemental composition was known, and they had been carefully classified according to solubility in various solvents. Proteins were known to be polymers of amino acids, and the different amino acids had, for the most part, been isolated and identified. Protein molecules were known to be very large in size, with molar masses ranging from several thousand to several million.

Knowledge of protein structure would answer many chemical and biological questions. But for a while the task of determining the actual structure of molecules of such colossal size appeared to be next to impossible. Then Linus Pauling, at the California Institute of Technology, attacked the problem by a new approach. Using X-ray diffraction techniques, Pauling and his collaborators painstakingly determined the bond angles and dimensions of amino acids and of dipeptides and tripeptides. After building accurate scale models of the dipeptides and tripeptides, they determined how these could be fitted into likely polypeptide configurations. Based on this work, Pauling and Robert Corey (1897–1971) proposed in 1951 that two different conformations—the *α-helix* and the *β-pleated sheet*—were the most probable stable polypeptide chain configurations of protein molecules. These two macromolecular structures are illustrated in Figure 30.1. Within a short time it was established that many proteins do have structures corresponding to those predicted by Pauling and Corey. This work was a very great achievement. Pauling received the 1954 Nobel prize in chemistry for this work on protein structure. Pauling's and Corey's work provided the inspiration for another great biochemical breakthrough—the concept of the double-helix structure for deoxyribonucleic acid, DNA (see Section 32.6).

Proteins are very large molecules. But just how many amino acid units must be present for a substance to be a protein? There is no universally agreed-upon answer to this question. Some authorities state that a protein must have a molar mass of at least 6000 or contain about 50 amino acid units. Smaller amino acid polymers, containing from 5 to 50 amino acid units, are classified as polypeptides and are not proteins. In reality, there is no clearly defined lower limit to the molecular size of proteins. The distinction is made to emphasize (1) that proteins usually serve structural or enzymatic functions, while polypeptides usually serve hormone-related functions, and (2) that the three-dimensional conformation of proteins is directly related to function, while the relationship is not so distinct with polypeptides.

In general, for a protein molecule to serve a specific biological function, it must have a closely defined overall shape. Chemists typically describe large proteins on several levels: (1) a primary structure, (2) a secondary structure, (3) a tertiary structure, and, for the most complex proteins, (4) a quaternary structure.

primary structure

The **primary structure** of a protein is established by the number, kind, and sequence of amino acid units composing the polypeptide chain or chains making up the molecule. The primary structure determines the alignment of side-chain characteristics, which, in turn, determines the three-dimensional shape into which the protein folds. In this sense the amino acid sequence is of primary importance in establishing protein shape.

Determining the sequence of the amino acids in even one protein molecule was a formidable task. The amino acid sequence of beef insulin was announced in 1955

(a) α-Helix

(c) Collagen

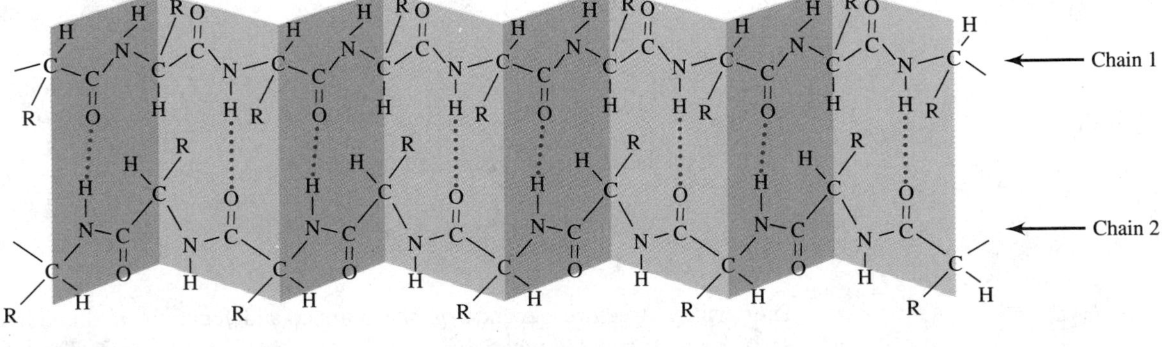

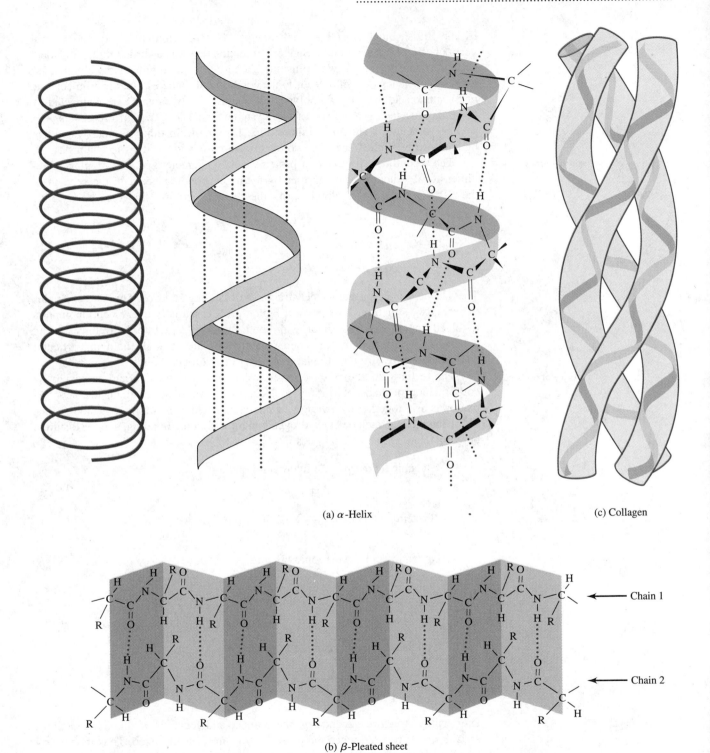

(b) β-Pleated sheet

FIGURE 30.1
Two common protein secondary structures.

by the British biochemist Frederick Sanger. This determination required several years of effort by a team under Sanger's direction. He was awarded the 1958 Nobel prize in chemistry for this work. Insulin is a hormone that regulates the blood-sugar level. A deficiency of insulin leads to the condition of diabetes. Beef insulin consists of 51 amino acid units in two polypeptide chains. The two chains are connected by disulfide linkages (—S—S—) of two cysteine residues at two different sites. The primary structure is shown in Figure 30.2. Insulins from other animals, including humans, differ slightly by one, two, or three amino acid residues in chain A.

secondary structure

The **secondary structure** of proteins can be characterized as a regular, three-dimensional structure held together by hydrogen bonding between the oxygen of the >C=O and the hydrogen of the H—N< groups in the polypeptide chains:

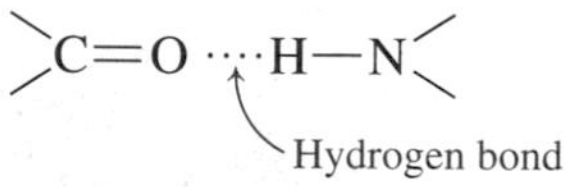

$$>C{=}O \cdots H{-}N<$$

Hydrogen bond

The α-helical and β-pleated-sheet structures of Pauling and Corey are two examples of secondary structure. As shown in Figure 30.1, essentially every peptide bond is involved in at least one hydrogen bond in these structures. Proteins having α-helical or β-pleated-sheet secondary structures are strongly held in particular conformations by virtue of the large number of hydrogen bonds.

tertiary structure

The **tertiary structure** of a protein refers to the distinctive and characteristic conformation, or shape, of a protein molecule. This overall three-dimensional conformation is held together by a variety of interactions between amino acid side chains. These interactions include (1) hydrogen bonding, (2) ionic bonding, and (3) disulfide bonding. Here are some examples

1. Glutamic acid–tyrosine hydrogen bonding

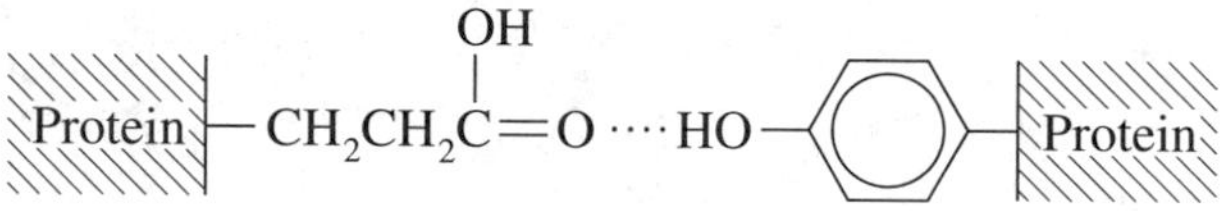

$$\text{Protein}{-}CH_2CH_2\overset{OH}{\overset{|}{C}}{=}O \cdots HO{-}C_6H_4{-}\text{Protein}$$

2. Glutamic acid–lysine ionic bonding

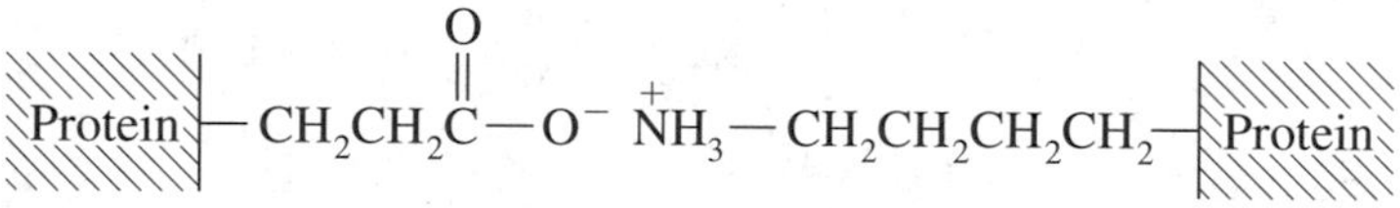

$$\text{Protein}{-}CH_2CH_2\overset{O}{\overset{\|}{C}}{-}O^- \ \overset{+}{N}H_3{-}CH_2CH_2CH_2CH_2{-}\text{Protein}$$

3. Cysteine–cysteine disulfide bonding

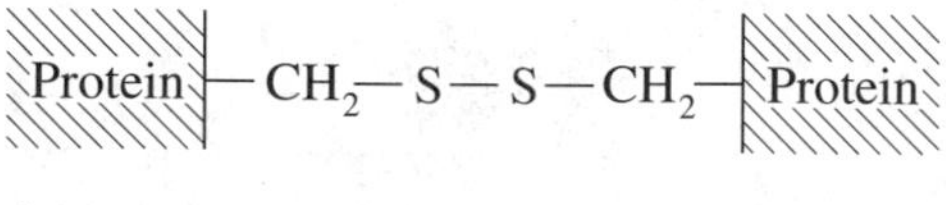

$$\text{Protein}{-}CH_2{-}S{-}S{-}CH_2{-}\text{Protein}$$

The tertiary structure depends on the number and location of these interactions, variables that are fixed when the primary structure is synthesized. Thus the tertiary structure depends on the primary structure. For example, there are three locations in the insulin molecule (Figure 30.2) where the primary sequence permits disulfide bonding. This in turn has an obvious bearing on the shape of insulin.

Hair is especially rich in disulfide bonds. These can be broken by certain reducing agents and restored by an oxidizing agent. This fact is the key to "cold"

```
                        15                         20     21
              ┌─────────Gln—Leu—Glu—Asn—Tyr—Cys—Asn
              │                                 |
             Tyr                                S
              |                                 |
             Leu                                S
              |                                 |    20
             Ser                             ┌Cys—Gly┐
              │         10                   Val      Glu
              └Cys—Val┐                       |        |
                |      Ser                   Leu      Arg
                S       |                     |        |
                |      Ala                   Tyr      Gly
                S       |                     |        |
  1            5|      ┘                   15 Leu      Phe
Gly—Ile—Val—Glu—Gln—Cys—Cys                   |        |
Chain A                 |                    Ala      Phe 25
                        S                     |        |
                        |                    Glu      Tyr
                        S                     |        |
  1            5        |        10           |        |
Phe—Val—Asn—Gln—His—Leu—Cys—Gly—Ser—His—Leu—Val       Thr
Chain B                                                |
                                                      Pro
                                                       |
                                                      Lys
                                                       |
                                                      Ala 30
```

FIGURE 30.2 Amino acid sequence of beef insulin.

permanent waving of hair. Some of the disulfide bonds are broken by applying a reducing agent to the hair. The hair is then styled with the desired curls or waves. These are then permanently set by using an oxidizing agent to reestablish the disulfide bonds at different points.

$$\vdash CH_2-S-S-CH_2 \dashv \xrightarrow[\text{agent}]{\text{reducing}}$$

bonds in normal hair

$$\vdash CH_2-S-H \quad H-S-CH_2 \dashv \xrightarrow[\text{agent}]{\text{oxidizing}} \vdash CH_2-S-S-CH_2 \dashv$$

broken bonds of "reduced" form — reestablished bonds of permanently waved hair

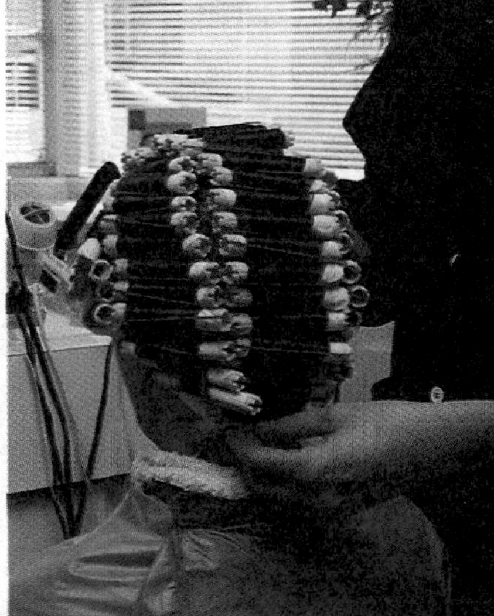

In a permanent wave, disulfide bonds in the hair are broken with reducing agents, set to a curling pattern, and then restored by an oxidizing agent.

If Table 30.1 is examined closely, it can be concluded that most of the amino acids have side chains that can form neither hydrogen, ionic, nor disulfide bonds. What then do the majority of amino acids do to hold together the tertiary structure of a protein? This is an important question and leads to an equally important answer. The uncharged relatively nonpolar amino acids form the center or core of most proteins. These amino acids have side chains that do not bond very well to water; they are like saturated hydrocarbons or lipids and are hydrophobic. When a protein is synthesized, the uncharged, nonpolar amino acids turn inward toward each other, excluding water and forming the core of the protein structure.

quaternary structure

A fourth type of structure, called **quaternary structure**, is found in some complex proteins. These proteins are made of two or more smaller protein subunits (polypeptide chains). Nonprotein components may also be present. The quaternary structure refers to the shape of the entire complex molecule and is determined by the way in which the subunits are held together by *noncovalent* bonds—that is, by hydrogen bonding, ionic bonding, and so on.

30.8 Some Examples of Proteins and Their Structures

Fibrous Proteins

fibrous protein

The **fibrous proteins** are an important class of proteins that contain highly developed secondary structures. As their name suggests, fibrous proteins have a "fiber-like" or elongated shape. Because secondary structures provide strength, these proteins tend to function in support roles. Several of these proteins are shown in Figure 30.3. The α-keratins depend on the α-helix for strength and are found in such diverse structures as hair, feathers, horns, and hooves. The silk protein, fibroin, is folded into a β-pleated-sheet secondary structure.

The most abundant protein in the animal kingdom, collagen, is a fibrous protein with a unique secondary structure (Figure 30.3). Collagen forms the bone matrix around which the calcium phosphate mineral can crystallize. Ligaments, tendons, and skin are composed of a large proportion of collagen. The structure of this protein allows it to provide a strong framework for each of these tissues. Collagen is a long, slender protein whose three strands are wrapped one around another as a rope would be woven, and, like a rope, the finished product is much stronger than a single strand. These triple helices are in turn stacked like cordwood to form collagen fibers. The resulting tough material can be aligned to form tendons or deposited as a network to form a bone or skin structure. Next time you take a stride, recognize the force exerted on your Achilles tendon and how your being able to function depends on the strength of collagen.

Globular Proteins

globular protein

The **globular proteins** share a similar compact shape. Although it may not be apparent at first, this roughly spherical structure allows each globular protein its own complex tertiary structure. Each unique structure is associated with a particular function in the biological world. Some proteins function to bind specific biochemicals. This function is important for biological transport (e.g., oxygen transport by hemoglobin) and for processes such as the immune response (e.g., immunoglobulins). Other globular proteins bind biochemicals and then catalyze a chemical reaction (e.g., enzymes). A catalyst is required for essentially all metabolic reactions.

We will consider several protein structures in detail. These proteins contain hundreds to thousands of atoms and look very complicated at first glance. However, scientists have learned to look for much simpler, major structural features. These features are often related to a protein's function. To help visualize these molecular relationships, we will use drawing techniques that emphasize major structural features.

Myoglobin—a binding protein Myoglobin is one of the simplest of the globular proteins (153 amino acid residues with a molar mass of about 17,000 g/mol). It functions as a *binding protein*, carrying oxygen for muscle tissue. The space-filling model in Figure 30.4(a) shows myoglobin to be a globular protein. To better visualize the details of myoglobin's structure, we must strip away the amino acid side chains and focus only on the covalent chain linking the amino acids as seen in Figure

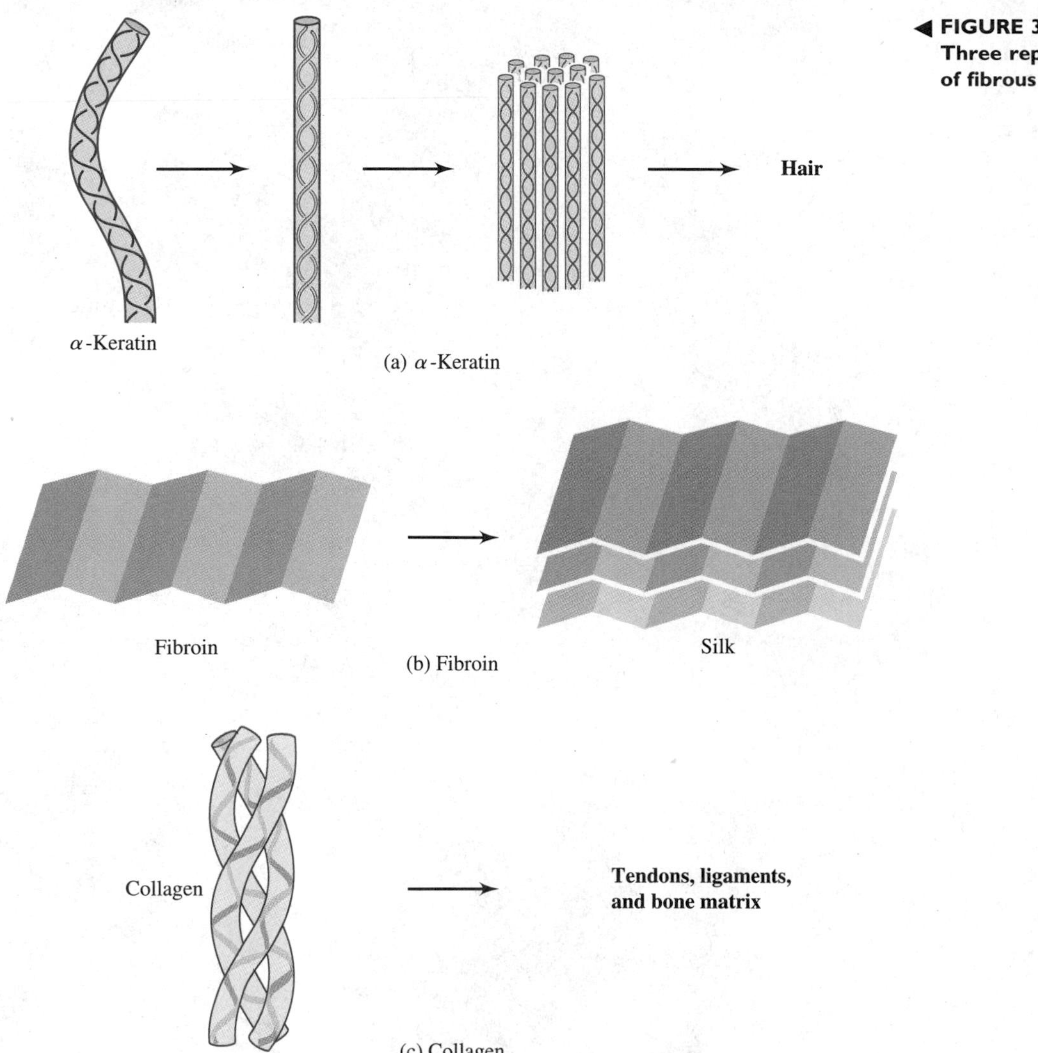

FIGURE 30.3
Three representative examples of fibrous proteins.

30.4(b). In this simplified view, myoglobin can be described as having a folded-sausage structure, where each section of the sausage represents a segment of an α-helix. The α-helical secondary structure provides stability for myoglobin. The folds of the myoglobin structure are held firmly in place by hydrogen bonds and other interactions between amino acid side chains. Within the center of this structure is an organically bound iron atom (in the heme group, shown in Figure 30.4(b) by a rectangular solid) that allows myoglobin to store and transport oxygen. Because of the heme group, myoglobin is a *conjugated* protein—that is, a protein that contains groups other than amino acid residues.

Carboxypeptidase A—an enzyme Carboxypeptidase A is a small enzyme (307 amino acid residues with a molar mass of 34,500 g/mol) that catalyzes protein digestion in the small intestine. The space-filling model shows carboxypeptidase A to be a globular protein (Figure 30.5a). The major structural features of carboxypep-

FIGURE 30.4 ▶
Representations of myoglobin.

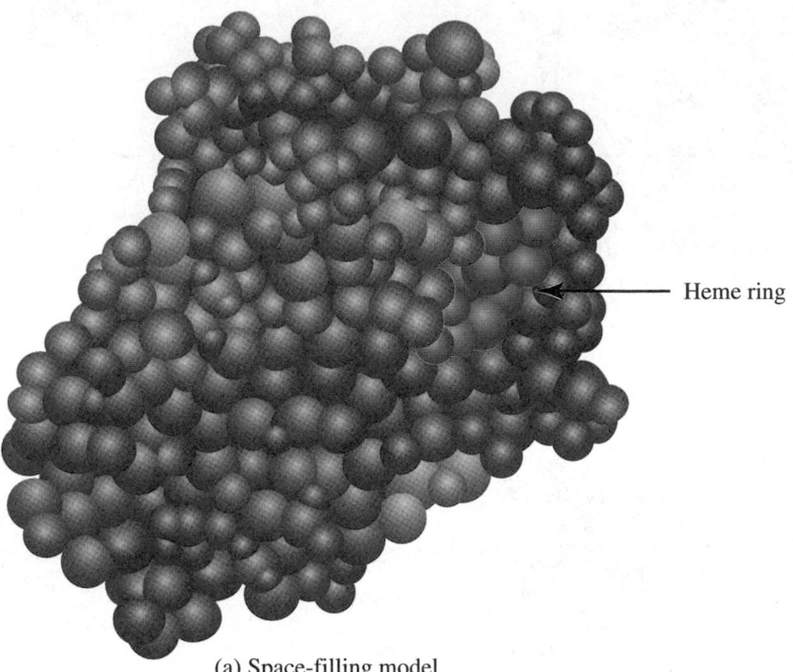

(a) Space-filling model

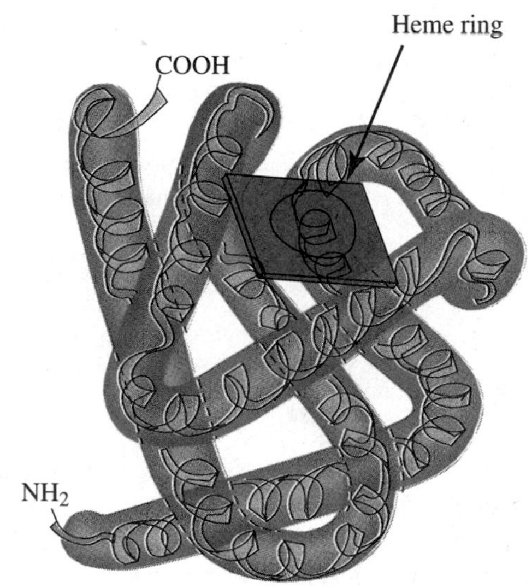

(b) Skeletal structure

tidase A are shown with a "ribbon structure" (Figure 30.5b). As we saw in myoglobin, this structure shows the position of the protein chain and leaves out the location of the amino acid side chains. An α-helix is shown with a curling line

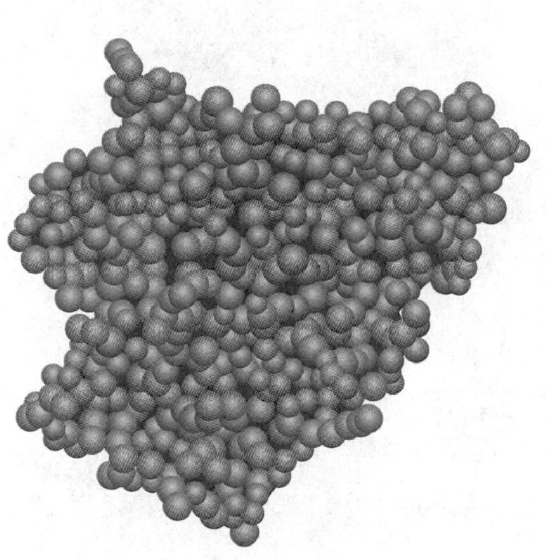

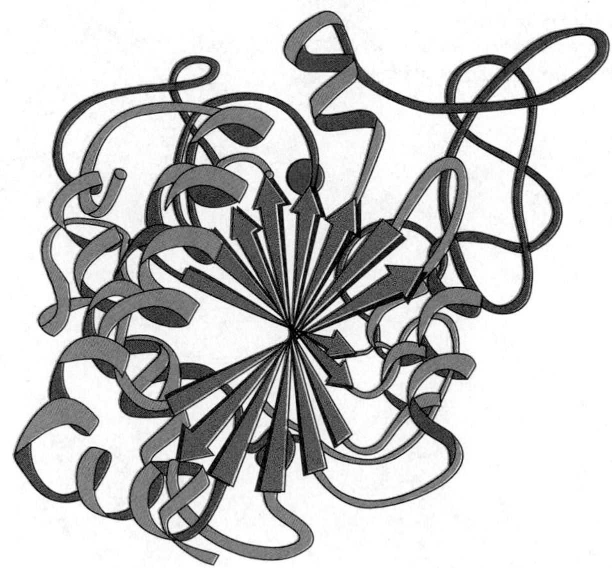

(a) Space-filling model

(b) Ribbon structure

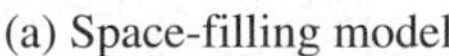

FIGURE 30.5 Representations of carboxypeptidase A.

while each strand of β-pleated sheet is a broader line with an arrowhead at its end to show which way the protein chain is running (from the amino end toward the carboxyl terminus).

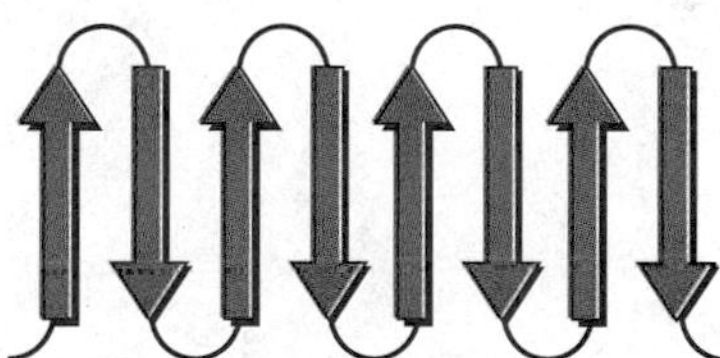

A β-pleated sheet provides a stable core for carboxypeptidase A. This sheet is often called a fan structure to describe the appearance of the twisted β-pleated sheet. A set of α-helices surround carboxypeptidase A's β-pleated sheet and further strengthen the protein structure.

Proteins share common skeletal structures just as various animal species have a common bone structure. Some proteins use the α-helix like myoglobin (see Figure 30.4); spiral tubes are twisted around to support the bulk of these proteins. In other proteins the β-pleated sheet is configured to provide a rigid protein framework. Some proteins have a fan of β-pleated sheet (like carboxypeptidase A) while for other molecules the β-pleated sheet has wrapped around on itself to form a barrel. The β-pleated sheets often act as the core of a protein with the α-helices on the outside.

Larger proteins are commonly folded into a number of compact globular units. These units are about the size of myoglobin or carboxypeptidase A and are called **protein domains**. In this way, a larger protein is like several myoglobin-sized proteins connected together by peptide bonds. Each protein domain has its own secondary structure that usually performs a discrete task in a protein's overall function.

protein domain

FIGURE 30.6 ▶ Representations of an immunoglobulin G protein.

(a) Space-filling molecule

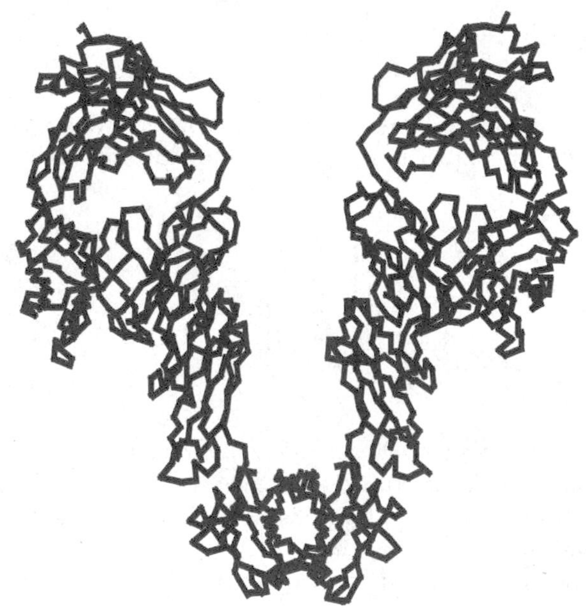

(b) Ribbon structure

Immunoglobulin G—a binding protein Immunoglobulins (antibodies) are large binding proteins (for immunoglobulin G, about 1320 amino acid residues with a molar mass of about 145,000 g/mol) that serve a critical biological function. These proteins bind molecules foreign to the body as a defense against disease. Figure 30.6 shows immunoglobulins to be complex molecules. They are composed of more than one polypeptide chain and have a quaternary structure as well as a primary, a

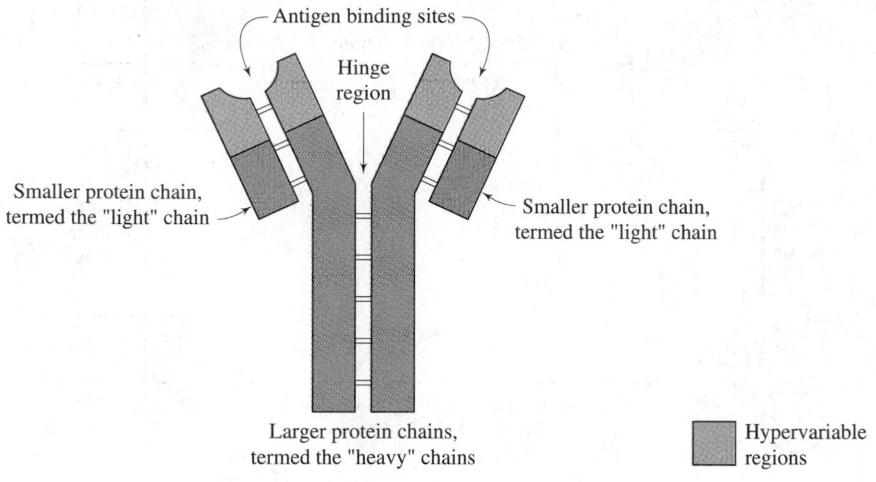

FIGURE 30.7
Diagrammatic shape of a typical immunoglobulin G protein.

secondary, and a tertiary structure. Each polypeptide chain has several protein domains, and β-pleated sheets form a stable core within each protein domain.

A schematic diagram of an immunoglobulin is often used to describe the binding function of these proteins (Figure 30.7). A successful immunological response requires that our bodies be prepared with a protein whose binding site is closely complementary to each of millions of different invading molecules (antigens). Just how our body accomplishes this feat can be described starting with the shape of a typical immunoglobulin G (Figure 30.7). Antigen binding regions are located at the ends of two arms. These are hinged together so that the distance between binding sites can vary with the size of the invading particle. The immunoglobulin protein chains are endowed with special hypervariable regions; every different immunoglobulin has a distinct amino acid sequence in its antigen binding sites and, therefore, has sites of unique size and shape. The body ensures that there will be immunoglobulins for many different antigens by producing millions of proteins with different hypervariable regions. Note that two different protein chains make up the antigen binding site. Since each chain has a hypervariable region that can be changed independently, the body has the ability to produce upwards of 10^{10} (ten billion) immunoglobulins with different binding sites, more than enough to handle exposure to invading particles.

Hemoglobin—a binding protein Quaternary structure is important in proteins that are involved in the control of metabolic processes. For example, hemoglobin (574 amino acid residues with a molar mass of 64,500 g/mol), the oxygen transport protein of the blood, is composed of four subunits (see Figure 30.8). Each subunit is similar to myoglobin in that a set of helical segments surround the oxygen-binding heme group. Of course, the important structural difference is that hemoglobin has a quaternary structure. This leads to an equally important functional difference. Whereas myoglobin always binds and releases oxygen in the same way, hemoglobin changes its oxygen-binding characteristics depending on the available O_2. When hemoglobin binds oxygen on one subunit, the protein's conformation changes to

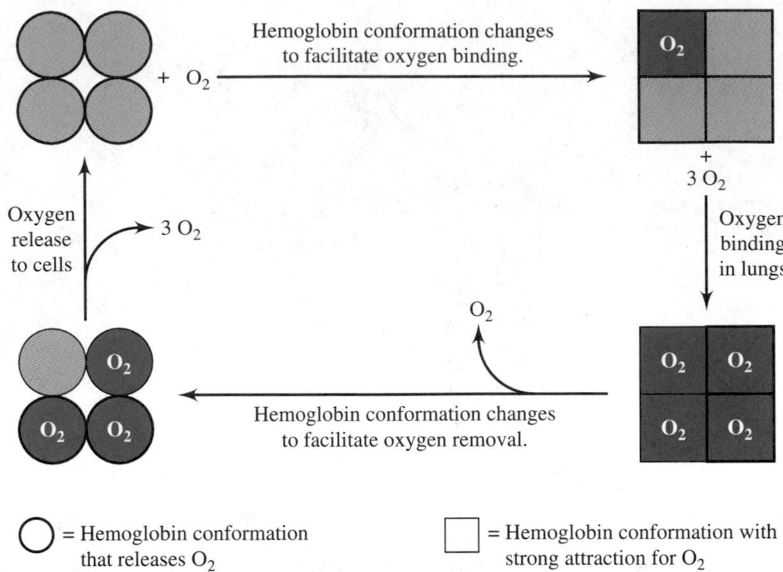

FIGURE 30.8 ▶
A schematic representation of the oxygen–hemoglobin binding process. The circles and the squares represent two different conformations of the hemoglobin molecule. Oxygen binding or release causes hemoglobin to change its conformation.

facilitate the binding of three additional oxygen molecules (see Figure 30.9). The oxygen binding is said to be *cooperative*; that is, the binding of an oxygen at one site promotes oxygen binding on the other three sites. In an oxygen-rich environment, hemoglobin becomes saturated with oxygen.

Conversely, the loss of one oxygen from hemoglobin facilitates the release of oxygen from the other sites. As hemoglobin moves to the oxygen-needy body tissues, its oxygen is cooperatively released. The presence of a quaternary structure allows the binding or removal of one oxygen molecule to control the binding or removal of three other oxygen molecules. Hence the oxygen-transport effectiveness of hemoglobin is multiplied by the quaternary structure.

Even minor alterations in primary structure may have drastic effects on the three-dimensional structural function of a protein. A graphic example of this fact is provided by sickle-cell anemia. This crippling genetic disease is due to red blood cells assuming a sickle shape. Sickled cells have impaired vitality and function. Sickle-cell anemia has been traced to a small change in the primary structure, or amino acid residue sequence, of hemoglobin. Normal hemoglobin molecules contain four polypeptide subunits—two identical α-polypeptide chains and two identical β-polypeptide chains. Sickle-cell and normal β-polypeptide chains of hemoglobin each contain 146 amino acid residues and differ only in the residue at the sixth position. In the sickle-cell hemoglobin chain, a glutamic acid residue has been replaced by a valine residue at the sixth position. This change is sufficient, under some circumstances, to cause hemoglobin to aggregate into long filaments—a different quaternary structure. Large amounts of hemoglobin are present in red blood cells, and the changed quaternary structure causes sickling of the cell and greatly diminishes cell vitality (see Figure 30.10). This cell affliction has led to the premature deaths of many affected individuals.

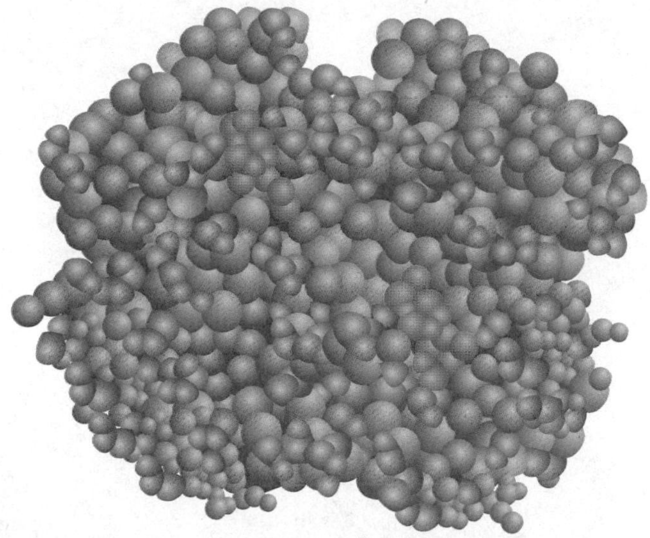

(a) Space-filling model

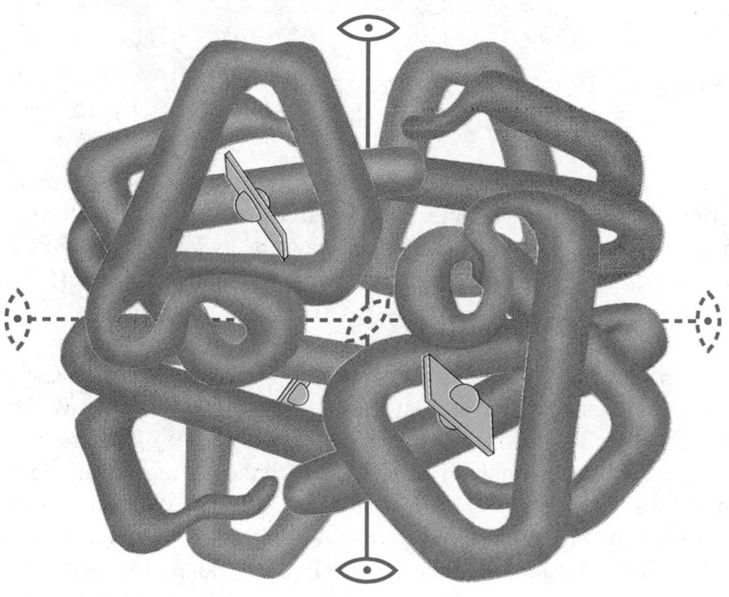

(b) Skeletal structure

▲
FIGURE 30.9
Representations of hemoglobin.

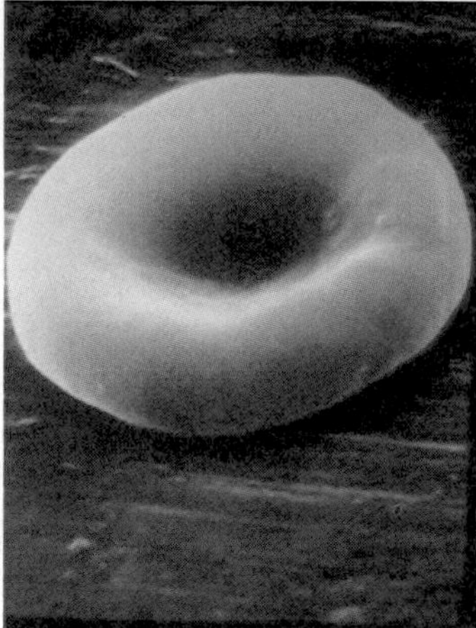

▲
FIGURE 30.10
Scanning electron micrograph of sickled red blood cell (top) and normal red blood cell (bottom).

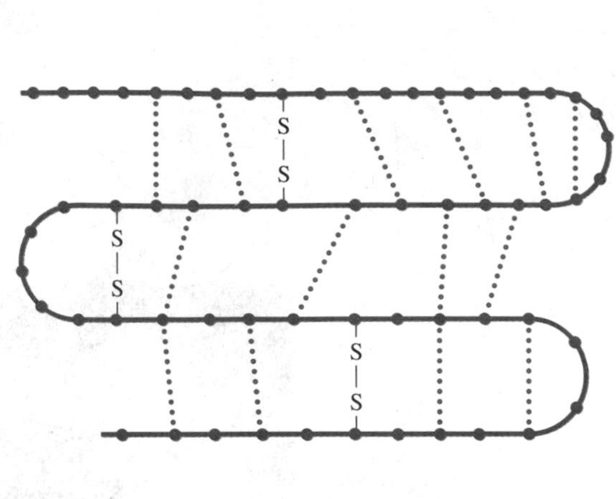

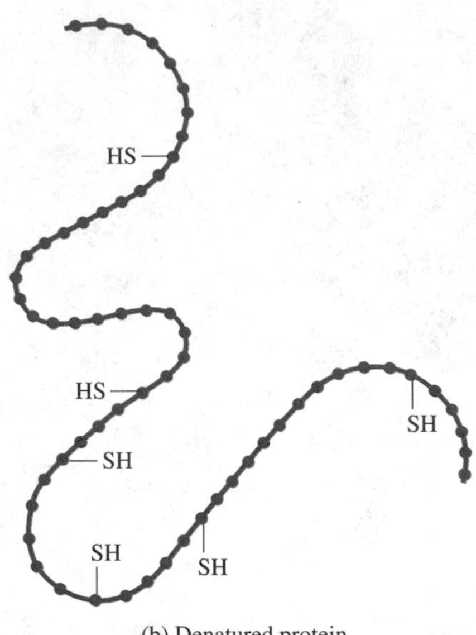

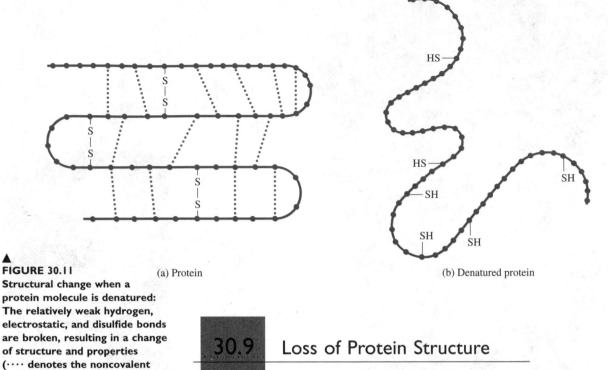

(a) Protein

(b) Denatured protein

FIGURE 30.11
Structural change when a protein molecule is denatured: The relatively weak hydrogen, electrostatic, and disulfide bonds are broken, resulting in a change of structure and properties (···· denotes the noncovalent bonds that stabilize a protein's natural conformation).

30.9 Loss of Protein Structure

Because protein structure is so important to life's functions, the loss of protein structure can be crucial. If a protein loses only its native three-dimensional conformation, the process is referred to as **denaturation**. In contrast, hydrolysis of peptide bonds ultimately converts proteins into their constitutent amino acids. Denaturation often precedes hydrolysis.

denaturation

Denaturation involves the alteration or disruption of the secondary, tertiary, or quaternary—but not primary—structure of proteins (see Figure 30.11). Because a protein's function depends on its natural conformation, biological activity is lost with denaturation. This process may involve changes ranging from the subtle and reversible alterations caused by a slight pH shift to the extreme alterations involved in tanning a skin to form leather.

Environmental changes may easily disrupt natural protein structure, which is held together predominantly by noncovalent, relatively weak bonds. As gentle an act as pouring a protein solution can cause denaturation. Purified proteins must often be stored under ice-cold conditions because room temperature denatures them. It is not surprising that a wide variety of chemical and physical agents can also denature proteins. To name a few: strong acids and strong bases, salts (especially those of heavy metals), certain specific reagents such as tannic acid and picric acid, alcohol and other organic solvents, detergents, mechanical action such as whipping, high temperature, and ultraviolet radiation. Denatured proteins are generally less soluble than native proteins and often coagulate or precipitate from solution. Cooks have taken advantage of this for many years. When egg white, which is a concentrated solution of egg albumin protein, is stirred vigorously (as with an egg beater), an

Chemistry in Action

Age and Memory: A Protein Connection

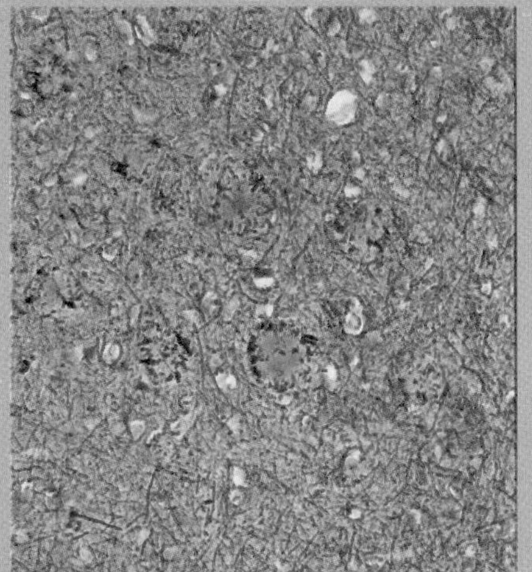

Alzheimer brain showing senile plaque.

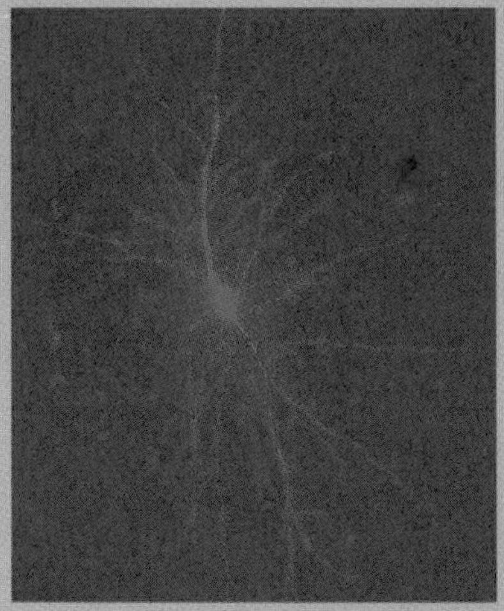

Normal neuron from the cortex of a human brain.

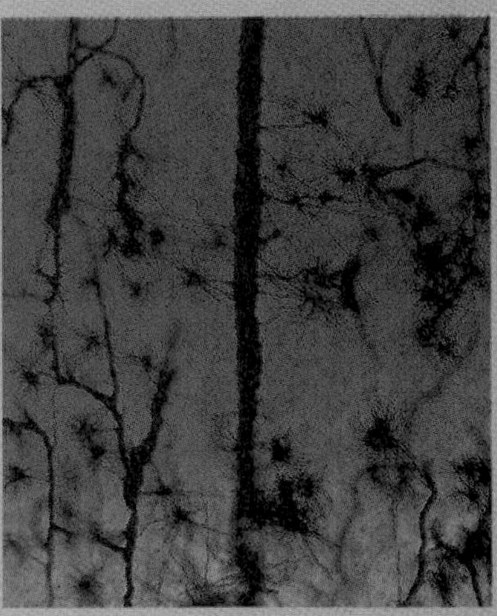

Neuron with "tangles."

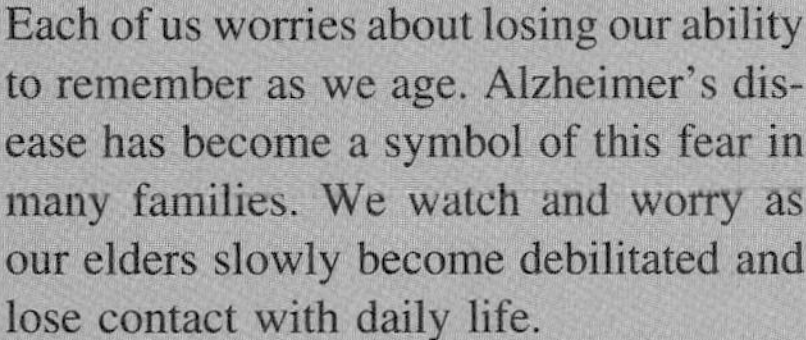

Each of us worries about losing our ability to remember as we age. Alzheimer's disease has become a symbol of this fear in many families. We watch and worry as our elders slowly become debilitated and lose contact with daily life.

On a biochemical level, Alzheimer's disease is caused by protein malfunctions. This disease affects many older people (17–20 million worldwide), causing increasing memory loss and dementia. Alzheimer's disease has complex biochemical roots, but it is now recognized that this disease results from the death of brain cells (neurons) and is associated with a buildup of "plaques" and "tangles." These changes reflect a serious problem with specific neuronal proteins.

"Senile plaques" are found throughout the brains of patients who suffer from Alzheimer's disease. These plaques are composed of dead and dying brain tissue and contain a large amount of a neurotoxic protein, β-amyloid. Understanding why this buildup occurs is central to understanding Alzheimer's disease.

Alzheimer's disease can be caused by an abnormal β-amyloid protein. In some types of Alzheimer's disease, genetic mutations of the β-amyloid protein are found. A minimal change of one amino acid is enough to cause a toxic buildup of β-amyloid protein. In other types of Alzheimer's disease, the β-amyloid protein may be incorrectly formed from its larger precursor protein. In turn, this can lead to the wrong three-dimensional β-amyloid shape and a buildup of this neurotoxin.

"Neurofibrillary tangles" are made up of a denatured protein called *tau.* The tau protein in Alzheimer patients has too many attached phosphate groups. The protein gains too large a negative charge, which causes it to denature and precipitate. The resulting "tangles" block normal transport of nutrients through the neurons.

It appears that changes associated with two neuronal proteins, β-amyloid and tau, are enough to disrupt the precise balance required in nerve metabolism. Cell death follows and the Alzheimer patient slowly loses mental capacity. As we discover more details of this disease, we will hopefully reach a point where it will become treatable. Treatment may be only a matter of correcting neuronal protein synthesis.

unsweetened meringue forms; the albumin denatures and coagulates. A cooked egg solidifies partially because egg proteins including albumin are denatured by heat.

In clinical laboratories the analysis of blood serum for small molecules such as glucose and uric acid is hampered by the presence of serum protein. This problem is resolved by first treating the serum with an acid to denature and precipitate the protein. The precipitate is removed, and the protein-free liquid is then analyzed.

▲
Left: **Raw egg white is concentrated protein.** ***Middle:*** **In a cooked egg, the proteins are denatured by heat.** ***Right:*** **If egg whites are beaten, the protein denatures and coagulates.**

Loss of protein structure also occurs with hydrolysis of the peptide bonds to produce free amino acids. This chemical reaction destroys the protein's primary structure. Proteins can be hydrolyzed by boiling in a solution containing a strong acid such as HCl or a strong base such as NaOH. At ordinary temperatures, proteins can be hydrolyzed using enzymes (see Chapter 31). These molecules, called proteolytic enzymes, are themselves proteins that function to catalyze or speed up the hydrolysis reaction. The essential reaction of hydrolysis is the breaking of a peptide linkage and the addition of the elements of water:

$$\sim\!N(H)\!-\!CH(R)\!-\!C(=O)\!-\!N(H)\!-\!CH(R)\!-\!C(=O)\!\sim \; + \; H_2O \longrightarrow \; \sim\!N(H)\!-\!CH(R)\!-\!C(=O)\!-\!OH \; + \; H_2N\!-\!CH(R)\!-\!C(=O)\!\sim$$

Peptide linkage in protein chain

(∿∿ = the rest of the protein chain)

Any molecule with one or more peptide bonds, from the smallest dipeptide to the largest protein molecule, can be hydrolyzed. During hydrolysis, proteins are broken down into smaller and smaller fragments until all component amino acids are liberated.

Dietary protein must have its structure completely destroyed before it can provide nutrition for the body. Thus digestion involves both the process of denaturation and hydrolysis. Stomach acid causes most proteins to denature. Then proteolytic enzymes in the stomach and the small intestine hydrolyze the proteins to smaller and smaller fragments until the free amino acids are formed and can be absorbed through the intestinal membranes into the bloodstream.

30.10 Tests for Proteins and Amino Acids

Many tests have been devised for detecting and distinguishing among amino acids, peptides, and proteins. Some examples are described here.

Xanthoproteic Reaction Proteins containing a benzene ring—for example, the amino acids phenylalanine, tryptophan, and tyrosine—react with concentrated nitric acid to give yellow reaction products. Nitric acid on skin produces a positive xanthoproteic test, as skin proteins are modified by this reaction.

Biuret Test A violet color is produced when dilute copper(II) sulfate is added to an alkaline solution of a peptide or protein. At least two peptide bonds must be present, as the color change occurs only when peptide bonds can surround the Cu^{2+} ion. Thus amino acids and dipeptides do not give a positive biuret test.

Ninhydrin Test Triketohydrindene hydrate, generally known as *ninhydrin*, is an extremely sensitive reagent for amino acids:

O
OH
OH
O

ninhydrin

All amino acids, except proline and hydroxyproline, give a blue solution with ninhydrin. Proline and hydroxyproline produce a yellow solution. Less than 1 μg (10^{-6}g) of an amino acid can be detected with ninhydrin.

Chromatographic Separation

Complex mixtures of amino acids are readily separated by thin layer, paper, or column chromatography. In chromatographic methods the components of a mixture are separated by means of differences in their distributions between two phases. Separation depends on the relative tendencies of the components to remain in one phase or the other. In *thin-layer chromatography* (TLC), for example, a liquid and a solid phase are used. The procedure is as follows: A tiny drop of a solution containing a mixture of amino acids (obtained by hydrolyzing a protein) is spotted on a strip (or sheet) coated with a thin layer of dried alumina or some other adsorbant. After the spot has dried, the bottom edge of the strip is put into a suitable solvent. The solvent ascends the strip (by diffusion), carrying the different amino acids upward at different rates. When the solvent front nears the top, the strip is removed from the solvent and dried. The locations of the different amino acids are established by spraying the strip with ninhydrin solution and noting where colored spots appear. The pattern of colored spots is called a chromatogram. The identities of the amino acids in an unknown mixture can be established by comparing the chromatogram of the mixture with a chromatogram produced by known amino acids. A typical chromatogram of amino acids is shown in Figure 30.12.

Since proteins tend to denature during thin-layer chromatography, they are often separated by column chromatography. In this technique a solution containing a mixture of proteins (liquid phase) is allowed to percolate through a column packed with beads of a suitable polymer (solid phase). Separation of the mixture is accomplished as the proteins partition between the solid and liquid phases. The separated proteins move at different rates and are collected as they leave the column. Unlike amino acid chromatography, mild conditions must be maintained to limit protein denaturation. In the past it was not uncommon for a protein column chromatography

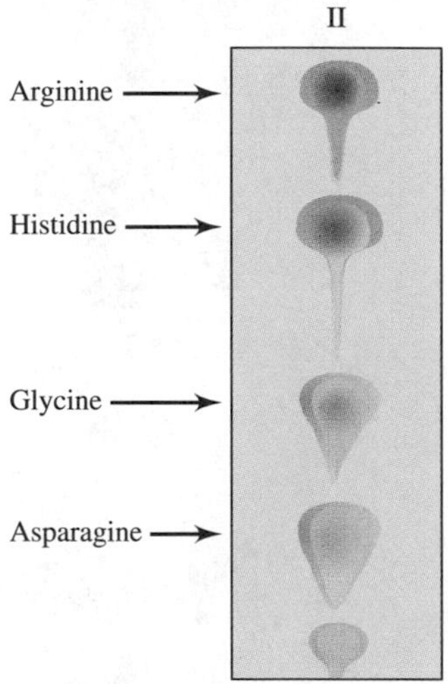

FIGURE 30.12 Chromatogram showing separation of selected amino acids. On slide I, spotted on the chromatographic strip, is an amino acid mixture containing arginine, histidine, glycine, and asparagine. On slide II, is the developed chromatogram showing the separated amino acids after treatment with ninhydrin. The solvent is a 250:60:250 volume ratio of 1-butanol–acetic acid–water.

to take from 6 to 12 hours. Recent developments in the field of high-performance liquid chromatography (HPLC) have shortened this time to minutes.

Electrophoresis

Perhaps the most precise separation technique available to the protein chemist is electrophoresis. This technique separates proteins based on both a difference in size and a difference in charge. Electrophoresis is based on the principle that charged particles move in an electric field. Since different proteins contain differing numbers of positively and negatively charged amino acids, proteins will separate in an electric field (see Figure 30.13). Sometimes proteins with an overall positive charge are separated from proteins with an overall negative charge (Figure 30.13a). But, most often, conditions are chosen so that all proteins have either a net positive or a net negative charge. Proteins with a greater charge are attracted more strongly to an electrode and move faster during electrophoresis (see Figure 30.13b). The faster moving proteins separate from the slower moving proteins.

Protein electrophoresis commonly is carried out in a gel matrix. This gel may be a natural polymer like starch or agarose or, it may be a purely synthetic polymer like polyacrylamide. No matter which gel is used, the proteins are forced to move through a sieve-like material. They experience friction; the larger the protein, the more friction and the slower the protein will tend to move. Thus the speed at which a protein moves depends on both charge *and* size during standard electrophoresis. Tens of proteins may be separated precisely using one gel electrophoresis.

This technique has proved not only valuable, but also versatile. Several modifications of the standard procedure are used routinely.

1. *SDS (sodium dodecyl sulfate) electrophoresis:* In this procedure, a negatively charged detergent (SDS) is added to the protein solution and acts to mask differences in charges between proteins. The electrophoretic separation depends primarily on size differences. SDS electrophoresis is often used to determine the approximate molar mass of proteins.
2. *Isoelectric focusing*: During this type of electrophoresis, proteins are forced to move through a pH gradient. The proteins stop moving when they reach their *isoelectric point* pH because at this pH the protein carries no net charge (see Section 30.5). The electrophoresis is completed once all the proteins have been "focused" at their isoelectric points. This technique separates proteins based on the number and kind of titratable amino acid side chains that they contain.

Today, two dimensional electrophoresis is not uncommon. This procedure multiplies the benefits of an electrophoretic separation by running one separation in the y-direction followed by a second electrophoresis in the x-direction using the same gel slab. Figure 30.14 shows a separation of all proteins from a bacterium (*E. coli*). The proteins were first separated by molar mass using SDS electrophoresis in the y-direction. Each band was then separated in the x-direction using isoelectric focusing, a technique that separates proteins based on charge. Over 1000 proteins have been separated on this one gel!

At beginning of electrophoresis

At a later time

Positive (+) electrode

+ Protein

− Protein

+ Protein

− Protein

Negative (−) electrode

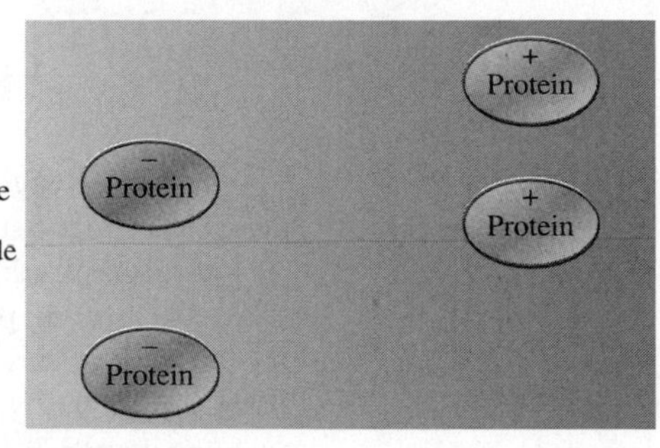

(a) Separating negatively charged proteins from positively charged proteins

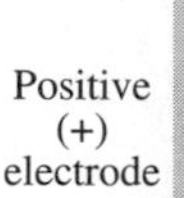

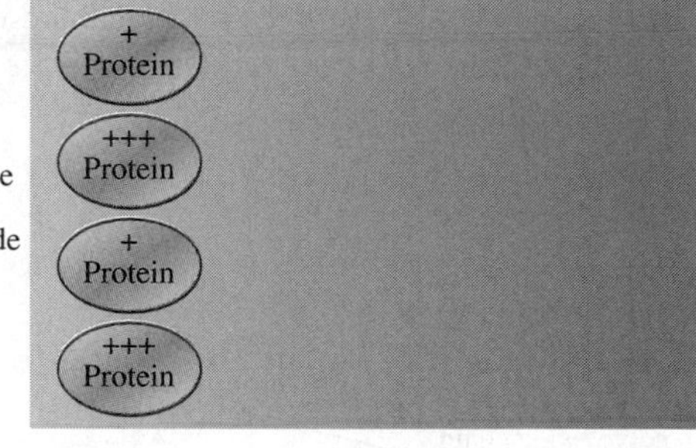

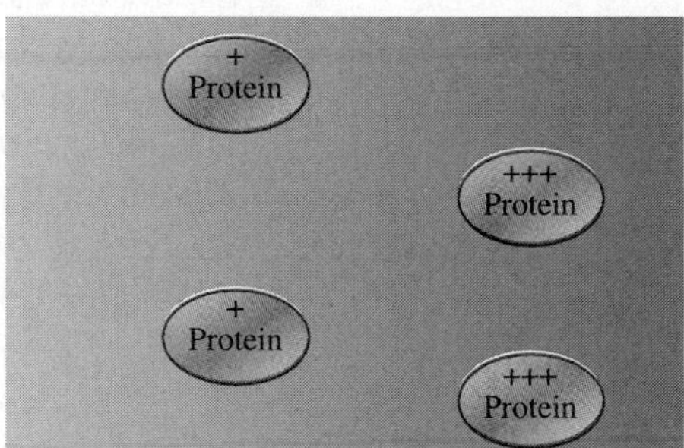

(b) Separating the more positively charged proteins from the less positively charged ones.

▲ **FIGURE 30.13**
A schematic representation of two forms of electrophoresis.

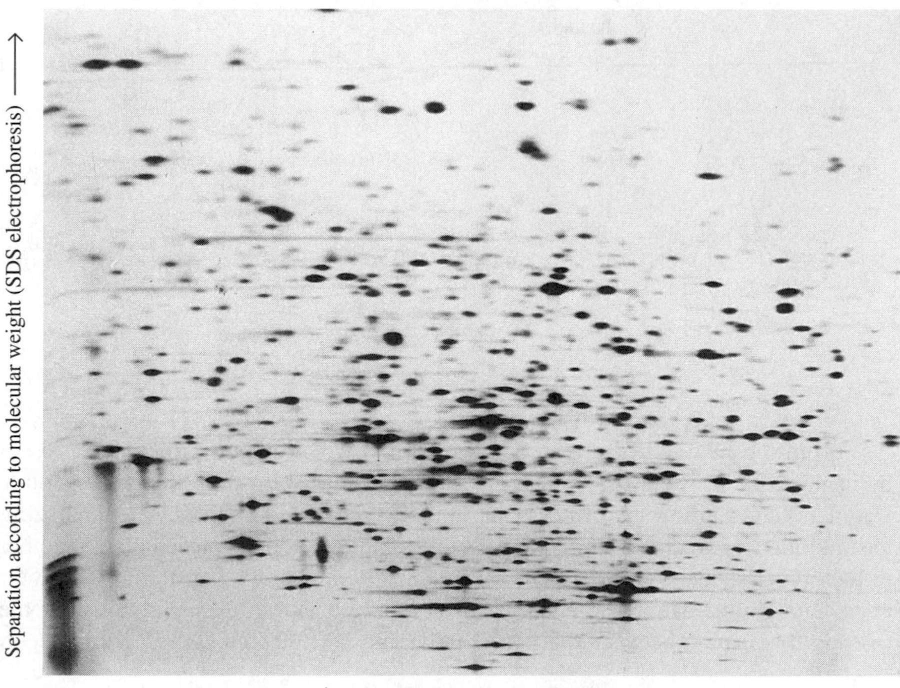

◀ **FIGURE 30.14**
Two-dimensional electrophoresis of proteins from the *E. coli* bacterium. Separation by molar mass has been used in the *y*-direction and separation by charge has been used in the *x*-direction.

30.11 Determination of the Primary Structure of Polypeptides

Sanger's Reagent The pioneering work of Frederick Sanger gave us the first complete primary structure of a protein, insulin, in 1955. He used specific enzymes to hydrolyze insulin into smaller peptides and amino acids, then separated and identified the hydrolytic products by various chemical reactions.

Sanger's reagent, 2-4-dinitrofluorobenzene (DNFB), reacts with the α-amino group of the N-terminal amino acid of a polypeptide chain. The carbon–nitrogen bond between the amino acid and the benzene ring of Sanger's reagent is more resistant to hydrolysis than are the remaining peptide linkages. Thus, when the substituted polypeptide is hydrolyzed, the terminal amino acid remains with the dinitrobenzene group and can be isolated and identified. The remaining peptide chain is hydrolyzed to free amino acids in the process. This method marked an important step in the determination of the amino acid sequence in a protein. The reaction of Sanger's reagent is illustrated in the following equations:

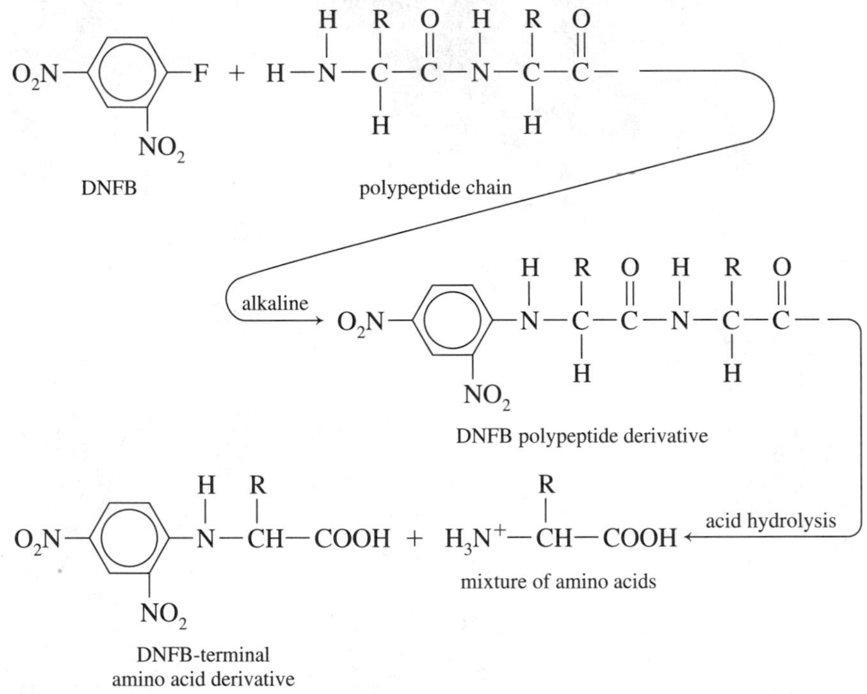

$$O_2N\text{—}C_6H_3(NO_2)\text{—}F + H\text{—}\underset{|}{\overset{H}{N}}\text{—}\underset{H}{\overset{R}{C}}\text{—}\overset{O}{\overset{\|}{C}}\text{—}\overset{H}{N}\text{—}\underset{H}{\overset{R}{C}}\text{—}\overset{O}{\overset{\|}{C}}\text{—}$$

DNFB — polypeptide chain

$$\xrightarrow{\text{alkaline}} O_2N\text{—}C_6H_3(NO_2)\text{—}\overset{H}{N}\text{—}\underset{H}{\overset{R}{C}}\text{—}\overset{O}{\overset{\|}{C}}\text{—}\overset{H}{N}\text{—}\underset{H}{\overset{R}{C}}\text{—}\overset{O}{\overset{\|}{C}}\text{—}$$

DNFB polypeptide derivative

$$\xrightarrow{\text{acid hydrolysis}} O_2N\text{—}C_6H_3(NO_2)\text{—}\overset{H}{N}\text{—}\overset{R}{CH}\text{—}COOH + H_3N^{+}\text{—}\overset{R}{CH}\text{—}COOH$$

DNFB-terminal amino acid derivative — mixture of amino acids

From the DNFB hydrolysis, Sanger learned which amino acids were present in insulin. By less drastic hydrolysis, he split the insulin molecule into peptide fragments consisting of two, three, four, or more amino acid residues. After analyzing vast numbers of fragments utilizing the N-terminal method, he pieced them together in the proper sequence by combining fragments with overlapping structures at their ends, finally elucidating the entire insulin structure. As an example, consider the overlap that occurs between the hexapeptide and heptapeptide shown here:

Gly-Glu-Arg-Gly-Phe-Phe		Hexapeptide
	Gly-Phe-Phe-Tyr-Thr-Pro-Lys	Heptapeptide
Gly-Glu-Arg-Gly-Phe-Phe-Tyr-Thr-Pro-Lys		Decapeptide

The three residues Gly-Phe-Phe at the end of the hexapeptide match the three residues at the beginning of the heptatpeptide. By using these three residues in common, the structure of the decapeptide shown, which occurs in chain B of insulin (residues 20–29), is determined. (See Figure 30.2.)

Edman Degradation More recently, the Edman degradation method has been developed to split off amino acids one at a time from the N-terminal end of a polypeptide chain. In this procedure the reagent, phenylisothiocyanate, is first added to the N-terminal amino group. The N-terminal amino acid–phenylisothiocyanate addition product is then cleaved from the polypeptide chain. The resultant substituted phenylthiohydantoin is isolated and identified. The shortened polypeptide chain is then ready to undergo another Edman degradation. By repeating this set of reactions, one can determine the amino acid sequence of a polypeptide directly. A machine, the protein sequenator, which carries out protein sequencing by Edman degradation automatically, is now available.

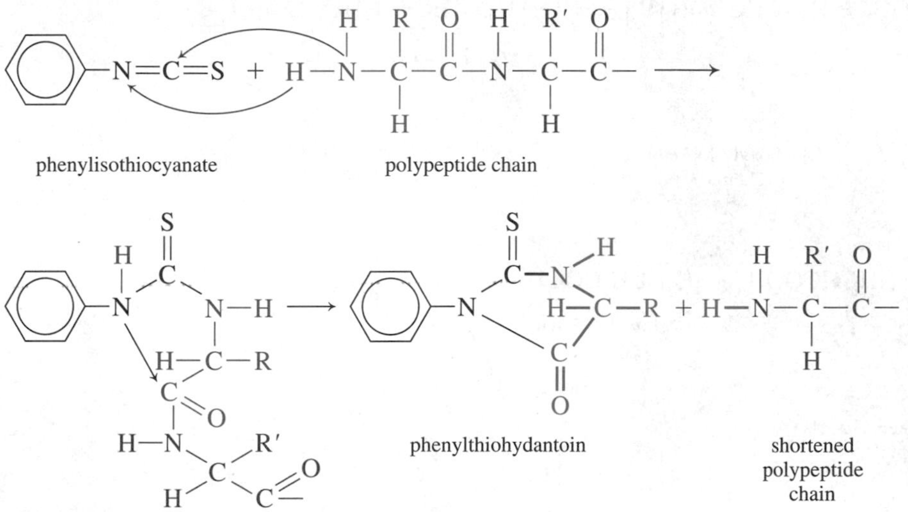

$$C_6H_5{-}N{=}C{=}S + H{-}N(H){-}CH(R){-}C(=O){-}N(H){-}CH(R'){-}C(=O){-} \longrightarrow$$

phenylisothiocyanate polypeptide chain

$$\longrightarrow \text{phenylthiohydantoin} + H{-}N(H){-}CH(R'){-}C(=O){-}$$

phenylthiohydantoin

shortened polypeptide chain

30.12 Synthesis of Peptides and Proteins

Since each amino acid has two functional groups, it is not difficult to form dipeptides or even fairly large polypeptide molecules. As mentioned in Section 30.6, glycine and alanine react to form two different dipeptides, glycylalanine and alanylglycine. If threonine, alanine, and glycine react, six tripeptides, each made up of three different amino acid units are obtained. By reacting a mixture of several amino acids, polypeptides are produced. This clarifies the fact that even a small protein like insulin cannot be produced by simply reacting a mixture of the required amino acids. When such a large polypeptide is synthesized, it is formed by joining amino acids one by one in the proper sequence. Within cells, the proper sequence is maintained by careful genetic control (see Section 32.10).

When a polypeptide chain of known primary structure is synthesized *in vitro* (synthesis in "glass" without the aid of living tissue), the process must be controlled so that only one particular amino acid is added at each stage. Remarkable progress has been made in developing the necessary techniques. Amino acids are used in which either the amino or carboxylic acid group has been inactivated or blocked with a suitable *blocking agent.* The blocked amino or carboxylic acid group is reactivated by removing the blocking agent in the next stage of the synthesis. For example, if the tripeptide Thr-Ala-Gly is to be made, amino-blocked threonine is reacted with carboxyl-blocked alanine to make the doubly blocked Thr-Ala dipeptide. The carboxyl-blocked alanine end of the dipeptide can then be reactivated and reacted with carboxyl-blocked glycine to make the blocked Thr-Ala-Gly tripeptide. The free tripeptide can then be obtained by removing the blocking agents from both ends. Using $\boxed{B}$ to designate the blocking agents, this synthesis is represented in schematic form as follows:

$$\underset{\text{amino-blocked threonine}}{\boxed{B}\,H_2N\underset{\substack{|\\ H-C-OH\\ |\\ CH_3}}{C}HCOOH} + \underset{\text{carboxyl-blocked alanine}}{H_2N\underset{\substack{|\\ CH_3}}{C}HCOOH\,\boxed{B}} \longrightarrow \underset{\text{double-blocked Thr-Ala dipeptide}}{\boxed{B}\,H_2N\underset{\substack{|\\ H-C-OH\\ |\\ CH_3}}{C}H\overset{\substack{O\\ \|}}{C}-NH\underset{\substack{|\\ CH_3}}{C}HCOOH\,\boxed{B}}$$

$$\underset{\text{amino-blocked Thr-Ala dipeptide}}{\boxed{B}\,H_2N\underset{\substack{|\\ H-C-OH\\ |\\ CH_3}}{C}H\overset{\substack{O\\ \|}}{C}-NH\underset{\substack{|\\ CH_3}}{C}HCOOH} + \underset{\text{carboxyl-blocked glycine}}{H_2NCH_2COOH\,\boxed{B}} \longrightarrow$$

$$\underset{\text{double-blocked Thr-Ala-Gly tripeptide}}{\boxed{B}\,H_2N\underset{\substack{|\\ H-C-OH\\ |\\ CH_3}}{C}H\overset{\substack{O\\ \|}}{C}-NH\underset{\substack{|\\ CH_3}}{C}H\overset{\substack{O\\ \|}}{C}-NHCH_2COOH\,\boxed{B}} \longrightarrow \text{Thr-Ala-Gly tripeptide}$$

Longer polypeptide chains of specified amino acid sequence can be made by this general technique. The procedure is very tedious. The synthesis of insulin required an effort equivalent to one person working steadily for several years.

Fortunately, it is not necessary to prepare an entire synthetic protein polypeptide chain by starting at one end and adding amino acids one at a time. Previously prepared shorter polypeptide chains of known structure can be joined to form a long polypeptide chain. This method was used in making synthetic insulin. The insulin molecule, shown in Figure 30.2, consists of two polypeptide chains bonded together by two disulfide bonds. A third disulfide bond, between two amino acids in one chain, makes a small loop in that chain. Once the two chains had been assembled from fragments, a seemingly formidable problem remained—how to form the disulfide bonds in the correct positions. As it turned out, it was necessary only to bring the two chains together, with the cysteine side chains in the reduced condition, and to treat them with an oxidizing agent. Disulfide bonds formed at the right places, and biologically active insulin molecules were obtained.

Production of a cloned protein. The technician is holding a container with cultured cells in a nutrient medium.

In the mid-1960s, a machine capable of automatically synthesizing polypeptide chains of known amino acid sequence was designed by R. Bruce Merrifield of Rockefeller University. The starting amino acid is bonded to a plastic surface (polystyrene bead) in the reaction chamber of the apparatus. Various reagents needed for building a chain of predetermined structures are automatically delivered to the reaction chamber in a programmed sequence. Twelve reagents and about 100 operations are needed to lengthen the chain by a single amino acid residue. But the machine is capable of adding residues to the chain at the rate of six a day. Such a machine makes it possible to synthesize complex molecules like insulin in a few days. The first large polypeptide (pancreatic ribonuclease), containing 124 amino acid residues, was synthesized by this method in 1969. More recently, a human growth hormone (HGH) containing 188 amino acid residues was synthesized using this technique.

Over the past 15 years, techniques have been developed that use biological systems to create and produce new proteins. These procedures start with genetic material that codes for the protein of interest. Then this material is properly modified and introduced into rapidly growing cells that are treated to overproduce this foreign protein. The result can be a harvest of much needed protein, such as the human insulin used by diabetics or the human growth hormone used to treat dwarfism. These procedures constitute a part of genetic engineering and will be discussed in more detail in Chapter 32. The advantages of these new techniques can be realized by comparing the rate of biological protein synthesis with that achieved by the machine designed by Merrifield. If a bacterium such as *E. coli* is used to produce a human protein such as growth hormone, one amino acid residue can be added every 0.01 second; a growth hormone molecule can be produced every 10–20 seconds. In contrast, the same growth hormone would take days to produce in the chemistry laboratory.

Concepts in Review

1. List five foods that are major sources of proteins.
2. List the elements that are usually contained in proteins.
3. Write formulas for and know the abbreviations for the common amino acids.
4. Write the zwitterion formula of an amino acid and know how it behaves in a dilute acidic and a dilute basic solution.

5. Understand why a protein in solution at its isoelectric pH does not migrate in an electrolytic cell.
6. Combine amino acids into polypeptide chains.
7. Understand how peptide chains are named and numbered.
8. Understand what is meant by the N-terminal and C-terminal residues of a peptide chain.
9. Briefly explain the primary, secondary, and tertiary structure of a protein.
10. Explain the meaning of the statement, "The primary structure helps establish a protein's three-dimensional structure."
11. Define quaternary structure and explain its importance in the function of hemoglobin.
12. Understand the bonding in the α-helical and β-pleated-sheet structure of a protein.
13. Give an example of a structural protein.
14. Explain the role cysteine plays in the structure of polypeptides and proteins.
15. Describe how the α-helix and β-pleated sheet are important in the structure of proteins.
16. List the most important protein functions.
17. Relate the structure of collagen to its biological function.
18. Discuss the role of the immunoglobulins in the body's defense mechanisms and identify important structural elements of these proteins.
19. Describe Sanger's work on determining the primary structure of proteins.
20. Reconstruct a peptide chain from a knowledge of its hydrolysis products.
21. Explain how Edman degradation allows the primary structure to be determined.
22. Describe chromatographic methods of separating amino acids.
23. Describe how electrophoresis is used to separate protein.
24. State the physical evidence observed in a positive reaction in the (a) xanthoproteic reaction, (b) biuret test, and (c) ninhydrin test.
25. Know what is meant by denaturation and the various methods by which proteins can be denatured.

Key Terms

The terms listed here have been defined within this chapter. Section numbers are referenced in parenthesis for each term.

conjugated protein (30.1)
denaturation (30.9)
essential amino acid (30.3)
fibrous protein (30.8)
globular protein (30.8)
isoelectric point (30.5)
peptide linkage (30.6)
polypeptide (30.6)
primary structure (30.7)
protein domain (30.8)
quaternary structure (30.7)
secondary structure (30.7)
simple protein (30.1)
tertiary structure (30.7)
zwitterion (30.5)

Questions

Questions refer to tables, figures, and key words and concepts defined within the chapter. A particularly challenging question or exercise is indicated with an asterisk.

1. Why are the common amino acids in proteins called α-amino acids?

2. What elements are present in amino acids and proteins?

3. Why are proteins from some foods of greater nutritional value than others?

4. Write the names of the amino acids that are essential to humans.

5. Why are amino acids amphoteric? Why are they optically active?

6. What can you say about the number of positive and negative charges on a protein molecule at its isoelectric point?

7. Explain what is meant by (a) the primary structure, (b) the secondary structure, (c) the tertiary structure, and (d) the quaternary structure of a protein.

8. What special role does the sulfur-containing amino acid cysteine have in protein structure?

9. How do myoglobin and hemoglobin differ in structure and function?

10. Differentiate between an α-helix and a β-pleated sheet structure of a protein.

11. Explain how hydrolysis of a protein differs from denaturation.

12. Which amino acids give a positive xanthoproteic test?

13. What is the visible evidence observed in a positive reaction for the following tests?
(a) xanthoproteic reaction
(b) biuret test
(c) ninhydrin test

14. Briefly describe separation of proteins by column chromatography.

15. **(a)** What is thin-layer chromatography?
(b) Describe how amino acids are separated using this technique.
(c) What reagent is commonly used to locate the amino acids in the chromatogram?

16. Which of these statements are correct? Rewrite each incorrect statement to make it correct.

(a) Proteins, like fats and carbohydrates, are primarily for supplying heat and energy to the body.

(b) Proteins differ from fats and carbohydrates in that they contain a large amount of nitrogen.

(c) A complete protein is one that contains all the essential amino acids.

(d) Except for glycine, amino acids that are found in proteins have the L-configuration.

(e) All amino acids have an asymmetric carbon atom and are therefore optically active.

(f) The amide linkages by which amino acids are joined together are called peptide bonds.

(g) Two different dipeptides can be formed from the amino acids glycine and phenylalanine.

(h) The compound Ala-Phe-Tyr has two peptide bonds and is therefore known as a dipeptide.

(i) A zwitterion is a dipolar ion form of an amino acid.

(j) The primary structure of a protein is the α-helical or β-pleated sheet form that it takes.

(k) The tertiary structure of a protein determines that protein's primary structure.

(l) The amino acid residues in a peptide chain are numbered beginning with the C-terminal amino acid.

(m) Insulin contains two polypeptide chains, one with 21 amino acids and the other with 30 amino acids.

(n) Collagen is an example of a binding protein.

(o) α-Helices cluster into a "twisted-sheet" conformation to form part of many protein skeletons.

(p) A protein structure with hinged arms provides binding flexibility for the immunoglobulins.

(q) When a protein is denatured, the polypeptide bonds are broken, liberating the amino acids.

(r) Irreversible coagulation or precipitation of proteins is called denaturation.

(s) Sanger's reagent reacts with polypeptide chains, isolating the N-terminal amino acid.

(t) Alzheimer's disease is caused by protein malfunctions.

(u) In Alzheimer's patients the tau protein may become denatured due to too many attached phosphate groups.

(v) Alzheimer's disease may result from a lack of the β-amyloid protein.

Paired Exercises

These exercises are paired. Each odd-numbered exercise is followed by a similar even-numbered exercise. Answers to the even-numbered exercises are given in Appendix V.

17. Write the structural formulas for the D and L forms of alanine

$$\underset{\displaystyle\mathrm{NH_2}}{\mathrm{CH_3\underset{|}{C}HCOOH}}$$

Which form is found in proteins?

18. Write the structural formulas for the D and L forms of serine

$$\mathrm{HOCH_2\underset{\underset{\displaystyle NH_2}{|}}{C}HCOOH}$$

Which form is found in proteins?

19. A 100.0 g sample of a food product is analyzed and found to contain 6.0 g of nitrogen. If protein contains an average of 16% nitrogen, what percentage of the food is protein?

20. A 250.0 g sample of hamburger is analyzed and found to contain 5.2 g of nitrogen. If protein contains an average of 16% nitrogen, how many grams of protein are contained in the hamburger?

21. Write the structural formula representing the following amino acid at its isoelectric point:

$$\mathrm{CH_3\underset{\underset{\displaystyle OH}{|}}{C}H{-}\underset{\underset{\displaystyle NH_2}{|}}{C}HCOOH}$$

threonine

22. Write the structural formula representing the following amino acid at its isoelectric point:

$$\mathrm{NH_2\underset{\underset{\displaystyle O}{\|}}{C}CH_2\underset{\underset{\displaystyle NH_2}{|}}{C}HCOOH}$$

asparagine

23. For the amino acid

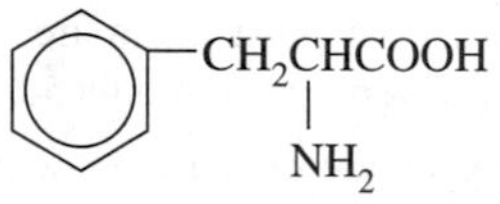

$$\mathrm{C_6H_5{-}CH_2\underset{\underset{\displaystyle NH_2}{|}}{C}HCOOH}$$

phenylalanine

write:
(a) the zwitterion formula
(b) the formula in 0.1 M H_2SO_4
(c) the formula in 0.1 M NaOH

24. For the amino acid

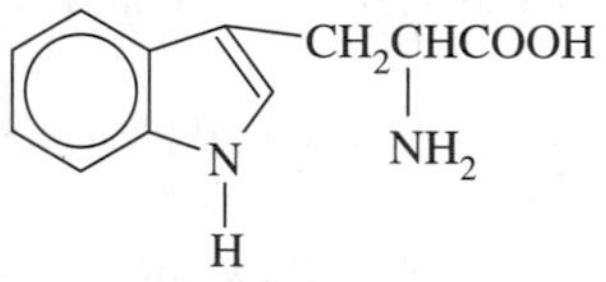

(indole ring with N–H) $\mathrm{{-}CH_2\underset{\underset{\displaystyle NH_2}{|}}{C}HCOOH}$

tryptophan

write:
(a) the zwitterion formula
(b) the formula in 0.1 M H_2SO_4
(c) the formula in 0.1 M NaOH

25. Write ionic equations to show how alanine

$$\mathrm{CH_3\underset{\underset{\displaystyle NH_2}{|}}{C}HCOOH}$$

acts as a buffer toward
(a) H^+ (b) OH^-

26. Write ionic equations to show how leucine

$$\mathrm{(CH_3)_2CHCH_2\underset{\underset{\displaystyle NH_2}{|}}{C}HCOOH}$$

acts as a buffer toward
(a) H^+ (b) OH^-

27. Write the structural formula representing the following amino acid at its isoelectric point:

$$\mathrm{CH_3SCH_2CH_2\underset{\underset{\displaystyle NH_2}{|}}{C}HCOOH}$$

methionine

28. Write the structural formula representing the following amino acid at its isoelectric point:

$$\mathrm{(CH_3)_2CH\underset{\underset{\displaystyle NH_2}{|}}{C}HCOOH}$$

valine

29. Write out the full structural formula of the two dipeptides containing

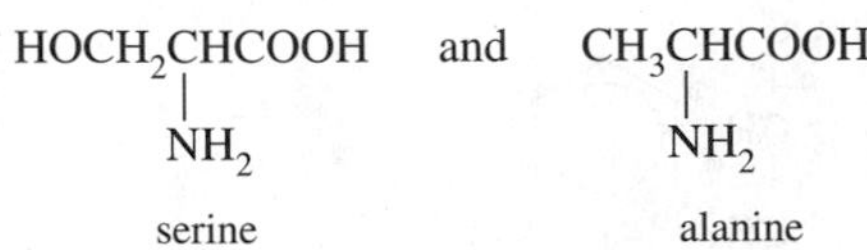

serine alanine

Indicate the location of the peptide bonds.

30. Write out the full structural formula of the two dipeptides containing

$CH_2(NH_2)COOH$ and $CH_3CH(OH)CH(NH_2)COOH$

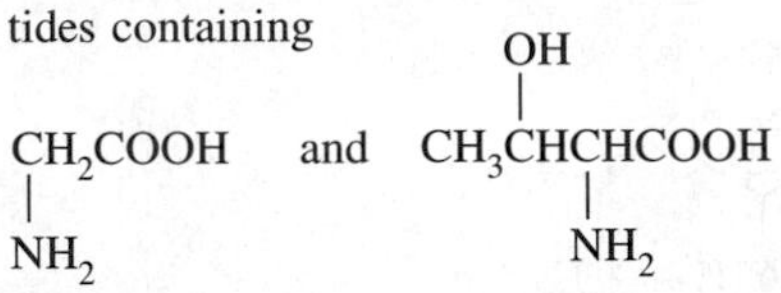

glycine threonine

Indicate the location of the peptide bonds.

31. Use the following amino acids

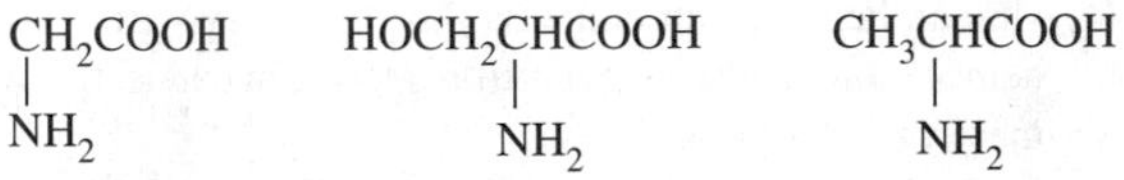

$CH_2(NH_2)COOH$ $HOCH_2CH(NH_2)COOH$ $CH_3CH(NH_2)COOH$

glycine serine alanine

to write structures for
(a) glycylglycine
(b) alanylglycylserine
(c) glycylserylglycine

32. Use the following amino acids

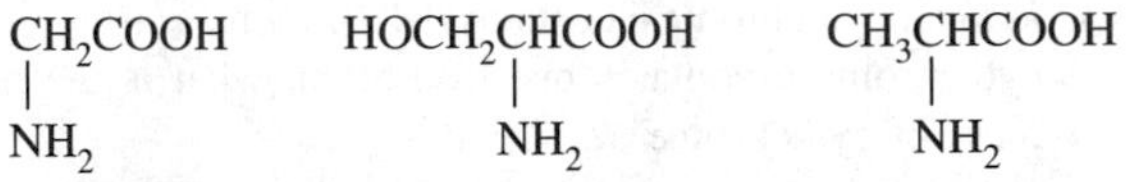

$CH_2(NH_2)COOH$ $HOCH_2CH(NH_2)COOH$ $CH_3CH(NH_2)COOH$

glycine serine alanine

to write structures for
(a) alanylalanine
(b) serylglycylglycine
(c) serylglycylalanine

33. Using amino acid abbreviations, write all the possible tripeptides that contain one unit each of glycine, phenylalanine, and leucine.

34. Using amino acid abbreviations, write all the possible tripeptides that contain one unit each of tyrosine, aspartic acid, and alanine.

35. What part of the following structure bonds when a β-pleated sheet forms? Identify the type of bonds that form.

$-\overset{H}{C}H\overset{O}{\overset{\|}{C}}\underset{H}{N}\overset{CH_3}{C}H-$

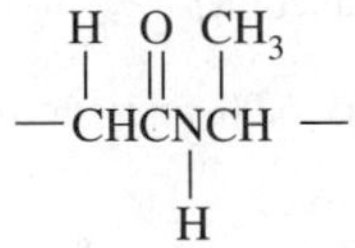

36. What part of the following structure bonds when an α-helix forms? Identify the type of bonds that form.

$-\overset{H}{C}H\overset{O}{\overset{\|}{C}}\underset{H}{N}\overset{CH_3}{C}H-$

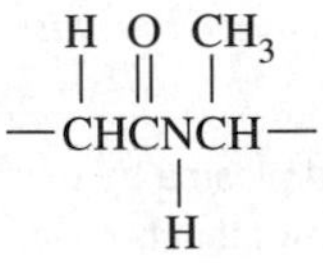

37. At pH = 7, draw the bond that might form between two serine side chains ($HOCH_2$—) to hold together a tertiary protein structure.

38. At pH = 7, draw the bond that might form between lysine side chain ($NH_2CH_2CH_2CH_2CH_2$—) and an aspartic acid side chain ($HOOCCH_2$—) to hold together a tertiary protein structure.

39. Would this compound

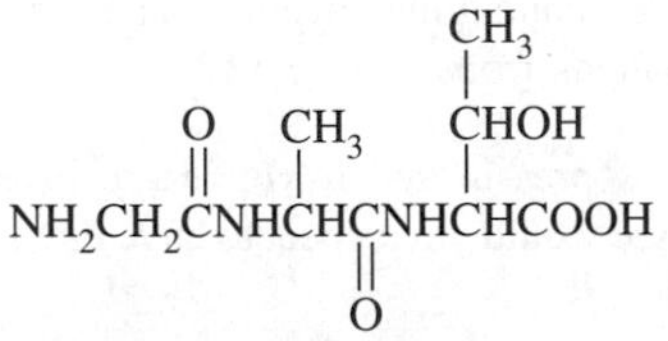

$NH_2CH_2C(=O)NHCH(CH_3)C(=O)NHCH(CH(OH)CH_3)COOH$

react with
(a) Sanger's reagent?
(b) concentrated HNO_3 to give a positive xanthoproteic test?
(c) ninhydrin?

40. Would this compound

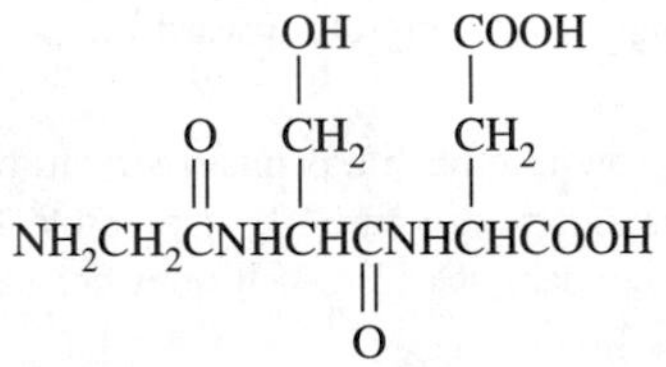

$NH_2CH_2C(=O)NHCH(CH_2OH)C(=O)NHCH(CH_2COOH)COOH$

react with
(a) Sanger's reagent?
(b) concentrated HNO_3 to give a positive xanthoproteic test?
(c) ninhydrin?

41. Write the products for the complete hydrolysis of

```
                 C6H5 (phenyl)
                   |       COOH
                   |        |
 CH3              CH2      CH2
  |                |        |
NH2CHCNHCHCNHCHCOOH
      ||      ||
      O       O
```

42. Write the products for the complete hydrolysis of

```
                                   OH
              HOOC                 |
                |                C6H4 (ring)
               CH2                 |
                |                  |
               CH2                 |
 CH3            |                 CH2
  |             |                  |
NH2CHCNHCHCNHCHCOOH
      ||      ||
      O       O
```

43. Human cytochrome c contains 0.43% by mass iron. If each cytochrome c contains one iron atom, what is the molar mass of cytochrome c?

44. Human hemoglobin contains 0.33% by mass iron. If each hemoglobin contains four iron atoms, what is the molar mass of hemoglobin?

45. What is the amino acid sequence of a heptapeptide, which upon hydrolysis yields the tripeptides Gly-Phe-Leu, Phe-Ala-Gly, and Leu-Ala-Tyr? The heptapeptide contains one residue each of Gly, Leu, and Tyr and two residues each of Ala and Phe.

46. What is the amino acid sequence of a heptapeptide, which upon hydrolysis yields the tripeptides Phe-Gly-Tyr, Phe-Ala-Ala, and Ala-Leu-Phe? The heptapeptide contains one residue each of Gly, Leu, and Tyr and two residues each of Ala and Phe.

Additional Exercises

These exercises are not paired or labeled by topic and provide additional practice on concepts covered in this chapter.

47. A nonapeptide is obtained from blood plasma by treatment with the enzyme trypsin. Analysis shows that both terminal amino acids are arginine. Total hydrolysis of this peptide yields Gly, Ser, 2 Arg, 2 Phe, 3 Pro. Partial hydrolysis gives Phe-Ser, Phe-Arg, Arg-Pro, Pro-Pro, Pro-Gly-Phe, Ser-Pro-Phe. What is the amino acid sequence of the nonapeptide?

48. **(a)** At what pH will arginine not migrate to either electrode in an electrolytic cell?

(b) In what pH range will it migrate toward the positive electrode?

49. Is it possible for a protein to have a primary structure if it does not have a secondary structure? Is it possible for a protein to have a secondary structure if it does not have a primary structure? Explain.

50. Ribonuclease is a protein that is slightly smaller than myoglobin. How many protein domains would you predict for the structure of ribonuclease? Explain your answer.

51. Threonine has two asymmetric centers. Write Fischer projection formulas for its stereoisomers.

52. How are the hypervariable regions in immunoglobulin chains important to this protein's function?

53. The isoelectric point for alanine is 6.00 and for lysine is 9.74.

(a) Write out the structure for alanine as it would be in a buffer with a pH of 9.0.

(b) Write out the structure for lysine as it would be in a buffer with a pH of 9.0.

(c) Would the charge on lysine be positive, negative or neutral when it is in a buffer with a pH of 9.0?

54. How many dipeptides containing glycine can be written using the 20 amino acids from Table 30.1?

55. The structures of vasopressin and oxytocin are given in Table 30.3. Which one would you expect to have the higher isoelectric point? Why?

56. Thin layer chromatography is used to separate amino acids from one another. The separation is based on the relative polarity of the amino acids. Which of these amino acids would you expect to be the most polar? Why?

leucine, alanine, glutamic acid

Answers to Practice Exercises

30.1 α-Amino acid: CH_2COOH with NH_2 on the CH_2 carbon (H_2NCH_2COOH)

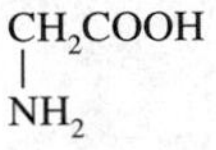

β-Amino acid: CH_2CH_2COOH with NH_2 on the first CH_2 carbon ($H_2NCH_2CH_2COOH$)

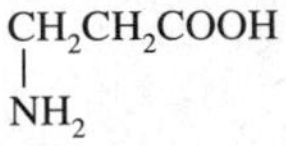

30.2 Leu (leucine), Lys (lysine), Trp (tryptophan)

30.3 $(CH_3)_2CHCH(NH_2)COO^-$

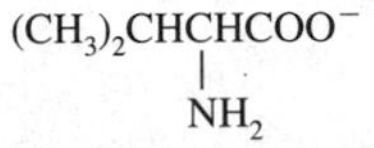

30.4 $CH_2(NH_2)C(=O)-NHCH[CH_2CH(CH_3)_2]C(=O)-NHCH(CH_2COOH)C(=O)-NHCH(CH_2OH)C(=O)-NHCH(CH_2SH)COOH$

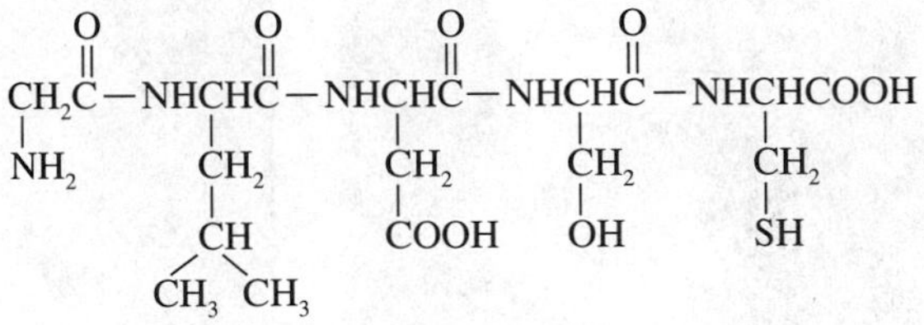

Enzymes

31

Chapter 31

Perhaps you have seen laundry detergents advertised as "containing enzymes." Enzymes are added to detergents to clean especially hard-to-remove spots. In living organisms, enzymes also take on the "tough" metabolic tasks—in fact, life cannot continue in the absence of enzymes.

Enzymes are important because they can accelerate a chemical reaction by 1 million to 100 million times. Imagine what would happen if some of our tasks were accelerated by that amount! A 2-hour daily homework assignment would be completed in about one-thousandth of a second. A flight from Los Angeles to New York would take about one-hundredth of a second. The building of Hoover Dam, a monumental task requiring 5 years of earth moving and complex steel and concrete work, would require only about 30 seconds.

Scientists now understand the general characteristics of enzymes. And, as we will see in this chapter, these molecules achieve almost miraculous results by following some very basic chemical principles.

31.1 Molecular Accelerators

enzyme

Enzymes are the catalysts of biochemical reactions. Enzymes catalyze nearly all the myriad reactions that occur in living cells. Uncatalyzed reactions that require hours of boiling in the presence of a strong acid or a strong base can occur in a fraction of a second in the presence of the proper enzyme at room temperature and nearly neutral pH. The catalytic functions of enzymes are directly dependent on their three-dimensional structures. It was believed until quite recently that all enzymes were proteins. However, research under way since about 1980 has shown that certain ribonucleic acids (RNAs, see Chapter 32) can also function as enzymes.

Louis Pasteur was one of the first scientists to study enzyme-catalyzed reactions. He believed that living yeasts or bacteria were required for these reactions, which he called *fermentations*—for example, the conversion of glucose to alcohol by yeast. In 1897, Eduard Büchner (1860–1917) made a cell-free filtrate that contained enzymes prepared by grinding yeast cells with very fine sand. The enzymes in this filtrate converted glucose to alcohol, thus proving that the presence of living cells was not required for enzyme activity. For this work Büchner received the Nobel prize in chemistry in 1907.

Enzymes are essential to life. The critical biochemical reactions occur too slowly in the absence of enzymes for life to be maintained. The typical biochemical reaction occurs more than a million times faster when catalyzed by an enzyme. For example, we know that the reduced carbons of carbohydrates can react with oxygen to produce carbon dioxide and energy. Yet the sucrose in the sugar bowl at home never reacts significantly with the oxygen in the air. These sucrose molecules must overcome an energy barrier (activation energy) before reaction can occur (see Figure 31.1). In the sugar bowl, the energy barrier is too large. But in the cell, enzymes lower the activation energy and enable sucrose to react rapidly enough to provide the energy needed for life processes.

◀ Chapter Opening Photo: Yeast provides the enzymes necessary to make bread rise.

FIGURE 31.1 ▶
A typical reaction-energy profile: The lower activation energy in the cell is due to the catalytic effect of enzymes.

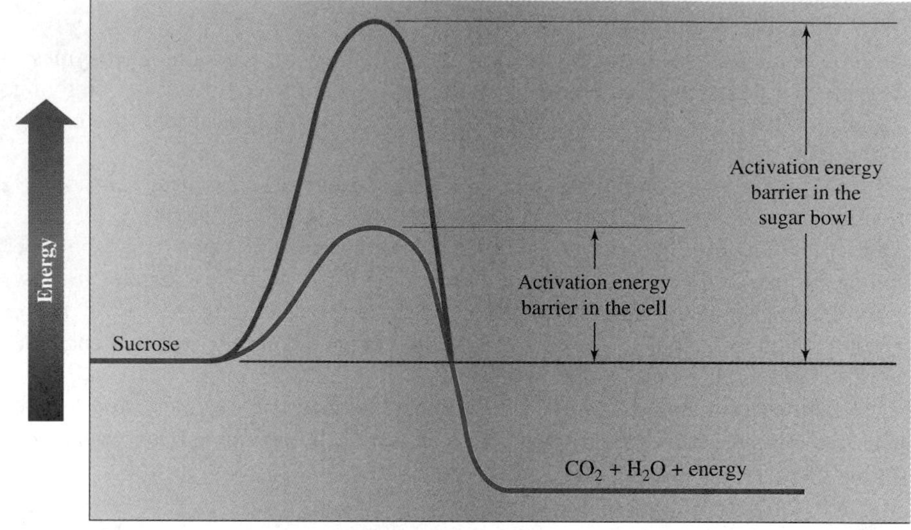

Each organism contains thousands of enzymes. Some are simple proteins consisting only of amino acid units. Others are conjugated and consist of a protein part, or *apoenzyme,* and a nonprotein part, or *coenzyme.* Both parts are essential, and a functioning enzyme consisting of both the protein and nonprotein parts is called a *holoenzyme*:

$$\text{Apoenzyme} + \text{Coenzyme} = \text{Holoenzyme}$$

Often the coenzyme is derived from a vitamin, and one coenzyme may be associated with many different enzymes.

For some enzymes, an inorganic component such as a metal ion (e.g., Ca^{2+}, Mg^{2+}, or Zn^{2+}) is required. This inorganic component is an *activator.* From the standpoint of function, an activator is analogous to a coenzyme, but inorganic components are not called coenzymes.

Another remarkable property of enzymes is their specificity of reaction; that is, a certain enzyme catalyzes the reaction of a specific type of substance. For example, the enzyme maltase catalyzes the reaction of maltose and water to form glucose. Maltase has no effect on the other two common disaccharides, sucrose and lactose. Each of these sugars requires a specific enzyme—sucrase to hydrolyze sucrose, lactase to hydrolyze lactose. These reactions are indicated by the following equations:

$$\underset{\text{maltose}}{C_{12}H_{22}O_{11}} + H_2O \xrightarrow{\text{maltase}} \underset{\text{glucose}}{C_6H_{12}O_6} + \underset{\text{glucose}}{C_6H_{12}O_6}$$

$$\underset{\text{sucrose}}{C_{12}H_{22}O_{11}} + H_2O \xrightarrow{\text{sucrase}} \underset{\text{glucose}}{C_6H_{12}O_6} + \underset{\text{fructose}}{C_6H_{12}O_6}$$

$$\underset{\text{lactose}}{C_{12}H_{22}O_{11}} + H_2O \xrightarrow{\text{lactase}} \underset{\text{glucose}}{C_6H_{12}O_6} + \underset{\text{galactose}}{C_6H_{12}O_6}$$

The substance acted on by an enzyme is called the *substrate*. Sucrose is the substrate of the enzyme sucrase. Enzymes have been named by adding the suffix *-ase* to the root of the substrate name. Note the derivations of maltase, sucrase, and lactase from maltose, sucrose, and lactose. Many enzymes, especially digestive enzymes, have common names such as pepsin, rennin, trypsin, and so on. These names have no systematic significance.

In the International Union of Biochemistry (IUB) System, enzymes are assigned to one of six classes, the names of which clearly describe the nature of the reaction they catalyze. Each of the classes has several subclasses. In this system the name of the enzyme has two parts; the first gives the name of the substrate, and the second, ending in *-ase*, indicates the type of reactions catalyzed by all enzymes in the group. The six main classes of enzymes are

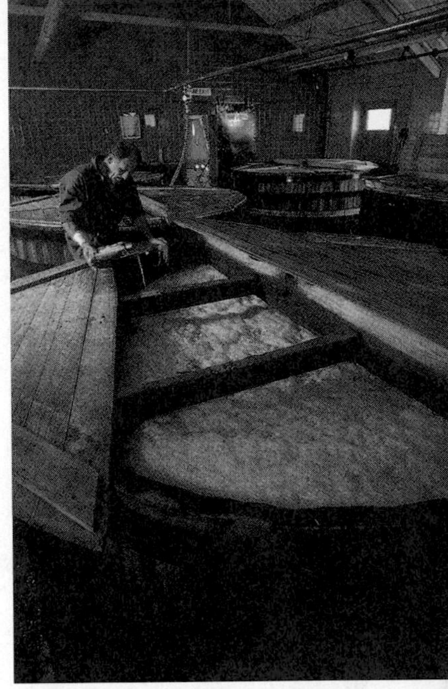

A worker adds yeast to a vat in a whiskey distillery.

1. *Oxidoreductases:* Enzymes that catalyze the oxidation–reduction between two substrates
2. *Transferases:* Enzymes that catalyze the transfer of a functional group between two substrates
3. *Hydrolases:* Enzymes that catalyze the hydrolysis of esters, carbohydrates, and proteins (polypeptides)
4. *Lyases:* Enzymes that catalyze the removal of groups from substrates by mechanisms other than hydrolysis
5. *Isomerases:* Enzymes that catalyze the interconversion of stereoisomers and structural isomers
6. *Ligases:* Enzymes that catalyze the linking of two compounds with the breaking of a pyrophosphate bond in adenosine triphosphate (ATP, see Chapter 32)

Because the systematic name is usually long and often complex, working or practical names are used for enzymes. For example, adenosine triphosphate creatine phosphotransferase is called creatine kinase, and acetylcholine acylhydrolase is called acetylcholine esterase.

31.2 Rates of Chemical Reactions

Enzymes catalyze biochemical reactions and thus increase the rate of these chemical reactions—but how does this process take place? To answer this question, we must first consider some general properties of a chemical reaction.

Every chemical reaction starts with at least one reactant and finishes with a minimum of one product. As the reaction proceeds, the reactant concentration decreases and the product concentration increases. Often these changes are plotted as a function of time, as shown in Figure 31.2 for the hypothetical conversion of reactant A into product B.

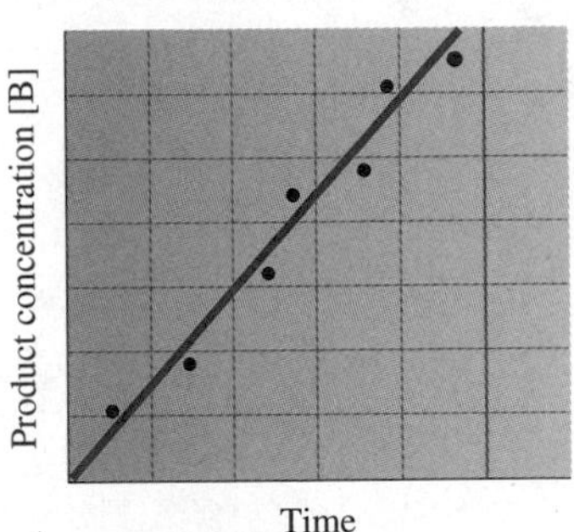

FIGURE 31.2
The change in product concentration [B] as a function of time. The reaction rate is determined by measuring the slope of this line.

$$A \longrightarrow B$$

A reaction rate is defined as a change in concentration with time—the rate at which the reactants of a chemical reaction disappear and the products form.

transition state

The reactant must pass through a high-energy **transition state** to be converted into a product. This transition state is an unstable structure with characteristics of

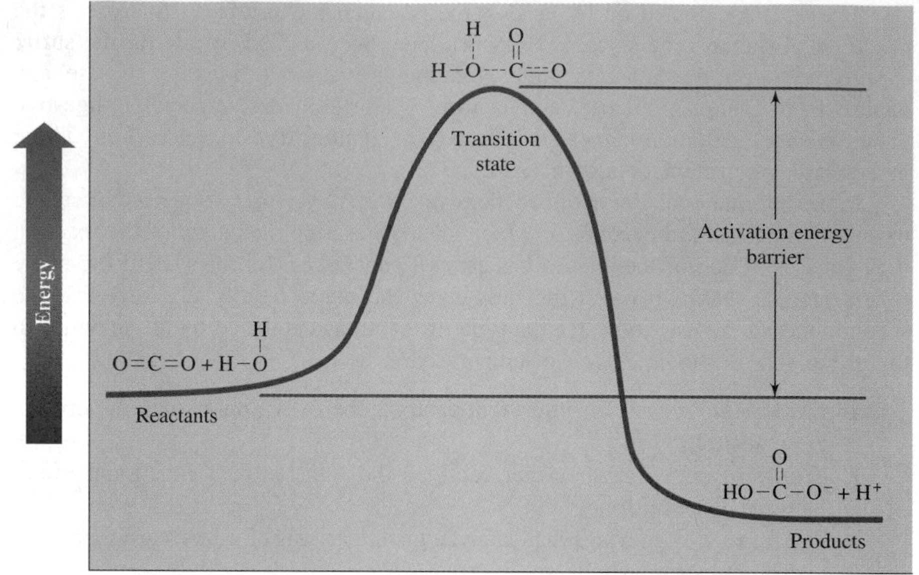

FIGURE 31.3 ▶ An energy profile for the reaction between water and carbon dioxide: The dashed lines in the transition state formula indicate bonds being formed and broken.

both the reactant and the product. The energy necessary to move a reactant to the transition state is termed the activation energy. The larger this energy barrier, the slower the reaction rate.

For example, carbon dioxide can be reacted with water to yield hydrogen carbonate ion.

$$CO_2 + H_2O \longrightarrow HCO_3^- + H^+$$

Yet, as shown in Figure 31.3, energy is required to form the transition state. Once the transition state is created, the remainder of the process proceeds easily. As with almost all chemical reactions, reaching the transition state is difficult and limits the rate at which reactants are converted to products.

There are three common ways to increase a reaction rate:

1. *Increasing the reactant concentration:* When the reactant concentrations are made larger, the number of reactant molecules with the necessary activation energy also increases. For simple reactions in the absence of a catalyst, the reaction rate increases with reactant concentration.
2. *Increasing the reaction temperature:* An increase in temperature generally means that each reactant molecule becomes more energetic. A larger fraction of the reactants have the activation energy necessary to be converted to products, and the reaction rate increases.
3. *Adding a catalyst:* A catalyst lowers the activation energy by allowing a new, lower-energy transition state. Since this new process has a lower activation energy, more reactants have the energy to become products. The reaction rate increases.

Biological systems can rarely change reactant concentration or temperature upon demand. Thus, to alter reaction rates, life has evolved a set of superb catalysts—the enzymes.

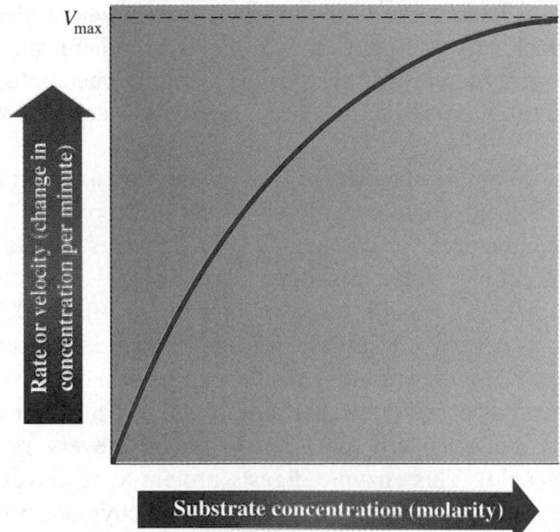

FIGURE 31.4
A Michaelis–Menten plot showing the rate of an enzyme-catalyzed reaction as a function of substrate concentration. The lower left portion of the graph marks the approximate area where an enzyme responds best to concentration changes.

31.3 Enzyme Kinetics

In 1913, two German researchers, Leonor Michaelis (1875–1949) and Maud Menten (1879–1960) measured enzyme-catalyzed reaction rates as a function of substrate (reactant) concentration. They observed that most enzyme-catalyzed reactions show an increasing rate with increasing substrate concentration *but* only to a specific maximum velocity, V_{max}. A graph like that in Figure 31.4 is often called a Michaelis–Menten plot.

Scientists have learned much about the nature of enzymes by studying the Michaelis–Menten plot. First, because the rate approaches a maximum, it can be concluded that enzymes have a limited catalytic ability. Once these enzymes are operating at a maximum, a further increase in substrate (reactant) concentration does not change the reaction rate. By analogy, an enzyme shuttles reactants to the transition state (and on to products) like an usher seating patrons at a theater. No matter how many people arrive at the entrance, a conscientious usher can only lead the ticket holders to their places at a set pace. Similarly, each specific enzyme has a set, maximum catalytic rate.

Scientists believe that each specific enzyme's catalytic abilities are tailored to fit a specific metabolic need. Some enzymes work well at low substrate concentrations, whereas others require much higher concentrations before they operate efficiently. Some enzymes catalyze reactions very quickly while others fit a biochemical need by reacting much more slowly. Several examples will illustrate these concepts.

There are two common mammalian enzymes that begin glucose metabolism, hexokinase and glucokinase. Both enzymes catalyze the same reaction, an addition of a phosphate group to glucose. These two enzymes differ in their effective concentration range; at low glucose concentrations, only hexokinase catalyzes the reaction while at higher glucose concentrations both hexokinase and glucokinase react. This difference means that hexokinase has first priority to use the available glucose. This is important because hexokinase is found in all tissues where glucose is used for energy, while glucokinase is found only in the liver, where glucose is stored as glycogen. When energy is needed, hexokinase starts the metabolism of glucose. If there is some excess glucose, the glucokinase enzyme starts converting this sugar to glycogen.

turnover number

Another measure of enzyme effectiveness is the **turnover number**. This measures how many molecules one enzyme can react or "turn over" in a given time span. Let's compare the turnover numbers for two enzymes. Catalase is responsible for destroying the cellular toxin, hydrogen peroxide, before it can do biological damage. This enzyme has a large turnover number of 10,000,000/s; that is, one catalase enzyme can destroy 10 million hydrogen peroxide molecules per second. This very large turnover number minimizes the danger posed by hydrogen peroxide. In contrast, the enzyme chymotrypsin has a much smaller turnover number of about 0.2/s (two reactant molecules are converted to products every 10 seconds by one chymotrypsin molecule). This enzyme digests protein as it moves slowly through the small intestine. Since the digestive process is slow, chymotrypsin is not required to react as quickly as catalase. In many cases it appears that an enzyme's turnover number matches the speed required for a biological process.

Example 31.1 Two enzymes react with the same substrate under identical conditions. Enzyme A is found to have a turnover number of 1500/s whereas enzyme B shows a turnover number of 500/s. Compare the catalytic efficiency of the two enzymes.

Solution The turnover number is a measure of catalytic efficiency. Under identical conditions, the larger the turnover number, the more efficient an enzyme is as a catalyst. Because enzyme A has a larger turnover number than enzyme B, enzyme A is the more efficient catalyst.

Practice 31.1

Under optimal conditions, the digestive enzyme pepsin has a turnover number of about 30/min while a second digestive enzyme, trypsin, has a turnover number of 12/min. Both of these enzymes digest proteins. Which would you judge to be the more efficient?

Industrial-Strength Enzymes

Not only are enzymes important in biology, they are becoming increasingly important in industry. Enzymes offer two major advantages to manufacturing processes and in commercial products: First, enzymes cause very large increases in reaction rates even at room temperature; second, enzymes are relatively specific and can be used to target selected reactants. Perhaps the biggest disadvantage to industrial enzymes is their relative short supply (and therefore higher cost as compared to traditional chemical treatments). Recent developments in biotechnology offer supplies of less expensive enzymes through genetic engineering.

Enzymes have long been used in food processing. The citrus industry has recently perfected a process to remove peel from oranges or grapefruits by using the enzyme pectinase. The pectinase penetrates the peel in a vacuum infusion process. There it dissolves the albedo (the white stringy material) that attaches the peel to the fruit. When the fruit is removed from the solution, the skin can be peeled easily by machine or hand. The industry is now marketing pre-peeled citrus to hospitals, airlines, and restaurants.

The enzyme pectinase can effectively peel an orange, leaving the fruit in perfect condition.

The enzyme lactase is used on an industrial scale to convert lactose to glucose and galactose. This change gives people who are lactose-intolerant the ability to consume milk products. Among the great number of different molecules in milk, lactase only reacts with lactose. Since the glucose and galactose formed are sweeter than lactose and differ in solubility, lactase treatment produces milk with different properties. Ice cream made from lactase-treated milk, for example, tends to be creamier.

About 25% of all industrial enzymes are used to convert cornstarch into syrups that are equivalent in sweetness and in calories to ordinary table sugar. More than 5 billion pounds of such syrups are produced annually. The process uses three enzymes: The first, α-amylase, catalyzes the liquefaction of starch to dextrins; the second, a glucoamylase, catalyzes the breakdown of dextrins to glucose; the third, glucose isomerase, converts glucose to fructose.

$$\text{Starch} \xrightarrow{\alpha\text{-amylase}} \text{Dextrins}$$

$$\text{Dextrins} \xrightarrow{\text{glucoamylase}} \text{Glucose}$$

$$\text{Glucose} \xrightarrow[\text{isomerase}]{\text{glucose}} \text{Fructose}$$

The product is a high-fructose syrup equivalent in sweetness to sucrose. One of these syrups, sold commercially since 1968, contains by dry weight about 42% fructose, 50% glucose, and 8% other carbohydrates.

Industrial enzymes offer solutions to environmental pollution problems for some manufacturers. For example, the paper industry, like other industries that use chemicals, is concerned with minimizing processes that produce potentially hazardous waste. Paper is produced from wood chips by first digesting the cellulose structure with calcium sulfite and then bleaching the pulp with chlorine to obtain a bright white paper. An excess of chlorine must be used because the pulp is not completely broken down. This excess creates a significant disposal problem as chlorine is environmentally hazardous. Recent developments in biotechnology offer a potential solution. The enzymes needed to complete wood fiber digestion (cellulase and hemicellulase) have been produced in large quantities via genetic engineering. With such enzymes to finish degrading the wood pulp, paper manufacturers may be able to markedly decrease the amount of chlorine used as bleach.

Consumer goods are increasingly impacted by enzyme technology. Many detergents are better cleansing agents because they contain enzymes; fully 40% of all industrially produced enzymes are used in this way. Most of these enzymes are specific for proteins, which are some of the most difficult molecules to remove from clothing. Proteins in grass stains, blood, milk, sweat, and so on dry and cross-link to the cloth fibers. Soap (even soap with hot water) is often ineffective at removing these stains while enzymes can specifically degrade these proteins. As a side benefit, detergents with enzymes can operate at lower wash temperatures, saving on water heating bills.

Clothing manufacturers are finding uses for newly available enzymes. More and more denim products are enzyme-treated to replace stonewashing, a process in which the material is washed with pumice to soften the fabric's appearance and remove some of the dye. Because this abrasion may weaken the fabric as well, some manufacturers now use "biostoning." The denim is treated with the enzyme cellulase, which changes the fabric's appearance without weakening the fabric structure.

The ability to produce large quantities of purified enzymes has raised a number of potential medical applications. Several enzymes are used to dissolve blood clots in the treatment of diseases such as lung embolism (clot in the lung), stroke (clot in the brain), and heart attack (clot in the heart). These enzymes are classified as plasminogen activators. They are proteinases (protein enzymes) that search through the hundreds of different proteins in the blood stream to find plasminogen. Once found, they convert it to plasmin, which then degrades the clot. Intravenous treatment of a patient with a plasminogen activator can dissolve a clot in 30–60 minutes.

31.5 Enzyme Active Site

Catalysis takes place on a small portion of the enzyme structure called the *enzyme active site.* Often this is a crevice or pocket on the enzyme representing only 1–5% of the total surface area. For example, Figure 31.5 shows a space-filling model of the enzyme hexokinase, which catalyzes a first step in the breakdown of glucose to provide metabolic energy. Notice that the active site is located in a crevice. Glucose enters this site and is bound. The enzyme then must change shape before the reaction takes place. Thus, although catalysis occurs at the small active site, the entire three-dimensional structure of the enzyme is important.

By examining values such as the turnover number and the effective substrate concentration range, scientists have gained a basic understanding of what takes place at an enzyme active site. To function effectively, an enzyme must attract and bind the substrate. Once the substrate is bound, a chemical reaction is catalyzed. This two-step process is described by the following general sequence. Enzyme (E) and substrate (S) combine to form an enzyme–substrate intermediate (E–S). This intermediate decomposes to give the product (P) and regenerate the enzyme:

$$\text{E} + \text{S} \xrightarrow[\text{binding}]{} \text{E–S} \xrightarrow[\text{catalysis}]{} \text{E} + \text{P}$$

For the hydrolysis of maltose by the enzyme maltase, the sequence is

$$\underset{\text{E}}{\text{Maltase}} + \underset{\text{S}}{\text{Maltose}} \xrightarrow[\text{binding}]{} \underset{\text{E–S}}{\text{Maltase–Maltose}}$$

$$\underset{\text{E–S}}{\text{Maltase–Maltose}} + \text{H}_2\text{O} \xrightarrow[\text{catalysis}]{} \underset{\text{E}}{\text{Maltase}} + 2\ \underset{\text{P}}{\text{Glucose}}$$

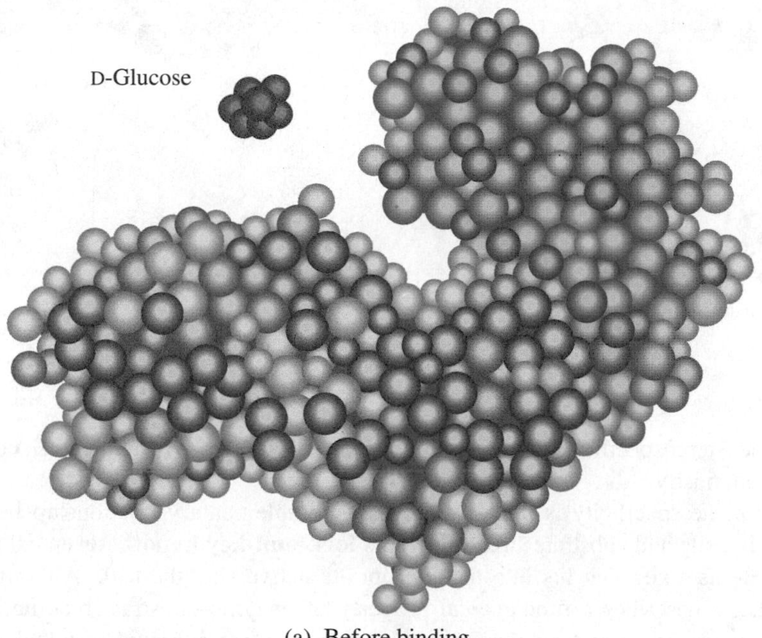

(a) Before binding

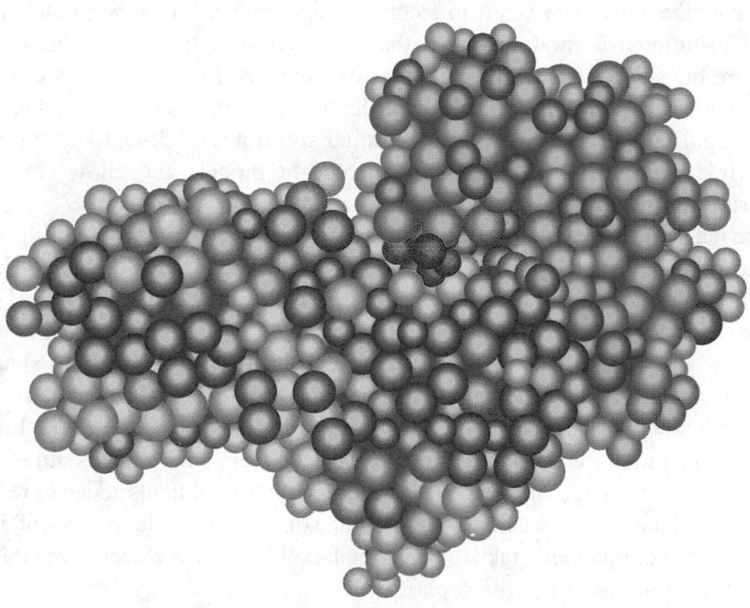
(b) After binding

◀ **FIGURE 31.5**
A space-filling model of the enzyme hexokinase (a) before and (b) after it binds to the substrate D-glucose. Note the two protein domains for this enzyme, which are colored differently.

Each different enzyme has its own unique active site whose shape determines, in part, which substrates can bind. Enzymes are said to be stereospecific; that is, each enzyme catalyzes reactions for only a limited number of different reactant structures. For example, maltase binds to maltose (two glucose units linked by an α-1,4-glycosidic bond) but not to lactose (galactose coupled to glucose). In fact, this enzyme can even distinguish maltose from cellobiose (glucose coupled to glucose by a β-1,4-linkage), in which only the glycosidic linkage is different.

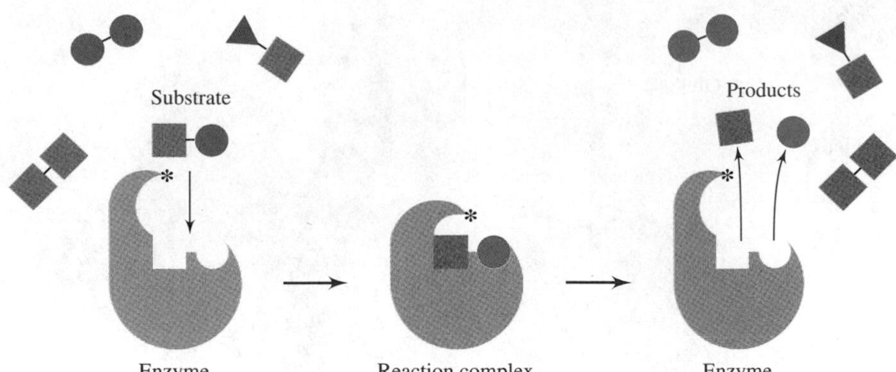

FIGURE 31.6 ▶ Enzyme–substrate interaction illustrating both the lock-and-key hypothesis and the induced-fit model. The correct substrate (■ — ●) fits the active site (lock-and-key hypothesis). This substrate also causes an enzyme conformation change that positions a catalytic group (∗) to cleave the appropriate bond (induced-fit model).

Enzyme stereospecificity is a very important means by which the cell controls its biochemistry.

lock-and-key hypothesis

Enzyme specificity is partially due to a complementary relationship between the active site and substrate structures. This **lock-and-key hypothesis** envisions the substrate as a key that fits into the appropriate active site, the lock. Although this hypothesis describes a fundamental property of enzyme–substrate binding, it has been known for some time that enzyme active sites are not rigid as a lock would be. Instead, the active site bends to a certain degree when the appropriate substrate binds. The **induced-fit model** proposes that the active site adjusts its structure in order to prepare the substrate–enzyme complex for catalysis. Enzyme stereospecificity is thus explained in terms of an active site having a somewhat flexible shape. This shape is rigid enough to exclude very dissimilar substrates (lock-and-key hypothesis) but flexible enough to accommodate (induced-fit model) and allow catalysis of appropriate substrates (see Figure 31.6).

induced-fit model

The fact that an enzyme is flexible can help to explain enzyme-binding specificity and also how the enzyme converts reactants into products. *An enzyme is a dynamic catalyst.* As the enzyme attracts the substrate into the active site, the enzyme's shape and the reactant's shape both begin to change. Note the change in hexokinase structure when glucose is bound (Figure 31.5). This alteration of enzyme shape aids the transformation of the reactant into the product. Figure 31.6 schematically depicts an enzyme where a shape change leads to catalysis. This only exemplifies the numerous alterations that can be brought about by enzymes, for example, bonds broken or formed, charges moved, and new molecular substituents added or removed.

An actual metabolic reaction will illustrate some of the basic features of enzyme catalysis. For example, let's take the enzyme hexokinase, which catalyzes the transformation of glucose to glucose-6-phosphate.

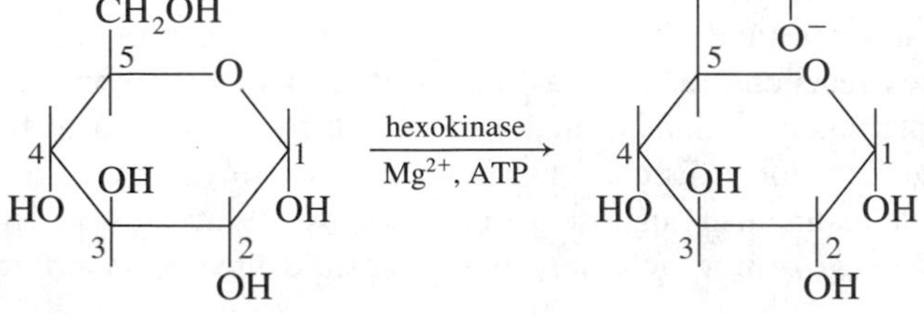

$$\text{glucose } (\text{C-6: } CH_2OH) \xrightarrow[\text{Mg}^{2+},\ \text{ATP}]{\text{hexokinase}} \text{glucose-6-phosphate } (\text{C-6: } CH_2OPO_3^{2-})$$

glucose — glucose-6-phosphate

Like many enzymes, hexokinase requires a second reactant. In this case, adenosine-5′-triphosphate (ATP) supplies the phosphate group that is transferred to the glucose. ATP is a very important molecule within the cell because it often transfers energy as well as phosphate groups from enzyme to enzyme (see Chapter 32). Additionally, the metal ion, Mg^{2+}, is required as an activator in this reaction. With all of these components present, the enzyme can start the catalytic process.

When glucose binds, the hexokinase shape changes to bring the ATP close to the carbon 6 of the glucose, thus forcing the transfer of phosphate to this specific carbon. Hexokinase speeds this reaction in several ways. First, this enzyme acts to bring the reactants close together, a process termed **proximity catalysis**. Second, hexokinase positions the reactants so the proper bonds will break and form. (The enzyme ensures that a phosphate is added to carbon 6 of glucose and not to one of the other carbons.) This is often termed the **productive binding hypothesis** because reactants are bound/oriented in such a way that products result. Hexokinase is a successful catalyst—glucose reacts with ATP 10 billion (10^{10}) times faster than it would in the absence of this enzyme. Increases in reaction rate of this magnitude are essential to life. In fact, without enzymes, cellular reactions are too slow to keep the cells alive.

proximity catalysis

productive binding hypothesis

Biotechnologists have made use of our understanding of enzyme catalysis to design (with nature's help) completely new enzymes. Antibodies (immunoglobulin proteins) are produced that bind tightly to a molecule like in the transition state. When these antibodies bind reactant molecules, the strong attractive forces "strain" the reactants as illustrated in Figure 31.7. The reactant molecule is impelled to change shape to fit the binding site. Catalysis occurs, and these antibodies act as enzymes! Scientists have termed this mode of catalysis the **strain hypothesis**, and it is thought to be important in many natural enzymes as well as these "antibody enzymes." In the not-too-distant future, such artificial enzymes may serve important industrial and medical applications.

strain hypothesis

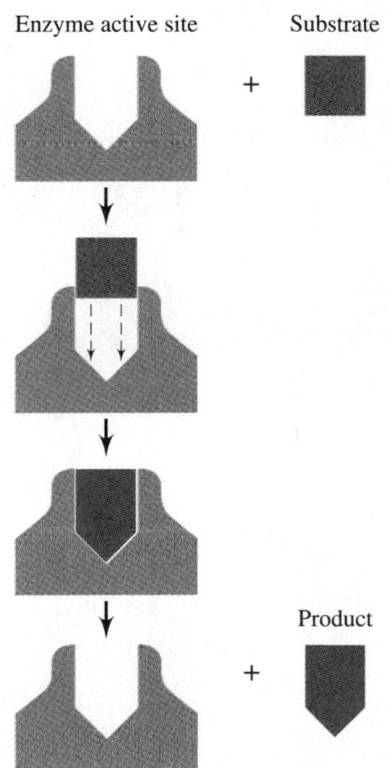

FIGURE 31.7
Schematic showing the substrate being forced toward the product shape by enzyme binding.

31.6 Temperature and pH Effects on Enzyme Catalysis

Essentially any change that will affect protein structure will affect an enzyme's catalytic function. If an enzyme is denatured, its activity will be lost. Thus strong acids and bases, organic solvents, mechanical action, and high temperature are examples of treatments that may decrease an enzyme-catalyzed rate of reaction.

Even slight changes in the pH can have profound effects on enzyme catalysis. Remember that some of the amino acids that make up enzymes have side chains whose charge depends on pH. For example, side chains with carboxylic acid functional groups may be either neutral or negatively charged; those with amino groups may be either neutral or positively charged. Thus as the pH is changed, the charges on an enzyme, or more specifically at the active site of an enzyme, also change. Enzyme catalysis is affected. For each enzyme there is an optimal pH; shifts to more acidic or more basic conditions decrease the enzyme activity as shown in Figure 31.8.

Because enzymes are so sensitive to small pH changes, our bodies have developed elaborate mechanisms to control the amount of acid and base present in cellular fluids. The kidneys and lungs share the responsibility for maintaining blood acid and base levels. If the blood pH shifts outside of the narrow range 6.8–7.8, death

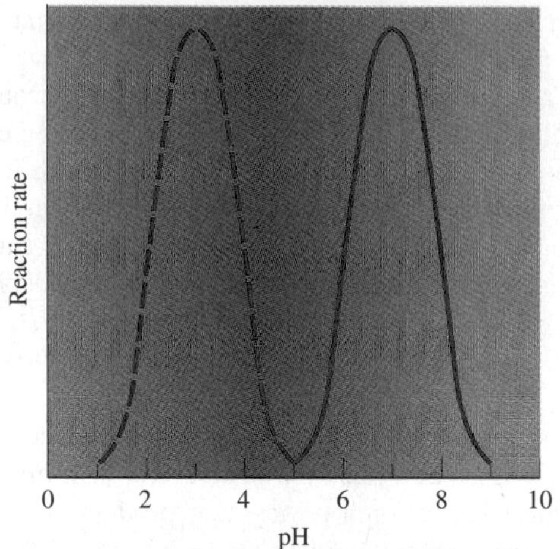

FIGURE 31.8 ▶
A plot of the enzyme-catalyzed rate as a function of pH: primitive bacterial enzyme (- - -); typical human enzyme (—).

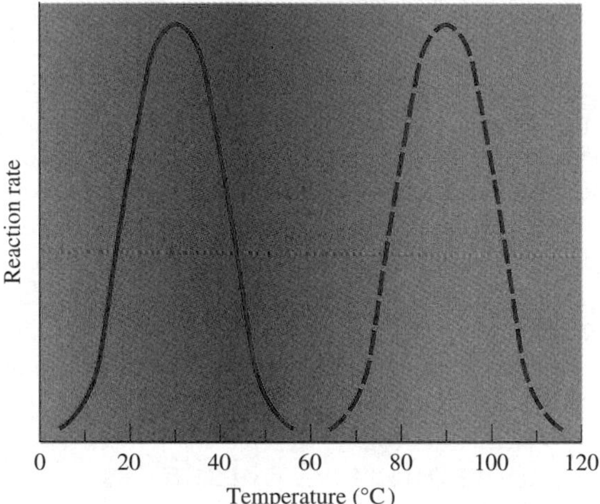

FIGURE 31.9 ▶
A plot of the temperature dependence of an enzyme-catalyzed reaction: typical human enzyme (—); primitive bacterial enzyme (- - -).

often results. The body's metabolic processes can no longer maintain life partially because of a pH-induced change in enzyme activities.

Body temperature is also carefully controlled partially because enzyme activities are particularly temperature-sensitive. A typical plot of temperature versus enzyme activity is shown in Figure 31.9. At very low temperatures, few reactant molecules have enough energy to overcome the activation energy barrier and the reaction occurs slowly. As the temperature increases, more reactants have the necessary activation energy and the rate increases. But a point is reached where the high temperature causes enzyme denaturation, and the rate of reaction starts to decrease. Thus both low and high temperatures cause slow reaction rates, and, as with pH, there is an optimal temperature for each enzyme. Not surprisingly, the optimal temperature for many human enzymes is approximately body temperature (about 37°C).

Not all enzymes operate optimally at around the physiological temperature (37°C) and pH (7). In fact, chemists have long sought enzymes that will have optimal activity at higher temperatures and more acidic pHs, conditions that favor industrial reactions. Such enzymes have been found in organisms that live in hot

This pharmacist is dispensing AZT, the AIDS drug that impedes virus reproduction.

springs. These bacteria are considered to be some of the most primitive of life forms; they may have become adapted to conditions that prevailed on earth over 1 billion years ago. Their enzymes show optimal activity at temperatures up to 100°C and pH values as low as 3.

31.7 Enzyme Regulation

Since enzymes are vital to life, enzyme catalysis is under careful cellular control. The cell uses a variety of mechanisms to change the rate of substrate conversion to product. Sometimes a new group of atoms is covalently joined to the enzyme in a process called *covalent modification.* In other cases, another molecule is noncovalently bound to the enzyme. The protein structural change that results may cause a decrease in enzyme activity, **enzyme inhibition**, or an increase in activity, **enzyme activation**.

enzyme inhibition
enzyme activation

Hexokinase is an example of an enzyme that is under cellular control. Because this enzyme begins the breakdown of glucose to yield cellular energy, the hexokinase-catalyzed reaction is often not needed when cellular energy levels are high. Control occurs in the following way: (1) if glucose is no longer needed for cellular energy, the glucose-6-phosphate concentration increases; (2) this compound binds to the hexokinase enzyme and inhibits the enzyme activity. Because glucose-6-phosphate is a product of the reaction, this form of enzyme control is called **product inhibition**. Hexokinase responds to the overall cellular energy state and does not use more glucose than needed.

product inhibition

Feedback inhibition and **feedforward activation** are two other common forms of enzyme control. To understand these mechanisms, visualize the various cellular processes as assembly lines with a different enzyme at each step from raw material to finished product. Feedback inhibition affects enzymes at the beginning of the molecular assembly line. A final product acts as "feedback" and inhibits an enzyme from using too many molecular starting materials. In contrast, feedforward activation

feedback inhibition
feedforward activation

often controls enzymes at the end of the molecular assembly line. If there is an excess of starting materials, these molecules will "feedforward" and activate enzymes, which in turn cause the whole process to move faster. As with other control mechanisms, both feedback inhibition and feedforward activation serve to coordinate enzyme processes within the cell.

A variety of drug therapies make use of enzyme control to selectively affect target cells. Careful drug design can create a molecule that binds to only one type of enzyme. By binding, the drug blocks normal catalysis and causes enzyme inhibition.

AZT (3′-azido-3′-deoxythymidine) was the first drug approved in the United States for treatment of the viral disease, acquired immune deficiency syndrome (AIDS). This drug is structurally similar to a reactant needed to make genetic material for the AIDS virus. When AZT binds to a specific enzyme, the formation of new viral genetic material is inhibited. Virus reproduction is impeded and the progression of AIDS is slowed.

This example of drug therapy illustrates a very important principle: Because biological processes depend on enzymes, enzyme control often has a major impact on life.

Concepts in Review

1. Briefly explain why living cells need catalysts.
2. List the six main classes of enzymes and their functions.
3. Compare a simple, uncatalyzed reaction with an enzyme-catalyzed process.
4. Define the symbol V_{max} and relate it to the Michaelis–Menten plot.
5. Discuss the relationship between the reaction rate and the enzyme concentration.
6. Describe an enzyme active site.
7. State what is meant by the specificity of an enzyme.
8. List several ways that an enzyme can facilitate the conversion of substrate to product.
9. Summarize the effects of pH and temperature on an enzyme-catalyzed reaction.
10. Describe how inhibition and activation are important in enzyme control.
11. Discuss the potential role of enzymes in medicine and industry.

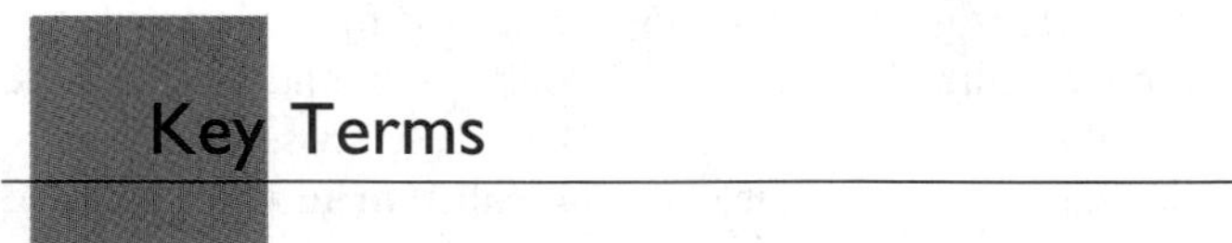

Key Terms

The terms listed here have been defined within this chapter. Section numbers are referenced in parenthesis for each term.

enzyme (31.1)
enzyme activation (31.7)
enzyme inhibition (31.7)
feedback inhibition (31.7)
feedforward activation (31.7)
induced-fit model (31.5)
lock-and-key hypothesis (31.5)
product inhibition (31.7)
proximity catalysis (31.5)
productive binding hypothesis (31.5)
strain hypothesis (31.5)
transition state (31.2)
turnover number (31.3)

Questions

Questions refer to tables, figures, and key words and concepts defined within the chapter. A particularly challenging question or exercise is indicated with an asterisk.

1. What is activation energy and how is this energy affected by enzymes?

2. What is the general role of enzymes in the body?

3. Distinguish between a coenzyme and an apoenzyme.

4. Give the names of enzymes that catalyze the hydrolysis of (a) sucrose, (b) lactose, and (c) maltose.

5. What are the six general classes of enzymes?

6. A catalyst increases the rate of a chemical reaction. List two other means of increasing reaction rates.

7. Differentiate between the lock-and-key hypothesis and the induced-fit model for enzymes.

8. List three ways that substrate binding to the active site helps the reactants convert to products.

9. How does an enzyme inhibitor differ from an enzyme substrate?

10. Which of the following statements are correct? Rewrite the incorrect ones to make them correct.

(a) Enzymes are proteins.

(b) Almost all enzymes increase the activation energy for a reaction.

(c) An enzyme active site is where catalysis occurs.

(d) The product of an enzyme-catalyzed reaction is called the substrate.

(e) Doubling the reactant concentration always doubles the rate of an enzyme-catalyzed reaction.

(f) Hydrolases are enzymes that catalyze the hydrolysis of esters, carbohydrates, and amino acids.

(g) Adding a catalyst lowers the activation energy and increases the reaction rate because more reactants have the energy to become products.

(h) The turnover number is a measure of an enzyme's catalytic ability.

(i) The lock-and-key hypothesis requires a flexible enzyme.

(j) The productive binding hypothesis states that an enzyme binds to and orients the reactants so that products can form most easily.

(k) Lyases are enzymes that catalyze the removal of groups from substrates by mechanisms other than hydrolysis.

(l) Feedforward activation increases the rate for enzymes at or close to the end of a metabolic process ("assembly line").

(m) Denaturation decreases an enzyme-catalyzed reaction rate.

(n) Enzyme treatment enables manufacturers to avoid more harsh chemical and physical processes.

Paired Exercises

These exercises are paired. Each odd-numbered exercise is followed by a similar even-numbered exercise. Answers to the even-numbered exercises are given in Appendix V.

11. An enzyme reacts 0.005 *M* substrate every 3.5 min. What is the reaction rate in units of molar per second?

12. An enzyme reacts 0.02 mol/L of substrate every 8 min. What is the reaction rate in units of molar per second?

13. Lysozyme, an enzyme that cleaves the cell wall of many bacteria, has a turnover number of 0.5/s. How many reactants can be converted to products by lysozyme in 1 min?

14. Pepsin, a digestive enzyme found in the stomach, has a turnover number of 1.2/s for a specific protein substrate. How many proteins can be digested by three pepsin molecules in 5 min?

15. Two enzymes are being studied; the first enzyme uses *n*-butanol, $C_4H_{10}O$, as a substrate while the second enzyme uses 2-methyl-2-propanol, $C_4H_{10}O$, as a substrate. Based on the lock-and-key hypothesis, how might the shapes for the enzyme active sites differ?

16. Two enzymes are being studied; the first enzyme uses butanoic acid, $C_4H_8O_2$, as a substrate while the second enzyme uses acetic acid, $C_2H_4O_2$, as a substrate. Based on the lock-and-key hypothesis, how might the shapes for the enzyme active sites differ?

17. Why does an enzyme-catalyzed reaction rate decrease at low temperatures?

18. Why does an enzyme-catalyzed reaction rate decrease at high temperatures?

19. If the V_{max} for an enzyme is decreased, would you predict activation or inhibition? Briefly explain.

20. If the turnover number for an enzyme is increased, would you predict activation or inhibition? Briefly explain.

21. A scientist studies two enzymes that catalyze the same reaction. Enzyme A has a turnover number of 225/s, while enzyme B has a turnover number of 120/s. The scientist concludes that enzyme B is more effective than enzyme A. Do you agree? Briefly explain.

22. A scientist studies two enzymes that catalyze the same reaction. Enzyme A has a turnover number of 0.05/s while enzyme B has a turnover number of 9.8×10^{-1}/s. The scientist concludes that enzyme B is more effective than enzyme A. Do you agree? Briefly explain.

23. The amino acid glutamine, is produced by a metabolic pathway in the liver. As the concentration of glutamine goes up, the metabolic pathway slows down. Is this control feedforward activation or feedback inhibition? Briefly explain.

24. The amino acid glutamine, is used by a liver metabolic pathway to make urea. As the concentration of glutamine goes up, the metabolic pathway speeds up. Is this control feedforward activation or feedback inhibition? Briefly explain.

Additional Exercises

These exercises are not paired or labeled by topic and provide additional practice on concepts covered in this chapter.

25. Chymotrypsin has a turnover number for a glycine-containing substrate of 0.05/s and a turnover number for an L-tyrosine–containing substrate of 200/s. For which substrate is chymotrypsin a more efficient catalyst? Briefly explain.

26. Feedback inhibition is an important form of enzyme regulation. Based on what you know about this control mechanism, how might a different process, "feedback activation," cause regulatory problems for a cell?

27. Explain the meaning of V_{max} by drawing a Michaelis–Menten plot.

28. The enzyme lactase is used in the food industry to modify milk products. What chemical change does lactase catalyze?

29. Name an important similarity between the enzymes used by paper manufacturers (for wood fiber digestion) and the enzyme used by clothing manufacturers (for "biostoning").

Answer to Practice Exercise

31.1 Under these conditions, pepsin is the more efficient because it has the larger turnover number, converting 30 reactant molecules to products per minute.

Nucleic Acids and Heredity

Chapter 32

The plight of Doctor Frankenstein's monster touches a chord in all of us. A scientist has given this monster life but cannot control his creation. The monster is "unnatural" and the story leads to tragedy.

The advent of genetic engineering has raised a similar specter. Genetic engineers work with the molecules that code life—the nucleic acids. By changing the code, new life forms can be produced. Already, bacteria have been altered to make needed human proteins. Recently, both cows and goats have been genetically engineered to produce a human protein in their milk.

The scientists involved in these initial programs have followed careful protocols and have produced valued medicines. However, the day may come when we can decide whether humans should be made smarter or stronger via genetic engineering. How this decision will be made and what it will be is the topic of ongoing heated discussions and much controversy; it has tremendous potential ramifications—and not only in the scientific community. The power to even consider such decisions and to possibly open Pandora's box in ways that were once only the stuff of science fiction is the result of our understanding of the biochemistry of nucleic acids.

32.1 Molecules of Heredity—A Link

The question of how hereditary material duplicates itself was for a long time one of the most baffling problems of biology. For many years biologists attempted in vain to solve this problem and to answer the question "Why are the offspring of a species undeniably of that species?" Many thought the chemical basis for heredity lay in the structure of the proteins. But no one was able to provide evidence showing how protein could reproduce itself. The answer to the heredity problem was finally found in the structure of the nucleic acids.

nucleoprotein
nucleic acid

The unit structure of all living things is the cell. Suspended in the nuclei of cells are chromosomes, which consist largely of proteins and nucleic acids. A simple protein bonded to a nucleic acid is called a **nucleoprotein**. There are two types of **nucleic acids**—those that contain the sugar deoxyribose and those that contain the sugar ribose. Accordingly, they are called deoxyribonucleic acid (DNA) and ribonucleic acid (RNA). Although many of us think of DNA as a recent discovery, it was actually discovered in 1869 by the Swiss physiologist Friedrich Miescher (1844–1895), who extracted it from the nuclei of cells.

32.2 Bases and Nucleosides

Nucleic acids are complex chemicals that combine several different classes of smaller molecules. As with many complex structures, it is easier to understand the whole by first studying its component parts. We begin our examination of nucleic acids by learning about a critical part of these molecules, two classes of heterocyclic bases called the purines and the pyrimidines:

◀ **Chapter Opening Photo: The science of genetics began when Gregor Mendel studied pea plants in the mid-nineteenth century.**

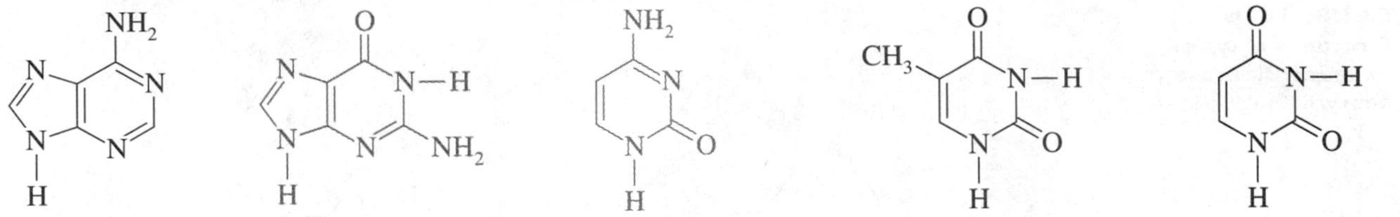

NH$_2$ … adenine (6-aminopurine)

guanine (2-amino-6-oxypurine)

cytosine (2-oxy-4-aminopyrimidine)

thymine (2,4-dioxy-5-methylpyrimidine)

uracil (2,4-dioxypyrimidine)

FIGURE 32.1 Purine and pyrimidine bases found in living matter.

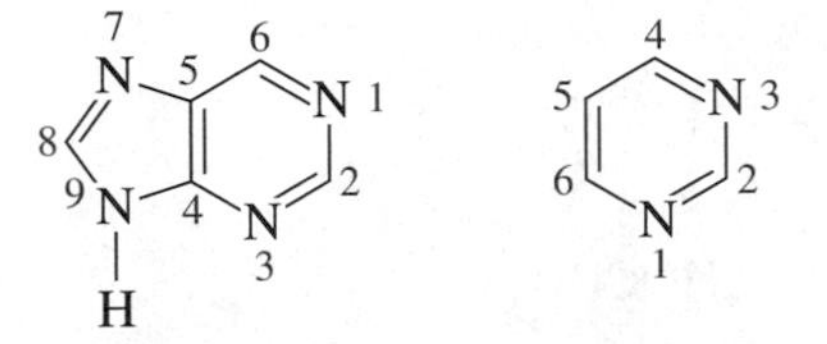

purine, $C_5H_4N_4$ pyrimidine, $C_4H_4N_2$

These parent compounds are related in structure, the pyrimidine being a six-membered heterocyclic ring while the purine contains both a five- and six-membered ring. The nitrogen atoms cause these compounds to be known as heterocycles (the rings are made up of more than just carbon atoms) and also as bases. Like the ammonia nitrogen, these heterocycles react with hydrogen ions to make a solution more basic.

There are five major bases commonly found in nucleic acids—two purine bases (adenine and guanine) and three pyrimidine bases (cytosine, thymine, and uracil). Figure 32.1 gives one stable form for each compound. Note that the bases differ one from another in their ring substituents. Each base has a lower-most nitrogen, which is bonded to a hydrogen as well as two carbons. This specific —NH shares chemical similarities with an alcohol (—OH) group. Just as two sugars can be linked when an alcohol of one monosaccharide reacts with a second monosaccharide (see Chapter 28), so a purine or pyrimidine can be bonded to a sugar by a reaction with the —NH group.

nucleoside

A **nucleoside** is formed when either a purine or pyrimidine base is linked to a sugar molecule, usually D-ribose or D-2′-deoxyribose.

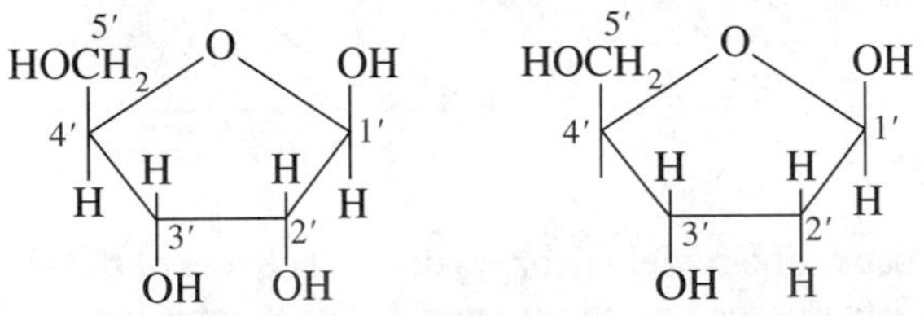

D-ribose   D-2′-deoxyribose

The base and sugar are bonded together between carbon 1′ of the sugar and either the purine nitrogen at position 9 or the pyrimidine nitrogen at position 1 by splitting out a molecule of water. Typical structures of nucleosides are shown in Figure 32.2. A prime is added to the position number to differentiate the sugar numbering system from the purine or pyrimidine numbering system.

The name of each nucleoside emphasizes the importance of the base to the chemistry of the molecule. Thus adenine and D-ribose react to yield adenosine,

FIGURE 32.2 ▶ Structures of typical ribonucleosides and deoxyribonucleosides.

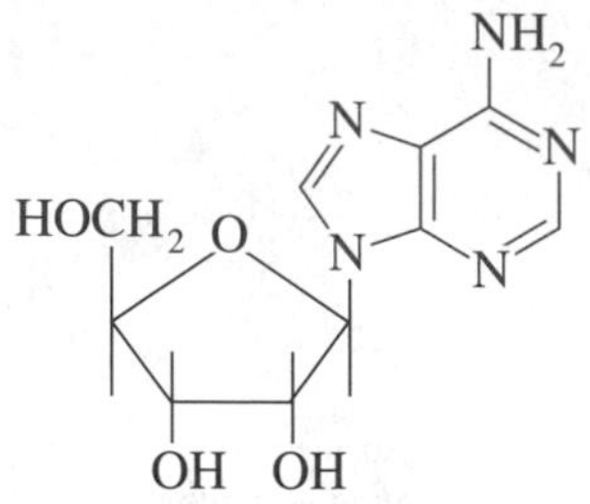

TABLE 32.1 Composition of Ribonucleosides and Deoxyribonucleosides

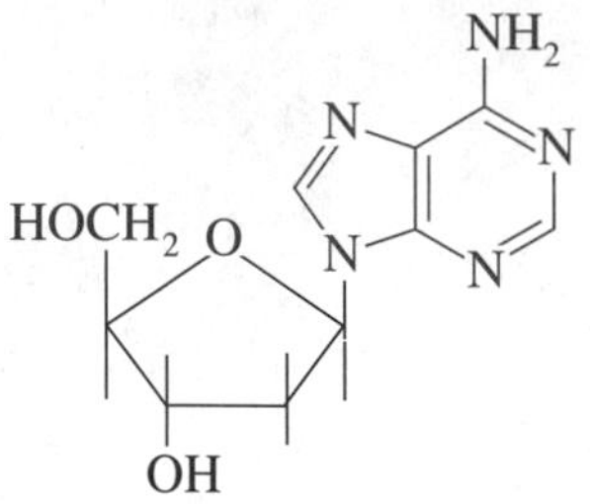

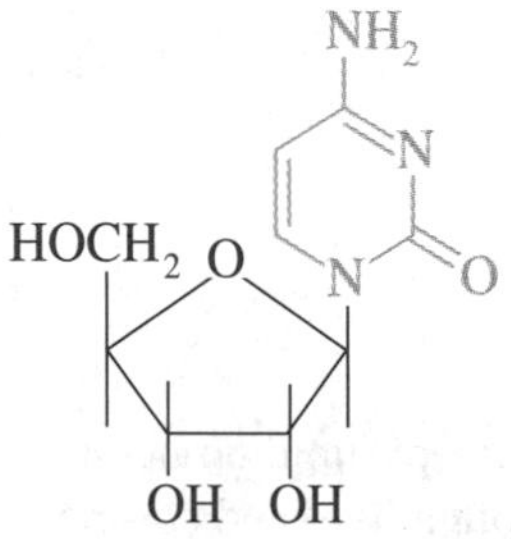

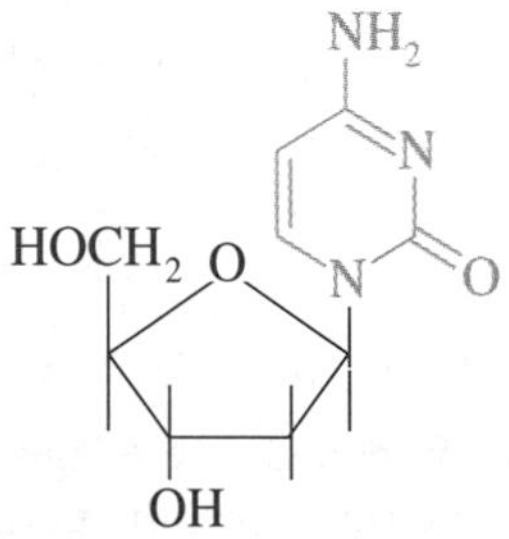

Name	Composition	Abbreviation
Adenosine	Adenine–ribose	A
Deoxyadenosine	Adenine–deoxyribose	dA
Guanosine	Guanine–ribose	G
Deoxyguanosine	Guanine–deoxyribose	dG
Cytidine	Cytosine–ribose	C
Deoxycytidine	Cytosine–deoxyribose	dC
Thymidine	Thymine–ribose	T
Deoxythymidine	Thymine–deoxyribose	dT
Uridine	Uracil–ribose	U
Deoxyuridine	Uracil–deoxyribose	dU

whereas cytosine and D-2′-deoxyribose yield deoxycytidine. The root of the nucleoside name derives from the purine or pyrimidine name. The compositions of the common ribonucleosides and the deoxyribonucleosides are given in Table 32.1.

32.3 Nucleotides: Phosphate Esters

nucleotide

A more complex set of biological molecules is formed by linking phosphate groups to nucleosides. Phosphate esters of nucleosides are termed **nucleotides**. These molecules consist of a purine or a pyrimidine base linked to a sugar, which in turn is bonded to at least one phosphate group.

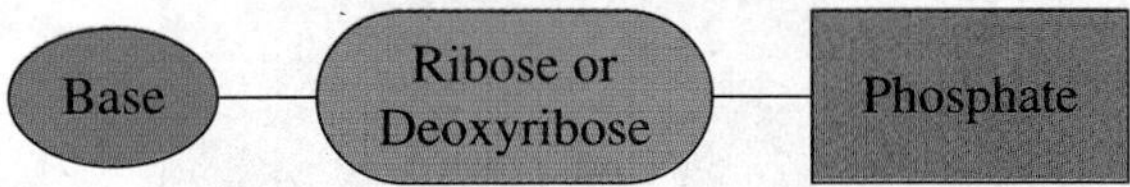

The ester may be a monophosphate, a diphosphate, or a triphosphate. When two or more phosphates are linked together, a high-energy phosphate anhydride bond is formed (see Section 32.4). The ester linkage may be to the hydroxyl group of position 2′, 3′, or 5′ of ribose or to position 3′ or 5′ of deoxyribose. Examples of nucleotide structures are shown in Figure 32.3.

Nucleotide abbreviations start with the corresponding nucleoside abbreviation (see Table 32.1). The letters MP (monophosphate) can be added to any of these to designate the corresponding nucleotide. Thus GMP is guanosine monophosphate. A lowercase d is placed in front of GMP if the nucleotide contains the deoxyribose sugar (dGMP). When the letters such as AMP or GMP are given, it is generally understood that the phosphate group is attached to position 5′ of the ribose unit (5′-AMP). If attachment is elsewhere, it will be designated, for example, as 3′-AMP.

Two other important adenosine phosphate esters are adenosine diphosphate (ADP) and adenosine triphosphate (ATP). Note that the letters DP are used for diphosphate and TP for triphosphate. In these molecules the phosphate groups are linked together. The structures are similar to AMP except that they contain two and three phosphate residues, respectively (see Figure 32.4). All the nucleosides form mono-, di-, and triphosphate nucleotides.

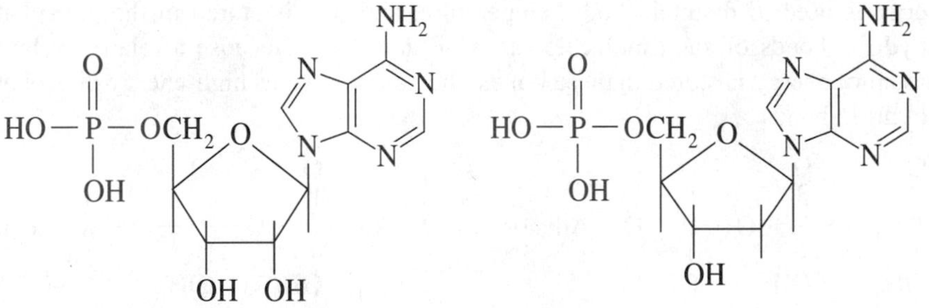

FIGURE 32.3
Examples of nucleotides.

adenosine-5′-monophosphate
(AMP)

deoxyadenosine-5′-monophosphate
(dAMP)

FIGURE 32.4
Structures of ADP and ATP.

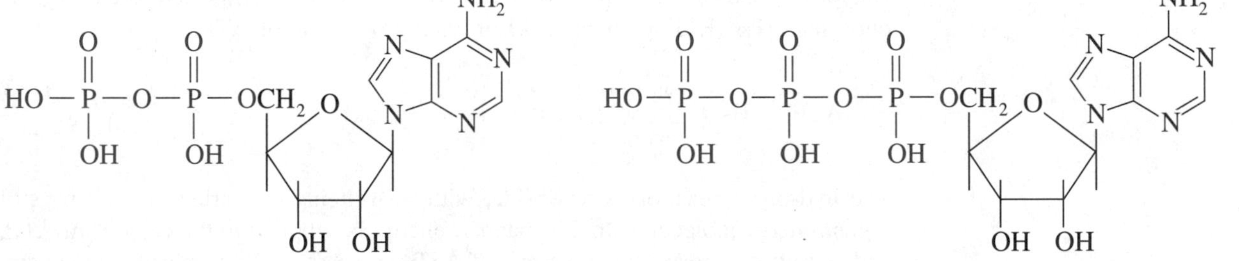

adenosine-5′-diphosphate
(ADP)

adenosine-5′-triphosphate
(ATP)

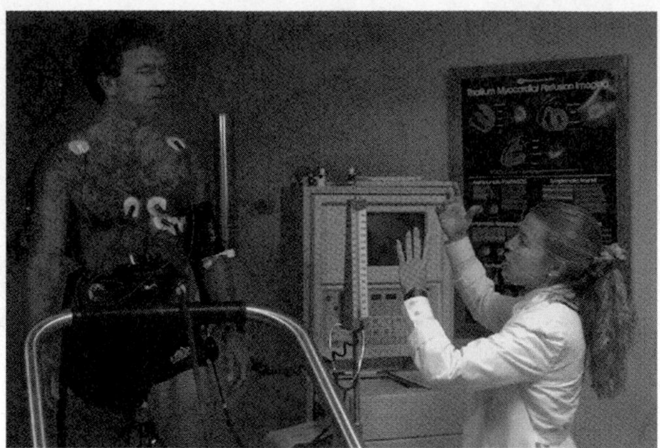

Muscle movements, including heart beats, are dependent on energy from ATP. ▶

32.4 High-Energy Nucleotides

The nucleotides have a central role in the energy transfers in many metabolic processes. ATP and ADP are especially important in these processes. The role of these two nucleotides is to store and release energy to the cells and tissues. The source of energy is the foods we eat, particularly carbohydrates and fats. Energy is released as the carbons from these foods are oxidized (see Chapter 34). Part of this energy is used to maintain body temperature, and part is stored in the phosphate anhydride bonds of such molecules as ADP and ATP. Because a relatively large amount of energy is stored in these bonds, they are known as high-energy phosphate anhydride bonds.

$$HO-\overset{\overset{\displaystyle O}{\|}}{\underset{\underset{\displaystyle OH}{|}}{P}}\sim O-\overset{\overset{\displaystyle O}{\|}}{\underset{\underset{\displaystyle OH}{|}}{P}}\sim O-\overset{\overset{\displaystyle O}{\|}}{\underset{\underset{\displaystyle OH}{|}}{P}}-O-\text{Adenosine}$$

High-energy phosphate anhydride bonds
(ATP)

$$HO-\overset{\overset{\displaystyle O}{\|}}{\underset{\underset{\displaystyle OH}{|}}{P}}\sim O-\overset{\overset{\displaystyle O}{\|}}{\underset{\underset{\displaystyle OH}{|}}{P}}-O-\text{Adenosine}$$

High-energy anhydride bond
(ADP)

Metabolic energy is released during the hydrolysis of high-energy phosphate anhydride bonds in ADP and ATP. In the hydrolysis, ATP forms ADP and inorganic phosphate (P_i) yielding about 35 kJ of energy per mole of ATP:

$$\text{ATP} + H_2O \underset{\text{energy storage}}{\overset{\text{energy utilization}}{\rightleftharpoons}} \text{ADP} + P_i + {\sim}35 \text{ kJ}$$

The hydrolysis reaction is reversible, with ADP being converted to ATP by still higher energy molecules. In this manner, energy is supplied to the cells from ATP, and energy is stored by the synthesis of ATP. Processes such as muscle movement, nerve sensations, vision, and even the maintenance of our heartbeats are all dependent on energy from ATP.

32.5 Polynucleotides; Nucleic Acids

Starting with two nucleotides, a dinucleotide is formed by splitting out a molecule of water between the —OH of the phosphate group of one nucleotide and the —OH on carbon 3′ of the ribose or deoxyribose of the other nucleotide. Then another and another nucleotide can be added in the same manner until a polynucleotide chain is formed. Each nucleotide is linked to its neighbors by phosphate ester bonds (see Section 24.13).

Two series of polynucleotide chains are known, one containing D-ribose and the other D-2′-deoxyribose. One polymeric chain consists of the monomers AMP, GMP, CMP, and UMP, and is known as a polyribonucleotide. The other chain contains the monomers dAMP, dGMP, dCMP, and dTMP, and is known as a polydeoxyribonucleotide:

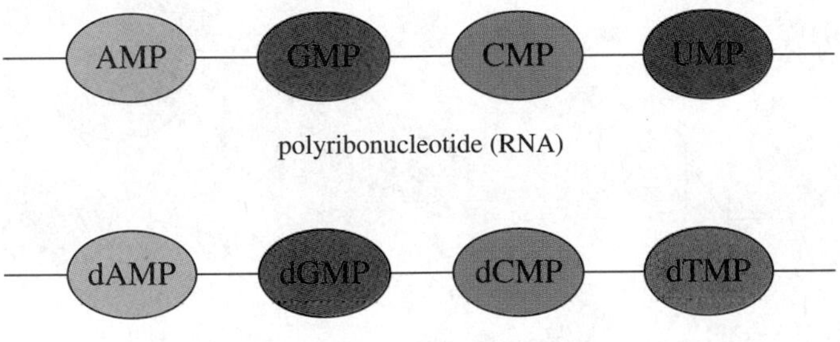

ribonucleic acid (RNA)

deoxyribonucleic acid (DNA)

The nucleic acids DNA and RNA are polynucleotides. **Ribonucleic acid (RNA)** is a polynucleotide that upon hydrolysis yields ribose, phosphoric acid, and the four purine and pyrimidine bases adenine, guanine, cytosine, and uracil. **Deoxyribonucleic acid (DNA)** is a polynucleotide that yields D-2′-deoxyribose, phosphoric acid, and the four bases adenine, guanine, cytosine, and thymine. Note that RNA and DNA contain one different pyrimidine nucleotide: RNA contains uridine, whereas DNA contains thymidine. A segment of a ribonucleic acid chain is shown in Figure 32.5. As will be described later, RNA and DNA also commonly differ in function: DNA serves as the storehouse for genetic information; RNA aids in expressing genetic characteristics.

32.6 Structure of DNA

Deoxyribonucleic acid (DNA) is a polymeric substance made up of the four nucleotides dAMP, dGMP, dCMP, and dTMP. The size of the DNA polymer varies with the complexity of the organism; more complex organisms tend to have larger DNAs. For example, simple bacteria like *Escherichia coli* have about 8 million nucleotides in their DNA while a human DNA contains up to 500 million nucleotides. The order in which these nucleotides occur varies in different DNA molecules, and it is within this order that the genetic information in a cell is stored.

Scientists have tried to understand DNA structure by determining the nucleotide

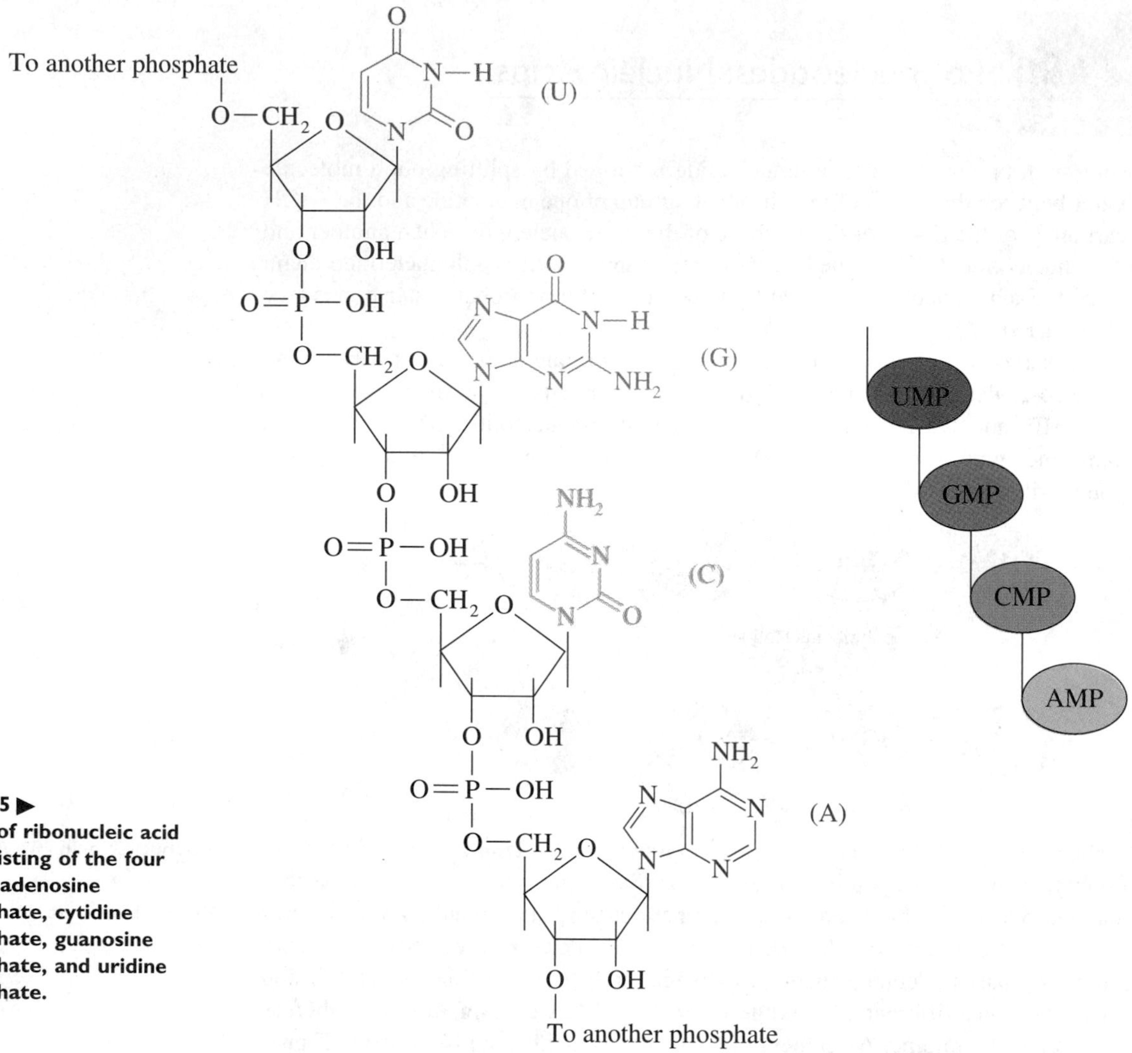

To another phosphate

(U)

(G)

(C)

(A)

To another phosphate

UMP
GMP
CMP
AMP

FIGURE 32.5 ▶
A segment of ribonucleic acid (RNA) consisting of the four nucleotides adenosine monophosphate, cytidine monophosphate, guanosine monophosphate, and uridine monophosphate.

composition of this molecule from many different sources. For a long time it was thought that the four nucleotides occurred in equal amounts in DNA. However, more refined analyses showed that the amounts of purine and pyrimidine bases varied in different DNA molecules. Surprisingly, careful consideration of these data also showed that the ratios of adenine to thymine and guanine to cytosine were always essentially 1:1. This observation served as an important key to unraveling the structure of DNA. The analyses of DNA from several species are shown in Table 32.2.

A second important clue to the special configuration and structure of DNA came from X-ray diffraction studies. Most significant was the work of Maurice Wilkins (b. 1916) of Kings College of London. Wilkins's X-ray pictures implied that the nucleotide bases were stacked one on top of another like a stack of saucers. From his work as well as that of others, the American biologist James D. Watson (b. 1928) and British physicist Francis H.C. Crick (b. 1916), working at Cambridge

TABLE 32.2 Relative Amounts of Purines and Pyrimidines in Samples of DNA

Source	Adenine	Thymine	Ratio A/T	Guanine	Cytosine	Ratio G/C
Beef thymus	29.0	28.5	1.02	21.2	21.2	1.00
Beef liver	28.8	29.0	0.99	21.0	21.1	1.00
Beef sperm	28.7	27.2	1.06	22.2	22.0	1.01
Human thymus	30.9	29.4	1.05	19.9	19.8	1.00
Human liver	30.3	30.3	1.00	19.5	19.9	0.98
Human sperm	30.9	31.6	0.98	19.1	18.4	1.04
Hen red blood cells	28.8	29.2	0.99	20.5	21.5	0.96
Herring sperm	27.8	27.5	1.01	22.2	22.6	0.98
Wheat germ	26.5	27.0	0.98	23.5	23.0	1.02
Yeast	31.7	32.6	0.97	18.3	17.4	1.05
Vaccinia virus	29.5	29.9	0.99	20.6	20.0	1.03
Bacteriophage T_2	32.5	32.6	1.00	18.2	18.6	0.98

University, designed and built a scale model of a DNA molecule. In 1953, Watson and Crick announced their now famous double-stranded helical structure for DNA. This was a milestone in the history of biology, and in 1962 Watson, Crick, and Wilkins were awarded the Nobel prize in medicine and physiology for their studies of DNA.

The structure of DNA, according to Watson and Crick, consists of two polymeric strands of nucleotides in the form of a double helix, with both nucleotide strands coiled around the same axis (see Figure 32.6). Along each strand are alternate phosphate and deoxyribose units with one of the four bases adenine, guanine, cytosine, or thymine attached to deoxyribose as a side group. The double helix is held together by hydrogen bonds extending from the base on one strand of the double helix to a complementary base on the other strand. The structure of DNA has been likened to a ladder that has been twisted into a double helix, with the rungs of the ladder kept perpendicular to the twisted railings. The phosphate and deoxyribose units alternate along the two railings of the ladder, and two nitrogen bases form each rung of the ladder.

In the Watson–Crick model of DNA, the two polynucleotide strands fit together best when a purine base is adjacent to a pyrimidine base. Although this allows for four possible base pairings, A–T, A–C, G–C, and G–T, only the adenine–thymine and guanine–cytosine base pairs can effectively hydrogen-bond together. Under normal conditions, an adenine on one polynucleotide strand is paired with a thymine on the other strand; a guanine is paired with a cytosine (see Figure 32.7). This pairing results in A:T and G:C ratios of 1:1, as substantiated by the data in Table 32.2. The hydrogen bonding of complementary base pairs is shown in Figure 32.8. Note that if the sequence of one strand is known, the sequence of the other strand can be determined. The two DNA polymers are said to be *complementary* to each other. As will be discussed in Section 32.7, the cell can chemically "read" one strand in order to synthesize its complementary partner.

The double helix is an important part of DNA structure. It does not explain how the very large DNA can be packed into a cell or the even smaller cell nucleus. For example, a human DNA molecule can be extended to almost 10 cm in length and yet is contained in a nucleus with a diameter about a hundred thousand times

FIGURE 32.6 ▶
Right: **Double-stranded helical structure of DNA (....denotes a hydrogen bond between adjoining bases).** ***Left:*** **Space filling model of DNA.**

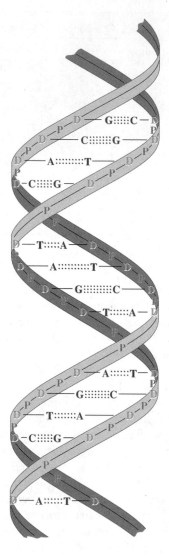

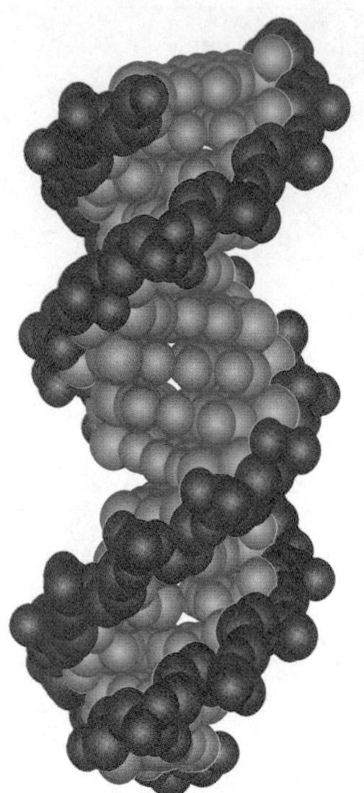

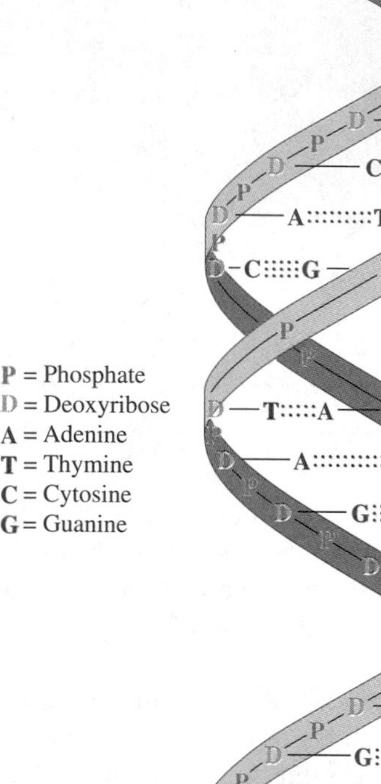

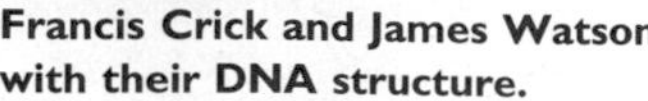
Francis Crick and James Watson with their DNA structure.

smaller. To begin the necessary packing, the DNA is looped around small aggregates of positively charged histone proteins. Hundreds of these aggregates are associated with each DNA molecule so that the DNA is foreshortened and has the appearance of a string of pearls. Further condensation is achieved by wrapping this structure into a tight coil called a solenoid, as shown in Figure 32.9. Finally, the solenoid nucleoprotein complex is wound around a protein scaffold within the nucleus. By following this complete procedure, human DNA can be taken to a length of about 10 μm in condensed chromosomes. Like a high-density computer disk, DNA takes up only a little space relative to the large amount of genetic information it contains.

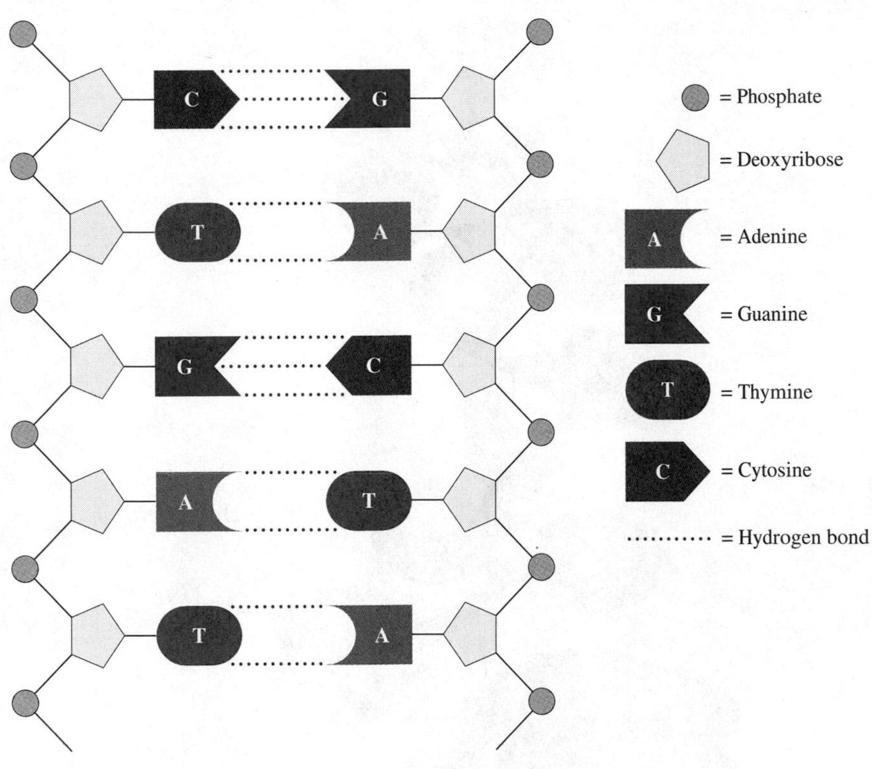

FIGURE 32.7
Schematic diagram of a DNA segment showing phosphate, deoxyribose, and complementary base pairings held together by hydrogen bonds.

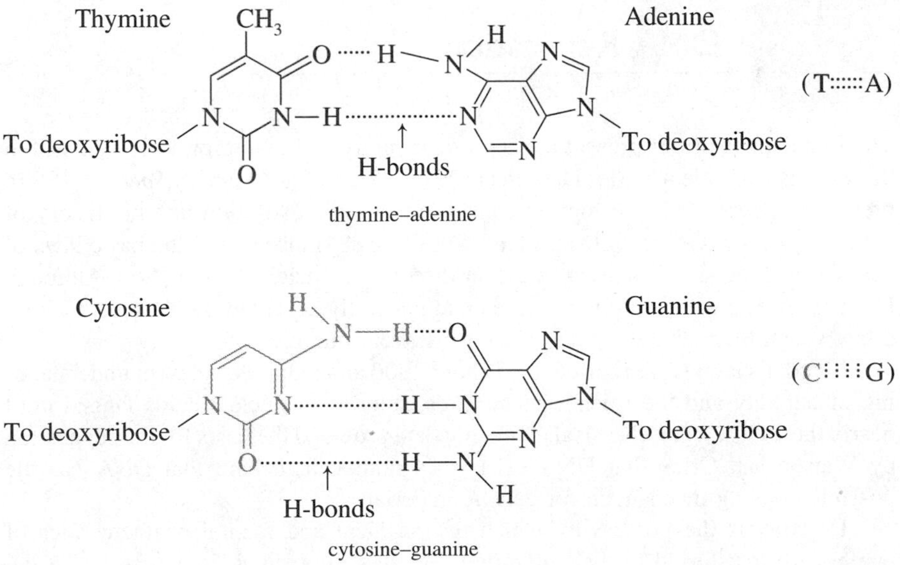

Thymine — Adenine

CH_3 … O ····· H—N(H) … N—H ················· N

To deoxyribose — To deoxyribose

H-bonds

thymine–adenine

(T::::::A)

Cytosine — Guanine

H—N—H ····· O … N ················· H—N … O ················· H—N—H

To deoxyribose — To deoxyribose

H-bonds

cytosine–guanine

(C⁝⁝⁝⁝G)

FIGURE 32.8
Hydrogen bonding between the complementary bases thymine and adenine (T : : : : A) and cytosine and guanine (C ⁝ ⁝ ⁝ ⁝ G). Note that one pair of bases has two hydrogen bonds and the other pair has three hydrogen bonds.

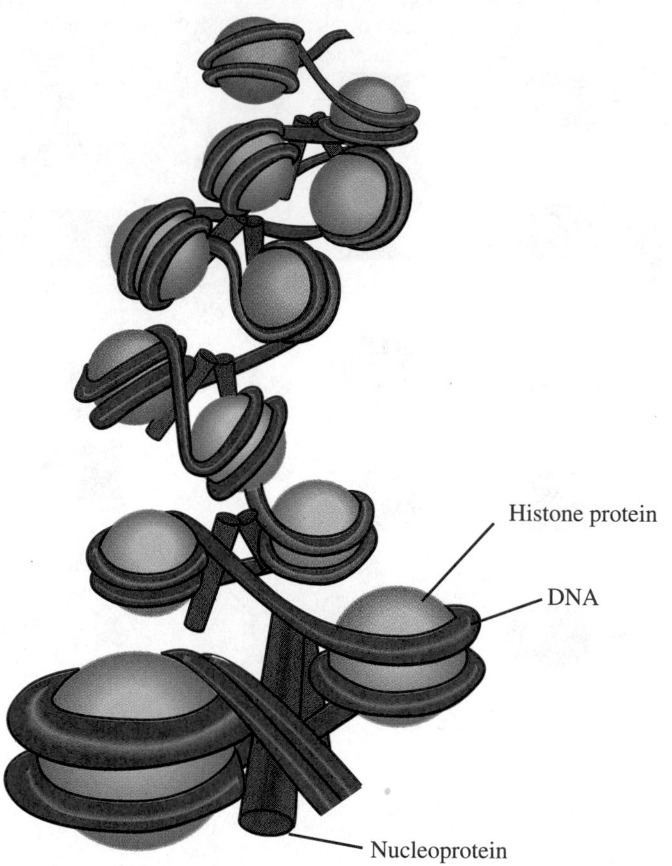

FIGURE 32.9 ▶
Condensed form of DNA. After the DNA polymer is wrapped twice around the histone protein aggregates (shown as round balls), it is coiled around a central axis to form a solenoid structure.

32.7 DNA Replication

The foundations of our present concepts of heredity and evolution were laid within the span of a decade. Charles Darwin (1809–1882), in *The Origin of Species* (1859), presented evidence supporting the concept of organic evolution and his theory of natural selection. Gregor Johann Mendel (1822–1884) discovered the basic laws of heredity in 1866, and Friedrich Miescher discovered nucleic acid in 1869. Although Darwin's views were widely discussed and generally accepted by biologists within a few years, Mendel's and Miescher's work went unnoticed for many years.

Mendel's laws were rediscovered about 1900 and led to our present understanding of heredity and the science of genetics. Interest in nucleic acids lagged until nearly the 1950s, when chemical and X-ray data provided the basis for the suggestion by Watson and Crick that DNA exists in a double helix and that DNA has the possible copying mechanism for genetic material.

Heredity is the process by which the physical and mental characteristics of parents are transferred to their offspring. For this to occur, it is necessary for the material responsible for genetic transfer to be able to make exact copies of itself. The polymeric DNA molecule is the chemical basis for heredity. The genetic information needed for transmittal of a species' characteristics is coded along the polymeric chain. Although the chain is made from only four different nucleotides, the information content of DNA resides in the sequence of these nucleotides.

The **genome** is the sum of all hereditary material contained in a cell. Within the eucaryotic genome are chromosomes, which are long, thread-like bodies composed of nucleic acids and proteins that contain the fundamental units of heredity, called genes. A **gene** is a segment of the DNA chain that controls formation of a molecule of RNA. In turn, many RNAs determine the amino acid sequences for specific polypeptides or proteins. One gene commonly directs the synthesis of only one polypeptide or protein molecule. The cell has the capability of producing a multitude of different proteins because each DNA molecule contains a large number of different genes.

genome

gene

For life to continue relatively unchanged, genetic information must be reproduced exactly each time a cell divides. **Replication**, as the name implies, is the biological process for duplicating the DNA molecule. The DNA structure of Watson and Crick holds the key to replication; because of the complementary nature of DNA's nitrogen bases, adenine bonds only to thymine and guanine only to cytosine. Nucleotides with complementary bases can hydrogen-bond to each single strand of DNA and hence be incorporated into a new DNA double helix. Every double-stranded DNA molecule that is produced contains one template strand and one newly formed, complementary strand. This form of DNA synthesis, known as *semiconservative* replication, is illustrated in Figure 32.10.

replication

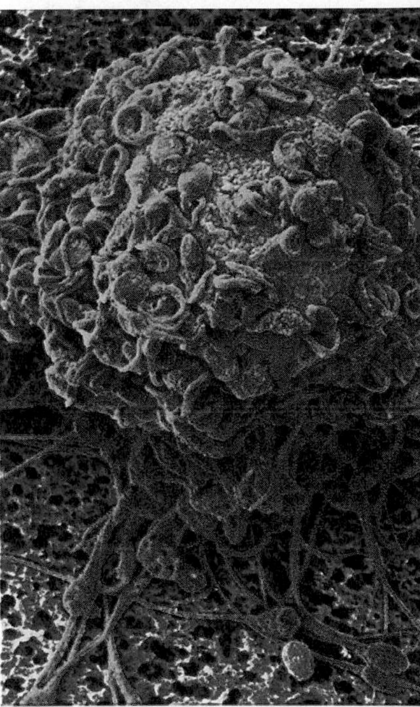

Although many sperm may approach an egg, only one penetrates the egg during fertilization.

Replication is one of the most complicated enzyme-catalyzed processes in life. Enzymes are required to unwind the DNA before replication and to repackage the DNA after synthesis. The two template strands are copied differently. One daughter strand grows directly toward the point at which the templates are unwinding while the other daughter strand is synthesized away from this point. In this way, the same enzyme-catalyzed reaction can be used to synthesize both strands. However, the one daughter strand that has been created in small fragments must be connected before replication is complete. So, while one DNA strand is formed by continuous synthesis, the other new strand is formed by a repetition of fragment synthesis followed by a coupling reaction (see Figure 32.11).

Other processes "proofread" the new polymers to check for errors. Amazingly, replication is so carefully coordinated that a mistake is passed on to the new strands only once in about a billion times. Of course, this important fact means that new cells retain the genetic characteristics of their parents.

DNA and Cell Division

The DNA content of cells doubles just before the cell divides, and one-half of the DNA goes to each daughter cell. After cell division is completed, each daughter cell contains DNA and the full genetic code that was present in the original cell. This process of ordinary cell division is known as **mitosis** and occurs in all the cells of the body except the reproductive cells.

mitosis

As we have indicated before, DNA is an integral part of the chromosomes. Each species carries a specific number of chromosomes in the nucleus of each of its cells. The number of chromosomes varies with different species. Humans have 23 pairs, or 46 chromosomes. The fruit fly has 4 pairs, or 8 chromosomes. Each

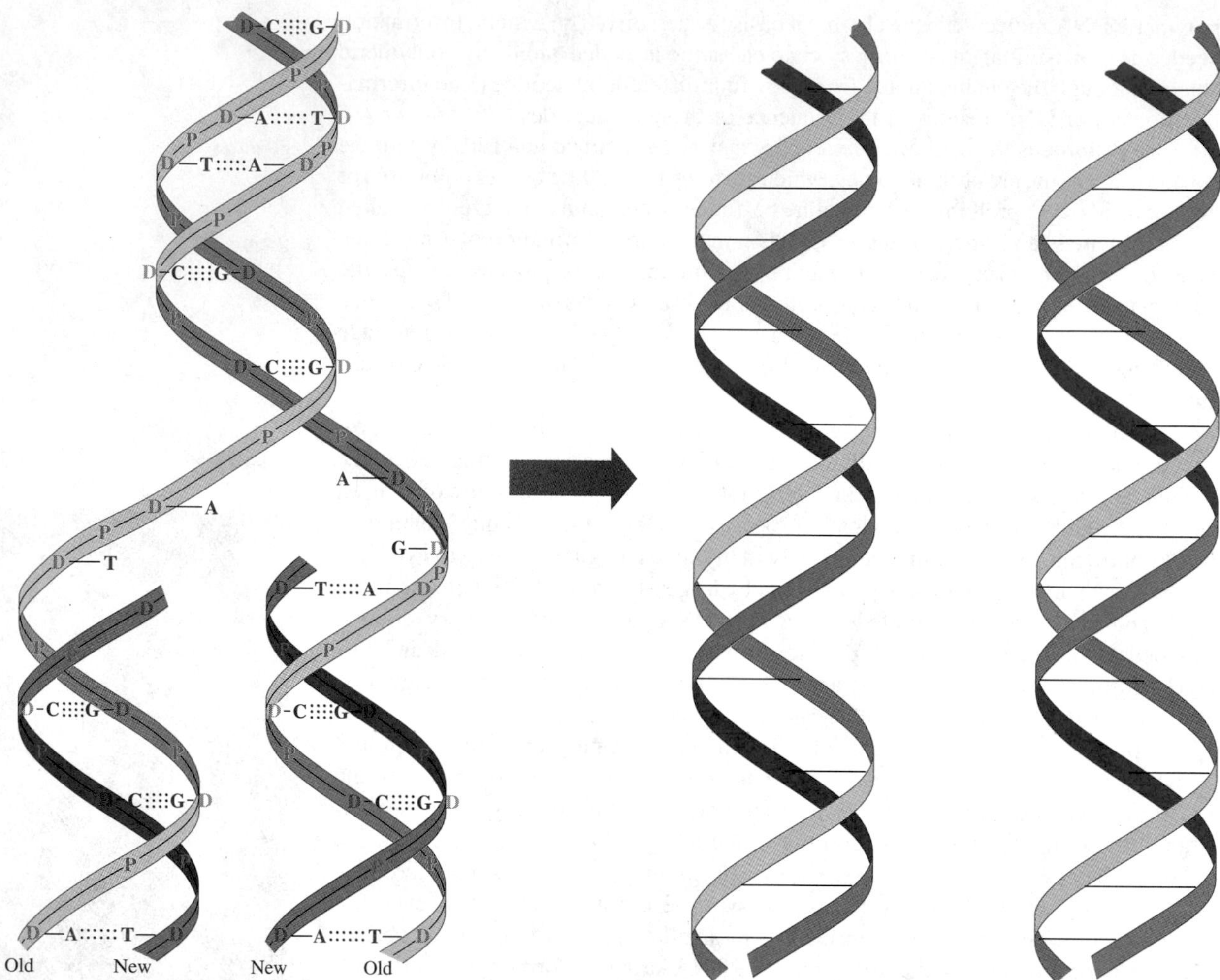

▲
FIGURE 32.10
Semiconservative replication of DNA: The two helices unwind, separating at the hydrogen bonds. Each strand then serves as a template, recombining with the proper nucleotides to form a new double-stranded helix. The newly synthesized DNA strands are shown in red.

chromosome contains DNA molecules. Mitosis produces cells with the same chromosomal content as the parent cell. However, in the sexual reproductive cycle, cell division occurs by a different process known as meiosis.

In sexual reproduction, two cells, the sperm cell from the male and the egg cell (or ovum) from the female, unite to form the cell of the new individual. If reproduction took place with mitotic cells, the normal chromosome content would double when two cells united. However, in **meiosis** the cell splits in such a way as to reduce the number of chromosomes to one-half of the number normally present (23 in humans). The sperm cell carries half of the chromosomes from its original cell, and the egg cell also carries half of the chromosomes from its original cell. When the sperm and the egg cells unite during fertilization, the cell once again contains the correct number of chromosomes and the hereditary characteristics of the species. Thus the offspring derives half its genetic characteristics from the father and half from the mother.

meiosis

Chemistry in Action

Fingerprints in the Genes

Just as for a set of fingerprints, no two individuals' DNA sequences are exactly the same. More than 100 years ago, detectives first started to use fingerprints from a crime scene to identify criminals. Today, forensic chemists use biochemistry to unlock the DNA sequence to identify possible criminals.

Because the human genome contains approximately 3 billion base pairs, it is not practical for forensic chemists to examine the entire genome sequence. Instead, they identify randomly selected pieces of DNA. If enough DNA is available (e.g., from a 1-cm diameter blood stain), this genetic material can be broken into pieces and separated on a gel by electrophoresis. The gel is then treated with radioactive molecules that bind to specific sites and serve as tags on the DNA fragments. These tags form a pattern on the gel known as a DNA "fingerprint."

A second technique uses the enzyme DNA polymerase to replicate the DNA from a very small sample (only several nanograms of DNA are needed). The replication process multiplies the amount of DNA available until there are enough DNA pieces to separate and visualize in gel electrophoresis. Again, a pattern is formed that can be treated like a fingerprint.

Unfortunately, DNA fingerprinting uses only a small segment of the total genome. Although each person has a unique total genome, there is an extremely small chance that two people will have the same DNA sequence for a particular segment. So, scientists must estimate the possibility of a match by chance, and courtroom decisions often hinge on complex statistical reasoning.

Still, DNA fingerprinting, a technique only available since 1990, continues to make courtroom history. Since most body fluids contain cells that carry DNA, DNA "fingerprints" can often be found at crime scenes, and many "DNA" convictions and exonerations have occurred because of this. Many experts believe that DNA fingerprinting will become more and more widely used in court proceedings in the future.

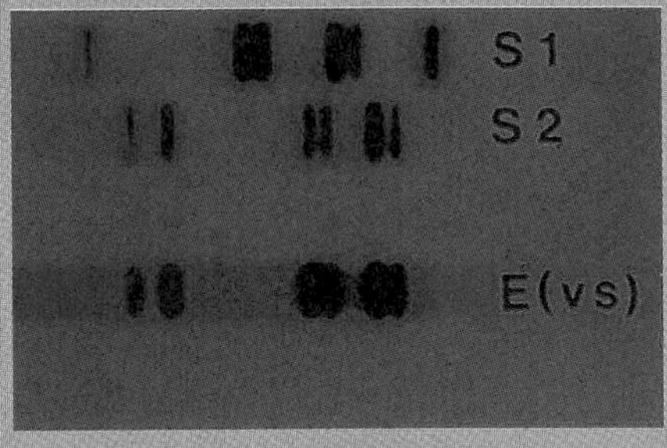

DNA fingerprint matching. Note that Suspect S2 matches evidence blood sample E(vs).

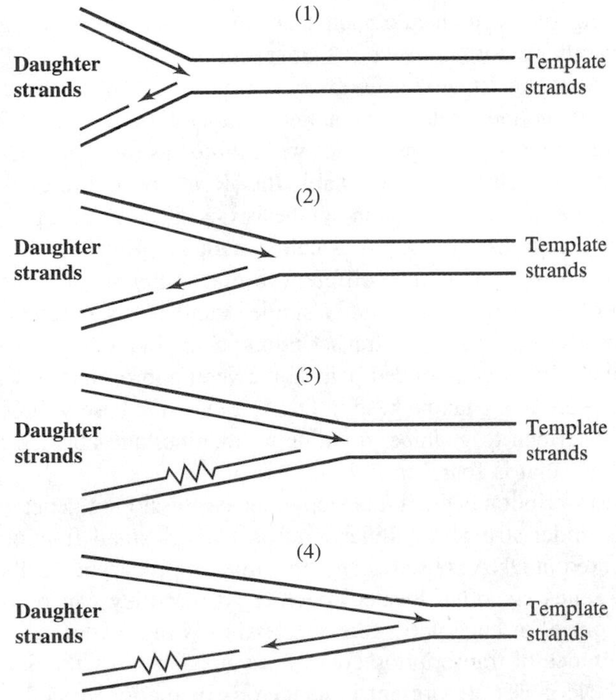

FIGURE 32.11 The basic replication process. Arrows indicate the direction of DNA synthesis. Between schematics (1) and (2), the template DNA strands unwind, and some synthesis occurs for both daughter strands. Moving to schematic (3), the DNA fragments are connected on one daughter strand (—WW—) while DNA synthesis continues on the other strand and the template strands unwind further. DNA synthesis again takes place for both daughter strands in schematic (4).

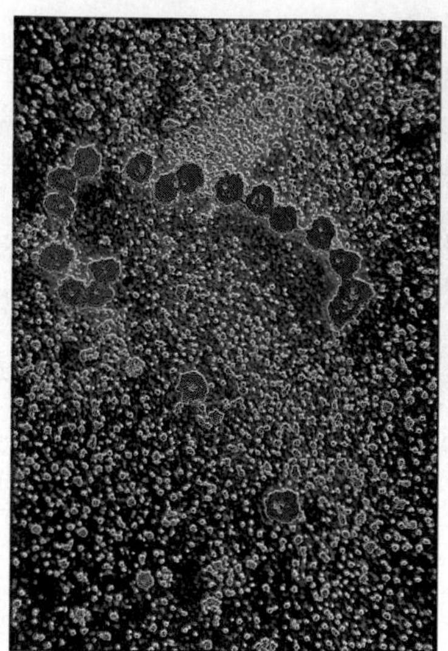

These ribosomes are translating mRNA strands to produce proteins.

32.8 RNA: Genetic Transcription

One of the main functions of DNA is to direct the synthesis of ribonucleic acids (RNAs). RNA differs from DNA in the following ways: (1) it consists of a single polymeric strand of nucleotides rather than a double helix; (2) it contains the pentose D-ribose instead of D-2′-deoxyribose; (3) it contains the pyrimidine base uracil instead of thymine; and (4) some types of RNA have a significant number of modified bases in addition to the common four. RNA also differs functionally from DNA. Whereas DNA serves as the storehouse of genetic information, RNA is used to process this information into proteins. Three types of RNA are needed to produce proteins: ribosomal RNA (rRNA), messenger RNA (mRNA), and transfer RNA (tRNA).

More than 80% of the cellular RNA is ribosomal RNA. It is found in the ribosomes, where it is associated with protein in proportions of about 60–65% protein to 30–35% rRNA. Ribosomes are the sites for protein synthesis.

Messenger RNA carries genetic information from DNA to the ribosomes. It is a template made from DNA and carries the code that directs the synthesis of proteins. The size of mRNA varies according to the length of the polypeptide chain it will encode.

The primary function of tRNA is to bring amino acids to the ribosomes for incorporation into protein molecules. Consequently, there exists at least one tRNA for each of the 20 amino acids required for proteins. Transfer RNA molecules have a number of structural features in common. The end of the chain of all tRNA molecules terminates in a CCA nucleotide sequence to which is attached the amino acid to be transferred to a protein chain. The primary structure of tRNA allows extensive folding of the molecule such that complementary bases are hydrogen-bonded to each other to form a structure that appears like a cloverleaf. The cloverleaf model of tRNA has an anticodon loop consisting of seven unpaired nucleotides. Three of these nucleotides make up an anticodon (see Figure 32.12). The anticodon is complementary to, and hydrogen-bonds with, three bases on an mRNA. The other two loops in the cloverleaf structure enable the tRNA to bind to the ribosome and other specific enzymes during protein synthesis (see Section 32.11).

transcription

The making of RNA from DNA is called **transcription**. The verb *transcribe* literally means to copy, often into a different format. When the nucleotide sequence of one strand of DNA is transcribed into a single strand of RNA, genetic information is copied from DNA to RNA. This transcription occurs in a complementary fashion and depends upon hydrogen-bonded pairing between appropriate bases (see Figure 32.13). Where there is a guanine base in DNA, a cytosine base will occur in RNA. Cytosine is transcribed to guanine, thymine to adenine, and adenine to uracil (the thymine-like base that is found in RNA).

Because transcription is the initial step in the expression of genetic information, this process is under stringent cellular control. Only a small fraction of the total information stored in DNA is used at any one time. In procaryotic cells (see Section 34.5), related genes are often located together so that they can be transcribed in concert. The control of eucaryotic gene expression is more complex.

oncogene

The importance of transcription control is emphasized by the discovery of the **oncogenes**. These genes are present in cancerous or malignant cells and code for proteins that control cell growth. To the surprise of many investigators, it was found

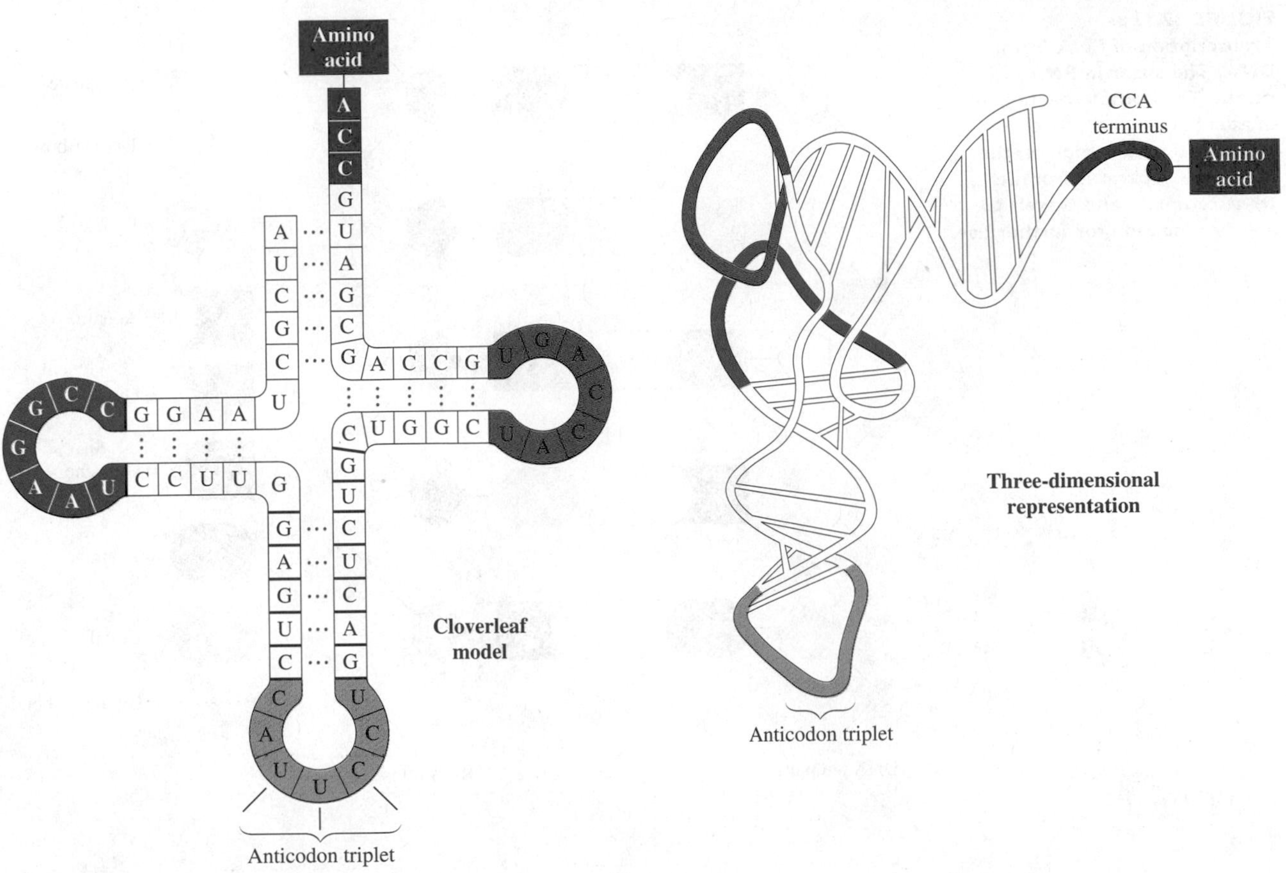

FIGURE 32.12
Representations of tRNA. The anticodon triplet (UUC) located at the lower loop is complementary to GGA (which is the code for glycine) on mRNA.

that these same oncogenes are present in many normal mammalian cells. The question arose, "Why do some cells become cancerous and some remain normal?" The answer: Cancerous cells lose control of oncogene transcription. Control is lost when the DNA structure is altered by, for example, a viral infection or a chemical carcinogen. The oncogenes are then transcribed too often, cell growth is uncontrolled—and a cancer develops.

Following transcription, the RNA molecules that are produced from DNA are often modified before they are put to use. Ribosomal RNA and tRNA molecules are formed as larger precursors and are then trimmed to the correct size. After transcription, many of the bases in tRNA are modified by methylation, saturation of a double bond, or by isomerization of the ribose-base linkage. This posttranscriptional modification or processing changes the information content of the RNA.

Eucaryotic mRNA may undergo considerable alteration before the message that it carries is ready to guide protein synthesis. These changes may include the elimination of portions of the mRNA molecule, the splicing together of two or more mRNA molecules, or the addition of new bases to one end of the mRNA. During this processing, the message encoded in the mRNA molecule is apparently refined so that it can be correctly read by the ribosome.

Figure 32.14 summarizes the processes by which genetic information is transferred within a cell. These steps culminate in the conversion of a coded nucleotide message into a protein amino acid sequence.

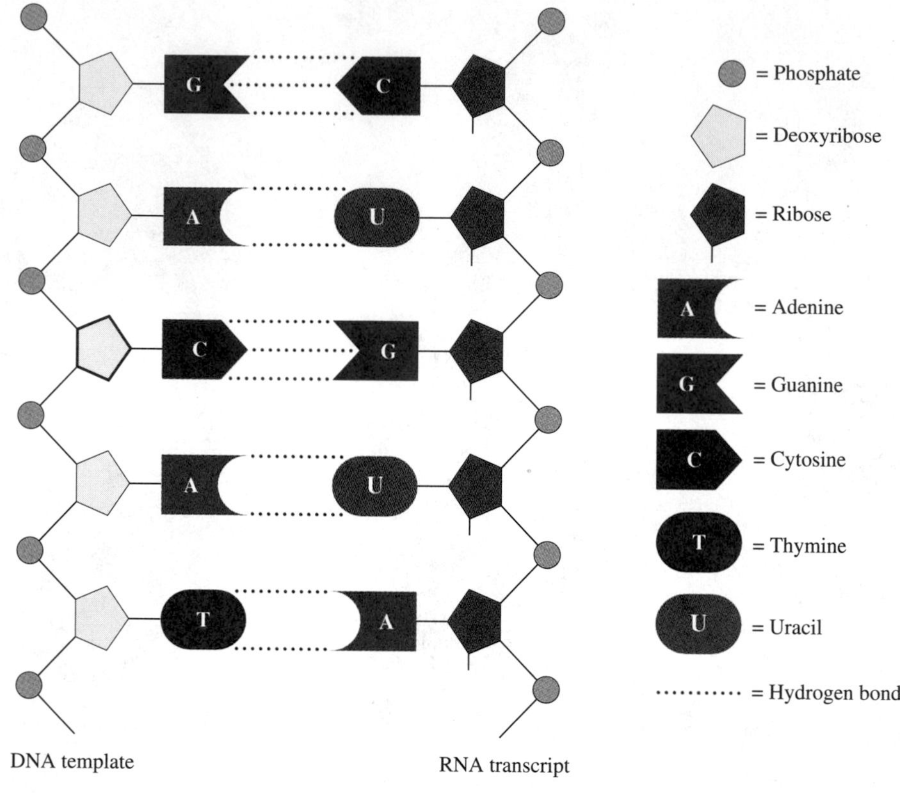

FIGURE 32.13 ▶ Transcription of RNA from DNA. The sugar in RNA is ribose. The complementary base of adenine is uracil. After transcription is complete, the new RNA separates from its DNA template and travels to another location for further use.

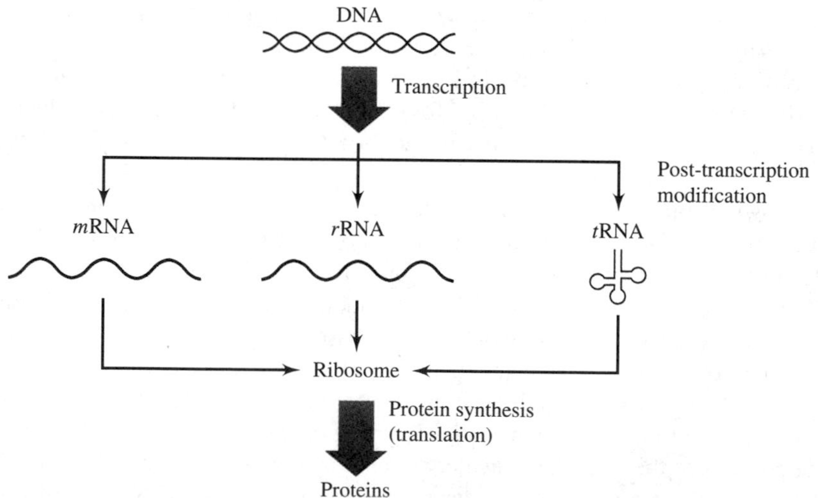

FIGURE 32.14 ▶ A flow diagram representing the processing of cellular genetic information.

32.9 The Genetic Code

For a long time after the structure of DNA was elucidated, scientists struggled with the problem of how the information stored in DNA could specify the synthesis of so many different proteins. Since the backbone of the DNA molecule contains a regular structure of repeated and identical phosphate and deoxyribose units, the key to the code had to lie with the four bases—adenine, guanine, cytosine, and thymine.

The code, using only the four nucleotides A, G, C, and T, must be capable of coding at least for the 20 amino acids that occur in proteins. If each nucleotide coded one amino acid, only four amino acids could be represented. If the code used two nucleotides to specify an amino acid, 16 (4×4) combinations would be possible—still not enough. Using three nucleotides, we can have 64 ($4 \times 4 \times 4$) possible combinations—which is more than enough to specify the 20 common amino acids in proteins. It has now been determined that each code word requires a sequence of three nucleotides. The code is therefore a triplet code. Each triplet of three nucleotides is called a **codon**, and, in general, each codon specifies one amino acid. Thus, to describe a protein containing 200 amino acid units, a gene containing at least 200 codons, or 600 nucleotides, is required.

codon

In the sequence of biological events, the code from a gene in DNA is first transcribed to a coded RNA, which, in turn, is used to direct the synthesis of a protein. The 64 possible codons for mRNA are given in Table 32.3. In this table a

TABLE 32.3 The Genetic Code for Messenger RNA

First nucleotide	Second nucleotide U	C	A	G	Third nucleotide
U	Phe	Ser	Tyr	Cys	U
U	Phe	Ser	Tyr	Cys	C
U	Leu	Ser	TC[a]	TC[a]	A
U	Leu	Ser	TC[a]	Trp	G
C	Leu	Pro	His	Arg	U
C	Leu	Pro	His	Arg	C
C	Leu	Pro	Gln	Arg	A
C	Leu	Pro	Gln	Arg	G
A	Ile	Thr	Asn	Ser	U
A	Ile	Thr	Asn	Ser	C
A	Ile	Thr	Lys	Arg	A
A	Met	Thr	Lys	Arg	G
G	Val	Ala	Asp	Gly	U
G	Val	Ala	Asp	Gly	C
G	Val	Ala	Glu	Gly	A
G	Val	Ala	Glu	Gly	G

[a]Termination or nonsense codons.

◀ This table shows the sequence of nucleotides in the triplet codons of messenger RNA that specify a given amino acid. For example, UUU or UUC is the codon for Phe, UCU is the codon for Ser; CAU or CAC is the codon for His.

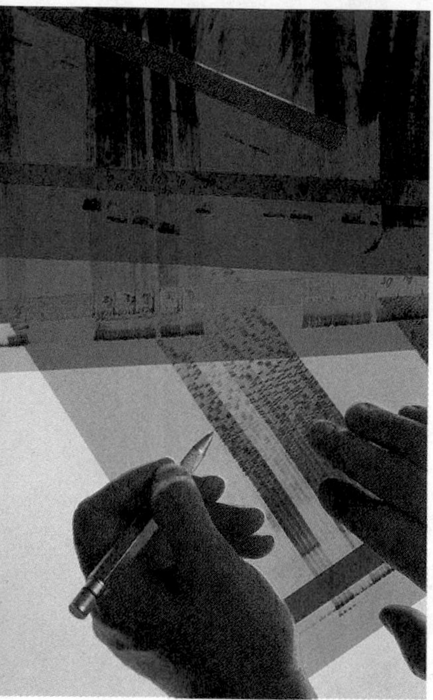

Scientist studies a gene map as part of the Human Genome Project.

three-letter sequence (first nucleotide–second nucleotide–third nucleotide) specifies a particular amino acid. For example, the codon CAC (cytosine–adenine–cytosine) is the code for the amino acid histidine (His). You will note that three codons in the table, marked TC, do not encode any amino acids. These are called *nonsense* or *termination codons*. They act as signals to indicate where the synthesis of a protein molecule is to end. The other 61 codons identify 20 amino acids. Methionine and tryptophan have only one codon each. For the other amino acids, the code is redundant; that is, each amino acid is specified by at least two, and sometimes by as many as six, codons.

It is believed that the genetic code is a universal code for all living organisms; that is, the same nucleotide triplet specifies a given amino acid regardless of whether that amino acid is synthesized by a bacterial cell, a pine tree, or a human being.

32.10 Genes and Medicine

It has been known for many years that most human traits can be traced to the genetic makeup of the individual. Also known was the fact that some diseases (e.g., sickle cell anemia and Tay-Sach's disease) occurred mainly within certain ethnic groups, clearly indicating a genetic connection for these maladies. This knowledge pointed to the need to learn more about the human genome. By the early 1980s techniques were available to "read" genetic material—that is, to determine the sequence of bases in the strings of DNA-comprising genes.

A complete sequencing or "mapping" of the human genome was desirable, but was a daunting task—given that this genome contains about three billion base pairs. The time that would be needed by a single laboratory to sequence or decode the entire genome was estimated to be about 1000 years! However, after much discussion and correspondence among many of the world's leading scientists, the Human Genome Project was organized. This project got under way in 1988 and is being carried forward in many laboratories in several countries with the object of obtaining nothing less than a complete map of the human genome.

The first five years of the project were used to locate genetic markers—"bench-marks" on the different human chromosome pairs. And as these genetic markers have been identified, the genes associated with a variety of diseases have been located. Genes have been found that trigger muscular dystrophy, Huntington's disease, cystic fibrosis, and some breast cancers, to name a few.

Recent research has turned to sequencing the chromosomes. This involves breaking a selected segment of the genome into pieces, sequencing each piece, and then matching overlapping sequences—like fitting a giant jigsaw puzzle together. Computer programs for fitting the pieces together have facilitated the project. It is now estimated that the basic library of human genetics, the human genome, may be known as early as the year 2001.

As more genetic information became available, a new method, known as gene therapy, emerged for treating certain diseases. This method involves inserting new genetic information into the genome of a cell and thereby changing its traits. Because the new information is transmitted to daughter cells, gene therapy is able to create "permanent" changes in the way these cells function.

Genetic diseases are the most obvious targets for gene therapy, and clinical trials are underway for several diseases. For example, cystic fibrosis is a lethal inherited disease that causes a buildup of mucus in the lungs. (More than 30,000 patients are presently afflicted in the United States.) This disease is caused by a defective (nonfunctional) gene for chloride membrane transport. Using genetic engineering techniques, a functional gene responsible for chloride membrane transport is identified, isolated, and replicated. Then it is introduced into the genome of a suitable virus, and is allowed to infect the patient's lung cells. The infection splices the genetic information from the virus into the host lung cell's genome. Thus as the virus infection occurs, a functional gene for chloride transport is spliced into the lung cell genome. The major problems with this treatment are found in choosing the correct virus and in maximizing the transfer of genetic information.

Only about 20% of gene therapies involve genetic diseases. The majority of clinical trials are focused on finding a cure for a variety of cancers. Several approaches are being explored. One strategy is to bring about better control of the oncogenes with the ultimate goal of "reversing" carcinogenesis—that is, to cause tumors to revert to normal tissue. Other strategies involve making cancer cells easier for the immune system to recognize and destroy and/or causing them to be more sensitive to particular chemotherapies.

A significant number of clinical trials are also focused on AIDS. One promising tactic involves introducing a marker into human cells that causes the body to strengthen its immune response to the HIV virus. Another tactic involves introducing genes that will interfere with the HIV viral life cycle.

It is expected that, as the Human Genome Project provides more detailed sequences, gene therapy will broaden in scope to include a greater variety of diseases. Although this prospect offers great potential and cause for hope, scientists warn that many obstacles need to be overcome before gene therapy comes into routine use by physicians.

32.11 Biosynthesis of Proteins

The biosynthesis of proteins is extremely complex, and the following is only a cursory description of the overall process. The production of a polypeptide using an mRNA template is called **translation**. This term is used because it literally means a change from one language to another. The genetic code is *translated* into the primary structure of a polypeptide or a protein.

translation

The biosynthesis of proteins begins when messenger RNA leaves the cellular nucleus and travels to the cytoplasm. Each mRNA is then bound to five or more ribosomes, the bodies responsible for protein synthesis. Amino acids must also be transferred to the ribosomes. To accomplish this step, cellular energy in the form of ATP is used to couple amino acids to tRNAs to form aminoacyl–tRNA complexes that can bind to the ribosomes.

$$\mathrm{R{-}CH(NH_2){-}\overset{\displaystyle O}{\overset{\|}{C}}{-}OH} + \mathrm{tRNA} + \mathrm{ATP} \xrightarrow[\text{synthetase}]{\text{aminoacyl–tRNA}}$$

(Products are given on the next page.)

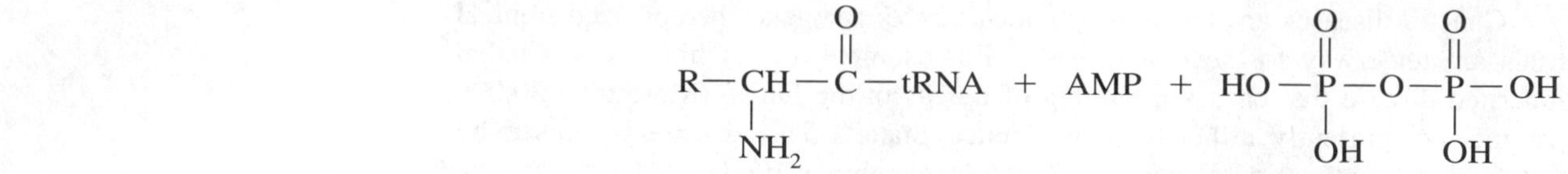

$$\underset{\text{aminoacyl–tRNA}}{\underset{\displaystyle NH_2}{\overset{}{R-\underset{|}{CH}}}-\overset{O}{\overset{\|}{C}}-tRNA} + AMP + HO-\underset{\underset{\displaystyle OH}{|}}{\overset{\overset{\displaystyle O}{\|}}{P}}-O-\underset{\underset{\displaystyle OH}{|}}{\overset{\overset{\displaystyle O}{\|}}{P}}-OH$$

A different specific enzyme is utilized for binding each of the 20 amino acids to a corresponding tRNA. Although only 20 amino acids are involved, there are about 60 different tRNA molecules in the cells. Thus the function of tRNA is to bring amino acids to the ribosome synthesis site.

Initiation The next step in the process is the initiation of polypeptide synthesis. Two codons signal for the start of protein synthesis, AUG and GUG (AUG being the more common). As shown in Table 32.3, these codons also code for incorporation of the amino acids methionine and valine. Most mRNAs have more than one AUG and GUG, and the ribosome must choose at which codon to begin. It appears that the ribosome uses information in addition to the AUG or GUG triplet to choose the correct starting codon for protein synthesis.

Once the correct starting point has been identified, a special initiator tRNA binds to the ribosome. This specific tRNA carries an *N*-formyl methionine in procaryotic cells (e.g., bacterial cells; see Section 34.5):

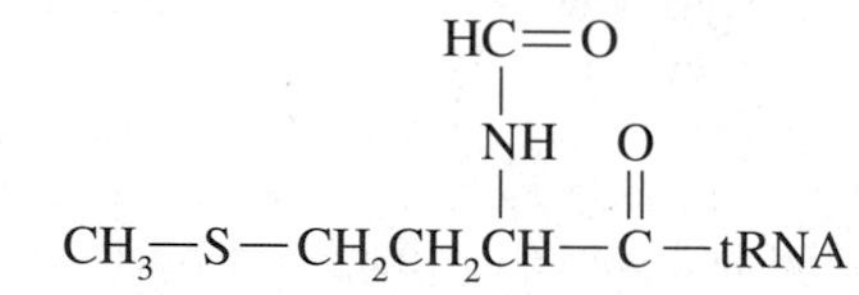

$$CH_3-S-CH_2CH_2\overset{\overset{\displaystyle HC=O}{|}}{\underset{}{\overset{\overset{}{NH}}{\overset{|}{CH}}}}-\overset{O}{\overset{\|}{C}}-tRNA$$

N-formyl methionine–tRNA

Because of the attached formyl group (CHO), a second amino acid will react only with the carboxyl group of methionine, the correct direction for protein synthesis. In eucaryotic cells, the initiator tRNA carries a methionine group.

Elongation The next stage involves the elongation or growth of the peptide chain, which is assembled one amino acid at a time. After the initiator tRNA is attached to the mRNA codon, the elongation of the polypeptide chain involves the following steps (see Figure 32.15):

1. The next aminoacyl–tRNA enters the ribosome and becomes attached to mRNA through the hydrogen bonding of the tRNA anticodon to the mRNA codon.
2. The peptide bond between the two amino acids is formed by the transfer of the amino acid from the initial aminoacyl–tRNA to the incoming aminoacyl–tRNA. In this step, which is catalyzed by the enzyme peptidyl transferase, the carboxyl group of the first amino acid separates from its tRNA and forms a peptide bond with the free amino group of the incoming aminoacyl–tRNA.
3. The tRNA carrying the peptide chain (now known as a peptidyl–tRNA) moves over in the ribosome, the free tRNA is ejected, and the next aminoacyl–tRNA enters the ribosome.

The peptide chain is transferred to the incoming amino acid, and the sequence is repeated over and over as the mRNA moves through the ribosome, just like a tape delivering its message. In each step the entire peptide chain is transferred to the incoming amino acid.

The initiation and elongation of polypeptide chains requires cellular energy. The primary source of this energy is the nucleotide guanosine-5′-triphosphate (GTP). At various steps in the growth of the protein chain, a high-energy GTP phosphate anhydride bond is hydrolyzed, yielding GDP and a phosphate group. The energy that is released drives protein synthesis. This reaction is analogous to that involving ATP (see Section 32.4).

Termination The termination of the polypeptide chain occurs when a "nonsense" or termination codon appears. In normal cells there are no tRNAs that have complementary anticodons to the termination codons. Because there is no new aminoacyl–tRNA to bind to the ribosome, the peptidyl–tRNA is hydrolyzed, and the free polypeptide (protein) is released. All of these amazing, coordinated steps are accomplished at a high rate of speed—about 1 min for a 146 amino acid chain of human hemoglobin and 10–20 s for a 300–500 amino acid chain in the bacterium *Escherichia coli (E. coli)*. This mechanism of protein synthesis is illustrated in Figure 32.15.

32.12 Changing the Genome: Mutations and Genetic Engineering

From time to time, a new trait appears in an individual that is not present in either parents or ancestors. These traits, which are generally the result of genetic or chromosomal changes, are called **mutations**. Some mutations are beneficial, but most are harmful. Because mutations are genetic, they may be passed on to the next or future generations.

mutation

Mutations can occur spontaneously or can be caused by chemical agents or various types of radiation such as X rays, cosmic rays, and ultraviolet rays. The agent that causes the mutation is called a **mutagen**. Exposure to mutagens may produce changes in the DNA of the sperm or ova. The likelihood of such changes is increased by the intensity and length of exposure to the mutagen. Mutations may then show up as birth defects in the next generation. Common types of genetic alterations include the substitution of one purine or pyrimidine for another during DNA replication. Such a substitution is a change in the genetic code and causes misinformation to be transcribed from the DNA. A mutagen may also alter genetic material by causing a chromosome or chromosome fragment to be added or removed.

mutagen

In recent years, laboratory techniques for controlling genetic change have been developed. These techniques, collectively known as genetic engineering, have already been responsible for considerable progress in medicine and biology. The genetic engineer is able to insert specific genes into the genome of a host cell and thus program it to produce new and different proteins that may be useful to humankind.

Genetic engineering has been made possible by several basic advances in nucleic acid biochemistry. First, scientists have gained the ability to isolate, identify, and then synthesize multiple copies of specific genetic messages. A key to this process is the *DNA polymerase chain reaction*. DNA polymerase is an enzyme that replicates DNA when supplied a starting fragment (the primer) and a DNA strand to copy (the template). By supplying the appropriate primer, the polymerase can be induced

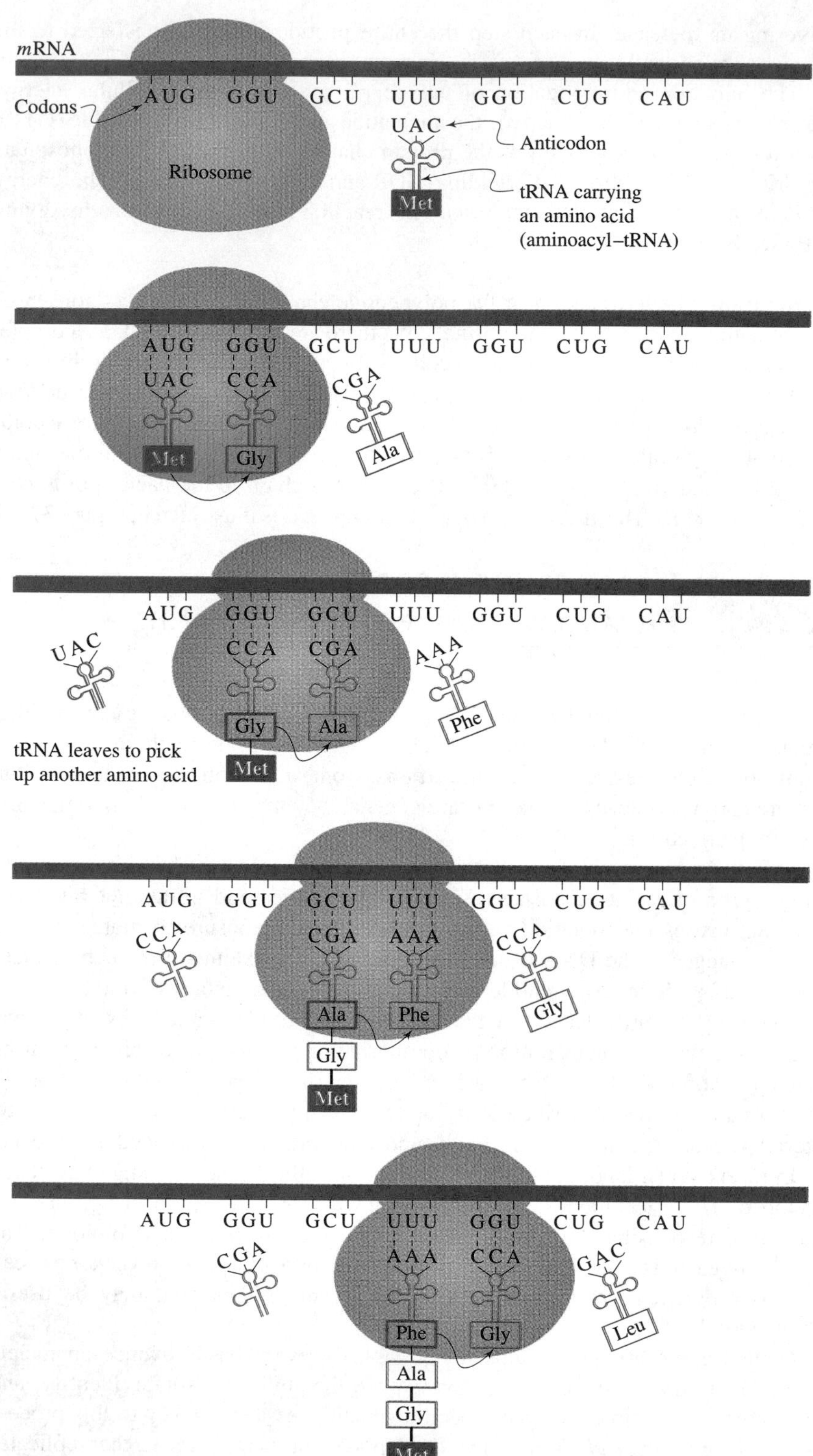

FIGURE 32.15 ▶ Biosynthesis of proteins: mRNA from DNA enters and complexes with the ribosomes. tRNA carrying an amino acid (aminoacyl–tRNA) enters the ribosome and attaches to the mRNA at its complementary anticodon. The peptide chain elongates when another aminoacyl–tRNA enters the ribosome and attaches to the mRNA. The peptide bond is then formed by the transfer of the peptide chain from the initial to the incoming aminoacyl–tRNA. The sequence is repeated until a termination codon appears in mRNA.

to synthesize a specific gene, even in the presence of a wide variety of other genetic material. What makes this reaction especially valuable is that the polymerase can use the newly synthesized DNA as a template for further replication. Thus, after 20 synthetic cycles, almost a million copies of the original gene can be produced. Like a radioactive chain reaction, the polymerase chain reaction leads to an explosion in the numbers of copied genes.

A second important process in genetic engineering is the insertion of genetic material into a "foreign" genome. Special enzymes, the *restriction endonucleases*, provide a key step during gene insertion. These enzymes split double-stranded DNA at very specific locations.

These genetically engineered tomatoes are disease-resistant and ripen slowly so that they can be picked later in the growing season.

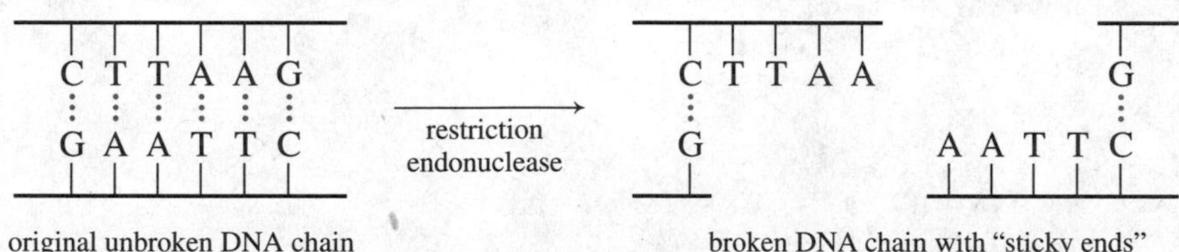

Often the newly formed break has what is termed "sticky ends"; that is, one end of the break can "stick" to the end of another break via hydrogen bonding between complementary bases. When both a gene and a foreign genome are processed in this way, they bond together. Then, with the aid of other enzymes (ligases), the gene can be covalently bonded into place.

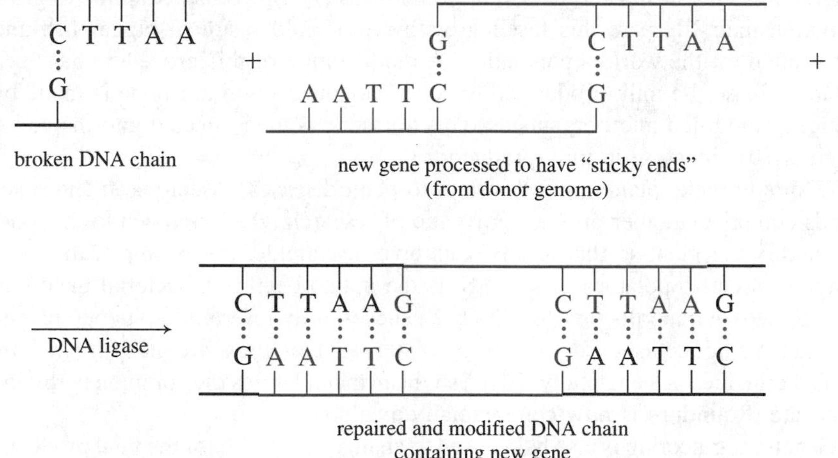

The resultant modified and repaired foreign genome now contains a new gene. The result of this process is a form of recombinant DNA. The term **recombinant DNA** refers to DNA whose genes have been rearranged to contain new or different hereditary information.

recombinant DNA

Agriculture is an industry being greatly impacted by genetic engineering. Planners estimate that the amount of arable land worldwide will not change much but population will continue to increase. Genetic engineering can improve land use by, for example, increasing yield and decreasing crop losses due to insects.

Increased yields can be achieved in a variety of ways. Presently, much effort is going into aquaculture; fish are genetically engineered to produce larger amounts

Bovine growth hormone injected into cows causes a 10% increase in milk production. ▶

of the fish growth hormone. These fish grow much faster than wild types (coho salmon up to 37 times faster; catfish up to 60% faster). Since fish continue to grow throughout their lifespan, this faster growth rate should result in bigger fish and more protein for the world's population. A modification of this procedure has been used to increase the milk production in cows. Bovine growth hormone is made by genetically modified microorganisms. This hormone is then injected into the cows, causing a 10% increase in milk production.

Efforts to make plants more resistant to plant diseases, insect pests, and other hazards comprise another promising avenue of research. One approach to this goal is to modify the plant so that it makes its own insecticide. For example, the most damaging insect to potatoes in the U.S. is the potato beetle. A bacterial gene that codes for a toxin that kills this pest has been successfully inserted into potato plants. A related technique is used to modify plants so that they are not affected by specific herbicides. A genetically altered soybean that tolerates the common herbicide glyphosate (Roundup) is now commercially available.

Genetic engineering is also being used to change the quality of the final products. Slow-ripening tomatoes are now being shipped to market. Because they ripen only slowly, the tomatoes can be picked later in the growing season; they should be better tasting when purchased by the consumer. An important oil-producing plant, rapeseed (canola), has been genetically altered to produce up to 40% lauric acid, an important component of many detergents. In the near future, this same plant may be producing another oil that is used to produce a nylon.

The techniques used in genetic engineering are becoming routine. It is clear that many benefits can be drawn from this technology. It is equally clear that there may be unknown negative consequences of these procedures. Genetic engineering remains controversial because it is difficult to foresee the future consequences of such a major innovation.

Concepts in Review

1. Write the structural formulas for the two purine and three pyrimidine bases found in nucleotides.
2. Distinguish between ribonucleotides and deoxyribonucleotides.
3. List the compositions, abbreviations, and structures for the ten nucleotides in DNA and RNA.
4. Write the structural formulas for ADP and ATP.
5. Identify where energy is stored in ADP and ATP.
6. Write a structural formula of a segment of a polynucleotide that contains four nucleotides.
7. Describe the double-helix structure of DNA according to Watson and Crick.
8. Explain the concept of complementary bases.
9. Describe and illustrate the replication process of DNA.
10. Understand how heredity factors are stored in DNA molecules.
11. Distinguish between mitosis and meiosis.
12. Describe how oncogenes are related to cancer.
13. Describe the genetic code.
14. State the functions of the three different kinds of RNA.
15. Describe the transcription of the genetic code from DNA to RNA.
16. Describe the biosynthesis of proteins.
17. Understand how mutations are caused.
18. Briefly describe genetic engineering.
19. Explain some of the advantages of gene therapy.
20. Describe DNA fingerprinting.

Key Terms

The terms listed here have been defined within this chapter. Section numbers are referenced in parenthesis for each term.

codon (32.9)
deoxynucleic acid (DNA) (32.5)
gene (32.7)
genome (32.7)
meiosis (32.7)
mitosis (32.7)
mutagen (32.12)
mutation (32.12)
nucleic acid (32.1)
nucleoprotein (32.1)
nucleoside (32.2)
nucleotide (32.3)
oncogene (32.8)
recombinant DNA (32.12)
ribonucleic acid (RNA) (32.5)
replication (32.7)
transcription (32.8)
translation (32.11)

Questions

Questions refer to tables, figures, and key words and concepts defined within the chapter. A particularly challenging question or exercise is indicated with an asterisk.

1. Write the names and structural formulas for the five nitrogen bases found in nucleotides.

2. What is the difference between a nucleoside and a nucleotide?

3. What are the principal structural differences between DNA and RNA?

4. What is the major function of ATP in the body?

5. Briefly describe the structure of DNA as proposed by Watson and Crick.

6. What is meant by the term *complementary bases*?

7. What is the genetic code?

8. Why are at least three nucleotides needed for one unit of the genetic code?

9. Starting with DNA, briefly outline the biosynthesis of proteins.

10. How does a codon differ from an anticodon?

11. Explain the role of *N*-formyl methionine in procaryotic protein synthesis.

12. What is a mutation?

13. What is meant by a "DNA fingerprint"?

14. Which of these statements are correct? Rewrite each incorrect statement to make it correct.

(a) Adenine and guanine are both purine bases and are found in both DNA and RNA.

(b) The ratio of adenine to thymine and guanine to cytosine in DNA is about 1:1.

(c) DNA and RNA are responsible for transmitting genetic information from parent to daughter cells.

(d) DNA is a polymer made from nucleotides.

(e) Codons are combinations of the base units in a tRNA molecule.

(f) The double-helix structure of DNA is held together by peptide linkages.

(g) Amino acids are linked to tRNA by an ester bond.

(h) The nucleotide adenosine monophosphate contains adenine, D-ribose, and a phosphate group.

(i) The letters ATP stand for adenine triphosphate.

(j) Thymine and uracil are both complementary bases to adenine.

(k) Messenger RNA is a transcribed section of DNA.

(l) The ratio of adenine to guanine and thymine to cytosine is 1:1 in DNA.

(m) Genetic information is based on the nucleotide sequence in DNA.

(n) Humans have 46 pairs of chromosomes.

(o) In mitosis the sperm cell and the egg cell, each with 23 chromosomes, unite to give a new cell containing 46 chromosomes.

(p) The genetic code consists of triplets of nucleotides; each triplet codes an amino acid.

(q) Transfer RNA carries the code for the synthesis of proteins.

(r) ADP and ATP contain high-energy phosphate anhydride bonds.

(s) On hydrolysis, DNA yields ribose, phosphoric acid, and the bases adenine, guanine, cytosine, and thymine.

(t) DNA has a double-helix conformation, whereas all RNAs have a single-helix conformation.

Paired Exercises

These exercises are paired. Each odd-numbered exercise is followed by a similar even-numbered exercise. Answers to the even-numbered exercises are given in Appendix V.

15. Identify the compounds represented by the following letters: A, AMP, dADP, UTP.

16. Identify the compounds represented by the following letters: G, GMP, dGDP, CTP.

17. Write structural formulas for the substances represented by
(a) A
(b) AMP
(c) CDP
(d) dGMP

18. Write structural formulas for the substances represented by
(a) U
(b) UMP
(c) CTP
(d) dTMP

19. Draw the structure for a three-nucleotide piece of single-stranded DNA using the bases, C, T, and A.

20. Draw the structure for a three-nucleotide piece of single-stranded RNA using the bases, G, U, and A.

21. Show by structural formulas the hydrogen bonding between adenine and uracil.

22. Show by structural formulas the hydrogen bonding between guanine and cytosine.

23. What is the role of RNA in the genetic process?

24. What is the role of DNA in the genetic process?

25. Differentiate between replication and transcription.

26. Differentiate between transcription and translation.

27. How does the function of tRNA differ from that of mRNA?

28. How does the function of mRNA differ from that of rRNA?

29. There are 146 amino acid residues in the β-polypeptide chain of hemoglobin. How many nucleotides are needed to code for the amino acids in this chain?

30. There are 573 amino acid residues in the phosphoglycerate kinase enzyme. How many nucleotides are needed to code for the amino acids in this chain?

31. A segment of a DNA strand consists of TCAATACCCGCG.
(a) What is the nucleotide order in the complementary mRNA?
(b) What is the anticodon order in the tRNA?
(c) What is the sequence of amino acids coded by the DNA?

32. A segment of a DNA strand consists of GCTTAGACCTGA.
(a) What is the nucleotide order in the complementary mRNA?
(b) What is the anticodon order in the tRNA?
(c) What is the sequence of amino acids coded by the DNA?

33. Describe the bond (e.g., the atoms involved, the functional groups that react) that forms during transcription to link nucleotides together.

34. Describe the bond (e.g., the atoms involved, the functional groups that react) that forms during translation to link amino acids together.

35. Briefly describe the events that occur during translation termination.

36. Briefly describe the events that occur during translation initiation.

37. Describe what the anticodon will be in tRNA if the codon in mRNA is as follows:
(a) GUC
(b) AGG
(c) UUU
(d) CCA

38. Describe what the anticodon will be in tRNA if the codon in mRNA is as follows:
(a) CGC
(b) ACA
(c) GAU
(d) UUC

Additional Exercises

These exercises are not paired or labeled by topic and provide additional practice on concepts covered in this chapter.

39. Explain why the ratio of thymine to adenine in DNA is 1:1, but the ratio of thymine to guanine is not necessarily 1:1.

40. In RNA does the guanine content have to be equal to the cytosine content? Explain. Do they have to be equal in DNA? Explain.

41. Mutations in DNA can occur when one base is substituted for another or when an extra base is inserted into the DNA strand. Explain why an insertion mutation is apt to be more harmful than a substitution mutation.

42. The sequence of bases in a segment of mRNA is UUUCAUAAG. Answer the following questions:
(a) What amino acids would be coded for by this segment of mRNA?
(b) What is the sequence of DNA that would produce the given segment of mRNA?

43. Describe the use of DNA polymerase in DNA fingerprinting.

44. What is the Human Genome Project?

Nutrition

33

Chapter 33

> Jack Sprat could eat no fat,
> his wife could eat no lean,
> yet twixt the two of them,
> they licked the platter clean.
>
> —*Mother Goose*

For hundreds of years it has been known that diet is important to good health and that diets may need to vary from one individual to another. Yet only in recent years have we been able to understand nutritional needs on a molecular level. In the nursery rhyme, Jack Sprat was clearly on a low-fat diet. Today, we can point to evidence that fat intake (especially saturated fat) is related to heart disease. Fat molecules accumulate on the blood vessel walls, which leads to hardening of the arteries.

Nutrition is a science of practical importance. Every person is concerned with diet. As a child you may have heard "Finish your milk" or "Please eat your vegetables." Do you remember being told that chocolate will cause complexion problems? How many of your friends are watching their weight?

An understanding of nutrition and digestion is closely coupled with an understanding of biochemistry. As more is learned about how diet impacts health, we can make better choices concerning which foods to purchase and eat.

33.1 Nutrients

The need for choosing the right diet has given rise to the science of nutrition, the study of nutrients—how they are digested, absorbed, metabolized, and excreted. **Nutrients** are components of the food we eat that provide for body growth, maintenance, and repair. Milk is a food, but the calcium from that milk is a nutrient. We eat meat, a food, for protein, a nutrient.

nutrient

As you study biochemistry, you learn about molecules that are required for life. In future chapters, you will learn how the cell uses and produces these molecules. Unfortunately, our cells are not self-sufficient. They require a constant input of energy, a source of carbon and nitrogen, as well as a variety of minerals and special molecules. Cells can synthesize many molecules, but they need starting materials. Nutrients, when digested and absorbed, provide the building blocks that allow cells to make carbohydrates, lipids, proteins, and nucleic acids, as well as provide the energy for life.

Nutritionists divide nutrients into six broad classes: (1) carbohydrates, (2) lipids, (3) proteins, (4) vitamins, (5) minerals, and (6) water. The first three classes—carbohydrates, lipids, and proteins—are the major sources of the building materials, replacement parts, and energy needs of the cells. These nutrients are used in relatively large amounts and are known as **macronutrients**. The next three classes of nutrients—vitamins, minerals, and water—have functions other than as sources of energy or building materials. Vitamins are components of some enzyme systems that are vital to the cells. Minerals are required to maintain specific concentrations of certain

macronutrient

◀ **Chapter Opening Photo: A healthy assortment of fresh vegetables is necessary to provide adequate nutrition.**

inorganic ions in the cellular and extracellular fluids. They are also utilized in a variety of other ways—for example, calcium and phosphorus in bones and teeth and iron in hemoglobin. Vitamins and minerals are needed in relatively small amounts and are termed **micronutrients**. Water is absolutely essential to every diet, because most biochemical reactions occur in aqueous solution.

micronutrient

Nutrients supply the needs of the body's many different cells. For example, red blood cells must produce hemoglobin and thus need iron. Muscle cells produce muscle fibers and need much protein. As they do their jobs, cells wear out and must be replaced. Red blood cells, on an average, are renewed every 6 weeks, whereas the cells lining the digestive tract must be replaced every 3 days.

The six classes of nutrients provide the basis for a healthy diet. This chapter will describe briefly (1) the relationship between diet and nutrition, (2) some characteristics of each nutrient category, and (3) the processes by which some nutrients are digested.

33.2 Diet

diet

Diet is the food and drink that we consume. Our choice of meals determines the nutrients available to our bodies. Thus our health depends directly on our diet.

Unfortunately, some dietary choices are not clear. Controversy surrounds many food selections. Are "fast foods" unhealthy? Should we avoid food additives? What are "natural" foods? To eat intelligently we must understand the role of the food nutrients in maintaining good health.

Nutritionists have the difficult problem of deciding the kinds of nutrients that should be in a diet. They often establish dietary need by correlating physical well-being with nutrient consumption. James Lind (1716–1794), a physician in the British Navy during the 18th century, was one of the first to use this approach. In a study of scurvy, a disease that afflicted sailors on long voyages, he placed seamen who suffered from scurvy on various diets, some of which contained citrus fruits. By observing changes in the conditions of the seamen, Lind was able to conclude that citrus fuits provide a nutrient that prevents scurvy. This and later work eventually led to the requirement of limes and lemons in the diets of the British Merchant Marine (1765) and the British Navy (1795). Scurvy is now recognized as a deficiency disease caused by a lack of vitamin C, ascorbic acid.

recommended dietary allowance (RDA)

As scientists like Lind discovered and studied more nutrients, they also found that the minimum quantities needed for good health varied from person to person. Thus, rather than establishing uniform minimum requirements, nutritionists have established a standard called the recommended dietary allowance. The **recommended dietary allowance (RDA)** for a nutrient is an average value that has been shown to maintain health for large groups of people. Although there is considerable variation in the dietary needs of individuals, RDAs are useful yardsticks in judging healthful diets. Table 33.1 shows the RDAs for a variety of nutrients.

TABLE 33.1 Recommended Daily Dietary Allowances[a] (revised 1989)

		Weight		Height			Fat-soluble vitamins			Water-soluble vitamins							Minerals					
	Age (yrs)	(kg)	(lb)	(cm)	(in.)	Protein (g)	Vitamin A (μg RE)[b]	Vitamin D (μg)[c]	Vitamin E (mg α-TE)[d]	Vitamin C (mg)	Thiamin (mg)	Riboflavin (mg)	Niacin (mg)	Vitamin B6 (mg)	Folic acid (μg)	Vitamin B12 (μg)	Ca (mg)	P (mg)	Mg (mg)	Fe (mg)	Zn (mg)	I (mg)
Infants	0.0–0.5	6	13	61	24	13	375	7.5	3	30	0.3	0.4	5	0.3	25	0.3	400	300	40	6	5	40
	0.5–1.0	9	20	71	28	14	375	10	4	35	0.4	0.5	6	0.6	35	0.5	600	500	60	10	5	50
Children	1–3	13	29	90	35	16	400	10	6	40	0.7	0.8	9	1.0	50	0.7	800	800	80	10	10	70
	4–6	20	44	112	44	24	500	10	7	45	0.9	1.1	12	1.1	75	1.0	800	800	120	10	10	90
	7–10	28	62	132	52	28	700	10	7	45	1.0	1.2	13	1.1	100	1.4	800	800	120	10	10	120
Males	11–14	45	99	157	62	45	1000	10	10	50	1.3	1.5	17	1.7	150	2.0	1200	1200	70	12	15	150
	15–18	66	145	176	69	59	1000	10	10	60	1.5	1.8	20	2.0	200	2.0	1200	1200	400	12	15	150
	19–24	73	160	177	70	58	1000	10	10	60	1.5	1.7	19	2.0	200	2.0	1200	1200	350	10	15	150
	25–50	79	174	176	70	63	1000	5	10	60	1.5	1.7	19	2.0	200	2.0	800	800	350	10	15	150
	51+	77	170	173	68	63	1000	5	10	60	1.2	1.4	15	2.0	200	2.0	800	800	350	10	15	150
Females	11–14	46	101	157	62	46	800	10	8	50	1.1	1.3	15	1.4	150	2.0	1200	1200	280	15	12	150
	15–18	55	120	163	64	49	800	10	8	60	1.1	1.3	15	1.5	180	2.0	1200	1200	300	15	12	150
	19–24	58	128	164	65	46	800	10	8	60	1.1	1.3	15	1.6	180	2.0	1200	1200	280	15	12	150
	25–50	63	138	163	64	50	800	5	8	60	1.0	1.3	15	1.6	180	2.0	800	800	280	15	12	150
	51+	65	143	160	63	50	800	5	8	60	1.0	1.2	13	1.6	180	2.0	800	800	280	10	12	150
Pregnant						50	800	10	10	70	1.5	1.6	17	2.2	400	2.2	1200	1200	320	30	15	175
Lactating						55	1200	10	11	90	1.6	1.8	20	2.1	260	2.6	1200	1200	350	15	17	200

Reference: *Recommended Dietary Allowances,* 10th ed. Food and Nutrition Board, National Research Council–National Academy of Sciences.

[a] The allowances are intended to provide for individual variations among most normal persons as they live in the United States under usual environmental stresses. Diets should be based on a variety of common foods in order to provide other nutrients for which human requirements have been less well defined.

[b] Retinol equivalents; 1 retinol equivalent = 1 μg retinol or 6 μg β-carotene.

[c] As cholecaliciferol; 10 μg cholecalciferol = 400 IU of vitamin D.

[d] α-Tocopherol equivalents; 1 mg α-tocopherol = 1 α-TE.

"Fast foods" make up a large portion of the daily diet for many people.

33.3 Energy in the Diet

An important component of every diet is the energy allowance that derives primarily from the energy-containing nutrients, the carbohydrates, lipids, and proteins. These molecules are a rich source of reduced carbons. As we will see in Chapter 34, almost all of the energy for life is derived from reactions in which cells oxidize carbon compounds.

The dietary energy allowance varies with activity, body size, age, and sex. Thus a 65-kg male office worker requires a diet that furnishes 2700 kcal/day, whereas a man of similar weight working as a carpenter requires about 3000 kcal/day. Women generally use less energy than men, and energy use decreases with age. Nutritionists express energy in units of kilocalories (kcal) or its equivalent, the large Calorie (Cal). Remember, there are 4.184 kJ per kilocalorie (or Calorie).

The balance between energy needs and the energy allowance is of vital importance. Calorie deficiency leads to a condition called *marasmus*, which affects many of the world's poor, particularly children. Marasmus is a wasting disease due to starvation. People suffering from this disease have limited diets that often consist of bulky, carbohydrate-containing foods—for example, sweet potatoes (32% carbohydrate, 1% fat, 2% protein). These foods are not energy-rich (about 160 kcal for a medium-size sweet potato). Children, with their small stomachs, have difficulty consuming enough to satisfy their energy requirements. It is estimated that 15–20% of the people in underdeveloped countries suffer from malnutrition due to insufficient calories.

In contrast, many people in developed countries consume far more calories than they need for health and well-being. Food is available in abundance, much of it rich in energy. For example, a fast-food lunch might consist of a hamburger (560 kcal), french fries (220 kcal), and a chocolate shake (380 kcal), which would supply nearly one-half of the daily energy allowance for an adult male. Excess accumulated calories lead to a condition called *obesity*, which is characterized by an overabundance of fatty tissue and by many attendant health problems.

Each pound of body fat contains about 3500 kcal of energy, or between 1 and 2 days' total RDA for energy. Because fat contains so much energy, it is normally accumulated only slowly. Unfortunately, for the same reason, fat is also very difficult to lose. To lose 1 lb of fat, a 65-kg man would have to swim nonstop for about 10 hours, play tennis continuously for about 8 hours, or run for 4 hours. Exercise alone is generally not sufficient to cure obesity without a change in diet as well.

Thus many people in the affluent, developed countries find it necessary to choose a restricted-calorie diet to lose weight. The difficulties associated with selecting a restricted-calorie diet have led to the creation of numerous fad diets that fail to provide sound nutrition. Nutritionists counsel that successful and nutritionally sound dieting can be accomplished by following these simple guidelines: (1) Reduce calorie intake by only a moderate amount (a 500-kcal/day reduction causes a loss of about 1 lb of fat per week) and be prepared to continue the diet for a long time; (2) carefully select foods so that the diet contains adequate amounts of all nutrients. The primary function of any diet is to maintain good health.

33.4 Carbohydrates in the Diet

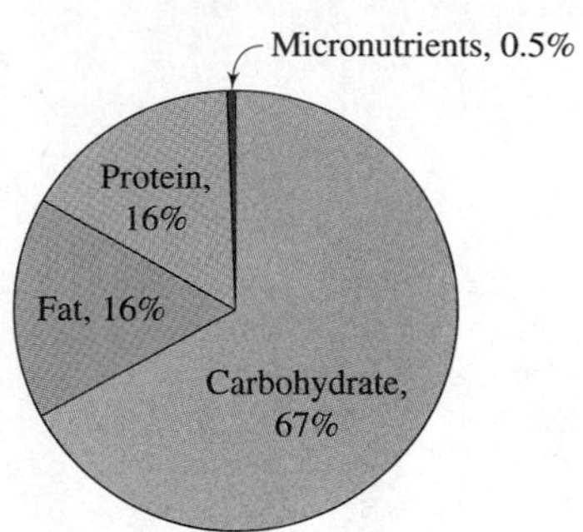

FIGURE 33.1
Relative masses of nutrients in a typical diet.

Carbohydrates are the major nutrient in most human diets (see Figure 33.1). They are justly described as macronutrients because they are a major source of usable, reduced carbon atoms and, therefore, of dietary energy. About half of the average daily calorie requirement derives from carbohydrates. These molecules are perhaps the most easily metabolized (see Chapter 35) of the energy-supplying nutrients and can be used for energy under both aerobic and anaerobic conditions. Additionally, carbons from carbohydrates are used to build other cellular molecules—amino acids, nucleic acids, as well as other carbohydrates. Excess dietary carbohydrate is most often converted to fat.

Dietary carbohydrates are primarily the polysaccharides starch and cellulose, the disaccharides lactose and sucrose, and the monosaccharides glucose and fructose. Seeds are the most common source of starch, grains being about 70% by mass starch, while dried peas and beans contain about 40% starch. A second major source of starch is tuber and root crops such as potatoes, yams, and cassava. The disaccharide lactose is an important component of milk, and sucrose is usually consumed as refined sugar (derived from sugar beets or sugar cane). The monosaccharides are often found in fruits.

The polysaccharides, also called *complex carbohydrates*, are difficult to digest because of their complex structures (see Section 28.15). Starch is digested only slowly, enabling the body to control distribution of this energy nutrient. Cellulose is not digested by humans; however, it is a major source of *dietary fiber*. As cellulose passes through the digestive tract, it absorbs water and provides dietary bulk. This bulk acts to prevent constipation and diverticulosis, a weakening of the intestinal walls. Many nutritionists recommend a daily fiber intake of 15–30 g, supplied by such foods as whole wheat bread (1.8 g/slice) and bran cereals (7.5 g/0.5 cup).

Although no RDA has been set for carbohydrate, many nutritionists recommend that a minimum of about 1000 kcal/day be derived from this source. In the American diet, most of these calories are derived from starch, although in recent years an increasing amount has come from sucrose.

The high sucrose content of many modern diets is due primarily to the large amounts of sucrose in commercially prepared foods. In 1900, an American consumed an average of about 20 lb of sucrose annually in prepared foods and beverages; by 1989, this figure had risen to about 65 lb annually. This large increase was caused mainly by (1) increased consumption of prepared foods and (2) an increased percentage of sucrose added to these foods by the manufacturers in attempts to gain larger shares of the market. (It is well known that many people, especially children, have a preference for sweet foods.) Although the consumption of other sweeteners such as high-fructose corn syrup and aspartame has risen, the consumption of sucrose has remained high.

The increase in sucrose consumption has troubled nutritionists for several reasons. First, sucrose is a prime factor in the incidence of dental caries. Because it is readily used by oral bacteria, sucrose promotes growth of the microorganisms that cause tooth decay. Second, ingested sucrose is rapidly hydrolyzed to monosaccharides, which are promptly absorbed from the intestine, leading to a rapid increase in blood-sugar levels. Wide variations in the blood-sugar level may stress the body's hormonal system. Finally, sucrose is said to provide "empty calories." This means that sucrose supplies metabolic energy (calories) but lacks other nutrients. Nutritionists usually recommend starch over sucrose as a major source of dietary carbohydrate.

33.5 Fats in the Diet

Fat is a more concentrated source of dietary energy than carbohydrate. Not only does fat contain more carbons per unit mass, but the carbons in fats are more reduced. As an energy source, fats provide about 9 kcal/g, whereas carbohydrates provide only about 4 kcal/g. As we will see in Chapter 36, energy can only be obtained from fat when oxygen is present; fat metabolism is strictly aerobic in humans. The carbons from fats can be used to synthesize amino acids, nucleic acids, and other fats, but our bodies cannot achieve a net synthesis of carbohydrate from fat. Thus fats tend not to be as versatile a nutrient as carbohydrates.

Fats contribute much of the dietary energy of many foods. For example, french fried potatoes contain about 18% fat, yet this fat provides about 40% of the calories in this food. A cup of whole milk has 170 kcal, while a cup of skim (nonfat) milk has 80 kcal. Many nutritionists counsel that the best way to reduce calorie intake is to eat foods containing less fat. Although fats are macronutrients, it is recommended that fat intake should not exceed 25–30% of the daily energy allowance.

In a diet, both the kind and the amount of fat are important. Fatty acids from meat and dairy products are relatively saturated, whereas those from plant sources are generally more unsaturated. Because there is a probable link between high consumption of saturated fats and atherosclerosis, the U.S. Department of Agriculture and other agencies concerned with nutrition have recommended that American diets should contain about equal portions of polyunsaturated and saturated fatty acids. Two ways of increasing polyunsaturated fats in the diets are to (1) cook with vegetable oils such as corn or canola oil and (2) use soft margarine, which usually contains more unsaturated fats than hard or stick margarine or butter.

Polyunsaturated fats generally also contain the three essential fatty acids—linoleic acid, linolenic acid, and arachidonic acid:

$$CH_3(CH_2)_4CH{=}CHCH_2CH{=}CH(CH_2)_7CO_2H$$

linoleic acid

$$CH_3CH_2CH{=}CHCH_2CH{=}CHCH_2CH{=}CH(CH_2)_7CO_2H$$

linolenic acid

$$CH_3(CH_2)_4CH{=}CHCH_2CH{=}CHCH_2CH{=}CHCH_2CH{=}CH(CH_2)_3CO_2H$$

arachidonic acid

Each of these nutrients has been shown to relieve the deleterious physiological changes, such as poor growth, skin lesions, kidney damage, and impaired fertility, that result from a totally fat-free diet. One essential biochemical function of these fatty acids is as precursors for prostaglandin synthesis.

TABLE 33.2 Common Dietary Amino Acids

Essential		Nonessential	
Isoleucine	Tryptophan	Alanine	Glutamine
Leucine	Valine	Arginine	Glycine
Lysine		Asparagine	Histidine
Methionine		Aspartic acid	Proline
Phenylalanine		Cysteine	Serine
Threonine		Glutamic acid	Tyrosine

33.6 Proteins in the Diet

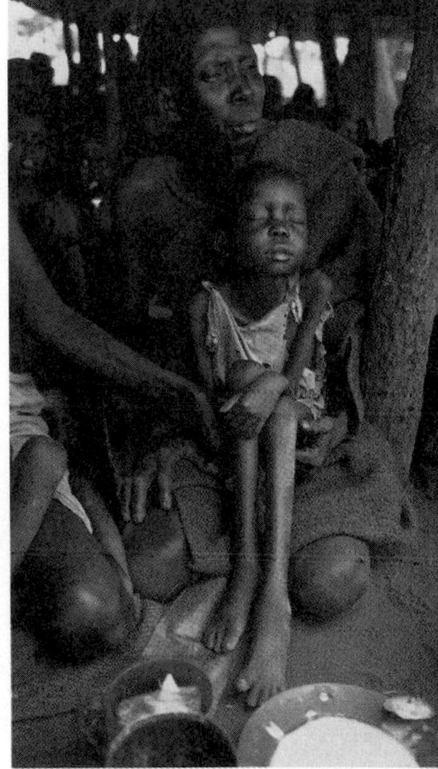

Kwashiorkor (protein deficiency) is a serious problem for children in many poverty-ridden areas of the world.

The third macronutrient is protein with an average energy yield of about 4.2 kcal/g. This energy is derived primarily from reduced carbons. However, recall that proteins also contain about 16% nitrogen; in fact, protein is the primary source of nitrogen in our diets. As discussed in Chapter 36, perhaps the most important function of dietary protein is to provide for synthesis of nitrogen-containing molecules such as nucleic acids, enzymes and other proteins, nerve transmitters, and many hormones. Because these materials are critical for human growth, protein malnutrition, or *kwashiorkor*, is a particularly serious problem. Unlike carbohydrates and fats (excluding the essential fatty acids), a specific RDA has been established for dietary protein (see Table 33.1).

Kwashiorkor can occur even when calorie intake is sufficient and thus is an especially insidious form of malnutrition. In many poverty-ridden areas of the world, the only reliable source of protein for children is mother's milk. After weaning, the children eat a protein-poor grain diet. Although the calorie intake is sufficient, these children show the stunted growth, poor disease resistance, and general body wasting that are characteristic of kwashiorkor.

Proteins are obtained from animal sources such as meat, milk, cheese, and eggs, and from plant sources such as cereals, nuts, and legumes (peas, beans, and soybeans). Animal proteins have nutritive values that are generally superior to those of vegetable proteins in that they supply all of the 20 amino acids that the body uses. In contrast, a single vegetable source often lacks several amino acids. This deficiency is critical, because humans cannot synthesize a group of amino acids called the *essential amino acids* (Table 33.2). These eight essential amino acids must be obtained from the diet.

Nutritionists often speak of a dietary source (or sources) of complete protein. A **complete protein** is one that supplies all of the essential amino acids. Animal products are, in general, sources of complete proteins. However, by either choice or necessity, animal protein is seldom consumed by a significant fraction of the world's population. Besides the moral, ethical, and religious reasons given for limiting consumption of animal protein, these proteins are relatively expensive sources of amino acids. Thus most of the people of the underdeveloped countries subsist primarily on vegetable protein. A constant danger inherent in a vegetarian

complete protein

diet is that the source of vegetable protein may be deficient in several of the essential amino acids; that is, the protein may be incomplete. For this reason nutritionists recommend that several sources of vegetable protein be included with each meal in a vegetarian diet. As an example, soybeans, which are rich in lysine, might supplement wheat, which is lysine deficient.

33.7 Vitamins

vitamin

Vitamins are a group of naturally occurring organic compounds that are essential for good nutrition and must be supplied in the diet. Whereas the energy supplying nutrients are digested and metabolized extensively, vitamins are often used after only minimal modification. Some of the vitamins necessary for humans are listed in Table 33.3. Note that vitamins are often classified according to their solubility, those that are fat soluble and those that are water soluble. The structural formulas of several vitamins are shown in Figure 33.2.

A prolonged lack of vitamins in the diet leads to vitamin deficiency diseases such as beriberi, pellagra, pernicious anemia, rickets, and scurvy. Left uncorrected, a vitamin deficiency ultimately results in death. Even when supplementary amounts of vitamins are provided, impaired growth due to a vitamin deficiency may be irreversible. For example, it is difficult to correct the distorted bone structures resulting from a childhood lack of vitamin D. Thus it is especially important that children receive sufficient vitamins for proper growth and development.

In a vegetarian diet it is important to include several sources of protein, such as beans and rice, to produce a complete protein.

Vitamins are required in only small amounts (see the RDAs in Table 33.1) and are classed as micronutrients. For example, a typical diet might include 250 g of carbohydrate (a macronutrient) and only 2 mg of vitamin B_6. However, the biochemistry of life cannot continue without vitamins. Each vitamin serves at least one specific purpose for an organism. The water-soluble compounds are generally involved in cellular metabolism of the energy-supplying nutrients. For example, thiamin is required to achieve a maximum energy yield from carbohydrates, and pyridoxine is of central importance in protein metabolism (see Chapter 35 and 36). Niacin and riboflavin are key components in almost all cellular redox reactions. The fat-soluble vitamins often serve very specialized functions. Vitamin D acts as a regulator of calcium metabolism. One function of vitamin A is to furnish the pigment that makes vision possible, while vitamin K enables blood clotting to occur normally.

Because some vitamin functions are not well understood, miraculous properties have been ascribed to these substances, such as vitamins C and E. In the absence of conclusive scientific studies, it is difficult to judge the merits of some of these claims.

Although vitamins are required only in small quantities, their natural availability is also low. Nutritionists caution that a diet should be balanced to include adequate sources of vitamins. As you can see from Table 33.3, the dietary sources of vitamins are varied. In general, fruits, vegetables, and meats are rich sources of the water-soluble vitamins; and eggs, milk products, and liver are good sources of the fat-soluble vitamins.

A seemingly balanced diet may be deficient in vitamins due to losses incurred in food processing, storing, and cooking. As much as 50–60% of the water-soluble vitamins in vegetables can be lost during cooking. Vitamin C can be destroyed by

TABLE 33.3 Some of the Most Important Vitamins

Vitamin	Important dietary sources	Some deficiency symptoms
FAT SOLUBLE		
Vitamin A (Retinol)	Green and yellow vegetables, butter, eggs, nuts, cheese, fish liver oil	Poor teeth and gums, night blindness
Vitamin D (Ergocalciferol, D_2; cholecalciferol, D_3)	Egg yolk, milk, fish liver oils; formed from provitamin in the skin when exposed to sunlight	Rickets (low blood-calcium level, soft bones, distorted skeletal structure)
Vitamin E (α-Tocopherol)	Meat, egg yolk, wheat germ oil, green vegetables; widely distributed in foods	Not definitely known in humans
Vitamin K (Phylloquinone, K_1; menaquinone, K_2)	Eggs, liver, green vegetables; produced in the intestines by bacterial reactions	Blood is slow to clot (antihemorrhagic vitamin)
WATER SOLUBLE		
Vitamin B_1 (Thiamin)	Meat, whole-grain cereals, liver, yeast, nuts	Beriberi (nervous system disorders, heart disease, fatigue)
Vitamin B_2 (Riboflavin)	Meat, cheese, eggs, fish, liver	Sores on the tongue and lips, bloodshot eyes, anemia
Vitamin B_6 (Pyridoxine)	Cereals, liver, meat, fresh vegetables	Skin disorders (dermatitis)
Vitamin B_{12} (Cyanocobalamin)	Meat, eggs, liver, milk	Pernicious anemia
Vitamin C (Ascorbic acid)	Citrus fruits, tomatoes, green vegetables	Scurvy (bleeding gums, loose teeth, swollen joints, slow healing of wounds, weight loss)
Niacin (Nicotinic acid and amide)	Meat, yeast, whole wheat	Pellagra (dermatitis, diarrhea, mental disorders)
Biotin (Vitamin H)	Liver, yeast, egg yolk	Skin disorders (dermatitis)
Folic acid	Liver, wheat germ, yeast, green leaves	Macrocytic anemia, gastrointestinal disorders

exposure to air. Removal of the outside hull from grains drastically decreases their B vitamin content. Thus, when polished rice became a dietary staple in the Orient, beriberi (the thiamin-deficiency disease) grew to epidemic proportions.

33.8 Minerals

A number of inorganic ions are needed for good health. Like vitamins, minerals are classified as micronutrients. Those that must be ingested in relatively large amounts, the *major elements*, include sodium, potassium, calcium, magnesium, chloride, and phosphate (phosphorus). Elements in a second group, the *trace ele-*

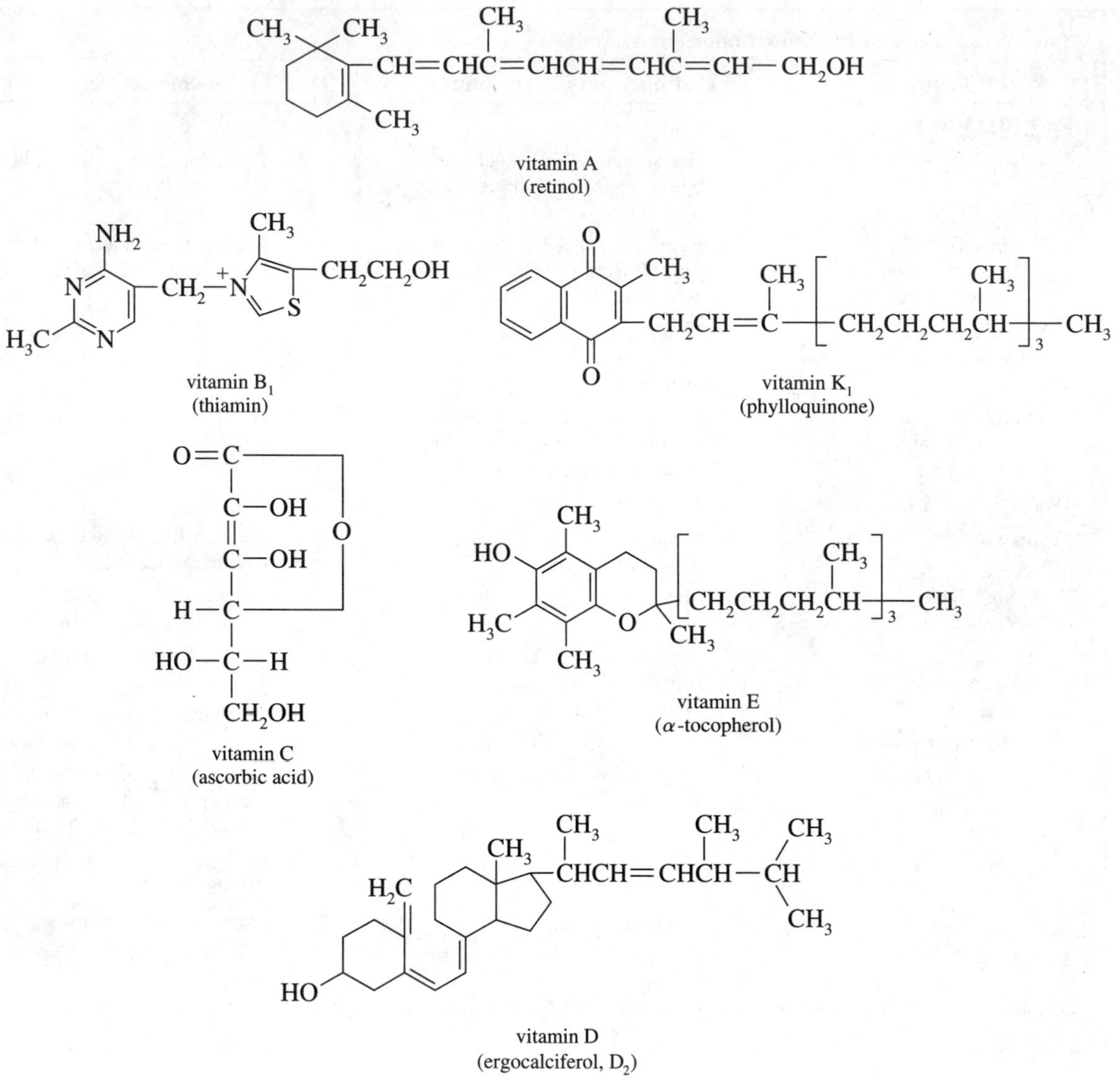

vitamin A
(retinol)

vitamin B_1
(thiamin)

vitamin K_1
(phylloquinone)

vitamin C
(ascorbic acid)

vitamin E
(α-tocopherol)

vitamin D
(ergocalciferol, D_2)

▲
FIGURE 33.2
The structures of selected vitamins.

ments, are required in much smaller amounts. As scientific studies of the body's mineral requirements proceed, the list of needed trace elements constantly lengthens. A recent compilation of these trace elements is given in Table 33.4.

Mineral nutrients differ from organic nutrients in that the body, in general, uses minerals in the ionic form in which they are absorbed. Although these elements are required for good health, they can also be toxic if ingested in quantities that are too large. Nutritionists have established RDAs for some of these minerals and warn against excess intake.

The major minerals—sodium, potassium, and chloride—are responsible for maintaining the appropriate salt levels in body fluids. Many enzyme reactions require an optimal salt concentration of 0.1–0.3 *M*. In addition, individual elements serve specific functions; nerve transmission, for example, requires a supply of extracellular

TABLE 33.4 Required Trace Elements

Element	Function	Human deficiency signs
Fluorine	Structure of teeth, possibly of bones; possible growth effect	Increased incidence of dental caries; possibly risk factor for osteoporosis
Silicon	Calcification; possible function in connective tissue	Not known
Vanadium	Not known	Not known
Chromium	Efficient use of insulin	Relative insulin resistance, impaired glucose tolerance, elevated serum lipids
Manganese	Mucopolysaccharide metabolism, superoxide dismutase enzyme	Not known
Iron	Oxygen and electron transport	Anemia
Cobalt	Part of vitamin B_{12}	Only as vitamin B_{12} deficiency
Nickel	Interaction with iron absorption	Not known
Copper	Oxidative enzymes; interaction with iron; cross-linking of elastin connective protein	Anemia, changes of ossification; possibly elevated serum cholesterol
Zinc	Numerous enzymes involved in energy metabolism and in transcription and translation	Growth depression, sexual immaturity, skin lesions, depression of immunocompetence, change of taste acuity
Arsenic	Not known	Not known
Selenium	Glutathione peroxidase; interaction with heavy metals	Endemic heart problems conditioned by selenium deficiency
Molybdenum	Xanthine, aldehyde, and sulfide oxidase enzymes	Not known
Iodine	Constituent of thyroid hormones	Goiter, depression of thyroid function, cretinism

sodium and intracellular potassium. Although there is no RDA for these two elements, 2–4 g/day represents an average NaCl consumption. But because high levels of NaCl can contribute to high blood pressure, many nutritionists advise using salt in moderation.

Calcium and magnesium serve many roles in the body, including being required by some enzymes. Fully 90% of all body calcium and an important percentage of the magnesium are found in the bones and teeth. Calcium also is needed for nerve transmission and blood clotting.

Trace elements are similar to vitamins in that (1) they are required in small amounts and (2) food contains only minute quantities of them. Usually a normal diet contains adequate quantities of all trace elements. The exact function of many trace elements remains unknown.

The most notable exception to the preceding generalizations is iron. This element is part of hemoglobin, the oxygen-binding protein of the blood. (Iron is also a critical component of some enzymes.) Relatively large amounts of iron are required for hemoglobin replenishment. Unfortunately, many foods are not rich enough in

Water has many functions in keeping our bodies healthy.

iron to provide the necessary RDA. This is especially true for the higher iron RDA for women. Therefore, nutritionists sometimes recommend a daily iron dietary supplement.

33.9 Water

Water is the solvent of life. As such it carries nutrients to the cells, allows biochemical reactions to proceed, and carries waste from the cells. Water makes up approximately 55–60% of the total body mass.

Our bodies are constantly losing water via urine (700–1400 mL/day), feces (150 mL/day), sweat (500–900 mL/day), and expired air (400 mL/day). If dehydration is to be prevented, this water output must be offset by water intake. In the normal diet, liquids provide about 1200–1500 mL of water per day. The remaining increment is derived from food (700–1000 mL/day) and water formed by biochemical reactions (metabolic water; 200–300 mL/day). Water losses vary and depend on the intake and the activity of the individual.

A proper water balance is maintained by control of water intake and water excretion. The kidney is the center for control of water output. Water intake is controlled by the thirst sensation. We feel thirsty when one or both of the two following events occur: (1) As the body's water stores are depleted, the salt concentration rises. Specific brain cells monitor this salinity and initiate the thirst feeling. (2) As water is drawn away from the salivary glands, the mouth feels dry; this event also leads to the sensation of thirst.

33.10 Nutrition Content Labeling

Nearly all foods in our supermarkets have been processed to some degree. And loss of nutrients is a problem with some kinds of food processing. In general, the more extensive the processing, the greater the loss. Processing can be quite extensive for such items as luncheon meats, TV dinners, and breakfast cereals.

In an effort to aid the consumer, nutritionists have established procedures for monitoring the nutritional values of processed foods. These values are reported on a package panel termed *Nutrition Facts* (see Figure 33.3). Much can be learned about food nutrients by examining such panels. First, some general observations: (1) Many people's daily energy requirement centers around a 2000 kcal (Cal) or a 2500 kcal (Cal) diet; the Nutrition Facts calculations are based on these approximations. (2) Fats are subdivided to recognize the important health difference between saturated and unsaturated fatty acids; cholesterol is given its own category. (3) Carbohydrates are categorized as sugars and fiber with the remainder being complex carbohydrates such as starch. (4) Total protein is always listed. (5) These nutritional facts are based on a standard serving size (set by the federal government).

Figure 33.3 compares the Nutrition Facts for two breakfast cereals. Notice that the serving size is the same (about 50 g) although the volume of one cereal is smaller. While both cereals are primarily carbohydrates, the "granola" cereal contains

NUTRITION FACTS
Serving Size 1 cup (50g)
Servings Per Container about 7

Amount Per Serving	1 Cup Cereal	Cereal with 1 Cup Vitamins A&D Skim Milk
Calories	180	270
Calories from Fat	15	15
		% Daily Value*
Total Fat 1.5g†	2%	2%
Saturated Fat 0g	0%	0%
Polyunsaturated Fat 1g		
Monounsaturated Fat 0g		
Cholesterol 0mg	0%	1%
Sodium 0mg	0%	5%
Total Carbohydrate 41g	14%	18%
Dietary Fiber 6g	26%	26%
Sugars 0g		
Protein 5g		
Vitamin A	**	10%
Vitamin C	**	4%
Calcium	2%	30%
Iron	8%	10%
Vitamin D	**	25%
Thiamin	6%	10%
Riboflavin	4%	20%
Niacin	15%	15%
Phosphorus	20%	45%
Magnesium	15%	20%

†Amount in 1 Cup Cereal. One cup skim milk contributes an additional 90 calories, less than 5mg cholesterol, 125mg sodium, 12g total carbohydrate (12g sugars), and 8g protein.
*Percent Daily Values are based on a 2,000 calorie diet. Your daily values may be higher or lower depending on your calorie needs.
**Contains less than 2% of the daily value of these nutrients.

	Calories:	2000	2500
Total Fat	Less than	65g	80g
Sat Fat	Less than	20g	25g
Cholesterol	Less than	300mg	300mg
Sodium	Less than	2,400mg	2,400mg
Total Carbohydrate		300g	375g
Dietary Fiber		25g	30g

(a) "Shredded wheat" cereal

NUTRITION FACTS
Serving Size ½ cup (48g)
Servings Per Container about 9

Amount Per Serving	
Calories 210	
Calories from Fat 70	
	% Daily Value*
Total Fat 8g	12%
Saturated Fat 3.5g	17%
Polyunsaturated Fat 1g	
Monounsaturated Fat 3g	
Cholesterol 0mg	0%
Sodium 15mg	1%
Potassium 210mg	6%
Total Carbohydrate 33g	11%
Other Carbohydrate 18g	
Dietary Fiber 4g	14%
Sugars 12g	
Protein 5g	
Vitamin A	0%
Vitamin C	0%
Calcium	4%
Iron	6%
Thiamin	10%
Phosphorus	10%
Magnesium	10%
Copper	15%

*Percent Daily Values are based on a 2,000 calorie diet. Your daily values may be higher or lower depending on your calorie needs.

	Calories:	2000	2500
Total Fat	Less than	65g	80g
Sat Fat	Less than	20g	25g
Cholesterol	Less than	300mg	300mg
Sodium	Less than	2,400mg	2,400mg
Total Carbohydrate		300g	375g
Dietary Fiber		25g	30g
Potassium		3,500mg	3,500mg

Calories per gram:
Fat 9 • Carbohydrate 4 • Protein 4

(b) "Granola" cereal

FIGURE 33.3
Two examples of Nutrition Facts panels.

about 25% (12 g out of 48 g) sugar. In contrast, the "shredded wheat" cereal contains only complex carbohydrates with a slightly higher Percent Daily Value of fiber. Both cereals are relatively low in fat, but the "granola" cereal contains more. Since common grains contain about the same percentage protein, both cereals contain 5 g of protein. The "shredded wheat" cereal manufacturer has opted to include nutritional information after adding 1 cup of skim milk. An additional food item such as milk can significantly change nutrient levels. Both cereals list vitamins and minerals; the list may vary depending on what is included in the food item. Finally, the recommended nutrient amounts for a healthful daily diet are given at the bottom of these nutritional facts (based on either 2000 Cal or 2500 Cal diets).

The Nutrition Facts panel gives us the amount of nutrients suggested for a 2000 kcal or a 2500 kcal diet. The Percent Daily Values are based on these suggestions. For example, the percent total carbohydrate in 48 g of granola cereal is readily calculated for a 2000 kcal diet (Figure 33.3) thus:

$$\frac{33\text{ g carbohydrate}}{300\text{ g carbohydrate}} \times 100 = 11\%$$

The nutrient amounts given at the bottom of the Nutrition Facts panel have been chosen with these important guidelines in mind: (1) Carbohydrate should be the bulk of a diet; (2) saturated fat should be a small proportion of the total fat intake; (3) consumption of both sodium and cholesterol should be minimized. We can estimate our nutrient intake and also learn more about nutrition from the Nutrition Facts panel.

33.11 Food Additives

Various chemicals are often added to foods during processing. In fact, more than 3000 of these *food additives* have been given the "generally recognized as safe" (GRAS) rating by the U.S. Food and Drug Administration.

The purpose of these food additives varies. Some additives enhance the nutritional value (e.g., when a food is vitamin enriched). Other additives serve as preservatives. For example, sodium benzoate is used to inhibit bacterial growth, and BHA (butylated hydroxyanisole) and BHT (butylated hydroxytoluene) are used as antioxidants. Still other additives may be used to improve the appearance and flavor of a food—for example, emulsifiers, thickeners, anticaking agents, flavors, flavor enhancers, nonsugar sweeteners, and colors (see Table 33.5).

There has been and continues to be a great deal of controversy concerning the use of food additives. Due to the nature and complexity of the subject, this debate will no doubt continue into the foreseeable future. Controversy centers around the problem of balancing the benefit derived from an additive against the risk to consumers. The use of at least some additives is necessary in the preparation of many foods. To discover that there is a risk and to assess the degree of risk for an additive requires long, difficult, and expensive research. As a case in point, salting and smoking were used in curing meats for centuries before there was any knowledge that these processes might involve a hazard. Research has shown that both processes involve risks to at least some consumers: Salt aggravates certain cardiovascular

TABLE 33.5 Some Common Food Additives

Food additive	Purpose
Sodium benzoate Calcium lactate Sorbic acid	Prevention of food spoilage (antimicrobials)
BHA (Butylated hydroxyanisole) BHT (Butylated hydroxytoluene) EDTA (Ethylenediaminetetraacetic acid)	Prevention of changes in color and flavor (antioxidants)
Calcium silicate Silicon dioxide Sodium silicoaluminate	Keeps powders and salt free-flowing (anticaking agents)
Carrageenan Lecithin	Aids even distribution of suspended particles (emulsifiers)
Pectin Propylene glycol	Imparts body and texture (thickeners)
MSG (Monosodium glutamate) Hydrolyzed vegetable protein	Supplements or modifies taste (flavor enhancers)

conditions, and smoke produces carcinogens in the meat. Yet in the eyes of many consumers, these risks are outweighed by the benefits to be had from salted and smoked meats. On the other hand, there would be few consumers indeed who would wish to have a compound known to be very carcinogenic used as an additive in their ice cream, even though that compound could improve the flavor of the ice cream remarkably!

Owing to the technical nature of the task, the consumer must rely largely on the judgment and integrity of the professional people who have been assigned the responsibility of protecting our food supply. Consumers should also inform themselves as fully as possible concerning the nature, purpose, and possible hazards of the additives that are used or proposed for use in our foods and, as responsible citizens, make sure that our government provides adequate support to the professionals charged with protecting our food.

33.12 A Balanced Diet

To summarize the preceding sections, the macronutrients (carbohydrates, fats, and proteins) comprise the majority of our diets. These nutrients supply the molecules that are needed for energy, growth, and maintenance. A second group of nutrients—vitamins, minerals, and water—is not used for energy but is nevertheless essential to our existence. Vitamins provide organic molecules that cannot be made in the body, and minerals provide the inorganic ions needed for life. Water is the solvent in which most of the chemical reactions essential to life occur. Vitamins and minerals

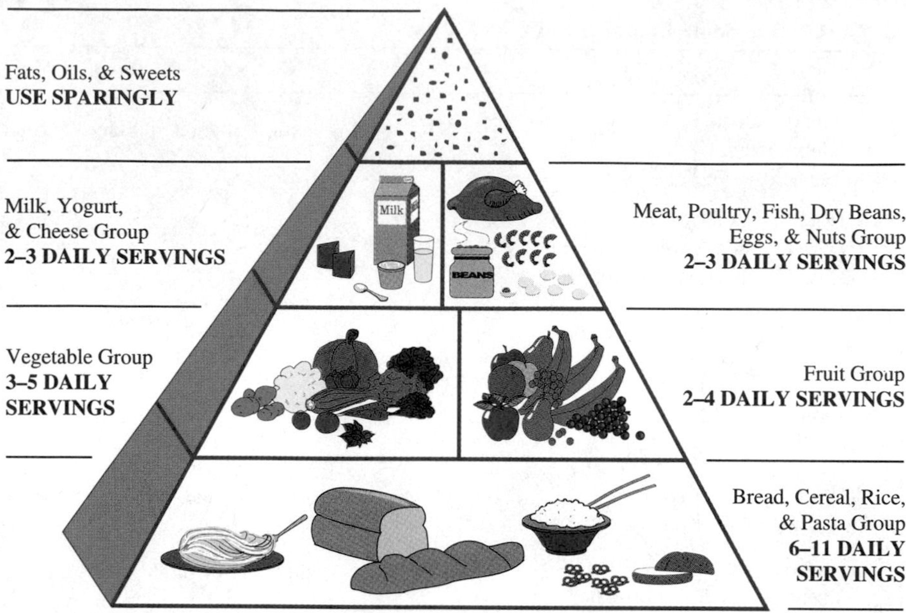

FIGURE 33.4 ▶ The food pyramid: a guide to a balanced diet.

are required only in small amounts, from a few micrograms to a few milligrams per day. But water must be consumed in large quantities, 2–3 L/day.

For health and well-being, each of six groups of nutrients must be in our diet. With the variety of foods that are available, how can we make sure that our diets contain enough of all the needed nutrients? How can we make sure that our diet is balanced?

To answer these questions, nutritionists have divided foods into five groups; (1) milk products; (2) vegetables; (3) fruits; (4) cereal products; and (5) meats, poultry, fish, beans, eggs, and nuts. Each group is a good source of one or more nutrients. To obtain a balanced mixture of carbohydrate, fat, vitamins, and minerals, a diet should contain food from several classes. Figure 33.4 is a pictorial guide to a balanced diet.

A balanced diet must include several food groups, even though each food may contain many nutrients. For example, compare the three breakfasts in Figure 33.5, each of which supplies about 600 kcal of food energy. By choosing only doughnuts (cereal food groups) and coffee, the consumer would gain some nutrients but only in small quantities. A breakfast of cold cereal (cereal group) and milk (milk group), toast (cereal group) with margarine (fats) and jelly (sweets), and coffee includes more food groups. Still more food groups are found in a breakfast of orange juice (fruit group), a fried egg (meat group), pancakes (cereal group) with margarine (fats) and syrup (sweets), milk (milk group), and coffee. Many other breakfasts could be chosen instead, yet an important generalization can be drawn from these three examples. The nutritional value of a meal improves if at least several food groups are included. The third breakfast in this illustration is the most balanced. Although the consumer often does not know the nutrient content of a specific food, overall nutrition can be ensured by choosing a balanced diet.

After the food composing a balanced diet is eaten, it must be digested, absorbed, and transported in order for the proper nutrients to reach the cells. Eating puts food

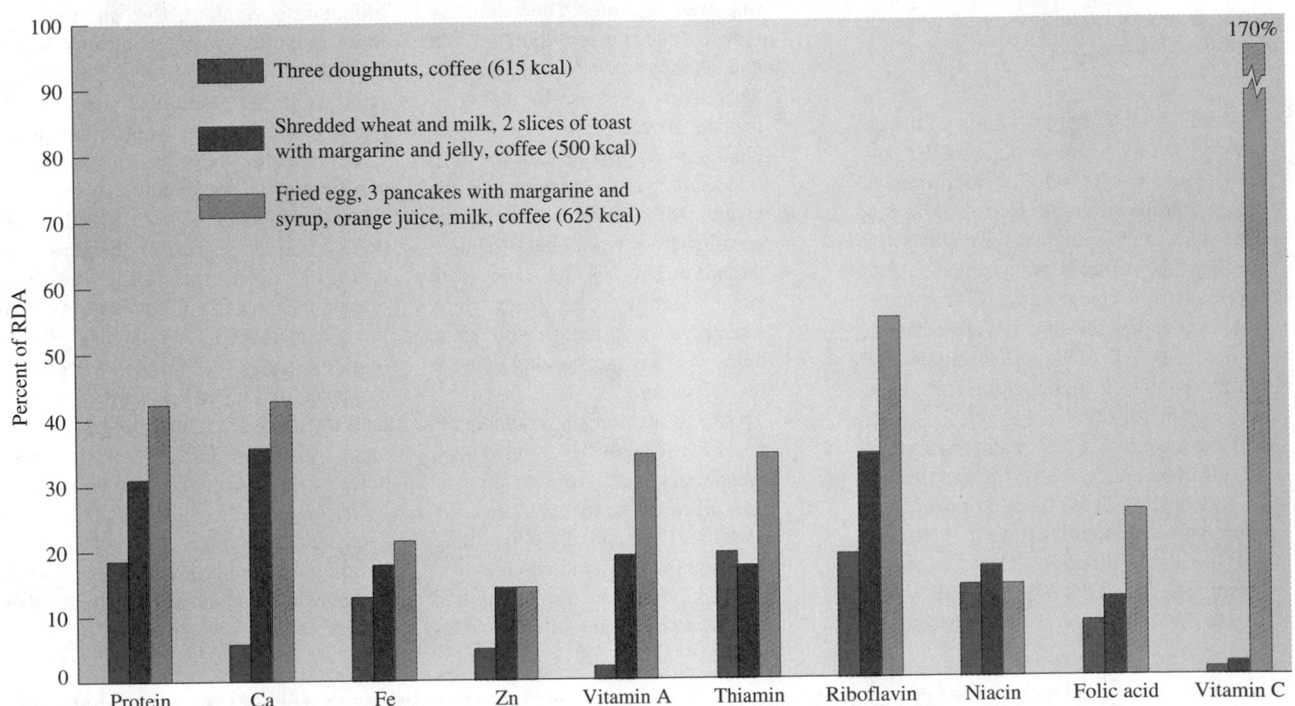

FIGURE 33.5
A comparison of nutritional values for three breakfasts. (No RDA has been established for carbohydrates and fats.)

Ice Cream: Food for Thought

Can food designers keep the desirable qualities of ice cream while trying to make it a healthier food?

For many people there is nothing quite as tasty as a serving of cool, smooth, rich ice cream. The texture is satin smooth and it literally melts in your mouth; you experience a slow increase in richness as one ingredient melds with another. And, although ice cream comes from the freezer, the temperature never seems to be too cold. All of these sensations and more combine to make eating ice cream a simple pleasure.

Like most foods, ice cream is a mixture of nutrients, and some of these nutrients are less healthy than others. For example, scientists recommend that we limit our fat and sugar consumption. This presents a dilemma for ice cream makers: How are they to decrease the fat and sugar content of ice cream while retaining eating enjoyment? This dilemma is shared by many food manufacturers. We can literally take a food apart into its separate nutrients. We can remove unhealthy components. But, can we put it back together again and have it taste as good? In essence, food scientists are being asked to be food designers. Ice cream manufacturing is a good example of this process.

Ice cream is a simple food that is composed of only a small number of major ingredients:

1. Water (between 60 and 70%) is the largest single ingredient, mostly in the form of small ice crystals that impart the cool sensation. To retain a smooth texture, these crystals must be very small (no larger than 30–45 μm in diameter).
2. Fats are another component of ice cream (the minimum for ice cream is 10% milk fat from cream in the United States). Fats are broken up into small globules (1–10 μm) and provide the creamy texture for the finished product. Fats are important for mouth feel because they reduce the sensation of cold and, as they warm to body temperature, slowly release the flavor of the ice cream.
3. Nonfat milk solids provide the third important basic component of ice cream. Milk proteins (casein and proteins from whey) help stabilize the mixture of fats and water. Milk sugar (lactose) contributes some sweetness and, along with salts, helps to lower the freezing point of the ice cream.
4. To these ice cream components are added emulsifiers, thickeners, sweeteners, and flavorings. Emulsifiers allow the fat to mix thoroughly with the water and also allow more air to be incorporated into the mixture. Chemicals such as mono- and diacylglycerols are often used as emulsifiers. Thickeners, such as carrageenan, guar gum, alginates, and pectins, ensure that ice cream components do not separate in transport or storage. Sweeteners include sucrose and corn syrup and often comprise from 12 to 20% of the ice cream. Finally, there are the flavorings.

Decreasing the sugar content has proven to be the easiest part of ice cream redesign. Nonnutritive sweeteners like aspartame substitute for sucrose or corn syrup. The milk sugar lactose can be broken down to its component monosaccharides to increase ice cream sweetness. Both changes retain sweetness while decreasing the amount of sugars needed.

In contrast, lowering the fat content of ice cream has proved difficult. Fats provide the creamy texture for ice cream. Perhaps more importantly, fats control the release of ice cream flavor so that the consumer can literally "savor" each bite. At this writing there is no ideal fat substitute. Some manufacturers have opted for nonfat polymers that form a suspension of small globules (several microns in diameter); these globules provide a creamy texture for the ice cream. Both carbohydrates (e.g., cellulose, modified starches, and pectins) and proteins (egg albumen and proteins from whey) have been used as substitutes. Unfortunately, these substitutes don't "melt in your mouth"; they don't release the ice cream flavors slowly, letting them blend as full-fat ice cream does. Other manufacturers have opted for lower calorie modified fats, "designer fats (e.g., olestra)." These molecules give the character of fat but are harder to digest (and thus provide fewer calories). Again, these substitutes yield a creamy product but lack the proper flavor release.

Ice cream can be made healthier. However, since we often choose food based not only on the nutritional value but also on an enjoyment factor, ice cream redesign is not a simple task.

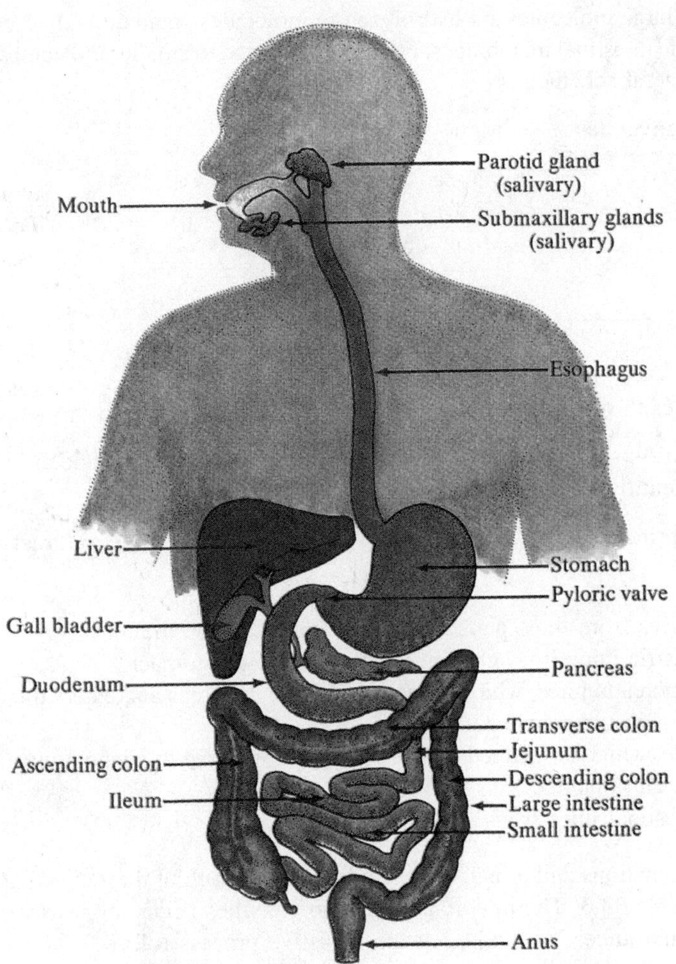

FIGURE 33.6
The human digestive tract.

into the alimentary canal (which includes the mouth, esophagus, stomach, small intestine, and large intestine). In a sense this canal is an extension of our external environment; that is, until food is broken down and processed, it cannot enter our internal environment to reach the cells. After digestion and absorption, the nutrients are transported to the cells by the blood and lymph systems. The liver has a vital role in controlling nutrient levels in the blood and neutralizing toxic substances.

33.13 Human Digestion

The human digestive tract is shown diagrammatically in Figure 33.6. Although food is broken up mechanically by chewing and by a churning action in the stomach, digestion is a chemical process. **Digestion** is a series of enzyme-catalyzed reactions

digestion

by which large molecules are hydrolyzed to molecules small enough to be absorbed through the intestinal membranes. Foods are digested to smaller molecules according to this general scheme:

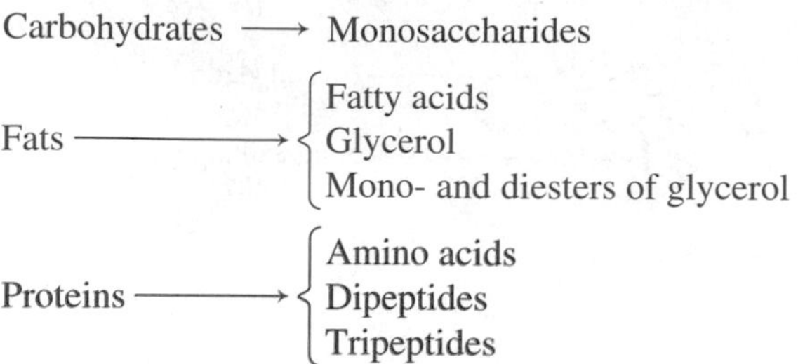

Food passes through the human digestive tract (Figure 33.6) in this sequence:

Mouth ⟶ esophagus ⟶ stomach ⟶ small intestine (duodenum, jejunum, and ileum) ⟶ large intestine

Five principal digestive juices (or fluids) enter the digestive tract at various points:

1. Saliva from three pairs of salivary glands in the mouth
2. Gastric juice from glands in the walls of the stomach
3. Pancreatic juice, which is secreted by the pancreas and enters the duodenum through the pancreatic duct
4. Bile, which is secreted by the liver and enters the duodenum via a duct from the gall bladder
5. Intestinal juice from intestinal mucosal cells

Detailed accounts of the various stages of digestion are to be found in biochemistry and physiology textbooks.

The main functions and principal enzymes found in each of these fluids are summarized in Table 33.6. The important digestive enzymes occur in gastric, pancreatic, and intestinal juices. An outline of the digestive process follows.

Salivary Digestion Food is chewed (masticated) and mixed with saliva in the mouth, and the hydrolysis of starch begins. The composition of saliva depends on many factors—age, diet, condition of teeth, time of day, and so on. Normal saliva is about 99.5% water. Saliva also contains mucin (a glycoprotein); a number of mineral ions such as K^+, Ca^{2+}, Cl^-, PO_4^{3-}, SCN^-; and one enzyme, salivary amylase (ptyalin). The pH of saliva ranges from slightly acidic to slightly basic, with the optimal pH about 6.6–6.8.

Mucin acts as a lubricant and facilitates the chewing and swallowing of food. The enzyme salivary amylase catalyzes the hydrolysis of starch to maltose:

$$\text{Starch} + \text{Water} \xrightarrow[\text{amylase}]{\text{salivary}} \text{Maltose}$$

Salivary amylase is inactivated at a pH of 4.0, so it has very little time to act before the food reaches the highly acidic stomach juices.

Saliva is secreted continuously, but the rate of secretion is greatly increased by the sight and odor, or even the thought, of many foods. The mouth-watering effect of the sight or thought of pickles is familiar to many of us. This is an example of a *conditioned reflex.*

TABLE 33.6 Digestive Fluids

Fluid (volume) produced daily)	Source	Principal enzymes and/or function
Saliva (1000–1500 mL)	Salivary glands	Lubricant, aids chewing and swallowing; also contains salivary amylase (ptyalin), which begins the digestion of starch
Gastric juice (2000–3000 mL)	Glands in stomach wall	Pepsin, gastric lipase; pepsin catalyzes partial hydrolysis of proteins to proteoses and peptones in the stomach
Pancreatic juice (500–800 mL)	Pancreas	Trypsinogen, chymotrypsinogen, procarboxypeptidase (converted after secretion to trypsin, chymotrypsin, and carboxypeptidase, respectively, which continue protein digestion); amylopsin (α-amylase, a starch-digestive enzyme), steapsin (a lipase)
Bile (500–1000 mL)	Liver	Contains no enzymes but does contain bile salts, which aid digestion by emulsifying lipids; serves to excrete cholesterol and bile pigments derived from hemoglobin
Intestinal juice	Intestinal mucosal cells	Contains a variety of finishing enzymes: sucrase, maltase, and lactase for carbohydrates; aminopolypeptidase and dipeptidase for final protein breakdown; intestinal lipase; nucleases and phosphatase for hydrolysis of nucleic acids

Gastric Digestion When food is swallowed, it passes through the esophagus to the stomach. In the stomach, mechanical action continues; food particles are reduced in size and are mixed with gastric juices until a material of liquid consistency, known as *chyme*, is obtained.

Gastric juice is a clear, pale yellow, acidic fluid having a pH of about 1.5–2.5. It contains hydrochloric acid; the mineral ions Na^+, K^+, Cl^-; some phosphates; and the digestive enzymes pepsin and lipase. The flow of gastric juice is accelerated by conditioned reflexes and by the presence of food in the stomach. The secretion of the hormone *gastrin* is triggered by food entering the stomach. This hormone, which is produced by the gastric glands, is absorbed into the bloodstream and returned to the stomach wall where it stimulates the secretion of additional gastric juice. Control of gastric secretion by this hormone is an example of one of the many chemical control systems that exist in the body.

The chief digestive function of the stomach is the partial digestion of protein. The principal enzyme of gastric juice is pepsin, which digests protein. The enzyme is secreted in an inactive form called pepsinogen, which is activated by hydrochloric acid to pepsin. Pepsin catalyzes the hydrolysis of proteins to fragments called proteoses and peptones, which are still fairly large molecules. Pepsin splits the peptide bonds adjacent to only a few amino acid residues, particularly tyrosine and phenylalanine:

$$\text{Protein} + \text{Water} \xrightarrow{\text{pepsin}} \text{Proteoses} + \text{Peptones}$$

The second enzyme in the stomach, gastric lipase, is a fat-digesting enzyme. Its action in the stomach is slight because the acidity is too high for lipase activity.

Food may be retained in the stomach for as long as 6 hours. It then passes through the pyloric valve into the duodenum.

Intestinal Digestion The next section of the digestive tract, the small intestine, is where most of the digestion occurs. The stomach contents are first made alkaline by secretions from the pancreatic and bile ducts. The pH of the pancreatic juice is 7.5–8.0, and the pH of bile is 7.1–7.7. The shift in pH is necessary because the enzymes of the pancreatic and intestinal juices are active only in an alkaline medium. Enzymes that digest all three kinds of food—carbohydrates, fats, and proteins—are secreted by the pancreas. Pancreatic secretion is stimulated by hormones that are secreted into the bloodstream by the duodenum and the jejunum.

The enzymes occurring in the small intestine include pancreatic amylases (diastase) that hydrolyze most of the starch to maltose, and carbohydrases (α-amylase, maltase, sucrase, and lactase) that complete the hydrolysis of disaccharides to monosaccharides. The proteolytic enzymes trypsin and chymotrypsin attack proteins, proteoses, and peptones, hydrolyzing them to dipeptides. Then the peptidases—carboxypeptidase, aminopeptidase, and dipeptidase—complete the hydrolysis of proteins to amino acids. Pancreatic lipases catalyze the hydrolysis of almost all fats. Fats are split into fatty acids, glycerol, and mono- and diesters of glycerol by these enzymes.

The liver is another important organ in the digestive system. A fluid known as bile is produced by the liver and stored in the gall bladder, a small organ located on the surface of the liver. When food enters the duodenum, the gall bladder contracts, and the bile enters the duodenum through a duct that is also used by the pancreatic juice. In addition to water, the major constituents of bile are bile acids (as salts), bile pigments, inorganic salts, and cholesterol. The bile acids are steroid monocarboxylic acids, two of which are shown here:

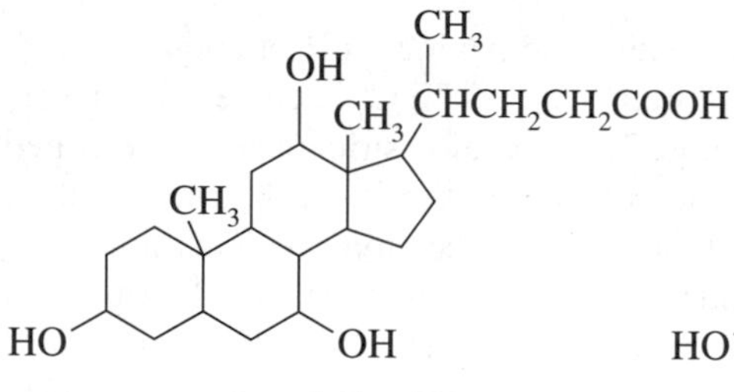

CH_3 | OH CH_3 $CHCH_2CH_2COOH$ CH_3 HO OH

cholic acid

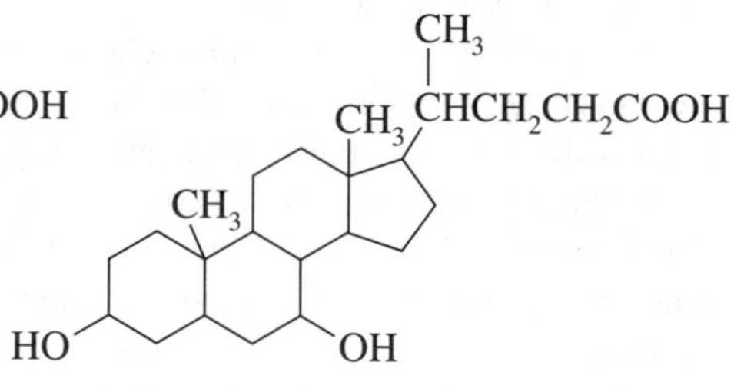

CH_3 | CH_3 $CHCH_2CH_2COOH$ CH_3 HO OH

chenodeoxycholic acid

The bile acids are synthesized in the liver from cholesterol, which is also synthesized in the liver. The presence of bile in the intestine is important for the digestion and absorption of fats. When released into the duodenum, the bile acids emulsify the fats, allowing them to be hydrolyzed by the pancreatic lipases. About 90% of the bile salts are reabsorbed in the lower part of the small intestine and are transported back to the liver and used again.

Most of the digested food is absorbed from the small intestine. Undigested and indigestible material passes from the small intestine to the large intestine, where it is retained for varying periods of time before final elimination as feces. Additional chemical breakdown, sometimes with the production of considerable amounts of gases, is brought about by bacteria (or rather by bacterial enzymes) in the large intestine. For a healthy person this additional breakdown is not important from the standpoint of nutrition since absorption of nutrients does not occur from the large intestine. However, large amounts of water, partly from digestive juices, are absorbed from the large intestine so that the contents become more solid before elimination as feces.

33.14 Absorption

For digested food to be utilized in the body, it must pass from the intestine into the blood and lymph systems. The process by which digested foods pass through the membrane linings of the small intestine and enter the blood and lymph is called **absorption**. Absorption is complicated, and we will consider only an overview of it.

absorption

After the food you have eaten is digested, the body must absorb billions upon billions of nutrient molecules and ions into the bloodstream. The absorption system is in the membranes of the small intestine, which, upon microscopic inspection, are seen to be wrinkled into hundreds of folds. These folds are covered with thousands of small projections called *villi*. Each projection is itself covered by many minute folds called *microvilli*. The small intestine's wrinkles, folds, and projections increase the surface area available for absorption. The average small intestine is about 4 m (13.3 ft) long and is estimated to contain about 8360 m^2 (90,000 ft^2) of absorbing surface.

The inner surface of the small intestine is composed of mucosal cells that produce many enzymes, such as disaccharidases, aminopeptidases, and dipeptidases, needed to complete the digestive process. As digestion is completed, the resulting nutrient molecules are absorbed by the mucosal cells and transferred to the blood and lymph systems. Water-soluble nutrients such as monosaccharides, glycerol, short-chain fatty acids, amino acids, and minerals enter directly into the bloodstream. Fat-soluble nutrients such as long-chain fatty acids and monoacylglycerols first enter the lymph fluid and then enter the bloodstream where the two fluids come together.

An important factor in the absorption process is that the membranes of the intestine are selectively permeable; that is, they prevent the passage of most large molecules but allow the passage of smaller molecules. For example, polysaccharides, disaccharides, and proteins are not ordinarily absorbed, but generally monosaccharides and amino acids are.

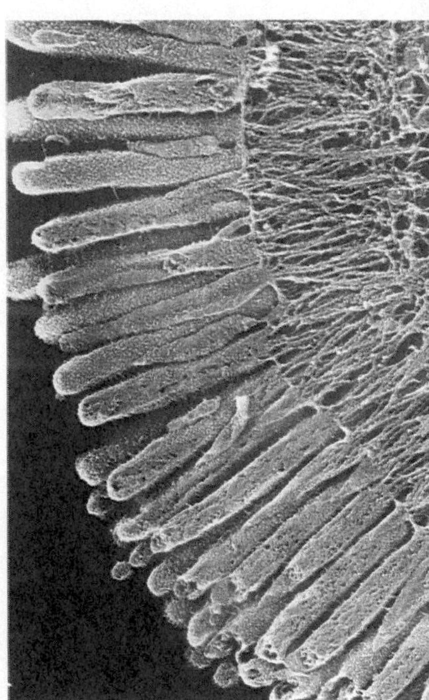

Absorption of digested food occurs in the small intestine through the villi and microvilli shown here.

33.15 Liver Function

The liver is the largest organ in the body and performs several vital functions. Two of these functions are the regulation of the concentrations of organic nutrients in the blood and the removal of toxic substances from the blood.

The concentration of blood sugar (glucose) is controlled and maintained by processes that occur in the liver. After absorption, excess glucose and other monosaccharides are removed from the blood and converted to glycogen in the liver. The liver is the principal storage organ for glycogen. As glucose is used in other cells, the stored liver glycogen is gradually hyrolyzed to maintain the appropriate blood-glucose concentration. Liver function is under sensitive hormonal control, as are most vital body functions. This regulation will be discussed further in Chapter 35.

A second major function of the liver is the detoxification of harmful and potentially harmful substances. This function apparently developed as higher vertebrates appeared in the evolutionary time scale. The liver is able to deal with most of the toxic molecules that occur in nature. For example, ethanol is oxidized in the liver, and nitrogenous metabolic waste products are converted to urea for excretion.

Organic chemists have learned to synthesize new substances that have no counterparts in the biological world. They are used as industrial chemicals, insecticides, drugs, and food additives. When these potentially toxic substances are ingested, even in small amounts, the body is faced with the difficult challenge of metabolizing or destroying substances that are unlike any found in nature.

The liver is able to meet this challenge and deal with most of these foreign molecules through an oxidation system located in the endoplasmic reticulum. Bound to these intracellular membranes are enzymes that can catalyze reactions between oxygen and the foreign molecules. As these molecules are oxidized, they become more polar and water soluble. Finally, the oxidation products of the potential toxins are excreted in the urine or bile fluid.

For example, most automobile antifreeze solutions contain ethylene glycol, a toxic substance. Even though this compound does not occur naturally, the liver can metabolize ethylene glycol. When small amounts are ingested, the following chemical changes occur:

$$\underset{\text{ethylene glycol}}{\mathrm{H{-}CH(OH){-}CH(OH){-}H}} \xrightarrow{\text{oxidation}} \underset{\text{glyoxal}}{\mathrm{H{-}CH(OH){-}C({=}O){-}H}} \xrightarrow{\text{oxidation}} \underset{\text{glycolate}}{\mathrm{H{-}CH(OH){-}C({=}O){-}O^-}} \xrightarrow{\text{oxidation}}$$

$$\underset{\text{glyoxalate}}{\mathrm{H{-}C({=}O){-}C({=}O){-}O^-}} \xrightarrow{\text{oxidation}} \underset{\text{formate}}{\mathrm{H{-}C({=}O){-}O^-}} + \underset{\text{bicarbonate}}{\mathrm{HO{-}C({=}O){-}O^-}}$$

Thus ethylene glycol is converted by oxidation to two more polar (charged) acid anions that are easily eliminated from the body.

Unfortunately, oxidation is not effective for some compounds. Halogenated hydrocarbons, which are particularly inert to oxidation, accumulate in fatty tissue or in the liver itself. Examples of halogenated hydrocarbons are carbon tetrachloride, hexachlorobenzene, DDT, dioxins, and polychlorinated biphenyls (PCBs). Some compounds become more toxic after oxidation. For example, polycyclic hydrocarbons (which can be formed when food is barbecued) become carcinogenic upon partial oxidation. Methanol becomes particularly toxic because it is converted by oxidation to formaldehyde. One of the most serious dangers of environmental pollution lies in the introduction of compounds that the liver cannot detoxify.

In a sense the liver is the final guardian along the pathway by which nutrients pass to the cells. This pathway starts with a balanced diet. Once foods are digested and absorbed, the liver adjusts nutrient levels and removes potential toxins. The blood can then provide nutrients for the cellular biochemistry that constitutes life.

Concepts in Review

1. Distinguish a nutrient from a food.
2. Describe the difference between a minimum dietary requirement and a recommended dietary allowance.
3. Summarize the importance of nutrients in metabolism.
4. Briefly discuss the major functions of dietary protein.
5. Discuss the dangers of kwashiorkor.
6. Explain why fats are a more concentrated source of metabolic energy than carbohydrates.
7. List some of the dangers of a high-sucrose diet.
8. Briefly describe the importance of dietary fiber.
9. List the essential fatty acids and essential amino acids.
10. Discuss a major danger inherent in a vegetarian diet.
11. Summarize some of the functions of the water-soluble vitamins.
12. Discuss some of the functions of the fat-soluble vitamins.
13. State the general results of vitamin deficiencies.
14. List the major mineral elements.
15. Compare the similarities between trace mineral elements and vitamins.
16. Briefly discuss the importance of water in the diet.
17. List four purposes served by food additives.
18. Understand the information given in the Nutrition Facts panel.
19. List the five principal digestive juices and where they originate in the body.
20. List the principal enzymes of the various digestive juices.
21. Give the main digestive functions of each of the five principal digestive juices.
22. List the classes of products formed when carbohydrates, fats, and proteins are digested.
23. Briefly describe the absorption of nutrients from the digestive tract into the lymph and blood.
24. Discuss the role of the liver in maintaining blood-glucose levels.
25. Explain how the liver metabolizes potentially toxic compounds.

Key Terms

The terms listed here have been defined within this chapter. Section numbers are referenced in parenthesis for each term.

absorption (33.14)
complete protein (33.6)
diet (33.2)
digestion (33.13)
macronutrient (33.1)
micronutrient (33.1)
nutrient (33.1)
recommended dietary allowance (RDA) (33.2)
vitamin (33.7)

Questions

Questions refer to tables, figures, and key words and concepts defined within the chapter. A particularly challenging question or exercise is indicated with an asterisk.

1. How does a food differ from a nutrient?
2. What is meant by the term *energy allowance*?
3. How does marasmus differ from kwashiorkor?
4. What is meant by the statement that candy provides empty calories?
5. What structural features do the essential fatty acids have in common?
6. What is meant by the term *essential fatty acid*?
7. List the essential amino acids.
8. How do animal proteins differ nutritionally from vegetable proteins?
9. Which vitamins are water soluble? Which are fat soluble?
10. List three functions that can be attributed to vitamins.
11. Distinguish the major elements from the trace elements.
12. List two major biological functions for calcium.
13. What percentage of the average water consumption comes from solid foods? How much from liquids?
14. List five common categories of food additives.
15. What are the five principal digestive juices?
16. Which federal agency is responsible for regulating the use of food additives?
17. What enzymes are in each of the digestive juices?
18. What is chyme?
19. What is the digestive function of the liver?
20. How do the intestinal mucosal cells aid digestion?
21. How does the liver metabolize toxic compounds?
22. Which of these statements are correct? Rewrite each incorrect statement to make it correct.
    (a) Water is an essential nutrient.
    (b) Most vegetable proteins are complete proteins.
    (c) Oleic acid is an essential fatty acid.
    (d) A dietary supply of glycine is not needed.
    (e) Of the three classes of energy nutrients, an RDA has been established only for proteins.
    (f) Vitamins are inorganic nutrients.
    (g) Most foods contain a variety of nutrients.
    (h) Calcium is an important trace element.
    (i) The functions of many trace elements are not well understood.
    (j) The main purpose of digestion is to hydrolyze large molecules to smaller ones that can be absorbed through the intestinal membranes.
    (k) Most of the digestion of food occurs in the stomach.
    (l) Gastric juice contains hydrochloric acid and has a pH of 1.5–2.5.
    (m) Digestion in the small intestine occurs in an alkaline medium.
    (n) The function of bile acids is to emulsify carbohydrates; this allows them to be hydrolyzed to monosaccharides.
    (o) One function of the liver is to detoxify toxic compounds.

Paired Exercises

These exercises are paired. Each odd-numbered exercise is followed by a similar even-numbered exercise. Answers to the even-numbered exercises are given in Appendix V.

23. A sirloin steak (85 g) has the following nutritional composition:

Protein	23 g
Carbohydrates	0
Fats	19 g
Sodium	52 mg
Potassium	297 mg
Magnesium	23 mg
Iron	2.5 mg
Zinc	4.7 mg
Calcium	9.0 mg
Vitamin A	15 μg
Vitamin C	0
Thiamin	92 μg
Riboflavin	218 μg
Niacin	3.2 mg
Vitamin B_6	330 μg
Folic acid	7 μg
Vitamin B_{12}	2 μg

(a) What is the mass of macronutrients contained in this food?
(b) What is the mass of minerals?
(c) What is the mass of vitamins?

24. A battered and fried chicken breast (280 g) has the following nutritional composition:

Protein	70 g
Carbohydrates	25 g
Fats	37 g
Sodium	770 mg
Potassium	564 mg
Magnesium	68 mg
Iron	3.5 mg
Zinc	2.7 mg
Calcium	56 mg
Vitamin A	56.5 μg
Vitamin C	0
Thiamin	322 μg
Riboflavin	408 μg
Niacin	29.5 mg
Vitamin B_6	1.2 mg
Folic acid	16 μg
Vitamin B_{12}	0.82μg

(a) What is the mass of macronutrients contained in this food?
(b) What is the mass of minerals?
(c) What is the mass of vitamins?

25. Fat yields about 9 kcal/g.
(a) How much energy is available from fat in sirloin steak from Exercise 23?
(b) If the total energy content is 270 kcal, what percentage of energy is derived from the fat?

26. Fat yields about 9 kcal/g.
(a) How much energy is available from fat in chicken breast from Exercise 24?
(b) If the total energy content is 728 kcal, what percentage of energy is derived from the fat?

27. For younger women, the RDA for calcium is 1.20 g. What percentage of the RDA is supplied by the sirloin steak in Exercise 23?

28. For older women, the RDA for calcium is 0.80 g. What percentage of the RDA is supplied by the chicken breast from Exercise 24?

29. The RDA for vitamin B_{12} is 2.0 μg for adults. What percentage of the RDA is supplied by the sirloin steak in Exercise 23?

30. The RDA for vitamin B_{12} is 2.0 μg for adults. What percentage of the RDA is supplied by the chicken breast from Exercise 24?

31. List three classes of nutrients that are considered macronutrients.

32. List two classes of nutrients that are considered micronutrients.

33. List three classes of nutrients that do *not* commonly supply energy for the cell.

34. List three classes of nutrients that commonly supply energy for the cell.

35. Why is starch important in the diet?

36. Why is cellulose important in the diet?

37. Is saliva (pH 6.6–6.8) considered to be slightly acidic, slightly basic, or neutral?

38. Is pancreatic juice (pH 7.5–8.0) considered to be slightly acidic, slightly basic, or neutral?

39. In what parts of the digestive system are proteins digested?

40. In what parts of the digestive system are carbohydrates digested?

41. Pancreatic amylase is a digestive enzyme found in the small intestine. What nutrient class does this enzyme digest?

42. Pepsin is a digestive enzyme found in the stomach. What nutrient class does this enzyme digest?

Additional Exercises

These exercises are not paired or labeled by topic and provide additional practice on concepts covered in this chapter.

43. Milk is a food. List four nutrients that can be obtained from milk.

44. You are told that a new diet will cause you to lose 9 kg (20 lb) of fat in 1 week. Is this reasonable? Briefly explain.

45. In what ways are the trace elements similar to vitamins? How do they differ?

46. Explain why a tablespoon of butter (a fatty food) approximately doubles the calorie content of a medium-size baked potato (a carbohydrate food).

47. Given the Nutrition Facts panel in the adjacent column from cream of chicken soup:

(a) What percentage of total energy is provided by the fat?

(b) How many grams of protein will be provided by two servings of this food?

(c) For a person on a reduced-calorie diet (1000 kcal/day) what percentage of the total energy is supplied by one serving?

(d) What percentage of the total carbohydrates are fiber?

48. Why must the food of higher animals be digested before it can be utilized?

49. Galactosemia is an inherited condition in some babies. It is a deficiency in the enzyme that catalyzes the conversion of galactose to glucose. Symptoms are lack of appetite, weight loss, diarrhea, and jaundice. What staple of a baby's diet must be changed to correct this problem?

NUTRITION FACTS
Serv. Size 1/2 cup (125g)
Servings about 2 1/2

Amount Per Serving	
Calories 100	Fat cal 50
	% Daily Value*
Total Fat 5g	8%
Saturated fat 2g	9%
Cholesterol 10mg	3%
Sodium 860mg	36%
Total carbohydrate 10g	3%
Fiber 2g	9%
Sugars 1g	
Protein 3g	
Vitamin A 10% •	Vitamin C 0%
Calcium 0%	• Iron 2%

*Percent Daily Values are based on a 2,000 calorie diet.

50. One large egg contains 270 mg of cholesterol. How does this compare with the daily recommended amount of cholesterol? (See the Nutrition Facts panel in Figure 33.3.)

Bioenergetics

34

Chapter 34

◀ Finding adequate sources of energy is a constant challenge for all living organisms, including this bear.

If you are traveling through the midwestern United States on a warm night in early summer, you might notice small yellow lights dancing through the fields. This beautiful sight is evidence that the fireflies, or lightning bugs, have returned. Their flickering light does not arise from a small fire or electrical discharge but rather from a chemical reaction. Bioenergetics are at work here.

The monarch butterfly is a striking insect with a wing span of several inches. As it flies, it is buffeted by the wind, first one direction and then another. Clearly, the monarch is light and not a powerful flier. Yet each year (in one leg of a complex, multigeneration migration) these small creatures fly from southern Canada to winter in central Mexico, a distance of over 2000 miles. Chemical reactions in the insects' cells make energy available for this arduous trip. Bioenergetics are at work again.

While you have been reading these paragraphs, fully 20% of the oxygen you breathed has been used by your brain. Chemical processes have reacted this oxygen with blood glucose, yielding the energy needed to comprehend your reading. This is another example of bioenergetics—the chemical processes directly related to energy needs and uses in life.

34.1 Energy Changes in Living Organisms

bioenergetics

One of the basic requirements for life is a source of energy. **Bioenergetics** is the study of the transformation, distribution, and utilization of energy by living organisms. Bioenergetics includes the radiant energy of sunlight used in photosynthesis, electrical energy of nerve impulses, mechanical energy of muscle contractions, heat energy liberated by chemical reactions within cells, and potential energy stored in "energy-rich" chemical bonds. The major source of biological energy includes chemical reactions occurring inside cells. Bioenergetics also includes the transport of needed reactants through cell membranes into the cell and the return of waste products to the surrounding fluids.

The bioenergetics of a cell can be compared to the energetics of a manufacturing plant (see Figure 34.1). In both cases, raw materials are delivered and finished products (as well as waste) are removed. Energy is delivered in one form and used in another.

Manufacturing plant energetics	Animal cell bioenergetics	Plant cell bioenergetics
Energy delivered as electricity	Energy delivered as reduced carbon atoms	Energy delivered as sunlight
Electricity converted to mechanical energy	Energy in reduced carbon atoms converted to high-energy phosphate bonds	Sunlight converted to chemical energy (first, as reduced carbon atoms and, second, as high-energy phosphate bonds)
Mechanical energy converts raw materials to finished products	High-energy phosphate bonds broken to do the work of the cell	High-energy phosphate bonds broken to do the work of the cell

FIGURE 34.1
A typical manufacturing plant converts electrical energy to mechanical energy and uses this energy to change raw materials to finished products.

This chapter will cover bioenergetics in the following order:

1. The delivery of oxygen to the cell and the removal of waste products
2. Two energy transformations that occur inside the cell, biological oxidation–reduction (redox) and the formation of high-energy phosphate bonds (phosphorylation)
3. The process by which sunlight is converted to chemical energy (photosynthesis)

34.2 Oxygen/Carbon Dioxide Exchange

Oxygen Delivery

Human muscle cells, like many other cells, must have an adequate oxygen supply. About 5 g of oxygen (4 L at 25°C and 1 atm) is required for every minute of strenuous activity by a young adult. The necessary oxygen is supplied by oxygenated arterial blood, and waste carbon dioxide is removed by oxygen-deficient venous blood.

Normally, the partial pressure of oxygen is about 100 torr in the lungs and 35 torr in the cells. Because gases move spontaneously from regions of higher pressure to regions of lower pressure, oxygen moves toward the cells. But the body must move oxygen at rates far faster than can be attained by simple diffusion. Red blood cells, which contain the oxygen-transport protein hemoglobin (Hb) (see Section 30.8), increase the oxygen-transport rate by 80–90 times that obtained by simple diffusion.

The amount of oxygen bound by hemoglobin changes as the partial pressure of oxygen changes. Figure 34.2 graphically shows the oxygen-binding characteristics of hemoglobin in the form of a steeply sloped **S**-shaped curve. The solid line represents the behavior of hemoglobin when the body is resting. The steep portion of the curve is in the 35–40 torr partial pressure range of oxygen found in the fluid surrounding the cell. In this steep portion of the curve, a small change in the

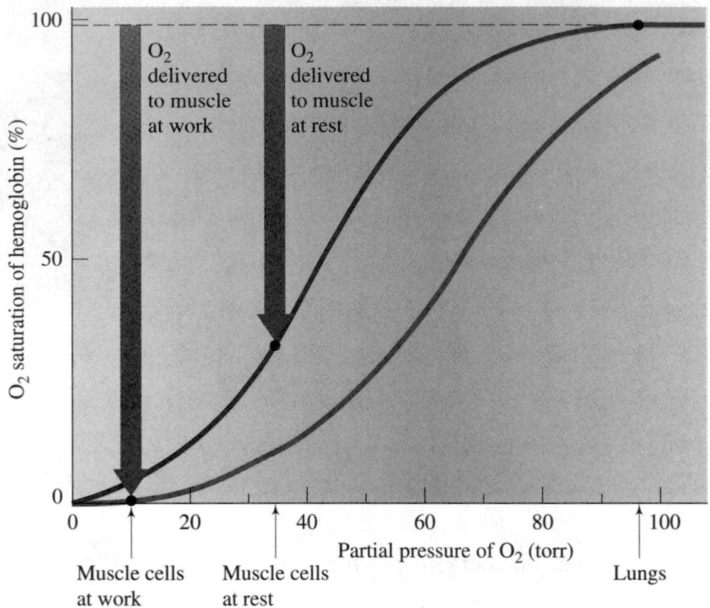

FIGURE 34.2 ▶ The oxygen-binding curve for hemoglobin under approximate physiological conditions (at rest,———; doing strenuous work,———). As the partial pressure of oxygen is increased, more O_2 binds to hemoglobin.

partial pressure of oxygen causes a relatively large change in the oxygen bound to hemoglobin. The lowering of the partial pressure of oxygen as hemoglobin moves from the lungs to resting muscle tissue causes the release of about 70% of the hemoglobin-bound oxygen.

When strenuous work begins, the muscle cells must have more oxygen to generate the needed energy. The rate of oxygen flow to the cells is speeded up by an increase in the pulse and respiration rates. But more subtle, and very effective, biochemical changes also occur.

First, in hard-working muscle tissue where bioenergetics proceeds rapidly, the partial pressure of oxygen may drop to 10 torr. This decrease in P_{O_2} from 35 torr to 10 torr is relatively small. But because the hemoglobin–oxygen-binding curve is steep, the amount of additional oxygen released is large. Hemoglobin may release up to 90% of its oxygen under these conditions.

Second, hydrogen ions are formed as the muscles work (see Section 34.3). These ions bind to oxygenated hemoglobin (HbO_2) and stimulate the release of additional oxygen.

$$H^+ + HbO_2 \longrightarrow HbH^+ + O_2 \quad (1)$$

This acid effect is represented graphically by a shift of the oxygen-binding curve toward higher partial pressures of oxygen, as shown in Figure 34.2. This shift, shown by the dashed line, means that more O_2 is released in the muscles. When muscle cells increase bioenergetics and need more oxygen, the acid produced by this activity triggers the release of more oxygen from HbO_2.

Carbon Dioxide Removal

When oxygen is used in the muscle cells, carbon dioxide is produced. This metabolic product must be removed from the cells. The partial pressure of carbon dioxide is

Movement utilizes chemical energy. The field of bioenergetics involves the study of these energy interrelationships at a cellular level.

highest in the muscle cells, decreases in the venous blood, and is lowest in the lungs. Thus carbon dioxide tends to move from the cells toward the lungs. But as with oxygen, carbon dioxide must be moved at a faster rate than can be attained by simple diffusion.

For the most part the rapid, efficient removal of carbon dioxide is effected by two methods. In the first method hemoglobin acts as a carrier molecule. Carbon dioxide reacts with amino groups on the hemoglobin molecule to form carbamino ion groups and H^+ ions:

$$\text{Hb—NH}_2 + \text{CO}_2 \longrightarrow \underset{\text{carbamino ion}}{[\text{Hb—NH—CO}_2]^-} + \text{H}^+ \qquad (2)$$

In the second method of removal, carbon dioxide reacts with water to form highly soluble hydrogen carbonate and hydrogen ions:

$$\text{CO}_2 + \text{H}_2\text{O} \rightleftharpoons \text{H}_2\text{CO}_3 \rightleftharpoons \text{H}^+ + \text{HCO}_3^- \qquad (3)$$

Reactions (2) and (3) occur in the fluid surrounding the muscles. The hydrogen ions produced by these reactions stimulate the further release of oxygen in the muscle tissue by the reaction shown in equation (1).

The hemoglobin carbamino ions, hydrogen carbonate ions, and excess hydrogen ions are transported to the lungs via the venous blood. There the exhalation of carbon dioxide enables reactions (2) and (3) to reverse, as shown in equations (4) and (5):

$$[\text{Hb—NH—CO}_2]^- + \text{H}^+ \longrightarrow \text{Hb—NH}_2 + \text{CO}_2 \qquad (4)$$

$$\text{HCO}_3^- + \text{H}^+ \rightleftharpoons \text{H}_2\text{CO}_3 \rightleftharpoons \text{H}_2\text{O} + \text{CO}_2 \qquad (5)$$

These reversible reactions serve not only to eliminate CO_2, but also to consume the excess H^+ ions that were formed in the muscle tissue.

As shown in reactions (1) to (5), acid (H^+) has an important role in both the delivery of O_2 and the removal of CO_2. In muscle, CO_2 and H^+ are formed.

This acid (H^+) causes HbO_2 to release the additional O_2 needed to continue the bioenergetics. In the lungs, CO_2 is exhaled and the blood becomes less acidic, causing Hb to bind more O_2. The overall interrelationships of oxygen, carbon dioxide, and acid can be summarized in this way:

Energy use at the muscle	Gas balance restored at the lungs
O_2 used	O_2 inhaled
CO_2 formed	CO_2 exhaled
Acidity increases	Acidity decreases
More O_2 released from HbO_2	More O_2 binds to Hb

34.3 Hydrogen Ion Production and Control

Acid (H^+) is usually produced as the muscles perform work. Two waste products of cellular bioenergetics—carbonic acid and lactic acid—are the primary sources of this acidity. Both substances produce hydrogen ions in solution:

$$CO_2 + H_2O \rightleftharpoons H_2CO_3 \rightleftharpoons H^+ + HCO_3^- \quad (6)$$

$$\underset{\text{lactic acid}}{CH_3CH(OH)COOH} \rightleftharpoons H^+ + \underset{\text{lactate ion}}{CH_3CH(OH)COO^-} \quad (7)$$

To prevent a toxic buildup of acid (abnormally low pH), the excess H^+ ions must be removed from the cells. The lungs provide a short-term partial solution to the problem of increased acidity in the cells and surrounding fluid. When CO_2 is exhaled from the lungs, reaction (6) is reversed and the H^+ ion concentration is reduced in the bloodstream. During hard work, the rate of removal of carbon dioxide from the lungs is accelerated by an increase in the respiration and pulse rate. Thus, as the breathing rate increases, the blood acidity decreases.

The liver and kidneys also have an important role in controlling the pH of body fluids. The liver converts lactic acid to glucose and thereby helps to prevent toxic acid buildup. The kidneys transfer water and selected ions from the bloodstream to the urine. Hydrogen ions and hydrogen carbonate ions are among those transferred. Because this is a relatively slow process, the kidneys are mainly concerned with the long-term maintenance of correct body fluid pH.

34.4 Metabolism

As we have seen, to do work, cells must take in extra O_2. This oxygen reacts with nutrients such as carbohydrates or fatty acids. Energy is released as the reduced carbons of the nutrients are converted by a series of reactions into the oxidized carbons of carbon dioxide and lactic acid. These reactions comprise an important fraction of all the biochemical reactions.

The sum of all chemical reactions that occur within a living organism is defined as **metabolism**. Many hundreds of different chemical reactions occur in a typical cell. To help make sense of this myriad of reactions, biochemists have subdivided metabolism into two contrasting categories—*anabolism* and *catabolism.* **Anabolism** is the process by which simple substances are synthesized (built up) into complex substances. **Catabolism** is the process by which complex substances are broken down into simpler substances. Anabolic reactions usually involve carbon reduction and consume cellular energy, whereas catabolic reactions usually involve carbon oxidation and produce energy for the cell.

metabolism

anabolism

catabolism

34.5 Metabolism and Cell Structure

Cells segregate many of their metabolic reactions into specific, subcellular locations. The simple **procaryotes**—cells without internal membrane-bound bodies—have a minimum amount of spatial organization (see Figure 34.3). The anabolic processes of DNA and RNA synthesis in these cells are localized in the nuclear material, whereas most other metabolic reactions are spread throughout the cytoplasm.

procaryote

In contrast, metabolic reactions in the cells of higher plants and animals are often segregated into specialized compartments. These cells, the **eucaryotes,** contain internal, membrane-bound bodies called **organelles** (see Figure 34.3). It is within the organelles that many specific metabolic processes occur.

eucaryote

organelle

In the eucaryotic cell, most of the DNA and RNA syntheses are localized in the nucleus. Anabolism of proteins takes place in the ribosomes, whereas that of carbohydrates and lipids occurs primarily in the cytoplasm.

There are a variety of specialized catabolic organelles within a eucaryotic cell. The lysosome contains the cell's digestive emzymes, and the peroxisome is the site of oxidative reactions that form hydrogen peroxide. Perhaps the most important catabolic organelle is the mitochondrion (plural, mitochondria). This membrane-bound body provides most of the energy for a typical cell. The energy is released by catabolic processes, which oxidize carbon-containing nutrients. Mitochondria consume most of the O_2 that is inhaled and produce most of the CO_2 that is exhaled by the lungs.

34.6 Biological Oxidation–Reduction

The ultimate source of biological energy on earth is sunlight. Plants capture light energy and transform it to chemical energy by a process called *photosynthesis* (see Section 34.9). This chemical energy is stored in the form of reduced carbon atoms in carbohydrate molecules. It is important to understand that the energy contained in carbohydrates, lipids, and proteins originally came from sunlight.

Humans, as well as other animals, draw most of their energy from foodstuffs that contain reduced carbons (see Chapter 33). For example, after we eat a meal, nutrients such as carbohydrates and fats are transported to our cells. The carbons in these compounds are in a reduced state and thus contain stored energy.

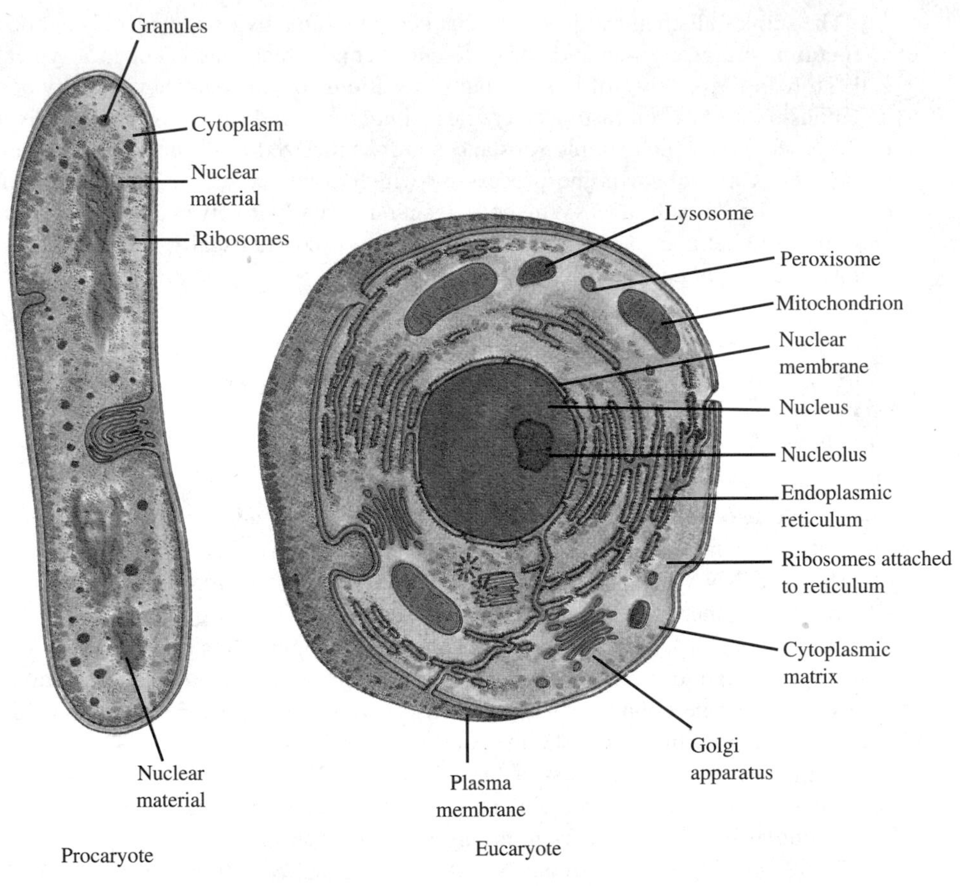

FIGURE 34.3 ▶
A schematic representation of a procaryotic and a eucaryotic cell. The procaryote lacks much of the organized structure and the numerous organelles found in the eucaryote.

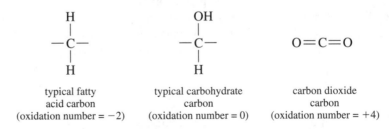

H–C(–H)–	HO–C(–H)–	O=C=O
typical fatty acid carbon (oxidation number = −2)	typical carbohydrate carbon (oxidation number = 0)	carbon dioxide carbon (oxidation number = +4)

Through the cell's metabolism these reduced carbons are oxidized, step by step, and are eventually converted to carbon dioxide.

Figure 34.4 is an energy diagram summarizing the energy flow through metabolism. Such a diagram will be used several times in this chapter. The black arrows trace progress from one chemical to the next. Because each chemical has its own special energy level, it is presented at a particular height with respect to the energy axis (the higher the level, the more energy). When energy levels change, energy must be released or absorbed as shown by the red arrows. Thus, in Figure 34.4, photosynthesis causes carbons to move to a higher energy level as they are reduced

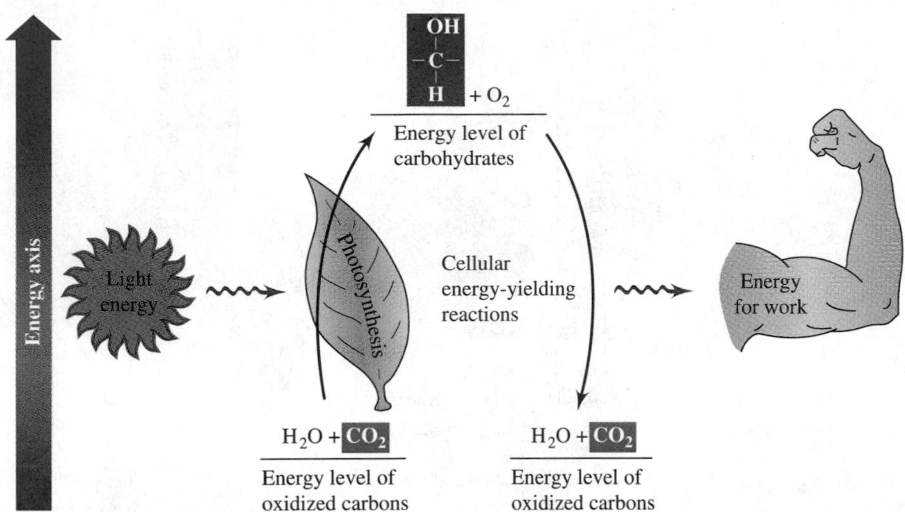

FIGURE 34.4
Energy flow through metabolism using the important energy nutrients, carbohydrates. Reduction of carbon stores energy while oxidation of carbon releases energy.

to carbohydrate carbons. The carbohydrate carbons are then oxidized to a lower energy level with the release of energy to do work.

In eucaryotic cells, specific organelles are present that specialize in redox reactions. The *mitochondria* (see Figure 34.5), often called the powerhouses of the cell, are the sites for most of the catabolic redox reactions. *Chloroplasts* (see Figure 34.5) are organelles found in higher plants and contain an electron transport system that is responsible for the anabolic redox reactions in photosynthesis (see Section 34.9).

To move electrons from one place to another (often outside of the mitochondrion or chloroplast), the cell uses a set of redox coenzymes. The redox coenzymes facilitate reactions by acting as temporary storage places for electrons. The three most common redox coenzymes (nicotinamide adenine dinucleotide, NAD^+; nicotinamide adenine dinucleotide phosphate, $NADP^+$; and flavin adenine dinucleotide, FAD) are shown in Figure 34.6. Humans synthesize NAD^+ and $NADP^+$ from the vitamin niacin while FAD is made from the vitamin riboflavin. In each case, the vitamin provides the reaction center of the coenzyme.

Recall from Section 31.1 that a coenzyme is an organic compound that is used and reused to help an enzyme-catalyzed reaction.

The addition or removal of electrons occurs in only one portion of each of these complex molecules. The nicotinamide ring is the reactive component within NAD^+ or $NADP^+$:

$$\text{(pyridinium ring, }{-}\overset{+}{N}{-}R\text{, with H and }{-}\overset{O}{\overset{\|}{C}}{-}NH_2\text{)} + 2\,e^- + H^+ \rightleftharpoons \text{(dihydropyridine ring, }{-}N{-}R\text{, with H H and }{-}\overset{O}{\overset{\|}{C}}{-}NH_2\text{)}$$

NAD^+ or $NADP^+$
(oxidized form)

NADH or NADPH
(reduced form)

(R represents the remainder of each molecule.)

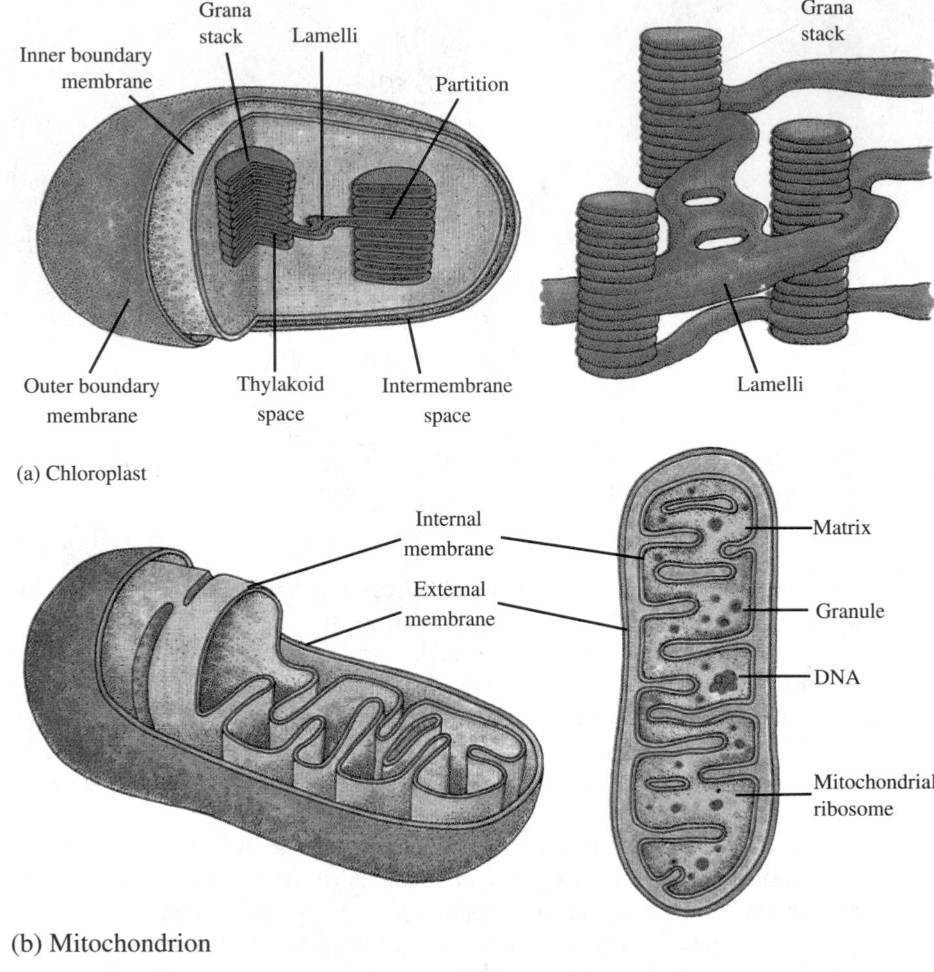

FIGURE 34.5 ▶ Schematic representation of (a) a chloroplast and (b) a mitochondrion.

For FAD, the flavin ring is the reactive component:

$$\text{FAD (oxidized form)} + 2\,e^- + 2\,H^+ \rightleftharpoons \text{FADH}_2 \text{ (reduced form)}$$

FAD
(oxidized form)

$FADH_2$
(reduced form)

(R represents the remainder of each molecule.)

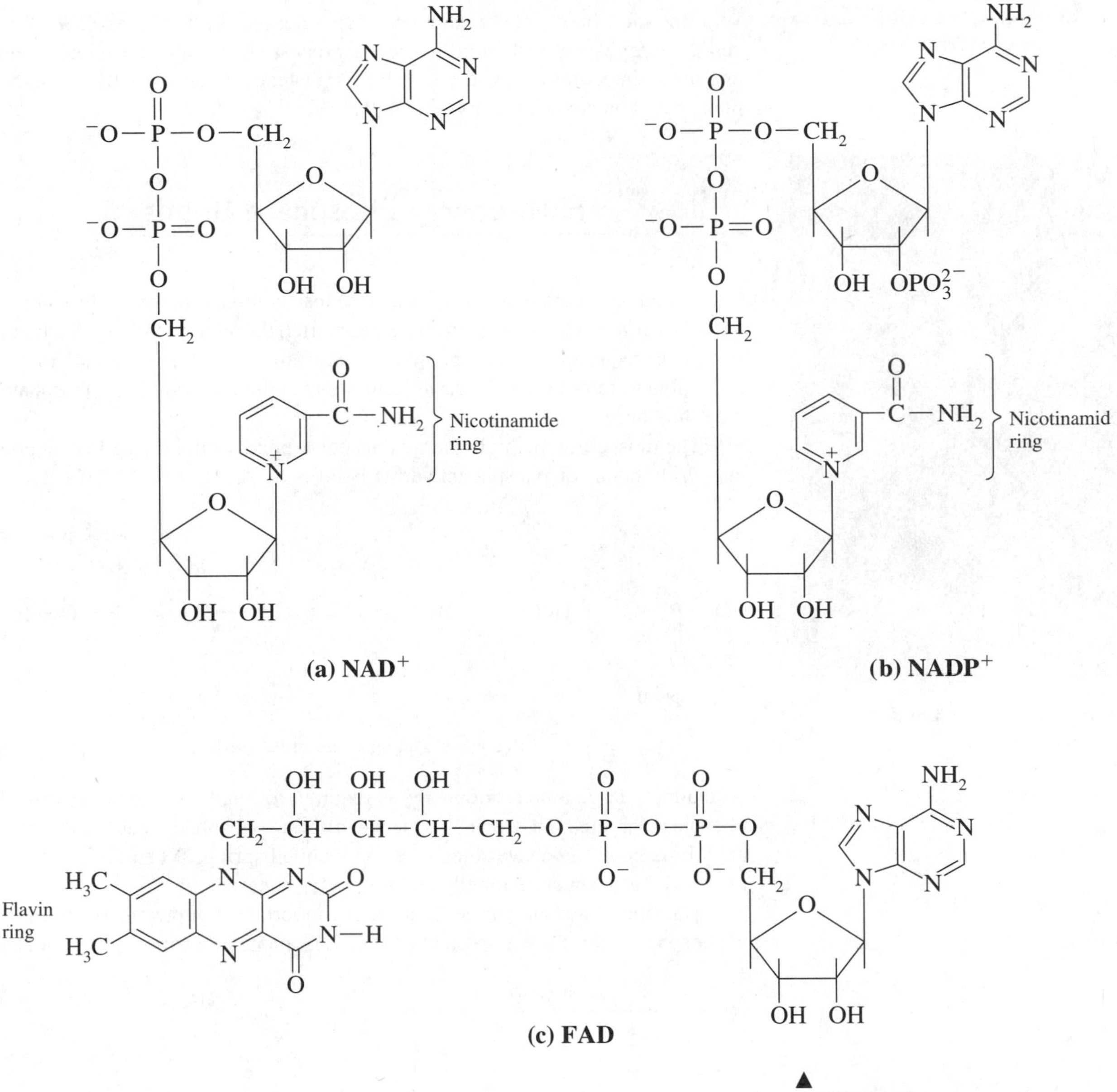

FIGURE 34.6 Structures of the redox coenzymes (a) nicotinamide adenine dinucleotide (NAD^+), (b) nicotinamide adenine dinucleotide phosphate ($NADP^+$), and (c) flavin adenine dinucleotide (FAD).

A very important function of these redox coenzymes is to carry electrons to the mitochondrial electron-transport system. As the coenzymes are oxidized, molecular oxygen is reduced:

$$\underset{\text{reduced coenzyme}}{2\ FADH_2} + O_2 \xrightarrow[\text{electron transport}]{\text{mitochondrial}} \underset{\text{oxidized coenzyme}}{2\ FAD} + 2\ H_2O$$

$$\underset{\text{reduced coenzyme}}{2\ NADH} + 2\ H^+ + O_2 \xrightarrow[\text{electron transport}]{\text{mitochondrial}} \underset{\text{oxidized coenzyme}}{2\ NAD^+} + 2\ H_2O$$

Remember, the mitochondria are the powerhouses of the cell!

With the movement of electrons, energy is released. In fact, over 85% of a typical cell's energy is derived from this redox process. However, the released energy is not used immediately by the cell but is instead stored, usually in high-energy phosphate bonds such as those in ATP.

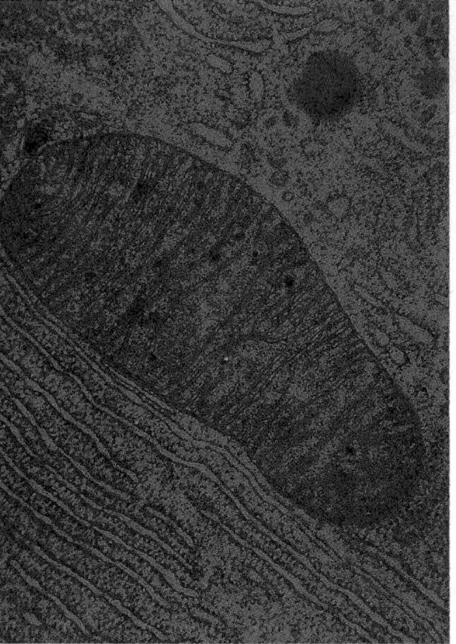

Photomicrograph of a mitochondrion.

34.7 High-Energy Phosphate Bonds

Cells need an energy delivery system. Most cellular energy is produced in the mitochondria, but this energy must be transported throughout the cell. Such a delivery system is required to carry relatively large amounts of energy and to be easily accessible to cellular reactions. Molecules that contain high-energy phosphate bonds meet this need.

The most common high-energy phosphate bond within the cell is the phosphate anhydride bond (or phosphoanhydride bond):

$$^{-}O-\overset{\overset{O^-}{|}}{\underset{\underset{O}{\|}}{P}}-OH + HO-\overset{\overset{O^-}{|}}{\underset{\underset{O}{\|}}{P}}-OR + 35\text{ kJ} \rightleftharpoons {}^{-}O-\overset{\overset{O^-}{|}}{\underset{\underset{O}{\|}}{P}}\sim O-\overset{\overset{O^-}{|}}{\underset{\underset{O}{\|}}{P}}-O-R + H_2O$$

phosphate — phosphate — diphosphate anhydride (Phosphate anhydride bond indicated by ~)

(R represents the remainder of each molecule.)

A relatively large amount of energy is required to bond the two negatively charged phosphate groups. The repulsion between these phosphates causes the phosphate anhydride bond to behave somewhat like a coiled spring. When the bond is broken, the phosphates separate rapidly and energy is released.

The phosphate anhydride bond is an important component of the nucleotide triphosphates, the most important of which is adenosine triphosphate (ATP):

$$\overbrace{^{-}O-\overset{\overset{O^-}{|}}{\underset{\underset{O}{\|}}{P}}\sim O-\overset{\overset{O^-}{|}}{\underset{\underset{O}{\|}}{P}}\sim O-\overset{\overset{O^-}{|}}{\underset{\underset{O}{\|}}{P}}}^{\text{triphosphate}}-OCH_2-\text{ribose (OH, OH)}-\text{adenine}$$

triphosphate; NH_2; N; adenine; ribose; OH OH

adenosine triphosphate (ATP)

Adenosine triphosphate plays an important role in all cells, from the simplest to the most complex. It functions by storing and transporting the energy in its high-energy phosphate bonds to the places in the cell where energy is needed. ATP is the common intermediary in energy metabolism.

The cell realizes several advantages by storing energy in ATP. First, the stored energy is easily accessible to the cell; it is readily released by a simple hydrolysis reaction yielding adenosine diphosphate (ADP) and an inorganic phosphate ion (P_i):

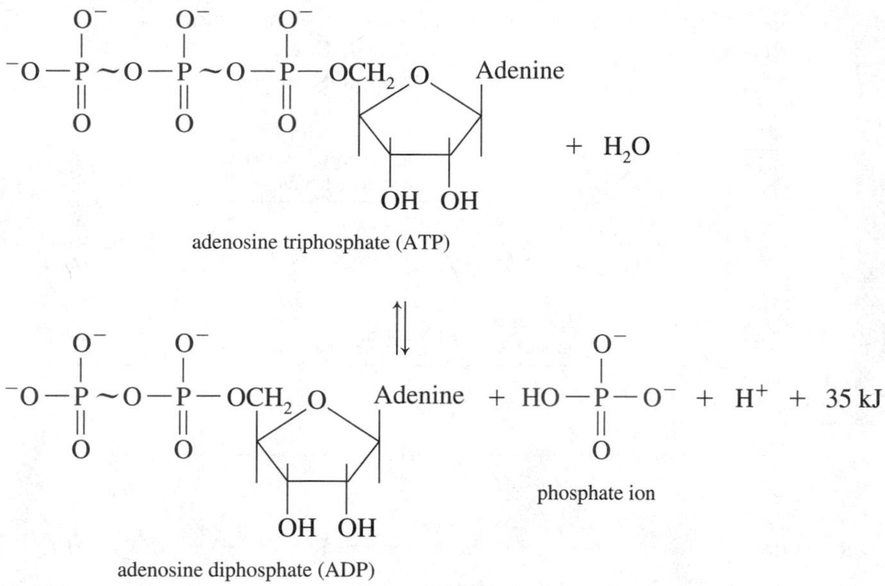

$$^{-}O-\underset{O}{\overset{O^-}{\overset{|}{\underset{||}{P}}}}\sim O-\underset{O}{\overset{O^-}{\overset{|}{\underset{||}{P}}}}\sim O-\underset{O}{\overset{O^-}{\overset{|}{\underset{||}{P}}}}-OCH_2\text{(ribose ring with O; OH OH)}\text{—Adenine} \quad + \; H_2O$$

adenosine triphosphate (ATP)

$$\rightleftharpoons$$

$$^{-}O-\underset{O}{\overset{O^-}{\overset{|}{\underset{||}{P}}}}\sim O-\underset{O}{\overset{O^-}{\overset{|}{\underset{||}{P}}}}-OCH_2\text{(ribose ring with O; OH OH)}\text{—Adenine} \; + \; HO-\underset{O}{\overset{O^-}{\overset{|}{\underset{||}{P}}}}-O^- \; + \; H^+ \; + \; 35\text{ kJ}$$

adenosine diphosphate (ADP) phosphate ion

Second, ATP serves as the common energy currency for the cell. Energy from catabolism of many different kinds of molecules is stored in ATP. For example, the energy obtained from oxidation of carbohydrates, lipids, and proteins is stored in ATP. The oxidation of each of these nutrients requires many different enzyme-catalyzed reactions. To make *direct* use of this energy, the cell would need a separate series of reactions for each different energy source. Instead, the cell channels most of the energy derived from redox reactions into the high-energy bonds of ATP. This process is analogous to an economic system that values all goods and services in terms of a common currency such as the dollar. Buying and selling within the system is thus greatly simplified. In the cell, energy utilization is greatly simplified by converting stored energy to ATP, the common energy currency.

34.8 Phosphorylation

To this point we have considered two forms for chemical storage of biological energy: reduced carbon atoms and high-energy phosphate bonds. It is vital that the cell be able to convert one form of stored energy to the other.

Figure 34.7 summarizes biological energy conversion. In the first process, a variety of cellular reactions oxidize the carbons of nutrients, converting these molecules from a high energy level to a lower energy state. The energy released (red arrow) is used to create high-energy phosphate bonds. Finally, when a cellular reaction needs energy, or work must be accomplished, molecules with high-energy phosphate bonds are recycled to low-energy molecules plus inorganic phosphate.

Energy is stored in phosphate anhydride bonds through two biological processes—*substrate-level phosphorylation* and *oxidative phosphorylation.* **Substrate-level phosphorylation** is the process whereby energy derived from oxidation is used to form high-energy phosphate bonds on various biochemical molecules (substrates) (Figure 34.8).

substrate-level phosphorylation

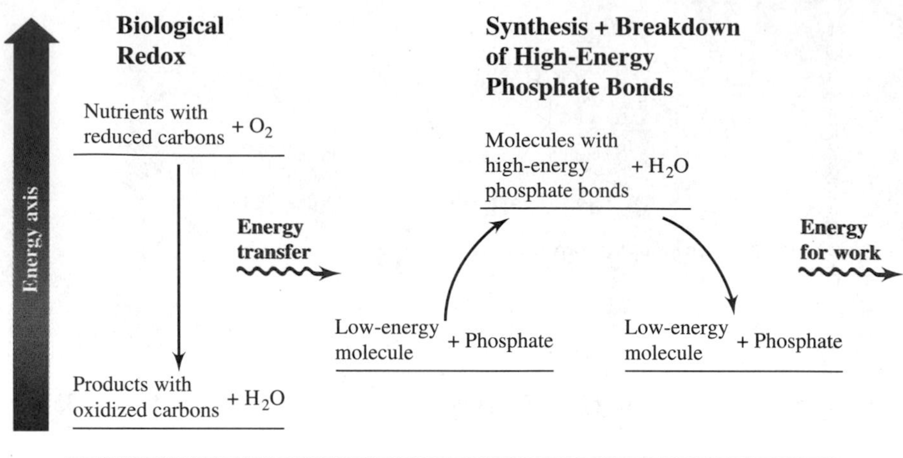

FIGURE 34.7 ▶ **Energy flow from nutrients with reduced carbons (energy-yielding nutrients) to high-energy phosphate bonds that are used to do work.**

$$\underset{\text{substrate}}{\mathrm{R{-}OH}} + \underset{\text{phosphate ion}}{\mathrm{HO{-}\overset{\overset{\textstyle O^-}{|}}{\underset{\underset{\textstyle O}{\|}}{P}}{-}O^-}} + \underset{\text{(derived from redox)}}{\text{Energy}} \longrightarrow \underset{\text{phosphorylated substrate}}{\mathrm{R{-}O{\sim}\overset{\overset{\textstyle O^-}{|}}{\underset{\underset{\textstyle O}{\|}}{P}}{-}O^-}} + H_2O$$

(R represents the remainder of the substrate molecule.)

In a succeeding reaction, the phosphorylated substrate often transfers the phosphate to ADP and forms ATP.

$$\underset{\text{phosphorylated substrate}}{\mathrm{R{-}O{\sim}\overset{\overset{\textstyle O^-}{|}}{\underset{\underset{\textstyle O}{\|}}{P}}{-}O^-}} + \underset{\text{adenosine diphosphate (ADP)}}{\mathrm{{}^-O{-}\overset{\overset{\textstyle O^-}{|}}{\underset{\underset{\textstyle O}{\|}}{P}}{\sim}O{-}\overset{\overset{\textstyle O^-}{|}}{\underset{\underset{\textstyle O}{\|}}{P}}{-}OCH_2\text{(ribose, OH OH)}{-}Adenine}} + H^+ \longrightarrow$$

$$\underset{\text{substrate}}{\mathrm{R{-}OH}} + \underset{\text{adenosine triphosphate (ATP)}}{\mathrm{{}^-O{-}\overset{\overset{\textstyle O^-}{|}}{\underset{\underset{\textstyle O}{\|}}{P}}{\sim}O{-}\overset{\overset{\textstyle O^-}{|}}{\underset{\underset{\textstyle O}{\|}}{P}}{\sim}O{-}\overset{\overset{\textstyle O^-}{|}}{\underset{\underset{\textstyle O}{\|}}{P}}{-}OCH_2\text{(ribose, OH OH)}{-}Adenine}}$$

This process is called substrate-level phosphorylation because ADP gains a phosphate from a cellular substrate.

Substrate-level phosphorylation is found most commonly in the catabolism of carbohydrates—that is, glycolysis (see Chapter 35). This process accounts for only a small amount of a resting cell's total energy production and does not require

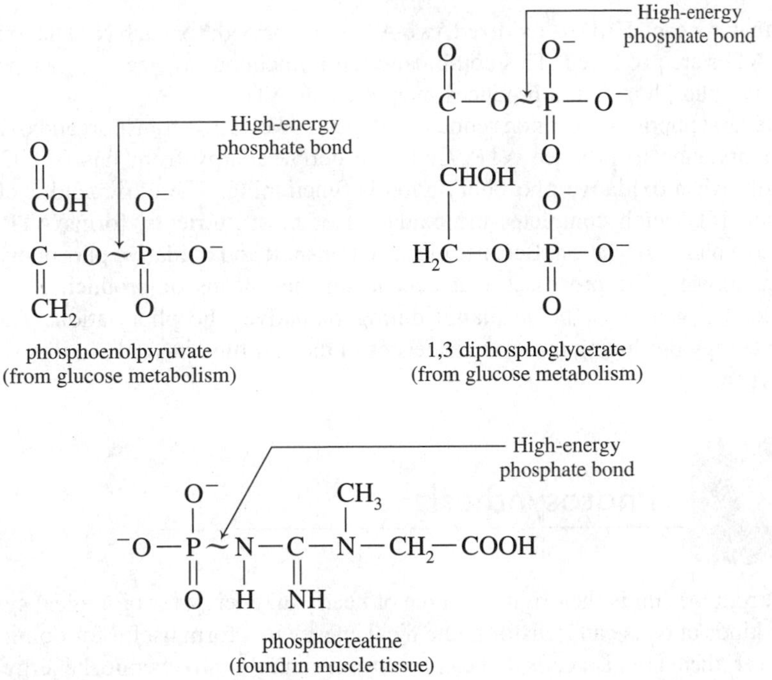

FIGURE 34.8
Some biological molecules that contain high-energy phosphate bonds (phosphorylated substrates).

oxygen. Under anaerobic conditions (e.g., when the bloodstream cannot deliver enough O_2 to hardworking muscles), substrate-level phosphorylation may be the cell's principal means of forming ATP.

oxidative phosphorylation

Oxidative phosphorylation is a process that directly uses energy from redox reactions to form ATP. This process occurs in the mitochondria and depends on the mitochondrial electron-transport system. The enzyme-catalyzed oxidation and phosphorylation reactions are coupled in such a way that energy released by the oxidation of a coenzyme is used to form ATP. The overall process is indicated by these equations:

$$FADH_2 + \tfrac{1}{2}O_2 \xrightarrow[]{\text{mitochondrial electron transport}} FAD + H_2O$$

energy

$$2\,ADP + 2\,P_i \xrightarrow[\text{oxidative phosphorylation}]{} 2\,ATP + 2\,H_2O$$

$$NADH + H^+ + \tfrac{1}{2}O_2 \xrightarrow[]{\text{mitochondrial electron transport}} NAD^+ + H_2O$$

energy

$$3\,ADP + 3\,P_i \xrightarrow[\text{oxidative phosphorylation}]{} 3\,ATP + 3\,H_2O$$

Note that, for each $FADH_2$ oxidized, two ATPs are formed; for each NADH oxidized, three ATPs are produced. This combination of mitochondrial electron transport and oxidative phosphorylation produces most cellular ATP.

Several important reaction sequences depend on electron transport and oxidative phosphorylation to produce ATP. Cells can derive energy from fats (see Chapter 36) only when oxidative phosphorylation is functioning. The citric acid cycle (see Chapter 35), which completes the oxidation of most nutrients, forms ATP using oxidative phosphorylation. Because electron transport and oxidative phosphorylation require oxygen, the processes that depend on this means of producing ATP are aerobic. Oxygen must be available during oxidative phosphorylation. Thus the major energy-producing reaction sequences in the cell function only in the presence of oxygen.

34.9 Photosynthesis

Light from the sun is the original source of nearly all energy for biological systems. Many kinds of cells can transform chemical energy to a form useful for doing work. However, there are also cells that can transform sunlight into chemical energy. Such cells use **photosynthesis**, a process by which energy from the sun is converted to chemical energy that is stored in chemical bonds.

photosynthesis

Photosynthesis is performed by a wide variety of organisms, both eucaryotic and procaryotic. Besides the higher plants, photosynthetic eucaryotes include multicellular green, brown, and red algae and unicellular organisms such as euglena. Photosynthetic procaryotes include the green and purple bacteria and the blue-green algae. Although the photosynthetic importance of higher plants is usually emphasized, it has been estimated that more than half of the world's photosynthesis is carried out by unicellular organisms.

Green algae shows spiral chloroplasts within the cytoplasm in the cells.

Photosynthesis in higher plants is a complex series of reactions in which carbohydrates are synthesized from atmospheric carbon dioxide and water:

$$6\ CO_2 + 6\ H_2O + 2820\ \text{kJ} \longrightarrow \underset{\text{glucose}}{C_6H_{12}O_6} + 6\ O_2$$

Sunlight provides the large energy requirement for this process. An important side benefit of photosynthesis is the generation of oxygen, which is crucial to all aerobic metabolism.

The 1961 Nobel prize in chemistry was awarded to the American chemist Melvin Calvin (1911–1992), of the University of California at Berkeley, for his work on photosynthesis. Calvin and his co-workers used radioactive carbon tracer techniques to discover the details of the complicated sequence of chemical reactions that occur in photosynthesis.

Photosynthesis traps light energy by reducing carbons. For eucaryotes the necessary electron-transfer reactions are segregated in the chloroplast (see Figure 34.5). Like the mitochondrion, the chloroplast contains an electron-transport system within its internal membranes. Unlike the mitochondrial system, which oxidizes coenzymes to liberate energy, the chloroplast electron-transport system reduces coenzymes with an input of energy:

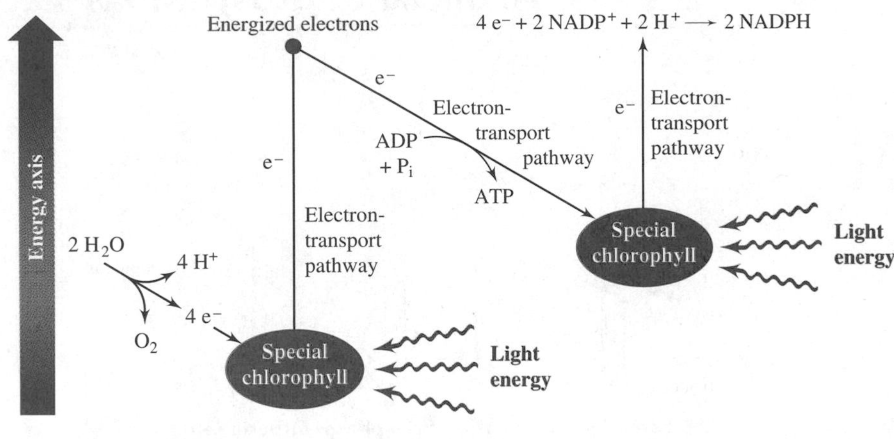

FIGURE 34.9 A schematic diagram showing the movement of electrons from water to $NADP^+$ in the photosynthetic electron transport pathway: Note that light energy causes the electrons to become more energetic so that they can reduce $NADP^+$.

$$NADP^+ + H_2O + \text{Energy} \xrightarrow[\text{electron transport}]{\text{chloroplast}} NADPH + H^+ + \tfrac{1}{2}\,O_2$$

The chloroplasts capture light energy and place it in chemical storage.

The photosynthetic mechanism is complex, but it can be divided into two general components—the *dark reactions* and the *light reactions.* The dark reactions produce glucose from carbon dioxide, reduced coenzymes, and ATP. No light is needed and, in nature, these reactions continue during the night.

The light reactions of photosynthesis form the ATP and NADPH needed to produce glucose. The mechanism for capturing light energy is unique to the photosynthetic process. Although much research has been devoted to this topic, not all of the details are clear. In general, light is absorbed by colored compounds (pigments) located in the chloroplasts. The most abundant of these pigments is chlorophyll. Once the light energy is absorbed, it is transferred to specific molecules (probably special chlorophylls) that lose electrons. These energized electrons travel through the chloroplast electron-transport system, as shown in Figure 34.9. Two events follow in quick succession: First, the electrons lost by these special chlorophylls are moved to higher energy levels until they can reduce molecules of the coenzyme $NADP^+$. Second, the special chlorophylls that lost electrons now regain them. Water is the electron donor, giving up electrons and producing oxygen gas (and hydrogen ions) in the process. The overall redox reaction moves four electrons from two water molecules to produce two molecules of NADPH.

This greenhouse creates ideal conditions for photosynthesis.

$$2\,H_2O + 2\,NADP^+ \xrightarrow{\text{light}} 2\,NADPH + O_2 + 2\,H^+$$

Photosynthesis uses light energy to force electrons to higher energy levels, as shown in Figure 34.9. As noted in the preceding paragraph, these energetic electrons are used to reduce $NADP^+$. But also notice that, when the electron loses energy in the middle of this electron-transport process, the released energy is used to make ATP. Thus the light reactions of photosynthesis supply both the NADPH and ATP needed to make glucose.

Chemistry in Action

Artificial Photosynthesis

We use huge quantities of energy in our modern society and are constantly looking for more. Our petroleum supplies are being depleted, and scientists are looking for a cleaner source of energy. Where are they looking? They are looking toward the sun, the same source of energy used by plants during photosynthesis. Sunlight is a readily available energy source *if* an inexpensive way to trap this energy can be found.

Solar cells (photovoltaic cells) can be used to power your calculator. However, this approach to trapping the sun's energy requires expensive, carefully produced components. Thus far, solar cells are an expensive way to trap the sun's energy. An alternative approach is to copy photosynthesis—to use sunlight to drive a chemical reaction.

Imagine drawing some tap water and generating enough hydrogen and oxygen to power your cooking and heating. The simple artificial photosynthetic cell shown in the photograph does just that. Like natural photosynthesis, artificial photosynthesis requires several components (summarized in the diagram):

1. There must be a means of absorbing the light energy (a photosensitive electrode).
2. Electrons must be energized and moved (electrodes and wires move electrons from one solution to another).
3. The energized electrons must be used to make high-energy chemicals (hydrogen and oxygen gases are often produced).

An artificial photosynthetic cell is much cheaper and easier to build than a photovoltaic cell. In fact, most households have the wires, glass jars, and water necessary to build an artificial photosynthetic cell. What is missing is a photosensitive electrode that does not corrode over time. A silicon electrode, for example, is converted to silicon dioxide as it absorbs light in an aqueous solution. Photosynthesis runs for a brief time and then stops because of electrode corrosion.

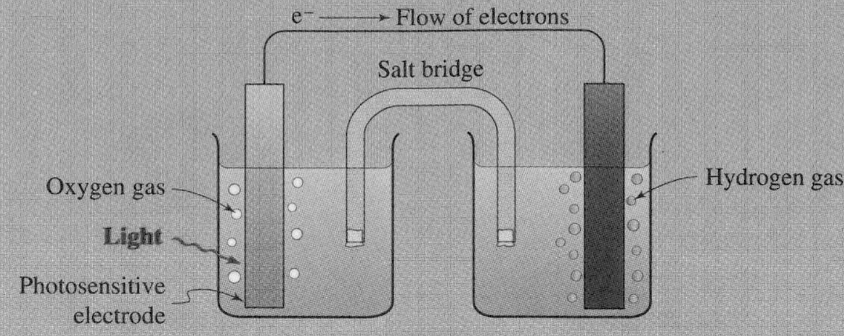

Schematic diagram of artificial photosynthetic cell.

Photograph of an artificial photosynthetic cell generating hydrogen gas.

Scientists appear to be close to providing an inexpensive solution to the electrode corrosion problem. Titanium dioxide (the white pigment in many paints) forms a noncorroding electrode but only absorbs ultraviolet light (a small fraction of the total sunlight). Engineers have borrowed an idea from natural photosynthesis; by adding colored pigments, the amount of absorbed light can be increased. Titanium dioxide can be coated with colored dyes to absorb up to 90% of all sunlight.

Using inexpensive materials, artificial photosynthesis can capture and convert about 20% of the sun's energy. This may seem to be a small percentage, but natural photosynthesis achieves only about a 3% conversion. Furthermore, consider the amount of energy that strikes the earth's surface every day. Even 20% represents a huge amount of useable energy. Today, the biggest problem with artificial photosynthesis lies with the photosensitive electrode. However, a commercial device is already in production in the United States. Within the next 20 years, artificial photosynthesis may trigger a revolution in the cost of energy.

Concepts in Review

1. Describe the importance of hemoglobin in oxygen transport.
2. Explain how oxygen transport is coordinated with carbon dioxide production.
3. Explain how an anabolic process differs from a catabolic process.
4. List three major classes of biochemical substances that provide metabolic energy.
5. Explain how reduced carbon atoms are important in the production of cellular energy.
6. Explain the difference between a procaryotic and a eucaryotic cell.
7. List four subcellular organelles and their functions.
8. Define the role of the mitochondria in metabolism.
9. Briefly describe electron transport.
10. Discuss the function of NAD^+, $NADP^+$, and FAD in biological processes.
11. Explain the importance of the high-energy phosphate bond in metabolism.
12. Give two advantages for using ATP as a common energy currency for the cell.
13. Contrast substrate-level phosphorylation with oxidative phosphorylation.
14. Compare the role of the mitochondrion with that of the chloroplast.
15. Outline the principal steps in the overall process of photosynthesis.

Key Terms

The terms listed here have been defined within this chapter. Section numbers are referenced in parenthesis for each term.

anabolism (34.4)
bioenergetics (34.1)
catabolism (34.4)
eucaryote (34.5)
metabolism (34.4)
organelle (34.5)
oxidative phosphorylation (34.8)
photosynthesis (34.9)
procaryote (34.5)
substrate-level phosphorylation (34.8)

Questions

Questions refer to tables, figures, and key words and concepts defined within the chapter. A particularly challenging question or exercise is indicated with an asterisk.

1. Explain how increased carbon dioxide production can create a more acidic environment in muscle tissue.
2. Show a chemical equation for the reaction that produces a bond between hemoglobin and carbon dioxide.
3. Why are fats and carbohydrates good sources of cellular energy?
4. Give a general structure for the most common high-energy phosphate bond found in the cell. With what compound is it generally associated?
5. Why is ATP known as the "common energy currency" of the cell?
6. What is an oxidation–reduction coenzyme?

7. Draw the ring structure portion of NAD^+ that becomes reduced during metabolism.
8. How does oxidative phosphorylation differ from substrate-level phosphorylation?
9. How do eucaryotes differ from procaryotes?
10. In what part of the eucaryotic cell does oxidative phosphorylation occur?
11. Compare the structural similarities between choroplasts and mitochondria.
12. What role do chloroplast pigments serve in photosynthesis?
13. Give the overall reaction for photosynthesis in higher plants.
14. Which of these statements are correct? Rewrite each incorrect statement to make it correct.
   (a) Slowed breathing, induced by a drug overdose, might cause the blood pH to become more acidic.
   (b) Anabolic processes are those in which complex biological substances are broken down into simpler substances.
   (c) Oxidation of reduced carbon atoms provides energy for the cell.
   (d) Typical carbohydrate carbons supply more energy to the cell than typical fatty acid carbons.
   (e) Many reduced carbons are oxidized to carbon dioxide in the mitochondria.
   (f) NAD^+ and $NADP^+$ are coenzymes that react as oxidizing agents in some metabolic processes.
   (g) Carbon dioxide provides much cellular energy.
   (h) A phosphate anhydride bond is considered to be a high-energy phosphate bond.
   (i) Mitochondrial electron transport serves to oxidize NADH.
   (j) Oxidative phosphorylation occurs in the mitochondria.
   (k) The chloroplast produces carbohydrates.
   (l) Chlorophyll absorbs light during photosynthesis.
   (m) NAD^+ is reduced during photosynthesis.

Paired Exercises

These exercises are paired. Each odd-numbered exercise is followed by a similar even-numbered exercise. Answers to the even-numbered exercises are given in Appendix V.

15. Write the equation that shows carbon dioxide reacting with water. As more carbon dioxide is made by muscle cells, how does the surrounding solution's pH change? As the pH changes, does oxygen bind more tightly to hemoglobin? Briefly explain.
16. Write the equation that shows carbon dioxide reacting with water. As more carbon dioxide is exhaled from the lungs, how does the surrounding solution's pH change? As the pH changes, does oxygen bind more tightly to hemoglobin? Briefly explain.
17. Many cells convert glucose to carbon dioxide. Is this an anabolic or a catabolic process? Briefly explain.
18. Many cells convert acetate to long-chain fatty acids. Is this an anabolic or a catabolic process? Briefly explain.
19. Your friend states, "The chloroplasts in this procaryotic cell are undergoing photosynthesis." What is wrong with this statement?
20. Your friend states, "The chloroplasts in this eucaryotic cell are undergoing oxidative phosphorylation." What is wrong with this statement?
21. Explain why the chemical changes in the mitochondria are said to be catabolic.
22. Explain why the chemical changes in the chloroplasts are said to be anabolic.
23. List three important characteristics of an anabolic process.
24. List three important characteristics of a catabolic process.
25. In mitochondrial electron transport, how many moles of oxygen gas are reduced to water by 2.38 mol of $FADH_2$?
26. In mitochondrial electron transport, how many moles of oxygen gas are reduced to water by 0.67 mol of NADH?
27. In mitochondrial chemical processes, how many moles of electrons will reduce 11.75 mol of NAD^+ to NADH?
28. In mitochondrial chemical processes, how many moles of electrons will reduce 0.092 mol of FAD to $FADH_2$?

29. How many high-energy phosphate anhydride bonds are contained in the following molecule?

```
                             O
                             ||
                    CH2—O—P—O⁻
   O       O         |       |
   ||      ||        |       O⁻
⁻O—P—O—P—O—CH
   |       |         |      O⁻      O
   O⁻      O⁻        |      |       ||
                    CH2—O—P—O—P—O⁻
                             ||      |
                             O       O⁻
```

30. How many high-energy phosphate anhydride bonds are contained in the following molecule?

```
                                         O
                                         ||
                                CH2—O—P—O⁻
   O       O       O             |       |
   ||      ||      ||            |       O⁻
⁻O—P—O—P—O—P—O—CH
   |       |       |             |       O⁻
  ⁻O       O⁻      O⁻            |       |
                                CH2—O—P—O⁻
                                         ||
                                         O
```

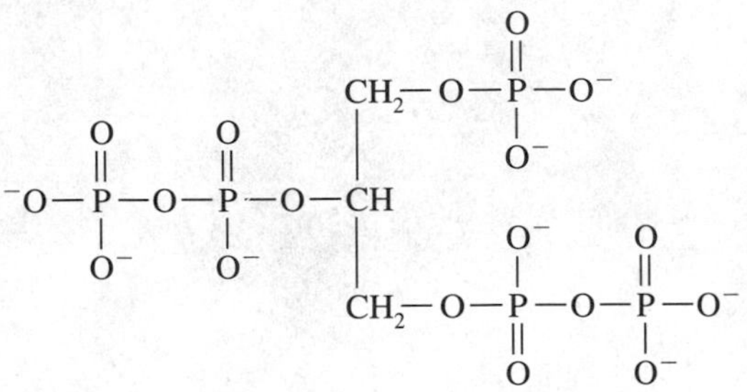

31. The equation for conversion of ADP to ATP shows an energy requirement of 35 kJ. How much energy would be required to convert 0.55 mol of ADP to ATP?

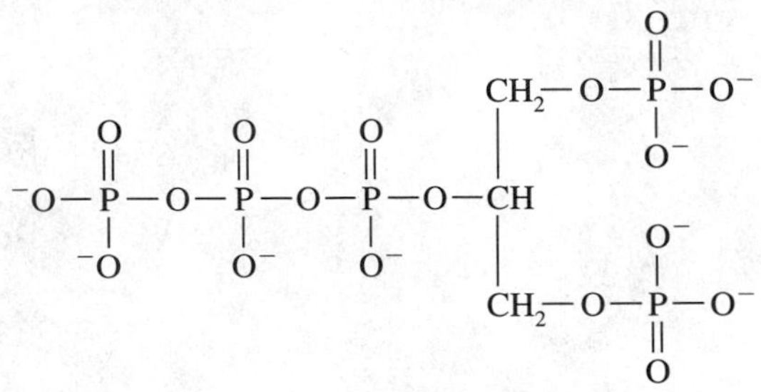

32. The equation for conversion of ATP to ADP shows an energy release of 35 kJ. How much energy would be released from 1.65 mol of ATP?

33. The following compound is formed during glucose catabolism:

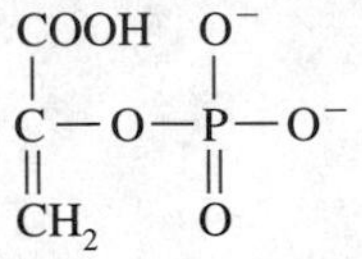

```
COOH  O⁻
 |    |
 C—O—P—O⁻
 ||   ||
 CH2  O
```

How many ATPs can be formed from one molecule of this compound during substrate-level phosphorylation? Explain.

34. The following compound is formed during glucose catabolism:

```
 O      O⁻
 ||     |
 C—O—P—O⁻
 |      ||
 |      O
CHOH
 |        O⁻
 |        |
CH2—O—P—O⁻
          ||
          O
```

How many ATPs can be formed from one molecule of this compound during substrate-level phosphorylation? Explain.

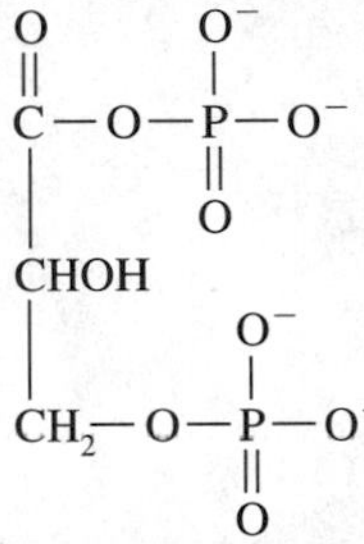

35. How many ATPs would be formed from 4 NADH and 2 $FADH_2$ using mitochondrial electron transport and oxidative phosphorylation?

36. How many ATPs would be formed from 2 NADH and 3 $FADH_2$ using mitochondrial electron transport and oxidative phosphorylation?

Additional Exercises

These exercises are not paired or labeled by topic and provide additional practice on concepts covered in this chapter.

37. How does rapid breathing (hyperventilation) change the blood pH? Explain.

38. Based on the function of the mitochondria, explain why these organelles are composed of about 90% membrane by mass.

39. Could a higher plant cell survive with chloroplasts but no mitochondria? Explain.

40. What structural similarities, if any, exist beween ATP and NAD^+?

41. One of your student colleagues has just told you that photosynthesis is a process by which carbon atoms are oxidized. Do you agree? Briefly explain.

Carbohydrate Metabolism

35

Chapter

Have you ever stopped to think about why carbohydrates are a staple of our world? Carbohydrates surround us on every side. Our diets consist mainly of carbohydrates, much of the building materials in our homes are carbohydrates, and most of the plant kingdom consists of carbohydrates. But *why* is this class of molecules so prevalent? Biochemists might answer this question in a number of ways, but one of the most important answers is in the air around us. Carbohydrates can be made from carbon dioxide, water, and sunlight by a process called *photosynthesis*; that is, the constituents for the process that results in carbohydrate metabolism are incredibly abundant.

Of course, a solid carbohydrate like sugar, which is sweet to the taste, has very little in common with gaseous carbon dioxide, which is tasteless and can be hazardous in high concentrations. But a series of chemical reactions can convert carbon dioxide and water to sugar. Photosynthesis is an example of carbohydrate metabolism and illustrates the fact that metabolism can achieve remarkable conversions. Our world depends on such metabolic processes as photosynthesis.

35.1 Metabolic Pathways

Carbohydrates are rich sources of energy for most organisms. The energy is released when a cell subjects a carbohydrate to a series of chemical reactions called a *metabolic pathway*. Despite the wide diversity of cell types, carbohydrate metabolic pathways vary little from cell to cell and from organism to organism.

Sunlight provides the energy that is stored in carbohydrates:

$$6\ CO_2 + 6\ H_2O + 2820\ \text{kJ} \xrightarrow[\text{chlorophyll}]{\text{enzymes}} C_6H_{12}O_6 + 6\ O_2 \qquad (1)$$

Equation (1) summarizes the production of glucose (a carbohydrate) and oxygen by the endothermic process called *photosynthesis*. Note that, in the overall transformation, carbon atoms from carbon dioxide are reduced and oxygen atoms from water are oxidized. The light energy is used to cause a net movement of electrons from water to carbon dioxide. As the carbons become more reduced, more energy is stored. Photosynthesis also supplies free oxygen to the atmosphere.

Equation (2) represents the oxidation of glucose and corresponds to the reversal of the overall photosynthesis reaction. Electrons are moved from carbohydrate carbons to oxygen. Energy stored in the reduced carbon atoms in glucose is released and can then be used by the cell to do work.

$$C_6H_{12}O_6 + 6\ O_2 \xrightarrow{\text{enzymes}} 6\ CO_2 + 6\ H_2O + 2820\ \text{kJ} \qquad (2)$$

Glucose can be burned in oxygen in the laboratory to produce carbon dioxide, water, and heat. But in the living cell, the oxidation of glucose does not proceed directly to carbon dioxide and water. Instead, like photosynthesis, the overall process proceeds by a series of enzyme-catalyzed intermediate reactions. These intermediate steps channel some of the liberated energy into uses other than heat production. Specifically, a portion of the energy is stored in the chemical bonds of ATP.

◀ **Chapter Opening Photo: An aerobics class is one way to increase the metabolism of carbohydrates.**

Even relaxing and watching TV requires metabolic energy.

35.2 Exercise and Energy Metabolism

Working or playing, studying or watching TV—everything we do requires metabolic energy. Sometimes we need the energy quickly, sometimes it can be delivered more slowly. This energy is liberated via catabolic pathways; different pathways liberate energy at different rates.

Carbohydrate catabolism is designed for relatively quick energy release, so this form of catabolism is activated during strenuous muscular exercise. When a muscle contracts, energy is consumed. Muscle contraction, as with almost all mechanical work, uses ATP. But, just like ready cash, ATP is in short supply. Muscle tissue can contract for no more than several seconds before the supply of high-energy phosphate bonds is depleted.

After the initial contraction, the muscle cells look for other energy sources. Muscle glycogen is the next available source. This polymer breaks down to glucose, which is oxidized to replenish the ATP supply. Because glucose oxidation is an involved process, muscle contraction must proceed at a slower rate. Unfortunately, even this energy supply is only useful for about 2 minutes of work. Muscles rapidly deplete their glycogen stores and rapidly build up lactic acid.

If this were the end of the story, our work schedule would always be very short. In fact, there are animals that respond in this way. For example, an alligator can run quickly, thrash its tail powerfully while capturing a prey, but only for several minutes. The alligator's exertion must be followed by a rest of several hours as the muscles recuperate their energy stores. Humans (and other mammals) are not so limited because their livers replenish the muscle glucose supply and remove lactic acid. Muscle contractions can continue with help from the liver, but the rate of contraction is slowed further as muscle cells wait on glucose transport.

Muscle cells choose their energy metabolism depending on the duration of strenuous exercise. On a personal basis we experience these choices daily, but the experiences of others are more convenient to cite. Take a track meet, for example. A world-class sprinter can run 100 m in about 10 s, a rate of 10 m/s. To run longer, athletes must slow their muscle contractions. In the 1000-m race, a world-class athlete will run for about 130 s, traveling at an average speed of about 7–8 m/s. The liver replenishes glucose stores in the muscles; this process is slower and medium distances are run at a slower pace. Marathon races are even slower. Runners move at a rate of about 5 m/s so that their muscles can continue to work for the 2+ hours necessary to cover the more than 26-mile (42-km) race. Muscle metabolism is carefully coordinated with the demands we place on our muscles. When more than one tissue is involved, the bloodstream is the intermediary in coordinating these demands.

35.3 Blood Glucose: The Center of Carbohydrate Metabolism

Glucose is the key monosaccharide in carbohydrate metabolism and is transported between tissues via the blood. The blood-glucose concentration is maintained within well-defined limits (70–90 mg/100 mL of blood for a healthy person before eating)

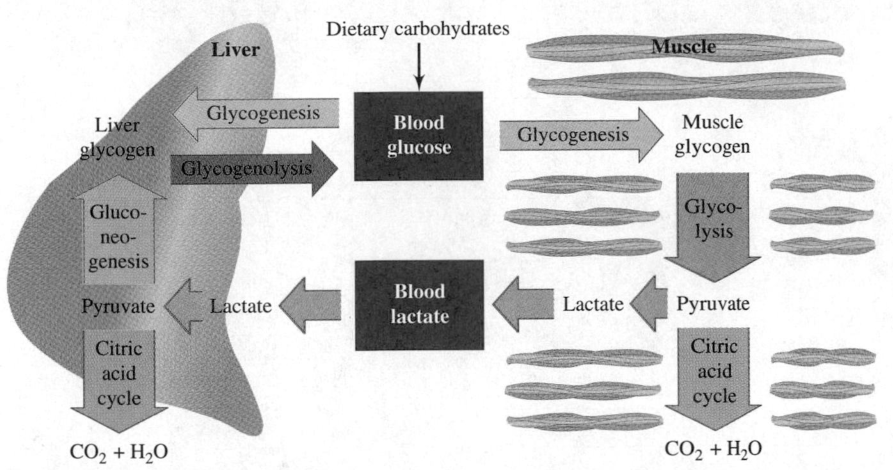

FIGURE 35.1
Overview of carbohydrate metabolism.

by the liver, which coordinates glucose metabolism. As muscles or other tissues use more glucose, the liver increases glucose output to the blood. After a meal, the liver removes excess glucose from the blood. Blood glucose is a metabolite pool that affects, and is affected by, each cell's carbohydrate metabolism (see Figure 35.1).

When blood glucose is in excess, it is converted to glycogen in the liver and in muscle tissue. Glycogen is a storage polysaccharide; it can be quickly hydrolyzed to replace depleted glucose supplies in the blood. The synthesis of glycogen from glucose is called **glycogenesis**; the hydrolysis, or breakdown, of glycogen to glucose is known as **glycogenolysis** (see Figure 35.1).

glycogenesis
glycogenolysis

Energy (stored as ATP) is obtained from glucose via several metabolic pathways. Under anaerobic (oxygen not required) conditions, a general sequence of reactions known as the *Embden–Meyerhof pathway* oxidizes glucose. When the final product of this process is lactic acid, the pathway is known as glycolysis. This reaction sequence is especially important for humans (and other animals) during strenuous activity when muscle tissue generates much lactic acid. Part of the lactic acid produced is sent to the liver and is converted to glucose in the *gluconeogenesis* pathway and then to glycogen. The rest is converted to carbon dioxide and water via the citric acid cycle, an aerobic (free or respiratory oxygen required) sequence. The *citric acid cycle*, also known as the Krebs cycle or the tricarboxylic acid cycle, is much more efficient at ATP production than is the glycolysis pathway. Consequently, the citric acid cycle is the major vehicle for obtaining metabolic energy from carbohydrates.

The diagram in Figure 35.1 presents a general overview of carbohydrate metabolism. The entire process requires specific enzymes at each reaction stage and is controlled by chemical regulatory compounds. The net result of these reactions is a conversion of carbohydrate to CO_2 and H_2O.

35.4 Anaerobic Sequence

In the absence of oxygen, glucose in living cells can be converted to a variety of end products including lactic acid (in muscle) and alcohol (in yeast). A similar conversion occurs in many different cells. At least a dozen reactions, many different

Chemistry in Action

Carbo-loading to Improve Athletic Performance

A Typical Depletion-Taper Precompetition Program for Glycogen-Loading

Day	Exercise	Diet
1	90-min period at maximum work	Mixed diet (50% carbohydrate)
2–3	Gradual tapering of time and intensity of work	Mixed diet (50% carbohydrate)
4–5	Tapering of time and intensity of work continues	Mixed diet (70% carbohydrate)
6	Complete rest	Mixed diet (70% carbohydrate)
7	Day of competition	High-carbohydrate pre-event diet

The day is at hand. Athletes from around the world arrive at the stadium. As they warm up, they picture the work that lies ahead of them. The marathoners are preparing to run about 26 miles in a little over 2 hours of time. One of the premier events of the Summer Olympics is about to begin. These athletes have been training for this event for years. Their bodies show evidence of consistent training—a high percentage of lean body mass and well-developed leg muscles for running. Perhaps as important, these athletes have prepared their body's metabolism to endure the marathon run.

In their pursuit of excellence, endurance athletes are constantly striving for the "winning edge." A new schedule of road work or even a new design for running shoes may provide this winning edge. Diet has always been an important part of training. As scientists have learned more about exercise, they have recognized the importance of carbohydrates in a training diet. Today a carbohydrate-loading diet is an accepted means toward gaining the winning edge. Perhaps you are familiar with athletes who "carbo-load" before events that require endurance. These competitors are trying to increase their body's glycogen stores knowing that carbohydrates provide important metabolic energy.

How effective is a carbohydrate-loading diet? To answer this question, we need to focus on the relationship between diet, metabolism, and exercise. A typical carbohydrate-loading diet is a timed sequence of dietary change coupled with changes in the exercise regimen. An example of such a diet is given in the table.

Diet has an important effect on glycogen stores. A low or carbohydrate-free diet does not provide for the replacement of glycogen stores. After several days on a carbohydrate-free diet (or about 1 day of complete starvation), the liver's glycogen stores are depleted. A 50% carbohydrate diet does not maintain glycogen stores for endurance athletes in training. On the glycogen-loading program, athletes in endurance training deplete their liver glycogen from the first day through the third day.

Glycogen is needed for work—for exercise. So the body attempts to maintain a supply of muscle glycogen. Even after 8 days on a carbohydrate-free diet (or 5 days of total starvation), muscle-glycogen levels have only decreased by about one-third. However, any loss of glycogen decreases the amount of work that can be done. Shifting from a normal diet to carbohydrate-free diet for as little as 3 days can halve the time muscles will work before exhaustion. On this diet, the liver no longer has much glycogen to contribute. After about 1 hour of work, people on carbohydrate-free diets become hypoglycemic. Athletes on the glycogen-loading program must taper their exercise from the second day until competition as their glycogen stores are depleted.

Increased glycogen stores can increase athletic performance. By exercising on a daily basis and by depleting the body's glycogen stores, an athlete prepares for carbo-loading. A switch to a carbohydrate-rich diet for about 3 days (from the fourth day through the seventh day on the above program) increases both the liver and muscle glycogen levels. In fact, these tissues are stimulated to store more than the normal amount of glycogen; they are "loaded" with glycogen. With the extra glycogen comes the ability to work longer. Some studies report an almost tripling of the work time available to the muscles before exhaustion. Carbo-loading before endurance atheletics follows sound biochemical principles.

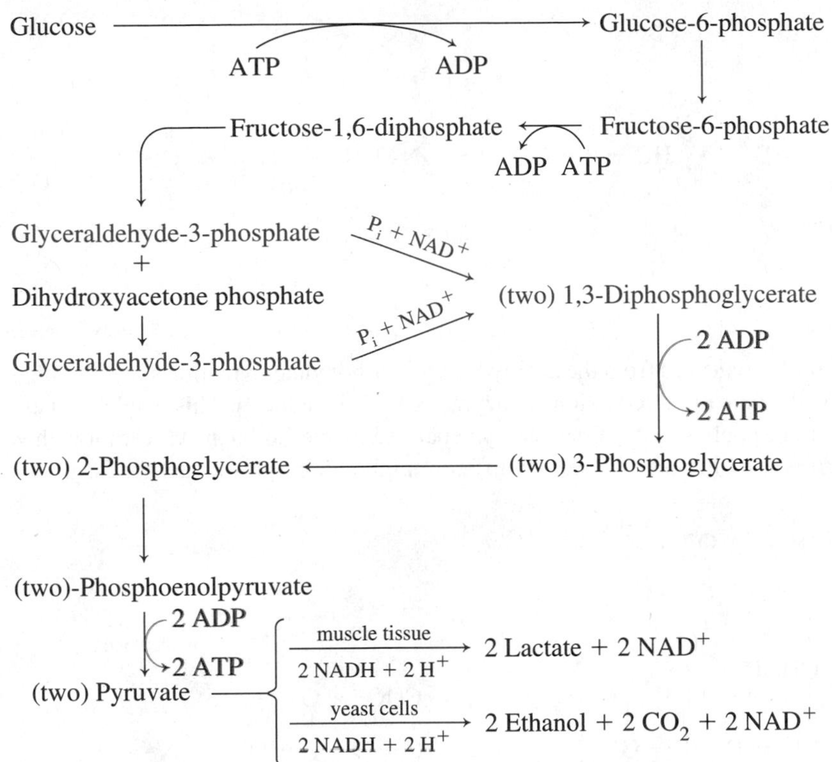

FIGURE 35.2 Conversion of glucose to pyruvate via the Embden–Meyerhof pathway (anaerobic sequence). Glycolysis proceeds further, forming lactic acid. Formation of ATP is denoted in green. The only oxidation–reduction reaction in this pathway is shown in pink.

enzymes, ATP, and inorganic phosphate (P_i) are required. Such a sequence of reactions from a particular reactant to end products is called a metabolic pathway.

Embden–Meyerhof pathway

The anaerobic conversion of glucose to pyruvate is known as the **Embden–Meyerhof pathway** (see Figure 35.2). The sequence is a catabolic one in which glucose is oxidatively degraded:

$$\underset{\substack{\text{D-glucose} \\ \text{(sum of oxidation numbers} \\ \text{for six carbons = 0)}}}{C_6H_{12}O_6} \xrightarrow[\text{pathway}]{\text{Embden–Meyerhof}} \underset{\substack{\text{pyruvate} \\ \text{(sum of oxidation numbers} \\ \text{for six carbons = +4)}}}{2\,CH_3\overset{\overset{\displaystyle O}{\|}}{C}{-}COO^-}$$

In glucose the sum of the oxidation numbers of the six carbon atoms is zero. After processing one molecule of glucose to two molecules of pyruvate by the pathway, the sum of the oxidation numbers of the six carbon atoms is +4. This makes it evident that an overall oxidation of carbon must occur in the Embden–Meyerhof pathway.

As with most catabolic processes, the Embden–Meyerhof pathway produces energy for the cell. Carbon atoms are oxidized; energy is released and stored in the form of ATP. The pathway uses the process of substrate-level phosphorylation (see Section 34.8)—the energy released from carbon oxidation is used to form a high-energy substrate–phosphate bond, which in turn is used to form ATP. It is interesting to note that there is only one oxidation–reduction reaction in the Embden–Meyerhof pathway—the conversion of glyceraldehyde-3-phosphate to 1,3-diphosphoglycerate (marked in red on Figure 35.2):

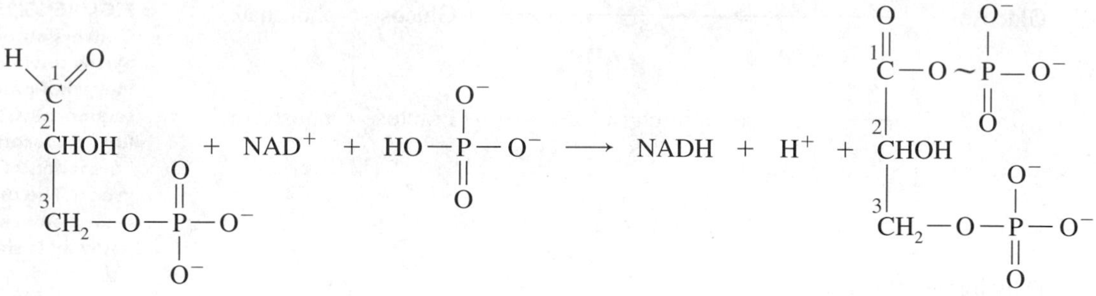

glyceraldehyde-3-phosphate $+ NAD^+ +$ $HO-PO_3^{2-}$ $\longrightarrow$ $NADH + H^+ +$ 1,3 diphosphoglycerate

Carbon 1 is oxidized from the aldehyde to the carboxylate oxidation state. Simultaneously, the oxidation–reduction coenzyme NAD^+ is reduced. This single oxidation–reduction supplies most of the energy generated by the Embden–Meyerhof pathway, which is used to make these two different high-energy phosphate bonds:

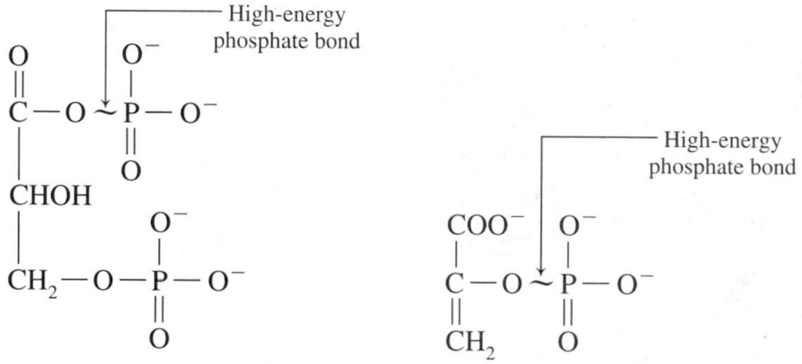

High-energy phosphate bond

1,3-diphosphoglycerate

High-energy phosphate bond

phosphoenolpyruvate

These high-energy bonds are found on intermediate compounds or pathway substrates. Substrate-level phosphorylation is complete when these substrates transfer their high-energy phosphate bonds to ADP, forming ATP (the ATP formation is shown in green on Figure 35.2).

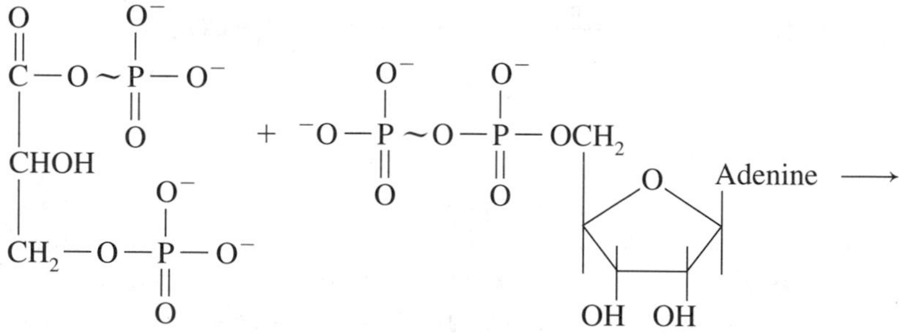

1,3-diphosphoglycerate + adenosine diphosphate (ADP) $\longrightarrow$

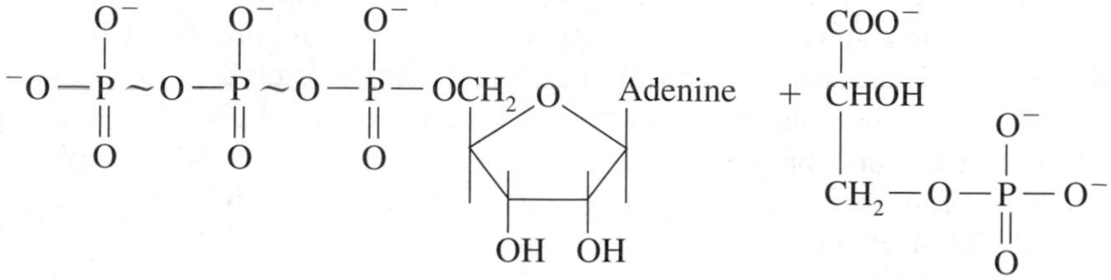

adenosine triphosphate (ATP) + 3-phosphoglycerate

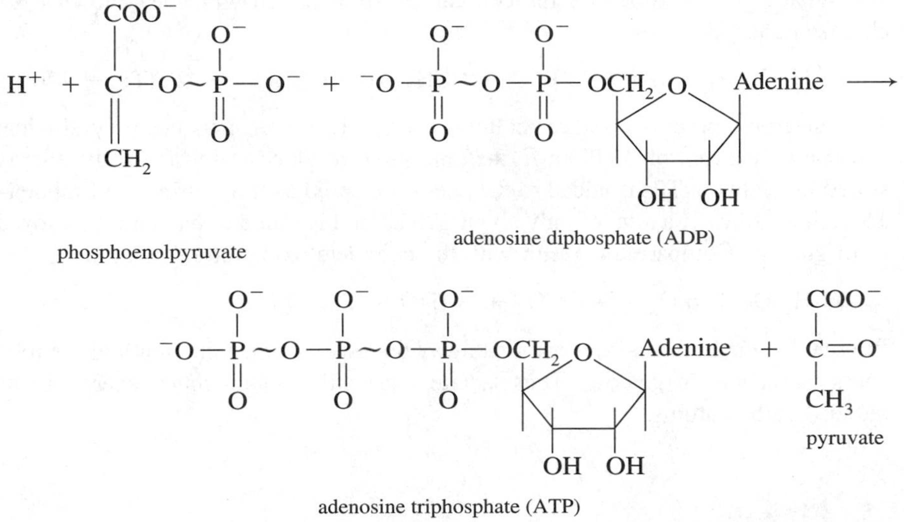

$$H^+ + \underset{\substack{\| \\ CH_2}}{\overset{\substack{COO^- \\ |}}{C}}-O\sim\overset{\substack{O^- \\ |}}{\underset{\substack{\| \\ O}}{P}}-O^- + {}^-O-\overset{\substack{O^- \\ |}}{\underset{\substack{\| \\ O}}{P}}\sim O-\overset{\substack{O^- \\ |}}{\underset{\substack{\| \\ O}}{P}}-OCH_2\text{(ribose, OH OH)}\text{Adenine} \longrightarrow$$

phosphoenolpyruvate

adenosine diphosphate (ADP)

$$^-O-\overset{\substack{O^- \\ |}}{\underset{\substack{\| \\ O}}{P}}\sim O-\overset{\substack{O^- \\ |}}{\underset{\substack{\| \\ O}}{P}}\sim O-\overset{\substack{O^- \\ |}}{\underset{\substack{\| \\ O}}{P}}-OCH_2\text{(ribose, OH OH)}\text{Adenine} + \underset{\substack{| \\ CH_3}}{\overset{\substack{COO^- \\ |}}{C}}=O$$

adenosine triphosphate (ATP)

pyruvate

It is important to note that the Embden–Meyerhof pathway oxidizes carbon and produces ATP in the absence of molecular oxygen. This pathway provides for *anaerobic* energy production. For this reaction sequence to continue, the coenzyme NADH must be recycled back to NAD^+; that is, an additional oxidation–reduction reaction is needed to remove electrons from NADH. In the absence of oxygen, pyruvate is used as an electron acceptor. In human muscle cells, pyruvate is reduced directly to lactate:

$$CH_3\underset{\substack{\| \\ O}}{C}COO^- + NADH + H^+ \rightleftharpoons CH_3\underset{\substack{| \\ OH}}{C}HCOO^- + NAD^+$$

pyruvate　　　　lactate

In yeast cells, pyruvate is converted to acetaldehyde, which is then reduced to ethanol:

$$CH_3\underset{\substack{\| \\ O}}{C}COO^- + H^+ \xrightarrow[\text{cells}]{\text{in yeast}} CH_3\overset{\substack{O \\ \|}}{C}-H + CO_2$$

pyruvate　　　　acetaldehyde

$$CH_3\overset{\substack{O \\ \|}}{C}-H + NADH + H^+ \rightleftharpoons CH_3CH_2OH + NAD^+$$

acetaldehyde　　　　ethanol

In each case, NADH is reoxidized to NAD^+, and a carbon atom from pyruvate is reduced. When lactate is the final product of anaerobic glucose catabolism, the pathway is termed **glycolysis**. As the Embden–Meyerhof pathway produces equal amounts of pyruvate and NADH, there is just enough pyruvate to recycle all the NADH. This is a good example of an important general characteristic of metabolism: Chemical reactions in the cell are precisely balanced so that there is never a large surplus or a large deficit of any metabolic product. If such a situation does occur, the cell may die.

glycolysis

What glycolysis does for the cell can be summarized with the following net chemical equation:

$$C_6H_{12}O_6 + 2\ ADP + 2\ P_i \longrightarrow 2\ CH_3CH(OH)COO^- + 2\ ATP + 150\ kJ$$

This anaerobic process produces cellular energy. However, it is not very efficient because only 2 mol of ATP are formed per mole of glucose. In fact, if the energy stored in the two ATPs is added to the energy released as heat during the Embden–Meyerhof pathway, a total of only about 220 kJ/mol is found to have been removed from glucose. Compare this result with the complete oxidation of glucose:

$$C_6H_{12}O_6 + 6\ O_2 \longrightarrow 6\ CO_2 + 6\ H_2O + 2820\ kJ$$

We see that the Embden–Meyerhof pathway releases less than one-tenth of the total energy available in glucose. Thus lactate must still contain much energy in its reduced carbon atoms.

35.5 Citric Acid Cycle (Aerobic Sequence)

As discussed previously, only a small fraction of the energy that is potentially available from glucose is liberated during the anaerobic conversion to lactate (glycolysis). Consequently, lactate remains valuable to the cells. The lactate formed may be (1) circulated back to the liver and converted to glycogen at the expense of some ATP or (2) converted back to pyruvate in order to enter the citric acid cycle.

$$\underset{\text{lactate}}{CH_3\overset{\overset{\displaystyle OH}{|}}{C}HCOO^-} + NAD^+ \rightleftharpoons \underset{\text{pyruvate}}{CH_3\overset{\overset{\displaystyle O}{\|}}{C}COO^-} + NADH + H^+$$

Pyruvate is the link between the anaerobic sequence (Embden–Meyerhof pathway) and the aerobic sequence (citric acid cycle). Pyruvate itself does not enter into the citric acid cycle. It is converted to acetyl coenzyme A (acetyl-CoA), a complex substance that, like ATP, is of great importance in metabolism:

$$CH_3\overset{\overset{\displaystyle O}{\|}}{C}-COO^- + \underset{\text{coenzyme A}}{CoASH} + NAD^+ \longrightarrow \underset{\text{acetyl-CoA}}{CH_3\overset{\overset{\displaystyle O}{\|}}{C}-SCoA} + NADH + CO_2$$

This important reaction depends on the availability of several vitamins. In addition to niacin (needed for synthesis of NAD^+), the enzyme that catalyzes this reaction requires riboflavin and thiamine. This is one of the few metabolic reactions that requires thiamine, but because this reaction is key to obtaining large amounts of energy from carbohydrate, thiamine is a vitamin of major importance. Also, our bodies need the vitamin pantothenic acid to synthesize coenzyme A.

Acetyl coenzyme A consists of an acetyl group bonded to a coenzyme A group. Coenzyme A contains the following units: adenine, ribose, diphosphate, pantothenic acid, and thioethanolamine. Coenzyme A is abbreviated as CoASH or CoA. Acetyl coenzyme A is abbreviated as acetyl-CoA or acetyl-SCoA.

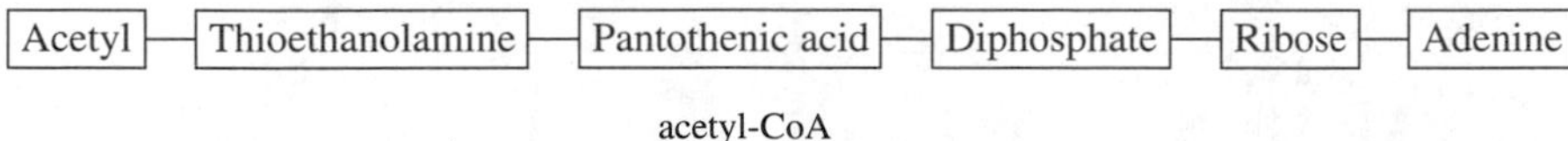

acetyl-CoA

The acetyl group is the group that is actually oxidized in the citric acid cycle. This group is attached to the large carrier molecule as a thioester—that is, by an ester linkage in which oxygen is replaced by sulfur:

$$\underbrace{CH_3-\overset{\overset{\displaystyle O}{\|}}{C}}_{\text{acetyl group}}-\underbrace{S}_{\text{thioester linkage}}-\underbrace{CoA}_{\text{carrier group}}$$

Acetyl-CoA not only is a key component of carbohydrate metabolism, but it also, as we will see in Chapter 36, serves to tie the metabolism of fats and that of certain amino acids to the citric acid cycle.

citric acid cycle

The **citric acid cycle** was elucidated by Hans A. Krebs (1900–1981), a British biochemist; thus it is also called the Krebs cycle. For his studies in intermediary metabolism, Krebs shared the 1953 Nobel prize in medicine and physiology with Fritz A. Lipmann (1899–1986), an American biochemist who discovered coenzyme A.

Krebs showed the citric acid cycle to be a series of eight reactions in which the acetyl group of acetyl-CoA is oxidized to carbon dioxide and water, and where many reduced coenzymes (both NADH and $FADH_2$) are formed. These reduced coenzymes then pass electrons through the electron-transport system and, in the process, ATP is produced. The citric acid cycle occurs in the mitochondria, which are the primary sites for the production of cellular energy.

The sequence of reactions involved in the citric acid cycle is shown in Figure 35.3. It is important to note that this cycle produces little usable cellular energy directly (only one GTP, or guanosine triphosphate, convertible to ATP). However, many ATPs are produced because of two other processes—electron transport and oxidative phosphorylation (see Section 34.8). Electron transport recycles the large number of reduced coenzymes formed by the citric acid cycle. As a consequence, oxidative phosphorylation can produce ATP from ADP and phosphate. Because electron transport requires oxygen and the citric acid cycle depends upon electron transport, the citric acid cycle is termed aerobic.

All three processes—the citric acid cycle (plus the initial conversion of pyruvate to acetyl-CoA), electron transport, and oxidative phosphorylation—team up to produce energy for the cell. All three take place in the mitochondria. Overall, these three processes result in the oxidation of 1 pyruvate ion to 3 carbon dioxide molecules and the formation of a maximum of 15 ATP molecules.

$$CH_3\overset{\overset{\displaystyle O}{\|}}{C}-COO^- \longrightarrow 3\,CO_2$$

A large quantity of energy (1260 kJ) becomes available for the cell when these three carbon atoms are fully oxidized. Most of this energy is used to reduce coenzymes:

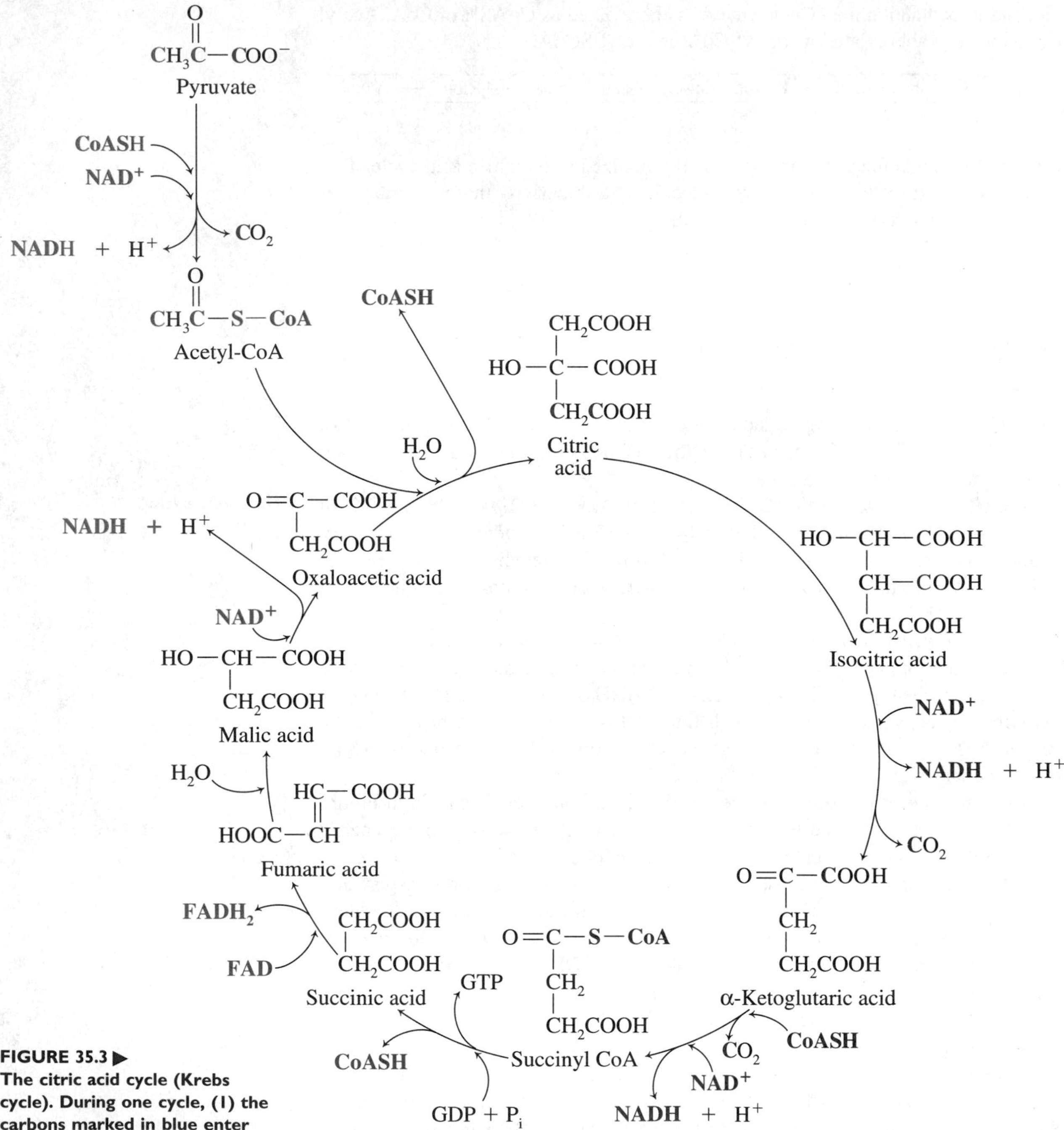

FIGURE 35.3 ▶
The citric acid cycle (Krebs cycle). During one cycle, (1) the carbons marked in blue enter the cycle and (2) the carbons marked in red are lost as CO_2. Coenzymes are marked in green.

$$4\ NAD^+ + FAD + 10\ H^+ + 10e^- \longrightarrow 4\ NADH + 4\ H^+ + FADH_2$$

These coenzymes, in turn, yield a total of 14 ATP molecules via mitochondrial electron transport and oxidative phosphorylation. Each mole of ATP stores approximately 35 kJ of energy. If we include the single GTP (convertible to ATP) that is formed in the citric acid cycle, the cell obtains about 462 kJ from each mole of pyruvate that is oxidized. By using an aerobic process, the cell produces 30 ATP molecules from the 2 lactates after gaining only 2 ATPs using the anaerobic pathway—that is, glycolysis. The presence of oxygen yields a large energy bonus for the cell.

Energy summation:
Glucose
↓ **glycolysis** — **2 ATP**
2 Lactate
↓ **citric acid cycle, e^- transport, oxidative phosphorylation** — **+30 ATP**
6 CO_2 — **32 ATP**

Myocardial infarction and stroke are two injuries that are especially serious because they deprive rapidly metabolizing tissue of oxygen. When the heart muscle loses at least part of its normal blood supply, myocardial infarction results; a similar occurrence in the brain results in a stroke. In both cases, a lack of blood circulation means at least part of the tissue loses its normal oxygen supply. Suddenly these cells lose most of their capacity to generate ATP; from a glucose molecule these cells can only gain 2 ATPs, where previously they could garner over 30. Because the heart and brain are very active tissues, the anaerobic process of glycolysis cannot support continued cell viability. Permanent tissue damage results from only minutes of oxygen deprivation.

35.6 Gluconeogenesis

A continuous supply of glucose is needed by the body, especially for the brain and the nervous system. But the amount of glucose and glycogen present in the body is sufficient to last for only about 4 hours at normal metabolic rates. Thus there is a metabolic need for a pathway that produces glucose from noncarbohydrate sources. The formation of glucose from noncarbohydrate sources is called **gluconeogenesis**.

gluconeogenesis

Most of the glucose formed during gluconeogenesis comes from lactate, certain amino acids, and the glycerol of fats. In each case, these molecules are converted to Embden–Meyerhof pathway intermediates (see Figure 35.2). As shown previously, lactate is converted to pyruvate. Amino acids first undergo *deamination* (loss of amino groups) and then are converted to pyruvate. Glycerol is converted to glyceraldehyde-3-phosphate (see Figure 35.4). Most of the steps in the Embden–Meyerhof pathway are reversed to transform these pathway intermediates into glucose.

Gluconeogenesis takes place primarily in the liver and also in the kidneys. These organs have the enzymes that catalyze reversal of the Embden–Meyerhof

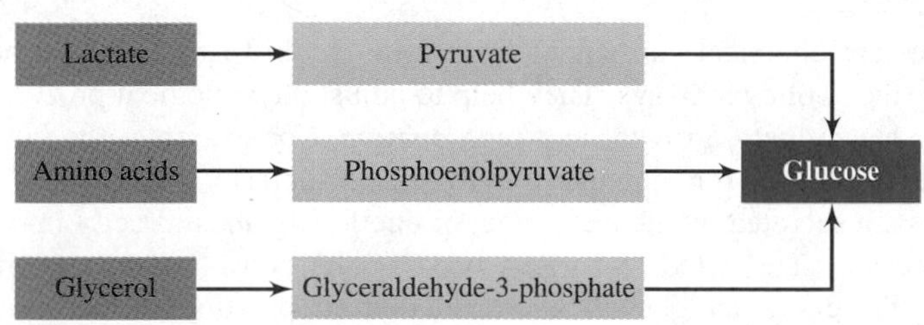

FIGURE 35.4
An overview of gluconeogenesis. All transformations except lactate to pyruvate require a series of reactions.

Long distance runners use gluconeogenesis to provide a source of glucose when more readily available supplies are exhausted.

pathway. Because of this capability the liver is the organ primarily responsible for maintaining normal blood-sugar levels.

35.7 Overview of Complex Metabolic Pathways

When we examine a metabolic pathway such as the Embden–Meyerhof pathway (Figure 35.2) or the citric acid cycle (Figure 35.3), the sheer complexity may be puzzling and overwhelming. We are tempted to ask why so many reactions are used by a cell to achieve its goal. This is a question that biochemists have asked for many years. It appears that a physiological design that includes a number of reactions per pathway achieves several vital objectives for the cell.

First, many of the chemicals that are formed in the middle of a pathway, *pathway intermediates*, are used in other metabolic processes within the cell.

$$A \rightleftarrows B \rightleftarrows C \rightleftarrows D$$ Pathway I (B, C = pathway intermediates)

$$B \rightleftarrows X \rightleftarrows Y \rightleftarrows Z$$ Pathway II

$$C \rightleftarrows I \rightleftarrows J \rightleftarrows K$$ Pathway III

These intermediates are like interchangeable machine parts. Once such parts are made, they can be used in more than one machine. And once pathway intermediates are formed, they can be used in several pathways and serve a variety of metabolic functions in the cell.

Second, having multiple-step pathways helps the cell to handle metabolic energy efficiently. For many pathways the total energy available is much greater than the cell can handle in a single reaction. As an example, a one-step complete oxidation of glucose yields enough energy to cook a cell, figuratively speaking. To avoid such disasters the cell extracts only a little energy from glucose at each chemical reaction (Figure 35.5). The quantity of energy released in each step is small enough to be handled by the cell. Thus, in order for a cell to extract the maximum energy, a metabolic pathway must have a number of steps.

Metabolic reaction sequences seem needlessly complex because we see only a small part of the total cellular function. Although the processes must be studied separately, it is important to remember that they are all interrelated.

35.8 Hormones

hormones

Hormones are chemical substances that act as control agents in the body, often regulating metabolic pathways. They help to adjust physiological processes such as digestion, metabolism, growth, and reproduction. For example, the concentration of glucose in the blood is maintained within definite limits by the actions of hormones. Hormones are secreted by the endocrine, or ductless, glands directly into the bloodstream and are transported to various parts of the body to exert specific control functions. The endocrine glands include the thyroid, parathyroid, pancreas, adrenal,

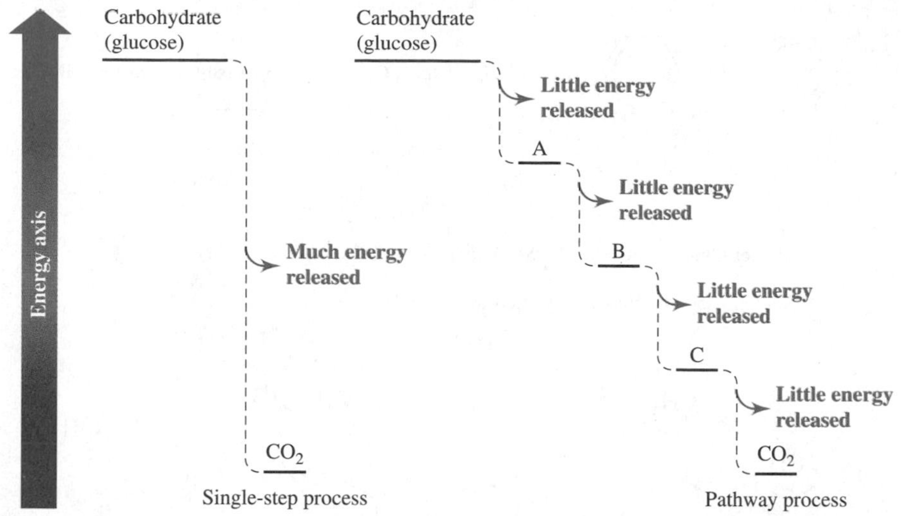

FIGURE 35.5 A single-step oxldation process compared with a multiple-step process: In the pathway, A, B, and C represent hypothetical pathway intermediates.

pituitary, ovaries, testes, placenta, and certain portions of the gastrointestinal tract. A hormone produced by one species is usually active in some other species. For example, the insulin used to treat diabetes mellitus in humans may be obtained from the pancreas of animals slaughtered in meat packing plants.

Hormones are often referred to as the chemical messengers of the body. They do not fit into any single chemical structural classification. Many are proteins or polypeptides, some are steroids, others are phenol or amino acid derivatives; examples are shown in Figure 35.6. Because a lack of any hormone can produce serious physiological disorders, many hormones are produced synthetically or are extracted from their natural sources and made available for medical use.

Like the vitamins, hormones are generally needed in only minute amounts. Concentrations range from 10^{-6} M to 10^{-12} M. Unlike vitamins, which must be supplied in the diet, the necessary hormones are produced in the body of a healthy person. A number of hormones and their functions are listed in Table 35.1.

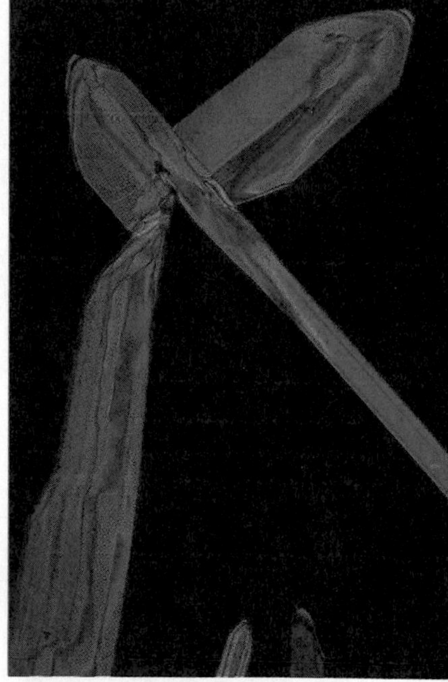

A crystal of thyroxine. This hormone is produced in the thyroid gland and controls metabolic rate.

35.9 Blood Glucose and Hormones

An adequate blood-glucose level must be maintained to ensure good health. To achieve this goal, hormones regulate and coordinate metabolism in specific organs. The hormones control selected enzymes, which, in turn, regulate the rates of reaction in the appropriate metabolic pathways. Let's examine some of the physiological mechanisms that maintain proper blood glucose levels.

Glucose concentrations average about 70–90 mg/100 mL of blood under normal fasting conditions—that is, when no nourishment has been taken for several hours. Most people are in a normal fasting condition before eating breakfast. After the ingestion of carbohydrates, the glucose concentration rises above the normal level, and a condition of *hyperglycemia* exists. If the concentration of glucose rises still

FIGURE 35.6 ▶ Structure of selected hormones. Thyroxine is produced in the thyroid gland; oxytocin is a polypeptide produced in the posterior lobe of the pituitary gland; glucagon is a polypeptide produced in the pancreas; testosterone and estradiol are steroid hormones produced in the testes and the ovaries, respectively; epinephrine is produced in the adrenal glands.

HO–(I,I-phenyl)–O–(I,I-phenyl)–$CH_2CHCOOH$ (NH_2)

thyroxine

Cys-Tyr-Ile-Gln-Asn-Cys-Pro-Leu-Gly-NH_2 (Cys–S—S–Cys)

oxytocin

1 His-Ser-Gln-Gly-Thr-Phe-Thr-Ser-Asp-Tyr-Ser-Lys-Tyr-Leu-Asp-Ser-Arg-Arg- (10, 15)
20 Ala-Gln-Asp-Phe-Val-Gln-Tyr-Leu-Met-Asn-Thr (29)

glucagon

testosterone (CH_3, CH_3, OH, O)

estradiol (CH_3, OH, OH)

epinephrine (OH, OH, $HOCHCH_2NHCH_3$)

further, the renal threshold for glucose is eventually reached. The *renal threshold* is the concentration of a substance in the blood above which the kidneys begin to excrete that substance into the urine. The renal threshold for glucose is about 140–170 mg/100 mL of blood. Glucose excreted by the kidneys can be detected in the urine by a test for reducing sugars (e.g., the Benedict test; see Section 28.13). When the glucose concentration of the blood is below the normal fasting level, hypoglycemia exists (see Figure 35.7).

Glucose concentration in the blood is under the control of various hormones. These hormones act as checks on one another and establish an equilibrium condition called *homeostasis*—that is, self-regulated equilibrium. Three hormones—insulin, epinephrine (adrenalin), and glucagon—are of special significance in maintaining glucose concentration within the proper limits. Insulin, secreted by the islets of Langerhans in the pancreas, acts to reduce blood-glucose levels by increasing the rate of glycogen formation. Epinephrine from the adrenal glands and glucagon from the pancreas increase the rate of glycogen breakdown (glycogenolysis) and thereby increase blood-glucose levels. These opposing effects are summarized as follows:

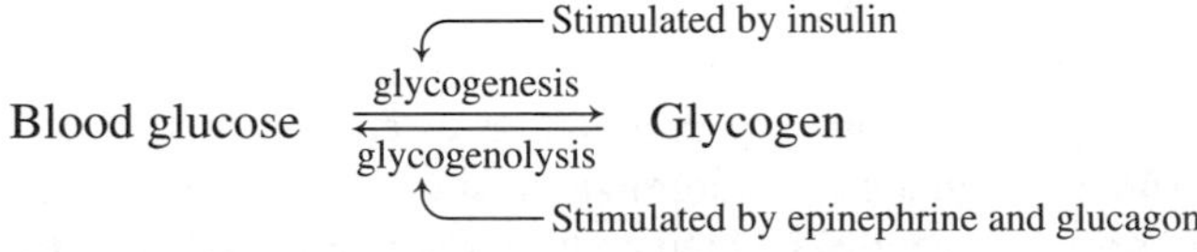

During the digestion of a meal rich in carbohydrates, the blood-glucose level of a healthy person rises into the hyperglycemic range. This stimulates insulin secretion, and the excess glucose is converted to glycogen, thereby returning the glucose level to normal. A large amount of ingested carbohydrates can over-stimulate insulin production and thereby produce a condition of mild hypoglycemia. This in turn triggers the secretion of additional epinephrine and glucagon, and the blood-glucose levels are again restored to normal. The body is able to maintain the normal fasting level of blood glucose for long periods of time without food by drawing on

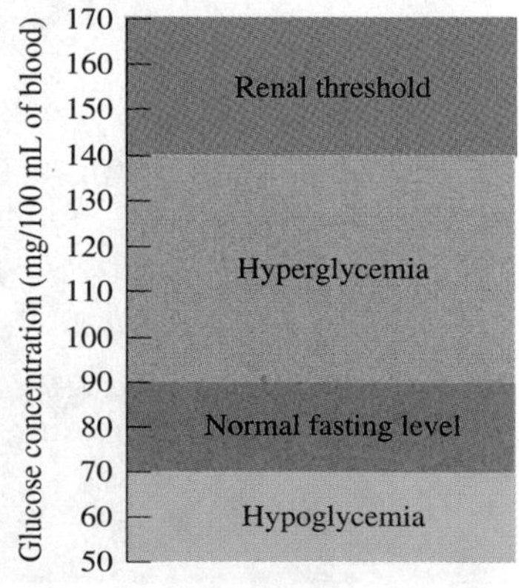

▲ FIGURE 35.7 Conditions related to concentration of glucose in the blood.

TABLE 35.1 Selected Hormones and Their Functions

Hormone	Source	Principal functions
Insulin	Pancreas	Controls blood-glucose level and storage of glycogen
Glucagon	Pancreas	Stimulates conversion of glycogen to glucose; raises blood-glucose level
Oxytocin	Pituitary gland	Stimulates contraction of the uterine muscles and secretion of milk by the mammary glands
Vasopressin	Pituitary gland	Controls water excretion by the kidneys; stimulates constriction of the blood vessels
Growth hormone	Pituitary gland	Stimulates growth
Adrenocorticotrophic hormone (ACTH)	Pituitary gland	Stimulates the adrenal cortex, which, in turn, releases several steroid hormones
Prolactin	Pituitary gland	Stimulates milk production by mammary glands after birth of a baby
Epinephrine (adrenalin)	Adrenal glands	Stimulates rise in blood pressure, accelerates the heartbeat, decreases secretion of insulin, and increases blood glucose
Cortisone	Adrenal glands	Helps control carbohydrate metabolism, salt and water balance, formation and storage of glycogen
Thyroxine and triiodothyronine	Thyroid gland	Increases the metabolic rate of carbohydrates and proteins
Calcitonin	Thyroid gland	Prevents the rise of calcium in the blood above the required level
Parathyroid hormone	Parathyroid gland	Regulates the metabolism of calcium and phosphate in the body
Gastrin	Stomach	Stimulates secretion of gastric juice
Secretin	Duodenum	Stimulates secretion of pancreatic juice
Estrogen	Ovaries	Stimulates development and maintenance of female sexual characteristics
Progesterone	Ovaries	Stimulates female sexual characteristics and maintains pregnancy
Testosterone	Testes	Stimulates development and maintenance of male sexual characteristics

liver glycogen, muscle glycogen, and finally on body fat as glucose replacement sources. Thus, in a normal person, neither hyperglycemia nor mild hypoglycemia has serious consequences since the body is able to correct these conditions. However, either condition, if not corrected, can have very serious consequences. Since the brain is heavily dependent on blood glucose for energy, hypoglycemia affects the brain and the central nervous system. Mild hypoglycemia may result in impaired vision, dizziness, and fainting spells. Severe hypoglycemia produces convulsions and unconsciousness; if prolonged, it may result in permanent brain damage and death.

Hyperglycemia may be induced by fear or anger because the rate of epinephrine secretion is increased under emotional stress. Glycogen hydrolysis is thereby speeded up and glucose-concentration levels rise sharply. This whole sequence readies the individual for the strenuous effort of either fighting or fleeing as the situation demands.

Insulin monitors are available to assist diabetics in watching their blood sugar levels.

Diabetes mellitus is a serious metabolic disorder characterized by hyperglycemia, glycosuria (glucose in the urine), frequent urination, thirst, weakness, and loss of weight. Prior to 1921, diabetes often resulted in death. In that year Frederick Banting and Charles Best, working at the University of Toronto, discovered insulin and devised methods for extracting the hormone from animal pancreases. For his work on insulin, Banting, with J. J. MacLeod, received the Nobel prize in medicine and physiology in 1923. Insulin is very effective in controlling diabetes. It must be given by injection, because, like any other protein, it would be hydrolyzed to amino acids in the gastrointestinal tract.

People with mild or borderline diabetes may show normal fasting blood-glucose levels, but they are unable to produce sufficient insulin for prompt control of ingested carbohydrates. As a result, their blood glucose rises to an abnormally high level and does not return to normal for a long period of time. Such a person has a decreased tolerance for glucose, which may be diagnosed by a glucose-tolerance test. After fasting for at least 12 hours, blood and urine specimens are taken to establish a beginning reference level. The person then drinks a solution containing 100 g of glucose (amount for adults). Blood and urine specimens are then collected at 0.5-, 1-, 2-, and 3-hour intervals and tested for glucose content. In a normal situation the blood-glucose level returns to normal after about 3 hours. Individuals with mild diabetes show a slower drop in glucose levels, but the glucose level in a severe diabetic remains high for the entire 3 hours. Responses to a glucose tolerance test are shown in Figure 35.8.

In 1978, the synthesis of insulin, identical in structure to that made by the human pancreas, was announced. In 1982, this insulin became the first genetically engineered human protein to receive FDA approval for sale. Today many diabetics have the option of choosing to use human insulin rather than the animal insulin obtained as a by-product of meat packing.

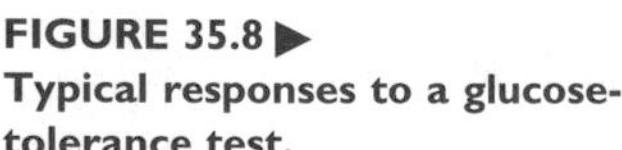

FIGURE 35.8 ▶
Typical responses to a glucose-tolerance test.

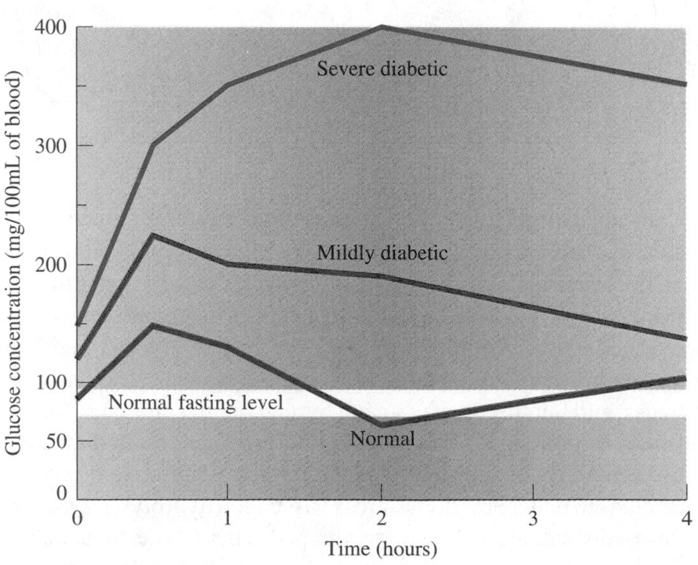

Concepts in Review

1. Briefly explain why carbohydrates are known as energy-storage molecules.
2. Briefly explain why enzymes are of vital importance in metabolism.
3. Explain why the Embden–Meyerhof pathway is catabolic.
4. State the purpose of the final reactions in anaerobic glucose catabolism.
5. Explain why the Embden–Meyerhof pathway is termed *anaerobic*.
6. Explain the role of glycogen in maintenance of blood-glucose levels.
7. Explain the role of epinephrine, insulin, and glucagon in the control of blood-glucose concentration.
8. Briefly describe how substrate-level phosphorylation is involved in the Embden–Meyerhof pathway.
9. Give an overall description of the Embden–Meyerhof pathway (anaerobic) for the metabolism of glucose.
10. Give an overall description of the citric acid cycle (aerobic).
11. Discuss the relationship between electron transport, oxidative phosphorylation, and the citric acid cycle.
12. Compare the amounts of energy liberated by the anaerobic and aerobic metabolic pathways of glucose.
13. Describe the function of acetyl-CoA in carbohydrate metabolism.
14. List three compounds that are used to make glucose during gluconeogenesis.
15. Explain the function of hormones in the body.
16. List the blood-glucose levels that are considered to be normal, hyperglycemic, and hypoglycemic.
17. Explain the renal threshold.
18. Predict what might happen to the blood-glucose level if a large overdose of insulin is taken.
19. Describe the glucose-tolerance test and how it ties in with the condition of diabetes mellitus.

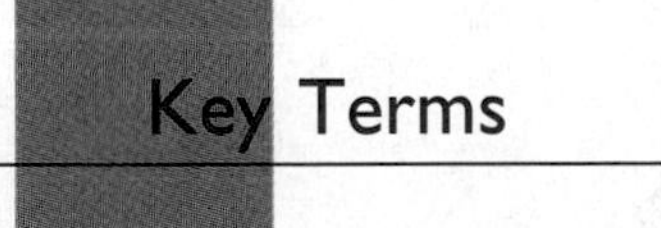

Key Terms

The terms listed here have been defined within this chapter. Section numbers are referenced in parenthesis for each term.

citric acid cycle (35.5)
Embden–Meyerhof pathway (35.4)
gluconeogenesis (35.6)
glycogenesis (35.3)
glycogenolysis (35.3)
glycolysis (35.4)
hormones (35.8)

Questions

Questions refer to tables, figures, and key words and concepts defined within the chapter. A particularly challenging question or exercise is indicated with an asterisk.

1. Why are carbohydrates considered to be energy-storage molecules?

2. List two general ways in which glucose catabolism differs from photosynthesis.

3. How much energy would be released if 3 mol of glucose were burned to form carbon dioxide and water?

4. Why are enzymes important in metabolism?

5. What form of chemical energy is first used for muscle contraction? What is used next?

6. How is the liver important in muscle contraction?

7. What is the purpose of the final reactions in anaerobic glucose metabolism?

8. How many high-energy phosphate bonds are directly formed in the citric acid cycle? List and identify the reduced coenzymes that are produced in this cycle.

9. What is acetyl-CoA and what is its function in metabolism?

10. What are the general functions of hormones and where are they produced?

11. **(a)** What is the range of glucose concentration in blood under normal fasting conditions?
(b) What blood-glucose concentrations are considered to be hyperglycemic? Hypoglycemic?

12. What is meant by the renal threshold?

13. Explain how the body maintains blood-glucose concentrations within certain definite limits despite wide variations in the rates of glucose intake and utilization.

14. Which of these statements are correct? Rewrite each incorrect statement to make it correct.
(a) Carbon dioxide is a final product of glucose catabolism.
(b) In general, only very few metabolic reactions require enzymes.
(c) Hormonal control often changes an enzyme's function.
(d) Glycolysis refers to the process of forming glycogen.
(e) The Embden–Meyerhof pathway uses oxidative phosphorylation to form ATP.
(f) Lactic acid is one product of gluconeogenesis.
(g) ATP is produced directly in the citric acid cycle.
(h) The Embden–Meyerhof metabolic pathway is an anaerobic sequence of reactions.
(i) In the Embden–Meyerhof metabolic pathway, 1 mol of glucose is converted to 2 mol of lactic acid in muscle tissue.
(j) The citric acid cycle is much more efficient in energy production than is the Embden–Meyerhof pathway.
(k) Hormones are regulatory agents that are secreted into the stomach and intestine to control metabolism.
(l) Hypoglycemia can affect the brain due to low blood glucose level.
(m) When the blood-glucose level exceeds the renal threshold, glucose is eliminated through the kidneys into the urine.

Paired Exercises

These exercises are paired. Each odd-numbered exercise is followed by a similar even-numbered exercise. Answers to the even-numbered exercises are given in Appendix V.

15. Draw the structure of the carbohydrate that is produced during glycogenolysis.

16. Draw the structure of the carbohydrate that is produced during glycogenesis.

17. Would you define glycogenesis as anabolic or catabolic? Briefly explain.

18. Would you define glycogenolysis as anabolic or catabolic? Briefly explain.

19. Is the ATP production that is associated with the citric acid cycle substrate-level phosphorylation or oxidative phosphorylation? Briefly explain.

20. Is the ATP production that is associated with the Embden–Meyerhof pathway substrate-level phosphorylation or oxidative phosphorylation? Briefly explain.

21. Circle the high-energy phosphate bond in the following molecule:

$$\begin{array}{l} O \quad\quad O \\ \| \quad\quad\;\; \| \\ C-O-P-O^- \\ | \quad\quad\;\;\; | \\ | \quad\quad\;\; O^- \\ CHOH \\ | \quad\quad\quad\; O \\ | \quad\quad\quad\; \| \\ CH_2-O-P-O^- \\ \quad\quad\quad\quad | \\ \quad\quad\quad\; O^- \end{array}$$

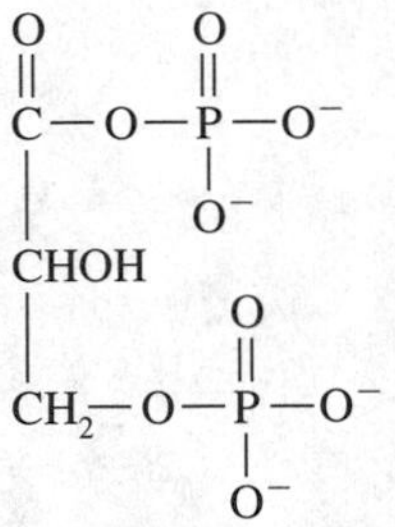

22. Circle the high-energy phosphate bond in the following molecule:

$$\begin{array}{l} O \\ \| \\ C-O^- \;\; O \\ | \quad\quad\quad \| \\ C-O-P-O^- \\ \| \quad\quad\quad | \\ CH_2 \quad\;\; O^- \end{array}$$

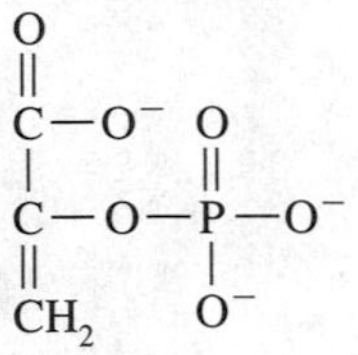

23. Explain why the Embden–Meyerhof pathway is considered to be anaerobic even though the oxidation of glucose occurs.

24. Explain why the citric acid cycle is considered to be aerobic even though no molecular oxygen (O_2) is used in this cycle.

25. How many moles of lactate will be formed from 2.8 mol of glucose through glycolysis?

26. How many moles of ATP will be formed from 0.85 mol of glucose through glycolysis?

27. What are the end products of the anaerobic catabolism of glucose in yeast cells?

28. What are the end products of the anaerobic catabolism of glucose in muscle tissue?

29. Is the Embden–Meyerhof pathway catabolic or anabolic? Briefly explain.

30. Is the gluconeogenesis pathway catabolic or anabolic? Briefly explain.

31. Fill in the coenzyme reactant and product for the following reaction:

$$\begin{array}{l} CH_2COOH \\ | \\ CH_2COOH \end{array} + \_\_ \longrightarrow \begin{array}{l} HC-COOH \\ \| \\ CH-COOH \end{array} + \_\_$$

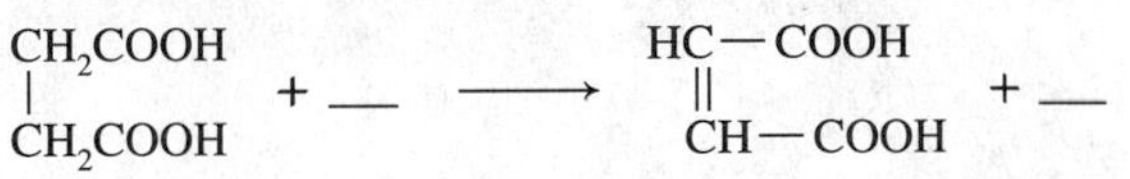

32. Fill in the coenzyme reactant and product for the following reaction:

$$\begin{array}{l} HO-CHCOOH \\ \quad\quad\;\; | \\ \quad\quad CH_2COOH \end{array} + \_\_ \longrightarrow \begin{array}{l} O=C-COOH \\ \quad\quad | \\ \quad CH_2-COOH \end{array} + \_\_ + H^+$$

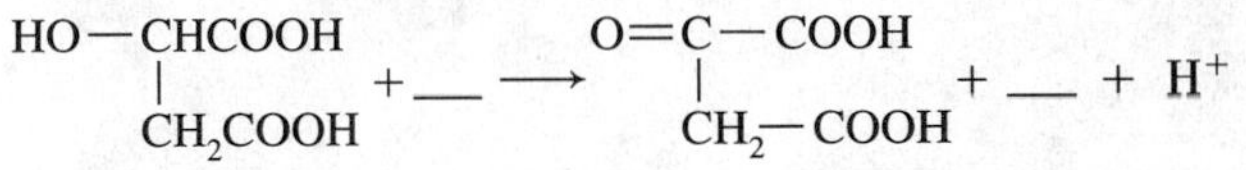

33. Why are electron transport and oxidative phosphorylation needed when the cell uses the citric acid cycle to produce energy?

34. Why are electron transport and oxidative phosphorylation not needed when the cell uses the glycolysis pathway to produce energy?

35. In what way(s) are hormones like vitamins?

36. In what way(s) do hormones differ from vitamins?

37. Exercise and a low-carbohydrate diet may cause a blood-glucose concentration of 65 mg/100 mL of blood. Is this hyperglycemic or hypoglycemic? Briefly explain.

38. After a meal, a diabetic may have a blood-glucose concentration of 350 mg/100 mL of blood. Is this hyperglycemic or hypoglycemic? Briefly explain.

Additional Exercises

These exercises are not paired or labeled by topic and provide additional practice on concepts covered in this chapter.

39. Why was the citric acid cycle not involved in the metabolism of life forms before the evolution of photosynthesis?

40. How does the function of hormones in the body differ from that of enzymes?

41. Why is the hormone insulin not effective when taken orally?

42. Predict what might happen to blood-glucose concentrations if a large overdose of insulin were taken by accident.

43. Why is epinephrine sometimes called the emergency or crisis hormone?

44. Write the structure of lactate. Would you predict that this molecule could provide cellular energy? Briefly explain.

45. In our bodies, what would be the most efficient way of using the energy produced: a multireaction system or a single-step reaction? Explain.

46. How does the use of pyruvate differ in aerobic catabolism and anaerobic catabolism?

Metabolism of Lipids and Proteins

36

One of the world's most pervasive nutritional problems is protein deficiency, known as kwashiorkor. It especially afflicts children and, when untreated, has a mortality rate of between 30 and 90%. These young people suffer from growth retardation, anemia, liver damage, and often appear bloated because of excess water absorption.

In more affluent societies, nutritional problems are often associated with a high intake of saturated fat—stroke and heart disease are closely correlated to lipid intake.

These disparate nutritional problems point toward an important similarity between protein and lipid metabolism. Our biochemical processes have needs for specific amino acids (proteins) and lipids. No matter how much food is available for our diet, we must also be concerned with meeting requirements for selected nutrients.

36.1 Metabolic Energy Sources

The ability to produce cellular energy is a vital characteristic of every cell's metabolism. As we have seen, carbohydrates are one of the major sources of cellular energy. The other two are lipids and proteins.

Of all the lipids, fatty acids are the most commonly used for cellular energy. Each fatty acid contains a long chain of reduced carbon atoms that can be oxidized to yield energy.

$$CH_3CH_2CH_2CH_2CH_2CH_2CH_2CH_2CH_2CH_2CH_2CH_2CH_2CH_2CH_2COOH$$

palmitic acid

The average oxidation number of the carbon atoms in fatty acids is about -2 compared with 0 in carbohydrates. Thus, when catabolized (oxidized), fatty acids yield more energy per carbon atom than do carbohydrates.

Proteins (amino acids) are also a source of reduced carbon atoms that can be catabolized to provide cellular energy. In addition, amino acids provide the major pool of usable nitrogen for cells. Proteins and amino acids also perform diverse functions. Some of these functions will be considered later in this chapter.

36.2 Fatty Acid Oxidation (Beta Oxidation)

Fats are the most energy-rich class of nutrients. Most of the energy from fats is derived from their constituent fatty acids. Palmitic acid derived from fat yields 39.1 kJ per gram when burned to form carbon dioxide and water. By contrast, glucose yields only 15.6 kJ per gram. Of course, fats are not actually burned in the body simply to produce heat. They are broken down in a series of enzyme-catalyzed reactions that also produce useful potential chemical energy in the form of ATP. In complete biochemical oxidation, the carbon and hydrogen of a fat ultimately are combined with oxygen (from respiration) to form carbon dioxide and water.

In 1904, Franz Knoop, a German biochemist, established that the catabolism of fatty acids involved a process whereby their carbon chains are shortened by two

◀ Chapter Opening Photo: These runners will metabolize lipids and proteins as well as carbohydrates by the end of the "Beat the Bridge Run" in Seattle, Washington.

carbon atoms at a time. Knoop knew that animals do not metabolize benzene groups to carbon dioxide and water. Instead, the benzene nucleus remains attached to at least one carbon atom and is eliminated in the urine as a derivative of either benzoic acid or phenylacetic acid.

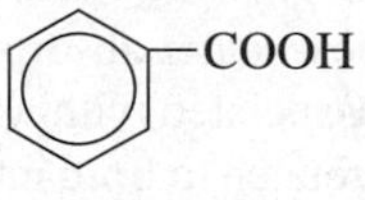

C_6H_5-COOH (benzoic acid) $\quad$ $C_6H_5-CH_2COOH$ (phenylacetic acid)

benzoic acid 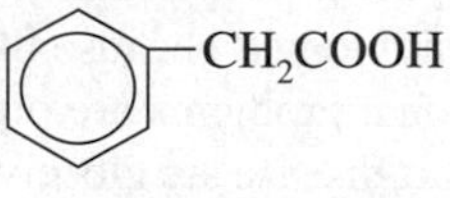 phenylacetic acid

Accordingly, Knoop prepared a homologous series of straight-chain fatty acids with a phenyl group at one end and a carboxyl group at the other end. He then fed these benzene-tagged acids to test animals. Phenylaceturic acid was identified in the urine of the animals that had eaten acids with an even number of carbon atoms; hippuric acid was present in the urine of the animals that had consumed acids with an odd number of carbon atoms:

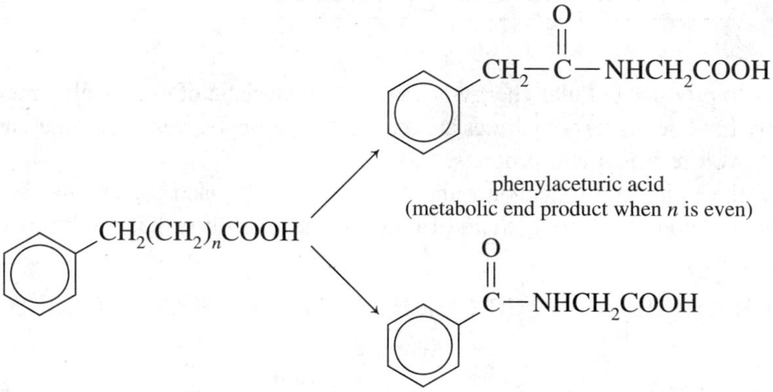

$$C_6H_5-CH_2(CH_2)_nCOOH \longrightarrow C_6H_5-CH_2-\overset{O}{\overset{\|}{C}}-NHCH_2COOH$$

phenylaceturic acid
(metabolic end product when n is even)

$$C_6H_5-CH_2(CH_2)_nCOOH \longrightarrow C_6H_5-\overset{O}{\overset{\|}{C}}-NHCH_2COOH$$

hippuric acid
(metabolic end product when n is odd)

These results indicated a metabolic pathway for fatty acids in which the carbon chain is shortened by two carbon atoms at each stage.

Knoop's experiments were remarkable for their time. They involved the use of tagged molecules and served as prototypes for modern research that utilizes isotopes to tag molecules.

Knoop postulated that the carbon chain of a fatty acid is shortened by successive removals of acetic acid units. The process involves the oxidation of the β-carbon atom and cleavage of the chain between the α and β carbons. A six-carbon fatty acid would produce three molecules of acetic acid; thus

First reaction sequence

$$CH_3CH_2CH_2\overset{\beta}{C}H_2\overset{\alpha}{C}H_2COOH \longrightarrow CH_3CH_2CH_2COOH + CH_3COOH$$

This C is oxidized. (β carbon); Chain is cleaved here (between β and α)

caproic acid → butyric acid + acetic acid

Second reaction sequence

$$CH_3\overset{\beta}{C}H_2\overset{\alpha}{C}H_2COOH \longrightarrow CH_3COOH + CH_3COOH$$

This C is oxidized (β carbon); Chain is cleaved here (between β and α)

The general validity of Knoop's theory of β-carbon atom oxidation has been confirmed. However, the detailed pathway for fatty acid oxidation was not established until about 50 years after his original work. The sequence of reactions involved, like those of the Embden–Meyerhof and citric acid pathways, is another fundamental metabolic pathway. **Beta oxidation**, or the *two-carbon chop*, is accomplished in a series of reactions whereby the first two carbon atoms of the fatty acid chain become the acetyl group in a molecule of acetyl-CoA.

beta oxidation

The catabolism proceeds in this manner: A fatty acid reacts with coenzyme A (CoASH) to form an activated thioester. The energy needed for this step of the catabolism is obtained from ATP.

Step 1 Activation: Formation of thioester with CoA

$$\underset{\text{fatty acid}}{RCH_2CH_2CH_2\overset{\overset{O}{\|}}{C}{-}OH} + CoASH + ATP \longrightarrow \underset{\text{CoA thioester of a fatty acid}}{RCH_2CH_2CH_2\overset{\overset{O}{\|}}{C}{-}SCoA} + AMP + \underset{\substack{\text{inorganic}\\ \text{phosphate}}}{2\,P_i}$$

The activated thioester next undergoes four more steps in the reaction sequence involving *oxidation, hydration, oxidation*, and *cleavage* to produce acetyl-CoA and an activated thioester shortened by two carbon atoms. The cleavage reaction requires an additional molecule of CoA.

Step 2 Oxidation: Dehydrogenation at carbons 2 and 3 (α and β carbons)

$$RCH_2CH_2CH_2\overset{\overset{O}{\|}}{C}{-}SCoA + FAD \longrightarrow RCH_2CH{=}CH\overset{\overset{O}{\|}}{C}{-}SCoA + FADH_2$$

Step 3 Hydration: Conversion to secondary alcohol

$$RCH_2CH{=}CH\overset{\overset{O}{\|}}{C}{-}SCoA + H_2O \longrightarrow RCH_2\overset{\overset{OH}{|}}{C}HCH_2\overset{\overset{O}{\|}}{C}{-}SCoA$$

Step 4 Oxidation: Dehydrogenation of carbon 3 (β carbon) to a keto group

$$RCH_2\overset{\overset{OH}{|}}{C}HCH_2\overset{\overset{O}{\|}}{C}{-}SCoA + NAD^+ \longrightarrow RCH_2\overset{\overset{O}{\|}}{C}CH_2\overset{\overset{O}{\|}}{C}{-}SCoA + NADH + H^+$$

Step 5 Carbon-chain cleavage: Reaction with CoA to produce acetyl-CoA and activated thioester of a fatty acid shortened by two carbons

$$RCH_2\overset{\overset{O}{\|}}{C}CH_2\overset{\overset{O}{\|}}{C}{-}SCoA + CoASH \longrightarrow RCH_2\overset{\overset{O}{\|}}{C}{-}SCoA + \underset{\text{acetyl-CoA}}{CH_3\overset{\overset{O}{\|}}{C}{-}SCoA}$$

The shortened-chain thioester repeats the reaction sequence of oxidation, hydration, oxidation, and cleavage to shorten the carbon chain further and produce another

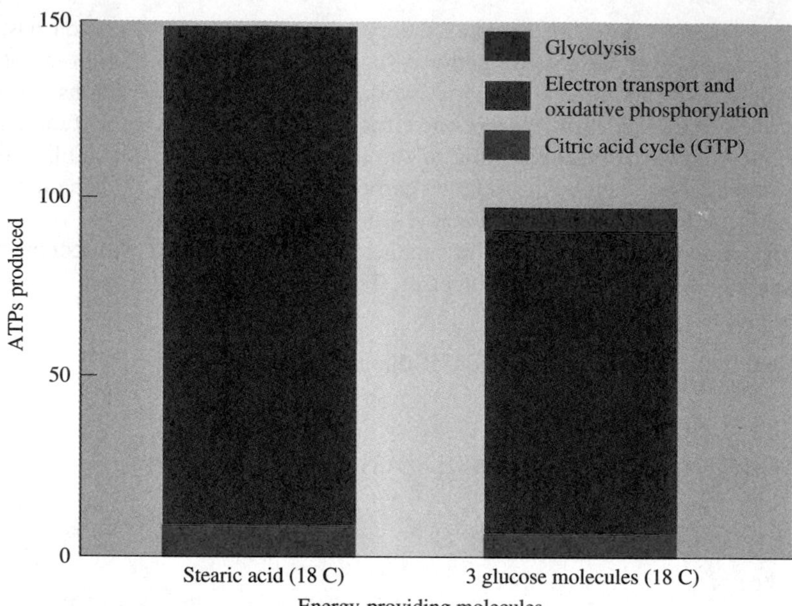

FIGURE 36.1 ▶ A comparison of the ATPs produced from 18 carbons of one stearic acid molecule and 18 carbons of the three glucose molecules.

molecule of acetyl-CoA. Thus, for example, eight molecules of acetyl-CoA can be produced from one molecule of palmitic acid.

As in the metabolic pathways for glucose, each reaction in the fatty acid oxidation pathway is enzyme catalyzed. No ATP is directly produced during fatty acid catabolism. Instead, ATP is formed when the reduced coenzymes, $FADH_2$ and NADH, are oxidized by the mitochondrial electron transport system in concert with oxidative phosphorylation. Fatty acid oxidation is aerobic because the products, $FADH_2$ and NADH, can only be reoxidized when oxygen is present.

In general, fatty acid catabolism yields more energy than can be derived from the breakdown of glucose (see Figure 36.1). For example, the reduced coenzymes derived from stearic acid (18 carbons) will yield 139 ATPs via electron transport and oxidative phosphorylation. Nine additional ATPs can be obtained from the 9 GTPs formed in the citric acid cycle while 1 ATP is used to start the beta oxidation process. Thus the 18 carbons of stearic acid yield a total of 147 ATPs or about 8.2 ATPs per carbon atom. In contrast, glucose can yield between 36 and 38 ATPs as its six carbons are completely oxidized to carbon dioxide. About 6 ATPs per carbon atom are gained from glucose as compared with about 8 ATPs per carbon atom from a fatty acid. Because fatty acid carbons are, in general, more reduced than glucose carbons, fatty acids are a more potent source of energy and yield more ATP molecules during metabolism.

Not surprisingly, the energy-storage molecule of choice in the human body is the fatty acid. On the average, a 70-kg male adult carries about 15 kg of fat (as triacylglycerols) but only about 0.22 kg of carbohydrate (as glycogen). Fat is such an effective way to store energy that obese people could exist for about 1 year without food. Unfortunately, fat is not the best energy-supply molecule for all tissues. For example, the brain normally derives all of its energy needs from glucose, using about 60% of all glucose metabolized by an adult at rest. Thus fatty acids are not a universal energy source although they are our most concentrated supply of energy.

36.3 Fat Storage and Utilization

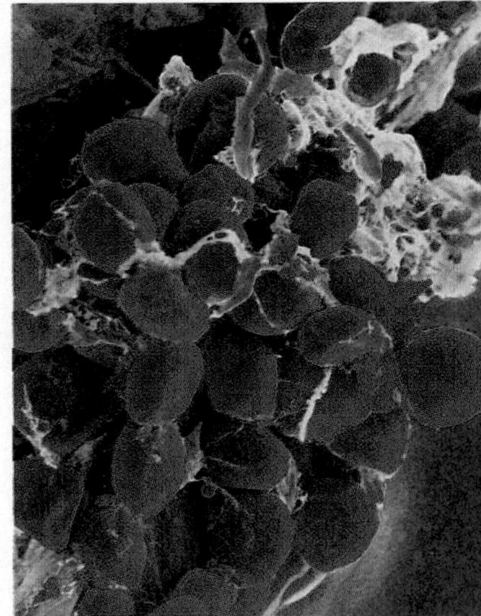

Photomicrograph of fat cells. These fat cells provide a major reserve of potential energy.

Fats (triacylglycerols) are stored primarily in adipose tissue, which is widely distributed in the body. Fat tends to accumulate under the skin (subcutaneous fat), in the abdominal region, and around some internal organs, especially the kidneys. Fat is deposited around internal organs as a shock absorber, or cushion. Subcutaneous fat acts as an insulating blanket. It is developed to an extreme degree in mammals such as seals, walruses, and whales that live in cold water.

Fat is the major reserve of potential energy. It is metabolized continuously. Stored fat does not remain in the body unchanged; there is a rapid exchange between the triacylglycerols of the plasma lipoproteins and the triacylglycerols in the adipose tissue. The plasma lipoprotein–bound triacylglycerols are broken down by the enzyme lipoprotein lipase that is found on the walls of all capillaries, and the resulting free fatty acids are transported into the adipose cells. When the body needs energy from fat, adipose cell enzymes hydrolyze triacylglycerols, and the fatty acids are exported to other body tissues. This vital process is under careful hormonal control. For example, a part of the "fight or flight" response caused by the hormone epinephrine (adrenalin) is an increased fatty acid output from the adipose tissue. Conversely, when there is more energy available in the diet than the body needs, the excess energy is used to make body fat. Continued eating of more food than the body can use results in obesity.

36.4 Biosynthesis of Fatty Acids (Lipogenesis)

lipogenesis

The biosynthesis of fatty acids from acetyl-CoA is called **lipogenesis**. Acetyl-CoA can be obtained from the catabolism of carbohydrates, fats, or proteins. Fatty acids, in turn, can be combined with glycerol to form triacylglycerols, which are stored in adipose tissue. Consequently, lipogenesis is the pathway by which all three of the major classes of nutrients are ultimately converted to body fat.

Is lipogenesis just the reverse of fatty acid oxidation (beta oxidation)? By analogy with carbohydrate metabolism (compare glycolysis with gluconeogenesis, see Chapter 35), we might expect this to be the case. However, fatty acid biosynthesis is not simply a reversal of fatty acid oxidation. The following are the major differences between the two pathways:

1. Fatty acid catabolism occurs in the mitochondria, but fatty acid anabolism (lipogenesis) occurs in the cytoplasm.
2. Lipogenesis requires a set of enzymes that are different from the enzymes used in the catabolism.
3. In the anabolic pathway (lipogenesis), the growing fatty acid chain is linked to a special acyl carrier protein (ACP—SH). The ACP—SH acts as a handle to transfer the growing chain from one enzyme to another through the series of enzyme-catalyzed reactions in the pathway. Coenzyme A is the carrier in fatty acid catabolism.
4. A preliminary set of reactions, involving malonyl-CoA, occurs for each two-carbon addition cycle in the synthesis. Malonyl is a three-carbon group and

has no counterpart in the catabolic pathway. Malonyl-CoA is synthesized from acetyl-CoA and carbon dioxide in the presence of the enzyme acetyl-CoA carboxylase, ATP, and the vitamin biotin.

$$\underset{\text{acetyl-CoA}}{CH_3\overset{O}{\overset{\|}{C}}-SCoA} + CO_2 \xrightarrow[\text{acetyl-CoA carboxylase}]{\text{ATP, biotin}} \underset{\text{malonyl-CoA}}{HO\overset{O}{\overset{\|}{C}}CH_2\overset{O}{\overset{\|}{C}}-SCoA}$$

The biosynthesis of a fatty acid occurs by addition of successive two-carbon-atom increments to a lengthening chain starting with acetyl-CoA. Each incremental addition follows this five-step pathway or reaction sequence.

Step 1 Acetyl-CoA and malonyl-CoA are linked to separate acyl carrier proteins:

$$\underset{\text{acetyl-CoA}}{CH_3\overset{O}{\overset{\|}{C}}-SCoA} + \underset{\text{malonyl-CoA}}{HO\overset{O}{\overset{\|}{C}}CH_2\overset{O}{\overset{\|}{C}}-SCoA} + \underset{\text{acyl protein carrier}}{2\ ACP-SH} \longrightarrow$$

$$\underset{\text{acetyl-ACP}}{CH_3\overset{O}{\overset{\|}{C}}-SACP} + \underset{\text{malonyl-ACP}}{HO\overset{O}{\overset{\|}{C}}CH_2\overset{O}{\overset{\|}{C}}-SACP} + 2\ CoASH$$

Step 2 Acetyl-ACP and malonyl-ACP condense, with loss of carbon dioxide (decarboxylation):

$$\underset{\text{acetyl-ACP}}{CH_3\overset{O}{\overset{\|}{C}}-SACP} + \underset{\text{malonyl-ACP}}{HO\overset{O}{\overset{\|}{C}}CH_2\overset{O}{\overset{\|}{C}}-SACP} \longrightarrow$$

$$\underset{\text{acetoacetyl-ACP}}{CH_3\overset{O}{\overset{\|}{C}}CH_2\overset{O}{\overset{\|}{C}}-SACP} + CO_2 + ACP-SH$$

The three steps that follow are approximate reversals of three steps in fatty acid beta oxidation (Section 36.2).

Step 3 Reduction: Hydrogenation of carbon 3 (β-keto group)

$$CH_3\overset{O}{\overset{\|}{C}}CH_2\overset{O}{\overset{\|}{C}}-SACP + NADPH + H^+ \longrightarrow$$

$$\underset{\beta\text{-hydroxybutyryl-ACP}}{CH_3\overset{OH}{\overset{|}{C}H}CH_2\overset{O}{\overset{\|}{C}}-SACP} + NADP^+$$

Step 4 Dehydration: Formation of a double bond between carbons 2 and 3

$$CH_3\overset{OH}{\overset{|}{C}H}CH_2\overset{O}{\overset{\|}{C}}-SACP \longrightarrow \underset{\text{crotonyl-ACP}}{CH_3CH{=}CH\overset{O}{\overset{\|}{C}}-SACP} + H_2O$$

Step 5 Reduction: Hydrogenation of carbons 2 and 3

$$CH_3CH{=}CH\overset{\overset{\displaystyle O}{\|}}{C}{-}SACP + NADPH + H^+ \longrightarrow$$

$$\underset{\text{butyryl-ACP}}{CH_3CH_2CH_2\overset{\overset{\displaystyle O}{\|}}{C}{-}SACP} + NADP^+$$

This completes the first cycle of the synthesis; the chain has been lengthened by two carbon atoms. Biosynthesis of longer-chain fatty acids proceeds by a series of such cycles, each lengthening the carbon chain by an increment of two carbon atoms. The next cycle would begin with the reaction of butyryl-ACP and malonyl-ACP, leading to a six-carbon chain, and so on. This synthesis commonly produces palmitic acid (16 carbons) as its end product. The synthesis of palmitic acid from acetyl-CoA and malonyl-CoA requires cycling through the series of steps seven times. The condensed equation for the formation of palmitic acid is

$$\underset{\text{acetyl-CoA}}{CH_3\overset{\overset{\displaystyle O}{\|}}{C}{-}SCoA} + 7\ \underset{\text{malonyl-CoA}}{HO\overset{\overset{\displaystyle O}{\|}}{C}CH_2\overset{\overset{\displaystyle O}{\|}}{C}{-}SCoA} + 14\ NADPH + 14\ H^+ \longrightarrow$$

$$\underset{\text{palmitic acid}}{CH_3(CH_2)_{14}COOH} + 7\ CO_2 + 6\ H_2O + 8\ CoASH + 14\ NADP^+$$

Nearly all naturally occurring fatty acids have even numbers of carbon atoms. A sound reason for this fact is that both the catabolism and the synthesis proceed by two-carbon increments.

In conclusion, it should be noted that the metabolism of fats has some features in common with that of carbohydrates. The acetyl-CoA produced in the catabolism of both carbohydrates and fatty acids can be used as a raw material for making other substances and as an energy source. When acetyl-CoA is oxidized via the citric acid cycle, more potential energy can be trapped in ATP. The ATP in turn serves as the source of energy needed for the production of other substances, including the synthesis of carbohydrates and fats.

36.5 Amino Acid Metabolism

Amino acids serve an important and unique role in cellular metabolism; they are the building blocks of proteins and also provide most of the nitrogen for other nitrogen-containing compounds.

Amino acid metabolism differs markedly from the biochemistry of carbohydrates or fatty acids. Amino acids always contain nitrogen, and the chemistry of this element presents unique problems for the cell. In addition, there is no structure common to all the carbon skeletons of amino acids. The carbohydrates can share a common metabolic pathway, as can fatty acids, because they share common structures. But the carbon skeletons of amino acids vary widely, and the cell must use a different metabolic pathway for almost every amino acid. Thus the metabolism of the carbon structures of amino acids is complex. In the sections that follow, a

Chemistry in Action

Ketone Bodies—A Stress Response

Extreme circumstances such as long-term, continuous exercise; starvation; and untreated diabetes mellitus cause a change in metabolism. In such cases the energy obtainable from glucose is severely limited. In response, the body produces small molecules from fatty acids, called *ketone bodies*, as glucose substitutes. These molecules include acetoacetic acid and two of its derivatives, β-hydroxybutyric acid and acetone.

$$CH_3\overset{\overset{\displaystyle O}{\|}}{C}CH_2COOH$$

acetoacetic acid

$$CH_3\overset{\overset{\displaystyle OH}{|}}{C}HCH_2COOH$$

β-hydroxybutyric acid

$$CH_3\overset{\overset{\displaystyle O}{\|}}{C}CH_3$$

acetone

Ketone bodies are highly soluble in blood and are able to provide cellular energy via beta oxidation and the citric acid cycle. When blood glucose becomes less available, some of the body's tissues can adapt to using ketone bodies as substitute energy-supply molecules. For example, the brain shifts its metabolism to use about 70% ketone bodies after about 40 days of minimal food intake.

An increase in ketone-body concentration is an emergency response to an extreme circumstance. This response can have dangerous side effects. For example, most ketone bodies are carboxylic acids. When their concentration increases in the blood, the acidity rises, causing a condition called *ketosis*, or *ketoacidosis*. Ketosis can affect respiration and cause a general deterioration of normal body function.

brief overview of amino acid metabolism will be presented together with a more detailed examination of nitrogen metabolism.

36.6 Metabolic Nitrogen Fixation

Nitrogen is an important component of many biochemicals. In addition to being a component of amino acids, nitrogen is found in nucleic acids, hemoglobin, and many vitamins. Every cell must have a continuous nitrogen supply. This supply might seem easy to obtain because the atmosphere contains about 78% free (elemental) nitrogen. Unfortunately, most cells cannot use elemental nitrogen.

Elemental nitrogen on earth exists as N_2 molecules, the two atoms being bonded together by a strong and stable triple bond. On an industrial scale, high temperature (400–500°C), high pressure (several hundred atmospheres), and a catalyst are required to react nitrogen gas with the reducing agent, hydrogen, to form ammonia. In the biosphere only a few organisms have the metabolic machinery necessary to use the abundant atmospheric nitrogen. These are procaryotes including *Azobacter* species, *Clostridium pasteurianum*, and *Rhizobium* species. By converting elemental nitrogen to compounds, these organisms make nitrogen available to the rest of the biological world.

nitrogen fixation

The conversion of diatomic nitrogen to a biochemically useful form is termed **nitrogen fixation**. Nitrogen is fixed by several methods, including (1) soil bacteria, (2) lightning, (3) Haber synthesis of ammonia, and (4) high-temperature processes such as combustion reactions. Of the procaryotes that are able to catalyze nitrogen fixation, *Rhizobium* bacteria deserve special attention. These bacteria flourish on the roots of legumes, such as clover and soybeans, in a symbiotic relationship. The bacteria have degenerated essentially into nitrogen-fixing machines that are maintained by the plants in the root nodules. The legumes even provide a special hemoglobin protein to assist the bacteria in their task.

Currently, in many university and industrial genetic-engineering laboratories, scientists are attempting to transfer the nitrogen-fixation capability of procaryotes to higher plants. To accomplish this transfer, the appropriate genes must be removed from procaryotic cells and spliced into the DNA of higher plants. This research is particularly challenging and, if successful, would have potentially great rewards. Success would mean that grains and other commercially important crops could be grown in poor soils using little or no nitrogen fertilizer. This achievement would create a vast increase in the world's food production capability.

Higher plants use nitrogen compounds primarily to produce proteins, which, in turn, enter the animal food chain. Both plant and animal proteins are important human nutrients.

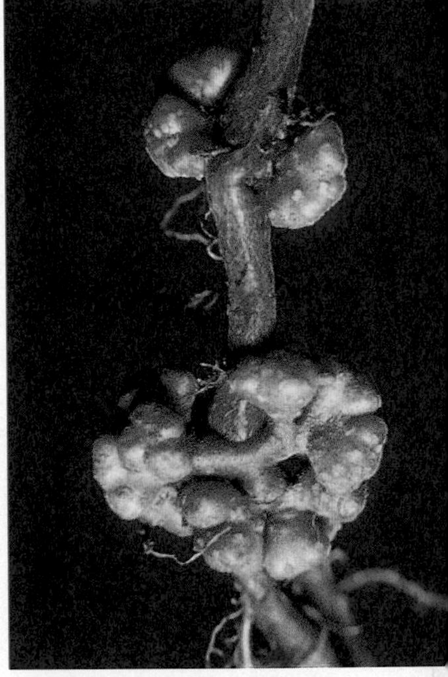

Legume root nodules containing nitrogen-fixing bacteria.

36.7 Amino Acids and Metabolic Nitrogen Balance

Protein is digested and absorbed to provide the amino acid dietary requirements (see Chapter 33). Once absorbed, an amino acid can be used in one of these ways; it can be

1. Incorporated into a protein
2. Utilized in the synthesis of other nitrogenous compounds such as nucleic acids
3. Deaminated to a keto acid, which either can be used to synthesize other compounds or can be oxidized to carbon dioxide and water to provide energy

Absorbed amino acids enter the amino acid pool. The **amino acid pool** is the total supply of amino acids available for use throughout the body. The amount of amino acids in the pool is maintained in balance with other cellular nitrogen pools (see Figure 36.2).

amino acid pool

One particularly important nitrogen pool is composed of the proteins in all the body's tissues. Amino acids continually move back and forth between the amino acid pool and the tissue proteins. In other words, our body proteins are constantly being broken down and resynthesized. The rate of turnover varies with different proteins. In the liver and other active organs, protein molecules may have a half-life of less than 1 week, whereas the half-life of some muscle proteins is about 6 months.

In a healthy, well-nourished adult, the amount of nitrogen excreted is equal to the amount of nitrogen ingested. The nitrogen pools within the body remain constant (see Figure 36.2), and such a person is said to be in **nitrogen balance**.

nitrogen balance

The amount of nitrogen consumed is more than that excreted for a growing child. Dietary protein feeds into the amino acid pool, which is used to make many new nitrogen-containing molecules. The nitrogen pools within the body expand, and the child is in **positive nitrogen balance**.

positive nitrogen balance

A fasting or starving person excretes more nitrogen than is ingested; such a person is in **negative nitrogen balance**. Tissue protein breaks down to supply amino

negative nitrogen balance

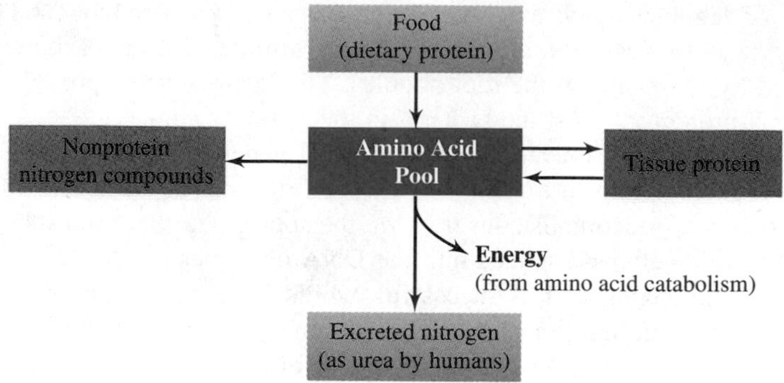

FIGURE 36.2 ▶ Major biological nitrogen pools related to the central amino acid pool.

acids (see Figure 36.2). These are used as an energy source and nitrogen is excreted. The body's nitrogen pools are depleted during negative nitrogen balance.

Nitrogen balance depends on the transition between nitrogen pools. If these pathways are disturbed, severe health problems may develop. For example, persons suffering from phenylketonuria (PKU) have lost the ability to catabolize phenylalanine. If this amino acid is in food protein, it will build up in the body's amino acid pool until it spills over into toxic side reactions. Babies who suffer from PKU develop extensive mental retardation. Fortunately, these symptoms can be avoided by removing most of the phenylalanine from the diet. Products like the artificial sweetener aspartame (Nutrasweet), which contain large amounts of phenylalanine also carry a warning label for people with the PKU genetic defect.

As shown schematically in Figure 36.2, amino acids are central to many different metabolic processes. They are the raw material for protein synthesis (see Section 32.10). They can be used to make other nitrogen-containing molecules. In addition, amino acids can supply metabolic energy if the amino acid nitrogen is first removed (or transferred to another molecule).

36.8 Amino Acids and Nitrogen Transfer

Amino acids are important in metabolism as carriers of usable nitrogen. If an amino acid is not directly incorporated into tissue proteins, its nitrogen may be incorporated into a variety of other molecules such as amino acids, nucleic acid bases, the heme of hemoglobin, and some lipids. In general, when an amino acid is to be used for some purpose other than protein synthesis, the amino acid carbon skeleton is separated from the amino acid nitrogen.

transamination

A process called transamination is responsible for most of the nitrogen transfer to and from amino acids. **Transamination** is the exchange of an oxygen atom for an amino group between an α-keto acid and an α-amino acid:

$$\underset{\alpha\text{-amino acid}}{\underset{\displaystyle |\atop NH_2}{RCH}-COOH} + \underset{\alpha\text{-keto acid}}{R'\overset{O}{\overset{\|}{C}}-COOH} \rightleftharpoons R\overset{O}{\overset{\|}{C}}-COOH + R'\underset{\displaystyle |\atop NH_2}{CH}-COOH$$

Transamination can involve many different molecules, with each different transamination requiring a different enzyme (transaminase). For example, one enzyme (glutamic-pyruvic transaminase) catalyzes the conversion of pyruvic acid to L-alanine.

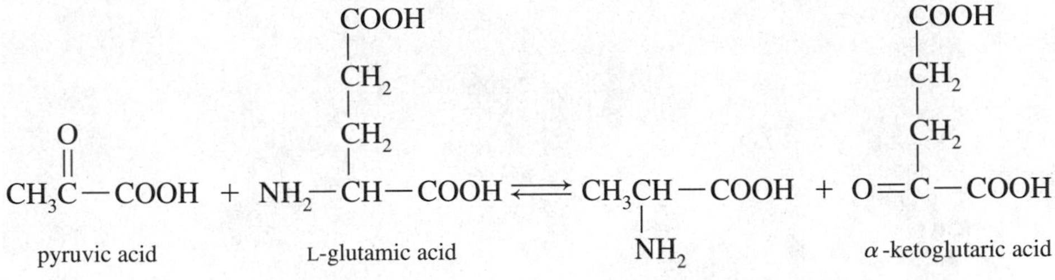

$$\underset{\text{pyruvic acid}}{CH_3\overset{\overset{O}{\|}}{C}-COOH} + \underset{\text{L-glutamic acid}}{NH_2-\overset{\overset{\overset{COOH}{|}}{\overset{CH_2}{|}}}{\overset{CH_2}{|}}}{CH}-COOH} \rightleftharpoons \underset{\text{L-alanine}}{CH_3\underset{\underset{NH_2}{|}}{CH}-COOH} + \underset{\alpha\text{-ketoglutaric acid}}{O=\overset{\overset{\overset{COOH}{|}}{\overset{CH_2}{|}}}{\overset{CH_2}{|}}}{C}-COOH}$$

A different enzyme catalyzes the production of L-aspartic acid from oxaloacetic acid:

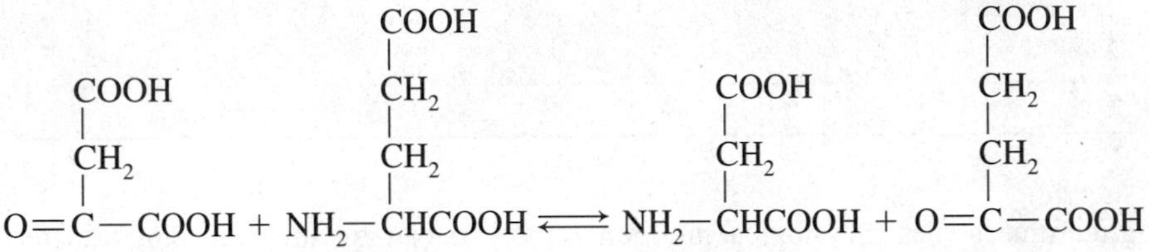

$$\underset{\text{oxaloacetic acid}}{O=\overset{\overset{COOH}{|}}{\overset{CH_2}{|}}}{C}-COOH} + \underset{\text{L-glutamic acid}}{NH_2-\overset{\overset{\overset{COOH}{|}}{\overset{CH_2}{|}}}{\overset{CH_2}{|}}}{C}HCOOH} \rightleftharpoons \underset{\text{L-aspartic acid}}{NH_2-\overset{\overset{COOH}{|}}{\overset{CH_2}{|}}}{C}HCOOH} + \underset{\alpha\text{-ketoglutaric acid}}{O=\overset{\overset{\overset{COOH}{|}}{\overset{CH_2}{|}}}{\overset{CH_2}{|}}}{C}-COOH}$$

Note that both of these reactions involve the conversion of L-glutamic acid to α-ketoglutaric acid. In fact, most transaminations use L-glutamic acid. Thus this amino acid plays a central role in cellular nitrogen transfer.

Transamination is the first step in the conversion of the carbon skeletons of amino acids to energy-storage compounds. Amino acids that are used to produce glucose and glycogen are termed **glucogenic amino acids**. Most amino acids are glucogenic (see Table 36.1), but some amino acids are converted to acetyl-CoA. When fed to starving animals, these amino acids cause an increase in the rate of ketone body formation and are therefore called **ketogenic amino acids**. Only leucine is completely ketogenic, but a few amino acids can be converted to either glucose or acetyl-CoA and are both ketogenic and glucogenic.

glucogenic amino acid

ketogenic amino acid

The amino acid pool of Figure 36.2 can now be described in more detail. At the center of this pool is L-glutamic acid, and other amino acids can either add or remove nitrogen from this central compound. L-Glutamic acid can also accept a second nitrogen atom and form L-glutamine:

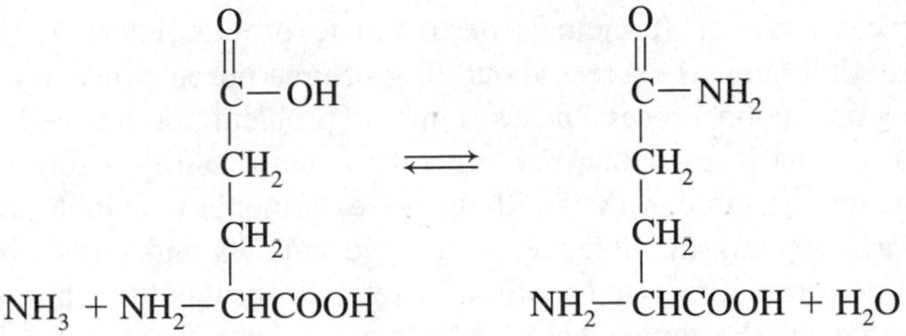

$$NH_3 + \underset{\text{L-glutamic acid}}{NH_2-\overset{\overset{\overset{\overset{O}{\|}}{C-OH}}{\overset{|}{CH_2}}}{\overset{|}{CH_2}}}{\overset{|}{C}}HCOOH} \rightleftharpoons \underset{\text{L-glutamine}}{NH_2-\overset{\overset{\overset{\overset{O}{\|}}{C-NH_2}}{\overset{|}{CH_2}}}{\overset{|}{CH_2}}}{\overset{|}{C}}HCOOH} + H_2O$$

Although this reaction can be considered as the simple addition of ammonia to yield an amide, the actual cellular reactions are more complex. The product,

TABLE 36.1 Classification of Amino Acids as Sources of Energy-Storage Molecules

Glucogenic	Ketogenic and glucogenic	Ketogenic
Alanine	Isoleucine	Leucine
Arginine	Lysine	
Aspartic acid	Phenylalanine	
Asparagine	Tyrosine	
Cysteine		
Glutamic acid		
Glutamine		
Glycine		
Histidine		
Methionine		
Proline		
Serine		
Threonine		
Tryptophan		
Valine		

L-glutamine, serves in biological nitrogen transfer, the amide nitrogen being transferable in a number of cellular reactions.

It is worthwhile to consider the synthesis of L-glutamine in more detail. Ammonia is a base and therefore is toxic to the cell. When ammonia forms an amide bond to L-glutamic acid, the nitrogen becomes less basic and also nontoxic. Thus L-glutamine serves as a safe package for transporting nitrogen. In the human body, L-glutamine is the major compound for transferring nitrogen from one cell to another via the bloodstream.

36.9 Nitrogen Excretion and the Urea Cycle

Nitrogen—unlike carbon, hydrogen, and oxygen—is often conserved for reuse by the cell. Nitrogen excretion does occur, however, when there is an excess of this element or when the carbon skeletons of nitrogen-containing compounds are needed for other purposes. Two examples from human nutrition arise: (1) when we consume more protein than is needed (an excess of nitrogen-containing molecules) or (2) when we experience starvation (protein is destroyed to provide energy). Under normal conditions, adult humans excrete about 20 g of urea nitrogen per day.

The nitrogen elimination process poses a major problem for the cell. The simplest excretion product is ammonia, but ammonia is basic and therefore toxic. Fish can excrete ammonia through their gills because ammonia is soluble and is swept away by water passing through the gills. Land animals and birds excrete nitrogen in less toxic forms. Birds and reptiles excrete nitrogen as the white solid uric acid, a derivative of the purine bases. Mammals excrete the water-soluble compound urea. Both of these compounds contain a high percentage of nitrogen in a nontoxic form.

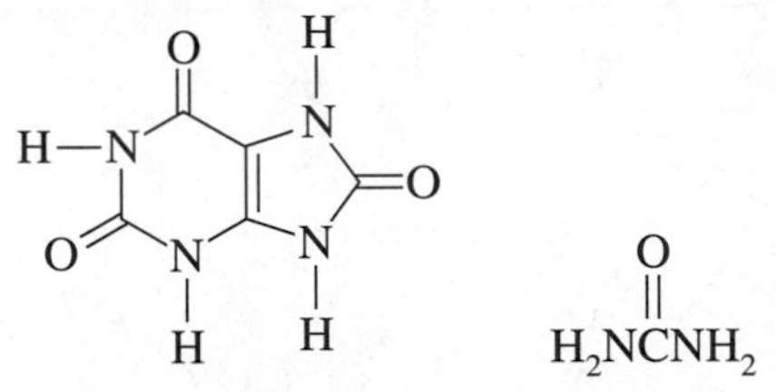

uric acid

$$H_2N\overset{O}{\overset{\|}{C}}NH_2$$

urea

Urea synthesis in mammals follows a pathway called the urea cycle (Figure 36.3), which takes place in the liver. Ammonium ion is first produced from L-glutamic acid in an oxidation–reduction reaction:

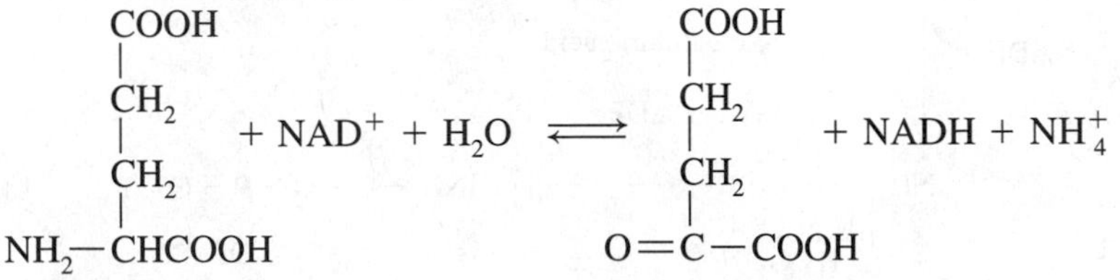

$$\underset{\text{L-glutamic acid}}{HOOC-CH_2-CH_2-CH(NH_2)COOH} + NAD^+ + H_2O \rightleftharpoons \underset{\alpha\text{-ketoglutaric acid}}{HOOC-CH_2-CH_2-C(=O)-COOH} + NADH + NH_4^+$$

The ammonium ion quickly reacts with the hydrogen carbonate ion and ATP to form carbamoyl phosphate:

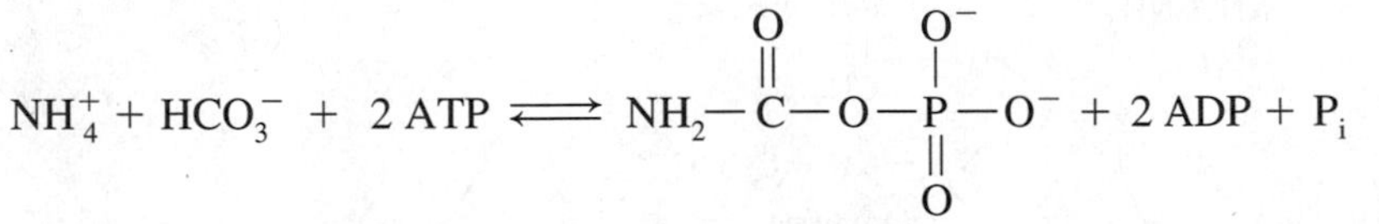

$$NH_4^+ + HCO_3^- + 2\,ATP \rightleftharpoons \underset{\text{carbamoyl phosphate}}{NH_2-\overset{O}{\overset{\|}{C}}-O-\overset{O^-}{\overset{|}{\underset{O}{\underset{\|}{P}}}}-O^-} + 2\,ADP + P_i$$

Finally, carbamoyl phosphate enters the urea cycle:

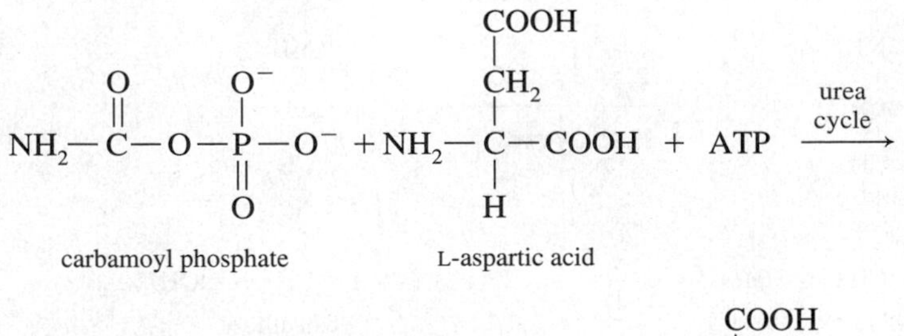

$$\underset{\text{carbamoyl phosphate}}{NH_2-\overset{O}{\overset{\|}{C}}-O-\overset{O^-}{\overset{|}{\underset{O}{\underset{\|}{P}}}}-O^-} + \underset{\text{L-aspartic acid}}{NH_2-\overset{COOH-CH_2}{\overset{|}{\underset{H}{\underset{|}{C}}}}-COOH} + ATP \xrightarrow{\text{urea cycle}}$$

$$\underset{\text{urea}}{NH_2-\overset{O}{\overset{\|}{C}}-NH_2} + \underset{\text{fumaric acid}}{HOOC-CH=CH-COOH} + P_i$$

Like the intermediate compounds of the citric acid cycle, the urea cycle intermediates do not appear in the overall reactions and serve only a catalytic function in the production of urea. Note that the urea cycle intermediates are α-amino acids.

FIGURE 36.3 ▶
Urea cycle.

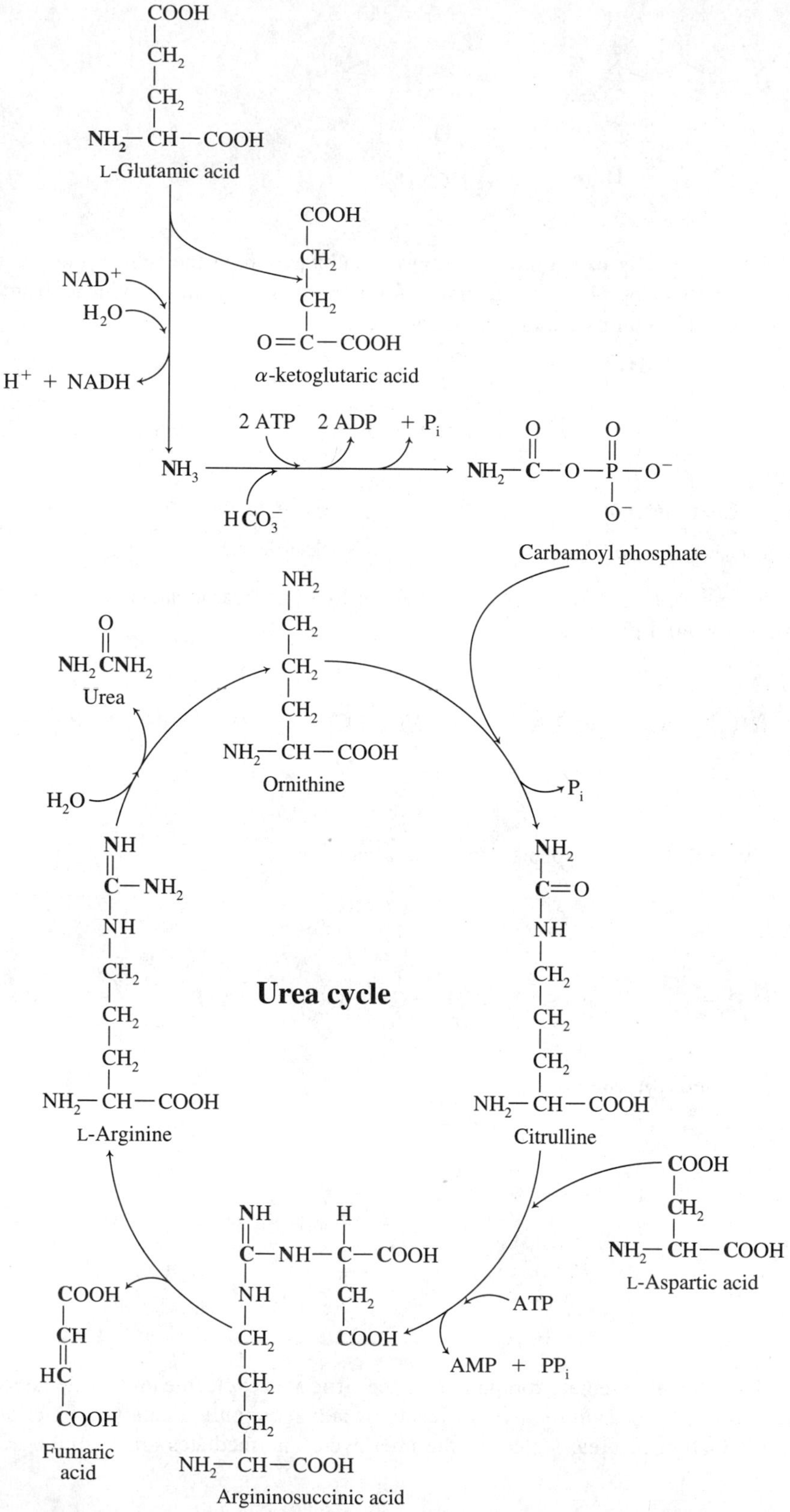

$HOOC-CH_2-CH_2-CH(NH_2)-COOH$
L-Glutamic acid

NAD^+, H_2O → $H^+ + NADH$

$HOOC-CH_2-CH_2-C(=O)-COOH$
α-ketoglutaric acid

$NH_3 + HCO_3^-$ —(2 ATP → 2 ADP + P_i)→ $NH_2-C(=O)-O-P(=O)(O^-)-O^-$
Carbamoyl phosphate

$NH_2-CH_2-CH_2-CH_2-CH(NH_2)-COOH$
Ornithine

$NH_2C(=O)NH_2$
Urea

H_2O

P_i

$NH_2-C(=O)-NH-CH_2-CH_2-CH_2-CH(NH_2)-COOH$
Citrulline

Urea cycle

$HOOC-CH_2-CH(NH_2)-COOH$
L-Aspartic acid

ATP → AMP + PP_i

$HN=C(NH-CH(COOH)-CH_2-COOH)-NH-CH_2-CH_2-CH_2-CH(NH_2)-COOH$
Argininosuccinic acid

$HOOC-CH=CH-COOH$
Fumaric acid

$HN=C(NH_2)-NH-CH_2-CH_2-CH_2-CH(NH_2)-COOH$
L-Arginine

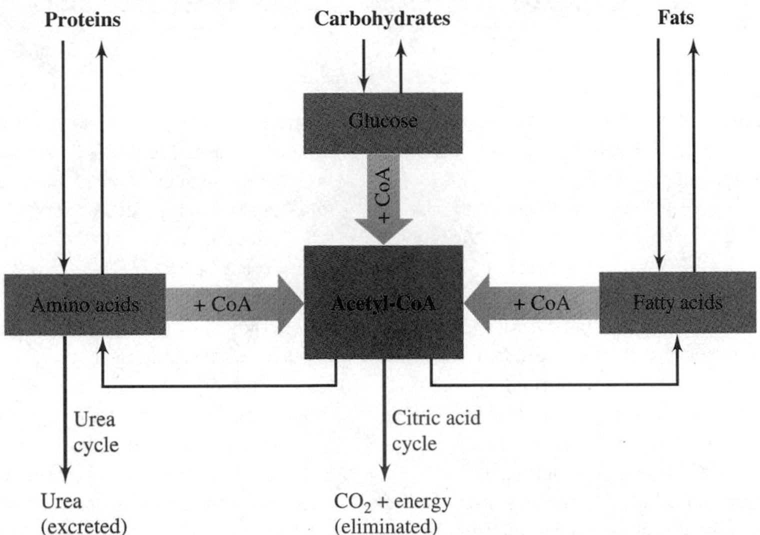

FIGURE 36.4 Simplified diagram showing acetyl-CoA at the hub of protein, carbohydrate, and fat metabolism.

These are amino acids that are rarely used in protein synthesis. Their primary role is in the formation of urea.

Also, it is important to recognize that the cell must expend energy, ATP, to produce urea. Formation of a nontoxic nitrogen excretion product is essential. In fact, impairment of the urea cycle is one of the major problems of liver cirrhosis caused by alcoholism. As liver function is impaired, more nitrogen is excreted as toxic ammonia.

Finally, let's look at the sources of the nitrogen that is excreted as urea. One nitrogen atom in each urea molecule comes from L-glutamic acid. The other nitrogen atom comes from L-aspartic acid, which may have gained its nitrogen from L-glutamic acid by transamination. Thus the amino acid that is central to nitrogen-transfer reactions is also the major contributor to nitrogen excretion.

36.10 Acetyl-CoA, a Central Molecule in Metabolism

As we think back through metabolism, we can identify some especially important compounds, molecules that are central to that portion of biochemistry. For example, glucose is the central compound in carbohydrate metabolism; glutamic acid is central in amino acid metabolism. There is one compound that is at the hub of all common metabolic processes; acetyl-CoA is central in the metabolism of carbon compounds (Figure 36.4). This molecule is a critical intermediate in the processes that form and break down both fats and amino acids. In addition, essentially all compounds that enter the citric acid cycle must first be catabolized to acetyl-CoA. In this section we will examine (1) the characteristics of acetyl-CoA that make it a central metabolic compound and (2) the advantages of a centralized metabolism for the cell.

Recall that acetyl-CoA consists of a small two-carbon unit (acetyl group) bonded by thioester linkage to a large organic coenzyme molecule, coenzyme A:

Chemistry in Action

An "Obese" Protein and Weight Loss

Obesity is a problem that is prevalent in the more affluent countries of the world. For example, between one-quarter and one-third of adults in the United States are overweight. With extra weight comes increased health risks and, sometimes social disfavor. We want to feel better and look better. So, in the United States alone, about 30 billion dollars are spent per year on weight loss programs. Unfortunately, most weight lost is regained over time; our bodies act as if they have a natural equilibrium weight.

Recently, a protein has been discovered that seems to establish the normal body weight for each individual. This discovery is the culmination of about 40 years of research. In the mid-1950s, a mutant mouse was discovered that grew to about double the normal weight—an obese mouse. Later it was learned that this obesity resulted from an inactive gene. When this gene was active, it coded for a protein called the obese protein. Subsequently, this gene has been found to reside in the adipose tissue of humans as well as mice.

The obese protein acts as a hormone; as more fat is accumulated, more obese protein is released into the bloodstream. The target organ for the obese protein is unknown although there is some evidence that it may be the brain. Wherever its destination, a high concentration of obese protein signals that the body has a large amount of fat reserves. The body's response is twofold. First, appetite is decreased. Second, metabolism is increased to burn fat. Coupled together, these two effects act as a feedback to control body weight. As fat accumulates, the obese protein serves to stop further accumulation.

Thinner people with low fat reserves will have a correspondingly lower obese protein concentration. This lower concentration leads to a larger appetite and a less active metabolism. The obese protein helps the body stay in a "normal" weight range, neither getting fatter nor thinner.

Doctors who treat obesity are particularly excited about this protein. Imagine injecting a medication that tells the body to decrease eating and to burn up more fat—just the responses for which weight loss clinics have searched. By adjusting the concentration of obese protein, a new normal body weight can be selected. However, there are still major hurdles to be overcome before this protein achieves its potential. For example, scientists point out that human obesity has many causes; it may be that the obese protein will not cure all obesity. Yet, the results are so encouraging that a biotechnology company has purchased the right to use a genetically engineered obese protein gene.

$$\underbrace{CH_3-\overset{\overset{\displaystyle O}{\|}}{C}}_{\text{acetyl group}}-\underset{\uparrow\ \text{thioester linkage}}{S}-\boxed{\text{CoA}}$$

This structure makes for an almost ideal central metabolic molecule with the following major advantages:

1. *Potential use in a wide variety of syntheses.* The small size and simple structure of the two-carbon acetyl fragment enables this molecule to be used to build a variety of diverse structures. Complex molecules with very different shapes and functions such as long-chain fatty acids, amino acids, and steroid hormones are synthesized from acetyl-CoA.
2. *Special reactivity of bonds.* The thioester causes both carbons in the acetyl fragment to be specially reactive. These carbons are "primed" to form new bonds as the acetyl-CoA enters various metabolic pathways.

3. *Structure that is recognizable by a wide variety of enzymes.* Coenzyme A acts as a kind of handle for the various enzymes that catalyze reactions of the acetyl group. Because many enzymes bind tightly to CoA, the acetyl group of acetyl-CoA can be involved in a great number of diverse reactions.

Acetyl-CoA can be compared with ATP as a central metabolic molecule. Remember that ATP (1) has a potential use in a wide variety of syntheses, (2) has a special reactivity in its phosphate anhydride bonds, and (3) has a structure that is recognizable and used by a wide variety of enzymes. For these important reasons, ATP serves as the "common energy currency" for the cell. In analogous fashion, acetyl-CoA might be termed the "common carbon currency" for the cell.

A consideration of these central metabolic compounds raises an important question, "How does centralization aid the cell?" We have seen numerous examples of centralization in biochemistry. Not only are there important central metabolites such as ATP and acetyl-CoA, but pathways such as the citric acid cycle that centralize the cellular metabolic machinery. A general answer to this question is that centralization improves the efficiency of metabolism and ensures that biochemistry can be under careful metabolic control.

Greater efficiency results from designing central metabolic pathways to handle a variety of different nutrients. For example, the citric acid cycle completes carbon oxidation for all energy-supplying nutrients that are first converted to the central metabolite, acetyl-CoA.

Greater control results from the dependence of many "feeder" pathways on a single central process. Control of a central path like the citric acid cycle will affect in a coordinated way the metabolism of a variety of nutrients. As scientists have learned more about metabolism, it has become clear that centralization is an important attribute of the chemistry of life.

Concepts in Review

1. Briefly explain what characteristic of fatty acids allows them to provide large amounts of metabolic energy.
2. Briefly describe Knoop's experiments on fatty acid oxidation and degradation and the conclusions derived from them.
3. Explain what is meant by beta oxidation and beta cleavage in relation to the metabolism of fatty acids.
4. State the purpose of ketone body production and indicate a dangerous side effect.
5. Briefly describe the biosynthesis of fatty acids using palmitic acid as an example.
6. List three major differences between beta oxidation and fatty acid synthesis.
7. Briefly describe how amino acid metabolism differs from carbohydrate and fatty acid metabolism.
8. List the possible metabolic fates of amino acids in humans.
9. Describe the major purpose of transamination.
10. Briefly explain the importance of L-glutamic acid and L-glutamine in nitrogen metabolism.

11. Explain how metabolism of proteins (amino acids) is tied into that of carbohydrates and fats.
12. Explain how a lack of essential amino acids in the diet affects the nitrogen balance.
13. Describe the purpose of the urea cycle.
14. Summarize the importance of acetyl-CoA in metabolism.
15. Discuss the major advantages of a centralized metabolism.

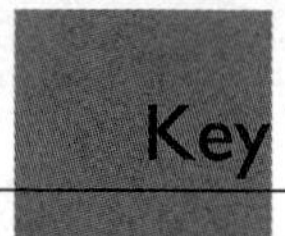

Key Terms

The terms listed here have been defined within this chapter. Section numbers are referenced in parenthesis for each term.

amino acid pool (36.7)
beta oxidation (36.2)
glucogenic amino acid (36.8)
ketogenic amino acid (36.8)
lipogenesis (36.4)
negative nitrogen balance (36.7)
nitrogen balance (36.7)
nitrogen fixation (36.6)
positive nitrogen balance (36.7)
transamination (36.8)

Questions

Questions refer to tables, figures, and key words and concepts defined within the chapter. A particularly challenging question or exercise is indicated with an asterisk.

1. What major characteristic of a fatty acid allows it to serve as an energy-storage molecule?
2. Briefly describe Knoop's experiment on fatty acid oxidation and catabolism.
3. What is meant by the terms *beta oxidation* and *beta cleavage* in relation to fatty acid catabolism?
4. How are ketone bodies important in energy metabolism?
5. Define *ketosis*.
6. Aside from being a food reserve, what are the two principal functions of body fat?
7. Is fatty acid synthesis simply the reverse of beta oxidation? Briefly explain.
8. In what two ways does the obese protein decrease fat accumulation in the body?
9. In general terms, how is the citric acid cycle involved in obtaining energy from fats?
10. Briefly describe why soybeans are a crop that enriches the soil.
11. Briefly describe why L-glutamic acid is considered to be the central amino acid of the amino acid pool.
12. In general, what are the metabolic fates of amino acids?
13. Why is L-glutamine a better nitrogen-transport molecule than ammonia?
14. In what compound is nitrogen excreted by (a) fish and (b) birds?
15. Give the structures of urea and uric acid.
16. How many ATPs are used to produce urea?
17. List three reasons why acetyl-CoA makes a good central intermediate in metabolism.
18. Which of these statements are correct? Rewrite each incorrect statement to make it correct.
    (a) The average carbon in palmitic acid is relatively oxidized, and this fatty acid thus contains little stored metabolic energy.
    (b) Carbohydrates provide more metabolic energy than fatty acids on a per gram basis.
    (c) Based on his experiments, Knoop postulated that fatty acids are broken down via beta oxidation—that is, by a "two carbon chop."
    (d) Fatty acid oxidation is an anaerobic reaction sequence.
    (e) The beta oxidation pathway directly produces 38 ATPs.
    (f) Acetyl-CoA is a very important product of beta oxidation.
    (g) When energy from glucose is severely limited, the body produces and uses ketone bodies to provide cellular energy.
    (h) Triacylglycerols stored in adipose tissue are continuously exchanged with triacylglycerols of the plasma lipoproteins.

(i) The catabolism of fatty acids is known as lipogenesis.
(j) Fatty acid anabolism occurs in the cytoplasm.
(k) An acyl carrier protein is used in both beta oxidation and fatty acid synthesis.
(l) Malonyl-CoA is synthesized from acetyl-CoA.
(m) The biosynthesis of fatty acids occurs by successive additions of two-carbon increments.
(n) During a dietary deficiency of protein, tissue protein acts as an emergency source to maintain a balanced amino acid pool.
(o) Most amino acids are glucogenic.
(p) Higher plant cells have the enzymes needed for nitrogen fixation.
(q) A positive nitrogen balance means more nitrogen is excreted than is consumed.
(r) A ketogenic amino acid can cause an increase in the rate of ketone body formation.
(s) Transamination allows formation of amino acids from α-keto acids.
(t) Acetyl-CoA can be produced from carbohydrates, fatty acids, and proteins.
(u) Acetyl-CoA consists of an acetyl group and a coenzyme A molecule linked through a phosphate ester bond.
(v) Mammals excrete nitrogen in the form of urea.
(w) Compared with other nitrogen excretion compounds, urea is especially toxic.

Paired Exercises

These exercises are paired. Each odd-numbered exercise is followed by a similar even-numbered exercise. Answers to the even-numbered exercises are given in Appendix V.

19. Circle the two carbons that will be found in acetyl-CoA after one pass through the beta oxidation pathway by butyric acid, $CH_3CH_2CH_2COOH$.

21. Write the structure and mark the β carbon in palmitic acid.

23. How many molecules of acetyl-CoA can be formed from lauric acid using the beta oxidation pathway?

25. One $FADH_2$ is produced for each pass through the beta oxidation pathway. How many of these reduced coenzymes will be formed from lauric acid?

27. Each $FADH_2$ can yield two ATPs and each NADH can yield three ATPs. Given that one NADH and one $FADH_2$ is made for each pass through the beta oxidation pathway, how many ATPs will be formed from myristic acid?

29. How is the acyl carrier protein (ACP) similar to coenzyme A (CoA)?

31. How might a low-protein diet cause a negative nitrogen balance?

33. A diet high in L-leucine was observed to cause an increase in blood levels of

acetoacetic acid

Is L-leucine a glucogenic or ketogenic amino acid? Briefly explain.

20. Circle the two carbons that will remain after two passes through the beta oxidation pathway by caproic acid, $CH_3CH_2CH_2CH_2CH_2COOH$.

22. Write the structure and mark the β carbon in myristic acid.

24. How many molecules of acetyl-CoA can be formed from palmitic acid using the beta oxidation pathway?

26. One NADH is produced for each pass through the beta oxidation pathway. How many of these reduced coenzymes will be formed from palmitic acid?

28. Each $FADH_2$ can yield two ATPs, and each NADH can yield three ATPs. Given that one NADH and one $FADH_2$ is made for each pass through the beta oxidation pathway, how many ATPs will be formed from lauric acid?

30. How is the acyl carrier protein (ACP) different from coenzyme A (CoA)?

32. How might a rapid growth spurt cause a positive nitrogen balance?

34. A diet high in L-aspartic acid was observed to cause no change in blood levels of

$$CH_3\overset{\displaystyle OH}{\overset{|}{C}}HCH_2COOH$$

β-hydroxybutyric acid

Is L-aspartic acid a glucogenic or ketogenic amino acid? Briefly explain.

35. Write the structural formulas of the compounds produced by transamination of the following amino acids:
(a) L-alanine (b) L-serine

36. Write the structural formulas of the compounds produced by transamination of the following amino acids:
(a) L-aspartic acid (b) L-phenylalanine

37. What portion of the carbamoyl phosphate molecule

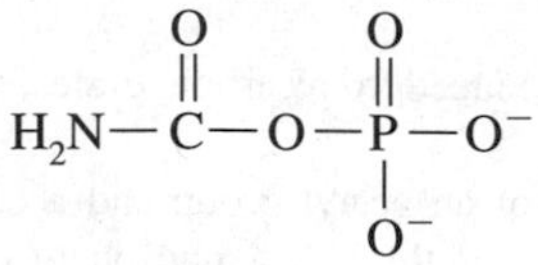

$$H_2N-\overset{\overset{\displaystyle O}{\|}}{C}-O-\underset{\underset{\displaystyle O^-}{|}}{\overset{\overset{\displaystyle O}{\|}}{P}}-O^-$$

is incorporated into urea?

38. What portion of the L-aspartic acid molecule

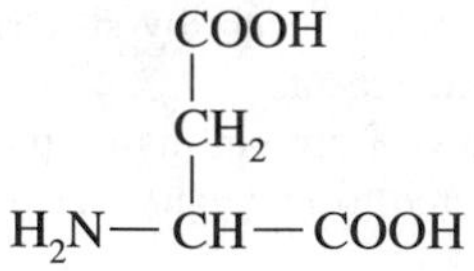

$$\begin{array}{c} COOH \\ | \\ CH_2 \\ | \\ H_2N-CH-COOH \end{array}$$

is incorporated into urea?

Additional Exercises

These exercises are not paired or labeled by topic and provide additional practice on concepts covered in this chapter.

39. How is it possible to accumulate fatty tissue (or adipose tissue) even though very little fat is included in the diet?

40. Give the name and structure of a ketone body that does *not* contain a ketone functional group.

41. Beta oxidation is used by the cell to produce energy, yet no ATPs are formed in this pathway. Briefly explain this seeming contradiction.

42. Based on your knowledge of catabolic pathways, why is the ATP yield from a six-carbon fatty acid greater than the ATP yield from a six-carbon hexose (glucose)?

43. Starting with acetyl-CoA and malonyl-CoA, write the condensed equation for the lipogenesis of myristic acid, $CH_3(CH_2)_{12}COOH$.

44. Starting with L-glutamic acid and

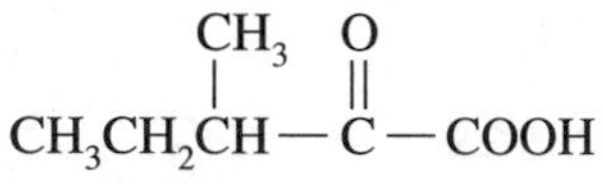

$$CH_3CH_2\overset{\overset{\displaystyle CH_3}{|}}{C}H-\overset{\overset{\displaystyle O}{\|}}{C}-COOH$$

write an equation showing the transamination to isoleucine.

45. You have been invited to a lecture entitled "The Role of Malonyl CoA." Would you expect the lecture to be about anabolism of fats or catabolism of fats or both? Briefly explain.

46. In Step 2 of beta oxidation of a fatty acid, dehydrogenation occurs. Is the fatty acid oxidized or reduced? Briefly explain.

47. The energy content of fats is 9 kcal/g. Calculate the energy content of 1.0 mole of palmitic acid.

Appendix I Mathematical Review

1. Multiplication Multiplication is a process of adding any given number or quantity to itself a certain number of times. Thus, 4 times 2 means 4 added two times, or 2 added together four times, to give the product 8. Various ways of expressing multiplication are

$$ab \qquad a \times b \qquad a \cdot b \qquad a(b) \qquad (a)(b)$$

Each of these expressions means a times b, or a multiplied by b, or b times a.

When $a = 16$ and $b = 24$, we have $16 \times 24 = 384$.

The expression °F = (1.8 × °C) + 32 means that we are to multiply 1.8 times the Celsius degrees and add 32 to the product. When °C equals 50,

$$°\text{F} = (1.8 \times 50) + 32 = 90 + 32 = 122°\text{F}$$

The result of multiplying two or more numbers together is known as the *product.*

2. Division The word *division* has several meanings. As a mathematical expression, it is the process of finding how many times one number or quantity is contained in another. Various ways of expressing division are

$$a \div b \qquad \frac{a}{b} \qquad a/b$$

Each of these expressions means a divided by b.

When $a = 15$ and $b = 3$, $\frac{15}{3} = 5$.

The number above the line is called the *numerator*; the number below the line is the *denominator.* Both the horizontal and the slanted (/) division signs also mean "per." For example, in the expression for density, the mass per unit volume:

$$\text{density} = \text{mass/volume} = \frac{\text{mass}}{\text{volume}} = \text{g/mL}$$

The diagonal line still refers to a division of grams by the number of milliliters occupied by that mass. The result of dividing one number into another is called the *quotient.*

3. Fractions and Decimals A fraction is an expression of division, showing that the numerator is divided by the denominator. A *proper fraction* is one in which the numerator is smaller than the denominator. In an *improper fraction,* the numerator is the larger number. A decimal or a decimal fraction is a proper fraction in which the denominator is some power of 10. The decimal fraction is determined by carrying out the division of the proper fraction. Examples of proper fractions and their decimal fraction equivalents are shown in the accompanying table.

Proper fraction		Decimal fraction		Proper fraction
$\frac{1}{8}$	=	0.125	=	$\frac{125}{1000}$
$\frac{1}{10}$	=	0.1	=	$\frac{1}{10}$
$\frac{3}{4}$	=	0.75	=	$\frac{75}{100}$
$\frac{1}{100}$	=	0.01	=	$\frac{1}{100}$
$\frac{1}{4}$	=	0.25	=	$\frac{25}{100}$

4. Addition of Numbers with Decimals To add numbers with decimals, we use the same procedure as that used when adding whole numbers, but we always line up the decimal points in the same column. For example, add 8.21 + 143.1 + 0.325:

$$\begin{array}{r} 8.21 \\ +143.1 \\ +\ \ 0.325 \\ \hline 151.635 \end{array}$$

When adding numbers that express units of measurement, we must be certain that the numbers added together all have the same units. For example, what is the total length of three pieces of glass tubing: 10.0 cm, 125 mm, and 8.4 cm? If we simply add the numbers, we obtain a value of 143.4, but we are not certain what the unit of measurement is. To add these lengths correctly, first change 125 mm to 12.5 cm. Now all the lengths are expressed in the same units and can be added:

$$\begin{array}{r} 10.0 \text{ cm} \\ 12.5 \text{ cm} \\ \underline{8.4 \text{ cm}} \\ 30.9 \text{ cm} \end{array}$$

5. Subtraction of Numbers with Decimals To subtract numbers containing decimals, we use the same procedure as for subtracting whole numbers, but we always line up the decimal points in the same column. For example, subtract 20.60 from 182.49:

$$\begin{array}{r} 182.49 \\ \underline{-\ \ 20.60} \\ 161.89 \end{array}$$

6. Multiplication of Numbers with Decimals To multiply two or more numbers together that contain decimals, we first multiply as if they were whole numbers. Then, to locate the decimal point in the product, we add together the number of digits to the right of the decimal in all the numbers multiplied together. The product should have this same number of digits to the right of the decimal point.

Multiply 2.05×2.05 (total of four digits to the right of the decimal):

$$\begin{array}{r} 2.05 \\ \underline{\times 2.05} \\ 1025 \\ \underline{4100} \\ 4.2025 \end{array} \quad \text{(four digits to the right of the decimal)}$$

Remember: When one of the numbers is a measurement, the answer must be adjusted to the correct number of significant figures.

Here are more examples:

$14.25 \times 6.01 \times 0.75 = 64.231875$ (six digits to the right of the decimal)
$39.26 \times 60 = 2355.60$ (two digits to the right of the decimal)

7. Division of Numbers with Decimals To divide numbers containing decimals, we first relocate the decimal points of the numerator and denominator by moving them to the right as many places as needed to make the denominator a whole number. (Move the decimal of both the numerator and the denominator the same amount and in the same direction.) For example,

$$\frac{136.94}{4.1} = \frac{1369.4}{41}$$

The decimal point adjustment in this example is equivalent to multiplying both numerator and denominator by 10. Now we carry out the division normally, locating the decimal point immediately above its position in the dividend:

$$\begin{array}{r} 33.4 \\ 41\overline{)1369.4} \\ \underline{123} \\ 139 \\ \underline{123} \\ 16\,4 \\ \underline{16\,4} \end{array} \qquad \frac{0.441}{26.25} = \frac{44.1}{2625} = \begin{array}{r} 0.0168 \\ 2625\overline{)44.1000} \\ \underline{26\,25} \\ 17\,850 \\ \underline{15\,750} \\ 2\,1000 \\ \underline{2\,1000} \end{array}$$

The foregoing examples are merely guides to the principles used in performing the various mathematical operations illustrated. There are, no doubt, shortcuts and other methods, and the student will discover these with experience. Every student of chemistry should learn to use a calculator for solving mathematical problems. The use of a calculator will save many hours of doing tedious longhand calculations. After solving a problem, the student should check for errors and evaluate the answer to see if it is logical and consistent with the data given.

8. Algebraic Equations Many mathematical problems that are encountered in chemistry fall into the following algebraic forms. Solutions to these problems are simplified by first isolating the desired term on one side of the equation. This rearrangement is accomplished by treating both sides of the equation in an identical manner until the desired term is isolated.

(a) $a = \frac{b}{c}$

To solve for *b*, multiply both sides of the equation by *c*:

$$a \times c = \frac{b}{\cancel{c}} \times \cancel{c}$$

$$b = a \times c$$

To solve for *c*, multiply both sides of the equation by $\frac{c}{a}$:

$$\cancel{a} \times \frac{c}{\cancel{a}} = \frac{b}{\cancel{c}} \times \frac{\cancel{c}}{a}$$

$$c = \frac{b}{a}$$

(b) $\frac{a}{b} = \frac{c}{d}$

To solve for *a*, multiply both sides of the equation by *b*:

$$\frac{a}{\cancel{b}} \times \cancel{b} = \frac{c}{d} \times b$$

$$a = \frac{c \times b}{d}$$

To solve for *b*, multiply both sides of the equation by $\frac{b \times d}{c}$:

$$\frac{a}{\cancel{b}} \times \frac{\cancel{b} \times d}{c} = \frac{\cancel{c}}{\cancel{d}} \times \frac{b \times \cancel{d}}{\cancel{c}}$$

$$b = \frac{a \times d}{c}$$

(c) $a \times b = c \times d$

To solve for *a*, divide both sides of the equation by *b*:

$$\frac{a \times \cancel{b}}{\cancel{b}} = \frac{c \times d}{b}$$

$$a = \frac{c \times d}{b}$$

(d) $\frac{b - c}{a} = d$

To solve for *b,* first multiply both sides of the equation by *a:*

$$\frac{\not{a}(b - c)}{\not{a}} = d \times a$$

$$b - c = d \times a$$

Then add *c* to both sides of the equation:

$$b - \not{c} + \not{c} = d \times a + c$$

$$b = (d \times a) + c$$

When $a = 1.8$, $c = 32$, and $d = 35$,

$$b = (35 \times 1.8) + 32 = 63 + 32 = 95$$

9. Exponents, Powers of 10, Expression of Large and Small Numbers In scientific measurements and calculations, we often encounter very large and very small numbers—for example, 0.00000384 and 602,000,000,000,000,000,000,000. These numbers are troublesome to write and awkward to work with, especially in calculations. A convenient method of expressing these large and small numbers in a simplified form is by means of exponents or powers of 10. This method of expressing numbers is known as **scientific or exponential notation.**

scientific or exponential notation

exponent

An **exponent** is a number written as a superscript following another number. Exponents are often called *powers* of numbers. The term *power* indicates how many times the number is used as a factor. In the number 10^2, 2 is the exponent, and the number means 10 squared, or 10 to the second power, or $10 \times 10 = 100$. Three other examples are

$$3^2 = 3 \times 3 = 9$$
$$3^4 = 3 \times 3 \times 3 \times 3 = 81$$
$$10^3 = 10 \times 10 \times 10 = 1000$$

For ease of handling, large and small numbers are expressed in powers of 10. Powers of 10 are used because multiplying or dividing by 10 coincides with moving the decimal point in a number by one place. Thus, a number multiplied by 10^1 would move the decimal point one place to the right; 10^2, two places to the right; 10^{-2}, two places to the left. To express a number in powers of 10, we move the decimal point in the original number to a new position, placing it so that the number is a value between 1 and 10. This new decimal number is multiplied by 10 raised to the proper power. For example, to write the number 42,389 in exponential form, the decimal point is placed between the 4 and the 2 (4.2389), and the number is multiplied by 10^4; thus, the number is 4.2389×10^4:

$$42{,}389 = 4.2389 \times 10^4$$
4 3 2 1

The exponent of 10 (4) tells us the number of places that the decimal point has been moved from its original position. If the decimal point is moved to the left, the exponent is a positive number; if it is moved to the right, the exponent is a negative number. To express the number 0.00248 in exponential notation (as a power of 10), the decimal point is moved three places to the right; the exponent of 10 is -3, and the number is 2.48×10^{-3}.

$$0.00248 = 2.48 \times 10^{-3}$$
1 2 3

Study the following examples.

$$
\begin{aligned}
1237 &= 1.237 \times 10^3 \\
988 &= 9.88 \times 10^2 \\
147.2 &= 1.472 \times 10^2 \\
2{,}200{,}000 &= 2.2 \times 10^6 \\
0.0123 &= 1.23 \times 10^{-2} \\
0.00005 &= 5 \times 10^{-5} \\
0.000368 &= 3.68 \times 10^{-4}
\end{aligned}
$$

Exponents in multiplication and division The use of powers of 10 in multiplication and division greatly simplifies locating the decimal point in the answer. In multiplication, first change all numbers to powers of 10, then multiply the numerical portion in the usual manner, and finally add the exponents of 10 algebraically, expressing them as a power of 10 in the product. In multiplication, the exponents (powers of 10) are added algebraically.

$$10^2 \times 10^3 = 10^{(2+3)} = 10^5$$
$$10^2 \times 10^2 \times 10^{-1} = 10^{(2+2-1)} = 10^3$$

Multiply: $40{,}000 \times 4200$

Change to powers of 10: $4 \times 10^4 \times 4.2 \times 10^3$

Rearrange: $4 \times 4.2 \times 10^4 \times 10^3$

$16.8 \times 10^{(4+3)}$

16.8×10^7 or 1.68×10^8 (Answer)

Multiply: 380×0.00020

$3.80 \times 10^2 \times 2.0 \times 10^{-4}$

$3.80 \times 2.0 \times 10^2 \times 10^{-4}$

$7.6 \times 10^{(2-4)}$

7.6×10^{-2} or 0.076 (Answer)

Multiply: $125 \times 284 \times 0.150$

$1.25 \times 10^2 \times 2.84 \times 10^2 \times 1.50 \times 10^{-1}$

$1.25 \times 2.84 \times 1.50 \times 10^2 \times 10^2 \times 10^{-1}$

$5.325 \times 10^{(2+2-1)}$

5.33×10^3 (Answer)

In division, after changing the numbers to powers of 10, move the 10 and its exponent from the denominator to the numerator, changing the sign of the exponent. Carry out the division in the usual manner and evaluate the power of 10. The following is a proof of the equality of moving the power of 10 from the denominator to the numerator:

$$1 \times 10^{-2} = 0.01 = \frac{1}{100} = \frac{1}{10^2} = 1 \times 10^{-2}$$

In division, change the sign(s) of the exponent(s) of 10 in the denominator and move the 10 and its exponent(s) to the numerator. Then add all the exponents of 10 together. For example,

$$\frac{10^5}{10^3} = 10^5 \times 10^{-3} = 10^{(5-3)} = 10^2$$

$$\frac{10^3 \times 10^4}{10^{-2}} = 10^3 \times 10^4 \times 10^2 = 10^{(3+4+2)} = 10^9$$

10. Significant Figures in Calculations The result of a calculation based on experimental measurements cannot be more precise than the measurement that has the greatest uncertainty. (See Section 2.5 for additional discussion.)

Addition and Subtraction The result of an addition or subtraction should contain no more digits to the right of the decimal point than are contained in the quantity that has the least number of digits to the right of the decimal point.

Perform the operation indicated and then round off the number to the proper number of significant figures:

$$\begin{array}{rl} 142.8 & \text{g} \\ 18.843 & \text{g} \\ \underline{36.42} & \text{g} \\ 198.063 & \text{g} \\ 198.1 & \text{g (Answer)} \end{array} \qquad \begin{array}{rl} 93.45 & \text{mL} \\ \underline{-18.0} & \text{mL} \\ 75.45 & \text{mL} \\ 75.5 & \text{mL (Answer)} \end{array}$$

Multiplication and Division In calculations involving multiplication or division, the answer should contain the same number of significant figures as the measurement that has the least number of significant figures. In multiplication or division the position of the decimal point has nothing to do with the number of significant figures in the answer. Study the following examples:

	Round off to
$2.05 \times 2.05 = 4.2025$	4.20
$18.48 \times 5.2 = 96.096$	96
$0.0126 \times 0.020 = 0.000252$ or $1.26 \times 10^{-2} \times 2.0 \times 10^{-2} = 2.520 \times 10^{-4}$	2.5×10^{-4}
$\dfrac{1369.4}{41} = 33.4$	33
$\dfrac{2268}{4.20} = 540.$	540.

11. Dimensional Analysis Many problems of chemistry can be readily solved by dimensional analysis using the factor-label or conversion-factor method. Dimensional analysis involves the use of proper units of dimensions for all factors that are multiplied, divided, added, or subtracted in setting up and solving a problem. Dimensions are physical quantities such as length, mass, and time, which are expressed in such units as centimeters, grams, and seconds, respectively. In solving a problem, we treat these units mathematically just as though they were numbers, which gives us an answer that contains the correct dimensional units.

A measurement or quantity given in one kind of unit can be converted to any other kind of unit having the same dimension. To convert from one kind of unit to another, the original quantity or measurement is multiplied or divided by a conversion factor. The key to success lies in choosing the correct conversion factor. This general method of calculation is illustrated in the following examples.

Suppose we want to change 24 ft to inches. We need to multiply 24 ft by a conversion factor containing feet and inches. Two such conversion factors can be written relating inches to feet:

$$\frac{12 \text{ in.}}{1 \text{ ft}} \quad \text{or} \quad \frac{1 \text{ ft}}{12 \text{ in.}}$$

We choose the factor that will mathematically cancel feet and leave the answer in inches. Note that the units are treated in the same way we treat numbers, multiplying or dividing as required. Two possibilities then arise to change 24 ft to inches:

$$24\ \cancel{\text{ft}} \times \frac{12\text{ in.}}{1\ \cancel{\text{ft}}} \quad \text{or} \quad 24\text{ ft} \times \frac{1\text{ ft}}{12\text{ in.}}$$

In the first case (the correct method), feet in the numerator and the denominator cancel, giving us an answer of 288 in. In the second case, the units of the answer are $\text{ft}^2/\text{in.}$, the answer being 2.0 $\text{ft}^2/\text{in.}$ In the first case, the answer is reasonable since it is expressed in units having the proper dimensions. That is, the dimension of length expressed in feet has been converted to length in inches according to the mathematical expression

$$\cancel{\text{ft}} \times \frac{\text{in.}}{\cancel{\text{ft}}} = \text{in.}$$

In the second case, the answer is not reasonable since the units ($\text{ft}^2/\text{in.}$) do not correspond to units of length. The answer is therefore incorrect. The units are the guiding factor for the proper conversion.

The reason we can multiply 24 ft times 12 in./ft and not change the value of the measurement is that the conversion factor is derived from two equivalent quantities. Therefore, the conversion factor 12 in./ft is equal to unity. When you multiply any factor by 1, it does not change the value:

$$12\text{ in.} = 1\text{ ft} \quad \text{and} \quad \frac{12\text{ in.}}{1\text{ ft}} = 1$$

Convert 16 kg to milligrams. In this problem it is best to proceed in this fashion:

$$\text{kg} \longrightarrow \text{g} \longrightarrow \text{mg}$$

The possible conversion factors are

$$\frac{1000\text{ g}}{1\text{ kg}} \quad \text{or} \quad \frac{1\text{ kg}}{1000\text{ g}} \qquad \frac{1000\text{ mg}}{1\text{ g}} \quad \text{or} \quad \frac{1\text{ g}}{1000\text{ mg}}$$

We use the conversion factor that leaves the proper unit at each step for the next conversion. The calculation is

$$16\ \cancel{\text{kg}} \times \frac{1000\ \cancel{\text{g}}}{1\ \cancel{\text{kg}}} \times \frac{1000\text{ mg}}{1\ \cancel{\text{g}}} = 1.6 \times 10^7\text{ mg}$$

Many problems can be solved by a sequence of steps involving unit conversion factors. This sound, basic approach to problem solving, together with neat and orderly setting up of data, will lead to correct answers having the right units, fewer errors, and considerable saving of time.

12. Graphical Representation of Data A graph is often the most convenient way to present or display a set of data. Various kinds of graphs have been devised, but the most common type uses a set of horizontal and vertical coordinates to show the relationship of two variables. It is called an x–y graph because the data of one variable are represented on the horizontal or x axis (abscissa) and the data of the other variable are represented on the vertical or y axis (ordinate). See Figure I.1.

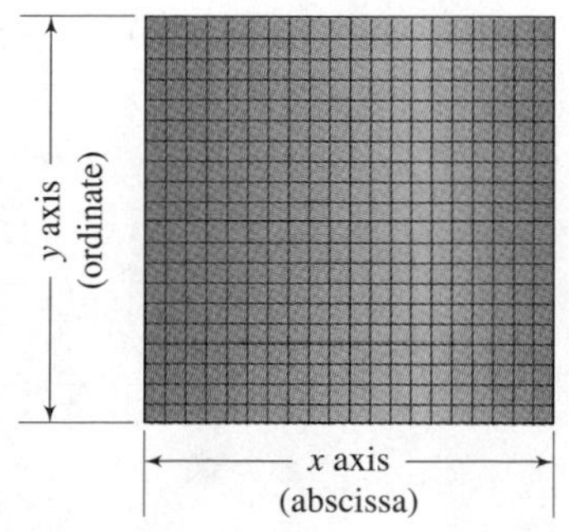

FIGURE I.1

As a specific example of a simple graph, let us graph the relationship between Celsius and Fahrenheit temperature scales. Assume that initially we have only the information in the following table:

°C	°F
0	32
50	122
100	212

On a set of horizontal and vertical coordinates (graph paper), scale off at least 100 Celsius degrees on the x axis and at least 212 Fahrenheit degrees on the y axis. Locate and mark the three points corresponding to the three temperatures given and draw a line connecting these points (see Figure I.2).

Here is how a point is located on the graph: Using the 50°C–122°F data, trace a vertical line up from 50°C on the x axis and a horizontal line across from 122°F on the y axis and mark the point where the two lines intersect. This process is called *plotting*. The other two points are plotted on the graph in the same way. [*Note*: The number of degrees per scale division was chosen to give a graph of convenient size. In this case, there are 5 Fahrenheit degrees per scale division and 2 Celsius degrees per scale division.]

The graph is Figure I.2 shows that the relationship between Celsius and Fahrenheit temperature is that of a straight line. The Fahrenheit temperature corresponding to any given Celsius temperature between 0 and 100° can be determined from the graph. For example, to find the Fahrenheit temperature corresponding to 40°C, trace a perpendicular line from 40°C on the x axis to the line plotted on the graph. Now trace a horizontal line from this point on the plotted line to the y axis and read the corresponding Fahrenheit temperature (104°F). See the dashed lines in Figure I.2. In turn, the Celsius temperature corresponding to any Fahrenheit temperature between 32 and 212° can be determined from the graph by tracing a horizontal line from the Fahrenheit temperature to the plotted line and reading the corresponding temperature on the Celsius scale directly below the point of intersection.

The mathematical relationship of Fahrenheit and Celsius temperatures is expressed by the equation $°F = 1.8 \times °C + 32$. Figure I.2 is a graph of this equation. Because the graph is a straight line, it can be extended indefinitely at either end. Any desired Celsius temperature can be plotted against the corresponding Fahrenheit temperature by extending the scales along both axes as necessary.

FIGURE I.2

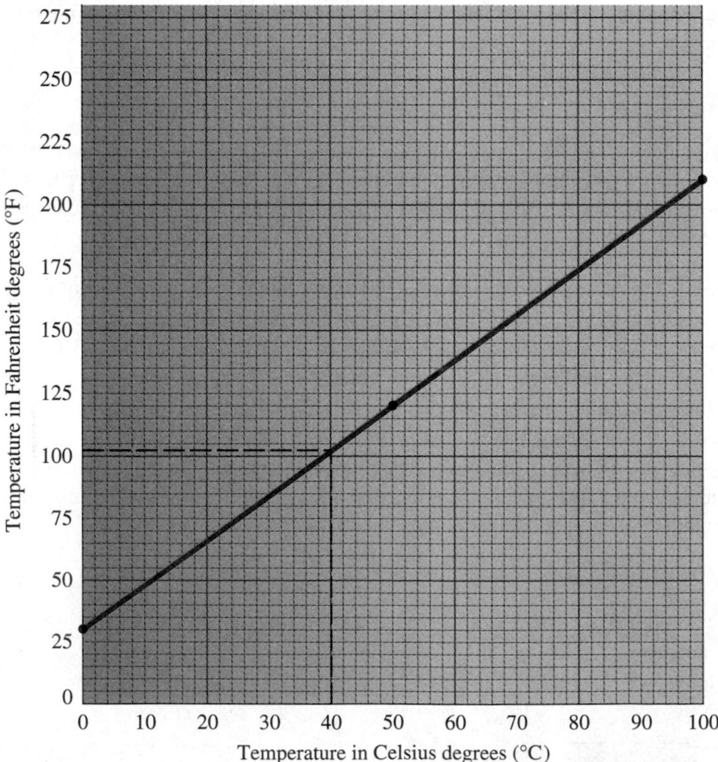

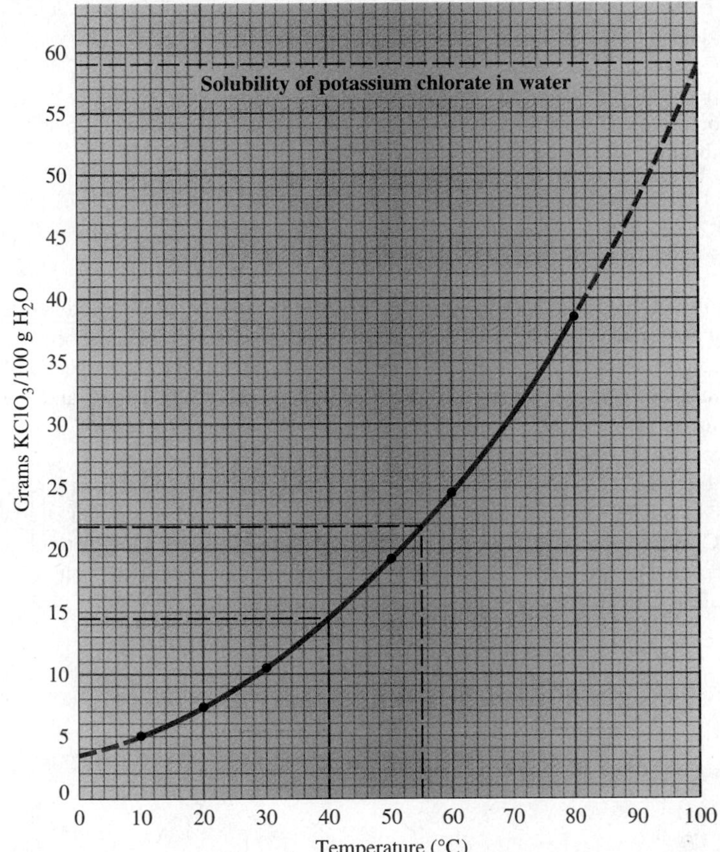

Temperature (°C)	Solubility (g $KClO_3$/100 g water)
10	5.0
20	7.4
30	10.5
50	19.3
60	24.5
80	38.5

◀ **FIGURE I.3**

Figure I.3 is a graph showing the solubility of potassium chlorate in water at various temperatures. The solubility curve on this graph was plotted from the data in the table next to the graph.

In contrast to the Celsius–Fahrenheit temperature relationship, there is no simple mathematical equation that describes the exact relationship between temperature and the solubility of potassium chlorate. The graph in Figure I.3 was constructed from experimentally determined solubilities at the six temperatures shown. These experimentally determined solubilities are all located on the smooth curve traced by the unbroken-line portion of the graph. We are therefore confident that the unbroken line represents a very good approximation of the solubility data for potassium chlorate over the temperature range from 10 to 80°C. All points on the plotted curve represent the composition of saturated solutions. Any point below the curve represents an unsaturated solution.

The dashed-line portions of the curve are *extrapolations*; that is, they extend the curve above and below the temperature range actually covered by the plotted solubility data. Curves such as this one are often extrapolated a short distance beyond the range of the known data, although the extrapolated portions may not be highly accurate. Extrapolation is justified only in the absence of more reliable information.

The graph in Figure I.3 can be used with confidence to obtain the solubility of $KClO_3$

at any temperature between 10 and 80°C, but the solubilities between 0 and 10°C and between 80 and 100°C are less reliable. For example, what is the solubility of $KClO_3$ at 55°C, at 40°C, and at 100°C?

First draw a perpendicular line from each temperature to the plotted solubility curve. Now trace a horizontal line to the solubility axis from each point on the curve and read the corresponding solubilities. The values that we read from the graph are

55°C	22.0 g $KClO_3$/100 g water
40°C	14.2 g $KClO_3$/100 g water
100°C	59 g $KClO_3$/100 g water

Of these solubilities, the one at 55°C is probably the most reliable because experimental points are plotted at 50°C and at 60°C. The 40°C solubility value is a bit less reliable because the nearest plotted points are at 30°C and 50°C. The 100°C solubility is the least reliable of the three values because it was taken from the extrapolated part of the curve, and the nearest plotted point is 80°C. Actual handbook solubility values are 14.0 and 57.0 g of $KClO_3$/100 g water at 40°C and 100°C, respectively.

The graph in Figure I.3 can also be used to determine whether a solution is saturated or unsaturated. For example, a solution contains 15 g of $KClO_3$/100 g of water and is at a temperature of 55°C. Is the solution saturated or unsaturated? *Answer*: The solution is unsaturated because the point corresponding to 15 g and 55°C on the graph is below the solubility curve; all points below the curve represent unsaturated solutions.

Appendix II Vapor Pressure of Water at Various Temperatures

Temperature (°C)	Vapor pressure (torr)	Temperature (°C)	Vapor pressure (torr)
0	4.6	26	25.2
5	6.5	27	26.7
10	9.2	28	28.3
15	12.8	29	30.0
16	13.6	30	31.8
17	14.5	40	55.3
18	15.5	50	92.5
19	16.5	60	149.4
20	17.5	70	233.7
21	18.6	80	355.1
22	19.8	90	525.8
23	21.2	100	760.0
24	22.4	110	1074.6
25	23.8		

Appendix III Units of Measurement

Physical Constants

Constant	Symbol	Value
Atomic mass unit	amu	1.6606×10^{-27} kg
Avogadro's number	N	6.022×10^{23}/mol
Gas constant	R (at STP)	0.08205 L atm/K mol
Mass of an electron	m_e	9.11×10^{-31} kg 5.486×10^{-4} amu
Mass of a neutron	m_n	1.675×10^{-27} kg 1.00866 amu
Mass of a proton	m_p	1.673×10^{-27} kg 1.00728 amu
Speed of light	c	2.997925×10^{8} m/s

SI Units and Conversion Factors

Length

SI unit: meter (m)

1 meter	=	1.0936 yards
1 centimeter	=	0.3937 inch
1 inch	=	2.54 centimeters (exactly)
1 kilometer	=	0.62137 mile
1 mile	=	5280 feet
	=	1.609 kilometers
1 angstrom	=	10^{-10} meter

Mass

SI unit: kilogram (kg)

1 kilogram	=	1000 grams
	=	2.20 pounds
1 pound	=	453.59 grams
	=	0.45359 kilogram
	=	16 ounces
1 ton	=	2000 pounds
	=	907.185 kilograms
1 ounce	=	28.3 g
1 atomic mass unit	=	1.6606×10^{-27} kilogram

Volume

SI unit: cubic meter (m^3)

1 liter	=	10^{-3} m^3
	=	1 dm^3
	=	1.0567 quarts
1 gallon	=	4 quarts
	=	8 pints
	=	3.785 liters
1 quart	=	32 fluid ounces
	=	0.946 liter
1 fluid ounce	=	29.6 milliliters

Temperature

SI unit: kelvin (K)

$0 \text{ K} = -273.15°\text{C}$

$= -459.67°\text{F}$

$\text{K} = °\text{C} + 273.15$

$$°\text{C} = \frac{°\text{F} - 32}{1.8}$$

$°\text{F} = 1.8(°\text{C}) + 32$

$°\text{F} = 1.8(°\text{C} + 40) - 40$

Energy

SI unit: joule (J)

1 joule	=	1 kg m^2/s^2
	=	0.23901 calorie
1 calorie	=	4.184 joules

Pressure

SI unit: pascal (Pa)

1 pascal	=	1 kg/m s^2
1 atmosphere	=	101.325 kilopascals
	=	760 torr (mm Hg)
	=	14.70 pounds per square inch (psi)

Appendix IV Solubility Table

	F^-	Cl^-	Br^-	I^-	O^{2-}	S^{2-}	OH^-	NO_3^-	CO_3^{2-}	SO_4^{2-}	$C_2H_3O_2^-$
H^+	*S*	*S*	*S*	*S*	*S*	*s*	*S*	*S*	*s*	*S*	*S*
Na^+	*S*	*S*	*S*	*S*	*S*	*S*	*S*	*S*	*S*	*S*	*S*
K^+	*S*	*S*	*S*	*S*	*S*	*S*	*S*	*S*	*S*	*S*	*S*
NH_4^+	*S*	*S*	*S*	*S*	—	*S*	*S*	*S*	*S*	*S*	*S*
Ag^+	*S*	*I*	*I*	*I*	*I*	*I*	—	*S*	*I*	*I*	*I*
Mg^{2+}	*I*	*S*	*S*	*S*	*I*	*d*	*I*	*S*	*I*	*S*	*S*
Ca^{2+}	*I*	*S*	*S*	*S*	*I*	*d*	*I*	*S*	*I*	*I*	*S*
Ba^{2+}	*I*	*S*	*S*	*S*	*s*	*d*	*s*	*S*	*I*	*I*	*S*
Fe^{2+}	*s*	*S*	*S*	*S*	*I*	*I*	*I*	*S*	*s*	*S*	*S*
Fe^{3+}	*I*	*S*	*S*	—	*I*	*I*	*I*	*S*	*I*	*S*	*I*
Co^{2+}	*S*	*S*	*S*	*S*	*I*	*I*	*I*	*S*	*I*	*S*	*S*
Ni^{2+}	*s*	*S*	*S*	*S*	*I*	*I*	*I*	*S*	*I*	*S*	*S*
Cu^{2+}	*s*	*S*	*S*	—	*I*	*I*	*I*	*S*	*I*	*S*	*S*
Zn^{2+}	*s*	*S*	*S*	*S*	*I*	*I*	*I*	*S*	*I*	*S*	*S*
Hg^{2+}	*d*	*S*	*I*	*I*	*I*	*I*	*I*	*S*	*I*	*d*	*S*
Cd^{2+}	*s*	*S*	*S*	*S*	*I*	*I*	*I*	*S*	*I*	*S*	*S*
Sn^{2+}	*S*	*S*	*S*	*s*	*I*	*I*	*I*	*S*	*I*	*S*	*S*
Pb^{2+}	*I*	*I*	*I*	*I*	*I*	*I*	*I*	*S*	*I*	*I*	*S*
Mn^{2+}	*s*	*S*	*S*	*S*	*I*	*I*	*I*	*S*	*I*	*S*	*S*
Al^{3+}	*I*	*S*	*S*	*S*	*I*	*d*	*I*	*S*	—	*S*	*S*

Key: *S* = soluble in water
s = slightly soluble in water
I = insoluble in water (less than 1 g/100 g H_2O)
d = decomposes in water

Appendix V Answers to Even-Numbered Questions and Exercises

Chapter 2

2. 7.6 cm
4. The most dense (mercury) at the bottom and the least dense (glycerin) at the top. In the cylinder the solid magnesium would sink in the glycerin and float on the liquid mercury.
6. 0.789 g/mL < ice < 0.91 g/mL
8. $D = m/V$ specific gravity $= \dfrac{d_{\text{substance}}}{d_{\text{water}}}$
10. A lean person weighs more in water than an obese person and has a higher density.
12. **Rule 1.** When the first digit after those you want to retain is 4 or less, that digit and all others to its right are dropped. The last digit retained is not changed.
Rule 2. When the first digit after those you want to retain is 5 or greater, that digit and all others to the right of it are dropped and the last digit is increased by one.
14. The correct statements are: a, c, d, e, g, h, i, j, l, n, p, q.
16. (a) mg (d) nm
(b) kg (e) Å
(c) m (f) μL
18. (a) not significant (d) significant
(b) significant (e) significant
(c) not significant (f) significant
20. (a) 40.0 (3) (c) 129,042 (6)
(b) 0.081 (2) (d) 4.090×10^{-3} (4)
22. (a) 8.87 (c) 130. (1.30×10^2)
(b) 21.3 (d) 2.00×10^6
24. (a) 4.56×10^{-2} (c) 4.030×10^1
(b) 4.0822×10^3 (d) 1.2×10^7
26. (a) 28.1 (d) 2.010×10^3
(b) 58.5 (e) 2.49×10^{-4}
(c) 4.0×10^1 (f) 1.79×10^3
28. (a) $\frac{1}{4}$ (b) $\frac{5}{8}$ (c) $1\frac{2}{3}$ or $\frac{5}{3}$ (d) $\frac{8}{9}$
30. (a) 1.0×10^2 (b) 4.6 mL (c) 22
32. (a) 4.5×10^8 Å (e) 6.5×10^5 mg
(b) 1.2×10^{-6} cm (f) 5.5×10^3 g
(c) 8.0×10^6 mm (g) 468 mL
(d) 0.164 g (h) 9.0×10^{-3} mL
34. (a) 117 ft (d) 4.3×10^4 g
(b) 10.3 mi (e) 75.7 L
(c) 7.4×10^4 mm^3 (f) 1.3×10^3 m^3
36. 50. ft/s
38. 0.102 km/s
40. 5.0×10^2 s
42. 3×10^4 mg
44. 3.0×10^3 hummingbirds to equal the mass of a condor
46. $2800
48. 57 L
50. 160 L
52. 4×10^5 m^2
54. 5 gal
56. 113°F Summer!
58. (a) 90.°F (c) 546 K
(b) −22.6°C (d) −300 K
60. −11.4°C = 11.4°F
62. 3.12 g/mL
64. 1.28 g/mL
66. 3.40×10^2 g
68. 7.0 lb
70. Yes, 116.5 L additional solution
72. −15°C > 4.5°F
74. *B* is 14 mL larger than *A*.
76. 76.9 g
78. 3.57×10^3 g
80. The container must hold at least 50 mL.
82. The gold bar is not pure gold.
84. 0.842 g/mL

Chapter 3

2. (a) Attractive forces among the ultimate particles of a solid (atoms, ions, or molecules) are strong enough to hold these particles in a fixed position within the solid and thus maintain the solid in a definite shape. Attractive forces among the ultimate particles of a liquid (usually molecules) are sufficiently strong to hold them together (preventing the liquid from rapidly becoming a gas) but are not strong enough to hold the particles in fixed positions (as in a solid).
(b) The ultimate particles in a liquid are quite closely packed (essentially in contact with each other) and thus the volume of the liquid is fixed at a given temperature. But, the ultimate particles in a gas are relatively far apart and essentially independent of each other. Consequently, the gas does not have a definite volume.
(c) In a gas the particles are relatively far apart and are easily compressed, but in a solid the particles are closely packed together and are virtually incompressible.
4. mercury and water
6. Three phases are present.
8. A system containing only one substance is not necessarily homogeneous.
10. 30 Si g/1 g H. There are more Si atoms than H atoms.
12. P H Mg N Ag
Al K Na Ni Pu
14. sodium tin mercury
potassium silver lead
iron tungsten
antimony gold
16. In an element all atoms are alike, while a compound contains two or more elements which are chemically combined. Compounds may be decomposed into simpler substances while elements cannot.
18. 7 metals 1 metalloid 2 nonmetals
20. aurum
22. A *compound* is two or more elements chemically combined in a definite proportion by mass. Its properties differ from those of its components. A *mixture* is the physical combining of two or more substances (not necessarily elements). The composition may vary, the substances retain their properties and may be separated by physical means.
24. characteristic physical and chemical properties
26. Cations are positive, anions are negative.
28. Homogeneous are one phase, heterogeneous have two or more phases.

30. (a) H_2 (c) HCl (e) NO
32. Sponge iron rusts easily. During this process hydrogen is generated and collected. The sponge iron can be regenerated from the rust formed and the process repeated indefinitely.
34. Charcoal—destructive distillation of wood.
Bone black—destructive distillation of bones or waste.
Carbon black—residue from burning natural gas.
36. Correct statements are: c, f, g, j, m, o, q, t, u, w, x, y.
38. (a) magnesium, bromine
(b) carbon, chlorine
(c) hydrogen, nitrogen, oxygen
(d) barium, sulfur, oxygen
(e) aluminum, phosphorus, oxygen
40. (a) $AlBr_3$ (c) $PbCrO_4$
(b) CaF_2 (d) C_6H_6
42. (a) 1 atom Al, 3 atoms Br
(b) 1 atom Ni, 2 atoms N, 6 atoms O
(c) 12 atoms C, 22 atoms H, 11 atoms O
44. (a) 2 atoms (d) 5 atoms
(b) 2 atoms (e) 17 atoms
(c) 9 atoms
46. (a) 2 atoms H (d) 4 atoms H
(b) 6 atoms H (e) 8 atoms H
(c) 12 atoms H
48. (a) element (c) element
(b) compound (d) mixture
50. (a) mixture (c) mixture
(b) element (d) compound
52. (a) HO (b) C_2H_6O (c) $Na_2Cr_2O_7$
54. No. The only common liquid elements (at least at room temperature) are mercury and bromine.
56. 75% solids
58. 420 atoms
60. 40 atoms H
62. (a) magnesium, manganese, molybdenum, mendelevium, mercury
(b) carbon, phosphorus, sulfur, selenium, iodine, astatine, boron
(c) sodium, potassium, iron, silver, tin, antimony
64. (a) As temperature decreases, density increases.
(b) 1.28 g/L, 1.17 g/L, 1.08 g/L

Chapter 4

2. solid
4. Water disappears. Gas appears above each electrode and as bubbles in solution.
6. A new substance is always formed during a chemical change, but never formed during physical changes.
8. The hot pack contains a solution of sodium acetate or sodium thiosulfate. A small crystal is added by squeezing a corner of the bag or bending a small metal activator. The solution crystallizes and heat is released to the surroundings. To reuse it, the pack is heated in boiling water until the crystals dissolve. Then, it is slowly cooled and stored until needed.
10. Potential energy is the energy of position. Kinetic energy is the energy matter possesses due to its motion.
12. The correct statements are: a, f, h, i.
14. (a) chemical (c) physical (e) chemical
(b) physical (d) chemical (f) physical
16. The copper wire, like the platinum wire, changed to a glowing red color when heated. Upon cooling, a new substance, black copper(II) oxide, had appeared.
18. Reactant: water
Products: hydrogen, oxygen
20. the transformation of kinetic energy to thermal energy
22. (a) + (d) +
(b) − (e) −
(c) −
24. 2.2×10^3 J
26. 5.03×10^{-2} J/g °C
28. 5 °C
30. 29.1 °C
32. 45.7 g coal
34. 656 °C
36. 16.7 °C
38. 6:06 and 54 s
40. at the same rate
42. The mercury and sulfur react to form a compound since the properties of the product are different from the properties of either reactant.

Chapter 5

2. A neutron is about 1840 times heavier than an electron.
4. An atom is electrically neutral. An ion has a charge.
6. a Wintergreen Lifesaver
8. SIRA technique is stable isotope ratio analysis.
10. The correct statements are: a, b, c, f, g.
12. The correct statements are: b, d.
14. (a) The nucleus of the atom contains most of the mass.
(b) The nucleus of the atom is positively charged.
(c) The atom is mostly empty space.
16. The nucleus of an atom contains nearly all of its mass.
18. Electrons: Dalton—Electrons are not part of his model.
Thomson—Electrons are scattered throughout the positive mass of matter in the atom.
Rutherford—Electrons are located out in space away from the central positive mass.
Positive matter: Dalton—No positive matter in his model.
Thomson—Positive matter is distributed throughout the atom.
Rutherford—Positive matter is concentrated in a small central nucleus.
20. The isotope of C with a mass of 12 is an exact number.
22. Three isotopes of hydrogen have the same number of protons and electrons but differ in the number of neutrons.
24. (a) 80 protons; +80 (b) Hg
26. 40
28. (a) 27 protons, 32 neutrons
(b) 15 protons, 16 neutrons
(c) 74 protons, 110 neutrons
(d) 92 protons, 143 neutrons
30. 24.31 amu
32. 6.716 amu
34. 1.0×10^{15}:1
36. (a) isotopes
(b) adjacent to each other on the periodic table
38. (a) The (+) particles are much lighter mass particles.
(b) negative

40. the number of protons and electrons
42. $^{60}Q = 50\%$ $^{63}Q = 50\%$
44. 6.03×10^{24} atoms
46.

Atomic number	Mass number	Symbol	Protons	Neutrons
(a) 8	16	O	8	8
(b) 28	58	Ni	28	30
(c) 80	199	Hg	80	119

Chapter 6

2. Charges on their ions must be equal and opposite in sign.
4. Seaborg is still alive. Elements can only be named for a person after he is dead according to IUPAC rules.
6. The correct statements are: a, b, c, e, f, h, i, k, n, p, q.
8. (a) BaO (d) $BeBr_2$
   (b) H_2S (e) Li_4Si
   (c) $AlCl_3$ (f) Mg_3P_2
10. Cl^- OH^- CO_3^{2-}
    Br^- S^{2-} HCO_3^-
    F^- SO_4^{2-} $C_2H_3O_2^-$
    I^- HSO_4^- ClO_3^-
    CN^- HSO_3^- MnO_4^-
    O^{2-} CrO_4^{2-} $C_2O_4^{2-}$
12. $(NH_4)_2SO_4$ NH_4Cl $(NH_4)_3AsO_4$ $NH_4C_2H_3O_2$ $(NH_4)_2CrO_4$
    $CaSO_4$ $CaCl_2$ $Ca_3(AsO_4)_2$ $Ca(C_2H_3O_2)_2$ $CaCrO_4$
    $Fe_2(SO_4)_3$ $FeCl_3$ $FeAsO_4$ $Fe(C_2H_3O_2)_3$ $Fe_2(CrO_4)_3$
    Ag_2SO_4 $AgCl$ Ag_3AsO_4 $AgC_2H_3O_2$ Ag_2CrO_4
    $CuSO_4$ $CuCl_2$ $Cu_3(AsO_4)_2$ $Cu(C_2H_3O_2)_2$ $CuCrO_4$
14. (a) carbon dioxide (f) dinitrogen tetroxide
    (b) dinitrogen oxide (g) diphosphorus pentoxide
    (c) phosphorus pentachloride (h) oxygen diflouride
    (d) carbon tetrachloride (i) nitrogen trifluoride
    (e) sulfur dioxide (j) carbon disulfide
16. (a) potassium oxide (e) sodium phosphate
    (b) ammonium bromide (f) aluminum oxide
    (c) calcium iodide (g) zinc nitrate
    (d) barium carbonate (h) silver sulfate
18. (a) $SnBr_4$ (d) $Hg(NO_2)_2$
    (b) Cu_2SO_4 (e) TiS_2
    (c) $Fe_2(CO_3)_3$ (f) $Fe(C_2H_3O_2)_2$
20. (a) $HC_2H_3O_2$ (d) H_3BO_3
    (b) HF (e) HNO_2
    (c) HClO (f) H_2S
22. (a) phosphoric acid (e) hypochlorous acid
    (b) carbonic acid (f) nitric acid
    (c) iodic acid (g) hydroiodic acid
    (d) hydrochloric acid (h) perchloric acid
24. (a) Na_2CrO_4 (h) $Co(HCO_3)_2$
    (b) MgH_2 (i) NaClO
    (c) $Ni(C_2H_3O_2)_2$ (j) $As_2(CO_3)_5$
    (d) $Ca(ClO_3)_2$ (k) $Cr_2(SO_3)_3$
    (e) $Pb(NO_3)_2$ (l) $Sb_2(SO_4)_3$
    (f) KH_2PO_4 (m) $Na_2C_2O_4$
    (g) $Mn(OH)_2$ (n) KSCN
26. (a) calcium hydrogen sulfate
    (b) arsenic(III) sulfite
    (c) tin(II) nitrite
    (d) iron(III) bromide
    (e) potassium hydrogen carbonate
    (f) bismuth(III) arsenate
    (g) iron(II) bromate
    (h) ammonium monohydrogen phosphate
    (i) sodium hypochlorite
    (j) potassium permanganate
28. (a) FeS_2 (e) $Mg(OH)_2$
    (b) $NaNO_3$ (f) $Na_2CO_3 \cdot 10\ H_2O$
    (c) $CaCO_3$ (g) C_2H_5OH
    (d) $C_{12}H_{22}O_{11}$
30. *-ide:* Suffix is used to indicate a binary compound except for hydroxides, cyanides, and ammonium compounds.
    -ous: Used in acids to indicate that the polyatomic anion contains the *-ite* suffix; also used for the lower ionic charge of a multivalent metal.
    hypo: Used as a prefix in acids or salts when the polyatomic ion contains less oxygen than that of *-ous* acid or the *-ite* salt.
    per: Used as a prefix in acids or salts when the polyatomic ion contains more oxygen than that of the *-ic* acid or the *-ate* salt.
    -ite: The suffix of a salt derived from an *-ous* acid.
    -ate: The suffix of a salt derived from an *-ic* acid.
    Roman numerals indicate the charge on the metal cation.
32. (a) $50e^-$, 50p (b) $48e^-$, 50p (c) $46e^-$, 50p
34. $Li_3Fe(CN)_6$ $AlFe(CN)_6$ $Zn_3[Fe(CN)_6]_2$
36. ammonium oxide zinc chloride
    ammonium carbonate zinc acetate
    ammonium chloride carbonic acid
    ammonium acetate acetic acid
    zinc oxide hydrochloric acid
    zinc carbonate

Chapter 7

2. A mole of gold has a higher mass than a mole of potassium.
4. A mole of gold atoms contains more electrons than a mole of potassium atoms.
6. 6.022×10^{23}
8. (a) 6.022×10^{23} atoms (d) 16.00 g
   (b) 6.022×10^{23} molecules (e) 32.00 g
   (c) 1.204×10^{24} atoms
10. Choosing 100 g of a compound allows us to simply drop the % sign and use grams for each percent.
12. calculation based on body mass in mg additive/kg mass/day
14. An empirical formula gives the smallest whole number ratio of the atoms present in a compound. The molecular formula represents the actual number of atoms of each element in a molecule of the compound. It may be the same as the empirical formula or may be a multiple of the empirical formula.
16. The correct statements are: a, d, f, g, j, k, m.
18. (a) 40.00 (d) 96.09 (g) 180.2
    (b) 275.8 (e) 146.3 (h) 368.4
    (c) 152.0 (f) 122.1 (i) 244.2
20. (a) 0.625 mol NaOH
    (b) 0.275 mol Br_2
    (c) 7.18×10^{-3} mol $MgCl_2$
    (d) 0.462 mol CH_3OH

(e) 2.03×10^{-2} mol Na_2SO_4
(f) 5.97 mol ZnI_2

22. (a) 0.0417 g H_2SO_4 (c) 0.122 g Ti
(b) 11 g CCl_4 (d) 8.0×10^{-7} g S

24. (a) 1.05×10^{24} molecules Cl_2
(b) 1.6×10^{23} molecules C_2H_6
(c) 1.64×10^{23} molecules CO_2
(d) 3.75×10^{24} molecules CH_4

26. (a) 3.271×10^{-22} g Au
(b) 3.952×10^{-22} g U
(c) 2.828×10^{-23} g NH_3
(d) 1.795×10^{-22} g $C_6H_4(NH_2)_2$

28. (a) 0.886 mol S
(b) 42.8 mol NaCl
(c) 1.05×10^{24} atoms Mg
(d) 9.47 mol Br_2

30. (a) 6.022×10^{23} molecules NH_3
(b) 6.022×10^{23} N atoms
(c) 1.807×10^{24} H atoms
(d) 2.41×10^{24} atoms

32. (a) 6.0×10^{24} atoms O
(b) 5.46×10^{24} atoms O
(c) 5.0×10^{18} atoms O

34. (a) 1.27 g Cl (b) 9.07 g H (c) 23.0 g I

36. (a) 47.97% Zn, 52.02% Cl
(b) 18.17% N, 9.153% H, 31.16% C, 41.51% O
(c) 12.26% Mg, 31.24% P, 56.48% O
(d) 21.21% N, 6.104% H, 24.27% S, 48.45% O
(e) 23.09% Fe, 17.37% N, 59.53% O
(f) 54.39% I, 45.61% Cl

38. (a) 47.55% Cl (c) 83.46% Cl
(b) 34.05% Cl (d) 83.63% Cl

40. 24.2% C 4.04% H 71.72% Cl

42. (a) $KClO_3$ (b) $KHSO_4$ (c) Na_2CrO_4

44. (a) CuCl (d) K_3PO_4
(b) $CuCl_2$ (e) $BaCr_2O_7$
(c) Cr_2S_3 (f) PBr_8Cl_3

46. V_2O_5

48. The empirical formula is CH_2O. The molecular formula is $C_6H_{12}O_6$.

50. 5.88 g Na

52. 5.54×10^{19} m

54. (a) 8×10^{16} drops
(b) 8×10^{6} mi^3

56. 10.3 mol H_2SO_4

58. (a) H_2O
(b) CH_3OH

60. 8.66 g Li

62. There is not sufficient S present.

64. 4.77 g O

66. (a) CCl_4 (c) C_6Cl_6
(b) C_2Cl_6 (d) C_3Cl_8

68. 2.4×10^{22} atoms Cu

70. 8×10^{-15} mol people

72. 32 g Mg

74. carbon

Chapter 8

2. the number of moles of each of the chemical species in the reaction

4. The correct statements are: a, d, e, f, h, i, j, l, n, o, p, q.

6. (a) $H_2 + Br_2 \longrightarrow 2\ HBr$
(b) $4\ Al + 3\ C \xrightarrow{\Delta} Al_4C_3$
(c) $Ba(ClO_3)_2 \xrightarrow{\Delta} BaCl_2 + 3\ O_2$
(d) $CrCl_3 + 3\ AgNO_3 \longrightarrow Cr(NO_3)_3 + 3\ AgCl$
(e) $2\ H_2O_2 \longrightarrow 2\ H_2O + O_2$

8. (a) combination (d) double displacement
(b) combination (e) decomposition
(c) decomposition

10. (a) $2\ MnO_2 + CO \longrightarrow Mn_2O_3 + CO_2$
(b) $Mg_3N_2 + 6\ H_2O \longrightarrow 3\ Mg(OH)_2 + 2\ NH_3$
(c) $4\ C_3H_5(NO_3)_3 \longrightarrow 12\ CO_2 + 10\ H_2O + 6\ N_2 + O_2$
(d) $4\ FeS + 7\ O_2 \longrightarrow 2\ Fe_2O_3 + 4\ SO_2$
(e) $2\ Cu(NO_3)_2 \longrightarrow 2\ CuO + 4\ NO_2 + O_2$
(f) $3\ NO_2 + H_2O \longrightarrow 2\ HNO_3 + NO$
(g) $2\ Al + 3\ H_2SO_4 \longrightarrow Al_2(SO_4)_3 + 3\ H_2$
(h) $4\ HCN + 5\ O_2 \longrightarrow 2\ N_2 + 4\ CO_2 + 2\ H_2O$
(i) $2\ B_5H_9 + 12\ O_2 \longrightarrow 5\ B_2O_3 + 9\ H_2O$

12. (a) $2\ H_2O \longrightarrow 2\ H_2 + O_2$
(b) $HC_2H_3O_2 + KOH \longrightarrow KC_2H_3O_2 + H_2O$
(c) $2\ P + 3\ I_2 \longrightarrow 2\ PI_3$
(d) $2\ Al + 3\ CuSO_4 \longrightarrow 3\ Cu + Al_2(SO_4)_3$
(e) $(NH_4)_2SO_4 + BaCl_2 \longrightarrow 2\ NH_4Cl + BaSO_4$
(f) $SF_4 + 2\ H_2O \longrightarrow SO_2 + 4\ HF$
(g) $Cr_2(CO_3)_3 \xrightarrow{\Delta} Cr_2O_3 + 3\ CO_2$

14. (a) $Cu + FeCl_3(aq) \longrightarrow$ no reaction
(b) $H_2 + Al_2O_3(s) \xrightarrow{\Delta}$ no reaction
(c) $2\ Al + 6\ HBr(aq) \longrightarrow 3\ H_2(g) + 2\ AlBr_3(aq)$
(d) $I_2 + HCl(aq) \longrightarrow$ no reaction

16. (a) $SO_2 + H_2O \longrightarrow H_2SO_3$
(b) $SO_3 + H_2O \longrightarrow H_2SO_4$
(c) $Ca + 2\ H_2O \longrightarrow Ca(OH)_2 + H_2$
(d) $2\ Bi(NO_3)_3 + 3\ H_2S \longrightarrow Bi_2S_3 + 6\ HNO_3$

18. (a) $C + O_2 \xrightarrow{\Delta} CO_2$
(b) $2\ Al(ClO_3)_3 \xrightarrow{\Delta} 9\ O_2 + 2\ AlCl_3$
(c) $CuBr_2 + Cl_2 \longrightarrow CuCl_2 + Br_2$
(d) $2\ SbCl_3 + 3\ (NH_4)_2S \longrightarrow Sb_2S_3 + 6\ NH_4Cl$
(e) $2\ NaNO_3 \xrightarrow{\Delta} 2\ NaNO_2 + O_2$

20. (a) exothermic (b) endothermic

22. (a) $2\ Al + 3\ I_2 \xrightarrow{H_2O} 2\ AlI_3 + \text{heat}$
(b) $4\ CuO + CH_4 + \text{heat} \longrightarrow 4\ Cu + CO_2 + 2\ H_2O$
(c) $Fe_2O_3 + 2\ Al \longrightarrow 2\ Fe + Al_2O_3 + \text{heat}$

24. 58 O on each side

26. A balanced equation tells us
(a) the type of atoms/molecules involved in the reaction.
(b) the relationship between quantities of the substances in the reaction.

A balanced equation gives no information about
(a) the time required for the reaction.
(b) odor or colors which may result.

28. Mg above Zn on the activity series

30. (a) $4\ K + O_2 \longrightarrow 2\ K_2O$
(b) $2\ Al + 3\ Cl_2 \longrightarrow 2\ AlCl_3$
(c) $CO_2 + H_2O \longrightarrow H_2CO_3$
(d) $CaO + H_2O \longrightarrow Ca(OH)_2$

32. (a) $Zn + H_2SO_4 \longrightarrow H_2 + ZnSO_4$
(b) $2\ AlI_3 + 3\ Cl_2 \longrightarrow 2\ AlCl_3 + 3\ I_2$
(c) $Mg + 2\ AgNO_3 \longrightarrow Mg(NO_3)_2 + 2\ Ag$
(d) $2\ Al + 3\ CoSO_4 \longrightarrow Al_2(SO_4)_3 + 3\ Co$

34. (a) $AgNO_3(aq) + KCl(aq) \longrightarrow AgCl(s) + KNO_3(aq)$
(b) $Ba(NO_3)_2(aq) + MgSO_4(aq) \longrightarrow Mg(NO_3)_2(aq) + BaSO_4(s)$
(c) $H_2SO_4(aq) + Mg(OH)_2(aq) \longrightarrow 2\ H_2O(l) + MgSO_4(aq)$
(d) $MgO(s) + H_2SO_4(aq) \longrightarrow H_2O(l) + MgSO_4(aq)$
(e) $Na_2CO_3(aq) + NH_4Cl(aq) \longrightarrow$ no reaction

36. chlorophyll — photosynthesis
carotenoids — absorb high energy oxygen and release it appropriately
anthocyanins — diversity in leaf color

38. Leaves containing all pigments:

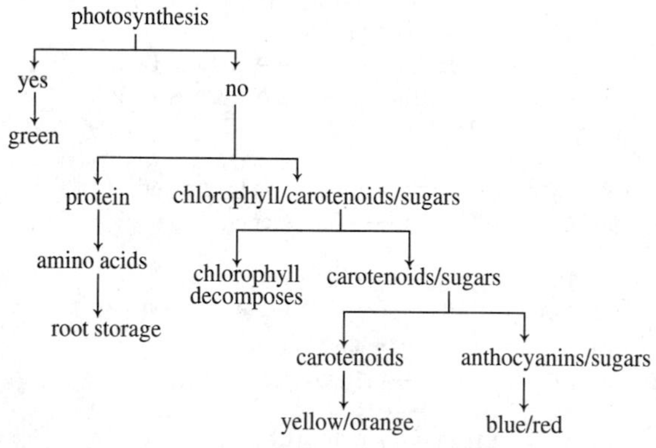

40. CO_2, CH_4, H_2O They act to trap the heat near the surface of the earth.

42. One-half the CO_2 released remains in the air.

Chapter 9

2. (a) correct (b) incorrect (c) correct (d) incorrect (e) correct (f) incorrect

4. (a) 25.0 mol $NaHCO_3$
(b) 3.85×10^{-3} mol $ZnCl_2$
(c) 16 mol CO_2
(d) 4.3 mol C_2H_5OH

6. (a) 1.31 g $NiSO_4$
(b) 3.60 g $HC_2H_3O_2$
(c) 373 g Bi_2S_3
(d) 1.35 g $C_6H_{12}O_6$
(e) 18 g K_2CrO_4

8. HCl

10. (a) $\frac{3\text{ mol } CaCl_2}{1\text{ mol } Ca_3(PO_4)_2}$ (b) $\frac{6\text{ mol } HCl}{2\text{ mol } H_3PO_4}$ (c) $\frac{3\text{ mol } CaCl_2}{2\text{ mol } H_3PO_4}$ (d) $\frac{1\text{ mol } Ca_3(PO_4)_2}{2\text{ mol } H_3PO_4}$ (e) $\frac{6\text{ mol } HCl}{1\text{ mol } Ca_3(PO_4)_2}$ (f) $\frac{2\text{ mol } H_3PO_4}{6\text{ mol } HCl}$

12. 15.5 mol CO_2

14. 4.20 mol HCl

16. 19.7 g $Zn_3(PO_4)_2$

18. 117 g H_2O
271 g Fe

20. (a) 0.500 mol Fe_2O_3 (b) 12.4 mol O_2 (c) 6.20 mol SO_2 (d) 65.6 g SO_2 (e) 0.871 mol O_2 (f) 332 g FeS_2

22. (a) H_2S is the limiting reactant and $Bi(NO_3)_3$ is in excess.
(b) H_2O is the limiting reactant and Fe is in excess.

24. (a) 16.5 g CO_2
(b) 59.9 g CO_2
(c) 6.0 mol CO_2, 8.0 mol H_2O, and 4.0 mol O_2

26. sulfur

28. 95.0% yield

30. 77.8% CaC_2

32. A subscript is used to indicate the number of atoms in a formula. It cannot be changed without changing the identity of the substance. Coefficients are used only to balance atoms in chemical equations. They may be changed as needed to achieve a balanced equation.

34. (a) 380 g C_2H_5OH
370 g CO_2
(b) 480 mL C_2H_5OH

36. 65 g O_2

38. 1.0 g Ag_2S

40. (a) 2.0 mol Cu, 2.0 mol $FeSO_4$, and 1.0 mol $CuSO_4$
(b) 15.9 g Cu, 38.1 g $FeSO_4$, 6.0 g Fe, and no $CuSO_4$

42. (a) 3.2×10^2 g C_2H_5OH
(b) 1.10×10^3 g $C_6H_{12}O_6$

44. 3.7×10^2 kg Li_2O

46. 13 tablets

48. The mask protects the parts of the machine that are not to be etched away.

50. Micromachines could be used as smart pills, drug reservoirs, or mini computers.

Chapter 10

2. A second electron may enter an orbital already occupied if its spin is opposite the electron already in the orbital, and if all other orbitals of the same sublevel contain an electron.

4. Both 1*s* and 2*s* orbitals are spherical in shape and located symmetrically around the nucleus. The radius of 2*s* is larger than the 1*s*.

6. 1*s*, 2*s*, 2*p*, 3*s*, 3*p*, 4*s*, 3*d*, 4*p*

8. The Bohr orbit has an electron traveling a specific path while an orbital is a region of space where the electron is most probably found.

10. *s* orbital

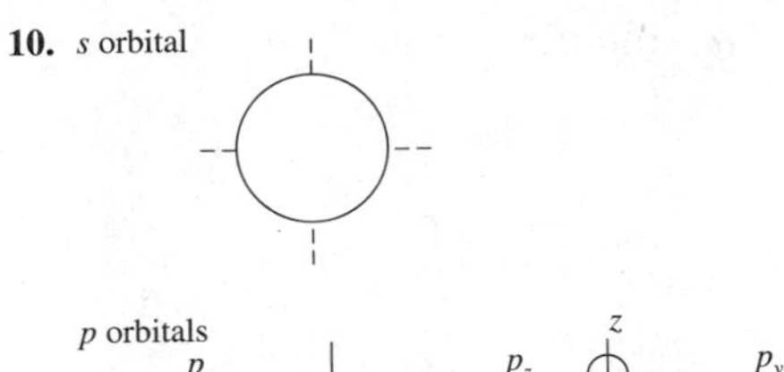

p orbitals

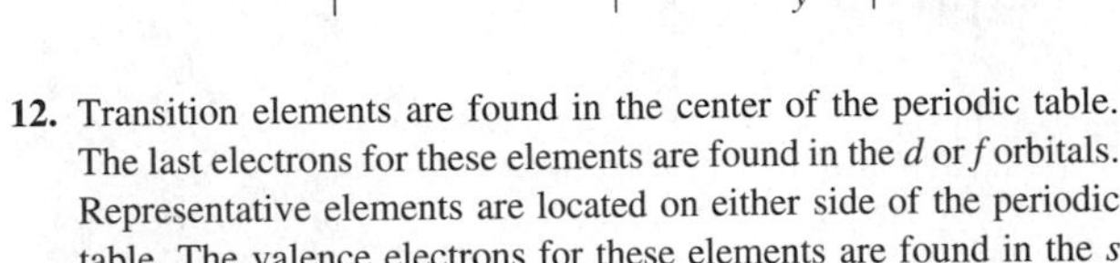

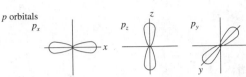

12. Transition elements are found in the center of the periodic table. The last electrons for these elements are found in the *d* or *f* orbitals. Representative elements are located on either side of the periodic table. The valence electrons for these elements are found in the *s* and/or *p* orbitals.

14.

Atomic number	**Symbol**
8	O
16	S
34	Se
52	Te
84	Po

All these elements have an outermost electron structure of s^2p^4.

16. 32; the 6th period has this number of elements.

18. Ar and K; Co and Ni; Te and I; Th and Pa; U and Np; Es and Fm; Md and No.

20. The correct statements are: a, c, e, f, g, i, and n.

22. (a) F 9 protons (c) Br 35 protons
(b) Ag 47 protons (d) Sb 51 protons

24. (a) Cl $1s^22s^22p^63s^23p^5$
(b) Ag $1s^22s^22p^63s^23p^64s^23d^{10}4p^65s^14d^{10}$
(c) Li $1s^22s^1$
(d) Fe $1s^22s^22p^63s^23p^64s^23d^6$
(e) I $1s^22s^22p^63s^23p^64s^23d^{10}4p^65s^24d^{10}5p^5$

26. Each line corresponds to a change from one orbit to another.

28. 32 electrons

30.

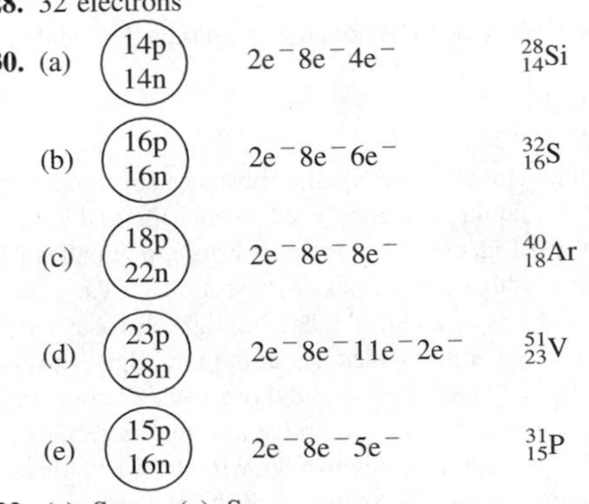

(a) 14p 14n $2e^-8e^-4e^-$ $^{28}_{14}Si$
(b) 16p 16n $2e^-8e^-6e^-$ $^{32}_{16}S$
(c) 18p 22n $2e^-8e^-8e^-$ $^{40}_{18}Ar$
(d) 23p 28n $2e^-8e^-11e^-2e^-$ $^{51}_{23}V$
(e) 15p 16n $2e^-8e^-5e^-$ $^{31}_{15}P$

32. (a) Sc (c) Sn
(b) Zr (d) Cs

34.

	Atomic number	**Electron structure**
(a)	9	$[He]2s^22p^5$
(b)	26	$[Ar]4s^23d^6$
(c)	31	$[Ar]4s^23d^{10}4p^1$
(d)	39	$[Kr]5s^24d^1$
(e)	52	$[Kr]5s^24d^{10}5p^4$
(f)	10	$[He]2s^22p^6$

36. (a) 13p 14n $2e^-8e^-3e^-$ $^{27}_{13}Al$
(b) 22p 26n $2e^-8e^-10e^-2e^-$ $^{48}_{22}Ti$

38. It is in the fourth energy level because the 4*s* orbital is at a lower energy level than the 3*d* orbital.

40. Noble gases each have filled *s* and *p* orbitals in the outer energy level.

42. All the elements in a Group have the same number of outer-shell electrons.

44. All of these elements have an s^2d^{10} electron configuration in their outermost energy levels.

46. (a) and (f)
(e) and (h)

48. 7, 33 since they are in the same group.

50. (a) nonmetal, (b) metal, (c) metal, (d) metalloid

52. Period 6, lanthinide series

54. Group VIIA contains 7 valence electrons.
Group VIIB contains 2 electrons in the outermost level and 5 electrons in an inner *d* orbital.

56. Nitrogen has more valence electrons on more energy levels. More varied electron jumps are possible.

58. (a) 100% (d) 24%
(b) 100% (e) 20%
(c) 19%

60. $\frac{1.5 \times 10^8}{1}$

62. In the scanning-probe microscope, a probe is placed near the surface of a sample and a parameter such as voltage is measured. The signals are translated electronically into a topographic image of the object. In a light microscope the image is formed by light reflecting from the object.

64. In the atomic force microscope, the probe measures the electric forces between electrons in the molecule, while in scanning-tunneling the actual movement of the electrons is measured.

66. Transition elements are found in Groups IB–VIIB and VIII.

68. 8, 16, 34, 52, 84. All have 6 electrons in their outer shell.

70. 35 would be in VIIA and 37 would be in IA.

72. (a) $[Rn]7s^26d^{10}5f^{14}7p^5$
(b) 7 valence electrons
(c) F, Cl, Br, I, At
(d) halogen family, Period 7

74. Most gases are located in the upper right part of the periodic table (H is the exception). They are nonmetals. Liquids show no pattern. Neither do solids, except that the vast majority of solids are metals.

76. Oil can be traced by added labeled fullerenes.

Chapter 11

2. More energy is required for neon because it has a very stable outer electron structure consisting of an octet of electrons in filled orbitals (noble gas electron structure).
4. The first ionization energy decreases from top to bottom because, in successive alkali metals, the outermost electron is farther away from the nucleus and is shielded from the positive nucleus by additional electron shells.
6. Barium has a lower ionization energy than beryllium.
8. (a) K > Na (d) I > Br
   (b) Na > Mg (e) Zr > Ti
   (c) O > F
10. Atomic size increases down the column since each successive element has an additional energy level.
12. Cs· Ba: Tl: ·Pb: ·Po: ·At: :Rn:

    Each of these is a representative element and has the same number of electrons in its outer shell as its periodic group number.
14. Valence electrons are the electrons found in the outermost energy level of an atom.
16. An aluminum ion has a +3 charge because it has lost 3 electrons in acquiring a noble gas electron structure.
18. The correct statements are: b, c, g, h, j, k, n, and o.
20. A bromine atom is smaller since it has one less electron than the bromine ion in its outer shell.
22.

	+	−		+	−
(a)	H	Cl	(d)	I	Br
(b)	Li	H	(e)	Mg	H
(c)	C	Cl	(f)	O	F

24. (a) covalent (c) covalent
    (b) ionic (d) ionic
26. (a) $F + 1e^- \rightarrow F^-$ (b) $Ca \rightarrow Ca^{2+} + 2e^-$
28. (a) Ca: + ·Ö: ⟶ CaO
    (b) Na· + :Br: ⟶ NaBr
30. Si (4) N (5) P (5) O (6) Cl (7)
32. (a) Chloride ion, none
    (b) Nitrogen atom, gain $3e^-$ or lose $5e^-$
    (c) Potassium atom, lose $1e^-$
34. (a) A magnesium ion, Mg^{2+}, is larger than an aluminum ion, Al^{3+}.
    (b) The Fe^{2+} ion is larger than the Fe^{3+} ion.
36. (a) SbH_3, Sb_2O_3
    (b) H_2Se, SeO_3
    (c) HCl, Cl_2O_7
    (d) CCl_4, CO_2
38. $BeBr_2$, beryllium bromide
    $MgBr_2$, magnesium bromide
    $RaBr_2$, radium bromide
40. (a) Ga: (b) $[Ga]^{3+}$ (c) $[Ca]^{2+}$
42. (a) covalent (c) covalent
    (b) ionic (d) covalent
44. (a) covalent (b) covalent (c) ionic
46. (a) :Ö::Ö: (b) :Br:Br: (c) :I:I:
48. (a) :S:H with H below S
    (b) :S::C::S:
    (c) H:N:H with H below N
    (d) [H:N:H with H above and below N]$^+$ [:Cl:]$^-$
50. (a) [:I:]$^-$
    (b) [:S:]$^{2-}$
    (c) [:O:C::O with :O: below C]$^{2-}$
    (d) [:O:Cl:O: with :O: below Cl]$^-$
    (e) [:O:N::O: with :O: below N]$^-$
52. (a) nonpolar (b) nonpolar (c) polar
54. (a) 2 electron pairs, linear
    (b) 4 electron pairs, tetrahedral
    (c) 4 electron pairs, tetrahedral
56. (a) tetrahedral (b) pyramidal (c) tetrahedral
58. (a) tetrahedral (b) bent (c) bent
60. potassium
62. (a) Zn (b) Be (c) Ne
64. Lithium has a +1 charge after the first electron is removed. It takes more energy to overcome that charge than to remove an e^- from helium.
66. $SnBr_2$, $GeBr_2$
68. A covalent bond results from the sharing of a pair of electrons between two atoms, while an ionic bond involves the transfer of one electron or more from one atom to another.
70. N, O, F, Cl
72. It is possible for a molecule to be nonpolar even though it contains polar bonds.
74. (a) 105° (c) 109.5°
    (b) 107° (d) 109.5°
76. 77 K is the boiling point of liquid N_2. If a substance superconducts at this temperature, liquid N_2 can be used to cool the material.
78. Current material used in superconductors is brittle, nonmalleable, and does not carry a high current per cross-sectional area.
80. Normally, the LCD acts as a mirror reflecting light. It has a series of layers, however. When the molecules at the top align with the lines etched on the first layer of glass and those at the bottom align with the grooves on the bottom layer of glass, the molecules in between form a twisted spiral (trying to align with those near them). If a current is applied to specific segments of the etched glass, the plates become charged and the spirals of molecules are attached

to the charged plate destroying the arrangement. The pattern of reflected light is changed and a number appears.

82. (a) Both use the p orbitals for bonding. B uses one s and $2p$ while N uses $3p$ orbitals.
(b) BF_3 is trigonal planar; NF_3 is pyramidal.
(c) BF_3 has no lone pairs; NF_3 has one lone pair.
(d) BF_3 has 3 very covalent bonds; NF_3 has 3 covalent bonds.

84. Each element in a particular group has the same number of valence electrons.

86. C_2Cl_4 :Cl: :Cl: / :Cl:C=C:Cl:

Chapter 12

2. The air pressure inside the balloon is greater than the air pressure outside the balloon.
4. 1 torr = 1 mm Hg.
6. 1 atm corresponds to 4 L.
8. The piston would move downward.
10. O_2, H_2S, HCl, F_2, CO_2
12. Rn, F_2, N_2, CH_4, He, H_2, decreasing molar mass
14. (a) pressure (b) volume (c) temperature (d) number of moles
16. A gas is least likely to behave ideally at low temperatures.
18. Equal volumes of H_2 and O_2 at the same T and P:
(a) have equal numbers of molecules (Avogadro's law)
(b) mass O_2 = 16 × mass H_2
(c) moles O_2 = moles H_2
(d) average kinetic energies are the same (T same)
(e) rate H_2 = 4 (rate O_2) (Graham's law of effusion)
(f) density O_2 = 16 (density H_2)
20. $N_2(g) + O_2(g) \longrightarrow 2NO(g)$
1 vol + 1 vol $\longrightarrow$ 2 vol
22. Conversion of oxygen to ozone is an endothermic reaction.
24. Heating a mole of N_2 gas at constant pressure has the following effects:
(a) Density will decrease.
(b) Mass does not change.
(c) Average kinetic energy of the molecules increases.
(d) Average velocity of the molecules will increase.
(e) Number of N_2 molecules remains unchanged.
26. The correct statements are: b, d, f1, f4, h, i, n, o, p, q, t.
28. (a) 715 torr (b) 953 mbar (c) 95.3 kPa
30. (a) 0.082 atm (b) 55.92 atm (c) 0.296 atm (d) 0.0066 atm
32. (a) 132 mL (b) 615 mL
34. 711 mm Hg
36. (a) 6.17 L (b) 8.35 L
38. 7.8×10^2 mL
40. 33.4 L
42. 681 torr
44. 1.45×10^3 torr
46. 1.19 L C_3H_8
48. 28.0 L N_2
50. 1.33 g NH_3
52. (a) 11 L H_2S (b) 14.7 L H_2S (c) 31.8 L H_2S
54. 2.69×10^{22} molecules CH_4
56. (a) 0.179 g/L He (b) 2.50 g/L C_4H_8
58. (a) 3.17 g/L Cl_2 (b) 1.46 g/L Cl_2
60. 19 L Kr
62. 72.8 L
64. (a) 5.6 mol NH_3 (b) 0.640 L NO (c) 1.1×10^2 g O_2
66. 153 L SO_2
68. The can will explode.
70. (a) 5 L Cl_2 (b) 5.0 L NH_3 (c) 5.9 L SO_3
72. (a) CH_3
(b) C_2H_6
(c)

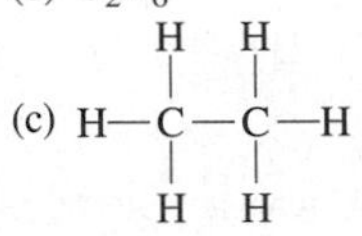

74. 430 mL CO_2
76. (a) 0.18 mol air
(b) 5.2 g air
(c) 0.3 g air
78. 1.0×10^2 atm
80. 1.5×10^3 torr
82. 65 atm
84. 6.1 L
86. 7.39×10^{21} molecules, 2.22×10^{22} atoms
88. (a) 34 mol
(b) 121 g H_2
90. 44.6 mol Cl_2
92. −78°C
94. 1.64×10^2 g/mol
96. 0.13 mol N_2
98. 9.0 atm
100. (a) Helium effuses twice as fast as CH_4.
(b) The gases meet 66.7 cm from the helium end.
102. (a) 10.0 mol CO_2; 3.0 mol O_2; no CO
(b) 29 atm
104. Some ammonia gas dissolves in the water squirted into the flask, lowering the pressure inside the flask. The atmospheric pressure outside is greater than the pressure inside the flask and thus pushes water from the beaker up the tube and into the flask.
106. (a) 1.1×10^2 torr CO_2; 13 torr H_2
(b) 120 torr
108. Air enters the room.

Chapter 13

2. H_2S, H_2Se, and H_2Te are gases
4. − / O / H H / +
6. Prefixes preceding the word *hydrate* are used to indicate the number of molecules of water present in the formulas.
8. about 70°C
10. case (b)
12. vapor pressure remains unchanged
14. (a) 88°C (b) 78°C (c) 16°C
16. melting point, 0°C; boiling point, 100°C (at 1 atm pressure);

colorless; odorless; tasteless; heat of fusion, 335 J/g (80 cal/g); heat of vaporization, 2.26 kJ/g (540 cal/g); density = 1.0 g/mL (at 4°C); specific heat = 4.184 J/g °C

18. If you apply heat to an ice-water mixture, the heat energy is absorbed to melt the ice, rather than to warm the water, so the temperature remains constant until all the ice has melted.

20. Ice floats in water because it is less dense than water. Ice sinks in ethyl alcohol because it is more dense than the alcohol.

22. Ethyl alcohol exhibits hydrogen bonding; ethyl ether does not.

24. Ammonia exhibits hydrogen bonding; methane does not.

26. $H_2NCH_2CH_2NH_2$

28. (a) mercury, acetic acid, water, toluene, benzene, carbon tetrachloride, methyl alcohol, bromine

(b) Highest is mercury; lowest is bromine.

30. In a pressure cooker, the temperature at which water boils increases above its normal boiling point, because the water vapor (steam) formed by boiling cannot escape. This results in an increased pressure over the water, and consequently, an increased boiling temperature.

32. As temperature increases, molecular velocities increase.

34. Ammonia

36. HF has a higher boiling point because of the strong H-bonding in HF.

38. 34.6°C

40. The expected temperature would be 4°C at the bottom of the lake.

42. endothermic

44. The correct statements are: a, b, c, f, h, l, m, o, p, s, t, w

46. $[HClO_4, Cl_2O_7]$ $[H_2CO_3, CO_2]$ $[H_3PO_4, P_2O_5]$

48. $[Ca(OH)_2CaO]$ $[KOH, K_2O]$ $[Ba(OH)_2, BaO]$

50. (a) $Li_2O + H_2O \longrightarrow 2\ LiOH$

(b) $2\ KOH \xrightarrow{\Delta} K_2O + H_2O$

(c) $Ba + 2\ H_2O \longrightarrow Ba(OH)_2 + H_2$

(d) $Cl_2 + H_2O \longrightarrow HCl + HClO$

(e) $SO_3 + H_2O \longrightarrow H_2SO_4$

(f) $H_2SO_3 + 2\ KOH \longrightarrow K_2SO_3 + 2\ H_2O$

52. (a) magnesium ammonium phosphate hexahydrate

(b) iron(II) sulfate heptahydrate

(c) tin(IV) chloride pentahydrate

54. (a) Distilled water has been vaporized by boiling and recondensed.

(b) Natural waters are generally not pure, but contain dissolved minerals and suspended matter, and can even contain harmful bacteria.

56. 0.262 mol $FeI_2 \cdot 4\ H_2O$

58. 1.05 mol H_2O

60. 48.7% H_2O

62. $FePO_4 \cdot 4\ H_2O$

64. 5.5×10^4 J

66. 42 g

68. The system will be at 0°C. It will be a mixture of ice and water.

70. (a) 0.784 g H_2O (b) 0.447 g H_2O (c) 0.167 g H_2O

72. Eventually the water will lose enough energy to change from a liquid to a solid (freeze).

74. (a) From 0°C to 40°C solid *X* warms until at 40°C it begins to melt. The temperature remains at 40°C until all of *X* is melted. After that, liquid *X* will warm steadily to 65°C where it will begin to boil and remain at 65°C until all the liquid becomes vapor. Beyond 65°C the vapor will warm steadily until 100°C.

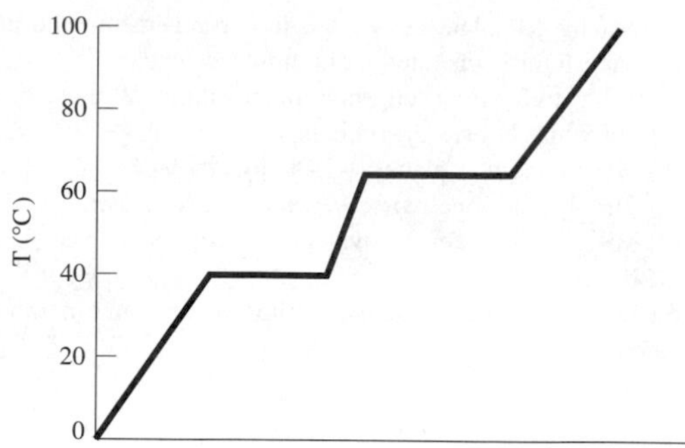

(b) 37,000 J

76. 75°C at 270 torr

78. $MgSO_4 \cdot 7\ H_2O$ $Na_2HPO_4 \cdot 12\ H_2O$

80. chlorine

82. When organic pollutants in water are oxidized by dissolved oxygen, there may not be sufficient dissolved oxygen to sustain marine life.

84. $Na_2\ zeolite(s) + Mg^{2+}(aq) \longrightarrow Mg\ zeolite(s) + 2Na^+(aq)$

86. Humectants are polar compounds while emollients are nonpolar compounds.

88. The triangle theory of sweetness states that "sweet" molecules contain three specific sites that produce the proper structure to attach to the taste buds and trigger the "sweet" response.

90. 1.6×10^5 cal

92. 3.1 kcal

94. 2.30×10^6 J

96. 40.2 g H_2O

98. 6.97×10^{18} molecules/s

100. (a) 40.0 mL O_2

(b) 20.0 mL O_2 unreacted

Chapter 14

2. 4.5 g NaF

4. For lithium and sodium halides, solubility is

$$F^- < Cl^- < Br^- < I^-$$

For potassium halides, solubility is

$$Cl^- < Br^- < F^- < I^-$$

6. KNO_3

8. $6 \times 10^2\ cm^2$

10. The dissolving process involves solvent molecules attaching to the solute ions or molecules. This rate decreases as more of the solvent molecules are already attached to solute molecules. As the solution becomes saturated, the number of unused solvent molecules decreases. Also, the rate of recrystallization increases as the concentration of dissolved solute increases.

12. The solution level in the thistle tube will fall.

14. It is not always apparent which component in a solution is the solute.

16. yes

18. Hexane and benzene are both nonpolar molecules.

20. Air is considered to be a solution because it is a homogeneous mixture of several gaseous substances and does not have a fixed composition.

22. The solubility of gases in liquids is greatly affected by the pressure of a gas above the liquid; little effect for solids in liquids.

24. In a saturated solution, the net rate of dissolution is zero.

26. 16 moles HNO_3/L of solution

28. The champagne would spray out of the bottle.

30. Water molecules can pass through in both directions.

32. A lettuce leaf immersed in salad dressing containing salt and vinegar will become limp and wilted as a result of osmosis.

34. The correct statements are: a, b, f, h, j, k, l, n, p, r, s, t, v, x, y, z

36. Since there are 2 equivalents per mole of H_2SO_4, 18 *M* is equivalent to 36 N.

38. (a) 1 *M* NaOH 1 L
(b) 0.6 *M* $Ba(OH)_2$ 0.83 L
(c) 2 *M* KOH 0.50 L
(d) 1.5 *M* $Ca(OH)_2$ 0.33 L

40. The vapor pressure of the solution equals the vapor pressure of the pure solvent at the freezing point. See Figure 14.8.

42. The presence of the methanol lowers the freezing point of the water.

44. The molarity of a 5 molal solution is less than 5 *M*.

46. When the paper on a scratch and sniff label is scratched or pulled open the fragrance is released into the air.

48. Three types of microencapsulation systems are:
(a) water diffuses through capsule forming a sølution which diffuses out again
(b) mechanical
(c) thermal

50. Reasonably soluble: (c) $CaCl_2$ (d) $Fe(NCl_3)_3$
Insoluble: (a) PbI_2 (b) $MgCO_3$ (e) $BaSO_4$

52. (a) 7.41% $Mg(NO_3)_2$ (b) 6.53% $NaNO_3$

54. 544 g solution

56. (a) 8.815% NaOH (b) 15% $C_6H_{12}O_6$

58. 37.5 g K_2CrO_4

60. 33.6% NaCl

62. 22% C_6H_{14}

64. (a) 2.5 *M* HCl
(b) 0.59 *M* $BaCl_2 \cdot 2H_2O$
(c) 2.19×10^{-3} *M* $Al_2(SO_4)_3$
(d) 0.172 *M* $Ca(NO_3)_2$

66. (a) 3.5×10^{-4} mol NaOH
(b) 16 mol $CoCl_2$

68. (a) 13 g HCl
(b) 8.58 g $Na_2C_2O_4$

70. (a) 3.91×10^4 mL
(b) 7.82×10^3 mL

72. (a) 0.250 L
(b) 1.81 *M* NaCl

74. (a) 20. mL 15 *M* NH_3
(b) 69 mL 18 *M* H_2SO_4

76. (a) 1.83 *M* H_2SO_4
(b) 0.30 *M* H_2SO_4

78. (a) 33.3 mL of 0.250 *M* Na_3PO_4
(b) 1.10 g $Mg_3(PO_4)_2$

80. (a) 0.13 mol Cl_2
(b) 16 mol HCl
(c) 1.3×10^2 mL of 6 *M* HCl
(d) 3.2 L Cl_2

82. (a) $\left(\frac{98.08 \text{ g } H_2SO_4}{\text{eq}}\right)$ $\left(\frac{56.11 \text{ g KOH}}{\text{eq}}\right)$
(b) $\left(\frac{49.00 \text{ g } H_3PO_4}{\text{eq}}\right)$ $\left(\frac{23.95 \text{ g LiOH}}{\text{eq}}\right)$
(c) $\left(\frac{63.02 \text{ g } HNO_3}{\text{eq}}\right)$ $\left(\frac{40.00 \text{ g NaOH}}{\text{eq}}\right)$

84. (a) 5.55 *N* H_3PO_4
(b) 0.250 *N* $HC_2H_3O_2$
(c) 1.25 *N* NaOH

86. (a) 11.94 mL NaOH (b) 10.09 mL NaOH

88. (a) 5.5 *m* $C_6H_{12}O_6$ (b) 0.25 *m* I_2

90. (a) 0.544 *m* (b) 2.7°C (c) 81.5°C

92. 163 g/mol

94. 97 g 10% NaOH solution

96. $C_8H_4N_2$

98. 0.14 *M*

100. (a) 4.5 g NaCl
(b) 450. mL H_2O must be evaporated.

102. 210. mL solution

104. 6.72 *M* HNO_3

106. 540. mL water to be added

108. To make 250. mL of 0.625 *M* KOH, take 31.3 mL of 5.00 *M* KOH and dilute with water to a volume of 250. mL.

110. 2.08 *M* HCl

112. 12.0 g $Mg(OH)_2$

114. 6.2 *m* H_2SO_4
5.0 *M* H_2SO_4

116. (a) 2.9 *m*
(b) 101.5°C

118. (a) 8.04×10^3 g $C_2H_6O_2$
(b) 7.24×10^3 mL $C_2H_6O_2$
(c) −4.0°F

120. 0.46 L HCl

122. (a) 7.7 L H_2O must be added
(b) 0.0178 mol
(c) 0.0015 mol

124. Mix together 667 mL 3.00 *M* HNO_3 and 333 mL 12.0 *M* HNO_3 to get 1000. mL of 6.00 *M* HNO_3

126. 2.84 g $Ba(OH)_2$ is formed.

128. (a) 0.011 mol
(b) 14 g
(c) 3.2×10^2 mL solution
(d) 1.5%

130. 8.9×10^{-3} *M*

Chapter 15

2. An electrolyte must be present in the solution for the bulb to glow.

4. First the orientation of the polar water molecules about the Na^+ and Cl^- ions is different. Second more water molecules will fit around Cl^-, since it is larger than the Na^+ ion.

6. tomato juice

8. Arrhenius: $HCl + NaOH \longrightarrow NaCl + H_2O$
Brønsted-Lowry: $HCl + KCN \longrightarrow HCN + KCl$
Lewis: $AlCl_3 + NaCl \longrightarrow AlCl_4^- + Na^+$

10. acids, bases, salts

12. Hydrogen chloride dissolved in water conducts an electric current. HCl does not ionize in benzene.

14. CH_3OH is a nonelectrolyte; NaOH is an electrolyte. This indicates that the OH group in CH_3OH must be covalently bonded to the CH_3 group.

16. The dissolving of NaCl is a dissociation process, while the dissolving of HCl in water is an ionization process.

18. Ions are hydrated in solution because there is an electrical attraction between the charged ions and the polar water molecules.

20. (a) $[H^+] = [OH^-]$
(b) $[H^+] > [OH^-]$
(c) $[OH^-] > [H^+]$

22. HCl is much more soluble in the polar solvent, water, than the nonpolar solvent, benzene.

24. The correct statements are: a, c, f, i, j, k, l, n, p, r, t, u

26. Colloids are prepared by two methods:
(a) dispersion (b) condensation

28. *Adsorption* refers to the adhesion of particles to a surface while *absorption* refers to the taking in of one material by another.

30. Dialysis is the process of removing dissolved solutes from colloidal dispersion by the use of a dialyzing membrane.

32. Acidic shampoo breaks hydrogen bonds and salt bridges in the hair leaving only disulfide bonds. In a depilatory, a basic solution is used which breaks all types of bonds (H, salt bridges, disulfide) and the hair dissolves.

34. (a) $H_2SO_4 - HSO_4^-$; $H_2C_2H_3O_2^+ - HC_2H_3O_2$
(b) Step 1: $H_2SO_4 - HSO_4^-$; $H_3O^+ - H_2O$
Step 2: $HSO_4^- - SO_4^{2-}$; $H_3O^+ - H_2O$
(c) $HClO_4 - ClO_4^-$; $H_3O^+ - H_2O$
(d) $H_3O^+ - H_2O$; $CH_3OH - CH_3O^-$

36. (a) $NaOH(aq) + HBr(aq) \longrightarrow NaBr(aq) + H_2O(l)$
(b) $KOH(aq) + HCl(aq) \longrightarrow KCl(aq) + H_2O(l)$
(c) $Ca(OH)_2(aq) + 2\ HI(aq) \longrightarrow CaI_2(aq) + 2\ H_2O(l)$
(d) $Al(OH)_3(s) + 3\ HBr(aq) \longrightarrow AlBr_3(aq) + 3\ H_2O(l)$
(e) $Na_2O(s) + 2\ HClO_4(aq) \longrightarrow 2\ NaClO_4(aq) + H_2O(l)$
(f) $3\ LiOH(aq) + FeCl_3(aq) \longrightarrow Fe(OH)_3(s) + 3\ LiCl(aq)$

38. (a) $NaHCO_3$—salt (e) RbOH—base
(c) $AgNO_3$—salt (f) K_2CrO_4—salt
(d) HCOOH—acid

40. (a) 0.75 *M* Zn^{2+}, 1.5 *M* Br^-
(b) 4.95 *M* SO_4^{2-}, 3.30 *M* Al^{3+}
(c) 0.682 *M* NH_4^+, 0.341 *M* SO_4^{2-}
(d) 0.0628 *M* Mg^{2+}, 0.126 *M* ClO_3^-

42. (a) 4.9 g Zn^{2+}, 12 g Br^-
(b) 8.90 g Al^{3+}, 47.5 g SO_4^{2-}
(c) 1.23 g NH_4^+, 3.28 g SO_4^{2-}
(d) 0.153 g Mg^{2+}, 1.05 g ClO_3^-

44. (a) $[K^+] = 1.0\ M$, $[Ca^{2+}] = 0.5\ M$, $[Cl^-] = 2.0\ M$
(b) No ions are present in the solution.
(c) 0.67 *M* Na^+, 0.67 *M* NO_3^-

46. (a) 0.147 *M* NaOH
(b) 0.964 *M* NaOH
(c) 0.4750 *M* NaOH

48. (a) $H_2S(g) + Cd^{2+}(aq) \longrightarrow CdS(s) + 2\ H^+(aq)$
(b) $Zn(s) + 2H^+(aq) \longrightarrow Zn^{2+}(aq) + H_2(g)$
(c) $Al^{3+}(aq) + PO_4^{3-}(aq) \longrightarrow AlPO_4(s)$

50. (a) 2 *M* HCl
(b) 1 *M* H_2SO_4

52. 1.57×10^3 mL of 0.245 *M* HCl

54. 7.0% NaCl in sample

56. 0.936 L H_2

58. (a) 7.0 (b) 0.30 (c) 4.00

60. (a) 4.30 (b) 10.47

62. 22.7 mL of 0.325 N H_2SO_4

64. 57 g/eq

66. 3.0×10^3 mL

68. The ionization of the acetic acid solution increases the particle concentration above that of the alcohol solution.

70. A hydronium ion is a hydrated hydrogen ion.

72. (a) 100°C pH = 6.0
25°C pH = 7.0
(b) H^+ concentration is higher at 100°C.
(c) The water is neutral at both temperatures.

74. 0.201 *M* HCl

76. 0.673 g KOH

78. 13.9 L of 18.0 *M* H_2SO_4

80. 0.3586 *N* NaOH

82. 1.2×10^2 mL

84. 0.100 *N* NaOH, 0.0500 *N* HCl

86. The molarity and normality of an acid solution will be the same when the acid has one ionizable hydrogen. The molarity and the normality of a base solution will be the same when the base has one ionizable hydroxide. The normality and molarity of a salt (ionic compound) solution will be the same when the salt contains one ionizable cation with a charge of +1.

88. 1.1, acidic

90. (a) $2\ NaOH(aq) + H_2SO_4(aq) \longrightarrow Na_2SO_4(aq) + 2\ H_2O(l)$
(b) 1.0×10^2 mL NaOH
(c) 0.71 g Na_2SO_4

92. 12 *M* HNO_3

94. (c)

Chapter 16

2. The reaction is endothermic because the increased temperature increases the concentration of product present at equilibrium.

4. The yield of NH_3 would be greater if it were carried out in a 1-L vessel.

6. Stronger than $HC_2H_3O_2$ are benzoic, cyanic, formic, hydrofluoric and nitrous acids. Weaker than $HC_2H_3O_2$ are carbonic, hydrocyanic, and hypochlorous acids.

8. (a) Ag_2CrO_4
(b) Ag_2CrO_4 has a greater molar solubility than $BaCrO_4$.

10. The figure should be modified to show the energy level of the product as lower than the energy level of the reactants.

12. increase since the number of collisions increases

14. increase since the number of molecules increases

16. The equilibrium shifts to the right, yielding a higher percent ionization.

18. The pH of the water can be different at different temperatures.

20. In HNO_3 solution, acetate ions react with H^+ to form $HC_2H_3O_2$ causing more $AgC_2H_3O_2$ to dissolve. More $AgC_2H_3O_2$ would dissolve in HCl than in pure water.

22. A buffer solution contains a weak acid or base plus a salt of that weak acid or base. When a small amount of a strong acid (H^+) is added to this buffer solution, the H^+ reacts with anions of the salt, thus neutralizing the added acid. When a strong base, OH^-, is added it reacts with un-ionized acid to neutralize the added base. As a result, in both cases, the approximate pH of the solution is maintained.

24. Both molecules bind oxygen. The myoglobin molecule binds oxygen differently than hemoglobin. Since the affinity between oxygen and myoglobin is higher than that between oxygen and hemoglobin, the hemoglobin will release oxygen to myoglobin for storage.

26. The correct statements are: c, d, g, i, j, k, l, n, p, q, s, t, u, v, z

28. (a) $H_2O(l) \overset{100°C}{\rightleftharpoons} H_2O(g)$

(b) $SO_2(l) \rightleftharpoons SO_2(g)$

30. (a) $[NH_3]$, $[O_2]$, and $[N_2]$ will be increased. $[H_2O]$ will be decreased. Reaction shifts left.
(b) The addition of heat will shift the reaction to the left.

32.

(a)	left	I	I	D
(b)	left	I	I	D
(c)	no change	N	N	N
(d)	?	?	I	I

34.

Reaction	Increase temperature	Increase pressure	Add catalyst
(a)	right	left	no change
(b)	left	left	no change
(c)	left	left	no change

36. (a) left (b) right (c) right

38. (a) $K_{eq} = \dfrac{[H^+][ClO_2^-]}{[HClO_2]}$

(b) $K_{eq} = \dfrac{[H^+][C_2H_3O_2^-]}{[HC_2H_3O_2]}$

(c) $K_{eq} = \dfrac{[NO]^4[H_2O]^6}{[NH_3]^4[O_2]^5}$

40. (a) $K_{sp} = [Fe^{3+}][OH^-]^3$
(b) $K_{sp} = [Sb^{5+}]^2[S^{2-}]^5$
(c) $K_{sp} = [Ca^{2+}][F^-]^2$
(d) $K_{sp} = [Ba^{2+}]^3[PO_4^{3-}]^2$

42. (a) pH is decreased. (b) pOH is increased. (c) OH^- is decreased. (d) K_W remains unchanged.

44. (a) $Ca(CN)_2$, basic (b) $BaBr_2$, neutral (c) $NaNO_2$, basic (d) NaF, basic

46. (a) $NH_4^+(aq) + H_2O(l) \rightleftharpoons H_3O^+(aq) + NH_3(aq)$
(b) $SO_3^{2-}(aq) + H_2O(l) \rightleftharpoons OH^-(aq) + HSO_3^-(aq)$

48. (a) $OCl^-(aq) + H_2O(l) \rightleftharpoons OH^-(aq) + HOCl(aq)$
(b) $ClO_2^-(aq) + H_2O(l) \rightleftharpoons OH^-(aq) + HClO_2(aq)$

50. When excess base gets into the bloodstream it reacts with H^+ to form water. Then H_2CO_3 ionizes to replace H^+, thus maintaining the approximate pH of the solution.

52. (a) $5.7 \times 10^{-6}\ M$
(b) 5.24
(c) $2.3 \times 10^{-3}\%$ ionized

54. 7×10^{-10}

56. (a) 3.72 (b) 4.23 (c) 4.72

58. 7.3×10^{-6}

60. $[OH^-] = 1.0\ M$, pOH = 0.00, pH = 14.00, $[H^+] = 1 \times 10^{-14}\ M$

62. (a) pH = 11.4, pOH = 2.60
(b) pH = 4.73, pOH = 9.8
(c) pH = 9.1, pOH = 4.92

64. (a) $[OH^-] = 2.5 \times 10^{-6}$
(b) $[OH^-] = 1.1 \times 10^{-13}$

66. (a) $[H^+] = 2.2 \times 10^{-9}$
(b) $[H^+] = 1.4 \times 10^{-11}$

68. (a) 1.2×10^{-23} (b) 2.6×10^{-13} (c) 1.81×10^{-18} (d) 5.13×10^{-17}

70. (a) $1.6 \times 10^{-2}\ M$ (b) $1.2 \times 10^{-4}\ M$

72. (a) 0.50 g Ag_2SO_4
(b) 7.0×10^{-4} g $Mg(OH)_2$

74. Precipitation occurs.

76. 2.5×10^{-12} mol AgBr will dissolve.

78. 4.74

80. Change in pH = 4.74 − 4.72 = 0.02 units in the buffered solution. Initial pH = 4.74.

82. (a) 3.16 mol HI
(b) 0.57 mol I_2
(c) $K_{eq} = 57$

84. $K_{eq} = 29$

86. Hypochlorous acid: $K_a = 3.5 \times 10^{-8}$
Propanoic acid: $K_a = 1.3 \times 10^{-5}$
Hydrocyanic acid: $K_a = 4.0 \times 10^{-10}$

88. (a) Precipitation occurs.
(b) Precipitation occurs.
(c) No precipitation occurs.

90. $BaSO_4$ precipitates first.

92. $K_{eq} = 1.1 \times 10^4$

94. 8.0 M

96. $K_{sp} = 4.00 \times 10^{-28}$

98. (a) The temperature could have been cooler.
(b) The humidity in the air could have been higher.
(c) The air pressure could have been greater.

100. $K_{eq} = 1$

102. $K_{eq} = 3$

104. (a) After an initial increase, $[OH^-]$ will be neutralized and equilibrium shifts to the right.
(b) $[H^+]$ will be reduced (reacts with OH^-). Equilibrium shifts to the right.
(c) $[NO_2^-]$ increases as equilibrium shifts to the right.
(d) $[HNO_2]$ decreases as equilibrium shifts to the right.

106. 1.1 g

Chapter 17

2. (a) Al (b) Ba (c) Ni
4. (a) $2\ Al + Fe_2O_3 \longrightarrow Al_2O_3 + 2\ Fe + \text{heat}$
   (b) Al is more active than Fe.
   (c) No
   (d) Yes
6. (a) Oxidation occurs at the anode.
   $2\ Cl^-(aq) \longrightarrow Cl_2(g) + 2e^-$
   (b) Reduction occurs at the cathode.
   $Ni^{2+}(aq) + 2\ e^- \longrightarrow Ni(s)$
   (c) The net chemical reaction is
   $$Ni^{2+}(aq) + 2\ Cl^-(aq) \xrightarrow[\text{energy}]{\text{electrical}} Ni(s) + Cl_2(g)$$
8. (a) It would not be possible to monitor the voltage produced, but the reactions in the cell would still occur.
   (b) If the salt bridge were removed, the reaction would stop.
10. $Ca^{2+} + 2\ e^- \longrightarrow Ca$ cathode reaction, reduction
    $2\ Br^- \longrightarrow Br_2 + 2\ e^-$ anode reaction, oxidation
12. Since lead dioxide and lead(II) sulfate are insoluble, it is unnecessary to have salt bridges in the cells of a lead storage battery.
14. cathode
16. A salt bridge permits movement of ions in the cell. This keeps the solution neutral with respect to the charged particles (ions) in the solution.
18. Patina is the blue-green product of the atmospheric corrosion of copper. It forms a coating on the surface that can protect the copper from further oxidation.
20. The correct statements are: a, c, e, g, j, k, m, p, q, r, s
22. (a) $K\underline{Mn}O_4$ +7
    (b) $\underline{I}_2$ 0
    (c) $\underline{N}H_3$ −3
    (d) $K\underline{Cl}O_3$ +5
    (e) $K_2\underline{Cr}O_4$ +6
    (f) $K_2\underline{Cr}_2O_7$ +6
24. (a) $\underline{O}_2$ 0
    (b) $\underline{As}O_4^{3-}$ +5
    (c) $Fe(\underline{O}H)_3$ −2
    (d) $\underline{I}O_3^-$ +5
26.

	Changing element	Type of reaction
(a)	S	oxidation
(b)	N	reduction
(c)	S	oxidation
(d)	Fe	oxidation

28. Equation (1):
    (a) As is oxidized, Ag^+ is reduced.
    (b) Ag^+ is the oxidizing agent, AsH_3 the reducing agent.
    Equation (2):
    (a) Br is oxidized, Cl is reduced.
    (b) Cl_2 is the oxidizing agent, NaBr the reducing agent.
30. (a) $3\ Cl_2 + 6\ KOH \longrightarrow KClO_3 + 5\ KCl + 3\ H_2O$
    (b) $3\ Ag + 4\ HNO_3 \longrightarrow 3\ AgNO_3 + NO + 2\ H_2O$
    (c) $3\ CuO + 2\ NH_3 \longrightarrow N_2 + 3\ Cu + 3\ H_2O$
    (d) $3\ PbO_2 + 2\ Sb + 2\ NaOH \longrightarrow 3\ PbO + 2\ NaSbO_2 + H_2O$
    (e) $5\ H_2O_2 + 2\ KMnO_4 + 3\ H_2SO_4 \longrightarrow 5\ O_2 + 2\ MnSO_4 + K_2SO_4 + 8\ H_2O$
32. (a) $6\ H^+ + ClO_3^- + 6\ I^- \longrightarrow 3\ I_2 + Cl^- + 3\ H_2O$
    (b) $14\ H^+ + Cr_2O_7^{2-} + 6\ Fe^{2+} \longrightarrow 2\ Cr^{3+} + 6\ Fe^{3+} + 7\ H_2O$
    (c) $2\ H_2O + 2\ MnO_4^- + 5\ SO_2 \longrightarrow 4\ H^+ + 2\ Mn^{2+} + 5\ SO_4^{2-}$
    (d) $6\ H^+ + 5\ H_3AsO_4 + 2\ MnO_4^- \longrightarrow 5\ H_3AsO_4 + 2\ Mn^{2+} + 3\ H_2O$
    (e) $8\ H^+ + Cr_2O_7^{2-} + 3\ H_3AsO_3 \longrightarrow 2\ Cr^{3+} + 3\ H_3AsO_4 + 4\ H_2O$
34. (a) $H_2O + 2\ MnO_4^- + 3\ SO_3^{2-} \longrightarrow 2\ MnO_2 + 3\ SO_4^{2-} + 2\ OH^-$
    (b) $2\ H_2O + 2\ ClO_2 + 2\ OH^- + SbO_2^- \longrightarrow 2\ ClO_2^- + Sb(OH)_6^-$
    (c) $8\ Al + 3\ NO_3^- + 18\ H_2O + 5\ OH^- \longrightarrow 3\ NH_3 + 8\ Al(OH)_4^-$
    (d) $4\ OH^- + 2\ H_2O + P_4 \longrightarrow 2\ HPO_3^{2-} + 2\ PH_3$
    (e) $2\ Al + 6\ H_2O + 2\ OH^- \longrightarrow 2\ Al(OH)_4^- + 3\ H_2$
36. (a) The oxidizing agent is $KMnO_4$.
    (b) The reducing agent is HCl.
    (c) $3.01 \times 10^{24}\ \dfrac{\text{electrons}}{\text{mol KMnO}_4}$
38. 20.2 L Cl_2
40. 66.2 mL of 0.200 *M* $K_2Cr_2O_7$ solution
42. 91.3% KI
44. 5.56 mol H_2
46. The electrons lost by the species undergoing oxidation must be gained (or attracted) by another species which then undergoes reduction.
48. Sn^{4+} can only be an oxidizing agent.
    Sn^0 can only be a reducing agent.
    Sn^{2+} can be both oxidizing and reducing agents.
50. Equations (a) and (b) represent oxidations.
52. (a) $F_2 + 2\ Cl^- \longrightarrow 2\ F^- + Cl_2$
    (b) $Br_2 + Cl^- \longrightarrow NR$
    (c) $I_2 + Cl^- \longrightarrow NR$
    (d) $Br_2 + 2\ I^- \longrightarrow 2\ Br^- + I_2$
54. $4\ Zn + NO_3^- + 10\ H^+ \longrightarrow 4\ Zn^{2+} + NH_4^+ + 3\ H_2O$
56. (a) Pb
    (b) Ag
    (c) Pb (anode)
    (d) Ag (cathode)
    (e) Electrons flow from the lead through the wire to the silver.
    (f) Positive ions flow through the salt toward the negatively charged strip of silver; negative ions flow toward the positively charged strip of lead.

Chapter 18

2. Alpha particles are much heavier.
4. Contributions to the early history of radioactivity include
   (a) Henri Becquerel—discovered radioactivity.
   (b) Marie and Pierre Curie—discovered polonium and radium.
   (c) Wilhelm Roentgen—discovered x rays and developed the technique for producing them.
   (d) Earnest Rutherford—discovered alpha and beta particles, established the link between radioactivity and transmutation and produced the first successful man-made transmutation.

(e) Otto Hahn and Fritz Strassmann—were first to produce nuclear fission.

6. *Isotope* is used with reference to atoms of the same element that contain different masses. Nuclide infers any isotope of any atom.

8.

	Charge	Mass	Nature of particles	Penetrating power
Alpha	+2	4 amu	He nucleus	low
Beta	−1	$\frac{1}{1837}$ amu	electron	moderate
Gamma	0	0	electromagnetic radiation	high

10. A disintegration series is a series of α and β emissions leading to production of a stable nuclide.

12. $^{232}_{90}Th \xrightarrow{-\alpha} {}^{228}_{88}Ra \xrightarrow{-\beta} {}^{228}_{89}Ac \xrightarrow{-\beta} {}^{228}_{90}Th \xrightarrow{-\alpha}$

$^{224}_{88}Ra \xrightarrow{-\alpha} {}^{220}_{86}Rn \xrightarrow{-\alpha} {}^{216}_{84}Po \xrightarrow{-\alpha}$

$^{212}_{82}Pb \xrightarrow{-\beta} {}^{212}_{83}Bi \xrightarrow{-\beta} {}^{212}_{84}Po \xrightarrow{-\alpha} {}^{208}_{82}Pb$

14. $^{211}_{83}Bi \longrightarrow {}^{4}_{2}He + {}^{207}_{81}Tl \qquad {}^{207}_{81}Tl \longrightarrow {}^{0}_{-1}e + {}^{207}_{82}Pb$

16. They were the first to report nuclear fission.

18. The fission reaction in a nuclear reactor and in an atomic bomb are essentially the same. The difference is that the fissioning is uncontrolled in the bomb. In a nuclear reactor, the fissioning rate is controlled.

20. The mass defect is the difference between the mass of an atom and the sum of the masses of the number of protons, neutrons, and electrons in that atom. The energy equivalent of this mass defect is known as the nuclear binding energy.

22. (a) High levels of radiation can cause nausea, vomiting, diarrhea, and death.
(b) Long exposure to low levels of radiation can weaken the body and cause malignant tumors.
(c) Radiation can damage DNA molecules in the body, causing mutations.

24. A radioactive "tracer" is a radioactive material whose presence is traced by a Geiger counter or some other detecting device.

26. Radioactivity could be used to locate a leak in an underground pipe by using a water soluble tracer element. For example, dissolve the tracer in water and pass the water through the pipe. Test the ground along the path of the pipe with a Geiger counter until radioactivity from the leak is detected. Then dig.

28. The correct statements are: c, e, f, g, j, k, n, o, q, r, t.

30.

		Protons	Neutrons	Nucleons
(a)	$^{235}_{92}U$	92	143	235
(b)	$^{82}_{35}Br$	35	47	82

32. Its atomic number increases by one, and its mass number remains unchanged.

34. Equations for alpha decays:
(a) $^{192}_{78}Pt \longrightarrow {}^{4}_{2}He + {}^{188}_{76}Os$
(b) $^{210}_{84}Po \longrightarrow {}^{4}_{2}He + {}^{206}_{82}Pb$

36. (a) $^{239}_{93}Np \longrightarrow {}^{0}_{-1}e + {}^{239}_{94}Pu$
(b) $^{90}_{38}Sr \longrightarrow {}^{0}_{-1}e + {}^{90}_{39}Y$

38. $^{30}_{15}P \longrightarrow {}^{30}_{14}S + {}^{0}_{+1}e$

40. (a) $^{66}_{29}Cu \longrightarrow {}^{66}_{30}Zn + {}^{0}_{-1}e$
(b) ${}^{0}_{-1}e + {}^{7}_{4}Be \longrightarrow {}^{7}_{3}Li$
(c) $^{27}_{13}Al + {}^{4}_{2}He \longrightarrow {}^{30}_{14}Si + {}^{1}_{1}H$
(d) $^{85}_{37}Rb + {}^{1}_{0}n \longrightarrow {}^{82}_{35}Br + {}^{4}_{2}He$

42. 3 half-lives; 1/8

44. (a) 5.4×10^{11} J/mol (b) 0.199% mass loss

46. 11,340 years old

48. (a) 0.0424 g/mol (b) 3.8×10^{12} J/mol

50. (a) Geiger counter: Radiation passes through a thin glass window into a chamber filled with argon gas and containing two electrodes. Some of the argon ionizes, sending a momentary electrical impulse between the electrodes to the detector. This signal is amplified electronically and read out on a counter or as a series of clicks.
(b) Scintillation counter: Radiation strikes a scintillator, which is composed of molecules that emit light in the presence of ionizing radiation. A light-sensitive detector counts the flashes and converts them into a digital readout.
(c) Film badge: Radiation penetrates a film holder. The silver grains in the film darken when exposed to radiation. The film is developed at regular intervals.

52. Fission is the process of splitting a large nucleus into two roughly equal mass pieces. Fusion is the process of combining two relatively small nuclei to form a single larger nucleus.

54. (a) $^{235}_{92}U + {}^{1}_{0}n \longrightarrow {}^{143}_{54}Xe + 3{}^{1}_{0}n + {}^{90}_{38}Sr$
(b) $^{235}_{92}U + {}^{1}_{0}n \longrightarrow {}^{102}_{39}Y + 3{}^{1}_{0}n + {}^{131}_{53}I$
(c) $^{14}_{7}N + {}^{1}_{0}n \longrightarrow {}^{1}_{1}H + {}^{14}_{6}C$

56. $^{236}_{92}U \longrightarrow {}^{90}_{38}Sr + 3{}^{1}_{0}n + {}^{143}_{54}Xe$

58. (a) $^{87}_{37}Rb \longrightarrow {}^{0}_{-1}e + {}^{87}_{38}Sr$
(b) $^{87}_{38}Sr \longrightarrow {}^{0}_{+1}e + {}^{87}_{37}Rb$

60. 7680 g

62. (a) 0.500 g left (c) 0.0625 g
(b) 0.250 g left (d) 9.77×10^{-4} g

64. 2.267×10^{4} years

66. 1.6397×10^{-13} J

Chapter 19

2. Carbon has the least metallic character in Group IVA. It is not lustrous or malleable.

4. The process of reduction is used to convert an ore to a free metal.

6. In a solution alloy, the metals are combined in the molten state and the components mix uniformly and randomly. In an intermetallic compound the alloy is homogeneous and of fixed composition.
solution alloy examples: brass, pewter, nitinol
intermetallic compound examples: dental amalgam, Co_5Sm, Cr_3Pt

8. Hydrogen is considered to be a family of one because its properties and reactions are unique and do not fit with those of other elements.

10. Fr ionization energy would most likely be less than that of Cs.

12. Answers will vary: regulation of pH, intercellular osmotic pressure, regulate water balance, etc.

14. (a) Precipitation of Mg^{2+} as $Mg(OH)_2$ from seawater:
$Mg^{2+} + 2\,OH^- \longrightarrow Mg(OH)_2(s)$
(b) Conversion of $Mg(OH)_2$ to $MgCl_2$:
$Mg(OH)_2(s) + 2HCl(aq) \longrightarrow MgCl_2(aq) + H_2O(l)$
(c) Electrolysis of $MgCl_2$:
$MgCl_2(aq) \xrightarrow{\text{electrolysis}} Mg(s) + Cl_2(g)$

16. Answers will vary.
Mg: present in chlorophyll of green plants; associated with the production and use of adenosine triphosphate (ATP), the central energy molecule of the cell.
Ca: found in inorganic salts in human bones and teeth; involved in muscle contraction and hormone regulation.

18. When these elements are heated in a flame some ground-state electrons are promoted to higher energy levels. When they fall back to ground state the absorbed energy is emitted as the characteristic light.

20. A mordant is a substance that binds to both cloth and dye molecules, adhering the dye to the cloth.

22. Answers will vary: lipstick, lotions, Silly Putty, car polish

24. Beginning with the atmosphere, nitrogen is fixed by bacterial action, combustion, or chemical fixation. In the soil, nitrogen is converted to nitrates, which are absorbed by higher plants and animals and converted to organic compounds. Eventually, the nitrogen is returned to the soil in the form of urea or feces or through bacterial decomposition of dead plants and animals. Some remains in the soil and the rest returns to the atmosphere as free nitrogen.

26. Free nitrogen is so inert because it is triple bonded and has a very high bond dissociation energy. This property makes nitrogen useful as a nonoxidizing atmosphere for preservation of food, wine, or other articles. It is also used as a liquid coolant to freeze things because of its low boiling point (−196°C).

28. Sulfur compounds are less ionic and more covalent than oxygen compounds.

30. F > Cl > Br > I as oxidizing agents. This is because F is most electronegative and therefore has the greatest capacity to remove an electron from another substance. As electronegativity decreases down the column so does the capacity to oxidize other substances.

32. Pig iron is the initial product in steel making and contains several impurities. In steel, these impurities have been removed or lowered to a controlled level.

34. Aluminum and magnesium both form oxides that adhere tightly to the surface of the metal, protecting it from further corrosion. Iron forms rust, which flakes off the surface, allowing further corrosion of the metal.

36. Answers will vary: wires in electrical system; alloys for coins and jewelry; a trace element essential for life; cooking utensils, etc.

38. The correct statements are: b, c, d, e, h, i j, m, o, p, r, s.

40. (a) Br nonmetal (c) Pd metal
(b) As metalloid (d) H nonmetal

42. (a) highest electronegativity N, Cl
(b) smallest atomic radius N
(c) smallest ionization energy Ca
(d) greatest metallic character Ca

44.

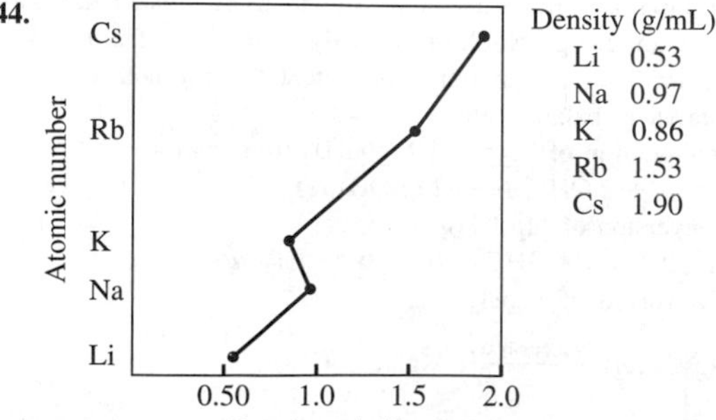

The density of Fr is probably less than 2.0 g/mL. According to density, Fr would sink. However, Fr would be so reactive with water that it would explode upon contact.

46.

oxide	BeO
hydroxide	$Mg(OH)_2$
acetate	$Ca(C_2H_3O_2)_2$
hydrogen carbonate	$Sr(HCO_3)_2$
carbonate	$BaCO_3$

48. (a) :N:::N: (c) $\left[\text{:Ö::N:Ö: / :Ö:}\right]^-$
(b) $:\ddot{\text{N}}:^{3-}$

50. (a) A lack of copper in the diet could affect the synthesis of hemoglobin, the development of connective tissue, the production of melanin, etc.
(b) A lack of zinc in the diet could affect certain enzymes in the digestive and respiratory systems.
(c) A lack of iron in the diet could affect the formation of hemoglobin. Iron is also required for the liver, spleen, and bone marrow.
(d) A lack of calcium in the diet could affect proper formation of teeth and bones, and can lead to osteoporosis. Calcium is also needed for muscle contraction, hormone regulation, and blood coagulation.

52.

$\cdot\ddot{\text{X}}: + \cdot\ddot{\text{X}}: \longrightarrow :\ddot{\text{X}}:\ddot{\text{X}}:$	covalent (share e^-)
$\text{H}\cdot + \cdot\ddot{\text{X}}: \longrightarrow \text{H}:\ddot{\text{X}}:$	covalent (share e^-)
$\text{Na}\cdot + \cdot\ddot{\text{X}}: \longrightarrow [\text{Na}]^+ [:\ddot{\text{X}}:]^-$	ionic (transfer e^-)

54. Hydrogen has an oxidation number of −1 in a hydride:
$2\ K + H_2 \longrightarrow 2\ KH$

56. (a) If carbon is burned in a limited supply of O_2, CO is formed.
(b) With an excess of O_2, CO_2 is formed.

58. $\left(\frac{300{,}000\ \text{L}\ H_2O}{1\ \text{hr}}\right)\left(\frac{1\ \text{kg}}{1\ \text{L}}\right)\left(\frac{1000\ \text{g}}{1\ \text{kg}}\right)\left(\frac{0.20\ \text{g Cl}}{1{,}000{,}000\ \text{g}\ H_2O}\right) = \frac{60\ \text{g Cl}}{\text{hr}}$

60. magnesium, strontium, or barium
All these elements are in Group IIA of the periodic table and contain the same outer shell electron structure as calcium.

62. Fe_2O_3 is formed. Fe_2O_3 is insoluble in water and alcohol, but soluble in acids.

Chapter 20

2. The carbon atom has only two unshared electrons, making two covalent bonds logical, but in CH_4, carbon forms four equivalent bonds. Promoting one 2s electron to the empty 2p orbital would make four bonds possible, but without hybridization, we could not explain the fact that all four bonds in CH_4 are identical, and the bond angles are equal (109.5°).

4. (a) A molecule of ethane, C_2H_6, contains seven sigma bonds.
(b) A molecule of butane, C_4H_{10}, contains thirteen sigma bonds.
(c) A molecule of 2-methylpropane also contains thirteen sigma bonds.

6. The correct statements are: a, b, c, d, g, i, j, l, m, p, s.

8. Lewis structures:

(a) CH_4
```
  H
H:C:H
  H
```
(b) C_3H_8
```
  H H H
H:C:C:C:H
  H H H
```
(c) C_5H_{12}
```
  H H H H H
H:C:C:C:C:C:H
  H H H H H
```

10. Compounds (b), (e), and (f) are identical. The others are different.

12. The formulas in Exercise 10 contain the following numbers of methyl groups:
(a) 4 (b) 4 (c) 4 (d) 4 (e) 4 (f) 4

14. hexane $CH_3CH_2CH_2CH_2CH_2CH_3$
```
CH3CH2CH2CHCH3
            |
           CH3

CH3CH2CHCH2CH3     CH3           CH3
      |             |             |
     CH3       CH3CH2CCH3     CH3CHCHCH3
                    |             |
                   CH3           CH3
```

16. (a) CH_2Cl_2, one
```
   H
   |
H—C—Cl
   |
   Cl
```
(b) C_3H_7Br, two
```
   H  H  H             H  H  H
   |  |  |             |  |  |
H—C—C—C—Br        H—C—C—C—H
   |  |  |             |  |  |
   H  H  H             H  Br H
```
(c) $C_3H_6Cl_2$, four
```
   H  H  H             H  H  H
   |  |  |             |  |  |
H—C—C—C—Cl        H—C—C—C—Cl
   |  |  |             |  |  |
   H  H  Cl            H  Cl H

   H  H  H             H  Cl H
   |  |  |             |  |  |
Cl—C—C—C—Cl       H—C—C—C—H
   |  |  |             |  |  |
   H  H  H             H  Cl H
```
(d) $C_4H_8Cl_2$, nine
```
   H  H  H  H            H  H  H  H
   |  |  |  |            |  |  |  |
H—C—C—C—C—Cl       H—C—C—C—C—Cl
   |  |  |  |            |  |  |  |
   H  H  H  Cl           H  H  Cl H

   H  H  H  H            H  H  H  H
   |  |  |  |            |  |  |  |
H—C—C—C—C—Cl      Cl—C—C—C—C—Cl
   |  |  |  |            |  |  |  |
   H  Cl H  H            H  H  H  H

   H  H  Cl H            H  Cl Cl H
   |  |  |  |            |  |  |  |
H—C—C—C—C—H        H—C—C—C—C—H
   |  |  |  |            |  |  |  |
   H  H  Cl H            H  H  H  H

      H                     H
      |                     |
   H—C—Cl                H—C—H
   H  |  H               H  |  H
   |  |  |               |  |  |
H—C—C—C—Cl          H—C—C—C—Cl
   |  |  |               |  |  |
   H  H  H               H  Cl H

      H
      |
   H—C—H
   H  |  H
   |  |  |
H—C—C—C—Cl
   |  |  |
   H  H  Cl
```

18. IUPAC names
(a) chloroethane
(b) 1-chloro-2-methylpropane
(c) 2-chlorobutane
(d) methylcyclopropane
(e) 2,4-dimethylpentane

20.
```
(a)                   CH3
                      |
    CH3CH2CHCH2CHCH3
            |
           CH2CH3

(b) CH3CH2CH2CHCH2CH2CH3
               |
              C(CH3)3

                                  CH3
                                  |
      CH3     CH2CH3     CHCH3
       |          |          |
(c) CH3CHCH2CCH2CH2CHCHCH2CH3
                  |             |
                 CH3           CH3

     CH3 CH2CH3
      |   |
(d) CH3—C—CHCH2CH2CH2CH2CH3
      |
     CH3
```

22. (a) 3-methyl-5-ethyloctane
```
CH3CH2CHCH2CHCH2CH2CH3
        |     |
       CH3   CH2CH3
```
Ethyl should be named before methyl (alphabetical order). The numbering is correct. The correct name is 5-ethyl-3-methyloctane.

(b) 3,5,5-triethylhexane
```
                 CH2CH3
                   |
CH3CH2CHCH2CCH3
        |      |
       CH2    CH2CH3
        |
       CH3
```
The name is not based on the longest carbon chain (7 carbons). The correct name is 3,5-diethyl-3-methylheptane.

(c) 4,4-dimethyl-3-ethylheptane
```
              CH3
               |
CH3CH2CH—CCH2CH2CH3
         |     |
    CH3CH2    CH3
```
Ethyl should be named before dimethyl (alphabetical order). The correct name is 3-ethyl-4,4 dimethylheptane.

24. $CH_3CH_2CH_2CH_2CH_2CH_2Cl$ $CH_3CH_2CH_2CH_2CHClCH_3$
$CH_3CH_2CH_2CHClCH_2CH_3$

26. (a) $CH_3CH_2CH_3 + Br_2 \xrightarrow{h\nu} CH_3CH_2CH_2Br + CH_3CHBrCH_3 + HBr$

(b) $CH_3CH_2CH_3 + 5\ O_2 \xrightarrow{\Delta} 3\ CO_2 + 4\ H_2O$

28. The formula for dodecane is $C_{12}H_{26}$

30. Data: $\frac{1\text{ gal}}{60\text{ miles}}$; 60 miles traveled; $\frac{19\text{ mol }C_8H_{18}}{\text{gal}}$

$$2\ C_8H_{18} + 25\ O_2 \longrightarrow 16\ CO_2 + 18\ H_2O$$

$$\frac{1\text{ gal}}{60\text{ miles}} \times 60\text{ miles} = 1\text{ gal gasoline used}$$

$$19\text{ mol }C_8H_{18} \times \frac{16\text{ mol }CO_2}{2\text{ mol }C_8H_{18}} = 1.5 \times 10^2\text{ mol }CO_2$$

$$PV = nRT \qquad V = \frac{nRT}{P}$$

$$V = \frac{1.5 \times 10^2\text{ mol }CO_2 \times 0.0821\text{ L-atm} \times 293\text{ K}}{1\text{ atm} \qquad \text{mol K}} = 3.6 \times 10^3\text{ L }CO_2$$

32. (a) elimination
(b) substitution
(c) addition

34. Propane is the most volatile. It has the lowest boiling point of the three alkanes.

36. (a) sp^3 hybrid orbitals in carbon
(b) Structural isomers are not possible because only one carbon atom is present.
(c) dichlorodifluoromethane
(d) The closest classification for Freon-12 (CF_2Cl_2) in Table 20.1 is alkyl halide.

Chapter 21

2. 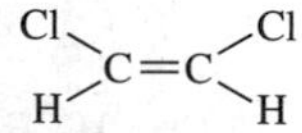 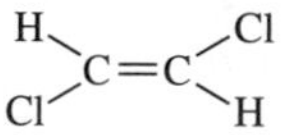 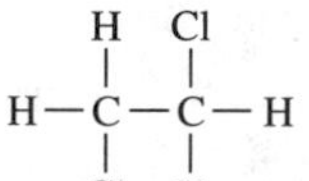 Cl(H)C=C(Cl)H   H(Cl)C=C(Cl)H   H–C(H)(Cl)–C(Cl)(H)–H

cis-1, 2-dichloroethene   trans-1, 2-dichloroethene   1, 2-dichloroethane

We are able to get cis-trans isomers of ethene because there is no rotation around the carbon–carbon double bond, so the two structures shown are not the same. With 1, 2-dichloroethane, there is free rotation of the carbon–carbon single bond. In the structure shown, exchanging the chlorine on the right with either hydrogenatom on that carbon appears to make a different isomer, but rotation makes any of the three positions equivalent.

4. Acetylene presents two different explosion hazards:
(a) It can form an explosive mixture with oxygen or air.
(b) When highly compressed or liquified, acetylene may decompose violently, either spontaneously or from a slight shock.

6. During the 10-year period from 1935 to 1945, the major source of aromatic hydrocarbons shifted from coal tar to petroleum due to the rapid growth of several industries that used aromatic hydrocarbons as raw material. These industries include drugs, dyes, detergents, explosives, insecticides, plastics, and synthetic rubber. Since the raw material needs far exceeded the aromatics available from coal tar, another source had to be found, and processes were developed to make aromatic compounds from alkanes in petroleum. World War II, which occurred in this period, put high demands on many of these industries, particularly explosives.

8. The correct statements are: a, f, g, h, i, j, k, m, n, o, s, v, w, x, y, z.

10. (a) propane (b) propene (c) propyne

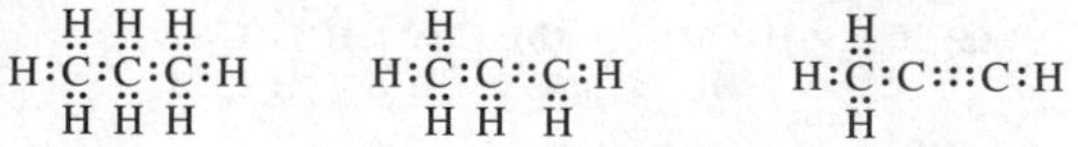

H:C:C:C:H (each C with H above and below)   H:C::C:C:H (H's above/below)   H:C:C:::C:H (H above and below first C)

12. (a) C_3H_5Cl

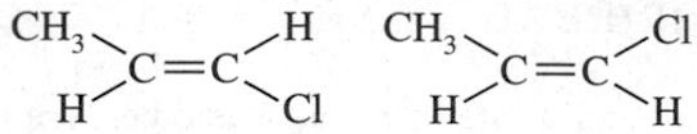

$CH_3(H)C=C(H)Cl$ $CH_3(H)C=C(Cl)H$ $CH_3CCl{=}CH_2$ $CH_2ClCH{=}CH_2$

(b) chlorocyclopropane

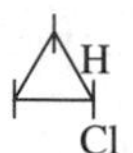

14. (a) 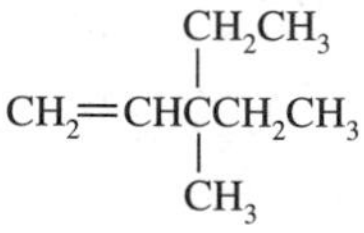 $CH_2{=}CHC(CH_2CH_3)(CH_3)CH_2CH_3$

(b) 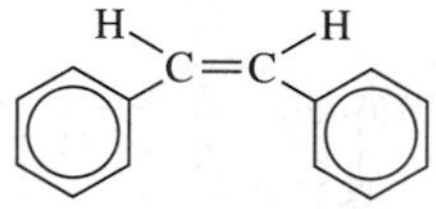 H(C₆H₅)C=C(H)(C₆H₅)

(c) $CH{\equiv}CCHCH_3$ with phenyl on CH

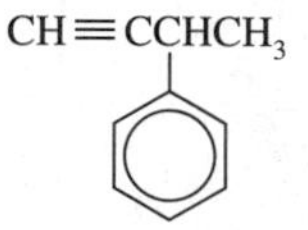

(d) 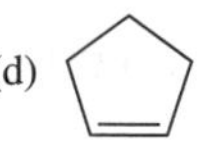

(e) 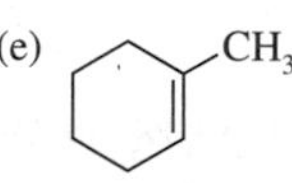

(f) 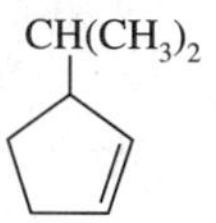 $CH(CH_3)_2$ on cyclopentene

16. (a) 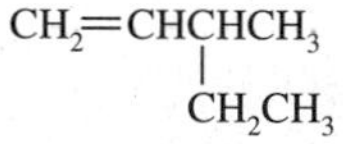 $CH_2{=}CHCHCH_3$ with CH_2CH_3 on CH. Longest chain contains five carbon atoms. Correct name: 3-methyl-1-pentene

(b) 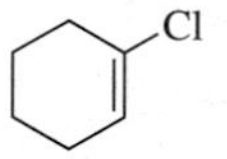 1-chlorocyclohexene (Cl on ring). The C=C bond in cyclohexene is numbered so that substituted groups have the smallest numbers. Correct name: 1-chlorocyclohexene

(c) $CH_3CH_2CH_2CH{=}CHCH_3$ Numbering was started from the wrong end of the structure. Correct name: 2-hexene

18. (a) cis-4-methyl-2-hexene (b) 2, 3-dimethyl-2-butene
(c) 3-isopropyl-1-pentene

20. all the pentynes, C_5H_8

1-pentyne: $CH_3CH_2CH_2C{\equiv}CH$ 2-pentyne: $CH_3CH_2C{\equiv}CCH_3$ 3-methyl-1-butyne: $CH_3CH(CH_3)C{\equiv}CH$

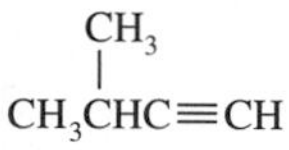

22. Only structure (c) will show cis-transisomers:
$CH_2ClCH{=}CHCH_2Cl$

24. (a) 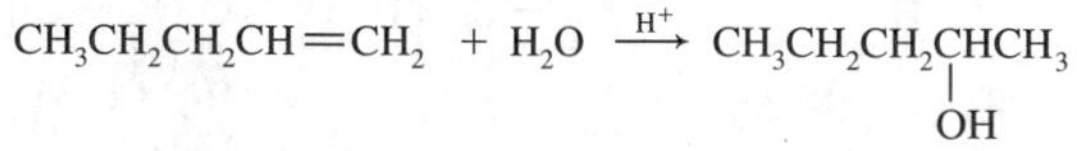 $CH_3CH_2CH_2CH{=}CH_2 + H_2O \xrightarrow{H^+} CH_3CH_2CH_2CH(OH)CH_3$

(b) $CH_3CH_2CH{=}CHCH_3 + HBr \longrightarrow CH_3CH_2CHBrCH_2CH_3 + CH_3CH_2CH_2CHBrCH_3$

(c) $CH_2{=}CHCl + Br_2 \longrightarrow CH_2BrCHClBr$

(d) 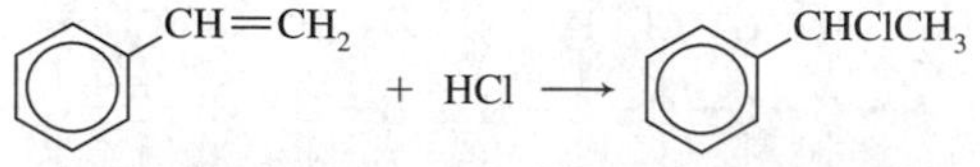 $C_6H_5CH{=}CH_2 + HCl \longrightarrow C_6H_5CHClCH_3$

(e) 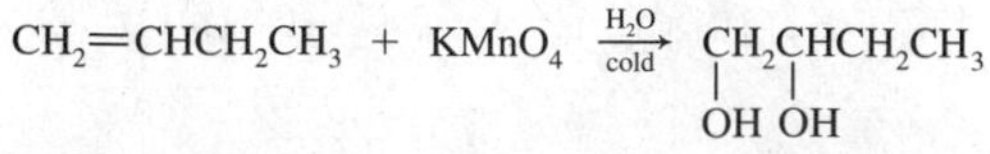 $CH_2{=}CHCH_2CH_3 + KMnO_4 \xrightarrow[\text{cold}]{H_2O} CH_2(OH)CH(OH)CH_2CH_3$

26. (a) $CH_3C{\equiv}CH + H_2(1\ mol) \xrightarrow[1\ atm]{Pt,\ 25°C} CH_3CH{=}CH_2$

(b) $CH_3C{\equiv}CCH_3 + Br_2(2\ mol) \longrightarrow CH_3CBr_2CBr_2CH_3$

(c) two-step reaction:

$$CH_3C{\equiv}CH + HCl \longrightarrow CH_3CCl{=}CH_2 \xrightarrow{HCl} CH_3CCl_2CH_3$$

28. When cyclopentene, [cyclopentene], reacts with

(a) Cl_2 the product is [1,2-dichlorocyclopentane: Cl, Cl]

(b) HBr the product is [bromocyclopentane: Br]

(c) H_2, Pt the product is [cyclopentane]

(d) H_2O, H^+ the product is [cyclopentanol: OH]

30. (a) $CH_3C{\equiv}CCH_2CH_3 + Cl_2(1\ mole) \longrightarrow CH_3CCl{=}CClCH_2CH_3$

$$CH_3CCl{=}CClCH_2CH_3 + H_2 \xrightarrow[1\ atm]{25°C} CH_3CHClCHClCH_2CH_3$$

(b) $CH_3C{\equiv}CCH_2CH_3 + HCl \longrightarrow CH_3CCl{=}CHCH_2CH_3$

$CH_3CCl{=}CHCH_2CH_3 + HCl \longrightarrow CH_3CCl_2CH_2CH_2CH_3$

Can also yield $CH_3CH_2CCl_2CH_2CH_3$ at the same time.

(c) $CH_3C{\equiv}CCH_2CH_3 + 2\ Cl_2 \longrightarrow CH_3CCl_2CCl_2CH_2CH_3$

32. (a) [benzene ring with CH_3]

(b) [benzene ring with CH_2CH_3]

(c) [benzene ring with OH]

(d) [anthracene]

34. (a) [benzene ring with Cl, Cl, NO_2]

(b) [benzene ring with NO_2, NO_2]

(c) [benzene ring–CH(CH_3)–benzene ring]

(d) [benzene ring with $CH{=}CH_2$]

36. (a) trichlorobenzenes

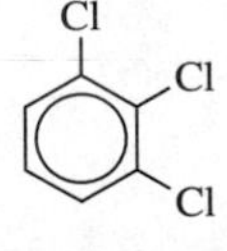

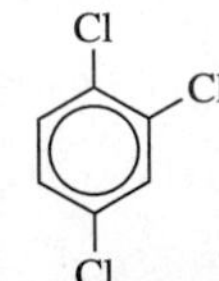

[1,3,5-trichlorobenzene: Cl, Cl, Cl]

1, 2, 3-trichlorobenzene   1, 2, 4-trichlorobenzene   1, 3, 5-trichlorobenzene

(b) the benzene derivatives of formula C_8H_{10}:

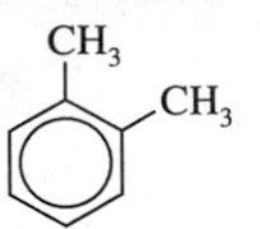

1, 2-dimethylbenzene or o-xylene

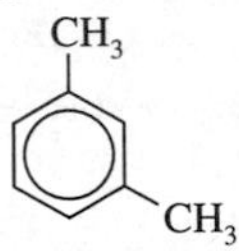

1, 3-dimethylbenzene or m-xylene

[benzene ring with CH_3, CH_3 para]

1, 4-dimethylbenzene or p-xylene

[benzene ring with CH_2CH_3]

ethylbenzene

38. The isomers that can be written substituting a third chlorine atom on *o*-dichlorobenzene are:

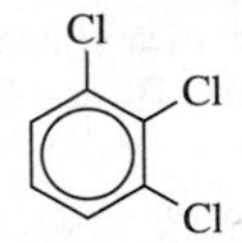

1, 2, 3-trichlorobenzene

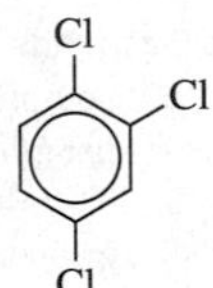

1, 2, 4-trichlorobenzene

40. (a) styrene

(b) *m*-nitrotoluene

(c) 2, 4-dibromobenzoic acid

(d) isopropylbenzene

(e) 2, 4, 6-tribromophenol

42. (a) [benzene] + CH_3CHCH_3 (with Cl on central C) $\xrightarrow{AlCl_3}$ [benzene ring with $CHCH_3$ bearing CH_3] + HCl

isopropylbenzene

(b) [benzene ring with CH_3] + $KMnO_4$ $\xrightarrow{H_2O}$ [benzene ring with COOH]

benzoic acid

44. Two tests can be used:

(a) Baeyer test—hexene will decolorize $KMnO_4$ solution; cyclohexane will not.

(b) In the absence of sunlight, hexene will react with and decolorize bromine; cyclohexane will not.

46. (a) $\overset{+}{C}H_3$

(b) $CH_3CH_2\overset{+}{C}H_2$

(c) $CH_3\overset{+}{C}(CH_3)CH_3$

(d) $CH_3CH_2CH_2CH_2\overset{+}{C}H_2$

48. The reaction mechanism by which benzene is brominated in the presense of $FeBr_3$:

$FeBr_3 + Br_2 \longrightarrow FeBr_4^- + Br^+$

benzene + Br^+ ⟶ (H, Br)-substituted benzenium ion (+)

(H, Br)-substituted benzenium ion (+) + $FeBr_4^-$ ⟶ bromobenzene + $FeBr_3$ + HBr

50. (a) $CH_3CHClCH_2CH_3 \xrightarrow{-HCl} CH_2{=}CHCH_2CH_3 +$
$CH_3CH{=}CHCH_3$

(b) $CH_2ClCH_2CH_2CH_2CH_3 \xrightarrow{-HCl} CH_2{=}CHCH_2CH_2CH_3$

(c) chlorocyclohexane (Cl) $\xrightarrow{-HCl}$ cyclohexene

52. Baeyer test: Add $KMnO_4$ solution to each sample. The $KMnO_4$ will lose its purple color with 1-heptene. There will be no reaction (no color change) with 1-heptane.

54. cyclopentane Other possibilities are:

methylcyclobutane (CH_3), ethylcyclopropane (CH_2CH_3), dimethylcyclopropane (CH_3 CH_3)

Chapter 22

2. Alkenes are almost never made from alcohols because the alcohols are almost always the higher value material. This is because recovering alkenes from hydrocarbon sources in an oil refinery (primarily catalytic cracking) is a relatively cheap process.

4. Oxidation of primary alcohols yields aldehydes. Further oxidation yields carboxylic acids. Examples are:

$$CH_3CH_2OH + [O] \longrightarrow CH_3\overset{\overset{\displaystyle O}{\|}}{C}{-}H + H_2O$$

$$CH_3\overset{\overset{\displaystyle O}{\|}}{C}{-}H + [O] \longrightarrow CH_3\overset{\overset{\displaystyle O}{\|}}{C}{-}OH$$

6. (a) Methanol is a poisonous liquid capable of causing blindness or death if taken internally. Exposure to methanol vapors is also very dangerous.

(b) Ethanol can act as a food, a drug, and a poison. The body can metabolize small amounts of ethanol to produce energy; thus it is a food. It depresses brain functions so that activities requiring skill and judgment are impaired; thus it is a drug. With very large consumption, the depression of brain function can lead to unconsciousness and death; thus it is a poison.

8. Low molar mass ethers present two hazards.

(a) They are very volatile and their highly flammable vapors form explosive mixtures with air.

(b) They also slowly react with oxygen in the air to form unstable explosive peroxides.

10. The correct statements are: a, b, e, g, h, i, j, m, o, q, s.

12. (a) $CH_3CH(OH)CH_2CH_3$

(b) $CH_3C(CH_3)(OH)CH_2CH_3$

(c) $CH_3CH(OH)CH_3$

(d) cyclopentane–OH

(e) $CH_3CH_2CH_2CH_2CH_2SH$

(f) $CH_3CH_2CH(CH_2CH_3)CH(OH)CH_2CH_3$

14. (a) C_3H_8O

	Alcohols	Ethers
(a) C_3H_8O	$CH_3CH_2CH_2OH$	$CH_3OCH_2CH_3$
	$CH_3CH(OH)CH_3$	
(b) C_4H_8O	$CH_3CH_2CH_2CH_2OH$	$CH_3OCH_2CH_2CH_3$
	$CH_3CH_2CH(OH)CH_3$	$CH_3OCH(CH_3)_2$
	$(CH_3)_2CHCH_2OH$	$CH_3CH_2OCH_2CH_3$
	$(CH_3)_3C{-}OH$	

16. The primary alcohols in Question 14 are:
$CH_3CH_2CH_2OH$ $CH_3CH_2CH_2CH_2OH$
$(CH_3)_2CHCH_2OH$
The secondary alcohols are:
$CH_3CH(OH)CH_3$ $CH_3CH_2CH(OH)CH_3$
The tertiary alcohol is: $(CH_3)_3C{-}OH$

18. (a) ethanol (ethyl alcohol)
(b) 2-phenylethanol
(c) 3-methyl-3-pentanol
(d) 1-methylcyclopentanol
(e) 3-pentanol
(f) 1,2-propanediol
(g) 4-ethyl-2-hexanol

20. (a) (CH_3 on cyclopentene ring) methylcyclopentene

(b) $CH_3CH{=}CHCH_3$ 2-butene

(c) $CH_3C(CH_3){=}CHCH_2CH_3$ 2-methyl-2-pentene

22. (a) cyclopentane–OH cyclopentanol

(b) cyclopentane with CH_3 and OH 1-methylcyclopentanol

(c) $CH_3CH(CH_3)CH(OH)CH(CH_3)CH_3$ 2,4-dimethyl-3-pentanol

24. Oxidation of a secondary alcohol:

$$\underset{\displaystyle\quad\;\,|\atop\displaystyle\quad\;\,OH}{\overset{\displaystyle\;\,CH_3\atop\displaystyle\;\;|}{CH_3CHCHCH_2CH_3}} \longrightarrow \underset{\displaystyle\quad\;\,\|\atop\displaystyle\quad\;\,O}{\overset{\displaystyle\;\,CH_3\atop\displaystyle\;\;|}{CH_3CCHCHCH_3}}$$

ketone

26. (a) $CH_3CH{=}CHCH_3 + H_2O \xrightarrow{H^+} CH_3CH_2\underset{\displaystyle |\atop\displaystyle OH}{C}HCH_3$

(b) $CH_3CH_2CH_2CH{=}CH_2 + H_2O \xrightarrow{H^+} CH_3CH_2CH_2\underset{\displaystyle |\atop\displaystyle OH}{C}HCH_3$

(c) $CH_3\underset{\displaystyle |\atop\displaystyle CH_3}{C}{=}CHCH_3 + H_2O \xrightarrow{H^+} CH_3\overset{\displaystyle OH\atop\displaystyle |}{\underset{\displaystyle |\atop\displaystyle CH_3}{C}}CH_2CH_3$

28. (a) $CH_3\underset{\displaystyle |\atop\displaystyle OH}{C}HCH_2CH_3$ 2-butanol

(b) $CH_3\underset{\displaystyle |\atop\displaystyle OH}{C}H\overset{\displaystyle CH_2CH_3\atop\displaystyle |}{C}HCH_2CH_3$ 3-ethyl-2-pentanol

(c) cyclopentane–OH cyclopentanol

30. (a) $2\ CH_3CH_2OH + 2\ Na \longrightarrow 2\ CH_3CH_2O^-Na^+ + H_2$

(b) $CH_3CH_2CH_2CH_2OH \xrightarrow[H_2SO_4]{K_2Cr_2O_7} CH_3CH_2CH_2\overset{\displaystyle O\atop\displaystyle \|}{C}{-}H$

(c) cyclopentyl–$CH{=}CH_2$ $\xrightarrow[H_2SO_4]{H_2O}$ cyclopentyl–$\underset{\displaystyle |\atop\displaystyle OH}{C}HCH_3$

(d) $CH_3CH_2\overset{\displaystyle O\atop\displaystyle \|}{C}{-}OCH_3 + NaOH$

$\longrightarrow CH_3CH_2\overset{\displaystyle O\atop\displaystyle \|}{C}{-}O^-Na^+ + CH_3OH$

32. (a) benzene ring with OH (top) and NO_2 (para)

(b) benzene ring with OH, flanked by H_3C and CH_3

(c) benzene ring with OH and OH (ortho)

34. (a) *p*-dihydroxybenzene (hydroquinone)
(b) 2,4-dimethylphenol
(c) 2,4-dinitrophenol
(d) *m*-hexylphenol

36. Order of decreasing solubility in water: [a (highest), d, c, b (lowest)]
(a) $CH_3CH(OH)CH(OH)CH_2OH$
(d) $CH_3CH(OH)CH_2CH_2OH$
(c) $CH_3CH_2CH_2CH_2OH$
(b) $CH_3CH_2OCH_2CH_3$

38. The six isomeric ethers, $C_5H_{12}O$, are:

$CH_3OCH_2CH_2CH_2CH_3$
methyl *n*-butyl ether
(1-methoxybutane)

$CH_3O\overset{\displaystyle CH_3\atop\displaystyle |}{C}HCH_2CH_3$
methyl sec-butyl ether
(2-methoxybutane)

$CH_3OCH_2\overset{\displaystyle CH_3\atop\displaystyle |}{C}HCH_3$
methyl isobutyl ether
(1-methoxy-2-methylpropane)

$CH_3O\overset{\displaystyle CH_3\atop\displaystyle |}{\underset{\displaystyle |\atop\displaystyle CH_3}{C}}CH_3$
methyl *t*-butyl ether
(2-methoxy-2-methylpropane)

$CH_3CH_2OCH_2CH_2CH_3$
ethyl *n*-propyl ether
(1-ethoxypropane)

$CH_3CH_2O\overset{\displaystyle CH_3\atop\displaystyle |}{C}HCH_3$
ethyl isopropyl ether
(2-ethoxypropane)

40. $CH_3CH_2OCH_2CH_3 + 6\ O_2 \longrightarrow 4\ CO_2 + 5\ H_2O$

42. These are possible combinations of reactants to make the following ethers by the Williamson syntheses:

(a) $CH_3CH_2CH_2OCH_2CH_2CH_3$

$CH_3CH_2CH_2ONa + CH_3CH_2CH_2Cl$

(b) $H\overset{\displaystyle CH_3\atop\displaystyle |}{\underset{\displaystyle |\atop\displaystyle CH_3}{C}}OCH_2CH_2CH_3$

$H\overset{\displaystyle CH_3\atop\displaystyle |}{\underset{\displaystyle |\atop\displaystyle CH_3}{C}}ONa + CH_3CH_2CH_2Cl$

or $CH_3CH_2CH_2ONa + CH_3\underset{\displaystyle |\atop\displaystyle Cl}{C}HCH_3$

44.

OH OH
cis-1,2-cyclopentanediol

OH OH
trans-1,2-cyclopentanediol

46. The common phenolic structure in the catecholamines is catechol, *o*-dihydroxybenzene:

benzene ring with OH and OH (ortho)

48. Methyl alcohol is converted to formaldehyde by an oxidation reaction:

$$CH_3OH \xrightarrow{[O]} H\overset{\displaystyle O\atop\displaystyle \|}{C}H$$

50. $p\text{-}CH_3C_6H_4OH + NaOH \longrightarrow p\text{-}CH_3C_6H_4O^-Na^+ + H_2O$

$CH_3CH_3 + Cl_2 \xrightarrow{\text{light}} CH_3CH_2Cl + HCl$

$p\text{-}CH_3C_6H_4O^-Na^+ + CH_3CH_2Cl \longrightarrow p\text{-}CH_3C_6H_4OCH_2CH_3 + NaCl$

52. Differentiation between phenols and alcohols:
Phenols are acidic compounds; alcohols are not acidic.
Phenols react with NaOH to form salts; alcohols do not react with NaOH.
The OH group is bonded to a benzene ring in phenols and to a nonbenzene carbon atom in alcohols.

54. Only compound (a) will react with NaOH:

$C_6H_5OH + NaOH \longrightarrow C_6H_5O^-Na^+ + H_2O$

Chapter 23

2. Aldehydes and ketones have lower boiling points than alcohols of similar molar mass because they do not form hydrogen bonds, since they contain no —OH groups.

4. The correct statements are: b, c, f, h, i, j, l, m, o, q, r, t, v.

6. (a) ethanal, acetaldehyde (b) butanal (c) benzaldehyde (d) 2-chloro-5-isopropylbenzaldehyde (e) 3-hydroxybutanal

8. (a) 2-butanone, methyl ethyl ketone (MEK)
(b) 3,3-dimethylbutanone, methyl *t*-butyl ketone
(c) 2,5-hexanedione
(d) 1-phenyl-2-propanone, methyl benzyl ketone

10. (a) $HOCH_2CH_2C(=O)-H$

(b) $CH_3CH_2C(=O)CH(CH_3)CH_2CH_3$

(c) cyclohexanone

(d) $CH_3CH(Cl)CH_2CH(Cl)CH_2CH(Cl)C(=O)-H$

(e) $CH_3CH=CHCH_2C(=O)-H$

12. Higher boiling point
(a) pentanal (b) benzyl alcohol (c) 2-hexanone (d) 1-butanol

14. Higher aqueous solubility
(a) acetaldehyde (b) 2,4-pentanediol (c) propanal

16. (a) 1-propanol:

$CH_3CH_2CH_2OH \xrightarrow[H_2SO_4]{K_2Cr_2O_7} CH_3CH_2C(=O)-H$ or CH_3CH_2COOH

$CH_3CH_2CH_2OH + O_2 \xrightarrow[\Delta]{Cu} CH_3CH_2C(=O)-H$

(b) 2,3 dimethyl-2 butanol:

$CH_3CH(CH_3)C(OH)(CH_3)CH_3 + O_2 \xrightarrow[\Delta]{Ag}$ No reaction (3° alcohol) with either oxidizing agent

18. (a) An aldehyde group, $-C(H)=O$, must be present to give a positive Fehling test.

(b) The visible evidence for a positive Fehling test is the formation of brick-red Cu_2O, which precipitates during the reaction.

(c) $CH_3C(H)=O + 2\,Cu^{2+} \xrightarrow[H_2O]{NaOH} CH_3COONa + Cu_2O(s)$ (brick-red)

20. (a) propanal $\longrightarrow$ propanoic acid + $Cu_2O(s)$
(b) acetone $\longrightarrow$ no reaction
(c) 3-methylpentanal $\longrightarrow$ 3-methylpentanoic acid + $Cu_2O(s)$

22. (a) 3-pentanone:

$2\,CH_3CH_2C(=O)CH_2CH_3 \xrightarrow[NaOH]{\text{dilute}} CH_3CH_2C(OH)(CH_2CH_3)-CH(CH_3)-C(=O)CH_2CH_3$

(b) propanal:

$2\,CH_3CH_2C(H)=O \xrightarrow[NaOH]{\text{dilute}} CH_3CH_2CH(OH)CH(CH_3)C(H)=O$

24. (a) $CH_3CH_2C(H)=O + 2\,CH_3CH_2CH_2OH \overset{\text{dry HCl}}{\rightleftharpoons} CH_3CH_2C(H)(OCH_2CH_2CH_3)-OCH_2CH_2CH_3 + H_2O$

(b) cyclohexanone $+ CH_3OH \overset{H^+}{\rightleftharpoons}$ 1-methoxycyclohexanol (ring C bearing OH and OCH_3)

(c) $CH_3CH_2CH_2CH(OCH_3)_2 \underset{H^+}{\overset{H_2O}{\rightleftharpoons}} CH_3CH_2CH_2C(H)=O + 2\,CH_3OH$

26. (a) $C_6H_5CHO + HCN \xrightarrow{OH^-} C_6H_5CH(OH)CN$

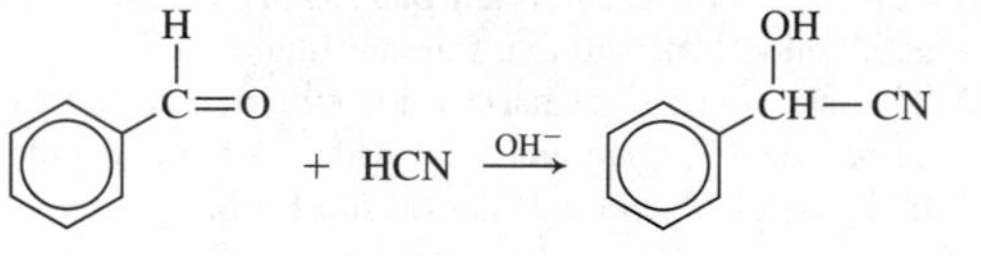

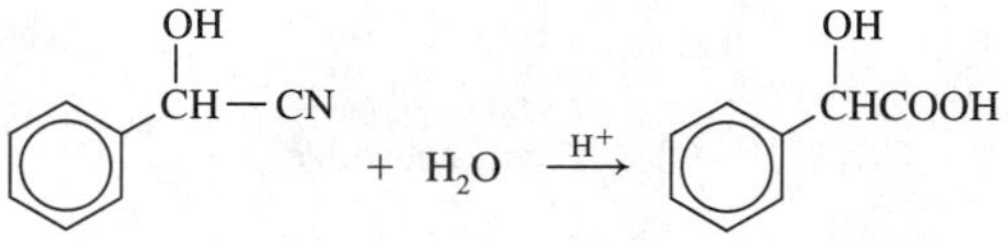

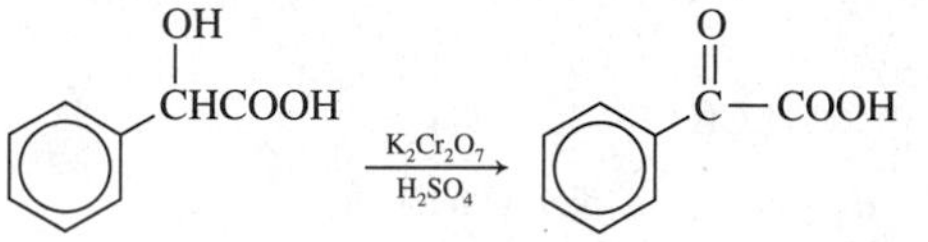

(b) $C_6H_5CH(OH)CN + H_2O \xrightarrow{H^+} C_6H_5CH(OH)COOH$

(c) $C_6H_5CH(OH)COOH \xrightarrow[H_2SO_4]{K_2Cr_2O_7} C_6H_5COCOOH$

28. $CH_3C(=O)H$ ethanal

30. The secondary alcohol in compound I is oxidized to the ketone (compound II).

32. $CH_3CH{=}O + HCN \longrightarrow CH_3CH(OH)CN \xrightarrow[H^+]{H_2O} CH_3CH(OH)COOH$

34. Pyruvic acid is changed to lactic acid by a reduction reaction.

36. (a) 3-pentanone: 3-pentanol $CH_3CH_2CH(OH)CH_2CH_3$

(b) methyl ethyl-ketone: 2-butanol $CH_3CH(OH)CH_2CH_3$

(c) 4-phenyl-2-butanone: 4-phenyl-2-butanol $C_6H_5CH_2CH_2CH(OH)CH_3$

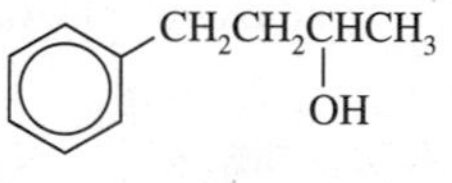

38. Benzaldehyde isomers of formula $C_9H_{10}O$:

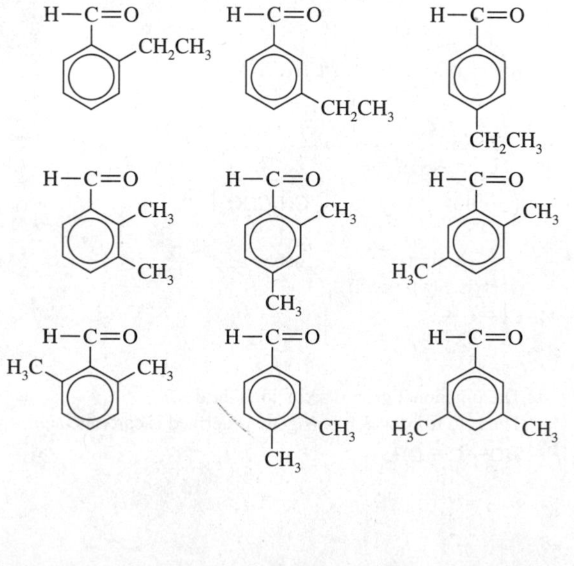

H—C=O on a benzene ring with: 2-CH_2CH_3; 3-CH_2CH_3; 4-CH_2CH_3; 2,3-diCH_3; 2,4-diCH_3; 2,5-diCH_3 (H_3C at 5); 2,6-diCH_3 (H_3C, CH_3); 3,4-diCH_3; 3,5-diCH_3 (H_3C, CH_3)

Chapter 24

2. The butyric acid solution would be expected to have the more objectionable odor because salts normally exhibit little or no odor since they are ionic and therefore have low volatility. For example, a dilute solution of acetic acid (vinegar) has considerable odor, but sodium acetate does not.

4. (a) The major difference between fats and oils is that fats are solids at room temperature, oils are liquids. The fatty acids in the molecules are mostly saturated in fats; more unsaturated in oils.
(b) Soaps are the sodium salts of high molar mass fatty acids. Syndets are synthetic detergents and occur in several different forms. They have cleansing action similar to soaps, but have different structural and solubility characteristics, such as being soluble in hard water. Some syndets also contain long hydrocarbon chains.
(c) Hydrolysis is the breaking apart of an ester in the presence of water to form an alcohol and a carboxylic acid. Mineral acids or digestive enzymes are used to speed the process. Saponification breaks the ester apart using sodium hydroxide to form alcohols and salts of carboxylic acids.

6. The principal advantage of synthetic detergents over soap is that the syndets do not form insoluble precipitates with the ions in hard water (Ca^{2+}, Mg^{2+}, Fe^{3+}).

8. methyl salicylate (pain relieving liniments)
phenyl salicylate (protective coating for pills)

10. The correct statements are: a, b, f, g, h, i, j, k, l, m, n, p, q.

12. (a) *o*-chlorobenzoic acid (b) oleic acid (c) *m*-toluic acid (d) 2-hydroxybutanoic acid (e) ammonium propanoate

14. (a) $HOOC{-}COOH$
(b) $CH_3CH_2CH_2CH_2COOH$
(c) benzene ring with COOH and COOH (ortho)
(d) $CH_3CH_2CH{=}CHCH_2CH{=}CHCH_2CH{=}CH(CH_2)_7COOH$
(e) benzene ring with COONa and NH_2 (para)
(f) $CH_3CH_2COONH_4$
(g) $CH_3CH(OH)CH_2COOH$

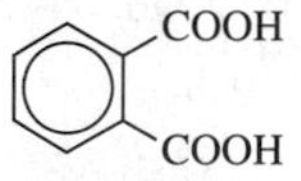

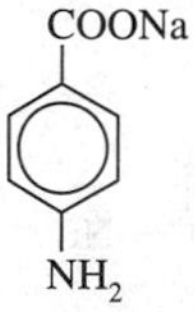

16. Decreasing pH means increasing acidity:

$KOH < NH_3 < KBr < C_6H_5OH < HCOOH < HBr$

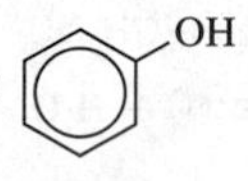

18. (a) methyl methanoate methyl formate
(b) propyl benzoate propyl benzoate
(c) ethyl propanoate ethyl propionate

20. (a) $CH_3\overset{O}{\overset{\|}{C}}-OCH_2CH_2CH_3$

(b) $\overset{O}{\overset{\|}{C}}-OCH_3$

(c) $CH_3CH_2CH_2CH_2CH_2\overset{O}{\overset{\|}{C}}-OCH_2CH_3$

22. (a) 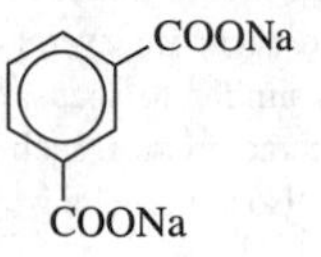 COONa, COONa sodium-*m*-phthalate

(b) HOOCCH$_2$CH$_2$COOH succinic acid

(c) $CH_3(CH_2)_7CH{=}CH(CH_2)_7COONa + H_2O$ sodium oleate

(d) $CH_3CH_2CH_2COOH$ butanoic acid

(e) 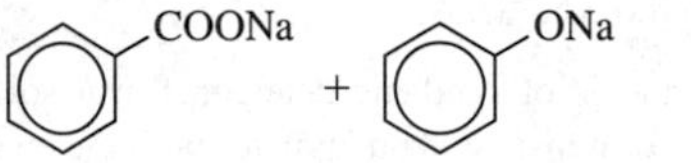 COONa + ONa

sodium benzoate sodium phenolate

24. (a) methanol and propanoic acid: $CH_3CH_2\overset{O}{\overset{\|}{C}}-OCH_3$

(b) 1-octanol and acetic acid: $CH_3\overset{O}{\overset{\|}{C}}-OCH_2(CH_2)_6CH_3$

(c) ethanol and butanoic acid: $CH_3CH_2CH_2\overset{O}{\overset{\|}{C}}-OCH_2CH_3$

26. (a) isopropyl formate: $CH_3\underset{OH}{\underset{|}{C}}HCH_3 + HCOOH$

(b) diethyl adipate: $CH_3CH_2OH + COOH-(CH_2)_4-COOH$

(c) benzyl benzoate: CH_2OH + $COOH$

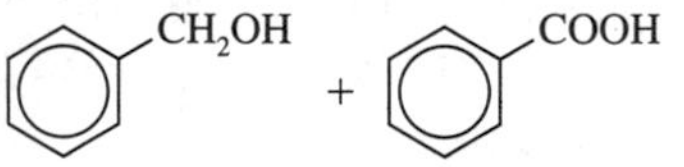

28. (a) 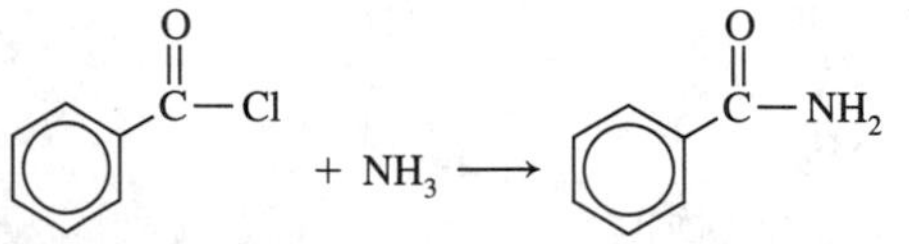 $\overset{O}{\overset{\|}{C}}-Cl + NH_3 \longrightarrow \overset{O}{\overset{\|}{C}}-NH_2$

(b) $CH_3CH_2COOH + SOCl_2 \longrightarrow CH_3CH_2\overset{O}{\overset{\|}{C}}-Cl$

(c) $CH_3CH_2\overset{O}{\overset{\|}{C}}-Cl + CH_3OH \longrightarrow CH_3CH_2\overset{O}{\overset{\|}{C}}-OCH_3$

(d) $CH_3CH_2CH_2C{\equiv}N + H_2O \xrightarrow{H^+} CH_3CH_2CH_2COOH$

(e) 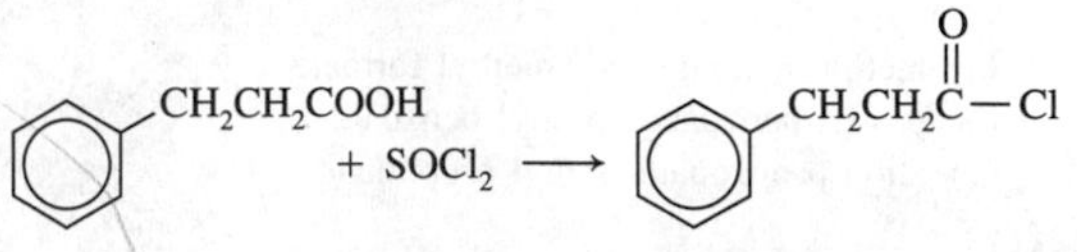 $CH_2CH_2COOH + SOCl_2 \longrightarrow CH_2CH_2\overset{O}{\overset{\|}{C}}-Cl$

30. (a) Benzoic acid and ethyl benzoate: benzoic acid is an odorless solid; ethyl benzoate is a fragrant liquid.

(b) Succinic acid and fumaric acid: fumaric acid has a carbon–carbon double bond and will readily add and decolorize bromine; succinic acid will not decolorize bromine.

32. (a) 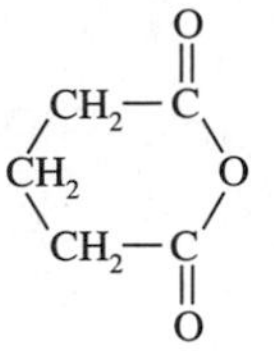 $CH_2-\overset{O}{\overset{\|}{C}}$, CH_2, O, $CH_2-\underset{O}{\underset{\|}{C}}$ glutaric anhydride

(b) $COOCH_2CH_3-CH_2-COOCH_2CH_3$ diethyl malonate

(c) 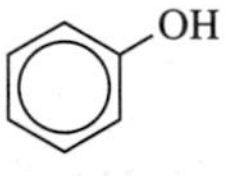 OH + CH_3COOH

phenol acetic acid

34. (a) cis-oleic acid: $CH_3(CH_2)_7(H)C{=}C(H)(CH_2)_7COOH$

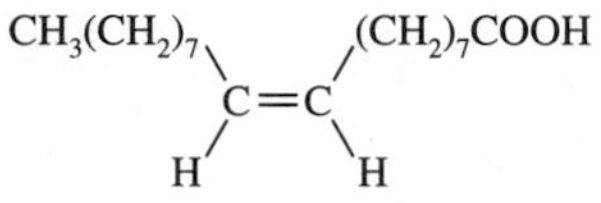

(b) cis, cis-linoleic acid: $CH_3(CH_2)_4(H)C{=}C(H)-CH_2-(H)C{=}C(H)(CH_2)_7COOH$

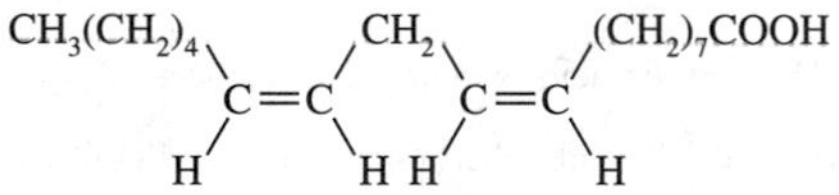

36. Sodium lauryl sulfate would be more effective as a detergent in hard water than sodium propyl sulfate because the hydrocarbon chain is only three carbons long in the latter, not long enough to dissolve grease well.

38. Only (c), $CH_3(CH_2)_{10}CH_2O(CH_2CH_2O)_7CH_2CH_2OH$, would be a good detergent in water. It is nonionic.

40. (a) $CH_3O-\overset{O}{\overset{\|}{P}}(OCH_3)-OCH_3 + 3\,H_2O$

(b) $CH_3O-\overset{O}{\overset{\|}{P}}(OH)-O\overset{CH_3}{\overset{|}{C}}HCH_2CH_3$

42. 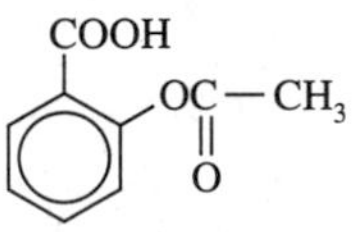 COOH, $O\underset{O}{\underset{\|}{C}}-CH_3$ + CH_3COOH

aspirin (acetylsalicylic acid) acetic acid

44. $Cl-\underset{O}{\underset{\|}{C}}-Cl$

The functional groups are acid chlorides.
The acid from which phosgene is derived is carbonic acid, $HO-\underset{O}{\underset{\|}{C}}-OH$.

46. 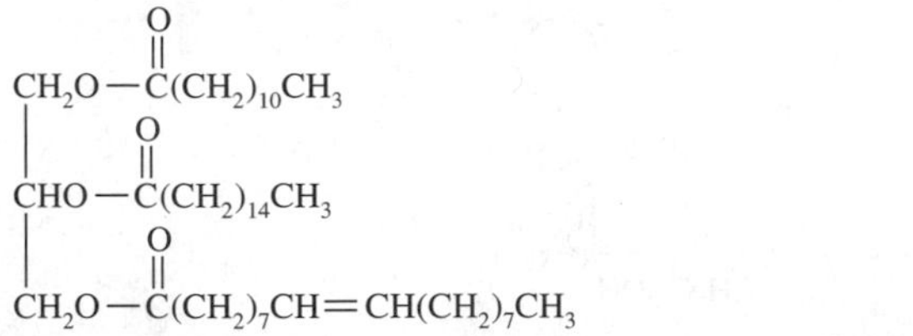
$$\begin{array}{l} CH_2O-\overset{O}{\overset{\|}{C}}(CH_2)_{10}CH_3 \\ | \\ CHO-\overset{O}{\overset{\|}{C}}(CH_2)_{14}CH_3 \\ | \\ CH_2O-\overset{O}{\overset{\|}{C}}(CH_2)_7CH=CH(CH_2)_7CH_3 \end{array}$$

There would be two other triacylglycerols containing all three of these acids. Each of the three acids can be the middle carbon.

48. 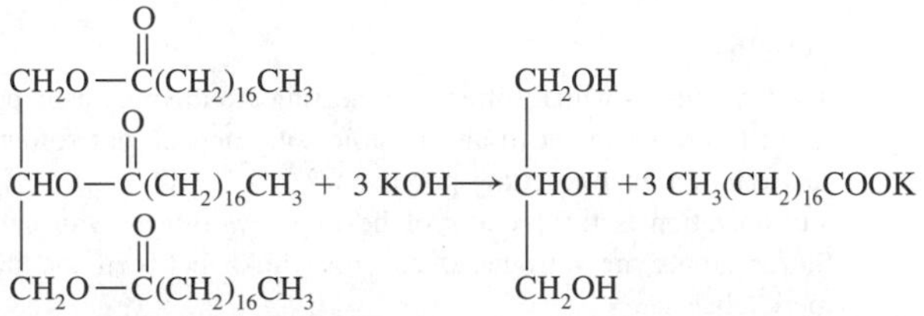
$$\begin{array}{l} CH_2O-\overset{O}{\overset{\|}{C}}(CH_2)_{16}CH_3 \\ | \\ CHO-\overset{O}{\overset{\|}{C}}(CH_2)_{16}CH_3 \\ | \\ CH_2O-\overset{O}{\overset{\|}{C}}(CH_2)_{16}CH_3 \end{array} + 3\ KOH \longrightarrow \begin{array}{l} CH_2OH \\ | \\ CHOH \\ | \\ CH_2OH \end{array} + 3\ CH_3(CH_2)_{16}COOK$$

The solubility in water will change considerably. Glyceryltristearate is insoluble in water but the products glycerol and potassium stearate (a salt) are soluble in water.

50. 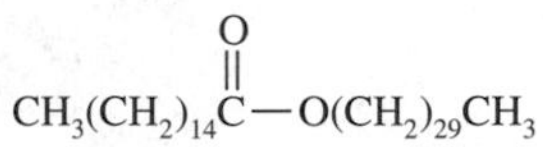 $CH_3(CH_2)_{14}\overset{O}{\overset{\|}{C}}-O(CH_2)_{29}CH_3$

52. 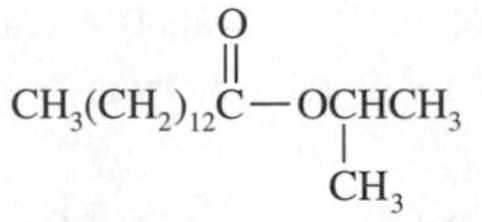 $CH_3(CH_2)_{12}\overset{O}{\overset{\|}{C}}-O\underset{CH_3}{\underset{|}{C}}HCH_3$

54. Smell both samples. Butanoic acid has an unpleasant rancid odor; ethyl butanoate has a pleasant odor of pineapple.

56. The statement is false. When methyl propanoate is hydrolyzed, propanoic acid and methanol are formed.

58. The common functional group that all three compounds (aspirin, acetaminophen, and ibuprofen) have in common is an aromatic ring.

Chapter 25

2. Amides: Unsubstituted amides (except formamide) are solids at room temperature. Many are odorless and colorless. Low molar-mass amides are water soluble. Solubility in water decreases as the molar mass increases. Amides are neutral compounds. The NH_2 group is capable of hydrogen bonding.

Amines: Low molar-mass amines are flammable gases with an ammonia-like odor. Aliphatic amines up to six carbon atoms are water soluble. Many amines have a "fishy" odor and many have very foul odors. Aromatic amines occur as liquids and solids. Soluble aliphatic amines give basic solutions. Aromatic amines are less soluble in water and less basic than aliphatic amines. The NH_2 group is capable of hydrogen bonding.

4. (a) Heterocyclic compounds are those in which all the atoms in the ring are not alike.

(b) The number of heterocyclic rings in each of the compounds is:

(i) purine, 2; (ii) ampicillin, 2; (iii) methadone, 0; (iv) nicotine, 2.

6. The correct statements are: h, j, m, o, p.

8. (a) 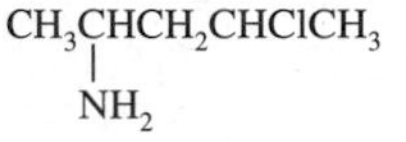 $CH_3\underset{NH_2}{\underset{|}{C}}HCH_2CHClCH_3$

(b) 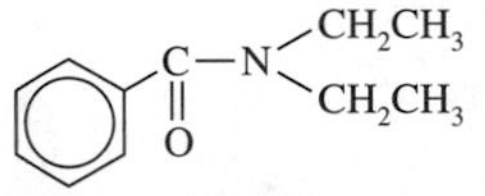 $C_6H_5-\overset{O}{\overset{\|}{C}}-N(CH_2CH_3)_2$ (benzene ring bonded to C=O, with N bearing two CH_2CH_3 groups)

(c) benzene ring with NH_2 and Cl in meta positions

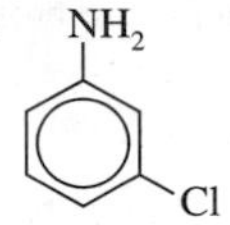

10. (a) 1-amino-3-methylbutane

(b) formamide

(c) *N*-phenylpropanamide

12. Increasing solubility in water

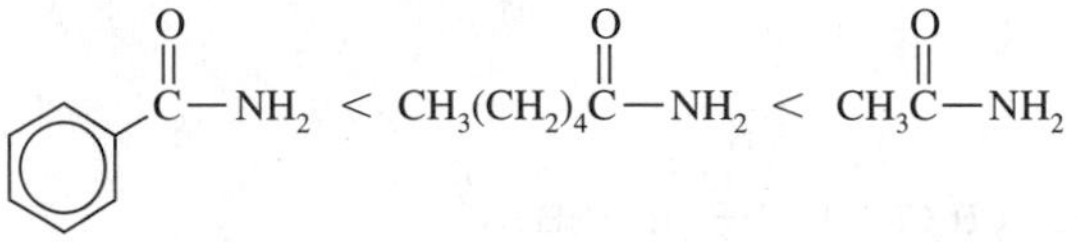

$$C_6H_5\overset{O}{\overset{\|}{C}}-NH_2 < CH_3(CH_2)_4\overset{O}{\overset{\|}{C}}-NH_2 < CH_3\overset{O}{\overset{\|}{C}}-NH_2$$

14. Several possibilities:

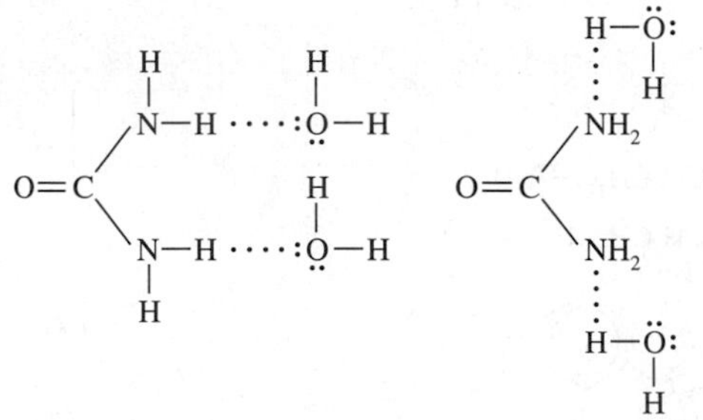

$O=C(NH_2)_2$ with each N–H hydrogen bonded to the O of a water molecule: $N-H \cdots :\ddot{O}-H$ (H on O); or each NH_2 nitrogen hydrogen bonded to the H of a water molecule: $H-\ddot{O}: \cdots NH_2$ (H on O).

16. (a) neither acid nor base

(b) base

(c) base

(d) neither acid nor base

(e) neither acid nor base

(f) acid

18. Organic products

(a) 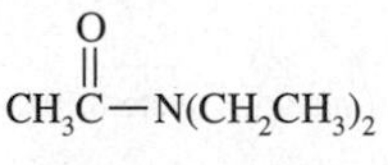 $CH_3\overset{O}{\overset{\|}{C}}-N(CH_2CH_3)_2$

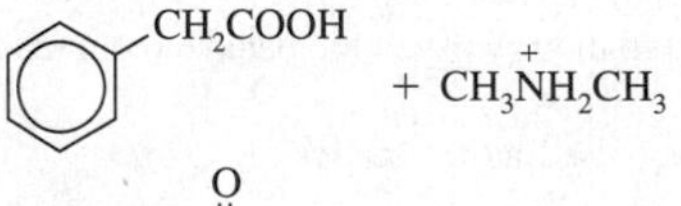

(b) 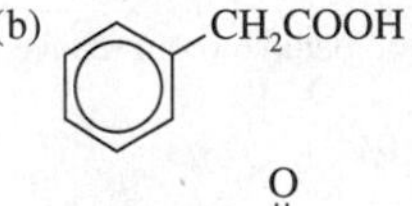 $C_6H_5CH_2COOH + CH_3\overset{+}{N}H_2CH_3$

(c) $CH_3\underset{CH_3}{\underset{|}{C}}HCH_2\overset{O}{\overset{\|}{C}}-NHCH_3$

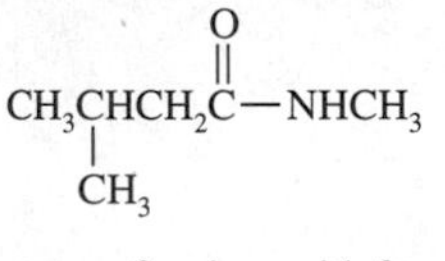

20. Structures of amines with formula C_3H_9N:

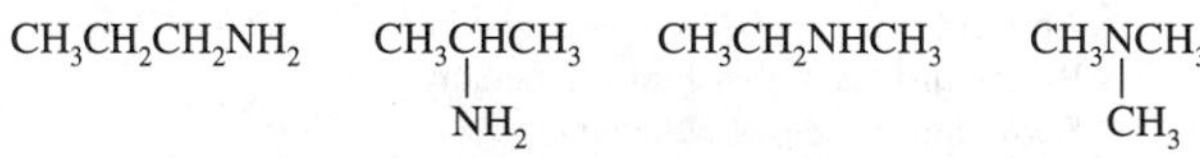

$CH_3CH_2CH_2NH_2$	$CH_3\underset{NH_2}{\underset{|}{C}}HCH_3$	$CH_3CH_2NHCH_3$	$CH_3\underset{CH_3}{\underset{|}{N}}CH_3$
1°	1°	2°	3°

22. (a) secondary

(b) tertiary

(c) secondary

24. The 2-aminopropane solution in 1.0*M* KOH would have the more objectionable odor because it would be in the form of the free amine, while in the acid solution the amine would form a salt that will have little or no odor.

26. (a) *N*-ethylaniline

(b) 2-aminobutane

(c) diphenylamine

(d) ethylammonium bromide
(e) pyridine
(f) *N*-ethylacetamide

28. (a) $CH_3CH_2CH_2CH_2NCH_2CH_2CH_2CH_3$ (with $CH_2CH_2CH_2CH_3$ on N)

(b) $\overset{+}{N}H_2Cl^-$ (with CH_3 on N, attached to a benzene ring)

(c) $CH_3CH_2NH_3^+Cl^-$

(d) $CH_3CH_2CH_2CHCH_2OH$ (with NH_2 on CH)

(e) $CH_3CH_2CH_2C{-}CH{-}NH{-}C_6H_5$ (with CH_3 above C, CH_3 below C and CH_3 below CH)

30. (a) $CH_3CH_2\overset{O}{\overset{\|}{C}}{-}NH_2$

(b) $CH_3CH_2CH_2CH_2NH_2$

(c) $H_3C{-}C_6H_4{-}NH_2$

(d) piperidine ring N${-}\overset{O}{\overset{\|}{C}}{-}CH_3$

32. *Amphetamines:* stimulate the central nervous system; used to treat depression, narcolepsy, and obesity
Tranquilizers: used to modify psychotic behavior and relieve pressure and anxiety
Antibacterial anents: used as antibiotics

34. ${-}\overset{O}{\overset{\|}{C}}{-}NH_2$ amide

Some classes of biochemicals that contain an amide are:
Antibacterial agents such as ampicillin
B-vitamins such as niacin (nicotinamide)
Barbiturates such as Nembutal
Tranquilizers such as Valium (diazepam)

36. thiophene${-}CH_2{-}\overset{O}{\overset{\|}{C}}{-}NH{-}$ (β-lactam/dihydrothiazine ring system)${-}CH_2O{-}\overset{O}{\overset{\|}{C}}CH_3$

amide, amide, ester

38. $H_2N{-}CH_2CH_2CH_2CH_2{-}NH_2$
Putrescine has two primary amine bonds.

40.

$H_2NCH\overset{O}{\overset{\|}{C}}{-}NHCH\overset{}{C}{-}OCH_3$ (with CH_2COOH on the first CH; $CH_2C_6H_5$ on the second CH; $\overset{\|}{O}$ on the last C)

amine, amide

Chapter 26

2. Those polymers which soften on reheating are thermoplastic polymers; those which set to an infusible solid and do not soften on reheating are thermosetting polymers.

4. Vulcanization is the process of heating raw rubber with sulfur. Sulfur atoms are introduced as cross links between the polymeric chains.

6. The correct statements are: a, b, d, f, i, k.

8. polystyrene: $-\!\!(CH_2{-}CH)_n-$ (with C_6H_5 on CH) C_8H_8

molar mass of 1 unit: 8(12.01 g/mol) + 8(1.008 g/mol) = 104.1 g/mol

$(3000 \text{ units})\left(\frac{104.1 \text{ g/mol}}{\text{unit}}\right) = 3 \times 10^5 \text{ g/mol} = \text{molar mass}$

10. (a) Saran $-\!(CH_2CCl_2)_n-$
(b) Orlon $-\!(CH_2CH)_n-$ (with CN on CH)
(c) Teflon $-\!(CF_2CF_2)_n-$
(d) Polystyrene $-\!(CH_2CH)_n-$ (with C_6H_5 on CH)
(e) Lucite $-\!(CH_2{-}C)_n-$ (with CH_3 and $\overset{}{C}OCH_3$, $\overset{\|}{O}$ on C)

12. polyethylene free radical (2 units) starting with RO • free radical
$ROCH_2CH_2CH_2CH_2\bullet$

14. (a) ethylene $-\!(CH_2{-}CH_2)_n-$
(b) chloroethene $-\!(CH_2CHCl)-$
(c) 1-butene $-\!(CH_2CH)_n-$ (with CH_2CH_3 on CH)

16. Two possible ways in which arylonitrile can polymerize to form Orlon are:

$-\!(CH_2CHCH_2CHCH_2CHCH_2CH)_n-$ (with CN on each CH)

$-\!(CH_2CH{-}CHCH_2CH_2CH{-}CHCH_2)_n-$ (with CN on each CH)

18. Gutta percha (all trans)

$-\!(CH_2(CH_3)C{=}CH{-}CH_2{-}CH_2(CH_3)C{=}CH{-}CH_2{-}CH_2(CH_3)C{=}CH{-}CH_2)_n-$

20. Yes, styrene—butadiene rubber contains carbon–carbon double bonds and is attacked by the ozone in smog, causing "age hardening" and cracking.

22. (a) synthetic natural rubber: $CH_2{=}C(CH_3)CH{=}CH_2$

(b) neoprene rubber: $CH_2{=}CCl{-}CH{=}CH_2$

24. (a) mass percent of oxygen in Dacron:

$$\left(\!-C(=O)-C_6H_4-C(=O)-OCH_2CH_2O-\!\right)_n \quad \text{molar mass} = 192.2\ \text{g/mol}$$

$$\frac{64.00\ \text{g/mol O}}{192.2\ \text{g/mol}} \times 100 = 33.30\%\ \text{O}$$

(b) mass percent of fluorine in Teflon:

$\left(CF_2CF_2\right)_n$ molar mass = 100.0 g/mol

$$\frac{76.00\ \text{g/mol F}}{100.0\ \text{g/mol}} \times 100 = 76.00\%\ \text{F}$$

26. (a) Isotactic polypropylene $\left(CH_2CH(CH_3)CH_2CH(CH_3)CH_2CH(CH_3)CH_2CH(CH_3)\right)_n$

(b) Another form of polypropylene

$\left(CH_2CH(CH_3)CH_2CH(CH_3)CH_2CH(CH_3)CH_2CH(CH_3)\right)_n$ (methyl groups alternating above and below the chain)

(other structures are possible)

28. Kevlar polymer:

$$\left(-C(=O)-C_6H_4-C(=O)-NH-C_6H_4-NH-C(=O)-C_6H_4-C(=O)-NH-C_6H_4-NH-\right)_n$$

30. Foam or sponge rubber materials achieve the foam by incorporating chemicals that release a gas within the material during the polymerization or the molding process. For example, in spongy polyurethane, water added during the polymerization reacts with the isocyanate to produce carbon dioxide, which causes the polymer to foam.

32. A glyptol polyester would more likely be thermosetting, because the glycerol is trifunctional, and would thus allow cross linking between chains in forming ester linkages.

34. The latex polymer belongs to the vinyl class of polymers.

$$\left(CH_2-CH(C(=O)OCH_2CH_3)-CH_2-CH(C(=O)OCH_2CH_3)-CH_2-CH(C(=O)OCH_2CH_3)\right)_n$$

36. (a) From 2,3-dimethyl-1,3-butadiene, $CH_2{=}C(CH_3){-}C(CH_3){=}CH_2$

this polymer can be made $\left(CH_2C(CH_3){=}C(CH_3)CH_2\right)_n$

(b) If produced by the free-radical mechanism, a random mixture of cis and trans connections are made. It is possible, using catalysts, for the reaction to proceed by an ionic mechanism which will give a stereochemically controlled polymer.

38. $CH_2{=}CH_2$ ethylene $CF_2{=}CF_2$ tetrafluoroethylene

$\left(CH_2-CH_2-CF_2-CF_2-CH_2-CH_2-CF_2-CF_2\right)_n$

(many other structures are possible)

40. (a) Quiana is a polyamide

(b) monomers

$HOOC(CH_2)_6COOH$ and $H_2N-C_6H_{10}-CH_2-C_6H_{10}-NH_2$

Chapter 27

2. When the axes of two pieces of polaroid film are parallel, the maximum brightness of light passes through both. When one piece is rotated 90° the polaroid appears black, indicating very little light passing through.

4. Enantiomers are nonsuperimposable mirror-image isomers. Diastereomers are stereoisomers that are not enantiomers (not mirror-image isomers).

6. The correct statements are: a, e, g, i, j.

8. Diastereoisomers do not have identical physical properties, so the differences form a basis for chemical or physical separation. Differences of boiling point, freezing point, and solubilities are most commonly used.

10. (a) a wood screw, (c) the letter g, (d) this textbook

12. Number of asymmetric carbon atoms

(a) 0 (b) 0 (c) 3 (d) 2

14. (a), (b), (c) will show optical activity.

16. Fructose, which has three asymmetric carbon atoms, will have eight possible stereoisomers. This can be determined from 2^n. $2^3 = 8$.

18. The two projection formulas (A) and (B) are the same compound, for it takes two changes to make (B) identical to (A).

(A)	(B)	1st change in (B) (H and CH_3)	2nd change in (B) (F and CH_3)
top H; Br–C–F; bottom CH_3	top CH_3; Br–C–H; bottom F	top H; Br–C–CH_3; bottom F	top H; Br–C–F; bottom CH_3

20. (+)-alanine: top COOH; H_2N–C–H; bottom CH_3 (−)-alanine: top COOH; H–C–NH_2; bottom CH_3

(b) top NH_2; H–C–COOH; bottom CH_3

(a) top CH_3; H_2N–C–H; bottom COOH

(c) top COOH; H–C–CH_3; bottom NH_2

(e) top COOH; CH_3–C–H; bottom NH_2

(d) top CH_3; H–C–NH_2; bottom COOH

(f) top H; CH_3–C–NH_2; bottom COOH

22. (a) 2,3-dichlorobutane

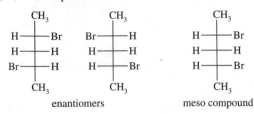

CH_3	CH_3	CH_3
H—C—Cl	Cl—C—H	H—C—Cl
Cl—C—H	H—C—Cl	H—C—Cl
CH_3	CH_3	CH_3

enantiomers (first two); meso compound (third)

(b) 2,4-dibromopentane

CH_3	CH_3	CH_3
H—C—Br	Br—C—H	H—C—Br
H—C—H	H—C—H	H—C—H
Br—C—H	H—C—Br	H—C—Br
CH_3	CH_3	CH_3

enantiomers (first two); meso compound (third)

(c) 3-hexanol

CH_2CH_3	CH_3CH_2
H—C—OH	HO—C—H
$CH_2CH_2CH_3$	$CH_3CH_2CH_2$

enantiomers

24. All the stereoisomers of 3,4-dichloro-2-methylpentane:

A	B	C	D
$CH(CH_3)_2$	$CH(CH_3)_2$	$CH(CH_3)_2$	$CH(CH_3)_2$
H—C—Cl	Cl—C—H	H—C—Cl	Cl—C—H
H—C—Cl	Cl—C—H	Cl—C—H	H—C—Cl
CH_3	CH_3	CH_3	CH_3

Compounds A and B and C and D are pairs of enantiomers. There are no meso compounds. Pairs of diastereomers are A and C, A and D, B and C, and B and D.

26. The four stereoisomers of 2-chloro-3-hexene:

cis	cis	trans	trans
CH_3	CH_3	CH_3	CH_3
H—C—Cl	Cl—C—H	H—C—Cl	Cl—C—H
H—C	H—C	H—C	H—C
‖	‖	‖	‖
H—C	H—C	C—H	C—H
CH_2CH_3	CH_2CH_3	CH_2CH_3	CH_2CH_3

The two cis compounds are enantiomers and the two trans compounds are enantiomers.

28. (a) $CH_3CH_2CHBr_2$ (i) $CH_3CHBrCH_2Br$ (ii)

$CH_3CBr_2CH_3$ (iii) $CH_2BrCH_2CH_2Br$ (iv)

(b) (ii) is chiral

CH_2Br	CH_2Br
H—C—Br	Br—C—H
CH_3	CH_3

enantiomers

(i), (iii), and (iv) are achiral; there are no meso compounds.

30. Assume (+)-2-chlorobutane is CH_3—CH_2—C(H)(Cl)—CH_3 (H—C—Cl, with CH_3CH_2 above and CH_3 below).

All possible isomers formed when (+)-2-chlorobutane is further chlorinated to dichlorobutane are:

A	B	C	D	E
CH_3	CH_2Cl	CH_3	CH_3	CH_3
CH_2	CH_2	CH_2	H—C—Cl	Cl—C—H
H—C—Cl	H—C—Cl	Cl—C—Cl	H—C—Cl	H—C—Cl
CH_2Cl	CH_3	CH_3	CH_3	CH_3

Compounds A, B, and E would be optically active; C does not have an asymmetric carbon atom; D is a meso compound.

32. If 1-chlorobutane and 2-chlorobutane were obtained by chlorinating butane, and then distilled, they would be separated into the two fractions, because their boiling points are different. 1-chlorobutane has no asymmetric carbon, so would not be optically active. 2-chlorobutane would exist as a racemic mixture (equal quantities of enantiomers) because substitution of Cl for H on carbon-2 gives equal amounts of the two enantiomers. Distillation would not separate the enantiomers because their boiling points are identical. The optical rotation of the two enantiomers of the 2-chlorobutane fraction would exactly cancel, and thus would not show optical activity.

34. Compound (d) is meso.

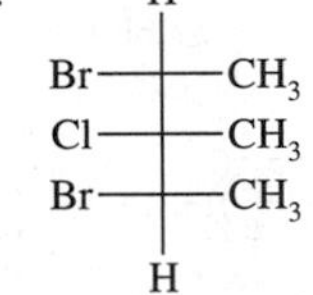

36. (c) is chiral.

(Two cyclopropane rings, each carrying H and Cl substituents, drawn as mirror images: H, H, Cl, Cl, H on the first; H, Cl, H, H, Cl on the second)

enantiomers

38. $(CH_3)_2CHCH_2$—C_6H_4—C(H)(CH_3)—COOH ← asymmetric carbon

HOOC—C(H)(CH_3)—C_6H_4—$CH_2CH(CH_3)_2$

40. (a) caraway (ring structure with CH_3, =O, H substituents, and a C(H)=CH_2 group; asymmetric carbon indicated)

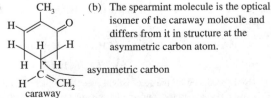

(b) The spearmint molecule is the optical isomer of the caraway molecule and differs from it in structure at the asymmetric carbon atom.

42. Ephedrine has two asymmetric carbons and can have four stereoisomers. This number is calculated using 2^n. $2^2 = 4$.

44. (−) placed in front of a name is used to indicate the rotation of plane-polarized light to the left, and (+) for rotation to the right. Therefore, we can write (−)-methorphan for levomethorphan and (+)-methorphan for dextromethorphan.

There is no obvious correlation between the structures of enantiomers and the direction in which they rotate plane-polarized light.

Chapter 28

2. The notations D and L in the name of a carbohydrate specify the configuration on the last asymmetric carbon. If that configuration is the same as D-glyceraldehyde, then the carbohydrate is designated as D. If it is the same as L-glyceraldehyde, it is designated as L.

4. Galactosemia is the inability of infants to metabolize galactose. The galactose concentration increases markedly in the blood and also appears in the urine. Galactosemia causes vomiting, diarrhea, enlargement of the liver, and often mental retardation. If not recognized a few days after birth it can lead to death.

6. A carbohydrate forms a five-member or six-member heterocyclic ring (one oxygen atom, the rest carbon atoms). If it forms a five-member ring, it is termed a furanose, after the compound furan. If it forms a six-membered ring it is termed a pyranose, after the compound pyran.

8. The cyclic forms of monosaccharides are hemicetals.

10. (a) sucrose: sugar beets and sugar cane
(b) lactose: milk
(c) maltose: sprouting grain and partially hydrolyzed starch

12. Invert sugar is sweeter than sucrose because it is a 50–50 mixture of fructose and glucose. Glucose is somewhat less sweet than sucrose, but fructose is much sweeter, so the mixture is sweeter.

14. In the Benedict test, both concentrated and dilute glucose solutions will give a precipitate. However, the concentrated glucose solution will produce a redder precipitate whereas the dilute glucose solution will appear more greenish-yellow.

16. The correct statements are: c, e, g, h, i, l, n, p, q, r, s, v, w, x.

18. (a) D-glyceraldehyde L-glyceraldehyde

```
D-glyceraldehyde        L-glyceraldehyde
  H—C=O                   H—C=O
    |                       |
  H—C—OH                 HO—C—H
    |                       |
   CH2OH                   CH2OH
```

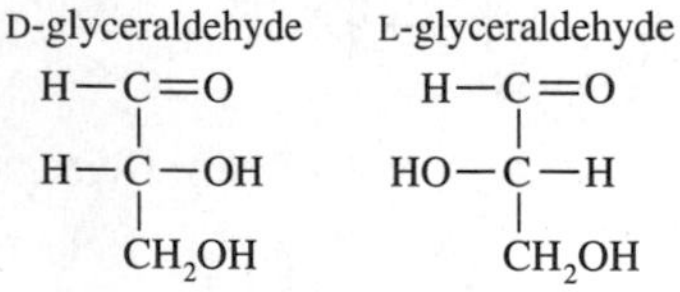

(b) If D-glyceraldehyde is reacted with hydrogen in the presence of a platinum catalyst, the product will be:

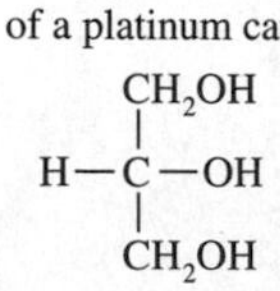

```
   CH2OH
    |
  H—C—OH
    |
   CH2OH
```

20. The epimer could differ at carbon 2 *or* carbon 3 *or* carbon 4 *or* carbon 5. One possible answer is:

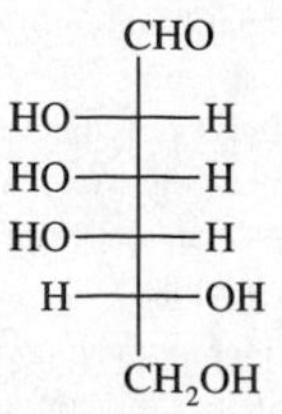

```
     CHO
HO ——+—— H
HO ——+—— H
HO ——+—— H
 H ——+—— OH
    CH2OH
```

22.

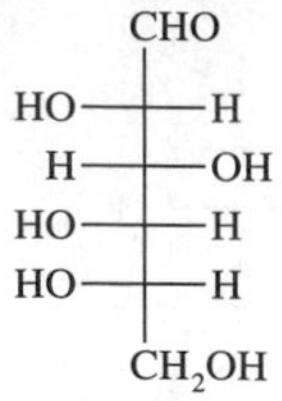

```
     CHO               CH2OH               CHO
HO ——+—— H              |             H ——+—— H
 H ——+—— OH            C=O           HO ——+—— H
HO ——+—— H         H ——+—— OH        HO ——+—— H
HO ——+—— H        HO ——+—— H             CH2OH
    CH2OH         HO ——+—— H
                      CH2OH
  L-glucose         L-fructose        L-2-deoxyribose
```

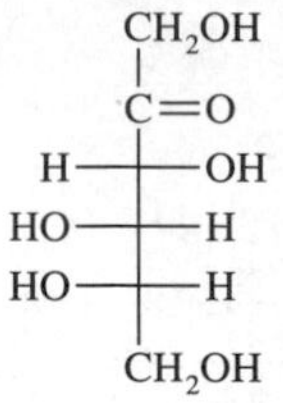

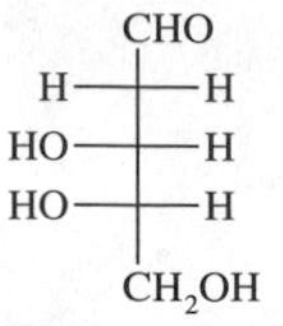

24. Either the Fischer projection formulas or Haworth formulas are satisfactory.

β-D-glucopyranose α-D-galactopyranose 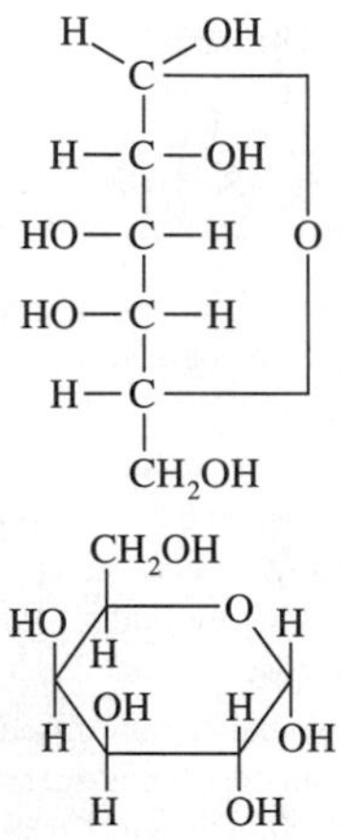 β-D-mannopyranose

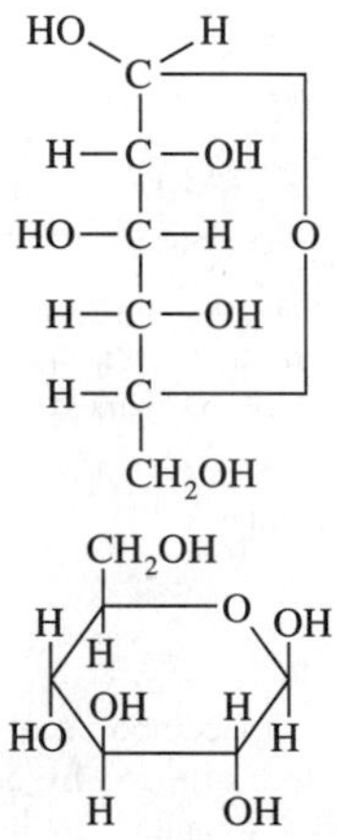

```
β-D-glucopyranose     α-D-galactopyranose    β-D-mannopyranose
 HO   H                H   OH                 HO   H
   \ /                  \ /                     \ /
    C ———┐               C ———┐                  C ———┐
    |    |               |    |                  |    |
  H—C—OH |             H—C—OH |               HO—C—H  |
    |    |               |    |                  |    |
 HO—C—H  O            HO—C—H  O               HO—C—H  O
    |    |               |    |                  |    |
  H—C—OH |            HO—C—H  |                H—C—OH |
    |    |               |    |                  |    |
  H—C————┘             H—C————┘                H—C————┘
    |                    |                       |
   CH2OH                CH2OH                   CH2OH
```

Haworth formulas:

β-D-glucopyranose: ring C5 bears CH_2OH (up) and H; ring O; C1: OH (up), H (down); C2: H (up), OH (down); C3: OH (up), H (down); C4: H (up), HO (down).

α-D-galactopyranose: ring C5 bears CH_2OH (up) and H; ring O; C1: H (up), OH (down); C2: H (up), OH (down); C3: OH (up), H (down); C4: HO (up), H (down).

β-D-mannopyranose: ring C5 bears CH_2OH (up) and H; ring O; C1: OH (up), H (down); C2: HO (up), H (down); C3: OH (up), H (down); C4: H (up), HO (down).

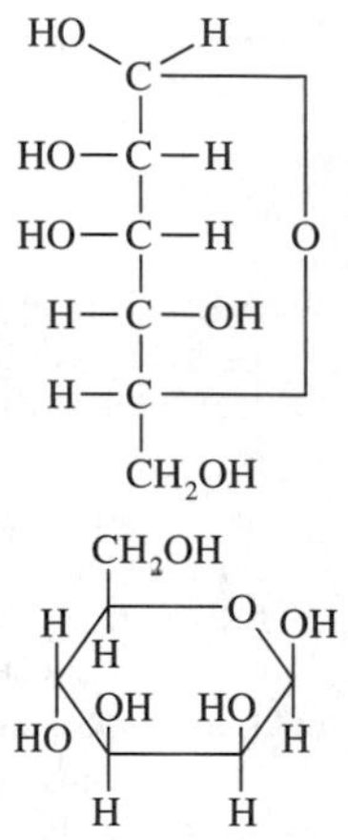

26. The structure of L-glyceraldehyde is

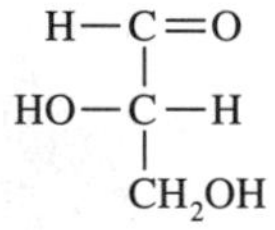

```
  H—C=O      1st carbon oxidation number = +1
    |
 HO—C—H      2nd carbon oxidation number =  0
    |
   CH2OH     3rd carbon oxidation number = −1
```

Carbon 3 has the most negative oxidation number and is the most reduced.

28. The Kiliani-Fischer synthesis of D-ribose starts with the proper D-triose, D-glyceraldehyde.

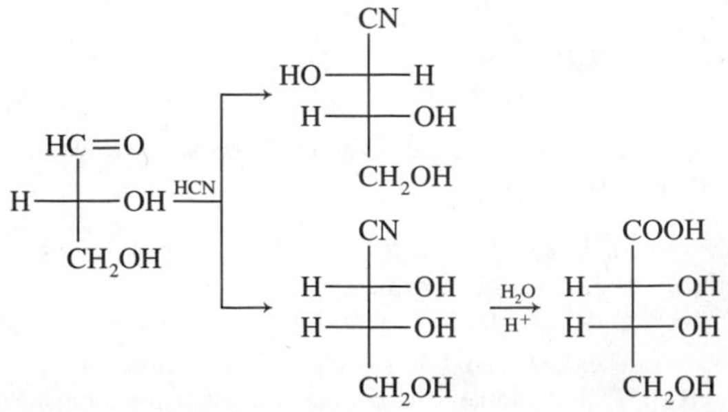

```
                              CN
                       HO ——+—— H
                 ┌──→   H ——+—— OH
  HC=O           │         CH2OH
 H ——+—— OH  HCN │
    CH2OH        │            CN                     COOH
                 │      H ——+—— OH    H2O     H ——+—— OH
                 └──→   H ——+—— OH    ——→     H ——+—— OH
                           CH2OH      H+         CH2OH
```

$$\begin{array}{c} HC{=}O \\ H-C-OH \\ H-C-OH \\ CH_2OH \end{array} \xrightarrow{HCN} \begin{array}{c} CN \\ HO-C-H \\ H-C-OH \\ H-C-OH \\ CH_2OH \end{array} \quad + \quad \begin{array}{c} CN \\ H-C-OH \\ H-C-OH \\ H-C-OH \\ CH_2OH \end{array}$$

(The first structure above is preceded by an arrow labelled Na(Hg).)

$$\begin{array}{c} CN \\ H-C-OH \\ H-C-OH \\ H-C-OH \\ CH_2OH \end{array} \xrightarrow[H^+]{H_2O} \begin{array}{c} COOH \\ H-C-OH \\ H-C-OH \\ H-C-OH \\ CH_2OH \end{array} \xrightarrow{Na(Hg)} \begin{array}{c} HC{=}O \\ H-C-OH \\ H-C-OH \\ H-C-OH \\ CH_2OH \end{array}$$

D-ribose

30. No, D-2-deoxygalactose is not the same as D-2-deoxyglucose. D-galactose differs from D-glucose at carbon 4, so replacement of the carbon 2 OH with an H does not make these two sugars identical.

32. The monosaccharide composition of:
(a) lactose: one glucose and one galactose unit
(b) amylopectin: many glucose units
(c) cellulose: many glucose units
(d) sucrose: one glucose and one fructose unit

34. Both maltose and isomaltose are disaccharides composed of two glucose units. The glucose units in maltose are linked by an α-1,4-glycosidic bond while the glucose units of isomaltose are linked by an α-1,6-glycosidic bond.

36. Both maltose and isomaltose will show mutarotation. Both disaccharides contain a hemiacetal structure that will open, allowing mutarotation.

38. [Haworth structure: maltose — labels CH_2OH, O, OH, OH, O, CH_2OH, O, OH, OH (circled), OH, OH]

maltose

[Haworth structure: lactose — labels CH_2OH, HO, O, OH, O, CH_2OH, O, OH, OH (circled), OH, OH]

lactose

The circled hemiacetal structures allow these two disaccharides to be reducing sugars.

40. The systematic name for maltose is α-D-glucopyranosyl-(1,4)-α-D-glucopyranose. The systematic name for lactose is β-D-galactopyranosyl-(1,4)-α-D-glucopyranose.

42. (a) D-ribose and D-2-deoxyribose. The D-2-deoxyribose has no OH group on the number 2 carbon, only 2 hydrogen atoms. Both are five-carbon sugars.

(b) Amylose and amylopectin. Amylose is a straight-chain polysaccharide; amylopectin has branched chains and more monomer units per molecule. Both are large polysaccharides composed of α-D-glucose units.

(c) Lactose and isomaltose. These sugars are disaccharides. Lactose is composed of one galactose unit and one glucose unit while isomaltose is composed of two glucose units. The monosaccharide units in lactose are linked by a β-1,4-glycosidic bond whereas the units in isomaltose are linked by an α-1,6-glycosidic bond. Both disaccharides also contain hemiacetal structures.

44.

$$\begin{array}{c} COOH \\ HO-C-H \\ HO-C-H \\ H-C-OH \\ H-C-OH \\ COOH \end{array}$$

mannaric acid

46.

$$\begin{array}{cccc} HC{=}O & HC{=}O & HC{=}O & HC{=}O \\ H-C-OH & HO-C-H & H-C-OH & HO-C-H \\ H-C-OH & H-C-OH & HO-C-H & HO-C-H \\ HO-C-H & HO-C-H & HO-C-H & HO-C-H \\ CH_2OH & CH_2OH & CH_2OH & CH_2OH \\ \text{I} & \text{II} & \text{III} & \text{IV} \end{array}$$

Structures I and II will form the same osazone. Structures III and IV will form the same osazone.

48. The Haworth formula for β-D-glucopyranosyl-(1,4)-α-D-galactopyranose is

[Haworth structure — labels CH_2OH, CH_2OH, O, H, H, O, H, O, H, OH, H, OH, H, HO, H, H, OH, H, OH, H, OH]

The Haworth formula for β-D-galactopyranosyl-(1,6)-β-D-mannopyranose is

[Haworth structure — labels CH_2OH, CH_2, HO, O, H, O, O, OH, H, H, OH, H, OH, HO, H, H, HO, H, H, OH, H, H]

50. Aspartame supplies many fewer calories than sucrose. In addition, oral bacteria cannot use aspartame as efficiently as sucrose and will form fewer dental caries.

52. Mucopolysaccharides serve a protective, lubricant function by absorbing water and taking on a slimy, spongy consistency. The acidic nature of mucopolysaccharides indicates that these polymers have many negative charges at physiological pH. Since negative charges repel each other, the polymer chains will move far apart making large holes to be filled by water, much like a sponge.

54. High fructose corn syrup is produced by breaking down some corn starch polymers to D-glucose monomers, which are then converted to D-fructose, a very sweet monosaccharide.

56. maltose

58.

$$\begin{array}{c} HC{=}O \\ H\text{—}\!\!+\!\!\text{—}OH \\ HO\text{—}\!\!+\!\!\text{—}H \\ CH_2OH \end{array}$$

These compounds are not epimers because they differ at more than one chiral carbon.

60. The β-1,4-glycosidic linkage in cellulose allows this polysaccharide to form fibers that are not digestable by humans. Starch (amylose and amylopectin) is composed of polysaccharides that have α-1,4-glycosidic linkages and α-1,6-glycosidic linkages; these polysaccharides can be digested.

62. No, the classmate should not be believed. Although D-glucose and D-mannose are related as epimers, it is pairs of *enantiomers* that

Chapter 29

2. Although caproic acid, $CH_3(CH_2)_4COOH$, has the same number of polar bonds as stearic acid, $CH_3(CH_2)_{16}COOH$, caproic acid has a shorter, nonpolar hydrocarbon chain and, therefore, is more water soluble.

4. The three essential fatty acids are linoleic, linolenic, and arachidonic acids. Diets lacking these fatty acids lead to impaired growth and reproduction, and skin disorders such as eczema and dermatitis.

6. Aspirin relieves inflammation by blocking the conversion of arachidonic acid to prostaglandins.

8. A membrane lipid must be (a) partially hydrophobic to act as a barrier to water and (b) partially hydrophilic, so that the membrane can interact with water along its surface.

10. The four classes of eicosanoids are prostaglandins, prostacyclins, thromboxanes, and leukotrienes.

12. Atherosclerosis is the deposition of cholesterol and other lipids on the inner walls of the large arteries. These deposits, called plaque, accumulate, making the arterial passages narrower and narrower. Blood pressure increases as the heart works to pump sufficient blood through the restricted passages. This may lead to a heart attack, or the rough surface can lead to coronary thrombosis.

14. Dietary fish oils provide fatty acids that inhibit the formation of thromboxanes—compounds that participate in blood clotting.

16. Because the interior of a lipid bilayer is very hydrophobic, molecules with hydrophilic character cross the lipid bilayer with difficulty.

18. Both facilitated diffusion and active transport catalyze movement of compounds through membranes. Active transport requires an input of energy whereas facilitated diffusion does not.

20. The correct statements are: d, e, g, j, l, m and n.

22.

$CH_2O—\overset{O}{\overset{\|}{C}}(CH_2)_{14}CH_3$ palmitic acid

$CHO—\overset{O}{\overset{\|}{C}}(CH_2)_{14}CH_3$ palmitic acid

$CH_2O—\overset{O}{\overset{\|}{C}}(CH_2)_7CH{=}CH(CH_2)_7CH_3$ oleic acid

There is one other possible triacylglycerol with the same components.

24. Yes, a triacylglycerol that contains three units of stearic acid would be more hydrophobic than a triacylglycerol that contains three units of myristic acid. Larger molecules tend to be more hydrophobic than smaller molecules, and a triacylglycerol with three stearic acid units is larger than a triacylglycerol with three myristic acid units.

26.

CH_2OH
$CHOH$
CH_2OH

$CH_3(CH_2)_7CH{=}CH(CH_2)_7COOH$

$CH_3(CH_2)_{14}COOH$

$CH_3(CH_2)_{16}COOH$

28.

$CH_2O—\overset{O}{\overset{\|}{C}}(CH_2)_{16}CH_3$

$CHO—\overset{O}{\overset{\|}{C}}(CH_2)_{16}CH_3$

$CH_2O—\overset{O}{\overset{\|}{P}}—OCH_2CH_2\overset{+}{N}(CH_3)_3$, with O^- on P

30.

$\overset{OH}{|}$
$CHCH{=}CH(CH_2)_{12}CH_3$

$CHNH—\overset{O}{\overset{\|}{C}}(CH_2)_{16}CH_3$

$CH_2O—\overset{O}{\overset{\|}{P}}—OCH_2CH_2\overset{+}{N}H_3$, with O^- on P

32.

$\overset{OH}{|}$
$CHCH{=}CH(CH_2)_{12}CH_3$

$CHNH—\overset{O}{\overset{\|}{C}}(CH_2)_7CH{=}CH(CH_2)_7CH_3$

CH_2O— (sugar ring with CH_2OH, OH, O, OH, OH)

or

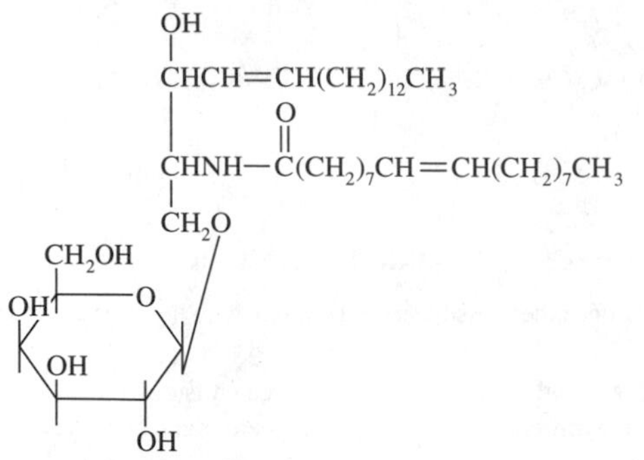

$$\begin{array}{l} \text{OH} \\ | \\ \text{CHCH}{=}\text{CH(CH}_2)_{12}\text{CH}_3 \\ | \qquad\qquad \text{O} \\ | \qquad\qquad \| \\ \text{CHNH}{-}\text{C(CH}_2)_7\text{CH}{=}\text{CH(CH}_2)_7\text{CH}_3 \\ | \\ \text{CH}_2\text{O}{-}\text{(galactose: CH}_2\text{OH, OH, OH, OH)} \end{array}$$

34. hydrophilic exterior (—COOH)

HO—C(CH$_2$)$_{12}$CH$_3$ (with C=O)

hydrophobic interior (—(CH$_2$)$_{12}$CH$_3$)

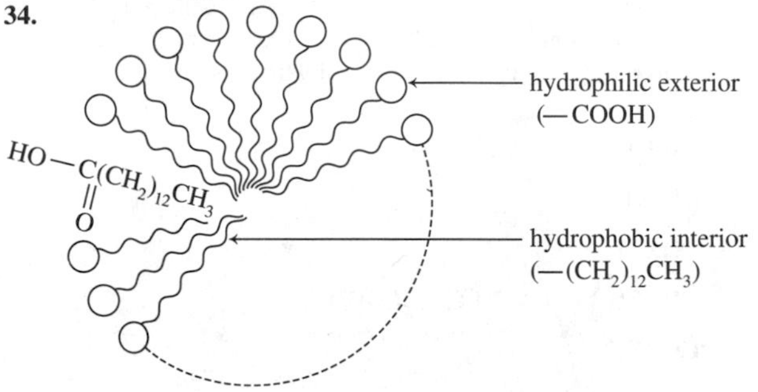

36. HDL (high density lipoprotein) differs from LDL (low density lipoprotein) in that
(a) LDL has a lower density than HDL
(b) LDL delivers cholesterol to peripheral tissues whereas HDL scavenges cholesterol and returns it to the liver.

38. (a) Both compounds have two long hydrophobic chains.
(b) For both compounds, the primary alcohol can react further with either acids (to form esters) or sugars (to form acetals).

40. Phosphate ion will move from a region of low concentration to a region of high concentration as it moves from a 0.1 *M* solution across a membrane to 0.5 *M* solution. This process requires energy and can be accomplished by active transport.

42. Thromboxanes act as vasoconstrictors and stimulate platelet aggregation. Leukotrienes have been associated with many of the symptoms of an allergy attack (e.g., an asthma attack).

44. Both olestra and natural fats contain fatty acid units. However, in natural fats, the fatty acid units are linked to glycerol while in olestra the fatty acid units are linked to sucrose.

46. Olestra is not shaped like a natural fat and is not attacked by digestive enzymes. Thus, olestra passes through the stomach and intestines undigested.

Chapter 30

2. All amino acids and proteins contain carbon, hydrogen, oxygen, and nitrogen. Sulfur is contained in some of the amino acids, and thus in most proteins.

4. The amino acids that are essential to humans are isoleucine, leucine, lysine, methionine, phenylalanine, threonine, tryptophan, and valine.

6. At its isoelectric point, a protein molecule must have an equal number of positive and negative charges.

8. The sulfur-containing amino acid, cysteine, has the special role in protein structure of creating disulfide bonding between polypeptide chains, which helps control the shape of the molecule.

10. The α-helix forms a tube composed of a spiraling polypeptide chain whereas the β-pleated sheet forms a plane composed of polypeptide chains aligned roughly parallel to each other.

12. Amino acids containing a benzene ring give a positive xanthoproteic test (formation of yellow reaction products). Among the common amino acids, these would include phenylalanine, tryptophan, and tyrosine.

14. Protein column chromatography uses a column packed with polymer beads (solid phase) through which a protein solution (liquid phase) is passed. Proteins separate based on differences in how they interact with the solid phase. The proteins move through the column at different rates and can be collected separately.

16. The correct statements are: b, c, d, f, g, i, m, p, r, s, t and u.

18. D-serine; L-serine (form commonly found in proteins)

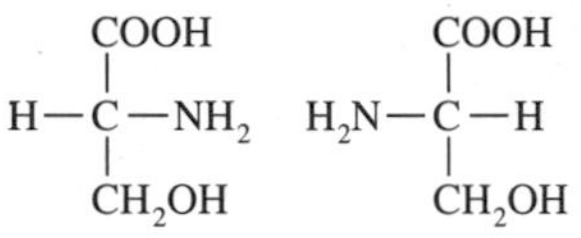

$$\begin{array}{c} \text{COOH} \\ | \\ \text{H}{-}\text{C}{-}\text{NH}_2 \\ | \\ \text{CH}_2\text{OH} \end{array} \qquad \begin{array}{c} \text{COOH} \\ | \\ \text{H}_2\text{N}{-}\text{C}{-}\text{H} \\ | \\ \text{CH}_2\text{OH} \end{array}$$

20. 32 g

22. $\text{NH}_2\overset{\text{O}}{\overset{\|}{\text{C}}}{-}\text{CH}_2{-}\underset{\text{NH}_3^+}{\underset{|}{\text{CH}}}{-}\text{COO}^-$

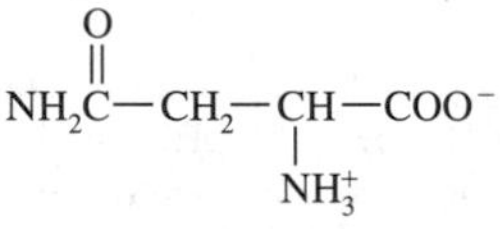

24. For tryptophan:

(a) Zwitterion formula: indole ring (N—H) with side chain $-\text{CH}_2\underset{\text{NH}_3^+}{\underset{|}{\text{C}}}\text{HCOO}^-$

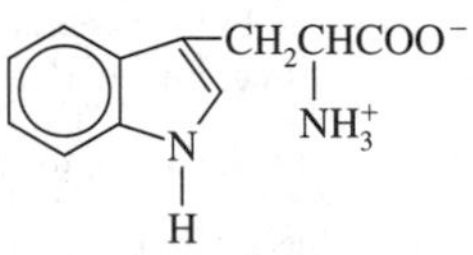

(b) formula in 0.1 *M* H_2SO_4: indole ring (N—H) with side chain $-\text{CH}_2\underset{\text{NH}_3^+}{\underset{|}{\text{C}}}\text{HCOOH}$

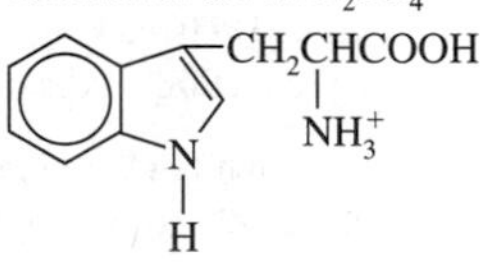

(c) formula in 0.1 *M* NaOH: indole ring (N—H) with side chain $-\text{CH}_2\underset{\text{NH}_2}{\underset{|}{\text{C}}}\text{HCOO}^-$

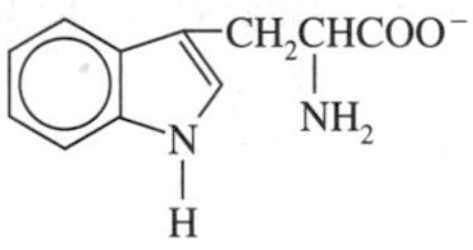

26. Ionic equations showing how leucine acts as a buffer toward:

(a) H^+

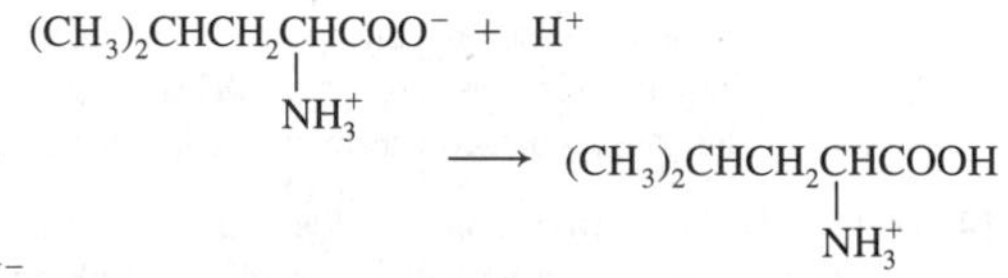

$$(\text{CH}_3)_2\text{CHCH}_2\underset{\text{NH}_3^+}{\underset{|}{\text{C}}}\text{HCOO}^- + \text{H}^+ \longrightarrow (\text{CH}_3)_2\text{CHCH}_2\underset{\text{NH}_3^+}{\underset{|}{\text{C}}}\text{HCOOH}$$

(b) OH^-

$$(\text{CH}_3)_2\text{CHCH}_2\underset{\text{NH}_3^+}{\underset{|}{\text{C}}}\text{HCOO}^- + \text{OH}^- \longrightarrow (\text{CH}_3)_2\text{CHCH}_2\underset{\text{NH}_2}{\underset{|}{\text{C}}}\text{HCOO}^- + \text{H}_2\text{O}$$

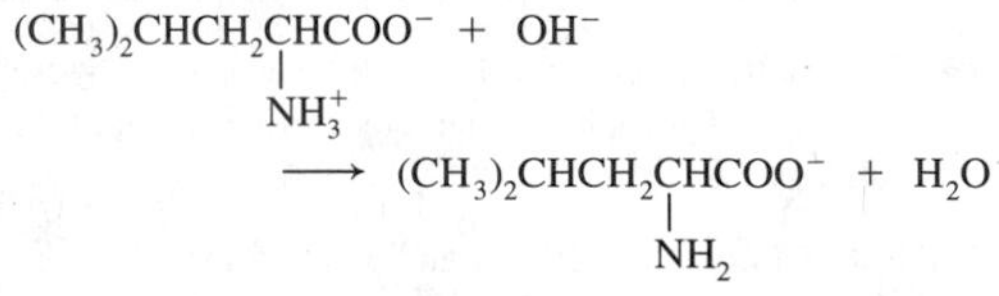

28. 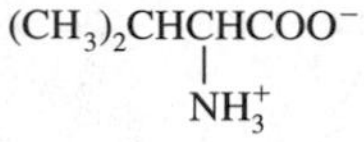 $(CH_3)_2CHCHCOO^-$
$\quad\quad\quad\quad\quad |$
$\quad\quad\quad\quad NH_3^+$

30.

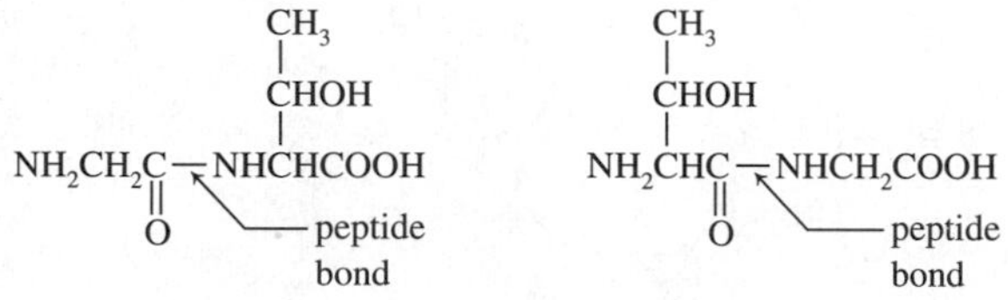

CH_3 / | / $CHOH$ / | / $NH_2CH_2C—NHCHCOOH$ / ‖ / O (peptide bond)

CH_3 / | / $CHOH$ / | / $NH_2CHC—NHCH_2COOH$ / ‖ / O (peptide bond)

32. (a) 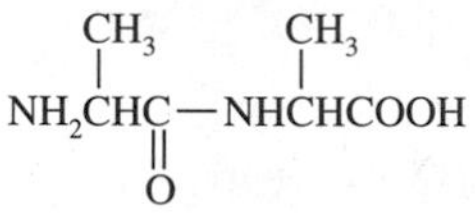
$CH_3 \quad CH_3$
$NH_2CHC—NHCHCOOH$
‖
O

(b) 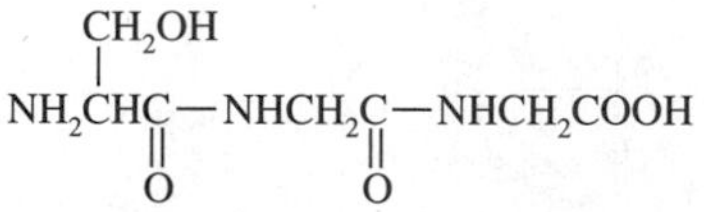
CH_2OH
$NH_2CHC—NHCH_2C—NHCH_2COOH$
‖ ‖
O O

(c) 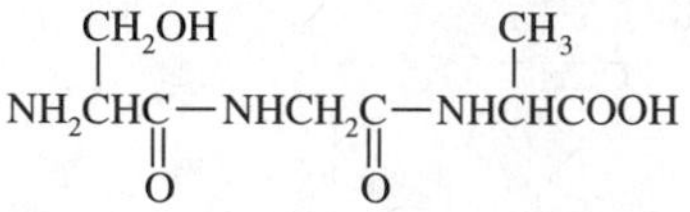
$CH_2OH \quad CH_3$
$NH_2CHC—NHCH_2C—NHCHCOOH$
‖ ‖
O O

34. Tyr-Asp-Ala Tyr-Ala-Asp Asp-Tyr-Ala
Asp-Ala-Tyr Ala-Tyr-Asp Ala-Asp-Tyr

36. 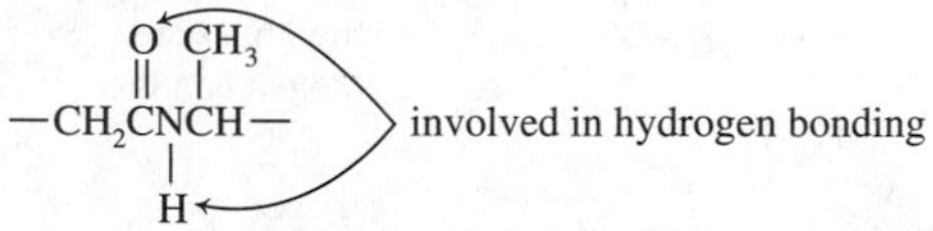
O CH_3
‖ |
$—CH_2CNCH—$ involved in hydrogen bonding
|
H

38. 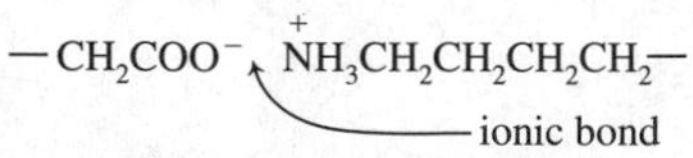 $—CH_2COO^-$ $\overset{+}{N}H_3CH_2CH_2CH_2CH_2—$
ionic bond

40. The tripeptide, Gly-Ser-Asp, will
(a) react with Sanger's reagent
(b) not react to give a positive xanthoproteic test
(c) react with ninhydrin

42. 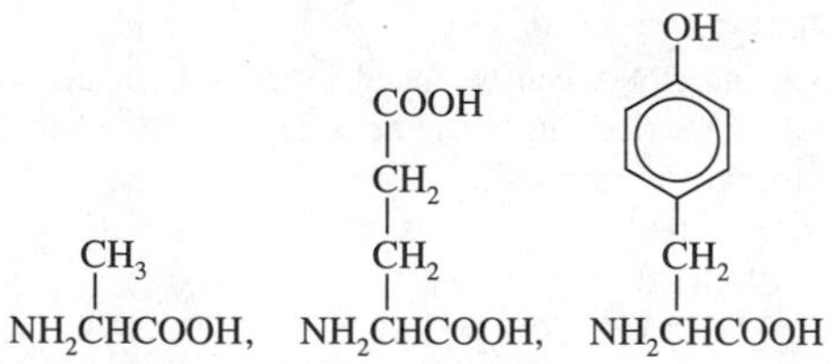
CH_3 / | / $NH_2CHCOOH$,

COOH / | / CH_2 / | / CH_2 / | / $NH_2CHCOOH$,

OH / (benzene ring) / | / CH_2 / | / $NH_2CHCOOH$

44. $6.8 \times 10^4 \dfrac{g}{mole}$

46. Phe-Ala-Ala-Leu-Phe-Gly-Tyr

48. (a) Arginine will not migrate to either electrode in an electrolytic cell at a pH of 10.8.
(b) Arginine will migrate toward the positive electrode at pH greater than 10.8, that is, more basic than its isoelectric point.

50. A small protein like ribonuclease (or myoglobin) would also have a small number of protein domains. Small proteins (e.g. myoglobulin, ribonuclease) commonly fold into one globular unit (one domain) while larger proteins will fold into more than one globular unit (more than one domain).

52. The immunoglobulin hypervariable regions allow the body to produce millions of different immunoglobulins, each with its distinct amino acid sequence and unique antigen binding site.

54. thirty-nine

56. Glutamic acid is the only one of these three amino acids with polar bonds in its side chain. Thus, glutamic acid will be the most polar.

Chapter 31

2. Enzymes serve as catalysts for chemical reactions in the body.

4. Sucrase catalyzes the hydrolysis of sucrose; lactase catalyzes the hydrolysis of lactose; maltase catalyzes the hydrolysis of maltose.

6. The reaction rate often can be increased (1) by increasing the reactant (substrate) concentration and (2) by raising the temperature.

8. When substrates bind to an active site, they may be converted more easily to products because (1) the substrates are close together (proximity catalysis); (2) the substrates are oriented to best react (productive binding hypothesis); (3) when the substrates bind to the active site, they change shape to be more like products (strain hypothesis).

10. The correct statements are: a, c, f, g, h, j, k, l, m, n.

12. 4×10^{-7} *M*/s

14. 1×10^3 reactants converted to products by three pepsin molecules

16. The first enzyme would have a narrow active site which is long enough to fit the four carbon carboxylic acid, butanoic acid ($CH_3CH_2CH_2COOH$). The second enzyme would have a narrow active site which would be shorter because it need only fit the two carbon carboxylic acid, acetic acid (CH_3COOH).

18. An enzyme-catalyzed rate decreases at high temperatures because the enzyme loses its natural shape (denatures).

20. If the turnover number for an enzyme is increased, the enzyme can convert more reactants to products per unit time; the enzyme is a better catalyst. Any process that increases an enzyme's catalytic abilities is termed activation.

22. Yes. Enzyme B is more effective than enzyme A based on a comparison of turnover numbers (9.8×10^{-1}/s for enzyme B vs. 0.05/s for enzyme A).

24. Glutamine is a starting material (reactant) for the liver metabolic pathway that produces urea. Thus, glutamine control of this pathway must be "feedforward". Since an increase in glutamine concentration causes the metabolic pathway to speed up, this control must be feedforward activation.

26. Feedback inhibition means that when a large amount of product has been formed, the beginning of a process will slow down. Thus, feedback inhibition protects against overproduction. "Feedback activation" means that when a large amount of product is formed, the beginning of the process will accelerate. A state of overproduction will worsen and might lead to cell death.

28. The enzyme lactase catalyzes hydrolysis of the disaccharide lactose to produce two monosaccharides, D-glucose and D-galactose.

Chapter 32

2. A nucleoside is a purine or pyrimidine base linked to a sugar molecule, usually ribose or deoxyribose. A nucleotide is a purine or pyrimidine base linked to a ribose or deoxyribose sugar which in turn is linked to a phosphate group.

4. The major function of ATP in the body is to store chemical energy, and to release it when called upon to carry out many of the complex reactions that are essential to most of our life processes.
6. Complementary bases are the pairs that hydrogen bond ("fit") to each other between the two helixes of DNA.
8. Since there are only four different bases to make up the code, one nucleotide could only specify four possible amino acids; two nucleotides could specify 16 amino acids; three nucleotides could specify 64 amino acids. Since there are at least 20 amino acids needed, three nucleotides are required.
10. A codon is a triplet of three nucleotides, and each codon specifies one amino acid. The cloverleaf model of transfer RNA has an anticodon loop consisting of seven unpaired nucleotides. Three of these make up the anticodon, which is complementary to, and hydrogen-bonds to the codon on mRNA.
12. From time to time a new trait appears in an individual that is not present in either parents or ancestors. These traits, which are generally the result of genetic or chromosomal changes, are called mutations.
14. The correct statements are: a, b, d, g, h, j, k, m, p, r.
16. (a) G, guanosine
    (b) GMP, guanosine-5′-monophosphate
    (c) dGDP, deoxyguanosine-5′-diphosphate
    (d) CTP, cytidine-5′-triphosphate
18. (a) U

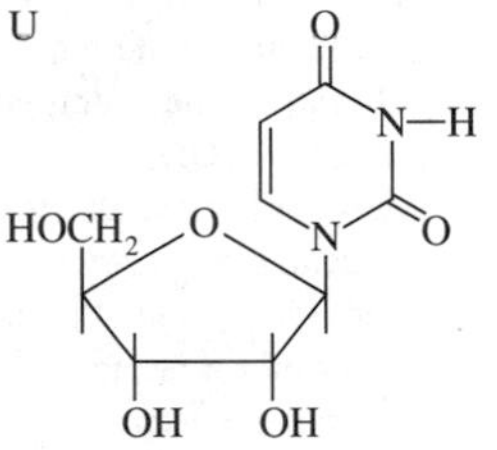

O, N—H, HOCH$_2$, O, N, O, OH, OH

(b) UMP

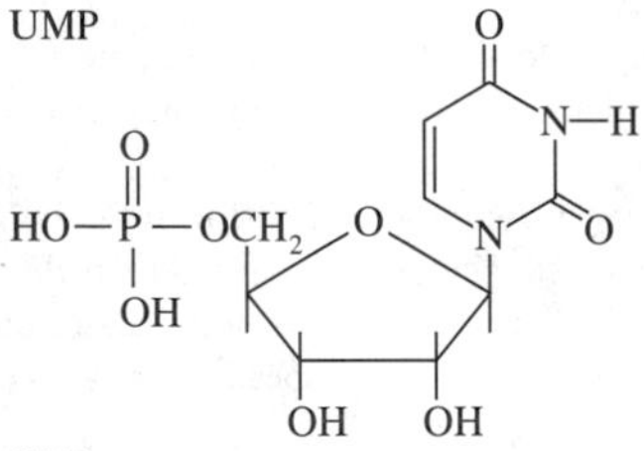

O, N—H, O, HO—P—OCH$_2$, O, N, O, OH, OH, OH

(c) CTP

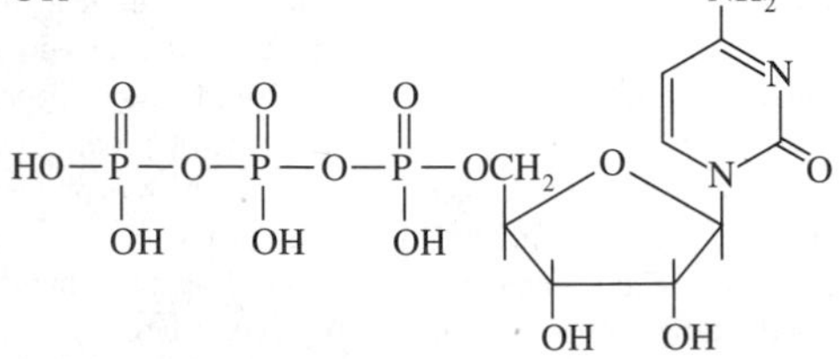

NH_2, N, O, O, O, HO—P—O—P—O—P—OCH$_2$, O, N, O, OH, OH, OH, OH, OH

(d) dTMP

O, H_3C, N—H, O, HO—P—OCH$_2$, O, N, O, OH, OH

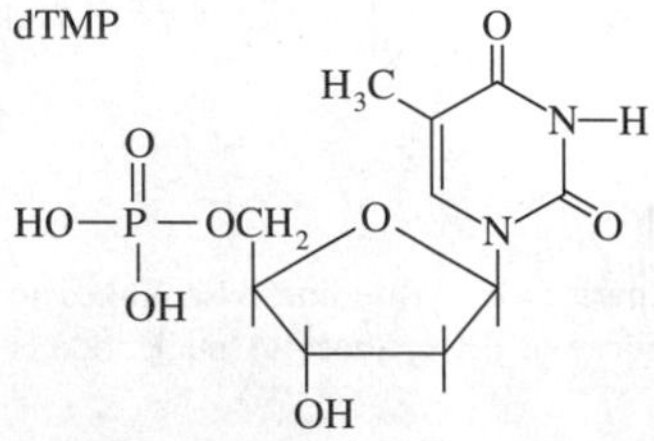

20. There are several possible sequences for the three-nucleotide single-stranded RNA. One possible structure follows:

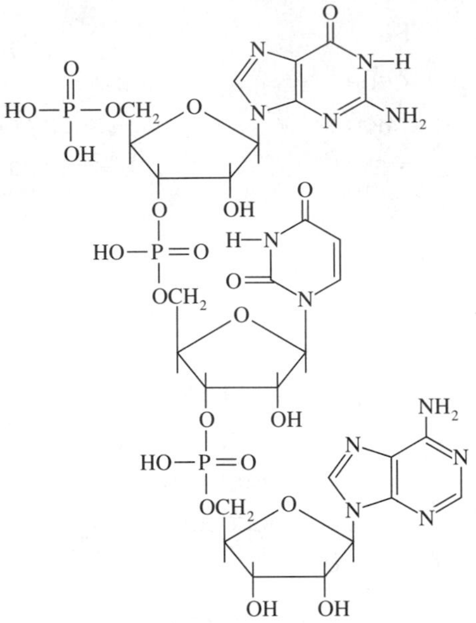

O, N, N—H, O, HO—P—OCH$_2$, O, N, N, NH_2, OH, O, O, OH, HO—P=O, H—N, O, OCH$_2$, N, O, O, OH, NH_2, N, N, HO—P=O, OCH$_2$, O, N, N, OH, OH

22. H, O······H—N, N, N—H ···· N, N, N, N, N—H···O, N, H, H, H (dotted lines are hydrogen bonds)

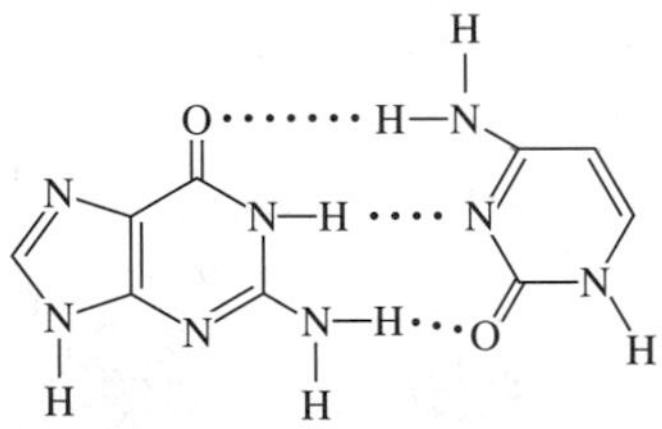

24. DNA is considered to be the genetic substance of life, because it contains the sequence of bases that carries the code for genetic characteristics.
26. Transcription is the process of making RNA using a DNA template, whereas translation is a process for making protein using an mRNA template.
28. Both mRNA and rRNA can be found in the ribosomes. rRNA serves as part of the ribosome structure, whereas mRNA serves as a template for protein synthesis.
30. 1719
32. (a) CGAAUCUGGACU
    (b) GCUUAGACCUGA
    (c) Arg-Ile-Trp-Thr
34. Translation makes a polymer of amino acids by forming amide bonds to connect the amino acids to each other. The amide bond combines an amine with a carboxylic acid.
36. Translation initiation occurs when the ribosome reaches a special AUG or GUG codon along the mRNA. Since there is commonly more than one AUG or GUG codon, the ribosome must use other information to choose the special AUG or GUG. This codon is the starting point for protein synthesis and is bound by either a special tRNA carrying *N*-formyl methionine (in procaryotes) or a tRNA carrying methionine (in eucaryotes).
38. (a) GCG (c) CUA
    (b) UGU (d) AAG
40. In RNA the guanine content does not have to be equal to the cytosine content, because RNA is a single strand. Its complements

are on the DNA template, not on the RNA. They have to be equal in DNA, because it is a double helix, with the complements together.

42. (a) Phe-His-Lys
(b) AAAGTATTC

44. The Human Genome Project is a cooperative effort by leading scientific laboratories to sequence the entire human genome (approximately 3 billion base pairs).

Chapter 33

2. The "energy allowance" represents the recommended dietary energy supply needed to maintain health.

4. Candy contains primarily sucrose, a source of calories, but little or no other nutrients. Thus, candy is said to provide "empty calories."

6. Essential fatty acids are required for normal growth and development.

8. Animal proteins are generally a source of all twenty common amino acids, whereas vegetable proteins may be deficient in one or several of the amino acids.

10. Vitamin D: functions as a regulator of calcium metabolism; Vitamin K: enables blood clotting to occur normally; Vitamin A: functions to furnish the pigment that makes vision possible. Many vitamins act as coenzymes.

12. Calcium is a major constituent of bones and teeth and is also important in nerve transmission among other functions.

14. Some common categories of food additives are (1) nutrients; (2) preservatives; (3) anticaking agents; (4) emulsifiers; (5) thickeners; (6) flavor enhancers; (7) colors; (8) nonsugar sweeteners.

16. The U.S. Food and Drug Administration is responsible for regulating the use of food additives.

18. Chyme is the material of liquid consistency consisting of food particles reduced to a small size and mixed with gastric juices in the stomach.

20. Intestinal mucosal cells produce many enzymes needed to complete digestion (e.g., disaccharidases, aminopeptidases, and dipeptidases).

22. The correct statements are: a, d, e, g, i, j, l, m, o.

24. (a) 132 g
(b) 1464 mg
(c) 31.5 mg

26. (a) 3×10^2 kcal
(b) 4×10^1 %

28. 7.0%

30. 41%

32. The two micronutrient classes are minerals and vitamins.

34. The three classes of nutrients that commonly supply energy for the cells are proteins, carbohydrates, and lipids.

36. Cellulose is important as dietary fiber. Although it is not digested, cellulose absorbs water and provides dietary bulk which helps maintain a healthy digestive tract.

38. Pancreatic juice is slightly basic.

40. Carbohydrates are digested in the mouth and small intestine.

42. Pepsin digests proteins.

44. 9 kg of body fat represents about 80,000 kcal of energy. Complete starvation for about 30 days would cause a net loss of 9 kg of fat. Thus, this new diet is unreasonable.

46. Fat (9 kcal/g) contains more than twice as much energy as carbohydrate (4 kcal/g). Thus, a tablespoon of butter will greatly increase the calorie content of a medium size baked potato.

48. The foods of higher animals must be digested before they can be utilized because foods are primarily large molecules. Foods must be broken down to much smaller molecules in order to pass through the intestinal walls into the blood and lymph systems where they can be utilized in the metabolic process of animals.

50. Nutritionists recommend less than 300 mg of dietary cholesterol per day. One large egg (270 mg cholesterol) represents almost the total daily recommended amount (about 90% of the RDA).

Chapter 34

2. $R{-}NH_2 + CO_2 \longrightarrow [R{-}NH{-}CO_2]^- + H^+$

4.

$$O^- {-} \overset{\overset{\displaystyle O}{\|}}{\underset{\underset{\displaystyle O^-}{|}}{P}} \sim O {-} \overset{\overset{\displaystyle O}{\|}}{\underset{\underset{\displaystyle O^-}{|}}{P}} {-} O {-} R$$

It is generally associated with adenosine triphosphate (ATP).

6. An oxidation–reduction coenzyme is a reusable organic compound which helps an enzyme carry out an oxidation–reduction reaction.

8. Oxidative phosphorylation uses the mitochondrial electron transport system to directly produce ATP from oxidation–reduction reactions. Substrate-level phosphorylation does not use the electron transport system, but, instead, involves transfer of a phosphate group from a substrate to ADP to form ATP.

10. Oxidative phosphorylation takes place in the mitochondria.

12. Chloroplast pigments trap light to provide energy for photosynthesis.

14. The correct statements are: a, c, e, f, h, i, j, k and l.

16. $CO_2(g) + H_2O(l) \rightleftharpoons H_2CO_3(aq) \rightleftharpoons H^+ + HCO_3^-(aq)$
The reaction between carbon dioxide and water is reversible. As carbon dioxide is exhaled from the lungs, carbonic acid is converted to carbon dioxide and the surrounding solution's pH increases. Since the hydrogen ion concentration decreases, hemoglobin binds to oxygen more tightly. The following reaction is reversed:
$H^+ + HbO_2 \rightleftharpoons HbH^+ + O_2$

18. The conversion of acetate to long-chain fatty acids is anabolic because:
(a) a smaller molecule (acetate) is converted to a larger molecule (long-chain fatty acids)
(b) the carbons from acetate become reduced (on the average) as they are converted to long-chain fatty acids

20. Chloroplasts carry out photosynthesis, not oxidative phosphorylation.

22. The chemical changes in the chloroplast are said to be anabolic because:
(a) smaller carbon dioxide molecules are converted to larger glucose molecules
(b) as this transformation takes place, the carbons become progressively more reduced
(c) this process requires an input of energy (from sunlight)

24. Three important characteristics of a catabolic process are (1) complex substances are broken down into simpler substances; (2) carbons are often oxidized; (3) cellular energy is often produced.

26. 0.34 mol O_2

28. 0.18 mol e^-

30. two

32. 58 kJ

34. Since the phosphorylated substrate 1,3-diphosphoglycerate contains only one high energy phosphate bond, substrate-level phosphorylation can produce only one ATP.

36. 12 ATP

38. The mitochondria carry out much of the oxidation–reduction reactions for the cell. Because these reactions involve electron movement, they are more easily controlled in a nonaqueous environment. Thus, mitochondria contain 90 percent membrane by mass.

40. Both ATP and NAD^+ contain a ribose that (1) is linked to an adenine at carbon one, and (2) is linked to a phosphate at carbon five.

Chapter 35

2. Glucose catabolism breaks down glucose to the smaller carbon dioxide in an oxidative process. This produces energy for the cell and is catabolic. Photosynthesis produces glucose from carbon dioxide via a reductive path. Light energy is required and this is an anabolic process.

4. Enzymes are needed so that biochemical reactions will proceed fast enough to keep the cell alive.

6. The liver replenishes muscle glucose (via the blood) by (a) releasing glucose from liver glycogen stores and (b) by converting lactic acid back into glucose.

8. One high energy phosphate bond (in GTP) is formed directly in the citric acid cycle. In addition, one $FADH_2$ and three NADHs are produced.

10. Hormones are chemical substances that act as control or regulatory agents in the body. Hormones are secreted by the endocrine, or ductless glands directly into the bloodstream and are transported to various parts of the body to exert specific control functions.

12. The renal threshold is the concentration of a substance in the blood above which the kidneys begin to excrete that substance.

14. The correct statements are: a, c, h, i, j, l, m.

16.

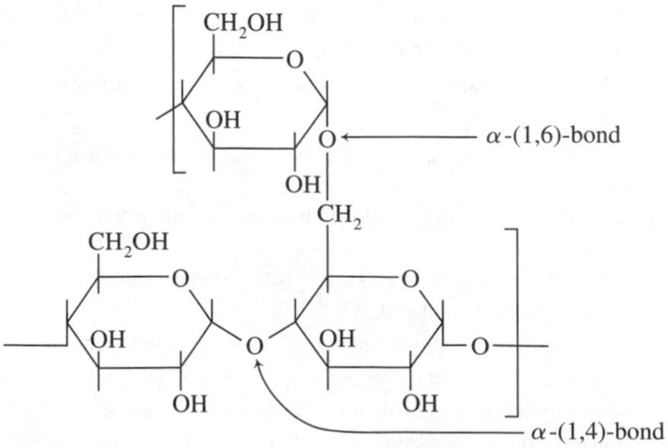

CH_2OH, O, OH, O ← α-(1,6)-bond, OH, CH_2

CH_2OH, O, OH, O, OH, O, OH, OH ← α-(1,4)-bond

18. Glycogenolysis is catabolic because it starts with larger precursors (glycogen molecules) and forms smaller products (glucose molecules).

20. The Embden-Meyerhof pathway uses substrate-level phosphorylation to produce ATP. That is, following carbon oxidation a phosphate group is transferred from a substrate to ADP, forming ATP.

22.

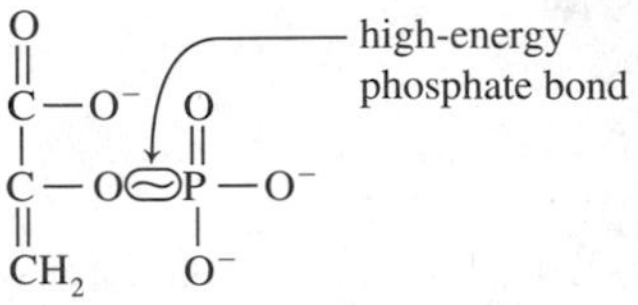

$$\begin{array}{l} O \\ \| \\ C-O^- \quad O \\ | \qquad\quad \| \\ C-O\sim P-O^- \\ \| \qquad\quad | \\ CH_2 \qquad O^- \end{array}$$

high-energy phosphate bond (arrow to the C–O–P bond)

24. Although no molecular oxygen (O_2) is used in the citric acid cycle, reduced coenzymes, NADH and $FADH_2$ are produced. These reduced coenzymes must be oxidized by electron transport so that the citric acid cycle can continue. And, it is electron transport that uses molecular oxygen. Thus, the citric acid cycle depends on the presence of molecular oxygen and is considered to be aerobic.

26. 1.7 mol ATP

28. The end product of the anaerobic catabolism of glucose in muscle tissue is lactate.

30. The gluconeogenesis pathway is considered to be anabolic because (a) smaller, noncarbohydrate precursors such as lactate are converted to the larger product, glucose, and (b) as this conversion takes place, the carbons become progressively more reduced.

32.

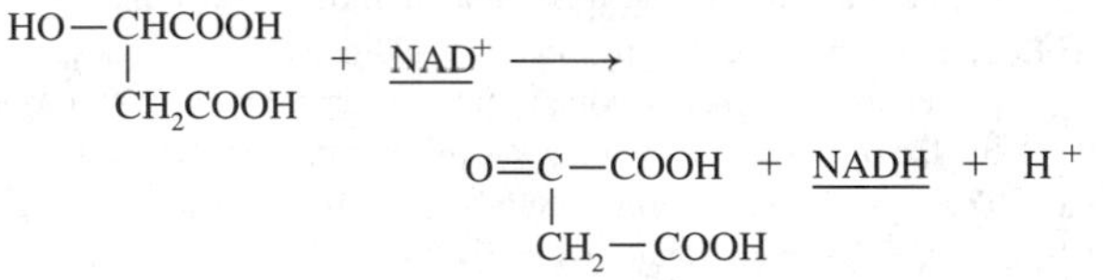

$$\begin{array}{l} HO-CHCOOH \\ \qquad\;\; | \\ \qquad CH_2COOH \end{array} + \underline{NAD^+} \longrightarrow \begin{array}{l} O=C-COOH \\ \quad\;\; | \\ \quad CH_2-COOH \end{array} + \underline{NADH} + H^+$$

34. The glycolysis pathway uses substrate-level phosphorylation to produce ATP and thus does not need electron transport and oxidative phosphorylation.

36. Unlike vitamins, which must be supplied in the diet, hormones are produced in the body.

38. The normal blood glucose concentration is 70–90 mg/100 mL of blood under fasting conditions. Hyperglycemia occurs when the blood glucose concentration rises above this range. Thus, 350 mg glucose/100 mL blood is hyperglycemic.

40. Hormones function as chemical messengers. They are chemical substances that act as control or regulatory agents in the body. Enzymes are catalysts for specific reactions, allowing these reactions to occur faster and under milder conditions than would otherwise be possible.

42. If a large overdose of insulin was taken by accident, the blood glucose concentration would drop to a low level, probably in the hypoglycemic range. This could result in fainting, convulsions, and unconsciousness.

44.

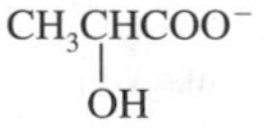

$$\begin{array}{l} CH_3CHCOO^- \\ \qquad\; | \\ \qquad OH \end{array}$$

lactate

All three carbons of lactate are more reduced than the carbon in CO_2. Thus, lactate can provide more cellular energy upon further oxidation of the carbons.

46. In aerobic catabolism, pyruvate is converted to carbon dioxide yielding many ATPs while consuming molecular oxygen. In anaerobic catabolism, pyruvate is used to recycle NADH back to NAD^+.

Chapter 36

2. Knoop prepared a homologous series of straight-chain fatty acids with a phenyl group at one end and a carboxyl group at the other end. He fed each of these acids to test animals and analyzed the urine of the animals for metabolic products of the acid. Those animals that had been fed acids with an even number of carbons excreted phenylaceturic acid. Those animals that had been fed acids with an odd number of carbon atoms excreted hippuric acid. The experiment showed that the metabolism of the acid must occur by shortening the chain two carbon atoms at a time.

4. Ketone bodies are relatively soluble in the bloodstream and can be transported to energy-deficient cells in time of need. Ketone bodies supply energy via β-oxidation and the citric acid cycle.

6. Fat acts as a cushion or shock absorber around internal organs and as an insulating blanket.

8. The obese protein signals the body to (a) increase catabolism of fats and (b) decrease the appetite for food.

10. Soybeans enrich the soil because their roots serve as host to the symbiotic, nitrogen-fixing bacteria.

12. (a) incorporation into a protein
(b) utilization in the synthesis of other nitrogenous compounds such as nucleic acids
(c) deamination to a keto acid

14. Nitrogen is excreted as (a) ammonia by fish and (b) uric acid by birds.

16. three ATPs

18. The correct statements are: c, f, g, h, j, l, m, n, o, r, s, t, v.

20. $(CH_3CH_2)CH_2CH_2CH_2COOH$ (the CH_3CH_2 group is circled)

22. $CH_3(CH_2)_{10}{}^{*}CH_2CH_2COOH$

24. 8 acetyl-CoA

26. 7 NADH

28. 25 ATP

30. (a) ACP is a protein whereas CoA is a coenzyme
(b) ACP is used in fatty acid synthesis whereas CoA is used in the β-oxidation pathway.

32. A rapid growth spurt commonly means the body is using ingested amino acids to build new proteins. Thus, less nitrogen will be excreted and the nitrogen balance should become more positive.

34. L-aspartic acid is not a ketogenic amino acid because a diet high in L-aspartic acid caused no increase in blood levels of the ketone body β-hydroxybutyric acid. (Other ketone bodies would not increase their blood levels also.) Because L-aspartic acid is not a ketogenic amino acid, it must be a glucogenic amino acid.

36. (a) $HOOCCH_2C(=O)COOH$

(b) 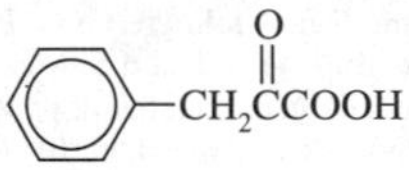 $C_6H_5{-}CH_2C(=O)COOH$

38. $HOOC{-}CH_2{-}CH(NH_2){-}COOH$ (the H_2N group is circled)

40. β-hydroxybutyric acid:

$CH_3CH(OH)CH_2COOH$

42. The six-carbon fatty acid yields more ATP than the six-carbon hexose during catabolism primarily because more acetyl-CoA molecules are formed and can feed into the citric acid cycle, electron transport, and oxidative phosphorylation.

44.
$$CH_3CH_2CH(CH_3){-}C(=O){-}COOH + \underset{\text{L-glutamic acid}}{HOOCCH_2CH_2CH(NH_2)COOH} \xrightarrow{\text{enzyme}}$$
$$\underset{\text{isoleucine}}{CH_3CH_2CH(CH_3)CH(NH_2)COOH} + \underset{\alpha\text{-ketoglutaric acid}}{HOOCCH_2CH_2C(=O)COOH}$$

46. In step 2, the fatty acid carbons at the alpha and beta positions change their oxidation numbers from -2 to -1; they are oxidized.

PHOTO CREDITS

FRONT MATTER: p. i, David Cavagnaro/Peter Arnold, p. ii, Ian Murphy/Tony Stone Images; p. vi, Courtesy of Merck and Company; p. xii, Courtesy of IBM/Tom Way. **CHAPTER 1:** p. 1, Murray Alcosser/Image Bank; p. 3, Courtesy of Merck and Company (2 photos); p. 4, Lori Ritzenthaler/Ex Libris; p. 5, Courtesy of William Buehler; p. 7, Rita Amaya; p. 8, Steve Leonard/Tony Stone Images; p. 9, Richard Nowitz/Phototake; p. 10, Stephen Marks/Image Bank. **CHAPTER 2:** p. 11, Murray Alcosser/Image Bank; p. 13, Andy Caulfield/Image Bank; p. 14, Courtesy of NASA; p. 18, Left: Joseph Drivas/Image Bank; Right: CNRI/Science Photo Library/Photo Researchers; p. 21, Ex Libris/Lori Ritzenthaler; p. 28, Rita Amaya (4 photos); p. 34, Lori Ritzenthaler/Ex Libris; p. 37, Will and Deni McIntyre/Photo Researchers. **CHAPTER 3:** p. 46, Michel Viard/Peter Arnold; p. 49, Edward S. Ross/Phototake; p. 53, Courtesy of IBM/Tom Way; p. 54, The Granger Collection; p. 55, Courtesy of International Academy of Science; pp. 56-57, Rita Amaya; p. 61, Courtesy of General Electric; p. 62, Rita Amaya (3 photos). **CHAPTER 4:** p. 68, Fritz Henle/Photo Researchers; p. 71, Leonard Lessin/Peter Arnold (2 photos); p. 72, Yoav Levy/Phototake; p. 75, Chris McCooey/Tony Stone Images; p. 76, Consumer Reports (3 photos); p. 78, Lee F. Snyder/Photo Researchers; p. 79, Lori Ritzenthaler/Ex Libris. **CHAPTER 5:** p. 83, Kent Wood/Photo Researchers; p. 85–86, The Granger Collection; p. 87, Julie Kranhold/Ex Libris; p. 88, Courtesy of IBM/Almaden Research Center; p. 89, The Granger Collection; p. 90, Peter Fowler/Science Photo Library/Photo Researchers; p. 94, Lori Ritzenthaler/Ex Libris. **CHAPTER 6:** p. 100, Antonio Rosario/Image Bank; p. 103, Jason Grow/San Jose Mercury News; p. 106, Richard Megna/Fundamental Photographs; p. 108, Courtesy of Sargent-Welch, a VWR Corporation; p. 119, Lori Ritzenthaler/Ex Libris. **CHAPTER 7:** p. 124, Michael Skott/Image Bank; p. 126, The Granger Collection; p. 130, Rita Amaya; p. 135, Julie Kranhold/Ex Libris. **CHAPTER 8:** p. 147, Richard Megna/Fundamental Photographs; p. 150, Yoav Levy/Phototake; p. 153, Lori Ritzenthaler/Ex Libris; p. 154, David Cavagnaro/Peter Arnold; p. 157, Yoav Levy/Phototake; p. 159, Richard Megna/Fundamental Photographs; p. 160, Michael Dalton/Fundamental Photographs; p. 161, Rita Amaya; p. 162, Andy Sacks/Tony Stone Images; p. 163, Rita Amaya (2 photos); p. 164, Dave Bjorn/Tony Stone Images. **CHAPTER 9:** p. 171, Courtesy of Merck and Company; p. 177, Courtesy of NASA; p. 181, Manfred Kage/Peter Arnold; p. 183, Gazelle Technologies/© 1991 Preferred Stock. **CHAPTER 10:** p. 193, Tim Brown/Tony Stone Images; p. 195, Courtesy of The National Institute of Standards and Technology; pp. 196-197, The Granger Collection; p. 198, Courtesy of IBM/Almaden Research Center; p. 199, Craig Wells/Tony Stone Images; p. 201, Left: Carolina Biological Supply/Phototake; Middle: Phototake; Right: Laurence Livermore Laboratory/Science Photo Library/Photo Researchers; p. 203, Courtesy of IBM/Almaden Research Center. **CHAPTER 11:** p. 214, Andrea Pistolesi/Image Bank; p. 219, Courtesy of Bancroft Library/University of California, Berkeley; p. 224, IBM Research/Peter Arnold; p. 231, AP/Wide World Photos; p. 233, Left: Courtesy of Bharat Bhushan/Ohio State University; p. 233, Right: Courtesy of IBM/Almaden Research Center. **CHAPTER 12:** p. 251, Ray Nelson/Phototake; p. 258, The Granger Collection; p. 263, Rita Amaya (3 photos); p. 265, The Granger Collection; p. 270, John Allison/Peter Arnold; p. 277, Stephen Frink/Tony Stone Images; p. 285, Peggy and Yoram Kahana/Peter Arnold; p. 286, Courtesy of NASA. **CHAPTER 13:** p. 294, Courtesy of NASA; p. 298, Lori Ritzenthaler/Ex Libris; p. 299, Yoav Levy/Phototake; p. 302, Courtesy of Matthew Senay and David Jewitt/University of Hawaii; p. 304, Rita Amaya; p. 308, Lori Ritzenthaler/Ex Libris; p. 312, Paul Silverman/Fundamental Photographs; p. 313, Sara Hotchkiss/SKA; p. 317, Courtesy of Mitch Kaufman/Southern California Edison. **CHAPTER 14:** p. 324, Warren Bolster/Tony Stone Images; p. 326, Richard Megna/Fundamental Photographs; p. 327, Kip Peticolas/Fundamental Photographs; p. 330, Lori Ritzenthaler/Ex Libris; p. 335, Richard Megna/Fundamental Photographs; p. 341, Courtesy of Sargent Welch/A VWR Corporation; pp. 346 and 348, Lori Ritzenthaler/Ex Libris; p. 349, Craig Newbauer/Peter Arnold; p. 352, Dennis Kunkel/Phototake (3 photos). **CHAPTER 15:** p. 361, G. Brad Lewis/Tony Stone Images; p. 369, Charles C. Place/The Image Bank; pp. 370 and 377, Rita Amaya; p. 378, Lori Ritzenthaler/Ex Libris; p. 379, Left: Rita Amaya; Right: Leonard Lessin/Peter Arnold; p. 382, Tim Bieber/The Image Bank; p. 388, Rita Amaya; p. 390, Photo Researchers. **CHAPTER 16:** p. 397, Jeff Hunter/Image Bank; p. 399, Lori Ritzenthaler/Ex Libris; p. 405, Grant Heilman/Grant Heilman Photography; p. 407, Rita Amaya; p. 420, Lori Ritzenthaler/Ex Libris; p. 421, David M. Phillips/Photo Researchers; p. 424, Lori Ritzenthaler/Ex Libris. **CHAPTER 17:** p. 432, Courtesy of Golden Gate Bridge, Highway, and Transportation District; p. 442, Courtesy of Corning Glass; p. 446, Peticolas/Megna/Fundamental Photographs; p. 450, Richard Nowitz/Phototake; p. 453, Rohan/Tony Stone Images; p. 454, Courtesy of Delco Corporation. **CHAPTER 18:** p. 460, Rainbow/Dan McCoy; p. 462, The Granger Collection; p. 470, Kevin Schafer/Peter Arnold; p. 471, Steve Proehl/The Image Bank; p. 472, Left: Rennie Van Munchow/Phototake; Right: Yoav Levy/Phototake; p. 473, Peticolas/Megna/Fundamental Photographs; p. 477, Catherine Pouedras/Photo Researchers; p. 481, Jim Olive/Peter Arnold; p. 483, Tom Carroll/Phototake. **CHAPTER 19:** p. 488, Larry Ulrich/Tony Stone Images; p. 491, Left: Paul Silverman/Fundamental Photographs; Right: Julie Kranhold/Ex Libris; p. 492, Tom Carroll/Phototake; p. 493, Richard Megna/Fundamental Photographs; p. 494, Michael Melford/Image Bank; p. 497, Don Smetzer/Tony Stone Images; p. 499, Jeff Hunter/Image Bank; p. 501, Gerhard Gscheidle/Peter Arnold; p. 502, J. B. Diederich/Contact Press/PNI; p. 504, Donovan Reese/Tony Stone Images; p. 508, Keith Wood/Tony Stone Images; p. 511, Dawn Goss/Phototake; p. 515, Bob Torrez/Tony Stone Images; p. 517, Andy Caulfield/Image Bank; p. 520, Butch Martin/Image Bank; p. 521, Chuck Davis/Tony Stone Images. **CHAPTER 20:** p. 525, L. D. Gordon/Image Bank; p. 527, Adeo/Image Bank; p. 531, Left: Mike Abrahams/Tony Stone Images; Middle: George Hunter/Tony Stone Images; Right: Jurgen Vogt/Image Bank; p. 545, Lowell Georgia/Photo Researchers; p. 549, J. L. Bohin/Photo Researchers; p. 551, Leonard Lessin/Peter Arnold; p. 554, Hans Pfletschinger/Peter Arnold. **CHAPTER 21:** p. 560, Gary Gladstone/Image Bank; p. 571, Paul H. Axelsen, M.D./Phototake; p. 574, Jerry Mason/Photo Researchers; p. 579, Andy Cox/Tony Stone Images; p. 584, Lori Ritzenthaler/Ex Libris; p. 585, David Hoffman/Tony Stone Images; p. 587, Helga Lade/Peter Arnold; p. 588, Julie Kranhold/Ex Libris. **CHAPTER 22:** p. 597, Alvis Upitis/Image Bank; p. 611, David Austen/Tony Stone Images; p. 612, Fernand Ivaldi/Tony Stone Images; p. 613, Courtesy of Chris Craft/Monosol Division; p. 617, Phil Schermeister/Tony Stone Images; p. 619, Mauritius GMBH/Phototake; p. 622, Alan Levenson/Tony Stone Images. **CHAPTER 23:** p. 632, Julie Kranhold/Ex Libris; p. 636, Coco McCoy/Rainbow/PNI; p. 641, Julie Kranhold/Ex Libris; p. 645, Malcom Linton/Black Star/PNI. **CHAPTER 24:** p. 656, Alain Altair/Image Bank; p. 658, George D. Dodge/Bruce Coleman/PNI; p. 663, Alan Reininger/Contact Press/PNI; p. 665, Jose Azel/Aurora/PNI; p.671, Ann Cecil/Photo 20-20/PNI; p. 674, Julie Kranhold/Ex Libris; p. 679, Tom Prettyman/PhotoEdit/PNI. **CHAPTER 25:** p. 691, Tony Molenaar/Image Bank; p. 699, Dugald Bremner/Tony Stone Images; p. 700, Infocus Intl/Image Bank. **CHAPTER 26:** p. 714, Steve Bly/Tony Stone Images; p. 716, Courtesy of IBM; p. 717, Julie Kranhold/Ex Libris; p. 719, Michael W. Nelson/Uniphoto South/PNI; p. 721, Left: Rainbow/PNI; Right: Steve Winter/Black Star/PNI; p. 722, Alvis Upitis/Image Bank; p. 725, Hank Morgan/Rainbow/PNI; p. 728, Srulik Haramaty/Phototake/PNI. **CHAPTER 27:** p. 733, David Gnizak/Phototake; p. 735, Rita Amaya; p. 736, PNI; p. 737, Dan McCoy/Rainbow/PNI; p. 739, Julie Kranhold/Ex Libris; p. 740, Left and Right: Julie Kranhold/Ex Libris; p. 745, Mark Burnside/Tony Stone Images/PNI. **CHAPTER 28:** p. 754, James Martin/Tony Stone Images; p. 756, Left and Right: Kristen Brochmann/Fundamental Photographs; p. 757, Tony Olmos/Black Star/PNI; p. 759, Ed Reschke/Peter Arnold; p. 768, Darrell Guillin/Tony Stone Images/PNI; p. 771, Bob Krist/Black Star/PNI; p. 772, Photo Network/PNI; p. 775, Julie Kranhold/Ex Libris; p. 779, Julie Kranhold/Ex Libris; p. 781, Roland Birke/Peter Arnold; p. 783, Biophoto Associates/Photo Researchers; p. 785, Phototake/PNI. **CHAPTER 29:** p. 791, Daniel J. Cox/Tony Stone Images; p. 795, Barts Medical Library/PNI; p. 797, Nicholas DeVore III/Bruce Coleman/PNI; p. 799, David Young-Wolff/PhotoEdit/PNI; p. 802, Chris Johns/Tony Stone Images/PNI; p. 805, Phil Schermeister/Tony Stone Images/PNI; p. 807, Julie Kranhold/Ex Libris; p. 809, Dan McCoy/Rainbow/PNI. **CHAPTER 30:** p. 816, Owen Franken/Stock Boston/PNI; p. 818, Manfred Kage/Peter Arnold; p. 819, Aaron Haupt/Stock Boston/PNI; p. 820, Contact Press/PNI/Carlos Humberto; p. 827, Joseph Nettis/Stock Boston/PNI; p. 831, Julie Kranhold/Ex Libris; p. 839, Stanley Flegler/Visuals Unlimited; p. 841, Left: Martin Rotker/Phototake; Middle: Dr. Dennis Kunkel/Phototake; Right: Carolina Biological Supply/Phototake; p. 842, Ex Libris/Julie Kranhold; p. 845, Courtesy of Patrick O'Farrell/UCSF; p. 849, James Holmes/Celltech Ltd/Science Photo Library/Photo Researchers. **CHAPTER 31:** p. 856, C.J. Allen/Stock Boston; p. 859, Catherine Karnow/Woodfin Camp/PNI; p. 863, Courtesy of Hunt, Rook & Scheibel; p. 869, Mark Ludak/Impact Visuals/PNI. **CHAPTER 32:** p. 873, Ann Shamel/Graphistock/PNI; p. 878, John Coletti/Stock Boston/PNI; p. 885, David Scharf/Peter Arnold; p. 887, Leonard Lessin/Peter Arnold; p. 888, Omikron/S.S./Photo Researchers; p. 892, Charles Thatcher/Tony Stone Images; p. 897, Gary Wagner/Stock Boston/PNI; p. 898, Peter Menzel/Stock Boston/PNI. **CHAPTER 33:** p. 902, Tis Hunter/Tony Stone Images; p. 906, Comstock; p. 909, Betty Press/Woodfin Camp/PNI; p. 910, Triolo Burke/FoodPix/PNI; p. 914, Alan Oddie/PhotoEdit/PNI; p. 919, Julie Kranhold/ExLibris; p. 920, Tony Freeman/PhotoEdit/PNI; p. 925, Fawcett/Hirokawa/Heuser/Science Source/Photo Researchers. **CHAPTER 34:** p. 931, Ron Sanford/Black Star/PNI; p. 933, Larry Goldstein/Tony Stone Images; p. 935, Comstock; p. 942, K.R. Porter/Science Source/Photo Researchers; p. 946, Ed Reschke/Peter Arnold; p. 947, Kevin Morris/Tony Stone Images/PNI; p. 948, Courtesy of Nathan S. Lewis, California Institute of Technology. **CHAPTER 35:** p. 952, Steve Dunwell/Image Bank; p. 954, Chad Ehlers/Tony Stone Images/PNI; p. 956, Bob Daemmrich/Stock Boston/PNI; p. 964, Peter Southwick/Stock Boston/PNI; p. 965, Dr. Dennis Kunkel/Phototake/PNI; p. 968, Rick Lance/Phototake/PNI. **CHAPTER 36:** p. 972, Jim Corwin/Stock Boston/PNI; p. 977, P. Motta/Science Photo Library/Photo Researchers; p. 981, Cabisco/Visuals Unlimited; p. 988, Rob Nelson/Black Star/PNI.

INDEX

Entries in boldface refer to key terms in the text.

NAMES, FORMULAS AND CHARGES OF COMMON IONS

Positive Ions (Cations)

Charge	Name	Formula
1+	Ammonium	NH_4^+
	Copper(I)	Cu^+
	(Cuprous)	
	Hydrogen	H^+
	Potassium	K^+
	Silver	Ag^+
	Sodium	Na^+
2+	Barium	Ba^{2+}
	Cadmium	Cd^{2+}
	Calcium	Ca^{2+}
	Cobalt(II)	Co^{2+}
	Copper(II)	Cu^{2+}
	(Cupric)	
	Iron(II)	Fe^{2+}
	(Ferrous)	
	Lead(II)	Pb^{2+}
	Magnesium	Mg^{2+}
	Manganese(II)	Mn^{2+}
	Mercury(II)	Hg^{2+}
	(Mercuric)	
	Nickel(II)	Ni^{2+}
	Tin(II)	Sn^{2+}
	(Stannous)	
	Zinc	Zn^{2+}
3+	Aluminum	Al^{3+}
	Antimony(III)	Sb^{3+}
	Arsenic(III)	As^{3+}
	Bismuth(III)	Bi^{3+}
	Chromium(III)	Cr^{3+}
	Iron(III)	Fe^{3+}
	(Ferric)	
	Titanium(III)	Ti^{3+}
	(Titanous)	
4+	Manganese(IV)	Mn^{4+}
	Tin(IV)	Sn^{4+}
	(Stannic)	
	Titanium(IV)	Ti^{4+}
	(Titanic)	
5+	Antimony(V)	Sb^{5+}
	Arsenic(V)	As^{5+}

Negative Ions (Anions)

Charge	Name	Formula
1–	Acetate	$C_2H_3O_2^-$
	Bromate	BrO_3^-
	Bromide	Br^-
	Chlorate	ClO_3^-
	Chloride	Cl^-
	Chlorite	ClO_2^-
	Cyanide	CN^-
	Fluoride	F^-
	Hydride	H^-
	Hydrogen carbonate	HCO_3^-
	(Bicarbonate)	
	Hydrogen sulfate	HSO_4^-
	(Bisulfate)	
	Hydrogen sulfite	HSO_3^-
	(Bisulfite)	
	Hydroxide	OH^-
	Hypochlorite	ClO^-
	Iodate	IO_3^-
	Iodide	I^-
	Nitrate	NO_3^-
	Nitrite	NO_2^-
	Perchlorate	ClO_4^-
	Permanganate	MnO_4^-
	Thiocyanate	SCN^-
2–	Carbonate	CO_3^{2-}
	Chromate	CrO_4^{2-}
	Dichromate	$Cr_2O_7^{2-}$
	Oxalate	$C_2O_4^{2-}$
	Oxide	O^{2-}
	Peroxide	O_2^{2-}
	Silicate	SiO_3^{2-}
	Sulfate	SO_4^{2-}
	Sulfide	S^{2-}
	Sulfite	SO_3^{2-}
3–	Arsenate	AsO_4^{3-}
	Borate	BO_3^{3-}
	Phosphate	PO_4^{3-}
	Phosphide	P^{3-}
	Phosphite	PO_3^{3-}